Water Worlds in the Solar System

"Either write something worth reading or do something worth writing"

Benjamin Franklin— (Inventor of lightning arrester, discoverer of Gulf Stream, and an able administrator)

Water Worlds in the Solar System

Antony Joseph
Formerly Chief Scientist at CSIR-National Institute of Oceanography, India

For access to more content please use this link
https://www.youtube.com/watch?v=dzyaTjntN4A&t=10s

Elsevier
Radarweg 29, PO Box 211, 1000 AE Amsterdam, Netherlands
The Boulevard, Langford Lane, Kidlington, Oxford OX5 1GB, United Kingdom
50 Hampshire Street, 5th Floor, Cambridge, MA 02139, United States

Copyright © 2023 Elsevier Inc. All rights reserved.

No part of this publication may be reproduced or transmitted in any form or by any means, electronic or mechanical, including photocopying, recording, or any information storage and retrieval system, without permission in writing from the publisher. Details on how to seek permission, further information about the Publisher's permissions policies and our arrangements with organizations such as the Copyright Clearance Center and the Copyright Licensing Agency, can be found at our website: www.elsevier.com/permissions.

This book and the individual contributions contained in it are protected under copyright by the Publisher (other than as may be noted herein).

Notices
Knowledge and best practice in this field are constantly changing. As new research and experience broaden our understanding, changes in research methods, professional practices, or medical treatment may become necessary.

Practitioners and researchers must always rely on their own experience and knowledge in evaluating and using any information, methods, compounds, or experiments described herein. In using such information or methods they should be mindful of their own safety and the safety of others, including parties for whom they have a professional responsibility.

To the fullest extent of the law, neither the Publisher nor the authors, contributors, or editors, assume any liability for any injury and/or damage to persons or property as a matter of products liability, negligence or otherwise, or from any use or operation of any methods, products, instructions, or ideas contained in the material herein.

ISBN: 978-0-323-95717-5

For Information on all Elsevier publications visit our website at
https://www.elsevier.com/books-and-journals

Publisher: Candice G Janco
Acquisitions Editor: Peter Llewellyn
Editorial Project Manager: Andrae Akeh
Production Project Manager: R. Vijay Bharath
Cover Designer: Matthew Limbert

Typeset by Aptara, New Delhi, India

Dedication

To Professor Stephen Hawking, World-renowned
theoretical physicist, who shaped modern
cosmology and inspired millions
(January 8, 1942–March 14, 2018).

Other Books by Dr. Antony Joseph through 2016

Tsunamis: Detection, Monitoring, and Early-Warning Technologies
Elsevier / Academic Press, New York, 448 p. (2011)

Measuring Ocean Currents: Tools, Technologies, and Data
Elsevier, New York, 426 p. (2013)

Investigating Seafloors and Oceans: From Mud Volcanoes to Giant Squid
Elsevier, New York, 581 p. (2016)

Contents

Foreword   xix
Preface   xxiii
Acknowledgments   xxvii

1. Solar/planetary formation and evolution

- 1.1 Planet formation   1
  - 1.1.1 Terrestrial planet formation   6
  - 1.1.2 Giant planet formation   7
- 1.2 Asteroids, meteorites, and chondrites   9
- 1.3 Giant-impact theory on the origin of Earth's Moon   14
  - 1.3.1 Single giant impact theory   15
  - 1.3.2 Multiple giant impact theory   22
  - 1.3.3 The concept of lunar magma ocean (LMO) of global dimensions   29
- 1.4 Influence of Moon-forming impacts on the environmental conditions on the early Earth   31
- 1.5 Earth's internal structure, development, orbit, and rotation   32
  - 1.5.1 Influence of collisions   32
  - 1.5.2 Features of Earth's core   32
  - 1.5.3 Earth's paleo-rotation and revolution—day: ~21 h; year: ~13 months and ~400 days   33
  - 1.5.4 Earth's inclination and orbit   34
- 1.6 Water and frost line in the astrophysical environments   34
  - 1.6.1 Water in the protoplanetary disk of the Sun   35
  - 1.6.2 Frost line   36
  - 1.6.3 Water stored on the surface and in the ground of modern Earth   38
- 1.7 Water-abundant celestial bodies in the Solar System—brief overview   39
- 1.8 Importance of understanding Earth's oceans in the search for life in extraterrestrial ocean worlds—NASA's ocean worlds exploration program   40
- 1.9 Importance of radiogenic heating and tidal dissipation in the generation and sustenance of extraterrestrial subsurface ocean worlds   44
- 1.10 Shedding light on extraterrestrial bodies—role of astronomical research   45
- References   46
- Bibliography   53

2. Geological timeline of significant events on Earth

- 2.1 An era from 4.5 to 4 billion years ago when the entire Earth was a "Fire Ball"   55
- 2.2 Importance of greenhouse gases in the atmosphere of the early Earth   55
- 2.3 Genesis of water on Earth   57
  - 2.3.1 Water on Earth through mantle evolution   59
  - 2.3.2 Water brought to Earth by comets and asteroids   59
- 2.4 Indispensability of water, biologically important chemical elements, and energy to sustain life as we know it   62
- 2.5 Formation of liquid water oceans on Earth about 3.8 billion years ago   65
- 2.6 Importance of deuterium to hydrogen ratio of water   67
- 2.7 Roles of Earth's Moon and Sun in generating tides—influences of local bathymetry and shoreline boundary on modifying tidal range and tidal pattern   69
  - 2.7.1 General characteristics of tidal oscillations   70
  - 2.7.2 Topographical influences on tidal range and tidal pattern   72
  - 2.7.3 Tidal bore—wall of tumbling and foaming water waves in some geometrically special water bodies during a spring tide flood tide   73
  - 2.7.4 Tidal currents—their role in mixing of ocean waters   75
  - 2.7.5 Implications of coastal tides and tidal bores   76

2.8	Appearance of microbes on Earth about 3.7 billion years ago	77
2.9	Stromatolites appearing on Earth about 3.5 billion years ago	77
2.10	Initiation of plate tectonics on Earth between 3.5 and 3.3 billion years ago	79
2.11	The great oxidation event ~2.4–2.0 billion years ago—an event that led to the banded iron formations and the rise of oxygen in Earth's atmosphere	81
2.12	An era when the entire Earth became fully covered with thick ice ~750–635 million years ago—"Snowball Earth" hypothesis	83
2.13	Multiple mass extinction events on Earth—important for understanding life	83
	2.13.1 Ordovician–Silurian extinction: ~440 million years ago	83
	2.13.2 Late Devonian extinction: ~365 million years ago	85
	2.13.3 Permian–Triassic extinction: ~253 million years ago	87
	2.13.4 Triassic–Jurassic extinction: ~201 million years ago	87
	2.13.5 The K–Pg extinction: ~66 million years ago: extinction of dinosaurs from Earth and subsequent appearance of modern humans' distant ancestors	88
2.14	Carbonate–silicate cycle and its role as a dynamic climate buffer	90
2.15	Occurrence of a sharp global warming ~56 million years ago	92
	2.15.1 Consequences	92
	2.15.2 Causes	92
2.16	Volcano eruptions on land causing atmospheric cooling and those happening underwater causing abnormal atmospheric warming	97
2.17	Synthesis of marine proxy temperature data across the Paleocene–Eocene thermal maximum	99
2.18	Fate of excess carbon released during the Paleocene–Eocene thermal maximum event	100
References		100
Bibliography		105

3. Beginnings of life on Earth

3.1	Origins of life and potential environments—multiple hypotheses on chemical evolution preceding biological evolution	115
	3.1.1 Lightning in the early atmosphere and the consequent production of amino acids—Miller–Urey "prebiotic soup" experiment	116
	3.1.2 Chemical processes at submarine hydrothermal vents	120
	3.1.3 Life brought to Earth from elsewhere in space	128
3.2	Biological evolution	129
	3.2.1 Discovery of DNA and its sequencing—the intriguing story of combined efforts by a group of scientists from different disciplines	131
	3.2.2 Role of National Human Genome Research Institute (NHGRI) in supporting development of new technologies for DNA sequencing	141
	3.2.3 Discovery of RNA and its sequencing—a combined effort by a group of researchers	141
	3.2.4 Genome sequencing	147
	3.2.5 Dark DNA	147
	3.2.6 Categorization of all living organisms into two major divisions: the cellular and the viral "empires" and three primary cellular domains—archaea, bacteria, and eukarya	148
3.3	Origins of life on Earth—importance of organic molecules	150
3.4	Life and living systems—interpretations	152
3.5	Why do a few million years or more are necessary for evolution from prebiotic chemical phase to biological phase?	153
3.6	Understanding the evolution of life	153
3.7	Influence of thermodynamic disequilibrium on life	155
3.8	Extraterrestrial life in the Solar System—implications of Kumar's hypothesis	156
3.9	Looking for possibility of extraterrestrial life in the Solar System—deriving clues from early Earth's conducive atmosphere for beginning of abundant life colonizing the Earth	159
References		160
Bibliography		166

4. Biosignatures—The prime targets in the search for life beyond Earth

4.1	Life	167
4.2	Use of fossil lipids for life-detection	169
4.3	Biosignatures	169
	4.3.1 Biosignatures of microorganisms	170
	4.3.2 Chemical biosignatures	170
	4.3.3 Morphological biosignatures	172
4.4	Serpentinization—implications for the search for biosignatures	174

4.5	Biosignatures versus bioindicators	175
4.6	Life and biomarkers	175
	4.6.1 Biomarker	175
	4.6.2 The search for life on Mars	176
	4.6.3 A potential biomarker identified on Venus	177
4.7	Identification of biosignature in Antarctic rocks	177
4.8	Existence of biosignatures under diverse environmental conditions	178
4.9	Characterizing extraterrestrial biospheres through absorption features in their spectra	179
4.10	Means of studying biosignatures	180
	4.10.1 Identification of stromatolites using portable network graphics analysis of layered structures captured in digital images	180
	4.10.2 Characterization of molecular biosignatures using time-of-flight secondary ion mass spectrometry	181
4.11	Detecting biosignature gases on extrasolar terrestrial planets	186
4.12	False positives and false negatives	189
4.13	Potential biosignatures—molecules that can be produced under both biological and nonbiological mechanisms but selectively/uniquely attributable to the action of biology	190
4.14	Atmospheric chemical disequilibrium (a generalized biosignature)—a proposed method for detecting extraterrestrial biospheres	192
4.15	Identification of amino acids in Murchison meteorite and Atarctic micrometeorites	192
4.16	Major challenges lurking in the study of extrasolar biosignature gases	194
	References	194
	Bibliography	200

5. Extremophiles—Organisms that survive and thrive in extreme environmental conditions

5.1	Relevance of astrobiology	201
5.2	Habitability	201
5.3	Importance of liquid water in maintaining habitability on celestial bodies	205
5.4	Habitability of extremophilic and extremotolerant bacteria under extreme environmental conditions	206
5.5	Why do extremophiles survive in extreme environments? Application of exopolymers derived from extremophiles in the food, pharmaceutical, and cosmetics industries	209
5.6	Microbial life on and inside rocks	211
5.7	Microbial life beneath the seafloor	214
5.8	Microbial life in Antarctic ice sheet	216
5.9	The year-2021 discovery of sessile benthic community far beneath an Antarctic ice shelf	217
5.10	Microbial life at the driest desert in the world	218
5.11	Tardigrades—Important extremophiles useful for investigating life's tolerance limit beyond earth	218
	5.11.1 Temperature tolerance in tardigrades	219
	5.11.2 Desiccation tolerance in tardigrades	220
	5.11.3 Radiation tolerance in tardigrades	221
	5.11.4 Dormancy strategies in tardigrades	223
	5.11.5 Ability of tardigrades to cope with high hydrostatic pressure	224
	5.11.6 Effect of extreme environmental stresses on tardigrades' DNA	224
5.12	Role of tardigrades as potential model organisms in space research	225
5.13	Discovery of a living Bdelloid Rotifer from 24,000-year-old Arctic permafrost	226
5.14	Archaea—single-celled microorganisms with no distinct nucleus—constituting a third domain in the phylogenetic tree of life	227
	5.14.1 The intriguing history of the discovery of archaea	227
	5.14.2 General features of archaea	230
	5.14.3 Unique feature of archaea	230
	5.14.4 Diverse sizes and shapes exhibited by archaea	230
	5.14.5 Extremophile archaea—halophiles, thermophiles, alkaliphiles, and acidophiles	231
	5.14.6 Extreme halophilic and hyperthermophilic archaea	231
	5.14.7 Implications of studies on archaea for the search for life on extraterrestrial worlds	235
5.15	How do extremophiles survive and thrive in extreme environmental conditions—clues from study of the DNA	236
5.16	Revival of panspermia concept encouraged by the discovery of survival limits of tardigrades in high-speed impacts	237
	5.16.1 Panspermia concept	237
	5.16.2 Ability of tardigrades to survive high-speed impact shocks	239
	References	242
	Bibliography	251

6. **Salinity tolerance of inhabitants in thalassic and athalassic saline and hypersaline waters & salt flats**

   6.1 Salinities in thalassic and athalassic water bodies—survival and growth of living organisms in saline and hypersaline environments   255

   6.2 Classification of organisms as osmo-regulators and osmo-conformers   260

   6.3 Life in thalassic water bodies   260
       6.3.1 Life in the oceans, seas, gulfs, and bays   260
       6.3.2 Life in thalassic brackish water bodies   272

   6.4 Life in athalassic brackish water bodies   276
       6.4.1 Life in the Caspian Sea   276
       6.4.2 Life in lake texoma   278

   6.5 Life in athalassic hypersaline water bodies   280
       6.5.1 Life in the Dead Sea—situated between Israel and Jordan   281
       6.5.2 Life in Great Salt Lake in the state of Utah in the United States   287
       6.5.3 Meagre microbial life in the hypersaline Don Juan Pond in Antarctica—the most saline water body on Earth   289
       6.5.4 The fauna of Athalassic salt lake at Sutton & Saline Pond at Patearoa in New Zealand   290
       6.5.5 Importance of the genus bacillus   291
       6.5.6 Microbial life in the hypersaline lake Assal in Djibouti in the Horn of Africa—the most saline hypersaline lake outside the Antarctic continent   291

   6.6 Life in lithium chloride-dominated hypersaline salt flats & ponds belonging to the lithium triangle zone (Argentina, Bolivia, and Chile) of South America   292
       6.6.1 Effects of lithium on microbial cells   293
       6.6.2 Microbial diversity with the presence of fungi, algae, and bacteria in the lithium-rich hypersaline environment of Salar del Hombre Muerto Salt Flat, Argentina   293
       6.6.3 Presence of bacteria and archaea in the lithium-rich hypersaline environment of Salar de Uyuni, Bolivia—the largest salt flat on Earth   295
       6.6.4 Life in lithium chloride-dominated hypersaline environments of Atacama hypersaline lakes & salt flats in Chile   298
       6.6.5 Importance of Atacama desert in astrobiological study of Mars   302

References   302
Bibliography   309

7. **Terrestrial analogs & submarine hydrothermal vents—their roles in exploring ocean worlds, habitability, and life beyond earth**

   7.1 Terrestrial analogs—their importance in understanding the secrets of extraterrestrial worlds   311
       7.1.1 Ice on Earth—analog for possible microbial life on extraterrestrial icy ocean worlds   311
       7.1.2 The NASA OCEAN project—an ocean-space analog   313
       7.1.3 Microbial communities colonizing on terrestrial submarine hydrothermal vents and volcanic rocks—analogs for life on early Earth and possible life on *Enceladus* & *Europa*   313
       7.1.4 Terrestrial lava tubes—analogs of underground tunnels on Earth's moon and Mars   314
       7.1.5 Ukrainian rocks—terrestrial analogs for botanical studies in simulation experiments involving possible growth of plants on Earth's moon   317
       7.1.6 Atacama desert in Chile—a terrestrial analog of Mars to examine its habitability conditions   318
       7.1.7 International efforts to map the distribution of extremophiles across the globe   318
       7.1.8 Extreme acidic environment of Rio Tinto basin—terrestrial analog of Mars to understand its microbial survival during its evolution   319
       7.1.9 Antarctic Ross Desert—terrestrial analog of Mars to understand its ecosystem and habitability conditions for different stages in its evolution   321
       7.1.10 Pingos on Earth—tools for understanding permafrost geomorphology on Mars   322

7.1.11	Spotted Lake (a hypersaline sulfate lake), British Columbia, Canada—a terrestrial analog of saline water bodies of ancient *Mars*: astrobiological implications	328
7.1.12	Hypersaline springs on Axel Heiberg Island, Canadian High Arctic—a unique analog to putative subsurface aquifers on *Mars*	330
7.1.13	Subglacial Lake Vostok in Antarctica and Mariana Trench in the Pacific Ocean—terrestrial analogs of Jupiter's Moon *Europa*	333
7.1.14	Methanogens and ecosystems in terrestrial volcanic rocks—terrestrial analogs to assess the plausibility of life on Saturn's Moon *Enceladus*	333
7.1.15	Bubbles bursting in Earth's oceans—terrestrial analogs in studying the transport of organics from *Enceladus*'s subsurface ocean to its surface and along its south polar jet	334
7.1.16	Haughton Crater in the Canadian Arctic Desert—a terrestrial analog for the study of craters on *Mars* and Saturn's Moon *Titan*	335
7.1.17	Terrestrial Salt Diapirs & Dust Devils—terrestrial analogs to study cantaloupe terrain & geyser-like plumes on the surface of Neptune's moon *Triton*	335
7.1.18	Earth's biosphere—terrestrial analog in the search for life on *Exoplanets*	336
7.2	Role of submarine hydrothermal vents in the emergence and persistence of life on Earth—their astrobiological implications	336
7.2.1	General features of terrestrial submarine hydrothermal vents	336
7.2.2	Rich microbial and faunal ecosystems harbored by submarine hydrothermal vents	344
7.2.3	Role of geothermal energy in driving deep-sea hydrothermal vent ecosystem	348
7.3	Discovery of indications of hydrothermal vents adorning the subsurface seafloors of *Enceladus* and *Europa*—possibility of life & habitability	348
7.3.1	Direct evidence for submarine hydrothermal vents on *Enceladus*	349
7.3.2	Submarine hydrothermal activity on *Europa*—inferences gleaned from telescopic observations and laboratory experiments	350
References		**350**
Bibliography		**358**

8. Surface environment evolution for Venus, Earth, and Mars—the planets which began with the same inventory of elements

8.1	Relevance of examining the surface environment evolution for Venus, Earth, and Mars	359
8.2	Specialties in the geologic activities of the three planets—Stagnant, Episodic, and Mobile Lid Regimes	359
8.3	Size and composition of Venus, Earth, and Mars	362
8.3.1	Earth's size and composition & differences with Mars	362
8.3.2	Venus and Earth—twins in terms of size & composition	363
8.3.3	How large is Mars & what is it made of?	364
8.4	Atmospheres of Venus, Earth, and Mars	365
8.4.1	Earth's atmosphere	365
8.4.2	Venus' atmosphere and clouds	366
8.4.3	Atmospheric composition—Mars versus Earth	368
8.4.4	Influence of mineral dust on Martian weather	368
8.5	Dust devils and vortices	370
8.5.1	Dust devil on Earth	370
8.5.2	Dust devil on Mars	371
8.5.3	Vortex on Venus	373
8.6	Distances of Venus, Earth, and Mars from Sun, and their current surface temperatures	374
8.6.1	Gradually growing distance between Earth and Sun over time	375
8.6.2	Distance of Mars from Sun and Earth & temperature differences	376
8.6.3	Venus' distance from Sun & its brightness and temperature profile	376
8.7	Volcanism and surface features of Venus, Earth, and Mars	377
8.7.1	Volcanism of Venus	377
8.7.2	Earth's volcanism	378
8.7.3	Volcanism on Mars—differences with terrestrial volcanic styles	379
8.8	Lava tubes on Earth and Mars	380

8.9 History of water on Mars and Venus 384
    8.9.1 Water once flowed on Mars—indications of hydrated minerals and clay, possible ocean, and vast river plains on martian surface 384
    8.9.2 Venus' dry surface 391
8.10 Magnetic fields of Earth, Venus, and Mars—intrinsic & induced magnetic fields 391
    8.10.1 Earth's magnetic field—intrinsic magnetic field 391
    8.10.2 Venus' unusual magnetic field—induced magnetosphere 391
    8.10.3 Induced magnetosphere of Mars—comparison with Venus and Titan 392
References 393
Bibliography 397

9. Lunar explorations—Discovering water, minerals, and underground caves and tunnel complexes

9.1 Orbiter- and lander-based explorations of Earth's Moon 399
9.2 A peep into the early history of the study of the origin of Earth's Moon 400
9.3 Origin of Earth's Moon—evidences from lunar observations 402
9.4 Differing preliminary views on the existence of water on Earth's Moon 403
9.5 Experimental search for clues to the existence of water on Earth's Moon 405
9.6 United States' Apollo missions culminating in the landing of Man on Earth's Moon 407
9.7 The year-2007 discovery of water molecules trapped/chemically bound in lunar rock samples 411
9.8 Discovery of adsorbed hydrogen and hydroxyl in volatile content of lunar volcanic glasses—additional clues to the presence of water on Earth's Moon 412
9.9 Identifying presence of water on Earth's Moon through spacecraft-borne spectroscopic measurements 414
9.10 The year-2009 confirmation of the presence of water on Earth's Moon by India's Moon Impact Probe (MIP) and America's Moon Mineralogy Mapper (M^3) carried by India's Chandrayaan-1 Lunar Probe 417
    9.10.1 Providing complete coverage of the Moon's polar regions—acquiring images of peaks and craters and conducting chemical and mineralogical mapping of the entire lunar surface 418
    9.10.2 Probing the poles in search of ice and water and providing confirmation of regolith hydration everywhere across the lunar surface 419
9.11 Permanently shadowed regions (cold traps) of the lunar poles—initial inferences 420
9.12 Role of NASA's LRO and LCROSS in confirming presence of water in the southern lunar crater *Cabeus* 420
9.13 Discovery of frozen water in permanently shadowed craters and poles of Earth's Moon 422
9.14 Water in the lunar interior 424
    9.14.1 Detection of water existing deeper within the lunar crust and mantle through analyses of melt inclusions that originated from deep within the Moon's interior 425
    9.14.2 Indications of past presence of water in the lunar interior (within the lunar mantle) from lunar convective core dynamo considerations 427
    9.14.3 Identification of water in the lunar interior through remote sensing of impact crater *Bullialdus* 429
9.15 Origin of water on Earth's Moon 430
9.16 The last phase of Chandrayaan-1 mission—the lunar probe lost and found 431
9.17 Discovery of subsurface empty lava tubes and caves on Earth's Moon—potential shelters for human settlement? 432
    9.17.1 Sinuous lunar rilles 432
    9.17.2 Underground caves and lava tubes on Earth's Moon—formation processes, strength, and durability 432
    9.17.3 Underground caves and tunnel complexes as potential shelters for lunar habitats 433
    9.17.4 Technical considerations in planning the use of lava tubes to house manned lunar bases 434
9.18 Lunar samples—what all tales do they tell the world about Earth's Moon? 434
    9.18.1 Protection of returned lunar samples from terrestrial contamination 434
    9.18.2 Fine-tuning the models for the origin of Earth's Moon 435
    9.18.3 First proposing and then disproving the "Dry Moon Paradigm" 436
    9.18.4 Understanding how and when large basins on the near side of the Earth's Moon were created 436
    9.18.5 Application of Apollo lunar samples in plant biology experiments 436

9.19 China's year-2019 landmark success in placing a lander and rover on Earth's Moon's far side for scientific data collection 437
    9.19.1 Importance of exploring lunar far side 437
    9.19.2 Establishing a relay satellite at the "halo orbit" around Lagrange point 2 of the Earth–Moon system—the first step before sending a data collection system to the far side of Earth's Moon 439
    9.19.3 The Chang'e-4 mission lander and rover 442
    9.19.4 Unveiling of Earth's Moon's far side shallow subsurface structure by China's Chang'E-4 lunar penetrating radar 443
9.20 Cotton seeds carried to Earth's Moon by China's Chang'e-4 probe—the first-ever to sprout on Earth's nearest neighbor 445
References 446
Bibliography 452

10. Liquid water lake under ice in Mars's southern hemisphere—Possibility of subsurface biosphere and life

10.1 Mars' glorious past 453
10.2 Mars—once a water-rich planet whose surface dried later 454
10.3 Favorable position of Mars in Sun's habitable zone 456
10.4 Water-bearing minerals on Mars, its hydrologic history, and its potential for hosting life 458
10.5 Evidence for hydrated sulfates on Martian surface—Biological implication 461
10.6 Day–night variations in liquid interfacial water in Martian surface 463
10.7 Clues leading to existence of water at Mars' subsurface 464
10.8 Searching for presence of liquid water at the base of the Martian polar caps using ground-penetrating radar on *Mars Express* spacecraft 465
10.9 The year-2018 discovery of an underground liquid water lake in Mars' southern hemisphere 466
10.10 Indications suggestive of Mars' subsurface harboring a vast microbial biosphere 468
    10.10.1 Prokaryotic life in Earth's deep subsurface offers clues to Mars' subsurface microbial biosphere 468
    10.10.2 Model studies suggestive of subsurface Martian biosphere 469
    10.10.3 Arguments in support of a deep biosphere on Mars 469
    10.10.4 Early Mars' potential as a better place for the origin of life compared to early Earth 470
10.11 Search for biosignatures and habitability on early Mars—*ExoMars* mission 471
    10.11.1 *ExoMars* mission to Mars—An astrobiology program of the European Space Agency (ESA) and the Russian Space Agency Roscosmos 471
    10.11.2 Prospects of lava tube caves on Mars as an extant habitable environment 473
    10.11.3 Mars' anticipated habitability potential 475
    10.11.4 Mars landing site selection criteria 475
10.12 When could life have probably arisen on early Mars? Clues from early Earth environments and biosignatures 476
10.13 Search for life on Mars under the first NASA Mars Scout mission—The Phoenix mission 478
10.14 Search for traces of life on Mars—NASA's biology experiments 479
10.15 Exploring Mars under United States' Viking Lander missions—Discovering volcanoes, lava plains, giant canyons, craters, wind-formed features, and seasonal dust storms 480
10.16 Looking for signs of primitive life on Mars 481
    10.16.1 Arguments favoring possibility of life on Mars 481
    10.16.2 Probable ancient life on Mars: Chemical arguments and clues from Martian meteorites 482
    10.16.3 UV is not always fatal—Examples from studies of haloarchaea and spore-forming bacteria 484
10.17 Exploring Mars under *Rover mission* to assess past environmental conditions for suitability for life 488
10.18 Mangalyaan—India's Mars Orbiter Mission (MOM) spacecraft—Exploring Mars' surface features, morphology, mineralogy, and atmosphere 489
10.19 Imaging of Mars' surface, its dynamical events, and its Moons *Phobos* and *Deimos* using Mars Color Camera (MCC) onboard MOM 491
    10.19.1 Atmospheric optical depth estimation in Valles Marineris—A huge canyon system 492

10.19.2 Imaging of Mars' twin Moons—*Phobos* and the far side of *Deimos* 493
10.19.3 Morphology study of *Ophir Chasma* canyon 494
10.19.4 Automatic extraction, monitoring, and change detection of area under polar ice caps on Mars 494
10.19.5 Application of reflectance data derived from a differential radiometer and the MCC on-board MOM in studying the mineralogy of Martian surface 494
10.20 Solar forcing on the Martian atmosphere and exosphere 495
10.21 Emission of thermal infrared radiation from Mars 496
10.22 D/H ratio estimation of Martian atmosphere/exosphere 497
10.23 Success achieved by MOM 498
10.24 Perchlorates on Mars 499
10.24.1 Use of perchlorates as an energy source—Example by Antarctic microbes 499
10.24.2 Effects of perchlorate salts on the function of a cold-adapted extreme halophile from Antarctica—No significant inhibition of growth and enzyme activity 500
10.24.3 Bacteriocidal effect of UV-irradiated perchlorate 500
10.24.4 Bacteriocidal effect of perchlorate salts under Martian analog conditions 502
10.24.5 Interactions of other Martian soil components 503
10.24.6 Implications of perchlorate detection on Mars 505
10.25 Hydrogen peroxide—Negatively impacting the habitability of Mars? 506
10.26 Exploring organic substances in the Martian soil 506
10.27 A ray of hope in Mars' horizon—Possibility of halophilic life on Mars 510
References 511
Bibliography 521

11. **Could near-Earth watery asteroid Ceres be a likely ocean world and habitable?**

11.1 Near-Earth asteroids *Ceres* and *Vesta*—why did they become targets of NASA discovery-class mission, *Dawn*? 523
11.2 An overview of the dwarf planet *Ceres* 526
11.3 Indications favoring the presence of water on *Ceres* 528
11.4 Application of reflectance spectroscopy for remote detection of water molecules on *Ceres* 529
11.5 Clues leading to the presence of water on the surface of *Ceres* 530
11.6 Inferring presence of water in the subsurface of *Ceres* 532
11.7 The year-2018 discovery of a seasonal water cycle on *Ceres* 532
11.8 Hint of an ocean hiding below the surface of *Ceres* 533
11.9 Possibility of hydrothermal geochemistry to take place on *Ceres* 535
11.10 Presence of organics on and inside *Ceres* 536
11.11 *Ceres*—a candidate ocean world in the asteroid belt 537
11.12 Habitability potential of *Ceres* 537
References 539
Bibliography 544

12. **An ocean and volcanic seafloor hiding below the icy crust of Jupiter's Moon *Europa*—Plumes of water vapor rising over 160 km above its surface**

12.1 Planet Jupiter and its water-bearing moons 545
12.1.1 Jupiter—the largest and the most massive planet in the solar system 545
12.1.2 NASA's Galileo mission to Jupiter and its mysterious moons 548
12.1.3 Searching for water vapor in Jupiter's atmosphere 548
12.1.4 Magnetic field production and surface topography of *Ganymede* 549
12.1.5 Water and oxygen exospheres of *Ganymede* 550
12.2 An overview of *Europa* 551
12.3 Jupiter's icy moons *Europa*, *Ganymede*, and *Callisto* containing vast quantities of liquid water 553
12.4 Oxygen, water, and sodium chloride on *Europa* 556
12.5 Understanding the mysteries of *Europa* 557
12.6 An approx. 100 km deep subsurface salty ocean hiding beneath *Europa*'s ice shell 559
12.7 Indirect methods for detecting an ocean hidden within *Europa* 560

12.7.1 Radar-based active detection—practical constraint 561
  12.7.2 Radar-based passive detection—an ingenious method 561
12.8 Direct methods for detecting *Europa*'s Ocean—plans in the pipeline 564
12.9 Lakes & layer of liquid water in & beneath *Europa*'s hard icy outer shell 565
12.10 Europa adorning a volcanic seafloor—the year 2021 numerical model prediction 566
12.11 Plumes of water vapor erupting from *Europa*'s surface 567
12.12 Saucer-shaped sills of liquid water in *Europa*'s ice shell 573
12.13 *Europa Clipper Mission* aimed at *Europa*'s exploration 573
12.14 Prospects of extant life on *Europa* 574
12.15 Future exploration of Europa in the pipeline 575
References 576

13. **Salty ocean and submarine hydrothermal vents on Saturn's Moon *Enceladus*—Tall plume of gas, jets of water vapor & organic-enriched ice particles spewing from its south pole**

13.1 Saturn and its glaring similarities and dissimilarities with Earth 583
13.2 Exploring Saturn and its two major moons (*Enceladus* and *Titan*)—importance of Cassini spacecraft mission 584
13.3 Presence of a global subsurface ocean on *Enceladus*—inferences from gravitational field measurements and forced physical wobbles in its rotation 586
13.4 Liquid water ocean hiding below the icy crust of *Enceladus*—inference drawn from plumes of water vapor & salty ice grain jutting out from the surface of *Enceladus* 587
13.5 Maintaining liquid oceans inside cold planets and moons—role of tidal heating 594
13.6 Eruptions in the vicinity of Enceladus' south pole—role of tidally driven lateral fault motion at its south polar rifts 596
13.7 Presence of macromolecular organic compounds in the subglacial water-ocean of *Enceladus* 597
13.8 Mechanisms driving the cryovolcanic plume emission from the warm fractures in *Enceladus* 602
13.9 Hydrothermal vents on the seafloor of *Enceladus*—possibility for harboring an ecosystem based on microbial populations 604
13.10 Resemblance of *Enceladus*'s organics-rich ocean to earth's primitive prebiotic ocean—a favorable scenario for life's emergence on *Enceladus* 606
13.11 Science goals and mission concept for future exploration of *Enceladus* 608
13.12 Implementable mission concepts to further explore Enceladus in the near future—Enceladus life finder (ELF) mission 610
References 612
Bibliography 615

14. **Hydrocarbon lakes and seas & internal ocean on *Titan*—Resemblance with primitive earth's prebiotic chemistry**

14.1 Titan—an earth-like system in some ways 617
14.2 Pre-Cassini mission knowledge of Titan 618
  14.2.1 Titan's dense atmosphere 618
  14.2.2 Size and shape of aerosol particles in Titan's atmosphere—results from voyager 1 and 2 missions 619
  14.2.3 Understanding the mechanisms of aerosol particle building in Titan's atmosphere 620
  14.2.4 Chemical transition of simple organic molecules into aerosol particles in Titan's atmosphere 621
  14.2.5 Understanding particle size distribution in Titan's hazy atmosphere 622
  14.2.6 Gaining Insight on the vertical distribution of Titan's atmospheric haze 622
  14.2.7 Greenhouse and antigreenhouse effects on Titan 622
  14.2.8 Inferring clues on Titan's surface 622
14.3 Role of Cassini spacecraft and Cassini–Huygens probe in understanding Titan better 623
  14.3.1 Cassini spacecraft mission to explore planet Saturn and some of its icy moons 623
  14.3.2 The European space agency's Huygens probe to explore Titan's hazy atmosphere and its surface 623
  14.3.3 Landing of Huygens probe on Titan's surface 625
14.4 Gaining better understanding of Titan based on data gleaned from Cassini spacecraft mission 625
  14.4.1 Titan's clouds, storms, and rain 626
  14.4.2 Detection of tall sand dunes in the equatorial regions of Titan 627

14.4.3 Understanding Titan's ionosphere using Cassini plasma spectrometer (CAPS)   629
14.4.4 Confirming the existence of strong winds in Titan's atmosphere   629
14.4.5 Knowing more on an unusual atmosphere surrounding Titan   630
14.4.6 Making the first direct identification of bulk atmospheric nitrogen and its abundance on Titan   631
14.4.7 Obtaining evidence for formation of Tholins in Titan's upper atmosphere and understanding the process   631
14.4.8 Detection of benzene (C_6H_6) in Titan's atmosphere   632
14.4.9 Direct measurements of carbon-based aerosols in Titan's atmosphere and deciphering their chemical composition   632
14.4.10 Understanding the role of nitrogen and methane in generating the orange blanket of haze in Titan's atmosphere   635
14.4.11 Examining the contribution of polycyclic aromatic hydrocarbons (PAHs) in producing organic haze layers in Titan's atmosphere   637
14.4.12 Deciphering the particle size distribution in Titan's hazy atmosphere   639
14.4.13 Investigating the role of organic haze in Titan's atmospheric chemistry   639
14.4.14 Understanding the processes responsible for the evolution of aerosols in Titan's atmosphere   640

14.5 Organic compounds on Titan's surface   642
14.6 Hydrocarbon reservoirs, seas, lakes, and rivers on Titan   643
14.6.1 Hydrocarbon seas on Titan's surface   643
14.6.2 Likelihood of finding transient liquid water environments on Titan's surface   645
14.6.3 Subsurface hydrocarbon reservoirs on Titan   645
14.6.4 Hydrocarbon lakes on Titan and their astrobiological significance   646
14.6.5 Ammonia-enriched salty liquid–water inner-ocean hiding far beneath Titan's frozen surface   648
14.6.6 Rivers and drainage networks on Titan—comparison and contrast with those on Earth and Mars   649

14.7 Presence of a salty liquid water inner ocean hiding far beneath Titan's frozen surface—evidence gleaned from a deformable interior, Schumann resonance, and cryovolcanism   652
14.8 Gaining insight into Titan's overall similarities and dissimilarities to Earth   654
14.9 Titan's resemblance to prebiotic Earth   655
14.10 Possibility of finding biomolecules on Titan   656
14.11 Practicability of photosynthesis on Titan's surface   657
14.12 Active cycling of liquid methane and ethane on Titan   657
14.13 Cassini spacecraft's retirement in September 2017 after successful 13 years' orbiting around the Saturn system   657
14.14 Is existence of life on Titan possible? Significance of carbon-rich and oxygen-loaded fullerenes in introducing oxygen to Titan's surface chemistry   659
14.15 Viability of a nonwater liquid capable of sustaining life on Titan   660
14.16 Science goals and mission concept for future exploration of Titan   661
References   663

15. A likely ocean world fostering a rare mixing of CO and N_2 ice molecules on Neptune's Moon *Triton*

15.1 General features of *Triton*   673
15.2 Origin of *Triton*—its uniqueness among all large moons in the solar system   674
15.3 *Triton*'s surface temperature and pressure   675
15.4 Chemical composition of *Triton*'s atmosphere & surface   675
15.5 *Triton*'s nitrogen deposits—interesting consequences   676
15.6 *Triton*'s surface—among the youngest surfaces in the solar system   677
15.7 Spectral features of *Triton*'s water–ice   678
15.8 Ridges on *Triton*   679
15.9 Dust devils-like tall plumes of gas and dark material rising through *Triton*'s atmosphere   679
15.10 Likely existence of a 135–190 km thick inner ocean at a depth of ~20–30 km beneath *Triton*'s icy surface   681
15.11 The Year-2019 discovery of *Triton* fostering a rare mixing of CO and N_2 ice molecules—shedding more light on *Triton*'s geysers   683

15.12	*Triton* in the limelight as a high-priority target under NASA's "Ocean Worlds" program	685		
15.13	Forthcoming mission to map Triton, characterize its active processes, and determine the existence of the predicted subsurface ocean—TRIDENT Mission	685		
References		686		
Bibliography		688		

16. Subsurface ocean of liquid water on Pluto

- 16.1 Pluto and its five moons 691
- 16.2 Pluto's highly eccentric orbit—an oddity in the general scheme 694
- 16.3 Long-period perturbations in the chaotic motion of Pluto 696
- 16.4 Pluto's complex crater morphology 697
- 16.5 Controversy over Pluto's planetary status—recent arguments 697
- 16.6 Studies on Pluto system and Pluto's physical and geological features prior to the launch of "New Horizons" spacecraft 698
  - 16.6.1 Dynamics of Pluto and its largest moon Charon 698
  - 16.6.2 Three-body orbital resonant interactions among Pluto's small moons Styx, Nix, and Hydra 700
  - 16.6.3 Insolation and reflectance changes on Pluto 701
  - 16.6.4 Pluto's atmosphere 702
  - 16.6.5 Presence of a subsurface ocean on Pluto—inference based on numerical studies 703
- 16.7 "New Horizons"—the spacecraft that brilliantly probed the distant Pluto from close quarters 704
  - 16.7.1 Powering the New Horizons spacecraft 704
  - 16.7.2 New Horizons spacecraft and its science payloads 706
  - 16.7.3 Discoveries by New Horizons during its long travel to Pluto and beyond 707
  - 16.7.4 New Horizons spacecraft performing the first ever flyby of Pluto 708
- 16.8 Understanding Pluto and its moons through the eyes of *New Horizons* spacecraft 709
  - 16.8.1 Shedding light on the consequences of Pluto's high orbital eccentricity and high obliquity 709
  - 16.8.2 Achieving confirmation on Pluto's small Moons' rotation, obliquity, shapes, and color & the absence of a predicted ring system surrounding Pluto 710
  - 16.8.3 Runaway Albedo effect on Pluto 711
  - 16.8.4 Knowing more on Pluto's insolation and reflectance in the light of new data from New Horizons mission 712
  - 16.8.5 Ice-laden "Heart-Shaped" region on Pluto's surface—formation and stability of Sputnik Planitia crater 713
  - 16.8.6 Gaining insights on Pluto's SP crater basin and its surroundings 715
  - 16.8.7 Subtle topography of ice domes and troughs of cellular plains within Sputnik Planitia crater 716
  - 16.8.8 Latitudinal variations of solar energy flux on Pluto—theoretical investigations 720
- 16.9 The current state of Pluto's atmosphere 722
- 16.10 Indirect detection of subsurface ocean of liquid water inside Pluto—in support of pre-New Horizons numerical studies 723
- 16.11 Distinct topographic signatures on Pluto's "Near-Side" and "Far-Side" 725
  - 16.11.1 Deep depression enclosing Sputnik Planitia ice sheet & north-south running complex ridge-trough system along 155° meridian 726
  - 16.11.2 Bladed terrain on Pluto's "Near-Side" 726
  - 16.11.3 Bladed terrain on Pluto's "Far-Side" 727
- 16.12 Is life possible on Pluto? 729
- 16.13 Expectations from future explorations of Pluto 730
- References 731
- Bibliography 735

17. Hunting for environments favorable to life on planets, moons, dwarf planets, and meteorites

- 17.1 A new frontier in planetary simulation 737
- 17.2 Role of tidally heated oceans of giant planets' moons in supporting an environment favorable to life 738
- 17.3 Prospects of life on *Mars* and Jupiter's Moon *Europa* 738
  - 17.3.1 Role of biomineralization in providing bacteria with an effective UV screen on Mars 738

17.3.2	Redox gradients may support life and habitability on Mars	739
17.3.3	Probable life on Mars—speculations driven by biochemistry	739
17.3.4	Best possible hideout on Mars—a string of lava tubes in the low-lying Hellas Planitia	740
17.3.5	Europa—one of most promising places in the solar system where possible extra-terrestrial life forms could exist	741

17.4 In the hope of finding evidence of past life on Mars—arrival of NASA's "Mars 2020 Perseverance Rover" at Mars 742

17.5 Looking for signs of primitive life on Jupiter & Saturn and their Moons 744

- 17.5.1 Prospects of life on Europa 745
- 17.5.2 Strategies for detection of life on Europa 746
- 17.5.3 Anticipation of an Earth-like chemical balance of Europa's ocean 747
- 17.5.4 Proposal for probing subglacial ocean of enceladus in search of life—orbiter and lander missions 747
- 17.5.5 Possibility of life on Enceladus 748
- 17.5.6 The anticipated nature of life on Saturn's Moon Titan 751
- 17.5.7 Possibility of amino acids production in Titan's haze particles 753

17.6 Astrobiological potential of the dwarf planet Pluto 753

17.7 Is there a prospect for life to arise on the asteroid belt resident dwarf planet ceres? 755

17.8 Recent and upcoming missions to the extraterrestrial worlds in the solar system in search of ingredients for life 757

- 17.8.1 Upcoming missions to Earth's Moon 757
- 17.8.2 New missions to Mars initiated in 2021 758
- 17.8.3 Titan's exploration planned to launch in 2026—NASA's dragonfly mission 759

17.9 Astrobiologists' thoughts on collection of samples from plumes emitted by Enceladus and Europa 762

References 764
Bibliography 769

Chemical names and their chemical formulae 773
Short forms and their expansion 775
Definition/Meaning 781
Index 807

Foreword

Water Worlds in the Solar System, Exploring Prospects of Extraterrestrial Habitability & Life is a very useful and interesting book for scientists, teachers, and students from various disciplines of science. Dr. Antony Joseph has explored broad and complex range of topics extensively and intensively from the core of the Earth and bottom of the oceans to the vast expanses of some of the Solar System bodies, critically exploring geological, geophysical, physical, geochemical, astronomical, astrophysical, biological, and biochemical aspects. This book has opened up a great new horizon for exploration of water and the possibility of life on some extraterrestrial celestial bodies in our Solar System. Several high-quality illustrations are provided. He has also added some videos in the e-book. Furthermore, text- and picture-specific web sites are provided, which give additional resources and extends the chapters of the text to the resources of *World Wide Web*. A large variety of topics have been introduced and presented in this book in such a simple and engrossing style that even nonspecialist people who have an interest in the topics of life and extraterrestrial water worlds can understand and enjoy reading this book. When I went through the manuscript of this book, I found it to be a thought-provoking and overwhelmingly impressive multidisciplinary endeavor of its own kind.

Dr. Antony Joseph is a scientist who had a long oceanographic career associated with CSIR-National Institute of Oceanography (CSIR-NIO), Goa, India, from February 1978 to 31 July 2012. He participated in numerous oceanographic cruises onboard research vessels. He had opportunities to see several types of remarkable marine creatures; and examine the intricacies of the seafloor, some of the oceanographic processes, and several valuable living and nonliving resources.

Apart from authoring the present book Dr. Joseph has authored three other oceanographic books published by Elsevier, New York. He has to his credit chapters in edited books, articles in Encyclopedias, and research papers. He has presented several research papers at various Science & Technology Conferences conducted in India and abroad. He was also a faculty member at the Indian Academy of Scientific and Innovative Research during his career at CSIR-NIO. That is why this book is very friendly to students, teachers, and researchers.

His vast interdisciplinary knowledge and broad range of interests, making difficult subjects easy to understand; and his excitement about various topics from different disciplines of science have made it a spellbound multidisciplinary book useful for the purpose of both teaching and research. He addresses all the topics from various disciplines so extensively. My admiration for this book is extraordinary and stupendous. I congratulate him for authoring such a multidisciplinary book.

The author begins with addressing the very interesting and elegant aspects of Solar/Planetary formation and evolution. It makes sense that most large planets in our Solar System stay near the ecliptic plane. Our Solar System is believed to be about 4.6 billion years old. It is thought to have arisen from an amorphous cloud of gas and dust in space. The original cloud was spinning, and this spin caused it to flatten out into a disk shape, rather than a spherical shape. The Sun and planets are believed to have formed out of this disk, which is why, today, the planets still orbit in a single plane around our Sun.

In consonance with the title, the book provides a brief overview of the water-abundant celestial bodies in the Solar System which could possibly be habitable for life. Life, as we know it, is based on carbon chemistry operating in an aqueous environment. Life is supposed to have originated in the Earth's oceans' water when chemical composition of the primordial ocean and the atmosphere was very different from what it is today. It is a consensus belief among scientists that the origin of life and the associated biological evolution was preceded by chemical evolution, which resulted in the production of organic molecules—the building blocks of life. Thousands and millions of molecules were assembled into structures exhibiting key properties of living things: metabolism, respiration, reproduction, and the transfer of genetic information. It is highly likely that life arose in water and that the necessary building blocks were organic compounds. The origin of water on the Earth has been described in much details in a very interesting manner in this book.

The panspermia hypothesis states that the seeds of life exist all over the universe and can be propagated through space from one location to another. For millennia, this idea has been a topic of philosophical debate. However, due to lack of any validation, it remained merely speculative until few decades ago. It is only with the recent discoveries and advances from different fields of research that panspermia has been given serious scientific consideration. Most of the major barriers against the acceptance of panspermia have been demolished when it has been shown that microorganisms can survive the high impact and velocity experienced during the ejection from one planet, the journey through space, and the impact process onto another world.

"Search for life" researchers are optimistic about the possible existence of life on some other planets and moons (e.g., Mars, Jupiter's icy moon Europa, Saturn's icy moons Enceladus and Titan) and even some extra-solar celestial bodies. While discoveries of liquid water bodies on a couple of distant extraterrestrial planets and moons are encouraging news, the Year-2018 discovery of subsurface liquid water lake on our nearest planet *Mars* is a good reason for extra cheers to every extraterrestrial enthusiast. The year-2021 discovery of macro- and mega-benthic communities far beneath an Antarctic ice shelf sheds optimism in terms of its astrobiological implications.

A biosignature is any substance, element, molecule or feature that can be used as evidence for past or present life. Biosignatures can also be organic molecules made by life. Another types of biosignatures are stromatolites which are microbial mats in shallow waters creating layered structures of minerals. Fossilized versions of microbial mats from billions of years ago are found today. Study of life here on Earth and knowing about the kinds of biosignatures will help us a lot as we keep questioning ourselves if we're alone in the universe and whether or not we might find signs of life on other celestial bodies. Dr. Antony has described in details about biosignatures in a separate chapter of his current book.

Extinction events in Earth's history, and Palaeocene–Eocene Thermal Maximum (PETM) event that occurred ~56 million years ago (Mya) resulting into the estimated global temperature increase in the range of 5–8°C, has also been described very well in this book. Serpentinization, which is the process of formation of serpentine minerals and considered to be responsible for CH_4 generation, contributed the bricks and mortar for many prebiotic reactions that could have led to the first proto-organisms. The process of serpentinization, as observed along mid-ocean ridge and producing hydrothermal fluids, gaseous methane and hydrogen, has received much attention by the astrobiologists.

Moon-forming impacts might have played a crucial role in the origin of Earth's atmosphere and ocean's water as well as some of the environmental conditions of the early Earth, which might have influenced the origin and evolution of life. Scientists now have reliable information that it is the cosmological evolution which has guided matter from simplicity to complexity; from inorganic to organic; and the origin and evolution of life which is a natural result of the evolution of that matter. How the warping of cosmic fabric of space-time can guide origin and evolution of life is described in Kumar's hypothesis which affirms that the organization of DNA molecule—the blueprint of life—is correlated to the astronomical organization of planets and moons in the Solar System through an elegant mathematical relationship. Life is supposed to have started immediately after the original accretion of the Earth, with a period of chemical evolution in a reducing primitive atmosphere of the newly formed Earth which resulted in the formation of small organic molecules that constitutes the building blocks essential for life; namely, amino acids, nucleic acids.

Evidence for biological activity can be derived from carbon isotopes because a high $^{12}C/^{13}C$ ratio is characteristic of biogenic carbon. Microfossil record only extends to ~3.5 BYA and the chemofossil record to ~3.8 BYA. Besides this, evidence of life on the Earth is manifestly preserved in the rock record as well. Given the temporal limits of rock record (~4 billion years), the detrital zircons as old as 4.38 billion years have been documented. Zircons, serving as time capsules, can capture and preserve their environment. There are reports on $^{12}C/^{13}C$ of graphite preserved in 4.1 billion years old zircon grains imaged by transmission X-ray microscopy. The carbon contained in the zircon has a characteristic signature—a specific $^{12}C/^{13}C$ ratio—that indicates the presence of photosynthetic life. So, the most recent research reported argue that the graphite's ^{12}C-rich isotopic signature may be evidence for the origin of life on the Earth by 4.1 BYA, based on examination of zircons discovered from Jack Hills, West Australia. The discovery indicates that life may have begun shortly after the Planet formed 4.54 BYA.

Scientists are of the view that atoms, stars and galaxies self-assembled out of the fundamental particles produced by *Big Bang*. Different combinations of the atoms compose both the inanimate and the living worlds. Life is a complex system of atoms and molecules undergoing natural selection. The search for life on other extraterrestrial worlds is often based on the strategy of finding the water. The research further suggests that life in the Universe could be abundant. Simple life on the Earth appears to have formed quickly, but it likely took many millions of years to evolve complex organisms. Keeping in view of the core subject matter of this book being "Water Worlds in the Solar System," a brief overview of the water-abundant celestial bodies in the Solar System is given adequate attention, highlighting the importance of understanding Earth's oceans in the search for microbial life in extraterrestrial ocean worlds. Dr. Antony Joseph has written the present comprehensive and scholarly reference book on complex topics addressed in 17 chapters.

Surprisingly, the question of the origin of Earth's Moon remained unresolved even after the scientifically productive Apollo missions, and it was only in the early 1980s that a model emerged—the Giant-Impact hypothesis—that eventually gained the support of most lunar scientists. It is now well recognized that during the formation of the Solar System, the Earth and the other terrestrial planets (i.e., Mercury, Venus, Earth, and Mars that are composed primarily of silicate rocks or metals) accreted in a gas-free environment, mostly from volatile-depleted planetesimals which were already differentiated into metallic cores and silicate mantles.

There are several celestial bodies in our Solar System where water is abundant and in most cases the surface water is in the form of frozen solid ice, and the liquid water is stored in subsurface oceans. It is now known that Earth's Moon possesses water not only in its poles and regolith, but also in its interior. Lunar scientists think that the caves and large tunnel complexes discovered on Earth's Moon could be harboring some form of life; they can even be potential sites for human settlement.

Earth's neighboring planet Mars was once rich in water. In the past, water had caused substantial erosion of the Martian surface, including extensive channeling and associated transport and deposition of material. It was found that water is present at the poles and in permafrost regions of Mars. Therefore, scientists think that this body of water could be a possible habitat for microbial life.

Near-Earth asteroid *Ceres* has given several indications favoring the presence of water. Jupiter's icy moons contain vast quantities of liquid water. Ganymede and Europa are known to possess oxygen and water. There are scientific reasons to believe that a liquid-water ocean lies hidden within Europa's hard ice shell. Search-for-life researchers consider that there are prospects of extant life on Europa. Apart from Jupiter's moons Ganymede and Europa, two of Saturn's moons *Enceladus* and *Titan* are also known to possess conducive environments to support microbial life. There is striking resemblance of Enceladus's organics-rich ocean to Earth's primitive prebiotic ocean, indicating a favorable scenario for life's emergence on Enceladus. The researchers have postulated a strong possibility of finding biomolecules on Titan. Neptune's moon *Triton* and the dwarf planet Pluto orbiting in the outer periphery of the Solar System are also endowed with the ocean of liquid water.

Microbes living at submarine hydrothermal vents do not depend on sunlight or oxygen. They grow around the vents and feed upon hydrogen sulfide gas (H_2S) and dissolved minerals to produce organic material through the process of chemosynthesis. There are some extraterrestrial ocean worlds, in which submarine hydrothermal vents are present on the seafloor (e.g., Saturn's icy moon *Enceladus*). When considering the habitability of an ocean world, especially one far from the Sun where photosynthesis is unlikely to be effective, a key issue is the nature of the underlying seafloor and water–rock interactions (e.g., submarine hydrothermal vent environment). Although the full diversity of settings observed on the Earth may not be repeated on any single other ocean world, an advantage provided by our home planet—Earth—is the opportunity to conduct a wealth of analogue studies spanning a broad range of conditions that may be relevant to other ocean worlds in terms of seafloor pressure, maximum water temperatures at those pressures, and interactions with different rock types. It is pertinent to note that the Earth's mid-ocean ridges (where most of the hydrothermal vents are located) remain 80% unexplored.

NASA's Ocean Worlds Exploration Program (OWEP) envisages to explore ocean worlds in the outer Solar System bodies that could possess subsurface oceans to assess their habitability and to seek biosignatures of simple extraterrestrial life. The various projects undertaken by the United States' NASA, European Space Agency (ESA) and some universities and space research organizations from various countries to search for planets that can sustain life has led to an increased curiosity in the 21st century about life beyond Earth.

When it comes to the topic of life, based on DNA sequence comparisons and structural and biochemical comparisons, all living organisms are categorize into three primary domains; namely, Bacteria, Archaea, and Eukarya (or Eukaryotes). Both Bacteria and Archaea are prokaryotes, single-celled microorganisms with no nuclei, and Eukarya includes humans and all other animals, plants, fungi, and single-celled protists—all organisms whose cells have nuclei to enclose their DNA apart from the rest of the cell. The fossil record indicates that the first living organisms were prokaryotes, and eukaryotes evolved billion years later.

Because life on Earth depends on a variety of biochemical respiratory chains involving electron transport, it has been argued that the presence of electrochemical gradients is one of the most critical prerequisites for life's initiation forming intense selection pressures on initial evolution. Genetic analysis of the respiratory chains in the Archaea, Eukarya, and Bacteria indicates that terminal oxidases are linked to oxygen, nitrate, sulfate, and sulfur which were all present in the last common ancestor (LCA) of living organisms. Note that oxidase is an enzyme which promotes the transfer of a hydrogen atom from a particular substrate to an oxygen molecule, forming water or hydrogen peroxide.

It is believed that our LCA must have evolved on a world in which these redox (oxidation and reduction considered together as complementary processes) gradients were present in large enough quantities to evolve and to be maintained, and in which metastable, energetic compounds could diffuse across redox boundaries. Because the efficiency of biological systems tends to improve over time as a result of the processes of random mutation and natural selection as propounded by Charles Darwin in 1859, primitive metabolic electron transport systems likely could not extract energy as efficiently from the redox couples as do those of modern organisms. Based on these observations it is argued that life is more likely to have evolved on the planet that had the broadest dynamical range of electrochemical species early in its history.

Oxygenic photosynthesis is Earth's dominant metabolism, having evolved to harvest the largest expected energy source at the surface of most terrestrial habitable zone planets. Using CO_2 and H_2O-molecules that are expected to be abundant and widespread on habitable terrestrial planets, oxygenic photosynthesis is plausible as a significant planetary process with a global impact. It is evident that the production of oxygen or oxidizing power radically changed Earth's surface and atmosphere during the Proterozoic Eon, making it less reducing than the conditions prevalent during the Archean. In addition to ancient rocks, our reconstruction of Earth's redox evolution is informed by our knowledge of biogeochemical cycles catalyzed by extant biota.

All life is organized as cells. Physical compartmentation from the environment and self-organization of self-contained redox reactions are the most conserved attributes of living things, hence inorganic matter with such attributes would be life's most likely forebear. It has been widely suggested that life based around carbon, hydrogen, oxygen, and nitrogen is the only plausible biochemistry, and specifically that terrestrial biochemistry of nucleic acids, proteins, and sugars is likely to be "universal." It is argued that many chemistries could be used to build living systems, and that it is the nature of the liquid in which life evolves that defines the most appropriate chemistry.

Water (a good solvent) plays an important role is sustaining life as we know it. For temperate, terrestrial planets, the presence of water is of great importance as an indicator of habitable conditions. Water is a good solvent, primarily because it is liquid at high temperature, and secondarily because of its chemical properties (polar, high dielectric constant). In a water molecule, two charges are present with a negative charge in the middle, and a positive charge at the ends. Thus, a water molecule (which is a polar inorganic compound) possesses an electric dipole moment, with a negatively charged end and a positively charged end. Polarity underlies a number of physical properties including surface tension, solubility, and melting and boiling points. Water is the most abundant substance on Earth and the only common substance to exist as a solid, liquid, and gas on Earth's surface. It is also the third most abundant molecule in the universe. Liquid water is both a crucial source of oxygen and a useful solvent for the generation of biomolecules.

Life is supposed to be a material system that undergoes reproduction, mutation, and natural selection. Probably the simplest, but not the only, proof of life is to find something that is alive. There are only two properties that can determine if an object is alive: metabolism (an organism's life functions, biomass increase, and reproduction) and motion. All living things require some level of metabolism. Life is basically a chemical system that is able to transfer its molecular information via self-replication and to evolve via mutations. Mutation is the change in the structure of a gene (DNA), resulting in a variant form which may be transmitted to subsequent generations.

Studies in extreme ambient conditions have changed the previous paradigm that life can only be found on pleasant Earth-like planets. In the last few decades, substantial changes have occurred regarding what scientists consider the limits of habitable environmental conditions. Our knowledge about extreme environments and the limits of life on Earth has greatly improved in recent decades. These advances have been fundamental to the development of astrobiology, which studies the origin, evolution, and distribution of life in the Universe. Numerous planetary bodies and moons in our Solar System have been suggested to sustain life (e.g., Mars, Europa, Enceladus).

Submarine seafloor hydrothermal vents are considered to have an important role in the emergence and persistence of life not only on Earth but also on worlds beyond Earth. Simulations of hydrothermal vent to study origin of life on the early *Earth*, Early *Mars*, and the *Icy Moons* of the outer planets have been done in the Laboratory. Discovery of the indications of hydrothermal vents adorning the subsurface seafloors of *Enceladus* and *Europa* is suggestive of the possibility of life & habitability.

Relevant information about a wide variety of terrestrial and extraterrestrial topics has been included in this book. New descendants of scientists are paying more attention to multidisciplinary approaches of research. We have got to acquire more knowledge, and encourage publication of more such multidisciplinary books. I believe that this book is an excellent piece of work worthy of finding a place in every library.

Prof. (Dr.) Arbind Kumar
Head and Director (Retired)
P. G. Department of Zoology,
Patna Science College Campus,
Patna University,
Patna, Bihar, India
September 28, 2022

Preface

Ever since the times of the Polish astronomer Nicolaus Copernicus (1473–1543) who discovered the theory of Sun-centric model of Solar System (i.e., placing the Sun at the center of the Solar System); Giordano Bruno (1548–1600)—Italian philosopher, astronomer, and mathematician, known for his theories of the infinite universe and the multiplicity of worlds; Johannes Kepler (1571–1630)—German astronomer, mathematician, astrologer, and natural philosopher who explained planetary motion for the first time, and well-known for formulating the famous *Fundamental Laws of Planetary Motion*; and Galileo Galilei (1564–1642)—Italian astronomer, physicist, and engineer who made a great contribution to astronomy by construction of the telescope and discovering Jupiter's 4 largest moons (Io, Europa, Ganymede, and Callisto) known as the "Galilean moons", and played a key role in the scientific revolution, matters concerning the Solar System and the Universe have been the subjects of continuing interest, speculation, and enquiry. These are also among the most challenging of all scientific problems. They are, perhaps to a unique degree, interdisciplinary, having attracted the attention of astronomers, physicists, mathematicians, geologists, geophysicists, chemists, biochemists, biologists, microbiologists, oceanographers, and even philosophers.

Prof. Stephen Hawking (1942–2018)—a brilliant British theoretical physicist and cosmologist since the world-renowned "relativity" theorist Dr. Albert Einstein—believed that in the distant future our planet Earth may not be able to sustain human life and hence, humans would need to travel in zero-gravity environments to colonize other planets and their moons. Prof. Hawking noted that a new space program should be humanity's top priority "with a view to eventually colonizing suitable planets for human habitation". Hawking warned that "We have reached the point of no return. Our Earth is becoming too small for us, global population is increasing at an alarming rate and we are in danger of self-destructing". Thus goes the vital importance of extraterrestrial exploration.

The historic landing of man on Earth's Moon on July 20, 1969—described as *"One giant leap for mankind"*—encouraged planetary scientists to thoroughly probe what lies on the surface and delve deep into what hides inside of the extraterrestrial planets and their moons.

Several marvelous things that we know now about the extraterrestrial worlds were mysteries in the past. Scientific theories are validated through observations that produce data such as, for example, temperature, chemical composition, density, gravity field, size, time, pattern, appearance, and so forth. Voluminous data are needed for analysis and hypotheses testing.

The mystery that shrouds the making of the extraterrestrial worlds is revealed by integrating simple concepts of chemistry, physics, meteorology, geology, oceanography, biology, and microbiology with computer simulations, laboratory analyses, and the data from the myriad of space missions (e.g., Lunar missions; Mars missions; Viking missions; Cassini mission; New Frontiers program, which includes the *New Horizons* mission to Pluto and the Kuiper Belt, Juno to Jupiter, and OSIRIS-REx to the asteroid Bennu) and Earth-based telescopic and spectroscopic observations.

The laws of physics and chemistry are the same on Earth and beyond Earth. Even biology is likely to be no different according to the present knowledge. Astrobiologists look at the extraterrestrial worlds under the assumption that the anticipated life there (extinct or extant) obeys the rules of terrestrial biology (water-based life). Astrobiologists consider that because water plays such an essential role in life on Earth, the presence of water is vital in the search of other habitable planets, dwarf planets, and moons.

The hydrocarbon lakes and seas discovered on Saturn's moon *Titan* is considered to have an intimate resemblance with the primitive Earth's prebiotic chemistry. Although the possible viability of a non-water liquid (liquid hydrocarbon; i.e., liquid methane and liquid ethane) capable of sustaining life on Titan is in the minds of astrobiologists, no evidence has yet been found.

If we examine the nature of studies carried out on Earth as well as beyond Earth, it is quite evident that observations and modeling are key to understanding the processes taking place on Earth and beyond Earth. Whereas in-situ and remote observations (land-based and satellite-based) and theoretical modeling are employed in terrestrial studies (e.g., oceanography, limnology, meteorology, seismology, etc.), Earth-based telescopic and spectroscopic observations and spacecraft-based astronomical observations and modeling are primarily employed in extraterrestrial studies. Quite often, well-tested models used in terrestrial studies are copied and used in extraterrestrial studies (with appropriate scale modifications). As an example, several of the theoretical model calculations applicable to terrestrial oceans will apply to similar mixing behavior within other ocean worlds as well, but the importance of these processes will scale with the depth of the ocean, the topography of the seafloor, the drivers for poleward currents, and the rotational velocity of the ocean world under consideration. The year-2021 discovery of multiple subglacial liquid water bodies below the south pole of Mars unveiled by the new data obtained from Mars Advanced Radar for Subsurface and Ionosphere Sounding (MARSIS) by the application of a method of analysis, based on signal processing procedures usually applied to terrestrial polar ice sheets, is a clear example of the utility and importance of the application of proven terrestrial research knowledge to unravel a hidden secret of Mars. In another example, three potential sea levels during tsunamis that struck the primordial Martian Ocean were determined by modeling tsunami propagation and the emplacement of thumbprint terrain by comparing the geomorphologic characteristics of the Martian deposits with the predictions of well-validated terrestrial models (scaled to Mars) of tsunami wave height, propagation direction, runup elevation, and distance.

Whether terrestrial studies or extraterrestrial studies, the results gleaned from model-based studies need to be rigorously tested by many additional measurements. Whereas sampling studies are routine in terrestrial oceanography, limnology, and geology, the only sample-based studies of the extraterrestrial worlds till date are based on meteorites that fell on Earth and the lunar samples collected by the astronauts. Landers and rovers are presently deployed on Earth's Moon and the planet Mars; however, samples collected (using these devices) from these celestial bodies will arrive on Earth years later. It is clear that the science and the methodologies used in both terrestrial and extraterrestrial studies are similar in nature.

"Do conditions favorable to the formation and sustenance of life occur elsewhere in the Solar system?" is a query that every space enthusiast throws. The *New Frontiers* program (a series of space exploration missions being conducted by NASA) has significantly transformed our understanding of the Solar System, uncovering the inner structure and composition of Jupiter's turbulent atmosphere, discovering the icy secrets of the dwarf planet Pluto's landscape, revealing mysterious objects in the Kuiper belt, and exploring a near-Earth asteroid for the building blocks of life. Through this process many things that have been considered as mysteries (considered to be beyond comprehension) in the past become marvels (astonishment-causing things on the strength of understanding).

A wealth of scientific evidences is now pouring in to suggest that oceans are not unique to Earth. We are now coming to realize that "ocean worlds" are all around us. Some planets and several planetary moons have subsurface seas and saltwater bodies (subsurface lakes, water bodies sandwiched between ice layers) that are of great interest for, among other reasons, their possible habitability. It may be recalled that the overarching goal of an Ocean Worlds exploration program as defined by the Roadmaps to Ocean Worlds (ROW) group of NASA's Outer Planets Assessment Group (OPAG) is to "identify ocean worlds, characterize their oceans, evaluate their habitability, search for life, and ultimately understand any life we find."

Recent reports of liquid water existing elsewhere in the Solar System (e.g., the year-2009 confirmation of the presence of water on Earth's Moon by India's Moon Impact Probe (MIP) and America's Moon Mineralogy Mapper (M^3) carried by India's Chandrayaan-1 Lunar Probe, which was launched in 2008; the year-2018 discovery of a sizable salt-laden liquid water lake under ice on the southern polar plain of the planet Mars; the year-2018 detection of signs of presence of water deep inside Jupiter's Great Red Spot) provide added impetus to vigorous extraterrestrial explorations.

A couple of potential extraterrestrial ocean worlds are presently known. Mars long ago was warmer and wetter, having possessed significant bodies of water, as evidenced by dry lake beds and river valleys on its surface. Whether any surface liquid water remains on Mars today has long been debated. However, as a pleasant surprise, in 2018, the European Space Agency's *Mars Express* orbiter found the first evidence for stable subsurface liquid water on Mars, below the planet's south polar ice cap. It was an exciting discovery, a lake below the red Martian soil that could, possibly, support life. Now, in a year-2021 report, additional data from *Mars Express* have not only confirmed that lake, but also found three more smaller lakes in close proximity to the first one.

Whereas Mars is categorized as an extinct ocean world, it has been found that subsurface oceans exist on some other planets' moons (e.g., Jupiter's moon *Europa* and Saturnian moons *Enceladus* and *Titan*). These are presently categorized as the confirmed present-day ocean worlds. Apart from these confirmed extraterrestrial ocean worlds, evidences have been found, suggesting that subsurface water exists also on some other moons (e.g., Jupiter's moons *Ganymede* and *Callisto*).

Various kinds of extraterrestrial exploration missions resulted in the discovery of several other extraterrestrial ocean worlds (e.g., potential candidate ocean worlds such as the dwarf planet *Pluto*, Uranus's moons *Triton*, *Miranda*, and *Ariel*; Saturn's moon *Dione*; and a dwarf planet *Ceres* in the asteroid belt), some of which are expected to be habitable environments likely to foster life. New models and analyses in conjunction with data returned from upcoming missions promise to significantly advance our knowledge of how life originated in our Solar System.

How do extraterrestrial oceans, lakes, and rivers differ from or compare with Earth's water bodies? Whereas Earth's water bodies consist of salt-water oceans, freshwater rivers and lakes, and intermediate saline estuaries and brackish water bodies, the known liquid bodies on the extraterrestrial worlds are considerably more complex in terms of chemical structure and location relative to the solid crust of the respective celestial bodies. The presence of a sufficient amount of ammonia (NH_3) in the known extraterrestrial oceans plays an important role in preventing freezing of these oceans. For example, it has been suggested that the presence of antifreeze such as ammonia in Triton's subsurface ocean help to prevent its solidification.

From physical oceanography perspective, the two major differences between Earth's oceans and the extraterrestrial oceans in the Solar System pertain to the interface of the water surface and the water temperature. On Earth's oceans, the water surface is in contact with a highly compressible and flexible atmosphere, because of which the ocean tides are free to rise and fall in consonance with the tide-generating forces induced by the gravitational pull of the neighboring celestial bodies and Earth's rotation about its own axis. However, in contrast, the surface of the chilly extraterrestrial oceans beyond Mars are hard, kilometers-thick icy crust, because of which the ocean tides are NOT free to rise and fall in consonance with the tide-generating forces. This gives rise to tidal strain and heating (apart from radiogenic heating, if any), which help to keep the subsurface ocean water in liquid form. It is this powerful stress that gives rise to the observed high-speed, tall (estimated to be about 200–300 km in height) jets of salt-laden, chemicals-rich water particles gushing out from the tiger-stripes (linear cracks) near the south pole of Saturn's icy moon *Enceladus*. It has been found that the tiger stripe region on Enceladus is warmer than the rest of its surface, indicating that the observed relative warming is induced by tidal strain. Furthermore, the water particle plume emanating from the tiger stripe region is found to exhibit Enceladus's tidal periodicity. It is particularly interesting that Saturn's moon *Titan*'s liquid hydrocarbon (ethane and methane) surface-seas and lakes have been found to resemble primitive Earth's primordial ocean.

One of the hypotheses with regard to the origin of life on Earth revolves around submarine hydrothermal vents. Such vents are ubiquitous on Earth's seafloors. Surprisingly, it has been found that the seafloors of Jupiter's moon *Europa* and Saturn's moon *Enceladus* are hydrothermally active. In fact, a clear understanding of terrestrial submarine hydrothermal vents is vital to understand such vents in the extraterrestrial worlds. Astrobiologists in search of signs of life or biochemicals required for life are pinning great hopes on such vents.

Earth remains the only inhabited world known so far, but scientists are finding that the Universe abounds with the chemistry of life. The origin of life appears to be closely tied to the formation and early evolution of the Solar System. Key questions deal with the source of abiotic organic material on the early Earth, the nature of interstellar organic material and its relationship to the observed organic compounds in the outer solar system, and the possible origin of life on Mars early in its history. Amino acids are essential to life, right from the primitive life-forms such as prokaryotes to the substantially more complex life-form, known as *Homo Sapiens* (we the humans). Scientists think that primitive life-forms could merely be the fore-runners of more advanced life-forms.

When there exists a vast Universe, there is no logical reason why life should be limited to Earth. Origin-of-life researchers consider that wherever life appears and whatever form it takes, it is likely to be governed by the same rules of evolution, ecological guilds and niches, and so on. Progress in cosmological investigations, supported by tremendous technological advancements achieved in recent decades in the fields of space observations, have brought to light the truth of some of the ancient visionaries' farsighted thinking (which proved to be "revelations" centuries later) that there is a strong possibility of existence of life elsewhere in the Universe. It may be recalled that a cosmological theoretician of the late medieval era and the Renaissance—Giordano Bruno (1548–1600)—postulated the possibility that, apart from Earth, the planets of even other stars (exo-planets) might foster life of their own (a revolutionary philosophical position known as *cosmic pluralism*), which is strongly favored by many present-day scientists, but was vehemently frowned upon in the 16th century. Bruno had to pay heavily for his visionary ideas (including cosmic pluralism) and for supporting Nicolaus Copernicus's heliocentrism (the astronomical model in which Earth and the other planets in the Solar System revolve around the Sun), which is opposed to geo-centrism (which placed the Earth at the center) and became the first martyr to the cause of free thought, at the hands of the condemnably intolerant and arrogant Roman Catholic Church of that era, when it enjoyed sweeping military and economic powers in Europe. Urban VIII, the Roman Catholic Pope, ordered and implemented Bruno's assassination in a most brutal and humiliating manner.

Turning the reader's attention to the present book, understanding the working of the worlds beyond Earth requires that we acquire a reasonably good knowledge about the working of our own planet Earth. It is in this context that this book is designed. Comprising 17 chapters, the first 7 chapters of this book focus on those subject matters that would help the reader in better understanding and appreciating the life- and habitability-related issues pertaining to Earth. The subsequent 10 chapters are designed to investigate the extraterrestrial worlds in the Solar System in general (their atmospheres, if any; surface conditions; physical nature; chemical components; subsurface scenarios), with special emphasis on identifying water, ocean worlds and other water bodies in the Solar System, characterizing them, evaluating the prospects of their habitability, searching for signs of extinct and extant life, and ultimately trying to understand any life we happen to find there.

The Solar System extraterrestrial bodies covered in this book in terms of availability of water include (in the order of decreasing proximity to Earth) Earth's Moon (*Chapter 9*), planet Mars (*Chapter 10*), Near-Earth Asteroid *Ceres* (*Chapter 11*), Jupiter's moons *Europa* (*Chapter 12*), Saturn's moons *Enceladus* (*Chapter 13*) and *Titan* (*Chapter 14*), Neptune's moon *Triton* (*Chapter 15*), and the dwarf planet Pluto (*Chapter 16*) that hugs the periphery of the Solar System. It is heartening that space missions are providing new details about some icy, ocean-bearing extraterrestrial bodies (e.g., Saturn's moons *Enceladus* and *Titan,* Neptune's moon *Triton,* dwarf planet *Pluto*), further heightening the scientific interest of this and other "ocean worlds" in our Solar system and beyond. The last chapter (*Chapter 17*) in the present book is dedicated to address the haunting issue of "environments favorable to life on planets, moons, dwarf planets, and meteorites".

An appropriate number of attractive, easily interpretable, and information-rich figures, which synthesizes the contents given in the text are provided so that the reader gets a visually comprehensible understanding of the information provided in the book. Notably, some chapters provide videos (available in the e-book) providing lively demonstration of certain aspects discussed in this book. For example, *Chapter 2* includes three videos, providing a lively demonstration of Earth's eccentricity (deviation of Earth's orbit from circularity), Earth's obliquity (Earth's axial tilt relative to its orbital plane), and Earth's precession (a gravity-induced slow and continuous change in the orientation of Earth's rotational axis, i.e., wobble). Likewise, in the context of addressing various curious subjects related to the beginnings of life on Earth, particularly multiple hypotheses on chemical evolution preceding biological evolution, *Chapter 3* provides a lively demonstration of the Miller-Urey "prebiotic soup" experiment through a video. In another example, in the context of addressing the significance of terrestrial extremophiles in our search of extraterrestrial habitability from an astrobiology perspective, *Chapter 5* provides a lively demonstration of extremophiles and their cell division through four interesting videos.

Having worked for more than three and a half decades in the field of oceanographic science and technology, and occasionally cruising in the vast oceans; coupled with my love for the wonderful oceans and the often complex (and even exotic) life within them gave me the insight and the impetus to pen this book, which covers various aspects involving water, subsurface/internal oceans, submarine hydrothermal vents, biologically important chemicals etc. in the extraterrestrial worlds in the Solar System. It is my hope that coverage of reasonably vast interdisciplinary topics in this book will be sufficient to prompt the readers to give wings to their fertile imagination about the incredibly vast worlds beyond Earth.

Dr. Antony Joseph

Acknowledgments

Several of my friends and well-wishers supported me while preparing the manuscript of this book. Dr. P. Mehra, Mr. K. Vijaykumar, and Mr. K. Sudheesh (all from CSIR—NIO) provided me help whenever needed. I wish to express my profound gratitude to bio-medical scientist and Prof. (Dr.) Arbind Kumar (Patna University)—discoverer of an elegant mathematical relationship between astronomy and biology (now known as Kumar's hypothesis)—who kindly agreed to write the foreword for this book. I thank my wife Lissy and daughter Rini, who silently supported me in this endeavor. I am greatly indebted to my elder daughter Dr. Runa who painstakingly carried out quality enhancement to some illustrations in this book at a very short notice.

After publishing my first Elsevier book, *Tsunamis*, my brothers Tomy and James prompted and urged me to continue writing. My co-brother Aprem and his daughter Resmi were always in the forefront, providing me moral support and appreciating my passionate efforts in exposing at least a miniscule drop of the knowledge of the vast extraterrestrial water-bearing worlds in the Solar System for the benefit of all those curious minds who are eager to read about such exotic putative habitable worlds.

I particularly acknowledge all those anonymous reviewers (from different disciplines) for their insightful feedback and constructive suggestions, which greatly helped me in improving the content of this book.

Dr. Antony Joseph

Chapter 1

Solar/planetary formation and evolution

1.1 Planet formation

A planet is an astronomical body orbiting a star. The plane of Earth's orbit around the Sun is known as the ecliptic. The planets in the Solar System orbit the Sun in the paths which are known as elliptical orbit. Each planet has its own orbit around the Sun and the direction in which all the planets orbit around the Sun are the same (in a counter-clockwise direction, when viewed from above the Sun's north pole). These orbits were well explained by the celebrated astronomer Kepler. With the exception of Mercury (which is the smallest planet in the Solar System, and whose orbit is inclined to the ecliptic by 7°) and the dwarf planet Pluto (whose orbit is inclined to the ecliptic by more than 17°), most large planets in our Solar System stay near the ecliptic plane. In fact, most major planets in our Solar System stay within 3° of the ecliptic. While most of the planets in the Solar System revolve around the Sun in the same direction and in virtually the same plane, as just mentioned, the small differences in the orbital inclinations with reference to the ecliptic are believed to stem from collisions that occurred late in the planets' formation.

The realization that most of the planets in the Solar System are orbiting around the Sun in a plane surface passing through the Sun's equator (the equatorial plane of the Sun) led the 18th century scientists Kant, Laplace and others to propose that planets in the Solar System formed from the protoplanetary disks of gas and dust. This hypothesis is known as the *"Nebular"* or *"Disk instability"* Hypothesis. Note that nebula is a distinct body of interstellar clouds (which can consist of cosmic dust, hydrogen, helium, molecular clouds; possibly as ionized gases). The nebular theory of Solar System formation states that our Solar System formed from the gravitational collapse of a giant interstellar gas cloud—the solar nebula (note that Nebula is the Latin word for cloud).

The idea that the Solar System originated from a nebula was first proposed in 1734 by Swedish scientist and theologian Emanual Swedenborg (Baker, 1983) (Fig. 1.1). Immanuel Kant (Fig. 1.2),—German philosopher and one of the central Enlightenment thinkers—who was familiar with Swedenborg's work, developed the theory further and published it in his *Universal Natural History and Theory of the Heavens* (1755). This treatise—dedicated to King Frederick II of Prussia—was an anonymous publication, arranged with the publisher Johann Friedrich Petersen. Given its grand scope and its targeted dedication, Kant clearly hoped that it would attract widespread attention from more powerful European figures and establish for himself a prominent scholarly reputation.

Watkins (2012) has described the story of publishing Kant's treatise thus: "In general terms, Kant's aim in the *Universal Natural History and Theory of the Heavens* was to show that the main elements of the entire observable universe—which include the constitution and regular motions not only of the Sun, the Earth, and the other planets, but also that of the moons, comets, and even other solar systems—can all be explained on the basis of three assumptions: (i) a certain initial state—a chaos in which matters endowed with different densities are distributed throughout space in the form of various indeterminate nebula; (ii) Newtonian mechanical principles—primarily attractive and repulsive forces, coupled with the law of universal gravitation; and (iii) the motions that these matters would have initiated and the states that they would eventually come to be in due to these motions and mechanical laws. In this way, Kant intended to lay bare the basic structure that governs the universe."

In his treatise, Kant argued that gaseous clouds (nebulae) slowly rotate, gradually collapsing and flattening due to gravity and forming stars and planets. A similar but smaller and more detailed model was proposed by Pierre-Simon Laplace in his treatise Exposition du system du monde (Exposition of the system of the world), which he released in 1796. Laplace theorized that the Sun originally had an extended hot atmosphere throughout the Solar System, and that this "protostar cloud" cooled and contracted. As the cloud spun more rapidly, it threw off material that eventually condensed to form the planets. According to the *Nebular* hypothesis, the Sun and planets formed together out of a rotating cloud of gas (the "solar nebula") and gravitational instabilities in the gas disk condense into

2 Water worlds in the solar system

FIG. 1.1 Emanuel Swedenborg (1688–1772)—a learned astronomer, who first proposed in 1734 the idea that the Solar System originated from a nebula (an interstellar cloud of dust and gas consisting of hydrogen and helium; possibly as ionized gases). Portrait by Carl Frederik von Breda. *(Image source: https://en.wikipedia.org/wiki/Emanuel_Swedenborg.)*

FIG. 1.2 Immanuel Kant—one of the proponents of the *"Nebular"* or *"Disk instability"* Hypothesis to explain the formation of Solar System from the protoplanetary disks of gas and dust; Portrait by Johann Gottlieb Becker, 1768. *(Image Source: https://en.wikipedia.org/wiki/Immanuel_ Kant#:~:text=In%201755%2C%20Kant%20received%20a,insight%20 into%20the%20coriolis%20force.)*

planets (Kant, 1755a, 1755b; Laplace, 1796). Laplace (1796) stated thus: *"It is astonishing to see all the planets move around the Sun from west to east, and almost in the same plane; all the satellites move around their planets in the same direction and nearly in the same plane as the planets; finally, the Sun, the planets, and all the satellites that have been observed rotate in the direction and nearly in the plane of their orbits…another equally remarkable phenomenon is the small eccentricity of the orbits of the planets and the satellites…we are forced to acknowledge the effect of some regular cause since chance alone could not give a nearly circular form to the orbits of all the planets."*

The Laplacian nebular model was widely accepted during the 19th century, but it had some rather pronounced difficulties. The main issue was angular momentum distribution between the Sun and planets, which the nebular model could not explain. In addition, Scottish scientist James Clerk Maxwell (1831–1879) asserted that different rotational velocities between the inner and outer parts of a ring could not allow for condensation of material (https://www.universetoday.com/38118/how-was-the-solar-system-formed/).

The Laplacian nebular model was also rejected by astronomer Sir David Brewster (1781–1868), who stated that: "those who believe in the Nebular Theory consider it as certain that our Earth derived its solid matter and its atmosphere from a ring thrown from the Solar atmosphere, which afterward contracted into a solid terraqueous sphere, from which the Moon was thrown off by the same process… [Under such a view] the Moon must necessarily have carried off water and air from the watery and aerial parts of the Earth and must have an atmosphere." (https://www.universetoday.com/38118/how-was-the-solar-system-formed/.)

By the early 20th century, the Laplacian model had fallen out of favor, prompting scientists to seek out new theories. However, it was not until the 1970s that the modern and most widely accepted variant of the nebular hypothesis—the solar nebular disk model (SNDM)—emerged. Credit for this goes to Soviet astronomer Victor Safronov and his book *Evolution of the protoplanetary cloud and formation of the Earth and the planets* (1972). According to the Safronov (nebular) hypothesis, planets form by the following multistage process requiring growth by 45 orders of magnitude in mass through many different physical processes (http://www.scholarpedia.org/article/Planetary_formation_and_migration):

1. Dust grains condense out from cooling gas disk and settle to the disk midplane
2. Dust coalesces into small (km-sized) solid bodies (planetesimals, i.e., bodies big enough that they are unaffected by gas)
3. Planetesimals collide and grow into "planetary cores"
4. Cores of intermediate and giant planets accrete gas envelopes before the gaseous disk disperses

Dust grains grow by colliding with one another and sticking together by electrostatic forces. Small particles also physically embed themselves in larger aggregates during high-speed collisions. The motion of small dust grains is closely coupled to that of the gas, and turbulence causes dust to diffuse over large distances leading to substantial radial and vertical mixing of material within the disk. Particles larger than 1 mm develop significant velocities relative to the gas because gas orbits the star somewhat more slowly than a solid body due to an outward pressure gradient in the disk. This velocity differential causes particles to migrate radially toward the star, and particles also settle vertically toward the midplane of the disk. Growth of bodies in the size range of centimeter to meter must be rapid, or else much of the solid material in the disk would evaporate when it enters the hot regions closer to the star. Alternatively, planetesimals (minute bodies which could come together with many others under gravitation to form a planet) might form via the gravitational collapse of regions containing dense concentrations of solid particles (the Goldreich-Ward mechanism). Both models face substantial challenges.

The formation of planets requires growth through at least 12 orders of magnitude in spatial scale, from micron-sized particles of dust and ice up to bodies with radii of thousands or tens of thousands of km.

The initial reservoir of solid materials for planet formation is micron-sized particles of rocky or icy dust. The dynamics of dust within a disk is dominated by gravity from the star and aerodynamic forces from the gas, including turbulence. In contrast, gravitational interactions between small bodies are very weak (the escape velocity from a 1-m-diameter rock is less than 0.1 cm/s). Aerodynamic forces remain dominant until bodies grow to 1–100 km in size. Such bodies, referred to as *planetesimals*, are massive enough that their gravitational interactions are significant, while the small value of their surface-to-volume ratio means they are only weakly affected by aerodynamic forces (http://www.scholarpedia.org/article/Planetary_formation_and_migration).

In Safronov's book, almost all major problems of the planetary formation process were formulated and many were solved. For example, the SNDM model has been successful in explaining the appearance of accretion disks around young stellar objects. Various simulations have also demonstrated that the accretion of material in these disks leads to the formation of a few Earth-sized bodies. Thus, the origin of terrestrial planets is now considered to be an almost solved problem. While originally applied only to the Solar System, the SNDM was subsequently thought by theorists to be at work throughout the Universe, and has been used to explain the formation of many of the exoplanets that have been discovered throughout our galaxy. With the recent advances in technology, the large amount of observational evidences related to newly formed and currently forming stars now pouring in is a clear testimony to the validity of the *Disk instability Hypothesis* and its variants.

Lissauer (1987) has outlined a unified scenario for Solar System formation consistent with astrophysical constraints. According to this scenario, Jupiter's core could have grown by runaway accretion of planetesimals to a mass sufficient to initiate rapid accretion of gas in times of order of 5×10^5–10^6 years, provided the surface density of solids in its accretion zone was at least 5–10 times greater than that required by minimum mass models of the protoplanetary disk. After Jupiter had accreted large amounts of nebular gas, it could have gravitationally scattered the planetesimals remaining nearby into orbits which led to escape from the Solar System. Most of the planetesimals in the Mars-asteroids accretion zone could have been perturbed into Jupiter-crossing orbits by resonances with Jupiter and/or interactions with bodies scattered inward from Jupiter's accretion zone; such Jupiter-crossing orbits would have subsequently led to ejection from the Solar System. However, removal of excess mass from sunward of 1 AU would have been much more difficult. The inner planets and the asteroids can be accounted for in this picture if the surface density of the solar nebula was relatively uniform (decreasing no more rapidly than $r^{-1/2}$) out to Jupiter's orbit. The total mass of the protoplanetary disk could have been less than one-tenth of a solar mass provided the surface density dropped off more steeply than r^{-1} beyond the orbit of Saturn. The outer regions of the nebula would still have contained enough solid matter to explain the growth of Uranus and Neptune in 5×10^6–10^8 years, together with the coincident ejection of comets to the Oort cloud. Lissauer (1987)'s study indicated that the formation of such a protoplanetary disk requires significant transport of mass and angular momentum, and is consistent with viscous accretion disk models of the solar nebula.

Astronomical observations have shown that protoplanetary disks are dynamic objects through which mass is transported and accreted by the central star. This transport causes the disks to decrease in mass and cool over time, and such evolution is expected to have occurred in our own solar nebula. Age dating of meteorite constituents shows that their creation, evolution, and accumulation occupied several Myr, and over this time disk properties would evolve significantly. Moreover, on this timescale, solid particles decouple from the gas in the disk and their evolution follows a different path. It is in this context that we must understand how our own solar nebula evolved and what effects this evolution had on the primitive materials contained within it (Ciesla and Cuzzi, 2006).

It makes sense that most large planets in our Solar System stay near the ecliptic plane. Our Solar System is believed to be about 4.6 billion years old. It is thought to have arisen from an amorphous cloud of gas and dust in space. The original cloud was spinning, and this spin caused it to flatten out into a disk shape, rather than a spherical shape. The Sun and planets are believed to have formed out of this disk, which is why, today, the planets still orbit in a single plane around our Sun.

As just mentioned, planetary systems are formed from rotating protoplanetary disks, which are the evolved phase of circumstellar disks produced during the collapse of a proto-stellar cloud with some angular momentum. A standard model of such a protoplanetary disk is that of a steady state disk in vertical hydrostatic equilibrium, with gas and dust fully mixed and thermally coupled (Kenyon and Hartmann, 1987). Such a disk is flared, not flat, but still geometrically thin in the sense defined by Pringle (1981). The disk intercepts a significant amount of radiation from the central star, but other heating sources (e.g., viscous dissipation) can be more important. If dissipation due to mass accretion is high, it becomes the main source of heating (Sasselov and Lecar, 2000). In a simpler language, a protoplanetary disk is a rotating circumstellar disk (a torus, pancake or ring-shaped accumulation of matter composed of gas, dust, planetesimals, asteroids, or collision fragments in orbit around a star) of dense gas and dust surrounding a young newly formed star. Although nobody has seen the protoplanetary disk that surrounded our Sun in its infancy, observations of recently formed and currently forming protoplanetary disk around new stars shed light on how the protoplanetary disk around Sun might have looked like.

Because new stars are being formed, and they are studied in minute detail by planetary scientists at several observatories, a clear picture of protoplanetary disk is now available. For example, Fig. 1.3A illustrates the protoplanetary disk surrounding the young star Elias 2-27, located some 450 light years away ("*Spirals with a Tale to Tell.*" www. eso.org.). This beautiful image, captured with the Atacama Large Millimeter/submillimeter Array (ALMA) features a protoplanetary disk surrounding the young stellar object Elias 2-27, some 450 light years away. ALMA has discovered and observed plenty of protoplanetary disks, but this disk is special as it shows two distinct spiral arms, almost like a tiny version of a spiral galaxy. Previously, astronomers noted compelling spiral features on the surfaces of protoplanetary disks, but it was unknown if these same spiral patterns also emerged deep within the disk where planet formation takes place. ALMA, for the first time, was able to peer deep into the mid-plane of a disk and discovered the clear signature of spiral density waves. Nearest to the star, ALMA found a flat disk of dust, which extends to what would approximately be the orbit of Neptune in our own Solar System. Beyond that point, in the region analogous to our Kuiper Belt, ALMA detected a narrow band with significantly less dust, which may be an indication for planet in formation. Springing from the outer edge of this gap are the two sweeping spiral arms that extend more than 10 billion kilometers away from their host star. The discovery of

FIG. 1.3A Protoplanetary disk surrounding the young star Elias 2-27, located some 450 light years away ("*Spirals with a Tale to Tell.*" www.eso.org.). This beautiful image, captured with the Atacama Large Millimeter/submillimeter Array (ALMA) features a protoplanetary disk surrounding the young stellar object Elias 2-27, some 450 light years away. ALMA has discovered and observed plenty of protoplanetary disks, but this disk is special as it shows two distinct spiral arms, almost like a tiny version of a spiral galaxy. *(Source: https://www.eso.org/public/images/potw1640a/ Author: B. Saxton (NRAO/AUI/NSF); ALMA (ESO/NAOJ/NRAO) Reproduced from: https://commons.wikimedia.org/wiki/File:Spirals_with_a_Tale_to_Tell.jpg.)*

spiral waves at these extreme distances may have implications on the theory of planet formation. Fig. 1.3B is an example of protoplanetary disk. Shown here is the Atacama Large Millimeter Array (ALMA) image of the protoplanetary disk surrounding the young star HL Tauri. These new ALMA observations reveal substructures within the disk that have never been seen before and even show the possible positions of planets forming in the dark patches within the system. Fig. 1.3C is an example of debris disk. Shown here is Hubble Space Telescope observation of the debris ring around Fomalhaut planet. Taking account of the features of several newly formed protoplanetary disk actually observed, artists have created images of protoplanetary disks of their concept. Fig. 1.3D is an artist's concept of a protoplanetary disk, where particles of dust and grit collide and accrete forming planets or asteroids. Cosmic dust grains, rotating around the primitive Sun, coalesced to form planetesimals, and then larger bodies (e.g., planets) through gravitation, giving rise to the Solar System about 4.6 billion years ago (Taylor and Norman, 1990).

As just mentioned, when it comes to the formation of our Solar System, the most widely accepted view is the Nebular Hypothesis and its variants. In essence, this theory states that the Sun, the planets, and all other objects in the Solar System formed from nebulous material billions of years ago. Originally proposed to explain the origin of the Solar System, this theory has gone on to become a widely accepted view of how all the star-systems came to be.

FIG. 1.3B An example of protoplanetary disk. Shown here is the Atacama Large Millimeter Array (ALMA) image of the protoplanetary disk surrounding the young star HL Tauri. These new ALMA observations reveal substructures within the disk that have never been seen before and even show the possible positions of planets forming in the dark patches within the system. *(Source: https://en.wikipedia.org/wiki/Protoplanetary_disk#/media/File:HL_Tau_protoplanetary_disk.jpg.)*

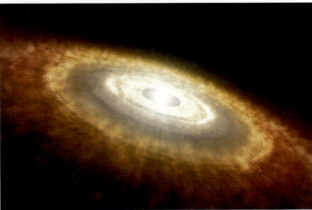

FIG. 1.3D (Top) Artist's concept of a protoplanetary disk, where particles of dust and grit collide and accrete forming planets or asteroids. *Source: NASA; http://origins.jpl.nasa.gov/stars-planets/ra4.html Author: NASA. Copyright: This file is in the public domain in the United States because it was solely created by NASA. NASA copyright policy states that "NASA material is not protected by copyright unless noted." (Source: https://commons.wikimedia.org/wiki/File:Protoplanetary-disk.jpg)* (Bottom) Artist's impression of the early Solar System, where collision between particles in an accretion disk led to the formation of planetesimals and eventually planets. *Credit: NASA/JPL-Caltech. (Source: https://www.universetoday.com/38118/how-was-the-solar-system-formed/.)*

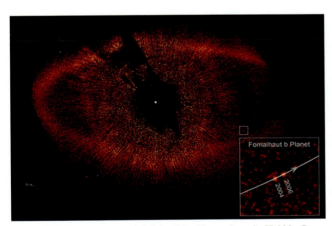

FIG. 1.3C An example of debris disk. Shown here is Hubble Space Telescope observation of the debris ring around Fomalhaut planet. The inner edge of the disk may have been shaped by the orbit of Fomalhaut b, at lower right. *Author: NASA, ESA, P. Kalas, J. Graham, E. Chiang, E. Kite (University of California, Berkeley), M. Clampin (NASA Goddard Space Flight Center), M. Fitzgerald (Lawrence Livermore National Laboratory), and K. Stapelfeldt and J. Krist (NASA Jet Propulsion Laboratory) This file is in the public domain because it was created by NASA and ESA. NASA Hubble material (and ESA Hubble material prior to 2009) is copyright-free and may be freely used as in the public domain without fee, on the condition that only NASA, STScI, and/or ESA is credited as the source of the material. This license does not apply if ESA material created after 2008 or source material from other organizations is in use. The material was created for NASA by Space Telescope Science Institute under Contract NAS5-26555, or for ESA by the Hubble European Space Agency Information Centre. (Source: https://commons.wikimedia.org/wiki/File:Fomalhaut_with_Disk_Ring_and_extrasolar_planet_b.jpg.)*

The nebular hypothesis of Solar System formation describes how protoplanetary disks are thought to evolve into planetary systems. According to this hypothesis, the Sun and all the planets of our Solar System began as a giant cloud of molecular gas and dust. Then, about 4.6 billion years ago, something happened that caused the cloud to collapse. This could have been the result of a passing star, or shock waves from a supernova (), but the end result was a gravitational collapse at the center of the cloud.

From this collapse, pockets of dust and gas began to collect into denser regions. As the denser regions pulled in more and more matter, conservation of momentum caused it to begin rotating, while increasing pressure caused it to heat up. Most of the material ended up in a ball at the center while the rest of the matter flattened out into disk that circled around it. While the ball at the center formed the

Sun, the rest of the material would form into the protoplanetary disk.

It has been argued that electrostatic and gravitational interactions may cause the dust and ice grains in the disk to accrete into planetesimals. The accretion process competes against the stellar wind, which drives the gas out of the system. Gravity (accretion) and internal stresses (viscosity) pull material into the center of a star. Planetesimals constitute the building blocks of both terrestrial and giant planets (Lissauer et al., 2009; D'Angelo et al., 2014).

Although the nebular theory is widely accepted, there are still problems with it that astronomers have not been able to resolve. For example, there is the problem of tilted axes. According to the nebular theory, all planets around a star should be tilted the same way relative to the ecliptic. But as we have learned, the inner planets and outer planets have radically different axial tilts.

Whereas the inner planets range from almost 0° tilt, others (like Earth and Mars) are tilted significantly (23.4° and 25°, respectively), outer planets have tilts that range from Jupiter's minor tilt of 3.13°, to Saturn and Neptune's more pronounced tilts (26.73° and 28.32°), to Uranus' extreme tilt of 97.77°, in which its poles are consistently facing toward the Sun.

It would be worth realizing that just when we think we have a satisfactory explanation, there remain those troublesome issues it just can't account for. However, between our current models of star and planet formation, and the birth of our Universe, we have come a long way. As we learn more about neighboring star systems and explore more of the cosmos, our models are likely to mature further (https://www.universetoday.com/38118/how-was-the-solar-system-formed/).

1.1.1 Terrestrial planet formation

As indicated earlier, it has been hypothesized that the planets formed by accretion from the protoplanetary disk, in which dust and gas gravitated together and coalesced to form ever larger bodies. Only metals and silicates having relatively higher density and melting point could exist closer to the Sun, and these would eventually form the terrestrial planets (Mercury, Venus, Earth, and Mars). Because metallic elements only comprised a very small fraction of the solar nebula, the terrestrial planets could not grow very large.

In the absence of direct observations or suitable laboratory experiments, much of what we know about terrestrial planet formation comes from computer simulations. Terrestrial-planet formation has been studied extensively using statistical models based on the coagulation equation to study the early stages of growth and N-body simulations to model later stages when the number of large bodies is small. Once planetesimals have formed, their subsequent evolution is dominated by mutual gravitational interactions and collisions as they orbit the central star. Colliding planetesimals typically merge to form a larger body with some mass escaping as small fragments. Planetesimals also undergo numerous close encounters, which alter their orbits but not their masses. At an early stage, *runaway growth* takes place, in which large bodies typically grow more rapidly than small ones due to differences in their orbital eccentricities and inclinations. Typical time scales for the runaway growth phase are of the order of 10^5 years. Runaway growth is followed by *oligarchic growth*, in which a relatively small number of large bodies grow at similar rates until they have swept up most of the smaller planetesimals. Collisions and radioactive decay heat the large bodies until they melt, causing dense elements such as iron to sink to the center to form a core overlain by a rocky mantle. Oligarchic growth generates a population of 100–1000 lunar-to-Mars-sized *planetary embryos*, probably in 1 million years or less. Subsequent collisions between these embryos lead to the final assembly of the terrestrial planets, on a time scale of up to 100 million years. Earth's Moon is thought to have formed about 40 million years after the start of the Solar System from debris placed into orbit about the Earth when it collided with a Mars-sized planetary embryo. A substantial fraction of the Earth's mass is thought to have been accreted via large impacts, so requiring such a cataclysmic even to form Earth's Moon is in principle not a problem, though only a small fraction of giant impacts would lead to the formation of a moon with the properties of the present Moon of the Earth (http://www.scholarpedia.org/article/Planetary_formation_and_migration).

A classic theory of Solar System formations is the "Kyoto Model." In the 1970s, Chushiro Hayashi led a research group at Kyoto University which proposed a fundamental creation scenario. To this day, the Kyoto Model is still used and expanded on for planet formation theories.

The time scale for lunar formation, along with other time scales such as that for asteroids to become large enough to differentiate, is derived by applying radionuclide chronometers to samples of rock. Such *cosmochemistry* evidence is becoming increasingly important, and provides a growing number of constraints on the formation of the early Solar System. Planets typically acquire mass from a range of distances within a protoplanetary disk, although the mixture is different for each object, leading to a unique chemical composition. It is likely that Earth acquired most of its water and other volatile materials from relatively cold regions of the Sun's protoplanetary disk such as the asteroid belt (http://www.scholarpedia.org/article/Planetary_formation_and_migration).

Simulations of terrestrial-planet formation are able to reproduce the basic architecture (a small number of terrestrial planets with low-eccentricity orbits) of the inner Solar System from plausible initial conditions. The stochastic nature of planetary accretion, however, means that a precision comparison between the Solar system and theoretical models is not possible. The number and masses of terrestrial planets are predicted to vary from one planetary system to another due to differences in the amount of solid material available and the presence or absence of giant planets, as well as the highly stochastic nature of planet formation. The presence of a giant planet probably frustrates terrestrial-planet formation in neighboring regions of the disk, leading to the absence of terrestrial planets in these regions or the formation of an asteroid belt. These predictions will be tested by ongoing and future space missions designed to search for extrasolar terrestrial planets, such as COROT and Kepler (http://www.scholarpedia.org/article/Planetary_formation_and_migration).

Once a planet is formed, planetary orbits may be modified as a result of interactions with the gas disks, or with other planets, stars, or small bodies present in the system. Such modifications can result in planetary migration. According to the *Nebula Hypothesis*, 4.6 billion years ago, the Solar System was formed by the gravitational collapse of a giant molecular cloud spanning several light years. Note that a molecular cloud is a type of interstellar cloud, the density and size of which permit the formation of molecules, most commonly molecular hydrogen (H_2). This is in contrast to other areas of the interstellar medium that contain predominantly ionized gas (atoms or molecules which have one or more orbital electrons stripped, thus attaining positive electrical charge or, rarely, an extra electron attached, thus attaining negative electrical charge). The gas that formed the Solar System was slightly more massive than the present Sun. Most of the mass concentrated in the center, forming the Sun, and the rest of the mass flattened into a protoplanetary disk, out of which all of the current planets, moons, asteroids, and other celestial bodies in the Solar System formed.

1.1.2 Giant planet formation

It appears that the *Frost Line* has played an important role in the giant planet formation. Note that frost line is the boundary surface/distance in the solar nebula from the central protostar where it is cold enough for volatile compounds such as water, ammonia, methane, carbon dioxide, and carbon monoxide to condense into solid ice grains; simply stated, the frost line is a boundary line located between the orbits of Mars and Jupiter where material is cool enough for volatile icy compounds to remain solid. In the current Solar System, the frost line is at about 5 astronomical unit (AU), which is a bit closer than Jupiter, so currently all the rocky planets are inside the frost line, and all the gas giants are beyond the frost line (Note that an astronomical unit (AU) is the average distance between Earth and the Sun, which is about 93 million miles or 150 million kilometers. Astronomical units are usually used to measure distances within our Solar System). This would seem to imply that it is the frost line that determines whether a rocky or gas planet will form. What is the relationship between frost line and the asteroid belt in the solar system? Martin and Livio (2012) have proposed that asteroid belts may tend to form in the vicinity of the frost line, due to nearby giant planets disrupting planet formation inside their orbit (more information on frost line is supplied in Section 1.6.2 in this chapter).

Stevenson and Lunine (1988) reported a model for enhancing the abundance of solid material in the region of the solar nebula in which Jupiter formed, by diffusive redistribution and condensation of water vapor. In this model, a turbulent nebula is assumed with temperature decreasing roughly inversely with the radial distance from the center, and time scales set by turbulent viscosities taken from recent nebular models. The diffusion equation in cylindrical coordinates is solved in the limit that the sink of water vapor is condensation within a narrow radial zone approximately 5 AU from the center. Most of the water vapor is extracted from the terrestrial planet-forming zone. This "cold finger" solution is then justified by analytic solution of the diffusion equation in the condensation zone itself and inward and outward of that zone. The length scale over which most of the diffusively transported water vapor condenses is calculated to be ~0.4 AU, provided the solids are not redistributed, and the surface density of ice in the formation zone of Jupiter is enhanced by as much as 75. It was found that the enhancement in surface density of solids is sufficient to trigger rapid accretion of planetesimals into a solid core along the lines of the model of Lissauer (1987), and hence produce Jupiter by nucleated instability on a time scale of approximately 10^5 to 10^6 years.

Giant planets are qualitatively distinct from terrestrial planets in that they possess significant gaseous envelopes. The giant planets (Jupiter, Saturn, Uranus, and Neptune) formed beyond the *Frost Line*. The ices that formed these planets were more plentiful than the metals and silicates that formed the terrestrial inner planets, allowing them to grow massive enough to capture large atmospheres of hydrogen and helium. The leftover debris that never became planets congregated in regions such as the Asteroid Belt, Kuiper Belt, and Oort Cloud.

In the Solar System, the *gas giants* (Jupiter and Saturn) are predominantly composed of hydrogen and helium gas, although these planets are enriched in elements heavier than helium. The *ice giants* (Uranus and Neptune) have lesser, but still substantial (several Earth masses) gas envelopes. The existence of these envelopes provides a critical constraint: giant planets must form relatively quickly, before the gas in the protoplanetary disk is dissipated. Observations of protoplanetary disks around stars in young clusters pin the gas-disk lifetime in the 3–10 million years range (http://www.scholarpedia.org/article/Planetary_formation_and_migration).

There exists hardly any core accretion simulation that can successfully account for the formation of Uranus or Neptune within the observed 2–3 Myr lifetimes of protoplanetary disks. Because solid accretion rate is directly proportional to the available planetesimal surface density, one way to speed up planet formation is to take a full accounting of all the planetesimal-forming solids present in the solar nebula. By combining a viscously evolving protostellar disk with a kinetic model of ice formation, which includes not just water but methane, ammonia, CO and minor ices, Dodson-Robinson et al. (2009) calculated the solid surface density of a possible giant planet-forming solar nebula as a function of heliocentric distance and time. Their results can be used to provide the starting planetesimal surface density and evolving solar nebula conditions for core accretion simulations, or to predict the composition of planetesimals as a function of radius. Dodson-Robinson et al. (2009) found three effects that favor giant planet formation by the core accretion mechanism: (1) a flow that brings mass from the inner solar nebula to the giant planet-forming region, (2) the fact that the ammonia and water ice lines should coincide, according to recent lab results from Collings et al. (2004), and (3) the presence of a substantial amount of methane ice in the trans-Saturnian region. Dodson-Robinson et al. (2009)'s results show higher solid surface densities than assumed in the core accretion models of Pollack et al. (1996), by a factor of 3–4 throughout the trans-Saturnian region. Dodson-Robinson et al. (2009) have discussed the location of ice lines and their movement through the solar nebula, and provide new constraints on the possible initial disk configurations from gravitational stability arguments.

There are currently two principal schools of thought regarding the formation of giant planets, which are exemplified by the *disk instability* model and the *core accretion* model. Owen (2004) pointed out that the predictions of the disk instability model are not consistent with the present composition of Jupiter's atmosphere as revealed by the Galileo Probe Mass Spectrometer (GPMS). Owen (2004) found that the core accretion model can explain the observed atmospheric composition of Jupiter if appropriate planetesimals were available to enhance the heavy elements. Hence core accretion seems to provide a better explanation for the formation of Jupiter, the most massive of our own giant planets. According to Owen (2004), disk instabilities may well be the source of giant planets in other planetary systems.

The standard theory for the formation of gas giants, *core accretion*, is a two-stage process whose first stage closely resembles the formation of terrestrial planets. A core with a mass of the order of 10 Earth masses forms in the disk by numerous collisions between planetesimals. Typically, there is not enough solid material to form bodies this massive in the inner region of a protoplanetary disk. Prior to further discussing this topic, a term that deserves particular mention is *snow line* (also known as *frost line*). The "snow line" (actually, it is a surface) divides the outer, cold, ice-rich region of the protoplanetary disk from the inner, steamy hot zone. Outward gas motions across the snow line result in the condensation of water into ice grains, which can collide, stick and grow until they dynamically decouple from the flow and are left behind. In this way, the snow line defines the inner edge of a cold trap in which the density of solids may have been sufficiently enhanced as to help speed the growth of planetesimals and, ultimately, of the cores of the giant planets (Stevenson and Lunine, 1988).

At larger orbital radii, beyond the *snow line (frost line)*, the temperature is low enough that ices as well as rocky materials can condense. This extra solid material, together with the reduced gravity of the central star, allows large solid cores to form in the outer regions of a disk. Initially a core is surrounded by a low-mass atmosphere, which grows steadily more massive as the gas cools and contracts onto the core. Eventually the core exceeds a "critical core mass," beyond which a hydrostatic envelope cannot be maintained. Determining an accurate time scale for reaching the critical core mass is very difficult, in part because the rate at which the gas cools depends upon how transparent the envelope is. The transparency varies dramatically with the amount of dust present, which is extremely uncertain. Once the core mass is exceeded, gas begins to flow onto the core. It is slow at first but increases rapidly as the planet becomes more massive. Growth ceases when the supply of gas is terminated, either because the planet opens a gap in the disk or because the disk gas dissipates (http://www.scholarpedia.org/article/Planetary_formation_and_migration).

Among various models available for attempting to explain the giant-planet formation (e.g., gravitational-instability model), core accretion is generally considered to be a more plausible model currently. First, theoretical calculations suggest that although young protoplanetary disks may be massive enough to be unstable, they are unlikely to cool rapidly enough to fragment (except perhaps at very large radius). Second, the core-accretion model naturally explains the existence of ice-giant planets like Neptune (although the time scale for formation of the ice giants is worryingly long if they formed at their present locations). Finally, the observed correlation between the frequency of extrasolar planets and the metallicity of their host stars is qualitatively explicable as a consequence of core accretion: if the disk is enriched in solids, a critical-mass core can form more readily. It is unclear whether this correlation can be explained by the gravitational-instability model. Against this, the inferred core mass of Jupiter (which can be estimated by comparing the measured multipoles of the gravitational field with theoretical structure models) is lower than simple estimates based on core accretion. More subtle observational constraints—such as the abundance of different elements measured in Jupiter's atmosphere by the

Galileo probe—are also in conflict with at least the simplest models of giant-planet formation. These problems suggest that a full understanding of giant-planet formation has yet to be attained. Observations of the frequency of giant planets in extrasolar planetary systems with very different properties to the Solar System promise to provide valuable new constraints (http://www.scholarpedia.org/article/Planetary_formation_and_migration).

According to Malhotra (1993):

- Early in the history of the Solar System there was debris left over between the planets
- Ejection of this debris by Neptune caused its orbit to migrate outward
- If Pluto were initially in a low-eccentricity, low-inclination orbit outside Neptune it is inevitably captured into 3:2 resonance with Neptune
- Once Pluto is captured its eccentricity and inclination grow as Neptune continues to migrate outward
- Other objects may be captured in the resonance as well

In general, giant-planets form, as in the solar system, at 5–10 AU. Note that the frost line is located at a distance of 5 AU from the Sun. Some process (e.g., close encounters between planets, tidal forces from a companion star [Kozai-Lidov oscillations], resonant capture during disk migration, secular chaos [as in delivery of meteorites to Earth]) excites their eccentricities to $e > 0.99$ so pericenter is $q = a(1 - e) < 0.1$ AU $= 20$ R$_\odot$. Tidal friction then damps the eccentricity.

Some of the moons of Jupiter, Saturn, and Uranus are believed to have formed from smaller, circumplanetary analogues of the protoplanetary disks (Canup and Ward, 2008; D'Angelo and Podolak, 2015). The formation of planets and moons in geometrically thin, gas- and dust-rich disks is the reason why the planets are arranged in an ecliptic plane. Tens of millions of years after the formation of the Solar System, the inner few AU of the Solar System likely contained dozens of Moon- to Mars-sized bodies that were accreting and consolidating into the terrestrial planets that we now see. The Earth's moon likely formed after a Mars-sized protoplanet obliquely impacted the proto-Earth ~30 million years after the formation of the Solar System.

In a recent study, Helled and Bodenheimer (2014) investigated the formation of Uranus and Neptune, according to the core-nucleated accretion model, considering formation locations ranging from 12 to 30 AU from the Sun, and with various disk solid-surface densities and core accretion rates. They showed that in order to form Uranus-like and Neptune-like planets in terms of final mass and solid-to-gas ratio, very specific conditions are required. Helled and Bodenheimer (2014) also show that when recently proposed high solid accretion rates are assumed, along with solid surface densities about 10 times those in the minimum-mass solar nebula, the challenge in forming Uranus and Neptune at large radial distances is no longer the formation timescale, but is rather finding agreement with the final mass and composition of these planets. In fact, these conditions are more likely to lead to gas-giant planets. Scattering of planetesimals by the forming planetary core is found to be an important effect at the larger distances. Helled and Bodenheimer (2014)'s study emphasizes how (even slightly) different conditions in the protoplanetary disk and the birth environment of the planetary embryos can lead to the formation of very different planets in terms of final masses and compositions (solid-to-gas ratios), which naturally explains the large diversity of intermediate-mass exoplanets.

1.2 Asteroids, meteorites, and chondrites

During the formation of the Solar System, the early formation of the gas giant, Jupiter, affected the subsequent development of inner solar system and is considered to be responsible for the existence of Solar System's Main Asteroid Belt (MAB) located between the orbits of Mars and Jupiter (Fig. 1.4), and the small size of Mars (Taylor, 1996). Asteroids are the bits and pieces left over from the initial agglomeration of the inner planets that include Mercury, Venus, Earth, and Mars. Asteroids whose orbits bring them relatively close to the Earth (perihelon distances

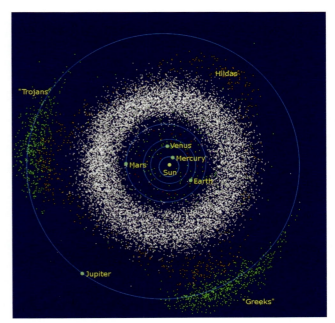

FIG. 1.4A The inner Solar System, from the Sun to Jupiter. Also includes the asteroid belt (the white donut-shaped cloud), the Hildas (the orange "triangle" just inside the orbit of Jupiter), the Jupiter trojans (green), and the near-Earth asteroids. The group that leads Jupiter are called the "Greeks" and the trailing group are called the "Trojans" (Murray and Dermott, Solar System Dynamics, p. 107) *(Image source: https://commons.wikimedia.org/wiki/File:InnerSolarSystem-en.png) Copyright information: This work has been released into the public domain by its author, Mdf at English Wikipedia. This applies worldwide.*

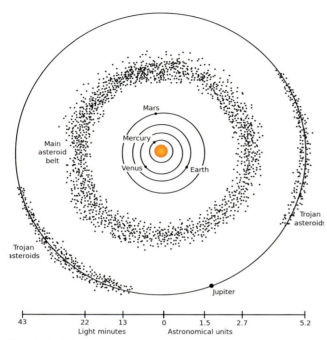

FIG. 1.4B Diagram of the Solar System's asteroid belt. *(Source: https:// commons.wikimedia.org/wiki/File:Asteroid_Belt.svg) © Public Domain.* Description note given with this diagram reads as follows: This file is in the public domain in the United States because it was solely created by NASA. NASA copyright policy states that "NASA material is not protected by copyright unless noted."

of less than 1.3 astronomical unit [AU]) are known as Near Earth Asteroids (NEAs). Most of the NEAs originate in the MAB and are perturbed inward through either collision between asteroids or the gravitational influence of Jupiter.

Jewitt et al. (2007) describe that aerodynamic drag causes 100-m scale ice-rich planetesimals from the outer disk to spiral inward on very short timescales (perhaps 1 AU per century). Some are swept up by bodies undergoing accelerated growth just outside the snow line. Others cross the snow line to quickly sublimate, enhancing the local water vapor abundance. Lyons and Young (2004) used a photochemical model of the solar nebula to investigate the time evolution of oxygen isotopes that occurs due to self-shielding during CO photodissociation, and to predict isotope values for initial water in the nebula. Coupled with on-going dissociation from ultraviolet photons, these freeze-out and sublimation processes lead to isotopic fractionation anomalies in water (Lyons and Young, 2005) that have already been observed in the meteorite record (Krot et al., 2005).

According to Lyons and Young (2005), changes in the chemical and isotopic composition of the solar nebula with time are reflected in the properties of different constituents that are preserved in chondritic meteorites. The aqueous alteration process in CM chondrites appears to have been largely isochemical if the bulk meteorites are considered as the reacting systems, although depletion patterns and isotopic anomalies indicate open-system behavior for a few highly mobile components. CR-group carbonaceous chondrites are among the most primitive of all chondrite types and must have preserved solar nebula records largely unchanged. Lyons and Young (2005) analyzed the oxygen and magnesium isotopes in a range of the CR constituents of different formation temperatures and ages, including refractory inclusions and chondrules of various types. The results provide new constraints on the time variation of the oxygen isotopic composition of the inner (<5 AU) solar nebula—the region where refractory inclusions and chondrules most likely formed. A chronology based on the decay of short-lived ^{26}Al ($t_{1/2}$ ~0:73 Myr) indicates that the inner solar nebula gas was ^{16}O-rich when refractory inclusions formed, but less than 0.8 Myr later, gas in the inner solar nebula became ^{16}O-poor, and this state persisted at least until CR chondrules formed ~1–2 Myr later. Lyons and Young (2005) suggest that the inner solar nebula became ^{16}O-poor because meter-sized icy bodies, which were enriched in ^{17}O and ^{18}O as a result of isotopic self-shielding during the ultraviolet photodissociation of CO in the protosolar molecular cloud or protoplanetary disk, agglomerated outside the snow line, drifted rapidly toward the Sun, and evaporated at the snow line. This led to significant enrichment in ^{16}O-depleted water, which then spread through the inner solar system. Astronomical studies of the spatial and temporal variations of water abundance in protoplanetary disks may clarify these processes (Lyons and Young, 2005).

Ciesla and Cuzzi (2006) reported a model which tracks how the distribution of water changes in an evolving disk as the water-bearing species experience condensation, accretion, transport, collisional destruction, and vaporization. Because solids are transported in a disk at different rates depending on their sizes, the motions will lead to water being concentrated in some regions of a disk and depleted in others. These enhancements and depletions are consistent with the conditions needed to explain some aspects of the chemistry of chondritic meteorites and formation of giant planets. The levels of concentration and depletion, as well as their locations, depend strongly on the combined effects of the gaseous disk evolution, the formation of rapidly migrating rubble, and the growth of immobile planetesimals. Understanding how these processes operate simultaneously is critical to developing our models for meteorite parent body formation in the Solar System and giant planet formation throughout the galaxy. Ciesla and Cuzzi (2006) have reported examples of evolution under a range of plausible assumptions and demonstrated how the chemical evolution of the inner region of a protoplanetary disk is intimately connected to the physical processes which occur in the outer regions. In the protoplanetary disk, opacity due to grains inhibited radiative cooling and raised the mid-plane kinetic temperature. Growth and migration of solids in the disk would have caused the opacity to change, so moving the snow line. For a fraction of a million years, it may have pushed in to the orbit of Mars or even closer (Sasselov and Lecar, 2000; Ciesla and Cuzzi, 2006), meaning that asteroids in the main belt, between Mars and Jupiter, could have incorporated water ice upon formation. Indeed, small samples of certain asteroids, available to us in the form of meteorites, contain hydrated minerals that probably formed when buried ice melted and chemically reacted with surrounding refractory materials (Kerridge et al., 1979; McSween, 1979a, 1979b).

According to the so-called "Nice model" of the early Solar System, at one time Jupiter and Saturn passed through a 2:1 orbital resonance resulting in large perturbations which destabilized many asteroids and "ejected" Uranus and Neptune to their current orbits. Resonances of a different kind impact asteroid population in the main asteroid belt. Jupiter's gravitational pull has a great effect on the asteroid belt. Rather than objects commonly existing at these resonances, repeated encounters with Jupiter ejects the asteroids onto another orbit. Due to this ejection process, there are gaps, called Kirkwood Gaps, which exist at the main orbital resonances with Jupiter, named after Daniel Kirkwood, who observed and explained the nature of the gaps in 1857.

Comets and asteroids offer clues to the chemical mixture from which the planets formed some 4.6 billion years ago. Asteroid science is a fundamental topic in planetary science and is key to furthering our understanding of planetary formation and the evolution of the Solar System. Ground-based observations and spacecraft missions have provided a wealth of new data in recent years, and forthcoming spacecraft missions promise further exciting results. Burbine (2016) has presented a comprehensive introduction to asteroid science, summarizing the astronomical and geological characteristics of asteroids.

To know the composition of the primordial mixture from which the planets formed, it would be necessary to determine the chemical constituents of the leftover debris from this formation process—the comets and asteroids. Comets and asteroids are rocky materials that originated from two different locations in the Solar system. Whereas a comet is a chunk of solid body, usually around 1–10 km across and made of ices, dust and rock originating from the outer Solar System (Uranus-Neptune region and the Kuiper Belt region), an asteroid is a hydrated rock in orbit located generally between Mars and Jupiter. Sometimes these rocky materials get bounced toward Earth. Visiting an asteroid will provide valuable mission experience and prepare us for the next steps—possibly for the first humans to step on Mars.

Many primitive meteorites originating from the asteroid belt once contained abundant water that is now stored as OH in hydrated minerals. Alexander et al. (2012) estimated the hydrogen isotopic compositions in 86 samples of primitive meteorites that fell in Antarctica and compared the results to those of comets and Saturn's moon, *Enceladus*. Water in primitive meteorites was less deuterium-rich than that in comets and Enceladus, implying that, in contradiction to recent models of the dynamical evolution of the Solar system, the parent bodies of primitive meteorites cannot have formed in the same region as comets. The results also suggest that comets are not the principal source of Earth's water.

The Earth's surface is composed of three layers—the *crust*, *mantle* and *core*. This layered structure can be compared to that of a boiled egg. The crust, the outermost layer, is rigid and very thin compared with the other two. Beneath the oceans, the crust varies little in thickness, generally extending only to about 5 km. The thickness of the crust beneath continents is much more variable but averages about 30 km; under large mountain ranges, such as the Alps or the Sierra Nevada, however, the base of the crust can be as deep as 100 km. Like the shell of an egg, the Earth's crust is brittle and can break (https://pubs.usgs.gov/gip/dynamic/inside.html). The mantle is the second layer of Earth that begins at ≤100 km under the surface and extends up to 2900 km. With the center of Earth around 6380 km from the surface, the only way to study material from such immense depths is through volcanic eruptions and magma samples. While we know about the formation and composition of the crust (the outermost layer), very little is known about the mantle and the core, which are located below the crust.

Researchers have now analyzed a meteorite that could hold clues about the composition of the mantle and offer insights into how Earth was formed. Meteorites are pieces of rock or metal that have fallen to the Earth's surface from outer space as meteors. Extraterrestrial organic matter in meteorites potentially retains a unique record of synthesis and chemical/thermal modification by parent body, nebular and even presolar processes (Alexander et al., 2007).

The planet Earth was formed from the similar material that constitutes present-day asteroids which is mostly made up of the mineral olivine. Note that olivine is a magnesium iron silicate with the chemical formula $(Mg, Fe)_2SiO_4$. Olivine being the primary component of the Earth's upper mantle, it is a common mineral in Earth's subsurface, but weathers quickly on the surface. Therefore, it is important to study olivine at high pressure and high temperature to understand its behavior. Olivine breaks down into bridgmanite (a magnesium-silicate mineral, $MgSiO_3$, the most abundant mineral on earth, making up around 70% of the lower mantle) and magnesiowüstite (a mineral composing 20% of the lower mantle. It is a cubic phase of composition $(Mg,Fe)O$) in the Earth's lower mantle which is one of the most important reactions that largely controls the physical and chemical properties of the Earth's interior. This breakdown may occur where the olivine remains in the solid state or may also form by melting of the olivine. The breakdown assemblage of bridgmanite and magnesiowüstite formed by both of these mechanisms has been reported in few Martian meteorites. Recently, this breakdown assemblage by the solid-state has been reported in the Suizhou meteorite (see Xiande et al., 2006). Studies carried out by Xiande et al. (2006) revealed that the Suizhou is a unique chondrite with specific and unusual shock-related mineralogical features. However, no such breakdown assemblage formed by melting has been found in meteorites originated from the asteroid belt.

In a recent incident of meteorite fall on Earth, a meteorite, which belonged to the asteroid belt located between the orbits of Mars and Jupiter, fell near a village (Kamargaon) in Assam, India, in 2015. This particular kind of meteorite is found in the asteroid belt—formed by accumulation of solid particles during the formation of planets. These materials are at times pulled out from the belt due to collision and gravitational forces. These meteorites have survived high-pressure and high-temperature events during their formation and fall on Earth due to the planet's gravitational pull. Tiwari et al. (2021) analyzed this shocked meteorite—one that has gone through high-pressure and high-temperature conditions due to an impact event—to conclude that it has a similar chemical composition as found in Earth's lower mantle. Tiwari et al. (2021) used a high-resolution electron microscope to image and scan the meteorite and conduct a set of complex analyses on a nanometer scale to find evidence of the complex chemical reaction that forms the Earth's mantle. These investigators found that olivine breaks down into bridgmanite and magnesiowustite in the Earth's lower mantle, which is one of the most important reactions that largely control the properties in the Earth's interior.

Tiwari et al. (2021) for the first time, reported the possible occurrence of bridgmanite and magnesiowüstite formed by incongruent melting of olivine in an ordinary chondrite. Tiwari et al. (2021) propose that this assemblage may have formed at pressure and temperature of ~25 GPa and ~2500°C. According to them, these observations suggest that the dissociation of olivine in the natural systems can also take place by the melting of olivine. The findings state that Earth's mantle was formed from a similar material that constitutes the Assam meteorite, which is mostly made up of a substance known as olivine. This is the first time that researchers have found compositions in a meteorite that is found when olivine is melted at high temperatures and pressures, confirming that the chemical found in the mantle is also present in the asteroid belt. Olivine is a rock-forming mineral found in dark-colored igneous rocks (igneous rocks form when hot, molten rock crystallizes and solidifies. The melt originates deep within the Earth near active plate boundaries or hot spots, then rises toward the surface) and has a very high crystallization temperature compared to other minerals. It is considered an important mineral in Earth's mantle. The samples found in the meteorite are similar to those observed on plate tectonics and could prove useful in studying earthquakes and volcanic activities. Scientists are now looking to prove the breakdown of olivine through lab experiments.

To give another recent example of meteorite fall on Earth, what was described as part of a space rock (a 30-pound chunk of iron meteorite) created a dramatic fireball over Uppsala, Sweden on November 7, 2020. It was found that a half-melted hunk of iron-rich rock found in Uppsala, Sweden, is part of a meteorite. According to the Swedish Museum of Natural History, this meteorite was once part of a larger space rock, probably weighing more than 9 tons, that created a dramatic fireball over Uppsala on Nov. 7, 2020. It is the first time that any meteorite fragments linked to an observed fireball have been recovered in Sweden for 66 years. Iron meteorites are the second-most common kind of meteorite that land on Earth, after stony meteorites. They originate in the cores of planets and asteroids, which means they can hold clues to the formation of the Solar System. Some iron-rich meteorites have been found to harbor minerals not found on Earth. Other types of meteorites contain complex organic compounds, perhaps hinting at how the building blocks of life originally landed on this planet.

Chondritic meteorites are asteroidal fragments that retain records of the first few million years of Solar System history. Many primitive meteorites originating from the asteroid belt once contained abundant water that is now stored as OH in hydrated minerals. Carbonaceous chondrites, generally considered to be the most primitive surviving materials from the early Solar System, form a distinctive group in terms of bulk Mg/Si, Ca/Si, and Al/Si ratios. The carbonaceous chondrites can be subdivided into five groups (CI, CM, CR, CO, and C) based on a number of petrologic and chemical criteria. Petrographic observations indicate that most carbonaceous chondrites have been processed, either by thermal metamorphism in the case of CO and C chondrites or by low-temperature aqueous alteration in the case of CI, CM, and CR chondrites.

CI chondrites, sometimes C1 chondrites, are a group of rare stony meteorites belonging to the carbonaceous chondrites. Samples have been discovered in France, Canada, India, and Tanzania. CM chondrites are a group of chondritic meteorites which resemble their type specimen, the Mighei meteorite. The CM is the most commonly recovered group of the 'carbonaceous chondrite' class of meteorites, though all are rarer in collections than ordinary chondrites. The CR chondrites are breccias consisting of two major components: the large layered chondrules and the matrix (+ dark inclusions). The overall degree of hydration in CR chondrites varies among meteorites, with Al Rais being the most hydrated. Carbonaceous chondrites or C chondrites are a class of chondritic meteorites comprising at least 8 known groups and many ungrouped meteorites. They include some of the most primitive known meteorites. The C chondrites represent only a small proportion (4.6%) of meteorite falls.

Thermal metamorphism resulted in Fe/Mg exchange between chondrules, olivine and pyroxene grains, and matrix, changes in the compositions of metal grains, and textural integration. Aqueous alteration probably produced hydrated phyllosilicate matrix phases and resulted in alteration of chondrules and replacement and vein filling by secondary carbonates and sulfates. The changes incurred during these processes appear to have been largely isochemical. However, if certain constituents behaved as open-system components, volatile elements or compounds may have been depleted during metamorphism, and isotopic patterns may have been changed during aqueous alteration. According to McSween (1979), the recognition of two different types of postaccretional processes resulting in petrological modifications necessitates a reinterpretation of the classification system for carbonaceous chondrites. The bulk hydrogen and nitrogen isotopic compositions of CI chondrites (a group of rare stony meteorites belonging to the carbonaceous chondrites) suggest that they were the principal source of Earth's volatiles (Alexander et al., 2012). Chondrites (the most primitive meteorites) are fragments of main-belt asteroids and are divided into several classes that are subdivided into numerous groups. The classes of chondrites include ordinary chondrites (OC); Rumuruti chondrites (RC); enstatite chondrites (EC); and carbonaceous chondrites (CC).

In a survey of the elemental and isotopic compositions of insoluble organic matter (IOM) from 75 carbonaceous, ordinary and enstatite chondrites, Alexander et al. (2007) found dramatic variations within and between chondrite classes. Based on available information at present, planetary scientists tentatively think that all IOM compositional variations are the result of parent body processing of a common precursor. It is presently uncertain whether the IOM formed in the interstellar medium or the outer Solar System, although the former is preferred in the studies of Alexander et al. (2007).

On the basis of their reflectance spectra and other evidences, the OCs and CCs have been linked to the S- and C-complex asteroids, respectively (Alexander et al., 2012). According to Alexander et al. (2012), it is possible that a number of chondrite groups contributed to Earth's volatile budget, but perhaps the simplest explanation is that most of the hydrogen and nitrogen (as well as other volatiles) was accreted in CI-like material, along with ~10% contributions to both elements from material with a solar isotopic, but elementally fractionated, composition.

Interestingly, water in primitive meteorites was less deuterium-rich than that in comets and Enceladus, implying that, in contradiction to recent models of the dynamical evolution of the Solar System, the parent bodies of primitive meteorites cannot have formed in the same region as comets.

Advances in the discovery and characterization of asteroids over the past decade have revealed an unanticipated underlying structure that points to a dramatic early history of the inner Solar System. The asteroids in the main asteroid belt have been discovered to be more compositionally diverse with size and distance from the Sun than had previously been known. This implies substantial mixing through processes such as planetary migration and the subsequent dynamical processes (DeMeo and Carry, 2014).

According to Levison et al. (2009)'s investigations, "the observed diversity of the asteroid belt is not a direct reflection of the intrinsic compositional variation of the protoplanetary disk. The dark captured bodies, composed of organic-rich materials, would have been more susceptible to collisional evolution than typical main-belt asteroids. Their weak nature makes them a prodigious source of micrometeorites—sufficient to explain why most are primitive in composition and are isotopically different from most macroscopic meteorites."

According to DeMeo and Carry (2014), "Advances in the discovery and characterization of asteroids over the past decade have revealed an unanticipated underlying structure that points to a dramatic early history of the inner Solar System. The asteroids in the main asteroid belt have been discovered to be more compositionally diverse with size and distance from the Sun than had previously been known. This implies substantial mixing through processes such as planetary migration and the subsequent dynamical processes." The main asteroid belt, which inhabits a relatively narrow annulus ~2.1–3.3 AU from the Sun, contains a surprising diversity of objects ranging from primitive ice–rock mixtures to igneous rocks. The standard model used to explain this assumes that most asteroids formed in situ from a primordial disk that experienced radical chemical changes within this zone.

Determining the source(s) of hydrogen, carbon, and nitrogen accreted by Earth is important for understanding the origins of water and life; and for constraining dynamical processes that operated during planet formation. According to Alexander et al. (2012), "The deuterium/hydrogen (D/H) values of water in carbonaceous chondrites are distinct from those in comets and Saturn's moon Enceladus, implying that they formed in a different region of the solar system, contrary to predictions of recent dynamical models. The D/H values of water in carbonaceous chondrites also argue against an influx of water ice from the outer solar system, which has been invoked to explain the nonsolar oxygen isotopic composition of the inner solar system." It may be noted that although chondrites share many similarities with comets, they do not now contain ice.

The United Nations Office for Outer Space Affairs (UNOOSA) (http://www.unoosa.org/oosa/en/ourwork/topics/

neos/index.html) defines a near-Earth object (NEO) as "an asteroid or comet which passes close to the Earth's orbit. In technical terms, a NEO is considered to have a trajectory which brings it within 1.3 astronomical units of the Sun and hence within 0.3 astronomical units, or approximately 45 million kilometers, of the Earth's orbit. NEOs generally result from objects that have experienced gravitational perturbations from nearby planets, moving them into orbits that allow them to come close to the Earth."

As just indicated, NEOs have been nudged by the gravitational attraction of nearby planets into orbits that allow them to enter the Earth's neighborhood. Composed mostly of water ice with embedded dust particles, comets originally formed in the cold outer planetary system while most of the rocky asteroids formed in the warmer inner Solar System between the orbits of Mars and Jupiter. The scientific interest in comets and asteroids is due largely to their status as the relatively unchanged remnant debris from the Solar System formation process some 4.6 billion years ago. The giant outer planets (Jupiter, Saturn, Uranus, and Neptune) formed from an agglomeration of billions of comets and the left-over bits and pieces from this formation process are the comets we see today.

There are three main types of NEA (COSMOS—The SAO Encyclopedia of Astronomy):

(i) Aten Asteroids—Earth crossing asteroids with semi-major axes smaller than 1 AU
(ii) Apollo Asteroids—Earth crossing asteroids with semimajor axes larger than 1 AU
(iii) Amor Asteroids—Earth approaching asteroids with orbits that lie between the Earth and Mars

Astronomers have so far detected approximately 18,000 NEOs, out of a population thought to number in the millions. With expected lifetimes of around 10 million years, the ultimate fate of NEAs may be gravitational ejection from the Solar System or collision with one of the terrestrial planets (note that a terrestrial planet [also called telluric planet, or rocky planet] is a planet that is composed primarily of silicate rocks or metals. Within the Solar System, the terrestrial planets are the inner planets closest to the Sun, i.e., Mercury, Venus, Earth, and Mars). In particular, NEAs that come especially close to the Earth and which are large enough to threaten civilization are labeled "Potentially Hazardous Asteroids."

Asteroids tens to hundreds of meters in diameter constitute the most immediate impact hazard to human populations. Studies carried out by Bland and Artemieva (2006) predict that "iron meteorites $>5 \times 10^4$ kg (2.5 m) arrive at the Earth's surface approx. once every 50 years. Iron bodies a few meters in diameter (10^5–10^6 kg), which form craters ~100 m in diameter, will strike the Earth's land area every 500 years. Larger bodies will form craters 0.5 km in diameter every 20,000 years, and craters 1 km in diameter will be formed on the Earth's land area every 50,000 years. Tunguska events (low-level atmospheric disruption of stony bolides $>10^8$ kg) may occur every 500 years. Bodies capable of producing hazardous tsunami (~200 m diameter projectiles) should strike the Earth's surface every ~100,000 years." Past collisions with such objects are evident on local (the Tunguska event) and global (the mass extinction at the Cretaceous–Tertiary boundary) scales.

The year 2013 Chelyabinsk event, in which a meteoroid exploded over Chelyabinsk, Russia, is a recent example of small asteroid hazard. When the meteorite entered Earth's atmosphere, it created a bright fire ball. Minutes later, a shock wave blasted out nearby windows, injuring hundreds of people. The meteoroid weighed around 10,000 tons, but only about 2000 tons of debris were recovered, which meant something happened in the upper atmosphere that caused it to disintegrate. Tabetah and Melosh (2018) summarized the 2013 Chelyabinsk event thus, "The entry and subsequent break-up of the ~17–20 m diameter Chelyabinsk meteoroid deposited approximately 500 kT of trinitrotoluene (TNT) equivalent energy to the atmosphere, causing extensive damage that underscored the hazard from small asteroid impacts. The break-up of this meteoroid was characterized by intense fragmentation that dispersed most of the original mass. In modeling of the entry, the apparent mechanical strength of the meteoroid during fragmentation, ~1–5 MPa, is two orders of magnitude lower than the mechanical strength of the surviving meteorites, ~330 Mpa." Tabetah and Melosh (2018) implemented a two-material computer code that allowed them to fully simulate the exchange of energy and momentum between the entering meteoroid and the interacting atmospheric air. Their simulations revealed a previously unrecognized process in which the penetration of high-pressure air into the body of the meteoroid greatly enhanced the deformation and facilitated the break-up of meteoroids similar to the size of the meteoroid that exploded over Chelyabinsk.

It may be noted that iron meteoroids are much smaller and denser, and even relatively small ones tend to reach the Earth's surface. For example, in the Canyon Diablo impact event (Artemieva and Pierazzo, 2009), an iron object 50 m in diameter impacted the Earth's surface after breaking up in the atmosphere, and this impact event created a Meteor Crater. It is likely that the Earth will encounter another of these objects in the future, and therefore several monitoring programs have been established to warn of possible danger.

1.3 Giant-impact theory on the origin of Earth's Moon

The Moon is Earth's only natural satellite. At about one-quarter the diameter of Earth, it is the largest natural satellite in the Solar System relative to the size of its planet, the fifth largest satellite in the Solar System overall, and is larger than any known dwarf planet.

Our own Earth's Moon, and the question of how it was formed, has long been a source of fascination and wonder. Moon has long held on to one of its most sought-after secrets: when it was formed, and how. The formation of the Moon has remained something of a puzzle. Various theories have been put forward over the centuries but none of them has been conclusively proved until recently; many, in fact, have fallen out of favor because of various inconsistencies.

Earth was formed at roughly the same time as the Sun and other planets in the Solar System some 4.6 billion years ago when the Solar system coalesced from a giant, rotating cloud of gas and dust known as the "Solar nebula." As the nebula collapsed under the force of its own gravity, it spun faster and flattened into a disk. Most of the material in that disk was then pulled toward the center to form the Sun. Other particles within the disk collided and stuck together to form ever-larger bodies, including Earth. It was believed that Earth started off as a waterless mass of rock. However, in recent years, new analyses of minerals trapped within ancient microscopic crystals suggest that there was liquid water already present on Earth during its first 500 million years.

Processes governing the evolution of planetesimals (minute planets; bodies which could come together with many others under gravitation to form planets) are critical to understanding how rocky planets are formed, how water is delivered to them, the origin of planetary atmospheres, how cores and magnetic dynamos develop, and ultimately, which planets have the potential to be habitable. Theoretical advances and new data from observations of asteroids and meteorites, coupled with spacecraft missions such as *Rosetta* and *Dawn*, have led to major advances in this field over the last decade (see Elkins-Tanton and Weiss, 2017). Theories about the Earth's formation, features about the young Earth, and the early life on this wonderful planet continue to spring surprises to the curious minds. Among the planets in the Solar System, there are a category of planets known as "terrestrial planets." This category of planets is also known as "telluric planets," or "rocky planets." These planets are composed primarily of silicate rocks or metals. Within the Solar System, these categories of planets are the inner planets closest to the Sun, that is, Mercury, Venus, Earth, and Mars.

There is a consensus opinion among planetary scientists that the Sun and its planets formed from a rotating disk of gas and grains largely made of molecular hydrogen and helium, known as the proto-solar nebula. This event is believed to have occurred about 4.6 billion years ago.

In the study of the origin of Earth's Moon, two theories have been advanced under the Giant-Impact concept; (i) Single Giant Impact Theory, and (ii) Multiple Giant Impact Theory. The latter theory has been developed to correct some of the inadequacies and inconsistencies between theory and observations noted in the former theory (particularly, the isotopic signatures from the Earth and the Impactor).

1.3.1 Single giant impact theory

The currently accepted theory for how the Earth gained its Moon is the so-called Big Whack theory, sometimes called the Big Splash (Fig. 1.5), or the Theia Impact (more popularly known as the "Giant-Impact" theory) put forward first by William Hartmann and Davis in 1975 and supported by Cameron and Ward in 1976 (Hartmann and Davis, 1975; Cameron and Ward, 1976). Note that NASA's Spitzer Space Telescope found evidence that a high-speed collision of this sort occurred a few thousand years ago around a young star, called HD 172555, still in the early stages of planet formation. The star is about 100 light years from Earth.

According to the Giant-Impact theory, an impact of another celestial body (an Impactor) on the proto-Earth (i.e., an early-stage fluid-Earth) approximately 4.6 billion years ago, in the Hadean eon (about 20 to 100 million years after the Solar System coalesced) set the initial conditions for the formation and evolution of the Earth–Moon system. Preliminary calculations by Cameron (1985) have shown that the single-impact theory is plausible. According to the Giant-Impact hypothesis, toward the end of the planetary accumulation process, the proto-Earth collided with a

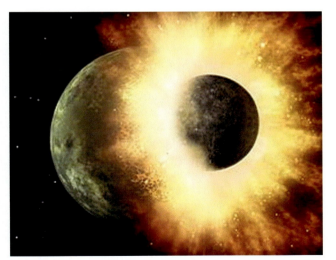

FIG. 1.5 Artist's depiction of a collision between two planetary bodies. According to the Single Giant Impact Theory, such an impact between Earth and a Mars-sized object likely formed Earth's Moon. Note that NASA's Spitzer Space Telescope found evidence that a high-speed collision of this sort occurred a few thousand years ago around a young star, called HD 172555, still in the early stages of planet formation. The star is about 100 light years from Earth. *Author: NASA/JPL-Caltech This file is in the public domain in the United States because it was solely created by NASA. NASA copyright policy states that "NASA material is not protected by copyright unless noted." Source: https://commons.wikimedia.org/wiki/File:Artist%27s_concept_of_collision_at_HD_172555.jpg https://en.wikipedia.org/wiki/Giant-impact_hypothesis#/media/File:Artist's_concept_of_collision_at_HD_172555.jpg.*

planetary body (an Impactor) having a substantial fraction of the proto-Earth mass. At the velocity characteristic of the collision (11–15 km/s), most rocky materials got vaporized.

As just mentioned, the most widely accepted hypothesis with regard to the story of origin of our planet Earth is that a few billion years ago, a hypothetical Mars-sized object happened to form in the inner Solar System, right in the orbital path of a proto-planet that would eventually become Earth. Note that a proto-planet is a large body of matter in orbit around the Sun or a Star and believed to be developing into a planet. The two eventually collided—known as "the giant impact" (Fig. 1.5)—and that large-scale and violent blow sent masses of debris out into space, some of which formed our Moon (read, Earth's Moon). Most planetary scientists believe that the Moon formed from catastrophic collision between the proto-Earth and a planet-sized body, which has been given the name "Theia." Theia is considered to be Earth proper, which officially came to be at around 4.5 billion years ago.

Earth formed by accretion (i.e., growth by the gradual accumulation of additional layers or matter) of Moon-to-Mars-size embryos coming from various heliocentric distances. Taking meteorites as a guide, most models assume that the Earth must have formed from a heterogeneous assortment of embryos with distinct isotopic compositions (Pahlevan and Stevenson, 2007; Mastrobuono-Battisti et al., 2015; Kaib and Cowan, 2015).

Billions of years ago, a messy disk of debris encircled the Sun, which would gradually clump together to form bigger and bigger orbiting objects. Eventually, these objects would get so big that when they collided because of the instability of their orbits, they would form rocky planets like Earth, Mars, and Venus. Current thinking on planetary formation states that one of these objects is the proto-planet, and the other is the colliding catalyst that completes its transition into a proper planet. If the colliding object is small, the newly formed planet would be showered with its shattered remains like meteorites. But planetary scientists are inclined to believe that if the colliding object is big enough, these shattered remains can orbit the newly formed planet for some time, like Saturn's rings, before eventually being swallowed up by it. But questions such as what if the two colliding objects were equal in size; without a clear proto-planet and colliding partner, who does what etc. remained unanswered.

In an attempt to unravel the mystery that shroud the formation of Earth some 4.6 billion years ago, Lock and Stewart (2017) mathematically modeled what happened when one very big and hot spinning object slams into a similar spinning mass at a very high angular momentum. They found that the most violent and evenly matched collisions would produce "synestias"—peculiar haloes of vaporized rock spinning around a molten core. Scientists think that about 4.6 billion years ago, synestia formation was a painfully common occurrence as the Solar System formed from the giant disk of dust, gas, and debris that orbited the young Sun. Lock and Stewart (2017) proposed that after a century or so a synestia will have lost so much heat that it would condense back into a solid. The researchers suspect that most planets that exist today could have been synestias at some point in their history.

Earth's Moon—Earth's closest celestial body—is among the strangest planetary bodies in the Solar System. Its orbit lies unusually far away from Earth, with a surprisingly large orbital tilt. Note that Earth's Moon's orbit around Earth is elliptical. At perigee—its closest approach—the moon comes as close as 225,623 miles (363,104 km). At apogee—the farthest away it gets—the moon is 252,088 miles (405,696 km) from Earth. On average, the distance from Earth to the moon is about 238,855 miles (384,400 km). Planetary scientists have struggled to piece together a scenario that accounts for these and other related characteristics of the Earth–Moon system. Ćuk et al. (2016) have suggested that the impact that formed the Earth's Moon also caused calamitous changes to Earth's rotation and the tilt of its spin axis.

Planetary research suggests that the impact sent Earth spinning much faster, and at a much steeper tilt, than it does today. In the several million years since that impact, complex interactions between Earth, its Moon, and Sun have smoothed out many of these changes, resulting in the Earth–Moon system that we see today. In this scenario, the remaining anomalies in the Moon's orbit are considered to be relics of the Earth–Moon system's explosive past.

Ćuk et al. (2016) experimented with many different scenarios. But the most successful ones involved a Moon-forming impact that sent Earth spinning extremely fast—as much as twice the rate predicted by other models. According to Ćuk et al. (2016)'s studies, both factors—a highly tilted, fast spinning Earth and an outwardly-migrating Moon—contributed to establishing the Moon's current weird orbit. The new-born Moon's orbit most likely tracked Earth's equator, tilted at a steep 60–80° angle that matched Earth's tilt immediately after the giant impact (Fig. 1.5).

Dauphas (2017) sought to explain the similarity in the composition of Earth and Moon rocks, based on a study of the abundance of oxygen isotopes—the yardstick for determining a rock's origin. Note that an isotope is any of two or more forms of a chemical element, having the same number of protons in the nucleus (i.e., the same atomic number), but having different numbers of neutrons in the nucleus. The atomic number is identical to the charge number of the nucleus. The atomic number uniquely identifies a chemical element. The atomic mass number is the total number of protons and neutrons in the nucleus of an atom. Isotopes of a single element possess almost identical properties.

By studying high-precision measurements, Dauphas (2017) showed that the Earth, the Moon and meteorites with a high concentration of the mineral enstatite have almost indistinguishable isotopic compositions. Therefore, the Earth most likely accreted from an isotopically homogenous reservoir. However, they diverged in their chemical evolution owing to subsequent fractionation by nebular and planetary processes, as argued by Dauphas et al. (2015).

Although the hypothesis of lunar origin by a single giant impact, as believed by several researchers, can explain some aspects of the Earth–Moon system, Rufu et al. (2017) found that it is difficult to agree with (reconcile) giant-impact models with the compositional similarity of the Earth and Moon without violating angular momentum constraints. According to Rufu et al. (2017), hitherto existing dynamical models suggest a chaotic mixing between the terrestrial bodies. Probably, the dynamical models do not incorporate important physical processes in the proto-planetary disk. Either way, assuming this is true, this will only strengthen Rufu et al. (2017)'s conclusion by increasing the success rate of forming the Moon, but it may be harder to prove whether several impacts would be needed to form the Moon or only one is enough.

Rufu et al. (2017)'s study, based on their dynamical solution, suggests that billions of years ago, a number of small objects collided with Earth at high velocity, over a period of millions of years. Such small collisions could mine more material from Earth than a single large collision. They carried out numerical simulations, which suggest that the debris agglomerated into small moonlets that, in turn, merged to become the Moon. They concluded that, assuming efficient merger of moonlets, a multiple-impact scenario can account for the formation of the Earth–Moon system with its present properties.

The newly proposed theory by Rufu et al. (2017) runs counter to the commonly held "giant impact" paradigm that the Moon is a single object that was formed following a single giant collision between a small Mars-like planet and the ancient Earth. In its last stages of the growth, Earth experienced many giant impacts with other bodies. Each of these impacts contributed more material to the proto-Earth, until it reached its current size. The tidal forces from Earth could cause moons to slowly migrate outward (the current Moon is slowly doing that at a pace of about 1 cm a year).

Using a computer software, named Highly Eccentric Rotating Concentric U (potential) Layers Equilibrium Structure (HERCULES), that mathematically solves for the equilibrium structure of planets as a series of overlapping constant-density spheroids, and a smoothed particle hydrodynamics software, Lock and Stewart (2017) showed that Earth-like bodies display a dramatic range of morphologies during rocky planet formation and evolution (from highly vaporized, rapidly rotating planetary structures).

For any rotating planetary body, there is a thermal limit beyond which the rotational velocity at the equator intersects the Keplerian orbital velocity. Beyond this corotation limit (CoRoL), a hot planetary body forms a special structure, which Lock and Stewart (2017) named a "synestia," with a corotating inner region connected to a disk-like outer region. Based on model studies of giant impacts and planet formation, Lock and Stewart (2017) showed that typical rocky planets are substantially vaporized multiple times during their formation (i.e., in their infant- and growth-stages), and are likely to form synestias. For the expected angular momentum of growing planets, a large fraction of postimpact bodies will exceed the CoRoL and form synestias. It was found that the common occurrence of hot, rotating states during accretion has major implications for planet formation and the properties of the final planets. In particular, the structure of postimpact bodies influences the physical processes that control accretion, core formation, and internal evolution. Synestias also lead to new mechanisms for moon formation. Finally, according to Lock and Stewart (2017)'s studies, the wide variety of possible structures for terrestrial bodies also expands the mass-radius range for rocky exo-planets.

As already indicated, a giant doughnut-shaped object called a synestia consists of a mass of hot, vaporized rock spinning around a molten mass left over from a planetary collision. Scientists believe that about 4.6 billion years ago, synestia formation was a painfully common occurrence as the Solar System formed from the giant disk of dust, gas, and debris that orbited the young Sun. Lock and Stewart (2017) argue that in the high-energy planetary collision events, some of the gas and debris would fly off so fast as to go into orbit. The centrifugal force operating on the spinning mass would form a donut because each point in the mass would be spinning at the same angular velocity about the core. It has been argued that the early Earth is likely to have suffered such an impact and formed a synestia that lasted about a hundred years, though such bodies would be longer lived around larger planets. If this did indeed occur as hypothesized, then it could explain how the Moon formed and why it is so similar in composition to the Earth. Perhaps at some point in the distant past, two bodies collided to form a synestia out of which the Earth formed from the molten core and the Moon from the orbiting mass. Planetary scientists believe that synestia might be found in other celestial systems once astronomers start looking for them alongside rocky planets and gas giants.

A term frequently used in the development of the Giant-Impact hypothesis is "Roche lobe," after the French mathematician and astronomer Edouard Roche (Fig. 1.6). Édouard Albert Roche—who taught at the University of Montpellier for 34 years—worked in the field of celestial mechanics, which required considerable mathematical dexterity (skill), which Roche apparently possessed in abundance. And in the field of eponymity (attaching name), he was an unusual triple-threat, contributing his name (*Roche*) to the *Roche lobe*, the *Roche sphere*, and the *Roche limit*—which have

FIG. 1.6 French mathematician Edouard Roche (1820–1883), well-known for the invention of the Roche limit, the Roche sphere, and the Roche lobe—which have been found useful in the investigation of celestial bodies in a binary system such as stars and their planets and planets and their moons. *(photograph, 1850s (academie-sciences.fr)) (Source: https://www.lindahall.org/edouard-roche/.)*

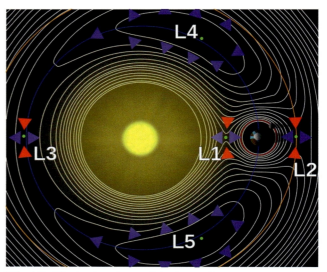

FIG. 1.7 The Roche spheres of the Sun (yellow) and the Earth. The gravitational forces of the two bodies are neutralized at the Lagrange points L1 to L5 (Wikimedia commons). *(Source: https://www.lindahall.org/edouard-roche/.)*

been found useful in the investigation of celestial bodies in a binary system such as stars & their planets and planets & their moons.

With regard to the interesting topics constituting *Roche lobe*, *Roche sphere*, and *Roche limit*, the below discussion may be worthwhile. When one draws a set of equipotential curves for the gravitational potential energy for a small test mass in the vicinity of two orbiting celestial bodies in a binary system (e.g., Sun and its planets; Earth and its Moon), there is a critical curve shaped like a figure-8 (see Fig. 1.7) which can be used to portray the gravitational domain of each celestial body in motion. If the figure-8 is rotated around the line joining the two celestial bodies in this binary system, two lobes are produced—known as *Roche lobes*. The term "Roche lobe" is used to describe a distinctively shaped region surrounding a celestial body in a binary system. This teardrop-shaped space defines the region in which material is bound to the celestial body by gravity.

The point where the two gravitational pulls just balance is called a *Lagrange point*, named after Joseph-Louis Lagrange, a former Scientist of the Day. There are five Lagrange points for the Sun-Earth system, and they were identified some 80 years before Roche, by Joseph-Louis Lagrange.

The point at which the Roche lobes of the two celestial bodies touch is called the inner *Lagrangian point*. If a celestial body in a close binary system evolves to the point at which it fills its Roche lobe, calculations predict that material from this celestial body will overflow both onto the companion celestial body (via the L1 point) and into the environment around the binary system.

The *Roche lobe* becomes all the more important when the two orbiting bodies are gaseous, as with a binary star. If the stars are close enough, gas will be pulled from each star toward the other, and the tear-shaped envelope of gas is the *Roche lobe*. The diagram shows a binary star system, showing the *Roche lobes* of a red giant star and a white dwarf.

The *Roche sphere* is the volume around a celestial body in a binary system in which, if you were to release a particle within this volume, it would fall back onto the surface of that celestial body. However, a particle released above the Roche lobe of either celestial body will, in general, occupy the circumbinary region that surrounds both celestial bodies. The Roche sphere also applies to two-body systems, such as the Sun and the Earth, or the Earth and the Moon. The Sun's Roche sphere is the region where the Sun's gravitational sphere predominates. Outside the Roche sphere, the Earth's gravitational field takes over.

In celestial mechanics, the *Roche limit*, also called *Roche radius*, is the distance from a celestial body within which a second celestial body, held together only by its own force of gravity, will disintegrate due to the first body's tidal forces exceeding the second body's gravitational self-attraction.

The Roche limit is applicable to any two mutually orbiting bodies (e.g., the Earth and the Moon). The Roche limit is the minimum distance that the Moon (or any satellite held together only by gravity) can be from the Earth (or any large body) without breaking up. If the Moon comes any closer than the Roche limit, it will break apart because of tidal forces. Thus, no planet can have a moon that lies within the Roche limit—it can have only rings. The size of the Roche limit depends on the mass of the two bodies. For the Earth–Moon system, the Roche limit is about 6000 miles, center to center. So, if the Moon were pushed toward the Earth, as it approached the Roche limit it would gradually deform into a potato-shape, and when it reached the Roche limit, it would no longer be able to hold itself together, and it would break apart like instant potato flakes, spreading out slowly into a ring. Of course, while this was happening (i.e., when the Moon was falling apart), people on the Earth would be experiencing tides that were thousands of feet high.

Fig. 1.8 provides an illustration of the fate of a celestial body, which was initially located well outside the Roche limit of another celestial body and approaching it under the influence of its force of gravity—initially spherical in shape, turning to potato shape as it approaches the *Roche limit*, and finally disintegrating when it begins to cross the *Roche limit*.

Because the Roche limit in the Earth–Moon binary system lies so close to the Earth, it has not affected our relationship to the Moon. But in the case of very massive planets, it does come into play, and the reason why Saturn has a ring and not a moon at that location is a direct result of Saturn's Roche limit, as Roche himself pointed out.

Laying stress on the importance of gravitational torques in understanding the origin of the Earth–Moon system, the Giant-Impact hypothesis investigators assumed that the proto-Earth-Impactor collision took place with a fairly large impact parameter (defined as the perpendicular distance between the path of a projectile and the center of a potential field created by an object that the projectile is approaching), so that the combined angular momentum of the two bodies was at least as great as that now in the Earth–Moon system.

Returning to the main theme of this Section, initially it was hypothesized that the pressure gradients in the vapor cloud (resulting from the collision between the proto-Earth and the Impactor) have played a crucial role in accelerating a relatively small fraction of the vapor into orbit about the proto-Earth, forming a disk (Cameron and Ward, 1976; Cameron, 1985). The dissipation of this disk was then thought to spread matter radially so that the Moon could collect together gravitationally beyond the Roche lobe (Ward and Cameron, 1978; Thompson and Stevenson, 1983). The Giant-Impact hypothesis investigators have subsequently realized that gravitational torques are much more important than gas pressure gradients in planetary-scale collisions. Thus, the initially proposed "gas pressure gradients" hypothesis was abandoned.

Surprisingly, the question of the origin of Earth's Moon remained unresolved even after the scientifically productive Apollo missions, and it was only in the early 1980s that a model emerged—the Giant-Impact hypothesis—that eventually gained the support of most lunar scientists. It is now well recognized that during the formation of the Solar System, the Earth and the other terrestrial planets (i.e., Mercury, Venus, Earth, and Mars that are composed primarily of silicate rocks or metals) accreted in a gas-free environment, mostly from volatile-depleted planetesimals (i.e., solid objects thought to exist in protoplanetary disks (Fig. 1.3A, B, D) and debris disks (Fig. 1.3C) and believed to have formed in the Solar System about 4.6 billion years ago) which were already differentiated into metallic cores and silicate mantles.

In this scenario the proto-Earth, shortly after its formation from the solar nebula (gaseous cloud from which the Sun and planets formed by condensation), was struck a glancing blow by an Impactor (a body the size of Mars). Prior to the impact, both the proto-Earth and the Impactor already had undergone differentiation into core and mantle. The titanic collision ejected a cloud of fragments, which

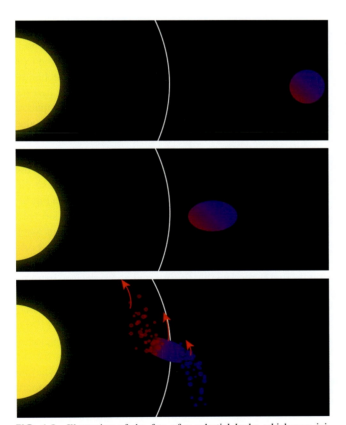

FIG. 1.8 Illustration of the fate of a celestial body, which was initially located well outside the Roche limit of another celestial body and approaching it under the influence of its force of gravity—initially spherical in shape, turning to potato shape as it approaches the *Roche limit*, and finally disintegrating when it begins to cross the *Roche limit*. (Top) Diagram showing a planet outside the Sun's Roche limit (white arc) (Wikimedia commons). (Middle) Diagram showing the same planet deforming as it is brought closer to the Roche limit (white arc) (Wikimedia commons). (Bottom) Diagram imagining a planet at the Sun's Roche limit (white arc), breaking apart from the Sun's gravitational force (Wikimedia commons). *(Source: https://www.lindahall.org/edouard-roche/.)*

aggregated into a full or partial ring around Earth and then coalesced into a proto-Moon. The ejected matter consisted mainly of mantle material from the Impactor (i.e., the colliding body) and the proto-Earth, and it experienced enormous heating from the collision. As a result, the proto-Moon that formed was highly depleted in volatiles, and relatively depleted in iron. Computer modeling of the collision shows that, given the right initial conditions, an orbiting cloud of debris as massive as the Moon could indeed have formed.

At a time when the single-impact hypothesis for forming the Moon has gained some favorable attention, Benz et al. (1986) reported the results of a series of three-dimensional numerical simulations of a hypothetical "Giant Impact" between the proto-Earth and an object about 0.1 of its mass. The tool (numerical method) used for the simulations, to explore the conditions in which a major planetary collision may have been responsible for the formation of the Earth–Moon system, was the so-called "smooth particle hydrodynamics" (SPH) method. Cameron (1997) neatly explained the SPH method thus: "In this method the domain of the computation is not divided into spatial cells, as in ordinary hydrodynamics, but rather the matter in the domain of computation is divided into smooth overlapping spheres. The density of a sphere has a radial distribution that is bell-shaped, with highest density at the center and a sharp outer edge. The particles carry individual internal energies, but such properties as density and pressure are collectively determined by the overlap between the density distributions of the particles; these are meaningful only when typically, a few tens of particles contribute to the overlap." While carrying out the SPH method of investigations, all particles were given the same mass (but could vary in composition), and their smoothing lengths were given fixed values (Cameron, 1997). Simulations were carried out with both the proto-Earth and the Impactor set to a temperature of 4000 K, presumed to have been developed by a prior history of collisions in the accumulations of these bodies (Cameron, 1997).

The SPH method was first introduced in the astronomical context by Lucy (1977), and was shown, especially by Gingold and Monaghan (1979, 1981, 1982, 1983; Monaghan and Gingold, 1983), to give very good results in many different applications. The SPH method is a fully Lagrangian one in which the trajectories of a finite number of mass points are followed in time. Each of these particles is assumed to have its mass spread out in space according to a given distribution called the kernel. For computational convenience both objects (proto-Earth and the Impactor) were assumed to be composed of granite. They studied the effects on the outcome of collision of varying the impact parameter, the initial internal energy, and the relative velocity. Their results showed that if the impact parameter is large enough so that the center of the Impactor approximately grazes the limb of the proto-Earth, the Impactor is not completely destroyed. Instead, part of the Impactor forms a clump in a large elliptical orbit about the Earth. Benz et al. (1986) found that this clump does not collide with the proto-Earth, because the effect of angular momentum transfer due to the rotation of the deformed Earth, modified the ballistic trajectory. However, according to Benz et al. (1986), because the orbit of the clump comes close to the Earth (within the *Roche limit*) the clump will be destroyed and spread out to form a disk around the proto-Earth. Benz et al. (1986)'s results indicated that the amount of angular momentum in the Earth–Moon system thus obtained fell short of the observed amount. Benz et al. (1986) found that this deficiency would be eliminated if the mass of the Impactor were somewhat greater than the one assumed by them. The scenario for making the Moon from a single-impact event was supported by their simulations.

As indicated above, the results of the simulations reported by Benz et al. (1986) showed that substantial amounts of mass could be injected into an orbit about the Earth. In Benz et al. (1987), these simulations have been made more realistic by introducing iron cores into the two planetary bodies (proto-Earth and the Impactor). They found that the iron cores play a distinct and important role in the mechanics of mass injection into orbit around the Earth. They investigated mainly what happens when the mass of the Impactor varies. A limited number of "high"-velocity cases were also considered. A total of nine new simulations were run by these investigators for various combinations of parameters. They also explored a range of mass ratio parameters of the Impactor relative to the target and identified the mass ratio interval in which the results were found to be promising for lunar formation. Benz et al. (1987)'s model included self-gravity in a completely self-consistent and rigorous way. Note that self-gravity is the dominant force when a body exceeds a size of about 50–100 km across. It is the self-gravity that is responsible for internal compression, increasing internal density and making vaporization of material more difficult. For both granite and iron, Benz et al. (1987) used a modified version of the so-called "Tillotson (1962) equation of state" (the ANEOS equation of state developed by Thompson and Lauson (1984), which includes a two-phase medium treatment in a more correct thermodynamic way). This analytical equation of state includes treatment of the melt liquid-vapor and solid-vapor transitions (Benz et al., 1987). Their results confirmed the ability of the code to compute collision problems accurately, in particular, the transformation of kinetic energy into internal energy and then the release of that energy as kinetic energy of an expanding cloud. Benz et al. (1987)'s modified simulations, as indicated above, revealed that the optimum conditions were obtained when the mass ratio of the Impactor to the proto-Earth was 0.136.

In another set of a series of modified computer simulations, Benz et al. (1989) investigated the hypothetical

Giant-Impact, which is believed to have culminated in the origin of the Earth–Moon system, by varying the equation of state and other related parameters. Their results reaffirmed the previous principal conclusions: the collisions produced a disk of rocky material in orbit, with most of the material derived from the impacting object. Benz et al. (1989)'s investigations also confirmed the conclusions of the previous studies that gravitational torques, and not pressure gradients, inject the orbiting mass. A particularly interesting finding gleaned from Benz et al. (1989)'s investigations was that the debris from the destroyed impacting object tends to form a straight rotating bar which is very effective in transferring angular momentum. Benz et al. (1989)'s investigations further indicated that "If the material near the end of the bar extends well beyond the *Roche lobe*, it may become unstable against gravitational clumping."

It has been noted that the method successively employed by various investigators, to explore the conditions in which a major planetary collision may have been responsible for the formation of the Earth–Moon system, has been successively improved variants of the smoothed particle hydrodynamics method (SPH). In another study, Cameron and Benz (1991) run 41 simulations of the planetary collisions, exploring some of the relevant parameter space. The parameters varied in their investigations include the mass ratio of the Impactor to the proto-Earth (values of 0.14, 0.16, and 0.25), the angular momentum in the collision (in units of the present angular momentum of the Earth–Moon system, from slightly more than 1 to slightly less than 2), and the velocity at infinity (0, 5, and 7 km/s).

According to Cameron and Benz (1991), their investigations present us with "two possible scenarios for the formation of the Moon: either directly in the collision or after dissipation and evolution of the disk." Their studies revealed that these collisions do "very extensive damage to the proto-Earth. The iron core of the Impactor plunges through the mantle of the proto-Earth and forms a very hot layer (several tens of thousands of degrees) around the surface of the proto-Earth core; this will later provide a heat source to drive vigorous mantle convection. The outer part of the mantle of the proto-Earth is also strongly heated, with rock decomposition products in gaseous form with an initial photospheric surface (i.e., outer shell from which light is radiated) at a temperature of 16,000 K or higher and a radius more than 20% greater than that of the Earth. This hot mantle surface should lead to the hydrodynamic escape of the Earth's primordial atmosphere and must be taken explicitly into account in any calculations of the evolution of the orbiting disk."

To circumvent the drawbacks found in the simulations reported in Cameron and Benz (1991), Cameron (1997) initiated yet another series of simulations of the Giant-Impact. Whereas a total of 3008 particles (all of the same mass) were used for each simulation reported in the earlier studies, in the new simulations reported by Cameron (1997) "the total was raised to 10,000 particles, which were equally divided between the proto-Earth and the Impactor, but in general the masses of the particles in these two bodies were unequal. The runs used several different mass ratios of Impactor to proto-Earth and varied the angular momentum in the collision. They were all started with zero velocity at infinity, and the results have been sufficiently definitive that it was clear that this parameter did not need to be varied." Unlike in previous simulation studies addressed above, in which the initial temperature was taken as 4000 K, this parameter was lowered to 2000 K in Cameron (1997)'s study to prevent extensive evaporation of rock vapor from the two planetary surfaces prior to collisions.

Cameron (1997) noted that the results reported in Cameron and Benz (1991) can be trusted to give good approximations to the dynamics of the collisions themselves and to the conditions in the interiors of the bodies in which extensive overlap of the particles occurred, but the particles in external orbits or that escaped were usually isolated from contact and thus could not respond to pressure gradients that would ordinarily affect their motions if they were in a gaseous state.

Cameron (1997)'s modified simulation study arrived at a "new and quite simple picture of the consequences of a Giant-Impact. Wherever the surface of the proto-Earth is hit hard by the impact, a very hot magma is produced. From this hot surface, rock vapor evaporates and forms a hot extended atmosphere around the proto-Earth. The atmosphere is partly hydrostatically and partly centrifugally supported. The mean temperature is in excess of 4000 K out to about 8 Earth-radii and is in excess of 2000 K out to about 20 Earth-radii." According to Cameron (1997), "The above description would apply to just about any Giant Impact involving an Impactor with at least 10% of an Earth mass. A candidate Moon-forming Giant Impact must also possess at least the present value of the Earth–Moon angular momentum, which places constraints on the Impactor." It may be noted that previous studies have shown that the Impactor needs to have at least 14% of an Earth mass to swallow up the Impactor iron core and avoid getting too much iron in the Moon. Cameron (1997) suggested that, "apart from this constraint it appears from the present simulations that any division of mass between the proto-Earth and the Impactor can produce a promising set of conditions." Whereas the previous simulations of the hypothetical Giant Impact appear to have characterized the internal effects of a Giant Impact on the proto-Earth, Cameron (1997)'s simulation has characterized the external environment of the proto-Earth following a Giant Impact. Cameron (1997) has suggested that the most promising direction for future simulations appears to involve Impactor/proto-Earth mass ratios around 0.3–0.5.

The giant-impact hypothesis has been favored for the formation of the Moon (Canup and Asphaug, 2001). Supporting evidences include:

- Earth's spin and the Moon's orbit have similar orientations (Mackenzie, 2003)
- The Earth–Moon system contains an anomalously high angular momentum. Meaning, the momentum contained in the Earth's rotation, the Moon's rotation, and the Moon revolving around the earth is significantly higher than the other terrestrial planets. A giant impact may have supplied this excess momentum.
- Moon samples indicate that the Moon was once molten down to a substantial, but unknown, depth (during the crystallization of the Lunar Magma Ocean over time, the first solids to form (e.g., olivine) are denser than the surrounding magma, thus sink toward the interior). This may have required more energy than predicted to be available from the accretion of a body of the Moon's size. An extremely energetic process, such as a giant impact, could provide this energy.
- The Moon has a relatively small iron core. This gives the Moon a lower density than the Earth. Computer models of a giant impact of a Mars-sized body with the Earth indicate the impactor's core would likely penetrate the Earth and fuse with its own core. This would leave the Moon with less metallic iron than other planetary bodies.
- The Moon is depleted in volatile elements (the group of chemical elements and chemical compounds that can be readily vaporized) compared to the Earth. Vaporizing at comparably-lower temperatures, they could be lost in a high-energy event, with the Moon's smaller gravity unable to recapture them while the Earth did.
- There is evidence in other star systems of similar collisions, resulting in debris disk.
- Giant collisions are consistent with the leading theory of the formation of the Solar System.
- The stable-isotope ratios of lunar and terrestrial rock are identical, implying a common origin (Wiechert et al., 2001).

Despite the above arguments, there remain several questions concerning the models of the giant-impact hypothesis (Clery, 2013). The energy of such a giant impact is predicted to have heated Earth to produce a global magma ocean, and evidence of the resultant planetary differentiation of the heavier material sinking into Earth's mantle has been documented (Rubie et al., 2007). However, there is no self-consistent model that starts with the giant-impact event and follows the evolution of the debris into a single Moon. Other remaining questions include when the Moon lost its share of volatile elements and why Venus—which experienced giant impacts during its formation—does not host a similar moon.

Even after much studies as indicated above, the origin of Earth's Moon remained a subject of considerable debate, as evidenced by emergence of several new ideas such as a hit-and-run giant impact scenario (Reufer et al., 2012), forming the Moon from terrestrial silicate-rich material (De Meijer et al., 2013) and so forth, as well as refinements of previously proposed models (Canup, 2012; Stewart and Ćuk, 2012). However, regardless of which model is more applicable to the lunar origin, one common feature appears to be processing of the Moon-forming material in a magma disk at very high temperatures corresponding to several thousands of Kelvins. It is hypothesized that the accretion of the Moon was followed by a Lunar Magma Ocean (LMO) phase, crystallization of which led to differentiation of the Moon, resulting in the planetary body as we know it today. The original depth, timing and the type of crystallization involved in the LMO have been a topic of considerable interest and debate in the lunar science community (see Anand et al., 2014).

Another attractive feature of the impact model is that it provides an explanation for the low bulk density of the Moon that is related to its small core, and hence its low iron/silicon ratio. Impact simulations show that the dense iron metal component is preferentially captured by Earth as the iron cores of both impactor and target merge (Canup, 2012; Ćuk and Stewart, 2012a, 2012b). In a glancing-blow impact, however, the major fraction of the Moon is made up of material from the impactor's mantle, not Earth, thus requiring that the impactor was compositionally similar to Earth. This is not impossible, but neither is it likely, given the randomness of planetary accretion and the timing of core formation on different planetary bodies (Carlson, 2019). To address this issue, more recent giant-impact models explore still more energetic collisions (Canup, 2012; Ćuk and Stewart, 2012a, 2012b). The most energetic of these (Lock et al., 2018) suggests that the outer layers of Earth expanded as a silicate vapor cloud of sufficient diameter to allow Moon formation from the condensed vapor as it cooled. These more energetic impacts mix Earth's mantle and the impactor to the point where any isotopic distinctions between target and impactor before the impact would be no longer resolvable.

1.3.2 Multiple giant impact theory

From the results gleaned from numerous studies carried out in the recent past, it is quite clear that Earth's Moon's origin remains enigmatic (difficult to interpret or understand; mysterious). A leading theory, which initially became favored for the formation of the Earth's Moon posits a scenario in which a Mars-sized planetesimal (Impactor) named Theia impacted the late-stage accreting Earth (Herwartz et al., 2014). The ejected material produced an Earth-orbiting disk, which later gravitationally accreted to a single Moon (Cameron, 2001; Canup, 2004a, 2004b; Reufer et al., 2011). The single impact theory initially had little impact of its own. Cameron's models had left off after a single giant impact, when a debris cloud from which the moon would arise formed around Earth. Cameron and her colleagues extended the modeling from debris cloud to finished moon. Cameron (2001)'s numerical model estimated that more than 40% of the Moon-forming disk material was derived from Theia. Other impact simulations study carried out by Canup (2004) indicated that the Impactor contributes more than 70% to the disk's mass.

Based on the scenario embodied in the single impact theory (i.e., the Impactor contributes a considerable percentage to the disk's mass), Earth's Moon should bear isotopic signatures from both the Earth and the Impactor (the body believed to have collided with the protoEarth). High-precision measurements show that the Earth, its Moon and enstatite meteorites (a diverse group of strange rocks that contain little or no oxidized iron, a rare occurrence in the Solar System) have almost indistinguishable isotopic compositions (Qin et al., 2010; Javoy et al., 2010; Warren, 2011; Zhang et al., 2012; Dauphas et al., 2015; Young et al., 2016; Dauphas and Schauble, 2016). Models have been proposed that in a way tend to accept (reconcile) the Earth–Moon similarity with the inferred heterogeneous nature of Earth-forming material, but these models either require specific geometries for the Moon-forming impact (Cuk and Stewart, 2012; Canup, 2012) or can explain only one aspect of the Earth–Moon similarity (that is the oxygen isotope, ^{17}O) (Pahlevan and Stevenson, 2007; Mastrobuono-Battisti et al., 2015; Kaib and Cowan, 2015).

Efforts to confirm that the impact had indeed taken place had centered on measuring the ratios between the isotopes of oxygen, titanium, silicon and others. These ratios are known to vary throughout the Solar System, but their close similarity between Earth and its Moon conflicted with theoretical models of the collision that indicated that Earth's Moon would form mostly from Theia, and thus would be expected to be compositionally different from Earth.

Herwartz et al. (2014) analyzed fresh basalt samples from three Apollo landing sites and compared them with several samples of Earth's mantle. The team initially used lunar samples which had arrived on Earth via meteorites, but as these samples had exchanged their isotopes with water from Earth, fresher samples were sought. These were provided by NASA from the Apollo 11, 12 and 16 missions.

High-precision measurements of isotopes indicate that Earth and its Moon are isotopically similar in oxygen, that is, oxygen isotopic compositions have been found to be identical between terrestrial and lunar samples (Wiechert et al., 2001) ($^{17}O/^{16}O$ and $^{18}O/^{16}O$ within 12±3 ppm (Wiechert et al., 2001; Herwartz et al., 2014)), titanium ($^{50}Ti/^{47}Ti$ within ±4 ppm (Zhang et al., 2012)) and prelate veneer (thin covering) tungsten ($^{182}W/^{184}W$ (Kruijer et al., 2015)). Isotopic equilibration with a hot protoplanetary atmosphere is efficient for oxygen but insufficient to explain the similarity in more refractory elements (a class of metals that are extraordinarily resistant to heat and wear) such as titanium (Pahlevan and Stevenson, 2007; Zhang et al., 2012).

However, it remained uncertain whether more refractory elements, such as titanium (Ti), show the same degree of isotope homogeneity as oxygen in the Earth–Moon system.

Oxygen makes up 90% of rocks' volume and 50% of their weight. More than 99.9% of Earth's oxygen is ^{16}O (where the superscript 16 represent the atomic weight of this oxygen atom), so called because each oxygen atom contains eight protons and eight neutrons. But there also are small quantities of heavier oxygen isotopes: ^{17}O, which have one extra neutron, and ^{18}O, which have two extra neutrons in the nucleus of their atoms. Earth, Mars and other planetary bodies in our Solar System each has a unique ratio of ^{17}O to ^{16}O—each one a distinctive "fingerprint." According to Hamilton et al. (2016), "it is clear from the sulfur isotope record and other geochemical proxies that the production of oxygen or oxidizing power radically changed Earth's surface and atmosphere during the Proterozoic Eon, pushing it away from the more reducing conditions prevalent during the Archean. In addition to ancient rocks, our reconstruction of Earth's redox evolution is informed by our knowledge of biogeochemical cycles catalyzed by extant biota." According to Meadows (2017), "Oxygenic photosynthesis is Earth's dominant metabolism, having evolved to harvest the largest expected energy source at the surface of most terrestrial habitable zone planets. Using CO_2 and H_2O-molecules that are expected to be abundant and widespread on habitable terrestrial planets—oxygenic photosynthesis is plausible as a significant planetary process with a global impact."

In the zoo of elements that show well-documented isotopic anomalies at a bulk planetary scale (Clayton, 1993; Dauphas et al., 2002; Trinquier et al., 2007, 2009), highly refractory titanium, with large nucleosynthetic anomalies on ^{50}Ti, is the most promising to assess the degree of homogeneity in the Earth–Moon system (Leya et al., 2008). In a study, Junjun Zhang of the University of Chicago and her colleagues made mass spectroscopic studies of the titanium isotopic compositions of 5 terrestrial samples, 37 bulk chondrites, and 24 lunar samples brought back by the 1970s Apollo missions, (8 whole rocks, 6 ilmenite separates, 1 pyroxene separate, and nine soil samples). Zhang et al. (2012) found that the $^{50}Ti/^{47}Ti$ ratios in lunar samples measured by mass spectrometry were essentially identical to those of the Earth's, within about four parts per million, which is only 1/150 of the isotopic range documented in meteorites. Thus, samples from meteorites, the closest stand-in to the colliding body available, are quite distinct from those of Earth. The isotopic homogeneity of this highly refractory element suggests that lunar material was derived from the proto-Earth mantle, an origin that could be explained by efficient impact ejection, by an exchange of material between the Earth's magma ocean and the protolunar disk, or by fission from a rapidly rotating postimpact Earth. Note that "fission" is Darwin (1879)'s notion that an early stage fluid-Earth began rotating so rapidly that it flung off a mass of material that formed the Moon. Thus, the idea that the Moon formed from Earth dates back to at least 1879 (Darwin, 1879; Brush, 1986) with the suggestion that the Moon was flung out by centrifugal force from a fast-spinning Earth. However, planetary-scale fission of this nature likely requires much more energy than can be provided by spin alone.

There is a priori (i.e., from logical point of view rather than from observation or experience) no reason to expect that Theia should have the same isotopic composition as the proto-Earth. Although the idea that Theia and the proto-Earth had identical compositions cannot be definitely ruled out (Wiechert et al., 2001), this idea seems to be contrived (deliberately created) and requires special circumstances for an embryo to have the same titanium, oxygen, and tungsten isotopic compositions as a growing planet.

According to Zhang et al. (2012), the isotopic homogeneity between the Earth and its Moon for a highly refractory element such as titanium has important implications for the giant-impact scenario. One possibility to explain this homogeneity is that the majority of the Moon-forming material came from the proto-Earth instead of the Impactor. However, this contradicts predictions of state-of-the-art giant single-impact simulations, which indicate that at most 60% of the Moon-forming material could originate from the proto-Earth (Reufer et al., 2011). Although several explanations have been put forward to reconcile with the single-impact theory, none seems to have succeeded in providing a definite answer to Earth's Moon's origin. Zhang et al. (2012)'s results place a new constraint to the Earth–Moon evolution system.

The formation of Earth's Moon has remained something of a puzzle. A leading theory proposes a cataclysmic (i.e., large-scale and violent) impact involving a Mars-sized object and a young Earth. But there are some inconsistencies with this scenario. As explained earlier, while the hypothesis of lunar origin by a single giant impact can explain some aspects of the Earth–Moon system, it is difficult to reconcile giant impact models with the compositional similarity of the Earth and its Moon without violating angular momentum constraints. Furthermore, successful giant single impact scenarios require highly specific initial conditions (for example, a collision with an object of a particular size traveling at a defined velocity and hitting Earth at a specific angle) such that they have a low probability of occurring. Furthermore, in a typical impact, different proportions of that object would have ended up in the Earth and the Moon, leaving a detectable difference between the bodies. But various chemical analyses of Earth's Moon's makeup, taken from samples returned by astronauts, reveal that it is nearly identical to that of Earth. In other words, there is no trace of the large body that supposedly hit Earth, and the theories, say the researchers, turn out to be improbable.

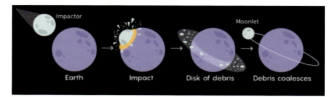

FIG. 1.9 Moonlet formation as a result of an impact on Earth: simplistic representation of the giant-impact hypothesis. (*Source: https://commons.wikimedia.org/wiki/File:Moon_-_Giant_Impact_Hypothesis_-_Simple_model.png.*)

Any scientific hypothesis for the formation of Earth's Moon must explain three key points: the low density of the Moon, the current angular momentum of the Earth–Moon system, and the similarities in isotope ratios between Earth and its Moon. Today's scientific community tends to favor the Giant-Impact Hypothesis precisely because it addresses these three features. However, collisions large enough to form Earth's Moon with a single impact are rare, appearing in only 2% of computer simulations. Instead, faster smaller collisions are more likely, giving credence to the Multiple-Impact Hypothesis.

As its name suggests, the Multiple-Impact Hypothesis theorizes that there were several smaller impacts throughout the course of Earth's formation. Each of these impacts discharged rock, creating an accretion disk—a rotating disk of debris matter—around the Earth. With low relative orbital velocities, particles from the accretion disk would gently clump together, growing larger like a snowball rolling down a mountain. Eventually all the debris would coalesce into a rigid-body satellite called a moonlet (see Fig. 1.9 for a visual depiction of this process). This entire process is estimated to take about 50 years—a relatively short period compared to the millions of years it took the Earth to form.

Rufu et al. (2017) recall that, "Deriving more disk material from the proto-Earth occurs in impact scenarios with increased angular momentum (Canup, 2012; Ćuk and Stewart, 2012a, 2012b) beyond the present value, that is later dissipated by an orbital resonance or an associated limit cycle (Wisdom and Tian, 2015). Studies of planetary accretion (Jacobson and Morbidelli, 2014; Kaib and Cowan, 2015) have shown that the equal sized impactors are extremely rare unless assuming a very early event, which is inconsistent with the recent Moon formation timing estimates (Jacobson et al., 2014)."

A study carried out by Rufu et al. (2017) considered a multi-impact hypothesis for Earth's Moon's formation (Ringwood, 1989). In this scenario, the proto-Earth experiences a sequence of collisions by medium to large size bodies ($0.01 - 0.1 M_\oplus$) where $M_\oplus$ represents Earth-mass (Fig. 1.10A, D). Small satellites form from the impact-generated disks (Fig. 1.10B, E) and migrate outward controlled by tidal interactions, faster at first, and slower as the body retreats away from the proto-Earth (Fig. 1.10C).

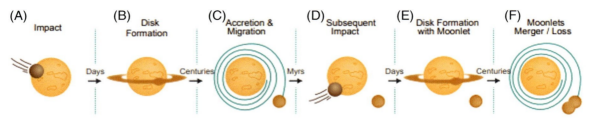

FIG. 1.10 Lunar formation in the multiple impact scenario. Moon to Mars sized bodies impact the proto-Earth (A) forming a debris disk (B). Due to tidal interaction, accreted moonlets migrate outward (C). Moonlets reach distant orbits before the next collision (D), and the subsequent debris disk generation (E). As the moonlet-proto-Earth distance grows, the tidal acceleration slows, moonlets enter their mutual Hill radii. The moonlet interactions can eventually lead to moonlet loss or merger (F). The time scale between these stages is estimated from previous works (Canup, 2004; Ida et al., 1997; Raymond et al., 2009). *(Source: Rufu, R., O. Aharonson, and H.B. Perets (2017), A multiple impact origin for the Moon, Nat. Geosci., 10(2):89–94. DOI: 10.1038/ngeo2866.)*

The slowing migration causes the moonlets to enter their mutual Hill radii and eventually coalesce to form the final Moon (Jutzi and Asphaug, 2011) (Fig. 1.10F). In this fashion, Earth's Moon forms as a consequence of a variety of multiple impacts in contrast to a more precisely tuned single impact. A similar scenario using smaller (0.001 –0.01$M_\oplus$), high velocity, late accreting impactors was previously suggested (Ringwood, 1989), but supporting calculations were not provided.

Rufu et al. (2017)'s study based on numerical simulations run on a computer cluster, suggests that a more plausible chain of events might involve a number of run-ins with smaller objects. A number of smaller collisions might better explain what happened several billion years ago, when the Solar System was taking shape. Such smaller bodies would have been more prevalent in the system, and thus collisions with the smaller objects would have been more likely. Small, high-velocity collisions could also mine more material from Earth than a single, large one. Numerous smaller collisions would have produced smaller moonlets that would have eventually coalesced into the single Moon we have today. Nearly a thousand computer simulations carried out by Rufu et al. (2017) produced results challenging the theory that Earth's Moon was formed by a single massive collision; rather, the investigation shows, the more likely explanation is that Earth's Moon could instead be the product of a succession of a variety of smaller collisions. More specifically, multiple impacts produced many moonlets, which coalesced into our solitary Moon. In this scenario, each collision forms a debris disk around the proto-Earth that then accretes to form a moonlet. How the smaller moonlets produced in the numerical simulations might have coalesced to form our Moon is a question yet to be fully understood. To answer this question, Rufu et al. (2017) argue that the moonlets tidally advance outward, and may coalesce to form the Moon. These investigators' results revealed that sublunar moonlets are a common result of impacts expected onto the proto-Earth in the early Solar System and found that the planetary rotation is limited by impact angular momentum drain.

N-body simulations of terrestrial planet accretion (Agnor et al., 1999) show that the final angular momentum of the Earth's system is a result of several impacts. The largest impactor is not necessarily the last one. That study shows that majority of single collisions with Earth cannot form the present Moon because the impact angular momentum is insufficient relative to that of the current value. Similar to the angular momentum of the Earth–Moon system, Rufu et al. (2017) argue that the Moon's mass is also the result of contributions from several last impactor.

Apart from the logical argument of a number of different bodies having collided with Earth over a period of millions of years with a larger range of angles (from head-on to near grazing impacts), and thus resulting in the formation of multiple moonlets which ultimately coalesced to form the Moon, the different chemical signatures of Earth, its Moon, and the Impactor must play an important role in justifying the impact theory concerning the origin of the Earth's Moon. In the multi-impact scenario, the 'compositional crisis' described above is mitigated by two effects. First, because the mass and angular momentum of the present Earth–Moon system provide constraints on the sum of multiple impacts, rather than a single impact, the additional freedom in impact geometries enables mining more material from Earth than in the conventional scenario. Second, the oxygen signature distribution of the cumulative sum of multiple moonlets will have a reduced variance, increasing the probability of the Earth–Moon similarity compared to that from a single event (Rufu et al., 2017).

In a different argument, Rufu et al. (2017) has shown that an off-axis energetic impact onto a rotating target produces a debris disk that is sourced from both impactor and target material. The cores are seen to merge, and the heating is concentrated in the upper planetary mantle (Fig. 1.11). It is interesting to note that in the accretionary simulations, disks of non-negligible mass are well extended beyond the *Roche limit*, with more than 60% of their mass outside the Roche radius.

Rufu et al. (2017) probed a broad phase space by simulating impact scenarios with ranging values of the

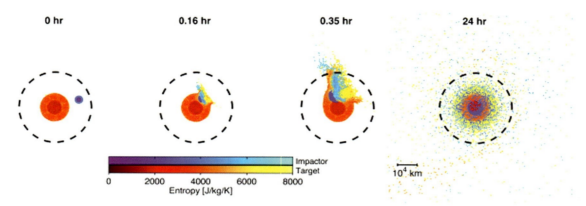

FIG. 1.11 Impact simulation. Several snapshots of one of the simulations with initial conditions of impactor mass ratio $\gamma = 0.025$, speed $V_{imp} = 2V_{esc}$, direction angle $\beta = 30°$ (relative to the line connecting the centers at contact) and planetary rotation $\omega = 0.5\omega_{max}$. The color bars represent the entropy of the impactor and target. All projections are on the equatorial plane with one hemisphere removed. The impactor core is shown over the target. The *Roche limit* is represented by the dashed line. *(Source: Rufu, R., O. Aharonson, and H.B. Perets (2017), A multiple impact origin for the Moon, Nat. Geosci., 10(2):89–94. DOI: 10.1038/ngeo2866.)*

impactor's velocity, mass, angle, and initial rotation of the target. They used Smoothed Particle Hydrodynamics (SPH) to simulate impacts in the gravity dominated regime, using the astrophysical code Gadget2 (Springel, 2005). The spatial distribution of each particle was defined by a spline density weighting function, known as the kernel, and a characteristic radius, known as the smoothing length. They performed over 800 simulations consisting of 10^5 SPH particles to span the parameter space and a smaller number of high-resolution simulations with 10^6 particles to better define the characteristics of the disk. For ease of comparison with previous work (Canup, 2008; Ćuk and Stewart, 2012), all impacts were assumed to occur in the equatorial plane of the target.

To escape from a celestial body, an object must exceed the escape velocity v, given by the expression, $v = \sqrt{\frac{2Gm}{R}}$, where G is the universal constant of gravitation; m is the mass of the celestial body; and R is nominally the distance from the center of the celestial body to the top of the atmosphere, if there is any (Jones, 2004). From the above expression, it is seen that escape velocity decreases with altitude and is equal to the square root of 2 (or about 1.414) times the velocity necessary to maintain a circular orbit at the same altitude. Based on this expression, the escape velocity from the Earth is 11.2 km/s. For comparison, from the Earth's Moon it is 2.4 km/s, and from Mars and Venus it is 5.0 and 10.4 km/s, respectively (Tepfer and Leach, 2006). A planet (or moon) cannot long retain an atmosphere if the planet's (moon's) escape velocity is low enough to be near the average velocity of the gas molecules making up the atmosphere.

Projection of material in excess of escape velocity could result from the impacts of comets, meteorites, and asteroids (Melosh, 1988; Vickery and Melosh, 1987), either acting directly on the site of impact or on particles already in the atmosphere. According to Burchell et al. (2003), "Because the ejection through the atmosphere is brief (order of seconds) heating of the ejected material by atmospheric drag is also minimal. The conclusion is that lightly shocked material, subject to only modest temperature increases, can be expelled into interplanetary space." For such ejected material, the mechanics of transfer to another body, via a heliocentric orbit, has been modeled by several researchers (e.g., Gladman et al., 1996; Gladman, 1997).

Rufu et al. (2017)'s hydrodynamic simulations indicated that "The high angle impacts ($\beta = 45°$) almost always produce a disk because the angular momentum of the impactor, even for relatively low velocities near the escape velocity, v, is sufficient to impart enough angular momentum to the ejecta to form a disk. The head-on and low-angle impacts require a high velocity to eject material to orbit. Moreover, high-angle (grazing) impacts result in little mixing between the planet and the disk with the impactor contributing a substantial fraction to the disk. Low-angle impacts produce disks with higher degree of compositional similarity to Earth."

Most terrestrial rocks have similar ratios of $^{182}W/^{184}W$, in which W refers to Wolfram (Tungsten). Note that Tungsten, or wolfram, is a chemical element with the symbol W and atomic number 74. Tungsten is a rare metal found naturally on Earth almost exclusively as compounds with other elements. It was identified as a new element in 1781 and first isolated as a metal in 1783. Rufu et al. (2017)'s investigations indicated that debris disks resulting from medium to large size impactors ($0.01–0.1 M_\oplus$) have sufficient angular momentum and mass to accrete a sublunar-size moonlet. Rufu et al. (2017) recall that "Freedom in impact geometry and velocity allows mining more material from Earth, and the sum of such impact-generated moonlets may naturally lead to the current values of the Earth–Moon system. Most terrestrial rocks have similar ratios of $^{182}W/^{184}W$, however excess in these ratios was found for Kostomuksha komatiites rocks dated at 2.8 billion year ago (Touboul et al., 2012), suggesting an unmixed mantle reservoir. Evidence of an unmixed mantle was also observed in noble gas samples (Mukhopadhyay, 2012). Efficient mantle mixing is predicted for the single impact high angular momentum scenarios (Nakajima and Stevenson, 2015) erasing any primordial heterogeneity that predated Earth's Moon formation. Multiple smaller impacts promote preservation of primordial heterogeneity of Earth's mantle, and potentially, also contribute to that of the Moon (Jutzi and Asphaug, 2011; Robinson et al., 2016)." Rufu et al. (2017) concluded that, assuming efficient merger of moonlets, a multiple impact scenario can account for the formation of the Earth–Moon system with its present properties.

As demonstrated by Rufu et al. (2017), although the Earth–Moon system has previously been suggested to have formed through a single giant collision, in which the Moon accreted from the impact-generated debris disk, such giant impacts are rare, and during its evolution the Earth experienced many more smaller impacts, producing smaller satellites (moonlets) that potentially coevolved. In the multiple-impact hypothesis of lunar formation, the current Moon was produced from the mergers of several smaller satellites (moonlets), each formed from debris disks produced by successive large impacts. Moonlet-creating impacts are estimated to have occurred on average every 6 million years, over the course of which a pre-existing moonlet's orbit drifts further away from the Earth due to tidal migration (note that our current Moon spirals away from Earth about 3.82 cm/year).

Citron et al. (2018) argued that in the million years (Myrs) between impacts, as the pre-existing moonlet orbits further from Earth due to tidal migration, a subsequent impact would produce a newer moonlet, which would orbit Earth with a smaller semimajor axis. A two-moonlet system would thus evolve with a newer moonlet orbiting closer to Earth and an older moonlet orbiting further away from Earth (see Fig. 1.12A).

In a study, Citron et al. (2018) examined the likelihood that pre-existing moonlets survive subsequent impact events, and explored the dynamics of Earth–Moonlet systems that contain two moonlets generated Myrs apart. While in theory the Multiple-Impact Hypothesis holds up, would impacts with Earth actually produce stable moonlets in real life? With the advent of fast processing computing power, Citron et al. (2018) ran thousands of simulations—each with slight parameter adjustments—to determine which outcomes are most likely.

The first simulations tested how a subsequent impact with Earth would affect the orbit of a pre-existing moonlet. Each simulation began with the pre-existing moonlet orbiting Earth at around 17 Earth radii (~100,000 km)

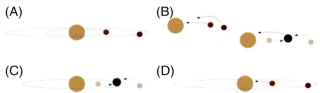

FIG. 1.12 Possible outcomes of an initial (A) two moonlet system. The two moonlets could (B) infall individually or as a merged unit. The two moonlets could (C) merge into one larger stable moonlet. One of the two moonlets (D) could infall. Note that the starting inclinations need not be coplanar (as assumed in (A)), and that the final inclination of the moonlets could change. To build the Moon from multiple moonlet-generating impacts, the probability of mergers and outer moonlet survival (where the outer moonlet could be a product of a previous merger) must be high. (*Source: Citron, R.I., H.B. Perets, and O. Aharonson (2018), The role of multiple giant impacts in the formation of the Earth–Moon system, Astrophys. J., 862:5 (11pp) © 2018. The American Astronomical Society. All rights reserved.*)

away from Earth—the expected tidal drift outward over the course of the estimated 6 million years between impacts. For comparison, the Moon is located at a distance of about 60 Earth radii today.

In each simulation, the incoming colliding body—the impactor—was assigned a random mass, velocity, impact angle, initial orbit, and impact parameter. After 5000 simulations, Citron et al. (2018) found that impactors of an Earth–Moonlet system are unlikely to affect the orbit of the pre-existing moonlet. Almost 90% of the simulations resulted in a stable pre-existing moonlet. The other 10% resulted in the moonlet either collapsing into Earth or ejecting from the system, both of which were likely caused by very fast massive impacts.

The second round of simulations tested the likelihood that a pre-existing outer moonlet and newly formed inner moonlet merge. In these simulations, the outer moonlet was assigned a random mass and distance from the earth, while the inner moonlet was assigned a random mass but an initial distance from the earth of 3.8 Earth radii just outside the *Roche limit*, below which tidal forces would tear the moonlet apart.

Citron et al. (2018) found mixed results. In short, the two moonlets merged about 70% of the time for prograde–prograde systems—systems in which both moonlets orbit the Earth in the same direction. The other 30% of the time, either both moonlets infell individually or as a merged unit (see Fig. 1.12B), only the inner moonlet infell (see Fig. 1.12D), or, in rare cases, both moonlets were ejected from the system.

Citron et al. (2018) demonstrated that pre-existing moonlets can tidally migrate outward, remain stable during subsequent impacts, and later merge with newly created moonlets to form one stable larger moonlet (see Fig. 1.12C) or re-collide with the Earth. This newly-merged larger moonlet would take the place of the outer moonlet, as a subsequent impact would then form a new inner moonlet.

As this process continues occurring again and again, the outer moonlet grows larger and larger until it evolves into the Moon as we know it.

Citron et al. (2018) argue that the abundance of giant impacts (each capable of producing a debris disk and moonlet(s)) in the late stages of planetary formation suggests that multiple-moon systems may have been a common occurrence. Their simulations show that the Earth may have had several past moons. According to Citron et al. (2018), "If prior moonlets merged, the Moon could have formed from a sequence of giant impacts, which could explain the similarity in isotopic composition to the Earth. And in the context of typical giant impacts during planetary formation, it may be more likely for several small, fast impactors to eject sufficient material into proto-Earth orbit to form the Moon than a single impact with finely tuned parameters. Sequences of impacts that result in 1–4 moonlet mergers are possible, particularly if the outer moonlet is larger than the inner moonlet and at an intermediate distance from the proto-Earth, and the likelihood of moonlet mergers could increase if subsequent moonlets are preferentially generated near the Laplace plane of pre-existing moonlets." Citron et al. (2018)'s simulations suggest that moonlets are also likely to infall and impact the proto-Earth, which could have consequences for early Earth evolution (Malamud et al., 2018). Citron et al. (2018)'s investigation implies that the formation of Earth's Moon from the mergers of several moonlets could be a natural by-product of the Earth's growth through multiple impacts.

During the last stages of the terrestrial planet formation, planets grew mainly through giant-impacts with large planetary embryos. The Earth's Moon was suggested to form through one of these impacts. However, since the proto-Earth has experienced many giant-impacts, several moons (and also the final Moon) are naturally expected to form through/as-part-of a sequence of multiple (including smaller scale) impacts. Each impact potentially forms a sub-lunar mass moonlet that interacts gravitationally with the proto-Earth and possibly with previously formed moonlets. Such interactions result in either moonlet–moonlet mergers, moonlet ejections or in-fall of moonlets on the Earth. The latter possibility, leading to low-velocity moonlet-Earth collisions has been explored by Malamud et al. (2018) for the first time. They made use of smooth particle hydrodynamical (SPH) simulations and considered a range of moonlet masses, collision impact-angles and initial proto-Earth rotation rates. They found that grazing/tidal-collisions are the most frequent (see Fig. 1.13) and produce comparable fractions of accreted-material and debris. The latter typically clump in smaller moonlets that can potentially later interact with other moonlets. Malamud et al. (2018)'s investigations suggest that other collision geometries are rarer. Note that Malamud et al. (2018) have considered only impacts that occur in the proto-Earth's equatorial plane.

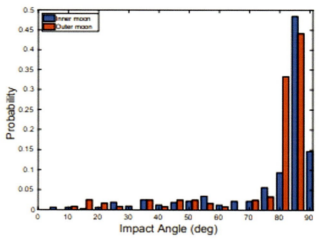

FIG. 1.13 The distribution of moonlet-Earth impact angles (0°—head-on impact; 90°—extremely grazing impact) based on analysis of data from Citron et al. (2018) study. As can be seen most collisions are extremely grazing, but the overall distribution includes a wide range of rarer head-on and intermediate impact angles. *(Source: Malamud, U., Perets, H.B., Schafer, C., and Burger, C. (2018), Moonfalls: collisions between the Earth and its past moons, arXiv:1805.00019 © 2018 The Authors Uri Malamud, and Hagai B. Perets: Department of Physics, Technion Israeli Intitute of Technology, Technion City, 3200003 Haifa, Israel Christoph Schäfer: Institut für Astronomie und Astrophysik, Eberhard Karls Universität Tübingen, Auf der Morgenstelle 10, 72076 Tübingen, Germany Christoph Burger: Department of Astrophysics, University of Vienna, Türkenschanzstraße 17, 1180 Vienna, Austria (Source: https://arxiv.org/pdf/1805.00019.pdf).*

Malamud et al. (2018)'s findings are: "Head-on collisions do not produce much debris and are effectively perfect mergers. Intermediate impact angles result in debris mass-fractions in the range of 2–25% where most of the material is unbound. Retrograde collisions produce more debris than prograde collisions, whose fractions depend on the proto-Earth initial rotation rate. Moonfalls can slightly change the rotation-rate of the proto-Earth. Accreted moonfall material is highly localized, potentially explaining the isotopic heterogeneities in highly siderophile elements in terrestrial rocks, and possibly forming primordial supercontinent topographic features." Malamud et al. (2018) are optimistic that their results can be used for simple approximations and scaling laws and applied to n-body studies of the formation of the Earth and its current Moon.

Note that siderophile elements are those elements that preferentially partition into metal phases during cosmochemical and geo-chemical processes. The siderophile elements include the Highly Siderophile Elements (HSE; Re, Os, Ir, Ru, Pt, Rh, Au, Pd), which are characterized by low-pressure metal–silicate distribution coefficients of more than 10^4; the Moderately Siderophile Elements (MSE; including Mo, W, Fe, Co, Ni, P, Cu, Ga, Ge, As, Ag, Sb, Sn, Tl, Bi), with low-pressure metal–silicate partition coefficients that are typically more than 10, but less than 10^4; and the Slightly Siderophile Elements (SSE; Mn, Cr, and V). The

SSE, along with the highly chalcophile elements (HCE: S, Se, Te, Pb), and elements that otherwise exhibit atmophile, lithophile, or chalcophile affinity, can become siderophiles at specific temperatures (T), pressures (P), compositions (x), or oxygen fugacity (fO_2). An example of this phenomenon is the composition of Earth's core (Day, 2018).

1.3.3 The concept of lunar magma ocean (LMO) of global dimensions

Prior to the Apollo landings on Earth's Moon, the prevailing view of the formation of terrestrial planets (planets that are composed primarily of silicate rocks or metals; i.e., Mercury, Venus, Earth, and Mars, which are closest to the Sun) was that planets grew by the gentle accumulation of asteroid-sized planetesimals (Urey, 1951), so that high temperatures only occurred locally in some of the larger impacts (see Carlson, 2019). It has been found that, to a large extent, the lunar highlands crust is composed of anorthosite, which consists predominantly of a single mineral, plagioclase feldspar. A handful of small mineral grains led to the suggestion that the entire lunar highlands were dominated by anorthosite (Wood et al., 1970; Smith et al., 1970). In molten rock (magma) of composition similar to that of the bulk Moon, anorthite plagioclase (a calcium-aluminum silicate), has a lower density than the magma, and so would float. An obvious way to form an anorthositic crust is to assume that a large portion of Earth's Moon was at one time molten, with the anorthositic crust forming by flotation above the crystallizing magma in much the same way that an iceberg floats on the ocean (Carlson, 2019). This idea led to what became known as the "magma ocean" model for early lunar differentiation (Fig. 1.14). The magma ocean model implies that some energetic process involving high temperatures was involved in Moon formation (Carlson, 2019).

Carlson (2019) recalls that based on the elevation of the highlands relative to the flat plains of basaltic lavas that infilled the major impact basins on the Moon's Earth-facing side (originally misidentified by early astronomers as seas, leading them to be called mare basins), the assumption of an anorthositic composition for the highlands implied a highlands crust some 25 km thick (Wood et al., 1970). The current best estimate for lunar nearside crustal thickness is ~30 km based on lunar gravity data measured more than 40 years later (Wieczorek et al., 2013).

Early models of the Moon-forming impact involved a glancing collision of the proto-Earth with an object roughly the size of Mars (Canup and Asphaug, 2001). Such an impact could place enough material into Earth orbit to form Earth's Moon. The rapid reassembly of this material into a single body would result in a largely molten Moon, consistent with the evidence for a Lunar Magma Ocean (LMO) (Carlson, 2019). At this point, it would be worthwhile to briefly revisit the LMO hypothesis for lunar differentiation following accretion.

FIG. 1.14 Diagram of the lunar magma ocean model. Convection (arrows) within the magma ocean circulates the magma. Crystallization of dense olivine and pyroxene (green) and buoyant plagioclase (gray) separates the magnesium-rich minerals of the interior from the anorthositic crust. The thickness of the lunar crust is exaggerated by about a factor of 10 for display. *(Source: Carlson, R.W. (2019), Analysis of lunar samples: Implications for planet formation and evolution, Science, 365 (6450), 240-243. DOI: 10.1126/science.aaw7580) The article from which this figure is reproduced is distributed under the terms of the Science Journals Default License Website: https://science.sciencemag.org/content/365/6450/240?intcmp=trendmd-sci.*

Anand et al. (2014) have beautifully summarized the LMO hypothesis thus: "As originally proposed (Smith et al., 1970; Wood et al., 1970) the model starts with the Moon fully or partially molten (Fig. 1.15) and, through fractional crystallization, the first minerals to crystallize from LMO were higher density ultramafic minerals such as Mg-rich olivine and pyroxenes that sank toward the bottom of the LMO. The crystallization sequence progressed such that, by about 80% crystallization, less-dense feldspar-dominated mineral assemblages began to crystallize. The residual LMO liquid at this stage was very Fe–Ti-rich and, therefore, denser than the crystallizing feldspar, which could essentially float at or near the surface of the LMO, where it eventually became the main component of the primary lunar crust (Smith et al., 1970; Wood et al., 1970; Taylor and Jjakes, 1974; Ringwood and Kesson, 1976). Not only was the residual LMO liquid dense, it was also highly enriched in incompatible elements such as potassium (K), rare earth elements (REEs) and phosphorus (P) (KREEP for short). Thus, the final 'layers' to crystallize from the residual liquid were KREEP-rich (termed urKREEP; [Warren and Wasson, 1979]) and Fe–Ti-rich. It is worth noting that the depth of the LMO is debated to be between 200 and 1000 km (Nakamura et al.,

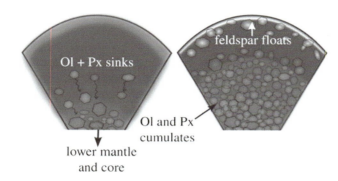

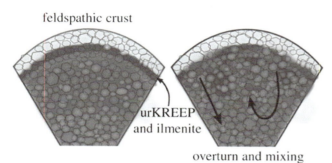

FIG. 1.15 A cartoon showing a possible LMO scenario (depth of melting not specified) (after PSRD graphic, G.J. Taylor). The model proceeds through crystallization of Mg-rich minerals (*Ol*+Px) which sink, then of low-density Ca-rich minerals (feldspar) which float, and finally of KREEP and Fe-Ti rich (ilmenite) layers. Note that this diagram is not to scale. *(Source: Anand, M., Tartèse, R., and Barnes, J.J. (2014), Understanding the origin and evolution of water in the Moon through lunar sample studies, Philos. Trans. R. Soc. A, 372:20130254) © 2014 The Authors. Published by the Royal Society under the terms of the Creative Commons Attribution License http://creativecommons.org/licenses/by/3.0/, which permits unrestricted use, provided the original author and source are credited. Website: https://royalsocietypublishing.org/doi/10.1098/rsta.2013.0254.*

Elardo et al., 2011). However, it is not known as to how much mixing of interior material was achieved by the overturn event."

As Carlson (2019) explained, "Because KREEP is rich in radioactive elements such as uranium, thorium, and potassium, whose concentrations can be measured from orbit because they emit gamma rays (Lawrence et al., 1998), we now have maps of the lunar surface that show that KREEP is concentrated dominantly on the Earth-facing side of the Moon, in the region of the major mare basalt-filled impact basins (Lawrence et al., 1998; Jolliff et al., 2000; Prettyman et al., 2006) (Fig. 1.16)." Lawrence et al. (1998) have shown that Lunar Prospector gamma-ray spectrometer spectra along with counting rate maps of thorium, potassium, and iron delineate large compositional variations over the lunar surface. Thorium and potassium are highly concentrated in and around the nearside western Maria and less so in the South Pole–Aitken basin. Counting rate maps of iron gamma-rays show a surface iron distribution that is in general agreement with other measurements from Clementine and the Lunar Prospector neutron detectors.

Carlson (2019) interpreted the restricted areal distribution of KREEP on the lunar surface thus: "The restricted areal distribution of KREEP could reflect the presence of a KREEP layer beneath the whole crust that was only excavated to the surface by impacts large enough to penetrate through the crust. However, the largest impact basin on the Moon, the South Pole–Aitken basin, shows only slightly elevated abundances of thorium (Fig. 1.16). Alternatively,

1973; Taylor and Jjakes, 1974; Hess and Parmentier, 2001) or even whole Moon melting (Mueller et al., 1988; Canup and Asphaug, 2001; Longhi, 2006). The dominant mode of LMO crystallization also remains contentious as to whether it was fractional, equilibrium or a mixture of both (Wood et al., 1970; Snyder et al., 1995; Elkins-Tanton et al., 2011a, 2011b). The estimated time for complete crystallization of the LMO varies between approximately 10 and 220 Ma (Meyer et al., 2010). In such a scenario having a dense Fe–Ti material on top of less-dense Mg-rich material is gravitationally unstable, such that the newly crystallized LMO cumulate pile (Mg-rich olivine and pyroxene layers toward the bottom while KREEP and Fe–Ti layers toward the top) re-organized itself via an 'overturn event' (Ringwood and Kesson, 1976; Hess and Parmentier, 1995). In a nutshell, the dense, KREEP-rich, Fe-Ti-rich material sank through the LMO cumulate pile and probably triggered the onset of Mg-rich basaltic magmatism (i.e., onset of Mg-suite and alkali-suite magmatism (Shearer and Papike, 2005, 2006;

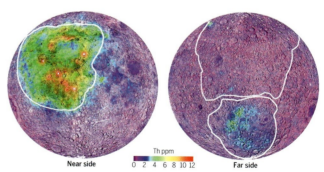

FIG. 1.16 Map of the distribution of thorium (Th) on the lunar surface. Data from the Lunar Prospector and Clementine missions (Gillis et al., 2004) allow division of the lunar crust into three compositionally distinct terranes, outlined here in white. On the near side is the Th-rich Procellarum-KREEP Terrane (PKT); the far side is dominated by the Th-poor Feldspathic Highlands Terrane (top) penetrated by the impact that formed the South Pole–Aitken basin (bottom) that exposed slightly more Th-rich material, but nowhere near as Th-rich as in the PKT. [Figure provided by Bradley Jolliff updated and modified from the figure shown in (Jolliff et al., 2000)]. *(Source: Carlson, R.W. (2019), Analysis of lunar samples: Implications for planet formation and evolution, Science, 365 (6450), 240-243. DOI: 10.1126/science.aaw7580). The article from which this figure is reproduced is distributed under the terms of the Science Journals Default License Website: https://science.sciencemag.org/content/365/6450/240?intcmp=trendmd-sci.*

both the apparently asymmetric distribution of KREEP and the thicker lunar crust on the far-side than on the near-side (Wieczorek et al., 2013) have led to proposals that the asymmetry originated in the magma ocean era by either gravitational or thermal interaction with Earth (Roy et al., 2014; Quillen et al., 2019). The concentration of radioactive KREEP into a smaller portion of the Moon also could have provided a heat source to explain both mare basalt volcanism that continued for more than a billion years after the Moon solidified (Laneuville et al., 2013) and the higher number of large impact basins facing Earth than on the lunar farside (Miljković et al., 2013)." Although the above description provides a glimpse of the lunar magma ocean, according to Carlson (2019), the simple magma ocean model described above almost certainly underestimates the complexity of the process.

1.4 Influence of Moon-forming impacts on the environmental conditions on the early Earth

There is a hypothesis that the Earth and its Moon were essentially dry immediately after the formation of the Moon—by a giant impact on the proto-Earth—and only much later they gained volatiles through accretion of wet material delivered from beyond the asteroid belt. Albarède (2009) examined evidence for this hypothesis and tried to justify this view, having been supported by uranium–lead (U–Pb) and iodine–xenon (I–Xe) chronologies, which show that water delivery peaked ~100 million years after the isolation of the Solar System. Note that uranium–lead dating, abbreviated U–Pb dating, is one of the oldest and most refined of the radiometric dating schemes. It can be used to date rocks that formed and crystallized from about 1 million year to over 4.5 billion years ago with routine precisions in the 0.1–1% range.

It is generally believed that much of the Earth's mantle was melted in the Moon-forming impact. Sleep et al. (2014) have addressed the history of the terrestrial aftermath of the Moon-forming impact, their description of which runs thus, "Gases that were not partially soluble in the melt, such as water and CO_2, formed a thick, deep atmosphere surrounding the postimpact Earth. This atmosphere was opaque to thermal radiation, allowing heat to escape to space only at the runaway greenhouse threshold of approximately 100 W/m². The duration of this runaway greenhouse stage was limited to approximately 10 million year (Myr) by the internal energy and tidal heating, ending with a partially crystalline uppermost mantle and a solid deep mantle. At this point, the crust was able to cool efficiently and solidified at the surface. After the condensation of the water ocean, approximately 100 bar of CO_2 remained in the atmosphere, creating a solar-heated greenhouse, while the surface cooled to approximately 500 K. Almost all this CO_2 had to be sequestered by subduction into the mantle by 3.8 Ga, when the geological record indicates the presence of life and hence a habitable environment." It may be noted that a "runaway greenhouse effect" is a process in which a net positive feedback between surface temperature and atmospheric opacity to thermal radiation increases the strength of the greenhouse effect on a planet until its oceans boil away. Note also that the present atmospheric pressure at the Earth's sea level is 1 bar.

According to Sleep et al. (2014), the deep CO_2 sequestration into the mantle could be explained by a rapid subduction of the old oceanic crust, such that the top of the crust would remain cold and retain its CO_2. Sleep et al. (2014) argue that "Kinematically, these episodes would be required to have both fast subduction (and hence seafloor spreading) and old crust. Hadean oceanic crust that formed from hot mantle would have been thicker than modern crust, and therefore only old crust underlain by cool mantle lithosphere could subduct. Once subduction started, the basaltic crust would turn into dense eclogite, increasing the rate of subduction. The rapid subduction would stop when the young partially frozen crust from the rapidly spreading ridge entered the subduction zone." Note that eclogite is a metamorphic rock containing granular minerals, typically garnet (a precious stone consisting of a deep red vitreous silicate mineral) and pyroxene (a group of important rock-forming inosilicate minerals found in many igneous and metamorphic rocks).

Earth scientists think that the environmental conditions on the early Earth can be deciphered from a sparse geological record combined with physics and molecular phylogeny (the branch of biology that deals with phylogenesis; i.e., the evolutionary development and diversification of a species or group of organisms, or of a particular feature of an organism). Zahnle et al. (2007) have addressed the first several hundred million years of Earth's history. According to them, "The Moon-forming impact left Earth enveloped in a hot silicate atmosphere that cooled and condensed over ~1000 years. As it cooled the Earth degassed its volatiles into the atmosphere. It took another ~2 Myrs for the magma ocean to freeze at the surface. The cooling rate was determined by atmospheric thermal blanketing. Tidal heating by the new Moon was a major energy source to the magma ocean. After the mantle solidified geothermal heat became climatologically insignificant, which allowed the steam atmosphere to condense, and left behind a ~100 bar, ~500 K CO_2 atmosphere. Thereafter cooling was governed by how quickly CO_2 was removed from the atmosphere. If subduction were efficient this could have taken as little as 10 million years. In this case the faint young Sun suggests that a lifeless Earth should have been cold and its oceans white with ice. But if carbonate subduction were inefficient the CO_2 would have mostly stayed in the atmosphere, which would have kept the surface near ~500 K for many tens of millions of years. Hydrous minerals are harder to subduct than carbonates and there is a good chance that the Hadean mantle was dry. Hadean heat flow was locally high enough to ensure that any ice cover would have been thin (<5 m) in places." Note that the Hadean is a geologic eon of the Earth, which began with the formation of the Earth about 4.6 billion years ago and ended 4 billion years ago.

Zahnle et al. (2007) have suggested that plate tectonics as it works now was inadequate to handle typical Hadean heat flows of 0.2–0.5 W/m². In its place, these researchers have hypothesized a convecting mantle capped by a ~100 km deep basaltic mush (a soft, wet, pulpy mass) that was relatively permeable to heat flow. Recycling and distillation of hydrous basalts (hard, dense, dark igneous rock) produced granitic rocks very early, which is consistent with preserved >4 Ga detrital zircons (particles of zircon minerals derived from pre-existing zircon rock crystals through processes of weathering and erosion). Zahnle et al. (2007) have argued that if carbonates in oceanic crust subducted as quickly as they formed, Earth could have been habitable as early as 10–20 Myrs after the Moon-forming impact.

According to Sleep (2010), "the Earth began hot after the moon-forming impact and cooled to the point where liquid water was present in ~10 million years. Subsequently, a few asteroid impacts may have briefly heated surface environments, leaving only thermophile survivors in kilometer-deep rocks. A warm 500 K, 100 bar CO_2 greenhouse persisted until subducted oceanic crust sequestered CO_2 into the mantle. It is not known whether the Earth's surface lingered in a ~70°C thermophile environment well into the Archean or cooled to clement or freezing conditions in the Hadean." This investigator thinks that the recently discovered ~4.3 Ga rocks near Hudson Bay (Canada) may have formed during the warm greenhouse. According to Sleep (2010), "Overall, mantle derived rocks, especially kimberlites and similar CO_2-rich magmas, preserve evidence of subducted upper oceanic crust, ancient surface environments, and biosignatures of photosynthesis."

According to Zahnle et al. (2010), the Moon-forming impact made Earth for a short time absolutely uninhabitable, and it set the boundary conditions for Earth's subsequent evolution. These researchers hypothesize that if life indeed began on Earth, as opposed to having migrated here from an external world, it would have done so after the Moon-forming impact. According to them, what took place before the Moon's formation determined the bulk properties of the Earth and probably determined the overall compositions and sizes of its atmospheres and oceans. Zahnle et al. (2010) argue that one interesting consequence of the Moon-forming impact is that the mantle is devolatized (i.e., volatile substances are removed from the solid mantle), so that the volatiles (i.e., substance that are easily evaporated at normal temperatures) subsequently fell out in a kind of condensation sequence (i.e., the conversion of a vapor or gas to a liquid). This process likely resulted in the volatiles getting concentrated toward the surface so that, for example, the oceans were likely salty from the very start. Zahnle et al. (2010) argue that the Moon-forming impact might have played a crucial role in the origin of Earth's atmosphere and ocean as well as some of the environmental conditions of the early Earth, which might have influenced the origin of life. They have also pointed out that an atmosphere generated by impact degassing (i.e., removal of dissolved gases) would tend to have a composition reflective of the impacting bodies (rather than the mantle), and these are almost without exception strongly reducing and volatile-rich. Because Earth is the one known example of an inhabited planet and to current knowledge the likeliest site of the one known origin of life, Zahnle et al. (2010)'s arguments assume great significance.

1.5 Earth's internal structure, development, orbit, and rotation

There are several factors that influenced the gradual development of Earth over a period of time in its long history. The prominent among them are briefly addressed below.

1.5.1 Influence of collisions

Just like other planets and moons in the Solar System, Earth was a target of frequent hits/collisions by asteroids and comets in the early history of (primitive) Earth, when the Sun was faint and young. Frequent collisions rendered the conditions on early Earth literally hellish. Radioactive materials in the rock and increasing pressure deep within the Earth generated enough heat to melt the planet's interior, causing some chemicals to rise to the surface and form water, while others became the gases of the atmosphere. Recent evidence suggests that Earth's crust and oceans may have formed within about 200 million years after the planet took shape.

Gravity increased as Earth grew larger. The collision of the dust, ice and rock created energy that raised Earth's temperature, making it hot enough to melt, sinking the densest materials to Earth's core and leaving the less dense material to harden as crust and mantle on Earth's surface.

1.5.2 Features of Earth's core

Earth's core is about 4400 miles (7100 km) wide, slightly larger than half the Earth's diameter and about the same size as Mars. The outermost 1400 miles (2250 km) of the core are liquid, while the inner core is solid. That solid core is about four-fifths as big as Earth's Moon, at some 1600 miles (2600 km) in diameter. Geologists think the temperature of Earth's outer core is about 6700–7800°F (3700–4300°C) and that the inner core may reach 12,600°F (7000°C)—hotter than the surface of the Sun. Above the core is Earth's mantle, which is about 1800 miles (2900 km) thick. The mantle is not completely stiff but can flow slowly. Earth's crust floats on the mantle much as a piece of wood floats on water. The slow motion of rock in the mantle shuffles continents around and causes earthquakes, volcanoes and the formation of mountain ranges.

Above the mantle, Earth has two kinds of crust. The dry land of the continents consists mostly of granite and other light silicate minerals, while the ocean floors are made up mostly of a dark, dense volcanic rock called basalt. Continental crust averages some 25 miles (40 km) thick, although it can be thinner or thicker in some areas. Oceanic crust is usually only about 5 miles (8 km) thick. Water fills in low areas of the basalt crust to form the world's oceans.

Earth gets warmer toward its core. At the bottom of the continental crust, temperatures reach about 1800°F (1000°C), increasing about 3°F per mile (1°C per km) below the crust. The surface of the primitive Earth was molten to a depth of 1000 km (Davies, 1990). The light elements had disappeared into space but various gases were retained on the surface by gravitation. As the temperature decreased, the surface of the magma ocean gradually solidified. Water vapor, carbon oxide, nitrogen gas began to cover the Earth's surface, forming the primitive atmosphere. Water vapor gave rise to clouds which turned into rain, feeding rivers and oceans. Dissolved metal ions from rocks entered into the primitive ocean. Lightnings and volcanic eruptions often occurred. Small and large meteorites also bombarded the early Earth. These events and light from the Sun were conducive to creating simple organic compounds and small biomolecules.

Earth is the third planet from the Sun. While Earth orbits the Sun, the planet is simultaneously spinning around an imaginary line called an axis that runs through the core, from the North Pole to the South Pole. It takes Earth 23.934 h to complete a rotation on its axis and 365.26 days to complete an orbit around the Sun——our days and years on Earth are defined by these gyrations. Earth has a diameter of roughly 8000 miles (13,000 km) and is mostly round because gravity generally pulls matter into a ball. But Earth's spin causes it to be squashed at its poles and swollen at the equator, making the true shape of the Earth an "oblate spheroid."

1.5.3 Earth's paleo-rotation and revolution— day: ~21 h; year: ~13 months and ~400 days

Earth's rotation or Earth's spin is the rotation of planet Earth around its own axis, as well as changes in the orientation of the rotation axis in space. Earth rotates eastward, in prograde motion. As viewed from the north pole star Polaris, Earth turns counter-clockwise.

Earth's rotation is slowing slightly with time; thus, a day was shorter in the past. This is due to the tidal effects the Moon has on Earth's rotation. Over millions of years, Earth's rotation has been slowed significantly by tidal acceleration through gravitational interactions with the Moon.

Thus, angular momentum is slowly transferred to the Moon at a rate proportional to r^{-6}, where r is the orbital radius of the Moon. This process has gradually increased the length of the day to its current value, and resulted in the Moon being tidally locked with Earth.

It has long been known that animals and plants reflect environmental rhythms in their tissues. Periodic growth features, such as the annual rings of trees, sheep's horns, and sea shells, have been used both for calculating the ages of organisms and their growth rates (Scrutton, 1978).

Studies carried out by Zahnle and Walker (1987) indicate that the semidiurnal atmospheric thermal tide would have been resonant with free oscillations of the atmosphere when the day was ~21 h long, c. 600 Ma ago. Very large atmospheric tides would have resulted, with associated surface pressure oscillations in excess of 10 mbar in the tropics. Near resonance the Sun's gravitational torque on the atmospheric tide - accelerating Earth's rotation - would have been comparable in magnitude to the decelerating lunar torque upon the oceanic tides. The balance of the opposing torques may have long maintained a resonant ~21 h day, perhaps for much of the Precambrian (the earliest aeon of the earth's history, preceding the Cambrian period (approx. 500 million years ago, when the first complex animals were evolving and the Phanerozoic aeon (the "time of ancient life")). Because the timescale of lunar orbital evolution is not directly affected, a constant daylength would result in fewer days/month. The hypothesis is shown not to conflict with the available (stromatolitic) evidence. Escape from the resonance could have followed a relatively abrupt global warming, such as that occurring at the end of the Precambrian. Alternatively, escape may simply have followed a major increase in the rate of oceanic tidal dissipation, brought about by the changing topography of the world's oceans. Zahnle and Walker (1987) integrated the history of the lunar orbit with and without a sustained resonance, finding that the impact of a sustained resonance on the other orbital parameters of the Earth–Moon system would not have been large.

Over the past decade the analysis of sedimentary cyclic rhythmites of tidal origin (i.e., stacked thin beds or laminae usually of sandstone, siltstone, and mudstone that display periodic variations in thickness reflecting a strong tidal influence on sedimentation), has provided information on Earth's paleo-rotation and the evolving lunar orbit for Precambrian time (before 540 Ma). Depositional environments of tidal rhythmites range from estuarine to tidal delta, with a wave-protected, distal ebb tidal delta setting being particularly favorable for the deposition and preservation of long, detailed rhythmite records. The potential sediment load of nearshore tidal currents and the effectiveness of the tide as an agent of sediment entrainment and deposition are related directly to tidal range (or maximum tidal height) and consequent water-current speed. Hence the thickness of successive laminae deposited by tidal currents can be a proxy tidal record, with paleotidal and paleorotational values being determined by analysis of measured records of lamina and cycle thickness. The validity of the findings can be investigated by testing the primary, observed values for internal self-consistency through application of the laws of celestial mechanics. Paleo-tidal and paleo-rotational values provided by late Neoproterozoic (~620 Ma) tidal rhythmites in South Australia are validated by these tests and indicate 13.1 ± 0.1 synodic (lunar) months/year, and 400 ± 7 solar days/year, a length of day of 21.9 ± 0.4 h. Concentrated study of Precambrian tidal rhythmites promises to illuminate the evolving dynamics of the early Earth–Moon system and may permit the lunar orbit to be traced back to near the time of the Moon's origin (Williams, 1990, 2000).

This gradual rotational deceleration is empirically documented by estimates of day lengths obtained from observations of tidal rhythmites and stromatolites; a compilation of these measurements (Williams, 2000) found that the length of the day has increased steadily from about 21 h at 600 Myr ago (Zahnle and Walker, 1987) to the current 24-h value. By counting the microscopic lamina that form at higher tides, tidal frequencies (and thus day lengths) can be estimated, much like counting tree rings, though these estimates can be increasingly unreliable at older ages (Scrutton, 1978).

Atomic clocks show that a modern-day is longer by about 1.7 ms than a century ago (McCarthy and Seidelmann, 2009), slowly increasing the rate at which UTC is adjusted by leap seconds. Analysis of historical astronomical records shows a slowing trend; the length of a day increased about 2.3 ms per century since the eighth century BCE (Richard, 2003). Scientists reported that in 2020 Earth has started spinning faster, after consistently slowing down in the decades before. Because of that, engineers worldwide are discussing a "negative leap second" and other possible timekeeping measures (Sarah, 2021).

1.5.4 Earth's inclination and orbit

Earth's axis of rotation is tilted in relation to the ecliptic plane—an imaginary surface passing through Earth's orbit around the Sun. This means the Northern and Southern hemispheres will sometimes point toward or away from the Sun depending on the time of year, and this changes the amount of light the hemispheres receive, resulting in the changing seasons.

Earth happens to orbit the Sun within the so-called "Goldilocks zone," where temperatures are just right to maintain liquid water on our planet's surface. Earth's orbit is not a perfect circle, but rather a slightly oval-shaped ellipse, similar to the orbits of all the other planets in the Solar System. Because Earth's orbit around the Sun is elliptical, or slightly oval-shaped, there is one point in the orbit where Earth is closest to the Sun, and another where Earth

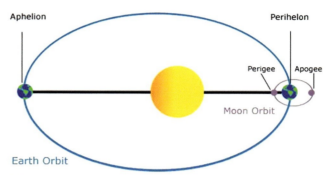

FIG. 1.17 A diagram of Earth's elliptical orbit around the Sun *(Image credit: NOAA). (Source: https://www.space.com/54-earth-history-composition-and-atmosphere.html) Video E1 shows Earth's elliptical motion around the Sun in action.*

is farthest from the Sun. The closest point occurs in early January, and the far point happens in early July (July 7, 2007). However, this proximity has a much smaller effect on the temperatures we experience on Earth's surface than does the tilt of Earth's axis.

Statistics about Earth's orbit, according to NASA is as follows (Choi, 2021):

- Average distance from the Sun: 92,956,050 miles (149,598,262 km)
- Perihelion (closest approach to the Sun): 91,402,640 miles (147,098,291 km)
- Aphelion (farthest distance from the Sun): 94,509,460 miles (152,098,233 km)
- Length of solar day (single rotation on its axis): 23.934 h
- Length of year (single revolution around the Sun): 365.26 days
- Equatorial inclination to orbit: 23.4393°

Fig. 1.17 shows a diagram of Earth's elliptical orbit around the Sun. Video E1 shows Earth's elliptical motion around the Sun in action.

1.6 Water and frost line in the astrophysical environments

Leonardo da Vinci stated thus: "*Water is the driving force of all nature.*" Leonardo's claim is only a slight exaggeration. After hydrogen and helium, oxygen is the third most cosmically abundant element (H/O ~1200 by number). Helium is chemically inert leaving hydrogen atoms free to combine to make H_2, or with oxygen to form H_2O. Water is the second most abundant molecule (after H_2) in those astrophysical environments where the temperature is below the thermal dissociation limit (~2000 K). These environments include almost the entire protoplanetary disk of the Sun. Therefore, it should be no surprise that water played an important role over a wide range of heliocentric distances in the Solar system from the region of the terrestrial planets out to the Kuiper belt and beyond (Jewitt et al., 2007).

Water is important for its obvious role as the enabler of life but more generally as the most abundant volatile molecule in the Solar system, containing about half of the condensible mass in solids. In its solid phase, water strongly influences the opacity of the protoplanetary disk and may determine how fast, and even whether, gas giant planets form. Water ice is found or suspected in a wide range of small-body populations, including the asteroids, the giant planet Trojan librators, Centaurs, comets and Kuiper belt objects. In addition to ice, there is mineralogical evidence for the past presence of liquid water in certain meteorites and, by inference, in their parent main-belt asteroids. The survival and evolution of liquid and solid water in small bodies has been discussed by Jewitt et al. (2007).

The study of water in small Solar system bodies involves perspectives that are astronomical, geochemical and physical. As is common in science, the study of one broad subject is breaking into many smaller subfields. It is important to address issues concerning water in the Solar system's small bodies in a language accessible to the general audience as well as astronomers and planetary scientists.

1.6.1 Water in the protoplanetary disk of the Sun

Astronomical observations have shown that protoplanetary disks are dynamic objects through which mass is transported and accreted by the central star. This transport causes the disks to decrease in mass and cool over time, and such evolution is expected to have occurred in our own solar nebula. Age dating of meteorite constituents shows that their creation, evolution, and accumulation occupied several Myr, and over this time disk properties would evolve significantly. Moreover, on this timescale, solid particles decouple from the gas in the disk and their evolution follows a different path. It is in this context that we must understand how our own solar nebula evolved and what effects this evolution had on the primitive materials contained within it.

Water is important in the solar nebula both because it is extremely abundant and because it condenses out at 5 AU (the distance of the *snow line* from the Sun), allowing all three phases of H_2O to play a role in the composition and evolution of the Solar System. Hayashi (1981) prescribed a "minimum-mass solar nebula," which contained just enough material to make the planets. This prescription, which has been widely used in constructing scenarios for planet formation, proposed that ice will condense when the temperature falls below 170 K (the "snow line"). In Hayashi's model, the proposed condensation occurred at 2.7 AU. It is usually assumed that the cores of the giant planets (Jupiter, Saturn, Uranus, and Neptune) form beyond the snow line. The snow line, in Hayashi's model, is where the temperature of a black body that absorbed direct sunlight and reradiated as much as it absorbed would be 170 K. Since Hayashi, there have been a series of more detailed models of the absorption by dust of the stellar radiation and of accretional heating, which alter the location of the snow line.

At low temperatures, water is thermodynamically stable as a solid; it condenses as frost or ice. Ice grains have a strong influence on the mean radiative opacity (Pollack et al., 1994) and so can change the temperature structure in the protoplanetary disk. In turn, the disk viscosity is related to the temperature of the gas. Hence, the freeze-out and distribution of icy grains together have a surprisingly strong impact on the transport of energy, mass and angular momentum and so influence the overall structure and evolution of the protoplanetary disk (Ciesla and Cuzzi, 2006).

Ciesla and Cuzzi (2006) reported a model which tracks how the distribution of water changes in an evolving disk as the water-bearing species experience condensation, accretion, transport, collisional destruction, and vaporization. Because solids are transported in a disk at different rates depending on their sizes, the motions will lead to water being concentrated in some regions of a disk and depleted in others. These enhancements and depletions are consistent with the conditions needed to explain some aspects of the chemistry of chondritic meteorites and formation of giant planets. The levels of concentration and depletion, as well as their locations, depend strongly on the combined effects of the gaseous disk evolution, the formation of rapidly migrating rubble, and the growth of immobile planetesimals. Understanding how these processes operate simultaneously is critical to developing models for meteorite parent body formation in the Solar System and giant planet formation throughout the galaxy. Ciesla and Cuzzi (2006) have presented examples of evolution under a range of plausible assumptions and demonstrated how the chemical evolution of the inner region of a protoplanetary disk is intimately connected to the physical processes which occur in the outer regions.

In a study, Kimberly et al. (1998) modeled the inward radial drift of ice particles from 5 AU. They linked the drift results to the outward diffusion of vapor in one overall model based on the two-dimensional diffusion equation, and numerically evolved the global model over the lifetime of the nebula. Kimberly et al. (1998) found that while the inner nebula is generally depleted in water vapor, there is a zone in which the vapor is enhanced by 20–100%, depending on the choice of ice grain growth mechanisms and rates. Kimberly et al. (1998) found that this enhancement peaks in the region from 0.1 to 2 AU and gradually drops off out to 5 AU. Because this result is somewhat sensitive to the choice of nebular temperature profile, these investigators examined representative hot (early) and cool (later) conditions during the quiescent phase of nebular evolution. It was found that variations in the pattern of vapor depletion and enhancement due to the differing temperature profiles vary only slightly from that given above. Such a pattern of vapor enhancement and depletion in the nebula is consistent with the observed radial dependence of water of hydration bands in asteroid spectra and the general trend of asteroid surface darkening. According to Kimberly et al. (1998), this pattern of water vapor abundance will also cause variations in the carbon-oxygen ratio (C:O ratio), shifting the ratio more in favor of C in zones of relative depletion, affecting local and perhaps even global nebular chemistry, the latter through quenching and radial mixing processes.

Jewitt et al. (2007) noted that water is important for its obvious role as the enabler of life but more generally as the most abundant volatile molecule in the Solar system, containing about half of the condensable mass in solids. In its solid phase, water strongly influences the opacity of the protoplanetary disk and may determine how fast, and even whether, gas giant planets form. Water ice is found or suspected in a wide range of small-body populations, including the asteroids, the giant planet Trojan librators, Centaurs, comets and Kuiper belt objects. In addition to ice, there is mineralogical evidence for the past presence of liquid water in certain meteorites and, by inference, in their parent mainbelt asteroids. Jewitt et al. (2007) have discussed the survival and evolution of liquid and solid water in small bodies.

At low temperatures, water is thermodynamically stable as a solid; it condenses as frost or ice. According to Pollack et al. (1994), ice grains have a strong influence on the mean radiative opacity. Therefore, they can change the temperature structure in the protoplanetary disk. In turn, the disk viscosity is related to the temperature of the gas. Hence, the freeze-out and distribution of icy grains together have a surprisingly strong impact on the transport of energy, mass and angular momentum and so influence the overall structure and evolution of the protoplanetary disk (Ciesla and Cuzzi, 2006).

Astronomical observations have shown that protoplanetary disks are dynamic objects through which mass is transported and accreted by the central star. This transport causes the disks to decrease in mass and cool over time, and such evolution is expected to have occurred in our own Solar nebula. Age dating of meteorite constituents shows that their creation, evolution, and accumulation occupied several Myr, and over this time disk-properties would evolve significantly. Moreover, on this timescale, solid particles decouple from the gas in the disk, and their evolution follows a different path. It is in this context that we must understand how our own solar nebula evolved and what effects this evolution had on the primitive materials contained within it. Ciesla and Cuzzi (2006) reported a model which tracks how the distribution of water changes in an evolving disk as the water-bearing species experience condensation, accretion, transport, collisional destruction, and vaporization. Because solids are transported in a disk at different rates depending on their sizes, the motions will lead to water being concentrated in some regions of a disk and depleted in others.

Small samples of certain asteroids, available to us in the form of meteorites, contain hydrated minerals that probably formed when buried ice melted and chemically reacted with surrounding refractory materials (Kerridge et al., 1979; McSween, 1979a, 1979b). Short-lived radionuclides were the most probable heat source in the early Solar System. Cosmo-chemical evidence for ^{26}Al (half-life 0.71 Myr) is firmly established (Lee et al., 1976). Additionally, ^{60}Fe (half-life ~1.5 Myr) may have provided supplemental heating (Mostefaoui et al., 2004).

1.6.2 Frost line

In astronomy or planetary science, the *frost line*, also known as the *snow line* or *ice line*, is the particular distance in the solar nebula from the central proto-star where it is cold enough for volatile compounds such as water, ammonia, methane, carbon dioxide, and carbon monoxide to condense into solid ice grains. The "snow line" (actually, it is a surface) divides the outer, cold, ice-rich region of the protoplanetary disk from the inner, steamy hot zone.

It was Chushiro Hayashi (see Hayashi, 1981) who prescribed a "minimum-mass solar nebula," which contained just enough material to make the planets. The Minimum Mass Solar Nebula (MMSN) is the protoplanetary disk of solar composition that has the amount of metals necessary to build the eight planets of the solar system (and the asteroid belts). From the masses and compositions of the planets, a density of solids is derived at several locations of the disk. Then, the solar composition is restored by adding gas, and a smooth protoplanetary disk density profile is derived (Crida, 2009). Hayashi's prescription, which has been widely used in constructing scenarios for planet formation, proposed that ice will condense when the temperature falls below 170 K (the "snow line"). In Hayashi's model, the condensation process occurred at 2.7 AU. The leading explanation for the delivery of water to Earth is that water-bearing, asteroid-like bodies from beyond the snow line were gravitationally scattered inward and accreted by the planet during its growth (e.g., Morbidelli et al. 2000).

The snow line is one of the key properties of protoplanetary disks that determine the water content of terrestrial planets in the habitable zone. Its location is determined by the properties of the star, the mass accretion rate through the disk, and the size distribution of dust suspended in the disk. Mulders et al. (2015) calculated the snow-line location from recent observations of mass accretion rates and as a function of stellar mass. By taking the observed dispersion in mass accretion rates as a measure of the dispersion in initial disk mass, Mulders et al. (2015) found that stars of a given mass will exhibit a range of snow-line locations.

It is usually assumed that the cores of the giant planets (e.g., Jupiter) form beyond the snow line (Sasselov and Lecar, 2000). The snow line, in Hayashi's model, is where the temperature of a black body that absorbed direct sunlight and reradiated as much as it absorbed would be 170 K. Since Hayashi, there have been a series of more detailed models of the absorption by dust of the stellar radiation and of accretional heating, which alter the location of the snow line. Sasselov and Lecar (2000) attempted a "self-consistent" model of a T Tauri disk in the sense that these investigators used dust properties and calculated surface temperatures that matched observed disks. Jewitt et al. (2007) recall that "Outward gas motions across the snow line result in the condensation of water into ice grains, which can collide, stick and grow until they dynamically decouple from the flow and are left behind. In this way, the snow line defines the inner edge of a cold trap in which the density of solids may have been sufficiently enhanced as to help speed the growth of planetesimals and, ultimately, of the cores of the giant planets (Stevenson and Lunine, 1988). Aerodynamic drag causes 100-m scale ice-rich planetesimals from the outer disk to spiral inward on very short timescales (perhaps 1 AU per century). Some are swept up by bodies undergoing accelerated growth just outside the snow line. Others cross the snow line to quickly sublimate, enhancing the local water vapor abundance. Coupled with on-going dissociation from ultraviolet photons, these freeze-out and sublimation processes lead to isotopic fractionation anomalies in water (Lyons and Young, 2005) that have already been observed in the meteorite record (Krot et al., 2005)." It may be noted that changes in the chemical and isotopic composition of the solar nebula with time are reflected in the properties of different constituents that are preserved in chondritic meteorites. CR carbonaceous chondrites are among the most primitive of all chondrite types and must have preserved solar nebula records largely unchanged (Krot et al., 2005).

As indicated earlier, Stevenson and Lunine (1988) reported a model for enhancing the abundance of solid material in the region of the solar nebula in which Jupiter formed, by diffusive redistribution and condensation of water vapor. They assumed a turbulent nebula with temperature decreasing roughly inversely with the radial distance from the center, and time scales set by turbulent viscosities taken from recent nebular models. They solved the diffusion equation in cylindrical coordinates in the limit that the sink of water vapor is condensation within a narrow radial zone approximately 5 AU from the center. Most of the water vapor is extracted from the terrestrial planet-forming zone. This "cold finger" solution is then justified by analytic solution of the diffusion equation in the condensation zone itself and inward and outward of that zone. Stevenson and Lunine (1988) calculated that the length scale over which most of the diffusively transported water vapor condenses is ~0.4 AU, provided the solids are not redistributed, and the surface density of ice in the formation zone of Jupiter is enhanced by as much as 75. Stevenson and Lunine (1988) found that the enhancement in surface density of solids is sufficient to trigger rapid accretion of planetesimals into a solid core along the lines of the model of Lissauer (1987), and hence produce Jupiter by nucleated instability on a time scale of approximately 10^5–10^6 years.

Jewitt et al. (2007) noted that "In the protoplanetary disk, opacity due to grains inhibited radiative cooling and raised the mid-plane kinetic temperature. Growth and migration of solids in the disk would have caused the opacity to change, so moving the snow line. For a fraction of a million years, it may have pushed in to the orbit of Mars or even closer (Sasselov and Lecar, 2000; Ciesla and Cuzzi, 2006), meaning that asteroids in the main belt, between Mars and Jupiter, could have incorporated water ice upon formation."

The lower temperature in the nebula beyond the frost line makes many more solid grains available for accretion into planetesimals and eventually planets. The frost line therefore separates terrestrial planets from giant planets in the Solar System (Kaufmann, 1987). However, giant planets have been found inside the frost line around several other stars (so-called hot Jupiters). They are thought to have formed outside the frost line, and later migrated inward to their current positions (D'Angelo et al., 2010). Earth, which lies less than a quarter of the distance to the frost line but is not a giant planet, has adequate gravitation for keeping methane, ammonia, and water vapor from escaping it. Methane and ammonia are rare in the Earth's atmosphere only because of their instability in an oxygen-rich atmosphere that results from life forms (largely green plants) whose biochemistry suggests plentiful methane and ammonia at one time, but of course liquid water and ice, which are chemically stable in such an atmosphere, form much of the surface of Earth.

Each volatile substance has its own snow line (e.g., carbon monoxide (Qi et al., 2013), nitrogen (Dartois et al., 2013), and argon (Öberg and Wordsworth, 2019)), so it is important to always specify which material's snow line is meant. A tracer gas may be used for materials that are difficult to detect; for example, diazenylium for carbon monoxide. The compositions of planets are linked to the chemical conditions in the solar nebula. Because the chemical conditions change across the nebula, a planet's composition provides clues to its formation locations, and therefore to its dynamical past (Dodson-Robinson et al., 2009; Öberg et al., 2011; Ciesla et al., 2015). Of all known giant planets, Jupiter presents the most well-constrained composition because of in situ measurements by the *Galileo* and *Juno* missions (Niemann et al., 1996; Bolton et al., 2017). In a study Öberg and Wordsworth (2019) found that Jupiter's atmosphere is enriched in C, N, S, P, Ar, Kr, and Xe with respect to solar abundances by a factor of ~3. Gas giant envelopes are mainly enriched through the dissolution of solids in the atmosphere, and this constant enrichment factor is puzzling since several of the above elements are not expected to have been in the solid phase in Jupiter's feeding zone; most seriously, Ar and the main carrier of N, N_2 only condense at the very low temperatures, 21–26 K, associated with the outer solar nebula. Öberg and Wordsworth (2019) proposed that a plausible solution to the enigma of Jupiter's uniform enrichment pattern is that Jupiter's core formed exterior to the N_2 and Ar snowlines, beyond 30 AU, resulting in a solar composition core in all volatiles heavier than Ne. During envelope accretion and planetesimal bombardment, some of the core mixed in with the envelope, causing the observed enrichment pattern. Öberg and Wordsworth (2019) showed that this scenario naturally produces the observed atmosphere composition, even with substantial pollution from N-poor pebble and planetesimal accretion in Jupiter's final feeding zone. Öberg and Wordsworth (2019) noted that giant core formation at large nebular radii is consistent with recent models of gas giant core formation through pebble accretion, which requires the core to form exterior to Jupiter's current location to counter rapid inward migration during the core- and envelope-formation process.

Different volatile compounds have different condensation temperatures at different partial pressures (thus different densities) in the proto-star nebula, so their respective frost lines will differ. The actual temperature and distance for the snow line of water ice depend on the physical model used to calculate it and on the theoretical solar nebula model:

- 170 K at 2.7 AU (Hayashi, 1981)
- 143 K at 3.2 AU to 150 K at 3 AU (Podolak and Zucker, 2010)
- 3.1 AU (Martin and Livio, 2012)
- ≈150 K for μm-size grains and ≈200 K for km-size bodies (D'Angelo and Podolak, 2015)

Podolak and Zucker (2010) computed the temperature of ice grains in a proto-stellar disk for a series of disk models. They calculated the region of stability against sublimation for small ice grains composed of either pure ice or "dirty" ice. They showed that in the optically thin photosphere of the disk the gas temperature must be around 145 K for ice grains to be stable. This is much lower than the temperature of 170 K that is usually assumed.

The radial position of the condensation/evaporation front varies over time, as the nebula evolves. Occasionally, the term "snow line" is also used to represent the present distance at which water ice can be stable (even under direct sunlight). This current snow line distance is different from the formation snow line distance during the formation of the Solar System, and approximately equals 5 AU (Jewitt et al., 2007). After formation of the Solar System, the ice got buried by infalling dust and it has remained stable a few meters below the surface. If ice within 5 AU is exposed, for example, by a crater, then it sublimates on short timescales. However, out of direct sunlight ice can remain stable on the surface of asteroids (and Earth's Moon and Mercury) if it is located in permanently shadowed polar craters, where temperature may remain very low over the age of the Solar System (e.g., 30–40 K on Earth's Moon).

Observations of the asteroid belt, located between Mars and Jupiter, suggest that the water snow line during formation of the Solar System was located within this region. The outer asteroids are icy C-class objects whereas the inner asteroid belt is largely devoid of water. This implies that when planetesimal formation occurred the snow line was located at around 2.7 AU from the Sun (Martin and Livio, 2012). For example, the dwarf planet Ceres lies almost exactly on the lower estimation for water snow line during the formation of the Solar System. Ceres appears to have an icy mantle and may even have a water ocean below the surface (McCord and Sotin, 2005; O'Brien et al., 2015).

Martin and Livio (2012) have proposed that asteroid belts may tend to form in the vicinity of the frost line, due to nearby giant planets disrupting planet formation inside their orbit. By analyzing the temperature of warm dust found around some 90 stars, they concluded that the dust (and therefore possible asteroid belts) was typically found close to the frost line. The underlying mechanism may be the thermal instability of snow line on the timescales of 1000–10,000 years, resulting in periodic deposition of dust material in relatively narrow circumstellar rings (Owen, 2020).

The formation mechanisms of the ice giants Uranus and Neptune, and the origin of their elemental and isotopic compositions, have long been debated. The density of solids in the outer protosolar nebula is too low to explain their formation, and spectroscopic observations show that both planets are highly enriched in carbon, very poor in nitrogen, and the ices from which they originally formed might have had deuterium-to-hydrogen ratios lower than the predicted cometary value, unexplained properties that were observed in no other planets. A scenario discussed by Ali-Dib et al. (2014) generalizes a well-known hypothesis that Jupiter formed on an ice line (water snow line) for the two ice giants, and might be a first step toward generalizing this mechanism for other giant planets.

1.6.3 Water stored on the surface and in the ground of modern Earth

Earth is a watery place. But just how much water exists on, in, and above our planet? About 29.2% of Earth's surface is land consisting of continents and islands. The remaining 70.8% is covered with water, mostly by oceans, seas, gulfs, and other salt-water bodies, but also by lakes, rivers, and other freshwater, which together constitute Earth's hydrosphere. Much of Earth's polar regions are covered in ice. Earth's surface water is any body of water found on its surface, including both the saltwater in the ocean and the freshwater in rivers, streams, and lakes. A body of surface water can persist all year long or for only part of the year. Rivers are a major type of surface water. Earth's surface water is a key component to its hydrologic cycle.

Apart from these sources, water also exists in the air as water vapor, in icecaps and glaciers, in the ground as soil moisture and in aquifers. Water is never sitting still. Thanks to the water cycle, our planet's water supply is constantly moving from one place to another and from one form to another. Things would get pretty stale without the water cycle!

There are three types of surface water: perennial, ephemeral, and man-made. Perennial, or permanent surface water persists throughout the year and is replenished with groundwater when there is little precipitation. Ephemeral, or semipermanent surface water exists for only part of the year. Ephemeral surface water includes small creeks, lagoons, and water holes. Man-made surface water is found in artificial structures, such as dams and constructed wetlands.

Even though one may only notice water on the Earth's surface, there is much more freshwater stored in the ground than there is in liquid form on the surface. In fact, some of the water one sees flowing in rivers comes from seepage of groundwater into river beds. Water from precipitation continually seeps into the ground to recharge aquifers, while at the same time water in the ground continually recharges rivers through seepage.

1.7 Water-abundant celestial bodies in the Solar System—brief overview

There are several celestial bodies in our Solar System, where water is known to exist in its various forms. For example, in the year 2009, presence of water on Earth's Moon (Earth's nearest celestial body) was confirmed first by India's Moon Impact Probe (MIP) and 10 months later reconfirmed by America's Moon Mineralogy Mapper (M^3) carried by India's Chandrayaan-1 Lunar Probe (Crotts, 2011), which orbited and surveyed the Moon (Clark, 2009; Pieters et al., 2009; Sunshine et al., 2009). This landmark discovery was a major boost to America's longstanding lunar exploration program and India's heartthrob success with its very first probe beyond Earth. India's Chandrayaan-1 mission provided also confirmation that regolith (the layer of loosely arranged solid material such as dust, broken rocks, and other related materials covering the bedrock of a planet, and present on Earth, Earth's Moon, Mars, some asteroids, and other terrestrial planets and moons) hydration exists everywhere across the lunar surface. It is now known that Earth's Moon possesses water not only in its poles and regolith (Bussey et al., 2011) but also in its interior (Colaprete et al.,

2010; Fisher et al., 2017; Li et al., 2018). Lunar scientists think that the caves and large tunnel complexes discovered on Earth's Moon could be harboring some form of life; they can even be potential sites for human settlement.

Earth's neighboring planet Mars was once rich in water. The earliest observations of Mars by spacecraft showed that, in the past, water had caused substantial erosion of the Martian surface, including extensive channeling and associated transport and deposition of material (Carr, 1996). In a study, Christensen (2006) found that water is present at the poles and in permafrost regions of Mars. In another study, Stoker et al. (2010) reported identification of near-surface ground ice at Mars' far northern hemisphere. This region on Mars is suspected to exhibit periodic presence of liquid water as orbital dynamics change the regional climate. Some regions on Mars (e.g., McLaughlin Crater and Arabia Terra) show evidence of groundwater activity (Andrews-Hanna et al., 2010). Based on survey results, Orosei et al. (2018) detected a huge reservoir of liquid water buried under the base of a thick slab of polar ice in the Planum Australe region (centered at 193°E, 81°S) in the southern hemisphere of Mars. The subsurface reservoir spotted in the Planum Australe region on Mars appears to be a sizable salt-laden liquid-water lake under ice (resembling a subglacial lake on Earth). Although the water in the Martian lake is below the normal freezing point, it remains liquid presumably because of the presence of high levels of dissolved salts and the pressure of the ice above. Subglacial lakes provide sufficient protection from harmful radiation. Therefore, scientists like to think that this body of water could be a possible habitat for microbial life.

Several indications have been found favoring the presence of water on the near-Earth asteroid *Ceres* (Russell et al., 2013b). Apart from these credible indications, clues leading to the presence of water in the subsurface of Ceres, and a seasonal water cycle on Ceres have also been discovered (Schröder et al., 2017).

Jupiter's icy moons contain vast quantities of liquid water. For example, the exospheres (the outermost region of a planet's or moon's atmosphere) of Ganymede (Jupiter's moon) are known to possess water and oxygen (Dijkstra et al., 2018). Likewise, Europa (another moon of Jupiter) possesses oxygen and water (Heggy et al., 2017). Furthermore, Europa is known for water vapor-plume eruptions from its warmer regions (Lorenz, 2016). Credible scientific reasons have been found to believe that a liquid-water ocean lies hidden within Europa's ice shell (Kargel et al., 2000; Lorenz, 2016). It has been found that lakes and a layer of liquid water lie in and beneath Europa's hard icy outer shell (Schmidt et al., 2011; Airhart, 2011). Search-for-life researchers consider that there are prospects of extant life on Europa (Kargel et al., 2000; Richard, 2010).

Apart from Jupiter's moons Ganymede and Europa, two of Saturn's moons (*Enceladus* and *Titan*) are also known to possess conducive environments to support microbial life. Enceladus has recently become famous for its tall jets of water vapor and other gases, and organic-enriched sodium-salt-rich ice grains spewing from the tiger-stripes-like cracks on its South Pole (McKay et al., 2008; Porco et al., 2017). Clear scientific evidences have been found to believe that Enceladus is in possession of a global subsurface ocean (Kivelson et al., 2000; MacKenzie et al., 2016). Furthermore, it has been found that Enceladus is home for macromolecular organic compounds in its subglacial water-ocean (Postberg et al., 2018). Interestingly, hydrothermal vents are present on the seafloor of Enceladus (Waite et al., 2017)—throwing the possibility for harboring an ecosystem based on microbial populations, akin to those found on the submarine hydrothermal vents existing on Earth's ocean floors. It has been found that there is striking resemblance of Enceladus's organics-rich ocean to Earth's primitive prebiotic ocean (the celebrated Russian biochemist Alexander Oparin's "primordial soup" which likely existed on Earth four billion years ago), indicating a favorable scenario for life's emergence on Enceladus (Deamer and Damer, 2017; Kahana et al., 2019).

Saturn's another moon, *Titan*, is home for several hydrocarbon lakes and an internal salt-water ocean (Stofan et al., 2007; Hayes et al., 2008; Hayes, 2016; Hörst, 2017). These water bodies resemble the primitive Earth's natural laboratory of prebiotic chemistry (Neish et al., 2006, 2008, 2009, 2018). Titan is known to witness frequent hydrocarbon rainfall on to its surface, akin to water rainfall on to Earth's surface (Graves et al., 2008; Schneider et al., 2012). The search-for-life researchers have postulated a strong possibility of finding biomolecules on Titan (McKay, 2016).

Likewise, Neptune's moon *Triton* is thought to be a likely ocean world, with a likely existence of a 135–190 km thick inner ocean at a depth of ~20–30 km beneath its icy surface (Gaeman et al., 2012).

Finally, the dwarf planet Pluto, orbiting in the outer periphery of the Solar System, is also endowed with a subsurface ocean of liquid water (Johnson et al., 2016).

From the above discussion, it is clear that there are several celestial bodies in our Solar System, where water is abundant. However, in most cases, the surface water is in the form of frozen solid ice, and the liquid water is stored in subsurface oceans, which are located at considerably large depths below the surface of these celestial bodies.

1.8 Importance of understanding Earth's oceans in the search for life in extraterrestrial ocean worlds—NASA's ocean worlds exploration program

Earth is the only planet yet where life is known to exist. Analogue studies of life in the Earth's oceans and other habitable niches provide our only anchor for extrapolations to other ocean-bearing worlds. Hendrix et al. (2019) have defined an "ocean world" as a body with a current liquid ocean (not necessarily global). All bodies in our Solar System that plausibly can have, or are known to have an ocean, are considered in this framework. The Earth is a well-studied ocean world that we use as a reference ("ground truth") and point of comparison. The ice giant planets are not included as ocean worlds.

By definition, ocean worlds provide a solvent, and the materials that planets form from can be expected to contain the necessary nutrients. Earth is an ocean world, in which liquid water is essential to sustain life as we know it. An often-posed question is whether the dark, alien subsurface/inner oceans of the outer Solar System could be habitable for simple life forms, and if so, what would their biochemistry might be. Because life on Earth is hypothesized to have emerged either in a "primordial soup" of the primitive ocean or in a submarine hydrothermal vent environment (both of which are essentially oceanic in nature), it is only logical to consider that exploring extraterrestrial ocean worlds could help answer the question of how life arose on Earth and whether it exists anywhere else in the Solar System. It may also be possible to find prebiotic chemistry occurring, which could provide clues to how life started on Earth. Another school of thought is that any life detected at the remote ocean worlds in the outer Solar System would likely have formed and evolved along an independent path from life on Earth, giving us a deeper understanding of the potential for life in the universe.

Earth's oceans are mostly composed of warm, less salty, water near the surface over cold, saltier, water in the ocean depths. These two regions don't mix except in certain special areas. The oceanic surface currents, the movement of the ocean in the surface layer, are driven mostly by the wind. In certain areas near the polar oceans, the colder surface water also gets saltier due to evaporation or sea ice formation. In these regions, the surface water becomes denser enough to sink to the ocean depths. This pumping of surface water into the deep ocean forces the deep water to move horizontally until it can find an area on the world where it can rise back to the surface and close the water current loop. This usually occurs in the equatorial ocean, mostly in the Pacific and Indian Oceans. This very large, slow water current loop is called the thermohaline conveyor belt circulation (Fig. 1.18); "thermohaline" because it is caused by temperature (thermo) and salinity (haline) variations. Sunlight falling on the surface of Earth drives ocean circulation through atmospheric circulation. Frictional forces transfer energy from winds at the base of the atmosphere, driving circulation of the uppermost ocean. Heating of the ocean surface from sunlight also helps to drive the deep thermohaline circulation on the Earth. Heating near the equator drives processes of evaporation and precipitation across the air–sea interface; meanwhile, cooling effects at

FIG. 1.18 Earth's oceans' thermohaline "conveyor belt" circulation. This simplified illustration shows this "conveyor belt" circulation which is driven by the difference in heat and salinity. Records of past climate suggest that there is some chance that this circulation could be altered by the changes projected in many climate models. *(Source: https://commons.wikimedia.org/wiki/File:Ocean_circulation_conveyor_belt.jpg) Copyright information: This work is in the public domain in the United States because it is a work prepared by an officer or employee of the United States Government as part of that person's official duties under the terms of Title 17, Chapter 1, Section 105 of the US Code.*

the poles lead to freezing of relatively fresh water-ice at the surface, leaving behind relatively salty, hence dense, ocean waters that sink, leading to deep ocean ventilation—that is, mixing of the entire deep ocean system on timescales that are extremely short (order 1000–2000 years) when compared with the age of the planet. This thermohaline circulation and global climate change (i.e., long-term fluctuations in temperature, precipitation, wind, and all other aspects of the Earth's climate) are considered to have an intimate connection.

Earthly oceans' thermohaline conveyor belt circulation is a part of the large-scale ocean circulation that is driven by global density gradients created by surface heat and freshwater fluxes. Basic characteristics of oceans (e.g., salinity [i.e., the total concentration of dissolved salts], and density, which changes with water temperature) are critical for ultimately understanding habitability. The hydrostatic pressure, p (given by the expression $p = h\rho g$) increases with depth (h) and will vary as a function of both the water density (ρ), and the gravitational acceleration (g) of the ocean world in question. The water density varies as a function of the physical properties (e.g., compressibility) of the dominant liquid phase. In an ocean world that has active ocean–atmosphere interactions, as on the Earth, significant variations in salinity can be established through evaporation–precipitation cycles.

According to Hendrix et al. (2019), changes in salinity are likely to play at least as important a role in driving physical circulation on other ocean worlds as does sunlight in driving circulation in the Earth's oceans. But understanding the chemical composition of that ocean will also represent an extremely effective way of understanding what the dominant lithologies (ultramafic/chondritic, basaltic, silicic) are at the underlying seafloor. This, in turn, will provide immediate insights into the geological evolution of that ocean world and what water–rock interactions are dominant, which are critical to understanding the planet's habitability. Changes in salinity and/or composition across the ocean world may also help reveal where any major seafloor inputs arise.

Apart from the thermohaline circulation—the Great Ocean Conveyor Belt—illustrated in Fig. 1.18, which is driven primarily by solar insolation and salinity, there are other ocean circulations including tidal currents. On the Earth, within currents flowing north or south, toward or away from the equator, the fluid flow is decoupled from the rigid underlying seafloor. As a parcel of water travels to higher latitudes, the rotational velocity of the underlying (eastward moving) seabed decreases and, hence, there is a relative motion of the ocean compared with the seafloor that is progressively toward the east at higher latitude (or to the west for currents flowing toward the equator). The exception arises when deep ocean currents intercept the seafloor mid-ocean ridges and both topographic steering and significant vertical mixing result (Toole and Warren, 1993; Scott et al., 2001).

Tidal forces operating on a given celestial body is primarily determined by the mass and the distance of the other celestial bodies which are gravitationally tied to it. For example, the tidal force operating on Earth is contributed primarily by its Moon than by the Sun (contribution from Sun is only approximately half of that from the Moon), although Sun is considerably more massive than Earth's moon. Likewise, the tidal stress acting on a celestial body is also determined by its obliquity and the eccentricity of its orbit. In astronomy, obliquity, also known as axial tilt, is the angle between a celestial body's rotational axis and its orbital axis, or, equivalently, the angle between its equatorial plane and orbital plane. Earth's obliquity oscillates between 22.1° and 24.5° on a 41,000-year cycle. The eccentricity of a celestial body's orbit is a measure of the amount by which its orbit deviates from a circle; it is found by dividing the distance between the focal points of the ellipse by the length of the major axis.

It is a recognized fact that tidal motions in topographically complex water bodies undergo several intricacies. Investigations by several celebrated men of science (e.g., Sir Isaac Newton [1642–1727], David Bernoulli (1700–1782), Marquis de Laplace (1749–1827), and Lord Kelvin (1824–1907)) have revealed that while the origin of tidal rhythm is primarily astronomical, tidal motions are heavily influenced by secondary bathymetric and topographical factors. Interesting scientific explanations on tidal oscillations may be found in Young (1823), Airy (1845), Darwin (1898), Harris (1907), Doodson (1927), Doodson and Warburg (1941), Munk and Cartwright (1966), Garrett and Munk (1971), Pugh (1987), and Pugh and Woodworth (2014). Although astronomically induced tidal variations are in general less than a meter in range (i.e., difference in sea level elevation between successive high water and low water) in the open ocean, when these tidal crests and troughs traverse into shallow waters, against landmasses and into confining channels, considerably large (sometimes dangerous) variations (e.g., tidal bore) occur in the tidal range. A discussion on topographical influences on tidal ranges and tidal patterns is provided in Joseph (2016).

In Earth's oceans, there are regions where invisible colossal waves below sea surface (known as *internal waves (IWs)*) are generated when favorable conditions occur. IWs, which are hidden below the sea surface and entirely between density layers within the interior of the oceans, are giant waves which can reach towering heights (up to a few hundred meters) with profound effects on the Earth's climate and on ocean ecosystems. Typical speeds for IWs are around 1 m/s. They have been described as "the lumbering giants of the ocean." They move fairly slowly relative to surface ocean waves but are very large in amplitude, carry considerable energy, and can ripple through the interior of the oceans over vast distances. Oceanic IWs are usually generated as tidal currents flow over a steep topography such as a ridge, a seamount, or a shelf break in the stratified ocean (see Baines, 2007). They are also found near islands, in straits, and along continental shelves. A discussion on the influence of abrupt seafloor topographical changes on IW generation is provided in Joseph (2016).

IWs can generate powerful turbulence, leading to strong vertical mixing of ocean waters. Because of their size and behavior, such as strong vertical and horizontal currents, and the turbulent mixing caused by their breaking, IWs affect a panoply of ocean processes. They carry nutrients from ocean depths to the upper layers, even up to the surface, to enrich marine organisms; enhance sediment and pollutant transport; and affect acoustic transmission, that is, propagation of sound waves (see Williams et al., 2001). The excellent vertical mixing produced by IWs is beneficial to the marine flora and fauna as a result of enhanced nutrient distribution in areas which are otherwise barren in terms of nutrients. Unlike tides (which are of astronomical origin), IWs are not always of astronomical origin although strong tidal currents are often needed to trigger their generation. Furthermore, IWs are not generally coherent with the movements of the Moon and the Sun. Seafloors of extraterrestrial ocean worlds are also likely to have similar geological and geophysical characteristics and features as those of Earth's seafloors and, therefore, a clear understanding of Earth's seafloors and their physical, geological, chemical, and biological influences would be beneficial in understanding extraterrestrial ocean worlds.

Life on the Earth utilizes, as sources of energy, light in the visible to near-infrared (IR) wavelength range and the chemical energy released in specific (mostly oxidation-reduction) chemical reactions. Because all life on the Earth derives energy either from sunlight or from the reaction energy of chemical redox pairs (either present in the environment or produced by cells from energy in the environment), identification of a redox gradient is highly relevant to any search for life in ocean worlds. The fundamental nature of planetary material results in a partitioning of oxidants and reductants at the planetary surface and interior, respectively. As on the Earth, interfaces that constitute boundaries between the surface and interior may be ideal habitats. On an ocean world, liquid water can extend these boundaries by transporting oxidants and reductants, placing them in close enough proximity to fuel life-sustaining processes. Solid surfaces can also provide a location for the concentration of nutrients, an attachment point for cells, and a stable habitat for life (potentially) in a habitable subsurface ocean.

As indicated above, based on numerous spectroscopic and model-based investigations, and discovery of plumes erupting from *Europa* and *Enceladus*, it is now known that some icy moons in the Solar System harbor hidden oceans beneath their icy crust. The confirmed ocean worlds in the Solar System so far are Jupiter's moons *Europa*, *Ganymede*, and *Callisto*, and Saturn's moons *Enceladus* and *Titan*. Europa and Enceladus stand out as ocean worlds with evidence for communication between the ocean and the surface, as well as the potential for interactions between the oceans and a rocky seafloor, which is important in terms of habitability considerations. Furthermore, Europa and Enceladus have the highest priority because their icy shells are thinner than the others (Europa's is less than 10 km; Enceladus' is about 40 km) and there is credible evidence that their oceans are in contact with the rocky mantle (plumes of salty water particles and other elements and compounds emanating from their surface), which could provide both energy and chemicals for life to form. Enceladus' ice crust has fractures at the south pole that allow ice and gas from the ocean to escape to space, where it has been sampled by mass spectrometers aboard the Cassini Saturn orbiter with tantalizing results (Postberg et al., 2018). Europa is also known to have active presence of tall and short plumes consisting of water particles and chemical molecules. Using analysis of radar data from NASA's Cassini spacecraft, it has been found that *Titan* has an ice layer that is 200 km thick, and an ocean somewhere in the range of 225–300 km thick; and no evidence for active plumes or ice volcanism have been observed.

The subsurface oceans of *Ganymede* and *Callisto* are also expected to be covered by relatively thick ice shells, making exchange processes with the surface more difficult, and with no obvious surface evidence of the oceans. Because oceans at Ganymede and Callisto are deeper and there is no evidence of communication between liquid water and the surface and/or a silicate core, these oceans should be better understood before exploring them as potentially habitable. This lack of knowledge limits their ability to support more of the Ocean Worlds science objectives, and, thus, they are lower in priority from the other known ocean worlds (Hendrix et al., 2019).

Icy worlds speckle our solar system. From Jupiter's moon Europa to Saturn's moon Enceladus, scientists have been investigating these alien worlds, discovering subsurface oceans hidden under their icy crusts. A host of other bodies in the outer Solar System, including the dwarf planet Pluto located in the outer periphery of our Solar System, are also inferred by a single type of observation or by theoretical modeling to have subsurface oceans. In subsequent "ocean worlds search" initiatives, researchers have turned their gaze to the moons orbiting Uranus, searching for secret oceans. During Voyager 2's flyby of the Uranus system in 1986, it sent back to Earth the first close-up views of the planet's five largest moons—Miranda, Ariel, Umbriel, Titania, and Oberon (Fig. 1.19A and B). Those images revealed that the moons' surfaces also display some of the classic signs of cryovolcanism (eruption of volatiles as in a cryovolcano), such as fresh un-cratered material and ridges, valleys, and folds. It may be noted that cryovolcanism is a physical sign of liquid water erupting through an ocean world and freezing on its surface.

As with Europa and Enceladus, a subsurface ocean is one way to create those cryovolcanism-like signs of recent geologic activity. Based on credible indications, scientists have inferred that several moons of Uranus could be sloshing with oceans hiding just below the surface. For example, extraterrestrial celestial bodies such as the dwarf planet *Pluto*, Uranus's moons *Triton*, *Miranda*, and *Ariel*, Saturn's moon *Dione*, and a dwarf planet (*Ceres*) in the asteroid belt (located between the orbits of the planets Mars and Jupiter), are considered to be potential candidate ocean worlds, based on hints from limited spacecraft observations (Hendrix et al., 2019). In a recent study, Cartier (2020) reported findings supporting earlier studies with regard to the existence of subsurface oceans on some of Uranus's moons. However, it has been cautioned that subsurface oceans in the Uranus system are likely to be farther beneath the surface than those in the Jupiter system simply because Uranus's moons are colder and so the icy shell is likely thicker.

There are some extraterrestrial ocean worlds, in which submarine hydrothermal vents are present on the seafloor (e.g., Saturn's icy moon *Enceladus*). When considering the habitability of an ocean world, especially one far from the Sun where photosynthesis is unlikely to be effective, a key issue is the nature of the underlying seafloor and water–rock interactions (e.g., submarine hydrothermal vent environment). Although the full diversity of settings observed on the Earth may not be repeated on any single other ocean world, an advantage provided by our home planet—Earth—is the opportunity to conduct a wealth of analogue studies spanning a broad range of conditions that may be relevant to other ocean worlds in terms of seafloor pressure, maximum water temperatures at those pressures, and interactions with different rock types (reactions that can both result in and be diagnosed from ocean chemical compositions). It is pertinent to note that the Earth's mid-ocean ridges (where most of the hydrothermal vents are located) remain 80% unexplored (Beaulieu et al., 2015). Because tectonic activities are taking place on all of the presently known and potential ocean worlds, it is likely that submarine hydrothermal vents could be present on the seafloors of these ocean worlds.

As just indicated, there are several—if not many—already confirmed ocean worlds and potential ocean worlds in our Solar System. All such worlds are targets for vigorous studies in the quest for understanding the distribution of life in the Solar System. There are several "candidate ocean worlds" in our Solar System that exhibit hints of possible oceans, and worlds that may theoretically harbor oceans but about which not enough is currently known to definitively determine whether an ocean exists or not. As a philosophy, it is critical to consider all of these worlds to understand the origin and development of oceans and life in different worlds. There could be several aspects with regard to life in extraterrestrial ocean worlds that we can learn about from studying the Earth.

Hendrix et al. (2019) reported that once an ocean has been demonstrated to exist in any suspected/assumed extraterrestrial ocean world, an immediate priority will be to understand the physics and chemistry of that ocean to gain knowledge of how the system functions and how that, in turn, informs our assessment of its potential habitability.

FIG. 1.19 Uranus and its five major moons (potential candidate ocean worlds) are depicted in this montage of images acquired by the Voyager 2 spacecraft. The moons, from largest to smallest as they appear here, are Ariel, Miranda, Titania, Oberon, and Umbriel (Image: © NASA/JPL). (*Source: https://www.space.com/uranus-moons-hiding-secret-oceans?utm_source=notification.*)

Such characterization permits identification of appropriate future life detection experiments, if warranted.

NASA's Ocean Worlds Exploration Program (OWEP) envisages to explore ocean worlds in the outer Solar System that could possess subsurface oceans to assess their habitability and to seek biosignatures of simple extraterrestrial life. The US House Appropriations Committee approved the bill on May 20, 2015 and directed NASA to create the OWEP. The "Roadmaps to Ocean Worlds" (ROW) was initiated in 2016, and was presented in January 2019 (Hendrix et al., 2019). Oceanographers, biologists and astrobiologists are part of the team developing the strategy roadmap. The planning also considers implementing planetary protection measures to avoid contaminating extraterrestrial habitable environments with resilient stowaway bacteria (i.e., bacteria cunningly hiding, without being seen) on their landers.

The formal program is being implemented within the agency by supporting the Europa Clipper orbiter mission to Europa (the report accompanying the bill directs NASA to create an "Ocean Worlds Exploration Program" of which the Europa mission is part), and the Dragonfly mission to Titan. The program is also supporting concept studies for a proposed Europa Lander, and concepts to explore Uranus's moon *Triton* (Hendrix et al., 2019). The Europa Clipper orbiter and lander, with the currently envisaged launch of the orbiter in 2025 and the potential Europa Lander shortly after, are expected to be important milestones in fulfilling the enormously challenging goals of the OWEP.

1.9 Importance of radiogenic heating and tidal dissipation in the generation and sustenance of extraterrestrial subsurface ocean worlds

Scientists are of the view that energy sources are perhaps the single most fundamental requirements for the maintenance of a present-day ocean on an otherwise frozen world (see Hendrix et al., 2019). The identification of extraterrestrial ocean worlds, therefore, requires identification of possible energy sources. Two sources of heating occurring in a celestial body are (1) radiogenic heating, and (2) tidal-stress-induced heating.

Radiogenic heating occurs as a result of the release of heat energy (thermal energy) from radioactive decay (i.e., spontaneous nuclear disintegrations) (see Alisa and Allaby, 1999) during the production of radiogenic nuclides (radiogenic isotopes).

Tidal heating of a planet and its moon(s) occurs through the tidal friction processes between these celestial bodies: orbital and rotational energy is dissipated as heat in either the surface ocean or interior of a planet or its moon(s). When an object is in an elliptical orbit, the tidal forces acting on it are stronger near periapsis (the point of closest approach) than near apoapsis (the point of greatest separation). Thus, the deformation of the body due to tidal forces varies over the course of its orbit, generating internal friction which heats its interior. This energy gained by the object comes from its gravitational energy, so over time in a two-body system, the initial elliptical orbit decays into a circular orbit. It is known that a planet's or its moon(s)' orbital or rotational properties play a role in determining the tidal stress and tidal dissipation experienced by these celestial bodies.

It is known that sustained tidal heating occurs when the elliptical orbit is prevented from circularizing due to additional gravitational forces from other bodies that keep tugging the object back into an elliptical orbit. In this more complex system, gravitational energy still is being converted to thermal energy. Tidal dissipation depends on the heat transport mechanism. For strong tidal forcing, an equilibrium between heat transport by convection and heat production by tidal dissipation can be obtained that determines the tidal dissipation (Clausen and Tilgner, 2015). The tidal dissipation is usually parameterized by the tidal quality/dissipation factor Q. Whereas Earth has $Q \approx 12$ (Murray and Dermott, 1999), the quality factor of other terrestrial planets in our solar system lie in the range $10 \leq Q \leq 190$ (Goldreich and Soter, 1966). Rocky bodies tend to have Q-values near 100, giant planets and stars have Q-values near 10^6.

As an example, tidal heating of Jupiter's moon Io (which is the most geologically active body in the Solar System) occurs through the tidal friction processes between Jupiter and Io. This is the heating mechanism of Io. Whereas the major heating source of Earth and its moon is radioactive heating, the heating source on Io is tidal heating. Jupiter's large moons Europa and Ganymede are in an orbital resonance with it. Io is the innermost of this set of resonant moons, and their interactions maintain its orbit in an eccentric (elliptical) state. The varying distance between Jupiter and Io continually changes the degree of distortion of Io's shape and flexes its interior, frictionally heating it. The friction-induced heating drives strong volcanic activities on the surface of Io.

As indicated in Hendrix et al. (2019), both radiogenic heating (e.g., for Europa, Ganymede, Callisto, and Titan) and tidal energy (e.g., for Europa, Enceladus) play a role in sustaining oceans (Hussmann et al., 2006). Tidal heating is a process that is critical to sustaining ocean worlds. The tidal dissipation factor (Q) of a planet is critical to driving the dynamics and heating in these systems. This factor is a complex function of the interior structure of the planet. The more we understand about ice giant interiors, the more we can learn about potential heat sources, to sustain oceans within the moons.

Available energy sources can be identified through either modeling or direct observation (or ideally a combination of the two). Theoretical modeling is an invaluable tool for predicting which bodies can sustain ocean worlds. Some models anticipated oceans on icy moons (e.g., Europa) long before such oceans were ever actually detected (Lewis, 1971; Consolmagno and Lewis, 1978). However, modeling alone may sometimes lead to misleading results. In the case of Enceladus, theoretical models indicated that insufficient conversion of tidal energy to heat should occur within the Moon (Meyer and Wisdom, 2007; Roberts and Nimmo, 2008).

Observations by the Cassini spacecraft have demonstrated that such heating does, in fact, occur, but the Moon emits an order of magnitude more energy (at the present time) than theoretical models predicted (Spencer et al., 2006; Howett et al., 2011). Identifying the sources of Enceladus' energy and its transient versus long-lived nature is an open and active area of research.

For the largest moons, remnant radiogenic heating may be sufficient to maintain an internal ocean, depending on the initial radiogenic content of the rock component, and the state of the overlying ice shell (Hussmann et al., 2006; Schubert et al., 2010). For smaller bodies (e.g., Enceladus), dissipation of tidal energy is critical (Schubert et al., 2010). Dissipation of tidal energy requires the presence of a parent planet or moons with sufficient gravitational energy to deform the body. Pluto and Charon lack a source of such energy, so the energetics that permit a long-lived ocean within these bodies are still in question (Nimmo and Spencer, 2015).

In addition, a body's orbit and/or rotation must be favorable to tidal dissipation, possibly through a high eccentricity (e.g., Europa) (Sotin et al., 2009), libration, or obliquity (e.g., Triton) (Nimmo and Spencer, 2015). Pluto and Charon lack a source of such energy because their orbits have evolved toward circularization (i.e., zero eccentricity). However, these two requirements are insufficient to ensure internal oceans, as the planet or moon must be able to convert available tidal energy to heat (Tobie et al., 2005). This is demonstrated by Mimas (the smallest and innermost of Saturn's major moons), which, despite its high eccentricity, dissipates little tidal energy, likely because its interior has remained cold since shortly after its formation (McKinnon, 2010). Note that Mimas is not quite big enough to hold a round shape, so it is somewhat ovoid in shape. The complex feedback between the orbital/rotational evolution of potential ocean worlds and their internal structure requires careful theoretical modeling (Ojakangas and Stevenson, 1989; Showman et al., 1997; Hussmann and Spohn, 2004).

Tidal modulation of the shell may also contribute to the ocean convection and transport dynamics (Tyler, 2008). As on the Earth, tidal processes may also lead to important mixing within an ocean—even in an icy ocean world. This will be highly dependent on the gravitational interaction of any ocean world and its adjacent moons. The same processes seen on the Earth are also important for Europa, for example, and may be for Enceladus and Titan, too (Tyler, 2009, 2011, 2014).

Similar mixing behavior within other ocean worlds could also affect such global calculations noted earlier, but the importance of these processes will scale with the depth of the ocean, the topography of the seafloor, the drivers for poleward currents, and the rotational velocity of the ocean world under consideration (Hendrix et al., 2019).

1.10 Shedding light on extraterrestrial bodies—role of astronomical research

The quest for understanding the formation and evolution of the Sun and our Solar System is everlasting, and, this is true for deciphering the origin and evolution of other Stars as well as the distant Galaxies. Systematic studies to understand stellar objects, including stars and planets, started more than 2600 years back, when Pythagoras (an ancient Greek philosopher who was born in 569 BCE) suggested that "Earth is a sphere and all other heavenly bodies are perfect spheres." Having been influenced by the philosophies of Pythagoras, another ancient Greek philosopher Plato, who was born in 428 BCE (one of the world's best known and most widely read and studied philosophers) suggested that the heavenly bodies appear to move in circles around the Earth. This led to the notion of an earth-centric universe, suggested by Aristotle, who was born in 384 BCE (another ancient Greek philosopher and scientist, one of the greatest intellectual figures of Western history), who also proposed that the stars define the edge of the Universe. Claudius Ptolemy (a Greco-Roman mathematician, astronomer, geographer and astrologer in first century AD) synthesized and consolidated the idea of an Earth-centric universe. He also introduced various geometrical parameters to explain the motion of planets.

Ever since the 16th century AD mathematician and astronomer Nicolaus Copernicus proposed the theory of Sun-centric model of the Solar System (Fig. 1.20), there has been an accelerated interest among several thinkers over a period of time to understand what lies within and beyond Earth. The 16th and the 17th century scientists, under innumerable obstacles and harsh religious opposition, have laid the foundation stone for understanding the vast expanse of the Universe in which Earth is a mere minuscule object. Copernicus conceived the idea that Earth and the other planets orbit around the Sun—a scientific principle known as helio-centrism. Giordano Bruno was a great visionary who supported the Copernican system and also foresaw the possibility of life lurking external to Earth, and much beyond. The Italian physicist, mathematician, astronomer, tireless experimenter, and inventor Galileo Galilei succeeded in courageously overturning the hitherto accepted notion of geo-centrism (the notion that the other planets and the Sun orbit around Earth) as a result of his invention of telescope and telescopic measurements of extraterrestrial planetary bodies. Galileo achieved great strides in space discoveries,

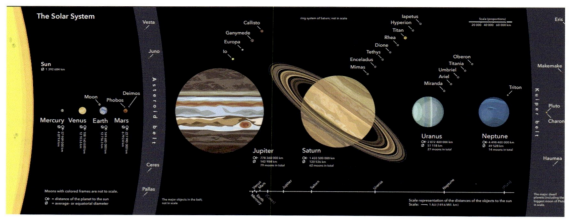

FIG. 1.20 Solar system, which consists of the Sun; the eight official planets, at least three "dwarf planets" including Pluto, more than 130 moons of the planets (not shown here), a large number of small bodies (the comets and asteroids), and the interplanetary medium. Note that Pluto has been removed from the list of major planets in the solar system; it is presently recognized as "dwarf planet." *(Image Source: https://en.wikipedia.org/wiki/File:Solar-System.pdf.)*

and contributed immensely to the scientific revolution that the seventeenth century witnessed.

Among several of Galileo's contributions to observational astronomy (e.g., Venus's disk occasionally appearing in the shape of a crescent; observation and analysis of sunspots) the discovery of the four largest moons of planet Jupiter (Io, Europa, Ganymede, and Callisto) (named the *Galilean moons* in honor of Galileo, who observed them in 1610) received much attention.

In the field of astronomy, Johannes Kepler—a great mathematician and astronomer—was the first person to theoretically explain planetary motion. Kepler's Rudolphine Tables provided data for calculating planetary positions for any past or future date. Kepler's three laws on planetary motion—the famous *Fundamental Laws of Planetary Motion*—constitute an essential study material for every physics student. With the advances in astrophysical studies and postulates, coupled with observational capabilities for studying stellar and planetary objects, we now have a better understanding about the origin and early evolution of the Sun and the Solar System.

The 20th century saw great strides in space missions that have now spilled over to the 21st century with accelerated vigor and surprisingly great technological leaps. Spacecraft visits literally hugging many extra-terrestrial targets have already transformed/refined some of the early notions; revealing the diversity of Solar System bodies and displaying active planetary processes at work (see Burns, 2010). Extensive studies carried out by several investigators over more than a decade suggest that formation of the Sun and our Solar system was triggered by an exploding star, a supernova, that is now accepted by the scientific community. Space probe missions targeting exo-planets (i.e., planets which orbit stars outside the Solar System) have already become a reality.

The various projects undertaken by the United States' National Aeronautics and Space Administration (NASA), European Space Agency (ESA)—an intergovernmental organization of 22 member states dedicated to the exploration of space, and some universities and space research organizations from various countries to search for planets and their moons that can sustain life has led to an increased curiosity in the 21st century about life beyond Earth.

References

Agnor, C.B., Canup, R.M., Levison, H.F., 1999. On the character and consequences of large impacts in the late stage of terrestrial planet formation. Icarus 142, 219–237.

Airhart, M., 2011. Scientists Find Evidence for "Great Lake" on Europa and Potential New Habitat for Life. In: Brush, S.G. (Ed.), Origin of the Moon (Lunar and Planetary Institute, 1986). Jackson School of Geosciences, The University of Texas at Austin, Houston, TX, pp. 3–15.

Airy, G.B., 1845. Tides and waves. Encyclopaedia Metropolitana, vol. 3, Burgersdijk & Niermans, Biddington, London, pp. 241–396.

Albarède, F., 2009. Volatile accretion history of the terrestrial planets and dynamic implications. Nature 461, 1227–1233.

Alexander, C.M.O.D., Fogel, M., Yabuta, H., Cody, G.D., 2007. The origin and evolution of chondrites recorded in the elemental and isotopic compositions of their macromolecular organic matter. Geochim. Cosmochim. Acta 71 (17), 4380–4403.

Alexander, C.M.O., Bowden, R., Fogel, M.L., Howard, K.T., Herd, C.D.K., Nittler, L.R., 2012. The provenances of asteroids, and their contributions to the volatile inventories of the terrestrial planets. Science 337, 721–723.

Ali-Dib, M., Mousis, O., Petit, J-M., Lunine, J.I., 2014. The measured compositions of Uranus and Neptune from their formation on the CO ice line. Astrophys. J. 793 (9), 7.

Alisa, A., Allaby, M., 1999. Radiogenic heating. In: Allaby, M. (Ed.). A Dictionary of Earth Sciences, Oxford University Press, Oxford ox2 6dp, pp. 635–647. ISBN: 978-0-19-921194-4.

Anand, M., Tartèse, R., Barnes, J.J., 2014. Understanding the origin and evolution of water in the Moon through lunar sample studies. Philos. Trans. R. Soc. A 372, 20130254.

Andrews-Hanna, J.C., Zuber, M.T., Arvidson, R.E., Wiseman, S.M., 2010. Early Mars hydrology: Meridiani playa deposits and the sedimentary record of Arabia Terra. J. Geophys. Res. 115, E06002.

Artemieva, N., Pierazzo, E., 2009. The Canyon Diablo impact event: Projectile motion through the atmosphere. Meteorit. Planet. Sci. 44 (1), 25–42.

Baines, P.G., 2007. Internal tide generation by seamounts. Deep-Sea Res. I 54, 1486–1508.

Baker, G.L. (1983), Emanuel Swenborg – an 18th century cosomologist, The Physics Teacher, pp. 441–446.

Beaulieu, S.E., Baker, E.T., German, C.R., 2015. Where are the undiscovered hydrothermal vents on oceanic spreading ridges? Deep Sea Res. II 121, 202–212.

Benz, W., Cameron, A.G., Melosh, H.J., 1989. The origin of the moon and the single-impact hypothesis III. Icarus 81, 113–131. doi:10.1016/0019-1035(89)90129-2.

Benz, W., Slattery, W.L., Cameron, A.G.W., 1986. The origin of the Moon and the single impact hypothesis I. Icarus 66, 515–535.

Benz, W., Slattery, W.L., Cameron, A.G.W., 1987. The origin of the Moon and the single impact hypothesis II. Icarus 71, 30–45.

Bland, P.A., Artemieva, N., 2010. The rate of small impacts on Earth. Meteorit. Planet. Sci. 41 (4), 607–631. doi:10.1111/j.1945-5100.2006.tb00485.x.

Brush, S.G., 1986. Origin of the Moon. Lunar and Planetary Institute, pp. 3–15.

Bolton, S.J., Adriani, A., Adumitroaie, V., et al., 2017. Jupiter's interior and deep atmosphere: the initial pole-to-pole passes with the Juno spacecraft. Science 356 (6340), 821–825. doi:10.1126/science.aal2108.

Burbine, T.H., 2016. Asteroids: Astronomical and Geological Bodies. Cambridge University Press, Cambridge, England, ISBN 978-1-107-09684-4.

Burchell, M.J., Galloway, J.A., Bunch, A.W., Brandão, P.F.B., 2003. Survivability of bacteria ejected from icy surfaces after hypervelocity impact. Origins Life Evol. Biosphere 33 (1), 53–74.

Burns, J.A., 2010. The four hundred years of planetary science since Galileo and Kepler. Nature 466, 575–584. doi:10.1038/nature09215.

Bussey, D.B.J., Neish, C.D., Spudis, P., Marshall, W., Thomson, B.J., Patterson, G.W., Carter, L.M., 2011. The nature of lunar volatiles as revealed by Mini-RF observations of the LCROSS impact site. J. Geophys. Res.: Planets (E01005), 116. doi:10.1029/2010JE003647.

Cameron, A.G.W., 1985. Formation of the prelunar accretion disk. Icarus 62, 319–327.

Cameron, A.G.W., 1997. The origin of the Moon and the single impact hypothesis V. Icarus 126, 126–137.

Cameron, A.G.W., 2001. From interstellar gas to the Earth–Moon system. Meteorit. Planet. Sci. 36, 9–22.

Cameron, A.G.W., Benz, W., 1991. The origin of the Moon and the single impact hypothesis IV. Icarus 92, 204–216.

Cameron, A.G.W., Ward, W.R., 1976. The origin of the Moon, Proc. Lunar Planet. Sci. Conf. 7th, 120–122.

Canup, R.M., 2004a. Dynamics of lunar formation. Annu Rev Astron Astr 42, 441–475.

Canup, R.M., 2004b. Simulations of a late lunar-forming impact. Icarus 168, 433–456.

Canup, R.M., 2008. Lunar-forming collisions with pre-impact rotation. Icarus 196, 518–538.

Canup, R.M., 2012. Forming a Moon with an Earth-like composition via a giant impact. Science 338, 1052–1055.

Canup, R.M., Asphaug, E, 2001. Origin of the Moon in a giant impact near the end of the Earth's formation. Nature 412, 708–712. doi:10.1038/35089010.

Canup, R.M., Ward, W.R., 2008. Origin of Europa and the Galilean Satellites. University of Arizona Press, Tucson, Arizona. p. 59. arXiv:0812.4995.

Carlson, R.W., 2019. Analysis of lunar samples: implications for planet formation and evolution. Science 365 (6450), 240–243. doi:10.1126/science.aaw7580.

Carr, M.H., 1996. Water on Mars. Oxford Univ. Press, New York.

Cartier, K.M.S., 2020. Do Uranus's moons have subsurface oceans? Eos 101. https://doi.org/10.1029/2020EO152056.

Choi, C.Q. (2021), Planet Earth: facts about our home planet, Space.com; https://www.space.com/54-earth-history-composition-and-atmosphere.html).

Christensen, P.R., 2006. Water at the poles and in permafrost regions of Mars. Elements (2), 151–155.

Ciesla, F.J., Cuzzi, J., 2006. The evolution of the water distribution in a viscous protoplanetary disk. Icarus 181 (1), 178–204.

Ciesla, F.J., Mulders, G.D., Pascucci, I., Apai, D, 2015. Volatile delivery to planets from water-rich planetesimals around low-mass stars. Astrophys. J. 804 (1), 9.

Citron, R.I., Perets, H.B., Aharonson, O., 2018. The role of multiple giant impacts in the formation of the Earth–Moon system. Astrophys. J. 862 (5), 11.

Clark, R.N., 2009. Detection of adsorbed water and hydroxyl on the Moon. Science 326 (5952), 562–564. doi:10.1126/science.1178105.

Clausen, N., Tilgner, A., 2015. Dissipation in rocky planets for strong tidal forcing. Astron. Astrophys. (A&A) 584, A60. doi:10.1051/0004-6361/201526082.

Clayton, R.N., 1993. Oxygen isotopes in meteorites. Annu. Rev. Earth Planet. Sci. 21, 115–149.

Clery, D., 2013. Impact theory gets whacked. Science 342 (6155), 183–185.

Colaprete, A., Schultz, P., Heldmann, J., Wooden, D., et al., 2010. Detection of water in the LCROSS ejecta plume. Science 330, 463–468.

Collings, M.P., Anderson, M.A., Chen, R., Dever, J.W., Viti, S., Williams, D.A., McCoustra, M.R.S., 2004. A laboratory survey of the thermal desorption of astrophysically relevant molecules. Mon. Not. R. Astron. Soc. 354, 1133–1140.

Consolmagno, G.J., Lewis, S.J., 1978. The evolution of icy satellite interiors and surfaces. Icarus 34, 280–293.

Crida, A., 2009. Minimum mass Solar Nebulae and planetary migration. Astrophys. J. 698 (1).

Crotts, A., 2011. Water on the Moon, I. Historical overview. Astron. Rev. 6 (7), 4–20. doi:10.1080/21672857.2011.11519687.

Ćuk, M., Hamilton, D.P., Lock, S.J., Stewart, S.T., 2016. Tidal evolution of the Moon from a high-obliquity, high-angular-momentum Earth. Nature. doi:10.1038/nature19846.

Ćuk, M., Stewart, S.T., 2012a. Early Solar System Impact Bombardment II (Lunar and Planetary Institute, Houston, Texas, #4006).

Ćuk, M., Stewart, S.T., 2012b. Making the moon from a fast-spinning earth: a giant impact followed by resonant despinning. Science 338, 1047–1052.

D'Angelo, G., Durisen, R.H., Lissauer, J.J., 2010. Giant planet formation. In: Seager, S. (Ed.), Exoplanets. University of Arizona Press, Tucson, AZ, pp. 319–346.

D'Angelo, G., Podolak, M., 2015. Capture and evolution of planetesimals in circumjovian disks. Astrophys. J. 806 (1), 29.

D'Angelo, G., Weidenschilling, S.J., Lissauer, J.J., Bodenheimer, P., 2014. Growth of Jupiter: enhancement of core accretion by a voluminous low-mass envelope. Icarus 241, 298–312 arXiv:1405.7305.

Dartois, E., Engrand, C., Brunetto, R., Duprat, J., Pino, T., Quirico, E., Remusat, L., Bardin, N., Briani, G., Mostefaoui, S., Morinaud, G., Crane, B., Szwec, N., Delauche, L., Jamme, F., Sandt, Ch., Dumas, P., 2013. Ultra carbonaceous Antarctic micrometeorites, probing the Solar System beyond the nitrogen snow-line. Icarus 224 (1), 243–252.

Darwin, G.H., 1879. On the precession of a viscous spheroid and on the remote history of the Earth. Phil. Trans. Roy. Soc. London 170, 447–538.

Darwin, G.H., 1898. The Tides. W.H. Freeman and Company, San Francisco and London, pp. 378p.

Dauphas, N., 2017. The isotopic nature of the Earth's accreting material through time. Nature 541, 521–524. doi:10.1038/nature20830.

Dauphas, N., Chen, J., Papanastassiou, D, 2015. Testing Earth–Moon isotopic homogenization with calcium-48. Lunar Planet. Sci. Conf. XXXXVI, 2436.

Dauphas, N., Marty, B., Reisberg, L., 2002. Molybdenum evidence for inherited planetary scale isotope heterogeneity of the protosolar nebula. Astrophys. J. 565, 640–644.

Dauphas, N., Schauble, E.A., 2016. Mass fractionation laws, mass-independent effects, and isotopic anomalies. Annu. Rev. Earth Planet. Sci. 44 (1). doi:10.1146/annurev-earth-060115-012157.

Davies, G.F., 1990. Heat and mass transport in the early Earth. In: Newsom, H.E, Jones, J.R (Eds.), Origin of the Earth. Oxford University Press, New York, pp. 175.

Day, J, 2018. Siderophile elements. In: White, W.M. (Ed.), Encyclopedia of Geochemistry. Encyclopedia of Earth Sciences Series. Springer, Cham. https://doi.org/10.1007/978-3-319-39312-4_234.

Deamer, D., Damer, B., 2017. Can life begin on Enceladus? A perspective from hydrothermal chemistry. Astrobiology 17 (9), 834–839.

De Meijer, R.J., Anisichkin, V.F., van Westrenen, W., 2013. Forming the Moon from terrestrial silicate-rich material. Chem. Geol. 345, 40–49. doi:10.1016/j.chemgeo.2012.12.015.

DeMeo, F.E., Carry, B., 2014. Solar System evolution from compositional mapping of the asteroid belt. Nature 505 (7485), 629–634. doi:10.1038/nature12908.

Dijkstra, H.E., van der Sanden, G.A.H.F., Peters, S., Zepper, J., Branchetti, M., van Westrenen, W., Foing, B.H., 2018. GLACE mission concept: Ganymede's life and curious exploration mission, 49th Lunar and Planetary Science Conference 2018.

Doodson, A.T., 1927. The analysis of tidal observations. Philos. Trans. R. Soc. Lond. A227, 223–279.

Doodson, A.T., Warburg, H.D., 1941. Admiralty Manual of Tides. HMSO, London, pp. 270.

Dodson-Robinson, S.E., Willacy, K., Bodenheimer, P., Turner, N.J., Beichman, C.A., 2009. Ice lines, planetesimal composition and solid surface density in the solar nebula. Icar 200 (2), 672–693.

Elardo, S.M., Draper, D.S., Shearer, C.K., 2011. Lunar magma ocean crystallization revisited: bulk composition, early cumulate mineralogy, and the source regions of the highlands Mg-suite. Geochim. Cosmochim. Acta 75, 3024–3045. doi:10.1016/j.gca.2011.02.033.

Elkins-Tanton, L.T., Burgess, S., Yin, Q.-Z., 2011a. The lunar magma ocean: Reconciling the solidification process with lunar petrology and geochronology. Earth Planet. Sci. Lett. 304, 326–336.

Elkins-Tanton, L.T., Grove, T.L., 2011b. Water (hydrogen) in the lunar mantle: results from petrology and magma ocean modelling. Earth Planet. Sci. Lett. 307, 173–179.

Elkins-Tanton, L., Weiss, B. (Eds.), 2017. Planetesimals, Early Differentiation and Consequences for Planets. Cambridge University Press, Cambridge, United Kingdom.

Fisher, E.A., et al., 2017. Evidence for surface water ice in the lunar polar regions using reflectance measurements from the Lunar Orbiter Laser Altimeter and temperature measurements from the Diviner Lunar Radiometer Experiment. Icarus 292, 74–85.

Gaeman, J., Hier-Majumder, S., Roberts, J.H., 2012. Sustainability of a subsurface ocean within Triton's interior. Icarus 220 (2), 339–347.

Garrett, C.J.R., Munk, W.H., 1971. The age of the tide and the 'Q' of the oceans. Deep-Sea Res. 18, 493–503.

Gingold, R.A., Monaghan, J.J., 1979. Binary fission in damped rotating polytropes, II. Mon. Not. R. Astron. Soc. 188, 39–44.

Gingold, R.A., Monaghan, J.J., 1981. The collapse of a rotating non-axisymmetric isothermal cloud. Mon. Not. R. Astron. Soc. 197, 461–475.

Gingold, R.A., Monaghan, J.J., 1982. Kernel estimates as a basis for general particle methods in hydrodynamics. J. Comput. Phys. 46, 429–453.

Gingold, R.A., Monaghan, J.J., 1983. On the fragmentation of differentially rotating clouds. Mon. Not. R. Astron. Soc. 204, 715–733.

Gladman, B., 1997. Destination: Earth. Martian meteorite delivery. Icarus 130, 228–246.

Gladman, B.J., Burns, J.A., Duncan, M., Lee, P., Levison, H.F., 1996. The exchange of impact ejecta between terrestrial planets. Science 271, 1378–1392.

Goldreich, P., Soter, S., 1966. Q in the Solar System. Icarus 5, 375.

Graves, S.D.B., McKay, C.P., Griffith, C.A., Ferri, F., Fulchignoni, M., 2008. Rain and hail can reach the surface of Titan. Planet. Space Sci. 56, 346–357.

Hamilton, T.L., et al., 2016. The role of biology in planetary evolution: cyanobacterial primary production in low-oxygen Proterozoic oceans. Environ. Microbiol. 18 (2), 325–340. doi:10.1111/1462-2920.13118.

Harris, R.A., 1907. Manual of Tides. In: Ockerbloom, J.M. (Ed.), Part V. U.S. Coast and Geodetic Survey. U.S. G.P.O., Washington, pp. 1894–1907, OBP copyrights and licenses.

Hartmann, W.K., Davis, D.R., 1975. Satellite-sized planetesimals and lunar origin. Icarus 24, 504–515.

Hayashi, C., 1981. Structure of the Solar Nebula, growth and decay of magnetic fields and effects of magnetic and turbulent viscosities on the Nebula. Prog. Theor. Phys. Suppl. 70, 35–53.

Hayes, A., Aharonson, O., Callahan, P., Elachi, C., Gim, Y., Kirk, R., Lewis, K., Lopes, R., Lorenz, R., Lunine, J., et al., 2008. Hydrocarbon lakes on Titan: distribution and interaction with a porous regolith. Geophys. Res. Lett. 35 (9).

Hayes, A.G., 2016. The lakes and seas of Titan. Annu. Rev. Earth Planet. Sci. 44 (1), 57–83. doi:10.1146/annurev-earth-060115-012247.

Heggy, E., Scabbia, G., Bruzzone, L., Pappalardo, R.T., 2017. Radar probing of Jovian icy moons: understanding subsurface water and structure detectability in the JUICE and Europa missions. Icarus 285, 237–251.

Helled, R., Bodenheimer, P., 2014. The formation of Uranus and Neptune: challenges and implications for intermediate-mass exoplanets. Astrophys. J. 789 (1), 69.

Hendrix, A.R., Hurford, T.A., Barge, L.M., Bland, M.T., Bowman, J.S., Brinckerhoff, W., Buratti, B.J., Cable, M.L., Castillo-Rogez, J., Collins, G.C., Diniega, S., German, C.R., Hayes, A.G., Hoehler, T., Hosseini, S., Howett, C.J.A., McEwen, A.S., Neish, C.D., Neveu, M., Nordheim, T.A.,

Patterson, G.W., Patthoff, A., Phillips, D., C., Rhoden, A., Schmidt, B.E., Singer, K.N., Soderblom, J.M., Vance, S.D. 2019., The NASA roadmap to ocean worlds. Astrobiology, 19, 1–27. doi:10.1089/ast.2018.1955.

Herwartz, D., Pack, A., Friedrichs, B., Bischoff, A., 2014. Identification of the giant impactor Theia in lunar rocks. Science 344, 1146–1150.

Hess, P.C., Parmentier, E.M., 1995. A model for the thermal and chemical evolution of the Moon's interior: implications for the onset of mare volcanism. Earth Planet. Sci. Lett. 134, 501–514. doi:10.1016/0012-821X(95)00138-3.

Hess, P.C., Parmentier, E.M., 2001. Thermal evolution of a thicker KREEP liquid layer. J. Geophys. Res. 106 (E11), 28 023–28,032.

Hörst, S.M, 2017. Titan's atmosphere and climate. J. Geophys. Res. Planets 122 (3), 432–482.

Howett, C.J.A., Spencer, J.R., Pearl, J., Segura, M., 2011. High heat flow from Enceladus' south polar region measured using 10–600 cm^{-1} Cassini/CIRS data. J. Geophys. Res. 116, E03003.

Hussmann, H., Sohl, F., Spohn, T., 2006. Subsurface oceans and deep interiors of medium-sized outer planet satellites and large trans-neptunian objects. Icarus 185, 258–273.

Hussmann, H., Spohn, T., 2004. Thermal-orbital evolution of Io and Europa. Icarus 171, 391–410.

Jacobson, S.A., Morbidelli, A., 2014. Lunar and terrestrial planet formation in the grand tack scenario. Philos. Trans. R. Soc. Lond. A: Math. Phys. Eng. Sci. 372 (2024).

Jacobson, S.A., et al., 2014. Highly siderophile elements in earth's mantle as a clock for the moon-forming impact. Nature 508, 84–87.

Javoy, M., et al., 2010. The chemical composition of the Earth: enstatite chondrite models. Earth Planet. Sci. Lett. 293, 259–268.

Jewitt, D., Chizmadia, L., Grimm, R., Prialnik, D., 2007. Water in the small bodies of the Solar System. In: Reipurth, B., Jewitt, D., Keil, K. (Eds.), Protostars and Planets V. University of Arizona Press, Tucson, pp. 951 pp., p. 863-878.

Johnson, B.C., Bowling, T.J., Trowbridge, A.J., Freed, A.M., 2016. Formation of the Sputnik Planum basin and the thickness of Pluto's subsurface ocean. Geophys. Res. Lett. 43, 10068–10077.

Jolliff, B.L., Gillis, J., Haskin, L., Korotev, R., Wieczorek, M., 2000. Major lunar crustal terranes: Surface expressions and crust-mantle origins. J. Geophys. Res. 105, 4197–4216. doi:10.1029/1999JE001103.

Jones, B.W., 2004. Life in the Solar System and Beyond. Praxis, Chichester, UK.

Joseph, 2016. Investigating Seafloors and Oceans: From Mud Volcanoes to Giant Squid. Elsevier Science & Technology Books Publishers, New York, pp. 581. ISBN: 978-0-12- 809357-3.

Jutzi, M., Asphaug, E, 2011. Forming the lunar farside highlands by accretion of a companion moon. Nature 476, 69–72.

Kahana, A., Schmitt-Kopplin, P., Lancet, D, 2019. Enceladus: First observed primordial soup could arbitrate origin-of-life debate. Astrobiology 19 (10), 1263–1278.

Kaib, N.A., Cowan, N.B., 2015. The feeding zones of terrestrial planets and insights into Moon formation. Icarus 252, 161–174.

Kant, I., 1755a. Theoretical Philosophy. Cambridge University Press, Cambridge, England, p. 496.

Kant, I., 1755b. Universal Natural History and Theory of the Heavens. Johann Friedrich Petersen, Konigsberg, Kaliningrad, Russia.

Kargel, J.S., Kaye, J.Z., Head, J.W., III. Marion, G.M., Sassen, R., Crowley, J.K., Ballesteros, O.P., Grant, S.A., Hogenboom, D.L, 2000. Europa's crust and ocean: origin, composition, and the prospects for life, Icarus, 148 (1), 226–265.

Kaufmann, W.J., 1987. Discovering the Universe. W.H. Freeman and Company, 41 Madison Avenue New York, NY, United States, p. 94. ISBN 978-0-7167-1784-3.

Kenyon, S.J., Hartmann, L., 1987. Spectral energy distributions of T Tauri stars: Disk flaring and limits on accretion. ApJ, 323, 714.

Kerridge, J.F., Mackay, A.L., Boynton, W.V., 1979. Magnetite in CI carbonaceous meteorites: origin by aqueous activity on a planetesimal surface. Science 205, 395–397.

Kimberly, E.C., William, D.S., Jonathan, I.L., 1998. Distribution and evolution of water ice in the Solar Nebula: implications for Solar System body formation. Icarus 135 (2), 537–548.

Kivelson, M.G., Khurana, K.K., Russell, C.T., Volwerk, M., Walker, R.J., Zimmer, C., 2000. Galileo magnetometer measurements: a stronger case for a Subsurface ocean at Europa. Science 289 (5483), 1340–1343. https://doi.org/10.1126/science.289.5483.1340.

Krot, A.N., Hutcheon, I., Yurimoto, H., Cuzzi, J., McKeegan, K., et al., 2005. Evolution of oxygen isotopic composition in the inner solar nebula. Astrophys. J. 622, 1333–1342.

Kruijer, T.S., Kleine, T., Fischer-Gödde, M., Sprung, P., 2015. Lunar tungsten isotopic evidence for the late veneer. Nature 520, 534–537.

Laneuville, M., Wieczorek, M.A., Breuer, D., Tosi, N., 2013. Asymmetric thermal evolution of the Moon. J. Geophys. Res. Planets 118, 1435–1452. doi:10.1002/jgre.20103.

Laplace 1796, Exposition du système du monde.

Lawrence, D.J., Feldman, W.C., Barraclough, B.L., Binder, A.B., Elphic, R.C., Maurice, S., Thomsen, D.R., 1998. Global elemental maps of the moon: the Lunar Prospector gamma-Ray spectrometer. Science 281, 1484–1489.

Lee, T., Papanastassiou, D.A., Wasserburg, G.J, 1976. Demonstration of ^{26}Mg excess in Allende and evidence for ^{26}Al. Geophys. Res. Lett. 3 (1), 41–44.

Levison, H.F., Bottke, W.F., Gounelle, M., et al., 2009. Contamination of the asteroid belt by primordial trans-Neptunian objects. Nature 460 (7253), 364–366. doi:10.1038/nature08094.

Lewis, J.S., 1971. Satellites of the outer planets: their physical and chemical nature. Icarus 15, 174–185.

Leya, I., Schönbächler, M., Wiechert, U., Krähenbühl, U., Halliday, A.N., 2008. Titanium isotopes and the radial heterogeneity of the solar system. Earth Planet. Sci. Lett. 266, 233–244.

Li, S., Lucey, P.G., Milliken, R.E., Hayne, P.O., Fisher, E., Williams, J-P., Hurley, D.M., Elphic, R.C., 2018. Direct evidence of surface exposed water ice in the lunar polar regions. Proc. Natl. Acad. Sci. USA 115 (36), 8907–8912. https://doi.org/10.1073/pnas.1802345115.

Lissauer, J.J., 1987. Timescales for planetary accretion and the structure of the protoplanetary disk. Icarus 69, 249–265.

Lissauer, J.J., Hubickyj, O., D'Angelo, G., Bodenheimer, P., 2009. Models of Jupiter's growth incorporating thermal and hydrodynamic constraints. Icarus 199 (2), 338–350. arXiv:0810.5186.

Lock, S.J., Stewart, S.T., 2017. The structure of terrestrial bodies: impact heating, corotation limits, and synestias. J. Geophys. Res.: Planets. doi:10.1002/2016JE005239.

Lock, S.J., Stewart, S.T., Petaev, M.I., Leinhardt, Z., Mace, M.T., Jacobsen, S.B., Cuk, M., 2018. The origin of the Moon within a terrestrial synestia. J. Geophys. Res. Planets 123, 910–951.

Longhi, J., 2006. Petrogenesis of picritic mare magmas: constraints on the extent of early lunar differentiation. Geochim. Cosmochim. Acta 70, 5919–5934. doi:10.1016/j.gca.2006.09.023.

Lorenz, R.D., 2016. Europa ocean sampling by plume flythrough: astrobiological expectations. Icarus 267, 217–219.

Lucy, L.B., 1977. A numerical approach to the testing of the fission hypothesis. Astron. J. 82, 1013–1024.

Lyons, J.R., Young, E.D., 2004. Evolution of oxygen isotopes in the Solar Nebula, 35th Lunar and Planetary Science Conference, March 15-19, 2004. League City, Texas abstract no.1970.

Lyons, J.R., Young, E.D., 2005. CO self-shielding as the origin of oxygen isotope anomalies in the early solar nebula. Nature 435, 317–320.

Mackenzie, D., 2003. The Big Splat, or How The Moon Came To Be. John Wiley & Sons, Hoboken, New Jersey. ISBN 978-0-471-15057-2.

MacKenzie, S.M., Caswell, T.E., Phillips-Lander, C.M., Stavros, E.N., Hofgartner, J.D., Sun, V.Z., Powell, K.E., Steuer, C.J., O'Rourke, J.G., Dhaliwal, J.K., Leung, C.W.S., Petro, E.M., Wynne, J.J., Phan, S., Crismani, M., Krishnamurthy, A., John, K.K., DeBruin, K., Budney, C.J., Mitchell, K.L., 2016. THEO concept mission: Testing the Habitability of Enceladus's Ocean. Adv. Space Res. 58 (6), 1117–1137.

Malamud, U., Perets, H.B., Schafer, C., Burger, C., 2018. Moonfalls: collisions between the Earth and its past moons. arXiv 1805, 00019.

Malhotra, R., 1993. The origin of Pluto's peculiar orbit. Nature 365, 819.

Martin, R.G., Livio, M., 2012. On the evolution of the snow line in protoplanetary discs. Mont. Not. R. Astron. Soc.: Lett. 425 (1). doi:10.1111/j.1745-3933.2012.01290.x.

Mastrobuono-Battisti, A., Perets, H.B., Raymond, S.N.A, 2015. Primordial origin for the compositional similarity between the Earth and the Moon. Nature 520, 212–215.

McCarthy, D.D., Seidelmann, K.P., 2009. Time: From Earth Rotation to Atomic Physics. John Wiley & Sons, Weinheim, Germany, p. 232. ISBN 978-3-527-62795-0.

McCord, T.B., Sotin, C., 2005. Ceres: evolution and current state. J. Geophys. Res.: Planets 110 (E5), E05009–EO5014.

McKay, C.P., 2016. Titan as the abode of life. Life 6, 8.

McKay, C.P., Porco, C.C., Altheide, T., Davis, W.L., Kral, T.A., 2008. The possible origin and persistence of life on Enceladus and detection of biomarkers in the plume. Astrobiology 8, 909–919. doi:10.1089/ast.2008.0265.

McKinnon, W.B., 2010. Why can't Mimas be more like Enceladus? [abstract #P31B–1534], 2010 AGU Fall Meeting. American Geophysical Union, Washington, DC.

McSween Jr., H.Y., 1979a. Are carbonaceous chondrites primitive or processed? A review. Geochim. Cosmochim. Acta 43, 1761–1770.

McSween Jr., H.Y., 1979b. Alteration in CM carbonaceous chondrites inferred from modal and chemical variations in matrix. Geochim. Cosmochim. Acta 43 (11), 1767–1770 1761-1765.

Meadows, V.S., 2017. Reflections on O_2 as a biosignature in exoplanetary atmospheres. Astrobiology 17 (10), 1022–1052.

Melosh, H.J., 1988. The rocky road to panspermia. Nature 21, 687–688.

Meyer, J., Elkins-Tanton, L.T., Wisdom, J., 2010. Coupled thermal-orbital evolution of the early Moon. Icarus 208, 1–10. doi:10.1016/j.icarus.2010.01.029.

Meyer, J., Wisdom, J., 2007. Tidal heating in Enceladus. Icarus 188, 535–539.

Miljković, K., Wieczorek, M.A., Collins, G.S., Laneuville, M., Neumann, G.A., Melosh, H.J., Solomon, S.C., Phillips, R.J., Smith, D.E., Zuber, M.T., 2013. Asymmetric distribution of lunar impact basins caused by variations in target properties. Science 342, 724–726. 10.1126/science.

Monaghan, J.J., Gingold, R.A., 1983. Shock simulation by the particle method SPH. J. Comput. Phys. 52 (2), 374–389.

Morbidelli, A., Chambers, J., Lunine, J.I., et al., 2000. Source regions and time scales for the delivery of water to Earth. Meteor. Planet. Sci. 35 (6), 1309–1320.

Mostefaoui, S., Lugmair, G.W., Hoppe, P., El Goresy, A, 2004. Evidence for live ^{60}Fe in meteorites. New Astron. Rev. 48, 155–159.

Mueller, S., Taylor, G.J., Phillips, R.J., 1988. Lunar composition: a geophysical and petrological synthesis. J. Geophys. Res. 93, 6338–6352. doi:10.1029/JB093iB06p06338.

Mukhopadhyay, S., 2012. Early dierentiation and volatile accretion recorded in deep-mantle neon and xenon. Nature 486, 101–104.

Mulders, G.D., Ciesla, F.J., Min, M., Pascucci, I., 2015. The snow line in viscous disks around low-mass stars: Implications for water delivery to terrestrial planets in the habitable zone. Astrophys. J. 807 (9), 7.

Munk, W.H., Cartwright, D.E., 1966. Tidal spectroscopy and prediction. Philos. Trans. R. Soc. Lond. A259, 533–581.

Murray, C., Dermott, S., 1999. Solar System Dynamics. Cambridge University Press, Cambridge, England.

Nakajima, M., Stevenson, D.J., 2015. Melting and mixing states of the earth's mantle after the moon-forming impact. Earth Planet. Sci. Lett. 427, 286–295.

Nakamura, Y., Lammlein, D., Latham, G., Ewing, M., Dorman, J., Press, F., Toksöz, N., 1973. New seismic data on the state of the deep lunar interior. Science 181, 49–51. doi:10.1126/science.181.4094.49.

Neish, C.D., Lorenz, R.D., 2014. Elevation distribution of Titan's craters suggests extensive wetlands. Icarus 228, 27–34.

Neish, C.D., Lorenz, R.D., O'Brien, D.P.the Cassini RADAR Team, 2006. The potential for prebiotic chemistry in the possible cryovolcanic dome Ganesa Macula on Titan. Int. J. Astrobiol. 5, 57–65.

Neish, C.D., Lorenz, R.D., Turtle, E.P., Barnes, J.W., Trainer, M.G., Stiles, B., Kirk, R., Hibbitts, C.A., Malaska, M.J., 2018. Strategies for detecting biological molecules on Titan. Astrobiology 18 (5), 571–585.

Neish, C.D., Molaro, J.L., Lora, J.M., Howard, A.D., Kirk, R.L., Schenk, P., Bray, V.J., Lorenz, R.D., 2016. Fluvial erosion as a mechanism for crater modification on Titan. Icarus 270, 114–129.

Neish, C.D., Somogyi, A., Imanaka, H., Lunine, J.I., Smith, M.A., 2008. Rate measurements of the hydrolysis of complex organic macromolecules in cold aqueous solutions: implications for prebiotic chemistry on the early earth and Titan. Astrobiology 8, 273–287.

Neish, C.D., Somogyi, A., Lunine, J.I., Smith, M.A., 2009. Low temperature hydrolysis of laboratory tholins in ammonia-water solutions: implications for prebiotic chemistry on Titan. Icarus 201, 412–421.

Niemann, H.B., Atreya, S.K., Carignan, G.R., et al., 1996. The Galileo probe mass spectrometer: composition of Jupiter's atmosphere. Science 272, 846–849.

Nimmo, F., Spencer, J.R., 2015. Powering Triton's recent geological activity by obliquity tides: implications for Pluto geology. Icarus 246, 2–10.

Öberg, K.I., Qi, C., Wilner, D.J., Andrews, S.M., 2011. The effects of snowlines on C/O in planetary atmospheres. Astrophys. J. Lett. 743 (1), 152–157.

Öberg, K.I., Wordsworth, R., 2019. Jupiter's composition suggests its core assembled exterior to the N_2 Snowline. Astron. J. 158 (5), 194–203.

O'Brien, D.P., Travis, B.J., Feldman, W.C., Sykes, M.V., Schenk, P.M., Marchi, S., Russell, C.T., Raymond, C.A., 2015. The potential for volcanism on Ceres due to crustal thickening and pressurization of a subsurface ocean, 46th Lunar and Planetary Science Conference, 2831.

Ojakangas, G.W., Stevenson, D.J., 1989. Thermal state of an ice shell on Europa. Icarus 81, 220–241.

Orosei, R., Lauro, S.E., Pettinelli, E., Cicchetti, A., Coradini, M., Cosciotti, B., Di Paolo, F., Flamini, E., Mattei, E., Pajola, M., Soldovieri, F.,

Cartacci, M., Cassenti, F., Frigeri, A., Giuppi, S., Martufi, R., Masdea, A., Mitri, G., Nenna, C., Noschese, R., Restano, M., Seu, R., 2018. Radar evidence of subglacial liquid water on Mars. Science 361 (6401), 490–493. doi:10.1126/science.aar7268.

Owen, J.E., 2020. Snow-lines can be thermally unstable. Mont. Not. R. Astron. Soc. 495 (3), 3160–3174.

Owen, T., 2006. A new constraint on the formation of Jupiter. In: Udry, S., Benz, W., Steiger, R.von (Eds.), Planetary Systems and Planets in Systems. ISSI Scientific Reports Series, ESA/ISSI, pp. 51–566. ISBN 978-92-9221-935-2.

Pahlevan, K., Stevenson, D.J., 2007. Equilibration in the aftermath of the lunar-forming giant impact. Earth Planet. Sci. Lett. 262, 438–449.

Pieters, C.M., Goswami, J.N., Clark, R.N., Annadurai, M., Boardman, J., Buratti, B., Combe, J.P., Dyar, M.D., Green, R., Head, J.W., Hibbitts, C., Hicks, M., Isaacson, P., Klima, R., Kramer, G., Kumar, S., Livo, E., Lundeen, S., Malaret, E., McCord, T., Mustard, J., Nettles, J., Petro, N., Runyon, C., Staid, M., Sunshine, J., Taylor, L.A., Tompkins, S., Varanasi, P., 2009. Character and spatial distribution of OH/H_2O on the surface of the Moon seen by M^3 on Chandrayaan-1. Science 326 (5952), 568–572. doi:10.1126/science.1178658.

Podolak, M., Zucker, S., 2010. A note on the snow line in protostellar accretion disks. Meteor. Planet. Sci. 39 (11), 1859.

Pollack, J., Hollenbach, D., Beckwith, S., Simonelli, D., Roush, T., Fong, W., 1994. Composition and radiative properties of grains in molecular clouds and accretion disks. Astrophys. J. 421, 615–639.

Pollack, J.B., Hubickyj, O., Bodenheimer, P., Lissauer, J.J., Podolak, M., Greenzweig, Y., 1996. Formation of the giant planets by concurrent accretion of solids and gas. Icarus 124, 62–85.

Porco, C.C., Dones, L., Mitchell, C., 2017. Could it be snowing microbes on Enceladus? Assessing conditions in its plume and implications for future missions. Astrobiology 17, 876–901.

Postberg, F., et al., 2018. Macromolecular organic compounds from the depths of Enceladus. Nature 558 (7711), 564–568.

Prettyman, T.H., Hagerty, J.J., Elphic, R.C., Feldman, W.C., Lawrence, D.J., McKinney, G.W., Vaniman, D.T., 2006. Elemental composition of the lunar surface: analysis of gamma ray spectroscopy data from Lunar Prospector. J. Geophys. Res. Planets 111, E12007. doi:10.1029/2005JE002656.

Pringle, J.E., 1981. Accretion discs in astrophysics. Annu. Rev. Astron. Astrophys. 19, 137–160.

Pugh, D.T., 1987. Tides, Surges and Mean Sea-Level: A Handbook for Engineers and Scientists. John Wiley and Sons, New York, pp. 472.

Pugh, D., Woodworth, P., 2014. Sea level Science: Understanding Tides, Surges, Tsunamis and Mean Sea Level Changes. Cambridge University Press, Cambridge, UK. ISBN 978-1-107-02819-7. xii+395 pp.

Qi, C., Oberg, K.I., Wilner, D.J., d'Alessio, P., Bergin, E., Andrews, S.M., Blake, G.A., Hogerheijde, M.R., van Dishoeck, E.F, 2013. Imaging of the CO snow line in a solar nebula analog. Science 341 (6146), 630–632.

Qin, L., Alexander, C.M.D., Carlson, R.W., Horan, M.F., Yokoyama, T, 2010. Contributors to chromium isotope variation of meteorites. Geochim. Cosmochim. Acta 74, 1122–1145.

Quillen, A.C., Martini, L., Nakajima, M., 2019. Near/far side asymmetry in the tidally heated Moon. Icarus 329, 182–196. doi:10.1016/j.icarus.2019.04.010.

Reufer, A., Meier, M.M.M., Benz, W., Wieler, R., 2011. Obtaining higher target material proportions in the giant impact by changing impact parameters and impactor compositions, Lunar Planet. Sci. Conf. 42nd Abstr., 1136.

Reufer, A., Meier, M.M.M., Benz, W., Wieler, R., 2012. A hit-and-run giant impact scenario. Icarus 221, 296–299. doi:10.1016/j.icarus.2012.07.021.

Richard, G., 2010. Transport rates of radiolytic substances into Europa's ocean: implications for the potential origin and maintenance of life. Astrobiology 10 (3), 275–283.

Richard, S.F., 2003. Historical eclipses and Earth's rotation. Astron. Geophys. 44 (2). doi:10.1046/j.1468-4004.2003.44222.x 2.22–2.27.

Ringwood, A., 1989. Flaws in the giant impact hypothesis of lunar origin. Earth Planet. Scie. Lett. 95, 208–214.

Ringwood, A.E., Kesson, S.E. 1976. A dynamic model for mare basalt petrogenesis, In: Proceedings of the Seventh Lunar Science Conference (Pergamon), pp. 1697–1722. Houston, TX. Lunar and Planetary Institute.

Roberts, J.H., Nimmo, F., 2008. Tidal heating and the long-term stability of a subsurface ocean on Enceladus. Icarus 194, 675–689.

Robinson, K.L., et al., 2016. Water in evolved lunar rocks: evidence for multiple reservoirs. Geochim. Cosmochim. Acta 188, 244–260.

Roy, A., Wright, J.T., Sigurðsson, S., 2014. Earthshine on a young Moon: explaining the lunar farside highlands. Astrophys. J. 788, L42. doi:10.1088/2041-8205/788/2/L42.

Rubie, D.C., Nimmo, F., Melosh, H.J., 2007. Formation of Earth's Core A2 – Schubert. Elsevier, Gerald. Amsterdam, pp. 51–90. ISBN 978-0444527486.

Rufu, R., Aharonson, O., Perets, H.B., 2017. A multiple-impact origin for the Moon. Nat. Geosci. 10, 89–94 10.1038/ngeo2866.

Russell, C.T., Raymond, C.A., Jaumann, R., MCSween, H.Y., De Sanctis, M.C., Nathues, A., Prettyman, T.H., Ammannito, E., Reddy, V., Preusker, F., O'Brien, D.P., Marchi, S., Denevi, B.W., Buczkowski, D.L., Pieters, C.M., MCCord, T.B., Li, J.-Y., Mittlefehldt, D.W., Combe, J.-P., Williams, D.A., Hiesinger, H., Yingst, R.A., Polanskey, C.A., Joy, S.P., 2013b. Dawn completes its mission at 4 Vesta. Meteor. Planet. Sci., 1–14. doi:10.1111/maps.12091.

Sarah, K., 2021. The Earth is spinning faster now than at any time in the past half century. The Telegraph.

Sasselov, D.D., Lecar, M., 2000. On the snow line in dusty protoplanetary disks. Astrophys. J. 528, 995–998.

Schmidt, B., Blankenship, D., Patterson, W., Schenk, P., 2011. Active formation of 'chaos terrain' over shallow subsurface water on Europa. Nature 479 (7374), 502–505.

Schneider, T., Graves, S.D.B., Schaller, E.L., Brown, M.E., 2012. Polar methane accumulation and rainstorms on Titan from simulations of the methane cycle. Nature 481, 58–61.

Schröder, S.E., Mottola, S., Carsenty, U., Ciarniello, M., Jaumann, R., Li, J.-Y., Longobardo, A., Palmer, E., Pieters, C., Preusker, F., Raymond, C.A., Russell, C.T., 2017. Resolved spectrophotometric properties of the Ceres surface from Dawn Framing Camera images. Icarus 288, 201–225.

Schubert, G., Hussmann, H., Lainey, V., Matson, D.L., McKinnon, W.B., Sohl, F., Sotin, C., Tobie, G., Turini, D., van Hoolst, T., 2010. Evolution of icy satellites. Space Sci. Rev. 153, 447–484.

Scott, J.R., Marotzke, J., Adcroft, A, 2001. Geothermal heating and its influence on the meridional overturning circulation. J. Geophys. Res. 106, 141–154.

Scrutton, C.T, 1978. Periodic growth features in fossil organisms and the length of the day and month. In: Brosche, P., Sündermann, J. (Eds.), Tidal Friction and the Earth's Rotation. Springer, Berlin Heidelberg, pp. 154–196. doi:10.1007/978-3-642-67097-8_12. ISBN 9783540090465.

Shearer, C.K., Papike, J.J., 2005. Early crustal building processes on the Moon: models for the petrogenesis of the magnesian suite. Geochim. Cosmochim. Acta 69, 3445–3461. doi:10.1016/j.gca.2005.02.025.

Shearer, C.K., et al., 2006. Thermal and magmatic evolution of the Moon. Rev. Mineral. Geochem. 60 (1), 365–518.

Showman, A.P., Stevenson, D.J., Malhotra, R., 1997. Coupled orbital and thermal evolution of Ganymede. Icarus 129, 367–383.

Sleep, N.H., 2010. The Hadean-Archaean environment. Cold Spring Harbor Perspect. Biol. 2, a002527.

Sleep, N.H., Zahnle, K.J., Lupu, R.E., 2014. Terrestrial aftermath of the Moon-forming impact. Philos. Trans. R. Soc. Math. Phys. Eng. Sci. 372, 20130172.

Smith, J.V., Anderson, A.T., Newton, R.C., Olsen, E.J., Wyllie, P.J., 1970. Petrologic history of the moon inferred from petrography, mineralogy and petrogenesis of Apollo 11 rocks, Proc. Apollo 11 Lunar Science Conf., *Houston, TX, 5–8 January 1970*. New York, NY. Pergamon Press, pp. 897–925.

Snyder, G.A., Neal, C.R., Taylor, L.A., Halliday, A.N., 1995. Processes involved in the formation of magnesian-suite plutonic rocks from the highlands of the Earth's Moon. J. Geophys. Res. 100, 9365–9388. doi:10.1029/95JE00575.

Sotin, C., Tobie, G., Wahr, J., McKinnon, W.B., 2009. Tides and tidal heating on Europa. In: Pappalrdo, R.T., McKinnon, W.B., Khurana, K. (Eds.), Europa. University of Arizona Press, Tucson, pp. 85–118.

Spencer, J.R., et al., 2006. Cassini encounters Enceladus: background and the discovery of a south polar hot spot. Science 311, 1401–1405.

Springel, V., 2005. The cosmological simulation code gadget-2. Mon. Not. R. Astron. Soc. 364, 1105–1134.

Stevenson, D.J., Lunine, J.I., 1988. Rapid formation of Jupiter by diffusive redistribution of water vapor in the solar nebula. Icarus 75, 146–155.

Stewart, S.T., Ćuk, M., 2012. Making the Moon from a fast-spinning Earth: a giant impact followed by resonant despinning. Science 338, 1047–1052. doi:10.1126/science.1225542.

Stofan, E.R., Elachi, C., Lunine, J.I., Lorenz, R.D., Stiles, B., Mitchell, K.L., Ostro, S., Soderblom, L., Wood, C., Zebker, H., Wall, S., Janssen, M., Kirk, R., Lopes, R., Paganelli, F., Radebaugh, J., Wye, L., Anderson, Y., Allison, M., Boehmer, R., Callahan, P., Encrenaz, P., Flamini, E., Francescetti, G., Gim, Y., Hamilton, G., Hensley, S., Johnson, W.T.K., Kelleher, K., Muhleman, D., Paillou, P., Picardi, G., Posa, F., Roth, L., Seu, R., Shaer, S., Vetrella, S., West, R., 2007. The lakes of Titan. Nature 445, 61–64.

Stoker, C.R., Zent, A., Catling, D.C., Douglas, S., Marshall, J.R., Archer, D., Clark, B., Kounaves, S.P., Lemmon, M.T., Quinn, R., Renno, N., Smith, P.H., Young, S.M.M., 2010. Habitability of the phoenix landing site. J. Geophys. Res. Planets 115, 1–24.

Sunshine, J.M., Farnham, T.L., Feaga, L.M., Groussin, O., Merlin, F., Milliken, R.E., A'Hearn, M, 2009. Temporal and spatial variability of lunar hydration as observed by the Deep Impact spacecraft. Science 326 (5952), 565–568. doi:10.1126/science.1179788.

Tabetah, M., Melosh, J., 2017. Air penetration enhances fragmentation of entering meteoroids. Meteorit. Planet. Sci. 53 (3). doi:10.1111/maps.13034.

Taylor, S.R., 1996. Origin of the terrestrial planets and the moon. J. R. Soc. West Aust. 79 (1), 59–65.

Taylor, S.R., Jjakes, P, 1974. The geochemical evolution of the moon, Proc. Fifth Lunar Sci. Conf. Houston, TX, 18–22 March 1974. Houston, TX. Lunar and Planetary Institute, pp. 1287–1305.

Taylor, S.R, Norman, M.D., 1990. Accretion of differentiated planetesimals to the Earth. In: Newsom, H.E., Jones, J.R. (Eds.), Origin of the Earth. Oxford University Press, New York, pp. 29.

Tepfer, D., Leach, S., 2006. Plant seeds as model vectors for the transfer of life through space. Astrophys. Space Sci. 306, 69–75. doi:10.1007/s10509-006-9239-0.

Thompson, A.C., Stevenson, D.J., 1983. Two-phase gravitational instabilities in thin disks with application to the origin of the Moon. Lunar Planet. Sci. XIV, 787–788.

Thompson, S.L., and H.S. Lauson (1984), Improvement in the chart D radiation-hydrodynamic code. III. Revised analytic equations of state. SC-RR-71 0714.

Tillotson, J.H. (1962), Metallic equations of state for hypervelocity impacts, General Atomic Report GA3216, July 1962.

Tiwari, K., Ghosh, S., Miyahara, M., Ray, D., 2021. Shock-induced incongruent melting of olivine in Kamargaon L6 chondrite. J. Geophys. Res. 48 (12), e2021GL093592, https://doi.org/10.1029/2021GL093592.

Tobie, G., Mocquet, A., Sotin, C., 2005. Tidal dissipation within large icy satellites: application to Europa and Titan. Icarus 177, 534–549.

Toole, J.M., Warren, B.A., 1993. A hydrographic section across the subtropical South Indian Ocean. Deep Sea Res. 40, 1973–2019.

Touboul, M., Puchtel, I.S., Walker, R.J., 2012. ^{182}W evidence for long-term preservation of early mantle di-erentiation products. Science 335, 1065–1069.

Trinquier, A., Birck, J.L., Allègre, C.J., 2007. Widespread 54Cr heterogeneity in the inner solar system. Astrophys. J. 655, 1179–1185.

Trinquier, A., et al., 2009. Origin of nucleosynthetic isotope heterogeneity in the solar protoplanetary disk. Science 324, 374–376.

Tyler, R., 2008. Strong ocean tidal flow and heating on moons of the outer planets. Nature 456, 770–772.

Tyler, R., 2011. Tidal dynamical considerations constrain the state of an ocean on Enceladus. Icarus 211 (1), 770–779. doi:10.1016/j.icarus.2010.10.007.

Tyler, R., 2014. Comparative estimates of the heat generated by ocean tides on icy satellites in the outer Solar System. Icarus 243, 358–385.

Tyler, R.H., 2009. Ocean tides heat Enceladus. Geophys. Res. Lett. 36 (15), L15205. doi:10.1029/2009GL038300.

Urey, H.C., 1951. The origin and development of the Earth and other terrestrial planets. Geochim. Cosmochim. Acta 1, 209–277. doi:10.1016/0016-7037(51)90001-4.

Vickery, A.M., Melosh, H.J., 1987. The large crater origin of the SNC meteorites. Science 237, 738–743.

Waite, J.H., Glein, C.R., Perryman, R.S., Teolis, B.D., Magee, B.A., Miller, G., Grimes, J., Perry, M.E., Miller, K.E., Bouquet, A., 2017. Cassini finds molecular hydrogen in the Enceladus plume: evidence for hydrothermal processes. Science 356, 155–159.

Ward, W.R., Cameron, A.G.W., 1978. Disc evolution within the Roche limit. Lunar Planet. Sci. IX, 1205–1207.

Warren, P.H., 2011. Stable-isotopic anomalies and the accretionary assemblage of the Earth and Mars: a subordinate role for carbonaceous chondrites. Earth Planet. Sci. Lett. 311, 93–100.

Warren, P.H., Wasson, J.T., 1979. The origin of KREEP. Rev. Geophys. 17 (1), 73–88. doi:10.1029/RG017i001p00073.

Watkins, E., 2012. Universal natural history and theory of the heavens or essay on the constitution and the mechanical origin of the whole universe according to Newtonian principles (1755). Kant: Natural Science. Published online by Cambridge University Press, Cambridge, England, pp. 182–308. https://doi.org/10.1017/CBO9781139014380.007.

Wiechert, U., et al., 2001. Oxygen isotopes and the Moon-forming Giant Impact. Science 294 (12), 345–348.

Wieczorek, M.A., et al., 2013. The crust of the Moon as seen by GRAIL. Science 339 (6120), 671–675.

Williams, G.E., 1990. Tidal rhythmites: key to the history of the Earth's rotation and the lunar orbit. J. Phys. Earth 38, 475–491.

Williams, G.E., 2000. Geological constraints on the Precambrian history of Earth's rotation and the Moon's orbit. Rev. Geophys. 38 (1), 37–59.

Williams, K.L., Henyey, F.S., Rouseff, D., Reynolds, S.A., Ewart, T., 2001. Internal wave effects on high-frequency acoustic propagation to horizontal arrays—experiment and implications to imaging. IEEE J. Oceanic Eng. 26, 102–112.

Wisdom, J., Tian, Z., 2015. Early evolution of the Earth–Moon system with a fast-spinning earth. Icarus 256, 138–146.

Wood, J.A., et al., 1970. Lunar anorthosites. Science 167 (3918), 602–604. doi:10.1126/science.167.3918.602.

Xiande, X., Chen, M., 2006. Evaluation of Shock Stage for Suizhou Meteorite. In: Suizhou Meteorite: Mineralogy and Shock Metamorphism. Springer, Berlin Heidelberg, ISSN 2194-3176; ISSN 2194-3184.

Young, T., 1823. Tides, Eighth ed. Encyclopaedia Britannica, vol. 21. Little & Brown, Boston.

Young, E.D., Kohl, I.E., Warren, P.H., Rubie, D.C., Jacobson, S.A., Morbidelli, A., 2016. Oxygen isotopic evidence for vigorous mixing during the Moon-forming giant impact. Science 351 (6272), 493. doi:10.1126/science.aad0525 2016.

Zahnle, K., Arndt, N., Cockell, C., Halliday, A., Nisbet, E., Selsis, F., Sleep, N.H., 2007. Emergence of a habitable planet. Space Sci. Rev. 129, 35–78.

Zahnle, K., Walker, J.C., 1987. A constant daylength during the Precambrian era? Precamb. Res. 37 (2), 95–105.

Zahnle, K., Schaefer, L., Fegley, B., 2010. Earth's earliest atmospheres. Cold Spring Harb Perspect Biol. 2, a004895. doi:10.1101/cshperspect.a004895.

Zhang, J., Dauphas, N., Davis, A.M., Leya, I., Fedkin, A.V., 2012. The proto-Earth as a significant source of lunar material. Nat. Geosci. 5 (4), 251–255. doi:10.1038/ngeo1429.

Bibliography

Alfvèn, H., 1954. On the Origin of the Solar System. Oxford University Pres, Oxford.

Alfvèn, H., Arrhenius, G, 1976. Evolution of the Solar System. U.S. Government Printing Office, Washington, D.C. NASA SP-345.

Clark, B.C., 1988. Primeval procreative comet pond. Orig Life Evol Biosph 18 (3), 209–238. doi:10.1007/BF01804671.

Endress, M., Zinner, E., Bischoff, A., 1996. Early aqueous activity on primitive meteorite parent bodies. Nature 379 (6567), 701–703. doi:10.1038/379701a0.

Herzog, G.F., Anders, E., Alexander, E.C.Jr, Davis, P.K., Lewis, R.S, 1973. Iodine-129/xenon-129 age of magnetite from the orgueil meteorite. Science 180 (4085), 489–491. doi:10.1126/science.180.4085.489.

Lauretta, D.S., et al., 2019. The unexpected surface of asteroid (101955) Bennu. Nature 568 (7750), 55–60. doi:10.1038/s41586-019-1033-6.

Lewis, R.S., Anders, E, 1975. Condensation time of the solar nebula from extinct I in primitive meteorites. Proc. Natl. Acad. Sci. U S A. 72 (1), 268–273. doi:10.1073/pnas.72.1.268.

Schöler, H.F., Nkusi, G., Niedan, V.W., Müller, G., Spitthoff, B, 2005. Screening of organic halogens and identification of chlorinated benzoic acids in carbonaceous meteorites. Chemosphere 60 (11), 1505–1512. doi:10.1016/j.chemosphere.2005.02.069.

Urey, H.C., 1966. A review of evidence for biological material in meteorites. Life Sci. Space Res. 4, 35–59.

White, L.F., Tait, K.T., Langelier, B., Lymer, E.A., Černok, A., Kizovski, T.V., Ma, C., Tschauner, O., Nicklin, R.I., 2020. Evidence for sodium-rich alkaline water in the Tagish Lake parent body and implications for amino acid synthesis and racemization. Proc. Natl. Acad. Sci. U S A 117 (21), 11217–11219. doi:10.1073/pnas.2003276117.

Chapter 2

Geological timeline of significant events on Earth

2.1 An era from 4.5 to 4 billion years ago when the entire Earth was a "Fire Ball"

The period after the formation of Earth has been compartmentalized into different eons (e.g., Hadean, Archean, etc.). Note that the Hadean Eon is named after the mythological underworld ruler—Hades—because during most of the Hadean period (4.5–4 billion years ago) the surface of the Earth must have been like our image of Hell (Fig. 2.1). The Hadean Eon began when the planet Earth first began to form, about 4.6 billion (4600 million) years ago and ended, as defined by the International Chronostratigraphic Chart-2015 (ICS-2015), 4 billion (4000 million) years ago. The Archean Eon began about 4 billion years ago with the formation of Earth's crust and extended to the start of the Proterozoic Eon 2.5 billion years ago; the latter is the second formal division of Precambrian time. Thus, the Archean Eon is a geologic eon, 4000–2500 million years ago (4–2.5 billion years ago), that followed the Hadean Eon and preceded the Proterozoic Eon. During the Archean eon, the Earth's crust cooled enough that rocks and continental plates began to form.

The Proterozoic eon (from 2500 million years ago [Ma] to 541 Ma) was a period of time during which several different events took place, eventually helping to shape the Earth as we know it today. During this eon, life began to evolve into more complex organisms.

2.2 Importance of greenhouse gases in the atmosphere of the early Earth

The greenhouse effect is the trapping of the Sun's warmth in a planet's lower atmosphere, due to the greater transparency of the atmosphere to visible radiation from the Sun than to infrared radiation emitted from the planet's surface. Irish physicist John Tyndall is commonly credited with discovering the greenhouse effect, which underpins the science of climate change. Starting in 1859, he published a series of studies on the way greenhouse gases including carbon dioxide trapped heat in the Earth's atmosphere. Greenhouse effect is essential for our survival because it maintains the temperature of our planet within a habitable range. Le Treut et al. (2007) have described "greenhouse effect" thus: "The Sun powers Earth's climate, radiating energy at very short wavelengths, predominantly in the visible or near-visible (e.g., ultraviolet) part of the spectrum. Roughly one-third of the solar energy that reaches the top of Earth's atmosphere is reflected directly back to space. The remaining two-thirds are absorbed by the surface and, to a lesser extent, by the atmosphere. To balance the absorbed incoming energy, the Earth must, on average, radiate the same amount of energy back to space. Because the Earth is much colder than the Sun, it (Earth) radiates at much longer wavelengths, primarily in the infrared part of the spectrum. Much of this thermal radiation emitted by the land and ocean is absorbed by the atmosphere, including clouds, and reradiated back to Earth. This is called the greenhouse effect. The glass walls in a greenhouse reduce airflow and increase the temperature of the air inside. Analogously, but through a different physical process, the Earth's greenhouse effect warms the surface of the planet. Without the natural greenhouse effect, the average temperature at Earth's surface would be below the freezing point of water. Thus, Earth's natural greenhouse effect makes life as we know it possible. However, human activities, primarily the burning of fossil fuels and clearing of forests, have greatly intensified the natural greenhouse effect, causing global warming."

A greenhouse gas is a gas in an atmosphere that absorbs and emits radiant energy within the thermal infrared range. This process is the fundamental cause of the greenhouse effect ("IPCC AR4 SYR Appendix Glossary"). Fig. 2.2 shows an idealized model of the natural greenhouse effect.

The common examples of greenhouse gases in Earth's atmosphere are water vapor (H_2O), carbon dioxide (CO_2), methane (CH_4), nitrous oxide (N_2O), ozone (O_3), and any fluorocarbons (FCs) such as CFC-12 (CCl_2F_2) and HCFC-22 ($CHClF_2$).

A discussion of Earth's early climate is so important and touches on so many interesting problems—such as the search for inhabitable extraterrestrial planets, what it can tell us about current greenhouse warming, what it can tell

FIG. 2.1 Artist's impression of Earth's surface landscape in the Hadean eon (4.6–4 billion years ago), when the surface of the Earth must have been like our image of Hell; Tim Bertelink - Own work; From Wikimedia Commons, the free media repository. *(Source: https://commons.wikimedia.org/wiki/File:Hadean.png) Image Author: Tim Bertelink. Copyright information: This file is licensed under the Creative Commons Attribution-Share Alike 4.0 International license.*

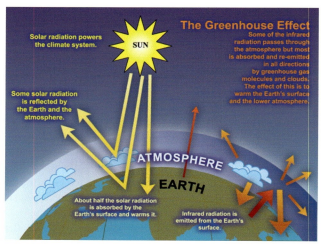

FIG. 2.2 **An idealized model of the natural greenhouse effect.** *(Source: Le Treut et al., 2007).*

us about early life on Earth. The geological record clearly shows that the Earth was very warm when it was young. Indeed, there is no persuasive evidence that there was *any* permanent ice on the Earth for at least the first 2.2 billion years of its existence. A mighty greenhouse effect suspected to have occurred in the early history of Earth would almost certainly have involved a high concentration of methane in the atmosphere, for two reasons: (1) methane is a much more efficient greenhouse gas than carbon dioxide, and (2) for the greenhouse effect to have been due solely or mostly to carbon dioxide would have required a level of CO_2 so high that it would have produced unusual rock chemistry that we simply do not see in the geological record.

Presumably, this methane would have come from bacteria. Bacteria that produce methane (methanogenic bacteria) are very common on Earth. Free oxygen is poisonous to methanogenic bacteria, so the only places they can live now are places where oxygen gas cannot reach them, such as inside animal organs or under layers of thick river sludge. But oxygen gas is produced by plants and photosynthetic bacteria, and there is reasonable evidence that photosynthetic bacteria evolved after methanogenic bacteria, thus once upon a time leaving all of planet Earth open to colonization by methanogenic bacteria. This would surely have led to a lot of methane in the Earth's atmosphere.

With the rise of oxygen-producing bacteria, and hence the filling of Earth's atmosphere with a gas that is poisonous to methanogens, the methane-producing bacteria would have been beaten into a slow retreat and the amount of methane in the atmosphere would have plummeted—thus drastically reducing the greenhouse effect and bringing on Earth's first ice age. This is the reason that the rise of oxygen in Earth's atmosphere and the time of the first real ice age essentially coincide. Naturally, this *coinciding* is not thought to be a *coincidence*.

Since that time, roughly 2.3 billion years ago, the Earth has gyrated chaotically between prolonged ice ages and warmer intervals. One reason for this is *positive feedback*: as the climate becomes cooler, more of the Earth becomes covered with snow and ice, which strongly reflects sunlight back into space rather than absorbing it. This serves to make the Earth even colder, which causes heavier snowfall, which increases the reflection. And going in the opposite direction, if an ice age starts to break then there is less ice and snow on the ground each year, which means the dark, exposed ground can absorb more sunlight and heat the Earth more generously, which helps bring it farther out of the ice age, and so there is even less snow on the ground.

The gradually increasing luminosity of the Sun has been balanced by, among other things, Earth's plants. Increased sunlight means increased photosynthesis, and that means less CO_2 in the atmosphere because plants consume CO_2 in the process of photosynthesis. Less CO_2 means less greenhouse effect, hence a cooling effect. There are many other factors that affect the Earth's climate, including slow cycles in the Earth's distance from the Sun, continental drift, outbreaks of volcanism, etc.

The main source of global climate change is human-induced changes in atmospheric composition (Karl and Trenberth, 2003). Although greenhouse gases are blamed for the current global warming, there is a need for some degree of greenhouse gases in our environment, without which our ecosystem would not be possible. It has been estimated that without greenhouse gases, the average temperature of Earth's surface would be about −18°C (about 0°F) ("NASA GISS: Science Briefs: Greenhouse Gases: Refining the Role of Carbon Dioxide"), rather than the present average of 15°C (59°F) (see Karl and Trenberth, 2003; Le Treut et al., 2007). Note that conversion of temperature (T) between degrees Fahrenheit (°F) and degrees Celsius (°C) is given by the expression: $T(°F) = (T(°C) \times 1.8) + 32$.

Awareness and a partial understanding of most of the interactive processes in the Earth system that govern climate and climate change predate the Intergovernmental Panel on Climate Change (IPCC), often by many decades. A deeper understanding and quantification of these processes and their incorporation in climate models have progressed rapidly since the IPCC First Assessment Report in 1990 (Le Treut et al., 2007).

Using a radiative–convective climate model, Kasting and Ackerman (1986) investigated the possible consequences of very high carbon dioxide concentrations in the Earth's early atmosphere. According to them, the early atmosphere would apparently have been stable against the onset of a runaway greenhouse (i.e., the complete evaporation of the oceans) for carbon dioxide pressures up to at least 100 bars. They think that "A 10–20-bar carbon dioxide atmosphere, such as may have existed during the first several hundred million years of the Earth's history, would have had a surface temperature of approximately 85–110° C." Kasting and Ackerman (1986) reckon that "the early stratosphere should have been dry, thereby precluding the possibility of an oxygenic prebiotic atmosphere caused by photo-dissociation of water vapor followed by escape of hydrogen to space. Earth's present atmosphere also appears to be stable against a carbon dioxide-induced runaway greenhouse." Note that stratosphere is the layer of the Earth's atmosphere above the troposphere, extending to about 50 km above the Earth's surface (the lower boundary of the mesosphere).

Over the last few decades, geochemical research has demonstrated that abiotic methane (CH_4), formed by chemical reactions which do not directly involve organic matter, occurs on Earth in several specific geologic environments. It can be produced by either high-temperature magmatic processes (e.g., melting in the mantle, transport to within the volcano, cooling and crystallization, assimilation of surrounding rocks, magma mixing, degassing etc.) in volcanic and geothermal areas, or via low-temperature (<100°C) gas–water–rock reactions in continental settings, even at shallow depths. The isotopic composition of carbon (C) and hydrogen (H) is a first step in distinguishing abiotic from biotic (including either microbial or thermogenic) CH_4 (Etiope and Sherwood Lollar, 2013). Although it has been traditionally assumed that abiotic CH_4 is mainly related to mantle-derived or magmatic processes, a new generation of data is showing that low-temperature synthesis related to gas–water–rock reactions is more common than previously thought. Etiope and Sherwood Lollar (2013) have reviewed the major sources of abiotic CH_4 and the primary approaches for differentiating abiotic from biotic CH_4, including novel potential tools such as clumped isotope geochemistry. They proposed a diagnostic approach for differentiation. Note that in geology, igneous differentiation, or magmatic differentiation, is an umbrella term used for the various processes by which magmas undergo bulk chemical change during the partial melting process, cooling, emplacement, or eruption.

Several researchers (e.g., Nisbet, 2000; Nisbet and Sleep, 2001; Schulte et al., 2006) think that geochemical/hydrothermal methane had to be abundant on early Earth. This is because methane (CH_4) was a by-product of the same successful recipe responsible for the sequestration (removal) of atmospheric CO_2. As seawater diffused downward through fractured ocean crust, it reacted with mantle host rocks at high temperatures and transformed into a hydrothermal fluid that became enriched in a variety of compounds and depleted in others, depending on the subsurface reaction conditions and the nature of the leached rocks (Konn et al., 2015). The result was emitted in the form of thick, smoke-like underwater plumes (hydrothermal vents) distributed ubiquitously in the terrestrial World Oceans.

It would be interesting to note how some of the greenhouse gases (especially water vapor and carbon dioxide) reached the atmosphere of the early Earth. Sleep et al. (2014) have summarized this process thus: "Much of the Earth's mantle was melted in the Moon-forming impact. Gases that were not partially soluble in the melt, such as water and CO_2, formed a thick, deep atmosphere surrounding the postimpact Earth. This atmosphere was opaque to thermal radiation, allowing heat to escape to space only at the runaway greenhouse threshold of approximately 100 W/m^2. The duration of this runaway greenhouse stage was limited to approximately 10 Myr by the internal energy and tidal heating, ending with a partially crystalline uppermost mantle and a solid deep mantle. At this point, the crust was able to cool efficiently and solidified at the surface. After the condensation of the water ocean, approximately 100 bar of CO_2 remained in the atmosphere, creating a solar-heated greenhouse, while the surface cooled to approximately 500 K. Almost all this CO_2 had to be sequestered by subduction into the mantle by 3.8 Ga, when the geological record indicates the presence of life and hence a habitable environment. The deep CO_2 sequestration into the mantle could be explained by a rapid subduction of the old oceanic crust, such that the top of the crust would remain cold and retain its CO_2. Kinematically, these episodes would be required to have both fast subduction (and hence seafloor spreading) and old crust. Hadean oceanic crust that formed from hot mantle would have been thicker than modern crust, and therefore only old crust underlain by cool mantle lithosphere could subduct. Once subduction started, the basaltic crust would turn into dense eclogite, increasing the rate of subduction. The rapid subduction would stop when the young partially frozen crust from the rapidly spreading ridge entered the subduction zone." Note that eclogite is a metamorphic rock containing granular minerals, typically garnet and pyroxene.

2.3 Genesis of water on Earth

Water is one of the most important molecules for sustaining of life as we know it. It would, therefore, be of interest to examine the origin of water on Earth, where diverse lifeforms thrive. What do we know about the origin of the Earth's water? Different phases of Earth's accretion, in its infant and subsequent stages of growth, are considered to have played an important role in the formation of water on Earth. Is it more likely that the origin of the Earth's water can be attributed to icy comets that struck the young Earth or from material released from the Earth's interior during volcanic activity? According to James C.G. Walker of the University of Michigan, "During accretion, the kinetic energy of the colliding planetesimals was converted into thermal energy, so the earth grew extremely hot as it came together. The material forming the earth was probably too hot for ice to have been a major carrier of water. Most of the water was probably present originally as water trapped in clay minerals or as separate hydrogen (in hydrocarbons) and oxygen (in iron oxides), rather than as ice."

Earth grew through collisions with Moon-sized to Mars-sized planetary embryos from the inner Solar System, but it also accreted material from greater heliocentric distances, including carbonaceous chondrite-like bodies, the likely source of Earth's water and highly volatile species. Understanding when and how this material was added to Earth is critical for constraining the dynamics of terrestrial planet formation and the fundamental processes by which Earth became habitable. However, earlier studies inferred very different timescales for the delivery of carbonaceous chondrite-like bodies, depending on assumptions about the nature of Earth's building materials.

One hypothesis claims that Earth accreted (gradually grew by accumulation of) icy planetesimals about 4.5 billion years ago, when it was 60–90% of its current size (Peslier et al., 2017). In this scenario, Earth was able to retain water in some form throughout accretion and major impact events. This hypothesis is supported by similarities in the abundance and the isotope ratios of water between the oldest known carbonaceous chondrite meteorites and meteorites from Vesta, both of which originate from the Solar System's asteroid belt (Sarafian et al., 2014). It is also supported by studies of osmium isotope ratios, which suggest that a sizeable quantity of water was contained in the material that Earth accreted early on (Drake, 2005). Measurements of the chemical composition of lunar samples collected by the Apollo 15 and 17 missions further support this, and indicate that water was already present on Earth before the Moon was formed (Cowen, 2013).

One problem with this hypothesis is that the noble gas isotope ratios of Earth's atmosphere are different from those of its mantle, which suggests they were formed from different sources (Owen et al., 1992; Dauphas, 2003). It may be noted that the origin of the terrestrial atmosphere is one of the most puzzling enigmas in the planetary sciences. Dauphas (2003) suggested that two sources contributed to the formation of the terrestrial atmosphere—fractionated nebular gases and accreted cometary volatiles. According to Dauphas (2003), "during terrestrial growth, a transient gas envelope was fractionated from nebular composition. This transient atmosphere was mixed with cometary material. The fractionation stage resulted in a high Xe/Kr ratio, with xenon being more isotopically fractionated than krypton. Comets delivered volatiles having low Xe/Kr ratios and solar isotopic compositions. The resulting atmosphere had a near-solar Xe/Kr ratio, almost unfractionated krypton delivered by comets, and fractionated xenon inherited from the fractionation episode. The dual origin therefore provides an elegant solution to the long-standing 'missing xenon' paradox." Dauphas (2003) demonstrated that such a model could explain the isotopic and elemental abundances of Ne, Ar, Kr, and Xe in the terrestrial atmosphere.

To explain the observation that the noble gas isotope ratios of Earth's atmosphere are different from those of its mantle (which suggests they were formed from different sources), a so-called "late veneer" theory has been proposed in which water was delivered much later in Earth's history, after the Moon-forming impact. However, the current understanding of Earth's formation allows for less than 1% of Earth's material accreting after the Moon formed, implying that the material accreted later must have been very water-rich. Models of early Solar System dynamics have shown that icy asteroids could have been delivered to the inner Solar System (including Earth) during this period if Jupiter migrated closer to the Sun (Gomes et al., 2005).

Yet a third hypothesis, supported by evidence from molybdenum (Mo) isotope ratios, suggests that the Earth gained most of its water from the same interplanetary collision that caused the formation of the Moon. For example, Budde et al. (2019) showed that the Mo isotopic composition of Earth's primitive mantle falls between those of the noncarbonaceous and carbonaceous reservoirs, and that this observation allowed these investigators to quantify the accretion of carbonaceous chondrite-like material to Earth independently of assumptions about its building blocks. As most of the Mo in the primitive mantle was delivered by late-stage impactors, their data demonstrated that Earth accreted carbonaceous bodies late in its growth history, probably through the Moon-forming impact. Budde et al. (2019) reckon that this late delivery of carbonaceous material probably resulted from an orbital instability of the gas giant planets, and it demonstrates that Earth's habitability is strongly tied to the very late stages of its growth.

One factor in estimating when water appeared on Earth is that water is continually being lost to space. H_2O molecules in the atmosphere are broken up by photolysis, and the resulting free hydrogen atoms can sometimes escape Earth's gravitational pull. When the Earth was younger and less massive, water would have been lost to space more easily. Lighter elements such as hydrogen and helium are expected to leak from the atmosphere continually, but isotopic ratios of heavier noble gases (historically also the inert gases such as helium [He], neon [Ne], argon [Ar], krypton [Kr], xenon [Xe], and the radioactive radon [Rn]) in the modern atmosphere suggest that even the heavier elements in the early atmosphere were subject to significant losses (Pepin, 1991). In particular, xenon is useful for calculations of water loss over time. Not only is it a noble gas (and therefore is not removed from the atmosphere through chemical reactions with other elements), but comparisons between abundances of its nine stable isotopes in the modern atmosphere reveal that the Earth lost at least one ocean of water early in its history, between the Hadean and Archean eras (Zahnle et al., 2019).

Several theories have been put forward to explain the presence of water on Earth. Some such theories are addressed below.

2.3.1 Water on Earth through mantle evolution

It is believed that early in the evolution of "terrestrial planets" (i.e., planets that are earth-like in nature; composed primarily of silicate rocks or metals; i.e., Mercury, Venus, Earth, and Mars) energetic impact, radioactive decay, and core formation may have created one or more whole or partial silicate mantle magma oceans. The time to mantle solidification and then to clement surface conditions allowing liquid water is highly dependent upon heat flux from the planetary surface through a growing primitive atmosphere. Elkins-Tanton (2008) modeled the time to clement conditions for whole and partial magma oceans on the Earth and Mars, and the resulting silicate mantle volatile compositions. Included in this researcher's calculations were partitioning of water and carbon dioxide between solidifying mantle cumulate mineral assemblages, evolving liquid compositions, and a growing atmosphere. Elkins-Tanton (2008) found that "small initial volatile contents (0.05 wt.% H_2O, 0.01 wt.% CO_2) can produce atmospheres in excess of 100 bars, and that mantle solidification is 98% complete in less than 5 million years (Myr) for all magma oceans investigated on both Earth and Mars, and less than 100,000 years for low-volatile magma oceans. Subsequent cooling to clement surface conditions occurs in 5–10 Ma, underscoring the likelihood of serial magma oceans and punctuated clement conditions in the early planets." Another finding reported by Elkins-Tanton (2008) is that "Cumulate mantles are volatile-bearing and stably stratified following solidification, inhibiting the onset of thermal convection but allowing for further water and carbon emissions from volcanoes even in the absence of plate tectonics." Based on the just-mentioned findings, Elkins-Tanton (2008) concluded that models produce a new hypothetical starting point for mantle evolution in the terrestrial planets.

There exists a belief among several planetary scientists that terrestrial planets, with silicate mantles and metallic cores, are likely to obtain water and carbon compounds during accretion. Elkins-Tanton (2011) examined the conditions that allow early formation of a surface water ocean (simultaneous with cooling to clement surface conditions), and the timeline of degassing the planetary interior into the atmosphere. According to Elkins-Tanton (2011), "The greatest fraction of a planet's initial volatile budget is degassed into the atmosphere during the end of magma ocean solidification, leaving only a small fraction of the original volatiles to be released into the atmosphere through later volcanism. Rocky planets that accrete with water in their bulk mantle have two mechanisms for producing an early water ocean: First, if they accrete with at least 1–3 mass% of water in their bulk composition, liquid water may be extruded onto the planetary surface at the end of magma ocean solidification. Second, at initial water contents as low as 0.01 mass% or lower, during solidification a massive supercritical fluid and steam atmosphere is produced that collapses into a water ocean upon cooling. The low water contents required for this process indicate that rocky super-Earth exoplanets may be expected to commonly produce water oceans within tens to hundreds of millions of years of their last major accretionary impact, through collapse of their atmosphere." Note that super-Earth exoplanets are defined as "being larger and more massive than Earth, but smaller and less massive than Neptune (about 17 times Earth's mass)."

In a continuation of related studies, Elkins-Tanton (2012) found that theory and observations point to the occurrence of magma ponds or magma oceans in the early evolution of terrestrial planets and in many early-accreting planetesimals (bodies which could come together with many others under gravitation to form planets). Elkins-Tanton (2012)'s review paper focuses on evidence for magma oceans on planetesimals and planets and on research concerning the processes of compositional differentiation in the silicate magma ocean, distribution and degassing of volatiles, and cooling. According to Elkins-Tanton (2012), "the apparent ubiquity of melting during giant accretionary impacts suggests that silicate and metallic material may be processed through multiple magma oceans before reaching solidity in a planet. The processes of magma ocean formation and solidification, therefore, strongly influence the earliest compositional differentiation and volatile content of the terrestrial planets, and they form the starting point for cooling to clement, habitable conditions and for the onset of thermally driven mantle convection and plate tectonics."

2.3.2 Water brought to Earth by comets and asteroids

Astronomers now recognize four regions to our Solar System. Zone 1 hosts the inner planets (terrestrial planets): Mercury, Venus, Earth, and Mars. Farther out lies zone 2; the outer planets: Jupiter, Saturn, Uranus, and Neptune. Pluto is a member of zone 3, the icy trans-Neptunian. Zone

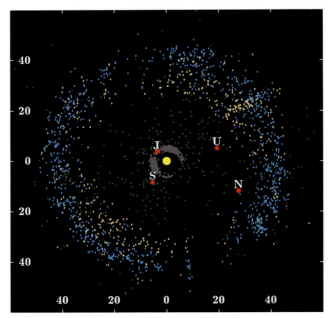

FIG. 2.3 The *Kuiper Belt*, the ring of frigid objects beyond Neptune's orbit. It is believed that buried water bodies may be abundant in the Kuiper Belt. *(Image Source: https://commons.wikimedia.org/wiki/File:Kuiper_belt_plot_objects_of_outer_solar_system.png). Copyright information: This file is licensed under the Creative Commons Attribution-Share Alike 3.0 Unported license.*

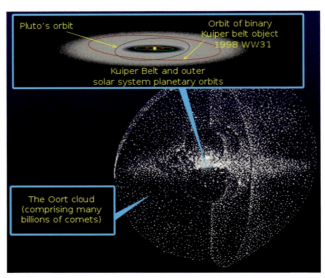

FIG. 2.4 **Kuiper Belt and Oort Cloud.** *(Source: https://commons.wikimedia.org/wiki/File:Kuiper_belt_-_Oort_cloud-en.svg) Image Author: NASA This SVG image was created by Medium69. Cette image SVG a été créée par Medium69. Credit: William Crochot. Copyright information: This file is in the public domain in the United States because it was solely created by NASA. NASA copyright policy states that "NASA material is not protected by copyright unless noted."*

4 lies beyond the Kuiper Belt (Figs. 2.3 and 2.4)—the ring of frigid objects beyond Neptune's orbit. The Kuiper belt, occasionally called the Edgeworth–Kuiper belt, is a circumstellar disk in the outer Solar System, extending from the orbit of Neptune to approximately 50 AU from the Sun. The Kuiper belt is similar to the asteroid belt, but is far larger—20 times as wide and 20–200 times as massive. The Oort Cloud (Fig. 2.4) may extend up to 1 light year from the Sun. Whereas Kuiper Belt was discovered in 1992, Oort Cloud was first imagined in 1950.

Objects in the *Kuiper belt* are believed to be a source of microscopic grains drifting toward the Sun and are thus present among the interplanetary dust particles (IDPs) collected in the Earth stratosphere. At least three dwarf planets are located in the Kuiper belt: *Pluto*, *Haumea* and *Makemake*. Also, some of the Solar System's moons are thought to have originated there, such as Neptune's *Triton* and Saturn's *Phoebe*.

Comets and asteroids are rocky materials that originate from two different locations in the Solar system. Whereas a comet is a chunk of solid body, usually around 1–10 km across and made of ices, dust and rock originating from the outer Solar System (Uranus-Neptune region and the Kuiper Belt region), an asteroid is a hydrated rock in orbit generally between Mars and Jupiter. Sometimes these rocky materials get bounced toward Earth.

It may be noted that asteroids are leftovers from the formation of our Solar System about 4.6 billion years ago. They are found in the asteroid belt (see Fig. 1.4)—a region of space between the orbits of Mars and Jupiter where most of the asteroids in our Solar System are found orbiting the Sun. Early on, the birth of Jupiter prevented any planetary bodies from forming in the gap between Mars and Jupiter, causing the small objects that were there to collide with each other and fragment into the asteroids seen today. The asteroid belt probably contains millions of asteroids. The asteroid belt is estimated to contain between 1.1 and 1.9 million asteroids larger than 1 km (0.6 mile) in diameter, and millions of smaller ones. One theory that astronomers have put forward regarding the existence of asteroid belt is that 4.6 billion years ago, when our Solar System was being formed, a tenth planet (ninth planet, since the removal of *Pluto* from the list of planets and its categorization as "dwarf planet") tried to form between Mars and Jupiter. However, Jupiter's gravitational forces were too strong, so the material was unable to form a planet.

In a recent incidence, Asteroid *Florence* (Fig. 2.5), a large Near-Earth asteroid (NEA), passed safely by Earth on September 1, 2017 (https://www.nasa.gov/feature/jpl/large-asteroid-to-safely-pass-earth-on-sept-1). Note that Near-Earth asteroids (NEAs) are asteroids whose orbits are close to Earth's orbit. Some NEAs' orbits intersect Earth's orbit so they pose a collision danger. NEAs only survive in their orbits for 10 million to 100 million years. In another instance, a small asteroid designated *2012 TC4* (Fig. 2.6)

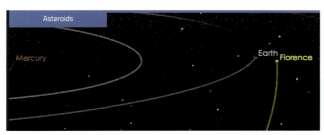

FIG. 2.5 Asteroid florence, a large near-Earth asteroid, which passed safely by Earth on Sept. 1, 2017, at a distance of about 4.4 million miles, (7.0 million kilometers, or about 18 Earth-Moon distances). *(Source: https://www.nasa.gov/feature/jpl/large-asteroid-to-safely-pass-earth-on-sept-1).*

In the primordial Solar System, the most plausible sources of the water accreted by the Earth were in the outer asteroid belt, in the giant planet regions, and in the Kuiper Belt. The bulk of the water presently on Earth was carried by a few planetary embryos, originally formed in the outer asteroid belt and accreted by the Earth at the final stage of its formation.

Based on dynamical models of primordial evolution of Solar System bodies, Morbidelli et al. (2000) argued that it is plausible that Earth accreted water all along its formation. They reckon that comets as well as asteroids from the Jupiter-Saturn region were the first water deliverers, when the Earth was less than half its present mass.

Meteorites (pieces of rock or metal that have fallen to the Earth's surface from outer space as meteors represent the state of matter at the time of planetary accretion. Over 90 per cent of meteorites are of rock while the remainder consist wholly or partly of iron and nickel. Many primitive meteorites originating from the asteroid belt once contained abundant water that is now stored as OH in hydrated minerals. Alexander et al. (2012) estimated the hydrogen isotopic compositions in 86 samples of primitive meteorites that fell in Antarctica and compared the results to those of comets and Saturn's moon, *Enceladus*. Water in primitive meteorites was less deuterium-rich than that in comets and *Enceladus*, implying that comets were not the principal source of Earth's water.

Observations of comets (see Fig. 2.7) further complicate the picture. As indicated earlier, as comets approach the Sun, the ice in comets sublimates to water vapor.

FIG. 2.6 Depiction of the safe flyby of asteroid 2012 TC4 as it passes under Earth on Oct. 12, 2017. While scientists cannot yet predict exactly how close it will approach, they are certain it will come no closer than 4200 miles (6800 km) from Earth's surface. *Image credit: NASA/JPL-Caltech (Source: https://www.jpl.nasa.gov/news/news.php?feature=6906).*

passed very close to Earth on October 12, 2017 (https://www.jpl.nasa.gov/news/news.php?feature=6906).

It was long thought that Earth's water did not originate from the planet's region of the protoplanetary disk. Instead, it was hypothesized water and other volatiles must have been delivered to Earth from the outer Solar System later in its history. There is evidence that water was delivered to Earth by impacts from icy planetesimals similar in composition to asteroids in the outer edges of the asteroid belt (Pepin, 1991).

FIG. 2.7 A comet with stars in the background. *(Source: http://cse.ssl.berkeley.edu/SegwayEd/lessons/findplanets/find_comet/WhatsAcomet.html).*

Spectroscopic studies of this water vapor have revealed D/H ratios of $(310 \pm 40) \times 10^{-6}$, substantially higher than that of terrestrial water (Bockelée-Morvan et al., 1995). The contribution of cometary water to terrestrial oceans should thus be small (<10%).

2.4 Indispensability of water, biologically important chemical elements, and energy to sustain life as we know it

Till date, Earth is the only place in the Universe where life is known to exist, and terrestrial life is "water-based" life. For this reason, the search for life on other worlds (extraterrestrial life, if any) is often characterized by the strategy of "follow the water." According to McKay (2016), "Water provides a convenient shorthand for the range of attributes that support life on Earth, including environmental processes associated with cycling of water, biochemistry of carbon in the medium of water, and the ecological systems that water-based life creates in water-cycling environments. Many of these attributes of the Earth, as an abode for life, are taken as given once the presence of liquid water is established." Although "search for life" researchers are optimistic about the possible existence of life (read, microbial life) in some other planets and moons (e.g., *Mars*, Jupiter's icy moon *Europa*, Saturn's icy moons *Enceladus* and *Titan*) and even some extrasolar celestial bodies, no traces of extinct or extant life have ever been found yet.

All life is organized as cells. Physical compartmentation from the environment and self-organization of self-contained redox reactions (chemical reactions in which the oxidation states of atoms are changed; see *Definitions* for details) are the most conserved attributes of living things, hence inorganic matter with such attributes would be life's most likely forebear (Martin and Russell, 2003). Although it has been widely suggested that life based around carbon, hydrogen, oxygen, and nitrogen is the only plausible biochemistry, and specifically that terrestrial biochemistry of nucleic acids, proteins, and sugars is likely to be "universal," the emerging view is that this notion is not an inevitable conclusion from our knowledge of chemistry. Bains (2004) argued that many chemistries could be used to build living systems, and that it is the nature of the liquid in which life evolves that defines the most appropriate chemistry. Bains (2004) further argued that fluids other than water could be abundant on a cosmic scale and could therefore be an environment in which nonterrestrial biochemistry could evolve. The chemical nature of these liquids could lead to quite different biochemistries, a hypothesis discussed in the context of the proposed "ammonochemistry" of the internal oceans of the Galilean satellites and a more speculative "silicon biochemistry" in liquid nitrogen. According to Bains (2004), these different chemistries satisfy the thermodynamic drive for life through different mechanisms, and so will have different chemical signatures than terrestrial biochemistry.

As already indicated, water (a good solvent) plays an important role is sustaining life as we know it. For temperate, terrestrial planets, the presence of water is of great importance as an indicator of habitable conditions. Water is a good solvent, primarily because it is liquid at high temperature, and secondarily because of its chemical properties (polar, high dielectric constant). In a water molecule, two charges are present with a negative charge in the middle, and a positive charge at the ends (Fig. 2.8). In the water molecule, the oxygen and hydrogen atoms share electrons in covalent bonds. There is a total of 10 protons (eight protons in an oxygen atom and one proton in a hydrogen atom) and 10 electrons (eight electrons in an oxygen atom and one electron in a hydrogen atom) so the water molecule as a whole is neutral. However, the electron cloud model shows that electrons are not shared equally in the water molecule. Electrons are a bit more attracted to the oxygen atom than they are to the hydrogen atoms. This makes the water molecule slightly negative at the oxygen end and slightly positive at the hydrogen end. When a neutral molecule has a positive area at one end and a negative area at the other, it is a polar molecule. Thus, a water molecule (which is a polar inorganic compound) possesses an electric dipole moment, with a negatively charged end and a positively charged end. Water molecules attract one another based on the attraction between the positive end of one water molecule and the negative end of another. Polarity underlies a number of physical properties including surface tension, solubility, and melting and boiling points. Water is described as the "universal solvent" and the "solvent of life." It is the most abundant substance on Earth and the only common substance to exist as a solid, liquid, and gas on Earth's surface. It is also the third most abundant molecule in the universe. Liquid water is both a crucial source of oxygen and a useful solvent for the generation of biomolecules.

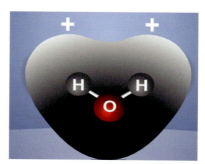

FIG. 2.8 In a water molecule, two charges are present with a negative charge in the middle, and a positive charge at the ends. *(Source: https://www.middleschoolchemistry.com/multimedia/chapter5/lesson1#polar_water_molecule).*

According to McKay (1991), life is a material system that undergoes reproduction, mutation, and natural selection. Cleland and Chyba (2002) have suggested that life might be like water, hard to define phenomenologically, but easy to define at the fundamental level. But life is like fire, not water—it is a process, not a pure substance. McKay (2004) thinks that such definitions are grist for philosophical discussion, but they neither inform biological research nor provide a basis for the search for life on other worlds. Our lack of data is reflected in our attempts to define life. Koshland (2002) lists seven features of life: (1) program (DNA), (2) improvization (response to environment), (3) compartmentalization, (4) energy, (5) regeneration, (6) adaptability, and (7) seclusion (chemical control and selectivity).

Probably the simplest, but not the only, proof of life is to find something that is alive. According to McKay (2004), there are only two properties that can determine if an object is alive: metabolism (the chemical processes that occur within a living organism in order to maintain life) and motion (metabolism is used here to include an organism's life functions, biomass increase, and reproduction). All living things require some level of metabolism to remain viable against entropy (note that entropy is a terminology indicating lack of order or predictability; gradual decline into disorder). Movement (either microscopic or macroscopic) in response to stimuli or in the presence of food can be a convincing indicator of a living thing.

From a scientific point of view, life is basically a chemical system that is able to transfer its molecular information via self-replication and to evolve via mutations (see Brack et al., 1999). Note that mutation is the changing of the structure of a gene, resulting in a variant form which may be transmitted to subsequent generations, caused by the alteration of single base units in DNA, or the deletion, insertion, or rearrangement of larger sections of genes or chromosomes.

Life on Earth uses proteins as the primary structural molecule. Using ~20 amino acids to form proteins, life can construct a combinatorially (i.e., involving combinations) large number of possible proteins. Due both to their internal interactions and especially to their interaction with water in both hydrophilic (i.e., having a tendency to mix with, dissolve in, or be wetted by water) and hydrophobic (i.e., tending to repel or fail to mix with water) ways, individual proteins fold into specific shapes. These shapes form the basis of structural molecules used by Earth life (McKay, 2016).

Life, as we know it, is based on carbon chemistry operating in an aqueous environment. On Earth, organisms, even microorganisms, live in communities. Within such communities, organisms exchange matter and genes. Arguably, communities of microorganisms can survive and thrive in conditions that would be difficult for any of the individuals in the community to grow in isolation. Communities can more efficiently cycle and recycle nutrients and can enhance genetic resilience. The liquid-water environment on Earth provides the medium for an organism to come into physical contact; as well, it mobilizes material released from cells (McKay, 2016).

On Earth, the vast majority of life-forms ultimately derive their energy from sunlight. The only other source of primary productivity known is chemical energy, and there are only two ecosystems known, both methanogen-based (Stevens and McKinley, 1995; Chapelle et al., 2002), that rely exclusively on chemical energy (i.e., they do not use sunlight or its product, oxygen). Photosynthetic organisms can use sunlight at levels even below the level of sunlight at the orbit of the dwarf planet Pluto, orbiting in the outer periphery of our Solar System (Ravens et al., 2000); therefore, energy is not the limitation for life (read, microbial life). Hydrogen, oxygen, carbon, nitrogen, sulfur, and phosphorus are the elements of life, and they are abundant in the Solar System. Indeed, the Sun and the outer Solar System have more than 10,000 times the carbon content of the bulk of Earth (McKay, 1991).

On Earth, life is widespread because habitable liquid water is widespread. Even in the driest desert on Earth, the Atacama Desert of Chile, there is occasional liquid water. Terrestrial life uses a chemistry dependent upon DNA and RNA (described as the software of life (McKay, 2004)), and the chemical synthesis of a critical component of this—the 5-carbon sugar, ribose—has only been stabilized in the presence of calcium borate minerals that are highly soluble in water (Ricardo et al., 2004). Because life on Earth exists whenever liquid water, organics, and energy coexist, understanding the chemical components of the liquid water and ice-grains found on extraterrestrial worlds could indicate whether life is potentially present on such worlds. According to the present knowledge, emergence of life requires liquid water, biologically important chemical elements, and some form of energy. Biologically important chemical elements are hydrogen (H), oxygen (O), carbon (C), nitrogen (N), sulfur (S) and phosphorous (P) (abbreviated as "HOCNSP"). Liquid water, under the right atmospheric conditions, is till date considered being the key to life as we know it.

In addition to liquid water and a supply of suitable chemical elements, organisms on Earth require a source of energy for metabolism, growth, maintenance, and reproduction. This energy is derived primarily from sunlight through photosynthesis (in certain bacteria and the bacterially derived chloroplasts of plants) or photorespiration (in archaea living in salty environments); a very small proportion is derived from geothermally driven chemical disequilibria. The chemical products of this activity span a range of oxidation potentials that drive a cascade of intermediate oxidation-reduction (redox) reaction pairs (detailed figure given in Gaidos et al., 1999). Organisms on Earth exploit these thermodynamically favored reactions as energy sources, and combinations of oxidants and reductants support various metabolic life-styles (Gaidos et al., 1999). Before the evolution of photosynthesis, organisms on Earth could not use sunlight directly and hence depended on abiotic sources of chemical energy. According to several researchers on the early Earth (e.g., Navarro-Gonzalez et al., 1998; Gaidos et al., 1999), abiotic sources of chemical energy would come in the form of disequilibrium concentrations of redox reactants driven by hydrothermal activity, solar ultraviolet radiation, electrical discharges, and impacts. All such redox pairs would be depleted by abiotic reactions and biological activity, so without external energy sources such as sunlight or geothermal energy, chemical equilibrium will ultimately terminate all life (Parkinson et al., 2007). Similarly, this thermodynamic condition will impose some formidable constraints for life elsewhere in the Solar System that is based on chemical energy (Gaidos et al., 1999).

Each report of liquid water existing elsewhere in the Solar System has reverberated through the international press and excited the imagination of humankind. This is because in the past few decades we have come to realize that where there is liquid water on Earth, virtually no matter what the physical conditions, there is life (Rothschild and Mancinelli, 2001). Liquid water is deemed necessary for life to evolve and develop because it allows the diffusion of complex molecules. Regarding the importance of liquid water concerning life, Claudi (2017) stated thus, "For what concerns liquid water, it has some important characteristics that make it the best solvent for life: a large dipole moment, the capability to form hydrogen bonds, to stabilize macromolecules, to orient hydro–phobic–hydrophilic molecules, etc." Note that in chemistry, a dipole refers to the separation of charges within a molecule between two covalently bonded atoms. A dipole moment is a measurement of the separation of two oppositely charged charges.

There is a belief among a class of origin-of-life scientists that life on Earth probably originated from the evolution of reduced organic molecules in liquid water. Although the source of this organic matter is debatable, there is a general view among origin-of-life scientists that the primitive Earth's atmosphere or hydrothermal vents could have acted as the source of this organic matter. Another school of thought is that a large fraction of prebiotic organic molecules might have been brought to Earth by extra-terrestrial meteoritic and cometary dust grains decelerated by the atmosphere. Proponents of extra-terrestrial life are of the view that any celestial body holding essential molecules containing HOCNSP atoms dissolved in liquid water at the right environmental condition (e.g., favorable ambient temperature, atmospheric pressure, humidity etc.) at a permanent basis can evolve and support life in the same manner as the primitive Earth allowed evolution and sustenance of life, if the said celestial bodies are free from harmful radiations (e.g., ultraviolet rays) and other life-threatening materials (e.g., poisonous matter).

Water is considered as an essential ingredient to life whether on Earth or elsewhere in the Solar system and even beyond. Claudi (2017) reinforces this argument thus, "Water is an abundant compound in our galaxy, it can be found in different environments, from cold dense molecular clouds to hot stellar atmospheres (e.g., Cernicharo and Crovisier, 2005; Lammer et al., 2009). Water is liquid at a large range of temperatures and pressures and it is a strong polar–nonpolar solvent. This dichotomy (entirely contrasting things) is essential for maintaining stable biomolecular and cellular structures (Des Marais et al., 2002). Furthermore, liquid water has a great heat capacity that makes it able to tolerate a heat shock. In addition, the most common solid form of water has a 'specific weight,' γ (weight of a substance per unit volume in absolute units; $\gamma = \rho g$, where ρ is the density and g is the standard gravity) lighter than that of its liquid form. This allows ice to float on a liquid ocean safeguarding the underlying liquid water. All those characteristics let grow the probability that life, once it emerges, could survive and evolve."

It is a well-known fact that each recent report of liquid water existing elsewhere in the Solar System (e.g., the year-2018 confirmation of the presence of water deposits in the darkest and coldest parts of the Earth's Moon's Polar Regions (Li et al., 2018) using data collected by NASA's Moon Mineralogy Mapper instrument borne by India's Chandrayan-I spacecraft, which was launched in 2008; the year-2018 discovery of a sizable salt-laden lake under ice on the southern polar plain of the planet Mars (Orosei et al., 2018); the year-2018 detection of signs of presence of water deep inside Jupiter's Great Red Spot (Bjoraker et al., 2018)) has reverberated through the international press and the social media, and excited the imagination of humankind. Such excitement arose primarily from the fact that in the past few decades we have come to realize that where there is liquid water on Earth, virtually no matter what the physical conditions, there is life. Scientists are optimistic that the water spots recently discovered on Earth's Moon and the planet Mars could be possible habitats for microbial life.

Surprisingly, life has been found to have a brave history of developing and flourishing very well in water that is very acidic, alkaline or is a strong brine solution. It has also survived and flourished in water at temperatures above 100°C. Life at and around hydrothermal vents is a glaring example of evolution, survival, and growth at high temperatures. There are some extremophiles that can survive without water for some time, but not indefinitely. Absence of water for long time results in the death of even extremophiles.

According to McKay (2004), the practical approach to the search for life is to determine what life needs. The simplest list is probably: energy, carbon, liquid water, and a few other elements such as oxygen, nitrogen, sulfur, and phosphorus (McKay, 1991). Life requires energy to maintain itself against entropy, as does any self-organizing open system. In the memorable words of Schrödinger (1945), "It feeds on negative entropy."

The importance of water molecules becomes obvious when we realize that, of all the elements in the Universe that form molecules, water (H_2O) is the combination of the first and second most abundant elements (i.e., elements in the HOCNSP group) in the Universe. According to Raymond et al. (2004, 2007) and Elkins-Tanton (2012), water should be a common feature of all the rocky planets. A general thinking in the scientific community is that some water-rich materials from beyond the snowline (boundary between a snow-covered and snow-free surface) are injected into the more water-poor material of the feeding zones of rocky planets in the circumstellar habitable zone (CHZ). Thus, it is likely that Venus and Mars both started out, like Earth, hot from accretional energy and impacts and wet from impacts of large hydrous (5–20% water) asteroids from the outer asteroid belt (Morbidelli et al., 2000, 2012) and other "wet" accretionary material (Drake, 2005). Scientists reckon that terrestrial planets (i.e., planets that are earth-like in nature, meaning those that are similar in composition [silicate rocks or metals] and structure to that of the Earth) in planetary systems other than the Solar System are also likely to start with variable, non-negligible amounts of water. Note that within the Solar System, the terrestrial planets are the inner planets closest to the Sun, that is, Mercury, Venus, Earth, and Mars.

It is believed that after the heavy bombardment in the early history of Earth, life could have emerged with some probability on initially wet rocky planets in the CHZ. However, due to large impacts and unstable abiotic volatile evolution with no tendency to maintain habitability, almost all life could have gone extinct early—with the rare exceptions of life that has undergone unusually rapid evolution. According to Chopra and Lineweaver (2016), the surface temperature and existence of liquid water at, or near, the surface of a planet or its moon could be predominantly due to Gaian regulation (i.e., the emergence of life's abilities to modify its environment and regulate initially abiotic positive feedback mechanisms rather than abiotic negative feedback). Liquid water on the surface of the planet (particularly old planets) would then not just be a prerequisite for life but a biosignature (Gorshkov et al., 2004). According to Luger and Barnes (2015), existence of liquid water on the surface of a planet may be a better biosignature than oxygen. Thus, the measurement of exoplanetary surface temperatures compatible with liquid water could be an important part of future search for extraterrestrial life. According to Lovelock and Kaplan (1975) and Krissansen-Totton et al. (2016), remote detection of atmospheric chemical equilibrium may soon develop into a mature science of remote biodetection.

Inside our own Solar System, the planetary moons of Jupiter and Saturn—particularly *Europa* and *Enceladus*—tickle the imagination because of the presence of water. There exists a general belief among planetary scientists and origin of life scientists that where there is water, the chance of life is higher, although there are exceptions (for example, habitability on the surface of celestial bodies suffering from harmful radiation and toxins). It has been shown by Wadsworth and Cockell (2017) that any environment where there is concentration of perchlorates, such as in the Martian brines, will produce unhabitable environments on account of the bacteriocidal properties of UV-irradiated perchlorates. Hostile geochemical properties such as these suggest that the mere presence of liquid water seeps, thought to be good locations to search for life, do not necessarily imply environments fit for life.

2.5 Formation of liquid water oceans on Earth about 3.8 billion years ago

The origin of liquid water on Earth is the subject of a body of research in the fields of planetary science, astronomy, and astrobiology. Earth is unique among the rocky planets in the Solar System in that it is the only planet known to have oceans of liquid water on its surface. Liquid water, which is necessary for life as we know it, continues to exist on the surface of Earth because the planet is at a distance, known as the habitable zone, far enough from the Sun that it does not lose its water to the runaway greenhouse effect (he complete evaporation of the oceans), but not so far that low temperatures cause all water on the planet to freeze.

The origin of copious amount of water vapor on Earth goes back to the time of the Earth's formation 4. 6 billion years ago, when the hot Earth was forming through the accumulation of smaller objects, called planetesimals. There are basically three possible sources for the water vapor on the early Earth. It could have (1) separated out from the rocks that make up the bulk of the Earth; (2) arrived as part of a late-accreting veneer of water-rich meteorites, similar to the carbonaceous chondrites that we see today; or (3) arrived as part of a late-accreting veneer of icy planetesimals, that is, comets.

What do we know about the origin of the earth's oceans? Is it more likely that they derive from icy comets that struck the young earth or from material released from the earth's interior during volcanic activity? According to Owen (1999), the composition of the ocean offers some clues as to its origin. If all the comets contain the same kind of water ice that have been examined in Comets Halley and Hyakutake—the only ones whose water molecules have been studied in detail—then comets cannot have delivered all the water in the earth's oceans because the ice in the comets contains twice as many atoms of deuterium (a heavy isotope of hydrogen) to each atom of ordinary hydrogen as we find in seawater. At the same time, the meteorites could not have delivered all of the water, because then the earth's atmosphere would contain nearly 10 times as much xenon (an inert gas) as it actually does. Meteorites all carry this excess xenon. Although nobody has yet measured the concentration of xenon in comets, recent laboratory experiments on the trapping of gases by ice forming at low temperatures suggest that comets do not contain high concentrations of the xenon. A mixture of meteoritic water and cometary water would not work either, because this combination would still contain a higher concentration of deuterium than is found in the oceans.

Hence, according to Owen (1999), the best model for the source of the oceans at the moment is a combination of water derived from comets and water that was caught up in the rocky body of the earth as it formed. This mixture satisfies the xenon problem. It also appears to solve the deuterium problem—but only if the rocky material out near the earth's present orbit picked up some local water from the solar nebula (the cloud of gas and dust surrounding the young Sun) before they accreted to form the Earth. Some new laboratory studies of the manner in which deuterium gets exchanged between hydrogen gas and water vapor have indicated that the water vapor in the local region of the solar nebula would have had about the right (low) proportion of deuterium to balance the excess deuterium seen in comets.

There are several planetary scientists who think that volatiles (elements and compounds, including water, that easily vaporize at low temperatures) were released from the solid phase as the earth accreted. Thus, the earth and its oceans and atmosphere grew together. During accretion, the kinetic energy of the colliding planetesimals was converted into thermal energy, so the earth grew extremely hot as it came together. The material forming the earth was probably too hot for ice to have been a major carrier of water. Most of the water was probably present originally as water trapped in clay minerals or as separate hydrogen (in hydrocarbons) and oxygen (in iron oxides), rather than as ice.

Since the end of the period of accretion, more than four billion years ago, there has been a continual exchange of volatile material—including water—between the surface of the earth and the planet's interior (that is, between the crust and the mantle). Volcanoes release water and carbon dioxide to the atmosphere and ocean. Subduction of sediments rich in volatiles takes place at deep ocean trenches. The sinking of oceanic crust at subduction zones carries water and carbon dioxide back into the mantle. These processes can all be seen at work today. In short, icy cometary material probably has not been important in providing water for the earth's oceans, but there is little sure knowledge in this field (https://www.scientificamerican.com/article/what-do-we-know-about-the/).

According to a theory, the ocean formed from the escape of water vapor and other gases from the molten rocks of the Earth to the atmosphere surrounding the cooling planet. Whatever may be the source, water on the early Earth remained in the form of vapor until the Earth cooled below the boiling point of water (100°C). At this time, about 3.8 billion years ago, the water condensed and rain began to fall—and continued to fall for centuries. As the water drained into the great hollows in the Earth's surface, the primeval ocean came into existence. The forces of gravity prevented the water from leaving the planet (https://oceanservice.noaa.gov/facts/why_oceans.html). Most scientists agree that the atmosphere and the ocean accumulated gradually over millions and millions of years with the continual 'degassing' of the Earth's interior.

There is geological evidence (e.g., discovery of remnants of pillow lavas in volcanic rock) that helps constrain the time frame for liquid water existing on Earth. Pillow lavas (Fig. 2.9) are lavas that contain characteristic pillow-shaped structures that are attributed to the extrusion of the lava underwater, or subaqueous extrusion. Pillow lavas in volcanic rock are characterized by thick sequences of discontinuous pillow-shaped masses, commonly up to 1 m in diameter. A sample of pillow basalt (a type of rock formed during an underwater eruption) was recovered from the Isua Greenstone Belt and provides evidence that water existed on Earth 3.8 billion years ago (Pinti and Nicholas, 2014). In the Nuvvuagittuq Greenstone Belt, Quebec, Canada, rocks dated at 3.8 billion years old by one study (Cates and Mojzsis, 2007) and 4.28 billion years old by another (O'Neil et al., 2012) show evidence of the presence of water at these ages. If oceans existed earlier than this, any geological evidence either has yet to be discovered or has since been destroyed by geological processes such as crustal recycling. More recently, in August 2020, researchers reported that sufficient water to fill the oceans may have always been on the Earth since the beginning of the planet's formation (Piani, 2020).

FIG. 2.9 Pillow lava on the ocean floor of Hawaii. Pillow lava rocks on the slope off Hawaii form when magma oozes from below. *(Source: https://commons.wikimedia.org/wiki/File:Nur05018-Pillow_lavas_off_Hawaii.jpg) Copyright info: This image is in the public domain because it contains materials that originally came from the US National Oceanic and Atmospheric Administration, taken or made as part of an employee's official duties.*

Unlike rocks, minerals called zircons are highly resistant to weathering and geological processes and so are used to understand conditions on the very early Earth. Mineralogical evidence from zircons has shown that liquid water and an atmosphere must have existed 4.404 ± 0.008 billion years ago, very soon after the formation of Earth (Wilde et al., 2001). But this period is in conflict with an era from 4.5 to 4 billion years ago when the entire earth was a "Fire Ball." How could liquid water have existed 4.4 billion years ago at a time when the entire earth was a fire ball?

This presents somewhat of a paradox, as the cool early Earth hypothesis suggests temperatures were cold enough to freeze water between about 4.4 billion and 4.0 billion years ago. Other studies of zircons found in Australian Hadean rock point to the existence of plate tectonics as early as 4 billion years ago. If true, that implies that rather than a hot, molten surface and an atmosphere full of carbon dioxide, early Earth's surface was much as it is today.

Furthermore, the "cool early Earth hypothesis" posits that for part of the Hadean geological eon, at the beginning of Earth's history, it had a modest influx of bolides (large meteors which explode in the atmosphere) and a cool climate, allowing the presence of liquid water. This would have been after the extreme conditions of Earth's earliest history between 4.6 and 4.4 billion years (Ga) ago, but before the Late Heavy Bombardment of 4.1–3.8 Ga ago.

In the history of Earth, initially there existed just one continent (Pangea, also called Pangaea) and one ocean (Panthalassa). Panthalassa, also known as the Panthalassic Ocean or Panthalassan Ocean, was the super-ocean that surrounded the super-continent Pangaea (Fig. 2.10), the latest in a series of supercontinents in the history of Earth. Panthalassa was fully assembled by the Early Permian Epoch (some 299 million to about 273 million years ago). Panthalassa is also referred to as the Paleo-Pacific ("old Pacific") or Proto-Pacific because the Pacific Ocean is a direct continuation of Panthalassa. During the Paleozoic–Mesozoic transition (about 250 Million years ago) Panthalassa occupied almost 70% of Earth's surface.

The supercontinent (Pangea) began to break apart about 200 million years ago, during the Early Jurassic Epoch (201 million to 174 million years ago), eventually forming the modern continents and the Atlantic and Indian oceans. Pangea's existence was first proposed in 1912 by German meteorologist Alfred Wegener as a part of his theory of continental drift.

The Panthalassic Ocean floor has completely disappeared because of the continuous subduction along the continental margins on its circumference (Isozaki, 2014). On the earth, plate tectonics has caused oceanic water to mix considerably with material from its interior; such contamination probably did not occur on Mars, where plate tectonics does not seem to occur. Therefore, apart from studying more comets, water on Mars needs to be examined, in which case

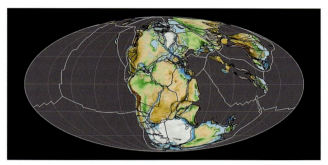

FIG. 2.10 Map of the supercontinent Pangaea 200 million years ago (Ma) *Author: Fama Clamosa Made using GPlates and data sets listed below: Amante, C. and Eakins, B. W. 2009. ETOPO1 1 Arc-Minute Global Relief Model: Procedures, Data Sources and Analysis. NOAA Technical Memorandum NESDIS NGDC-24, 19. Matthews, K.J., Maloney, K.T., Zahirovic, S., Williams, S.E., Seton, M., and Müller, R.D. (2016). Global plate boundary evolution and kinematics since the late Paleozoic, Global and Planetary Change, 146, 226-250. DOI: 10.1016/j.gloplacha.2016.10.002. Müller, R.D., Seton, M., Zahirovic, S., Williams, S.E., Matthews, K.J., Wright, N.M., Shephard, G.E., Maloney, K.T., Barnett-Moore, N., Hosseinpour, M., Bower, D.J. & Cannon, J. 2016. Ocean Basin Evolution and Global-Scale Plate Reorganization Events Since Pangea Breakup, Annual Review of Earth and Planetary Sciences, vol. 44, pp. 107. DOI: 10.1146/annurev-earth-060115-012211. (Source: https://commons.wikimedia.org/wiki/File:Pangaea_200Ma.jpg) Copyright info: This file is licensed under the Creative Commons Attribution-Share Alike 4.0 International license.*

we have another chance to investigate the sources described above. These investigations (and other related studies) are currently under way. This is an active area of research.

2.6 Importance of deuterium to hydrogen ratio of water

Deuterium (or hydrogen-2, also known as heavy hydrogen) is one of two stable isotopes of hydrogen (the other being protium, or hydrogen-1). The nucleus of a deuterium atom, called a deuteron, contains one proton and one neutron, whereas the far more common protium has no neutrons in the nucleus. Deuterium has a natural abundance in Earth's oceans of about one atom in 6420 of hydrogen. Thus, deuterium accounts for approximately 0.0312% by mass of all the naturally occurring hydrogen in the oceans. The abundance of deuterium changes slightly from one kind of natural water to another (see Vienna Standard Mean Ocean Water).

The name deuterium is derived from the Greek deuteros, meaning "second," to denote the two particles composing the nucleus (O'Leary, 2012). Deuterium was discovered and named in 1931 by Harold Urey. When the neutron was discovered in 1932, this made the nuclear structure of deuterium obvious, and Urey won the Nobel Prize in 1934 "for his discovery of heavy hydrogen." Soon after deuterium's discovery, Urey and others produced samples of "heavy water" in which the deuterium content had been highly concentrated.

Deuterium is destroyed in the interiors of stars faster than it is produced. Other natural processes are thought to produce only an insignificant amount of deuterium. Nearly all deuterium found in nature was produced in the Big Bang that is considered to have occurred 13.8 billion years ago, as the basic or primordial ratio of hydrogen-1 to deuterium (about 26 atoms of deuterium per million hydrogen atoms) has its origin from that time. This is the ratio found in the gas giant planets, such as Jupiter. The analysis of deuterium–protium ratios in comets found results very similar to the mean ratio in Earth's oceans (156 atoms of deuterium per million hydrogen atoms). This reinforces theories that much of Earth's ocean water is of cometary origin (Hersant et al., 2001; Hartogh et al., 2011). However, Alexander et al. (2012)'s study suggests that comets are not the principal source of Earth's water. The deuterium–protium ratio of the comet 67P/Churyumov-Gerasimenko, as measured by the Rosetta space probe, is about three times that of Earth water. This figure is the highest yet measured in a comet (Altwegg et al., 2015).

It would be worthwhile to shed some light on the Deuterium to Hydrogen ratio (D/H ratio) because this ratio will be mentioned frequently in several subsequent chapters in this book. It may be noted that hydrogen isotopes (deuterium, D; and standard hydrogen, H) provide unique insight into the origin of water in planetary bodies (see Robert, 2001, 2006; Alexander et al., 2012; Marty, 2012; Halliday, 2013). Deuterium atom has an extra neutron. In a static solar nebula, radial temperature gradients, combined with equilibrium and kinetic factors, mean that water D/H values should have increased with increasing formation distance from the Sun (Drouart et al., 1999; Mousis et al., 2000). Given its dynamic evolution, the spatial and temporal variations in ice compositions in the solar nebula are likely to have been more complex. Nevertheless, objects that formed in the same source regions and at similar times should have accreted ice with similar hydrogen isotopic compositions.

The Solar System consists of reservoirs containing water with an extremely wide range of D/H ratios. The use of the D/H ratio demonstrates a clear connection between the Solar System and interstellar water. Water molecules originating from different places in the Solar System have different amounts of deuterium. In general, things formed closer to the Sun have less deuterium than things formed further out. Thus, the variation in D/H ratios partly reflects the primordial gradient of water and other volatiles (i.e., most easily vaporized materials) through the Solar System as a function of distance from the Sun (note that the reference distance called 1 AU is the Earth-Sun distance, which is approximately 150 million km). A deuterium/hydrogen (D/H) ratio of $(149 \pm 3) \times 10^{-6}$ has been estimated for the bulk Earth (Lécuyer et al., 1998), compared with a solar ratio of $(20 \pm 4) \times 10^{-6}$ deduced from solar wind implanted into lunar soils (Geiss and Gloecker, 1998).

The net result of accretion from these several reservoirs is that the water on Earth had essentially the deuterium/hydrogen ratio (D/H ratio) typical of the water condensed in the outer asteroid belt. This is in agreement with the observation that the D/H ratio in the oceans is very close to the mean value of the D/H ratio of the water inclusions in carbonaceous chondrites (CCs). Note that chondritic meteorites are asteroidal fragments that retain records of the first few million years of Solar System history. The bulk water in carbonaceous chondrite meteorites (a class of chondritic meteorites comprising at least 8 known groups and many ungrouped meteorites, representing only a small proportion (4.6%) of meteorite falls) has a D/H ratio similar to that of Earth's water (Lécuyer et al, 1998; Alexander et al., 2012), suggesting that these meteorites might be responsible for bringing water to the terrestrial planets (Fig. 2.11).

Hydrogen isotopic compositions are available for the water in some comets and Saturn's icy moon *Enceladus*. The D/H values of the OH in hydrated silicates in chondrites are likely to have been modified during alteration from the water compositions at the time of accretion. The chondrites also accreted organic matter (Alexander et al., 2007), and the hydrogen in it further complicates the determination of the original isotopic compositions of their water. Alexander et al. (2012) reported the bulk hydrogen, carbon, and nitrogen elemental and isotopic compositions of 86 samples of chondrites and adopted two approaches to use them for estimating or placing limits on the original hydrogen isotopic compositions of their water.

Alexander et al. (2012) found that *the six measured Oort cloud comets and Enceladus have water D/H values that are enriched in deuterium by roughly a factor of 2 compared to Earth and more than an order of magnitude relative to the initial solar composition. The one measured Jupiter-family comets (JFC) has a water hydrogen isotopic composition*

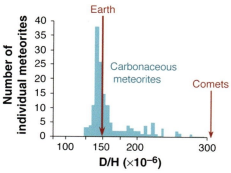

FIG. 2.11 Water from meteors. Distribution of the hydrogen isotopic ratio in carbonaceous meteorites compared with Earth and comets. According to this distribution, water on Earth seems mostly derived from a meteoritic source *(Source: Robert, F. (2001), The origin of water on Earth, Science, 293, 1056-1058; DOI: 10.1126/science.1064051).*

that is similar to Earth's (Hartogh et al., 2011). This suggests that there may not be a monotonic rise in the D/H values of icy bodies with formation distance from the Sun. Nevertheless, the D/H value of this JFC is still enriched in deuterium by almost an order of magnitude compared to the initial solar composition. Consequently, one would still expect that water accreted by asteroids that formed in the asteroid belt (e.g., S-complex asteroids) would have lower D/H values than objects that formed in the giant planet region and beyond (e.g., C-complex asteroids) (Alexander et al., 2012).

Based on detailed studies, Alexander et al. (2012) provided the below summary on D/H ratio: "The deuterium/hydrogen (D/H) values of water in carbonaceous chondrites are distinct from those in comets and Saturn's moon Enceladus, implying that they formed in a different region of the solar system, contrary to predictions of recent dynamical models. The D/H values of water in carbonaceous chondrites also argue against an influx of water ice from the outer solar system, which has been invoked to explain the nonsolar oxygen isotopic composition of the inner solar system. The bulk hydrogen and nitrogen isotopic compositions of CI chondrites suggest that they were the principal source of Earth's volatiles."

The D/H value for ocean water on Earth is known very precisely to be $(1.5576 \pm 0.0005) \times 10^{-4}$ (Hagemann et al., 1970). This value represents a mixture of all of the sources that contributed to Earth's reservoirs, and is used to identify the source or sources of Earth's water. The ratio of deuterium to hydrogen may have increased over the Earth's lifetime as the lighter isotope is more likely to leak to space in atmospheric loss processes. However, no process is known that can decrease Earth's D/H ratio over time (Catling, 2017). This loss of the lighter isotope is one explanation for why Venus has such a high D/H ratio, as that planet's water was vaporized during the runaway greenhouse effect and subsequently lost much of its hydrogen to space (Donahue et al., 1982). Because Earth's D/H ratio has increased significantly over time, the D/H ratio of water originally delivered to the planet was lower than at present. This is consistent with a scenario in which a significant proportion of the water on Earth was already present during the planet's early evolution (Genda, 2016). Earth's current deuterium to hydrogen ratio matches ancient eucrite chondrites, which originate from the asteroid Vesta in the outer asteroid belt (Sarafian et al., 2014). Note that eucrites (Fig. 2.12) are achondritic stony meteorites, many of which originate from the surface of the asteroid 4 Vesta and as such are part of the HED meteorite clan. They are the most common achondrite group with well over 100 distinct finds at present. CI, CM, and eucrite chondrites are believed to have the same water content and isotope ratios as ancient icy protoplanets from the outer asteroid belt that later delivered water to Earth (Morbidelli et al., 2000).

FIG. 2.12 Stannern meteorite eucrite. *(Source: https://commons.wikimedia.org/wiki/File:EUKRIT-METEORIT_VON_STANNERN_0041.jpg) Copyright info: This file is licensed under the Creative Commons Attribution-Share Alike 3.0 Unported license.*

2.7 Roles of Earth's Moon and Sun in generating tides—influences of local bathymetry and shoreline boundary on modifying tidal range and tidal pattern

An observer who spends a day on a beach witnesses a low-frequency rhythmic rising and lowering of sea level known as "tidal oscillation" or simply "tide." Tides have exerted such a great influence on the lives of coastal dwellers and beach visitors that the word itself may be found sprinkled throughout the poetry and folklore of many maritime regions. Indeed, so awe-inspiring are tides that they have inspired the creation of several myths. Despite many direct and indirect influences of tidal rhythms on mankind, little was known about what caused them. This led to many strange notions and mythologies among ancient observers. Meanwhile, progress in science has slowly and steadily begun to unravel the hidden secrets of these wonderful tidal rhythms. Over time, the theory of tidal rhythms has received a sound mathematical footing from physicists, astronomers, and mathematicians including: Sir Isaac Newton (1642–1727), David Bernoulli (1700–1782), Marquis de Laplace (1749–1827), and Lord Kelvin (1824–1907).

Investigations by these celebrated men of science have revealed that the origin of tidal rhythm (Fig. 2.13) is primarily astronomical with secondary bathymetric and topographical influences. To Newton we owe the artifice of the equilibrium tide, one that would exist in the absence of inertia on a world that is fully covered by water of infinite depth. In 1778, Laplace established the fluid dynamic equations of motion under gravity on a rotating sphere and described the tide-producing force as the residual differential force acting on a fluid particle in the ocean after the force at the center

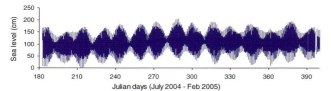

FIG. 2.13 Tidal oscillations observed at Takoradi harbor in Ghana (West Africa) at the Central East Atlantic Coast for Climate Change Research Studies under the aegis of IOC-UNESCO. The sea level gauge and the deployment mechanism have been designed by the Instrumentation Division of CSIR-NIO in India and established at Takoradi harbor (Ghana) by Antony Joseph and Prakash Mehra of CSIR-NIO, Goa, India. Local support and data retrieval were accomplished by Joseph T. Odametey and Emmanuel K. Nkebi of the Survey Department, Accra, Ghana. (Reference: Joseph et al., 2006).

of the Earth is subtracted. It is now known that tidal rhythms are in reality produced by a combination of several forces, namely;

- Force of gravitational attraction of the Moon and the Sun on Earth
- Centrifugal force produced by the revolution of the Earth-Moon system about their common center-of-mass
- Centrifugal force produced by the revolution of Earth around the Sun
- Rotation of the Earth about its own axis

Thus, tides are forced oscillations generated by the attractions of the Moon and Sun, and they have the same periods as the motion of the Sun and Moon relative to the Earth. Every 14th day at full moon or new moon, the attraction forces of the Sun and Moon reinforce one another. These conditions give rise to "spring tide" during which the tidal range (i.e., the difference between successive high water and low water elevations) is the largest in a fortnight.

2.7.1 General characteristics of tidal oscillations

Sea level oscillations on an approximately twice daily (semidiurnal) or daily (diurnal) basis are a worldwide phenomenon routinely observed at continental coasts and shores of islands. At most coastal or island locations, the interval between successive high waters (or successive low waters) is about 12 h and 25 min, which is half the time of the Moon's apparent revolution around the Earth. This type of tidal oscillation is called *semidiurnal tide*.

The fact that different oceans, basins, gulfs, and straits can respond to the different periods of the tide-generating forces allows a variety of tidal patterns to be developed. In some water bodies, the interval between successive high waters (or successive low waters) is about 24 h. This type of tidal oscillation is called diurnal tide. The East China Sea is an example of such a water body, suppressing semidiurnal tides and selectively promoting diurnal tides. The dominant tidal constituents are the diurnal constituents, K_1, O_1, P_1, Q_1, and S_1, with periods of 23.93, 25.82, 24.07, 26.87, and 24.00 h, respectively, and the semidiurnal constituents M_2, S_2, N_2, and S_2, with periods of 12.42, 12.00, 12.66, and 11.97 h, respectively.

In fact, mixed and diurnal tides predominate in parts of the Pacific. The most regular rhythmic motions in the sea level in several parts of the world's oceans have periods in the range of approximately 12–24 h.

A parameter known as tidal form factor (F), defined by the ratio of the two main diurnal constituents (K1 and O1) and semidiurnal constituents (M2 and S2) provides a quantitative measure of the general characteristic of tidal oscillations at a given location (i.e., whether the tides are semidiurnal, diurnal, or mixed).

$$F = \frac{(H_{K1} + H_{O1})}{(H_{M2} + H_{S2})}$$

Tidal regimes where F is below 0.25 are normally said to be semidiurnal, F between 0.25 and 1.5 is said to be mixed and mainly semidiurnal, F between 1.50 and 3 is mixed and mainly diurnal, while F greater than 3 is diurnal (see Pugh and Woodworth, 2014). A map providing a quantitative measure of F in the world's oceans is shown in Fig. 2.14.

Assuming that the Earth were fully covered by water (this, for simplicity of explanation), the full rotation of the solid Earth on its own axis (once in 24 h), combined with the "heaping" action resulting from the horizontal flow of water toward two diametrically opposite regions on the Earth representing the positions of the maximum gravitational attraction and centrifugal forces of the combined lunar and solar influences, makes each point on the ocean surface pass through two maximum levels (tidal bulges) and two minimum levels (tidal depressions) for each daily

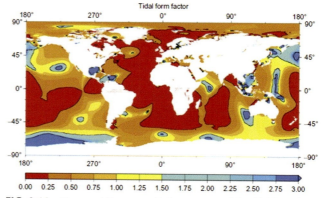

FIG. 2.14 Map providing a quantitative measure of the tidal form factor in the world's oceans. The amplitudes used to make this map were taken from the Technical University of Denmark DTU-10 global tide model (see Cheng and Andersen, 2010). (*Courtesy of Dr. Philip Woodworth; Formerly, Leader, PSMSL*).

rotation of Earth. This causes two high waters and two low waters in a day in a given location, giving rise to the semidiurnal tidal rhythm. This is not applicable at amphidromic points, whose dynamics are entirely different. While the equilibrium high tides are generated as a result of the heaping action, the equilibrium low tides are created by a compensating maximum withdrawal of water from the null force regions around the Earth, which lies midway between these two tidal humps. Note that an amphidromic point, also called a tidal node, is a geographical location which has zero tidal amplitude for one harmonic constituent of the tide. The tidal range (the peak-to-peak amplitude, or height difference between high tide and low tide) for that harmonic constituent increases with distance from this point.

The Moon, in making a full revolution around the Earth once each month, passes from a position of maximum angular distance north of the equatorial plane of the Earth to a position of maximum angular distance south of the Earth's equatorial plane and vice versa. The effect of this angular distance of the Moon relative to the equatorial plane of the Earth (i.e., declination) is to produce inequalities in the heights of successive high waters (and successive low waters) at a given point because this point passes below the two tidal bulges as a result of the rotation of the Earth about its own axis.

A point P which experiences a higher high tidal level (HHT) will experience a lower high tidal level (LHT) approximately 12 h later when the Earth's rotation on its own axis has brought this point to a diametrically opposite point P′ where the height of the ellipsoidal tidal bulge of water is lesser than that at P. This causes the successive tidal ranges (i.e., the difference in levels between two successive high- and low-waters) to be different. This inequality in the range of two successive tides in a day is called the diurnal inequality of the tide.

Diurnal inequalities (see Fig. 2.15) vary markedly with seasons. The strongest diurnal inequality is possible when spring tides (maximum tidal ranges) occur during the solstices when both celestial bodies are near their maximum declination and acting together. For example, in June and December, the maximum diurnal inequalities of the year are observed because of the occurrence of the maximum solar declinations during these months. However, in March and September, solar declination becomes zero (Equinox), causing the minimum diurnal inequalities of the year. From a purely theoretical point of view, the diurnal inequality arising from solar declination must vanish on the equator. However, lunar declination can give rise to diurnal inequality even on Equinox. Similarly, the diurnal inequality in any location must be zero when both the Moon and the Sun are on the equatorial plane of the Earth. However, as a result of the varying depths and boundaries, the real ocean responses to the tidal forcing are too complicated for these ideal situations to occur in practice.

The interference of the solar tidal forces with the lunar tidal forces (which, due to the relative proximity between Earth and Moon, are about 2.2 times stronger than the solar tidal forces) causes the regular variation of the tidal range between spring tide (maximum tidal range) and neap tide (minimum tidal range). Spring tides occur when the Earth, Moon, and Sun are in line, which is at new moon and full moon. The synodic period from new moon to new moon is 29.5 days, and the time from one spring tide to the next is 14.8 days.

Neap tides occur when the positions of the Moon and the Sun relative to the Earth are at right angles to each other, which is at half-moon. During these times, the tide-generating forces caused by the Moon and the Sun counteract each other so that the resulting tide-generating force is reduced. Thus, the observed high tides are lower than average, and the low tides are higher than average. Such tides of diminished range are called neap tides.

There are two spring tides and two neap tides in a month, and one spring tide is usually larger than the other (see Fig. 2.15) because of the elliptical orbit of the Moon's motion around Earth. Within a lunar synodic period (the time required for the Moon to return to the same or approximately the same position relative to the Sun as seen by an observer on the Earth), the two sets of spring tides are usually of different amplitudes. This difference is due to the varying distance of the Moon from the Earth as a result of the elliptical path of the Moon around the Earth. At lunar perigee (the nearest approach of the Moon to the Earth), which occurs once each month, the lunar tide-raising force will be larger and therefore the spring tide during this period is the largest. Approximately 2 weeks later, when the Moon is farthest from the Earth (i.e., at apogee), the lunar tide-raising force will be smaller, and the spring tide during this period is smaller than that at lunar perigee.

One complete cycle from perigee to perigee takes 27.6 days. At perigee, the tidal forces are approximately

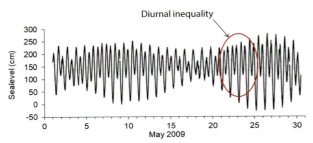

FIG. 2.15 Regular variation of tidal range (observed at Ratnagiri, west coast of India) between two spring tides (which experience maximum tidal ranges) and two neap tides (which experience minimum tidal range) in a month. The data have been acquired from the real/near-real time reporting internet-accessible sea level station network designed and established by the Instrumentation Division of CSIR-NIO in India (graph prepared by Prakash Mehra). *(Source: Joseph, 2016).*

15% larger than the mean tidal forces. At apogee, the lunar forces are approximately 15% less than the mean tidal forces. Variations in the distance between the Earth and the Moon from a minimum to a maximum in 13.8 days cause fortnightly modulations in the tidal amplitudes.

Very high spring tides occur when the Moon and Sun are overhead at the equator of the Earth, near the Mar. 21 and Sep. 23 equinoxes, when day and night are of equal length.

The combinations of astronomical forcing that define spring and neap cycles and diurnal inequalities are further modified by local bathymetry and shoreline boundary influences. All of these factors combine to produce tidal envelopes that vary from location to location. The result is site-specific tidal signatures, which can be classified as semidiurnal, diurnal, or mixed.

In the Sun-Earth system, the Earth follows an elliptical path relative to the Sun. When the Earth is closest to the Sun (perihelion), about Jan. 2 of each year, the solar contribution in the tidal pattern at a place will be enhanced. However, when the Earth is farthest from the Sun (aphelion), around Jul. 2, the solar contribution in the tidal pattern at a place will be reduced. Variation of the distance between the Earth and the Sun from a minimum to a maximum in 6 months causes a yearly modulation in the tidal amplitudes.

Among several interesting characteristics of tidal oscillations, the spring tide enjoys a special status in a biblical sense—namely, the Easter spring. The high spring tides of Easter weekend are inevitable because Easter Sunday is defined in the western Christian calendar in terms of the lunar cycle and the date of the spring equinox. Easter is fixed as the first Sunday after the full moon that happens on, or next after, the Mar. 21 spring equinox. It may fall on any one of the 35 days from Mar. 22 to Apr. 25 (see Pugh and Woodworth, 2014). Thus, Easter Sunday varies from year to year, depending on the time of spring tide after the Mar. 21 equinox.

In addition to semidiurnal and diurnal tides, there are also long-period tides. Surprisingly, periods of approximately 2 weeks (fortnightly) to approximately 19 years (Metonic cycle) exist. Tidal oscillations are periodic in a complex manner as mentioned because they are related to the complex motions of the Earth, the Moon, and the Sun.

A fourth group of tides, not of direct astronomic origin, consists of the shallow-water tides that result from the mutual interaction of the semidiurnal and diurnal tides and several other constituents of the astronomical tides having specific periods. Tidal oscillations of all such periods are superimposed on the observed sea level oscillations. This all-weather, worldwide, rhythmic rising and lowering of sea level is known as tidal oscillation or simply as tide. Fig. 2.13 shows tidal oscillations observed at Takoradi harbor in Ghana (West Africa) at the Central East Atlantic Coast.

2.7.2 Topographical influences on tidal range and tidal pattern

In the open ocean, the tidal range (i.e., difference in sea level elevation between successive high water and low water) is generally on the order of tens of centimeters up to a meter. Likewise, in some semienclosed seas (e.g., the Mediterranean Sea and the Baltic Sea), the tidal range for both semidiurnal and diurnal tides is quite small. A detailed analysis of the characteristics of tides in several partially enclosed seas has been reported by Pugh and Woodworth (2014). Despite relatively small tidal amplitudes, the tidal currents in many of these semienclosed seas are very strong (2–3 m/s). The reasons for this have been found through analyses as reported by Tsimplis et al. (1995) and Tsimplis (1997).

Although astronomically induced tidal variations are in general less than a meter in range in the open ocean, when these tidal crests and troughs traverse into shallow waters, against landmasses and into confining channels, noticeable variations occur in the tidal range. In some other water bodies, particularly in bays and adjacent seas, the tidal range could be exceedingly large because their geometrical shape (bathymetry and topography) may favor tidal amplification. (This is called geometrical amplification.) It is known that the tidal range may be locally amplified further by several factors including the proximity of the period of natural resonant oscillation of the water body in the bay and estuary and the tidal period (see Clancy, 1968). Broad, funnel-shaped estuaries (with the broader portion open to the sea) cause exceptionally large tidal ranges in their head region, which is relatively much narrower and shallower than the mouth region that faces the sea. The Gulf of Kachchh and the Gulf of Khambhat in NW India are just two of several examples where such tidal amplification occurs, as the tide propagates up the gulf because of its funnel shape, with the mouth of the funnel opening to the sea. Fig. 2.16 shows the tidal pattern at the mouth and head of the Gulf of Kachchh on the west coast of India, illustrating the amplification. Geometrical amplification, as in the examples mentioned, takes place as the same volume of seawater that enters the mouth of the gulf under continuous tidal forcing is constrained to traverse up the estuary along progressively diminishing vertical cross-sectional area of the gulf, thereby compelling the water mass at a progressively increasing distance from the gulf mouth to spring up in order to conserve mass, momentum, and energy. However, frictional forces play an important role in influencing the quantum of amplification that really occurs as the tide propagates along the gulf toward its head region. The largest known tides occur in the Bay of Fundy (see Fig. 2.17) where spring tidal ranges up to 15 m have been measured.

The reverse is true (i.e., tidal range diminishes) in a broad water body if it opens to the sea via a narrow inlet. One of several examples to illustrate such a situation (geometrical

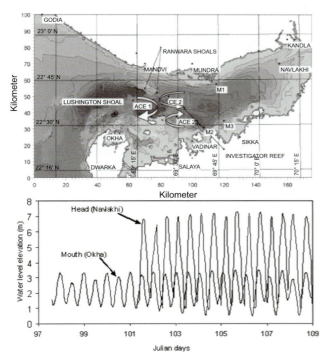

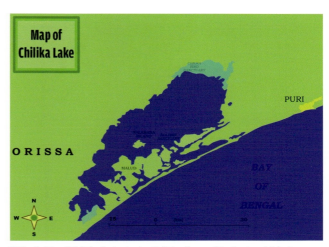

FIG. 2.16 (Top) Gulf of Kachchh on the north-eastern Arabian Sea. *(From Babu, M.T., Vethamony, P., Desa, E., 2005.Modelling tide-driven currents and residual eddies in the Gulf of Kachchh and their seasonal variability: amarine environmental planning perspective. Ecol. Model. 184, 299–312; Elsevier. © 2004 Elsevier B.V. All rights reserved). (Bottom) Measured tides at the mouth (Okha) and the head (Navlakhi) of the Gulf of Kutchch in India, illustrating the amplification of tides as it propagates up the gulf. Amplification occurs as a result of geometrical amplification and resonance (bottom figure prepared by Prakash Mehra, CSIR-NIO, Goa, based on measurements carried out by CSIR-NIO).*

FIG. 2.17 **Bay of Fundy where the largest known tides are found in the world.** *(Source: https://commons.wikimedia.org/wiki/File:Wpdms_nasa_topo_bay_of_fundy_-_en.jpg). Author: Decumanus at English Wikipedia. Licensing info: This file is licensed under the Creative Commons Attribution-Share Alike 3.0 Unported license. Subject to disclaimers."*

attenuation) is the Chilika Lake in the Odisha State of India (see Fig. 2.18). Chilika Lake is the largest brackish water lake in Asia. It is approximately 65 km long, its width varying from 20 km in the middle to 7 km at the ends. The lake is connected to the Bay of Bengal through a 25-km long narrow

FIG. 2.18 "Chilika Lake in the Odisha State of India, in which the water body opens to the sea via a narrow inlet resulting in geometrical attenuation of tidal range in the lake, which is 61 km long. The maximum width is 15 km. *Image Source: https://commons.wikimedia.org/wiki/File:Chilika_lake.png. Author: User:gppande / User_talk:gppande (I created this map taking help from demis website.) Licensing info: This file is licensed under the Creative Commons Attribution-Share Alike 4.0 International, 3.0 Unported, 2.5 Generic, 2.0 Generic and 1.0 Generic license."*

inlet channel. In this case, the narrow tail of the water body opens to the sea. As a result, the tidal range in some locations within the lake diminishes up to one-seventh of that observed in the adjacent open sea. This is an example of geometrical attenuation. There are several water bodies of complex geometrical shape (e.g., the Kochi backwaters shown in Fig. 2.19 in which the tidal range progressively decreases from their mouth to their head as a result of geometrical attenuation (see Fig. 2.20). The increasing prominence of the Msf tide (tidal constituent having period of 14.77 days) toward the head region, resulting from shallow water processes, is notable. In some gulfs and bays, tidal oscillations are triggered by the tides of the open ocean as a forced standing wave to which the *Corioli's force* (deflecting force of the Earth's rotation) adds a swinging cross-component, and both oscillations together result in a rotating tide.

If an estuary is long, narrow, and shallows rapidly upstream, the rising tide (i.e., *flood tide*) tends to be steepened when it meets the river water. This happens because as a progressive wave enters shallow water, its speed decreases. Because the trough is shallower than the crest, its retardation is greater, resulting in a steepening of the wave front. Therefore, in many tidal rivers, the duration of the rise is considerably less than the duration of the fall.

2.7.3 Tidal bore—wall of tumbling and foaming water waves in some geometrically special water bodies during a spring tide flood tide

A tidal bore is a sharp rise in free-surface water elevation propagating upstream in an estuarine system or narrow bay

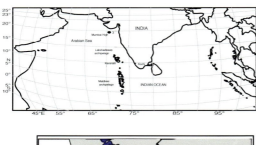
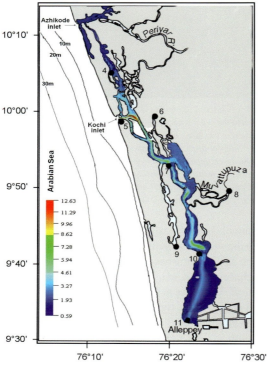
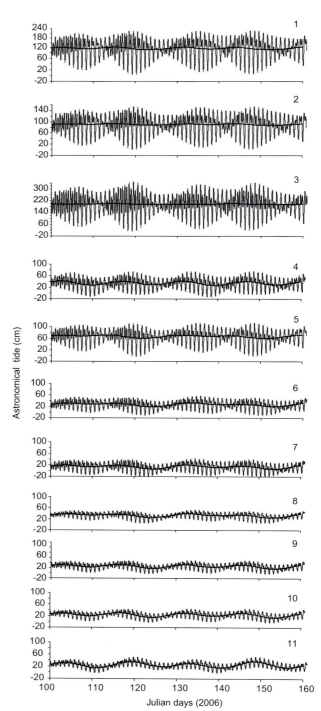
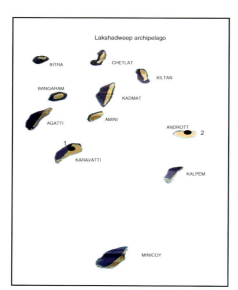

FIG. 2.19 Map of Kochin backwaters and the adjoining Lakshadweep group of islands located west of the South Indian State of Kerala, India. *From: Joseph, A., Balachandran, K.K., Mehra, P., Prabhudesai, R.G., Kumar, V., Agarvadekar, Y., Revichandran, C., Dabholkar, N., 2009. Amplified Msf tides at Kochi backwaters on the southwest coast of India. Curr. Sci. 97 (6), 776–784.*

FIG. 2.20 Progressive decrease of tidal range and increasing dominance of Msf tide in the Kochi backwaters in India, from its mouth to its head, as a result of geometrical attenuation. *From: Joseph, A., Balachandran, K.K., Mehra, P., Prabhudesai, R.G., Kumar, V., Agarvadekar, Y., Revichandran, C., Dabholkar, N., 2009. Amplified Msf tides at Kochi backwaters on the southwest coast of India. Curr. Sci. 97 (6), 776–784.*

during a flood tide (i.e., as the tidal flow turns to rising) against the direction of the river's or bay's water current during *spring tide*. A tidal bore is literally a wall of tumbling and foaming water waves roaring up an estuary with high speed and thundering noise. The origin of the word "bore" is believed to derive from the Icelandic "bara" (billow), indicating a potentially dangerous phenomenon—that is, a tidal bore with a breaking roller (see Coates, 2007).

In fluid dynamical term, a bore is a fast, nonlinear, gravity wave. Because the water behind the front of the nonlinear wave has larger depth than the water ahead, the back travels faster than the front and the wave ultimately breaks (see Stoker, 1957; Lighthill, 1978). A low barometric pressure and a positive surge on top of a spring tide are most effective for producing big bores. A tidal bore is in fact the most spectacular form of tidal distortion, seen in only a few estuaries. Most field occurrences showed well-defined undulations behind the leading wave, often described as an undular bore process (see Koch and Chanson, 2008). A tidal bore is a typical hydraulic phenomenon, frequently observed in some shallow estuaries during spring tide when the tidal ranges are appreciably large (exceeding 4–6 m) and the flood tide is confined to a narrow funnel-shaped estuary. As described above, this amplifies the tide wave in a process, discussed earlier, called geometrical amplification. In such a situation, the steepened flood tide becomes near-vertical in shape at the crest and the tide roars up the estuary as a wall of tumbling and foaming water producing a tidal bore. Simply stated, the bore is more or less the front of a tide wave where the water surface changes abruptly and surges forward, often as a group of steep waves. It may be noted that the bore height is not the same as the tidal height; in fact, the bore rides on the tide. So, the bore height is actually the surplus height above the instantaneous tidal height. For further details on tidal bore, see Joseph (2016).

2.7.4 Tidal currents—their role in mixing of ocean waters

The periodic rise and fall of tides, in concert with astronomically induced tide-generating forces and topographically induced interactions among various tidal constituents, are associated with a periodic horizontal flow of water, known as tidal current. Tidal currents are strongest in restricted areas such as funnel-shaped estuaries with wide mouths and narrow heads, shallow areas, and over the sills (concordant intrusions of rocks). A decrease of cross-sectional area of the flow channel produces faster flows. In the open sea, where the direction of flow is not restricted by any barriers, the tidal current is rotary; that is, it flows continuously, with the direction changing through 360° during the tidal period. The tendency for the rotation in direction has its origin in the deflecting force of the Earth's rotation, known as *Corioli's force*. The current speed usually varies throughout the tidal cycle, passing through two maximums in approximately opposite directions and two minimums about halfway between the maximums in time and direction. Rotary current, depicted by a series of arrows representing the speed and direction of the current, is usually known as a current rose. A line joining the extremities of the radius vectors will form a curve roughly approximating an ellipse. The cycle is completed in one-half tidal day or in a whole tidal day, according to whether the tidal current is of the semidiurnal or the diurnal type. A current of the mixed type will give a curve of two unequal loops each tidal day. Because of the elliptical pattern formed by the envelope joining the ends of the arrows, it is also referred to as a tidal current ellipse (for details, see Joseph, 2013).

Tidal currents in the open ocean are generally weak. However, tidal currents in coastal water bodies belong to the most vigorous flow phenomena in the sea. Nonlinear instabilities give rise to a large variety of "secondary currents" (Grant et al., 1962) to the topographically "frozen" quasi-two-dimensional residual eddies (Zimmerman, 1978, 1980). Topographic obstacles in the path of large currents give rise to production of vortex streets in the wakes of such obstacles, and the vortices are carried along by the current. Any existing quasi-two-dimensional turbulence in tidal currents must have a pronounced influence on the physical transport processes in tidal areas and hence on the total physical environment (Veth and Zimmerman, 1981).

In estuaries or straits or where the direction of water flow is more or less restricted to certain channels, the tidal current is *reversing*, that is, it flows alternately in approximately opposite directions with a short period of little or no current, called *slack water*, at each reversal of the current. During the flow in each direction, the speed varies from zero at the time of slack water to a maximum, called *strength*, about midway between the slacks. The water current movement from the sea toward shore or upstream is the *flood current*, and the water current movement away from shore or downstream is the *ebb current*. Tidal currents may be of the semidiurnal, diurnal, or mixed type, depending on the type of tide at the location. Offshore rotary currents that are purely semidiurnal repeat the elliptical pattern each tidal cycle of 12 h and 25 min duration. If there is considerable diurnal inequality (i.e., if there is considerable difference in the two successive tidal ranges), the arrows representing the instantaneous tidal current vector describe a set of two ellipses of different sizes during a period of 24 h and 50 min (one lunar day). The difference between the sizes of the two ellipses is dependent on the diurnal inequality. In a completely diurnal rotary current, the smaller ellipse disappears and only one ellipse is produced in duration of 24 h and 50 min.

The magnitude of the tidal current speed varies with the tidal range (i.e., the difference between successive high-tide and low-tide elevations). Thus, the stronger *spring* and *perigean* currents occur, respectively, near the times of new and full moon and near the times of the moon's perigee. The weaker *neap* and *apogean* currents occur, respectively, at the times of *neap* and *apogean* tides. An important role of the strong tidal currents is to produce intense tidal mixing, which, by pumping nutrients into the surface layers, is one of the ultimate causes of a location's biological richness.

Tidal currents in ridge valleys are particularly interesting. For instance, Garcia-Berdeal et al. (2006) have reported fascinating vertical structures of time-dependent water currents in the axial valley at Endeavour Segment of the Juan de Fuca Ridge, which lies along the divergent boundary between the Pacific and Juan de Fuca plates, approximately 400–500 km off the coasts of Washington and Oregon.

Diurnal tidal current motions are trapped to the ridge and experience an amplification of their clockwise rotary component that is attributed to the generation of anticyclonic vorticity by vortex squashing over the ridge. Whereas the across-valley water currents diminish with increasing depth, the along-valley currents are accelerated and intensified with depth. Alignment of the current flow with the along-valley direction yields a velocity vector that spirals with depth.

2.7.5 Implications of coastal tides and tidal bores

Coastal tides and tidal bores have profound navigational, societal, and ecological implications. For example, they significantly affect the flow of water into and out of estuaries. Tidal streams in restricted channels too exhibit a regular pattern of ebb and flood. The tidal current that is directed toward land or runs up an estuary (the tidal mouth of a large river, where the tide meets the stream) is known as a flood current because it floods the land. Conversely, the tidal current that is directed toward the sea or runs down an estuary is known as an ebb current (current moving away from the land). Coastal tidal oscillations are an important phenomenon accounting for a significant amount of the ocean's energy on the shelf. Their signals dominate the sea level and current spectra for frequencies of order 1 cycle per day (cpd) or greater. Conversely, their high energy implies that shelf currents can be strongly influenced by tidal oscillations, both through tidal rectification and tidal friction. Tidal currents also play an important role in mixing, material dispersion, and sediment transport.

Information on the nature of tidal oscillations at a place is important in many ways. For example, it plays a significant role in the efficient use of nautical charts in shallow water bodies. To the mariner, the importance of tidal information increases very quickly as the land approaches, reaching critical level for deep-draft ships in shallow ports, straits, and channels. Navigators approaching a coast and steaming into harbor need information on the tidally influenced changing water depth in these areas in relation to the bottom of their ships. Tidal information pertaining to a given coastal water body, coupled with water depth information available from the nautical charts for this water body, helps the navigator to safely cruise over complex navigation channels. Moreover, it is advantageous for ships to sail in the direction of the tidal current flow, in consonance with the proverbial analogy of "swimming with the tide."

Ocean currents that change under the influence of tidal oscillations exert strong direct or indirect influences on the distribution and behavior of some organisms. For example, the patchiness of phytoplankton in the vicinity of river mouths is coupled with the periodicity of the tidal cycle. Phytoplankton aggregations occur at frontal zones that form during flood tide when river- and tidal-flows are in opposition (see Dustan and Pinckney, 1989). Furthermore, migration patterns of some types of fish are considered to be related to tidal streams.

The intertidal area between highest and lowest water levels generally shows high biological productivity and contains a very rich and diverse range of species. The most common intertidal species are crabs, periwinkles, barnacles, and mussels. These intertidal species often develop in intense clusters to create microhabitats in which more moisture is retained during exposure at low tide. Valuable seaweeds are often found in select intertidal areas. The many reasons for this high level of biological productivity include the regular availability of nutrients during each tidal cycle in water that is shallow enough for photosynthesis. However, it is also a region of great environmental stress. The potential for high productivity can also be realized by those species that have highly adapted survival mechanisms. Different coastal conditions require different mechanisms.

Lunar periodicity plays an important role in the occurrence of crustaceans. For example, Chatterji et al. (1994) found lunar periodicity exhibiting a significant influence on the occurrence of edible crabs (*Portunus pelagicus*, *Charybdis cruciata*, and *Portunus sanguinolentus*). They found that whereas high density of these crabs was recorded in the trawl catches during full moon and new moon, low value was found during intermediate phase.

Some animals are more sensitive to tidal bore sounds than humans (see Warfield, 1973; Fay, 1988). When a bore closes in, its rumbling noise disorients some species. In the Baie du Mont Saint Michel, sheep have been outrun and drowned by the tidal bore. In Alaska, deer have tried unsuccessfully to outrun the bore (see Molchan-Douthit, 1998). In each case, the animals panicked with the deafening noise, although they could outrun the bore front. As described, a key feature of a tidal bore is its rumble noise that can be heard from far away.

Tidal bores do have a significant effect on the natural channels and their ecology. Bores are usually associated with a massive mixing of the estuarine waters that stirs the organic matter and creates some rich fishing grounds. Its occurrence is essential to many ecological processes and to the survival of unique ecosystems. The tidal bore is also an integral part of the cultural heritage of many regions: for instance, the Qiantang River bore in China, the Severn River bore in the UK, and the Dordogne River in France.

Chanson (2011) has addressed in some detail the environmental, ecological, and cultural impacts of tidal bores. According to his studies, tidal bores have a significant impact on ecosystems. For example, tidal bore-affected estuaries are the natural habitat, as well as the feeding zone and breeding grounds, of several forms of wildlife. The evidence connects both scientific and anecdotal observations. In Brazil, the bore sets organic matter into suspension and the estuarine zone is the feeding grounds of piranhas (a class of freshwater fishes that inhabit South American rivers, floodplains, lakes and reservoirs). In Alaska and in France, several birds of prey fish behind the tidal bore front and next to the banks: for example, bald eagles in Alaska and buzzards in France. Visual observations in Alaska and France showed a number of fish being ejected above the bore roller by the flow turbulence. For example, Chanson (2011) saw this several times in the Dordogne River. In Alaska, eagles catch fish jumping off and projected upward above the tidal bore roller. Several large predators feed immediately behind the tidal bore during its upstream progression. These include beluga whales in Alaska (at Turnagain Inlet), seals in the Baie du Mont Saint Michel, sharks in Queensland (the Styx River and Broadsound), and crocodiles in northern Australia and Malaysia (the Daly and Batang Lupar Rivers).

2.8 Appearance of microbes on Earth about 3.7 billion years ago

The earliest life forms we know of were microscopic organisms (microbes) that left signals of their presence in rocks about 3.7 billion years ago. It is well known that in the beginning of Earth's history, its surface was extremely hot, because the Earth is the product of a collision between two planets—a collision that also created Earth's Moon. Most of the heat within the very young Earth was lost quickly to space while the surface was still quite hot. As the very young Earth cooled, its surface passed monotonically (either entirely nonincreasing, or entirely nondecreasing) through every temperature regime between silicate vapor to liquid water; and perhaps, even to ice, eventually reaching equilibrium with sunlight (Sleep et al., 2001). In general, warm ocean conditions prevailed for hundreds of millions of years (Hren et al., 2009), although it is probable that Earth's surface may have experienced a number of cold spells (Ashkenazy et al., 2013), as well as several major impacts (Bada et al., 1994). During this period, microbes diversified, achieved higher degrees of functional complexity, and proceeded to colonize all surface and subsurface habitats available to them. As they spread, microorganisms developed an ever more important capacity to influence environments and affect the regulation of planetary feedback mechanisms—two factors that may have contributed greatly to life's enduring persistence on Earth (Chopra and Lineweaver, 2016).

Sleep et al. (2001) have summarized the initiation of clement surface conditions on Earth thus, "Inevitably the surface passed through a time when the temperature was around 100°C at which modern thermophile organisms live. How long this warm epoch lasted depends on how long a thick greenhouse atmosphere can be maintained by heat flow from the Earth's interior, either directly as a supplement to insolation, or indirectly through its influence on the nascent carbonate cycle. In both cases, the duration of the warm epoch would have been controlled by processes within the Earth's interior where buffering by surface conditions played little part. A potentially evolutionarily significant warm period of between 10^5 and 10^7 years seems likely, which nonetheless was brief compared to the vast expanse of geological time."

Understanding the role of biology in planetary evolution remains an outstanding challenge to geo-biologists. Progress toward unraveling this puzzle for Earth is hindered by the scarcity of well-preserved rocks from the Archean (4.0–2.5 Gyr ago) and Proterozoic (2.5–0.5 Gyr ago) Eons. In addition, the microscopic life that dominated Earth's biota for most of its history left a poor fossil record, consisting primarily of lithified microbial mats, rare microbial body fossils and membrane-derived hydrocarbon molecules that are still challenging to interpret (Hamilton et al., 2016).

2.9 Stromatolites appearing on Earth about 3.5 billion years ago

Stromatolites (Fig. 2.21)—a laminated usually mounded sedimentary fossil formed from alternating layers of cyanobacteria (microbes that live primarily in seawater/shallow water), calcium carbonate, trapped silicate sediment and fossilized algal mats—are produced over geologic time by the trapping, binding, or precipitating of sediment by groups of micro-organisms, primarily cyanobacteria (Fig. 2.22). Stromatolites are formed over the years by mats (1–10 mm in thickness) of microorganisms (cynobacteria among others) found in shallow, mainly marine waters. The microorganisms precipitate mineral particles, which makes the mat to thicken, but only the upper part survives. Most stromatolites display characteristically layered structures. Only the layers are visible to the naked eye. Stromatolites are commonly defined as "organo-sedimentary structures built by microbes". Apart from cyanobacteria, stromatolites are associated with a diverse range of microbial communities and environments (e.g., Reid et al., 2000; Cady et al., 2003; Allwood et al., 2007). Massive formations of stromatolites showed up along shorelines all over the world about 3.5 billion years ago. They were the earliest visible

FIG. 2.21 (Top) Fossilized stromatolite in Strelley Pool chert, about 3.4 billion years old (Duda et al., 2016), from Pilbara Craton, Western Australia. *(Source: https://commons.wikimedia.org/wiki/File:Stromatolithe_Pal%C3%A9oarch%C3%A9en_-_MNHT.PAL.2009.10.1.jpg) Copyright info: This file is licensed under the Creative Commons Attribution-Share Alike 4.0 International license. (bottom) Stromatolites growing in Hamelin Pool Marine Nature Reserve, Shark Bay in Western Australia. (Source: https://commons.wikimedia.org/wiki/File:Stromatolites_in_Sharkbay.jpg) Copyright info: This file is licensed under the Creative Commons Attribution-Share Alike 3.0 Unported license.*

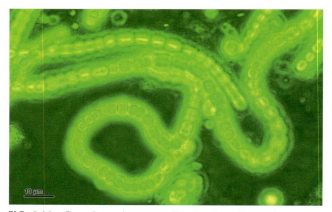

FIG. 2.22 **Cyanobacteria: responsible for the buildup of oxygen in the Earth's atmosphere.** *(Source: By Doc. RNDr. Josef Reischig, CSc. - Author's archive, CC BY-SA 3.0, https://commons.wikimedia.org/w/index.php?curid=31550579) Image author: Doc. RNDr. Josef Reischig, CSc. Copyright information: This file is licensed under the Creative Commons Attribution-Share Alike 3.0 Unported license.*

manifestation of life on Earth and dominated the scene for more than two billion years. Stromatolites are considered to be biological mileposts that encode the role played by ancient micro-organisms in the evolution of life on Earth.

The discovery of stromatolites and microfossils in 3.5-Ga-old sedimentary rock formations is evidence for the existence of phototrophic prokaryotes at that time. Carbon isotopic values for sedimentary organic carbon strongly suggest autotrophic CO_2 fixation, and the existence of large deposits of sedimentary sulfate is consistent with a photosynthesis dependent on reduced sulfur compounds for reducing power (Olson and Pierson, 1986).

Cyanobacteria are an ancient but advanced form of life. Cyanobacteria are microbes that live primarily in seawater. Oceanic cyanobacteria (also called *blue-green algae*) are believed to have been the first organisms on Earth to perform oxygenic photosynthesis. In the process of oxygenic photosynthesis, in the presence of sunlight, the carbon dioxide (CO_2) gas molecules inhaled by the cyanobacteria are broken up into its constituent molecules, namely; carbon (C) molecules and oxygen (O_2) molecules. It may be noted that organic carbon molecules are the building blocks of life. The carbon (C) molecules and the O_2 gas molecules, unlocked from the CO_2 gas molecules in the process of photosynthesis, take different pathways. While the carbon molecules are used for body building (i.e., growth) of the various organisms, some of the oxygen gas molecules released during photosynthesis get dissolved in seawater and some of the rest of it escape into the atmosphere. Thus, oxygenic photosynthesis stimulated growth of marine organisms (e.g., cyanobacteria). When they died, they drifted down to the seafloor and gradually stored that same carbon in deep marine sediments.

On the basis of photobiological, evolutionary, paleontological, paleoenvironmental and physiological arguments, García-Pichel (1998) proposed a time course for the role of solar ultraviolet radiation (UVR, wavelengths below 400 nm) in the ecology and evolution of cyanobacteria in which three main periods can be distinguished. According to García-Pichel (1998), an initial stage, before the advent of oxygenic photosynthesis, when high environmental fluxes of UVC (wavelengths below 280 nm) and UVB (280–320 nm) may have depressed the ability of proto-cyanobacteria to develop large populations or restricted them to UVR refuges. A second stage lasting between 500 and 1500 Ma (million years), started with the appearance of true oxygen-evolving cyanobacteria and the concomitant formation of oxygenated (micro) environments under an oxygen-free-atmosphere. García-Pichel (1998) suggested that in this second stage, the age of UV, the overall importance of UVR must have increased substantially, since the incident fluxes of UVC and UVB remained virtually unchanged, but additionally the UVA portion of the spectrum (320–400 nm) suddenly became biologically injurious and extremely reactive oxygen species must have formed wherever oxygen and UVR spatially coincided.

In García-Pichel (1998)'s proposal, the last period began with the gradual oxygenation of the atmosphere and the formation of the stratospheric ozone shield. The physiological stress due to UVC all but disappeared and the effects of UVB were reduced to a large extent. García-Pichel (1998) drew evidence in support of this dynamics from the phylogenetic distribution of biochemical UV-defense mechanisms among cyanobacteria and other microorganisms. García-Pichel (1998) used the specific physical characteristics of UVR and oxygen exposure in planktonic, sedimentary and terrestrial habitats to explore the plausible impact of UVR in each of the periods on the ecological distribution of cyanobacteria.

Harvesting light to produce energy and oxygen (photosynthesis) is the signature of all land plants. This ability was co-opted from cyanobacteria. Today these bacteria, as well as microscopic algae, supply oxygen to the atmosphere and churn out fixed nitrogen in Earth's vast oceans. Microorganisms may also have played a major role in atmosphere evolution before the rise of oxygen. Under the dimmer light of a young Sun cooler than today's, certain groups of anaerobic bacteria may have been pumping out large amounts of methane, thereby keeping the early climate warm and inviting. It is, therefore, not surprising that the evolution of Earth's atmosphere is linked tightly to the evolution of its biota (see Kasting and Siefert, 2002).

The emergence of oxygenic photosynthesis in ancient cyanobacteria (the only microorganisms capable of oxygenic photosynthesis) represents one of the most impressive microbial innovations in Earth's history, and oxygenic photosynthesis is the largest source of O_2 in the atmosphere today. Thus, the study of microbial metabolisms and evolution provides an important link between extant (i.e., still in existence; surviving) biota and the clues from the geologic record (see Hamilton et al., 2016).

Hamilton et al. (2016) studied the physiology of cyanobacteria, their co-occurrence with anoxygenic phototrophs in a variety of environments and their persistence in low-oxygen environments, including in water columns as well as mats, throughout much of Earth's history. They also examined insights gained from both the rock record and cyanobacteria presently living in early Earth analogue ecosystems and synthesized current knowledge of these ancient microbial mediators in planetary redox evolution. Their analysis supports the hypothesis that anoxygenic photosynthesis, including the activity of metabolically versatile cyanobacteria, played an important role in delaying the oxygenation of Earth's surface ocean during the Proterozoic Eon (the period ranging from 2.5 billion years old to 541 million years old. During this time, most of the central parts of the continents had formed and the plate tectonic process had started.).

Although the oxygen releasing oceanic cyanobacteria evolved into multicellular forms more than 2.7 billion years ago, there was a long lag time—hundreds of millions of years—before Earth's atmosphere first gained significant amounts of oxygen, some 2.4 billion to 2.3 billion years ago. Before the Great Oxygenation Event (GOE) in the history of Earth (Flannery and Walter, 2012), any free oxygen produced by the cyanobacteria was chemically captured by dissolved iron or organic matter. The GOE was the point in time when these oxygen sinks became saturated, because of which the excess oxygen became fully available to escape into the atmosphere.

As indicated earlier, stromatolites are still being formed in lagoons in Australasia (see Fig. 2.21). Decomposition of stromatolites removes oxygen from seawater, and in turn, from the atmosphere. It may be noted that while features of some stromatolites are suggestive of biological activity, others possess features that are more consistent with abiotic (nonbiological) precipitation (Grotzinger and Rothman, 1996). According to Lepot et al. (2008), finding reliable ways to distinguish between biologically formed and abiotic stromatolites is an active area of research in geology.

The oldest known fossils, in fact, are cyanobacteria from Archean rocks of Western Australia. It is surprising that stromatolite-building communities date back some 3.5 billion years when Earth's environments were too hostile to support life as we know it today. The great importance of stromatolites stems from the fact that they encode the environmental conditions that supported and led to the preservation of this evidence for early life; and reveal the role that ancient micro-organisms played in the evolution of life on Earth (see Allwood et al., 2009).

Paleontologists (scientists who study fossils as a way of getting information about the history of life on Earth and the structure of rocks) consider that studies of stromatolites will be helpful in recognizing Earth's oldest environments (see Noffke, 2009). At this time, when planetary scientists are attaching much emphasis to the search for signs of microbial life on Mars and several other extraterrestrial celestial bodies, study of stromatolites (if any) from such celestial bodies is likely to attract much attention in the years to come.

2.10 Initiation of plate tectonics on Earth between 3.5 and 3.3 billion years ago

Plate tectonics is a scientific theory describing the large-scale motion of the plates making up the Earth's lithosphere since tectonic (building) processes began on Earth between 3.5 and 3.3 billion years ago. It has been hypothesized that after an initial hot period, Earth's surface temperatures in the late Hadean (period in the range 4.6–4 billion years ago) may have been mild beneath an atmosphere endowed with greenhouse gases over an ocean-dominated planetary surface. The first crust was mafic, which consisted of dark-colored, mainly ferromagnesian minerals such as pyroxene and olivine and it internally melted repeatedly to produce a different kind of rocks known as felsic rocks, which are rich in silicon, sodium,

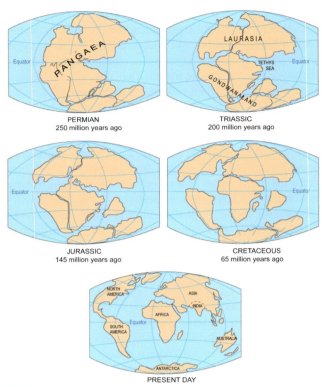

FIG. 2.23 Diagrams showing the breakup of the supercontinent Pangaea, which figured prominently in the theory of continental drift, the forerunner to the theory of plate tectonics, and the evolution of Earth's oceans. *(Courtesy of US Geological Survey, http://pubs.usgs.gov/publications/text/historical.html).*

potassium, calcium, aluminum, and lesser amounts of iron and magnesium. This crust was destabilized during the Late Heavy Bombardment (LHB) or lunar cataclysm period. It may be noted that LHB is a hypothesized event thought to have occurred approximately 4.1–3.8 billion years (Ga) ago. According to the hypothesis, during this interval, a disproportionately large number of asteroids collided with the early terrestrial planets in the inner Solar System, including Mercury, Venus, Earth and Mars (Claeys and Morbidelli, 2011). These came from both postaccretion and planetary instability-driven populations of impactors (Bottke and Norman, 2017). Plate tectonics probably started soon after the heavy bombardment event and had produced voluminous continental crust by the mid Archean (geologic eon, 4–2.5 billion years ago), but ocean volumes were sufficient to submerge much of this crust.

The initiation of plate tectonics on Earth is a critical event in our planet's history. The process of plate tectonics has continuously reshaped the landscape on Earth, pushing mountain ranges up between colliding continental plates, and opening ocean basins as landmasses slowly pull apart (Fig. 2.23). Sleep (2007) considers that plate-like processes became evident as soon as there was a geological record that might have preserved them. According to him, "Transform faults, oceanic crust, rifting, continents with thick chemically buoyant lithosphere, and subduction were all present in the Archean. As on the modern Earth, vertical tectonic processes dominated in many Archean regions. The high temperature of the interior allowed for broad regions of hot mobile continental crust." According to Sleep (2007), plate processes are complex and involve poorly understood mechanical properties of partly molten rock and faults. Poor geological preservation cloaks tectonic processes on the early Earth.

While the first proto-subduction occurred on Earth about 4 billion years ago, global tectonics on Earth occurred about 3 billion years ago. This time lag of approximately 1 billion year between these two events suggests that plates and plate boundaries on Earth became widespread over a period of approximately 1 billion years. Albarède (2009) suggested that introduction of water into the terrestrial mantle triggered plate tectonics, which may have been crucial for the emergence of life. The oceans provided ample water to Earth and are thought to have played a prominent role in the origin of life and its sustenance on Earth.

According to Sleep (2007), about a billion years in the future, the Earth's interior will become too cold for plate tectonics. Then the Earth will become a one-plate planet like Mars. Arndt and Nisbet (2012) have suggested that "in the Hadean and early Archean, hydrothermal systems around abundant komatiitic volcanism may have provided suitable sites to host the earliest living communities and for the evolution of key enzymes. Evidence from the Isua Belt, Greenland, suggests life was present by 3.8 Gya, and by the mid-Archean, the geological record both in the Pilbara in Western Australia and the Barberton Greenstone Belt in South Africa shows that microbial life was abundant, probably using anoxygenic photosynthesis. By the late Archean, oxygenic photosynthesis had evolved, transforming the atmosphere and permitting the evolution of eukaryotes." Note that komatiite is a type of ultramafic mantle-derived volcanic rock defined as having crystallized from a lava of at least 18 wt% MgO. Lower temperature mantle melts such as basalt and picrite have essentially replaced komatiites as an eruptive lava on the Earth's surface.

Based on a study, Bercovici and Ricard (2014) suggested that when sufficient lithospheric damage (which promotes shear localization and long-lived weak zones) combines with transient mantle flow and migrating proto-subduction, it leads to the accumulation of weak plate boundaries and eventually to fully formed tectonic plates driven by subduction alone. In Bercovici and Ricard (2014)'s studies, they simulated this process using a grain evolution and damage mechanism with a composite rheology, which is compatible with field and laboratory observations of polycrystalline rocks, coupled to an idealized model of pressure-driven lithospheric flow in which a low-pressure zone is equivalent to the suction of convective down-wellings. These researchers have suggested that "In the simplest case, for Earth-like conditions, a few successive rotations of the driving pressure field yield relic damaged weak zones that are inherited by the lithospheric flow to form a nearly perfect plate, with passive spreading and strike-slip margins that persist and localize further, even though flow is driven only by subduction. But for hotter surface conditions, such as those on Venus, accumulation and inheritance of damage is negligible; hence only subduction zones survive and plate tectonics does not spread, which corresponds to observations. After plates have developed, continued changes in driving forces, combined with inherited damage and weak zones, promote increased tectonic complexity, such as oblique subduction, strike-slip boundaries that are subparallel to plate motion, and spalling of minor plates."

2.11 The great oxidation event ~2.4–2.0 billion years ago—an event that led to the banded iron formations and the rise of oxygen in Earth's atmosphere

The evolution and extinction of life are tied intimately to the oxygen state of the atmosphere and the ocean on a global scale. The composition of the Earth's earliest atmosphere is not known with certainty. However, the bulk was likely dinitrogen (N_2), and carbon dioxide (CO_2), which are also the predominant carbon- and nitrogen-bearing gases produced by volcanism today. These are relatively inert gases. The Sun shone at about 70% of its current brightness 4 billion years ago, but there is strong evidence that liquid water existed on Earth at the time. A warm Earth, in spite of a faint Sun, is known as the faint young Sun paradox (Kasting, 1993). Either carbon dioxide levels were much higher at the time, providing enough of a greenhouse effect to warm the Earth, or other greenhouse gases were present. The most likely such gas is methane (CH_4), which is a powerful greenhouse gas and was produced by early forms of life known as methanogens. Scientists continue to research how the Earth was warmed before life arose (Shaw, 2008). An atmosphere of N_2 and CO_2 with trace amounts of H_2O, CH_4, carbon monoxide (CO), and hydrogen (H_2), is described as a weakly reducing atmosphere. Such an atmosphere contains practically no oxygen.

Due to the intimate association between life, ocean, and atmosphere chemistry on Earth, biogeochemical signatures preserved within the sedimentary record enable study of the early atmosphere. Paleosols (fossil soils), detrital grains (particles of rock derived from pre-existing rock through processes of weathering and erosion), and red beds (red-colored sandstones that are coated with hematite) are evidence of low-level oxygen (Catling, 2017). The 2.45-Byr-old weathering profile developed on early Proterozoic mafic volcanics located near Cooper Lake, Ontario, Canada, was examined by Utsunomiya et al. (2003) geochemically and mineralogically for a better understanding of the atmospheric oxygen evolution. They found that paleosols (fossil soils) older than 2.4 billion years old have low iron concentrations that suggest anoxic weathering (Utsunomiya et al., 2003). Hofmann et al. (2009) carried out multiple S and Fe isotope analyses of rounded pyrite grains from 3.1 to 2.6 Ga (billion years ago) conglomerates (lithified sedimentary rock consisting of rounded fragments greater than 2 mm in diameter) of southern Africa. Their analysis indicates the detrital origin of those pyrite grains, which supports anoxic surface conditions in the Archean. It was found that detrital grains found in sediments older than 2.4 billion years old contain minerals that are stable only under low oxygen conditions (Hofmann et al., 2009). Red beds indicate that there was enough oxygen to oxidize iron to its ferric state (Eriksson and Cheney, 1992).

The basic premise adopted in the study of formation and evolution of oxygen on Earth is that the Earth was without a primordial atmosphere and that its secondary atmosphere has arisen primarily from local heating and volcanic action associated with continent building. Because no oxygen can be derived in this way, the initial formation of oxygen from photochemical dissociation of water vapor is found to provide the primitive oxygen in the atmosphere. Because of the Urey self-regulation of this process by shielding H_2O vapor with O_2, O_3, and CO_2, primitive oxygen levels cannot exceed O_2 ~0.001 present atmospheric level (P.A.L.) (Berkner and Marshall, 1965). The rise of oxygen from the primitive levels can only be associated with photosynthetic activity, which in turn depends upon the range of ecologic conditions at any period. When oxygen passes ~0.01 P.A.L., the ocean surfaces are sufficiently shadowed to permit widespread extension of life to the entire hydrosphere. Likewise, a variety of other biological opportunities arising from the metabolic potentials of respiration are opened to major evolutionary modification when oxygenic concentration rises to this level. It is generally agreed that the atmosphere contained little or no free oxygen initially and that oxygen concentrations increased markedly near 2.0 billion years ago, but the precise timing of and reasons for its rise remain unexplained (Kasting, 1993).

The initial increase of O_2 in the atmosphere, its delayed build-up in the ocean, its increase to near-modern levels in the sea and air much later, and its cause-and-effect relationship with life are among the most compelling stories in Earth's history. The Great Oxidation Event (GOE), also called the Great Oxygenation Event, was a time period when the Earth's atmosphere and the shallow ocean first experienced a rise in oxygen, approximately 2.4–2.0 Ga (billion years ago) during the Paleoproterozoic era, an era spanning the time period from 2500 to 1600 million years ago (Lyons et al., 2014). It was caused by cyanobacteria doing photosynthesis. The modern atmosphere contains abundant oxygen, making it an oxidizing atmosphere (Wiechert, 2002). The rise in oxygen is attributed to photosynthesis by cyanobacteria, which are thought to have evolved as early as 3.5 billion years ago (Baumgartner et al., 2019).

While the Great Oxidation Event is generally thought to be a result of oxygenic photosynthesis by ancestral cyanobacteria, the presence of cyanobacteria in the Archean before the Great Oxidation Event is a highly controversial topic (Catling and Zahnle, 2020). Structures that are claimed to be fossils of cyanobacteria exist in rock as old as 3.5 billion years old (Schopf, 2006). These include microfossils of supposedly cyanobacterial cells and macrofossils called stromatolites, which are interpreted as colonies of microbes, including cyanobacteria, with characteristic layered structures.

Cyanobacteria are among the most diverse prokaryotic phyla, with morphotypes (any of a group of different types of individuals of the same species in a population) ranging from unicellular to multicellular filamentous forms, including those able to terminally (i.e., irreversibly) differentiate in form and function. These diverse growth strategies have enabled cyanobacteria to inhabit almost every terrestrial and aquatic habitat on Earth. It has been suggested that cyanobacteria raised oxygen levels in the atmosphere around 2.45–2.32 billion years ago during the Great Oxidation Event (GOE), hence dramatically changing life on the planet. The period around 2.45–2.32 billion years ago is attributed to the existence of cyanobacteria, based on the assumption that cyanobacteria were responsible for the accumulation of atmospheric oxygen levels, referred to as GOE (Blankenship, 2002; Bekker et al., 2004; Kopp et al., 2005; Allen and Martin, 2007; Frei et al., 2009). Schirrmeister et al. (2013) estimated divergence times of extant cyanobacterial lineages under Bayesian relaxed clocks for a dataset of 16S rRNA sequences representing the entire known diversity of this phylum. Note that "relaxed clock" models have been developed to allow the molecular rate to vary among species. The first methods were developed under the penalized-likelihood and maximum-likelihood frameworks (Sanderson, 1997; Yang and Yoder, 2003). In Bayesian clock dating, such models are integrated into the analysis as the prior on rates. Note that 16S and Internal Transcribed Spacer (ITS) ribosomal RNA (rRNA) sequencing are common amplicon sequencing methods used to identify and compare bacteria or fungi present within a given sample. Amplicon sequencing is a highly targeted approach that enables researchers to analyze genetic variation in specific genomic regions. The ultra-deep sequencing of PCR products (amplicons) allows efficient variant identification and characterization. This method uses oligonucleotide probes designed to target and capture regions of interest, followed by next-generation sequencing (NGS). In molecular biology, an amplicon is a piece of DNA or RNA that is the source and/or product of amplification or replication events. It can be formed artificially, using various methods including polymerase chain reactions or ligase chain reactions, or naturally through gene duplication. Schirrmeister et al. (2013) tested whether the evolution of multicellularity overlaps with the GOE, and whether multicellularity is associated with significant shifts in diversification rates in cyanobacteria. Schirrmeister et al. (2013)'s results indicate an origin of cyanobacteria before the rise of atmospheric oxygen. The evolution of multicellular forms coincides with the onset of the GOE and an increase in diversification rates. These results suggest that multicellularity could have played a key role in triggering cyanobacterial evolution around the GOE.

Evidence for the Great Oxidation Event is provided by a variety of petrological and geochemical markers. Geological, isotopic, and chemical evidence suggests that biologically produced molecular oxygen (dioxygen, O_2) started to accumulate in Earth's atmosphere and changed it from a weakly reducing atmosphere practically free of oxygen into an oxidizing atmosphere containing abundant oxygen. The event is inferred to have been caused by cyanobacteria producing the oxygen, which stored enough chemical energy (Schmidt-Rohr, 2020) to enable the subsequent development of multicellular life forms (Schirrmeister et al., 2013).

Banded iron formations are composed of thin alternating layers of chert (a fine-grained form of silica) and iron oxides, magnetite and hematite. Extensive deposits of this rock type are found around the world, almost all of which are more than 1.85 billion years old and most of which were deposited around 2.5 billion years ago. The iron in banded iron formation is partially oxidized, with roughly equal amounts of ferrous and ferric iron (Trendall and Blockley, 2004). Deposition of banded iron formation requires both an anoxic deep ocean capable of transporting iron in soluble ferrous form, and an oxidized shallow ocean where the ferrous iron is oxidized to insoluble ferric iron and precipitates onto the ocean floor (Cox et al., 2013). The reconstruction of oceanic paleo-redox conditions on Earth is essential for investigating links between biospheric oxygenation and major periods of biological innovation and extinction, and for unraveling feedback mechanisms associated with paleoenvironmental change. The deposition of banded iron formation before 1.8 billion years ago suggests the ocean was in a persistent ferruginous state (containing iron), but deposition was episodic (consisting of a series of separate events) and there may have been significant intervals of euxinia (Canfield and Poulton, 2011). Note that euxinia or euxinic conditions occur when water is both anoxic and sulfidic. This means that there is no oxygen (O_2) and a raised level of free hydrogen sulfide (H_2S). Deposition of banded iron formation ceased when conditions of local euxinia on continental platforms and shelves began precipitating iron out of upwelling ferruginous (containing iron oxides) water as pyrite (a shiny yellow mineral consisting of iron disulfide and typically occurring as intersecting cubic crystals) (Lyons et al., 2009; Canfield and Poulton, 2011).

2.12 An era when the entire Earth became fully covered with thick ice ~750–635 million years ago—"Snowball Earth" hypothesis

The "Snowball Earth" hypothesis proposes that during one or more of Earth's icehouse climates, Earth's surface became entirely or nearly entirely frozen, sometime earlier than 650 Mya (million years ago) during the Cryogenian period. It is also believed that the Snowball Earth existed during the Neoproterozoic era (about 750–635 million years ago). Over a period of time, due to some as yet unclear reasons, the geological, geophysical, and meteorological features of Earth underwent a multitude of changes. As a result, the Earth went through a difficult time when the entire Earth became fully covered with thick ice. Thus, the snowball phase of the Earth continued for a long time. It is believed that freezing of Earth is a consequence of a lowering of greenhouse gases (e.g., carbon dioxide (CO_2), methane (CH_4)).

However, with passage of time, the "Snowball Earth" gave way to a more comfortable Earth. Emergence of greenhouse gases, which we blame now for the current undesirable global warming scenario, is responsible for this desirable change from the "Snowball Earth" to a more comfortable normal Earth. In the present era, rising global atmospheric temperatures is blamed for the often-noticed erratic weather patterns (e.g., heavy rains and floods, thunderstorms, heat waves). A worrying trend is that "periodicity and severity" had increased. According to the Intergovernmental Panel on Climate Change (IPCC) fifth assessment report on climate change, it is likely that the frequency of heavy precipitation will increase in the 21st century over many regions.

It has been hypothesized that without sufficient greenhouse gases in the atmosphere, the early Earth would have become a permanently frozen planet because the "young Sun" of that era was less luminous than it is today. However, Bada et al. (1994) showed that bolide (a large meteor which explodes in the atmosphere) impacts between about 4.0 and 3.6 billion years ago could have episodically melted an ice-covered early ocean. The era in which the change took place from the "Snowball Earth" to a rather normal Earth is debatable because if bolide impacts as argued by Bada et al. (1994) episodically melted an ice-covered early ocean, a "Snowball Earth" would not have existed during the Neoproterozoic era.

2.13 Multiple mass extinction events on Earth—important for understanding life

Mass extinctions are episodes of accelerated extinction of variable magnitude that affect widespread taxa and cause at least temporary declines in their diversity. Although such episodes are often difficult to identify and characterize precisely in the fossil record, it is clear that they have been frequent throughout the history of complex life. More than 99% of all organisms that have ever lived on Earth are extinct (Greshko, 2019). As new species evolve to fit the ever-changing ecological niches, older species fade away. But the rate of extinction is far from constant.

Mass extinctions are defined as "any substantial increase in the amount of extinction (lineage termination) suffered by more than one geographically wide-spread higher taxon during a relatively short interval of geologic time, resulting in an at least temporary decline in their standing diversity" (Sepkoski, 1986, p. 278). In the last 500 million years, life has had to recover from five major catastrophic blows. At least a handful of times in the last 500 million years, 75 to more than 90% of all species on Earth have disappeared in a geological blink of an eye in catastrophes we call mass extinctions. Though mass extinctions are deadly events, they open up the planet for new forms of life to emerge (Greshko, 2019). As a prelude to the discussion of Earth and the early life on this planet, it would be interesting to skim through some important periods in Earth's geological past, and recollect what really happened during those periods.

2.13.1 Ordovician–Silurian extinction: ~440 million years ago

The end Ordovician (Hirnantian) extinction was the first of the five big Phanerozoic extinction events, and the first that involved metazoan-based communities. The first mass extinction on Earth occurred in a period when organisms such as corals and shelled brachiopods filled the world's shallow waters but hadn't yet ventured onto land. Life itself was beginning to spread and diversify, having first emerged around 3.7 billion years ago. But about 440 million years ago, a climatic shift caused sea temperatures to change, and the majority of life in the ocean died.

The terminal Ordovician was marked by the first of five great mass extinction events, when up to 86% of the marine species became extinct. The Ordovician period, from 485 to 444 million years ago, was a time of dramatic changes for life on Earth. Over a 30-million-year stretch, species diversity blossomed, but as the period ended, the first known mass extinction struck.

According to Sheehan (2001), a brief glacial interval produced two pulses of extinction. The massive glaciation locked up huge amounts of water in an ice cap that covered parts of a large south polar landmass. The icy onslaught may have been triggered by the rise of North America's Appalachian Mountains. The large-scale weathering of these freshly uplifted rocks sucked carbon dioxide out of the atmosphere and drastically cooled the planet. As a result, sea levels plummeted by hundreds of feet.

The first pulse of extinction was at the beginning of the glaciation, when sea-level decline drained epicontinental (those areas of sea overlying the continental shelf) seaways, produced a harsh climate in low and mid-latitudes, and initiated active, deep-oceanic currents that aerated the deep oceans and brought nutrients and possibly toxic material up from oceanic depths.

Creatures living in shallow waters would have seen their habitats cool and shrink dramatically, dealing a major blow. Whatever life remained recovered haltingly (stopping often) in chemically hostile waters. Following that initial pulse of extinction, surviving faunas adapted to the new ecologic setting.

According to Sheehan (2001), the glaciation ended suddenly, and sea level rose. Once sea levels started to rise again, marine oxygen levels fell, which in turn caused ocean waters to more readily hold onto dissolved toxic metals, and another pulse of extinction occurred. The Ordovician–Silurian Extinction event killed an estimated 85% of all species. The event took its hardest toll on marine organisms such as corals, shelled brachiopods, eel-like creatures called conodonts, and the trilobites (fossil marine arthropods that occurred abundantly during the Paleozoic era, with a carapace over the forepart, and a segmented hind part divided longitudinally into three lobes, hence the name trilobite).

The second extinction marked the end of a long interval of ecologic stasis (an Ecologic-Evolutionary Unit). Recovery from the event took several million years, but the resulting fauna had ecologic patterns similar to the fauna that had become extinct. Other extinction events that eliminated similar or even smaller percentages of species had greater long-term ecologic effects (Sheehan, 2001).

The Ordovician saw major diversification in marine life abruptly terminated by the Late Ordovician mass extinction (LOME). At the end of the Ordovician period, a rapid onset of mass glaciation covered the southern supercontinent, Gondwana (Fig. 2.24). According to Barash (2014), the continental glaciation on Gondwana determined by its position in the South Pole area; the cooling; the hydrodynamic changes through the entire water column in the World Ocean; and the corresponding sea level fall, which was responsible for the reduction of shelf areas and shallow-water basins, that is, the main ecological niche of the Ordovician marine biota, were main prerequisites of the stress conditions. Barash (2014) suggest that similar to other mass extinction events, these processes were accompanied by volcanism, impact events, a corresponding reduction of the photosynthesis and bioproductivity, the destruction of food chains, and anoxia (an absence of oxygen). The appearance and development of terrestrial plants and micro-phytoplankton, which consumed atmospheric carbon dioxide, and thus, diminishing the greenhouse effect and promoting the transition of the climatic system to the glacial mode, played a unique role in that period (Barash, 2014). Glaciation on this scale locked away high percentages of the world's liquid water and dramatically lowered global sea levels, which stripped away vital habitats from many species, destroying food chains and decreasing reproductive success (Barash, 2014).

According to Harper et al. (2014), the Ordovician extinctions comprised two discrete pulses, both linked in different ways to an intense but short-lived glaciation at the South Pole. The first, occurring at, or just below, the Normalograptus extraordinarius graptolite Biozone (an interval comprising the initial pulse of the end-Ordovician extinction; see Kröger, 2007), mainly affected nektonic and planktonic species together with those living on the shallow shelf and in deeper water whereas the second, within the N. persculptus graptolite Biozone, was less focused, eradicating faunas across a range of water depths. In all, about 85% of marine species were removed. Proposed kill mechanisms for the first phase have included glacially induced cooling, falling sea level and chemical recycling in the oceans, but a general consensus is lacking. The second phase is more clearly linked to near-global anoxia associated with a marked transgression during the Late Hirnantian. Note that the Hirnantian is the final internationally recognized stage of the Ordovician Period of the Paleozoic Era. The

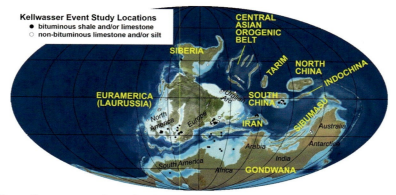

FIG. 2.24 Map indicating the southern supercontinent, Gondwana. At the end of the Ordovician period, a rapid onset of mass glaciation covered this supercontinent. *(Source: Carmichael et al., 2019), Paleogeography and paleoenvironments of the Late Devonian Kellwasser event: A review of its sedimentological and geochemical expression, Global Planet. Change, 183: 102984 (https://ars.els-cdn.com/content/image/1-s2.0-S0921818118306258-ga1_lrg.jpg).*

Hirnantian was of short duration, lasting about 1.4 million years, from 445.2 to 443.8 Ma (million years ago). Most recently, however, new drivers for the extinctions have been proposed, including widespread euxinia together with habitat destruction caused by plate tectonic movements, suggesting that the end Ordovician mass extinctions were a product of the coincidence of a number of contributing factors. Moreover, when the deteriorating climate intensified, causing widespread glaciation, a tipping point was reached resulting in catastrophe (Harper et al., 2014).

According to Bond and Grasby (2020), in the Ordovician–Silurian mass extinction event, around 85% of species were eliminated in two pulses 1 m.y. apart. The first pulse, in the basal Hirnantian, has been linked to cooling and Gondwanan glaciation. The second pulse, later in the Hirnantian, is attributed to warming and anoxia. Previously reported mercury (Hg) spikes in Nevada (USA), South China, and Poland implicate an unknown large igneous province (LIP) in the crisis, but the timing of Hg loading has led to different interpretations of the LIP-extinction scenario in which volcanism causes cooling, warming, or both (Bond and Grasby, 2020).

It is not known exactly what triggered these events. One theory is that the cooling process may have been caused by the formation of the North American Appalachian Mountains. Large-scale erosion of these mountainous silicate rocks is associated with the removal of the greenhouse gas carbon dioxide from the atmosphere. Not all scientists agree with this, however. Alternative theories suggest that toxic metal may have dissolved into ocean waters during a period of oxygen depletion, wiping out marine life, according to National Geographic. Other scientists suggest that a gamma-ray burst from a supernova ripped an enormous hole in the ozone layer, allowing deadly ultraviolet radiation to kill life below, according to APS News, and another theory suggests that volcanism was the cause, according to a study.

Bond and Grasby (2020) reported close correspondence between Hg, Mo, and U anomalies, declines in enrichment factors of productivity proxies, and the two LOME (Late Ordovician Mass Extinction) pulses at the Ordovician–Silurian boundary stratotype (Dob's Linn, Scotland). These support an extinction scenario in which volcanogenic greenhouse gases caused warming around the Katian–Hirnantian boundary that led to expansion of a pre-existing deep-water oxygen minimum zone, productivity collapse, and the first LOME pulse. Note that the Katian is the second stage of the Upper Ordovician. It is preceded by the Sandbian and succeeded by the Hirnantian Stage. The Katian began 453 million years ago and lasted for about 7.8 million years until the beginning of the Hirnantian 445.2 million years ago. During the Katian climate cooled which started the Late Ordovician glaciation. Renewed volcanism in the Hirnantian stimulated further warming and anoxia and the second LOME pulse.

According to Bond and Grasby (2020), rather than being the odd-one-out of the "Big Five" extinctions with origins in cooling, the LOME is similar to the others in being caused by volcanism, warming, and anoxia.

2.13.2 Late Devonian extinction: ~365 million years ago

The origin of the Late Devonian biotic crisis is a subject of continuing debate. The Late Devonian extinction consisted of several extinction events in the Late Devonian Epoch, which collectively represent one of five largest mass extinction events in the history of life on Earth. Overall, 19% of all families and 50% of all genera became extinct. Although various causes have been proposed, including bolide impacts, oceanic overturn (a large system of ocean currents, like a conveyor belt, driven by differences in temperature and salt content—the water's density—heralding and driving climate shifts), sea-level changes, and global climate change (Copper, 1986; Geldsetzer et al., 1987; McGhee, 1991; Claeys et al., 1992), none has gained general acceptance. Algeo et al. (1995) propose that evolutionary innovations among vascular land plants (land plants with tissues for conducting water, minerals, and the products of photosynthesis throughout the plant) were the ultimate cause of both the Late Devonian biotic crisis and a variety of coeval (contemporary) sedimentologic and geochemical anomalies. Often referred to as the "age of fish," the Devonian period saw the rise and fall of many prehistoric marine species. Although by this time animals had begun to evolve on land, the majority of life swam through the oceans. That was until vascular plants, such as trees and flowers, likely caused a second mass extinction (Algeo et al., 1995). The main lines of evidence supporting this hypothesis are (1) close temporal relations between Late Devonian paleo-botanic developments and major episodes of oceanic anoxia and mass extinction, and (2) a model that successfully links these paleo-botanic developments to the Late Devonian biotic crisis and coeval sedimentologic and geochemical anomalies through changes in pedogenic (relating to or denoting processes occurring in soil or leading to the formation of soil) rates and processes.

Note that paleo-botany is the branch of botany dealing with the recovery and identification of plant remains from geological contexts, and their use for the biological reconstruction of past environments (paleogeography), and the evolutionary history of plants, with a bearing upon the evolution of life in general. The emergence of paleobotany as a scientific discipline can be seen in the early 19th century, especially in the works of the German paleontologist Ernst Friedrich von Schlotheim, the Czech (Bohemian) nobleman and scholar Kaspar Maria von Sternberg, and the French botanist Adolphe-Théodore Brongniart (Cleal et al., 2005).

As plants evolved roots, they inadvertently transformed the land they lived on, turning rock and rubble into soil. This nutrient-rich soil then ran into the world's oceans, causing algae to bloom on an enormous scale. These blooms essentially created giant "dead zones," which are areas where algae strips oxygen from the water, suffocating marine life and wreaking havoc on marine food chains. Species that were unable to adapt to the decreased oxygen levels and lack of food died. One sea monster that was wiped from the world's oceans was a 33-foot-long (10 m) armored fish called *Dunkleosteus*. A fearsome predator, this giant fish had a helmet of bone plates that covered its entire head and created a fang-like cusp on its jaw.

The Frasnian–Famennian (F–F) global event, one of the five largest biotic crises of the Phanerozoic, has been inconclusively linked to rapid climatic perturbations promoted in turn by volcanic cataclysm, especially in the Viluy large igneous province (LIP) of Siberia. Conversely, the triggers of four other Phanerozoic mass extinction intervals have decisively been linked to LIPs, owing to documented mercury anomalies, shown as the diagnostic proxy.

The theory put forward by Algeo et al. (1995) is debated, and some scientists believe that volcanic eruptions were responsible for the decrease in oxygen levels in the ocean. For example, Racki et al. (2018) reported multiple mercury (Hg) enrichments in the two-step late Frasnian (Kellwasser) crisis interval from paleo-geographically distant successions in Morocco, Germany, and northern Russia. The distinguishing signal, >1 ppm Hg in the domain of closing Rheic Ocean (an ocean which separated two major paleocontinents, Gondwana and Laurussia (Laurentia-Baltica-Avalonia)), was identified in different lithologies immediately below the F–F boundary and approximately correlated with the onset of the main extinction pulse. This key Hg anomaly, comparable only with an extreme spike known from the end-Ordovician extinction, was not controlled by increased bioproductivity in an anoxic setting. Racki et al. (2018) suggested, therefore, that the global chemo-stratigraphic pattern near the F–F boundary records a greatly increased worldwide Hg input, controlled by the Center Hill eruptive pulse of the Eovariscan volcanic acme (most highly developed volcano), but likely not manifested exclusively by LIP(s). Consequently, all five major biotic crises of the Phanerozoic have now been more reliably linked to volcanic cataclysms (i.e., large-scale and violent volcanic events (Racki et al., 2018).

According to Carmichael et al. (2019), The Late Devonian (383–359 Ma) was a time of prolonged climate instability with catastrophic perturbation of global marine ecosystems at the Frasnian–Famennian (F–F) and the Devonian–Carboniferous (D–C) boundaries. Paschall et al. (2019) described the D–C transition thus: "The Devonian–Carboniferous transition (359 Ma) was a time of extreme climate and faunal change and is associated with the end-Devonian biodiversity crisis. The transition is characterized by transgressive/regressive cycles, which culminated in the onset of widespread ocean anoxia (the Hangenberg Black Shale event) and a remarkable sea-level fall close to the Devonian–Carboniferous boundary (the Hangenberg Sandstone event); together these are known as the Hangenberg Crisis. The Hangenberg Crisis has been documented around the globe, but the trigger mechanisms for its onset remain unknown. The Pho Han Formation on Cat Ba Island in northeastern Vietnam preserves the Devonian–Carboniferous transition and Hangenberg Crisis in a sediment-starved basinal facies on the South China carbonate platform. Although the Hangenberg Black Shale event is generally preserved as a discrete anoxic interval in Devonian–Carboniferous boundary sections of North America and Europe, the Pho Han Formation records sustained dysoxic/anoxic conditions from the Famennian (Upper Devonian) through the Tournasian (early Carboniferous), with severe anoxia (approaching euxinia) throughout the Hangenberg Black Shale event interval (as determined by trace element proxies, increased total organic carbon, and framboidal pyrite distributions). There is also significant mercury enrichment corresponding to the Hangenberg Crisis in the Pho Han Formation. The isolated paleogeography of the region suggests that the mercury is most likely sourced from distal volcanic emissions. It is therefore possible that large-scale volcanic activity acted as a trigger mechanism for the Hangenberg Crisis and biodiversity drop at the Devonian–Carboniferous transition, similar to other major mass extinction events."

With regard to the Frasnian–Famennian (F–F) interval Carmichael et al. (2019) stated thus, "The causes and mechanisms of anoxia and extinction at the Frasnian–Famennian (F–F) interval are not clearly delineated, and alternative explanations for virtually every aspect of this interval are still intensely debated. In many (but not all) locations, the F–F interval is characterized by two dark, organic-rich lithologies: the Lower and Upper Kellwasser beds (as originally described in Germany) that represent a stepwise ocean anoxia and extinction sequence. The Upper and Lower Kellwasser anoxia event beds are often collectively termed the Kellwasser Event, and the termination of this sequence is within the Upper Kellwasser Event at the F–F boundary. Current knowledge is limited by significant sampling bias, as most previous studies sampled epicontinental seaways or passive continental shelves, primarily from localities across Europe and North America. Together these formed a single equatorial continent with a rising mountain chain during the Late Devonian. Our understanding of the Kellwasser Event is thus based on data and observations from a restricted set of paleoenvironments that may not represent the complete range of Late Devonian environments and oceanic conditions. In the last decade, new methodologies and research in additional paleoenvironments around the world confirm that the Kellwasser Event was global in scope, but also that its expression varies with both paleoenvironment and paleogeography."

It has been hard to nail down the cause for the late Devonian extinction pulses, but volcanism is a possible trigger: Within a couple million years of the Kellwasser event, a large igneous province called the Viluy Traps erupted 240,000 cubic miles of lava in what is now Siberia. The eruption would have spewed greenhouse gases and sulfur dioxide, which can cause acid rain. Asteroids may also have contributed. Sweden's 32-mile-wide Siljan crater, one of Earth's biggest surviving impact craters, formed about 377 million years ago (Carmichael et al., 2019).

Carmichael et al. (2019) have suggested that studying the many differing geochemical and lithological expressions of the Kellwasser Event using (a) a wide variety of paleoenvironments, (b) a multiproxy approach, and (c) placement of results into the broader context of Late Devonian marine biodiversity patterns is vital for understanding the true scope of ocean anoxia, and determining the causes of the marine biodiversity crisis at the F–F boundary.

2.13.3 Permian–Triassic extinction: ~253 million years ago

The Permian–Triassic extinction event, also known as the End-Permian Extinction and colloquially as the Great Dying, formed the boundary between the Permian and Triassic geologic periods, as well as between the Paleozoic and Mesozoic eras, approximately 253 million years ago. The Permian–Triassic extinction event, often referred to as the "Great Dying," is the largest to ever hit Earth. It wiped out some 90% of all the planet's species and decimated the reptiles, insects and amphibians that roamed on land. What caused this catastrophic event was a period of rampant volcanism. At the end of the Permian period, the part of the world we now call Siberia erupted in explosive volcanoes. This released a large amount of carbon dioxide into the atmosphere, causing a greenhouse effect that heated up the planet. As a result, weather patterns shifted, sea levels rose and acid rain beat down on the land.

The scientific consensus is that the causes of extinction were elevated temperatures and in the marine realm widespread oceanic anoxia and ocean acidification due to the large amounts of carbon dioxide that were emitted by the eruption of the Siberian Traps. In the ocean, increased levels of carbon dioxide dissolved into the water, poisoning marine life and depriving them of oxygen-rich water, according to the Sam Noble Museum in Oklahoma. At the time, the world consisted of one supercontinent called Pangaea, which some scientists believe contributed to a lack of movement in the world's oceans, creating a global pool of stagnant water that only perpetuated carbon dioxide accumulation. Rising sea temperatures also reduced oxygen levels in the water.

Corals were a group of marine life forms that were among the worst affected—it took 14 million years for the ocean reefs to rebuild to their former glory. The Permian extinction was characterized by the elimination of over 95% of marine and 70% of terrestrial species (https://www.livescience.com/mass-extinction-events-that-shaped-Earth.html).

2.13.4 Triassic–Jurassic extinction: ~201 million years ago

The Triassic period was the first period of the Mesozoic era and occurred between 251.9 million and 201.3 million years ago. It followed the great mass extinction at the end of the Permian period and was a time when life outside of the oceans began to diversify.

At the beginning of the Triassic, most of the continents were concentrated in the giant C-shaped supercontinent known as Pangaea (see Fig. 2.10). Climate was generally very dry over much of Pangaea with very hot summers and cold winters in the continental interior. A highly seasonal monsoon climate prevailed nearer to the coastal regions. Although the climate was more moderate farther from the equator, it was generally warmer than today with no polar ice caps. Late in the Triassic, seafloor spreading in the Tethys Sea led to rifting between the northern and southern portions of Pangaea, which began the separation of Pangaea into two continents, Laurasia and Gondwana (see Fig. 2.24), which would be completed in the Jurassic period (Bagley, 2014).

The oceans had been massively depopulated by the Permian extinction when as many as 95% of extant marine genera were wiped out by high carbon dioxide levels. Fossil fish from the Triassic period are very uniform, which indicates that few families survived the extinction. The mid- to late Triassic period shows the first development of modern stony corals and a time of modest reef building activity in the shallower waters of the Tethys near the coasts of Pangaea (Bagley, 2014).

The Triassic period erupted in new and diverse life, and dinosaurs began to populate the world. Unfortunately, numerous volcanoes also erupted at that time. Although it remains unclear exactly why this fourth mass extinction occurred, scientists think that massive volcanic activity occurred in an area of the world now covered by the Atlantic Ocean. Similar to the Permian extinction, volcanoes released enormous amounts of carbon dioxide, driving climate change and devastating life on Earth. Global temperatures increased, ice melted, and sea levels rose and acidified. As a result, many marine and land species became extinct; these included large prehistoric crocodiles and some pterosaurs. Note that pterosaurs were flying reptiles of the extinct clade. They existed during most of the Mesozoic: from the late Triassic to the end of the Cretaceous. Pterosaurs are the earliest vertebrates known to have evolved powered flight.

The cause of the end-Triassic extinction is a matter of considerable debate. As just indicated, many scientists contend that this event was caused by climate change and rising sea levels resulting from the sudden release of large amounts of carbon dioxide. Only the phylloceratid ammonoids survived the extinction event and later gave rise to the cephalopods in the Jurassic period. Several families of gastropods, brachiopods, marine reptiles, and bivalves also became extinct.

There are alternative theories explaining this mass extinction, which suggest that rising carbon dioxide levels released trapped methane from permafrost, which would have resulted in a similar series of events. In this extinction event, the species made extinct is estimated to be 80%.

2.13.5 The K–Pg extinction: ~66 million years ago: extinction of dinosaurs from Earth and subsequent appearance of modern humans' distant ancestors

The most famous of all the mass extinction events is the Cretaceous–Paleogene extinction—better known as the short period of time during which the dinosaurs vanished. The event is sometimes also known as the K–T extinction, and geologists call it the "K–Pg" extinction because the letter "C" is shorthand for a previous geological period called the Cambrian. The "K" is from the German word "Kreide," which means "Cretaceous."

Note that the (K–Pg) extinction event was a mass extinction of some three-quarters of the plant and animal species on Earth that occurred over a geologically short period of time, approximately 66 million years ago, having ended the 180 million-year reign of the dinosaurs on Earth. The Cenozoic era (covering the period from 66 million years ago to the present day) is notorious in Earth's history because this era has seen the extinction of the nonavian dinosaurs (members of the dinosaur descendants other than birds and their immediate ancestors). The Cretaceous–Paleogene extinction event is the most recent mass extinction and the only one definitively connected to a major asteroid impact. Some 76% of all species on the planet, including all nonavian dinosaurs, went extinct.

There are a lot of theories about why the K–T extinction occurred, but a widely accepted theory is that one day about 66 million years ago, an asteroid about 10–15 km across slammed/ crash-landing into the waters off of what is now Mexico's Yucatán Peninsula at a terrific speed of approximately 45,000 miles an hour. The massive impact—which left a large crater (about 180–200 km in diameter), known as the Chicxulub crater—flung huge volumes of dust, debris, and sulfur into the atmosphere, bringing on severe global cooling. Wildfires ignited any land within 900 miles of the impact, and a huge tsunami rippled outward from the impact. Overnight, the ecosystems that supported nonavian dinosaurs began to collapse.

As just indicated, an approximately 15-km diameter asteroid is implicated in the Cretaceous/Paleogene (K/Pg) mass extinction, where it struck in shallow water off of the Yucatan Peninsula. Based on a model-based study, Range et al. (2018) reported what is, to their knowledge, the first global simulation of the Chicxulub impact tsunami. These investigators used the axisymmetric iSALE-2D hydrocode to simulate the initial formation of the Chicxulub impact tsunami. The hydrocode results at 10 min postimpact were merged with 66 Ma bathymetry into a shallow-water model—the Geophysical Fluid Dynamics Laboratory (GFDL) Modular Ocean Model Version 6 (MOM6)—to trace the tsunami's propagation throughout the world ocean. The paleo-bathymetry came from a combination of the PALEOMAP dataset, where present-day crust was used to reconstruct 66 Ma crust, and the Muller et al. dataset (see Muller et al., 2008a,b; Muller et al., 2013; Muller and Seton, 2015), which uses basin-age depth relations to construct the 66 Ma bathymetry. The impact tsunami spread quickly out of the Gulf of Mexico into the Atlantic and through the Central American seaway into the Pacific within the first 24 h. Wave reflection and refraction create a more complex tsunami propagation pattern by 48 h postimpact. Flow velocities exceeded 20 cm/s along shorelines worldwide and may have disturbed sediments over 6000 km from the impact origin. In open-ocean areas in the Tethyian region, the South Atlantic, the North Pacific, and the Indian Ocean appear geographically protected from high flow velocities. Compared to the December 26, 2004 Indian Ocean tsunami, one of the largest tsunamis in the modern record, the impact tsunami was approximately 2600 times more energetic. Range et al. (2018) are hopeful that these records may be able to indicate if the impact tsunami model propagation and magnitude is consistent with various upper-Cretaceous sediments found around the world. This model suggests that the bolide impact not only had major effects on the global atmosphere and biosphere, it also created a tsunami of such magnitude that its effect is felt across much of the world ocean.

A study carried out by Kinsland et al. (2021) indicate that ancient "mega-ripples" as tall as five-story buildings are hiding deep under Louisiana (located in the vicinity of the Gulf of Mexico), and their unique geology indicates that they formed in the immediate aftermath of the asteroid strike that killed the nonavian dinosaurs. The inference that a huge tsunami rippled outward from the impact was obtained when the energy corporation "Devon Energy" took a three dimensional (3-D) seismic survey of Iatt Lake. Note that a seismic survey entails creating loud sound waves (often made with explosives or big thumps) and placing surface detectors around the area that can capture the sound waves, which are reflected when they hit various underground rock

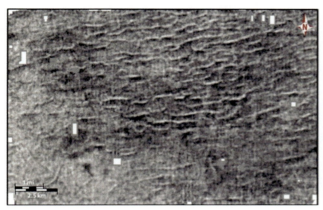

FIG. 2.25 A black-and-white seismic image of the mega-ripples, created by study coresearcher Kaare Egedahl for his master's thesis. The seismic image covers an area of about 11 by 7 miles (18 by 11 km). *(Image credit: Kinsland, G.L. et al. (2021), Earth Planet. Sci. Lett., Elsevier); Kaare Egedahl). (Source: Geggel, L., 2021) Copyright © 2021 Elsevier B.V. or its licensors or contributors.*

layers. Data from these reflected sound waves (seismic waves) allow researchers to make maps of the underground geology. Study coresearcher Kaare Egedahl, then a master's student of petroleum geology at the University of Louisiana at Lafayette, took the "Devon Energy" petroleum industry seismic dataset and created a seismic image of the subterranean area (see Fig. 2.25). It was found that the seismic image was so different from the conventional image, which one would expect to see in deposits laid down by the sea or by rivers.

Kinsland et al. (2021) found that the mega-ripples are asymmetrical, which shows the direction the water was flowing when the mega-ripples were created. In this case, it was found that the long, asymmetrical side of the mega-ripples have a south-southeast-facing slope, which points back to the Chicxulub impact crater, and going backward from the mega-ripples leads one right in Chicxulub. Fig. 2.26 shows the map illustrating the Chicxulub impact crater (red arrow) and the location of the newly discovered mega-ripples (red star) that were likely left by a tsunami caused when the asteroid hit 66 million years ago. It was found that the mega-ripples have an average wavelength (straight line distance from one crest to the next) of 1968 feet (600 m). That, combined with their 52-foot-high amplitude, makes them "the largest ripples documented on Earth" (Kinsland et al., 2021).

The study indicates that the 52-foot-tall (16 m) mega-ripples are about 5000 feet (1500 m) under the Iatt Lake area, in north central Louisiana, and date to the end of the Cretaceous period 66 million years ago, when that part of the state was underwater. The mega-ripples' size and orientation suggest that they formed after the giant space rock, known as the Chicxulub asteroid, slammed into the Yucatán Peninsula, leading to the Chicxulub impact tsunami, whose

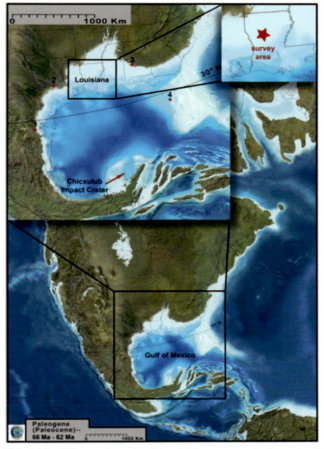

FIG. 2.26 This map shows the Chicxulub impact crater (red arrow) and the location of the newly discovered mega-ripples (red star) that were likely left by a tsunami caused when the asteroid hit 66 million years ago. *(Image credit: Kinsland, GL. et al. (2021), Earth Planet. Sci. Lett., Elsevier); Original base map by Ron Blakey/Colorado Plateau Geosystems; Nina Zamanialavijeh). (Source: Geggel, L., 2021) Copyright © 2021 Elsevier B.V. or its licensors or contributors.*

waves then rushed into shallower waters and created the mega-ripple marks on the seafloor.

The mega-ripples indicate that after the space rock hit Earth 66 million years ago, a tsunami rushed across the Gulf of Mexico and then shoaled and broke offshore as it reached the abrupt shallowing of the Gulf of Mexico within what is now central Louisiana. The resulting pulses of water flowing north-northeast over the shelf area produced the asymmetric mega-ripples which are imaged within the seismic data. Moreover, these mega-ripples are at the top of the Cretaceous/Paleogene geological boundary dating to 66 million years ago, and lie beneath a layer of debris that were kicked up in the aftermath of the Chicxulub impact.

It is an observed fact that ripples left by water waves on a sandy beach are short-lived. So how did the mega-ripples persist for 66 million years? Kinsland et al. (2021) explain that after the tsunami created the mega-ripples, they remained underwater. They were deep enough underwater that when storms swept through the Gulf of Mexico, the mega-ripples remained undisturbed. Then, the mega-ripples were buried by shale—in essence, a sedimentary rock made of mud mixed with clay and mineral fragments—over a period of about 5 million years, during the Paleocene epoch (66 million to 56 million years ago). Later, that shale was covered by even younger sediments.

In a nutshell, large-scale mega-ripples have been recognized in a petroleum industry 3-D seismic horizon near the Cretaceous/Paleogene (K–Pg) boundary. These features occur at the top of the Cretaceous/Paleogene Boundary Deposit (KPBD) which is a "cocktail" of mass transport deposits and debris widely recognized as resulting from the impact of a large bolide 66 million years ago (Ma) creating the Chicxulub impact crater on the north-western corner of the Yucatan Peninsula of Mexico. Kinsland et al. (2021) examined the seismic data and associated well-logs and concluded that the features are mega-ripples caused by a tsunami resulting from the impact. These mega-ripples are preserved as a result of having formed below storm wave base and being buried by Paleocene deep water shales. This association suggests that the Chicxulub impact is the single cause for the KPBD, the mega-ripples, and the end of the Mesozoic.

It is believed that global warming fuelled by volcanic eruptions at the Deccan Flats in India may have aggravated the event. Some scientists even argue that some of the Deccan Flats eruptions could have been triggered by the impact.

What followed the impact were months of blackened skies caused by debris and dust being hurled into the atmosphere. This prevented plants from absorbing sunlight, and they died out en masse (in a group; all together) and broke down the dinosaurs' food chains. It also caused global temperatures to plummet, plunging the world into an extended cold winter. Scientists estimate that most extinctions on Earth at the time would have occurred in just months after the impact. However, many species that could fly, burrow or dive to the depths of the oceans survived. For example, the only true descendants of the dinosaurs living today are modern-day birds—more than 10,000 species are thought to have descended from impact survivors.

It is noteworthy that the Cenozoic era is equally famous in the sense that the *Homo* (i.e., the genus that comprises the species *Homo sapiens*, which includes modern humans, as well as several extinct species classified as ancestral to or closely related to modern humans, most notably *Homo erectus*) began to appear on Earth subsequent to the extinction of dinosaurs.

The temporal match between the ejecta layer (which resulted from the asteroid impact) and the onset of the extinctions and the agreement of ecological patterns in the fossil record with modeled environmental perturbations (for example, darkness and cooling) led Schulte et al. (2010) to conclude that the Chicxulub impact triggered the most devastating mass-extinction event in Earth's history and abruptly ended the age of the dinosaurs.

Analyses of fossil records reveal that several other major groups suffered considerable, although not complete, species-level extinction. For marine phytoplankton (which are the major drivers of ocean productivity), darkness and suppression of photosynthesis were likely major killing mechanisms. On land, the Chicxulub impact resulted in the loss of diverse vegetation, destruction of diverse forest communities, abrupt mass extinction, and ecosystem disruption.

The Eocene Epoch (lasting from 56 to 33.9 million years ago) saw the replacement of older mammalian orders by modern ones. Hoofed animals (i.e., those animals having horny covering encasing their foot, as the ox and horse) first appeared, including the famous Eohippus (dawn horse). Early bats, rabbits, beavers, rats, mice, carnivorous mammals, and whales also evolved during the Eocene Epoch. The vegetation of the Eocene was fairly modern; the climate was warm.

2.14 Carbonate–silicate cycle and its role as a dynamic climate buffer

Atmospheric gaseous constituents play an important role in determining the surface temperatures and habitability of a planet. Since the classic work of Urey (1952) it has been often postulated that, over geologic time scales, the level of atmospheric carbon dioxide (CO_2) is greatly affected by, if not controlled by, the transformation of silicate rocks to carbonate rocks by weathering and sedimentation and re-transformation back to silicate rocks by metamorphism (alteration of the composition or structure of a rock by heat, pressure, or other natural agency) and magmatism (the motion or activity of magma) (Holland, 1978; Budyko and Ronov, 1979; Mackenzie and Pigott, 1981). Thus, the carbonate–silicate geochemical cycle describes the transformation of silicate rocks to carbonate rocks by weathering and sedimentation at Earth's surface and the re-transformation of carbonate rocks back into silicates by metamorphism and magmatism.

The carbonate–silicate cycle impacts the global carbon cycle, as carbon dioxide is removed from the Earth's surface through the burial of weathered minerals in deep ocean sediments and returned to the atmosphere through metamorphism and volcanism (the phenomenon of eruption of molten rock onto the surface of the Earth or a solid-surface planet or moon, where lava, pyroclastics and volcanic gases erupt through a break in the surface called a vent). However, this process is far from being a closed loop. Walker et al. (1981) suggested that the partial pressure of carbon dioxide in the atmosphere is buffered, over geological time scales, by a negative feedback mechanism in which the rate of weathering of silicate minerals (followed by deposition of carbonate minerals) depends on surface temperature, and the surface temperature, in turn, depends on carbon dioxide partial pressure through the greenhouse effect. In Earth's history generally the formation of carbonates significantly outpaces the formation of silicates, effectively removing carbon dioxide from the atmosphere. Because carbon dioxide is a potent greenhouse gas, the carbonate–silicate cycle is suspected to initiate ice ages by creating a negative feedback on the global temperature with a typical time scale of a few million years that is capable of countering water vapor and carbon dioxide short-term positive feedback on global temperature (Walker et al., 1981).

Berner et al. (1983) constructed a computer model that considers the effects of several processes on the carbon dioxide level of the atmosphere, and the calcium, magnesium, and bicarbonate (HCO_3) levels of the ocean, providing details on the various chemical reactions involved. Results obtained from their computer model indicated that the CO_2 content of the atmosphere is highly sensitive to changes in seafloor spreading rate and continental land area, and to a much lesser extent, to changes in the relative masses of calcite and dolomite. Note that calcite (chemical formula: $CaCO_3$) is a rock-forming mineral. It is extremely common and found throughout the world in sedimentary, metamorphic, and igneous rocks. Some geologists consider it to be a "ubiquitous mineral"—one that is found everywhere. Calcite is the principal constituent of limestone and marble. Likewise, dolomite (calcium magnesium carbonate), whose chemical formula is $CaMg(CO_3)_2$ is also a common rock-forming mineral. Dolomite is the primary component of the sedimentary rock known as dolostone and the metamorphic rock known as dolomitic marble.

From the above discussions it is quite evident that the crustal Urey cycle of CO_2 involving silicate weathering and metamorphism acts as a dynamic climate buffer. Sleep and Zahnle (2001) have neatly summarized the mechanism of Urey cycle acting as a dynamic climate buffer thus: "In this cycle, warmer temperatures speed silicate weathering and carbonate formation, reducing atmospheric CO_2 and thereby inducing global cooling. Over long periods of time, cycling of CO_2 into and out of the mantle also dynamically buffers CO_2. In the mantle cycle, CO_2 is outgassed at ridge axes and island arcs, while subduction of carbonatized oceanic basalt and pelagic sediments returns CO_2 to the mantle. Negative feedback is provided because the amount of basalt carbonatization depends on CO_2 in seawater and therefore on CO_2 in the air. On the early Earth, processes involving tectonics were more vigorous than at present, and the dynamic mantle buffer dominated over the crustal one. The mantle cycle would have maintained atmospheric and oceanic CO_2 reservoirs at levels where the climate was cold in the Archean unless another greenhouse gas was important. Reaction of CO_2 with impact ejecta and its eventual subduction produce even lower levels of atmospheric CO_2 and small crustal carbonate reservoirs in the Hadean. Despite its name, the Hadean climate would have been freezing unless tempered by other greenhouse gases."

Using a global climate model and a parameterization of the carbonate–silicate cycle, Edson et al. (2012) explored the effect of the location of the substellar point on the atmospheric CO_2 concentration and temperatures of a tidally locked terrestrial planet, using the present Earth continental distribution as an example. They found that the substellar point's location relative to the continents is an important factor in determining weathering and the equilibrium atmospheric CO_2 level. Edson et al. (2012)'s study indicated that "Placing the substellar point over the Atlantic Ocean results in an atmospheric CO_2 concentration of 7 ppmv and a global mean surface air temperature of 247 K, making ~30% of the planet's surface habitable, whereas placing it over the Pacific Ocean results in a CO_2 concentration of 60,311 ppmv and a global temperature of 282 K, making ~55% of the surface habitable." Note that in the above discussion, ppmv stands for "Parts Per Million by Volume."

The carbonate–silicate geochemical cycle plays a large part in the carbon cycle, because the equilibrium point of the Carbonate–Silicate cycle dictates the pace of carbon release from the lithosphere (Edson et al., 2012). The Carbonate–Silicate cycle involves several chemical reactions that occur in different environments. In the atmosphere, gaseous carbon dioxide (CO_2) dissolves in rainwater, forming natural carbonic acid (H_2CO_3). This weak acid weathers silicate rocks on continents, slowly dissolving the rock and releasing aqueous minerals. These dissolved minerals are eventually carried by water to the ocean, where they are used by living organisms such as foraminifera (single-celled organisms, characterized by an external shell for protection, that live in the open ocean, along the coasts and in estuaries, either floating in the water column (planktonic) or living on the sea floor (benthic)), radiolarians (unicellular predatory protists (single-celled organisms of the kingdom Protista, such as a protozoan or simple alga) encased in elaborate globular shells usually made of silica and pierced with holes), coccolithopores (unicellular, eukaryotic phytoplankton belonging either to the kingdom Protista, according to Robert Whittaker's Five kingdom classification, or clade Hacrobia, according to the newer biological classification system), and diatoms (a major group of algae, specifically microalgae, having a cell wall of silica, and found in the oceans, waterways and soils of the world) to create shells of $CaCO_3$ (calcite) or SiO_2 (opal). When these organisms die, many shells are re-mineralized but some shells fall all the way to the seafloor and are buried. The cycle is completed when the seafloor is subducted and carbonate minerals recombine with silicate minerals under temperatures above 300°C to reform calcium silicates and release gaseous CO_2 through volcanism.

2.15 Occurrence of a sharp global warming ~56 million years ago

There was a time interval in the history of Earth when it experienced a series of sudden and extreme global warming events (hyperthermals) superimposed on a long-term warming trend. The first and largest of these global warming event that occurred ~56 million years ago (Mya) is known as the Paleocene–Eocene Thermal Maximum (PETM) (see McInerney and Wing, 2011; Dunkley et al., 2013). The PETM is characterized by a massive input of carbon, ocean acidification and an increase in global temperature of about 5°C within a few thousand years. Although various explanations for the PETM have been proposed, a satisfactory model that accounts for the source, magnitude and timing of carbon release at the PETM and successive hyperthermals remains elusive.

The PETM—also called Initial Eocene Thermal Maximum (IETM)—is a relatively short-lived (in geological terms) transient climatic event of maximum temperature lasting approximately 100,000–200,000 years during the late Paleocene and early Eocene epochs (roughly 55 million years ago). The PETM period is believed to be the most rapid and extreme natural global warming event of the last 56 million years. The global temperature increase was estimated to have been in the range 5–8°C.

2.15.1 Consequences

The PETM phenomenon gave rise to far-reaching ecological consequences, having witnessed rapid global warming of both the oceans and the continents. Earth's warming in the range 5–8°C, over a few thousand years at the onset of the PETM, adversely affected life both in the sea and on the land in different ways. For example, many species of benthic organisms in the deep sea went extinct (e.g., benthic foraminifera) and parts of the deep ocean became anoxic (oxygen depleted). On land, the water cycle strengthened, leading to both floods and droughts. It is estimated that it took about 150,000 years for Earth's climate to naturally recover from this "fever" and regain some sort of equilibrium.

In terms of weather change, sea surface temperature (SST) and continental air temperature increased by more than 5°C during the transition into the PETM. It is believed that SST in the high-latitude Arctic may have been as high as 23°C (73°F). This high temperature is comparable to that of modern subtropical and warm-temperate seas.

Just prior to the PETM, Earth was very different than it is at present. For example, the Polar Regions were devoid of ice sheets. Because of the warm atmospheric temperature that prevailed then in Antarctica, the coastlines of Antarctica witnessed flourishing temperate- or even subtropical-forests. Likewise, the Arctic Canada resembled the swamplands of modern Florida. Furthermore, the deep oceans were about 10°C warmer than the present. These examples indicate that the warm climate zones were all shifted pole-ward.

The PETM was characterized by several scientifically interesting episodes. Most important among these are (1) a large negative excursion in stable carbon isotope ratios, (2) a prominent shoaling of the global ocean carbonate compensation depth (CCD), and (3) a widespread positive temperature anomaly (Zachos et al., 2005; Sluijs et al., 2007a). Note that "Carbonate compensation depth" is the depth in the oceans below which the rate of supply of calcite (calcium carbonate) lags behind the rate of solvation (i.e., interaction of solvent with dissolved molecules), such that no calcite is preserved. Shells of animals therefore dissolve and carbonate particles may not accumulate in the sediments on the sea floor below this depth. As a result, the PETM has been the focus of considerable research effort since it was first recognized over 20 years ago (Thomas, 1989; Kennett and Stott, 1991). In this context, the characteristic carbon isotope excursion and associated lithologic (rock characteristics) and biotic changes have been documented at numerous localities across the globe and in disparate environments (see reviews of Sluijs et al., 2007a; McInerney and Wing, 2011).

Another noteworthy consequence of the PETM phenomenon is that while a chain of events during the PETM event aided the removal of carbon from the ancient atmosphere it also led to oxygen starvation in some parts of the deep sea—analogous to the "dead zones" that form today in areas like the Gulf of Mexico. The PETM event was also characterized by ocean acidification (Zachos et al., 2005), resulting from a massive input of carbon dioxide into the oceans.

2.15.2 Causes

It is believed that during the PETM event, thousands of petagrams (note that 1 petagram is equivalent to 10^{12} kg) of carbon were released into the ocean-atmosphere system. Such excessive release of carbon gave rise to considerable changes in the carbon cycle, climate, ocean chemistry, and marine and continental ecosystems (see McInerney and Wing, 2011).

There are two major schools of thought associated with the origin of PETM event: (1) sudden release of methane hydrates from ocean sediments (methane burp hypothesis), and (2) a sudden release of carbon dioxide and/or methane.

2.15.2.1 Volcanic eruptions and seaquakes ~56 million years ago—Greenland and North America drifting away from Europe, resulting in the formation of North Atlantic Ocean

At the onset of the PETM when the atmosphere was beginning to warm up excessively, Greenland and North America were drifting away from Europe, creating the North Atlantic effect. Such greenhouse gases trap heat radiated off of the surface of the Earth, forming a type of insulation around the planet. Studies pertaining to incessant large-scale volcano eruptions associated with the opening of the North Atlantic have shown that the PETM was triggered by greenhouse gas release during magma interaction with basin-filling carbon-rich sedimentary rocks proximal to the embryonic plate boundary between Greenland and Europe (Storey et al., 2007).

Based on the analysis of 56-million-year-old sediments (sediment core and borehole data analysis), it is now known that there was a huge release of "new" carbon into the atmosphere and oceans at the time of the PETM. Yet where this carbon came from has always been disputed. Carbon can be emitted as carbon dioxide (CO_2) or methane (CH_4) and both are greenhouse gases. Using a combination of new geochemical measurements and a novel numerical global climate model, Gutjahr et al. (2017) found that the PETM was associated with a total input of more than 10,000 petagrams (i.e., 10^{16} kg) of carbon from massive carbon emissions from Earth's interior. This led them to identify volcanism associated with the North Atlantic Igneous Province (Svensen et al., 2004; Storey et al., 2007), rather than carbon from a surface reservoir, as the main driver of the PETM. Gutjahr et al. (2017)'s study showed that PETM was associated with a geologically rapid doubling of atmospheric CO_2 in less than 25,000 years, with volcanoes squarely to blame.

A sudden release of methane hydrates from ocean sediments can be triggered also by seaquakes in the gas hydrate deposit regions. For example, there have been instances of fire breaking out over the sea surface and thick smoke columns having been ejected into the atmosphere over the Japan Sea, below which gas hydrate deposits have been identified (see Joseph, 2011).

2.15.2.2 Methane hydrates emission

During the Paleocene–Eocene Thermal Maximum (PETM), ~56 Mya, thousands of petagrams of carbon were released into the ocean-atmosphere system with attendant changes in the carbon cycle, climate, ocean chemistry, and marine and continental ecosystems. The period of carbon release is thought to have lasted <20 ka, the duration of the whole event was ~200 ka, and the global temperature increase was 5–8°C. Terrestrial and marine organisms experienced large shifts in geographic ranges, rapid evolution, and changes in trophic ecology, but few groups suffered major extinctions with the exception of benthic foraminifera. The PETM provides valuable insights into the carbon cycle, climate system, and biotic responses to environmental change that are relevant to long-term future global changes (McInerney and Wing, 2011).

Extreme global warmth and an abrupt negative carbon isotope excursion during the Paleocene–Eocene Thermal Maximum (PETM) have been attributed to a massive

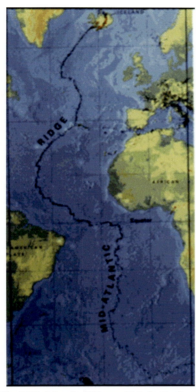

FIG. 2.27 **The Mid-Atlantic Ridge, which is the longest mountain range in the world that is almost entirely submerged below the sea surface.** When Greenland and North America started drifting away from Europe about 66 million years ago, ultimately creating the North Atlantic Ocean, a string of volcanic activity, which occurred along what is now the Mid-Atlantic Ridge, gave rise to the emission of massive amount of carbon dioxide (CO_2) from these volcanoes. This period, known as Paleocene–Eocene thermal maximum (PETM), witnessed the most rapid and extreme natural global warming event (about 5°C rises in atmospheric temperature) of the last 66 million years, driven by the greenhouse effect triggered by the carbon dioxide emission from the volcanoes *(Image Source: USGS / wiki)*.

Ocean and a string of volcanic activity along what is now the Mid-Atlantic Ridge (Fig. 2.27). It would be interesting to note that the PETM volcanism largely took place under water and at a slower pace. The most rapid and extreme natural global warming event of the last 65 million years was driven by massive carbon dioxide (CO_2) emissions from volcanoes during the just-mentioned formation of the North Atlantic Ocean. While it has long been suggested that the PETM event was caused by the injection of carbon into the ocean and atmosphere, the ultimate trigger, the source of this carbon, and the total amount released, have up to now all remained elusive.

As already noted, a sudden release of methane hydrates from ocean sediments can be triggered by a massive volcanic eruption. There are many reasons that large volcanic eruptions have such far-reaching effects on global climate. Volcanic eruptions produce major quantities of carbon dioxide (CO_2), a gas known to contribute to the greenhouse release of methane hydrate from sediments on the continental slope. However, the magnitude of the warming (5–6°C and rise in the depth of the Carbonate Compensation Depth (CCD) (> 2 km) indicate that the size of the carbon addition was larger than can be accounted for by the methane hydrate hypothesis. Additional carbon sources associated with methane hydrate release (e.g., pore-water venting and turbidite oxidation) are also insufficient. Although the PETM event is commonly thought to have been driven primarily by the destabilization of carbon from surface sedimentary reservoirs such as methane hydrates, fingers have been raised against such a view (see Dickens et al., 1995; Higgins and Schrag, 2006; McInerney and Wing, 2011; DeConto et al., 2012). Dickens et al. (1995) have suggested that the fate of CH_4 in oceanic hydrates must be considered in developing models of the climatic and paleoceanographic regimes that operated during the "Latest Paleocene Thermal Maximum" (LPTM), which occurred roughly 55 million years ago.

As just mentioned, apart from carbon dioxide (CO_2) emission to the atmosphere, a massive methane (CH_4) release from marine gas hydrate reservoirs have also been blamed for the occurrence of the PETM event. A rapid increase of water temperatures is capable of triggering a massive thermal dissociation of gas hydrate reservoirs beneath the seafloor. For example, the notorious Bermuda Triangle imbroglio has been attributed to the excessive warming of the Gulf Stream warm current.

According to Katz et al. (2001), with the available data, neither thermal dissociation nor mechanical disruption of sediments can be identified unequivocally as the triggering mechanism for methane release. They suggested that further documentation with high-resolution benthic foraminifera isotopic records and with seismic profiles tied to borehole data is needed to clarify whether erosion, thermal dissociation, or a combination of these two was the triggering mechanism for the PETM methane release.

To establish constraints on thermal dissociation, Katz et al. (2001) modeled heat flow through the sediment column and showed the effect of the temperature change on the gas hydrate stability zone through time. In addition, they provided seismic evidence tied to borehole data for methane release along portions of the US continental slope.

Higgins and Schrag (2006) found that the oxidation of at least 5000 Gt C of organic carbon is the most likely explanation for the observed geochemical and climatic changes during the PETM, for which there are several potential mechanisms. They found that production of thermogenic CH_4 and CO_2 during contact metamorphism associated with the intrusion of a large igneous province into organic rich sediments is capable of supplying large amounts of carbon, but is inconsistent with the lack of extensive carbon loss in metamorphosed sediments, as well as the abrupt onset and termination of carbon release during the PETM. A global conflagration (i.e., an extensive fire which destroys a great deal of land or property) of Paleocene peatlands (land consisting largely of peat—an accumulation of partially decayed vegetation or organic matter) highlights a large terrestrial carbon source, but massive carbon release by fire seems unlikely as it would require that all peatlands burn at once and then for only 10–30 ky. In addition, this hypothesis requires an order of magnitude increase in the amount of carbon stored in peat.

According to Higgins and Schrag (2006), the isolation of a large epicontinental seaway (i.e., those areas of sea or ocean overlying the continental shelf) by tectonic uplift associated with volcanism or continental collision, followed by desiccation and bacterial respiration of the aerated organic matter is another potential mechanism for the rapid release of large amounts of CO_2. In addition to the oxidation of the underlying marine sediments, the desiccation of a major epicontinental seaway would remove a large source of moisture for the continental interior, resulting in the desiccation and bacterial oxidation of adjacent terrestrial wetlands.

2.15.2.3 Orbitally triggered (Milankovitch cycles) decomposition of soil organic carbon in polar permafrost

Cycles of various kinds play key roles in Earth's short-term weather and long-term climate. DeConto et al. (2012) used an astronomically calibrated cyclostratigraphic record (i.e., records pertaining to astronomically forced climate cycles within sedimentary successions) from central Italy (Galeotti et al., 2010) to show that the Early Eocene hyperthermals (extreme global warming events) occurred during orbits with a combination of high eccentricity and high obliquity in Earth's orbital chronology (i.e., orbital triggering). Corresponding climate–ecosystem–soil simulations accounting for rising concentrations of background greenhouse gases (Beerling et al., 2011) and orbital forcing show that the magnitude and timing of the PETM and subsequent hyperthermals can be explained by the orbitally triggered decomposition of soil organic carbon in circum-Arctic and Antarctic thawing (the process of ice and snow becoming liquid or soft as a result of warming up) terrestrial permafrost. This massive carbon reservoir had the potential to repeatedly release huge quantity of carbon to the atmosphere–ocean system, once a long-term warming threshold had been reached just before the PETM. According to DeConto et al. (2012), replenishment of permafrost soil carbon stocks following peak warming probably contributed to the rapid recovery from each event, while providing a sensitive carbon reservoir for the next hyperthermal. As background temperatures continued to rise following the PETM, the areal extent of permafrost steadily declined, resulting in an incrementally smaller available carbon pool and smaller hyperthermals at each successive orbital forcing maximum.

DeConto et al. (2012) found that a mechanism linking Earth's orbital properties with release of soil carbon from permafrost provides a unifying model accounting for the salient features of the hyperthermals.

Buis (2020) recalls that a century ago, Serbian scientist Milutin Milankovitch hypothesized that the long-term, collective effects of changes in Earth's position relative to the Sun are a strong driver of Earth's long-term climate, and are responsible for triggering the beginning and end of glaciation periods (Ice Ages). Specifically, Milankovitch examined how variations in three types of Earth orbital movements affect how much solar radiation (known as insolation) reaches the top of Earth's atmosphere as well as where the insolation reaches. These cyclical orbital movements, which became known as the *Milankovitch cycles*, cause variations of up to 25% in the amount of incoming insolation at Earth's mid-latitudes (the areas of our planet located between about 30° and 60° north and south of the equator).

The Milankovitch cycles include (Buis, 2020):

1. The shape of Earth's orbit, known as *eccentricity*;
2. The angle Earth's axis is tilted with respect to Earth's orbital plane, known as *obliquity*; and
3. The direction Earth's axis of rotation is pointed, known as *precession*.

Buis (2020) recalls that Earth's annual pilgrimage around the Sun is not perfectly circular, but it is pretty close. Over time, the pull of gravity from our Solar system's two largest gas giant planets, Jupiter and Saturn, causes the shape of Earth's orbit to vary from nearly circular to slightly elliptical.

Eccentricity: *Eccentricity* measures how much the shape of Earth's orbit departs from a perfect circle. These variations affect the distance between Earth and the Sun. Fig. 2.28 provides an illustration of changes in eccentricity of Earth (100,000-year cycle), exaggerated so that the effects can be clearly seen. Earth's orbit shape varies between 0.0034 (almost a perfect circle) to 0.058 (slightly elliptical). Video E1 provides a lively illustration of what is shown in Fig. 2.28.

Buis (2020) explains that eccentricity is the reason why our seasons are slightly different lengths, with summers in the Northern Hemisphere currently about 4.5 days longer than winters, and springs about 3 days longer than autumns. As eccentricity decreases, the length of our seasons gradually evens out.

The difference in the distance between Earth's closest approach to the Sun (known as perihelion), which occurs on or about January 3 each year, and its farthest departure from the Sun (known as aphelion) on or about July 4, is currently about 5.1 million kilometers (about 3.2 million miles), a variation of 3.4%. That means each January, about 6.8% more incoming solar radiation reaches Earth than it does each July (Buis, 2020).

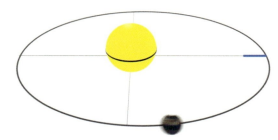

Fig. 2.28 Illustration of changes in eccentricity of Earth (100,000-year cycle), exaggerated so that the effects can be clearly seen. Earth's orbit shape varies between 0.0034 (almost a perfect circle) to 0.058 (slightly elliptical). Credit: Climate.NASA.Gov. Video E1 presents live demonstration. (Source: Buis, A. (2020), Milankovitch (orbital) cycles and their role in Earth's climate, NASA's Jet Propulsion Laboratory Report, Global Climate Change, Vital Signs of the Planet; https://climate.nasa.gov/news/2948/milankovitch-orbital-cycles-and-their-role-in-earths-climate/).

When Earth's orbit is at its most elliptic, about 23% more incoming solar radiation reaches Earth at our planet's closest approach to the Sun each year than does at its farthest departure from the Sun. *Currently, Earth's eccentricity is near its least elliptic (most circular) and is very slowly decreasing, in a cycle that spans about 100,000 years* (Buis, 2020), that is, the cyclicity of Earth's eccentricity variation is 100,000 years.

The total change in global annual insolation due to the eccentricity cycle is very small. Because variations in Earth's eccentricity are fairly small, they are a relatively minor factor in annual seasonal climate variations (Buis, 2020).

Obliquity: The angle between Earth's axis of rotation and Earth's orbital plane as it orbits around the Sun (i.e., Earth's axial tilt angle) is known as *obliquity*. Obliquity is responsible for the existence of seasons occurring on Earth. Over the last million years, Earth's obliquity has varied between 22.1° and 24.5°. The greater Earth's axial tilt angle, the more extreme our seasons are, as each hemisphere receives more solar radiation during its summer, when the hemisphere is tilted toward the Sun, and less during winter, when it is tilted away. Larger tilt angles favor periods of deglaciation (the melting and retreat of glaciers and ice sheets). These effects are not uniform globally—higher latitudes receive a larger change in total solar radiation than areas closer to the equator. Fig. 2.29 illustrates the changes in Earth's obliquity over a period of 41,000 years (i.e., the cyclicity of Earth's obliquity variation is 41,000 years). Video E2 provides a live demonstration of the changes in Earth's obliquity over a period of 41,000 years.

Earth's axis is currently tilted 23.4°, or about half way between its extremes, and this angle is very slowly decreasing in a cycle that spans about 41,000 years. It was last at its maximum tilt about 10,700 years ago and will reach

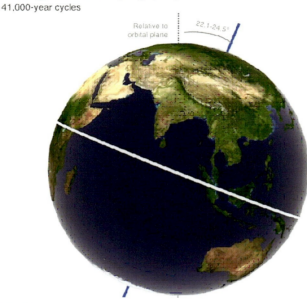

FIG. 2.29 Illustration of the changes in Earth's obliquity over a period of 41,000 years (i.e., the cyclicity of Earth's obliquity variation is 41,000 years). *Credit: NASA/JPL-Caltech Video E2 presents live demonstration. (Source: Buis, A. (2020), Milankovitch (orbital) cycles and their role in Earth's climate, NASA's Jet Propulsion Laboratory Report, Global Climate Change, Vital Signs of the Planet; https://climate.nasa.gov/news/2948/milankovitch-orbital-cycles-and-their-role-in-earths-climate/).*

FIG. 2.30 Illustration of Earth's 26,000-year cycle axial precession (wobble). *Credit: NASA/JPL-Caltech Video E3 presents live demonstration of Earth's axial precession (wobble) over a period of 26,000 years. (Source: Buis, A. (2020), Milankovitch (orbital) cycles and their role in Earth's climate, NASA's Jet Propulsion Laboratory Report, Global Climate Change, Vital Signs of the Planet; https://climate.nasa.gov/news/2948/milankovitch-orbital-cycles-and-their-role-in-earths-climate/).*

its minimum tilt about 9800 years from now. As obliquity decreases, it gradually helps make our seasons milder, resulting in increasingly warmer winters, and cooler summers that gradually, over time, allow snow and ice at high latitudes to build up into large ice sheets. As ice cover increases, it reflects more of the Sun's energy back into space, promoting even further cooling (Buis, 2020).

Precession: As Earth rotates, it wobbles slightly upon its axis (Fig. 2.30), like a slightly off-center spinning toy top. This wobble is due to tidal forces caused by the gravitational influences of the Sun and Moon that cause Earth to bulge at the equator, affecting its rotation. The trend in the direction of this wobble relative to the fixed positions of stars is known as *axial precession*. The cycle of axial precession spans about 25,771.5 years (Buis, 2020). Video E3 provides a lively demonstration of Earth's Precession cycle shown in Fig. 2.30.

Axial precession makes seasonal contrasts more extreme in one hemisphere and less extreme in the other. Currently perihelion occurs during winter in the Northern Hemisphere and in summer in the Southern Hemisphere. This makes Southern Hemisphere summers hotter, and moderates the Northern Hemisphere seasonal variations. But in about 13,000 years, axial precession will cause these conditions to flip, with the Northern Hemisphere seeing more extremes in solar radiation and the Southern Hemisphere experiencing more moderate seasonal variations (Buis, 2020).

Apsidal precession: There's also *apsidal precession*. Not only does Earth's axis wobble, but Earth's entire orbital ellipse also wobbles irregularly, primarily due to its interactions with Jupiter and Saturn. The cycle of apsidal precession spans about 112,000 years. Apsidal precession changes the orientation of Earth's orbit relative to the elliptical plane (Buis, 2020).

The combined effects of axial and apsidal precession result in an overall precession cycle spanning about 23,000 years on average (Buis, 2020).

The small changes set in motion by *Milankovitch cycles* operate separately and together to influence Earth's climate over very long timespans, leading to larger changes in our climate over tens of thousands to hundreds of thousands of years. Milankovitch combined the cycles to create a comprehensive mathematical model for calculating differences in solar radiation at various Earth latitudes along with corresponding surface temperatures. The model is sort of like a *climate time machine* (https://climate.nasa.gov/interactives/climate-time-machine): it can be run backward and forward to examine past and future climate conditions (Buis, 2020).

Milankovitch assumed changes in radiation at some latitudes and in some seasons are more important than others to the growth and retreat of ice sheets. In addition, it was his belief that obliquity was the most important of the three cycles for climate, because it affects the amount of insolation in Earth's northern high-latitude regions during summer (the relative role of precession versus obliquity is still a matter of scientific study) (Buis, 2020).

Milankovitch calculated that Ice Ages occur approximately every 41,000 years. Subsequent research confirms that they did occur at 41,000-year intervals between one and three million years ago. But about 800,000 years ago, the cycle of Ice Ages lengthened to 100,000 years, matching Earth's eccentricity cycle (Buis, 2020).

Milankovitch's work was supported by other researchers of his time, and he authored numerous publications on his hypothesis. But it was not until about 10 years after his death in 1958 that the global science community began to take serious notice of his theory. Using deep-sea sediment cores Hays et al. (1976) found that, in agreement with Milankovitch cycles, for 500,000 years, major climatic changes have followed variations in obliquity and precession, with Ice Ages occurring when Earth was undergoing different stages of orbital variation.

Hays et al. (1976) summarized their findings thus:

1. Three indices of global climate have been monitored in the record of the past 450,000 years in Southern Hemisphere ocean-floor sediments.
2. Over the frequency range 10^{-4}–10^{-5} cycle per year, climatic variance of these records is concentrated in three discrete spectral peaks at periods of 23,000, 42,000, and approximately 100,000 years. These peaks correspond to the dominant periods of the earth's solar orbit, and contain respectively about 10%, 25%, and 50% of the climatic variance.
3. The 42,000-year climatic component has the same period as variations in the obliquity of the earth's axis and retains a constant phase relationship with it.
4. The 23,000-year portion of the variance displays the same periods (about 23,000 and 19,000 years) as the quasi-periodic precession index.
5. The dominant, 100,000-year climatic component has an average period close to, and is in phase with, orbital eccentricity. Unlike the correlations between climate and the higher-frequency orbital variations (which can be explained on the assumption that the climate system responds linearly to orbital forcing), an explanation of the correlation between climate and eccentricity probably requires an assumption of nonlinearity.
6. It is concluded that changes in the earth's orbital geometry are the fundamental cause of the succession of Quaternary ice ages.
7. A model of future climate based on the observed orbital-climate relationships, but ignoring anthropogenic effects, predicts that the long-term trend over the next several thousand years is toward extensive Northern Hemisphere glaciation.

Several other projects and studies have also upheld the validity of Milankovitch's work, including research using data from ice cores in Greenland and Antarctica that has provided strong evidence of Milankovitch cycles going back many hundreds of thousands of years. In addition, his work has been embraced by the National Research Council of the US National Academy of Sciences (Buis, 2020).

Scientific research to better understand the mechanisms that cause changes in Earth's rotation and how specifically Milankovitch cycles combine to affect climate is ongoing. But the theory that they drive the timing of glacial-interglacial cycles is well accepted (Buis, 2020).

2.16 Volcano eruptions on land causing atmospheric cooling and those happening underwater causing abnormal atmospheric warming

It is well-known that volcanic eruptions taking place on land (Fig. 2.31) inject enormous quantities of ash and aerosol (a suspension of fine solid or liquid particles dispersed in air) clouds into the atmosphere and produce large quantities of carbon dioxide. For example, the AD 1783–1784 Laki eruption in south–east Iceland produced about 14 km^3 of basalt (thin, black, fluid lava) during more than 8 months of activity. The ash cloud caused a dense haze across Europe that dimmed the Sun over a considerable region covering even far beyond Iceland. In addition to ash, the eruptive cloud consisted primarily of vast quantities of sulfur dioxide (SO_2), hydrogen chloride (HCl), and hydrogen fluoride gases (HF). It has been found that the formation of atmospheric sulfur aerosols has a more substantial effect on global temperatures than simply the volume of ash produced during an eruption. Sulfate aerosols appear to take several years to settle out of the atmosphere, which is one of the reasons their effects are so widespread and long lasting.

There are several instances in which volcano eruptions have caused atmospheric cooling (https://www.scientificamerican.com/article/how-do-volcanoes-affect-w/). However, such a cooling effect was presumably first reported by the well-known American inventor and statesman Benjamin Franklin. It was also Benjamin Franklin who made what may have been the first connection between volcanoes and global climate. In 1784, Benjamin Franklin observed that during the summer of 1783, the climate was abnormally cold, both in Europe and the United States. The ground froze early, the first snow stayed on the ground without melting, the winter was more severe than usual, and there seemed to be "a constant fog over all Europe, and a great part of North America." What Benjamin Franklin

98 Water worlds in the solar system

FIG. 2.31 Mosaic of some of the eruptive structures formed during volcanic activity: a Plinian eruption column, Hawaiian pahoehoe flows, and a lava arc from a Strombolian eruption. *(Source: https://commons.wikimedia.org/wiki/File:Lava_forms.jpg) Image author: יקי (in Hebrew). Copyright information: This file is licensed under the Creative Commons Attribution-Share Alike 3.0 Unported license.*

observed was indeed the result of the Laki volcanic eruption activity.

Laki is a volcanic fissure in the western part of Vatnajökull National Park, Iceland. The system erupted violently over an 8-month period between June 1783 and February 1784 from the Laki fissure and the adjoining volcano Grímsvötn, pouring out an estimated 42 billion tons or 14 km³ (3.4 cu mi) of basalt lava and clouds of poisonous hydrofluoric acid and sulfur dioxide compounds that contaminated the soil, leading to the death of over 50% of Iceland's livestock population, and the destruction of the vast majority of all crops. This led to a famine which then killed approximately 69% of the island's human population (Karlsson, 2000). The lava flows also destroyed 21 villages.

The Laki eruption and its aftermath caused a drop in global temperatures, as 120 million tons of sulfur dioxide was spewed into the Northern Hemisphere. This caused crop failures in Europe and may have caused droughts in North Africa and India.

According to Karen Harpp, record atmospheric temperature lows in the United States of America occurred during the winter of 1783–1784 in association with the Laki eruptions in Iceland (https://www.scientificamerican.com/article/how-do-volcanoes-affect-w/). In fact, the temperature decreased about 1°C in the Northern Hemisphere overall. For comparison, the global temperature of the most recent Ice Age (about 11,700 years ago, when glaciers covered huge parts of the planet Earth) was only about 5°C below the current average.

In terms of atmospheric temperature change, volcano eruptions of various kinds have been reported to have caused contradictory consequences (i.e., cooling effects in some instances and warming effects in some other instances). It is interesting, perhaps ironic, to note that while volcanic eruptions that occurred on land gave rise to abnormal cooling, volcanic eruptions that occurred underwater gave rise to abnormal atmospheric warming (e.g., the PETM event that occurred ~56 million years ago with an abnormal increase of 5–9°C in atmospheric temperature). An example of a volcanic eruption that occurred on land is the just-mentioned 1783-Laki eruptions. An example of a chain of volcanic eruptions that took place in water is that occurred on the Mid-Atlantic ridge during the opening of the North Atlantic.

These seemingly contradictory effects presented by volcanic eruptions occurring on land and underwater can be attributed to the nature of ejecta ultimately reaching the atmosphere. For example, in the case of volcanic eruptions occurring on land, the ash clouds reach the stratosphere (the layer of gases surrounding the Earth at a height of between 15 and 50 km). These thick and vast clouds cause a dense haze that dim the Sun. Furthermore, the eruptive cloud consisting primarily of vast quantities of sulfur aerosols reflect the sunlight away from Earth's surface thereby preventing its entry on the Earth's surface. Although the carbon dioxide greenhouse gas is one of the ejecta materials associated with volcanic eruption, its warming effect is over-ridden by the dominant cooling effect caused by the combined effect of thick and vast ash clouds and the aerosols. This explains the observed cooling effect experienced in association with volcanic eruptions occurring on land.

On the other hand, volcanic eruptions occurring underwater (e.g., from oceanic ridges) give rise to atmospheric warming because, in this case, the volcanic ejecta such as ash clouds and aerosol-forming sulfur dioxide (SO_2) (that are responsible for atmospheric cooling) get dissolved in the water layer lying over the volcano and, therefore, these materials are prevented from reaching the atmosphere. However, the carbon dioxide gas (CO_2) ejecta that are responsible for atmospheric warming (by virtue of greenhouse effect) largely get access to the atmosphere, although a certain percentage of CO_2 gas gets dissolved in seawater, contributing to ocean acidification. This mechanism allows volcanic eruptions occurring underwater to generate abnormal atmospheric warming such as that occurred in association with PETM.

2.17 Synthesis of marine proxy temperature data across the Paleocene–Eocene thermal maximum

Dramatic warming and upheaval of the carbon system at the end of the Paleocene Epoch have been linked to massive dissociation of sedimentary methane hydrate. However, testing the Paleocene–Eocene thermal maximum hydrate dissociation hypothesis has been hindered by the inability of available proxy records to resolve the initial sequence of events. The cause of the Paleocene–Eocene thermal maximum carbon isotope excursion remains speculative, primarily due to uncertainties in the timing and duration of the Paleocene–Eocene thermal maximum (Thomas et al., 2002).

Thomas et al. (2002) reported high-resolution stable isotope records based on analyses of single planktonic and benthic foraminiferal shells from Ocean Drilling Program Site 690 (Weddell Sea, Southern Ocean), demonstrating that the initial carbon isotope excursion was geologically instantaneous and was preceded by a brief period of gradual surface-water warming. Both of these findings support the thermal dissociation of methane hydrate as the cause of the Paleocene–Eocene thermal maximum carbon isotope excursion. Furthermore, the data reveal that the methane-derived carbon was mixed from the surface ocean downward, suggesting that a significant fraction of the initial dissociated hydrate methane reached the atmosphere prior to oxidation.

In a study, Lunt et al. (2010) proposed changes in ocean circulation as a trigger mechanism for the large coupled climate and carbon cycle perturbations at the Paleocene–Eocene Thermal Maximum (PETM, ca. 55 Ma). According to them, an abrupt warming of oceanic intermediate waters could have initiated the thermal destabilization of sediment-hosted methane gas hydrates and potentially triggered sediment slumps and slides. In an ensemble of fully coupled atmosphere-ocean general circulation model (AOGCM) simulations of the late Paleocene and early Eocene, Lunt et al. (2010) identified such a circulation-driven enhanced intermediate-water warming. Lunt et al. (2010)'s study indicated that this warming is largely focused on the equatorial and South Atlantic and is driven by a significant reduction in deep-water formation from the Southern Ocean. According to Lunt et al. (2010), this scenario is consistent with altered PETM circulation patterns inferred from benthic carbon isotope data and the intensity of deep-sea carbonate dissolution in the South Atlantic. The linkage between intermediate-water warming and gas hydrate destabilization could provide an important feedback in the establishment of peak PETM warmth (Lunt et al., 2010).

It is now generally recognized that the PETM is clearly a planetary-scale perturbation of the Earth's carbon cycle, biosphere and climate. However, there are several investigators who think that there remains considerable uncertainty over the size (e.g., Cui et al., 2011, 2012; Sluijs et al., 2012) and source (DeConto et al., 2012) of the PETM carbon release. Consequently, the exact magnitude of greenhouse gas forcing arising from this carbon release is poorly constrained. Whereas the extent of the PETM carbon cycle perturbation has received considerable attention (e.g., Dickens, 2003; Zachos et al., 2005), there has been no compilation and thorough reassessment of the available PETM temperature proxy data. When quoted, the PETM sea surface temperature anomaly is usually stated as a range, typically "4–8°C" (Thomas et al., 2002), or more recently "5–9°C" (Zeebe et al., 2009). The robustness, interpretation, quantity and geographic spread of the underlying data are, however, rarely commented upon. Even recent "PETM" climate modeling studies have used a range of published PETM paleo-temperature proxy data as if they were relatively uncomplicated, unbiased and without inherent uncertainty (Lunt et al., 2010).

Constraining the greenhouse gas forcing, climatic warming and estimates of climate sensitivity across ancient large transient warming events is a major challenge to the paleoclimate research community. It is generally recognized that the paleo-temperature proxy data are both spatially limited and subject to varying degrees of uncertainty and potential biases. With a view to presenting a full synthesis and interpretation of all the available data, Jones et al. (2013) used the results of an early Eocene fully coupled Atmosphere-Ocean General Circulation Model (AOGCM).

Jones et al. (2013)'s compilation of PETM proxy data is derived from near-surface dwelling planktic foraminifera, sea surface-derived organic lipid biomarkers tumor-derived exosomal (TEX_{86}) and benthic foraminifera, mostly from deep-sea drilling cores recovered in intermediate to abyssal water depths (~1000–3000 m). Jones et al. (2013) treated these data, collected over a period of 20 years prior to 2013, with a consistent methodology for the identification of pre-PETM and PETM intervals based on carbon isotope stratigraphies (the study of the composition, relative positions, etc., of rock strata in order to determine their geological history), the application of uniform paleo-temperature calibrations and a more robust and transparent assessment of uncertainties associated with temporal averaging and calibrations. Jones et al. (2013) also discussed major uncertainties in these paleo-temperature estimates that cannot yet be reliably quantified, including the preservation-state of foraminiferal carbonate and the application of TEX_{86} calibrations from the modern oceans to ancient warm-climates states.

Jones et al. (2013)'s study presented a detailed reassessment of the existing marine paleo-temperature estimates across the PETM hyperthermal event. They applied a consistent and up-to-date set of calibrations to all published data. This provided the basis for a more rigorous comparison of the PETM temperature anomaly between records published over the past two decades prior to 2013, from widely distributed locations.

Jones et al. (2013)'s investigations suggest that it is unlikely that the global mean surface temperature anomaly across the PETM exceeded 5°C. These researchers are optimistic that if this assessment of the global temperature anomaly proves to be relatively robust then, with refined controls on the PETM carbon cycle perturbation and carbon input, a more accurate estimate of early Paleogene climate sensitivity is within reach.

On the basis of the above-mentioned comparison and taking into account the patterns of intermediate-water warming Jones et al. (2013) estimated that the global mean surface temperature anomaly for the PETM is within the range of 4–5°C. These investigators hoped that a clear presentation of the proxy data and its limitations will justify and encourage future efforts to gather reliable PETM paleo-temperature proxy data across both marine and terrestrial environments and from geographically disparate localities.

2.18 Fate of excess carbon released during the Paleocene–Eocene thermal maximum event

An approximately 170,000-year-long period of global warming about 56 million years ago, during the ill-famous PETM event, has been attributed to the release of an enormous quantity of carbon into the oceans, atmosphere, and biosphere (see Zeebe et al., 2009). However, the fate of this excess carbon at the end of the PETM event is poorly constrained. Several investigators (e.g., Dickens et al., 1997; Bains et al., 2000; Dickens, 2001; Zeebe et al., 2009) have attributed the drawdown of atmospheric carbon dioxide to an increase in the weathering of silicates or to increased rates of organic carbon burial. In an attempt to provide clarity on the fate of the excess carbon released, Bowen and Zachos (2010) developed constraints on the rate of carbon drawdown based on rates of carbon isotope change in well-dated marine and terrestrial sediments spanning the event. They found that the rate of recovery is an order of magnitude more rapid than that expected for carbon drawdown by silicate weathering alone. Their results imply that more than 2000 Petagram (Pg) of carbon were sequestered as organic carbon over 30,000–40,000 years at the end of the PETM. Their findings suggest that the accelerated sequestration of organic carbon could be a result of the regrowth of carbon stocks in the biosphere or shallow lithosphere that were released at the onset of the event.

References

Albarède, F., 2009. Volatile accretion history of the terrestrial planets and dynamic implications. Nature 461, 1227–1233.

Alexander, C.M.O.D., Fogel, M., Yabuta, H., Cody, G.D., 2007. The origin and evolution of chondrites recorded in the elemental and isotopic compositions of their macromolecular organic matter. Geochim. Cosmochim. Acta 71 (17), 4380–4403.

Alexander, C.M., Bowden, R., Fogel, M.L., Howard, K.T., Herd, C.D., Nittler, L.R., 2012. The provenances of asteroids, and their contributions to the volatile inventories of the terrestrial planets. Science 337 (6095), 721–723. doi:10.1126/science.1223474.

Algeo, T.J., Berner, R.A., Maynard, J.B., Scheckler, S.E., 1995. Late Devonian oceanic anoxic events and biotic crises: "Rooted" in the evolution of vascular land plants? GSA Today 5 (3).

Allen, J.F., Martin, W., 2007. Evolutionary biology: out of thin air. Nature 445 (7128), 610–612.

Allwood, A.C., Walter, M.R., Burch, I.W., Kamber, B.S., 2007. 3.43 billion-year-old stromatolite reef from the Pilbara Craton of Western Australia: Ecosystem-scale insights to early life on Earth. Precamb. Res. 158, 198–227. doi:10.1016/j.precamres.2007.04.013.

Allwood, A.C., Grotzinger, J.P., Knoll, A.H., Burch, I.W., Anderson, M.S., Coleman, M.L., Kanika, I., 2009. Controls on development and diversity of early Archean stromatolites. Proc. Natl. Acad. Sci. USA 106 (24), 9548–9555. doi:10.1073/pnas.0903323106.

Altwegg, K., Balsiger, H., Bar-Nun, A., Berthelier, J.J., Bieler, A., Bochsler, P., et al., 2015. Cometary science. 67P/Churyumov-Gerasimenko, a Jupiter family comet with a high D/H ratio. Science 347 (6220), 1261952.

Arndt, N.T., Nisbet, E.G., 2012. Processes on the Young Earth and the habitats of early life. Annu. Rev. Earth Planet Sci. 40, 521–549.

Ashkenazy, Y., Gildor, H., Losch, M., Macdonald, F.A., Schrag, D.P., Tziperman, E., 2013. Dynamics of a Snowball Earth ocean. Nature 495, 90–93.

Bada, J.L., Bigham, C., Miller, S.L., 1994. Impact melting of frozen oceans on the early Earth: implications for the origin of life. Proc. Natl. Acad. Sci. U S A 91, 1248–1250.

Bagley, M., 2014. Triassic period facts: climate, animals & plants. Live Science. https://www.livescience.com/mass-extinction-events-that-shaped-Earth.html.

Bains, S., Norris, R.D., Corfield, R.M., Faul, K.L., 2000. Termination of global warmth at the Palaeocene/Eocene boundary through productivity feedback. Nature 407, 171–174.

Bains, W., 2004. Many chemistries could be used to build living systems. Astrobiology 4, 137–167.

Barash, M.S., 2014. Mass extinction of the marine biota at the Ordovician–Silurian Transition due to environmental changes. Oceanology 54 (6). doi:10.1134/S0001437014050014.

Baumgartner, R.J., Van Kranendonk, M.J., Wacey, D., Fiorentini, M.L., Saunders, M., Caruso, S., Pages, A., Homann, M., Guagliardo, P., 2019. Nano–porous pyrite and organic matter in 3.5-billion-year-old stromatolites record primordial life. Geology 47 (11), 1039–1043.

Beerling, D.J., Fox, A., Stevenson, D.S., Valdes, P.J., 2011. Enhanced chemistry-climate feedbacks in past greenhouse worlds. Proc. Natl. Acad. Sci. USA 108, 9770–9775.

Bekker, A., et al., 2004. Dating the rise of atmospheric oxygen. Nature 427 (6970), 117–120.

Bercovici, D., Ricard, Y., 2014. Plate tectonics, damage and inheritance. Nature 508, 513–516.

Berkner, L.V., Marshall, L.C., 1965. On the origin and rise of oxygen concentration in the Earth's atmosphere. J. Atmos. Sci. 22 (3), 225–261.

Berner, R.A., Lasaga, A.C., Garrels, R.M., 1983. The carbonate–silicate geochemical cycle and its effect on atmospheric carbon dioxide over the past 100 million years. Am. J. Sci. 283, 641–683. doi:10.2475/ajs.283.7.641.

Bjoraker, G.L., Wong, M.H., de Pater, I., Hewagama, T., Ádámkovics, M., Orton, G.S., 2018. The gas composition and deep cloud structure of Jupiter's Great Red Spot. Astron. J. 156 (3), 101–116.

Blankenship, R.E., 2002. Molecular Mechanisms of Photosynthesis. Blackwell Science, Oxford.

Bockelée-Morvan, D., Brooke, T.Y., Crovisier, J., 1995. On the origin of the 3.2–3.6 μm emission features in comets. Icarus 116, 18–39.

Bond, D.P.G., Grasby, S.E., 2020. Late Ordovician mass extinction caused by volcanism, warming, and anoxia, not cooling and glaciation. Geology 48 (8), 777–781. https://doi.org/10.1130/G47377.1.

Bottke, W.F., Norman, M.D., 2017. The Late Heavy Bombardment. Annu. Rev. Earth Planet. Sci. 45 (1), 619–647. doi:10.1146/annurev-earth-063016-020131.

Bowen, G.J., Zachos, J.C., 2010. Rapid carbon sequestration at the termination of the Palaeocene–Eocene thermal maximum. Nat. Geosci. 3, 866–869.

Brack, A., Fitton, B., Raulin, F., 1999. Exobiology in the Solar System and the Search for Life on Mars. ESA Publications Division, ESTEC, Noordwijk, The Netherlands.

Budyko, M.I., Ronov, A.B., 1979. Chemical evolution of the atmosphere in the Phanerozoic. Geochem. Int. 16(3), 1–9.

Buis, A., 2020. Milankovitch (orbital) cycles and their role in Earth's climate, NASA's Jet Propulsion Laboratory Report, Global Climate Change, Vital Signs of the Planet, https://climate.nasa.gov/news/2948/milankovitch-orbital-cycles-and-their-role-in-earths-climate/

Cady, S.L., Farmer, J.D., Grotzinger, J.P., Schopf, J.W., Steele, A., 2003. Morphological biosignatures and the search for life on Mars. Astrobiology 3 (2), 351–368. doi:10.1089/153110703769016442.

Canfield, D.E., Poulton, S.W., 2011. Ferruginous conditions: a dominant feature of the ocean through Earth's history. Elements 7 (2), 107–112. doi:10.2113/gselements.7.2.107.

Carmichael, S.K., Waters, J.A., Königshof, P., Suttner, T.J., Kido, E., 2019. Paleogeography and paleoenvironments of the Late Devonian Kellwasser event: a review of its sedimentological and geochemical expression. Global Planet. Change 183, 102984.

Cates, N.L., Mojzsis, S.J., 2007. Pre-3750 Ma supracrustal rocks from the Nuvvuagittuq supracrustal belt, northern Québec. Earth Planet. Sci. Lett. 255 (1–2), 9–21.

Catling, D.C., Kasting, J.F., 2017. Atmospheric Evolution on Inhabited and Lifeless Worlds. Cambridge University Press, Cambridge. doi:10.1017/9781139020558 ISBN 978-1-139-02055-8.

Catling, D.C., Zahnle, K.J., 2020. The Archean atmosphere. Sci. Adv. 6 (9), eaax1420. doi:10.1126/sciadv.aax1420.

Cernicharo, J., Crovisier, J., 2005. Water in space: the water world of ISO. Space Sci. Rev. 119, 29.

Chanson, H., 2011. Current knowledge in tidal bores and their environmental, ecological and cultural impacts. Environ. Fluid Mech. 11, 77–98.

Chapelle, F.H., O'Neil, K., Bradley, P.M., Methé, B.A., Ciufo, S.A., et al., 2002. A hydrogen-based subsurface microbial community dominated by methanogens. Nature 415, 312–315.

Chatterji, A., Ansari, Z.A., Ingole, B.S., Sreepada, R.A., Kanti, A., Parulekar, A.H., 1994. Effect of lunar periodicity on the abundance of crabs from the Goa coast. Indian J. Mar. Sci. 23, 180–181.

Cheng, Y., Andersen, O.B., 2010. Improvement in global ocean tide model in shallow water regions, Ocean Surface Topography Science Team Meeting, Lisbon, October 18–22. Poster, SV.1-68, 45. http://www.space.dtu.dk/english/Research/Scientific_data_and_models/Global_Ocean_Tide_Model.

Chopra, A., Lineweaver, C.H., 2016. The case for a Gaian bottleneck: the biology of habitability. Astrobiology 16 (1), 7–22. doi:10.1089/ast.2015.1387.

Claeys, P., Casier, J.-G., Margolis, S.V., 1992. Microtektites and mass extinctions: evidence for a Late Devonian asteroid impact. Science 257, 1102–1104.

Claeys, P., Morbidelli, A., Gargaud, M., Amils, R., Quintanilla, J.C., Cleaves II, H.J., Irvine, W.M., Pinti, D.L., Viso, M., 2011. Late Heavy Bombardment. Encyclopedia of Astrobiology. Springer Berlin Heidelberg, pp. 909–912. doi:10.1007/978-3-642-11274-4_869.

Clancy, E.P., 1968. The Tides Pulses of the Earth. Anchor Books, New York, NY, pp. 228.

Cleal, C.J., Lazarus, M., Townsend, A., 2005. Illustrations and illustrators during the 'Golden Age' of palaeobotany: 1800–1840. In: Bowden, A.J., Burek, C.V., Wilding, R. (Eds.), History of Palaeobotany: Selected Essays. Geological Society of London, London, pp. 41. ISBN 9781862391741.

Claudi, R., 2017. Exoplanets: Possible biosignatures, *Proceedings of Science*, Frontier Research in Astrophysics âAS II, 23-28 May 2016, Mondello (Palermo), Italy arXiv:1708.05829v1 [astro-ph.EP] 19 Aug 2017.

Cleland, C.E., Chyba, C.F., 2002. Defining life. Orig. Life Evol. Biosph. 32, 387–393.

Coates, R., 2007. The genealogy of eagre 'tidal surge in the river Trent. Eng. Lang. Linguist 11 (3), 507–523.

Copper, P., 1986. Frasnian/Famennian mass extinction and cold-water oceans. Geology 14, 835–839.

Cox, G.M., Halverson, G.P., Minarik, W.G., Le, H.D.P., Macdonald, F.A., Bellefroid, E.J., Strauss, J.V., 2013. Neoproterozoic iron formation: An evaluation of its temporal, environmental and tectonic significance. Chem. Geol. 362, 232–249.

Cui, Y., Kump, L.R., Ridgwell, A.J., Charles, A.J., Junium, C.K., Diefendorf, A.F., K.H.Freeman, N.M.Urban, Harding, I.C., 2011. Slow release of fossil carbon during the Palaeocene–Eocene thermal maximum. Nat. Geosci. 4, 481–485.

Cui, Y., Kump, L.R., Ridgwell, A.J., Charles, A.J., Junium, C.K., Diefendor, A.F., Freeman, K.H., Urban, N.M., Harding, I.C., 2012. Constraints on hyperthermals—reply. Nat. Geosci. 5, 231–232.

Dauphas, N., 2003. The dual origin of the terrestrial atmosphere. Icarus 165 (2), 326–339. doi:10.1016/S0019-1035(03)00198-2.

DeConto, R.M., Galeotti, S., Pagani, M., Tracy, D., Schaefer, K., Zhang, T., Pollard, D., Beerling, D.J., 2012. Past extreme warming events linked to massive carbon release from thawing permafrost. Nature 484, 87–91.

Des Marais, D.J., Harwit, M.O., Jucks, K.W., Kasting, J.F., Lin, D.N.C., Lunine, J.I., Schneider, J., Seager, S., Traub, W.A., Woolf, N.J., 2002. Remote sensing of planetary properties and biosignatures on extrasolar terrestrial planets. Astrobiology 2, 153.

Dickens, G.R., O'Neil, J.R., Rea, D.K., Owen, R.M., 1995. Dissociation of oceanic methane hydrate as a cause of the carbon isotope excursion at the end of the Paleocene. Paleoceanography 10, 965–971.

Dickens, G.R., Castillo, M.M., Walker, J.C.G., 1997. A blast of gas in the latest Paleocene, simulating first-order effects of massive dissociation of oceanic methane hydrate. Geology 25, 259–262.

Dickens, G.R., 2001. In: Western North Atlantic Palaeogene And Cretaceous Palaeoceanography 9. Kroon, K., Norris, R.D., Klaus, A. (Eds.), 9. Geological Society Publishing House, pp. 293–305.

Dickens, J., 2003. Rethinking the global carbon cycle with a large, dynamic and microbially mediated gas hydrate capacitor. Earth Planet. Sci. Lett. 213, 169–183.

Drake, M.J., 2005. Origin of water in the terrestrial planets. Meteorit Planet Sci. 40, 519–527.

Drouart, A., Dubrulle, B., Gautier, D., Robert, F., 1999. Structure and transport in the solar nebula from constraints on deuterium enrichment and giant planets formation. Icarus 140 (1), 129–155. doi:10.1006/icar.1999.6137.

Duda, J-P., Van Kranendonk, M.J., Thiel, V., Ionescu, D., Strauss, H., Schäfer, N., Reitner, J., 2016. A rare glimpse of Paleoarchean life: geobiology of an exceptionally preserved microbial mat facies from the 3.4 Ga Strelley Pool formation, Western Australia. PLOS One 11 (1), e0147629.

Dunkley, J.T., et al., 2013. Climate model and proxy data constraints on ocean warming across the Paleocene–Eocene Thermal Maximum. Earth Sci. Rev. 125, 123–145.

Dustan, P., Pinckney Jr., J.L., 1989. Tidally induced estuarine phytoplankton patchiness. Limnol. Oceanogr. 34 (2), 410–419.

Edson, A.R., Kasting, J.F., Pollard, D., Lee, S., Bannon, P.R., 2012. The Carbonate–Silicate cycle and CO_2/climate feedbacks on tidally locked terrestrial planets. Astrobiology 12 (6), 562–571. doi:10.1089/ast.2011.0762.

Elkins-Tanton, L.T., 2008. Linked magma ocean solidification and atmospheric growth for Earth and Mars. Earth Planet Sci Lett 271, 181–191.

Elkins-Tanton, L.T., 2011. Formation of early water oceans on rocky planets. Astrophys. Space Sci. 332, 359–364.

Elkins-Tanton, L.T., 2012. Magma oceans in the inner Solar System. Annu. Rev. Earth Planet. Sci. 40, 113–139.

Eriksson, P.G., Cheney, E.S., 1992. Evidence for the transition to an oxygen-rich atmosphere during the evolution of red beds in the lower proterozoic sequences of southern Africa. Precamb. Res. 54 (2–4), 257–269.

Etiope, G., Sherwood Lollar, B., 2013. Abiotic methane on Earth. Rev. Geophys. 51, 276–299.

Fay, R.R., 1988. Hearing in Vertebrates: A Psychophysics Databook. Hill-Fay Associates, Winnetka, IL.

Flannery, D.T., Walter, R.M., 2012. Archean tufted microbial mats and the Great Oxidation Event: new insights into an ancient problem. Austr. J. Earth Sci. 59 (1), 1–11. doi:10.1080/08120099.2011.607849.

Frei, R., Gaucher, C., Poulton, S.W., Canfield, D.E., 2009. Fluctuations in Precambrian atmospheric oxygenation recorded by chromium isotopes. Nature 461 (7261), 250–253.

Gaidos, E.J., et al., 1999. Life in ice-covered oceans. Science 284 (5420), 192–200.

Galeotti, S., et al., 2010. Orbital chronology of Early Eocene hyperthermals from the Contessa Road section, central Italy. Earth Planet. Sci. Lett. 290, 192–200.

García-Pichel, F., 1998. Solar ultraviolet and the evolutionary history of cyanobacteria. Origins Life Evol. B 28, 321–347.

Garcia-Berdeal, I., Hautala, S.L., Thomas, L.N., Johnson, H.P, 2006. Vertical structure of time-dependent currents in a mid-ocean ridge axial valley. Deep-Sea Res. I 53, 367e386.

Geiss, J., Gloecker, G., 1998. Abundances of deuterium and helium-3 in the protosolar cloud. Space Sci. Rev. 84 (1), 239–250.

Geldsetzer, H.H.J., Goodfellow, W.D., McLaren, D.J., Orchard, M.J., 1987. Sulfur-isotope anomaly associated with the Frasnian-Famennian extinction, Medicine Lake, Alberta, Canada. Geology 15, 393–396.

Gorshkov, V.G., Makarieva, A.M., Gorshkov, V.V., 2004. Revising the fundamentals of ecological knowledge: the biota-environment interaction. Ecol. Complexity 1, 17–36.

Grant, H.L., Stewart, R.W., Moillet, A., 1962. Turbulence spectra from a tidal channel. J. Fluid Mech. 12, 241e268.

Greshko, M., 2019. What are mass extinctions, and what causes them? Natl. Geogr., https://www.nationalgeographic.com/science/article/mass-extinction.

Grotzinger, J.P., Rothman, D.H., 1996. An abiotic model for stromatolite morphogenesis. Nature 383, 423–425. doi:10.1038/383423a0.

Gutjahr, M., Ridgwell, A., Sexton, P.F., Anagnostou, E., Pearson, P.N., Pälike, H., Norris, R.D., Thomas, E., Foster, G.L., 2017. Very large release of mostly volcanic carbon during the Palaeocene–Eocene thermal maximum. Nature 548, 573–577. doi:10.1038/nature23646.

Halliday, A.N., 2013. The origins of volatiles in the terrestrial planets. Geochim. Cosmochim. Acta 105, 146.

Hamilton, T.L., et al., 2016. The role of biology in planetary evolution: cyanobacterial primary production in low-oxygen Proterozoic oceans. Environ. Microbiol. 18 (2), 325–340. doi:10.1111/1462-2920.13118.

Harper, D.A.T., Hammarlund, E.U., Rasmussen, C.M.Ø., 2014. End Ordovician extinctions: a coincidence of causes. Gondwana Res. 25 (4), 1294–1307.

Hartogh, P., Lis, D.C., Bockelée-Morvan, D, de Val-Borro, M., Biver, N., Küppers, M., et al., 2011. Ocean-like water in the Jupiter-family comet 103P/Hartley 2. Nature 478 (7368), 218–220. doi:10.1038/nature10519.

Hays, J.D., Imbrie, J., Shackleton, N.J., 1976. Variations in the earth's orbit: pacemaker of the ice ages. Science 194, 1121–1132.

Hersant, F., Gautier, D., Hure, J., 2001. A two-dimensional model for the primordial nebula constrained by D/H measurements in the Solar System: Implications for the formation of Giant Planets. Astrophys. J. 554 (1), 391–407.

Higgins, J.A., Schrag, D.P., 2006. Beyond methane: towards a theory for the Paleocene–Eocene thermal maximum. Earth Planet. Sci. Lett. 245, 523–537.

Hofmann, A., Bekker, A., Rouxel, O., Rumble, D., Master, S., 2009. Multiple sulphur and iron isotope composition of detrital pyrite in Archaean sedimentary rocks: A new tool for provenance analysis. Earth Planet. Sci. Lett. 286 (3–4), 436–445.

Holland, 1978. The Chemistry of the Atmospheres and Oceans. Wiley Intersci., New York, pp. 351.

Hren, M.T., Tice, M.M., Chamberlain, C.P., 2009. Oxygen and hydrogen isotope evidence for a temperate climate 3.42 billion years ago. Nature 462, 205–208.

Jones, T.D., Lunt, D.J., Schmidt, D.N., Ridgwell, A., Sluijs, A., Valdes, P.J., Maslin, M., 2013. Climate model and proxy data constraints on ocean warming across the Paleocene–Eocene thermal maximum. Earth-Science Reviews 125, 123–145.

Joseph, A., Odametey, J.T., Nkebi, E.K., Pereira, A., Prabhudesai, R.G., Mehra, P., Rabinovich, A.B., Kumar, V., Prabhudesai, S., Woodworth, P.L., 2006. The 26 December 2004 Sumatra tsunami recorded on the coast of West Africa. Afr. J. Mar. Sci. 28 (3&4), 705–712.

Joseph, A., 2011. Tsunamis: Detection, Monitoring, and Early-Warning Technologies. *Elsevier/Academic Press*, New York, pp. 448. doi:10.1016/B978-0-12-385053-9.10001-8. ISBN: 978-0-12-385053-9.

Joseph, A., 2013. Measuring Ocean Currents: Tools, Technologies, and Data. *Elsevier Science & Technology Books Publishers*, New York, pp. 426. ISBN: 978-0-12-415990-7.

Joseph, 2016. Investigating Seafloors and Oceans: From Mud Volcanoes to Giant Squid. Elsevier Science & Technology Books Publishers, New York, pp. 581. ISBN: 978-0-12- 809357-3.

Karl, T.R., Trenberth, K.E., 2003. Modern global climate change. Science 302 (5651), 1719–1723.

Kasting, J.F., Ackerman, T.P., 1986. Climatic consequences of very high CO_2 levels in Earth's early atmosphere. Science 234, 1383–1385.

Kasting, J., 1993. Earth's early atmosphere. Science 259 (5097), 920–926. doi:10.1126/science.11536547.

Kasting, J.F., Siefert, J.L., 2002. Life and the evolution of Earth's atmosphere. Science 296 (5570), 1066–1068.

Katz, M.E., Cramer, B.S., Mountain, G.S., Katz, S., Miller, K.G., 2001. Uncorking the bottle: What triggered the Paleocene/Eocene thermal maximum methane release? Paleoceanography. doi:10.1029/2000 PA000615.

Kennett, J.P., Stott, L.D., 1991. Abrupt deep-sea warming, palaeoceanographic changes and benthic extinctions at the end of the Paleocene. Nature 353 (6341), 225–229.

Kinsland, G.L., Egedahl, K., Strong, M.A., Ivy, R., 2021. Chicxulub impact tsunami megaripples in the subsurface of Louisiana: Imaged in petroleum industry seismic data. Earth Planet. Sci. Lett. 570 (2021), 117063.

Koch, C., Chanson, H., 2008. Turbulent mixing beneath an Undular Bore Front. J. Coast. Res. 24 (4), 999–1007. http://dx.doi.org/10.2112/06-0688.1.

Koshland, D.E., 2002. The seven pillars of life. Science 295, 2215–2216.

Krissansen-Totton, J., Bergsman, D.S., Catling, D.C., 2016. On detecting biospheres from thermodynamic disequilibrium in planetary atmospheres. Astrobiology 16, 39–67.

Konn, C., Charlou, J.L., Holm, N.G., Mousis, O., 2015. The production of methane, hydrogen, and organic compounds in ultramafic-hosted hydrothermal vents of the Mid-Atlantic Ridge. Astrobiology 15, 381–399.

Kopp, R.E., Kirschvink, J.L., Hilburn, I.A., Nash, C.Z., 2005. The Paleoproterozoic snowball Earth: a climate disaster triggered by the evolution of oxygenic photosynthesis. Proc. Natl. Acad. Sci. USA 102 (32), 11131–11136.

Kröger, B., 2007. Concentrations of juvenile and small adult cephalopods in the Hirnantian cherts (Late Ordovician) of Porkuni, Estonia. Acta Palaeontol. Polon. 52 (3), 591–608.

Lammer, H., Bredehöft, J.H., Cousteins, A., Khodachenko, M.L., Kaltenegger, L., Grasset, O., Prieur, D., Raulin, F., Ehrenfreund, P., Yamauchi, M., Wahlund, J.E., Griessmeier, J.M., Stangl, G., Cockell, C.S., Kulikov, Yu.N., Grenfell, J.L., Rauer, H., 2009. What makes a planet Habitable? Astron. Astroph. Rev. 17, 181.

Le Treut, H., Somerville, R., Cubasch, U., Ding, Y., Mauritzen, C., Mokssit, A., Peterson, T., Prather, M, 2007. Historical overview of climate change science. In: Solomon, S., Qin, D., Manning, M., Chen, Z., Marquis, M., Averyt, K.B., Tignor, M., Miller, H.L. (Eds.), Climate Change 2007: The Physical Science Basis. Contribution of Working Group I to the Fourth Assessment Report of the Intergovernmental Panel on Climate Change. Cambridge University Press, Chapter 1, Pages 94–127, https://www.ipcc.ch/site/assets/uploads/2018/03/ar4-wg1-chapter1.pdf.

Lécuyer, C., Gillet, P., Robert, F., 1998. The hydrogen isotope composition of seawater and the global water cycle. Chem. Geol. 145 (3), 249–261.

Lepot, K., Benzerara, K., Brown, G.E., Philippot, P., 2008. Microbially influenced formation of 2.7 billion-year-old stromatolites. Nat. Geosci. 1 (2), 118–121.

Lighthill, J., 1978. Waves in Fluids. Cambridge University Press, Cambridge, pp. 504.

Lovelock, J.E., Kaplan, I.R., 1975. Thermodynamics and the recognition of alien biospheres [and discussion]. Philos. Trans. R. Soc. Lond. B Biol. Sci. 189, 167–181.

Luger, R., Barnes, R., 2015. Extreme water loss and abiotic O_2 buildup on planets throughout the habitable zones of M dwarfs. Astrobiology 15, 1–26.

Lunt, D.J., Valdes, P.J., Dunkley, T., Jones, A., Ridgwell, A.M., Haywood, D.N., Schmidt, R.Marsh, Maslin, M., 2010. CO_2-driven ocean circulation changes as an amplifier of Paleocene–Eocene thermal maximum hydrate destabilization. Geology 38 (10), 875–878.

Lyons, T.W., Anbar, A.D., Severmann, S., Scott, C., Gill, B.C., 2009. Tracking euxinia in the ancient ocean: A multiproxy perspective and proterozoic case study. Annu. Rev. Earth Planetary Sci. 37 (1), 507–534.

Lyons, T.W., Reinhard, C.T., Planavsky, N.J., 2014. The rise of oxygen in Earth's early ocean and atmosphere. Nature 506 (7488), 307–315.

Mackenzie, F.T., Pigott, J.P., 1981. Tectonic controls of Phanerozoic sedimentary rock cycling. Geology Soc. (London) J. 138, 183–191.

Martin, W., Russell, M.J., 2003. On the origins of cells: a hypothesis for the evolutionary transitions from abiotic geochemistry to chemoautotrophic prokaryotes, and from prokaryotes to nucleated cells. Philos. Trans. R. Soc. B: Biol. Sci. 358 (1429), 59–83. doi:10.1098/rstb.2002.1183.

Marty, B., 2012. The origins and concentrations of water, carbon, nitrogen and noble gases on Earth. Earth Planet. Sci. Lett. 56, 56–66.

McGhee Jr., G.R., 1991. Extinction and diversification in the Devonian brachiopoda of New York State: no correlation with sea level? Historical Biology 5, 215–227.

McInerney, F.A., Wing, S.L., 2011. The Paleocene–Eocene thermal maximum: a perturbation of carbon cycle, climate, and biosphere with implications for the future. Annu. Rev. Earth Planet. Sci. 39, 489–516.

McKay, C.P., 1991. Urey Prize lecture: Planetary evolution and the origin of life. Icarus 91, 92–100.

McKay, C.P., 2004. What is life—and how do we search for it in other worlds? PLoS Biol 2 (9), e302. https://doi.org/10.1371/journal.pbio.0020302.

McKay, C.P., 2016. Titan as the abode of life. Life 6, 8.

Molchan-Douthit, M., 1998. Alaska Bore Tales. National Oceanic and Atmospheric Administration. Anchorage, USA. Revised pp. 2.

Morbidelli, A., Chambers, J., Lunine, J.I., Petit, J.M.,, Robert, F., Valsecchi, G.B., Cyr, K.E., 2000. Source regions and timescales for the delivery of water to the Earth. Meteorit. Planet. Sci. 35 (6), 1309–1320.

Morbidelli, A., Lunine, J.I., O'Brien, D.P., Raymond, S.N., Walsh, K.J., 2012. Building terrestrial planets. Annu. Rev. Earth Planet Sci. 40, 251–275.

Mousis, O., et al., 2000. Constraints on the formation of comets from D/H ratios measured in H_2O and HCN. Icarus 148 (2), 513–525. doi:10.1006/icar.2000.6499.

Muller, R.D., Sdrolias, M., Gaina, C., Roest, W.R., 2008a. Age, spreading rates, and spreading asymmetry of the world's ocean crust. Geochem. Geophys. Geosyst. 9 (4), Q04006. doi:10.1029/2007GC001743.

Muller, R.D., Sdrolias, M., Gaina, C., Steinberger, B., Heine, C., 2008b. Long-term sea level fluctuations driven by ocean basin dynamics. Science 319 (5868), 1357–1362.

Muller, R.D., Dutkiewicz, A., Seton, M., Gaina, C., 2013. Seawater chemistry driven by supercontinent assembly, break-up and dispersal. Geology 41, 907–910. doi:10.1007/978-94-007-6644-0_84-1.

Muller, R.D., and Seton, M., 2015. Paleophysiography of ocean basins. *Encyclopedia of Marine Geosciences*.

Navarro-Gonzalez, R., Molina, M.J., Molina, L.T., 1998. Nitrogen fixation by volcanic lightning in the early Earth. Geophys. Res. Lett. 25, 3123–3126.

Nisbet, E., 2000. The realms of Archaean life. Nature 405, 625–626.

Nisbet, E.G., Sleep, N.H., 2001. The habitat and nature of early life. Nature 409, 1083–1091. doi:10.1038/35059210.

Noffke, N., 2009. The criteria for the biogenicity of microbially induced sedimentary structures (MISS) in Archean and younger, sandy deposits. Earth-Sci. Rev. doi:10.1016/j.earscirev.2008.08.00.

O'Leary, D., 2012. The deeds to deuterium. Nat. Chem. 4 (3), 236.

Olson, J.M., Pierson, B.K., 1986. Photosynthesis 3.5 thousand million years ago. Photosynth. Res. 9, 251–259.

O'Neil, J., Carlson, R.W., Paquette, J-L., Francis, D, 2012. Formation age and metamorphic history of the Nuvvuagittuq Greenstone Belt. Precamb. Res. 220–221, 23–44.

Owen, 1999. What do we know about the origin of the earth's oceans? Is it more likely that they derive from icy comets that struck the young earth or from material released from the earth's interior during volcanic activity? Sci. Am. https://www.scientificamerican.com/article/what-do-we-know-about-the/.

Parkinson, C., et al., 2007. Enceladus: Cassini observations and implications for the search for life. Astron. Astrophys. 463 (1). doi:10.1051/0004-6361:20065773.

Paschall, O., Carmichael, S.K., Königshof, P., Waters, J.A., Ta, P.H., Komatsu, T., Dombrowski, A., 2019. The Devonian–Carboniferous boundary in Vietnam: sustained ocean anoxia with a volcanic trigger for the Hangenberg Crisis? Global Planet. Change 175, 64–81.

Pepin, R.O., 1991. On the origin and early evolution of terrestrial planet atmospheres and meteoritic volatiles. Icarus 92 (1), 2–79.

Piani, L., 2020. Earth's water may have been inherited from material similar to enstatite chondrite meteorites. Science 369 (6507), 1110–1113.

Pinti, D.L., Nicholas, A., 2014. Oceans, Origin of. Encyclopedia of Astrobiology. Springer, Berlin Heidelberg, pp. 1–5. doi:10.1007/978-3-642-27833-4_1098-4.

Orosei, R., Lauro, S.E., Pettinelli, E., Cicchetti, A., Coradini, M., Cosciotti, B., Di Paolo, F., Flamini, E., Mattei, E., Pajola, M., Soldovieri, F., Cartacci, M., Cassenti, F., Frigeri, A., Giuppi, S., Martufi, R., Masdea, A., Mitri, G., Nenna, C., Noschese, R., Restano, M., Seu, R., 2018. Radar evidence of subglacial liquid water on Mars. Science 361 (6401), 490–493. doi:10.1126/science.aar7268.

Pugh, D., Woodworth, P., 2014. Sea level Science: Understanding Tides, Surges, Tsunamis and Mean Sea Level Changes. Cambridge University Press, Cambridge, UK. ISBN 978-1-107-02819-7. xii+395 pp.

Racki, G., Rakociński, M., Marynowski, L., Wignall, P.B., 2018. Mercury enrichments and the Frasnian-Famennian biotic crisis: a volcanic trigger proved? Geology 46 (6), 543–546. https://doi.org/10.1130/G40233.1.

Raymond, S.N., Quinn, T., Lunine, J.I., 2004. Making other Earths: dynamical simulations of terrestrial planet formation and water delivery. Icarus 168, 1–17.

Raymond, S.N., Quinn, T., Lunine, J.I., 2007. High-resolution simulations of the final assembly of Earth-like planets. 2. Water delivery and planetary habitability. Astrobiology 7, 66–84.

Reid, R.P., Visscher, P.T., Decho, A.W., Stolz, J.F., Bebout, B.M., Dupraz, C., MacIntyre, I.G., Pearl, H.W., Pinckney, J.L., Prufert-Bebout, L., Steppe, T.F., Des Marais, D.J., 2000. The role of microbes in accretion, lamination and early lithification of modern marine stromatolites. Nature 406, 989–991. doi:10.1038/35023158.

Robert, F., 2001. The origin of water on Earth. Science 293, 1056–1058. doi:10.1126/science.1064051.

Robert, F., 2006. Solar System Deuterium/Hydrogen ratio. In: Lauretta, D.S., McSween, H.Y., Jr. (Eds.). Meteorites and the Early Solar System II, 2006. Univ. of Arizona Press, Tucson, AZ, pp. 341–352.

Rothschild, L.J., Mancinelli, R.L., 2001. Life in extreme environments. Nature 409, 1092–1101. doi:10.1038/35059215.

Sanderson, M.J., 1997. A nonparametric approach to estimating divergence times in the absence of rate constancy. Mol. Biol. Evol. 14, 1218–1232.

Schmidt-Rohr, K., 2020. Oxygen is the high-energy molecule powering complex multicellular life: fundamental corrections to traditional bioenergetics. ACS Omega 5, 2221–2233. http://dx.doi.org/10.1021/acsomega.9b03352.

Schirrmeister, B.E., de Vos, J.M., Antonelli, A., Bagheri, H.C., 2013. Evolution of multicellularity coincided with increased diversification of cyanobacteria and the Great Oxidation Event. Proc. Natl. Acad. Sci. 110 (5), 1791–1796.

Schopf, J.W., 2006. Fossil evidence of Archaean life. Philos. Trans. R. Soc. B: Biol. Sci. 361 (1470), 869–885. doi:10.1098/rstb.2006.1834.

Schrödinger, E., 1945. *What is life?* Cambridge University Press, Cambridge, pp. 91.

Schulte, M., Blake, D., Hoehler, T., McCollom, T., 2006. Serpentinization and its implications for life on the early Earth and Mars. Astrobiology 6, 364–376.

Schulte, P., et al., 2010. The Chicxulub asteroid impact and mass extinction at the Cretaceous–Paleogene boundary. Science 327, 1214–1218.

Sepkoski Jr., J.J., 1986. Phanerozoic overview of mass extinction. In: Raup, D.M., Jablonski, D. (Eds.). *Patterns and Processes in the History of Life*. Dahlem Workshop Reports (Life Sciences Research Reports), 36. Springer, Berlin, Heidelberg, pp. 277–295. https://doi.org/10.1007/978-3-642-70831-2_15.

Shaw, G.H., 2008. Earth's atmosphere—Hadean to early Proterozoic. Geochemistry 68 (3), 235–264. doi:10.1016/j.chemer.2008.05.001.

Sheehan, P.M., 2001. The late Ordovician mass extinction. Annu. Rev. Earth Planet. Sci. 29, 331–364.

Sleep, N.H., Zahnle, K., 2001. Carbon dioxide cycling and the climate of ancient Earth. J. Geophys. Res. Planets 106, 1373–1399.

Sleep, N.H., Zahnle, K., Neuhoff, P.S., 2001. Initiation of clement surface conditions on the earliest Earth. Proc. Natl. Acad. Sci. USA 98, 3666–3672.

Sleep, N.H., 2007. Plate tectonics through time. Treatise Geophys. 9, 145–169.

Sleep, N.H., Zahnle, K.J., Lupu, R.E., 2014. Terrestrial aftermath of the Moon-forming impact. Philos. Trans. R. Soc. Math. Phys. Eng. Sci. 372, 20130172.

Sluijs, A., Bowen, G.J., Brinkhuis, H., Lourens, L.J., Thomas, E., 2007. The Palaeocene–Eocene thermal maximum super greenhouse: biotic and geochemical signatures, age models and mechanisms of global change. In: Williams, M., Haywood, A.M., Gregory, J., Schmidt, D.N. (Eds.), Deep-time Perspectives on Climate Change: Marrying the Signal from Computer Models and Biological Proxies. The Micropalaeontological Society Special Publication, London, pp. 323–350.

Sluijs, A., Zachos, J.C., Zeebe, R.E., 2012. Constraints on hyperthermals. Nat. Geosci. 5, 231.

Stevens, T.O., McKinley, J.P., 1995. Lithoautotrophic microbial ecosystems in deep basalt aquifers. Science 270, 450–454.

Stoker, J.J., 1957. Water Waves. Interscience, New York, NY, pp. 567.

Storey, M., Duncan, R.A., Swisher III, C.C., 2007. Paleocene–Eocene thermal maximum and the opening of the Northeast Atlantic. Science 316 (5824), 587–589. doi:10.1126/science.1135274.

Svensen, H., et al., 2004. Release of methane from a volcanic basin as a mechanism for initial Eocene global warming. Nature 429, 542–545.

Thomas, E., 1989. Development of Cenozoic deep-sea benthic foraminiferal faunas in Antarctic waters. Geol. Soc. Lond. Special Publ. 47, 283–296.

Thomas, D., Zachos, J., Bralower, T., Thomas, E., Bohaty, S., 2002. Warming the fuel for the fire: evidence for the thermal dissociation of methane hydrate during the Paleocene–Eocene thermal maximum. Geology 30 (12), 1067–1070.

Trendall, A.F., Blockley, J.G., 2004. Precambrian iron-formation. In: Eriksson, P.G., Altermann, W., Nelson, D.R., Mueller, W.U., Catuneanu, O. (Eds.). Evolution of the Hydrosphere and Atmosphere, 12. Developments in Precambrian Geology, pp. 359–511. doi:10.1016/S0166-2635(04)80007-0.

Tsimplis, M.N., Proctor, R., Flather, R.A., 1995. A two-dimensional tidal model for the Mediterranean Sea. J. Geophys. Res. 100, 16223–16239. http://dx.doi.org/10.1029/95JC01671.

Tsimplis, M.N., 1997. Tides and sea level variability at the strait of Euripus. Estuar. Coast. Shelf Sci. 44, 91–101. http://dx.doi.org/10.1006/ecss.1996.0128.

Urey, H.C., 1952. The Planets, Their Origin and Development. Yale University Press, New Haven, pp. 245.

Utsunomiya, S., Murakami, T., Nakada, M., Kasama, T., 2003. Iron oxidation state of a 2.45 Byr-old paleosol developed on mafic volcanics. Geochim. Cosmochim. Acta 67 (2), 213–221.

Veth, C., Zimmerman, J.T.F., 1981. Observations of quasi-two-dimensional turbulence in tidal currents. J. Physical Oceanogr. 11, 1425e1430.

Wadsworth, J., Cockell, C.S., 2017. Perchlorates on Mars enhance the bacteriocidal effects of UV light. Nat. Sci. Rep. 7 (4662). doi:10.1038/s41598-017-04910-3.

Warfield, D., 1973. Gay, E.Methods of Animal ExperimentationIV. Academic Press, London, pp. 43–143.

Walker, J.C.G., Hays, P.B., Kasting, J.F., 1981. A negative feedback mechanism for the long-term stabilization of Earth's surface temperature. J. Geophys. Res.: Oceans 86 (C10), 9776–9782.

Wiechert, U.H., 2002. GEOLOGY: Earth's early atmosphere. Science 298 (5602), 2341–2342. doi:10.1126/science.1079894.

Yang, Z., Yoder, A.D., 2003. Comparison of likelihood and Bayesian methods for estimating divergence times using multiple gene loci and calibration points, with application to a radiation of cute-looking mouse lemur species. Syst. Biol. 52, 705–716.

Zachos, J.C.U. Röhl, Schellenberg, S.A., Sluijs, A., Hodell, D.A., Kelly, D.C., E.Thomas, M.Nicolo, Raffi, I., Lourens, L.J., McCarren, H., Kroon, D., 2005. Rapid acidification of the ocean during the Paleocene–Eocene Thermal Maximum. Science 308, 1611–1615.

Zahnle, K.J., Marko, G., Catling, D.C., 2019. Strange messenger: A new history of hydrogen on Earth, as told by Xenon. Geochim. Cosmochim. Acta 244, 56–85.

Zeebe, R.E., Zachos, J.C., Dickens, G.R., 2009. Carbon dioxide forcing alone insufficient to explain Palaeocene–Eocene thermal maximum warming. Nat. Geosci. 2, 576–580.

Zimmerman, J.T.F., 1978. Topographic generation of residual circulation by oscillatory (tidal) currents. Geophys. Astrophys. Fluid Dyn. 11, 35e47.

Zimmerman, J.T.F., 1980. Vorticity transfer by tidal currents over an irregular topography. J. Mar. Res. 38, 601e630.

Bibliography

Abbot, D.S., Huber, M., Bousquet, G., Walker, C.C., 2009. High-CO_2 cloud radiative forcing feedback over both land and ocean in a global climate model. Geophys. Res. Lett. 36.

Abbot, D.S., Tziperman, E., 2008. Sea ice, high-latitude convection, and equable climates. Geophys. Res. Lett. 35.

Agnor, C.B., Canup, R.M., Levison, H.F., 1999. On the character and consequences of large impacts in the late stage of terrestrial planet formation. Icarus 142, 219–237.

Anagnostou, E., et al., 2016. Changing atmospheric CO_2 concentration was the primary driver of early Cenozoic climate. Nature 533, 380–384.

Anand, P., Elderfield, H., Conte, M., 2003. Calibration of Mg/Ca thermometry in planktonic foraminifera from a sediment trap time series. Paleoceanography 18 (2), 1050.

Anderson, J.B., et al., 2011. Progressive Cenozoic cooling and the demise of Antarctica's last refugium. Proc. Natl. Acad. Sci. USA 108, 11356–11360.

Aziz, H.A., et al., 2008. Astronomical climate control on paleosol stacking patterns in the upper Paleocene-lower Eocene Willwood Formation, Bighorn Basin, Wyoming. Geology 36, 531–534.

Badro, J., Brodholt, J.P., Piet, H., Siebert, J., Ryerson, F.J., 2015. Core formation and core composition from coupled geochemical and geophysical constraints. Proc. Natl Acad. Sci.USA 112, 12310–12314.

Bains, S., Corfield, R.M., Norris, R.D., 1999. Mechanisms of climate warming at the end of the Paleocene. Science 285, 724–727.

Bains, W., 2004. Many chemistries could be used to build living systems. Astrobiology 4 (2), 137–167. doi:10.1089/153110704323175124.

Barboni, M., Boehnke, P., Keller, B., Kohl, I.E., Schoene, B., Young, E.D., McKeegan, K.D., 2017. Early formation of the Moon 4.51 billion years ago. Sci. Adv. 3 (1), e1602365. doi:10.1126/sciadv.1602365.

Barker, S., Greaves, M., Elderfield, H., 2003. A study of cleaning procedures used for foraminiferal Mg/Ca paleothermometry. Geochem. Geophys. Geosyst. 4, 8407.

Bell, D.R., Rossman, G.R., 1992. Water in Earth's mantle: The role of nominally anhydrous minerals. Science 255, 1391–1397.

Bemis, B., Spero, H., Bijma, J., Lea, D., 1998. Reevaluation of the oxygen isotopic composition of planktonic foraminifera: experimental results and revised paleotemperature equations. Paleoceanography 13 (2), 150–160.

Berner, R.A., Lasaga, A.C., Garrels, R.M., 1983. Carbonate–Silicate geochemical cycle and its effect on atmospheric carbon dioxide over the past 100 million years. Am. J. Sci. 283, 7.

Berner, R., 2004. A model for calcium, magnesium and sulfate in seawater over Phanerozoic time. Am. J. Sci. 304 (5), 438–453.

Bijl, P.K., Schouten, S., Sluijs, A., Reichart, G.-J., Zachos, J.C., Brinkhuis, H., 2009. Early Palaeogene temperature evolution of the southwest Pacific Ocean. Nature 461, 776–779.

Bijl, P.K., Houben, A.J.P., Schouten, S., Bohaty, S.M., Sluijs, A., Reichart, G.J., Sinninghe Damsté, J.S., Brinkhuis, H., 2010. Transient middle Eocene atmospheric CO_2 and temperature variations. Science 330, 819–821.

Bijl, P.K., Pross, J., Warnaar, J., Stickley, C.E., Huber, M., Guerstein, R., Houben, A.J.P., A.Sluijs, H.Visscher, Brinkhuis, H., 2011. Environmental forcings of Paleogene Southern Ocean dinoflagellate biogeography. Paleoceanography 26, PA1202.

Bijl, P.K., Bendle, J.A.P., Bohaty, S.M., Pross, J., Schouten, S., Tauxe, L., Stickley, C.E., McKay, R.M., Röhl, U., Olney, M., Sluijs, A., Escutia, C., Brinkhuis, H., Expedition 318 Scientists, 2013. Eocene cooling linked to early flow across the Tasmanian Gateway. Proc. Natl. Acad. Sci. 110, 9645–9650.

Binzel, R.P., 1991. Urey prize lecture: physical evolution in the solar system—present observations as a key to the past. Icarus 100 (1992), 274–287.

Boato, G., 1954. The isotopic composition of hydrogen and carbon in the carbonaceous chondrites. Geochim. Cosmochim. Acta 6, 209–220.

Bornemann, A., et al., 2014. Persistent environmental change after the Paleocene–Eocene thermal maximum in the eastern North Atlantic. Earth Planet. Sci. Lett. 394, 70–81.

Bottke, W.F., Walker, R.J., Day, J.M., Nesvorny, D., Elkins-Tanton, L., 2010. Stochastic late accretion to Earth, the Moon, and Mars. Science 330, 1527–1530.

Bowen, G.J., Beerling, D.J., Koch, P.L., Zachos, J.C., Quattlebaum, T.A., 2004. Humid climate state during the Palaeocene–Eocene thermal maximum. Nature 432, 495–499.

Bowen, Zachos, 2010. Rapid carbon sequestration at the termination of the Paleocene–Eocene thermal maximum. Nat. Geosci. 3 (12). doi:10.1038/ngeo1014.

Brady, P.V., 1991. The effect of silicate weathering on global temperature and atmospheric CO_2. J. Geophys. Res. 96, 18101–18106.

Bralower, T.J., Thomas, E., Zachos, J.C., 1995. Late Paleocene to Eocene paleoceanography of the equatorial Pacific Ocean: stable isotopes recorded at Ocean Drilling Program Site 865, Allison Guyot. Paleoceanography 19, 841–865.

Bralower, T.J., et al., 1997. High-resolution records of the late Paleocene thermal maximum and circum-Caribbean volcanism: is there a causal link? Geology 25, 963–965.

Broecker, W., Yu, J., 2011. What do we know about the evolution of Mg to Ca ratios in seawater? Paleoceanography 26, PA3203.

Brown, M.E., 2012. The compositions of Kuiper Belt Objects. Annu. Rev. EarthPlanet. Sci. 40, 467–494.

Budde, G., Burkhardt, C., Kleine, T., 2019. Molybdenum isotopic evidence for the late accretion of outer Solar System material to Earth. Nat. Astron. 3 (8), 736–741.

Buffett, B.Á., Zatsepina, O.Y., 2000. Formation of gas hydrate from dissolved gas in natural porous media. Mar. Geol 164, 69–77.

Buffett, B., Archer, D., 2004. Global inventory of methane clathrate: sensitivity to changes in the deep ocean. Earth Planet. Sci. Lett. 227, 185–199.

Bush, M.B., Rivera, R., 1998. Pollen dispersal and representation in a neotropical rain forest. Glob. Ecol. Biogeogr. 7, 379–392.

Canup, R.M., Asphaug, E., 2001. Origin of the Moon in a giant impact near the end of the Earth's formation. Nature 412, 708–712.

Canup, R.M., 2012. Forming a Moon with an Earth-like composition via a giant impact. Science. doi:10.1126/science.1226073.

Cao, L., et al., 2009. The role of ocean transport in the uptake of anthropogenic CO_2. Biogeosciences 6, 375–390.

Chapman, M.R., 2010. Seasonal production patterns of planktonic foraminifera in the NE Atlantic Ocean: implications for paleotemperature and hydrographic reconstructions. Paleoceanography 25 (1), PA1101.

Charles, A.J., et al., 2011. Constraints on the numerical age of the Paleocene–Eocene boundary. Geochem. Geophys. Geosyst. 12, Q0AA17.

Catling, D.C., 2017. Atmospheric Evolution on Inhabited and Lifeless Worlds. Cambridge University Press, pp. 180.

Chun, C.O.J., Delaney, M.L., Zachos, J.C., 2010. Paleoredox changes across the Paleocene–Eocene thermal maximum, Walvis Ridge (ODP Sites 1262, 1263, and 1266): evidence from Mn and U enrichment factors. Paleoceanography 25 (4), PA4202.

Cicerone, R.J., Oremland, R.S., 1988. Biogeochemical aspects of atmospheric methane. Global Biogeochem. Cycles 2, 299–327.

Clennell, M.B., Hovland, M., Booth, J.S., Henry, P., Winters, J.W., 1999. Formation of natural gas hydrates in marine sediments, 1, Conceptual model of gas hydrate growth conditioned by host sediment properties. J. Geophys. Res. 104 (22) 985–23,003.

Clyde, W.C., Gingerich, P.D., 1998. Mammalian community response to the latest Paleocene thermal maximum: An isotaphonomic study in the northern Bighorn Basin, Wyoming. Geology 26, 1011–1014.

Coggon, R.M., Teagle, D.A.H., Smith-Duque, C.E., Alt, J.C., Cooper, M.J., 2010. Reconstructing past seawater Mg/Ca and Sr/Ca from mid-ocean ridge flank calcium carbonate veins. Science 327, 1114–1117.

Coggon, R.M., Teagle, D.A.H., Dunkley Jones, T., 2011. Comment on "What do we know about the evolution of Mg to Ca ratios in seawater?" by Wally Broecker and Jimin Yu. Paleoceanography 26, PA3224.

Colbourn, G., Ridgwell, A., Lenton, T.M., 2015. The time scale of the silicate weathering negative feedback on atmospheric CO_2. Glob. Biogeochem. Cycles 29, 583–596.

Cowen, R., 2013. Common source for Earth and Moon water. Nature. doi:10.1038/nature.2013.12963.

Cramer, B.S., Wright, J.D., Kent, D.V., Aubry, M.-P., 2003. Orbital climate forcing of $\delta^{13}C$ excursions in the late Paleocene–early Eocene (chrons C24n-C25n). Paleoceanography 18 (1097). http://dx.doi.org/10.1029/2003PA000909.

Cramer, B.S., Miller, K.G., Barrett, P.J., Wright, J.D., 2011. Late Cretaceous–Neogene trends in deep ocean temperature and continental ice volume: reconciling records of benthic foraminiferal geochemistry

(δ^{18}O and Mg/Ca) with sea level history. J. Geophys. Res. 116 (C12023). doi:10.1029/2011JC007255.

Creech, J.B., Baker, J.A., Hollis, C.J., Morgans, H.E.G., Smith, E.G.C., 2010. Eocene sea temperatures for the mid-latitude southwest Pacific from Mg/Ca ratios in planktonic and benthic foraminifera. Earth Planet. Sci. Lett. 299 (3–4), 483–495.

Cui, Y., Kump, L.R., Ridgwell, A.J., Charles, A.J., Junium, C.K., Diefendorf, A.F., K.H.Freeman, N.M.Urban, Harding, I.C., 2011. Slow release of fossil carbon during the Palaeocene–Eocene thermal maximum. Nat. Geosci. 4 (7), 481–485.

Cui, Y., Kump, L.R., 2015. Global warming and the end-Permian extinction event: proxy and modeling perspectives. Earth Sci. Rev. 149, 5–22.

Cyr, K.E., Sears, W.D., Lunine, J.I., 1998. Distribution and evolution of water ice in the solar nebula: Implications for Solar System body formation. Icarus 135, 537–548.

Dauphas, N., 2003. The dual origin of the terrestrial atmosphere. Icarus 165 (2), 326–339.

Dauphas, N., Burkhardt, C., Warren, P.H., Fang-Zhen, T., 2014. Geochemical arguments for an Earth-like Moon-forming impactor. Phil. Trans. R. Soc. Lond. A 372, 20130244.

Davies, R., Cartwright, J., Rana, J., 1999. Giant hummocks in deep-water marine sediments: Evidence for large-scale differential compaction and density inversion during early burial. Geology 27, 907–910.

Deloule, E., Doukhan, J.C., Robert, F., 1998. Interstellar hydroxyl in meteorite chondrules: Implications for the origin of water in the inner solar system. Geochim. Cosmochim. Acta 62, 3367–3378.

Delsemme, A.H., 1999. The deuterium enrichment observed in recent comets is consistent with the cometary origin of seawater. Planet. Space Sci. 47, 125–131.

DeLucia, E.H., et al., 1999. Net primary production of a forest ecosystem with experimental CO_2 enrichment. Science 284, 1177–1179.

Des Marais, D.J., Harwit, M.O., Jucks, K.W., Kasting, J.F., Lin, D.N.C., Lunine, J.I., Schneider, J., Seager, S., Traub, W.A., Woolf, N.J., 2002. Remote sensing of planetary properties and biosignatures on extrasolar terrestrial planets. Astrobiology 2, 153.

Demicco, R.V., Lowenstein, T.K., Hardie, L., Spencer, R.J., 2005. Model of seawater composition for the Phanerozoic. Geology 33 (11), 877–880.

Dickens, G.R., Quinby-Hunt, M.S., 1994. Methane hydrate stability in seawater. Geophys. Res. Lett. 21, 2115–2118.

Dickens, G.R., O'Neil, J.R., Rea, D.K., Owen, R.M., 1995. Dissociation of oceanic methane hydrate as a cause of the carbon isotope excursion at the end of the Paleocene. Paleoceanography 10, 965–971.

Dickens, G.R., Castillo, M.M., Walker, J.C.G., 1997. A blast of gas in the latest Paleocene: Simulating first-order effects of massive dissociation of oceanic methane hydrate. Geology 25, 259–264.

Dickens, G.R., Paull, C.K., Wallace, P., 1997. Direct measurement of in situ methane quantities in a large gas-hydrate reservoir. Nature 384, 426–428.

Dickens, G.R., 2000. Methane oxidation during the late Palaecoene thermal maximum. Bull. Soc. Geol. Fr. 171, 37–49.

Dickens, G.R., 2001. Carbon addition and removal during the late Paleocene thermal maximum: basic theory with a preliminary treatment of the isotope record at Ocean Drilling Program Site 1051, Blake Nose, *Western North Atlantic Palaeogene and Cretaceous Palaeoceanography*. Geol. Soc. Spec. Publ. 183D. Kroon, R. D. Norris, A. Klaus 2001, 293–305.

Dickens, G.R., 2003. Rethinking the global carbon cycle with a large, dynamic and microbially mediated gas hydrate capacitor. Earth Planet. Sci. Lett. 213 (3–4), 169–183.

Dickson, J., 2004. Echinoderm skeletal preservation: calcite-aragonite seas and the Mg/Ca ratio of Phanerozoic oceans. J. Sediment. Res. 74 (3), 355–365.

Dickson, A.J., Cohen, A.S., Coe, A.L., 2012. Seawater oxygenation during the Paleocene–Eocene thermal maximum. Geology 40, 639–642.

Dillon, W.P., Grow, J.A., Paull, C.K., 1980. Unconventional gas hydrate seals may trap gas off southeast U.S. Oil Gas J 78, 124–130.

Donahue, T.M., Hoffman, J.H., Hodges, R.R., Watson, A.J., 1982. Venus was wet: a measurement of the ratio of deuterium to hydrogen. Science 216 (4546), 630–633.

Dones, L., 1996. Simulations of the discovery of Centaurs and Kuiper Belt objects. AAS Bull. 28, 1081.

Drake, M.J., 2005. Origin of water in the terrestrial planets. Meteor. Planet. Sci. 40 (4), 519–527.

Duncan, M.J., Levison, H.F., 1997. Scattered comet disk and the origin of Jupiter family comets. Science 276, 1670–1672.

Duncan, M.J., Quinn, T.R., Tremaine, S., 1987. The formation and extent of the solar system comet cloud. Astron. J. 94, 1330–1338.

Dunkley, J.T., Ridgwell, A., Lunt, D.J., Maslin, M.A., Schmidt, D.N., Valdes, P.J., 2010. A Paleogene perspective on climate sensitivity and methane hydrate instability. Philos. Trans. R. Soc. A 368, 2395–2415.

Dunkley, J.T., Lunt, D.J., Schmidt, D.N., Ridgwell, A., Sluijs, A., Valdes, P.J., Maslin, M., 2013. Climate model and proxy data constraints on ocean warming across the Paleocene–Eocene thermal maximum. Earth Sci. Rev. 125, 123–145.

Eberle, J.J., Fricke, H.C., Humphrey, J.D., Hackett, L., Newbrey, M.G., Hutchison, J.H., 2010. Seasonal variability in Arctic temperatures during early Eocene time. Earth Planet. Sci. Lett. 296 (3–4), 481–486.

Edgar, K.M., Bohaty, S.M., Gibbs, S.J., Sexton, P.F., Norris, R.D., Wilson, P.A., 2013. Symbiont 'bleaching' in planktonic foraminifera during the Middle Eocene climatic optimum. Geology 41 (1), 15–18.

Edwards, N.R., Marsh, R., 2005. Uncertainties due to transport-parameter sensitivity in an efficient 3-D ocean-climate model. Clim. Dyn. 24, 415–433.

Ehhalt, D.H., 1974. The atmospheric cycle of methane. Tellus 26, 58–70.

Elderfield, H., Ferretti, P., Greaves, M., Crowhurst, S., McCave, I.N., Hodell, D., Piotrowski, A.M., 2012. Evolution of ocean temperatures and ice volume through the Mid-Pleistocene climate transition. Science 337 (704), 704–709.

Erez, J., Luz, B., 1983. Experimental paleo-temperature equation for planktonic foraminifera. Geochim. Cosmochim. Acta 47, 1025–1031.

Ershov, E.D., Yakushev, V.S., 1992. Experimental research on gas hydrate decomposition in frozen rocks. Cold Reg. Sci. Technol 20, 147–156.

Evans, D., Müller, W., 2012. Deep time foraminifera Mg/Ca paleothermometry: nonlinear correction for secular change in seawater Mg/Ca. Paleoceanography 27, PA4205.

Ewing, M., Worzel, J.L., Burk, C.A., 1969. Regional aspects of deep-water drilling in the Gulf of Mexico, east of the Bahama Platform and on the Bermuda Rise. Init. Rep. Deep Sea Drill. Program 1, 624–640.

Farley, K.A., Eltgroth, S.F., 2003. An alternative age model for the Paleocene–Eocene thermal maximum using extraterrestrial He-3. Earth Planet. Sci. Lett. 208, 135–148.

Fitoussi, C., Bourdon, B., 2012. Silicon isotope evidence against an enstatite chondrite Earth. Science 335, 1477–1480.

Foster, G.L., 2008. Seawater pH, pCO_2 and $[CO_3^{2-}]$ variations in the Caribbean Sea over the last 130 kyr: a boron isotope and B/Ca study of planktic forminifera. Earth Planet. Sci. Lett. 271, 254–266.

Foster, G.L., et al., 2013. Inter-laboratory comparison of boron isotope analyses of boric acid, seawater and marine $CaCO_3$ by MC-ICPMS and NTIMS. Chem. Geol. 358, 1–14.

Foster, G.L., Rae, J.W.B., 2016. Reconstructing ocean pH with boron isotopes in foraminifera. Annu. Rev. Earth Planet. Sci. 44, 207–237.

Franchi, I.A., Wright, I.P., Pllllnger, C.T., 1993. Constraints on the formation conditions of iron meteorites based on concentrations and isotopic compositions of nitrogen. Geochim. Cosmochim. Acta 57, 3105–3121.

Francis, J.E., 1996. Antarctic palaeobotany: clues to climate change. Terra Antartica 3, 135–140.

Franklin, F., Lecar, M., 2000. On the transport of bodies within and from the asteroid belt. Meteorit. Planet. Sci. 35, 331–340.

Gamboa, L.A., Buffler, R.T., Barker, P.F., 1983. Seismic stratigraphy and geologic history of the Rio Grande Basin. Initial Rep. Deep Sea Drill. Program 72, 481–497.

Genda, H., 2016. Origin of Earth's oceans: An assessment of the total amount, history and supply of water. Geochem. J. 50 (1), 27–42.

Gibbs, S.J., et al., 2016. Ocean warming, not acidification, controlled coccolithophore response during past greenhouse climate change. Geology 44, 59–62.

Gibson, T.G., Bybell, L.M., Owens, J.P., 1993. Latest Paleogene lithologic and biotic events in neritic deposits of southwest New Jersey. Paleoceanography 8, 495–514.

Giusberti, L., et al., 2007. Mode and tempo of the Paleocene–Eocene thermal maximum in an expanded section from the Venetian pre-Alps. Geol. Soc. Am. Bull. 119, 391–412.

Gomes, R., Levison, H.F., Tsiganis, K., Morbidelli, A., 2005. Origin of the cataclysmic Late Heavy Bombardment period of the terrestrial planets. Nature 435 (7041), 466–469.

Gornitz, V., Fung, I., 1994. Potential distribution of methane hydrates in the world's oceans. Global Biogeochem. Cycles 8, 335–347.

Gradstein, F.M., Bukry, D., Habib, D., Renz, O., Roth, P.H., Schmidt, R.R., Weaver, F.M., Wind, F.H., 1978. Biostratigraphic summary of DSDP Leg 44: Western North Atlantic Ocean, Initial Rep. Deep Sea Drill. Program 44, 657–678.

Greenwood, D.R., Archibald, S.B., Mathewes, R.W., Moss, P.T., 2005. Fossil biotas from the Okanagan Highlands, southern British Columbia and northeastern Washington State: climates and ecosystems across an Eocene landscape. Can. J. Earth Sci. 42, 167–185.

Hagemann, R., Nief, G., Roth, E., 1970. Absolute isotopic scale for deuterium analysis of natural waters. Absolute D/H ratio for SMOW. Tellus 22 (6), 712–715.

Hasiuk, F., Lohmann, K., 2010. Application of calcite Mg partitioning functions to the reconstruction of paleocean Mg/Ca. Geochem. Cosmochem. Acta 74 (23), 6751–6763.

Hatzikiriakos, S.G., Englezos, P., 1993. The relationship between global warming and methane gas hydrates in the Earth. Chem. Eng. Sci. 48, 3963–3969.

Hay, W.W., et al., 1999. Alternative global Cretaceous paleogeography. Spec. Publ. Geol. Soc. Am. 332, 1–47.

Henehan, M.J., et al., 2013. Calibration of the boron isotope proxy in the planktonic foraminifera *Globigerinoides ruber* for use in palaeo-CO_2 reconstruction. Earth Planet. Sci. Lett. 364, 111–122.

Henry, P., Thomas, M., Clennell, M.B., 1999. Formation of natural gas hydrates in marine sediments, 2, Thermodynamic calculations of stability conditions in porous sediments. J. Geophys. Res. 104, 23005–23022.

Herwartz, D., Pack, A., Friedrichs, B., Bischoff, A., 2014. Identification of the giant impactor Theia in lunar rocks. Science 344 (6188), 1146. doi:10.1126/science.1251117.

Higgins, J.A., Schrag, D.P., 2006. Beyond methane: Towards a theory for the Paleocene–Eocene Thermal Maximum. Earth Planet. Sci. Lett. 245 (3–4), 523–537.

Higgins, J.A., Schrag, D.P., 2012. Records of Neogene seawater chemistry and diagenesis in deep-sea carbonate sediments and pore fluids. Earth Planet. Sci. Lett. 357–358, 386–396.

Hollis, C.J., Taylor, K.W.R., Handley, L., Pancost, R.D., Huber, M., Creech, J.B., B.R.Hines, E.M.Crouch, Morgans, H.E.G., Crampton, J.S., Gibbs, S., Pearson, P.N., Zachos, J.C., 2012. Early Paleogene temperature history of the Southwest Pacific Ocean: reconciling proxies and models. Earth Planet. Sci. Lett. 349–350, 53–66.

Hönisch, B., et al., 2003. The influence of symbiont photosynthesis on the boron isotopic composition of foraminifera shells. Mar. Micropaleontol. 49, 87–96.

Hönisch, B., et al., 2012. The geological record of ocean acidification. Science 335, 1058–1063.

Horita, J., Zimmermann, H., Holland, H., 2002. Chemical evolution of seawater during the Phanerozoic: implications from the record of marine evaporates. Geochem. Cosmochim. Acta 66 (21), 3733–3756.

Hovland, M., Lysne, D., Whiticar, M.J., 1995. Gas hydrate and sediment gas composition, Hole 892A. Proc. Ocean Drill. Program Sci. Results 146, 151–161.

Huber, M., Caballero, R., 2011. The early Eocene equable climate problem revisited. Clim. Past. 7, 603–633.

Hulsbos, R.E., 1987. Eocene benthic foraminifers from the upper continental rise off New Jersey, Deep Sea Drilling Project Site 605, DSDP Site 605. Initial Rep. Deep Sea Drill. Program 93, 525–533.

Hyndman, R.D., Davis, E.E., Wright, J.A., 1979. The measurement of marine geothermal heat flow by a multi-penetration probe with digital acoustic telemetry and in situ thermal conductivity. Mar. Geophys. Res. 4, 181–193.

Inglis, G.N., et al., 2020. Global mean surface temperature and climate sensitivity of the early Eocene Climatic Optimum (EECO), Paleocene–Eocene Thermal Maximum (PETM), and latest Paleocene. Clim. Past 16, 1953–1968. https://doi.org/10.5194/cp-16-1953-2020.

Isozaki, Y., 2014. Memories of pre-Jurassic lost oceans: how to retrieve them from extant lands. Geosci. Canada 41 (3), 283–311.

Jansa, L.F., 1981. Mesozoic carbonate platforms and banks of the eastern North American margin. Mar. Geol. 44, 97–117.

Jenkyns, H.C., 1980. Cretaceous anoxic events—from continents to oceans. J. Geol. Soc. Lond. 137, 171–188.

John, C.M., Bohaty, S.M., Zachos, J.C., Sluijs, A., Gibbs, S., Brinkhuis, H., Bralower, T.J., 2008. North American continental margin records of the Paleocene–Eocene thermal maximum: implications for global carbon and hydrological cycling. Paleoceanography 23. doi:10.1029/2007PA001465.

John, E.H., Pearson, P.N., Birch, H., Coxall, H.K., Wade, B.S., Foster, G.L., 2013. Warm ocean processes and carbon cycling in the Eocene. Philos. Trans. R. Soc. A.

Kaiho, K., et al., 1996. Latest Palaeocene benthic foraminiferal extinction and environmental changes at Tawanui, New Zealand. Paleoceanography 11, 447–465.

Kaplan, J.O., et al., 2003. Climate change and Arctic ecosystems: 2. Modeling, paleodata-model comparisons, and future projections. J. Geophys. Res. 108 (D19), 8171. http://dx.doi.org/10.1029/2002JD002559.

Karlsson, G., 2000. Iceland's 1100 Years. The History of a Marginal Society. Hurst & Co, London, pp. 181.

Katz, D.L., Cornell, D., Kobayashi, R., Poetmann, F.H., Vary, J.A., Elenblass, J.R., Weinaug, C.F., 1959. Handbook of Natural Gas Engineering802. McGraw-Hill, New York.

Katz, M.E., Miller, K.G., 1991. Early Paleogene benthic foraminiferal assemblages and stable isotopes in the Southern Ocean. Proc. Ocean Drill. Program Sci. Results 114, 481–512.

Katz, M.E., Pak, D.K., Dickens, G.R., Miller, K.G., 1999. The source and fate of massive carbon input during the latest Paleocene thermal maximum. Science 286, 1531–1533.

Katz, M., Katz, D., Wright, J., Miller, K., Pak, D., Shackleton, N., Thomas, E., 2003. Early Cenozoic benthic foraminiferal isotopes: species reliability and interspecies correction factors. Paleoceanography 18 (2), 1024.

Keating-Bitonti, C.R., Ivany, L.C., Affek, H.P., Douglas, P., Samson, S.D., 2011. Warm, not super-hot, temperatures in the early Eocene subtropics. Geology 39 (8), 771–774.

Kelly, D., Bralower, T., Zachos, J., Premoli-Silva, I., Thomas, E., 1996. Rapid diversification of planktonic foraminifera in the tropical Pacific (ODP Site 865) during the late Paleocene thermal maximum. Geology 24 (5), 423–426.

Kelly, D., Bralower, T., Zachos, J., 1998. Evolutionary consequences of the latest Paleocene thermal maximum for tropical planktonic foraminifera. Palaeogeogr. Palaeoclimatol. Palaeoecol. 141 (1), 139–161.

Kelly, D.C., Zachos, J.C., Bralower, T.J., Schellenberg, S.A., 2005. Enhanced terrestrial weathering/runoff and surface ocean carbonate production during the recovery stages of the Paleocene–Eocene thermal maximum. Paleoceanography 20 (4).

Kelly, D.C., Nielsen, T.M.J., McCarren, H.K., Zachos, J.C., Röhl, U., 2010. Spatiotemporal patterns of carbonate sedimentation in the South Atlantic: implications for carbon cycling during the Paleocene–Eocene thermal maximum. Palaeogeogr. Palaeoclimatol. Palaeoecol. 293, 30–40.

Kennett, J.P., Stott, L.D., 1990. Proteus and Proto-oceanus: ancestral Paleogene oceans as revealed from Antarctic stable isotopic results: ODP Leg 113. Proc. Ocean Drill. Program Sci. Results 113, 865–880.

Kennett, J.P., Stott, L.D., 1991. Abrupt deep-sea warming, paleoceanographic changes and benthic extinctions at the end of the Palaeocene. Nature 353, 225–229.

Kiehl, J.T., et al., 1998. The National Center for Atmospheric Research Community Climate Model: CCM3. J. Clim. 11, 1131–1149.

Kim, H.C., Bishnoi, P.R., Heidmann, R.A., Rizvi, S.S.H., 1987. Kinetics of methane hydrate decomposition. Chem. Eng. Sci. 42, 1645–1653.

Kim, J.-H., Schouten, S., Hopmans, E.C., Donner, B., Damste, J.S.S., 2008. Global sediment core-top calibration of the TEX_{86} paleothermometer in the ocean. Geochim. Cosmochim. Acta 72 (4), 1154–1173.

Kim, J.-H., Meer, J.v.d, Schouten, S., Helmke, P., Willmott, V., Sangiorgi, F., Koç, N., Hopmans, E.C., Damsté, J.S.S., 2010. New indices and calibrations derived from the distribution of crenarchaeal isoprenoid tetraether lipids: implications for past sea surface temperature reconstructions. Geochim. Cosmochim. Acta 74 (16), 4639–4654.

Klochko, K., Kaufman, A.J., Yao, W.S., Byrne, R.H., Tossell, J.A., 2006. Experimental measurement of boron isotope fractionation in seawater. Earth Planet. Sci. Lett. 248, 276–285.

Koch, P.L., Zachos, J.C., Gingerich, P., 1992. Correlation between isotope records in marine and continental carbon reservoirs near the Palaeocene/Eocene boundary. Nature 358, 319–322.

Kozdon, R., Kelly, D.C., Kita, N.T., Fournelle, J.H., Valley, J.W., 2011. Planktonic foraminiferal oxygen isotope analysis by ion microprobe technique suggests warm tropical sea surface temperatures during the Early Paleogene. Paleoceanography 26 (3), PA3206.

Kvenvolden, K.A., 1995. A review of the geochemistry of methane in natural gas hydrate. Org. Geochem. 23, 997–1008.

Lammer, H., Kasting, J.F., Chassefière, E., Johnson, R.E., Kulikov, Y.N., Tian, F., 2008. Atmospheric escape and evolution of terrestrial planets and satellites. Space Sci. Rev. 139, 399–436.

Lang, T.H., Wise Jr, S.W., 1987. Neogene and Paleocene-Maestrichtian calcareous nannofossil stratigraphy, DSDP Sites 604 and 605, upper continental rise off New Jersey: sedimentation rates, hiatuses, and correlations with seismic stratigraphy, Leg 93. Initial Rep. Deep Sea Drill. Program 93, 661–684.

Larcher, W., Winter, A., 1981. Frost susceptibility of palms: experimental data and their interpretation. Principes 25, 143–152.

Laskar, J., et al., 2004. A long-term numerical solution for the insolation quantities of the Earth. Astron. Astrophys. 428, 261–285.

Lear, C.H., Elderfield, H., Wilson, P.A., 2000. Cenozoic deep-sea temperatures and global ice volumes from Mg/Ca in benthic foraminiferal calcite. Science 287, 269–272.

Lear, C., Rosenthal, Y., Slowey, N., 2002. Benthic foraminiferal Mg/Ca-paleothermometry: a revised core-top calibration. Geochim. Cosmochim. Acta 66 (19), 3375–3387.

Le Quéré, C., et al., 2016. Global carbon budget 2016. Earth Syst. Sci. Data 8, 605–649.

Li, S., Lucey, P.G., Milliken, R.E., Hayne, P.O., Fisher, E., Williams, J-P., Hurley, D.M., Elphic, R.C., 2018. Direct evidence of surface exposed water ice in the lunar polar regions. Proc. Natl. Acad. Sci. USA 115 (36), 8907–8912. https://doi.org/10.1073/pnas.1802345115.

Lissauer, J.J., Barnes, J.W., Chambers, J.E., 2012. Obliquity variations of a moonless Earth. Icarus 217, 77–87.

Liu, Z., Pagani, M., Zinniker, D., DeConto, R., Huber, M., Brinkhuis, H., Shah, S.R., Leckie, R.M., Pearson, A., 2009. Global cooling during the Eocene–Oligocene climate transition. Science 323, 1187–1190.

Lord, N.S., Ridgwell, A., Thorne, M.C., Lunt, D.J., 2016. An impulse response function for the 'long tail' of excess atmospheric CO_2 in an Earth system model. Global Biogeochem. Cycles 30, 2–17.

Lowenstein, T., Timofeeff, M., Brennan, L., Hardie, L., Demicco, R., 2001. Oscillations in Phanerozoic seawater chemistry: evidence from fluid inclusions. Science 294, 1086–1088.

Lu, G.Y., Keller, G., 1993. The Paleocene Eocene Transition in the Antarctic Indian-Ocean - inference from Planktic foraminifera. Mar. Micropaleontol. 21, 101–142.

Lu, G., Keller, G., Adatte, T., Ortiz, N., Molina, E., 1996. Long-term (10^5) or short-term (10^3) d 13 C excursion near the Palaeocene–Eocene transition: evidence from the Tethys. Terra Nova 8, 347–355.

Lunt, et al., 2010. CO_2-driven ocean circulation changes as an amplifier of Paleocene–Eocene thermal maximum hydrate destabilization. Geology 38 (10), 875–878.

Lunt, D.J., Dunkley Jones, T., Heinemann, M., Huber, M., LeGrande, A., Winguth, A., C.Loptson, J.Marotzke, Tindall, J., Valdes, P., Winguth, C., 2012. A model-data comparison for a multi-model ensemble of early Eocene atmosphere-ocean simulations: EoMIP. Clim. Past 8, 1229–1273.

Lyle, M., Wilson, P.A., Janecek, T., et al., 2002, Proceeding of the Ocean Drilling Program, Initial Reports, 199. doi:10.2973/odp.proc. ir.199.113.2002.

Ma, Z., et al., 2014. Carbon sequestration during the Palaeocene–Eocene thermal maximum by an efficient biological pump. Nat. Geosci. 7, 382–388.

Markwick, P.J., 2007. The palaeogeographic and palaeoclimatic significance of climate proxies for data-model comparisons. In: Williams,

M., Haywood, A.M., Gregory, F.J., Schmidt, D.N. (Eds.), Deep-Time Perspectives on Climate Change, The Micropalaeontological Society Special Publications. The Geological Society, London, pp. 251–312.

Martínez-Boti, M.A., et al., 2015. Boron isotope evidence for oceanic carbon dioxide leakage during the last deglaciation. Nature 518, 219–222.

Matsuyama, I., Mitrovica, J.X., Manga, M., Perron, J.T., Richards, M.A., 2006. Rotational stability of dynamic planets with elastic lithospheres. J. Geophys. Res. Planets 111, E02003.

Matsuyama, I., Nimmo, F., 2007. Rotational stability of tidally deformed planetary bodies. J. Geophys. Res. 112, E11003.

McCarren, H., Thomas, E., Hasegawa, T., Rohl, U., Zachos, J.C., 2008. Depth dependency of the Paleocene–Eocene carbon isotope excursion: paired benthic and terrestrial biomarker records (Ocean Drilling Program Leg 208, Walvis Ridge). Geochem. Geophys. Geosyst. 9, Q10008.

McInerney, F.A., Wing, S.L., 2011. The Paleocene–Eocene thermal maximum - a perturbation of carbon cycle, climate, and biosphere with implications for the future. Annu. Rev. Earth Planet. Sci. 39, 489–516.

McIver, R.D., 1982. Role of naturally occurring gas hydrates in sediment transport. AAPG Bull 66, 789–792.

Montadert, L., Roberts, D.G., 1979. Initial Reports of the Deep Sea Drilling Project. U.S. Government Printing Office, Washington, pp. 73–123.

Morbidelli, A., et al., 2000. Source regions and timescales for the delivery of water to the Earth. Meteor. Planet. Sci. 35 (6), 1309–1329.

Mosbrugger, V., Utescher, T., 1997. The coexistence approach—a method for quantitative reconstructions of Tertiary terrestrial palaeoclimate data using plant fossils. Palaeogeogr. Palaeoclimatol. Palaeoecol. 134, 61–86.

Mountain, G.S., Miller, K.G., 1992. Seismic and geologic evidence for early Paleogene deep-water circulation in the western North Atlantic. Paleoceanography 7, 423–439.

Murray, C.D., Dermott, S.F., 2000. Solar System Dynamics. Cambridge University Press 1652000.

Nadeau, 2010 A., and McGhee, R.A. (2015), A simple method for calculating a planet's mean annual insolation by latitude, Preprint at https://arxiv.org/abs/1510.04542.

Nimmo, F., O'Brien, D., Kleine, T., 2010. Tungsten isotopic evolution during late-stage accretion: constraints on Earth–Moon equilibration. Earth Planet. Sci. Lett. 292, 363–370.

Norris, R.D., 1996. Symbiosis as an evolutionary innovation in the radiation of Paleocene planktic foraminifera. Paleobiology 22, 461–480.

Norris, R.D., et al., 1998. *Proceedings of the Ocean Drilling Program, Initial Reports*, 171B, 749, Ocean Drill. Program. College Station, Tex.

Norris, R.D., Röhl, U., 1999. Carbon cycling and chronology of climate warming during the Palaeocene/Eocene transition. Nature 401, 775–778.

Norris, R.D., 2001. Global warming 55 million years ago. Ocean Sciences at the New Millenium. Natl. Sci. Found.

Nunes, F., Norris, R.D., 2006. Abrupt reversal in ocean overturning during the Palaeocene/Eocene warm period. Nature 439, 60–63.

Owen, T., Bar-Nun, A., Kleinfeld, I., 1992. Possible cometary origin of heavy noble gases in the atmospheres of Venus, Earth and Mars. Nature 358 (6381), 43–46.

Pagani, M., Caldeira, K., Berner, R., Beerling, D., 2009. The role of terrestrial plants in limiting atmospheric CO_2 decline over the past 24 million years. Nature 460, 85–88.

Pak, D.K., Miller, K.G., 1992. Paleocene to Eocene benthic foraminiferal isotopes and assemblages: Implications for deep water circulation. Paleoceanography 7, 405–422.

Panchuk, K., Ridgwell, A., Kump, L.R., 2008. Sedimentary response to Paleocene–Eocene thermal maximum carbon release: a model-data comparison. Geology 36, 315–318.

Pardo, A., Keller, G., Molina, E., Canudo, J.I., 1997. Planktic foraminiferal turnover across the Paleocene–Eocene transition at DSDP site 401, Bay of Biscay, North Atlantic. Mar. Micropaleontol. 29, 129–158.

Paull, C.K., Ussier III, W., Borowski, W.S., Speiss, F.N., 1995. Methane-rich plumes on the Carolina continental rise: associations with gas hydrates. Geology 23, 89–92.

Paull, C.K., Buelow, W.J., Ussler III, W., Borowski, W.S., 1996. Increased continental-margin slumping frequency during sea-level lowstands above gas hydrate-bearing sediments. Geology 24, 143–146.

Payne, J.L., et al., 2010. Calcium isotope constraints on the end-Permian mass extinction. Proc. Natl. Acad. Sci. USA 107, 8543–8548.

Peale, S.J., 1977. In: Planetary Satellites, Burns, J.A. (Ed.), Univ. Arizona Press 1977, pp. 87–111.

Pearson, P., Ditchfield, P., Singano, J., Harcourt-Brown, K., Nicholas, C., Olsson, R., Shackleton, N., Hall, M., 2001. Warm tropical sea surface temperatures in the Late Cretaceous and Eocene epochs. Nature 413 (6855), 481–487.

Pearson, P.N., van Dongen, B.E., Nicholas, C.J., Pancost, R., Schouten, S., Singano, J., Wade, B.S., 2007. Stable warm tropical climate through the Eocene Epoch. Geology 35 (3), 211–214.

Pearson, P., 2012. Oxygen isotopes in foraminifera: overview and historical review. Paleontol. Soc. Pap. 18, 1–38.

Pearson, P.N., Nicholas, C.J., 2014. Layering in the Paleocene/Eocene boundary of the Millville core is drilling disturbance. Proc. Natl. Acad. Sci. USA 111, E1064–E1065.

Pearson, P.N., Thomas, E., 2015. Drilling disturbance and constraints on the onset of the Paleocene–Eocene boundary carbon isotope excursion in New Jersey. Clim. Past. 11, 95–104.

Penman, D.E., Hönisch, B., Zeebe, R.E., Thomas, E., Zachos, J.C., 2014. Rapid and sustained surface ocean acidification during the Paleocene–Eocene thermal maximum. Paleoceanography 29, 357–369.

Penman, D.E., et al., 2016. An abyssal carbonate compensation depth overshoot in the aftermath of the Palaeocene–Eocene Thermal Maximum. Nat. Geosci. 9, 575–580.

Peslier, A.H., Schönbächler, M., Busemann, H., Karato, S.-I., 2017. Water in the Earth's interior: distribution and origin. Space Sci. Rev. 212 (1–2), 743–810.

Poole, I., Cantrill, D., Utescher, T., 2005. A multi-proxy approach to determine Antarctic terrestrial palaeoclimate during the Late Cretaceous and Early Tertiary. Palaeogeogr. Palaeoclimatol. Palaeoecol. 222, 95–121.

Pross, J., Contreras, L., Bijl, P.K., Greenwood, D.R., Bohaty, S., Schouten, S., Bendle, J.A., Röhl, U., Tauxe, L., Raine, J., Huck, C.E., van de Flierdt, T., Jamieson, S.S.R., Stickley, C.E., van de Schootbrugge, B., Escutia, C., Brinkhuis, H., Scientists, I.O.D.P.E., 2012. Persistent near-tropical warmth on the Antarctic continent during the early Eocene epoch. Nature 488, 73–77.

Rampino, M.R., 2013. Peraluminous igneous rocks as an indicator of thermogenic methane release from the North Atlantic Volcanic Province at the time of the Paleocene–Eocene Thermal Maximum (PETM). Bull. Volcanol. 75, 1–5.

Ridgwell, A., 2001. Glacial–Interglacial Perturbations in the Global Carbon Cycle. Univ. East Anglia PhD thesis.

Ridgwell, A., et al., 2007. Marine geochemical data assimilation in an efficient Earth system model of global biogeochemical cycling. Biogeosciences 4, 87–104.

Ridgwell, A., Hargreaves, J.C., 2007. Regulation of atmospheric CO_2 by deep-sea sediments in an Earth system model. Global Biogeochem. Cycles 21, GB2008.

Ries, J.B., 2010. Review: geological and experimental evidence for secular variation in seawater Mg/Ca (calcite-aragonite seas) and its effects on marine biological calcification. Biogeosciences 7 (9), 2795–2849.

Roberts, C.D., LeGrande, A.N., Tripati, A.K., 2011. Sensitivity of seawater oxygen isotopes to climatic and tectonic boundary conditions in an early Paleogene simulation with GISS ModelE-R. Paleoceanography 26, PA4203. doi:10.1029/2010PA002025.

Röhl, U., Bralower, T.J., Norris, R.D., Wefer, G., 2000. A new chronology for the late Paleocene thermal maximum and its environmental implications. Geology 28, 927–930.

Röhl, U., Brinkhuis, H., Sluijs, A., Fuller, M., Exon, N.F., Malone, M., Kennett, J.P. (Eds.), Geophysical Monograph Series 151. The Cenozoic Southern Ocean: Tectonics, Sedimentation, and Climate Change between Australia and Antarctica2004. American Geophysical Union, pp. 113–125.

Röhl, U., Westerhold, T., Bralower, T.J., Zachos, J.C., 2007. On the duration of the Paleocene–Eocene thermal maximum (PETM). Geochem. Geophys. Geosyst. 8, Q12002.

Rohling, E.J., et al., 2012. Making sense of palaeoclimate sensitivity. Nature 491, 683–691.

Ronov, A., Khain, V., Balukhovsky, S., 1989. Atlas of Lithological-Paleogeographical Maps of the World: Mesozoic and Cenozoic of Continents and Oceans Barsukov, V.L., Laviorov, N.P. (Eds.), Moscow Editorial Publishing Group VNII Zarubezh-Geologia, Moscow.

Royer, D.L., Osborne, C.P., Beerling, D.J., 2002. High CO_2 increases the freezing sensitivity of plants: implications for paleoclimatic reconstructions from fossil floras. Geology 30, 963–966.

Rubie, D.C., et al., 2015. Accretion and differentiation of the terrestrial planets with implications for the compositions of early-formed Solar System bodies and accretion of water. Icarus 248, 89–108.

Rubie, D.C., et al., 2016. Highly siderophile elements were stripped from Earth's mantle by iron sulfide segregation. Science 353, 1141–1144.

Rudge, J.F., Kleine, T., Bourdon, B., 2010. Broad bounds on Earth's accretion and core formation constrained by geochemical models. Nat. Geosci. 3, 439–443.

Saint-Marc, P., 1987. Biostratigraphic and paleoenvironmental study of Paleocene benthic and planktonic foraminifers, DSDP Site 605. Initial Rep. Deep Sea Drill. Program 93, 539–547.

Sanyal, A., Bijma, J., Spero, H., Lea, D.W., 2001. Empirical relationship between pH and the boron isotopic composition of *Globigerinoides sacculifer*: implications for the boron isotope paleo-pH proxy. Paleoceanography 16, 515–519.

Sarafian, A.R., Nielsen, S.G., Marschall, H.R., McCubbin, F.M., Monteleone, B.D., 2014. Early accretion of water in the inner solar system from a carbonaceous chondrite-like source. Science 346 (6209), 623–626.

Saunders, A.D., 2016. Two LIPs and two Earth-system crises: the impact of the North Atlantic Igneous Province and the Siberian Traps on the Earth-surface carbon cycle. Geol. Mag. 153, 201–222.

Schaefer, K., Zhang, T., Bruhwiler, L., Barrett, A.P., 2011. Amount and timing of permafrost carbon release in response to climate warming. Tellus B 63, 165–180.

Schaller, M.F., Fung, M.K., Wright, J.D., Katz, M.E., Kent, D.V., 2016. Impact ejecta at the Paleocene–Eocene boundary. Science 354, 225–229.

Schouten, S., Hopmans, E., Schefuss, E., Sinninghe Damsté, J.S., 2002. Distributional variations in marine crenarchaeotal membrane lipids: a new tool for reconstructing ancient sea water temperatures? Earth Planet. Sci. Lett. 204 (1–2), 265–274.

Schouten, S., Hopmans, E., Forster, A., van Breugel, Y., Kuypers, M., Sinninghe Damsté, J.S., 2003. Extremely high sea-surface temperatures at low latitudes during the middle Cretaceous as revealed by archaeal membrane lipids. Geology 31 (12), 1069–1072.

Schouten, S., Hopmans, E., Sinninghe Damsté, J.S., 2013. The organic geochemistry of glycerol dialkyl glycerol tetraether lipids: a review. Org. Geochem. 54, 19–61.

Schmidt, R.R., 1978. Calcareous nannoplankton from the western North Atlantic DSDP Leg 44. Initial Rep. Deep Sea Drill Program 44, 703–729.

Schmitz, B., Speijer, R.P., Aubry, M.-P., 1996. Latest Paleocene benthic extinction event on the southern Tethyan shelf (Egypt): foraminiferal stable isotopic ($\delta^{13}C$, $\delta^{18}O$) records. Geology 24, 347–350.

Schönbächler, M., Carlson, R., Horan, M., Mock, T., Hauri, E., 2010. Heterogeneous accretion and the moderately volatile element budget of Earth. Science 328, 884–887.

Schubert, B.A., Jahren, A.H., 2013. Reconciliation of marine and terrestrial carbon isotope excursions based on changing atmospheric CO_2 levels. Nat. Commun. 4, 1653.

Schuur, E.A.G., et al., 2008. Vulnerability of permafrost carbon to climate change: implications for the global carbon cycle. Bioscience 58, 701–714.

Seth, A., Morbidelli, J.A., Raymond, S.N., O'Brien, D.P., Walsh, K.J., Rubie, D.C., 2014. Highly siderophile elements in Earth's mantle as a clock for the Moon-forming impact. Nature 508 (7494), 84. doi:10.1038/nature13172.

Sexton, P.F., Wilson, P.A., Pearson, P.N., 2006. Microstructural and geochemical perspectives on planktic foraminiferal preservation: "glassy" versus "frosty". Geochem. Geophys. Geosyst. 7, Q12P19.

Sexton, P., Wilson, P., Norris, R., 2006. Testing the Cenozoic multisite composite $\delta 18O$ and $\delta 13C$ curves: new monospecific Eocene records from a single locality, Demerara Rise (Ocean Drilling Program Leg 207). Paleoceanography 21 (2).

Shackleton, N.J., Kennett, J.P., 1975. Paleotemperature history of the Cenozoic and the initiation of Antarctic glaciation: oxygen and carbon isotope analyses in DSDP sites 277, 279 and 281. Initial Rep. Deep Sea Drill Program 29, 743–755.

Shackleton, N.J., Hall, M.A., Boersma, A., 1984. Oxygen and carbon isotope data from Leg 74 foraminifers. Initial Rep. Deep Sea Drill Proj 74, 599–612.

Shellito, C.J., Sloan, L.C., Huber, M., 2003. Climate model sensitivity to atmospheric CO_2 levels in the Early-Middle Paleogene. Palaeogeogr. Palaeoclimatol. Palaeoecol. 193, 113–123.

Shi, G., 1992. Radiative forcing and greenhouse effect due to the atmospheric trace gasses. Sci. China B 35, 217–229.

Siebert, J., Corgne, A., Ryerson, F.J., 2011. Systematics of metal–silicate partitioning for many siderophile elements applied to Earth's core formation. Geochim. Cosmochim. Acta 75, 1451–1489.

Siegler, M.A., Miller, R.S., Keane, J.T., Laneuville, M., Paige, D.A., Matsuyama, I., Lawrence, D.J., Crotts, A., Poston, M.J., 2016. Lunar true polar wander inferred from polar hydrogen. Nature 531 (7595), 480. doi:10.1038/nature17166.

Sijp, W.P., England, M.H., Huber, M., 2011. Effect of deepening of the Tasman Gateway on the global ocean. Paleoceanography 26, PA4207. doi:10.1029/2011PA002143.

Sloan, L.C., Walker, J.C.G., Moore Jr., T.C., Rea, D.K, 1992. Possible methane-induced polar warming in the early Eocene. Nature 357, 320–322.

Sluijs, A., Schouten, S., Pagani, M., Woltering, M., Brinkhuis, H., Damste, J.S., Dickens, G.R., Huber, B.T., Reichart, G.-J., Stein, R., Matthiessen, J., Lourens, L.J., Pedentchouk, N., Backman, J., Moran, K., the Expedition 302 Scientists, 2006. Subtropical Arctic Ocean temperatures during the Palaeocene/Eocene thermal maximum. Nature 441, 610–613.

Sluijs, A., Brinkhuis, H., Schouten, S., Bohaty, S., John, C., Zachos, J.C., Reichart, G.-J., Sinninghe Damsté, J.S., Crouch, E.M., Dickens, G.R., 2007. Environmental precursors to rapid light carbon injection at the Palaeocene/Eocene boundary. Nature 450, 1218–1221.

Sluijs, A., Bowen, G.J., Brinkhuis, H., Lourens, L.J., Thomas, E., 2007. Deep Time Perspectives on Climate Change: Marrying the Signal from Computer Models and Biological Proxies, Williams, M., Haywood, A., Gregory, J., Schmidt, D. N. (Eds.), Geological Society of London, TMS Special Publication, pp. 267–293.

Sluijs, A., Röhl, U., Schouten, S., Brumsack, H.-J., Sangiorgi, F., Sinninghe Damsté, J.S., Brinkhuis, H., 2008. Arctic late Paleocene–early Eocene paleoenvironments with special emphasis on the Paleocene–Eocene thermal maximum (Lomonosov Ridge, Integrated Ocean Drilling Program Expedition 302). Paleoceanography 23, PA1S11.

Sluijs, A., Schouten, S., Donders, T.H., Schoon, P.L., Röhl, U., Reichart, G.-J., F.Sangiorgi, J.-H.Kim, Damsté, J.S.S., Brinkhuis, H., 2009. Warm and wet conditions in the Arctic region during Eocene thermal maximum 2. Nat. Geosci. 2 (11), 777–780.

Sluijs, A., Dickens, G.R., 2012. Assessing offsets between the $\delta^{13}C$ of sedimentary components and the global exogenic carbon pool across Early Paleogene carbon cycle perturbations. Global Biogeochem. Cycles 26, GB4005.

Stassen, P., Thomas, E., Speijer, R.P., 2012. Integrated stratigraphy of the Paleocene–Eocene thermal maximum in the New Jersey Coastal Plain: towards understanding the effects of global warming in a shelf environment. Paleoceanography 27, PA4210. doi:10.1029/2012PA002323.

Stassen, P., Speijer, R.P., Thomas, E., 2014. Unsettled puzzle of the Marlboro clays. Proc. Natl. Acad. Sci. USA 111, E1066–E1067.

Stern, S.A., 1996. On the collisional environment, accretion time scales, and architecture of the massive, primordial Kuiper Belt. Astron. J. 112, 1203–1210.

Stern, S.A., Colwell, J.E., 1997. Collisional erosion in the primordial Edgeworth-Kuiper Belt and the generation of the 30-50 AU Kuiper gap. Astrophys. J. 490, 879–885.

Stocker, T.F., et al., 2013. IPCC. *Climate Change 2013: The Physical Science Basis*. Contribution of Working Group I to the Fifth Assessment Report of the Intergovernmental Panel on Climate Change. Cambridge Univ. Press 2013.

Stoll, H.M., 2005. Limited range of interspecific vital effects in coccolith stable isotopic records during the Paleocene–Eocene thermal maximum. Paleoceanography 20, PA1007.

Storey, M., Duncan, R.A., Swisher III, C.C., 2007. Paleocene–Eocene thermal maximum and the opening of the northeast Atlantic. Science 316, 587–589.

Stott, L.D., Sinha, A., Thiry, M., Aubry, M.-P., Berggren, W.A., 1996. Global $\delta^{13}C$ changes across the Paleocene–Eocene boundary: criteria for terrestrial-marine correlations. In: Knox, R.W.O.B., Corfield, R.M., Dunay, R.E. (Eds.), Correlation of the Early Paleogene in Northwest Europe. Geological Society London, London, pp. 381–399 Special Publication.

Svensen, H., Planke, S., Corfu, F., 2010. Zircon dating ties NE Atlantic sill emplacement to initial Eocene global warming. J. Geol. Soc. Lond. 167, 433–436.

Tarnocai, C., et al., 2009. Soil organic carbon pools in the northern circumpolar permafrost region. Glob. Biogeochem. Cycles 23, GB2023. http://dx.doi.org/10.1029/2008GB003327.

Thomas, D., Bralower, T., Zachos, J., 1999. New evidence for subtropical warming during the late Paleocene thermal maximum: stable isotopes from Deep Sea Drilling Project Site 527, Walvis Ridge. Paleoceanography 14 (5), 561–570.

Thomas, E., Shackleton, N.J., 1996. The Palaeocene–Eocene benthic foraminiferal extinction and stable isotope anomalies. Geol. Soc. Lond. Special Publ. 101, 401–441.

Thomas, E., 1998. The biogeography of the late Paleocene benthic foraminiferal extinction. In: Aubry, M.-P., Lucas, S., Berggren, W.A. (Eds.), Late Paleocene–Early Eocene Biotic and Climatic Events in the Marine and Terrestrial Records. Columbia University Press, pp. 214–243.

Thomas, E., Zachos, J.C., Bralower, T.J., 2000. Deep-sea environments on a warm earth: latest Paleocene–early Eocene. In: Huber, B., MacLeod, K., Wing, S. (Eds.), Warm Climates in Earth History. Cambridge University Press, pp. 132–160.

Thomas, D.J., Zachos, J.C., Bralower, T.J., Thomas, E., Bohaty, S., 2002. Warming the fuel for the fire: Evidence for the thermal dissociation of methane hydrate during the Paleocene–Eocene thermal maximum. Geology 30, 1067–1070.

Thomas, E., 2003. Extinction and food at the seafloor: a high-resolution benthic foraminiferal record across the Initial Eocene thermal maximum, Southern Ocean Site 690. Special Papers-Geological Society of America, 319–332.

Thomas, D.J., Bralower, T.J., Jones, C.E., 2003. Neodymium isotopic reconstruction of late Paleocene-early Eocene thermohaline circulation. Earth Planet. Sci. Lett. 209, 309–322.

Thompson, S.L., Pollard, D, 1995. A global climate model (GENESIS) with a land-surface-transfer scheme (LSX). Part I: present-day climate. J. Clim. 8, 732–761.

Thompson, S.L., Pollard, D., 1997. Greenland and Antarctic mass balances for present and doubled atmospheric CO_2 from the GENESIS Version-2 Global Climate Model. J. Clim. 10, 871–900.

Tindall, J., Flecker, R., Valdes, P., Schmidt, D.N., Markwick, P., Harris, J., 2010. Modelling oxygen isotope distribution of ancient seawater using a coupled ocean-atmosphere GCM: implications for reconstructing early Eocene climate. Earth Planet. Sci. Lett. 292, 265–273.

Tjoelker, M.G., Oleksyn, J., Reich, P.B., 2001. Modelling respiration of vegetation: evidence for a general temperature-dependent Q10. Glob. Change Biol. 7, 223–230.

Touboul, M., Puchtel, I.S., Walker, R.J., 2015. Tungsten isotopic evidence for disproportional late accretion to the Earth and Moon. Nature 520, 530–533. doi:10.1038/nature14355.

Tripati, A.K., Elderfield, H., 2004. Abrupt hydrographic changes in the equatorial Pacific and subtropical Atlantic from foraminiferal Mg/Ca indicate greenhouse origin for the thermal maximum at the Paleocene–Eocene boundary. Geochem. Geophys. Geosyst. 5, Q02006.

Tripati, A., Elderfield, H., 2005. Deep-sea temperature and circulation changes at the Paleocene–Eocene thermal maximum. Science 308 (5730), 1894–1898.

Truswell, E.M., Macphail, M.K., 2009. Polar forests on the edge of extinction: what does the fossil spore and pollen evidence from East Antarctica say? Aust. Syst. Bot. 22, 57–106.

Turcotte, D.L., et al., 1981. Role of membrane stresses in the support of planetary topography. J.Geophys. Res. 86, 3951–3959.

Turner, S.K., Sexton, P.F., Charles, C.D., Norris, R.D., 2014. Persistence of carbon release events through the peak of early Eocene global warmth. Nat. Geosci. 7, 748–751.

Turner, S.K., Ridgwell, A., 2016. Development of a novel empirical framework for interpreting geological carbon isotope excursions, with implications for the rate of carbon injection across the PETM. Earth Planet. Sci. Lett. 435, 1–13.

van Sickel, W.A., Kominz, M.A., Miller, K.G., Browning, J.V., 2004. Late Cretaceous and Cenozoic sea-level estimates: backstripping analysis of borehole data, onshore New Jersey. Basin Res. 16, 451–465.

Ward, B.B., Kilpatrick, K.A., Novelli, P.C., Scranton, M.I., 1987. Methane oxidation and methane fluxes in the ocean surface layer and deep anoxic waters. Nature 327, 226–229.

Westerhold, T., Röhl, U., McCarren, H.K., Zachos, J.C., 2009. Latest on the absolute age of the Paleocene–Eocene Thermal Maximum (PETM): new insights from exact stratigraphic position of key ash layers +19 and −17. Earth Planet. Sci. Lett. 287, 412–419.

Wetherill, G.W., 1992. An alternative model for the formation of the asteroids. Icarus 100, 307–325.

Wieczorek, R., Fantle, M.S., Kump, L.R., Ravizza, G., 2013. Geochemical evidence for volcanic activity prior to and enhanced terrestrial weathering during the Paleocene Eocene Thermal Maximum. Geochim. Cosmochim. Acta 119, 391–410.

Wilde, S.A., Valley, J.W., Peck, W.H., Graham, C.M., 2001. Evidence from detrital zircons for the existence of continental crust and oceans on the Earth 4.4 nGyr ago. Nature 409 (6817), 175–178.

Willemann, R.J., 1984. Reorientation of planets with elastic lithospheres. Icarus 60, 701–709.

Wilson, D.S., Luyendyk, B.P., 2009. West Antarctic paleotopography estimated at the Eocene-Oligocene climate transition. Geophys. Res. Lett. 36, L16302.

Wilson, D.S., et al., 2011. Antarctic topography at the Eocene-Oligocene boundary. Palaeogeogr. Palaeoclimatol. Palaeoecol. 335–336, 24–34.

Winguth, A., Shellito, C., Shields, C., Winguth, C., 2010. Climate response at the Paleocene–Eocene thermal maximum to greenhouse gas forcing—a model study with CCSM3. J. Clim. 23, 2562–2584.

Wright, J.D., Schaller, M.F., 2013. Evidence for a rapid release of carbon at the Paleocene–Eocene thermal maximum. Proc. Natl. Acad. Sci. USA 110, 15908–15913.

Wright, J.D., Schaller, M.F., 2014. Reply to Pearson and Nicholas, Stassen et al., and Zeebe et al.: Teasing out the missing piece of the PETM puzzle. Proc. Natl. Acad. Sci. USA 111, E1068–E1071.

Yoder, J., Schollaert, S., O'Reilly, J., 2002. Climatological phytoplankton chlorophyll and sea surface temperature patterns in continental shelf and slope waters off the northeast U.S. coast. Limnol. Oceanogr. 47 (3), 672–682.

Yu, J., Elderfield, H., 2008. Mg/Ca in the benthic foraminifera Cibicidoides wuellerstorfi and Cibicidoides mundulus: temperature versus carbonate ion saturation. Earth Planet. Sci. Lett., 1–11.

Zachos, J.C., Lohmann, K.C., Walker, J.C.G., Wise, S.W., 1993. Abrupt climate change and transient climates during the Paleogene: a marine perspective. J. Geol 101, 191–213.

Zachos, J.C., Stott, L.D., Lohmann, K.C., 1994. Evolution of early Cenozoic marine temperatures. Paleoceanography 9, 353–387.

Zachos, J.C., Pagani, M., Sloan, L.C., Thomas, E., Billups, K., 2001. Trends, rhythms, and aberrations in global climate 65 Ma to present. Science 292, 686–693.

Zachos, J.C., Wara, M.W., Bohaty, S., Delaney, M.L., Petrizzo, M.R., Brill, A., Bralower, T.J., Premoli-Silva, I., 2003. A transient rise in tropical sea surface temperature during the Paleocene–Eocene thermal maximum. Science 302 (5650), 1551–1554.

Zachos, J.C., Röhl, U., Schellenberg, S.A., Sluijs, A., Hodell, D.A., Kelly, D.C., Thomas, E., Nicolo, M., Raffi, I., Lourens, L.J., McCarren, H., Kroon, D., 2005. Rapid acidification of the ocean during the Paleocene–Eocene thermal maximum. Science 308 (5728), 1611–1615.

Zachos, J.C., Bohaty, S.M., John, C.M., McCarren, H., Kelly, D.C., Nielsen, T., 2007. The Palaeocene–Eocene carbon isotope excursion: constraints from individual shell planktonic foraminifer records. Philos. Trans. R. Soc. A 365 (1856), 1829–1842.

Zachos, J.C., Dickens, G.R., Zeebe, R.E., 2008. An early Cenozoic perspective on greenhouse warming and carbon-cycle dynamics. Nature 451, 279–283.

Zahnle, K.J., Kasting, J.F., Pollack, J.B., 1988. Evolution of a steam atmosphere during Earth's accretion. Icarus 74, 62–97.

Žarić, S., Donner, B., Fischer, G., Mulitza, S., Wefer, G., 2005. Sensitivity of planktic foraminifera to sea surface temperature and export production as derived from sediment trap data. Mar. Micropaleontol. 55, 75–105.

Zeebe, R.E., Wolf-Gladrow, D.A., Bijma, J., Hönisch, B., 2003. Vital effects in foraminifera do not compromise the use of delta B-11 as a paleo-pH indicator: evidence from modelling. Paleoceanography 18, 1043.

Zeebe, R.E., Zachos, J., Dickens, G.R., 2009. Carbon dioxide forcing alone insufficient to explain Paleocene–Eocene thermal maximum warming. Nature Geosci 2, 576–580.

Zeebe, R.E., Dickens, G.R., Ridgwell, A., Sluijs, A., Thomas, E, 2014. Onset of carbon isotope excursion at the Paleocene–Eocene thermal maximum took millennia, not 13 years. Proc. Natl. Acad. Sci. USA 111, E1062–E1063.

Zeebe, R.E., Ridgwell, A., Zachos, J.C., 2016. Anthropogenic carbon release rate unprecedented during the past 66 million years. Nat. Geosci. 9, 325–329.

Zhang, T., Heginbottom, J.A., Barry, R.G., Brown, J., 2000. Further statistics on the distribution of permafrost and ground ice in the northern hemisphere. Polar Geogr 24, 126–131.

Chapter 3

Beginnings of life on Earth

3.1 Origins of life and potential environments—multiple hypotheses on chemical evolution preceding biological evolution

Alexander Oparin (1894–1980)—Russian biochemist—suggested that simple molecules (e.g., methane (CH_4), ammonia (NH_3)) in the early Earth, reacted to form small biomolecules and complex biopolymers (e.g., nucleoside, nucleotide, peptide, polynucleotide) which then evolved into multimolecular functional systems, and finally "life."

As early as 1922, at a meeting of the Russian Botanical Society, Oparin had first introduced his concept of a primordial organism arising in a brew of already-formed organic compounds. He asserted the following tenets (http://physicsoftheuniverse.com/scientists_oparin.html).

- There is no fundamental difference between a living organism and lifeless matter, and the complex combination of manifestations and properties so characteristic of life must have arisen in the process of the evolution of matter.
- The infant Earth had possessed a strongly reducing atmosphere, containing methane, ammonia, hydrogen, and water vapor, which were the raw materials for the evolution of life.
- As the molecules grew and increased in complexity, new properties came into being and a new colloidal-chemical order was imposed on the simpler organic chemical relations, determined by the spatial arrangement and mutual relationship of the molecules.
- Even in this early process, competition, speed of growth, struggle for existence, and natural selection determined the form of material organization, which has become characteristic of living things.

Oparin showed how organic chemicals in solution may spontaneously form droplets and layers, and outlined a way in which basic organic chemicals might form into microscopic localized systems (possible precursors of cells) from which primitive living things could develop. He suggested that different types of coacervates (spherical aggregation of lipid molecules making up a colloidal inclusion which is held together by hydrophobic forces) might have formed in the Earth's primordial ocean and, subsequently, been subject to a selection process, eventually leading to life.

Clearly, Oparin effectively extended Charles Darwin (1809–1882)'s theory of evolution backward in time to explain how simple organic and inorganic materials might have combined into more complex organic compounds, which could then have formed primordial organisms.

In 1924, Oparin officially put forward his influential theory that life on Earth developed through gradual chemical evolution of carbon-based molecules in a "primordial soup" (see Oparin, 1924). Subsequently, the British biologist J.B.S. Haldane independently proposed a similar theory (see Haldane, 1929), not having knowledge about Oparin's theory (which was published in Russian language). Of all the hypotheses that were suggested (Farely, 1977), few were as fruitful as those of Oparin. As a result, Oparin became notable for his contributions to the theory of the origin of life on Earth, and particularly for the "primordial soup" theory of the evolution of life from carbon-based molecules.

Oparin's proposal that life developed effectively by chance, through a progression from simple to complex self-duplicating organic compounds, initially met with strong opposition, but has since received experimental support (such as the famous 1953 experiments of Stanley Miller and Harold Urey at the University of Chicago), and has been accepted by the scientific community as a legitimate hypothesis.

Life on Earth began more than 3 billion years ago, evolving from the most basic of microbes into a dazzling array of complexity over time. But how did the first organisms on the only known home to life in the Universe develop from the primordial soup proposed by the origin of life researchers? In evolutionary biology, *abiogenesis*, or informally the origin of life (OoL), is the natural process by which life has arisen from nonliving matter, such as simple organic compounds. Unlike in religious beliefs, in which no evidence is sought by the believer for what he/she believes in, nothing that unambiguously passes the test of experimental evidence is treated as "truth" in science. Thus, mere assumption or hypothesis of a "spontaneous origin of life from nonliving matter" at the end of the "prebiotic era" needed to be tested before accepting it as a scientific truth. Essential to the spontaneous origin of life was the availability of organic molecules as building blocks.

Based on cues gleaned from experimental findings, a general model for the beginnings of life on Earth is emerging. It starts, almost immediately after the original accretion of the Earth, with a period of chemical evolution which resulted in the formation of small organic molecules that are essential for life. These molecules were synthesized from the gases that made up the primitive atmosphere. This first atmosphere, which surrounded the newly formed Earth, is believed to have been reducing in nature. That is, it contained no free oxygen (O) and was composed mainly of hydrogen molecules (H_2) and hydrogen-rich gases such as methane (CH_4), ammonia (NH_3), and water (H_2O) vapor. The reducing primitive atmosphere of the newly formed Earth was in possession of all the ingredients necessary for the production/formation of organic molecules that constitutes the building block of life; namely, amino acids.

As time passed, under the influence of primitive planetary environmental conditions, the molecules in these gases were activated by energy sources such as lightning discharges and ultraviolet radiation from the Sun. The original molecules broke apart, and their atoms recombined to make new, larger, even more complex molecules. Finally, thousands and millions of molecules were assembled into structures exhibiting *key properties of living things*: *metabolism, respiration, reproduction, and the transfer of genetic information*. In simple language, this is one of the stories explaining the beginning of the prebiotic synthesis of organic matter; meaning the chemical reactions that preceded the development of life.

A number of hypotheses have been proposed regarding the origins of life on Earth. Significance of hydrothermal vents in the origin of life has already been addressed in Section 3.2.4. The most important multiple hypotheses beyond submarine hydrothermal systems in the emergence of life are addressed below.

3.1.1 Lightning in the early atmosphere and the consequent production of amino acids—Miller–Urey "prebiotic soup" experiment

One candidate mechanism associated with the origin of life involves lightning in the early atmosphere and the consequent production of amino acids that, when combined in long chains, provided the basic constituents of life. Electric sparks can generate amino acids and sugars from an atmosphere loaded with water vapor, methane, ammonia and hydrogen, as was shown in the famous Miller–Urey experiment reported in 1953, suggesting that lightning might have helped create the key building blocks of life on Earth in its early eons. Over millions of years, larger and more complex molecules could form. The importance of amino acids in the context of origin of life stems from the fact that amino acids are essential to life. Amino acids are biologically important organic compounds composed of amine ($-NH_2$) and carboxylic acid ($-COOH$) functional groups, along with a side chain specific to each amino acid. The key elements of an amino acid are carbon, hydrogen, oxygen, and nitrogen, though other elements are found in the side chains of certain amino acids. Although research since then has revealed the early atmosphere of Earth was actually hydrogen-poor, scientists have suggested that volcanic clouds in the early atmosphere might have held methane, ammonia and hydrogen and been filled with lightning as well.

In laboratory studies, origin-of-life researchers used electric discharge apparatus to simulate conditions in the primitive atmosphere of the Earth that may have produced the chemicals of life.

Miller–Urey "prebiotic soup" experiment: There was likely a wide variety of molecules in the prebiotic chemical inventory, so the first formidable challenge was to find an abiotic mechanism by which "useful" building blocks were selected from a complex mixture. The Miller–Urey experiment (Hill and Nuth, 2003) or Miller experiment (Balm et al., 1991) was a chemical experiment that simulated the conditions thought at the time (1952) to be present on the early Earth and tested the chemical origin of life under those conditions.

In a fundamental experiment, known as the Miller–Urey "prebiotic soup" experiment (see Miller, 1953; Fig. 3.1) incorporating simple chemical molecules thought to have existed in the early terrestrial atmosphere (i.e., water vapor (H_2O), hydrogen (H_2), methane (CH_4), and ammonia (NH_3)), it was found that a large variety of more

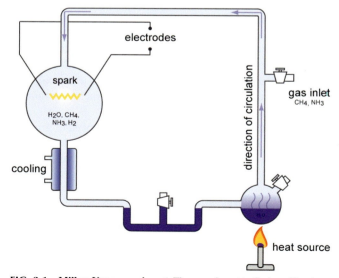

FIG. 3.1 Miller–Urey experiment. The experiment to find out if amino acids could have formed in the early Earth. *(Source: https://commons.wikimedia.org/wiki/File:MUexperiment.png)* Copyright info: This file is licensed under the Creative Commons Attribution 2.5 Generic license.

complicated biogenic molecules (e.g., several amino acids, hydroxyl acids, and other molecules) had formed spontaneously, thereby strongly supporting the lightning theory. The summary of the Miller–Urey "prebiotic soup" experiment (see Video_E4) is as follows (The Community College Consortium for Biosciences Credentials (c^3bc) #TC-23761-12-60-A-37):

- The experiment used water vapor (H_2O), methane (CH_4), ammonia (NH_3), and hydrogen (H_2) to simulate early Earth's reducing atmosphere.
- The chemicals were all sealed inside flasks connected in a loop, with one flask half-full of liquid water and another flask containing a pair of electrodes.
- The liquid water was heated to induce evaporation, sparks were fired between the electrodes to simulate lightning, and then the atmosphere was cooled again so that the water could condense and trickle back into the first flask in a continuous cycle.
- At the end of 1 week of continuous operation, Miller and Urey observed that as much as 10–15% of the carbon within the system was now in the form of organic compounds including the common 20 amino acids now found in living systems.

Note that biogenic molecules or biomolecules are molecules that are present in living organisms, including large macromolecules such as proteins, polysaccharides, lipids, and nucleic acids, as well as small molecules such as primary metabolites, secondary metabolites, and so forth. In fact, experimental studies on the origin of life kick-started with the classic Miller–Urey experiment.

Miller (1957) had shown that the mechanism of synthesis of amino acids is the Strecker condensation, where hydrogen cyanide (HCN) and aldehydes formed by the electrical discharges condense with each other in the presence of ammonia, forming the amino nitriles that upon hydrolysis yield the amino acids. A similar mechanism can account for the synthesis of the reported hydroxyl acids. A study by Miller and Van Trump (1981) has shown that both amino acids and hydroxyl acids can be synthesized even at high dilutions of HCN and aldehydes in a primitive hydrosphere.

The experiment at the time supported Alexander Oparin's and J.B.S. Haldane's hypothesis that putative conditions on the primitive Earth favored chemical reactions that synthesized more complex organic compounds from simpler inorganic precursors. Considered to be the classic experiment investigating abiogenesis, it was performed in 1952 by Stanley Miller, supervised by Harold Urey at the University of Chicago, and published the following year (Miller, 1953; Miller and Urey, 1959; Lazcano and Bada, 2004).

After Miller's death in 2007, scientists examining sealed vials (small containers, typically cylindrical and made of glass, used especially for holding liquid medicines) preserved from the original experiments were able to show that there were actually well over 20 different amino acids produced in Miller's original experiments. That is considerably more than what Miller originally reported, and more than the 20 that naturally occur in the genetic code. More recent evidence suggests that Earth's original atmosphere might have had a composition different from the gas used in the Miller experiment, but prebiotic experiments continue to produce racemic mixtures of simple-to-complex compounds—such as cyanide—under varying conditions (Bada, 2013). Note that a racemic mixture is a mixture that has equal amounts of left- and right-handed enantiomers of a chiral molecule. The first known racemic mixture was racemic acid, which Louis Pasteur found to be a mixture of the two enantiomeric isomers of tartaric acid.

Much after the Darwinian era, the great success achieved through the Miller–Urey experiment, which succeeded in the synthesis of biogenic molecules under the early Earth's climatic, physical, and chemical conditions, and kickstarted the origin of life studies (see Miller, 1986), is considered to be an important milestone in the origin of life research. The Miller–Urey experiment has been repeated and extended many times since, and it has been found that similar results are obtained with other dissociating and ionizing mechanisms, including ultraviolet (UV) radiation and X-rays.

Urey (1952) developed a model of Oparin's reducing primordial terrestrial atmosphere. Following this development, Miller (1953) showed that a number of protein amino acids and a diverse assortment of other small organic molecules of biochemical significance could be made in the laboratory under environmental conditions thought to be representative of the Hadean or early Archean Earth. Since then, a wide variety of organic compounds of biochemical significance have been experimentally formed from simple molecules such as water, methane, ammonia, and HCN (Miller, 1987).

In continuation of the results obtained from Stanley Miller's 1953 experiment, considerable amounts of work on chemical evolution and the origin of life on the Earth have been carried out (e.g., Alexander and Bacq, 1960; de Duve, 1991; Pleasant and Ponnamperuma, 1980). Since then, several other steps in the synthesis of organic molecules have been carried out and experiments simulating prebiotic conditions have yielded sugars, fatty acids, and nucleic acid bases (Calvin, 1975; Keefe et al., 1995; Robertson and Miller, 1995).

Observations of Jupiter and Saturn had shown that they contained ammonia and methane, and large amounts of hydrogen were inferred to be present there (this inference was confirmed later, and it is now known that hydrogen is the major atmospheric component of these planets). These "chemically reducing" atmospheres of the giant planets were regarded as captured remnants of the solar nebula; and by analogy the atmosphere of the early Earth was assumed to have been similar. Thus, the experimental conditions in the Miller–Urey experiment were assumed to simulate those on the primitive Earth. This assumption was found to have been realistic as attested in a subsequent study, which supports an early "reducing" atmosphere.

The significance of these experiments to the origin of life has been debated, with no firm conclusions emerging, except that the generation of biogenic molecules under astronomical conditions is not particularly difficult, and can be achieved in a variety of ways. Therefore, it is interesting that essentially all the starting gases of the various Miller–Urey experiments are now found in the interstellar gas, under conditions that are radically different from those in planetary atmospheres and oceans, with densities lower by at least 15 orders of magnitude (see Thaddeus, 2006).

Some of the end products of the Miller–Urey experiment are found in space as well; for example, formaldehyde and cyanoacetylene. Note that cyanoacetylene is a major nitrogen-containing product of the action of an electric discharge on a mixture of methane and nitrogen. It reacts with simple inorganic substances in aqueous solution to give products including aspartic acid, asparagine, and cytosine (Sanchez et al., 1966). It can be reasonably suspected that many more end-products of the Miller–Urey experiment are currently lurking just below our current level of sensitivity, amino acids especially.

In a confirmatory test, the group of Bada reanalyzed samples from Miller's 1950s spark experiments, simulating water vapor-rich volcanic eruptions. Such eruptions would have released reducing gases. In these samples, amino acids were more varied than in the classical Miller experiment, and yields were comparable or even higher, indicating that even if the Earth's atmosphere had been neutral, localized prebiotic synthesis could have been effective (Johnson et al., 2008).

Chondrites (i.e., primitive material from the solar nebulae) are generally believed to be the building blocks of the Earth and other rocky planets, asteroids, and moons. During and after planet formation, gases escape (i.e., out-gassing occurs) from the chondritic material due to high temperature and pressure. Systematic, detailed calculations on what these gases must have been show that they are mainly the highly reducing hydrogen, methane, and ammonia; the same gases as in the Miller–Urey-type experiments (Schaefer and Fegley, 2007).

There was likely a wide variety of molecules in the prebiotic chemical inventory, so the first formidable challenge was to find an abiotic mechanism by which "useful" building blocks were selected from a complex mixture. In a fundamental experiment, known as the Miller–Urey "prebiotic soup" experiment (see Miller, 1953), consisting of injecting electrical discharges acting for a week in a "spark discharge apparatus" into which a mixture of water (H_2O) vapor, hydrogen (H_2), methane (CH_4), and ammonia (NH_3) had been injected into an "Electric discharge apparatus (see Miller, 1953; Oro, 1963b)" to undergo chemical reaction, it was found that a large variety of more complicated biogenic molecules (e.g., several amino acids, hydroxyl acids, and other molecules) had formed spontaneously, thereby strongly supporting the above theories. Spontaneous formation of amino acids through Miller–Urey "prebiotic soup" experiment is particularly noteworthy because amino acids are essential to life. The chemicals used in Miller–Urey "prebiotic soup" experiment were simple molecules thought to have existed in the early terrestrial atmosphere.

Note that biogenic molecules are molecules that are present in living organisms, including large macro-molecules such as proteins, polysaccharides, lipids, and nucleic acids (DNA and RNA; discussed in Section 2.2 *Biological Evolution*), as well as small molecules such as primary metabolites, secondary metabolites, and so forth. In fact, experimental studies on the origin of life kick-started with the classic Miller–Urey experiment. DNA needs proteins in order to form, and proteins require DNA to form, so how could these have formed without each other? The answer may be RNA, which can store information like DNA, serve as an enzyme like proteins, and help create both DNA and proteins. Later DNA and proteins succeeded this "RNA world," because they are more efficient.

RNA still exists and performs several functions in organisms, including acting as an on-off switch for some genes. The question still remains how RNA got here in the first place. And while some scientists think the molecule could have spontaneously arisen on Earth, others say that was very unlikely to have happened. Nucleic acids other than RNA have been suggested as well, such as Peptide Nucleic Acid (PNA) or TNA, which are more esoteric (i.e., intended for or likely to be understood by only a small number of people with a specialized knowledge or interest).

Note that PNA is an artificially synthesized polymer similar to DNA or RNA (see Nielsen et al., 1991). Threose nucleic acid (TNA) is an artificial genetic polymer in which the natural five-carbon ribose sugar found in RNA has been replaced by an unnatural four-carbon threose sugar (Schöning et al., 2000). Invented by Albert Eschenmoser as part of his quest to explore the chemical etiology of RNA (Eschenmoser, 1999), TNA has become an important synthetic genetic polymer (XNA) due to its ability to efficiently base pair with complementary sequences of DNA and RNA (Schöning et al., 2000).

As already noted, genetic information storage and processing rely on just two polymers, DNA and RNA. Whether their role reflects evolutionary history or fundamental functional constraints is unknown. Using polymerase evolution and design, Pinheiro et al. (2012) have shown that genetic information can be stored in and recovered from six alternative genetic polymers (XNAs) based on simple nucleic acid architectures not found in nature. Pinheiro et al. (2012) also selected XNA aptamers, which bind their targets with high affinity and specificity, demonstrating that beyond heredity, specific XNAs have the capacity for Darwinian evolution and for folding into defined structures. Thus, heredity and evolution, two hallmarks of life, are not limited to DNA and RNA but are likely to be emergent properties of polymers capable of information storage.

Much after the Darwinian era, the great success achieved through the Miller–Urey experiment, which succeeded in the synthesis of biogenic molecules under the early Earth's climatic, physical, and chemical conditions, and kick-started the origin of life studies (see Miller, 1986), is considered to be an important milestone in the origin of life research. The Miller–Urey experiment has been repeated and extended many times since, and it has been found that similar results are obtained with other dissociating and ionizing mechanisms, including ultraviolet (UV) radiation and X-rays.

Urey (1952) developed a model of Oparin's reducing primordial terrestrial atmosphere. Following this development, Miller (1953) showed that a number of proteins, amino acids, and a diverse assortment of other small organic molecules of biochemical significance could be made in the laboratory under environmental conditions thought to be representative of the Hadean Earth (period in the range 4.6–4 billion years ago) or early Archean Earth (period in the range 4–2.5 billion years ago). Since then, a wide variety of organic compounds of biochemical significance have been experimentally formed from simple molecules such as water, methane, ammonia, and HCN (Miller, 1987).

In continuation of the results obtained from Stanley Miller's 1953 experiment, considerable amounts of work on chemical evolution and the origin of life on the Earth have been carried out (e.g., Alexander and Bacq, 1960; Pleasant and Ponnamperuma, 1980; de Duve, 1991). Since then, several other steps in the synthesis of organic molecules have been carried out and experiments simulating prebiotic conditions have yielded sugars, fatty acids, and nucleic acid bases (Calvin, 1975; Keefe et al., 1995; Robertson and Miller, 1995).

In a confirmatory test, the group of Bada re-analyzed samples from Miller's 1950s spark experiments, simulating water vapor-rich volcanic eruptions. Such eruptions would have released reducing gases. In these samples, amino acids were more varied than in the classical Miller–Urey experiment, and yields were comparable or even higher, thereby indicating that even if the Earth's atmosphere had been neutral, localized prebiotic synthesis could have been effective (Johnson et al., 2008).

Subsequent to the Miller–Urey experiment that produced amino acids and other biogenic molecules under the early Earth's simulated highly reducing atmospheric conditions, Tian et al. (2005) measured the production of organic molecules through UV photolysis under the highly reducing atmospheric conditions of the early Earth, and concluded that at 1010 kg/year it "would have been orders of magnitude greater than the rate of either the synthesis of organic compounds in hydrothermal systems or the exogenous delivery of organic compounds to early Earth." Although a chemically "reducing" atmosphere of methane and ammonia is extremely vulnerable to destruction by UV sunlight, researchers such as Kuhn and Atreya (1979), Kasting et al. (1983), and Tian et al. (2005) point out that, based on recent developments, a reducing atmosphere is more stable than previously believed.

The Miller–Urey-type experiments have demonstrated that the generation of biogenic molecules under astronomical conditions is not particularly difficult, and can be achieved in a variety of ways. Therefore, it is not surprising that essentially all the starting gases of the various Miller–Urey experiments are now found in the interstellar gases, under conditions that are radically different from those in planetary atmospheres and oceans (see Thaddeus, 2006). Some of the end products of the Miller–Urey experiment are found in space as well; for example, formaldehyde and cyano-acetylene. Note that cyano-acetylene is a major nitrogen-containing product of the action of an electric discharge on a mixture of methane and nitrogen. It reacts with simple inorganic substances in aqueous solution to give products including aspartic acid, asparagine, and cytosine (Sanchez et al., 1966). In the light of these findings, it can be reasonably suspected that many more end-products of the Miller–Urey experiment are currently lurking just below our current level of sensitivity, amino acids especially.

Oparin proposed a long period of chemical abiotic synthesis of organic compounds as a necessary precondition for the appearance of the first life-forms. The first forms of life would then have been anaerobic heterotrophic microorganisms.

According to Oparin's "ocean scenario" (see Oparin, 1938, 1957), the primitive oceans became the ultimate repository for the great variety of chemicals and biochemicals thought to have been produced in the young Earth's primitive anaerobic atmosphere through the action of ultraviolet radiation and electrical discharge upon water vapor, carbon dioxide, nitrogen, and other gases. In this way, the primitive ocean became a "soup" of energy-rich biochemicals. Interactions among these chemicals produced ever more complicated structures, which eventually (somehow) turned into complex living (self-replicating) cellular entities (Oparin, 1957).

Research on the origin of life is highly heterogeneous. After a peculiar historical development, it still includes strongly opposed views which potentially hinder progress. In the 1st Interdisciplinary Origin of Life Meeting, early-career researchers gathered to explore the commonalities between theories and approaches, critical divergence points, and expectations for the future. In a study, Preiner et al. (2020), found that even though classical approaches and theories—for example, bottom-up and top-down, RNA world vs. metabolism-first—have been prevalent in origin of life research, they are ceasing to be mutually exclusive and they can and should feed integrating approaches. Preiner et al. (2020) have focused on pressing questions and recent developments that bridge the classical disciplines and approaches, and highlighted expectations for future endeavors in origin of life research.

3.1.2 Chemical processes at submarine hydrothermal vents

A second mechanism associated with the origin of life on Earth concerns chemical processes at submarine volcanic vents (hydrothermal vents). Superheated water and minerals spewing from the hydrothermal vents create black smokers and white smokers, with some stacks reaching 30 feet (10 m) in height. Microorganisms feed on the chemicals from these vents and in turn support higher lifeforms. The deep-sea vent theory suggests that life may have begun at submarine hydrothermal vents spewing key hydrogen-rich molecules. Their rocky nooks could then have concentrated these molecules together and provided mineral catalysts for critical reactions. Even now, these vents, rich in chemical and thermal energy, sustain vibrant ecosystems. According to the thermal vent model, organic molecules sink to the seafloor around a thermal vent, and polymerize under conditions of high pressure and temperature (Mulkidjanian et al., 2012). The polymers formed would then move away from the thermal vent. The synthesis of glycine peptides with montmorillonite (a clay mineral) under trench-like hydrothermal conditions (5–100 MPa pressure; 150°C temperature) has been reported by Ohara et al. (2007) who obtained up to 10-mers of glycine.

How the primitive Earth cooked up proteins is a chemical mystery. These molecules—vital to biological functions—are made of long strands of hundreds of amino acids, but researchers are unclear how even some of the shortest amino acid chains, called peptides (polymers of amino acids), formed prior to the dawn of living organisms.

How and where did life on Earth originate? To date, various environments have been proposed as plausible sites for the origin of life. However, discussions have focused on a limited stage of chemical evolution, or emergence of a specific chemical function of proto-biological systems. It remains unclear what geochemical situations could drive all the stages of chemical evolution, ranging from condensation of simple inorganic compounds to the emergence of self-sustaining systems that were evolvable into modern biological ones. In a review, Kitadai and Maruyama (2018) have summarized the reported experimental and theoretical findings for prebiotic chemistry relevant to this topic, including availability of biologically essential elements (N and P) on the Hadean Earth, abiotic synthesis of life's building blocks (amino acids, peptides, ribose, nucleobases, fatty acids, nucleotides, and oligonucleotides), their polymerizations to biomacromolecules (peptides and oligonucleotides), and emergence of biological functions of replication and compartmentalization. It is indicated from the overviews that completion of the chemical evolution requires at least eight reaction conditions of (1) reductive gas phase, (2) alkaline pH, (3) freezing temperature, (4) fresh water, (5) dry/dry-wet cycle, (6) coupling with high energy reactions, (7) heating-cooling cycle in water, and (8) extraterrestrial input of life's building blocks and reactive nutrients. The necessity of these mutually exclusive conditions clearly indicates that life's origin did not occur at a single setting; rather, it required highly diverse and dynamic environments that were connected with each other to allow intra-transportation of reaction products and reactants through fluid circulation. Future experimental research that mimics the conditions of the proposed model are expected to provide further constraints on the processes and mechanisms for the origin of life.

Recent experiments have demonstrated how a volcanic gas, carbonyl sulfide (COS), may have been instrumental in the "prebiotic" build-up of peptides. Carbonyl sulfide is known to fume out of volcanoes today and was likely present in the planet's fiery past. Orgel and colleagues formed peptides by adding COS to a watery solution containing various amino acids at room temperature. About 7% of the amino acids formed pairs and triplets. This peptide yield increased to as high as 80% when the researchers added metal ions to the solution.

Bioactive peptides are organic compounds formed of amino acids covalently attached by peptide bonds. However, some of them may exist freely in nature but most bioactive peptides are encrypted into the parent protein structure, which can be extracted or produced through processes like enzyme synthesis, chemical synthesis or microbial synthesis (Chauhan and Kanwar, 2020). These results lend credence to a theory that life arose near underwater volcanic vents, which to this day support thriving, self-contained ecosystems.

Because carbonyl sulfide (COS) breaks down quickly in water, it is speculated that chains of amino acids most likely formed on rocks near the COS source. Whether life could have blossomed on this ocean bed of peptides is not yet known. However, it is likely that other planets with volcanic activity might have this sort of chemistry.

Hot, acidic hydrothermal fluids produced porous, "beehive-like" structures, rich in Fe and Mg minerals (Russell and Hall, 1997; Martin and Russell, 2003). The thermodynamic driving force came from the chemical potential of the gases discharged by the vents (Russell et al., 2013a). Such vents, similar to the seafloor "black smokers" of today (see Rona et al., 1986), are thought to have been common in the Archean period (i.e., early period in Earth's history), and life at those depths would have been shielded from the strong ultraviolet radiation that existed at that time due to the absence of an ozone layer. According to Russell and Arndt (2005), the first viable protocells—probably endolithic autotrophs exhibiting the rudiments of autonomy, energy conversion, and reproduction, but lacking much of the complexity of modern-day archaea—could have relied on physicochemical attributes of the vents' porous network and circulating fluids for many of their functions.

According to Vago et al. (2017), an argument favoring the role of submarine hydrothermal vents in association with the origin of life on Earth is that "by ~4.4 Ga ago, the ocean waters surrounding the innumerable submarine vents spewing out a rich cocktail of reduced compounds could have attained temperatures less than 80°C (Zahnle et al., 2007; Sleep, 2010). This is important because this value can be considered as an upper limit for the survival of complex organic molecules."

Based on numerous studies, Vago et al. (2017) highlighted the importance of serpentinization process in the generation of the first proto-organisms thus, "Serpentinization, which is considered to be responsible for CH_4 generation (Holm et al., 2015), contributed the bricks and mortar for many prebiotic reactions that, in time, could have led to the first proto-organisms (Russell and Hall, 1997; Kelley et al., 2005; Miller and Cleaves, 2006; Kasting, 2009; Russell et al., 2010, 2014; Grosch and Hazen, 2015; Saladino et al., 2016; Sojo et al., 2016)."

3.1.2.1 Significance of hydrothermal vents in the origin of life

The earliest known life-forms are putative fossilized microorganisms, found in hydrothermal vent precipitates, that may have lived as early as 4.28 Gya (billion years ago), relatively soon after the oceans formed 4.41 Gya, and not long after the formation of the Earth 4.54 Gya (Dodd et al., 2017).

All life is organized as cells. Physical compartmentation from the environment and self-organization of self-contained redox reactions are considered to be the most conserved attributes of living things, hence it has been argued that inorganic matter with such attributes would be life's most likely forebear. However, "Where did life begin?" is a question for which there is still no consensus answer. The environment that could have fostered the beginning of life is still debated. With several hypotheses in play, the race is n to replicate the conditions that allowed life to emerge. In 1977, the first deep sea hydrothermal vent (Fig. 3.2)

FIG. 3.2 "Black smoker" hydrothermal vent on East Pacific Rise 21°N Copyright info: This image is in the public domain in the United States because it only contains materials that originally came from the United States Geological Survey, an agency of the United States Department of the Interior. For more information, see the official USGS copyright policy. (Source: https://commons.wikimedia.org/wiki/File:BlackSmoker.jpg)

was discovered in the East Pacific Rise mid-oceanic ridge. Fig. 3.3 shows sampling of hot rocks from a "Beehive Smoker" at the Mid-Atlantic Ridge. Named "black smokers," the vents emit geothermally heated water up to 400°C, with high levels of sulfides that precipitate on contact with the cold ocean to form the black smoke. Several black smokers were discovered since then, and all of them were located on mid-oceanic ridges and triple junctions (see Joseph, 2016). This was followed in 2000 by the discovery of a new type of alkaline deep-sea hydrothermal vent found a little off axis from mid-ocean ridges. The first field, known as the "Lost City" (Fig. 3.4), was discovered on the seafloor of Atlantis Massif mountain in the mid-Atlantic (Fig. 3.5). While deep sea hydrothermal vents have been suggested as the birthplace of life, alkaline vents, like those

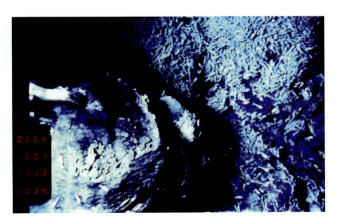

FIG. 3.3 Sampling of hot rocks from a "Beehive Smoker" at the Mid-Atlantic Ridge (TAG site; 21°N; depth approx. 4000m). (Source: Stetter, K.O. (2006), Hyperthermophiles in the history of life, Philos. Trans. R. Soc. Lond. B Biol. Sci., 361(1474): 1837–1843. doi: 10.1098/rstb.2006.1907) Copyright info: © 2006 The Royal Society.

FIG. 3.4 A 5-foot-wide flange, or ledge, on the side of a chimney in the "Lost City" Field is topped with dendritic carbonate growths that form when mineral-rich vent fluids seep through the flange and come into contact with the cold seawater. Note that alkaline vents such as the "Lost City" Field do not form chimneys, as in black smokers (and indeed do not normally "smoke" at all) but rather are labyrinthine networks of interconnected micropores bounded by thin inorganic walls, through which hydrothermal fluids (and ocean waters) percolate. Copyright info: This work is in the public domain in the United States because it is a work prepared by an officer or employee of the United States Government as part of that person's official duties under the terms of Title 17, Chapter 1, Section 105 of the US Code. *(Source: https://commons.wikimedia.org/wiki/File:Lost_City_(hydrothermal_field)02.jpg)*

found at "the Lost City" field in the mid-Atlantic began to receive particular attention.

It has been found that active deep-sea hydrothermal vent environments have much resemblance to the primordial Earth, in which a "prebiotic soup" or "primeval broth" covered the Earth. This theory, popularly known as "prebiotic soup theory," generally attributed to Haldane (1929) and Oparin (1952), embraces the concept that life arose from such a primeval broth of organic compounds that is considered to have permeated and pervaded the primordial oceans on early Earth. The theory received tremendous support from Miller's (1953) demonstration that organic molecules could be obtained by the action of simulated lightning on a mixture of the gases methane (CH_4), ammonia (NH_3) and hydrogen (H_2), which were thought at that time to represent Earth's earliest atmosphere. The organic compounds that were measured included hydrogen cyanide (HCN), aldehydes, amino acids, oil and tar. Additional amino acids were produced by Strecker synthesis through the hydrolysis of the reaction products of HCN, ammonium chloride and aldehydes, and in later experiments polymerization of HCN produced the nucleic acid bases adenine and guanine (Bada, 2004). The best-known theories for the origin of organic compounds are based on the notion of an "organic soup" that was generated either by lightning-driven reactions in the early atmosphere of the Earth as indicated above, or by delivery of organic compounds from space by comets (de Duve, 1995) to the Earth. Although there has been no debate about the occurrence of organic molecules in comets, this mechanism still produces organic soup, but without the help of lightning (Martin et al., 2008).

The chemistry of life has been believed to be the chemistry of reduced organic compounds. Organic compounds can be produced easily in the reducing conditions, but hardly in nonreducing conditions (Islam et al., 2001). Nowadays, primitive earth atmosphere is supposed to be nonreducing or mildly reducing. Submarine hydrothermal systems (SHSs) preserve, however, reducing environments even in the present oxidizing Earth environments. For this and

FIG. 3.5 Map showing the location of the "Lost City" hydrothermal vent field on the seafloor of Atlantis Massif mountain in the mid-Atlantic (30° 07' N; 42° 07' W). Copyright info: This file is licensed under the Creative Commons Attribution-Share Alike 3.0 Unported license. *(Sources: https://commons.wikimedia.org/wiki/File:World_location_map_(equirectangular_180).svg https://en.wikipedia.org/wiki/Lost_City_Hydrothermal_Field)*

some other reasons, SHSs have been regarded as models of hellish primeval ocean where chemical evolution toward the origin of life occurred (Holm, 1992). They are possible sites for the abiotic formation of bio-organic compounds in the present and past terrestrial environments (Corliss et al., 1979; Milller and Van Trump, 1981; Schulte and Shock, 1995; Shock, 1996).

When the discovery of a submarine hydrothermal vent was first reported in 1977 (see Lonsdale, 1977), and more such vents continued to be discovered from various submarine ridge sites, the prevailing hypotheses on the source of life's reduced carbon started to change. Hydrothermal vents revealed a vast and previously unknown domain of chemistry on the Earth. It has been realized that the hot (approximately 360°C) sulfide chimneys of the vent systems are primordial environments that are reminiscent of the primordial Earth, with reactive gases, dissolved elements, and thermal and chemical gradients that operate over spatial scales of centimeters to meters. As (Martin et al., 2008) recalls, "this discovery had an immediate impact on hypotheses about the origin of life, because it was recognized that the vent systems were chemically reactive environments that constituted suitable conditions for sustained prebiotic syntheses (Baross and Hoffman, 1985)."

The quantum of incessant research that went into understanding various aspects of submarine hydrothermal vents is surprisingly large. For example, Corliss et al. (1981) carried out an analysis of a hypothesis concerning the relationship between submarine hot springs and the origin of life on Earth. A diverse set of observations from Archean fossil-bearing rocks, modern submarine hydrothermal systems, experimental and theoretical work on the abiotic synthesis of organic molecules and primitive organized structures, and on water-rock interactions suggests that submarine hot springs were the site for the synthesis of organic compounds leading to the first living organisms on Earth.

Baross and Hoffman (1985) discussed the role of submarine hydrothermal vents and associated gradient environments as sites for the origin and evolution of life. They proposed that a multiplicity of physical and chemical gradients as indicated earlier provided the necessary multiple pathways for the abiotic synthesis of chemical compounds, origin and evolution of "precells" and "precell" communities and, ultimately, the evolution of free-living organisms. According to them, this hypothesis is consistent with the tectonic, paleontological, and degassing history of the Earth; and with the use of thermal energy sources in the laboratory to synthesize amino acids and complex organic compounds.

Holm (1992) explained the reasons for proposing hydrothermal systems as plausible environments for the origin of life. Shock (1992) examined the chemical environments in submarine hydrothermal systems. Hennet et al. (1992) examined abiotic synthesis of amino acids under hydrothermal conditions and the origin of life. These investigators suspected that such amino acid synthesis could perhaps be a perpetual phenomenon.

Hydrocarbons—molecules critical to life—are being generated by the simple interaction of seawater with the hot volcanic rocks or mantle rocks under hydrothermal vents (e.g., under axial vents or the "Lost City" hydrothermal vent field in the mid-Atlantic Ocean). Hydrothermal vents pour large amount of heat and chemicals into the world's oceans and there is an abundance of life that thrives in such conditions. According to Martin et al. (2008), hydrothermal vents unite microbiology and geology to breathe new life into research into one of biology's most important questions—what is the origin of life? There is a school of thought, which advocates that life started from carbon dioxide (CO_2), and life began in hydrothermal vents.

The Lost City hydrothermal field (LCHF)—a completely new type of vent system discovered in 2000, which is characterized by carbonate chimneys that rise 60 m above the ultramafic (i.e., silica-poor) seafloor (Kelley et al., 2001, 2005)—is considered to be particularly relevant to our understanding of the origins of life. The ultramafic underpinnings of the Lost City system have a similar chemical composition to lavas that erupted into the primordial oceans on early Earth (de Wit, 1998; Kelley et al., 2005). Consequently, the LCHF provides insights into past mantle geochemistry and presents a better understanding of the chemical constraints that existed during the evolutionary transition from geochemical to biochemical processes. Furthermore, according to Martin et al. (2008), alkaline pH is an important property of vents that should be considered when contemplating the biochemical origins of life.

There were two critical differences in the ocean chemistry in the Hadean and Archean, around 4 billion years ago (Pinti, 2005): oxygen was absent (Bekker et al., 2004; Kasting, 2013); and the CO_2 concentration in the oceans was substantially higher (although there is little consensus on how much higher; see Russell and Arndt, 2005; Sleep, 2010; Arndt and Nisbet, 2012). The higher CO_2 concentration in Hadean oceans should have increased carbon availability (modern alkaline hydrothermal vents are often carbon limited, from carbonate precipitation and removal by living cells (Proskurowski et al., 2008; Bradley et al., 2009) and lowered the pH of the oceans, probably to around pH 5–7 (Arndt and Nisbet, 2012). That could have produced pH gradients of 5 or 6 pH units between the alkaline hydrothermal fluids and acidic oceans. While mixing could prevent such steep gradients being juxtaposed across single barriers, laminar flow in elongated hydrothermal pores does make it feasible for sharp gradients of several pH units to exist across distances of a few micrometers (Herschy et al., 2014).

The chimneys at the LCHF are mainly composed of carbonates, rather than of iron mono-sulfide (FeS) minerals, which because of their catalytic properties play a central part in our thinking about biochemical origins (Russell and Hall, 1997; Cody, 2004; Martin and Russell, 2007). An exciting property of "Lost City" hydrothermal vent effluent is its alkalinity (pH 9–11), which produces a pH gradient at the vent–ocean interface. It is feasible that the naturally "chemiosmotic" nature of alkaline hydrothermal vents in the Hadean eon, when chimney interiors had a pH of 9–10 and the outer walls of chimneys were bathed in ocean fluids with a pH of 5–6, were essential for the origin of free-living cells (Russell et al., 1994; Russell and Hall, 1997; Martin and Russell, 2007). Note that in biology, "chemiosmosis" refers to the process of moving ions (e.g., protons) to the other side of the membrane, resulting in the generation of an electrochemical gradient that can be used to drive Adenosine triphosphate (ATP) synthesis. Note that ATP is an organic compound and hydrotrope (a compound that solubilizes hydrophobic compounds in aqueous solutions by means other than micellar solubilization) that provides energy to drive many processes in living cells, such as muscle contraction, nerve impulse propagation, condensate dissolution, and chemical synthesis. The gradient mentioned above also incites the ions to return passively with the help of the proteins embedded in the membrane. By passively, it means that the ions will move from an area of higher concentration to an area of lower concentration. This process is similar to osmosis where water molecules move passively. In the case of chemiosmosis, though, it involves the ions moving across the membrane; in osmosis, it is the water molecules. Nevertheless, both processes require a gradient. In osmosis, this is referred to as an osmotic gradient. The differences in the pressures between the two sides of the membrane drive osmosis. As for chemiosmosis, the movement of ions is driven by an electrochemical gradient, such as a proton gradient. Not only is chemiosmosis similar to osmosis, it is also similar to other forms of passive transport, such as facilitated diffusion. It employs a similar principle. The ions move downhill. Also, the molecules are transferred to the other side of the membrane with the help of membrane proteins. Membrane proteins help the ions to move across since the membrane is not readily permeable to ions, basically because of its bilipid feature. These proteins in the membrane facilitate their movement by acting as a temporary shuttle or by serving as a channel or a passageway. Chemiosmosis uses membrane proteins to transport specific ions. Furthermore, it does not require chemical energy (e.g., ATP) as opposed to an active transport system that does. In chemiosmosis, the formation of an ion gradient leads to the generation of potential energy that is sufficient to drive the process (for details, see Mitchell, 1961; Berg et al., 2013; https://www.biologyonline.com/dictionary/chemiosmosis).

A diverse bacterial assemblage populates the vent exteriors, where sulfur-oxidizing and methane-oxidizing bacteria use the interface of oxygenated seawater with H_2- and CH_4-rich hydrothermal fluid. Sulfate-reducing firmicutes have also been identified; these organisms might serve as a link between the high-temperature, anaerobic (requiring an absence of free oxygen) vent interiors and the seawater-bathed chimney exteriors. The vent effluent is devoid of oxygen and harbors only anaerobes (organisms requiring an absence of free oxygen), although aerobes (microorganisms which grow in the presence of air or requires oxygen for growth) occur where there is contact with ocean water. Hydrothermal vents have breathed fresh life into a century-old concept regarding the origin of life. This concept is known today as autotrophic (requiring only carbon dioxide or carbonates as a source of carbon and a simple inorganic nitrogen compound for metabolic synthesis of organic molecules (such as glucose)) origins and posits that life started from CO_2, that the first organisms were autotrophs (organisms that manufacture their own food from inorganic substances, such as carbon dioxide and ammonia) and that these autotrophs obtained their reduced carbon from CO_2 and other simple C_1 compounds, using H_2 as the main electron donor (Baross and Hoffman, 1985).

Martin et al. (2008) examined the possible role of hydrothermal vents in the origin of life. They found that there are striking parallels between the chemistry of the H_2–CO_2 redox couple that is present in hydrothermal systems and the core energy metabolic reactions of some modern prokaryotic autotrophs. The biochemistry of these autotrophs might, in turn, harbor clues about the kinds of reactions that initiated the chemistry of life (Martin et al., 2008). Note that "metabolism" is the chemical reactions in the body's cells that change food into energy. Our bodies need this energy to do everything from moving to thinking to growing. Specific proteins in the body control the chemical reactions of metabolism. Thousands of metabolic reactions happen at the same time—all regulated by the body—to keep our cells healthy and working.

According to Martin et al. (2008), "a real or virtual sojourn to active deep-sea hydrothermal vent environments is also a visit to primordial Earth—active hydrothermal systems existed as soon as liquid water accumulated on the Earth more than 4.2 billion years ago. It is possible that present-day hydrothermal vent microorganisms harbor relict physiological characteristics that resemble the earliest microbial ecosystems on the Earth. It is also possible that geochemical processes of carbon reduction in hydrothermal systems represent the same kind of energy-releasing chemistry that gave rise to the first biochemical pathways. Life need not have evolved this way, but the mere prospect that it could have is reason enough to probe these environments further."

There are some origin-of-life researchers who believe that it is possible that life (read microbial life) emerged and became widespread on Earth prior to the Archean (4–2.5 billion years ago). Using thermal models Abramov and Mojzsis (2009) argued that life would have persisted in subsurface niches during an eon when Earth's surface was frequently and heavily hit by comets and asteroids of all sizes (this eon is known as the late heavy bombardment period spanning from about 4 to 3.8 billion years ago). Such findings strengthen the hypothesis that widespread hydrothermal activity facilitated life's emergence and early diversification (Pace, 1997). It may be noted that hydrothermal activity is responsible for the production of subsurface biomes (large naturally occurring communities of organisms occupying a major habitat) for chemotrophic (deriving energy from the oxidation of organic (chemo-organotrophic) or inorganic (chemo-lithotrophic) compounds) hyperthermophilic ("superheat-loving" bacteria and archaea, which are found within high-temperature environments, representing the upper temperature border of life) communities.

It is interesting to note that carbonaceous morphological remains of subsurface biofilms have been discovered in hydrothermal precipitates produced by meteorite impacts (Hode et al., 2008). The observation that fossil biosignatures have been found in hydrothermal deposits (e.g., Reysenbach and Cady, 2001; Konhauser et al., 2003) is an indication that ancient microbial life could have survived in hydrothermal niches. This interpretation has important implications in the search for ancient and extra-terrestrial life (e.g., Farmer and Des Marais, 1999). In the light of these observations and inferences, the recently reported finding of rock outcrops on Mars could be interpreted as a consequence of hydrothermal activity on Mars (see Squyres et al., 2008; Allen and Oehler, 2008).

Almost uncontested to date, is the view that both atmosphere and ocean would have remained anoxic (oxygen-depleted) until the great oxidizing event postulated at 2.4 billion years ago (Konn et al., 2015). It is well known that chemolithotrophic microbial communities commonly colonize hydrothermal vents. Note that chemolithotrophic microbial communities are microbial communities possessing an energy metabolism that allows them, in the absence of light, to use the oxidation of inorganic substances as a source of energy for cell biosynthesis and maintenance (for details, see Kelly, 1981). Hydrothermal fields at the seafloor, arising as a result of subsurface hydrothermal circulation, are considered to be the foci of submarine oases of life. Fig. 3.6 shows the Mid Ocean Ridge (MOR) system, illustrating the presently known and sampled hydrothermal sites on Earth. According to Konn et al. (2015), hydrothermal circulation occurs when seawater percolates downward through fractured ocean crust. The heated seawater is transformed into hydrothermal fluid through reaction with the host rock at temperatures that can exceed 400°C and exhaled forming thick smoke-like plumes of black metal sulfides often termed "black smokers." During its transit through the oceanic crust, seawater composition is mainly modified by phase separation and water-rock interactions but is also influenced by microbiological processes and magmatic degassing. As a result, fluids become enriched in a variety of compounds and depleted in some others. In spite of their similar appearance, high-temperature hydrothermal vent fluids exhibit a wide range of temperatures and chemical compositions depending on subsurface reaction conditions and nature of the leached rocks (basalts, ultramafic rocks, felsic rocks which are enriched in the lighter elements such as silicon, oxygen, aluminum, sodium, and potassium). Notably, alteration of ultramafic rocks (i.e., magmatic rocks with very low silica content and rich in minerals such as hypersthene, augite, and olivine) is associated with hydrogen release, which leads to high reducing conditions in these environments. In turn, it has been suggested that these reducing conditions would be favorable for the abiogenic production of methane and other organic molecules. One of the major implications for abiogenic synthesis is the origin of life. Konn et al. (2015) recall that, "life may have appeared on Earth in the earliest Archean or even before in the Hadean (Russell and Hall, 1997; Rosing, 1999; Korenaga, 2013). Hydrothermal activity is relevant to Hadean and Archean Earth, as it began as soon as water condensed to form oceans, and some kind of plate tectonics (corresponding to crust formation) appeared 4.4 billion years ago (Wilde et al., 2001). Also, hydrothermal systems as well as ultramafic rocks were much more abundant on primitive Earth than today (Russell et al., 1988)."

According to Konn et al. (2015), "conditions at modern seafloor hydrothermal systems seem to be similar, to some extent, to early Earth's conditions and thus can be considered a place of primary focus in the search for the origin of life. Moreover, hydrothermal vents constitute very favorable environments for the start of life, as much in terms of protection against the sterilizing effect of giant impacts as in terms of scale. Microenvironments such as

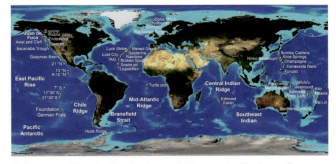

FIG. 3.6 **Global distribution of known hydrothermal vents.** Temperature and chemical anomalies hint that many more sites exist throughout the world's oceans. Data courtesy of D. Fornari and T. Shank, Woods Hole Oceanographic Institute, Massachusetts, USA. *(Source: Martin, W., Baross, J., Kelley, D., and Russell, M.J. (2008), Hydrothermal vents and the origin of life, Nat. Rev. Microbiol., 6:805–814. Article in Nature Reviews Microbiology • October 2008: DOI: 10.1038/nrmicro1991)* Copyright info: ©2008 Macmillan Publishers Limited. All rights reserved.

mineral surfaces favor adsorption, concentration of organics, and subsequent reactions. In addition, a hydrothermal mound provides some kind of protection (niches), physicochemical gradients, and nonequilibrium conditions that are required for the majority of macromolecules typical of the cell organization to persist as well as for the emergence of a living organism (Russell and Hall, 1997; Kompanichenko, 2009)."

Konn et al. (2015) argue that archaea methanogens, which are one of the most common microorganism groups found at hydrothermal vents, are of particular interest, because they synthesize methane (CH_4) from carbon dioxide (CO_2) and hydrogen (H_2) (Schoell, 1988; Takai et al., 2004; Brazelton et al., 2011; Nishizawa et al., 2014). Also, from CO_2 and H_2, acetogenic bacteria are able to generate acetate which can, in turn, be used as substrate by heterotrophic methanogens (e.g., Chapelle and Bradley, 1996).

In 2015, traces of carbon minerals interpreted as "remains of biotic life" were found in 4.1-billion-year-old rocks in Western Australia (Bell et al., 2015). It is, therefore, believed that the physical, geological, and meteorological conditions at the surface of the young (late Hadean and early Archean) Earth were suitable for the emergence and evolution of life. It was early in the Archean that life first appeared on Earth. Our oldest fossils dating to roughly 3.5 billion years ago consist of bacteria microfossils. Alkaline (high pH) hydrothermal systems are thought to be even more relevant to Archean hydrothermal vents, and according to Konn et al. (2015) "the Lost City hydrothermal field could provide particular insights into past mantle geochemistry and present a better understanding of the chemical constraints that existed during the evolutionary transition from geochemical to biochemical processes." It is heartening that in the quest of the origin of life, experimental work simulating alkaline hydrothermal vents and/or Hadean conditions is being done (Herschy et al., 2014; Yamaguchi et al., 2014).

Benner (2017) suggested that Darwinism can be detected from molecules in the Enceladus plumes, Jupiter's moons, and other planetary water lagoons. Fly-bys of Saturn's moon Enceladus have revealed liquid water, hydrogen gas, and small carbon compounds, which together could potentially support microbial cells. From a perspective from hydrothermal chemistry, Deamer and Damer (2017) consider that life beginning on Enceladus could be feasible. Deamer and Damer (2017) further recalls that a mechanism exists by which dilute solutions of organic compounds can be sufficiently concentrated to undergo chemical reactions, and particularly condensation leading to polymer synthesis, provided the condensation reactions are thermodynamically feasible. Furthermore, polymers resembling peptides and nucleic acids can be synthesized by condensation reactions, in which spontaneous hydrolysis reaction is favored.

Importance of terrestrial hydrothermal vents became all the more apparent and important with the discovery of evidence for the presence of hydrothermal vents on the floor of Saturn's icy moon Enceladus's subsurface ocean, and for exploring the astrobiological implications of Enceladus. Submarine hydrothermal vents are the only contemporary geological environment which may be called truly primeval; they continue to be a major source of gases and dissolved elements to the modern ocean as they were to the Archean ocean. Then, as now, they encompassed a multiplicity of physical and chemical gradients as a direct result of interactions between extensive hydrothermal activity in the Earth's crust and the overlying oceanic and atmospheric environments.

Submarine hydrothermal systems are possible sites where abiotic formation of organic compounds toward origin of life took place. As indicated earlier, scientific investigations of the origins of life in connection with a marine environment started at the end of the 1970s, when the 'black smokers' in the Pacific were discovered and the Red Sea deep hydrothermal brines were found to be a fruitful environment for abiotic synthesis of life precursors. For a while this research was categorized under the heading "chemistry," but in less than a decade the topic became fully integrated into the science of "oceanography." The Scientific Committee on Oceanographic Research (SCOR) initiated Working Group 91: Chemical Evolution and Origin of Life in Marine Hydrothermal Systems' (see Holm, 1992; Shock, 1990).

3.1.2.2 Functional resemblance of iron-sulfide membrane in alkaline hydrothermal
Vents to biological cell membrane

In the history of advancement of science, scientists have sometimes succeeded in predicting certain natural phenomena that could have been beyond one's imagination. Predicting the existence and properties of hitherto unknown deep-ocean alkaline hydrothermal systems is just one example of such predictions in the realm of science. Herschy et al. (2014) recall that "Russell and colleagues (Russell et al., 1988, 1989, 1993, 1994; Russell and Hall, 1997) predicted the existence and properties of deep-ocean alkaline hydrothermal systems more than a decade before their discovery, pointing out their suitability as natural electrochemical reactors capable of driving the origin of life. The discovery of Lost City hydrothermal field (Kelley et al., 2001, 2005) was remarkable in that its properties corresponded almost exactly to those postulated by Russell et al. (1993). As indicated earlier, Lost City is powered by a process called serpentinization, the exothermic reaction of ultramafic minerals from the upper mantle, in particular olivine, with water (Bach et al., 2006; Sleep et al., 2004; Martin et al., 2008; Russell et al. 2010). This reaction produces large volumes of H_2 (the presence of the mineral awaruite in some serpentinizing systems indicating as much as

200 mM (McCollom and Bach, 2009)) dissolved in warm (45–90°C) alkaline (pH 9–11) fluids containing magnesium hydroxides (Kelley et al., 2001, 2005). Alkaline vents do not form chimneys, as in black smokers (and indeed do not normally "smoke" at all) but rather are labyrinthine networks of interconnected micropores bounded by thin inorganic walls, through which hydrothermal fluids (and ocean waters) percolate." It is believed that the Lost City hydrothermal system represents a novel, natural laboratory for investigating linkages between geological, biological, and hydrothermal processes in a system dominated by moderate temperature serpentinization reactions.

Chemiosmotic coupling also known as chemiosmosis (i.e., movement of ions across a semipermeable membrane bound structure, down their electrochemical gradient) is universal. Practically all biological cells harness electrochemical proton gradients across membranes to drive adenosine triphosphate (ATP) synthesis, powering biochemistry. ATP is a nucleotide that consists of three main structures: the nitrogenous base (adenine); the sugar (ribose); and a chain of three phosphate groups serially bonded to ribose (Fig. 3.7). Found in all known forms of life, ATP is often referred to as the "molecular unit of currency" of intracellular energy transfer (Knowles, 1980); and is considered as the main source of energy reserve in all the living organisms. ATP is the "energy currency" of the cell, as it provides readily releasable energy in the bond between the second and third phosphate groups. The bond between the three phosphate molecules is very strong and large amount of energy is stored in this bond. Autotrophic cells (living thing that can make its own food from simple chemical substances such as carbon dioxide), including phototrophs (i.e., organisms that produce complex organic compounds [such as carbohydrates] and acquire energy using the energy from sunlight) and chemolithotrophs (i.e., organisms that are capable of using inorganic reduced compounds as a source of energy), also use proton gradients to power carbon fixation directly. The universality of chemiosmotic coupling suggests that it arose very early in evolution, but its origins are obscure (Herschy et al., 2014).

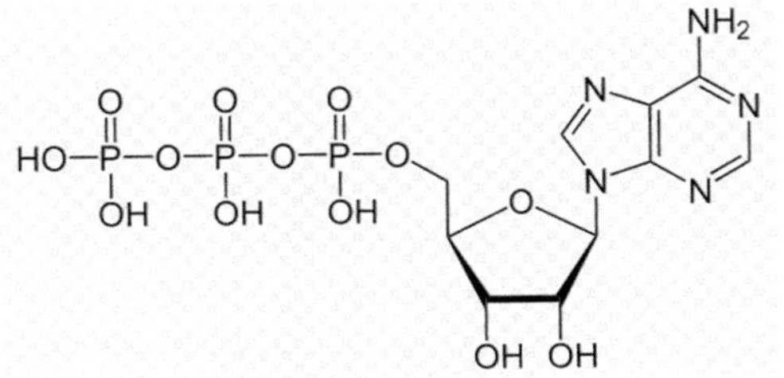

FIG. 3.7 Structure of adenosine triphosphate (ATP), protonated. (Source: https://commons.wikimedia.org/wiki/File:Adenosintriphosphat_protoniert.svg) Copyright info: This work is ineligible for copyright and therefore in the public domain because it consists entirely of information that is common property and contains no original authorship.

Before alkaline vents were actually discovered in the year 2000, Russell et al. (1993) suggested a mechanism by which life could have started at such vents. They proposed that the precipitation of a gelatinous iron-sulfide membrane is the necessary first step toward life. In cell biology, a vesicle is a structure within or outside a cell, consisting of liquid or cytoplasm enclosed by a lipid bilayer. According to Russell et al. (1993)'s hypothesis, membrane vesicles were inflated with alkaline, sulfide-bearing hydrothermal (<200°C) solution and grew on a submarine sulfide mound in acid iron-bearing Hadean seawater. Once a critical size had been reached (0.1–1 mm) vesicles would have budded contiguous self-similar daughters. Russell et al. (1993) assumed that the membrane was rendered insulating by the adsorption and/or oxidative precipitation of organic and organo-sulfur compounds. As a consequence of the naturally induced proton-motive (chemi-osmotic) force, and the activity of the iron mono-sulfide redox catalysts within the membrane, organic compounds would have accumulated within the vesicle. Osmotically driven growth therefore became more significant with time. Russell et al. (1993) argued that the geochemical environment envisaged as responsible for this first step toward life is consistent with that widely accepted for the early Earth.

Russell et al. (1993)'s ideas were updated in 2003. Accordingly, Martin and Russell (2003) suggested that life came from harnessing the energy gradients that exist when alkaline vent water mixes with more acidic seawater (the early oceans were thought to contain more carbon dioxide than now, and hence more acidic). This mirrors the way that cells harness energy. Cells maintain a proton gradient by pumping protons across a membrane to create a charge differential from inside to outside. Known as the proton-motive force, this can be equated to a difference of about 3 pH units. It is effectively a mechanism to store potential energy and this can then be harnessed when protons are allowed to pass through the membrane to phosphorylate adenosine diphosphate (ADP), making Adenosine triphosphate (ATP), which is the energy-carrying molecule found in the cells of all living things. ATP captures chemical energy obtained from the breakdown of food molecules and releases it to fuel other cellular processes.

It is known that submarine hydrothermal vents possess labyrinths (complicated irregular network of passages or paths) of micropores bounded by thin inorganic walls containing catalytic Fe(Ni)S minerals (Nitschke and Russell, 2009; Lane and Martin, 2012). Martin and Russell (2003)'s theory suggests that micropores in the hydrothermal vent chimneys provided templates for cells, with the same 3 pH unit difference across the thin mineral walls of the interconnected vent micropores that separate the vent and seawater. This energy, along with catalytic iron nickel sulfide minerals, allowed the reduction of carbon dioxide and production of organic molecules, then self-replicating molecules, and eventually true cells with their own membranes (see Brazil, 2017).

It has been found that alkaline hydrothermal vents provide a dynamic concentration mechanism known as thermophoresis (Braun and Libchaber, 2002; Baaske et al., 2007). Convection currents and thermal diffusion across the interconnected microporous matrix of alkaline vents produce thermal gradients that can concentrate organic molecules in the cooler regions. Herschy et al. (2014) recall that "The natural inorganic compartments in alkaline vents could facilitate not only the concentration of organics by thermophoresis, but also the beginnings of selection for metabolism (Branciamore et al., 2009; Koonin and Martin, 2005). The two processes combined could potentially drive the replication of simple organic vesicles composed of mixed amphiphiles enclosing primitive replicators within vent pores (Mauer and Monndard, 2011). Such vesicles are capable of growth and division, while retaining RNA (Hanczyc et al., 2003; Mansy et al., 2008) and are en route to the known end-point, modern cells with lipid membranes."

In all known autotrophic bacteria and archaea, carbon and energy metabolism are driven by electrochemical ion (generally proton) gradients across membranes; Peter Mitchell's (see Mitchell, 1959, 1961, 1966) chemiosmotic coupling (Maden, 1995; Stetter, 2006; Lane et al., 2010). Alkaline hydrothermal systems sustain natural proton gradients across the thin inorganic barriers of interconnected micropores within deep-sea vents. Herschy et al. (2014) suggest that in Hadean oceans, these inorganic barriers should have contained catalytic Fe(Ni)S minerals similar in structure to cofactors in modern metabolic enzymes, suggesting a possible abiotic origin of chemiosmotic coupling (Note that a cofactor is a nonprotein chemical compound or metallic ion that is required for an enzyme's activity as a catalyst. Cofactors can be considered "helper molecules" that assist in biochemical transformations. The rates at which these happen are characterized in an area of study called enzyme kinetics). The continuous supply of H_2 from vent fluids and CO_2 from early oceans offers further parallels with the biochemistry of ancient autotrophic cells, notably the acetyl-CoA (acetyl coenzyme A) pathway in archaea and bacteria. Note that the acetyl-CoA pathway utilizes carbon dioxide as a carbon source and often times, hydrogen as an electron donor to produce acetyl-CoA. Acetyl-CoA is a molecule that participates in many biochemical reactions in protein, carbohydrate and lipid metabolism. Its main function is to deliver the acetyl group to the citric acid cycle (Krebs cycle) to be oxidized for energy production. The acetyl-CoA pathway is the only exergonic pathway of carbon fixation, drawing on just H_2 and CO_2 as substrates to drive both carbon and energy metabolism (Fuchs and Stupperich, 1985; Ragsdale and Pierce, 2008; Ljungdahl, 2009); what Everett Shock has called "a free lunch you're paid to eat" (Shock et al., 1998). Despite being exergonic overall, the acetyl-CoA pathway is strictly dependent on chemiosmotic coupling in both methanogens and acetogens (Thauer et al., 2007; Poehlein et al., 2012).

The Last Universal Common Ancestor (LUCA) was the common ancestor of bacteria and archaea (Dagan et al., 2010). It has been suggested that several paradoxical issues concerning cell membrane and cell wall (Koga et al., 1998; Pereto et al., 2004), enzymes for lipid biosynthesis (Martin and Russell, 2003; Pereto et al., 2004; Lombard et al., 2012), biochemical pathways such as glycolysis (Say and Fuchs, 2010), and heme and quinone synthesis (Sousa et al., 2013) could be resolved if LUCA depended on natural (geochemically sustained) proton gradients to drive carbon and energy metabolism, a life-style demanding the membranes to be extremely leaky to protons (Lane and Martin, 2012; Sojo et al., 2016). This requirement for leaky membranes could explain the divergence in other traits that might have coevolved later with membranes (see Sojo et al., 2016) notably DNA replication (in which the replicon is attached to the membrane in most bacteria), and cell wall synthesis, which requires membrane-integral export machinery (Herschy et al., 2014). All these factors—H_2, CO_2, Fe(Ni) S catalysis and electrochemical proton gradients—point to one very specific environment on the early Earth as the cradle of life: alkaline hydrothermal vents (Martin et al., 2008).

3.1.3 Life brought to Earth from elsewhere in space

A third mechanism with regard to the origin of life on Earth is life originating from the carbon and hydrocarbons in comets and meteorites as they burned in the atmosphere. The craters on Moon were formed about 3.8–4.0 billion years ago through bombardment by meteorites and asteroids (Chyba and Sagan, 1992). At the same time, huge numbers of meteorite and asteroids would have hit the relatively larger Earth because of its proximity to Moon. Carbonaceous (C-)chondrite is considered to be the earliest type of meteorite, containing a "memory" of the primitive Solar System. C-chondrites are mainly composed of Mg-rich minerals including a hydrous silicate, serpentine. They also contain organic and bio-organic molecules (e.g., amino acids). When such meteorites rained down on the early Earth, the energy of collision would convert simple organic molecules to bio-organic compounds (Hashizume, 2012). In a "shock" experiment by Blank et al. (2001), simulating collision of meteorite and asteroids with Earth, amino acids were polymerized into oligo-peptides (mostly dimers and trimmers).

Sources of organic molecules on the early Earth can be divided into three categories: delivery by extraterrestrial objects; organic synthesis driven by impact shocks;

and organic synthesis by other energy sources (such as ultraviolet light or electrical discharges). According to Chyba and Sagan (1992), estimates of these sources for plausible end-member oxidation states of the early terrestrial atmosphere suggest that the heavy bombardment before 3.5 Gyr ago either produced or delivered quantities of organics comparable to those produced by other energy sources. Which sources of prebiotic organics were quantitatively dominant? This depends strongly on the composition of the early terrestrial atmosphere. In the event of an early strongly reducing atmosphere, production by atmospheric shocks seems to have dominated that due to electrical discharges. Chyba and Sagan (1992) proposed that endogenous, exogenous and impact-shock sources of organics could each have made a significant contribution to the origins of life. The school of thought that perhaps life did not begin on Earth at all, but was brought to Earth from elsewhere in space is part of a notion known as *panspermia*. Rocks regularly get blasted off Mars by cosmic impacts, and a number of Martian meteorites have been found on Earth that some researchers have controversially suggested brought microbes over here, potentially making us all Martians originally.

Other scientists have even suggested that life might have hitchhiked on comets from other star systems. However, even if this concept were true, the question of how life began on Earth would then only change to how life began elsewhere in space. According to the notion of terrestrial life having an extraterrestrial origin, the amino acid building blocks may not have formed at the submarine hydrothermal vents as suggested by some origin-of-life scientists, but instead may have rained down in comets and meteorites. Astronomers have identified many small organic molecules in space, which opens the possibility of peptide factories being seeded on places besides Earth.

3.2 Biological evolution

Heredity and evolution are the two hallmarks of life. Genetic information storage and processing rely on just two polymers—the nucleic acids deoxyribonucleic acid (DNA) and ribonucleic acid (RNA). Nucleic acids are biopolymers (natural polymers produced by the cells of living organisms), which are essential to all known forms of life. Nucleic acids are composed of nucleotides, which are organic molecules composed of three subunit molecules; namely, a nucleobase, a 5-carbon sugar (ribose or deoxyribose), and a phosphate group consisting of one to three phosphates, and a nitrogenous base. The two main classes of nucleic acids are DNA and RNA. If the sugar is ribose, the polymer is RNA; if the sugar is the ribose derivative deoxyribose, the polymer is DNA.

Nucleic acids are chemical compounds that serve as the primary information-carrying molecules in biological cells and make up the genetic material. Nucleic acids are found in abundance in all living things, where they create, encode, and then store information of every living cell of every life-form on Earth. In turn, they function to transmit and express that information inside and outside the cell nucleus to the interior operations of the cell and ultimately to the next generation of each living organism. The encoded information is contained and conveyed via the nucleic acid sequence, which provides the "ladder-step" ordering of nucleotides (Adenine (abbreviated to A), Thymine (abbreviated to T), Cytosine (abbreviated to C), Guanine (abbreviated to G), and Uracil (abbreviated to U)) within the molecules of RNA and DNA. Note that whereas the four nucleotides present in DNA are A, T, C, and G, those present in RNA are C, G, A, and U. Nucleotides play an especially important role in directing protein synthesis.

Simply stated, DNA and RNA provide the molecular basis for all life through their unique ability to store and propagate information. DNA and RNA are nucleic acids, which are made up of just four subunits, called nucleotides. Fig. 3.8 shows comparison of single-stranded RNA (left) with double-stranded DNA (right), showing the corresponding helices and nucleobases that each molecule employs. DNA is a molecule composed of two polynucleotide chains that coil around each other to form a double helix carrying genetic instructions for the development, functioning, growth and reproduction of all known organisms and many viruses.

Strings of nucleotides are bonded to form helical backbones—typically, one for RNA, two for DNA—and assembled into chains of base-pairs selected from the five primary (or canonical), nucleobases, which are: adenine (Fig. 3.9A), cytosine (Fig. 3.9B), guanine (Fig. 3.9C), thymine (Fig. 3.9D), and uracil (Fig. 3.9E). Thymine occurs only in DNA and uracil only in RNA. Using amino acids and the process known as protein synthesis, the specific sequencing in DNA of these nucleobase-pairs enables storing and transmitting coded instructions as genes. In RNA, base-pair sequencing provides for manufacturing new proteins that determine the frames and parts and most chemical processes of all life forms. Fig. 3.10 illustrates the double-helical structure of DNA, showing dimensional and molecular structure details.

Pray (2008) explains that "One of the ways that scientists have elaborated on Watson and Crick's model is through the identification of three different conformations of the DNA double helix. In other words, the precise geometries and dimensions of the double helix can vary. The most common conformation in most living cells (which is the one depicted in most diagrams of the double helix, and the one proposed by Watson and Crick) is known as B-DNA. There are also two other conformations: A-DNA, a shorter and wider form that has been found in dehydrated samples of DNA and rarely under normal physiological circumstances; and Z-DNA, a left-handed conformation.

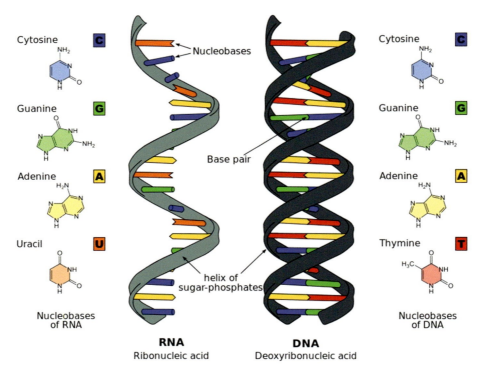

FIG. 3.8 Comparison of single-stranded RNA (left) with double-stranded DNA (right), showing the corresponding helices and nucleobases that each molecule employs. *(Source: https://commons.wikimedia.org/wiki/File:Difference_DNA_RNA-EN.svg)* Copyright info: This file is licensed under the Creative Commons Attribution-Share Alike 3.0 Unported license.

FIG. 3.9A (Left) Chemical structure of adenine in vector format. *(Source: https://commons.wikimedia.org/wiki/File:Adenine.svg)* Author: Pepemonbu Copyright info: This file is licensed under the Creative Commons Attribution-Share Alike 3.0 Unported license. (middle) Chemical Structure of Adenine in 3D balls format Color code: ■ Carbon, C: black □ Hydrogen, H: white ■ Oxygen, O: red ■ Nitrogen, N: blue Source: https://commons.wikimedia.org/wiki/File:Adenine-3D-balls.png Author: Vesprcom Copyright info: I, the copyright holder of this work, release this work into the public domain. This applies worldwide. (right) Chemical Structure of Adenine in 3D vdW format (Source: https://commons.wikimedia.org/wiki/File:Adenine-3D-vdW.png Author: Vesprcom Copyright info: I, the copyright holder of this work, release this work into the public domain. This applies worldwide.

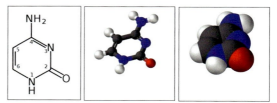

FIG. 3.9B (Left) Chemical structure of cytosine in vector format. *(Source: https://commons.wikimedia.org/wiki/File:Cytosine_chemical_structure.svg)* Author: Engineer gena Copyright info: I, the copyright holder of this work, release this work into the public domain. This applies worldwide. (middle) Chemical Structure of Cytosine in 3D balls format Color code: ■ Carbon, C: black □ Hydrogen, H: white ■ Oxygen, O: red ■ Nitrogen, N: blue Source: https://commons.wikimedia.org/wiki/File:Cytosine-3D-balls.png Author: Vesprcom Copyright info: I, the copyright holder of this work, release this work into the public domain. This applies worldwide. (right) Chemical Structure of Cytosine in 3D vdW format (Source: https://commons.wikimedia.org/wiki/File:Cytosine-3D-vdW.png Author: Vesprcom Copyright info: I, the copyright holder of this work, release this work into the public domain. This applies worldwide.

Z-DNA is a transient form of DNA, only occasionally existing in response to certain types of biological activity (Fig. 3.11). Z-DNA was first discovered in 1979, but its existence was largely ignored until recently. Scientists have since discovered that certain proteins bind very strongly to Z-DNA, suggesting that Z-DNA plays an important biological role in protection against viral disease (Rich and Zhang, 2003)."

Whereas DNA is found mostly in the cell nucleus, RNA is common in the cytoplasm. In the nucleus, the DNA code is "transcribed," or copied, into a messenger RNA (mRNA) molecule. Having briefly underlined the salient features of the wonderful biological molecule, known as DNA, it would be interesting and thought-provoking to examine how the secrets of this molecule have been unraveled/discovered and how difficult was the path to this discovery.

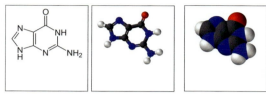

FIG. 3.9C (Left) Chemical structure of guanine in vector format. *(Source: https://commons.wikimedia.org/wiki/File:Guanin.svg)* Author: NEUROtiker Copyright info: This work is ineligible for copyright and therefore in the public domain because it consists entirely of information that is common property and contains no original authorship. (middle) Chemical Structure of Guanine in 3D balls format. Color code: ■ Carbon, C: black □ Hydrogen, H: white ■ Oxygen, O: red ■ Nitrogen, N: blue Source: https://commons.wikimedia.org/wiki/File:Guanine-3D-balls.png Author: Vesprcom Copyright info: I, the copyright holder of this work, release this work into the public domain. This applies worldwide. (right) Chemical Structure of Guanine in 3D vdW format (Source: https://commons.wikimedia.org/wiki/File:Guanine-3D-vdW.png Author: Vesprcom Copyright info: I, the copyright holder of this work, release this work into the public domain. This applies worldwide.

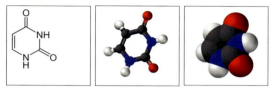

FIG. 3.9E (Left) Chemical structure of uracil in vector format. *(Source: https://commons.wikimedia.org/wiki/File:Uracil.svg)* Author: NEUROtiker Copyright info: This work is ineligible for copyright and therefore in the public domain because it consists entirely of information that is common property and contains no original authorship. (Middle) Chemical structure of uracil in 3D balls format. Color code: ■ Carbon, C: black □ Hydrogen, H: white ■ Oxygen, O: red ■ Nitrogen, N: blue. Source: https://commons.wikimedia.org/wiki/File:Uracil-3D-balls.png Author: Kemikungen Copyright info: I, the copyright holder of this work, release this work into the public domain. This applies worldwide. (right) Chemical Structure of Uracil in 3D vdW format (Source: https://commons.wikimedia.org/wiki/File:Uracil-3D-vdW.png) Author: Kemikungen Copyright info: I, the copyright holder of this work, release this work into the public domain. This applies worldwide.

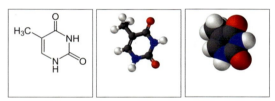

FIG. 3.9D (Left) Chemical structure of thymine in vector format. *(Source: https://commons.wikimedia.org/wiki/File:Thymin.svg)* Author: NEUROtiker Copyright info: This work is ineligible for copyright and therefore in the public domain because it consists entirely of information that is common property and contains no original authorship. (middle) Chemical Structure of Thymine in 3D balls format. Color code: ■ Carbon, C: black □ Hydrogen, H: white ■ Oxygen, O: red ■ Nitrogen, N: blue. Source: https://commons.wikimedia.org/wiki/File:Thymine-3D-balls.png Author: Vesprcom Copyright info: I, the copyright holder of this work, release this work into the public domain. This applies worldwide. (right) Chemical Structure of Thymine in 3D vdW format (Source: https://commons.wikimedia.org/wiki/File:Thymine-3D-vdW.png Author: Vesprcom Copyright info: I, the copyright holder of this work, release this work into the public domain. This applies worldwide

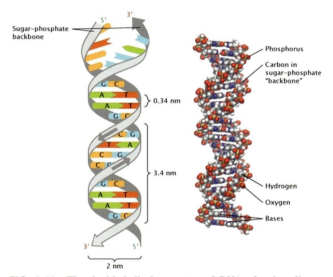

FIG. 3.10 **The double-helical structure of DNA, showing dimensional and molecular structure details.** The three-dimensional double helix structure of DNA, correctly elucidated by James Watson and Francis Crick. Complementary bases are held together as a pair by hydrogen bonds. Note that the precise geometries and dimensions of the double helix can vary. *(Source: Pray, L. (2008), Discovery of DNA structure and function: Watson and Crick, Nat. Educ., 1(1):100.)*

3.2.1 Discovery of DNA and its sequencing—the intriguing story of combined efforts by a group of scientists from different disciplines

DNA is a long linear polymer found in the nucleus of every biological cell and shaped like a double helix. DNA is the genetic information that all living creatures carry in the nucleus of each of their cells. Thus, DNA is like a blueprint of biological guidelines that a living organism must follow to exist and remain functional. DNA is self-replicating (i.e., making replicas of it).

A clear understanding of biological evolution was made possible only after the discovery of DNA, which hold the genetic blueprint to all known forms of life. For this reason, it would be interesting and thought-provoking to skim through the difficult pathways of the highly strenuous journey painstakingly undertaken by a group of determined scientists in the field of physics, biology, and medicine.

3.2.1.1 Isolating nucleic acid—Johannes Friedrich Miescher: the first scientist

Many people believe that American biologist James Watson and English physicist Francis Crick discovered DNA—a nucleic acid (i.e., an acid that is enclosed in the nucleus of every biological cell) in the 1950s. In reality, this is not

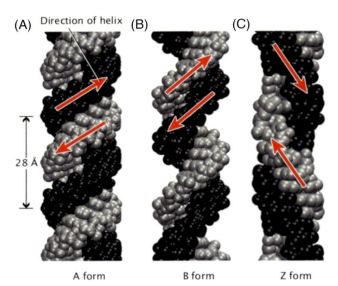

FIG. 3.11 **Three different conformations of the DNA double helix.** (A) A-DNA is a short, wide, right-handed helix. (B) B-DNA, the structure proposed by Watson and Crick, is the most common conformation in most living cells. (C) Z-DNA, unlike A- and B-DNA, is a left-handed helix. © 2014 Nature Education Adapted from Pierce, Benjamin. Genetics: A Conceptual Approach, 2nd ed. All rights reserved. *(Source: Pray, L. (2008), Discovery of DNA structure and function: Watson and Crick, Nat. Educ., 1(1):100.)*

FIG. 3.12 Johannes Friedrich Miescher (13 August 1844–26 August 1895)—a Swiss physician and biologist—who was the first scientist to isolate nucleic acid in 1869.

the case. Rather, DNA was first identified and isolated by Swiss physician and biologist Johannes Friedrich Miescher in 1869 (Fig. 3.12).

As part of his research on nucleic acids, Miescher devised different salt solutions, eventually producing one with sodium sulfate. The cells were filtered. Because centrifuges were not available at the time, the cells were allowed to settle to the bottom of a beaker. He then tried to isolate the nuclei free of cytoplasm (a thick solution that fills each cell and is enclosed by the cell membrane). Miescher had isolated various phosphate-rich chemicals from the nuclei of white blood cells in Felix Hoppe-Seyler's laboratory at the University of Tübingen, Germany (Dahm, 2008). Miescher was a perfectionist and a workaholic, and often worked very long hours to do the nuclein isolations. He subjected the purified nuclei to an alkaline extraction followed by acidification, resulting in the formation of a precipitate that Miescher called *nuclein*. The term "nuclein" was later changed to "nucleic acid" and eventually to "deoxyribonucleic acid," or "DNA." Miescher found that this contained phosphorus and nitrogen, but not sulfur.

Hoppe-Seyler's laboratory was one of the first in Germany to focus on tissue chemistry. At a time when scientists were still debating the concept of "cell," Hoppe-Seyler and his lab were isolating the molecules that made up cells. Miescher was given the task of researching the composition of lymphoid cells—white blood cells. These cells were difficult to extract from the lymph glands, but they were found in great quantities in the pus from infections.

Miescher collected bandages from a nearby clinic and washed off the pus. He experimented and isolated a new molecule—nuclein—from the cell nucleus. He determined that nuclein was made up of hydrogen, oxygen, nitrogen and phosphorus and there was a unique ratio of phosphorus to nitrogen. He was able to isolate nuclein from other cells and later used salmon sperm (as opposed to pus) as a source (Concept 15: DNA and proteins are key molecules of the cell nucleus, In: *DNA from the Beginning: An animated primer of 75 experiments that made modern genetics;* https://dnalc.cshl.edu/websites/dnaftb.html). Sensing the importance of his findings, Miescher wrote, "It seems probable to me that a whole family of such slightly varying phosphorous-containing substances will appear, as a group of nucleins, equivalent to proteins" (Wolf, 2003).

Although Miescher did most of his work in 1869, his paper on nuclein wasn't published until 1871. Nuclein was such a unique molecule that Hoppe-Seyler was sceptical and wanted to confirm Miescher's results before publication. Miescher's discovery (that DNA contained phosphorus and nitrogen, but not sulfur) was so unlike anything else at the time that Hoppe-Seyler repeated all Miescher's research himself before publishing it. Miescher's discovery paved the way for the identification of DNA as the carrier of inheritance. The significance of the discovery—first published in 1871 (Miescher, 1871)—was not at first apparent, and Albrecht Kossel (a German physiologist specializing in the physiological chemistry of the cell and its nucleus and of proteins) made the initial inquiries into the chemical structure of nuclein (Jones, 1953). Since the beginning of human history, people have wondered how traits are inherited from one generation to the next. Although children often look more like one parent than the other, most offspring seem to be a blend of the characteristics of both parents.

Centuries of breeding of domestic plants and animals had shown that useful traits—speed in horses, strength in oxen, and larger fruits in crops—can be accentuated (made more noticeable or prominent) by controlled mating. Miescher's discovery played an important part in the identification of nucleic acids as the carriers of inheritance. Miescher raised the idea that the nucleic acids could be involved in heredity (see Bryson, 2005) and even posited that there might be something akin to an alphabet that might explain how variation is produced.

Miescher was the first to identify DNA as a distinct molecule. Surprisingly, the story of genetics typically omits the original discovery of the molecular nature of DNA (see Lamm et al., 2000). As already indicated, in 1869, the young Swiss biochemist Friedrich Miescher discovered the molecule we now refer to as DNA, developing techniques for its extraction. Miescher, already having been suffering from weak hearing as a result of the typhoid he contracted in his childhood, left this world in 1895 from tuberculosis at the age of 51. Lamm et al. (2000) explain why Miescher's name is all but forgotten, and his role in the history of genetics is mostly overlooked. In a study, Lamm et al. (2000) have focused on the role of national rivalries and disciplinary turf wars (acrimonious—angry and bitter—dispute between rival groups) in shaping historical memory, and on how the story they tell shapes our understanding of the science. Lamm et al. (2000) have highlighted that Miescher could just as correctly be portrayed as the person who understood the chemical nature of chromatin (before the term existed), and the first to suggest how stereochemistry (the branch of chemistry concerned with the three-dimensional arrangement of atoms and molecules and the effect of this on chemical reactions) might serve as the basis for the transmission of hereditary variation. Note that "chromatin" is a substance within a chromosome (a threadlike structure of nucleic acids and protein found in the nucleus of most living cells, carrying genetic information in the form of genes) consisting of DNA and protein. The DNA carries the cell's genetic instructions. The major proteins in chromatin are histones, which help package the DNA in a compact form that fits in the cell nucleus.

Unfortunately, Miescher's notion of DNA as the carrier of inheritance remained under-studied for many decades because proteins, rather than DNA, were thought to hold the genetic blueprint to life. Then, in the decades following Miescher's discovery, other scientists—notably, Russian-American biochemist Phoebus Levene and Erwin Chargaff (an Austro-Hungarian-born American biochemist, Bucovinian Jew who emigrated to the United States during the Nazi era)—carried out a series of research efforts that revealed additional details about the DNA molecule, including its primary chemical components and the ways in which they joined with one another. For example, Phoebus Levene (1869–1940), who had discovered ribose sugar in 1909

FIG. 3.13 (Left) Portrait of Oswald T. Avery, cropped from a Rockefeller Institute for Medical Research staff photograph *(Source: https://commons.wikimedia.org/wiki/File:Oswald_T._Avery_portrait_1937.jpg)* Copyright info: This work is in the public domain in the United States because it is a work prepared by an officer or employee of the United States Government as part of that person's official duties under the terms of Title 17, Chapter 1, Section 105 of the US Code. (Middle) Colin MacCleod, geneticist *(Source: https://commons.wikimedia.org/wiki/File:ColinMacCleod.jpg)* Copyright info: This work is in the public domain in the United States because it is a work prepared by an officer or employee of the United States Government as part of that person's official duties under the terms of Title 17, Chapter 1, Section 105 of the US Code. (right) Maclyn McCarty with Francis Crick and James D. Watson. *(Source: https://commons.wikimedia.org/wiki/File:Maclyn_McCarty_with_Francis_Crick_and_James_D_Watson_-_10.1371_journal.pbio.0030341.g001-O.jpg)* Copyright info: This file is licensed under the Creative Commons Attribution 3.0 Unported license.

and deoxyribose sugar in 1929, suggested the structure of nucleic acid as a repeating tetramer (a polymer comprising four monomer units). He called the phosphate-sugar-base unit a nucleotide. Likewise, Erwin Chargaff contributed to the basic understanding of double helix structure of DNA. Chargaff noticed that DNA—whether taken from a plant or animal—contained equal amounts of adenine and thymine and equal amounts of cytosine and guanine.

Without the scientific foundation provided by these pioneers, Watson and Crick may never have reached their ground-breaking conclusion of 1953: that the DNA molecule exists in the form of a three-dimensional double helix (Pray, 2008).

The undesirable situation (of DNA remaining under-studied) changed after 1944 as a result of some experiments by Oswald Avery (Fig. 3.13 (left)), Colin MacLeod (Fig. 3.13 (middle)), and Maclyn McCarty (Fig. 3.13 (right)), demonstrating that purified DNA could change one strain of bacteria into another. The Avery–MacLeod–McCarty experiment was an experimental demonstration, reported in 1944 by Oswald Avery, Colin MacLeod, and Maclyn McCarty, that DNA is the substance that causes bacterial transformation (Avery et al., 1944), in an era when it had been widely believed that it was proteins that served the function of carrying genetic information (with the very word protein itself coined to indicate a belief that its function was primary). It was the culmination of research in the 1930s and early 20th century at the Rockefeller Institute for Medical Research to purify and characterize the "transforming principle" responsible for the transformation phenomenon first described in Griffith's experiment of 1928.

This was the first time that DNA was shown capable of transforming the properties of cells.

As just indicated, the acceptance of the genetic role of DNA began in 1944 with the publication of Avery et al. (1944) on the identification of the "transforming principle" in pneumococcal bacteria as DNA (Cobb, 2014). For much of the 1950s, the suggestion that DNA was the hereditary material in all organisms was accepted as a "working hypothesis" but nothing more. A key insight came in 1953, when Watson and Crick suggested that the sequence of bases on a DNA molecule contains "genetical information" (Watson and Crick, 1953). The issue then became how that information was turned into biological function—the nature of the genetic code and how it worked. The person initially responsible for focusing attention on this problem was the cosmologist George Gamow. One of the continuous concerns throughout this period was that it remained unclear how genes functioned.

3.2.1.2 DNA sequencing—contributions of Frederick Sanger, Francis Crick, and James D. Watson

DNA sequencing has played an important role in understanding biological evolution. Sequencing DNA means determining the order of the four chemical building blocks—called "bases"—that make up the DNA molecule. The sequence tells scientists the kind of genetic information that is carried in a particular DNA segment. The sequence data can highlight changes in a gene that may cause disease. As already noted, the canonical structure of DNA has four bases: adenine (A), thymine (T), cytosine (C), and guanine (G). DNA sequencing is the determination of the physical order of these bases in a molecule of DNA. DNA sequencing includes any method or technology that is used to determine the order of the four bases indicated above. DNA sequencing technology is helping scientists unravel questions that humans have been asking about organisms (individual animals, plants, or single-celled life forms) for centuries. Knowledge of DNA sequences has become indispensable for basic biological research.

In the DNA double helix (see Fig. 3.8), the four chemical bases always bond with the same partner to form "base pairs." For example, adenine (A) always pairs with thymine (T). Likewise, cytosine (C) always pairs with guanine (G). This pairing is the basis for the mechanism by which DNA molecules are copied when cells divide, and the pairing also underlies the methods by which most DNA sequencing experiments are done. In the fields of molecular biology and genetics, a "genome" is the complete set of genes or genetic material (i.e., complete DNA sequences) present in a cell or organism. The human genome contains about 3 billion base pairs that spell out the instructions for making and maintaining a human being.

The rapid speed of sequencing attained with modern DNA sequencing technology has been instrumental in the sequencing of complete DNA sequences, or genomes, of numerous types and species of life, including the human genome and other complete DNA sequences of many animal, plant, and microbial species. Sequencing is used in molecular biology to study genomes and the proteins they encode. Information obtained using sequencing allows researchers to identify changes in genes. The first DNA sequences were obtained in the early 1970s by academic researchers using laborious methods based on two-dimensional chromatography. Following the development of fluorescence-based sequencing methods with a DNA sequencer (Olsvik et al., 1993), DNA sequencing has become easier and orders of magnitude faster (Pettersson et al., 2009).

DNA sequencing may be used to determine the sequence of individual genes, larger genetic regions (i.e., clusters of genes or operons), full chromosomes, or entire genomes of any organism. DNA sequencing is also the most efficient way to indirectly sequence RNA or proteins (via their open reading frames). In fact, DNA sequencing has become a key technology in many areas of biology and other sciences such as medicine, forensics, and anthropology.

Returning to the intriguing story of the discovery of DNA and its sequencing, the foundation for sequencing proteins was first laid by the work of Frederick Sanger (Fig. 3.14) who by 1955 had completed the sequence of all the amino acids in insulin, a small protein secreted by the pancreas. This provided the first conclusive evidence that proteins were chemical entities with a specific molecular pattern rather than a random mixture of material suspended in fluid. Sanger's success in sequencing insulin spurred on X-ray crystallographers, including Francis Crick and James D. Watson, who by now were trying to understand how DNA directed the formation of proteins within a cell.

Although DNA had first been discovered in the late nineteenth century, it remained little studied for many decades. In part this was because proteins, rather than DNA, were considered to hold the genetic blueprint for organisms. As Sanger admitted in 1997, when he had started working in the Department of Biochemistry in the 1940s, he had "thought of DNA as an inert substance." Indeed, he continued, "the notion" that DNA contained "all the information for making a complete organism would have been thought of as science fiction" (Garcia-Sancho, 2006).

Although Sanger did not progress very far in understanding the function of insulin through sequencing, his sequencing methods generated great excitement among X-ray crystallographers investigating the three-dimensional structure of DNA. One of the first people inspired by Sanger's work was the British physicist and crystallographer Francis Crick (Fig. 3.15). Other researchers had made important but seemingly unconnected findings about the composition of DNA. Finally, it fell to Francis Crick

FIG. 3.14 (Top) Double Nobel Laureate Frederick Sanger, a pioneer of sequencing. Sanger is one of the few scientists who was awarded two Nobel prizes, one for the sequencing of proteins, and the other for the sequencing of DNA. (Copyright info: This image is a work of the National Institutes of Health, part of the United States Department of Health and Human Services, taken or made as part of an employee's official duties. As a work of the U.S. federal government, the image is in the public domain). *(Source: https://commons.wikimedia.org/wiki/File:Frederick_Sanger2.jpg) (bottom) Sanger in his laboratory, 1969. Credit: The Laboratory of Molecular Biology (LMB).*

FIG. 3.15 Nobel laureate British physicist and crystallographer Francis Crick in his office. (Source: https://commons.wikimedia.org/wiki/File:Francis_Crick_crop.jpg) Copyright info: This file is licensed under the Creative Commons Attribution 2.5 Generic license.

FIG. 3.16 Nobel laureate American geneticist and biophysicist Dr. James D. Watson, Chancellor, Cold Spring Harbor Laboratory. *(Source: https://en.wikipedia.org/wiki/File:James_D_Watson.jpg)* Copyright info: This image is a work of the National Institutes of Health, part of the United States Department of Health and Human Services, taken or made as part of an employee's official duties. As a work of the U.S. Federal Government, the image is in the public domain.

and American geneticist and biophysicist James Watson (Fig. 3.16) to unify these disparate findings into a coherent theory of genetic transfer. The British biochemist Alexander Todd had determined that the backbone of the DNA molecule contained repeating phosphate and deoxyribose sugar groups. Todd was awarded the 1957 Nobel Prize in chemistry for his work on the synthesis of nucleotides (the small units that make up the larger molecule of nucleic acids), the hereditary material of cells. The American biochemist Erwin Chargaff had found that while the amount of DNA and of its four types of bases—the purine bases adenine (A) and guanine (G), and the pyrimidine bases cytosine (C) and thymine (T)—varied widely from species to species, A and T always appeared in ratios of one-to-one, as did G and C. The New Zealand-born British biophysicist Maurice Wilkins and the English physical chemist Rosalind Franklin had obtained high-resolution X-ray images of DNA fibers that suggested a helical, corkscrew-like shape. Crick's discovery was also made possible by recent advances in model building, or the assembly of possible three-dimensional structures based upon known molecular distances and bond angles, a technique advanced by American biochemist Linus Pauling. Linus Pauling, then the world's leading physical chemist, had discovered the single-stranded alpha helix, the structure found in many proteins, prompting biologists to think of helical forms. Moreover, Pauling had pioneered the method of model building in chemistry by which Watson and Crick were to uncover the structure of DNA. Indeed, Crick and Watson feared that they would be upstaged by

Pauling, who proposed his own model of DNA in February 1953, although his three-stranded helical structure quickly proved erroneous.

The time, then, was ripe for Watson and Crick's discovery. Jerry Donohue, a visiting physical chemist from the United States who shared Watson and Crick's office for the year, pointed out that the configuration for the rings of carbon, nitrogen, hydrogen, and oxygen (the elements of all four bases) in thymine and guanine given in most textbooks of chemistry was incorrect. On February 28, 1953, Watson, acting on Donohue's advice, put the two bases into their correct form in cardboard models by moving a hydrogen atom from a position where it bonded with oxygen to a neighboring position where it bonded with nitrogen. While shifting around the cardboard cut-outs of the accurate molecules on his office table, Watson realized in a stroke of inspiration that A, when joined with T, very nearly resembled a combination of C and G, and that each pair could hold together by forming hydrogen bonds. If A always paired with T, and likewise C with G, then not only were Chargaff's rules (that in DNA, the amount of A equals that of T, and C that of G) accounted for, but the pairs could be neatly fitted between the two helical sugar-phosphate backbones of DNA, the outside rails of the ladder. The bases connected to the two backbones at right angles while the backbones retained their regular shape as they wound around a common axis, all of which were structural features demanded by the X-ray evidence. Similarly, the complementary pairing of the bases was compatible with the fact, also established by the X-ray diffraction pattern, that the backbones ran in opposite direction to each other, one up, the other down.

Fig. 3.17 shows a pencil sketch of the DNA double helix drawn by Crick in his notebook, way back in the year 1953.

It shows a right-handed helix and the nucleotides of the two antiparallel strands. Fig. 3.18 illustrates a more recent schematic representation of the structure of DNA, showing how its nucleotides are arranged. Each strand is composed of four complementary nucleotides—adenine (A), cytosine (C), guanine (G) and thymine (T)—with an A on one strand always paired with T on the other, and C always paired with G. Watson and Crick proposed that such a structure allowed each strand to be used to reconstruct the other, an idea central to the passing on of hereditary information between generations (Watson and Crick, 1953).

Watson and Crick published their findings in a paper, with the title "A Structure for Deoxyribose Nucleic Acid," in the British scientific weekly *Nature* on April 25, 1953, illustrated with a schematic drawing of the double helix by Crick's wife, Odile. A coin toss decided the order in which they were named as authors (Watson and Crick, 1953a). Foremost among the "novel features" of "considerable biological interest" they described was the pairing of the bases on the inside of the two DNA backbones: A=T and C=G. The pairing rule immediately suggested a copying mechanism for DNA: given the sequence of the bases in one strand, that of the other was automatically determined, which meant that when the two chains separated, each served as a template for a complementary new chain. Watson and Crick developed their ideas about genetic replication in a second

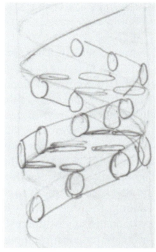

FIG. 3.17 Pencil sketch of the DNA double helix by Crick. It shows a right-handed helix and the nucleotides of the two antiparallel strands, 1953, Crick's notebook. Credit: Wellcome Library, file PP/CRI/H/1/16. *(Source: https://www.whatisbiotechnology.org/index.php/exhibitions/sanger/path)*

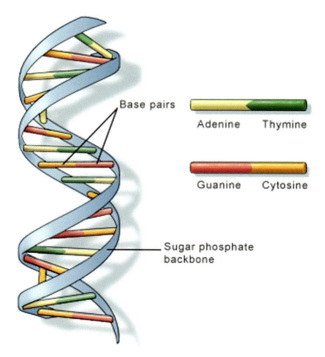

FIG. 3.18A Schematic diagram illustrating a more recent representation of the structure of DNA showing how its nucleotides are arranged. Credit: U.S. National Library of Medicine. The sugar phosphate backbone shown in this diagram is called a "strand." *(Source: https://www.whatisbiotechnology.org/index.php/exhibitions/sanger/path)*

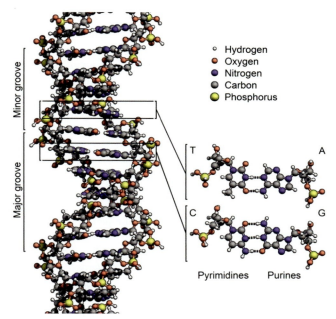

FIG. 3.18B The structure of the DNA double helix. The atoms in the structure are color-coded by element and the detailed structures of two base pairs are shown in the bottom right. Copyright info: This file is licensed under the Creative Commons Attribution-Share Alike 3.0 Unported license. *(Source: https://en.wikipedia.org/wiki/File:DNA_Structure%2BKey%2BLabelled.pn_NoBB.png)*

article in *Nature*, published on May 30, 1953 (Watson and Crick, 1953b).

As indicated above, in 1953, Francis Crick together with James Watson put forward what was to become a famous model for the molecular structure of DNA based on X-ray diffraction images produced by Rosalind Franklin and Maurice Wilkins. Watson, Crick, and Maurice Wilkins were awarded the 1962 Nobel Prize in Physiology or Medicine "for their discoveries concerning the molecular structure of nucleic acids and its significance for information transfer in living material." In subsequent years, it has been recognized that Watson and his colleagues did not properly attribute colleague Rosalind Franklin for her contributions to the discovery of the double helix structure (Stasiak, 2001; Maddox, 2003).

As just indicated, an enduring controversy has been generated by Watson and Crick's use of Rosalind Franklin's crystallographic evidence of the structure of DNA, which was shown to them, without her knowledge, by her estranged colleague, Maurice Wilkins, and by Max Perutz. Rosalind Franklin's evidence demonstrated that the two sugar-phosphate backbones lay on the outside of the molecule, confirmed Watson and Crick's conjecture that the backbones formed a double helix, and revealed to Crick that they were antiparallel (i.e., parallel but oriented in opposite directions). Thus, while each strand contained four complementary nucleotides: adenine (A), cytosine (C), guanine (G), and thymine (T), the two strands were oriented in opposite directions (antiparallel) so that adenine always joined thymines (A T) and cytosines were linked with guanines (C G). It was this structure, they argued, which enabled each strand to reconstruct the other and facilitated the passing on of hereditary information from parent to offspring (Watson and Crick, 1953).

Franklin's superb experimental work thus proved crucial in Watson and Crick's discovery. Yet, they gave her scant acknowledgment. Even so, Franklin bore no resentment toward them. She had presented her findings at a public seminar to which she had invited the two (https://profiles.nlm.nih.gov/spotlight/sc/feature/doublehelix). She soon left DNA research to study tobacco mosaic virus. She became friends with both Watson and Crick, and spent her last period of remission from ovarian cancer in Crick's house (Franklin died in 1958). Crick believed that he and Watson used her evidence appropriately, while admitting that their patronizing attitude toward her, so apparent in *The Double Helix* (a book authored by Watson in 1968), reflected contemporary conventions of gender in science (https://profiles.nlm.nih.gov/spotlight/sc/feature/doublehelix).

Crick and Watson recognized, at an early stage in their careers, that gaining a detailed knowledge of the three-dimensional configuration of the gene was the central problem in molecular biology. Without such knowledge, heredity and reproduction could not be understood. They seized on this problem during their very first encounter, in the summer of 1951, and pursued it with single-minded focus over the course of the next 18 months. This meant taking on the arduous intellectual task of immersing themselves in all the fields of science involved: genetics, biochemistry, chemistry, physical chemistry, and X-ray crystallography. Drawing on the experimental results of others (they conducted no DNA experiments of their own), taking advantage of their complementary scientific backgrounds in physics and X-ray crystallography (Crick) and viral and bacterial genetics (Watson), and relying on their brilliant intuition, persistence, and luck, the two showed that DNA had a structure sufficiently complex and yet elegantly simple enough to be the master molecule of life.

Crick and Watson had shown that in DNA, form is function: the double-stranded molecule could both produce exact copies of itself and carry genetic instructions. During the following years, Crick elaborated on the implications of the double-helical model, advancing the hypothesis, revolutionary then but widely-accepted since, that the sequence of the bases in DNA forms a code by which genetic information can be stored and transmitted.

Although recognized today as one of the seminal scientific papers of the twentieth century, Watson and Crick's original article in Nature was not frequently cited at first. Its true significance became apparent, and its circulation widened, only toward the end of the 1950s, when the structure of DNA they had proposed was shown to provide a mechanism for controlling protein synthesis, and when their conclusions were confirmed in the laboratory by Matthew Meselson, Arthur Kornberg, and others.

Researchers working on DNA in the early 1950s used the term "gene" to mean the smallest unit of genetic information, but they did not know what a gene actually looked like structurally and chemically, or how it was copied, with very few errors, generation after generation. Attitudes to DNA began to change in the wake of some experiments on pneumococcal bacteria carried out by Oswald Avery, Colin MacLeod and Maclyn McCarty in 1944. In 1944, Oswald Avery and colleagues had shown that DNA was the "transforming principle," the carrier of hereditary information, in pneumococcal bacteria. Nevertheless, many scientists continued to believe that DNA had a structure too uniform and simple to store genetic information for making complex living organisms. The genetic material, they reasoned, must consist of proteins, much more diverse and intricate molecules known to perform a multitude of biological functions in the cell. However, the findings of Oswald Avery and colleagues established that DNA could transform the properties of cells. As a result, a number of researchers began investigating the structure of DNA hoping that this would reveal how the molecule worked.

The discovery in 1953 of the double-helix, the twisted-ladder structure of deoxyribonucleic acid (DNA), by James Watson and Francis Crick marked a milestone in the history of science and gave rise to modern molecular biology, which is largely concerned with understanding how genes control the chemical processes within cells. In short order, their discovery yielded ground-breaking insights into the genetic code and protein synthesis. During the 1970s and 1980s, it helped to produce new and powerful scientific techniques, specifically recombinant DNA research, genetic engineering, rapid gene sequencing, and monoclonal antibodies, techniques on which today's multibillion-dollar biotechnology industry is founded. Major current advances in science, namely genetic fingerprinting and modern forensics, the mapping of the human genome, and the promise, yet unfulfilled, of gene therapy, all have their origins in Watson and Crick's inspired work. The double helix has not only reshaped biology, it has become a cultural icon, represented in sculpture, visual art, jewelry, and toys.

Following their elucidation of the structure of DNA, Crick and Watson began to investigate how DNA directed the formation of proteins within a cell. This they saw as fundamental to understanding how DNA dictated metabolism and other functional processes (Garcia-Sancho, 2012). On starting to deliberate the question, in October 1954 Crick attended a series of lectures given by Sanger about sequencing insulin. Inspired by Sanger's results with insulin, Crick began to develop a theory from the mid-1950s which argued that the arrangement of nucleotides in DNA determined the sequence of amino acids in proteins and that this in turn regulated how a protein folded into its final shape; it was this shape which determined the function of a protein. He further hypothesized that an intermediary molecule helped the DNA to specify the sequence of the amino acids in a protein (Crick, 1958).

In Crick's notebook (Wellcome Library, file: PP/CRI/H/1/6), he noted Sanger discussing three methods, including his own "jig-saw" method whereby proteins were broken into fragments and working out the sequence based on their overlaps. Another method mentioned is that of Edman which enabled sequences to be determined in their correct order.

Sanger's work on insulin not only helped shaped Crick's sequence hypothesis, but also provided an important experimental approach to test it. Before proceeding any further, however, Crick needed to find a way to demonstrate how a single mutant gene could alter the sequence of amino acids in the protein it coded for. He soon latched on to the idea of doing this by examining how an inherited genetic defect affected the sequence of amino acids in a protein (de Chadarevian, 1996).

Crick quickly turned to looking at sickle-cell anemia, a common genetic blood disorder. Just a few years before, in 1949, William Castle, an American hematologist, had spotted that hemoglobin, a protein molecule that exists in red blood cells and delivers oxygen to cells in the body, was shaped like a sickle in blood taken from patients with sickle-cell anemia. This, he believed, might be due to the protein being deprived of oxygen. The hemoglobin also differed from normal adult hemoglobin when subjected to electrophoretic tests and displayed unusual properties when investigated under polarized light. Subsequent research by Jim Neel, an American geneticist, suggested the abnormal hemoglobin was linked to an inherited genetic defect (Watson and Crick, 1953).

With regard to the question of whether the difference between normal hemoglobin and that taken from sickle-cell patients could be all about a likely difference in their number of amino acids, many were sceptical that the alteration of just one amino acid, out of approximately 300, could produce a molecule as lethal as sickle-cell hemoglobin. Techniques for sequencing the protein's amino acids, however, were not sufficiently sophisticated to settle the matter (Crick, 1958; Ingram, 2004). Vernon Ingram (Fig. 3.19), a German American postdoctoral protein biochemist, who was doing research in this area, fled to England as a refugee from Nazi Germany. He joined the MRC Unit in 1952. Meanwhile, it was suggested that Ingram deploy Sanger's latest fingerprinting techniques which he was already refining to characterize some other large protein fragments (Ingram, 2004).

Rather than sequencing the whole hemoglobin protein which, as Ingram recalled, was "a Herculean task," he

FIG. 3.19 Vernon Ingram, pictured here, fled to England as a refugee from Nazi Germany. He joined the MRC Unit in 1952. Credit: Ingram Family; Weatherall, 2010. *(Source: https://www.whatisbiotechnology.org/index.php/exhibitions/sanger/path)*

decided to cleave it into manageable peptide fragments, using a pancreatic enzyme called trypsin. With this he obtained 26 peptide fragments. Ingram, with some help from Sanger, then began separating the fragments, using paper electrophoresis and chromatography. His ultimate goal was to find a peptide fragment with a demonstrable electrophoretic difference. This involved "characterizing each peptide by its position on a two-dimensional map, a sheet of 'blotting paper'" (Ingram, 2004). After many hours of painstaking work, Ingram determined that the difference between normal and sickle-cell hemoglobin was down to the replacement of "only one of nearly 300 amino acids". Ingram's finding was a significant breakthrough. Not only did it confirm Crick's sequence hypothesis, but it was also the first time that anyone had managed to break the genetic code, the process by which cells translate information stored in DNA into proteins (de Chadarevian, 1996). Fig. 3.20 shows Ingram's sequencing results from normal hemoglobulin and sickle-cell hemoglobin.

A method to sequence proteins was the use of an enzyme to break up the nucleic acid into partial fragments and then to separate the nucleotides in the fragments. One of the most popular enzymes used for this work was ribonuclease T1, an enzyme discovered in 1957 by Kimiko Sato-Asano and Fujio Egami based at Nagoya University. The advantage of this enzyme was that it cut nucleic acids at a very specific point on its nucleotide chain, where guanine was present (Sanger and Dowding, 1996).

As early as 1955 John Kendrew, a budding protein crystallographer, believed Sanger's sequencing method could provide an important tool to work out the protein's three-dimensional structure which he believed could not be determined with X-ray crystallography alone. In addition to Kendrew, Sydney Brenner, a South African biologist, and Crick were keen to have Sanger join them in their work on sequencing some proteins they had produced with some mutant viruses. They hoped to demonstrate that the order of changes seen in a sequence of a mutated gene corresponded with that of the amino acids in the protein it coded for (de Chadarevian, 1996).

Creation of a new center, to be called the Laboratory of Molecular Biology (LMB), signaled a new alliance that had begun to emerge between protein crystallographers, molecular geneticists and protein chemists all working toward the development of a new discipline—molecular biology. Given the institution's strong biological orientation Sanger soon dropped his reservations about joining the Cavendish team (de Chadarevian, 1996). How DNA specified the structure of proteins lay at the heart of their research (Garcia-Sancho, 2010).

By now scientists had begun to understand how it was that DNA, a nucleic acid that is enclosed in the nucleus of the cell, could make proteins outside of a cell's nucleus, in the cytoplasm, the fluid beyond the cell's nucleus. This process involves a number of mechanisms. The first is the ribosome which exists on the outside a cell's nucleus and is responsible for protein synthesis in cells. Ribosome had first been discovered in 1955 by George E Palade, a Romanian-American cell biologist based at the Rockefeller Institute in New York. The second mechanism is ribonucleic acid (RNA), another nucleic acid. RNA is very similar to DNA in that it contains the same number of nucleotides, but it only has one strand.

DNA is inherently limited by its four natural nucleotides. Efforts to expand the genetic alphabet, by addition of an unnatural base pair, promise to expand the biotechnological applications available for DNA as well as to be an essential first step toward expansion of the genetic code. Leconte et al. (2008) conducted two independent screens of hydrophobic unnatural nucleotides to identify novel candidate base pairs that are well recognized by a natural DNA polymerase.

DNA is not only a repository of genetic information for life, it is also a unique polymer with remarkable properties: it associates according to well-defined rules, it can be assembled into diverse nanostructures of defined geometry, it can be evolved to bind ligands (molecules that bind to other (usually larger) molecules) and catalyze chemical reactions and it can serve as a supramolecular scaffold to arrange chemical groups in space. However, its chemical makeup is rather uniform and the physicochemical properties of the four canonical bases only span a narrow range. Much wider chemical diversity is accessible through solid-phase synthesis but oligomers are limited to <100 nucleotides and variations in chemistry can usually not be replicated and thus are not amenable to evolution. Recent advances in nucleic acid chemistry and polymerase engineering promise to bring the synthesis, replication and ultimately evolution of nucleic acid polymers with greatly

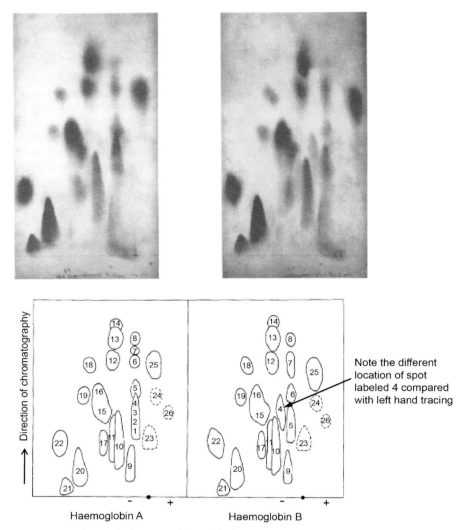

FIG. 3.20 This shows Ingram's sequencing results from normal hemoglobulin (labeled A on the left) and sickle-cell hemoglobin (labeled S on the right). At the bottom are tracings of the top chromatograms. Dotted lines indicate peptides that only became visible after heating the chromatogram. Credit: Ingram, 1958, Fig. 3. *(Source: https://www.whatisbiotechnology.org/index.php/exhibitions/sanger/path)*

expanded chemical diversity within our reach (see Loakes and Holliger, 2009).

Starting from pyranose nucleic acids, several series of modified nucleic acids with a six-membered carbohydrate moiety (mimic) have been synthesized and analyzed over a period of 20 years, and this work is summarized in (Herdewijn, 2010). The process starts with structural and conformational considerations, followed by synthetic efforts and a structural analysis, and ends up with a biological confirmation of the concept, demonstrating that these modified nucleic acids represent very valuable tools in chemistry and biology.

Oligonucleotide chemistry has been developed greatly over the past three decades, with many advances in increasing nuclease resistance, enhancing duplex stability and assisting with cellular uptake. Locked nucleic acid (LNA) is a structurally rigid modification that increases the binding affinity of a modified-oligonucleotide. In contrast, unlocked nucleic acid (UNA) is a highly flexible modification, which can be used to modulate duplex characteristics. In a tutorial review, Campbell and Wengel (2011) compared the synthetic routes to both of these modifications, contrasted the structural features, examined the hybridization properties of LNA and UNA modified duplexes, and discussed how they have been applied within biotechnology and drug research. LNA has found widespread use in antisense oligonucleotide technology, where it can stabilize interactions with target RNA and protect from cellular nucleases. The newly emerging field of siRNAs has made use of LNA and, recently, also UNA. These modifications are able to increase double-stranded RNA stability in serum and decrease off-target effects seen with conventional siRNAs. LNA and UNA are also emerging as versatile modifications for aptamers (oligonucleotide or peptide molecules that bind to a specific target molecule). Their application to known aptamer structures has opened up the possibility of future selection of LNA-modified aptamers. Each of these oligonucleotide technologies has the potential to become a new type of therapy to treat a wide variety of diseases, and LNA and UNA will no doubt play a part in future developments of therapeutic and diagnostic oligonucleotides (Campbell and Wengel, 2011).

3.2.2 Role of National Human Genome Research Institute (NHGRI) in supporting development of new technologies for DNA sequencing

DNA sequencing determines the order of the four chemical building blocks—called "bases"——that make up the DNA molecule. Since the completion of the Human Genome Project (an international scientific research project with the goal of determining the base pairs that make up human DNA, and of identifying, mapping and sequencing all of the genes of the human genome from both a physical and a functional standpoint), technological improvements and automation have increased speed and lowered costs to the point where individual genes can be sequenced routinely, and some labs can sequence well over 100,000 billion bases per year, and an entire genome can be sequenced for just a few thousand dollars.

Many of these new technologies were developed with support from the National Human Genome Research Institute (NHGRI) Genome Technology Program and its Advanced DNA Sequencing Technology awards. One of NHGRI's goals is to promote new technologies that could eventually reduce the cost of sequencing a human genome of even higher quality than is possible today and for less than $1000. (https://www.genome.gov/about-genomics/fact-sheets/DNA-Sequencing-Fact-Sheet#:~:text=Sequencing%20DNA%20means%20determining%20the, in%20a%20particular%20DNA%20segment.)

Since the completion of the Human Genome Project, technological improvements and automation have increased speed and lowered costs to the point where individual genes can be sequenced routinely, and some labs can sequence well over 100,000 billion bases per year, and an entire genome can be sequenced for just a few thousand dollars.

One new sequencing technology involves watching DNA polymerase molecules as they copy DNA—the same molecules that make new copies of DNA in our cells—with a very fast movie camera and microscope, and incorporating different colors of bright dyes, one each for the letters A, T, C, and G. This method provides different and very valuable information than what is provided by the instrument systems that are in most common use.

Another new technology in development entails the use of nanopores to sequence DNA. Nanopore-based DNA sequencing involves threading single DNA strands through extremely tiny pores in a membrane. DNA bases are read one at a time as they squeeze through the nanopore. The bases are identified by measuring differences in their effect on ions and electrical current flowing through the pore. Using nanopores to sequence DNA offers many potential advantages over current methods. The goal is for sequencing to cost less and be done faster. Unlike sequencing methods currently in use, nanopore DNA sequencing means researchers can study the same molecule over and over again (NHGRI website).

Researchers now are able to compare large stretches of DNA—1 million bases or more—from different individuals quickly and cheaply. Such comparisons can yield an enormous amount of information about the role of inheritance in susceptibility to disease and in response to environmental influences. In addition, the ability to sequence the genome more rapidly and cost-effectively creates vast potential for diagnostics and therapies (NHGRI website).

Although routine DNA sequencing in the doctor's office is still many years away, some large medical centers have begun to use sequencing to detect and treat some diseases. In cancer, for example, physicians are increasingly able to use sequence data to identify the particular type of cancer a patient has. This enables the physician to make better choices for treatments.

Researchers in the NHGRI-supported Undiagnosed Diseases Program use DNA sequencing to try to identify the genetic causes of rare diseases. Other researchers are studying its use in screening new-borns for disease and disease risk.

Moreover, The Cancer Genome Atlas project, which is supported by NHGRI and the National Cancer Institute, is using DNA sequencing to unravel the genomic details of some 30 cancer types. Another National Institutes of Health program examines how gene activity is controlled in different tissues and the role of gene regulation in disease. Ongoing and planned large-scale projects use DNA sequencing to examine the development of common and complex diseases, such as heart disease and diabetes, and in inherited diseases that cause physical malformations, developmental delay and metabolic diseases.

Comparing the genome sequences of different types of animals and organisms, such as chimpanzees and yeast, can also provide insights into the biology of development and evolution.

3.2.3 Discovery of RNA and its sequencing— a combined effort by a group of researchers

RNA—shaped like a single helix—helps carry out the blueprint's guidelines residing in the DNA. Three main types of RNA are involved in protein synthesis. They are messenger RNA (mRNA), transfer RNA (tRNA), and ribosomal RNA (rRNA). rRNA forms ribosomes, which are essential in protein synthesis.

Ribonucleic acid (RNA) is a molecule that is present in the majority of living organisms and viruses. It is made up of nucleotides, which are ribose sugars attached to nitrogenous bases and phosphate groups. The nitrogenous bases include adenine, guanine, uracil, and cytosine. RNA mostly exists in the single-stranded form, but there are special RNA viruses that are double-stranded. The RNA molecule can have a variety of lengths and structures. An RNA virus uses RNA instead of DNA as its genetic material and can cause many human diseases. Transcription is the process of RNA formation from DNA, and translation is the process of protein synthesis from RNA. The means of RNA synthesis and the way that it functions differs between eukaryotes and prokaryotes. Specific RNA molecules also regulate gene expression and have the potential to serve as therapeutic agents in human diseases (Wang and Farhana, 2021).

It would be worth passing through the difficult path to the discovery of this complex biological molecule. According to Cobb (2015), the complexity of what actually took place in the discovery of different types of RNA is much more in keeping with what we know about science—a series of different groups attack a problem, using slightly different techniques, seeing the problem from different angles, before eventually a breakthrough makes clear what was previously problematic. From this point of view, priority of publication is not the sole criterion for contributing to discovery. Jean Brachet in Belgium and Torbjörn Caspersson in Sweden, who in the 1940s had reported that RNA was found primarily in the cytoplasm, where protein synthesis took place, and that RNA levels increased in cells that were actively synthesizing proteins (Brachet, 1942; Caspersson, 1947). Crick also became convinced that protein synthesis did not directly involve chromosomal DNA, but instead took place in the cytoplasm and required RNA, although it was not at all clear how that process occurred, or what the form or the function of RNA was. The first hypothesis about how RNA fitted into gene function came from the Paris laboratory of André Boivin. In 1947, Boivin published a French-language article with Roger Vendrely in *Experientia* outlining his view; the idea was pithily expressed by the editor's English-language summary: "the macromolecular desoxyribonucleic acids govern the building of macro-molecular ribonucleic acids, and, in turn, these control the production of cytoplasmic enzymes" (Boivin and Vendrely, 1947).

Cobb (2015) reports that up until the middle of the 1950s, thinking about what was taking place in the cytoplasm during protein synthesis was blurred by lack of knowledge. Although RNA-rich structures called microsomal particles were identified in the cytoplasm in the 1950s, it was only in 1958 that they were baptized "ribosomes," during informal discussions at a conference (Roberts, 1958). Ribosomal RNA was the only form of RNA that had been clearly identified, and it was quite possible that this was the RNA intermediary between DNA and proteins that so many scientists assumed existed. Above all, there was no good evidence that any form of RNA existed without being bound up with a protein (Watson, 1963).

Watson (1963) considered that the ordered interaction of three classes of RNA controls the assembly of amino acids into proteins. Watson (1963) summarized his thoughts about RNA thus, "We can now have considerable confidence that the broad features of protein synthesis are understood. The involvement of RNA is very much more complicated than was imagined in 1953. There is not one functional RNA. Instead, protein synthesis demands the ordered interaction of three classes of RNA—ribosomal, soluble, and messenger. Many important questions, however, remain unanswered. For instance, there is no theoretical framework for the ribosomal subunits, nor for that matter, do we understand the functional significance of ribosomal RNA. Most satisfying is the realization that all the steps in protein replication will be shown to involve well-understood chemical forces. As yet we do not know all the details. For example, are the DNA base pairs involved in messenger RNA selection of the corresponding amino-acyl-sRNA? With luck, this will soon be known. We can thus have every expectation that future progress in understanding selective protein synthesis (and its consequences for embryology) will have a similarly well-defined and, when understood, easy-to-comprehend chemical basis."

Cobb (2015) reports Crick's idea on RNA thus, "In 1957, Francis Crick gave a talk at University College, London, as part of a Society of Experimental Biology symposium entitled "The Biological Replication of Macromolecules" (Crick, 1958). Published the next year, this lecture became famous for its description of what Crick called the central dogma, which outlined a hypothesis for the transfer of information inside the cell, and argued that it was not possible for information to be transferred from proteins to DNA. In an uncirculated 1956 document Crick drew a little diagram summarizing his view (Fig. 3.21; this was not included in the published version)." Crick was still hobbled by the lack of understanding about the nature and function of the ribosome. It seemed likely that the RNA in the ribosome was the template upon which the protein was made. Cobb (2015) notes that "to explain how each amino acid got to the ribosome, Crick hypothesized the existence of what he called "the adaptor": a small, highly unstable set of RNA molecules that would bring each amino acid to the ribosome in order to allow the ribosome to make the protein. Unknown to either side, Hoagland and Zamecnik were simultaneously identifying such an RNA species, which eventually became known as transfer RNA (Hoagland et al., 1958)."

Cobb (2015) further notes that the announcement of the discovery of messenger RNA (abbreviated to mRNA) and

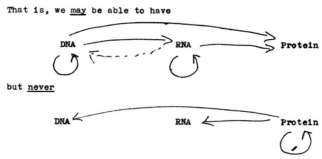

FIG. 3.21 Francis Crick's unpublished 1956 sketch of the central dogma. (Image: Wellcome Library, London.) *(Source: Cobb, M. (2015), Who discovered messenger RNA? Current Biology, 25 (13), R526-R532)* Copyright info: https://doi.org/10.1016/j.cub.2015.05.032Get rights and content (https://www.sciencedirect.com/science/article/pii/S0960982215006065) Under an Elsevier user license.

the cracking of the genetic code took place within weeks of each other in a climax of scientific excitement. Two types of RNA had been discovered in 1956. The first was mRNA, discovered by Elliot Volkin (Fig. 3.22) and Lazarus Astrachan () at the Oak Ridge National Laboratory (Volkin and Astrachan, 1956). It was found that the mRNA is like a tape that copies information from DNA and then carries that information to the ribosome, which reads it off and follows the instructions to make the appropriate protein. At the time of the discovery of mRNA, the just-mentioned "tape recorder metaphor" was a cutting-edge analogy, using the latest technological developments to explain a new biological phenomenon. Fig. 3.23 illustrates the relationship between DNA and mRNA in protein synthesis. Fig. 3.24 illustrates a modern method of DNA sequencing. Although

FIG. 3.22 Ken Volkin, one of the first to observe mRNA. (Image courtesy of Oak Ridge National Laboratory, U.S. Dept. of Energy.) *(Source: Cobb, M. (2015), Who discovered messenger RNA? Curr. Biol., 25 (13), R526-R532)* Copyright info: https://doi.org/10.1016/j.cub.2015.05.032Get rights and content (https://www.sciencedirect.com/science/article/pii/S0960982215006065) Under an Elsevier user license.

mRNA is of decisive importance to our understanding of gene function, no Nobel Prize was awarded for its discovery. The large number of people involved, the complex nature of the results, and the tortuous path that was taken over half a century ago, all show that simple claims of priority may not reflect how science works. Found inside a cell's nucleus, mRNA is responsible for carrying the genetic code to the ribosome to build a protein. Studies leading to the isolation of mRNA was announced in 1961 by Brenner et al.

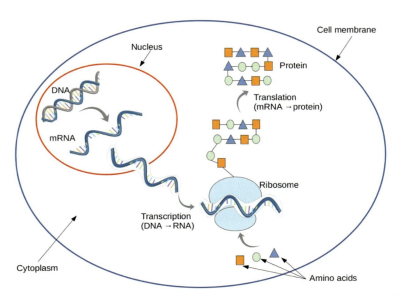

FIG. 3.23 Diagram showing the relationship between DNA and mRNA in protein synthesis. Adapted from illustrations from R. Hesketh, The War on Cancer, New York (2012) p.65, and K. Spencer Joyce, Woods Hole Oceanographic Institution. *(Source: https://www.whatisbiotechnology.org/index.php/exhibitions/sanger/path)*

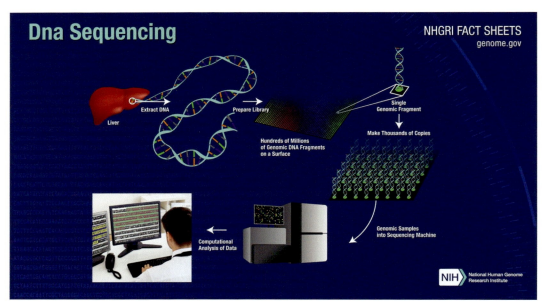

FIG. 3.24 A modern method of DNA sequencing; DNA sequencing fact sheet. *(Source: National Human Genome Research Institute. https://www.genome.gov/about-genomics/fact-sheets/DNA-Sequencing-Fact-Sheet#:~:text=Sequencing%20DNA%20means%20determining%20the,in%20a%20particular%20DNA%20segment.)*

(1961) and Gros et al. (1961). In a review study, Jacob and Monod (1961) put mRNA into a theoretical context, arguing for its role in gene regulation. Aside from the technical prowess involved, these papers were feats of the imagination, for they represented an entirely new way of thinking about gene function.

Although insight and hard thinking played a decisive role in developing this new view of life, this work built upon over a decade of research by many groups in the United States and Europe as they attempted to unravel how the genetic message gets from DNA to produce proteins. We can reconstruct what happened in these years not only by studying the papers that were produced, but also by examining the reminiscences of those who were involved, both in their memoirs (Gros, 1979; Crick, 1988; Jacob, 1988; Judson, 1996; Brenner, 2001), and in oral histories (Volkin, 1995), including talks by participants at the conference on the history of mRNA that took place in August 2014 as part of the Cold Spring Harbor Laboratory Genentech Center Conferences on the History of Molecular Biology and Biotechnology (Cobb, 2015).

Having tersely traversed the history of the difficult path to the discovery of mRNA, a simple view of the history of mRNA would claim that François Jacob and Jacques Monod named it, while Sydney Brenner, François Jacob and Matt Meselson subsequently isolated it. Cobb (2015) summarized the story of discovery of mRNA thus:

The answer to the question "who discovered mRNA?" depends on what you mean by "discovered." Many different groups have a claim, depending on which part of the mRNA story is being focused upon:

- The first person to argue that DNA produces RNA which in turn leads to protein synthesis was André Boivin, in 1947.
- The first suggestion that small RNA molecules move from the nucleus to the cytoplasm and associate with ribosomes where they drive protein synthesis was made by Raymond Jeener in 1950.
- The first reports of what we would now identify as mRNA were made by Al Hershey's group in 1953 and by Elliot "Ken" Volkin and Astrachan in 1956.
- The realization that mRNA might exist, with the functions we now ascribe to it, first came about through the insight of Sydney Brenner and Francis Crick, while François Jacob and Jacques Monod named mRNA and put it in a theoretical framework.
- The first unambiguous description of mRNA was jointly the work of Sydney Brenner, Francis Crick and Matt Meselson on the one hand, and of Jim Watson's team on the other (although the Brenner–Crick–Meselon group got their results first).
- Finally, the first people to prove the function of mRNA were Marshall Nirenberg and Heinrich Matthaei, although they did not frame their results in these terms.

RNA makes proteins using amino acids. There are 20 different types of amino acids that make up a protein's primary structure. Once a ribosome binds to an mRNA transcript, it starts decoding the mRNA codons and recruits tRNAs with the encoded amino acid. Codons are deciphered using the genetic code. In the genetic code, each codon represents a specific amino acid—for example, CUU codes for leucine and GGU codes for glycine.

Besides the most important function of RNA to make proteins, other critical cellular functions include modifying and restructuring of other RNAs and regulation of gene expressions during growth and development, and changing cellular environments.

Some RNAs also function as catalytic RNA to drive biochemical reactions; hence they are termed as ribozymes. The ribozymes also sometimes pair with auxiliary proteins to carry out their catalytic functions. The biochemical reactions catalyzed by ribozymes include protein synthesis, RNA splicing, and RNA cleavage.

Wang and Farhana (2021) explain that the primary function of RNA is to create proteins via translation. RNA carries genetic information that is translated by ribosomes into various proteins necessary for cellular processes. mRNA, rRNA, and tRNA are the three main types of RNA involved in protein synthesis. RNA also serves as the primary genetic material for viruses. Other functions include RNA editing, gene regulation, and RNA interference. These processes are carried out by a group of small regulatory RNAs, which include small nuclear RNA, microRNA, and small interfering RNA.

The three primary types of RNAs are briefly addressed below.

3.2.3.1 mRNA

To better understand the nucleic acid properties and discover relevant parameters for the chemical basis of molecular information encoding, nucleic acid structure has been dissected by systematic variation of nucleobase, sugar and backbone moieties (Eschenmoser, 1996; Nielsen, 1995; Schoning et al., 2000; Leconte et al., 2008; Loakes and Holliger, 2009; Herdewijn, 2010; Campbell and Wengel, 2011). These studies have revealed the profound influence of backbone, sugar and base chemistry on nucleic acid properties and function. Crucially, only a small subset of chemistries allows information transfer through base pairing with DNA or RNA, a prerequisite for crosstalk with extant biology. According to the present knowledge, mRNA is transcribed from DNA and contains the genetic blueprint to make proteins. Prokaryotic mRNA does not need to be processed and can proceed to synthesize proteins immediately. In eukaryotes, a freshly transcribed RNA transcript is considered a pre-mRNA and needs to undergo maturation to form mRNA. A pre-mRNA contains noncoding and coding regions known as introns and exons, respectively. During pre-mRNA processing, the introns are spliced, and the exons are joined together. A 5' cap known as 7-methylguanosine is added to the 5' end of the RNA transcript and the 3' end is polyadenylated. Polyadenylation refers to the process where a poly(A) tail, which is a sequence of adenine nucleotides, is added to the transcript. The 5' cap protects the mRNA from degradation, and the 3' poly(A) tail contributes to the stability of mRNA and aids it in transport.

Researchers are also studying mRNA as an anticancer treatment due to its ability to modify cells (Van Lint et al., 2013).

3.2.3.2 tRNA

The second type of RNA is transfer RNA (abbreviated to tRNA), originally known as soluble RNA or S-RNA, found by Paul Zamecnik, Mahlon Hoagland and other colleagues at Massachusetts General Hospital attached to Harvard University. Located in the cell's cytoplasm, tRNA helps transfer specific amino acids from the cytoplasm to the ribosome where they are joined in a specific order to make a protein.

Wang and Farhana (2021) explain that tRNAs are RNA molecules that translate mRNA into proteins. They have a cloverleaf structure that consists of a 3' acceptor site, 5' terminal phosphate, D arm, T arm, and anticodon arm. The primary function of a tRNA is to carry amino acids on its 3' acceptor site to a ribosome complex with the help of aminoacyl-tRNA synthetase. Aminoacyl-tRNA synthetases are enzymes that load the appropriate amino acid onto a free tRNA to synthesize proteins. Once an amino acid is bound to tRNA, the tRNA is considered an aminoacyl-tRNA. The type of amino acid on a tRNA is dependent on the mRNA codon, which is a sequence of three nucleotides that codes for an amino acid. The anticodon arm of the tRNA is the site of the anticodon, which is complementary to an mRNA codon and dictates which amino acid to carry. tRNAs also regulate apoptosis through acting as a cytochrome c scavenger (Raina and Ibba, 2014).

3.2.3.3 rRNA

rRNA forms ribosomes, which are essential in protein synthesis. A ribosome contains a large and small ribosomal subunit. In prokaryotes, a small 30S and large 50S ribosomal subunit make up a 70S ribosome. In eukaryotes, the 40S and 60S subunit form an 80S ribosome. The ribosomes contain an exit (E), peptidyl (P), and acceptor (A) site to bind aminoacyl-tRNAs and link amino acids together to create polypeptides.

3.2.3.4 Sequencing RNA

Sequencing RNA involves isolating the RNA, preparation of an RNA library, and utilization of next-generation sequencing technology. Skimming through the history of sequencing, sequencing nucleic acids initially seemed a much more formidable challenge to Sanger than proteins. One of the major obstacles arose because there was no suitable pure small nucleic acid available to experiment on. Another issue was the composition of nucleic acids. As nucleic acids were made up of just four subunits, nucleotides, he was concerned that he might struggle to break them down into sufficiently large fragments with enough of an overlap with other fragments. Such overlaps had been crucial in his determining the sequence in insulin. The fact that nucleic acids had just four components, however, meant that he might find the final analysis easier than when he had analyzed the 20 amino acids in insulin (Sanger, 1988).

It was Sanger who succeeded in first sequencing RNA while he was wrestling with nucleic acids trying to refine his sequencing techniques, which included experimenting with ways to label proteins and enzymes with a radioactive phosphorous isotope known as ^{32}P. Progress, however, was slow. He soon realized that one solution might be to focus on tRNA. Zamecnik and Hoagland's findings suggested it contained just 60 nucleotides, which seemed a manageable number to sequence (Finch, 2008). Sanger's "Insulin" laboratory notebook, 13/13, dated 1958 highlights the variable results Sanger was getting in his early experiments on S-RNA (later came to be known as transfer RNA (tRNA).

Sanger, who was putting all his energy into sequencing RNA, was aided in this by John Smith, another researcher who was also hard at work on the nucleic acid in the laboratory and was happy to teach Sanger some of his skills in fractionating nucleotides. In 1963 Sanger's work on RNA was given an extra boost when a doctoral student George Brownlee arrived, and opted to work on nucleic acid under his supervision (Sanger, 1988; Finch, 2008).

Much of Sanger's team's early effort was directed toward purifying tRNA from different species, including yeast and *Escherichia coli*. Most of this work was done by Smith, but progress was slow so Sanger soon focused on ways to develop a rapid and simple fraction technique for sequencing the nucleotides in RNA. The question was which method to explore. Many scientists then working on sequencing nucleic acids deployed methods similar to those used to sequence proteins.

Once the nucleic acid was partially digested, it was then commonly fractionated on ion exchange columns. The use of columns had largely replaced the two-dimensional paper fraction technique Sanger had used for insulin. Sanger, however, saw the use of columns as somewhat laborious. He much preferred a paper-based method. As he put it, "I still had a preference for paper techniques, especially for preliminary experiments, as they were quicker and, in general, gave more information—though of a qualitative rather than [a] quantitative nature." Some scientists were already experimenting with a paper-based method for fractionating single nucleotides with the help of ultraviolet light. Sanger, however, did not find this method very sensitive and remarked that he "found it impossible to see any distinct spots from partial digests of RNA" (Sanger and Dowding, 1996).

One technique, which Sanger thought the process might be improved, was attaching a radioactive label to RNA. This label would act as a probe to detect the nucleic acid or any fragments derived from it. He believed this would provide a more rapid approach than the paper-based methods which relied on the detection of nucleic acids based on their absorption of ultraviolet light. Sanger's reasoning was based on some experiments he had already conducted on enzymes and proteins with the radioactive phosphorous isotope ^{32}P. He believed that the same label had the potential to incorporate well into RNA, because every one of the nucleotides in RNA contained phosphorous atoms. Furthermore, ^{32}P could be easily detected in autoradiographs (Sanger et al, 1965; Sanger and Dowding, 1996; Sanger, 1988).

In 1965 Sanger was joined by another collaborator, Kjeld Marcker, a Danish postdoctoral researcher, and together they set about testing his radioactive sequencing approach. Their first efforts were directed toward tagging RNA with ^{32}P attached to an amino acid, methionine. Results from this research proved confusing, however, because methionine kept appearing as an extra spot in the paper electrophoresis read-outs. Further investigation revealed the amino acid had been potentially modified during the experiment. Following this, Sanger worked out a way to synthesize radioactive RNA by adding ^{32}P inorganic phosphate to some *Escherichia coli* and yeast that were growing in culture (Sanger, 1992; Finch, 2008). Sanger's RNA notebook (1/3) recorded the steps to produce radioactive RNA.

In tandem with the radioactive labeling work, Sanger began exploring different separation techniques to facilitate sequencing. By early 1965, Sanger and Brownlee, together with Bart Barrell, had successfully devised a two-dimensional partition method which used ionophoresis on cellulose acetate followed by ionophoresis on ion exchange paper (Sanger and Dowding, 1996).

Looking back on his working life, Sanger commented that he could not remember many moments of particular elation, but one that stuck out in his memory came from the time he had worked on the two-dimensional partition technique. He recalled "one occasion when Bart Barrell, who usually developed the day's autoradiographs first thing in the morning, came into my lab brandishing a beautiful sheet of film with clear, round, well-separated spots on it. This was certainly exciting after the streaky, unresolved pictures we had been getting before." The great advantage of the method was that it was quick. Furthermore, "it avoided a good deal of final analysis" (Sanger, 1988). An outline of the technique was published in the Journal of Molecular Biology in September 1965 (Sanger et al, 1965).

While the new system appeared to be robust for sequencing, Sanger's team was unable to test it on any tRNA, his original target, because they had not yet managed to purify the nucleic acid. The best they could manage was a test run with a sample of ribosomal RNA which was easy to prepare in radioactive form. While too large a molecule to provide any useful sequencing information, the ribosomal RNA confirmed the utility of the system (Sanger, 1988).

In the end the first RNA to be sequenced was alanine tRNA purified from yeast, achieved by Robert Holley and colleagues at Cornell University. This was the first determination of the nucleotide sequence of a nucleic acid and the culmination of 7 years' hard work. Three of those years had been spent in purifying the nucleic acid by using a countercurrent distribution system. The following 4 years had been devoted to sequencing, using a similar procedure to that adopted by Sanger for insulin, whereby the nucleic acid was first cut up into 16 small fragments with enzymes and then assembling them together like a puzzle. The RNA was found to contain 77 nucleotides (Holley et al, 1965; Kresge et al., 2005).

Two years after Holley's success, Sanger's team announced the successful sequencing of another short RNA, 5S ribosomal RNA from *Escherichia coli*. Much of the work had been carried out by Brownlee as part of his doctorate. The RNA contained 120 nucleotides. It was substantially larger than any tRNA sequenced so far, which ranged from 77 to 85 nucleotides (Brownlee et al, 1967). The sequencing of 5S ribosomal RNA was greatly aided by the ^{32}P label. Importantly, it also showed up well-defined spots on the autoradiographs, making it possible to identify individual nucleotides on the basis of their position and to work out their sequence order directly. Thereafter, ^{32}P became a standard tool for sequencing RNA, until it was displaced by fluorescent labeling (Sanger, 1992; Sanger and Dowding, 1996).

In addition to proving Sanger's radioactive sequencing approach effective, the results from 5S ribosomal RNA demonstrated the power of another fractionation system called homochromatography. Sanger devised this method to obtain longer fragments when separating products partially digested by T1 ribonuclease from one another. The method was a type of displacement chromatography. It rested on the displacement of oligonucleotides (small groups of nucleotides) fixed on some ion-exchange paper by some unlabeled oligonucleotides (Brownlee, 2015).

3.2.4 Genome sequencing

Genome is the complete set of genes or genetic material present in a cell or organism. Genome sequencing is figuring out the order of DNA nucleotides, or bases, in a genome—the order of As, Cs, Gs, and Ts that make up an organism's DNA. The human genome is made up of over 3 billion of these genetic letters.

Genome sequencing allows us to compare and contrast the DNA of different organisms and work out how they evolved in their own unique ways. At present, DNA sequencing on a large scale—the scale necessary for ambitious projects such as sequencing an entire genome—is mostly done by high-tech machines. Much as one's eye scans a sequence of letters to read a sentence, these machines "read" a sequence of DNA bases. A DNA sequence that has been translated from life's chemical alphabet into our alphabet of written letters might look like this:

AGTCCGCGAATACAGGCTCGGT

That is, in this particular piece of DNA, an adenine (A) is followed by a guanine (G), which is followed by a thymine (T), which in turn is followed by a cytosine (C), another cytosine (C), and so on.

3.2.5 Dark DNA

As just mentioned, genome sequencing allows us to compare and contrast the DNA of different organisms and work out how they evolved in their own unique ways. But in some cases, we are faced with a mystery. Hargreaves (2017) explains that some animal genomes seem to be missing certain genes, ones that appear in other similar species and must be present to keep the animals alive. These apparently missing genes have been dubbed "dark DNA." Microbiologists have managed to identify the chemical products that the instructions from the "missing" genes would create. This would only be possible if the genes were present somewhere in the genome, indicating that they weren't really missing but just hidden. And its existence could change the way we think about evolution. It is now known that GC-rich sequences (note that G and C molecules are two of the four "base" molecules that make up DNA) cause problems for certain DNA-sequencing technologies. This makes it more likely that the genes scientists were looking for were simply "hard to detect" rather than "missing." For this reason, scientists call the hidden sequence "dark DNA" as a reference to "dark matter," the stuff that we think makes up about 25% of the Universe but that we can't actually detect. The advent of rapid DNA sequencing methods has greatly accelerated biological and medical research and discovery (Behjati and Tarpey, 2013; Hargreaves, 2017).

Most textbook definitions of evolution state that it occurs in two stages: mutation followed by natural selection. Hargreaves (2017) recalls that DNA mutation is a common and continuous process, and occurs completely at random. Natural selection then acts to determine whether mutations are kept and passed on or not, usually depending on whether they result in higher reproductive success. In short, mutation creates the variation in an organism's DNA, natural selection decides whether it stays or if it goes, and so biases the direction of evolution.

Hargreaves (2017) explains that hotspots of high mutation within a genome mean genes in certain locations have a higher chance of mutating than others. This means that such hotspots could be an underappreciated mechanism that could also bias the direction of evolution, meaning natural selection may not be the sole driving force. The most exciting puzzle to solve is considered to be working out what effect dark DNA has had on biological evolution.

In certain organisms, the mutation hotspot may have made their adaptation to a certain environment possible. But on the other hand, the mutation may have occurred so quickly that natural selection hasn't been able to act fast enough to remove anything detrimental in the DNA. If true, this would mean that the detrimental mutations could prevent an organism from surviving outside its current environment. The discovery of such a weird phenomenon certainly raises questions about how genomes evolve, and what could have been missed from existing genome sequencing projects. Perhaps we need to go back and take a closer look (Hargreaves, 2017).

Because DNA is an informative macromolecule in terms of transmission from one generation to another, DNA sequencing is used in evolutionary biology to study how different organisms are related and how they evolved. In February 2021, scientists reported, for the first time, the sequencing of DNA from animal remains, a mammoth in this instance, which lived more than a million years ago. This is the oldest DNA sequenced to date (Callaway, 2021).

3.2.6 Categorization of all living organisms into two major divisions: the cellular and the viral "empires" and three primary cellular domains—archaea, bacteria, and eukarya

How do scientists study and classify life-forms? How can we understand the complex evolutionary connections between living organisms? According to Koonin (2010), "Comparative genomics, which involves analysis of the nucleotide sequences of genomes, shows that the known life-forms comprise two major divisions: the cellular and the viral 'empires.' The cellular empire consists of three domains: Archaea, Bacteria, and Eukarya. What are the evolutionary relationships between the two empires and the three domains? Comparative genomics sheds light on this key question by showing that the previous conception of the Tree of Life should be replaced by a complex network of treelike and netlike routes of evolution to depict the history of life. Even under this new perspective on evolution, the two empires and the three cellular domains remain distinct. Furthermore, comparative genomics suggests that eukaryotes are archaebacterial chimeras, which evolved as a result of, or at least under a strong influence of, an endosymbiotic event that gave rise to mitochondria."

3.2.6.1 Cells, viruses, and the classification of organisms

Koonin (2010) explains that All living organisms consist of elementary units called cells. Cells are membrane-enclosed compartments that contain genomic DNA (chromosomes), molecular machinery for genome replication and expression, a translation system that makes proteins, metabolic and transport systems that supply monomers for these processes, and various regulatory systems. Scientists have performed careful microscopic observations and other experiments to show that all cells reproduce by different forms of division. Cell division is an elaborate process that ensures faithful segregation of copies of the replicated genome into daughter cells. The best-characterized cells are the relatively large cells of animals, plants, fungi, and diverse unicellular organisms known as protists, such as amoebae or paramecia. These cells possess an internal cytoskeleton and a complex system of intracellular membrane partitions, including the nucleus, a compartment that encloses the chromosomes. These organisms are known as eukaryotes because they possess a true nucleus (karyon in Greek). In contrast, the much smaller cells of bacteria have no nucleus and are named prokaryotes.

Invention of new imaging methods in the 20th century, such as electron microscopy, enabled scientists to view tiny particles that are much smaller than cells—a second fundamental form of biological organism known as "virus." Viruses are obligate intracellular parasites. These selfish genetic elements typically encode some proteins essential for viral replication, but they never contain the full complement of genes for the proteins and RNAs required for translation, membrane function, or metabolism. Therefore, viruses exploit cells to produce their components (Koonin, 2010).

As already noted, DNA sequence is the basis of grouping organisms. A revolution occurred in 1977 when Carl Woese and his coworkers performed pioneering studies to compare the nucleotide sequences of a molecule that is conserved in all cellular life-forms: the small subunit of ribosomal RNA (known as 16S rRNA). By comparing the nucleotide sequences of the 16S rRNA, they were able to derive a global phylogeny of cellular organisms for the first time. This phylogeny overturned the eukaryote-prokaryote dichotomy by showing that the 16S rRNA tree neatly divided into three major branches, which became known as the three domains of (cellular) life: archaea, bacteria, and eukarya (Woese et al., 1990). Eukarya are also called eukaryotes. Both archaea and bacteria are prokaryotes, single-celled microorganisms with no nuclei. Eukarya includes humans and all other animals, plants, fungi, and single-celled protists (single-celled organisms of the kingdom Protista, such as a protozoan or simple alga)—all organisms whose cells have nuclei to enclose their DNA apart from the rest of the cell. The fossil record indicates that the first living organisms were prokaryotes, and eukaryotes arose a billion years later.

3.2.6.2 The cellular domains: archaea, bacteria, and eukarya

Because life on Earth depends on a variety of biochemical respiratory chains involving electron transport, it has been argued (see Kirschvink and Weiss, 2002) that the presence of electrochemical gradients is one of the most critical prerequisites for life's initiation and would form intense selection pressures during its initial evolution. Genetic analysis of the respiratory chains in the Archaea, Bacteria, and Eukarya (Castresana and Moreira, 1999) indicates that terminal oxidases linked to oxygen, nitrate, sulfate, and sulfur were all present in the last common ancestor (LCA) of living organisms. Note that oxidase is an enzyme which promotes the transfer of a hydrogen atom from a particular substrate to an oxygen molecule, forming water or hydrogen peroxide.

According to Koonin (2010), Woese's breakthrough was momentous for at least three reasons. First, he had traced the evolution of cellular life directly by comparing molecules that actually undergo evolutionary changes. Second, the detection of the 16S rRNA sequence conservation in all forms of cellular life provided the strongest possible support for Darwin's hypothesis of the common ancestry of life on Earth. These results provided strong evidence that the last universal common ancestor (LUCA) of all cellular life really existed, although we still know little about what this ancestor was like and how it lived. Finally, the three-domain structure of Woese's tree shows that evolutionary history is decoupled from biological organization.

It is believed that our LCA must have evolved on a world in which the redox (oxidation and reduction considered together as complementary processes) gradients were present in large enough quantities to evolve and to be maintained, and in which metastable, energetic compounds could diffuse across redox boundaries. Since the efficiency of biological systems tends to improve over time as a result of the processes of random mutation and natural selection (Darwin, 1859), primitive metabolic electron transport systems likely could not extract energy as efficiently from the redox couples as do those of modern organisms. Based on these observations, Kirschvink and Weiss (2002) argue that life is more likely to have evolved on the planet that had the broadest dynamical range of electrochemical species early in its history.

Results of gene sequencing in the 1980s supported the three-domain classification. Moreover, evolutionary biologists developed approaches to deduce the root position of the tree. Strikingly, they placed the LUCA between bacteria on one side and archaea together with eukaryotes on the other side, implying that archaea and eukaryotes share a common ancestor to the exclusion of bacteria (Gogarten et al., 1989; Brown and Doolittle, 1997). This finding emphasizes that similarity of cellular organization and common ancestry are two very different things (Koonin, 2010).

Koonin (2010) explains that "The discovery of Archaea as a distinct, new domain of cellular life stimulated extensive studies into the molecular biology of these microbes, many of which thrive in unusual, extremely hot or salty environments. From these studies, researchers learned that the three domains are indeed fundamentally different at several cell biological levels, and not just in universal genes like the 16S rRNA. How do the domains of life differ? Scientists identified two key distinctions related to the DNA replication system and the membrane. The replication system of archaea is largely unrelated to that of bacteria, but it is homologous to the replication machinery of eukaryotes. Conversely, the archaeal membrane and the proteins involved in its formation are unique, whereas bacteria and eukaryotes share homologous membranes. Thus, archaea and bacteria differ with respect to the origin of some of their central cellular systems, whereas eukaryotes seem to combine important features of both archaea and bacteria."

3.2.6.3 Viruses

A virus is a submicroscopic infectious agent that replicates only inside the living cells of an organism. Viruses infect all life forms, from animals and plants to microorganisms, including bacteria and archaea (Koonin et al., 2006). The origins of viruses in the evolutionary history of life are unclear. Viruses were discovered at the end of the nineteenth century as ultramicroscopic parasites of plants and animals, which passed through filters that held back bacteria. By the middle of the twentieth century, it became clear that viruses can replicate only within cells. However, the actual prominence of viruses in the biosphere and their role in the evolution of life were not revealed until the advances of metagenomics allowed for the massive sequencing of genes and genomes in environmental samples without the isolation of individual organisms. Viruses turn out to be the dominant biological entities on Earth. In the ocean, for example, viral particles outnumber cells by an order of magnitude. Viruses are also dominant in terms of genetic variety. A unique feature of viruses is that, in contrast with cellular life-forms—which all employ the same, classic strategy of DNA replication, transcription, and translation—viruses possess diverse genetic cycles. Viruses employ nearly all imaginable strategies of genome replication and expression: Some viruses have single-stranded or double-stranded RNA genomes that do not involve DNA in their replication, some have RNA genomes that use DNA as a replication intermediate, and some have genomes that are either single-stranded or double-stranded DNA molecules. New discoveries have revealed the previously unrealized prominence of the viral world. This second biological empire seems to be even more vast and diverse than the empire of cellular life-forms (Koonin, 2010).

Koonin (2010) explains that in comparison to cellular life-forms, viruses possess small genomes, ranging in size from between about 1000 and 1,000,000 nucleotides. Viruses typically lack many of the genes that are universal among the three domains of cellular life—in particular, genes for translation system components. However, a small core of viral "hallmark genes" have been discovered that are missing in cellular life-forms. These genes encode proteins essential for virus reproduction (e.g., polymerases, helicases, and core virus particle components). These hallmark genes are shared by an extremely diverse group of viruses with different replication strategies, although none of the genes is strictly universal among viruses. The discovery of the hallmark genes reveals the evolutionary unity of the viral empire (Koonin et al., 2006). According to Koonin (2010), clearly, viruses constitute a distinct, major biological "empire" that is distinct from the empire of cellular life-forms, and the viral empire seems to eclipse the latter in terms of genetic complexity (Raoult and Forterre, 2008).

3.3 Origins of life on Earth—importance of organic molecules

Between 4.6 billion and 4.0 billion years ago, there was probably no life on Earth. The planet's surface was at first molten and even as it cooled, it was getting pulverized by asteroids and comets. All that existed were simple chemicals. But about 3.8 billion years ago, the bombardment stopped, and life arose. Most scientists think the "last universal common ancestor"——the creature from which everything on the planet descends—appeared about 3.6 billion years ago.

The origin of life on Earth is a long-standing and controversial subject concerned with how the first known single-cell organisms called prokaryotes probably originated in the Archean period (4–2.5 billion years ago (BYA)) and about 3.8 BYA in the Earth's oceans when chemical composition of the ocean and the atmosphere was very different from what it is today (see Dostal et al., 2009). Microbial life has prevailed in the biosphere (the regions of the surface, atmosphere, and hydrosphere of the earth (or analogous parts of other celestial bodies) occupied by living organisms) since the Archean period at least; and, according to the most recent study, probably for the last 4.1 billion years (see Bell et al., 2015).

Throughout most of Earth's history, it has experienced a range of pressures, both dynamic pressure when the young Earth (proto-Earth) was heavily bombarded by particles of celestial origin and static pressure in subsurface environments that could have served as a refuge and where microbial life nowadays flourishes. Picard and Daniel (2013) have discussed the extent of high-pressure habitats in early and modern times and provided a short overview of microbial survival under dynamic pressures. They have summarized the current knowledge concerning the impact of microbial activity on biogeochemical cycles under pressures characteristic of the deep subsurface. Picard and Daniel (2013) have also evaluated the possibility that pressure can be a limiting parameter for life at depth and discussed the open questions and knowledge gaps that exist in the field of high-pressure geo-microbiology.

All known cosmic and geological conditions and laws of chemistry and thermodynamics allow that complex organic matter, considered to be the building blocks of life, could have formed spontaneously on a pristine planet Earth about 4 billion years ago. It is believed that simple gasses and minerals on the early Earth's surface and in its primitive oceans reacted and were eventually organized in supra-molecular aggregates and enveloped cells that evolved into primitive forms of life. Chemical evolution, which preceded all species of organisms that are still in existence, is now accepted as a fact (Follmann and Brownson, 2009). The first indication of primitive life on Earth occurred just over 3.5–3.0 billion (1 billion = 10^9) years ago (Alexander and Bacq, 1960; Fowler, 1975; Kerr, 1995); that is, life did not exist on this planet before this "biotic era." It is thus a consensus belief among scientists that the origin of life and the associated biological evolution was preceded by chemical evolution, which resulted in the production of organic molecules—the building blocks of life. While no-one knows exactly how life began, one thing is certain; these tiny particles—organic molecules—played an important role. A fundamental problem in origins of life research is how the first polymers with the properties of nucleic acids were synthesized and incorporated into living systems on the prebiotic Earth. Da Silva et al. (2015) showed that RNA-like polymers can be synthesized nonenzymatically from $5'$-phosphate mononucleosides in salty environments.

How did life on Earth begin? It has been one of modern biology's greatest mysteries: How did the chemical soup that existed on the early Earth lead to the complex molecules needed to create living, breathing organisms? In our own Solar system, proto-planet evolution was guarded and strengthened by an active prebiotic chemistry that brought about the emergence of life on Earth. Despite all the recent advances of evolutionary genomics, we still have to answer the most fundamental questions: How did cells evolve in the first place, what caused the fundamental differences between the two prokaryotic domains (archaea and bacteria), and what triggered the emergence of the complex organization of the eukaryotic cell? The search for life signatures requires as the first step the knowledge of planet atmospheres—main objective of future exoplanetary space explorations (Claudi, 2017). The development of life on Earth witnessed several stages in Earth's long history. The following sections endeavor to shed some light on these aspects briefly to prepare the ground for understanding the possibility of life beyond Earth.

While the chemical paths by which life emerged are far from consensual, it is highly likely that life arose in water and that the necessary building blocks were organic compounds. It appears that molecular size and complexity are necessary but not sufficient conditions for being relevant to life, as some very large carbon-based molecules are abiotic (Kahana et al., 2019). Life's origin indisputably necessitates an ample supply of organic molecules. While the origin of terrestrial organic compounds remains only partly deciphered, there is clear evidence for carbon-containing compounds wherever one probes in the Solar System. There are ample data on carbon-containing molecules up to atomic mass of 200 u in interstellar space (Kahana et al., 2019). Earth has frequently been hit by icy comets, whose origin is Kuiper belt located in the outer periphery of the Solar System. In icy comets, about 25% of the nucleus mass consists of carbon compounds (Greenberg, 1998).

An important class of organic molecules that constitutes the building block of life is amino acids. The importance of amino acids in the context of origin of life stems from the fact that amino acids are essential to life. Amino acids are biologically important organic compounds composed of amine ($-NH_2$) and carboxylic acid ($-COOH$) functional groups, along with a side chain specific to each amino acid. The key elements of an amino acid are carbon, hydrogen, oxygen, and nitrogen, though other elements are found in the side chains of certain amino acids. About 500 amino acids are known (Wagner and Musso, 1983), and can be classified in many ways.

Amino acids are natural molecules that form a very large network of molecules, known as proteins, by a process of polymerization; that is, by chemical binding to other molecules. Thus, proteins are simply amino acid polymers. In other words, amino acids are the building blocks of proteins; that is, the structural units (monomers) that make up proteins. They join together to form short polymer chains called *peptides* or longer chains called either *poly-peptides* or "proteins." These polymers are linear and un-branched, with each amino acid within the chain attached to two neighboring amino acids. The process of making proteins is called *translation* and involves the step-by-step addition of amino acids to a growing protein chain by a ribozyme (Wagner and Musso, 1983). Note that a ribozyme is a ribonucleic acid (RNA) molecule capable of acting as an enzyme. It may be noted that an enzyme is a substance produced by a living organism which acts as a catalyst to bring about a specific biochemical reaction.

A complex network of more than 200 essential proteins is found in today's most elementary biological cells. The formation of stable covalent chemical bonds between the relatively small molecules sets polymerization apart from other processes, such as crystallization, in which large numbers of molecules aggregate under the influence of weak inter-molecular forces. Sreenivasachary and Lehn (2005, 2008) have shown that polymers with the capacity for constitutional reorganization show a number of intriguing material properties, including self-healing and chemo-responsive characteristics (Sreenivasachary et al. 2006; Cordier et al. 2008).

Now, researchers claim that they have found the missing link. How did the chemistry of simple carbon-based molecules lead to the information storage of ribonucleic acid, or RNA? The RNA molecule must store information to code for proteins. Research carried out by Wolfenden et al. (2015) indicate that the close linkage between the physical properties of amino acids, the genetic code, and protein folding was likely the key factor in the evolution from building blocks to organisms when Earth's first life was emerging from the primordial soup. Wolfenden et al. (2015) found that the hydrophobicities of the 20 common amino acids are reflected in their tendencies to appear in interior positions in globular proteins and in deeply buried positions of membrane proteins. To determine whether these relationships might also have been valid in the warm surroundings where life may have originated, they examined the effect of temperature on the hydrophobicities of the amino acids as measured by the equilibrium constants for transfer of their side-chains from neutral solution to cyclohexane ($K_{w>c}$). Wolfenden et al. (2015) found that the hydrophobicities of most amino acids increase with increasing temperature. Because that effect is more pronounced for the more polar amino acids, the numerical range of $K_{w>c}$ values decreases with increasing temperature. There are also modest changes in the ordering of the more polar amino acids. However, those changes are such that they would have tended to minimize the otherwise disruptive effects of a changing thermal environment on the evolution of protein structure.

Earlier, the genetic code was found to be organized in such a way that—with a single exception (threonine)—the side-chain dichotomy polar/nonpolar matches the nucleic acid base dichotomy purine/pyrimidine at the second position of each coding triplet at 25°C. That dichotomy is preserved at 100°C. The accessible surface areas of amino acid side-chains in folded proteins are moderately correlated with hydrophobicity, but when free energies of vapor-to-cyclohexane transfer (corresponding to size) are taken into consideration, a closer relationship becomes apparent (Wolfenden et al., 2015).

The universal genetic code is the earliest point to which biological inheritance can be traced. Earlier work hinted at a relationship between the codon bases and the physical properties of the 20 amino acids that dictate the 3D conformations of proteins in solution. Carter looked at the way a molecule called "transfer RNA," or tRNA, reacts with different amino acids. In a study, Carter Jr. and Wolfenden (2015) examined the way a molecule called "transfer RNA," or tRNA, reacts with different amino acids, and showed that acceptor stems and anticodons, which are at opposite ends of the tRNA molecule, code, respectively, for size and polarity. These two distinct properties of the amino acid side-chains jointly determine their preferred locations in folded proteins. The early appearance of an acceptor stem code based on size, β-branching, and carboxylate groups might have favored the appearance of antiparallel peptides that have been suggested to have a special affinity for RNA.

Aminoacyl-tRNA synthetases recognize tRNA anticodon and 3′ acceptor stem bases. Synthetase urzymes acylate cognate tRNAs even without anticodon-binding domains, in keeping with the possibility that acceptor stem recognition preceded anticodon recognition. Representing tRNA identity elements with two bits per base, Carter Jr. and Wolfenden (2015) showed that the anticodon encodes the hydrophobicity of each amino acid side-chain as represented by its water-to-cyclohexane distribution coefficient, and this relationship holds true over the entire temperature range of liquid water. The acceptor stem codes preferentially for the surface area or size of each side-chain, as represented by its vapor-to-cyclohexane distribution coefficient. Carter Jr. and Wolfenden (2015) found that these orthogonal experimental properties are both necessary to account satisfactorily for the exposed surface area of amino acids in folded proteins. Moreover, the acceptor stem codes correctly for β-branched and carboxylic acid side-chains, whereas the anticodon codes for a wider range of such properties, but not for size or β-branching. According to Carter Jr. and Wolfenden (2015), these and other results suggest that genetic coding of 3D protein structures evolved in distinct stages, based initially on the size of the amino acid and later on its compatibility with globular folding in water. They found that one end of the tRNA could help sort amino acids according to their shape and size, while the other end could link up with amino acids of a certain polarity. In that way, this tRNA molecule could dictate how amino acids come together to make proteins, as well as determine the final protein shape. That is similar to what the adenosine triphosphate (ATP) enzyme does today, activating the process that strings together amino acids to form proteins. Note that today ATP does the job of linking amino acids into proteins, activated by an enzyme called aminoacyl tRNA synthetase.

The above two studies (Wolfenden et al., 2015; Carter Jr. and Wolfenden, 2015) suggest a way for RNA to control the production of proteins by working with simple amino acids that does not require the more complex enzymes that exist today. This link would bridge this gap in knowledge between the primordial chemical soup and the complex molecules needed to build life. Current theories say life on Earth started in an "RNA world," in which the RNA molecule guided the formation of life, only later taking a backseat to DNA, which could more efficiently achieve the same end result. Like the DNA, the RNA is a helix-shaped molecule that can store or pass on information (DNA is a double-stranded helix, whereas RNA is single-stranded). Many scientists think the first RNA molecules existed in a primordial chemical soup—probably pools of water on the surface of Earth billions of years ago.

The very first RNA molecules formed from collections of three chemicals: a sugar (called a ribose); a phosphate group, which is a phosphorus atom connected to oxygen atoms; and a base, which is a ring-shaped molecule of carbon, nitrogen, oxygen, and hydrogen atoms. RNA also needed nucleotides, made of phosphates and sugars.

The question is: how did the nucleotides come together within the soupy chemicals to make RNA? In a study, Patel et al. (2015) showed that a cyanide-based chemistry could make two of the four nucleotides in RNA and many amino acids. They demonstrated a plausible prebiotic scheme showing that the precursors of pyrimidine nucleotides formed from hydrogen cyanide can also form precursors of lipids and amino acids, providing significant evidence that early life may have emerged from a common chemistry on prebiotic Earth. This work has been heralded by Nobel-prize winning geneticist Jack Szostak as an important advance in understanding the origins of life.

3.4 Life and living systems—interpretations

Life is a difficult and contentious phenomenon to define. According to Claudi (2017), "Life has been described as a (thermodynamically) open system (Prigogine et al., 1972), which exploits gradients in its surroundings to create imperfect copies of itself, makes use of chemistry based on carbon, and exploits liquid water as solvent for the necessary chemical reactions (Owen, 1980; Des Marais et al., 2002)." A widely accepted simple definition of life is "life is that which replicates and evolves" (Nowak and Ohtsuki, 2008). Because living organisms are open systems, they must receive energy and materials from outside themselves, and are not therefore limited by the Second Law of Thermodynamics (which is applicable only to closed systems in which energy is not replenished). Based on the simple definition of life, its origin appears to entail a transition from the abiogenic availability of various organic compounds to chemical entities capable of producing their own copies. In this context, organic compounds are those that contain covalently bound combinations of carbon with one or more of the five other main life elements, hydrogen, nitrogen, oxygen, phosphorus, and sulfur (see Kahana et al., 2019). It appears that what defines life molecules is their capacity to undergo specific mutual interactions, leading to the emergence of complex networks (Toparlak and Mansy, 2018).

Life is usually considered to be a characteristic of organisms that exhibit certain biologic processes (such as chemical reactions or other events that results in a transformation), that are capable of growth (through metabolism) and reproduction. The abilities to ingest food and excrete waste are also sometimes considered to be requirements of life.

The two distinguishing features of living systems are sometimes considered to be complexity and organization. Some organisms can communicate, and many can adapt to their environment through internally generated changes, although these are not universally considered to be prerequisites for life. This is an approximate background about life.

3.5 Why do a few million years or more are necessary for evolution from prebiotic chemical phase to biological phase?

There exists a wide gap between chemical molecules and biological cells, and how that gap is bridged in the history of origin of life is a big question, the answer to which is still unclear. Based on available knowledge Vago et al. (2017) attempted to answer this very complicated and difficult question thus, "It is not the case that, once we had an interesting mix of organics, cellular organization took care of itself. This does not happen in the laboratory and most surely did not on early Earth (Schrum et al., 2010). We must, therefore, accept the need for an extended phase (perhaps a few million years—or more—we will never be sure) of prebiotic chemical evolution during which the various molecular building blocks generated and associated, underpinned by replication, to gradually progress from elements to system (Lazcano and Miller, 1996; Joyce, 2002; Orgel, 2004; Harold, 2014). We can perhaps call this a period of converging prebiotic chemistry." Once cells are evolved, there is a need for them to settle somewhere. To be able to disperse and settle in other environments, whether reached through open water or the subsurface, along fractures and fault zones, likely required a higher degree of sophistication, including proper membranes (Vago et al., 2017).

3.6 Understanding the evolution of life

"Why it took complex life so long to form" is a question that arouses much curiosity. The key building block of life is the cell, with its complex genetic and biochemical systems. Animals and plants are all made of cells, so cells had to evolve first. To make tissues and organs, cells need to multiply, specialize in function, and cooperate. The evolution of these basic building blocks and their integration took time. Larger organisms require even more specialized and integrated cellular systems. The fossil record tells us that this took billions of years.

Rampelotto (2010c) summarized the story of the evolution of life thus, "When life appeared on Earth after a period of heavy bombardment, this planet was hotter and more hostile, in comparison with nowadays (Lal, 2008). The early atmosphere did not contain oxygen until the 'great oxidation event,' 2.3–2.4 billion years ago. Because life started at least 3.5 billion years ago, it was exclusively anaerobic for at least 1.5 billion years. Heat resistance and anaerobic features are observed in many hyperthermophiles. Thus, the last universal common ancestor (LUCA) of all life on Earth has been suggested to be hyperthermophilic (Martin et al., 2008). Indeed, based upon phylogenetic studies, such extremophilic microbes form a cluster on the base of the tree of life (Stetter, 2007)." In the above description, note that hyperthermophiles are particularly extreme thermophiles for which the optimal temperatures are above 80°C. These organisms are found living in hydrothermal vents. For example, the Archaea Strain 121 and Methanopyrus kandleri, which are the most hyperthermophilic organisms reported, were isolated from such environments, as well as the living organism with the smallest genome known, the Archaea *Nanoarchaeum equitans* (Waters et al., 2003). In contrast to the classical Miller–Urey experiments to the origins of life, which depend upon the external sources of energy, such as simulated lightning or ultraviolet irradiation, microbes living at hydrothermal vents do not depend on sunlight or oxygen. They grow around the vents and feed upon hydrogen sulfide gas (H_2S) and dissolved minerals to produce organic material through the process of chemosynthesis.

Scientists now have reliable information that it is cosmological evolution which has guided matter from simplicity to complexity; and from inorganic to organic. Fossils document the increase in brain size of our ancestors. It has been noticed that life is hardly more than a combination of simple chemicals operating in complex ways; and the origin of life is a natural result of the evolution of that matter. Indeed, this new aspect of the research highlights the vision of Sri Aurobindo (August 15, 1872–December 5, 1950)—an Indian philosopher and yogi. He said, "There is an ascending evolution in nature which goes from stone to the plant, from the plant to the animal, and from animal to man. Because man is, for the moment, the last rung at the summit of the ascending evolution; he considers himself as the final stage in this ascension and believes there can be nothing on Earth superior than him. He is mistaken in that perception. In his physical nature, he is yet almost wholly an animal—a thinking and speaking animal—but still an animal in his material habits and instincts. Undoubtedly, nature cannot be satisfied with such an imperfect result; she endeavors to bring out a being who will be to man what man is to animal, a being who will remain a man in its external form, and yet whose consciousness will rise far above the mind and its slavery to ignorance" (see De, 2017). Indeed, we are all active participants in the cosmological relay race holding and passing the batons of evolution!

Evolution has played a role even in changing the Earth's atmosphere. For example, three billion years ago, there was no free oxygen in the atmosphere at all. Life was anaerobic, meaning that it did not need oxygen to live and grow. That all changed due to the evolution of *Cyanobacteria*, a group of single-celled, blue-green bacteria. Likewise, when we imagine what made early modern humans unique, it is tempting to imagine a caveperson, brandishing a lit torch,

FIG. 3.25 A model of the face of an adult female *Homo erectus*, one of the first truly human ancestors of modern humans, on display in the Hall of Human Origins in the Smithsonian Museum of Natural History in Washington, DC (humanorigins.si.edu, smithsonianmag.com, gurche.com). Reconstruction by John Gurche, Smithsonian Museum of Natural History, based on KNM ER 3733 and 992. *(Source: https://en.wikipedia.org/wiki/Homo_erectus#/media/File:Homo_erectus_adult_female_-_head_model_-_Smithsonian_Museum_of_Natural_History_-_2012-05-17.jpg)*

FIG. 3.26 Approximate reconstruction of a Neanderthal skeleton and artistic interpretation of the La Ferrassie 1 Neanderthal man (a male Neanderthal skeleton estimated to be 70–50,000 years old; discovered at the La Ferrassie site in France by Louis Capitan and Denis Peyrony in 1909) from the National Museum of Nature and Science, Tokyo, Japan; Author: Photaro—Own work *(Source: https://en.wikipedia.org/wiki/Neanderthal#/media/File:Skeleton_and_restoration_model_of_Neanderthal_La_Ferrassie_1.jpg)*

using fire to eke out a better living from a hostile environment. However, humans were not alone in their use of fire. Homo erectus (Fig. 3.25) and Neanderthals (Fig. 3.26) also used fire. Compared with modern Homo sapiens, which have only been around for the last 200,000 years, Homo erectus, or "upright man," had a long reign. The ancient ancestor of modern humans lived from 2 million years ago till about 100,000 years ago, possibly even 50,000 years ago. Neanderthals, members of a group of archaic humans, emerged at least 200,000 years ago and were replaced or assimilated by early modern human populations (*Homo sapiens*).

Now, new research suggests that part of what distinguished early humans from their close relatives was not their use of fire, but how the human lineage evolved in response to fire. Rosas et al. (2017), who studied the fossilized skeleton of Neanderthal child, obtained from a collection of the remains of 13 individuals (consisting of seven adults, three teenagers, and three younger children) from the El Sidron cave—located in Pilona, northern Spain—have found that the growth of our ancient relatives were no different than that of a modern-day human child. According to Antonio Rosas, from Spanish National Research Council (CSIC), "Discerning the differences and similarities in growth patterns between Neanderthals and modern humans helps us better define our own history. Modern humans and Neanderthals emerged from a common recent ancestor, and this is manifested in a similar overall growth rate." Applying pediatric growth assessment methods, the researchers found that this Neanderthal child is no different to a modern-day child. It was found that Neanderthals had a greater cranial (i.e., of the skull) capacity than today's humans. Likewise, Neanderthal adults had an intracranial volume of 1520 cm^3, while that of modern adult man is 1195 cm^3. The growth and development of this juvenile Neanderthal matches the typical characteristics of human ontogeny (the study of the entirety of an organism's lifespan), where there is a slow anatomical growth between weaning (accustoming to food other than its mother's milk) and puberty (the period during which adolescents reach sexual maturity and become capable of reproduction). In fact, the skeleton and dentition of this Neanderthal present a physiology which is similar to that of a sapiens of the same age, except for the thorax area (the part of the body of a mammal between the neck and the abdomen, including the cavity enclosed by the ribs, breastbone, and dorsal vertebrae, and containing the chief organs of circulation and respiration; the chest), which corresponds to a child between 5 and 6 years, in that it is less developed. The only divergent aspect in the growth of both species is the moment of maturation of the vertebral column. The researchers obtained some clues suggesting that the Neanderthal had slower brain growth.

In another example, it is interesting to note that *Homo sapiens* (i.e., we the humans) come in many different shapes, sizes, hues, and appearances. While we tend to notice differences that are easy to spot, other "stealth" variations in human populations are not necessarily observable from physical appearance alone. Now, new research into Arctic-dwelling Inuit populations points to a surprising origin for one such "stealth" trait—the ability to tolerate frigid temperatures.

Biology and paleontology provide an increasingly detailed picture of the evolution of life. Natural selection has resulted in a huge range of organisms living in a multitude of environments, and the geological record shows us how the Earth as a whole has changed over time. To understand the stage on which the evolution of life plays out, however, we must turn to astronomy.

3.7 Influence of thermodynamic disequilibrium on life

Life has significantly altered the Earth's atmosphere, oceans and crust. Dyke et al. (2010) provided estimates for the power generated by various elements in the Earth system. This includes, among other things, surface life generation of 264 TW of power (note that TW (Terawatt) is a metric measurement unit of power: $1 \text{ TW} = 10^{12}$ Watts), much greater than those of geological processes such as mantle convection at 12 TW. According to Dyke et al. (2010), this high power results from life's ability to harvest energy directly from the Sun. Life needs only utilize a small fraction of the generated free chemical energy for geochemical transformations at the surface, such as affecting rates of weathering and erosion of continental rocks, in order to affect interior, geological processes. Consequently, when assessing the effects of life on Earth, and potentially any planet with a significant biosphere, dynamical models that better capture the coupled nature of biologically-mediated surface and interior processes may be required (Dyke et al., 2010).

Photosynthetic life generates substantial quantity of geochemical free energy by performing photochemistry, thereby substantiating the notion that a geochemical composition far from chemical equilibrium can be a sign for strong biotic activity. The Earth's chemical composition far from chemical equilibrium is unique in our Solar System, and this uniqueness has been attributed to the presence of widespread life on the planet.

It is known that in moist convection, atmospheric motions transport water vapor from the Earth's surface to the regions where condensation occurs. This transport is associated with three other aspects of convection: the latent heat transport, the expansion work performed by water vapor, and the irreversible entropy production due to diffusion of water vapor and phase changes (Pauluis and Held, 2002a). Note that entropy is a thermodynamic quantity representing the unavailability of a system's thermal energy for conversion into mechanical work, often interpreted as the degree of disorder or randomness in the system. An analysis of the thermodynamic transformations of atmospheric water yields what is referred to as the entropy budget of the water substance, providing a quantitative relationship between these three aspects of moist convection. According to Pauluis and Held (2002a), "The water vapor transport can be viewed as an imperfect heat engine that produces less mechanical work than the corresponding Carnot cycle because of diffusion of water vapor and irreversible phase changes." Pauluis and Held (2002a) showed that diffusion of water vapor and irreversible phase changes can be interpreted as the irreversible counterpart to the continuous dehumidification resulting from condensation and precipitation. These researchers described moist convection where it acts more as an atmospheric dehumidifier than as a heat engine.

According to Kleidon (2012), "atmospheric composition is the only one aspect of the Earth system that is maintained far from a state of equilibrium." Another example of disequilibrium is the atmospheric water vapor content, which is mostly far from being saturated. If it were not for the continuous work being performed in the form of dehumidification by the atmospheric circulation (Pauluis and Held, 2002a, 2002b), then the atmosphere would gain moisture until it is completely saturated, as this is the state of thermodynamic equilibrium between the liquid and gaseous state of water.

It would be interesting to know as to what extent life has affected interior geological processes. To address this question, Dyke et al. (2010) formulated three models of geological processes: mantle convection, continental crust uplift, and erosion and oceanic crust recycling. According to Dyke et al. (2010), "These processes are characterized as nonequilibrium thermodynamic systems. Their states of disequilibrium are maintained by the power generated from the dissipation of energy from the interior of the Earth. Altering the thickness of continental crust via weathering and erosion affects the upper mantle temperature which leads to changes in rates of oceanic crust recycling and consequently rates of outgassing of carbon dioxide into the atmosphere."

Topographical gradients on land also reflect disequilibrium as erosion would act to deplete these gradients into a uniform equilibrium state, and life plays an important role in the processes that shape topographical gradients (Dietrich and Perron, 2006; Dyke et al., 2010). Dietrich and Perron (2006) elaborated on the role of topographical gradients and their impact of life thus, "Landscapes are shaped by the uplift, deformation and breakdown of bedrock and the erosion, transport and deposition of sediment. Life is important in all of these processes. Over short timescales, the impact of life is quite apparent: rock weathering, soil formation and erosion, slope stability and river dynamics are directly influenced by biotic processes that mediate chemical reactions, dilate soil, disrupt the ground surface and add strength with a weave of roots. Over geologic time, biotic effects are less obvious but equally important: biota affect climate, and climatic conditions dictate the mechanisms and rates of erosion that control topographic evolution."

A query that has been posed is whether, apart from the obvious influence of humans, the resulting landscape bears an unmistakable stamp of life? According to Dietrich and Perron (2006) the influence of life on topography is a topic that has remained largely unexplored. According to them "erosion laws that explicitly include biotic effects are needed to explore how intrinsically small-scale biotic processes can influence the form of entire landscapes, and to determine whether these processes create a distinctive topography."

The maintenance of some of the aforementioned disequilibrium states may seem to contradict the fundamental trend toward states of thermodynamic equilibrium, as formulated by the second law of thermodynamics. The second law of thermodynamics states that the total entropy of an isolated system can never decrease over time. The total entropy of a system and its surroundings can remain constant in ideal cases where the system is in thermodynamic equilibrium, or is undergoing a reversible process. Kleidon (2012) suggested that to understand why this is fully consistent and even to be expected from the laws of thermodynamics, it would be necessary to resort to the formulations of non-equilibrium thermodynamics and formulate Earth system processes and their interactions on this basis. It would be interesting to investigate how disequilibrium is generated and maintained without violating the second law of thermodynamics. Kleidon (2012) showed how free energy is generated from one thermodynamic gradient and transferred to another, causing disequilibrium in thermodynamic variables that are not directly related to heat and entropy.

To understand the generation and maintenance of disequilibrium—and the associated low entropy within the system—it may be noted that the Earth system is not an isolated system, so that the nature of its thermodynamic state is substantially different. A state of disequilibrium does not violate the laws of thermodynamics, but is rather a consequence of these. Kleidon (2012) reported a brief overview of the thermodynamics away from equilibrium to provide the basics of generating and maintaining disequilibrium and its relation to free energy generation within a system.

3.8 Extraterrestrial life in the Solar System—implications of Kumar's hypothesis

In biological evolution, proteins were formed by polymerization of amino acids; and aggregates—called *coacervates*—appeared in the water medium along with other organic compounds. Only after the origin of the genetic code, determined by the sequence of bases in nucleic acids, could the first self-reproducing molecule, self-perpetuating cells have arisen. Thus, organization of life first took place at the molecular level; for example, proteins and nucleic acids (DNA and RNA).

Earth is the only planet as we know so far that hosts life. So, the search for signs of life in the worlds beyond Earth is based on the assumption that extraterrestrial life has the fundamental characteristics of life as we know it on Earth. All life that is known to exist on Earth today and all life for which there is evidence in the geological record seems to be of the same form—one based on DNA genomes and protein enzymes.

A genome is an organism's complete set of DNAs, including all of its genes—the units of heredity. Each genome of an organism contains all of the information needed to build and maintain that organism. In humans, a copy of the entire genome, more than 3 billion DNA base pairs, is contained in the nucleus of all cells. Associated with the transmission of genetic information, DNA is often described as "the king of molecules."

Today, most living beings are composed of three basic ingredients: (1) proteins, which give a body its structure and carry out chemical reactions; (2) double-stranded DNA, which encodes genes, the units of heredity; and (3) RNA, a single-stranded molecule similar to DNA. It may be noted that DNA and RNA are made from building blocks called nucleotides. Note also that a nucleotide is a subunit of DNA or RNA that consists of a nitrogenous base, a phosphate molecule, and a sugar molecule. Among many other jobs, RNA carries the information for building proteins from a cell's genes to its protein factories.

RNA usually forms more irregular secondary structures than DNA with its double helix. Of the two, RNA is more versatile than DNA, capable of performing numerous, diverse tasks in an organism, but DNA is more stable and holds more complex information for longer periods of time.

Organization of life at the molecular level (i.e., proteins, DNA, and RNA) was followed by evolution of the first cell on the Earth; that is, prokaryotic cells, which did not have distinct organization of nucleus. Prokaryotic cells evolved into more complex cells termed eukaryotic cells, in which organization of nucleus took place (Avakian et al., 1970; Cavalier-Smith, 1975; Portelli, 1979; White, 1994). Subsequently, in the long passage of time, continued evolutionary processes resulted in the formation of all the higher forms of complex multicellular living organisms spearheaded by modern humans purely from the lower forms of unicellular living organisms (Darwin, 1859; Simpson, 1967). It is now a well-established fact that the genetic code has played a key role throughout the evolutionary process (Kerr, 1995; Weiner, 1995; Gibbons, 1995; Futuyma, 1995) and acted as an intra-cellular computer in the organization of the cell and nucleus (Portelli, 1976, 1979).

Current terrestrial life is built of monomers (repeating subunits of a polymer) such as amino acids, fatty acids, sugars, and nitrogenous bases. Amino acids link together to form proteins; fatty acids link to form lipids; sugars link to form carbohydrates; and nitrogenous bases combine with sugar and phosphate to make nucleotide monomers, which link to form RNA/DNA (Lineweaver and Chopra, 2012b). Thus, life on Earth emerged when available monomers linked together to make polymers. Amino acids and other organic monomers fall from the skies in carbonaceous chondrites. We have no reason to believe that the availability of these monomers is somehow unique to Earth or the Solar System. The flux of such organics was particularly high during the first billion years of the formation of Earth, and we have no reason to believe that this will not be the case during the formation of terrestrial planets in other planetary systems. The assortment of organic compounds found in carbonaceous chondrites and the probable universality of early heavy meteoritic bombardments indicate that new planetary systems should also be supplied with organic ingredients and be conducive to the synthesis of prebiotic molecules (Ehrenfreund and Charnley, 2000; Herbst and van Dishoeck, 2009; Tielens, 2013).

Life on Earth is made of hydrogen, oxygen, carbon, nitrogen, sulfur, and phosphorus: "HOCNSP" (Chopra et al., 2010). Hydrogen and oxygen together form water (H_2O). Thus, the ingredients essential for life include water, carbon, nitrogen, sulfur, and phosphorus.

Water functions as a solvent, allowing key chemical reactions to take place. Almost all the processes that make up life on Earth can be broken down into chemical reactions; and most of those reactions require a liquid to break down substances so they can move and interact freely. Liquid water is an essential requirement for life on Earth because it functions as a solvent. It is capable of dissolving substances and enabling key chemical reactions in animal, plant and microbial cells. Its chemical and physical properties allow it to dissolve more substances than most other liquids.

Many complex molecules are needed to perform the thousands of functions sustaining complex life. Carbon is the simple building block that organisms need to form organic compounds such as proteins, carbohydrates and fats. Carbon's molecular structure allows its atoms to form long chains, with each link leaving two potential bonds free to join with other atoms. It bonds particularly easily with oxygen, hydrogen and nitrogen. The free bonds can even join with other carbon atoms to form complex three-dimensional (3D) molecular structures, such as rings and branching trees.

Carbon is a fundamental component of organic compounds, but it can't do it alone. The complex proteins required for life are built up from smaller compounds called amino acids—simple organic compounds that contain nitrogen. Nitrogen is also needed to make DNA and RNA, the carriers of the genetic code for life on Earth. Liquid nitrogen is also put to use because it can maintain temperatures far below the freezing point of water. Many bacteria can convert nitrogen from the atmosphere into a form that is used in living cells.

Phosphorus is a key component of adenosine triphosphate (ATP), an organic substance that acts as life's molecular unit of currency. ATP transports chemical energy around the body's cells, powering nearly every cellular process that requires energy. Phosphorus is a vital element in cell membranes, the layer surrounding the inside of cells that controls the movement of substances in and out. And like nitrogen, phosphorus is necessary to create DNA and RNA. The phosphate group acts like glue in DNA, so the bodies of living organisms would not work without it.

Sulfur is part of most biochemical processes on Earth, and most enzymes cannot function without it. It is also a component of many vitamins and hormones. In the absence of oxygen and light, it is possible to use sulfur as an energy source. Bacteria that live under severe environmental conditions (called *extremophiles*) have been found to gain their energy for growth from sulfur and hydrogen alone.

HOCNSP are among the most abundant elements in the Universe (Pace, 1997; Lodders et al., 2009; Lineweaver and Chopra, 2012b). Because the elemental ingredients for life happen to be the most common elements in the Universe, it is not surprising that the molecular ingredients of life are also common. We expect all the ingredients of life as we know it (HOCNSP, water, amino acids, sugars, nucleic acids, HCN, and other organics) to be present and available on wet rocky planets throughout the Universe. An "ingredient bottleneck" seems implausible.

Based on the latest estimates, there exist about 8.7 ± 1.3 million different species of organisms on the Earth today. At the cellular and molecular levels, there is a unique master plan of organization common to all. For example, all organisms have the cells as their basic structural and functional unit. Likewise, all organisms have essentially the same genetic code made of DNA.

While we do not know the specifics of the prebiotic chemistry and geo-chemical environments necessary for life to emerge (e.g., Orgel, 1998; Stueken et al., 2013), the view that life will emerge with high probability on Earth-like planets is shared by many scientists. For instance, Darwin (1871) speculated that the environment, ingredients, and energy sources needed for the origin of life could be common, when he wrote about life emerging in "some *warm little pond* with all sort of ammonia and phosphoric salts, light, heat, electricity present, that a protein compound was chemically formed, ready to undergo still more complex changes."

de Duve (1995) has been a strong proponent of the view that the emergence of life is a cosmic imperative: "Life is either a reproducible, almost commonplace manifestation of matter, given *certain conditions*, or a miracle. Too many steps are involved to allow for something in between." Discarding miracles, de Duve leaves us with life as an "almost commonplace manifestation of matter." As far as we know, Darwin's *warm little pond* or de Duve's "certain conditions" are common on Earth-like planets and may be sufficient for the emergence of life. Biota throughout the Universe would emerge from chemistry through proto-biotic molecular evolution (e.g., Eigen and Winkler, 1992; Eschenmoser and Volkan Kisakurek, 1996; Orgel, 1998; Ward and Brownlee, 2000; Martin and Russell, 2007; Russell et al., 2013a).

There is some tension between this conclusion and the inability of synthetic biologists to produce life, despite having access to a wide variety of ingredients, environmental setups, and energy sources. According to Chopra and Lineweaver (2016), plausible explanations for this tension include (1) there is an emergence bottleneck (i.e., low probability for the emergence of life) due to a convoluted recipe whose requirements only rarely occur naturally; or (2) the recipe for life is simple—but we are not as imaginative or as resourceful as nature, so we have not replicated the recipe in the relatively short time we have been investigating the origin of life. Chopra and Lineweaver (2016) concluded that the idea of an emergence bottleneck is still plausible. However, the evidence for it is getting weaker as we find out more about the proto-biotic molecular evolution that led to the emergence of life on Earth (e.g., Benner, 2013; England, 2013; Nisbet and Fowler, 2014; Sousa and Martin, 2014). This weakness provides motivation for alternative explanations for the apparent paucity of life in the Universe.

In the background of increasingly growing curiosity about the origin of life and the possibility of life beyond Earth, it is heartening that one of the building blocks of life was spotted around three separate stars, each one very similar to the Sun when it was young. The molecule was found inside the warm cocoons of cosmic dust (the material produced by the fragmentation of asteroids and comets) and gas surrounding each star! Based on this finding, scientists arrived at two view-points about how life on Earth might have come about; (i) either life originated entirely on the surface of the Earth, or (ii) some of the building blocks were formed around the Sun before the Earth had even formed. Based on the discovery of biogenic molecules—known as the building blocks of life—inside the warm cocoons of cosmic dust and gas surrounding different stars, it looks likely that the second view-point is more reasonable. If so, these organic molecules became part of the comets in our Solar System. These comets may then have delivered the materials to our planet Earth, where they led to the rise of the very first life-forms. Some scientists think that it was comets that brought biomolecules to Earth in the early days of our Solar System.

A study by Lin and Loeb (2015) revealed that if life can travel between the stars, it would spread in a characteristic pattern that we could potentially identify. In their theory, clusters of life could form, grow, and overlap like bubbles in a pot of boiling water. Lin and Loeb (2015)'s model assumes that seeds from one living planet spread outward in all directions. If a seed reaches a habitable planet orbiting a neighboring star, it can take root there. Over time, the result of this process would be a series of life-bearing oases dotting the galactic landscape. The theoretical model result suggests that life could spread from host star to host star in a pattern similar to the outbreak of an epidemic.

According to Kumar's hypothesis (see Kumar, 2009; Joseph, 2016), the DNAs in the cells of all known life-forms (including the single-cell organism called prokaryote and all multicell organisms known as eukaryote including plants and we humans) on Earth are related to the astronomical organization of planets in the Solar System through an elegant mathematical relationship. The gravitational force, due to the curvature of four-dimensional space-time (propounded by the celebrated relativity scientist Dr. Albert Einstein), exerted by the Sun not only holds the members of the Solar System together but also causes them to take part in its true movement; that is, a movement relative to the adjacent stars and a revolution about the center of the Galactic System which itself is moving in the space. The Milky Way Galaxy is most significant to humans because it is home sweet home. Interestingly, our galaxy is a typical barred spiral, about 100,000 light-years across, much like billions of other galaxies in the Universe. Looking down on it from the top, there is a central bulge surrounded by four large spiral arms that wrap around it. Spiral galaxies make up about two-third of the galaxies in the Universe. The Milky Way does not sit still, but is constantly rotating; with the arms moving through space. The Sun and the Solar System travel with them. The Solar System travels at an average speed of 515,000 mph (828,000 km/h). Even at this rapid speed, the Solar System would take about 230 million years to travel all the way around the Milky Way (https://www.space.com/19915-milky-way-galaxy.html). It may be noted that, like the Galactic System, the stars are also constantly moving (this is a consequence of the expanding Universe).

Taking due consideration of the motions of the stars in the galaxies and those (motions) of the galaxies themselves, the "World Line" (i.e., the path of an object in the space-time continuum) of orbital motion of the Earth is a helix and geodesic (curved four-dimensional space-time). Kumar's hypothesis (propounded by Prof. [Dr.] Arbind Kumar) implies that there is a correlation between the helical "world line" of Earth's orbital motion and the helical structure of the DNA molecule, which is the first self-reproducing double-helical molecule to have formed in the beginning of evolution of life. This is the reason behind the existence of a mathematical relationship between organization of DNA molecule and Solar System, indicating a common underlying order and principle both in biological and astrophysical sciences, opening a new avenue for research.

This correlation also suggests that all natural self-assembly—from microscopic to macroscopic world (microcosm to macrocosm)—are influenced and controlled by gravitational curvature of space-time continuum as, for example, the self-assembly of Solar System and DNA double helical molecule. This emphasizes an underlying unifying relationship between seemingly two different disciplines of science. DNA and life originated and evolved in the gravitational curvature of the fabric of space-time caused by Sun's large mass and Earth's mass as proposed in Einstein's General Theory of Relativity. It is truly gratifying that Einstein's last prediction about gravitational curvature and gravitational waves was confirmed on February 11, 2016 after a century of his prediction.

Kumar's hypothesis combined with Einstein's General Theory of Relativity implies a close relationship between the organizational plan of DNA molecule, cells, tissues, organs and the whole body of the organisms during organic evolution and the organizational plan of the Solar System, galaxies, cluster of galaxies and the whole Universe during inorganic evolution. If intelligent life exists elsewhere in the cosmos, it is likely that the basic structural organization of genetic code and cell, tissue and organ system would be the same as on the Earth because of the universal gravity acting in the cosmic fabric of warped space-time continuum. This, in turn, implies that the DNA of all life-forms that may exist in the planets and their moons in the Solar System (although life beyond Earth is yet to be found) might also have the same basic structure. As just indicated, there are schools of thought postulating that life on Earth may have its origin in the cosmic dust.

Many scientists who believe in an underlying order in nature search for an ultimate, unifying principle in the physical sciences as well as in biological sciences. Kumar's hypothesis suggests gravitational force and waves due to warping of the cosmic fabric of space-time as a unifying force/entity between the biological science and physical/astrophysical/chemical sciences. It opens new vistas in our understanding of the origin, evolution and structural organization of the life and the Universe. It emphasizes that self-organization of self-reproducing polynucleotide chains of DNA molecule during the biochemical evolution might be governed by Sun's gravitational curvature of space-time and its true movement along with the Earth, resulting in the synthesis of genetic code containing message and language of life.

Many phenomena—involving stars, nebulae, galaxies, cluster of galaxies and black holes—occurring in the Universe seems to be analogous to the phenomena occurring in the living system. These analogies existing on the two different scales of space and time might appear unrelated initially. DNA, the blueprint of all living organisms on the Earth, is involved in the structural organizations and all vital functions such as protein synthesis and production of energy. It proposes that the Universe may have a blueprint for structural organizations and functions of the Universe.

3.9 Looking for possibility of extraterrestrial life in the Solar System—deriving clues from early Earth's conducive atmosphere for beginning of abundant life colonizing the Earth

Abundant life is believed to have colonized the Earth by approximately 3.8 billion years ago (BYA). Just prior to this period, something similar to modern photosynthesis developed and oxygen was produced. Over time, it transformed Earth's atmosphere to its current state. Some of the oxygen reacted to form ozone, which collected in a layer near the upper part of the atmosphere. By blocking the ultraviolet radiation, it allowed cells to colonize the surface of the ocean and ultimately the land. Fish, the earliest vertebrates, evolved in the oceans around 530 million years ago (MYA).

Only the most durable biosignatures that survive extensive metamorphosis (a conspicuous and relatively abrupt change in a living organism's body structure through cell growth and differentiation; e.g., the metamorphosis in the life of a butterfly) provide evidence. Some scientists concentrate on photosynthetic life because its high primary productivity has left a durable ancient record that has seen extensive study.

Today the origin of life is more than ever the subject of intensive research by biologists and anthropologists. As the early European explorers were soon followed by botanists, geographers, geologists, and surveyors who thoroughly and systematically explored the New World, cosmologists today are following the lead of the astronomical explorers of the last century, embarking on their own thorough and systematic exploration of the heavens. What were once fanciful stories are now scientific models; but whatever form they take, ideas about the origins of the Universe and of life on the planets both reflect and enrich the imagination of the people who generate them (see Luminet, 2011).

Science aims to discover what really happened in historical terms by means of theories supported by observations. Often considered to be antimyth, science has in fact created new stories about the origin of the Universe and life supported by planet Earth, and possibly about other planets and galaxies (e.g., the Big Bang model of the Universe, the theory of biological evolution, the ancestry of mankind, and the like).

There was a time when people believed that life exists only on Earth, and the rest of the Universe remains lifeless. However, such notions have begun to slowly vanish into thin air, and a rather revolutionary view that water (in its different manifestations such as vapor, liquid water, snow, ice) and life could exist also elsewhere in the Universe began to permeate into the minds of many intellectuals and several other open-minded nonspecialists who attach much value to observations and the logical deductions made based on experimental results. Many new findings show the remarkable potential that life has for adaptation and survival in extreme environments. All those results from different fields of science are guiding our perspectives and strategies to look for life in other Solar System objects as well as beyond, in extrasolar worlds (Cottin et al., 2017).

The problems of the origin, evolution, and structure of the Universe and those of life on Earth and outside Earth have been intriguing and exciting questions for scientists. The Renaissance of the 15th and 16th centuries led to an increased concern with the real world and a turning away from scholastic theology. It is now generally accepted that the origin of life on Earth was closely tied to the origin of the Solar System itself. Because of the umbilical relationship found between organic molecules and life on Earth, searching for clues to the origin of prebiotic organic molecules beyond Earth is a first step toward searching for clues to the origin of life there.

References

Abramov, O., Mojzsis, S.J., 2009. Microbial habitability of the Hadean Earth during the late heavy bombardment. Nature 459, 419–422. doi:10.1038/nature08015.

Alexander, P.A., Bacq, Z.M., 1960. Aspects of the Origin of Life (International Series of Monographs on Pure and Applied Biology6. Pergamon Press Oxford, London, New York, Paris.

Allen, C.C., Oehler, D.Z., 2008. A case for ancient springs in Arabia Terra: Mars. Astrobiology 8 (6), 1093–1112. doi:10.1089/ast.2008.0239.

Arndt, N., Nisbet, E., 2012. Processes on the young earth and the habitats of early life. Annu. Rev. Earth Planet. Sci. 40, 521–549.

Avakian, A.A., Tordzhian, I.Kh., Oparin, A.I., 1970. Submicroscopic organization of obligate anaerobic bacteria in the light of theories of the evolution of microorganisms. Dokl. Akad. Nauk. SSSR **193**, 1397–1399.

Avery, O.T., Colin, M.M., Maclyn, M., 1944. Studies on the chemical nature of the substance inducing transformation of pneumococcal types: induction of transformation by a deoxyribonucleic acid fraction isolated from pneumococcus type III. J. Exp. Med. 79 (2), 137–158. doi:10.1084/jem.79.2.137.

Baaske, P., Weinert, F.M., Duhr, S., et al., 2007. Extreme accumulation of nucleotides in simulated hydrothermal pore systems. Proc. Natl. Acad. Sci. USA 104, 9346–9351.

Bach, W., Paulick, H., Garrido, C.J., et al., 2006. Unraveling the sequence of serpentinization reactions: petrography, mineral chemistry, and petrophysics of serpentinites from MAR 15 °N (ODP Leg209, Site 1274). Geophys. Res. Lett. 33, L13306.

Bada, J.L., 2004. How life began on Earth: a status report. Earth Planet. Sci. Lett. 226, 1–15.

Bada, J.L., 2013. New insights into prebiotic chemistry from Stanley Miller's spark discharge experiments. Chem. Soc. Rev. 42 (5), 2186–2196.

Balm, S.P., Hare, J.P., Kroto, H.W., 1991. The analysis of comet mass spectrometric data. Space Sci. Rev. 56 (1–2), 185–189.

Baross, J.A., Hoffman, S.E., 1985. Submarine hydrothermal vents and associated gradient environments as sites for the origin and evolution of life. Orig. Life Evol. Biosphere 1, 327–345.

Behjati, S., Tarpey, P.S., 2013. What is next generation sequencing? Arch. Dis. Childhood: Educ. Pract. Ed. 98 (6), 236–238. doi:10.1136/archdischild-2013-304340.

Bekker, A., Holland, H.D., Wang, P.L., Rumble, D., Stein, H.J., Hannah, J.L., Coetzee, L.L., Beukes, N.J., 2004. Dating the rise of atmospheric oxygen. Nature 427, 117–120.

Bell, E.A., Boehnike, P., Harrison, T.M., et al., 2015. Potentially biogenic carbon preserved in a 4.1 billion-year-old zircon. Proc. Natl. Acad. Sci. U.S.A. 112, 14518–14521.

Benner, S.A., 2013. Keynote: planets, minerals and life's origin. Mineral Mag 77, 686.

Benner, S.A., 2017. Detecting Darwinism from molecules in the Enceladus plumes, Jupiter's moons, and other planetary water lagoons. Astrobiology 17, 840–851.

Berg, J.M., Tymoczko, J.L., Lubert Stryer 2012. *A Proton Gradient Across the Thylakoid Membrane Drives ATP Synthesis*, Nih.Gov; W H Freeman. https://www.ncbi.nlm.nih.gov/books/NBK22519/.

Blank, J.G., Miller, G.H., Ahrens, M.J., Winans, R.E., 2001. Experimental shock chemistry of aqueous amino acid solution and the cometary delivery of prebiotic compounds. Orig. Life Evol. Biosphere 31, 15–51.

Boivin, A., Vendrely, R., 1947. Sur le role possible des deux acides nucléiques dans la cellule vivante. Experientia 3, 32–34.

Brachet, J., 1942. La localisation des acides pentose nucléiques dans les tissus animaux et les oeufs d'amphibiens en voie de développement. Arch. Biol. 53, 207–257.

Bradley, A.S., Hayes, J.M., Summons, R.E., 2009. Extraordinary C-13 enrichment of diether lipids at the Lost City Hydrothermal Field indicates a carbon-limited ecosystem. Geochim. Cosmochim. Acta 73, 102–118.

Branciamore, S., Gallori, E., Szathmary, E., Czaran, T., 2009. The origin of life: chemical evolution of a metabolic system in a mineral honeycomb? J. Mol. Evol. 69, 458–469.

Braun, D., Libchaber, A., 2002. Trapping of DNA by thermophoretic depletion and convection. Phys. Rev. Lett. 89, 188103.

Brazelton, W.J., Mehta, M.P., Kelley, D.S., Baross, J.A., 2011. Physiological differentiation within a single-species biofilm fueled by serpentinization. mBio 2. doi:10.1128/mBio.00127-11.

Brazil, R., 2017. Hydrothermal vents and the origins of life, https://www.chemistryworld.com/features/hydrothermal-vents-and-the-origins-of-life/3007088.article.

Brenner, S., Jacob, F., Meselson, M., 1961. An unstable intermediate carrying information from genes to ribosomes for protein synthesis. Nature 190, 576–581.

Brenner, S., 2001. My Life in Science. BioMedCentral, London.

Brown, J.R., Doolittle, W.F., 1997. Archaea and the prokaryote-to-eukaryote transition. Microbiol. Mol. Biol. Rev. 61, 456–502.

Brownlee, G.B., Sanger, F., Barrell, B.G., 1967. Nucleotide sequence of 5S-ribosomal RNA from *Escherichia coli*. Nature 215, 735–736.

Brownlee, G.B., 2015. Fred Sanger Double Nobel Laureate: A Biography. Cambridge Univ. Press, Cambridge.

Bryson, B., 2005. A Short History of Nearly Everything. Broadway Books, New York, pp. 500.

Calvin, M., 1975. Chemical evolution. Am. Sci. **63**, 169–177.

Callaway, E., 2021. Million-year-old mammoth genomes shatter record for oldest ancient DNA - permafrost-preserved teeth, up to 1.6 million years old, identify a new kind of mammoth in Siberia. Nature 590 (7847), 537–538. doi:10.1038/d41586-021-00436-x.

Campbell, M.A., Wengel, J., 2011. Locked vs. unlocked nucleic acids (LNA vs. UNA): contrasting structures work towards common therapeutic goals. Chem. Soc. Rev. 40, 5680.

Carter Jr., C.W., Wolfenden, R., 2015. tRNA acceptor stem and anticodon bases form independent codes related to protein folding. Proc. Natl. Acad. Sci. 112 (24), 7489–7494.

Caspersson, T., 1947. The relations between nucleic acid and protein synthesis. Symp. Soc. Exp. Biol. 1, 127–151.

Castresana, J., Moreira, D., 1999. Respiratory chains in the last common ancestor of living organisms. J. Mol. Evol. 49, 453–460.

Cavalier-Smith, T., 1975. The origin of nuclei and of eukaryotic cells. Nature 256, 463–468.

Chapelle, F.H., Bradley, T.M., 1996. Microbial acetogenesis as a source of organic acids in ancient Atlantic coastal plain sediments. Geology 24, 925–928.

Chauhan, V., Kanwar, S.S., 2020. Bioactive peptides. M.L. Verma and A.K. Chandel (Eds.), Biotechnological Production of Bioactive Compounds, Imprint Elsevier, ISBN: 978-0-444-64323-0 (Copyright © 2020 Elsevier B.V. All rights reserve), 107–137. doi:10.1016/B978-0-444-64323-0.00004-7.

Chopra, A., Lineweaver, C.H., Brocks, J.J., Ireland, T.R., 2010. Palaeoecophylostoichiometrics searching for the elemental composition of the last universal common ancestor, Proceedings from Ninth Australian Space Science Conference. Sydney, Australia. National Space Society of Australia.

Chopra, A., Lineweaver, C.H., 2016. The case for a Gaian bottleneck: The biology of habitability. Astrobiology 16 (1), 7–22. doi:10.1089/ast.2015.1387.

Chyba, C., Sagan, C., 1992. Endogenous production, exogenous delivery and impact-shock synthesis of organic molecules: an inventory for the origins of life. Nature 355, 125–132.

Claudi, R., 2017. Exoplanets: Possible biosignatures, *Proceedings of Science*, Frontier Research in Astrophysics âAS II, 23-28 May 2016. Mondello (Palermo), Italy arXiv:1708.05829v1 [astro-ph.EP] 19 Aug 2017.

Cobb, M., 2014. Oswald Avery, DNA, and the transformation of biology. Curr. Biol. 24, R55–R60.

Cobb, M., 2015. Who discovered messenger RNA? Curr. Biol. 25 (13), R526–R532.

Cody, G.D., 2004. Transition metal sulfides and the origin of metabolism. Annu. Rev. Earth Planet. Sci. 32, 569–599.

Cordier, P., Tournilhac, F., Soulie-Ziakovic, C., Leibler, L., 2008. Self-healing and thermo-reversible rubber from supramolecular assembly. Nature **451**, 977–980.

Corliss, J.B., Baross, J.A, Hoffman, S.E., 1981. An hypothesis concerning the relationship between submarine hot springs and the origin of life on Earth. Oceanol. Acta., 59 No. Sp..

Corliss, J.B., Dymond, J., Gordon, L.I., Edmond, J.M., von Herzen, R.P., Ballard, R.D., Green, K., Williams, D., Bainbridge, A., Crane, K., van Andle, T.H., 1979. Submarine thermal sprirngs on the galapagos rift. Science 203 (4385), 1073–1083. doi:10.1126/science.203.4385.1073.

Cottin, H., Kotler, J.M., Bartik, K., Cleaves, H.J., Cockell, C.S., de Vera, J-P.P., Ehrenfreund, P., Leuko, S., Kate, I.L.T., Martins, Z., Pascal, R., Quinn, R., Rettberg, P., Westall, F., 2017. Astrobiology and the possibility of life on Earth and elsewhere? Space Sci. Rev. 209 (1-4), 1–42.

Crick, F., 1958. On protein synthesis. In: Sanders, F.K. (Ed.), The Biological Replication of Macromolecules, *Symposia of the Society of Experimental Biology*, XII. Cambridge, pp. 138–163.

Crick, F., 1988. What Mad Pursuit: A Personal View of Scientific Discovery Basic. Cambridge, Mass.

Dagan, T., Roettger, M., Bryant, D., Martin, W., 2010. Genome networks root the tree of life between prokaryotic domains. Genome Biol. Evol. 2, 379–392.

Dahm, R., 2008. Discovering DNA: Friedrich Miescher and the early years of nucleic acid research. Hum. Genet. 122 (6), 565–581. doi:10.1007/s00439-007-0433-0.

Darwin, C., 1859. On the Origin of Species by Means of Natural Selection, or the Preservation of Favored Races in the Struggle for Life. Cambridge University Press, Cambridge. Also http://www.literature.org/authors/darwin-charles/the-origin-of-species/.

Da Silva, L., Maurel, M.C., Deamer, D., 2015. Salt-promoted synthesis of RNA-like molecules in simulated hydrothermal conditions. J. Mol. Evol. 80, 86–97.

De, S., 2017. Understanding evolution of life. Navhind Times September 28, 10.

de Chadarevian, S., 1996. Sequences, conformation, information: biochemists and molecular biologists in the 1950s. J. History Biol. 29, 361–386.

Deamer, D., Damer, B., 2017. Can life begin on Enceladus? A perspective from hydrothermal chemistry. Astrobiology 17 (9), 834–839.

de Duve, C., 1995. Vital Dust: Life as a Cosmic Imperative. Basic Books, New York.

de Duve, C., 1991. Blueprint for a Cell: The Nature and Origin of Life, Burlington. Neil Patterson Publishers, Carolina Biological Supply Company, North Carolina, pp. 179.

de Wit, M.J., 1998. Early Archean processes: evidence from the South African Kaapvaal craton and its greenstone belts. Geol. Mijinbouw 76, 369–371.

Des Marais, D.J., Harwit, M., Jucks, K., Kasting, J.F., Lunine, J.I., Lin, D., Seager, S., Schneider, J., Traub, W., Woolf, N., 2002. Remote sensing of planetary properties and biosignatures on extrasolar terrestrial planets. Astrobiology (2), 153–181.

Dietrich, W.E., Perron, J.T., 2006. The search for a topographic signature of life. Nature 439, 411–418. doi:10.1038/nature04452.

Dodd, M.S., Papineau, D., Grenne, T., Slack, J.F., Rittner, M., Pirajno, F., O'Neil, J., Little, C.T.S., 2017. Evidence for early life in Earth's oldest hydrothermal vent precipitates. Nature 543 (7643), 60–64.

Dostal, J., Murphy, J.B., Nance, R.D., 2009. History of the Earth. Volume 2 of Earth Systems: History and Natural Variability, *Encyclopedia of Life Support Systems*. Developed Under the Auspices of the UNESCO. Eolss Publishers, Oxford, UK.

Dyke, J.G., Gans, F., Kleidon, A., 2010. Assessing life's effects on the interior dynamics of planet Earth using non-equilibrium thermodynamics. Earth Syst. Dyn. Discuss. 1, 191–246. doi:10.5194/esdd-1-191-2010.

Ehrenfreund, P., Charnley, S.B., 2000. Organic molecules in the interstellar medium, comets, and meteorites: a voyage from dark clouds to the early Earth. Annu. Rev. Astron. Astrophys. 38, 427–483.

Eigen, M., Winkler, R., 1992. Steps towards Life: A Perspective on Evolution. Oxford University Press, Oxford, UK.

England, J.L., 2013. Statistical physics of self-replication. J. Chem. Phys. 139, 1–8.

Eschenmoser, A., Volkan Kisakurek, M., 1996. Chemistry and the origin of life. Helv. Chim. Acta 79, 1249–1259.

Eschenmoser, A., 1999. Chemical etiology of nucleic acid structure. Science 284, 2118–2124.

Farely, J., 1977. The Spontaneous Generation Debate from Descartes to Oparin. Johns Hopkins University Press, Baltimore, pp. 225.

Farmer, J.D., Des Marais, D.J, 1999. Exploring for a record of ancient Martian life. J. Geophys. Res. Planets 104 (E11), 26,977-26,995. doi: 10.1029/1998JE000540.

Finch, J., 2008. A Nobel Laureate on Every Floor. Cambridge Univ. Press, Cambridge.

Follmann, H., Brownson, C., 2009. Darwin's warm little pond revisited: from molecules to the origin of life. Naturwissenschaften 96 (11), 1265–1292.

Fowler, W.A., 1975. A foundation for research. Science 188, 414–420.

Fuchs, G., Stupperich, E., 1985. Evolution of autotrophic CO_2 fixation. In: Schleifer, K.H., Stackebrandt, E. (Eds.), Evolution of Prokaryotes, FEMS Symposium No. 29. Academic Press, London, pp. 235–251.

Futuyma, D.J., 1995. The uses of evolutionary biology. Science 267, 41–42.

Garcia-Sancho, M., 2006. The rise and fall of the idea of genetic information (1948-2006). Genom. Soc. Policy 2/3, 16–36.

Garcia-Sancho, M., 2010. A new insight into Sanger's development of sequencing: from proteins to DNA, 1943-1977. J. History Biol. 43, 265 -23.

Garcia-Sancho, M., 2012. Biology, Computing and the History of molecular Sequencing: From Proteins to DNA, 1945-2000, Basingstoke. Euskal Herriko Unibertsitatea / Universidad del País Vasco, Spain.

Gibbons, A., 1995. The mystery of humanity's missing mutations. Science 267, 35–36.

Gogarten, J.P., et al., 1989. Evolution of the vacuolar H+-ATPase: implications for the origin of eukaryotes. PNAS 86, 6661–6665.

Greenberg, J.M., 1998. Making a comet nucleus. Astron. Astrophys. 330, 375–380.

Gros, F., Hiatt, H., Gilbert, W., Kurland, C.G., Risebrough, R.W., Watson, J.D., 1961. Unstable ribonucleic acid revealed by pulse labelling of Escherichia coli. Nature 190, 581–585.

Gros, F., 1979. The messenger. In: Lwoff, A., Ullmann, A. (Eds.), Origins of Molecular Biology: A Tribute to Jacques Monod. *Academic Press*, London, pp. 117–124.

Grosch, E.G., Hazen, R.M., 2015. Microbes, mineral evolution, and the rise of microcontinents—origin and coevolution of life with early earth. Astrobiology 15, 922–939.

Haldane, J.B.S., 1929. The origin of life. Rationalist Annu. 148, 3–10.

Hanczyc, M., Fujikawa, S., Szostak, J., 2003. Experimental models of primitive cellular compartments: encapsulation, growth, and division. Science 302, 618–622.

Hargreaves, A., 2017. Introducing 'dark DNA' – the phenomenon that could change how we think about evolution, https://theconversation.com/introducing-dark-dna-the-phenomenon-that-could-change-how-we-think-about-evolution-82867

Harold, F.M., 2014. Search of Cell History. University of Chicago Press, Chicago, IL, pp. 304. doi:10.7208/chicago/9780226174310.001.0001.

Hashizume, H., 2012. Role of clay minerals in chemical evolution and the origins of life. *Clay Minerals in Nature - Their Characterization, Modification and Application*. Open Access. doi:10.5772/50172.

Hennet, R.J-C., Holm, N.G., Engel, M.H., 1992. Abiotic synthesis of amino acids under hydrothermal conditions and the origin of life: a perpetual phenomenon? Naturwissenschaften 79, 361–365.

Herbst, E., van Dishoeck, E.F., 2009. Complex organic interstellar molecules. Annu. Rev. Astron. Astrophys. 47, 427–480.

Herdewijn, P., 2010. Nucleic acids with a six-membered 'carbohydrate' mimic in the backbone. Chem. Biodivers. 7, 1.

Herschy, B., Whicher, A., Camprubi, E., Watson, C., Dartnell, L., Ward, J., Evans, J.G., Lane, N., 2014. An origin-of-life reactor to simulate alkaline hydrothermal vents. J. Mol. Evol. 79, 213–227.

Hill, H.G., Nuth, J.A., 2003. The catalytic potential of cosmic dust: implications for prebiotic chemistry in the solar nebula and other protoplanetary systems. Astrobiology 3 (2), 291–304.

Hoagland, M.B., Stephenson, M.L., Scott, J.F., Hecht, L.I., Zamecnik, P., 1958. A soluble ribonucleic acid intermediate in protein synthesis. J. Biol. Chem. 231, 241–257.

Hode, T., Cady, S.L., von Dalwigk, I., Kristiansson, P., 2008. Evidence of ancient microbial life in a large impact structure and its implications for astrobiology: A case study. In: Seckbach, J., Walsh, M. (Eds.), From Fossils to Astrobiology. Springer Science Series Volume 12-Cellular Origin, Life in Extreme Habitats and Astrobiology, Berlin, pp. 249–273.

Holley, R.W., Agar, J., Everett, G.A., et al., 1965. Structure of a Ribonucleic Acid. Science 47/3664, 1462–1465.

Holm, N.G., 1992. Why are hydrothermal systems proposed as plausible environments for the origin of life? In: Holm (Ed.), Marine Hydrothermal Systems and the Origin of Life. Origins Life Evol. Biosphere 22, 1–242.

Holm, N.G., Oze, C., Mousis, O., Waite, J.H., Guilbert-Lepoutre, A., 2015. Serpentinization and the formation of H_2 and CH_4 on celestial bodies (Planets, Moons, Comets). Astrobiology 15, 587–600.

Ingram, V.M., 2004. Sickle-cell anaemia hemoglobin: the molecular biology of the first "molecular disease" - the crucial importance of serendipity. Genetics 167, 1–7.

Islam, M.D., Kaneko, T., Kobayashi, K., 2001. Determination of amino acids formed in supercritical water flow reactor simulating submarine hydrothermal systems. Anal. Sci. 17, 1631–1634.

Jacob, F., Monod, J., 1961. Genetic regulatory mechanisms in the synthesis of proteins. J. Mol. Biol. 3, 318–356.

Jacob, F., 1988. The Statue Within. Unwin Hyman, London.

Johnson, A.P., Cleaves, H.J., Dworkin, J.P., Glavin, D.P., Laczano, A., Bada, J.L., 2008. The Miller volcanic spark discharge experiment. Science 322, 404.

Jones, M.E., 1953. Albrecht Kossel, a biographical sketch. Yale J. Biol. Med. 26 (1), 80–97. PMC 2599350. PMID 13103145.

Joseph, 2016. Investigating Seafloors and Oceans: From Mud Volcanoes to Giant Squid. Elsevier Science & Technology Books Publishers, New York, pp. 581. ISBN: 978-0-12-809357-3.

Joyce, G.F., 2002. The antiquity of RNA-based evolution. Nature 418, 214–221.

Judson, H.F., 1996. The Eighth Day of Creation: Makers of the Revolution in Biology. Cold Spring Harbor Laboratory Press, Plainview.

Kahana, A., Schmitt-Kopplin, P., Lancet, D., 2019. Enceladus: first observed primordial soup could arbitrate origin-of-life debate. Astrobiology 19 (10), 1263–1278.

Kasting, J.F., Zahnle, K.J., Walker, J.C.G., 1983. Photochemistry of methane in the earth's early atmosphere. Precamb. Res. 20, 121–148.

Kasting, J.F., 2009. The primitive Earth. In: Wong, J.T., Lazcano, A. (Eds.), Prebiotic Evolution and Astrobiology. Landes Bioscience, Austin, TX, pp. 158.

Kasting, J.F., 2013. What caused the rise in atmospheric O_2? Chem. Geol. 362, 13–25.

Keefe, A.D., Newton, G.L., Miller, S.L., 1995. A possible prebiotic synthesis of pantetheine, a precursor to coenzyme A. Nature 373, 683–685.

Kelley, D.S., Karson, J.A., Blackman, D.K., Fruh-Green, G.L., Butterfield, D.A., Lilley, M.D., Olson, E., Schrenk, M.O., Roe, K., Lebon, G.T., Rivissigno, P., the AT3-60 Shipboard Party, 2001. An off-axis hydrothermal vent field near the Mid-Atlantic Ridge at 30°N. Nature 412, 145–149.

Kelley, D.S., Karson, J.A., Fruh-Green, G.L., Yoerger, D., Shank, T.M., Butterfield, D.A., Hayes, J.M., Schrenk, M.O., Olson, E., Proskurowski, G., Jakuba, M., Bradley, A., Larson, B., Ludwig, K., Glickson, D.,

Buckman, K., Bradley, A.S., Brazelton, W.J., Roe, K., Elend, M.J., Delacour, A., Bernasconi, S.M., Lilley, M.D., Baross, J.A., Summons, R.E., Sylva, S.P., 2005. A serpentinite-hosted ecosystem: the lost city hydrothermal field. Science 307, 1428–1434.

Kelly, D.P., 1981. Introduction to the chemolithotrophic bacteria. In: Starr, M.P., Stolp, H., Trüper, H.G., Balows, A., Schlegel, H.G. (Eds.), The Prokaryotes. Springer, Berlin, Heidelberg.

Kerr, R.A., 1995. Timing evolution's early bursts. Science 267, 33–34.

Kirschvink, J.L., Weiss, B.P., 2002. Mars, Panspermia, and the Origin of Life: Where Did It All Begin? (A Modified Version of the Carl Sagan Lecture Given by Joseph Kirschvink at the American Geophysical Union Meeting in December, 2001). Coquina Press.

Kitadai, N., Maruyama, S., 2018. Origins of building blocks of life: a review. Geosci. Front. 9 (4), 1117–1153.

Kleidon, A., 2012. How does the Earth system generate and maintain thermodynamic disequilibrium and what does it imply for the future of the planet? J.M.T. Thompson and J. Sieber (Eds.) Philos. Trans. R. Soc. A: Math. Phys. Eng. Sci, 370 (1962). https://doi.org/10.1098/rsta.2011.0316.

Knowles, J.R., 1980. Enzyme-catalyzed phosphoryl transfer reactions. Annu. Rev. Biochem. 49, 877–919. doi:10.1146/annurev.bi.49.070180.004305.

Koga, Y., Kyuragi, T., Nishihara, M., Sone, N., 1998. Did archaeal and bacterial cells arise independently from noncellular precursors? A hypothesis stating that the advent of membrane phospholipid with enantiomeric glycerophosphate backbones caused the separation of the two lines of descent. J. Mol. Evol. 46, 54–63.

Kompanichenko, V.N., 2009. Changeable hydrothermal media as potential cradle of life on a planet. Planet Space Sci. 57, 468–476.

Konhauser, K.O., Jones, B., Reysenbach, A.-L., Renaut, R.W., 2003. Hot spring sinters: keys to understanding Earth's earliest life forms. Can. J. Earth Sci. 40 (11), 1713–1724. doi:10.1139/e03-059.

Konn, C., Charlou, J.L., Holm, N.G., Mousis, O., 2015. The production of methane, hydrogen, and organic compounds in ultramafic-hosted hydrothermal vents of the Mid-Atlantic Ridge. Astrobiology 15, 381–399.

Koonin, E.V., Martin, W., 2005. On the origin of genomes and cells within inorganic compartments. Trends Genet. 21, 647–654.

Koonin, E.V., Senkevich, T.G., Dolja, V.V., 2006. The ancient Virus World and evolution of cells. Biol. Direct 19, 29.

Koonin, E.V., 2010. The two empires and three domains of life in the postgenomic age. Nat. Educ. 3 (9), 27.

Korenaga, J., 2013. Initiation and evolution of plate tectonics on Earth: theories and observations. Annu. Rev. Earth Planet Sci. 41, 117–151.

Kresge, N., Simoni, R.D, Hill, R.L., 2005. The purification and sequencing of alanine transfer ribonucleic acid: the work of Robert W Holley. J. Biol. Chem. 281/7, e7–e9.

Kuhn, W.R., Atreya, S.K., 1979. Ammonia photolysis and the greenhouse effect in the primordial atmosphere of the earth. Icarus 37, 207–213.

Kumar, A., 2009. Evolution and relation of microcosm and macrocosm. Biospectra 4 (2), 459–466.

Lal, A.K., 2008. Origin of life. Astrophys. Space Sci. 317, 267–278.

Lamm, E., Harman, O., Veigl, S.J., 2000. Before Watson and Crick in 1953 Came Friedrich Miescher in 1869. Genetics 215 (2), 291–296. https://doi.org/10.1534/genetics.120.303195.

Lane, N., Allen, J.F., Martin, W., 2010. How did LUCA make a living? Chemiosmosis in the origin of life. Bio Essays 32, 271–280.

Lane, N., Martin, W.F., 2012. The origin of membrane bioenergetics. Cell 151, 1406–1416.

Lazcano, A., Miller, S.L., 1996. The origin and early evolution of life: prebiotic chemistry, the pre-RNA world, and time. Cell 85, 793–796.

Lazcano, A., Bada, J.L., 2004. The 1953 Stanley L. Miller experiment: fifty years of prebiotic organic chemistry. Orig. Life Evol. Biospheres 33 (3), 235–242.

Leconte, A.M., Hwang, G.T., Matsuda, S., Capek, P., Hari, Y., Romesberg, F.E., 2008. Discovery, characterization, and optimization of an unnatural base pair for expansion of the genetic alphabet. J. Am. Chem. Soc. 130, 2336–2343.

Lin, H.W., Loeb, A., 2015. Statistical signatures of Panspermia in exoplanet surveys. Astrophys. J. Lett., 810 (1): 10.1088/2041-8205/810/1/l3.

Lineweaver, C.H., Chopra, A., 2012b. What can life on Earth tell us about life in the Universe? In: Seckbach, J. (Ed.), Genesis—In the Beginning: Precursors of Life, Chemical Models and Early Biological Evolution. Springer, Dordrecht, the Netherlands, pp. 799–815.

Ljungdahl, L.G., 2009. A life with acetogens, thermophiles, and cellulolytic anaerobes. Annu. Rev. Microbiol. 63, 1–25.

Loakes, D., Holliger, P., 2009. Polymerase engineering: towards the encoded synthesis of unnatural biopolymers. Chem. Commun. 40 (31), 4619–4631.

Lodders, K., Palme, H., Gail, H.-P., 2009. Abundances of the elements in the Solar System. In: Trumper, J. (Ed.), *Landolt-Bornstein, New Series, Astronomy and Astrophysics*, Vol. VI/4B, Chapter 4.4. Springer-Verlag, Berlin, pp. 560–630.

Lombard, J., Lopez-Garcıa, P., Moreira, D., 2012. The early evolution of lipid membranes and the three domains of life. Nat. Rev. Microbiol. 10, 507–515.

Lonsdale, P., 1977. Structural geomorphology of a fast-spreading rise crest: the East Pacific Rise near 3°25′S. Mar. Geophys. Res. 3, 251–293.

Luminet, J.-P., 2011. The Rise of Big Bang Models, From Myth to Theory to Observations, LaboratoireUnivers et Theories, CNRS-UMR 8102, Observatoire de Paris, France.

Maddox, B., 2003. The double helix and the 'wronged heroine'. Nature 421 (6921), 407–408.

Maden, B.E.H., 1995. No soup for starters? Autotrophy and the origins of metabolism. Trends Biochem. Sci. 20, 337–341.

Mansy, S.S., Schrum, J.P., Krishnamurthy, M., Tobe, S., Treco, D.A., et al., 2008. Template-directed synthesis of a genetic polymer in a model protocell. Nature 454, 122–125.

Martin, W.F., Russell, M.J., 2007. On the origin of biochemistry at an alkaline hydrothermal vent. Philos Trans R Soc Lond B Biol Sci 362, 1887–1925.

Martin, W., Russell, M.J., 2003. On the origins of cells: a hypothesis for the evolutionary transitions from abiotic geochemistry to chemoautotrophic prokaryotes, and from prokaryotes to nucleated cells. Philos. Trans. R. Soc. B: Biol. Sci. 358 (1429), 59–83. doi:10.1098/rstb.2002.1183.

Martin, W., Baross, J., Kelley, D., Russell, M.J., 2008. Hydrothermal vents and the origin of life. Nat. Rev. Microbiol. 6, 805–814.

Mauer, S.E., Monndard, P.A., 2011. Primitive membrane formation, characteristics and roles in the emergent properties of a protocell. Entropy 13, 466–484.

McCollom, T.M., Bach, W., 2009. Thermodynamic constraints on hydrogen generation during serpentinization of ultramafic rocks. Geochim. Cosmochim. Acta 73, 856–875.

Miescher, F., 1871. Ueber die chemische Zusammensetzung der Eiterzellen (On the chemical composition of pus cells). Med.-chem. Untersuchungen 4, 441–460.

Miller, S.L., 1953. A production of amino acids under possible primitive Earth conditions. Science 117, 528–529.

Miller, S.L., 1957. The mechanism of synthesis of amino acids by electric discharges. Biochem. Biophys. Acta 23, 480.

Miller, S.L., Urey, H.C., 1959. Organic compound synthesis on the primitive Earth. Science 130 (3370), 245–251.

Milller, S.L., and Van Trump J.E. (1981), The Strecker synthesis in the primitive ocean, In: Origin of Life, Y. Wolman (ed.), D. Reidel Publishing Company Dordrecht: Holland / Boston: U.S.A. London: England, 135–142.

Miller, S.L., 1986. Current status of the prebiotic synthesis of small molecules. Chem. Scripta 26B, 5–11.

Miller, S.L., 1987. Which organic compounds could have occurred on the prebiotic Earth? Cold Spring Harbor Symp. Quant. Biol. 52, 17–27.

Miller, S., Cleaves, H., 2006. Prebiotic chemistry on the primitive Earth. Genomics 1, 3–56.

Mitchell, P., 1959. The origin of life and the formation and organizing functions of natural membranes. In: Oparin, A.I. et al (Ed.), Proceedings of the first International Symposium on the Origin of Life on the Earth.

Mitchell, P., 1961. Coupling of phosphorylation to electron and hydrogen transfer by a chemi-osmotic type of mechanism. Nature 191 (4784), 144–148. https://doi.org/10.1038%2F191144a0.

Mitchell, P., 1966. Chemiosmotic coupling in oxidative and photo-synthetic phosphorylation. Biol Rev 41, 445–501.

Mulkidjanian, A.Y., Bychkov, A.Y., Dibrova, D.V., Galperin, M.Y., Koonin, E.V., 2012. Origin of first cells at terrestrial, anoxic geothermal fields. PNAS. http://www.pnas.org/cgi/doi/10.1073/pnas.1117774109. Accessed 2012 Jan 20.

Nielsen, P.E., Egholm, M., Berg, R.H., Buchardt, O., 1991. Sequence-selective recognition of DNA by strand displacement with a thymine-substituted polyamide. Science 254 (5037), 1497–1500. doi:10.1126/science.1962210.

Nielsen, P.E., 1995. DNA analogues with nonphosphodiester backbones. Annu. Rev. Biophys. Biomol. Struct. 24, 167–183.

Nishizawa, M., Miyazaki, J., Makabe, A., Koba, K., Takai, K., 2014. Physiological and isotopic characteristics of nitrogen fixation by hyperthermophilic methanogens: key insights into nitrogen anabolism of the microbial communities in Archean hydrothermal systems. Geochim. Cosmochim. Acta 138, 117–135.

Nitschke, W., Russell, M.J., 2009. Hydrothermal focusing of chemical and chemiosmotic energy, supported by delivery of catalytic Fe, Ni, Mo, Co, S and Se forced life to emerge. J. Mol. Evol. 69, 481–496.

Nisbet, E.G., Fowler, C.M.R., 2014. The early history of life. In: Palme, H., O'Neill, H. (Eds.), 2nd ed. Treatise on Geochemistry, 10 Elsevier, Amsterdam, pp. 1–42.

Nowak, M.A., Ohtsuki, H., 2008. Revolutionary dynamics and the origin of evolution. Proc. Natl. Acad. Sci. USA 105, 14924–14927.

Ohara, S., Kakegawa, T., Nakazawa, H., 2007. Pressure effects on the abiotic polymerization of glycine. Orig. Life Evol. Biosphere 37, 215–223.

Olsvik, O., Wahlberg, J., Petterson, B., Uhlén, M., Popovic, T., Wachsmuth, I.K., Fields, P.I., 1993. Use of automated sequencing of polymerase chain reaction-generated amplicons to identify three types of cholera toxin subunit B in *Vibrio cholerae* O1 strains. J. Clin. Microbiol. 31 (1), 22–25. doi:10.1128/JCM.31.1.22-25.1993.

Oparin, A.I., 1924. Proiskhodenie Zhizni, Moscow: Moscoksky Rabotichii. Proiskhodenie Zhizni, Moscow: Moscoksky Rabotichii, pp. 71. Transl., 1967 as appendix in Bernal, J.D., The Origin of Life, Cleveland: World, pp. 345.

Oparin, A.I., 1938. The Origin of Life. MacMillan, New York, pp. 270 (Orginal: (1924), *Proiskhozhdenie zhizny*. Moscow: Izd. Moskovski Rabochii.

Oparin, A.I., 1952. The Origin of Life. Dover, New York.

Oparin, A.I., 1957. The Origin of Life on Earth, 3rd ed. Oliver and Boyd, Edinburgh.

Orgel, L., 1998. The origin of life—a review of facts and speculations. Trends Biochem. Sci. 23, 491–495.

Orgel, L.E., 2004. Prebiotic chemistry and the origin of the RNA world. Crit. Rev. Biochem. Mol. Biol. 39, 99–123.

Oro, J., 1963b. Synthesis of organic compounds by electric discharges. Nature 197, 862–867.

Owen, T., 1980. Strategies for the Search for Life in the Universe, M.D. Papagiannis (Ed.), Reidel Publ. Co., Dordrecht, Holland, pp. 177.

Pauluis, O., Held, I.M., 2002a. Entropy budget of an atmosphere in radiative convective equilibrium. Part I: maximum work and frictional dissipation. J. Atmos. Sci. 59, 126–139.

Pauluis, O., Held, I.M., 2002b. Entropy budget of an atmosphere in radiative convective equilibrium. Part I: maximum work and frictional dissipation. J. Atmos. Sci. 59, 126–139.

Pace, N.R., 1997. A molecular view of microbial diversity and the biosphere. Science 276, 734–740. doi:10.1126/science.276.5313.734.

Patel, B.H., Percivalle, C., Ritson, D.J., Duffy, C.D., Sutherland, J.D., 2015. Common origins of RNA, protein, and lipid precursors in a cyanosulfidic protometabolism. Nat. Chem. 7 (4), 301–307.

Pereto, J., Lopez-Garcia, P., Moreira, D., 2004. Ancestral lipid biosynthesis and early membrane evolution. Trends Biochem. Sci. 29, 469–477.

Pettersson, E., Lundeberg, J., Ahmadian, A., 2009. Generations of sequencing technologies. Genomics 93 (2), 105–111. doi:10.1016/j.ygeno.2008.10.003.

Picard, A., Daniel, I., 2013. Pressure as an environmental parameter for microbial life—a review. Biophys Chem 183, 30–41.

Pinheiro, V.B., et al., 2012. Synthetic genetic polymers capable of heredity and evolution. Science 336 (6079), 341–344. doi:10.1126/science.1217622.

Pinti, D., 2005. The origin and evolution of the oceans. Lect. Astrobiol. 1, 83–112.

Pleasant, L.G., Ponnamperuma, C., 1980. Chemical evolution and the origin of life. Origins Life 10, 379–404.

Poehlein, A., Schmidt, S., Kaster, A-K., Goenrich, M., Vollmers, J., Thurmer, A., Bertsch, J., Schuchmann, K., Voigt, B., Hecker, M., Daniel, D., Thauer, R.K., Gottschalk, G., Muller, V., 2012. An ancient pathway combining carbon dioxide fixation with the generation and utilization of a sodium ion gradient for ATP synthesis. PLoSONE 7, e33439.

Portelli, C., 1976. DNA-histones, a computer model. Acta. Biotheor. 25, 130–152.

Portelli, C., 1979. The origin of life, A cybernetic and informational process. Acta Biotheor. 28, 19–47.

Pray, L., 2008. Discovery of DNA structure and function: Watson and Crick. Nat. Educ. 1 (1), 100.

Preiner, M., et al., 2020. The future of origin of life research: bridging decades-old divisions. Life (Basel) 10 (3), 20. doi:10.3390/life10030020.

Prigogine, I., Nicolis, G., Babloyants, A., 1972. Thermodynamics of evolution. Phys. Today 25, 25.

Proskurowski, G., et al., 2008. Abiogenic hydrocarbon production at lost city hydrothermal field. Science 319, 604–607.

Ragsdale, S.W., Pierce, E., 2008. Acetogenesis and the Wood-Ljungdahl pathway of CO_2 fixation. Biochim. Biophys. Acta 1784, 1873–1898.

Raina, M., Ibba, M., 2014. tRNAs as regulators of biological processes. Front. Genet. 5, 171.

Rampelotto, P.H., 2010c. Resistance of microorganisms to extreme environmental conditions and its contribution to astrobiology. Sustainability 2, 1602–1623. doi:10.3390/su2061602.

Raoult, D., Forterre, P., 2008. Redefining viruses: lessons from mimivirus. Nat. Rev. Microbiol. 6, 315–319. doi:10.1038/nrmicro1858.

Reysenbach, A.-L., Cady, S.L., 2001. Microbiology of ancient and modern hydrothermal systems. Trends Microbiol. 9, 79–86. doi:10.1016/S0966-842X(00)01921-1.

Rich, A., Zhang, S., 2003. Z-DNA: the long road to biological function. Nat. Rev. Genet. 4, 566–572.

Roberts, R.B., 1958. Microsomal Particles and Protein Synthesis. Pergamon, London.

Robertson, M.P., Miller, S.L., 1995. Prebiotic synthesis of 5-substituted uracils: a bridge between the RNA world and the DNA-protein world. Science 268, 702–705.

Rona, P.A., Klinkammer, G., Nelsen, T.A., Trefry, J.H., Elderfield, H., 1986. Black smokers, massive sulphides, and vent biota at the Mid-Atlantic Ridge. Nature 321, 33–37.

Rosas, A., et al., 2017. The growth pattern of Neandertals, reconstructed from a juvenile skeleton from El Sidrón (Spain). Science. science.sciencemag.org/cgi/doi … 1126/science.aan6463.

Rosing, M.T., 1999. ^{13}C-depleted carbon microparticles in >3700-Ma sea-floor sedimentary rocks from west Greenland. Science 283, 674–676.

Russell, M.J., Hall, A.J., Cairns-Smith, A.G., Braterman, P.S., 1988. Submarine hot springs and the origin of life. Nature 336, 117.

Russell, M.J., Hall, A.J., Turner, D., 1989. In-vitro growth of iron sulphide chimneys: possible culture chambers for origin-of-life experiments. Terra Nova 1, 238–241.

Russell, M.J., Daniel, R., Hall, A.J., 1993. On the emergence of life via catalytic iron-sulphide membranes. Terra Nova 5, 343. doi:10.1111/j.1365-3121.1993.tb00267.x.

Russell, M.J., Daniel, R.M., Hall, A.J., Sherringham, J., 1994. A hydrothermally precipitated catalytic iron sulphide membrane as a first step toward life. J. Mol. Evol. 39, 231–243.

Russell, M.J., Hall, A.J., 1997. The emergence of life from iron monosulphide bubbles at a submarine hydrothermal redox and pH front. J. Geol. Soc. London 154, 377–402.

Russell, M.J., Arndt, N.T., 2005. Geodynamic and metabolic cycles in the Hadean. Biogeosciences 2, 97–111.

Russell, L.M., et al., 2010. Carbohydrate-like composition of submicron atmospheric particles and their production from ocean bubble bursting. Proc. Natl. Acad. Sci. USA 107, 6652–6657.

Russell, M.J., Nitschke, W., Branscomb, E., 2013a. The inevitable journey to being. Philos. Trans. R. Soc. Lond. B Biol. Sci. 368, 1–19.

Russell, M.J., Barge, L.M., Bhartia, R., Bocanegra, D., Bracher, P.J., Branscomb, E., Kidd, R., McGlynn, S., Meier, D.H., Nitschke, W., Shibuya, T., Vance, S., White, L., Kanik, I., 2014. The drive to life on wet and icy worlds. Astrobiology 14, 308–343.

Saladino, R., Pilat-Lohinger, E., Palomba, E., Harrison, J., Rull, F., Muller, C., Strazzulla, G., Brucato, J.R., Rettberg, P., Capria, M.T., 2016. AstRoMap European Astrobiology Roadmap. Astrobiology 16, 201–243.

Sanchez, R.A., Ferris, J.P., Orgel, L.E., 1966. Cyanoacetylene in prebiotic synthesis. Science 154, 784–785.

Sanger, F., Brownlee, G.B., Barrell, B.G., 1965. A two-dimensional fractionation procedure for radioactive nucleotides. J. Mol. Biol. 13/2, 373–398.

Sanger, F., 1988. Sequences, sequences, sequences. Annu. Rev. Biochem. 57, 1–29.

Sanger, F., 1992. A life of research on the sequences of proteins and nucleic acids: Dr Fred Sanger in conversation with George Brownlee. Biochem. Soc. Arch., London: Imperial College.

Sanger, F., Dowding, M. (Eds.), 1996. Selected papers of Frederick Sanger (with commentaries), London.

Say, R.F., Fuchs, G., 2010. Fructose 1,6-bisphosphate aldolase/phosphatase may be an ancestral gluconeogenic enzyme. Nature 464, 1077–1081.

Schaefer, L., Fegley Jr., B., 2007. Outgassing of ordinary chondritic material and some of its implications for the chemistry of asteroids, planets, and satellites. Icarus 186, 462–483.

Schoell, M., 1988. Multiple origins of methane in the Earth. Chem Geol 71, 1–10.

Schoning, K.U., et al., 2000. Chemical etiology of nucleic acid structure: the α-threofuranosyl-(3′→2′) oligonucleotide system. Science 290 (5495), 1347–1351. doi:10.1126/science.290.5495.1347.

Schrum, J.P., Zhu, T.F., Szostak, J.W., 2010. The origins of cellular life. Cold Spring Harb. Perspect. Biol. 2, a002212.

Schulte, M.D., Shock, E.L., 1995. Thermodynamics of strecker synthesis in hydrothermal systems. Orig. Life Evol. Biosphere 25, 161–173.

Stasiak, A., 2001. Rosalind Franklin. EMBO Rep, National Institutes of Health, 2 (3), 181. doi:10.1093/embo-reports/kve037.

Stetter, K.O., 2006. Hyperthermophiles in the history of life. Philos. Trans. R. Soc. Lond. B Biol. Sci. 361 (1474), 1837–1843. doi:10.1098/rstb.2006.1907.

Shock, E.L., 1992. Chemical environments in submarine hydrothermal systems. In: Holm NG (Ed.), Marine Hydrothermal Systems and the Origin of Life, a special issue of Orig. Life Evol. Biosphere, 22: 67–107.

Shock, E.L., 1996. Hydrothermal systems as environments for the emergence of life. In: Bock, G.R., Goode, J.A. (Eds.), Evolution of Hydrothermal Ecosystems on Earth (and Mars?). John Wiley & Sons, New York.

Shock, E.L., McCollom, T.M., Schulte, M.D., 1998. Thermophiles: The Keys to Molecular Evolution and The Origin of Life? Wiegel, J., Adams, M.W.W. (Eds.), Taylor and Francis, London, pp. 59–76.

Shock, E., 1990. Geochemical constraints on the origin of organic compounds, In: Hydrothermal Systems. Orig Life 20, 331–367.

Simpson, G.G., 1967. The Meaning of Evolution. Yale University Press, New Haven, CT.

Sleep, N.H., Meibom, A., Fridriksson, T., Coleman, R.G., Bird, D.K., 2004. H_2-rich fluids from serpentinization: geochemical and biotic implications. Proc. Natl Acad. Sci. USA 101, 12818–12823.

Sleep, N.H., 2010. The Hadean-Archaean environment. Cold Spring Harb Perspect. Biol. 2 (6), a002527.

Sojo, V., Herschy, B., Whicher, A., Camprubí, E., Lane, N., 2016. The origin of life in alkaline hydrothermal vents. Astrobiology 16, 181–197.

Sousa, F.L., Martin, W.F., 2014. Biochemical fossils of the ancient transition from geoenergetics to bioenergetics in prokaryotic one carbon compound metabolism. Biochim Biophys Acta 1837, 964–981.

Sousa, F.L., et al., 2013. Early bioenergetic evolution. Philos. Trans. R. Soc. Lond. B Biol. Sci. 368 (1622) 20130088.

Squyres, S.W., Arvidson, R.E., Ruff, S., Gellert, R., Morris, R.V., Ming, D.W., Crumpler, L., Farmer, J.D., Des Marais, D.J., Yen, A., McLennan, S.M., Calvin, W., Bell III, J.F., Clark, B.C., Wang, A., McCoy, T.J., Schmidt, M.E., de Souza Jr., P.A., 2008. Detection of silica-rich deposits on Mars. Science 320, 1063–1067. doi:10.1126/science.1155429.

Sreenivasachary, N., Lehn, J.M., 2005. Gelation-driven component selection in the generation of constitutional dynamic hydrogels based on guanine-quartet formation. Proc. Natl. Acad. Sci. 102, 5938–5943.

Sreenivasachary, N., Hickman, D.T., Sarazin, D., Lehn, J.M., 2006. DyNAs: constitutional dynamic nucleic acid analogues. Chem.-Eur. J. 12, 8581–8588.

Sreenivasachary, N., Lehn, J.M., 2008. Structural selection in G-quartet-based hydrogels and controlled release of bioactive molecules. Chem.-Asian J. **3**, 134–139.

Stetter, K.O., 2007. Hyperthermophilic life on Earth—and on Mars? In: Pudritz, R., Higgs, P., Stone, J. (Eds.), Planetary Systems and the Origins of Life. Cambridge University Press, Cambridge, UK.

Stueken, E.E., Anderson, R.E., Bowman, J.S., Brazelton, W., Colangelo-Lillis, J., Goldman, A.D., Som, S.M., Baross, J.A., 2013. Did life originate from a global chemical reactor? Geobiology 11, 101–126.

Takai, K., Nealson, K.H., Horikoshi, K., 2004. *Methanotorris formicicus* sp nov., a novel extremely thermophilic, methane-producing archaeon isolated from a black smoker chimney in the Central Indian Ridge. Int. J. Syst. Evol. Microbiol. 54, 1095–1100.

Thaddeus, P., 2006. The prebiotic molecules observed in the interstellar gas. Philos. Trans. R. Soc. B 361, 1681–1687.

Thauer, R.K., Kaster, A-K., Seedorf, H., Buckel, W., Hedderich, R., 2007. Methanogenic archaea: ecologically relevant differences in energy conservation. Nat. Rev. Microbiol. 6, 579–591.

Tian, F., Toon, O.B., Pavlov, A.A., De Sterck, H., 2005. A hydrogen-rich early Earth atmosphere. Science 308, 1014–1017.

Tielens, A.G.G.M., 2013. The molecular universe. Rev. Mod. Phys. 85, 1021–1081.

Toparlak, O.D., Mansy, S.S., 2018. Progress in synthesizing protocells. Exp. Biol. Med. (Maywood) 244, 304–313.

Urey, H.C., 1952. The Planets, Their Origin and Development. Yale University Press, New Haven, pp. 245.

Vago, J.L., et al., 2017. Habitability on early Mars and the search for biosignatures with the ExoMars Rover. Astrobiology 17 (6-7), 471–510.

Volkin, E., Astrachan, L., 1956. Phosphorous incorporation in *Escherichia coli* ribonucleic acid after infection with bacteriophage T2. Virology 2, 149–161.

Volkin, E., 1995. What was the message? TIBS 20, 206–209.

Van Lint, S., Heirman, C., Thielemans, K., Breckpot, K., 2013. mRNA: From a chemical blueprint for protein production to an off-the-shelf therapeutic. Hum. Vaccin Immunother. 9 (2), 265–274.

Wagner, I., Musso, H., 1983. New naturally occurring amino acids. Angew. Chem. Int. Ed. Engl. 22 (22), 816–828.

Wang, D., Farhana, A., 2021. Biochemistry, RNA Structure. StatPearls Publishing LLC, Tampa, Florida, United States.

Ward, P.D., Brownlee, D., 2000. Rare Earth: Why Complex Life is Uncommon in the Universe. Copernicus Books, New York.

Waters, E., Hohn, M.J., Ahel, I., Graham, D.E., Adams, M.D., Barnstead, M., Beeson, K.Y., Bibbs, L., Bolanos, R., Keller, M., Kretz, K., Lin, X., Mathur, E., Ni, J., Podar, M., Richardson, T., Sutton, G.G., Simon, M., Soll, D., Stetter, K.O., Short, J.M., Noordewier, M., 2003. The genome of *Nanoarchaeum equitans*: insights into early archaeal evolution and derived parasitism. Proc. Natl. Acad. Sci. USA 100, 12984–12988.

Watson, J.D., Crick, F.H.C., 1953a. Molecular structure of nucleic acids: a structure for deoxyribose nucleic acid. Nature 171 (4356), 737–738.

Watson, J.D., Crick, F.H.C., 1953b. Genetical implications of the structure of deoxyribose nucleic acid. Nature 171 (1953), 964–967.

Watson, J.D., Crick, F.H., 1953. The structure of DNA. Cold Spring Harb. Symp. Quant. Biol. 18, 123–131. doi:10.1101/SQB.1953.018.01.020.

Watson, J.D., 1963. Involvement of RNA in the synthesis of proteins: the ordered interaction of three classes of RNA controls the assembly of amino acids into proteins. Science 140 (3562), 17–26.

Weiner, J., 1995. Evolution made visible. Science 267, 30–33.

White, S.H., 1994. The evolution of proteins from random amino acid sequences: II. Evidence from the statistical distributions of the lengths of modern protein sequences. J. Mol. Evol. 38, 383–394.

Wilde, S.A., Valley, J.W., Peck, W.H., Graham, C.M., 2001. Evidence from detrital zircons for the existence of continental crust and oceans on the Earth 4.4 Gyr ago. Nature 409, 175–178.

Woese, C.R., Kandler, O., Wheelis, M.L., 1990. Towards a natural system of organisms: proposal for the domains archaea, bacteria, and eucarya. PNAS 87, 4576–4579.

Wolf, G., 2003. Friedrich Miescher: the man who discovered DNA. Chem. Heritage 21 (10-11), 37–41.

Wolfenden, R., Lewis Jr., C.A., Yuan, Y., Carter Jr., C.W., 2015. Temperature dependence of amino acid hydrophobicities. Proc. Natl. Acad. Sci.. doi:10.1073/pnas.1507565112.

Yamaguchi, A., Yamamoto, M., Takai, K., Ishii, T., Hashimoto, K., Nakamura, R., 2014. Electrochemical CO_2 reduction by Ni-containing iron sulfides: how is CO_2 electrochemically reduced at bisulfide-bearing deep-sea hydrothermal precipitates? Electrochim Acta 141, 311–318.

Zahnle, K., Arndt, N., Cockell, C., Halliday, A., Nisbet, E., Selsis, F., Sleep, N.H., 2007. Emergence of a habitable planet. Space Sci. Rev. 129, 35–78.

Bibliography

Allison, A.C., 2004. Two lessons from the interface of genetics and medicine. Genetics 166, 1591–1599.

Astrachan, L., Volkin, E., 1958. Properties of ribonucleic acid turnover in T2-infected *Escherichia coli*. Biochim. Biophys. Acta 29, 536–544.

Davis, T.H., 2004. Biography of Vernon M Ingram. PNAS 101/40, 14323–14325.

de Chadarevian, S., 1999. Protein sequencing and the making of molecular genetics. Trends Biochem Sci. *(TIBS)* 24(5), 203–206. doi: 10.1016/s0968-0004(99)01360-2.

Eck, R.V., 1961. Non-randomness in amino-acid "alleles". Nature 191, 1284–1285.

Hoagland, M.B., Zamecnik, P.C., Stephenson, M.L., 1957. Intermediate reactions in protein biosynthesis. Biochim. Biophys. Acta 24, 215–216.

Ingram, V.M., 1957. Gene mutations in human haemoglobin: The chemical difference between normal and sickle-cell haemglobulin. Nature 180, 326–328.

Ingram, V.M., 1958. Abnormal human haemoglobins: The comparison of normal human and sickle-cell haemoglobins by 'fingerprinting'. Biochim. Biophys. Acta 28, 539–545.

Ingram, V.M., 1989. A case of sickle-cell anaemia. Biochim. Biophys. Acta 1000, 147–150.

Ingram, V.M., 2001. The Sickle-cell Story, 2001.

Watson, J.D., Crick, F.H.C., 1953. Molecular structure of nucleic acids. Nature 4356/171, 737–738.

Weatherall, D.J., 2010. Molecular medicine, the road to the better integration of medical sciences in the twenty-first century. R. Soc. Notes Rec. 64/1, S5–S15.

Chapter 4

Biosignatures—The prime targets in the search for life beyond Earth

4.1 Life

A defining characteristic of terrestrial life is its metabolic versatility and adaptability and it is reasonable to expect that this is universal. Different physiologies operate for carbon acquisition, the garnering of energy and the storage and processing of information. As well as having a range of metabolisms, organisms build biomass suited to specific physical environments, habitats and their ecological imperatives. This overall "metabolic diversity" manifests itself in an enormous variety of accompanying product molecules (i.e., natural products). The whole field of organic chemistry grew from their study and now provides tools to link metabolism (i.e., physiology) to the occurrence of biomarkers specific to, and diagnostic for, particular kinds of metabolism (Summons et al., 2008a).

Another characteristic of living thing, also likely to be pervasive, is that an enormous diversity of large molecules is built from a relatively small subset of universal precursors. These include the four bases of DNA, 20 amino acids of proteins and two kinds of lipid building blocks. The building blocks of lipids are one glycerol molecule and at least one fatty acid, with a maximum of three fatty acids. Glycerol is a sugar alcohol with three OH groups. It acts as a backbone for fatty acids to bond. Fatty acids are made up of a long hydrocarbon with carboxyl group, which is represented as COOH. When the total number of carbons exceeds four, carboxylic acid is known as a fatty acid. To become a lipid, a combination must occur between fatty acids and glycerol.

A protein is a naturally occurring, extremely complex substance that consists of amino acid residues joined by peptide bonds. Proteins are large biomolecules and macromolecules that comprise one or more long chains of amino acid residues. Proteins perform a vast array of functions within organisms, including catalyzing metabolic reactions, DNA replication, responding to stimuli, providing structure to cells and organisms, and transporting molecules from one location to another. Proteins differ from one another primarily in their sequence of amino acids, which is dictated by the nucleotide sequence of their genes, and which usually results in protein folding into a specific 3D structure that determines its activity (Fig. 4.1). Proteins are present in all living organisms and include many essential biological compounds such as enzymes, hormones, and antibodies. In terms of in biochemistry, a protein is an organic compound that contains carbon, hydrogen, oxygen, nitrogen, and in some cases sulfur and are made up of small molecules called amino acids (note that all amino acids contain an amine group [NH_2] and a carboxyl group [COOH]). Proteins make up the tissue of living organisms. There are 20 different types of amino acids commonly found in the proteins of living things. All amino acid molecules have the same basic structure; with only the side chain differing from one molecule to another. Of the 20 types of amino acids in proteins, each type of amino acid possesses different chemical properties. A protein molecule is made from a long chain of these amino acids, each linked to its neighbor through a covalent peptide bond. Note that a peptide is a compound consisting of two or more amino acids linked in a chain, the carboxyl group of each amino acid being joined to the amino group of the next by a bond of the type –OC–NH–.

Given that a defining characteristic of terrestrial life is its metabolic versatility and adaptability, it is reasonable to expect that life is universal. The quest for extraterrestrial life is an old and inevitable ambition of modern science. In its practical implementation, the core challenge is to reduce this grand vision into bite-sized pieces of research. In the search for organic remnants of past life, it is enormously helpful to have a paradigm to guide exploration. This begins with assessing habitability (Grotzinger, 2014). Interestingly, the hitherto observed manifestations of life are as multitude as the number of organisms and life-forms themselves. For example, among different organisms, different physiologies operate for carbon acquisition, the garnering of energy and the storage and processing of information.

Our search for signs of life in the extraterrestrial worlds is based on the assumption that extraterrestrial life shares

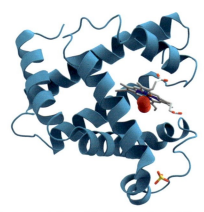

FIG. 4.1 A representation of the 3D structure of the protein myoglobin showing turquoise α-helices. This protein was the first to have its structure solved by X-ray crystallography. Toward the right-center among the coils, a prosthetic group called a heme group (shown in gray) with a bound oxygen molecule (red). *(Source: https://commons.wikimedia.org/wiki/File:Myoglobin.png) Author: AzaToth Copyright info: I, the copyright holder of this work, release this work into the public domain. This applies worldwide.*

fundamental characteristics with life on Earth, in that it requires liquid water as a solvent and has a carbon-based chemistry (see, e.g., Brack, 1993; Des Marais et al., 2002). According to Kaltenegger et al. (2010), life on the base of a different chemistry is not yet seriously considered by astrobiologists (although there is an emerging school of thought that life on the base of a different chemistry could be possible on Titan) because the vast possible life-forms produce signatures in their atmosphere that are so far unknown. Therefore, astrobiologists assume that extraterrestrial life is similar to life on Earth in its use of the same input and output gases, which it exists out of thermodynamic equilibrium (Lovelock, 1975).

The fact that evidence of complex microbial life is to be found on Earth stretching back some 3800 million years—to within about 700 million years of its formation—shows that it emerged very early and that once established, it proved to be adept at evolving and adapting to a wide range of changing physical and chemical conditions. The discovery of an active biosphere several hundred meters below the Earth's surface is witness to this incredible diversity and adaptability of life.

An understanding of the co-evolution of life and its physical and chemical settings relies on the ability to decipher evidence of life preserved in the rock and sediment records. While any phenomenon produced by life (modern or ancient) can be considered a biosignature (cf. Steele et al., 2005), the main challenge in ancient and extraterrestrial life detection is determining whether the phenomenon (or suite of phenomena) can be uniquely attributed to life. Taphonomic changes (i.e., changes taking place during the process of fossilization) inevitably alter the chemical and structural fidelity of all biosignatures over time.

A characteristic of living thing, also likely to be pervasive, is that an enormous diversity of large molecules that have been found to occur on Earth is built from a relatively small subset of universal precursors. According to McKay (2016), one would know a great deal about life on Earth even if all that one knew were that life utilized the following: These include a specific five nucleotide bases including the four bases of DNA, 20 amino acids of proteins, two kinds of lipid building blocks, and polysaccharides made up mostly from a few specific simple sugars. The information in DNA is stored as a code made up of four chemical bases: adenine (A), guanine (G), cytosine (C), and thymine (T). Human DNA consists of about 3 billion bases, and more than 99% of those bases are the same in all people. The premier example of biochemical selectivity is chirality. This property is most well known in the amino acids used in proteins. Life on Earth uses only the L version of amino acids in proteins, not the mirror image, the D version.

The conundrum (a confusing and difficult problem or question) of finding a "definition" for life can be sidestepped by asking how people actually identify examples of life, and using this as the basis for life detection strategies. Bains (2014) has illustrated how astrobiologists actually select things that are living, from things that are not living, with a simple exercise, and used this as the starting point to develop four characteristics that underlie their decisions: (1) highly distinctive structure (physical or chemical), (2) dynamic behavior (physical or chemical), (3) multiple instances of life forming a "natural group" and that the structural and dynamic characteristics of the group are independent of the details of the substrate on which life is growing. Bains (2014) showed that all these characteristics derive the role of a code in the dynamic maintenance and propagation of life; and argued that evolution is neither a useful nor a practical way of identifying life; and concluded with some specific ways that these general categories of the observable properties of life can be detected.

Molecules associated with living organisms are organic. These include nucleic acids, fats, sugars, proteins, enzymes and hydrocarbon fuels. All organic molecules contain carbon, nearly all contain hydrogen, and many also contain oxygen. According to Claudi (2017), "The chemicals produced by life on Earth are hundreds of thousands (estimated from plant natural products, microbial natural products, and marine natural products), but only a subset of hundreds are volatile enough to enter the atmosphere at more than trace concentrations. Among these, only a few handfuls accumulate to high enough levels to be remotely detectable for astronomical purposes and defined as biosignatures. Apart from oxygen, these biosignature (and bioindicator) gases range from highly abundant gases in Earth's atmosphere that are either already existing or predominantly produced by geochemical or photochemical processes (N_2, Ar, CO_2, and H_2O) to those that are relatively abundant and attributed to life (N_2O, CH_4, and H_2S)."

Earth is the only planet yet in the universe known to possess life. Life, as we know it, is based on carbon chemistry operating in an aqueous environment. Living organisms process chemicals, make copies of themselves, are autonomous and evolve in concert with the environment. All these characteristics are driven by, and operate through, carbon chemistry. The carbon chemistry of living systems is an exact branch of science.

Earth boasts several million described species, living in habitats ranging from the bottom of the deepest ocean to a few miles up into the atmosphere. Scientists think far more species remain that have yet to be described to science. Scientists have yet to precisely nail down exactly how our primitive ancestors first showed up on Earth, although most believe that a chemical soup on the planet gave rise to the building blocks of living organisms. Another theory suggests that life first evolved on the nearby planet Mars, which could once have been habitable, then traveled to Earth on meteorites hurled from the Red Planet by impacts from other space rocks.

4.2 Use of fossil lipids for life-detection

In biology and biochemistry, a lipid is a macro biomolecule of biological origin that is soluble in nonpolar solvents ("lipids," doi:10.1351/goldbook. L03571). Nonpolar solvents are typically hydrocarbons used to dissolve other naturally occurring hydrocarbon lipid molecules that do not (or do not easily) dissolve in water. Lipid comprises a group of naturally occurring molecules that include fats, waxes, sterols, fat-soluble vitamins (such as vitamins A, D, E, and K), monoglycerides, diglycerides, triglycerides, and phospholipids. The functions of lipids include storing energy, signaling, and acting as structural components of cell membranes (Fahy et al., 2009; Subramaniam et al., 2011).

The importance of lipids arises from the fact that fossil lipids are used for life-detection (Eigenbrode, 2008) so that they are recognized as universal biomarkers of life, including extraterrestrial life (see Georgiou and Deamer, 2014). Eigenbrode (2008) has described the importance of lipids thus, "The geological preservation of lipids from the cell membranes of organisms bestows a precious record of ancient life, especially for the Precambrian eon (>542 million years ago) when Earth life was largely microbial. All organisms produce lipids that, if the lipids survive oxidative degradation, become molecular fossils entrained with information on biological diversity, environmental conditions, and postdepositional alteration history." As with most biosignatures, the molecular fossil record that is indigenous (of the same place) and syngenetic (of the same age) to host rocks can be compromised by the introduction from and reaction with foreign or younger materials (e.g., petroleum or endolithic life). Deciphering the resulting complex pool of organic signals requires tests for the provenance of molecular fossils and the overall quality of the geobiological record itself. Eigenbrode (2008) has reviewed the basis for the very existence of a molecular fossil record from lipid biochemistry to mechanisms of organic-matter preservation and geochemical alteration. In that review, a systematic approach to resolving the provenance of molecular fossils and historical qualities of the record has been presented in a case study of an early Earth record, clearly demonstrating the value of geological context and the integration of independent geobiological parameters, which are critical to the detection and understanding of the ecological processes responsible for records.

In a study, Brocks (1999) found that molecular fossils of biological lipids are preserved in 2700-million-year-old shales (soft finely stratified sedimentary rock that formed from consolidated mud or clay and can be split easily into fragile plates) from the Pilbara Craton, Australia. Molecular fossils provide an indication of the type of microbes represented by the fossils. For example, 2α-methylhopanes are characteristic of cyanobacteria. In Brocks (1999)'s sequential extraction of adjacent samples it was observed that these hydrocarbon biomarkers are indigenous and identical to the Archean shales, greatly extending the known geological range of such molecules.

4.3 Biosignatures

Thomas-Keprta et al. (2002) defined a biosignature to be a physical and/or chemical marker of life that does not occur through random, stochastic interactions or through directed human intervention. Biosignature is something (some clue) that provides scientific evidence of past or present life. In a broad sense, biosignature (also called *chemical fossil,* or *molecular fossil*) is any substance or phenomenon that provides unambiguous signal/scientific evidence of past or present life. In a broad sense, a biosignature is any substance—such as an element, isotope, molecule, a chemical compound, or cellular component—or phenomenon that indicates or suggests the presence of a biological process indicative of past or present life. In the literature, "biomarker" is occasionally used as synonymous for "biosignature," but traditionally the term refers only to organic molecules. Therefore, "molecular biosignatures" has been suggested as a more specific synonym (Summons et al., 2008b).

Life exploits the specificity inherent in the spatial, that is, the three-dimensional qualities of organic chemicals (stereo-chemistry). These characteristics then lead to some readily identifiable and measurable generic attributes that would be diagnostic as biosignatures.

Microbes have the potential to become fossils and serve as "biosignatures" for researchers seeking signs of early life in the remnants of ancient dried-up lakes, geologic formations, or subsurface environments on the worlds beyond Earth. In the hunt for life in an extraterrestrial environment, discovery of a biosignature will help astronomers and astrobiologists learn to recognize that a celestial body or its atmosphere has been influenced by life.

4.3.1 Biosignatures of microorganisms

According to Cady et al. (2003), biosignatures of microorganisms fall into one of three categories: (1) bona fide cellular fossils (cf. Cady, 2002) and carbonaceous remnants of microbial cells and their extra-cellular matrices (Cady, 2001), that is, cellular fossils that preserve organic remains of microbes and their extracellular matrices; (2) bioinfluenced (microbially influenced) fabrics and sedimentary structures (for details, see Westall, 2008, 2012; Davies et al., 2016), which provide a macroscale imprint of the presence of microbial biofilms that can be more readily identified, for example, laminated stromatolites (Grotzinger and Knoll, 1999); and (3) organic chemofossils preserved in the geological record (for details, see Parnell et al., 2007; Summons et al., 2008b) that can be either primary biomolecules or diagenetically altered compounds known as biomarkers.

4.3.2 Chemical biosignatures

Chemical biosignatures may consist of biomarkers or nonrandom mixtures of lipids or other compounds that could not have been assembled by abiogenetic processes. The most obvious of these are repeating structural subunits, reflecting the biosynthetic assemblage of lipids. Biosignatures may eventually be a critical component in recognizing extraterrestrial life. Taxon-specific biomarkers (TSBs) are recognized as complex biosynthetic molecules (biomarkers) that are utilized or synthesized by one specific group of organisms. Thus, they are signature compounds with demonstrated efficacy for tracing evolutionary history and the early development of the Earth's biosphere (Moldowan and Jacobson, 2010).

A dramatic secular change (i.e., a linear long-term trend) in Earth's history has been the impact of life on the diversity of minerals. Hazen et al. (2008) estimated that, over the past 4.56 billion years, the number of different minerals has increased from about a dozen to more than 4300 known types. Although only a small number of these can be considered biominerals (i.e., chemical fossils), even their use as definitive evidence for life remains, justifiably, problematic (e.g., Golden et al., 2004; Altermann et al., 2009). In any case, biology has altered the relative abundances of different groups of minerals (most notably since the oxidation of the atmosphere), expanded the range of compositional variants (which include solid solutions and minor and trace element variations), affected the kinetics of mineral formation (hence the degree of ordering and density/type of defect microstructure), and created distinctive morphological habits. The emergence of key microbial metabolic innovations throughout Earth's history and development of bioskeletons during the Phanerozoic Eon (the current geologic eon in the geologic time scale, covering 541 million years to the present, and the one during which abundant animal and plant life has existed) resulted in the biomineralization mechanisms that persist today (cf. Ehrlich and Newman, 2009). Collectively, the diverse metabolic and behavioral activities of life have created and sustained chemical gradients in geochemically dynamic environments, which has led to an abundance of mineral varieties distributed over scales that range from microenvironments around, and within, cells to regional-sized terrains.

Earthly life is based on carbon chemistry operating in an aqueous environment. Summons et al. (2008b) have addressed the potential for detecting extra-terrestrial life using molecular organic chemistry although they have also acknowledged that other kinds of chemistry may constitute "life." All living organisms on Earth process chemicals, make copies of themselves, are autonomous, and evolve in concert with the environment. All these characteristics are driven by, and operate through, carbon chemistry which is an exactly known branch of science, and we have detailed knowledge of the basic metabolic and reproductive machinery of living organisms. We can recognize the residual biochemicals long after life has expired and otherwise lost most life-defining features. Carbon chemistry affords a robust tool for identifying extant (living, or recently dead) and extinct life on today's Earth, in the sedimentary record of the early Earth and, potentially, throughout the Universe. Note that in biology, extinction is the termination of an organism or of a group of organisms (taxon), normally a species. The moment of extinction is generally considered to be the death of the last individual of the species, although the capacity to breed and recover may have been lost before this point.

Among several potential isotopic biomarkers (e.g., H, N, O, and S), perhaps the most important biomarker—signature of life—common to all life as we know it at present include the preferential isotopic fractionation of carbon. It may be noted that atoms are the basic unit of natural elements such as carbon. Although the atom is the smallest unit having the properties of its element, atoms are composed of subatomic particles, among which protons, neutrons, and electrons are the most important. All atoms of a given element have the same number of protons in the nucleus of the atoms, but some atoms may have more neutrons than other atoms of the same element. These different atomic forms are called isotopes of the element, and in nature there is usually a mixture of isotopes. The basic (purely inorganic) carbon is ^{12}C.

The heavy isotope ^{13}C is incorporated into a plant during photosynthesis. However, the animal/creature that consumed the plant, and might have later got fossilized with the passage of time, preserves the ^{13}C signature.

In the context of identification of biosignatures, it is important to differentiate between "inorganic" carbon and "organic" carbon. Inorganic carbon is the carbon extracted from ores and minerals, as opposed to organic carbon found in nature through plants and living things, and their fossilized forms. The ratio of carbon-13 (denoted as ^{13}C) to carbon-12 (denoted as ^{12}C) isotopes in plant tissues is a good indicator of the photosynthetic pathway in many terrestrial plants.

Organic carbon is an element, which is so important to living organisms. Organic carbon is found in organic compounds which are found mainly in living things. Organic compounds make up the cells and other structures of organisms. They also carry out life processes. Carbon possesses unique properties, such as an exceptional ability to bind with a wide variety of other elements, including hydrogen, oxygen, and nitrogen. Interestingly, carbon atoms can also form stable bonds with other carbon atoms (forming single, double, or even triple bonds). Claudi (2017) described the importance of carbon thus, "considering life like a phenomenon with a non–zero probability to happen as soon as the conditions are satisfied it appears, we have to consider all the conditions that maximize this probability. In this framework, carbon is the only atom with which it is possible to form molecules with a total number of atoms up to 13 (e.g., $HC_{11}N$) thus allowing the formation of very complex molecules. Carbon is also very easy to oxidize (CO_2) and reduce (CH_2)."

Note that each atom of every carbon element (irrespective of "inorganic" carbon or "organic" carbon) has six protons and six electrons (i.e., every carbon atom's atomic number is 6). Most carbon atoms have six neutrons. Because the atomic weight is largely determined by the mass of the protons plus neutrons, this isotope is called carbon-12 (denoted as ^{12}C), where the number 12 denotes the atomic weight (i.e., each atom having six neutrons and six protons, thus having an atomic mass 12); it is the most common form of carbon, accounting for about 99% of the carbon in nature. Most of the remaining 1% of carbon consists of atoms of the isotope carbon-13 (denoted as ^{13}C), with 7 rather than 6 neutrons; thus this isotope is heavier than ^{12}C. A third isotope, ^{14}C (i.e., each atom having eight neutrons and six protons in its nucleus, thus having an atomic mass 14), is present in the environment in minute quantities but is not very stable. Consequently, it may decay spontaneously, giving off radiation, and thus it is a radioactive isotope. Note that ^{12}C and ^{13}C are considered to be stable isotopes. The carbon element naturally exists as a mix of primarily three isotopes. Although a total of 15 isotopes of carbon are known, ^{13}C and ^{12}C isotopes are more important in biosignature investigations. Note that although the isotopes of an element contain equal numbers of protons (and therefore an equal number of electrons) but different numbers of neutrons in their nuclei, and hence differ in relative atomic mass, the chemical properties of its isotopes remain the same because the chemical properties of any atom is determined by the number of electrons in the outer shell of the atom.

Stable isotopes are measured on a mass spectrometer, an instrument that separates atoms on the basis of their mass differences. Initially, the plant material is combusted, and the carbon dioxide given off is analyzed by the mass spectrometer for the ratio of ^{13}C to ^{12}C isotopes. This ratio is compared to the ratio of ^{13}C to ^{12}C in an internationally accepted standard and is expressed as the difference between the sample and the standard minus 1. This number is multiplied by 1000 and expressed as a "per mil." In plant matter, this number is always negative. The more negative the ratio of ^{13}C to ^{12}C, the less ^{13}C is present (Ehleringer and Osmond, 1989).

The $^{13}C/^{12}C$ ratio in sedimentary organic carbon, as compared to the same isotopic ratio for coexisting carbonate carbon, provides a remarkably consistent signal of biological carbon fixation which has been shown to stretch back over some 3500 million years of the fossil record of microbial life on Earth. The origin of this fractionation in favor of ^{12}C in the biological pathway of carbon fixation derives for the most part from a kinetic isotope effect imposed on the first CO_2-fixing enzymatic carboxylation reaction of the photosynthetic pathway.

One of the aims of the European Space Agency's ExoMars mission (launched on 14 March 2016 and arrived on 19 October 2016) was to seek evidence of organic compounds of biological and nonbiological origin at the Martian surface. One of the instruments in the Pasteur payload is a Life Marker Chip (Sims et al., 2012) that utilizes an immunoassay approach to detect specific organic molecules or classes of molecules. The scientific aims of the instrument are to detect organics in the form of biomarkers that might be associated with extinct life, extant life or abiotic sources of organics. Therefore, it is necessary to define and prioritize specific molecular targets for antibody development. Target compounds have been selected to represent meteoritic input, fossil organic matter, extant (i.e., living, or recently dead) organic matter, and contamination. Once organic molecules are detected on Mars, further information is likely to derive from the detailed distribution of compounds rather than from single molecular identification. This will include concentration gradients beneath the surface and gradients from generic to specific compounds. The choice of biomarkers is informed by terrestrial biology but is wide ranging, and nonterrestrial biology may be evident from unexpected molecular distributions. One of the most important requirements is to sample where irradiation and oxidation are minimized, either by drilling or by using naturally excavated exposures. Analyzing regolith samples will allow for the search of both extant and fossil biomarkers, but sequential extraction would be required to optimize the analysis of each of these in turn (Parnell et al., 2007).

The detection of biomarkers plays a central role in our effort to establish whether there is, or was, life beyond Earth. In a review, Röling et al. (2015) addressed the importance of considering mineralogy in relation to the selection of locations and biomarker detection methodologies with characteristics most promising for exploration. They also addressed relevant mineral-biomarker and mineral-microbe interactions. The local mineralogy on a particular planet reflects its past and current environmental conditions and allows a habitability assessment by comparison with life under extreme conditions on Earth. The type of mineral significantly influences the potential abundances and types of biomarkers and microorganisms containing these biomarkers. The strong adsorptive power of some minerals aids in the preservation of biomarkers and may have been important in the origin of life. On the other hand, this strong adsorption as well as oxidizing properties of minerals can interfere with efficient extraction and detection of biomarkers. Differences in mechanisms of adsorption and in properties of minerals and biomarkers suggest that it will be difficult to design a single extraction procedure for a wide range of biomarkers. While on Mars, samples can be used for direct detection of biomarkers such as nucleic acids, amino acids, and lipids, on other planetary bodies remote spectrometric detection of biosignatures has to be relied upon. Claudi (2017) considered different classes of spectral features coming from life metabolic reactions working on Earth. The interpretation of spectral signatures of photosynthesis can also be affected by local mineralogy. Röling et al. (2015) made an earnest attempt for identifying current gaps in our knowledge and indicated how they may be filled to improve the chances of detecting biomarkers on Mars and beyond.

4.3.3 Morphological biosignatures

The rock record contains a rich variety of sedimentary surface textures on siliciclastic sandstone, siltstone and mudstone bedding planes. In contrast to chemical sedimentary rocks, which are composed of mineral crystals, "detrital" or "clastic" sedimentary rocks are composed of rock fragments. Fig. 4.2 diagrammatically illustrates plausible Mars habitable environments suitable for hosting detrital organics on altered volcanics (phyllosilicates), chemotrophs near hydrothermal vents, and lithotrophs on volcanic particles during the Early- to Middle-Noachian eon. Note that the *Noachian* is an early time period on the planet Mars characterized by high rates of meteorite and asteroid impacts and the possible presence of abundant surface water. The absolute age of the Noachian period is uncertain but probably corresponds to the periods of 4100–3700 million years ago (Tanaka, 1986), during the interval known as the Late Heavy Bombardment (Carr and Head, 2010). Many of the large impact basins on Earth's Moon and Mars formed at this time. The Noachian Period is roughly equivalent to the Earth's Hadean and early Archean eons when the first life forms likely arose (Abramov and Mojzsis, 2009).

Some of the submarine hydrothermal settings may have been active long enough to witness the appearance of life (especially in the case of long-term hydrothermal activity); others could have hosted already flourishing microorganisms. Fig. 4.3 shows examples of some typical submarine hydrothermal vents. Under wet environments, microorganisms have a tendency to form biofilms (thin but robust

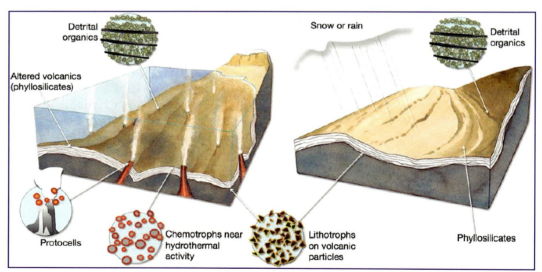

FIG. 4.2 **Diagrammatic illustration of plausible Mars habitable environments suitable for hosting detrital organics on altered volcanics (phyllosilicates), chemotrophs near hydrothermal vents, and lithotrophs on volcanic particles during the Early- to Middle-Noachian eon.** *(Source: Vago, J.L. et al. (2017), Habitability on early Mars and the search for biosignatures with the ExoMars Rover, Astrobiology, 17(6-7): 471-510) (Reference # 97) © Jorge L. Vago et al., 2017; Published by Mary Ann Liebert, Inc. This Open Access article is distributed under the terms of the Creative Commons License (http://creativecommons.org/licenses/by/4.0), which unrestricted use, distribution, and reproduction in any medium, provided the original work is properly credited.*

FIG. 4.3 (*Top left*) Deep-sea hydrothermal vents and the fauna dispersion around them found on the East Scotia Ridge in the Southern Ocean, hosting high-temperature black smokers up to 382.8°C and diffuse venting. *Source: Rogers et al. (2012), The discovery of new deep-sea hydrothermal vent communities in the Southern Ocean and implications for biogeography,* PLoS Biol. *10(1), e1001234, http://dx.doi.org/10.1371/journal.pbio.1001234. Copyright info:* © *2012 Rogers et al. This is an open-access article distributed under the terms of the Creative Commons Attribution License, which permits unrestricted use, distribution, and reproduction in any medium, provided the original author and source are credited.* (*top right*) A hydrothermal vent field along the Arctic Mid-Ocean Ridge, close to where a group of microorganisms called "Loki" (described as a missing link connecting the simple cells that first populated Earth to the complex cellular life that emerged roughly 2 BYA) was found in marine sediments. *(From: Image courtesy Lawrence Berkeley National Laboratory).* (*Bottom left*) High-temperature black smoker hydrothermal vents at South Cleft, Juan de Fuca Ridge, NE Pacific. Jul. 2000 ROPOS dive 542. *Source: NOAA PMEL EOI Program; http://www.pmel.noaa.gov/eoi/gallery/smoker-images.html. Copyright info: Unless otherwise noted (copyrighted material for example), information presented on the EOI Program World Wide Web site is considered public information and may be distributed freely.* (*Bottom right*) Hydrothermal vent found at Kairei Field at Rodriguez Triple Junction (RTJ) in the Indian Ocean east of Madagascar (for sake of clarity the portion of ROV Kaiko measuring the temperature of the vent fluid has been removed from the photograph). *Source: Hashimoto, et al. (2001), First hydrothermal vent communities from the Indian Ocean discovered,* Zool. Sci., *18 (5), 717–721 [Published By: Zoological Society of Japan].*

layers of viscous secretions adhering to solid surfaces and containing highly organized microbial communities of bacteria and other microorganisms), which are able to affect the accumulation of "detrital" sediments. Biofilms trapping by detrital sediments (sedimentary rocks that form from transported solid material) can result in the formation of macroscopic edifices (large structures). In recent years, an increasing number of these sedimentary surface textures have been attributed to surficial microbial mats at the time of deposition, resulting in the classification of these sedimentary structures as Microbially Induced Sedimentary Structures, or MISS (Noffke and Awramik, 2011; Noffke et al., 2013; Davies et al., 2016). According to Davies et al. (2016), MISS exhibits a pan-environmental and almost continuous record since the Archean Eon (4–2.5 billion years ago). During the Archean, the Earth's crust had cooled enough to allow the formation of continents and life started to form. Although the individual cells would be too small to distinguish, their compressed microbial colonies and biofilms would be much larger. Microbes accumulated on such sedimentary edifices are amenable to be recognized and studied with cameras and close-up imagers.

The earliest known fossils, found in Precambrian rocks (i.e., rocks in the earliest part of Earth's history), and still being formed in lagoons in Australasia, are known as stromatolites. Note that stromatolites are dome-shaped structures consisting of alternating layers of carbonate or silicate sediments and fossilized algal mats, produced over geologic time by the trapping, binding, or precipitating of sediment by groups of microorganisms, primarily cyanobacteria. Stromatolites are commonly defined as "organo-sedimentary structures built by microbes." Interestingly, stromatolites constitute essential beacons of information, recording snapshots of microbial communities and environments

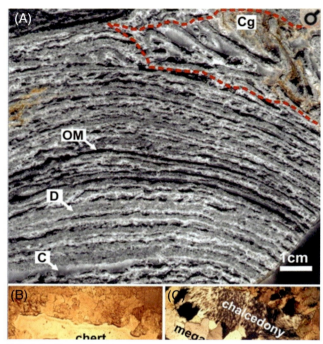

FIG. 4.4 **Encrusting/domical stromatolite fabrics.** (A) Polished slab, showing irregular wrinkly laminar fabric consisting of dolomite (D), chert (C), and organic laminae (OM). *(Source: Allwood et al. (2009), Controls on development and diversity of Early Archean stromatolites, Proc. Natl. Acad. Sci. USA., 106:9548–9555). Source: National Academy of Sciences of the United States of America.*

throughout Earth's history (Allwood et al., 2006, 2009, 2013). The 3430-million-year-old Strelley Pool Chert (SPC) (Pilbara Craton, Australia) is a sedimentary rock formation containing laminated structures of probable biological origin (stromatolites). Fig. 4.4 illustrates encrusting/domical stromatolite fabrics.

Allwood et al. (2006) managed to overcome many obstacles to interpretation of the fossils in the SPC because of the broad extent, excellent preservation and morphological variety of its stromatolitic outcrops—which provide comprehensive paleontological information on a scale exceeding other rocks of such age. They argued that the diversity, complexity and environmental associations of the stromatolites describe patterns that reflect the presence of organisms.

In another study, Allwood et al. (2009) found that the ≈3450-million-year-old Strelley Pool Formation in Western Australia contains a reef-like assembly of laminated sedimentary accretion structures (stromatolites) that have macroscale characteristics suggestive of biological influence. They found that by examining spatiotemporal changes in microscale characteristics it is possible not only to recognize the presence of probable microbial mats during stromatolite development, but also to infer aspects of the biological inputs to stromatolite morphogenesis. They suggested that the persistence of an inferred biological signal through changing environmental circumstances and stromatolite types indicates that benthic microbial populations adapted to shifting environmental conditions in early oceans.

MISS and stromatolites stem from the cooperative action of microbes; in particular phototrophs produce large amounts of extracellular polymeric substances in the biofilm. If the biofilm covers a large enough area experiencing similar conditions, often multiple organo-sedimentary structures can arise in regularly spaced groups (see Vago et al., 2017). Recognizing that abiotic processes may also create morphologically similar features, Davies et al. (2016) suggested a practical methodology for classifying such structures in the geological record. This methodology claims to offer a scientific approach that appreciates the low likelihood of conclusively identifying microbial structures from visual appearance alone, informing the search for true MISS in Earth's geological record and potentially on other planetary bodies such as Mars. In any case, Davies et al. (2016) noted that MISS should be treated with caution as they are a subset of "sedimentary surface textures" that include those of abiotic origin.

4.4 Serpentinization—implications for the search for biosignatures

Serpentinization is a process that takes place at depth in the seafloor, leading to significant changes in topography, focused microseismic activity as a result of continuous cracking, and significant heat generation (Fyfe, 1974; Allen and Seyfried, 2004). It may be noted that chemically, serpentinization is the hydration of the olivine $((Mg,Fe)_2SiO_4)$ and orthopyroxene minerals $\{(Mg,Fe,Ca)(Mg,Fe,Al)(Si,Al)_2O_6\}$ that mainly constitute the upper mantle. Hydrogen production by serpentinization proceeds most effectively in ultramafic rocks (i.e., silica-poor rocks). Serpentinization involves the hydrolysis and transformation of primary ferromagnesian minerals such as olivine and pyroxenes $((Mg,Fe)SiO_3)$ to produce hydrogen-rich fluids and a variety of secondary minerals over a wide range of environmental conditions. In the process of serpentinization, the low-temperature (150–400°C) hydrolysis and transformation of ultramafic rocks produces hydrogen gas (H_2). The hydrogen gas thus produced reacts with simple oxidized carbon compounds, such as carbon dioxide (CO_2) and carbon monoxide (CO), under reducing conditions to release methane gas (CH_4) and other organic molecules through Fischer–Tropsch-type synthesis. The continual and elevated production of hydrogen is capable of reducing carbon, thus initiating an inorganic pathway to produce organic compounds, particularly methane (Holm et al., 2015).

According to Konn et al. (2015), "field data associated with laboratory experiments (McCollom and Seewald, 2001) show that methane (CH_4), together with hydrogen (H_2), is a major emission by-product of serpentinization.

CH$_4$ out-gassing associated with intense H$_2$ output has consistently been observed in ultramafic settings such as in peridotites (a dense, coarse-grained plutonic rock (an igneous rock, such as granite, of holocrystalline granular texture regarded as having solidified at considerable depth below the surface) containing a large amount of olivine, believed to be the main constituent of the Earth's mantle) of the Oman ophiolite (Neal and Stanger, 1983), in serpentinized rocks of the Zambales ophiolite in Philippines (Abrajano et al., 1988), in serpentine seamount drilled during ODP Leg 125 in the Mariana Fore-arc (Haggerty, 1991), and along MORs (Charlou et al., 1991, 1993a, 1996a, 2010; Charlou and Donval, 1993; Kelley, 1996; Früh-Green and Kelley, 1998)." Note that MOR denotes mid-ocean ridge. Widespread serpentinization, which is considered to be responsible for CH$_4$ generation on the early Earth (Etiope and Sherwood Lollar, 2013; Holm et al., 2015), played a fundamental greenhouse role on early Earth.

Production of H$_2$ and H$_2$-dependent CH$_4$ in serpentinization systems has received significant interdisciplinary interest, especially with regard to the abiotic synthesis of organic compounds and the origins and maintenance of life in Earth's lithosphere (the rocky outer part of the Earth) and elsewhere in the Universe. Holm et al. (2015) have reviewed serpentinization with an emphasis on the formation of H$_2$ and CH$_4$ within the context of the mineralogy, temperature/pressure, and fluid/gas chemistry present in planetary environments. According to them, whether deep in Earth's interior or in Kuiper Belt Objects in space in the outer periphery of the Solar System, serpentinization is a feasible process to invoke as a means of producing astrobiologically indispensable H$_2$ capable of reducing carbon to organic compounds.

As just indicated, serpentinization is a weathering process in which ultramafic rocks react with water, generating a range of products, including serpentine and other minerals, in addition to H$_2$ and low-molecular-weight hydrocarbons that are capable of sustaining microbial life. Lipid biomarker analyses of serpentinite-hosted ecosystems hold promise as tools for investigating microbial activity in ancient Earth environments and other terrestrial planets such as Mars because lipids have the potential for longer term preservation relative to DNA, proteins, and other more labile (i.e., easily broken down or displaced) organic molecules. In a study, Newman et al. (2020) reported the first lipid biomarker record of microbial activity in the mantle section of the Samail Ophiolite, in the Sultanate of Oman, a site undergoing active serpentinization. They detected isoprenoidal (archaeal) and branched (bacterial) glycerol di-alkyl glycerol tetraether (GDGT) lipids, including those with 0–3 cyclopentane moieties (note that in organic chemistry, a moiety is a part of a molecule that is given a name because it is identified as a part of other molecules as well), and crenarchaeol (an isoprenoidal GDGT containing four cyclopentane and one cyclohexane moieties), as well as mono-ether lipids and fatty acids indicative of sulfate-reducing bacteria. Comparison of their geochemical data and 16S rRNA (a component of the prokaryotic ribosome 30S subunit) data from the Samail Ophiolite with those from other serpentinite-hosted sites identifies the existence of a common core serpentinization microbiome. In light of these findings, Newman et al. (2020) have also discussed the preservation potential of serpentinite lipid biomarker assemblages on Earth and Mars. According to Newman et al. (2020), continuing investigations of the Samail Ophiolite and other terrestrial analogues will enhance our understanding of microbial habitability and diversity in serpentinite-hosted environments on Earth and elsewhere in the Solar System.

4.5 Biosignatures versus bioindicators

Gases that are, or could be, indicative of biological processes, but can also be produced abiotically, or biosignature gases that can be transformed into other chemical species abiotically are considered bioindicators (see Kiang et al., 2007). As the name suggests, bioindicators are indicative of biological processes but can also be produced abiotically. It is their quantities, and detection along with other atmospheric species, and in a certain context (for instance the properties of the star and the planet) that points toward a biological origin (Kaltenegger et al., 2010). For example, on Earth, ozone gas (O$_3$) is photochemically produced by oxygen (O$_2$). Oxygen in high abundance is a promising bioindicator.

Through some examples Claudi (2017) explains the difference between biosignature and bioindicator thus, "Methane is considered a "chemical energy gradient" biosignature gas because it is generated by methanogen bacteria at the seafloor reducing the CO$_2$ available in the seawater due to the mixing with the atmosphere. In order to reduce the CO$_2$ these bacteria use the H$_2$ released by hot water coming from rocks (serpentinization). On the other hand, methane is also released volcanically from hydrothermal systems. Most of the methane found in the present atmosphere of Earth has a biological origin but a small fraction is produced abiotically in hydrothermal systems. Therefore, the detection of CH$_4$ alone cannot be considered as a sign of life, though its detection in an oxygen–rich atmosphere could be an indication of the presence of a biosphere."

4.6 Life and biomarkers

While discussing extant and extinct life, an important topic that warrants consideration is biomarker. The below sections briefly address life and biomarkers.

4.6.1 Biomarker

In recognizing that certain distinctive compounds isolable (i.e., capable of being isolated) from living systems had related fossil derivatives, organic geochemists coined the term "biological marker compound" or "biomarker" (Eglinton et al., 1964; Eglinton and Calvin, 1967) to describe them. In this terminology, biomarkers are defined as metabolites (intermediate or end products of metabolism) or biochemicals (i.e., compounds/substances that contain carbon and are found in living things; e.g., carbohydrates, proteins, lipids [fats], and nucleic acids) by which we can identify particular kinds of living organisms and their defining physiologies together with the molecular fossil derivatives through which we identify defunct counterparts. Thus, "biological marker," "biomarker," and "molecular biosignature" are used interchangeably.

We can recognize the residual biochemicals long after life has expired and otherwise lost most life-defining features. Carbon chemistry provides a tool for identifying extant and extinct life on Earth and, potentially, throughout the Universe. In recognizing that certain distinctive compounds isolable from living systems had related fossil derivatives, organic geochemists coined the term biological marker compound or biomarker (e.g., Eglinton et al., 1964) to describe them. In this terminology, biomarkers are metabolites or biochemicals by which we can identify particular kinds of living organisms as well as the molecular fossil derivatives by which we identify defunct counterparts. The terms biomarker and molecular biosignature are synonymous.

Detecting the presence of the ensemble of primary biomolecules (such as amino acids, proteins, nucleic acids, carbohydrates, some pigments, and intermediary metabolites) associated with active microorganisms in high abundance would be diagnostic of extant life, but unfortunately, they degrade quickly once microbes die (Vago et al., 2017). Lipids and other structural biopolymers, however, are biologically essential components (e.g., of cell membranes) known to be stable for billions of years when buried (Brocks, 1999; Georgiou and Deamer, 2014). According to Vago et al. (2017), it is the recalcitrant (i.e., capable of long persistence in the environment) hydrocarbon backbone that is responsible for the high-preservation potential of lipid-derived biomarkers relative to that of other biomolecules (Eigenbrode, 2008).

4.6.2 The search for life on Mars

Evidence of biogenic activity on any celestial body has profound scientific implications for our understanding of the origin of life on Earth and the presence and diversity of life within the Cosmos. Most of Earth's biological matter exists in the form of carbonaceous (i.e., containing carbon or its compounds) macromolecules stored within layered sedimentary rocks, which are orders of magnitude more abundant than that in living beings (Summons et al., 2011). If life existed on an ancient celestial body (e.g., Mars), its remains may also have accumulated in extensive, organic-rich sedimentary deposits (Vago et al., 2017).

The search for life on Mars is now zeroing in on indirect markers of life rather than extant life, as few reputable researchers think it reasonable that life-as-we-know-it could be found at the surface of Mars. According to Banfield et al. (2004), "if life ever existed, or still exists, on Mars, its record is likely to be found in minerals formed by, or in association with, microorganisms. An important concept regarding interpretation of the mineralogical record for evidence of life is that, broadly defined, life perturbs disequilibria that arise due to kinetic barriers and can impart unexpected structure to an abiotic system. Many features of minerals and mineral assemblages may serve as biosignatures even if life does not have a familiar terrestrial chemical basis. Biological impacts on minerals and mineral assemblages may be direct or indirect. Crystalline or amorphous biominerals, an important category of mineralogical biosignatures, precipitate under direct cellular control as part of the life cycle of the organism (shells, tests, phytoliths) or indirectly when cell surface layers provide sites for heterogeneous nucleation." Note that biominerals are natural composite materials based upon biomolecules (such as proteins) and minerals produced by living organisms via processes known as biomineralization, yielding materials with impressive mechanical properties such as bones, shells and teeth (Weiner and Lowenstam, 1989; Sigel et al., 2008; Cuif et al., 2011). Banfield et al. (2004) explain that biominerals also form indirectly as byproducts of metabolism due to changing mineral solubility. Mineralogical biosignatures include distinctive mineral surface structures or chemistry that arise when dissolution and/or crystal growth kinetics are influenced by metabolic by-products. Mineral assemblages themselves may be diagnostic of the prior activity of organisms where barriers to precipitation or dissolution of specific phases have been overcome.

Critical to resolving the question of whether life exists, or existed, on Mars is knowing how to distinguish biologically induced structure and organization patterns from inorganic phenomena and inorganic self-organization. This task assumes special significance when it is acknowledged that the majority of, and perhaps the only, material to be returned from Mars will be mineralogical (Banfield et al., 2004).

Life exploits the specificity inherent in the spatial, that is, the three-dimensional qualities of organic chemicals (stereochemistry). These characteristics then lead to some readily identifiable and measurable generic attributes that would be diagnostic as biosignatures. Measurable attributes of molecular biosignatures include (Summons et al., 2008b):

- Enantiomeric excess
- Diastereoisomeric preference
- Structural isomer preference
- Repeating constitutional subunits or atomic ratios
- Systematic isotopic ordering at molecular and intramolecular levels
- Uneven distribution patterns or clusters (e.g., C-number, concentration, $\delta^{13}C$) of structurally related compounds.

Cady et al. (2004) have provided a rationale for the advances in instrumentation and understanding needed to assess claims of ancient and extraterrestrial life made on the basis of morphological biosignatures. Morphological biosignatures consist of bona fide microbial fossils as well as microbially influenced sedimentary structures. To be recognized as evidence of life, microbial fossils must contain chemical and structural attributes uniquely indicative of microbial cells or cellular or extracellular processes. When combined with various research strategies, high-resolution instruments can reveal such attributes and elucidate how morphological fossils form and become altered, thereby improving the ability to recognize them in the geological record on Earth or other planets. Also, before fossilized microbially influenced sedimentary structures can provide evidence of life, criteria to distinguish their biogenic from nonbiogenic attributes must be established. This topic can be advanced by developing process-based models. A database of images and spectroscopic data that distinguish the suite of bona fide morphological biosignatures from their abiotic mimics will avoid detection of false-positives for life. The use of high-resolution imaging and spectroscopic instruments, in conjunction with an improved knowledge base of the attributes that demonstrate life, will maximize our ability to recognize and assess the biogenicity of extraterrestrial and ancient terrestrial life. Summons et al. (2008b) have detailed knowledge of the basic metabolic and reproductive machinery of living organisms.

4.6.3 A potential biomarker identified on Venus

As already indicated, a biomarker, or biological marker is a measurable indicator of some biological state or condition. In September 2020, it was announced that phosphine, a potential biomarker, had been detected in the atmosphere of Venus. There is no known abiotic source of phosphine on Venus that could explain the presence of the substance in the concentrations detected (Greaves et al., 2020). However, the detection of phosphine was suggested to be a possible false positive in October 2020 (Thompson, 2021). In January 2021 further research attributed the spectroscopic signal to that of sulfur dioxide (Lincowski et al., 2021), although later research has refuted the sulfur dioxide claim and confirmed the existence of phosphine (Greaves et al., 2021).

4.7 Identification of biosignature in Antarctic rocks

Deciphering biosignatures and evidence of microbial activity in ancient rock remains a central challenge in geobiology studies (e.g., Rosing, 1999; Fedo and Whitehouse, 2002; Lepland et al., 2005). When surface-derived, organic-bearing rocks are transferred to Earth's shallow interior, the combination of burial and deformation can ultimately make it impossible to distinguish a biological signature in relict carbonaceous compounds (Brasier et al., 2005). Consequently, determination of the degree of metamorphism beyond which life's signatures are no longer recognizable in ancient carbon is a research topic of considerable interest in early Earth and extra-terrestrial studies (e.g., Schopf and Kudryavtsev, 2009; Glikson et al., 2008; Oehler et al., 2009). For example, a recent approach focuses on the applicability and limitations of using Raman spectroscopy to characterize evidence of ancient life (e.g., van Zuilen et al., 2002; Schopf et al., 2005; McKeegan et al., 2007; Schiffbauer et al., 2007; van Zuilen et al., 2007; Marshall et al., 2007).

The cold, dry ecosystems of Antarctica are known to harbor traces left behind by microbial activity within certain types of rocks. Some indirect biomarkers of cryptoendolithic activity (i.e., ability of microorganisms to colonize the empty spaces or pores inside a rock) in the Antarctic cold desert zone have been described in the literature; primarily geophysical and geochemical bioweathering patterns macroscopically observed in sandstone rock. Wierzchos et al. (2003) have shown that in this extreme environment, minerals are biologically transformed, and as a result, Fe-rich diagenetic minerals (i.e., recombination or rearrangement of chemical or mineral constituents, resulting in a new product) in the form of iron hydroxide nanocrystals and biogenic clays are deposited around chasmoendolithic hyphae (long, branching filamentous structured organisms, growing in the interior of rocks, inhabiting fissures and cracks of the rock) and bacterial cells. Thus, when microbial life decays, these characteristic neocrystalized minerals act as distinct biomarkers of previous endolithic activity. Wierzchos et al. (2003) have suggested that the ability to recognize these traces may have potential astrobiological implications because the Antarctic Ross Desert is considered a terrestrial analogue of a possible ecosystem on early Mars.

The cryptoendolithic habitat of the Antarctic Dry Valleys has been considered a good analogy for past Martian ecosystems, if life arose on the planet. Yet cryptoendoliths are thought to favor the colonization of rocks that have a pre-existing porous structure, for example, sandstones. This may weaken their significance as exact analogues of potential rock-colonizing organisms on Mars, given our current understanding of the dominant volcanic nature of Martian geology. However, the production of oxalic acid, by these lichen-dominated communities, and its weathering potential indicate that it could be an aid in rock colonization, enabling endoliths (organisms that live inside rocks or in pores between mineral grains) to inhabit a wider variety of rock types. Utilizing ICP-AES and scanning electron microscope techniques, a study carried out by Blackhurst et al. (2005) investigated elemental and mineralogical compositions within colonized and uncolonized layers in individual sandstone samples. This was in order to determine if the weathering of mineral phases within the colonized layers causes an increase in the amount of pore space available for colonization. The results show that colonized layers are more weathered than uncolonized, deeper portions of the rock substrate. It was found that layers within uncolonized samples have uniform compositions. Differences between the colonized and uncolonized layers also occur to varying extents within colonized rocks of different mineralogical maturities. The results confirm that cryptoendoliths modify their habitat through the production of oxalic acid and suggest that over time this directly increases the porosity of their inhabited layer, potentially increasing the biomass it can support (Blackhurst et al., 2005).

In terms of possible detection of signs of extra-terrestrial life, a meteorite that reached Earth from Mars (the meteorite was found to have been encased in Antarctic ice for 13,000 years and lay exposed on the surface for ~500 years (McKay et al., 1996)) attracted considerable interest and attention not only among planetary scientists and astrobiologists but also among administrators and the general public. For example, quoting from Vago et al. (2017), "In 1996, David McKay and his colleagues published the first description of possible microbial signatures in extraterrestrial rocks, namely in a meteorite from Mars called ALH84001 (McKay et al., 1996). The subject was so delicate that President Bill Clinton announced the news in a press conference (Statement, 1996). The ensuing interest in the scientific world spurred a huge increase in astrobiological research and, in particular, the study of biosignatures."

4.8 Existence of biosignatures under diverse environmental conditions

Evidence of biogenic activity on any celestial body beyond Earth has profound scientific implications for our understanding of the origin of life on Earth and the presence and diversity of life within the Cosmos. Biosignatures (also sometimes called biomarkers) have been identified under diverse environmental conditions. For example, Parenteau et al. (2014) reported the production and early preservation of lipid biosignatures in iron hot-springs. It may be noted that the bicarbonate-buffered anoxic vent waters at Chocolate Pots hot springs in Yellowstone National Park are 51–54°C, pH 5.5–6.0, and are very high in dissolved Fe(II) at 5.8–5.9 mg/L. The aqueous Fe(II) is oxidized by a combination of biotic and abiotic mechanisms and precipitated as primary siliceous nanophase iron oxyhydroxides (ferrihydrite). Four distinct prokaryotic photosynthetic microbial mat types grow on top of these iron deposits. Lipids were used to characterize the community composition of the microbial mats, link source organisms to geologically significant biomarkers, and investigate how iron mineralization degrades the lipid signature of the community. The phospholipid and glycolipid fatty acid profiles of the highest-temperature mats indicate that they are dominated by cyanobacteria and green nonsulfur filamentous anoxygenic phototrophs (FAPs). It has been found that diagnostic lipid biomarkers of the cyanobacteria include midchain branched mono- and dimethyl-alkanes and, most notably, 2-methylbacteriohopanepolyol. Diagnostic lipid biomarkers of the FAPs (*Chloroflexus* and *Roseiflexus* spp.) include wax esters and a long-chain tri-unsaturated alkene. Surprisingly, the lipid biomarkers resisted the earliest stages of microbial degradation and diagenesis to survive in the iron oxides beneath the mats. Understanding the potential of particular sedimentary environments to capture and preserve fossil biosignatures is of vital importance in the selection of the best landing sites for future astrobiological missions to Mars. Parenteau et al. (2014)'s study explored the nature of organic degradation processes in moderately thermal Fe(II)-rich groundwater springs—environmental conditions that have been previously identified as highly relevant for Mars exploration.

Effective chemical identification of biosignatures requires access to well-preserved organic molecules. Because the Martian atmosphere is more tenuous (very weak) than Earth's, three important physical agents reach the surface of Mars with adverse effects for the long-term preservation of biomarkers (see Vago et al., 2017): (1) The UV radiation dose is higher than on our planet and will quickly damage exposed organisms or biomolecules. (2) UV-induced photochemistry is responsible for the production of reactive oxidant species that, when activated, can also destroy chemical biosignatures. The diffusion of oxidants into the subsurface is not well characterized and constitutes an important measurement that the mission must perform. Finally, (3) ionizing radiation penetrates into the uppermost meters of the planet's subsurface. This causes a slow degradation process that, operating over many millions of years, can alter organic molecules beyond the detection sensitivity of analytical instruments. Radiation effects are depth-dependent: the material closer to the surface is exposed to higher doses than that buried deeper.

Regarding the search for molecular biosignatures on Mars, the site must provide easy access to locations with reduced radiation accumulation in the subsurface. The presence of fine-grained sediments in units of recent exposure age would be very desirable (on Earth, organic molecules are better preserved in fine-grained sediments—which are more resistant to the penetration of biologically damaging agents, such as oxidants—than they are in porous, coarse materials). Young craters can provide the means to access deeper sediments, and studies on Earth suggest that fossil biomarkers can survive moderate impact heating (Parnell and Lindgren, 2006). In addition, impact-related hydrothermal fractures might have contributed to creating habitats for microbial life in the past.

4.9 Characterizing extraterrestrial biospheres through absorption features in their spectra

To characterize a planet's atmosphere and its potential habitability, planetary scientists look for absorption features in the emergent and transmission spectrum of the planet. The spectrum of the planet can contain signatures of atmospheric species, what creates its spectral fingerprint (Kaltenegger et al., 2010). According to them, both spectral regions (thermal infrared and visible to near-infrared) contain the signature of atmospheric gases that may indicate habitable conditions and, possibly, the presence of a biosphere: carbon dioxide (CO_2), water (H_2O), ozone (O_3), methane (CH_4), and nitrous oxide (N_2O) in the thermal infrared; and H_2O, O_3, oxygen (O_2), CH_4 and CO_2 in the visible to near-infrared. The presence or absence of these spectral features (detected individually or collectively) will indicate similarities or differences with the atmospheres of terrestrial planets, and its astrobiological potential. On Earth, some atmospheric species exhibiting noticeable spectral features in the planet's spectrum result directly or indirectly from biological activity: the main molecules are O_2, O_3, CH_4, and N_2O. In addition, CO_2 and H_2O are important as greenhouse gases in a planet's atmosphere and potential sources for high O_2 concentration from photosynthesis (Kaltenegger et al., 2010). According to Kaltenegger et al. (2010), it is relatively straightforward to remotely ascertain that Earth is a habitable planet, replete with oceans, a greenhouse atmosphere, global geochemical cycles, and life—if one has data with arbitrarily high signal-to-noise and spatial and spectral resolution. It is heartening that O_2, O_3, and CH_4 are good biomarker candidates that can be detected by a low-resolution (resolution <50) spectrograph.

It is well known that photosynthesis captures the carbon in CO_2 into biomass, releasing oxygen that, nowadays, is 20% by volume of the Earth's atmosphere. Léger et al. (1993) consider that oxygen is a robust biosignature. According to Walker (1977), less than 1ppm (parts per million) of atmospheric O_2 comes from abiotic processes. Because of the presence (on Earth) of a high quantity of a reactive gas such as O_2 with a short atmospheric lifetime, Owen (1980) suggested that searching for O_2 as a tracer of life is quite reasonable. According to Kasting and Catling (2003), without continual replenishment by photosynthesis in plants and bacteria, O_2 would be ten orders of magnitude less than present today in the Earth's atmosphere. According to Claudi (2017) any observer seeing oxygen in Earth's spectrum would know that some nongeological chemistry must be producing it.

Note that if the presence of biogenic gases such as O_2/O_3 + CH_4 may imply the presence of a massive and active biosphere, their absence does not imply the absence of life. Life existed on Earth before the interplay between oxygenic photosynthesis and carbon cycling produced an oxygen-rich atmosphere (Kaltenegger et al., 2010). It may be noted that oxygenic photosynthesis is a noncyclic photosynthetic electron chain where the initial electron donor is water and, as a consequence, molecular oxygen is liberated as a by-product. The use of water as an electron donor requires a photosynthetic apparatus with two reaction centers.

On Earth, photosynthetic organisms are responsible for the production of nearly all of the oxygen in the atmosphere. However, according to Kaltenegger et al. (2010), in many regions of the Earth, and particularly where surface conditions are extreme, for example in hot and cold deserts, photosynthetic organisms can be driven into and under substrates where light is still sufficient for photosynthesis. These communities exhibit no detectable surface spectral signature. The same is true of the assemblages of photosynthetic organisms at more than a few meters depth in water bodies. These communities are widespread and dominate local photosynthetic productivity (Kaltenegger et al., 2010). Such world is *terrestrial analogues* (see Section 7.1 for details) that would not exhibit a biological surface feature in the disc-averaged spectrum (see Cockell et al., 2009).

According to Domagal-Goldman et al. (2014), in any case, we have to pay attention to the retrieved spectra in order to unambiguously identify O_2 and other atmospheric gases, which would set the environmental context in which we are confident that the O_2 is not being geochemically or photochemically generated. The O_2 has a strong visible signature at 0.76 µm (the Frauenhofer A-band) observable with low/medium resolution and also a faint one at 0.69 µm, visible with high resolution.

It has been suggested that a very promising biosignature could be the dimethyl sulfide (DMS), a secondary metabolic by-product, but in that case, on Earth, we have to compel with small atmospheric concentration.

During its mission, flying toward Jupiter, the probe Galileo took ultra-violet (UV), visible, and near-infra red (NIR) spectra of Earth. Sagan et al. (1993) analyzed a spectrum of the Earth taken by the Galileo probe, searching for signatures of life. They found a large amount of oxygen (O_2) and the simultaneous presence of methane (CH_4) traces. They concluded that this co-presence of oxygen and methane is strongly suggestive of biology (read, life). It has been suggested that rocky planets, named super-Earths (i.e., planets of less than 10 Earth masses—the so-called 10 M_{Earth}), seems to be the really putative loci where to search for alien life.

From astrobiological perspective, Saturn's moon Titan is an important astronomical body in our Solar System. Owing to the presence of Titan's thick nitrogen–methane atmosphere, remote spectroscopic measurements are restricted to a discrete number of atmospheric "windows," where scattering and/or absorption are reduced (Lorenz and Mitton, 2002). For example, the VIMS (Visible and Infrared Mapping Spectrometer) instrument on Cassini has only been able to image the surface of Titan at seven atmospheric windows at wavelengths ranging between 0.94 and 5 mm (Brown et al., 2004). High spectral resolution is crucial for the remote identification of surface materials. Neish et al. (2018) have recalled that "The observation of key spectral features has provided essential information about the composition of many planetary bodies, including the identification of water ice on the Galilean satellites (Pilcher et al., 1972), carbonates on Mars (Ehlmann et al., 2008), and hydroxyl on the Moon (Clark, 2009; Pieters et al., 2009; Sunshine et al., 2009)." Neish et al. (2018) consider that with only a handful of wavelengths available for surface analysis, similar identifications may be impossible on Titan. The observations are further complicated by residual absorption and scattering within Titan's atmospheric windows. According to Neish et al. (2018), these limitations are present for both orbital and aerial platforms (such as a balloon or aircraft). This is true even though the amount of atmospheric absorption between an aerial platform and the surface is much less than that encountered by an orbiter.

According to Claudi (2017), the importance of super-Earth is that they are rocky planets with a rigid interface between the interior and the atmosphere and if they stay at the right distance from their host star, they could retain liquid water on their surface. This condition generally defines a planet as habitable (because all life on Earth, including extremophiles, requires liquid water to survive for long). Surface liquid water requires a suitable surface temperature. The surface temperature of planets with thin atmospheres is determined by the fraction of flux reaching the surface of the planet from the host star. Planets at the "right distance" are planets on orbits between the hot planet and the cold planet orbits, in particular they are in the well-known "Habitable Zone (HZ)" (Kasting et al., 1993; Selsis et al., 2007; Koppaparu et al., 2013), or the zone around a star where a rocky planet with a thin atmosphere, heated by its star, may have liquid water on its surface.

According to Kaltenegger et al. (2010), "For small planets (<0.5 M_{Earth}) close to inner edge of the habitable zone (<0.93 AU from the present Sun), there is a risk of abiotic oxygen detection, but this risk becomes negligible for big planets further away from their star. The fact that, on the Earth, oxygen and indirectly ozone are by-products of the biological activity does not mean that life is the only process able to enrich an atmosphere with these compounds." The question of the abiotic synthesis of biomarkers is crucial, but only very few studies have been dedicated to it (Leger et al. 1993; Rosenqvist and Chassefiere, 1995; Selsis et al., 2002; Lagrange et al., 2009).

4.10 Means of studying biosignatures

Enough confusion and controversy has surrounded the interpretation of various stromatolites, commonly defined as "organo-sedimentary structures built by microbes" (e.g., Walter, 1976; Grotzinger and Knoll, 1999; Riding and Awramik, 2000) that one can begin from a default assumption that it is not currently possible to distinguish biogenic from abiogenic structures reliably by simple visual examination. Yet the ability to separate biogenic from abiogenic structures is vital to interpreting any such layered structures that may be encountered on other planets. Further, in extraterrestrial environments, the interpretation of laminated structures will be even more challenging than it is on Earth (Cady et al., 2003). In the near term, layered extraterrestrial samples will be observed remotely with data collected in situ by a rover or lander. Key contextual clues may be overlooked or simply unobserved due to the limited capabilities of the rover or lander. Even if humans are present to observe the samples, analogies with terrestrial environments may be tenuous or nonexistent. Therefore, given the potential utility of such a tool, Wagstaff and Corsetti (2010) sought to test whether the assumption of inseparability holds when applying objective, quantitative analysis methods: is it possible to draw a confident dividing line between the two kinds of structures?

4.10.1 Identification of stromatolites using portable network graphics analysis of layered structures captured in digital images

As indicated earlier, stromatolites are commonly defined as "organo-sedimentary structures built by microbes." Information-theoretic methods have been used for detecting structural microbial biosignatures (biomarkers). In a study, Wagstaff and Corsetti (2010) carried out an evaluation of these methods. The first observations of extraterrestrial environments will most likely be in the form of digital images. Given an image of a rock that contains layered structures, is it possible to determine whether the layers were created by life (biogenic)? While conclusive judgments about biogenicity are unlikely to be made solely on the basis of image features, an initial assessment of the importance of a given sample can inform decisions about follow-up searches for other types of possible biosignatures (e.g., isotopic or chemical analysis). In a study, Wagstaff and Corsetti (2010) evaluated several quantitative measures that capture the degree of complexity in visible structures, in terms of compressibility (to detect order) and the entropy (spread) of their intensity distributions. Computing complexity inside a sliding analysis window yields a map of each of these features that indicates how they vary spatially across the sample. Wagstaff and Corsetti (2010) conducted experiments on both biogenic and abiogenic terrestrial stromatolites and on laminated structures found on Mars. The degree to which each feature separated biogenic from abiogenic samples (separability) was assessed quantitatively. None of the techniques provided a consistent, statistically significant distinction between all biogenic and abiogenic samples. Portable Network Graphics (PNG) analysis is a tool that has been employed for detecting structural microbial biosignatures. The PNG format has built-in transparency, but can also display higher color depths, which translates into millions of colors. This tool has been employed in Wagstaff and Corsetti (2010)'s studies, in which the PNG compression ratio provided the strongest distinction (2.80 in standard deviation units) and could inform future techniques. Increasing the analysis window size or the magnification level, or both, improved the separability of the samples. Finally, data from all four Mars samples plotted well outside the biogenic field suggested by the PNG analyses, although Wagstaff and Corsetti (2010) caution against a direct comparison of terrestrial stromatolites and Martian nonstromatolites.

4.10.2 Characterization of molecular biosignatures using time-of-flight secondary ion mass spectrometry

Microorganisms have acted as a strong geobiological force for the majority of Earth's history because of the multiple element cycles they drive, as well as their role in the formation and transformation of organic matter, gases, minerals, and rocks. Such biosignatures may consist of morphological, chemical, and isotopic traces and should ideally be applicable to both recent and ancient materials, thus providing a direct linkage between living systems and those from the geological past. Organic molecules have a long-standing history as molecular biosignatures [e.g., porphyrins (Treibs, 1934)]. Such biomarkers (Eglinton et al., 1964) carry information not only about the biological sources of organic matter and the particular metabolic pathways involved, but also about the physicochemical conditions during formation, deposition, and burial.

4.10.2.1 Advantage of time-of-flight secondary ion mass spectrometry over other techniques for obtaining biomarker information

Thiel and Sjövall (2011) recall that "Biomarkers are commonly identified using coupled gas chromatography/mass spectrometry (GC/MS) and coupled liquid chromatography/mass spectrometry (LC/MS). These techniques are well-established and widely distributed tools, but when studying small (submillimeter) geobiological systems, their spatial resolution is limited because large sample amounts must be used for extraction (for biomass, usually milligram amounts; for sediments, gram amounts). Homogenization (the process of making things uniform or similar) at the beginning of the extraction and chromatography (a technique for the separation of a mixture by passing it in solution or suspension through a medium in which the components move at different rates) procedure destroys all information on the distribution or localization of source-specific biomarkers. Thus, it has been infeasible so far to directly couple biomarker information to microscopic, histological, or petrological sample structures, as can be done for inorganic compounds with other techniques commonly used in geobiology [e.g., scanning electron microscopy (SEM), energy dispersive X-ray spectroscopy (EDX)]. Such coupling, however, is essential for obtaining biomarker information from typical structures of geobiological interest, for example, finely zoned mineral precipitates, biofilms and microbial mats, or even individual cells. Similarly, because the integrity of the sample cannot be maintained, it is difficult with extract-based techniques to recognize biomarkers that are abundant in only a small part of the sample."

Secondary-ion mass spectrometry (SIMS) is a technique used to analyze the composition of solid surfaces and thin films by sputtering the surface of the specimen with a focused primary ion beam and collecting and analyzing ejected secondary ions. The application of SIMS has tremendous value for the field of geobiology, representing a powerful tool for identifying the specific role of microorganisms in biogeochemical cycles. SIMS performs isotope and elemental analysis at microscale, enabling the investigation of the physiology of individual microbes within complex communities. Additionally, through the study of isotopic or chemical characteristics that are common in both living and ancient microbial communities, SIMS allows for direct comparisons of potential biosignatures derived from extant microbial cells and their fossil equivalents (Orphan and House, 2009).

Unlike SIMS in which molecular information about the sample has been inaccessible owing to the molecular damage caused by the high-energy ions colliding with the sample surface during analysis, Time-of-flight secondary ion mass spectrometry (ToF-SIMS) allows for analysis in the static SIMS regime, in which the analysis is completed before the incoming ions have significantly damaged the sample surface (Vickerman, 2001). ToF-SIMS is a technique designed to analyze the composition and spatial distribution of molecules and chemical structures on surfaces. These capabilities have generated much interest in its use in geobiology, in particular for the characterization of organic biomarkers (molecular biosignatures) at the microscopic level.

Thiel and Sjövall (2011) have discussed the strengths, weaknesses, and potential of ToF-SIMS for biomarker analyses with a focus on applications in geobiology, including biogeochemistry, organic geochemistry, geomicrobiology, and paleo-biology. After describing the analytical principles of ToF-SIMS, they have discussed issues of biomarker spectral formation and interpretation. Furthermore, key applications of ToF-SIMS to soft (microbial matter, cells), hard (microbial mineral precipitates), and liquid (petroleum) samples relevant in geobiology have been reviewed. They also examined the potential of ToF-SIMS in biomarker research and the current limitations and obstacles for which further development would be beneficial to the field.

By using mass spectrometry to analyze the atomic and molecular secondary ions that are emitted from a solid surface when bombarded with ions, one obtains detailed information about the chemical composition of the surface. A time-of-flight mass spectrometer is especially suitable for the analysis of secondary ions because of its high transmission, high mass resolution, and ability to detect ions of different masses simultaneously. By using a finely focused primary ion beam it is also possible to analyze microareas and generate surface images with a lateral resolution of 0.1 μm or less. Static time-of-flight secondary ion mass spectrometry (TOF-SIMS) allows monolayer imaging and local analysis of monolayers with high sensitivity, a wide mass range, high mass resolution, and high lateral resolution (Benninghoven, 1994).

Spectacular improvements in terms of sensitivity of TOF–SIMS imaging methods have allowed many biological applications to be successfully tested (Brunelle and Laprévote, 2009). A ToF-SIMS mass spectrum contains a wide range of molecularly specific ion peaks, which allows for detailed chemical analysis, for example, the identification of specific organic compounds (Benninghoven, 1994). Improved capabilities to obtain organic molecular information by the use of cluster primary ion sources has stimulated a growing use of ToF-SIMS for analysis of biological systems (Sjövall et al., 2004; Touboul et al., 2005b; Winograd and Garrison, 2010). Indeed, full mass spectra from organic (and inorganic) surfaces can be obtained with ToF-SIMS at a resolution of less than 1 μm, which is within the size range of typical microbial cells (Thiel and Sjövall, 2011). The concept of obtaining integrated molecular information at the microscopic level nevertheless offers interesting prospects for those involved in the study of geobiological systems. These prospects encompass, but are not limited to, the below benefits/advantages (Thiel and Sjövall, 2011):

- Localization of particular biomarkers/metabolites within cells, consortia, and complex microbial systems (e.g., biofilms, microbial mats)
- Definition of molecular biosignature patterns from unculturable microorganisms
- Assignment of orphan biomarkers, that is, biomarkers with no known biological source
- Study of the structure and composition of lipid membranes
- Definition of the distribution and function of microbial EPS (extracellular polymeric substances) in environmental samples
- Study of heterogeneous—inorganic and organic—bio- and geomaterials (biomineralization)
- Identification of the relevant microbial protagonists and ancient biogeochemical pathways in sedimentary rocks; and
- Exploration of biomarkers for early life on Earth in fluid inclusions.

Thiel and Sjövall (2011) have critically examined the current limitations of static ToF-SIMS in the study of molecular biosignatures and how these limits are being overcome.

With regard to the "microbial protagonists" indicated above, Johnson (1998) described that microorganisms may alter the availability of metals in the environment by a variety of means. While eukaryotic microorganisms can bring about certain of the various metal transformations, the major protagonists of metal cycling in aquatic ecosystems are prokaryotes (bacteria and archaea). The cycling of metals generally involves phase changes (usually between soluble and insoluble forms) which have major impacts on the biological availability of metals.

4.10.2.2 Generic scheme of a time-of-flight secondary ion mass spectrometry experiment

A generic scheme of a ToF-SIMS experiment is shown in Fig. 4.5. A beam of high-energy ions (primary ions) bombards the sample surface, resulting in the emission of secondary ions from the outermost molecular layers of the sample. Analysis of these ions with respect to mass yields a mass spectrum that contains chemical information about the sample surface. During the measurement, the primary ion beam is scanned over a selected analysis area, and individual mass spectra are recorded from each raster point within this area. The acquired data can then be used to produce (a) ion images, which show the signal intensity of selected secondary ions across the analysis area, and (b) mass spectra from selected regions of interest.

Thiel and Sjövall (2011) explain that to obtain molecularly specific secondary ion peaks, the data acquisition must be completed before the primary ions damage a significant fraction of the surface. This typically occurs at a primary ion dose density (PIDD) of 10^{12}–10^{13} ions/cm^2, which is termed the static limit and defines two important SIMS

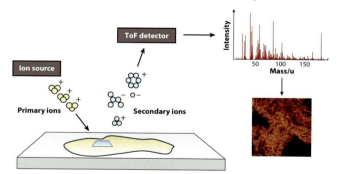

FIG. 4.5 **General representation of a time-of-flight secondary ion mass spectrometry (ToF-SIMS) experiment.** The bombardment of a surface by high-energy primary ions generated in an ion source results in the emission of secondary ions from the uppermost molecular layers of the sample surface. The secondary ions are analyzed with respect to mass-to-charge ratio (m/z; in ToF-SIMS typically $z = 1$ and $m/z = $ Mass[u]) in a ToF analyzer, producing a mass spectrum that provides chemical information about the sample surface. By scanning the primary ion beam over a preselected analysis area on the sample surface, separate spectra are acquired from each point, and ion images showing the lateral distribution of selected ion signals can be produced. *Source: Thiel, V., and P. Sjövall (2011), Using time-of-flight secondary ion mass spectrometry to study biomarkers, Annu. Rev. Earth Planet. Sci., 39(1): 125-156). Publisher: Annual Reviews, California https://www.annualreviews.org/doi/full/10.1146/annurev-earth-040610-133525 No info available on Copyright. Email IDs of the Source authors: Volker Thiel: vthiel@gwdg.de Peter Sjövall: peter.sjovall@sp.se.*

analysis regimes. In static SIMS, the analysis is completed at a PIDD below the static limit and is therefore capable of providing molecularly specific secondary ions, whereas in dynamic SIMS, the analysis is conducted at a higher PIDD on a molecularly damaged and continuously eroding surface, and only atomic or diatomic secondary ions contribute to the recorded mass spectra (see sidebar, Dynamic SIMS). This situation is, however, somewhat relaxed when C_{60}^+ (buckminsterfullerene) ions are used, as is discussed below.

Thiel and Sjövall (2011) recall that dynamic SIMS has been used for many exciting geobiological applications. Examples include studies of microbial metabolic pathways using isotopically labeled substrates (Dekas et al., 2009; Orphan et al., 2009), and of biosignatures in kerogen particles enclosed in Precambrian rocks (Oehler et al., 2010; Rasmussen et al., 2008; Wacey et al. 2008). For recent reviews on the use of dynamic SIMS/NanoSIMS in the study of biological materials, see Boxer et al. (2009), Herrmann et al. (2007), and Orphan and House (2009).

According to Thiel and Sjövall (2011), an important advantage of ToF-SIMS is the ability to perform SIMS analysis in the static regime. This capability arises because in ToF-SIMS, the primary ions are supplied as very short pulses, and for each pulse, all secondary ions (subject to analyzer transmission and detector efficiency factors) are detected and recorded, that is, a secondary ion spectrum is produced by a minimum number of primary ion collisions. A key advantage of ToF-SIMS compared with other SIMS techniques is therefore its capability to produce mass spectra with detailed molecular information (Thiel and Sjövall, 2011).

The area of analysis: During analysis, the primary ion beam is scanned over a predetermined area of the sample surface, acquiring data from a selected number of points (typically 128 × 128 or 256 × 256 pixels). This is primarily done to allow for the acquisition of secondary ion images, but it also has the important effect of spreading out the primary ion dose to a larger area and thereby avoiding analysis of damaged areas (Thiel and Sjövall, 2011).

Time-of-flight analysis: After collision of the primary ion with the sample surface, the emitted secondary ions are extracted and accelerated by electrostatic potentials into a ToF analyzer, where they travel at a constant kinetic energy to the detector. The kinetic energy of the ions is given by (Thiel and Sjövall, 2011):

$$E_{kin} = zeU_{ToF} = \frac{mv^2}{2}$$

where z is the ionic charge state, e is the electron charge, U_{ToF} is the electric potential inside the ToF analyzer (relative to the sample), m is the ion mass, and v is the ion velocity. The ions travel at different velocities depending on the mass-to-charge ratio (m/z) and hence arrive at the detector at different times. Keeping track of the time duration needed to travel through the ToF analyzer for each detected secondary ion and translating this flight time to mass according to the equation

above result in a mass spectrum, which shows the number of detected secondary ions as a function of mass ($z = 1$ for almost all secondary ions). The use of a pulsed primary ion source and a ToF analyzer allows the detection of essentially all secondary ions generated. This is in contrast to magnetic sector or quadrupole analyzers, in which mass selection is done by filtering away those ions that do not agree with the chosen transmission mass criteria (Thiel and Sjövall, 2011).

The IONTOF instruments use a reflectron-type ToF analyzer in which the ions are electrostatically reflected in an almost 180° angle before reaching the detector (Fig. 4.6). Thiel and Sjövall (2011) explain that "The extraction of

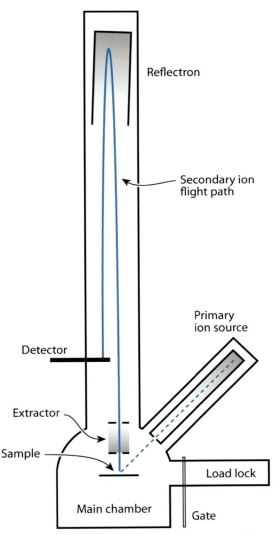

FIG. 4.6 Schematic drawing of a common type of time-of-flight secondary ion mass spectrometry (ToF-SIMS) instrument (IONTOF) GmbH Germany. *Source: Thiel, V., and P. Sjövall (2011), Using time-of-flight secondary ion mass spectrometry to study biomarkers, Annu. Rev. Earth Planet. Sci., 39(1): 125-156). Publisher: Annual Reviews, California https://www.annualreviews.org/doi/full/10.1146/annurev-earth-040610-133525 No info available on Copyright Email IDs of the Source authors: Volker Thiel: vthiel@gwdg.de Peter Sjövall: peter.sjovall@sp.se.*

secondary ions into the ToF analyzer is efficient only for secondary ions emitted at angles up to a certain value relative to the direction of the analyzer, which depends on the analyzer design and settings (Lee et al., 2008). For samples with surface roughness, the limited angle of acceptance results in a reduced signal intensity from areas on the surface where the surface normal is tilted away from the direction of the analyzer. This topographic effect is observed in ion images from rough sample surfaces, which typically show signal intensity variations that are related only to the topography of the sample surface and not to concentration variations. Topographic effects must therefore be considered carefully in the interpretation of ion images, and reliable concentration variations may be difficult to obtain from very rough surfaces." Another topographic effect noted by Thiel and Sjövall (2011) is "the peak shift that is produced in mass spectra from different areas of rough insulating samples. Positioned in the analyzer extraction field, such samples will have a varying electric potential depending on the surface height. This varying potential will, in turn, result in a slightly varying kinetic energy of the secondary ions in the ToF analyzer and, consequently, in shifts in the peak positions in the recorded mass spectrum." Thiel and Sjövall (2011) caution that these shifts will give rise to peak broadening, multiple peaks, or, double peaks in the total area mass spectrum; however, these can be removed by acquiring mass spectra from a selected region of interest with constant surface potential.

Continuous inspection of the sample surface in the analysis position is accomplished using a video camera and a lens system mounted on the main chamber. Additional components of the ToF-SIMS instrument include a low-energy electron gun that is used for charge compensation of insulating samples and an electron detector for detection of secondary electrons emitted during primary ion bombardment, which produces topographic images of the sample surface similar to SEM images.

Secondary ion formation: Thiel and Sjövall (2011) explain that the collision of primary ions and the resulting generation of secondary ions is the core process that determines the inherent limits in detection sensitivity and lateral resolution. The secondary ion yield, that is, the number of ions of a specific type that are emitted per incident primary ion, varies by several orders of magnitude depending on the analyte molecule and on the specific ion that is used to probe it. Furthermore, the secondary ion yield depends heavily on the properties of the primary ion beam.

Fig. 4.7 shows schematically a sample surface after the collision of two types of high-energy ions, obtained from a molecular dynamics simulation (Colla et al., 2000). In the case of an atomic primary ion, such as Ga^+ or Au^+ (Fig. 4.7A), the ion penetrates relatively deep into the sample and thereby distributes its kinetic energy in a relatively large volume below the sample surface. This induces a collision cascade in which chemical bonds are broken and the atoms and molecules undergo frequent collisions at varying energies. Particles located at the surface will be emitted if they undergo collisions of sufficient energy during the collision cascade, but deeper particles will remain in the sample (molecularly damaged). In the case of a cluster primary ion, such as C_{60}^+ or Au_4^+ (Fig. 4.7B), the situation is quite different in that most of the kinetic energy of the primary ion is deposited closer to the sample surface, partly owing to the higher mass of the projectile but also because of dissociation of the cluster into separate particles, each with a smaller amount of kinetic energy. According to Thiel and Sjövall (2011), this has several advantageous consequences for the analytical use of cluster ions as compared with atomic ions: (a) a dramatically increased sputter yield (number of

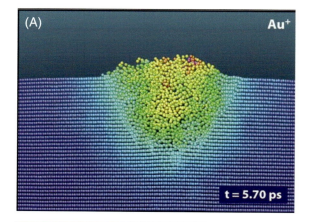

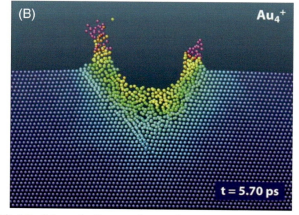

FIG. 4.7 **Schematic diagram of a sample surface after the impact of different primary ions.** (A) An atomic primary ion (e.g., Au^+) penetrates deep into the sample, inducing a collision cascade in the surface region. (B) A cluster primary ion (e.g., C_{60}^+, Au_4^+) deposits most of its kinetic energy close to the sample surface (from Colla et al., 2000). Copyright by the American Physical Society. Source: Thiel, V., and P. Sjövall (2011), Using time-of-flight secondary ion mass spectrometry to study biomarkers, Annu. Rev. Earth Planet. Sci., 39(1): 125-156). Publisher: Annual Reviews, California https://www.annualreviews.org/doi/full/10.1146/annurev-earth-040610-133525.

sputtered molecules per incident primary ion); (b) a reduced damage region that is limited to a shallow damage layer close to the surface; and (c) an increased yield of large organic ions with masses as high as ~2000 amu. Because of these positive effects, the capability to analyze organic materials, particularly biological materials, has improved dramatically since the development of cluster primary ion sources (Touboul et al. 2005a). The advantage of the cluster source for organic analysis is twofold: (a) the higher secondary ion yield of large organic fragment and molecular ions provides increased detection sensitivity and imaging capability for specific organic molecules, and (b) the reduced molecular damage in combination with a high sputter yield means that molecular information can be retained in the ToF-SIMS spectrum after sputtering (Colla et al., 2000; Garrison and Postawa, 2008; Winograd and Garrison, 2010).

According to Thiel and Sjövall (2011), a consequence of the surface sensitivity of ToF-SIMS is that the technique is very sensitive to small amounts of surface contamination, which may partially or completely cover the actual sample surface. To prepare a clean surface for ToF-SIMS analysis, a good strategy is to expose a fresh surface as immediately as possible before analysis, for example, by cutting or fracturing the sample.

Sjövall et al. (2004) explain that for biological samples (e.g., microbial mats), cryosectioning is a convenient and frequently used method to produce flat, clean surfaces for ToF-SIMS analysis. Thin sections are cut from the frozen sample using a sharp knife and placed on a substrate for subsequent drying and analysis.

Several investigators (e.g., Cannon et al., 2000; Lanekoff et al., 2011; Ostrowski et al., 2004) have reported the application of freeze fracturing of cell samples as a method to expose the cell interior for ToF-SIMS analysis. In this method, a cell suspension is pressed between two silicon substrates in a sandwich geometry and plunge frozen in liquid propane. Separation of the two silicon substrates induces a fracture through the thin ice layer and through the cells. Polystyrene spacer beads (5–10-μm diameter) are sometimes used to protect the cells while pressing together the silicon substrates and to define a fracture plane. If the sample is fractured inside the ToF-SIMS instrument, the cells can be analyzed either in the frozen hydrated state or in a freeze-dried state, whereas if the sample is fractured outside the instrument (in liquid nitrogen), exposure of the cold sample to the ambient atmosphere results in water condensation on the cell surfaces, and freeze drying is needed to expose the cell surfaces for analysis (Thiel and Sjövall, 2011).

Thiel and Sjövall (2011) have suggested that ion etching using a C_{60}^+ sputter source may be used to remove material from the sample surface to expose new structures for analysis. As C_{60}^+ projectiles deposit their energy very close to the surface, little damage is induced in the underlying layers, and the chemical integrity of the molecules under study is preserved (Kurczy et al., 2010). This technique can be used for different purposes, such as removal of contaminants and removal of the plasma membrane to expose the cell interior or to provide a complete 3D analysis of a single cell (Breitenstein et al., 2008; Fletcher et al., 2007).

Data interpretation: The tendency of ToF-SIMS to produce data too complex for a straightforward interpretation has increasingly urged analysts to use chemometrics to simplify data evaluation (Fletcher et al., 2006, Vaidyanathan et al., 2009). Interestingly, ions derived from lipids, namely fatty acids and phospholipids, were of particular diagnostic importance (Vaidyanathan et al., 2009).

Ionization and fragmentation of biomarkers—membrane lipids and fatty acids: Thiel and Sjövall (2011) explain that much of the published literature on biological ToF-SIMS imaging deals with the analysis of membrane-derived lipids (Brunelle and Laprévote, 2009; Johansson, 2006). Indeed, ToF-SIMS seems to favor to some degree lipid ions, and published studies on other types of organic compounds are relatively sparser. From a geobiological perspective, this is no disadvantage, as lipids are the compound class from which most biomarkers have been recruited so far. The typical mass range of ToF-SIMS (0–2000 amu) also favors lipid biomarker analysis. Basic information about the ToF-SIMS spectral patterns of lipids can be obtained from standard compounds (Heim et al., 2009; Ostrowski et al., 2005; Suo et al., 2007).

4.10.2.3 Demonstration of potential of time-of-flight secondary ion mass spectrometry for biomarker research

Thiel and Sjövall (2011) reported examples of three different applications relating to key areas of current biomarker research to demonstrate the potential of ToF-SIMS for such investigations. In these examples, emphasis has been placed on typical geobiological samples, including soft (e.g., microbial mats, cells), hard (e.g., microbial mineral precipitates), and liquid (e.g., fluid inclusions) materials.

Lipid biomarkers in soft microbial matter (biomass): From a geobiological perspective, membrane lipids are excellent biomarkers to specify organic matter with respect to source biota and related biogeochemical pathways, and have been amply studied using extract-based methods, in particular Liquid Chromatography/Mass Spectrometry (LC/MS) (Bradley et al., 2009; Pitcher et al., 2009; Schubotz et al., 2009; Sturt et al., 2004). Because intact polar membrane lipids easily degrade after the death of an organism, they are not found in ancient materials, but have been interpreted as proxies for the quantity of live microbial biomass in sediments (Lipp and Hinrichs, 2009; Zink et al., 2003). If not fully degraded, the alkyl moieties (parts of a molecule that is given a name because it is identified as a part of other molecules as well) of membrane lipids may further persist, either incorporated into macromolecular matter (kerogen),

A possible advantage of ToF-SIMS is thus offered by its capability to analyze extremely small sample volumes in environmental samples such as microbial consortia, or ideally, single cells. A ToF-SIMS experiment would also make sense when attempting to obtain information on the spatial distribution of lipids within cells and into the transport mechanisms of lipids between different cell compartments (Thiel and Sjövall, 2011).

Wanger et al. (2006) published a study on microbial coatings on water-bearing rock fractures from a South African gold mine. Fresh surfaces sampled at 2.8 km below the surface were examined for morphological and organic traits of the natural, deep subsurface biosphere. This study provided images of the physical and chemical environment of the sessile (permanently attached or established: not free to move about) microbial community, such as different cell types, organic film coatings on the fracture surface, and areas covered with microbial extracellular polymeric substances (Wanger et al., 2006).

Thiel's group used ToF-SIMS to study lipids in 10-μm-thick cryosections of calcifying microbial mats from the Black Sea (Thiel et al., 2007a). This study was aimed at assigning archaea-derived lipids to known phylogenetic groups of methanotrophic (methane-utilizing) archaea thriving in this anoxic marine environment. Using a Bi_3^+ source, the intensities of (quasi)molecular ions derived from isopranyl glycerol di- and tetraether-lipids were mapped on the sections, and the individual lipid fingerprints of different microbial consortia were determined. Subsequent examination of the areas studied with conventional fluorescence microscopy enabled association of the observed lipid patterns with distinctive morphological traits (filaments) of the candidate microbes. This helped to constrain the source of glycerol tetraether lipids to filamentous methanotrophic archaea of the phylogenetic ANME-1 group (ANME stands for anaerobic methane oxidizers). These lipids, or derivatives thereof, have frequently been observed in modern and ancient methane-rich settings (Birgel et al., 2006; Blumenberg et al., 2004; Rossel et al., 2008).

These lipids, or derivatives thereof, have frequently been observed in modern and ancient methane-rich settings (Birgel et al., 2006; Blumenberg et al., 2004; Rossel et al., 2008). With growing knowledge of the environmental requirements of the source organisms (Nauhaus et al., 2005), such occurrences in ancient sediments can be translated into better proxies for paleo-environmental reconstruction (Stadnitskaia et al., 2008).

Biomarkers in minerals and rocks: Rocks and mineral precipitates offer exciting targets for ToF-SIMS experiments, as analysis of pristine biomarkers could reveal thrilling results on the biogenicity and identity of particular morphological microstructures. In situ detection of biomarkers in ancient body fossils, biofilms, or microbialites (rock-like underwater structures that look like reefs but are made entirely of millions of microbes) with ToF-SIMS was first put into practice by Toporski et al. (2002), who studied biofilms (thin but robust layers of viscous secretion adhering to solid surfaces and containing a community of bacteria and other microorganisms) and small macrostructures (coprolite, tadpole) in the organic-rich (total organic carbon >8%) Late Oligocene (25.8 My old) Enspel Fossillagerstätte. Spectra from freshly fractured internal surfaces of rock particles using 25 keV primary Ga^+ ions revealed considerable organic material associated with these fossils (Thiel and Sjövall, 2011).

Guidry and Chafetz (2003) analyzed two suites of siliceous (containing or consisting of silica) precipitates from Yellowstone hot springs using light microscopy, SEM, and ToF-SIMS. Samples included mildly silicified (converted into or impregnated with silica) microbial mats from the surface of an active hot spring and diagenetically altered (changes and alterations that have taken place on biological material) siliceous sinters from a drill core, which were studied to examine the preservation potential of microbial biomarkers in these environments.

4.11 Detecting biosignature gases on extrasolar terrestrial planets

Biosignature gases are all those gases that are produced by life and that can accumulate in a planet atmosphere to detectable levels. According to Claudi (2017), "In the case of the bodies of the Solar System, the best way to investigate if they are bearing life on their surface is to go on-site and explore the planet and or its moons searching for evidences of life. This is made 'easy' by the vicinity of these planets. Venus and Mars are targets of this wandering of the human beings in the Solar System. This is not possible yet for extrasolar planets. In this case, and also in the former actually, the remote sensing is the 'easier' way of investigation."

Kaltenegger et al. (2010) have discussed how to read a planet's spectrum to assess its habitability and search for the signatures of a biosphere. After a decade rich in giant exoplanet detections, observation techniques have now advanced to a level where we now have the capability to find planets of less than 10 Earth masses (10 M_{Earth}—the so-called Super-Earths) that may potentially be habitable. How can we characterize those planets and assess whether they are habitable? These are challenges staring at the astrobiologists. Claudi (2017) reviewed the sites where to search for life and their main characteristics, what we should search for in their atmospheres (biosignatures) and why, and eventually the main methods for this quest. According to Kaltenegger et al. (2010), the new field of extrasolar planet search has shown an extraordinary ability to combine research by astrophysics, chemistry, biology and geophysics into a new and exciting interdisciplinary approach to understand our place in the universe.

Seager et al. (2013) had also suggested that detection of exoplanet atmospheric biosignature gases by remote sensing spectroscopy is usually taken as inevitable for the future of exoplanets. They argued that "This sentiment is being borne out with the discovery of increasing numbers of smaller and lower-mass planets each year. In addition, the development of larger and more sophisticated telescopes (such as the James Webb Space Telescope (JWST) (Gardner et al., 2006) and the giant 20 m to 40 m class ground-based telescopes continues to fuel the concept that the eventual detection and study of biosignature gases is a near certainty." In the context of the human struggle aimed at detecting biosignature gases on extrasolar terrestrial planets, an important question is "which gases should we search for?"

According to Seager et al. (2016), although a few biosignature gases are prominent in Earth's atmospheric spectrum (O_2, CH_4, N_2O), others have been considered as being produced at or able to accumulate to higher levels on exo-Earths (e.g., dimethyl sulfide (C_2H_6S) and methyl chloride (CH_3Cl). Life on Earth produces thousands of different gases (although most in very small quantities). Some might be produced and/or accumulate in an exo-Earth atmosphere to high levels, depending on the exo-Earth ecology and surface and atmospheric chemistry. According to Kiang et al. (2007), "So far, O_2 has been assumed as the best case for a biosignature gas in the search for life beyond our solar system, and the presence into the atmosphere of its photosynthetized product O_3 is considered as the evidence of the presence of life-forms producing oxygen."

Thousands of exoplanets are known to orbit nearby stars, and super-Earth exoplanets are being discovered with increasing frequency. The changing view of planets orbiting low-mass stars (M stars—red stars characterized primarily by surface temperature of 3000 kelvins or less—called also Type M stars) as potentially hospitable worlds for life and its remote detection was motivated by several factors, including the demonstration of viable atmospheres and oceans on tidally locked planets, and predictions of unusually strong spectral biosignatures. However, it must be borne in mind that M star habitable zone (HZ) terrestrial planets must survive a number of early trials in order to enjoy their many Gyr of stability. Their formation may be jeopardized by an insufficient initial disk supply of solids, resulting in the formation of objects too small and/or dry for habitability (Scalo et al., 2007). According to them, "During the first approximately 1 Gyr, atmospheric retention is at peril because of intense and frequent stellar flares and sporadic energetic particle events, and impact erosion, both enhanced, the former dramatically, for M star HZ semimajor axes." Scalo et al. (2007) concluded that "attempts at remote sensing of biosignatures and nonbiological markers from M star planets are important, not as tests of any quantitative theories or rational arguments, but instead because they offer an inspection of the residues from a Gyr-long biochemistry experiment in the presence of extreme environmental fluctuations."

Seager et al. (2013) studied biosignature gases on exoplanets with thin H_2 atmospheres and habitable surface temperatures, using a model atmosphere with photochemistry and a biomass estimate framework for evaluating the plausibility of a range of biosignature gas candidates. According to Seager et al. (2013), "In Sun-Earth-like UV radiation environments, H (and in some cases O) will rapidly destroy nearly all biosignature gases of interest. The lower UV fluxes from UV-quiet M stars would produce a lower concentration of H (or O) for the same scenario, enabling some biosignature gases to accumulate." They further suggest that "Most potential biosignature gases, such as dimethylsulfide and CH_3Cl, are therefore more favorable in low-UV, as compared with solar-like UV, environments. A few promising biosignature gas candidates, including NH_3 and N_2O, are favorable even in solar-like UV environments, as these gases are destroyed directly by photolysis and not by H (or O)."

According to Scalo et al. (2007), "Biological viability seems supported by unmatched very long-term stability conferred by tidal locking, small habitable zone (HZ) size, an apparent short-fall of gas giant planet perturbers, immunity to large astrosphere compressions, and several other factors, assuming incidence and evolutionary rate of life benefit from lack of variability." According to them, "The formation and retention of a thick atmosphere and a strong magnetic field as buffers for a sufficiently massive planet emerge as prerequisites for an M star planet to enter a long period of stability with its habitability intact." After a review of evidence concerning disks and planets associated with M stars, Scalo et al. (2007) evaluated M stars as targets for future HZ planet search programs.

M stars (also called M dwarfs or Red dwarfs) are relatively cool, red stars on the main sequence of M spectral type, having a surface temperature of less than 3600 K and an absorption spectrum dominated by molecular bands, especially titanium oxide. "Main sequence stars" are those stars that fuse hydrogen atoms to form helium atoms in their cores. About 90% of the stars in the Universe, including the Sun, are main sequence stars. These stars can range from about a tenth of the mass of the Sun to up to 200 times as massive. Stars start their lives as clouds of dust and gas. After condensation of mass and ignition of a star, it generates thermal energy in the dense core region through nuclear fusion of hydrogen atoms into helium. According to Kilston et al. (2007), "Stable, hydrogen-burning, M dwarf stars make up about 75% of all stars in the Galaxy. They are extremely long-lived, and because they are much smaller in mass than the Sun (between 0.5 and 0.08 M (Sun)), their temperature and stellar luminosity are low and peaked in the red." Because the luminosity of M dwarfs peak in the red, these stars are also called "Red dwarfs." These diminutive stars, much smaller and dimmer than our own Sun are not bright enough to see with the naked eye. Proxima Centauri, the nearest star to the Sun, is an M dwarf (Type M5), as are fifty of the sixty nearest stars.

Historically, scientists interested in the search for extrasolar life have shied away from studying M dwarfs. Because they put out relatively paltry amounts of light and heat, compared to the Sun, the general feeling among scientists was that they were unlikely to host habitable planets. Because the SETI Institute is building a massive radio telescope array to search for radio signals from extraterrestrial civilizations, a group of astronomers and biologists got together to reconsider the question; they want to know whether to include M dwarfs in their list of search targets (https://www.astrobio.net/alien-life/m-dwarfs-the-search-for-life-is-on/).

Kilston et al. (2007) re-examined what is known at present about the potential for a terrestrial planet forming within, or migrating into, the classic liquid-surface-water habitable zone close to an M dwarf star. According to Kilston et al. (2007), "Particularly in light of the claimed detection of the planets with masses as small as 5.5 and 7.5 M (Earth) orbiting M stars, there seems no reason to exclude the possibility of terrestrial planets. Tidally locked synchronous rotation within the narrow habitable zone does not necessarily lead to atmospheric collapse, and active stellar flaring may not be as much of an evolutionarily disadvantageous factor as has previously been supposed." Kilston et al. (2007) concluded that M dwarf stars may indeed be viable hosts for planets on which the origin and evolution of life can occur. According to them, "a number of planetary processes such as cessation of geothermal activity or thermal and nonthermal atmospheric loss processes may limit the duration of planetary habitability to periods far shorter than the extreme lifetime of the M dwarf star. Nevertheless, it makes sense to include M dwarf stars in programs that seek to find habitable worlds and evidence of life." Kilston et al.'s (2007) paper presents the summary conclusions of an interdisciplinary workshop (http://mstars.seti.org) sponsored by the NASA Astrobiology Institute and convened at the SETI Institute.

According to Seager et al. (2016), "Plans for the next generation of space-based and ground-based telescopes are fuelling the anticipation that a precious few habitable planet can be identified in the coming decade. Even more highly anticipated is the chance to find signs of life on these habitable planets by way of biosignature gases." According to Prof. Seager, "any kind of ab initio approach" to predicting what biosignature gases might be is so challenging that nearly all work done to date basically follows the "We know what Earth life produces, so what might Earth's products look like if transplanted to another, slightly different, Earth-like planet" (https://www.saraseager.com/research/exoplanet-biosignature-gases/). Note that "Earth-like planet" refers to a planet with about the same size and mass as Earth, with oceans and continents, a thin N_2-CO_2-O_2 atmosphere, and a radiation environment similar to that of Earth's. Gases studied in this context include oxygen, methyl halides, sulfur compounds, and some other gases. According to Prof. Seager, the main purpose of the ongoing extensive research in the realm of exoplanet biosignature gas aims to push the frontiers to consider biosignature gases on planets very different from Earth and also as many molecules as possible so that we do not miss our chance to identify gases that might be produced by life. Seager et al. (2013) have reported a novel approach to the subject of biosignature gases by systematically constructing a long list of volatile molecules in different categories, totaling about 14,000 molecules. To maximize our chances of recognizing biosignature gases, Seager et al. (2013) promoted the concept that all stable and potentially volatile molecules should initially be considered as viable biosignature gases.

The rapid rate of discoveries of exoplanets has expanded the scope of the science for possible remote detection of life beyond Earth. According to Schwieterman et al. (2018), "In the coming years and decades, advanced space- and ground-based observatories will allow an unprecedented opportunity to probe the atmospheres and surfaces of potentially habitable exoplanets for signatures of life. Life on Earth, through its gaseous products and reflectance and scattering properties, has left its fingerprint on the spectrum of our planet."

The Exoplanet Biosignatures Workshop Without Walls (EBWWW) held in 2016 engaged the international scientific community across diverse scientific disciplines, to assess the state of the science and technology in the search for life on exoplanets, and to identify paths for progress. According to Kiang et al. (2018), "Strong themes that emerged from the workshop were that biosignatures must be interpreted in the context of their environment, and that frameworks must be developed to link diverse forms of scientific understanding of that context to quantify the likelihood that a biosignature has been observed. Models are needed to explore the parameter space where measurements will be widespread but sparse in detail. Given the technological prospects for large ground-based telescopes and space-based observatories, the detection of atmospheric signatures of a few potentially habitable planets may come before 2030."

Seager et al. (2016) believe that to maximize our chances of recognizing biosignature gases, it is necessary to promote the concept that all stable and potentially volatile molecules should initially be considered as viable biosignature gases. Toward this end, Seager et al. (2016) reported a new approach to the subject of biosignature gases by systematically constructing lists of about 14,000 volatile molecules in different categories, in which about 2500 of these are HOCNSP compounds. These researchers have further shown that about one-fourth of HOCNSP molecules are known to be produced by life on Earth. Considering their accumulation and possible false positives on exoplanets with atmospheres and surface environments different from those of Earth, Seager et al. (2016) claim that "the list can be used to study classes of chemicals that might be potential biosignature gases. The list can also be used for terrestrial biochemistry applications." Apart from providing some examples, they have additionally provided an online community usage database to serve as a registry for volatile molecules including biogenic compounds. Claudi (2017) discussed the possible false positives that can be generated by geophysical and geochemical reactions, and indicated that on Earth, the only sure biosignatures netted off false positives are O_2 and N_2O.

Biosignature gas detection is one of the ultimate future goals for exoplanet atmosphere studies. An accepted method used for detection of evidence of habitable conditions and life on terrestrial-sized extrasolar planets is seeking spectroscopic evidence. In Des Marais et al. (2002)'s studies, the highest priority was given to the detection of oxygen (O_2) or its photolytic product ozone (O_3). Although liquid H_2O is not a bioindicator, this is considered essential to life. Substantial carbon dioxide (CO_2) gas indicates an atmosphere and oxidation state typical of a terrestrial planet (i.e., a planet, such as Mercury, Venus, Earth, and Mars, which is composed primarily of silicate rocks or metals). Likewise, abundant methane (CH_4) might require a biological source, although abundant CH_4 also can arise from a crust and upper mantle more reduced than that of Earth. It is believed that the range of characteristics of extrasolar rocky planets might far exceed that of the Solar System. Planetary size and mass are very important indicators of habitability and can be estimated in the mid-IR and potentially also in the visible to near-IR. In the context of putative extrasolar analogs, Des Marais et al. (2002) assessed known spectroscopic molecular band features of Earth, Venus, and Mars. In their studies, the preferred wavelength ranges were 7–25 microns (μm) in the mid-IR and 0.5 microns to approximately 1.1 microns in the visible to near-IR. Additional spectroscopic features merit study, for example, features created by other biosignature compounds in the atmosphere or on the surface; and features due to Rayleigh scattering. Des Marais et al. (2002) found that both the mid-IR and the visible to near-IR wavelength ranges offer valuable information regarding biosignatures and planetary properties; therefore, both merit serious scientific consideration.

Meadows et al. (2018) have described how environmental context can help determine whether oxygen (O_2) detected in extrasolar planetary observations is more likely to have a biological source. They have provided an in-depth, interdisciplinary example of O_2 biosignature identification and observation, which serves as the prototype for the development of a general framework for biosignature assessment. Photosynthetically generated O_2 is a potentially strong biosignature, and at high abundance, it was originally thought to be an unambiguous indicator for life. However, according to Meadows et al. (2018), "as a biosignature, O_2 faces two major challenges: (1) it was only present at high abundance for a relatively short period of Earth's history, and (2) we now know of several potential planetary mechanisms that can generate abundant O_2 without life being present. Consequently, our ability to interpret both the presence and absence of O_2 in an exoplanetary spectrum relies on understanding the environmental context."

Meadows et al. (2018) have examined the co-evolution of life with the early Earth's environment to identify how the interplay of sources and sinks may have suppressed O_2 release into the atmosphere for several billion years, producing a false negative for biologically generated O_2. According to them these studies suggest that planetary characteristics that may enhance false negatives should be considered when selecting targets for biosignature searches. Meadows et al. (2018) have reviewed the most recent knowledge of false positives for O_2, planetary processes that may generate abundant atmospheric O_2 without a biosphere. They have also provided examples of how future photometric, spectroscopic, and time-dependent observations of O_2 and other aspects of the planetary environment can be used to rule out false positives and thereby increase our confidence that any observed O_2 is indeed a biosignature. According to Meadows et al. (2018), these insights are expected to guide and inform the development of future exoplanet characterization missions.

4.12 False positives and false negatives

As already noted, a biosignature is an unambiguous signal of past life. The organic molecules listed under the category of biosignature include nucleic acids, lipids (any of a class of organic compounds that are fatty acids or their derivatives and are insoluble in water but soluble in organic solvents), sugars, proteins, and enzymes. All organic molecules contain carbon, nearly all contain hydrogen, and many also contain oxygen. Only those organic molecules and gases that result exclusively from biological processes can be categorized as biosignatures.

Most of the features described above as biosignatures actually are not unique by–products due to the presence of life. There are a lot of atmospheric and geophysical processes that are able to produce the same kind of molecules in detectable quantities. Thus, a difficulty that arises in the detection of biosignature is that those organic molecules and gases indicative of biological processes can also be produced abiotically. In this case, a biosignature molecule can turn out to be "false positive." Note that, by definition, "false positive" is an error in data reporting in which a test result improperly indicates presence of a condition, when in reality it is not present; while a "false negative" is an error in which a test result improperly indicates no presence of a condition (the result is negative), when in reality it is present. The activities of the Exoplanet Biosignatures Workshop Without Walls (EBWWW) held in 2016 has provided an encyclopedic review of known and proposed biosignatures and models used to ascertain them (Schwieterman et al., 2018); and an in-depth review of O_2 as a biosignature, rigorously examining the nuances of false positives and false negatives for evidence of life (Meadows et al., 2018) (see Kiang et al., 2018 for details).

The most prone to false positives are type-I biosignatures (i.e., primary metabolic biosignatures). In this case, in fact, geology uses the same redox gradient in order to produce the same molecules produced by life. According to Claudi (2017), "If the planet is geologically active, hot spots, volcanism, fumaroles and hot springs are the main actors that are able to produce CO, CO_2, CH_4, H_2O, N_2, H_2S. The last one is produced by volcanoes in large amount. Moreover, there are gases, like N_2 and H_2O that are by–products of life and that are present in considerable amount in the Earth's atmosphere." Claudi (2017) further noted that "The most important false positive is the abiotic production of oxygen. The main abiotic routes for oxygen are both coming from photochemical reactions due to the photodissociation of CO_2 and H_2O. The CO_2 photolysis is due to UV radiation coming from the host star with a wavelength of about 140 nm." Water was also present on Earth since the beginning. According to Segura et al. (2007), normally, the detection of the water vapor bands simultaneously with the O_2 band can rule out this abiotic mechanism. However, one should be careful, as the vapor pressure of H_2O over a high-albedo icy surface might be high enough to produce detectable H_2O bands (Kaltenegger et al., 2010).

According to Reinhard et al. (2017), though a great deal of work has centered on refining our understanding of "false positives" for remote life detection, much less attention has been paid to the possibility of false negatives, that is, cryptic (i.e., serving to camouflage them in their natural environment) biospheres that are widespread and active on a planet's surface but are ultimately undetectable or difficult to detect in the composition of a planet's atmosphere.

As an example of false positive, Seager et al. (2013) noted that "most gases produced by life that are fully hydrogenated forms of an element, such as CH_4 and H_2S, are not effective signs of life in an H_2-rich atmosphere because the dominant atmospheric chemistry will generate such gases abiologically, through photochemistry or geochemistry." Seager et al. (2013) are hopeful that "Suitable biosignature gases in H_2-rich atmospheres for super-Earth exoplanets transiting M stars could potentially be detected in transmission spectra with the *James Webb Space Telescope*."

Less than 1ppm of atmospheric O_2 comes from abiotic processes (Walker, 1977). Cyanobacteria and plants are responsible for this production by using the solar photons to extract hydrogen from water and using it to produce organic molecules from CO_2. This metabolism is called oxygenic photosynthesis. According to Meadows (2017), "Photosynthetic O_2 has long been considered particularly robust as a sign of life on a habitable exoplanet, due to the lack of known "false positives"—geological or photochemical processes that could also produce large quantities of stable O_2. O_2 has other advantages as a biosignature, including its high abundance and uniform distribution throughout the atmospheric column and its distinct, strong absorption in the visible and near-infrared." However, recent modeling work has shown that false positives for abundant oxygen or ozone could be produced by abiotic mechanisms, including photochemistry and atmospheric escape. Environmental factors for abiotic O_2 have been identified and will improve our ability to choose optimal targets and measurements to guard against false positives. "Most of these false-positive mechanisms are dependent on properties of the host star and are often strongest for planets orbiting M dwarfs."

Meadows (2017) further states, "In advance of the next generation of telescopes, thorough evaluation of potential biosignatures—including likely environmental context and factors that could produce false positives—ultimately works to increase our confidence in life detection."

4.13 Potential biosignatures—molecules that can be produced under both biological and nonbiological mechanisms but selectively/uniquely attributable to the action of biology

The term "biosignatures" is generally used to mean detectable species, or set of species, whose presence at significant abundance strongly suggests a biological origin. Although methane (CH_4) is an abundant constituent of the cold planetary atmospheres in the outer solar system, the methane found in the present atmosphere of the Earth has a biological origin. On Earth, methane is produced abiotically (i.e., without any biological intervention or influence) in hydrothermal systems where H_2 (produced from the oxidation of Fe by water) reacts with CO_2 in a certain range of pressures and temperatures. In the absence of atmospheric oxygen, abiotic methane could build up to detectable levels. Depending on the degree of oxidation of a planet's crust and upper mantle, such nonbiological mechanisms can also produce large amounts of CH_4 under certain circumstances. Therefore, the detection of methane alone cannot be considered a sign of life, while its detection in an oxygen-rich atmosphere would be difficult to explain in the absence of a biosphere. Therefore, the detection of CH_4 cannot be attributed unambiguously to life (see Kaltenegger et al., 2010).

Reduced gases and oxygen have to be produced concurrently to be detectable in the atmosphere, as they react rapidly with each other. Thus, the chemical imbalance traced by the simultaneous signature of O_2 and/or O_3 and of a reduced gas like CH_4 can be considered as a signature of biological activity (e.g. couple CH_4+O_2, or CH_4+O_3 (Lovelock, 1975)).

Kaltenegger et al. (2010) further mentions that "The case of NH_3 is similar to the one of CH_4. They are both released into Earth's atmosphere by the biosphere with similar rates but the atmospheric level of NH_3 is orders of magnitude lower due to its very short lifetime under UV irradiation. The detection of NH_3 in the atmosphere of a habitable planet would thus be extremely interesting, especially if found with oxidized species." According to Kaltenegger et al. (2010), "The detection of H_2O and CO_2, not as biosignatures themselves, are important in the search for signs of life because they are raw materials for life and thus necessary for planetary habitability."

An interesting biosignature is nitrous oxide (N_2O). According to Kaltenegger et al. (2010), N_2O is produced in abundance by life but only in negligible amounts by abiotic processes. Nearly all of Earth's N_2O is produced by the activities of anaerobic denitrifying bacteria. Nitrous oxide is difficult to be detected in a humid atmosphere (like in Earth's atmosphere) with low resolution because of the presence of the molecular band of water vapor while it would become more apparent in atmospheres with more N_2O or less H_2O vapor, or a combination of the two (Kaltenegger et al., 2010).

Measurable attributes of life include its complex physical and chemical structures and also its utilization of free energy and the production of biomass and wastes. Due to its unique characteristics, a biosignature can be interpreted as having been produced by living organisms; however, it is important that they not be considered definitive because there is no way of knowing in advance which ones are universal to life and which ones are unique to the peculiar circumstances of life on Earth.

Ocean–atmosphere chemistry on Earth has undergone dramatic evolutionary changes throughout its long history, with potentially significant ramifications for the emergence and long-term stability of atmospheric biosignatures. Reinhard et al. (2017) have summarized recent developments from geochemical proxy records and Earth system models that provide insight into the long-term evolution of the most readily detectable potential biosignature gases on Earth—oxygen (O_2), ozone (O_3), and methane (CH_4). According to them "the canonical O_2-CH_4 disequilibrium biosignature would perhaps have been challenging to detect remotely during Earth's ~4.5-billion-year history and that in general atmospheric O_2/O_3 levels have been a poor proxy for the presence of Earth's biosphere for all but the last ~500 million years." More broadly, Reinhard et al. (2017) emphasize that "internal oceanic recycling of biosignature gases will often render surface biospheres on ocean-bearing silicate worlds cryptic (i.e., serving to camouflage them in their natural environment), with the implication that the planets most conducive to the development and maintenance of a pervasive biosphere will often be challenging to characterize via conventional atmospheric biosignatures."

Measurable attributes of molecular biosignatures include (Summons et al., 2008b):

- Enantiomeric excess
- Diastereoisomeric preference
- Structural isomer preference
- Repeating constitutional subunits or atomic ratios
- Systematic isotopic ordering at molecular and intramolecular levels
- Uneven distribution patterns or clusters (e.g., C-number, concentration, $\delta^{13}C$) of structurally related compounds.

Summons et al. (2008b) have addressed details of the chemical and biosynthetic basis for these features, which largely arise as a consequence of construction from small, recurring subunits.

According to Cady and Noffke (2009), the main challenge for any search-for-life mission consists in determining whether a candidate observation (or better yet, a collection of observations) can be uniquely attributed to the action of biology. This caution was found to be valuable in the interpretation of possible signs of extraterrestrial microbial life on the Martian meteorite ALH84001. Although polycyclic aromatic hydrocarbons (PAHs) were detected in numerous fresh fracture surfaces of ALH84001 (McKay et al., 1996), and it was stated that the PAHs had a Martian origin (Clemett et al., 1998), it was suggested (Steele et al., 2000) that the presence of a filamentous organism observed on a fracture just beneath the fusion crust is proof of some terrestrial biogenic activity subsequent to ALH84001's fall to Earth. Subsequent detailed analysis of ALH84001 specimens (Stephan et al., 2003) revealed that the meteorite had been exposed to terrestrial contamination. Furthermore, a survey of the association of abiotic macromolecular carbon with magmatic minerals on several Martian meteorites (ranging in age from 4.2 Ga to 190 Ma) indicated that Martian magmas favor the precipitation of (abiotic) reduced carbon species during crystallization (Steele et al., 2012). On the basis of the above findings and reasonings, Vago et al. (2017) concluded that the claim that ALH84001 PAHs may result from the action of past Martian life is not sufficiently substantiated by the data.

It is interesting and impressive to note that as well as having a range of metabolisms, organisms build biomass suited to specific physical environments, habitats and their ecological imperatives. This overall "metabolic diversity" manifests itself in an enormous variety of accompanying natural products. The whole field of organic chemistry grew from the study of natural products and now provides tools to link metabolism (i.e., physiology) to the occurrence of biomarkers specific to, and diagnostic for, particular kinds of metabolism.

4.14 Atmospheric chemical disequilibrium (a generalized biosignature)—a proposed method for detecting extraterrestrial biospheres

Atmospheric chemical disequilibrium has been proposed as a method for detecting extraterrestrial biospheres from exoplanet observations (see Krissansen-Totton et al., 2016). Chemical disequilibrium is potentially a generalized biosignature because it makes no assumptions about particular biogenic gases or metabolisms. Kleidon (2012) had shown how this notion can be quantified using nonequilibrium thermodynamics. According to Kleidon (2012), generating and maintaining disequilibrium in a thermodynamic variable requires the extraction of power from another thermodynamic gradient, and the second law of thermodynamics imposes fundamental limits on how much power can be extracted. Krissansen-Totton et al. (2016) noted that "only on Earth is the chemical disequilibrium energy comparable to the thermal energy per mole of atmosphere. Earth's disequilibrium is biogenic, mainly caused by the coexistence of N_2, O_2 and liquid water instead of more stable nitrate. In comparison, the O_2-CH_4 disequilibrium is minor." They further noted that the purely gas phase disequilibrium in Earth's atmosphere is mostly attributable to O_2 and CH_4.

According to Claudi (2017), "The very first discussions on the search for life by atmospheric sensing focused on the idea that a system at thermodynamic disequilibrium shall be by itself an atmospheric biosignature (Lederberg, 1965; Lovelock, 1965). This disequilibrium (sign of life) could be testified by the simultaneous presence in the Earth's atmosphere of hydrocarbons and molecular oxygen (a redox disequilibrium)." Lippincot et al. (1967) identified methane (CH_4) as the hydrocarbon to be contrasted with oxygen (O_2) as sign of life in their first systematic thermodynamic equilibrium calculations on the atmospheres of the rocky planets of the Solar System. They found that CH_4 is out of thermodynamic equilibrium on Earth together with other gases (H_2, N_2O, and SO_2) but they all cannot be considered unambiguous signs of life (with the possible exception of N_2O) due to the production of these molecules by geochemical processes. Lovelock and Kaplan (1975) supported the idea of Lippincot et al. (1967) that the O_2–CH_4 disequilibrium is strong evidence for life. Subsequently, the role of CH_4 as a biosignature gas also seemed to be established (Lovelock and Kaplan, 1975; Sagan, 1975). In any case, a lot of arguments tackles the use of thermodynamic equilibrium as a life indicator (Seager, 2014).

Kleidon (2012) elaborated the history of how thermodynamic disequilibrium began to be considered as a sign of a habitable planet thus, "In the search for easily recognizable signs of planetary habitability, Lovelock (1965) suggested the use of the chemical disequilibrium associated with the composition of a planetary atmosphere as a sign for presence of widespread life on a planet. He argued that the Earth's high concentration of oxygen in combination with other gases, particularly methane, constitutes substantial chemical disequilibrium that would quickly be dissipated by chemical reactions if it were not continuously replenished by some processes. Since life dominates these exchange fluxes, he argued that life is the primary driver that generates and maintains this state of chemical disequilibrium in the Earth's atmosphere (see also Lovelock (1975))."

While Krissansen-Totton et al. (2016) suggested the possibility of detecting biospheres from chemical thermodynamic disequilibrium in planetary atmospheres; they cautioned that further work would be needed to establish whether thermodynamic disequilibrium is a practical exoplanet biosignature, requiring an assessment of false positives, noisy observations, and other detection challenges.

4.15 Identification of amino acids in Murchison meteorite and Atarctic micrometeorites

Amino acids are an important class of organic molecules essential for the development of primitive life. Therefore, identification of these biogenic molecules in meteorite and micrometeorites strengthens the notion of the possibility of life's existence there, may be in the most primitive form such as prokaryotes. As we have found on Earth, such primitive life-forms could merely be the fore-runners of more advanced life-forms.

Cosmic dust form into solidified droplets during their hypervelocity entry into the Earth's atmosphere and finally get spread over the Earth's surface. The Murchison meteorite is a meteorite that fell in Australia in 1969 near Murchison, Victoria. It belongs to a group of meteorites rich in organic compounds. Due to its mass and the fact that it was an observed fall, the Murchison meteorite is one of the most studied of all meteorites.

Murchison contains common amino acids such as glycine, alanine, and glutamic acid as well as unusual ones such as isovaline and pseudoleucine (Kvenvolden et al., 1970). A complex mixture of alkanes was isolated as well, similar to that found in the Miller–Urey experiment. In 1997, L-excesses also were found in a nonprotein amino acid, isovaline (Cronin and Pizzarello, 1997), suggesting an extraterrestrial source for molecular asymmetry in the Solar System. A specific family of amino acids called diamino acids was identified in the Murchison meteorite as well (Meierhenrich et al., 2004). By 2001, the list of organic materials identified in the meteorite was extended to polyols (Cooper et al., 2001).

The Murchison meteorite contained a mixture of left-handed and right-handed amino acids; most amino acids used by living organisms are left-handed in chirality, and most sugars used are right-handed. A team of chemists in Sweden demonstrated in 2005 that this homochirality could have been triggered or catalyzed, by the action of a left-handed amino acid such as proline (Córdova et al., 2005).

Several lines of evidence indicate that the interior portions of well-preserved fragments from Murchison are pristine. A 2010 study using high resolution analytical tools including spectroscopy, identified 14,000 molecular compounds, including 70 amino acids, in a sample of the Murchison meteorite (Schmitt-Kopplin et al., 2010). The limited scope of the analysis by mass spectrometry provides for a potential 50,000 or more unique molecular compositions, with the team estimating the possibility of millions of distinct organic compounds in the meteorite (Matson, 2010). It is surprising that a space rock such as the Murchison meteorite contains an organic molecular feast. Scientists consider that a large fraction of homochiral amino acids might have been delivered to the primitive Earth via meteorites and micrometeorites. Note that "homochirality" is a uniformity of chirality, or handedness. Objects are chiral when they cannot be superposed on their mirror images. For example, the left and right hands of a human are approximately mirror images of each other but are not their own mirror images, so they are chiral. In biology, 19 of the 20 natural amino acids are homochiral, being L-chiral (left-handed), while sugars are D-chiral (right-handed) (Lehninger et al., 2008).

Subsequent to the hypervelocity entry of cosmic dust droplets into the Earth's atmosphere, those droplets which fell on land surface vanish mostly, and sink into oblivion with the passage of time; or rains take some of them into the oceans through rivers. Most of the droplets so reached the oceans from land and those directly fell into the oceans' surface sink down and remain buried in deep-sea sediments to recover as cosmic spherules and micrometeorites, or for the posterity to ponder over the past.

During the 4-year (1850–1854) Arctic odyssey of H.M.S. *Investigator* expedition (conducted under command of Captain McClure, with the sole purpose of discovering a safe navigation route across the Arctic Ocean, from the Pacific Ocean to the Atlantic Ocean), which was riddled with the most vigorous ice-packs, the expedition team members had the most enjoyable and auspicious opportunity to witness three continuous days of meteor showers decorating a clear and cloudless sky (see Stein, 2015; p. 80). Cosmic spherules from the deep-sea sediments were retrieved for the first time during the HMS *Challenger* expedition (Murray and Renard, 1891). Both stony and iron types were found and their origin was assigned to meteoroids.

It is estimated that about 40,000 ± 20,000 tons of dust of all sizes rains on the Earth annually (Love and Brownlee, 1993). Materials raining into the Earth's atmosphere have two origins: particles released from comets during their perihelional passage, and debris generated by collisions of asteroids (Brownlee, 1981); and dust collection is the only earth-based technique for obtaining typical cometary solids.

While sample returns are very rare and expensive, to date only lunar samples are present as sample returns, although this scene has changed marginally with the important collections made during the Stardust Mission. In effect, cosmic dust, due to its large flux, is available in plenty in the Earth's materials and more importantly, is a unique way of actually studying asteroidal as well as cometary material. Furthermore, the dust accreted by the Earth offers the widest spectrum of solar system material than represented by conventional meteorites. It is expected that all the spherules >50 μm in size are asteroidal and these results provide information on the composition of what may be the most common main belt in asteroid type. The information provided by these cosmic spherules is considered to be better, and less biased than those provided by conventional meteorites of centimeter- and larger-size for investigation of asteroid composition.

The majority of the particles entering the Earth's surface become ferromagnetic, partly because they had to pierce the Earth's magnetosphere during their earth-ward traverse, and partly because of magnetic formation due to friction during atmospheric entry. The ferromagnetic property of these particles allows them to be easily attracted by simple hand held magnets. This aspect has been utilized in almost all deep-sea collections so far.

Many large collections of cosmic dust have been made in a variety of environments (including deep sea, beach sands, paleo-sediments, polar regions, stratosphere etc.), with a recent, larger concentration of efforts in the Polar Regions. These collections have enhanced our understanding of the types of materials that comprise the asteroids and manifold comets. However, it is also acknowledged that cosmic dust, more so large collections, provide greater insights into the types of materials that asteroids or comets comprise of, and in turn have the potential to provide clues to the early Solar System processes.

Cosmic dust has been isolated from beach sands (Einaudi and Marvin, 1967), Antarctic ice (Maurette et al., 1991, 1992; Koeberl and Hagen, 1989) from Greenland ice melt water lakes (Maurette et al., 1986, 1987) and from old, lithified sedimentary formations on land (Tailor and Brownlee, 1991). Most initial efforts were concentrated in recovering the dust from deep-sea sediments (Millard and Finkelman, 1970; Brownlee et al., 1997) in view of the low sedimentation rates in these regions and the relatively easier accessibility as compared with the Polar Regions.

A large collection of micrometeorites has been recently extracted from Antarctic old blue ice (Maurette et al., 1991). It was found that in the 50–100 μm size range, the carbonaceous micrometeorites represent 80% of the samples and contain 2% of carbon. It is believed that they might have brought more carbon to the surface of the primitive Earth than that involved in the present surficial biomass. It proved to be music to life-scientists that amino acids such as "amino isobutyric acid" have been identified in these Antarctic micrometeorites.

FIG. 4.8 (Left) Murchison meteorite at The National Museum of Natural History (Washington). *(Source: https://en.wikipedia.org/wiki/Murchison_meteorite#/media/File:Murchison_crop.jpg)* (Right) Pair of grains from the Murchison meteorite. *This image is a work of a United States Department of Energy (or predecessor organization) employee, taken or made as part of that person's official duties. As a work of the US Federal Government, the image is in the public domain. (Source: https://commons.wikimedia.org/wiki/File:Murchison-meteorite-stardust.jpg).*

Interestingly, enantiomeric excesses of L-amino acids have been detected in the Murchison meteorite (Fig. 4.8).

4.16 Major challenges lurking in the study of extrasolar biosignature gases

While looking for hints for the presence of extinct/extant life on extraterrestrial bodies in our very own Solar System is beset with innumerable hurdles, hunting for signatures of life on extrasolar planets (exoplanets) and their moons is all the more difficult. According to Kaltenegger et al. (2010), "The interpretation of observations of other planets with limited signal-to-noise ratio and spectral resolution as well as absolutely no spatial resolution, as envisioned for the first generation instruments, will be far more challenging and implies that we need to gather information on the planet environment to understand what we will see." In theory, spectroscopy can provide some detailed information on the thermal profile of a planetary atmosphere. This, however, requires a spectral resolution and a sensitivity that are well beyond the performance of a first-generation spacecraft. Despite the discovery of increasing numbers of smaller and lower-mass planets each year, and the development of larger and more sophisticated telescopes; according to Seager et al. (2013), "The topic of biosignature gases may remain a futuristic one unless a number of extreme challenges can be overcome. The biggest near-term challenge is to find a large enough pool of potentially habitable exoplanets accessible for follow-up atmospheric study." By the term "potentially habitable," Seager et al. (2013) mean "rocky planets with surface liquid water and not those with massive envelopes making any planet surface too hot for the complex molecules required for life." According to these investigators, a large pool of such planets is needed because there could be a large difference in the numbers of seemingly potentially habitable planets (based on their measured host stellar type, orbit, mass or size, and inferred surface temperature) and those that are inhabited by life that produces useful biosignature gases (which will be inferred from measured atmospheric spectra). By the term "useful biosignature gases" Seager et al. (2013) mean "those that can accumulate in the planet atmosphere, are spectroscopically active, and are not overly contaminated by geophysical false positives." Seager et al. (2013) suggest that a contemporary, related point to identifying "a large enough pool of exoplanets" is that even the fraction of small or low-mass planets that are potentially habitable—that is, with surface conditions suitable for liquid water—is not yet known. According to Seager et al. (2013), "The reason is that the factors controlling a given planet's surface temperatures are themselves not yet observed or known, including the atmosphere mass (and surface pressure), the atmospheric composition, and hence the concomitant greenhouse gas potency." Details on this topic are given in a review by Seager (2013).

Seager et al. (2013) are of the opinion that "A second major challenge for the study of biosignature gases is the capability of telescopes to robustly detect molecules in terrestrial exoplanet atmospheres. This challenge is continuously faced in today's hot Jupiter atmosphere studies (e.g., Seager, 2010), where many atmospheric molecular detections based on data from the Hubble Space Telescope (HST) or the Spitzer Space Telescope remain controversial." Deming et al. (2013) and references therein provide information on the just-mentioned controversy pertaining to atmospheric molecular detections.

Note that Hot Jupiters are a class of gas giant exoplanets that are inferred to be physically similar to Jupiter but that have very short orbital periods ($P < 10$ days). The close proximity to their stars and high surface-atmosphere temperatures resulted in their informal name "hot Jupiters."

According to Seager et al. (2013), "A third major challenge in the study of biosignature gases has to do with geological false positive signatures. These false positives are gases that are produced geologically and emitted by volcanoes or vents in the crust or ocean." These investigators suggest that geochemistry has the same chemicals to work with that life does, and therefore false positives are inevitable.

References

Abramov, O., Mojzsis, S.J., 2009. Microbial habitability of the Hadean Earth during the late heavy bombardment. Nature 459 (7245), 419–422. doi:10.1038/nature08015.

Abrajano, T.A., Sturchio, N.C., Bohlke, J.K., Lyon, G.L., Poreda, R.J., Stevens, C.M., 1988. Methane-hydrogen gas seeps, Zambales ophiolite, Philippines: deep or shallow origin? Chem. Geol. 71, 211–222.

Allen, D.E., Seyfried, W.E., 2004. Serpentinization and heat generation: constraints from Lost City and Rainbow hydrothermal systems. Geochim Cosmochim Acta 68, 1347–1354.

Allwood, A.C., Walter, M.R., Kamber, B.S., Marshall, C.P., Burch, I.W., 2006. Stromatolite reef from the Early Archaean era of Australia. Nature 441, 714–718.

Allwood, A.C., Grotzinger, J.P., Knoll, A.H., Burch, I.W., Anderson, M.S., Coleman, M.L., Kanik, I., 2009. Controls on development and diversity of Early Archean stromatolites. Proc. Natl. Acad. Sci. U S A 106, 9548–9555.

Allwood, A.C., Burch, I.W., Rouchy, J.M., Coleman, M., 2013. Morphological biosignatures in gypsum: diverse formation processes of messinian (~6.0 ma) gypsum stromatolites. Astrobiology 13, 870–886.

Altermann, W., Böhmer, C., Gitter, F., Heimann, F., Heller, I., Läuchli, B., Putz, C., 2009. Discussion about "Defining biominerals and organominerals. Sediment. Geol. 213, 150–151. doi:10.1016/j.sedgeo.2008.04.001.

Bains, W., 2014. What do we think life is? A simple illustration and its consequences. Int. J. Astrobiol. 13 (02), 101–111.

Banfield, J.F., Moreau, J.W., Chan, C.S., Welch, S.A., Little, B., 2004. Mineralogical biosignatures and the search for life on Mars. Astrobiology 1 (4). https://doi.org/10.1089/153110701753593856.

Benninghoven, A., 1994. Chemical analysis of inorganic and organic surfaces and thin films by static time-of-flight secondary ion mass spectrometry (TOF-SIMS). Angew. Chem. Int. Ed. 33, 1023–1043.

Birgel, D., Thiel, V., Hinrichs, K-U., Elvert, M., Campbell, K.A., et al., 2006. Lipid biomarker patterns of methane-seep microbialites from the Mesozoic convergent margin of California. Org. Geochem. 37, 1289–1302.

Blackhurst, R.L., Genge, M.J., Kearsley, A.T., Grady, M.M., 2005. Cryptoendolithic alteration of Antarctic sandstones: Pioneers or opportunists? J. Geophys. Res.: Planets 110 (E12), E12S24–E12S24.

Blumenberg, M., Seifert, R., Reitner, J., Pape, T., Michaelis, W., 2004. Membrane lipid patterns typify distinct anaerobic methanotrophic consortia. Proc. Natl. Acad. Sci. USA 101, 11111–11116.

Boxer, S.G., Kraft, M.L., Weber, P.K., 2009. Advances in imaging secondary ion mass spectrometry for biological samples. Annu. Rev. Biophys. 38, 53–74.

Brack, A., 1993. Liquid water and the origin of life. Origins of Life and Evolution of the Biosphere 23 (1), 3–10. Springer Nature, London, UK (global) Berlin, Germany (corporate). doi:10.1007/BF01581985.

Bradley, A.S., Fredricks, H., Hinrichs, K-U., Summons, R.E., 2009. Structural diversity of diether lipids in carbonate chimneys at the Lost City Hydrothermal Field. Org. Geochem 40, 1169–1178.

Brasier, M.D., Green, O.R., Lindsay, J.F., McLoughlin, N., Steele, A., Stoakes, C., 2005. Critical testing of Earth's oldest putative fossil assemblage from the ~3.5 Ga Apex chert, Chinaman Creek, Western Australia. Precamb. Res. 140, 55–102. doi:10.1016/j.precamres.2005.06.008.

Breitenstein, D., Rommel, C.E., Stolwijk, J., Wegener, J., Hagenhoff, B., 2008. The chemical composition of animal cells reconstructed from 2D and 3D ToF-SIMS analysis. Appl. Surf. Sci. 255, 1249–1256.

Brocks, J.J., Logan, G.A., Buick, R., Summons, R.E., 1999. Archean molecular fossils and the early rise of eukaryotes. Science 285, 1033–1036.

Brown, R.H., Baines, K.H., Bellucci, G., Bibring, J.P., Buratti, B.J., Capaccioni, F., Cerroni, P., Clark, R.N., Coradini, A., Cruikshank, D.P., Drossart, P., Formisano, V., et al., 2004. The Cassini visual and infrared mapping spectrometer (VIMS) investigation. Space Sci. Rev. 115, 111–168.

Brocks, J.J., 1999. Archean molecular fossils and the early rise of eukaryotes. Science 285, 1033–1036.

Brownlee, D.E., 1981. Extraterrestrial components in deep sea sediments. In: Emiliani, E. (Ed.). The Sea, 7. Wiley, New York, pp. 733–762.

Brownlee, D.E., Bates, B., Schramm, L., 1997. The elemental composition of stony cosmic spherules. Meteor. Planet. Sci. 32, 157–175.

Brunelle, A., Laprévote, O., 2009. Lipid imaging with cluster time-of-flight secondary ion mass spectrometry. Anal. Bioanal. Chem. 393, 31–35.

Cady, S.L., 2001. Paleobiology of the Archean Blum, P., ed., Archaea: Ancient Microbes, Extreme Environments, and the Origin of Life. Adv. Appl. Microbiol. 50, 1–35.

Cady, S.L., 2002. Formation and preservation of bona fide microfossils. Signs of Life: A Report Based on the April 2000 Workshop on Life-Detection Techniques: The National Academies Space Studies Board and Board on Life Sciences. Carnegie Institution of Washington, National Academies Press, Washington, D.C, pp. 149–155.

Cady, S.L., Farmer, J.D., Grotzinger, J.P., Schopf, J.W., Steele, A., 2003. Morphological biosignatures and the search for life on Mars. Astrobiology 3 (2), 351–368. doi:10.1089/153110703769016442.

Cady, S.L., Farmer, J.D., Grotzinger, J.P., Schopf, J.W., Steele A., 2004. Life on Mars, https://doi.org/10.1089/153110703769016442.

Cady, S., Noffke, N., 2009. Geobiology: evidence for earlylife on Earth and the search for life on other planets. GSAToday 19, 4–10.

Cannon, D.M., Pacholski, M.L., Winograd, N., Ewing, A.G., 2000. Molecule specific imaging of freeze-fractured, frozen-hydrated model membrane systems using mass spectrometry. J. Am. Chem. Soc. 122, 603–610.

Carr, M.H., Head, J.W., 2010. Geologic history of Mars. Earth Planet. Sci. Lett. 294 (3–4), 185–203. doi:10.1016/j.epsl.2009.06.042.

Charlou, J.L., Bougault, H., Appriou, P., Nelsen, T., Rona, P.A., 1991. Different TDM/CH_4 hydrothermal plume signatures: TAG site at 26°N and serpentinised ultrabasic diapir at 15°05′N on the Mid-Atlantic Ridge. Geochim Cosmochim Acta 55, 3209–3222.

Charlou, J.L., Bougault, H., Donval, J.P., Pellé, H., Langmuir, C.; the FAZAR scientific party, 1993a. Seawater CH_4 concentration over the Mid-Atlantic Ridge from the Hayes F.Z. to the Azores Triple Junction. Eos 74 (Supplement), 380.

Charlou, J.L., Bougault, H., Fouquet, Y., Donval, J.P., Douville, D., Radford-Knoery, J., Aalléa, M., Needham, H.D., Jean-Baptiste, P., Rona, P.A., Langmuir, C., German, C.R., 1996a. Methane degassing, hydrothermal activity and serpentinization between the fifteen-twenty Fracture Zone area and the Azores Triple Junction area (MAR), MAR Symposium, June 19–22, 1996, Reykjavik, Iceland. J. Conf. Abst. 1, 771–772.

Charlou, J.L., Donval, J.P., Konn, C., Ondreas, H., Fouquet, Y., Jean Baptiste, P., Fourré, E., 2010. High production and fluxes of H_2 and CH_4 and evidence of abiotic hydrocarbon synthesis by serpentinization in ultramafic-hosted hydrothermal systems on the Mid-Atlantic Ridge. In: Rona, P.A., Devey, C.W., Dyment, J., Murton, B.J. (Eds.), Diversity of Hydrothermal Systems on Slow-Spreading Ocean Ridges. American Geophysical Union, Washington, DC, pp. 265–296.

Charlou, J.L., Donval, J.P., 1993. Hydrothermal methane venting between 12°N and 26°N along the Mid-Atlantic Ridge. J. Geophys. Res. 98, 9625–9642.

Clark, R.N., Curchin, J.M., Hoefen, T.M., Swayze, G.A., 2009. Reflectance spectroscopy of organic compounds: 1. Alkanes. J. Geophys. Res. 114, E03001.

Claudi, R., 2017. Exoplanets: Possible biosignatures, *Proceedings of Science*, Frontier Research in Astrophysics âAS II, 23-28 May 2016. Mondello (Palermo), Italy arXiv:1708.05829v1 [astro-ph.EP] 19 Aug 2017.

Clemett, S.J., Dulay, M.T., Seb Gillette, J., Chillier, X.D.F., Mahajan, T.B., Zare, R.N., 1998. Evidence for the extraterrestrial origin of polycyclic aromatic hydrocarbons in the Martian meteorite ALH84001. Farad. Discuss. 109, 417–436.

Cockell, C.S., Kaltenegger, L., Raven, J.A., 2009. Cryptic photosynthesis—extrasolar planetary oxygen without a surface biological signature. Astrobiology 9 (7), 623–636.

Colla, T.J., Aderjan, R., Kissel, R., Urbassek, H.M., 2000. Sputtering of Au (111) induced by 16-keV Au cluster bombardment: spikes, craters, late emission, and fluctuations. Phys. Rev. B 62, 8487–8493.

Cooper, G., Kimmich, N., Belisle, W., Sarinana, J., Brabham, K., Garrel, L., 2001. Carbonaceous meteorites as a source of sugar-related organic compounds for the early Earth. Nature 414 (6866), 879–883.

Córdova, A., Engqvist, M., Ibrahem, I., Casas, J., Sundén, H., 2005. Plausible origins of homochirality in the amino acid catalyzed neogenesis of carbohydrates. Chem. Commun. (15), 2047–2049.

Cronin, J.R., Pizzarello, S., 1997. Enantiomeric excesses in meteoritic amino acids. Science 275 (5302), 951–955.

Cuif, J-P., Dauphin, Y., Sorauf, J.E., 2011. Biominerals and Fossils Through Time. Cambridge Univ. Press, Cambridge. ISBN 978-0-521-87473-1.

Davies, N.S., Liu, A.G., Gibling, M.R., Miller, R.F., 2016. Resolving MISS conceptions and misconceptions: a geological approach to sedimentary surface textures generated by microbial and abiotic processes. Earth Sci. Rev. 154, 210–246.

Dekas, A.E., Poretsky, R.S., Orphan, V.J., 2009. Deep-sea archaea fix and share nitrogen in methane-consuming microbial consortia. Science 326, 422–426.

Deming, D., Wilkins, A., McCullough, P., et al., 2013. Infrared transmission spectroscopy of the exoplanets HD 209458b and XO-1b using the wide field camera-3 on the *hubble space telescope*. Astrophys. J. 774 (2), 95.

Des Marais, D.J., Harwit, M., Jucks, K., Kasting, J.F., Lunine, J.I., Lin, D., Seager, S., Schneider, J., Traub, W., Woolf, N., 2002. Remote sensing of planetary properties and biosignatures on extrasolar terrestrial planets. Astrobiology (2), 153–181.

Domagal-Goldman, S.D., Segura, A., Claire, M.W., Robinson, T.D., Meadows, V.S., 2014. Abiotic ozone and oxygen in atmospheres similar to prebiotic Earth. Astrophys. J. 792, 90.

Eglinton, G., Scott, P.M., Belsky, T., Burlingame, A.L., Calvin, M., Cloud Jr., P.E., 1964. Hydrocarbons of biological origin from a one-billion-year-old sediment. Science 145, 263–264.

Eglinton, G., Scott, P.M., Belsky, T., Burlingame, A.L., Calvin, M., 1964. Hydrocarbons of biological origin from a one-billion-year-old sediment. Science 145, 263–264.

Eglinton, G., Calvin, M., 1967. Chemical fossils. Sci. Am. 216, 32–43.

Ehleringer, J.R., Osmond, C.B., 1989. Stable isotopes, In: Pearcy, R.W., Ehleringer, J.R. Mooney, H.A. Rundel, P.W. (Eds.), *Plant Physiological Ecology: FieldMethods and Instrumentation*, Chapman and Hall, New York, 1989.

Ehlmann, B.L., Mustard, J.F., Murchie, S.L., Poulet, F., Bishop, J.L., Brown, A.J., Calvin, W.M., Clark, R.N., Des Marais, D.J., Milliken, R.E., Roach, L.H., Roush, T.L., Swayze, G.L., Wray, J.J., 2008. Orbital identification of carbonate bearing rocks on Mars. Science 322, 1828–1832.

Ehrlich, H.L., Newman, D.K., 2009. Geomicrobiology, fifth ed. Taylor and Francis Group, CRC Press, Boca Raton, pp. 606.

Eigenbrode, J.L., 2008. Fossil lipids for life-detection: a case study from the early earth record. Space Sci. Rev. 135, 161–185.

Einaudi, M.T., Marvin, U.B., 1967. Black magnetic spherules from Pleistocene and recent beach sands. Geochim. Cosmochim. Acta 31, 1871–1872.

Etiope, G., Sherwood Lollar, B., 2013. Abiotic methane on Earth. Rev. Geophys. 51, 276–299.

Fahy, E., Subramaniam, S., Murphy, R.C., Nishijima, M., Raetz, C.R., Shimizu, T., Spener, F., van Meer, G., Wakelam, M.J., Dennis, E.A., 2009. Update of the LIPID MAPS comprehensive classification system for lipids. J. Lipid Res. 50 (S1), S9–S14. doi:10.1194/jlr.R800095-JLR200.

Fedo, C.M., Whitehouse, M.J., 2002. Metasomatic origin of quartz-pyroxene rock, Akilia, Greenland, and implications for Earth's earliest life. Science 296, 1448–1452. doi:10.1126/science.1070336.

Fletcher, J.S., Henderson, A., Jarvis, R.M., Lockyer, N.P., Vickerman, J.C., Goodacre, R., 2006. Rapid discrimination of the causal agents of urinary tract infection using ToF-SIMS with chemometric cluster analysis. Appl. Surf. Sci. 252, 6869–6874.

Fletcher, J.S., Lockyer, N.P., Vaidyanathan, S., Vickerman, J.C., 2007. TOF-SIMS 3D biomolecular imaging of Xenopus laevis oocytes using buckminsterfullerene (C60) primary ions. Anal. Chem. 79, 2199–2206.

Früh-Green, G.L., Kelley, D.S., 1998. Volatiles at MORs. Eos 79, 45.

Fyfe, W.S., 1974. Heats of chemical reactions and submarine heat production. Geophys. j. R. Astron. Soc. 37, 213–215.

Gardner, J.P., et al., 2006. The James Webb Space Telescope. Space Sci. Rev. 123, 485.

Garrison, B.J., Postawa, Z., 2008. Computational view of surface based organic mass spectrometry. Mass Spectrom. Rev. 27, 289–315.

Georgiou, C.D., Deamer, D.W., 2014. Lipids as universal biomarkers of extraterrestrial life. Astrobiology 14, 541–549.

Glikson, M., Duck, L.J., Golding, S.D., Hofmann, A., Bolhar, R., Webb, R., Baiano, J.C.F., Sly, L.I., 2008. Microbial remains in some earliest Earth rocks: comparison with a potential modern analogue. Precamb. Res. 164 (3-4), 187–200. doi:10.1016/j.precamres.2008.05.002.

Golden, D.C., Ming, D.W., Morris, R.V., Brearley, A.J., Lauer Jr., H.V., Treiman, A.H., Zolensky, M.E., Schwandt, C.S., Lofgren, G.E., McKay, G.A, 2004. Evidence for exclusively inorganic formation of magnetite in Martian meteorite ALH84001. Am. Mineral. 89, 681–695.

Greaves, J.S., Richards, A.M.S., Bains, W, 2020. Phosphine gas in the cloud decks of Venus. Nat. Astron. 5 (7), 655–664.

Greaves, J.S., et al., 2021. Reply to: No evidence of phosphine in the atmosphere of Venus from independent analyses. Nat. Astron. 5 (7), 636–639.

Grotzinger, J.P., Knoll, A.H., 1999. Stromatolites in Precambrian carbonates: evolutionary mileposts or environmental dipsticks?. Annu. Rev. Earth Planet. Sci. 27, 313–358. doi:10.1146/annurev.earth.27.1.313.

Grotzinger, J.P., 2014. Introduction to special issue – habitability, taphonomy, and the search for organic carbon on Mars. Science 343 (6169), 386–387. doi:10.1126/science.1249944.

Guidry, S.A., Chafetz, H.S., 2003. Depositional facies and diagenetic alteration in a relict siliceous hot-spring accumulation: examples from Yellowstone National Park, USA. J. Sediment. Res. 73, 806–823.

Haggerty, J.A., 1991. Evidence from fluid seeps atop serpentine seamounts in the Mariana Forearc: clues for emplacement of the seamounts and their relationship to forearc tectonics. Mar. Geol. 102, 293–309.

Hashimoto, J., Ohta, S., Gamo, T., Chiba, H., Yamaguchi, T., Tsuchida, S., Okudaira, T., Watabe, H., Yamanaka, T., Kitazawa, M., 2001. First hydrothermal vent communities from the Indian Ocean discovered. Zool. Sci. 18 (5), 717–721.

Hazen, R.M., Papineau, D., Bleeker, W., Downs, R.T., Ferry, J.M., McCoy, T.J., Sverjensky, D.A., Yang, H, 2008. Mineral evolution. Am. Mineral. 93, 1693–1720. doi:10.2138/am.2008.2955.

Heim, C., Sjövall, P., Lausmaa, J., Leefmann, T., Thiel, V., 2009. Spectral characterisation of eight glycerolipids and their detection in natural samples using time-of-flight secondary ion mass spectrometry. Rapid Commun. Mass Spectrom. 23, 2741–2753.

Herrmann, A.M., Ritz, K., Nunan, N., Clode, P.L., Pett-Ridge, J., et al., 2007. Nano-scale secondary ion mass spectrometry—a new analytical tool in biogeochemistry and soil ecology: a review article. Soil Biol. Biochem. 39, 1835–1850.

Holm, N.G., Oze, C., Mousis, O., Waite, J.H., Guilbert-Lepoutre, A., 2015. Serpentinization and the formation of H_2 and CH_4 on celestial bodies (Planets, Moons, Comets). Astrobiology 15, 587–600.

Johansson, B., 2006. ToF-SIMS imaging of lipids in cell membranes. Surf. Interface Anal. 38, 1401–1412.

Johnson, D.B., 1998. Microorganisms and the biogeochemical cycling of metals in aquatic environments. In: Langston, W.J., Bebianno, M.J. (Eds.), Metal Metabolism in Aquatic Environments. Springer, Boston, MA. https://doi.org/10.1007/978-1-4757-2761-6_3.

Kaltenegger, L., Selsis, F., Fridlund, M., Lammer, H., Beichman, C., Danchi, W., Eiroa, C., Henning, T., Herbst, T., Léger, A., Liseau, R., Lunine, J., Paresce, F., Penny, A., Quirrenbach, A., Rottgering, H., Schneider, J., Stam, D., Tinetti, G., White, G.J., 2010. Deciphering spectral fingerprints of habitable exoplanets. Astrobiology 10, 89.

Kasting, J.F., Whitmire, D.P., Reynolds, R.T., 1993. Habitable zones around main sequence stars. Icarus 101, 108.

Kasting, J.F., Catling, D., 2003. Evolution of a habitable planet. Ann. Rev. Astron. Astrophys. 41, 429.

Kelley, D.S., 1996. Methane-bearing fluids in the oceanic crust: gabbro-hosted fluid inclusions from the southwest Indian ridge. J. Geophys. Res. 101, 2943–2962.

Kiang, N.Y., Siefert, J., Govindjee, Blankenship, R.E, 2007. Spectral signatures of photosynthesis. I. Review of Earth organisms. Astrobiology 7, 7.

Kiang, N.Y., Domagal-Goldman, S., Parenteau, M.N., Catling, D.C., Fujii, Y., Meadows, V.S., Schwieterman, E.W., Walker, S.I., 2018. Exoplanet biosignatures: at the dawn of a new era of planetary observations. Astrobiology 18 (6), 619–629.

Kilston, S., Liu, M.C., Meikle, E., Reid, I.N., Rothschild, L.J., Scalo, J., Segura, A., Tang, C.M., Tiedje, J.M., Turnbull, M.C., Walkowicz, L.M, Weber, A.L., Young, R.E., 2007. A reappraisal of the habitability of planets around M dwarf stars. Astrobiology 7 (1), 30–65.

Kleidon, A., 2012. How does the Earth system generate and maintain thermodynamic disequilibrium and what does it imply for the future of the planet?. Philos. Trans. R. Soc. A: Math. Phys. Eng. Sci. https://doi.org/10.1098/rsta.2011.0316.

Koeberl, C., Hagen, E.H., 1989. Extraterrestrial spherules in glacial sediment from the Transantarctic Mountains, Antarctica: structure, mineralogy, and chemical composition. Geochim. Cosmochim. Acta 53, 937–944.

Koppaparu, R.K., Ramirez, R., Kasting, J.F., Eymet, V., Robinson, T.D., Mahadevan, S., Terrien, R.C., Domagal–Goldman, S., Meadows, V., Deshpande, R., 2013. Habitable zones around main–sequence stars: new estimates. Astroph. J. 765, 131.

Krissansen-Totton, J., Bergsman, D.S., Catling, D.C., 2016. On detecting biospheres from chemical thermodynamic disequilibrium in planetary atmospheres. Astrobiology 16 (1). doi:10.1089/ast.2015.1327.

Kurczy, M.E., Piehowsky, P.D., Willingham, D., Molyneaux, K.A., Heien, M.L., et al., 2010. Nanotome cluster bombardment to recover spatial chemistry after preparation of biological samples for SIMS imaging. J. Am. Soc. Mass Spectrom. 21, 833–836.

Kvenvolden, K.A., Lawless, J., Pering, K., Peterson, E., Flores, J., Ponnamperuma, C., Kaplan, I.R., Moore, C., 1970. Evidence for extraterrestrial amino-acids and hydrocarbons in the Murchison meteorite. Nature 228 (5275), 923–926.

Lagrange, A.-M., Gratadour, D., Chauvin, G., Fusco, T., Ehrenreich, D., Mouillet, D., Rousset, G., Rouan, D., Allard, F., Gendron, É., Charton, J., Mugnier, L., Rabou, P., Montri, J., Lacombe, F., 2009. A probable giant planet imaged in the β Pictoris disk. VLT/NaCo deep L'-band imaging. Astron. Astrophys. 493, L21–L25.

Lanekoff, I., Kurczy, M.E., Adams, K.L., Malm, J., Karlsson, R., et al., 2011. An in situ fracture device to image lipids in single cells using ToF-SIMS. Surf. Interface Anal. 43, 257–260.

Lederberg, J., 1965. Signs of life. Criterion-system of exobiology. Nature 207, 9.

Lee, J.L.S., Gilmore, I.S., Fletcher, I.W., Seah, M.P., 2008. Topography and field effects in the quantitative analysis of conductive surfaces using ToF-SIMS. Appl. Surf. Sci. 255, 1560–1563.

Leger, A., Pirre, M., Marceau, F.J., 1993. Search for primitive life on a distant planet: relevance of O_2 and O_3 detections. Astron. Astrophys. 277, 309.

Léger, A., Pirre, M., Marceau, F.J., 1993. Search for primitive life on a distant planet: relevance of O_2 and O_3 detections. Astron. Astrophys. 277, 309.

Lehninger, N., et al., 2008. Lehninger Principles of Biochemistry. Macmillan, Stuttgart, Germany, pp. 474.

Lepland, A., van Zuilen, M.A., Arrhenius, G., Whitehouse, M.J., Fedo, C.M., 2005. Questioning the evidence for Earth's earliest life—Akilia revisited. Geology 33, 77–79. doi:10.1130/G20890.1.

Lincowski, A.P., et al., 2021. Claimed detection of PH_3 in the clouds of Venus is consistent with mesospheric SO_2. Astrophys. J. Lett. 908 (2), L44.

Lipp, J.S., Hinrichs, K-U., 2009. Structural diversity and fate of intact polar lipids in marine sediments. Geochim. Cosmochim. Acta 73, 6816–6833.

Lippincott, E.R., Eck, R.V., Dayhoff, M.O., Sagan, C., 1967. Thermodynamic equilibria in planetary atmospheres. Astrophys. J. 147, 753.

Lorenz, R., Mitton, J., 2002. Lifting Titan's Veil. Cambridge University Press, Cambridge, United Kingdom, pp. 260.

Love, S.G., Brownlee, D.E., 1993. A direct measurement of the terrestrial mass accretion rate of cosmic dust. Science 256, 550–553.

Lovelock, J.E., 1965. A physical basis for life detection experiments. Nature 207, 568.

Lovelock, J.E., 1975. Thermodynamics and the recognition of alien biospheres. Proc. R. Soc. Lond. Ser. B Biol. Sci. 189 (1095), 167–180.

Lovelock, J.E., Kaplan, I.R., 1975. Thermodynamics and the recognition of alien biospheres. Proc. R. Soc. Lond. B 189, 167.

Marshall, C.P., Love, G.D., Snape, C.E., Hill, A.C., Allwood, A.C., Walter, M.R., Van Kranendonk, M.J., Bowden, S.A., Sylva, S.P., Summons, R.E., 2007. Structural characterization of kerogen in 3.4 Ga Archaean cherts from the Pilbara Craton, Western Australia. Precamb. Res. 155, 1–23. doi:10.1016/j.precamres.2006.12.014.

Matson, J., 2010. Meteorite that fell in 1969 still revealing secrets of the early Solar System. Sci. Am.

Maurette, M., Hammer, C., Brownlee, D.E., Reeh, N., 1986. Traces of cosmic dust in blue ice lake Greenland. Science 233, 869–872.

Maurette, M., Jehanno, C., Robin, E., Hammer, C., 1987. Characteristics and mass distribution of extraterrestrial dust from the Greenland ice cap. Nature 328, 699–702.

Maurette, M., Olinger, C., Michel-Levy, Ch.M., Kurat, G., Purchet, M., Brandstatter, F., Bourot-Denise, M., 1991. A collection of diverse micrometeorites recovered from 100 tonnes of Antarctic blue ice. Nature 351, 44–47.

Maurette, M., Immel, G., Perreau, M., Porchet, M., Vincent, C., Kurat, G., 1992. The 1991 EUROMET collection of micrometeorites at Cap Prudhomme, Antarcica: discussion of possible collection biases, Proceedings of the Lunar Planetary Science Conference 33rd, 859–860.

McKay, D.S., Gibson, E.K., Thomas-Keprta, K.L., Vali, H., Romanek, C.S., Clemett, S.J., Chillier, X.D.F., Maechling, C.R., Zare, R.N., 1996. Search for past life on Mars: possible relic biogenic activity in martian meteorite ALH84001. Science 273, 924–930.

McCollom, T.M., Seewald, J.S., 2001. A reassessment of the potential for reduction of dissolved CO_2 to hydrocarbons during serpentinization of olivine. Geochim. Cosmochim. Acta 65, 3769–3778.

McKay, C.P., 2016. Titan as the abode of life. Life 6, 8.

McKeegan, K.D., Kudryantsev, A.B., Schopf, J.W., 2007. Raman and ion microscopic imagery of graphitic inclusions in apatite from older than 3830 Ma Akiliasupracrustal rocks, west Greenland. Geology 35, 591–594. doi:10.1130/G23465A.1.

Meadows, V.S., 2017. Reflections on O_2 as a biosignature in exoplanetary atmospheres. Astrobiology 17 (10), 1022–1052.

Meadows, V.S., Reinhard, C.T., Arney, G.N., Parenteau, M.N., Schwieterman, E.W., Domagal-Goldman, S.D., Lincowski, A.P., Stapelfeldt, K.R., Rauer, H., DasSarma, S., Hegde, S., Narita, N., Deitrick, R., Lustig-Yaeger, J., Lyons, T.W., Siegler, N., Grenfell, J.L., 2018. Exoplanet biosignatures: understanding oxygen as a biosignature in the context of its environment. Astrobiology 18 (6), 630–662.

Meierhenrich, U.J., Bredehöft, J.H., Jessberger, E.K., Thiemann, W.H.-P., 2004. Identification of diamino acids in the Murchison meteorite. PNAS 101 (25), 9182–9186.

Millard, H.T., Finkelman, R.B., 1970. Chemical and mineralogical compositions of cosmic and terrestrial spherules from a marine sediment. J. Geophys. Res. 75, 2125–2133.

Moldowan, J.M., Jacobson, S.R., 2010. Chemical signals for early evolution of major taxa: biosignatures and taxon-specific biomarkers, 805-812. https://doi.org/10.1080/00206810009465112

Murray, J., Renard, A.F., 1891. Deep-sea deposits (based on the specimens collected during the voyage of HMS Challenger in the years 1872 to 1876). Report on the scientific results of the voyage of H.M.S. Challenger during the years 1873-76. John Menzies and Co., Endinburgh, United Kingdom, pp. 688.

Nauhaus, K., Treude, T., Boetius, A., Kruger, M., 2005. Environmental regulation of the anaerobic oxidation of methane: a comparison of ANME-I and ANME-II communities. Environ. Microbiol. 7, 98–106.

Neal, C., Stanger, G., 1983. Hydrogen generation from mantle source rocks in Oman. Earth Planet Sci. Lett. 66, 315–320.

Neish, C.D., Lorenz, R.D., Turtle, E.P., Barnes, J.W., Trainer, M.G., Stiles, B., Kirk, R., Hibbitts, C.A., Malaska, M.J., 2018. Strategies for detecting biological molecules on Titan. Astrobiology 18 (5), 571–585. https://doi.org/10.1089/ast.2017.1758.

Newman, S.A., et al., 2020. Lipid biomarker record of the serpentinite-hosted ecosystem of the Samail Ophiolite, Oman and implications for the search for biosignatures on Mars. Astrobiology 20 (7), 830–845. doi:10.1089/ast.2019.2066.

Noffke, N., Awramik, S.M., 2011. Stromatolites and MISS—differences between relatives. GSA Today 23, 4–9.

Noffke, N., Christian, D., Wacey, D., Hazen, R.M., 2013. Microbially induced sedimentary structures recording an ancient ecosystem in the ca. 3.48 billion-year-old dresser formation, Pilbara, Western Australia. Astrobiology 13, 1103–1124.

Oehler, D.Z., Robert, F., Walter, M.R., Sugitani, K., Allwood, A., Meibom, A., Mostefaoui, S., Selo, M., Thomen, A., and Gibson, E.K. (2009), NanoSIMS: Insights to biogenicity and syngeneity of Archaean carbonaceous structures, Precamb. Res., doi: 10.1016/j.precamres.2009.01.001.

Oehler, D.Z., Robert, F., Walter, M.R., Sugitani, K., Meibom, A., et al., 2010. Diversity in the Archean biosphere: new insights from NanoSIMS. Astrobiology 10, 413–424.

Orphan, V.J., House, C.H., 2009. Geobiological investigations using secondary ion mass spectrometry: microanalysis of extant and paleo-microbial processes. Geobiology 7, 360–372.

Orphan, V.J., Turk, K.A., Green, A.M., House, C.H., 2009. Patterns of 15N assimilation and growth of methanotrophic ANME-2 archaea and sulfate-reducing bacteria within structured syntrophic consortia revealed by FISH-SIMS. Environ. Microbiol. 11, 1777–1791.

Ostrowski, S.G., van Bell, C.T., Winograd, N., Ewing, A.G., 2004. Mass spectrometric imaging of highly curved membranes during *Tetrahymena* mating. Science 305, 71–73.

Ostrowski, S.G., Szakal, C., Kozole, J., Roddy, T.P., Xu, J., et al., 2005. Secondary ion MS imaging of lipids in picoliter vials with a buckminsterfullerene ion source. Anal. Chem. 77, 6190–6196.

Owen, T., 1980. Strategies for the Search for Life in the Universe. In: Papagiannis, M.D. (Ed.). Reidel Publ. Co., Dordrecht, Holland, pp. 177.

Parenteau, M.N., Jahnke, L.L., Farmer, J.D., Cady, S.L., 2014. Production and early preservation of lipid biomarkers in iron hot-springs. Astrobiology 14 (6), 502–521.

Parnell, J., Lindgren, P., 2006. Survival of reactive carbon through meteorite impact melting. Geology 34 (12), 1029–1032. doi.org/10.1130/G22731A.1.

Parnell, J., Cullen, D., Sims, M.R., Bowden, S., Cockell, C.S., Court, R., Ehrenfreund, P., Gaubert, F., Grant, W., Parro, V., Rohmer, M., Sephton, M., Stan-Lotter, H., Steele, A., Toporski, J., Vago, J., 2007. Searching for life on Mars: selection of molecular targets for ESA's aurora ExoMars mission. Astrobiology 7, 578–604.

Pieters, C.M., Goswami, J.N., Clark, R.N., Annadurai, M., Boardman, J., Buratti, B., Combe, J.-P., Dyar, M.D., Green, R., Head, J.W., Hibbitts, C., Hicks, M., Isaacson, P., Klima, R., Kramer, G., Kumar, S., et al., 2009. Character and spatial distribution of OH/H_2O on the surface of the moon seen by M3 on Chandrayaan-1. Science 326, 568–572.

Pilcher, C.B., Ridgway, S.T., McCord, T.B., 1972. Galilean satellites: identification of Water Frost. Science 178, 1087–1089.

Pitcher, A., Hopmans, E.C., Schouten, S., Sinninghe Damsté, J.S., 2009. Separation of core and intact polar archaeal tetraether lipids using silica columns: insights into living and fossil biomass contributions. Org. Geochem. 40, 12–19.

Rasmussen, B., Fletcher, I.R., Brocks, J.J., Kilburn, M.R., 2008. Reassessing the first appearance of eukaryotes and cyanobacteria. Nature 455, 1101–1104.

Reinhard, C.T., Olson, S.L., Schwieterman, E.W., Lyons, T.W., 2017. False negatives for remote life detection on ocean-bearing planets: lessons from the Early Earth. Astrobiology 17 (4), 287–297.

Riding, R.E., Awramik, S.M (Eds.), 2000. Microbial Sediments. Springer, Springer-Verlag, Berlin Heidelberg New York, pp. 329.

Rogers, A.D., Tyler, P.A., Connelly, D.P., Copley, J.T., James, R., Larter, R.D., Linse, K., Mills, R.A., Garabato, A.N., Pancost, R.D., Pearce, D.A., Polunin, N.V.C., German, C.R., Shank, T., Boersch-Supan, P.H., Alker, B.J., Aquilina, A., Bennett, S.A., Clarke, A., Dinley, R.J.J., Graham, A.G.C., Green, D.R.H., Hawkes, J.A., Hepburn, L., Hilario, A., Huvenne, V.A.I., Marsh, L., Ramirez-Llodra, E., Reid,

W.D.K., Roterman, C.N., Sweeting, C.J., Thatje, S., Zwirglmaier, K., 2012. The discovery of new deep-sea hydrothermal vent communities in the Southern Ocean and implications for biogeography. PLoS Biol. 10 (1), e1001234. http://dx. doi.org/10.1371/journal.pbio.1001234.

Röling, W.F., Aerts, J.W., Patty, C.H., ten Kate, I.L., Ehrenfreund, P., Direito, S.O., 2015. The significance of microbe-mineral-biomarker interactions in the detection of life on Mars and beyond. Astrobiology 15 (6), 492–507. doi:10.1089/ast.2014.1276.

Rosenqvist, J., Chassefiere, E., 1995. Inorganic chemistry of O_2 in a dense primitive atmosphere. Planet. Space Sci. 43, 3–10.

Rosing, M.T., 1999. 13-C depleted carbon microparticles in >3700 Ma seafloor sedimentary rocks from West Greenland. Science 283, 674–676. doi:10.1126/science.283.5402.674.

Rossel, P.E., Lipp, J.S., Fredricks, H.F., Arnds, J., Boetius, A., et al., 2008. Intact polar lipids of anaerobic methanotrophic archaea and associated bacteria. Org. Geochem. 39, 992–999.

Sagan, C., 1975. The recognition of extraterrestrial intelligence. Proc. R. Soc. Lond. B 189, 143.

Sagan, C., Thompson, W.R., Carlson, R., Gurnett, D., Hord, C., 1993. A search for life on Earth from the Galileo spacecraft. Nature 365, 715.

Scalo, J., Kaltenegger, L., Segura, A., Fridlund, M., Ribas, I., Kulikov, Y.N., Grenfell, J.L., Rauer, H., Odert, P., Leitzinger, M., Selsis, F., Khodachenko, M.L., Eiroa, C., Kasting, J., Lammer, H., 2007. M stars as targets for terrestrial exoplanet searches and biosignature detection. Astrobiology 7 (1), 85–166.

Schiffbauer, J.D., Yin, L., Bodnar, R.J., Kaufman, A.J., Meng, F., Hu, J., Shen, B., Yuan, X., Bao, H., Xiao, S, 2007. Ultrastructural and geochemical characterization of Archean-Paleoproterozoic graphite particles: Implications for recognizing traces of life in highly metamorphosed rocks. Astrobiology 7, 684–704. doi:10.1089/ast.2006.0098.

Schmitt-Kopplin, P., Gabelica, Z., Gougeon, R.D., Fekete, A., Kanawati, B., Harir, M., Gebefuegi, I., Eckel, G., Hertkorn, N., 2010. High molecular diversity of extraterrestrial organic matter in Murchison meteorite revealed 40 years after its fall. PNAS 107 (7), 2763–2768.

Schopf, J.W., Kudryavtsev, A.B., Agresti, D.G., Czaja, A.D., Wdowiak, T.J., 2005. Raman imagery: A new approach to assess the geochemical maturity and biogenicity of permineralized Precambrian fossils. Astrobiology (5), 333–371. doi:10.1089/ast.2005.5.333.

Schopf, J.W., Kudryavtsev, A.B., 2009. Confocal laser scanning microscopy and Raman imagery of ancient microscopic fossils. Precamb. Res. doi:10.1016/j.precamres.2009.02.007.

Schubotz, F., Wakeham, S.G., Lipp, J.S., Fredricks, H.F., Hinrichs, K-U., 2009. Detection of microbial biomass by intact polar membrane lipid analysis in the water column and surface sediments of the Black Sea. Environ. Microbiol. 11, 2720–2734.

Schwieterman, E.W., Kiang, N.Y., Parenteau, M.N., Harman, C.E., DasSarma, S., Fisher, T.M., Arney, G.N., Hartnett, H.E., Reinhard, C.T., Olson, S.L., Meadows, V.S., Cockell, C.S., Walker, S.I., Grenfell, J.L., Hegde, S., Rugheimer, S., Hu, R., Lyons, T.W., 2018. Exoplanet biosignatures: A review of remotely detectable signs of life. Astrobiology 18 (6), 663–708.

Seager, S., 2010. Exoplanet Atmospheres: Physical Processes. Princeton Univ. Press, Princeton, NJ.

Seager, S., 2013. Exoplanet habitability. Science 340 (6132), 577–581. doi:10.1126/science.1232226.

Seager, S., Bains, W., Hu, R., 2013. A biomass-based model to estimate the plausibility of exoplanet biosignature gases. Astrophys. J. 775, 104.

Seager, S., Bains, W., Hu, R., 2013. Biosignature gases in H_2-dominated atmospheres on rocky exoplanets. Astrophys. J. 777 (2), 95.

Seager, S., 2014. The future of spectroscopic life detection on exoplanets. PNAS 111, 12634.

Seager, S., Bains, W., Petkowski, J.J., 2016. Toward a list of molecules as potential biosignature gases for the search for life on exoplanets and applications to terrestrial biochemistry. Astrobiology 16 (6), 465–485.

Segura, A., Meadows, V.S., Kasting, J.F., Crisp, D., Cohen, M., 2007. Abiotic formation of O_2 and O_3 in high- CO_2 terrestrial atmospheres. Astron. Astrophys. 472, 665.

Selsis, F., Despois, D., Parisot, J.-P., 2002. Signature of life on exoplanets: Can Darwin produce false positive detections? Astron. Astrophys. 388, 985–991.

Selsis, F., Kasting, J.F., Levrard, B., Paillet, J., Ribas, I., Delfosse, X., 2007. Habitable planets around the star Gliese 581? Astron. Astrophys. 476, 1373.

Sigel, A., Sigel, H., Sigel, R.K.O., 2008. Biomineralization: From Nature to Application, Metal Ions in Life Sciences, Wiley, Hoboken, New Jersey, pp. 700.

Sims, M.R., Cullen, D., Rix, C.S., et al., 2012. Development status of the life marker chip instrument for ExoMars. Planet. Space Sci. 72 (1), 129–137. doi:10.1016/j.pss.2012.04.007.

Sjövall, P., Lausmaa, J., Johansson, B, 2004. Mass spectrometric imaging of lipids in brain tissue. Anal. Chem. 76, 4271–4278.

Stadnitskaia, A., Nadezhkin, D., Abbas, B., Blinova, V., Ivanov, M.K., Sinninghe Damsté, J.S., 2008. Carbonate formation by anaerobic oxidation of methane: evidence from lipid biomarker and fossil 16S rDNA. Geochim. Cosmochim. Acta 72, 1824–1836.

Steele, A., Goddard, D.T., Stapleton, D., Toporski, J.K.W., Peters, V., Bassinger, V., Sharples, G., Wynn-Williams, D.D., Mckay, D.S., 2000. Investigations into an unknown organism on the martian meteorite Allan Hills 84001. Meteorit. Planet Sci. 35, 237–241.

Steele, A., Beaty, D.W., Amend, J., Anderson, R., Beegle, L., Benning, L., Bhattacharya, J., Blake, D., Brinckerhoff, W., Biddle, J., Cady, S., Conrad, P., Lindsay, J., Mancinelli, R., Mungas, G., Mustard, J., Oxnevad, K., Toporski, J., and Waite, H. (2005), The Astrobiology Field Laboratory: JPL Document Review Services (Reference #CL#06-3307), Mars Exploration Program Analysis Group (MEPAG), Unpublished white paper, 72 p., http://mepag.jpl.nasa.gov/reports/index.html

Steele, A., McCubbin, F.M., Fries, M., Kater, L., Boctor, N.Z., Fogel, M.L., Conrad, P.G., Glamoclija, M., Spencer, M., Morrow, A.L., Hammond, M.R., Zare, R.N., Vicenzi, E.P., Siljestrom, S., Bowden, R., Herd, C.D.K., Mysen, B.O., Shirey, S.B., Amundsen, H.E.F., Treiman, A.H., Bullock, E.S., Jull, A.J.T., 2012. A reduced organic carbon component in martian basalts. Science 337, 212–215.

Stein, G.M., 2015. Discovering the North-West Passage: The Four-Year Arctic Odyssey of H.M.S. Investigator and the McClure Expedition. McFarland & Company, Inc., Publishers, Jefferson, North Carolina, pp. 376.

Stephan, T., Jessberger, E.K., Heiss, C.H., Rost, D, 2003. TOF-SIMS analysis of polycyclic aromatic hydrocarbons in Allan Hills 84001. Meteorit. Planet. Sci. 38, 109–116.

Sturt, H.F., Summons, R.E., Smith, K., Elvert, M., Hinrichs, K-U., 2004. Intact polar membrane lipids in prokaryotes and sediments deciphered by high-performance liquid chromatography/electrospray ionization multistage mass spectrometry—new biomarkers for biogeochemistry and microbial ecology. Rapid Commun. Mass Spectrom. 18, 617–628.

Subramaniam, S., Fahy, E., Gupta, S., Sud, M., Byrnes, R.W., Cotter, D., Dinasarapu, A.R., Maurya, M.R., 2011. Bioinformatics and systems biology of the lipidome. Chem. Rev. 111 (10), 6452–6490. doi:10.1021/cr200295k.

Summons, R.E., Albrecht, P., McDonald, G., Moldowan, J.M., 2008b. Molecular biosignatures. Space Sci. Rev. 135, 133–159.

Summons, R.E., Albrecht, P., McDonald, G., Moldowan, J.M., 2008. Molecular biosignatures. In: Botta, O., Bada, J.L., Gomez-Elvira, J., Javaux, E., Selsis, F., Summons, R. (Eds.). *Strategies of Life Detection*, Space Sciences Series of ISSI, 25. Springer, Boston, MA. https://doi.org/10.1007/978-0-387-77516-6_11.

Summons, R.E., Amend, J.P., Bish, D., Buick, R., Cody, G.D., Des Marais, D.J., Dromart, G., Eigenbrode, J.L., Knoll, A.H., Sumner, D.Y., 2011. Preservation of martian organic and environmental records: final report of the Mars biosignature working group. Astrobiology 11, 157–181.

Sunshine, J.M., Farnham, T.L., Feaga, L.M., Groussin, O., Merlin, F., Milliken, R.E., A'Hearn, M.F., 2009. Temporal and spatial variability of lunar hydration as observed by the deep impact spacecraft. Science 326, 565–568.

Suo, Z., Avci, R., Schweitzer, M.H., Deliorman, M., 2007. Porphyrin as an ideal biomarker in the search for extraterrestrial life. Astrobiology 7, 605–615.

Tanaka, K.L., 1986. The stratigraphy of Mars. J. Geophys. Res. 91 (B13), E139–E158. doi:10.1029/JB091iB13p0E139.

Taylor, S., Brownlee, D.E., 1991. Cosmic spherules in the geologic record. Meteoritics 26, 203–211.

Thiel, V., Heim, C., Arp, G., Hahmann, U., Sjövall, P., Lausmaa, J., 2007a. Biomarkers at the microscopic range: ToF-SIMS molecular imaging of Archaea-derived lipids in a microbial mat. Geobiology 5, 413–421.

Thiel, V., Sjövall, P., 2011. Using time-of-flight secondary ion mass spectrometry to study biomarkers. Annu. Rev. Earth Planet. Sci. 39 (1), 125–156.

Thomas-Keprta, K.L., Clemett, S.J., Bazylinski, D.A., Kirschvink, J.L., McKay, D.S., Wentworth, S.J., Vali, H., Gibson, E.K., Romanek, C.S., 2002. Magnetofossils from ancient Mars: a robust biosignature in the Martian meteorite ALH84001. Appl. Environ. Microbiol. 68, 3663–3672.

Thompson, M.A., 2021. The statistical reliability of 267 GHz JCMT observations of Venus: No significant evidence for phosphine absorption. Mon. Not. R. Astronom. Soc.: Lett. 501 (1), L18–L22.

Toporski, J.K.W., Steele, A., Westall, F., Avci, R., Martill, D.M., McKay, D.S., 2002. Morphological and spectral investigation of exceptionally well preserved bacterial biofilms from the Oligocene Enspel formation, Germany. Geochim. Cosmochim. Acta 66, 1773–1791.

Touboul, D., Brunelle, A., Halgand, F., De La Porte, S., Laprevote, O., 2005a. Lipid imaging by gold cluster time-of-flight secondary ion mass spectrometry: application to Duchenne muscular dystrophy. J. Lipid Res. 46, 1388–1395.

Touboul, D., Kollmer, F., Niehuis, E., Brunelle, A., Laprévote, O., 2005b. Improvement of biological time-of-flight-secondary ion mass spectrometry imaging with a bismuth cluster ion source. J. Am. Soc. Mass Spectrom. 16, 1608–1618.

Treibs, A., 1934. Chlorophyll- und Häminderivate in bituminösen Gesteinen, Erdölen, Erdwachsen und Asphalten. Ein Beitrag zur Entstehung des Erdöls. Justus Liebigs Ann. Chem. 510, 42–62.

Vago, J.L., et al., 2017. Habitability on early Mars and the search for biosignatures with the ExoMars Rover. Astrobiology 17 (6-7), 471–510.

Vaidyanathan, S., Fletcher, J.S., Jarvis, R.M., Henderson, A., Lockyer, N.P., et al., 2009. Explanatory multivariate analysis of ToF-SIMS spectra for the discrimination of bacterial isolates. Analyst 134, 2352–2360.

van Zuilen, M.A., Lepland, A., Arrhenius, G, 2002. Reassessing the evidence for the earliest traces of life. Nature 418, 627–630. doi:10.1038/nature00934.

van Zuilen, M.A., Chaussidon, M., Rollion-Bard, C., Marty, B, 2007. Carbonaceous cherts of the Barberton Greenstone Belt, South Africa; Isotopic, chemical, and structural characteristics of individual microstructures. Geochim. Cosmochim. Acta 71, 655–669. doi:10.1016/j.gca.2006.09.029.

Vickerman, J.C., 2001. ToF-SIMS—an overview. In: Vickerman, J.C., Briggs, D. (Eds.), ToF-SIMS: Surface Analysis by Mass Spectrometry. IM Publications, Chichester, UK, pp. 1–36.

Wacey, D., Kilburn, M.R., McLoughlin, N., Parnell, J., Stoakes, CA, et al., 2008. Use of NanoSIMS in the search for early life on Earth: ambient inclusion trails in a c. 3400 Ma sandstone. J. Geol. Soc. 165, 43–53.

Wagstaff, K.L., Corsetti, F.A., 2010. An evaluation of information-theoretic methods for detecting structural microbial biosignatures. Astrobiology 10 (4), 363–379.

Walker, J.C.G., 1977. Evolution of the Atmosphere. Macmillan, New York.

Walter, M.R., 1976. Stromatolites, Vol. 20, 1st edition, Elsevier, eBook ISBN: 9780080869322.

Wanger, G., Southam, G., Onstott, T.C., 2006. Structural and chemical characterization of a natural fracture surface from 2.8 kilometers below land surface: biofilms in the deep subsurface. Geomicrobiol. J. 23, 443–452.

Weiner, S., Lowenstam, H.A., 1989. On Biomineralization. Oxford University Press, Oxford [Oxfordshire] ISBN 978-0-19-504977-0.

Westall, F., 2008. Morphological biosignatures in early terrestrial and extraterrestrial materials. Space Sci. Rev. 135, 95–114.

Westall, F., 2012. The early Earth. In: Impey, C., Lunine, J., Funes, J. (Eds.), Frontiers of Astrobiology. Cambridge University Press, Cambridge, pp. 331.

Wierzchos, J., Ascaso, C., Sancho, L.G., Green, A., 2003. Iron-rich diagenetic minerals are biomarkers of microbial activity in Antarctic rocks. Geomicrobiol. J. 20, 15–24.

Winograd, N., Garrison, B.J., 2010. Biological cluster mass spectrometry. Annu. Rev. Phys. Chem. 61, 305–322.

Zink, K-G., Wilkes, H., Disko, U., Elvert, M., Horsfield, B., 2003. Intact phospholipids–microbial "life markers" in marine deep subsurface sediments. Org. Geochem. 34, 755–769.

Bibliography

Kleidon, A., 2012. How does the Earth system generate and maintain thermodynamic disequilibrium and what does it imply for the future of the planet?. Philos. Trans. R. Soc. A: Math. Phys. Eng. Sci. https://doi.org/10.1098/rsta.2011.0316.

Lovelock, J.E., 1965. A physical basis for life detection experiments. Nature 207, 568–570. doi:10.1038/207568a010.1038/207568a0.

Lovelock, J.E., Margulis, L., 1974. Atmospheric homeostasis by and for the biosphere: the Gaia hypothesis. Tellus 26, 2–10. doi:10.1111/j.2153-3490.1974.tb01946.x10.1111/j.2153-3490.1974.tb01946.x.

Lovelock, J.E., 1975. Thermodynamics and the recognition of alien biospheres [and Discussion]. Proc. R. Soc. Lond. B 189, 167–181. doi:10.1098/rspb.1975.005110.1098/rspb.1975.0051.

Poch, O., Frey, J., Roditi, I., Pommerol, A., Jost, B., Thomas, N., 2017. Remote sensing of potential biosignatures from rocky, liquid, or icy (exo)planetary surfaces. Astrobiology 17 (3), 231–252.

Watson, A.J., Lovelock, J.E., 1983. Biological homeostasis of the global environment: the parable of Daisyworld. Tellus 35B, 284–289. doi:10.1111/j.1600-0889.1983.tb00031.x10.1111/j.1600-0889.1983.tb00031.x.

Westall, F., Foucher, F., Bost, N., Bertrand, M., Loizeau, D., Vago, J.L., Kminek, G., Gaboyer, F., Campbell, K.A., Bréhéret, J.-G., Gautret, P., Cockell, C.S., 2015. Biosignatures on Mars: What, where, and how? Implications for the search for Martian life. Astrobiology 15, 998–1029.

Chapter 5

Extremophiles—Organisms that survive and thrive in extreme environmental conditions

5.1 Relevance of astrobiology

Is organic life a unique phenomenon on Earth? It was only during the past few decades that the science of studying life outside Earth began to advance. Today this scientific field is known as "astrobiology." In the context of planetary research, astrobiology (i.e., study of the origin, evolution, and distribution of life in the context of cosmic evolution; including habitability in the solar system and beyond) assumes great importance and relevance.

According to Rampelotto (2010), "One of the prominent goals of astrobiology is to discover life or signs of life on planets beyond Earth. Because currently, there is no direct evidence for life on another planet, our notions of habitability are necessarily constrained by our knowledge of life on Earth. Thus, one question emerges: Are there environments on Earth, which may resemble those we expect to find on other worlds? In this question lies the fundamental connection between the extreme environments on Earth and the search for life on other planets. When we look at our own planet's most challenging environments, we are really looking for clues to what may be the normal conditions on other planets. Therefore, we use our knowledge of the extremes of life on Earth to assess extraterrestrial environments and the plausibility that they can sustain life. Nowadays, based upon the increasing knowledge originated by studies of terrestrial extreme environments, numerous planetary bodies in our Solar System appear to have provided suitable conditions at some point in their history for the emergence of life. Among them, the two candidates that appear more susceptible to sustain life are Mars and Europa."

Cottin et al. (2017) have presented an interdisciplinary review of current research in astrobiology, covering the major advances and main outlooks in the field. They summarized the meaning and relevance of astrobiology thus, "Astrobiology is an interdisciplinary scientific field not only focused on the search of extraterrestrial life, but also on deciphering the key environmental parameters that have enabled the emergence of life on Earth. Understanding these physical and chemical parameters is fundamental knowledge necessary not only for discovering life or signs of life on other planets, but also for understanding our own terrestrial environment. Therefore, astrobiology pushes us to combine different perspectives such as the conditions on the primitive Earth, the physicochemical limits of life, exploration of habitable environments in the Solar System, and the search for signatures of life in exoplanets."

The astrobiology institute of the United Nation's National Aeronautics and Space Administration (NASA), which witnessed revolutionary discoveries, defines astrobiology as "the study of the origins, evolution, distribution, and future of life in the universe. This interdisciplinary field requires a comprehensive, integrated understanding of biological, geological, planetary, and cosmic phenomena. Astrobiology encompasses the search for habitable environments in our solar system and on planets around other stars; the search for evidence of prebiotic chemistry or life on solar system bodies such as Mars, Jupiter's moon Europa, and Saturn's moon Titan; and research into the origin, early evolution, and diversity of life on Earth. Astrobiologists address three fundamental questions: How does life begin and evolve? Is there life elsewhere in the Universe? What is the future of life on Earth and beyond?"

Taking account of the interdisciplinary nature of astrobiology, extensive contributions have been made by chemists, biologists, geologists, planetologists, and astrophysicists; and such contributions have brought about profound transformations in this interdisciplinary research field over a relatively short period of time.

5.2 Habitability

Habitability is a widely used term in the geoscience, planetary science, and astrobiology literature, but what does it mean? This was a question that troubled the minds of several researchers. Different investigators have looked at the term from different perspectives, and, therefore, it is difficult to provide a definitive definition to the term "habitability." For example, Westall et al. (2013) proposed to use the term "habitable" only for conditions necessary for the origin of life, the proliferation of life, and the survival of life. From their perspective, "Not covered by this term would be conditions necessary for prebiotic chemistry and conditions that would allow the recognition of extinct or hibernating life." According to Westall et al. (2013), "The term habitable is misleading when applied to conditions in which prebiotic chemistry can lead to the origin of life. The principal requirement of a prebiotic environment would be the simultaneous coexistence of the ingredients of life and the range of chemical and physicochemical reactions necessary to result in a protolife entity—a chemical reactor of sorts. It is entirely possible that an environment conducive to prebiotic chemical processes leading to the origin of life could have been spatially confined and toxic compared to the present terrestrial environment yet able to provide that first spark, with subsequent evolution occurring eventually in other locations. Similarly, conditions that would allow life to flourish are likely to be different from those in which the molecular building blocks formed."

Extraterrestrial habitability is a complex notion. A star's habitable zone (HZ) is the region where liquid water can exist on its planets' surface. This region changes depending on the star's temperature. Sun has a "habitable zone," a relatively narrow region around it where conditions are favorable for life (i.e., temperatures could be just right for liquid water to be present), and perhaps alien life to be present. Within the habitable or "Goldilocks" zone, temperatures are not too hot or too cold but just right for surface water to exist as a liquid. A habitable zone planet could have oceans, lakes, and rivers harboring liquid water.

Earth, where we live, is situated halfway between the middle of the Sun's habitable zone and its inner boundary. Likewise, Mars is situated halfway between the middle of the Sun's habitable zone and its outer boundary. Venus is situated just on the brink of the inner boundary of the Sun's habitable zone. Thus, in terms of habitability, Earth and Mars are the only two planets that are favorably positioned in the Solar system.

Cockell (2014) divided all environments in the Universe into three types: (1) uninhabitable, (2) uninhabited habitats, or (3) inhabited habitats (Cockell, 2011; Zuluaga et al., 2014). Habitable planets are the most appealing places. Originally defined as a planet's potential to hold life of any kind, a more "binary" definition was introduced by Cockell et al. (2016). According to them, if a habitat is a place that can support the activity of at least one known organism, then it follows that it is meaningless to speak of more or less habitable places. The assessment is binary—either an environment can support the activity of a given organism or it cannot. Thus, Cockell et al. (2016) defined habitability as the ability of an environment to support the activity of at least one known organism—yes or no (although in reality microbial colonies in nature are almost always multispecies). It is likely that environments such as entire planets might be capable of supporting more or less species diversity or biomass compared with that of Earth. Cockell et al. (2016) have suggested that "The investigation of the origin of life on Earth, its persistence on the planet since its emergence, and the search for evidence of life on other planetary bodies all require that we define what conditions life requires." Cockell et al. (2016) suggested that a clarity in understanding habitability can be obtained by defining *instantaneous habitability* as the conditions at any given time in a given environment required to sustain the activity of at least one known organism, and *continuous planetary habitability* as the capacity of a planetary body to sustain habitable conditions on some areas of its surface or within its interior over geological timescales. Our search for life beyond Earth may be thwarted by the short timescales over which planets may remain inhabited.

Cockell et al. (2016) also sought to distinguish between *surface liquid water worlds* (such as Earth) that can sustain liquid water on their surfaces and *interior liquid water worlds*, such as icy moons and terrestrial-type rocky planets with liquid water only in their interiors. This distinction is considered to be important because, while the former can potentially sustain habitable conditions for oxygenic photosynthesis that leads to the rise of atmospheric oxygen and potentially complex multicellularity and intelligence over geological timescales, the latter is unlikely to. Habitable environments do not need to contain life. If a planetary body hosts a habitat of any kind on any scale, it is a habitable planetary body. The volume of Earth that is inhabited is less than 0.5% of its total volume (Jones and Lineweaver, 2010a, 2010b), but we describe Earth as a habitable planet. According to Cockell et al. (2016), although the decoupling of habitability and the presence of life may be rare on Earth, it may be important for understanding the habitability of other planetary bodies.

On Earth, life is connected to the global environment. According to McKay (2016), "Habitability is determined by the physicochemical interactions of the fluids (water on Earth) in a specific environment, and by the solids present in that environment. Through physical processes such as sunlight and volcanism and their interaction with cycles of water, important physical conditions for life are established on the Earth. These include: (1) sources of chemical or light energy suitable for life; (2) nutrients; (3) liquid habitats; and (4) transport cycles of liquid moving nutrients and wastes."

Much of the efforts expended in extraterrestrial missions have been aimed at looking for traces of microbial life (past and present) and attempting to find a congenial atmospheric, surface, and subsurface conditions necessary to support life. Taking this aspect into account, astrobiologists have attempted to provide a review of habitability from an astrobiological perspective. According to Cockell et al. (2016), "In astrobiology, the habitability of an environment is the capacity of a particular physical space to support the activity of an organism, that is, to provide the set of resources and conditions required for its way of life."

It may be noted that in the standard scenario of planet formation, planets are formed from a proto-planetary disk that consists of gas and dust. The building blocks of solid planets are called planetesimals; they are formed by coagulation of dust (Kokubo and Ida, 2012), which could come together with many others under gravitation to form planets. Rocky planet formation is a process in which gas-rich proto-planetary disks evolve into dust disks in which planetesimals form and undergo growth into planetary embryos as they differentiate into iron-nickel-rich cores, silicate mantles, and crusts dominated by incompatible lithophiles (Elkins-Tanton, 2012; Morbidelli et al., 2012; Chambers, 2014; Hardy et al., 2015). Models and observations suggest that this sequence—the formation of terrestrial planets—is not a rare occurrence that needs special initial conditions. For most rocky planets in the circumstellar habitable zone (CHZ) to remain habitable, they may have to be inhabited: "habitability depends on inhabitance and the width of the habitable zone is difficult to characterize" (Goldblatt, 2015).

According to Chopra and Lineweaver (2016), just because Earth is at 1 Astronomical Unit (1 AU; 150 million kilometers distant from the Sun) and has been inhabited for ~4 billion years does not mean that there is a physics-based, biology-independent, computable continuous habitable zone. With thermal instability and increasing stellar luminosity, it is not clear that a physics-based continuously habitable zone even exists. There may be no range of orbital distances (or any region of multidimensional abiotic parameter space) for which the surface environments of initially wet rocky planets have sufficiently strong abiotic negative feedback to maintain habitability. If this is the case, purely abiotic computations of a continuously habitable zone may be misleading (Chopra and Lineweaver (2016). If life gets started on a celestial body, there are many potential ways in which life can regulate the mechanisms that create or maintain the temperatures and pressures needed for liquid water (Schneider and Boston, 1991; Schneider, 2004; Harding and Margulis, 2010). There is a school of thought that life on Earth evolved to become integrated into previously abiotic feedback systems that can modify or regulate surface temperature and the hydrological cycle (e.g., Lenton, 1998; Nisbet et al., 2012).

It is usually assumed that the circumstellar habitable zone (CHZ) is determined by abiotic physical parameters such as stellar mass and luminosity, planetary mass and atmospheric greenhouse gas composition, surface albedo, and sometimes clouds. More recently planetary spin, orbital eccentricity, obliquity, and initial water content have been added to the list of physical parameters (e.g., Gaidos et al., 2005; Gonzalez, 2005; Lammer et al., 2009; Gudel et al., 2014; Shields et al., 2016). Chopra and Lineweaver (2016) argued that these abiotic parameters can fleetingly (i.e., for a very short time) enable the emergence of life but cannot maintain habitable surface conditions with liquid water. As the early heavy bombardment subsides, strong selection pressure on life begins to regulate, control, and even dominate the mechanisms that create or maintain the temperatures and pressures at the surface of a celestial body that allow liquid water. If so, then biology (rather than physics or chemistry) can play the most important role in maintaining habitability.

On Earth, life began to modulate the greenhouse gas composition of the atmosphere as soon as life became widespread (Nisbet, 2002; Nisbet et al., 2012; Nisbet and Fowler, 2014; Johnson and Goldblatt, 2015). If we are able to find well-preserved, ~4.3 to ~3.8 billion-year-old rocks on Venus or Mars, then we may be able to identify isotopic anomalies produced by biotic actions, in a way analogous to how $^{12}C/^{13}C$ ratios are used to infer the existence of the earliest life in Isua (Fig. 5.1), Greenland (Ohtomo et al., 2014).

It may be noted that the Isua Greenstone Belt is an Archean greenstone belt in southwestern Greenland, aged between 3.8 and 3.7 billion years (Nutman et al., 2007). The belt contains variably metamorphosed mafic volcanic and sedimentary rocks, and is the largest exposure of Eoarchaean supracrustal rocks on Earth (Nutman and Friend, 2009). Due to its age and low metamorphic grade (Ramírez-Salazar et al., 2021) relative to many Eoarchaean rocks, the Isua Greenstone Belt has become a focus for investigations on the emergence of life (Nutman et al., 2016; Allwood et al., 2018) and the style of tectonics that operated on the early Earth (Furnes et al., 2007; Alexander et al., 2020).

The complexity of the stromatolites found at Isua, if they are indeed stromatolites, suggests that life on Earth was already sophisticated and robust by the time of their formation, and that the earliest life on Earth likely evolved over 4 billion years ago (Nutman et al., 2016). This conclusion is supported in part by the instability of Earth's surface conditions 3.7 billion years ago, which included intense asteroid bombardment (Allwood, 2016). The possible formation and preservation of fossils from this period indicate that life may have evolved early and prolifically in Earth's history (Allwood, 2016).

The surrounding rocks at Isua suggest that the stromatolites may have been deposited in a shallow marine environment (Nutman et al., 2016). While most rocks in the Isua Greenstone Belt are too metamorphically altered to preserve fossils, the area of stromatolite discovery may have preserved original sedimentary rocks and the fossils inside

FIG. 5.1 Image of the general location of the Isua Greenstone belt (Nuuk Region). *(Source: https://commons.wikimedia.org/wiki/File:Nuuk_Location.jpg) Author: A.bre.clare. Copyright info: This file is licensed under the Creative Commons Attribution-Share Alike 4.0 International license.*

them (Bolus, 2001). However, some geologists interpret the structures as the result of deformation and alteration of the original rock (Allwood, 2018).

In the book *Vital Dust*, de Duve (1995) presented the case that water and energy are common and abiogenesis may be a cosmic imperative. The most important constraint on the existence of life in the Universe may be whether life, after emerging and evolving into a biosphere, can evolve global mechanisms rapidly enough to mediate the positive and negative feedbacks of abiotic atmospheric evolution.

The search for habitable exoplanets inspires the question—how do habitable planets form? Planet habitability models traditionally focus on abiotic processes and neglect a biotic response to changing conditions on an inhabited planet. The Gaia hypothesis postulates that life influences the Earth's feedback mechanisms to form a self-regulating system, and hence that life can maintain habitable conditions on its host planet. If life has a strong influence, it will have a role in determining a planet's habitability over time.

The prerequisites and ingredients for life seem to be abundantly available in the Universe. However, the Universe does not seem to be teeming with life. The most common explanation for this is a low probability for the emergence of life (an emergence bottleneck), notionally due to the intricacies of the molecular recipe. Chopra and Lineweaver (2016) reported an alternative Gaian bottleneck explanation, according to which, "If life emerges on a planet, it only rarely evolves quickly enough to regulate greenhouse gases and albedo, thereby maintaining surface temperatures compatible with liquid water and habitability. Such a Gaian bottleneck suggests that (i) extinction is the cosmic default for most life that has ever emerged on the surfaces of wet rocky planets in the Universe and (ii) rocky planets need to be inhabited to remain habitable. In the Gaian bottleneck model, the maintenance of planetary habitability is a property more associated with an unusually rapid evolution of biological regulation of surface volatiles than with the luminosity and distance to the host star."

Whether isotopic anomalies evolved independently of life on Earth will be difficult to determine. According to Chopra and Lineweaver (2016), in the far future, we may be able to find evidence for biogenic isotopic anomalies on the initially wet rocky planets around most stars. Since life does not persist for long in the Gaian bottleneck model, it predicts a universe filled with isotopic or microscopic fossils from the kind of life that can evolve in ~1 billion years, not the fossils of larger multi-cellular eukaryotes or anything else that would take several billion years to evolve.

Chopra and Lineweaver (2016) hypothesized that the early evolution of biologically mediated negative feedback processes, or Gaian regulation as proposed by Lovelock and Margulis (1974), may be necessary to maintain habitability because of the strength, rapidity, and universality of abiotic positive feedbacks on the surfaces of rocky planets in traditional CHZs. Chopra and Lineweaver (2016) argued that the habitable surface environments of rocky planets usually become uninhabitable due to abiotic runaway positive feedback mechanisms involving surface temperature, albedo, and the loss of atmospheric volatiles. Because of the strength, rapidity, and universality of abiotic positive feedbacks in the atmospheres of rocky planets in traditional CHZs, biotic negative feedback or Gaian regulation may be necessary to maintain habitability.

Nicholson et al. (2018) reported the ExoGaia model—a model of simple "planets" host to evolving microbial biospheres. In this model, microbes interact with their host planet via consumption and excretion of atmospheric chemicals. Model planets orbit a "star" that provides incoming radiation, and atmospheric chemicals have either an albedo or a heat-trapping property. Planetary temperatures can therefore be altered by microbes via their metabolisms. Nicholson et al's. (2018) model results indicated that the underlying geochemistry plays a strong role in determining long-term habitability prospects of a planet. Nicholson et al. (2018) found five distinct classes of model planets, including clear examples of "Gaian bottlenecks"—a phenomenon whereby life either rapidly goes extinct leaving an inhospitable planet or survives indefinitely maintaining planetary habitability. These results suggest that life might play a crucial role in determining the long-term habitability of planets.

As already indicated, in our own solar system, HZ is the area roughly between the orbits of Venus and Mars. But for cooler stars, even a close-orbiting planet could harbor water. Discovery of exo-planets (i.e., planets that orbit stars outside the Solar System) in the HZ has been an important target in the field of exoplanetary research.

Just like in exoplanetary research, habitable zone has been an important target in the field of planetary research as well. It is worth emphasising that the existence of a planet in the HZ does *not* imply that life exists on the planet, or even that it will necessarily be able to support a biosphere. Most of the current assessments of habitable planets have tended to focus on superficial metrics which can be misleading, as rightly pointed out in Schulze-Makuch and Guinan (2016) and Tasker et al. (2017). These metrics evaluate the degree of similarity between certain, physically relevant, parameters of a given exoplanet and the Earth, and have led to unfortunate misconceptions that planets with higher values (of the similarity indices) are automatically more habitable.

5.3 Importance of liquid water in maintaining habitability on celestial bodies

Based on the present knowledge, the availability of liquid water is the first step to habitability. It may be noted that if a place is habitable, it does not mean that it is actually inhabited, just that the conditions could allow for the survival of some extremely hardy forms of life (extremophiles) that we know of on Earth. There are many scientists who think that several planets initially had some habitable regions but, through volatile evolution or other transient factors, lost their surface water and evolved away from habitable conditions {e.g., a runaway (i.e., running out of control) greenhouse or runaway glaciated planet}. Without significant abiotic negative feedback mechanisms, the surface environments of initially wet rocky planets are volatile and change rapidly without any tendency to maintain the habitability that they may have temporarily possessed as their early unstable surface temperatures transited through habitable conditions.

It is generally agreed that liquid water is not easy to maintain on a planetary surface. The initial inventory and the timescale with which water is lost to space due to a runaway greenhouse (i.e., temperatures too hot for life), or frozen due to ice-albedo positive feedback, are poorly quantified, but plausible estimates of future trajectories have been made. On Earth, dissociation of water vapor by UV radiation in the upper atmosphere is ongoing. Several scientists are of the view that such dissociation of water vapor will eventually (~1–2 billion years from now) lead to the loss of water from the bio-shell and the subsequent extinction of life on Earth (Caldeira and Kasting, 1992; Franck, 2000; Lenton and von Bloh, 2001; Franck et al., 2002; von Bloh et al., 2005).

Chopra and Lineweaver (2016) have addressed the issue of liquid water loss from planets and the general tendency to evolve away from habitability. In our solar system, Venus, Earth, and Mars are usually assumed to have started out in similar conditions: hot from accretion and wet from the impacts of aqueous bodies from beyond the snowline. However, atmospheric evolution of these planets diverged significantly (Kasting, 1988; Kulikov et al., 2007; Driscoll and Bercovici, 2013). It has been suggested that Venus has lost the vast majority of its water molecules (H_2O); Mars lost about 85% of its initial water content, and the rest froze into the polar ice-caps and subsurface permafrost (Kurokawa et al., 2014; Villanueva et al., 2015); and Earth was able to keep a larger fraction of its H_2O (Pope et al., 2012).

According to Chopra and Lineweaver (2016), the answer to the question "Why didn't Earth undergo a runaway greenhouse (i.e., temperatures too hot for life) like Venus or a runaway glaciation (i.e., lowering the temperature and/or water activity to levels not conducive to life) like Mars?" may have as much, or more, to do with life on Earth than with Earth's distance from the Sun. The biotic mechanisms of how this preservation has been achieved have been discussed in the context of the Gaia hypothesis by Harding and Margulis (2010). The early de-volatilization of Earth-like planets around M-stars (relatively cool, red stars, having surface temperature of less than 3600 K) due to an extended pre-main sequence period of high extreme UV flux (above the dissociation energy of water molecules, ~5 eV) could apply to some extent to Earth-like planets around more massive stars (Luger and Barnes, 2015; Tian, 2015).

The amount of water (and volatiles in general) deposited or devolatilized during the late accretion phase of rocky planet formation in the Universe is highly variable (Raymond et al., 2004, 2009) and can produce desert worlds (Abe et al., 2011), ocean worlds (Leger et al., 2003), and probably everything in between. Planetary scientists think that abiotic volatile evolution will be rapid, stochastic (having a random probability distribution or pattern that may be analyzed statistically but may not be predicted precisely), and hostage to the timing, mass, volatile content, and impact parameters of the largest impactors and the runaway feedbacks they could induce.

Chopra and Lineweaver (2016) argued that abiotic habitable zones are available initially and fleetingly (for a very short time) to wet planets within a wide range of orbital radii (~0.5 to ~2 AU) because of the thermal instability of their surfaces. Wide-ranging unstable temperatures could provide transitory abiotic habitable zones during the first half billion years after formation (Nisbet, 2002). There are two ways to influence the surface temperature of a planet: change the albedo or change the greenhouse gas content of the atmosphere (Kasting, 2012). The amount and the phases of the volatiles such as water (H_2O), carbon dioxide (CO_2), methane (CH_4) of rocky planet atmospheres control both the albedo and greenhouse warming. Albedo and greenhouse warming, in turn, control the amount and phases of the volatiles. Strong positive feedback cycles may lead to both (1) runaway greenhouse (i.e., temperatures too hot for life) with runaway loss of atmosphere (i.e., hydrogen loss and thus water loss) or (2) runaway glaciation (lowering the temperature and/or water activity to levels not conducive to life).

5.4 Habitability of extremophilic and extremotolerant bacteria under extreme environmental conditions

Exploring the diversity of microorganisms and understanding their mechanisms of adaptation permit the development of hypotheses regarding the conditions required for the origin and early diversification of life on Earth (Westall, 2005; Rampelotto, 2010). The research also expands our notions of the potential habitable environments able to sustain life beyond Earth. Indeed, our increasing knowledge about the biology of microbes in extreme habitats has led numerous scientists to raise the possibility of finding life in various planetary bodies within the Solar System, including Mars, Venus, and the moons of Jupiter (Io, Europa, Ganymede, and Callisto) and Saturn (Titan and Enceladus) (Cavicchioli, 2002; Horneck and Rettberg, 2007). Furthermore, a variety of studies in this field of research has given support to Panspermia—the hypothesis that life migrates naturally through space (Rampelotto, 2009).

In assessing the bacterial populations present in spacecraft assembly, spacecraft test, and launch preparation facilities, La Duc et al. (2007) isolated *extremophilic* bacteria (i.e., bacteria requiring severe conditions for growth) and *extremotolerant* bacteria (i.e., bacteria that are tolerant to extreme conditions). They employed several cultivation approaches to select for and identify bacteria that not only survive the nutrient-limiting conditions of clean room environments but can also withstand even more inhospitable environmental stresses. Due to their proximity to space-faring objects, these bacteria pose a considerable risk for forward contamination of extraterrestrial sites. Samples collected from four geographically distinct US National Aeronautics and Space Administration (NASA) clean rooms were challenged with UV-C irradiation, 5% hydrogen peroxide, heat shock, pH extremes (pH: 3.0 and 11.0), temperature extremes (4–65°C), and hypersalinity (25% salt {NaCl}) prior to and/or during cultivation as a means of selecting for extremotolerant bacteria.

Microscopy is one tool that is useful in distinguishing live and dead microbial populations. However, it is not a practical method for spacecraft and cleanroom surfaces due to the low abundance of microbes in these areas and high levels of debris, which promote false positive results from autofluorescence as well as issues with insufficient and nonspecific binding of strains (Moter and Gobel, 2000; La Duc et al., 2003; La Duc et al., 2007). Molecular methods, on the other hand, utilize universally common cellular compounds such as deoxyribonucleic acid (DNA) and adenosine triphosphate (ATP) for microbial detection and can be modified to distinguish the total and viable populations (Venkateswaran et al., 2003; La Duc et al., 2004; Nocker and Camper, 2006; Nocker et al., 2007). ATP is a key molecule found in living cells.

In a study, Hendrickson et al. (2017) used molecular assays to determine the total organisms (TO, dead, and live) and the viable organisms (VO, live). The ATP assay used in this study measured ATP using a commercially available ATP assay kit capable of measuring total and viable populations (Stanley, 1989; Selan et al., 1992). The TO was measured using ATP and quantitative polymerase chain reaction (qPCR) assays. The VO was measured using internal ATP, propidium monoazide (PMA)-qPCR, and flow cytometry (after staining for viable microorganisms) assays.

The phrase "Polymerase chain reaction" (PCR) was first used in a paper describing a novel enzymatic amplification of DNA (Saiki et al., 1985). Real-time PCR (quantitative PCR, qPCR) is now a well-established method for the detection, quantification, and typing of different microbial agents in the areas of clinical and veterinary diagnostics and food safety (Kralik, P., and M. Ricchi (2017). The most substantial milestone in PCR utilization was the introduction of the concept of monitoring DNA amplification in real time through monitoring of fluorescence (Holland et al., 1991; Higuchi et al., 1992). In real-time PCR (also denoted as quantitative PCR—qPCR), fluorescence is measured after each cycle and the intensity of the fluorescent signal reflects the momentary amount of DNA amplicons in the sample at that specific time. In initial cycles the fluorescence is too low to be distinguishable from the background. However, the point at which the fluorescence intensity increases above the detectable level corresponds proportionally to the initial number of template DNA molecules in the sample. This point is called the quantification cycle (Cq) and allows determination of the absolute quantity of target DNA in the sample according to a calibration curve constructed of serially diluted standard samples (usually decimal dilutions) with known concentrations or copy numbers (Yang and Rothman, 2004; Kubista et al., 2006; Bustin et al., 2009).

Duc et al. (2007) employed culture-independent approaches to measure viable microbial (ATP-based) and total bacterial (quantitative PCR-based) burdens. In their investigations, intracellular ATP concentrations suggested a viable microbial presence ranging from below detection limits to 10^6 cells/m^2. However, only 0.1 to 55% of these viable cells were able to grow on defined culture medium. It was found that isolated members of the *Bacillaceae* family were more physiologically diverse than those reported in previous studies, including thermophiles (Geobacillus), obligate anaerobes (Paenibacillus), and halotolerant, alkalophilic species (*Oceanobacillus* and *Exiguobacterium*). Nonspore-forming microbes (alpha- and beta-proteobacteria and actinobacteria) exhibiting tolerance to the selected stresses were also encountered. It was found that the multiassay cultivation approach employed by Duc et al. (2007) enhances the current understanding of the physiological diversity of bacteria housed in these clean rooms and leads the curious minds to ponder over the origin and means of translocation of thermophiles, anaerobes, and halotolerant alkalophiles into these environments.

Data on the survival of microbes in the space environment and modeling of the potential for transfer of life between celestial bodies, suggests that life could be more common than previously thought. It is interesting that what we previously thought of as insurmountable physical and chemical barriers to life, we now see as yet another niche harboring "extremophiles." This realization, coupled with new data on the survival of microbes in the space environment and modeling of the potential for transfer of life between celestial bodies, suggests that life could be more common than previously thought. Rothschild and Mancinelli (2001) have examined critically what it means to be an extremophile, and the implications of this for evolution, biotechnology, and especially the search for life in the Universe, forcefully substantiating Giordano Bruno's 16th-century revolutionary philosophical position on the possibility of life outside Earth. Note that Giordano Bruno was a great visionary who supported the Copernican system of celestial bodies and also foresaw the possibility of life lurking external to Earth, and much beyond. Giordano Bruno was sentenced to be burned to death by the Roman Inquisition for his heretical ideas, which he refused to recant. Bruno is, perhaps, chiefly remembered for the tragic death he suffered at the stake because of the tenacity with which he maintained his unorthodox ideas at a time when both the Roman Catholic and Reformed churches were reaffirming rigid Aristotelian and Scholastic principles in their struggle for the evangelization of Europe.

Various categories of extremophiles thrive in ice, boiling water, acid, the water core of nuclear reactors, salt crystals, and toxic waste and in a range of other extreme habitats that were previously thought to be inhospitable for life. Extremophiles include representatives of all three domains (bacteria, archaea, and eucarya); however, the majority are microorganisms, and a high proportion of these are archaea. Knowledge of extremophile habitats is expanding the number and types of extraterrestrial locations that may be targeted for exploration. In addition, contemporary biological studies are being fueled by the increasing availability of genome sequences and associated functional studies of extremophiles. This is leading to the identification of new biomarkers, an accurate assessment of cellular evolution, insight into the ability of microorganisms to survive in meteorites and during periods of global extinction, and knowledge of how to process and examine environmental samples to detect viable life forms. Cavicchioli (2004) has evaluated extremophiles and extreme environments in the context of astrobiology and the search for extraterrestrial life.

High-latitude polar deserts are among the most extreme environments on Earth. Polar deserts, despite the extreme low temperature and scarcity of water, are not devoid of life. Cockell and Stokes (2004) have described a large and previously unappreciated habitat for photosynthetic life under opaque rocks in the Arctic and Antarctic polar deserts. This habitat is created by the periglacial movement of the rocks, which allows some light to reach their underside.

The underside of rocks in climatically extreme deserts acts as a refugium for photosynthetic microorganisms (defined as "hypoliths") and their community (the "hypolithon") (Golubic et al., 1981). Photosynthetic microorganisms are capable of converting light energy (from the sun) into chemical energy through a process known as photosynthesis. They are also classified as photoautotrophs because they can make their own energy using inorganic material from their surroundings. At the underside of rocks, the organisms are protected from harsh ultraviolet radiation (Cockell et al., 2003) and wind scouring, and trapped moisture can provide them with a source of liquid water. Broady (1981) reported that sublithic (hypolithic) algae were found to be widely distributed under quartz stones at the Vestfold Hills, Antarctica. Smith et al. (2000) reported that quartz stone sublithic cyanobacterial communities are common throughout the Vestfold Hills, Eastern Antarctica (68° S 78° E), contributing biomass in areas otherwise devoid of any type of vegetation. Colonization also usually requires the rocks to be sufficiently translucent to allow for the penetration of light—all hypoliths reported so far have been found under quartz, which is one of the most common translucent rocks (Schlesinger et al., 2003).

Asteroid and comet-impacts on Earth are commonly viewed as agents of ecosystem destruction, be it on local or global scales. However, for some microbial communities, impacts may represent an opportunity for habitat formation as some substrates are rendered more suitable for colonization when processed by impacts. Cockell et al. (2002) have described how heavily shocked gneissic crystalline basement rocks exposed at the Haughton impact structure, Devon Island (Nunavut, Arctic Canada), are hosts to endolithic photosynthetic microorganisms in significantly greater abundance than lesser-shocked or unshocked gneisses (metamorphic rocks with a banded or foliated structure, typically coarse-grained and consisting mainly of feldspar, quartz, and mica). According to Cockell et al. (2002), two factors contribute to this enhancement: (1) increased porosity due to impact fracturing and differential mineral vaporization, and (2) increased translucence due to the selective vaporization of opaque mineral phases. Using biological ultraviolet radiation dosimetry (determination and measurement of the amount or dosage of radiation absorbed by a substance or living organism by means of a dosimeter), and by measuring the concentrations of photoprotective compounds, Cockell et al. (2002) demonstrated that a covering of 0.8 mm of shocked gneiss can provide substantial protection from ultraviolet radiation, reducing the inactivation of *Bacillus subtilis* spores by 2 orders of magnitude. The colonization of the shocked habitat represents a potential mechanism for pioneer microorganisms to invade an impact structure in the earliest stages of postimpact primary succession. Cockell et al. (2002) found that the communities are analogous to the endolithic communities associated with sedimentary rocks in Antarctica, but because they occur in shocked crystalline rocks, they illustrate a mechanism for the creation of microbial habitats on planetary surfaces that do not have exposed sedimentary units. This might have been the case on early Earth. According to Cockell et al. (2002), the data have implications for the microhabitats in which biological signatures might be sought on Mars.

Cockell and Stokes (2004) have described a large and previously unappreciated habitat for photosynthetic life under opaque rocks in the Arctic and Antarctic polar deserts. This habitat is created by the periglacial (i.e., at the edges of glacial areas) movement of the rocks, which allows some light to reach their underside. According to the studies of Cockell and Stokes (2004), the productivity of this ecosystem is at least as great as that of above-ground biomass and potentially doubles previous productivity estimates for the polar desert ecozone.

Cockell and Stokes (2004) examined 850 randomly selected opaque dolomitic rocks, without regard to the local patterns of periglacial rock sorting, on Cornwallis Island and Devon Island in the Canadian high Arctic. These are regions typical of extreme polar desert, where vegetation cover was measured to be 1.2% or less (Bliss et al., 1984). On Devon Island, 95% of rocks were found to be colonized, and on Cornwallis Island, 94% were colonized. Cockell and Stokes (2004) found that the photosynthetic organisms form a well-defined green band on the underside of the rocks. The mean (±s.d.) band width of these communities was 3.1 ± 1.9 cm and 3.0 ± 1.6 cm on Devon Island and Cornwallis, respectively. The Arctic hypoliths are dominated by cyanobacteria. Species found in Cockell and Stokes (2004)'s study include *Gloeocapsa* cf. atrata Kützing, *Gloeocapsa* cf. punctata Nägeli, *Gloeocapsa* cf. kuetzingiana Nägeli and *Chroococcidiopsis*-like cells; unicellular algal chlorophytes were also present.

Cockell and Stokes (2004) further investigated the colonization of well-developed polygons, which are just one manifestation of a diversity of linear and circular features caused by the long-term action of periglacial processes (Kessler and Werner, 2003). In the Arctic, Cockell and Stokes (2004) found colonization on 68% of rocks within polygons, with a mean (±s.d.) band width of 0.7 ± 0.8 cm, where the rocks are surrounded by fine soil sorted into their centers. At the edges, where the cracks around the rocks are larger, they found 100% colonization with a mean (±s.d.) band width of 3.6 ± 1.4 cm.

Cockell and Stokes (2004) studied similar polygonal terrains at Mars Oasis on Alexander Island in the Antarctic Peninsula, a location where hypolithic colonization occurs. The percentage colonization was 5% within polygons and 100% on the edges of polygons, with mean (±s.d.) band widths of 0.7 ± 0.1 and 2.1 ± 0.3 cm, respectively. Cockell and Stokes (2004) proposed that rock sorting by periglacial action, including that during freeze–thaw cycles, improves light penetration around the edges of rocks, one factor that might account for the widespread colonization of the underside of opaque rocks.

In Arctic and Antarctic ecology, a *hypolith* is a photosynthetic organism, and an extremophile, that lives underneath rocks in climatically extreme deserts such as Cornwallis Island and Devon Island in the Canadian high Arctic. The community itself is the hypolithon. In a study carried out in the Arctic and Antarctic, Thomas (2005) found that the productivity of the cyanobacteria and algae (*hypoliths*) that colonize the underside of the stones is strongly related to the pattern of the stones. Interestingly, the hypolith assemblages were in some cases as productive as lichens, bryophytes, and plants that resided nearby.

In another example pertaining to habitability of extremophilic and extremotolerant bacteria under extreme environmental conditions, life (*Sulfalobus acidocaldarius*) has been found to exist in low pH (3) and the high temperature (80°C) environments in Congress Pool, Norris Geyser Basin, Yellowstone National Park, USA (Fig. 5.2). Likewise, the pink filamentous *Thermocrinis ruber* (Fig. 5.3) thrives at high pH (8.8–8.3) and high temperature (95–83°C) environments at Octopus Spring (a hot-spring) in Yellowstone National Park. Taking into account the truly marvellous identification of organisms that thrive in extreme environments {e.g., hot vents on the seafloor (Fig. 5.4), cold gas seeps on the seafloor (Fig. 5.5)}, hot subseafloor sediments, oxygen minimum zones—dead zones (see Joseph, 2016). Macelroy (1974) named these lovers ("philos" to the Greeks) of extreme environments "extremophiles."

FIG. 5.2 Congress Pool, Norris Geyser Basin, Yellowstone National Park, USA, where Tom Brock originally isolated *Sulfalobusacidocaldarius*. The average pH is 3 and the average temperature is 80°C. Photo taken on September 20, 2000 *(Source: Rothschild and Mancinelli, 2001; Life in extreme environments, Nature, 409, 1092–1100)*.

FIG. 5.4 Deep-sea hydrothermal hot vents found on the East Scotia Ridge in the Southern Ocean, hosting high-temperature black smokers up to 382.8°C and diffuse venting. *(Source: Rogers et al., 2012; PLoS Biol., 10(1), e1001234)*.

FIG. 5.3 Octopus Spring, an alkaline (pH 8.8–8.3) hot-spring in Yellowstone National Park, USA, is situated several miles north of Old Faithful geyser. The water flows from the source at 95°C to an outflow channel, where it cools to a low of 83°C. About every 4–5 min a pulse of water surges from the source raising the temperature as high as 88°C. In this environment, the pink filamentous *Thermocrinisruber* thrives (lower right). Where the water cools to 75°C, growth of photosynthetic organisms is permitted. The inset on left shows the growth of a thermophilic cyanobacterium, *Synechococcus*, tracking the thermal gradient across the channel. At 65°C a more complex microbial mat forms with *Synechococcus* on the top overlaying other bacteria, including species of the photosynthetic bacterium *Chloroflexus* (upper right). The yellow object at 65°C was part of an experimental set-up. Photo taken on July 4, 1999. *(Source: Rothschild and Mancinelli, 2001; Life in extreme environments, Nature, 409, 1092–1100)*.

The term "extremes" include physical extremes (for example, temperature, radiation or pressure) and geochemical extremes (for example, desiccation, salinity, pH, oxygen species or redox potential). It could be argued that extremophiles should include organisms thriving in biological extremes (for example, nutritional extremes, and extremes of population density, parasites, prey, and so on (Rothschild and Mancinelli, 2001).

Although all hyperthermophiles are members of the archaea and bacteria, eukaryotes are common among the psychrophiles, acidophiles, alkaliphiles, piezophiles, xerophiles, and halophiles (which, respectively, thrive at low temperatures, low pH, high pH, and under extremes of pressure, desiccation and salinity (see http://www.astrobiology.com/ extreme.html foran overview). Extremophiles include multicellular organisms, and psychrophiles include vertebrates.

5.5 Why do extremophiles survive in extreme environments? Application of exopolymers derived from extremophiles in the food, pharmaceutical, and cosmetics industries

Scientists and industrial researchers study the incredible microbes known as *extremophiles* to tap into the extraordinary strength of their proteins for numerous applications. Extreme environments offer novel microbial biodiversity that produces varied and promising exopolysaccharides (EPSs). EPSs, forming a layer surrounding the cell, provide an effective protection against extreme environments such as high or low temperature and salinity, high pressure, radiation, or against possible predators.

EPSs are high molecular weight carbohydrate polymers that make up a substantial component of the extracellular polymers surrounding most microbial cells in the harsh environment. Many marine bacteria produce EPS as a strategy for growth, adhering to solid surfaces, and to survive adverse conditions. There is growing interest in isolating

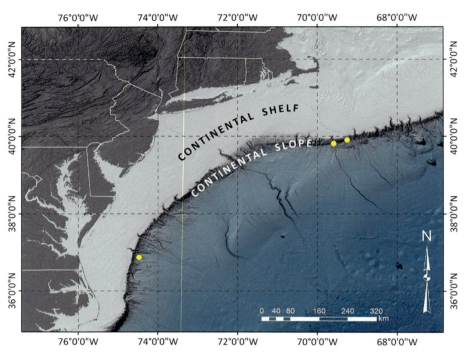

FIG. 5.5 Deep-ocean cold gas seeps (indicated with yellow circles) on the U.S. Atlantic coast between Cape Hatteras, NC, and Cape Ann, MA. *(Source: NOAA explorers discover deep-water gas seeps off U.S. Atlantic coast, December 19, 2012); Credit: NOAA.*

new EPS-producing bacteria from marine environments, particularly from extreme marine environments such as deep-sea hydrothermal vents characterized by high pressure and temperature and heavy metal presence (Poli et al., 2010).

EPSs make up a substantial component of the extracellular polymers surrounding most microbial cells in extreme environments such as Antarctic ecosystems, saline lakes, geothermal springs or deep-sea hydrothermal vents. The extremophiles have developed various adaptations, enabling them to compensate for the deleterious effects of extreme conditions, e.g. high temperatures, salt, low pH or temperature, high radiation. Among these adaptation strategies, EPS biosynthesis is one of the most common protective mechanisms. The unusual metabolic pathways revealed in some extremophiles raised interest in extremophilic microorganisms as potential producers of EPSs with novel and unusual characteristics and functional activities under extreme conditions (Nicolaus et al., 2010).

For example, it has been found that EPS is a major component of the exopolymer secreted by many marine bacteria to enhance survival and is abundant in sea ice brine channels (Nichols et al., 2005a). It is believed that EPS may have a cryoprotective role (having the capability to prevent the freezing of tissues, or to prevent damage to cells during freezing) in brine channels of sea ice, where extremes of high salinity and low temperature impose pressures on microbial growth and survival (Nichols et al., 2005a). It has been found that EPSs constitute a large fraction of the reduced carbon reservoir in the ocean and enhance the survival of marine bacteria by influencing the physicochemical environment around the bacterial cell. For this reason, microbial EPSs are abundant in the Antarctic marine environment, for example, in sea ice and ocean particles, where they may assist microbial communities to endure extremes of temperature, salinity, and nutrient availability (Nichols et al., 2005b).

Marine EPS-producing microorganisms have been isolated from several extreme niches such as the cold marine environments typically of Arctic and Antarctic sea ice, characterized by low temperature and low nutrient concentration, and the hypersaline environment found in a wide variety of aquatic and terrestrial ecosystems such as salt lakes and salterns (sets of pools in which seawater is left to evaporate to make salt). Most of their EPSs are heteropolysaccharides containing three or four different monosaccharides arranged in groups of 10 or less to form the repeating units (Poli et al., 2010).

Whereas the Antarctic marine environment is characterized by low-temperature extremes, deep-sea hydrothermal vent environments are characterized by high pressure, extreme high temperature inside and extreme low temperature outside, and heavy metals (Heavy metals are generally defined as metals with relatively high densities, atomic weights, or atomic numbers, for example, As, Cd, Pb, Cr, Cu, Hg, and Ni). The commercial value of microbial EPSs from these habitats has been established recently. The biotechnological potential of these biopolymers from hydrothermal vent environments as well as from Antarctic marine ecosystems remains largely untapped.

As noted above, extremophile microorganisms are found in several extreme marine environments, such as hydrothermal vents, hot springs, salty lakes and deep-sea floors. The ability of these microorganisms to support extremes of temperature, salinity and pressure demonstrate their great potential for biotechnological processes. Extremophilic microorganisms provide nonpathogenic products, appropriate for applications in the food, pharmaceutical, and cosmetics industries as emulsifiers (substances that stabilize an emulsion, in particular additives used to stabilize processed foods), stabilizers, gel agents, coagulants (substances that cause fluids to coagulate, that is, cause to change to a solid or semisolid state), thickeners and suspending agents. The commercial value of EPSs synthesized by microorganisms from extreme habitats has been established recently. By examining their structure and chemical-physical characteristics it is possible to gain insight into the commercial application of EPSs, and they are employed in several industries. Indeed, EPSs produced by microorganisms from extreme habitats show biotechnological promise ranging from pharmaceutical industries, for their immunomodulatory and antiviral effects, bone regeneration and cicatrizing capacity (capacity for scar formation at the site of a healing wound), to food-processing industries for their peculiar gelling and thickening properties. Moreover, some EPSs are employed as biosurfactants {surfactants (substances that tend to reduce the surface tension of liquids in which they are dissolved) of microbial origin} and in detoxification mechanisms (mechanisms of removing toxic substances) of petrochemical oil-polluted areas (Poli et al., 2010). Poli et al. (2010) have given an overview of current knowledge on EPSs produced by marine bacteria including symbiotic (involving mutually beneficial relationship or interaction between two different organisms living in close physical association) marine EPS-producing bacteria isolated from some marine annelid worms that live in extreme niches.

Hydrolases (a class of enzyme that commonly perform as biochemical catalysts that use water to break a chemical bond, which typically results in dividing a larger molecule into smaller molecules) including amylases (enzymes, found chiefly in saliva and pancreatic fluid, that convert starch and glycogen into simple sugars), cellulases (enzymes that convert cellulose into glucose), peptidases (enzymes which break down peptides into amino acids) and lipases (proteins, made by pancreas, which help the body to digest fats) from hyperthermophiles (particularly extreme thermophiles for which the optimal temperatures are above 80°C), psychrophiles (organisms that grow best at temperatures close to freezing), halophiles (organisms, especially microorganisms, that grow in or can tolerate saline conditions) and piezophiles (organisms with optimal growth under high hydrostatic pressure) have been investigated by several researchers. Extremozymes (enzymes, often created by archaea, which are known prokaryotic extremophiles, that can function under extreme environments) are adapted to work in harsh physical-chemical conditions and their use in various industrial applications such as the biofuel, pharmaceutical, fine chemicals, and food industries has increased. The understanding of the specific factors that confer the ability to withstand extreme habitats on such enzymes has become a priority for their biotechnological use. The most studied marine extremophiles are prokaryotes. Dalmaso, et al. (2015) have reviewed the most studied archaea and bacteria extremophiles and their hydrolases, and discussed their use for industrial applications.

It is heartening that commercial interests in deep-sea and hydrothermal vent systems and remote ocean areas are increasing rapidly. Blasiak et al. (2018) accessed 38 million records of genetic sequences associated with patents and created a database of 12,998 sequences extracted from 862 marine species. They identified >1600 sequences from 91 species associated with deep-sea and hydrothermal vent systems, reflecting commercial interest in organisms from remote ocean areas, as well as a capacity to collect and use the genes of such species. These researchers found that a single corporation registered 47% of all marine sequences included in gene patents, exceeding the combined share of 220 other companies (37%). Universities and their commercialization partners registered 12%. Actors located or headquartered in 10 countries registered 98% of all patent sequences, and 165 countries were unrepresented.

5.6 Microbial life on and inside rocks

The microbial way of life spans at least 3.8 billion years of evolution. Microbial organisms are pervasive, ubiquitous, and essential components of all ecosystems. The geochemical composition of Earth's biosphere has been molded largely by microbial activities. Yet, despite the predominance of microbes during the course of life's history, general principles and theory of microbial evolution and ecology are not well developed. Molecular phylogenetic surveys have revealed a vast array of new microbial groups. Many of these new microbes are widespread and abundant among contemporary microbiota and fall within novel divisions that branch deep within the tree of life. The breadth and extent of extant microbial diversity has become much clearer. A remaining challenge for microbial biologists is to better characterize the biological properties of these newly described microbial taxa. This more comprehensive picture will provide much better perspective on the natural history, ecology, and evolution of extant microbial life (DeLong and Pace, 2001).

Rocks offer a diversity of habitats to microorganisms and, each habitat can experience different microclimatic conditions, and each habitat may host different colonists.

The endolithic environment (i.e., the pore space in rocks) is a ubiquitous (found everywhere) microbial habitat. Photosynthesis-based endolithic communities inhabit the outer few millimeters to centimeters of rocks exposed to the surface. Such endolithic ecosystems have been proposed as simple, tractable models for understanding basic principles in microbial ecology. The results obtained from studies (see Walker and Pace, 2007) of selected endolithic communities in the Rocky Mountain region of the United States with culture-independent molecular methods indicate that endolithic ecosystems are seeded from a select, global metacommunity and form true ecological communities that are among the simplest microbial ecosystems known. Collectively, results of this study support the idea that patterns of microbial diversity found in endolithic communities are governed by principles similar to those observed in macro-ecological systems. According to Walker and Pace (2007), the photosynthesis-based endolithic communities inhabiting the outer centimeters of rocks exposed to the surface offer model systems for microbial ecology, geobiology, and astrobiology. In another study, Horath and Bachofen (2009) found that endolithic microorganisms colonize the pores in exposed dolomite rocks in the Piora Valley in the Swiss Alps. It was found that these microorganisms appear as distinct greyish-green bands about 1–8 mm below the rock surface. Based on environmental small subunit ribosomal RNA gene sequences, these investigators noticed a diverse community driven by photosynthesis.

According to Friedmann (1980), the interior of rocks has merited particular attention in the world's extreme deserts as a potential habitat for microorganisms, on account of the ameliorated environmental conditions that are often found in these habitats compared with their surfaces. There are several microorganisms that survive under rock-related (lithic) habitats (Friedmann, 1977; Friedmann, 1982; Smith et al., 2000; Westall and Folk, 2003; De los Rios et al, 2005). For instance, Westall and Folk (2003) investigated the rocks from the Isua Greenstone Belt (south-western Greenland) and found that these rocks contain endolithic microorganisms (i.e., inhabiting cracks in rocks). They found that the microorganisms include cyanobacteria, filamentous microorganisms such as fungal hyphae and possibly bacteria, as well as large, unidentified cells or spores. According to their report, most of the microorganisms appear to have been fossilized. It was found that the endoliths are evidently younger than the host rock, but must have infiltrated at different periods, most likely after the Inland Ice retreated (~8000 years ago). These investigators suspect that some of the previously described Isuan microorganisms probably represent recent, endolithic contamination.

In general, organisms living in or penetrating into stone/rock are known as endolithic organisms. Golubic et al. (1981) introduced the general term "lithobiontic" to include organisms on and within hard rock substrates. It

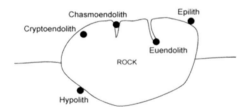

FIG. 5.6 Illustration of stone-related habitats (lithic habitats). Note that hypolithic habitats are those found on underside of rocks. Cryptoendolithic habitats are those found within the interior of rocks. Chasmoendolithic habitats are those found in fissures and cracks within rocks. Euendolithic habitats are those formed by active boring/ penetration by microorganisms. Epilithic habitats are those found on surface of rocks. *(Source: Cockell et al., 2005; Meteoritics & Planetary Science). Copyright holder: © The Meteoritical Society Publisher: Wiley-Blackwell on behalf of the Meteoritical Society.*

may be noted that lithic habitats include epilithic habitats (i.e., those found on surface of rocks), hypolithic habitats (i.e., those found on underside of rocks), cryptoendolithic habitats (i.e., those found within the interior of rocks), chasmoendolithic habitats (i.e., those found in fissures and cracks within rocks), and euendolithic habitats (i.e., those formed by active boring/penetration by microorganisms). This scenario is illustrated in Fig. 5.6. In the McMurdo Dry Valleys of Antarctica, microorganisms colonize the pore spaces of exposed rocks and are thereby protected from the desiccating environmental conditions on the surface. These cryptoendolithic communities have received attention in microscopy and culture-based studies but have not been examined by molecular approaches. de la Torre et al. (2003) surveyed the microbial biodiversity of selected cryptoendolithic communities by analyzing clone libraries of rRNA genes amplified from environmental DNA.

It has been found that rock porosity is not a factor in growth in the cases of epilithic (growing on the surface of rock) and hypolithic (living underneath rocks) organisms. The discovery of Precambrian euendolithic (formed by active boring/penetration into rocks) organisms (Campbell, 1982), shows that lithic (rock-related) habitats have probably been important throughout the history of life on Earth.

It is interesting to note that both land-based and sea-floor-based rocks are home to microorganisms. Direct and indirect evidence suggests that an extensive biosphere exists in the rocks below the seafloor. For example, Fisk et al (1998)'s survey of basalts of the Atlantic, Pacific, and Indian Oceans supports the hypothesis that bacteria have colonized much of the upper oceanic crust.

It is of particular interest that in environments where the surfaces of rocks are exposed to extremes of temperature, UV radiation, and desiccation, the interior of rocks can be an important refugium for life. According to Westall and Folk (2003), the opportunities that rock interiors afford as an escape from environmental extremes have made them the subject of interest as possible locations for life on early Earth. Wierzchos et al. (2003) found that iron-rich diagenetic minerals (minerals formed by water-rock interactions) are biomarkers of microbial activity in Antarctic rocks. These investigators went a step ahead and speculated that the protection from environmental extremes provided by rock interiors could be suggestive of the presence of life on the surface of other planets as well.

Asteroid and comet-impacts on Earth are commonly viewed as agents of ecosystem destruction, be it on local or global scales. However, for some microbial communities, impacts may represent an opportunity for habitat formation as some substrates are rendered more suitable for colonization when processed by impacts. For example, Cockell et al. (2002) have reported how heavily shocked gneissic (foliated) crystalline basement rocks exposed at the Haughton impact structure, Devon Island, Nunavut, Arctic Canada, are hosts to endolithic (living inside rock) photosynthetic (oxygen-producing) microorganisms in significantly greater abundance than lesser-shocked or unshocked gneisses (foliated metamorphic rock corresponding in composition to a feldspathic plutonic rock such as granite). Note that gneiss is a common distributed type of rock formed by high-grade regional metamorphic processes from pre-existing formations that were originally either igneous or sedimentary rocks. It is often foliated (composed of layers of sheet-like planar structures) (Fig. 5.7). The foliations are characterized by alternating darker and lighter colored bands, called "gneissic banding" (https://en.wikipedia.org/wiki/Gneiss). According to Cockell et al. (2002), two factors that contribute to an enhancement of endolithic photosynthetic microorganisms are: (1) increased porosity due to impact fracturing and differential mineral vaporization, and (2) increased translucence due to the selective vaporization of opaque mineral phases.

Impact structures are a rare habitat on Earth. However, where they do occur, they can potentially have an important influence on the local ecology. Impact-shocked rocks have become a habitat for endolithic microorganisms, and large, impact-shattered blocks of rock are used as resting sites by avifauna (the birds of a given region, considered as a whole). The colonization of the shocked habitat represents a potential mechanism for pioneer microorganisms to invade an impact structure in the earliest stages of postimpact primary succession. The communities are analogous to the endolithic communities associated with sedimentary rocks in Antarctica, but because they occur in shocked crystalline rocks, they illustrate a mechanism for the creation of microbial habitats on planetary surfaces that do not have exposed sedimentary units. Cockell et al. (2002) have suggested that this might have been the case on early Earth. However, it may be noted that some materials produced by an impact, such as melt sheet rocks, can make craters more biologically depauperate (render lacking in numbers or variety) than the area surrounding them.

Parnell et al. (2004) reported that hydrothermal gypsum deposits in the Haughton impact structure, Devon Island, Canada, contain microbial communities in an endolithic habitat within individual gypsum crystals. The crystals are transparent and, therefore, they allow transmission of light for photosynthesis, while affording protection from dehydration and wind. The relative ease with which microbial colonization may be detected and identified in impact-generated sulfate deposits at Haughton suggests that analogous settings on other planets might merit future searches for biosignatures. Parnell et al. (2004) argue that the proven occurrence of sulfates on the Martian surface suggests that sulfate minerals should be a priority target in the search for life on Mars.

Impact craters are a universal phenomenon on solid planetary surfaces. Cockell et al. (2002) have suggested that the data of potential biological relevance obtained from impact sites have implications for the microhabitats in which biological signatures might be sought on other planetary surfaces, particularly Mars.

Some of the types of habitats created in the immediate postimpact environment are not specific to the impact phenomenon. However, some others are specifically linked to processes of impact processing. Citing the Haughton impact structure in the Canadian High Arctic, Cockell et al. (2003b) have reported examples of how impact processing of target materials has created novel habitats that improve the opportunities for colonization.

Studies carried out by several researchers (e.g., Hickey et al., 1988; Grieve, 1988; Grieve and Pesonen, 1992; Cockell and Lee, 2002; Cockell et al., 2002; Cockell et al., 2003b; Fike et al., 2003; Cockell et al., 2004; Hoppert et al., 2004; Parnell et al., 2004; Cockell and Lim, 2005; Cockell et al., 2005) have shown that asteroid and comet impacts on planets and their moons cause profound changes to target rocks (Frisch and Thorsteinsson, 1978). For example, it has been suggested that impact-induced fractures and pore space is likely to increase interior space for colonization in many rock types, which can influence the availability of

FIG. 5.7 Sample of gneiss exhibiting "gneissic banding." (*Source: https://en.wikipedia.org/wiki/Gneiss#/media/File:Gneiss.jpg*) *Photographer: Siim Sepp. Date of the creation: 20th April, 2005. This rock is a property of museum of geology of University of Tartu.*

habitats for microorganisms. The interior of the cracks and fissures provides escape from desiccation and rapid temperature variations, and also protects against UV radiation damage (Cockell et al., 2002).

As already indicated, cryptoendolithic organisms live within the interstices of rocks, invading the pore spaces to reach the interior of the material. An increase in porosity caused by impact bulking and fracturing has been observed in many shocked rock lithologies in comparison to their unshocked forms (see Cockell et al., 2005). In a study, Saiz-Jimenez et al. (1990) found that the pore space must be interconnected in order to allow microorganisms to access the interior of the rock from the surface and allow them to spread within the rock matrix itself. Cockell et al. (2002) found that because impact bulking creates fractures and deformation features within target rocks, an increase in porosity is often accompanied by an increase in fracturing, which, in turn, allows microorganisms to move within the rock matrix either by growth along surfaces or by water transport. Fike et al. (2003) found that the interconnected fractures within the shocked material provide a habitat suitable for a wide diversity of nonphotosynthetic heterotrophic (i.e., requiring complex organic compounds of nitrogen and carbon, such as that obtained from plant or animal matter, for metabolic synthesis) microorganisms. However, photosynthetic endoliths require light penetration. Furthermore, impact alteration of rock translucence (the quality or state of being semitransparent) may change availability of habitat.

In particular, asteroid and comet impacts can have a profound influence on the habitats available for plants that grow on rocks (i.e., lithophytic microorganisms). Using evidence from the Haughton impact structure, Nunavut, Canadian High Arctic, Cockell et al. (2005) have described the role of impacts in influencing the nature of the lithophytic ecological niche. They observed that chasmoendolithic (colonizing fissures and cracks in the rock) habitats are commonly increased in abundance as a result of impact bulking. Investigations carried out by several researchers (e.g., Broady, 1979, 1981a; Budel and Wessels, 1991) indicated that the most widely characterized chasmoendoliths are cyanobacterial colonists. Cockell et al. (2005) have presented a synthetic understanding of the influence of asteroid and comet impacts on the availability and characteristics of rocky habitats for microorganisms. Because impact events are a universal phenomenon and would have been more frequent on early Earth than today, understanding their influence on habitats for microorganisms is a fundamental objective in microbiology (Cockell et al., 2005).

5.7 Microbial life beneath the seafloor

One of the startling discoveries with regard to life on Earth in the past few decades is that it can—and does—flourish beneath the ocean floor, in the planet's dark, dense, rocky crust. The only way to get there is by drilling through meters of sediment until hitting the rock; so, for obvious reasons, information on this ubiquitous but buried marine biosphere is still scarce.

Deep-sea sediments become apparently more hostile to life with increasing depth as temperature and pressure rise, and organic matter becomes increasingly recalcitrant (resistant). Demonstrations of high bacterial populations in deep sediments may thus appear enigmatic (mysterious). The great challenge that scientists face then is to be able to explain the continued presence of active bacterial populations in deep sediments that are over 10 million years old. The subseafloor biosphere is the largest prokaryotic habitat on Earth but also a habitat with the lowest metabolic rates. Modeled activity rates are very low, indicating that most prokaryotes may be inactive or have extraordinarily slow metabolism.

Wellsbury et al. (1997) showed that heating coastal surface marine sediments to simulate increasing temperature during burial produces an increase of over three orders of magnitude in acetate concentration and increases bacterial activity. It may be noted that volatile fatty acids, particularly acetate, are important intermediates in the anaerobic degradation of organic matter. Wellsbury et al. (1997) found that pore-water (i.e., water contained in pores in soil or rock) acetate concentration at two sites in the Atlantic Ocean increased at depths below about 150 m and was associated with a significant stimulation in bacterial activity. Comparing these acetate concentrations to in-situ temperatures confirmed that there was a notable generation of acetate associated with temperature increases during burial. This was supported by heating experiments with deep sediments. Thus, according to Wellsbury et al. (1997), acetate generation from organic matter during burial may explain the presence of a deep bacterial biosphere in marine sediments, and could underpin an even deeper and hotter biosphere than has previously been considered.

Scientists have found that diverse microbial communities and numerous energy-yielding activities occur in deeply buried sediments of the eastern Pacific Ocean. D'Hondt et al. (2004) reported that the rates of activities, cell concentrations, and populations of cultured bacteria vary consistently from one subseafloor environment to another. Parkes et al. (2005) reported results from two Pacific Ocean sites—margin and open ocean—both of which have deep, subsurface stimulation of prokaryotic processes associated with geochemical and/or sedimentary interfaces. At 90 m depth in the margin site, stimulation was such that prokaryote numbers were higher (about 13-fold) and activity rates higher than or similar to near-surface values. Parkes et al. (2005)'s analysis of high-molecular-mass DNA confirmed the presence of viable prokaryotes and showed changes in biodiversity with depth that were coupled to geochemistry, including a marked community change at the 90-m interface. It was found that deep sedimentary prokaryotes can have high activity, have changing diversity associated with interfaces and are active over geological timescales.

Oceanic crust encompasses the largest aquifer on Earth, with a liquid volume equal to approximately 2% of the ocean's volume (Johnson and Pruis, 2003). Oceanic crust harbors a substantial reservoir of microbial life which, according to Orcutt et al. (2013) may influence global-scale biogeochemical cycles. To date, knowledge of microbial life in basaltic crust is derived primarily from the study of crustal fluids from the warm ridge flank in the NE Pacific where anaerobic and thermophilic microbial lifestyles dominate (Cowen et al., 2003; Huber et al., 2006; Jungbluth et al., 2013) and from the examination of basaltic rocks collected from mid-ocean spreading ridges (Lysnes et al., 2004; Mason et al., 2009; Santelli et al., 2009). However, very little is known about the oceanic crustal aquifer in low-temperature ridge flank hydrothermal systems, where circulating fluids are predicted to be cold (<20°C) and oxygenated (Edwards et al., 2012). North Pond is a sedimented basin along the Mid-Atlantic ridge that is relatively well studied in terms of its geology and hydrology, and may serve as a model for thousands of similar sediment basins flanking slow-spreading mid-ocean ridges (Schmidt-Schierhorn et al., 2012). The sediment layer (<200 m thick) at North Pond provides an effective barrier to seawater-basalt exchange, while the surrounding ridges of exposed rocky outcrops facilitate the exchange of fluids between the deep ocean and the crustal aquifer. Previous work has shown that fluid flows rapidly in a horizontal direction through the porous oceanic crust beneath North Pond (Becker et al., 2001; Langseth and Yon, 1992), and oxygen profiles of the sediment column revealed that the deepest sediment contains higher concentrations of oxygen than shallower sediment, providing evidence that the basaltic aquifer is oxygenated (Orcutt et al. 2013; Ziebis et al., 2012). Less well constrained are what microbial communities reside in these fluids and how this microbial life impacts biogeochemical cycling in the crustal aquifer and overlying ocean.

Recent studies have added new details to our understanding of the nature of life way down under. Meyer et al. (2016) have offered the first description of an active microbial community buried in cold oceanic crust at North Pond, an isolated sediment pond on the western flank of the Mid-Atlantic Ridge. The samples used in the study were obtained from the CORK (Circulation Obviation Retrofit Kit) subseafloor observatory, installed at North Pond by the Integrated Ocean Drilling Program in 2011.

It may be noted that CORK subseafloor observatories (Fig. 5.8), installed through scientific ocean drilling programs (e.g., Integrated Ocean Drilling Program, IODP), allow access to subseafloor fluids and enable the characterization of microbial life in low-temperature ridge-flank hydrothermal systems (Wheat et al., 2011). These CORK observatories are installed in several locations, including in the warm (64°C), 3.5 million-year-old ridge flank of the Juan de Fuca Ridge, and in the cold (<20°C), 8 million-year-old ridge flank at North Pond. New CORK observatories in two drill holes were installed at North Pond in November 2011 during IODP Expedition 336 (Edwards et al., 2012a,b) and the first crustal fluids for microbiological analysis were collected from these boreholes six months later using a remotely operated vehicle (ROV). These CORK observatories are designed and constructed to minimize contamination and serve studies that utilize basaltic

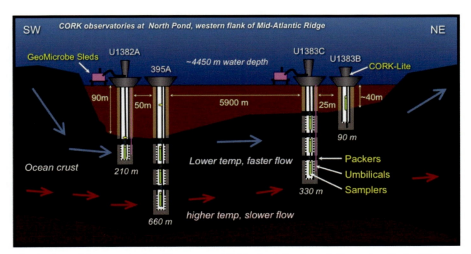

FIG. 5.8 A distinct and active bacterial community in cold oxygenated fluids circulating beneath the western flank of the Mid-Atlantic ridge; Map (A) and diagram (B) of CORK observatories at North Pond. Crustal fluids were extracted from IODP Holes U1382A and U1383C for microbial community analysis in this study *(Source: Edwards, K.J. et al. (2012), Design and deployment of borehole observatories and experiments during IODP Expedition 336, Mid-Atlantic Ridge flank at North Pond, In: Edwards, K.J., Bach, W., Klaus, A., and the Expedition 336 Scientists, Proc. IODP, 336: Tokyo (Integrated Ocean Drilling Program Management International, Inc.). doi:10.2204/iodp.proc.336.109.2012).*

crustal fluids, allowing for samplers to be deployed within the seafloor at multiple isolated depth intervals as well as at the CORK wellhead (Edwards et al., 2012a,b). Specifically, the CORK observatory in U1382A has a packer seal in the bottom of the casing and monitors/samples a single zone in uppermost oceanic crust extending from 90 to 210 mbsf (meters below seafloor). Hole U1383C was equipped with a three-level CORK observatory that spans a zone of thin basalt flows with intercalated limestone (~70–146 mbsf), a zone of glassy, thin basaltic flows and hyaloclastites (146–200 mbsf), and a lowermost zone (~200–331.5 mbsf) of more massive pillow flows with occasional hyaloclastites (volcanoclastic accumulations or breccia consisting of glass fragments formed by quench fragmentation of lava flow surfaces during submarine or subglacial extrusion) in the upper part (Edwards et al., 2012a). Both CORK observatories consist of fiberglass casing within the oceanic crust (Edwards et al., 2012a). Meyer et al. (2016) reported the initial microbiological and geochemical characterization of the crustal fluids from cold, oxygenated igneous crust at North Pond on the western flank of the Mid Atlantic Ridge. They examined the geochemical and microbial signatures of crustal fluids to resolve the extent of geochemical transformations during passage through the crust and the presence and activities of microbes living in the circulating subseafloor fluids. Combining genomic technologies with geochemical measurements, Meyer et al. (2016) examined crustal fluid samples retrieved in 2012 from 50 to 250 meters beneath the seafloor, under 4.5 kilometers of seawater.

The oceanic crust is anything but static: Seawater runs through its rocky crevices, creating a dynamic aquifer through which the entire volume of the ocean circulates every 200,000 years. Meyer et al. (2016) discovered that the microbial community in North Pond crustal samples was oxygenated, heterogeneous, and markedly distinct from that found in ocean bottom seawater.

The rock-hosted, oceanic crustal aquifer is one of the largest ecosystems on Earth, yet little is known about its indigenous microorganisms. Meyer et al. (2016) provided the first phylogenetic (i.e., relating to the evolutionary development and diversification of a species or group of organisms, or of a particular feature of an organism) and functional description of an active microbial community residing in the cold oxic (of a process or environment in which oxygen is involved or present) crustal aquifer. Using subseafloor observatories, they recovered crustal fluids and found that the geochemical composition is similar to bottom seawater, as are cell abundances. However, based on relative abundances and functional potential of key bacterial groups, the crustal fluid microbial community was found to be heterogeneous (diverse in character or content) and markedly distinct from seawater. Potential rates of autotrophy (the process of making food from inorganic substances, using photosynthesis) and heterotrophy (the quality or condition of being capable of utilizing only organic materials as a source of food) in the crust exceeded those of seawater, especially at elevated temperatures (25°C) and deeper in the crust. Together, these results reveal an active, distinct, and diverse bacterial community engaged in both heterotrophy and autotrophy in the oxygenated crustal aquifer, providing key insight into the role of microbial communities in the ubiquitous cold dark subseafloor biosphere.

Interestingly, in many cases, Meyer et al. (2016) found the same general group of bacteria in the crustal aquifer and in bottom seawater, but different species within that group. This finding suggests distinct differences in potential microbial activity between the two sites, such as more carbon fixation in the aquifer. Whereas prior work has focused on the hot, volcanic fluids at mid-ocean ridges and the subseafloor microbes that survive there, Meyer et al. (2016)'s investigations describe the subseafloor microbial community in a cold crustal aquifer site. It may be noted that the cold crustal aquifer is a different environment that is also globally important not just in terms of life, but biogeochemical cycling. According to the researchers, these investigations suggest that we are only starting to discover how things proceed there.

5.8 Microbial life in Antarctic ice sheet

The cold, arid (having little or no rain), remotely located and perennially ice-covered Antarctic has long been considered an analog to how life might persist in the barren, frozen landscape of Mars and Europa (Abyzov et al., 1998a; 1999). Although Antarctica has one of the harshest environments on Earth, it is evident from studies on Antarctic glacial and lake ice (Abyzov et al., 1998b; Priscu et al., 1998), permafrost (Gilichinsky et al., 2007), sea ice (Thomas and Dieckmann, 2002) and cryoconite holes (Christner et al., 2003a) that bacteria are prevalent in this environment. Bacteria entrapped in ice cores as old as 1800 years to 2 million years have been isolated and characterized (Christner et al., 2000; D'Elia et al., 2008).

Although glaciers are considered free of living organisms, there are areas like supraglacial water-filled depressions called cryoconite holes (Fig. 5.9). They can be 0.5 cm to a meter in diameter and up to 0.5 m deep and formed when dark granular matter is deposited onto the surface of the glacier, and the low albedo of the matter causes increased absorption of solar energy, which results in melting of the ice and subsequently formation of a water-filled hole with dark granular material at the base. They are considered as microbial hotspots, where carbon and nutrients help the growth of photosynthetic and heterotrophic microorganisms. Cryoconite holes were first described and named by Nils A.E. Nordenskiöld when he traveled on Greenland's icecap in 1870. During summer, cryoconite holes frequently

FIG. 5.9 Sampling a cryoconite on Longyearbreen glacier during field work of Arctic microbiology course, Svalbard. *(Source: https://commons.wikimedia.org/wiki/File:Noorteadlased_kr%C3%BCokoniidist_proovi_v%C3%B5tmas.jpg Author: Kertu Liis Krigul Copyright info: This file is licensed under the Creative Commons Attribution-Share Alike 5.0 International license).*

contain liquid water and thus provide a niche for cold-adapted microorganisms like bacteria, algae and animals like rotifers (Fontaneto et al., 2015) and tardigrades to thrive. Cryoconite typically settles and concentrates at the bottom of these holes creating a noticeable dark mass.

Study of ice core samples from Antarctica has revealed the presence of a diverse assemblage of prokaryotic and eukaryotic organisms including bacteria, yeasts, fungi, viruses, diatoms, and pollen (et al., 1998a; 1999). Microbiological analysis of Vostok ice core at a depth of 1500–2750 m revealed that bacteria were present in numbers ranging from 0.8 to 11×10^3 cells/mL (Karl et al., 1999). In a deeper section of the core at 3590 m, bacteria were observed at concentrations ranging from 2.8×10^3 to 3.6×10^4 cells/ml (Priscu et al., 1999). Bacteria have also been reported in samples from the sediment environment beneath the Antarctic ice sheet at concentrations of $\sim 10^7$ cells/g (Lanoil et al., 2009). There is now growing evidence that bacteria carry out DNA and protein syntheses at sub-zero temperatures (Carpenter et al., 2000; Christner, 2002). Bacterial metabolism in ice is important not only for the survival of bacteria, but also for interpreting the ice core glacio-chemical records (Sowers, 2001; Tung et al., 2005; Rohde et al., 2008).

India's National Centre for Polar and Ocean Research (NCPOR) {formerly National Centre for Antarctic and Ocean Research (NCAOR)} researchers have carried out several studies in Antarctica in various disciplines. For example, Antony et al. (2009) reported for the first time the isolation of a bacterium known as *Cellulosimicrobium cellulans* (which is known to be adapted to moderate temperature environments) from Antarctic snow collected from the Larsemann Hills region of East Antarctica. It was found that this strain demonstrated physiological traits that were markedly different from that of the mesophilic *C. cellulans* type strain DSM 43879^T. Note that a mesophile is an organism that grows best in moderate temperature, neither too hot nor too cold, typically between 20 and 45°C (68 and 113°F). The optimal temperature is 37°C. The term is mainly applied to microorganisms. As already indicated, organisms that prefer extreme environments are known as extremophiles.

The study carried out by Antony et al. (2009) indicated that the ability of *C. cellulans* to survive in Antarctic snow could be due to its modified physiological properties (cellular composition and metabolic capabilities) that distinguish it from its mesophilic counterpart. The difference in cell wall composition of this strain appears to be an adaptation for survival in the extreme cold environment.

Carbon utilization studies demonstrated that *C. cellulans* preferred complex carbon substrates over simple ones, thereby suggesting that it could play a potential role in carbon uptake in snow. Furthermore, the isolation of this strain from the perennially cold Antarctic provides new insight on the physiological adaptation of *C. cellulans* to extreme environments. The survival of bacteria in extremely cold environments is dependent on the successful occurrence of several adaptive features. The metabolic flexibility of *C. cellulans* in terms of its ability to rapidly utilize a wide array of simple carbon substrates like amino acids, amides, carboxylic acids, carbohydrates as well as complex substrates like polymers indicate that it could play a potentially significant role in the scavenging of simple and complex carbons in snow. Antony et al.'s (2009) study shows that this genus, isolated earlier only from mesophilic environments, is also capable of surviving in extreme cold conditions.

5.9 The year-2021 discovery of sessile benthic community far beneath an Antarctic ice shelf

Hard substrate sessile benthic community (sponges and other animals) have been discovered stuck to a boulder under 900 m of ice and 500 m of water in Antarctica (Fig. 5.10). The creatures were spotted by chance by an underwater camera, after researchers drilled through the Filchner-Ronne ice shelf on the south-eastern Weddell Sea in Antarctica to obtain a sediment core from the seabed. These borehole records are the result of images captured as part of geological and glaciological sampling which happened to record images of the seafloor life beneath the ice shelves. No sunlight reaches the animals, and they are thought to live on food carried hundreds of km from the nearest light. Photos and video footage of the boulder show that it is home to at least two types of sponge, one of which has a long stem that opens into a head. But other organisms, which could be tube worms or stalked barnacles, also appear to be growing on the rock.

FIG. 5.10 Marine organisms on a boulder on the sea floor beneath 900 metres (3,000ft) of Antarctic ice shelf. Photograph: Huw Griffiths/UKRI BAS. *(Source: Griffiths, H.J., Anker, P., Linse, K., Maxwell, J., Post, A.L., Stevens, C., Tulaczyk, S., and Smith, J.A. (2021), Breaking all the rules: The first recorded hard substrate sessile benthic community far beneath an Antarctic ice shelf, Front. Mar. Sci., 8:642040. doi: 10.3389/fmars.2021.642040) Copyright © 2021 Griffiths, Anker, Linse, Maxwell, Post, Stevens, Tulaczyk and Smith. This is an open-access article distributed under the terms of the Creative Commons Attribution License (CC BY). The use, distribution or reproduction in other forums is permitted, provided the original author(s) and the copyright owner(s) are credited and that the original publication in this journal is cited, in accordance with accepted academic practice. No use, distribution or reproduction is permitted which does not comply with these terms.*

The discovery is surprising because the areas beneath the floating ice shelves are hidden from daylight and often far from areas of primary productivity (Ingels et al., 2021). Under ice shelf assemblages are generally believed to resemble the communities of the oligotrophic (relatively poor in plant nutrients and containing abundant oxygen in the deeper parts) deep sea, subsisting on advected food particles (Ingels et al., 2018). This community is somewhere between 625 and 1500 km from the nearest region of photosynthesis. At least some of these animals might be carnivorous sponges. Southern Ocean species represent ~20% of all known carnivorous sponges and they are often found in oligotrophic bathyal regions (the zone of the sea between the continental shelf and the abyssal zone) or on isolated seamounts (Goodwin et al., 2017). According to Griffiths et al. (2021), the biological and physical attributes that allow this community to survive, despite our current theories, suggest that these communities are either better connected to the outside world than we can currently explain or that the organisms themselves represent highly specialized extreme oligotrophic adaptation.

5.10 Microbial life at the driest desert in the world

The Atacama Desert, located in the "rain shadow" between two mountain ranges—the Andes and the Chilean Coast Range—is a plateau in northern Chile in South America, covering a 1000-km (600-mi) strip of land located on the Pacific coast, west of the Andes mountains. It is the driest and the oldest desert on Earth, as well as the only true desert to receive less precipitation than the polar deserts. In 2004, the Yungay region in this hyperarid desert was established as the driest site of this desert and also close to the dry limit for life on Earth. Since then, much has been published about the extraordinary characteristics of this site and its pertinence as a Mars analog model. However, as a result of a more systematic search in the Atacama, Azua-Bustos et al. (2015) described a new site, María Elena South (MES), which is much drier than Yungay. According to Azua-Bustos et al. (2015), "The mean atmospheric relative humidity (RH) at MES was 17.3%, with the RH of its soils remaining at a constant 14% at the depth of 1 m, a value that matches the lowest RH measurements taken by the Mars Science Laboratory at Gale Crater." Remarkably, Azua-Bustos et al. (2015) found a number of viable bacterial species in the soil profile at MES using a combination of molecular-dependent and independent methods, unveiling the presence of life in the driest place on the Atacama Desert.

5.11 Tardigrades—Important extremophiles useful for investigating life's tolerance limit beyond earth

Despite the typical human notion that the Earth is a habitable planet, over three quarters of our planet is uninhabitable by us without assistance. The organisms that live and thrive in these "inhospitable" extreme environments are known by the name "extremophiles" and are found in all Domains of Life (Bacteria, Archaea, and Eucarya). Extremophiles thrive in ice, boiling water, acid, the water core of nuclear reactors, salt crystals, and toxic waste and in a range of other extreme habitats that were previously thought to be inhospitable for life. Extremophiles include representatives of all three domains of life just indicated; however, the majority are microorganisms, and a high proportion of these are Archaea. Over the past few decades, it has been found that many extreme environments are inhabited by extremophiles. According to von Hegner (2019), knowledge of their emergence, adaptability, and limitations seems to provide a guideline for the search of extra-terrestrial life, since some extremophiles presumably can survive in extreme environments such as the planet Mars, Jupiter's moon Europa, and Saturn's moon Enceladus. According to several planetary scientists, extremophiles, especially those thriving under multiple extremes, represent a key area of research for multiple disciplines, spanning from the study of adaptations to harsh conditions, to the biogeochemical cycling of elements.

Extremophile research also has implications for origin of life studies and the search for life on other planetary and celestial bodies. Merino et al. (2019) have reviewed the current state of knowledge for the biospace in which life operates on Earth and discussed it in a planetary context, highlighting knowledge gaps and areas of opportunity. Coker (2019) has described how extremophiles have adapted to live/thrive/survive in their niches, helped scientists unlock major scientific discoveries, advance the field of biotechnology, and inform us about the boundaries of life and where we might find it in the Universe.

In the last few decades of the twentieth century, numerous true extreme-loving organisms were found. Tardigrades ("water bears") (Fig. 5.11) are small invertebrate animals that survive a remarkable array of stresses. The name "water-bear" comes from the way they walk, reminiscent of a bear's gait. The name *Tardigradum* means "slow walker." Tardigrades are remarkable diminutive (extremely or unusually small) creatures of anywhere from 100 to 1000 microns in maximum length and live mainly in freshwater environments (Weronika and Lukasz, 2017). Tardigrades are also able to survive extreme environmental conditions such as low temperatures, vacuum, and radiation (Persson et al., 2011; Schill and Hengherr, 2018; Jönsson, 2019) and have survived exposure in space on the exterior of space vehicles (Jönsson et al., 2008; Rebecchi et al., 2009; Persson et al., 2011).

Tardigrades in the "tun" state (i.e., shedding almost all the water in its body, and then curling up into a dry husk), can survive temperatures from —253°C to 151°C, X-rays, vacuum and, when in perfluorocarbon, pressures up to 600 MPa, almost 6,000 times atmospheric pressure at sea level (Seki and Toyoshima, 1998). Tardigrades are one of nature's smallest animals. They are never more than 1.5 mm long, and can only be seen clearly with a microscope. The video (Habt_Worlds_Ch-5_Tardigrade_in_real_time) shows a video of the movements of tardigrade under the microscope (Source of video: Tardigrade; From Wikipedia, the free encyclopedia).

Tardigrades have podgy faces with folds of flesh. They have eight legs, with ferocious claws resembling those of great bears. Their mouth is also a serious weapon, with dagger-like teeth that can spear prey. Fossils of tardigrades have been dated to the Cambrian period (over 500 million years ago), when the first complex animals were evolving. In 1948, the Italian zoologist Tina Franceschi claimed that tardigrades found in dried moss from museum samples over 120 years old could be reanimated. After rehydrating a tardigrade, she observed one of its front legs moving. This finding has never been replicated. But it does not seem impossible. In 1995, dried tardigrades were brought back to life after 8 years.

Tardigrades are well-known for their ability to tolerate complete desiccation and very high and low temperatures (Wright et al., 1992; Wright, 2001; Møbjerg et al., 2011). They also belong to the most radiation tolerant animals on Earth. In a study, Jagadeesh et al. (2018) established a metric tool for distinguishing the potential survivability of active and cryptobiotic tardigrades on rocky-water and water-gas planets in our solar system and exoplanets.

When tardigrades start to dry out, they seem to make a lot of antioxidants. These are chemicals, like vitamins C and E, that soak up dangerously reactive chemicals. This may mop up harmful chemicals in the tardigrades' cells. The antioxidants may explain one of tardigrades' neatest abilities. If a tardigrade stays in its dry "tun" state for a long time, its DNA gets damaged. But after it reawakens it is able to quickly fix it. It is clear that the "tun" state is key to tardigrades' ability to cope with being dried out.

5.11.1 Temperature tolerance in tardigrades

Tardigrades can tolerate blazing heat. They can also tolerate very low temperatures. It seems tardigrades can actually tolerate ice forming within their cells. Either they can protect themselves from the damage caused by ice crystals, or they can repair it. Tardigrades may produce chemicals called ice nucleating agents. These encourage ice crystals to form outside their cells rather than inside, protecting the vital molecules. While we have some idea of how tardigrades cope with the cold, we have no idea how they cope with heat.

Many animals that have evolved to live in hot environments, such as hot springs and scorching deserts, produce chemicals called heat shock proteins. These act as chaperones (body guards) for proteins inside cells, helping them keep their shape. They also repair heat-damaged proteins.

Survival in microhabitats that experience extreme fluctuations in water availability and temperature requires special adaptations. To withstand such environmental conditions, tardigrades, as well as some nematodes and rotifers,

FIG. 5.11 A microscope image of Tardigrade. The tiny animal tardigrade (water bear), which is capable of surviving temperatures from —253°C to 151°C, X-rays, vacuum and, when in perfluorocarbon, pressures up to 600 MPa, almost 6,000 times atmospheric pressure at sea level. *(Source: https://commons.wikimedia.org/wiki/File:SEM_image_of_Milnesium_tardigradum_in_active_state_-_journal.pone.0045682.g001-2.png) Author: Schokraie, E., Warnken, U., Hotz-Wagenblatt, A., Grohme, M.A., Hengherr, S., et al. (2012), Comparative proteome analysis of Milnesium tardigradum in early embryonic state versus adults in active and anhydrobiotic state, PLoS ONE, 7(9): e45682. doi:10.1371/journal.pone.0045682 Copyright info: This file is licensed under the Creative Commons Attribution 2.5 Generic license.*

enter a completely desiccated state known as anhydrobiosis. Hengherr et al. (2009) examined the effects of high temperatures on fully desiccated (anhydrobiotic) tardigrades. They exposed nine species from the classes *Heterotardigrada* and *Eutardigrada* to temperatures of up to 110°C for 1 hour. Exposure to temperatures of up to 80°C resulted in a moderate decrease in survival. Exposure to temperatures above this resulted in a sharp decrease in survival, with no animals of the families *Macrobiotidae* and *Echiniscidae* surviving 100°C. However, Milnesium tardigradum (Milnesidae) showed survival of >90% after exposure to 100 °C; temperatures above this resulted in a steep decrease in survival. Vitrification (conversion into a glassy substance by heat and fusion) is assumed to play a major role in the survival of anhydrobiotic organisms during exposure to extreme temperatures, and consequently, the glass-transition temperature (T_g) is critical to high-temperature tolerance. In this study, Hengherr et al. (2009) provided the first evidence of the presence of a glass transition during heating in an anhydrobiotic tardigrade through the use of differential scanning calorimetry.

5.11.2 Desiccation tolerance in tardigrades

Desiccation is a biological state in which all the moisture from the body is completely dried out. The phenomenon of desiccation tolerance, also called "anhydrobiosis," involves the ability of an organism to survive the loss of almost all cellular water without sustaining irreversible damage.

In one of the key discoveries in 1922, a German scientist named H. Baumann found that when a tardigrade dries out, it retracts its head and its eight legs. It then enters a deep state of suspended animation that closely resembles death. Shedding almost all the water in its body, the tardigrade curls up into a dry husk, which Baumann called "Tönnchenform"—now commonly known as a "tun." Its metabolism (i.e., the chemical processes that occur within a living organism in order to maintain life) slows to 0.01% of the normal rate. It can stay in this state for decades, only re-animating when it comes into contact with water.

How tardigrades survive desiccation has remained a mystery for more than 250 years. To withstand desiccation, many invertebrates such as rotifers, nematodes, and tardigrades enter a state known as "anhydrobiosis," which is thought to require accumulation of compatible osmolytes, such as the nonreducing disaccharide trehalose to protect against dehydration damage. Trehalose—a disaccharide essential for several organisms to survive drying—is detected at low levels or not at all in some tardigrade species. Hengherr et al. (2008) studied the trehalose levels of eight tardigrade species comprising *Heterotardigrada* and *Eutardigrada* in five different states of hydration and dehydration. They found that although many species accumulate trehalose during dehydration, the data revealed significant differences between the species. Furthermore, although trehalose accumulation was found in species of the order *Parachela (Eutardigrada)*, it was not possible to detect any trehalose in the species *Milnesium tardigradum* and no change in the trehalose level has been observed in any species of *Heterotardigrada* investigated until this research. These results are considered to have contributed considerably in expanding our understanding of anhydrobiosis in tardigrades and, for the first time, demonstrate the accumulation of trehalose in developing tardigrade embryos, which have been shown to have a high level of desiccation tolerance.

As already indicated, certain organisms found across a range of taxa, including bacteria, yeasts, plants and many invertebrates such as nematodes and tardigrades are able to survive almost complete loss of body water, remaining in "anhydrobiosis" state for decades without apparent damage. When water again becomes available, they rapidly rehydrate and resume active life. Research in anhydrobiosis has focused mainly on sugar metabolism and stress proteins. Despite the discovery of various molecules which are involved in desiccation and water stress, knowledge of the regulatory network governing the stability of the cellular architecture and the metabolic machinery during dehydration is still fragmentary and not well understood. A combination of transcriptional, proteomic, and metabolic approaches with bioinformatics tools can provide a better understanding of gene regulation that underlie the biological functions and physiology related to anhydrobiosis. The development of this concept will raise exciting possibilities and techniques for the preservation and stabilization of biological materials in the dry state (Schill et al., 2009).

As just mentioned, upon desiccation, some tardigrades enter an ametabolic dehydrated state called anhydrobiosis and can survive a desiccated environment in this state. For successful transition to anhydrobiosis, some anhydrobiotic tardigrades require pre-incubation under high humidity conditions, a process called preconditioning, prior to exposure to severe desiccation. Although tardigrades are thought to prepare for transition to anhydrobiosis during preconditioning, the molecular mechanisms governing such processes remain unknown. In a study, Kondo et al. (2015) used chemical genetic approaches to elucidate the regulatory mechanisms of anhydrobiosis in the anhydrobiotic tardigrade, *Hypsibius dujardini*. They first demonstrated that inhibition of transcription or translation drastically impaired anhydrobiotic survival, suggesting that de novo gene expression is required for successful transition to anhydrobiosis in this tardigrade. They then screened 81 chemicals and identified 5 chemicals that significantly impaired anhydrobiotic survival after severe desiccation, in contrast to little or no effect on survival after high humidity exposure only. In particular, cantharidic acid, a selective inhibitor of protein phosphatase (PP) 1 and PP2A, exhibited the most profound inhibitory effects. Another PP1/PP2A inhibitor, okadaic acid, also significantly and specifically impaired anhydrobiotic survival, suggesting that PP1/PP2A activity plays an important role for anhydrobiosis in this species. This is considered to be the first report of the required activities of signaling molecules for desiccation tolerance in tardigrades. The identified inhibitory chemicals could provide novel clues to elucidate the regulatory mechanisms underlying anhydrobiosis in tardigrades.

Detection of trehalose (a sugar of the di-saccharide class produced by some fungi, yeasts, and similar organisms to survive drying) at low levels or not at all in some tardigrade species indicates that tardigrades possess potentially novel mechanisms for surviving desiccation. Boothby et al. (2017) showed that tardigrade-specific intrinsically disordered proteins (TDPs) are essential for desiccation tolerance. They found that TDP genes are constitutively expressed at high levels or induced during desiccation in multiple tardigrade species. TDPs are required for tardigrade desiccation tolerance, and these genes are sufficient to increase desiccation tolerance when expressed in heterologous systems. Note that in cell biology and protein biochemistry, heterologous expression means that a protein is experimentally put into a cell that does not normally make (i.e., express) that protein. TDPs form noncrystalline amorphous solids (vitrify) upon desiccation, and this vitrified state mirrors their protective capabilities against desiccation. Boothby et al. (2017)'s study identifies TDPs as functional mediators of tardigrade desiccation tolerance, expanding our knowledge of the roles and diversity of disordered proteins involved in stress tolerance.

5.11.3 Radiation tolerance in tardigrades

Tardigrades have a reputation of being extremely tolerant to extreme environmental conditions including tolerance to ionizing radiation while in a desiccated, anhydrobiotic state. In 1964, scientists exposed tardigrades to lethal doses of X-rays and found that they could survive. Later experiments showed they can also cope with excessive amounts of alpha, gamma, and ultraviolet radiation—even if they are not in the tun state. Radiation was one of the biggest threats facing the tardigrades sent into space in 2007. Those exposed to higher levels of radiation fared worse than those protected, but the mortality rate was not 100%.

Several studies have documented a very high tolerance of adult tardigrades to both low-LET (Linear Energy Transfer) radiation {X-rays: (May et al., 1964); gamma rays: (Jönsson et al., 2005; Horikawa et al., 2006; Beltrán-Pardo et al., 2015)}, high-LET radiation {alpha particles: (Horikawa et al., 2006); protons: (Nilsson et al., 2010)}, and UV radiation (Altiero et al., 2011; Horikawa et al., 2013).

Tardigrades are also the only animals so far that have survived the combined exposure to cosmic radiation, UV radiation, and vacuum under real space conditions (Jönsson et al., 2008). The dose-responses documented in these studies show that several tardigrade species are able to survive several thousands of Gray of ionizing radiation. For comparison, the dose at which 50% of humans die within 30 days is less than 5 Gy (Bolus, 2001). It is also clear that the tolerance is not restricted to the desiccated anhydrobiotic state in tardigrades, but is expressed also in active hydrated animals. Embryos are considerably more sensitive to radiation, particularly in the early stage of development (Beltrán-Pardo et al., 2015; Jönsson et al., 2013; Beltrán-Pardo et al., 2013; Horikawa et al., 2012).

Jönsson et al (2005) reported an investigation on radiotolerance in desiccated and hydrated specimens of the eutardigrade *Richtersius coronifer*. Jönsson et al (2005)'s study suggested that radiation tolerance in tardigrades is not due to biochemical protectants connected with the desiccated state. Rather, cryptobiotic tardigrades may rely on efficient mechanisms of DNA repair, the nature of which remained unknown during their studies.

In 2007, thousands of tardigrades were attached to a satellite and blasted into space. After the satellite had returned to Earth, scientists examined them and found that many of them had survived. Some of the females had even laid eggs in space, and the newly-hatched young were healthy.

Tardigrades represent one of the most desiccation and radiation-tolerant animals on Earth, and several studies have documented their tolerance in the adult stage. Studies on tolerance during embryological stages are rare, but differential effects of desiccation and freezing on different developmental stages have been reported, as well as dose-dependent effect of gamma irradiation on tardigrade embryos. Beltrán-Pardo et al. (2013) reported a study evaluating the tolerance of eggs from the *eutardigrade Milnesium* cf. *tardigradum* to three doses of gamma radiation (50, 200 and 500 Gy) at the early, middle, and late stage of development. They found that embryos of the middle and late developmental stages were tolerant to all doses, while eggs in the early developmental stage were tolerant only to a dose of 50 Gy, and showed a declining survival with higher dose. They also observed a delay in development of irradiated eggs, suggesting that periods of DNA repair might have taken place after irradiation induced damage. The delay was independent of dose for eggs irradiated in the middle and late stage, possibly indicating a fixed developmental schedule for repair after induced damage. These results show that the tolerance to radiation in tardigrade eggs changes in the course of their development. The mechanisms behind this pattern are unknown, but may relate to changes in mitotic activities (i.e., having to do with the presence of dividing and proliferating cells. As an example, cancer tissue generally has more mitotic activity than normal tissues) over the embryogenesis and/or to activation of response mechanisms to damaged DNA in the course of development.

As already indicated, tardigrades are highly tolerant to desiccation and ionizing radiation but the mechanisms of this tolerance are not well understood. Beltrán-Pardo et al. (2015) reported studies on dose responses of adults and eggs of the tardigrade *Hypsibius dujardini* exposed to gamma radiation. In adults, the LD50/48h for survival was estimated at ~4200 Gy, and doses higher than 100 Gy reduced both fertility and hatchability of laid eggs drastically. Note that the gray (symbol: Gy) is a derived unit of ionizing radiation dose in the International System of Units (SI). It is defined as the absorption of one joule of radiation energy per kilogram of matter. The observations suggest a judgment that 4.5 Gy (450 rad) absorbed dose in the bone marrow for energetic and therefore penetrating gamma rays giving reasonably uniform irradiation of the marrow could be regarded as the LD50 in circumstances where those irradiated were protected from thermal radiation and blast damage (Mole, 1984). Beltrán-Pardo et al. (2015) also evaluated the effect of radiation (doses 50 Gy, 200 Gy, 500 Gy) on eggs in the early and late embryonic stage of development, and observed a reduced hatchability in the early stage, while no effect was found in the late stage of development. Survival of juveniles from irradiated eggs was highly affected by a 500 Gy dose, both in the early and the late stage. Juveniles hatched from eggs irradiated at 50 Gy and 200 Gy developed into adults and produced offspring, but their fertility was reduced compared to the controls. Finally, they measured the effect of low temperature during irradiation at 4000 Gy and 4500 Gy on survival in adult tardigrades, and observed a slight delay in the expressed mortality when tardigrades were irradiated on ice. Because *H. dujardini* is a freshwater tardigrade with lower tolerance to desiccation compared to limno-terrestrial tardigrades, the high radiation tolerance in adults, similar to limno-terrestrial tardigrades, is unexpected and seems to challenge the idea that desiccation and radiation tolerance rely on the same molecular mechanisms. Beltrán-Pardo et al. (2015) suggested that the higher radiation tolerance in adults and late-stage embryos of *H. dujardini* (and in other studied tardigrades) compared to early stage embryos may partly be due to limited mitotic activity (having to do with the presence of dividing and proliferating cells. Cancer tissue generally has more mitotic activity than normal tissues.), since tardigrades have a low degree of somatic cell division (eutely), and dividing cells are known to be more sensitive to radiation.

Genomic DNA stores all genetic information and is indispensable for maintenance of normal cellular activity and propagation. Radiation causes severe DNA lesions, including double-strand breaks, and leads to genome instability and even lethality. Regardless of the toxicity of radiation, some organisms exhibit extraordinary tolerance against radiation. These organisms are supposed to possess special mechanisms to mitigate radiation-induced DNA damages. Extensive study using radio-tolerant bacteria suggested that effective protection of proteins and enhanced DNA repair system play important roles in tolerability against high-dose radiation. Recent studies using the tardigrade—an extremo-tolerant animal—provides new evidence that a tardigrade-unique DNA-associating protein, termed Dsup, suppresses the occurrence of DNA breaks by radiation in human-cultured cells.

The majority of previous studies on radiation tolerance in tardigrades have investigated limno-terrestrial (being or inhabiting a moist terrestrial environment that is subject to periods of both immersion and desiccation) species that are also highly tolerant to desiccation (Richtersius coronifer, Milnesium tardigradum, Ramazzottius varieornatus), the only exception being a study on the limnic (relating to bodies of water with low salt concentration, such as lakes and ponds) species Hypsibius dujardini (Beltrán-Pardo et al., 2015). A common view on why tardigrades are highly tolerant to radiation is that their tolerance relies on molecular mechanisms that have evolved to allow survival under extreme but natural environmental conditions, in dry or cold environments (Jönsson, 2003). This hypothesis predicts that the most desiccation tolerant tardigrades should also be the most radiation tolerant, regardless of evolutionary origin and taxonomic (concerned with the classification of things, especially organisms) affiliation. So far, all tardigrades evaluated for radiation tolerance belong to the class of *Eutardigrada*. In a first study of its kind, Jönsson et al. (2016) reported on tolerance to ionizing radiation in a marine hetero-tardigrade species, *Echiniscoides sigismundi*, and evaluated the dose-response to gamma irradiation. In Jönsson et al. (2016)'s studies, they used specimens of the marine tidal heterotardigrade Echiniscoides sigismundi collected from barnacle shells at Lynæs, Zealand, Denmark. Fig. 5.12 shows dose-response curves showing the proportion of active Echiniscoides sigismundi after different doses of gamma radiation, at activity assessments 24 h, 48 h, 72 h and 7 days after irradiation.

Jönsson et al. (2016)'s study provided the first evidence that tardigrades also of the class *Heterotardigrada* have a high tolerance to ionizing radiation, with some specimens surviving doses of gamma ray up to 4 kGy. High radiation tolerance therefore is not restricted to the class *Eutardigrada* but seems to have a wide taxonomic distribution within the phylum *Tardigrada*. However, in contrast to previous dose-response studies with gamma rays in eutardigrades, which exhibit LD50 values in the range of 3–5 kGy, 1–2 days after irradiation, calculated LD50 values of E. sigismundi were considerably lower. Already at 1 kGy the effect of irradiation on animal activity was considerable, while in the eutardigrades (a class of tardigrades without lateral appendices) Macrobiotus areolatus (May et al., 1964), Richtersius coronifer (Jönsson et al., 2005), Milnesium tardigradum (Horikawa et al., 2006), and Hypsibius dujardini (Beltrán-Pardo et al., 2015), doses of 3 kGy or higher were required

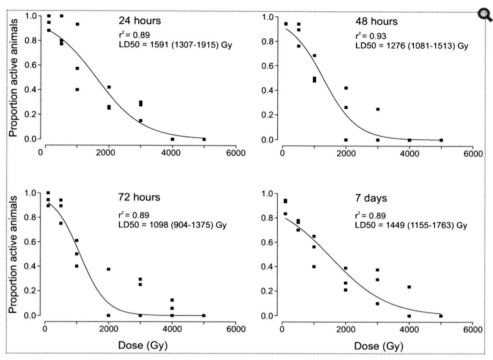

FIG. 5.12 shows dose-response curves showing the proportion of active Echiniscoides sigismundi after different doses of gamma radiation, at activity assessments 24h, 48h, 72h and 7 days after irradiation. *(Source:Jönsson et al., 2016; PLoSOne).*

to significantly reduce viability. Thus, under the current protocol, E. sigismundi seems to be more sensitive to low-LET ionizing radiation than the *eutardigrades* studied so far.

Jönsson et al. (2016) consider that the current study broadens our knowledge on radiation tolerance in tardigrades to include also the marine heterotardigrade E. sigismundi. According to them, an inclusion of nontidal marine tardigrade species and limno-terrestrial heterotardigrades (a class of tardigrades that have cephalic appendages and legs with four separate but similar digits or claws on each) in future studies would further contribute to the picture of the taxonomic and environmental distribution of radiation tolerance in tardigrades.

In a review, Hashimoto and Kunieda (2017) provided a brief summary of the current knowledge on extremely radio-tolerant animals, and presented novel insights from the tardigrade research, which expand our understanding on molecular mechanism of exceptional radio-tolerability.

5.11.4 Dormancy strategies in tardigrades

In a review Guidetti et al. (2011) analyzed the dormancy strategies of metazoans (all animals having the body composed of cells differentiated into tissues and organs and usually a digestive cavity lined with specialized cells) inhabiting "hostile to life" habitats, which have a strong impact on their ecology and in particular on the traits of their life history. Tardigrades are considered a model animal, being aquatic organisms colonizing terrestrial habitats. Tardigrades evolved a large variety of dormant stages that can be ascribed to diapause (a period of suspended development in an insect, other invertebrate, or mammal embryo, especially during unfavorable environmental conditions), cryptobiosis (a physiological state in which metabolic activity is reduced to an undetectable level without disappearing altogether), and cryobiosis (a metabolic state of life entered by an organism in response to adverse environmental conditions such as desiccation, freezing, and oxygen deficiency). In tardigrades, diapause and cryptobiosis can occur separately or simultaneously, consequently the adoption of one adaptive strategy is not necessarily an alternative to the adoption of the other. Encystment (the process of becoming enclosed by a cyst, which is a thin-walled hollow organ or cavity in an animal or plant, containing a liquid secretion; a sac, vesicle, or bladder) and cyclomorphosis (occurrence of cyclic or seasonal changes in the phenotype of an organism through successive generations) are characterized by seasonal cyclic changes in morphology (form, shape, or structure) and physiology (normal functions of living organisms and their parts) of the animals. They share several common features and their evolution is strictly linked to the molting process (the process of shedding the old outer covering of the body to make way for a new growth).

Biological bet hedging occurs when organisms suffer decreased fitness in their typical conditions in exchange for increased fitness in stressful conditions. Bet-hedging theory addresses how individuals should optimize fitness in varying and unpredictable environments by sacrificing mean fitness to decrease variation in fitness. So far, three main bet-hedging strategies have been described: conservative bet-hedging (play it safe), diversified bet-hedging (don't put all eggs in one basket) and adaptive coin flipping (choose a strategy at random from a fixed distribution) (for details, see Olofsson et al., 2009). A bet-hedging strategy with different patterns of egg hatching time has been observed in a tardigrade species. Four categories of eggs have been identified: (1) subitaneous, (2) delayed-hatching, (3) abortive, and (4) diapause resting eggs, which needs a stimulus to hatch (rehydration after a period of desiccation). Cryptobiotic tardigrades are able to withstand desiccation (anhydrobiosis) and freezing (cryobiosis) at any stage of their life-cycle. This ability involves a complex array of factors working at molecular (bioprotectans), physiological and structural levels. Animal survival and the accumulation of molecular damage are related to the time spent in the cryptobiotic state, to the abiotic parameters during the cryptobiotic state, and to the conditions during the initial and final phases of the process. Cryptobiosis evolved independently at least two times in tardigrades, in eutardigrades and in echiniscoids. Within each evolutionary line, the absence of cryptobiotic abilities is more related to selective pressures to local habitat adaptation than to phylogenetic relationships. The selective advantages of cryptobiosis (e.g. persistency in "hostile to life" habitats, reduction of competitors, parasites and predators, escaping in time from stressful conditions) could explain the high tardigrade species diversity and number of specimens found in habitats that dry out compared to freshwater habitats.

5.11.5 Ability of tardigrades to cope with high hydrostatic pressure

Apart from several adverse environmental conditions discussed so far, it has been found that tardigrades can also cope with extreme hydrostatic pressure that would squash most animals flat. When an animal is exposed to high hydrostatic pressure, its cellular membranes, proteins and DNA are damaged. High pressures solidify the fatty membranes around cells. At pressures of around 30 megapascals (MPa), proliferation and metabolism in microorganisms stops; at 300 MPa, most bacteria and multicellular organisms die. But, according to a study carried out by Seki and Toyoshima (1998), in perfluorocarbon at pressures as high as 600 MPa, tardigrades can survive in a dehydrated state. Seki, K.; Toyoshima, M. Preserving tardigarades under pressure. Nature 395(6705):853–854; 1998.

This is beyond anything they might encounter in nature. The deepest part of the sea is the Challenger Deep in the Mariana Trench in the Pacific Ocean, which goes down 10,994 m. There, the water pressure is around 100 MPa. Somehow the tardigrades survived six times that. At these crushing pressures, proteins and DNA are ripped apart. Cell membranes, which are composed of fat, become solid like butter in a fridge. Most microorganisms stop metabolising at 30 MPa, and bacteria cannot survive much beyond 300 MPa.

5.11.6 Effect of extreme environmental stresses on tardigrades' DNA

Extreme heat and cold, radiation and high pressures all have one thing in common: they damage DNA and other bits of the tardigrades' cells. Heat and cold both cause proteins to unfold, stick together, and stop working. Radiation tears up DNA and other crucial molecules.

According to Boothby et al. (2017), if all the stressors cause similar problems, may be the tardigrades only need a handful of tricks to survive them; but nobody knows for sure. Scientists postulate that there could certainly be some good reasons to think that overlapping strategies might be used to cope with some of these extremes. For instance, being dried out and being exposed to radiation both damage the tardigrades' DNA. So, it would make sense that tardigrades respond to these two conditions in a similar way, by making antioxidants and repairing the damaged DNA.

Freezing a tardigrade and drying it out both cause the same problem: not enough liquid water in the animal's cells. Unlike bacteria that live in boiling hot springs or other extreme sites, most tardigrades live in relatively unremarkable places. They tend to live in or near water, and there is nothing a tardigrade does like more than a good chunk of moss and lichen. Their lives are not even that exciting: while most creatures their size dart about frantically, tardigrades are sluggish. Yet despite their rather tedious lifestyle, they have evolved to cope with environments so extreme, which do not even exist on Earth.

There are other tardigrades that have a different story to reveal. For example, the oldest and most primitive group of tardigrades, the *Arthrotardigrada*, cannot survive extreme conditions or suspend their metabolism. These more vulnerable creatures offer a clue to why the other tardigrades got so tough. *Arthrotardigrada* only live in the ocean. It is only land-dwelling and fresh-water species that have the extreme survival skills. That suggests leaving the ocean was the key.

It has been postulated that one reason that marine tardigrades are not as good at surviving extremes is that they just don't need to be. Oceans are so big that they don't undergo rapid changes in temperature or salinity, and they certainly don't dry up overnight. By contrast, the land is dangerously changeable. Tardigrades need a thin layer of water around their bodies to breath, eat, mate and move around. But in many parts of the land, drought is a risk. The tardigrades that live in these places need to be able to cope when their

Extremophiles—Organisms that survive and thrive in extreme environmental conditions Chapter | 5 225

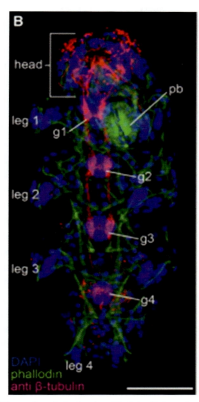

FIG. 5.13 The tardigrade body plan. *H. dujardini* specimen with DAPI-stained nuclei, phalloidin-stained muscles, and nervous system visualized using a β-tubulin antibody. g1–g4, ganglion 1 through ganglion 4; pb, pharyngeal bulb. The scale bar represents 20 mm. (*Source: Smith, F.W., Boothby, T.C., Giovannini, I., Rebecchi, L., Jockusch, E.L, and B. Goldstein (2016), The compact body plan of tardigrades evolved by the loss of a large body region, Current Biology, 26:224-229) Copyright © 2016 Elsevier Ltd All rights reserved.*

environments suddenly change. So, it makes sense that land-dwelling tardigrades would evolve a way to survive suddenly being dried out. It was a matter of survival. What is more, once they had it, the land tardigrades could exploit new habitats. Today they can be found in some of the driest places on Earth, where other animals cannot survive. Fig. 5.13 shows the tardigrade body plan.

The little critters seem adept at living in some of the harshest regions of Earth. They have been discovered 5546 m (18,196 ft) up a mountain in the Himalayas, in Japanese hot springs, at the bottom of the ocean and in Antarctica. They can withstand huge amounts of radiation, being heated to 150°C, and being frozen almost to absolute zero.

5.12 Role of tardigrades as potential model organisms in space research

Exposure of living organisms to open space requires a high level of tolerance to desiccation, cold, and radiation. Some organisms have a remarkable ability to survive the loss of all, or almost all, cellular water and enter into a state of suspended animation in which their metabolism comes reversibly to a standstill. This ability is known as "anhydrobiosis," which means "life without water" (desiccation tolerance). Among animals, only anhydrobiotic species can fulfil the requirements of high level of tolerance to desiccation. The invertebrate phylum *Tardigrada* includes many anhydrobiotic species, which are adapted to survive in very dry or cold environmental conditions. As a likely by-product of the adaptations for desiccation and freezing, tardigrades also show a very high tolerance to a number of other, unnatural conditions, including exposure to ionizing radiation. This makes tardigrades an interesting candidate for experimental exposure to open space. Jönsson (2007) has reviewed the tolerances that make tardigrades suitable for astrobiological studies and the reported radiation tolerance in other anhydrobiotic animals. Several studies have shown that tardigrades can survive gamma-irradiation well above 1 kilogray, and desiccated and hydrated (active) tardigrades respond similarly to irradiation. Thus, tolerance is not restricted to the dry anhydrobiotic state, and Jönsson (2007) discussed the possible involvement of an efficient, but yet undocumented, mechanism for DNA repair. Other anhydrobiotic animals (*Artemia*, *Polypedium*), when desiccated, show a higher tolerance to gamma-irradiation than hydrated animals, possibly due to the presence of high levels of the protective disaccharide trehalose in the dry state. Tardigrades and other anhydrobiotic animals provide a unique opportunity to study the effects of space exposure on metabolically inactive but vital metazoans.

Tardigrades are tiny (less than 1 mm in length) invertebrate animals that have the potential to survive travel to other planets because of their tolerance to extreme environmental conditions by means of a dry ametabolic state called anhydrobiosis. While the tolerance of adult tardigrades to extreme environments has been reported, there are few reports on the tolerance of their eggs. Horikawa et al. (2012) examined the ability of hydrated and anhydrobiotic eggs of the tardigrade *Ramazzottius varieornatus* to hatch after exposure to ionizing irradiation (helium ions), extremely low and high temperatures, and high vacuum. They previously reported that there was a similar pattern of tolerance against ionizing radiation between hydrated and anhydrobiotic adults. In contrast, anhydrobiotic eggs (50% lethal dose; 1690 Gy) were substantially more radioresistant than hydrated ones (50% lethal dose; 509 Gy). These researchers found that anhydrobiotic eggs also have a broader temperature resistance compared with hydrated ones. Over 70% of the anhydrobiotic eggs treated at either -196°C or +50°C hatched successfully, but all the hydrated eggs failed to hatch. After exposure to high-vacuum conditions (5.3×10^{-4} Pa to 6.2×10^{-5} Pa), Horikawa et al. (2012) compared the hatchability of the anhydrobiotic eggs to that of untreated control eggs.

5.13 Discovery of a living Bdelloid Rotifer from 24,000-year-old Arctic permafrost

Bdelloid rotifers are microscopic multicellular animals, known for their ability to survive extremely low temperatures (Shain et al., 2016). Previous reports suggested survival after six to ten years when frozen between −20° to 0°C (Newsham et al., 2006; Iakovenko et al., 2015; Shain et al., 2016). In some invertebrates, reproduction from an ovum without fertilization is a normal process. This specialty found in certain invertebrates is known as parthenogenesis. More specifically, parthenogenesis is a natural form of asexual reproduction in which growth and development of embryos occur without fertilization by sperm. In animals, parthenogenesis means development of an embryo from an unfertilized egg cell. Shmakova et al. (2021) have reported the survival of an obligate (i.e., requiring a suitable host to complete its life-cycle and perform reproduction) parthenogenetic bdelloid rotifer, recovered from north-eastern Siberian permafrost radiocarbon-dated to ∼24,000 years BP (Before Present). To be more specific, the Accelerator Mass Spectrometry (AMS) analysis dated the material as 23,960–24,485 years old (calibrated age of low-temperature combustion humin fraction, σ 95%; University of Arizona AMS Laboratory, sample AA109004). While simple organisms such as bacteria can often survive millennia in permafrost, bdelloid rotifer is an animal with a nervous system and brain.

This constitutes the longest reported case of rotifer survival in a frozen state.

The term metagenome refers to the genetic content of any group of microorganisms. Metagenome is also the recovery and complete sequencing of genetic material extracted directly from all environmental samples. Metagenomics is the study of genetic material recovered directly from environmental samples. To confirm that rotifers originated from the permafrost core, Shmakova et al. (2021) searched for their DNA sequences in a metagenome obtained from the same core fragment (NCBI SRA: SRR13615827). Their analysis demonstrated the presence of an actin gene fragment (two paired reads) belonging to a bdelloid rotifer. By morphological and molecular markers, Shmakova et al. (2021) found that the discovered rotifer belonged to the genus *Adineta*, and aligns with a contemporary *Adineta vaga* isolate collected in Belgium. To follow the process of freezing and recovery of the ancient rotifer, Shmakova et al. (2021) randomly selected 144 individuals from the strain SCL-15-7 and froze them at −15°C for one week. Surviving individuals were counted one-hour post-thawing, and the process of recovery was documented for one individual per plate. The movement of the recovered rotifer can be viewed at Video_R1_Rotifer, reproduced from Shmakova et al. (2021). Data on survival of the ancient Adineta sp. were compared to those of contemporary Adineta species from Svalbard, Alaska, Western and Southern Europe, tropical Asia and Africa, and North America (10 species, 404 individuals in total), frozen using the same protocol. These experiments demonstrated that the ancient rotifer withstands slow cooling and freezing (∼1°C/min) for at least seven days. Shmakova et al. (2021) also show that a clonal culture can continuously reproduce in the laboratory by parthenogenesis. In Shmakova et al. (2021)'s studies, initial cultures from the permafrost core were maintained for about one month and yielded, among other microscopic organisms, numerous living rotifers (Fig. 5.14A and Video).

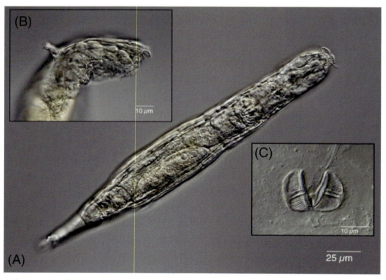

FIG. 5.14 Adineta sp. isolated from permafrost. *(Source: Shmakova, L., S. Malavin, N. Iakovenko, D. Shain, M. Plewka, E. Rivkina (2021), A living bdelloid rotifer from 24,000-year-old Arctic permafrost, Current Biology, 31 (11), R712-R713. DOI: https://doi.org/10.1016/j.cub.2021.05.077) Copyright © 2021 Elsevier Inc. Video. A rotifer recovered from 20,000-year-old permafrost (00:00–00:23) Adineta sp. SCL-15-7 in feeding locomotion. (00:24–06:10) Adineta sp. SCL-15-7 recovering from a week-long cryobiosis in the lab. (file name: Video_R1_Rotifer_Current_Biology_2021) Copyright © 2021 Elsevier Inc.*

Shmakova et al. (2021)'s studies indicated that the ancient rotifer is capable of surviving a relatively slow freezing process that allows ice crystals detrimental for cells to form (the duration of complete freezing of a well with a rotifer 45 ± 4 min). In combination with its occurrence in permafrost, this suggests that the discovered Adineta sp. has effective biochemical mechanisms of organ and cell shielding necessary to survive low temperatures. Shmakova et al. (2021) claim that their discovery is of interest not only for evolutionary biology but also for practical purposes of cryobiology (the branch of biology which deals with the properties of organisms and tissues at low temperatures) and biotechnology (the branch of biology involving the exploitation of biological processes for industrial and other purposes, especially the genetic manipulation of microorganisms for the production of antibiotics, hormones, etc.).

5.14 Archaea—single-celled microorganisms with no distinct nucleus—constituting a third domain in the phylogenetic tree of life

Archaea are single-celled microorganisms with no distinct nucleus—microorganisms that may have evolved as long ago as 4 billion years. The word "archaea" means "ancient things," and it was assumed that their metabolism reflected Earth's primitive atmosphere and the organisms' antiquity, but as new habitats were studied, more organisms were discovered. Archaea are considered to be a major part of Earth's life. They are part of the microbiota (the microorganisms of a particular site, habitat, or geological period) of all organisms. In the human microbiome (the genes of the 10-100 trillion symbiotic microbial cells harbored by each human being, primarily bacteria in the gut; see Turnbaugh et al., 2007; Ursell et al., 2012), they are important in the gut, mouth, and on the skin (Bang and Schmitz, 2015). Their morphological, metabolic, and geographical diversity permits them to play multiple ecological roles: carbon fixation; nitrogen cycling; organic compound turnover; and maintaining microbial symbiotic (denoting a mutually beneficial relationship between two or more different organisms living in close physical association) and syntrophic communities (i.e., cross-feeding communities in which the growth of one partner depends on the nutrients, growth factors, or substrates provided by the other partner) (Moissl-Eichinger et al., 2018). Several studies have shown that archaea exist not only in mesophilic (growing best at moderate temperatures, between 25°C and 40°C) and thermophilic (thriving at relatively high temperatures, between 41 and 122°C) environments but are also present, sometimes in high numbers, at low temperatures as well. For example, archaea are common in cold oceanic environments such as polar seas (López-García et al., 2001).

5.14.1 The intriguing history of the discovery of archaea

For much of the 20th century, prokaryotes were regarded as a single group of organisms and were classified based on their biochemistry, morphology and metabolism. In a highly influential 1962 paper, Roger Stanier and C.B. van Niel first established the division of cellular organization into prokaryotes and eukaryotes, defining prokaryotes as those organisms lacking a cell nucleus (Stanier and Van Niel, 1962; Pace, 2009). Since then it was generally assumed that all life shared a common prokaryotic ancestor (Pace, 2009; Oren, 2010).

Conventionally, the classification scheme long used by biologists about the basic structure of the tree of life was the two-domain classification—prokaryotes versus eukaryotes. In 1977, Carl Richard Woese and George E. Fox experimentally disproved this universally held hypothesis (Pace et al., 2012). In 1977, Carl Woese, studying the genetic sequences of organisms, developed a new comparison method that involved splitting the RNA into fragments that could be sorted and compared with other fragments from other organisms (Woese and Fox, 1977). The more similar the patterns between species, the more closely they are related (Howland, 2000).

Woese used his new ribosomal RNA (rRNA) comparison method to categorize and contrast different organisms. He compared a variety of species and happened upon a group of methanogens (methane-producing organisms possessing the capability to chemically reduce carbon dioxide to methane) with rRNA vastly different from any known prokaryotes or eukaryotes (Woese and Fox, 1977). These methanogens were much more similar to each other than to other organisms, leading Woese to propose the new domain of Archaea (Woese and Fox, 1977). Their experiments showed that the archaea were genetically more similar to eukaryotes than prokaryotes, even though they were more similar to prokaryotes in structure (Cavicchioli, 2011). This led to the conclusion that Archaea and Eukarya shared a common ancestor more recent than Eukarya and Bacteria (Cavicchioli, 2011). The development of the nucleus occurred after the split between Bacteria and this common ancestor (Woese et al., 1990; Cavicchioli, 2011).

Archaea were first classified separately from bacteria in 1977 by Carl Woese and George E. Fox based on their ribosomal RNA (rRNA) genes (Woese and Fox, 1977). Archaea were split off as a third domain because of the large differences in their ribosomal RNA structure. The particular molecule 16S rRNA is key to the production of proteins in all organisms. Because this function is so central to life, organisms with mutations in their 16S rRNA are unlikely to survive, leading to great (but not absolute) stability in the structure of this polynucleotide (a biopolymer composed

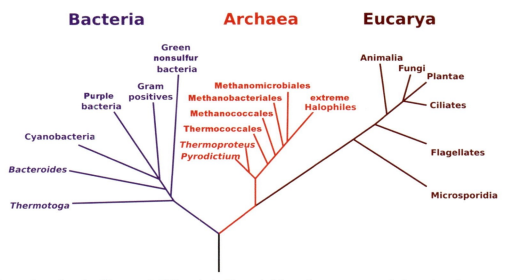

FIG. 5.15 Phylogenetic tree based on Woese et al. rRNA analysis. The vertical line at bottom represents the last universal common ancestor (LUCA) *Author: By Maulucioni - Own work, CC BY-SA 3.0, https://commons.wikimedia.org/w/index.php?curid=24740337. (Source: https://commons.wikimedia.org/wiki/File:PhylogeneticTree,_Woese_1990.PNG).*

of 13 or more nucleotide monomers covalently bonded in a chain) over generations. 16S rRNA is large enough to show organism-specific variations, but still small enough to be compared quickly.

Woese (July 15, 1928 – December 30, 2012) was an American microbiologist and biophysicist. Woese and Fox discovered a kind of microbial life that they called the "archaebacteria" (Woese and Fox, 1977). They reported that the archaebacteria comprised "a third kingdom" of life (Fig. 5.15) as distinct from bacteria as plants and animals (Woese and Fox, 1977).

As just indicated, when the single-celled microorganism now known as "Archaea" was first discovered, it was named "archaebacteria," which Woese (Fig. 5.16) later renamed "Archaea" when he realized that they are as distinct from bacteria as bacteria are from us. Woese is famous for defining the Archaea (a new domain of life) in 1977 by phylogenetic taxonomy of 16S ribosomal RNA, a technique he pioneered that revolutionized microbiology (Woese et al., 1990; Woese and Fox, 1977; Morell, 1997). He also originated the RNA world hypothesis in 1967, although not by that name (Woese, 1967). Woese held the Stanley O. Ikenberry Chair and was professor of microbiology at the University of Illinois at Urbana–Champaign (Noller, 2013; Goldenfeld and Pace, 2013). Woese turned his attention to the genetic code while setting up his lab at General Electric's Knolls Laboratory in the fall of 1960 (Sapp, 2009). Interest among physicists and molecular biologists had begun to coalesce around deciphering the correspondence between the twenty amino acids and the four-letter alphabet of nucleic acid bases in the decade following James D. Watson, Francis Crick, and Rosalind Franklin's discovery of the structure of DNA in 1953 (Nair, 2012). Woese published a series of papers on the topic.

It was found that the "archaebacteria" were distinct from bacteria, for instance in the sequences of their ribosomal RNA. Through this discovery of an entirely new group of organisms: Archaea, the hitherto prevailed conventional notion was upended. The discovery of archaea not only

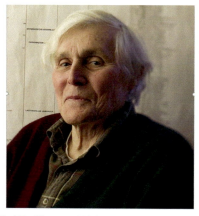

FIG. 5.16 Carl R. Woese (1928-2012), who propose the new domain of Archaea — a Third Domain — in the Phylogenetic Tree of Life. The portrait was taken by Don Hamerman in 2004 at the University of Illinois.

changed the shape of the tree of life; it also raised questions about the common ancestor at the base of that tree and offered clues about how it may have lived. Thus, a surprising biological discovery upended the long-held two-domain classification scheme used by most biologists—prokaryotes versus eukaryotes. That new discovery was an entirely new group of organisms: archaea. Initially thought to be bacteria, these single-celled microbes—many of which were first found in seemingly unlivable habitats such as the volcanic hot springs of Yellowstone National Park—were in fact so different at molecular and genetic levels as to constitute a separate, third domain beside bacteria and eukaryotes. Their discovery sparked a conceptual revolution in our understanding of the evolution of life.

Most of the archaea that had initially been isolated were extremophiles. These include hyperthermophilic (super-heat-loving—from 60°C upwards) microbes that thrive above 80°C and are typically found in habitats such as deep-ocean vents. Up to the 1970s, the consensus had been that most such habitats were hostile to life, but a handful of ground-breaking microbiologists changed that notion. Thomas Brock, for instance, began to isolate hyperthermophilic archaea, including the genus *Sulfolobus*, from hot springs in Yellowstone National Park, Wyoming. Later, German microbiologist Karl Stetter showed that many surprising habitats, even oil fields, teemed with microbial life.

Having defined Archaea as a new "urkingdom" (later domain) that were neither bacteria nor eukaryotes, Woese redrew the taxonomic tree. His three-domain system, based on phylogenetic relationships rather than obvious morphological similarities, divided life into 23 main divisions, incorporated within three domains: Bacteria, Archaea, and Eucarya (Woese et al., 1990).

Molecular biologist Patrick Forterre (2016) narrates in his highly educational, well-balanced, and exciting book entitled *Microbes from Hell* the intriguing history of the discovery of archaea—single-celled microorganisms with no distinct nucleus that may have evolved as long ago as 4 billion years. Forterre's book—written after several expeditions to areas of hydrothermal activity, both high up in the mountains and deep down on the ocean floor—brings us along on the discovery of life in extreme environments, a surprising and little-known world that thrived in the past and continues to thrive, and even evolve. Forterre—one of the world's leading experts on archaea and hyperthermophiles, or organisms that have evolved to flourish in extreme temperatures—offers a colorful, engaging account of this taxonomic upheaval. Blending tales of his own search for thermophiles with discussions of both the physiological challenges thermophiles face and the unique adaptations they have evolved to live in high-temperature environments, Forterre illuminates our developing the understanding of the relationship between archaea and the rest of Earth's organisms.

Forterre was fascinated by the ideas of Woese. Ever since Forterre read Woese's work on the identification of the archaea and its implications for the tree of life, he has wondered about a last universal common ancestor of all life. His book *Microbes from Hell* walks the reader through his fascinating journey to understand how life evolved. It was Forterre who, in the 1980s, found that certain archaea wind their DNA using reverse gyrase enzymes, which work differently from the gyrase found in bacteria (see Albers, 2016).

In the 1980s, Forterre began to analyze the hyperthermophilic archaea isolated by Stetter and Zillig, looking for reverse gyrase. The enzyme causes the DNA double helix to cross over on itself (supercoiling), and Forterre discovered that hyperthermophiles contain a form of it that induces positive supercoiling—adding extra twists. This enzyme, also found in hyperthermophilic bacteria, has not yet been seen in organisms growing at lower temperatures, leading to speculation that it might be one reason that hyperthermophiles can grow at such high temperatures. One theory is that the enzyme is important in sensing unpaired regions in hyperthermophile genomes, then initiating repair.

In 1999, Forterre and his technician (later wife) Évelyne Marguet joined the AMISTAD expedition of the French National Center for Scientific Research and the French Research Institute for Exploration of the Sea. The book "Microbes from Hell" narrates stories depicting science in action, including a submarine trip to collect samples in deep-sea hydrothermal vents. The aim of the AMISTAD expedition was to isolate hyperthermophilic archaea from the deep eastern Pacific Ocean. Marguet gives a rousing account of her 2,600-metre dive in the submersible *Nautile* to gather samples from "smokers." These rock chimneys form at geologically active sites on the sea floor, where superheated, metal-laden water is funneled from vents.

Back on the ship *Atalante*, Marguet was enthralled to see cells growing in cultures from her samples, and isolated several *Thermococcus* species. She also tried to isolate the first viruses from these archaea, using methods established by Zillig. She and Forterre discovered that *Thermococcus* strains produced a vast amount of membrane vesicles from cells containing plasmid DNA. Since then, large quantities of membrane vesicles (structures within or outside biological cells, consisting of liquid or cytoplasm enclosed by lipid bilayers) have been found, particularly in ocean water, and produced by eukaryotes and bacteria. They are thought to contribute to DNA transfer between species, so they may have a role in evolution.

In 1990, Woese and his colleagues proposed to divide life into three domains: bacteria, archaea, and eukaryotes. The concept has gradually been accepted, but Forterre—with microbiologists Wolfram Zillig and Otto Kandler, among others—was an early "believer."

5.14.2 General features of archaea

As already indicated, Archaea constitute a domain of single-celled organisms. These microorganisms lack cell nuclei. Archaea were initially classified as bacteria, receiving the name archaebacteria, but this term has fallen out of use. Archaea are microorganisms that are similar to bacteria in size and simplicity of structure but radically different in molecular organization. They are now believed to constitute an ancient group that is intermediate between the bacteria and eukaryotes. Archaeal cells have unique properties separating them from the other two domains, Bacteria, and Eukaryota. Archaea are further divided into multiple recognized phyla. Archaea and bacteria are generally similar in size and shape, although a few archaea have very different shapes, such as the flat and square cells of *Haloquadratum walsbyi* (Stoeckenius, 1981).

For a long time, archaea were seen as extremophiles that exist only in extreme habitats such as hot springs and salt lakes, but by the end of the 20th century, archaea had been identified in nonextreme environments as well. Today, they are known to be a large and diverse group of organisms abundantly distributed throughout nature (DeLong, 1998). This new appreciation of the importance and ubiquity of archaea came from using polymerase chain reaction (PCR) to detect prokaryotes from environmental samples (such as water or soil) by multiplying their ribosomal genes.

Despite the morphological similarity to bacteria, archaea possess genes and several metabolic pathways (linked series of chemical reactions occurring within a cell) that are more closely related to those of eukaryotes, notably for the enzymes (proteins that act as biological catalysts that accelerate chemical reactions) involved in transcription (the process of copying a segment of DNA into RNA) and translation (the process in which ribosomes in the cytoplasm or endoplasmic reticulum synthesize proteins after the process of transcription of DNA to RNA in the cell's nucleus). Note that morphology is a branch of biology dealing with the study of the form and structure of organisms and their specific structural features. This includes aspects of the outward appearance (shape, structure, color, pattern, size), that is, external morphology, as well as the form and structure of the internal parts such as bones and organs, i.e. internal morphology. Archaea use more diverse energy sources than eukaryotes, ranging from organic compounds such as sugars, to ammonia, metal ions or even hydrogen gas. The salt-tolerant *Haloarchaea* use sunlight as an energy source, and other species of archaea fix carbon, but unlike plants and cyanobacteria, no known species of archaea does both. Archaea reproduce asexually by binary fission, fragmentation, or budding. The first observed archaea were extremophiles, living in extreme environments such as hot springs and salt lakes with no other organisms. Improved molecular detection tools led to the discovery of archaea in almost every habitat, including soil, oceans, and marshlands. Archaea are particularly numerous in the oceans.

5.14.3 Unique feature of archaea

One property unique to archaea is the abundant use of ether-linked lipids in their cell membranes. Ether linkages are more chemically stable than the ester linkages found in bacteria and eukarya, which may be a contributing factor to the ability of many archaea to survive in extreme environments that place heavy stress on cell membranes, such as extreme heat and salinity. Comparative analysis of archaeal genomes has also identified several molecular conserved signature indels and signature proteins uniquely present in either all archaea or different main groups within archaea (Gao and Gupta, 2007; Gupta and Shami, 2011; Gupta et al., 2015). An insertion/deletion polymorphism, commonly abbreviated "indel," is a type of genetic variation in which a specific nucleotide sequence is present (insertion) or absent (deletion). Indels are widely spread across the genome. It may be noted that conserved signature inserts and deletions (CSIs) in protein sequences provide an important category of molecular markers for understanding phylogenetic relationships (Baldauf, 1993; Gupta, 1998). CSIs, brought about by rare genetic changes, provide useful phylogenetic markers that are generally of defined size and they are flanked on both sides by conserved regions to ensure their reliability. While indels can be arbitrary inserts or deletions, CSIs are defined as only those protein indels that are present within conserved regions of the protein (Gupta, 1998; Rokas and Holland, 2000; Gupta and Griffiths, 2002; Cutiño-Jiménez et al., 2010). The CSIs that are restricted to a particular clade or group of species, generally provide good phylogenetic markers of common evolutionary descent (Gupta, 1998).

Another unique feature of archaea, found in no other organisms, is methanogenesis (the metabolic production of methane). Methanogenic archaea play a pivotal role in ecosystems with organisms that derive energy from oxidation of methane, many of which are bacteria, as they are often a major source of methane in such environments and can play a role as primary producers. Methanogens also play a critical role in the carbon cycle, breaking down organic carbon into methane, which is also a major greenhouse gas (Deppenmeier, 2002).

5.14.4 Diverse sizes and shapes exhibited by archaea

Individual archaea range from 0.1 micrometers (μm) to over 15 μm in diameter and occur in various shapes, commonly as spheres, rods, spirals, or plates (Krieg, 2005). Other morphologies in the Crenarchaeota include irregularly shaped lobed cells in *Sulfolobus*, needle-like filaments that are less than half a micrometer in diameter in *Thermofilum*, and almost perfectly rectangular rods in T*hermoproteus* and *Pyrobaculum* (Barns and Burggraf, 1997). Archaea in the genus *Haloquadratum* such as *Haloquadratum walsbyi* are flat, square specimens that live in hypersaline pools (Walsby, 1980). These unusual shapes are probably maintained by both their cell walls and a prokaryotic cytoskeleton. Proteins related to the cytoskeleton components of other organisms exist in archaea (Hara et al., 2007), and filaments form within their cells (Trent et al., 1997), but in contrast with other organisms, these cellular structures are poorly understood (Hixon and Searcy, 1993). In *Thermoplasma* and *Ferroplasma* the lack of a cell wall means that the cells have irregular shapes, and can resemble amoebae (Golyshina et al., 2000).

5.14.5 Extremophile archaea—halophiles, thermophiles, alkaliphiles, and acidophiles

Archaea exist in a broad range of habitats, and are now recognized as a major part of global ecosystems (DeLong, 1998), and may represent about 20% of microbial cells in the oceans (DeLong and Pace, 2001). However, the first-discovered archaeans were extremophiles (Valentine, 2007). Extremophile archaea are members of four main physiological groups. These are the halophiles (salt-loving), thermophiles (thriving at relatively high temperatures, between 41 and 120°C), alkaliphiles (capable of survival in alkaline environments, growing optimally at a pH of 10), and acidophiles (thriving under highly acidic conditions) (Pikuta et al., 2007).

Halophile is an organism that needs high salt concentrations for growth. A widely used definition is that of Kushner and Kamekura (1988) who classify organisms depending on the salt concentration needed for optimum growth. Thus, nonhalophiles grow best in media containing less than 0.2 M salts while halophiles grow best in media containing from 0.2 to 5.2 M dissolved salts. Halophiles can be further divided into slightly halophilic (optimum growth between 0.2 and 0.5 M salt), moderately halophilic (0.5–2.5 M salt), and extremely halophilic (above 2.5 M salt). All three domains of life include halophilic microorganisms. Archaeal halophiles, all belonging to the Euryarchaeota, can be found among the methanogens and the members of the order Halobacteriales, that only includes extreme halophiles (Antón, 2011).

Fig. 5.17 shows cluster of cells of *Halobacterium* sp. strain NRC-1. Indeed, some archaea survive high temperatures, often above 100°C (212°F), as found in geysers (springs characterized by intermittent discharge of water ejected turbulently and accompanied by steam), black smokers, and oil wells. The genus *Halobacterium* ("salt" or

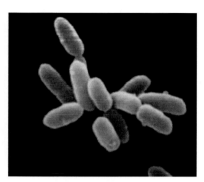

FIG. 5.17 Cluster of cells of Halobacterium sp. strain NRC-1 *Author: NASA - en:Image:Halobacteria.jpg (Taken from NCBI webpage on Halobacterium), Public Domain, https://commons.wikimedia.org/w/index.php?curid=2979987. (Source: https://en.wikipedia.org/wiki/Archaea).*

"ocean bacterium") consists of several species of Archaea with an aerobic metabolism that requires an environment with a high concentration of salt; many of their proteins will not function in low-salt environments. Other common habitats include very cold habitats and highly saline, acidic, or alkaline water, but archaea include mesophiles that grow in mild conditions, in swamps and marshland, sewage, the oceans, the intestinal tract of animals, and soils (DeLong, 1998).

5.14.6 Extreme halophilic and hyperthermophilic archaea

Extreme halophilic (Magrum et al., 1978) and hyperthermophilic Archaea (Stetter, 1996) have been discovered in acid mine drainage (Fig. 5.18). Acid mine drainage causes severe environmental problems in Rio Tinto, Spain. Note that Acid mine drainage, acid and metalliferous drainage (AMD), or acid rock drainage (ARD) is the outflow of acidic water from metal mines or coal mines. *Haloquadratum walsbyi* (Fig. 5.19) is a species of archaea that was discovered in a brine pool in the Sinai Peninsula of Egypt. It is noted for both its flat, square-shaped cells, and its unusual ability to survive in aqueous environments with high concentrations of sodium chloride and magnesium chloride (Oren et al., 1999; Bolhuis et al., 2006). The species' genus name *Haloquadratum* literally translates from Greek and Latin as "salt square."

Thermophiles grow best at temperatures above 45°C (113°F), in places such as hot springs, and hyperthermophilic archaea grow optimally at temperatures greater than 80°C (176°F) (Madigan and Martino, 2006). The archaea *Methanopyrus kandleri* Strain 116 can even reproduce at 122°C (252°F), the highest recorded temperature of any organism (Takai et al., 2008). Other archaea exist in very acidic or alkaline conditions (Pikuta et al., 2007). For example, one of the most extreme archaean acidophiles is *Picrophilus torridus*, which grows at pH 0 (Ciaramella

FIG. 5.18 A group of archaea discovered in acid mine drainage. Acid mine drainage causes severe environmental problems in the Rio Tinto, Spain. By Carol Stoker, NASA - ACD03-0051-13 from http://www.nasa.gov/centers/ames/news/releases/2003/03images/tinto/tinto.html, Copyright info: Public Domain. This file is in the public domain in the United States because it was solely created by NASA. NASA copyright policy states that "NASA material is not protected by copyright unless noted". (Source: https://commons.wikimedia.org/w/index.php?curid=4359761).

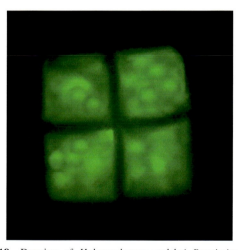

FIG. 5.19 Drawing of *Haloquadratum walsbyi* Permission: Public domain Author: By Rotational - Own work, Public Domain, https://commons.wikimedia.org/w/index.php?curid=4587362.

et al., 2005). This resistance to extreme environments has made archaea the focus of speculation about the possible properties of extraterrestrial life (Javaux, 2006). Some extremophile habitats are not dissimilar to those on Mars (Nealson, 1999), leading to the suggestion that viable microbes could be transferred between planets in meteorites (Davies, 1996).

Among the numerous archaea that had been isolated, most of them were extremophiles. These include hyperthermophilic microbes that thrive above 80°C and are typically found in habitats such as deep-ocean hydrothermal vents. Up to the 1970s, the consensus had been that most such habitats were hostile to life, but a handful of groundbreaking microbiologists changed that notion. Thomas Brock, for instance, began to isolate hyperthermophilic archaea, including the genus *Sulfolobus*, from hot springs in Yellowstone National Park, Wyoming. Later, German microbiologist Karl Stetter showed that many surprising habitats, even oil fields, teemed with microbial life.

Stetter (2006) provided the following information about hyperthermophilic (HT) organisms, "hyperthermophilic ("superheat-loving") bacteria and archaea are found within high-temperature environments, representing the upper-temperature border of life. They grow optimally above 80°C and exhibit an upper-temperature border of growth up to 113°C. Members of the genera, *Pyrodictium* and *Pyrolobus*, survive at least 1h of autoclaving. In their basically anaerobic environments, hyperthermophiles (HT) gain energy by inorganic redox reactions employing compounds like molecular hydrogen, carbon dioxide, sulfur and ferric and ferrous iron. Based on their growth requirements, HT could have existed already on the early Earth about 3.9Gyr ago. In agreement, within the phylogenetic tree of life, they occupy all the short deep branches closest to the root. The earliest archaeal phylogenetic lineage is represented by the extremely tiny members of the novel kingdom of Nanoarchaeota, which thrive in submarine hot vents. HT is very tough survivors, even in deep-freezing at −140°C. Therefore, during impact ejecta, they could have been successfully transferred to other planets and moons through the coldness of space."

Only superheat-loving microbes similar to the hyperthermophiles (HT; Stetter, 1992) would have been able to thrive and survive in an "early times of life" scenario, such as the times at the end of the heavy meteorite bombardment, when the surface of the early Earth must have been much hotter than today. Expelled by impacts, microbes could have spread in between the planets and moons of the early Solar System. In the year 1981, the first HT had been isolated, which exhibited unprecedented optimal growth temperatures above 80°C, where usual mesophilic and thermophilic bacteria are killed within seconds (Brock, 1978; Stetter et al., 1981; Zillig et al., 1981; Stetter, 1982). HT turned out to be very common in hot terrestrial and submarine environments. Their position within the universal phylogenetic tree of life provides further strong evidence of a hyperthermophilic last common ancestor.

In the 1980s, Forterre began to analyze the hyperthermophilic archaea isolated by Stetter and Zillig, looking for reverse gyrase. Note that reverse gyrase is an enzyme that induces positive supercoiling in closed circular DNA in vitro. It is unique to thermophilic organisms and found without exception in all microorganisms defined as hyperthermophiles, that is, those having optimal growth temperatures of 80°C and above (Lipscomb et al., 2017). Forterre found that an enzyme causes the DNA double helix to cross over on itself (supercoiling), and Forterre discovered that hyperthermophiles contain a form of it that induces positive supercoiling—adding extra twists. This enzyme, also found in hyperthermophilic bacteria, has not yet been seen in organisms growing at lower temperatures, leading to speculation that it might be one reason that hyperthermophiles can grow at such high temperatures. One theory is that the enzyme is important in sensing unpaired regions in hyperthermophile genomes, then initiating repair.

Stetter (2006) recalls that based on the pioneering work of Carl Woese, the small subunit ribosomal RNA (ss rRNA) is widely used in phylogenetic studies (Woese and Fox 1977; Woese et al., 1990). In Bacteria and Archaea, it consists of about 1500 bases. Owing to sequence comparisons, a phylogenetic tree is now available (Fig. 5.20). It shows a tripartite division of the living world into the bacterial, archaeal, and eukaryal domains. Within this tree, deep branches are evidenced for early separation. The separation of the bacteria from the stem shared by Archaea and Eukarya represents the deepest and earliest branching point. Short phylogenetic branches indicate a rather slow rate of evolution. Stetter (2006) explains that in contrast to the Eukarya, the bacterial and archaeal domains within the phylogenetic tree exhibit some extremely short and deep branches. Surprisingly, these are covered exclusively by HT, which therefore form a cluster around the phylogenetic root (Fig. 5.20, thick lineages). The deepest and shortest phylogenetic branches are represented by the Aquificales and Thermotogales within the bacteria and the Nanoarchaeota, Pyrodictiaceae and Methanopyraceae within the Archaea, indicating a slow rate of evolution. Now, several total genome sequences are already available. Within the sequences, as a rule, trees based on genes involved in information management (e.g. DNA replication, transcription, translation) parallel the ss

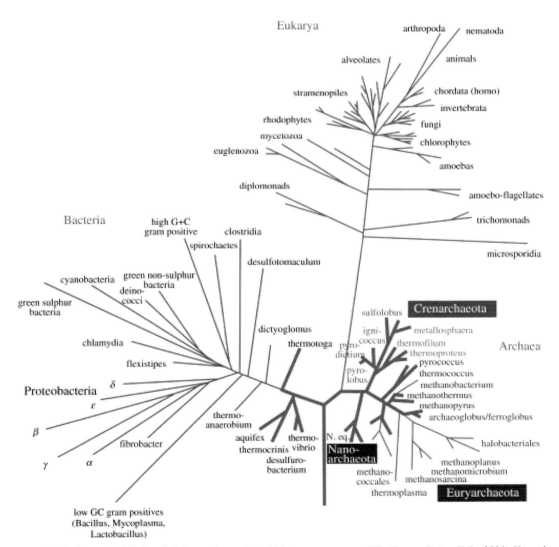

FIG. 5.20 Small subunit ribosomal RNA-based phylogenetic tree. The thick lineages represent HT. *(Source: Stetter, K.O. (2006), Hyperthermophiles in the history of life, Philos Trans R Soc Lond B Biol Sci., 361(1474): 1837–1843. doi: 10.1098/rstb.2006.1907) Copyright info: © 2006 The Royal Society.*

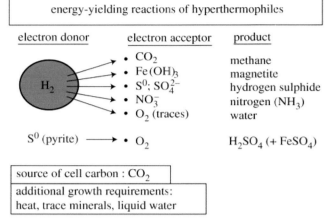

FIG. 5.21 Schematic of main energy-yielding reactions in chemolithoautotrophic HT. *(Source: Stetter, K.O. (2006), Hyperthermophiles in the history of life, Philos Trans R Soc Lond B Biol Sci., 361(1474): 1837–1843. doi: 10.1098/rstb.2006.1907) Copyright info: © 2006 The Royal Society.*

rRNA tree. However, within the genes of metabolism, based on the frequent lateral gene transfer, a network rather than a tree may reflect their phylogenetic relations (Doolittle, 1999).

With regard to the energy sources of HT, Stetter (2006) explains that "most species exhibit a chemolithoautotrophic mode of nutrition (Fig. 5.21). Anaerobic and aerobic types of respiration follow inorganic redox reactions (chemolithotrophic), and CO_2 is the only carbon source required to build-up organic cell material (autotrophic). Therefore, these organisms fix CO_2 by chemosynthesis and are designated chemolithoautotrophs. Molecular hydrogen serves as an important electron donor. Other electron donors are sulfide, sulfur, and ferrous iron. As in mesophilic respiratory organisms, in some HT, oxygen may serve as an electron acceptor. However, in contrast, oxygen-respiring HT are usually microaerophilic (requiring little free oxygen, or oxygen at a lower partial pressure than that of atmospheric oxygen) and therefore grow only at reduced oxygen concentrations. Anaerobic respiration types are the nitrate, sulfate, sulfur and carbon dioxide respirations. While chemolithoautotrophic HT produce organic matter, there is some HT that depend on organic material as energy and carbon sources. They are designated as heterotrophs." Note that chemolithoautotrophic means that these organisms obtain the necessary carbon for metabolic processes from carbon dioxide in their environment. According to Stetter (2006), several chemolithoautotrophic HT are opportunistic heterotrophs. Note that a heterotroph is an organism that cannot produce its own food, instead taking nutrition from other sources of organic carbon, mainly plant or animal matter. In the food chain, heterotrophs are primary, secondary and tertiary consumers, but not producers. Heterotrophs are able to use organic material alternatively to inorganic nutrients, whenever it is provided by the environment (e.g. by decaying cells). Heterotrophic HT gain energy either by aerobic or different types of anaerobic respiration, using organic material as electron donor or by fermentation (Stetter, 2006).

Pulschen et al. (2020) recall that although similar techniques have been applied to the study of halophilic archaea (Delpech et al., 2018; Bisson-Filho et al., 2018; Eun et al., 2018; Walsh et al., 2019; Li et al., 2019), our ability to explore the cell biology of thermophilic archaea has been limited by the technical challenges of imaging at high temperatures. *Sulfolobus* are the most intensively studied members of TACK archaea and have well-established molecular genetics (Wagner et al., 2009; Wagner et al., 2012; Zebec et al., 2014; Zhang et al., 2018). Note that TACK is a group of archaea (acronym for Thaumarchaeota, Aigarchaeota, Crenarchaeota, and Korarchaeota), the first groups discovered. TACK is a clade that is close to the branch that gave rise to the eukaryotes. Additionally, as Pulschen et al. (2020) recall, studies using *Sulfolobus* were among the first to reveal striking similarities between the cell biology of eukaryotes and archaea (Robinson et al., 2004; Duggin et al., 2008; Lindås et al., 2008; Samson et al. 2008; Lindås and Bernander, 2013; Takemata et al., 2019). However, it has not been possible to image *Sulfolobus* cells as they grow and divide. Pulschen et al. (2020) have reported the construction of a device, which they named Sulfoscope, which is a heated chamber on an inverted fluorescent microscope that enables live-cell imaging of thermophiles. By using thermostable fluorescent probes together with this system, Pulschen et al. (2020) were able to image *Sulfolobus acidocaldarius* cells live to reveal tight coupling between changes in DNA condensation, segregation, and cell division.

DNA reorganization is tightly coupled to cell division in hyper-thermophilic archaeon *Sulfolobus acidocaldarius*. Pulschen et al. (2020) have described the development of the "Sulfoscope," an imaging platform that makes it possible to image hyper-thermophilic archaeon *Sulfolobus* cell divisions live {Video S1 displaying cell division of hyper-thermophilic archaeon *Sulfolobus acidocaldarius* DMS639 stained with dye Nile Red (membrane) and SYBR safe (DNA) & Video S2 displaying field of cell of *sulfolobus acidcaldarius* DMS639 stained with dye nile red (membrane) and SYBR safe (DNA), wherein the dye retain their optical properties at high temperature and low pH, showing several examples of divisions; Reproduced from Pulschen et al. (2020), *Current Biology*, Elsevier Inc.}. Note that "SYBR Safe" is a cyanine dye used as a nucleic acid stain in molecular biology. This dye binds to DNA. The resulting DNA-dye-complex absorbs blue light (λ_{max} = 509 nm) and emits green light (λ_{max} = 524 nm).

Pulschen et al. (2020)'s first use of the Sulfoscope revealed a tight coupling between DNA reorganization, nucleoid separation, and membrane deformation during division. These events appear to occur in a defined order in the wild-type cell (both S. acidocaldarius and S. solfataricus). This begins as replicated chromosomes lose their diffuse organization as they separate and is followed by a relatively sudden DNA compaction, which coincides with the onset of furrowing. Thus, DNA compaction may help to ensure that the segregated chromosomes remain out of the way of the furrow as it closes. According to Pulschen et al. (2020), these findings are in line with observations in fixed cells of Sulfolobus (Popławski and Bernander, 1997) and are similar to DNA segregation behavior observed in fixed cells of Halobacterium salinarium (Herrmann and Soppa, 2002).

In one model, Endosomal Sorting Complex Required for Transport (ESCRT)-III proteins forms tubes, funnels, or spirals inside the membrane neck. By combining molecular genetics and live-cell imaging, Pulschen et al. (2020) were also able to use the Sulfoscope to define homolog-specific roles for the two ESCRT-III proteins (cdvB1 and cdvB2) that form part of the contractile division ring (Risa et al., 2019). According to Pulschen et al. (2020), although some of the differences in the findings in some different studies may reflect strain differences, live-cell imaging greatly aids the mechanistic study of dynamic processes like division. Note that in genetics, homolog is a gene related to a second gene by descent from a common ancestral DNA sequence. The term, homolog, may apply to the relationship between genes separated by the event of speciation or to the relationship between genes separated by the event of genetic duplication.

Pulschen et al. (2020) are optimistic that though they have used the Sulfoscope to reveal fundamental novel aspects of cell division, this type of high-temperature live-imaging platform can be used to shed light on other exciting areas of Sulfolobus cell biology, including "DNA re-modeling (Kalliomaa-Sanford et al., 2012; Takemata et al., 2019), swimming (Tsai et al., 2020), conjugation (Fröls et al., 2008; van Wolferen et al., 2016), viral infection (Bize et al., 2009; Liu et al., 2017), competition (Prangishvili et al., 2000; Ellen et al., 2011), and cell-cell fusion (Zhang et al., 2019)." In addition, according to Pulschen et al. (2020), this system can now also be applied to the study of other thermophilic microbes, from eukaryotes to bacteria (Cava et al., 2009; Berka et al., 2011). Although Pulschen et al. (2020) are optimistic that their work can open new avenues of research and lead to further developments in the field of live imaging of thermophiles (Horn et al., 1999; Charles-Orszag et al., 2020), the further development of thermophile cell biology will also depend on the establishment of thermostable fluorescent proteins (Visone et al., 2017; Frenzel et al., 2018) and the use of microfluidics, to allow for the tracking of cells over long periods across multiple generations, as currently performed in Haloarchaea (Abdul-Halim et al., 2020).

5.14.7 Implications of studies on archaea for the search for life on extraterrestrial worlds

As noted earlier, Woese argued that Bacteria, Archaea, and Eukaryotes represent separate lines of descent that diverged early on from an ancestral colony of organisms (Woese and Gupta, 1981; Woese, 1998). One possibility (Woese, 1998; Kandler, 1998) is that this occurred before the evolution of cells, when the lack of a typical cell membrane allowed unrestricted lateral gene transfer, and that the common ancestors of the three domains arose by fixation of specific subsets of genes (Woese, 1998; Kandler, 1998). It is possible that the last common ancestor of bacteria and archaea was a thermophile, which raises the possibility that lower temperatures are "extreme environments" for archaea, and organisms that live in cooler environments appeared only later (Gribaldo and Brochier-Armanet, 2006).

Archaea and bacteria have generally similar cell structure, but cell composition and organization set the archaea apart. Like bacteria, archaea lack interior membranes and organelles (Woese, 1994). Like bacteria, the cell membranes of archaea are usually bounded by a cell wall and they swim using one or more flagella (Thomas et al., 2001). Structurally, archaea are most similar to gram-positive bacteria. Most have a single plasma membrane and cell wall, and lack a periplasmic space; the exception to this general rule is *Ignicoccus*, which possess a particularly large periplasm that contains membrane-bound vesicles and is enclosed by an outer membrane (Rachel et al., 2002). Note that periplasm is a concentrated gel-like matrix in the space between the inner cytoplasmic membrane and the bacterial outer membrane called the periplasmic space in gram-negative bacteria. Using cryo-electron microscopy it has been found that a much smaller periplasmic space is also present in gram-positive bacteria. Gram-negative bacteria are bacteria that do not retain the crystal violet stain used in the Gram staining method of bacterial differentiation.

Because archaea and bacteria are no more related to each other than they are to eukaryotes, the term prokaryote may suggest a false similarity between them (Woese, 1994). However, structural and functional similarities between lineages often occur because of shared ancestral traits or evolutionary convergence. These similarities are known as a grade, and prokaryotes are best thought of as a grade of life, characterized by such features as an absence of membrane-bound organelles. Note that organelles are specialized structures that perform various jobs inside cells. The term literally means "little organs." In the same way organs, such as the heart, liver, stomach, and kidneys, serve specific functions to keep an organism alive, organelles serve specific functions to keep a cell alive.

Woese's research on Archaea is significant in its implications for the search for life on other celestial bodies. Now, most microbiologists believe Archaea are ancient, and may have robust evolutionary connections to the first organisms on Earth (Kelly et al., 2010). Organisms similar to those archaea that exist in extreme environments may have developed on other celestial bodies, some of which harbor conditions conducive to extremophile life (Stetter, 2006). Notably, Woese's elucidation of the tree of life shows the overwhelming diversity of microbial lineages: single-celled organisms represent the vast majority of the biosphere's genetic, metabolic, and ecologic niche diversity (Woese, 2006).

Woese wrote, "My evolutionary concerns center on the bacteria and the archaea, whose evolutions cover most of the planet's 4.5-billion-year history. Using ribosomal RNA sequence as an evolutionary measure, my laboratory has reconstructed the phylogeny of both groups, and thereby provided a phylogenetically valid system of classification for prokaryotes. The discovery of the archaea was in fact a product of these studies" ("Carl R Woese, Professor of Microbiology." University of Illinois at Urbana–Champaign).

5.15 How do extremophiles survive and thrive in extreme environmental conditions—clues from study of the DNA

Knowledge of extremophile habitats is expanding the number and types of extraterrestrial locations that may be targeted for exploration (Cavicchioli, 2002). Contemporary biological studies are being fuelled by the increasing availability of genome sequences and associated functional studies of extremophiles. According to Cavicchioli (2002), such studies are leading to the identification of new biomarkers, an accurate assessment of cellular evolution, insight into the ability of microorganisms to survive in meteorites and during periods of global extinction, and knowledge of how to process and examine environmental samples to detect viable life forms. Cavicchioli (2002) has evaluated extremophiles and extreme environments in the context of astrobiology and the search for extraterrestrial life.

Tardigrades are an outgroup (a group of organisms not belonging to the group whose evolutionary relationships are being investigated) to arthropods (invertebrate animals having an exoskeleton, a segmented body, and paired jointed appendages) in the Ecdysozoa (a group of protostome animals, including Arthropoda Nematoda, and several smaller phyla) and, as such, can provide insight into how gene functions have evolved among the arthropods and their close relatives. Tardigrades make up a phylum of microscopic ecdysozoan animals {a morphologically heterogeneous group of animals which have a cuticle (the dead skin at the base of a fingernail or toenail) and grow by molting}. Tardigrades share many characteristics with *C. elegans* and *Drosophila* that could make them useful laboratory models, but long-term culturing of tardigrades historically has been a challenge, and there have been few studies of tardigrade development.

In a study, Gabriel et al. (2007) showed that the tardigrade *Hypsibius dujardini* can be cultured continuously for decades and can be cryo-preserved. They reported that *H. dujardini* has a compact genome, a little smaller than that of *C. elegans* or *Drosophila*, and that sequence evolution has occurred at a typical rate. *H. dujardini* has a short generation time, 13-14 days at room temperature. Gabriel et al. (2007) found that the embryos of *H. dujardini* have a stereotyped cleavage pattern (a series of mitotic divisions whereby the enormous. volume of egg cytoplasm is divided into numerous smaller, nucleated cells) with asymmetric cell divisions, nuclear migrations, and cell migrations occurring in reproducible patterns. Gabriel et al. (2007) reported a cell lineage of the early embryo and an embryonic staging series. These investigators expect that their data can serve as a platform for using *H. dujardini* as a model for studying the evolution of developmental mechanisms.

In another study, Gabriel and Goldstein (2007) developed immunostaining methods for tardigrade embryos, and they used cross-reactive antibodies to investigate the expression of homologs of the pair-rule gene paired (Pax3/7) and the segment polarity gene engrailed in the tardigrade *Hypsibius dujardini*. Note that in genetics, homolog is a gene related to a second gene by descent from a common ancestral DNA sequence. The term, homolog, may apply to the relationship between genes separated by the event of speciation or to the relationship between genes separated by the event of genetic duplication. Gabriel and Goldstein (2007) found that in *H. dujardini* embryos, Pax3/7 protein localizes not in a pair-rule pattern but in a segmentally iterated pattern, after the segments are established, in regions of the embryo where neurons later arise. It was found that engrailed protein localizes in the posterior ectoderm of each segment before ectodermal segmentation is apparent. The Engrailed gene is thought to be a "selector" gene that controls the expression of other genes to confer a "posterior identity" on groups of cells that are related to each other by lineage. Together with previous results from others, Gabriel and Goldstein (2007)'s data support the conclusions that the pair-rule function of Pax3/7 is specific to the arthropods (invertebrate animals having an exoskeleton, a segmented body, and paired jointed appendages), that some of the ancient functions of Pax3/7 and Engrailed in ancestral bilaterians may have been in neurogenesis, and that Engrailed may have a function in establishing morphological boundaries between segments that is conserved at least among the Panarthropoda (a proposed animal clade containing the extant phyla Arthropoda, Tardigrada and Onychophora).

The origin and diversification of segmented metazoan (any of a group that comprises all animals having the body composed of cells differentiated into tissues and organs and usually a digestive cavity lined with specialized cells) body plans has fascinated biologists for over a century. Tardigrada is one of the three phyla of segmented animals; the other two being Euarthropoda, Onychophora in the superphylum Panarthropoda. This superphylum includes representatives with relatively simple, and representatives with relatively complex, segmented body plans. At one extreme of this continuum, euarthropods exhibit an incredible diversity of serially homologous segments (i.e., similar in position, structure, and evolutionary origin but not necessarily in function). At the other extreme, all tardigrades share a simple segmented body plan that consists of a head and four leg-bearing segments.

Although there are several physiological, morphological and ecological studies on tardigrades, only limited DNA sequence information is available. Therefore, Mali et al. (2010) explored the transcriptome (the sum total of all the messenger RNA molecules expressed from the genes of an organism) in the active and anhydrobiotic state ("life without water," referring to the remarkable ability of some organisms to survive the loss of all, or almost all, water and enter into a state of suspended animation in which their metabolism comes reversibly to a standstill) of the tardigrade *Milnesium tardigradum* which has extraordinary tolerance to desiccation and freezing. In their study, they reported the first overview of the transcriptome of *M. tardigradum* and its response to desiccation and discussed potential parallels to stress responses in other organisms.

Mali et al. (2010) sequenced a total of 9984 expressed sequence tags (ESTs) from two cDNA libraries from the eutardigrade *M. tardigradum* in its active and inactive, anhydrobiotic (tun) stage. Assembly of these ESTs resulted in 3283 putative unique transcripts, whereof approximately 50% showed significant sequence similarity to known genes. The resulting uni-genes were functionally annotated using the Gene Ontology (GO) vocabulary. A GO term enrichment analysis revealed several GOs that were significantly under-represented in the inactive stage. Furthermore, Mali et al. (2010) compared the putative uni-genes of *M. tardigradum* with ESTs from two other eutardigrade species that are available from public sequence databases, namely "Richtersius coronifer" and "Hypsibius dujardini." The processed sequences of the three tardigrade-species revealed similar functional content and the *M. tardigradum* dataset contained additional sequences from tardigrades not present in the other two.

Mali et al. (2010)'s study described novel sequence data from the tardigrade *M. tardigradum*, which significantly contributed to the available tardigrade sequence data and will help to establish this extraordinary tardigrade as a model for studying anhydrobiosis. Functional comparison of active and anhydrobiotic tardigrades revealed a differential distribution of Gene Ontology terms associated with chromatin structure and the translation machinery, which are under-represented in the inactive animals.

These findings imply a widespread metabolic response of the animals on dehydration. The collective tardigrade transcriptome data is expected to serve as a reference for further studies and support the identification and characterization of genes involved in the anhydrobiotic response.

Tardigrades are valuable research subjects for investigating how organisms and biological materials can survive extreme conditions. Methods to disrupt gene activity are essential to each of these efforts, but no such method existed for the Phylum *Tardigrada*. However, Tenlen et al. (2013) developed a protocol to disrupt tardigrade gene functions by double-stranded RNA-mediated RNA interference (RNAi). They showed that targeting tardigrade homologs of essential developmental genes by RNAi produced embryonic lethality, whereas targeting green fluorescent protein did not. They found that disruption of gene functions appears to be relatively specific by two criteria: targeting distinct genes resulted in distinct phenotypes that were consistent with predicted gene functions and by RT-PCR, RNAi reduced the level of a target mRNA and not a control mRNA. These studies represent the first evidence that gene functions can be disrupted by RNAi in the phylum *Tardigrada*. Tenlen et al. (2013)'s results form a platform for dissecting tardigrade gene functions for understanding the evolution of developmental mechanisms and survival in extreme environments.

With regard to body plans of tardigrade, the modular body plans of panarthropods make them a tractable model for understanding diversification of animal body plans more generally. Smith and Goldstein (2017) reviewed results of recent morphological and developmental studies of tardigrade segmentation. These results complement investigations of segmentation processes in other panarthropods and paleontological studies to illuminate the earliest steps in the evolution of panarthropod body plans.

5.16 Revival of panspermia concept encouraged by the discovery of survival limits of tardigrades in high-speed impacts

Tardigrades have proved to be wonderful microscopic extremophiles, which possess several special features. These special features made them targets in astrobiological studies and even encouraged the revival of the once forgotten philosophical concept known as "panspermia."

5.16.1 Panspermia concept

The panspermia hypothesis—a philosophical thought that life migrates naturally through space—states that the seeds of life exist all over the Universe and can be propagated through space from one location to another. Particular models of panspermia, such as lithopanspermia (Melosh, 1988), involve movement of rocks that contain life from one planetary surface to another, that is, launch into space from the surface of a planet on impact ejecta and subsequent arrival on a new body at high speed. Both ejection and arrival involve accelerations and shocks.

For millennia, the idea embodied in the panspermia concept has been a topic of philosophical debate (Raulin-Cerceau et al., 1998). However, due to lack of any validation, it remained merely speculative until few decades ago. It is only with the recent discoveries and advances from different fields of research that panspermia has been given serious scientific consideration (Rampelotto, 2009). Most of the major barriers against the acceptance of panspermia have been demolished when it has been shown that microorganisms can survive the high impact and velocity experienced during the ejection from one planet, the journey through space, and the impact process onto another world (Rampelotto, 2010).

According to Rampelotto (2010), this interest was revived in the late 70s primarily by the recognition of Martian meteorites here on Earth (Fritz et al., 2005) which prove beyond doubt that intact rocks can be transferred between the surfaces of planetary bodies in the Solar System. Petrographic analysis of the Martian meteorites and mathematical simulations of impact induced ejection demonstrated that these rocks experienced shocks from 5–10 GPa to 55 GPa (Nyquist et al., 2001), heating in the range from 40°C to 350°C (Shuster and Weiss, 2005), and acceleration on the order of 3.8×10^6 m/s^2 (Mastrapa et al., 2001).

Based on this data, mechanisms for the transfer of planetary material have been proposed. The most well-accepted mechanism, developed by Melosh (1988), indicates that materials can be expelled into interplanetary space under light shocks and modest temperature increases. In fact, measurements in the Allan Hills 84001 Martian meteorite (ALH84001 meteorite, for short) have shown that it was probably not heated over 40 °C since before it was ejected from Mars (Weiss et al., 2000). A discussion on ALH84001 meteorite is given in Chapter 10.16.2.

According to Rampelotto (2010), these results led to the question of whether living organisms have been transported between the planets of our Solar System by the same mechanism. The viable transfer from one planet to another requires microorganisms to survive the escape process from one planet, the journey through space, as well as the re-entry/impact process on another planet (Horneck et al., 2001). In this context, a variety of studies have been performed in order to simulate different aspects of lithopanspermia {movement of rocks that contain life from one planetary surface to another, that is, launch into space from the surface of a planet on impact ejecta and subsequent arrival on a new body (Melosh, 1988)}— which postulates that meteors are the transfer vehicles for life through space (Rampelotto, 2010).

Rampelotto (2010) recalls that further support to the theory of lithopanspermia has been given by simulation experiments in which model microbes are subjected to ultracentrifugation, hypervelocity, shock pressure, and heating in the range defined for the Martian meteorites found on Earth. These experiments simulate the physical forces that hypothetical endolithic microbes (organisms such as archaeon, bacterium, fungus, lichen, algae or amoeba that are able to acquire the necessary resources for growth in the inner part of a rock, mineral, coral, animal shells, or in the pores between mineral grains of a rock) would be subjected to during ejection from one planet and landing upon another. Previous simulation experiments have measured each of these stresses in an isolated manner and the results indicate that spores (reproductive cells capable of developing into new individuals without fusion with other reproductive cells) can survive each stress applied singly (Burchell et al., 2004; Stoffler et al., 2007). The analyses of the combined stresses can be most closely simulated in the laboratory via hypervelocity ballistics experiments. The results demonstrated that microbes could survive rapid acceleration to Mars escape velocities and subsequent impact into surfaces of different compositions (Horneck et al., 2008; Fajardo-Cavazos et al., 2009). Thus, there is a body of evidence suggesting that microbes can survive the conditions of interplanetary transfer from Mars to Earth or from any Mars-like planet to other habitable planets in the same Solar System (Rampelotto, 2010).

Rampelotto (2010) points out that the Earth-Mars system is not the only place where natural transfer may occur. The discovery of potentially habitable environments, such as the moons of Jupiter and Saturn (e.g., Io, Europa, Ganymede, Callisto, Titan, and Enceladus), expands the possibility of interplanetary transfer of life in the Solar System. Therefore, in recent years, most of the major barriers against the acceptance of panspermia have been demolished and this theory re-emerges as a promising field of research (Rampelotto, 2010).

Spaceflight experiments demonstrate that with minimal UV shielding, several types of microbes can survive for years at exposures to the harsh environment of space (Rettberg et al., 2002). Furthermore, it is estimated that, if shielded by 2 meters of meteorite, a substantial number of spores would survive after 25 million years in space (Horneck et al., 2002). Surprisingly, on Earth, microbes were brought back to life after 250 million years (Vreeland et al., 2000).

Studies in extreme ambient conditions have changed the previous paradigm that life can only be found on pleasant Earth-like planets. In the last few decades, substantial changes have occurred regarding what scientists consider the limits of habitable environmental conditions. Our knowledge about extreme environments and the limits of life on Earth has greatly improved in recent decades. These advances have been fundamental to the development of astrobiology, which studies the origin, evolution, and distribution of life in the Universe. Numerous planetary bodies and moons in our Solar System have been suggested to sustain life (e.g., Mars, Europa, Enceladus).

Detailed mineral compositions of Mars, obtained by landers and orbiters, have revealed that Mars presented a geochemically active environment (Gendrin et al., 2005; Bibring et al., 2007). The evidence indicates that Mars once had an environment similar to Earth, supporting the hypothesis that in the past, life may have flourished abundantly on early Mars (Schulze-Makuch et al., 2005; Yung et al., 2010). The discovery of trace amounts of methane in the Martian atmosphere (Formisano et al., 2004), which could perhaps be biotic (produced by extant life, such as methanogens (Atreya et al., 2007; Onstott et al., 2006) and the recent discovery of subsurface liquid-water lakes on Mars (Orosei et al., 2018; Lauro et al., 2021) have prompted the astrobiologists consider that life could have survived and adapted to the subsurface conditions as microorganisms do in extreme environments on Earth (Schulze-Makuch et al., 2008). The most proper terrestrial extreme environments analogous to those of Mars may be the Atacama Desert, the Antarctic Dry Valleys, the Rio Tinto region, and the deep basalt aquifers.

Due to the presence of a hydrothermally active subsurface ocean, Europa has been considered the most promising target in the search for extant extraterrestrial life within the Solar System. Life could exist within its subglacial ocean, perhaps subsisting in an environment similar to Earth's multitudes of deep-ocean hydrothermal vents or the Antarctic Lake Vostok. Furthermore, the ice layer may provide other potential habitats for life (Rampelotto, 2010).

Presence of hydrothermal vents on the subseafloor of Enceladus and the high-speed jets of water vapor and organic-rich particles gushing out from the tiger stripes on the southern region of Enceladus provides optimism to the astrobiologists that some form of life could be present on Enceladus.

5.16.2 Ability of tardigrades to survive high-speed impact shocks

The ability of tardigrades to survive extreme environmental conditions such as low temperatures, vacuum, and radiation (Persson et al., 2011; Schill and Hengherr, 2018; Jönsson, 2019) and their ability to survive exposure in space on the exterior of space vehicles (Jönsson et al., 2008; Rebecchi et al., 2009; Persson et al., 2011) have led to suggestions that tardigrades could be a vector for panspermia, that is, natural movement of life between celestial bodies within and beyond the Solar System (e.g., see Veras et al., 2018 for a review of panspermia).

It has long been proposed that ejecta (and fossilized material within) from giant impacts on Earth could have struck Earth's Moon as well and became preserved (Armstrong et al., 2002; Armstrong, 2010; Burchell et al., 2014, 2017). In a study, Armstrong (2010) determined an average lunar impact speed for terrestrial ejecta of some 2.5 km/s. However, it is the vertical component of impact speed that determines peak shock pressure; Armstrong (2010) has shown that this has a mean value of about 1.3 km/s. At vertical impact speeds of even 1 km/s, the peak shock pressure in lunar impacts was estimated by Armstrong (2010) to be 2 GPa (Traspas and Burchell, 2021).

Burchell et al. (2017) calculated the peak shock pressures for terrestrial material impacting Earth's Moon for a range of possible impactor and lunar surface material combinations, and shock pressures were found to be in the range of 2–5 GPa, which is consistent with those of Armstrong (2010), who pointed out, nonetheless, that 43% of impacts of terrestrial ejecta onto the lunar surface would be at speeds below 1 km/s. It was found that 29% of such impacts have a vertical impact speed of less than 0.5 km/s, and 10% at less than 0.1 km/s. These correspond to peak pressures of 0.5 and 0.02 GPa, respectively, which is well within tardigrade survival limits. Two qualifications, however, are required: (1) the degree of shock during ejection from Earth, and (2) the increase in temperature due to the shock impact. These points are discussed, for example, in the work of Halim et al. (2021); however, their simulations still show biomarkers potentially surviving in terrestrial ejecta that impact the Earth's Moon.

In an impressive experimental study, Traspas and Burchell (2021) investigated the ability of tardigrades to survive impact shocks in the kilometer per second and gigapascal range. This investigation tested whether tardigrades can survive in impacts typical of those that occur naturally in the Solar System. Traspas and Burchell (2021) found that tardigrades can survive impacts up to 0.9 km/s, which is equivalent to 1.14 GPa shock pressure, but cannot survive impacts above this. Traspas and Burchell (2021)'s study demonstrated the survival limits of tardigrades in high-speed impacts, signaling the possible implications for panspermia. It was once thought that Earth is a biogeographical island with respect to living creatures; they have been trapped, like a flightless bird on a remote island. However, Traspas and Burchell (2021)'s study indicates that although panspermia is hard, it is not impossible. Meteorite impacts on Earth typically arrive at speeds of more than 11 kilometers per second. On Mars, they collide at least at 8 kilometers per second. Traspas and Burchell (2021)'s study indicates that these impact-speeds are well above the threshold for tardigrades (if trapped in meteorites) to survive. However, these researchers argue that some parts of a meteorite impacting Earth or Mars would experience lower shock pressures that a tardigrade could live through.

This new experimental study, which places new limits on tardigrades' ability to survive impacts in space—and potentially seed life on other planets, also has implications for our ability to detect life on icy moons in the Solar System. As already indicated, Saturn's moon Enceladus, for example, ejects plumes of water into space from a subsurface ocean that could support life, as Jupiter's moon Europa. If the findings of the new study apply to potential life trapped in the plumes, a spacecraft orbiting Enceladus—at relatively low speeds of hundreds of meters per second—might sample and detect existing life without killing it. Traspas and Burchell (2021)'s study also places new limits on the panspermia theory, which suggests some forms of life could move between worlds, as stowaways (like persons who hide on a ship, aircraft, or other vehicles) on meteorites kicked up after an asteroid strikes a planet or moon. Eventually, the meteorite could impact another planet—along with its living cargo.

Traspas and Burchell (2021) have noted that as is the case for Earth's Moon, a similar scenario can be applied to Earth's neighboring planet Mars, that is, Martian impact ejecta striking Mars' moon Phobos. Chappaz et al. (2013) estimated that as much as 50 mg of Martian surface material lies within every 100 g of Phobosian regolith, of which 0.2 mg will have been deposited in the last 10 million years (Traspas and Burchell, 2021). Similarly, Ramsley and Head (2013) estimated that, over the past 500 Myr, around 250 ppm of Martian surface material has been deposited in the Phobosian regolith, and there is of order 6.5×10^8 kg of Martian ejecta in the Phobosian upper surface. The impact speed on Phobos is estimated to range from 1 to 4.5 km/s (Chappaz et al., 2013), which, if typical material parameters are assumed, is likely to produce peak shock pressures just above those that permit tardigrade survival. However, even in the event some of this material was lightly enough shocked to permit tardigrade survival, long-term exposure to solar and cosmic radiation would still have sterilized much of it (Kurosawa et al., 2019).

In recent times, tardigrades have attracted much attention among astrobiologists and the supporters of panspermia primarily because this organism has been found to possess an ability to survive raw exposure to space for at least short periods. Traspas and Burchell (2021) believed that tardigrades' resistance to shock pressures could likely be a limiting factor in their success or otherwise as vectors for panspermia. The ability of tardigrades to survive extreme conditions is linked to their ability to enter a "tun" state in which they dehydrate, expelling 90%+ of their water, and produce antioxidants, which allows their metabolic rate to fall to 0.01% of normal. It was in this tun state that the tardigrades were tested in the study carried out by Traspas and Burchell (2021).

It is interesting to note that the new research carried out by Traspas and Burchell (2021) was inspired by a 2019 Israeli mission called Beresheet, which attempted to land on Earth's Moon. The probe infamously included tardigrades on board that mission managers had not disclosed to the public, and the lander crashed with its passengers in tow, raising concerns about contamination (O'Callaghan, 2021). Traspas and Burchell (2021) wanted to find out whether tardigrades on a spacecraft could survive an impact akin to crash-landing on a celestial body such as Earth's Moon or the planet Mars—and they wanted to conduct their experiment ethically. So, after feeding about 20 tardigrades moss and mineral water, they put them into hibernation, a so-called "tun" state in which their metabolism decreases to 0.1% of their normal activity, by freezing them for 48 hours. These researchers then placed two to four at a time in a hollow nylon bullet and fired them at increasing speeds using a two-stage light gas gun, a tool in physics experiments that can achieve muzzle velocities far higher than any conventional gun. When shooting the bullets into a sand target several meters away, the researchers found the creatures could survive impacts up to about 900 meters per second (3240 kilometres per hour), and momentary shock pressures up to a limit of 1.14 gigapascals (GPa). Above those speeds, they just mush (O'Callaghan, 2021). It may be noted that the sand target can be likened to the regolith (sandy dust) found on the surface of Earth's Moon.

The tardigrade used by Traspas and Burchell (2021) in their study were the species *Hypsibius dujardini*, which were handled according to the ethical rules for invertebrates with the consent of the departmental ethics officer. Traspas and Burchell (2021) described their impact testing methodology thus, "The tardigrades were fed mineral water and moss (Fig. 5.22A and B). They were fired from a two-stage light gas gun

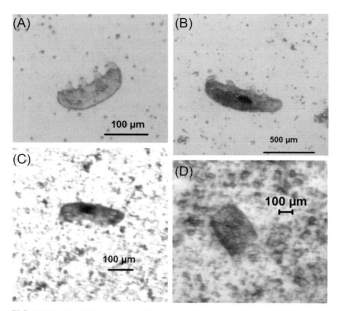

FIG. 5.22 (A, B) Example tardigrades before impact testing. Tardigrades ranged in size from 150 to 850 μm. (C) Tardigrade recovered after an impact at 0.728 km/s. (d) Tardigrade fragment from shot at 0.901 km/s. *(Source: Traspas, A., and M.J. Burchell (2021), Tardigrade survival limits in high-speed impacts —Implications for panspermia and collection of samples from plumes emitted by ice worlds, Astrobiology, https://doi.org/10.1089/ast.2020.2405 Open Access; Open Access license Website: https://www.liebertpub.com/doi/full/10.1089/ast.2020.2405).*

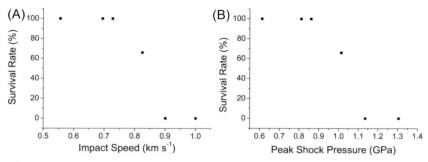

FIG. 5.23 (A, B) Results of impact experiments onto sand. (A) Tardigrade survival rate vs. impact speed. (B) Tardigrade survival rate vs. peak shock pressure. *(Source: Traspas, A., and M.J. Burchell (2021), Tardigrade survival limits in high-speed impacts —Implications for panspermia and collection of samples from plumes emitted by ice worlds, Astrobiology,https://doi.org/10.1089/ast.2020.2405 Open Access; Open Access license Website: https://www.liebertpub.com/doi/full/10.1089/ast.2020.2405).*

(Burchell et al., 1999; Hibbert et al., 2017) at sand targets in a vacuum chamber. Prior to shooting, two or three tardigrades were loaded into a water-filled shaft in a nylon sabot (the number was measured in each case). The sabot was then frozen for 48 h so that the tardigrades were in a tun state during the shot. The sabot was then placed in the gun and fired at normal incidence into the sand. The whole sabot impacted the target in each shot. Impact speeds were measured in each shot to better than ±1% using two laser light stations mounted transverse to the direction of flight and focused onto photodiodes. The signals from the photodiodes, combined with their known separation (499 mm), provided the speed."

Traspas and Burchell (2021) further explain that six shots were executed at speeds from 0.556 to 1.00 km/s. After each shot, the sand target was poured into a water column to separate the sand from other materials and isolate the tardigrades. The recovered tardigrades were then observed over time to discern whether they returned to a mobile state (i.e., an active state). The time to achieve this was noted. As a control, tests were made to freeze 20 tardigrades and then defrost them without their being fired with the gun. All 20 were revived successfully, and it took them 8–9 h to recover to a mobile state, with none requiring more than 9 h. Traspas and Burchell (2021) recall that "in an earlier study (Pasini et al., 2014), tardigrades were frozen in an ice target that was impacted. Survival of the tardigrades in the target after impact was then evaluated. However, in that study, even in unimpacted frozen control samples, about two-thirds of the tardigrades died. The current study thus represents an improvement in the overall handling of the samples (with 20 out of 20 control samples surviving). The experimental method also provides a more uniform shock to the samples during the experiments, as a result of their being mounted in the small interior volume of the sabot, rather than their being distributed throughout the target."

Traspas and Burchell (2021) have provided interesting results with regard to the survival rate from each shot (in Fig. 5.23A versus impact speed) and in Fig. 5.23B versus

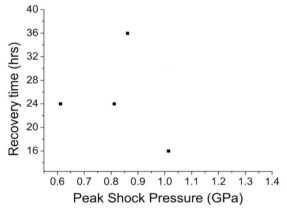

FIG. 5.24 Recovery time posshot for those tardigrades which survived to regain mobility. All recovery times greatly exceed the 8–9 h recovery time from just being frozen. *(Source: Traspas, A., and M.J. Burchell (2021), Tardigrade survival limits in high-speed impacts —Implications for panspermia and collection of samples from plumes emitted by ice worlds, Astrobiology,https://doi.org/10.1089/ast.2020.2405. Open Access; Open Access license. Website: https://www.liebertpub.com/doi/full/10.1089/ast.2020.2405).*

peak shock pressure. It can be seen that survival fell from 100% to 0% between 0.728 and 0.901 km/s (corresponding to 0.86–1.14 GPa). In the shots up to and including 0.825 km/s, intact tardigrades were recovered postshot (e.g., Fig. 5.22C), but in the higher-speed shots only fragments of tardigrades were recovered (e.g., Fig. 5.22D). Thus, Traspas and Burchell (2021) point out that shortly after the onset of lethality, the tardigrades were also physically broken apart as impact speed increased.

Traspas and Burchell (2021)'s study further indicates that for active tardigrades that were found postshot, the recovery times to achieve a fully mobile state were significantly greater than those for tardigrades in the frozen/defrosted control samples (8–9 h) (Fig. 5.24). This result suggests that the impact shock had a more significant effect than freezing alone.

The results obtained in Traspas and Burchell (2021)'s study suggest that peak shock pressures above 1.14 GPa will kill tardigrades. According to Traspas and Burchell (2021), it is, therefore, likely that the arrival of a tardigrade on Earth, for example by way of a meteorite impact, is not likely to be a viable means of a successful transfer even for such hardy organisms. There are other places in the Solar System, however, where biological material, during transfer, would encounter low shock pressures.

Traspas and Burchell (2021) have admitted that it is not clear whether the reproduction cycle can be undertaken by the survivors. This was not observed after any shot in their study, but the sample numbers were small and the samples kept isolated; thus, further study is needed. Similarly, collecting samples of tardigrade eggs, using them in the projectiles, and then assessing whether they can develop afterward would also be a fruitful area of study.

References

Abe, Y., Abe-Ouchi, A., Sleep, N.H., Zahnle, K.J., 2011. Habitable zone limits for dry planets. Astrobiology 11, 443–460.

Abdul-Halim, M.F., Schulze, S., DiLucido, A., Pfeiffer, F., Bisson Filho, A.W., Pohlschroder, M., 2020. Lipid anchoring of archaeosortase substrates and midcell growth in haloarchaea. MBio 11, e00349-20.

Abyzov, S.S., Mitskevich, I.N., Poglazova, M.N., Barkov, N.I., Lipenkov, V.Y, Bobin, N.E., Koudryashov, B.B., Pashkevich, V.M., 1998a. Antarctic ice sheet as a model in search of life on other planets. Adv. Space Res. 22, 363–368.

Abyzov, S.S., Mitskevich, I.N., Poglazova, M.N., 1998b. Microflora of deep glacier horizons of central Antarctica. Microbiology 67, 451–458.

Albers, S-V., 2016. Extremophiles: life at the deep end. Nature 538, 457. https://doi.org/10.1038/538457a.

Alexander, A., et al., 2020. A non–plate tectonic model for the Eoarchean Isua supracrustal belt. Lithosphere 12 (1), 166–179.

Allwood, A., 2016. Geology: evidence of life in Earth's oldest rocks. Nature, 537 (7621), 500–501. doi: 10.1038/nature19429.

Allwood, A., et al., 2018. Reassessing evidence of life in 3,700-million-year-old rocks of Greenland. Nature 563 (7730), 241–244. doi:10.1038/s41586-018-0610-4.

Altiero, T., Guidetti, R., Caselli, V., Cesari, M., Rebecchi, L., 2011. Ultraviolet radiation tolerance in hydrated and desiccated eutardigrades. J Zool Syst Evol Res 49, 104–110.

Antón, J. (2011), Halophile, Encyclopedia of Astrobiology, 2011 Edition (Eds.): Muriel Gargaud, Ricardo Amils, José Cernicharo Quintanilla, Henderson James (Jim) CleavesII, William M. Irvine, Daniele L. Pinti, Michel Viso.

Antony, R., Krishnan, K.P., Thomas, S., Abraham, W.P., Thamban, M., 2009. Phenotypic and molecular identification of Cellulosimicrobium cellulans isolated from Antarctic snow. Antonie Van Leeuwenhoek. doi:10.1007/s10482-009-9377-9.

Armstrong, J.C., Wells, L.E., Gonzales, G, 2002. Rummaging through Earth's attic for remains of ancient life. Icarus 160, 183–196.

Armstrong, J.C., 2010. Distribution of impact locations and velocities of Earth meteorites on the Moon. Earth Moon Planets 107, 43–54.

Atreya, S.K., Mahaffy, P.R., Wong, A.S., 2007. Methane and related trace species on Mars: origin, loss, implications for life, and habitability. Planet. Space Sci. 55, 358–369.

Azua-Bustos, A., Caro-Lara, L., Vicuña, R., 2015. Discovery and microbial content of the driest site of the hyperarid Atacama Desert, Chile. Environ Microbiol Rep 7, 388–394.

Baldauf, S.L., 1993. Animals and fungi are each other's closest relatives: congruent evidence from multiple proteins. Proc. Natl. Acad. Sci. 90 (24), 11558–11562.

Bang, C., Schmitz, R.A., 2015. Archaea associated with human surfaces: not to be underestimated. FEMS Microbiol. Rev. 39 (5), 631–648.

Barns, S., Burggraf, S., 1997. Crenarchaeota, The Tree of Life Web Project. Version 01 January 1997. http://tolweb.org/Crenarchaeota/9.

Becker, K., Bartetzko, A., Davis, E.E., 2001. Leg 174B Synopsis: Revisiting Hole 395A for logging and long-term monitoring of off-axis hydrothermal processes in young oceanic crust. Proc. ODP, Sci. Results 174B, 1–13.

Beltrán-Pardo, E., Jönsson, K.I., Wojcik, A., Haghdoost, S., Harms-Ringdahl, M., Bermúdez-Cruz, R.M., Bernal Villegas, J.E., 2013. Effects of ionizing radiation on embryos of the tardigrade Milnesium cf. tardigradum at different stages of development. PLoS One 8 (9), e72098. Epub 2013 Sep 6.

Beltrán-Pardo, E., Jönsson, K.I., Harms-Ringdahl, M., Haghdoost, S., Wojcik, A., 2015. Differences in tolerance to gamma radiation in the tardigrade Hypsibiusdujardini from embryo to adult correlate inversely with cellular proliferation. PLoS One 10 (7), e0133658. doi:10.1371/journal.pone.0133658.

Berka, R.M., Grigoriev, I.V., Otillar, R., Salamov, A., Grimwood, J., Reid, I., Ishmael, N., John, T., Darmond, C., Moisan, M.C., et al., 2011. Comparative genomic analysis of the thermophilic biomass-degrading fungi Myceliophthora thermophila and Thielavia terrestris. Nat. Biotechnol. 29, 922–927.

Bibring, J.P., Arvidson, R.E., Gendrin, A., Gondet, B., Langevin, Y., Le Mouelic, S., Mangold, N., Morris, R.V., Mustard, J.F., Poulet, F., Quantin, C., Sotin, C., 2007. Coupled ferric oxides and sulfates on the Martian surface. Science 317, 1206–1210.

Bisson-Filho, A.W., Zheng, J., Garner, E., 2018. Archaeal imaging: leading the hunt for new discoveries. Mol. Biol. Cell 29, 1675–1681.

Bize, A., Karlsson, E.A., Ekefjärd, K., Quax, T.E., Pina, M., Prevost, M.C., Forterre, P., Tenaillon, O., Bernander, R., Prangishvili, D., 2009. A unique virus release mechanism in the Archaea. Proc. Natl. Acad. Sci. USA 106, 11306–11311.

Blasiak, R., Jouffray, J.B., Wabnitz, C.C.C., Sundstrom, E., Osterblom, H., 2018. Corporate control and global governance of marine genetic resources. Sci. Adv. 4, p 6.

Bliss, L.C., Svoboda, J., Bliss, D.I., 1984. Polar deserts, their plant cover and plant production in the Canadian High Arctic. Holarctic Ecol 7, 305–324.

Bolhuis, H., Palm, P., Wende, A., Falb, M., Rampp, M., Rodriguez-Valera, F., Pfeiffer, F., Oesterhelt, D, 2006. The genome of the square archaeon Haloquadratum walsbyi: life at the limits of water activity. BMC Genomics 7 (1), 169. doi:10.1186/1471-2164-7-169.

Bolus, N.E., 2001. Basic review of radiation biology and terminology. J. Nucl. Med. Technol. 29 (2), 67–73.

Boothby, T.C., Tapia, H, Brozena, A.H., Piszkiewicz, S., Smith, A.E., Giovanninni, I., Rebecchi, L., Pielak, G.J., Koshland, D., Goldstein, B., 2017. Tardigrades use intrinsically disordered proteins to survive desiccation. Mol. Cell 65, 975–984.

Broady, P.A., 1979. The terrestrial algae of Signy Island, South Orkney Islands. British Antarctic Survey Science Reports 98, 1–117.

Broady, P.A., 1981. The ecology of sublithic terrestrial algae at the Vestfold Hills, Antarctica. Brit. Phycol. J. 16, 231–240.

Broady, P.A., 1981a. The ecology of chasmolithic algae at coastal locations of Antarctica. Phycologia 20, 259–272.

Brock, T.D., 1978 Thermophilic Microorganisms and Life at High Temperaturesxi. Springer, New York, NY, pp. 465.

Budel, B., Wessels, D.C.J., 1991. Rock inhabiting blue-green algae/cyanobacteria from hot arid regions. Algological Studies 64, 385–398.

Burchell, M.J., Cole, M.J., McDonnell, J.A.M, et al., 1999. Hypervelocity impact studies using the 2 MV Van de Graaff accelerator and two stage light gas gun of the University of Kent at Canterbury. Meas. Sci. Technol. 10, 41–50.

Burchell, M.J., Mann, J.R., Bunch, A.W., 2004. Survival of bacteria and spores under extreme shock pressures. Mon. Not. R. Astron. Soc. 352, 1273–1278.

Burchell, M.J., McDermott, K.H., Price, M.C., 2014. Survival of fossils under extreme shocks induced by hypervelocity impacts. Philos Trans A Math Phys Eng Sci 372. doi:10.1098/rsta.2013.0190.

Burchell, M.J., Harriss, K.H., Price, M.C., et al., 2017. Survival of fossilised diatoms and forams in hypervelocity impacts with peak shock pressures in the 1–19 GPA range. Icarus 290, 81–88.

Bustin, S.A., Benes, V., Garson, J.A., Hellemans, J., Huggett, J., Kubista, M., et al., 2009. The MIQE guidelines: minimum information for publication of quantitative real-time PCR experiments. Clin. Chem. 55, 611–622. doi:10.1373/clinchem.2008.112797.

Caldeira, K., Kasting, J.F., 1992. The life span of the biosphere revisited. Nature 360, 721–723.

Campbell, S.E., 1982. Precambrian endoliths discovered. Nature 299, 429–431.

Carpenter, E.J., Lin, S., Capone, D.G., 2000. Bacterial activity in South Pole snow. Appl. Environ. Microbiol. 66, 4514–4517.

Cava, F., Hidalgo, A., Berenguer, J., 2009. Thermus thermophilus as biological model. Extremophiles 13 (2009), 213–231.

Cavicchioli, R., 2002. Extremophiles and the search for extraterrestrial life. Astrobiology 2 (3), 281–292.

Cavicchioli, R., 2011. Archaea — timeline of the third domain. Nature Reviews. Microbiology 9 (1), 51–61.

Chambers, J., 2014. Planet formation. Treatise on Geochemistry 2, 55–72.

Chappaz, L., Melosh, H.J., Vaquero, M., et al., 2013. Transfer of impact ejecta from the surface of Mars to Phobos and Deimos. Astrobiology 13, 963–980.

Charles-Orszag, A., Lord, S.J., Mullins, R.D., 2020. High-temperature live-cell imaging of cytokinesis, cell motility and cell-cell adhesion in the thermoacidophilic crenarchaeon Sulfolobus acidocaldarius. bioRxiv. doi:10.1101/2020.02.16.951772.

Chopra, A., Lineweaver, C.H., 2016. The case for a gaian bottleneck: the biology of habitability. Astrobiology 16 (1), 7–22. doi:10.1089/ast.2015.1387.

Christner, B.C., Mosley-Thompson, E., Thompson, L.G., Zagorodnov, V., 2000. Recovery and identification of viable bacteria immured in glacial ice. Icarus 144, 479–485.

Christner, B.C., 2002. Incorporation of DNA and protein precursors into macromolecules by bacteria at −15°C. Appl. Environ. Microbiol. 68, 6435–6438.

Christner, B.C., Kvitko II, B.H., Reeve, J.N, 2003a. Molecular identification of bacteria and eukarya inhabiting an Antarctic cryoconite hole. Extremophiles 7, 177–183.

Ciaramella, M., Napoli, A., Rossi, M., 2005. Another extreme genome: how to live at pH 0. Trends Microbiol. 13 (2), 49–51.

Cockell, C.S., Lee, P., 2002. The biology of impact craters—a review. Biological Reviews 77, 279–310.

Cockell, C.S., Lee, P., Osinski, G., Horneck, H., Broady, P., 2002. Impact-induced microbial endolithic habitats. Meteorit. Planet. Sci. 37, 1287–1298.

Cockell, C.S., Rettberg, P., Horneck, G., Scherer, K., Stokes, D.M., 2003. Measurements of microbial protection from ultraviolet radiation in polar terrestrial microhabitats. Polar Biol 26, 62–69.

Cockell, C.S., Osinski, G.R., Lee, P., 2003b. The impact crater as a habitat: effects of impact processing of target materials. Astrobiology 3, 181–191.

Cockell, C.S., Pösges, G., Broady, P., 2004. Chasmoendolithic colonization of impact-shattered rocks—A geologically nonspecific habitat for microorganisms in the post-impact environment (abstract). Int. J. Astrobiol. 3, 71.

Cockell, C.S., Stokes, M.D., 2004. Widespread colonization by polar hypoliths. Nature 431, 414.

Cockell, C.S., Lim, D.S.S., 2005. Impact craters, water and microbial life. In: Tokano, T. (Ed.), Water and life on Mars. Springer-Verlag, Heidelberg, pp. 261–275.

Cockell, C.S., Lee, P., Broady, P., Lim, D.S.S., Osinski, G.R., Parnell, J., Koeberl, C., Pesonen, L., Salminen, J., 2005. Effects of asteroid and comet impacts on habitats for lithophytic organisms—A synthesis. Meteorit. Planet. Sci. 40 (12), 1901–1914.

Cockell, C.S., 2011. Vacant habitats in the Universe. Trends Ecol. Evol. 26, 73–80.

Cockell, C.S., 2014. Types of habitat in the Universe. Int. J. Astrobiol. 13, 158–164.

Cockell, C.S., Bush, T., Bryce, C., Direito, S., Fox-Powel, M., Harrison, J.P., Lammer, H., Landenmark, H., Martin-Torres, J., Nicholson, N., Noack, L., O'Malley-James, J., Payler, S.J., Rushby, A., Samuels, T., Schwendner, P., Wadsworth, J., Zorzano, M.P., 2016. Habitability: a review. Astrobiology 16 (1), 89–117.

Coker, J.A., 2019. Recent advances in understanding extremophiles. F1000Res 8: F1000 Faculty Rev-1917. doi: 10.12688/f1000research.20765.1. eCollection 2019.

Cottin, H., Kotler, J.M., Bartik, K., Cleaves, H.J., Cockell, C.S., de Vera, J-P.P., Ehrenfreund, P., Leuko, S., Kate, I.L.T., Martins, Z., Pascal, R., Quinn, R., Rettberg, P., Westall, F., 2017. Astrobiology and the possibility of life on Earth and elsewhere? Space Sci. Rev. 209 (1-4), 1–42.

Cowen, J.P., et al., 2003. Fluids from aging ocean crust that support microbial life. Science 299, 120–123.

Cutiño-Jiménez, A.M., et al., 2010. Evolutionary placement of Xanthomonadales based on conserved protein signature sequences. Molecular Phylogenetics Evolution 54 (2), 524–534. doi:10.1016/j.ympev.2009.09.026.

Dalmaso, G.Z., Ferreira, D., Vermelho, A.B., 2015. Marine extremophiles: a source of hydrolases for biotechnological applications. Mar Drugs 13 (4), 1925–1965. doi:10.3390/md13041925.

Davies, P.C., 1996. The transfer of viable microorganisms between planets Ciba Foundation Symposium. Novartis Foundation Symposia 202, 304–314 discussion 314–17.

de Duve, C., 1995. Vital Dust: Life as a Cosmic Imperative. Basic Books, New York, NY.

D'Elia, T., Veerapaneni, R., Rogers, S.O., 2008. Isolation of microbes from Lake Vostok accretion ice. Appl. Environ. Microbiol. 74, 4962–4965.

D'Hondt, S., Jørgensen, B.B., Miller, D.J., Batzke, A., Blake, R., Cragg, B.A., Cypionka, H., Dickens, G.R., Ferdelman, T., Hinrichs, K., Holm, N.G., Mitterer, R., Spivack, A., Wang, G., Bekins, B., Engelen, B., Ford, K., Gettemy, G., Rutherford, S.D., Sass, H., Skilbeck, C.G., Aiello, I.W., Guèrin, G., House, C.H., Inagaki, F., Meister, P., Naehr,

T., Niitsuma, S., Parkes, R.J., Schippers, A., Smith, D.C., Teske, A., Wiegel, J., Padilla, C.N., Acosta, J.L.S., 2004. Distributions of microbial activities in deep subseafloor sediments. Science 306, 2216–2221.

de la Torre, J.R., Goebel, B.M., Friedmann, E.I., Pace, N.R., 2003. Microbial diversity of cryptoendolithic communities from the McMurdo Dry Valleys, Antarctica. Appl. Environ. Microbiol. 69 (7), 3858–3867.

DeLong, E.F., 1998. Everything in moderation: archaea as 'non-extremophiles. Current Opinion Genetics Develop. 8 (6), 649–654.

DeLong, E.F., Pace, N.R., 2001. Environmental diversity of bacteria and archaea. Syst. Biol. 50 (4), 470–478.

De los Rios, A., Sancho, L.G., Grube, M., Wierzchos, J., Ascaso, C, 2005. Endolithic growth of two Lecidea lichens in granite from continental Antarctica detected by molecular and microscopy techniques. New Phytol. 165, 181–189.

Delpech, F., Collien, Y., Mahou, P., Beaurepaire, E., Myllykallio, H., Lestini, R., 2018. Snapshots of archaeal DNA replication and repair in living cells using super-resolution imaging. Nucleic. Acids. Res. 46, 10757–10770.

Deppenmeier, U., 2002. The unique biochemistry of methanogenesis. Prog. Nucleic Acid Res. Mol. Biol. 71, 223–283.

Doolittle, W.F., 1999. Phylogenetic classification and the universal tree. Science 284 (5423), 2124–2129.

Driscoll, P., Bercovici, D, 2013. Divergent evolution of Earth and Venus: influence of degassing, tectonics, and magnetic fields. Icarus 226, 1447–1464.

Duggin, I.G., McCallum, S.A., Bell, S.D., 2008. Chromosome replication dynamics in the archaeon Sulfolobus acidocaldarius. Proc. Natl. Acad. Sci. USA 105, 16737–16742.

Edwards, K.J., Fisher, A.T., Wheat, C.G., 2012a. The deep subsurface biosphere in igneous ocean crust: Frontier habitats for microbiological exploration. Front. Microbiol. 3, 1–11.

Edwards, K.J., et al., 2012b. Design and deployment of borehole observatories and experiments during IODP Expedition 336, Mid-Atlantic Ridge flank at North Pond and the Expedition 336 Scientists. In: Edwards, K.J., Bach, W., Klaus, A. (Eds.). Proc. IODP, 336. doi:10.2204/iodp.proc.336.109.2012 Tokyo (Integrated Ocean Drilling Program Management International, Inc.

Elkins-Tanton, L.T., 2012. Magma oceans in the inner Solar System. Annu. Rev. Earth Planet. Sci. 40, 113–139.

Ellen, A.F., Rohulya, O.V., Fusetti, F., Wagner, M., Albers, S.V., Driessen, A.J., 2011. The sulfolobicin genes of Sulfolobus acidocaldarius encode novel antimicrobial proteins. J. Bacteriol. 193, 4380–4387.

Eun, Y.J., Ho, P.Y., Kim, M., LaRussa, S., Robert, L., Renner, L.D., Schmid, A., Garner, E., Amir, A., 2018. Archaeal cells share common size control with bacteria despite noisier growth and division. Nat. Microbiol. 3, 148–154.

Fajardo-Cavazos, P., Langenhorst, F., Melosh, H.J., Nicholson, W.L., 2009. Bacterial spores in granite survive hypervelocity launch by spallation: Implications for Lithopanspermia. Astrobiology 9, 647–657.

Fike, D.A., Cockell, C.S., Pearce, D., Lee, P., 2003. Heterotrophic microbial colonization of the interior of impact-shocked rocks from Haughton impact structure, Devon Island, Nunavut, Canadian High Arctic. Int. J. Astrobiol. 1, 311–323.

Fisk, M.R., Giovannoni, S.J., Thorseth, I.H., 1998. Alteration of oceanic volcanic glass: textural evidence of microbial activity. Science 281, 978–979.

Fontaneto, D., Iakovenko, N., De Smet, W.H, 2015. Diversity gradients of rotifer species richness in Antarctica. Hydrobiologia 750 (1), 235–248. doi:10.1007/s10750-015-2258-5.

Formisano, V., Atreya, S., Encrenaz, T., Ignatiev, N., Giuranna, M., 2004. Detection of methane in the atmosphere of Mars. Science 306, 1758–1761.

Forterre, P., 2016. Microbes from Hell, Translated by Teresa Lavender Fagan. The University of Chicago Press, pp. 363 pages. https://press.uchicago.edu/ucp/books/book/chicago/M/bo20952527.html.

Franck, S., 2000. Habitable zone for Earth-like planets in the Solar System. Planet. Space Sci. 48, 1099–1105.

Franck, S., von Bloh, W., Bounama, C., Steffen, M., Schonberner, D., and Schellnhuber, H.-J. (2002), Habitable zones and the number of Gaia's sisters. In: The Evolving Sun and Its Influence on Planetary Environments, Vol. 269, edited by B. Montesinos, A. Gimenez, and E. F. Guinan, Astronomical Society of the Pacific, San Francisco.

Frenzel, E., Legebeke, J., van Stralen, A., van Kranenburg, R., Kuipers, O.P., 2018. In vivo selection of sfGFP variants with improved and reliable functionality in industrially important thermophilic bacteria. Biotechnol. Biofuels 11, 8.

Friedmann, E.I., 1977. Microorganisms in Antarctic desert rocks from dry valleys and Dufek Massif. Antarctic J. United States 12, 26–30.

Friedmann, E.I., 1980. Endolithic microbial life in hot and cold deserts. Origins Life Evolution Biosphere 10, 223–235.

Friedmann, E.I., 1982. Endolithic microorganisms in the Antarctic cold desert. Science 215, 1045–1053.

Frisch, T., Thorsteinsson, R., 1978. Haughton astrobleme: A midCenozoic impact crater, Devon Island, Canadian Arctic Archipelago. arct. 31, 108–124.

Fritz, J., Artemieva, N., Greshake, A., 2005. Ejection of Martian meteorites. Meteoritics Planet. Sci. 40, 1393–1411.

Fröls, S., Ajon, M., Wagner, M., Teichmann, D., Zolghadr, B., Folea, M., Boekema, E.J., Driessen, A.J., Schleper, C., Albers, S.V., 2008. UV-inducible cellular aggregation of the hyperthermophilic archaeon Sulfolobus solfataricus is mediated by pili formation. Mol. Microbiol. 70, 938–952.

Furnes, H., et al., 2007. A vestige of Earth's oldest ophiolite. Science 315 (5819), 1704–1707. doi:10.1126/science.1139170.

Gabriel, W.N., McNuff, R., Patel, S.K., Gregory, T.R., Jeck, W.R., Jones, C.D., Goldstein, B., 2007. The Tardigrade Hypsibiusdujardini, a New Model for Studying the Evolution of Development. Dev. Biol. 312, 545–559.

Gabriel, W.N., Goldstein, B., 2007. Segmental expression of Pax3/7 and engrailed homologs in tardigrade development. Develop. Genes Evol. 217, 421–433.

Gaidos, E.J., Deschenes, B., Dundon, L., Fagan, K., Menviel-Hessler, L., Moskovitz, N., Workman, M., 2005. Beyond the principle of plentitude: a review of terrestrial planet habitability. Astrobiology 5, 100–126.

Gao, B., Gupta, R.S., 2007. Phylogenomic analysis of proteins that are distinctive of Archaea and its main subgroups and the origin of methanogenesis. BMC Genomics 8, 86.

Gendrin, A., Mangold, N., Bibring, J.P., Langevin, Y., Gondet, B., Poulet, F., Bonello, G., Quantin, C., Mustard, J., Arvidson, R., LeMouélic, S., 2005. Sulfates in Martian layered terrains: The OMEGA/Mars Express view. Science 307, 1587–1591.

Gilichinsky, D.A., Wilson, G.S., Friedmann, E.I., McKay, C.P., Sletten, R.S., Rivkina, E.M., Vishnivetskaya, T.A., Erokhina, L.G., Ivanushkina, N.E., Kochkina, G.A., Shcherbakova, V.A., Soina, V.S., Spirina, E.V., Vorobyova, E.A., Fyodorov-Davydov, D.G., Hallet, B., Ozerskaya, S.M., Sorokovikov, V.A., Laurinavichyus, K.S., Shatilovich, A.V., Chanton, P., Ostroumov, V.E., Tiedje, J.M., 2007. Microbial populations in Antarctic permafrost: biodiversity, state, age and implication for astrobiology. Astrobiology 7, 275–311.

Goldblatt, C., 2015. Habitability of waterworlds: runaway greenhouses, atmospheric expansion, and multiple climate states of pure water atmospheres. Astrobiology 15, 362–370.

Goldenfeld, N., Pace, N.R., 2013. Retrospective: Carl R. Woese (1928-2012). Science 339 (6120), 661.

Golubic, S., Friedmann, I., Schneider, J., 1981. The lithobiontic ecological niche, with special reference to microorganisms. J. Sed. Petrol. 51 (2), 475–478.

Golyshina, O.V., Pivovarova, T.A., Karavaiko, G.I., Kondratéva, T.F., Moore, E.R., Abraham, W.R., et al., 2000. Ferroplasma acidiphilum gen. nov., sp. nov., an acidophilic, autotrophic, ferrous-iron-oxidizing, cell-wall-lacking, mesophilic member of the Ferroplasmaceae fam. nov., comprising a distinct lineage of the Archaea. Int. J. Systematic Evolutionary Microbiol. 50 (3), 997–1006.

Gonzalez, G., 2005. Habitable zones in the Universe. Orig Life, Evol Biosph. 35, 555–606.

Goodwin, C.E., Berman, J., Downey, R.V., Hendry, K.R., 2017. Carnivorous sponges (Porifera: Demospongiae: Poecilosclerida: Cladorhizidae) from the drake passage (Southern Ocean) with a description of eight new species and a review of the family Cladorhizidae in the Southern Ocean. Inverteb. Syst. 31, 37–64. doi:10.1071/is16020.

Gribaldo, S., Brochier-Armanet, C., 2006. The origin and evolution of Archaea: a state of the art. Philosophical Transactions Royal Society London. Series B, Biological Sci. 361 (1470), 1007–1022.

Grieve, R.A.F., 1988. The Haughton impact structure: summary and synthesis of the results of the HISS project. Meteoritics 23, 249–254.

Grieve, R.A.F., Pesonen, L.J., 1992. The terrestrial impact cratering record. Tectonophysics 216, 1–30.

Griffiths, H.J., Anker, P., Linse, K., Maxwell, J., Post, A.L., Stevens, C., Tulaczyk, S., Smith, J.A., 2021. Breaking all the rules: the first recorded hard substrate sessile benthic community far beneath an Antarctic ice shelf. Front. Mar. Sci. 8, 642040. doi:10.3389/fmars.2021.642040.

Gudel, M., Dvorak, R., Erkaev, N., Kasting, J.F., Khodachenko, M.L., Lammer, H., Pilat-Lohinger, E., Rauer, H., Ribas, I., and Wood, B.E., 2014. Astrophysical conditions for planetary habitability, In: Protostars and Planets VI, edited by H. Beuther, R.S. Klessen, C.P. Dullemond, and T. Henning, University of Arizona Press, Tucson, pp. 883–906.

Guidetti, R., Altiero, T., Rebecchi, L., 2011. On dormancy strategies in tardigrades. J. Insect Physiol. 57 (5), 567–576.

Gupta, R.S., 1998. Protein phylogenies and signature sequences: A reappraisal of evolutionary relationships among Archaebacteria, Eubacteria, and Eukaryotes. Microbiol. Molecular Biol. Rev. 62 (4), 1435–1491.

Gupta, R.S., Griffiths, E., 2002. Critical issues in bacterial phylogeny. Theoretical Population Biol. 61 (4), 423–434. doi:10.1006/tpbi.2002.1589.

Gupta, R.S., Shami, A., 2011. Molecular signatures for the Crenarchaeota and the Thaumarchaeota. Antonie Van Leeuwenhoek 99 (2), 133–157.

Gupta, R.S., Naushad, S., Baker, S., 2015. Phylogenomic analyses and molecular signatures for the class Halobacteria and its two major clades: a proposal for division of the class Halobacteria into an emended order Halobacteriales and two new orders, Haloferacales ord. nov. and Natrialbales ord. nov., containing the novel families Haloferacaceae fam. nov. and Natrialbaceae fam. Nov. Int. J. Systematic Evolut. Microbiol. 65 (Pt 3), 1050–1069.

Halim, S.H., Crawford, I.A., Collins, G.S., et al., 2021. Assessing the survivability of biomarkers within terrestrial material impacting the lunar surface. Icarus 354. doi:10.1016/j.icarus.2020.114026.

Hara, F., Yamashiro, K., Nemoto, N., Ohta, Y., Yokobori, S., Yasunaga, T., et al., 2007. An actin homolog of the archaeon Thermoplasma acidophilum that retains the ancient characteristics of eukaryotic actin. J. Bacteriol. 189 (5), 2039–2045.

Harding, S. and Margulis, L. 2010. Water Gaia: 3.5 thousand million years of wetness on planet Earth. In: Gaia in Turmoil: Climate Change, Biodepletion, and Earth Ethics in an Age of Crisis, edited by E. Crist and H.B. Rinker, MIT Press, Cambridge, MA, pp 41–60.

Hardy, A., Caceres, C., Schreiber, M.R., Cieza, L., Alexander, R.D., Canovas, H., Williams, J.P., 2015. Probing the final stages of protoplanetary disk evolution with ALMA. Astron. Astrophys. 583. doi:10.1051/0004-6361/201526504.

Hashimoto, T., Kunieda, T., 2017. DNA protection protein, a novel mechanism of radiation tolerance: Lessons from tardigrades. Life (Basel) 7 (2). Epub 2017 Jun 15.

Hendrickson, R., Lundgren, P., Babu, G., Mohan, M., Urbaniak, C., Benardini, J.N., Venkateswaran, K., 2017. Comprehensive measurement of microbial burden in nutrient-deprived cleanrooms, 47th International Conference on Environmental Systems, 16-20 July 2017. Charleston, South Carolina, ICES-2017-177.

Hengherr, S., Heyer, A.G., Köhler, H.R., Schill, R.O., 2008. Trehalose and anhydrobiosis in tardigrades—evidence for divergence in responses to dehydration. FEBS J. 275 (2), 281–288. Epub 2007 Dec 6.

Hengherr, S., Worland, M.R., Reuner, A., Brümmer, F., Schill, R.O., 2009. High-temperature tolerance in anhydrobiotic tardigrades is limited by glass transition. Physiol.Biochem. Zool. 82 (6), 749–755.

Hibbert, R.H., Cole, M.J., Price, M.C., et al., 2017. The hypervelocity impact facility at the University of Kent: recent upgrades and specialised facilities. Procedia Engineering 204, 208–214.

Hickey, L.J., Johnson, K.R., Dawson, M.R., 1988. The stratigraphy, sedimentology, and fossils of the Haughton formation: a post-impact crater-fill, Devon Island, N.W.T., Canada. Meteoritics 23, 221–231.

Higuchi, R., Dollinger, G., Walsh, P.S., Griffith, R, 1992. Simultaneous amplification and detection of specific DNA sequences. Biotechnology 10, 413–417.

Hixon, W.G., Searcy, D.G., 1993. Cytoskeleton in the archaebacterium Thermoplasma acidophilum? Viscosity increase in soluble extracts. Bio Syst. 29 (2–3), 151–160.

Holland, P.M., Abramson, R.D., Watson, R., Gelfand, D.H., 1991. Detection of specific polymerase chain reaction product by utilizing the 5′—3′ exonuclease activity of *Thermus aquaticus* DNA polymerase. Proc. Natl. Acad. Sci. U.S.A. 88, 7276–7280.

Hoppert, M., Flies, C., Pohl, W., Gunzl, B., Schneider, J., 2004. Colonization strategies of lithobiontic microorganisms on carbonate rocks. Environ. Geol. 46, 421–428.

Horath, T., Bachofen, R., 2009. Molecular characterization of an endolithic microbial community in dolomite rock in the central Alps (Switzerland). Microb. Ecol. 58 (2), 290–306.

Horikawa, D.D., Sakashita, T., Katagiri, C., Watanabe, M., Kikawada, T., Nakahara, Y, et al., 2006. Radiation tolerance in the tardigrade Milnesiumtardigradum. Int. J. Radiat. Biol. 82, 843–848. doi:10.1080/09553000600972956.

Horikawa, D.D., Yamaguchi, A., Sakashita, T., Tanaka, D., Hamada, N., Yukuhiro, F., et al., 2012. Tolerance of anhydrobiotic eggs of the Tardigrade Ramazzottiusvarieornatus to extreme environments. Astrobiology 12, 283–289. doi:10.1089/ast.2011.0669.

Horikawa, D.D., Cumbers, J., Sakakibara, I., Rogoff, D., Leuko, S., Harnoto, R., et al., 2013. Analysis of DNA repair and protection in the Tardigrade *Ramazzottiusvarieornatus* and *Hypsibiusdujardini* after exposure to UVC radiation. PLoS One 8, e64793. doi:10.1371/journal.pone.0064793.

Horn, C., Paulmann, B., Kerlen, G., Junker, N., Huber, H., 1999. In vivo observation of cell division of anaerobic hyperthermophiles by using a high-intensity dark-field microscope. J. Bacteriol. 181, 5114–5118.

Horneck, G., Stoeffler, D., Eschweiler, U., Hornemann, U., 2001. Bacterial spores survive simulated meteorite impact. Icarus 149, 285–290.

Horneck, G., Miliekowsky, C., Melosh, H.J., Wilson, J.W., Cucinotta, F.A., Gladman, B., 2002. Viable transfer of microorganisms in the solar system and beyond. In: Horneck, G., Baumstark-Khan, C. (Eds.). Astrobiology: The Quest for the Conditions of Life, 2002. Springer, Berlin.

Horneck, G., Rettberg, P., 2007. Complete Course in Astrobiology. Wiley-VCH Verlag GmbH & Co. KGaA, Weinheim.

Horneck, G., Stoffler, D., Ott, S., Hornemann, U., Cockell, C.S., Moeller, R., Meyer, C., de Vera, J.P., Fritz, J., Schade, S., Artemieva, N.A., 2008. Microbial rock inhabitants survive hypervelocity impacts on Mars-like host planets: First phase of Lithopanspermia experimentally tested. Astrobiology 8, 17–44.

Howland, J.L., 2000. The Surprising Archaea: Discovering Another Domain of Life, New York, Oxford University Press. ISBN-13: 978-0195111835; ISBN-10: 0195111834.

Huber, J.A., Johnson, H.P., Butterfield, D.A., Baross, J.A., 2006. Microbial life in ridge flank crustal fluids. Environ, Microbiol 8, 88–99.

Iakovenko, N.S., Smykla, J., Convey, P., Kašparová, E., Kozeretska, I.A., Trokhymets, V., Dykyy, I., Plewka, M., Devetter, M., Duriš, Z., et al., 2015. Antarctic bdelloid rotifers: diversity, endemism and evolution. Hydrobiologia 761, 5–43.

Ingels, J., Aronson, R.B., Smith, C.R., 2018. The scientific response to Antarctic ice-shelf loss. Nat. Clim. Chang. 8, 848–851. doi:10.1038/s41558-018-0290-y.

Ingels, J., Aronson, R.B., Smith, C.R., Baco, A., Bik, H.M., Blake, J.A., et al., 2021. Antarctic ecosystem responses following ice-shelf collapse and iceberg calving: science review and future research. Wiley Interdiscip. Rev. 12, e682. doi:10.1002/wcc.682.

Jagadeesh, M.K., Roszkowska, M., Kaczmarek, Ł., 2018. Tardigrade indexing approach on exoplanets. Life Sci Space Res (Amst) 19, 13–16.

Javaux, E.J., 2006. Extreme life on Earth — past, present and possibly beyond. Res. Microbiol. 157 (1), 37–48.

Johnson, H.P., Pruis, M., 2003. Fluxes of fluid and heat from the oceanic crustal reservoir. Earth and Planet. Sci. Lett. 216, 565–574.

Johnson, B., Goldblatt, C., 2015. The nitrogen budget of Earth. Earth Sci. Rev. 148, 150–173.

Jones, E.G., Lineweaver, C.H., 2010a. Pressure-temperature phase diagram of the Earth. ASP Conference Series 430, 145–151.

Jones, E.G., Lineweaver, C.H., 2010b. To what extent does terrestrial life "follow the water"? Astrobiology 10, 349–361.

Jönsson, K.I., 2003. Causes and consequences of excess resistance in cryptobiotic metazoans. Physiol. Biochem. Zool. 76, 429–435. doi:10.1086/377743.

Jönsson, K.I., Harms-Ringdahl, M., Torudd, J., 2005. Radiation tolerance in the eutardigrade*Richtersiuscoronifer*. Int. J. Radiat. Biol. 81 (9), 649–656. doi:10.1080/09553000500368453.

Jönsson, K.I., 2007. Tardigrades as a potential model organism in space research. Astrobiology 7 (5), 757–766.

Jönsson, K.I., Rabbow, E., Schill, R.O., Harms-Ringdahl, M., Rettberg, P., 2008. Tardigrades survive exposure to space in low Earth orbit. Curr. Biol. 18, R729–R731. doi:10.1016/j.cub.2008.06.048.

Jönsson, K.I., Beltran-Pardo, E., Haghdoost, S., Wojcik, A., Bermúdez-Cruz, R.M., Bernal Villegas, J.E., et al., 2013. Tolerance to gamma-irradiation in eggs of the tardigrade *Richtersiuscoronifer* depends on stage of development. J. Limnol. 72 (1s), 73–79.

Jönsson, K.I., Hygum, T.L., Andersen, K.N., Clausen, L.K., Møbjerg, N., 2016. Tolerance to gamma radiation in the marine heterotardigrade, echiniscoidessigismundi. PLoSOne 11 (12), e0168884.

Jönsson, K.I., 2019. Radiation tolerance in tardigrades: current knowledge and potential applications in medicine. Cancers 11. doi:10.3390/cancers11091333.

Joseph, A, 2016. Investigating Seafloors and Oceans: From Mud Volcanoes to Giant Squid. Elsevier Science & Technology Books, New York, NY, pp. 586. ISBN: 978-0-12-809357-3.

Jungbluth, S.P., Grote, J., Lin, H.-T., Cowen, J.P., Rappé, M.S., 2013. Microbial diversity within basement fluids of the sediment-buried Juan de Fuca Ridge flank. ISME J. 7, 161–172.

Kalliomaa-Sanford, A.K., Rodriguez- Castañeda, F.A., McLeod, B.N., Latorre-Roselló, V., Smith, J.H., Reimann, J., Albers, S.V., Barillà, D., 2012. Chromosome segregation in Archaea mediated by a hybrid DNA partition machine. Proc. Natl. Acad. Sci. USA 109, 3754–3759.

Kandler, O., 1998. The early diversification of life and the origin of the three domains: A proposal. In: Wiegel, J, Adams, WW (Eds.), *Thermophiles: The keys to molecular evolution and the origin of life*? Taylor and Francis, Athens, pp. 19–31.

Karl, D.M., Bird, D.F., Björkman, K., Houlihan, T., Shackelfor, R., Tupas, L., 1999. Microorganisms in the accreted ice of Lake Vostok, Antarctica. Science 286, 2144–2147.

Kasting, J.F., 1988. Runaway and moist greenhouse atmospheres and the evolution of Earth and Venus. Icarus 74, 472–494.

Kasting, J.F., 2012. How to Find a Habitable Planet. Princeton University Press, Princeton, NJ.

Kelly, S., Wickstead, B., Gull, K., 2010. Archaeal phylogenomics provides evidence in support of a methanogenic origin of the Archaea and a thaumarchaeal origin for the eukaryotes. Proc. Royal Society B: Biological Sci. 278 (1708), 1009–1018.

Kessler, M.A., Werner, B.T., 2003. Self-organization of sorted patterned ground. Science 299, 380–383.

Kokubo, E., Ida, S., 2012. Dynamics and accretion of planetesimals. Progress Theoretical Experimental Phys. 2012 (1), 01A308.

Kondo, K., Kubo, T., Kunieda, T., 2015. Suggested involvement of PP1/PP2A activity and de novo gene expression in anhydrobiotic survival in a tardigrade, hypsibiusdujardini, by chemical genetic approach. PLoS One 10 (12), e0144803. Epub 2015 Dec 21.

Kralik, P., Ricchi, M., 2017. A basic guide to real time PCR in microbial diagnostics: definitions, parameters, and everything. Front. Microbiol. https://doi.org/10.3389/fmicb.2017.00108.

Krieg, N, 2005. Bergey's Manual of Systematic Bacteriology. Springer, pp. 21–26 ISBN 978-0-387-24143-2.

Kubista, M., Andrade, J.M., Bengtsson, M., Forootan, A., Jonak, J., Lind, K., et al., 2006. The real-time polymerase chain reaction. Mol. Aspects Med. 27, 95–125. doi:10.1016/j.mam.2005.12.007.

Kulikov, Y.N., Lammer, H., Lichtenegger, H.I.M., Penz, T., Breuer, D., Spohn, T., Lundin, R., Biernat, H.K., 2007. A comparative study of the influence of the active young Sun on the early atmospheres of Earth, Venus, and Mars. Space Sci Rev. 129, 207–243.

Kurokawa, H., Sato, M., Ushioda, M., Matsuyama, T., Moriwaki, R., Dohm, J.M., Usui, T., 2014. Evolution of water reservoirs on Mars: constraints from hydrogen isotopes in martian meteorites. Earth Planet. Sci. Lett. 394, 179–185.

Kurosawa, K., Genda, H., Hyodo, R., et al., 2019. Assessment of the probability of microbial contamination for sample return from Martian moons II: the fate of microbes on Martian moons. Life Sci. Space Res. 23, 85–100.

Kushner, D.J., Kamekura, M., 1988. Physiology of halophilic eubacteria. In: Rodríguez-Valera, F (Ed.), Halophilic bacteria. CRC, Boca Raton.

La Duc, M.T., Nicholson, W., Kern, R., Venkateswaran, K, 2003. Microbial characterization of the Mars Odyssey spacecraft and its encapsulation facility. Environ. Microbiol. 5 (10), 977–985.

Duc, La, T., M., Kern, R.G., Venkateswaran, K, 2004. Microbial monitoring of spacecraft and associated environments. Microb. Ecol. 47, 150–158.

La Duc, M.T., Dekas, A., Osman, S., Moissl, C., Newcombe, D., Venkateswaran, K, 2007. Isolation and characterization of bacteria capable of tolerating the extreme conditions of clean room environments. Appl. Environ. Microbiol. 73 (8), 2600–2611.

Lammer, H., Bredehöft, J.H., Coustenis, A., Khodachenko, M.L., Kaltenegger, L., Grasset, O., Prieur, D., Raulin, F., Ehrenfreund, P., Yamauchi, M., Wahlund, J.-E., Griebmeier, J.-M., Stangl, G., Cockell, C.S., Kulikov, Y.N., Grenfell, J.L., Rauer, H., 2009. What makes a planet habitable? Astronomy Astrophysics Rev. 17, 181–249.

Langseth, G., Yon, P, 1992. Heat and fluid flux through sediment on the western flank of the mid-atlantic ridge: a hydrogeological study of North Pond. Geophys. Res. Lett. 19, 517–520.

Lanoil, B., Skidmore, M., Priscu, J.C., Han, S., Foo, W., Vogel, S.W., Tulaczyk, S., Engelhardt, H., 2009. Bacteria beneath the West Antarctic ice sheet. Environ. Microbiol 11, 609–615.

Lauro, S.E., Pettinelli, E., Caprarelli, G., et al., 2021. Multiple subglacial water bodies below the south pole of Mars unveiled by new MARSIS data. Nat Astron 5, 63–70. https://doi.org/10.1038/s41550-020-1200-6.

Leger, A., Selsis, F., Sotin, C., Guillot, T., Despois, D., Lammer, H., Ollivier, M., and Brachet, F., 2003. A new family of planets? ''Oceanplanets.'' In: Proceedings of the Conference on Towards Other Earths: DARWIN/TPF and the Search for Extrasolar Terrestrial Planets, edited by M. Fridlund and T. Henning, compiled by H. Lacoste, ESA SP-539, ESA Publications Division, Noordwijk, pp 253–259.

Lenton, T.M., 1998. Gaia and natural selection. Nature 394, 439–447.

Lenton, T.M., von Bloh, W., 2001. Biotic feedback extends the life span of the biosphere. Geophys. Res. Lett. 28, 1715–1718.

Li, Z., Kinosita, Y., Rodriguez-Franco, M., Nußbaum, P., Braun, F., Delpech, F., Quax, T.E.F., Albers, S.V., 2019. Positioning of the motility machinery in Halophilic Archaea. MBio 10, e00377 -19.

Lindås, A.C., Karlsson, E.A., Lindgren, M.T., Ettema, T.J., Bernander, R., 2008. A unique cell division machinery in the Archaea. Proc. Natl. Acad. Sci. USA 105, 18942–18946.

Lindås, A.C., Bernander, R., 2013. The cell cycle of archaea. Nat. Rev. Microbiol. 11, 627–638.

Lipscomb, G.L., Hahn, E.M., Crowley, A.T., Adams, M.W.W., 2017. Reverse gyrase is essential for microbial growth at 95°C. Extremophiles 21 (3), 603–608. doi:10.1007/s00792-017-0929-z.

Liu, J., Gao, R., Li, C., Ni, J., Yang, Z., Zhang, Q., Chen, H., Shen, Y., 2017. Functional assignment of multiple ESCRT-III homologs in cell division and budding in Sulfolobus islandicus. Mol. Microbiol. 105, 540–553.

López-García, P., López-López, A., Moreira, D., Rodríguez-Valera, F, 2001. Diversity of free-living prokaryotes from a deep-sea site at the Antarctic Polar Front. FEMS Microbiol. Ecol. 36 (2–3), 193–202.

Lovelock, J.E., Margulis, L., 1974. Atmospheric homeostasis by and for the biosphere: the Gaia hypothesis. Tellus 26, 1–2.

Luger, R., Barnes, R., 2015. Extreme water loss and abiotic O_2 buildup on planets throughout the habitable zones of M dwarfs. Astrobiology 15, 1–26.

Lysnes, K., et al., 2004. Microbial community diversity in seafloor basalt from the Arctic spreading ridges. FEMS Microbiol. Ecol. 50, 213–230.

Macelroy, R.D., 1974. Some comments on the evolution of extremophiles. Biosystems 6, 74–75.

Madigan, M.T., Martino, J.M., 2006. Brock Biology of Microorganisms, 11th ed. Pearson Prentice Hall, Upper Saddle River, NJ, pp. 136 ISBN 978-0-13-196893-6.

Magrum, L.J., Luehrsen, K.R., Woese, C.R., 1978. Are extreme halophiles actually "bacteria"? J. Molecular Evolution 11 (1), 1–8.

Mason, O.U., et al., 2009. Prokaryotic diversity, distribution, and insights into their role in biogeochemical cycling in marine basalts. ISME J. 3, 231–242.

Mastrapa, R.M.E., Glanzberg, H., Head, J.N., Melosh, H.J., Nicholson, W.L., 2001. Survival of bacteria exposed to extreme acceleration: Implications for panspermia. Earth Planet. Sci. Lett. 189, 1–8.

May, R.M., Maria, M., Guimard, J., 1964. Action différentielle des rayons x et ultraviolets sur le tardigrade Macrobiotusareolatus, a l'étatactif et desséché. Bull Biol FrBelg 98, 18.

McKay, C.P., 2016. Titan as the abode of life. Life 6, 8.

Melosh, H.J., 1988. The rocky road to panspermia. Nature 332, 687–688.

Merino, N., Aronson, H.S., Bojanova, D.P., Feyhl-Buska, J., Wong, M.L., Zhang, S., andGiovannelli, D., 2019. Living at the extremes: extremophiles and the limits of life in a planetary context. Front. Microbiol. 10, 780. doi:10.3389/fmicb.2019.00780.

Meyer, J.L., Jaekel, U., Tully, B.J., Glazer, B.T., Wheat, C.G., Lin, H-T., Hsieh, C-C., Cowen, J.P., Hulme, S.M., Girguis, P.R., Huber, J.A., 2016. A distinct and active bacterial community in cold oxygenated fluids circulating beneath the western flank of the Mid-Atlantic ridge. Sci. Rep. 6. doi:10.1038/srep22541 22541.

Møbjerg, N., Halberg, K.A., Jørgensen, A., Persson, D., Bjørn, M., Ramløv, H., et al., 2011. Survival in extreme environments—on the current knowledge of adaptations in tardigrades. Acta Physiol (Oxf) 202 (3), 409–420.

Moissl-Eichinger, C., Pausan, M., Taffner, J., Berg, G., Bang, C., Schmitz, R.A., 2018. Archaea are interactive components of complex microbiomes. Trends Microbiol. 26 (1), 70–85. doi:10.1016/j.tim.2017.07.004.

Mole, R.H., 1984. The LD50 for uniform low LET irradiation of man. Br. J. Radiol. 57 (677), 355–369. doi:10.1259/0007-1285-57-677-355.

Morbidelli, A., Lunine, J.I., O'Brien, D.P., Raymond, S.N., Walsh, K.J., 2012. Building terrestrial planets. Annu. Rev. Earth Planet. Sci. 40, 251–275.

Morell, V., 1997. Microbiology's scarred revolutionary. Science 276 (5313), 699–702.

Moter, A., Gobel, U.B., 2000. Fluorescence in situ hybridization (FISH) for direct visualization of microorganisms. J. Microbiol. Methods 41, 85–112.

Nair, P., 2012. Woese and Fox: Life, rearranged. Proc. National Acad. Sci. 109 (4), 1019–1021.

Nealson, K.H., 1999. Post-Viking microbiology: new approaches, new data, new insights. Origins Life Evolution Biosphere 29 (1), 73–93.

Newsham, K.K., Maslen, N.R., McInnes, S.J., 2006. Survival of Antarctic soil metazoans at −80°C for six years. CryoLetters 27, 291–294.

Nicolaus, B., Kambourova, M., Oner, E.T., 2010. Exopolysaccharides from extremophiles: from fundamentals to biotechnology. Environ. Technol. 31, 1145–1158.

Nichols, C.M., Bowman, J.P., Guezennec, J., 2005a. Effects of incubation temperature on growth and production of exopolysaccharides by an Antarctic sea ice bacterium grown in batch culture. Appl. Environ. Microbiol. 71 (7), 3519–3523. doi:10.1128/AEM.71.7.3519-3523.2005.

Nichols, C.A., Guezennec, J., Bowman, J.P., 2005b. Bacterial exopolysaccharides from extreme marine environments with special consideration of the Southern Ocean, sea ice, and deep-sea hydrothermal vents: a review. Mar. Biotechnol. (NY) 7 (4), 253–271. doi:10.1007/s10126-004-5118-2.

Nicholson, A.E., Wilkinson, D.M., Williams, H.T.P., Lenton, T.M., 2018. Gaian bottlenecks and planetary habitability maintained by evolving model biospheres: the ExoGaia model. Monthly Notices Royal Astronomical Soc. 477 (1), 727–740. https://doi.org/10.1093/mnras/sty658.

Nilsson, C.E.J., Jönsson, K.I., Pallon, J., 2010. Tolerance to proton irradiation in the eutardigrade Richtersiuscoronifer—a nuclear microprobe study. Int. J. Radiat. Biol. 86, 420–427. doi:10.3109/09553000903568001.

Nisbet, E.G., 2002. Fermor lecture: the influence of life on the face of the Earth: garnets and moving continents. Geol. Soc. Spec. Publ. 199, 275–307.

Nisbet, E.G., Fowler, C., Nisbet, R.E.R., 2012. The regulation of the air: a hypothesis. Solid Earth 3, 87–96.

Nisbet, E.G. and Fowler, C.M.R., 2014. The early history of life. In: Treatise on Geochemistry, Vol. 10, 2nd ed., edited by H. Palme and H. O'Neill, Elsevier, Amsterdam, pp 1–42.

Nocker, A., Camper, A.K., 2006. Selective removal of DNA from dead cells of mixed bacterial communities by use of ethidium monoazide. Appl. Environ. Microbiol. 72 (3), 1997–2004.

Nocker, A., Sossa-Fernandez, P., Burr, M.D., Camper, A.K., 2007. Use of propidium monoazide for live/dead distinction in microbial ecology. Appl. Environ. Microbiol. 73 (16), 5111–5117.

Nutman, A.P., Friend, C.R.L., Horie, K., Hidaka, H., 2007. The Itsaq Gneiss complex of Southern West Greenland and the construction of Eoarchean crust at convergent plate boundaries. Devel. Precambrian Geol. 15, 187–218. doi:10.1016/S0166-2635(07)15033-7.

Nutman, A.P., Friend, C.R.L., 2009. New 1:20,000 scale geological maps, synthesis and history of investigation of the Isua supracrustal belt and adjacent orthogneisses, southern West Greenland: A glimpse of Eoarchaean crust formation and orogeny. Precambrian Res. 172 (3), 189–211. doi:10.1016/j.precamres.2009.03.017.

Nutman, A.P., et al., 2016. Rapid emergence of life shown by discovery of 3,700-million-year-old microbial structures. Nature 537 (7621), 535–538. doi:10.1038/nature19355.

Nyquist, L.E., Bogard, D.D., Shih, C.Y., Greshake, A., Stöffler, D., Eugster, O., 2001. Ages and geologic histories of Martian meteorites. Space Sci. Rev. 96, 105–164.

O'Callaghan, J., 2021. Hardy water bears survive bullet impacts—up to a point, https://www.sciencemag.org/news/2021/05/hardy-water-bears-survive-bullet-impacts-point?utm_campaign=news_daily_2021-05-18&et_rid=730936342&et_cid=3778040. Accessed on September 19, 2022.

Ohtomo, Y., Kakegawa, T., Ishida, A., Nagase, T., Rosing, M.T., 2014. Evidence for biogenic graphite in early Archaean Isua metasedimentary rocks. Nat. Geosci. 7, 25–28.

Olofsson, H., Ripa, J., Jonzén, N., 2009. Bet-hedging as an evolutionary game: the trade-off between egg size and number. Proc. Royal Soc. B 276 (1669). https://doi.org/10.1098/rspb.2009.0500.

Onstott, T.C., McGown, D., Kessler, J., Lollar, B.S., Lehmann, K.K., Clifford, S.M, 2006. Martian CH4: sources, flux, and detection. Astrobiology 6, 377–395.

Orcutt, B.N., et al., 2013. Oxygen consumption rates in subseafloor basaltic crust derived from a reaction transport model. Nature Commun 4, 2539.

Oren, A., Ventosa, A., Gutierrez, M.C., Kamekura, M., 1999. Haloarcula quadrata sp. nov., a square, motile archaeon isolated from a brine pool in Sinai (Egypt). Int. J. Syst. Bacteriol. 49 (3), 1149–1155. doi:10.1099/00207713-49-3-1149.

Oren, A., 2010. Concepts about phylogeny of microorganisms–an historical perspective. In: Oren, Aharon, Thane Papke, R. (Eds.), Molecular Phylogeny of Microorganisms. Caister Academic Press, Norfolk, pp. 1–22. ISBN 9781904455677.

Orosei, R., Lauro, S.E., Pettinelli, E., Cicchetti, A., Coradini, M., Cosciotti, B., Di Paolo, F., Flamini, E., Mattei, E., Pajola, M., Soldovieri, F., Cartacci, M., Cassenti, F., Frigeri, A., Giuppi, S., Martufi, R., Masdea, A., Mitri, G., Nenna, C., Noschese, R., Restano, M., Seu, R., 2018. Radar evidence of subglacial liquid water on Mars. Science 361 (6401), 490–493. doi:10.1126/science.aar7268.

Pace, N.R., 2009. Problems with "Procaryote". J. Bacteriol. 191 (7), 2008–2010.

Pace, N.R., Sapp, J., Goldenfeld, N., 2012. Phylogeny and beyond: scientific, historical, and conceptual significance of the first tree of life. Proc. Nat. Acad. Sci. 109 (4), 1011–1018.

Parkes, R.J., Webster, G., Cragg, B.A., Weightman, A.J., Newberry, C.J., Ferdelman, T.G., Kallmeyer, J., Jørgensen, B.B., Aiello, I.W., Fry, J.C., 2005. Deep sub-seafloor prokaryotes stimulated at interfaces over geologic time. Nature 436, 390–394.

Parnell, J., Lee, P., Cockell, C.S., Osinski, G.R., 2004. Microbial colonization in impact-generated hydrothermal sulphate deposits, Haughton impact structure, and implications for sulphates on Mars. Int. J. Astrobiol. 3, 247–256.

Pasini, D.L.S., Price, M.C., Burchell, M.J., et al., 2014. Survival of the tardigrade Hypsibius dujardini during hypervelocity impact events up to 3.23 km s^{-1} [abstract 1780], 45th Lunar and Planetary Science Conference, Lunar and Planetary Institute. Houston.

Persson, D., Halberg, K.A., Jørgenson, A., et al., 2011. Extreme stress tolerance in tardigrades: surviving space conditions in low earth orbit. J Zool Syst Evol Res 49 (S1), 90–97.

Pikuta, E.V., Hoover, R.B., Tang, J., 2007. Microbial extremophiles at the limits of life. Critical Rev. Microbiol. 33 (3), 183–209.

Poli, A., Anzelmo, G., Nicolaus, B., 2010. Bacterial exopolysaccharides from extreme marine habitats: production, characterization and biological activities. Mar Drugs 8 (6), 1779–1802. doi:10.3390/md8061779.

Pope, E.C., Bird, D.K., Rosing, M.T., 2012. Isotope composition and volume of Earth's early oceans. Proc. Natl. Acad. Sci. USA. 109, 4371–4376.

Popławski, A., Bernander, R., 1997. Nucleoid structure and distribution in thermophilic Archaea. J. Bacteriol. 179, 7625–7630.

Prangishvili, D., Holz, I., Stieger, E., Nickell, S., Kristjansson, J.K., Zillig, W., 2000. Sulfolobicins, specific proteinaceous toxins produced by strains of the extremely thermophilic archaeal genus Sulfolobus. J. Bacteriol. 182, 2985–2988.

Priscu, J.C., Fritsen, C.H., Adams, E.E., Giovannoni, S.J., Paerl, H.W., McKay, C.P., Doran, P.T., Gordon, D.A., Lanoil, B.D., Pinckney, J.L., 1998. Perennial Antarctic lake ice: an oasis for life in a polar desert. Science 280, 2095–2098.

Priscu, J.C., Adams, E.E., Lyons, W.B., Voytek, M.A., Mogk, D.W., Brown, R.L., McKay, C.P., Takacs, C.D., Welch, K.A., Wolf, C.F., Kirshtein, J.D., 1999. Geomicrobiology of subglacial ice above Lake Vostok, Antarctica. Science 286, 2141–2144.

Pulschen, et al., 2020. Live imaging of a hyperthermophilic archaeon reveals distinct roles for two ESCRT-III homologs in ensuring a robust and symmetric division. Curr. Biol. 30 (14), 2852–2859.

Rachel, R., Wyschkony, I., Riehl, S., Huber, H., 2002. The ultrastructure of Ignicoccus: evidence for a novel outer membrane and for intracellular vesicle budding in an archaeon. Archaea 1 (1), 9–18.

Ramírez-Salazar, A., et al., 2021. Tectonics of the Isua supracrustal belt 1: P-T-X-d constraints of a poly-metamorphic terrane. Tectonics 40 (3), e2020TC006516. doi:10.1029/2020TC006516.

Rampelotto, P.H., 2009. Are we descendants of extraterrestrials? J. Cosmol. 1, 86–88.

Rampelotto, P.H., 2010. Resistance of microorganisms to extreme environmental conditions and its contribution to astrobiology. Sustainability 2 (6), 1602–1623. https://doi.org/10.3390/su2061602.

Ramsley, K.R., Head III, J.W., 2013. Mars impact ejecta in the regolith of Phobos: bulk concentration and distribution. Planet. Space Sci. 87, 115–129.

Raulin-Cerceau, F., Maurel, M.C., Schneider, J., 1998. From Panspermia to Bioastronomy: The evolution of the hypothesis of universal life. Orig. Life. Evol. Biosph. 28, 597–612.

Raymond, S.N., Quinn, T., Lunine, J.I., 2004. Making other Earths: dynamical simulations of terrestrial planet formation and water delivery. Icarus 168, 1–17.

Raymond, S.N., O'Brien, D.P., Morbidelli, A., Kaib, N.A., 2009. Building the terrestrial planets: constrained accretion in the inner Solar System. Icarus 203, 644–662.

Rebecchi, L., Altiero, T., Guidetti, R, et al., 2009. Tardigrade resistance to space effects: first results of experiments on the LIFE-TARSE mission on FOTON-M3. Astrobiology 9, 581–591.

Rettberg, P., Eschweiler, U., Strauch, K., Reitz, G., Horneck, G., Wänke, H., Brack, A., Barbier, B., 2002. Survival of microorganisms in space protected by meteorite material: Results of the experiment "EXOBIOLOGIE" of the PERSEUS mission. Adv. Space Res 30, 1539–1545.

Risa, G.T., Hurtig, F., Bray, S., Hafner, A.E., Harker-Kirschneck, L., Faull, P., Davis, C., Papatziamou, D., Mutavchiev, D.R., Fan, C., et al., 2019. Proteasome-mediated protein degradation resets the cell division cycle and triggers ESCRT-III-mediated cytokinesis in an archaeon. bioRxiv. doi:10.1101/774273.

Robinson, N.P., Dionne, I., Lundgren, M., Marsh, V.L., Bernander, R., Bell, S.D., 2004. Identification of two origins of replication in the single chromosome of the archaeon Sulfolobus solfataricus. Cell 116, 25–38.

Rogers, A.D., Tyler, P.A., Connelly, D.P., Copley, J.T., James, R., Larter, R.D., Linse, K., Mills, R.A., Garabato, A.N., Pancost, R.D., Pearce, D.A., Polunin, N.V.C., German, C.R., Shank, T., Boersch-Supan, P.H., Alker, B.J., Aquilina, A., Bennett, S.A., Clarke, A., Dinley, R.J.J., Graham, A.G.C., Green, D.R.H., Hawkes, J.A., Hepburn, L., Hilario, A., Huvenne, V.A.I., Marsh, L., Ramirez-Llodra, E., Reid, W.D.K., Roterman, C.N., Sweeting, C.J., Thatje, S., Zwirglmaier, K., 2012. The discovery of new deep-sea hydrothermal vent communities in the Southern Ocean and implications for biogeography. PLoS Biol. 10 (1), e1001234. doi:10.1371/journal.pbio.1001234.

Rohde, R.A., Price, B.P., Bay, R.C., Bramal, N.E., 2008. In situ microbial metabolism as a cause of gas anomalies in ice, Proc. Natl. Acad. Sci. USA, 105, 8667–8672.

Rokas, A., Holland, P.W.H., 2000. Rare genomic changes as a tool for phylogenetics. Trends Ecol. Evolution 15 (11), 454–459. doi:10.1016/S0169-5347(00)01967-4.

Rothschild, L., Mancinelli, R.L., 2001. Life in extreme environments. Nature 409, 1092–1100.

Saiki, R.K., Scharf, S., Faloona, F., Mullis, K.B., Horn, G.T., Erlich, H.A., et al., 1985. Enzymatic amplification of beta-globin genomic sequences and restriction site analysis for diagnosis of sickle cell anemia. Science 230, 1350–1354.

Saiz-Jimenez, C., Garcia-Rowe, J., Garcia del Cura, M.A., Ortega-Calvo, J.J., Roekens, E., van Grieken, R., 1990. Endolithic cyanobacteria in Maastricht Limestone. Sci. Total Environ. 94, 209–220.

Samson, R.Y., Obita, T., Freund, S.M., Williams, R.L., Bell, S.D., 2008. A role for the ESCRT system in cell division in archaea. Science 322, 1710–1713.

Santelli, C.M., Edgcomb, V.P., Bach, W., Edwards, K.J., 2009. The diversity and abundance of bacteria inhabiting seafloor lavas positively correlate with rock alteration. Environ. Microbiol. 11, 86–98.

Sapp, J.A., 2009. The New Foundations of Evolution: On the Tree of Life. Oxford University Press, New York, NY ISBN 978-0-199-73438-2.

Schill, R.O., Mali, B., Dandekar, T., Schnölzer, M., Reuter, D., Frohme, M., 2009. Molecular mechanisms of tolerance in tardigrades: new perspectives for preservation and stabilization of biological material. Biotechnol. Adv. 27 (4), 348–352.

Schill, R.O., and Hengherr, S., 2018. Environmental adaptations: desiccation tolerance. In: Water Bears: The Biology of Tardigrades, edited by R.O. Schill, Springer International Publishing, Cham, pp 273–293.

Schlesinger, W.H., et al., 2003. Community composition and photosynthesis by photoautotrophs under quartz pebbles, southern Mojave Desert. Ecology 84, 3222–3231.

Schmidt-Schierhorn, F., Kaul, N., Stephan, S., Villinger, H., 2012. Geophysical site survey results from North Pond (Mid-Atlantic Ridge). In: Edwards, K.J., Bach, W., Klaus, A. (Eds.), the Expedition 336 Scientists, *Proc. IODP*, 336. Tokyo (Integrated Ocean Drilling Program Management International, Inc. doi:10.2204/iodp.proc.336.107.2012.

Schneider, S.H., Boston, P.J., 1991. Scientists on Gaia. MIT Press, Cambridge, MA.

Schneider, S.H., 2004. Scientists Debate Gaia: The Next Century. MIT Press, Cambridge, MA.

Schulze-Makuch, D., Irwin, L.N., Lipps, J.H., LeMone, D., Dohm, J.M., Fairén, A.G., 2005. Scenarios for the evolution of life on Mars. J. Geophys. Res. 110, E12S23.

Schulze-Makuch, D., Fairén, A.G., Davila, A.F., 2008. The case for life on Mars. Int. J. Astrobiology 7, 117–141.

Schulze-Makuch, D., Guinan, E., 2016. Another Earth 2.0? Not so fast. Astrobiology 16 (11), 817–821.

Seki, K., Toyoshima, M., 1998. Preserving tardigrades under pressure. Nature 395 (6705), 853–854.

Selan, L., Berlutti, F., Passariello, C., Thaller, M.C., Renzini, G., 1992. Reliability of a bioluminescence ATP assay for detection of bacteria. J. Clin. Microbiol. 30 (7), 1739–1742.

Shain, D.H., Halldórsdóttir, K., Pálsson, F., Aðalgeirsdóttir, G., Gunnarsson, A., Jónsson, þ., Lang, S.A., Pálsson, H.S., Steinþórsson, S., Arnason, E, 2016. Colonization of maritime glacier ice by bdelloid Rotifera. Mol. Phylogenet. Evol. 98, 280–287.

Shields, A.L., Barnes, R., Agol, E., Charnay, B., Bitz, C., Meadows, V.S., 2016. The effect of orbital configuration on the possible climates and habitability of Kepler-62f. Astrobiology. doi:10.1089/ast.2015.1353.

Shmakova, L., Malavin, S., Iakovenko, N., Shain, D., Plewka, M., Rivkina, E., 2021. A living bdelloid rotifer from 24,000-year-old Arctic permafrost. Current Biol. 31 (11), R712–R713. https://doi.org/10.1016/j.cub.2021.04.077.

Shuster, D.L., Weiss, B.P., 2005. Martian surface paleotemperatures from thermochronology of meteorites. Science 309, 594–597.

Smith, M.C., Bowman, J.P., Scott, F.J., Line, M.A., 2000. Sublithic bacteria associated with Antarctic quartz stones. Antarctic Sci. 12, 177–184.

Smith, F.W., Goldstein, B., 2017. Segmentation in Tardigrada and diversification of segmental patterns in Panarthropoda. Arthropod Structure & Development 46, 328–340.

Sowers, T., 2001. The N_2O record spanning the penultimate deglaciation from the Vostok ice core. J. Geophys. Res. 106, 31903–31914.

Stanier, R.Y., Van Niel, C.B., 1962. The concept of a bacterium. Archiv für Mikrobiologie 42, 17–35.

Stanley, P.E., 1989. A review of bioluminescent ATP techniques in rapid microbiology. J. Bioluminescence Chemiluminescence 4 (1), 375–380.

Stetter, K.O., 1996. Hyperthermophiles in the history of life. Ciba Foundation Symp. 202, 1–10 discussion 11–8.

Stetter, K.O., 2006. Hyperthermophiles in the history of life. Philosophical Trans. Royal Society B: Biological Sci. 361 (1474), 1837–1843.

Stetter, K.O., Thomm, M., Winter, J., Wildgruber, G., Huber, H., Zillig, W., Janecovic, D., König, H., Palm, P., Wunderl, S., 1981. *Methanothermus fervidus*, sp. nov., a novel extremely thermophilic methanogen isolated from an Icelandic hot spring. Zbl. Bakt. Hyg., I. Abt. Orig. C2, 166–178.

Stetter, K.O., 1982. Ultrathin mycelia-forming organisms from submarine volcanic areas having an optimum growth temperature of 105°C. Nature 300, 258–260. doi:10.1038/300258a0.

Stetter, K.O., 1992. Life at the upper temperature border. In: Tran Thanh Van, J., Tran Thanh Van, K., Mounolou, J.C., Schneider, J., McKay, C. (Eds.), Frontiers of Life. Frontieres, Gif-sur-Yvette, pp. 195–219.

Stetter, K.O., 2006. Hyperthermophiles in the history of life. Philos Trans R Soc Lond B Biol Sci. 361 (1474), 1837–1843. doi:10.1098/rstb.2006.1907.

Stoeckenius, W., 1981. Walsby's square bacterium: fine structure of an orthogonal prokaryote. J. Bacteriol. 148 (1), 352–360.

Stoffler, D., Horneck, G., Ott, S., Hornemann, U., Cockell, C.S., Moeller, R., Meyer, C., de Vera, J.P., Fritz, J., Artemieva, N.A., 2007. Experimental evidence for the potential impact ejection of viable microorganisms from Mars and Mars-like planets. Icarus 186, 585–588.

Takai, K., Nakamura, K., Toki, T., Tsunogai, U., Miyazaki, M., Miyazaki, J., Hirayama, H., Nakagawa, S., Nunoura, T., Horikoshi, K., 2008. Cell proliferation at 122°C and isotopically heavy CH_4 production by a hyperthermophilic methanogen under high-pressure cultivation, Proceedings of the National Academy of Sciences of the United States of America, 105, 10949–10954.

Takemata, N., Samson, R.Y., Bell, S.D., 2019. Physical and functional compartmentalization of archaeal chromosomes. Cell 179, 165–179.e18.

Tasker, E., Tan, J., Heng, K., Kane, S., Spiegel, D., Brasser, R., Casey, A., Desch, S., Dorn, C., Hernlund, J., Houser, C., Laneuville, M., Lasbleis, M., Libert, A.-S., Noack, L., Unterborn, C., Wicks, J., 2017. The language of exoplanet ranking metrics needs to change. Nat. Astron. 1, 0042.

Tenlen, J.R., McCaskill, S., Goldstein, B., 2013. RNA interference can be used to disrupt gene function in tardigrades. Develop. Genes Evolution 223, 171–181.

Thomas, N.A., Bardy, S.L., Jarrell, K.F., 2001. The archaeal flagellum: a different kind of prokaryotic motility structure. FEMS Microbiol. Rev. 25 (2), 147–174.

Thomas, D.N., Dieckmann, G.S., 2002. Antarctic sea ice – a habitat for extremophiles. Science 295, 641–644.

Thomas, D.N., 2005. Photosynthetic microbes in freezing deserts. Trends Microbiol. 13 (3), 87–88.

Tian, F., 2015. Atmospheric escape from Solar System terrestrial planets and exoplanets. Annu. Rev. Earth Planet Sci. 43, 459–476.

Traspas, A., Burchell, M.J., 2021. Tardigrade survival limits in high-speed impacts —Implications for panspermia and collection of samples from plumes emitted by ice worlds. Astrobiology. doi:10.1089/ast.2020.2405.

Trent, J.D., Kagawa, H.K., Yaoi, T., Olle, E., Zaluzec, N.J., 1997. Chaperonin filaments: the archaeal cytoskeleton? Proc. Nat. Acad. Sci. United States of America 94 (10), 5383–5388.

Tsai, C.L., Tripp, P., Sivabalasarma, S., Zhang, C., Rodriguez-Franco, M., Wipfler, R.L., Chaudhury, P., Banerjee, A., Beeby, M., Whitaker, R.J., et al., 2020. The structure of the periplasmic FlaG-FlaF complex and its essential role for archaellar swimming motility. Nat. Microbiol. 5, 216–225.

Tung, H.C., Bramall, N.E., Price, P.B., 2005. Microbial origin of excess methane in glacial ice and implications for life on Mars. Proc. Natl. Acad. Sci. USA 102, 18292–18296.

Turnbaugh, P.J., Ley, R.E., Hamady, M., Fraser-Liggett, C.M., Knight, R., Gordon, J.I, 2007. The human microbiome project. Nature 449, 804–810.

Ursell, L.K., Metcalf, J.L., Parfrey, L.W., Knight, R., 2012. Defining the human microbiome. Nutr. Rev. 70 (Suppl 1), S38–S44. doi:10.1111/j.1753-4887.2012.00493.x.

Valentine, D.L., 2007. Adaptations to energy stress dictate the ecology and evolution of the Archaea. Nature Rev. Microbiol. 5 (4), 316–323.

van Wolferen, M., Wagner, A., van der Does, C., Albers, S.V., 2016. The archaeal Ced system imports DNA. Proc. Natl. Acad. Sci. USA 113, 2496–2501.

Venkateswaran, K., Hattori, N., La Duc, M.T., Kern, R., 2003. ATP as a biomarker of viable microorganisms in cleanroom facilities. J. Microbiol. Methods 52, 367–377.

Veras, D., Armstrong, D.J., Blake, J.A., et al., 2018. Dynamical and biological panspermia constraints within multiplanet exosystems. Astrobiology 18, 1106–1122.

Villanueva, G.L., Mumma, M.J., Novak, R.E., Hartogh, P., Encrenaz, T., Tokunaga, A., Khayat, A., Smith, M.D., 2015. Strong water isotopic anomalies in the martian atmosphere: probing current and ancient reservoirs. Science 348, 218–221.

Visone, V., Han, W., Perugino, G., Del Monaco, G., She, Q., Rossi, M., Valenti, A., Ciaramella, M., 2017. In vivo and in vitro protein imaging in thermophilic archaea by exploiting a novel protein tag. PLoS One 12, e0185791.

von Bloh, W., Bounama, C., Franck, S., 2005. Dynamic habitability of extrasolar planetary systems, In: A Comparison of the Dynamical Evolution of Planetary Systems, Eds R. Dvorak and S. Ferraz-Mello, Springer, Dordrecht, pp 287–300.

von Hegner, I., 2019. Extremophiles: a special or general case in the search for extra-terrestrial life? Extremophiles. doi:10.1007/s00792-019-01144-1.

Vreeland, R.H., Rosenzweig, W.D., Powers, D.W., 2000. Isolation of a 250-million-year-old halotolerant bacterium from a primary salt crystal. Nature 407, 897–900.

Wagner, M., Berkner, S., Ajon, M., Driessen, A.J., Lipps, G., Albers, S.V., 2009. Expanding and understanding the genetic toolbox of the hyperthermophilic genus Sulfolobus. Biochem. Soc. Trans. 37, 97–101.

Wagner, M., van, M., Wolferen, A.Wagner, Lassak, K., Meyer, B.H., Reimann, J., Albers, S.V., 2012. Versatile genetic tool box for the crenarchaeote sulfolobus acidocaldarius. Front. Microbiol. 3, 214.

Walker, J.J., Pace, N.R., 2007. Phylogenetic composition of Rocky Mountain endolithic microbial ecosystems. Appl. Environ. Microbiol. 73 (11), 3497–3504.

Walsby, A.E., 1980. A square bacterium. Nature 283 (5742), 69–71.

Walsh, J.C., Angstmann, C.N., Bisson-Filho, A.W., Garner, E.C., Duggin, I.G., Curmi, P.M.G., 2019. Division plane placement in pleomorphic archaea is dynamically coupled to cell shape. Mol. Microbiol. 112, 785–799.

Weiss, B.P., Kirschvink, J.L., Baudenbacher, F.J., Vali, H., Peters, N.T., MacDonald, F.A., Wikswo, J.P., 2000. A low temperature transfer of ALH84001 from Mars to Earth. Science 290, 791–795.

Wheat, C.G., et al., 2011. Fluid sampling from oceanic borehole observatories: design and methods for CORK activities (1990–2010). In: Fisher, A.T., Tsuji, T., Petronotis, K. (Eds.), and the Expedition 327 Scientists, Proc. IODP, 327: Tokyo (Integrated Ocean Drilling Program Management International, Inc.). doi:10.2204/iodp.proc.327.109.2011.

Wellsbury, P., Goodman, K., Barth, T., Cragg, B.A., Barnes, S.P., Parkes, R.J., 1997. Deep marine biosphere fuelled by increasing organic matter availability during burial and heating. Nature 388, 573–576.

Weronika, E., Lukasz, K., 2017. Tardigrades in space research—past and future. Orig Life Evol Biosph 47, 545–553.

Westall, F., Folk, R.L., 2003. Exogenous carbonaceous microstructures in Early Archaean cherts and BIFs from the Isua Greenstone Belt: Implications for the search for life in ancient rocks. Precambrian Res. 126, 313–330.

Westall, F., 2005. Early life on Earth: The ancient fossil record. In: Ehrenfreund, P., Irvine, W., Owen, T., Becker, L., Blank, J., Brucato, J., Colangeli, L., Derenne, S., Dutrey, A., Despois, D., Lascano, A., Robert, F. (Eds.). Astrobiology: Future Perspectives, 2005. Springer, New York, NY.

Westall, F., Loizeau, D., Foucher, F., Bost, N., Betrand, M., Vago, J., Kminek, G., 2013. Habitability on Mars from a microbial point of view. Astrobiology 13, 887–897.

Wierzchos, J., Ascaso, C., Sancho, L.G., Green, A., 2003. Iron-rich diagenetic minerals are biomarkers of microbial activity in Antarctic rocks. Geomicrobiol. J. 20, 15–24.

Woese, C., 1967. The Genetic Code: the Molecular basis for Genetic Expression. Harper & Row, New York, NY.

Woese, C.R., Fox, G.E., 1977. Phylogenetic structure of the prokaryotic domain: the primary kingdoms. Proc. Nat. Academy Sci. United States of America 74 (11), 5088–5090.

Woese, C.R., Gupta, R., 1981. Are archaebacteria merely derived 'prokaryotes'? Nature 289 (5793), 95–96.

Woese, C.R., Kandler, O., Wheelis, M.L., 1990. Towards a natural system of organisms: proposal for the domains Archaea, Bacteria, and Eucarya. Proc. Nat. Acad. Sci. United States of America 87 (12), 4576–4579.

Woese, C.R., 1994. There must be a prokaryote somewhere: microbiology's search for itself. Microbiol. Rev. 58 (1), 1–9.

Woese, C., 1998. The universal ancestor. Proc. Nat. Acad. Sci. United States of America 95 (12), 6854–6859.

Woese, C.R., 2006. How we do, don't and should look at bacteria and bacteriology. In: Dworkin, M., Falkow, S., Rosenberg, E., Schleifer, K.H., Stackebrandt, E. (Eds.), The Prokaryotes. New York, NY. https://doi.org/10.1007/0-387-30741-9_1.

Wright, J.C., Westh, P., Ramløv, H., 1992. Cryptobiosis in tardigrade. Biol Rev 67, 1–29.

Wright, J.C., 2001. Cryptobiosis 300 years on from van Leuwenhoek: what have we learned about tardigrades? ZoolAnz 240, 563–582.

Yang, S., Rothman, R.E., 2004. PCR-based diagnostics for infectious diseases: uses, limitations, and future applications in acute-care settings. Lancet Infect. Dis. 4, 337–348. doi:10.1016/S1473-3099(04)01044-8.

Yung, Y.L., Russell, M.J., Parkinson, C.D., 2010. The search for life on Mars. J. Cosmol. 5, 1121–1130.

Zebec, Z., Manica, A., Zhang, J., White, M.F., Schleper, C., 2014. CRISPR-mediated targeted mRNA degradation in the archaeon Sulfolobus solfataricus. Nucleic. Acids. Res. 42, 5280–5288.

Zhang, C., Phillips, A.P.R., Wipfler, R.L., Olsen, G.J., Whitaker, R.J., 2018. The essential genome of the crenarchaeal model Sulfolobus islandicus. Nat. Commun. 9, 4908.

Zhang, C., Wipfler, R.L., Li, Y., Wang, Z., Hallett, E.N., Whitaker, R.J., 2019. Cell structure changes in the hyperthermophilic crenarchaeon sulfolobus islandicus lacking the S-layer. MBio 10, e01589 -19.

Ziebis, W., et al., 2012. Interstitial fluid chemistry of sediments underlying the North Atlantic gyre and the influence of subsurface fluid flow. Earth Planet. Sci. Lett. 323–324, 79–91.

Zillig, W., Stetter, K.O., Schäfer, W., Janekovic, D., Wunderl, S., Holz, I., Palm, P., 1981. *Thermoproteales*: a novel type of extremely thermoacidophilic anaerobic archaebacteria isolated from Icelandic solfataras. Zbl. Bakt. Hyg., I. Abt. Orig. C2, 205–227.

Zuluaga, J.I., Salazar, J.F., Cuartas-Restrepo, P., Poveda, G., 2014. The habitable zone of inhabited planets. Biogeosci Discuss 11, 8443–8483.

Bibliography

Abbot, D.S., Cowan, N.B., Ciesla, F.J., 2012. Indication of insensitivity of planetary weathering behavior and habitable zone to surface land fraction. Astrophys. J. 756. doi:10.1088/0004-637X/756/2/178.

Albertsen, M., et al., 2013. Genome sequences of rare uncultured bacteria obtained by differential coverage binning of multiple metagenomes. Nat. Biotechnol. 31, 533–538.

Amann, R.I., Krumholz, L., Stahl, D.A., 1990. Fluorescent-oligonucleotide probing of whole cells for determinative, phylogenetic, and environmental studies in microbiology. J. Bacteriol. 172, 762–770.

Arakawa, K., Yoshida, Y., Tomita, M., 2016. Genome sequencing of a single tardigrade Hypsibiusdujardini individual. Sci Data 3, 160063. doi:10.1038/sdata.2016.63.

Bates, N.R., Michaels, A.F., Knap, A.H., 1996. Alkalinity changes in the Sargasso Sea: geochemical evidence of calcification? Marine Chem. 51, 347–358.

Blair, N., et al., 1985. Carbon isotopic fractionation in heterotrophic microbial metabolism. Appl. Environ. Microbiol. 50, 996–1001.

Boothby, T.C., Tenlen, J.R., Smith, F.W., Wang, J.R., Patanella, K.A., Osborne Nishimura, E., Tintori, S.C., Li, Q., Jones, C.D., Yandell, M., Messina, D.N., Glasscock, J., Goldstein, B., 2015. Evidence for extensive horizontal gene transfer from the draft genome of a tardigrade. Proc. Natl Acad. Sci. 112, 15976–15981.

Bruinsma, S., Forbes, J.M., Marty, J.C., Zhang, X., Smith, M.D., 2014. Long-term variability of Mars' exosphere based on precise orbital analysis of Mars Global Surveyor and Mars Odyssey. J. Geophys. Res. Planets 119, 210–218.

Caporaso, J.G., et al., 2012. Ultra-high-throughput microbial community analysis on the Illumina HiSeq and MiSeq platforms. ISME J. 6, 1621–1624.

Caspers, H., Parsons, T.R., Maita, Y., Lalli, C.M., 1984. A Manual of Chemical and Biological Methods for Seawater Analysis. Pergamon Press.

Chaufray, J.Y., Modolo, R., Leblanc, F., Chanteur, G., Johnson, R.E., Luhmann, J.G., 2007. Mars solar wind interaction: formation of the Martian corona and atmospheric loss to space. J. Geophys. Res. 112, E09009. doi:10.1029/2007JE002915.

Chaufray, J.Y., Bertaux, J.L., Leblanc. F., Quémerais, E., 2008. Observation of the hydrogen corona with SPICAM on Mars Express. Icarus 195 (2), 598–613.

Chaufray, J.Y., Leblanc, F., Quemerais, E., Bertaux, J.L., 2009. Martian oxygen density at the exobase deduced from O I 130.4-nm observations by SPICAM on Mars Express. J. Geophys. Res. (Planets) 114, E02006. doi:10.1029/2008JE003130.

Clausen, L.K.B., Andersen, K.N., Hygum, T.L., Jørgensen, A., Møbjerg, N., 2014. First record of cysts in the tidal tardigrade *Echiniscoidessigismundi*. Helgol,Mar Res 68, 531–537.

Clegg, J.S., 2005. Desiccation tolerance in encysted embryos of the animal extremophile, Artemia. Integr. Comp. Biol. 45, 715–724. doi:10.1093/icb/45.5.715.

Crowe, J.H., Higgins, R.P., 1967. The revival of Macrobiotusareolatus Murray (Tardigrada) from the cryptobiotic state. Trans. Am. Microsc. Soc. 86, 286–294.

Cowen, J.P., et al., 2012. Advanced instrument system for real-time and time-series microbial geochemical sampling of the deep (basaltic) crustal biosphere. Deep-Sea Res I 61, 43–56.

Dallas, L.J., Keith-Roach, M., Lyons, B.P., Jha, A.N., 2012. Assessing the impact of ionizing radiation on aquatic invertebrates: A critical review. Rad Res 177, 693–716.

Deming, J., Somers, L., Straube, W., Swartz, D., Macdonell, M., 1988. Isolation of an obligately barophilic bacterium and description of a new genus, Colwellia gen-nov. System. Appl. Microbiol. 10, 152–160.

Dennerl, K., et al., 2006. First observation of Mars with XMM Newton High resolution X-ray spectroscopy with RGS. Astron. Astrophys. 451, 709–722. doi:10.1051/0004-6361:20054253.

Dickson, A., Sabine, C., and Christian, J. (2007), Guide to Best practices for ocean CO2 measurements (PICES Science Report).

Dostal, J., Murphy, J.B., Nance, R.D., 2009. History of the Earth. Volume 2 of Earth Systems: History and Natural Variability, Encyclopedia of Life Support Systems. Developed Under the Auspices of the UNESCO. Eolss Publishers, Oxford.

Edgar, R.C., 2010. Search and clustering orders of magnitude faster than BLAST. Bioinformatics 26, 2460–2461.

Eren, A.M., Vineis, J.H., Morrison, H.G., Sogin, M.L., 2013. A filtering method to generate high quality short reads using Illumina paired-end technology. PLoS One 8, 6–11.

Feldman, P.D., et al., 2013. Rosetta-Alice observations of exospheric hydrogen and oxygen on Mars. Icarus 214, 394–399.

Filippidou, S., Wunderlin, T., Junier, T., Jeanneret, N., Dorador, C., Molina, V., Johnson, D.R., Junier, P., 2016. A combination of extreme environmental conditions favor the prevalence of endospore-forming firmicutes. Front. Microbiol. 7.

Fisher, A.T., Becker, K., 2000. Channelized fluid flow in oceanic crust reconciles heat-flow and permeability data. Nature 403, 71–74.

Fox, J.L., 2004. Response of the Martian thermosphere/ionosphere to enhanced fluxes of solar soft X-rays. J. Geophys. Res. 109, A11310.

Franck, S., Block, A., Bloh, W., Bounama, C., Garrido, I., Schellnhuber, H.J., 2001. Planetary habitability: is Earth commonplace in the Milky Way? Naturwissenschaften 88, 416–426.

Gaboyer, F., Le Milbeau, C., Bohmeier, M., Schwendner, P., Vannier, P., Beblo-Vranesevic, K., Rabbow, E., Foucher, F., Gautret, P., Guégan, R., Richard, A., Sauldubois, A., Richmann, P., Perras, A.K., Moissl-Eichinger, C., Cockell, C.S., Rettberg, P., Marteinsson, E.Monaghan, Ehrenfreund, P., Garcia-Descalzo, L., Gomez, F., Malki, M., Amils, R., Cabezas, P., Walter, N., Westall, F., 2017. Mineralization and preservation of an extremo-tolerant bacterium isolated from an early Mars analog environment. Sci. Rep. 7 (1).

Girguis, P.R., Childress, J.J., 2006. Metabolite uptake, stoichiometry, and chemoautotrophic function of the hydrothermal vent tubeworm Riftiapachyptila: Responses to environmental variations in substrate concentrations and temperature. J. Exp. Biol. 209, 3516–3528.

Gladyshev, E., Meselson, M., 2008. Extreme resistance of bdelloid rotifers to ionizing radiation. Proc. Nat. Acad. Sci. U.S.A 105, 5139–5144. doi:10.1073/pnas.0800966105.

Gross, V., Mayer, G., 2015. Neural development in the tardigrade Hypsibiusdujardini based on anti-acetylated α-tubulin immunolabeling. Evodevo 6, 12. Epub 2015 Apr 25.

Hallberg, K., Hedrich, S., Johnson, D, 2011. Acidiferrobacterthiooxydans, gen. nov. sp. nov.; an acidophilic, thermo-tolerant, facultatively anaerobic iron- and sulfur-oxidizer of the family Ectothiorhodospiraceae. Extremophiles 15, 271–279.

Han, Y., Perner, M., 2015. The globally widespread genus Sulfurimonas: versatile energy metabolisms and adaptations to redox clines. Front. Microbiol. 6, 989. doi:10.3389/fmicb.2015.00989.

Hashimoto, T., Horikawa, D.D., Saito, Y., Kuwahara, H., Kozuka-Hata, H., Shin-I, T, et al., 2016. Extremotolerant tardigrade genome and improved radiotolerance of human cultured cells by tardigrade-unique protein. Nat. Commun. 7, 12808. doi:10.1038/ncomms12808.

Heidemann, N.W.T., Smith, D.K., Hygum, T.L., Stapane, L., Clausen, L.K.B., Jørgensen, A, et al., 2016. Osmotic stress tolerance in semi-terrestrial tardigrades. Zool. J. Linn. Soc. 178 (4), 912–918.

Hering, L., Bouameur, J.E., Reichelt, J., Magin, T.M., Mayer, G., 2016. Novel origin of lamin-derived cytoplasmic intermediate filaments in tardigrades. Elife 5, e11117. doi:10.7554/eLife.11117.

Huang, Y., Gilna, P., Li, W., 2009. Identification of ribosomal RNA genes in metagenomic fragments. Bioinformatics 25, 1338–1340.

Huber, J.A., et al., 2007. Microbial population structures in the deep marine biosphere. Science 318, 97–100.

Huse, S.M., et al., 2008. Exploring microbial diversity and taxonomy using SSU rRNA hypervariable tag sequencing. PLos Genet. 4 (11), e1000255. doi:10.1371/journal.pgen.1000255.

Hyatt, D., Chen, G.-L., Larimer, F.W., Hauser, L.J., 2010. Prodigal: prokaryotic gene recognition and translation initiation site identification. BMC Bioinfor. 11, 119.

Hygum, T.L., Clausen, L.K.B., Halberg, K.A., Jørgensen, A., Møbjerg, N., 2016. Tun formation is not a prerequisite for desiccation tolerance in the marine tidal tardigrade *Echiniscoidessigismundi*. Zool. J. Linn. Soc. 178 (4), 907–911.

Iwasaki, T., 1964. Sensitivity of Artemia eggs to the γ-irradiation. I. Hatchability of encysted dry eggs. J. Radiat. Res. 5, 69–75.

Jahnke, L.L., Summons, R.E., Hope, J.M., Des Marais, D.J, 1999. Carbon isotopic fractionation in lipids from methanotrophic bacteria II: the effects of physiology and environmental parameters on the biosynthesis and isotopic signatures of biomarkers. Geochem. Cosmochim. Acta 63, 79–93.

Jehlička, J., Culka, A., Nedbalová, L., 2016. Colonization of snow by micro-organisms as revealed using miniature Raman spectrometers—Possibilities for detecting carotenoids of psychrophiles on Mars? Astrobiology 16 (12), 913–924.

Kallio, E., Chaufray, J.V., Modolo, R., Snowden, D., Winglee, R., 2011. Modeling of Venus, Mars and Titan. Space Sci. Rev. 162, 267–307.

Kasting, J.F., Whitmire, D.P., Reynolds, R.T., 1993. Habitable zones around main sequence stars. Icarus 101, 108–128.

Kopparapu, R.K., Ramirez, R., Kasting, J.F., Eymet, V., Robinson, T.D., Mahadevan, S., Terrien, R.C., Domagal-Goldman, S., Meadows, V., Deshpande, R., 2013. Habitable zones around main-sequence stars: new estimates. Astrophys. J. 765. doi:10.1088/0004-637X/765/2/131.

Koutsovoulos, G., Kumar, S., Laetsch, D.R., Stevens, L., Daub, J., Conlon, C., Maroon, H., Thomas, F., Aboobaker, A.A., Blaxter, M., 2016. No evidence for extensive horizontal gene transfer in the genome of the tardigrade Hypsibiusdujardini. Proc. Natl. Acad. Sci. USA. 113 (18), 5053–5058. Epub 2016 Mar 24.

Krasnopolsky, V.A., Gladstone, G.R., 1996. Helium on Mars: EUVE and PHOBOS data and implications for Mars' evolution. J. Geophys. Res. 101, 15765–15772.

Lammer, H., et al., 2003. Outgassing history and escape of the Martianatmosphere and water inventory. Space Sci. Rev. 174, 113–154.

Lang, S.Q., Butterfield, D.A., Lilley, M., Johnson, H.P., Hedges, J., 2006. Dissolved organic carbon in ridge-axis and ridge-flank hydrothermal systems. Geochem. Cosmochim. Acta 70, 3830–3842.

Lin, H.-T., Hsieh, C.C., Cowen, J.P., Rappé, M.S., 2015. Data report: dissolved and particulate organic carbon in the deep sediments of IODP Site U1363 near Grizzly Bear seamount. In: Fisher, A.T., Tsuji, T., Petronotis, K. (Eds.), and the Expedition 327 Scientists, Proc. IODP, 327: Tokyo (Integrated Ocean Drilling Program Management International, Inc.). doi:10.2204/iodp.proc.327.202.2015.

Lineweaver, C.H., Chopra, A., 2012. The habitability of our Earth and other Earths: astrophysical, geochemical, geophysical, and biological limits on planet habitability. Annu Rev Earth Planet Sci. 40, 597–623.

Markowitz, V.M., et al., 2006. The integrated microbial genomes (IMG) system. Nucleic. Acids. Res. 34, D344–D348.

Martin, C., Gross, V., Hering, L., Tepper, B., Jahn, H., de Sena Oliveira, I., Stevenson, P.A., Mayer, G., 2017. The nervous and visual systems of onychophorans and tardigrades: learning about arthropod evolution from their closest relatives. J. Comp. Physiol. A. Neuroethol Sens Neural Behav Physiol. Epub 2017 Jun 9.

Martin, A., McMinn, A., 2017. Sea ice, extremophiles and life on extraterrestrial ocean worlds. Int. J. Astrobiol., 1–16.

Mason, O.U., et al., 2007. The phylogeny of endolithic microbes associated with marine basalts. Environ. Microbiol. 9, 2539–2550.

Masui, N., Morono, Y., Inagaki, F., 2008. Microbiological assessment of circulation mud fluids during the first operation of riser drilling by the deep-earth research vessel Chikyu. Geomicrobiol. J. 25, 274–282.

Methé, B.A., et al., 2005. The psychrophilic lifestyle as revealed by the genome sequence of Colwelliapsychrerythraea 34H through genomic and proteomic analyses. Proc. Natl. Acad. Sci. USA 102, 10913–10918.

Mattimore, V., Battista, J.R., 1996. Radioresistance of *Deinococcusradiodurans*: functions necessary to survive ionizing radiation are also necessary to survive prolonged desiccation. J. Bacteriol. 178, 633–637.

Mayer, G., Kauschke, S., Rüdiger, J., Stevenson, P.A., 2013. Neural markers reveal a one-segmented head in tardigrades (water bears). PLoS One 8 (3), e59090. Epub 2013 Mar 13.

Mayer, G., Martin, C., Rüdiger, J., Kauschke, S., Stevenson, P.A., Poprawa, I., Hohberg, K., Schill, R.O., Pflüger, H.J., Schlegel, M., 2013. Selective neuronal staining in tardigrades and onychophorans provides insights into the evolution of segmental ganglia in panarthropods. BMC Evol. Biol. 13, 230. Epub 2013 Oct 24.

Mayer, G., Hering, L., Stosch, J.M., Stevenson, P.A., Dircksen, H., 2015. Evolution of pigment-dispersing factor neuropeptides in Panarthropoda: Insights from Onychophora (velvet worms) and Tardigrada (water bears). J. Comp. Neurol. 523 (13), 1865–1885. Epub 2015 Apr 7.

Meyers, P.A., 1994. Preservation of elemental and isotopic source identification of sedimentary organic matter. Chem. Geol. 114, 289–302.

Miller, C.S., Baker, B.J., Thomas, B.C., Singer, S.W., Banfield, J.F., 2011. EMIRGE: reconstruction of full-length ribosomal genes from microbial community short read sequencing data. Genome Biol. 12, R44.

Moissl-Eichinger, C., Cockell, C., Rettberg, P., Albers, S.-V., 2016. Venturing into new realms? Microorganisms in space. FEMS Microbiol. Rev. 40 (5), 722–737.

Morin, R., Hess, A., Becker, K., 1992. situ measurement of fluid flow in Holes 395A and 534A. Geophys. Res. Lett. 19, 509–512.

Nadkarni, M.A., Martin, F.E., Jacques, N.A., Hunter, N., 2002. Determination of bacterial load by real-time PCR using a broad-range (universal) probe and primers set. Microbiol 148, 257–266.

Nakagawa, S., et al., 2005. Distribution, phylogenetic diversity and physiological characteristics of epsilon-Proteobacteria in a deep-sea hydrothermal field. Environ. Microbiol. 7, 1619–1632.

Neir, A.O., et al., 1976. Composition and structure of the Martian atmosphere: preliminary results from Viking 1. Science 193, 786–788.

Nier, A.O., McElroy, M.B., 1976. Structure of the neutral upper atmosphere of Mars: results from Viking 1 and Viking 2. Science 194, 1298–1300.

Nisbet, E.G., Zahnle, K.J., Gerasimov, M.V., Helbert, J., Jaumann, R., Hofmann, B.A., Benzerara, K., Westall, F., 2007. Creating habitable zones, at all scales, from planets to mud micro-habitats, on Earth and on Mars. Space Sci. Rev. 129, 79–121.

Noack, L., Snellen, I., Rauer, H., 2017. Water in extra-solar planets and implications for habitability. Space Sci. Rev.

Nogi, Y., Hosoya, S., Kato, C., Horikoshi, K., 2004. Colwelliapiezophilasp. nov., a novel piezophilic species from deep-sea sediments of the Japan Trench. Int. J. Syst. Evol. Microbiol. 54, 1627–1631.

Orcutt, B.N., Sylvan, J.B., Knab, N.J., Edwards, K.J., 2011. Microbial ecology of the dark ocean above, at, and below the seafloor. Microbiol. Molecul. Biol. Rev. 75, 361–422.

Orcutt, B.N., et al., 2015. Carbon fixation by basalt-hosted microbial communities. Front. Microbiol. 6, 904.

Pernthaler, A., Pernthaler, J., Amann, R, 2002. Fluorescence in situ hybridization and catalyzed reporter deposition for the identification of marine bacteria. Appl. Environ. Microbiol. 68, 3094–3101.

Picard, A., Ferdelman, T., 2011. Linking microbial heterotrophic activity and sediment lithology in oxic, oligotrophic sub-seafloor sediments of the North Atlantic Ocean. Front. Microbiol. 2, 263.

Pirajano, F., 2009. Hydrothermal Processes and Mineral Systems. Springer Netherlands.

Porter, K., Feig, Y., 1980. The use of DAPI for identification and enumeration of bacteria and blue-green algae. Limnol. Oceanogr. 25, 943–948.

Quast, C., et al., 2013. The SILVA ribosomal RNA gene database project: Improved data processing and web-based tools. Nucleic. Acids. Res. 41, 590–596.

Seyfried, W. (1977), Seawater-basalt interaction from 25°–300°C and 1-500 bars: implications for the origin of submarine metal-bearing hydrothermal solutions and regulation of ocean chemistry, Ph.D. Thesis, Univ. Southern California, 216 pp.

Shields, A.L., Ballard, S., Johnson, J.A., 2016. The habitability of planets orbiting M-dwarf stars. Phys. Rep. 663, 1–38.

Smith, F.W., Boothby, T.C., Giovannini, I., Rebecchi, L., Jockusch, E.L., Goldstein, B., 2016. The Compact Body Plan of Tardigrades Evolved by the Loss of a Large Body Region. Curr. Biol. 26 (2), 224–229. Epub 2016 Jan 14.

Sogin, M.L., et al., 2006. Microbial diversity in the deep sea and the underexplored "rare biosphere. Proc. Nat. Acad. Sci. USA 103, 12115–12120.

Stahl, D., Amann, R, 1991. Development and application of nucleic acid probes in bacteria systematic. In: Stackebrandt, E., Goodfellow, M. (Eds.), Nucleic Acid Techniques in Bacterial Systematics. John Wiley and Sons.

Sylvan, J.B., Toner, B.M., Edwards, K.J., 2012. Life and death of deep-sea vents: Bacterial diversity and ecosystem succession on inactive hydrothermal sulphides. mBio 3, 1–10.

Sylvan, J., Pyenson, B., Rouxel, O., German, C.G., Edwards, K.J., 2012. Time-series analysis of two hydrothermal plumes at 9°50'N East Pacific Rise reveals distinct, heterogeneous bacterial populations. Geobiol 10, 178–192.

Takai, K., Horikoshi, K., 2000. Rapid detection and quantification of members of the archaeal community by quantitative PCR using fluorogenic probes. Appl Environ. Microbiol. 66, 5066–5072.

Tenlen, J.R., McCaskill, S., Goldstein, B., 2013. RNA interference can be used to disrupt gene function in tardigrades. Dev. Genes Evol. 223 (3), 171–181. doi:10.1007/s00427-012-0432-6.

Vago, J.L., Westall, F.Pasteur Instrument Teams, Landing Site Selection Working Group, and Other Contributors, Coates, A.J., Jaumann, R., Korablev, O., Ciarletti, V., Mitrofanov, I., Josset, J.-L., De Sanctis, M.C., Bibring, J.-P., Rull, F., Goesmann, F., Steininger, H., Goetz, W., Brinckerhoff, W., Szopa, C., Raulin, F., Westall, F., Edwards, H.G.M., Whyte, L.G., Fairén, A.G., Bibring, J.-P., Bridges, J., Hauber, E., Ori, G.G., Werner, S., Loizeau, D., Kuzmin, R.O., Williams, R.M.E., Flahaut, J., Forget, F., Vago, J.L., Rodionov, D., Korablev, O., Svedhem, H., Sefton-Nash, E., Kminek, G., Lorenzoni, L., Joudrier, L., Mikhailov, V., Zashchirinskiy, A., Alexashkin, S., Calantropio, F., Merlo, A., Poulakis, P., Witasse, O., Bayle, O., Bayón, S., Meierhenrich, U., Carter, J., García-Ruiz, J.M., Baglioni, P., Haldemann, A., Ball, A.J., Debus, A., Lindner, R., Haessig, F., Monteiro, D., Trautner, R., Voland, C., Rebeyre, P., Goulty, D., Didot, F., Durrant, S., Zekri, E., Koschny, D., Toni, A., Visentin, G., Zwick, M., van Winnendael, M., Azkarate, M., Carreau, C.The ExoMars Project Team, 2017. Habitability on early Mars and the search for biosignatures with the ExoMars Rover. Astrobiology 17 (6-7), 471–510.

Vilella, K., Kaminski, E., 2017. Fully determined scaling laws for volumetrically heated convective systems, a tool for assessing habitability of exoplanets. Phys. Earth Planetary Interiors 266, 18–28.

Watanabe, M., Sakashita, T., Fujita, A., Kikawada, T., Nakahara, Y., Hamada, N, et al., 2006. Estimation of radiation tolerance against high LET heavy ions in an anhydrobiotic insect, Polypedilumvanderplanki. Int. J. Radiat. Biol. 82, 835–842. doi:10.1080/09553000600979100.

Wheat, C.G., et al., 2004. Venting formation fluids from deep sea boreholes in a ridge flank setting: ODP Sites 1025 and 1026. Geochem. Geophys. Geosyst. 5. doi:10.1029/2004GC000710.

Wheat, C.G., et al., 2010. Subseafloor seawater-basalt-microbe reactions: Continuous sampling of borehole fluids in a ridge flank environment. Geochem. Geophys. Geosyst. 11. doi:10.1029/2010GC003057.

Wheat, C., et al., 2012. CORK-Lite: bringing legacy boreholes back to life. Sci. Drill. 14, 39–43.

Wickham, H. (2009), *ggplot2: elegant graphics for data analysis.*

Wright, J.C., 1989. Desiccation tolerance and water-retentive mechanisms in tardigrades. J. Exp. Biol. 142, 267–292.

Yagi, M., Leblanc, F., Chaufray, J.Y., Gonzalez-Galindo, F., Hess, S., Modolo, R., 2012. Mars exospheric thermal and nonthermal components: seasonal and local variations. Icarus 221, 682–693.

Yoshida, Y., Koutsovoulos, G., Laetsch, D.R., Stevens, L., Kumar, S., Horikawa, D.D., Ishino, K., Komine, S., Kunieda, T., Tomita, M., et al., 2017. Comparative genomics of the tardigrades Hypsibiusdujardini and Ramazzottiusvarieornatus. PLoS Biol. 15 (7), e2002266. Epub 2017 Jul 27.

Chapter 6

Salinity tolerance of inhabitants in thalassic and athalassic saline and hypersaline waters & salt flats

6.1 Salinities in thalassic and athalassic water bodies—survival and growth of living organisms in saline and hypersaline environments

Organisms that live in saline and hypersaline habitats are of great interest to ecologists and physiologists. The most saline locations on Earth contain salt, or sodium chloride, and also other minerals and contain these in much greater quantities than the ocean. Saline and hypersaline habitats are unusual habitats that confront potential colonizing organisms with extremely severe osmotic problems, but which nevertheless contain some characteristic inhabitants. These inhabitants, through the evolution of special mechanisms, have achieved a remarkable degree of osmotic independence from, or tolerance of, their unusual environment.

Cubillos et al. (2019) made the below observations with regard to the survival and growth of living organisms in saline and hypersaline environments: "The survival and growth of living organisms in any environment are controlled by biotic and abiotic factors (Kristjánsson and Hreggvidsson, 1995). From an anthropocentric point of view, extreme environments can be defined as habitats where one or more abiotic conditions extend beyond the typical physiological tolerances of higher organisms (Wiegel and Canganella, 2011), e.g., low or high temperatures, extremes of pH, high concentrations of metals or salts (Peoples and Bartlett, 2017). Organisms that can live in such hostile habitats are typically microbes, and are commonly known as extremophile or extremotolerant taxa (Rodriguez-Valera et al., 1985; Orellana et al., 2018). The most commonly used classification of extremophile is related to NaCl concentrations, where taxa can be classified as either non-halophilic, halophile or extremely halophile (Tiquia et al., 2007; Lee et al., 2018). Such microorganisms are commonly found in evaporation ponds, sea waters, saline salterns, and sediments, as well as other saline habitats (Maheshwari and Saraf, 2015; Lee et al., 2018). In particular, bacteria belonging to the Bacillales order are found in almost every environment on Earth from the stratosphere (Sass et al., 2008), through to NaCl-hypersaline environments, such as salterns (Garabito et al., 1998), saline soils (Zahran, 1997), freshwater lakes (Bagheri et al., 2012; Amoozegar et al., 2014, 2016), and even in alkaline lakes (Weisser and Trüper, 1985). Moreover, they have been isolated from brines with different chemical composition including $MgCl_2$ (Hallsworth et al., 2007; Sass et al., 2008), Na_2CO_3 (Jones et al., 1998), and NaCl/LiCl (Rodríguez-Contreras et al., 2013; Martínez et al., 2018). There is an increasing awareness that saline environments include both natural and artificial environments, and extend beyond those dominated by NaCl, with microbial life being reported at high salinities of other salts or even different pH as Na_2CO_3 (Imhoff et al., 1979; Banciu and Muntyan, 2015), $CaCl_2$ (Dickson et al., 2013), $MgCl_2$ (Yakimov et al., 2015), LiCl (Cubillos et al., 2018) or acid (Benison, 2008, 2019) and basic brines (Samylina et al., 2014)."

The term "thalassic" means inhabiting or growing in the sea. Following Bayly and Williams (1966) and Bayly (1967), the term *athalassic* is used to refer to (1) waters associated with land, irrespective of their salinity or position relative to the coastline, which have never been joined to the sea during geologically "Recent Times" (the ions contained in these waters have for the most part been supplied from rocks and soils by weathering, or from the sea via the atmosphere, or by a combination of both processes); (2) waters occupying basins with a geologically "Recent" connection with the sea which was permanently lost and in which any originally enclosed seawater evaporated to dryness before reinstatement as an aquatic environment (such waters are biologically athalassic even though a considerable portion of the ions may be of relic marine origin). Stated in simple words, situations in which a body of seawater permanently loses connection with the sea, and dries up before influx of fresh water restores a saline body of water, are regarded by Bayly (1967) and Bayly and Williams (1966) as athalassic in nature.

With reference to thalassic saline and hypersaline waters, on average, seawater in the world's oceans has a salinity of approximately 3.5%, or 35 PSU {35 parts per thousand (ppt)}, abbreviated as 35‰ or 35g/kg. This means that for every 1 L (1000 mL) of seawater there are 35 grams of salts (mostly, but not entirely, sodium chloride) dissolved in it. Thus, the average salinity of oceanic waters is 35 ppt (35‰). Note that the surface waters of the North Atlantic have a higher salinity than those of any other ocean, reaching values exceeding 37‰ in latitudes 20° to 30° N. Fig. 6.1 shows the map of the World Ocean surface salinity, taken by the Aquarius/SAC-D spacecraft, August–September 2011.

There are several athalassic water bodies on Earth. Inland bodies of saltwater such as the Dead Sea the, Caspian Sea, and the Great Salt Lake are distinct from the world's oceans. Despite their chemical diversity, athalassic waters possess a real biological unity (Bayly, 1972). Ionescu et al. (2012) recall that the Dead Sea is a terminal lake located on the border between Jordan, the Palestinian Authority and Israel, and is part of a larger geological system known as the Jordan Dead Sea Rift. The lake consists of a deeper northern basin (deepest point at ~725 m below sea level) and a southern basin, which has dried out but is kept shallow by continuous transfer of water from the northern basin as it is used for commercial mineral production. Until 1979 the Dead Sea was a meromictic lake (a lake that has layers of water that do not intermix) with hypersaline, anoxic, and sulfidic deep waters and a seasonally varying mixolimnion (the freely circulating upper layer of a meromictic lake) (Anati et al., 1987). The extensive evaporation in the absence of major water input led to an increase in the density of the upper water layer, which caused the lake to overturn in 1979 (Steinhorn et al., 1979). Since then, except after two rainy seasons in 1980 and 1992, the Dead Sea remained holomictic (having a uniform temperature and density from top to bottom at a specific time during the year, which allows the water body to completely mix) and has been characterized by a NaCl supersaturation and halite (rock salt) deposition on the lake bottom, with total dissolved salt concentrations reaching 347 g/L (compare this with the mean salinity of seawater, which is 35 g/L). Due to the continuous evaporation of the Dead Sea, Na^+ precipitates out as halite while Mg^{2+}, whose salts are more soluble, is further concentrated and has become the dominant cation (Oren, 2010). The salt concentration in the Dead Sea is a staggering 34.7%, 9.9 times saltier than ocean water, which is only about 3.5% salt. As a result, the Dead Sea is famous for its body buoyancy properties, as people who take an exploratory dip generally find themselves riding high on its waters. The Dead Sea is also the lowest point on Earth, and getting lower every year, as water that would ordinarily fill it by flowing in from the Jordan River has been diverted to quench the thirst of Israel, Jordan, and Palestine. Every year, the lake drops over a meter per year.

Diving studies carried out in the Dead Sea by a team led by Ionescu in 2011 (see Ionescu et al., 2012) indicated that this water body is adorned with several underground springs at its floor. Ionescu et al. (2012) found a new microbial ecosystem in the Dead Sea, tightly associated with underwater springs that emerge at the lake floor at depths between 10–30 m. Ionescu et al. (2012) found that compared to the Dead Sea water (Möller et al. 2003), underwater springs were significantly less saline and had a higher pH. Furthermore, the southern springs were more saline and had a lower pH than the northern ones. The fluctuating nature of the Dead Sea spring system does not permit the establishment of neither a constant salinity or a permanent gradient (Ionescu et al., 2012). Their study reveals that the ecosystem is diverse, and the presence of multiple major biogeochemical pathways is indicated. However, as demonstrated by their data, the springs do not serve as an

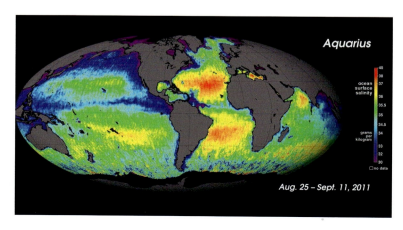

FIG. 6.1 Map of ocean salinity, 2011. The first map of ocean salinity taken by the Aquarius/SAC-D spacecraft, August–September 2011. GSFC—JPL-Caltech/NASA.

input of this diversity, but more likely as a source of nourishment for the native Dead Sea community. The dominant microorganisms in the ecosystem, as implicated by their microbial community and water chemistry data, are phototrophs (organisms that carry out photon capture—i.e., use the energy from light to carry out various cellular metabolic processes—to produce complex organic compounds and acquire energy), sulfide oxidizers (organisms having the capability to oxidize various reduced inorganic sulfur compounds with high efficiency to obtain electrons for their autotrophic growth), sulfate reducers (organisms which can perform anaerobic respiration utilizing sulfate as terminal electron acceptor, reducing it to hydrogen sulfide), nitrifiers (organisms that change ammonia or ammonium into nitrite or change nitrite into nitrate as part of the nitrogen cycle) and iron reducers. Iron-reducing bacteria (FeRB) play key roles in anaerobic metal and carbon cycling and carry out biogeochemical transformations that can be harnessed for environmental bioremediation (Luef et al., 2013).

The Caspian Sea (Fig. 6.2)—the world's largest inland body of water, which lies between Europe and Asia; east of the Caucasus, west of the broad steppe of Central Asia, south of the fertile plains of Southern Russia in Eastern Europe, and north of the mountainous Iranian Plateau of Western Asia—has a salinity of approximately 1.2% (12 g/L), about a third that of average seawater. Salinity changes from north to south, from 1.0 to 13.5 ppt (Szalay, 2017). By contrast, the North Atlantic Ocean has a salinity of 37 ppt.

The Great Salt Lake (Fig. 6.3) is located at the bottom of a 35,000-square-mile drainage basin that has a human population of more than 1.5 million. Great Salt Lake, the shrunken remnant of prehistoric Lake Bonneville, has no outlet. Because the lake does not have an outlet, water flows into the lake and then evaporates, leaving dissolved minerals behind as residue. On average, 2.9 million acre-feet of water enters the lake each year from the Bear, Weber, and Jordan Rivers. The inflow carries about 2.2 million tons of salt. About 4.3 billion tons of salt are in the lake. Dissolved salts accumulate in the lake by evaporation. A railroad causeway separates the lake into a north and south part. Most freshwater enters the south part, and water in the north part evaporates faster than it enters. The north part is always much saltier than the south part. The salinity of Great Salt Lake, normally about 23‰, decreases when its area increases. The chief constituent of the dissolved salts is sodium chloride, which is recovered in commercial quantities. The lake has been estimated to contain more than 5 million tonnes of sodium chloride in solution. Salinity south of the causeway of the Great Salt Lake has ranged from 6% to 27% (i.e., 60–270‰) over a period of 22 years (~2 to 8 times saltier than the ocean). The high salinity supports a mineral industry that extracts about 2 million tons of salt from the lake each year (https://pubs.usgs.gov/wri/wri994189/PDF/WRI99-4189.pdf). As it is endorheic (has no outlet besides evaporation), it has very high salinity, far saltier than seawater, which makes swimming similar to floating, and its mineral content is steadily increasing.

Don Juan Pond in Antarctica (Fig. 6.4) is a small and very shallow, flat-bottom, hypersaline lake in the

FIG. 6.2 (top) Location of the Caspian Sea in the world. *Author: Uwe Dedering; Licensed under the Creative Commons Attribution-Share Alike 3.0 Unported license. (Source: https://commons.wikimedia.org/wiki/File:Asia_laea_relief_location_map.jpg)* (bottom). The Caspian Sea as taken by the MODIS on the orbiting Terra satellite, June 2003. This file is in the public domain in the United States because it was solely created by NASA. *NASA copyright policy states that "NASA material is not protected by copyright unless noted". (Source: https://en.wikipedia.org/wiki/File:Caspian_Sea_from_orbit.jpg).*

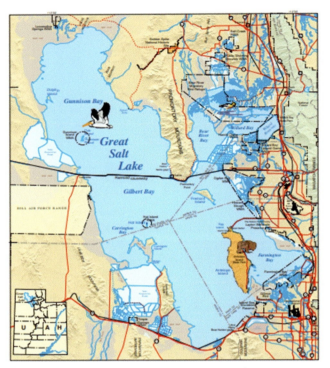

FIG. 6.3 Map of the Great Salt Lake. *Credit: USGS. (Source: Stephens, D.W., and J. Gardner (2007), Great Salt Lake, Utah; U.S. Geological Survey; Water-Resources Investigations Report 1999–4189; https://pubs.usgs.gov/wri/wri994189/PDF/WRI99-4189.pdf).*

western end of Wright Valley (South Fork), Victoria Land, Antarctica, 9 km (5.6 mi) west from Lake Vanda. With a salinity level of 33.8%, Don Juan Pond is the saltiest of the Antarctic lakes (Hammer, 1986; Grobljar, 2017). It has the second highest total dissolved solids on record, 1.3 times greater salinity than the Dead Sea. Salinity varies over time from 200 to 474 g/L, dominated by calcium chloride. It is the only Antarctic hypersaline lake that almost never freezes. It has been described as a groundwater discharge zone (Salty Antarctic pond could be a replica of Mars' subsurface water). The area around Don Juan Pond is covered with sodium chloride and calcium chloride salts that have precipitated as the water evaporated (Hammer, 1986; Oren, 2007). The just mentioned high salinity keeps the pond from completely freezing, even in Antarctic winters; allowing the pond to remain liquid even at temperatures as low as −50°C (−58°F) due to the interference of salts with the bonding of water molecules. The water thus remains fluid in what is among the coldest and driest locations on Earth (https://geographyandyou.com/ten-saline-locations-earth/).

Lake Assal (Fig. 6.5)—the most saline hypersaline lake outside the Antarctic Continent—is a crater lake in central-western Djibouti, a country located in the Horn of Africa. The Red Sea and the Gulf of Aden lie in its eastern border. Lake Assal is the most saline body of water on earth after

FIG. 6.4 (top left) Map of Antarctica indicating the Don Juan Pond in East Antarctica. *(Source: https://en.wikipedia.org/wiki/Don_Juan_Pond)* (top right) Satellite photo of Don Juan Pond *Author: NASA/Goddard Space Flight Center Scientific Visualization Studio, Landsat 7 Project Science Office; MODIS Rapid Response Team, NASA Goddard Space Flight Center (http://rapidfire.sci.gsfc.nasa.gov Copyright: This file is in the public domain in the United States because it was solely created by NASA. NASA copyright policy states that "NASA material is not protected by copyright unless noted". (Source: https://commons.wikimedia.org/wiki/File:DonJuanSTILL.0660_web.jpg)* (bottom) Don Juan Pond, McMurdo Dry Valleys, Antarctica, seen from the air. View is from the south. Image by Turckish D. December 2013. *(Source: https://commons.wikimedia.org/wiki/File:Don_Juan_Pond,_McMurdo_Dry_Valleys,_Antarctica.jpg) Author: Dturme: Copyright: This file is licensed under the Creative Commons Attribution-Share Alike 4.0 International license.*

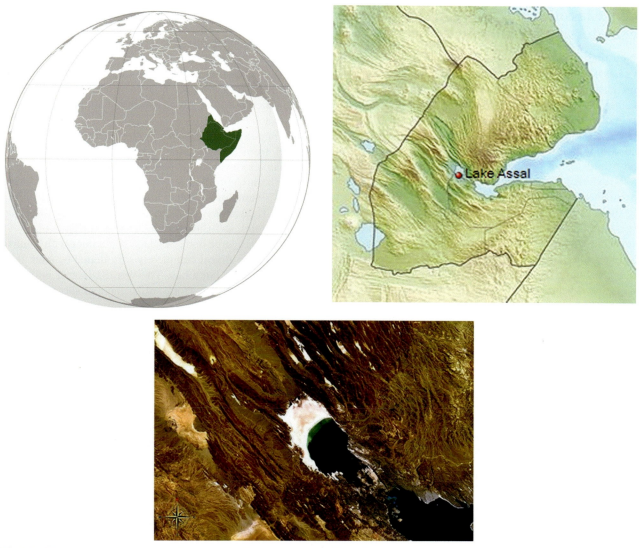

FIG. 6.5 The hypersaline Lake Assal in Djibouti in the Horn of Africa (top) Horn of Africa (orthographic projection). *(Source: https://commons.wikimedia.org/wiki/File:Horn_of_Africa_(orthographic_projection).svg) Copyright info: This file is licensed under the Creative Commons Attribution-Share Alike 3.0 Unported license.* (middle) Location of Lake Assal in Djibouti *(Source: https://commons.wikimedia.org/wiki/File:Djibouti_relief_location_map.jpg) Copyright info: This file is licensed under the Creative Commons Attribution-Share Alike 3.0 Unported license.* (bottom) Satellite view of Lake Assal *(Source: https://commons.wikimedia.org/wiki/File:Lake_Assal_NASA.jpg) Copyright info: This image is in the public domain because it is a screenshot from NASA's globe software World Wind using a public domain layer, such as Blue Marble, MODIS, Landsat, SRTM, USGS or GLOBE.*

Don Juan Pond with 34.8% average salt concentration (up to 40% at 20 m (66 ft) depth); higher than the 33.7% level in the Dead Sea. The dissolved salts include NaCl, KCl, $MgCl_2$, $CaCl_2$, $CaSO_4$, and $MgBr_2$, with NaCl dominating in Lake Assal. The surface concentration of salts is 276.5 g/L for Lake Assal.

Bayly (1967b) discussed the chemical composition of two athalassic saline water bodies located near central Otago {1) Salt Lake, Sutton, and 2) saline pond near Patearoa}. One water (dominated by Na and Cl) had a salinity of 15‰. These are the only regions of New Zealand in which the average annual rainfall is less than 30 in. Studies carried out by Bayly (1967b) found that the Salt Lake at Sutton in New Zealand is a chloride water and the pond at Patearoa in New Zealand is a (bi-)carbonate water. At Sutton the cationic order of dominance was reported to be Na > Mg > K > Ca and the anions were strongly dominated by chloride. In the pond at Patearoa the ionic orders of dominance were Na > K = Ca > Mg and HCO_3 > Cl > SO_4. Presumably in both cases an appreciable amount of sodium carbonate is present in the surrounding rocks and soils. This is almost certainly true of the Patearoa locality; the occurrence in this general area of sodium carbonate and pH values in excess of 8.4 was pointed out by Raeside et al. (1966). The soil type is characterized by the presence of numerous salt pans some of which are alkaline.

6.2 Classification of organisms as osmo-regulators and osmo-conformers

Many marine organisms have the same osmotic pressure as seawater. When the salt concentration of their surroundings changes, however, they must be able to adjust. Two means of contending with this situation are employed, and, depending on how they regulate the salt concentrations of their tissues, organisms are classified as osmo-regulators and osmo-conformers. The osmotic concentration of the body fluids of an osmo-conformer changes to match that of its external environment. However, an osmo-regulator controls the osmotic concentration of its body fluids, keeping them constant in spite of external alterations. Aquatic organisms that can tolerate a wide range of external ion concentrations (salinity) are called euryhaline; those that have a limited tolerance are called stenohaline. Even if aquatic organisms have an integument (a tough outer protective layer) that is relatively impermeable to water, as well as to small inorganic ions, their respiratory exchange surfaces are permeable. Hence, organisms occurring in water that has a lower solute concentration than their tissues (e.g., trout in mountain streams) will constantly lose ions to the environment as water flows into their tissues. In contrast, organisms in salty environments face a constant loss of water and an influx of ions (Michael B. Thompson; https://www.britannica.com/science/biosphere/Salinity#ref589476).

6.3 Life in thalassic water bodies

The Earth's oceans (Fig. 6.6) constitute a continuous body of saltwater that covers more than 70% of the Earth's surface. Geographers divide the ocean into five major basins: the Pacific, Atlantic, Indian, Arctic, and Southern. The five oceans of the earth are in reality one large interconnected water body.

In terms of geography, seas are smaller than oceans and are usually located where the land and ocean meet. By definition, a sea is a smaller part of an ocean and is typically partially enclosed by land. There are over 50 smaller seas scattered around the world.

Whereas oceans constitute very large bodies of water, bays, gulfs, and straits are types of waterbodies that are contained within a larger body of water near land. These three waterbodies are usually located at important points of human activities.

A bay is a small body of water or a broad inlet that is set off from a larger body of water generally where the land curves inward. The San Francisco Bay, off the coast in northern California, is a well-known bay in the United States. Examples of other bays include the Bay of Pigs (Cuba), Hudson Bay (Canada), Chesapeake Bay (Maryland and Virginia), and Bay of Bengal (near India).

A gulf is a large body of water, sometimes with a narrow mouth, that is almost completely surrounded by land. It can be considered a large bay. The world's largest gulf is the Gulf of Mexico, with a total surface area of about 1,554,000 square kilometers (600,000 square miles). It is surrounded by Mexico, the southern coast of the United States, and Cuba, and contains many bays, such as Matagorda Bay (Texas) and Mobile Bay (Alabama). Examples of other gulfs include the Gulf of California, Gulf of Aden (between the Red Sea and the Arabian Sea), and the Persian Gulf (between Saudi Arabia and Iran).

Marine science and ecosystems are crucial to understanding how to keep our oceans and marine life healthy and abundant. Ecosystems widely vary from fjords (long, narrow, deep inlets of the sea between high cliffs, as in Norway) in Scandinavia and Chile to common beaches to deep hydrothermal vents.

6.3.1 Life in the oceans, seas, gulfs, and bays

Oceans serve as Earth's largest habitat and the global reservoirs of life. The oceans are home to millions of Earth's

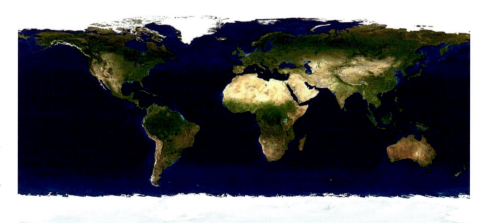

FIG. 6.6 True-color image of areas of the Earth that are ocean (blue), created by stitching together satellite imagery. Image: NASA, 2002. *(Source: https://www.geographyrealm.com/what-is-the-difference-between-a-sea-and-an-ocean/).*

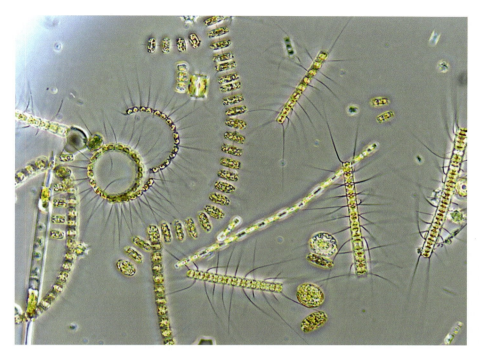

FIG. 6.7 **Mixed phytoplankton community NASA.** *Credits: University of Rhode Island/Stephanie Anderson. - NASA Earth Expeditions. Copyright: This file is in the public domain in the United States because it was solely created by NASA. NASA copyright policy states that "NASA material is not protected by copyright unless noted". (Source: https://commons.wikimedia.org/wiki/File:Mixed_phytoplankton_community_2.png).*

plants and animals—from tiny single-celled organisms to the gargantuan (enormously large) blue whale, the planet's largest living animal. The well-being, prosperity, and sustainability of the human enterprise rely on the functioning of Earth's oceans and life within it. The ocean represents our planet's largest habitat and supports more than half its species. From marine microbes to anemones, corals, algae, and whales—the sheer diversity of marine life is staggering.

6.3.1.1 Phytoplankton & algae

Photosynthesis in the ocean provides about half of Earth's oxygen. Phytoplankton (Fig. 6.7), also known as microalgae, are similar to terrestrial plants in that they contain chlorophyll and require sunlight in order to live and grow. Most phytoplankton are buoyant and float in the upper part of the ocean, where sunlight penetrates the water. The microscopic plants called phytoplankton that float at the surface and carry out this photosynthesis form the base of the ocean food web and, as such, feed most of the species in the sea. Other fodder for sea dwellers includes seaweed and kelp, which are types of algae, and seagrasses, which grow in shallower areas where they can catch sunlight.

Phytoplankton are very diverse, varying from photosynthesising bacteria to plant-like algae to armour-plated coccolithophores (Fig. 6.8). Important groups of phytoplankton include the diatoms, cyanobacteria and dinoflagellates, although many other groups are represented (Pierella et al., 2020).

FIG. 6.8 **Coccolithus pelagicus ssp. braarudii.** *Copyright: This file is licensed under the Creative Commons Attribution 2.5 Generic license. (Source: https://commons.wikimedia.org/wiki/File:Coccolithus_pelagicus.jpg).*

Most phytoplankton are too small to be individually seen with the unaided eye. However, when present in high enough numbers, some varieties may be noticeable as colored patches on the water surface due to the presence of chlorophyll within their cells and accessory pigments (such as *phycobiliproteins* or *xanthophylls*) in some species.

Algae (Fig. 6.9) are defined as a group of predominantly aquatic, photosynthetic, and nucleus-bearing organisms that lack the true roots, stems, leaves, and specialized multicellular reproductive structures of land plants. The largest and most complex marine algae are called seaweeds (Fig. 6.10). Seaweed species such as kelps provide essential nursery habitat for fisheries and other marine species and thus protect food sources; other species, such as planktonic algae, play a vital role in capturing carbon, producing up to 50% of Earth's oxygen.

Because of their importance in marine ecologies and for absorbing carbon dioxide, recent attention has been on cultivating seaweeds as a potential climate change mitigation strategy for bio-sequestration of carbon dioxide, alongside other benefits such as nutrient pollution reduction, increased habitat for coastal aquatic species, and reducing local ocean acidification (Duarte et al., 2017). The IPCC Special Report on the Ocean and Cryosphere in a Changing Climate recommends "further research attention" as a mitigation tactic (Bindoff et al., 2019).

6.3.1.2 Cephalopods, crustaceans, & other shellfish

A cephalopod is any member of the molluscan class Cephalopoda (Queiroz et al., 2020) such as a squid (Fig. 6.11), octopus (Fig. 6.12), cuttlefish, or nautilus. These

FIG. 6.9 A variety of algae growing on the sea bed in shallow waters. Copyright: This file is licensed under the Creative Commons Attribution-Share Alike 3.0 Unported license. *(Source: https://commons.wikimedia.org/wiki/File:NSW_seabed_1.JPG).*

FIG. 6.11 The squid *Euprymna scolopes* **swimming in the water column.** Image *Credit: Chris Frazee and Margaret McFall-Ngai. Copyright: © 2014 McFall-Ngai. This is an open-access article distributed under the terms of the Creative Commons Attribution License, which permits unrestricted use, distribution, and reproduction in any medium, provided the original author and source are credited. (Source: (2014) PLoS Biology Issue Image | Vol. 12(2) February 2014. PLoS Biol 12(2): ev12.i02. https://doi.org/10.1371/image.pbio.v12.i02. https://journals.plos.org/plosbiology/article/info%3Adoi%2F10.1371%2Fimage.pbio.v12.i02).*

FIG. 6.10 *Fucus serratus*—a seaweed—in its natural habitat. Copyright: This file is licensed under the Creative Commons Attribution-Share Alike 4.0 International license. *(Source: https://commons.wikimedia.org/wiki/File:Fucus_serratus_2015-09-08_ag_M0010140.jpg).*

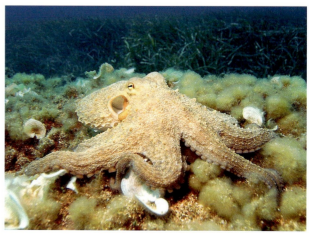

FIG. 6.12 *Octopus vulgaris* **(common octopus)** Author: albert kok. Copyright: This file is licensed under the Creative Commons Attribution-Share Alike 3.0 Unported, 2.5 Generic, 2.0 Generic and 1.0 Generic license. *(Source: https://commons.wikimedia.org/wiki/File:Octopus2.jpg).*

exclusively marine animals are characterized by bilateral body symmetry, a prominent head, and a set of arms or tentacles (muscular hydrostats) modified from the primitive molluscan foot. Fishers sometimes call cephalopods "inkfish," referring to their common ability to squirt ink. The squid species shown in Fig. 6.11 is a night-active predator in the shallow sand flats of the Hawaiian archipelago, uses the light produced by its luminous bacterial symbiont, *Vibrio fischeri*, as a camouflage. Much like the microbiota of the human gut, the squid's bioluminescent partner is acquired anew each generation and resides extracellularly along the surface of epithelial tissues. The relationship between *E. scolopes* and *V. fischeri* has been studied for over 25 years as a model for the establishment and maintenance of animal-bacterial symbioses (McFall-Ngai, 2014).

Like other cephalopods (active predatory molluscs of the large class Cephalopoda), an octopus is bilaterally symmetric with two eyes and a beaked mouth at the center point of the eight limbs. The soft body can radically alter its shape, enabling octopuses to squeeze through small gaps. They trail their eight appendages behind them as they swim. The siphon is used both for respiration and for locomotion, by expelling a jet of water. Octopuses have a complex nervous system and excellent sight, and are among the most intelligent and behaviorally diverse of all invertebrates. Octopuses inhabit various regions of the ocean, including coral reefs, pelagic waters, and the seabed; some live in the intertidal zone and others at abyssal depths.

Cuttlefish or cuttles are marine molluscs belonging to the class Cephalopoda, which also includes squid, octopuses, and nautiluses. Cuttlefish have a unique internal shell, the cuttlebone, which is used for control of buoyancy.

About 800 living species of cephalopods have been identified. Two important extinct taxa are the Ammonoidea (ammonites) and Belemnoidea (belemnites). Extant cephalopods range in size from the 10 mm (0.3 in) *Idiosepius thailandicus* to the 14 m (45.1 ft) colossal squid, the largest extant invertebrate.

Crabs, Starfish, Oysters, and Snails crawl and scoot along the ocean bottom. Cephalopods, crustaceans, and other shellfish play important roles in maintaining healthy ocean systems. Cephalopods such as the southern blue-ringed octopus and the vampire squid look other-worldly, Atlantic blue crab (Fig. 6.13) and American lobster (Fig. 6.14) are the focus of major fishing industries.

6.3.1.3 Ocean fishes

The world's oceans are estimated to be home to 20,000 species of fish. Ocean fishes come in all shapes, sizes, colors and live in drastically different depths and temperatures. Despite this diversity, the United Nations Food and Agriculture Organization reported in 2016 that 89.5% of fish stocks are fully fished or overfished.

FIG. 6.13 An Atlantic Blue Crab (*Callinectes sapidus*) in the permanent collection of The Children's Museum of Indianapolis. *(Source: https://commons.wikimedia.org/wiki/File:The_Childrens_Museum_of_Indianapolis_-_Atlantic_blue_crab.jpg) Copyright info: This file is licensed under the Creative Commons Attribution-Share Alike 3.0 Unported license.*

FIG. 6.14 (top) An Atlantic Lobster. *(Source: https://commons.wikimedia.org/wiki/File:Lobster.jpg) Author: U.S. National Oceanic and Atmospheric Administration Copyright info: his image is in the public domain because it contains materials that originally came from the U.S. National Oceanic and Atmospheric Administration, taken or made as part of an employee's official duties.* (bottom) A previously V-notched lobster is recaptured carrying eggs *(Source: https://commons.wikimedia.org/wiki/File:Homarus_americanus_eggs.jpg) Author: U.S. National Oceanic and Atmospheric Administration Copyright info: This image is in the public domain because it contains materials that originally came from the U.S. National Oceanic and Atmospheric Administration, taken or made as part of an employee's official duties.*

Fishes in the ocean are classified into various groups, depending on the oceanic zones where they prefer to live. For example, pelagic fish live in the pelagic zone

FIG. 6.15 An underwater picture taken in Moofushi Kandu, Maldives, showing a school of large pelagic predator fish (bluefin trevally) sizing up a school of small pelagic prey fish (anchovies). *Copyright info: This file is licensed under the Creative Commons Attribution-Share Alike 2.5 Italy license. (Source: https://en.wikipedia.org/wiki/Pelagic_fish#/media/File:Moofushi_Kandu_fish.jpg).*

FIG. 6.16 **Green turtle, Chelonia mydas is going for the air.** *Copyright info: This file is licensed under the Creative Commons Attribution-Share Alike 3.0 Unported license. (Source: https://commons.wikimedia.org/wiki/File:Chelonia_mydas_is_going_for_the_air_edit.jpg).*

of ocean—being neither close to the bottom nor near the shore—in contrast with demersal fish that do live on or near the bottom, and reef fish that are associated with coral reefs (Lal and Fortune, 2000). Marine pelagic fish can be divided into pelagic coastal fish and oceanic pelagic fish (McLintock, 2007). Coastal fish inhabit the relatively shallow and sunlit waters above the continental shelf, while oceanic fish inhabit the vast and deep waters beyond the continental shelf (even though they also may swim inshore) (Walrond, 2007).

Pelagic fish range in size from small coastal forage fish, such as herrings and sardines, to large apex predator oceanic fishes, such as bluefin tuna and oceanic sharks (Lal and Fortune, 2000). They are usually agile swimmers with streamlined bodies, capable of sustained cruising on long-distance migrations. Many pelagic fishes swim in schools weighing hundreds of tonnes. Others, such as the large ocean sunfish, are solitary (Lal and Fortune, 2000). Fig. 6.15 shows a school of large pelagic predator fish (bluefin trevally) sizing up a school of small pelagic prey fish (anchovies).

6.3.1.4 Sea turtles & reptiles

Sea turtles, sometimes called marine turtles (Avise and Hamrick, 1996), are reptiles of the order *Testudines* and of the suborder *Cryptodira*. The seven existing species of sea turtles are the green sea turtle, loggerhead sea turtle, Kemp's ridley sea turtle, olive ridley sea turtle, hawksbill sea turtle, flatback sea turtle, and leatherback sea turtle (Fisheries, NOAA. "Sea Turtles: NOAA Fisheries." www.nmfs.noaa.gov.). All six of the sea turtle species present in US waters (loggerhead, green sea turtle, hawksbill, Kemp's ridley, olive ridley, and leatherback) are listed as endangered and/or threatened under the Endangered Species Act. The seventh sea turtle species is the Flatback, which exists in the waters of Australia, Papua New Guinea and Indonesia. Sea turtles can be separated into the categories of hard-shelled (cheloniid) and leathery-shelled (dermochelyid) (Wyneken, 2001). While these highly migratory species periodically come ashore to either bask or nest, sea turtles spend the bulk of their lives in the ocean. Fig. 6.16 shows a green turtle, Chelonia mydas, going for the air.

Sea turtles have been around since the time of the dinosaurs, but all seven species in the world face potential extinction. *Oceana* (the largest international advocacy group dedicated entirely to ocean conservation) runs proven campaigns to protect sea turtles from death as bycatch, habitat degradation and other issues.

6.3.1.5 Marine mammals

Dolphins, whales, sea otters, and other marine mammals capture the imagination and demand the affection of people around the world. Many of these marine animals adhere to complex social systems and exhibit remarkable intelligence.

Dolphins (Fig. 6.17) are mammals not fish. Like every mammal, dolphins are warm-blooded. Unlike fish, who

FIG. 6.17 **A common dolphin.** Photo by NOAA (https://www.fisheries.noaa.gov/region/west-coast#southwest-science). *Copyright: This image is in the public domain because it contains materials that originally came from the U.S. National Oceanic and Atmospheric Administration, taken or made as part of an employee's official duties. (Source: https://commons.wikimedia.org/wiki/File:Comdolph.jpg).*

breathe through gills, dolphins breathe air using lungs. Dolphins must make frequent trips to the surface of the water to catch a breath. Dolphins range in size from the relatively small 1.7-m-long (5 ft 7 in) and 50-kg (110-pound) bodied Maui's dolphin to the 9.5 m (31 ft 2 in) and 10-tonne (11-short-ton) killer whale. Dolphins can sometimes leap about 30 feet (9.1 m). Several species of dolphins exhibit sexual dimorphism (the condition where the sexes of the same species exhibit different characteristics, particularly characteristics not directly involved in reproduction), in that the males are larger than females. They have streamlined bodies and two limbs that are modified into flippers (broad flat limb without fingers, used for swimming). Though not quite as flexible as seals, some dolphins can travel at speeds 29 km (18 mi) per hour for short distances (Grady et al., 2019). Dolphins use their conical shaped teeth to capture fast-moving prey. They have well-developed hearing which is adapted for both air and water and is so well developed that some can survive even if they are blind. Some species are well adapted for diving to great depths. They have a layer of fat, or blubber, under the skin to keep warm in the cold water.

Although dolphins are widespread, most species of them prefer the warmer waters of the tropic zones, but some, like the right whale dolphin, prefer colder climates. Dolphins feed largely on fish and squid, but a few, like the killer whale, feed on large mammals, like seals. Male dolphins typically mate with multiple females every year, but females only mate every two to three years. Calves are typically born in the spring and summer months and females bear all the responsibility for raising them. Mothers of some species fast and nurse their young for a relatively long period of time. Dolphins produce a variety of vocalizations, usually in the form of clicks and whistles.

Whales are a widely distributed and diverse group of fully aquatic placental marine mammals. Whales are fully aquatic, open-ocean creatures: they can feed, mate, give birth, suckle and raise their young at sea. Whales range in size from the 2.6 m (8.5 ft) and 135 kg (298 lb) dwarf sperm whale to the 29.9 m (98 ft) and 190 metric tons (210 short tons) blue whale, which is the largest known creature that has ever lived. The sperm whale is the largest toothed predator on Earth. Several whale species exhibit sexual dimorphism, in that the females are larger than males.

Whales evolved from land-living mammals, and must regularly surface to breathe air, although they can remain under water for long periods of time. Some species, such as the sperm whale can stay underwater for up to 90 min (Gray, 2013). They have blowholes (modified nostrils) located on top of their head, through which air is taken in and expelled (Fig. 6.18). They are warm-blooded, and have a layer of fat known as blubber under the skin. The whale is the largest mammal on Earth, and it takes a big camera to capture whales in all their majesty. Fig. 6.19 shows the tail and head portion of a whale while diving down and up.

FIG. 6.18 The blow of a blue whale. *(Source: https://en.wikipedia.org/wiki/Blue_whale#/media/File:Bluewhale1_noaa_crop.jpg) Copyright info: This image is in the public domain because it contains materials that originally came from the U.S. National Oceanic and Atmospheric Administration, taken or made as part of an employee's official duties.*

FIG. 6.19 Whale (top) tail portion while diving down). (bottom) head portion. *(Source: https://unsplash.com/images/animals/whale free whale images.).*

With streamlined fusiform (tapering at both ends) bodies and two limbs that are modified into flippers, whales can travel at speeds of up to 20 knots (37.04 km/h), though they are not as flexible or agile as seals. Whales produce a great variety of vocalizations, notably the extended songs of the

FIG. 6.20 Sea otter. *Copyright info: This file is licensed under the Creative Commons Attribution 2.0 Generic license. (Source: https://commons.wikimedia.org/wiki/File:Sea_otter_with_sea_urchin.jpg).*

FIG. 6.21 **Sea urchins like this purple sea urchin can damage kelp forests by chewing through kelp holdfasts.** *(Source: https://commons.wikimedia.org/wiki/File:Seaurchin_300.jpg) Copyright info: This image is in the public domain because it contains materials that originally came from the U.S. National Oceanic and Atmospheric Administration, taken or made as part of an employee's official duties.*

humpback whale. Although whales are widespread, most species prefer the colder waters of the northern and southern hemispheres, and migrate to the equator to give birth. Species such as humpbacks and blue whales are capable of traveling thousands of miles without feeding. Males typically mate with multiple females every year, but females only mate every two to three years. Calves are typically born in the spring and summer; females bear all the responsibility for raising them.

The sea otter (Fig. 6.20) is a marine mammal. Adult sea otters typically weigh between 14 and 45 kg (31 and 99 lb), making them the heaviest members of the weasel family (a family of carnivorous mammals), but among the smallest marine mammals. Unlike most marine mammals, the sea otter's primary form of insulation is an exceptionally thick coat of fur, the densest in the animal kingdom. Although it can walk on land, the sea otter is capable of living exclusively in the ocean.

The sea otter inhabits nearshore environments, where it dives to the sea floor to forage. It preys mostly on marine invertebrates such as sea urchins, various mollusks and crustaceans, and some species of fish. Its foraging and eating habits are noteworthy in several respects. First, its use of rocks to dislodge prey and to open shells makes it one of the few mammal species to use tools. In most of its range, it is a keystone species, controlling sea urchin (Fig. 6.21) populations that would otherwise inflict extensive damage to kelp forest (Fig. 6.22) ecosystems. Its diet includes prey species that are also valued by humans as food, leading to conflicts between sea otters and fisheries. Note that a keystone species is a species that has a disproportionately large effect on its natural environment relative to its abundance, a concept introduced in 1969 by the zoologist Robert T. Paine. Keystone species play a critical role in maintaining the structure of an ecological community, affecting many other organisms in an ecosystem and helping to determine

FIG. 6.22 **A diver in a kelp forest off the coast of California.** *(Source: https://commons.wikimedia.org/wiki/File:Diver_in_kelp_forest.jpg) Copyright info: This file is licensed under the Creative Commons Attribution 2.0 Generic license.*

the types and numbers of various other species in the community. Without keystone species, the ecosystem would be dramatically different or cease to exist altogether.

6.3.1.6 Sharks & rays

The earliest known sharks date back to more than 420 million years ago. Sharks (Fig. 6.23) are found in all seas and are common to depths up to 2000 m (6600 ft). They generally do not live in freshwater, although there are a few known exceptions, such as the bull shark and the river shark, which can be found in both seawater and freshwater (Allen, 1999). Sharks have a covering of dermal denticles (structurally minute teeth) that protect their skin from damage and parasites in addition to improving their fluid dynamics. They have numerous sets of replaceable teeth (Budker, 1971). Several species are apex predators (predators that are at the top of their food chain without natural predators). Select examples include the tiger shark, blue shark, great white shark, mako shark, thresher shark, and hammerhead shark.

FIG. 6.23 Sharks (top and bottom views of two different species). Copyright info: This file is licensed under the Creative Commons Attribution-Share Alike 4.0 International license. (Source: https://commons.wikimedia.org/wiki/File:TDpGUipa.jpg).

FIG. 6.24 (top) Frilled shark *chlamydoselachus anguineus* stuffed at Aquarium tropical du Palais de la Porte Dorée (Paris, France). *Source: https://commons.wikimedia.org/wiki/File:Chlamydoselachus_anguineus_head.jpg Copyright info: © Citron / CC BY-SA 3.0 This file is licensed under the Creative Commons Attribution-Share Alike 3.0 Unported license.* (bottom) Extended tips of the gill filaments of the frilled shark's six pairs of gills. *Source: https://commons.wikimedia.org/wiki/File:Frilled_shark_throat.jpg Copyright info: This file is licensed under the Creative Commons Attribution-Share Alike 2.5 Generic license.*

Sharks are caught by humans for shark meat or shark fin soup. Many shark populations are threatened by human activities. Since 1970, shark populations have been reduced by 71%, mostly from overfishing (Einhorn, 2021). Sharks have played a vital role in maintaining healthy oceans for hundreds of millions of years as a top predator. More than 450 species of sharks cruise the world's oceans, ranging in size from 8 inches to a whopping 40 feet long. But today, nearly one in four sharks and their relatives are threatened with extinction. A major cause is the demand for shark fins. Every year, fins from as many as 73 million sharks end up in the global fin trade.

Humans rarely encounter frilled sharks (Fig. 6.24 **(top)**), which prefer to remain in the oceans' depths, up to 5000 feet (1500 meters) below the surface. Considered living fossils, frilled sharks bear many physical characteristics of ancestors who swam the seas in the time of the dinosaurs (https://www.nationalgeographic.com/environment/article/ocean).

The frilled shark (*Chlamydoselachus anguineus*) and the southern African frilled shark (*Chlamydoselachus africana*) are the two extant species of shark in the family Chlamydoselachidae. The frilled shark is considered a living fossil, because of its primitive, anguilliform (eel-like) physical traits, such as a dark-brown color, amphistyly (the articulation of the jaws to the cranium), and a 2.0 m (6.6 ft)–long body, which has dorsal, pelvic, and anal fins located toward the tail. The common name, frilled shark, derives from the fringed appearance of the six pairs of gill slits at the shark's throat (Fig. 6.24 (bottom)). When hunting food, the frilled shark moves like an eel, bending and lunging to capture and swallow whole prey with its long and flexible jaws, which are equipped with 300 recurved, needle-like teeth (Ebert, 2003).

FIG. 6.25 An example for the family of stingrays Author: Albert Kok. *Copyright info: This image is in the public domain. (Source: https://commons.wikimedia.org/wiki/File:Stingray.jpg).*

Stingrays (Fig. 6.25) generally are not dangerous—in fact, they have a reputation for being gentle. They often burrow beneath the sand in the shallows and swim in the open water. Stingrays will usually only sting when disturbed or stepped on by unaware swimmers. Most of the time, one can avoid being stung by a stingray. However, there are instances in which humans have been killed by them. For example, the world was stunned with the news that on

September 4, 2006, a 44-year-old Australian conservationist and television personality Steeve Irwin was killed by the poisoned barb (a sharp projection near the end of the tail, which is angled away from the main point so as to make extraction difficult) of a huge sting ray while he was filming an underwater documentary in the Great Barrier Reef of Queensland. Irwin was brought to the surface unconscious. Steve Irwin died after being accidentally pierced in the chest. The stinger penetrated his thoracic wall, causing massive trauma (Crocodile Hunter, 2012; "Discovery Channel Mourns the Death of Steve Irwin." Animal.discovery.com).

6.3.1.7 Corals—reef builders in the oceans

Corals are living creatures that often live in compact colonies and build up coral reefs, the biggest of which is the Great Barrier Reef off the coast of Queensland, Australia. Corals and other invertebrates face major threats from climate change and destructive fishing such as dynamite fishing, bottom trawling and more.

Tiny marine creatures known as corals and other calcium-depositing animals build coral reefs usually on top of a rocky outcrop on the seafloor. Although coral is often mistaken for a rock or a plant, it is actually composed of tiny, fragile animals called *coral polyps*. Corals, which are considered to represent an early stage in the evolution of multi-cellular organisms, are marine invertebrates. Polyps are approximately cylindrical in shape and elongated at the axis of the vase-shaped body. Polyps can live individually (like many mushroom corals do) or in large colonies that comprise an entire reef structure (Fig. 6.26). The polyps may remain with their tentacles open or closed. But some of them have been seen with their tentacles folded toward their mouth and partially covering the mouth. Sometimes the polyps completely fold their tentacles toward their mouth, covering it totally. Polyps can live on their own, but are primarily associated with the spectacularly diverse limestone communities, or reefs, they construct. The polyp group includes the important reef builders that inhabit tropical oceans and secrete calcium carbonate to form a hard skeleton.

It is an almost universal attribute of polyps to reproduce asexually by the method of budding. This mode of reproduction may be combined with sexual reproduction (e.g., stony corals). Those polyps that reproduce asexually by the method of budding are entirely devoid of sexual organs. In an overwhelming majority of stony coral (*scleractinia*) species that reproduce sexually, there is ordinarily a synchronized release of eggs and sperm into the water during brief spawning events (Harrison et al., 1984).

An interesting and important species in the family of corals is the Brain Coral (Fig. 6.27). The name "brain coral" stems from the fact that the surfaces of its colonies are covered in a meandering, brain-like pattern of ridges, or walls, across its surface. In this species, the walls are relatively thin (Veron, 1993, 2000a,b) and are usually separated by short "valleys," although longer, meandering valleys do also sometimes occur (Dai and Horng, 2009). Like other corals, Brain Coral (*P. sinensis*) forms colonies that consist of many polyps. Each polyp of the Brain Coral secretes a hard skeleton, and in this species adjacent polyps share their skeleton walls. Colonies of this species display a wide range of colors, with some being quite dull in appearance while others are more brightly colored. The corallites (the skeletons of coral polyps) of the Brain Coral share common walls (Veron, 1993, 2000a,b; Dai and Horng, 2009), and the mouths of the polyps are aligned along the colony's

FIG. 6.26 **Alcyonium glomeratum or red sea fingers—A species of soft coral in the family *Alcyoniidae*.** (*Source: Alcyonium glomeratum or red sea fingers soft coral, 2008, Khurr-Fakkan, UAE Author: Yahia.Mokhtar Copyright info: Permission is granted to copy, distribute and/or modify this document under the terms of the GNU Free Documentation License, Version 1.2 or any later version published by the Free Software Foundation; with no Invariant Sections, no Front-Cover Texts, and no Back-Cover Texts. This work is licensed under the Creative Commons Attribution-ShareAlike 3.0 License. https://en.wikipedia.org/wiki/File:2008-01-26_Inshcape_1_Khurr-Fakkan_-_UAE_(2).jpg.*)

FIG. 6.27 **Brain coral (*Diploria labyrinthiformis*).** Picture taken by Jan Derk in 2005 on Bonaire. *Copyright info: This work has been released into the public domain by its author, Janderk at English Wikipedia. This applies worldwide. In some countries this may not be legally possible; if so: Janderk grants anyone the right to use this work for any purpose, without any conditions, unless such conditions are required by law. (Source: https://en.wikipedia.org/wiki/File:Brain_coral.jpg).*

valleys, with most valleys only having one mouth, but more elongated valleys sometimes having several (Veron, 1993; Dai and Horng, 2009).

The photosynthetic pigments of the coral's symbiotic algae give coral reefs their fascinating color. The term "coral" actually refers to the animals. Stony corals—or just hard corals—which consist of polyps that cluster in groups, are the primary reef builders in the oceans. *Polyps* are clones, each having the same genetic structure. Reefs are formed by skeletons of organisms bound together into reef mounds that grow upward from the seafloor. Each polyp generation grows on the skeletal remains of previous generations, forming a structure that has a shape characteristic of the species, but that is subject to environmental influences.

Often called "rainforests of the sea," shallow coral reefs form some of the most diverse ecosystems on Earth. They occupy less than 1% of the world's ocean surface, yet they provide a home for at least 25% of all marine species (Spalding and Grenfell, 1997; Spalding et al., 2001; Mulhall, 2009), including fish, mollusks, worms, crustaceans, echinoderms, sponges, tunicates, and other cnidarians (living species consisting of corals, hydras, jellyfish etc.) (Hoover, 2007). It has been estimated that over 4000 species of fish inhabit coral reefs (Spalding et al., 2001). According to reliable estimates, healthy reefs can produce up to 35 tons of fish per square kilometer each year, but damaged reefs produce much less (McClellan and John, 2008). About six million tons of fish are taken each year from coral reefs. Well-managed coral reefs have an annual yield of 15 tons of seafood on average per square kilometer.

The best-known types of reefs are tropical coral reefs which exist in most tropical waters. Although coral reefs are mostly found at depths where sunlight penetration occurs (i.e., photic zone), there are also coral reefs that exist in deep waters. In fact, the mesophotic (i.e., middle-light) coral reef ecosystems are the deepest of the light-dependent coral ecosystems. Coral reefs are found in the deep sea away from continental shelves, around oceanic islands, and as atolls (ring-shaped reefs, or chain of islands formed of coral). The vast majority of these islands are volcanic in origin. The few exceptions have tectonic origins where plate movements have lifted the deep ocean floor on the surface. Cold-water corals also exist on smaller scales in other areas. Seamounts are also favorite locations where corals thrive. A detailed discussion on corals and coral reefs has been provided by Joseph (2016). Coral are threatened by pollution, sedimentation, and global warming. Researchers are seeking ways to preserve fragile, ailing ecosystems such as Australia's Great Barrier Reef.

6.3.1.8 Starfish, jellyfish, & sea slugs

Starfish (Fig. 6.28) are marine invertebrates. They typically have a central disc and usually five arms, though some species have a larger number of arms. The aboral or upper

FIG. 6.28 Seastar (*Fromia monilis*). *Copyright info: This file is licensed under the Creative Commons Attribution-Share Alike 3.0 Unported license. (Source: https://commons.wikimedia.org/wiki/File:Fromia_monilis_(Seastar).jpg).*

surface may be smooth, granular or spiny, and is covered with overlapping plates. Many species are brightly colored in various shades of red or orange, while others are blue, grey or brown. Starfish have tube feet operated by a hydraulic system and a mouth at the center of the oral or lower surface. They are opportunistic feeders (exhibiting the practice of taking advantage of circumstances, guided primarily by self-interested motives) and are mostly predators on benthic invertebrates. Several species have specialized feeding behaviors including eversion (the process of turning inside-out) of their stomachs and suspension feeding. They have complex life cycles and can reproduce both sexually and asexually. Most can regenerate damaged parts or lost arms and they can shed arms as a means of defence. They occupy several significant ecological roles. Starfish, such as the ochre sea star (*Pisaster ochraceus*) and the reef sea star (*Stichaster australis*), have become widely known as examples of the keystone species concept in ecology.

The tropical crown-of-thorns starfish (*Acanthaster planci*) is a voracious predator of coral throughout the Indo-Pacific region, and the northern Pacific sea star is considered to be one of the world's 100 worst invasive species.

Despite regular discoveries about the ocean and its denizens (animal, or plant that lives or is found in a particular place), much remains unknown. More than 80% of the ocean is unmapped and unexplored, which leaves open the question of how many species there are yet to be discovered. At the same time, the ocean hosts some of the world's oldest creatures: Jellyfish (Fig. 6.29) have been around more than half a billion years, horseshoe crabs almost as long.

Jellyfish are mainly free-swimming marine animals with umbrella-shaped bells and trailing tentacles, although a few are anchored to the seabed by stalks rather than being mobile. The bell can pulsate to provide propulsion for highly efficient locomotion. The tentacles are armed with stinging cells and may be used to capture prey and defend against predators. Jellyfish have a complex life cycle; the

FIG. 6.29 **A Pacific sea nettle (*Chrysaora fuscescens*) at the Monterey Bay Aquarium in California, USA.** *Copyright info: This file is licensed under the Creative Commons Attribution-Share Alike 2.0 Generic license. (Source: https://commons.wikimedia.org/wiki/File:Jelly_cc11.jpg).*

medusa is normally the sexual phase, which produces planula larva that disperse widely and enter a sedentary polyp phase before reaching sexual maturity.

Jellyfish are found all over the world, from surface waters to the deep sea. Scyphozoans (the "true jellyfish") are exclusively marine, but some hydrozoans with a similar appearance live in freshwater. Large, often colorful, jellyfish are common in coastal zones worldwide. The medusae of most species are fast-growing, and mature within a few months, and die soon after breeding. However, the polyp stage in which it is attached to the seabed may be much more long-lived. Jellyfish have been in existence for at least 500 million years, and possibly 700 million years or more, making them the oldest multi-organ animal group.

Sea slug (Fig. 6.30) is a common name for some marine invertebrates with varying levels of resemblance to terrestrial slugs. Most creatures known as sea slugs are actually gastropods, that is, they are sea snails (marine gastropod mollusks) that over evolutionary time have either completely lost their shells, or have seemingly lost their shells due to having a greatly reduced or internal shell. The name "sea slug" is most often applied to nudibranchs, as well as to a paraphyletic set of other marine gastropods without obvious shells (Thompson, 1976).

Sea slugs have an enormous variation in body shape, color, and size. Most are partially translucent. The often bright colors of reef-dwelling species implies that these are under constant threat of predators, but the color can serve as a warning to other animals with regard to the sea slug's toxic stinging cells (nematocysts) or offensive taste. Like all gastropods, they have small, razor-sharp teeth, called radulas. Most sea slugs have a pair of rhinophores—sensory tentacles used primarily for the sense of smell—on their head, with a small eye at the base of each rhinophore. Many have feathery structures (cerata) on the back, often in a contrasting color, which act as gills. All species of genuine sea slugs have a selected prey animal on which they depend for food, including certain jellyfish, bryozoans, sea anemones, and plankton as well as other species of sea slugs (Reference: Byatt et al., 2001; "sea slug." The Columbia Encyclopedia, 6th ed., 2015. Encyclopedia.com. 10 Nov. 2015 http://www.encyclopedia.com).

6.3.1.9 Seabirds

From the emperor penguin to the blue-footed booby to the brown pelican, seabirds come in all shapes and sizes (some can't even fly) and play a vital role in ocean ecosystems. The emperor penguin (*Aptenodytes forsteri*) (Fig. 6.31) is the tallest and heaviest of all living penguin species and is

FIG. 6.30 **Sea slug (*Chelidonura varians*).** *Copyright info: This file is licensed under the Creative Commons Attribution 2.0 Generic license. (Source: https://commons.wikimedia.org/wiki/File:Chelidonura_varians.jpg).*

FIG. 6.31 **Two adult Emperor Penguins with a juvenile on Snow Hill Island, Antarctica.** *Copyright info: This file is licensed under the Creative Commons Attribution 2.0 Generic license. (Source: https://commons.wikimedia.org/wiki/File:Aptenodytes_forsteri_-Snow_Hill_Island,_Antarctica_adults_and_juvenile-8.jpg).*

endemic to Antarctica. The male and female are similar in plumage and size, reaching 100 cm (39 in) in length and weighing from 22–45 kg (49–99 lb). Feathers of the head and back are black and sharply delineated from the white belly, pale-yellow breast and bright-yellow ear patches.

Like all penguins it is flightless, with a streamlined body, and wings stiffened and flattened into flippers for a marine habitat. Its diet consists primarily of fish, but also includes crustaceans, such as krill, and cephalopods, such as squid. While hunting, the species can remain submerged around 20 min, diving to a depth of 535 m (1755 ft). It has several adaptations to facilitate this, including an unusually structured haemoglobin to allow it to function at low oxygen levels, solid bones to reduce barotrauma (injury to one's body because of changes in barometric or water pressure), and the ability to reduce its metabolism and shut down nonessential organ functions (https://en.wikipedia.org/wiki/Emperor_penguin).

The blue-footed booby (*Sula nebouxii*) (Fig. 6.32) is a marine bird native to subtropical and tropical regions of the eastern Pacific Ocean. It is one of six species of the genus Sula – known as boobies. It is easily recognizable by its distinctive bright blue feet, which is a sexually selected trait. Males display their feet in an elaborate mating ritual by lifting them up and down while strutting before the female. The female is slightly larger than the male and can measure up to 90 cm (35 in) long with a wingspan up to 1.5 m (5 ft) ("Blue-footed Booby Day." Galapagos Conservation Trust. 2010).

The natural breeding habitats of the blue-footed booby are the tropical and subtropical islands of the Pacific Ocean. It can be found from the Gulf of California south along the western coasts of Central and South America to Peru.

FIG. 6.33 A Brown Pelican in Florida. *Copyright info: This file is licensed under the Creative Commons Attribution-Share Alike 3.0 Unported license (Source: https://commons.wikimedia.org/wiki/File:Brown_Pelican21K.jpg).*

About half of all breeding pairs nest on the Galápagos Islands ("Blue-Footed Booby." *National Geographic*). Its diet mainly consists of fish, which it obtains by diving and sometimes swimming under water in search of its prey. It sometimes hunts alone, but usually hunts in groups (*Handbook of the Birds of the World Vol 1*. Lynx Edicions. 1992).

The brown pelican (*Pelecanus occidentalis*) (Fig. 6.33) is a bird of the pelican family, Pelecanidae, one of three species found in the Americas and one of two that feed by diving into water. It is found on the Atlantic Coast from New Jersey to the mouth of the Amazon River, and along the Pacific Coast from British Columbia to northern Chile, including the Galapagos Islands.

The brown pelican mainly feeds on fish, but occasionally eats amphibians (frogs, toads, newts, salamanders, and caecilians), crustaceans (crab, lobster, shrimp, or barnacle), and the eggs and nestlings of birds. It nests in colonies in secluded areas, often on islands, vegetated land among sand dunes, thickets of shrubs and trees, and mangroves. In 1903, Theodore Roosevelt set aside the first National Wildlife Refuge, Florida's Pelican Island, to protect the species from hunters.

6.3.1.10 Microbes living in subseafloor sediment layers

Despite the fact that only about 1% of the total marine primary production of organic carbon is available for deep-sea microorganisms, subseafloor sediments harbor over half of all prokaryotic cells on Earth (see Schippers et al., 2005). This estimation has been arrived at from numerous microscopic cell-counts in sediment cores collected under the Ocean Drilling Program (ODP). Most parts of the seafloor consist of a thick layer of sediment. This sediment layer functions as an abode for various kinds of microbes, particularly high-temperature-loving archaebacteria. Such organisms have shed a great deal of light on the complicated topic of the origin of life on planet Earth. The subseafloor hot

FIG. 6.32 Blue Footed Booby, Galapagos islands. *(Source: https://commons.wikimedia.org/wiki/File:Blue-footed-booby.jpg) Copyright info: This image is not in the Public Domain Author: Benjamint 444 Contact address.*

sediments are a fertile ground for several groups of researchers who undertake studies of thermophilic microbes.

Archaebacteria are a group of microorganisms considered to be an ancient form of life that evolved separately from the bacteria and blue-green algae, and they are sometimes classified as a kingdom. Hyperthermophilic (i.e., high-temperature-loving) archaebacteria, found in submarine hydrothermal areas (see Stetter, 1982; Segerer, et al., 1985; Stetter, 1986; Huber et al., 1989; Neuner et al., 1990), thrive at temperatures in the range 80–110°C, and they are unable to grow below 60°C. Various extremely thermophilic archaebacteria exhibit optimum growth at above 80°C. *Pyrodictium* is the most thermophilic of these organisms, growing at temperatures of up to 110°C and exhibiting optimum growth at about 105°C. All of these organisms grow by diverse types of anaerobic (nonoxygen) and aerobic (oxygen) metabolism (i.e., through a set of life-sustaining chemical transformations within the living cells) (see Stetter et al., 1986).

These high-temperature-loving bacteria represent life at the known upper temperature limit. Within the marine environment, hyperthermophilic archaebacteria have been found in shallow-water hydrothermal fields as well as in deep hot sediments and vents (see Stetter, 1982, 1986; Neuner et al., 1990). Sulfur-dependent archaebacteria can be assigned to two distinct branches: the aerobic, sulfur-oxidizing *Sulfolobales* (see Brock et al., 1972; Brierley and Brierley, 1973; Zillig et al., 1980; Woese et al., 1984) and the strictly anaerobic sulfur-reducing *Thermoproteales* (see Zillig et al., 1981; Stetter, 1982; Fischer et al., 1983). Archaebacteria are important as primary producers and consumers of organic matter within high-temperature ecosystems. Their distribution and possible modes of dissemination are at present largely unknown.

The deepest reaches of the ocean were once thought to be devoid of life, because no light penetrates beyond 1000 m (3300 feet). But then hydrothermal vents were discovered. These chimney-like structures allow tube worms, clams, mussels, and other organisms to survive not via photosynthesis but chemosynthesis, in which microbes convert chemicals released by the vents into energy (see Chapters 2 & 5 in Joseph, 2016). It was found that Earth's ocean floors support submarine hydrothermal vents, in whose vicinity organisms capable of living above 100°C thrive under anaerobic conditions (see Joseph, 2016). Bizarre fish with sensitive eyes, translucent flesh, and bioluminescent lures jutting from their heads lurk about in nearby waters, often surviving by eating bits of organic waste and flesh that rain down from above, or on the animals that feed on those bits (see Joseph, 2016).

6.3.2 Life in thalassic brackish water bodies

Salinity in thalassic brackish water bodies is much lower than that of ocean water (which averages 3.5%), as a result of abundant freshwater runoff from the surrounding land (rivers, streams, and alike). Consequently, life in brackish water bodies depends on the salinity sensitivity of various life-forms. Quite often, the fauna of thalassic brackish water bodies is a mixture of marine and freshwater species.

6.3.2.1 Fauna in the Baltic sea—world's largest inland brackish sea & an arm of the Atlantic ocean

The Baltic Sea (Fig. 6.34) is the world's largest inland brackish sea (Snoeijs-Leijonmalm and Andrén, 2017). This brackish water body is an arm of the Atlantic Ocean, enclosed by Denmark, Estonia, Finland, Germany, Latvia, Lithuania, Poland, Russia, Sweden and the North and Central European Plain. The sea stretches from 53°N to 66°N latitude and from 10°E to 30°E longitude. As a marginal sea of the Atlantic, water exchange between the two water bodies is limited (https://en.wikipedia.org/wiki/Baltic_Sea).

The Baltic Sea's salinity is much lower than that of ocean water. The open surface waters of the Baltic Sea "proper" generally have a salinity of 0.3 to 0.9%, which is border-line freshwater. The flow of freshwater into the sea from approximately two hundred rivers and the introduction of salt from the southwest build up a gradient of salinity in the Baltic Sea. The highest surface salinities, generally 0.7–0.9%, are in the south-western-most part of the Baltic, in the Arkona and Bornholm basins (the former

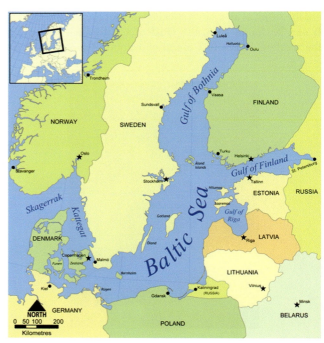

FIG. 6.34 Map of the Baltic Sea. *(Source: https://commons.wikimedia.org/wiki/File:Baltic_Sea_map.png) Copyright info: This file is licensed under the Creative Commons Attribution-Share Alike 3.0 Unported license.*

located roughly between southeast Zealand and Bornholm, and the latter directly east of Bornholm). Salinity gradually falls further east and north, reaching the lowest in the Bothnian Bay at around 0.3% (Viktorsson, 2018). As saltwater is denser than freshwater, the bottom of the Baltic Sea is saltier than the surface. This creates a vertical stratification of the water column, a halocline (a comparatively thin, typically horizontal layer within the water column, in which a property of the fluid varies greatly over a relatively short vertical distance), that represents a barrier to the exchange of oxygen and nutrients, and fosters completely separate maritime environments.

The Baltic Sea being a brackish water body, the fauna of the Baltic Sea is a mixture of marine and freshwater species. Among marine fishes are Atlantic cod (Fig. 6.35), Atlantic herring, European hake, European plaice, European flounder (Fig. 6.36), shorthorn sculpin and turbot, and examples of freshwater species include European perch, northern pike (Fig. 6.37), whitefish and common roach. Freshwater species may occur at outflows of rivers or streams in all coastal sections of the Baltic Sea. Otherwise marine species dominate in most sections of the Baltic, at least as far north as Gävle, where less than one-tenth are freshwater species. Further north the pattern is inverted. In the Bothnian Bay, roughly two-thirds of the species are freshwater. In the

FIG. 6.37 **Pike.** *(Source: https://commons.wikimedia.org/wiki/File:Esox_lucius_ZOO_1.jpg) Copyright info: This file is licensed under the Creative Commons Attribution-Share Alike 3.0 Unported, 2.5 Generic, 2.0 Generic and 1.0 Generic license.*

FIG. 6.35 **Atlantic Cod, *Gadus morhua*, taken through glass at Atlanterhavsparken, Ålesund, Norway.** Photo: Hans-Petter Fjeld (CC-BY-SA). *(Source: https://commons.wikimedia.org/wiki/File:Gadus_morhua_Cod-2b-Atlanterhavsparken-Norway.JPG) Copyright info: This file is licensed under the Creative Commons Attribution-Share Alike 2.5 Generic license.*

FIG. 6.36 **European Flounder from Oost Dyck bank in the Southern North Sea.** *(Source: https://en.wikipedia.org/wiki/European_flounder#/media/File:Platichthys_flesus_1.jpg) Attribution: © Hans Hillewaert Copyright info: This file is licensed under the Creative Commons Attribution-Share Alike 4.0 International license.*

FIG. 6.38 **The Atlantic red sea star *Asterias rubens*.** This starfish was sampled on the Belgian Continental Shelf. Photographed by Hans Hillewaert. *(Source: https://commons.wikimedia.org/wiki/File:Asterias_rubens.jpg) Copyright info: This file is licensed under the Creative Commons Attribution-Share Alike 4.0 International license. Attribution: © Hans Hillewaert.*

far north of this bay, saltwater species are almost entirely absent. For example, the common starfish (Fig. 6.38) and shore crab, two species that are very widespread along European coasts, are both unable to cope with the significantly lower salinity. Their range limit is west of Bornholm, meaning that they are absent from the vast majority of the Baltic Sea. Some marine species, such as the Atlantic cod and European flounder, can survive at relatively low salinities but need higher salinities to breed, which therefore occurs in deeper parts of the Baltic Sea (Nissling and Westin, 1997; Momigliano et al., 2018).

There is a decrease in species richness from the Danish belts to the Gulf of Bothnia. The Gulf of Bothnia is the northernmost arm of the Baltic Sea. The decreasing salinity along this path causes restrictions in both physiology and habitats (Lockwood et al., 1998). At more than 600 species of invertebrates, fish, aquatic mammals, aquatic birds and macrophytes, the Arkona Basin (roughly between southeast

Zealand and Bornholm) is far richer than other more eastern and northern basins in the Baltic Sea, which all have less than 400 species from these groups, with the exception of the Gulf of Finland with more than 750 species. However, even the most diverse sections of the Baltic Sea have far fewer species than the almost-full saltwater Kattegat, which is home to more than 1600 species from these groups. The lack of tidal propagation has affected the marine species as compared with the Atlantic.

6.3.2.2 Flora and fauna of the Bay of Bengal-connected Chilika Lake—a lake with a delicate salinity gradient between its different parts

Chilika Lake (Fig. 6.39) is a brackish water lagoon, located in the Odisha state on the east coast of India, at the mouth of the Daya River, flowing into the Bay of Bengal, covering an area of over 1100 km^2. It is the largest coastal lagoon in India and the largest brackish water lagoon in the world. The lake's salinity varies—from 0‰ in the northern sector, where there is complete freshwater, to 33‰ in the mouth, which is very close to complete saltwater. This delicate salinity gradient between different parts of the lake supports a wide variety of ecosystems.

The water spread area of the lake ranges between 1165 and 906 km^2 during the monsoon and summer respectively. A 32 km long, narrow, outer channel connects the lagoon to the Bay of Bengal. More recently a new mouth has been opened by the Chilika Development Authority (CDA) which has brought a new lease of life to the lagoon. The lake is home to a number of threatened species of plants and animals.

Birds: Chilika Lake is the largest wintering ground for migratory birds on the Indian sub-continent. The lake hosts over 160 species of birds in the peak migratory season. Birds from as far as the Caspian Sea, Lake Baikal, Aral Sea, and other remote parts of Russia, Kirghiz steppes of Kazakhstan, Central and southeast Asia, Ladakh, and Himalayas come here. These birds travel great distances; some of them possibly travel as much as 12,000 km to reach Chilika Lake (https://en.wikipedia.org/wiki/Chilika_Lake).

Migratory water fowl {e.g., Brown-headed Gull (Fig. 6.40)} arrive here from as far as the Caspian Sea, Baikal Lake and remote parts of Russia, Mongolia, Siberia, Iran, Iraq, Afghanistan and from the Himalayas.

Flora: Microalgae, marine seaweeds, sea grasses flourish in the brackish water of the Chilika Lake. Recent surveys revealed an overall 726 species of flowering plants belonging to 496 genera and 120 families. Fabaceae is the most dominant plant family followed by Poaceae (Fig. 6.41)

FIG. 6.40 A Brown-headed Gull (Chroicocephalus brunnicephalus) at the bird feeding point, Chilika, Odisha, India. *(Source: https://commons.wikimedia.org/wiki/File:Brown-headed_Gull_Chroicocephalus_brunnicephalus_3.jpg) Copyright info: This file is licensed under the Creative Commons Attribution-Share Alike 4.0 International license.*

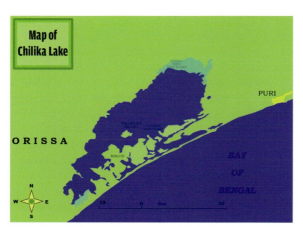

FIG. 6.39 Map of Lake Chilka (Puri, Orissa, India). *(Source: https://commons.wikimedia.org/wiki/File:Chilika_lake.png) Copyright info: This file is licensed under the Creative Commons Attribution-Share Alike 4.0 International, 3.0 Unported, 2.5 Generic, 2.0 Generic and 1.0 Generic license.*

FIG. 6.41 Flowering head of meadow foxtail belonging to the Poaceae family. *(Source: https://commons.wikimedia.org/wiki/File:Meadow_Foxtail_head.jpg) Copyright info: This file is licensed under the Creative Commons Attribution-Share Alike 3.0 Unported license. Attribution: Fir0002.*

FIG. 6.42 Dugong (*Dugong dugon*). (Source: https://commons.wikimedia.org/wiki/File:Dugong.JPG) Copyright info: This file is licensed under the Creative Commons Attribution-Share Alike 3.0 Germany license.

and Cyperaceae. The Fabaceae or Leguminosae, commonly known as the legume, pea, or bean family, are a large and economically important family of flowering plants. The flora is predominantly of aquatic and sub-aquatic plants. The species reported are leguminosae, poaceae, and cyperaceae; endemic cassipourea ceylanica; five species of seagrass, and more (https://en.wikipedia.org/wiki/Chilika_Lake). Especially the recovery of seagrass beds in recent years is a welcoming trend which may eventually result in re-colonization of endangered dugongs (Fig. 6.42). The dugong is a medium-sized marine mammal. It is one of four living species of the order Sirenia, which also includes three species of manatees. It is the only living representative of the once-diverse family Dugongidae.

<u>Fauna</u>: The lake is an ecosystem with large fishery resources, fish, and crab. As per the Chilika Development Authority's (CDA) updated data (2002), 323 aquatic species, which including 261 fish species {e.g., Helicopter catfish (Fig. 6.43)}, 28 prawn species and 34 crab species are reported, of which 65 species breed in the lake. 27 species are freshwater fishes and two genera of prawns. The remaining species migrate to the sea to breed. 21 species of herrings and sardines of the family *Clupeidae* are reported (https://en.wikipedia.org/wiki/Chilika_Lake).

Between 1998–2002, 40 fish species were recorded here for the first time and following the reopening of the lake mouth in 2000, six threatened species have reappeared, including:

1. Milk fish (Seba khainga).
2. Indo-Pacific tarpon (Panialehio).
3. Ten pounder (Nahama).
4. Bream (Kala khuranti).
5. Hilsa (Tenuealosa) ilisha (ilishi).
6. Mullet R. corsula (Kekenda).

Chilika lake is home to the only known population of Irrawaddy dolphins (Fig. 6.44) in India (Sutaria, 2005) and one of only two lagoons in the world that are home to this species (Ghosh and Pattnaik, 2008). In fact, the Irrawaddy dolphin (*Orcaella brevirostris*) is the flagship species of Chilika lake. The Irrawaddy dolphin is a euryhaline species of oceanic dolphin found in discontinuous subpopulations near sea coasts and in estuaries and rivers in parts of the Bay of Bengal and Southeast Asia. It closely resembles the Australian snubfin dolphin (of the same genus, *Orcaella*) and was not described as a separate species until 2005. It has a slate blue to a slate grey color and are a part of genus *Orcaella* which is also known as snubfin dolphins. Although found in much of the riverine and marine zones of South and Southeast Asia, the only concentrated lagoon populations are found in Chilika Lake and Songkhla Lake in southern Thailand (Brian and Perrin, 2007). Some Irrawaddy dolphins used to be sighted only along the inlet channel and in a limited portion of the central sector of the lake. After the opening of the new mouth at Satapada in the year 2000, they are now well distributed in the central and the southern sector of the lake (Ghosh and Pattnaik,

FIG. 6.43 Helicopter catfish (*Wallago attu*) – A common type of fish in Chilika lake. (Source: https://commons.wikimedia.org/wiki/File:Wallago_attu_1.jpg) Copyright info: This file is licensed under the Creative Commons Attribution-Share Alike 3.0 Unported license.

FIG. 6.44 (top) Irrawaddy dolphin (Orcaella brevirostris) of Chilika Lake, Odisha, India. (Source: https://commons.wikimedia.org/wiki/File:DSC_0764f.jpg) Copyright info: his file is licensed under the Creative Commons Attribution-Share Alike 4.0 International license. (bottom) Irrawaddy dolphin (Source: https://commons.wikimedia.org/wiki/File:Irrawaddy_dolphin-Orcaella_brevirostris_by_2eight.jpg) Copyright info: This file is licensed under the Creative Commons Attribution-Share Alike 3.0 Germany license. Flag of Germany.svg Attribution: Foto: Stefan Brending, Lizenz: Creative Commons by-sa-3.0 de.

FIG. 6.45 **Bottlenose dolphin.** *(Source: https://en.wikipedia.org/wiki/File:Tursiops_truncatus_01.jpg) Copyright info: This file is in the public domain in the United States because it was solely created by NASA. NASA copyright policy states that "NASA material is not protected by copyright unless noted".*

2008). The number of dolphins sighted has varied from 50 to 170. A 2006 census counted 131 dolphins and the 2007 census revealed 138 dolphins. Of the 138 dolphins, 115 were adults, 17 adolescents and six calves. 60 adults were spotted in the outer channel, followed by 32 in the central sector, and 23 in the southern sector (Das, 2008). It is classified as Endangered according to International Union for Conservation of Nature (IUCN).

A small population of Bottlenose dolphins (Fig. 6.45) also migrate into the lagoon from the sea (Forest and Environment Department, "Chilika"). Bottlenose dolphins inhabit warm and temperate seas worldwide, being found everywhere except for the Arctic and Antarctic Circle regions (Wells and Scott, 2009). Chilika fishermen say that when Irrawaddy dolphins and bottlenose dolphins meet in the outer channel, the former get frightened and are forced to return toward the lake (Sinha, 2004).

6.4 Life in athalassic brackish water bodies

As already indicated, brackish water, also sometimes termed brack water, is water occurring in a natural environment, having salinity less than that of seawater. It may result from mixing seawater (saltwater) with fresh water together, as in estuaries, or it may occur in brackish fossil aquifers. Certain human activities can produce brackish water, in particular civil engineering projects such as dikes and the flooding of coastal marshland to produce brackish water pools for freshwater prawn farming. Brackish water is also the primary waste product of the salinity gradient power process.

Technically, brackish water contains between 0.5 and 30 grams of salt per liter—more often expressed as 0.5 to 30 parts per thousand (‰), which is a specific gravity of between 1.0004 and 1.0226. Thus, "brackish" covers a range of salinity regimes and is not considered a precisely defined condition. It is characteristic of many brackish surface waters that their salinity can vary considerably over space or time. Water with a salt concentration greater than 30‰ is considered saline.

Some seas and lakes are brackish. The Caspian Sea is the world's largest lake and contains brackish water with a salinity about one-third that of normal seawater. The Caspian is famous for its peculiar animal fauna, including one of the few non-marine seals (the Caspian seal) and the great sturgeons, a major source of caviar (the pickled eggs of fish or shellfish, eaten as a delicacy).

6.4.1 Life in the Caspian Sea

The Caspian Sea (see Fig. 6.2)—the largest inland brackish water body in the world: bordered in the northeast by Kazakhstan, in the southeast by Turkmenistan, in the south by Iran, in the southwest by Azerbaijan, and in the northwest by Russia—has historically been considered a sea because of its size and its saline water, but it embodies many characteristics of lakes. While the Caspian Sea is not fresh water, its salty water is diluted by the inflow of fresh water, especially in the north. The Caspian Sea is named after the Caspi Tribe, which settled on its southwestern shore.

The Caspian Sea is a remnant of the ancient Paratethys Sea, part of the Tethys Ocean that existed 50 million to 60 million years ago. At that time, the Tethys Ocean was connected to the Atlantic and Pacific oceans, according to WorldLakes.org. Over millennia, continental platforms shifted, and the Tethys Ocean lost its connections to other oceans. Much of it evaporated during hot and dry periods, and eventually the Caspian Sea, the Black Sea, and the Aral Sea formed. The Caspian Sea is estimated to be about 30 million years old. The saltwater from the Tethys Sea remained and accounts for the Caspian Sea's salinity (Szalay, 2017).

The Caspian Sea is the world's largest lake and contains brackish water with a salinity about one-third of normal seawater. The Caspian is famous for its peculiar animal fauna, including one of the few non-marine seals (the Caspian seal) and the great sturgeons, a major source of caviar. The Caspian Sea is known for its biodiversity. In many areas, the shores are dotted with shallow saline pools in which birds, small fish, crustaceans, and invertebrates thrive. Birds are present throughout the year, and many species use the Caspian Sea as a migratory refuge. Nearly 2000 species and subspecies of animals live in and around the Caspian Sea, according to *Casp Info*—An EU funded project (which ended in 2010) aimed to strengthen the data management infrastructure around the Caspian Sea. About 400 of the animals living in and around the Caspian Sea are endemic to the area, including the Caspian gull, Caspian turn, spur-thighed tortoise, Horsfield's tortoise, Caspian white fish, Caspian salmon and Caspian seal, which is the only aquatic mammal in the area (Szalay, 2017).

More than 500 plant species are found in the Caspian Sea. The Volga River Delta in the North Caspian is home to a wide range of endemic or rare aquatic plants, according to the World Wildlife Fund. The vegetation in the Turkmenistan portion of the Caspian shores is considered impoverished. Nevertheless, there are some specialized salt-resistant plants such as shrubs and sagebrush. Blue-green algae (cyanobacteria) and diatoms constitute the greatest biomass concentrations, and there are several species of red and brown algae. The Caspian Sea is home to a wide range of fish species. Animal life—which has been affected greatly by changes in salinity—includes fish species such as sturgeon, herring, pike, perch, and sprat; several species of mollusks; and a variety of other organisms including sponges. Some 15 species of Arctic (e.g., the Caspian seal) and Mediterranean types complement the basic fauna. Some organisms have migrated to the Caspian relatively recently: barnacles, crabs, and clams, for example, have been transported by sea vessels, while gray mullets have been deliberately introduced by humans (https://www.britannica.com/place/Caspian-Sea/Marine-life).

6.4.1.1 Caspian seal

The Caspian seal (*Pusa caspica*) (Fig. 6.46) is one of the smallest members of the earless seal family and unique in that it is found exclusively in the brackish Caspian Sea. They are found not only along the shorelines, but also on the many rocky islands and floating blocks of ice that dot the Caspian Sea. In winter, and cooler parts of the spring and autumn season, these marine mammals populate the Northern Caspian. As the ice melts in the warmer season, they can be found on the mouths of the Volga and Ural Rivers, as well as the southern latitudes of the Caspian where cooler waters can be found due to greater depth.

Evidence suggests that the Caspian seals are descended from Arctic ringed seals that reached the area from the north during an earlier part of the Quaternary period (~2.59 million years ago to the present) and became isolated in the landlocked Caspian Sea when continental ice sheets melted.

Adults are about 126–129 cm (50–51 in) in length. Males are longer than females at an early age, but females experience more rapid growth until they reach ten years of age. Males can grow gradually until they reach an age of about 30 or 40 years (Wilson et al., 2014). Adults weigh around 86 kg (190 lb); males are generally larger and bulkier.

In the summer and winter, during mating season, Caspian seals tend to be gregarious (living in flocks or loosely organized communities), spending most of their time in large colonies/groups. At other times of the year, Caspian seals are solitary. During the summer, however, they make aggressive snorts (explosive sound made by the sudden forcing of breath through the nose) or use flipper waving to tell other seals to keep their distance. Little else is known about their behavior (Reeves et al., 2002).

In winter, and cooler parts of the spring and autumn season, these marine mammals populate the Northern Caspian. In the first days of April, spring migration to the southern part of the Caspian Sea begins with mature female seals and their pups, during this migration hungry seals eat the fish in the nets. Male mature seals stay in the northern Caspian Sea longer and wait until the moulting (shedding old hair to make way for a new growth) is completed.

Caspian seals are piscivorous (eating primarily fish). However, they eat a variety of food depending on season and availability. A typical diet for Caspian seals found in the northern Caspian Sea consists of crustaceans and various fish species. Caspian seals are shallow divers, typically diving 50 m (160 ft) for about 1 min, although scientists have recorded Caspian seals diving much deeper and for longer periods of time. After foraging during a dive, they rest at the surface of the water (Reeves et al., 2002).

6.4.1.2 Caspian sturgeon—the world's largest freshwater fish

The most famous and financially valuable fish in the region is the beluga sturgeon, sometimes called the European or Caspian sturgeon. The world's largest freshwater fish, the beluga sturgeon is known for its eggs, which are processed into caviar (pickle). The majority of the world's beluga caviar comes from the Caspian Sea. This has caused problems with overfishing. Dams have also destroyed much of their spawning grounds, and pesticides used in land agriculture have limited their fertility. The beluga sturgeon is now critically endangered, according to the World Wildlife Fund.

Among the wide range of fish species, sturgeons (the common name for the 27 species of fish belonging to the family *Acipenseridae*) are long-lived, late-maturing fishes with distinctive characteristics, such as a heterocercal (having unequal upper and lower lobes) caudal (tail-like) fin similar to those of sharks, and an elongated, spindle-like body that is smooth-skinned, scale-less, and armored with five lateral rows of bony plates called scutes. Several species can grow quite large, typically ranging 7–12 ft

FIG. 6.46 Caspian seal. *(Source: https://commons.wikimedia.org/wiki/File:Caspian_seal_03.jpg) Author: Aboutaleb Nadri Copyright info: This file is licensed under the Creative Commons Attribution 4.0 International license.* Attribution: *Mehr News Agency.*

FIG. 6.47 Beluga sturgeon fish found in the Caspian Sea. Author: Максим Яковлєв. *Copyright info: This file is licensed under the Creative Commons Attribution-Share Alike 4.0 International license. (Source: https://en.wikipedia.org/wiki/Beluga_(sturgeon)).*

FIG. 6.48 **Ossetra sturgeon (also known as Russian sturgeon) found in the Caspian Sea.** *(Source: https://commons.wikimedia.org/wiki/File:Waxdick_(Acipenser_gueldenstaedtii_)_-_crop.jpg. https://en.wikipedia.org/wiki/Russian_sturgeon) Copyright info: This file is licensed under the Creative Commons Attribution-Share Alike 3.0 Unported license.*

(2–3+1/2 m) in length. The largest sturgeon on record was a beluga female captured in the Volga estuary in 1827, measuring 7.2 m (24 ft) long and weighing 1571 kg (3463 lb). Most sturgeons are anadromous (migrating up rivers from the sea to spawn) bottom-feeders, which migrate upstream to spawn, but spend most of their lives feeding in river deltas and estuaries. Some species inhabit freshwater environments exclusively, while others primarily inhabit marine environments near coastal areas, and are known to venture into open ocean. The Caspian Sea is best known for sturgeons' caviar—roe (the mass of eggs contained in the ovaries of fish) from wild sturgeon in the Caspian Sea and Black Sea (Davidson and Jane, 2006) (Beluga, Ossetra, and Sevruga caviars). Fig. 6.47 shows Beluga sturgeon fish found in the Caspian Sea. Note that Caviar is considered a delicacy food, and is eaten as a garnish or a spread.

The beluga, also known as the beluga sturgeon or great sturgeon, is a species of anadromous fish in the sturgeon family (*Acipenseridae*) of order Acipenseriformes. It is found primarily in the Caspian and Black Sea basins, and formerly in the Adriatic Sea. Based on maximum size, it is the third-most-massive living species of bony fish.

Ossetra sturgeon (Fig. 6.48) found in the Caspian Sea weighs 50–400 pounds and can live up to 50 years. Also known as Russian sturgeon, the diamond sturgeon or Danube sturgeon, Ossetra sturgeon is a species of fish in the family *Acipenseridae.*

The sevruga sturgeon (also known as starry sturgeon, and stellate sturgeon) is a species of sturgeon, native to the Black, Azov, Caspian and Aegean Sea basins. It suffered local extinction and it is predicted that the remaining natural population will follow soon due to overfishing (Qiwei, 2010). Fig. 6.49 shows a starry sturgeon in a bazaar in Odessa, Ukraine

Several species of sturgeon are harvested for their roe (fully ripe internal egg masses in the ovaries, or the released external egg masses), which is processed into the luxury food caviar. This has led to serious overexploitation, which combined with other conservation threats, has brought most of the species to critically endangered status, at the edge of extinction.

As already indicated, the Caspian Sea has been famous for its sturgeon, a fish prized for its caviar, and the sea accounts for the great bulk of the world catch. During the

FIG. 6.49 **Starry sturgeon in a bazaar in Odessa, Ukraine.** *(Source: https://commons.wikimedia.org/wiki/File:Odesa_bazaar_(8)_Sturgeon.JPG) https://en.wikipedia.org/wiki/Starry_sturgeon Author: Jewgienij Bal Copyright info: This file is licensed under the Creative Commons Attribution-Share Alike 3.0 Unported license.*

long period (1929–1977) of water-level decline and consequent drying of the most favorable spawning grounds, the sturgeon population fell considerably. A number of measures, including prohibition of sturgeon fishing in the open sea and the introduction of aquaculture, have been undertaken to improve the situation.

6.4.2 Life in lake texoma

Lake Texoma (Fig. 6.50), a reservoir on the border between the US states of Texas and Oklahoma, is a rare example of a brackish lake that is neither part of an endorheic basin (body of water that does not flow into the sea) nor a direct arm of the ocean, though its salinity is considerably lower than that of many other bodies of water. Lake Texoma is formed by Denison Dam on the Red River in Bryan County, Oklahoma, and Grayson County, Texas, about 726 miles (1168 km) upstream from the mouth of the river. It is located at the confluence of the Red and Washita

FIG. 6.50 Map of Lake Texoma. Drawn from satellite imagery. *(Source: https://commons.wikimedia.org/wiki/File:Lake_Texoma_map.gif) Copyright info: This file is licensed under the Creative Commons Attribution-Share Alike 3.0 Unported license.*

Rivers. The reservoir was created by the damming of the Red River of the South, which (along with several of its tributaries) receives large amounts of salt from natural seepage from buried deposits in the upstream region. The salinity is high enough that striped bass, a fish normally found only in saltwater, has self-sustaining populations in the lake (Malewitz, 2013; U.S. Geological Survey Fact Sheet 170-197). Fig. 6.51 provides a glimpse of a small section of the Texoma Lake, covering the water body and the shoreline.

The Red River that formed Lake Texoma is a saltwater river due to salt deposits left over from a 250-million-year-old former sea that was in the current Texas-Oklahoma border region. As time passed, that sea evaporated, leaving salt deposits—mostly sodium chloride. Rock and silt eventually buried the deposits, but the salt continues to leach through natural seeps in tributaries above Lake Texoma, sending as much as 3450 tons of salt per day flowing down the Red River. Due to this phenomenon striped bass, a saltwater fish, thrive in Lake Texoma. Lake Texoma is home to the only self-sustaining population of striped bass in Texas ("1996–97 Stream Monitoring and Outreach Activities Fact Sheet").

Lake Texoma provides habitat for at least 70 species of fish. Those species popular for recreational fishing include largemouth, spotted, white, and striped bass; white crappie; and catfish. Downstream of the dam is a tailwater fishery that supports the species and the three species of local catfish. American gizzard shad, threadfin shad, and inland silverside are important forage species. Freshwater drum, common carp, gars, buffaloes, and river carpsucker also inhabit the lake.

The striped bass (*Morone saxatilis*), also called the Atlantic striped bass, striper, linesider, rock or rockfish, is found primarily along the Atlantic coast of North America. Striped bass found in the Gulf of Mexico are a separate strain referred to as Gulf Coast striped bass (Reference: Gulf Coast Striped Bass. Welaka National Fish Hatchery. Fws.gov (September 16, 2009)).

The striped bass is a typical member of the family *Moronidae* in shape, having a streamlined, silvery body marked with longitudinal dark stripes running from behind the gills to the base of the tail. Common mature size is 20 to 40 pounds (9–18 kg). The largest specimen recorded was 124 pounds (56 kg), netted in 1896. Striped bass are believed to live for up to 30 years (Froese and Pauly, 2008). The average size in length is 20 to 35 inches (50–90 cm) and approximately 5 to 20 pounds (2–9 kg).

Striped bass are native to the Atlantic coastline of North America from the St. Lawrence River into the Gulf of Mexico to approximately Louisiana. They are anadromous fish (fish born in freshwater who spend most of their lives in saltwater and return to freshwater to spawn) that migrate between fresh and saltwater. Spawning takes place in fresh water. Fig. 6.52 shows a researcher holding up a large striped bass, *Morone saxatilis*.

Though the population of striped bass was growing and repopulating in the late 1980s and throughout the 1990s, a study executed by the Wildlife and Fisheries Program at West Virginia University found that the rapid growth of the striped bass population was exerting a tremendous pressure on its prey (river herring, shad, and blueback herring). This pressure on their food source was putting their own population at risk due to the population of prey naturally not coming back to the same spawning areas (Hartman, 2003).

Striped bass are spawn in fresh water, and although they have been successfully adapted to freshwater habitat, they naturally spend their adult lives in saltwater (i.e., they are anadromous). Striped bass is of significant value for sport

FIG. 6.51 A glimpse of a small section of the Texoma Lake, covering the water body and the shoreline. *(Source: https://commons.wikimedia.org/wiki/File:Lake_Texoma.JPG) Copyright info: This file is licensed under the Creative Commons Attribution-Share Alike 3.0 Unported license.*

FIG. 6.52 A researcher holds up a large striped bass, Morone saxatilis. *(Source: https://commons.wikimedia.org/wiki/File:Researcher_with_striped_bass.jpg)* Copyright info: *This work is in the public domain in the United States because it is a work prepared by an officer or employee of the United States Government as part of that person's official duties under the terms of Title 17, Chapter 1, Section 105 of the US Code.*

fishing, and have been introduced to many waterways outside their natural range.

The lake was stocked with striped bass in the late 1960s, and has proven to be an excellent habitat for them. It is one of the seven US inland lakes where the striped bass reproduce naturally, instead of being farmed and released into the waters. The "stripers" feed on large schools of shad, and often reach sizes of 12 to 20 pounds (5.4 to 9.1 kg), with a lake record of 35.12 pounds (15.93 kg) caught on April 25, 1984.

In 2004, a blue catfish was pulled from the lake that weighed 121.5 pounds (55.1 kg), temporarily setting a world weight record for rod-and-reel-caught catfish (Lambeth, M., "2007 Oklahoma Catfish." Oklahoma Game & Fish. Intermedia Outdoors, Inc.). The fish was moved to a freshwater aquarium in Athens, Texas. More commonly, catfish in Lake Texoma weigh between 5 and 70 pounds (2.3 and 31.8 kg).

6.5 Life in athalassic hypersaline water bodies

The environment has a profound influence on the microbial population of an aquatic medium. Apart from availability of food, salinity is an important parameter that determines the survivability of life in any water body. The term salinity refers to the amount of dissolved salts that are present in water. Sodium and chloride are the predominant ions in seawater, and the concentrations of magnesium, calcium, and sulfate ions are also substantial. Naturally occurring waters vary in salinity from the almost pure water, devoid of salts, in snowmelt to the saturated solutions in salt lakes such as the Dead Sea. In general, salinity in the oceans is constant but is more variable along the coast where seawater is diluted with freshwater from runoff or from the emptying of rivers. This brackish water forms a barrier separating marine and freshwater organisms. The cells of organisms also contain solutions of dissolved ions, but the range of salinity that occurs in tissues is narrower than the range that occurs in nature. Although a minimum number of ions must be present in the cytoplasm for the cell to function properly, excessive concentrations of ions will impair cellular functioning. Organisms that live in aquatic environments and whose integument (tough outer protective layer) is permeable to water, therefore, must be able to contend with osmotic pressure. This pressure arises if two solutions of unequal solute concentration exist on either side of a semipermeable membrane such as the skin. Water from the solution with a lower solute concentration will cross the membrane diluting the more highly concentrated solution until both concentrations are equalized. If the salt concentration of an organism's body fluids is higher than that of the surrounding environment, the osmotic pressure will cause water to diffuse through the skin until the concentrations are equal unless some mechanism prevents this from happening.

Studies have shown that, in hypersaline media, the bacterial communities are particularly rich in halophilic and hyperhalophilic bacteria. A study (B. Elazari-Volcani, doctoral thesis, Hebrew University, Jerusalem, 1940) on the waters of the Dead Sea indicated this, and in 1891 Lortet (Cited in E. Mace, Trait practique de bacteriologie, p. 669, 1901. Bailliere Ed., Paris) had isolated strains of *Clostridium tetani* from mud of the same sea. This showed that the strict halophiles were not the only microorganisms present in these particular waters (Brisou et al., 1974).

It has been known for long that the strongly saline environment is primarily a domain of prokaryotes and the spectrum of eukaryotic species in highly saline biotopes (the region of a habitat associated with a particular ecological community) is rather restricted. Several studies (e.g., Imhoff et al., 1979; Trüper and Galinski, 1986; Imhoff, 1988, 2001, 2002) have found that the dominant primary producers are halophilic (capable of flourishing in a salty environment) and halotolerant (having ability to grow at salt concentrations higher than those required for growth) algae (simple, non-flowering, and typically aquatic plants of a large group that includes the seaweeds and many single-celled forms that contain chlorophyll but lack true stems, roots, leaves, and vascular tissue) and cyanobacteria (a phylum comprised of photosynthetic bacteria that live in aquatic habitats and moist soils) as well as anoxygenic phototrophic bacteria (a diverse collection of organisms that are defined by their ability to grow using energy from light without generating oxygen (George et al., 2020)). A variety of anoxygenic phototrophic bacteria has been isolated from different hypersaline habitats, such as marine salterns (Rodriguez-Valera et al., 1985; Caumette et al., 1988, 1991; Caumette, 1993), alkaline soda lakes in the Egyptian Wadi Natrun (Imhoff and Trüper, 1977, 1981; Imhoff et al., 1978), in Siberia and Mongolia (Bryantseva et al., 1999, 2000) and from Solar Lake (Sinai) (Cohen and Krumbein, 1977; Caumette et al., 1997) as reviewed in Imhoff (2001). Also, green sulfur bacteria have been observed in various saline environments mainly based on microscopic and macroscopic observations (Giani et al., 1989; Caumette, 1993; Oren, 1993). Note that the photosynthetic metabolism of green sulfur bacteria differs from that of Cyanobacteria, algae, and green plants in that water cannot serve as an electron-donating substrate and molecular oxygen is not generated. Species of the genus *Prosthecochloris* were obtained from marine and saline environments and are recognized as halotolerant and moderately halophilic organisms (Gorlenko, 1970; Imhoff, 2001, 2003; Vila et al., 2002; Alexander and Imhoff, 2006; Triadó-Margarit et al., 2010).

6.5.1 Life in the Dead Sea—situated between Israel and Jordan

In Arabic language the Dead Sea is called the Sea of Death. Although referred to as a sea, the Dead Sea is actually an extremely large landlocked Salt Lake located between Israel and Jordan, in Western Asia. The Dead Sea appears quite dead. In general, because of high salinity, there are no plants, fish, or any other visible life in the sea. Occasionally, when conditions are right, the sea blooms red with life. This happened in 1980 and 1992 (Frazer, 2011). The majority of the microbial lineages are known from environments of various salinities (Oren, 2002), usually in well-established microbial mats or in overlaying waters with a constant salinity. Hence, a full comparison between the microbial communities in this system to those in any other system is difficult (Ionescu et al., 2012).

6.5.1.1 High density of archaea in the spring waters in the lake bed

Due to its extreme salinity and high magnesium (Mg) concentration, the Dead Sea is characterized by a very low density of cells, most of which are Archaea. According to Ionescu et al. (2012), "The increased salinity and the elevated concentration of divalent ions make the Dead Sea an extreme environment that is not tolerated by most organisms. This is reflected in a generally low diversity and very low abundance of microorganisms. The microbiology of the lake has been subject for research since the 1930s when Benjamin Elazari-Volcani (Wilkansky at the time) isolated the first microorganisms from the sediment of the Dead Sea (Wilkansky, 1936). Besides Bacteria and Archaea (Elazari-Volcani, 1940), these isolates included algae (Elazari-Volcani, 1943a), protozoa (Elazari-Volcani, 1943b), and ciliates (Elazari-Volcani, 1944). Since then, several Bacterial and Archaeal isolates have been obtained in culture, both from the sediment and from the water body (Oren, 2010). The general cell abundance in the Dead Sea water is very low (<5 × 10^4 cells/mL; Ionescu et al. 2012's study), except for two blooms in 1980 and 1992, when after severe winters the upper meter of the water column was diluted by 15–30%, floods provided an input of phosphate, and the cell concentrations reached 20–35 × 10^6 cells/mL (Oren and Gurevich, 1995)."

According to Ionescu et al. (2012), the relatively high fraction of Archaea detected in the spring waters by collection along the path, which was much higher than in other groundwater aquifers and freshwater bodies (Keough et al., 2003; O'Connell et al., 2003), suggest that these Archaea may originate from two different possible sources. Shoreline pore water (groundwater held within a soil or rock, in gaps between particles) can be the first source as suggested by the fact that majority of the detected Archaeal sequences were from the halophilic group *Halobacteria*, which usually require a minimum salinity of 15%. The pore water of the Dead Sea shoreline is saturated with brine and contains organic matter from flood-carried debris, which provides suitable conditions for *Halobacteria*. Furthermore, no *Halobacteria* were detected in springs far from the sandy shoreline, i.e., at the bottom of a steep cliff.

Ionescu et al. (2012) further suggested that an additional or alternative origin of the Archaea could be a second water source. This is suggested by the fact that the majority of *Halobacteria* detected in the spring water fall within a cluster of organisms associated with deep hydrothermal vents and differ from the Halobacteraceae found in the spring's sediments. It has been shown before that this deep-sea group of the Halobacteria can thrive in less saline environments (Elshahed et al., 2004; Graças et al., 2011). The Dead Sea is located in a tectonically active environment (Lazar and Ben-Avraham, 2002) and ascending hydrothermal brines are often observed as thermal springs (e.g., Swarieh, 2000; Lazar and Ben-Avraham, 2002; Shalev et al., 2008).

Ionescu et al. (2012)'s study indicates that sulfide oxidation could be the key metabolism in this ecosystem, as suggested by the presence of a large variety of sulfide oxidizing bacteria. In the biofilms (thin but robust layers of viscous secretions adhering to solid surfaces and containing highly organized microbial communities of bacteria and other microorganisms) from the southern system 75–85% of the sequences are associated with known sulfide oxidizing bacteria. In the northern system the percentage is much lower (1.5–4.5%). Hydrogen sulfide (H_2S) was measured in significant concentrations in the underwater springs and could be often smelled from the freshly retrieved sediment cores and from shore springs. Oxygen is present in the Dead Sea waters (20–40 μM; Shatkay et al., 1993); thus, the process of aerobic sulfide oxidation is feasible. The sulfide oxidizing communities in the southern springs are different from those in the northern springs. The southern springs consist mainly of *Epsilonproteobacteria* and phototrophic sulfide oxidizers, while the *Thiotrichaceae* and *Acidithiobacillaceae* inhabit sediments and biofilms around the northern springs. The difference in sequence abundance between these spring sites suggests that sulfide plays a more important role in the southern system. This is also evident by the significantly larger size of white biofilms in the southern springs. Interestingly, sulfide oxidizers were not found in the water of all springs. Only a few of the organisms found in the spring-associated biofilms and sediments originate from the spring water. *Epsilonproteobacteria* are common among the non-phototrophic sulfide oxidizing bacteria in moderate saline environments (Benlloch et al., 2002) and were found to be the main players in salt gradient systems such as the deep sea Mediterranean brines (van der Wielen et al., 2005; La Cono et al., 2011).

6.5.1.2 Dense biofilms of diatoms, bacteria, and cyanobacteria surrounding underwater springs in the lake bed

Bacteria and cyanobacteria are the two types of prokaryotes that do not contain membrane-bound organelles such as nucleus, mitochondria, chloroplasts, etc. Chloroplast and mitochondria are two organelles found in the cell. The chloroplast is a membrane-bound organelle found only in algae and plant cells. Mitochondria are found in fungi, plants and animal such as eukaryotic cells. The main difference between chloroplast and mitochondria is their functions; chloroplasts are responsible for the production of sugars with the aid of sunlight in a process called photosynthesis whereas mitochondria are the powerhouses of the cell which break down sugar in order to capture energy in a process called cellular respiration. The main difference between bacteria and cyanobacteria is that the bacteria are mainly heterotrophs (organisms deriving their nutritional requirements from complex organic substances) while the cyanobacteria are autotrophs (organisms that can produce their own food using light, water, carbon dioxide, or other chemicals; because autotrophs produce their own food, they are sometimes called producers). Furthermore, bacteria do not contain chlorophyll while cyanobacteria contain chlorophyll-a.

In 2011, divers from Israel and Germany led by Ionescu (see Ionescu et al., 2012) braved the waters to see what might have been causing the concentric-ringed ripples observed near shore. Ionescu et al. (2012) discovered a complex system of underwater springs in the Dead Sea. The divers found freshwater springs, jetting into the bottom of the Dead Sea from inside craters. Found as deep as 100 feet from the surface, the springs lie at the base of craters as large as 50 feet wide and 65 feet deep. What makes this place biologically amazing was the life they found near the plumes, in a place hitherto considered to have been "dead." To everybody's surprise, it was found that the top of the springs' rocks is covered with green biofilms, which use both sunlight and sulfide—naturally occurring chemicals from the springs—to survive. It was further noticed that exclusively sulfide-eating bacteria coat the bottoms of the rocks in a white biofilm. Bacterial mats or biofilms have never been found in the Dead Sea before. Films of green photosynthetic bacteria can be seen on top of a rock and a film of white sulfide-oxidizing bacteria underneath it. Not only have the organisms evolved in such a harsh environment, Ionescu speculates that the bacteria can somehow cope with sudden fluxes in fresh water and saltwater that naturally occur as water currents shift around the springs. It may be noted that all known hard-core halophiles (salt-loving microbes) die if they are put in freshwater.

A more detailed exploration revealed that these springs harbor microbial communities with much higher diversity and cell density than reported for the Dead Sea, including dense biofilms covering sediments and rocks around the springs. In this study Ionescu et al. (2012) provided the first description of these habitats and the associated microbial communities. Based on comparative analyses of the community structure and geochemical reconstruction of the spring water sources, they proposed hypotheses about the main energy resources and metabolic pathways that drive these microbial ecosystems, as well as discussed, their possible origins and environmental adaptations.

Ionescu et al. (2012) explain that the underwater springs are located in the Darga area on the western coast of the Dead Sea, and are divided into two systems (Fig. 6.53).

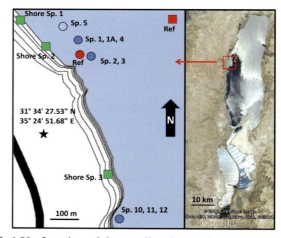

FIG. 6.53 Locations of the sampling sites on the west coast of the Dead Sea, showing the northern and southern spring (Sp) systems. Underwater springs with the corresponding reference site (Ref) are marked with blue and red circles, whereas shore springs together with their reference site are marked with green and red squares, respectively. The open-water reference site for the shore springs was used only for comparison of dissolved organic matter (DOM) and total dissolved nitrogen (TDN). The open blue circle is located in the center of an underwater spring upwelling and was sampled for DOM and TDN analysis. The contour lines on the left panel represent the yearly drop in the lake level and are a close approximation of the areal topography. The satellite image was created using Google Earth. *(Source: Ionescu et al. (2012), Microbial and chemical characterization of underwater fresh water springs in the Dead Sea, PLoS ONE, 7(6): e38319 Copyright info: © 2012 Ionescu et al. This is an open-access article distributed under the terms of the Creative Commons Attribution License, which permits unrestricted use, distribution, and reproduction in any medium, provided the original author and source are credited. Website: https://journals.plos.org/plosone/article?id=10.1371/journal.pone.0038319.*

Salinity tolerance of inhabitants in thalassic and athalassic saline and hypersaline waters & salt flats Chapter | 6 | 283

FIG. 6.54 Sketch of the northern spring system at the floor of the Dead Sea. The water seep shown on the slope of the sketch is found only in deeper parts of the southern system where water seeps through the sediment surface over a large area without defined boundaries. The shafts have steep, laminated walls (see Fig. 6.55A) and contain one or more springs (blue). Localized water sources are either directly visible on the shaft bottom (Fig. 6.55B) or are hidden within deeper cavities (Fig. 6.55C). In the southern spring system (not shown in the sketch) springs do not form shafts and are covered by cobbles (Fig. 6.55D). *(Source: Ionescu et al. (2012), Microbial and chemical characterization of underwater fresh water springs in the Dead Sea, PLoS ONE, 7(6): e38319 Copyright info: © 2012 Ionescu et al. This is an open-access article distributed under the terms of the Creative Commons Attribution License, which permits unrestricted use, distribution, and reproduction in any medium, provided the original author and source are credited. Website: https://journals.plos.org/plosone/article?id=10.1371/journal.pone.0038319.)*

The northern system (springs 1–5) consists of one or more springs at the bottom of deep steep-walled shafts (Fig. 6.54). Often several such shafts are connected and form a large system that extends from shallow (~10 m) to deeper (~30 m) waters. The diameter and depth of each shaft can reach up to 15 and 20 m, respectively. Fig. 6.55 provides examples of underwater springs in the northern and southern systems from the floor of the Dead Sea. The walls of the shafts are finely laminated (Fig. 6.55A). Groundwater emerges from either small seeps (~20 cm in diameter, Fig. 6.55B) or deeper shafts that are hidden within deeper cavities (Fig. 6.55C). The springs in the southern system (springs 10–12) do not form shafts. Instead, they are located at the bottom of steep walls descending directly from the water surface, and are covered by cobble (Fig. 6.55D). Uniquely in the southern system and in addition to the described springs, large water seeps without clear boundaries were also found, mostly at depths of more than 15 m. Detailed hydrological and geological maps of the area can be found in Laronne Ben-Itzhak and Gvritzman (2005).

Ionescu et al. (2012) explain that dense white biofilms covered sediments around the underwater springs at all sites. Fig. 6.56 shows different types of biofilms found near the underwater springs. The biofilms around the northern springs 1–5 formed small thin patches adjacent to the water outlet (Fig. 6.56A). In contrast, biofilms around the southern springs covered relatively large (2–10 m^2) patches of sediment next to areas where water seeped out without

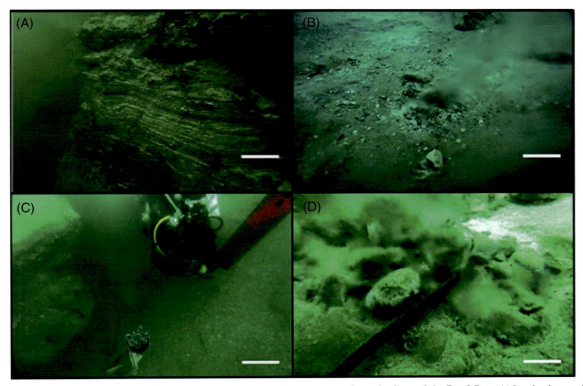

FIG. 6.55 Examples of underwater springs in the northern and southern systems from the floor of the Dead Sea. (A) Lamination on the walls of the shafts created by the springs in the northern system. (B) An example of a single water source out of several at the bottom of a shaft in the northern system. (C) An example of an in-shaft cavity from which water springs out. (D) Cobble covered spring in the southern system. Biofilms are visible on the cobble. Scale bar 0.2 m. *(Source: Ionescu et al. (2012), Microbial and chemical characterization of underwater fresh water springs in the Dead Sea, PLoS ONE, 7(6): e38319 Copyright info: © 2012 Ionescu et al. This is an open-access article distributed under the terms of the Creative Commons Attribution License, which permits unrestricted use, distribution, and reproduction in any medium, provided the original author and source are credited. Website: https://journals.plos.org/plosone/article?id=10.1371/journal.pone.0038319.)*

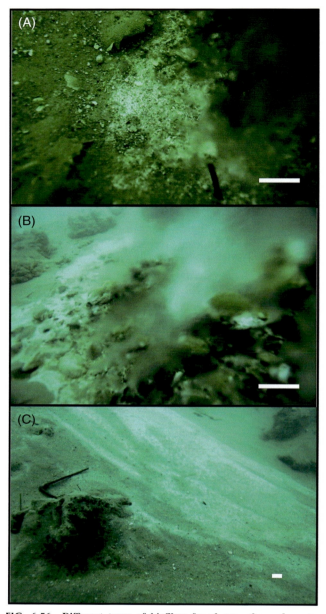

FIG. 6.56 **Different types of biofilms found near the underwater springs.** (A) Small patches of thin white biofilms covered sediments adjacent to the water source in springs 1–5. (B) Thick white biofilms covered sediments around spring 12, whereas top and bottom surfaces of rocks found within this spring were covered with green and white biofilms, respectively. (C) Large white biofilms covered slopes below springs 10 and 11 at depths ca. 20 m, although no water seepage was detected. Scale bar: 0.2 m. *(Source: Ionescu et al. (2012), Microbial and chemical characterization of underwater fresh water springs in the Dead Sea, PLoS ONE, 7(6): e38319 Copyright info: © 2012 Ionescu et al. This is an open-access article distributed under the terms of the Creative Commons Attribution License, which permits unrestricted use, distribution, and reproduction in any medium, provided the original author and source are credited. Website: https://journals.plos.org/plosone/article?id=10.1371/journal.pone.0038319.*

at depth ~20 m (Fig. 6.56C). Thickest white biofilms covering an area of several square meters were found around spring 12 (Fig. 6.56B). Microscopic analysis revealed that large, sulfur-storing filamentous bacteria, which are typical for sulfidic environments, were not present in the white biofilms.

Ionescu et al. (2012) found that in addition to white biofilms, rocks around the southern springs (particularly around spring 12) were covered by thick green biofilms. The white and green biofilms covered exclusively the lower and upper sides of the rocks, respectively. Microscopic observations of the green biofilms revealed the presence of diatoms and unicellular cyanobacteria (Fig. 6.57). Fig. 6.58A shows examples of absorption spectra of green biofilm samples from spring 12. The presence of unicellular cyanobacteria was confirmed by hyper-spectral imaging, which revealed high concentrations of chlorophyll a and phycocyanin, characteristic pigments of this functional group (Fig. 6.59).

Hyper-spectral imaging additionally revealed a high abundance of bacteriochlorophyll c, a pigment characteristic for green sulfur bacteria, which was tightly associated with chlorophyll a (Fig. 6.59B). Microscopic observations showed that this association had a specific spatial structure, with patches of cyanobacteria surrounded by or co-localized with green sulfur bacteria (Fig. 6.59C). Hyper-spectral imaging showed a peculiar association between the green sulfur bacteria and the cyanobacteria, with patches of cyanobacteria surrounded by or co-localized with green sulfur bacteria. A co-localization of cyanobacteria and green sulfur bacteria was not described before. Green sulfur bacteria are usually strictly anaerobic organisms and contain a quenching mechanism for protection against oxidation (Frigaard and Bryant, 2004). On the other hand, although cyanobacteria that are able to switch to anoxygenic photosynthesis are also known (Cohen et al., 1986; Oren et al., 2005), cyanobacteria are oxygenic phototrophs. The nature of the association found in Ionescu et al. (2012)'s study is unclear and will be the subject of future studies.

Ionescu et al. (2012) found that cell densities in the spring waters ranged between 7×10^5 and 10^7 cells/mL, and were between 10 to 100 times higher than in the ambient Dead Sea water. Bacteria made 30–50% of the total cell counts, whereas in ambient Dead Sea water where bacteria could not be detected (Bodaker et al., 2010). Both pyrosequencing and Automated Approach for Ribosomal Intergenic Spacer Analysis (ARISA) (see Fisher and Triplett, 1999) showed clearly that the microbial communities in the spring waters and spring sediments are different. The microbial communities in the green and white biofilms are closer to the spring water. Thus, the communities in the spring water and the biofilms differ from the Dead Sea communities but the sediments near the springs are colonized largely by normal Dead Sea microbial communities. All spring-associated communities were very different (maximum 10% similarity

clear boundaries (Fig. 6.56B). Large areas covered with biofilms were also found on sediments with no detectable water seepage, such as on slopes below springs 10 and 11

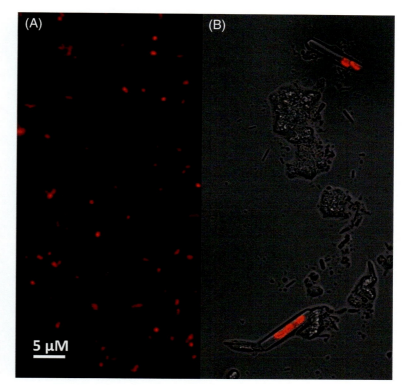

FIG. 6.57 Chlorophyll a autofluorescence confocal laser scanning microscopy of samples from the green biofilms of spring 12 showing small unicellular cyanobacteria (A) and diatoms (B). The images were acquired by Mr. Assaf Lowenthal. *(Source: Ionescu et al. (2012), Microbial and chemical characterization of underwater fresh water springs in the Dead Sea, PLoS ONE, 7(6): e38319 Copyright info: © 2012 Ionescu et al. This is an open-access article distributed under the terms of the Creative Commons Attribution License, which permits unrestricted use, distribution, and reproduction in any medium, provided the original author and source are credited. Website: https://journals.plos.org/plosone/article?id=10.1371/journal.pone.0038319.*

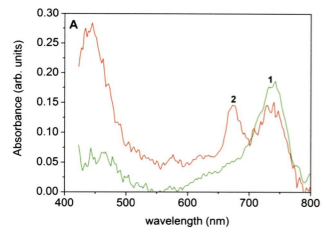

FIG. 6.58A **Examples of absorption spectra of green biofilm samples from spring 12.** Major peaks at 675 nm and 740 nm correspond to in vivo absorption maxima of chlorophyll a and bacteriochlorophyll c, respectively. *(Source: Ionescu et al. (2012), Microbial and chemical characterization of underwater fresh water springs in the Dead Sea, PLoS ONE, 7(6): e38319 Copyright info: © 2012 Ionescu et al. This is an open-access article distributed under the terms of the Creative Commons Attribution License, which permits unrestricted use, distribution, and reproduction in any medium, provided the original author and source are credited. Website: https://journals.plos.org/plosone/article?id=10.1371/journal.pone.0038319.*

based on species composition) from the residual Dead Sea communities described by Bodaker et al. (2010) as well as from the communities identified during the 1992 bloom linked to the dilution of the upper water layer of the Dead Sea (Rhodes et al., 2010).

Ionescu et al. (2012) found that compared to Bacteria, the diversity of Archaea detected in the spring-associated samples was much lower, though in many samples (especially in spring sediments) no Archea were detected. Most abundant phototrophs in the green biofilms were green sulfur bacteria (25%), consistent with the presence of bacteriochlorophyll c, and purple sulfur bacteria (10%). Purple non-sulfur bacteria and Chloroflexaceae were common in sediment samples, but not in biofilms and spring water.

Study carried out by Ionescu et al. (2012) provides the first description of dense microbial communities in the Dead Sea. It shows that these communities are exclusively linked to groundwater seepage at the lake floor. In some underwater springs, H_2S was detected in relatively high concentrations. This H_2S possibly originates from bacterial sulfate reduction that occurs along the flow path of the spring water, as shown for other sulfidic spring systems (Gavrieli et al., 2001; Macalady et al., 2007). This process requires organics and dissolved sulfate (SO_4). The latter

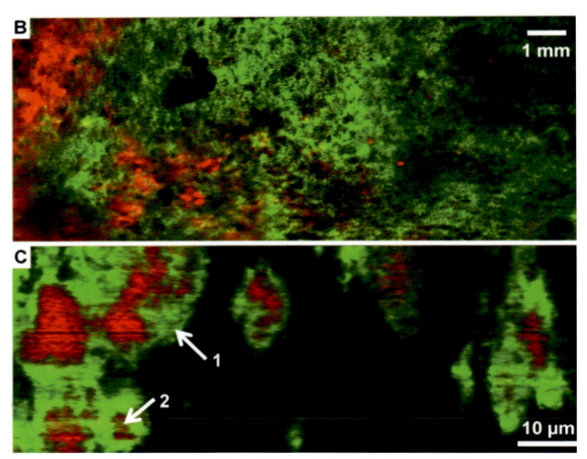

FIG. 6.59 (B and C) Examples of absorption spectra of green biofilm samples from spring 12. (B–C) Distributions of pigments in whole-biofilm samples (B) and inbiofilm samples under the microscope (C). Pigments characteristic for cyanobacteria (chlorophyll a and phycocyanin) are shown in red, whereas the pigment characteristic for green sulfur bacteria (bacteriochlorophyll c) is shown in green. Cyanobacteria were always co-localized with the green sulfur bacteria and never detected alone. *(Source: Ionescu et al. (2012), Microbial and chemical characterization of underwater fresh water springs in the Dead Sea, PLoS ONE, 7(6): e38319 Copyright info: © 2012 Ionescu et al. This is an open-access article distributed under the terms of the Creative Commons Attribution License, which permits unrestricted use, distribution, and reproduction in any medium, provided the original author and source are credited. Website: https://journals.plos.org/plosone/article?id=10.1371/journal.pone.0038319.*

may be provided to the hydro-biological system by mineral dissolution.

According to Ionescu et al. (2012), the freshwater input in the brackish underwater springs originates from the Judea Group Aquifers (JGA). The JGA waters are affected by a number of processes before they emerge at the Dead Sea floor. These include interaction with clay minerals, fine clastics and FeOOH complexes, dissolution or precipitation of different types of evaporates (e.g., aragonite, gypsum, anhydrite, halite, celestite, barite), bacterial degradation of organic matter (by sulfate reduction), and mixing with interstitial (forming, or occupying intervening spaces, especially very small ones) and ascending brines in the Dead Sea Group sediment. The sum of these processes results in water with high concentrations of dissolved ammonium and organic matter, sulfide, sulfate and phosphorus; all of which are necessary to support the microbial communities at the springs outlets in the Dead Sea. Microbial communities in the spring waters contained high cell densities. The majority of these cells were alive at the time of sampling (DeLong et al., 1999; Ravenschlag et al., 2000).

As already noted, the Dead Sea is a hypersaline lake where only a few types of organisms can grow. Recently, abundant and diverse microbial life was discovered in biofilms covering rocks and permeable sediments around underwater freshwater springs and seeps. Häusler et al. (2014a) used a newly developed salinity mini-sensor (spatial resolution 300 μm) to investigate the salinity environment around these biofilms in a flume that simulates an underwater spring. Compared with the hypersaline bulk water, salinity at the sediment surface decreased to zero at seeping velocities of 7 cm/s. At similar flow velocities, salinity above rocks decreased to 100–200 g/L at a distance of 300 μm from the surface. This depended on the position on the rock, and coincided with the locations of natural biofilms. The salinity reduction substantially diminished at flow velocities of 3.5 cm/s. Häusler et al. (2014a) suggest that locally decreased salinity due to freshwater input is one of the main factors that make areas around underwater freshwater springs and seeps in the Dead Sea more favorable for life. However, due to frequent fluctuations in the freshwater flow, the locally decreased salinity is unstable. Therefore, microorganisms that inhabit these environments must be capable of withstanding large and rapid salinity fluctuations (Häusler et al., 2014a).

Although cyanobacteria and diatom mats are ubiquitous in hypersaline environments, they have never been observed in the Dead Sea, one of the most hypersaline lakes on Earth. In another study, Häusler et al. (2014b) reported the discovery of phototrophic microbial mats at underwater freshwater seeps in the Dead Sea. They found that these mats are either dominated by diatoms or unicellular cyanobacteria and are spatially separated. Using in situ and ex situ O_2 microsensor measurements they showed that these organisms are photosynthetically active in their natural habitat. The diatoms, which are phylogenetically associated to the *Navicula* genus, grew in culture at salinities up to 40 % Dead Sea water (DSW) (14 % total dissolved salts, TDS). The unicellular cyanobacteria belong to the extremely halotolerant *Euhalothece* genus and grew at salinities up to 70 % DSW (24.5% TDS). As suggested by a variable O_2 penetration depth measured in situ, the organisms are exposed to drastic salinity fluctuations ranging from brackish to DSW salinity within minutes to hours. Häusler et al. (2014b) could demonstrate that both phototrophs are able to withstand such extreme short-term fluctuations. Nevertheless, while the diatoms recover better from rapid fluctuations, the cyanobacteria cope better with long-term exposure to DSW. Häusler et al. (2014b) conclude that the main reason for the development of these microbial mats is a local dilution of the hypersaline Dead Sea to levels allowing growth. According to Häusler et al. (2014b), the spatial distribution of these microbial mats in the seeping areas is a result of different recovery rates from short or long-term fluctuation in salinity.

6.5.2 Life in Great Salt Lake in the state of Utah in the United States

The Great Salt Lake (see Fig. 6.3), located in the northern part of the US state of Utah, is the largest saltwater lake in the Western Hemisphere. In an average year the lake covers an area of around 4400 km², but the lake's size fluctuates substantially due to its shallowness. For instance, in 1963 it reached its lowest recorded size at 2460 km², but in 1988 the surface area was at the historic high of 8500 km². The lake is irregular in shape and is about 121km long from north to south and 48 to 80km wide. The average depth of the lake is about 6m, but great seasonal variations occur. The lake is deepest in the spring, when it is fed by the melting snow from the nearby Wasatch Mountains to the east. Great Salt

FIG. 6.60 **A pair of birds found in the Great Salt Lake.** *Copyright info: © ESA 2000-2021 (Source: https://earth.esa.int/web/earth-watching/image-of-the-week/content/-/article/great-salt-lake-utah).*

Lake occupies a portion of what was, in Pleistocene times, the bed of the great Bonneville Lake. The present lake has no outlets and loses water naturally only by evaporation, which concentrates the dissolved salts carried into the lake by its tributaries and causes its salinity. Its shallow, warm waters cause frequent, sometimes heavy lake-effect snows from late fall through spring. Although it has been called "America's Dead Sea," the lake provides habitat for millions of native birds (Fig. 6.60), brine shrimp, shorebirds, and waterfowl, including the largest staging population of Wilson's phalarope in the world.

Hypersaline lakes typically support fewer species of aquatic organisms than freshwater lakes because only a few specialized species can withstand the stress of saline water. In 1899, the US Fisheries Service considered implanting oysters into the mouths of the larger rivers that enter the lake but decided that oysters could not be raised commercially. In 1986, killifish entered from Timpie Springs on the south end when the lake was at a salinity of about 5.5%. The lake is used extensively by millions of migratory and nesting birds.

The total number of species in Great Salt Lake depends on the salinity and is not accurately known because microscopic species have not been well studied. When the salinity is high (28% in Gunnison Bay), there may be as few as six species. At lower salinities (6%–9% in Gilbert Bay), the number of species has been estimated at 32. These organisms interact to form a complex and highly productive ecosystem.

6.5.2.1 Algae, bacteria, protozoa, and brine flies

Algae (25 species reported in 1979) provide the base for the food chain through photosynthesis. An alga with a red pigment lives in the north part of the lake and causes the purplish-pink coloration, especially when the salinity is greater than about 25 percent. The algae serve as food for brine shrimp and brine flies, which in turn are consumed by birds or by the shrimp harvested for the aquaculture industry.

Several species of purple salt-tolerant bacteria are present in the lake. Bacteria, protozoa (a phylum or grouping of phyla which comprises the single-celled microscopic animals), and brine flies recycle the organic material to release nutrients and keep energy flowing through the system.

Two species of brine flies (genus *Ephydra*) live in the lake. They are very numerous but do not bite. Adult flies congregate along the shoreline for breeding and occasionally reach densities of 370 million per mile of beach. Although they do not bite, they often land on people and may be annoying. The larval form of the brine fly feeds on detritus (dead organisms and fecal material, typically hosting communities of microorganisms that colonize and decompose them) and benthic algae and plays an important role in recycling nutrients within the lake. Numbers of these flies may reach 370 million per mile of shoreline during summer (Stephens and Gardner, 2007).

6.5.2.2 Brine shrimp

The invertebrate community of Great Salt Lake undergoes large annual variations in timing and abundance of life stages. Of the various animals, the most dominant and famous is the brine shrimp (*Artemia franciscana*). The variations are partly the result of changes in salinity, water temperature, nutrients, and algal populations. The young artemia, called nauplii, typically appear from overwintering eggs (hard-walled cysts containing an embryo in dormancy) from late February to early March when water temperatures are between 32- and 41-degrees Fahrenheit. Their appearance follows the winter bloom of phytoplankton that provides food for the developing nauplii (the first larval stage of many crustaceans, having an unsegmented body and a single eye) by about 1 month. Peak numbers of nauplii typically occur between mid-April and mid-May. The first generation of adults produced by the nauplii reproduces sexually and, if food availability and environmental conditions are suitable, will produce the second generation ovoviviparously (by live birth; producing young by means of eggs which are hatched within the body of the parent). If conditions are less favorable, oviparous reproduction produces cysts. Note that oviparous animals are female animals that lay their eggs, with little or no other embryonic development within the mother. This is the reproductive method of most fish, amphibians, most reptiles, and dinosaurs (including birds). Two or three generations of artemia (brine shrimp) are produced each year. The artemia graze the phytoplankton to near extinction by May. The lack of food causes adult females to begin oviparous cyst production. The maximum number of adults occurs about mid-May and then gradually declines until December, when water temperatures drop below about 37 degrees Fahrenheit and the adults die ((Stephens and Gardner, 2007).

Harvest of its best-known species, the brine shrimp (Fig. 6.61), annually supplies millions of pounds of food for the aquaculture industry worldwide. The shrimp likely arrived as cysts (eggs) on the feet of migratory birds. Brine

FIG. 6.61 Brine shrimp (Artemia franciscana) adults and cysts. Credit: USGS. (Source: Stephens, D.W., and J. Gardner (2007), Great Salt Lake, Utah; U.S. Geological Survey; Water-Resources Investigations Report 1999–4189; https://pubs.usgs.gov/wri/wri994189/PDF/WRI99-4189.pdf.

shrimp cysts and fecal pellets in a deep sediment core have been dated to be 15,000 years old. The body cuticle (the dead skin at the base of a fingernail or toenail) prevents water loss and salt entry. The shrimp drink large amounts of saltwater and the gut takes up the water (and some salt). The branchia, or gills, then excrete the salt back into the water. This is how the brine shrimp live in the extremely salty (hypersaline) environment of the lake (Stephens and Gardner, 2007).

Brine shrimp from Great Salt Lake have been commercially harvested as adults since 1950 and as cysts since 1952. The annual harvest is regulated by the Utah Division of Wildlife Resources and begins October 1. The harvest lasts about 1 to 4 months, depending on the number of cysts available. The cysts are marketed worldwide and used extensively in aquaculture as a food source for larval stages of giant prawns and fish. Reported cyst harvest from 1964 to 1978 ranged from 83 tons in 1966 to 265 pounds in 1968 to about 7400 tons during 1995–1996. (Reported amounts include total unprocessed biomass of which about half is cysts.) Historically, salinity of the lake has been the major factor associated with the abundance and distribution of artemia and the success of the cyst harvest. When salinity of Gilbert Bay varied from 6%–10% from 1982 to 1989, cyst production was poor and most harvesters moved to Gunnison Bay, the north part of the lake, where the salinity was 15%–17%. Since about 1990, reproducing populations of artemia have been limited to Gilbert Bay, although during the 1998–1999 season limited harvest efforts resumed in Gunnison Bay (Stephens and Gardner, 2007).

6.5.2.3 Migratory birds

Great Salt Lake supports from 2 to 5 million shorebirds, as many as 1.7 million Eared Grebes, and hundreds of thousands of waterfowl seasonally (during spring and fall

FIG. 6.62 Wilson's Phalarope (Phalaropus tricolor) female in breeding plumage: Original uploaded from Flickr account of Dominic Sherony. *(Source: https://commons.wikimedia.org/wiki/File:Phalaropus_tricolor_-_breeding_female.jpg) Copyright info: This file is licensed under the Creative Commons Attribution-Share Alike 2.0 Generic license.*

migration). Because of its importance to migratory birds, the lake was designated a part of the Western Hemisphere Shorebird Reserve Network in 1992. The lake is a major staging area for some shorebirds {sites that attract large concentrations (many thousands) of birds}, such as the Wilson's Phalarope. This bird is named after Scottish-American ornithologist Alexander Wilson (Szabo, 2013). Wilson's Phalaropes (Fig. 6.62) are small shorebirds with long legs, slender necks, and very thin, straight, long bills. They have sharply pointed wings. are grayish birds with cinnamon or rusty highlights especially on the neck. In the breeding season females are more colorful than males, with a dark line through the eye extending down the neck. The throat is white and the neck is washed rusty.

Wilson's phalaropes are unusually halophilic (salt-loving) and feed in great numbers when on migration on saline lakes such as Mono Lake in California, Lake Abert in Oregon, and the Great Salt Lake of Utah, often with red-necked phalaropes. When feeding, a Wilson's phalarope will often swim in a small, rapid circle, forming a small whirlpool. Their funny behavior of bobbing on the surface (making quick, short movements up and down), often spinning in circles, is thought to aid feeding by raising small food items from the bottom of shallow water, thereby bringing them (small food items) within reach of their slender bills. The bird will reach into the outskirts of the vortex with its bill, plucking small insects or crustaceans caught up therein.

The typical avian (relating to birds) sex roles are reversed in the three phalarope species. Females are larger and more brightly colored than males. The females pursue males, compete for nesting territory, and will aggressively defend their nests and chosen mates. Once the females lay their eggs, they begin their southward migration, leaving the males to incubate the eggs. Three to four eggs are laid in a ground nest near water. The young feed themselves. Nonbreeding (nonreproducing) birds are pale gray above, white below, without the strong facial markings of other phalarope species. Phalaropes are the only shorebirds that regularly swim in deep water. (https://www.allaboutbirds.org/guide/Wilsons_Phalarope/id).

In July 1986, approximately 387,000 Wilson's Phalarope were estimated in a 1-day aerial survey, and 600,000 were estimated on a single day in July 1991. Studies indicate that at least 5000 to 10,000 Snowy Plovers nest on the alkaline flats surrounding the lake. The current estimate for breeding American Avocets is 40,000 and for Black-necked Stilts 30,000. Hundreds of thousands of Eared Grebes stage on the lake, fattening on the abundant brine shrimp. One of the world's largest populations of White-faced Ibis nests in the marshes along the east side of the lake. The lake hosts the largest number of breeding California Gulls, including the world's largest recorded single colony. About 150,000 breeding adults have been counted in recent years. The American White Pelican colony on Gunnison Island ranks among the three largest colonies in North America (Stephens and Gardner, 2007).

The lake and its marshes provide a resting and staging area (a stopping place or assembly point en route to a destination) for the birds as well as an abundance of brine shrimp and brine flies that serve as food. Bird watching is seasonally popular at most of the wetlands along the east shore.

6.5.3 Meagre microbial life in the hypersaline Don Juan Pond in Antarctica—the most saline water body on Earth

Studies of lifeforms in the hypersaline (and/or brine) water of Don Juan Pond have been ambiguous until 1979 (Siegel et al., 1979). However, in a subsequent study, Siegel et al. (1983) found on the edge of this pond a mat of mineral and detritus cemented by organic matter, containing Oscillatoria and other cyanobacteria, unicellular forms, colonial forms rich in carotenoids, and diatoms. Oscillatoria is a genus of filamentous cyanobacterium which is named after the oscillation in its movement. Oscillatoria is an organism that reproduces by fragmentation. Oscillatoria forms long filaments of cells which can break into fragments called hormogonia. The hormogonia can grow into a new, longer filament. Each hormogonium consists of one or more cells and grow into a filament by cell division in one direction. Fig. 6.63 shows *Oscillatoria* filaments (Cyanobacteria: "blue-green algae").

Diatoms are a major group of algae, specifically microalgae, found in the oceans, waterways, and soils of the world. Diatoms are classified as eukaryotes, organisms with a membrane-bound cell nucleus, that separates them from the prokaryotes, archaea, and bacteria. Diatoms are a type of plankton called phytoplankton, the most common of the plankton types. Diatoms are classed as microalgae. Several systems for classifying the individual diatom species exist.

FIG. 6.63 *Oscillatoria* filaments (Cyanobacteria: "blue-green algae"). *(Source: https://commons.wikimedia.org/wiki/File:Oscillatoria_filaments.jpg) Copyright info: This file is licensed under the Creative Commons Attribution-Share Alike 4.0 International license.*

Diatoms are unicellular: they occur either as solitary cells or in colonies, which can take the shape of ribbons, fans, zigzags, or stars. Individual cells range in size from 2 to 200 micrometers (Grethe et al., 1996). In the presence of adequate nutrients and sunlight, an assemblage of living diatoms doubles approximately every 24 h by asexual multiple fission; the maximum life span of individual cells is about six days. Diatoms have two distinct shapes: a few (centric diatoms) are radially symmetric, while most (pennate diatoms) are broadly bilaterally symmetric. A unique feature of diatom anatomy is that they are surrounded by a cell wall made of silica (hydrated silicon dioxide), called a frustule. These frustules (the silicified cell walls of diatoms, consisting of two valves or overlapping halves) have structural coloration due to their photonic nanostructure, prompting them to be described as "jewels of the sea" and "living opals." Movement in diatoms primarily occurs passively as a result of both water currents and wind-induced water turbulence; however, male gametes (sperms) of centric diatoms have flagella, permitting active movement for seeking female gametes (eggs). Zygote is a fertilized egg cell that results from the union of a female gamete (egg or ovum) with a male gamete (sperm). Similar to plants, diatoms convert light energy to chemical energy by photosynthesis, although this shared autotrophy evolved independently in both lineages.

It was found that although Bacteria are rare, fungal filaments are not. Oscillatoria showed motility but only at temperatures <10°C. Acetone extracts of the mat and nearby muds yielded visible spectra similar to those of laboratory grown *O. sancta*, with 50- to 70-fold molar ratio of chlorophyll a to b. Although rare, tardigrades were also found. The algal mat had enzymatic activities characteristic of peroxidase, catalase, dehydrogenase, and amylase. Cellulose, chitin, protein, lipid and ATP were present. Previously, algae in the Wright Valley, in which the Don Juan Pond is located, have been described in melt water, not in the brine itself.

Wright Valley has been used as a near sterile Martian model. It obviously contains an array of hardy terrestrial organisms.

6.5.4 The fauna of Athalassic salt lake at Sutton & Saline Pond at Patearoa in New Zealand

Athalassic saline environment has attracted little attention in New Zealand, and it might be doubted whether such an environment could exist in New Zealand at all (Bayly (1967b). With reference to the athalassic saline waters in New Zealand, Marples (1962) stated thus: "At least one inland saline lake exists in this country, a small lake in Otago," and he further stated that this was a shallow, temporary, saline lake at Sutton. Sutton Salt Lake in Otago (Fig. 6.64) is New Zealand's only inland saltwater lake (Craw, 2021). It is located 10 km west of Sutton, in the Strath Taieri. New Zealand also has one coastal saltwater lake, Lake Grassmere, a lagoon with no direct outflow to the sea. The lake is filled only by rainwater and dries up during periods of warm weather and little rain. When full, the lake has salinity about one quarter to one third of that of seawater. The salinity is due to marine aerosols in rainwater which is concentrated by around 20000 as a result of evaporation and refilling cycles (Craw, 2021).

Sutton Salt Lake is also unique in that it lies in a region of windy cool-temperate maritime climate. This makes it different from most of the world's saline lakes, which usually form in arid continental landscapes. The total area of the lake, when the shallow depression is filled with rainwater, is nearly two hectares. The nearest coast is 50 kilometers from the lake (Craw, 2021).

The area around Sutton Lake gets 500 mm of rainfall annually. Coastal hills create a minor barrier to the rain, halving the amount of rainfall that the lake receives. Strong winds increase the lake's surface evaporation rate.

Bayly (1967b) discussed the fauna and chemical composition of two athalassic saline waters located near central Otago. One water (dominated by Na and Cl) had a

FIG. 6.64 Sutton Salt Lake, Otago, New Zealand. *(Source: https://commons.wikimedia.org/wiki/File:Sutton_Salt_Lake_11.jpg) Author: Rudolph89 Copyright info: This file is licensed under the Creative Commons Attribution-Share Alike 3.0 Unported license.*

salinity of 15‰ and contained the rotifer *Brachionus plicatilis*, the copepod *Microcyclops monacanthus*, the ostracod *Diacypris*, and larvae of the dipteran *Ephydrella*. The second water (dominated by Na and HCO_3) had a salinity of 6‰ and contained larvae of *Ephydrella* (adults of *E. novaezealandiae* occurred on the shore), and the hemipterans *Sigara arguta, Anisops wakefieldi* and *Anisops assimilis*. Bayly (1967b) further discussed in some detail the relationship between *Microcyclops monacanthus* and *M. arnaudi* (an Australian species), and the basis of their current taxonomic separation. At a minimum, the faunas of Australian and New Zealand athalassic saline waters have in common the following distinctive (non-cosmopolitan) forms: *Ephydrella, Diacypris*, and two closely related halobiont species of *Microcyclops*.

According to Bayly (1967b), the fauna at the Sutton Salt Lake consist of the following:

Rotatoria: *Brachionus plicatilis* Müller; Copepoda: *Microcyclops (Metacyclops) monacanthus* (Kiefer); Ostracoda: *Diacypris sp.*; Diptera: *Ephydrella sp.*; and the alga *Chora sp.* present in abundance.

The occurrence of salt pans and local areas of saline soil in Patearoa region is mentioned by Raeside et al. (1966). In this restricted part of New Zealand evaporation exceeds rainfall (Raeside et al., 1966).

According to Bayly, (1967b), the fauna at the saline pond near Patearoa consist of Diptera: *Ephydrella sp*(p) (larvae) and *E. novaezealandiae* Tonnoir and Mallock (adults); numerous larvae were present in the aquatic collections, and several adults were captured from the thick swarms flying around the margins of the pond.

6.5.5 Importance of the genus bacillus

Bacillus, (genus *Bacillus*) is any of a genus of rod-shaped, Gram-positive, aerobic, or (under some conditions) anaerobic bacteria widely found in soil and water. Bacillus species can be either obligate aerobes: oxygen-dependent; or facultative anaerobes: having the ability to continue living in the absence of oxygen. The term bacillus has been applied in a general sense to all cylindrical or rod-like bacteria. In bacteriology, Gram-positive bacteria are bacteria that give a positive result in the Gram stain test, which is traditionally used to quickly classify bacteria into two broad categories according to their type of cell wall. *Bacillus* frequently occurs in chains. A Gram-positive, aerobic, or facultative (capable of but not restricted to a particular function or mode of life) anaerobic microorganism with rod-shaped cells, Bacillus is among the most well-characterized microbial genera at biochemical, genomic, and proteomic (protein-related) levels (Alcaraz et al., 2010). The ubiquity of this microbial group has been well reported, due to its presence in a wide range of contrasting habitats including air, fresh and marine water, alkaline soils, geothermal heated soils, stone surfaces of ancient monuments, soda lakes, and cold environments (Logan and De Vos, 2015), highlighting the wide metabolic potential of Bacillus (Alcaraz et al., 2010). Bacillus can reduce themselves to oval endospores and can remain in this dormant state for years. The endospore of one species from Morocco is reported to have survived being heated to 420°C (Beladjal et al., 2018). Since the first description of dormancy in Bacillus subtilis (Cohn, 1872) and the recognition of Bacillus anthracis as the agent of anthrax (Koch, 1876), the importance of the genus Bacillus was recognized early in the timeline of modern science (Zeigler and Perkins, 2015).

Colonial morphology in Bacillus varies from circular to irregular, from entire edge to undulate, with sizes ranging from 1 to 5 mm; and colors from a creamy-gray to off-white (Logan and De Vos, 2015). Up until 2015, 142 Bacillus-species were described (Logan and De Vos, 2015), with species subclassified according to their ecophysiology as halophile, acidophiles, psychrophiles, alkalophiles, and thermophiles (Ravel and Fraser, 2005). Subsequently, more species were discovered, with 266 named species known. An alternative approach to classification of Bacillus species reflects their use by, and their impacts on human society, including environmental (*B. subtilis*), pathogenic (*B. anthracis*), and industrial uses (*B. licheniformis*) (Alcaraz et al., 2010).

Endospore formation is usually triggered by a lack of nutrients: the bacterium divides within its cell wall, and one side then engulfs the other. They are not true spores (i.e., not an offspring) (Reference: Bacterial Endospores." Cornell University College of Agriculture and Life Sciences, Department of Microbiology). Endospore formation originally defined the genus, but not all such species are closely related, and many species have been moved to other genera of the Bacillota (Madigan and Martinko, 2005). Only one endospore is formed per cell. The spores are resistant to heat, cold, radiation, desiccation, and disinfectants. Beyond their use in industry, the highly resistant endospores of *B. subtilis* have been widely studied as a model species in astrobiology (Nicholson et al., 2000). Genomic studies of Bacillus from different habitats (halophile, aquatic, pathogenic, deep-ocean, and soil) have revealed that 814 genes form the core genome of Bacillus and that 53 are related to sporulation and the competence core genome (Alcaraz et al., 2010). Further studies on Bacillus have been reported by Cubillos et al. (2019).

6.5.6 Microbial life in the hypersaline lake Assal in Djibouti in the Horn of Africa— the most saline hypersaline lake outside the Antarctic continent

Lake Assal (see Fig. 6.5) is a volcanic lake, and is located at the top of the Great Rift Valley that passes through the Danakil Desert, and is composed of two principal divisions. The first is a dry bed of salt white in color that is the result of the evaporation of water from the lake in earlier times, leaving behind the dry salt bed. The second portion is the very saline body of water in the lake. Lake Assal—lying in a closed depression—has an oval shape (length 19 km (12 mi) and width 6.5 km (4.0 mi) (Brisou et al., 1974). Lake Assal, 420 longitude East and 120 latitude North, is situated in the French territory of Afars and Issas, and can be reached only on foot by difficult tracks; it has a maximum depth of 40 m and is 156 m below the level of the Red Sea.

The watershed area of the lake measures 900 km^2 (350 sq mi). There is a residual surface runoff of fresh water into the lake from the sporadic rainfall from ephemeral streams; there are no defined streams in the catchment. Runoff from rainfall is in the form of flash floods which may last from several hours up to two or three days depending on the intensity of rainfall. The main source of supply, however, is subsurface geothermal springs.

The area in which lake Assal is located is also among the hottest regions on Earth, where temperatures can reach up to 50°C. Lake Assal lies 155 m (509 ft) below sea level in the Afar Triangle, making it the lowest point on land in Africa and the third-lowest point on Earth after the Sea of Galilee and the Dead Sea. No outflow occurs from the lake, and due to high evaporation, the salinity level of its waters is 10 times that of the sea, making it the third most saline body of water in the world behind Don Juan Pond and Gaet'ale Pond (Warren, 2006). Lake Assal is the world's largest salt reserve.

The Lake Assal region is wild and desert-like, and no fauna or flora can be seen in the syrupy waters of the lake. The high temperature of the water (33 to 34°C) favors evaporation, and on one side of the lake there is an immense mass of salt. The lake is fed by hot springs whose salinity is close to that of seawater. There is just a residual runoff of fresh water into the lake. It is believed that some freshwater runoff might be hidden under the present surface.

The lake water is rich in minerals but it does not support any kind of aqua fauna. In a study, Brisou et al. (1974) found that all of the heterotrophic bacteria isolated from this hypersaline lake were aerobes; no strictly anaerobic strains were found. Ninety percent of the strains were euryhalines (able to tolerate a wide range of salinity) and 10% were strict halophiles (growing only in high saline conditions). The extreme halophiles belonged to the species *Halobacterium trapanicum* and *Halococcus morrhuae*.

Brisou et al. (1974) recall that the waters of Lake Assal, which are very rich in mineral, containing 400 g of salts per liter and showing no apparent signs of life, have a rich population of common bacteria. These are aerobic forms. The genus *Bacillus* was the most abundant from surface and depth and much more frequent than in the other aquatic media (Denis et al., 1972). Contrary to what might have been expected, the strictly halophilic strains were in the minority and represented only 10% of the total, the great majority of the bacteria being tolerant to the salinity. From this picture, it is clear that the bacteria were, on the whole, common bacteria belonging to the same species which are currently isolated from large and small rivers and from seawater. The strictly halophilic strains belonged to already known species or to halophilic ecotypes of normally euryhaline species. The hyperhalophilic strains belonged mostly to the species Halobacterium *trapanicum*, which had been found by Elazari-Volcani (doctoral thesis) in the Dead Sea; one specimen of Halococcus morrhuae was also isolated.

Brisou et al. (1974) reported that "The moderate halophiles were distributed as follows: V. marinopraesens, P. petasitis, A. aquamarinus, F. amocontactum, and several halophilic ecotypes of euryhaline strains. The usual microbial indicators of pollution, such as the enterobacteria, Pseudomonas aeruginosa and Streptococcus faecalis, were not found, probably because the area around the lake is not frequented by man or animals. It would appear, finally, that even an environment as specific as lake water containing 40% mineral salts in an ambient temperature of 33–34°C results in the selection of only a few strictly halophilic strains (10%), with the large majority of specimens consisting of euryhaline halotolerant bacteria that are found in most aquatic media, but with a very definite preponderance of the genus *Bacillus*, which accounts for half of the total isolated strains."

6.6 Life in lithium chloride-dominated hypersaline salt flats & ponds belonging to the lithium triangle zone (Argentina, Bolivia, and Chile) of South America

Whereas sodium chloride (common salt) is the major source of salinity in most water bodies, there exists a group of peculiar athalassic water bodies where lithium chloride is the major source of salinity. Lithium chloride-dominated hypersaline salt flats & ponds belonging to the Lithium Triangle Zone (Argentina, Bolivia, and Chile) of South America are an example to such a situation. While the average salinity in the global ocean is 35 PSU (equivalent to per thousand or (o/oo) or to g/kg), the salinity of lithium chloride-rich Salar de Uyuni salt pond in Bolivia is between 132 and 177 PSU.

The bulk of lithium is found in natural brines, and half of all world reserves are found in three countries in South America (Argentina, Bolivia, and Chile; Del Barco and Foladori, 2014). Argentina, Bolivia, and Chile include three large salars (salt flats) which globally, represent the largest lithium reserves, and are commonly referred to as the "Lithium Triangle Zone." This zone holds the world's largest reserves of lithium, a key component in high energy density batteries. Natural or artificial hypersaline environments are not limited to high concentrations of NaCl, and under such conditions, specific adaptation mechanisms are necessary to permit microbial survival and growth. Microbial life inhabiting hypersaline environments belong to a limited group of extremophile or extremotolerant taxa and, therefore, it would be interesting to examine the biology of these unique environments. Studies carried out by Cubilloset al. (2019) revealed that two bacterial strains (isolates LIBR002 and LIBR003), classified to the *Bacillus* genera, are lithium-tolerant and that they are phylogenetically differentiated from those *Bacillus* associated with high NaCl concentration environments, and form a new clade from the Lithium Triangle Zone.

In nature, lithium is found in both a solid (in minerals) and liquid form, as brines in aquatic ecosystems (Wanger, 2011). The so-called Lithium Triangle Zone represents up to 85% of global reserves of soluble lithium (Martínez et al., 2018) and is formed by the Salar del Hombre Muerto (Argentina), Salar de Uyuni (Bolivia) and Salar de Atacama (Chile) (see Cubilloset al., 2019).

At a microbiological level, lithium has been classified as an antimicrobial compound (Lieb, 2004), associated with the stimulation of autolysis (Sugahara et al., 1983), sporulation in Bacillus-species (Warburg et al., 1985; Li et al., 2014) or growth inhibition in fungi (Kurita and Funabashi, 1984) and yeasts (Morris, 2004). Nevertheless, the capacity of some bacteria to accumulate lithium by teichoic acid polymers from solution has been reported (Tsuruta, 2005; Belfiore et al., 2018) (see Cubilloset al., 2019).

6.6.1 Effects of lithium on microbial cells

Cubillos et al. (2018) have pointed out that in the case of prokaryotes, it has been suggested that some fungal groups have resistance or tolerance mechanisms against lithium (de Assunção, 2012). Further other effects of lithium on microbial cells have been demonstrated including the stimulation of autolysis (Sugahara et al., 1983) and sporulation (Warburg et al., 1985), as well as growth inhibition (Takenishi and Takada, 1984). Conversely, lithium has been used as a growth medium for the isolation of Straphylococcus aureus, Listeria monocytogenes, and Bifidobacterium sp. (Cox et al., 1990; Lapierre et al., 1992; Minor and Marth, 1976), and evidence exists of its accumulation in bacteria, actinomycetes, fungi, and yeasts (Perkins and Gadd, 1993; Tsuruta, 2005). In addition, it has been shown that Micrococcus varians ssp. halophilus, a moderate halophilic bacterium, can grow in concentrations of up to 1.5 M of lithium chloride (Kamekura and Onishi, 1982), one of the highest values reported (Cubillos et al., 2018).

6.6.2 Microbial diversity with the presence of fungi, algae, and bacteria in the lithium-rich hypersaline environment of Salar del Hombre Muerto Salt Flat, Argentina

Part of the "Lithium Triangle of salars," Salar del Hombre Muerto (Fig. 6.65) in Argentina (Fig. 6.66) is one of the

FIG. 6.65 **Salar del Hombre Muerto, in North-Western Argentina.** A salar is a salt pan, created when water repeatedly evaporates from a shallow lake, leaving behind a crusty layer of salt minerals, which are brilliant white in this image. *(Source: https://en.wikipedia.org/wiki/File:Salar_del_Hombre_Muerto.tif) Copyright info: This file is in the public domain because it was created by NASA. NASA copyright policy states that "NASA material is not protected by copyright unless noted". (NASA copyright policy page or JPL Image Use Policy).*

world's most important sources of lithium. This lithium-rich salt flat covers an area of 600 square km (230 sq mi). Unlike many salt pans in the region in and near the world's driest desert—the Atacama Desert—Salar del Hombre Muerto salt pan receives enough rainfall to occasionally be covered by a thin layer of water. The water evaporates or percolates through the salt crust to form a layer of brine. The brine in Salar del Hombre Muerto is rich in lithium, an element used in a wide range of products from batteries to medication.

FIG. 6.66 **Argentina, South America, where Salar del Hombre Muerto is located.** *(Source: https://commons.wikimedia.org/wiki/File:Relief_Map_of_Argentina.jpg) Copyright info: This file is licensed under the Creative Commons Attribution-Share Alike 3.0 Unported license.*

FIG. 6.67 James's Flamingos found at the Salar del Hombre Muerto salt flat, Argentina. *(Image Source: https://commons.wikimedia.org/wiki/File:Flamingos_Laguna_Colorada.jpg) Copyright info: This file is licensed under the Creative Commons Attribution 2.0 Generic license.*

FIG. 6.69 Llamas grazing around the Salar del Hombre Muerto salt flat, Argentina. *(Image source: https://commons.wikimedia.org/wiki/File:Llamas,_Vernagt-Stausee,_Italy.jpg) Copyright info: This file is licensed under the Creative Commons Attribution-Share Alike 4.0 International license.*

FIG. 6.68 Bunch grass around the Salar del Hombre Muerto salt flat, Argentina, grazed by burros. *(Image Source: https://commons.wikimedia.org/wiki/File:Donkey_in_Clovelly,_North_Devon,_England.jpg) Copyright info: This work has been released into the public domain by its author, Arpingstone at English Wikipedia. This applies worldwide.*

The mine in the lower left corner of the image is extracting lithium from the brine by pumping it into solar ponds. The dry, sunny, windy environment quickly evaporates the water from the brine, leaving behind a more concentrated solution of lithium.

The mean temperatures range from 23°C (73°F) in summer to 8°C (46°F) in winter; the day-night variation is about 20–25°C (36–45°F) (Godfrey et al., 2013) and maximum temperatures at Salar del Hombre Muerto are about 28°C (82°F) (Lowenstein et al., 1998). The climate is arid; the 60–80 mm per year (2.4–3.1 in/year) (Garrett, 1998) precipitation originates mainly in the Amazon and comes to the salar during summer, but winter snowfall also occurs (Godfrey et al., 2003).

Algae in the perennial water surfaces draw flamingos (Fig. 6.67), and bunch grass around the salar is grazed by burros (Fig. 6.68), and llamas (Garrett, 1998) (Fig. 6.69), while copepods (Fig. 6.70) live in the Salar (Locascio de Mitrovich et al., 2000).

The Salar del Hombre Muerto is a salt flat with great microbial activity despite the existing extreme conditions such as high altitude, lack of water, low level of oxygen, high radiation, and high concentration of sodium and lithium chlorides. Despite these unfavorable conditions, Martínez et al. (2019) found microbial diversity with the presence of fungi, algae, and bacteria. From aqueous solutions and soil samples, a total of 238 bacteria were isolated and 186 of them were able to grow in the presence of salt. About 30% of the strains showed the ability to grow in solid medium proximally to a LiCl solution close to saturation (636 g/L). These isolates were characterized taking into account the morphology, Gram stain, ability to form biofilms and to produce pigments, and mainly according to the tolerance against lithium chloride. *Bacillus* was predominant among the most tolerant 26 microorganisms found, followed by *Micrococcus* and *Brevibacterium*. Members of the genera *Kocuria*, *Curtobacterium* and *Halomonas* were also represented among the bacteria with tolerance to 30 and 60 g/L of LiCl in defined liquid medium. All the capacities found in these microorganisms make them extremely interesting for biotechnological applications (Martínez et al., 2019).

FIG. 6.70 Copepods (a group of small crustaceans) living in the Salar del Hombre Muerto salt flat, Argentina (For the photo, the darkfield method was combined with polarization). *Image source: Copyright info: This file is licensed under the Creative Commons Attribution-Share Alike 4.0 International license.*

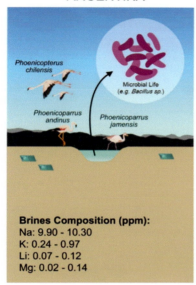

FIG. 6.71 The general characteristics of the Salar del Hombre Muerto salt flat in Argentina. *(Source: Cubillos et al. (2019), Insights into the microbiology of the Chaotropic Brines of Salar de Atacama, Chile, Front. Microbiol., https://doi.org/10.3389/fmicb.2019.01611) Copyright info: © 2019 Cubillos, Paredes, Yáñez, Palma, Severino, Vejar, Grágeda and Dorador. This is an open-access article distributed under the terms of the Creative Commons Attribution License (CC BY). The use, distribution or reproduction in other forums is permitted, provided the original author(s) and the copyright owner(s) are credited and that the original publication in this journal is cited, in accordance with accepted academic practice. No use, distribution or reproduction is permitted which does not comply with these terms.*

The general characteristics of the Salar del Hombre Muerto salt flat are illustrated in Fig. 6.71.

6.6.3 Presence of bacteria and archaea in the lithium-rich hypersaline environment of Salar de Uyuni, Bolivia—the largest salt flat on Earth

Salar de Uyuni (Fig. 6.72), together with about 75 other salt flats, situated in the Southwest of the Bolivian Altiplano, and covering an area of more than 10,000 km², is the largest salina-type evaporite basin on Earth (Rettig et al., 1980). Climatic conditions define Salar de Uyuni as a desert. The salar is a hydrologically closed basin that only receives water from rainfall and riverine inflow. Consequently, large amounts of evaporites accumulated over time inside the salar (Warren, 2010). The brine salinity of the solution within the porous halite crust is about 300 g/L (Rettig et al., 1980).

As just indicated, the most important hypersaline environment in Bolivia is the Uyuni salt flat. This salt flat presents certain unique chemical characteristics and composition on its surface such as a gradient of ion concentrations from south to north. Perez-Fernandez et al. (2016) carried out a study in order to describe the structure of the microbial communities and determine any possible correlations with abiotic factors. In this study, they extracted total DNA from rock salt samples obtained at different locations, and carried out 16S rDNA followed by Terminal Restriction Fragment Length Polymorphism (T-RFLP) analyses. Statistical analyses of the communities indicated that the highest diversity indices were found in the southern area, and the microbial communities were clustered in three groups for bacteria and in two groups for archaea. This variation could be explained by different concentrations of lithium and calcium, in addition to other abiotic variables on the surface crust. Perez-Fernandez et al. (2016)'s results indicate that even under extreme hypersaline conditions abiotic factors such as wind or geological activity may determine the composition of the resident microbiota.

According to Haferburg et al. (2017), in many regards the Salar de Uyuni offers a plethora of geological and hydrochemical superlatives. Brines of this athalasso-haline hypersaline environment are rich in lithium and boron. Haferburg et al. (2017) reported for the first time on the prokaryotic diversity of four brine habitats across the salar (salt flat). The brine is characterized by salinity values between 132 and 177 PSU, slightly acidic to near-neutral pH, and lithium and boron concentrations of up to 2.0 and 1.4 g/L, respectively. Hypersalinity in combination with intense UV irradiance, high lithium concentration and moderately acidic pH make these brines of the world's largest salt flat unique for studies on diversity and adaptation of polyextremophilic microorganisms.

Haferburg et al. (2017) reported that "The surface of the salar is almost perennially dry; intermittent flooding only occurs during the rainy season from December to April. Almost 90% of the annual precipitations (141 mm on av.) accumulate in this period. This leads at intervals to the transformation of the salar to a transitory hypersaline lake. The only permanent surface inflows occur South of Uyuni through the Rio Grande de Lipez and the Rio Colorado. A lacking outflow makes the Salar de Uyuni a gigantic endorheic basin that encloses stratified salt and mud layers up to a total depth of at least 121 m (Fornari et al., 2001)." Note that an "endorheic basin" is a drainage basin that normally retains water and allows no outflow to other external bodies of water, such as rivers or oceans, but drainage converges instead into lakes or swamps, permanent or seasonal, that equilibrate through evaporation. In this salt flat, extended deposits in combination with the high concentrations make the brine attractive for lithium extraction. With an amount of estimated 5.4 million tons, the Salar de Uyuni

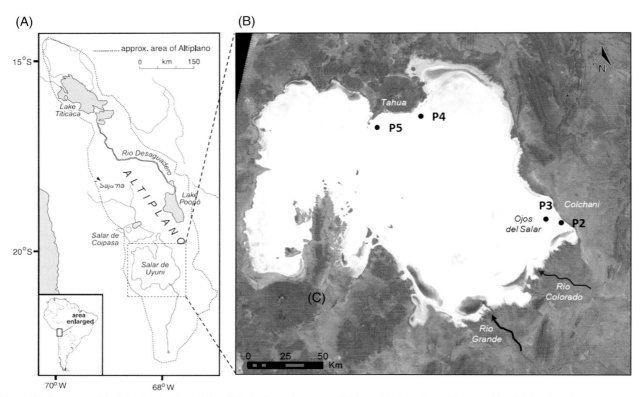

FIG. 6.72 Location of the Salar de Uyuni, Bolivia within the South American Altiplano (A). Overview of the salar, highlighting the brine pools selected for sampling (B). Pools of the four sampling sites are filled with clear (P2, P4 and P5) or turbid (P3) brine. *(Source: Haferburg, G., Gröning, J. A. D., Schmidt, N., Kummer, N. A., Erquicia, J. C., and Schlömann, M. (2017). Microbial diversity of the hypersaline and lithium-rich Salar de Uyuni, Bolivia. Microbiol. Res. 199, 19–28. doi: 10.1016/j.micres.2017.02.007) Website: https://www.sciencedirect.com/science/article/pii/S0944501317301532?via%3Dihub Copyright info: © 2017 Elsevier GmbH. {Elsevier}.*

is ranked one of the lithium-richest deposits in the world (An et al., 2012). During several sampling campaigns, moderately acidic brines of pH values between 4.62 and 5.84 were found in a section of the south-eastern shore of Salar de Uyuni (Hönninger et al., 2004; Schmidt, 2010).

Salar de Uyuni offers the opportunity to study microbial communities of polyextremophilic organisms in their resilience toward hyper-salinity, intense UV radiation, high lithium and boron concentrations, and in various pH ranges. Haferburg et al. (2017) have provided the below background information on the hitherto known microbial life in Salar de Uyuni salt flat in Olivia: "Extremely halophilic microorganisms occur in brine solutions of at least 10% (w/v) salt up to saturation (Oren, 2002; Bowers et al., 2009). Likewise, they can inhabit fluid inclusions of primary or secondary halite crystals and survive or even thrive for millions of years as salt-entrapped cell entities (McGenity et al., 2000). "The halobacteria's confusion to biology"—Larsen's famous lecture on the life of halophiles at the "borderland of physiological possibilities" is continuously repeated for good reasons (Becking, 1928; Larsen, 1973). Although some bacterial genera as, for example, *Salinibacter*, *Halomonas*, and *Salicola* are known to dwell in hypersaline environments as well, this type of extreme habitat is predominantly inhabited by members of the haloarchaea reaching partially enormous cell densities that exceed 10^7 cells/mL (Ventosa, 2006). Currently, the family Halobacteriaceae contains 47 formally described and validly named genera with a constant increase of new findings. Meanwhile methods based on next-generation sequencing became indispensable for phylogenetic and taxonomic studies on the composition of extremely halophilic communities (see, e.g., Ghai et al., 2011, Vavourakis et al., 2016). Besides the great amount of ecologically analyzable data, metagenomic community analysis retrieves good indications for the occurrence of potentially novel taxa (Sharpton et al., 2011).

In a study, Haferburg et al. (2017) reported for the first time the composition of microbial (prokaryotic) communities dwelling in near neutral and slightly acidic, lithium-rich, hypersaline brine pools at four different locations of Salar de Uyuni (see Fig. 6.73). The general characteristics of the Salar de Uyuni salt flat in Bolivia, South America, is shown in Fig. 6.74.

Haferburg et al. (2017) explain that salinity of the investigated brine pools amounted up to ca. 29–36%, hence reaching the upper range of brines from Andean salars that have been analyzed for prokaryotic diversity to date (Maturrano et al., 2006, Demergasso et al., 2008). Hypersaline brines

FIG. 6.73 Brine pools of sampling location P2 (A), P3 (B) and (C), P5 (D). Note the gas emanations at sampling point P3. *(Source: Haferburg, G., Gröning, J. A. D., Schmidt, N., Kummer, N. A., Erquicia, J. C., and Schlömann, M. (2017). Microbial diversity of the hypersaline and lithium-rich Salar de Uyuni, Bolivia. Microbiol. Res. 199, 19–28. doi: 10.1016/j.micres.2017.02.007) Website: https://www.sciencedirect.com/science/article/pii/S0944501317301532?via%3Dihub Copyright info: © 2017 Elsevier GmbH. {Elsevier}.*

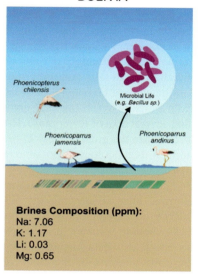

FIG. 6.74 **General characteristics of the Salar de Uyuni salt flat in Bolivia, South America.** *(Source: Cubillos et al. (2019), Insights into the microbiology of the Chaotropic Brines of Salar de Atacama, Chile, Front. Microbiol., https://doi.org/10.3389/fmicb.2019.01611) Copyright info: Copyright © 2019 Cubillos, Paredes, Yáñez, Palma, Severino, Vejar, Grágeda and Dorador. This is an open-access article distributed under the terms of the Creative Commons Attribution License (CC BY). The use, distribution or reproduction in other forums is permitted, provided the original author(s) and the copyright owner(s) are credited and that the original publication in this journal is cited, in accordance with accepted academic practice. No use, distribution or reproduction is permitted which does not comply with these terms.*

are known to harbor communities of haloarchaea with comparably high and bacteria with rather low diversity (Martínez-Murcia et al., 1995). Surprisingly bacterial taxa were completely missing and only halobacterial taxa were encountered at the investigated sites (Haferburg et al., 2017).

According to Haferburg et al. (2017), it seems possible that the comparably high lithium concentrations represent an important reason for the total absence of extremely halophilic bacteria such as, for example, *Salinibacter* sp. and *Halomonas* sp. The occurrence of these genera in various Andean salars with low lithium concentrations has previously been shown (Demergasso et al., 2008, Maturrano et al., 2006). The underlying resistance mechanism for growth of haloarchaea in presence of elevated Li⁺ concentration still awaits elucidation (Haferburg et al., 2017).

Haferburg et al. (2017) have suggested that the retrieval of sequences of methanogens and Archaeoglobi as obligate anaerobes at sampling location P3 can probably be explained by stirred-up anoxic sediments that were part of the brine sample. Hyper-thermophilic and slightly halophilic Archaeoglobi are known from sulfate-rich habitats (Burggraf et al., 1990, Huber et al., 1997). According to Haferburg et al. (2017), with a sulfate content of ca. 9 g/L the investigated brine seems to provide the appropriate concentration for an Archaeoglobi inhabited brine pool. The identification of the family Archaeoglobaceae as adventitious (happening as a result of an external factor or chance rather than design or inherent nature.) part of the microbial community within the brine pools of "Ojos del Salar" presents a further indication for the existence of a thermal spring in the Eastern region of the salt flat, because Archaeoglobaceae are commonly affiliated with high-temperature hydrothermal systems (Jungbluth et al., 2016).

Haferburg et al. (2017)'s study has shown that the four investigated brine pool communities of Salar de Uyuni consist solely of archaeal taxa (any unit used in the science of biological classification, or taxonomy. Taxa are arranged in a hierarchy from kingdom to subspecies, a given taxon ordinarily including several taxa of lower rank). However, these investigators recognize the great number of biases in DNA extraction and PCR methodology as possible reason for the total lack of bacterial sequences.

In Haferburg et al. (2017)'s study, community analysis was performed after sequencing the V3–V4 region of the 16S rRNA genes employing the Illumina MiSeq technology. The mothur software package was used for sequence processing and data analysis. The 16S rRNA-based metagenomic analysis of the brine pool samples from Salar de Uyuni revealed a clear predominance of the class Halobacteria. Sequences originating from the classes Archaeoglobi, Methanomicrobia and Nanohaloarchaea were less abundant. Within the family Halobacteriaceae 26 genera were found. *Halonotius* occurs as highly abundant genus in all brine habitats (Haferburg et al., 2017).

Metagenomic analysis revealed the occurrence of an exclusively archaeal community comprising 26 halobacterial genera including only recently identified genera such as Halapricum, Halorubellus and Salinarchaeum. It was found that *Halapricum* represents with a relative abundance of about 6% at sampling location P3—the second most dominant genus in the brine community.

Haferburg et al. (2017)'s investigation presented a first small insight into the composition of archaeal brine communities from the microbiologically previously uncharacterized ecosystem Salar de Uyuni. Their study of the microbial diversity in naturally lithium-rich brines from Salar de Uyuni, Bolivia (salinity ≤356 g/L), only revealed the presence of Archaea in its brines. These investigators are optimistic that future investigations of different habitats of the salar as, for exmaple, halite crystals with liquid inclusions, lacustrine (relating to or associated with lakes) sediments from remnants of paleolakes (ancient lakes, especially those that no longer exist) and salt/soil boundaries at Isla Incahuasi within the salar will extend our knowledge on the ecology of extreme halophiles. Additionally, special attention should be paid to the impact of lithium on the diversity of halobacteria in the context of the salar as the world's largest lithium reservoir. Note that halobacterium (plural halobacteria) is a genus in the family *Halobacteriaceae*. The genus Halobacterium consists of several species of Archaea with an aerobic metabolism which requires an environment with a high concentration of salt; many of their proteins will not function in low-salt environments.

6.6.4 Life in lithium chloride-dominated hypersaline environments of Atacama hypersaline lakes & salt flats in Chile

The Salar de Atacama, or the Chilean salt flat (Fig. 6.75) is a closed saline basin within the pre-Andean depression of the Atacama Desert located in northern Chile and covers approximately 2900 km^2 (Zúñiga et al., 1991; Demergasso et al., 2004). It is a geologically young, dynamic system; the second-largest salt flat in the world. The Salar de Atacama is the largest evaporitic basin in Chile. Note that an evaporite is a water-soluble sedimentary mineral deposit that results from concentration and crystallization by evaporation from an aqueous solution. The area is rich in lithium reserves, so much so that in the year 2008, the area provided 30 per cent of global lithium carbonate. Salar de Atacama is surrounded by mountains, and has no drainage outlets. In the east it is enclosed by the main chain of the Andes, while to the west lies a secondary mountain range of the Andes called Cordillera de Domeyko. Large volcanoes dominate the landscape.

Salar de Atacama has several permanent hypersaline lakes that receive waters from the Andes Range (Risacher and Alonso, 1996). Highly saline lakes in the extremely arid Atacama Desert located in northern Chile are characterized by high UV radiation, high salt concentrations and wide diurnal temperature variations. There are small ponds and a number of shallow lakes with high concentrations of salts (hypersaline lakes), which receive streams of fresh water from the subsurface (Zúñiga et al., 1991). The low

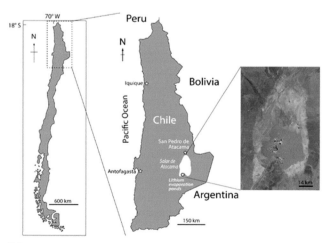

FIG. 6.75 Map of Salar de Atacama and satellite image of evaporation ponds of lithium brines concentration process (blue squares; source EarthExplorer, U.S. Geological Survey, https://earthexplorer.usgs.gov/). *(Source: Cubillos, C.F., Aguilar, P., Grágeda, M., and Dorador, C. (2018). Microbial communities from the world's largest lithium reserve, Salar de Atacama, Chile: life at high LiCl concentrations. J. Geophys. Res. Biogeosci. 123, 3668–3681. doi: 10.1029/2018JG004621) Website: https://agupubs.onlinelibrary.wiley.com/doi/full/10.1029/2018JG004621 Copyright info: © 2021 American Geophysical Union.*

precipitation together with an exceptionally high evaporation of 1800–3200 mm/year leads to a hyper-arid ecosystem (Risacher et al., 2003; Boschetti et al., 2007). Solar radiation is high, especially UV-B radiation, which is 20% high compared with that at sea level (Cabrera et al., 1995). Occasional floods do reach the saltpan, and flood waters carry gravel, sand, clay, and salt. Because the playa (salt flat) lacks drainage, water only leaves by evaporation. As it evaporates, salts remain behind and form crusts.

It was found that the pH (a measure of the acidity or basicity of aqueous or other liquid solutions) was only slightly alkaline and despite the shallow character of the lakes, low dissolved oxygen concentrations have been measured in Thiel et al. (2010)'s study (1.2 mg/L) and in previous studies (0.6 mg/L, Zúñiga et al., 1991; Boschetti et al., 2007). Risacher and Alonso (1996) found that sodium and chloride are the dominating ions, followed by sulfate.

The saline systems in Salar de Atacama include a number of different constituents: these include a liquid fraction, largely made up of chlorinated brines; a detrital fraction, consisting of sands, salts, and intertwined clays in the saline body; and a salt fraction, produced by the deposition of different salts resulting in areas dominated by carbonate, sulfate, and chlorides (Vila, 2010). Of these salts, lithium is important in the Salar de Atacama, which exhibits some of the highest reported concentrations of lithium in the world (Munk et al., 2016), with a mean concentration of 1500 ppm (1.5 g/L) of Li (Habashi, 1997).

The hypersaline lakes in the Atacama Desert represent unique and extreme habitats clearly dominated by various forms of microbial life. Like other hypersaline environments, the studied lakes of the Salar de Atacama (Laguna Chaxa and Laguna Tebenquiche) are inhabited by only a few higher organisms such as brine shrimps, some copepods and surrounding macrophytes (Zúñiga et al., 1991). Visually, the shallow Laguna Tebenquiche and Laguna Chaxa exhibit the presence of extensive red–purple-colored microbial mats on the surface of the lake sediments. Although the biological productivity in these lakes is expected to be high (Boschetti et al., 2007), the content of chlorophyll a was shown to be rather low in some studies (e.g., Demergasso et al., 2008), leading to the assumption of a considerable impact of anoxygenic phototrophic bacteria on the primary productivity in these habitats.

Thiel et al. (2010) recall that heterotrophic strains of moderately halophilic bacteria have been analyzed by numerical taxonomy (Prado et al., 1991; Valderrama et al., 1991) and chemotaxonomic analysis (Marquez et al., 1993) and dominated the isolation-based studies (Ramos-Cormenzana, 1993; Campos, 1997). The only phototrophic bacteria so far described in Laguna Tebenquiche (Salar de Atacama) were oxygenic cyanobacteria represented by Oscillatoria (Zúñiga et al., 1991). Recently, the bacterial diversity in water samples of Laguna Tebenquiche has been studied by ribosomal gene library analysis (Demergasso et al., 2008).

In order to specifically study the communities of phototrophic prokaryotes of these habitats, Thiel et al. (2010) used molecular genetic analyses with group-specific primers for functional genes (pufLM, fmoA), which target phototrophic bacterial communities. Because they represent a physiological group of polyphyletic origin (derived from more than one common evolutionary ancestor or ancestral group and therefore not suitable for placing in the same taxon), it is not possible to recover the diversity of phototrophic communities using 16S rRNA gene sequences. The pufLM genes encode for the light (L) and medium (M) subunit of the photosynthetic reaction center type II structural proteins of phototrophic purple bacteria including purple sulfur bacteria, purple non-sulfur bacteria and aerobic anoxygenic phototrophic bacteria, as well as Chloroflexaceae (Thiel et al., 2010).

Thiel et al. (2010) recall that "These genes have been used previously to access phototrophic bacteria in environmental samples (Oz et al., 2005; Hu et al., 2006) and were demonstrated to be suitable phylogenetic markers for purple sulfur bacteria (Tank et al., 2009). The fmoA gene encodes the monomer of the FMO protein, which binds Bchl a and is associated in a trimeric structure (Fenna et al., 1974). Its unique occurrence in green sulfur bacteria and the recently described 'Candidatus Chloracidobacterium thermophilum' (Bryant et al., 2007) makes fmoA an appropriate target to specifically analyze environmental communities of these bacteria (Alexander and Imhoff, 2006)."

Studies carried out by Thiel et al. (2010) indicated that highly diverse communities of anoxygenic phototrophic bacteria were present in both Laguna Chaxa and Laguna Tebenquiche, but significant differences were found between the two lakes according to pufLM T-RFLP analysis. The studied hypersaline lakes in the Salar de Atacama are unique habitats harboring highly diverse anoxygenic phototrophic bacterial communities, including numerous pufLM containing anoxygenic phototrophic bacteria. In addition, clear differences were also obvious between the subsamples of each of the two lakes. It was found that the communities have a significantly different composition in the two lakes as well as in subsamples of each of the lakes. Both pufLM T-RFLP and clone libraries revealed a high heterogeneity of the anoxygenic phototrophic bacterial communities within the different samples of both lakes (Thiel et al., 2010).

According to Thiel et al. (2010), the identification of a novel group of phototrophic Gammaproteobacteria, which inhabits and possibly dominates the extreme habitats of the Salar de Atacama, indicates that the available knowledge of the diversity of anoxygenic phototrophic bacteria is far from complete. Thiel et al. (2010) suggest that this novel lineage is represented by a phylogenetically diverse number of phylotypes possibly reflecting several species and even genera (differentiated at a 86% pufLM sequence similarity level as proposed by Tank et al. (2009) and may be specifically adapted to Chilean salt lakes, similar to a group of Bacteroidetes found in Laguna Tebenquiche (Demergasso et al., 2008; Dorador et al., 2009). Note that Bacteroidetes are a phylum of rod-shaped, Gram-negative bacteria that are commonly found in the environment, including in soil, sea water, and in the GI tract and on the skin of animals. Members of this genus are among the so-called good bacteria, because they produce favorable metabolites, including SCFAs, which have been correlated with reducing inflammation. The phylum Bacteroidota is composed of three large classes of Gram-negative, non-spore-forming, anaerobic or aerobic, and rod-shaped bacteria that are widely distributed in the environment, including in soil, sediments, and sea water, as well as in the guts and on the skin of animals.

Thiel et al. (2010) suggest that the presence of aerobic phototrophic Alpha- and Gammaproteobacteria in Laguna Chaxa found in their study indicates the coexistence of both aerobic and anaerobic anoxygenic phototrophic bacteria in the bacterial mats covering the shallow lake sediments and in its saline waters. In their studies, the molecular approach using photosynthesis genes specifically targeting anoxygenic phototrophic bacteria demonstrated a rich diversity of anoxygenic phototrophic Proteobacteria, including aerobic anoxygenic phototrophic Gammaproteobacteria in the Salar de Atacama. With regard to the microbial life in the Salar de Atacama, Thiel et al. (2010) recall that the pufM sequences related to those found in the Salar de Atacama were also present in a Tibetan hypersaline lake (Jiang et al., 2010), indicating a possible preference for high-salt conditions in these bacteria.

Green sulfur bacteria are quite likely a minor component of the phototrophic bacterial community in the Salar de Atacama, but may develop under appropriate conditions, as demonstrated by enrichment cultures (Thiel et al., 2010). Species of the genus *Prosthecochloris* are known as salt-dependent organisms from marine and hypersaline habitats and have been obtained by isolation and molecular methods from different saline habitats (Gorlenko, 1970; Imhoff, 2001, 2003; Vila et al., 2002; Alexander and Imhoff, 2006; Triadó-Margarit et al., 2010; Imhoff & Thiel, 2010). Detection of the halotolerant Prostecochloris indica (7% salt tolerance) in the Salar de Atacama correlates with the recognition of Prosthecochloris species being common as representatives of the Chlorobiaceae in saline and hypersaline habitats (Alexander et al., 2002; Triadó-Margarit et al., 2010). At the same time, they demonstrate low abundance and a possibly quite restricted phylogenetic diversity of this group in the salt lakes of the Salar de Atacama (Thiel et al., 2010).

Thiel et al. (2010)'s study indicated that highly variable conditions such as water variability, salt concentrations, and light conditions apparently shape the community structure in microhabitats of the salt lakes in Salar de Atacama. Their study demonstrated the great power of molecular methods targeting the photosynthesis-related functional genes pufLM in studying natural communities of anoxygenic phototrophic Proteobacteria in environmental samples.

Prior to the study reported by Cubillos et al. (2018), little was known regarding the microbiota associated with lithium evaporation ponds in Salar de Atacama. One report suggested that extremely saline (40% and 70% salinity) brines from the Salar de Atacama were devoid of life (Pedrós-Alió, 2004), and another report showed the presence of Archaea but not bacteria in this type of ponds (Demergasso et al., 2004). These microbial groups were also found in Cubillos et al. (2018)'s study, but these investigators found the presence of both domains Bacteria and Archaea, in hypersaline lithium brines of Salar de Atacama (salinity = 556 g/L). According to Cubillos et al. (2018), the different results regarding the presence of bacteria likely reflect methodological differences, given that the results of high-throughput sequencing depend strongly on primer choice (Klinworth et al., 2013).

As already indicated, Salar de Atacama is one of the largest global reservoirs of natural lithium brines (mean lithium concentration = 1500 ppm), enabling Chile to be a leading producer of lithium products. This large salt flat (3000 km^2), located in the Atacama Desert at 2300 m above sea level, is dominated by microorganisms. Cubillos et al. (2018) studied lithium as a modulator of microbial richness and diversity in brines representing natural conditions (34.7% salinity) and conditions prior to lithium production with a concentrated brine (55.6% salinity). They found that brines only supported a single archaeal family (Halobacteriaceae): natural brines included the archaeal genera *Halovenus*, *Natronomonas*, *Haloarcula*, and *Halobacterium*. Concentrated brines included the archaeal genera *Halovenus*, *Halobacterium*, and *Halococcus*. Cubillos et al. (2018) found that the most abundant bacterial families in natural brine were Rhodothermaceae and Staphylococcaceae. It was noticed that Xanthomonadaceae dominated the bacterial community in the concentrated brine. A comparison of the entire microbial community (Archaea and Bacteria) revealed that only seven operational taxonomic units were shared between samples, all of which were Archaea. Further, Cubillos et al. (2018)'s results showed that Bacteria were phylogenetically more diverse and richer in the concentrated brine, while archaeal diversity was maximized in the natural brine. The concentrated lithium brines of the Salar de Atacama represent one of the most saline environments described to date (dominated by LiCl). Cubillos et al. (2018) suggest that elevated concentrations of lithium could greatly modulate microbial diversity and give insights into the adaptive biology of microorganisms required to cope with extremely high concentrations of salts that extend beyond that of NaCl, which is a far more commonly studied salt.

Cubillos et al. (2018)'s research showed that concentrated brines support life and are dominated by hundreds of species of microorganisms. Due to saline stress these "extremophiles" have developed very special (and previously undescribed) strategies to survive in this lithium soup. Cubillos et al. (2018) claim that these results have implications beyond Earth: they have marked implications for our understanding of the potential for life on Mars, where liquid water is known to occur as brine. Cubillos et al. (2018) further suggest that although lithium production has clear economic importance, their results show that we should consider how we will preserve these unusual ecosystems that act as reservoirs of unique microbial life.

There is little knowledge regarding the composition of microbial communities present in lithium brines, which can be considered as an underexplored microbial habitat. Cubillos et al. (2018) carried out a study aimed at determining and to describing the diversity and composition of Archaea and Bacteria present in natural and concentrated, highly lithium-enriched brines from Salar de Atacama. They also aimed to clarify the role of lithium as a modulator of changes in microbial diversity associated with the concentration process of brines in the Salar de Atacama. Thus, Cubillos et al. (2018)'s results described the microbial community composition from one environment with high concentrations of non-NaCl salts, which may be key for approaches in the search for extraterrestrial life, for example, on other planets or celestial bodies.

In another study, Cubillos et al. (2019) found the presence of two bacterial strains (isolates LIBR002 and LIBR003) from one of the most hypersaline lithium-dominated man-made environments (total salinity 556 g/L; 11.7 M LiCl) reported to date. Both isolates were classified to the *Bacillus* genera, but displayed differences in 16S rRNA gene and fatty acid profiles. To determine osmoadaptation strategies in these microorganisms, Cubillos et al. (2019) characterized both isolates using morphological, metabolic and physiological attributes. Their characterization of bacterial isolates from a highly lithium-enriched environment has revealed that even at such extreme salinities with high concentrations of chaotropic solutes, scope for microbial life exists. Note that chaotropic solutes decrease the net hydrophobic effect (the observed tendency of nonpolar substances to aggregate in an aqueous solution and exclude water molecules) of hydrophobic regions because of a disordering of water molecules adjacent to the protein. This solubilises the hydrophobic region in the solution, thereby denaturing the protein. These conditions have previously been considered to limit the development of life, and Cubillos et al. (2019)'s work extends the window of life beyond high concentrations of $MgCl_2$, as previously reported, to LiCl. Cubillos et al. (2019) have suggested that their results can be used to further the understanding of salt tolerance, most especially for LiCl-dominated brines, and

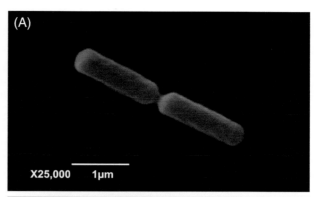

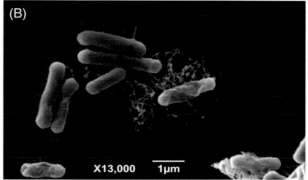

FIG. 6.76 Visualization of *Bacillus sp.*, strains from lithium brines by Scanning Electron Microscopy (SEM). (A) LIBR002 strain; (B) LIBR003 strain. *(Source: Cubillos, C.F., A. Paredes, C. Yáñez, J. Palma, E. Severino, D. Vejar, M. Grágeda, and C. Dorador (2019), Insights into the microbiology of the Chaotropic Brines of Salar de Atacama, Chile, Front. Microbiol., https://doi.org/10.3389/fmicb.2019.01611) Copyright info: Copyright © 2019 Cubillos, Paredes, Yáñez, Palma, Severino, Vejar, Grágeda and Dorador. This is an open-access article distributed under the terms of the Creative Commons Attribution License (CC BY). The use, distribution or reproduction in other forums is permitted, provided the original author(s) and the copyright owner(s) are credited and that the original publication in this journal is cited, in accordance with accepted academic practice. No use, distribution or reproduction is permitted which does not comply with these terms.*

likely have value as models for the understanding of putative extra-terrestrial (e.g., Martian) life.

Cubillos et al. (2019) evaluated growth in the presence of lithium of two lithium-tolerant bacterial isolates obtained from a hypersaline LiCl-dominated environment (total salinity 556 g/L; 11.7 M LiCl), (Cubillos et al., 2018). Furthermore, they compared the similarities of the 16S rRNA gene of their isolates with type strains of Bacillus associated with NaCl and from the Lithium Triangle Zone of South America.

Cubillos et al. (2019) reported the isolation of two bacterial species, LIBR002 strain and LIBR003 strain (Fig. 6.76) from one of the most hypersaline environments on Earth (total salinity = 556 g/L), where lithium chloride is the principal salt. Furthermore, they established the bacteriostatic (the agent prevents the growth of bacteria; i.e., it keeps them in the stationary phase of growth) effects of lithium on the microbial cells and they further suggested that the NaCl-tolerance, generation of spores, and fluidity change of membrane of Bacillus could be keys to survive in this poly-extreme environment, as reported elsewhere (Filippidou et al., 2019). The information detailed in Cubillos et al. (2019)'s study is likely relevant to the search for microbial life in planets and other celestial bodies with surficial salt-lakes or brines that provide a potential niche for microbial life. In consequence, these fragile salar ecosystems (salares), known as the Lithium Triangle Zone, must be protected, because they are base or support of large food chain from animals (flamingos), going through microscopic eukaryotes (ciliates and amoebas), and microbial life (as archaea and bacteria organisms).

Interestingly, Hurlbert and Chang (1983) reported that experimental exclusion of the Andean flamingo (*Phoenicoparrus andinus*) from shallow water areas of a Salt Lake in the Bolivian Andes caused large increases in the biomass of microorganisms inhabiting the surface sediments, especially a large diatom (*Surirella wetzeli*), amebas, ciliates (single-celled animals of a large and diverse phylum distinguished by the possession of short microscopic hair-like vibrating structure found in large numbers on the surface of certain cells, either causing currents in the surrounding fluid, or, in some protozoans and other small organisms, providing propulsion), and nematodes (roundworms). This is a conservative demonstration of the influences that water birds in general exert on the structure of aquatic ecosystems.

Cubillos et al. (2019) hypothesized that Salar de Hombre Muerto, Salar de Uyuni and Salar de Atacama support extreme, unique and unstudied microbial life capable of resisting the presence of various stressful agents including high concentrations of lithium and other salts, chaotropic activity (i.e., an activity that disrupts the structure of, and denatures, macromolecules such as proteins and nucleic acids; e.g. DNA and RNA), evaporation, and high radiation, providing a potential microbial niche. Furthermore, Cubillos et al. (2019) suggested that this information could be important for understanding physiology at the extreme boundary of life in terrestrial or even extra-terrestrial brines such as those found on Mars (Benison and Bowen, 2006).

6.6.5 Importance of Atacama desert in astrobiological study of Mars

Cubillos et al. (2018) recall that in recent decades, the Atacama Desert has been subject to an increased focus for astrobiological research, largely due to the presence of "Mars-like" soils (Bull et al., 2016; Navarro-González et al., 2003). It has been found that the "Mars-like" soils present in the Atacama Desert harbor active halophilic microbial communities (Ojha et al., 2015; Oren, 2013), specifically in areas with high concentrations of NaCl, such as halite rocks (Robinson et al., 2015). The presence of abundant deposits of magnesium and sodium salts (major cations), as well as of sulfates and chlorides (major anions) on Mars (Clark and van Hart, 1981; Forte et al., 2016); and the abundant presence of lithium in meteorites (Shima and Honda, 1963) and stars including the Sun (Grevesse, 1968), potentially present important chemical similarities with the brines studied in Cubillos et al. (2018)'s research.

Cubillos et al. (2018) recall that "the process of brine concentration via evaporation and subsequent harvesting of salts resulted in a change in the saline composition of the natural brine (high sodium/low lithium concentration) to a brine concentrated in lithium (low sodium/high lithium concentration). In addition, this change resulted in a marked shift in microbial diversity in initial/final brines of the process, with Archaea as the most phylogenetically diverse in natural brine, and Bacteria as the most phylogenetically diverse in concentrated lithium brine. However, Archaea was the only group shared in both."

Halites (chloride minerals, sometimes tinted by impurities, found in beds as an evaporite) are known (Davila et al., 2010) to have a characteristic feature in which at a specific relative humidity the crust is transformed to brines (water strongly impregnated with salt). According to Cubillos et al. (2018), their study results showing notably higher bacterial diversity in concentrated brines relative to natural brines, combined with the deliquesce effects (habit of becoming liquid by absorbing moisture from the air, as certain salts) of chloride in dry environments (like the Atacama Desert) in supporting an active microbial community in halite, have implications for potential life on Mars. According to Cubillos et al. (2018), it is not too difficult to believe that in deep brines (Orosei et al., 2018) and the chloride-rich deposits of Terra Sirenum (Glotch et al., 2010) of Mars could represent suitable habitat (similar to the concentrated lithium brines) for extant or extinct Bacteria. If present, Martian-microbial life probably displays a salt-out osmo-adaptation strategy, allowing flexibility for variation in ionic composition and water availability (Cubillos et al., 2018).

Consequently, Cubillos et al. (2018) have proposed that the brines of Salar de Atacama could represent a useful future model for studies examining biological resistance mechanisms under life-limiting conditions, as well as possible biological indicators for the search and probable development of extraterrestrial life.

References

Alcaraz, L.D., Moreno-Hagelsieb, G., Eguiarte, L.E., Souza, V., Herrera-Estrella, L., Olmedo, G., 2010. Understanding the evolutionary relationships and major traits of Bacillus through comparative genomics. BMC Genomics 11, 332. doi:10.1186/1471-2164-11-332.

Alexander, B., Andersen, J.H., Cox, R.P., Imhoff, J.F., 2002. Phylogeny of green sulfur bacteria on the basis of gene sequences of 16S rRNA and of the Fenna–Matthews–Olson protein. Arch. Microbiol. 178, 131–140.

Alexander, B., Imhoff, J.F., 2006. Communities of green sulfur bacteria in marine and saline habitats analyzed by gene sequences of 16S rRNA and Fenna–Matthews–Olson protein. Int. Microbiol. 9, 259–266.

Allen, T.B., 1999. The Shark Almanac. The Lyons Press, New York, NY ISBN 978-1-55821-582-5.

Amoozegar, M.A., Bagheri, M., Didari, M., Mehrshad, M., Schumann, P., Spröer, C., et al., 2014. *Aquibacillus halophilus* gen. nov., Sp. nov., A moderately halophilic bacterium from a hypersaline lake, and reclassification of *Virgibacillus koreensis* as *Aquibacillus koreensis* comb. nov. and *Virgibacillus albus* as *Aquibacillus albus* comb. nov. Int. J. Syst. Evol. Microbiol. 64, 3616–3623. doi:10.1099/ijs.0.065375-0.

Amoozegar, M.A., Bagheri, M., Makhdoumi, A., Nikou, M.M., Fazeli, S.A.S., Schumann, P., et al., 2016. *Oceanobacillus halophilus* sp. nov., a novel moderately halophilic bacterium from a hypersaline lake. Int. J. Syst. Evol. Microbiol. 66, 1317–1322. doi:10.1099/ijsem.0.000952.

An, J.W., Kang, D.J., Tran, K.T., Kim, M.J., Lim, T., Tran, T, 2012. Hydrometallurgy recovery of lithium from Uyuni salar brine. Hydrometallurgy 117–118, 64–70. doi:10.1016/j.hydromet.2012.02.008.

Anati, D.A., Stiller, M., Shasha, S., Gat, J.R., 1987. Changes in the thermo-haline structure of the Dead Sea? 1979–1984. Earth Planet. Sci. Lett. 84 (109), 121.

Avise, J.C., Hamrick, J.L., 1996. Conservation Genetics. Imprint: New York: Chapman & HallSpringer, c1996. ISBN 978-0412055812.

Bagheri, M., Didari, M., Amoozegar, M.A., Schumann, P., Sánchez-Porro, C., Mehrshad, M., et al., 2012. *Bacillus iranensis* sp. nov., a moderate halophile from a hypersaline lake. Int. J. Syst. Evol. Microbiol. 62, 811–816. doi:10.1099/ijs.0.030874-0.

Banciu, H.L., Muntyan, M.S., 2015. Adaptive strategies in the double-extremophilic prokaryotes inhabiting soda lakes. Curr. Opin. Microbiol. 25, 73–79. doi:10.1016/j.mib.2015.05.003.

Bayly, I.A.E., Williams, W.D., 1966. Chemical and biological studies on some saline lakes of south-east Australia. Aust. J. mar. Freshwat. Res. 17, 177–228.

Bayly, I.A.E., 1967. The general biological classification of aquatic environments with special reference to those of Australia. In: Weatherley, A.H. (Ed.), Studies on Australian Inland Waters and their Fauna. Australian National University Press, Canberra.

Bayly, I.A.E., 1967b. The fauna and chemical composition of some athalassic saline waters in New Zealand. N.Z. J. Mar. Freshwater Res. 1 (2), 105–117. doi:10.1080/00288330.1967.9515197.

Bayly, I.A.E., 1972. Salinity tolerance and osmotic behavior of animals in athalassic saline and marine hypersaline waters. Annual Review of Ecology and Systematics 3, 233–268.

Becking, L.G.M.B., 1928. On organisms living in concentrated brine. Tijdschr. Ned. Dierkund Ver. Ser. III 1, 6–9.

Beladjal, L., Gheysens, T., Clegg, J.S., Amar, M., Mertens, J., 2018. Life from the ashes: survival of dry bacterial spores after very high temperature exposure. Extremophiles: Life Under Extreme Conditions 22 (5), 751–759. doi:10.1007/s00792-018-1035-6.

Belfiore, C., Curia, M., Farías, M., 2018. Characterization of Rhodococcus sp. A5wh isolated from a high-altitude Andean lake to unravel the survival strategy under lithium stress. Rev. Argent. Microbiol. 50, 311–322. doi:10.1016/j.ram.2017.07.005.

Benison, K.C., Bowen, B.B., 2006. Acid saline lake systems give clues about past environments and the search for life on Mars. Icarus 183, 225–229. doi:10.1016/j.icarus.2006.02.018.

Benison, K.C., 2008. Life and death around acid-saline lakes. Palaios 23, 571–573. doi:10.2110/palo.2008.s05.

Benison, K.C., 2019. The physical and chemical sedimentology of two high-altitude acid salars in Chile: sedimentary processes in an extreme environment. J. Sediment. Res. 89, 147–167. doi:10.2110/jsr.2019.9.

Benlloch, S., López-López, A., Casamayor, E.O., Øvreås, L., Goddard, V, 2002. Prokaryotic genetic diversity throughout the salinity gradient of a coastal solar saltern. Environ. Microbiol. 4 (349), 360.

Bindoff, N.L., Cheung, W.W.L., Kairo, J.G., Arístegui, J., et al., 2019. Chapter 5. Changing Ocean, Marine Ecosystems, and Dependent Communities, IPCC Special Report on the Ocean and Cryosphere in a Changing Climate, 447–587.

Bodaker, I., Sharon, I., Suzuki, M.T., Feingersch, R., Shmoish, M., 2010. Comparative community genomics in the Dead Sea: an increasingly extreme environment. ISME. J. 4 (399), 407.

Boschetti, T., Cortecci, G., Barbieri, M., Mussi, M., 2007. New and past geochemical data on fresh to brine waters of the Salar de Atacama and Andean Altiplano, northern Chile. Geofluids 7, 33–50.

Bowers, K.J., Mesbah, N.M., Wiegel, J., 2009. Biodiversity of polyextremophilic bacteria: does combining the extremes of high salt, alkaline pH and elevated temperature approach a physico-chemical boundary for life? Saline Syst 5, 9.

Brian, D.S., Perrin, W., 2007. Conservation Status of Irrawaddy dolphins (*Orcaella Brevirostris*), 14th Meeting of the CMS Scientific Council. Bonn, 14–17.

Brierley, C.L., Brierley, J.A., 1973. Chemoautotrophic and thermophilic microorganism isolated from an acid hot spring. Can. J. Microbiol. 19 (2), 183–188.

Brisou, J., Courtois, D., Denis, F., 1974. Microbiological study of a hypersaline lake in French Somaliland. Appl. Microbiol. 27 (5), 819–822.

Brock, T.D., Brock, K.M., Belley, R.T., Weiss, R.L., 1972. *Sulfolobus*: a new genus of sulfur-oxidizing bacteria living at low pH and high temperature. Archs. Microbiol. 84 (1), 54–68.

Bryantseva, I., Gorlenko, V.M., Kompantseva, E.I., Imhoff, J.F., Süling, J., Mityushina, L., 1999. *Thiorhodospira sibirica gen. nov.*, sp. nov., a new alkaliphilic purple sulfur bacterium from a Siberian soda lake. Int. J. Syst. Bacteriol. 49, 697–703.

Bryantseva, I.A., Gorlenko, V.M., Kompantseva, E.I., Imhoff, J.F., 2000. Thioalkalicoccus limnaeus gen. nov., sp nov., a new alkaliphilic purple sulfur bacterium with bacteriochlorophyll b. Int J Syst Evol Micr. 50, 2157–2163.

Bryant, D.A., Costas, A.M.G., Maresca, J.A., et al., 2007. Candidatus Chloracidobacterium Thermophilum: an aerobic phototrophic acidobacterium. Science 317, 523–526.

Budker, P., 1971. The Life of Sharks, Columbia University Press, New York; 222 pages.

Bull, A.T., Asenjo, J.A., Goodfellow, M., Gómez-Silva, B, 2016. The Atacama Desert: Technical resources and the growing importance of novel microbial diversity. Annu. Rev. Microbiol. 70, 215–234.

Burggraf, S., Jannasch, H.W., Nicolaus, B., Stetter, K.O., 1990. *Archaeoglobus profundus* sp. nov.: represents a new species within the sulfate-reducing archaebacteria. Syst. Appl. Microbiol. 13, 24–28.

Byatt, A., Fothergill, A., Holmes, M., 2001. The Blue Planet: A Natural History of the Oceans. DK, Print. sea slugs, New York, NY.

Cabrera, S., Bozzo, S., Fuenzalida, H., 1995. Variation in UV radiation in Chile. J Photoch Photobio B 28, 137–142.

Campos, V, 1997. Microorganismos de ambientes extremos: Salar de Atacama, Chile. In: Gonzales, C (Ed.), El Altiplano: Ciencia y Conciencie de los Andes. Editirial Artegrama, Santiago, Chile, pp. 143–147.

Caumette, P., Baulaigue, R., Matheron, R., 1988. Characterization of *Chromatium salexigens* sp. nov., a halophilic Chromatiaceae isolated from Mediterranean Salinas. Syst Appl Microbio. 110, 284–292.

Caumette, P., Baulaigue, R., Matheron, R., 1991. *Thiocapsa halophila* sp. nov, a new halophilic phototrophic purple sulfur bacterium. Arch Microbiol. 155, 170–176.

Caumette, P., 1993. Ecology and physiology of phototrophic bacteria and sulfate reducing bacteria in marine salterns. Experientia 49, 473–481.

Caumette, P., Imhoff, J.F., Süling, J., Matheron, R., 1997. *Chromatium glycolicum* sp. nov., a moderately halophilic purple sulfur bacterium that uses glycolate as substrate. Arch. Microbiol. 167, 11–18.

Clark, B.C., van Hart, D.C., 1981. The salts of Mars. Icarus 45 (2), 370–378.

Cohen, Y., Krumbein, W.E., 1977. Solar Lake (Sinai). 2. Distribution of photosynthetic microorganisms and primary production. Limnol. Oceanogr. 22, 609–620.

Cohen, Y., Jørgensen, B.B., Revsbech, N.P., Poplawski, R., 1986. Adaptation to hydrogen sulfide of oxygenic and anoxygenic photosynthesis among cyanobacteria. Appl. Environ. Microbiol. 51 (398), 407.

Cohn, F., 1872. Untersuchungen über Bakterien. Beit. Biol. Pflanz. 1, 127–224.

Cox, L., Dooleyl, D., Nestlk, A., 1990. Effect of lithium chloride growth of Listeria spp. Int. J. Food Microbiol. 7, 311–325.

Craw, D., 2021. Sutton Salt Lake. University of Otago, Otago.

Cubillos, C.F., Aguilar, P., Grágeda, M., Dorador, C., 2018. Microbial communities from the world's largest lithium reserve, Salar de Atacama, Chile: life at high LiCl concentrations. J. Geophys. Res. Biogeosci. 123, 3668–3681. doi:10.1029/2018JG004621.

Cubillos, C.F., Paredes, A., Yáñez, C., Palma, J., Severino, E., Vejar, D., Grágeda, M., Dorador, C., 2019. Insights into the microbiology of the Chaotropic Brines of Salar de Atacama, Chile. Front. Microbiol. https://doi.org/10.3389/fmicb.2019.01611.

Dai, C.F., Horng, S., 2009. Scleractinia Fauna of Taiwan. II. The Robust Group. National Taiwan University Press, Taipei.

Das, S., 2008. Dolphins better off in Chilika — Survey reveals dip in death toll of Irrawaddy School. The Telegraph Calcutta, India: Calcutta. pp. Front page.

Davidson, I., Jane, T., 2006. The Oxford Companion to Food. Oxford University Press: Oxford, England, p. 150. ISBN 0-19-280681-5, ISBN 978-0-19-280681-9.

Davila, A.F., Duport, L.G., Melchiorri, R., Jänchen, J., Valea, S., de los Rios, A., Fairén, A.G., Möhlmann, D., McKay, C.P., Ascaso, C., Wierzchos, J., 2010. Hygroscopic salts and the potential for life on Mars. Astrobiology 10 (6), 617–628.

de Assunção, L.S., da Luz, J.M.R., da Silva, M.D.C.S., Vieira, P.A.F., Bazzolli, D.M.S., Vanetti, M.C.D., Kasuya, M.C.M., 2012. Enrichment of mushrooms: An interesting strategy for the acquisition of lithium. Food Chem. 134 (2), 1123–1127.

Del Barco, R., Foladori, G., 2014. Nanotechnology and Lithium: A window of opportunity for Bolivia? Advances and Challenges. Paper presented at the 6th International Scientific Conference on Economic and Social Development and 3rd Eastern European ESD Conference: Business Continuity Book of Proceedings: 283. Vienna, Austria.

DeLong, E.F., Taylor, L.T., Marsh, T.L., Preston, C.M, 1999. Visualization and enumeration of marine planktonic archaea and bacteria by using polyribonucleotide probes and fluorescent in situ hybridization. Appl. Environ. Microbiol. 65, 5554 5563.

Demergasso, C., Casamayor, E., Chong, G., Escudero, L., Pedrós-Alió, C., 2004. Distribution of prokaryotic genetic diversity in athalassohaline lakes of the Atacama Desert, Northern Chile. FEMS Microbiol. Ecol. 48, 57–69.

Demergasso, C., Escudero, L., Casamayor, E.O., Chong, G., Balague, V., Pedros-Alio, C., 2008. Novelty and spatio-temporal heterogeneity in the bacterial diversity of hypersaline Lake Tebenquiche (Salar de Atacama). Extremophiles 12, 491–504.

Denis, F., Brisou, J., Courtois, D., Niaussat, P., 1972. Microbiologie comparke de quelques milieux aquatiques. Rev. Epidemiol. Med. Soc. Sante Publ. 20, 195–203.

Dickson, J.L., Head, J.W., Levy, J.S., Marchant, D.R., 2013. Don Juan Pond, Antarctica: near-surface CaCl2-brine feeding Earth's most saline lake and implications for Mars. Sci. Rep. 3, 1166. doi:10.1038/srep01424.

Dorador, C., Meneses, D., Urtuvia, V., Demergasso, C., Vila, I., Witzel, K.P., Imhoff, J.F., 2009. Diversity of Bacteroidetes in high altitude saline evaporitic basins in northern Chile. J. Geophys. Res. 114, 35–44.

Duarte, C.M., Wu, J., Xiao, X., Bruhn, A., Krause-Jensen, D, 2017. Can seaweed farming play a role in climate change mitigation and adaptation? Frontiers in Marine Science 4. doi:10.3389/fmars.2017.00100.

Ebert, D.A., 2003. Sharks, Rays, and Chimaeras of California. University of California Press, Downtown Oakland, California, pp. 50–52. ISBN 0-520-23484-7.

Einhorn, C. (2021), Shark populations are crashing, with a 'very small window' to avert disaster, The New York Times

Elazari-Volcani, B., 1940. Algae in the bed of the Dead Sea. Nature 145, 975 –975.

Elazari-Volcani, B., 1943a. Bacteria in the bottom sediments of the Dead Sea. Nature 152, 301–302.

Elazari-Volcani, B., 1943b. A dimastig amoeba in the bed of the Dead Sea. Nature 152, 301–302.

Elazari-Volcani, B., 1944. A ciliate from the Dead Sea. Nature 154, 335 –335.

Elshahed, M.S., Najar, F.Z., Roe, B.A., Oren, A., Dewers, T.A., 2004. Survey of archaeal diversity reveals an abundance of halophilic Archaea in a low-salt, sulfide-and sulfur-rich spring. Appl. Environ. Microbiol. 70 (2230), 2239.

Fenna, R.E., Matthews, B.W., Olson, J.M., Shaw, E.K., 1974. Structure of a bacteriochlorophyll protein from green photosynthetic bacterium Chlorobium limicola– crystallographic evidence for a trimer. J. Mol. Biol. 84, 231–240.

Filippidou, S., Junier, T., Wunderlin, T., Kooli, W.M., Palmieri, I., Al-Dourobi, A., et al., 2019. Adaptive strategies in a poly-extreme environment: differentiation of vegetative cells in *Serratia ureilytica* and resistance to extreme conditions. Front. Microbiol. 10, 102. doi:10.3389/fmicb.2019.00102.

Fischer, F., Zillig, W., Stetter, K.O., Schreiber, G., 1983. Chemolithoautotrophic metabolism of anaerobic extremely thermophilic archaebacteria. Nature 301, 511–513.

Fisher, M.M., Triplett, E.W., 1999. Automated Approach for Ribosomal Intergenic Spacer Analysis of microbial diversity and its application to freshwater bacterial communities. Appl. Environ. Microbiol. 65 (10), 4630–4636.

Fornari, M., Risacher, F., Feraud, G., 2001. Dating of paleolakes in the central Altiplano of Bolivia. Palaeogeogr. Palaeoclimatol. Palaeoecol. 172, 269–282.

Forte, E., Dalle Fratte, M., Azzaro, M., Guglielmin, M., 2016. Pressurized brines in continental Antarctica as a possible analogue of Mars. Sci. Rep. 6, 33158.

Frazer, J., 2011. Fountains of life found at the bottom of the Dead Sea. Sci. Am. https://blogs.scientificamerican.com/artful-amoeba/fountains-of-life-found-at-the-bottom-of-the-dead-sea/.

Frigaard, N.U., Bryant, D.A., 2004. Seeing green bacteria in a new light: genomics-enabled studies of the photosynthetic apparatus in green sulfur bacteria and filamentous anoxygenic phototrophic bacteria. Arch. Microbiol. 182 (265), 276.

Froese, R., Pauly, D. (Eds.), 2008. Morone saxatilis, In: FishBase, World Wide Web electronic publication. (version 04/2008).

Garabito, M.J., Márquez, M., Ventosa, A., 1998. Halotolerant Bacillus diversity in hypersaline environments. Can. J. Microbiol. 44, 95–102. doi:10.1139/cjm-44-2-95.

Garrett, D.E., 1998. Lake or brine deposits. Borates: Handbook of Deposits, Processing, Properties, and Use; Elsevier, 227–253; eBook ISBN: 9780080500218; https://www.elsevier.com/books/borates/garrett/978-0-12-276060-0.

Garret, D.E., 2004. Handbook of Lithium and Natural Calcium Chloride, Their Deposits, Processing, Uses and Properties. Elsevier, Amsterdam.

Gavrieli, I., Yechieli, Y., Halicz, L., Spiro, B., Bein, A., 2001. The sulfur system in anoxic subsurface brines and its implication in brine evolutionary pathways: the Ca-chloride brines in the Dead Sea area. Earth Planet. Sci. Lett. 186 (199), 213.

George, D.M., Vincent, A.S., Mackey, H.R., 2020. An overview of anoxygenic phototrophic bacteria and their applications in environmental biotechnology for sustainable resource recovery. Biotechnology Reports 28, e00563.

Ghai, R., Pašić, L., Fernández, A.B., Martin-Cuadrado, A.B., Mizuno, C.M., McMahon, K.D., Papke, R.T., Stepanauskas, R., Rodriguez-Brito, B., Rohwer, F., Sánchez-Porro, C., Ventosa, A., Rodríguez-Valera, F., 2011. New abundant microbial groups in aquatic hypersaline environments. Sci. Rep. 1, 135.

Ghosh, A.K., Pattnaik, A.K., 2008. Chilika Lagoon Basin. Chilika Lagoon Experience and Lessons learned Brief. UNEP International Waters Learning Exchange and Resource Network, p. 115.

Giani, D., Seeler, J., Giani, L., Krumbein, W.E., 1989. Microbial mats and physicochemistry in a saltern in the Brittany (France) and in a laboratory scale saltern model. FEMS Microbiol. Ecol. 62, 151–162.

Glotch, T.D., Bandfield, J.L., Tornabene, L.L., Jensen, H.B., Seelos, F.P., 2010. Distribution and formation of chlorides and phyllosilicates in Terra Sirenum, Mars. Geophys. Res. Lett. 37, L16202.

Godfrey, L.V, Jordan, T.E, Lowenstein, T.K, Alonso, R.L, 2003. Stable isotope constraints on the transport of water to the Andes between 22° and 26°S during the last glacial cycle. Palaeogeogr. Palaeoclimatol. Palaeoecol. 194 (1–3), 299–317.

Godfrey, L.V., Chan, L.-H., Alonso, R.N., Lowenstein, T.K., McDonough, W.F., Houston, J., Li, J., Bobst, A., Jordan, T.E., 2013. The role of climate in the accumulation of lithium-rich brine in the Central Andes. Appl. Geochem. 38, 92–102.

Gorlenko, V.M., 1970. A new phototrophic green sulphur bacterium Prosthecochloris aestuarii nov. gen. nov. spec. Z. Allg. Mikrobiol. 10, 147–149.

Graças, D.A., Miranda, P.R., Baraúna, R.A., McCulloch, J.A., Ghilardi, R, 2011. Microbial diversity of an anoxic zone of a hydroelectric power station reservoir in Brazilian Amazonia. Microb. Ecol. 62 (853), 861.

Grady, J.M., Maitner, B.S., Winter, A.S., Kaschner, K., Tittensor, D.P., Record, S., Smith, F.A., Wilson, A.M., Dell, A.I., Zarnetske, P.L., Wearing, H.J., 2019. Metabolic asymmetry and the global diversity of marine predators. Science 363 (6425), eaat4220. doi:10.1126/science.aat4220.

Gray, R. (2013), How the sperm whale can hold its breath for 90 minutes, *Daily Telegraph*. ISSN 0307-1235

Grethe, R.H., Syvertsen, E.E., Steidinger, K.A., Tangen, K., 1996. Marine Diatoms. In: Tomas, Carmelo R. (Ed.), Identifying Marine Diatoms and Dinoflagellates. Academic Press, pp. 5–385 ISBN 978-0-08-053441-1.

Grevesse, N., 1968. Solar abundances of lithium, beryllium and boron. Sol. Phys. 5 (2), 159–180.

Habashi, F., 1997. Handbook of extractive metallurgy. Wiley-Vch, Germany.

Haferburg, G., Gröning, J.A.D., Schmidt, N., Kummer, N.A., Erquicia, J.C., Schlömann, M., 2017. Microbial diversity of the hypersaline and lithium-rich Salar de Uyuni, Bolivia. Microbiol. Res. 199, 19–28. doi:10.1016/j.micres.2017.02.007.

Hallsworth, J., Yakimov, M., Golyshin, P., Gillion, J., D'auria, G., de Lima Alves, F., et al., 2007. Limits of life in $MgCl_2$ -containing environments: chaotropicity defines the window. Environ. Microbiol. 9, 801–813. doi:10.1111/j.1462-2920.2006.01212.x.

Hammer, U.T., 1986. Saline Lake Ecosystems of the World. Published by Kluwer Academic Publishers Group, Dordrecht, p. 616, ISBN 10: 9061935350: ISBN 13: 9789061935353.

Hartman, K.J., 2003. Population-level consumption by Atlantic coastal striped bass and the influence of population recovery upon prey communities. Fisheries Management and Ecology 10 (5), 281–288.

Häusler, S., Noriega-Ortega, B.E., Polerecky, L., Meyer, V., de Beer, D., Ionescu, D., 2014a. Microenvironments of reduced salinity harbour biofilms in Dead Sea underwater springs. Environ Microbiol Rep 6 (2), 152–158.

Häusler, S., Weber, M., de Beer, D., Ionescu, D, 2014b. Spatial distribution of diatom and cyanobacterial mats in the Dead Sea is determined by response to rapid salinity fluctuations. Extremophiles 18 (6), 1085–1094.

Hönninger, G., Bobrowski, N., Palenque, E.R., Torrez, R., Platt, U., 2004. Reactive bromine and sulfur emissions at Salar de Uyuni, Bolivia. Geophys. Res. Lett. 31, L04101.

Hoover, J., 2007. Hawaii's Sea Creatures: A Guide to Hawaii's Marine Invertebrates, Mutual Publishing; Honolulu, HI 96816, United States; Revised edition. 5th printing; 366 pages; ISBN-10: 1566472202; ISBN-13: 978-1566472203.

Hu, Y.H., Du, H.L., Jiao, N.Z., Zeng, Y.H., 2006. Abundant presence of the gamma-like proteobacterial pufM gene in oxic seawater. FEMS Microbiol. Lett. 263, 200–206.

Huber, R., Kurr, M., Jannasch, H.W., Stetter, K.O., 1989. A novel group of abyssal methanogenic archaebacteria (*Methanopyrus*) growing at 110° C. Nature 342, 833–834.

Huber, H., Jannasch, H., Rachel, R., Fuchs, T., Stetter, K.O., 1997. *Archaeoglobus veneficus* sp. nov. a novel facultative chemolithoautotrophic hyperthermophilic sulfite reducer, isolated from abyssal black smokers. Syst. Appl. Microbiol. 20, 374–380.

Hurlbert, S.H., Chang, C.C.Y., 1983. Ornitholimnology: effects of grazing by the Andean flamingo (Phoenicoparrus andinus). Proc. Natl. Acad. Sci. U.S.A. 80, 4766–4769. doi:10.1073/pnas.80.15.4766.

Imhoff, J.F., Trüper, H.G., 1977. Ectothiorhodospira halochloris sp. nov. – a new extremely halophilic phototrophic bacterium containing bacteriochlorophyll b. Arch. Microbiol. 114, 115–121.

Imhoff, J.F., Trüper, H.G., 1981. Ectothiorhodospira abdelmalekii sp. nov., a new halophilic and alkaliphilic phototropic bacterium. Zbl Bakt Mik Hyg I. C2, 228–234.

Imhoff, J.F., Hashwa, F., Trüper, H.G., 1978. Isolation of extremely halophilic phototrophic bacteria from the alkaline Wadi Natrun, Egypt. Arch Hydrobiol 84, 381–388.

Imhoff, J.F., Sahl, H.G., Soliman, G.S.H., Truper, H.G., 1979. Wadi Natrun - chemical composition and microbial mass developments in alkaline brines of eutrophic desert lakes. Geomicrobiol. J. 1, 219–234.

Imhoff, J.F., 1988. Halophilic phototrophic bacteria. In: Rodriguez-Valera, F (Ed.), Halophilic Bacteria. CRC Press, Boca Raton, FL, pp. 85–108.

Imhoff, J.F., 2001. True marine and halophilic anoxygenic phototrophic bacteria. Arch. Microbiol. 176, 243–254.

Imhoff, J.F., 2002. Phototrophic anoxygenic bacteria in marine and hypersaline environments. In: Bitton, G (Ed.), Encyclopedia of Environmental Microbiology. John Wiley & Sons Ltd, New York, NY, pp. 2470–2489.

Imhoff, J.F., 2003. Phylogenetic Taxonomy of the family *Chlorobiaceae* on the basis of 16S rRNA and fmo (Fenna–Matthews–Olson protein) gene sequences. Int J Syst Evol Micr 53, 941–951.

Ionescu, D., Siebert, C., Polerecky, L., Munwes, Y.Y., Lott, C., Häusler, S., et al., 2012. Microbial and chemical characterization of underwater fresh water springs in the Dead Sea. PLoS One 7 (6), e38319. https://doi.org/10.1371/journal.pone.0038319.

Jiang, H.C., Deng, S.C., Huang, Q.Y., Dong, H.L., Yu, B.S, 2010. Response of aerobic anoxygenic phototrophic bacterial diversity to environment conditions in saline lakes and Daotang River on the Tibetan Plateau. Geomicrobiol J27, 400–408.

Jones, B.E., Grant, W.D., Duckworth, A.W., Owenson, G.G., 1998. Microbial diversity of soda lakes. Extremophiles 2, 191–200. doi:10.1007/s007920050060.

Joseph, 2016. Investigating Seafloors and Oceans: From Mud Volcanoes to Giant Squid. *Elsevier Science & Technology Books Publishers*, New York, NY, p. 581 ISBN: 978-0-12- 809357-3.

Joseph, 2016. Citadel-Building: Tiny Creatures in the Oceans—Corals", Chapter 7. Investigating Seafloors and Oceans: From Mud Volcanoes to Giant Squid. *Elsevier Science & Technology Books Publishers*, New York, NY, p. 581 ISBN: 978-0-12- 809357-3, pp. 377–435.

Jungbluth, S.P., Bowers, R.M., Lin, H.T., Cowen, J.P., Rappé, M.S., 2016. Novel microbial assemblages inhabiting crustal fluids within mid-ocean ridge flank subsurface basalt. ISME J. 10, 2033–2047.

Kamekura, M., Onishi, H., 1982. Cell-associates cations of the moserate halophile *Micrococcus varians* ssp. halophilus grown in media of high concentrations of LiCl, NaCl, KCl, RbCl, or CsCl. Can. J. Microbiol. 28, 155–161. doi:10.1139/m82-020.

Keough, B.P., Schmidt, T.M., Hicks, R.E., 2003. Archaeal nucleic acids in picoplankton from great lakes on three continents. Microb. Ecol. 46 (238), 248.

Klinworth, A., Pruesse, E., Schweer, T., Peplies, J., Quast, C., Horn, M., Glöckner, F., 2013. Evaluation of general 16S ribosomal RNA gene PCR primers for classical and next-generation sequencing-based diversity studies. Nucleic Acids Res. 41 (1), e1. https://doi.org/10.1093/nar/gks808.

Koch, R., 1876. Die Aetiologie der Milzbranded-Krankheit, begrundet auf die Entwicklungsgeschichete Bacillus Anthracis. Beit. Biol. Pflanz. 2, 277–310.

Kristjánsson, J.K., Hreggvidsson, G.O., 1995. Ecology and habitats of extremophiles. World J. Microbiol. Biotechnol. 11, 17–25. doi:10.1007/BF00339134.

Kurita, N., Funabashi, M., 1984. Growth-inhibitory effect on fungi of alkali cations and monovalent inorganic anions and antagonism among different alkali cations. Agric. Biol. Chem. 48, 887–893. doi:10.1080/00021369.1984.10866259.

La Cono, V., Smedile, F., Bortoluzzi, G., Arcadi, E., Maimone, G, 2011. Unveiling microbial life in new deep-sea hypersaline Lake Thetis. Part I: Prokaryotes and environmental settings. Environ. Microbiol. 13, 2250–2268.

Lal, B.V., Fortune, K., 2000. The Pacific Islands: An Encyclopedia. University of Hawaii Press, Honolulu, p. 8 ISBN 978-0-8248-2265-1.

Lapierre, L., Undeland, P., Cox, L.J., 1992. Lithium chloride-sodium propionate agar for the enumeration of bifidobacteria in fermented dairy products. J. Dairy Sci. 75 (5), 1192–1196.

Laronne Ben-Itzhak, L., Gvirtzman, H, 2005. Groundwater flow along and across structural folding: an example from the Judean Desert, Israel. J. Hydrol. 312 (51), 69.

Larsen, H., 1973. The halobacteria's confusion to biology. Antonie Leeuwenhoek 39, 383–396.

Lazar, M., Ben-Avraham, Z., 2002. First images from the bottom of the Dead Sea–Indications of recent tectonic activity. Israel J Earth Sci 51, 3–4 211.218.

Lee, C.J.D., McMullan, P.E., O'Kane, C.J., Stevenson, A., Santos, I.C., Roy, C., et al., 2018. NaCl-saturated brines are thermodynamically moderate, rather than extreme, microbial habitats. FEMS Microbiol. Rev. 42, 672–693. doi:10.1093/femsre/fuy026.

Li, H.R., Liu, W.M., Cheng, S.J., Jiang, Y., 2014. Effect of lithium on growth process of environmental microorganism by microcalorimetry and SEM. Adv. Mater. Res. 955-959, 445–449. doi:10.4028/www.scientific.net/amr.955-959.445.

Lieb, J., 2004. The immunostimulating and antimicrobial properties of lithium and antidepressants. J. Infect. 49, 88–93. doi:10.1016/j.jinf.2004.03.006.

Locascio de Mitrovich, C., Morrone, J.J., Menu-Marque, S., 2000. Distributional patterns of the South American species of boeckella (Copepoda: Centropagidae): A track analysis. Journal of Crustacean Biology 20 (2), 272.

Lockwood, A.P.M., Sheader, M., Williams, J.A., 1998. Life in Estuaries, Salt Marshes, Lagoons and Coastal Waters. In: Summerhayes, C.P., Thorpe, S.A. (Eds.), Oceanography: An Illustrated Guide2nd ed. Manson Publishing, London, p. 246 ISBN 978-1-874545-37-8.

Logan, N.A., De Vos, P., 2015. Bacillus. In: Whitman, W.B. (Ed.), Bergey's Manual of Systematics of Archaea and Bacteria. John Wiley & Sons, Hoboken, NJ.

Lorenzini, G.C., 2012. Estudo da transferência de oxigênio em cultivo de Bacillus megaterium. J. Appl. Microbiol. 114, 1378–1387.

Lowenstein, T.K., Jianren, L., Jeffrey, H., Teh-lung, K.U., Shangde, L., 1998. 80,000-year paleoclimate record from the arid Andes, Salar de Hombre Muerto, Argentina. Abstracts with Programs - Geological Society of America, Boulder, CO. 30 (7), 115–116 ISSN 0016-7592.

Luef, B., et al., 2013. Iron-reducing bacteria accumulate ferric oxyhydroxide nanoparticle aggregates that may support planktonic growth. ISME J. 7 (2), 338–350.

Macalady, J.L., Jones, D.S., Lyon, E.H., 2007. Extremely acidic, pendulous cave wall biofilms from the Frasassi cave system, Italy. Environ. Microbiol. 9 (1402), 1414.

Madigan, M., Martinko, J., 2005. Brock Biology of Microorganisms, 11th ed. Prentice Hall, Hoboken, New Jersey ISBN 978-0-13-144329-7.

Maheshwari, D.K., Saraf, M., 2015. Halophiles: Biodiversity and Sustainable Exploitation. Springer, Berlin.

Malewitz, J., 2013. Communities along Red River seek Feds' help. The Texas Tribune.

Marples, B.J., 1962. An Introduction to Freshwater Life in New Zealand. Whitcombe & Tombs, Christchurch.

Marquez, M.C., Quesada, E., Bejar, V., Ventosa, A., 1993. A chemotaxonomic study of some moderately halophilic Gram-positive isolates. J. Appl. Bacteriol. 75, 604–607.

Martínez-Murcia, A., Acinas, S.G., Rodríguez-Valera, F., 1995. Evaluation of prokaryotic diversity by restrictase digestion of 16S rDNA directly amplified from hypersaline environments. FEMS Microbiol. Ecol. 17, 247–255.

Martínez, F.L., Orce, I.G., Rajal, V.B., Irazusta, V.P., 2019. Salar del Hombre Muerto, source of lithium-tolerant bacteria. Environ. Geochem. Health 41, 529–543. doi:10.1007/s10653-018-0148-2.

Maturrano, L., Santos, F., Rosselló-Mora, R., Antón, J., 2006. Microbial diversity in Maras salterns: A hypersaline environment in the Peruvian Andes. Journal of Applied and Environmental Microbiology 72 (6), 3887–3895.

McClellan, K., and J. Bruno, 2008. Coral Degradation through Destructive Fishing Practices, Encyclopedia of Earth. Washington, DC: Environmental Information Coalition, National Council for Science and the Environment.

McFall-Ngai, M., 2014. Divining the essence of symbiosis: Insights from the squid-vibrio model. PLoS Biol. 12 (2), e1001783. https://doi.org/10.1371/journal.pbio.1001783.

McGenity, T.J., Gemmell, R.T., Grant, W.D., Stan-Lotter, H., 2000. Origins of halophilic microorganisms in ancient salt deposits. Environ. Microbiol. 2, 243–250.

McLintock, A.H.(ed.), 2007. *Fish, Marine*, Te Ara – *The Encyclopaedia of New Zealand*. Manatū Taonga, New Zealand; Ministry for Culture and Heritage; Publisher: R.E. Owen, Government Printer.

Minor, T.E., Marth, E.H, 1976. Staphylococci and their significance in foods. Elsevier Scientific Publishing, Amsterdam, Oxford.

Möller, P., Rosenthal, E., Dulski, P., Geyer, S., Guttman, Y., 2003. Rare earths and yttrium hydrostratigraphy along the Lake Kinneret–Dead Sea–Arava transform fault, Israel and adjoining territories. Appl. Geochem. 18, 1613–1628.

Momigliano, M., Denys, G.P.J., Jokinen, H., Merilä, J., 2018. *Platichthys solemdali* sp. nov. (Actinopterygii, Pleuronectiformes): A new flounder species from the Baltic Sea, *Front.* Mar. Sci. 5 (225). doi:10.3389/fmars.2018.00225.

Morris, E.O., 2004. Yeast growth. Brewing Science and Practice. Academic Press Inc, Washington, DC:, pp. 469–508. doi:10.1533/9781855739062.469.

Mulhall, M., 2009. Saving rainforests of the sea: an analysis of international efforts to conserve coral reefs. Duke Environ. Law Policy Forum 19, 321–351.

Munk, L.A., Hynek, S.A., Bradley, D., Jochens, H., 2016. Lithium brines: A global perspective. Econ. Geol. 18, 339–365.

Nacif, F., Lacabana, M., 2015. ABC Del Litio Sudamericano. *Atuel y Cara o Ceca*, Buenos Aires.

Navarro-González, R., Rainey, F.A., Molina, P., Bagaley, D.R., Hollen, B.J., de la Rosa, J., Small, A.M., Quinn, R.C., Grunthaner, F.J., Cáceres, L., Gomez-Silva, B., McKay, C., 2003. Mars-like soils in the Atacama Desert, Chile, and the dry limit of microbial life. Science 302 (5647), 1018–1021.

Neuner, A., Jannasch, H.W., Belkin, S., Stetter, K.O., 1990. Arch. Microbiol. 153, 205–207.

Nicholson, W.L., Munakata, N., Horneck, G., Melosh, H.J., Setlow, P., 2000. Resistance of bacillus endospores to extreme terrestrial and extraterrestrial environments. Microbiol. Mol. Biol. Rev. 64, 548–572. doi:10.1016/j.tplants.2014.05.005.

Nissling, L., Westin, A., 1997. Salinity requirements for successful spawning of Baltic and Belt Sea cod and the potential for cod stock interactions in the Baltic Sea. Marine Ecology Progress Series 152 (1/3), 261–271.

O'Connell, S.P., Lehman, R.M., Snoeyenbos-West, O., Winston, V.D., Cummings, D.E., 2003. Detection of Euryarchaeota and Crenarchaeota in an oxic basalt aquifer. FEMS Microbiol. Ecol. 44 (165), 173.

Ojha, L., Wilhelm, M.B., Murchie, S.L., McEwen, A.S., Wray, J.J., Hanley, J., Massé, M., Chojnacki, M., 2015. Spectral evidence for hydrated salts in recurring slope lineae on Mars. Nat. Geosci. 8 (11), 829–832.

Orellana, R., Macaya, C., Bravo, G., Dorochesi, F., Cumsille, A., Valencia, R., et al., 2018. Living at the frontiers of life: extremophiles in chile and their potential for bioremediation. Front. Microbiol. 9, 2309. doi:10.3389/fmicb.2018.02309.

Oren, A., 1993. Ecology of extremely halophilic microorganisms. In: Vreeland, RH, Hochstein, LJ (Eds.), The Biology of Halophilic Bacteria. CRC Press, Boca Raton, FL, pp. 25–54.

Oren, A., Gurevich, P., 1995. Dynamics of a bloom of halophilic archaea in the Dead Sea. Hydrobiologia 315, 149–158.

Oren, A., 2002. Halophilic microorganisms and their environments. Kluwer Academic Press, Dordrecht (Nl), p. 600 Seckbach, J.

Oren, A., Ionescu, D., Lipski, A., Altendorf, K, 2005. Fatty acid analysis of a layered community of cyanobacteria developing in a hypersaline gypsum crust. Arch Hydrobiol Suppl Algol Stud 117 (339), 347.

Oren, A., 2007. Salts and Brines. In: Whitton, B.A., Potts, M. (Eds.), The Ecology of Cyanobacteria: Their Diversity in Time and Space. Springer, Switzerland, p. 287 ISBN 9780306468551.

Oren, A., 2010. The dying Dead Sea: The microbiology of an increasingly extreme environment. Lakes Reservoirs: Res Manage 15 (215), 222.

Oren, A., 2013. Life in magnesium-and calcium-rich hypersaline environments: Salt stress by chaotropic ions. In: Seckbach, J., Oren, A., Stan-Lotter, H. (Eds.), Polyextremophiles: Life Under Multiple Forms of Stress. Springer, Dordrecht, pp. 217–232.

Orosei, R., Lauro, S.E., Pettinelli, E., Cicchetti, A., Coradini, M., Cosciotti, B., di Paolo, F., Flamini, E., Mattei, E., Pajola, M., Soldovieri, F., Cartacci, M., Cassenti, F., Frigeri, A., Giuppi, S., Martufi, R., Masdea, A., Mitri, G., Nenna, C., Noschese, R., Restano, M., Seu, R., 2018. Radar evidence of subglacial liquid water on Mars. Science 361 (6401), 490–493.

Oz., A., Sabehi, G., Koblizek, M., Massana, R., Beja, O, 2005. Roseobacter-like bacteria in Red and Mediterranean Sea aerobic anoxygenic photosynthetic populations. Appl Environ Microb 71, 344–353.

Pedrós-Alió, C., 2004. Trophic ecology of solar salterns. Halophilic microorganisms. Springer, Berlin, Heidelberg, pp. 33–48.

Peoples, L.M., Bartlett, D.H., 2017. Ecogenomics of Deep-Ocean Microbial Bathytypes. In: Chénard, C., Lauro, F. (Eds.), Microbial Ecology of Extreme Environments. Springer, Cham, pp. 7–50. doi:10.1007/978-3-319-51686-8_2.

Perez-Fernandez, C.A., Iriarte, M., Hinojosa-Delgadillo, W., Veizaga-Salinas, A., Cano, R.J., Rivera-Perez, J., et al., 2016. First insight into microbial diversity and ion concentration in the Uyuni salt flat, Bolivia. Caribb. J. Sci. 49, 57–75. doi:10.18475/cjos.v49i1.a6.

Perkins, J., Gadd, G.M., 1993. Accumulation and intracellular compartmentation of lithium ions in Saccharomyces cerevisiae. FEMS Microbiol. Lett. 107 (2-3), 255–260.

Pierella, K.J.J., Ibarbalz, F.M., Bowler, C., 2020. Phytoplankton in the Tara Ocean. Annual Review of Marine Science 12 (1), 233–265. doi:10.1146/annurev-marine-010419-010706.

Prado, B., Delmoral, A., Quesada, E., Rios, R., Monteolivasanchez, M., Campos, V., Ramos-Cormenzana, A., 1991. Numerical taxonomy of moderately halophilic gram-negative rods isolated from the Salar de Atacama. Chile, Syst Appl Microbiol 14, 275–281.

Qiwei, W. (2010), Acipenser stellatus, IUCN Red List of Threatened Species. 2010: e.T229A13040387. doi:10.2305/IUCN.UK.2010-1.RLTS.T229A13040387.en

Queiroz, K., Cantino, P.D., Gauthier, J.A., 2020. Phylonyms: A Companion to the PhyloCode. CRC Press, Boca Raton, Florida, p. 1843. ISBN 978-1-138-33293-5.

Raeside, J.P., Cutler, E.J.B., Miller, R.B., 1966. Soils and related irrigation problems of part of Maniototo Plains, Otago. N.Z. Soil Bur. Bull. 23, 68.

Ramos-Cormenzana, A., 1993. Ecology of moderately halophilic bacteria. In: Vreeland, RH, Hochstein, LJ (Eds.), Ecology of moderately halophilic bacteria. Biology of Halophilic Bacteria, 55–86.

Ravel, J., Fraser, C.M., 2005. Genomics at the genus scale. Trends Microbiol. 13, 95–97. doi:10.1016/j.tim.2005.01.004.

Ravenschlag, K., Sahm, K., Knoblauch, C., Jorgensen, B.B., Amann, R.I., 2000. Community structure, cellular rRNA content, and activity of sulfate-reducing bacteria in marine Arctic sediments. Appl. Environ. Microbiol. 66, 3592–3602.

Reeves, R., Stewart, B., Clapham, P., Powell, J, 2002. Guide to Marine Mammals of the World. Chanticlear Press, New York, NY.

Rettig, S.L., Jones, B.F., Risacher, F., 1980. Geochemical evolution of brines in the Salar of Uyuni, Bolivia. Chem. Geol. 30, 57–79.

Rhodes, M.E., Fitz-Gibbon, S.T., Oren, A., House, C.H., 2010. Amino acid signatures of salinity on an environmental scale with a focus on the Dead Sea. Environ. Microbiol. 12, 2613 2623.

Risacher, F., Alonso, H, 1996. Geochemistry of the Salar de Atacama. Part 2: water evolution. Revista Geológica de Chile 23, 123–134.

Risacher, F., Alonso, H., Salazar, C., 2003. The origin of brines and salts in Chilean salars: a hydrochemical review. Earth-Sci Rev 63, 249–293.

Robinson, C.K., Wierzchos, J., Black, C., Crits-Christoph, A., Ma, B., Ravel, J., Ascaso, C., Artieda, O., Valea, S., Roldán, M., Gómez-Silva, B., DiRuggiero, J., 2015. Microbial diversity and the presence of algae in halite endolithic communities are correlated to atmospheric moisture in the hyper-arid zone of the Atacama Desert. Environ. Microbiol. 17 (2), 299–315.

Rodriguez-Valera, F., Ventosa, A., Juez, G., Imhoff, J.F., 1985. Variation of environmental features and microbial populations with salt concentrations in a multi-pond saltern. Microb. Ecol. 11, 107–115.

Rodríguez-Contreras, A., Koller, M., de Sousa Dias, M.M., Calafell, M., Braunegg, G., Marqués-Calvo, M.S., 2013. Novel poly[(R)-3-hydroxybutyrate]-producing bacterium isolated from a bolivian hypersaline lake. Food Technol. Biotechnol. 51, 123–130. doi:10.1039/b816556d.

Rubin, S.S., Marin, I., Gomez, M.J., Morales, E.A., Zekker, I., San Martin, P., et al., 2017. Prokaryotic diversity and community composition in the Salar de Uyuni, a large scale, chaotropic salt flat. Environ. Microbiol. 19, 1–41. doi:10.1111/1462-2920.

Samylina, O.S., Sapozhnikov, F.V., Gainanova, O.Y., Ryabova, A.V., Nikitin, M.A., Sorokin, D.Y., 2014. Algo-bacterial communities of the Kulunda steppe (Altai Region, Russia) Soda Lakes. Microbiology 83, 849–860. doi:10.1134/S0026261714060162.

Sass, A.M., McKew, B.A., Sass, H., Fichtel, J., Timmis, K.N., McGenity, T.J., 2008. Diversity of Bacillus-like organisms isolated from deep-sea hypersaline anoxic sediments. Saline Systems 4 (8). doi:10.1186/1746-1448-4-8.

Schippers, A., Neretin, L.N., Kallmeyer, J., Ferdelman, T.G., Cragg, B.A., Parkes, R.J., Jørgensen, B.B., 2005. Prokaryotic cells of the deep sub-seafloor biosphere identified as living bacteria. Nature 433, 861–864.

Schmidt, N., 2010. Hydrogeological and hydrochemical investigations at the Salar de Uyuni (Bolivia) with regard to the extraction of lithium. FOG 26, 1–131.

Segerer, A., Stetter, K.O., Klink, F., 1985. Two contrary modes of chemolithotrophy in the same archaebacterium. Nature 313, 787–789.

Shalev, E., Levitte, D., Gabay, R., Zemach, E., 2008. Assessment of Geothermal Resources in Israel, the Ministry of National Infrastructures Geological Survey of Israel, Jerusalem, 23 p.

Sharpton, T.J., Riesenfeld, S.J., Kembel, S.W., Ladau, J., O'Dwyer, J.P., Green, J.L., Eisen, J.A., Pollard, K.S., 2011. PhylOTU: a high-throughput procedure quantifies microbial community diversity and resolves novel taxa from metagenomic data. PLoS Comput. Biol. 7, e1001061.

Shatkay, M., Anati, D.A., Gat, J.R., 1993. Dissolved oxygen in the Dead Sea - seasonal changes during the holomictic stage. Int J Salt Lake Res 2 (93), 110.

Shima, M., Honda, M., 1963. Isotopic abundance of meteoritic lithium. J. Geophys. Res. 68 (9), 2849–2854.

Siegel, B.Z., McMurty, G., Siegel, S.M., Chen, J., Larock, P., 1979. Life in the calcium chloride environment of Don Juan Pond, Antarctica. Nature 280 (5725), 828–829.

Sinha, R.K., 2004. The Irrawaddy Dolphins Orcaella of Chilika Lagoon, India. Journal of the Bombay Natural History Society. Mumbai, India: online edition: Environmental Information System (ENVIS), Annamalai University, Centre of Advanced Study in Marine Biology, Parangipettai - 608 502, Tamil Nadu, India. 101 (2), 244–251.

Snoeijs-Leijonmalm, P., Andrén, E., 2017. Why is the Baltic Sea so special to live in? In: Snoeijs-Leijonmalm, P., Schubert, H., Radziejewska, T. (Eds.), Biological Oceanography of the Baltic Sea. Springer, Dordrecht, pp. 23–84 ISBN 978-94-007-0667-5.

Spalding, M.D., Grenfell, A.M., 1997. New estimates of global and regional coral reef areas. Coral Reefs 16 (4), 225–230. http://dx.doi.org/10.1007/s003380050078.

Spalding, M., Ravilious, C., Green, E., 2001. World Atlas of Coral Reefs. University of California Press and UNEP/WCMC, Berkeley, CA ISBN: 0520232550.

Steinhorn, I., Assaf, G., Gat, J.R., Nishry, A., Nissenbaum, A., 1979. The dead sea: deepening of the mixolimnion signifies the overture to overturn of the water column. Science 206 (55), 57.

Stephens, D.W., and J. Gardner (2007), Great Salt Lake, Utah; U.S. Geological Survey; Water-Resources Investigations Report 1999–4189; https://pubs.usgs.gov/wri/wri994189/PDF/WRI99-4189.pdf

Stetter, K.O., 1982. Ultrathin mycelia-forming organisms from submarine volcanic areas having an optimum growth temperature of 105°C. Nature 300, 258–260.

Stetter, K.O., 1986. Thermophiles: General, Molecular and Applied Microbiology. Wiley, New York, London, Sidney, Toronto, pp. 39–74 Brock, T.D.

Sugahara, I., Hayashi, K., Kimura, T., Jinno, C., 1983. Studies on marine bacteria producing lytic enzymes-IX: effect of inorganic salts on the autolysis of bacterial cells capable of producing lytic enzyme. Bull. Fac. Fish. Mie Univ. 10, 33–39.

Sutaria, D., 2005. Irrawaddy dolphin — India (On-line). Whale and Dolphin Conservation Society. Accessed April 24, 2010 at http://www.wdcs.org/submissions_bin/consprojectirr.pdf.

Swarieh, A. (2000), Geothermal energy resources in Jordan, country update report. Proceedings World Geothermal Geothermal Congress.

Szabo, M.J., 2013. Wilson's phalarope. New Zealand Birds Online Miskelly, C.M.

Szalay, J. (2017), Caspian Sea: Largest Inland Body of Water, https://www.livescience.com/57999-caspian-sea-facts.html

Takenishi, M., Takada, H., 1984. Effect of lithium ions on growth of respiratory deficient mutant of Saccharomyces cerevisiae in medium supplied with galactose as a sole carbón source. Int. J. Antimicrob. Agents 12, 323–328.

Tank, M., Thiel, V., Imhoff, J.F., 2009. Phylogenetic relationship of phototrophic purple sulfur bacteria according to pufL and pufM genes. Int. Microbiol. 12, 175–185.

Thiel, V., Tank, M., Neulinger, S.C., Gehrmann, L., Dorador, C., Imhoff, J.F., 2010. Unique communities of anoxygenic phototrophic bacteria in saline lakes of Salar de Atacama (Chile): evidence for a new phylogenetic lineage of phototrophic Gammaproteobacteria from pufLM gene analyses. FEMS Microbiol. Ecol. 74 (3), 510–522.

Thompson, T.E., 1976Biology of opisthobranch molluscs1. Ray Society, Scion Pub Ltd; UK 207 p. 21 pls.

Tiquia, S.M., Davis, D., Hadid, H., Kasparian, S., Ismail, M., Sahly, R., et al., 2007. Halophilic and halotolerant bacteria from river waters and shallow groundwater along the Rouge River of southeastern Michigan. Environ. Technol. 28, 297–307. doi:10.1080/09593332808618789.

Triadó-Margarit, X., Vila, X., Abella, C.A., 2010. Novel green sulfur bacteria phylotypes detected in saline environments: ecophysiological characters versus phylogenetic taxonomy. Antonie Van Leeuwenhoek 97, 419–431.

Trüper, H.G., Galinski, E.A., 1986. Concentrated brines as habitats for microorganisms. Cell. Mol. Life Sci. 42, 1182–1187.

Tsuruta, T., 2005. Removal and recovery of lithium using various microorganisms. J. Biosci. Bioeng. 100, 562–566. doi:10.1263/jbb.100.562.

Valderrama, M.J., Prado, B., Del Moral, A., Rios, R., Ramos-Cormenzana, A., Campos, V., 1991. Numerical taxonomy of moderately halophilic Gram-positive cocci isolated from the Salar de Atacama (Chile). Microbiología (Madrid, Spain) 7, 35–41.

van der Wielen, P.W.J.J., Bolhuis, H., Borin, S., Daffonchio, D., Corselli, C., 2005. The enigma of prokaryotic life in deep hypersaline anoxic basins. Science 307 (121), 123.

Vavourakis, C.D., Ghai, R., Rodriguez-Valera, F., Sorokin, D.Y., Tringe, S.G., Hugenholtz, P., Muyzer, G., 2016. Metagenomic insights into the uncultured diversity and physiology of microbes in four hypersaline soda lake brines. Front. Microbiol. 7, 211.

Ventosa, A., 2006. Unusual microorganisms from unusual habitats: hypersaline environments. In: Logan, N.A., Lappin-Scott, H.M., Oyston, P.C.F. (Eds.), *Prokaryotic Diversity – Mechanisms and Significance* (Society for General Microbiology Symposium No 66). Cambridge University Press, Cambridge, pp. 223–253.

Veron, J.E.N., 1993. Corals of Australia and the Indo-Pacific. University of Hawaii Press, Honolulu, Hawaii.

Veron, J.E.N. (2000a), Corals of the World. Australian Institute Marine Science, Townswille. 1, XII+463 pp. 2, VIII+429 pp. 3, VIII +490.

Veron, J.E.N., 2000b. Corals of the World, vol. 3. Australian Institute of Marine Science, Townsville, Australia.

Viktorsson, L. (2018), Hydrogeography and oxygen in the deep basins, HELCOM.

Vila, X., Guyoneaud, R., Cristina, X.P., Figueras, J.B., Abella, C.A., 2002. Green sulfur bacteria from hypersaline Chiprana Lake (Monegros, Spain): habitat description and phylogenetic relationship of isolated strains. Photosynth. Res. 71, 165–172.

Vila, T., 2010Geología de los depósitos salinos andinos2. *Andean Geology*, Provincia de Antofagasta, Chile. https://doi.org/10.5027/andgeoV2n1-a05.

Walrond, C., 2007. Oceanic fish. Encyclopedia of New Zealand.

Wanger, T.C., 2011. The Lithium future-resources, recycling, and the environment. Conserv. Lett. 4, 202–206. doi:10.1111/j.1755-263X.2011.00166.x.

Warburg, R.J., Moir, A., Smith, D.A., 1985. Influence of alkali metal cations on the germination of spores of wild-type and gerD mutants of Bacillus subtilis. Microbiology 131, 221–230. doi:10.3386/w19846.

Warren, J.K., 2006. Evaporites: sediments, resources and hydrocarbons. Birkhäuser, p. 280 ISBN 978-3-540-26011-0.

Warren, J.K., 2010. Evaporites through time Tectonic, climatic and eustatic controls in marine and nonmarine deposits. Earth-Sci. Rev. 98, 217–268.

Weisser, J., Trüper, H.G., 1985. Osmoregulation in a New Haloalkaliphilic Bacillus from the Wadi Natrun (Egypt). Syst. Appl. Microbiol. 6, 7–11. doi:10.1016/S0723-2020(85)80003-5.

Wells, R.S., Scott, M., 2009. Common bottlenose dolphin (Tursiops truncatus). Encyclopedia of Marine Mammals 6, 249–255. doi:10.1016/B978-0-12-373553-9.00062-6.

Wiegel, J., Canganella, F., 2011. Extremophiles: from abyssal to terrestrial ecosystems and possibly. Naturwissenschaften 98, 253–279. doi:10.1038/npg.els.0000392.

Wilkansky, B., 1936. Life in the Dead Sea. Nature 138, 467 -467.

Wilson, S., Tariel, E., Masao, A., Paul, J., Simon, G., 2014. The role of canine distemper virus and persistent organic pollutants in mortality patterns of Caspian seals (*Pusa caspica*). PLoS One 9 (7), e99265.

Woese, C.R., Gupta, R., Hahn, C.M., Zillig, W., Tu, J., 1984. The phylogenetic relationships of three sulfur-dependent archaebacteria. Syst. Appl. Microbiol. 5 (1), 97–105.

Wyneken, J., 2001. The Anatomy of Sea Turtles. U.S Department of Commerce NOAA Technical Memorandum NMFS-SEFSC-470, pp. 1–172.

Yakimov, M.M., La Cono, V., Spada, G.L., Bortoluzzi, G., Messina, E., Smedile, F., et al., 2015. Microbial community of the deep-sea brine Lake Kryos seawater-brine interface is active below the chaotropicity limit of life as revealed by recovery of mRNA. Environ. Microbiol. 17, 364–382. doi:10.1111/1462-2920.12587.

Zahran, H.H., 1997. Diversity, adaptation and activity of the bacterial flora in saline environments. Biol. Fertil. Soils 25, 211–223. doi:10.1007/s003740050306.

Zeigler, D.R., Perkins, J.B., 2015. The Genus Bacillus. CRC Press, New York, NY.

Zillig, W., Stetter, K.O., Simon, W., Wolfgang, S., Harro, P., Ingrid, S., 1980. The Sulfolobus-"Caldariella" group: taxonomy on the basis of the structure of DNA-dependent RNA polymerases. Arch. Microbiol. 125 (3), 259–269.

Zillig, W., Stetter, K.O., Schaefer, W., Janekovic, D., Wunderl, S., Holz, I., Palm, P., 1981. Thermoproteales: a novel type of extremely thermoacidophilic anaerobic archaebacteria isolated from Icelandic solfataras. Zbl. Bakt. Hyg. I. Abt. Orig. C2 (3), 205–227.

Zúñiga, L.R., Campos, V., Pinochet, H., Prado, B., 1991. A limnological reconnaissance of lake Tebenquiche, Salar de Atacama, Chile. Hydrobiologia 210, 19–24.

Bibliography

Arnow, T., and Stephens, D. (1990), Hydrologic characteristics of the Great Salt Lake, Utah: 1847–1986: U.S. Geological Survey Water-Supply Paper 2332, p. 32.

Brisou, J., 1955. Microbiologie du milieu marin. Flammarion Ed., Paris.

Curry, D.R., Atwood, G., Mabey, D.R., 1984. Major levels of Great Salt Lake and Lake Bonneville: Map 73. Utah Geological and Mineral Survey.

Cline, J.D., 1969. Spectrophotometric determination of hydrogen sulfide in natural waters. Limnol. Oceanogr. 14, 454–458.

Dulski, P., 1994. Interferences of oxide, hydroxide and chloride analyte species in the determination of rare earth elements in geological samples by inductively coupled plasma-mass spectrometry. Fresenius J. Anal. Chem. 350, 194–203.

Dulski, P., 2001. Reference Materials for Geochemical Studies? New Analytical Data by ICP-MS and Critical Discussion of Reference Values. Geostandard Newslett 25 (87), 125.

Gibbons, N.E., 1970. Isolation, growth and requirements of halophilic bacteria. In: Norris, J.R., Ribbons, D.W. (Eds.), *Methods in microbiology*, Vol. 3 B. Academic Press Inc., New York, pp. 169–182.

Gvirtzman, H., Stanislavsky, E., 2000. Large-scale flow of geofluids at the Dead Sea Rift. J. Geochem. Explor. 69 (207), 211.

Gwynn, J.W., 1980. Great Salt Lake, A scientific, historical, and economic overview: Utah Geological and Mineral. Survey Bulletin 116, 400.

Häusler, S., Weber, M., de Beer, D., Ionescu, D, 2014. Spatial distribution of diatom and cyanobacterial mats in the Dead Sea is determined by response to rapid salinity fluctuations. Extremophiles 18 (6), 1085–1094.

Ishida, Y., 1970. Growth behavior of Halobacteria in relation to concentration of NaCl and temperature environments. Bull. Jap. Soc. of Sci. Fish 36, 397.

Kocur, M., Hodgkiss, W., 1973. Taxonomic status of the genus *Halococcus Schoop 1935*. Int. J. Syst. Bacteriol. 23, 151–156.

Morgan, D.L., 1947. The Great Salt Lake: Salt Lake City. University of Utah Press, Salt Lake City, Utah, p. 432.

Skerman, V.B.D., 1967. A guide to the identification of the genera of bacteria. The Williams & Wilkins Co., Baltimore.

Smith, N.R., Gordon, R.E., Clark, F.E., 1952. Aerobic spore forming bacteria. Agriculture Monograph. *U.S. Dept. of Agriculture*, Washington, DC.

Stephens, D.W., 1998. Salinity-induced changes in the aquatic ecosystem of Great Salt Lake. In: Pitman, J., Carroll, A. (Eds.), Modern and Ancient Lake Systems. Utah Geological Survey Guidebook, Utah, pp. 1–7, 26.

ZoBell, C.E., Upham, H.C., 1944. A list of marine bacteria including 60 new species. Bull. Scripps Inst. Oceanogr. 5, 239–253.

Chapter 7

Terrestrial analogs & submarine hydrothermal vents—their roles in exploring ocean worlds, habitability, and life beyond earth

7.1 Terrestrial analogs—their importance in understanding the secrets of extraterrestrial worlds

Extraterrestrial exploration carried out with a view to understanding the secrets of the unknown worlds takes advantage of some peculiar Earth sites which can serve as field facilities or Earth references. These field sites, which bear an analogy or similarity in some way to other planetary bodies of the Solar System, are called "terrestrial analogs." Terrestrial analogs can be used as a reference for extraterrestrial studies because they have particular similarities or characteristics that resemble those found on extraterrestrial planetary body/bodies. In the pursuit of knowing the unknown extraterrestrial worlds and the search for habitable environments there, terrestrial analogs have often been employed by planetary scientists to shed light on such worlds.

7.1.1 Ice on Earth—analog for possible microbial life on extraterrestrial icy ocean worlds

While there is no broadly accepted definition of life, one working definition of life has become influential in the origins-of-life community. According to one definition, "life is a self-sustained chemical system capable of undergoing Darwinian evolution" (Joyce, 1994). But according to several researchers (e.g., Fleischaker, 1990; Chyba and McDonald, 1995), however valuable this Darwinian definition may be for guiding laboratory experiments, it is unlikely to prove useful to a remote in situ search for life. According to Chyba et al. (2000), in a search for extraterrestrial life in our Solar System, we instead fall back on a less ambitious notion of "life as we know it," meaning life based on a liquid water solvent, a suite of "biogenic" elements (most famously carbon, but others as well), and a source of free energy. Chyba and Phillips (2001) have suggested that the availability of these ingredients on a given world would suggest life to be possible, so that further exploration may be warranted.

Ice on the Earth is a rich habitat. Properties of the external ice shell and its outer surface environment will not only influence the icy ocean world habitability but also determine the extent to which biosignatures may be expressed on the surface and impact the design of any future exploration (Chyba and Phillips, 2001; Figueredo et al., 2003; Hand et al., 2009). The idea of habitability was introduced by Dole (1964) and followed up by Sagan (1996) to refer to those planetary conditions suitable for human life. The word "habitability" has since then come to imply requirements both less stringent and less anthropocentric (i.e., regarding humankind as the central or most important element of existence, especially as opposed to God or animals), referring instead to the stability of liquid water at a world's surface (Chyba and Phillips, 2001). Given the importance of liquid water to sustain life, the historical emphasis on surface liquid water is easy to understand. First, life on Earth—still our sole example of a biology—utterly depends on liquid water (Blum, 1955; Chyba et al., 2000). Terrestrial microbes in an ice-rich environment exhibit a variable distribution, in which they are strongly concentrated in nutrient-rich regions near the ice-water interface (Madigan et al., 1997; Whitman et al., 1998). Should life arise in ocean worlds, then their ice shells may be similarly rich in habitable environments, given proper conditions. Of specific interest are those properties that affect transport processes through the ice shell, both thermal and physical, as they play a crucial role in the evolution and dynamics of the world. Models of the tidally heated, floating ice shell proposed for the Jovian (planet Jupiter's) moon Europa generally find shell thicknesses less than 30 km (McKinnon, 1999). Past parameterized convection models indicated that such shells are stable against convective overturn, which otherwise ostensibly/apparently leads to freezing of the ocean underneath. In order to investigate the convective instability in Europa's floating ice shell, McKinnon (1999) invoked grain sizes observed in terrestrial polar glaciers. Material transport between the surface and subsurface is potentially crucial for mediating energy and nutrient flow, as well as ocean pH, oxidants, and other factors that govern habitability of the shell itself (potentially rife with brine zones, water pockets, and habitable ice grain boundaries) and that of the ocean below. The overall thickness of the ice shell, and any spatial variability, will determine the propensity (an inclination or natural tendency to behave in a particular way) and timescales for ocean–surface interaction and what modes of transport are possible (McKinnon, 1999; Gaidos and Nimmo, 2000; Pappalardo and Barr, 2004; Katterhorn and Prockter, 2014). Gaidos and Nimmo (2000) have argued that some surface features of Europa, which hosts a subsurface water ocean, are formed by soft ice that is heated by viscous dissipation of tidal motion along faults. Figueredo et al. (2003) evaluated the astrobiological potential of the major classes of geologic units on Europa with respect to possible biosignatures preservation on the basis of surface geology observations. They found that some lineaments (linear features on a planet's surface, such as a fault) and impact craters may be promising sites for closer study despite the comparatively lower astrobiological potential of their classes.

The surface of Europa is peppered (covered) by topographic domes, interpreted as sites of intrusion and extrusion. Theoretical model based investigations carried out by Pappalardo and Barr (2004) revealed that the exclusion of impurities from warm upwellings allows sufficient buoyancy for icy plumes to create the observed surface topography, provided the ice shell has a small effective elastic thickness (~0.2–0.5 km) and contains low-eutectic-point impurities at the few percent level. Note that "eutectic-point" is the temperature at which a particular eutectic (homogeneous) mixture of substances freezes or melts; this single temperature is lower than the freezing/melting point of any of the constituents. Pappalardo and Barr (2004)'s model suggests that the ice shell may be depleted in impurities over time.

Europa has one of the youngest planetary surfaces in the Solar System, implying rapid recycling by some mechanism. Europa's dilational bands (a tabular zone of new crustal material that intruded between the progressively dilating walls of a tension fracture) resemble terrestrial mid-ocean spreading zones. Katterhorn and Prockter (2014) reported several lines of evidence that subduction may be recycling surface material into the interior of Europa's ice shell. Based on the interpretation of *Galileo spacecraft* images, they discovered a discrete tabular zone and interpreted this zone as a subduction-like convergent boundary that abruptly truncates older geological features and is flanked by potential cryo-lavas on the overriding ice. They proposed that Europa's ice shell has a brittle, mobile, plate-like system above convecting warmer ice. Hence, Europa may be the only Solar System body other than Earth to exhibit a system of plate tectonics.

Fractures in brittle conductive shells may provide a direct path for material exchange from the subsurface to the surface; however, downward material motion is poorly understood and perhaps unlikely, as in the case of Enceladus' south polar terrain (Glein et al., 2015). For thicker, more temperate ice shells, large-scale convective motion of the ice and entrained or endogenic (formed or occurring beneath the surface) liquids can provide a geologically rapid transport mechanism between the underlying ocean and the upper portions of the shell (McKinnon, 1999; Pappalardo and Barr, 2004; Schmidt et al., 2011; Peddinti and McNamara, 2015). Quasi-circular areas of ice disruption called "chaos terrains" are unique to Europa. Features such as Conamara Chaos stand above surrounding terrain and contain matrix domes. The buoyancy of material rising as either plumes of warm, pure ice called diapirs or convective cells in a thick (>10 km) shell is insufficient to produce the observed chaos heights, and no single plume can create matrix domes. Note that *diapir* is a type of geologic intrusion in which a more mobile and ductily deformable material is forced into brittle overlying rocks; Diapir consists of material buoyantly rising as either plumes of warm, pure ice, etc., on to the surface of a celestial body. Diapir often manifests itself as a geological structure consisting of mobile material that was forced into more brittle surrounding rocks, usually by the upward flow of material from a parent stratum. Diapirs may take the shape of domes, waves, mushrooms, teardrops, or dikes (large slabs of rocks that cut through another types of rock).

Schmidt et al. (2011) reported an analysis of archival data from Europa, guided by processes observed within Earth's subglacial volcanoes and ice shelves. The data suggest that "chaos terrains" form above liquid water lenses perched within the ice shell as shallow as 3 km. Their results suggest that ice-water interactions and freeze-out give rise to the diverse morphologies and topography of chaos terrains. *Thrace Macula* is one of the largest chaos regions on Europa, visible in global-scale images because of its large extent and low relative albedo; it likely formed above a melt lens that is currently liquid. Schmidt et al. (2011) further found that the sunken topography of *Thrace Macula* indicates that Europa is actively resurfacing over a lens comparable in volume to the Great Lakes in North America. For an archetypical (representing or constituting an original type after which other similar things are patterned) layered shell, like that of Europa with a brittle stagnant lid overlying a convecting ice mantle, a combination of tectonic processes (such as subsumption/subduction) (Katterhorn and Prockter, 2014) and convective overturn could produce a continuous chemical cycling of ocean-derived reductants to the surface and oxidized materials from the surface back into the underlying ocean.

Europa exhibits a deformed icy surface with salt deposits concentrated along the varied geological features. Two-way transport of salts from a liquid-water ocean beneath the ice shell to the surface, and vice versa, has been speculated. Peddinti and McNamara (2015) reported dynamical models that demonstrate the incorporation of newly frozen ice into convective plumes within the ice shell, caused by convection within the ice shell that drives dynamic topography along the ice-ocean boundary. Their study results reveal that the new ice that forms at the freezing front can be transported by the rising ice plumes toward the surface until it is blocked by a high-viscosity lid at the surface. Weakening of the lid by tidal or tectonic forces could then lead to the surface detection of ocean trace chemistry captured in the newly formed ice.

7.1.2 The NASA OCEAN project—an ocean-space analog

An Advanced Life Support (ALS) system with bioregenerative components may one day be required for long-term, deep space exploration, in extended missions to Mars or in establishing long-term bases on Earth's Moon. Intensive research programs on such ALS systems have been ongoing throughout the National Aeronautics and Space Administration (NASA) since 1988 (Chamberland, 1996). Recycling waste products during orbital (e.g., International Space Station) and planetary missions (e.g., lunar base, Mars transit mission, Martian base) will reduce storage and resupply costs. Wastes streams on the space station will include human hygiene water, urine, faeces, and trash. Longer-term missions will contain human waste and inedible plant material from plant growth systems used for atmospheric regeneration, food production, and water recycling. The feasibility of biological and physical-chemical waste recycling is being investigated as part of NASA's ALS Program (Garland et al., 1998). Notably, projects have been initiated at the John F. Kennedy Space Center (KSC), Ames Research Center (ARC), and the Lyndon B. Johnson Space Center (JSC). The KSC ALS work has been named the "Breadboard Project" because of its approach developing the components and combining them into a breadboard to understanding the bioregenerative ALS picture [also called a Controlled Ecological Life Support System (CELSS)] in smaller pieces, similar to an electronic "breadboard." The Breadboard Project has been involved for 7 years in the study of higher crops grown in a 113 m^3 chamber—the longest operating and largest such closed, controlled growth chamber in the world. This chamber has proven itself to be very successful in growing a wide variety of crops from seedlings to harvest and in helping researchers understand the complex biological cycle of such edible plants in closed, environmentally controlled environments. Because the system's ultimate use will be a more challenging environment, moving a specially designed piece of the system into extreme conditions was an important test. Engineers at KSC developed a compact, portable, functional plant module for testing in the world's only *fixed seafloor laboratory* at Key Largo, FL. The laboratory, called *MarineLab*, is operated out of the facilities of the Marine Resources Development Foundation in a lagoon of some 10 m depth. The project was called the OCEAN project (Ocean CELSS Experimental Analog NASA). The Breadboard Project at Kennedy Space Center in NASA was focused on the development of the bioregenerative life support components, crop plants for water, air, and food production, and bioreactors for recycling of wastes. The keystone of the Breadboard Project was the Biomass Production Chamber (BPC), which was supported by 15 environmentally controlled chambers and several laboratory facilities holding a total area of 2150 m^2. In supporting the ALS Program, the Project utilizes these facilities for large-scale testing of components and development of required technologies for human-rated test-beds at Johnson Space Center in NASA, in order to enable a Lunar and a Mars mission finally (Guo and Ai, 2001).

7.1.3 Microbial communities colonizing on terrestrial submarine hydrothermal vents and volcanic rocks—analogs for life on early Earth and possible life on *Enceladus* & *Europa*

Chemolithotrophic microbial communities (organisms that are able to use inorganic reduced compounds as a source of energy) commonly colonizing on and in the vicinity of terrestrial submarine hydrothermal vents are considered to represent analogs for life on early Earth and some other planets and their moons (Konn et al., 2015). Vent ecosystems depend almost exclusively on oxidants such as SO_4^{2-}, O_2, and CO_2 (Tunnicliffe, 1991; Lutz and Kennish, 1993; Gaidos et al., 1999). Vent ecosystem communities are organisms that utilize only inorganic and/or abiotic simple molecules for their carbon and energy sources so that they do not rely on other living organisms to feed, develop, and multiply (Lang et al., 2012). By examining insights gained from both the rock record and cyanobacteria (a group of photosynthetic bacteria, some of which are nitrogen-fixing, that live in a wide variety of moist soils and water either freely or in a symbiotic relationship with plants or lichen-forming fungi) presently living in early Earth analog ecosystems, it became possible to synthesize current knowledge of these ancient microbial mediators in planetary redox evolution

(Hamilton et al., 2016). On the Earth, we have seen the existence of life in submarine hydrothermal vents, and potential analogs on Jupiter's icy moon *Europa* have been suggested (e.g. Chyba, 2000).

Several theories for the origin of life on Earth would apply to Saturn's icy moon *Enceladus* (McKay et al., 2008). These are (1) origin in an organic-rich mixture, (2) origin in the redox gradient of a submarine vent, and (3) panspermia (the theory that life on the Earth originated from microorganisms or chemical precursors of life present in outer space and able to initiate life on reaching a suitable environment; according to this hypothesis, life exists throughout the Universe, distributed by space dust, meteoroids, asteroids, comets, planetoids, and also by spacecraft carrying unintended contamination by microorganisms). There are three known microbial ecosystems on Earth that do not rely on sunlight, oxygen, or organics produced at the surface and, thus, provide analogs for possible ecologies on Enceladus (McKay et al., 2008). Two of these ecosystems are found deep in volcanic rock, and the primary productivity is based on the consumption by methanogens (various archaeabacteria that are capable of producing methane through the process of methanogenesis) of hydrogen produced by rock reactions with water. The third ecosystem is found deep below the surface in South Africa and is based on sulfur-reducing bacteria consuming hydrogen and sulfate, both of which are ultimately produced by radioactive decay. Methane gas has been detected in the plume of high-speed particles gushing out from the "tiger stripe" cracks found in the vicinity of the South Pole region of Enceladus and may be biological in origin. An indicator of biological origin may be the ratio of nonmethane hydrocarbons to methane, which is very low (0.001) for biological sources but is higher (0.1–0.01) for nonbiological sources (McKay et al., 2008).

7.1.4 Terrestrial lava tubes—analogs of underground tunnels on Earth's moon and Mars

On Earth it has been found that when the lava-flows [an outburst of lava (molten rock) that moves during a nonexplosive effusive volcanic eruption] stops moving, it cools and hardens to form igneous rock (i.e., rock having solidified from lava). While lava can be up to 100,000 times more viscous than water, it can flow for a large distance before abating and solidifying (Pinkerton, 1995). Lava flows can occasionally form lava tubes, which appear to form because of rapid lava flow. Lava tubes, which are usually formed from extremely fluid smooth, unbroken lava (pahoehoe lava), manifest themselves as natural subsurface caves or caverns. Lava tubes typically form when the exterior surface of lava channels cools more rapidly and forms a strong crust over the subsurface lava (Léveillé and Saugata, 2010). The lava flow ultimately stops and drains out of the tube, leaving a conduit-shaped void located several feet underneath the surface.

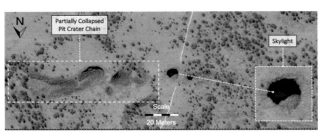

FIG. 7.1 Similarities between a pit crater chain and pit crater "skylight" at El Mapais, New Mexico. *(Source: NASA, Processing: Center for Planetary Science).* (Reproduced from Paris, A.J., E.T. Davies, L. Tognetti, and C. Zahniser (2019), Prospective lava tubes at Hellas Planitia: Leveraging volcanic features on Mars to provide crewed missions protection from radiation, *The Journal of The Washington Academy of Sciences,* arXiv:2004.13156 [astro-ph.EP]).

A line of "pit craters" near a dormant volcano tells the story of a lava tube that formed, drained, and then partially collapsed in one section or another—sometimes even leaving behind "skylight" holes (light-transmitting windows that form part of the roof space of a dwelling/building for daylighting and ventilation purposes) in the middle of the crater (see Fig. 7.1). On Earth, lava tubes are most easily spotted when they collapse, forming long furrows in the dirt. Partial collapses sometimes form chains of skylights that reveal hidden lava tubes that are mostly intact.

Paris et al. (2019) recall that "skylights" associated with lava tubes are dark, nearly rounded features that are hypothesized as entrances to lava tubes (Léveillé and Saugata, 2010). At this point, light from the Sun enters into the permanent darkness of the lava tube from above, forming a skylight (Green and Short, 1971). Many lava tubes on Earth, for instance, have been identified through the discovery of skylights. Researchers have speculated since the 1960s that lava tubes might exist on Earth's Moon and planet Mars, but in recent years Martian and lunar orbiters have beamed home images showing how common these formations likely are, both on Mars and on Earth's Moon (Letzter, 2020).

Earth's closest celestial body is its Moon; and although Venus is considered to be Earth's closest planet, occasionally Mars also happens to be Earth's closest planet. Sinuous collapse chains and skylights in lunar and Martian volcanic regions have often been interpreted as collapsed lava tubes (also known as pyro-ducts). This hypothesis has fostered a forty years debate among planetary geologists trying to define if analog volcano-speleogenetic processes (those processes which operate on or close to the surface of the Earth and which involve weathering, mass movement, fluvial, aeolian, glacial, periglacial, and coastal processes) acting on Earth could have created similar subsurface linear voids in extra-terrestrial volcanoes (Sauro et al., 2020). A clear understanding of the potential morphologies and dimensions of Martian and Lunar lava tubes remains elusive.

On Earth lava tubes are well-known thanks to speleological exploration (exploration of caves) and mapping in several shield volcanoes (a type of volcano named for its low profile, resembling a warrior's shield lying on the ground), with examples showing different genetic processes (inflation and over-crusting) and morphometric (size and shape) characters. In the case of Earth's Moon, initial indications on such underground complexes have been obtained from distant aerial photographs taken primarily during the United States' Apollo missions. On Earth's Moon subsurface cavities have been inferred from several skylights in Maria smooth plains and corroborated using gravimetry and radar sounder.

Many efforts have been expended in understanding Earth's Moon and the planet Mars. Underground empty lava tubes and caves—believed to have potential as shelters for human settlement on Earth's Moon and Mars—have been discovered during different stages of Lunar and Martian explorations.

Although it is still impossible to gather direct information on the interior of Martian and Lunar lava tube candidates, scientists have the possibility to investigate their surface expression through the analysis of collapses and skylight morphology, morphometry (quantitative analysis of form, a concept that encompasses size and shape) and their arrangement, and compare these findings with terrestrial analogs. Thus, with the lack of extensive field information from Earth's Moon and Mars, it became necessary to study terrestrial lava tubes as terrestrial analogs to gain a clearer insight into the formation process of the photographically observed extraterrestrial underground features (e.g., Greeley, 1971; Cruikshank and Wood, 1972; Peterson and Swanson, 1974). Fig. 7.2 shows a prototype rover creeping through a lava tube in Spain's Canary Island of Lanzarote, part of a training campaign to explore settings on Earth that could be similar to those on Earth's Moon

FIG. 7.3 **The El Calderon lava tubes in El Mapais, New Mexico (Prof. Antonio Paris, the Principal Investigator conducting research, is seen standing at the lava tube entrance).** *(Source: Paris, A.J., E.T. Davies, L. Tognetti, and C. Zahniser (2019),* Prospective lava tubes at Hellas Planitia: Leveraging volcanic features on Mars to provide crewed missions protection from radiation, *The Journal of The Washington Academy of Sciences,* arXiv:2004.13156 [astro-ph.EP]).

and Mars. Fig. 7.3 shows the El Calderon lava tubes in El Mapais, New Mexico (Prof. Antonio Paris, the Principal Investigator conducting research, is seen standing at the lava tube entrance).

Although there is an order of magnitude size difference between lunar and terrestrial lava tubes, the formational processes appear very similar (Coombs and Hawke, 1992). Note that an order-of-magnitude estimate of a variable, whose precise value is unknown, is an estimate rounded to the nearest power of ten. Lunar researchers found that lunar lava tubes seem to be much larger, with collapse sites 300–700 times the size of Earth's. A lava tube on Earth's Moon could easily contain a small city within its walls. Interestingly, in the Moon's lava tubes, openings can be as wide as a football field (Letzter, 2020). These lava tubes are truly enormous, and might offer safer habitats than the lunar or Martian surfaces.

As already indicated, lava tubes have been detected not only on Earth's Moon, but also on Mars. Pit craters have been discovered on Earth, Mars, as well as on the various moons in our solar system (including Earth's Moon). They are generally located in a series of ranged or offset chains; and in these instances, they are referred to as pit crater chains. When adjacent walls between pits in a pit crater chain collapse, they become troughs. A commonly invoked hypothesis to explain these troughs is that they are collapsed lava tubes, essentially tunnels formed underground by rivers of lava (Paris et al., 2019).

Earth analog studies carried out by Paris et al. (2019) shed much light on the feasibility of using lava tubes on Earth's Moon and the planet Mars for human settlement (despite the presence of harmful radiations on these two celestial bodies), in terms of the ability of lava tubes in substantially reducing (by ~82%) the intensity of harmful radiations inside of the lava tubes. Paris et al. (2019)

FIG. 7.2 **A prototype rover creeps through a lava tube in Spain's Canary Island of Lanzarote, part of a training campaign to explore settings on Earth that could be similar to those on Earth's Moon and Mars.** *(Image credit: ESA/Robbie Shone). (Source: Letzter, R. (2020), Entire cities could fit inside the moon's monstrous lava tubes, https://www.livescience.com/lava-tubes-mars-and-moon-habitable.html).*

reported their investigation to identify and study prospective lava tubes at *Hellas Planitia*, a plain located inside the large impact basin *Hellas* in the southern hemisphere of Mars, through the use of Earth analog structures. The search for lava tubes at *Hellas Planitia* was primarily due to the low radiation environment at this particular location. Several studies by NASA spacecraft have measured radiation levels in this region at ~342 μSievert per day (μSv/day), which is considerably less than other regions on the surface of Mars (~547 μSv/day). Notwithstanding, a radiation exposure of ~342 μSv/day is still sizably higher than what human beings are annually exposed to on Earth. After surveying 1500 images from MRO, this investigation has identified three candidate lava tubes in the vicinity of Hadriacus Mons as prospective sites for manned exploration. To complement this investigation, 30 in-situ radiation monitoring experiments have been conducted at three analog lava tubes (located at Mojave, California; El Malpais, New Mexico; and Flagstaff, Arizona). Paris et al. (2019) found that on average, the total amount of solar radiation detected outside the analog lava tubes was approximately ~0.470 μSv/h, while the average inside the analog lava tubes decreased by 82% to ~0.083 μSv/h. Paris et al. (2019)'s investigation, therefore, concluded that "terrestrial lava tubes can be leveraged for radiation shielding and, accordingly, that the candidate lava tubes on Mars (as well as known lava tubes on the lunar surface) can serve as natural radiation shelters and habitats for a prospective crewed mission to the planet."

The largest lava tubes on Earth are maximum ~40 m (130 feet) of width and height, like a very large motorway tunnel. On Mars collapsed lava tubes tend to be about 80 times larger than Earth's, with diameters of 130–1300 feet (40–400 m). The sheer scale of these extraterrestrial lava tubes is likely a result of low Martian and lunar gravity [it may be noted that the heights of material eruptions (plumes) from low-gravity moons such as Europa and Enceladus are in the range 100–200 km or more], as well as differences in how volcanoes operated on those bodies compared to Earth. To assess the size of lunar and Martian lava tubes, the researchers collected 3D laser scans of their counterparts—both collapsed and intact—on Earth. Then they collected all the available satellite images of collapsed lava tubes on Mars and Earth's Moon, and modeled the size of the intact tubes based on the relationships between collapsed and intact tunnels on Earth. The dataset reported by Sauro et al. (2020) indicates that Martian and lunar lava tubes are 1 to 3 orders of magnitude more voluminous than on Earth, and suggests that the same processes of inflation and over-crusting were active on Mars, while deep inflation and thermal entrenchment (a trench or system of trenches that provide a place of shelter) was the predominant mechanism of emplacement (being set in place) on Earth' Moon. Even with these outstanding dimensions (with total volumes exceeding 1 billion m^3), lunar tubes remain well within the roof stability threshold. The analysis carried out by Sauro et al. (2020) shows that aside of collapses triggered by impacts/tectonics, most of the lunar tubes could be intact, making Earth's Moon an extraordinary target for subsurface exploration and potential settlement in the wide protected and stable environments of lava tubes.

Models suggest that Martian lava tubes were more likely to have grown to the point of collapse when the planet was volcanically active, and might be harder to find intact. According to researchers, a greater proportion of lunar tubes are probably structurally sound, rendering the moon's lava tubes better candidates for exploration.

In the case of observed formation of lava tubes during the 1970–1971 period at Kilauea Volcano, Hawaii, it has been noted that as the supply rate of lava diminished during an eruption, the level of liquid in the tube dropped, leaving a void space between the top of the lava flow and the roof of the tube (Peterson and Swanson, 1974). At the end of an eruption, most of the lava was observed to drain out of the main tube to leave an open tunnel of varying dimensions (e.g., Macdonald et al., 1983). Coombs et al. (1989) found that, of the several hundred lava tubes known to exist on and around Kilauea, more than 90% are open or void; less than 30% of the tube has been filled in by later flows. Coombs and Hawke (1992) argued that the once empty lava tubes may later act as conduits for younger lava flows erupting from the common source vent. According to them, these later flows may partly or completely fill the older lava tube, depending on the lava supply rate and amount of material flowing through. It has been suggested that it is likely that similar occurrences may have happened during the formation of the lunar lava tubes as well.

Two of the largest of the intact lava tubes, that exist along the east and southwest rift zones of Kilauea Volcano on the Big Island of Hawaii, are Thurston lava Tube (Keanakahina), which was formed 350–500 years before present and an unnamed tube on the floor of Kilauea Caldera (a large volcanic crater, especially one formed by a major eruption leading to the collapse of the mouth of the volcano) that was formed during the 1919 eruption of Halemaumau. The respective average dimensions of these two lava tubes are 4.90 m wide × 2.20 m high and 8.60 m wide × 3.73 m high (Coombs and Hawke, 1992). Both of these lava tubes have maintained their structural integrity while constantly being shaken by local seismic tremors (see Klein et al., 1987). These examples pertaining to terrestrial lava tubes suggest that a long and varied seismic history has had minimal effect on the lava tubes in the Kilauea area and elsewhere on the island of Hawaii. It is possible then, that many of the lunar lava tubes have remained intact over the billions of years of the seismic shaking generated by meteorite impacts and tectonically originated moonquakes (Coombs and Hawke, 1992).

According to the researchers, lava tubes make appealing human habitats for a number of reasons, including protection from meteors that don't burn up as easily in the thin Martian and lunar atmospheres. They also likely contain useful chemicals, such as water ice and volatile chemicals that can be used to make fuel. A thick layer of rock overhead can also offer shielding against solar radiation. And skylights would still offer easy access to the surface. Although a lava tube could provide a shelter to thermal excursion, radiation and micro impacts, it is not easily accessible and the basaltic rocks of its interior can be razor-like sharp and the terrain very uneven, thereby throwing some engineering challenges in converting the lava tubes to habitable housings.

Before initiating any in-situ surveys (e.g., lander/rover-based surveys), satellites equipped with ground-penetrating radar or other remote-sensing technology should build detailed maps of tube formations underground. Lunar and Martian researchers are of the view that the process of discovering ideal sites for human sub-lunar or sub-Martian habitation will likely take a long time and involve many intermediate steps (Letzter, 2020).

If lunar lava tubes are suitable for housing permanent lunar bases, free from cosmic radiation and possessing a comfortable temperature condition [the constant temperature of around 20°C (Horz, 1985)], then the possibility for some form of life-forms existing there cannot be ruled out. Therefore, it would be worth searching for biosignatures inside lunar lava tubes and caves.

7.1.5 Ukrainian rocks—terrestrial analogs for botanical studies in simulation experiments involving possible growth of plants on Earth's moon

Because the amount of native lunar material is limited for botanical studies involving possible growth of plants on Earth's Moon, terrestrial analogs can be evaluated in simulation experiments to define pioneer plant cultivation. Microorganisms may be a key element in a precursory scenario of growing pioneer plants for extraterrestrial exploration. They can be used for plant inoculation to leach nutritional elements from regolith (loose unconsolidated rock and dust, including soil, which is a biologically active medium and a key component in plant growth), to alleviate lunar stressors, as well as to decompose both lunar rocks and the plant straw in order to form a proto-soil.

Zaets et al. (2011) examined bioleaching capacities of both French marigold (*Tagetes patula L.*) and the associated bacteria in contact with a lunar rock simulant (terrestrial anorthosite) using the model plant-bacteria microcosms (i.e., plant-bacteria regarded as encapsulating in miniature the characteristics of something much larger) under controlled conditions. Note that anorthosite is a granular plutonic igneous rock (a type of rock formed through the cooling and solidification of magma or lava) composed almost exclusively of a soda-lime feldspar (such as labradorite). Plutonic rocks are igneous rocks that are solidified from a melt at great depth. Magma rises, bringing minerals and precious metals such as gold, silver, molybdenum, lead, etc., with it, forcing its way into older rocks. Zaets et al. (2011) found that marigold accumulated K, Na, Fe, Zn, Ni, and Cr at higher concentrations in anorthosite compared to the podzol soil (an infertile acidic soil characterized by a white or grey subsurface layer resembling ash, from which minerals have been leached into a lower dark-colored stratum). Zaets et al. (2011) further found that plants inoculated with the consortium of well-defined species of bacteria accumulated higher levels of K, Mg, and Mn, but lower levels of Ni, Cr, Zn, Na, Ca, Fe, which exist at higher levels in anorthosite. Bacteria also affected the Ca/Mg and Fe/Mn ratios in the biomass of marigold grown on anorthosite. Despite their growth retardation, the inoculated plants had 15% higher weight on anorthosite than noninoculated plants. The data suggest that the bacteria supplied basic macro-and micro-elements to the model plant. Zaets et al. (2011) found that Ukrainian rocks from Korosten Pluton (Penizevitchi, Turchynka deposits, Zhytomyr oblast) provide a suitable test-bed for modeling bio-mobilization of plant-essential elements from the lunar-like rocks.

In another related study, temperature-programmed desorption (TPD) experiments carried out by (Hibbitts et al., 2011) showed that lunar-analog basaltic-composition glass is hydrophobic (tending to repel or fail to mix with water), with water–water interactions dominating over surface chemisorption (an adsorption process that involves a chemical reaction and that implies the formation of a covalent bond between the molecule and one or more atoms on the surface). While investigating the adsorption (the process by which a solid holds molecules of a gas or liquid or solute as a thin film) of molecular water onto lunar analog materials under ultra-high vacuum with the goal to better understand the thermal stability and evolution of water on the lunar surface, Hibbitts et al. (2011) found that lunar agglutinates (materials clumped together) will tend not to adsorb water (i.e., hold molecules of water as a thin film on the outside surface or on internal surfaces within the agglutinates) at temperatures above where water clusters and multilayer ice forms. Their studies further suggested that mineral surfaces will adsorb more water than mare or mature (glassy, agglutinate rich) surfaces and may explain the association of water with fresh feldspathic (containing feldspar, which is an abundant rock-forming mineral typically occurring as colorless or pale-colored crystals and consisting of aluminosilicates of potassium, sodium, and calcium) craters at high latitudes. They also determined the activation energies for the thermal desorption of water from these materials, and along with values from the literature, used to model the grain-to-grain migration of water within the lunar regolith.

7.1.6 Atacama desert in Chile—a terrestrial analog of Mars to examine its habitability conditions

As noted earlier, a biosignature is any substance—such as an element, isotope, or molecule—or phenomenon that provides scientific evidence of past or present life. Measurable attributes of life include its complex physical or chemical structures and its use of free energy and the production of biomass and wastes. Any traces of biosignatures that might have been present on the surface of celestial bodies are likely to be destroyed over time as a result of any one or more of a multitude of destructive factors (e.g., ultraviolet radiation, loss of atmosphere). In such cases, study of terrestrial analogs becomes helpful. A known terrestrial analog to Mars is the Atacama Desert in Chile, South America (Fig. 7.4), the dryness of which is found to be close to the dry limit for life on Earth.

Note that Atacama Desert is the driest desert in the world, as well as the only true desert to receive less precipitation than the polar deserts. Low water activity is the case for the Atacama Desert's similarity with the Mars surface. Through the study of Atacama Desert (a Mars terrestrial analog), Fairén et al. (2010) assessed and constrained the habitability conditions for each of the three important stages in the evolution of Mars (i.e., beginning with a water-rich epoch, followed by a cold and semi-arid era, and transitioning into present-day arid and very cold desert conditions), the geochemistry of the surface, and the likelihood for the preservation of organic and inorganic biosignatures. They claim that "the study of these analog environments provides important information to better understand past and current mission results as well as to support the design and selection of instruments and the planning for future exploratory missions to Mars."

7.1.7 International efforts to map the distribution of extremophiles across the globe

Physiologic, chemical, and biologic adaptations allow extremophiles to thrive in the most acidic, saline, hot, cold, and barophilic (relatively high pressure (up to 110 MPa)) environments on our planet. The secrets to their survival lie in the versatility and adaptability of their genomes (Dittami et al., 2012; Rampelotto, 2013; Dalmaso et al., 2015). According to Tighe et al. (2017), "The understanding of extremophiles—their genomes, molecular machinery, and how they interact with their environments—has potential health and research benefits for humanity. There are applications in bioremediation of polluted sites deemed too unbearable for most living organisms or as sources for novel therapeutics in medicine and potentially, an alternative process for biofuel or energy production. The metabolic mechanisms of these organisms are rather specialized and could inspire innovations in such diverse areas as synthetic biology and research into human survival in space. Many extreme environments offer relatively accessible proxies for the harsh environments found beyond Earth."

FIG. 7.4 Atacama Desert located between the Andes and Chilean Coast Range on the west coast (Pacific coast) of Chile, South America. It sits at a height [~2,500 m (8202 feet) above sea level] that prevents moisture reaching it from the Pacific or Atlantic Oceans. It receives less than 1mm of rainfall per year and some parts of the desert have no recorded rainfall at all. Image by NASA—NASA. *Copyright info: This image is in the Public Domain because it is a screenshot from NASA's globe software World Wind using a public domain layer, such as Blue Marble, MODIS, Landsat, SRTM, USGS or GLOBE.*

Taxonomic classification of extremophiles has been a pioneering field of study since the 1950s. Efforts to characterize extremophiles increased to such a degree that the International Society of Extremophiles was established in the 1990s and publishes the dedicated peer-reviewed journal, *Extremophiles*. In recent years, international efforts, such as the Earth Microbiome Project (EMP), have initiated large-scale endeavors to map the distribution of microorganisms (including extremophiles) across the globe (Tighe et al., 2017).

Researchers are inventing laboratory methods for handling the "extremophile" microbes that thrive in the world's most inhospitable environments and enjoy life at near-boiling or near-freezing temperatures, high pressures or environments steeped in salt, acids, alkalis, metals, or radioactivity. Scientists are particularly drawn to the novelty of the extremophiles that might harbor useful enzymes for industrial processes or antibiotics to save lives. There is an ongoing initiative known as the "Extreme Microbiome Project (XMP)"—a project launched by the Association of Biomolecular Resource Facilities Metagenomics Research Group (ABRF MGRG) that focuses on whole genome shotgun sequencing of extreme and unique environments using a wide variety of biomolecular techniques. The unique XMP consortium was founded in 2014 with the intention to create a comprehensive molecular profile of various extreme sites using novel culturing methods, long-read and short-read whole genome shotgun sequencing (instead of rRNA amplicon-based methods), improved RNA and DNA extraction methods, methylation tracing, and for future studies, metaproteomics (Tighe et al., 2017). Through this initiative, the researchers hope to find genes that might indicate how extremophiles survive and whether they might make compounds that could work as antibiotics.

Since the inception of the XMP, the consortium has begun collecting and analyzing data from 12 sites across the world, with more sites under consideration. Presently, the XMP is characterizing sample sites around the world with the intent of discovering new species, genes, and gene clusters. Sites include Lake Hillier in Western Australia, the "Door to Hell" crater in Turkmenistan, deep ocean brine lakes of the Gulf of Mexico, deep ocean sediments from Greenland, permafrost tunnels in Alaska, ancient microbial biofilms from Antarctica, Blue Lagoon Iceland, Ethiopian toxic hot springs, and the acidic hypersaline ponds in Western Australia (Tighe et al., 2017). Each site is sampled and analyzed as a complete stand-alone project. The selected sites are defined as extreme or "novel," based on such metadata as salinity, temperature, pressure, moisture, pH, or remoteness, with many sites falling into more than one category.

The ABRF MGRG works as a collaborative team to study these environments, using both traditional and novel methods. This includes a modified, nucleic acid-free sample collection; extraction of the DNA/RNA using methods to preserve nucleic acid length; culturing; microscopy; and multiple types of nucleic acid sequencing. Culturing methods are included to address the questions of viability and co-dependency, as well as the relationship to the detection of species using next-generation sequencing (NGS) and bioinformatics analyses (Tighe et al., 2017).

Studying extremophiles in the laboratory has reportedly thrown several challenges. Understandably, it was found that the extremophile microbes that are well-known to be tough enough to withstand extreme conditions also resisted scientists' attempts to break them open and recover their DNA.

Tighe et al. (2017) have reported genomic methods and microbiological technologies for profiling novel and extreme environments for the XMP. They have highlighted the details on the various methods they use to characterize the microbiome and metagenome of these complex samples. In particular, they reported data of a novel multienzyme extraction protocol that they developed, called Polyzyme or MetaPolyZyme.

According to Tighe et al. (2017), "These unique samples are often difficult to collect because of their remote site location and sometimes, even worse to extract their DNA and RNA as a result of reticent cells (cells hard to revealing themselves readily). Samples from harsh environments tend to have robust cell walls requiring special lysis (the disintegration of a cell by rupture of the cell wall or membrane) procedures. Consequently, one of the major goals of the MGRG was the development of a novel extraction method tailored to difficult samples that avoided beater beads (an effective method of cell lysis used to disrupt virtually any biological sample by rapidly agitating samples with a lysing matrix, sometimes referred to as grinding media or beads, in a bead beater) whenever possible to minimize unnecessary shearing of DNA. These protocols included substituting a novel multienzyme blend, called "Polyzyme," in place of lysozyme (an enzyme which catalyzes the destruction of the cell walls of certain bacteria, and occurs notably in tears and egg white) and further extraction of DNA to recover long fragment length DNA compatible with long sequencing strategies. Whereas culturing is not the primary focus of the projects, it does provide "minimum truth" in a sample and also requires multiple techniques, such as the following: 1) use of a multitude of microbial growth media and broths (including sample site enrichment media), 2) culturing of anaerobically and aerobically, 3) incubation at different times and temperatures, and 4) identification using full-length, 16S DNA sequencing and/or the Microbial Identification system (Biolog, Hayward, CA, USA)." As extremophilic samples are extraordinarily difficult to work with, they require new approaches that can be applied to other microbiome projects. RNA sequencing, coupled with functional and phylogenetic bioinformatics, provides results on the dynamics of the metatranscriptome (microbes within natural environments) at these sites, notably, the metabolic systems of extremophiles in situ.

7.1.8 Extreme acidic environment of Rio Tinto basin—terrestrial analog of Mars to understand its microbial survival during its evolution

The Rio Tinto, known by the Phoenicians as "River of Fire," because of its deep red color and high acidity, flows through the world's largest pyritic (iron sulfide) belt in south-western Spain. Bioleaching (the process of extracting metals from ores or waste by using microorganisms to oxidize the metals, producing soluble compounds) and interrelated acidophilic (growing best in acidic conditions) microorganisms have been reported from Río Tinto (López–Archilla et al., 1993). In a study, the microbial community composition and ecology of the acidic aquatic environment prevailing in the Tinto River have been reported by López–Archilla et al. (2001). Surprisingly, eukaryotic microbes are the principal contributors of biomass in this ecologically hostile river, which has a pH of 2 and contains much higher concentrations of heavy metals (Fe, Cu, Zn, As, Cr) than are typically found in fresh waters. Iron- and sulfur-oxidizing prokaryotes have also been identified in the Tinto River (Amils et al., 2002). Amaral-Zettler et al. (2002) showed that the Rio Tinto exhibits an unexpected degree of eukaryotic diversity and includes new lineages that these investigators have identified by sequence analysis of genes encoding small-subunit ribosomal RNAs. The diversity of these eukaryotes is much greater than that of prokaryotes, whose metabolism is responsible for the extreme environment.

High acidity and salinity do not always pose insurmountable challenges to microbial life on Earth. Acidic waters can support a phylogenetically (in a way that relates to the evolutionary development and diversification of a species or group of organisms) wide array of bacteria and microscopic eukaryotes (Amaral-Zettler et al., 2002); microorganisms can also accommodate to both short-term desiccation (removal of moisture) and (within limits) persistent hypersalinity. Such organisms, however, belong to specialized populations that have evolved to survive in highly acidic or saline environments. It is less clear that such conditions are suitable for the kinds of prebiotic chemical reactions commonly invoked to explain the origin of life.

It has been suspected that acidic environment in the worlds beyond Earth poses severe threat to the development and survival of life there. For example, a set of challenges to the development of Martian biology deals with water chemistry. Mineral assemblages similar to those in Eagle crater in Mars are found on Earth. Commonly these are associated with mine drainage, but in some cases, notably the Rio Tinto basin in Spain (Fernández-Remolar et al., 2003), mineral deposition predates mining activity. Precipitating waters tend to be acidic, sometimes strongly so. In addition, in the saline waters suggested by Meridiani geology (Meridiani Planum, Mars), water activity could have been at least episodically prohibitive.

Exploration by the NASA rover *Opportunity* has revealed sulfate- and hematite-rich sedimentary rocks exposed in craters and other surface features of Meridiani Planum, Mars. Modern, Holocene (the present epoch), and Plio-Pleistocene (geological pseudo-period, which begins about 5 million years ago and, drawing forward, combines the time ranges of the formally defined Pliocene and Pleistocene epochs—marking from about 5 Mya to about 12 kya) deposits of the Río Tinto provide at least a partial environmental analog to Meridiani Planum rocks, facilitating our understanding of Meridiani mineral precipitation and diagenesis, while informing considerations of Martian astrobiology (Fernández-Remolar et al., 2005).

Oxidation, thought to be biologically mediated, of pyritic ore bodies by groundwaters in the source area of the Río Tinto generates headwaters enriched in sulfuric acid and ferric iron. Seasonal evaporation of river water drives precipitation of hydronium jarosite ($KFe_3(SO_4)_2(OH)_6$) and schwertmannite ($Fe_8O_8(OH)_6(SO_4) \cdot nH_2O$), while (Mg, Al, Fe^{3+})-copiapite, coquimbite (hydrous ferric sulfate $Fe_2(SO_4)_3 \cdot 9H_2O$), gypsum ($CaSO_4 \cdot 2H_2O$), and other sulfate minerals precipitate nearby as efflorescences where locally variable source waters are brought to the surface by capillary action. Note that in chemistry, efflorescence is the migration of a salt to the surface of a porous material, where it forms a coating. The essential process involves the dissolving of an internally held salt in water, or occasionally in another solvent. During the wet season, hydrolysis of sulfate salts results in the precipitation of nanophase goethite. Fernández-Remolar et al. (2005) found that Holocene (the current geological epoch, which began approximately 11,650 years before present, after the last glacial period) and Plio-Pleistocene (geological pseudo-period, which begins about 5 million years ago and, drawing forward, combining the time ranges of the formally defined Pliocene and Pleistocene epochs—marking from about 5 Mya to about 12 kya) terraces show increasing goethite (a dark or yellowish-brown mineral consisting of hydrated iron oxide, occurring typically as masses of fibrous crystals) crystallinity and then replacement of goethite with hematite through time. Hematite (a common iron oxide compound with the formula, Fe_2O_3 and is widely found in rocks and soils) in Meridiani spherules (small spheres) also formed during diagenesis (physical and chemical processes that affect sedimentary materials after deposition and before metamorphism and between deposition and weathering), although whether this replaced precursor goethite (a dark or yellowish-brown mineral consisting of hydrated iron oxide, occurring typically as masses of fibrous crystals) or precipitated directly from groundwaters is not known. Note that Martian spherules are the abundant spherical hematite inclusions discovered by the Mars rover *Opportunity* at Meridiani Planum on the planet Mars in 2004. Note that an inclusion is any material trapped inside a mineral as it forms. That material could be a rock trapped inside another rock. It may be noted that diagenesis refers to sum of all processes, chiefly chemical, by which changes in a sediment are brought about after its deposition but before its final lithification (conversion to rock). Diagenesis is considered a relatively low-pressure, low-temperature alteration process. An example of diagenesis is the chemical alteration of a feldspar (an abundant rock-forming mineral typically occurring as colorless or pale-colored crystals and consisting of aluminosilicates of potassium, sodium, and calcium) to form a distinctly new mineral in its place, a clay mineral. According to Fernández-Remolar et al. (2005), the retention of jarosite [a basic hydrous sulfate of potassium and ferric iron (Fe-III) with a chemical formula of $KFe_3(SO_4)_2(OH)_6$] and other soluble sulfate salts suggests that water limited the diagenesis of Meridiani rocks.

As already noted, diverse prokaryotic and eukaryotic microorganisms inhabit acidic and seasonally dry Río Tinto environments. Organic matter does not persist in Río Tinto sediments, but biosignatures imparted to sedimentary rocks as macroscopic textures of coated microbial streamers (bacterial biofilms degenerated in the form of filamentous structures), surface blisters (small to large, broken or unbroken bubbles, which are under or within a coating) formed by biogenic gas, and microfossils preserved as casts and molds in iron oxides help to shape strategies for astrobiological investigation of Meridiani outcrops (Fernández-Remolar et al., 2005).

The geo-microbiological characterization of the water column and sediments of Río Tinto (Huelva, South-western Spain) have proven the importance of the iron and the sulfur cycles, not only in generating the extreme conditions of the habitat (low pH, high concentration of toxic heavy metals) but also in maintaining the high level of microbial diversity detected in the basin. It has been proven that the extreme acidic conditions of Río Tinto basin are not the product of 5000 years of mining activity in the area, but the consequence of an active underground bioreactor that obtains its energy from the massive sulfidic minerals existing in the Iberian (countries of Spain and Portugal) Pyrite Belt. Two drilling projects, MARTE (Mars Astrobiology Research and Technology Experiment) (2003–2006) and IPBSL (Iberian Pyrite Belt Subsurface Life Detection) (2011–2015), was developed and carried out to provide evidence of subsurface microbial activity and the potential resources that support these activities. The reduced substrates and the oxidants that drive the system appear to come from the rock matrix. These resources need only groundwater to launch diverse microbial metabolisms. The similarities between the vast sulfate and iron oxide deposits on Mars and the main sulfide bioleaching products found in the Tinto basin have given Río Tinto the status of a geochemical and mineralogical Mars terrestrial analog (Amils et al., 2014).

High-acidity conditions appear to be the main factor underlying the limited diversity of the microbial populations thriving in some natural systems environments, although temperature, ionic composition, total organic carbon, and dissolved oxygen are also considered to significantly influence their microbial life. This natural reduction in diversity driven by extreme conditions was reflected in several studies on the microbial populations inhabiting the various micro-environments present in such ecosystems. In a review, Méndez-García et al. (2015) reported a complete overview of the bacterial, archaeal, and eukaryotic diversity in these ecosystems, and included a thorough depiction of the metabolism and element cycling in acid mine drainage (AMD) habitats. These investigators also reviewed different metabolic network structures at the organismal level, which is necessary to disentangle the role of each member of the AMD communities described thus far.

Some micro-algae are adapted to extremely acidic environments in which toxic metals are present at high levels. However, little is known about how acidophilic (growing best in acidic conditions) algae evolved from their respective neutrophilic ancestors [leucocytes (a type of blood cell that is made in the bone marrow and found in the blood and lymph tissue) having lobed nucleus and a fine granular cytoplasm, which stains with neutral dyes] by adapting to particular acidic environments. To gain insights into this issue, Hirooka et al. (2017) determined the draft genome sequence of the acidophilic green alga *Chlamydomonas eustigma* and performed comparative genome and transcriptome analyses between *Ceustigma* and its neutrophilic relative *Chlamydomonas reinhardtii*. It was found that the arsenic detoxification genes have been multiplied in the genome. The features in *Ceustigma* that probably contributed to the adaptation to an acidic environment have also been found in other acidophilic green and red algae, suggesting the existence of common mechanisms in the adaptation to acidic environments.

Acid mine drainages are characterized by their low pH (i.e., high acidity) and the presence of dissolved toxic metallic species. Microorganisms survive in different microhabitats within the ecosystem, namely water, sediments, and bio-films. In a study, Mesa et al. (2017) surveyed the microbial diversity within all domains of life in the different microhabitats at Los Rueldos abandoned mercury underground mine (NW Spain), and predicted bacterial function based on community composition. It was found that sediment samples contained higher proportions of soil bacteria (AD3, Acidobacteria), as well as *Crenarchaeota* and *Methanomassiliicoccaceae* archaea. It was found that oxic (having enough oxygen in the tissues to sustain bodily functions) and hypoxic (not having enough oxygen in the tissues to sustain bodily functions) biofilm samples were enriched in bacterial iron oxidizers from the genus *Leptospirillum*, order Acidithiobacillales, class Betaproteobacteria, and archaea from the class Thermoplasmata. It was also found that water samples were enriched in Cyanobacteria and Thermoplasmata archaea at a 3–98% of the sunlight influence, whilst Betaproteobacteria, Thermoplasmata archaea, and Micrarchaea dominated in acid water collected in total darkness. Eukaryotes were detected in biofilms and open-air water samples.

7.1.9 Antarctic Ross Desert—terrestrial analog of Mars to understand its ecosystem and habitability conditions for different stages in its evolution

Mars has undergone three main climatic stages throughout its geological history, beginning with a water-rich epoch, followed by a cold and semi-arid era, and transitioning into the present-day arid and very cold desert conditions. These global climatic eras also represent three different stages of planetary habitability: an early, potentially habitable stage when the basic requisites for life as we know it were present (liquid water and energy); an intermediate extreme stage, when liquid solutions became scarce or very challenging for life; and the most recent stage during which conditions on the surface have been largely uninhabitable, except perhaps in some isolated niches. Our understanding of the evolution of Mars is now sufficient to assign specific terrestrial environments to each of these periods.

An organism that lives inside rock, coral, animal shells, or in the pores between mineral grains of a rock is an "endolith" (archaeon, bacterium, fungus, lichen, algae, or amoeba). Many are extremophiles, living in places long imagined inhospitable to life. In the frigid desert of the Antarctic dry valleys there are no visible life forms on the surface of the soil or rocks. Yet in certain rock types a narrow subsurface zone has a favorable microclimate and is colonized by microorganisms. Dominant are lichens (very small grey or yellow plants that spread over the surface of rocks, walls, and trees and do not have any flowers) of unusual organization. They survive not by physiological adaptation to lower temperatures, but by changing their mode of growth, being able to grow between the crystals of porous rocks. Their activity results in mobilization of iron compounds and in rock weathering with a characteristic pattern of exfoliation (the process of removing dead skin cells from the outer layer of the skin). This simple ecosystem lacks both higher consumers and predators (Friedmann, 1982).

Endolith are of particular interest to astrobiologists, who theorize that endolithic environments on Mars and other planets constitute potential refugia for extraterrestrial microbial communities (Wierzchos et al., 2011). "Cryptoendolithic" refers to one of the three subclasses in which "Endolithic" microorganisms are classified. Cryptoendolithic microorganisms are those able to colonize the empty spaces or pores inside a rock with the connotation of being hidden. The cold, dry ecosystems of Antarctica have been shown to harbor traces left behind by microbial activity within certain types of rocks, but only two indirect biomarkers of cryptoendolithic activity in the Antarctic cold desert zone seem to have been described thus far. These are the geophysical and geochemical bio-weathering patterns macroscopically observed in sandstone rock. Wierzchos et al. (2003) showed that in this extreme environment, minerals are biologically transformed, and as a result, Fe-rich diagenetic minerals (i.e., all minerals formed after deposition of a sediment and prior to its disintegration by weathering or alteration by metamorphism) in the form of iron hydroxide nanocrystals and biogenic clays are deposited around chasmo-endolithic (i.e., colonizing in fissures and cracks in the rock) hyphae (each of the branching filaments that make up the mycelium of a fungus) and bacterial cells. Thus, when microbial life decays, these minerals act as distinct biomarkers of previous endolithic activity. According to Wierzchos et al. (2003), the ability to recognize these traces may have potential astrobiological implications because the Antarctic *Ross Desert* is considered a terrestrial analog of a possible ecosystem on early Mars.

7.1.10 Pingos on Earth—tools for understanding permafrost geomorphology on Mars

Pingos are meso-scale (order 100-m-diameter) ice-cored mounds (Fig. 7.5)—typically conical in shape and growing and persisting only in permafrost environments—that develop through pressurized groundwater flow mechanisms and progressive freezing. These specific formation conditions make them helpful climatic and hydrologic markers. The potential of the ice core to host psychrophilic (i.e., extremophilic cold-loving) microorganisms makes them attractive as potential astrobiologic targets. These considerations make identifying pingos on Mars useful for piecing together Mars' climatic and hydrologic history and for providing potential biotic or prebiotic targets (Burr et al., 2009). According to Burr et al. (2009), pingos are commonly (although not universally) identified on Mars based on the axiom that if their formation mechanisms are similar, terrestrial and Martian pingos should have a similar shape. It has been shown that pingo size is not a function of gravity (see Cabrol et al., 2000; Burr et al., 2005).

FIG. 7.5 Pingos near Tuktoyaktuk, Northwest Territories, Canada. *Author: Emma Pike. (Source: https://commons.wikimedia.org/wiki/File:Pingos_near_Tuk.jpg) Copyright information: This work has been released into the public domain by its author, Emma Pike. This applies worldwide. In some countries this may not be legally possible; if so: Emma Pike grants anyone the right to use this work for any purpose, without any conditions, unless such conditions are required by law.*

7.1.10.1 General features of terrestrial pingos

Pingos found on Earth are formed when water, rising by hydraulic pressure through gaps in the permafrost, freezes, and uplifts a mound of ice covered by a layer of alluvium (loose, soil or sediment that has been eroded, reshaped by water in some form, and redeposited in a nonmarine setting). Pingos are dynamic and can pulse vertically. It has also been observed that the distinctive feature of these mounds is an ice core, and the creation is dependent on the expansion of pore-water in the soil of the mound. The analysis of distribution and characteristics suggest that pingos are extremely dynamic and individualistic and a wide range of heights, diameters, slopes, and spacing exist.

Pingos and their collapsed forms are found in periglacial terrain (i.e., terrain subject to repeated freezing and melting) and paleo-periglacial terrain on Earth. Pingos are widespread in permafrost regions of Earth's Northern Hemisphere, but in the Southern Hemisphere, the presence of pingos is not clearly confirmed. On Earth, pingos have been located at several freezing cold regions such as Svalbard (an archipelago in Norway), West Greenland, Alaska, Siberia, Tibet, Canadian Arctic and Mongolia. Kenji Yoshikawa (a research professor at the University of Alaska) has located pingos in several locations such as these, and examined their internal structures and hydrology (see Yoshikawa, 2013).

The two types of pingos are (i) open-system (hydraulic) pingos and (ii) closed-system (hydrostatic) pingos (Swift et al., 2015). The former is more common in valley bottoms and foot slopes in both discontinuous and continuous permafrost, and the latter occurs in lowland settings within the continuous permafrost zone. Burn (2020) has discussed open-system pingos in some detail. Mackay (1998) has addressed the issue of pingo growth and collapse.

Pingo formation in an "open system" takes place as follows (Rowley et al., 2015): In an open system, a pingo forms as a result of artesian pressure (natural pressure developed in water-bearing strata lying at an angle producing a constant supply of water with little or no pumping) induced from locally higher topographies. Fig. 7.6 shows the diagram of an artesian well. Like an artesian well, "open system" pingos commonly occur at the base of a hillslope where hydraulic pressures are high, allowing the groundwater to inject into the pingo. As the groundwater approaches the top of the pingo, it freezes and accumulates, causing the pingo to grow. Because groundwater is essential for the creation and growth of an open system pingo, these often occur in areas of discontinuous permafrost where groundwater can flow freely. Note that "artesian pressure" refers to or denotes the natural pressure producing a constant supply of water at the foot of a hill with little or no pumping, such as what happens when a well is bored perpendicularly into water-bearing strata lying at an angle on top of a hill. Fig. 7.7 shows three open-system pingos in a

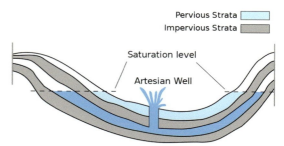

FIG. 7.6 Diagram of an artesian well. The pervious (permeable) strata allows water to pass through. The impervious strata does not allow water to pass through. *(Source: https://commons.wikimedia.org/wiki/File:Artesian_Well.svg) Copyright info: This file is licensed under the Creative Commons Attribution-Share Alike 2.0 Generic license.*

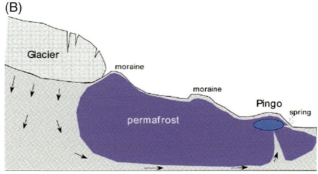

FIG. 7.7 (A) An oblique aerial photograph showing three open-system pingos in a mountain river valley, Spitsbergen, Norway. The two pingos on the left are collapsed and have central ponds. "Ny" pingo on the right shows both radial cracks from dilation and diagonal cracks trending from upper left to lower right. A glacier is located in the center background (flow direction is out of the photo), with two moraines in front of it. (B) A generalized schematic of the hydrological and glaciological situation in the photo in (A). Subglacial meltwater percolates through an unfrozen zone in the discontinuous permafrost to form a pingo. *(Source: Burr, D.M., Tanaka, K.L. and Yoshikawa, K. (2009), Pingos on Earth and Mars, Planetary and Space Science, 57 (5-6): 541–555. © 2008 Elsevier Ltd. All rights reserved.).*

mountain river valley, Spitsbergen, Norway. In Fig. 7.7B, Burr et al. (2009) provide a beautifully picturized generalized schematic of the hydrological and glaciological situation in which a pingo develops and grows in a low-lying

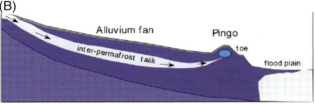

FIG. 7.8 (A) An oblique aerial photo showing an open-system pingo (arrow) at the toe of a fan (to left), Hulahula River, Brooks, Range, Alaska. (B) A generalized schematic showing the inferred formation mechanism for the pingo in (A). Groundwater flow occurs from higher elevation to lower elevation through an unfrozen zone ("talik") within the permafrost. The hydraulic head is provided by elevation and confinement within the permafrost. *(Source: Burr, D.M., Tanaka, K.L. and Yoshikawa, K. (2009), Pingos on Earth and Mars, Planetary and Space Science, 57 (5-6): 541–555. © 2008 Elsevier Ltd. All rights reserved.).*

region. Fig. 7.8 shows another open-system pingo located at the toe of a fan at Hulahula River, Alaska. Fig. 7.8B is a generalized schematic showing the inferred formation mechanism for the pingo in Fig. 7.8A. In this case, groundwater flow occurs from higher elevation to lower elevation through an unfrozen zone ("talik") within the permafrost (Burr et al., 2009). The hydraulic head required to sustain the flow is provided by elevation and confinement within the permafrost. Fig. 7.9 shows a pingo located adjacent to South Kunlun Fault, Kunlun Mountain, China), in a peculiar situation in which pingos develop along minor faults associated with major fault systems as a result of flow along the minor faults. In this case, artesian pressure (i.e., natural pressure producing a constant supply of water with little or no artificial intervention) of sub-permafrost water is 22 m higher than the ground surface. A generalized schematic (Burr et al., 2009) of the situation depicted in Fig. 7.9B explains the pingo formation mechanism in this case. Fig. 7.10A shows a swath of pingos about 1km in total length along a formerly submerged coastline on the river delta of Adventdalen, Spitsbergen. The overburden of pingo ice is composed mainly of fine silt and mud, but has a thin peat layer. The largest pingo is approximately 4.5 m high, 500 m long, and 140 m wide, and is situated approximately 25 km from the former shoreline (see Yoshikawa

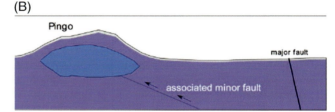

FIG. 7.9 (A) A ground photograph of a pingo located adjacent to South Kunlun Fault, Kunlun Mountain, China. Artesian pressure of subpermafrost water is 22 m higher than the ground surface. Polygonal terrain due to ice wedging is visible in the foreground. (B) A generalized schematic of the situation inferred for the photo in (A). Pingos develop along minor faults associated with major fault systems as a result of flow along the minor faults. *(Source: Burr, D.M., Tanaka, K.L. and Yoshikawa, K. (2009), Pingos on Earth and Mars, Planetary and Space Science, 57 (5-6): 541–555. © 2008 Elsevier Ltd. All rights reserved.).*

and Harada, 1995, for details). Fig. 7.10B depicts a generalized schematic of the hypothesized situation in the photo in Fig. 7.10A. The marine terrace is rebounding from glacial isostasy, and developing new permafrost. Note that "isostasy" is a concept that describes the response of the planet to a change in surface load. "Glacial isostasy" is concerned with the planet's response to the changing surface loads of ice and water during the waxing and waning of large ice sheets. In the present case under discussion, deglaciation (i.e., the disappearance of ice from a previously glaciated region) permits upwelling of artesian groundwater through unfrozen zones ("taliks") in the newly forming permafrost. New permafrost thickness is ~30 m.

In contrast to open system pingos (i.e., hydraulic pingos), closed system or hydrostatic pingos are found in regions of continuous permafrost (as reviewed in Mackay, 1979, 1998), which creates an impermeable (i.e., not allowing fluid to pass through) layer at depth (Burr et al., 2009). Most closed-system pingos form in drained "thaw lakes" (bodies of freshwater, usually shallow, that are formed in a depression formed by ice-rich permafrost becoming liquid or soft as a result of

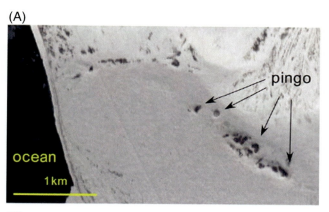
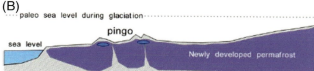

FIG. 7.10 (A) An aerial photo showing a swath of pingos (arrows) about 1 km in total length along a formerly submerged coastline on the river delta of Adventdalen, Spitsbergen. The overburden of pingo ice is composed mainly of fine silt and mud, but has a thin peat layer. The largest pingo is approximately 4.5 m high, 500 m long, and 140 m wide, and is situated approximately 25 km from the former shore line. (See Yoshikawa and Harada, 1995, for details.) (B) A generalized schematic of the hypothesized situation in the photo in (A). The marine terrace is rebounding from glacial isostasy, and developing new permafrost. Deglaciation permits upwelling of artesian groundwater through unfrozen zones ("taliks") in the newly forming permafrost. New permafrost thickness is ~30 m. (*Source: Burr, D.M., Tanaka, K.L. and Yoshikawa, K. (2009), Pingos on Earth and Mars, Planetary and Space Science, 57 (5-6): 541–555. © 2008 Elsevier Ltd. All rights reserved.*).

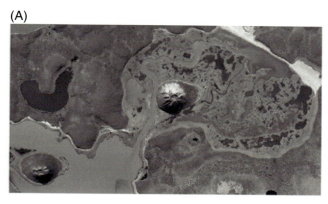
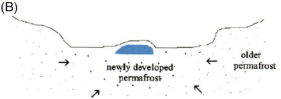

FIG. 7.11 (A) A black-and-white aerial photograph showing two closed-system pingos on the Tuktoyaktuk Peninsula, NWT, Canada. "Split" is located to the lower left and "Ibyuk" is approximately at image center. Dilational cracking is visible on both pingos, and thermal contraction polygons are visible on the surrounding terrain. "Ibyuk" sits within a drained thaw lake, and "Split" has been inundated by the Arctic Ocean (just off the image to the left). "Ibyuk" is 300 m in diameter for scale. (B) A generalized schematic depicting formation of a closed-system pingo (after Mackay, 1998). Arrows show the direction of permafrost aggradation. Pore water expulsion ahead of the aggrading permafrost causes subsurface pooling of the pore water and upward (hydrostatic) pressure. Freezing of this pore water creates the pingo ice (blue). ((*Source: Burr, D.M., Tanaka, K.L. and Yoshikawa, K. (2009), Pingos on Earth and Mars, Planetary and Space Science, 57 (5-6): 541–555. © 2008 Elsevier Ltd. All rights reserved.*).

warming up), in which the saturated lake basin sediments are exposed to the atmosphere. This exposure causes the saturated sediments to freeze, forming new permafrost. However, residual ponds within the lakes retard permafrost growth, resulting in locally thinner permafrost and lesser resistance to deformation. As the permafrost aggrade (i.e., alter the grade or level of a stream bed) downward, the pore water is expelled ahead of the freezing front. Blockage of this downward pore water flow by the continuous permafrost at depth redirects the flow inward from the basin edges. Injection of expelled water beneath the overlying permafrost, followed by freezing of the injected water, causes progressive doming of the permafrost overburden. This doming, commonly found beneath the least permafrost aggradation, is a result of the massive ice forming the core of the pingo.

Rowley et al. (2015) tersely described a "closed system" pingo thus, "In a closed system, a pingo forms as a result of hydrostatic pressure. Closed system pingos commonly form in drained shallow lake basins. Where surface water is present, a thermal gradient lowers the permafrost table, leaving an unfrozen layer termed a 'talik.' As a surface water basin drains, the residual water in the saturated soil is exposed to the atmosphere where it freezes. As residual pore water freezes, cryostatic pressure (i.e., the pressure exerted on soil when freezing occurs) pushes remaining unfrozen water in the talik toward the top of the pingo where it also freezes; eventually creating the inner ice core dominant in pingos."

Fig. 7.11 shows two closed-system pingos on the Tuktoyaktuk Peninsula, Northwest Territories (NWT), Canada (Burr et al., 2009). Fig. 7.11B is a generalized schematic depicting formation of a closed-system pingo (after Mackay, 1998).

7.1.10.2 Pingo-like forms (PLFs) identified on Mars

Pingos have been hypothesized to explain variously-shaped, meso-scale features found in Mars' mid-latitudes. Although Dundas et al. (2008) noted one PLF in the southern hemisphere, mapping of PLFs in the mid-latitudes has focused largely on Utopia Planitia. Centered near 50 N 120 E, Utopia

Planitia is an ancient multiring impact basin about 3300 km in diameter (McGill, 1989). It is believed that Utopia Planitia may have been inundated by a northern ocean (Baker et al., 1991; Parker et al., 1993; Head et al., 1999).

Couple of studies of pingo-like forms (PLFs) in Utopia Planitia on the planet Mars has indicated that Martian PLFs are the most viable candidates for pingos or collapsed pingos. PLFs are periglacial features that have been discovered but are usually not classed as pingos. This is because they are not large enough to be classified as pingos, or there is not enough evidence to class them as pingos (Burr et al., 2009). Apart from Martian PLFs, the feature found in Manannán crater on Europa shows some characteristics to pingos on Earth (Steinbrügge et al., 2020).

Importance of gaining sound understanding of terrestrial pingos lies in the fact that such knowledge, in conjunction with knowledge of Martian conditions, allows the assessment of PLFs on Mars. Since the year 2000, a series of visible-wavelength cameras on Mars spacecraft have provided high-resolution images of various hypothesized periglacial (subject to repeated freezing and thawing) features, including fields of inferred pingos. The Mars data include images and compositional spectral information from the Thermal Emission Imaging Spectrometer (THEMIS), the Thermal Emission Spectrometer (TES), the Compact Reconnaissance Imaging Spectrometer for Mars (CRISM), the Mars Orbiter Camera (MOC), the High-Resolution Imaging Science Experiment (HiRISE), and the High-Resolution Stereo Camera (HRSC), as well as topographic information from the Mars Orbiter Laser Altimeter (MOLA) (Burr et al., 2009).

It may be noted that Utopia Planitia is a large plain hosting an albedo feature within Utopia, the largest recognized impact basin on Mars and in the Solar System with an estimated diameter of 3300 km (McGill, 1989). Dundas and McEwen (2010) systematically searched for candidate pingos on Mars using images from the High-Resolution Imaging Science Experiment (HiRISE) camera. Because pingos are expected to develop surface fractures due to extension of the frozen ground over the ice core, they searched for fractured features and identified a variety of mounds. They found that these features are confined to the Martian mid-latitudes, in the bands where gullies [deep, narrow gorges with steep sides, formed by the action of water] are also most common. They found that the observed fractured mounds have a variety of morphologies and are likely of multiple origins. According to Dundas and McEwen (2010), "Isolated fractured mounds found on the floors of gullied craters in the southern hemisphere match the general morphologic characteristics of terrestrial pingos and are the best candidates for Martian pingos, but there is currently no direct evidence for presence of ice cores and it is difficult to produce the necessary water volumes, so these features should still be interpreted with caution.

Other fractured mounds appear more likely to be erosional remnants of an unusual mantling layer or possibly thermokarst structures. Flat-topped mounds in Utopia have some characteristics (fracture pattern and latitudinal distribution) consistent with pingos but differ in other aspects such as shape and setting." Note that thermokarst is a form of periglacial topography resembling karst (a landscape that is characterized by numerous caves, sinkholes, fissures, and underground streams), with hollows produced by the selective melting of permafrost.

The necessary elements for terrestrial-style pingo formation—namely, a sedimentary substrate, subsurface groundwater, and freezing conditions, as well as structural anomalies—have all been inferred in Utopia Planitia (Skinner and Tanaka, 2007). Some areas of the Utopia Planitia's surface seem to have been carved out by an ice cream scoop. This surface is thought to have formed by the degradation of an ice-rich permafrost (Sejourne et al., 2012). Many features that look like pingos on the Earth are found in Utopia Planitia (~35–50 N; ~80–115 E) (Soare et al., 2019). On November 22, 2016, NASA reported finding a large amount of underground ice in the Utopia Planitia region of Mars. The volume of water detected in the scalloped terrain (Fig. 7.12) in the Utopia Planitia region has been estimated to be equivalent to the volume of water in Lake Superior, which is the largest of the Great Lakes of North America, the world's largest freshwater lake by surface area, and the

FIG. 7.12 This vertically exaggerated view shows scalloped depressions in Mars' Utopia Planitia region, one of the area's distinctive textures that prompted researchers to check for underground ice, using ground-penetrating radar aboard NASA's Mars Reconnaissance Orbiter. More than 600 overhead passes with the spacecraft's Shallow Radar (SHARAD) instrument provided data for determining that about as much water as the volume of Lake Superior lies in a thick layer beneath a portion of Utopia Planitia. The vertical dimension is exaggerated fivefold in proportion to the horizontal dimensions, to make texture more apparent in what is a rather flat plain. (Source: https://en.wikipedia.org/wiki/File:PIA21136_-_Scalloped_Terrain_Led_to_Finding_of_Buried_Ice_on_Mars.jpg) Copyright information: This file is in the public domain in the United States because it was solely created by NASA. NASA copyright policy states that "NASA material is not protected by copyright unless noted." (See Template:PD-USGov, NASA copyright policy page or JPL Image Use Policy.).

third largest freshwater lake by volume. These scalloped depressions on the surface as seen in Fig. 7.12 are typically about 100–200 m wide. Scalloped terrain led to the discovery of a large amount of underground ice, enough water to fill Lake Superior. The foreground of this view covers ground about one mile (1.6 km) across. The perspective view is based on a three-dimensional terrain model derived from a stereo pair of observations by the High-Resolution Imaging Science Experiment (HiRISE) camera on the Mars Reconnaissance Orbiter. Similar scalloped depressions are found in portions of the Canadian Arctic, where they are indicative of ground ice. Scalloped topography (i.e., topography consisting of shallow, rimless depressions) is common in the mid-latitudes of Mars, between 45 and 60 north and south. It is particularly prominent in the region of Utopia Planitia (Morgenstern et al., 2007; Lefort et al., 2009) in the northern hemisphere of Mars and in the region of Peneus and Amphitrites Patera (Zanetti et al., 2009; Lefort, et al., 2010).

Burr et al. (2009) recall that early identification of PLFs in north-western Utopia focused on the floor of a crater near 65 N, 67 E, showing a set of roughly circular mounds 25–50 m in diameter (Soare et al., 2005). These mounds were hypothesized to be pingos on the basis of their size and morphology, which included radial cracks and summit depressions visible in Mars Orbiter Camera (MOC) image E03-00299. It may be recalled that radial cracking is observed on terrestrial pingos as dilation cracking; and summit depressions on terrestrial pingos result from melting or sublimation of the ice core and subsequent collapse. According to Burr et al. (2009), "associated features inferred to be periglacial in origin, namely polygonal cracking, and a curvilinear (channel-like) feature, led to interpretation of the crater floor as supporting an assemblage of features most consistent with formation by aqueous periglacial processes. The proposed terrestrial analog is the assemblage of features found in the Tuktoyaktuk Peninsula, Northwest Territories, Canada, which includes one of the largest concentrations of closed-system (hydrostatic) pingos on Earth (Mackay, 1998 and references therein)." Images from HiRISE (e.g., Fig. 7.13A) show the same suite of features inferred from the lower resolution MOC image, namely, mounds with radial cracks that in some cases extends onto the surrounding terrain (Fig. 7.13B) and/or summit depressions (Fig. 7.13C). Burr et al. (2009) recall that "in addition, polygonal cracking of the surrounding crater floor and bright curvilinear features are visible, previously hypothesized to be thermal contraction polygons and meltwater flow features, respectively (Soare et al., 2005). Thus, the higher resolution data are consistent with the earlier interpretation of the mounds as pingos. The water source for the pingos was hypothesized to have been atmospheric, tied to obliquity variations (Soare et al., 2005)."

Fig. 7.14 shows pingo-like features on Mars. On Mars, pingos have been hypothesized to explain various-shaped,

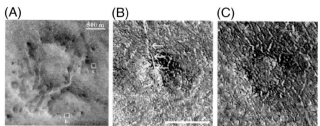

FIG. 7.13 (A) Subset of HiIRSE image PSP_008492_2450 (centered near 65.4 N, 67.3 E), showing the floor of an impact crater in Utopia Planitia with features hypothesized by Soare et al. (2005) to be pingos, thermal contraction polygons, and drainage features. White boxes show the locations of (B) and (C). (B) and (C) Subsets of HiRISE image PSP_008492_2450, showing hypothesized pingos with (B) radial cracking (white arrows), and (C) a summit depression. The 50-m scale bar on (B) applies to both images (B) and (C). *(Source: Burr, D.M., Tanaka, K.L. and Yoshikawa, K. (2009), Pingos on Earth and Mars, Planetary and Space Science, 57 (5-6): 541–555. Copyright info: © 2008 Elsevier Ltd. All rights reserved.)*

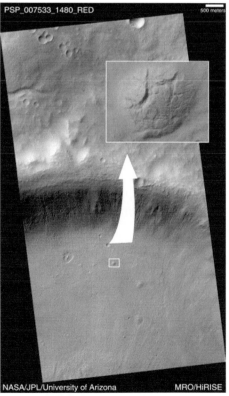

FIG. 7.14 **Pingo-like features on Mars.** The size of this mound (ca. 100 m diameter) and cracking pattern seems similar to Terrestrial pingos. This mound is located at the bottom of an impact crater where an ideally developed artesian condition could occur, if groundwater is present. Thus, if this is a pingo, it would be of the open system type. (PSP_007533_1480) Subsets of HiRISE image *(Image: NASA/JPL/University of Arizona). (Source: Yoshikawa, K. (2013), Glacial and Periglacial Geomorphology, In: Treatise on Geomorphology) Copyright info: © 2020 Elsevier Ltd. B.V. or its licensors or contributors.*

mesoscale features in the mid-latitudes. Although pingo-like forms (PLFs) have been noted in the Southern Hemisphere of Mars, mapping of PLFs in the mid-latitudes has focused on Utopia Planitia. According to Yoshikawa (2013), the reason that the Martian landforms, which are most likely to be pingos, are clustered in Utopia Planitia may be due to general conditions in this northern mid-latitude basin. The basin experienced repeated aqueous inflow during the Amazonian epoch (a geologic system and time period on the planet Mars, thought to have begun around 3 billion years ago and continues to the present day, characterized by low rates of meteorite and asteroid impacts and by cold, hyperarid conditions broadly similar to those on Mars today). Such surface aqueous flow would provide near-surface groundwater; and near-surface groundwater is necessary for pingo formation. Candidate pingos have not been hypothesized in large, southern midlatitude basins that experienced sedimentary infilling and aqueous inflow. A limited number of PLFs have been identified in small (approximately 10 km diameter) craters in the southern midlatitudes (Fig. 7.14). Using the MOC and Thermal Emission Imaging Spectrometer (THEMIS) images, more different types of morphology have been identified.

Based on a survey of Martian PLFs, Burr et al. (2009) found that the regional context in the Utopia Planitia is favorable to terrestrial-style pingo formation and that "the observed features span a range of forms that are similar to the range of terrestrial pingos undergoing growth and collapse." However, according to them, some questions remain about these Utopia Planitia features, including the water source for the hypothesized pingos, the timing of their formation, and the flat-topped morphology observed in some cases. Of the Martian landforms that have been hypothesized in the literature to be pingos, those in Utopia Planitia are the most likely candidates to be terrestrial-style pingos or pingo scars. PLFs reported to date elsewhere on Mars are less likely pingo candidates or are better explained by some other mechanism (Burr et al., 2009).

According to Burr et al. (2009), the reason that the landforms most likely to be pingos on Mars are clustered in Utopia basin may be due to general conditions in this northern mid-latitude basin. They recall that "Utopia basin has been a spatially extensive site for sediment accumulation throughout the Amazonian Period. Except for a single reported example of a pingo forming in bedrock (Muller, 1959), all terrestrial pingos form in unconsolidated sediments. So, sediment accumulation appears generally required for pingo formation. In addition, Utopia basin also experienced repeated aqueous inflow during the Amazonian. Such surface aqueous flow would provide near-surface groundwater, which is also generally necessary for pingo formation."

Interestingly, candidate pingos have not been hypothesized in large, southern mid-latitude basins that experienced sedimentary infilling and aqueous inflow. According to Burr et al. (2009)'s extensive literature survey, "A limited number of PLFs have been identified in small (~10-km diameter) craters in the southern mid-latitudes (Dundas et al., 2008). Additional data analysis is required to determine the extent of PLFs in the southern hemisphere. Such future research will help to determine the veracity of this seeming north–south asymmetry in PLF occurrence and, if true, its climatologic, hydrologic, and sedimentologic implications."

7.1.11 Spotted Lake (a hypersaline sulfate lake), British Columbia, Canada—a terrestrial analog of saline water bodies of ancient *Mars*: astrobiological implications

Hypersaline brines have salinities ranging from 35 g/L to more than 400 g/L. Brines can also be highly chaotropic [disrupting the structure of, and denatures, macromolecules such as proteins and nucleic acids (e.g. DNA and RNA)], or membrane destabilizing (see Fox-Powell et al., 2016). Strong chaotropes (disorder-makers; one of two or more solutes in the same solution that can disrupt the hydrogen bonding network between water molecules and reduce the stability of the native state of proteins) such as Ca^{2+} and Mg^{2+}, when not countered by a suitable kosmotrope (order-maker; stabilizing ion), prove incredibly hostile to life. The water activity (aw) of a food is the ratio between the vapor pressure of the food itself, when in a completely undisturbed balance with the surrounding air media, and the vapor pressure of distilled water under identical conditions. The habitability of a saline environment relies heavily on water activity (aw), a function of the ionic composition and concentration of the brine (see Baldwin, 1996; Ha and Chan, 1999). Although hypersaline environments impose severe stresses on micro-organisms, such as high osmotic pressures and potentially low water activity level, (aw ~0.75) (Grant, 2004), some specialized life-forms can persist under severe osmotic stress and low water activity in hypersaline environments (see Stevenson et al., 2015). For example, analysis carried out by Pontefract et al. (2017) suggests that hypersaline-associated species occupy niches characterized foremost by differential abundance of Archaea, uncharacterized Bacteria, and Cyanobacteria. They have also discussed potential biosignatures in this environment.

Pontefract et al. (2017) investigated Spotted Lake (British Columbia, Canada) (Fig. 7.15), a hypersaline lake with extreme levels of sulfate salts as a model of the conditions thought to be associated with ancient Mars. These investigators provided the first characterization of microbial structure in Spotted Lake sediments through meta-genomic

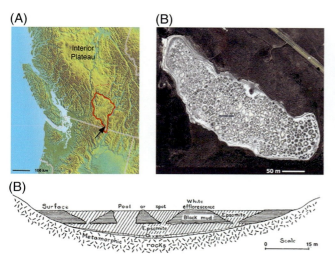

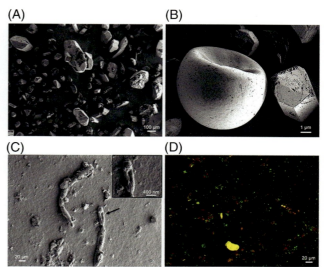

FIG. 7.15 Spotted Lake. (A) Spotted Lake (black arrow) is located on the edge of the Thompson Plateau (red line). (B) Its hundreds of brine pools are seasonally connected during periods of higher water levels, and separated during periods of low water input and evaporation. (B) Imagery 2014 DigitalGlobe, Map data 2014 Google. (C) During mining of nearby Hot Lake, it was discovered that spots represent the bases of inverted cones or cylindrical eposmite (the natural form of Epsom salt, $Mg_2SO_4 \cdot 7H_2O$, found as a crust in caves and lake deposits) masses that connect to a more basal horizontal bed underlain by gypsum. Reprinted from Fig. 7.4 of Jenkins (1918) with permission from the American Journal of Science. (Source: Pontefract A, Zhu TF, Walker VK, Hepburn H, Lui C, Zuber MT, Ruvkun G and Carr CE (2017) Microbial Diversity in a Hypersaline Sulfate Lake: A Terrestrial Analog of Ancient Mars, Front. Microbiol., 8:1819. doi: 10.3389/fmicb.2017.01819) Copyright info: 2017 Pontefract, Zhu, Walker, Hepburn, Lui, Zuber, Ruvkun and Carr. This is an open-access article distributed under the terms of the Creative Commons Attribution License (CC BY). The use, distribution or reproduction in other forums is permitted, provided the original author(s) or licensor are credited and that the original publication in this journal is cited, in accordance with accepted academic practice. No use, distribution or reproduction is permitted which does not comply with these terms.

FIG. 7.16 Scanning electron microscopy (SEM) micrographs. Representative micrographs show (A) the mineral substrate consisting of sulfate salts; (B) a brine shrimp egg (one of many hundreds seen in other micrographs), revealing the presence of higher-order biology in the system; (C) presence of in situ microbes within the soil samples; (D) Sediment-derived microbes, visualized with SYTO 9 and propidium iodide. (Source: Pontefract, A., Zhu. T.F., Walker, V.K., Hepburn, H., Lui, C., Zuber, M.T., Ruvkun, G., and Carr, C.E. (2017), Microbial Diversity in a Hypersaline Sulfate Lake: A Terrestrial Analog of Ancient Mars, Front. Microbiol., 8:1819. doi: 10.3389/fmicb.2017.01819) Copyright info: 2017 Pontefract, Zhu, Walker, Hepburn, Lui, Zuber, Ruvkun and Carr. This is an open-access article distributed under the terms of the Creative Commons Attribution License (CC BY). The use, distribution or reproduction in other forums is permitted, provided the original author(s) or licensor are credited and that the original publication in this journal is cited, in accordance with accepted academic practice. No use, distribution or reproduction is permitted which does not comply with these terms.

sequencing, and reported a bacteria-dominated community with abundant Proteobacteria, Firmicutes, and Bacteroidetes, as well as diverse extremophiles.

The high concentration of sulfates within Spotted Lake makes this one of the most hypersaline environments in the world—yet microbial life is both abundant and diverse, indicating that hyper-salinity in and of itself is not a barrier to microorganisms. Rather, habitability is likely more dependent on the water activity (a_w) and chaotropicity (the entropic disordering of lipid bilayers and other biomacromolecules, caused by substances dissolved in water) of the brine, controlled by the ionic strength and composition of the solution. In comparison to other hypersaline environments, Spotted Lake is notable for its high sequence diversity, reflected in the high abundance of "unclassified bacteria" present, as well as its low levels of Archaea (Halobacteria) (Fig. 7.16).

Pontefract et al. (2017) recall that high salt concentrations, such as those found in the Spotted Lake, preserve biosignatures and allow organic compounds, and even entire cells, to be preserved on geologic time scales (e.g., Vreeland et al., 2000; Aubrey et al., 2006). Furthermore, organisms have also been shown to exist in fluid inclusions trapped in rapidly forming salt crystals, and viable isolates have been obtained from inclusions that are on the order of 10^5 years old (e.g., Mormile et al., 2003; Fendrihan et al., 2006).

Pontefract et al. (2017) recall that "Orbital and in situ observations of Mars have revealed that extensive water flows, as well as saline and acidic fluids, were once present on the planet's surface (Tosca et al., 2008). Ancient Mars transitioned from wet to dry during the Hesperian (beginning 3.7 Ga), a time of ephemeral (lasting a very short time) lakes, resulting in the widespread deposition of sulfate and chloride salts observed today on the Martian surface (Wanke et al., 2001; Clark et al., 2005; Crisler et al., 2012; Goudge et al., 2016). Magnesium sulfate salts ($MgSO_4 \cdot nH_2O$) are common on Mars and are distributed globally, with some sediments containing 10–30% sulfate by weight (Vaniman et al., 2004; Gendrin et al., 2005). The presence of hydrated magnesium sulfates within the rim of Columbia Crater is ascribed to the existence of a paleolake, which at times must have been hypersaline in nature (Wray et al., 2011). Targets for future life-detection missions (to Mars) include such salty environments that could have once been habitable, and are relevant today because of their potential to retain water and generate liquid water brines (McEwen et al., 2011; Möhlmann and Thomsen, 2011; Chevrier and Valentin, 2012; Karunatillake et al., 2016)."

On Mars, as noted above, evidence for the past presence of saline bodies of water is prevalent and resulted in the widespread deposition of sulfate and chloride salts. Hypersaline environments have been shown to preserve biological material on timescales exceeding that of nonsaline systems, thus, the presence of paleo-sulfate lakes on Mars is of particular interest for exobiology, as these deposits may retain evidence of previous microbial habitation on the planet (Pontefract et al., 2017).

Pontefract et al. (2017) have suggested that with its high sulfate levels and seasonal freeze-thaw cycles, Spotted Lake is an analog for ancient paleolakes on Mars in which sulfate salt deposits may have offered periodically habitable environments, and could have concentrated and preserved organic materials or their biomarkers over geologic time. According to Pontefract et al. (2017), it is plausible that salt precipitation and drying-out events on Mars could result in the entrapment of high molarity fluids that still may maintain transient water activity levels high enough to support microbial activity. Any organisms in such brines would have to cope with osmotic stress, desiccation, and freeze-thaw stresses, as well as cosmic irradiation if within 1–2 m of the surface. Pontefract et al. (2017) suggest that Spotted Lake-analogous sulfate-rich closed-basin paleolakes on Mars would represent excellent locations to search for preserved organic material and associated biomarkers, which would be concentrated through evaporative processes, entombed in sulfates, and preserved within lake-bed deposits, conserving signatures of ancient life that could persist over geologic time.

7.1.12 Hypersaline springs on Axel Heiberg Island, Canadian High Arctic—a unique analog to putative subsurface aquifers on *Mars*

Terrestrial cryo-environments/cryogenic environments (i.e., low-temperature environments in which the temperature range is below the point at which permanent gases begin to liquefy) in polar regions are characterized by extremely low temperatures, limited water availability, and a thick permafrost layer. Diverse suites of cryophilic (preferring or thriving at low temperatures) microorganisms have successfully colonized these environments and play key ecological roles in carbon and nutrient cycling (Margesin and Miteva, 2011). These extreme environments, characterized by low

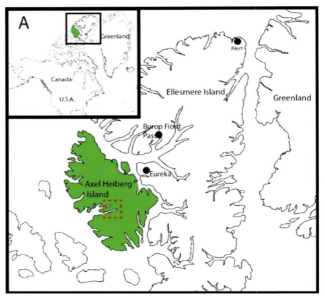

FIG. 7.17 **Geographic location of Gypsum Hill spring, Canadian high Arctic.** (A) Axel Heiberg Island (green) and Ellesmere Island, boxed area shown in inset. Expedition Fiord, Axel Heiberg is outlined by the red box. Burop Fiord Pass, Ellesmere Island is indicated. Landsat-7 image courtesy of the U.S. Geological Survey. *(Source: Sapers, H.M., Ronholm, J., Raymond-Bouchard, I., Comrey, R., Osinski, G.R. and Whyte, L.G. (2017), Biological characterization of microenvironments in a hypersaline cold spring Mars analog, Front. Microbiol., 8:2527. doi: 10.3389/fmicb.2017.02527) Copyright info: 2017 Sapers, Ronholm, Raymond-Bouchard, Comrey, Osinski and Whyte. This is an open-access article distributed under the terms of the Creative Commons Attribution License (CC BY). The use, distribution or reproduction in other forums is permitted, provided the original author(s) or licensor are credited and that the original publication in this journal is cited, in accordance with accepted academic practice. No use, distribution or reproduction is permitted which does not comply with these terms.*

temperatures and often high salinity are exceptional analogs for putative habitable environments beyond Earth (Fairén et al., 2010; McKay et al., 2012).

Axel Heiberg Island, in the high Canadian Arctic, hosts a series of perennial cold springs (Pollard et al., 1999; Omelon et al., 2001, 2006; Battler et al., 2013) (Fig. 7.17) dominated by unique microbial communities (Lay et al., 2012, 2013; Lamarche-Gagnon et al., 2015) usually associated with deep subsurface or submarine environments. It has been hypothesized that the springs originate from sub-permafrost salt aquifers and rise to the surface through the permafrost (Andersen, 2002); however, the source aquifer has not been defined and the subsurface hydrogeologic system is largely unknown. The hypersaline cold springs on Axel Heiberg Island are among the only known perennial springs flowing through thick permafrost on Earth (Andersen, 2002) and comprise a unique opportunity to study microbial diversity in cryo-environments.

The current atmospheric pressure on Mars precludes the formation of standing water at the surface; that is, at present, the surface conditions on Mars are inconsistent with the formation of standing bodies of water. Liquid water on Mars is in the form of subsurface eutectic brines in a spatially restricted hydrogeological cycle existing in thick permafrost (e.g., Martínez and Renno, 2013). Putative evidence of transient liquid water lies in observations of high-latitude seepage gullies (Malin and Edgett, 2000), recurring slope lineae (RSL) containing evidence of hydrated salts indicating briny water (Ojha et al., 2015), and somewhat more controversially, Martian slope streaks (Bhardwaj et al., 2017).

The physiochemical parameters that characterize the cryo-environments on Axel Heiberg Island, Canada, and specifically, the Gypsum Hill spring system represents a terrestrial analog for putatively habitable subsurface briny aquifers on Mars (Malin and Edgett, 2000; Andersen, 2002; McKay et al., 2012; Battler et al., 2013). The extensive diapirism characterizing the Expedition Fiord area of Axel Heiberg Island creates a series of chemical tails (long hydrophobic chains) coupling supra and sub-permafrost reservoirs allowing for continual hypersaline fluid circulation and perennial spring activity (Pollard et al., 1999; Andersen, 2002; Harrison and Jackson, 2014).

The Gypsum Hill springs, one such system on Axel Heiberg Island, at nearly 80 N, is the only known nonvolcanic, hypersaline, sulfidic, perennial cold spring system on Earth. Gypsum Hill is one of at least 6 identified perennial saline spring systems on Axel Heiberg Island. Gypsum Hill is located in Expedition Fiord approximately 3 km from the White and Thompson Glaciers on the NE side of Expedition river. The system is comprised of 30–40 small outlets concentrated over an area of approximately 3000 m^2 in a narrow band (~300 m long × 30 m wide) that perennially discharge a combined average of 15–20 L/s of anoxic [mean oxidoreduction potential (ORP) ~325 mV] brines (7.5–8% salts) at the base of Expedition Diapiar (Gypsum Hill).

Detailed studies characterizing the microbial community of Gypsum Hill spring (Perreault et al., 2007, 2008; Niederberger et al., 2009) indicate that the springs' microbial community is primarily sustained by chemolithoautotrophic primary production performed by sulfur-oxidizing bacteria, with little to no evidence of phototrophic metabolism despite being surface exposed and in continuous illumination during the Arctic summer. To date, ecosystems of this type have been found only in permanently dark hydrothermal vents and sulfidic groundwater (Perreault et al., 2008). A study of streamers (plants that stream like a flag) growing in snow-covered run-off channels forming during the winter months identified extremely limited microbial diversity again dominated by sulfur-oxidizing bacteria (Niederberger et al., 2009).

The microbial community and metabolic profile of surface waters from the Gypsum Hill springs suggests that the community may be derived from subsurface communities inoculated into the hypersaline fluids upwelling through the underlying Expedition diapir as opposed to seeding by surface-associated microorganisms (e.g., Perreault et al., 2008). Alternatively, the community may be evolved from the source water reservoir, having undergone selection during the residence time of the fluids in the subsurface. The spatial geometry of microbial mats and the physical coupling of diverse metabolisms, niche partitioning, and symbiosis allows mat communities to thrive in extreme environments (Sapers et al., 2017).

During fieldwork in the summer of 2013, Sapers et al. (2017) observed a flood plain hosting red and green microbial mats associated with the Gypsum Hill springs (Fig. 7.18). It was pointed out that an outstanding question is whether summer flood plains may support ephemeral (lasting for a very short time), seasonal microbial communities colonizing the flood plain and how much diversity these seasonal communities share with the source pool. According to Sapers et al. (2017), such a question is important in understanding the role of perennial fluid flow in permafrost environments in sustaining and/or initiating microbial activity. Sapers et al. (2017) analyzed the bacterial community in visually distinct mats in a proximal flood plain formed during the summer months and compared the results with the sediment-associated bacterial community in Gypsum Hill outlet to assess the potential contribution of surface microorganisms to the spring community. The data

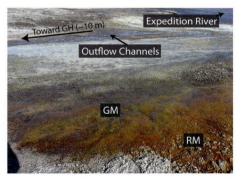

FIG. 7.18 Visually distinct red and green mats forming the Gypsum Hill flood plain in Canadian high Arctic. The red (RM) and green (GM) microbial mats sampled in this study are shown relative to the Gypsum Hill (GH) outflow channels proximal to Expedition River. Note the intermixing of the microbial mats with the red mats dominating the areas furthest away from the GH outflow channels. *(Source: Sapers, H.M., Ronholm, J., Raymond-Bouchard, I., Comrey, R., Osinski, G.R. and Whyte, L.G. (2017), Biological characterization of microenvironments in a hypersaline cold spring Mars analog, Front. Microbiol., 8:2527. doi: 10.3389/fmicb.2017.02527) Copyright info: 2017 Sapers, Ronholm, Raymond-Bouchard, Comrey, Osinski and Whyte. This is an open-access article distributed under the terms of the Creative Commons Attribution License (CC BY). The use, distribution or reproduction in other forums is permitted, provided the original author(s) or licensor are credited and that the original publication in this journal is cited, in accordance with accepted academic practice. No use, distribution or reproduction is permitted which does not comply with these terms.*

highlighted compositional differences at high taxonomic levels and provided insight to plausible mechanisms of niche differentiation.

Sapers et al. (2017) used Scanning Electron Microscopy (SEM) to evaluate cell-substrate associations in the mat samples. In their studies, imaging of GM samples revealed the presence of completely intact and undamaged diatoms as well as a massive biofilm (Fig. 7.19A and B) while specific cell-substrate associations were not observed. Sapers et al. (2017) have explained that despite several limitations, such as limited samples and the inability to fix immediately in the field, electron microscopy did reveal differences between the red and green mat samples. The dominant bacterial taxon, *Thiomicrospira*, identified through 16S rRNA targeted-amplicon sequencing in the green mat sample is not known to form biofilms (Niederberger et al., 2009) and does not have a known coccoid morphology. Sapers et al. (2017), therefore, reckon that the observed biofilm could be formed by algal cells.

Sapers et al. (2017) found that the bacterial diversity, presence of diatoms, cyanobacteria, and coloration described for the first time in the surficial mats associated with the Gypsum Hill summer flood plain is distinct from the source pool. This provides support for the hypothesis that the communities in the Gypsum Hill spring are sourced from the subsurface (Sapers et al., 2017). In Sapers et al. (2017)'s studies, it was found that the "mat communities are distinctly different from both the spring outlet and the snow-covered winter streamers suggesting temporal and spatial heterogeneity in community composition. The red mat bacterial community in particular is the first microbial community to be isolated from hypersaline cold spring systems that are not dominated by T*hiomicrospira*, but rather by *Marinobacter* (40.6%)." Sapers et al. (2017) have been able to link macro-scale visual differences to distinct microbial communities that show very different micro-scale cell-substrate associations. These investigators further found that the microbial communities of the red and green mats differ also from the reported microbial communities of the Gypsum Hill outlet (Perreault et al., 2007). Sapers et al. (2017)'s data suggests that minor variations in chemistry, even between propinquitous (in propinquity; nearby, close at hand) sites, can have significant implications for community structure. According to Sapers et al. (2017), detailed studies of emerging, transiently habitable niches in Mars analog environments will aid our understanding of the potential for microbial life, and its detection, in putatively habitable environments beyond Earth.

While results from Sapers et al. (2017)'s studies are hint at a surficial community active only during the summer months, these researchers are optimistic that future metagenomics [study of a collection of genetic material (genomes) from a mixed community of organisms recovered directly from environmental samples] and metatranscriptomic (study of gene expression of microbes within natural environments, i.e., the metatranscriptome) analyses would lead to a better understanding of the metabolic potential and active members inhabiting these mat communities. They further reckon that a deeper assessment of the mat cyanobacterial diversity would allow for a more thorough comparison to other mat communities and replicate analyses are required to better understand seasonal variability and stability.

On Axel Heiberg Island, the perennial cold spring such as Gypsum Hill represent the surface expression of subsurface briny aquifers flowing through thick permafrost. Sapers et al. (2017)'s results suggest that the communities within these springs are distinct from the transient surficial communities that form in flood plains during the summer

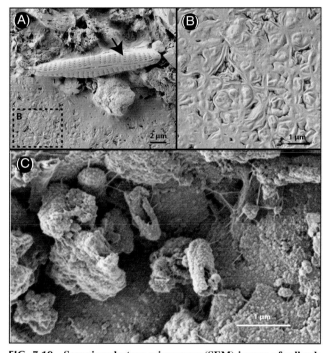

FIG. 7.19 Scanning electron microscopy (SEM) images of cell-substrate interaction observed in Gypsum Hill spring, Canadian high Arctic. (A) Intact diatom (black arrow) and a desiccated coccoid biofilm (box B) dominated the GM samples. (B) The texture of the biofilm in GM samples is a result of desiccated and puckered coccoid cells. (C) Rod shaped cells adhered to the substrate in RM samples. Notice the rough surface textures of the cells consistent with biomineralization processes. *(Source: Sapers, H.M., Ronholm, J., Raymond-Bouchard, I., Comrey, R., Osinski, G.R. and Whyte, L.G. (2017), Biological characterization of microenvironments in a hypersaline cold spring Mars analog, Front. Microbiol., 8:2527. doi: 10.3389/fmicb.2017.02527) Copyright info: 2017 Sapers, Ronholm, Raymond-Bouchard, Comrey, Osinski and Whyte. This is an open-access article distributed under the terms of the Creative Commons Attribution License (CC BY). The use, distribution or reproduction in other forums is permitted, provided the original author(s) or licensor are credited and that the original publication in this journal is cited, in accordance with accepted academic practice. No use, distribution or reproduction is permitted which does not comply with these terms.*

months. According to Sapers et al. (2017), "understanding the geomicrobiology and microbial activity in hypersaline cold springs like Gypsum Hill shed insight into the potential habitability of putative subsurface Martian brines. The development of the mat communities demonstrates the potential for ephemeral environments linked to seasonal melt events, and/or flooding to host diverse and unique microbial communities."

The finding that the surficial complex microbial communities in the hypersaline springs on Axel Heiberg—a unique analog to putative subsurface aquifers on Mars—exist in close proximity to perennial springs is encouraging in the search for life on Mars. The Martian subsurface represents the longest-lived potentially habitable environment on Mars and a better understanding of the microbial communities on Earth that thrive in analog conditions will help direct future life detection missions (Sapers et al. (2017).

7.1.13 Subglacial Lake Vostok in Antarctica and Mariana Trench in the Pacific Ocean—terrestrial analogs of Jupiter's Moon *Europa*

The cold, arid, remotely located, and perennially ice-covered Antarctic has long been considered an analog to how life might persist in the barren, frozen landscape of Mars and Jupiter's moon Europa (Abyzov et al., 1998;1999). Due to the presence of a subsurface ocean, Europa has been considered to be one of the most promising targets in the search for extant extraterrestrial life within the Solar System. Deep ocean basins such as the Mariana Trench are good Earth analogs for the cold, high-pressure ocean of Europa.

Many of the best terrestrial analogs for potential Europan habitats are in the Arctic and Antarctica. For example, a possible analog to the planet Jupiter's moon Europa's subsurface environment is *Lake Vostok* in Antarctica—the largest and deepest known subglacial lake and likely the oldest in Antarctica (Siegert et al., 2001). Lake Vostok (meaning "Lake East") is the largest of Antarctica's almost 400 known subglacial lakes. Lake Vostok (Fig. 7.20) in Antarctica is located beneath Russia's Vostok Station under the surface of the central East Antarctic Ice Sheet. Similarities can be related with physicochemical conditions, such as dryness, low water activity, mineralogy, or chemical composition of the environment. A subsurface ocean under an ice layer applies approximately in the case of Lake Vostok and Europa (https://link.springer.com/referenceworkentry/).

Karl et al. (1999) found that numerous bacteria live in ice core at depths exceeding 1500 m in the Lake Vostok. D'Elia et al. (2008) isolated microbes from Lake Vostok accretion ice, and the assembly of microbes that was found in the Lake Vostok accretion ice samples indicates that the lake sustains a diverse population of microorganisms and potentially a complex ecosystem, although the concentrations of microbes are expected to be lower than those in

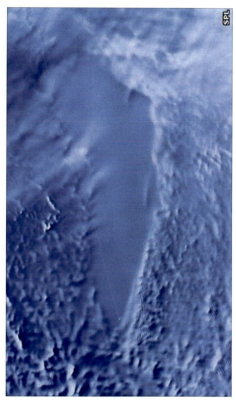

FIG. 7.20 RADARSAT image of Lake Vostok in Antarctica. *Image by Goddard Space Flight Center—NASA. Copyright info: The file containing this figure is in the public domain in the United States because it was solely created by NASA. NASA copyright policy states that "NASA material is not protected by copyright unless noted." (See Template:PD-USGov, NASA copyright policy page or JPL Image Use Policy Public Domain, https://commons.wikimedia.org/w/index.php?curid=12841168.*

most environments on Earth. Rampelotto (2010) considers that life could exist within Europa's sub-glacial ocean, perhaps subsisting in an environment similar to Earth's deep-ocean hydrothermal vents or the Antarctic Lake Vostok. No wonder, in studies examining the utility of analog environments in preparations for a Europa lander mission, it was found that Antarctic and Arctic analog sites are useful in the demonstration and rehearsal of engineering functions such as sample acquisition from an icy surface, as well as in the exercise of the scientific protocols needed to identify organic, inorganic and possible biological impurities in ice (see Lorenz et al., 2011).

7.1.14 Methanogens and ecosystems in terrestrial volcanic rocks—terrestrial analogs to assess the plausibility of life on Saturn's Moon *Enceladus*

To assess the plausibility of life on Saturn's icy moon *Enceladus*, planetary researchers look at the fundamental requirements of life on Earth (McKay et al., 2008) and consider studies done to assess the possibility of life on other planets with ice-covered oceans such as Europa (Gaidos et al., 1999). The jets of icy particles and water vapor issuing from the south pole of Enceladus (Porco et al., 2006) are evidence for activity driven by some geophysical energy source. The vapor has also been shown to contain simple organic compounds (Postberg et al., 2018), and the south polar terrain is bathed in excess heat coming from below (Parkinson et al., 2007).

Based on the inferences gleaned from several studies that a possible liquid water environment exists beneath the south polar cap of Enceladus, McKay et al. (2008) argued that such an environment may be conducive to life. McKay et al. (2008)'s arguments connecting the possibility for a conducive environment for existence of life on Enceladus based on the existence of life on Earth are the following: "Several theories for the origin of life on Earth would apply to Enceladus. These are (1) origin in an organic-rich mixture, (2) origin in the redox gradient of a submarine vent, and (3) panspermia.

There are three microbial ecosystems on Earth that do not rely on sunlight, oxygen, or organics produced at the surface and, thus, provide analogs for possible ecologies on Enceladus (McKay et al., 2008). According to McKay et al. (2008), "Two of these ecosystems are found deep in volcanic rock, and the primary productivity is based on the consumption by methanogens of hydrogen produced by rock reactions with water. The third ecosystem is found deep below the surface in South Africa and is based on sulfur-reducing bacteria consuming hydrogen and sulfate, both of which are ultimately produced by radioactive decay. Methane has been detected in the plume of Enceladus and may be biological in origin. An indicator of biological origin may be the ratio of nonmethane hydrocarbons to methane, which is very low (0.001) for biological sources but is higher (0.1–0.01) for nonbiological sources."

It has been found that to date, methanogenic archaea are the only known microorganisms that are capable of performing biological methane (CH_4) production in the absence of oxygen (see Liu and Whitman, 2008; Taubner et al., 2015). On Earth, methanogens are found in a wide range of pH (4.5–10.2), temperatures (<0–122°C), and pressures (0.005–759 bar) (Taubner et al., 2015) that overlap with conditions predicted in Saturn's tiny moon Enceladus' subsurface ocean, that is, temperatures between 0 and above 90°C (Hsu et al., 2015), pressures of 40–100 bar (Hsu et al., 2015), a pH between 8.5–10.5 (Hsu et al., 2015) and 10.8–13.5 (Glein et al., 2015), and a salinity in the range of Earth's oceans. Taubner et al. (2018) showed that methanogens can produce CH_4 under Enceladus-like conditions, and that the estimated H_2 production rates on this icy moon can potentially be high enough to support autotrophic [requiring only carbon dioxide or carbonates as a source of carbon and a simple inorganic nitrogen compound for metabolic synthesis of organic molecules (such as glucose)], hydrogenotrophic (converting hydrogen to other compounds as part of their metabolism) methanogenic (producing methane as a by-product of energy metabolism) life.

7.1.15 Bubbles bursting in Earth's oceans—terrestrial analogs in studying the transport of organics from *Enceladus*'s subsurface ocean to its surface and along its south polar jet

Formation of aerosol from bubbles, which are covered by an organic microlayer, bursting is a well-studied process in Earth's oceans (deLeeuw et al., 2011). On Earthly water bodies bubbles are generated mostly by breaking of waves (Wilson et al., 2015). In Earth's oceans, bubbles generated by both "surface waves" and "internal waves" (see Joseph, 2016) not only produce aerosols, but are also very efficient in "harvesting" organic molecules from the deeper oceanic environment by collecting these substances on their surfaces while ascending. The increase of relative organic concentrations observed near the surface of Earth's oceans by 2 to 3 orders of magnitude (Russell et al., 2010; Burrows et al., 2014; Jayarathne et al., 2016) suggests a selective transport of organic matter from the bulk seawater to the water table and then from the microlayer to atmospheric aerosols (Schmitt-Kopplin et al., 2012; Gantt and Meskhidze, 2013; Wilson et al., 2015). Bubbles ascend through tens of kilometers of ocean before reaching the surface. Organic-bearing sea spray serves as highly efficient nucleation cores of ice clouds over polar waters on Earth (Wilson et al., 2015) and is found preferentially in the smallest aerosols between 40 nm and 250 nm in size (Leck and Bigg, 2008; deLeeuw et al., 2011), whereas larger aerosols with sizes between 500 nm and 1000 nm are either organics mixed with salt or are entirely without organics (Gantt and Meskhidze, 2013).

Surprisingly large plume of gas, tall jets of water vapor and organic-enriched salty ice particles spewing from the "Tiger Stripes" in the South Pole of Saturn's tiny moon *Enceladus* is an interesting event that appeals every space enthusiast and even lay people. In assessing conditions in the *Enceladus*'s jetting-plume and implications for future spacecraft missions to *Enceladus*, study of the role of bubbles in transporting organics from Enceladus's subsurface ocean to the ice moon's surface and further along the jet through the water table located inside water-filled cracks in Enceladus's ice crust became important. Several volatile gases have been detected in substantial concentrations in the plume of Enceladus (CO_2, CH_4, NH_3, and H_2) (Waite. Jr. et al., 2017) that will inevitably create bubbles when rising though the ocean, which then burst at the water table (Postberg et al., 2018). No wonder, a terrestrial analog that is used to explain the transport of organic materials and salt particles from below the water-filled cracks (tiger stripes) on the South Polar Region on Enceladus is air bubbles.

In order to deduce the mechanism that enables the presence of an organic enriched layer at the Enceladean water table located inside the tiger stripes in the Enceladean ice crust, Postberg et al. (2018) invoked a scenario, well known from ice cloud formation over polar waters on Earth (Wilson et al., 2015). There, organic aerosols of mostly biogenic origin (de Leeuw et al., 2011) thrown up by bubble bursting serve as highly efficient nucleation seeds. When bubbles burst on Earth's ocean, an organic-free sea spray forms in parallel with pure organic aerosols and mixed-phase organic-bearing sea spray (Gaston et al., 2011; Gantt and Meskhidze, 2013). In a similar process, ascending gas bubbles from the subsurface ocean—inferred to be present below the tiger stripes on Enceladus—efficiently transport organic material (Porco et al., 2017) into the water-filled cracks in the Enceladean south polar ice crust. Organics ultimately concentrate in a thin organic layer on top of the water table located inside the icy vents. When gas bubbles present in the Enceladean tiger stripe burst, they form aerosols made of insoluble organic material that later serve as efficient condensation cores for an icy crust from water vapor thereby forming high mass organic cations (HMOC)-type particles found in the tall jets of water vapor and other gases & ice grains spewing from Enceladus' South Pole.

The organic mass fraction of sea-spray aerosol has been consistently shown to be inversely related to aerosol size (Keene et al., 2007; Facchini et al., 2008; Gantt and Meskhidze, 2013) and is mostly water-insoluble (Facchini et al., 2008; deLeeuw et al., 2011). The purely organic endmembers (organic materials disseminated in rocks) are found preferentially in the smallest aerosols (Leck and Bigg, 2008; deLeeuw et al., 2011).

Under low gravity conditions of *Enceladus* (relative to those of Earth), one would expect larger gas bubbles and therefore somewhat larger film drops. Mass spectra showing organic species with an increasing number of carbon atoms—the so-called High Mass Organic Cations (HMOC) likely representing molecules with 7–15 carbon atoms—are generated by ice grains with radii around 1μm. Insoluble organic nucleation cores, a few hundred of nm in size, would naturally explain the high organic content in HMOC grains. Like in Earth's oceans, organic substances can accumulate efficiently on the bubble walls (Porco et al, 2017), thus probing the oceanic organic inventory at depth.

These kinds of knowledge gained from the study of aerosol formation process from bubble bursting in Earth's oceans have been successfully used by planetary scientists in the fascinating discovery of the presence of complex organic compounds in *Enceladus*'s subglacial water ocean (see Postberg et al., 2018).

7.1.16 Haughton Crater in the Canadian Arctic Desert—a terrestrial analog for the study of craters on *Mars* and Saturn's Moon *Titan*

To identify an appropriate sampling site on Saturn's moon *Titan* in an attempt for detecting biological molecules on this moon, Neish et al. (2018) considered a relevant terrestrial analog: Haughton crater in the Canadian Arctic. The 39 Ma old Haughton impact structure is a well-preserved 23-km diameter crater in a polar desert, with little to no obscuring vegetation (Osinski et al., 2005; Tornabene et al., 2005). Thus, it is an excellent analog for the study of craters on worlds that have experienced moderate amounts of erosion, such as *Mars* or *Titan*.

7.1.17 Terrestrial Salt Diapirs & Dust Devils—terrestrial analogs to study cantaloupe terrain & geyser-like plumes on the surface of Neptune's moon *Triton*

Triton is the largest of Neptune's 13 moons. This moon is unusual because it is the only large moon in our Solar System that orbits in the opposite direction of its planet's rotation—a retrograde orbit. Scientists think Triton is a Kuiper Belt Object captured millions of years ago by Neptune's gravity.

In a terrestrial analog example pertaining to Triton, an infamous pattern called "cantaloupe terrain" (an organized cellular pattern of noncircular dimples) on the surface of Triton shows up some similarity with terrestrial "salt diapirs" exhibiting some kind of patterns (e.g., polygon) on the surface of the Earth. These terrestrial surface patterns are natural regularities defined by stones, ground cover, or topography that assume forms such as circles, stripes, and polygons in water-laden soil on Earth that undergoes repeated (seasonal or daily) freeze-thaw cycles. Some of these patterns stretch over square kilometers while others less than a square meter. The pattern ground on Earth results from the free convection of water. The onset of convection is induced by that strange character of water that is densest above melting point (densest at 4°C (277 K)). The analogy may be of interest in spite of the fact that water obviously cannot cause the small-scale structure on Triton because of its low surface temperature (37–39 K). Planetary scientists consider that, in drawing a comparison with Earth, the "circulating fluid" in the case of Triton's cantaloupe terrain may be nitrogen, and the "soil" could be a mixture of ammonia/methane/water-ice/clathrate (Ilés-Almar, 1992).

An example of the closest terrestrial analog to the geyser-like plumes (carrying clouds of ice and dark particles into the atmosphere) observed on Triton, may be "dust devils" developed on Earth's surface, which are atmospheric vortices originating in the unstable layer close to the ground. Patches of unfrosted ground near the subsolar point on

Triton could act as sites for dust devil formation because they heat up relative to the surrounding nitrogen frost (Ingersoll and Tryka, 1990). Details on these interesting phenomena observed on Triton are given in Chapter 15.9.

7.1.18 Earth's biosphere—terrestrial analog in the search for life on *Exoplanets*

In the context of studying signs of life on exoplanets, Schwieterman et al. (2018) turned to Earth's biosphere, both in the present and through geologic time, for analog signatures that will aid in the search for life elsewhere in the Universe. Considering the insights gained from modern and ancient Earth, and the broader array of hypothetical exoplanet possibilities, they compiled a comprehensive overview of our current understanding of potential exoplanet biosignatures, including gaseous, surface, and temporal biosignatures.

Importance of terrestrial analogs in studying extraterrestrial biosignatures is beyond dispute. Meinert et al. (2016) demonstrated that numerous prebiotic molecules can be formed in an interstellar-analog sample containing a mixture of simple ices of water, methanol, and ammonia.

7.2 Role of submarine hydrothermal vents in the emergence and persistence of life on Earth—their astrobiological implications

Submarine hydrothermal vents, as the name suggests, are naturally forming structures protruding out over fissures on the seafloors. In most cases, the vents expel a fluid that was geothermally heated to extreme temperatures when seeping through the Earth's crust. A submarine hydrothermal vent was discovered for the first time in 1977 at a depth of 2500 m along the Galapagos Rift spreading center (Lonsdale, 1977).

Some of the mysteries about the origins of life still persist. There are theories that life originated in hydrothermal vents which exist at a depth of 4–5 km in the ocean. Several countries are engaged in conducting oceanographic studies in general, and some organizations have focused their attention to the study of submarine hydrothermal vents. For example, India will soon scour the ocean bed to unravel the mysteries of the origins of life as scientists are set to travel up to 6000 m below the sea surface under a deep ocean mission (DOM). Initially, the mission will entail scientists traveling to a depth of 500 m to test various technologies being developed for the purpose before taking a deeper dive into the unknown. It is completely dark at the depth of 4–5 km, but there are living organisms. How is life born at that depth, how does life survive? The deep ocean mission will also help us understand this.

7.2.1 General features of terrestrial submarine hydrothermal vents

Submarine hydrothermal vents usually occur on divergent plate boundaries, where tectonic plates are moving apart. Such vents are akin to hot springs on the seafloor. As tectonic plates of oceanic crust move apart, the crust is stretched, and in places it breaks, forming cracks and fissures within it. Seawater then percolates into these cracks and seeps deep into the crust, where it comes into close physical contact either with the underlying magma chambers or with the deeper mantle and is heated. As the heated seawater circulates through the crust beneath the seafloor, it picks up dissolved gases and minerals/chemicals and gains heat until there is enough heat for the fluids to rise buoyantly and vent back into the ocean. In other words, the high temperature of the fluid makes it buoyant, and the superheated water eventually begins to move back up to the seabed where it is expelled through a vent. When the hot fluid erupts from the seafloor, it starts to mix with the colder surrounding seawater. As the warm fluids mix with cold seawater, the chemicals separate from the vent fluids (precipitate), and solidify, sometimes piling up into impressive, short & tall, solid structures, mounds, spires, and chimneys of minerals.

As already indicated, underwater volcanoes at spreading ridges and plate boundaries produce hot springs known as hydrothermal vents. Submarine hydrothermal vents occur on the seafloor spreading zones and have a global distribution (Baker and German, 2004). Such vents are commonly found near volcanically active places, areas where tectonic plates are moving apart, and where new crust is being formed. Hydrothermal activity, accounting for water temperature elevations ranging from a few degrees above ambient to 350–400°C, is known to occur at seafloor spreading centers worldwide (Juniper et al., 1990).

7.2.1.1 *Magma-chambers-fed black smoker vents & white smoker vents*

Most submarine hydrothermal vent fields studied until the late 1990s are characterized by basalt-hosted black smoker sulfide structures on or very near the ridge axis. For this reason, such vents are known as "axial vents." On the basis of the color of the ejecta (mineral-laden fluid and gas) emitted from submarine hydrothermal vents, they are often classified into two categories, namely: "black smokers" and "white smokers." The mineral-laden fluid is emitted either as a warm (5–100°C), diffuse flow from seabed cracks or as plumes of superheated water (250–400°C) from chimney-like structures.

"Black smokers" (Fig. 7.21) are chimneys formed from deposits of iron sulfide, which is black. These chimneys are located directly above magma chambers that are found 1–3 km beneath the seafloor (Kelley et al., 2002). Black smoker chimneys emit hot (up to 405°C), chemically

FIG. 7.21 A black smoker chimney venting mineral-laden water at temperatures of 350 to 400°C and velocities of 1 to 5 m/s. Photo by the Woods Hole Oceanographic Institution. *(Source: Lutz, R.A., and M.J. Kennish (1993), Ecology of deep-sea hydrothermal vent communities: A review, Reviews of Geophysics, 31(3), 211-242). Copyright 1993 by the American Geophysical Union.*

modified seawater (Von Damm et al., 2003). Beneath the fissured seafloor, downwelling seawater comes into close contact with the magma chamber during its circulation from the ocean floor, before moving through the crust to re-emerge at the vents.

The light appearance of "white smokers" chimneys (Fig. 7.22) is due to the minerals carried, which can include deposits of barium, calcium, and silicon, which are white. When these minerals precipitate, they appear white. White smokers typically occur at lower temperatures.

At spreading zones, "magma chambers" that contain molten rock (800–1,200°C) discharge lavas onto the ocean floor over time periods that range from <10 years between eruptions to >50,000 years between eruptions (Hammond, 1997). The close proximity of heat-laden magma chambers to the seafloor, in conjunction with tectonic plate movement, deformation, and cooling of newly created oceanic crust, results in convective circulation of dense, cold seawater through the cracked and fissured upper portions of the lithosphere (Sleep et al., 1983; Edmond, 1986; Rona, 1987; Eberhart et al., 1988; Haymon, 1989; Haymon et al., 1991; Sleep, 1991). This circulation promotes the formation of numerous hot springs that emanate from vents along seafloor spreading centers. Most high-temperature hydrothermal activity takes place within the neovolcanic zone [zones of recent volcanic activity; from the Tertiary period (up to about 65 million years ago) to the present time (contrasted with palaeovolcanic)] along axial summit grabens (elongated blocks of the earth's crust lying between two faults and displaced downwards relative to the blocks on either side, as in a rift valley) or calderas (large volcanic craters, especially those formed by major eruptions leading to the collapse of the mouth of the volcanos) of the mid-ocean ridge crest (Macdonald, 1982; Haymon and Macdonald, 1985; Haymon, 1989; Haymon et al., 1991).

Because the black smoker systems are fuelled by volcanoes, black smoker fluids commonly contain high concentrations of magmatic CO_2 (4–215 mmol per kg), H_2S (3–110 mmol per kg) and dissolved H_2 (0.1–50 mmol per kg), with varying amounts of CH_4 (0.05–4.5 mmol per kg) that is formed both through biogenic (produced or brought about by living organisms) and abiogenic (not produced or brought about by living organisms; of inorganic origin) processes (Kelley et al., 2002). A range of temperatures exist, from the hot interior of black smokers to the interface with cold (2°C), oxygenated sea-water (Fig. 7.23B). Effluent at black smokers is typically acidic (pH 2–3) and rich in dissolved transition metals (Von Damm, 1995), such as Fe(II) and Mn(II) (Martin et al., 2008).

It has been found that hot black-smoke-vomiting submarine hydrothermal vents are mostly found along the axis of mid-oceanic ridges (long, narrow submarine hilltop, mountain ranges), which are seafloor spreading zones and have a global distribution. Ridges are found in all the oceans, running to a total length of approximately 75,000 km. Naturally, most of the vent systems discovered thus far are located on the axis of ridges. However, vent systems have been discovered at almost all seafloor locations that have been studied in detail (Baker and German, 2004).

The dissolved gases and metals in black smokers and the associated diffuse flow systems produced as a result of eruptions fuel microbial communities that serve as the base of the food chain in these ecosystems and, consequently host dense and diverse biological communities (Delaney et al., 1998; Embley and Lupton, 2004). Some of the archaea in black smokers can replicate at temperatures up to 121°C (Kashefi and Lovley, 2003), which is currently thought to

FIG. 7.22 (left) These white chimneys are around 50 cm tall and release a fluid that is around 103°C. *Copyright info: NOAA Image Library via Flickr (CC BY 2.0)* (right) The vents release hot fluid that was superheated in the Earth's crust and has mixed with dissolved gases and minerals. *Copyright info: NOAA Photo Library via Flickr (CC BY 2.0 [inline image] https://creativecommons.org/licenses/by/2.0/.*

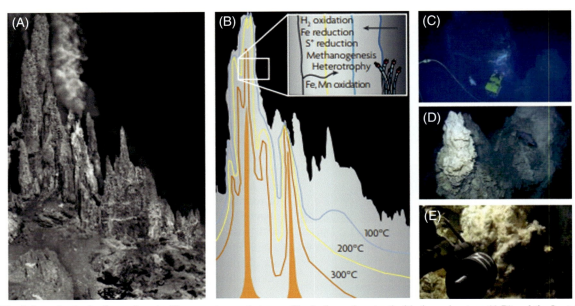

FIG. 7.23 **Submarine hydrothermal vents.** There are two main types of hydrothermal vents: the black smoker type (A,B) and the Lost City type (C–E). (A) A black smoker in the Faulty Towers complex in the Mothra hydrothermal field on the Endeavour Segment of the Juan de Fuca Ridge. The tallest chimney rises 22 m above the sea-floor. The "furry" appearance of the chimneys reflects the fact that the chimney walls are encrusted in dense communities of tube worms, scale worms, palm worms, sulphide worms and limpets. (B) The outer surface of black smoker chimneys is bathed in a mixture of 2°C, oxygenated sea-water and warm vent fluid that escapes from within the structure. The inner walls that form the boundary of the central up-flow conduits commonly exceed 300°C, and temperatures are fixed by a steady supply of rapidly rising, strongly reducing vent fluid. Intermediate conditions exist as gradients between these extremes. (C) Microbial sampling at the Lost City hydrothermal field. White areas are active or recently active sites of venting. (D) The top left part of the Nature Tower. Actively venting edifices are composed of aragonite ($CaCO_3$) and brucite ($Mg(OH)_2$). The grey–brown material also contains carbonate, but is richer in calcite that has recrystallized from aragonite. (E) A close up of a 75°C, diffusely venting carbonate chimney showing a titanium water sampler. Dense colonies of filamentous bacteria thrive in the high pH, CH_4-, and H_2-rich fluids. Parts [A,B] courtesy of D. Kelley and J. Deloney, University of Washington, USA. Images [C–E] courtesy of D. Kelley, Institute for Exploration, University of Rhode Island, USA, and the National Ocean and Atmospheric Administration Office of Ocean Exploration. *(Source: Martin, W., Baross, J., Kelley, D., and Russell, M.J. (2008), Hydrothermal vents and the origin of life, Nat Rev Microbiol, 6:805–814. Article in Nature Reviews Microbiology October 2008: DOI: 10.1038/nrmicro1991 2008 Macmillan Publishers Limited. All rights reserved.)*

be the upper-temperature limit of life. The thermal regime and permeability field in the new oceanic crust (both of which are believed to be strongly linked to magmatic and tectonic activity on the ridge crest) are the parameters that govern hydrothermal flow, and consequently the distribution of organisms, at seafloor spreading centers (Lutz and Kennish, 1993).

Three general types of hydrothermal vents have been recognized based on temperature: (1) diffuse vents, emitting low-temperature, clear waters up to 30°C; (2) white smoker vents, releasing milky fluids with temperatures ranging from 200 to 330°C; and (3) black smoker vents, discharging jets of water blackened by sulfide precipitates at temperatures between 300 and 400°C (Macdonald et al., 1980; Macdonald, 1983; Rona, 1984, 1988; Haymon and Macdonald, 1985; Rona et al., 1986). In low-temperature diffuse-venting systems, as exemplified by Galapagos-style vents, heated fluids emerge unrestricted from cracks and fissures around basalt pillows.

Note that pillow basalt is a type of Basalt in which lava erupts underwater or flows into the sea. Contact with the water quenches the surface and the lava forms a distinctive pillow shape, through which the hot lava breaks to form another pillow. This "pillow" texture is very common in underwater basaltic flows and is diagnostic of an underwater eruption environment when found in ancient rocks. Pillows typically consist of a fine-grained core with a glassy crust and have radial jointing. The size of individual pillows varies from 10 cm up to several meters.

Black smokers, in contrast, are often constrained by sulfide-rich chimney structures that rise up to 50 m in height (Lutz and Kennish, 1993). Gaill and Hunt (1991) described the fourth kind of smoker, the basal mound variety, characterized by sulfide-cemented biogenic tubes. The most common type of basal mound smoker found on the East Pacific Rise at 13 N is built of Alvinella tubes cemented together by iron sulfide associated with silica. By comparison, the most commonly occurring basal mound smoker at the hydrothermal vent region of the Juan de Fuca Ridge (46 N) in the northeast Pacific is composed of zinc sulfide-encrusted vestimentiferan worm tubes with traces of lead and arsenic sulfides. Iron, zirconium, lead, and arsenic sulfides compose the periphery of these mounds, as is the case in white smokers (Lutz and Kennish, 1993).

In the case of terrestrial submarine hydrothermal vents, the emerging water can approach 350°C and is prevented from boiling only by the crushing pressure of the deep water. However, the water cools rapidly as it mixes with the frigid ocean around it. The water's carrying capacity (defined as the maximum sustainable population and socioeconomic scale that the water environment can support in a specific region for some period of time without obvious adverse effect on the local water environment) drops and many of the dissolved minerals instantly precipitate out. This produces a great cloudy plume of dark particles, which gives these vents their nickname "black smokers." The effluent at black smokers is rich in dissolved transition metals, such as Fe^{2+} and Mn^{2+}, and contains high concentrations of magmatic CO_2, H_2S, and dissolved H_2, with varying amounts of CH_4 (Kelley et al., 2002). The dissolved gases and metals in black smokers fuel the microbial communities serving as the base of the food chain in these ecosystems (Takai et al., 2001). Rona et al. (1987) investigated serpentinite outcrops near the 15200 Fracture Zone at the Mid-Atlantic Ridge (MAR) and found evidence of hydrothermal activity influenced by serpentinization reactions. Serpentinization delivers, and has always delivered, a substantial amount of H_2 as a source of electrons for primary production in submarine ecosystems. No wonder, submarine hydrothermal vents are geochemically reactive habitats that harbor rich microbial communities.

7.2.1.2 Serpentinite-hosted carbonate chimneys—Lost City carbonate structures

The *Lost City* hydrothermal vent structures (vent field), taller than any seen before, were serendipitously discovered on Dec. 4, 2000, in the mid-Atlantic Ocean near the Mid-Atlantic Ridge (MAR), south of the Azores, during a US National Science Foundation (NSF)-funded expedition. The field was named *Lost City* with passing reference to the mythical "Lost City of Atlantis" mentioned in the prestigious literary work *The Republic*, authored by the Greek philosopher Plato around 375 BC. Also, this Hydrothermal Field sits on top of a submerged seamount "Atlantis," so named because it was discovered during an expedition aboard the research vessel *Atlantis*, operated by the Woods Hole Oceanographic Institution (WHOI) in the US, and led by Scripps Institution of Oceanography's Donna Blackman, Washington's Kelley and Duke University's Jeffrey Karson.

The *Lost City* hydrothermal vent field is formed in a very different way than the ocean-floor vents studied since the late 1970s. This new class of hydrothermal vents apparently forms where circulating seawater reacts directly with mantle rocks, as opposed to where seawater interacts with basaltic rocks from magma chambers beneath the seafloor. Note that as the Earth cooled after its formation, the heavier, denser materials sank to the center and the lighter materials rose to the top. Because of this phenomenon, the crust is made of the lightest materials (rock-basalts and granites) and the core consists of heavy metals (nickel and iron). The mantle is the layer of the earth located below the lighter crust and above the denser core. Uniquely, at the *Lost City* hydrothermal vent field location, the process of serpentinization (in which seawater encounters peridotite rock from the Earth's mantle) generates calcium-rich water that reacts with carbon in seawater to create tall chimney structures.

The "Lost City" Hydrothermal Field (LCHF) at 30 N—a completely new type of vent system discovered in late 2000, as indicated earlier—is an off-axis, moderate temperature (40–91°C), high-pH (9–10.8), serpentinite-hosted vent system, located at a water depth of 750–850 m near the summit of a 4000-m mountain named the Atlantis Massif (Fig. 7.24) that sits on 1.5–2-million-year-old crust, at a water depth of ~750 m (Kelley et al., 2001; Blackman et al., 2002).

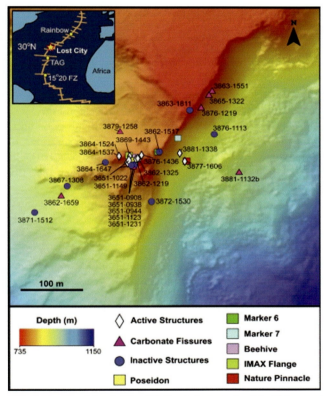

FIG. 7.24 **The Atlantis Massif is located approximately 15 km west of the spreading axis of the Mid-Atlantic Ridge.** The Lost City Field is located on the southern face of the massif on variably altered and deformed mantle rocks with lesser gabbroic material that is 1.5 My in age (Blackman et al., 2002). The massif rises from a water depth of 6000 m to 700 m over a horizontal distance of 20 km. Markers denote specific features within the field. Locations of most carbonate samples examined in this study are classified (active, inactive, fissure) based on field observations. Samples 3873-1233 and 3881-1338 are beyond the western and southern boundaries of this map, respectively. *(Source: Ludwig, K.A., D.S. Kelley, D.A. Butterfield, B.K. Nelson, and G. Fruh-Green (2006), Formation and evolution of carbonate chimneys at the Lost City Hydrothermal Field, Geochimica et Cosmochimica Acta, 70: 3625–3645) © 2006 Elsevier Inc. All rights reserved.*

FIG. 7.25 An underwater spire: the top few feet of an actively venting carbonate chimney at Lost City. Credit: NSF, NOAA, University of Washington. Image courtesy: National Science Foundation. (Source: Johnson, D.E. (2019), Protecting the lost city hydrothermal vent system: All is not lost, or is it?, Marine Policy, 107: 103593) 2019 Johnson, D.E Published by Elsevier Ltd. This is reproduced from an open access article under the CC BY-NC-ND license (http://creativecommons.org/licenses/BY-NC-ND/4.0/).

Possessing intricate and delicate structures (see Fig. 7.25), and spewing out high temperature (90°C), metal-poor, alkaline (pH 9–11) fluids containing high concentrations of nonbiogenic hydrogen and methane (Johnson, 2019), the "Lost City" is considered to be the oldest among the hitherto known hydrothermal systems.

A carbonate chimney is unlike the familiar "magma-chambers-fed" hydrothermal vent systems in a number of ways. For example, whereas the magma-chambers-fed hydrothermal vent systems are mostly located on or close to the axis of submarine ridges (those found at "triple junctions" are exceptions), the mountains of the Lost City-like systems are tens of kilometers off-axis, rarely contain volcanic rocks, and are formed by sustained fault activity that has lasted for millions of years (Kelley et al., 2001, 2005; Karson et al., 2006).

Carbonate vents are formed by a process known as serpentinization. Seabed rock, in particular, olivine (magnesium iron silicate) reacts with water and produces large volumes of hydrogen. In the Lost City, when the warm alkaline fluids (45–90°C and pH 9–11) are mixed with seawater, they create white calcium carbonate chimneys 30–60 m tall. In contrast to high-temperature black smoker-type chimneys consisting of Cu–Fe sulfides, the Lost City chimneys are composed of brucite and calcium carbonate (aragonite, calcite), which form when serpentinization fluids (SF) mix with seawater (SW) (Ludwig et al., 2006). Nascent Lost City chimneys are dominated by aragonite and brucite, whereas older structures are dominated by calcite. During aging, brucite undersaturated in seawater dissolves and aragonite recrystallizes to calcite (Ludwig et al., 2006):

$$Ca^{2+} + OH^- + HCO_3^- = CaCO_3 + H_2O$$
SF   SF   SW

$$Mg^{2+} + 2OH^- = Mg(OH)_2$$
SW   SF

The carbonate chimney vents are nearly 100 percent carbonate, the same material as limestone in caves, and range in color from a clean white to cream or grey, in contrast to black smoker vents that are a darkly mottled mix of sulfide minerals. Furthermore, the height attained by some of the structures (for example, the mighty 180-foot vent scientists named Poseidon) is considerably larger than the previously studied vents that reach 80 feet or less. Hydrothermal circulation appears to be driven by seawater that permeates into the deeply fractured surface and transforms olivine (magnesium iron silicate) in the mantle rocks into a new mineral, serpentine, in a process called serpentinization.

Fluid circulation within the massif is driven by convection that dissipates heat from the underlying mantle rocks, and perhaps, in part, by exothermic chemical reactions between the circulating fluids and host rocks. These rocks have different compositions compared with those of submarine volcanoes, because they are dominated by the magnesium- and iron-rich mineral olivine and because they have lower silica concentrations. This geochemical setting results in a highly alkaline (pH 9–11) effluent and a combination of extreme conditions that have not previously been observed in the marine environment, including venting of 40–91°C hydrothermal fluids with high concentrations of dissolved H_2, CH_4, and other low-molecular-mass hydrocarbons, but almost no dissolved CO_2 (Kelley et al., 2005; Proskurowski et al., 2006, 2008).

Mixing of warm, high-pH fluids (that reach the seafloor due to hydrothermal venting from the mantle rocks deep below the seafloor) with seawater results in carbonate precipitation and the growth of tall chimneys above the surrounding seafloor. The ^{14}C radio-isotopic dating indicates that hydrothermal activity has been ongoing for at least 30,000 years (Früh-Green et al., 2003). However, recent uranium–thorium dating indicates that venting has been active for ~100,000 years (Ludwig et al., 2005). These mantle rocks (ultramafic rocks) are sites of an important set of geochemical reactions named serpentinization, and have been producing geological H_2 for as long as there has been water on the Earth.

Unlike ridge-axis (also known as *axial*) hydrothermal vents, those hydrothermal vents responsible for the generation of carbonate chimneys are located away from the ridge axis (such vents are known as "off-axis vents") (Williams et al., 1979). At off-axis vents, seawater invades the warm (100°C) to hot (200°C) oceanic crust through cracks and crevasses where the chemical reactions of serpentinization take place. It is believed that serpentinization has probably been ongoing ever since there were oceans on Earth (Sleep et al., 2004). Off-axis vents are located several kilometers away from the spreading zone. Their exhalate has also circulated through the crust, where it can be heated up to ~200°C (Früh-Green et al., 2003; Proskurowski et al., 2006), but their waters do not come into close contact with the magma chamber.

Kelley et al. (2001) have reported the discovery of the LCHF and important differences of LCHF geochemistry compared with black smokers. Although off-axis carbonate vents are radically different to black smokers, our current understanding of these systems is based on research into only one system: the LCHF (Kelley et al., 2005; Ludwig et al., 2006). Because Lost City systems are profoundly different from black smokers, it is important to contrast the two (for an in-depth comparison of Lost City systems and black smokers, see Kelley et al., 2002).

The LCHF, hosted on 1.5 Ma crust, is actively venting carbonate chimneys towering up to 60 m above the seafloor, making them the tallest vent structures known (Kelley et al., 2001; Kelley et al., 2005). The chemistry of the chimneys and vent fluids is controlled by serpentinization reactions between seawater and the underlying peridotite (a dense, coarse-grained plutonic rock containing a large amount of olivine, believed to be the main constituent of the earth's mantle). Note that plutonic rocks are igneous rocks that solidified from a melt at great depth. Magma rises, bringing minerals and precious metals such as gold, silver, molybdenum, and lead with it, forcing its way into older rocks.

Unlike the diverse mineral suites that typify black smoker chimneys, the mineralogy of the Lost City deposits is simple. Active carbonate structures are predominantly composed of aragonite (calcium carbonate), calcite (the most stable form of calcium carbonate crystal), and brucite (hydrated magnesium hydroxide). Calcite, minor brucite, and variable amounts of aragonite dominate inactive chimneys. Mixing of <40–91°C calcium-rich vent fluids with seawater results in the precipitation of these minerals. The resultant deposits range from tall, graceful pinnacles to fragile flanges and delicate precipitates that grow outward from fissures in the bedrock. Exposed ultramafic rocks are prevalent along the MidAtlantic, Arctic, and Indian Ocean ridge networks and, according to Ludwig et al. (2006), it is likely that other Lost City-type systems exist. The brilliant study carried out by Ludwig et al. (2006) provides the first detailed mineralogical and chemical analyses of carbonate chimneys within the serpentinite-hosted LCHF. Their results show that the massive edifices form through complex fluid-chemical mineralogical reactions over tens of thousands of years.

Compared to other studied sites, the hydrothermal geochemistry of Lost City Hydrothermal Field is controlled primarily by serpentinization reactions that produce pH 9–11, <40–91°C fluids enriched in H_2 (up to 15 mmol/kg) and CH_4 (1–2 mmol/kg), with 1 mmol/kg of CO_2 (Kelley et al., 2005; Proskurowski et al., 2006). These fluids are low in most metals and silica, and are enriched in calcium (up to 30 mmol/kg). Mixing of these fluids with seawater results in the precipitation of large carbonate chimneys. Results presented in Ludwig et al. (2006)'s study, based on an analysis of samples collected across the LCHF (Fig. 7.24) during two cruises in 2000 and 2003 (Kelley et al., 2001, 2005), provide the first detailed examination of the geologic, tectonic, and chemical processes that produce and influence the Lost City carbonate structures over time.

According to Ludwig et al. (2006)'s description, "within the field and in the surrounding areas, "active," "inactive," and "fissure-filling" carbonate formations are observed. Most of the carbonate structures grow directly from serpentinite bedrock (Fig. 7.26A). Splintered debris and talus of extinct chimneys surround the base of several of the large structures (Fig. 7.24) and accumulate on the slope or in saddles between chimneys. Fluid percolation through extinct carbonate deposits and talus forms cross-cutting veins and small chimneys, and leads to variable cementation of the carbonate talus. Diffuse venting occurs from numerous chimneys and flange deposits within the core of the field (Fig. 7.26). Radiocarbon isotopic analyses on a subset of active and inactive chimneys and vein material collected in 2000 indicate that hydrothermal activity has spanned at least 30,000 years (Fruh-Green et al., 2003)." Note that in the above descriptions, "saddle" refers to the low part lying between two higher chimneys; and "talus" refers to an outward sloping and accumulated heap or mass of rock fragments of any size or shape derived from and lying at the base of a cliff or very steep, rocky slope, and formed chiefly by gravitational falling, rolling, or sliding.

It has been found that actively venting chimneys within the LCHF emit 40–91°C fluids and range in size from a fledgling, <1 m tall structures to mature chimneys that grow up to 60 m in height. It has also been found that the bases and significant portions of some of the larger active structures are well lithified (i.e., transformed into stone). These sections grade upward and outward into more porous carbonate where diffuse venting is concentrated (Fig. 7.26B).

The results of Ludwig et al. (2006)'s study indicate that the chemical composition of carbonate deposits and vent fluids within the Lost City field and its extended history of venting are likely the product of extensive faulting within the Atlantis Massif and long-lived serpentinization reactions in the subsurface. Complex networks of carbonate veins and fissure-filling deposits within the serpentinized basement rocks correspond to the progressive funneling of hydrothermal fluids from depth to the surface. According to Ludwig et al. (2006), these veins are reminiscent of ophicalcite (crystalline limestone or marble spotted with greenish

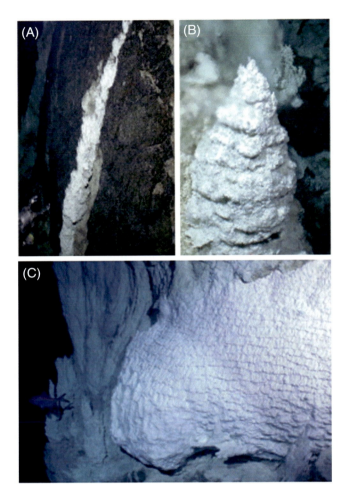

FIG. 7.26 Fissure-filling deposits and actively venting structures at Lost City. (A) Carbonate forms in a fissure in serpentinite bedrock on the east side of the field. (B) The 80 cm-tall Beehive structure emits pH 10.7 shimmering (shining with a soft tremulous or fitful light) fluids at 91°C. (C) A diffusely venting shingled (clad with shingles like in a tiled roof) wall of carbonate on the north face of the Nature Tower; a 30 m tall edifice on the east side of the field. The wreck fish in the left portion of the photo is approximately 1 m in length. *(Source: Ludwig, K.A., D.S. Kelley, D.A. Butterfield, B.K. Nelson, and G. Fruh-Green (2006), Formation and evolution of carbonate chimneys at the Lost City Hydrothermal Field, Geochimica et Cosmochimica Acta, 70: 3625–3645) Copyright info: © 2006 Elsevier Inc. All rights reserved.*

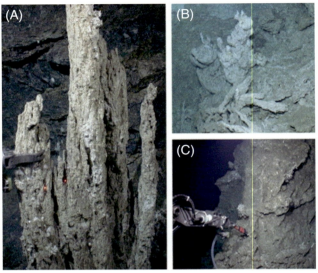

FIG. 7.27 Inactive structures at Lost City. (A) Inactive chimney composed of well-lithified carbonate with a dark brown, pitted exterior, and white interior. The two red laser points shown from the submersible Alvin are 10 cm apart. (B) Tumbled, inactive chimney debris form a talus pile on the south east corner of the Poseidon edifice. (C) Dark, knobby exterior walls populated by worms and deep-sea corals characterize old, inactive structures. *(Source: Ludwig, K.A., D.S. Kelley, D.A. Butterfield, B.K. Nelson, and G. Fruh-Green (2006), Formation and evolution of carbonate chimneys at the Lost City Hydrothermal Field, Geochimica et Cosmochimica Acta, 70: 3625–3645) Copyright info: © 2006 Elsevier Inc. All rights reserved.*

serpentine, like a serpent or snake) deposits in ancient ophiolite (a section of Earth's oceanic crust and the underlying upper mantle that has been uplifted and exposed above sea level and often emplaced onto continental crustal rocks) deposits.

Another interesting observation is that the actively venting structures are characteristically snow white, friable, and porous. However, some of the actively venting carbonate material is significantly discolored bright orange-red or pale pistachio green on both the exterior and interior of the structures. The cause of these discolorations remains unexplained, but Ludwig et al. (2006) reckon that the observed discolorations may reflect biological influences.

Credible indications have been found to believe that some of the chimneys have been clearly inactive for some time. These chimneys are characterized by their dark grey, brown, or black exterior, and white, ivory, or grey interior. The outer walls and rinds (the tough outer skin portions) of these older structures are characteristically rough and pitted or knobby in texture and are colonized by corals and worms (Fig. 7.27). These formations are typically massive and extremely well-lithified and little evidence remains of the original conduits for fluid flow.

Poseidon, which is the largest of the edifices (Kelley et al., 2001, 2005), hosts a gigantic pillar, towering 60 m above the seafloor and is 15 m in diameter at its top where fluids vent at temperatures up to 80°C. The "Beehive" structure on Poseidon vents the highest temperature fluids measured at Lost City (91.4°C). Most of its base is composed of massive carbonate with several buttresses resembling "drapery" formations (hanging in loose folds, akin to curtain clothing) commonly observed in caves. According to Ludwig et al. (2006), these basal structures are likely relict carbonate formations that have melded together.

Ludwig et al. (2006) made the following observations regarding the mineralogy (a subject of geology specializing

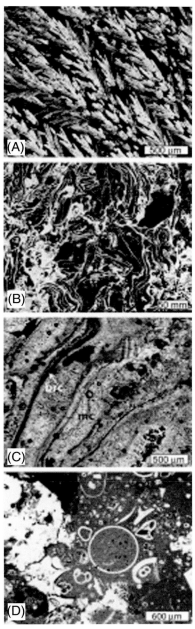

FIG. 7.28 **Representative photomicrographs of nascent, mature, and old vent structures at Lost City.** (A) Delicate, needle-like aragonite crystals form in the direction of fluid flow (toward the upper left corner of the photo). This image is from sample 3876-1436, an actively venting flange on the "Beehive" structure. (B) Sinuous (having many curves and turns), well-defined fluid flow paths (black) are preserved in carbonate (sample 3862-1659) filling a fissure. This deposit represents a mature system (Stage III), and is located on the side of the Poseidon edifice. (C) Relict, micritic calcite (mc)-filled fluid flow paths in the inactive structure 3867-1308. Brucite (brc) lines the channel walls. (D) Abundant fossils and shells of marine organisms are preserved in the outer walls of the inactive, old structure (sample 3651-0908). *(Source: Ludwig, K.A., D.S. Kelley, D.A. Butterfield, B.K. Nelson, and G. Fruh-Green (2006), Formation and evolution of carbonate chimneys at the Lost City Hydrothermal Field, Geochimica et Cosmochimica Acta, 70: 3625–3645) Copyright info: © 2006 Elsevier Inc. All rights reserved.*

in the scientific study of the chemistry, crystal structure, and physical properties of minerals and mineralized artifacts) and petrography (the branch of science concerned with the composition and properties of rocks) of the Lost City vent structure, "In young, actively venting flanges, nascent fluid flow channels form between blades of aragonite (Fig. 7.28A). In more mature, but still active chimneys, sinuous flow paths are well-formed (Fig. 7.28B) and aragonite, brucite, and calcite are present within the structures. In older deposits that are no longer venting, microchannels

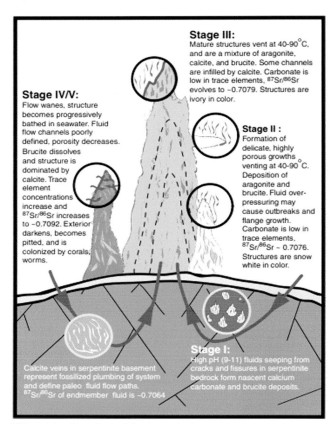

FIG. 7.29 **Growth model schematic for carbonate deposits at the Lost City Hydrothermal Field.** Five stages of development are defined by petrographic and geochemical differences. *(Source: Ludwig, K.A., D.S. Kelley, D.A. Butterfield, B.K. Nelson, and G. Fruh-Green (2006), Formation and evolution of carbonate chimneys at the Lost City Hydrothermal Field, Geochimica et Cosmochimica Acta, 70: 3625–3645) Copyright info: © 2006 Elsevier Inc. All rights reserved.*

are progressively in-filled with micritic calcite, and only trace brucite remains (Fig. 7.28C). In well-lithified structures, marine fossils are incorporated into the chimney walls (Fig. 7.28D). These fossils include shells of planktonic marine foraminifera, gastropods, echinoderm spines, sponge spicules, and pteropods."

Ludwig et al. (2006)'s petrographic and geochemical analyses of a representative suite of samples define five stages of formation (Fig. 7.29). These stages include the transformation from nascent, fragile, and highly porous active chimneys that are dominated by aragonite and brucite, to highly lithified, inactive deposits dominated by calcite and extinct deposits that lack well-defined flow channels. In these later stages, brucite dissolves resulting in decreased concentrations of magnesium (Mg). Concomitant with mineralogical and physical changes during aging, the chimneys tend to become enriched in seawater-derived trace metals, and strontium (Sr) isotope ratios increase toward seawater values.

7.2.2 Rich microbial and faunal ecosystems harbored by submarine hydrothermal vents

Both magma-chambers-fed hydrothermal vents and serpentinite-hosted carbonate hydrothermal vents contain various carbon compounds, including graphite, CH_4, and CO_2.

7.2.2.1 Fascinating life supported by magma-chambers-fed vents

Life is typically sparse on the deep seafloor, where organisms endure high pressure, near-freezing temperatures, and pitch-black darkness. But unique ecosystems teem with unusual animal species at hydrothermal vent sites, and the assortment of animals surrounding them are referred to as hydrothermal vent communities. When first discovered in 1977, the sight of hydrothermal vent teeming with a thriving community of fascinating life around it was a complete surprise. No wonder, deep-sea hydrothermal vents have since then been described as the "Oases on the ocean floor" (see Lutz, 1984).

Most bacteria (a type of biological cells constituting a large domain of prokaryotic microorganisms) and archaea (microorganisms that are similar to bacteria in size and simplicity of structure but radically different in molecular organization) cannot survive in the superheated hydrothermal fluids of the chimneys or "black smokers." However, hydrothermal microorganisms are able to thrive just outside the hottest waters, in the temperature gradients that form between the hot venting fluid and cold seawater. These microbes are the foundation for life in hydrothermal vent ecosystems. Instead of using light energy to turn carbon dioxide into sugar like plants do, these microbes harvest chemical energy from the minerals and chemical compounds that spew from the vents—a process known as *chemosynthesis*. These compounds—such as hydrogen sulfide, hydrogen gas, ferrous iron, and ammonia—lack carbon. The microbes produce and release new compounds after chemosynthesis, some of which are toxic, but others can be taken in nutritionally by other organisms. For example, at hydrothermal vents, vent bacteria oxidize hydrogen sulfide, add carbon dioxide and oxygen, and produce sugar, sulfur, and water: $CO_2 + 4H_2S + O_2 \rightarrow CH_2O + 4S + 3H_2O$. Other bacteria make organic matter by reducing sulfide or oxidizing methane (Smithsonian Ocean Team). Our knowledge of chemosynthetic communities is relatively new, brought to light by ocean exploration when humans first observed a vent on the deep ocean floor in 1977 and found a thriving community where there was no light.

Magma-chambers-fed vents harbor rich ecosystems, the energy source of which stems mainly from mid-ocean ridge volcanism (Corliss et al., 1979; Spiess et al., 1980). As just mentioned, hydrothermal vent systems can support life without input from photosynthesis and they harbor fascinating life with symbiotic relationships that involve lithoautotrophic microorganisms that use chemical energy to support metazoans (all animals having the body composed of cells differentiated into tissues and organs and usually a digestive cavity lined with specialized cells). Note that lithoautotrophs or chemolithoautotrophs are microbes that derive energy from reduced compounds of mineral origin. Lithoautotrophs are a type of lithotrophs (a diverse group of organisms using an inorganic substrate to obtain reducing equivalents for use in biosynthesis or energy conservation via aerobic or anaerobic respiration) with autotrophic (capable of manufacturing complex organic nutritive compounds from simple inorganic sources such as carbon dioxide, water, and nitrates, using energy from the sun) metabolic pathways. Lithoautotrophs are exclusively microbes; macrofauna do not possess the capability to use mineral sources of energy (Martin et al., 2008).

Benthic communities associated with deep-sea hydrothermal vents have been studied intensively since their initial discovery in 1977. Studies of the many active and inactive hydrothermal vents found since then have radically altered views of biological and geological processes in the deep sea. It has been found that biological processes, such as rates of metabolism and growth, in vent organisms are comparable to those observed in organisms from shallow-water ecosystems. An abundant energy source is provided by chemosynthetic bacteria that constitute the primary producers sustaining the lush communities at the hydrothermal sites. As recounted by Grassle (1986), the spectacular nature of the communities was manifested by specimens of vent mussels, clams, tube worms, crabs, and other fauna. Active hydrothermal vent fields in the eastern Pacific, harboring very dense megafaunal (the large or giant animals of an area, habitat, or geological period) and macrofaunal (animals that are 1 cm or more long but smaller than an earthworm) populations at some vent sites, have been found as far north as 49 42′N on the Explorer Ridge and as far south as 22 S on the East Pacific Rise (Ballard et al., 1981; Lutz, 1984, 1988; Crane, 1985; Crane et al., 1985; Tunnicliffe et al., 1986; Southward, 1991; Tunnicliffe, 1991). The high chemosynthetic primary production at hydrothermal vents provides nutrition directly to luxuriant populations of clams (Fig. 7.30), mussels, tube worms, and a variety of grazers, while supporting indirectly a diverse community of scavengers (Fig. 7.31), detritus feeders, and predators (Turner, 1985).

The distribution and abundance of species in chemosynthetically based animal communities at hydrothermal vents have become the focus of considerable research (Hessler and Smithey, 1983; Grassle, 1986; Grassle et al., 1986; Juniper et al., 1990; Van Dover and Hessler, 1990). The number of new vent species, genera, and families grew appreciably during the 1980s with more than 20 new families, over 90 new genera, and nearly 300 new species being

FIG. 7.30 **Giant white clams (Calyptogena magnifica) living in clusters within the crevices between basalt pillows in the area known as Clam Acres at 21 N on the East Pacific Rise.** Other vent fauna include a clump of tube worms (Riftia pachyptila), galatheid crabs (Munidopsis subsquamosa), limpets, and serpulids. Note manipulator arm of the submersible Alvin collecting a clam approximately 25 cm in length. Photo by R. A. Lutz. *(Source: Lutz, R.A., and M.J. Kennish (1993), Ecology of deep-sea hydrothermal vent communities: A review, Reviews of Geophysics, 31(3), 211-242). Copyright info: Copyright 1993 by the American Geophysical Union.*

FIG. 7.31 **Porcelain white brachyuran crabs (Bythograea thermydron) resting on basaltic rock adjacent to a thicket of tube worms (Riftia pachyptila) at the Genesis hydrothermal vent.** Photo by R. A. Lutz. *(Source: Lutz, R.A., and M.J. Kennish (1993), Ecology of deep-sea hydrothermal vent communities: A review, Reviews of Geophysics, 31(3), 211-242). Copyright info: Copyright 1993 by the American Geophysical Union.*

registered from hydrothermal vent environments (Grassle, 1985, 1986, 1989; Tunnicliffe, 1991, 1992).

Active hydrothermal vent activity may span years or decades, as has been shown by independent studies of heat loss and rates of dissolution of giant white clam (*Calyptogena magnifica*) shells commonly found in massive death assemblages at previously active vent localities and by direct observations of community changes through time (Macdonald et al., 1980; Turner and Lutz, 1984; Grassle, 1985, 1986; Lutz et al., 1985, 1988; Fustec et al.,

1987; Tunnicliffe, 1991). Campbell et al. (1988) proposed a time scale of decades for the lifetime of a vent field. Somewhat in contrast to the preceding estimates, Converse et al. (1984) calculated a longevity of hundreds to thousands of years for vent activity using fluid flow rates (Lutz and Kennish, 1993).

It has been suggested that the environmental unpredictability, transient nature, and high biological production associated with hydrothermal systems favor continuous reproduction, rapid and constant recruitment, accelerated growth, high fecundity (fertility, the natural capability to produce offspring, measured by the number of eggs, seed set, etc.), tolerance to change, and rapid response of organisms to the exploitation of resources (Tunnicliffe, 1990; Gaill and Hunt, 1991).

Konn et al. (2015) made the following observations about the life abounding on hydrothermal vents, "Chemolithotrophic microbial communities commonly colonize hydrothermal vents and may represent analogs for life on early Earth and other planets. Chemolithotrophic organisms, by definition, utilize only inorganic and/or abiotic simple molecules for their carbon and energy sources so that they do not rely on other living organisms to feed, develop, and multiply (Lang et al., 2012). To date, the maximum temperature for some of such organisms to grow is 122°C (Takai et al., 2008)."

7.2.2.2 Organisms supported by Lost City and Lost City-Type hydrothermal systems

Subseafloor mixing of reduced hydrothermal fluids with seawater is believed to provide the energy and substrates needed to support deep chemolithoautotrophic life in the hydrated oceanic mantle (i.e., serpentinite). As already noted, hydrothermal venting and the formation of carbonate chimneys in the Lost City hydrothermal field (LCHF) are driven predominantly by serpentinization reactions and cooling of mantle rocks, resulting in a highly reducing, high-pH environment with abundant dissolved hydrogen and methane. Detailed mapping of this hydrothermal vent field as well as chemical and microbial analyses (Kelley et al., 2005) reveal that an extraordinary array of micro- and macro-organisms inhabit the giant white carbonate chimneys of Lost City. For example, it has been found that the four known vents (IMAX, Poseidon, Seeps, and Nature) of the Lost City marine hydrothermal vent field support a low biomass but a high species diversity (Kelley et al., 2005). The serpentinite-hosted Lost City hydrothermal field is a remarkable submarine ecosystem in which geological, chemical, and biological processes are intimately interlinked, and some scientists postulate this to be a contemporary analog of conditions where life may have originated. Studies carried out by Kelley et al. (2005) found that a low diversity of microorganisms related to methane-cycling Archaea thrive in the warm porous interiors of the edifices. Their studies revealed that macrofaunal communities show a degree of species diversity at least as high as that of black smoker vent sites along the Mid-Atlantic Ridge, but they lack the high biomasses of chemosynthetic organisms that are typical of volcanically driven systems.

Actively venting chimneys host a microbial community with a relatively high proportion of methanogenic archaea (the Lost City Methanosarcinales), methanotrophic bacteria, and sulfur-oxidizing bacteria, whereas typical sulfate-reducing bacteria are rare (Brazelton et al., 2006, 2010; Schrenk et al., 2013). Geochemical evidence for significant microbial sulfate reduction in basement lithologies and distinct microbial communities in Lost City vent fluids and chimneys suggest that subsurface communities may be different from those in chimney walls (Brazelton et al., 2006, 2010; Delacour et al., 2008). The lack of modern seawater bicarbonate and low $CO_2/^3He$ in Lost City fluids clearly indicate bicarbonate removal before venting, but it remains unclear if bicarbonate removal occurred by "dark" microbial carbon fixation in a serpentinization-fueled deep biosphere, by carbonate precipitation, or both (Proskurowski et al., 2008).

For the Lost City hydrothermal vent field, the access and benefit sharing of marine genetic resources associated with novel microbial communities could be very significant. For example, Brazelton et al. (2006) recorded biofilms of archaea (Lost City Methanoscarcinates) from the oxygen-free interior zones of chimneys and entirely different bacterial assemblage on chimney outer walls. In their studies, phylogenetic and terminal restriction fragment length polymorphism analyses of 16S rRNA genes in fluids and carbonate material from the Lost City hydrothermal field (LCHF) site indicated the presence of organisms similar to sulfur-oxidizing, sulfate-reducing, and methane-oxidizing bacteria as well as methanogenic and anaerobic methane-oxidizing Archaea. Note that 16S rRNA (16S ribosomal RNA) is the RNA component of the 30S small subunit of a prokaryotic ribosome. The 16S rRNA gene is the DNA sequence corresponding to rRNA encoding bacteria, which exists in the genome of all bacteria.

With reference to the bacteria and Archaea mentioned above, Brazelton et al. (2006) made the following observation: "The presence of these metabolic groups indicates that microbial cycling of sulfur and methane may be the dominant biogeochemical processes active within this ultramafic rock-hosted environment." Whereas 16S rRNA gene sequences grouping within the Methylobacter and Thiomicrospira clades (groups of organisms believed to comprise all the evolutionary descendants of a common ancestor) were recovered from a chemically diverse suite of carbonate chimney and fluid samples, 16S rRNA genes corresponding to the Lost City Methanosarcinales phylotype were found exclusively in high-temperature chimneys. Furthermore, a phylotype of anaerobic methanotrophic

Archaea (ANME-I) was restricted to lower-temperature, less vigorously venting sites. Brazelton et al. (2006) reckon that "a hyperthermophilic habitat beneath the LCHF may be reflected by 16S rRNA gene sequences belonging to Thermococcales and uncultured Crenarchaeota identified in vent fluids." These investigators argue that "The finding of a diverse microbial ecosystem supported by the interaction of high-temperature, high-pH fluids resulting from serpentinization reactions in the subsurface provides insight into the biogeochemistry of what may be a pervasive process in ultramafic subseafloor environments."

Metagenomics (study of genetic material recovered directly from environmental samples; the broad field may also be referred to as environmental genomics, ecogenomics, or community genomics) and environmental sequencing of ribosomal RNA carried out by Martin et al (2008) have shown that microbial communities in actively venting carbonate chimneys in the LCHF are dominated by a novel phylotype (an observed similarity used to classify a group of organisms by their phenetic relationship; this phenetic similarity, particularly in the case of asexual organisms, may reflect the evolutionary relationships) of anaerobic (requiring an absence of free oxygen) methanogens (microorganisms that produce methane as a metabolic by-product in hypoxic conditions) from the Methanosarcinales order (Schrenk et al., 2004; Kelley et al., 2005; Brazelton et al., 2006). It was found that these methanogens can use several organic compounds, some of which have been implicated in anaerobic methane oxidation (AMO) in both hydrothermal sediments (Orphan et al., 2002; Teske et al., 2002) and methane seeps (Boetius et al., 2000; Aloisi et al., 2002). In chimneys that have little or no active venting, the Lost City Methanosarcinales (LCMs) group is replaced by a single phylotype of the anaerobic methanotrophic clade ANME-1.

Where the LCHF carbonate chimneys are bathed in >80°C hydrothermal fluid, the Lost City Methanosarcinales (LCMs) group forms dense biofilms that are tens of micrometers thick and comprise ~100% of the archaeal community (Brazelton et al., 2006). Recently, methyl-coenzyme M reductase (mcrA) gene sequences that correspond to both LCMs and ANME-1 have also been recovered from LCHF carbonate chimneys (Brazelton et al., 2006). At LCHF, biofilms that are composed of organisms related to the ANME-3 group are in direct contact with serpentinization-derived CH_4 at temperatures in excess of 90°C (Kelley et al., 2005; Brazelton et al., 2006). According to Martin et al. (2008)'s interpretation, the finding that CH_4 and associated short hydrocarbons in the effluent of LCHF are not formed by biological activity, but instead are of geochemical origin. Proskurowski et al. (2008) suggest that LCMs and ANME-1 are probably oxidizing CH_4 at LCHF and that they are doing so in the presence of abundant environmental H_2.

It is pertinent to note that Lost City is a marine ecosystem characterized by extreme conditions, supporting anaerobic thermophile communities, where the impacts of climate change may be least felt (Johnson et al., 2018). Studies carried out by Blasiak et al. (2018) indicate that the alkaline environment of Lost City is unique, and may support specially adapted extremophile microbes that could offer marine genetic resources not found elsewhere. According to Johnson (2019), a principal threat to the Lost City ecosystem is indirect impact of deep-sea mining from possible plumes and discharges, if exploitation for polymetallic sulfides is sanctioned at adjacent or neighboring locations.

The suggestion that the Lost City is a unique marine hydrothermal vent field is recognized by the Convention on Biological Diversity as being within an "Ecologically or Biologically Significant Marine Area" and has been mooted as a possible World Heritage site (Johnson, 2019). Only discovered in 2000 (Boetius, 2005; Kelley et al., 2007), the geochemical and micro-biological processes at this section of the MAR are still the subject of ongoing scientific research. For example, the goals of the IODP Expedition 357 were to investigate serpentinization processes and microbial activity in the shallow subsurface of highly altered ultramafic (relating to or denoting igneous rocks composed chiefly of mafic minerals) and mafic (relating to or denoting a group of dark-colored, mainly ferromagnesian minerals such as pyroxene and olivine) sequences that have been uplifted to the seafloor along a major detachment fault zone (Früh-Green et al., 2018). According to Früh-Green et al. (2018), a major achievement of IODP Expedition 357 was to obtain microbiological samples along a west–east profile, which will provide a better understanding of how microbial communities evolve as ultramafic and mafic rocks are altered and emplaced on the seafloor. New technologies were also developed for the seabed drills to enable biogeochemical and microbiological characterization of the environment. In 2014, a Convention on Biological Diversity (CBD) Regional Workshop for the North-West Atlantic included Lost City as a feature within a Hydrothermal Vent Fields Ecologically or Biologically Significant Marine Area (EBSA) description (Johnson, 2019). It is expected that the UN Decade of Ocean Science for Sustainable Development (2021–2030) provides an opportunity to better understand key sites such as Lost City.

Serpentinized mantle rocks constitute a major component of oceanic plates, subduction zones, and passive margins. Serpentinization systems have existed throughout most of Earth's history, and it has been suggested that mixing of serpentinization fluids with Archean seawater produced conditions conducive to abiotic synthesis and the emergence of life on Earth (Sousa et al., 2013). The exposure of mantle rocks to seawater during the breakup of Pangaea fuelled chemolithoautotrophic microbial communities at the Iberia Margin (an active tectonic zone along and offshore of the coast of Portugal; the Southwest Iberian Margin (SIM) is believed to be an incipient subduction zone and the source of the great Lisbon earthquake of 1755), possibly before the onset of seafloor spreading. Lost City-type serpentinization systems have been discovered at mid-ocean ridges, in forearc settings of subduction zones, and at continental margins. It appears that, wherever they occur, they can support microbial life, even in deep subseafloor environments.

While Klein et al. (2015) were examining the samples obtained from cores drilled from the Iberian continental margin (located along the Azores-Gibraltar Fracture Zone off the coast of Spain and Portugal) in 1993 (the samples came from rock 760 m below the current seafloor, which would have been 65 m below the early un-sedimented ocean floor), they came across some unusual looking veins in the samples, composed of minerals also found at the Lost City hydrothermal system. This discovery was intriguing to them because this mineral assemblage is only formed when the alkaline hydrothermal fluids are mixed with the acidic seawater. Within these veins, dated to 120 million years ago, Klein and his team found inclusion of fossilized microbes. It was suggested that the desiccating properties of the mineral brucite ($Mg(OH)_2$) might explain the preservation of organic molecules from the microbes. These included amino acids, proteins and lipids which were identified by confocal Raman spectroscopy. Analysis of extracted samples confirmed unique lipid biomarkers for sulfate-reducing bacteria and archaea, which are also found in the Lost City hydrothermal vents system. Scanning Electron Microscopy (SEM) imaging showed carbon inclusions which seemed to look like micro-colonies of micro-organisms. Note that SEM is a form of microscopy in which a focused beam of accelerated electrons is scanned across the surface of a specimen, generating a number of signals that yield information about its morphology, elemental composition, and, when outfitted with appropriate detectors, crystalline microstructure or other features. The presence of the microbes mentioned above is an indication that life is possible in seafloor environments in hydrothermal systems, that were probably present and active throughout most of the early Earth. Notably, the sub-seafloor represents another more protected environment.

Klein et al. (2015) provided biogeochemical, micropaleontological, and petrological constraints on a sub-seafloor habitat at the passive Iberia Margin, where mixing of oxidized seawater and strongly reducing hydrothermal serpentinization fluids at moderate temperatures created conditions capable of supporting microbial activity. They found that this mixing zone is inhabited by bacteria and archaea and is comparable to the active Lost City hydrothermal field at the Mid-Atlantic Ridge. Their results highlight the potential of magma-poor passive margins to host Lost City-type hydrothermal systems that support microbial activity in subseafloor environments. At the Iberia Margin, Klein et al. (2015) detected a combination of bacterial diether lipid biomarkers and archaeal tetraethers analogous to those found in carbonate chimneys at the active Lost City hydrothermal field. Klein et al. (2015) argued that because equivalent systems have likely existed throughout most of Earth's history in a wide range of oceanic environments, fluid mixing may have provided the substrates and energy to support a unique subseafloor community of microorganisms over geological timescales.

Other active and fossil seafloor hydrothermal systems similar to Lost City exist in a range of seafloor environments, including the Mid-Atlantic Ridge (Lartaud et al., 2011), New Caledonia (Cox et al., 982), and the Mariana forearc (Fryer et al., 1990; Ohara et al., 2012). Bathymodiolus mussels are, in some places, associated with these systems, suggesting that active serpentinization is supporting not only microbial chemosynthetic ecosystems but also macrofaunal communities (Ohara et al., 2012). However, biological processes in the subseafloor of these Lost City-type systems are poorly understood (Klein et al., 2015).

7.2.3 Role of geothermal energy in driving deep-sea hydrothermal vent ecosystem

Geothermal energy rather than solar energy drives the deep-sea hydrothermal vent ecosystem. The reaction of seawater with crustal rocks at high temperatures produces reduced inorganic compounds that discharge from the hot springs and provide energy for free-living and/or symbiotic chemosynthetic bacteria (alternatively referred to as chemoautotrophs or chemolithotrophs) which form the base of the food chain in these unique habitats (Jannasch and Wirsen, 1979; Jannasch, 1983, 1984, 1985; Jannasch and Mottl, 1985; Jannasch et al., 1985). By oxidizing sulfides, especially hydrogen sulfide (H_2S), as well as other reduced substrates such as hydrogen (H_2), iron (Fe^{2+}), or manganese (Mn^{2+}) released from vents, the microbes obtain energy to synthesize organic compounds from carbon dioxide (CO_2) in seawater; this process is known as autotrophy (Cavanaugh, 1985a, b; Turner, 1985). Two vent animals that have developed mechanisms for detoxifying sulfide, enabling them to inhabit vent environments containing potentially lethal concentrations of sulfide, include the vesicomyid clam *Calyptogena magnifica* and the tube worm *Riftia pachyptila* (see Lutz and Kennish, 1993).

7.3 Discovery of indications of hydrothermal vents adorning the subsurface seafloors of *Enceladus* and *Europa*—possibility of life & habitability

Vance et al. (2007) have shown that in the case of icy worlds (e.g., the planet Saturn's moon *Enceladus*) the time-varying hydrostatic head of a tidally forced ice shell may drive hydrothermal fluid flow through the seafloor, which can generate moderate but potentially important heat through viscous interaction with the matrix of porous seafloor rock.

7.3.1 Direct evidence for submarine hydrothermal vents on *Enceladus*

Water-rich plumes of vapor and ice particles with sodium salts erupting from the warm fractures near the south pole of Saturn's icy moon *Enceladus* suggest the presence of a liquid water reservoir in the interior (Schmidt et al., 2008; Waite Jr et al., 2009; Postberg et al., 2009, 2011). Investigations combining measurements obtained from *Cassini* spacecraft (which orbited Saturn and its moons *Enceladus* and *Titan*) and experimental results show that some of the observed plume materials erupting from Enceladus's south polar "tiger stripes" are associated with ongoing hydrothermal activity in the interior (Hsu et al., 2015). Nanometre-sized silica particles with a confined size range detected by the Cassini Cosmic Dust Analyser are found to have originated from Enceladus (Hsu et al., 2015). Supported by the results of hydrothermal experiments, it is indicated that these particles originated from nano-silica colloids that formed when silica saturation was reached upon cooling of hydrothermal fluids (Hsu et al., 2015). According to Sekine et al. (2015), the presence of these particles provides tight constraints on the particular conditions of the interior ocean; that is, the presence of high-temperature reactions (≥~90°C), moderate salinity (≤~4%), and alkaline seawater (pH=8.5–10.5) (Hsu et al., 2015). Studies carried out by Iess et al. (2014) and McKinnon (2015) indicate that the products of ongoing hydrothermal reactions would have been transported upwards from an interior ocean located at a depth of ~30 km beneath the surface at Enceladus' south pole. According to Matson et al. (2012), hydrothermal reaction products that reached the bottom of the tiger stripes would have then been ejected into the plume. McKay et al. (2008) discussed the possible origin and persistence of life on Enceladus and detection of biomarkers in the plume.

As just indicated, an interesting revelation with respect to Enceladus's subsurface ocean is the indication of ongoing hydrothermal activities within Enceladus (Hsu et al., 2015). Mayhew et al. (2013) have reported hydrogen generation from low-temperature water-rock reactions. To sustain the formation of silica nanoparticles, the composition of Enceladus' core needs to be similar to that of carbonaceous chondrites. Note that several groups of carbonaceous chondrites, notably the CM (carbonaceous chondrite) and CI (also known as C1 chondrites; a group of rare stony meteorites belonging to the carbonaceous chondrites) groups, contain high percentages (3%–22%) of water, as well as organic compounds. They are composed mainly of silicates, oxides, and sulfides, with the minerals olivine and serpentine being characteristic. C chondrites contain a high proportion of carbon (up to 3%), which is in the form of graphite, carbonates, and organic compounds, including amino acids. In addition, they contain water and minerals that have been modified by the influence of water. In a study, Sekine et al. (2015) found high-temperature water-rock interactions and hydrothermal environments in the chondrite-like core of Enceladus. The recent discovery of silica nanoparticles derived from Enceladus shows the presence of ongoing hydrothermal reactions in the interior. Sekine et al. (2015) reported results from detailed laboratory experiments to constrain the reaction conditions. They showed that the presence of hydrothermal reactions would be consistent with NH_3- and CO_2-rich plume compositions. They further suggested that high reaction temperatures (>50°C) are required to form silica nanoparticles whether Enceladus' ocean is chemically open or closed to the icy crust. Sekine et al. (2015) argued that such high temperatures imply either that Enceladus formed shortly after the formation of the Solar System or that the current activity was triggered by a recent heating event. Under the required conditions, hydrogen production would proceed efficiently, which could provide chemical energy for chemoautotrophic life.

In a subsequent study, Waite et al. (2017) examined the data from Cassini spacecraft, which orbited Enceladus, and found molecular hydrogen in the Enceladus plume. This discovery is an evidence of hydrothermal processes taking place in Enceladus's subsurface ocean (Waite et al., 2017). In another study, Choblet et al. (2017) explained the geophysical mechanism involved in powering prolonged hydrothermal activity inside Enceladus. It has been suggested that molecular hydrogen, detected in the plumes and likely reflecting hydrothermal activity at the core (Waite et al., 2017), along with other reducing gases such as methane and ammonia could contribute to further chemical modifications.

The observed strong indications of ongoing hydrothermal activities within Enceladus are particularly interesting because, as already described in this chapter, submarine hydrothermal vents are abundantly found on Earth's ocean floors. Because several kinds of terrestrial hydrothermal vent organisms that thrive under low-temperature and high-pressure environmental conditions do not require sunlight or oxygen to survive and thrive, there exists a strong possibility that Enceladus's dark and cold subsurface ocean should also be harboring organisms similar to those found on and in the vicinity of Earth's submarine hydrothermal vents.

Several investigators proposed that the presence of tall jets of water vapor and organic-enriched salty ice particles spewing from the tiger stripes on the south pole of Enceladus (McKay et al., 2014), taken together with the presence of submarine hydrothermal vents adorning Enceladus's subsurface seafloor, could be indications of strong possibility for habitability of Enceladus. It could be that the complex organic molecules identified in Enceladus's subsurface ocean may have evolved in seafloor hydrothermal systems by mineral catalysis, and these organic molecules could be primordial. Williams et al. (2002) investigated smectite (a clay mineral which undergoes reversible expansion on absorbing water) incubation of organic molecules in seafloor hydrothermal systems. Williams et al. (2005) showed that smectite-type clays can protect and promote the development of diverse organic compounds that may be precursors to biomolecules. Smectite provides a safe haven for the synthesis of organic molecules, essentially like a "primordial womb." Their experiments simulated seafloor hydrothermal conditions. Hydrothermal sites are excellent sources of nutrients and chemical energy needed to support metabolism.

According to Deamer and Georgiou (2015), hydrothermal conditions are conducive to the origin of cellular life. It is currently well recognized that there is a deep subsurface ocean on Enceladus, at least 10 km thick under the south polar terrain and covered by a layer of ice tens of kilometers thick but perhaps thinner at the south pole, and that Enceladus has abundant tidal heat (Lainey et al., 2012) that is sufficient to maintain a long-lived liquid ocean. According to Deamer and Damer (2017), a good way to sharpen the argument that life can plausibly begin in hydrothermal conditions is to compare hydrothermal vents with an alternative site, the "warm little pond" first envisaged by Charles Darwin (Darwin, 1871). Bouquet et al. (2015) found possible evidence for a methane source in Enceladus' ocean. Taubner et al. (2018) showed that biological methane production under putative Enceladus-like conditions is viable. Kahana et al. (2019) proposed that the first observed primordial soup could arbitrate origin-of-life debate on Enceladus.

7.3.2 Submarine hydrothermal activity on *Europa*—inferences gleaned from telescopic observations and laboratory experiments

There are several lines of evidence for the presence of a subsurface ocean within Jupiter's icy moon *Europa*; however, its oceanic chemistry and geochemical cycles are largely unknown. The recent observations by large telescopes show that sulfur ions and sulfur dioxide (SO_2) occurring on the surface of Europa are implanted from Io (the third largest of Jupiter's moons) and accumulate as sulfuric acids in Europa's trailing hemisphere. This suggests that a large amount of sulfate could have been supplied into Europa's subsurface ocean over geological timescales. The telescope observations also suggest that chloride salts appear on the "chaotic terrains" of Europa, suggesting that the primary oceanic anion may be chloride despite a supply of sulfate into the ocean. These observations imply the presence of possible sinks of exogenic (occurring on the surface) sulfate within the ocean. Tan et al. (2021) have reported the results of laboratory experiments on hydrothermal sulfate reduction under the pressure conditions that correspond to Europa's seafloor. Using a Dickson-type experimental system, they obtained the reaction rate of sulfate reduction at a pressure of 100 MPa and temperature of 280°C for various pH levels (pH 2–7). They found strong pH dependence and little pressure dependence of the reaction rate. It was found that sulfate reduction proceeds effectively at fluid pH < 6, whereas it is kinetically inhibited at fluid pH ~7. These results suggest that, if hydrothermal fluid pH is <6, hydrothermal sulfate reduction within Europa can be a sink of exogenic sulfate within the ocean in addition to precipitation of sulfate salts. Such acidic fluid pH may be achieved if hydrothermal activity is hosted by basaltic rocks. Tan et al. (2021) suggest the importance of the thermal evolution of the rocky interior for both the ocean chemistry and sulfur cycles of Europa.

References

Abyzov, S.S., Mitskevich, I.N., Poglazova, M.N., Barkov, N.I., Lipenkov, V.Y, Bobin, N.E., Koudryashov, B.B., Pashkevich, V.M., 1998. Antarctic ice sheet as a model in search of life on other planets. Adv. Space Res. 22, 363–368.

Abyzov, S.S., Mitskevich, I.N., Poglazova, M.N., Barkoti, N.I., Lipenkos, V.Y., Bobin, N.E., Kudryashoti, B.B., Pashkevich, V.M., 1999. Antarctic ice sheet as an object for solving some methodological problems of exobiology. Adv. Space Res. 23, 371–376.

Aloisi, G.I., et al., 2002. CH4-consuming microorganisms and the formation of carbonate crusts at cold seeps. Earth Planet. Sci. Lett. 203, 195–203.

Amaral-Zettler, L.A., Gómez, F., Zettler, E., Keenan, B.G., Amils, R., Sogin, M.L., 2002. Microbiology: eukaryotic diversity in Spain's River of Fire. Nature 417 (6885), 137.

Amils, R., González-Toril, E., Fernández-Remolar, D., Gómez, F., Rodríguez, N., Durán, C., 2002. Interaction of the sulfur and iron cycles in the Tinto River ecosystem. Rev Environ Sci Biotechnol 1, 299–309.

Amils, R., Fernández-Remolar, D., 2014. Río Tinto: a geochemical and mineralogical terrestrial analogue of Mars. Life (Basel) 4 (3), 511–534. http://doi.org/10.3390/life4030511.

Andersen, D.T., 2002. Cold springs in permafrost on Earth and Mars. J. Geophys. Res. 107, 5015–5017. http://doi.org/10.1029/2000JE001436.

Aubrey, A., Cleaves, H.J., Chaimers, J.H., Skelley, A.M., Mathies, R.A., Grunthaner, F.J., et al., 2006. Sulfate minerals and organic compounds on Mars. Geology 34, 357–360. http://doi.org/10.1130/G22316.1.

Baker, V.R., Strom, R.G., Gulick, V.C., Kargel, J.C., Komatsu, G., Kale, V.S., 1991. Ancient oceans, ice sheets and the hydrological cycle on Mars. Nature 352, 589–594.

Baker, E.T. & German, C.R. (2004), *Mid-Ocean Ridges: Hydrothermal Interactions between the Lithosphere and Oceans* (eds. German, C., Lin, J. & Parson, L.M.): 245–266 (American Geophysical Union, Washington, DC).

Baldwin, R.L., 1996. How Hofmeister ion interactions affect protein stability. Biophys. J. 71, 2056–2063.

Ballard, R.D., Francheteau, J., Juteau, T., Rangan, C., Normack, W., 1981. East Pacific Rise at 21 N: The volcanic, tectonic, and hydrothermal processes of the central axis. Earth Planet. Sci. Lett. 55, 1–10.

Battler, M.M., Osinski, G.R., Banerjee, N.R., 2013. Mineralogy of saline perennial cold springs on Axel Heiberg Island, Nunavut, Canada and implications for spring deposits on Mars. Icarus 224, 364–381. http://doi.org/10.1016/j.icarus.2012.08.031.

Bhardwaj, A., Sam, L., Martín-Torres, F.J., Zorzano, M.-P., Fonseca, R.M., 2017. Martian slope streaks as plausible indicators of transient water activity. Sci. Rep. 7, 7074. http://doi.org/10.1038/s41598-017-07453-9.

Blackman, D.K., et al., 2002. Geology of the Atlantis massif, (Mid-Atlantic Ridge 30 N): implications for the evolution of an ultramafic oceanic core complex. Mar. Geophys. Res. 23, 443–469.

Blasiak, R., Jouffray, J.B., Wabnitz, C.C.C., Sundstrom, E., Osterblom, H., 2018. Corporate control and global governance of marine genetic resources. Sci. Adv. 4, 6.

Blum, H.F., 1955. Time's Arrow and Evolution. Harper, New York, NY.

Boetius, A., et al., 2000. A marine microbial consortium apparently mediating anaerobic oxidation of methane. Nature 407, 623–626.

Boetius, A., 2005. Lost city life. Science 307, 1420–1422.

Bouquet, A., Mousis, O., Waite, J.H., Picaud, S., 2015. Possible evidence for a methane source in Enceladus' ocean. Geophys. Res. Lett. 42, 1334–1339. http://doi.org/10.1002/2014GL063013.

Brazelton, W.J., Schrenk, M.O., Kelley, D.S., Baross, J.A., 2006. Methane and sulfur metabolizing microbial communities dominate the Lost City hydrothermal field ecosystem. Appl. Environ. Microbiol. 72, 6257–6270.

Brazelton, W.J., et al., 2010. Archaea and bacteria with surprising microdiversity show shifts in dominance over 1,000-year time scales in hydrothermal chimneys. Proc Natl Acad Sci USA 107 (4), 1612–1617.

Burn, C.R., 2020. Permafrost Landscape Features, In: Reference Module in Earth Systems and Environmental Sciences, Elsevier ScienceDirect. Online Reference Collection.

Burr, D.M., Soare, R.J., Wan Bun Tseung, J.M., Emery, J.P., 2005. Young (late Amazonian), near-surface, ground ice features near the equator, Athabasca Valles, Mars. Icarus 178 (1), 56–73.

Burr, D.M., Tanaka, K.L., Yoshikawa, K., 2009. Pingos on Earth and Mars. Planet. Space Sci. 57 (5-6), 541–555.

Burrows, S.M., et al., 2014. A physically based framework for modeling the organic fractionation of sea spray aerosol from bubble film Langmuir equilibria. Atmos. Chem. Phys. 14, 13601–13629.

Cabrol, N.A., Grin, E.A., Pollard, W.H., 2000. Possible frost mounds in an ancient martian lake bed. Icarus 145, 91–107.

Campbell, A.C., et al., 1988. Chemistry of hot springs on the Mid-Atlantic Ridge. Nature 335, 514–519.

Cavanaugh, C., 1985a. Symbiosis of chemoautotrophic bacteria and marine invertebrates from hydrothermal vents and reducing sediments. In: Jones, M.L. (Ed.). Hydrothermal Vents of the Eastern Pacific: An Overview, 6. *Bull. Biol. Soc. Wash.*, pp. 373–388. Imprint: Vienna, Va.: Published for the Biological Society of Washington by INFAX Corp., 1985; 547 p.

Cavanaugh, C.M., 1985b. Symbiosis of chemoautotrophic bacteria in marine invertebrates Ph.D. thesis Harvard Univ., Cambridge, Mass.

Chamberland, D., 1996. The NASA OCEAN Project — An Ocean-Space Analog. Life Support Biosph. Sci. 2 (3-4), 183–190.

Chevrier, V.F., Valentin, E.R., 2012. Formation of recurring slope lineae by liquid brines on present-day Mars. Geophys. Res. Lett. 39, L21202. http://doi.org/10.1029/2012GL054119.

Choblet, G., et al., 2017. Powering prolonged hydrothermal activity inside Enceladus. Nat. Astronomy 1, 841–847.

Chyba, C.F, McDonald, G.D., 1995. The origin of life in the Solar System; Current issues. Annu Rev Earth Planet Sci. 23, 215–249.

Chyba, C.F, Whitmire, D.P, and Reynolds, R. (2000), In: Protostars and Planets IV. Mannings V, Boss AP, Russell SS, Ed. Tucson: Univ. of Arizona Press, pp. 1365–1393.

Chyba, C., 2000. Correction: energy for microbial life on Europa. Nature 406, 368. https://doi.org/10.1038/35019159.

Chyba, C.F., Phillips, C.B., 2001. Possible ecosystems and the search for life on Europa. Proc Natl Acad Sci U S A. 98 (3), 801–804.

Clark, B.C., Morris, R.V., McLennan, S.M., Gellert, R., Jolliff, B., Knoll, A.H., et al., 2005. Chemistry and mineralogy of outcrops at Meridiani Planum. Earth Planet Sci. Lett. 240, 73–94. http://doi.org/10.1016/j.epsl.2005.09.040.

Converse, D.R., Holland, H.D., Edmond, J.M., 1984. Flow rates in the axial hot springs on the East Pacific Rise (21 N): implications for the heat budget and the formation of massive sulfide deposits. Earth Planet. Sci. Lett. 69, 159–175.

Coombs, C.R., Hawke, B.R, Wilson, L., 1989. Kauhako crater and channel, Kalaupapa, Molokai: a terrestrial analog to lunar sinuous rilles. Proc. Lunar Planet. Sci. Conf. 20th, 195–206.

Coombs, C.R., Hawke, B.R., 1992. A search for intact lava tubes on the Moon: possible lunar base habitats. In: Mendell, W.W. (Ed.). The Second Conference on Lunar Bases and Space Activities of the 21st Century, 3166. NASA Conferences Publication, Houston, TX (United States), pp. 219–229.

Corliss, J.B., Dymond, J., Gordon, L.I., Edmond, J.M., von Herzen, R.P., Ballard, R.D., Green, K., Williams, D., Bainbridge, A., Crane, K., van Andle, T.H., 1979. Submarine thermal sprirngs on the galapagos rift. Science 203 (4385), 1073–1083. http://doi.org/10.1126/science.203.4385.1073.

Crane, K., 1985. The distribution of geothermal fields on the midocean ridge. In: Jones, M.L. (Ed.). Hydrothermal Vents of the Eastern Pacific: An Overview, 6. Bull. Biol. Soc. Wash., pp. 3–18. Imprint: Vienna, Va.: Published for the Biological Society of Washington by INFAX Corp., 1985; 547 p.

Crane, K., Aikman III, F., Embley, R., Hammond, S., Malahoff, A., Lupton, J., 1985. The distribution of geothermal fields on the Juan de Fuca Ridge. J. Geophys. Res. 90, 727–744.

Crisler, J.D., Newville, T.M., Chen, F., Clark, B.C., Schneegurt, M.A., 2012. Bacterial growth at the high concentrations of magnesium sulfate found in martian soils. Astrobiology 12, 98–106. http://doi.org/10.1089/ast.2011.0720.

Cruikshank, D.P., Wood, C.A., 1972. Lunar rilles and Hawaiian volcanic features: possible analogues. Moon 3, 412–447.

Dalmaso, G.Z., Ferreira, D., Vermelho, A.B., 2015. Marine extremophiles: a source of hydrolases for biotechnological applications. Mar Drugs 13 (4), 1925–1965. http://doi.org/10.3390/md13041925.

Darwin, C., 1871. Letter no. 7471. Darwin Correspondence Project. Cambridge University Library, Cambridge. Available online at http://www.darwinproject.ac.uk/DCP-LETT-7471.

Deamer, D.W., Georgiou, C.D., 2015. Hydrothermal conditions and the origin of cellular life. Astrobiology 15, 1091–1095.

Deamer, D., Damer, B., 2017. Can Life Begin on Enceladus? a perspective from hydrothermal chemistry. Astrobiology 17 (9), 834–839.

Delacour, A., Früh-Green, G.L., Bernasconi, S.M., Kelley, D.S., 2008. Sulfur in peridotites and gabbros at Lost City (30 N, MAR): Implications for hydrothermal alteration and microbial activity during serpentinization. Geochim. Cosmochim. Acta 72 (20), 5090–5110.

Delaney, J.R., et al., 1998. The quantum event of oceanic crustal accretion: impacts of diking at mid-ocean ridges. Science 281, 222–230.

D'Elia, T., Veerapaneni, R., Rogers, S.O., 2008. Isolation of microbes from Lake Vostok accretion ice. Appl. Environ. Microbiol. 74, 4962–4965.

deLeeuw, G., et al., 2011. Production flux of sea spray aerosol. Rev. Geophys. 49, RG2001.

Dittami, S.M., Tonon, T., 2012. Genomes of extremophile crucifers: new platforms for comparative genomics and beyond. Genome Biol 13, 166.

Dole, S.H., 1964. Habitable Planets for Man. Blaisdell, New York, NY.

Dundas, C.M., Mellon, M.T., McEwen, A.S., Lefort, A., Keszthelyi, L.P., Thomas, N.the HiRISE Team, 2008. HiRISE observations of fractured mounds: possible Martian pingos. Geophys. Res. Lett. 35 (4), L04201.

Dundas, C.M., McEwen, A.S., 2010. An assessment of evidence for pingos on Mars using HiRISE. Icarus 205 (1), 244–258.

Eberhart, G.L., Rona, P.A., Honnorez, J., 1988. Geologic controls of hydrothermal activity in the Mid-Atlantic Ridge rift valley: tectonics and volcanics. Mar. Geophys. Res. 10, 233–259.

Edmond, J.M., 1986. Hydrothermal activity in the North Atlantic. In: Vogt, P.R., Tucholke, B.E. (Eds.), *The Geology of North America*, vol. M, *The Western North Atlantic Region*. Geological Society of America, Boulder, pp. 173–188.

Embley, R.W. & Lupton, J.E. (2004), In: The Subseafloor Biosphere at Mid-Ocean Ridges (eds Wilcock, W.S.D., DeLong, E.F., Kelley, D.S., Baross, J.A. & Cary, S.C.) 75–97 (American Geophysical Union, Washington, DC).

Facchini, M.C., et al., 2008. Primary submicron marine aerosol dominated by insoluble organic colloids and aggregates. Geophys. Res. Lett. 35, L17814.

Fairén, A.G., Davila, A.F., Lim, D., Bramall, N., Bonaccorsi, R., Zavaleta, J., Uceda, E.R., Stoker, C., Wierzchos, J., Dohm, J.M., Amils, R., Andersen, D., McKay, C.P., 2010. Astrobiology through the ages of Mars: the study of terrestrial analogues to understand the habitability of mars. Astrobiology 10 (8), 821–843.

Fendrihan, S., Legat, A., Pfaffenhuemer, M., Gruber, C., Weidler, G., Gerbi, F., et al., 2006. Extremely halophilic archaea and the issue of long-term microbial survival. Rev. Env. Sci. Bio/Tech. 5, 203–218. http://doi.org/10.1007/s11157-006-0007-y.

Fernández-Remolar, D.C., Rodriguez, N., Gómez, F., Amils, R., 2003. Geological record of an acidic environment driven by iron hydrochemistry: the Tinto River system. J. Geophys. Res. 108. doi:10.1029/2002JE001918).

Fernández-Remolar, D., Morris, R.V., Gruener, J.E., Amils, R., Knoll, A.H., 2005. The Río Tinto Basin, Spain: mineralogy, sedimentary geobiology, and implications for interpretation of outcrop rocks at Meridiani Planum, Mars. Earth Planet Sci Lett 240, 149–167.

Figueredo, P., Greeley, R., Neuer, S., Irwin, L., Schulze-Makuch, D., 2003. Locating potential biosignatures on Europa from surface geology observations. Astrobiology 3, 851–861.

Fleischaker, G.R., 1990. Origins of life: an operational definition. Orig Life Evol Biosph 20, 127–137.

Fox-Powell, M.G., Hallsworth, J.E., Cousins, C.R., Cockell, C.S., 2016. Ionic strength is a barrier to the habitability of Mars. Astrobiology 16, 427–442. http://doi.org/10.1089/ast.2015.1432.

Friedmann, E.I., 1982. Endolithic microorganisms in Antarctic cold desert. Science 215, 1045–1053.

Fruh-Green, G.L., Kelley, D.S., Bernasconi, S.M., Karson, J.A., Ludwig, K.A., Butterfield, D.A., Boschi, C., Proskurowski, G., 2003. 30,000 years of hydrothermal activity at the Lost City vent field. Science 301, 495–498.

Frűh-Green, G.L., Orcutt, B.N., Roumejon, S., Lilley, M.D., Morono, Y., Cotterill, C., Green, S., Escartin, J., John, B.E., McCaig, A.M., Cannat, M., Menez, B., Schwarzenbach, E.M., Williams, M.J., Morgan, S., Lang, S.Q., Schrenk, M.O., Brazelton, W.J., Akizawa, N., Boschi, C., Dunkel, K.G., Quemeneur, M., Whattam, S.A., Mayhem, L., Harris, M., Bayrakci, G., Behrmann, J.H., Herrero-Bervera, E., Hesse, K., Liu, H.Q., Sandaruwan Ratnayake, A., Twing, K., Weis, D., Zhao, R., Bilenker, L., 2018. Magmatism, serpentinization and life: insights through drilling the Atlantis Massif (IODP Expedition 357). Lithos 323, 137–155.

Fryer, P., et al., 1990. Conical Seamount: SeaMARC II, Alvin Submersible, and Seismic-Reflection Studies. Proc Ocean Drill Program Initial Rep 125, 69–80.

Fustec, A., Desbruy•res, D., Juniper, S.K., 1987. Deep-sea hydrothermal vent communities at 13 N on the East Pacific Rise: Microdistribution and temporal variations. Biol. Oceanogr. 4, 121–164.

Gaidos, E.J., Nealson, K.H., Kirschvink, J.L., 1999. Life in ice-covered oceans. Science 284 (5420), 1631–1633.

Gaidos, E.J., Nimmo, F., 2000. Planetary science: tectonics and water on Europa. Nature 405, 637.

Gaill, F., Hunt, S., 1991. The biology of annelid worms from high temperature hydrothermal vent regions. Rev. Aquat. Sci. 4, 107–137.

Gantt, B., Meskhidze, N., 2013. The physical and chemical characteristics of marine primary organic aerosol: a review. Atmos. Chem. Phys. 13, 3979–3996.

Garland, J.L., Alazraki, M.P., Atkinson, C.F., Finger, B.W., 1998. Evaluating the feasibility of biological waste processing for long term space missions. Acta Hortic 469, 71–78. http://doi.org/10.17660/actahortic.1998.469.6.

Gaston, C.J., et al., 2011. Unique ocean-derived particles serve as a proxy for changes in ocean chemistry. J. Geophys. Res. 116, D18310.

Gendrin, A., Mangold, N., Bibring, J.P., Langevin, Y., 2005. sulfates in Martian layered terrains: the OMEGA/Mars express view. Science 307, 1587–1591. http://doi.org/10.1126/science.1109087.

Glein, C.R., Baross, J.A., Waite, J.H., 2015. The pH of Enceladus' ocean. Geochim.Cosmochim.Acta 162, 202–219. http://doi.org/10.1016/j.gca.2015.04.017.

Goudge, T.A., Fassett, C.I., Head, J.W., Mustard, J.F., Aureli, K.L., 2016. Insights into surface runoff on early Mars from paleolake basin morphology and stratigraphy. Geol. Soc. Am. 44, 419–422. http://doi.org/10.1130/G37734.1.

Grant, W.D., 2004. Life at low water activity. Phil. Trans. R. Soc. B Biol. Sci. 359, 1249–1267. http://doi.org/10.1098/rstb.2004.1502.

Grassle, J.F., 1985. Hydrothermal vent animals: Distribution and biology. Science 229, 713–717.

Grassle, J.F., 1986. The ecology of deep-sea hydrothermal vent communities. Adv. Mar. Biol. 23, 301–362.

Grassle, J.F., Humphris, S.E., Rona, P.A., Thompson, G., Van Dover, C.L., 1986. Animals at Mid-Atlantic Ridge hydrothermal vents (abstract). Eos Trans. AGU 67, 1022.

Grassle, J.F., 1989. Species diversity in deep-sea communities. Trends Ecol. Evol. 4, 12–15.

Greeley, R., 1971. Lunar Hadley rille: considerations of its origin. Science 172, 722–725.

Green, J., Short, N.M, 1971. Volcanic Landforms and Surface Features: A Photographic Atlas and Glossary. Springer-Verlag, New York, Heidelberg, and Berlin, pp. 1–18.

Guo, S.S. and Ai, W.D., 2001. Prospect of the Advanced Life Support Program Breadboard Project at Kennedy Space Center in USA. Space Med. Med. Eng. (Beijing) 14 (2), 149–153.

Ha, Z., Chan, C.K., 1999. The water activities of $MgCl_2$, $Mg(NO_3)_2$, $MgSO_4$, and their mixtures. Aerosol. Sci. Tech. 31, 154–169. http://doi.org/10.1080/027868299304219.

Hamilton, T.L., Bryant, D.A., Macalady, J.L., 2016. The role of biology in planetary evolution: cyanobacterial primary production in low-oxygen Proterozoic oceans. Environ Microbiol 18 (2), 325–340.

Hammond, S.R., 1997. Offset caldera and crater collapse on Juan de Fuca ridge-flank volcanoes. Bull. Volcanol. 58, 617–627.

Hand, K., Chyba, C.F., Priscu, J.C., Carlson, R.W., Nealson, K.H, 2009. Astrobiology and the potential for life on Europa. In: Pappalardo, R.T., McKinnon, W.B., Khurana, K.K. (Eds.), Europa. University of Arizona Press, Tucson, pp. 589–630.

Harrison, J.C., Jackson, M.P.A, 2014. Exposed evaporite diapirs and minibasins above a canopy in central Sverdrup Basin, Axel Heiberg Island, Arctic Canada. Basin Res. 26, 567–596. http://doi.org/10.1111/bre.12037.

Haymon, R.M., Macdonald, K.C., 1985. The geology of deep-sea hot springs. Am. Sci. 73, 441–449.

Haymon, R.M., 1989. Hydrothermal processes and products on the Galapagos Rift and East Pacific Rise. In: Winterer, E.L., Hussong, D.M., Decker, R.W. (Eds.), *The Geology of North America*, vol. N, *The Eastern Pacific Ocean and Hawaii*. Geological Society of America, Boulder, pp. 173–188.

Haymon, R.M., Fornari, D.J., Edwards, M.H., Carbotte, S., Wright, D., Macdonald, K.C., 1991. Hydrothermal vent distribution along the East Pacific Rise crest (9 09′-54′N) and its relationship to magmatic and tectonic processes on fast-spreading mid-ocean ridges. Earth Planet. Sci. Lett. 104, 513–534.

Head III, J.W., Hiesinger, H., Ivanov, M.A., Kreslavsky, A., Pratt, S., Thomson, B.J, 1999. Possible ancient oceans on Mars: evidence from Mars Orbiter Laser Altimeter data. Science 286, 2134–2137.

Hessler, R.R., Smithey. Jr, W.M., 1983. The distribution and community structure of megafauna at the Galapagos Rift hydrothermal vents. In: Rona, P.A., Bostrom, K., Laubier, L., Smith, K.L. Jr. (Eds.). Hydrothermal Processes at Seafloor Spreading Centers, NATO Conf. Ser. 4, 12. Plenum, New York, NY, pp. 735–770.

Hibbitts, C.C., Grieves, G.A., Poston, M.J., Dyar, M.D., Alexandrov, A.B., Johnson, M.A., Orlando, T.M., 2011. Thermal stability of water and hydroxyl on the surface of the Moon from temperature-programmed desorption measurements of lunar analog materials. Icarus 213 (1), 64–72.

Hirooka, S., Hirose, Y., Kanesaki, Y., Higuchi, S., Fujiwara, T., Onuma, R., Era, A., Ohbayashi, R., Uzuka, A., Nozaki, H., et al., 2017. Acidophilic green algal genome provides insights into adaptation to an acidic environment. Proc Natl Acad Sci U S A 114 (39), E8304–E8313.

Horz, F., 1985. Lava tubes: Potential shelters for habitats. In: Mendell, W.W. (Ed.), Lunar Bases and Space Activities of the 21st Century. Lunar and Planetary Institute, Houston, pp. 405–412.

Hsu, H.W., Postberg, F., Sekine, Y., Shibuya, T., Kempf, S., Horányi, M., Juhász, A., Altobelli, N., Suzuki, K., Masaki, Y., et al., 2015. Ongoing hydrothermal activities within Enceladus. Nature 519 (7542), 207–210.

Iess, L., et al., 2014. The gravity field and interior structure of Enceladus. Science 344, 78–80.

Ilés-Almar, E., 1992. A possible terrestrial analogy of the "Cantaloupe Terrain" on Triton's surface, American Astronomical Society, 24th DPS Meeting, id.18.11-P; Bulletin of the American Astronomical Society, Vol. 24, 968.

Ingersoll, A.P., Tryka, K.A., 1990. Triton's plumes: the dust devil hypothesis. Science 250 (4979), 435–437.

Jannasch, H.W., Wirsen, C.O., 1979. Chemosynthetic primary production at East Pacific seafloor spreading centers. Bioscience 29, 592–598.

NATO Conf. Ser. 4 , H.W.Jannasch, 1983. Microbial processes at deep-sea hydrothermal vents. In: Rona, P.A., Bostr6m, K., Laubier, L., Smith, K.L. (Eds.). Hydrothermal Processes at Seafloor Spreading Centers, 12. Plenum, New York, NY, pp. 677–709 NATO Conf. Ser. 4.

Jannasch, H.W., 1984. Chemosynthesis: the nutritional basis for life at deep-sea vents. Oceanus 27, 73–78.

Jannasch, H.W., 1985. The chemosynthetic support of life and the microbial diversity at deep-sea hydrothermal vents. Proc. R. Soc. London, Ser. B 225, 277–297.

Jannasch, H.W., Mottl, M.J., 1985. Geomicrobiology of deep-sea hydrothermal vents. Science 229, 717–725.

Jannasch, H.W., Wirsen, C.O., Nelson, D.C., Robertson, L.A., 1985. *Thiomicrospira crunogena* sp. nov., A colorless sulfur oxidizing bacterium from a deep-sea hydrothermal vent. Int. J. Syst. Bacteriol. 35, 422–424.

Jayarathne, T., et al., 2016. Enrichment of saccharides and divalent cations in sea spray aerosol during two phytoplankton blooms. Environ. Sci. Technol. 50, 11511–11520.

Johnson, D., Ferreira, M.A., Kenchington, E., 2018. Climate change is likely to severely limit the effectiveness of deep-sea ABMTs in the North Atlantic. Mar. Policy 87, 111–122.

Johnson, D.E., 2019. Protecting the lost city hydrothermal vent system: All is not lost, or is it? Mar. Policy 107, 103593.

Joseph, 2016. Investigating Seafloors and Oceans: From Mud Volcanoes to Giant Squid. *Elsevier Science & Technology Books Publishers*, New York, NY, pp. 581 ISBN: 978-0-12- 809357-3.

Joyce, G.F. (1994), In: Origins of Life: The Central Concepts. Deamer, D.W, Fleischaker, G.R. Eds. Boston: Jones & Bartlett; 1994. Pp. xi–xii.

Juniper, S.K., Tunnicliffe, V., Desbruybres, D., 1990. Regional-scale features of northeast Pacific, East Pacific Rise, and Gulf of Aden vent communities. In: McMurray, G.R. (Ed.), Gorda Ridge: A Seafloor Spreading Center in the United States' Exclusive Economic Zone. Springer-Verlag, New York, NY, pp. 965–967.

Kahana, A., Schmitt-Kopplin, P., Lancet, D., 2019. Enceladus: First observed primordial soup could arbitrate origin-of-life debate. Astrobiology 19 (10), 1263–1278.

Karl, D.M., Bird, D.F., Björkman, K., Houlihan, T., Shackelfor, R., Tupas, L., 1999. Microorganisms in the accreted ice of Lake Vostok, Antarctica. Science 286, 2144–2147.

Karson, J.A., Früh-Green, G.L., Kelley, D.S., Williams, E.A., Yoerger, D.R., Jakuba, M., 2006. Detachment shear zone of the Atlantis Massif core complex, Mid-Atlantic ridge, 30 N. Geochem. Geophys. Geosyst. 7, Q06016.

Karunatillake, S., Wray, J.J., Gasnault, O., McLenna, S.M., Rogers, A.D., Squyres, S.W., et al., 2016. Sulfates hydrating bulk soil in the Marian low and middle latitudes. Geophys. Res. Lett. 41, 7987–7996. http://doi.org/10.1002/2014GL061136.

Kashefi, K., Lovley, D.R., 2003. Extending the upper temperature limit for life. Science 301, 934.

Katterhorn, S.A., Prockter, L.M., 2014. Evidence for subduction in the ice shell of Europa. Nat. Geosci. 7, 762–767.

Keene, W.C., et al., 2007. Chemical and physical characteristics of nascent aerosols produced by bursting bubbles at a model air–sea interface. J. Geophys. Res. 112, D21202.

Kelley, D.S., Karson, J.A., Blackman, D.K., Fruh-Green, G.L., Butterfield, D.A., Lilley, M.D., Olson, E., Schrenk, M.O., Roe, K., Lebon, G.T., Rivissigno, P., 2001. An off-axis hydrothermal vent field near the Mid-Atlantic Ridge at 30 N. Nature 412, 145–149.

Kelley, D.S., Baross, J.A., Delaney, J.R., 2002. Volcanoes, fluids, and life at mid-ocean ridge spreading centers. Annu. Rev. Earth Planet Sci. 30, 385–491.

Kelley, D.S., Karson, J.A., Frűh-Green, G.L., Yoerger, D., Shank, T.M., Butterfield, D.A., Hayes, J.M., Schrenk, M.O., Olson, E., Proskurowski, G., Jakuba, M., Bradley, A., Larson, B., Ludwig, K., Glickson, D., Buckman, K., Bradley, A.S., Brazelton, W.J., Roe, K., Elend, M.J., Delacour, A., Bernasconi, S.M., Lilley, M.D., Baross, J.A., Summons, R.E., Sylva, S.P., 2005. A Serpentinite-hosted ecosystem: the lost city hydrothermal field. Science 307, 1428–1434.

Kelley, D.S., Frűh-Green, G.L., Karson, J.A., Ludwig, K.A., 2007. The lost city hydrothermal field revisited. Oceanography 20, 90–99.

Klein, F.W., Koyanaki, R.Y., Nakata, J.S., and Tanigawa, W.R., 1987. The seismicity of Kilauea's magma system. In: Volcanism in Hawaii, Edited by Robert W. Decker, Thomas L. Wright, and Peter H. Stauffer; pp. 1019-1185; U.S. Geol. Surv. Professional Paper 1350; 1667 pp., https://pubs.usgs.gov/pp/1987/1350/.

Klein, F., Humphris, S.E., Guo, W., Schubotz, F., Schwarzenbach, E.M., Orsi, W.D., 2015. Fluid mixing and the deep biosphere of a fossil Lost City-type hydrothermal system at the Iberia Margin. Proc. Natl Acad. Sci. USA 112 (39), 12036–12041. https://doi.org/10.1073/pnas.1504674112.

Konn, C., Charlou, J.L., Holm, N.G., Mousis, O., 2015. The production of methane, hydrogen, and organic compounds in ultramafic-hosted hydrothermal vents of the Mid-Atlantic Ridge. Astrobiology 15, 381–399.

Lainey, V., Karatekin, O., Desmars, J., Charnoz, S., Arlot, J.-E., Emelyanov, N., Le Poncin-Lafitte, C., Mathis, S., Remus, F., Tobie, G., 2012. Strong tidal dissipation in Saturn and constraints on Enceladus' thermal state from astrometry. Astrophys. J. 752, 14.

Lamarche-Gagnon, G., Comery, R., Greer, C.W., Whyte, L.G., 2015. Evidence of in situ microbial activity and sulphidogenesis in perennially sub-0 C and hypersaline sediments of a high Arctic permafrost spring. Extremophiles 19, 1–15. https://doi.org/10.1007/s00792-014-0703-4.

Lang, S.Q., Früh-Green, G.L., Bernasconi, S.M., Lilley, M.D., Proskurowski, G., Méhay, S., Butterfield, D.A., 2012. Microbial utilization of abiogenic carbon and hydrogen in a serpentinite-hosted system. Geochim. Cosmochim. Acta. 92, 82–99.

Lartaud, F., et al., 2011. Fossil evidence for serpentinization fluids fueling chemosynthetic assemblages. Proc. Natl. Acad. Sci. USA 108 (19), 7698–7703.

Lay, C.-Y., Mykytczuk, N.C.S., Niederberger, T.D., Martineau, C., Greer, C.W., Whyte, L.G., 2012. Microbial diversity and activity in hypersaline high Arctic spring channels. Extremophiles 16, 177–191. https://doi.org/10.1007/s00792-011-0417-9.

Lay, C.-Y., Mykytczuk, N.C.S., Yergeau, E., Lamarche-Gagnon, G., Greer, C.W., Whyte, L.G., 2013. Defining the functional potential and active community members of a sediment microbial community in a high-arctic hypersaline subzero spring. Appl. Environ. Microbiol. 79, 3637–3648. https://doi.org/10.1128/AEM.00153-13.

Leck, C., Bigg, E.K., 2008. Comparison of sources and nature of the tropical aerosol with the summer high arctic aerosol. Tellus B 60, 118–126.

Lefort, A., Russell, P.S., Thomas, N., McEwen, A.S., Dundas, C.M., Kirk, R.L., 2009. Observations of periglacial landforms in Utopia Planitia with the High Resolution Imaging Science Experiment (HiRISE). J. Geophys. Res. 114 (E4). https://doi.org/10.1029/2008JE003264.

Lefort, A., Russell, P.S., Thomas, N., 2010. Scalloped terrains in the Peneus and Amphitrites Paterae region of Mars as observed by HiRISE. Icarus 205 (1), 259. https://doi.org/10.1016/j.icarus.2009.06.005.

Letzter, R. (2020), Entire cities could fit inside the moon's monstrous lava tubes, https://www.livescience.com/lava-tubes-mars-and-moon-habitable.html

Léveillé, R.J., Saugata, D., 2010. Lava tubes and basaltic caves as astrobiological targets on Earth and Mars: a review. Planet. Space Sci. 58 (4), 592–598.

Liu, Y., Whitman, W.B., 2008. Metabolic, phylogenetic, and ecological diversity of the methanogenicarchaea. Ann. N. Y. Acad. Sci. 1125, 171–189. https://doi.org/10.1196/annals.1419.019.

López–Archilla, A.I., Marín, I., Amils, R., 1993. Bioleaching and interrelated acidophilic microorganisms from Río Tinto, Spain. Geomicrobiol. J. 11, 223–233.

López–Archilla, A.I., Marín, I., Amils, R., 2001. Microbial community composition and ecology of an acidic aquatic environment: the Tinto River, Spain. Microb. Ecol. 41, 20–35.

Lonsdale, P., 1977. Structural geomorphology of a fast-spreading rise crest: The East Pacific Rise near 3 25′S. Mar. Geophys. Res. 3, 251–293.

Lorenz, R.D., Gleeson, D., Prieto-Ballesteros, O., Gomez, F., Hand, K., Bulat, S., 2011. Analog environments for a Europa lander mission. Adv. Space Res. 48 (4), 689–696.

Ludwig, K.A., Kelley, D.S., Butterfield, D.A., Nelson, B.K., Fruh-Green, G., 2006. Formation and evolution of carbonate chimneys at the Lost City Hydrothermal Field. Geochim. Cosmochim. Acta 70, 3625–3645.

Lutz, R.A., 1984. Deep-sea hydrothermal vents: Oases on the ocean floor. In: Calhoun, D. (Ed.), 1985 Yearbook of Science and the Future. Encyclopaedia Britannica, Chicago, IL, pp. 226–242.

Lutz, R.A., Fritz, L.W., Rhoads, D.C., 1985. Molluscan growth at deep-sea hydrothermal vents. In: Jones, M.L. (Ed.). Hydrothermal Vents of the Eastern Pacific: An Overview, 6. Bull. Biol. Soc. Wash., pp. 199–210. Imprint: Vienna, Va.: Published for the Biological Society of Washington by INFAX Corp., 1985; 547 p.

Lutz, R.A., 1988. Dispersal of organisms at deep-sea hydrothermal vents' A review. Oceanol. Acta 8, 23–29.

Lutz, R.A., Kennish, M.J., 1993. Ecology of deep-sea hydrothermal vent communities: a review. Rev. Geophys. 31 (3), 211–242.

Macdonald, K.C., Becker, K., Spiess, F.N., Ballard, R.D., 1980. Hydrothermal heat flux of the "black smoker" vents on the East Pacific Rise. Earth Planet. Sci. Lett. 48, 1–7.

Macdonald, K.C., 1982. Mid-ocean ridges' fine scale tectonic, volcanic, and hydrothermal processes within the plate boundary zone. Annu. Rev. Earth Planet. Sci. 10, 155–190.

Macdonald, K.C., 1983. Crustal processes at spreading centers. Rev. Geophys. 21, 1441–1454.

Macdonald, G.A., Abott, A.T., Peterson, F.L., 1983. Volcanoes in the Sea. Univ. of Hawaii, Honolulu, pp. 517.

Mackay, J.R., 1979. Pingos of the Tuktoyaktuk Peninsula area, Northwest Territories. Geogr. Phys. Quat. 23 (1), 3–61.

Mackay, J.R., 1998. Pingo growth and collapse, Tuktoyaktuk Peninsula area, western Arctic coast, Canada: a long-term field study. Géographie physique et Quaternaire 52, 271–323.

Madigan, M.T., Martinko, J.M., Parker, J., 1997. Brock Biology of Microorganisms. Prentice–Hall, Upper Saddle River, NJ.

Malin, M.C., Edgett, K.S., 2000. Evidence for recent groundwater seepage and surface runoff on Mars. Science 288, 2330–2335. https://doi.org/10.1126/science.288.5475.2330.

Margesin, R., Miteva, V., 2011. Diversity and ecology of psychrophilic microorganisms. Res. Microbiol. 162, 346–361. https://doi.org/10.1016/j.resmic.2010.12.004.

Martin, W., Baross, J., Kelley, D., Russell, M.J., 2008. Hydrothermal vents and the origin of life. Nat Rev Microbiol 6, 805–814.

Martínez, G.M., Renno, N.O., 2013. Water and brines on Mars: current evidence and implications for MSL. Space Sci. Rev. 175, 29–51. https://doi.org/10.1007/s11214-012-9956-3.

Matson, D.L., Castillo-Rogez, J.C., Davies, A.G., Johnson, T.V., 2012. Enceladus: a hypothesis for bringing both heat and chemicals to the surface. Icarus 221, 53–62.

Mayhew, L.E., Ellison, E.T., McCollom, T.M., Trainor, T.P., Templeton, A.S., 2013. Hydrogen generation from low-temperature water-rock reactions. Nat. Geosci. 6, 478–484. https://doi.org/10.1038/ngeo1825.

McEwen, A.S., Ohja, L., Dundas, C.M., Mattson, S.S., Byrne, S., Wray, J.J., et al., 2011. Seasonal flows on warm martian slopes. Science 333, 740–743. https://doi.org/10.1126/science.1204816.

McGill, G.E., 1989. Buried topography of Utopia, Mars: Persistence of a giant impact depression. J. Geophys. Res. 94, 2753–2759.

McKay, C.P., Porco, C.C., Altheide, T., Davis, W.L., Kral, T.A., 2008. The possible origin and persistence of life on Enceladus and detection of biomarkers in the plume. Astrobiology 8 (5), 909–919.

McKay, C., Mykytczuk, N., Whyte, L., 2012. Life in ice on other worlds. In: Miller, R., Whyte, L. (Eds.), Polar Microbiology: Life in a Deep Freeze. ASM Press, Washington, DC, pp. 290–304. https://doi.org/10.1128/9781555817183.

McKay, C.P., Anbar, A.D., Porco, C., Tsou, P., 2014. Follow the plume: the habitability of Enceladus. Astrobiology 14 (4), 352–355.

McKinnon, W.B., 1999. Convective instability in Europa's floating ice shell. Geophys Res Lett 26, 951–954.

McKinnon, W.B., 2015. Effect of Enceladus's rapid synchronous spin on interpretation of Cassini gravity. Geophys. Res. Lett. 42, 2137–2143.

Meinert, C., Myrgorodska, I., deMarcellus, P., Buhse, T., Nahon, L., Hoffmann, S.V, DHendecourt, L.L.S., Meierhenrich, U.J., 2016. Ribose and related sugars from ultraviolet irradiation of interstellar ice analogs. Science 352, 208–212.

Méndez-García, C., Peláez, A.I., Mesa, V., Sánchez, J., Golyshina, O.V., Ferrer, M., 2015. Microbial diversity and metabolic networks in acid mine drainage habitats. Front. Microbiol. 6, 475. https://doi.org/10.3389/fmicb.2015.00475.

Mesa, V., Gallego, J.L.R., González-Gil, R., Lauga, B., Sánchez, J., Méndez-García, C., Peláez, A.I., 2017. Bacterial, archaeal, and eukaryotic diversity across distinct microhabitats in an acid mine drainage. Front. Microbiol. 8, 1756.

Möhlmann, D., Thomsen, K., 2011. Properties of cryobrines on Mars. Icarus 212, 123–130. https://doi.org/10.1016/j.icarus.2010.11.025.

Morgenstern, A., Hauber, E., Reiss, D., van Gasselt, S., Grosse, G., Schirrmeister, L., 2007. Deposition and degradation of a volatile-rich layer in Utopia Planitia, and implications for climate history on Mars. J. Geophysical Res.: Planets 112 (E6), E06010. https://doi.org/10.1029/2006JE002869.

Mormile, M.R., Biesen, M.A., Gutierrez, M.C., Ventosa, A., Pavlovich, J.B., Onstott, T.C., et al., 2003. Isolation of *Halobacterium salinarum* retrieved directly from halite brine inclusions. Environ. Microbiol. 5, 1094–1102. https://doi.org/10.1046/j.1462-2920.2003.00509.x.

in GermanTranslated by , F.Muller, 1959. Observations on pingos in German. In: Sinclair, D.A. (Ed.). Meddelelser om Grønland, 153. National Research Council of Canada, Ottawa, pp. 117 Translated byTT-1073, 1963.

Neish, C.D., Lorenz, R.D., Turtle, E.P., Barnes, J.W., Trainer, M.G., Stiles, B., Kirk, R., Hibbitts, C.A., Malaska, M.J., 2018. Strategies for Detecting Biological Molecules on Titan. Astrobiology 18 (5), 571–585. https://doi.org/10.1089/ast.2017.1758.

Niederberger, T.D., Perreault, N., Lawrence, J.R., Nadeau, J.L., Mielke, R.E., Greer, C.W., et al., 2009. Novel sulfur-oxidizing streamers thriving in perennial cold saline springs of the Canadian high Arctic. Environ. Microbiol. 11, 616–629. https://doi.org/10.1111/j.1462-2920.2008.01833.x.

Ohara, Y., et al., 2012. A serpentinite-hosted ecosystem in the Southern Mariana Forearc. Proc Natl Acad Sci USA 109 (8), 2831–2835.

Ojha, L., Wilhelm, M.B., Murchie, S.L., McEwen, A.S., Wray, J.J., Hanley, J., et al., 2015. Spectral evidence for hydrated salts in recurring slope lineae on Mars. Nat. Geosci. 8, 829–832. https://doi.org/10.1038/ngeo2546.

Omelon, C.R., Pollard, W.H., Marion, G.M., 2001. Seasonal formation of ikaite (caco3 · 6h 2o) in saline spring discharge at Expedition Fiord, Canadian High Arctic: assessing conditional constraints for natural crystal growth. Geochim. Cosmochim. Acta 65, 1429–1437. https://doi.org/10.1016/S0016-7037(00)00620-7.

Omelon, C.R., Pollard, W.H., Andersen, D.T., 2006. A geochemical evaluation of perennial spring activity and associated mineral precipitates at Expedition Fjord, Axel Heiberg Island, Canadian High Arctic. Appl. Geochem. 21, 1–15. https://doi.org/10.1016/j.apgeochem.2005.08.004.

Orphan, V.J., House, C.H., Hinrichs, K.U., McKeegan, K.D., DeLong, E.F., 2002. Direct phylogenetic and isotopic evidence for multiple groups of Archaea involved in the anaerobic oxidation of methane. Geochim. Cosmochim. Acta 66, A571.

Osinski, G.R., Lee, P., Parnell, J., Spray, J.G., 2005. A case study of impact-induced hydrothermal activity: the Haughton impact structure, Devon Island, Canadian High Arctic. Meteorit Planet Sci. 40, 1859–1877.

Pappalardo, R.T., Barr, A.C., 2004. The origin of domes on Europa: the role of thermally induced compositional diapirism. Geophys. Res. Lett. 31. https://doi.org/10.1029/2003GL019202.

Paris, A.J., Davies, E.T., Tognetti, L., Zahniser, C., 2019. Prospective lava tubes at Hellas Planitia: Leveraging volcanic features on Mars to provide crewed missions protection from radiation. J. Washington Acad. Sci. arXiv:2004.13156 [astro-ph.EP].

Parker, T.J., Gorsline, D.S., Saunders, R.S., Pieri, D.C., Schneeberger, D.M., 1993. Coastal geomorphology of the Martian northern plains. J. Geophys. Res. 98 (E6), 11061–11078.

Parkinson, C.D., Liang, M.-C., Hartman, H., Hansen, C.J., Tinetti, G., Meadows, V., Kirschvink, J.L., Yung, Y.L., 2007. Enceladus: Cassini observations and implications for the search for life (Research Note). Astron. Astrophys. 463, 353–357. https://doi.org/10.1051/0004-6361:20065773.

Peddinti, D.A., McNamara, A.K., 2015. Material transport across Europa's Ice Shell. Geophys Res Lett 42, 4288–4293.

Perreault, N., Andersen, D.T., Pollard, W.H., Greer, C.W., Whyte, L.G., 2007. Characterization of the prokaryotic diversity in cold saline perennial springs of the Canadian high Arctic. Appl. Environ. Microbiol. 73, 1532–1543. https://doi.org/10.1128/AEM.01729-06.

Perreault, N., Greer, C.W., Andersen, D.T., Tille, S., Lacrampe-Couloume, G., Lollar, B.S., et al., 2008. Heterotrophic and autotrophic microbial populations in cold perennial springs of the high arctic. Appl. Environ. Microbiol. 74, 6898–6907. https://doi.org/10.1128/AEM.00359-08.

Peterson, D.W., Swanson, D.A., 1974. Observed formation of lava tubes during 1970-1971 at Kilauea Volcano, Hawaii. Studies Speleol. 2 (Part 6), 209–222.

Pinkerton, H., 1995. Rheological properties of basaltic lavas at sub-liquidus temperatures: laboratory and field measurements on lavas from Mount Etna. J. Volcanol. Geotherm. Res. 68 (4), 307–323.

Pollard, W., Omelon, C., Andersen, D., McKay, C., 1999. Perennial spring occurrence in the Expedition Fiord area of western Axel Heiberg Island, Canadian High Arctic. Can. J. Earth Sci. 36, 105–120. https://doi.org/10.1139/e98-097.

Pontefract, A., Zhu, T.F., Walker, V.K., Hepburn, H., Lui, C., Zuber, M.T., Ruvkun, G., Carr, C.E., 2017. Microbial Diversity in a Hypersaline Sulfate Lake: A Terrestrial Analog of Ancient Mars. Front. Microbiol. 8, 1819. https://doi.org/10.3389/fmicb.2017.01819.

Porco, C.C., et al., 2006. Cassini observes the active south pole of Enceladus. Science 311, 1393–1401.

Porco, C.C., Dones, L., Mitchell, C., 2017. Could it be snowing microbes on Enceladus? Assessing conditions in its plume and implications for future missions. Astrobiology 17, 876–901.

Postberg, F., et al., 2009. Sodium salts in E-ring ice grains from an ocean below the surface of Enceladus. Nature 459, 1098–1101.

Postberg, F., Schmidt, J., Hillier, J., Kempf, S., Srama, R., 2011. A saltwater reservoir as the source of a compositionally stratified plume on Enceladus. Nature 474, 620–622.

Postberg, F., Khawaja, N., Abel, B., Choblet, G., Glein, C.R., Gudipati, M.S., Henderson, B.L., Hsu, H-W., Kempf, S., Klenner, F., Moragas-Klostermeyer, G., Magee, B., Nölle, L., Perry, M., Reviol, R., Schmidt, J., Srama, R., Stolz, F., Tobie, G., Trieloff, M., Waite, J.H., 2018. Macromolecular organic compounds from the depths of Enceladus. Nature 558, 564–568.

Proskurowski, G., Lilley, M.D., Kelley, D.S., Olson, E.J., 2006. Low temperature volatile production at the Lost City Hydrothermal Field, evidence from a hydrogen stable isotope geothermometer. Chem. Geol. 229, 331–343.

Proskurowski, G., et al., 2008. Abiogenic hydrocarbon production at lost city hydrothermal field. Science 319, 604–607.

Rampelotto, P.H., 2010. Resistance of microorganisms to extreme environmental conditions and its contribution to astrobiology. Sustainability 2 (6), 1602–1623. https://doi.org/10.3390/su2061602.

Rampelotto, P.H, 2013. Extremophiles and extreme environments. Life (Basel) 3, 482–485.

Rona, P.A., 1984. Hydrothermal mineralization at seafloor spreading centers. Earth Sci. Rev. 20, 1–104.

Rona, P.A., Klinkammer, G., Nelsen, T.A., Trefry, J.H., Elderfield, H., 1986. Black smokers, massive sulphides, and vent biota at the Mid-Atlantic Ridge. Nature 321, 33–37.

Rona, P.A., 1987. Oceanic ridge crest processes. Rev. Geophys. 25, 1089–1114.

Rona, P.A., Windenfalk, L., Bostrom, K., 1987. Serpentinized ultramafics and hydrothermal activity at the Mid-Atlantic Ridge Crest near 15N. J. Geophys. Res. 92, 1417–1427.

Rona, P.A., 1988. Hydrothermal mineralization at oceanic ridges. Can. Mineral. 26, 431–465.

Rowley, T., R. Giardino, R.G. Aguilar, and J.D. Vitek (2015), Periglacial processes and landforms in the critical zone, Developments in Earth Surface Processes; 19:397-447. DOI:10.1016/B978-0-444-63369-9.00013-6.

Russell, L.M., et al., 2010. Carbohydrate-like composition of submicron atmospheric particles and their production from ocean bubble bursting. Proc. Natl Acad. Sci. USA 107, 6652–6657.

Sagan, C. (1996), In: *Circumstellar Habitable Zones*. Doyle, L.R. Ed, Menlo Park, CA: Travis House; pp. 3–14.

Sapers, H.M., Ronholm, J., Raymond-Bouchard, I., Comrey, R., Osinski, G.R., Whyte, L.G., 2017. Biological characterization of microenvironments in a hypersaline cold spring Mars analog. Front. Microbiol. 8, 2527. https://doi.org/10.3389/fmicb.2017.02527.

Sauro, F., Pozzobon, R., Massironi, M., Berardinis, P.D., Santagata, T., Waele, J.D., 2020. Lava tubes on Earth, Moon and Mars: A review on their size and morphology revealed by comparative planetology. Earth Sci. Rev. 209, 103288.

Schmidt, J., Brilliantov, N., Spahn, F., Kempf, S., 2008. Slow dust in Enceladus' plume from condensation and wall collisions in tiger stripe fractures. Nature 451, 685–688.

Schmidt, B., Blankenship, D., Patterson, G., Schenk, P., 2011. Active formation of chaos terrain over shallow subsurface water on Europa. Nature 479, 502–505.

Schmitt-Kopplin, P., et al., 2012. Dissolved organic matter in sea spray: a transfer study from marine surface water to aerosols. Biogeosciences 9, 1571–1582.

Schrenk, M.O., Kelley, D.S., Bolton, S., Baross, J.A., 2004. Low archaeal diversity linked to sub-seafloor geochemical processes at the Lost City Hydrothermal Field, Mid-Atlantic Ridge. Environ. Microbiol. 6, 1086–1095.

Schrenk, M.O., Brazelton, W.J., Lang, S.Q., 2013. Serpentinization, carbon, and deep life. Rev Mineral Geochem 75 (1), 575–606.

Schwieterman, E.W., Kiang, N.Y., Parenteau, M.N., Harman, C.E., DasSarma, S., Fisher, T.M., Arney, G.N., Hartnett, H.E., Reinhard, C.T., Olson, S.L., Meadows, V.S., Cockell, C.S., Walker, S.I., Grenfell, J.L., Hegde, S., Rugheimer, S., Hu, R., Lyons, T.W., 2018. Exoplanet biosignatures: A review of remotely detectable signs of life. Astrobiology 18 (6), 663–708.

Sejourne, A., et al., 2012. Evidence of an eolian ice-rich and stratified permafrost in Utopia Planitia, Mars. Icarus 60, 248–254.

Sekine, Y., Shibuya, T., Postberg, F., Hsu, H.W., Suzuki, K., Masaki, Y., Kuwatani, T., Mori, M., Hong, P.K., Yoshizaki, M., et al., 2015. High-temperature water-rock interactions and hydrothermal environments in the chondrite-like core of Enceladus. Nat. Commun. 6, 8604.

Siegert, M.J., Ellis-Evans, J.C., Tranter, M., Mayer, C., Petit, J., Salamatin, A., Priscu, J.C., 2001. Physical, chemical, and biological processes in Lake Vostok and other Antarctic subglacial lakes. Nature 414, 603–609.

Skinner Jr., J.A., Tanaka, K.L, 2007. Evidence for and implications of sedimentary diapirism and mud volcanism in the southern Utopia highland–lowland boundary plain, Mars. Icarus 186, 41–59.

Sleep, N.H., Morton, J.L., Burns, L.E., Wolery, T.J., 1983. Geophysical constraints on the volume of hydrothermal flow at ridge axes. In: Rona, P.A., Bostrtm, K., Laubier, L., Smith, K.L. Jr. (Eds.). *Hydrothermal Processes at Seafloor Spreading Centers, NATO Conf.* Ser. 4, 12. Plenum, New York, NY, pp. 53–69.

Sleep, N.H., 1991. Hydrothermal circulation, anhydrite precipitation, and thermal structure of ridge axes. J. Geophys. Res. 96, 2375–2387.

Sleep, N.H., Meibom, A., Fridriksson, T., Coleman, R.G., Bird, D.K., 2004. H_2-rich fluids from serpentinization: geochemical and biotic implications. Proc. Natl Acad. Sci. USA. 101, 12818–12823.

Smithsonian Ocean Team, The microbes that keep hydrothermal vents pumping; https://ocean.si.edu/ecosystems/deep-sea/microbes-keep-hydrothermal-vents-pumping.

Soare, R.J., Burr, D.M., Wan Bun Tseung, J-M., 2005. Pingos and a possible periglacial landscape in Utopia Planitia. Icarus 174 (2), 373–382.

Soare, E., et al., 2019. Possible (closed system) pingo and ice-wedge/thermokarst complexes at the mid latitudes of Utopia Planitia, Mars. Icarus. https://doi.org/10.1016/j.icarus.2019.03.010.

Sousa, F.L., et al., 2013. Early bioenergetic evolution. Philos. Trans. R Soc. Lond. B Biol. Sci. 368 (1622), 20130088.

Southward, E.C., 1991. Three new species of Pogonophora, including two vestimentiferans, from hydrothermal sites in the Lau Back-Arc Basin (southwest Pacific Ocean). J. Nat. Hist. 25, 859–881.

Spiess, F.N., et al., 1980. East Pacific rise: hot springs and geophysical experiments. Science 207, 1421–1433.

Steinbrügge, G., Voigt, J.R.C., Wolfenbarger, N.S., Hamilton, C.W., Soderlund, K.M., Young, D.A., Blankenship, D.D., Vance, S.D., Schroeder, D.M., 2020. Brine migration and impact-induced cryovolcanism on Europa. Geophys. Res. Lett. 47, e2020GL090797. https://doi.org/10.1029/2020GL090797.

Stevenson, A., Cray, J.A., Williams, J.P., Santos, R., Sahay, R., Neuenkirchen, N., et al., 2015. Is there a common water-activity limit for the three domains of life? ISME 9, 1333–1351. https://doi.org/10.1038/ismej.2014.219.

Swift, D.A. et al. 2015. Ice and snow as land-forming agents, In book: Snow and Ice-Related Hazards, Risks, and Disasters (Second Edition) pp.165-198; Publisher: Elsevier.

Takai, K., Komatsu, T., Inagaki, F., Horikoshi, K., 2001. Distribution of archaea in a black smoker chimney structure. Appl. Environ. Microbiol. 67, 3618–3629.

Takai, K., Nakamura, K., Toki, T., Tsunogai, U., Miyazaki, M., Miyazaki, J., Hirayama, H., Nakagawa, S., Nunoura, T., Horikoshi, K., 2008. Cell proliferation at 122°C and isotopically heavy CH_4 production by a hyperthermophilic methanogen under high-pressure cultivation. Proc. Natl. Acad. Sci. USA 105, 10949–10954.

Tan, S., Sekine, Y., Shibuya, T., Miyamoto, C., Takahashi, Y., 2021. The role of hydrothermal sulfate reduction in the sulfur cycles within Europa: laboratory experiments on sulfate reduction at 100 MPa. Icarus 357, 114222.

Taubner, R.S., Schleper, C., Firneis, M.G., Rittmann, S.K.M.R., 2015. Assessing the ecophysiology of methanogens in the context of recent astrobiological and planetological studies. Life 5, 1652–1686. https://doi.org/10.3390/life5041652.

Taubner, R.S., Pappenreiter, P., Zwicker, J., Smrzka, D., Pruckner, C., Kolar, P., Bernacchi, S., Seifert, A.H., Krajete, A., Bach, W., Peckmann, J., Paulik, C., Firneis, M.G., Schleper, C., Rittmann, S.K.R., 2018. Biological methane production under putative Enceladus-like conditions. Nat. Commun. 9 (1), 748.

Teske, A., et al., 2002. Microbial diversity of hydrothermal sediments in the Guaymas Basin: evidence for anaerobic methanotrophic communities. Appl. Environ. Microbiol. 68, 1994–2007.

Tighe, S., et al., 2017. Genomic methods and microbiological technologies for profiling novel and extreme environments for the Extreme Microbiome Project (XMP). J. Biomol. Tech. 28 (1), 31–39.

Tornabene, L.L., Moersch, J.E., Osinski, G.R., Lee, P., Wright, S.P., 2005. Spaceborne visible and thermal infrared lithologic mapping of impact-exposed subsurface lithologies at the Haughton impact structure, Devon Island, Canadian High Arctic: applications to Mars. Meteorit Planet Sci. 40, 1835–1858.

Tosca, N.J., Knoll, A.H., McLennan, S.M., 2008. Water activity and the challenge for life on early Mars. Science 320, 1204–1207. https://doi.org/10.1126/science.1155432.

Tunnicliffe, V., Botros, M., De Burgh, M.E., Dinet, A., Johnson, H.P., Juniper, S.K., McDuff, R.E., 1986. Hydrothermal vents of Explorer Ridge, northeast Pacific. Deep Sea Res 33, 401–412.

Tunnicliffe, V., 1990. Observations on the effects of sampling on hydrothermal vent habitat and fauna of Axial Seamount, Juan de Fuca Ridge. J. Geophys. Res. 95 12,961-12,966.

Tunnicliffe, V., 1991. The biology of hydrothermal vents: ecology and evolution, *Oceanogr*. Mar. Biol. Ann. Rev. 29, 319–407.

Tunnicliffe, V., 1992. Hydrothermal-vent communities of the deep sea. Am. Sci. 80, 336–349.

Turner, R.D., Lutz, R.A., 1984. Growth and distribution of molluscs at deep-sea vents and seeps. Oceanus 27, 54–62.

Turner, R.D., 1985. Notes on mollusks of deep-sea vents and reducing sediments. Am. Malacol. Bull., 23–34.

Vance, S., Harnmeijer, J., Kimura, J., Hussmann, H., Demartin, B., Brown, J.M., 2007. Hydrothermal systems in small ocean planets. Astrobiology 7 (6), 987–1005.

Van Dover, C.L., Hessler, R.R., 1990. Spatial variation in faunal composition of hydrothermal vent communities on the East Pacific Rise and Galapagos spreading center. In: Mc-Murray, G.R. (Ed.), Gorda Ridge: A Seafloor Spreading Center in the United States' Exclusive Economic Zone. Springer-Verlag, New York, NY, pp. 253–264.

Vaniman, D.T., Bish, D.L., Chipera, S.J., Fialips, C.I., Carey, J.W., Feldman, W.C., 2004. Magnesium sulphate salts and the history of water on Mars. Nature 431, 663–665. https://doi.org/10.1038/nature02973.

Von Damm, K.L. (1995), In: *Physical, Chemical, Biological, and Geological Interactions within Seafloor Hydrothermal Systems* (Eds Humphris, S., Zierenberg, R., Mullineau, L. & Thomson R.) 222–247 (American Geophysical Union, Washington, DC).

Von Damm, K.L., et al., 2003. Extraordinary phase separation and segregation in vent fluids from the southern East Pacific Rise. Earth Planet Sci. Lett. 206, 265–378.

Vreeland, R.H., Rosenzweig, W.D., Powers, D.W., 2000. Isolation of a 250 million-year-old halotolerant bacterium from a primary salt crystal. Nature 407, 897–900. https://doi.org/10.1038/35038060.

Waite Jr, J.H., et al., 2009. Liquid water on Enceladus from observations of ammonia and 40Ar in the plume. Nature 460, 487–490.

Waite Jr, J.H., et al., 2017. Cassini finds molecular hydrogen in the Enceladus plume: evidence for hydrothermal processes. Science 356, 155–159.

Wanke, H., Bruckner, J., Dreibus, G., Rieder, R., Ryabchikov, I., 2001. Chemical composition of rocks and soils at the Pathfinder site. Space Sci. Rev. 96, 317–330. https://doi.org/10.1023/A:1011961725645.

Whitman, W.B., Coleman, D.C., Wiebe, W.J., 1998. Prokaryotes: the unseen majority. Proc Natl Acad Sci USA 95, 6578–6583.

Wierzchos, J., Ascaso, C., Sancho, L.G., Green, A., 2003. Iron-rich diagenetic minerals are biomarkers of microbial activity in Antarctic rocks. Geomicrobiol. J. 20, 15–24.

Wierzchos, J., Camara, B., De Los Rios, A., Davila, A.F., Sanchaz Almazo, M., Artieda, O., Wierzchos, K., Gomez-Silva, B., McKay, C., Ascaso, C., 2011. Microbial colonization of Ca-sulfate crusts in the hyperarid core of the Atacama Desert: Implications for the search for life on Mars. Geobiology 9 (1), 44–60. https://doi.org/10.1111/j.1472-4669.2010.00254.x. PMID 20726901.

Williams, D.L., Green, K., van Andel, T.H., Von Herzen, R.P., Dymond, J.R., Crane, K., 1979. The hydrothermal mounds of the Galapagos Rift: Observations with DSRV Alvin and detailed heat flow studies. J. Geophys. Res. 84, 7467–7484.

Williams, L.B., Canfield, B., Holloway, J.R., Williams, P., 2002. Smectite incubation of organic molecules in seafloor hydrothermal systems, The V.M. Goldschmidt Conference. Davos, A837.

Williams, L.B., Canfield, B., Voglesonger, K.M., et al., 2005. Organic molecules formed in a "primordial womb. Geology 33, 913–916.

Wilson, T.W., et al., 2015. A marine biogenic source of atmospheric ice-nucleation particles. Nature 525, 234–238.

Wray, J.J., Miliken, R.E., Dundas, C.M., Swayze, G.A., Andrews-Hanna, J.C., Baldridge, A.M., et al., 2011. Columbus crater and other possible groundwater-fed paleolakes of Terra Sirenum, Mars. J. Geophys. Res. 116, E01001. https://doi.org/10.1029/2010JE003694.

Yoshikawa, K., Harada, K., 1995. Observations on Nearshore pingo growth, Adventdalen, Spitsbergen. Permafrost Periglac. Processes 6, 361–372.

Editor in Chief , K.Yoshikawa, 2013. Pingos. In: Shroder, J., Giardino, R., Harbor, J. (Eds.). Treatise on Geomorphology, 8. Academic Press, San Diego, CA, pp. 274–297 Glacial and Periglacial Geomorphology.

Zaets, I., Burlak, O., Rogutskyy, I., Vasilenkoa, A., Mytrokhyn, O., et al., 2011. Bioaugmentation in growing plants for lunar bases. Adv. Space Res. 47, 1071–1078.

Zanetti, M., Hiesinger, H., Reiss, D., Hauber, E., Neukum, G., 2009. Scalloped Depression Development on Malea Planum and the Southern Wall of the Hellas Basin, Mars. Lunar Planetary Sci. 40, 2178.

Bibliography

Früh-Green, G.L., et al., 2003. 30,000 years of hydrothermal activity at the Lost City Vent Field. Science 301, 495–498.

Ludwig, K.A., Kelley, D.S., Shen, C., Cheng, H., Edwards, R.L., 2005. U/Th geochronology of carbonate chimneys at the Lost City hydrothermal field. Eos Trans. AGU 86, V51B–1487.

Tunnicliffe, V., 1991. The biology of hydrothermal vents: Ecology and evolution, *Oceanogr*. Mar. Biol. 29, 319–407.

Chapter 8

Surface environment evolution for Venus, Earth, and Mars— the planets which began with the same inventory of elements

8.1 Relevance of examining the surface environment evolution for Venus, Earth, and Mars

The three inner planets in the Solar System—Venus, Earth, and Mars—began with the same inventory of elements. Three things have—more than anything—contributed to their similarities and differences: (1) The size of the planets and their ability to keep hold of their atmospheres; (2) their distance from the faint young Sun; (3) chance. The latter may seem simplistic, but the last major collisions during planetary formation—e.g., the one that gave rise to Earth's Moon—primed obliquities, and rotation rates, may have decided the fate of plate tectonics being turned on or not. It would be useful to have a go at this thorny comparative assessment. It helps to open people's minds to what may be there to explore.

The presence of surface water and life, and plate tectonics make Earth unique in our Solar System. By contrast, Venus, the planet closest to the Earth in terms of bulk properties, is characterized by a lack of surface water and life, significantly higher surface temperature and pressure, and a different mode of tectonics. These observations have motivated decades of studies aimed at the question of what leads to the differences between Earth and Venus. Modeling studies that address this question, as well as similar issues arising with comparisons of Earth and Mars, often assume that the profound differences between the current states of Earth and Venus reflect differences in planetary size, position to the Sun, and material parameters (e.g. strength of near-surface rock) (Lenardic et al., 2016).

Mars is about half the size of Earth by diameter and has a much thinner atmosphere, with an atmospheric volume less than 1% of Earth's. The atmospheric composition is also significantly different: primarily carbon dioxide-based, while Earth's is rich in nitrogen and oxygen.

An alternative view is that the current states and tectonic regimes of Earth and Venus represent two equally possible solutions to the dynamical evolution of a planet with the same size, bulk composition, solar proximity, and material characteristics. Rather than being distinct consequences of physical or chemical conditions, Earth and Venus may instead represent an inherent "bi-stability" in the dynamic state of terrestrial planets (Lenardic et al., 2016).

8.2 Specialties in the geologic activities of the three planets—Stagnant, Episodic, and Mobile Lid Regimes

Critical to the ideas of thermal and lid-state evolution is the hypothesis that the tectonic state of a planet can change over its geologically active lifetime. Growing evidence suggests that planetary tectonic regimes do transition. The Earth is the only body in the Solar System for which significant observational constraints are accessible to such a degree that they can be used to discriminate between competing models of Earth's tectonic evolution. It is a natural tendency to use observations of the Earth to inform more general models of planetary evolution. The solid portion of Earth was formed from accretion of material and debris formed in the primitive Solar System. Earth's early evolution included the differentiation of its interior and the development of a primordial atmosphere. However, our understating of Earth's evolution is far from complete.

The ultimate tectonic state of a planet is a result of a balance between the coupling of the plates and the mantle beneath, and also the buoyancy forces driving convective motion. These two factors are critically sensitive to how a planet's thermal state evolves through time; buoyancy forces are strongly coupled to the temperature drop across the convecting mantle and induced lithospheric stresses to the internal viscosity, and thus mantle temperature (O'Neill et al., 2016).

According to O'Neill et al. (2016), the tectonic regime of a planet depends critically on the contributions of basal and internal heating to the planetary mantle, and how these evolve through time. Weller and Lenardic (2018) explain that "In a stagnant-lid regime, the cold and stiff outermost rock layer does not participate in mantle overturn, nor does it exhibit significant horizontal surface motions. In contrast to the stagnant-lid regime, plate tectonics (as manifest on Earth), is characterized by the horizontal motions of strong surface plates. Surface motion is accommodated by localized failure along relatively narrow plate boundary zones. The critical difference between a stagnant-lid and a plate tectonic regime, in terms of a planet's thermal state, is that the cold surface plates of plate-tectonics participate in mantle overturn and the associated cooling of the planetary interior. As a result, plate tectonics is considered to be an example of a mobile-lid style of mantle convection (also referred to as active-lid convection). In contrast to observations of both Earth and Mars, it has been suggested that Venus has been, and perhaps still is, operating in an episodic-lid regime (e.g., Turcotte, 1993; Fowler and O'Brien, 1996; Moresi and Solomatov, 1998). The episodic regime is highly dynamic, characterized by periods of extreme quiescence (akin to stagnant-lid) punctuated with rapid episodes of surface overturn and mobility (Moresi and Solomatov, 1998)." Mobile-lids are identified by active yielding of the surface, with appreciable horizontal motions and interaction with the deep interior (e.g., subducting slabs) (Weller and Lenardic, 2018).

Weller and Lenardic (2018) further explain that a general survey of geologic activity of the three main terrestrial planets (Venus, Earth, and Mars) in the Solar System suggests that the Earth is currently unique in that it operates within a plate tectonic regime. In contrast to the Earth, both Mars and Venus exhibit very different tectonic states. Observations suggest that Mars may operate within what is termed a single plate mode of tectonics, or a stagnant-lid regime (e.g., Nimmo and Stevenson, 2000).

O'Neill et al. (2016) used visco-plastic mantle convection simulations, with evolving core–mantle boundary temperatures, and radiogenic heat decay, to explore how these factors affect tectonic regime over the lifetime of a planet. The simulations demonstrated that (1) hot, mantle conditions, coming out of a magma ocean phase of evolution, can produce a "hot" stagnant-lid regime, whilst a cooler post magma ocean mantle may begin in a plate tectonic regime; (2) planets may evolve from an initial hot stagnant-lid condition, through an episodic regime lasting 1–3 Gyr, into a plate-tectonic regime, and finally into a cold, senescent stagnant lid regime after ~10 Gyr of evolution, as heat production and basal temperatures wane; and (3) the thermal state of the post magma ocean mantle, which effectively sets the initial conditions for the subsolidus mantle convection phase of planetary evolution, is one of the most sensitive parameters affecting planetary evolution—systems with exactly the same physical parameters may exhibit completely different tectonics depending on the initial state employed. O'Neill et al. (2016)'s estimates of the early Earth's temperatures suggest Earth may have begun in a hot stagnant lid mode, evolving into an episodic regime throughout most of the Archaean, before finally passing into a plate tectonic regime. The implication of O'Neill et al. (2016)'s results is that, for many cases, plate tectonics may be a phase in planetary evolution between hot and cold stagnant states, rather than an end-member.

Geologists have long debated the timing of the onset of plate tectonics on Earth (e.g., O'Neill et al., 2007b; Condie and Pearse, 2008). There has been a consensus that as Earth cools, tectonic activity will wane, and eventually Earth will settle into a cold, stagnant-lid regime, similar to Mars today (e.g., Nimmo and Stevenson, 2000, O'Neill et al., 2007a). However, there is no such consensus on what form tectonics might have taken during the Earth's deep geologic past.

According to Weller et al. (2018), "A key observation and open question in the Earth and Planetary Sciences is that the Earth is seemingly unique in that it exhibits plate tectonics and a buffered climate allowing liquid water to exist at the surface over its geologic lifetime. While we know that plate-tectonics is currently in operation on the Earth, the timing of its onset, the length of its activity, and its prevalence outside the Earth are far from certain. Recent work suggests that the Earth has not always been within a plate-tectonic regime, and that it has evolved over time. Multiple lines of geochemical and geologic evidence, as well as geophysical models of planetary evolution, suggest that the Earth initiated in a stagnant-lid (one plate-planet), followed by an "adolescent" episodic-lid (alternating between stagnant and mobile-lids), before settling into a "mature" modern style of plate-tectonics (mobile-lid) [e.g., Debaille et al., 2013; O'Neill and Debaille, 2014; O'Neill et al., 2016; Lenardic et al., 2016; Weller and Lenardic, 2018]."

A major change in Earth's geodynamics occurred ~ 3 billion years (Ga) ago, likely related to the onset of modern and continuous plate tectonics. However, the question of how Earth functioned prior to this time is poorly constrained. Using anomaly in the isotope of a chemical element Neodymium (symbol Nd)—a rare-earth element—in a 2.7 Ga old tholeiitic lava flow from the Abitibi Greenstone Belt, Debaille et al (2013) provided insight into the early-formed mantle heterogeneities of Earth. Using a numerical modeling approach, Debaille et al (2013) showed that convective mixing is inefficient in absence of mobile-lid plate tectonics. According to Debaille et al (2013), the preservation of a ^{142}Nd anomaly until 2.7 Ga ago can be explained if throughout the Hadean and Archean, Earth was characterized by a stagnant-lid regime, possibly with sporadic and short subduction episodes. The major change in geodynamics observed around ~3 Ga ago can then reflect the transition from stagnant-lid plate tectonics to modern mobile-lid plate tectonics. Solving the paradox of a convective but poorly-mixed mantle has implications not only for Archean Earth, but also for other planets in the Solar System such as Mars.

In recent years, there has been growing geodynamic and geochemical evidence that suggests that plate tectonics may not have operated on the early Earth, with both the timing of its onset and the length of its activity far from certain. Recently, the potential of tectonic bi-stability (multiple stable, energetically allowed solutions) has been shown to be dynamically viable, both from analytical analysis and through numeric experiments in two and three dimensions. This indicates that multiple tectonic modes may operate on a single planetary body at different times within its temporal evolution. It also allows for the potential that feedback mechanisms between the internal dynamics and surface processes (e.g., surface temperature changes driven by long-term climate evolution), acting at different thermal evolution times, can cause terrestrial worlds to alternate between multiple tectonic states over giga-year timescales. The implication within this framework is that terrestrial planets have the potential to migrate through tectonic regimes at similar "thermal evolution times" (e.g., points where they have a similar bulk mantle temperature and energies), but at very different "temporal times" (time since planetary formation). It can be further shown that identical planets at similar stages of their evolution may exhibit different tectonic regimes due to random variations.

Weller and Lenardic (2018) have discussed constraints on the tectonic evolution of the Earth and presented a novel framework of planetary evolution that moves toward probabilistic arguments based on general physical principles, as opposed to particular rheologies (deformations and flows of matter), and incorporates the potential of tectonic regime transitions and multiple tectonics states being viable at equivalent physical and chemical conditions.

Weller and Lenardic (2018) recall that while the nature of tectonics that the early Earth exhibited is hotly debated, an important aspect in planetary evolution that has long been in consensus is that as the Earth cools the driving energy for plate-tectonics will wane, and the Earth will begin to move into stagnant-lid regime similar to observations for current day Mars (e.g., Nimmo and Stevenson, 2000). The implication is that the lid-state of a planet can change over time. Recently, this idea has been bolstered through several studies exploring the sensitivity of mantle convection and lid-states to changes in internal temperature, through internal heating and/or long-term climatic effects (O'Neill et al., 2007; O'Neill et al., 2016; Lenardic et al., 2008; Landuyt and Bercovici, 2009; Foley et al., 2012; Lenardic and Crowley, 2012; Stein et al., 2013; Weller et al., 2015; Lenardic et al., 2016a; Weller and Lenardic, 2016; Weller et al., 2016).

Based on Earth's most robust record, several studies based both on geochemical and geodynamic methods have argued for stagnant to episodic behavior in the Archean through the Precambrian (e.g., Debaille et al., 2013; O'Neill et al., 2013; O'Neill and Debaille, 2014). Several disparate sources provide reasonably compelling evidence for significant changes in the linked internal/external processes of the Earth over time.

Multiple lines of significant, if indirect, evidence for fluctuations in the tectonics of a case study planet, the Earth, are available. The data indicates a relatively quiescent early Earth (post late heavy bombardment), punctuated by periods of extreme activity (from ~3 Ga to ~1 Ga), with perhaps a modern style of plate tectonics emerging within roughly the last Gyr. The key question then becomes—is this temporal pathway of evolution (stagnant/near stagnant to episodic to mobile) representative of general planetary evolution? (Weller and Lenardic, 2018).

Weller et al. (2015) and O'Neill et al. (2016) showed that the thermal evolution of planetary convective systems with high levels of internal heating strongly favor early (hot) stagnant-lid states. However, as radiogenic heating is tapped, the hot stagnant-lid can yield through an intermediary episodic-lid, into a mobile-lid regime. With a further decrease in radiogenic heating, the mobile-lid transitions back into a (now) cold stagnant-lid (summarized in Fig. 8.1).

In the mobile-lids cases, the surface velocity is near that of the interior velocity. Stagnant-lids by contrast show highly limited (e.g., 'resurfacing' times greater than the planet's lifetime) to no surface motions, with no active yielding or communication with the interior. Surface velocities in these cases will be far less than internal velocities (<0.1, often <0.01, internal velocities). For the same parameter values, stagnant-lids have thicker boundary layers, lower heat flux, and higher internal temperatures. Both results are defined from statistically steady-state conditions. Episodic regimes oscillate strongly between both end-member states, with surface velocities of an O(10) increase from mobile-lid values (Weller and Lenardic, 2018).

Plate tectonics, except possibly on a limited scale, is not in evidence on Venus (Schubert and Sandwell, 1995). About 20% of Venus has experienced severe tectonic disruption, possibly as a prelude to plains emplacement (Basilevsky and Head, 1995). Mars also doesn't have plate tectonics. After its formation, Mars was a searing mass of molten rock that eventually cooled to form a static crust around a rocky mantle, yet it is unclear how hot Mars' insides are today.

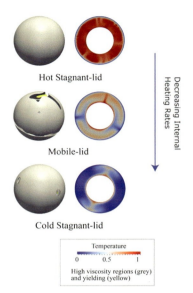

FIG. 8.1 **Effects of internal heating on tectonic regimes.** Internal heating is varied from high to none. Shown are viscosity plots (left— grey shells: high viscosity "plates," yellow bands: regions of yielding) and nondimensional thermal profiles (right) from the Core Mantle Boundary (CMB) to surface. The system exists as a hot stagnant-lid for high internal heating rates, a mobile-lid for intermediate internal heating rates, and in a cold stagnant-lid for low internal heating rates (modified after Weller et al., 2015). *(Source: Weller, M.B., and Lenardic, A., (2018), On the evolution of terrestrial planets: Bi-stability, stochastic effects, and the nonuniqueness of tectonic states, Geoscience Frontiers, 9(1): 91-102 [**Elsevier**])* Copyright info: Reproduced from Elsevier journal.

8.3 Size and composition of Venus, Earth, and Mars

Earth, Venus, and Mars are terrestrial planets that are located within the Sun's Habitable Zone (aka. "Goldilocks Zone"). Venus is often referred to as "Earth's Twin" (or "sister planet"), and for good reason. Despite some rather glaring differences, not the least of which is their vastly different atmospheres, there are enough similarities between Earth and Venus that many scientists consider the two to be closely related. In short, they are believed to have been very similar early in their existence, but then evolved in different directions.

Unlike other planet's in our Solar System, the majority of Earth's surface is covered in liquid water. In fact, about 70.8% of the surface is covered by oceans, lakes, rivers, and other sources, with much of the continental shelf below sea level. In addition, Earth's terrain varies greatly from place to place, regardless of whether or not it is above or below sea level.

The submerged surface has mountainous features, as well as undersea volcanoes, oceanic trenches, submarine canyons, oceanic plateaus and abyssal plains. The remaining portions of the surface are covered by mountains, deserts, plains, plateaus, and other landforms. Over long periods, the surface undergoes reshaping due to a combination of tectonic activity and erosion. A bird's eye view of the similarities and differences among Earth, Venus, and Mars is presented in the below sections.

8.3.1 Earth's size and composition & differences with Mars

Earth is the third planet from the Sun, and the fifth-largest planet in the Solar System. It is smaller than the four gas giants (giant, low-density, planets composed mainly of hydrogen, helium, methane, and ammonia in either gaseous or liquid state)—Jupiter, Saturn, Uranus, and Neptune—but larger than the three other rocky planets, Mercury, Mars, and Venus. Earth has a diameter of roughly 8000 miles (13,000 km) and is mostly round because gravity generally pulls matter into a ball. But the spin of our home planet causes it to be squashed at its poles and swollen at the equator, making the true shape of the Earth an "oblate spheroid." In terms of size, Mars is about half the size of Earth by diameter.

Scientists think Earth was formed at roughly the same time as the Sun and other planets in the Solar System some 4.6 billion years ago when the Solar System coalesced from a giant, rotating cloud of gas and dust known as the *solar nebula*. As the nebula collapsed under the force of its own gravity, it spun faster and flattened into a disk. Most of the material in that disk was then pulled toward the center to form the Sun. Other particles within the disk collided and stuck together to form ever-larger bodies, including Earth. It was thought that because of the asteroids and comets flying around started colliding with Earth, conditions on early Earth may have been hellish.

Earth's interior is divided into layers based on their chemical or physical properties, consisting of a core in the center, a mantle located above the core, and an outer crust. Whereas Earth's core region (inner core & outer core) consists of nickel and iron, its mantle and outer crust are composed of silicate rock and minerals. The diameter of Earth's core is about 4400 miles (7100 km), slightly larger than half the Earth's diameter and about the same size as Mars. The outermost 1400 miles (2250 km) of the core (outer core) are liquid, while the inner core is solid. That solid core is about four-fifths as big as Earth's Moon, at some 1600 miles (2600 km) in diameter. Earth's crust floats on the liquid mantle much as a piece of wood floats on liquid water.

As just indicated, Earth's core is divided between an inner solid and an outer liquid core. The liquid outer core also rotates in the opposite direction as the planet, producing a dynamo effect that is believed to be the source of Earth's magnetosphere.

It was initially believed that Earth started off as a waterless mass of rock. However, in recent years, new analyses of minerals trapped within ancient microscopic crystals suggest that there was liquid water already present on Earth during its first 500 million years. Radioactive materials in the rock and increasing pressure deep within the Earth generated enough heat to melt the planet's interior, causing some chemicals to rise to the surface and form water, while others became the gases of the atmosphere. Recent evidence suggests that Earth's crust and oceans may have formed within about 200 million years after the planet took shape.

Earth's crust is made up of several elements: oxygen (46.6% by weight), silicon (27.7%), aluminium (8.1%), iron (5%), calcium (3.6%), sodium (2.8%), potassium (2.6%), and magnesium (2.1%). The layers of Earth are crust (5 to 70 km thick), mantle (~2,900 km thick), and outer core (~2,200 km thick).

Earth's core consists mostly of iron and nickel and potentially smaller amounts of lighter elements, such as sulfur and oxygen. The mantle is made of iron and magnesium-rich silicate rocks. (The combination of silicon and oxygen is known as silica, and minerals that contain silica are known as silicate minerals).

Holland et al. (2009) reported the discovery of primordial krypton (Kr) in samples derived from Earth's mantle and show it to be consistent with a meteorite or fractionated solar nebula source. These researchers analyzed the krypton and xenon found in upper-mantle gases leaking from the Bravo Dome gas field in New Mexico. They found that the two noble gases have isotopic signatures characteristic of early Solar System material similar to meteorites instead of the modern atmosphere and oceans. According to Holland et al. (2009), the high-precision Kr and Xe isotope data together suggest that Earth's interior acquired its volatiles from accretionary material similar to average carbonaceous chondrites and that the noble gases in Earth's atmosphere and oceans are dominantly derived from later volatile capture rather than impact degassing or outgassing of the solid Earth during its main accretionary stage.

Mars and Earth are very different planets when it comes to temperature, size, and atmosphere, but geologic processes on the two planets are surprisingly similar. Surface temperature (average) on Earth is 57°F (14°C), the average surface temperature on Mars is -81°F (-63°C). We now know that though Mars may currently be very cold, very dry, and very inhospitable, this was not always the case. What is more, we have come to see that even in its current form, Mars and Earth actually have a lot in common. On Mars we see volcanoes, canyons, and impact basins much like the ones we see on Earth. Many of the same physical land features we see on Earth also exist on Mars.

8.3.2 Venus and Earth—twins in terms of size & composition

Venus' proximity to Earth has made it a prime target for early interplanetary exploration. Venus was the first planet beyond Earth visited by a spacecraft (Mariner 2 in 1962), and the first to be successfully landed on (by Venera 7 in 1970). Venus's thick clouds render observation of its surface impossible in visible light, and the first detailed maps did not emerge until the arrival of the Magellan orbiter in 1991.

The Venusian surface was a subject of speculation until some of its secrets were revealed by planetary science in the 20th century. Venera landers in 1975 and 1982 returned images of a surface covered in sediment and relatively angular rocks (Mueller, 2014). The surface was mapped in detail by Magellan in 1990–1991. The ground shows evidence of extensive volcanism, and the sulfur in the atmosphere may indicate that there have been recent eruptions (Bullock and Grinspoon, 2001). Plans have been proposed for rovers or more complex missions, but they are hindered by Venus's hostile surface conditions.

Venus is one of the four terrestrial planets in the Solar System, meaning that it is a rocky body like Earth. Venus is the closest planet to the Earth, not only in distance but also in size, mass, structures, and compositions. Venus and Earth are often called twins because they are similar in size, mass, density, bulk composition, and gravity (Lopes and Gregg, 2004). Venus and Earth are much larger than the other terrestrial planetary bodies of our Solar System. With a radius about 5% smaller, the uncompressed density of Venus works out to be nearly the same as Earth's (relative density: 0.950), implying a similar structure and composition for the body of both planets.

While little direct information exists about Venus' seismology, its similarity in size and density to Earth suggests that it has a similar internal structure—consisting of a core, mantle, and crust. Like that of Earth, the Venusian core is at least partially liquid because the two planets have been cooling at about the same rate.

Venus is actually only a little bit smaller than Earth, with a mass 81.5% of Earth's. The interior of Venus is made of a metallic iron core that is roughly 2400 miles (6000 km) wide. Venus' molten rocky mantle is roughly 1200 miles (3000 km) thick. Venus' crust is mostly basalt, and is estimated to be 6 to 12 miles (10–20 km) thick, on average. The diameter of Venus is 12,103.6 km (7,520.8 mi)—only 638.4 km (396.7 mi) less than Earth's. In brief, Venus is roughly 0.9499 the size of Earth and 0.815 as massive. In terms of volume, the two planets are almost neck and neck, with Venus possessing 0.866 as much volume as Earth.

Past exploration missions reveal that Venus is hellishly hot, devoid of oceans, and apparently lacking plate tectonics. When and why did Venus and Earth's evolutionary paths diverge? Did Venus ever host habitable conditions? These fundamental and unresolved questions drive the need for vigorous new exploration of Venus. The answers are central to understanding Venus in the context of terrestrial planets and their evolutionary processes. Critically, Venus provides important clues to understanding our planet Earth—does hot, dry Venus represent the once and future Earth?

Volcanoes provide a time-dependent source of volatiles to the atmosphere, a process shared by the Earth and Venus. Although variations in the volcanic flux do exist, heat transport by the creation and subduction of the lithosphere on the Earth provides a steady and reliable means to transport heat from the interior (Turcotte and Schubert, 1982). The formation of immense volcanic provinces on the Earth, possibly associated with large buoyant plumes of magma within the mantle, may also be associated with larger impulses of volatiles to the atmosphere (Coffin, 1994). On Venus, it appears that transport of heat from the interior has been accomplished in the recent geologic past by the formation of the basaltic plains, and later by the large volcanoes that grew on top of them. Massive volcanic edifices do exist, however, perched above volcanic plains that cover 80% of the planet (Head et al., 1992).

According to Bullock and Grinspoon (2001)'s description, radar images from the Magellan mission show that the surface of Venus has been geologically active on a global scale, yet its sparse impact cratering record is almost pristine. This geologic record on Venus is consistent with an epoch of rapid plains emplacement 600 to 1100 Myr ago. According to Bullock and Grinspoon (2001), the average crater density on Venus, as revealed by detailed radar images from the Magellan spacecraft, implies that the surface of the planet is 600 –1100 Myr old (McKinnon et al., 1997; Phillips et al., 1992; Schaber et al., 1992). The nature and rates of the planetary resurfacing processes are recorded in the styles and distribution of modified craters. A distinctive feature of Venus' crater population is that only a small number of them are apparently modified by volcanism (Phillips et al., 1991; Schaber et al., 1992). Schaber et al. (1992) report that out of a total of 912 identified impact craters, only 4–7% are partially embayed by volcanic lavas, while 33% are tectonically modified. These facts alone suggest that the surface is young and has undergone little geological processing since it was emplaced. The implication is that an older surface may have been almost entirely wiped out during an epoch of rapid resurfacing followed by a quiescent period and the collection of random impacts from space ever since (Bullock et al., 1993; Schaber et al., 1992).

Weller et al. (2018) have pointed out that based on observations, Venus resurfaced by vast volcanic plains that cover ∼ 80% of the surface, which are thought to have been emplaced in the last 300–1000 Myr (Schaber et al., 1992; Strom et al., 1994; McKinnon et al., 1997), perhaps 'catastrophically' (Schaber 1992; Strom et al., 1994). These observations, along with inferences of limited large scale shortening, are consistent with suggestions of an episodic-lid regime.

Current and future efforts to locate and characterize planetary systems beyond our Solar System (e.g., the Kepler mission and the Transiting Exoplanet Survey Satellite) are aimed at Earth-size planets in the "habitable zones" of their parent stars. Precisely because it may have begun so like

FIG. 8.2 Mars is about half the size of Earth in diameter. *(Image credit: NASA). (Source: Sharp, T. (2012), How Big is Mars? Size of Planet Mars, Space.com; https://www.space.com/16871-how-big-is-mars.html) Copyright info: Unknown.*

Earth, yet evolved to be so different, Venus is the planet most likely to yield new insights into the conditions that determine whether a Venus-sized exoplanet can sustain long-lived habitability.

8.3.3 How large is Mars & what is it made of?

Mars, the fourth planet from the Sun, is the second smallest planet in the Solar System; only Mercury is smaller. Mars (diameter 6790 km) is only slightly more than half (about 53%) the size of Earth (diameter 12750 km; Fig. 8.2). Mars' mass is 6.42×10^{23} kilograms, about 10 times less than Earth. This affects the force of gravity. Gravity on Mars is 38% of Earth's gravity. The circumference of Mars around the equator is about 13,300 miles (21,343 km), but from pole-to-pole Mars is only 13,200 miles (21,244 km) around. This shape is called an oblate spheroid.

Mars is the "Red Planet" for a very good reason: its surface is made of a thick layer of oxidized iron dust and rocks of the same color. In other words, Mars is a "Rusty" planet. The dust that covers the surface of Mars is fine like talcum powder (Fig. 8.3). Beneath the layer of dust, the

FIG. 8.3 Dusty, glass-rich sand dunes covering much of the Martian surface (False color image) *(Image credit: NASA/JPL/University of Arizona). (Source: Sharp, T. (2017a), What is Mars Made Of? Composition of Planet Mars, Space.com. https://www.space.com/16895-what-is-mars-made-of.html) Copyright info: Unknown.*

Martian crust consists mostly of volcanic basalt rock. The soil of Mars also holds nutrients such as sodium, potassium, chloride, and magnesium. The crust is between 6 and 30 miles (10 and 50 km) thick, according to NASA (Sharp, 2017b).

Mars is home to both the highest mountain and the deepest, longest valley in the Solar System. Olympus Mons is roughly 17 miles (27 km) high, about three times as tall as Mount Everest. Olympus Mons is also one of the largest volcanoes in the solar system. It is about 370 miles (600 km) in diameter.

The Valles Marineris system of valleys—named after the Mariner 9 probe that discovered it in 1971—can go as deep as 6 miles (10 km) and runs east-west for roughly 2500 miles (4000 km), about one-fifth of the distance around Mars and close to the width of Australia

Despite appearances, Mars is not a sphere. Because the planet rotates on its axis (every 24.6 h), it bulges at the equator (as do Earth and other planets). At its equator, Mars has a diameter of 4222 miles (6794 km), but from pole to pole, the diameter is 4196 miles (6752 km).

Mars' crust is thought to be one piece. Unlike Earth, the red planet has no tectonic plates that ride on the mantle to reshape the terrain. Since there is little to no movement in the crust, molten rock flowed to the surface at the same point for successive eruptions, building up into the huge volcanoes that dot the Martian surface.

Landslides on Mars exhibit features such as steep collapse, extreme deposit thinning, and long runout. De Blasio and Crosta (2017) investigated the flow dynamics of Martian landslides particularly in Valles Marineris, where landslides are among the largest and longest. They observed that landslides in Valles Marineris share a series of features with terrestrial landslides fallen onto glaciers. The presence of suspected glacial and periglacial morphologies from the same areas of Valles Marineris, and the results of remote sensing measurements suggest the presence of ice under the soil and into the rock slopes. Thus, De Blasio and Crosta (2017) explored with numerical simulation the possibility that such landslides have been lubricated by ice. To establish a plausible rheological (concerned with the flow of matter, primarily in a liquid or gas state, but also as "soft solids" or solids under conditions in which they respond with plastic flow rather than deforming elastically in response to an applied force) model for these landslides, De Blasio and Crosta (2017) introduced two possible scenarios. One scenario assumes ice only at the base of the landslide, the other inside the rock-soil. A numerical model was extended here to include ice in these two settings, and the effect of lateral widening of the landslide. It was found that only if the presence of ice is included in the calculations, do results reproduce reasonably well both the vertical collapse of landslide material in the scarp area, and the extreme thinning and runout in the distal area, which are evident characteristics of large landslides in Valles Marineris. The calculated velocity of landslides (often well in excess of 100 m/s and up to 200 m/s at peak) compares well with velocity estimates based on the run-up of the landslides on mounds. De Blasio and Crosta (2017) conclude that ice may have been an important medium of lubrication of landslides on Mars, even in equatorial areas such as Valles Marineris. The calculated velocity of landslides compares well with velocity estimates based on the run-up of the landslides on mounds.

Difference in color between Earth and Mars is noteworthy, in that Mars is the "Red Planet." Mars' redness is for a very good reason: its surface is made of a thick layer of oxidized iron dust and rocks of the same color. Maybe another name for Mars could be "Rusty." But the ruddy surface does not tell the whole story of the composition of this world. The Mars rovers have used gas chromatography, mass spectrometry, and laser spectrometry to determine the composition of Martian soil. Mars regolith is mostly silicon dioxide and ferric oxide, with a fair amount of aluminum oxide, calcium oxide, and sulfur oxide. Almost 70% of Earth's surface is covered by liquid water. In contrast, Mars now has no liquid water on its surface and is covered with bare rock and dust. Beneath the layer of dust, the Martian crust consists mostly of volcanic basalt rock. The soil of Mars also holds nutrients such as sodium, potassium, chloride, and magnesium. The crust is between 6 and 30 miles (10 and 50 km) thick.

8.4 Atmospheres of Venus, Earth, and Mars

Throughout their evolution, planetary atmospheres are strongly influenced by the radiation and particle emissions from their host star. Different studies have shown that the Sun's radiation in the extreme ultraviolet (EUV) part of the solar spectrum was higher in the past (Ribas et al., 2005; Tu et al., 2015), and thus, the planetary atmospheres are exposed to varying external conditions. Tu et al. (2015) have shown that the star's initial rotation rate and its rotational evolution play an important role for the EUV flux enhancement and the evolution of the atmospheres of terrestrial planets.

8.4.1 Earth's atmosphere

Ideas about atmospheric composition and climate on the early Earth have evolved considerably since the early 1960s. It is generally agreed that the atmosphere contained little or no free oxygen initially and that oxygen concentrations increased markedly near 2.0 billion years ago. A better understanding of past atmospheric evolution is important to understanding the evolution of life and to predicting whether Earth-like planets might exist elsewhere in the galaxy (Kasting, 1993).

According to the present knowledge, Earth's atmosphere is composed of about 78% nitrogen, 21% oxygen, 0.9% argon, and 0.1% other gases. Trace amounts of carbon dioxide, methane, water vapor, and neon are some of the other gases that make up the remaining 0.1%. Heavy noble gases in the atmosphere could have been acquired during the initial accretion process or may have accumulated later through gravitational volatile capture. Noble gas isotopes are key tracers of both the origin of volatiles found within planets and the processes that control their eventual distribution between planetary interiors and atmospheres. No other planet in the Solar system has an atmosphere loaded with free oxygen, which is vital to one of the other unique features of Earth: life. It is usually conceded that the atmospheric greenhouse effect must have been higher in the past to offset reduced solar luminosity, but the levels of atmospheric carbon dioxide and other greenhouse gases required remain speculative.

Water vapor, carbon dioxide, and other gases in the atmosphere trap heat from the Sun, warming Earth. Without this so-called "greenhouse effect," Earth would probably be too cold for life to exist. The air surrounding the Earth becomes thinner farther from the surface. Roughly 100 miles (160 km) above Earth, the air is so thin that satellites can zip through the atmosphere with little resistance. Still, traces of atmosphere can be found as high as 370 miles (600 km) above the planet's surface. Sunlight heats Earth's surface, causing warm air to rise into the troposphere. This air expands and cools as air pressure decreases, and because this cool air is denser than its surroundings, it sinks down and gets warmed by the Earth again. Earth-orbiting satellites have shown that the upper atmosphere actually expands during the day and contracts at night due to heating and cooling.

Above the troposphere, some 30 miles (48 km) above the Earth's surface, is the stratosphere. The still air of the stratosphere contains the ozone layer, which was created when ultraviolet light caused trios of oxygen atoms to bind together into ozone molecules. Ozone prevents most of the Sun's harmful ultraviolet radiation from reaching Earth's surface, where it can damage and mutate life.

Holland et al. (2009)'s study shows that krypton (Kr) and xenon (Xe) trapped in the upper mantle have isotopic signatures characteristic of early Solar System material similar to meteorites rather than those of the modern atmosphere and oceans. Thus, noble gases trapped within the young Earth did not contribute to Earth's later atmospheric composition. Holland et al. (2009) found that the gases which formed the Earth's atmosphere—as well as its oceans—did not come from inside the Earth but from comets and meteorites hitting Earth during the "Late Heavy Bombardment" period. Holland et al. (2009) tested volcanic gases to uncover the new evidence.

According to the theory of the Late Heavy Bombardment (LHB), the inner Solar System was pounded by a sudden rain of solar system debris only 700 million years after it formed, which likely had monumental effects on the nascent Earth. So far, the evidence for this event comes primarily from the dating of lunar samples, which indicates that most impact melt rocks formed in this very narrow interval of time. But Holland et al. (2009)'s research on the origin of Earth's atmosphere may lend credence to this theory as well. The new research found a clear meteorite signature in volcanic gases. This finding suggests that the volcanic gases could not have contributed in any significant way to the Earth's atmosphere. Therefore, the atmosphere and oceans must have come from somewhere else, possibly from a late bombardment of gas- and water-rich materials similar to comets. It, therefore, appears that noble gases (historically also the inert gases; the six naturally occurring noble gases are helium, neon, argon, krypton, xenon, and the radioactive radon) trapped within the young Earth did not contribute to Earth's later atmosphere. It is worthwhile to note that until now, no one has had instruments capable of looking for these subtle signatures in samples from inside the Earth—but now we can do exactly that.

Water vapor, carbon dioxide and other greenhouse gases in the atmosphere trap heat from the Sun, warming Earth. Without this so-called "greenhouse effect," Earth would probably be too cold for life to exist, although a runaway greenhouse effect led to the hellish conditions of Venus' current surface.

8.4.2 Venus' atmosphere and clouds

The atmosphere of the planet Venus is the layer of gases surrounding it. The amount of haze in the Venus middle atmosphere is about 10 times that found in Earth's stratosphere after the most recent major volcanic eruptions, and the thermal energy required for this injection on Venus is greater by about an order of magnitude than the largest of these recent Earth eruptions and about as large as the Krakatoa eruption of 1883 (Esposito, 1984; Bullock and Grinspoon, 2001).

As just indicated, Venus is bathed in a thick, reactive atmosphere. Venus has a much more massive atmosphere than its terrestrial sibling, with a surface pressure about 90 times higher and a composition predominantly of carbon dioxide. It is composed primarily of carbon dioxide (a powerful greenhouse gas) and is much denser and hotter than that of Earth. The present climate of Venus is controlled by an efficient carbon dioxide–water vapor greenhouse effect and by the radiative properties of its global cloud cover. Both the greenhouse effect and clouds are sensitive to perturbations in the abundance of atmospheric water vapor and sulfur gases. It is speculated that the atmosphere of Venus up to around 4 billion years ago was more like that of the Earth with liquid water on the surface. A runaway greenhouse effect may have been caused by the evaporation of the surface water and subsequent rise of the levels of other greenhouse gases (Kasting, 1988).

Planetary-scale processes involving the release, transport, and sequestering of volatiles affect the abundances of atmospheric water vapor and sulfur gases over time, driving changes in climate (Bullock and Grinspoon, 2001). The temperature at the surface is 740 K (467°C, 872°F). Venus has thus the credit of being the hottest planet in our Solar System, even though Mercury is closer to the Sun. With scorching temperatures, Venus has a hellish atmosphere.

The atmosphere of Venus is heavier than any other planet [the atmospheric pressure of Venus is 93 bar (1350 psi)], leading to a surface pressure that is about 93 times that of Earth—somewhat similar to the pressure that exists 930 m deep in the ocean. The very top layer of Venus' clouds zips around the planet every four Earth days, much faster than the planet's sidereal day of 243 days, propelled by hurricane-forced winds traveling roughly 224 mph (360 kph). This super-rotation of Venus's atmosphere, some 60 times faster than Venus itself rotates, may be one of Venus' biggest mysteries.

Bullock and Grinspoon (2001) developed a numerical model of the climate evolution of Venus. Atmospheric temperatures are calculated using a one-dimensional two-stream radiative–convective model that treats the transport of thermal infrared radiation in the atmosphere and clouds. These radiative transfer calculations are the first to utilize high-temperature, high-resolution spectral databases for the calculation of infrared absorption and scattering in Venus' atmosphere. Bullock and Grinspoon (2001) used a chemical/microphysical model of Venus' clouds to calculate changes in cloud structure that result from variations in atmospheric water and sulfur dioxide. It was found that atmospheric abundances of water, sulfur dioxide, and carbon dioxide change under the influence of the exospheric escape of hydrogen, outgassing from the interior, and heterogeneous reactions with surface minerals.

Bullock and Grinspoon (2001)'s models show that intense volcanic outgassing of sulfur dioxide and water during this time would have resulted in the formation of massive sulfuric acid/water clouds and the cooling of the surface for 100–300 Myr. The thick clouds would have subsequently given way to high, thin water clouds as atmospheric sulfur dioxide was lost to reactions with the surface. Surface temperatures approaching 900 K would have been reached 200–500 Myr after the onset of volcanic resurfacing. Evolution to current conditions would have proceeded due to loss of atmospheric water at the top of the atmosphere, ongoing low-level volcanism, and the reappearance of sulfuric acid/water clouds. Bullock and Grinspoon (2001)'s study revealed that the maintenance of sulfuric acid/water clouds on Venus today requires sources of outgassed sulfur active in the past 20–50 Myr, in contrast with the 1.9 Myr as determined from geochemical arguments alone (Fegley and Prinn, 1989).

As just noted, apart from carbon dioxide, the atmosphere of Venus is perpetually shrouded in thick, yellowish clouds of sulfuric acid that trap heat, causing a runaway greenhouse effect. Because the Venusian atmosphere supports opaque clouds of sulfuric acid, making optical Earth-based and orbital observation of the surface is impossible. Consequently, information about the topography has been obtained exclusively by radar imaging (Basilevsky and Head, 2003). Aside from carbon dioxide, the other main component is nitrogen and only trace amounts of water.

The winds supporting super-rotation blow at a speed of 100 m/s (≈360 km/h or 220 mph) (Svedhem et al., 2007) or more. Winds move at up to 60 times the speed of the planet's rotation, while Earth's fastest winds are only 10% to 20% rotation speed (Normile, 2010). On the other hand, the wind speed becomes increasingly slower as the elevation from the surface decreases, with the breeze barely reaching the speed of 10 km/h (2.8 m/s) on the surface (DK Space Encyclopedia: Atmosphere of Venus p 58). Near the poles are anticyclonic structures called polar vortices. Each vortex is double-eyed and shows a characteristic S-shaped pattern of clouds (Piccioni et al., 2007). Above there is an intermediate layer of mesosphere which separates the troposphere from the thermosphere (Svedhem et al., 2007; Bertaux et al., 2007). The thermosphere is also characterized by strong circulation, but very different in its nature—the gases heated and partially ionized by sunlight in the sunlit hemisphere migrate to the dark hemisphere where they recombine and down-well (Bertaux et al., 2007).

Despite the harsh conditions on the surface, the atmospheric pressure and temperature at about 50 km to 65 km above the surface of the planet is nearly the same as that of the Earth, making its upper atmosphere the most Earth-like area in the Solar System, even more so than the surface of Mars. Due to the similarity in pressure and temperature and the fact that breathable air (21% oxygen, 78% nitrogen) is a lifting gas on Venus in the same way that helium is a lifting gas on Earth, the upper atmosphere has been proposed as a location for both exploration and colonization (Landis, 2003).

The clouds also carry signs of meteorological events known as gravity waves (waves moving through a stable layer of the atmosphere), caused when winds blow over geological features, causing rises and falls in the layers of air. The winds at the planet's surface are much slower, estimated to be just a few miles per hour. On Earth's atmosphere, thunderstorm updrafts will produce gravity waves as they try to punch into the tropopause.

There exist unusual stripes in the upper clouds of Venus. These stripes are dubbed "blue absorbers" or "ultraviolet absorbers" because they strongly absorb light in the blue and ultraviolet wavelengths. These are soaking up a huge amount of energy—nearly half of the total solar energy the planet absorbs. As such, they seem to play a major role in keeping Venus as hellish as it is. Weller et al. (2018) have pointed out that based on observations, Venus is a world that has both a thick 92 bar atmosphere (note that Earth's atmospheric pressure at mean sea level is 1 bar), comprised of 96.5% CO_2.

Venus has a similar size, composition, and gravity to Earth, but its dense carbon dioxide atmosphere, sulfuric acid clouds, and surface temperatures above 450°C make it inhospitable to life as we know it. Conditions are more moderate a few dozen kilometers up, in Venus's clouds, and examples on Earth of microbes that can survive in clouds and in acidic environments had made life on Venus seem plausible.

Not so, according to a study carried out by Hallsworth et al. (2021), whose analysis of Venusian sulfuric acid cloud droplets found that their water content is two orders of magnitude below the known limit for life, set by a common household fungus here on Earth. The researchers drew on measurements of temperature, pressure, and moisture made by probes sent to Venus in the 1970s and 80s to calculate the water activity—a value analogous to relative humidity—of droplets at different temperatures and sulfuric acid concentrations.

The recent suggestion of phosphine in Venus's atmosphere has regenerated interest in the idea of life in clouds. However, such analyses usually neglect the role of water activity, which is a measure of the relative availability of water, in habitability. In a study, Hallsworth et al. (2021) computed the water activity within the clouds of Venus and other Solar System planets from observations of temperature and water-vapor abundance. Hallsworth et al. (2021) found water-activity values of sulfuric acid droplets, which constitute the bulk of Venus's clouds, of ≤0.004, two orders of magnitude below the 0.585 limit for known extremophiles. Considering other planets, ice formation on Mars imposes a water activity of ≤0.537, slightly below the habitable range, whereas conditions are biologically permissive (>0.585) at Jupiter's clouds (although other factors such as their composition may play a role in limiting their habitability). By way of comparison, Earth's troposphere conditions are, in general, biologically permissive, whereas the atmosphere becomes too dry for active life above the middle stratosphere (Hallsworth et al., 2021).

Hallsworth et al. (2021)'s results make it seem impossible that Earth-like organisms could survive on Venus. However, scientists' understanding of Venus's clouds and their chemistry remain limited. New missions should help with that, but they probably won't make life on Venus seem more probable.

8.4.3 Atmospheric composition—Mars versus Earth

In contrast with Earth, Mars has a much thinner atmosphere, with an atmospheric volume less than 1% of Earth's. The atmospheric composition is also significantly different: primarily carbon dioxide-based, while Earth's is rich in nitrogen and oxygen. The atmosphere of Mars has evolved. Evidence on the surface suggest that Mars was once much warmer and wetter. The atmosphere of Mars is about 100 times thinner than Earth's, and it is 95% carbon dioxide. Mars' atmospheric composition, according to a NASA fact sheet, is as shown below (Sharp, 2017b):

- Carbon dioxide: 95.32%.
- Nitrogen: 2.7%.
- Argon: 1.6%.
- Oxygen: 0.13%.
- Carbon monoxide: 0.08%.
- Also, minor amounts of: water, nitrogen oxide, neon, hydrogen-deuterium-oxygen, krypton, and xenon.

Mars is a planet that shows climate change on a large scale. The climate of Mars comes from a variety of factors, including its ice caps, water vapor and dust storms. At times, giant dust storms can blanket the entire planet and last for months, turning the sky hazy and red.

Understanding if life could have ever existed in such conditions is one of the hot topics of Mars exploration, and for the ESA–Roscosmos ExoMars mission. The ExoMars Trace Gas Orbiter is capable of sniffing out the composition of the planet's trace gases—which make up less than 1% by volume of a planet's atmosphere—in minute amounts. Although making up a very small amount of the overall atmospheric inventory, methane in particular holds key clues to the planet's current state of activity.

On Earth, living organisms release much of the planet's methane. It is also the main component of naturally occurring hydrocarbon gas reservoirs, and a contribution is also provided by volcanic and hydrothermal activity. Because of the key role natural biology plays in Earth's methane production, confirming the existence of methane on Mars, and distinguishing between its potential sources, is a top priority of the ExoMars Trace Gas Orbiter. Fig. 8.4 shows comparison of atmospheres of Mars and Earth. The planets in this graphic are not to scale. Mars atmospheric values are as measured by NASA's Curiosity rover.

8.4.4 Influence of mineral dust on Martian weather

Mineral dust is responsible for the highest measured aerosol optical thickness (AOT) on Earth (the three largest mineral dust source regions on Earth are: eastern Asia, the Arabian Peninsula, and the Sahara Desert). Many of the processes that create aerosols on Earth, such as sulfate emissions from volcanic eruptions and fossil fuel combustion, sea salt from spray, and smoke and soot from fires, do not occur on Mars (although volcanic eruptions and possibly sea salt spray contributed to atmospheric aerosols in the distant

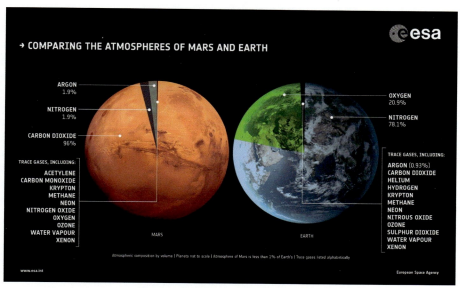

FIG. 8.4 Comparison of atmospheres of Mars and Earth. The planets in this graphic are not to scale. Mars atmospheric values are as measured by NASA's Curiosity rover. *(Source: https://www.esa.int/ESA_Multimedia/Images/2018/04/Comparing_the_atmospheres_of_Mars_and_Earth)* Copyright info: ©ESA.

past). Nevertheless, mineral dust is a major forcing mechanism in the Martian atmosphere, and it is intimately linked with interannual variability in the Martian climate (e.g., Zurek et al. 1992; Read and Lewis, 2004). Dust on Mars has long been known from Earth-based telescopic observations. Martin and Zurek (1993) summarized observations of "yellow clouds" dating as far back as 1873. These phenomena were correctly attributed to lofted dust, but determining details of their formation, development, and dissipation required a closer inspection from orbiting spacecraft.

With regard to the dust on Mars, Fenton et al. (2016) recall that "Lifting may be accomplished by winds linked to large-scale weather systems or atmospheric tides (Wang et al., 2003; Hinson and Wang, 2010; Hinson et al., 2012; Wang and Richardson, 2015), local mesoscale gusts or topographic flows (Spiga and Lewis, 2010; Mulholland et al., 2015), or on much smaller scales by convective motions (Spiga and Forget, 2009), such as the DDs (Balme and Greeley, 2006; Greeley et al., 2010)." Note that DDs refer to Dust Devils.

According to Fenton et al. (2016), the primary effect of Martian dust is to provide local heating to the atmosphere through absorption of solar shortwave radiation. Dust in the atmosphere also absorbs, scatters and re-radiates radiation at longer wavelengths, such as thermal infrared emission originating from the surface (Smith, 2004). Fenton et al. (2016) further note that dust aerosols have an additional potential feedback as nuclei for cloud ice particles, which in turn impact atmospheric radiative heating and cooling (Montmessin et al., 2004; Wilson et al., 2008; Madeleine et al., 2012a; Hinson et al., 2014; Navarro et al., 2014; Steele et al., 2014a; 2014b), although, unlike on Earth, it seems likely that there will always be a sufficient supply of small dust particles on Mars to nucleate the relatively thin water ice clouds that have been observed (Heavens et al., 2010; Madeleine et al. 2012b). Clouds, in turn, may further increase the complexity of the climate feedbacks, by accelerating the removal of dust from the atmosphere by scavenging smaller particles, thereby enhancing the sedimentation rate (Madeleine et al. 2012a; Navarro et al., 2014).

Fenton et al. (2016) explain that "The net effect is to warm the atmosphere where it is most dusty and in daylight, and to cool the surface below regions with very high dust opacity. If atmospheric dust loading varies from place-to-place, this may introduce or steepen horizontal temperature gradients within the atmosphere that are, in turn, linked to winds. Winds in the atmosphere both advect dust and may lift more from the surface. In a heavily dust-laden atmosphere, the effect is to warm the atmosphere relative to the surface, which increases its static stability and tends to ultimately reduce both vertical convection during the day and large-scale wave-like instabilities, both reducing the likelihood of dust lifting from the surface and ultimately leading to the slow decay of planet-encircling dust events (e.g., Cantor, 2007). In this way, atmospheric dust provides complex positive and negative feedbacks to the Martian climate system."

Diniega et al. (2021) recall that within the present-day Mars climate, wind and frost/ice are the dominant drivers, resulting in large avalanches of material down icy, rocky, or sandy slopes; sediment transport leading to many scales of aeolian bedforms and erosion; pits of various forms and patterned ground; and substrate material carved out from under subliming ice slabs. Due to the ability to collect correlated observations of surface activity and new landforms with relevant environmental conditions with spacecraft on or around Mars, studies of Martian geomorphologic activity are uniquely positioned to directly test surface-atmosphere interaction and landform formation/evolution models outside of Earth. In a study, Diniega et al. (2021) have outlined the currently observed and interpreted surface activity occurring within the modern Mars environment, and tie this activity to wind, seasonal surface CO_2 frost/ice, sublimation of subsurface water ice, and/or gravity drivers.

8.5 Dust devils and vortices

A dust devil (DD) (Fig. 8.5) is a strong, well-formed, and relatively short-lived whirlwind, ranging from small (half a meter wide and a few meters tall) to large (more than 10 m wide and more than 1 km tall) in size. The primary vertical motion is upward. They are comparable to tornadoes in that both are a weather phenomenon involving a vertically oriented rotating column of wind.

DDs form when a pocket of hot air near the surface rises quickly through cooler air above it, forming an updraft. If conditions are just right, the updraft may begin to rotate. As the air rapidly rises, the column of hot air is stretched vertically, thereby moving mass closer to the axis of rotation, which causes intensification of the spinning effect by conservation of angular momentum. The secondary flow in the DD causes other hot air to speed horizontally inward to the bottom of the newly forming vortex. As more hot air rushes in toward the developing vortex to replace the air that is rising, the spinning effect becomes further intensified and self-sustaining. A dust devil, fully formed, is a funnel-like chimney through which hot air moves, both upwards and in a circle.

FIG. 8.5 (left) A dust devil in Arizona, developed on 10 June 2005. *(Source: https://commons.wikimedia.org/wiki/File:Dust_devil.jpg (Copyright info: This file is in the **public domain** in the United States because it was solely created by NASA. NASA copyright policy states that "NASA material is not protected by copyright **unless noted**".)* (right) A dust devil in Ramadi, Iraq. *(Source: https://commons.wikimedia.org/wiki/File:Iraqi_Dust_Devil.jpg) (Copyright info: This work has been released into the public domain by its author, Ultratone85 at English Wikipedia. This applies worldwide).*

As the hot air rises, it cools, loses its buoyancy, and eventually ceases to rise. As it rises, it displaces air which descends outside the core of the vortex. This cool air returning acts as a balance against the spinning hot-air outer wall and keeps the system stable (Ludlum, 1997). The spinning effect, along with surface friction, usually will produce a forward momentum. The DD is able to sustain itself longer by moving over nearby sources of hot surface air (Andrea, 2021). As available hot air near the surface is channeled up the dust devil, eventually surrounding cooler air will be sucked in. Once this occurs, the effect is dramatic, and the dust devil dissipates in seconds. Usually, this occurs when the dust devil is not moving fast enough (depletion) or begins to enter a terrain where the surface temperatures are cooler.

8.5.1 Dust devil on Earth

Dust devils (DDs) are common on Earth. Dust devils appearing on Earth are atmospheric vortices originating in the unstable layer close to the ground. Dust devils form if there is a very unstable layer of air. Such layers occur in deserts because often there is a large temperature difference between the ground and the air only a short distance above it.

The three largest mineral dust source regions on Earth are eastern Asia, the Arabian Peninsula and the Sahara Desert. Severe DDs have caused damage and even deaths in the past. One such DD struck the Coconino County Fairgrounds in Flagstaff, Arizona, on September 14, 2000, causing extensive damage to several temporary tents, stands and booths, as well as some permanent fairgrounds structures. Fig. 8.5 shows dust devils in Arizona, USA, and Ramadi, Iraq. Several injuries were reported, but there were no fatalities. Based on the degree of damage left behind, it is estimated that the dust devil produced winds as high as 75 mph (120 km/h). On May 19, 2003, a dust devil lifted the roof off a two-story building in Lebanon, Maine, causing it to collapse and kill a man inside ("Man Dies In Windstorm." The New York Times. May 21, 2003). In East El Paso, Texas, in 2010, three children in an inflatable jump house were picked up by a dust devil and lifted over 10 feet (3 m), traveling over a fence and landing in a backyard three houses away. In Commerce City, Colorado in 2018, a powerful dust devil hurtled two portapotties (chemical toilets used for collecting human excreta in a holding tank and using chemicals to minimize odors) into the air (*Two children killed after bouncy castle is swept into air by "dust devil" in central China*, South China Morning Post, April 1, 2019).

DDs, even small ones (on Earth), can produce radio noise and large electrical fields. A dust devil picks up small dirt and dust particles. As the particles whirl around, they become electrically charged through contact or frictional charging (triboelectrification). The whirling charged particles also create a magnetic field that fluctuates between 3 and 30 times each second (Houser et al., 2003).

These electric fields may assist the vortices in lifting material off the ground and into the atmosphere. Field experiments indicate that a dust devil can lift 1 gram of dust per second from each square meter (10 lb/s from each acre) of ground over which it passes. A large dust devil measuring about 100 m (330 ft) across at its base can lift about 15 metric tonnes (17 short tons) of dust into the air in 30 minutes. Giant DD storms that sweep across the world's deserts contribute 8% of the mineral dust in the atmosphere each year during the handful of storms that occur. In comparison, the significantly smaller dust devils that twist across the deserts during the summer lift about three times as much dust, thus having a greater combined impact on the dust content of the atmosphere. When this occurs, they are often called sand pillars (Kok and Renno, 2006).

DDs have been implicated in several aircraft accidents (Lorenz, 2005). While many incidents have been simple taxiing problems, a few have had fatal consequences. DDs are also considered major hazards among skydivers and paragliding pilots as they can cause a parachute or a para-glider to collapse with little to no warning, at altitudes considered too low to cut away, and contribute to the serious injury or death of parachutists.

Terrestrial dust devils have not yet been observed directly in satellite imagery until 2016. Reiss (2016) reported about the first terrestrial dust devil observations with visible and thermal satellite data on an alluvial fan in the Taklimakan desert, China. DDs were first recognized in high-resolution visible image data using Google Earth. Further inspection of medium resolution image data (Aster and Landsat 7 and 8) revealed that dust devils in this area are numerous and large. In addition, several larger terrestrial dust devils are resolved in thermal Landsat 8 images.

The continued monitoring of the amount of dust lofted by daytime convective turbulence, mainly by DDs from orbital platforms opens new avenues of research that, informed by the extensive studies performed on Mars, could prove to be of use to the field of terrestrial climate science (Fenton et al., 2016).

8.5.2 Dust devil on Mars

Large dust devils have been observed on the planet Mars, where they can rise to a height of about 6 km in its thin atmosphere. Dust devils on Mars were first photographed by the Viking orbiters in the 1970s. DDs on Mars are whirlwinds resulting from sunlight warming the ground, causing convective rising of air. Observations of these dust devils over the Martian environment provide significant information about wind directions and interaction between the surface and atmosphere.

Arya et al. (2015a) reported that the MCC onboard MOM detected a major dust devil prograding between Oct 20 & Oct 28, 2014 in Mars' northern and equatorial region. The dust front was seen spreading eastward along Acidalia to central Noachies, to the Hellas, then swirling westward near Argyre and ultimately fading out along Vallies Marineris.

Spaceborne observations of lofted dust began with the first weather satellites in the 1960s. Until recently, most such phenomena have been associated with dust hazes and smog that have been transported far from their source regions (e.g., United States 1964), but new data have highlighted a complex interplay between dust emission and daytime dry convective turbulence in the planetary boundary layer (Fenton et al. 2016).

Fenton et al. (2016) recall that the best first-look at Martian dust from space is that from Mariner 9, which entered orbit 5 around Mars in November 1971, in the midst of one of the most intense planet-encircling dust storms on record. Features related to dust entrained by subkilometer-scale daytime convective turbulence, such as DDs, were not expected to be resolved in images from 1970s-era spacecraft; as a result of which they were only identified many years later after careful inspection of these data sets (Thomas and Gierasch, 1985).

Martian DDs are typically a kilometre in diameter and altitude of 10 km. Fig. 8.6 shows a towering dust devil casts a serpentine shadow over the Martian surface. This image was acquired by the High Resolution Imaging Science Experiment (HiRISE) camera on NASA's Mars Reconnaissance Orbiter (MRO). The devil is 800m in height and 30m wide.

Dust devils on Mars were first detected in images taken by the Viking orbiters. Columnar, cone-shaped, and funnel-shaped clouds rising 1 to 6 km above the surface of Mars have been identified in Viking Orbiter images. They are interpreted as dust devils, confirming predictions of their occurrence on Mars and giving evidence of a specific form of dust entrainment (Thomas and Gierasch, 1985). In 1997, the Mars Pathfinder lander detected a dust devil passing over it ("Martian Dust Devils Caught." Climate Research USA. Ruhr-Universität Bochum. March 21, 2000). On Mars, dust devils are frequently observed with orbital image data

FIG. 8.6 A towering dust devil casts a serpentine shadow over the Martian surface in this image acquired by the High Resolution Imaging Science Experiment (HiRISE) camera on NASA's Mars Reconnaissance Orbiter. The devil is 800m in height and 30m wide. *(Source: https://commons.wikimedia.org/wiki/File:The_Serpent_Dust_Devil_on_Mars_PIA15116.jpg) (Copyright info: This file is in the public domain in the United States because it was solely created by NASA. NASA copyright policy states that "NASA material is not protected by copyright unless noted".).*

(Thomas and Gierasch, 1985; Cantor et al., 2006; Stanzel et al., 2008; Towner, 2009; Greeley et al., 2010; Reiss et al., 2014). Dust devils are important boundary layer processes for energy transport and their contribution to the atmosphere's dust load (Fenton et al., 2016; Klose et al., 2016). There was no evidence suggesting dust devils cause or lead to initiation of dust storms. Model-derived tangential wind speeds of large vortices were >20 m/s at 20 m above the surface. Dust flux calculations suggest that dust devils are a contributor to the background dust opacity observed through northern spring and summer (Cantor et al., 2006).

In Stanzel et al. (2008)'s studies, a total of 205 dust devils were detected in 23 High Resolution Stereo Camera (HRSC) images taken between January 2004 and July 2006 with the ESA Mars Express orbiter, in which average dust devil heights were ~660 m and average diameters were ~230 m. For the first time, dust devil velocities were directly measured from orbit, and range from 1 to 59 m/s. It was found that the observed dust devil directions of motion are consistent with data derived from a General Circulation Model (GCM). In some respects, HRSC dust devil properties agree favorably with data from the NASA Mars Exploration Rover Spirit dust devil analyses. The spatial distribution of the active dust devils detected by HRSC supports the conjecture that the ascending branch of the Hadley circulation is responsible for the increase in dust devil activity, especially observed during southern summer between 50° and 60° S latitude. Combining the dust-lifting rate of 19 kg/km^2/sol derived from the Spirit observations with the fewer in number but larger in size dust devils from various other locations observed by HRSC, Stanzel et al. (2008) suggested that dust devils make a significant contribution to the dust entrainment into the atmosphere and to the Martian dust cycle.

Dust devils and convective vortices are common on Aeolis Mons (volcanoes) relative to the nearby Aeolis Palus (Lemmon, 2017). Reiss et al. (2014) derived the horizontal motion (speed and direction) of dust devils from time-delayed Mars Reconnaissance Orbiter (MRO) coordinated image data sets of the Compact Reconnaissance Imaging Spectrometer for Mars (CRISM) to the Context Camera (CTX) and/or the HiRISE acquired between 2008 and 2011. In their study, in total, 47 dust devils were observed in 15 regions with diameters ranging from 15 to 280 m with an average diameter of 100 m and heights from 40 to 4400 m. Horizontal speeds of 44 dust devils range from 4 to 25 m/s with average speeds of 12 m/s. The majority of dust devils were observed in the northern hemisphere (79%), mainly in Amazonis Planitia (67.5% from the northern hemisphere dust devils). It was found that seasonal occurrence of dust devils in the northern hemisphere is predominant in early and mid-spring (76%). Reiss et al. (2014) compared their measured dust devil horizontal speeds and directions of motion to the monthly climatologies (wind speed and direction) released in the Mars Climate Database (MCD) derived from General Circulation Model (GCM) predictions. They found that there is a broad agreement between dust devil horizontal speeds and MCD wind speed predictions within the Planetary Boundary Layer (PBL) as well as dust devil directions of motion and MCD predicted wind directions occurring within the PBL. It was noticed that comparisons between dust devil horizontal speeds and MCD near-surface wind speed predictions at 10 m height above the surface do not correlate well: dust devils move about a factor of 2 faster than MCD near-surface wind predictions. The largest offsets between dust devil motion and MCD predictions were related to three dust devils occurring near the Phoenix landing site when the lander was still active. The offsets could be explained by a regional weather front passing over the Phoenix landing site. In general, the good agreement between dust devil horizontal speeds and directions of motion, and ambient wind speeds and directions predicted within the PBL through GCM, show that dust devils on Mars move with ambient winds in the PBL, hence faster than near-surface winds.

The impact of dust aerosols on the climate and environment of Earth and Mars is complex and forms a major area of research. A difficulty arises in estimating the contribution of small-scale dust devils to the total dust aerosol. This difficulty is due to uncertainties in the amount of dust lifted by individual dust devils, the frequency of dust devil occurrence, and the lack of statistical generality of individual experiments and observations. In a study, Klose et al. (2016) reviewed results of observational, laboratory, and modeling studies and provided an overview of dust devil dust transport on various spatio-temporal scales as obtained with the different research approaches. Methods used for the investigation of dust devils on Earth and Mars vary. For example, while the use of imagery for the investigation of dust devil occurrence frequency is common practice for Mars, this is less so the case for Earth. Modeling approaches for Earth and Mars are similar in that they are based on the same underlying theory, but they are applied in different ways. Insights into the benefits and limitations of each approach suggest potential future research focuses, which can further reduce the uncertainty associated with dust devil dust entrainment. Klose et al. (2016) have discussed the potential impacts of dust devils on the climates of Earth and Mars on the basis of research results.

Over the past several decades, orbital observations of lofted dust have revealed the importance of mineral aerosols as a climate forcing mechanism on both Earth and Mars. Increasingly detailed and diverse data sets have provided an ever-improving understanding of dust sources, transport pathways, and sinks on both planets, but the role of dust in modulating atmospheric processes is complex and not always well understood. Fenton et al. (2016) have presented a review of orbital observations of entrained dust on Earth and Mars, particularly that produced by the dust-laden structures produced by daytime convective turbulence called "dust devils." On Earth, dust devils are thought to contribute only a small fraction of the atmospheric dust budget; accordingly, there are not yet any published accounts of their occurrence from orbit. In contrast, dust devils on Mars are thought to account for several tens of percent of the planet's atmospheric dust budget; the literature regarding Martian dust devils is quite rich. Because terrestrial dust devils may temporarily contribute significantly to local dust loading and lowered air quality, Fenton et al. (2016) have suggested that Martian dust devil studies may inform future studies of convectively-lofted dust on Earth.

Fenton et al. (2016) explain that as on Earth, Martian dust devils form most commonly when the insolation (the amount of solar radiation reaching a given area) reaches its daily and seasonal peak and where a source of loose dust is plentiful. However, this pattern is modulated by variations in weather, albedo (the proportion of the incident light or radiation that is reflected by the surface), or topography, which produce turbulence that can either enhance or suppress dust devil formation. For reasons not well understood, when measured from orbit, Martian dust devil characteristics (dimensions, and translational and rotational speeds) are often much larger than those measured from the ground on both Earth and Mars. Studies connecting orbital observations to those from the surface are needed to bridge this gap in understanding.

Martian dust devils have been used to remotely probe conditions in the planetary boundary layer (PBL) [e.g., convective boundary layer (CBL) depth, wind velocity]; the same could be done in remote locations on Earth. The CBL is composed of structured turbulent eddies, containing tens-of-meter-scale vortices that form most commonly in narrow updrafts at the intersections of three or more kilometer-scale convection cells (e.g., Willis and Deardorff, 1979; Hess and Spillane, 1990; Kanak et al., 2000). When dust-laden, these vortices become visible to the eye as dust devils (DDs). Nonrotating gusts may also entrain dust, likely occurring most commonly along upwelling sheets where two convection cells meet.

According to Fenton et al. (2016), Martian dust devils appear to play a major role in the dust cycle, waxing and waning (increase and decrease) in relative importance and spatial patterns of occurrence with the planet's orbital state. Orbital studies of terrestrial dust devils would provide a basis for comparative planetology that would broaden the understanding of these dusty vortices on both planets (Fenton et al., 2016).

Dust-devils processes represent the drops in transient daytime pressure (Kahanpaa, et al., 2018). Small whirlwinds on Mars, known as vortices, are not uncommon. They can be tens of meters wide and several kilometers high, with wind speeds upwards of 100 km/h. Singh and Arya (2019) recall that the Mars Science Laboratory (MSL) rover has also observed dust devils simultaneously by imaging and by meteorological measurements. Knowledge of the threshold transient wind velocity from MSL helps in parametrizing the amount of dust lifted by dust devil in numerical models of the Martian atmosphere.

As already indicated, dust devils are whirlwinds that result from solar warming of the ground, prompting convective air to rise into the atmosphere. The dust devils on Mars are relatively in larger number. Understanding dust devils is important for exploration of Mars. Dust related studies are main concern for the current and upcoming Martian missions. The dust devils on Mars observed on November 7, 2016 using Mars color camera (MCC) developed at the Space Applications Centre of the Indian Space Research Organisation (ISRO). In this study, the altitude of dust devil has been estimated using shadow method and sun-sensor geometry. The shadow method works well if the Sun is not too high in the sky and with images that have very high spatial resolution (Reiss et al., 2014). Many dust devils have been detected by missions on Mars, such as the Curiosity rover and InSight lander. If these vortices lift up Martian dust, they can form so-called dust devils (O'Callaghan, 2021).

Martian dust devils can be up to fifty times as wide and ten times as high as terrestrial dust devils, and large ones may pose a threat to terrestrial technology sent to Mars (Smith and Nilton, 2001). On November 7, 2016, five such dust devils ranging in heights of 0.5 to 1.9 km were imaged in a single observation by the Mars Orbiter Mission in the Martian southern hemisphere (Singh and Arya, 2019).

Mission members monitoring the Spirit rover on Mars reported on March 12, 2005, that a lucky encounter with a dust devil had cleaned the solar panels of that robot. Power levels dramatically increased and daily science work was anticipated to be expanded (David, 2005). A similar phenomenon (solar panels mysteriously cleaned of accumulated dust) had previously been observed with the Opportunity rover, and dust devils had also been suspected as the cause. The electrical activity associated with dust devils is widely thought to generate lightning on Mars, but this has not yet been conclusively detected (Harrison et al., 2016). NASA's latest Mars rover, Perseverance, encountered 100 dust devils in its first 90 days on Mars after landing in February. 2021 (O'Callaghan, 2021).

8.5.3 Vortex on Venus

Several planets in the Solar System, including Earth, have been found to possess hurricane-like vortices, where clouds and winds rotate rapidly around the poles. Some of these take on strange shapes, such as the hexagonal structure on Saturn, but none of them are as variable or unstable as the southern polar vortex on Venus (https://sci.esa.int/web/venus-express/-/54062-1-shape-shifting-polar-vortices).

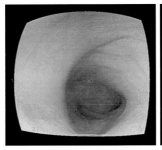 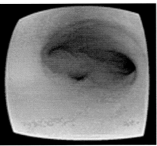

FIG. 8.7 **Southern polar vortex on Venus.** This image sequence shows thermal infrared radiation (at a wavelength of 3.8 microns) emitted by the cloud-tops at the southern polar region of Venus, as viewed by the VIRTIS imaging spectrometer on Venus Express. The scale has been inverted: white regions show cooler cloud, likely to be at higher altitude. *Copyright: ESA/VIRTIS-VenusX/INAF-IASF/LESIA-Obs. de Paris (G. Piccioni, INAF-IASF) (Source: https://sci.esa.int/web/venus-express/-/48598-the-southern-polar-vortex-on-venus).*

While Mars is well known for immense dust storms that cover the entire planet, Venus has an incredibly thick and fast-moving atmosphere that can form permanent vortices at its poles. At the south pole of Venus is a large vortex (Fig. 8.7) the size of Europe swirling in the atmosphere. This vortex appears to have been around for a long time and is a result of some strange properties on the planet. The atmosphere on Venus moves faster than the planet, reaching speeds of up to 250 miles (400 km) per hour—60 times faster than the planet rotates, according to the European Space Agency.

Venus is well known for its remarkable, super-rotating upper atmosphere, which sweeps around the planet once every four Earth days. This is in stark contrast to the rotation of the planet itself—the length of the day—which takes a comparatively laborious 243 Earth days.

The existence of the polar vortices on Venus has been known for many years, but high-resolution infrared measurements obtained by the Visible and Infrared Thermal Imaging Spectrometer (VIRTIS) instrument on Venus Express have revealed that the southern vortex is far more complex than previously believed.

After its arrival at the planet in 2006, Venus Express was able to conduct the most detailed survey of cloud motions in the atmosphere of Venus, with the surprising discovery that the planet's high-level winds have become faster over time.

By tracking the movements of distinct cloud features in the cloud tops some 70 km above the planet's surface over a period of 10 Venusian years (6 Earth years), scientists were able to monitor patterns in the long-term global wind speeds.

In 2006, average cloud-top wind speeds between latitudes 50° on either side of the equator were clocked at roughly 300 km/h. However, detailed cloud tracking studies revealed that these already remarkably rapid winds are becoming even faster, increasing to 400 km/h over the course of the mission. The reason for this dramatic increase is unknown.

On top of this long-term increase in the average wind speed, studies have also revealed regular variations linked to the local time of day and the altitude of the Sun above the horizon, as well as the rotation period of Venus.

There have also been dramatic variations in the average wind speeds observed between consecutive orbits of Venus Express around the planet. In some cases, wind speeds at low latitudes varied such that clouds completed one journey around the planet in 3.9 days, while on other occasions they took 5.3 days.

The new observations show that the center of the vortex has a highly variable shape and internal structure, and its morphology is constantly changing on timescales of less than 24 h, as a result of differential rotation. This fast-moving feature is all the more surprising since its center of rotation is offset from the geographical South Pole.

The images show that the core of the dynamic southern vortex can take almost any shape, so although it often looks like an "S", it may become completely irregular, or even chaotic, in appearance. These rapid shape changes indicate complex weather patterns, which are strongly influenced by the fact that the center of the vortex does not coincide with the geographical pole.

Images from the Venus Monitoring Camera and from the VIRTIS instrument show that the speeds of the zonal winds change with latitude, so that the vortex is continually being pulled and stretched. Although the mean zonal wind is retrograde (blowing from east to west), its speed decreases toward the pole.

The center of rotation drifts right around the pole over a period of 5–10 Earth days. Its average displacement from the South Pole is about 3° of latitude or several hundred kilometers.

Although its highly elliptical orbit means that Venus Express flies too close to the planet's North Pole for detailed imaging, it is likely that both vortices have similar structures and behave in a similar way (https://sci.esa.int/web/venus-express/-/54062-1-shape-shifting-polar-vortices).

8.6 Distances of Venus, Earth, and Mars from Sun, and their current surface temperatures

The habitable zone is the area around a star where it is not too hot and not too cold for liquid water to exist on the surface of surrounding planets. In the Solar System, the distance of a planet from the Sun plays an important role in providing a habitable environment to a planet. Apart from greenhouse effect and volcanic activities, the surface temperature of a planet in the Solar System is largely determined by its distance from the Sun. The below Sections address the distances of Venus, Earth, and Mars from Sun, and their current surface temperatures.

8.6.1 Gradually growing distance between Earth and Sun over time

Sky watchers have been trying to gauge the Sun-Earth distance for thousands of years. In the third century BC, Aristarchus of Samos incorrectly estimated the Sun to be 20 times farther away than Earth's Moon.

By the late 20th century, astronomers had a much better grip on this fundamental cosmic metric—what came to be called the astronomical unit. In fact, thanks to radar beams pinging off various Solar-System bodies and to tracking of interplanetary spacecraft, the Sun-Earth distance has been pegged with remarkable accuracy. The current value stands at 149,597,870.696 km (https://www.newscientist.com/article/dn17228-why-is-the-earth-moving-away-from-the-sun/#ixzz7ClClHH6q). Thus, Earth's current average distance to the Sun is about 150×10^6 km (150 million km); i.e., 1 Astronomical Unit (AU). However, Earth's distance from Sun is gradually increasing as suggested by various scientific investigations.

Having such a precise yardstick as indicated above allowed Krasinsky and Brumberg (2004) to calculate that the Sun and Earth are gradually moving apart. It is not much; just 15 cm per year. From the analysis of all available radiometric measurements of distances between the Earth and the major planets (including observations of Martian landers and orbiters over 1971–2003 with the errors of few meters) the positive secular trend in the Astronomical Unit (AU) is estimated as $\frac{d}{dt}AU = 15 \pm 4 (m/cy)$, The given uncertainty is the 10 times enlarged formal error of the least-squares estimate and so accounts for possible systematic errors of measurements and deficiencies of the mathematical model. Krasinsky and Brumberg (2004) have discussed the reliability of this estimate as well as its physical meaning. Their study indicated that a priori most plausible attribution of this effect to the cosmological expansion of the Universe turns out inadequate. A model of the observables developed in the frame of the relativistic background metric of the uniform isotropic Universe showed that the corresponding dynamical perturbations in the major planet motions are completely cancelled out by the Einstein effect of dependence of the rate of the observer's clock (that keeps the proper time) on the gravitational field, though separately values of these two effects are quite large and attainable with the accuracy achieved. Another tentative source of the secular rate of AU is the loss of the solar mass due to the solar wind and electromagnetic radiation but it amounts in $\frac{d}{dt}AU$ only to 0.3 m/cy. Excluding other explanations that seem exotic (such as secular decrease of the gravitational constant) at present there is no satisfactory explanation of the detected secular increase of AU, at least in the frame of the considered uniform models of the Universe (Krasinsky and Brumberg, 2004).

In another study, Miura et al. (2009) argued that the Sun and Earth are literally pushing each other away due to their tidal interaction. Miura et al. (2009) reported an idea and the order-of-magnitude estimations to explain the reported secular increase of the Astronomical Unit (AU) by Krasinsky and Brumberg (2004). The idea proposed by Miura et al. (2009) is analogous to the tidal acceleration in the Earth-Moon system, which is based on the conservation of the total angular momentum; and Miura et al. (2009) applied this scenario to the Sun-planets system. Assuming the existence of some tidal interactions that transfer the rotational angular momentum of the Sun and using reported value of the positive secular trend in the astronomical unit, $\frac{d}{dt}AU = 15 \pm 4 (m/cy)$ the suggested change in the period of rotation of the Sun is about 21 ms/cy in the case that the orbits of the eight planets have the same "expansion rate." This value is sufficiently small, and at present it seems there are no observational data which exclude this possibility. Miura et al. (2009) investigated the effects of the change in the Sun's moment of inertia as well. It is pointed out that the change in the moment of inertia due to the radiative mass loss by the Sun may be responsible for the secular increase of AU, if the orbital "expansion" is happening only in the inner planets system. Although the existence of some tidal interactions is assumed between the Sun and planets, concrete mechanisms of the angular momentum transfer remain to be done as future investigations.

Tidal interaction is the same process that is gradually driving Earth's Moon's orbit outward: Tides raised by the Moon in Earth's oceans are gradually transferring Earth's rotational energy to lunar motion. As a consequence, each year Earth's Moon's orbit expands by about 4 cm and Earth's rotation slows by 0.000017 s.

Likewise, Miura et al. (2009) assume that Earth's mass is raising a tiny but sustained tidal bulge in the Sun. They calculated that, thanks to Earth, the Sun's rotation rate is slowing by 3 milliseconds per century (0.00003 s per year). According to their explanation, the distance between the Earth and Sun is growing because the Sun is losing its angular momentum.

Tidal acceleration is an effect of the tidal forces between an orbiting natural satellite and the primary object that it orbits. The acceleration causes a gradual recession of a natural satellite in a prograde orbit away from the primary, and a corresponding slowdown of the primary's rotation. The process eventually leads to tidal locking, usually of the smaller first, and later the larger body. The Earth–Moon system is the best studied case where this process is driving the Moon's orbit outward: Tides raised by the Moon in Earth's oceans are gradually transferring Earth's rotational energy to lunar motion. As a consequence, each year the Moon's orbit expands by about 4 cm and Earth's rotation slows by 0.000017 s (Allen E Hall, Owner at Aerospace Eng & Mfg Consultant; https://www.quora.com/

Is-the-distance-from-the-earth-to-the-sun-changing). The Sun and Earth are literally pushing each other away due to their tidal interaction.

Earth's mass is raising a tiny but sustained tidal bulge in the Sun. This causes the Sun's rotation rate to slow by 3 milliseconds per century (0.00003 s per year). As already indicated, the distance between the Earth and Sun is growing 15 cm per year because the Sun is losing its angular momentum. This and the expansion of space at 75 km per second per megaparsec will move Earth's orbit 40–50 million miles farther from the Sun by the time the Sun turns into a red giant. The planets' orbits are actually chaotic over billions of years.

It has been suggested that if the Earth and the outer planets survive until the Sun's dying last gasps cast off nearly half its mass as a planetary nebula, the planets may drift off into dark space if disturbed by a large passing mass and roam the darkness alone. Most likely for all eternity or at least until the end of time.

Earth's gravity causes small bulge in Sun at the Sun's equator. The Sun's rotation drags this bulge in front of the orbiting Earth because the Sun's rotation is faster than Earth's orbital period. This dragging ahead of the Earth bleeds the Sun's angular momentum and transfers this energy to the Earth by pulling the Earth off the center of its orbit so that it moves farther from the Sun.

8.6.2 Distance of Mars from Sun and Earth & temperature differences

Mars has a very eccentric orbit; that is, it deviates from a perfect circle more than any other planet's orbit. At its farthest distance (aphelion), Mars is 154 million miles (249 million km) from the Sun. At its closest (perihelion), Mars is 128 million miles (206 million km) distant. On average, the distance to Mars from the Sun is 142 million miles (229 million km), according to NASA. Mars revolves around the Sun in 687 Earth days, which represents a Martian year.

The distance to Mars from Earth is constantly changing. Like a pair of cars on a racetrack, Mars and Earth orbit the Sun at different speeds. Earth has an inside lane and moves around the Sun more quickly. Furthermore, both have elliptical orbits, rather than perfect circles.

In theory, the closest the planets could come together would be when Mars is at its closest point to the Sun (perihelion) and Earth is at its farthest point (aphelion). In that situation, the planets would be 33.9 million miles (54.6 million km) from each other. But that has never happened in recorded history. The closest known approach was 34.8 million miles (56 million km) in August 2003. Fig. 8.8 shows the ever-changing distance between Earth and Mars.

In their race around the Sun, Earth on its inside track laps Mars every 26 months. This close approach provides

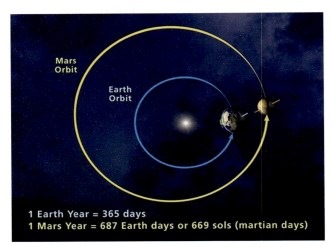

FIG. 8.8 The ever-changing distance between Earth and Mars. *(Image credit: NASA). (Source: https://www.space.com/16875-how-far-away-is-mars.html).*

an opportunity—a launch window—to send spacecraft to the Red Planet. Rather than pointing the spacecraft at Mars, engineers aim it in a wide orbit around the Sun. The Sun's gravity gives the spacecraft a boost—called a gravity assist or slingshot effect—saving time and fuel. The spacecraft's orbit then intersects with Mars.

8.6.3 Venus' distance from Sun & its brightness and temperature profile

Venus is the second planet from the Sun. Its distance from Sun is 108.2 million km, or 0.7 AU. Because Venus travels in an elliptical orbit around the Sun, its distance from the Sun varies throughout its year from 66,782,000 miles (107,476,000 km) to 67,693,000 miles (108,942,000 km). As the brightest natural object in Earth's night sky after the Moon, Venus can cast shadows and can be, on rare occasions, visible to the naked eye in broad daylight. When seen from Earth, Venus is brighter than any other planet or even any star in the night sky because of its highly reflective clouds and its closeness to Earth. With conditions on Venus that could be described as infernal, the ancient name for Venus—Lucifer ("morning star" or, as an adjective, "light-bringing")—seems to fit. Although Venus is not the planet closest to the Sun, its dense atmosphere traps heat in a runaway version of the greenhouse effect that warms Earth, thereby rendering Venus to the status of being the hottest planet in the Solar system. As a result, temperatures on Venus reach 880°F (471°C), which is more than hot enough to melt lead. Spacecraft have survived only a few hours after landing on the planet before being destroyed.

Although the planet's surface is like a red-hot furnace, conditions are very different at an altitude of 125 km, where Venus Express revealed a very frigid layer with a

8.7 Volcanism and surface features of Venus, Earth, and Mars

Volcanism (or volcanicity) is any of various processes and phenomena associated with the surficial discharge of molten rock, pyroclastic fragments, or hot water and steam, including volcanoes (mountains or hills, typically conical, having craters or vents through which lava, rock fragments, hot vapor, and gas are or have been erupted from the earth's crust), geysers (springs characterized by intermittent discharge of water ejected turbulently and accompanied by steam, as in hydrothermal vents), and fumaroles (vents in the surface of the Earth or other rocky planet from which hot gases and vapors are emitted, without any accompanying liquids or solids). Volcanism involves eruption of molten rock (magma) onto the surface of the Earth or a solid-surface planet or moon, where lava, pyroclastics (fragments of rock erupted by a volcano) and volcanic gases erupt through a break in the surface called a vent. It includes all phenomena resulting from and causing magma within the crust or mantle of the body, to rise through the crust and form volcanic rocks on the surface. Volcanic activity can influence climate in a number of ways at different timescales. Individual volcanic eruptions can release large quantities of sulfur dioxide and other aerosols into the stratosphere, reducing atmospheric transparency and thus the amount of solar radiation reaching Earth's surface and troposphere. A recent example is the 1991 eruption in the Philippines of Mount Pinatubo, which had measurable influences on atmospheric circulation and heat budgets. The 1815 eruption of Mount Tambora on the island of Sumbawa had more dramatic consequences, as the spring and summer of the following year (1816, known as "the year without a summer") were unusually cold over much of the world. New England and Europe experienced snowfalls and frosts throughout the summer of 1816 (https://www.britannica.com/science/climate-change/Evidence-for-climate-change#ref994262). Although volcanism is best known on Earth, there is evidence that it has been important in the development of the other terrestrial planets—Mercury, Venus, and Mars—as well as some natural satellites such as Earth's Moon and Jupiter's moon Io.

8.7.1 Volcanism of Venus

The impact cratering record on Venus is unique among the terrestrial planets. Strom et al. (1994) reported that fully 84% of the craters on Venus are in pristine condition, and only 12% are fractured. Remarkably, only 2.5% of the craters and crater-related features are embayed by lava, although intense volcanism and tectonism have affected the entire planet. Furthermore, the spatial and hypsometric (height) distribution of the craters is consistent with a completely random one, including stochastic variations. Monte Carlo simulations of equilibrium resurfacing models result in a minimum of 17

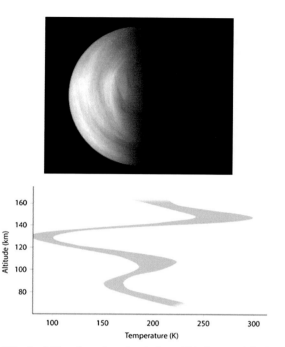

FIG. 8.9 (top) Venus' southern hemisphere. This image of the Venus southern hemisphere illustrates the terminator—the transitional region between the dayside (left) and nightside of the planet (right). The south pole is near the terminator, just above the center of the image. The complex atmosphere that surrounds the planet is also clearly visible. *Copyright: ESA/MPS, Katlenburg-Lindau, Germany (Source: https://sci.esa.int/web/venus-express/-/50886-venus-terminator)* (bottom) The temperature profile along the terminator for altitudes of 70–160 km above the surface of Venus. *Copyright: ESA/AOES (Source: https://sci.esa.int/web/venus-express/-/50885-terminator-temperature-profile).*

temperature of around −175°C. The unexpected cold layer is far chillier than any part of Earth's atmosphere, despite Venus being much closer to the Sun. Surprisingly, Venus Express discovered a surprisingly cold region high in the planet's atmosphere, where conditions may be frigid enough for carbon dioxide to freeze out as ice or snow.

Since the temperature at some heights dips below the freezing temperature of carbon dioxide, the main constituent of the atmosphere, carbon dioxide ice might be able to form there—possibly forming clouds of ice or snow particles. The data also show that the cold layer above the terminator (the dividing line between the day and night sides of the planet) is sandwiched between two comparatively warm layers. The temperature profiles on the hot dayside and cool night side at altitudes above 120 km are extremely different, so the terminator is affected by conditions on both sides (Fig. 8.9). The night side may be playing a greater role at one altitude and the dayside might be playing a larger role at other altitudes.

This situation is unique to Venus, because the temperature profile along the terminator in the atmospheres of Earth or Mars is very different (https://sci.esa.int/web/venus-express/-/54066-5-snow-on-venus).

times more embayed craters than observed, or unobserved nonrandom crater distributions for resurfacing areas between 0.03% and 100% of the planet's surface. Strom et al. (1994)'s study indicated that Venus experienced a global resurfacing event about 300 m.y. ago followed by a dramatic reduction of volcanism and tectonism. This global resurfacing event ended abruptly (<10 m.y.). The present crater population has accumulated since then and remains largely intact. Thermal history models suggest that similar global resurfacing events probably occurred episodically in the past. Monte Carlo simulations carried out by Strom et al. (1994) indicate that only about 4%–6% of the planet has been volcanically resurfaced since the global event, and that the lava production rate has been no more than 0.01–0.15 km^3/yr during this time. This rate is significantly less than the current rate of intraplate volcanism on Earth (0.33–0.5 km^3/yr). Most of Venus' recent volcanism occurs in the Beta-Atla-Themis region, and most of the recent tectonism is associated with the major global-scale tectonic disruption zones that lie within and connect the equatorial highlands. The approximately 33% of the planet's surface bounded by latitudes 30°N and 30°S, longitudes 60° and 300°E contains twice as many heavily fractured craters and 1.4 times more lava-embayed craters as the planetary average. This region includes most of the major tectonic belts in the equatorial region. Strom et al. (1994)'s study further indicate that the effects of recent geologic activity are much less than those of the earlier global resurfacing event, when the record of all the early heavy bombardment and much of the later light bombardment was erased from the surface by massive volcanism and tectonic activity. Strom et al. (1994) point out that episodic regional resurfacing events that had global effects also occurred on Earth (e.g., the mid-Cretaceous superplume) and probably on Mars. On Mars they may have triggered the catastrophic releases of water that formed the outflow channels.

Tectonically, much of the Venus surface resembles the terrestrial continents, as expected on the basis of their similar strength profiles. One area of Venus, referred to as Nuwa Campus, is similar in size and shape to the Tarim Basin, an exotic continental terrane in NW China. Despite these similarities, detailed mapping of these two terranes reveals important differences consistent with a thinner crustal thickness on Venus: if Earth's continents are analogous to drifting icebergs, then the Venus crust would be a sea of pack ice (Ghail et al., 2018).

Roughly two-thirds of the Venusian surface is covered by flat, smooth plains that are marred by thousands of volcanoes, some of which are still active today, ranging from about 0.5 to 150 miles (0.8 to 240 km) wide, with lava flows carving long, winding canals that are up to more than 3000 miles (5,000 km) in length.

Six mountainous regions make up about one-third of the Venusian surface. One mountain range, called Maxwell, is about 540 miles (870 km) long and reaches up to some 7 miles (11.3 km) high, making it the highest feature on the planet.

Venus also possesses a number of surface features that are unlike anything on Earth. For example, Venus has coronae, or crowns—ring-like structures that range from roughly 95 to 1300 miles (155 to 2100 km) wide. Scientists believe these formed when hot material beneath the planet's crust rose up, warping the planet's surface. Venus also has tesserae, or tiles—raised areas in which many ridges and valleys have formed in different directions. Currently, Venus shows no clear evidence of Earth-like plate tectonic activity or surface conditions.

8.7.2 Earth's volcanism

On Earth, volcanism occurs in several distinct geologic settings. Most of these are associated with the boundaries of the enormous rigid plates that make up the lithosphere—the crust and upper mantle. The majority of active terrestrial volcanoes (roughly 80%) and related phenomena occur where two tectonic plates converge and one overrides the other, forcing it down into the mantle to be reabsorbed. Long curved chains of islands known as island arcs form at such subduction zones (https://www.britannica.com/science/volcanism). Movement of molten rock in the mantle, caused by thermal convection currents, coupled with gravitational effects of changes on the Earth's surface (erosion, deposition, even asteroid impact and patterns of post-glacial rebound) drive plate tectonic motion and ultimately volcanism.

An explosive eruption is a volcanic eruption of the most violent type. A notable example is the 1980 eruption of Mount St. Helens (Fig. 8.10 left) and an early stage of the July 12, 2009 eruption of Sarychev volcano, located

FIG. 8.10 (left) Mount St. Helens explosive eruption on July 22, 1980. *(Source: https://commons.wikimedia.org/wiki/File:MSH80_st_helens_eruption_plume_07-22-80.jpg) Copyright info: This image is in the public domain in the United States because it only contains materials that originally came from the United States Geological Survey, an agency of the United States Department of the Interior.* (right) An early stage of the July 12, 2009 eruption of Sarychev volcano, seen from space *(Source: https://en.wikipedia.org/wiki/File:Sarychev_Volcano_edit.jpg) Copyright info: This file is in the public domain in the United States because it was solely created by NASA. NASA copyright policy states that "NASA material is not protected by copyright unless noted."*

in the Kuril Islands, Russia. Mount St. Helens, located in Washington State, is the most active volcano in the Cascade Range, and it is the most likely of the contiguous U.S. volcanoes to erupt in the future. Sarychev volcano is a young, highly symmetrical stratovolcanic cone (described in the below paragraph). The height of the plume during the 2009 eruption was estimated at 12 to 18 km (7.5 to 11.2 mi). Such eruptions result when sufficient gas has dissolved under pressure within a viscous magma such that expelled lava violently froths into volcanic ash when pressure is suddenly lowered at the vent. Sometimes a lava plug will block the conduit to the summit, and when this occurs, eruptions are more violent. Explosive eruptions can send rocks, dust, gas and pyroclastic material up to 20 km (12 mi) into the atmosphere at a rate of up to 100,000 tonnes per second, traveling at several hundred meters per second. This cloud may then collapse, creating a fast-moving pyroclastic flow of hot volcanic matter.

A stratovolcano, also known as a composite volcano, is a conical volcano built up by many layers (strata) of hardened lava and tephra. Unlike shield volcanoes, stratovolcanoes are characterized by a steep profile with a summit crater and periodic intervals of explosive eruptions and effusive eruptions, although some have collapsed summit craters called calderas. The lava flowing from stratovolcanoes typically cools and hardens before spreading far, due to high viscosity. The magma forming this lava is often felsic, having high-to-intermediate levels of silica (as in rhyolite, dacite, or andesite), with lesser amounts of less-viscous mafic magma. Extensive felsic lava flows are uncommon, but have travelled as far as 15 km (9.3 mi). Volcanoes of the explosive type make up many of the islands of a single arc or the inner row of islands of a double arc. All such islands that border the Pacific basin are built up from the seafloor, usually by the extrusion of basaltic and andesitic magmas. Note that andesitic magma is a high-temperature magma (~1,200°C) characterized by flowing lava, and it is made up of about 45–55% silica (SiO_2) by weight.

A second major site of active volcanism is along the axis of the oceanic ridge system, where the plates move apart on both sides of the ridge and magma wells up from the mantle, creating new ocean floor along the trailing edges of both plates. Virtually all of this volcanic activity occurs underwater. In a few places the oceanic ridges are sufficiently elevated above the deep seafloor that they emerge from the ocean, and subaerial volcanism occurs. Iceland is the best-known example. The magmas that are erupted along the oceanic ridges are basaltic in composition.

A relatively small number of volcanoes occur within plates far from their margins. Some, as exemplified by the volcanic islands of Hawaii that lie in the interior of the Pacific Plate, are thought to occur because of plate movement over a "hot spot" from which magmas can penetrate to the surface. These magmas characteristically generate a chain of progressively older volcanoes that mark the direction of past motion of the plate over a particular hot spot. The active volcanoes of the East African Rift Valley also occur within a plate (the African Plate), but they appear to result from a different mechanism—possibly the beginning of a new region of plates moving apart (https://www.britannica.com/science/volcanism).

8.7.3 Volcanism on Mars—differences with terrestrial volcanic styles

The geological evolution of Mars is largely a story of volcanic activity. About 70% of the planet's crust was resurfaced by basaltic volcanism, but a significant fraction of that volcanic material is from unknown sources. Volcanic activity, or volcanism, has played a significant role in the geologic evolution of Mars (Head, 2007). Scientists have known since the Mariner 9 mission in 1972 that volcanic features cover large portions of the Martian surface. These features include extensive lava flows, vast lava plains, and the largest known volcanoes in the Solar System (Masursky et al., 1973; Carr, 1973). Carr et al. (1973) constructed a geologic map of Mars largely on the basis of photographic evidence. Four classes of units are recognized: (1) primitive cratered terrain, (2) sparsely cratered volcanic eolian plains, (3) circular radially symmetric volcanic constructs such as shield volcanoes, domes, and craters, and (4) tectonic erosional units such as chaotic and channel deposits. It was found that grabens (elongated blocks of the earth's crust lying between two faults and displaced downwards relative to the blocks on either side, as in a rift valley) are the main structural features; compressional and strike-slip features are almost completely absent. Most grabens are part of a set radial to the main volcanic area, Tharsis. Martian volcanic features range in age from Noachian (>3.7 billion years) to late Amazonian (< 500 million years), indicating that the planet has been volcanically active throughout its history (Michalski and Bleacher, 2013), and some speculate it probably still is so today. Both Earth and Mars are large, differentiated planets built from similar chondritic materials. Many of the same magmatic processes that occur on Earth also occurred on Mars, and both planets are similar enough compositionally that the same names can be applied to their igneous rocks and minerals.

Because the lower gravity of Mars generates less buoyancy forces on magma rising through the crust, the magma chambers that feed volcanoes on Mars are thought to be deeper and much larger than those on Earth (Wilson and Head, 1994). If a magma body on Mars is to reach close enough to the surface to erupt before solidifying, it must be big. Consequently, eruptions on Mars are less frequent than on Earth but are of enormous scale and eruptive rate when they do occur. Somewhat paradoxically, the lower gravity of Mars also allows for longer and more widespread lava flows. Lava eruptions on Mars may be unimaginably huge. A vast lava-flow (the size of the state of Oregon) has recently been described in western Elysium Planitia. The flow is believed to have been emplaced turbulently over the span of several weeks and is thought to be one of the youngest lava flows on Mars (Jaeger et al., 2010).

Michalski and Bleacher (2013) found that several irregularly shaped craters located within Arabia Terra on Mars represent a new type of highland volcanic construct and together constitute a previously unrecognized Martian igneous province. Similar to terrestrial supervolcanoes, these low-relief paterae (broad, shallow bowl-shaped features on the surface) possess a range of geomorphic features related to structural collapse, effusive volcanism and explosive eruptions. Extruded lavas contributed to the formation of enigmatic highland ridged plains in Arabia Terra. Outgassed sulfur and erupted fine-grained pyroclastics from these calderas probably fed the formation of altered, layered sedimentary rocks and fretted terrain found throughout the equatorial region. The discovery of a new type of volcanic construct in the Arabia volcanic province fundamentally changes the picture of ancient volcanism and climate evolution on Mars. Michalski and Bleacher (2013) have suggested that other eroded topographic basins in the ancient Martian highlands that have been dismissed as degraded impact craters should be reconsidered as possible volcanic constructs formed in an early phase of widespread, disseminated magmatism on Mars.

There are several volcanoes (Mons) on Mars. The "Aster clouds" are thought to form under weak atmospheric static stability and weak background flow, and are probably related to the local up-slope winds associated with the Mons. These clouds are observed during mid to late northern summer on western side of Olympus Mons. Mars color camera (MCC) onboard MOM observed ASTER clouds over Olympus Mons [a shield volcano 624 km (374 mi) in diameter, 25 km (16 mi) high, and is rimmed by a 6 km (4 mi) high scarp. A caldera 80 km (50 mi) wide is located at the summit of Olympus Mons] and Elysium Mons [a volcano on Mars located in the volcanic province Elysium, at 25.02°N 147.21°E, in the Martian eastern hemisphere], that have unique morphology and sometimes forms rays around the central disk of the Mountain. The Aster clouds are thought to form under weak atmospheric static stability and weak background flow, and are probably related to the local up-slope winds associated with the Mons. These clouds are observed during mid to late northern summer on western side of Olympus Mons.

MCC has also observed Lee-Wave cloud over Ascraeus Mons. Lee wave clouds are a regular train of clouds aligned orthogonal to the prevailing wind. Mountain waves (lee waves, or gravity waves) result from a parcel of air being forced up due to a topographic high, condensing out as a cloud, then dropping back down. Other than Lee-wave Clouds, the MCC onboard MOM observed ASTER clouds over Olympus and Elysium Mons, that have unique morphology and sometimes forms rays around the central disk. These clouds are observed during mid to late northern summer.

Besides, the shadow of clouds casted on the Martian surface have also been used to estimate the cloud height and one such patch of cloud is estimated to be at an altitude of about 35–38 km., which is abode of CO_2 clouds (LPSC 2015). It is worth mentioning that unlike Earth, there are two types of clouds in Martian atmosphere viz. water-ice clouds and CO_2 ice clouds. The later occurs mostly 20–25 km above the Mars surface.

8.8 Lava tubes on Earth and Mars

Lava tubes, found on Earth and detected on Earth's Moon, have been identified also on Mars. For example, images, taken by the High-Resolution Stereo Camera (HRSC) on board ESA's Mars Express, show elongated depressions on the Martian surface on Pavonis Mons (the central volcano of the three 'shield' volcanoes that comprise Tharsis Montes), which researchers believe are the surface manifestations of lava tubes that lie beneath; channels originally formed by hot, flowing lava that forms a crust as the surface cools. Lava continues to flow beneath this hardened surface, but when the lava production ends and the tunnels empty, the surface collapses, forming elongated depressions. ESA's Mars Express spacecraft obtained these images using the HRSC during orbit 902 with a ground resolution of approximately 14.3 m per pixel. The images were acquired in the region of Pavonis Mons, at approximately 0.6° South and 246.4° East. The context map is centerd on Pavonis Mons, one of the three volcanoes called Tharsis Montes (the others being Arsia and Ascreus Montes, aligned with Pavonis in a line nearly 1500 km long). The dramatic features visible in the color image are located on the south-west flank of Pavonis Mons. The long, continuous lava tube in the northwest of the color image extends over 59 km and ranges from approximately 1.9 km to less than 280 m wide. Pit chains, strings of circular depressions thought to form as a result of collapse of the surface, are also visible within the color image. In the northeast, there is a clear distinction between the brighter terrain at higher elevations and darker material located down slope. In the southwest, the lava tubes appear to be covered by subsequent lava flows.

Fig. 8.11 shows a dark pit—caved-in roof of a lava tube running beneath—which forms the lava tube's entrance on the Martian surface; captured by NASA's HiRISE Camera on board the Mars Reconnaissance Orbiter from an orbit about 260 km (160 miles) above the Martian surface. According to MRO team member Ross Beyer on the HiRISE website, "Fortunately, HiRISE is sensitive enough to actually see things in this otherwise dark pit. Since HiRISE turned by

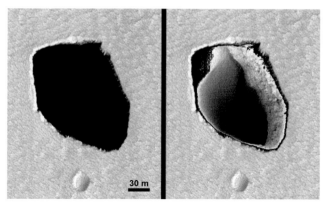

FIG. 8.11 A dark pit—caved-in roof of a lava tube running beneath—which forms the lava tube's entrance on the Martian surface; captured by NASA's high-resolution HiRISE Camera on board the Mars Reconnaissance Orbiter from an orbit about 260 km (160 miles) above the Martian surface. In this cut-out, the "normal" view of the HiRISE image is on the left, while the right shows what happens when the brightness of the pixels inside the pit is enhanced. *Credit: NASA/JPL/University of Arizona. (Source: https://phys.org/news/2020-02-caved-in-roof-lava-tube-good.html).*

almost 30° to capture this image, we can see the rough eastern wall of the pit. The floor of the pit appears to be smooth sand and slopes down to the southeast."

Lava tubes below the surface of Mars, as well as those below the lunar surface, can be identified by locating pit craters in the vicinity of known ancient lava flows. Paris et al. (2019) described a pit crater thus: "A pit crater is a circular or elliptical depression shaped by the collapsing or sinking of the surface lying above a void or hollow cavity, rather than by the eruption of a lava vent or volcano. Pit craters generally lack a raised rim, uplifting, ejecta blankets associated with impact craters, or radial patterns of lava flows discharging downslope from volcanic calderas. There are generally two types of pit craters—atypical and bowl-shaped. Atypical pit craters exhibit a distinctive set of morphologies and characteristics that set them apart from the commonly observed bowl-shaped pit craters. Instead of bowls, the interior of atypical pit craters is cylindrical or bell-shaped with vertical to overhanging walls that extend down to their floors without forming talus slopes. In some instances, atypical pit craters show evidence of vents, fissures, and caverns/caves that could be intact." Note that a caldera is a large volcanic crater, especially one formed by a major eruption leading to the collapse of the mouth of the volcano.

Pit craters have been discovered on Earth, Mars, as well as on the various moons in our solar system (including Earth's Moon). They are generally located in a series of ranged or offset chains; and in these instances, they are referred to as pit crater chains. When adjacent walls between pits in a pit crater chain collapse, they become troughs. A commonly invoked hypothesis to explain these troughs is that they are collapsed lava tubes, essentially tunnels formed underground by rivers of lava (Paris et al., 2019).

The largest lava tubes on Earth are maximum ~40 m [130 feet] of width and height, like a very large motorway tunnel. On Mars collapsed lava tubes tend to be about 80 times larger than Earth's, with diameters of 130 to 1300 feet (40 to 400 m). The sheer scale of these extraterrestrial lava tubes is likely a result of low Martian gravity, as well as differences in Martian lava tubes. For the sake of comparison, the researchers collected 3D laser scans of their counterparts—both collapsed and intact—on Earth. Then they collected all the available satellite images of collapsed lava tubes on Mars, and modeled the size of the intact tubes based on the relationships between collapsed and intact tunnels on Earth. In the context of the consequences of low gravity, it may be noted that the heights of material eruptions (plumes) from low-gravity moons such as Europa and Enceladus are in the range 100–200 km or more. The dataset reported by Sauro et al. (2020) indicates that Martian lava tubes are 1 to 3 orders of magnitude more voluminous than on Earth, and suggests that the same processes of inflation and over-crusting were active on Mars.

Dark pits on Mars are fascinating—probably because they provide mysteries and possibilities. Could anything be inside? Or perhaps this could be a place where humans could set up a base that would provide shelter from Mars' harsh environment. A line of "pit craters" near a dormant volcano tells the story of a lava tube that formed, drained and then partially collapsed in one section or another—sometimes even leaving behind "skylight" holes in the middle of the crater. On Mars several deep skylights have been identified on lava flows with striking similarities with terrestrial cases (Sauro et al., 2020). By comparing literature and speleological (pertaining to caves) data from terrestrial analogs and measuring lunar and Martian collapse chains on satellite images and digital terrain models (DTMs), Sauro et al. (2020) sheds light on tube size, depth from surface, and several other morphometric (quantitative analysis of form, encompassing size and shape) parameters pertaining to the lava tubes on the three different planetary bodies.

If a future rover mission were to land nearby, this pit might be worth a look—from a safe distance around the rim. The *Hellas Planitia* lava tubes are considered to be among the safest places for Martian explorers to camp out. On Mars, with its lower gravity, simulations suggest that the hollowed-out tubes would be much larger in size than those found on Earth. Although the Martian surface is a radiation hot zone, the Martian lava tubes might offer safety. There are other potential advantages to life in the tubes. It might be possible to shore them up (i.e., support them from falling by placing something under or against it), seal them off, and pressurize them and warm them up to create liveable environments. Like human-made shelters, the tubes would also offer protection from micrometeorites, temperature fluctuations and potentially dangerous substances in the Martian surface dust (Letzter, 2020; Murdock, 2020). The candidate lava tubes, moreover, can serve as important locations for

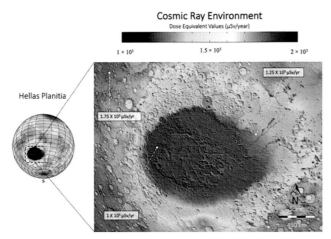

FIG. 8.12 An expanded view of the radiation environment at Hellas Planitia. *(Source: Center for Planetary Science) Reproduced from Paris, A.J., E.T. Davies, L. Tognetti, and C. Zahniser (2019), Prospective lava tubes at Hellas Planitia: Leveraging volcanic features on Mars to provide crewed missions protection from radiation, The Journal of The Washington Academy of Sciences, arXiv:2004.13156 [astro-ph.EP].*

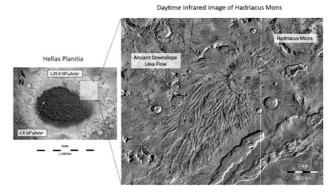

FIG. 8.13 An expanded view of Hadriacus Mons along the edge of Hellas Planitia. *(Source: NASA and Center for Planetary Science). Reproduced from Paris, A.J., E.T. Davies, L. Tognetti, and C. Zahniser (2019), Prospective lava tubes at Hellas Planitia: Leveraging volcanic features on Mars to provide crewed missions protection from radiation, The Journal of The Washington Academy of Sciences, arXiv:2004.13156 [astro-ph.EP].*

direct observation and study of Martian geology and geomorphology, as well as potentially uncovering any evidence for the development of microbial life early in the natural history of Mars.

Volcanic formations at an impact basin in Mars' southern hemisphere, known as *Hellas Planitia* offer a few protective advantages: NASA probes have shown that the most intense radiation environments on Mars are at the poles. But *Hellas Planitia* lies closer to the equator (Fig. 8.12). Radiation level is an important consideration on Mars. Micro-sieverts (μSv) is a unit of radiation exposure. According to data from NASA's MARIE experiment, radiation levels in this region have been measured at $\sim 1.25 \times 10^5$ μSv/year (~ 342.46 μSv/day), which is considerably lower than that at the higher topographic elevations on Mars, which were measured at $\sim 2 \times 10^5$ μSv/year. Notwithstanding, a radiation exposure of 1.25×10^5 μSv/year is still sizably higher than what human beings in developed nations are annually exposed to on Earth. And of all Martian environments, the impact basin is among the most low-lying at about 23,464 feet (7152 m) deep, implying that at this impact basin more of Mars' thin atmosphere lies overhead. It has been estimated that about 50% less radiation reaches the basin floor than higher-elevation regions of Mars. Explorers could expect about 342 μSv/day in the basin, compared with 547 μSv/day elsewhere on Mars. Although this amount of radiation exposure is relatively small on the Martian surface, it is much higher than what is typically considered safe.

The Context Camera (CTX) deployed on NASA's Mars Reconnaissance Orbiter (MRO) observes Mars's surface at ~ 6 m/pixel in swaths of 30 km across and up to ~ 160 km in length. MRO also observes the Martian surface earlier in the day; thus, more pit craters can be observed with partially sunlit floors (Walden et al., 2014). A partially sunlit floor allows planetary scientists to identify specific pit crater characteristics such as skylights, individual boulders, dusty, or rocky surface textures, overhanging rims, wall and floor morphologies, and bedrock stratification (Walden et al., 2014).

Paris et al. (2019) surveyed 1500 images taken by the MRO and identified three candidate lava tubes as possible sites for manned exploration. It was found that *Hadriacus Mons* region is having low radiation levels and is also home for prospective lava tubes and, therefore, this region could be a prime candidate for manned exploration. Paris et al. (2019) describe that the lava tubes near Hadriacus Mons—an ancient, low-relief volcanic mountain located along the north-eastern edge of Hellas (Fig. 8.13) —could be used as natural radiation shelters and habitats for a crewed mission to Mars. According to Paris et al. (2019), "These natural caverns have roofs estimated to be tens of meters thick, which would provide the crew protection from not only exposure to too much radiation, but also the bombardment of micrometeorites, exposure to dangerous soil perchlorates due to long-term dust storms, and extreme temperature fluctuations [Underground towns on the Moon and Mars: Future human habitats could be hidden in lava tubes, *Spaceflight Insider* (available at http://www.spaceflightinsider.com)]. Moreover, although the exact conditions of the interior of Martian lava tubes will remain unknown until they are actually explored, planetary scientist are of the consensus that they represent prime locations for direct observation of pristine Martian bedrock, where keys critical to understanding the natural history of this planet will be found."

Paris et al. (2019)'s investigations revealed that the $\sim 4{,}500$-meter collapsed lava pit (Fig. 8.13A), which is positioned between two partially collapsed sinuous (having many curves and turns) pit crater chains, depicts two surface areas that have not collapsed (Fig. 8.14A and B). Imagery analysis indicates the area of the northern un-collapsed

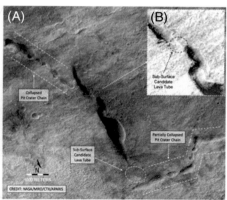

FIG. 8.14 **Candidate lava tubes south of Hadriacus Mons.** *(Credit: Center for Planetary Science). Reproduced from Paris, A.J., E.T. Davies, L. Tognetti, and C. Zahniser (2019), Prospective lava tubes at Hellas Planitia: Leveraging volcanic features on Mars to provide crewed missions protection from radiation, The Journal of The Washington Academy of Sciences, arXiv:2004.13156 [astro-ph.EP].*

surface is ∼600 m long and ∼300 m wide (Fig. 8.14A), and that the area of the southern un-collapsed surface is ∼900 m long and ∼600 m wide (Fig. 8.14B). Further analysis of the ∼4,500-meter collapsed lava pit with its contrast range limited to low-end radiance values indicate that the pit crater walls encompassing both un-collapsed areas are structurally intact—ruling out a lava bridge—indicating that a lava tube below the surface is plausible, since these parts of the pit crater chain have not collapsed (Paris et al., 2019).

Another candidate lava tube identified in Paris et al. (2019)'s investigations is located southwest of Hadriacus Mons at latitude -36.870° and longitude 89.498°E. The ∼25-meter collapsed pit, which appears to indicate the entrance to a candidate cavern and/or cave, is preceded by a partially collapsed trough (Fig. 8.15A). Paris et al.

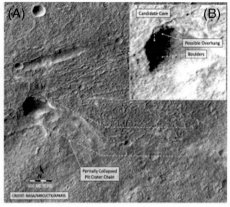

FIG. 8.15 **Candidate lava tubes south of Hadriacus Mons.** *(Credit: Center for Planetary Science). Reproduced from Paris, A.J., E.T. Davies, L. Tognetti, and C. Zahniser (2019), Prospective lava tubes at Hellas Planitia: Leveraging volcanic features on Mars to provide crewed missions protection from radiation, The Journal of The Washington Academy of Sciences, arXiv:2004.13156 [astro-ph.EP].*

(2019) found that the collapsed trough is sinuous, and it could be associated with the presence of a lava tube below the surface. According to Paris et al. (2019)'s interpretation, the same collapsed pit with its contrast range limited to low-end radiance values depicts the presence of boulders that more than likely arose from the collapse of the western wall section, which could be the entrance to a lava tube (Fig. 8.15B). It was further found that the region of the candidate lava tube is characterized by numerous sinuous collapsed pit crater chains ranging in size—providing further evidence that one or more lava tubes could exist below the surface. Paris et al. (2019) found that, interestingly, this proposed lava tube is analogous to Giant Ice Cave in El Mapais, NM (Fig. 8.15), which is also characterized by a partially collapsed sinuous trough that leads into the entrance of a lava tube.

Assuming that the observed clues point to the existence of real lava tubes in Hellas Planitia on Mars, Paris et al. (2019) visited similar terrestrial analog sites in the American Southwest and conducted 30 analog radiation experiments to test the idea of feasibility of lava tubes as radiation shields. Though cosmic radiation on Earth's surface is much lower than on Mars, some of those particles do make it to Earth's surface. The experiments were conducted during solar noon at terrestrial lava tubes (Fig. 8.16) located at California's Mojave Aiken tube, Arizona's Lava River Cave, and New Mexico's Big Skylight, Giant Ice Cave and Junction Cave; using Geiger counter capable of detecting Gamma and X-rays (ionizing radiation) down to 10 keV thru the window, 40 keV minimum through the case, and with a Gamma sensitivity of 10,000 µSv/h. Paris et al. (2019) explain that "Because radiation from the sun strikes the surface of Earth at different angles, conducting the experiments at solar noon (when the sun is directly overhead) allowed the surface area where the experiments were conducted to receive the most electromagnetic energy from the Sun possible (Nobel, 2009). Furthermore, a Sun at 90° above the surface of each lava tube prevented the radiation from becoming too scattered and diffused." Although Earth's atmosphere and magnetic field protect us from most ionizing radiation from the Sun, streams of ionizing radiation still reach the surface of Earth. Occasionally, during a solar particle event, larger amounts of ionizing radiation strike the surface of the Earth, which can also be detected by the Geiger counter (Paris et al., 2019).

Getting down to the nitty-gritty of the experiments (i.e., the most important aspects/practical details of the experiments), Paris et al. (2019) explain that preceding each experiment, the Geiger counter device was turned on for a minimum of 1 h to establish a baseline reading. The experiment was then conducted in two parts—measuring the amount of solar radiation outside the lava tube and then comparing it with the amount of solar radiation measured inside the lava tube. In total, six one-hour readings were

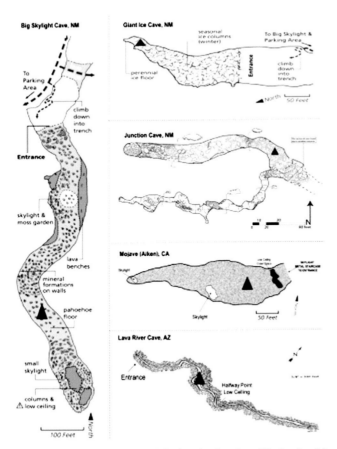

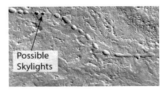

FIG. 8.17 An image from the Mars Reconnaissance Orbiter showing a possible lava tube on Mars near the southern flank of the Martian volcano Arsia Mons. *(Image credit: Mars Reconnaissance Orbiter/NASA). (Source: Letzter, R. (2020), Entire cities could fit inside the moon's monstrous lava tubes, Live Science, https://www.livescience.com/lava-tubes-mars-and-moon-habitable.html).*

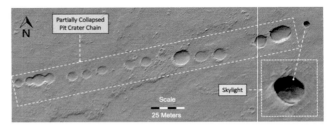

FIG. 8.18 Similarities between a pit crater chain and pit crater "skylight" at South of Arsia Mons on Mars. *(Source: NASA, Processing: Center for Planetary Science). (Reproduced from Paris, A.J., E.T. Davies, L. Tognetti, and C. Zahniser (2019), Prospective lava tubes at Hellas Planitia: Leveraging volcanic features on Mars to provide crewed missions protection from radiation, The Journal of The Washington Academy of Sciences, arXiv:2004.13156 [astro-ph.EP]).*

FIG. 8.16 Lava tube maps and the interior locations (black triangle) where each radiation experiment was conducted. *(Map Sources: Mojave lava tube sketched by Prof. Antonio Paris, others National Park Service). Reproduced from Paris, A.J., E.T. Davies, L. Tognetti, and C. Zahniser (2019), Prospective lava tubes at Hellas Planitia: Leveraging volcanic features on Mars to provide crewed missions protection from radiation, The Journal of The Washington Academy of Sciences, arXiv:2004.13156 [astro-ph.EP].*

completed for each lava tube (three exterior and three interior) for a total of 30 observation hours. Paris et al. (2019) found that on average, the total amount of radiation detected during solar noon on the exterior of all five lava tubes was ~0.471 μSv/h while the average radiation detected inside the lava tubes was ~0.083 μSv/h (i.e., 5.67 times reduction). Furthermore, during the experiments, Paris et al. (2019) observed a significant drop in temperature inside the terrestrial lava tubes, which likewise would be the case for lava tubes on Mars. Thus, comparing measurements of radiation inside and outside, Paris et al. (2019) found a significant radiation-shielding effect. Extrapolating their results to Mars, they calculated that living in a Hellas lava tube, people might experience just about 61.64 μSv/day.

According to recent research findings, Mars is pockmarked with absolutely massive lava tubes, with ceilings as high as the Empire State Building (see Letzter, 2020).

Fig. 8.17 shows an image from the Mars Reconnaissance Orbiter, indicating a possible lava tube on Mars near the southern flank of the Martian volcano Arsia Mons. Fig. 8.18 shows similarities between a pit crater chain and pit crater "skylight" at South of Arsia Mons on Mars.

8.9 History of water on Mars and Venus

Liquid water is an important ingredient in the search for habitability of any celestial body. Therefore, the existence of extinct or extant liquid water is of considerable importance to astrobiologists and life researchers. Venus and Mars being the closest neighbours to Earth, gaining an understanding on the history of water on these two planets is of considerable interest.

8.9.1 Water once flowed on Mars—indications of hydrated minerals and clay, possible ocean, and vast river plains on martian surface

Although Mars' atmosphere used to be thick enough for water to run on the surface, today that water is either scarce or nonexistent. Mars is too cold for liquid water to exist for any length of time, but features on the surface suggest that water once flowed on Mars. Today, water exists in the form of ice in the soil, and in sheets of ice in the polar ice caps.

8.9.1.1 Hydrated minerals and clay on Martian surface

The polar ice caps are Mars' largest known reservoir of water. The water stored there is thought to document the evolution of Mars's water from the wet Noachian period, which ended about 3.7 billion years ago, to the present. At present, no liquid water is known to exist on Martian surface.

Besides seeing features that were signs of past surface water, researchers found other types of evidence for past water. For example, minerals detected in many locations needed water to form. The OMEGA/Mars Express hyperspectral imager identified hydrated sulfates on light-toned layered terrains on Mars. Likewise, outcrops in Valles Marineris, Margaritifer Sinus, and Terra Meridiani show evidence for kieserite, gypsum, and poly-hydrated sulfates. This identification has its basis in vibrational absorptions between 1.3 and 2.5 micrometers. These minerals constitute direct records of the past aqueous activity on Mars (Gendrin et al., 2005). Bibring et al. (2006) reported that global mineralogical mapping of Mars by the Observatoire pour la Mineralogie, l'Eau, les Glaces et l'Activité (OMEGA) instrument on the European Space Agency's Mars Express spacecraft provided new information on Mars' geological and climatic history. Bibring et al. (2006)'s study indicated that phyllosilicates formed by aqueous alteration very early in the planet's history (the "phyllocian" era) is present in the oldest terrains; and sulfates were formed in a second era (the "theiikian" era) in an acidic environment. However, beginning about 3.5 billion years ago, the last era (the "siderikian") is dominated by the formation of anhydrous ferric oxides in a slow superficial weathering, without liquid water playing a major role across the planet (Bibring et al., 2006). An instrument onboard Mars Odyssey orbiter mapped the distribution of water in the shallow surface.

Several lines of evidence suggest that the mean global temperatures on early Mars were well below the freezing point of pure water. Modeling studies, carried out by Fairén et al. (2009), of freezing and evaporation of Martian fluids with a chemical composition resulting from the weathering of basalts as reflected in the chemical compositions found at Mars landing sites suggests that a significant fraction would remain in the liquid state at temperatures well below 273 K.

Water on Mars has been explained by invoking controversial and mutually exclusive solutions based on warming the atmosphere with greenhouse gases (the "warm and wet" Mars) or on local thermal energy sources acting in a global freezing climate (the "cold and dry" Mars). Both have critical limitations and none has been definitively accepted as a compelling explanation for the presence of liquid water on Mars. Replacing these two hypotheses, Fairén (2010) came up with a third hypothesis, termed the "cold and wet" Mars hypothesis. Fairén (2010) examined the hypothesis that cold, saline and acidic liquid solutions have been stable on the subzero surface of Mars for relatively extended periods of time, completing a hydrogeological cycle in a water-enriched but cold planet. Fairén (2010) developed computer simulations to analyze the evaporation processes of a hypothetical Martian fluid with a composition resulting from the acid weathering of basalt. This model is based on orbiter- and lander-observed surface mineralogy of Mars, and is consistent with the sequence and time of deposition of the different mineralogical units. Fairén (2010)'s simulations indicated that the hydrological cycle would have been active only in periods of dense atmosphere, as having a minimum atmospheric pressure is essential for water to flow, and relatively high temperatures (over $\sim$245 K) are required to trigger evaporation and snowfall; minor episodes of limited liquid water on the surface could have occurred at lower temperatures (over $\sim$225 K). During times with a thin atmosphere and even lesser temperatures (under $\sim$225 K), only transient liquid water can potentially exist on most of the Martian surface. Assuming that surface temperatures have always been maintained below 273 K, Mars can be considered a "cold and wet" planet for a substantial part of its geological history.

8.9.1.2 Mars once possessed a Primordial ocean—Indications

The possibility that a large ocean once occupied the Martian northern plains is one of the most important and controversial hypotheses to have originated from the exploration of Mars. It is believed that nearly a third of the surface of Mars was covered by an ocean of liquid water (the Mars ocean hypothesis) early in the planet's geologic history (Clifford and Parker, 2001; Rodriguez et al., 2015). According to Baker et al. (1991), a variety of anomalous geomorphological features on Mars can be explained by a conceptual scheme involving episodic ocean and ice-sheet formation. The formation of valley networks early in Mars' history is evidence for a long-term hydrological cycle, which may have been associated with the existence of a persistent ocean. Cataclysmic flooding, triggered by extensive Tharsis volcanism, subsequently led to repeated ocean formation and then dissipation on the northern plains, and associated glaciation in the southern highlands until relatively late in Martian history.

Several physical features in the present geography of Mars suggest the past existence of a primordial ocean. In the 1980s, Viking spacecraft images revealed two possible ancient shorelines near the Martian pole, each thousands of kilometers long with features like those found in Earth's coastal regions. The shorelines—Arabia and the younger Deuteronilus—date from between 2 and 4 billion years ago. In 1987, John E. Brandenburg published the hypothesis of a primordial Mars ocean he dubbed Paleo-Ocean (Brandenburg, 1987). The ocean hypothesis is important because the existence of large bodies of liquid water in the past would have had a significant impact on ancient

FIG. 8.19 An artist's impression of ancient (billions of years ago) Mars and its oceans based on geological data. It is based on Mars Orbiter Laser Altimeter (MOLA) data. The elevations have been updated so the shore lines will closely approximate their ancient locations. Also, any mountains less than two billion years old have been removed. *Copyright info: This file is licensed under the Creative Commons Attribution-Share Alike 3.0 Unported license. (Source: https://commons.wikimedia.org/wiki/File:AncientMars.jpg).*

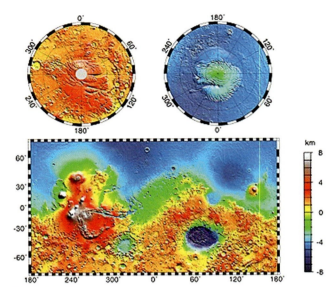

FIG. 8.20 The blue region of low topography in the Martian northern hemisphere is hypothesized to be the site of a primordial ocean of liquid water (Brandenburg, J.E. (1987), The Paleo-Ocean of Mars, *MECA Symposium on Mars: Evolution of its Climate and Atmosphere*, Lunar and Planetary Institute. pp. 20–22.). *Copyright info: Public domain (Source: https://en.wikipedia.org/wiki/Mars_ocean_hypothesis#/media/File:MarsTopoMap-PIA02031_modest.jpg).*

Martian climate, habitability potential and implications for the search for evidence of past life on Mars. The Mars Orbiter Laser Altimeter (MOLA), which accurately determined in 1999 the altitude of all parts of Mars, found that the watershed for an ocean on Mars would cover three-quarters of the planet (Smith, 1999). Fig. 8.19 depicts an artist's impression of ancient (billions of years ago) Mars and its oceans based on geological data. In Fig. 8.20, the blue region of low topography in the Martian northern hemisphere is hypothesized to be the site of a primordial ocean of liquid water.

In a study, Clifford and Parker (2001) examined the hydraulic and thermal conditions that gave rise to the elevated source regions of the Late Hesperian outflow channels and explored their implications for the evolution of the Martian hydrosphere. They found that if the outflow channel floodwaters were derived from a subpermafrost aquifer, then it implies that, throughout the planet's first billion years of evolution, as much as one-third of its surface was covered by standing bodies of water and ice. Following the development of the global dichotomy, the bulk of this water would have existed as an ice-covered ocean in the northern plains. Clifford and Parker (2001) demonstrated that the progressive crustal assimilation of this early surface reservoir of H_2O (punctuated by possible episodes of less extensive flooding) was a natural consequence of the planet's subsequent climatic and geothermal evolution—potentially cycling the equivalent of a km-deep global ocean of water through the atmosphere and subsurface every $\sim 10^9$ years. In response to the long-term decline in planetary heat flow, the progressive cold-trapping of H_2O into the growing cryosphere (those portions of a celestial body's surface where water is in solid form, including sea ice, lake ice, river ice, snow cover, glaciers, ice caps, ice sheets, and frozen ground) is expected to have significantly depleted the original inventory of groundwater—a development that could well explain the apparent decline in outflow channel activity observed during the Amazonian. Although primarily a theoretical analysis, Clifford and Parker (2001)'s findings appear remarkably consistent with the geomorphic and topographic evidence that Mars once possessed a primordial ocean and that a substantial relic of that body continues to survive as massive ice deposits within the northern plains. Confirmation of the presence of such deposits, combined with the potential detection of a global-scale groundwater system, would provide persuasive support for the validity of this analysis.

About four billion years ago, the young planet would have had enough water to cover its entire surface in a liquid layer about 140 m deep, but it is more likely that the liquid would have pooled to form an ocean occupying almost half of Mars's northern hemisphere, and in some regions reaching depths greater than 1.6 km.

Mars' primordial ocean, dubbed Paleo-Ocean (Brandenburg, 1987) and Oceanus Borealis (Baker et al.,

1991) would have filled the basin Vastitas Borealis in the northern hemisphere, a region which lies 4–5 km (2.5–3 miles) below the mean planetary elevation, at a time period of approximately 4.1–3.8 billion years ago. Evidence for this ocean includes geographic features resembling ancient shorelines, and the chemical properties of the Martian soil and atmosphere (Villanueva et al., 2015). Early Mars would have required a denser atmosphere and warmer climate to allow liquid water to remain at the surface (Read and Lewis, 2004; Fairén, 2010; Fairén et al., 2011).

Study carried out by Villanueva et al. (2015) showed that a primitive ocean on Mars held more water than Earth's Arctic Ocean, and covered a greater portion of the planet's surface than the Atlantic Ocean does on Earth. These investigators used ESO's Very Large Telescope, along with instruments at the W.M. Keck Observatory and the NASA Infrared Telescope Facility, to monitor the atmosphere of the planet and map out the properties of the water in different parts of Mars's atmosphere over a 6-year period. These new maps are the first of their kind.

Villanueva et al. (2015)'s study provides a solid estimate of how much water Mars once had, by determining how much water was lost to space. With this work, we can better understand the history of water on Mars. The new estimate is based on detailed observations of two slightly different forms of water in Mars's atmosphere. One is the familiar form of water, consisting of two hydrogen atoms and one oxygen, H_2O. The other is HDO, or semiheavy water, a naturally occurring variation in which one hydrogen atom is replaced by a heavier form, called deuterium. As the deuterated form is heavier than normal water, it is less easily lost into space through evaporation. So, the greater the water loss from the planet, the greater the ratio of HDO to H_2O in the water that remains.

Villanueva et al. (2015) distinguished the chemical signatures of the two types of water using ESO's Very Large Telescope in Chile, along with instruments at the W.M. Keck Observatory and the NASA Infrared Telescope Facility in Hawaii. By comparing the ratio of HDO to H_2O, Villanueva et al. (2015) measured by how much the fraction of HDO has increased and thus determined how much water has escaped into space. This in turn allowed the amount of water on Mars at earlier times to be estimated.

In the just mentioned study, the team mapped the distribution of H_2O and HDO repeatedly over nearly six Earth years—equal to about three Mars years—producing global snapshots of each, as well as their ratio. The maps reveal seasonal changes and microclimates, even though modern Mars is essentially a desert.

Villanueva et al. (2015)'s results show that atmospheric water in the near-polar region was enriched in HDO by a factor of seven relative to Earth's ocean water, implying that water in Mars's permanent ice caps is enriched eightfold. Mars must have lost a volume of water 6.5 times larger than the present polar caps to provide such a high level of enrichment. The volume of Mars's early ocean must have been at least 20 million cubic kilometers.

Based on the surface of Mars today, a likely location for this water would be the Northern Plains, which have long been considered a good candidate because of their low-lying ground. An ancient ocean there would have covered 19% of the planet's surface—by comparison, the Atlantic Ocean occupies 17% of the Earth's surface.

According to Villanueva et al. (2015), with Mars losing that much water, the planet was very likely wet for a longer period of time than previously thought, suggesting the planet might have been habitable for longer. It is possible that Mars once had even more water, some of which may have been deposited below the surface. Because the new maps reveal microclimates and changes in the atmospheric water content over time, they may also prove to be useful in the continuing search for underground water.

Many craters and other depressions on Mars show deltas that resemble those on Earth. In addition, if a lake lies in a depression, channels entering it will all stop at the same altitude. Such an arrangement is visible around places on Mars that are supposed to have contained large bodies of water, including around a possible ocean in the north.

8.9.1.3 Vast river plains on Martian surface

There are indications to suspect that surface liquid water could have existed for periods of time due to geothermal effects, chemical composition or asteroid impacts (Newsom, 1980; McKay and Davis, 1991; Newsom et al., 1996; Abramov and Kring, 2005; Fairén et al., 2009; Newsom, 2010). Many features of the Martian landscape are thought to have been formed by liquid water flow.

In summer 1965, the first close-up images from Mars showed a cratered desert with no signs of water. However, over the decades, as more parts of the planet were imaged with better cameras on more sophisticated orbital spacecrafts, Mars showed evidence of past river valleys, lakes and present ice in glaciers and in the ground.

Networks of gullies that merge into larger channels imply erosion by a liquid agent, and resemble ancient riverbeds on Earth. Enormous channels, 25 km wide and several hundred meters deep, appear to direct flow from underground aquifers in the Southern uplands into the Northern lowlands (Rodriguez et al., 2015). Much of the northern hemisphere of Mars is located at a significantly lower elevation than the rest of the planet (the Martian dichotomy), and is unusually flat.

Studies indicate a much higher density of stream channels than formerly believed. Regions on Mars with the most valleys are comparable to what is found on the Earth. In the research carried out by Rodriguez et al., 2015, the team developed a computer program to identify valleys by searching for U-shaped structures in topographical data.

The large amount of valley networks strongly supports rain on the planet in the past. The global pattern of the Martian valleys could be explained with a big northern ocean in the past. A large ocean in the northern hemisphere in the past would explain why there is a southern limit to valley networks; the southernmost regions of Mars, farthest from the water reservoir, would get little rainfall and would develop no valleys. In a similar fashion, the lack of rainfall would explain why Martian valleys become shallower from north to south.

Scientists consider that the climate of early Mars could have supported a complex hydrological system and possibly a northern hemispheric ocean covering up to one-third of the planet's surface (Parker et al., 1989; Baker et al., 1991; Head et al., 1999; Clifford and Parker, 2001; Fairén et al., 2003). This notion has been repeatedly proposed (Parker et al., 1989; Baker et al., 1991; Head et al., 1999; Clifford and Parker, 2001; Fairén et al., 2003) and challenged (Malin and Edgett, 1999; Ghatan and Zimbelman, 2006) over the past two decades, and remains one of the largest uncertainties in Mars research. DiAchille and Hynek (2010) used global databases of known deltaic deposits, valley networks (Hynek et al., 2010) and present-day Martian topography to test for the occurrence of an ocean on early Mars. The distribution of ancient Martian deltas delineates a planet-wide equipotential surface within and along the margins of the northern lowlands. Hynek et al. (2010) suggest that the level reconstructed from the analysis of the deltaic deposits may represent the contact of a vast ocean covering the northern hemisphere of Mars around 3.5 billion years ago. This boundary is broadly consistent with palaeo-shorelines suggested by previous geomorphologic, thermophysic and topographic analyses, and with the global distribution and age of ancient valley networks. Hynek et al. (2010)'s findings lend credence to the hypothesis that an ocean formed on early Mars as part of a global and active hydrosphere.

A study of deltas on Mars, carried out by DiAchille and Hynek (2010), revealed that seventeen of them are found at the altitude of a proposed shoreline for a Martian ocean. This is what would be expected if the deltas were all next to a large body of water (DiBiasse et al., 2013). Note that a river delta is a landform created by deposition of sediment that is carried by a river as the flow leaves its mouth and enters slower-moving or stagnant water. This occurs where a river enters an ocean, sea, estuary, lake, reservoir, or (more rarely) another river that cannot carry away the supplied sediment. It has been suggested that the Hypanis Valles fan complex is a delta with multiple channels and lobes, which formed at the margin of a large, standing body of water. That body of water was a northern ocean. This delta is at the dichotomy boundary between the northern lowlands and southern highlands near Chryse Planitia (Fawdon et al., 2018).

8.9.1.4 Open-lake systems on ancient Mars

As already indicated, Mars is currently cold and hyperarid; liquid water is not stable at its surface. However, orbital and rover observations of features including valley networks, sedimentary fans and ancient lake beds indicate the planet once had rainfall and a warmer, wetter climate (Craddock and Howard, 2002; Grotzinger et al., 2015; Wordsworth et al., 2021). Observations from orbital spacecraft have shown that Jezero crater on Mars contains a prominent fan-shaped body of sedimentary rock deposited at its western margin. Water is thought to have filled much of Jezero crater, which has a diameter of 45 km. The ancient lake that once sat in Jezero crater on Mars flooded billions of years ago, transporting large boulders through a river delta and depositing fine-grained clay that could potentially preserve signs of ancient life. Mangold et al. (2021) identified three parts of a rock formation, called Kodiak butte, at the opening of the lake. At the top, there are large boulders, the biggest of which is 1.5 m wide and 1 m high, that suggest the flow of water into the lake sped up enough at one point that it could carry the rocks over tens of kilometers. Below the boulders, they found a build-up of sediment that points to a steady and consistent river flow before the boulder-carrying floods hit the crater. Previous studies have proposed that Jezero crater hosted an open-lake system with the water level at an elevation of −2395 m (Fassett and Head III, 2005); this inference is derived from the observations that the inlet valley (feeding the western fan) and the breaching valley (that dissects the eastern rim of the crater) have approximately the same elevation of −2395 m.

Orbital images showed geomorphic expressions of two sedimentary fan structures (western and northern) at the edges of the Jezero crater (Fassett and Head III, 2005; Ehlmann et al., 2008). Images of a prominent butte (an isolated flat-topped hill) located ∼1 km south of the main fan deposit, which is informally named Kodiak, record ancient sedimentary processes at Jezero crater. The geomorphic expressions were inferred to be river delta deposits that formed in an ancient lake basin during the Late Noachian or Early Hesperian epochs on Mars (∼3.6–3.8 Ga) (Fassett and Head III, 2005; Ehlmann et al., 2008; Goudge et al., 2015; Goudge et al., 2017; Mangold et al., 2020). The *Perseverance* rover landed in Jezero crater in February 2021. Spectroscopic observations from orbit have detected phyllosilicates and carbonates, minerals indicative of past aqueous environments (Ehlmann et al., 2008; Goudge et al., 2015; Horgan, et al., 2020). Rover investigations on the surface could provide insight into the evolution of Jezero's ancient lake system and the timescale of liquid water residence on the surface.

Mangold et al. (2021) analyzed images taken by the rover in the three months after landing. It was found that the fan has outcrop faces that were invisible from orbit, which

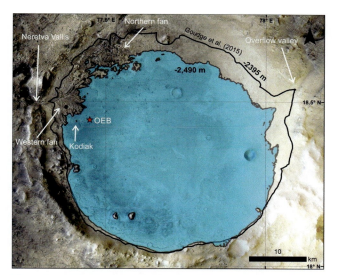

FIG. 8.21 **Inferred paleolake level inside Jezero crater at the time of Kodiak sediment deposition.** Blue shading indicates assumed lake level filled to the −2490 m gray contour following the uppermost elevation deduced from deltaic architecture at Kodiak. The red star indicates Octavia E. Butler (OEB) landing site of the Perseverance rover. The black outline of the implied earlier minimum water stand, corresponding to the overflow valley breach (Goudge et al., 2017), is shown for comparison. Rocks present on the crater floor might not have been emplaced during the period of lake activity. Both western and northern fans are above the inferred lake surface and the basin appears closed, 100 m below the breach to the east (labeled overflow valley). Background from the Context Camera (CTX) mosaic. *(Source: Mangold et al. (2021), Perseverance rover reveals an ancient delta-lake system and flood deposits at Jezero crater, Mars, Science, DOI: 10.1126/science.abl405 https://www.science.org/doi/10.1126/science.abl4051).*

record the hydrological evolution of Jezero crater. Mangold et al. (2021) interpret the presence of inclined strata in these outcrops as evidence of deltas that advanced into a lake. In contrast, the uppermost fan strata are composed of boulder conglomerates, which imply deposition by episodic high-energy floods. According to Mangold et al. (2021), this sedimentary succession indicates a transition, from a sustained hydrologic activity in a persistent lake environment, to highly energetic short-duration fluvial flows.

Mangold et al. (2021)'s investigations indicate that the ancient lake level was ~100 m below the inferred open-system lake level (Fig. 8.21). Thus, Jezero lake was closed (no outlet river) at the time of the delta progradation at Kodiak, which is a hydrological system conducive to short-term fluctuations in the lake level. Nevertheless, the overall stratigraphy indicates progradation of the western delta system and long-term lake level regression. It is believed that the lake was about 35 km wide and about 900 km^2 in area. There is evidence of rainfall on ancient Mars (Craddock and Howard, 2002). However, it is unclear if the water came from glacial lakes upstream or was it just rain? It is also unknow how old the lake is or when it dried up, nor whether the water was fresh or salty, which could impact the types of potential life it may have sustained.

8.9.1.5 Evidence of tsunami waves striking in the primordial Martian ocean—caused by asteroids hitting Mars millions of years apart

The geologic shape of what were once shorelines through Mars' northern plains convinces scientists that two large meteorites—hitting the planet millions of years apart—triggered a pair of mega-tsunamis. These gigantic waves forever scarred the Martian landscape and yielded evidence of cold, salty oceans conducive to sustaining life. A large team of scientists described how some of the surface in Ismenius Lacus quadrangle on Mars was altered by two tsunamis. The tsunamis were caused by asteroids striking the primordial Martian ocean. Both were thought to have been strong enough to create 30 km diameter craters. According to a study carried out by Rodriguez et al. (2016), about 3.4 billion years ago, a big meteorite impact triggered the first tsunami wave. This wave was composed of liquid water. It formed widespread backwash channels to carry the water back to the ocean. The first tsunami picked up and carried boulders the size of cars or small houses. The backwash from the wave formed channels by rearranging the boulders. The second came in when the ocean was 300 m lower. The second carried a great deal of ice which was dropped in valleys. Calculations show that the average height of the waves would have been 50 m, but the heights would vary from 10 m to 120 m. Numerical simulations show that in this particular part of the ocean two impact craters of the size of 30 km in diameter would form every 30 million years. In the millions of years between the two meteorite impacts and their associated mega-tsunamis, Mars went through frigid climate change, where water turned to ice. According to Rodriguez et al. (2016)'s study, the ocean level receded from its original shoreline to form a secondary shoreline, because the climate had become significantly colder. The implication here is that a great northern ocean may have existed for millions of years. One argument against an ocean has been the lack of shoreline features. These features may have been washed away by these tsunami events. The parts of Mars studied in this research are Chryse Planitia and northwestern Arabia Terra. These tsunamis affected some surfaces in the Ismenius Lacus quadrangle and in the Mare Acidalium quadrangle (Rodriguez et al., 2016). Fig. 8.22 shows an example of backwash channels in northwestern Arabia Terra on Mars, displaying streamlined bars generated by a tsunami wave that struck an ancient Martian Ocean. The impact that created the crater Lomonosov has been identified as a likely source of tsunami waves (Costard et al., 2017a,b).

Recently, the identification of lobate deposits, which appear to originate from within the plains and onlap the plains margin, have been interpreted as potential tsunami deposits associated with the existence of a former ocean (Rodriguez et al., 2016; Costard et al., 2017a,b). Rodriguez et al. (2016) argued that the deposits they identified were

FIG. 8.22 Example of backwash channels in north-western Arabia Terra on Mars, displaying streamlined bars generated by a tsunami wave that struck an ancient Martian Ocean (Part of HiRISE image ESP_028537_2270, 25 cm/pixel.). *(Source: Rodriguez et al. (2016), Tsunami waves extensively resurfaced the shorelines of an early Martian ocean, Scientific Reports, 6: 25106-1788.*

formed by two separate ocean impact events that occurred ~3.6–3.4 Ga, and that the location and morphological characteristics of the second event suggested a retreating shoreline and a significantly colder climate. Costard et al. (2017a,b) compared the geomorphologic characteristics of the Martian deposits with the predictions of well-validated terrestrial models (scaled to Mars) of tsunami wave height, propagation direction, runup elevation, and distance, for three potential sea levels. These deposits appear to have originated from one or more impact-generated tsunamis in a Martian northern ocean.

The possibility that a large ocean once occupied the Martian northern plains is one of the most important and controversial hypotheses to have originated from the exploration of Mars. Costard et al. (2017a,b) mapped lobate deposits, which appear and are potential tsunami deposits associated with the existence of a former ocean. They identified the most probable crater sources of the proposed tsunami deposits from a numerical modeling. They tested the ability of the tsunami hypothesis to explain the origin and location of the previously enigmatic "thumbprint terrain" of northern Arabia Terra which our models suggest were formed by the interaction of the refracted tsunami waves with those reflected by the shore and offshore islands/obstacles. The age of the deposits is determined by the latest crater dating techniques. It provides evidence of the existence of Martian ocean as recently as Early Amazonian (~3 Ga), which has implications for understanding the volatile inventory Mars, its hydrologic and climatic evolution, and the potential for the origin of survival of life.

Costard et al. (2017a,b) did a comparison between the geomorphologic characteristics of the Martian thumbprint terrain, determined from a previous GIS mapping, with the predictions of tsunami wave height, propagation direction, runup elevation, and distance, for three potential sea levels, using their Volcflow (http://lmv.univ-bpclermont.fr/volcflow/) tsunami models, scaled to Mars. Three impact craters in Acidalia Planitia were considered (Fig. 8.23) based on their apparent age/state of degradation and locations,

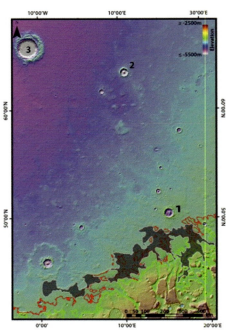

FIG. 8.23 The locations of the three impact craters in Acidalia Planitia considered as the potential sources of the tsunamis on Mars. The observed distribution of Thumbprint terrains and terminal lobate deposits in Arabia Terra are depicted in grey. The -3940 m (red line) and -3760 m (green line) topographic contour lines represent the minimum and maximum elevations of the ocean shorelines considered in Costard et al. (2017a)'s numerical simulations. Mars Orbiter Laser Altimeter (MOLA) elevations are superimposed on a shaded relief map with Cassini projection. Courtesy of NASA/JPL/USGS. Crater 1: 50.52°N/16.39°E, crater 2: 63.70°N/10.98°E, crater 3: 64.88°N/09.15°W. *(Source: Costard, F. et al. (2017a), Modelling investigation of tsunamis on Mars, Lunar and Planetary Science XLVIII, The Woodlands, Texas: Lunar and Planetary Institute. p. 1171. https://www.hou.usra.edu/meetings/lpsc2017/pdf/1171.pdf).*

relative to the proposed shorelines. The crater #3 seems to be the best single candidate because the inundated areas and backwashed areas from the modeling show the best spatial correlation to the observed Thumbprint terrains (TT) distribution.

Costard et al. (2017a,b)'s simulations indicate that, when the initial tsunami collided with the complex topography along the "dichotomy boundary," it gave rise to multiple reflected and refracted waves that propagated back out to sea creating the depositional conditions responsible for the characteristic arcuate pattern of the TT. It was found that the tsunami hypothesis can explain the origin and location of the previously enigmatic TT of northern Arabia Terra which Costard et al. (2017a,b)'s models suggests were formed by the interaction of the refracted tsunami waves with those reflected by the shore and offshore islands/obstacles.

Costard et al. (2017a,b) claim that the detection of potential impact-generated tsunami deposits on Mars provides evidence of the existence of Martian ocean as recently as Early Amazonian, which has implications for understanding the volatile inventory Mars, its hydrologic and climatic evolution, and the potential for the origin of survival of life.

8.9.2 Venus' dry surface

Scientists have speculated that liquid water existed on the surface of Venus before the runaway greenhouse effect heated the planet. Assuming the process that delivered water to Earth was common to all the planets near the habitable zone, it has been estimated that liquid water could have existed on its surface for up to 600 million years during and shortly after the Late Heavy Bombardment, but this figure can vary from as little as a few million years to as much as a few billion (Atkinson, 2010; Michael, 2016). Recent studies from September 2019 concluded that Venus may have had surface water and a habitable condition for around 3 billion years and may have been in this condition until 700 to 750 million years ago. If correct, this would have been an ample amount of time for the formation of life, and for microbial life to evolve to become aerial.

Venus' surface is extremely dry. During its evolution, ultraviolet rays from the Sun evaporated water quickly, keeping the planet in a prolonged molten state. There is no liquid water on its surface today because the scorching heat created by its ozone-filled atmosphere would cause water to immediately boil away.

8.10 Magnetic fields of Earth, Venus, and Mars—intrinsic & induced magnetic fields

Magnetic fields in any given celestial body is developed and sustained by the motion of charged particles in a loop, either inside or outside of that celestial body.

8.10.1 Earth's magnetic field—intrinsic magnetic field

As already indicated, Earth's core is about 4400 miles (7100 km) wide. The outermost 1400 miles (2250 km) of Earth's core are liquid. This part of Earth's core is responsible for the planet's magnetic field. Earth's magnetic field is generated by currents flowing in Earth's outer core (ionized lava rotating in a plane approximately parallel to Earth's equatorial plane, as a result of Earth's rotation about its axis).

It is Earth's magnetic field, which helps to deflect harmful charged particles shot from the Sun. When charged particles from the Sun get trapped in Earth's magnetic field, they smash into air molecules above the magnetic poles, causing them to glow. This phenomenon is known as the aurorae, the northern and southern lights.

The magnetic poles are always on the move, with the magnetic North Pole accelerating its northward motion to 24 miles (40 km) annually since tracking began in the 1830s. It is estimated that Earth's magnetic North Pole will likely exit North America and reach Siberia in a matter of decades. Earth's magnetic field is changing in other ways, too.

A few times in every million years or so, the field completely flips, with the North and the South poles swapping places. The magnetic field can take anywhere from 100 to 3000 years to complete the flip. The strength of Earth's magnetic field decreased by about 90% when a field reversal occurred in ancient past. The drop in the strength of the Earth's magnetic field makes the planet more vulnerable to solar storms and radiation, which could significantly damage satellites as well as communication and electrical infrastructure.

Globally, the magnetic field has weakened 10% since the 19th century, according to NASA. As just indicated, in Earth's history, reversal of magnetic fields has been happening approximately every few million years (for details, see Joseph, 2016). An inevitability of magnetic field reversal is that during a certain period, Earth's magnetic field will become zero, as a result of which Earth's magnetosphere will dwindle and vanish, giving rise to harmful radiation from the Sun penetrating the Earth's atmosphere and causing health hazards such as accelerated cancer risk. It is heartening that such an event is a long way in the future and scientists are optimistic that future technologies may be developed to avoid huge damage.

8.10.2 Venus' unusual magnetic field—induced magnetosphere

Whereas Earth takes 24 h to complete one full rotation on its axis, Venus takes 243 Earth days to rotate on its axis. Venus' rotation on its axis is by far the slowest of any of the major planets. Because of this sluggish spin, its molten metal core cannot generate a magnetic field similar to Earth's. The magnetic field of Venus is 0.000015 times that of Earth's magnetic field.

Allen et al. (2021) recall that the interactions between planets and stellar winds from their host star are a fundamental aspect in space physics. Venus and Mars, as opposed to the other planets in our Solar System, do not have intrinsic magnetic fields. Solar Orbiter mission, a cooperative mission between NASA and the European Space Agency (ESA), is making some new discoveries about our next-door planetary neighbor Venus. An analysis of Solar Orbiter data collected during its first flyby of Venus shows that the planet's unique magnetic environment, which is not generated by the planet, is robust enough to accelerate particles to millions of miles per hour. Venus' densely packed atmosphere is responsible for the planet's unusual induced magnetosphere. Venus' induced magnetosphere is caused when the Sun's solar wind interacts with Venus' ionosphere, producing a planetary magnetic field that the solar wind drags behind the planet.

The first flyby of Venus by Solar Orbiter permitted a survey of the energetic particles within and around the Venusian system. Venus and Mars, as opposed to the other planets in our Solar System, do not have intrinsic magnetic fields. Unlike Mars, Venus does have a substantial ionosphere, formed by the photoionization of its thick atmosphere, that supports current systems generated by the variable solar wind. This induced magnetic field, in turn, acts to establish the near-Venus plasma environment (see reviews by Luhmann, 1986; Phillips & McComas, 1991; Bertucci et al., 2011; Dubinin et al. 2011; Futaana et al., 2017). Unlike Earth, which has an intrinsic magnetic field from the sloshing, molten material inside its outer core, Venus generates its magnetic field from the interaction of the Sun's solar wind with the planet's ionosphere, the atmospheric region filled with charged atoms. Those charged atoms create electric currents. As the solar wind drapes over Venus, it interacts with those currents to produce a full-fledged magnetosphere around the planet. It is to be noted that the induced magnetosphere of Venus is "an unstable system," twisting and undulating with the solar wind. Allen et al. (2021) have provided a detailed insight into the physics of the problem.

Although scientists knew about this unusual magnetosphere from Venus missions flown from the 1960s to the 1980s, there are still many unknowns. The solar wind, for example, drags a planet's magnetosphere behind the planet to form a "tail," known as a magnetotail. Allen et al. (2021) explain that the interaction between the solar wind and Venus establishes several distinct regions. The deflection of the supersonic and super-Alfvénic solar wind from the induced magnetosphere creates a bow shock that is about a tenth of the size of Earth's bow shock (Slavin et al., 1979) and is far weaker than at Earth (Russell et al., 1979). The shocked solar wind then forms the magneto-sheath, comprised of both solar wind originating ions as well as picked-up ions from the Venusian atmosphere (Gröller et al., 2010). The magneto-sheath is characterized by hot dense plasma (Phillips and McComas, 1991) and a highly turbulent fluctuating magnetic field consisting of mirror-mode waves (Volwerk et al., 2008) with different spectral and scaling properties (Vörös et al., 2008).

Magnetic fields accelerate charged particles, such as electrons and protons. But does an induced magnetosphere accelerate particles the same way—and to the same speeds—as an intrinsic one? Solar Orbiter found that Venus' magnetic field extends at least 188,000 miles behind the planet—about the distance from Earth to the Moon. That's small relative to Earth's magnetotail, which extends more than twice that distance. Despite the magnetic field's size and instability, Solar Orbiter found it was accelerating particles to over 5 million mph that far from the planet.

The team identified multiple mechanisms propelling the particles, all of which are found in magnetospheres such as Earth's. Waves traveling along the magnetic field lines, for example, sometimes matched the rate that charged particles twirled around those lines, giving the particles an extra push. In other cases, turbulence in the magnetic field or particles trapped within current sheets—electric currents confined to a surface—passed on enough energy that particles could come flying out at up to almost 7 million mph.

8.10.3 Induced magnetosphere of Mars—comparison with Venus and Titan

Venus and Mars, as opposed to the other planets in our Solar System, do not have intrinsic magnetic fields. Unlike Mars, Venus does have a substantial ionosphere, formed by the photoionization of its thick atmosphere, that supports current systems generated by the variable solar wind. This induced magnetic field, in turn, acts to establish the near-Venus plasma environment. Mars and Venus do not have a global magnetic field and, as a result, solar wind interacts directly with their ionospheres and upper atmospheres (Dubinin et al., 2011). Neutral atoms ionized by solar UV, charge exchange and electron impact, are extracted and scavenged by solar wind, providing a significant loss of planetary volatiles. There are different channels and routes through which the ionized planetary matter escapes from the planets. Processes of ion energization driven by direct solar wind forcing and their escape are intimately related. Forces responsible for ion energization in different channels are different and, correspondingly, the effectiveness of escape is also different. Dubinin et al. (2011) have provided a review of the classification of the energization processes and escape channels on Mars and Venus and also their variability with solar wind parameters. Dubinin et al. (2011) attempted to distinguish between classical pickup and "mass-loaded" pickup processes, energization in boundary layer and plasma sheet, polar winds on unmagnetized planets with magnetized ionospheres and enhanced escape flows from localized auroral regions in the regions filled by strong crustal magnetic fields.

Bertucci et al. (2011) reported their investigation of the main features of the induced magnetospheres of Mars, Venus, and Titan. Their investigation revealed that all three celestial bodies form a well-defined induced magnetosphere (IM) and magnetotail as a consequence of the interaction of an external wind of plasma with the ionosphere and the exosphere of these objects. Bertucci et al. (2011)'s study indicates that in all three, photoionization seems to be the most important ionization process. In all three, the IM displays a clear outer boundary characterized by an enhancement of magnetic field draping and mass-loading, along with a change in the plasma composition, a decrease in the plasma temperature, a deflection of the external flow, and, at least for Mars and Titan, an increase of the total density. Bertucci et al. (2011)'s study further indicated that, the magnetotail geometries of all these three celestial objects follow the orientation of the upstream magnetic field and flow velocity under quasi-steady conditions. Exceptions to this are fossil fields observed at Titan and the near Mars regions where crustal fields dominate the magnetic topology. Bertucci et al. (2011)'s study indicated that the magnetotails also concentrate the escaping plasma flux from these three objects and similar acceleration mechanisms are thought to be at work. In the case of Mars and Titan, global reconfiguration of the magnetic field topology (reconnection with the crustal sources and exits into Saturn's magneto-sheath, respectively) may lead to important losses of plasma. Finally, an ionospheric boundary related to local photoelectron signals may be, in the absence of other sources of pressure (crustal fields) a signature of the ultimate boundary to the external flow (Bertucci et al., 2011).

References

Abramov, O., Kring, D., 2005. Impact-induced hydrothermal activity on early Mars. J. Geophys. Res. 110 (E12), E12S09.

Allen, R.C., et al., 2021. Energetic ions in the Venusian system: insights from the first solar orbiter flyby. Astronomy Astrophys 656 (A7): 11p., DOI. https://doi.org/10.1051/0004-6361/202140803.

Andrea, T., 2021. How Do Dust Devils Form? Sci. Am, Retrieved 26 May 2021.

Arya, A.S., Rudravaram, P.R., Singh, R.B., Sur, K., et al. (2015a), Mars color camera onboard Mars Orbiter Mission: initial observations & results, 46th Lunar and Planetary Science Conference, The Woodlands, Texas, USA, March 16–20, 2015.

Atkinson, N., 2010. Was Venus Once a Waterworld? *Universe Today*, https://www.universetoday.com/67240/was-venus-once-a-waterworld/.

Baker, V.R., Strom, R.G., Gulick, V.C., Kargel, J.S., Komatsu, G., Kale, V.S., 1991. Ancient oceans, ice sheets and the hydrological cycle on Mars. Nature 352 (6336), 589–594.

Balme, M., Greeley, R., 2006. Dust devils on Earth and Mars. Rev. Geophys. 44, RG3003. http://doi.org/10.1029/2005RG000188.

Basilevsky, A.T., Head, J.W., 1995. Global stratigraphy of Venus: Analysis of a random sample of thirty-six test areas. Earth Moon Planets 66, 285–336.

Basilevsky, A.T., Head, J.W., 2003. The surface of Venus. Rep. Prog. Phys. 66 (10), 1699–1734.

Bertaux, J-L., Vandaele, A-C., Korablev, O., Villard, E., Fedorova, A., Fussen, D., Quémerais, E., Belyaev, D., et al., 2007. A warm layer in Venus' cryosphere and high-altitude measurements of HF, HCl, H_2O and HDO. Nature 450 (7170), 646–649.

Bertucci, C., Duru, F., Edberg, N., Fraenz, M., Martinecz, C., Szego, K., Vaisberg, O., 2011. The induced magnetospheres of Mars, Venus, and Titan. Space Sci Rev 162, 113–171. http://doi.org/10.1007/s11214-011-9845-1.

Bibring, J., et al., 2006. Global mineralogical and aqueous history derived from OMEGA observations. Science 312 (5772), 400–404.

Brandenburg, J.E., 1987. The Paleo-Ocean of Mars, *MECA Symposium on Mars: Evolution of its Climate and Atmosphere*. Lunar Planetary Institute, Houston, TX; Smithsonian Institution's National Air and Space Museum, July 17-19, 1986, Washington, DC., 20–22.

Bullock, M.A., Grinspoon, D.H., Head, J.W., 1993. Venus resurfacing rates: Constraints provided by 3-D Monte Carlo simulations. Geophys. Res. Lett. 20, 2147–2150.

Bullock, M.A., Grinspoon, D.H., 2001. The recent evolution of climate on Venus. Icarus 150 (1), 19–37.

Cantor, B.A., K.M., K., Edgett, K.S., 2006. Mars Orbiter Camera observations of Martian dust devils and their tracks (September 1997 to January 2006) and evaluation of theoretical vortex models. JGR 111, E12.

Cantor, B.A., 2007. MOC observations of the 2001 Mars planet-encircling dust storm. Icarus 186 (1), 60–96. http://doi.org/10.1016/j.icarus.2006.08.019.

Carr, M.H., Masursky, H., and Saunders, R.S. (1973), A generalized geologic map of Mars, 78 (20): 4031-4036

Carr, M.H., 1973. Volcanism on Mars. J. Geophys. Res. 78 (20), 4049–4062.

Clifford, S.M., Parker, T.J., 2001. The evolution of the Martian hydrosphere: Implications for the fate of a primordial ocean and the current state of the northern plains. Icarus 154 (1), 40–79.

Coffin, M.F., 1994. Large igneous provinces: Crustal structure, dimensions, and external consequences. Rev. Geophys. 32, 1–36.

Condie, K.C., Pearse, V., 2008. *When Did Plate Tectonics Begin on Planet Earth?* The Geological Society of America, Boulder, CO, p. 295.

Costard, F., Séjourné, A., Kelfoun, K., Clifford, S., Lavigne, F., Di Pietro, I., Bouley, S., 2017. Modelling investigation of tsunamis on Mars, *Lunar and Planetary Science XLVIII*. Lunar and Planetary Institute, The Woodlands, TX, p. 1171.

Costard, F., Séjourné, A., Kelfoun, K., Clifford, S., Lavigne, F., Pietro, I.D., Bouley, S., 2017b. Modeling tsunami propagation and the emplacement of thumbprint terrain in an early Mars ocean. JGR Planet 122 (3), 633–649.

Craddock, R.A., Howard, A.D., 2002. The case for rainfall on a warm, wet early Mars. J. Geophys. Res. 107, 5111.

David, L., 2005. Spirit gets a dust devil once-over. Space.com.

Debaille, et al., 2013. Stagnant-lid tectonics in early Earth revealed by ^{142}Nd variations in late Archean rocks. Earth Planet. Sci. Lett. 373, 83–92.

De Blasio, F.V., Crosta, G.B., 2017. Modelling Martian landslides: dynamics, velocity, and paleoenvironmental implications. Eur. Phys. J. Plus 132 (468). https://doi.org/10.1140/epjp/i2017-11727-x.

DiAchille, G., Hynek, B., 2010. Ancient ocean on Mars supported by global distribution of deltas and valleys. Nat. Geosci. 3 (7), 459–463.

DiBiasse, L.A., Scheingross, J., Fischer, W., Lamb, M., 2013. Deltic deposits at Aeolis Dorsa: Sedimentary evidence for a standing body of water on the northern plains of Mars. J. Geophysical Res.: Planets 118 (6), 1285–1302.

Diniega, S., et al., 2021. Modern Mars' geomorphological activity, driven by wind, frost, and gravity. Geomorphology 380, 107627.

Dubinin, E., Fraenz, M., Fedorov, A., Lundin, R., Edberg, N., Duru, F., Vaisberg, O., 2011. Ion energization and escape on Mars and Venus. Space Sci. Rev. 162, 173–211. http://doi.org/10.1007/s11214-011-9831-7.

Ehlmann, B.L., Mustard, J.F., Fassett, C.I., Schon, S.C., Head III, J.W., Des Marais, D.J., Grant, J.A., Murchie, S.L., 2008. Clay minerals in delta deposits and organic preservation potential on Mars. Nat. Geosci. 1, 355–358.

Esposito, L.W., 1984. Sulfur dioxide: Episodic injection shows evidence for active Venus volcanism. Science 223 (4640), 1072–1074.

Fairén, A.G., et al., 2003. Episodic flood inundations of the northern plains of Mars. Icarus 165, 53–67.

Fairén, A.G., Davila, A.F., Gago-Duport, L., Amils, R., McKay, C.P., 2009. Stability against freezing of aqueous solutions on early Mars. Nature 459 (7245), 401–404.

Fairén, A.G., 2010. A cold and wet Mars. Icarus 208 (1), 165–175.

Fairén, A.G., et al., 2011. Cold glacial oceans would have inhibited phyllosilicate sedimentation on early Mars. Nat. Geosci. 4 (10), 667–670.

Fassett, C.I., Head III, J.W., 2005. Fluvial sedimentary deposits on Mars: Ancient deltas in a crater lake in the Nili Fossae region. Geophys. Res. Lett. 32, L14201.

Fawdon, P., et al., 2018. Hypanis Valles Delta: The last high-stand of a sea on early Mars, 49th Lunar and Planetary Science Conference 2018 (LPI Contrib. No. 2083).

Fegley, B., Prinn, R.G., 1989. Estimation of the rate of volcanism on Venus from reaction rate measurements. Nature 337, 55–58.

Fenton, L., Reiss, D., Lemmon, M., Marticorena, B., Lewis, S., Cantor, B., 2016. Orbital observations of dust lofted by daytime convective turbulence. Space Sci. Rev. 203 (1), 89–142.

Foley, B.J., Bercovici, D., Landuyt, W., 2012. The conditions for plate tectonics on super-Earths: inferences from convection models with damage. Earth Planet. Sci. Lett. 331, 281.

Fowler, A.C., O'Brien, S.B.G., 1996. A mechanism for episodic subduction on Venus. J. Geophys. Res. 101, 4755–4763.

Futaana, Y., Wieser, G.S., Barabash, S., Luhmann, J.G., 2017. Venus goals, objectives, and investigations. Space Sci Rev 212, 1453–1509. http://doi.org/10.1007/s11214-017-0362-8.

Gendrin, A., et al., 2005. Sulfates in martian layered terrains: the OMEGA/Mars express view. Science 307 (5715), 1587–1591. http://doi.org/10.1126/science.1109087.

Ghail, R., et al., 2018. Venus crustal tectonics analogous to jostling pack ice, Lunar and Planetary Science Conference. The Woodlands, TX.

Ghatan, G.J., Zimbelman, J.R., 2006. Paucity of candidate coastal constructional landforms along proposed shorelines on Mars: Implications for a northern lowlands-filling ocean. Icarus 185, 171–196.

Goudge, T.A., Mustard, J.F., Head, J.W., Fassett, C.I., Wiseman, S.M., 2015. Assessing the mineralogy of the watershed and fan deposits of the Jezero crater paleolake system. Mars. J. Geophys. Res. 120, 775–808.

Goudge, T.A., Milliken, R.A., Head, J.W., J.F., Mustard, C.I.F., 2017. Sedimentological evidence for a deltaic origin of the western fan deposit in Jezero crater, Mars and implications for future exploration. Earth Planet. Sci. Lett. 458, 357–365.

Greeley, R., Waller, D.A., Cabrol, N.A., Landis, G.A., Lemmon, M.T., Neakrase, L.D.V., Pendeleton Hoffer, M., Thompson, S.D., Whelley, P.L., 2010. Gusev crater, Mars: Observations of three dust devil seasons. J. Geophys. Res. 115, E00F02. http://doi.org/10.1029/2010JE003608.

Gröller, H., Shematovich, V.I., Lichtenegger, H.I.M., Lammer, H., Pfleger, M., Kulikov, Y.N., Macher, W., Amerstorfer, U.V., Biernat, H.K., 2010. Venus' atomic hot oxygen environment. J. Geophys. Res. 115, E12017. http://doi.org/10.1029/2010JE003697.

Grotzinger, J.P., et al., 2015. Deposition, exhumation, and paleoclimate of an ancient lake deposit, Gale crater, Mars. Science 350 aac7575.

Hallsworth, J.E., Koop, T., Dallas, T.D., et al., 2021. Water activity in Venus's uninhabitable clouds and other planetary atmospheres. Nature Astronomy 5, 665–675. https://doi.org/10.1038/s41550-021-01391-3.

Harrison, R.G., Barth, E., Esposito, F., Merrison, J., Montmessin, F., Aplin, K.L., Borlina, C., Berthelier, J.J., Déprez, G., Farrell, W.M., Houghton, I.M.P., 2016. Applications of electrified dust and dust devil electrodynamics to Martian atmospheric electricity. Space Sci. Rev. 203 (1), 299–345.

Head, J.W., Crumpler, L.S., Aubele, J.C., Guest, J.E., Saunders, R.S., 1992. Venus volcanism: Classification of volcanic features and structures, associations, and global distribution from Magellan data. J. Geophys. Res. 97 (13) 153–13,198.

Head, J.W., et al., 1999. Possible ancient oceans on Mars: Evidence from Mars orbiter laser altimeter data. Science 286, 2134–2137.

Head, J.W., 2007. The Geology of Mars: New Insights and Outstanding Questions. In: Chapman, M. (Ed.), The Geology of Mars: Evidence from Earth-Based Analogs. Cambridge University Press, Cambridge, p. 10.

Heavens, N.G., Benson, J.L., Kass, D.M., Kleinbohl, A., Abdou, W.A., McCleese, D.J., Richardson, M.I., Schofield, J.T., Shirley, J.H., Wolkenberg, P.M., 2010. Water ice clouds over the Martian tropics during northern summer. Geophys. Res. Lett. 37. http://doi.org/10.1029/2010gl044610.

Hess, G.D., Spillane, K.T., 1990. Characteristics of dust devils in Australia. J. Appl. Meteorol 29, 498–506.

Hinson, D.P., Wang, H., 2010. Further observations of regional dust storms and baroclinic eddies in the northern hemisphere of Mars. Icarus 206, 290–305. http://doi.org/10.1016/j.icarus.2009.08.019.

Hinson, D.P., Wang, H., Smith, M.D., 2012. A multi-year survey of dynamics near the surface in the northern hemisphere of Mars: Short-period baroclinic waves and dust storms. Icarus 219, 307–320. http://doi.org/10.1016/j.icarus.2012.03.001.

Hinson, D.P., Asmar, S.W., Kahan, D.S., Akopian, V., Haberle, R.M., Spiga, A., Schofield, J.T., Kleinböhl, A., Abdou, W.A., Lewis, S.R., Paik, M., Maalouf, S.G., 2014. Initial results from radio occultation measurements with the Mars Reconnaissance Orbiter: a nocturnal mixed layer in the tropics and comparisons with polar profiles from the Mars Climate Sounder. Icarus 243 (2014), 91–103. http://doi.org/10.1016/j.icarus.2014.09.019.

Holland, G., Cassidy, M., Ballentine, C.J., 2009. Meteorite Kr in Earth's mantle suggests a late accretionary source for the atmosphere. Science 326 (5959), 1522–1525. http://doi.org/10.1126/science.1179518.

Horgan, B.H., Anderson, R.B., Dromart, G., Amador, E.S., Rice, M.S., 2020. The mineral diversity of Jezero crater: evidence for possible lacustrine carbonates. Icarus 339, 113526.

Houser, J.G., Farrell, W.M., Metzger, S.M., 2003. ULF and ELF magnetic activity from a terrestrial dust devil. Geophys. Res. Lett. 30 (1). doi:10.1029/2001GL014144.

Hynek, B.M., Beach, M., Hoke, M.R.T., 2010. Updated global map of Martian valley networks and implications for climate and hydrologic processes. J. Geophys. Res doi:10.1029/2009JE003548.

Jaeger, W.L., Keszthelyi, L.P., Skinner Jr., J.A., Milazzo, M.P., McEwen, A.S., Titus, T.N., Rosiek, M.R., Galuszka, D.M., Howington-Kraus, E., Kirk, R.L., 2010. Emplacement of the youngest flood lava on Mars: A short, turbulent story. Icarus 205 (1), 230–243.

Joseph, 2016. Investigating Seafloors and Oceans: From Mud Volcanoes to Giant Squid. *Elsevier Science & Technology Books Publishers*, New York, NY, p. 581 ISBN: 978-0-12- 809357-3.

Kahanpää, H., Lemmon, M.T., Reiss, D., Raack, J., Mason, E., et al., 2018. Martian dust devils observed simultaneously by imaging and by meteorological measurements, Forty-Ninth Lunar and Planetary Science Conference. Woodlands, TX March 19–23, 2018.

Kanak, K.M., Lilly, D.K., Snow, J.T., 2000. The formation of vertical vortices in the convective boundary layer. Q. J. R. Meteorol. Soc. 126, 2789–2810. http://doi.org/10.1002/qj.49712656910.

Kasting, J.F., 1993. Earth's early atmosphere, Science, 259 (5097), 920–926.

Kasting, J.F., 1988. Runaway and moist greenhouse atmospheres and the evolution of Earth and Venus. Icarus 74 (3), 472–494.

Klose, et al., 2016. Dust devil sediment transport: From lab to field to global impact. Space Science Rev 203, 377–426.

Kok, J.F., Renno, N.O., 2006. Enhancement of the emission of mineral dust aerosols by electric forces. Geophys. Res. Lett. 33, L19S10.

Krasinsky, G.A., Brumberg, V.A., 2004. Secular increase of astronomical unit from analysis of the major planet motions, and its interpretation. Celestial Mechanics Dynamical Astronomy 90, 267–288.

Landis, G.A., 2003. Colonization of Venus. AIP Conf. Proc 654 (1), 1193–1198.

Landuyt, W., Bercovici, D., 2009. Variations in planetary convective via the effect of climate on damage. Earth Planet. Sci. Lett. 277, 29–37.

Lenardic, A., Jellinek, A., Moresi, L.-N., 2008. A climate change induced transition in the tectonic style of a terrestrial planet. Earth Planet. Sci. Lett. 271, 34–42.

Lenardic, et al., 2016. The Solar System of Forking Paths: Bifurcations in Planetary Evolution and the Search for Life-Bearing Planets in Our Galaxy. Astrobiology 16 (7), 9.

Lenardic, A., Crowley, J.W., 2012. On the notion of well-defined tectonic regimes for terrestrial planets in this solar system and others. Astrophysical J. 755 (2), 132.

Lenardic, A., Crowley, J.W., Jellinek, A.M., Weller, M.B., 2016. The solar system of forking paths: bifurcations in planetary evolution and the search for life bearing planets in our Galaxy. Astrobiology 16 (7). doi:10.1089/ast.2015.1378.

Lemmon, M.T., 2017. Dust devil activity at the Curiosity Mars rover field site, Conference: Lunar and Planetary Science XLVIII (2017). Woodlands, TX, XLVIII.

Letzter, R. (2020), These lava tubes could be the safest place for explorers to live on Mars, https://www.livescience.com/radiation-mars-safe-lava-tubes.html.

Lopes, R.M.C., Gregg, T.K.P., 2004. Volcanic Worlds: Exploring the Solar System's Volcanoes. Springer Publishing, Springer Berlin, Heidelberg, XXIV, p. 236. Hardcover ISBN: 978-3-540-00431-8; Softcover ISBN: 978-3-642-05586-7. Jointly published with Praxis Publishing, UK.

Lorenz, R., 2005. Dust devil hazard to aviation: a review of US air accident reports. J. Meteorol. 28 (298), 178–184.

Ludlum, D.M., 1997. National Audubon Society Field Guide to North American Weather. Knopf ISBN 978-0-679-40851-2.

Luhmann, J.G., 1986. The solar wind interaction with Venus. Space Sci. Rev. 44, 241–306. http://doi.org/10.1007/BF00200818.

Madeleine, J.-B., Forget, F., Millour, E., Navarro, T., Spiga, A., 2012a. The influence of radiatively active water ice clouds on the Martian climate. Geophys. Res. Lett. 39, L23202. http://doi.org/10.1029/2012GL053564.

Madeleine, J.-B., Forget, F., Spiga, A., Wolff, M.J., Montmessin, F., Vincendon, M., Jouglet, D., Gondet, B., Bibring, J.-P., Langevin, Y., Schmitt, B., 2012b. Aphelion water-ice cloud mapping and property retrieval using the OMEGA imaging spectrometer onboard Mars Express. J. Geophys. Res. Planets 117, E00J07. http://doi.org/10.1029/2011JE003940.

Malin, M., Edgett, K., 1999. Oceans or seas in the martian northern lowlands: High resolution imaging tests of proposed coastlines. Geophys. Res. Lett. 26, 3049–3052.

Mangold, N., Dromart, G., Ansan, V., Salese, F., Kleinhans, M.G., Massé, M., Quantin-Nataf, C., Stack, K.M., 2020. Fluvial regimes, morphometry, and age of Jezero crater paleolake inlet valleys and their exobiological significance for the 2020 Rover mission landing site. Astrobiology 20, 994–1013.

Mangold, N., Gupta, S., Gasnault, O., Dromart, G., Tarnas, J.D., Sholes, S.F., Horgan, B., Quantin-Nataf, C., Brown, A.J., Williford, K.H., et al., 2021. Perseverance rover reveals an ancient delta-lake system and flood deposits at Jezero crater, Mars. Science. doi:10.1126/science.abl4051.

Martin, L.J., Zurek, R.W., 1993. An analysis of the history of dust activity on Mars. J. Geophys. Res. 98 (3221). http://doi.org/10.1029/92JE02937.

Masursky, H., Masursky, H., Saunders, R.S., 1973. An overview of geological results from Mariner 9. J. Geophys. Res. 78 (20), 4009–4030.

McKinnon, W.B., Zahnle, K.J., Ivanov, B.A., Melosh, H.J., 1997. Cratering on Venus: models and observations. In: Bougher, S.W., Hunten, D.M., Phillips, R.J. (Eds.), Venus II. Univ. of Arizona Press, Tucson, pp. 969–1014.

McKay, C., Davis, W., 1991. Duration of liquid water habitats on early Mars. Icarus 90 (2), 214–221.

Michalski, J.R., Bleacher, J.E., 2013. Supervolcanoes within an ancient volcanic province in Arabia Terra, Mars. Nature 502 (7469), 46–52.

Michael, J.W., 2016. Was Venus the first habitable world of our Solar System? Geophys. Res. Lett. 43 (16), 8376–8383.

Miura, T., Arakida, H., Kasai, M., Kuramata, S., 2009. Secular increase of the astronomical unit: a possible explanation in terms of the total angular momentum conservation law. Earth Astronomical Soc. Japan 61 (6), 1247–1250. https://doi.org/10.1093/pasj/61.6.1247.

Montmessin, F., Forget, F., Rannou, P., Cabane, M., Haberle, R.M., 2004. Origin and role of water ice clouds in the Martian water cycle as inferred from a general circulation model. J. Geophys. Res. Planets 109, E10004. https://doi.org/10.1029/2004JE002284.

Moresi, L., Solomatov, V., 1998. Mantle convection with a brittle lithosphere: thoughts on the global tectonic style of the Earth and Venus. Geophys. J. Int. 133, 669–682.

Mueller, N. (2014), Venus Surface and Interior, In: Tilman, Spohn, Breuer, Doris, Johnson, T.V. (eds.). *Encyclopedia of the Solar System* (3rd ed.). Oxford: Elsevier Science & Technology. ISBN 978-0-12-415845-0.

Mulholland, D.P., Spiga, A., Listowski, C., Read, P.L., 2015. An assessment of the impact of local processes on dust lifting in Martian climate models. Icarus 252, 212–227. https://doi.org/10.1016/j.icarus.2015.01.017.

Murdock, J. (2020), Lava tubes on Mars could be 'natural shelters for first humans on the Red Planet, https://www.newsweek.com/mars-lava-tubes-red-planet-crewed-missions-nasa-hellas-planitia-1503432

Navarro, T., Madeleine, J.-B., Forget, F., Spiga, A., Millour, E., Montmessin, F., Määttänen, A., 2014. Global climate modeling of the Martian water cycle with improved microphysics and radiatively active water ice clouds. J. Geophys. Res. Planets 119, 1479–1495. https://doi.org/10.1002/2013JE004550.

Newsom, H., 1980. Hydrothermal alteration of impact melt sheets with implications for Mars. Icarus 44 (1), 207–216.

Newsom, H., et al., 1996. Impact crater lakes on Mars. J. Geophys. Res. 101 (E6), 14951.

Newsom, H. (2010), Heated Lakes on Mars, In: Cabrol, N., and E. Grin (eds.). *Lakes on Mars*, Elsevier. New York, NY.

Nimmo, F., Stevenson, D.J., 2000. Influence of early plate tectonics on the thermal evolution and magnetic field of Mars. J. Geophys. Res. 105, 11969–11979.

Nobel, P.S., 2009, Solar Radiation. In: Physicochemical and Environmental Plant Physiology (4th Edition), Imprint: Academic Press, ISBN: 978-0-12-374143-1; DOI https://doi.org/10.1016/B978-0-12-374143-1.X0001-4 (available at https://www.sciencedirect.com/topics/biochemistry-genetics-and-molecular-biology/solarradiation).

Normile, D., 2010. Mission to probe Venus's curious winds and test solar sail for propulsion. Science 328 (5979), 677.

O'Callaghan, J. (2021), NASA's Perseverance rover has been hit by 100 'dust devils' on Mars, New Scientist

O'Neill, C., et al., 2007a. Conditions for the onset of plate tectonics on terrestrial planets and moons. EPSL 261, 20–32.

O'Neill, C., et al., 2007b. Episodic Precambrian subduction. EPSL 262, 552–562.

O'Neill, C., Lenardic, A., Condie, K., 2013. Earth's punctuated evolution: cause and effect. Special Publications, London, p. 389. *University Geological Society*. https://doi.org/10.1144/SP389.4.

O'Neill, C., Debaille, 2014. The evolution of Hadean–Eoarchaean geodynamics. Earth Planet. Sci. Lett. 406, 49–58.

O'Neill, C., et al., 2016. A window for plate tectonics in terrestrial planet evolution? Phys. Earth Planet. Inter. 255, 80–92.

Paris, A.J., Davies, E.T., Tognetti, L., Zahniser, C., 2019. Prospective lava tubes at Hellas Planitia: leveraging volcanic features on Mars to provide crewed missions protection from radiation. J. Washington Acad. Sci., 105 (3), 13–36.

Parker, T.J., Saunders, R.S., Schneeberger, D.M., 1989. Transitional morphology in the west Deuteronilus Mensae region of Mars: Implications for modification of the lowland/upland boundary. Icarus 82, 111–145.

Phillips, J.L., McComas, D.J., 1991. The magnetosheath and magnetotail of Venus. Space Sci Rev 55, 1–80. https://doi.org/10.1007/BF00177135.

Phillips, R.J., Arvidson, R.E., Boyce, J.M., Campbell, D.B., Guest, J.E., Schaber, G.G., Soderblom, L.A., 1991. Impact craters on Venus: Initial analysis from Magellan. Science 252, 288–297.

Phillips, R.J., Raubertas, R.F., Arvidson, R.E., Sarker, I.C., Herrick, R.R., Izenberg, N.R., Grimm, R.E., 1992. Impact craters and Venus resurfacing history. J. Geophys. Res. 97 15,923–15,948.

Piccioni, G., Drossart, P., Sanchez-Lavega, A., Hueso, R., Taylor, F.W., Wilson, C.F., Grassi, D., Zasova, L., 2007. South-polar features on Venus similar to those near the north pole. Nature 450 (7170), 637–640.

Read, P.L., Lewis, S.R., 2004 The Martian Climate Revisited: Atmosphere and Environment of a Desert Planet 2004. Springer Praxis Books, Berlin; New York.

Reiss, D., Erkeling, G., Spiga, A., 2014. The horizontal motion of dust devils on Mars derived from CRISM and CTX/HiRISE observations. Icarus 227, 8–20.

Reiss, D., 2016. First observations of terrestrial dust devils in orbital image data: Comparison with dust devils in Amazonis Planitia, Mars, 47th Lunar and Planetary Science Conference, 2016.

Ribas, I., Guinan, E.F., Güdel, M., Audard, M., 2005. Evolution of the solar activity over time and effects on planetary atmospheres. I. High-energy irradiances (1–1700 Å). Astrophys. J. 622, 680–694.

Rodriguez, J.A.P., Kargel, J.S., Baker, V.R., Gulick, V.C., et al., 2015. Martian outflow channels: How did their source aquifers form, and why did they drain so rapidly? Sci. Rep. 5, 13404.

Rodriguez, J., et al., 2016. Tsunami waves extensively resurfaced the shorelines of an early Martian ocean. Sci. Rep. 6 25106-1788.

Russell, C.T., Elphic, R.C., Slaven, J.A., 1979. Pioneer magnetometer observations of the Venus bow shock. Nature 282, 815–816. https://doi.org/10.1038/282815a0.

Sauro, F., Pozzobon, R., Massironi, M., Berardinis, P.D., Santagata, T., Waele, J.D., 2020. Lava tubes on Earth, Moon and Mars: a review on their size and morphology revealed by comparative planetology. Earth Sci. Rev. 209, 103288.

Schaber, G.G., Strom, R.G., Moore, H.J., Soderblom, L.A., Kirk, R.L., Chadwick, D.J., Dawson, D.D., Gaddis, L.R., Boyce, J.M., Russel, J., 1992. Geology and distribution of impact craters on Venus: What are they telling us? J. Geophys. Res. 97 (E8) 13,257–13,302.

Schubert, G., Sandwell, D.T., 1995. A global survey of possible subduction sites on Venus. Icarus 117, 173–196.

Sharp, T., 2012. How Big is Mars? size of planet mars. Space.com. https://www.space.com/16871-how-big-is-mars.html.

Sharp, T., 2017a. What is Mars Made Of? composition of Planet Mars. Space.com. https://www.space.com/16895-what-is-mars-made-of.html.

Sharp, T., 2017b. Mars' Atmosphere: Composition, Climate & Weather. Space.com.

Singh, R., Arya, A.S., 2019. Martian dust devils observed by Mars Colour Camera onboard Mars Orbiter Mission, 50th Lunar and Planetary Science Conference, 2019 (LPI Contrib. No. 2132).

Slavin, J.A., Elphic, R.C., Russell, C.T., Wolfe, J.H., Intriligator, D.S., 1979. Position and shape of the Venus bow shock: Pioneer Venus Orbiter observations. Geophys. Res. Lett. 6, 901–904. https://doi.org/10.1029/GL006i011p00901.

Smith, D.E, 1999. The global topography of Mars and implications for surface evolution. Science 284 (5419), 1495–1503.

Smith, P., Nilton, R., 2001. Studying Earth dust devils for possible Mars mission. *Uni Sci News*. Retrieved December 1, 2006.

Smith, M.D., 2004. Interannual variability in TES atmospheric observations of Mars during 1999–2003. Icarus 167 (1), 148–165. https://doi.org/10.1016/j.icarus.2003.09.010.

Spiga, A., Forget, F., 2009. A new model to simulate the Martian mesoscale and microscale atmospheric circulation: Validation and first results. J. Geophys. Res. 114. https://doi.org/10.1029/2008je003242.

Spiga, A., Lewis, S.R., 2010. Martian mesoscale and microscale wind variability of relevance for dust lifting. Mars 5, 146–158. https://doi.org/10.1555/mars.2010.0006.

Stanzel, C., Pätzold, M., Williams, D., Whelley, P.L., Greeley, R., Neukum, G., Team, HCo-I, HRSC., C-I.T, 2008. Dust devil speeds, directions of motion and general characteristics observed by the Mars Express High Resolution Stereo Camera. Icarus 197 (1), 39–51. https://doi.org/10.1016/j.icarus.2008.04.017.

Steele, L.J., Lewis, S.R., Patel, M.R., 2014a. The radiative impact of water ice clouds from a reanalysis of Mars Climate Sounder data. Geophys. Res. Lett. 41, 4471–4478. https://doi.org/10.1002/2014GL060235.

Steele, L.J., Lewis, S.R., Patel, M.R., Montmessin, F., Forget, F., Smith, M.D., 2014b. The seasonal cycle of water vapour on Mars from assimilation of thermal emission spectrometer data. Icarus 237, 97–115. https://doi.org/10.1016/j.icarus.2014.04.017.

Stein, C., Lowman, J.P., Hansen, U., 2013. The influence of mantle internal heating on lithospheric mobility: implications for super-Earths. Earth Planet. Sci. Lett. 361, 448–459.

Strom, R.G., Schaber, G.G., Dawson, D.D., 1994. The global resurfacing of Venus 99 (E5), 10899–10926.

Svedhem, H., Titov, D.V., Taylor, F.V., Witasse, O., 2007. Venus as a more Earth-like planet. Nature 450 (7170), 629–632.

Thomas, P., Gierasch, P.J., 1985. Dust devils on Mars. Science 230 (4722), 175–177.

Towner, M.C., 2009. Charateristics of large Martian dust devils using Mars Odyssey Thermal Emission Imaging System visual and infrared images, J. Geophys. Res., 114, E02010, doi:10.1029/2008JE003220.

Tu, L., Johnstone, C.P., Güdel, M., Lammer, H., 2015. The extreme ultraviolet and X-ray Sun in Time: High-energy evolutionary tracks of a solar-like star. Astron. Astrophys. 577, L3. https://doi.org/10.1051/0004-6361/201526146.

Turcotte, D.L., Schubert, G., 1982. Geodynamics. Wiley, New York, NY.

Turcotte, D.L., 1993. An episodic hypothesis for Venusian tectonics. J. Geophys. Res. 98, 17061–17068.

Villanueva, G., Mumma, M., Novak, R., Käufl, H., Hartogh, P., Encrenaz, T., Tokunaga, A., Khayat, A., Smith, M., 2015. Strong water isotopic anomalies in the martian atmosphere: Probing current and ancient reservoirs. Science 348 (6231), 218–221.

Volwerk, M., Zhang, T.L., Delva, M., Vörös, Z., Baumjohann, W., Glassmeier, K.-H., 2008. First identification of mirror mode waves in Venus' magnetosheath? Geophys. Res. Lett. 35, L12204. https://doi.org/10.1029/2008GL033621.

Vörös, Z., Zhang, T.L., Leubner, M.P., Volwerk, M., Delva, M., Baumjohann, W., Kudela, K., 2008. Magnetic fluctuations and turbulence in the Venus magnetosheath and wake. Geophys. Res. Lett. 35, L11102. https://doi.org/10.1029/2008GL033879.

Walden, B.E., Billings, T.L., York, C.L., Gillett, S.L., Herbert, M.V., 1998. Utility of lava tubes on other worlds. Using in situ Resources for Construction of Planetary Outposts, Vol. 1, p. 16.

Wang, M.I.R, Wilson, R.J., Ingersoll, A.P., Toigo, A.D., Zurek, R.W., 2003. Cyclones, tides, and the origin of a cross-equatorial dust storm on Mars. Geophys. Res. Lett. 30. doi:10.1029/2002GL016828.

Wang, H., Richardson, M.I., 2015. The origin, evolution, and trajectory of large dust storms on Mars during Mars years 24–30 (1999–2011). Icarus 251, 112–127. https://doi.org/10.1016/j.icarus.2013.10.033.

Weller, M.B., Lenardic, A., O'Neill, C., 2015. The effects of internal heating and large-scale climate variations on tectonic bi-stability in terrestrial planets. Earth Planet. Sci. Lett., 85–94.

Weller, M.B., Lenardic, A., 2016. The energetics and convective vigor of mixed-mode heating: scaling and implications for the tectonics of exoplanets. Geophys. Res. Lett. 43, 6. doi:10.1002/2016GL069927.

Weller, M.B., Lenardic, A., Moore, W.B., 2016. Scaling relationships and physics for mixed heating convection in planetary interiors: isoviscous spherical shells. Journal of Geophysical Research: Solid Earth 121, 20. doi:10.1002/2016JB013247.

Weller, M.B., Lenardic, A., Jellinek, M., 2018. Life potential on early Venus connected to climate and geologic history, Lunar and Planetary Science Conference. Woodlands,TX.

Weller, M.B., Lenardic, A., 2018. On the evolution of terrestrial planets: Bi-stability, stochastic effects, and the non-uniqueness of tectonic states. Geosci. Front. 9 (1), 91–102.

Willis, G.E., Deardorff, J.W., 1979. Laboratory observations of turbulent penetrative convection planforms. J. Geophys. Res. 84, 295–302. https://doi.org/10.1029/JC084iC01p00295.

Wilson, L., Head, J.W., 1994. Mars: Review and analysis of volcanic eruption theory and relationships to observed landforms. Rev. Geophys. 32 (3), 221–263.

Wilson, R.J., Lewis, S.R., Montabone, L., Smith, M.D., 2008. Influence of water ice clouds on Martian tropical atmospheric temperatures. Geophys. Res. Lett. 35. https://doi.org/10.1029/2007GL032405.

Wordsworth, R., Knoll, A.H., Hurowitz, J., Baum, M., Ehlmann, B.L., Head, J.W., Steakley, K., 2021. A coupled model of episodic warming, oxidation and geochemical transitions on early Mars. Nat. Geosci. 14, 127–132.

Zurek, R.W., Barnes, J.R., Haberle, R.M., Pollack, J.B., Tillman, J.E., Leovy, C.B., 1992. Dynamics of the atmosphere of Mars. In: Kieffer, H.H., Jakosky, B.M., Snyder, C.W., Matthews, M.S. (Eds.). Mars, 1992. University of Arizona Press, Tucson, pp. 835–933.

Bibliography

Debaille, V., O'Neill, C., Brandon, A.D., Haenecour, P., Yin, Q., Mattielli, N., Treiman, A.H., 2013. Stagnant-lid tectonics in early Earth revealed by 142Nd variations in late Archean rocks. Earth Planet. Sci. Lett. 373, 83–92.

O'Neill, C., Lenardic, A., Moresi, L., Torsvik, T., Lee, C.A., 2007. Episodic Precambrian subduction. Earth Planet. Sci. Lett. 262, 552–562.

O'Neill, C., Debaille, V., 2014. The evolution of Hadean-Eoarchaean geodynamics. Earth Planet. Sci. Lett. 406, 49–58.

O'Neill, C., Lenardic, A., Weller, M., Moresi, L., Quenette, S., Zhang, S., 2016. A window for plate tectonics in terrestrial planet evolution? Phys. Earth Planet. Inter. 255, 80–92.

Chapter 9

Lunar explorations— Discovering water, minerals, and underground caves and tunnel complexes

9.1 Orbiter- and lander-based explorations of Earth's Moon

Earth's Moon has captured a special place in the hearts and minds of the humans for several reasons; first, this beautiful celestial body faintly illuminates the Earth's surface through its scattered sunlight during several cloud-free nights, most prominently on the Full Moon (the phase of the Moon in which its whole disc is illuminated); second, its appearance changing night after night from crescent (any time less than half the Moon is lit by the Sun) to gibbous (any time more than half the Moon is lit by the Sun) and back again provides a fascinating sight to every Moon-watcher having an artistic mind; third, the Moon exerts a major influence in generating an all-weather, worldwide, semidiurnal/ diurnal, rhythmic rising and lowering of sea level known as *tide*, which, apart from a plethora of oceanographic consequences, provides a fulfilling experience to every observer, who spends a day on a beach (only approximately 50% of the Moon's influence is exerted by the Sun); and fourth, the Moon generates biological rhythms (periodic biological fluctuations) among organisms, especially in the context of reproductive cycles of animals, particularly marine animals. Earth's Moon has been with us since before humans laid foot on Earth. Any person you have ever known or heard of has seen it shine bright in the night sky. Yet, researchers can only speculate about its creation.

Earth's Moon has always been one of the closest sources of cosmic curiosity across the world. Apart from the beginners such as the Soviet Union, the United States of America, the European Union, and Japan, relatively new entrants including India, China, and Turkey are currently involved in a wide range of missions to the lunar surface. Between 1959 and 2018, there have been twenty successful soft-landings on the lunar surface, including the most prominent one in July 1969, carried out by the United States' NASA, which culminated in taking the first human to Earth's Moon. However, the origin of the Earth–Moon System remained intractable until recently. Robotic exploration continues to deliver profound answers to the multitude of curious questions targeted about the secrets of our Universe by visiting far-off destinations, providing reconnaissance and collecting scientific data. The United States, the Soviet Union and China are the three nations which have successfully landed their spacecrafts on the surface of Earth's Moon. USA is the only country to have ever placed people on the Moon until today. Russia (the USSR), Japan, China, and the European Space Agency (ESA) have all made visits to Earth's Moon via probes. An Israeli nonprofit organization SpaceIL attempted to send a lander named Beresheet to Earth's Moon in 2019, but it crashed after it entered the lunar orbit, leaving the debris on the lunar surface. India's first lunar mission, called Chandrayaan-1, orbited Earth's Moon in 2008 and helped confirm the presence of water on our nearest celestial body.

The US National Aeronautics and Space Administration (NASA) has carried out several lunar landings, including the one in July 1969 that took the first human, Neil Armstrong, to the Moon. Before this, the US had sent unmanned lunar probes to conduct scientific studies. Its latest probe the Lunar Atmosphere and Dust Environment Explorer (LADEE) mission began on September 7, 2013, and ended on April 18, 2014.

Between 1959 and 2018, there have been twenty successful soft-landings on the lunar surface (of 44 attempts; 45% success rate). In 2013, China became the third country to land a spacecraft on the Moon. On 3 January 2019, China's Chang'E-4 (CE-4) successfully landed on the eastern floor of Von Kármán crater within the South Pole–Aitken Basin, becoming the first spacecraft in history to land on the Moon's far side. India's second mission to the Moon, Chandrayaan-2 was launched on 22nd July 2019 from Satish Dhawan Space Center, Sriharikota. Millions of Indians and several others watched the Vikram Moon lander's final heart-stopping descent in the early hours of 7 September 2019. However, a problem occurred during the final stage—known as the "hovering" stage—and the lander made a "hard landing." The lander was about 2.1 km (1.3 miles) from the lunar surface when it lost contact with scientists. New pictures from a NASA spacecraft showed the targeted landing site; but they were taken at dusk, and the pictures were unable to locate the lander. India would have become the fourth country to have landed on the lunar surface if the lander of its Moon mission Chandrayaan-2 had soft-landed successfully near its previously unexplored South Pole.

9.2 A peep into the early history of the study of the origin of Earth's Moon

Unlike the moons of several other planets in the Solar System, the planet Earth and its Moon operate as a coupled system, in the sense that the two bodies of the Earth–Moon system revolve together about their common center-of-mass, or common center-of-gravity with a period of 27.32 days, defined as one sidereal month. These two astronomical bodies are held together by gravitational attraction, but are simultaneously kept apart by an equal and opposite centrifugal force produced by their individual revolutions around the common center-of-mass.

Because the Earth is approximately 81 times more massive than its Moon, the common center-of-mass of the Earth–Moon system actually lies within the Earth at a distance of approximately 1718 km beneath the Earth's surface, on the side toward the Moon and along a line connecting the center-of-masses of the Earth and the Moon (Darwin, 1898). Because the common center-of-mass happens to lie below the surface of Earth, it appears (inaccurately though) that the Moon is orbiting around the Earth. A particularly interesting astronomical phenomenon that exists in the Earth–Moon system is "tidal locking" (also called "gravitational locking," "captured rotation" and "spin–orbit locking"). It may be noted that "tidal locking" occurs when an orbiting astronomical body always has the same face toward the object it is orbiting. This is known as synchronous rotation: *the tidally locked body takes just as long to rotate around its own axis as it does to revolve around its partner*. Thus, while the Earth and its Moon revolve together about their common center-of-mass, the same side of the Moon (known as Moon's "near side") faces the Earth. The other side of the Moon is known as "far side." The far side is also sometimes called (incorrectly though) the "dark side" just because that side is hidden from the line of sight of an observer on Earth although the far side comes facing the Sun during the daily rotation of the Earth–Moon system. It was learned only recently that the dwarf planet Pluto's largest and the nearest moon *Charan* may be comparable to Earth's Moon in respect of tidal locking (see *Chapter 16* for details). Likewise, Mars and one of its moons "Deimos" are tidally locked, so one would always get to view the same side of Deimos from Mars (Arya et al., 2015).

Since Galileo's discovery (in 1610) of four small bodies orbiting the planet Jupiter (known as *Galilean moons*), it has been known that the Earth's Moon is only one of several moons in the Solar System. One might, therefore, expect that selenogony (the study of the origin of Earth's Moon) would be only a special case of the theory of formation of moons, and indeed one of the most popular theories ("binary accretion" or "sister" hypothesis) treats Earth's Moon this way (Brush, 1984). But many scientists have thought that Earth's Moon deserves its own special hypothesis: first because it is unusual in being a single and relatively large companion of its primary (i.e., Earth), unlike the Jovian and Saturnian systems, which could be described as miniature planetary systems; second, because we have much more detailed information about it (i.e., Earth's Moon) and thus presumably provides an excellent opportunity to construct a more reliable quantitative theory.

Having said this, considering how conspicuous and starkly beautiful Earth's Moon is, and what a unique object it is in the Solar System, it is surprising how little scientific thought was given to its origin before the present epoch (Wood, 1984). Even the rapid rise of scientific inquiry in the Renaissance period of the Scientific Revolution (1450–1630) failed to induce enthusiasm in investigating the origin of the Earth's Moon. Brush (1984) recalls that "One of the first attempts to explain the Moon's formation in the framework of the new heliocentric astronomy of the 17th century is found in Descartes' *Le Monde*. This work was apparently written around 1630, but withheld from publication because of the condemnation of the heliocentric system in the notorious Trial of Galeleo in 1633; but it was published posthumously in 1664 (Descartes, 1664)." It is surprising to note that this was the scenario at the end of the Scientific Revolution! Wood (1984) further recalls that "The Moon appears only as a minor detail in the grand World System of authors such as Immanuel Kant (1755). The only widely cited reference on lunar origin from this period is George Darwin's (1879) treatise on fission of the Moon from the Earth." Lunar investigators attempted to fit theories on the origin of Earth's Moon to the available information, and the question of the Moon's formation became a part of the attempt to explain the observed properties of the Solar system. At first, the approach was largely founded on a mathematical examination of the dynamics of the Earth–Moon system.

The initially proposed Lunar origin theories such as coaccretion (the notion that the Moon and Earth were formed together from a primordial cloud of gas and dust), fission (the notion that an early stage fluid-Earth began rotating so rapidly that it flung off a mass of material that formed the Moon), and capture (the notion that the Moon formed elsewhere in the solar system and was later trapped by the strong gravitational field of Earth) failed on several grounds, including their inability to explain the large angular momentum (i.e., the quantity of rotation of a body, which is the product of its mass, velocity, and radius) of the present Earth–Moon system. Note that the Earth's Moon moves in its approximately geocentric orbit in the same prograde sense that the Earth rotates. The amount of angular momentum in the Earth–Moon system is substantial, 3.45×10^{41} rad g cm^2/s. This is the sum of the rotational angular momentum of the Earth and the orbital angular momentum of the Moon (Wood, 1984).

As noted above, a long-standing difficulty with the fission model (i.e., formation of Earth's Moon by fission from a rapidly rotating Earth following a giant impact (Ćuk and Stewart, 2012)) is that the angular momentum of the Earth–Moon system concentrated into the proto-Earth would leave the Earth spinning insufficiently rapidly to allow fission (Melosh, 2009). However, it has recently been suggested that this angular momentum could have been dissipated through resonances among the Moon, the Earth's core and the Sun. Further work remains to be done to assess this possibility. Zhang et al. (2012) suggest that, "Alternatively, an icy impactor that formed beyond the snow line would have delivered water, but would have delivered minimal amounts of rock-forming elements such as titanium to the Earth–Moon system. The water–steam disk created by such an impact may have promoted oxygen isotope equilibration between Earth's mantle and the protolunar disk." The feasibility of such a scenario remains to be evaluated from a dynamical point of view. According to Zhang et al. (2012), in all instances, the isotopic homogeneity of refractory elements (a group of metallic elements that are highly resistant to heat and wear; for example, tungsten, molybdenum, niobium, tantalum and rhenium) in the Earth–Moon system is a fundamental new constraint to lunar-formation theories. With regard to Zhang et al. (2012)'s suggestion of oxygen isotope equilibration between Earth's mantle and the protolunar disk, it is pertinent to remember that 45% of the lunar soil is composed of oxygen and that on Earth's Moon, oxygen isotopes are essentially in the same ratio as on Earth (Arlin Crotts (Columbia University), Water on The Moon, I. Historical Overview).

Although Mayer had noted (Mayer, 1848; 1851) that tidal action would progressively increase the Moon's distance from Earth, his idea did not attract much attention at the time (Lindsay, 1973), and no one seems to have pursued the cosmogonic implications. However, rigorous analysis of careful observations over a period of more than 200 years gradually revealed that, because of tidal effects, the rotations of both the Moon and Earth are indeed slowing and the Moon is slowly receding from Earth. Brush (1984) has described the consequence arising from the tidal action increasing the Moon's distance from Earth thus: "If the average Earth–Moon distance has been continuously increasing in the past, either at a constant rate or as a result of a force varying with distance in a known manner, we can extrapolate backward in time to reconstruct the earlier history of the Moon's orbit. This leads us to an epoch when the Moon would have been inside the 'Roche limit': the distance at which the Earth's tidal force would break up a body held together by gravitational forces. At that point (if not before) the extrapolation becomes invalid and we must introduce a specific hypothesis about the Moon's earlier motion, and indeed about whether she even existed in her present form." Studies then turned back to consider the state of the system when the Moon was closer to Earth. Throughout the 17th, 18th, and 19th centuries, lunar investigators examined different theories on lunar origin in an attempt to find one that would agree with the observations. Edouard Roche's 1848 formula for the tidal stability limit of a moon (Ward and Cameron, 1978) plays a crucial role in many modern theories, including the currently accepted "Giant-Impact" theory on the origin of the Earth–Moon system. For example, the disk structure, including its entropy, density and vapor fraction are important in determining the constituents and efficiency of moonlet accretion in the "multiple impact theory," which is an advanced version of the "Giant-Impact" theory (See Chapter 1.3: Giant-Impact Theory on the Origin of Earth's Moon). The importance of Roche limit is that tidal forces within the Roche limit prevent accretion, therefore the inner disk material must spread beyond the Roche limit before it can aggregate. The accretion of material outside the Roche limit is efficient, whereas the accretion of the inner disk is self-limiting because the newly spawned moonlets (described under Chapter 1.3.2: The "multiple impact theory") at the boundary will confine the disk and cause disk mass to fall onto Earth (Salmon and Canup, 2014). The current location of Earth's Moon is (1.2–20) R_{Roche} (see Rufu et al., 2017).

By the mid-20th century, scientists had imposed additional requirements for a viable lunar-origin theory. Of great importance is the observation that the Moon is much less dense than Earth, and the only likely reason is that the Moon contains significantly less iron. Such a large chemical difference argued against a common origin for the two bodies. Independent-origin theories, however, had their own problems.

In the mid-20th century, the prospect of water on the Moon was adopted by a more competent and respected scientist, Harold Urey, recipient of the 1934 Nobel Prize in Chemistry for discovering deuterium (heavy isotope of hydrogen), and noted for contributions to uranium isotopic enrichment, cosmochemistry, meteoritics and isotopic analysis, and the idea that life might have arisen from the chemistry of Earth's primordial atmosphere (Crotts, 2011). According to Wood (1984), "It appears that the first person who really cared about the origin of the Moon was Harold Urey (1952). The Moon was a crucial element of Urey's World System, ….." Wood (1984) further stated that Urey's great interest in the Earth's Moon and his scientific prestige (as stated above) made him a powerful advocate of lunar exploration in the period when United States' National Aeronautics and Space Administration (NASA) was founded and potential space missions were being defined.

Crotts (2011) recalls that, "In the early days of the space program, Urey was decisive in persuading NASA to scientifically explore the Moon. He began writing about the Moon in 1952, in his book *The Planets*, arguing that to understand planetary origins we should study the Moon, the 'Rosetta Stone of the Solar System'—a primitive world that might hold samples indicative of the early Earth. He based this argument in part on the idea that the Moon and planets perhaps formed by accretion of cold material like dust (Newell, 1972). In such a scenario he imagined water playing important roles." Note that Rosetta Stone is a black basalt stone found in 1799 that bears an inscription, having given the first clue to the decipherment of Egyptian hieroglyphics (mysterious symbols or writing). It is often considered that without the Rosetta Stone, it is likely that Egyptian hieroglyphics would still be a mystery. Rosetta Stone is thus considered as something that gives a clue to understanding.

Wood (1984) recalls that "By 1964, with four different US programs of lunar exploration at various stages of execution, scientific interest in the Moon had become intense. During January 20–21 of that year, just 5 months before the first successful Ranger flight to the Moon, a conference on 'The Dynamics of the Earth–Moon System' was held at the Institute for Space Studies (New York City) of the NASA Goddard Space Flight Center. Discussions centered on the origin of the Moon via Earth fission or capture, and the subsequent orbital evolution of the Moon. This was (to my knowledge) the first conference that focused on the question of the origin of the Moon."

There was widespread expectation that the Apollo exploration of the Moon would settle the question of its origin; in fact, this had been cited frequently as one of the scientific goals of the Apollo program. Unfortunately, even the most advanced space exploration adventure at the time, leading to the successful landing of man on Earth's Moon on July 20, 1969 could not shed adequate light to explain the origin of the Earth–Moon system. The only clear reading on lunar origin to come from the Apollo data was that Urey's concept of a cold, primitive Moon had been wrong. Unfortunately, as Apollo scientists set to work studying the geological evolution of Earth's Moon, the question of its origin receded into the background (Wood, 1984).

Twenty years after the just-mentioned 1964-conference on "The Dynamics of the Earth–Moon System," the question has been re-opened by another conference on "The Origin of the Moon," held at Kona, Hawaii (October 14–16, 1984). However, how the Earth's Moon was formed could not be deciphered. Several independent investigators showed that coaccretion, the model that had been most widely accepted by lunar scientists until then (at least at a subconscious level), could not account for the angular momentum content of the Earth–Moon system. With this alternative seemingly removed, many in the lunar community turned to collisional ejection as the model that appeared most plausible among those remaining (Wood, 1984). Thus, it became clear at the Kona conference that a major shift of confidence has occurred among lunar scientists toward the collisional ejection model.

9.3 Origin of Earth's Moon—evidences from lunar observations

An origin of the Earth's Moon by a Giant Impact is presently the most widely accepted theory of lunar origin. According to the current knowledge, the origin of Earth's Moon by a single massive impact with a Mars-sized Impactor named "Theia" explains the angular momentum, orbital characteristics and the unique nature of the Earth–Moon coupled system. Such an origin is consistent with the major lunar observations: its exceptionally large size relative to the host planet, the high angular momentum of the Earth–Moon system, the extreme depletion of volatile elements, and the delayed accretion, quickly followed by the formation of a global crust and mantle. The impact produced a protolunar cloud. Fast accretion of the Moon from the dense cloud ensured an effective transformation of gravitational energy into heat and widespread melting. A "Magma Ocean" of global dimensions formed, and upon cooling, an anorthositic crust (granular intrusive igneous rock composed almost exclusively of a soda-lime feldspar) and a mafic mantle (silicate mineral or igneous rock rich in magnesium and iron) were created by gravitational separation (Geiss and Rossi, 2013).

The Giant Impact lunar formation hypothesis, growing into favor in the 1970s and 1980s, explained why so little remained of volatile elements (lead, etc.) or compounds, such as water. They boiled away into space. Thus, with hydrogen evaporated and the Moon's core largely lost inside Earth, only intermediate-mass elements remain abundant, starting with oxygen, which binds tightly to some heavier elements such as silicon. The average temperature in prelunar material reached several thousand Kelvins, at which any mineral would melt and tend to release its dissolved gas.

Although much progress has been made in the relentless efforts to understand the origin of Earth's Moon, there is still no definitive answer to the question "how did the moon really form?" The consensus is that impacts are the likely precursors to the Moon's formation, but the question is *how many*? The simulations show that it is *possible*—at least, in an idealized digital computer world—that billions of years ago, multiple impacts created several moonlets which eventually merged to form the Moon. Although the Multiple-Impact Hypothesis is just that—a hypothesis—hopefully, with future research, we will come to a better understanding of how the moon formed. While we continue to look *back* in time to try to better understand the origins of our moon, it is also important to look *up* at night and appreciate the Moon and its beautiful glow (Azoulay, 2019).

The Lunar Prospector (LP) gamma-ray spectrometer (GRS) has acquired global maps of elemental composition of Earth's Moon. It has long been known that such maps will significantly improve our understanding of lunar formation and evolution (The NASA Lunar Exploration Science Working Group (LExSWG), A Planetary Science Strategy for the Moon, JSC-25920 (1992)). For example, one long-standing issue of lunar formation concerns the bulk composition of the Moon. There are suggestions from Apollo, Galileo, and Clementine data that the Moon is enriched in refractory elements (Al, U, and Th) and FeO compared to Earth (Taylor and Esat, 1996). If so, then lunar origin models that assume that most of the Moon's material comes from Earth's mantle would not be correct (see Lawrence et al., 1998). According to Lawrence et al. (1998), another question is the compositional variability and evolution of the lunar highlands, which contain KREEP-rich materials [potassium (K), rare earth elements (REE), and phosphorus (P)]. KREEP-rich rocks are thought to have formed at the lunar crust-mantle boundary as the final product of the initial differentiation of the Moon. The distribution of these rocks on the lunar surface therefore gives information about how the lunar crust has evolved over time (Lawrence et al., 1998). Other issues that can be addressed in the future using GRS data include: (i) identifying and delineating basaltic units in the maria; (ii) determining the composition of ancient or "cryptic" mare units found in the highlands using Clementine data (Schultz and Spudis, 1983), and searching for more of these units using mainly the Fe and Ti data; (iii) identifying and delineating highland petrologic units; and (iv) searching for anomalous areas with unusual elemental compositions that might be indicative of deposits with resource potential (Lawrence et al., 1998).

9.4 Differing preliminary views on the existence of water on Earth's Moon

From the perspective of planetary environments, liquid water is the essential requirement for life and serves as a surrogate indicator for life. Water, among other constituents, is essential for the synthesis of amino acids. Therefore, existence of water (in the form of water vapor, liquid water, or ice floes) beyond Earth is considered to be one of the essential prerequisites in the search for life there, even for extremophiles.

Although there are microbes that survive and thrive in space, hot sediments below the seafloor, and under adverse environmental conditions of hydrothermal vents in the absence of oxygen; water is one of the most essential substances needed to sustain most life-forms. If one considers the future possibility of man inhabiting on Earth's Moon and planet Mars, as well as other extra-terrestrial planets and their moons, then existence of water and sufficiently congenial atmospheric conditions are necessary for his survival and comfortable living in such celestial bodies. It would be interesting to explore the complex nature of the presence of water in its various forms in different celestial bodies.

By mid-19th century, astronomers strongly suspected that Earth's Moon was largely dry and airless, based on the absence of any observable weather. Based on actual telescopic examination and by inference from absence of clouds, respectable scientists realized that significant amounts of water on the Earth's Moon's surface would rapidly sublime into the vacuum (Dana, 1846). It was also found that there are no streams, lakes or seas on Earth's Moon. There exists a hypothesis that the Moon and the Earth were essentially dry immediately after the formation of the Moon—by a single giant impact or multiple impacts on the proto-Earth—and only much later gained volatiles through accretion of wet material delivered from beyond the asteroid belt.

Volatiles, and water in particular, have been thought to be unstable on the lunar surface because of the rapid removal of constituents of the lunar atmosphere by solar radiation, solar wind, and gravitational escape. It is generally presumed that gases of low molecular weight escape very rapidly from Earth's Moon. As a consequence, it has been assumed that volatile substances, such as water, which possess short relaxation times for escape, do not exist there. The limiting factor in removal of a volatile from Earth's Moon, however, is actually the evaporation rate of the solid phase, which will be collected at the coldest points on the lunar surface. Urey (1952) recognized that there may be depressions in which the Sun never shines, and in which some condensed volatile substances could be present. However, Urey (1952) concluded that no solid or liquid water could exist on the Moon for more than very short periods of time.

In 1961 Watson et al. (1961a) showed how water vapor might stick in such cold traps for geologically long times. Water, even more than many heavier substances such as sulfur dioxide, carbon dioxide or hydrogen chloride, would stick longer in these polar cold traps (Watson et al., 1961b). In polar areas of Earth's Moon, the maximum temperatures reached in some permanently shaded areas are well below the temperature required to retain water ice for billions of years, and cold enough to hold other volatiles for shorter periods (Hodges, Jr., 1980). The idea that ice and other trapped volatiles exist in permanently shadowed regions near the lunar poles (originally proposed by Watson et al. (1961a, 1961b), was re-examined by Arnold (1979) in the light of the vast increase of our lunar knowledge. The stability of the traps and the trapping mechanism have been verified in Arnold (1979)'s studies. Arnold's 1979 study on the subject of ice in the lunar polar regions, which concluded that there is a reasonable likelihood that large quantities of volatiles, particularly H_2O, are trapped in permanently shadowed regions at high lunar latitudes, has aroused a good deal of interest and some controversy. For example, Lanzerotti et al. (1981) used the laboratory measurements of the erosion of H_2O ice to estimate the loss rates of possible "trapped" volatiles in the cold lunar polar regions. They concluded that a significant accumulation of water ice is unlikely to occur.

In 1961, 8 years before man first landed on the Moon in 1969 through the successful Apollo-11 Lunar mission carried out by NASA, three farsighted scientists (Watson, Murray, and Brown) proposed a brilliant visionary idea that ice and other trapped volatiles might be existing in the permanently shadowed regions near the lunar poles (see Watson et al., 1961). Watson et al. (1961b) presented a detailed theory of the behavior of volatiles on the lunar surface based on solid-vapor kinetic relationships, and showed that water is far more stable there than the noble gases or other possible constituents of the lunar atmosphere. Numerical calculations carried out by them indicate that the amount of water lost from the Moon (since the present surface conditions were initiated) is only a few grams per square centimeter of the lunar surface. They further found that the amount of ice eventually detected in lunar "cold traps" thus will provide a sensitive indication of the degree of chemical differentiation of the moon.

The argument put forward by them to justify their prediction was based on their theoretical models of water stability on the lunar surface, which indicated that hydrogen atoms ultimately derived from the solar wind can sustain stable ice in polar cold traps. Pictures of the Moon obtained through Ranger 7 in 1964 and Ranger 9 in 1965, particularly the close-up photos of the crater Alphonsus indicating crevasses, influenced Harold Urey to suspect that the Earth's Moon's surface harbors rivers and channels of flowing liquid. Although it was not possible to decide whether such crevasses are the result of lava flows or the evaporation of massive amounts of water from beneath the surface, Urey believed that the water interpretation is the more likely of the two (see Crotts, 2011).

Turning the tables on those who advocated the presence of water on the surface of Earth's Moon, it was quite shocking that the mineral and rock samples collected from the Moon during initial Lunar Mission programs ("Apollo" programs) were found to have been devoid of indigenous water (Epstein and Taylor, 1973, 1974). The severe depletion of volatile elements in the Moon, including water, in lunar rock samples (Taylor et al., 2006) is thought to have occurred during a "giant impact event" between a proto-Earth and a Mars-sized object (Hartmann and Davis, 1975; Canup and Asphaug, 2001), although Earth may have received additional input of volatiles by impacts after Moon formation (Owen and Bar-Nun, 1995).

Subsequent to the examination of the data obtained from the Ranger 9 impact on the Moon in 1965, Lunar scientists were rather skeptical in regard to all sources of water on the Moon. The first samples returned from the Moon exhibited almost no signs of hydrated minerals. The evidence from Apollo against water on the Moon was varied and manifest, and there was little in favor. Notwithstanding the initial understanding gleaned from the Apollo-11 mission that our Moon is totally devoid of water, 10 years after man's historical landing on the Moon, Watson et al. (1961)'s visionary idea was re-examined by another scientist Arnold (see Arnold, 1979), in the light of the vast increase of our lunar knowledge. Arnold examined the stability of the traps (where volatiles including water could be preserved) and the trapping mechanism, according to which there are likely to be four potential sources of lunar water (Arnold, 1979):

1. solar wind reduction of iron oxide (FeO) in the regolith
2. water-containing meteoroids
3. cometary impact, and
4. (the least certain) degassing of the interior

With reference to (1), solar wind comprises of flux of particles, chiefly protons and electrons together with nuclei of heavier elements in smaller numbers, that are accelerated by the high temperatures of the solar corona, or outer region of the Sun, to velocities large enough to allow them to escape from the Sun's gravitational field. Note that the outer surface layer of the Moon is termed the regolith; it is a loose layer or debris blanket probably, up to 10 m deep in places, continually churned up by the impacts of micrometeorites (P. Moore, In: *Encyclopedia of Geology*, 2005). Solar wind protons are implanted directly into the top 100 nm of the lunar near-surface region, but can either quickly diffuse out of the surface or be retained, depending upon surface temperature and the activation energy, U, associated with the implantation site. In a study, Farrell et al. (2015) explored the distribution of activation energies upon implantation and the associated hydrogen-retention times; this for comparison with recent observation of OH on the lunar surface.

With reference to (2) and (3), beyond the Sun every major Solar System body shows the presence of water, as do many lesser ones, including all comets and many asteroids.

Asteroids and comets, which contain water, undoubtedly strike the Moon and presumably add water vapor to its atmosphere, temporarily. If those water molecules enter the ultracold craters' permanently shadowed regions near the poles, they will stick (see Crotts, 2011). With reference to (4), Ustunisik et al. (2011) suggested that if one accepts that outgassing 3–4 billion years ago included water that might stick within the regolith for geologically long times, or that more recent outgassing might contain water vapor, this might explain the observed variation in hydration signal across the polar regions. In another study, Feoktistova et al. (2018) investigated the possibility of existence of the hydrogen-containing volatile compounds, similar to those found in the Cabeus crater, in the area of the proposed landing ellipses of the Luna-Glob mission. They found that "the existence of water ice and other hydrogen-containing substances is possible only in the presence of a shielding layer of regolith. The time of existence of such deposits does not exceed several tens or hundreds of years for a layer of regolith with a thickness of 0.4 m and several thousand years for a layer of regolith 1 m thick."

In Arnold's investigations, two important destructive mechanisms have also been identified, namely: (1) photo-dissociation of water (H_2O) molecules adsorbed on the sunlit surface, and (2) sputtering or decomposition of trapped H_2O by solar wind particles. Arnold (1979) suggested that the question of the presence of H_2O molecules in the traps remains open; it can be settled by experiment. Since Arnold's investigations, further research aimed at obtaining the much sought-after answer to the curious question of the presence (or otherwise) of water on Earth's Moon continued for a long time by a battalion of scientists, and finally, as indicated in the below paragraphs, Watson et al. (1961)'s visionary idea was proved to be impeccably correct.

Surprisingly, the argument put forward by Watson et al. (1961) based on their theoretical models indicating that hydrogen atoms derived from the solar wind can sustain stable ice in polar cold traps received indirect support much later through studies carried out by Eke et al. (2009), who showed that hydrogen is present either in the form of a volatile compound or as solar wind protons implanted into small regolith grains (the layer of unconsolidated solid material covering the bedrock) on Earth's Moon. Based on a study, Eke et al. (2009) reported an analysis of the Lunar Prospector epithermal neutron data, providing an improved map of the distribution of hydrogen near to the lunar poles. This was achieved using a specially developed pixon image reconstruction algorithm to de-convolve the instrumental response of the Lunar Prospector's neutron spectrometer from the observed data, while simultaneously suppressing the statistical noise. The results showed that these data alone require the hydrogen to be concentrated into the cold traps of Earth's Moon at up to 1 wt% water-equivalent hydrogen. This combination of localization and high concentration suggested that hydrogen is present either in the form of a volatile compound or as solar wind protons implanted into small regolith grains.

9.5 Experimental search for clues to the existence of water on Earth's Moon

Earth's Moon is depleted in all volatile elements compared with Earth (Vaniman et al., 1991). Volatiles, and water in particular, have been thought to be unstable on the lunar surface because of the rapid removal of constituents of the lunar atmosphere by solar radiation, solar wind, and gravitational escape (Watson et al., 1961). Water was believed to have been brought to Earth's Moon by comets and asteroids and was presumed to have been formed by the reduction of ferrous oxide (FeO) in lunar materials by solar wind hydrogen, and some juvenile water may have been released from the lunar interior over billions of years (Watson et al., 1961; Arnold, 1979). Studies of the transport of such water over the lunar surface after its release indicate that 20–50% should be retained as frozen water ice within permanently shaded craters near both poles (Watson et al., 1961; Arnold, 1979; Hodges, Jr., 1980; Ingersoll et al., 1992; Butler, 1997).

On the surface of Earth's Moon, the existence of a broken-up surface layer (regolith) was known long before the first soft landings, though its properties were not clearly understood. The regolith is a thin deposit containing a record of billions of years of the history of the Moon, and of the inner Solar System and, therefore, a rich subject for study.

According to Morgan and Shemansky (1991), the presence of sodium and potassium on the Moon implies that other more abundant species should be present. Volatile molecules such as H_2O are significantly more abundant than sodium in any of the external atmospheric sources. Source mechanisms which derive atoms from the surface should favor abundant elements in the Lunar regolith (Morgan and Shemansky, 1991). A preferential loss process for oxygen on the far side of the Moon has been examined by Morgan and Shemansky (1991), in which ionization by electron capture in surface collisions leads to escape through acceleration in the local electric field. Morgan and Shemansky (1991) found that the episodic nature of cometary insertion may allow formation of ice layers which act as a stabilized source of OH.

For examining the presence of water in lunar regolith by identifying hydrogen molecule, hydrogen is assumed to be in the form of water molecules. A unique identification of chemical species enriched in hydrogen and a characterization of their spatial distribution are possible through measurement of neutron flux spectra. For example, Feldman et al. (1993) found that remotely sensed neutrons and gamma rays are useful in delineating the abundance of ground ice and its distribution with depth. Likewise, Feldman et al. (1997) simulated neutron leakage flux spectra to explore whether the nature of volatile deposits residing within the permanently shaded craters near the poles of the planet Mercury can be identified using state-of-the-art space neutron detection techniques deployed aboard a flyby mission to Mercury. They found that the most sensitive method for discriminating between H_2O (ice)-rich and S-rich deposits involves measurement of the flux of epithermal neutrons. In the curious search for clues to the existence of water on Earth's Moon, Feldman et al. (1998) used maps of epithermal- and fast-neutron fluxes measured by the *Lunar Prospector* to search for deposits enriched in hydrogen at both lunar poles.

Feldman et al. (1998) noted that there are three general energy (E) ranges that correspond to low or thermal, intermediate or epithermal, and high or fast neutrons, which reflect the neutron production and energy moderation process. Fast neutrons are formed at high energies as the result of interactions between galactic cosmic rays and the nuclear constituents of the regolith. They transfer energy (E) to nuclei in the regolith at a nearly constant fractional rate during subsequent, primarily elastic scattering collisions. This process results in an epithermal flux spectrum that is proportional to E^{-1}. As their energy approaches that of the thermal motion of regolith nuclei, they begin to absorb energy as fast as they lose it and thereby develop a Maxwellian velocity distribution. The thermal-neutron population builds in amplitude until the rate of injection from the fast-neutron population equals the rate of absorption due to thermal capture reactions plus the rate of loss to space.

Feldman et al. (1998) found that the dominant effect of hydrogen (here assumed to be in the form of water molecules) is to monotonically decrease the intensity of epithermal and fast neutrons. In contrast, the effect of hydrogen on the thermal population is not monotonic, producing a maximum between about 3 and 5 wt% of H_2O. Also, all the spectra slope downward toward lower energies in the epithermal range because of the absorption of neutrons by regolith nuclei.

In a study, seeking evidence for water ice at the lunar poles, Feldman et al. (1998) observed depressions in epithermal fluxes close to permanently shaded areas at both poles of the Moon. It was found that the peak depression at the North Pole is 4.6% below the average epithermal flux intensity at lower latitudes, and that at the South Pole it is 3.0% below the low-latitude average. However, no measurable depression in fast neutrons was seen at either pole. Feldman et al. (1998) argued that the observed data are consistent with deposits enriched in hydrogen in the form of water ice that are covered by as much as 40 cm of desiccated regolith within permanently shaded craters near both poles.

In a different kind of search for the possibility of hydrogen-enriched deposits (e.g., water) on Earth's Moon, Crider and Vondrak (2000) investigated the solar wind as a source for the deposits of hydrogen at the lunar poles. Monte Carlo simulations are often used to model the probability of different outcomes in a process that cannot easily be predicted due to the intervention of random variables (the use of random variables is most common in probability and statistics, where they are used to quantify outcomes of random occurrences). Monte Carlo methods, or Monte Carlo experiments, are a broad class of computational algorithms that rely on repeated random sampling to obtain numerical results. The underlying concept is to use randomness to solve problems that might be deterministic in principle.

In the search for possible presence of water ice at the lunar poles, Feldman et al. (1998) created an interesting Monte Carlo model that simulated the migration of atmospheric particles through the lunar exosphere. Making the model general enough to incorporate any physical process that might affect the particles, they developed a tool that would estimate the number and form of particles that reach and stick to lunar cold traps. In their model, each particle was allowed to follow a series of ballistic trajectories as it hops around the surface of the Moon. They traced the path of the particle until it was removed from the system by photo-processes such as ionization or dissociation, by thermal escape, or by reaching a cold trap. Accumulating statistics on the outcomes for various input particles, they determined the amount and form of hydrogen that was able to migrate to the lunar cold traps over time for a typical solar wind input flux. In this exercise, Crider and Vondrak (2000) found that although the fraction of hydrogen delivered to the Moon that ultimately reached the poles was small, a slow steady source such as the solar wind might have provided enough hydrogen over 83 Myr to account for the observed deposits. Furthermore, they found that enrichment in the deuterium-hydrogen ratio (for short, D/H ratio) occurred by the migration process. The amount of fractionation was found to have been dependent on the molecular form of the migrating hydrogen. Interestingly, atomic deuterium/hydrogen was found to have been enriched by a factor of 4 over the delivered fraction by the migration process.

It may be noted that the Earth's Moon is generally thought to have formed and evolved through a single or a series of catastrophic heating events (resulting from collisions with other celestial bodies wandering/loitering in space), during which most of the highly volatile elements were lost. Hydrogen, being the lightest of all elements, is believed to have been completely lost during these "heating events" period. With this background notion, Saal et al. (2008) made use of considerable advances in secondary ion mass spectrometry to obtain improved limits on the indigenous volatiles contents (e.g., carbon dioxide (CO_2), water (H_2O), fluorine (F), sulfur (S), and chlorine (Cl)) of the most primitive basalts in the Moon—the lunar volcanic glasses. Although the pre-eruptive water content of the lunar volcanic glasses could not be precisely constrained, numerical modeling of diffusive degassing of the very-low-titanium (Ti) glasses provided a best estimate of 745 parts per million (p.p.m.) water, with a minimum of 260 p.p.m. at the 95% confidence level. Saal et al. (2008)'s results indicate that, contrary to prevailing ideas, the bulk Moon might not be entirely depleted in highly volatile elements, including water. Thus, Saal et al. (2008) concluded that the presence of water must be considered in models constraining the Moon's formation and its thermal and chemical evolution.

9.6 United States' Apollo missions culminating in the landing of Man on Earth's Moon

On May 25, 1961, the US President John F. Kennedy announced the goal of sending astronauts to Earth's Moon before the end of the decade. Eight years later at 9:32 a.m. EDT on July 16, 1969, that dream became a reality as the swing arms moved away and a plume of flame signaled the lift-off of the Apollo 11 carrying astronauts Neil A. Armstrong, Michael Collins, and Buzz Aldrin from Kennedy Space Center Launch Complex 39A to the Moon. The image in Fig. 9.1 shows the Apollo 11 crew "suiting up" for countdown demonstration test. Fig. 9.2 shows the three astronauts in spacesuits without helmets sitting in front of a large photo of the Moon.

The United States' Apollo program was designed to land humans on Earth's Moon and bring them safely back to Earth. The photograph in Fig. 9.3 shows the Saturn V launch vehicle (SA-506) for the Apollo 11 mission lift-off

FIG. 9.2 Three astronauts in space suits without helmets sitting in front of a large photo of the Moon. The Apollo 11 lunar landing mission crew, pictured from left to right, Mission commander Neil A. Armstrong, Command Module pilot Michael Collins, and Lunar Module pilot Edwin "Buzz" E. Aldrin, Jr. *(Source: NASA; NASA photo ID: S69-31739); Public Domain in the United States.*

FIG. 9.1 This image is of the Apollo 11 crew "suiting up" for countdown demonstration test. *(Source: https://www.nasa.gov/multimedia/imagegallery/image_feature_1134.html) Image Credit: NASA, Last Updated: Aug. 4, 2017, Editor: NASA Content Administrator.*

FIG. 9.3 This photograph shows the Saturn V launch vehicle (SA-506) for the Apollo 11 mission lift-off at 8:32 a.m. CDT, July 16, 1969, from launch complex 39A at the Kennedy Space Center. Apollo 11 was the first manned lunar landing mission with a crew of three astronauts: Mission commander Neil A. Armstrong, Command Module pilot Michael Collins, and Lunar Module pilot Edwin "Buzz" E. Aldrin, Jr. It placed the first humans on the surface of Earth's Moon on July 20 and returned them back to Earth on July 24. *(Source: https://www.nasa.gov/centers/marshall/history/apollo_11_140716.html) Image credit: NASA Last Updated: Aug. 4, 2017 Editor: Lee Mohon.*

FIG. 9.4 Saturn V carrying Apollo 11 rises past the launch tower camera. At 9:32 a.m. EDT, the swing arms move away and a plume of flame signals the lift-off of the Apollo 11 Saturn V space vehicle and astronauts Neil A. Armstrong, Michael Collins and Edwin E. Aldrin, Jr. from Kennedy Space Center Launch Complex 39A. *(Source: NASA; restored by Michel Vuijlsteke - Great Images in NASA Description); Public Domain in the United States.*

FIG. 9.5 Launch of Apollo 11, the first Lunar landing mission, on July 16, 1969. The massive Saturn V rocket lifted off from NASA's Kennedy Space Center with astronauts Neil A. Armstrong, Michael Collins, and Edwin "Buzz" Aldrin at 9:32 a.m. EDT. *(Source: https://www.nasa.gov/multimedia/imagegallery/image_feature_359a.html) Image Credit: NASA Last Updated: Aug. 4, 2017 Editor: NASA Content Administrator.*

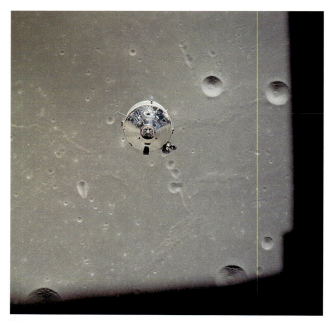

FIG. 9.6 Apollo 11 Command/Service Module Columbia in lunar orbit, photographed from the Lunar Module Eagle; The Apollo 11 Command and Service Modules (CSM) are photographed from the Lunar Module (LM) in lunar orbit during the Apollo 11 lunar landing mission. The lunar surface below is in the north central Sea of Fertility. The coordinates of the center of the picture are 51° east longitude and 1° north latitude. About half of the crater Taruntius G is visible in the lower left corner of the picture. Part of Taruntius H can be seen at lower right. *(Source: NASA - NASA Images at the Internet Archive (Image link)); Copyright info: Public Domain, https://commons.wikimedia.org/w/index.php?curid=28490842.*

at 8:32 a.m. CDT, July 16, 1969, from launch complex 39A at the Kennedy Space Center. Fig. 9.4 shows Saturn V carrying Apollo 11 rising past the launch tower camera. The image in Fig. 9.5 shows the launch of Apollo 11, the first Lunar landing mission, on July 16, 1969. The massive Saturn V rocket lifted off from NASA's Kennedy Space Center with astronauts: Mission commander Neil A. Armstrong, Command Module pilot Michael Collins, and Lunar Module pilot Edwin "Buzz" Aldrin, Jr. at 9:32 a.m. EDT.

After being sent toward the Moon by the Saturn V's upper stage, the astronauts separated the spacecraft from it and traveled for 3 days until they entered into lunar orbit (see Fig. 9.6). Neil Armstrong became the first to step onto the lunar surface 6 h after landing on July 21 at 02:56:15 UTC; Aldrin joined him about 20 min later. They spent about two and a quarter hour together outside the spacecraft, and collected 47.5 pounds (21.5 kg) of lunar material to bring back to Earth. Michael Collins piloted the command module Columbia alone in lunar orbit while his colleagues were on the Moon's surface. Armstrong and Aldrin spent just under a day on the lunar surface before re-joining Columbia in lunar orbit.

The photograph of the Earth seen from Apollo 11 just after leaving Earth orbit (translunar injection) is indeed fascinating (see Fig. 9.7). Armstrong and Aldrin then moved into the lunar module Eagle (see Fig. 9.8) and landed in the

FIG. 9.7 **Earth seen from Apollo 11 just after leaving Earth orbit (translunar injection).** This view of Earth showing clouds over water was photographed from the Apollo 11 spacecraft following translunar injection. *(Source: NASA- http://spaceflight.nasa.gov/gallery/images/apollo/apollo11/html/as11-36-5299.html (direct link) (JSC link)); Copyright info: Public Domain in the United States.*

FIG. 9.9 **(Top image) This photograph of the Lunar Module at Tranquility Base was taken by Neil Armstrong during the Apollo 11 mission, from the rim of Little West Crater on the lunar surface.** Armstrong's shadow and the shadow of the camera are visible in the foreground. When he took this picture, Armstrong was clearly standing above the level of the Lunar Module's footpads. Darkened tracks lead leftward to the deployment area of the Early Apollo Surface Experiments Package (EASEP) and rightward to the TV camera. This is the furthest distance from the lunar module traveled by either astronaut while on the moon. *(Source: NASA: https://www.nasa.gov/image-feature/lunar-module-at-tranquility-base Image Credit: NASA; Last Updated: Aug. 4, 2017; Editor: Sarah Loff (Bottom image) A panorama of the top image) (Source: https://www.hq.nasa.gov/alsj/a11/a11pan1111231EvMHR.jpg).*

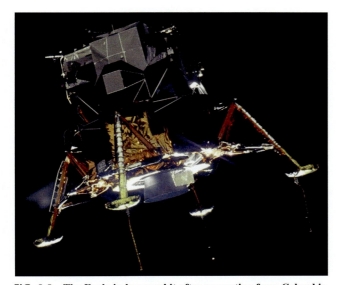

FIG. 9.8 **The Eagle in lunar orbit after separating from Columbia.** The Apollo 11 Lunar Module Eagle, in a landing configuration was photographed in lunar orbit from the Command and Service Module Columbia. Inside the module were Commander Neil A. Armstrong and Lunar Module Pilot Buzz Aldrin. The long rod-like protrusions under the landing pods are lunar surface sensing probes. Upon contact with the lunar surface, the probes sent a signal to the crew to shut down the descent engine. *(Source: NASA; https://www.nasa.gov/multimedia/imagegallery/image_feature_1161.html); Public Domain in the United States.*

"Sea of Tranquility" (see Fig. 9.9). The landing was broadcast on live TV to a worldwide audience. Armstrong stepped onto the lunar surface and described the event as "one small step for [a] man, one giant leap for mankind." Apollo 11 effectively ended the Space Race and fulfilled a US national goal proposed in 1961 by US President John F. Kennedy: "before this decade is out, of landing a man on the Moon and returning him safely to the Earth" (Richard, 2001).

Apollo 11, the first of six successful lunar missions, marked the first-time humans set foot on another planetary surface. Apollo 11 was the first manned lunar landing mission with a crew of three astronauts, having landed the first two humans on Earth's Moon. Mission commander Neil Armstrong and pilot Buzz Aldrin, both Americans, landed the lunar module Eagle on July 20, 1969, at 20:18 UTC.

Six of the missions (Apollos 11, 12, 14, 15, 16, and 17) achieved this goal. Apollos 7 and 9 were Earth orbiting missions to test the Command and Lunar Modules, and did not return lunar data. Apollo 8 and Apollo 10 tested various components while orbiting the Moon, and returned photography of the lunar surface. Apollo 13 did not land on the Moon due to a malfunction, but also returned photographs. The six missions that landed on the Moon returned a wealth

FIG. 9.10 **The Apollo 11 crew await pickup by a helicopter from the USS Hornet, prime recovery ship for the historic lunar landing mission.** The fourth man in the life raft is a United States Navy underwater demolition team swimmer. All four men are wearing biological isolation garments. The Apollo 11 Command Module Columbia with astronauts Neil Armstrong, Michael Collins, and Buzz Aldrin aboard splashed down at 11:49 a.m. CDT, July 24, 1969, about 812 nautical miles southwest of Hawaii and only 12 nautical miles from the USS Hornet. *(Source: https://www.nasa.gov/multimedia/imagegallery/image_feature_1429.html) Image Credit: NASA Last Updated: Aug. 4, 2017 Editor: NASA Content Administrator.*

FIG. 9.11 **Within the Mobile Quarantine Facility, Apollo 11 astronauts (left to right) Michael Collins, Edwin E. Aldrin Jr. and Neil A. Armstrong relax following their successful lunar landing mission.** They spent two-and-one-half days in the quarantine trailer enroute from the USS Hornet, prime recovery ship, to the Lunar Receiving Laboratory at the Manned Spacecraft Center in Houston. The Hornet docked at Pearl Harbor where the trailer was transferred to a jet aircraft for the flight to Houston. *(Source: https://www.nasa.gov/content/apollo-11-astronauts-relax-following-successful-mission) Image Credit: NASA Last Updated: Aug. 4, 2017 Editor: Sarah Loff.*

of scientific data and almost 400 kg of lunar samples. Experiments included soil mechanics, meteoroids, seismic, heat flow, lunar ranging, magnetic fields, and solar wind experiments (https://nssdc.gsfc.nasa.gov/planetary/lunar/apollo.html).

After staying for a total of about 21.5 h on the lunar surface, the two astronauts used Eagle's upper stage to lift off from the lunar surface and re-join Collins in the command module. They jettisoned (abandoned) Eagle before they performed the maneuvers that blasted them out of lunar orbit on a trajectory back to Earth. They returned to Earth and landed in the Pacific Ocean on July 24 (see Fig. 9.10).

To prevent the possibility of any alien contamination polluting the Earth, it was necessary to quarantine the three astronauts. Fig. 9.11 shows the three astronauts relaxing within the Mobile Quarantine Facility, following their successful lunar landing mission. Fig. 9.12 shows President Barack Obama chats with Apollo 11 astronauts, from left, Buzz Aldrin, Michael Collins and Neil Armstrong, Monday, July 20, 2009, in the Oval Office of the White House in Washington, on the 40th anniversary of the Apollo 11 lunar landing.

FIG. 9.12 **Former US President Barack Obama chatting with Apollo 11 astronauts, from left, Buzz Aldrin, Michael Collins and Neil Armstrong, Monday, July 20, 2009, in the Oval Office of the White House in Washington, on the 40th anniversary of the Apollo 11 lunar landing.** *(Source: https://www.nasa.gov/multimedia/imagegallery/image_feature_1422.html) Image Credit: NASA/Bill Ingalls Last Updated: Aug. 4, 2017 Editor: NASA Content Administrator.*

9.7 The year-2007 discovery of water molecules trapped/chemically bound in lunar rock samples

It is now realized that water plays a critical role in the evolution of planetary bodies (Albaréde, 2009), and determination of the amount and sources of lunar water has profound implications for our understanding of the history of the Earth–Moon system. In the following sections, the present author attempts to take the readers of this Book through the winding trajectories of some of the relentless scientific investigations carried out by a select group of scientists toward understanding the presence of water in our Moon.

As already noted, Earth's Moon is thought to have formed from a disc of debris left either after a "giant impact" at an angle of 45° or more—a powerful sideswipe—to a still-forming Earth (proto-Earth) by a "planetary embryo" called *Theia*, almost 4.5 billion years ago as argued by Dauphas (2017), or after a number of smaller, high-velocity collisions of several smaller celestial bodies with a proto-Earth about 4.5 billion years ago, very early in Earth's history when the Solar system was taking shape, as argued by Rufu et al. (2017). Some studies have shown that the Moon formed nearly 100 million years after the start of the Solar system (see Seth et al., 2014). Scientists have long assumed that the heat from an impact of that size would cause hydrogen and other volatile elements to boil off into space, meaning the Moon must have started off completely dry.

An interesting question is whether the collision with a Mars-size celestial body known as *Theia* or several other smaller celestial bodies removed any water that the early Earth might have contained. After the collision—perhaps tens of millions of year later—small asteroids likely hit the Earth, including ones that may have been rich in water (see Young et al., 2016). In this context, it is important to realize that collisions of growing bodies (example, proto-Earth) are thought to have occurred very frequently back then, primarily because of orbit instabilities, although Mars is considered to have avoided large collisions.

The possibility that water has existed on Earth's Moon for varying lengths of time, both in liquid and in solid form, and both beneath the surface and on the surface, has been widely discussed during 1950s. The "Ranger program" was a series of unmanned space missions by the United States in the 1960s whose objective was to obtain the first close-up images of the surface of Earth's Moon. NASA again launched five successful Lunar Orbiter missions to Earth's Moon from 1966 through 1967 to map the surface before the Apollo landings. During the initial photographic picture-based investigations of Earth's Moon in the 1960s, surface features resembling terrestrial meandering rivers, with their oxbow loops, have been noticed in the Orbiter 4 & 5 and the Ranger pictures of the Moon (Urey, 1967a). Harold Urey turned out to be an ardent proponent of a watery Moon (Urey, 1967b). But Urey was not alone in this thought. Unfortunately, those features were in reality lunar sinuous rilles (curved paths having the shape and features of a mature river, and are commonly thought to be the remains of collapsed lava tubes or extinct lava flows), having been mistaken for flowing rivers on the Moon. On realizing the blunder, Urey suffered sharp criticism for his openness to ideas of water on the Moon (Crotts, 2011). Urey relented, abandoning any idea of lunar water.

One of the scientific discoveries resulting from the Apollo missions was the pervasive waterless nature of Earth's Moon and its rocks. The initial studies of the returned samples were pristine and showed no evidence of aqueous alteration, and water analyses were below detection limits. The subsequent 40 years of lunar sample analysis have only supported and strengthened the idea that indigenous water was nearly absent from the Moon's interior (McCubbin et al., 2010a). Initial observations of the samples brought back by Apollo 11 astronauts showed that lunar magmatic rocks were devoid of water-bearing minerals. This result was later on found to have been a result of a lack of sufficiently accurate and proper instrumentation and techniques to dissect contents of the lunar samples. For the past 50 years or so, Earth's Moon has been described as nearly devoid of indigenous water; however, evidence for water both on the lunar surface and within the lunar interior have recently emerged, calling into question this long-standing lunar dogma. The bulk water content of Earth's Moon was estimated to be less than 1 ppb (Taylor et al., 2006a), which would make the Moon at least six orders of magnitude drier than the interiors of Earth (Bolfan-Casanova, 2005; Williams et al., 2001) and Mars (McCubbin et al., 2010). This extremely low water content is in keeping with the pervasive volatile-element depletion signature recorded in all lunar materials, because hydrogen is the most volatile of the elements. The exact cause for this volatile-element depletion is still under question; however, many have argued that it stems from the high temperatures associated with the Moon-forming giant-impact event at ~4.5 Ga (Canup, 2004; Lucey et al., 2006; Saal et al., 2008; Albarede, 2009).

When water was first discovered in lunar samples in 2007, it was very surprising because from the time Apollo astronauts brought lunar samples from July 1969 onward, scientists thought that the Moon contained virtually no water. In 2006 water's presence anywhere on the Moon was in doubt, at least in some scientists' minds. The general disbelief among lunar scientists with regard to the presence of water on Earth's Moon is beautifully portrayed by Crotts (2011), who stated thus: "I heard it said at the time that some would not believe lunar water existed until they drank it from a glass." This skepticism expressed by some lunar scientists can be likened to that described in the biblical story of "Doubting Thomas"—a skeptic who refused to believe without direct personal experience—(one of the 12 disciples of Jesus Christ), who said, when the other disciples told him that Jesus Christ appeared to them in his absence after his crucifixion and death, "Except I shall see in his hands the print of the nails, and put my finger into the print of the nails, and thrust my hand into his side, I will not believe" (Gospel of John chapter 20). No wonder, in April that year NASA announced a plan to probe water on the Moon almost as directly as pouring a glass. LCROSS (Lunar CRater Observation and Sensing Satellite) would slam into a permanently shadowed lunar crater and dig out tons of regolith, and hopefully water (Crotts, 2011).

Ironically, the water discovered in lunar samples brought by the Apollo heroes was not liquid water, but water molecules trapped in volcanic glasses or chemically bound in mineral grains inside lunar rocks. Thus, it was realized that the water that exists on Earth's Moon came from inside the Moon. It is in the rocks. Despite skepticism about hydrated Apollo samples, lunar scientists began to believe by 2008 that surprisingly large amounts of water (H_2O)/hydroxyl (OH) could be locked in some lunar minerals. This optimism derives from better instrumentation and techniques to dissect contents internal to rock samples, clarifying that some minerals have prodigious amounts of water, primarily in samples from Apollo. Crotts (2011) recalls that "previously one would vaporize a sample for analysis, ionize it with an electron beam, then run these charged particles through electromagnetic fields in a mass analyzer to measure the charge/mass ratio for the atoms, molecules and molecular fragments that result. A newer technique is SIMS or 'secondary ion mass spectrometry' using an ion gun to blast tiny subsamples from material being studied, then sending the resulting shrapnel of ions into the mass analyzer. The ion gun can be focused to a small spot that is scanned across the sample, analyzing each particle blast and constructing a map of each constituent across the sample face. Since rocks often contain a jumble of tiny mineral inclusions or grains, this is a powerful way to study each mineral separately. (Electron beams carry more charge per unit kinetic energy, so may distort the charge distribution across the sample in troublesome ways.) One can study compositional variations on scales of microns, smaller than most grains."

The Moon is generally thought to have formed and evolved through a single or a series of catastrophic heating events, during which most of the highly volatile elements were lost. Hydrogen, being the lightest element, is believed to have been completely lost during this period. In a first, Saal et al. (2007) made use of considerable advances in secondary ion mass spectrometry to obtain improved limits on the indigenous volatile (CO_2, H_2O, F, S, Cl) contents of the most primitive basalts in the Moon—the lunar volcanic glasses (picritic glasses). Picrite composed of olivine and pyroxene from deep in the Moon, and glasses quenched by sudden cooling at the lunar surface. The samples were in fact tiny glass spheres from fire fountains: lava droplets spewing into space, solidifying, then hitting the ground as beads. Expanding gas propelled these eruptions. The gases erupting from the Moon was volcanic in composition: high in sulfur dioxide and water, even if low in carbon dioxide or monoxide. The samples were fire fountain beads (Saal et al., 2008), typically a few tenth of a millimeter in diameter, from Apollo 15: the green glass (indicative of olivine), and Apollo 17's famous orange soil (ilmenite and olivine) (Sato, 1979). By slicing the beads in half and pecking an ion microprobe every 15 microns across their diameter, Saal and collaborators showed that volatiles were concentrated on the interior of the beads: volatiles had leaked out (presumably when the droplets were flying through the vacuum), not leaked in (due to contamination). The amounts were large: 115–576 parts per million of sulfur (presumably associated with SO_2) and 4–46 ppm of water. There was little chlorine (0.06–2 ppm) but significant fluorine (4–40 ppm) (Saal et al., 2007). Surprisingly, there was almost no appreciable carbon dioxide, which is expected in many theoretical models (Saal et al., 2008). Modeling how much of each gas is lost based on their concentration profile across the beads, Saal et al. (2008) extrapolated that 260–745 ppm of water were originally present in the beads, and up to about 700 ppm sulfur. This is a huge number of volatiles, comparable to the amount found in basalt extruded from mid-ocean ridges on Earth (Dixon et al., 2002). Saal et al. (2008)'s studies have revealed that although the preeruptive water content of the lunar volcanic glasses cannot be precisely constrained, numerical modeling of diffusive degassing of the very-low-Ti glasses provides a best estimate of 745 p.p.m. water, with a minimum of 260 p.p.m. at the 95% confidence level. Saal et al. (2008)'s results indicate that, contrary to prevailing ideas, the bulk Moon might not be entirely depleted in highly volatile elements, including water. Thus, Saal et al. (2008) suggested that the presence of water must be considered in models constraining the Moon's formation and its thermal and chemical evolution. It was found that rocks originating from some areas in the lunar interior contain much more water than rocks from other places. Furthermore, the hydrogen isotopic composition of lunar water also varies from region to region, much more dramatically than on Earth.

9.8 Discovery of adsorbed hydrogen and hydroxyl in volatile content of lunar volcanic glasses—additional clues to the presence of water on Earth's Moon

One of the several ways of finding the existence and origin of the Moon's water is to examine the adsorbed (taken in or soaked up by chemical or physical action) hydrogen (H) and hydroxyl (OH—a molecule consisting of one oxygen atom (O) and one hydrogen atom (H)) in the volatile content of melt inclusions found in samples brought back from the Apollo space craft missions to the Moon.

In a study carried out by McCubbin et al. (2010), hydroxyl (as well as fluoride and chloride) was analyzed by secondary ion mass spectrometry in apatite [$Ca_5(PO_4)_3(F,Cl,OH)$] from three different lunar samples in order to obtain quantitative constraints on the abundance of water in the lunar interior. This work confirmed that hundreds to thousands of ppm water (of the structural form hydroxyl) is present in apatite from the Moon. Moreover, two of the studied samples likely had water preserved from magmatic processes, which would qualify the water as being indigenous to the Moon. The presence of hydroxyl in apatite from a number of different types of lunar rocks indicated that water may be ubiquitous within the lunar interior, potentially as early as the time of lunar formation. The water contents analyzed for the lunar apatite indicate minimum water contents of their lunar source region to range from 64 ppb to 5 ppm H_2O. This lower limit range of water contents is at least two orders of magnitude greater than the previously reported value for the bulk Moon, and the actual source region water contents could be significantly higher.

It may be realized that the water in the Earth's Moon is a tracer of the processes that operated in the hot, partly silicate gas, partly magma disk that surrounded the Earth after the giant impact(s) with the Earth that resulted in the formation of the Earth's Moon (more specifically, the Earth–Moon system). Basically, during this (these) giant impact(s), whatever happened to the Moon also happened to the Earth. Therefore, the source of the Moon's water has important implications for determining the source of Earth's water, which is vital to life. With regard to the existence of water on Earth's Moon, there are two options: either, water was inherited by the Moon from Earth during the Moon-forming impact, or it was added to the Moon later by impingement of comets or asteroids. It might also be a combination of these two processes. A recent review of hundreds of chemical analyses of the Moon rocks indicates that the amount of water in the Moon's interior varies regionally—revealing clues about how water originated and was redistributed in the Moon.

Examination of the following lunar materials shed much light on the presence of water on Earth's Moon:

1. Volcanic glass—the amorphous product of rapidly cooling magma. Like all types of glass, it is a state of matter intermediate between the close-packed, highly ordered array of a crystal and the highly disordered array of gas.
2. Melt inclusions—tiny dots of volcanic glass trapped within crystals called olivine, which prevent water escaping during an eruption and enable researchers to get an idea of what the inside of the Moon is like.
3. Apatite—hydrogen-rich calcium phosphate mineral found in samples of the ancient lunar crust.
4. Plagioclase feldspar—dominant mineral in the ancient crustal lunar highlands.
5. Anorthosites—intrusive igneous rock characterized by its composition: mostly plagioclase feldspar, with a minimal mafic component. Pyroxene, ilmenite, magnetite, and olivine are the mafic minerals most commonly present.

Detection of indigenous hydrogen in a diversity of lunar materials, including volcanic glass (Saal et al., 2008), melt inclusions (Hauri et al., 2011), apatite (Boyce et al., 2010; McCubbin et al., 2010), and plagioclase feldspar (Hui et al., 2013) suggests that water (H_2O), which is abundant in hydrogen (H), played a role in the chemical differentiation of the Moon. Water contents measured in plagioclase feldspar have been used to predict that 320 p.p.m. water initially existed in the lunar magma ocean (Hui et al., 2013) whereas measurements in apatite, found as a minor mineral in lunar rocks, representing younger potassium-enriched melt predict a bulk Moon with <100 ppm water (Mills et al., 2016).

From a lunar magma with the composition expected for the bulk Moon, the dense magnesium silicates olivine and pyroxene would have been the first minerals to crystallize (Walker et al., 1975; Snyder et al. 1992; Shearer et al., 2006; Wieczorek et al., 2013; Elardo et al., 2011). According to Carlson (2019), "Settling of these crystals to the bottom of the magma ocean left the remaining magma richer in calcium and aluminum, leading to crystallization of plagioclase. Olivine and pyroxene crystallization also drove the remaining magma to higher concentrations of iron, which served to further increase its density. By the time the magma became saturated in plagioclase, as it crystallized, its buoyancy with respect to the magma caused the plagioclase to float to the surface of the magma ocean to form the highlands crust. As the crust formed, it trapped beneath it the remaining magma that was enriched in all elements that were not incorporated into the minerals that had crystallized by that time. These elements include potassium (K), the rare earth elements (REE), and phosphorus (P), leading to the acronym KREEP for the rocks formed from this residual magma (Warren and Wasson, 1979)."

One of the important materials in the study of lunar water is KREEP. Sample-based estimates of water content of KREEP-rich magmas from measurements of OH-, F, and Cl in lunar apatites suggest a low water concentration in the KREEP component with 2–140 p.p.m. magmatic water (McCubbin et al., 2010). Using these data Elkins-Tanton and Grove (2011) predicted that the bulk water content of the magma ocean would have been <10 p.p.m. In contrast, Hui et al. (2013) estimated water contents of 320 p.p.m. for the bulk Moon and 1.4 wt. % for the evolved melt residuum of the magma ocean (i.e., urKREEP) from water measured in lunar anorthosite (Mills et al., 2016). As Evans et al.

(2014) have noted; the anorthositic composition of the lunar highlands and KREEP terrane of the lunar nearside (Jolliff et al., 1999) support the past existence of a large-scale lunar magma ocean leading to fractional crystallization and compositional stratification of the Moon (Wood et al., 1970; Wood, 1972; Warren, 1985).

Although past studies of the Earth's Moon's formation and its chemical and thermal evolution have focused on a bulk Moon with minimal volatile content and the absence of water (Shearer, 2006), geochemical analyses of very low titanium (Ti) glasses and lunar melt inclusions present compelling evidence that water concentrations of at least 260 p.p.m. and up to 6000 p.p.m. were present in some regions of the lunar interior prior to 3 Ga (Saal et al., 2008; Hauri et al., 2011). Furthermore, analysis of the electrical conductivity of the Moon suggests a reduced viscosity layer above the core–mantle boundary consistent with an enriched wet region of approximately 100–200 km (Karato, 2013; Khan et al., 2013). If water was a constituent of the bulk Moon, fractional crystallization of all but the deep mantle would have resulted in a sequestered source of interior water. Ultimately, in this scenario, a postmagma ocean-Moon would contain separate subcrustal and deep mantle reservoirs, which may be consistent with both localized water-rich reservoirs (≥260 ppm) (Saal et al., 2008) and a bulk magma ocean containing under 100 ppm water (Elkins-Tanton and Grove, 2011).

According to Mills et al. (2016), geochemical surveys by the Lunar Prospector (Jolliff et al., 2011) and Diviner Lunar Radiometer Experiment (Glotch et al., 2010; Jolliff et al., 2011) indicating the global significance of evolved igneous rocks suggest that the formation of these granites removed water from some mantle source regions, helping to explain the existence of mare basalts with <10 ppm water, but must have left regions of the interior relatively wet as seen by the water content in volcanic glass and melt inclusions. Although these early-formed evolved melts were water-rich, their petrogenesis (branch of science dealing with the origin and formation of rocks) supports the conclusion that the Moon's mantle had <100 p.p.m. water for most of its history.

Greenwood et al. (2011) have reported ion microprobe measurements of water and hydrogen isotopes in the hydrous mineral apatite, derived from crystalline lunar mare basalts (interpreted as products of partial melting of mantle cumulates produced during the early lunar differentiation) and highlands rocks collected during the Apollo missions. Note that ion microprobes use a focused beam of ions to sputter ions from a small rock sample into a mass spectrometer. The ratio of deuterium to hydrogen (i.e., D/H ratio) can indicate the source of the water or trace magmatic processes in the lunar interior. Water has now been detected in apatite in many different lunar rock types. Greenwood et al. (2011) found significant water in apatite from both mare and highlands rocks, indicating a role for water during all phases of the Moon's magmatic history. According to them, variations of hydrogen isotope ratios in apatite suggest sources for water in lunar rocks could come from the lunar mantle, solar wind protons and comets. Greenwood et al. (2011) concluded that a significant delivery of cometary water to the Earth–Moon system occurred shortly after the Moon-forming impact.

In another study, Saal et al. (2013) examined the isotopic composition of the hydrogen trapped in the melt inclusions. Note that melt inclusion is a small parcel or "blobs" of melt(s) that is entrapped by crystals growing in magma and eventually forming igneous rocks. Melt inclusions tend to be microscopic in size and can be analyzed for volatile contents that are used to interpret trapping pressures of the melt at depth. In principle, the isotopic composition of hydrogen is used as a fingerprint in order to understand the origin of that hydrogen. Saal et al. (2013) carried out in-situ measurements of the isotopic composition of hydrogen dissolved in primitive volcanic glass and olivine-hosted melt inclusions recovered from the Moon by the Apollo 15 and 17 missions. Using a multicollector ion microprobe, the researchers measured the amount of deuterium in the samples compared to the amount of regular hydrogen.

As indicated earlier, deuterium is an isotope of hydrogen with an extra neutron (compared to the normal hydrogen). Water molecules originating from different places in the Solar system have different amounts of deuterium. In general, things formed closer to the Sun have less deuterium than things formed farther out. Saal and his colleagues found that the deuterium/hydrogen ratio in the melt inclusions was relatively low and matched the ratio found in carbonaceous chondrites—meteorites originating in the asteroid belt (located near planet Jupiter) and considered to be among the oldest objects in the Solar System. That means the source of the water on the Moon is primitive meteorites, not comets, as some scientists thought. In any case, Saal et al. (2013) found that the melt inclusions on Earth's Moon have plenty of water.

9.9 Identifying presence of water on Earth's Moon through spacecraft-borne spectroscopic measurements

Lunar Prospector results indicated that potential water on the Moon likely rests a large fraction of a meter underground, covered by dry soil. Later probes imply similar results. Despite indirect indications on the possible presence of water on the Earth's Moon, conclusive evidence was lacking. The search for water on the surface of the presumably anhydrous Moon had remained an unfulfilled quest for 40 years. However, researchers remained optimistic, and continued with various studies with the hope of finding direct indications of the presence of water on our Moon someday. Finally, scientists succeeded in challenging the prevailing notion of an anhydrous Moon. Data from the Visual and Infrared Mapping Spectrometer (VIMS) on Cassini during its flyby of the Moon in 1999 show a broad absorption at 3 micrometers (μs) due to adsorbed water and near 2.8 μs attributed to hydroxyl in the sunlit surface on the Moon. The amounts of water indicated in the spectra depend on the type of mixing and the grain sizes in the rocks and soils but could be 10–1000 parts per million and locally higher. Water in the polar regions may be water that has migrated to the colder environments there. Trace hydroxyl is observed in the anorthositic highlands at lower latitudes (Clark, 2009). Infrared spectroscopic measurements of the lunar surface from spacecraft provide unambiguous evidence for the presence of hydroxyl (OH) or water (see Lucey, 2009).

In a joyous note, and curiously enough, the Deep Impact spacecraft found the entire surface to be hydrated during some portions of the day. Sunshine et al. (2009) found that hydroxyl (OH) and water (H_2O) absorptions in the near infrared were strongest near the North Pole of the Moon and are consistent with <0.5 wt% H_2O. It was also found that hydration varied with temperature, rather than cumulative solar radiation, but no inherent absorptivity differences with composition were observed. In Sunshine et al. (2009)'s studies, comparisons between data collected 1 week (a quarter lunar day) apart showed a dynamic process with diurnal changes in hydration that were greater for mare basalts (approximately 70%) than for highlands (approximately 50%). Interestingly, this hydration loss and return to a steady state occurred entirely between local morning and evening, requiring a ready daytime source of water-group ions, which is consistent with a solar wind origin. Note that the lunar maria (singular: mare) are large, dark, basaltic plains on Earth's Moon, formed by ancient asteroid impacts. They were dubbed maria (Latin for "seas"), by early astronomers who mistook them for actual seas.

The reason for the observed diurnal variations in the distribution of water and hydroxyl on the illuminated Moon remained unanswered for quite a long time in the history of lunar research. Hibbitts et al. (2011) investigated the adsorption of molecular water onto lunar analogue materials under ultra-high vacuum with the goal of better understanding the thermal stability and evolution of water on the lunar surface. Temperature-programmed desorption (TPD) experiments showed that lunar-analogue basaltic-composition glass is hydrophobic, with water–water interactions dominating over surface chemisorption. This suggests that lunar agglutinates (pyroclastic igneous rock formed from partly fused volcanic bombs) will tend not to adsorb water at temperatures above where water clusters and multilayer ice forms. The basalt JSC-1A lunar mare analogue, which is a complex mixture of minerals and glass, adsorbs (hold as a thin film on the outside surface or on internal surfaces within the material) water above 180 K with an adsorption profile that extends to 400 K, showing evidence for a continuum of water adsorption sites. It was found that bancroft albite adsorbs more water, more strongly, than JSC-1A, with a well-defined desorption (the release of an adsorbed substance from a surface) peak near 225 K. This suggests that mineral surfaces will adsorb more water than mare or mature (glassy, agglutinate rich) surfaces and may explain the association of water with fresh feldspathic (containing feldspar) craters at high latitudes. Hibbitts et al. (2011) determined the activation energies for the thermal desorption of water from these materials, and along with values from the literature, used to model the grain-to-grain migration of water within the lunar regolith. These models suggested that a combination of recombinative desorption of hydroxyl along with molecular desorption of water and its subsequent migration within and out of the regolith may explain the observed diurnal variations in the distribution of water and hydroxyl on the illuminated Moon.

In another study, the Moon Mineralogy Mapper (M^3) spectrometer, a NASA instrument on Chandrayaan-1, India's first mission to the Moon, detected absorption features near 2.8–3.0 micrometer (μm) on the uppermost surface of the Moon (see Clark, 2009; Pieters et al., 2009; Sunshine et al., 2009). The M^3 spectrometer measured visible and near-infrared wavelengths, which contain highly diagnostic absorptions due to minerals as well as hydroxyl (OH) and water (H_2O). Absorptions occur as solar radiation passes through multiple randomly oriented particles in the upper 1–2 mm of soil. Reflectance spectra exhibit these combined absorptions from all particles. As soils evolve in the lunar environment, individual grains develop silicate glass coatings that contain nanophase metallic iron (npFeO).

For silicate bodies, the observed absorption features are typically attributed to hydroxyl- and/or water-bearing materials. On Earth's Moon, the feature is seen as a widely distributed absorption that appears strongest at cooler high latitudes (i.e., toward Lunar poles) and at several fresh feldspathic craters. The general lack of correlation of this feature in sunlit M^3 data with neutron spectrometer hydrogen abundance data suggests that the formation and retention of hydroxyl and water are ongoing surficial processes. Pieters et al. (2009) argued that hydroxyl/water production processes may feed polar cold traps and make the lunar regolith a candidate source of volatiles for human exploration. In any case, Pieters et al. (2009)'s measurements, acquired by M^3, showed small amounts of OH/H_2O on the surface of the Moon.

In a different kind of study by Clark (2009), data from the Visual and Infrared Mapping Spectrometer (VIMS) on Cassini during its flyby of the Moon in 1999 showed a broad absorption at 3 μm due to adsorbed water and near 2.8 μm attributed to hydroxyl in the sunlit surface on the Moon. It was found that the amounts of water indicated in the spectra depend on the type of mixing and the grain sizes in the rocks and soils, but could be 10–1000 parts per million and locally higher. It was argued that water in the Polar Regions of the Earth's Moon may be water that has migrated to the colder environments there. Trace hydroxyl was observed in the anorthositic (a granular plutonic igneous rock composed almost exclusively of a soda-lime feldspar) highlands at lower latitudes.

The Moon was thought to have been generally anhydrous, yet the Deep Impact spacecraft found the entire surface to be hydrated during some portions of the day. In a study, Sunshine et al. (2009) found that hydroxyl (OH) and water (H_2O) absorptions in the near infrared are strongest near the North Pole and are consistent with <0.5 wt% H_2O. Hydration varied with temperature, rather than cumulative solar radiation, but no inherent absorptivity differences with composition were observed. However, comparisons between data collected 1 week (a quarter lunar day) apart indicated a dynamic process with diurnal changes in hydration that were greater for mare basalts (approximately 70%) than for highlands (approximately 50%). This hydration loss and return to a steady state occurred entirely between local morning and evening, requiring a ready daytime source of water-group ions, which is consistent with a solar wind origin. Note that "mare" is one of the darker areas on Earth's Moon. The mare formed billions of years ago. Large meteorites impacted the surface of the moon and broke up the crust. Later lavas formed by melting of rock within the moon due to the decay of radioactive elements. Mare basalts are igneous rocks from the mare.

In yet another study of the M^3 spectrometer data pertaining to Earth's Moon, McCord et al. (2011) reported the analysis of two absorption features near 3 μm in the lunar reflectance spectrum, and interpreted them as being due to OH and H_2O. They proposed that solar wind proton-induced hydroxylation (a chemical process that introduces a hydroxyl group into an organic compound; the degree of hydroxylation refers to the number of OH groups in a molecule) is the creation process, and its products could be a source for other reported types of hydrogen-rich material and water. The irregular and damaged fine-grained lunar soil seems especially adapted for trapping solar wind protons and forming hydroxyl (OH) owing to abundant dangling oxygen (O) bonds. The M^3 spectrometer data reveal that the strengths of the two absorptions are correlated and widespread, and both are correlated with lunar composition, but in different ways. Feldspathic material (containing feldspar, consisting of aluminosilicates of potassium, sodium, and calcium) seems richer in OH. These results seem to rule out water from the lunar interior and cometary infall as major sources responsible for the presence of water on the lunar surface. McCord et al. (2011) suggested that there appear to be correlations of apparent band strengths with time of day and lighting conditions. They also found that thermal emission from Earth's Moon reduces the apparent strengths of the M^3 absorptions, and its removal is not yet completely successful. Furthermore, according to their studies, many of the lunar physical properties are themselves intercorrelated, and so separating these dependencies on the absorptions is difficult, due to the incomplete M^3 spectrometer data set. According to McCord et al. (2011), this process should also operate on other airless silicate surfaces, such as the planet *Mercury* and the asteroid *Vesta*.

It must be noted, and even appreciated, that several remote sensing discoveries of hydroxyl and water on the lunar surface have considerably reshaped our view of the distribution of water and related compounds on airless bodies such as the Moon (Pieters et al. 2009; Sunshine et al., 2009; Clark, 2009). The exact origin of this surface water is still unclear (McCord et al., 2011), but it has been suggested that hydroxyl in the lunar regolith can result from the implantation of hydrogen ions by the solar wind (Pieters et al. 2009; Ichimura et al., 2012).

In a different study, Liu et al. (2012) reported Fourier transform infrared spectroscopy and secondary ion mass spectrometry analyses of Apollo lunar samples that revealed the presence of significant amounts of hydroxyl in glasses formed in the lunar regolith by micro-meteorite impacts. Hydrogen isotope compositions of these glasses suggested that some of the observed hydroxyl is derived from solar wind sources. Liu et al. (2012)'s findings imply that ice in polar cold traps could contain hydrogen atoms ultimately derived from the solar wind, as predicted by early theoretical models of water stability on the lunar surface (Watson et al., 1961). Liu et al. (2012) suggested that a similar mechanism may contribute to hydroxyl on the surfaces of other airless terrestrial bodies where the solar wind directly interacts with the surface, such as the planet *Mercury* and the asteroid *4-Vesta*.

The heat flux incident upon the surface of an airless planetary body is dominated by solar radiation during the day, and by thermal emission from topography at night. Motivated by the close relationship between this heat flux, the surface temperatures, and the stability of volatiles, Rubanenko and Aharonson (2017) examined the effect of the slope distribution on the temperature distribution and hence prevalence of cold-traps, where volatiles may accumulate over geologic time. They developed a thermo-physical model accounting for insolation, reflected and emitted radiation, and subsurface conduction, and used it to examine several idealized representations of rough topography. They showed how subsurface conduction alters the temperature distribution of bowl-shaped craters compared to predictions given by past analytic models. Rubanenko and Aharonson (2017) modeled the dependence of cold-traps on crater geometry and quantified the effect that while deeper depressions cast more persistent shadows, they are often too warm to trap water ice due to the smaller sky fraction and increased reflected and re-emitted radiation from the walls. In order to calculate the temperature distribution outside craters, Rubanenko and Aharonson (2017) considered rough random surfaces with a Gaussian slope distribution. Using their derived temperatures and additional volatile stability models, they estimated the potential area fraction of stable water ice on Earth's Moon. For example, surfaces with slope RMS ~15° located near the poles have been found to have a ~10% exposed cold-trap area fraction. It was found that in the subsurface, the diffusion barrier created by the overlaying regolith increases this area fraction to ~40%. Additionally, some buried water ice was shown to remain stable even beneath temporarily illuminated slopes, making it more readily accessible to future lunar excavation missions. Finally, due to the exponential dependence of stability of ice on temperature, it is possible to constrain the maximum thickness of the unstable layer to a few decimeters.

9.10 The year-2009 confirmation of the presence of water on Earth's Moon by India's Moon Impact Probe (MIP) and America's Moon Mineralogy Mapper (M³) carried by India's Chandrayaan-1 Lunar Probe

Chandrayaan-1 (lit: moon-traveler, or moon vehicle) was India's first unmanned lunar probe (Fig. 9.13). India's first successful lunar mission, Chandrayaan-1, put a spacecraft in orbit around the Moon in 2008 and then later sent a probe hurtling toward the Moon's south pole, where it deliberately crashed and released material that got analyzed by the orbiter's scientific instruments, helping to confirm the presence of water on Earth's Moon. It was launched by the Indian Space Research Organisation (ISRO) in October 2008 (Fig. 9.14), and operated until August 2009. The mission included a lunar orbiter and an impactor. The mission was a major boost to India's space program, as India researched and developed its own technology in order to explore the Moon. The vehicle was successfully inserted into lunar orbit on 8 November 2008.

On 14 November 2008, the Moon Impact Probe separated from the Chandrayaan orbiter at 20:06 and struck the south pole in a controlled manner, making India the fourth

FIG. 9.14 PSLV-C11/Chandrayaan-1 Lifting off from Second Launch Pad of Satish Dhawan Space Centre, Sriharikota Range on 22 October 2008, 06:24:53. (Source: https://www.isro.gov.in/pslv-c11-chandrayaan-1/pslv-c11-chandrayaan-1-gallery) Author: Indian Space Research Organisation This file is a copyrighted work of the Government of India, licensed under the Government Open Data License - India (GODL) All users are provided a worldwide, royalty-free, nonexclusive license to use, adapt, publish (either in original, or in adapted and/or derivative forms), translate, display, add value, and create derivative works (including products and services), for all lawful commercial and noncommercial purposes, and for the duration of existence of such rights over the data or information. The user must acknowledge the provider, source, and license of data by explicitly publishing the attribution statement, including the DOI (Digital Object Identifier), or the URL (Uniform Resource Locator), or the URI (Uniform Resource Identifier) of the data concerned. (Source: https://commons.wikimedia.org/wiki/File:PSLV-C11_launch2.jpg).

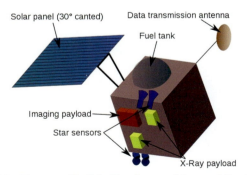

FIG. 9.13 Diagram of India's Chandrayaan-1 lunar orbiter. (Source: https://nssdc.gsfc.nasa.gov/nmc/spacecraftDisplay.do?id=2008-052A) Author: Reproduction by User: Watercred of an image from NASA This file is made available under the Creative Commons CC0 1.0 Universal Public Domain Dedication The person who associated a work with this deed has dedicated the work to the public domain by waiving all of their rights to the work worldwide under copyright law, including all related and neighboring rights, to the extent allowed by law. You can copy, modify, distribute and perform the work, even for commercial purposes, all without asking permission. (Source: https://commons.wikimedia.org/wiki/File:Chandrayaan-1.svg).

country to place its flag on the Moon. The probe impacted near Shackleton Crater at 20:31 ejecting underground soil that could be analyzed for the presence of lunar water ice (see Arlin, 2011).

The remote sensing lunar satellite had a mass of 1380 kg (3042 lb) at launch and 675 kg (1488 lb) in lunar orbit (because of the reduced gravity of the Moon). The spacecraft was orbiting around the Moon at a height of 100 km from the lunar surface for chemical, mineralogical and photo-geologic mapping of the Moon. The spacecraft carried 11 scientific instruments built in India, USA, UK, Germany, Sweden, and Bulgaria.

9.10.1 Providing complete coverage of the Moon's polar regions—acquiring images of peaks and craters and conducting chemical and mineralogical mapping of the entire lunar surface

The Chandrayaan-1 lunar orbiter carried high resolution remote sensing equipment for visible, near infrared, and soft and hard X-ray frequencies. The Chandrayaan-1 mission performed high-resolution remote sensing of the Earth's Moon in visible, near infrared (NIR), low energy X-rays and high-energy X-ray regions. One of the objectives was to prepare a three-dimensional atlas (with high spatial and altitude resolution) of both near and far side of the moon. It also aimed at conducting chemical and mineralogical mapping of the entire lunar surface for distribution of mineral and chemical elements such as magnesium, aluminum, silicon, calcium, iron, and titanium as well as high atomic number elements such as Radon, Uranium, and Thorium with high spatial resolution. The mineral content on the lunar surface was mapped with the Moon Mineralogy Mapper (M^3, for short) (Fig. 9.15), a NASA instrument on board the orbiter. The presence of iron was reiterated and changes in rock and mineral composition have been identified. The Oriental Basin region of the Moon was mapped, and it indicates abundance of iron-bearing minerals such as pyroxene.

ISRO announced in January 2009 the completion of the mapping of the Apollo Moon missions landing sites by the orbiter, using multiple payloads. Six of the sites have been mapped including landing sites of Apollo 15 and Apollo 17.

On 26 November, the indigenous Terrain Mapping Camera, which was first activated on 29 October 2008, acquired images of peaks and craters (Fig. 9.16). This came as a surprise to ISRO officials because the Moon was believed to have been consisting mostly of craters. Interesting data on lunar polar areas was provided by Lunar Laser Ranging Instrument (LLRI) and High Energy X-ray Spectrometer (HEX) of ISRO as well as Miniature Synthetic Aperture Radar (Mini-SAR) of the USA. The LLRI covered both the lunar poles and additional lunar regions of interest,

FIG. 9.15 Left side of the Moon Mineralogy Mapper that was located on the Chandrayaan-1 lunar orbiter. *(Primary Source: http://m3.jpl.nasa.gov/images/M3LeftSide.jpg) Author: NASA This file is in the public domain in the United States because it was solely created by NASA. NASA copyright policy states that "NASA material is not protected by copyright unless noted." Secondary Source: (https://commons.wikimedia.org/wiki/File:Moon_Mineralogy_Mapper_left.jpg).*

HEX made about 200 orbits over the lunar poles and Mini-SAR provided complete coverage of both North and South Polar Regions of the Moon.

Another ESA payload—Chandrayaan-1 imaging X-ray Spectrometer (C1XS)—detected more than two dozen weak solar flares during the mission duration. The Bulgarian payload called Radiation Dose Monitor (RADOM) was

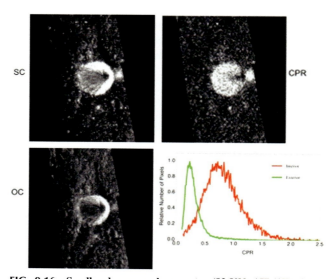

FIG. 9.16 **Small polar anomalous crater (83.8°N, 157.1°W; 8 km diameter) on the floor of the large crater Rozhdestvensky (85.2 N, 155.4 W; 177 km diameter).** Clockwise from upper left are the same-sense circular image (SC), CPR, CPR histogram, and the opposite sense circular (OC) images. Elevated CPR is found inside but not outside the crater rim; this crater interior is in permanent shadow (Noda et al., 2008). These properties are consistent with volume scattering in this crater interior. *(Source: Spudis, P.D., et al., 2010; Initial results for the north pole of the Moon from Mini-SAR, Chandrayaan-1 mission, Geophys. Res. Lett., 37, L06204, doi: http://doi.org/10.1029/2009GL042259).*

activated on the day of the launch itself and worked until the mission's end.

ISRO said scientists from India and participating agencies expressed satisfaction on the performance of Chandrayaan-1 mission as well as the high quality of data sent by the spacecraft (https://en.wikipedia.org/wiki/Chandrayaan-1). The Chandrayaan-1 payload has enabled scientists to study the interaction between the solar wind and a celestial body like the Moon without a magnetic field. In its 10-month orbit around Earth's Moon, Chandrayaan-1's X-ray Spectrometer (C1XS) detected titanium, confirmed the presence of calcium, and gathered the most accurate measurements yet of magnesium, aluminum, and iron on the lunar surface.

9.10.2 Probing the poles in search of ice and water and providing confirmation of regolith hydration everywhere across the lunar surface

The Polar Regions are of special interest as they might contain ice (Bhandari, 2005). The Mini-SAR onboard Chandrayan-I imaged many of the permanently shadowed regions that exist at both poles of the Moon. On March 2010, it was reported that the Mini-SAR on board the Chandrayaan-1 had discovered more than 40 permanently darkened craters near the Moon's north pole, which are hypothesized to contain an estimated 600 million metric tons of water-ice. It was found that the estimated amount of water ice potentially present is comparable to the quantity estimated from the previous mission of Lunar Prospector's neutron data. Furthermore, other NASA instruments onboard Chandrayaan-1 (the Moon Mineralogy Mapper (M^3) discovered water molecules in the Moon's polar regions, while water vapor was detected by NASA's Lunar CRater Observation and Sensing Satellite, or LCROSS.

The Mini-RF instruments on ISRO's Chandrayaan-1 and NASA's Lunar Reconnaissance Orbiter (LRO) obtained S-band (12.6 cm) synthetic aperture radar images of the impact site at 150 and 30 m resolution, respectively. These observations show that the floor of Cabeus has a circular polarization ratio (CPR) comparable to or less than the average of nearby terrain in the southern lunar highlands. Furthermore, <2% of the pixels in Cabeus crater have CPR values greater than unity. This observation is not consistent with presence of thick deposits of nearly pure water ice within a few meters of lunar surface, but it does not rule out the presence of small (<~10 cm), discrete pieces of ice mixed in with the regolith (Bussey et al., 2011).

Among the several achievements made by the Chandrayaan-1 lunar mission, undoubtedly, the greatest achievement is considered to be the discovery of the widespread presence of water molecules in the lunar soil. On 18 November 2008, the Moon Impact Probe was released from Chandrayaan-1 at a height of 100 km (62 mi). During its 25-min descent, Chandra's Altitudinal Composition Explorer (CHACE) recorded evidence of water in 650 mass spectra readings gathered during this time. In subsequent investigations, the Moon Mineralogy Mapper (M^3)—an imaging spectrometer—on Chandrayaan-1 detected water ice on Earth's Moon. On 25 September 2009, ISRO announced that the Moon Impact Probe (MIP), another instrument on board Chandrayaan-1, had discovered water on the Moon just before impact and had discovered it 3 months before NASA's M^3. However, the heart-throb announcement of this joyous discovery—something the lunar scientists have been eagerly searching for, but also possibly something they had no idea existed—was not made until NASA confirmed it.

The M^3 probe detected absorption features near 2.8–3.0 μm on the surface of the Moon. For silicate bodies, such features are typically attributed to hydroxyl (OH)- and/or water (H_2O)-bearing materials. On Earth's Moon, the feature is seen as a widely distributed absorption that appears strongest at cooler high latitudes and at several fresh feldspathic craters. The general lack of correlation of this feature in sunlit M^3 data with neutron spectrometer H-abundance data suggests that the formation and retention of OH and H_2O is an ongoing surficial process. Many scientists think that OH/H_2O production processes may feed polar cold traps and make the lunar regolith a candidate source of volatiles for human exploration.

It may be noted that the lunar scientists had been discussing the possibility of water repositories for decades. They are now increasingly confident that the decades-long debate is over. It is now generally agreed that Earth's Moon, in fact, has water in all sorts of places; not just locked up in minerals, but scattered throughout the broken-up surface, and, potentially, in blocks or sheets of ice at depth. The results from the Chandrayaan-1 mission are also offering a wide array of "watery signals."

Although Vilas et al. (2008) used images from Galileo's flyby of the Moon (on the way to Jupiter) and identified spectral reflectance signature near the lunar South pole and the South Pole-Aitken Basin, as noted earlier, to the great surprise of lunar scientists, confirmation of the hitherto suspected regolith hydration came from another spacecraft—India's first lunar probe, Chandrayaan-1 (Crotts, 2011). To confirm this surprising result, investigators called upon data from two other crafts having flown by the Moon: Deep Impact heading to comet Tempel 1 and Cassini spacecraft on its way to Saturn. M^3 was a spectrometer operating in the visual to infrared wavelength range, from 0.4 to 3.0 microns, splitting light into 260 wavelength bins or "colors." At any moment M^3 would view a rectangle of Moon about 40 km by 0.07 km and split this image into 600 individual spectra. An instant later M^3 would view the next 0.07 km-wide strip passing beneath Chandrayaan-1 and slowly build an image of the entire lunar surface (requiring about 300 parallel swaths) in each of the 260 colors (Crotts, 2011). The amazing result from M^3 was that hydration seemed ubiquitous across the lunar surface, stronger near the poles (the 3-micron cut-off, however, made characterizing the nature of hydration, H$_2$O versus OH, difficult). This was so unexpected that earlier 3-microns images taken by Deep Impact and Cassini were used to confirm it: hydration is everywhere across the lunar surface, varying in longitude depending on the angle of the illumination of the Sun in a way which seemed to imply that the solar wind might be responsible for making it. However, M^3 found variation toward the poles (where the Sun barely shines) in the opposite sense: some places near the poles showed a concentration of OH (or H$_2$O) as large as about 0.07%, whereas near the equator concentrations were typically 0.002%. Deep Impact and Cassini supported this (without resolving the H$_2$O versus OH issue) (Pieters et al., 2009). Crotts (2011) most beautifully described the euphoria created by the Chandrayaan-1/M^3 thus: "The Chandrayaan-1/M^3 hydration result in 2009 created much excitement in the U.S and India. Here was a detection of lunar hydration by a NASA effort that received the public endorsement of the agency, and also demonstrated scientific success for India's first probe beyond Earth, despite its demise due to overheating and power failure. This was the first time many were aware lunar water (or at least hydroxyl) had been detected. In fact, Chandrayaan-1 deployed a second craft, the Moon Impact Probe (MIP), 10 months before the M^3 announcement, and it too found water (but this result was not revealed until the M^3 publication)."

9.11 Permanently shadowed regions (cold traps) of the lunar poles—initial inferences

It takes the Moon the equivalent of 27.3 Earth days to complete a single rotation on its axis, the same amount of time it takes to complete a single orbit around Earth. This is the reason why the same side of the Moon always faces the Earth. This side of the Moon is known as "near side" of the Moon. Whereas the length of a day (the time between two noons or sunsets) is 24 h on Earth, the length of a day by the same definition on Earth's Moon is equivalent to 1 month on Earth (i.e., roughly two Earth-weeks continuous darkness and two Earth-weeks continuous sunshine in a given location on Earth's Moon).

A possible accumulation of volatiles, including water frost and ice, in the permanently shadowed regions of the lunar-poles has been discussed for decades. The Lunar Prospector Neutron Spectrometer (LP-NS) directly measured H on the Moon and found a higher abundance of H in association with the permanently shadowed regions of both poles, implying that the lunar poles could be potential cold traps for volatiles, some of which could be linked to solar-wind hydrogen. Components of the neutron energy spectrum measured by instruments on the Lunar Prospector (LP) spacecraft have been optimistically interpreted as indicating massive deposits of water ice or other hydrogenic minerals in the polar regions of Earth's Moon. Hodges Jr. (2002) reported thus, "About midway through the Lunar Prospector (LP) mission a widely distributed press release announced that neutron detectors on the LP spacecraft had provided evidence that 'each polar region may contain as much as three billion metric tons of water ice' (NASA Ames Research Center release 98–158, September 3, 1998). Almost simultaneously, the basis for this claim was published by Feldman et al. (1998a) in a preliminary analysis of 6 months of Lunar Prospector neutron spectrometer (LPNS) data obtained from a 100 km polar orbit. A subsequent study by Feldman et al. (2000a) of neutron data from the entire LP mission, which includes a period of measurements at a nominal altitude of 30 km, led to a modified conclusion that the data require about 6×10^7 metric tons of hydrogen distributed in the regolith near the north pole, and roughly 3.8×10^8 metric tons of hydrogen deposited more densely in permanently shaded craters near the south pole." Detailed analysis by Hodges Jr. (2002) revealed that "What is clear is that the existence of polar water ice or hydride deposits has not been proven by analyses of Lunar Prospector data." According to Hodges Jr. (2002) "The important point is that the LPNS data do not uniquely identify hydrogen deposits." Hodges Jr. (2002) noted abnormally high SiO$_2$ content could mimic hydrogen. Nonetheless, one can detect hydrogen and use different energies to decide how deep it is in the soil. For example, Lawrence et al. (2006) found that "the average hydrogen abundance near both lunar poles is 100–150 ppm and is likely buried by 10 ± 5 cm of dry lunar soil, a result that is consistent with previous studies. The localized hydrogen abundance for small (<20 km) areas of permanently shaded regions remains highly uncertain and could range from 200 ppm H up to 40 wt% H$_2$O in some isolated regions."

9.12 Role of NASA's LRO and LCROSS in confirming presence of water in the southern lunar crater *Cabeus*

Hydrogen has been inferred to occur in enhanced concentrations within the permanently shadowed regions and, hence, the coldest areas of the lunar poles. Several remote observations have indicated that water ice may be present in permanently shadowed craters of Earth's Moon. The Lunar CRater Observation and Sensing Satellite (LCROSS) mission was designed to provide direct evidence, in particular, to detect hydrogen-bearing volatiles directly. The Lunar Reconnaissance Orbiter (LRO) and LCROSS mission, launched on June 18, 2009, as part of the shared Lunar Precursor Robotic Program, was conceived as a low-cost means of exploring the presence of water ice in a permanently shadowed crater (*Cabeus*) near the south pole of the Moon (*"NASA - LCROSS: Mission Overview". Nasa.gov.*).

Neutron flux measurements of the Moon's south polar region from the Lunar Exploration Neutron Detector (LEND) on the LRO spacecraft were used to select the optimal impact site for LCROSS. LEND data show several regions where the epithermal (having energy above that of thermal agitation) neutron flux from the surface is suppressed, which is indicative of enhanced hydrogen content (Mitrofanov et al., 2010). These regions are not spatially coincident with permanently shadowed regions of the Moon.

LCROSS was designed to collect and relay data from the impact and debris plume resulting from the launch vehicle's spent Centaur upper stage striking the crater *Cabeus*. The LCROSS science instrument payload, provided by NASA's Ames Research Center (ARC), consisted of a total of nine instruments: one visible, two near infrared, and two mid-infrared cameras; one visible and two near-infrared spectrometers; and a photometer. A data handling unit (DHU) would collect the information from each instrument for transmission back to LCROSS Mission Control. Both LRO and LCROSS were together launched and operated by NASA immediately after discovery of lunar water by India's Chandrayaan-1. The LRO-LCROSS mission was the first American mission to the Moon in over 10 years.

On its final approach to the Earth's Moon, the Shepherding Spacecraft and Centaur booster rocket separated on October 9, 2009, at 01:50 UTC. The Centaur upper stage acted as a heavy impactor to create a debris plume that would rise above the lunar surface. On 9 October 2009, LCROSS sent a kinetic impactor (a spent Centaur rocket) to strike Cabeus crater, on a mission to search for water ice and other volatiles expected to be trapped in lunar polar soils. This impactor struck the persistently shadowed region within the lunar south pole crater *Cabeus*, ejecting debris, dust, and vapor. The LCROSS struck one of the coldest of the cryogenic regions, where subsurface temperatures are estimated to be 38 Kelvin. Large areas of the lunar polar regions are currently cold enough to cold-trap water ice as well as a range of both more volatile and less volatile species (Paige et al., 2010). Diviner Lunar Radiometer Experiment surface-temperature maps reveal the existence of widespread surface and near-surface cryogenic regions that extend beyond the boundaries of persistent shadow. According to Paige et al. (2010), "The diverse mixture of water and high-volatility compounds detected in the LCROSS ejecta plume is strong evidence for the impact delivery and cold-trapping of volatiles derived from primitive outer solar system bodies."

As its detached upper-stage launch vehicle collided with the surface, instruments on the trailing LCROSS Shepherding Spacecraft monitored the impact and ejecta. Schultz et al. (2010) found that the faint impact flash in visible wavelengths and thermal signature imaged in the mid-infrared together indicate a low-density surface layer. They further found that the evolving spectra reveal not only OH within sunlit ejecta but also other volatile species. As the Shepherding Spacecraft approached the surface, it imaged a 25–30-m-diameter crater and evidence of a high-angle ballistic ejecta plume still in the process of returning to the surface—an evolution attributed to the nature of the impactor (Schultz et al., 2010).

The LRO Diviner instrument detected a thermal emission signature 90 s after the LCROSS Centaur impact and on two subsequent orbits. According to Hayne et al. (2010), "The impact heated a region of 30–200 m^2 to at least 950 Kelvin, providing a sustained heat source for the sublimation of up to ~300 kg of water ice during the 4 min of LCROSS postimpact observations. Diviner visible observations constrain the mass of the sunlit ejecta column to be ~10^{-6}–10^{-5} kg per square meter, which is consistent with LCROSS estimates used to derive the relative abundance of the ice within the regolith."

The Lyman Alpha Mapping Project (LAMP) ultraviolet spectrograph onboard the LRO observed the plume generated by the LCROSS impact as far-ultraviolet emissions from the fluorescence of sunlight by molecular hydrogen and carbon monoxide, plus resonantly scattered sunlight from atomic mercury, with contributions from calcium and magnesium. Gladstone et al. (2010) found that "The observed light curve is well simulated by the expansion of a vapor cloud at a temperature of ~1000 kelvin, containing ~570 kg of carbon monoxide, ~140 kg of molecular hydrogen, ~160 kg of calcium, ~120 kg of mercury, and ~40 kg of magnesium."

Following 4 min after impact of the Centaur upper stage, a second "shepherding" spacecraft, which carried nine instruments, including cameras, spectrometers, and a radiometer, flew through this debris plume, collecting and relaying data back to Earth before it struck the lunar surface to produce a second debris plume. The impact velocity was projected to be 9000 km/h (5600 mph) or 2.5 km/s. Fig. 9.17 provides an illustration of the LCROSS Centaur rocket stage and shepherding spacecraft as they approach impact with the lunar South Pole on October 9, 2009.

Because of data bandwidth issues, the exposures were kept short, which made the plume difficult to see in the images in the visible spectra. This resulted in the need for image processing to increase clarity. The infrared camera also captured a thermal signature of the booster's impact. On 13 November 2009, NASA reported that multiple lines of evidence show water was present in both the high-angle vapor plume and the ejecta curtain created by the LCROSS

FIG. 9.17 An illustration of the LCROSS Centaur rocket stage and shepherding spacecraft as they approach impact with the lunar South Pole on October 9, 2009. *Source: NASA -http://lcross.arc.nasa. gov/index.htm (http://lcross.arc.nasa.gov/images/LCROSS_Centaur_Sep. jpg) Public Domain "This is an artist's illustration of the Lunar Crater Observation and Sensing Satellite (LCROSS) Centaur rocket stage and shepherding spacecraft as they approach impact with the lunar south pole on October 9, 2009." LCROSS spacecraft separates from Centaur stage Author: NASA.*

Centaur impact. As of November 2009, the concentration and distribution of water and other substances required more analysis. Additional confirmation came from an emission in the ultraviolet spectrum that was attributed to hydroxyl fragments, a product from the break-up of water by sunlight. Analysis of the spectra indicate that a reasonable estimate of the concentration of water in the frozen regolith is on the order of 1%. Evidence from other missions suggests that this may have been a relatively dry spot, as thick deposits of relatively pure ice appear to present themselves in other craters.

A later, more definitive, analysis carried out by Colaprete et al. (2010) found that near-infrared absorbance attributed to water vapor and ice, and ultraviolet emissions attributable to hydroxyl radicals support the presence of water in the debris. According to them, the maximum total water vapor and water ice within the instrument field of view was 155 ± 12 kg. Colaprete et al. (2010) reported that "Given the estimated total excavated mass of regolith that reached sunlight, and hence was observable, the concentration of water ice in the regolith at the LCROSS impact site is estimated to be 5.6 ± 2.9% by mass. In addition to water, spectral bands of a number of other volatile compounds were observed, including light hydrocarbons, sulfur-bearing species, and carbon dioxide." According to

Mitrofanov et al. (2010), "The LCROSS impact site inside the *Cabeus* crater demonstrates the highest hydrogen concentration in the lunar south polar region, corresponding to an estimated content of 0.5–4.0% water ice by weight, depending on the thickness of any overlying dry regolith layer. The distribution of hydrogen across the region is consistent with buried water ice from cometary impacts, hydrogen implantation from the solar wind, and/or other as yet unknown sources."

As already indicated, the LCROSS mission was designed to search for evidence of water in a permanently shadowed region near the lunar south pole. In this mission, an instrumented Shepherding Spacecraft followed a kinetic impactor and provided—from a nadir perspective—the only images of the debris plume. With independent observations of the visible debris plume from a more oblique view, the angles and velocities of the ejecta from this unique cratering experiment are better constrained. Strycker et al. (2013) reported the first visible observations of the LCROSS ejecta plume from Earth, thereby ascertaining the morphology of the plume to contain a minimum of two separate components, placing limits on ejecta velocities at multiple angles, and permitting an independent estimate of the illuminated ejecta mass. Strycker et al. (2013)'s mass estimate implies that the lunar volatile inventory in the Cabeus permanently shadowed region includes a water concentration of 6.3 ± 1.6% by mass.

The research findings summarized above indicate that the LRO-LCROSS mission was successful in confirming water in *Cabeus*. Together, LCROSS and LRO form the vanguard of NASA's return to the Moon, and seemingly influenced United States government decisions on whether or not to colonize the Moon.

9.13 Discovery of frozen water in permanently shadowed craters and poles of Earth's Moon

As noted earlier, in 1961, 8 years prior to man's landmark first landing on the Moon in July 1969, three farsighted scientists (Watson, Murray, and Brown) predicted the possibility of the existence of ice and other trapped volatiles in the permanently shadowed regions near the lunar poles (see Watson et al., 1961). In a piece of heart-throbbing and much-awaited news to the entire world, this prediction was proved right after a long gap of about 50 years when evidence for existence of water in the North Pole of the Moon was reported in 2010 by Spudis and coresearchers based on data obtained from Mini-synthetic aperture radar (SAR)—S-band imaging radar—deployed on India's Chandrayaan-1 mission to Earth's Moon (see Spudis et al., 2010). Likewise, in a simultaneous investigation, Colaprete and coresearchers discovered the presence of water in the persistently shadowed region within the lunar South Pole crater *Cabeus* (see Colaprete et al., 2010). Some of the studies unraveling the existence of water on Earth's Moon are briefly addressed in the following paragraphs.

From a scientific perspective, volatiles deposited on the Moon—as a result of bombardment by water-bearing objects such as comets and meteorites, and the continuous impingement of solar wind hydrogen on the surface of the airless Moon—might migrate to and get collected in permanently shadowed dark cold traps near the lunar poles, where they would be stable over geologic time (Arnold, 1979). Modeling of the thermal conditions of permanently dark areas near the poles (Vasavada et al., 1999) indicates that they are significantly colder than 100 K. Likewise, early data from the Lunar Reconnaissance Orbiter (LRO) Diviner experiment indicate cold trap temperatures as low as 25 K. At such very low temperatures, water–ice and other volatiles are stable and once deposited in these cold traps and covered by a thin (few cm) layer of regolith are largely protected from processes of removal.

Because these cold traps receive no direct solar illumination, and emit little radiation, most of these cold traps are difficult to observe from the Earth. However, radar can identify deposits of frozen volatiles because under certain conditions, they produce a unique backscatter signature, arising from multiple internal reflection and/or coherent backscatter produced by low-loss material, such as water–ice (Ostro and Shoemaker, 1990). For example, a high (>1.0) ratio of same-sense to opposite-sense circular polarization and high reflectivity has been detected by radar from the ice caps on the Galilean satellites of Jupiter (Hapke, 1990; Ostro and Shoemaker, 1990), the residual south polar ice cap of Mars (Muhleman et al., 1991) and the inside of permanently shadowed polar craters on Mercury (Harmon and Slade, 1992; Butler et al., 1993).

Note that in radar literature, the ratio of the power of the received signal in the same sense (SC; same sense circular) to that of the opposite sense (OC; opposite sense circular) as transmitted (SC/OC) is known as Circular Polarization Ratio (CPR). It is interesting to note that the Clementine bistatic radar experiment (Nozette et al., 1996, 1997, 2001) revealed elevated CPR in the South Polar Region at bistatic angles close to zero, suggesting the presence of patchy ice deposits in the permanently dark areas of *Shackleton* crater.

Independently, the neutron spectrometer on the Lunar Prospector mission measured low epithermal neutron flux at both poles, indicating elevated amounts of near-surface hydrogen near the poles (Feldman et al., 1998, 2000). Combined with study of permanently shadowed areas (Margot et al., 1999), the neutron signal has been attributed to the presence of water ice in dark polar craters (Lawrence et al., 2006; Elphic et al., 2007).

Spudis et al. (2010) reported polarimetric radar data for the surface of the North Pole of the Moon acquired with the Mini-SAR experiment onboard India's Chandrayaan-1 Lunar mission spacecraft. Mini-SAR is an S-band (wavelength = 12.6 cm) imaging radar designed to map the polar terrain and to collect information about the scattering properties of illuminated and permanently dark areas of the lunar poles near optimum (45°) viewing geometry (Spudis et al., 2009). The Mini-SAR instrument utilizes a hybrid polarity architecture (Raney, 2007) which transmits circular polarization and receives two orthogonal linear polarizations coherently. This configuration allows reconstruction of the four Stokes parameters of the backscattered field that are equivalent to those which would be measured by a conventional radar astronomical telescope that transmitted and received exclusively in circular polarizations.

Between mid-February and mid-April, 2009, the Mini-SAR imaging radar mapped more than 95% of the areas pole-ward of 80° latitude at a resolution of 150 m. The North Polar Region displays backscatter properties typical for the Moon, with CPR values in the range of 0.1–0.3, increasing to over 1.0 for young primary impact craters. Analyses of the CPR data for anomalous polar craters (Fig. 9.18) reveal them to be unique. Several craters found near the north pole of the Moon show a different pattern (Figs. 9.18 and 9.19). While most of the floor of the crater *Rozhdestvensky* (85.2° N, 155.4° W; 177 km diameter) is not in permanent darkness, the interiors of small (5–15 km diameter) craters found on its floor are in permanent darkness (Margot et al., 1999; Bussey et al., 2005; Noda et al., 2008).

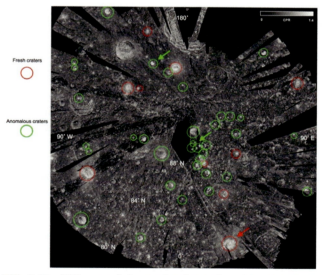

FIG. 9.18 CPR map of the north polar region of Earth's Moon. Normal appearing fresh craters are indicated by red circles while anomalous craters having high CPR in their interiors are shown by green circles. Although both crater types are found throughout the map area, anomalous craters (green) are concentrated at the higher latitudes and have at least some portion of their floors in permanent shadow. (*Source: Spudis, P.D., et al., 2010; Initial results for the north pole of the Moon from Mini-SAR, Chandrayaan-1 mission, Geophys. Res. Lett., 37, L06204, doi: http://doi.org/10.1029/2009GL042259*).

FIG. 9.19 These images show a very young lunar crater on the side of the moon that faces away from Earth, as viewed by NASA's Moon Mineralogy Mapper on the Indian Space Research Organization's Chandrayaan-1 spacecraft. On the left is an image showing brightness at shorter infrared wavelengths. On the right, the distribution of water-rich minerals (light blue) is shown around a small crater. Both water- and hydroxyl-rich materials were found to be associated with material ejected from the crater. (Credits: ISRO/NASA/JPL-Caltech/USGS/Brown Univ. Sources: https://en.wikipedia.org/wiki/Chandrayaan-1#/media/File:Chandrayaan1_Spacecraft_Discovery_Moon_Water.jpg, http://www.nasa.gov/topics/moonmars/features/moon20090924.html (Public domain)).

A few non-polar craters have been identified that also exhibit high interior and lower exterior CPR values. The spatial distribution of lunar polar hydrogen deposits after SELENE, 2009 (available at http://lunarscience2009.arc.nasa.gov/node/73) are all consistent with the presence of water ice in these craters. The recent discovery of significant amounts of mobile water in the higher latitudes of the Moon (Pieters et al., 2009) may provide one source for the polar ice inferred from these measurements.

The observed higher CPR values in the high-latitude regions of the Earth's Moon likely reflect surface roughness associated with these fresh features. In contrast, some craters in this region show elevated CPR in their interiors, but not exterior to their rims. Almost all of these features are in permanent sun-shadow and correlate with proposed locations of polar ice modeled on the basis of Lunar Prospector neutron data. These relations are consistent with deposits of water ice in these craters.

Apart from Mini-SAR S-band imaging radar, several other remote sensing devices have been used for exploration of water on Earth's Moon, the observations from which have indicated that frozen water (water ice) may be present in permanently shadowed craters of the Moon. For example, the Lunar Crater Observation and Sensing Satellite (LCROSS) mission was designed to provide direct evidence of the presence of water on the Moon.

Ingenious and well-planned exploration techniques have been carried out to accomplish the objective of discovering water on the Moon. As part of such exploration, on 9 October 2009, a spent Centaur rocket struck the persistently shadowed region (named "LCROSS impact site") within the lunar South Pole crater *Cabeus*, resulting in the ejection of debris, dust, and vapor. This material was observed by a second "shepherding" spacecraft, which carried nine instruments, including cameras, spectrometers, and a radiometer. Studies carried out by Colaprete et al. (2010) have indicated that near-infrared absorbance attributed to water vapor and ice; and ultraviolet emissions attributable to hydroxyl radicals support the presence of water in the debris. They estimated that the maximum total water vapor and water ice within the instrument field of view was 155 ± 12 kg. Given the estimated total excavated mass of regolith that reached sunlight, and hence was observable, the concentration of water ice in the regolith at the LCROSS impact site is estimated to be 5.6 ± 2.9% by mass. In addition to water, spectral bands of a number of other volatile compounds were observed, including light hydrocarbons, sulfur-bearing species, and carbon dioxide.

As already indicated, Circular Polarization Ratio (CPR) mosaics from Mini-SAR on Chandrayaan-1 and Mini-RF on LRO have been used to study craters near to the lunar North Pole. The look direction of the detectors strongly affects the appearance of the crater CPR maps. Rectifying the mosaics to account for parallax also significantly changes the CPR maps of the crater interiors. Eke et al. (2014) showed that the CPRs of crater interiors in un-rectified maps are biased to larger values than crater exteriors, because of a combination of the effects of parallax and incidence angle. Using the LOLA Digital Elevation Map (DEM), they studied the variation of CPR with angle of incidence. It was found that for anomalous craters, the CPR interior to the crater increases with both incidence angle and distance from the crater center. Furthermore, central crater CPRs have been found to be similar to those in the crater exteriors. Eke et al. (2014) suggested that CPR does not appear to correlate with temperature within craters. Furthermore, the anomalous polar craters have diameter-to-depth ratios that are lower than those of typical polar craters. These results strongly suggest that the high CPR values in anomalous polar craters are not providing evidence of significant volumes of water ice. Rather, according to Eke et al. (2014), anomalous craters are of intermediate age, and maintain sufficiently steep sides that sufficient regolith does not cover all rough surfaces.

9.14 Water in the lunar interior

As already indicated, during the initial years of lunar exploration and research, it has been thought that the Moon is highly depleted in volatiles such as water, and indeed published direct measurements of water in lunar volcanic glasses have never exceeded 50 parts per million (p.p.m.). For many years, researchers believed that the rocks from the Moon were "bone dry," and that any water detected in the Apollo samples had to be mere contamination from Earth.

Although data from orbital spacecraft revealed the presence of water on the lunar surface, the water so detected was thought to be a thin layer formed from solar wind hitting the surface of the airless Moon.

However, thanks to persistent scientific investigations carried out by planetary geology scientists from various countries, much progress has been made in subsequent years in better understanding the quantum of water in our Moon. For example, widespread presence of surface-cited water as well as evidence for the presence of subsurface water ice in the lunar polar regions (Pieters et al., 2009; Spudis et al., 2010; Colaprete et al., 2010) have established that Moon is not "bone dry" as hypothesized earlier. First signatures for presence of water intrinsic to the Moon came from studies of lunar volcanic glasses collected at the Apollo 15 and 17 landing sites (Saal et al., 2008). This together with reports of indigenous volatiles in lunar apatite (Boyce et al., 2010; McCubbin et al., 2010a; Greenwood et al., 2011; Sarbadhikari et al., 2013; Tartèseand Anand, 2013; Tartèse et al., 2013) and in olivine hosted melt inclusions within Apollo 17 volcanic glasses (Hauri et al., 2011), suggest that the volatiles in lunar interior could have influenced the lunar magmatic processes.

Presence of water could also lead to lowering of viscosity in the lunar mantle. Water intrinsic to the Moon has also been inferred from studies of laboratory reflectance spectra of lunar anorthosites (Hui et al., 2013) and Chandrayaan-1 spectral reflectance data for different regions on the Moon (Bhattacharya et al., 2013; Klima et al., 2013).

The Apollo 15 and Apollo 17 volcanic glasses contained 4–46 p.p.m. of water (Saal et al., 2008). A value of 745 p.p.m. was estimated for the pre-eruptive water content, taking into account diffusive degassing and subsequent glass formation (Saal et al., 2008). Olivine hosted melt inclusions in Apollo 17 volcanic glasses contain 615–1410 p.p.m. of water, 50–78 p.p.m. of fluorine, and 1.5–3.0 p.p.m. of chlorine (Hauri et al., 2011). High pressure–temperature experimental petrology coupled with geochemical studies (Delano, 1980; Longhi, 1992) suggests a deep mantle source (≥ 500 km) for the Apollo 15 and 17 volcanic glasses. The apatites in lunar basalts belonging to low-Ti, high-Ti, and high-Al groups also host water over a wide range from few hundred to several thousands of p.p.m. (Boyce et al., 2010; McCubbin et al., 2010; Greenwood et al., 2011; Sarbadhikari et al., 2013; Tartèseand Anand, 2013; Tartèse et al., 2013). These studies suggest water content ranging from 70 to 240 p.p.m. in the parent magmas of lunar basalts.

It is believed that presence and distribution of water and other volatiles in the lunar interior could have played a key role in the early evolution of the Moon. Sarbadhikari et al. (2016) reported abundance of water along with fluorine and chlorine in apatite present in the Apollo 15 lunar basalt 15555, which is considered to be the primitive end member of the low-Ti mare basalt suite. Apatites are rare in this basalt and are devoid of significant spatial variation in volatile content. Considering a late-stage crystallization of apatite, Sarbadhikari et al. (2016) inferred 100–160 p.p.m. water, 80–90 p.p.m. fluorine and 10–20 p.p.m. chlorine in the parent magma of 15555. It was found that the inferred water content is much lower than that reported earlier for the parent magma of lunar volcanic glasses, as well as in melt inclusions trapped within the glasses that sampled much deeper regions of Moon. It has been argued that this difference suggests a nonuniform distribution of water and other volatiles in lunar mantle source regions, which could have significantly influenced early thermo-chemical evolution of the Moon.

It would be interesting to skim through various studies that scientists have carried out in establishing the presence of water in our Moon's interior. The following sections seek to achieve this modest objective in terms of brief descriptions of some such studies.

9.14.1 Detection of water existing deeper within the lunar crust and mantle through analyses of melt inclusions that originated from deep within the Moon's interior

It would be difficult to reach the Moon's interior and examine the presence of water there. Therefore, an ingenious method that scientists have developed to examine the presence (or otherwise) of water content in the Moon's interior was to analyze the melt inclusions (a microscopic size parcel of melt(s) that is entrapped by crystals growing in magma and eventually forming igneous rocks) that originated from deep within the Moon's interior, but ultimately reached the Moon's surface through some natural processes such as volcanic eruption, or crater-forming celestial impacts. In planetary geological science, water that originates from deep within a planet's interior is known as magmatic water.

It must be recalled that Earth's Moon was once considered dry compared with Earth. However, laboratory analyses of igneous components of lunar samples have suggested that the Moon's interior is not entirely anhydrous (Saal et al., 2008; McCubbin 2011). Water (H_2O) and hydroxyl (OH) have also been detected from the lunar surface, but these have been attributed to nonindigenous sources, such as interactions with the solar wind (Pieters et al., 2009; Clark, 2009; Sunshine et al., 2009).

In a first of its kind of magmatic water detection, Hauri et al. (2011) reported in situ measurements of magmatic water in lunar melt inclusions, which originated from primitive lunar magma, by virtue of being trapped within olivine crystals before volcanic eruption, and therefore did not experience posteruptive degassing. Hauri et al. (2011) found that the lunar melt inclusions they examined contained 615–1410 p.p.m. water and high correlated amounts of fluorine (50–78 p.p.m.), sulfur (612–877 p.p.m.), and chlorine (1.5–3.0 p.p.m.). Hauri et al. (2011) found that these volatile contents are very similar to primitive terrestrial mid-ocean ridge basalts. Their findings indicated that some parts of the lunar interior contain as much water as Earth's upper mantle.

Understanding the story of lunar apatite has implications beyond determining how much water is locked inside lunar rocks and soil. Many scientists theorize that Earth's Moon formed when a giant impact tore free a large chunk of Earth more than 4 billion years ago. If this "giant impact" model is correct, the Moon would have become completely molten, and lighter elements such as hydrogen should have bubbled to the surface and escaped into space. Hydrogen (H) is the lightest element, and it is a key component of water (H_2O). A proto-Moon formed from the Earth as a result of a giant celestial impact should be completely free of hydrogen post-impact because the intense heat generated during the impact should have totally expelled all of the originally present hydrogen molecules into space. According to the predominant theory of how did Earth's Moon originally form, hydrogen and other volatile elements should not be present at all in lunar rocks. On this basis, for decades, scientists believed that Earth's Moon was almost entirely devoid of water from its very genesis. The majority of lunar samples are in fact very dry and missing lighter elements. However, the discovery of hydrogen-rich mineral apatite within lunar rocks in 2010 seemed to hint at a waterier past. It may be noted that the mineral apatite is the most widely used medium for estimating the amount of water in lunar rocks.

It is ironical that in the background of the generally accepted "giant impact" model of Earth's Moon's formation, hydrogen-rich apatite crystals have been found in a whole host of lunar samples. This presented a paradox for scientists. To explain away the riddle, scientists had to agree that somehow, despite the Moon's fiery beginning, some water and other volatiles might have remained, though perhaps not as much as apatite initially implied. Lunar scientists had years of believing in a dry moon, and now they have some evidence that the old dry model of Earth's Moon wasn't perfect. However, they are aware of the need to be cautious and look carefully at each piece of evidence before they decide that rocks on the Moon are as wet as those on Earth.

Scientists originally assumed that information obtained from a small sample of apatite could predict the original water content of a large body of magma, or even the entire Moon, but a new study indicates that apatite may, in fact, be deceptive. Lunar volcanic glasses collected at the Apollo 17 landing site have been found to contain fluorine and chlorine along with water (Hauri et al., 2011). Boyce et al. (2014) have found that the amount of water present in the Earth's Moon may have been over-estimated by scientists studying the mineral apatite. Boyce et al. (2014)'s study outlined the problems associated with accurately retrieving volatile content in melts due to compatible behavior of fluorine in apatite, which can lead to significant volatile zoning.

When water is present as molten rock cools, apatite can form by incorporating hydrogen atoms into its crystal structure. However, hydrogen will be included in the newly crystallizing mineral only if apatite's preferred building blocks, fluorine (F) and chlorine (Cl), have been mostly exhausted. Boyce et al. (2014) suspected that the estimate of large water content within lunar apatite results from a quirk in the crystallization process rather than a water-rich lunar environment. To clear their suspicion, they created a computer model to accurately predict how apatite would have crystallized from cooling bodies of lunar magma early in the moon's history. Their simulations revealed that the unusually hydrogen-rich apatite crystals observed in many lunar rock samples may not have formed within a water-rich environment, as was originally expected. This discovery has overturned the long-held assumption that the hydrogen in apatite is a good indicator of overall lunar water content. This discovery also indicates that the amount of water present in Earth's Moon may have been overestimated by scientists studying the mineral apatite.

Boyce et al. (2014) argue that early-forming apatite is so fluorine-rich that it vacuums all the fluorine out of the magma, followed by chlorine. Apatite that forms later doesn't see any fluorine or chlorine and becomes hydrogen-rich because it has no choice. Therefore, when fluorine and chlorine become depleted, a cooling body of magma will shift from forming hydrogen-poor apatite to forming hydrogen-rich apatite, with the latter not accurately reflecting the original water content in the magma.

In disagreement with Boyce et al. (2014)'s theory of occurrence of significant volatile zoning in apatite, Sarbadhikari et al. (2016) reported the identification of lunar apatite devoid of significant volatile zoning in mare basalt 15555, a slowly cooled micro-gabbro. This basalt, considered to be the primitive end member of the low-Ti mare basalt suite (Walker et al., 1977), is composed of olivine, pyroxene, plagioclase and accessory phases, such as tridymite, ilmenite, chromite, ulvöspinel, troilite, glass, Fe–Ni metal, and small fluorapatite grains (Walker et al., 1977), and have a crystallization age of ~3.3 Ga (Papanastassiou and Wasserburg, 1973). High-pressure experiments together with geochemical characteristics of 15555 indicated cosaturation of olivine-orthopyroxene-spinel in the parent magma, at ~8.5 kbar pressure, that is, 150–200 km below the lunar surface, in a closed system (Walker et al., 1977).

In Sarbadhikari et al. (2016)'s studies, they analyzed a polished thick section of the lunar basalt 15555. Apatite generally occurs in association with clinopyroxene, plagioclase, silica, ilmenite and glass, and was identified by taking back-scattered electron and X-ray elemental (Ca, P) images with a Cameca SX-100 electron microprobe. Two relatively large apatite grains (B1: ~30 μm and B2: ~20 μm) were selected for this study. Sarbadhikari et al. (2016) used a Cameca secondary ion mass spectrometer (nanoSIMS50) to analyze the apatite grains for their volatile composition. In their experiments, the thick section of 15555 was re-polished to remove a few micron layers to conduct isotope studies on fresh surfaces of apatite. Each grain and its surrounding areas were extensively presputtered with a high intensity ($\sim$1 nA) Cs+ beam for 15–20 min, to expose clean, pristine grain surfaces. A 16 keV Cs+ primary beam of $\sim$50 pA and diameter 0.5 μm was used to raster 40 μm × 40 μm area covering the apatite grain surface. Pressure in the analysis chamber was maintained at $\sim 10^{-10}$ torr to minimize OH$^-$ background. Secondary ions, ^{16}O ^{1}H$^-$, ^{18}O$^-$, ^{19}F$^-$, ^{31}P$^-$, and ^{35}Cl$^-$, were collected simultaneously using five different electron multipliers. Magnetic field stability was controlled using an NMR probe that ensured minimal field drift ($\sim$10 p.p.m.) during analyses.

Sarbadhikari et al. (2016) analyzed data from specific regions of interest in each lunar apatite, away from fractures, cracks and grain boundaries. It was found that water content in the two apatite grains ranged from 2200 to 2850 p.p.m. (apatite B1) and 3400–3750 p.p.m. (apatite B2). Abundances of fluorine (F) in the two grains were found to have been 2.7–3.1 wt% (B1) and 2.7–2.8 wt% (B2). The values for chlorine (Cl) content were 0.14–0.15 wt% (B1) and 0.085–0.125 wt% (B2). The dark red patches, present inconsistently from center to near the edge of the apatite grain, host water in the range of 2200–2450 p.p.m., and the yellowish-red regions represent water content of 2500–2850 p.p.m.

As indicated earlier, Boyce et al. (2014) had pointed out that the simplified approach used in previous studies of lunar apatite for inferring lunar volatile content may not be exact as the volatiles OH, F, and Cl prefer specific site occupancy rather than behaving like incompatible elements during apatite crystallization. In such a case, one expects systematic spatial variation of volatile content in individual apatite grains from core to rim and this was indeed seen in large (100–400 μm) lunar apatites (McCubbin et al., 2010; Greenwood et al., 2011). Even though minor variation in water content is present in the two small apatite grains from 15555, there is no discernible trend suggestive of volatile compositional gradation or zoning. The distributions of Cl and F in this apatite are also not suggestive of compositional zoning. The apparent high water-contents in this apatite in some narrow zones are not intrinsic and, according to Sarbadhikari et al. (2016), are artifacts due to the presence of internal cracks and/or grain edges. According to Sarbadhikari et al. (2016), the nearly uniform distribution of volatiles in the two grains makes it highly unlikely that this could result from preferred site-occupancy or exchange equilibration, and suggest that the volatiles entered into apatite structure guided primarily by their respective partition coefficients.

It may be recalled that studies of Apollo 15 and Apollo 17 lunar volcanic glasses, that sampled a deeper region ($\geq$500 km) (Delano, 1980; Longhi, 1992) of the lunar interior, relative to the mare basalts, yielded a value of 745 p.p.m. for the abundance of water in the parent magmas of these lunar volcanic glasses (Saal et al., 2008). Melt inclusions trapped in Apollo 17 volcanic glasses yielded water content in the range of 615–1410 p.p.m. (Hauri et al., 2011). These values are much higher than those for the parent magmas of the Apollo 15 mare basalt 15555 that originated at a depth of 150–200 km below the lunar surface (Walker et al., 1977). This suggests a nonuniform distribution of water and other volatiles in the lunar interior. Sarbadhikari et al. (2016) proposed that this finding needs to be taken into account in any estimation of the bulk water content of Earth's Moon.

Presence of water in lunar mantle could have significantly affected the early evolution of Earth's Moon and in particular, helped in sustaining a lunar core dynamo for an extended duration (Evans et al., 2014). High water content would also lower the solidus–liquidus temperature of the deeper mantle source of volcanic glasses in comparison to the upper mantle source of the mare basalts. This can influence a change in thermo-chemical processes, for example, differential degree of melting, in different mantle source regions during the early evolutionary stages of the Moon. It has been suggested that further studies of melt inclusions trapped in lunar volcanic glasses and in lunar basalts are needed for a better understanding of the distribution of volatiles in the lunar interior.

9.14.2 Indications of past presence of water in the lunar interior (within the lunar mantle) from lunar convective core dynamo considerations

An interesting feature about the Sun, several planets, and their moons is a magnetic field surrounding them. A kind of dynamo process is believed to be responsible for large planetary magnetic fields. Interestingly, the past several years has witnessed dramatic developments in the study of planetary magnetic fields, including a wealth of new data, together with major improvements in theoretical modeling effort of the dynamo process. Stevenson (2003) described the dynamo process supposedly leading to the generation of planetary magnetic fields thus, "These dynamos arise from thermal or compositional convection in fluid regions of large radial extent. The relevant electrical conductivities range from metallic values to values that may be only about 1% or less that of a typical metal, appropriate to ionic fluids and semiconductors. In all planets, the Coriolis force is dynamically important, but slow rotation may be more favorable for a dynamo than fast rotation. The maintenance and persistence of convection appears to be easy in gas giants and ice-rich giants, but is not assured in terrestrial planets because the quite high electrical conductivity of iron-rich cores guarantees a high thermal conductivity (through the Wiedemann–Franz law), which allows for a large core heat flow by conduction alone. In this sense, high electrical conductivity is unfavorable for a dynamo in a metallic core. Planetary dynamos mostly appear to operate with an internal field $\sim(2\rho\Omega/\sigma)^{1/2}$ where ρ is the fluid density, Ω is the planetary rotation rate, and σ is the conductivity (SI units)." Stevenson (2003) further mentioned that the presence or absence of a dynamo in a terrestrial body appears to depend mainly on the thermal histories and energy sources of these bodies, especially the convective state of the silicate mantle and the existence and history of a growing inner solid core. According to this investigator, induced fields observed in planetary bodies is suggestive of strong likelihood of water oceans in these bodies.

The recent reanalyses of Apollo-era lunar samples for magnetization have identified the likely existence of an internally generated magnetic field on the Moon between $\sim$4.2 and 3.56 Ga (Garrick-Bethell et al., 2009; Shea et al., 2012; Suavet et al., 2013). It has been noticed that the Earth's magnetism is similar to that which surrounds a circular coil of wire when direct current (DC) electricity flows through it, with the plane of the coil approximately coinciding with the Earth's equatorial plane. Scientists believe that the earth's magnetism is driven by a dynamic action involving a circular motion of ionized (i.e., electrically charged) molten rock (i.e., lava) in the earth's outer core (i.e., mantle convection) along and approximately parallel to the Earth's equatorial plane. This hypothesis (i.e., existence of a convecting liquid-iron-rich terrestrial core dynamo) supports the existence of the earth's magnetic poles in the vicinity of the earth's geographic poles. Note that earth's geographic poles are two points (one in the Arctic region and the other in the Antarctic region), that connects an imaginary line around which the earth spins.

It may be noted that the magnetic field of the Earth is primarily a dipolar field that exists in two antipodal polarity states, conventionally named as normal and reverse, which are equally likely, stable and alternates through geological time. Contrary to the Sun, where the magnetic field reverses regularly about every 11 year, the pattern of geomagnetic field reversals on Earth is aperiodic, and the rate of reversals has varied over time (see Lowrie and Kent, 2004). The duration and dynamics of geomagnetic field reversals are crucial requirements for theoretical models of the geo-dynamo, and experimental constraints on the geomagnetic reversal process can only be obtained through detailed paleomagnetic investigations in suitable stratigraphic sequences.

Recent re-examination of Apollo samples and the lunar geophysical data suggests that Earth's Moon contains at least some regions with high water content. Although it is unclear whether the measured water concentrations in the samples derived from the Moon's interior are representative of the entire lunar mantle or limited to a water-enriched reservoir (Hauri et al., 2011), the existence and subsequent enrichment of water in the lunar interior could well have had a significant effect on early lunar thermo-chemical evolution (Hirth and Kohlstedt, 1995; Shearer, 2006; Karato, 2010). In particular, water may have aided the cooling of the early Moon (Neumann et al., 1996; Hood and Zuber, 2000), in addition to influencing the expression of surface features such as impact basins.

It has been found that the mechanism of lunar magnetic field generation is the same (i.e., convecting liquid iron-rich core dynamo) as that for Earth. However, unlike the Earth (Bercovici and Karato, 2003), deep reservoirs of water within the Moon could possibly exist near the core–mantle boundary and, if so, could decrease lower mantle viscosity, possibly facilitating a prolonged and conceivably vigorous lunar dynamo (Evans et al., 2014). It has been argued that, given bounds on the lunar core size (Runcorn, 1996; Weber et al., 2011), it may be implausible for lunar mantle convection from secular cooling to generate such a long-lasting dynamo via core convection (Suavet et al., 2013; Laneuville et al., 2014) without some of the several factors such as (1) mediation by a thermal blanket (Stegman et al., 2003; Zhang et al., 2013), (2) impact-induced rotation changes (Le Bars et al., 2011), or, (3) possibly, a water-rich layer near the core–mantle boundary.

Under the lunar magma ocean model, water would be progressively enriched within compatible elements during solidification, and a fraction may have been retained by the lunar mantle (Warren and Wasson, 1979; Elkins-Tanton and Grove, 2011). Although the cooling history of the Moon is still ambiguous, there is some evidence, including recovered lunar alkalic igneous rocks, indicative of a late, rapid cooling scenario for the shallow Moon, which may have been preceded by, or overlapped with, a slow cooling phase of the deep Moon (Longhi and Ashwal, 1985; Jolliff et al., 1999; McCallumand Schwartz, 2001); such a scenario would be consistent with a convective core dynamo.

In a study of the Moon's interior water, Evans et al. (2014) anticipated that if water was transported or retained preferentially in the deep interior, it would have played a significant role in transporting heat out of the deep interior and reducing the lower mantle temperature. They addressed the influence of water on lunar evolution by incorporating an attenuating strain rate as a proxy for decreased viscosity (Hirth and Kohlstedt, 1995; Shearer, 2006; Kohlstedt, 2006) for potential wet regions in the lunar interior. Moon's interior water has two main effects (Evans et al., 2014): (1) it may change the duration of a dynamo for a homogeneous mantle by up to 0.4 Gyr allowing for a convective core dynamo to be active during the most recently recorded magnetic field sample age of 3.56 Ga and (2) in the lower mantle, it provides a mechanism to enhance the amount of heat advected away from the core, thereby strengthening the dynamo field intensity. From experimental studies of the Earth's upper mantle (300 MPa), the presence of a small amount of water (~40 wt p.p.m.) can result in a viscosity reduction by a factor in excess of 100 (Hirth and Kohlstedt, 1995; Kohlstedt, 2006).

Evans et al. (2014) examined the influence of compositional stratification and water via protonic weakening (Kohlstedt, 2006)—proton diffusion into nominally anhydrous minerals—on the deep interior. Evans et al. (2014) utilized a finite element thermo-chemical evolution model and investigated changes in temperature, modes of heat transport, core–mantle boundary heat flux, and surface magnetic field intensity. They then used the core–mantle boundary heat flux to provide constraints on the duration of core dynamo activity.

In examining the role of water on the thermal evolution of the Moon's interior, Evans et al. (2014) found that water enrichment at depth is likely to further decrease the temperature of the lower mantle over time and, depending on the water concentration and regional extent, shorten or lengthen the duration of a possible core dynamo. With a range of density and temperature profiles, the addition of water in the deep interior facilitates the transport of heat out of the deep mantle and provides a stronger core–mantle boundary (CMB) heat flux.

From model results, Evans et al. (2014) found that the existence of water within the lunar mantle can result in a CMB heat flux that is able to sustain a core dynamo beyond 3.56 Ga. Kohlstedt (2006) had found that enriched water in the lower mantle acts as a catalyst for transporting heat out of the deep mantle due to the reduced viscosity from protonic weakening. Investigations by Evans et al. (2014) indicate that, water, if enriched in the lower mantle, could have influenced core dynamo timing by over 1.0 Gyr and enhanced the vigor of a lunar core dynamo. The cases that Evans et al. (2014) examined demonstrated that with the consideration of a chemically layered mantle or water-enriched lower mantle, the core heat flux is no longer a limiting factor in sustaining a convective core dynamo even beyond the period currently indicated by the Apollo samples.

9.14.3 Identification of water in the lunar interior through remote sensing of impact crater *Bullialdus*

Magmatic lunar volatiles—evidence for water indigenous to the lunar interior—have not previously been detected remotely. Under this background history, in 2009, NASA's Moon Mineralogy Mapper (M^3), aboard the Indian Space Research Organisation's Chandrayaan-1 spacecraft fully imaged the lunar impact crater *Bullialdus*. This crater is within 25° latitude of the Moon's equator and so not in a favorable location for the solar-wind to produce significant surface water. The rocks in the central peak of this crater

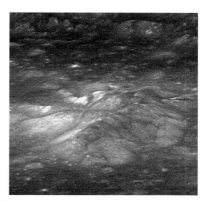

FIG. 9.20 Central peak of the lunar impact crater Bullialdus rising above the crater floor with the crater wall in the background. *Credit: NASA/GSFC/Arizona State University (Source: Johns Hopkins University Applied Physics Laboratory. "Scientists detect magmatic water on moon's surface." ScienceDaily. ScienceDaily, 27 August 2013. www.sciencedaily.com/releases/2013/08/130827091355.htm).*

are of a type called "norite" that usually crystallizes when magma ascends but gets trapped underground instead of erupting at the surface as lava. *Bullialdus* crater is not the only location where this rock type is found, but the exposure of these rocks combined with generally low regional water abundance enabled scientists to quantify the amount of internal water in these rocks.

Based on detailed analysis of spectroscopic data from M^3, Klima et al. (2013) found that the central peak of the Moon's *Bullialdus Crater* (Fig. 9.20) is significantly enhanced in hydroxyl (OH)—a molecule consisting of one oxygen atom and one hydrogen atom—relative to its surroundings. These researchers further suggested that the observed strong and localized hydroxyl absorption features are inconsistent with a surficial origin; instead, they are consistent with hydroxyl bound to magmatic minerals that were excavated from depth by the impact that formed the *Bullialdus Crater*. Furthermore, estimates of thorium concentration in the central peak based on data from the Lunar Prospector Orbiter indicated an enhancement in incompatible elements, in contrast to the compositions of water-bearing lunar samples (McCubbin et al., 2011). Klima et al. (2013) claimed that the hydroxyl-bearing material was excavated from a magmatic source that is distinct from that of samples analyzed thus far. Although there have been some measurements of internal water in lunar samples, until now this form of native lunar water has not been detected from orbit. According to Klima, the discovery represents an exciting contribution to the rapidly changing understanding of lunar water (Johns Hopkins University Applied Physics Laboratory. "Scientists detect magmatic water on moon's surface." ScienceDaily. ScienceDaily, 27 August 2013. <www.sciencedaily.com/releases/2013/08/130827091355.htm>). Because this impressive research confirms earlier lab analyses of Apollo samples, it has been suggested that this kind of research will help broaden our understanding of how this water originated and where it might exist in the lunar mantle.

9.15 Origin of water on Earth's Moon

Several researchers have attempted to understand the origin of water on Earth's Moon. In these attempts, several possibilities have been looked at. One possible source was thought to be comets. It may be noted that comets, like meteorites, are known to carry water and other volatiles, but most comets formed in the far reaches of the Solar System in a formation called the *Oort Cloud*, more than 1000 times more distant than Neptune. Because comets formed so far away from the Sun, they tend to have high deuterium/hydrogen ratios—much higher ratios than in the Moon's interior.

To find the origin of the Moon's water, scientists examined the trapped volcanic glass, referred to as a melt inclusion. The surrounding olivine crystals prevent water from escaping during an eruption, providing researchers an idea of what the inside of the Moon is like. According to the investigations carried out by Saal et al. (2013), the water found on Earth's Moon, like that on Earth, came from small meteorites called carbonaceous chondrites in the first 100 million years or so after the Solar system formed. Evidence discovered within samples of Moon's dust returned by the lunar crews of Apollo-15 and 17 dispels the theory that comets delivered the molecules.

Interestingly, it was found that the D/H ratio in the Earth's water and in the water retrieved from specks of volcanic glass trapped in crystals within the Earth's Moon dust match the ratio found in the chondrites. However, the proportions are far different from those in comet water. It may be remembered that solar wind implantation, cosmic-ray spallation, and magmatic degassing can all potentially modify the water content and D/H ratios. Although the effect of solar wind implantation is negligible in the samples studied by Saal et al. (2013), cosmic ray spallation and degassing processes might have significantly affected the water and D/H ratios of the lunar glasses.

To determine the ratios that would currently be found deep in the Moon's interior, Saal et al. (2013) modeled the loss of gasses from inside melt inclusions and the influence of degassing on the deuterium. The researchers also had to take into account the impact of cosmic rays—high-energy rays that carry charged particles—on the water trapped inside the inclusions. The interaction produces more deuterium than hydrogen. In total, the effects proved to be small for the melt inclusions, and the ratios remained consistent with those of the chondrites. In short, after consideration of cosmic-ray spallation and degassing processes, Saal et al. (2013)'s results demonstrate that lunar magmatic water (i.e., water that originates from deep within the Moon's interior) has an isotopic composition that is indistinguishable from that of the bulk water in carbonaceous chondrites and similar to that of terrestrial water, implying a common origin for the water contained in the interiors of Earth and its Moon. The carbonaceous chondrites-bearing meteorites originated in the asteroid belt near Jupiter and are thought to be among the oldest objects in the Solar System. That means the source of water on Earth's Moon is primitive meteorites. By showing that water on the Moon and Earth came from the same source, this new study offers yet more evidence that the Moon's water has been there all along, or nearly so.

It may be recalled that several observations from the United States' NASA spacecrafts and Indian Space Research Organization's (ISRO's) Chandrayan-1 lunar mission, as well as new research on samples from the Apollo missions have shown that Earth's Moon actually has water, both on its surface and beneath. Scientists have studied the amount of water within lunar rocks returned during the Apollo missions, as well as the water ice (frozen water) body trapped inside of the polar craters of the Moon.

In the context of finding the origin of water content found on Earth's Moon, it may be recalled that an interesting study carried out by Robert (2001) had shown that water on Earth is mostly derived from a meteoritic source. In fact, recent research has found that as much as 98% of the water on Earth comes from primitive meteorites. Combining these two findings and those by Saal et al. (2013) provide a clear indication that the carbon-bearing chondrites were a common source for the water in the Earth and its Moon alike, and perhaps the entire inner solar system.

It has been suggested that the easiest way to explain away the argument for a common source for water on Earth and that on its Moon is that the water was already present on the early Earth, and part of this water was transferred to the Moon. The similarity of both pre-eruptive H_2O content and magmatic D/H ratios of the lunar glasses to terrestrial magmas as shown in the study of Saal et al. (2013) supports the notion that the water from Earth and its Moon have a common origin. It has been argued that the Moon must have received its water during or shortly after its accretion, before the formation of a robust lunar lithosphere ≤100 My after the generation of the Moon.

The simplest scenario consistent with the observations of Saal et al. (2013) is that Earth was wet at the time of the Moon-forming impact, as predicted by dynamic models (Morbidelli et al., 2000; Walsh et al., 2012), and that the water was not completely lost during this event. Furthermore, Saal et al. (2013)'s results provide evidence that Earth's water budget and isotopic composition at the time of the giant impact were broadly similar to what they are today. The hydrogen isotopic similarity suggests that chemical exchange of even the most volatile elements between the molten Earth and the proto-lunar disc could have been pervasive and extensive, even at the very high temperatures expected after a giant impact. This could have been aided by the presence of a high-temperature convective atmospheric envelope surrounding Earth and the proto-lunar disc as the Moon solidified (Pahlevan et al., 2011). Alternatively, according to Saal et al. (2013), it is conceivable that a portion of the lunar interior escaped the widespread melting and degassing expected in the aftermath of a giant impact and simply inherited water from the proto-Earth. The latter alternative is consistent with the hypothesis that the Moon began with a 200- to 300-km-thick outer shell near melting conditions and a relatively cold interior (Solomon and Chaiken, 1976). This hypothesis has received support from recent gravity gradiometry (an instrument for measuring the gradient of gravity) observations by the Gravity Recovery and Interior Laboratory (GRAIL) (Andrews-Hanna et al., 2013). Any dynamic model proposed for the formation of the Earth–Moon system must meet the constraints imposed by the presence of H_2O with an isotopic composition similar to that of terrestrial water in the lunar interior. Saal et al. (2013)'s findings also have implications for the origin of water ice in permanently shadowed lunar craters, which has been attributed to solar wind implantation and to cometary and meteoritic impacts. It is conceivable that at least part of this water could have originated from magmatic degassing during lunar volcanic eruptions.

The argument that water was already present on the early Earth and was transferred to the Moon is not necessarily inconsistent with the idea that the Moon was formed by a giant impact with the early Earth, but presents a problem. If the Moon is made from material that came from Earth, it makes sense that the water in both would share a common source. However, there is still the question of how that water was able to survive such a violent collision.

These new findings suggest that Earth may have had water at the time of that impact, and some of that water might have been transferred to the Moon. This means that although the Moon's water bears a chemical signature (i.e., D/H ratio) similar to that on meteorites, the Moon's water would not have come directly from meteorites but was already present on Earth 4.5 billion years ago, when the collision/collisions, as already indicated in previous Sections in this chapter, sent material from Earth to form its Moon. By showing that water on the Moon and on Earth came from the same source, this new study offers yet more evidence that the Moon's water has been there all along. Water found in ancient Moon rocks might have actually originated from the proto-Earth and even survived the Moon-forming event.

It may be argued that the impact somehow didn't cause all the water to be lost, but we don't know what that process would be. According to researchers, there are some important processes we don't yet understand about how planets and their moons are formed. Saal et al. (2013)'s study suggests that even highly volatile elements may not be lost completely during a giant impact. Researchers would need to go back to the drawing board and discover more about what giant impacts do, and they also need a better handle on volatile inventories in the Moon.

It is now generally accepted that the water found in ancient Moon rocks might have actually originated from the proto-Earth and even survived the Moon-forming event. Several researchers investigated the amount of water present in the mineral apatite, a calcium phosphate mineral found in samples of the ancient lunar crust. Based on a plethora of such studies, the Moon, including its interior, is presently believed to be much wetter than was envisaged during the Apollo era. It has been found that the ancient lunar rocks contain appreciable amounts of water locked into the crystal structure of apatite. Researchers have also measured the hydrogen isotopic signature of the water in these lunar rocks to identify the potential source(s) for the water. Such studies confirmed several other earlier studies that the water locked into the mineral apatite in the lunar rocks has an isotopic signature very similar to that of the Earth and some carbonaceous chondrite meteorites.

It is worth noting that the Moon's chemical composition differs from Earth's. It is enriched by a factor of two to three in refractory elements (those that condense first from a high-temperature gas) such as aluminum, calcium, and titanium; however, some of the most easily vaporized (that is, volatile) elements, such as sodium and potassium, are rare (Chaussidon, 2008). With regard to water on Earth's Moon, analyses of lunar volcanic glasses show that they are rich in volatile elements and water. If parts of the lunar mantle contain as much water as Earth's, does this imply that the water has a common origin? There are still several issues, a reliable answer to which is yet to be found.

Looking from another angle, the remarkable consistency between the hydrogen composition of lunar samples and water-reservoirs of the Earth strongly suggests that there is a common origin for water in the Earth–Moon system. The study of water in Earth's Moon is still quite new, and many rocks have not yet been studied for water. Planetary geology researchers have a new set of Apollo samples from NASA that they will be studying, looking for additional clues about the early life of Earth and its Moon.

9.16 The last phase of Chandrayaan-1 mission—the lunar probe lost and found

After the successful completion of all the major mission objectives, the Chandrayaan-1 orbit has been raised to 200 km during May 2009. The spacecraft made more than 3400 orbits around Earth's Moon and the mission was concluded when the communication with the spacecraft was lost on August 29, 2009.

After suffering from several technical issues including failure of the star sensors in orbit after 9 months of operation, and poor thermal shielding, Chandrayaan-1 stopped sending radio signals at 1:30 a.m. IST on 29 August 2009. Note that the star sensor is a device used for determining the spacecraft's orientation in space. Afterward, the orientation of Chandrayaan-1 was determined using a back-up procedure using a two-axis Sun sensor and taking a "bearing" from an Earth station. This was used to update the three-axis gyroscopes, which enabled spacecraft operations. The second failure, detected on 16 May, was attributed to excessive radiation from the Sun. Shortly after these failures, the ISRO officially declared the mission over. Chandrayaan-1 operated for 312 days as opposed to the intended 2 years but the mission is claimed to have achieved 95% of its planned objectives. Among its many achievements was the discovery of the widespread presence of water molecules in lunar soil.

Surprisingly, Chandrayaan-1, which was considered lost more than 7 years after it shut down, has been located by NASA on July 2, 2016. NASA used ground-based radar systems to relocate Chandrayaan-1 in its lunar orbit (Faith, 2017; Agle, 2017). The spacecraft was found orbiting the Moon some 200 km above the lunar surface.

9.17 Discovery of subsurface empty lava tubes and caves on Earth's Moon—potential shelters for human settlement?

Among the major discoveries made by planetary missions, complex channel structures formed by a wide range of fluids such as water and lavas have proven to be surprisingly common.

9.17.1 Sinuous lunar rilles

A rille is a fissure or narrow channel on the Moon's surface. Sinuous lunar rilles are winding, meandering, long, narrow depressions/channels or valleys on the lunar surface, having many curves and turns, unlike normal rilles that are either straight or very greatly curved (large radius of curvature). Sinuous rilles on the surface of Earth's Moon fall under such complex channel structures. Since Apollo-11 lunar mission, several lava channels have been identified on the lunar surface (e.g., Oberbeck et al., 1969; Cruikshank and Wood, 1972; Schaber et al., 1976; Greeley and King, 1977; Masursky et al., 1978). Lunar rilles, once hypothesized to have formed by water action, now constitute an important lava channel type. Most unusual among lava channels is the great variety of channels produced by large-scale, low-viscosity lava eruptions.

Various theories have been advanced with regard to the origin of lunar sinuous rilles. Some scientists have proposed that the rilles represent water-eroded streams, or valleys carved by ice-melted water under a ruble-covered permafrost. The basic argument in favor of water erosion lies in the claimed morphological similarity between sinuous rilles on the Moon and terrestrial rivers, particularly with respect to meandering oxbow bends or goosenecks. Identification of gigantic flood channels in continental regions and submarine channels on the continental shelves provide new channel categories to an already rich inventory of terrestrial channels. However, Gornitz (1973) presented evidence to show that lunar rille geometry differs considerably from terrestrial rivers.

Numerous studies carried out by several researchers over more than a decade (e.g., Hulme, 1973; Wilson and Head, 1981; Coombs et al., 1987) have shown that lunar rilles formed as a result of the extrusion of hot, fluid, low-viscosity basaltic magma (see Coombs and Hawke, 1992). Studies carried out by Hulme (1973, 1982), Wilson and Head (1980, 1981), and several others have shown that lunar sinuous rilles are products of low-viscosity, high-temperature basaltic lava flows, and that these features are very similar to terrestrial basaltic lava channels and tubes.

The majority of lunar sinuous rilles occur in the maria region on the lunar near side, at mare-highland boundaries, but a few begin in the highlands. Most are concentrated around the margins of Mare Imbrium and Oceanus Procellarum (Gornitz, 1973). The origin of these channels must be explained by eruption parameters, probable high discharge rates, high temperatures, long duration, and the possible involvement of very fluid silicates or lavas of exotic composition. Most of the observed rilles seem to originate in craters or irregular depressions. Some of the rilles tend to become narrower and shallower with increasing distance from their presumed source. Sinuous rilles can be over 200 km long and 4–5 km wide, but more typical dimensions are lengths of 30–40 km and widths of less than one km (Gornitz, 1973).

9.17.2 Underground caves and lava tubes on Earth's Moon—formation processes, strength, and durability

Although sinuous rilles on the lunar surface have been identified even before the Apollo-11 mission, which succeeded in landing man on the Moon, their importance in the context of this book lies in the fact that some of the lunar rilles may have evolved into underground lava tubes when segments of the channels roofed over (e.g., Oberbeck et al., 1969; Greeley, 1971; Cruikshank and Wood, 1972; Hulme, 1973; Coombs et al., 1987). An underground lava tube may form when an active basaltic lava stream develops a continuous crust. Depending on the viscosity, temperature, supply rate, rate of flow, and the rheology of the lava (i.e., whether the lava is in liquid state, "soft solids" state, or plastic-flow-solid state), a lava tube may form by a variety of mechanisms (see Cruikshank and Wood, 1972; Greeley, 1987). Wilson and Head (1981) reckon that, under the conditions of lunar basaltic eruptions (lower gravity field, no atmosphere), such processes would have produced lunar lava channels and associated tubes at least an order of magnitude greater in size than those found on Earth.

In a study, Coombs and Hawke (1992) conducted a survey of all available Lunar Orbiter and Apollo photographs in order to locate possible intact lava tubes. In their survey, the criteria used to identify tube candidates were (1) the presence of an un-collapsed, or roofed, segment or, preferably, a series of segments along a sinuous rille; (2) the presence of un-collapsed segments between two or more elongate depressions that lie along the trend of the rille; and (3) the presence of an un-collapsed section between an irregular-shaped depression, or source vent, and the rest of the channel. Other volcanic features such as endogenic depressions, crater chains, and other types of rilles were also examined as some may be associated with a lava tube.

Coombs and Hawke (1992) estimated the maximum tube widths by projecting the walls of adjacent rille segments along the roofed-over segments. Measuring tube lengths was straightforward; direct measurements were possible from the Lunar Orbiter and Apollo photographs. However, measuring roof thickness was rather tricky; it had to be estimated following the crater-geometry argument presented by Horz (1985) whereby the largest impact crater superposed on an un-collapsed roof was assumed to yield a minimum measure of roof thickness.

The strength and durability of existing lunar lava tubes have been topics that attracted much attention among lunar researchers. In this context, questions have been raised as to whether or not the tube roofs are structurally stable enough to withstand prolonged meteoroid impact and sufficiently thick enough to provide protection from cosmic radiation (see Horz, 1985). However, it has been pointed out that several empty lava tubes remained intact during the billions of years of meteoritic bombardment and seismic shaking to which they have been subjected since their formation. Oberst and Nakamura (1988) have examined the seismic risk for a lunar base but have not dealt specifically with the effect of moonquakes on lunar lava tubes.

An interesting argument put forward as the major reason for the structural stability of underground caves and empty lava tubes is the geometric shape of their roofs. For example, studies carried out by Oberbeck et al. (1972) showed that the arc shape (a portion of the circumference of a circle or an ellipse) of the roofs of lunar underground caves and empty lava tubes is responsible for their observed structural stability.

India's Chandrayaan-1 imaged a lunar rille, formed by an ancient lunar lava flow, with an un-collapsed segment indicating the presence of a lunar lava tube, a type of large cave below the lunar surface (Arya et al., 2011). As indicated earlier, typically a rille can be up to several kilometers wide and hundreds of kilometers in length. However, the term has also been used loosely to describe similar structures on a number of planets in the Solar System, including Mars, Venus, and on a number of moons. All bear a structural resemblance to each other. Interestingly, the tunnel, which was discovered near the lunar equator, is an empty volcanic tube, measuring about 2 km (1.2 mi) in length and 360 m (1180 ft) in width. According to Arya et al. (2011), this could be a potential site for human settlement on Earth's Moon. Earlier, Japanese Lunar orbiter SELENE (Kaguya) also recorded evidence for other caves on the Moon (Nadia, 2016).

9.17.3 Underground caves and tunnel complexes as potential shelters for lunar habitats

Even before America's Apollo-11 lunar mission, the prospect of using the natural underground cavities formed by drained intact lunar tubes for housing manned lunar bases has been speculated by lunar researchers. For example, Brown and Finn (1962) evaluated the prospect of permanent lunar base, with special reference to biological problem areas. Henderson (1962) examined setting lunar housing in place for human settlement. Coombs and Hawke (1992) reported survey of lunar sinuous rilles and other volcanic features in an effort to locate intact lava tubes that could be used to house an advanced lunar base. Horz (1985) examined the possible application of lava tubes as potential shelters for habitats and noted that "the lunar lava tubes would be ideal for locating the lunar base because they (1) require little construction and enable a habitat to be placed inside with a minimal amount of building or burrowing; (2) provide natural environmental control; (3) provide protection from natural hazards (i.e., cosmic rays, meteorites and micrometeorite impacts, impact crater ejecta); and (4) provide an ideal natural storage facility for vehicles and machinery."

Coombs and Hawke (1992) considered several factors as part of the assessment of the suitability of the identified tube segments for housing an advanced lunar base. These include the usefulness of the locality (i.e., presence of scientific targets near the base site, geologic diversity, etc.), whether or not the site would be readily available (minor excavation and construction) for habitation, and its location. Other important considerations in Coombs and Hawke (1992)'s studies were the presence of potential lunar resources (e.g., basalt or pyroclastics) in the region and ease of access to these deposits. Lunar pyroclastic deposits are known to be associated with some source vents for the lunar sinuous rilles and lava tubes (Coombs et al., 1987). The black spheres that dominate some regional pyroclastic deposits are known to be rich in ilmenite (Heiken et al., 1974; Pieters et al., 1973, 1974; Adams et al., 1974). It is believed that the ilmenite-rich pyroclastics could be a source of Ti, Fe, and O. Coombs and Hawke (1992) consider that pyroclastics and regolith found in the vicinity of some of the tube candidates may be a good source for sulfur as well as other volatile elements. Coombs and Hawke (1992) further consider that "sulfur could be used as a propellant, as a fertilizer, and in industrial chemistry as suggested by Vaniman et al. (1988). The volcanic material may also be used as construction materials. Big pieces of rock may be utilized as bricks while small pyroclastic debris may be incorporated in cement compounds or broken down into individual elements." Other considerations involved in the selection of base sites included the presence of empty tube complex in a very flat region, which ensures ease of mobility, and appropriate tube widths and roof thicknesses in terms of structural stability.

Coombs and Hawke (1992) identified several advantages to using intact lunar lava tubes as the site for a manned lunar base, the most important of which is the protection provided by the natural tube roof from cosmic radiation. Furthermore, the protected area offered by the intact tubes would provide storage facilities, living quarters, and space for industrial production. The constant temperature of around −20°C (Horz, 1985) available inside the empty underground complexes is conducive to many projects and experiments, and could be altered to maintain a controlled environment for a variety of experiments as well as comfortable living conditions (Coombs and Hawke, 1992).

Coombs and Hawke (1992) reckon that "Unused or uninhabitable portions of lava tubes would also provide an additional disposal facility for solid waste products generated from the manned lunar base. Biological and industrial (i.e., mining, construction) waste may be safely discarded within these structures without diminishing the vista of the lunar surface. This method of waste disposal may provide an alternative to the crater filling, burial, or hiding-in-the-shade methods proposed by Ciesla (1988)."

The Lunar Reconnaissance Orbiter has imaged over 200 pits craters on the lunar surface. Many of these lunar pits craters show "skylights" into subsurface voids or caverns, ranging in diameter from about 5m to more than 900m, although some of these are likely to be postflow features rather than volcanic skylights. Moreover, there is observational evidence from orbiting spacecrafts to infer there are lava tubes along the Marius Hills, Hadley Rille and Mare Serenitatis regions of the Moon (Coombs and Hawke, 1992). The lunar pits are nearly 80 m deep. The temperature is expected to be a pleasantly warm weather of approx. minus 20°C to minus 25°C compared to surface temperatures reaching 130°C during the daytime and up to minus 150°C in the night-time (Horz, 1985; Díaz-Flores, A. et al., 2021). Therefore, lava tubes on the Moon (as well as on Mars), once sealed off, could be warmed up and pressurized with a breathable atmosphere (Cain, 2018).

9.17.4 Technical considerations in planning the use of lava tubes to house manned lunar bases

In a study, Coombs and Hawke (1992) pointed out the strong possibility of there being an intact open tube system on the lunar surface that could be incorporated into future plans for an advanced, manned lunar base. However, they also underpinned a major problem to consider when planning the use of a lava tube to house the first manned lunar base. That major problem is the difficulty in confirming, absolutely, that a tube does in fact exist in a particular spot and determining what its exact proportions are. Efforts are currently underway to determine a method for identifying evacuated, intact, lava tubes on Earth's Moon. According to Coombs and Hawke (1992), "such methods might include initial gravity and seismic surveys with later drilling and/or 'lunarnaut' and rover reconnaissance, or a portable radar system." Coombs and Hawke (1992) further suggested that "The construction of highly detailed geologic and topographic maps for the lunar areas also would greatly enhance the efforts to accurately determine the locations, dimensions, and existence of the lunar lava tubes. Thus, until a tube or tube system is positively identified on the lunar surface, mission planners should not rely on the presence of a lava tube when designing the first manned lunar habitat."

9.18 Lunar samples—what all tales do they tell the world about Earth's Moon?

Planetary scientists seek to understand how planets formed and evolved to their present states. Because Earth is geologically active (i.e., Earth is subject to frequent mountain-building processes, volcanic flows, earthquakes, canyon-creating processes, plate motions, erosion (by wind and water), and so on), much of the terrestrial record from the time of planet formation has been overwritten so many times that it is now hard to separate the role of ancient events from more recent ones. In contrast, many features of the Moon have been preserved from this formative era. Some can be studied remotely, but others were revealed only after the lunar samples returned during the Apollo and Luna missions were analyzed in terrestrial laboratories (see Carlson, 2019).

9.18.1 Protection of returned lunar samples from terrestrial contamination

The initial investigations of the lunar surface have been carried out based on analyses of soil samples that have been returned from the United States' Apollo landing sites and Soviet lunar stations. Note that Luna 2, originally named the Second Soviet Cosmic Rocket, was the sixth Soviet attempt to send a probe crashing into Earth's Moon. But it was the first successful attempt for any nation, making the Luna 2 probe the first human-made object to reach the surface of another celestial body. Thus, the first human-made object to touch Earth's Moon was the Soviet Union's Luna 2, on 13 September 1959. Luna 2 deposited Soviet emblems on the lunar surface. While on the lunar surface, the panoramic television system was operated. Lunar samples were obtained by means of an extendable drilling apparatus. The ascent stage carried the collected lunar samples in a sealed capsule.

The United States' Apollo 11 was the first crewed mission to land on the Moon, on 20 July 1969. Six of these sites (Apollo 11, 12, 15, and 17; Luna 16 and 24) are entirely or mainly mare sites (large meteorites-impacted darker surface areas on the Moon, initially mistaken as sea, and hence the name "maria"). However, Apollo 16 and Luna 20 landed in highland areas near maria, and Apollo 14 sampled the Fra Mauro formation, a high area of unusual chemistry surrounded by Oceanus Procellarum (Langevin and Arnold, 1977). The three Soviet lunar sample return missions (Luna 16, 20, and 24) from 1970 to 1976 brought back a total of 327 g of lunar soil. The six Apollo lunar landing missions in 1969–1972 returned 381,700 g of rock and soil. Apollo won the samples race (Crotts, 2011). Nearly all our present knowledge derives from the analyses of the samples returned from the Moon, and secondarily from observations made on the surface or from lunar orbit.

Anand et al. (2014) stated thus on the protection of returned lunar samples from terrestrial contamination: "Ever since lunar samples collected during Apollo and Luna missions were brought to Earth, they have been stored and curated in specially designed ultra-clean curatorial facilities in order to minimize any terrestrial contamination. Although it may not be possible to rule out some surface contamination, it is highly unlikely that terrestrial contamination would have been introduced inside the rock samples (and furthermore into the apatite lattice) that are curated under such strict inert conditions."

9.18.2 Fine-tuning the models for the origin of Earth's Moon

The commonly accepted view of the growth of the Earth is that it proceeded over several tens of million years by collision and accretion of Moon- to Mars-sized planetary embryos (Chambers and Wetherill, 1998). During this stage of chaotic growth, radial mixing of material from different regions of the disk took place. The analysis of lunar samples returned to Earth by the Apollo and Luna missions changed our view of the processes involved in planet formation. The data obtained on lunar samples brought to light the importance during planet growth of highly energetic collisions that lead to global-scale melting. This violent birth determines the initial structure and long-term evolution of planets (Carlson, 2019).

In a study, Zhang et al. (2012) found that the majority of lunar samples have titanium isotopic compositions identical to terrestrial samples within the level of uncertainties. However, several samples show measurable departures from terrestrial composition, with ε^{50}Ti deficits reaching -0.23 ± 0.04 for Apollo 15556 (low-Ti basalt). Leyaa et al. (2008) had found homogeneous titanium isotopic composition between the Earth and Moon but their uncertainties were up to 15 times larger than those of Zhang et al. (2012), so the negative anomalies that Zhang et al. (2012) detected could not have been resolved. Trinquier et al. (2009) measured only one lunar sample and found terrestrial isotopic composition (within ± 0.32 ε-unit using a Neptune MC-ICPMS and ± 0.13 using an Axiom MC-ICPMS). However, the variability in ε^{50}Ti values of lunar rocks documented here poses the question of the representativeness of a single sample measurement.

A model proposed recently to explain the small mass of Mars is that the terrestrial planets were formed in a narrow annulus of matter truncated at 1 AU by the inward then outward migration of Jupiter (Walsh et al., 2011). Accordingly, the proto-Earth mantle and *Theia* (the Mars size celestial body that bombarded with the proto-Earth and gave rise to the birth of Earth's Moon) may have been made from similar materials and could have shared the same isotopic compositions (Zhang et al., 2012). An argument against this idea is given by the identical tungsten isotopic compositions of lunar metals and Earth's mantle. Touboul et al. (2007) inferred late formation and prolonged differentiation of Earth's Moon from W (tungsten) isotopes in lunar metals. König et al. (2011) estimated the Earth's tungsten budget during mantle melting and crust formation. König et al. (2011)'s results indicate that interpretations of the tungsten isotope results are difficult because ^{182}W/^{184}W variations are produced by the decay of the short-lived nuclide ^{182}Hf (isotope of hafnium) and there is still significant uncertainty on the Hf/W ratios of the terrestrial and lunar mantles. Natural hafnium (^{72}Hf) consists of five stable isotopes and one very long-lived radioisotope, ^{174}Hf, with a half-life of 2×10^{15} years. In addition, there are 30 other known radionuclides, the most stable of which is ^{182}Hf (CRC Handbook of Chemistry and Physics (85th ed.). Boca Raton, Florida: CRC Press). With this caveat (limitation) in mind, it is likely that the mantles of Theia and the proto-Earth would have followed different ^{182}Hf-^{182}W paths and their tungsten isotopic compositions at the time of impact were probably different (Zhang et al., 2012).

Several investigations have explored more complicated evolution scenarios than the simple magma ocean model, including the consequences of density-driven overturn of different layers in the lunar interior after magma ocean crystallization (Ringwood and Kesson, 1976; Hess and Parmentier, 1995; Elkins Tanton et al., 2002). Nonetheless, the basic magma ocean model explains many characteristics of the lunar samples (Carlson, 2019). Later Apollo missions that landed on, or close to, the highlands returned substantial amounts of anorthosite (an igneous rock type that is known, but uncommon, on Earth).

Some mare basalts have compositions similar to terrestrial basalts, but others have distinctively high contents of titanium (Shearer et al., 2006; Neal and Taylor, 1992). These unusual lunar lava compositions most likely reflect a wide range in the composition of the rocks in the lunar interior that melted to create these lava flows (Shearer et al., 2006; Walker et al., 1975; Elardo et al., 2011). This type of compositional diversity in the lunar interior is predicted by the magma ocean model, as it fits the sequence of dense mineral accumulation that would occur as the magma ocean cooled and crystallized (Snyder et al., 1992).

9.18.3 First proposing and then disproving the "Dry Moon Paradigm"

Analyses of lunar samples have enabled us to gain some clues on the presence or otherwise of water content on the surface and interior of Earth's Moon. For nearly 40 years after return to the Earth of the first lunar samples in 1969, the Moon's interior has been considered dry. Observations of the samples brought back by Apollo 11 astronauts showed that lunar magmatic rocks were devoid of water-bearing minerals, and neither those that form when wet magmas crystallize, such as amphibole or biotite, nor those that form in response to surface weathering, such as clays, were found. In addition, small grains of metallic iron, which should have rusted rapidly if lunar rocks had ever interacted with water, were observed in many igneous rocks. Yet, some observations, such as the existence of vesicular basalts and pyroclastic deposits, nevertheless hinted for the presence of some volatiles in some portions of the lunar interior, but these were nowhere to be found. This gave rise to the "Dry Moon Paradigm" (Tartèse, 2015). However, this scenario soon changed as a result of better analyses using advanced techniques.

In a study, McCubbin et al. (2010) analyzed apatites (mineral, consisting of calcium phosphate with some fluorine, chlorine, and other elements) from three different lunar samples for hydroxyl by using secondary ion mass spectrometry (SIMS). The hydroxyl analyses were subsequently used to infer information about magmatic water contents and the water contents of the magmatic source regions (where possible). The samples investigated by McCubbin et al. (2010) span three distinct rock types, and it was found that they all have KREEP-like incompatible trace element signatures. The samples investigated included an Apollo 14 high-Al basalt, a clast-bearing impact melt rock,

and an olivine-gabbro cumulate lunar meteorite (Northwest Africa 2977). McCubbin et al. (2010) compared the magmatic water contents inferred from this study to previous investigations of water in mare volcanic materials to summarize the current state of knowledge regarding water in the lunar interior.

9.18.4 Understanding how and when large basins on the near side of the Earth's Moon were created

The density and chemical differences concerning the Earth and its Moon are accounted for by deriving the Moon from the mantle of the Impactor (Taylor, 1996). However, the discrepancy between the impact records pertaining to the Earth and its Moon in the time period, 4.0–3.5 Ga (during which the heavy and declining lunar bombardments are believed to have taken place) calls for a re-evaluation of the cause and localization of the late lunar bombardment. A relative crater count timescale was established and calibrated by radiometric dating (i.e., dating by use of radioactive decay) of rocks returned from six Apollo landing regions and three Luna landing spots. Fairly well calibrated are the periods ≈4 Ga to ≈3 Ga BP (before present) and ≈0.8 Ga BP to the present. Radiometric dating and stratigraphy (branch of geology concerned with the study of rock layers and layering) have revealed that many of the large basins on the near side of the Moon were created by impacts about 4.1–3.8 Ga ago. The lunar timescale is not only used for reconstructing lunar evolution, but it serves also as a standard for chronologies of the terrestrial planets, including Mars and possibly early Earth (Geiss and Rossi, 2013).

9.18.5 Application of Apollo lunar samples in plant biology experiments

The bioavailability of regolith for plant nutrition and its putative safety was tested following the delivery of lunar regolith. Botanical studies indicated that the lunar material from Apollo 11 and 12 outposts could provide mineral nutrients for germinating seeds, for liverworts growth, and plant tissue culture development (Walkinshaw et al., 1970; Weete and Walkinshaw, 1972; Johnston et al., 1975). The general conclusion was that the lunar rock used as a substrate to grow plants had low bioavailability and needed mineral additives. Because the amount of native lunar material is limited, terrestrial analogues can be evaluated in simulation experiments to define pioneer plant cultivation. Zaets et al. (2011) found that Ukrainian rocks from Korosten Pluton (Penizevitchi, Turchynka deposits, Zhytomyr oblast) provide a suitable test-bed for modeling biomobilization of plant-essential elements from the lunar-like rocks. The Penizevitchi anorthosite in addition to intermediate plagioclase, a low-calcium pyroxene and olivine, all contain minor quantities of ilmenite ($FeTiO_3$), orthoclase ($K[AlSi_3O_8]$), biotite ($K(Mg, Fe)_3[AlSi_3O_{10}](OH,F)_2$), and apatite ($Ca_5[PO_4]_3(F,OH, Cl)$) (Mytrokhyn et al., 2003). The Turchynka type anorthosite is composed of plagioclase, pyroxene of low-calcium content, and olivine (Mytrokhyn et al., 2003, 2008). As compared to an average composition of the lunar anorthosite, Ukrainian anorthosites contain a bit more SiO_2, Na_2O, FeO, MgO, TiO_2, and less CaO and Al_2O_3. It reflects more "alkaline" composition of plagioclase and a higher content of mafic minerals (Ashwal, 1993). There are other rock deposits on the Earth, such as the Stillwater (MA, USA) and the Skaergaard Intrusion (East Greenland) complexes, which are counterparts to lunar anorthosites (Bérczi et al., 2008).

Apart from the importance of plants in space-flight missions, plants also hold a uniquely powerful position in the historical record of lunar biology experiments. Indeed, plants were prominent test organisms during the Apollo era and were used extensively during early evaluations of the biological impact of lunar samples. As discussed below, experiments on plant interactions with lunar regolith (sandy soil on the Moon's surface layers) were key to those scientific conclusions which indicated that the returned lunar samples were not overtly dangerous to the terrestrial biosphere and that they did not contain toxic elements or harmful alien life-form contaminants. Moreover, plant experiments with lunar return samples showed that plants could potentially derive nutrients from lunar regolith (Ferl and Paul, 2010). But while those Apollo-era experiments showed that terrestrial life-forms could safely interact with the regoliths of the lunar environment and that lunar samples were safe to bring back to Earth, they did not answer all the necessary questions of plant responses to growth in lunar soils.

Ferl and Paul (2010) have provided a review of the interaction of Apollo lunar samples in plant biology experiments, with a view to collating and contextualizing the experiments of the Apollo era that placed plants directly in contact with lunar regolith. In their review, Ferl and Paul (2010) have attempted to illuminate what is actually known about the science of plants grown in lunar soils and to expose what is yet unknown. The result is hoped to be a knowledge base that would allow the next era of lunar exploration to draw appropriately from previous data while understanding the limitations of those data in the design of future experiments.

Ironically, or coincidentally, the United States' Apollo-11 lunar mission that successfully landed man on Earth's Moon on 21st July 1969 for the first time in the history of the world, and the plan of bringing to Earth the lunar samples was under active consideration by the Lunar mission planners and scientists, a techno-novel "The Andromeda Strain"—a story line dealing specifically with the return to Earth of an alien form that is devastating to life on Earth—was published in 1969 (Crichton, 1969). Its movie adaptation was released in 1971. Although the Lunar scientists might have already preplanned to take all the necessary precautions to avoid a terrible possible calamity as outlined in the novel, its publication triggered the lunar mission planners and scientists to take extra care to protect terrestrial life from the introduction of alien life-forms from the Moon.

The Lunar Receiving Laboratory (LRL) was the primary focal point for all initial biological experiments with lunar regolith and especially the returned samples from the early Apollo missions (McLane et al., 1967; Kemmerer et al., 1969; Mangus and Larsen, 2004). The need to protect terrestrial life from the introduction of alien life-forms from the Moon was strongly felt. Therefore, the biology science policy surrounding the lunar missions and the lunar samples was one of preventing what is commonly referred to as "back contamination"—the idea that samples returning from an extra-terrestrial environment such as the Moon might harbor alien organisms that are pathogenic or otherwise harmful to terrestrial life-forms, or substances that could cause mutation or pathogenesis.

9.19 China's year-2019 landmark success in placing a lander and rover on Earth's Moon's far side for scientific data collection

On 3 January 2019, the Chinese National Space Administration (CNSA) of the People's Republic of China (PRC) successfully landed its spacecraft Chang'E-4 (CE-4) on the eastern floor of Von Kármán crater within the South Pole-Aitken Basin (Fig. 9.21), the Moon's largest and oldest impact crater, becoming the first spacecraft in history to land on Earth's Moon's far side. Chang'e-4 spacecraft's soft landing at the far side of Earth's Moon inaugurated mankind's first close observation of what is also known as the "dark side of the Moon."

9.19.1 Importance of exploring lunar far side

The far side of Earth's Moon, for its special location, is of unique peculiarity that the near side cannot match. On one hand, the far side shields all kinds of radio waves emitted from the Earth, thus becomes the best place for cosmic radio spectrum detection. On the other hand, the original information of the Moon is hidden in the largest, deepest and eldest South-Pole Aitken (SPA). Such information is crucial for the study of the history, evolution, composition and components of the deep-layer of both the Moon and the Earth system.

The well-known South-Pole Aitken (SPA) basin is located in the southern part of the far side, with the central area at latitude 40°–60°S, longitude around 180°. The

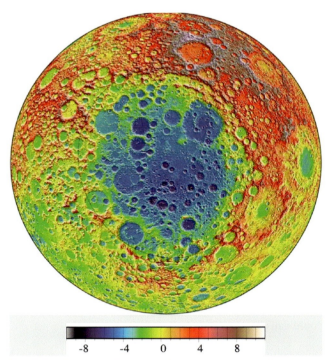

FIG. 9.21 Elevation diagram of the South-Pole Aitken Basin (scale in km). This LOLA (Lunar Orbiter Laser Altimeter) image on NASA's LRO (Lunar Reconnaissance Orbiter) mission centers on the SPA (South Pole-Aitken) basin, the largest impact basin on the Moon (diameter = 2600 km), and one of the largest impact basins in the Solar System. The distance from its depths to the tops of the highest surrounding peaks is over 15 km, almost twice the height of Mount Everest on Earth. *(Image credit: NASA/GSFC). (Source: https://directory.eoportal.org/web/eoportal/satellite-missions/c-missions/chang-e-4).*

SPA basin is of high scientific interest because it is the largest impact basin on the Moon (diameter = 2600 km), and one of the largest impact basins in the Solar System. The distance from its depths to the tops of the highest surrounding peaks is over 15 km, almost twice the height of Mount Everest on Earth. SPA is interesting for a number of reasons. To begin with, large impact events can remove surficial materials from local areas and bring material from beneath the impact craters to, or closer to, the surface. The larger the crater, the deeper the material that can be exposed. As SPA is the deepest impact basin on the Moon, more than 8 km deep, the deepest lunar crustal materials should be exposed here. In fact, the Moon's lower crust may be revealed in areas within SPA: something not found anywhere else on the Moon

How SPA is formed remains controversial and deserves further research. Soft landing on the SPA as well as roving exploration of it are of great scientific significance mainly in the following two aspects, namely; (i) Planetary formation and evolution, and (ii) Ideal observation site for low-frequency radio signals (https://directory.eoportal.org/web/eoportal/satellite-missions/c-missions/chang-e-4):

(i) Planetary formation and evolution:
- The study of SPA may benefit the discovery of material composition of Lunar crust and mantle. So, it opens an important window to the study of the deep-layer material composition of the Moon.
- SPA is a basin (its altitude is 13 km lower than its surrounding highlands) and of thin crust. Whether in the passive or active modes that bring out the Lunar mare basalt, there should have emerged large amount of basalt in SPA. However, currently obtained data cannot effectively prove that the basin has abundant basalt. On the other hand, absence of basalt may indicate something happened in the process of Lunar thermal evolution and differentiation in early times.
- Comparing the craters in SPA with the Lunar mare we can see that the degradation situation of SPA is not obvious. Also, no crater with Lunar rays has been discovered. Therefore, the formation, evolution, topography and chemical characteristics of craters in the SPA are apparently different from those of other terrains.
- Measuring roughly 2500 km in diameter and 13 km deep, the South Pole-Aitken Basin is the single-largest impact basin on the Moon and one of the largest in the Solar System (Williams, 2018). Apart from its large size, this basin is of great interest to scientists because in recent years, it has been discovered that the region contains vast amounts of water ice. These are thought to be the results of impacts by meteors and asteroids which left water ice that survived because of how the region is permanently shadowed. Without direct sunlight, water ice in these craters has not been subject to sublimation and chemical dissociation.

(ii) Ideal observation site for low-frequency radio signals (Ye et al., 2017)
- The astronomical observation of radio waves is one of the most effective methods to study and understand the Universe. At present, most portion of the spectrum has been detected, such as ultraviolet wave (in 1890s), radio wave (wavelength less than meters, in 1930s), X-ray (in 1940s), infrared and millimeter wave (in 1950s), Gamma-ray (in 1960s). But no myriametric wave (<30 MHz) has been detected yet. The detection of myriametric wave is of much importance for all-sky imaging obtained by continuous sky scanning of discrete radio source, cosmic dark times study (21 cm radiation in dark times), solar physics, space weather, extreme-high-energy cosmic ray and neutrino study (Jester and Falcke, 2009).
- Interfered by ionosphere and Earth radio waves, it is impossible to detect myriametric wave on the Earth. In earlier times, wave detection satellites are

RAE-A/B (NASA). RAE-A was launched in 1968 and operated in near-Earth orbit. Its scientific objective was to detect the intensity of cosmic ray (0.2–20 MHz). But it was interfered by radio waves in Earth orbit. RAE-B was launched in 1973 and was injected into the lunar orbit, whose scientific objective was to detect the long-wavelength radio waves (working frequency 25 kHz–13.1 MHz). It demonstrated that the lunar far side is ideal for myriametric wave detection (Alexander et al., 1975).

- At present, low-frequency radio detection was mainly achieved via spacecraft operating in circumlunar orbit by some countries but none of them has done this on the Lunar surface.

Since the 1960s, several missions have explored the SPA basin region from orbit, including the Apollo 15, 16, and 17 missions, the LRO (Lunar Reconnaissance Orbiter) and India's Chandrayaan-1 orbiter. This last mission (which was launched in 2008) also involved sending the Moon Impact Probe to the surface to trigger the release of material, which was then analyzed by the orbiter.

The exploration of Change'4 will further promote people's understanding of the far side of the Moon. With comprehensive analysis and study on the near side exploration data, more general understanding about the Earth's Moon will be obtained and the reliability of a theoretical system will be increased.

9.19.2 Establishing a relay satellite at the "halo orbit" around Lagrange point 2 of the Earth–Moon system—the first step before sending a data collection system to the far side of Earth's Moon

It is important to understand what lies on the Moon's far side, and how the far side differs (or otherwise) from the Moon's near side. However, it is impossible to establish any sort of direct communication between any point on the far side of the Moon and Earth. If some means are not established for such communications (between the far side of the Moon and Earth), placing a lander or rover on the Moon's far side will be meaningless. To overcome this technical difficulty, in preparation to placing a Lander on the Far-Side of Earth's Moon, China deployed a relay satellite, *Queqiao*, in May 2018 at a "strategically important location" above the Moon's far side (Xu, 2018; Wall, 2018). It would be curious to comprehend what the "strategically important location above the Moon's far side" is.

It may be noted that a mechanical system with three objects, say the Sun, Earth, and Earth's Moon, constitutes a three-body problem. For the Sun-Earth–Moon system, the Sun's mass is so dominant that it can be treated as a fixed object and the Earth–Moon system treated as a two-body system from the point of view of a reference frame orbiting the Sun with that system. One of the contributions of an 18th century mathematician Joseph-Louis Lagrange was to plot contours of equal gravitational potential energy for systems where the third mass was very small compared to the other two. Below is a sketch of such equipotential contours for a system like the Earth–Moon system. The equipotential contour that makes a figure-8 around both masses is important in assessing scenarios were one partner loses mass to the other. These equipotential loops form the basis for the concept of the *Roche lobe* (for details, see Chapter 1).

The 18th century mathematicians Leonhard Euler and Joseph-Louis Lagrange discovered that there are five special points in this rotating reference frame where a gravitational equilibrium could be maintained. That is, an object placed at any one of these five points in the rotating frame would stay there, with the effective forces with respect to this frame cancelling. Such an object would then orbit the Sun, maintaining the same relative position with respect to the Earth–Moon system. These five points were named Lagrange points and numbered from L_1 to L_5. Fig. 9.22 illustrates the Lagrange points of the Earth–Moon system.

More technical names for the Lagrange points are Earth–Moon Liberation (EML) points. The three-body problem is famous in both mathematics and physics circles, and mathematicians in the 1950s finally managed an elegant proof that it is impossible to solve. However, approximate solutions can be very useful, particularly when the masses of the three objects differ greatly.

From purely theoretical point of view, the Lagrange points L_1, L_2 and L_3 would not appear to be so useful because they are unstable equilibrium points. Like balancing a pencil on its point, keeping a spacecraft there is theoretically possible, but any perturbing influence will drive it out of equilibrium. However, in practice, these Lagrange points have proven to be very useful indeed because a spacecraft can be made to execute a small orbit about one of these Lagrange points with a very small expenditure of energy. In

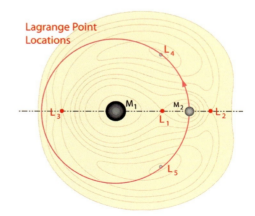

FIG. 9.22 Lagrange points of the Earth–Moon system. *(Source: http://hyperphysics.phy-astr.gsu.edu/hbase/Mechanics/lagpt.html).*

fact, surrounding each actual Lagrange point is an extended zone where a spacecraft can conveniently park itself in an orbit that requires little fuel to maintain. Therefore, they can be used by spacecraft to "hover."

When one hears about a spacecraft orbiting at a certain Lagrange point, it really means the probe is traveling within or near one of these extended, three-dimensional islands of orbits. Thus, the Lagrange points have provided useful places to "park" a spacecraft for observations. These orbits around L_1 and L_2 are often called "halo orbits." L_3 is on the opposite side of the Earth from the Moon, so is not so easy to use.

Note that a "halo" orbit (a term first advocated by the American national and NASA scientist Robert W. Farquhar in 1966) is a periodic, three-dimensional orbit near one of the L_1, L_2, or L_3 Lagrange points in the three-body problem of orbital mechanics. Although a Lagrange point is just a point in empty space, its peculiar characteristic is that it can be orbited by a Lissajous orbit or a halo orbit. These can be thought of as resulting from an interaction between the gravitational pull of the two planetary bodies and the Coriolis and centrifugal force on a spacecraft. Halo orbits exist in any three-body system, for example, the Sun–Earth–orbiting system or the Earth–Moon–orbiting system. Continuous "families" of both Northern and Southern halo orbits exist at each Lagrange point. Because halo orbits tend to be unstable, station-keeping may be required to keep a satellite on the orbit. Most satellites in halo orbit serve scientific purposes, for example as space telescopes.

The Lagrange points L_4 and L_5 constitute stable equilibrium points, so that an object (say, a spacecraft) placed there would be in a stable orbit with respect to the Earth and Moon. With small departures from L_4 or L_5, there would be an effective restoring force to bring this spacecraft back to the stable point. The L_4 and L_5 points make equilateral triangles with the Earth and Moon.

By the same logic, the Earth-Sun system also has five Lagrange points, each with an extended region where spacecraft can orbit with little fuel (https://astronomy.com/magazine/ask-astro/2013/10/lagrangian-points). As indicated earlier, of the five Lagrange points, three are unstable and two are stable. The unstable Lagrange points—labeled L_1, L_2, and L_3—lie along the line connecting the two large masses. The islands of stability get bigger farther from the Sun and also for more massive planets. The size of these islands varies. The ones associated with Earth are roughly 500,000 miles (800,000 km) wide. Each planet in the Solar System has its own Lagrange points. Scientists place probes that monitor the Sun at the L_1 point, and craft that survey the infrared and microwave sky at the L_2 point. The second Lagrange point (L_2 point) is rapidly establishing itself as a pre-eminent location for advanced space probes. The L_2 point is considered to be an orbital "sweet-spot." L_2 is located 1.5 million kilometers directly "behind"

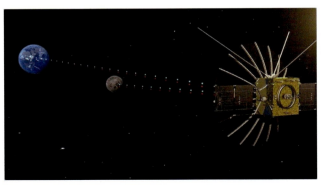

FIG. 9.23 An illustration of China's "umbrella" antenna Queqiao relay satellite located near the Lagrange point L2 of the Earth–Moon binary system, facing the far side of the Moon. *(Image credit: China National Space Administration (CNSA)/Chinese Academy of Sciences (CAS). (Source: https://directory.eoportal.org/web/eoportal/satellite-missions/c-missions/chang-e-4#foot28%29).*

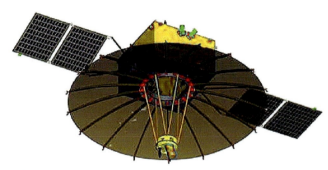

FIG. 9.24 Illustration of the deployed Chang'-4 relay satellite. *(Image credit: CAST). (Source: https://directory.eoportal.org/web/eoportal/satellite-missions/c-missions/chang-e-4#foot28%29).*

the Earth as viewed from the Sun (http://www.esa.int/Science_Exploration/Space_Science/Herschel/L2_the_second_Lagrangian_Point#:~:text=Lagrangian%20points%20are%20locations%20in,as%20viewed%20from%20the%20Sun).

It is quite clear that the Earth–Moon L_2 point is located about 65,000 km above the far side of Earth's Moon (for all practical purposes, in the range 60,000–80,000 km away; halo orbit). A relay satellite remaining at the Lagrange point L_2 or orbiting around L_2 is sufficiently far away from the far side hemisphere of the Moon (Fig. 9.23). As indicated earlier, Chang'e-4 relay satellite (Fig. 9.24) was launched in May 2018 in preparation for a later landing on the far side of the Moon. Fig. 9.25 shows the flight profile of the relay satellite, from Earth to a location near the Lagrange point L_2 of the Earth–Moon binary system. Note that the relay satellite "Queqiao," is not stationary at the Lagrange point L_2 of the Earth–Moon binary system, rather it is orbiting around the Lagrange point L_2 in the halo orbit around the L_2 point of the Earth–Moon system.

Fig. 9.26 provides a static demonstration of the Lissajous/halo orbit of the "Queqiao" Chang'e-4 relay

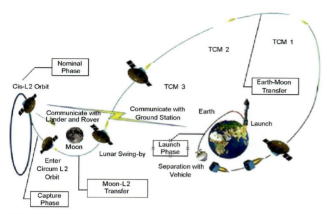

FIG. 9.25 Flight profile of the relay satellite, from Earth to a location near the Lagrange point L2 of the Earth–Moon binary system, facing the far side of Earth's Moon. *(Image credit: CAST). (Source: https://directory.eoportal.org/web/eoportal/satellite-missions/c-missions/chang-e-4#foot28%29).*

FIG. 9.26 A static demonstration of the Lissajous/halo orbit of the Queqiao Chang'e-4 relay satellite (orbiting around the Earth–Moon Lagrange point L2), which is used to communicate with the lander and rover operating from the far side of the Earth's Moon and download the data to a ground station on Earth (live demonstration is provided in the video M1_Halo-Orbit_Relay_Satellite). *(Source: https://directory.eoportal.org/web/eoportal/satellite-missions/c-missions/chang-e-4#foot28%29).*

FIG. 9.27 Artist's rendering of China's Chang'e-4 relay satellite, launched in May 2018, and lander and rover located on the lunar far side. *(Image credit: CAS). (Source: https://directory.eoportal.org/web/eoportal/satellite-missions/c-missions/chang-e-4#foot28%29).*

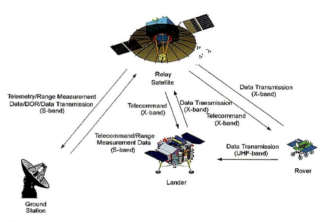

FIG. 9.28 Relay communication link profile of China's Chang'e-4 communication system in its totality, including telecommunication between the lander and rover operating from the lunar far side. *(Image credit: CNSA, CAST). (Source: https://directory.eoportal.org/web/eoportal/satellite-missions/c-missions/chang-e-4#foot28%29).*

satellite orbiting around the Lagrange point L_2, located above and facing the far side of Earth's Moon), which is used to communicate with the lander and rover operating from the far side of the Earth's Moon and download the data to a ground station on Earth (live demonstration is provided in the Video M1_Halo-Orbit_Relay_Satellite). Fig. 9.27 shows artist's rendering of China's Chang'e-4 relay satellite, launched in May 2018, and lander and rover located on the lunar far side. Fig. 9.28 shows relay communication link profile of China's Chang'e-4 communication system in its totality, including telecommunication between the lander and rover operating from the lunar far side.

Because the far side of Earth's Moon never faces the Earth, communications with the spacecraft is facilitated by the "Queqiao" relay satellite launched in May 2018, and inserted into a "halo" orbit around the second Earth–Moon Lagrange point in June 2018. From this vantage point between 65,000–85,000 km beyond the Moon, the Queqiao relay satellite achieves constant line-of-sight with both the Chang'e-4 spacecraft and Chinese ground stations in China, at Kashi and Jiamusi, Namibia and Argentina.

With its special environmental and complex geological history, the far side of Earth's Moon is a hot spot for scientific and space exploration. The Chinese Chang'e-4 mission to the hidden side (far side) of Earth's Moon has generated quite a lot of interest globally. China had planned the mission well by overcoming the obstacle to communicate from

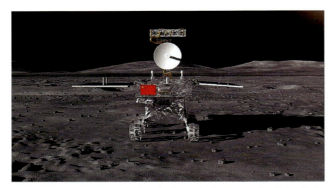

FIG. 9.29 Illustration of China's Chang'e-4 rover in the South Pole-Aitken Basin on the far side of the moon. *(Image credit: CNSA, CASC, Ref. 10). (Source: https://directory.eoportal.org/web/eoportal/satellite-missions/c-missions/chang-e-4).*

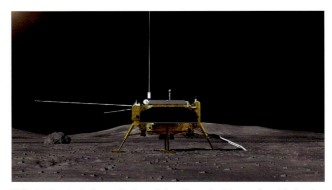

FIG. 9.30 Artist's rendering of the Chang'e-4 lunar far side lander, released in August 2018. *(Image credit: CNSA, CASC).* Visible on the newly released image of the lander are the antennas for the LFS (Low Frequency Spectrometer), which will take advantage of the uniquely quiet electromagnetic environment offered by the far side of the moon. *(Source: https://directory.eoportal.org/web/eoportal/satellite-missions/c-missions/chang-e-4).*

the far side (hidden side) of the Moon by sending a relay satellite for Chang'e-4 to the "halo orbit" of the Earth–Moon Lagrange Point L_2, and then launch the Chang'e-4 lunar lander and rover to the Aitken Basin of the south pole region later (launching a relay satellite took place in May 2018). From the Earth–Moon Lagrange point 2 (L_2 point) it is easy to connect both ways to Earth and the hidden side of the Moon.

9.19.3 The Chang'e-4 mission lander and rover

The Chang'e-4 mission consists of a relay orbiter, a lander and a rover, the primary purpose of which is to explore the geology of the South Pole-Aitken Basin. Chang'e-4 is the backup to the Chang'e-3 mission which put a lander and rover on Mare Imbrium in late 2013. The relay satellite "Queqiao" is responsible for relaying signals between Chang'e-4 and ground control. Following that success, the lunar craft have been repurposed for a pioneering landing on the Moon's far side to communicate via the Chang'e-4 Relay (Queqiao) satellite (Ye et al., 2017). Fig. 9.29 is an illustration of China's Chang'e-4 rover in the South Pole-Aitken Basin on the far side of the Moon. Designed and built by China Aerospace Science and Technology Corp (CAST), Yutu 2 has six wheels, two foldable solar panels, and a radar dish and weighs 140 kg. It is 1.5 m long, 1 m wide and 1.1 m high. The launch mass of Chang'e-4 spacecraft is ~3780 kg; the lander has a mass of ~1200 kg, while the rover has a mass of 140 kg. The rover (Fig. 9.29) is capable of avoiding some obstacles. If there are obstacles in front of it, it can stop and plan a new route on its own. It can also climb some slopes and cross some rocks. Fig. 9.30 is an artist's rendering of the Chang'e-4 lunar far side lander, released in August 2018.

The four scientific instruments mounted on the Yutu 2 rover for the purpose of conducting scientific investigations are: a panoramic camera, a lunar penetrating radar, a visible and near-infrared imaging spectrometer and the Swedish-developed Advanced Small Analyzer for Neutrals.

Chang'e-4's lander and the rover are programmed to operate only during the lunar day (2 weeks on Earth); they are in a dormant state during the lunar night (2 weeks on Earth), when the temperature falls below −180°C, and there is no sunlight to provide power to the probe. The data collected during the working session will be used to help scientists reveal the origins of substances beneath the South Pole-Aitken Basin.

Note that as a result of the tidal locking effect, Earth's Moon's revolution cycle is the same as its rotation cycle, and the same side of the Moon always faces Earth. This is the reason that a lunar day equals 2 weeks on Earth, and a lunar night is the same length. The Chang'e-4 probe switched to dormant mode during the lunar night due to a lack of solar power. The rover and the lander carried a radioisotope heat source, which helped keep the probe warm during the lunar night. The lander was also equipped with an isotope thermoelectric cell and dozens of temperature data collectors to measure the temperatures on the surface of the Moon during the lunar night. Used for the first time in a Chinese spacecraft, the isotope thermoelectric generation technology to transform heat into power on Chang'e-4 is a prototype for future deep-space exploration. NASA's Curiosity rover also adopts this power technology, freeing it from the sunshine, sand and dust restrictions that have affected its predecessors Opportunity and Spirit. It is a technology that we must master if we want to go to the Moon's polar regions or farther than Jupiter into deep space, where solar power cannot be used as the primary power source.

Great significance is attached to soft-landing exploration on the Lunar far side because at Lunar far side, numerous highland terrains are distributed all over, including the highest peak, which is up to 10.9 km in height. In the highland area, craters and mountains are widely spread. Chang'e-4, which remains operational, has enabled scientists to learn more about Earth's Moon, deepening knowledge of the early history of the extraterrestrial body and the Solar System. The mission's Yutu 2 rover, the second of its kind made by China and the world's first to reach the far side, has become the second-longest operational rover on Earth's Moon. China's first lunar rover, which worked on the Moon for 972 days, far outliving its designed life span of 3 months. Yutu 2 broke the previous record for the longest operational lunar rover—321 Earth days—which was held for more than 43 years by the former Soviet Union's Lunokhod 1. It was carried by the Luna 17 spacecraft and landed on the Moon on November 17, 1970.

9.19.4 Unveiling of Earth's Moon's far side shallow subsurface structure by China's Chang'E-4 lunar penetrating radar

Chang'e-4 (CE-4)'s successful autonomous landing on the far side of Earth's Moon was one of the most technologically impressive feats China's space program has accomplished to date, given the uneven landscape in the target location and the need for a relay satellite to facilitate communication with Earth. When countries raced to the Earth's Moon, China decided to explore a section that has been left ignored for decades- the far side (the side not visible to Earth). The Chinese National Space Administration (CNSA) in January 2019 successfully landed its probe Chang'e-4, named after the goddess of the Moon, on the side that always faces away from Earth. The second successful landing came on the shoulders of Chang'e-3 probe that had reached the lunar surface on December 14, 2013. While China also landed its third probe Chang'e-5 in December 2020 on the near side of the Moon, the probe on the far side is still transmitting, exceeding its expected 3 months operational period on Earth's Moon. The Chang'e-5, which landed on Mons Rumker area of the huge volcanic plain Oceanus Procellarum, known as the "Ocean of Storms" created a new record by returning a sample from the lunar surface for the first time since 1976. The last lunar sample was collected by Soviet Union's Luna mission.

The CE-4 Lunar Penetrating Radar (LPR) is a dual-frequency Ground Penetrating Radar (GPR) system, operating at 60 MHz (low frequency) and 500 MHz (high frequency), with a frequency bandwidth of 40–80 MHz and 250–750 MHz, respectively (Fang et al., 2014; Su et al., 2014). The two monopole antennas have a length of 1.15 m and are spaced 0.8 m from each other. These antennas are installed in the back of the rover and suspended 0.6 m above the ground. The high-frequency system is equipped with one transmitting and two receiving bowtie antennas (CH2A and CH2B), with a nominal central frequency of 500 MHz and 250–750-MHz bandwidth. The antennas are located at the bottom of the Yutu-2 rover, ~0.3 m above the ground (contactless), and are separated 0.16 m from each other (Fang et al., 2014). LPR data were collected along the radar route, while the rover was traveling. The time intervals between two adjacent traces are 1.536 s at 60 MHz and 0.6636 s at 500 MHz. Because of the strong coupling of the antennas with both the metallic rover and the lunar surface, which generates severe disturbances on the transmitted and received signals, low-frequency data have not been analyzed at this stage, as they need to be properly calibrated and accurately checked before confirming their reliability (Li et al., 2018). Furthermore, because the first receiving antenna (CH2A) in the high-frequency system is affected by a strong cross-talk, only the data collected by the second antenna (CH2B) have been analyzed. The high-frequency raw data were found to be of good quality and were processed according to the standard GPR procedure.

There are four science instruments each on the lander and the rover. The instruments on lander include the landing camera, the terrain camera and the low frequency spectrometer. The fourth one, a lunar lander neutrons and dosimetry was provided by Germany. The rover has instruments such as a panoramic camera, the lunar penetrating radar, the visible and near-infrared imaging spectrometer, and the advanced small analyzer for neutrals, contributed by Sweden.

Li et al. (2020) have reported the observations made by the dual-frequency Lunar Penetrating Radar (LPR) onboard the Yutu-2 rover during the first two lunar days. They processed and interpreted only the high-frequency LPR data. As in the CE-3 mission, the low-frequency antennas are installed on the back of the CE-4 rover; unfortunately, this leads to electromagnetic coupling with the rover's metallic body (Angelopoulos et al., 2014), resulting in strong disturbances that largely overlap the signals coming from the subsurface. These disturbances could easily be misinterpreted as subsurface reflectors (Li et al., 2018), and thus, more data should be acquired to effectively mitigate/remove these artifacts from the radargram.

The overall terrain of Von Kármán crater, where the Yutu-2 rover landed, is relatively flat, and its floor was flooded by Imbrian-aged mare basalts that appear dark in optical images (Pasckert et al., 2018). Secondary craters are widespread over the floor of Von Kármán, as indicated by their spatial patterns (e.g., chains or clusters with herringbone-shaped morphology) (Xiao, 2016; Huang et al., 2018). The CE-4 spacecraft landed on the bright distal ejecta from surrounding large craters, in an area with a low abundance of rocks and boulders on the surface. Li et al. (2020) recall that "The stratigraphic architecture of the lunar subsurface is the result of a complex and long process of emplacement and modification of the deposited materials (Hiesinger and Head, 2006). Such architecture can be inferred from remote sensing observations (Huang et al., 2018) or can be directly imaged using geophysical techniques such as seismic or electromagnetic wave prospections (Reynolds, 2011). Both methods were successfully tested during the Apollo missions, providing ground-breaking information about the Lunar interior. In the past 20 years, spaceborne/

rover-deployed Ground Penetrating Radar (GPR) has progressively become the most suitable geophysical technique (tool) to investigate planetary subsurface stratigraphy."

To convert the two-way travel time into depth, an estimation of the electromagnetic wave velocity is necessary, which Li et al. (2020) calculated by applying the hyperbola calibration method. Using seven well-defined point-source hyperbolic events, which are only present within the first section (between 30 and 90 ns), they computed the average wave velocity at different time delays. Such velocity progressively decreases and reaches a constant value, $v = 0.16$ m/ns, at about 70 ns. Given the lack of any other information, Li et al. (2020) assumed that this velocity does not appreciably change with depth and that it does not vary across the entire radar section. On the basis of this value, the thickness of the homogeneous material at the top of the radar section (0–150 ns) is about 12 m, whereas the underlying material (150–500 ns) is about 28 m thick. Li et al. (2020) used the set of retrieved velocity values to estimate the material bulk density in the first 7 m of the radar section.

Li et al. (2020) found a ($FeO + TiO_2$) content of $9 \pm 4\%$ and $11 \pm 4\%$, in good agreement with the range, 11.4–15.9%, measured by Kaguya Multiband Imager (MI) (Ohtake et al., 2008) at this site. This compatibility is quite notable, as the two techniques probe different portions of the lunar regolith: Kaguya data refer to the surface material, whereas LPR data refer to the large volume of material located to a depth of 6 m.

Li et al. (2020) described the salient geological features of the study area thus, "The shallower unit (depth interval, 0–12 m), unit 1, looks rather uniform, with sporadic large rocks distributed along the radar section. The underlying unit (depth interval, 12–24 m), unit 2, is composed of an upper part containing a large amount of randomly (almost evenly) distributed 0.2–1-m-wide rocks and a deeper part that is much more inhomogeneous in terms of both rock distribution and size. In the latter, three blocky regions, with boulder dimensions of 1–3, 0.3–1, and 1 m, alternate with zones of fine materials. In the bottom unit (24–40 m), unit 3, the boulder density distribution decreases considerably, and the rocky material is essentially localized on the top of this interval. Below 28 m, a lens of rather uniform and transparent material is clearly visible; this lens is probably made of very fine-grained material with no large rocks. Beneath this, there is a zone with relative uniform distribution of small rocks overlaying a completely transparent layer and thus again probably made of very fine materials."

Combining the information provided by the radargram, the tomographic image, and the quantitative analysis (estimated wave velocity and loss tangent), Li et al. (2020) concluded thus, "the subsurface internal structure at the landing site is essentially made by low-loss, highly porous granular materials embedding boulders of different sizes. Given such a strong geological constraint, the most plausible interpretation is that the sequence is made of a layer of regolith

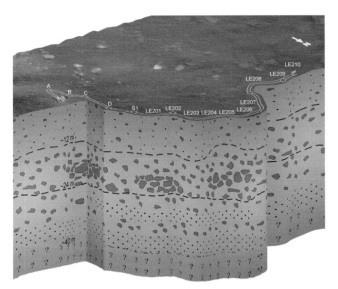

FIG. 9.31 Schematic representation of the subsurface geological structure at the CE-4 landing site inferred from Lunar Penetrating Radar (LPR) observations. *(Image credit: CLEP/CRAS/NAOC) (Sources: Li, C. et al. (2020), The Moon's farside shallow subsurface structure unveiled by Chang'E-4 Lunar Penetrating Radar, Sci. Adv., 6 (9): eaay6898. doi: http://doi.org/10.1126/sciadv.aay6898, https://directory.eoportal.org/web/eoportal/satellite-missions/c-missions/chang-e-4#foot28%29).*

overlaying a sequence of ejecta deposits from various craters (Fig. 9.31), which progressively accumulated after the emplacement of the mare basalts on the floor of Von Kármán crater. It has been found that the subsurface can be divided into three units: Unit 1 (up to 12 m) consists of lunar regolith; unit 2 (depth range, 12–24 m) consists of coarser materials with embedded rocks; and unit 3 (depth range, 24–40 m) contains alternating layers of coarse and fine materials. The layer of regolith (unit 1) is rather homogenous both laterally and vertically, and is mostly composed of fine materials. It developed from the uppermost portion of the ejecta deposits, which were thicker than 12 m and were delivered to this area by multiple impact craters, mostly Finsen, Von Kármán L, and Von Kármán L' craters (Huang et al., 2018). Unit 2 (depth, 12–24 m) is characterized by large rocks and boulders that are interbedded with thin layers of fine materials."

According to Li et al. (2020), ejecta deposition never occurs via simple carpeting, but it is accompanied by horizontal shearing and mixing, excavation, and subsurface structural disturbances (Melosh, 1989), especially in areas beyond the continuous ejecta deposits of the source crater. Fine materials can be produced by the internal shearing of ejecta deposits, which reduces particle sizes, and the fine layers can also be formed due to regolith development during the intervals between different impact events. Li et al. (2020) reckon that Unit 3 (24–40 m), which contains alternating layers of coarse and fine materials, can be interpreted as a combination of ejecta deposits, which were delivered by various craters older than Finsen and regolith developed during the impact intervals. Finsen is an Eratothenian-aged crater (<3.2 Ga), and many older craters have contributed to deliver ejecta deposits on the crater floor of Von Kármán before that age. The lack of a detectable radar signal below 40 m did not allow Li et al. (2020) to make any definite conclusion about the properties of the materials at the base of the ejecta, although we can speculate that the granular materials extend beneath such a depth; accordingly, the upper contact of the mare basal layer must be deeper than 40 m.

The CE-4 Lunar Penetrating Radar (LPR) images provide clear information about the structure of the subsurface, which is primarily made of low-loss, highly porous, granular materials with embedded boulders of different sizes; the images also indicate that the top of the mare basal layer should be deeper than 40 m. These results represent the first high-resolution image of a lunar ejecta sequence ever produced and the first direct measurement of its thickness and internal architecture. Li et al. (2020)'s study shows that the extensive use of the LPR could greatly improve our understanding of the history of lunar impact and volcanism and could shed new light on the comprehension of the geological evolution of Earth Moon's far side.

9.20 Cotton seeds carried to Earth's Moon by China's Chang'e-4 probe—the first-ever to sprout on Earth's nearest neighbor

Although astronauts have cultivated plants on the International Space Station, and rice and Arabidopsis were grown on China's Tiangong-2 space lab, those experiments were conducted in low-Earth orbit, at an altitude of about 400 km. The environment on Earth's Moon, 380,000 km from Earth, is more complex. None had such experience before, and simulating the lunar environment, such as microgravity and cosmic radiation, on Earth, was difficult.

An experiment that saw the first-ever plant sprouting on Earth's Moon in January 2019 was born in a natural disaster that devastated China's cotton-industry almost three decades ago. The seed was one of the best varieties developed by the Institute of Cotton Research (ICR) of the Chinese Academy of Agricultural Sciences. After making the first-ever soft landing on the far side of Earth's Moon, China's Chang'e-4 mission pioneered the first mini biosphere experiment on the Moon. One of the cotton seeds carried to the Moon by China's Chang'e-4 probe is the first ever to sprout on the Moon, according to scientists of a mini biosphere experiment ("Moon sees first cotton-seed sprout," Space Daily, 16 January 2019,http://www.spacedaily.com/reports/Moon_sees_first_cotton_seed_sprout_999.html;

"Moon sees first cotton-seed sprout," XinhuaNet, 16 January 2019, URL: http://www.xinhuanet.com/english/2019-01/15/c_137745432.htm)

A canister on the Chang'e-4 lander contained seeds of cotton, rapeseed, potato and Arabidopsis (a small flowering plant related to cabbage and mustard), as well as fruit fly eggs and some yeast, to form a simple mini biosphere. The reason for choosing these species specifically was that potatoes could be a major source of food for future space travelers. The growth period of Arabidopsis is short and easy to observe. Yeast could play a role in regulating carbon dioxide and oxygen in the mini biosphere, and the fruit fly would be the consumer of the photosynthesis process. Researchers used biological technology to render the seeds and eggs dormant during the 2 months when the probe went through the final checks in the launch center and journey of more than 20 days through space. After Chang'e-4 landed on the far side of the Moon on 3 January 2019, the ground control center instructed the probe to water the plants to start the growing process. A tube directs natural light on the surface of the Moon into the canister to allow the plants to grow.

The cylinder canister, made from special aluminum alloy materials, is 198 mm tall, with a diameter of 173 mm and a weight of 2.6 kg. It also holds water, soil, air, two small cameras and a heat control system. More than 170 pictures have been taken by the cameras and sent back to Earth. Images from the probe showed that only a cotton sprout was growing. Although the sprout couldn't survive the extremely cold lunar night, Li, head of ICR, believes it could help acquire knowledge for building a base and long-term residence on Earth's Moon. The cotton seeds were selected for the experiment because of their outstanding performance on Earth. The seeds belonged to a transgenic insect-resistant cotton variety developed in China and named CCRI 41. On Jan. 12, 2019, Chang'e 4 sent back to Earth the last photo of the bio test load showing that tender shoots have come out and the plants are growing well inside the sealed test can (Fig. 9.32). It is the first time that humans conducted a biological growth and cultivation experiment on the surface of the Moon. Unfortunately, on February 1, 2019, China's Chang'e 4 probe, having made the first-ever soft landing on Earth's Moon's far side, found

FIG. 9.32 At 8 p.m. on Jan. 12, 2019, Chang'e 4 sent back to Earth the last photo of the bio test load showing that tender shoots have come out and the plants are growing well inside the sealed test can. It is the first-time humans conducted a biological growth and cultivation experiment on the surface of Earth's Moon. *(Image credit: China Daily).* *(Source: https://directory.eoportal.org/web/eoportal/satellite-missions/c-missions/chang-e-4).*

that the temperature of the lunar surface dropped to as low as minus 190°C, colder than expected. This was the first time Chinese scientists have received first-hand data about the temperatures on the surface of the Moon during the lunar night. The experiment has ended. The organisms will gradually decompose in the totally enclosed canister, and will not affect the lunar environment, according to CNSA (China National Space Administration).

References

Adams, B., Pieters, C.M., McCord, T.B., 1974. Orange glass: evidence for regional deposits of pyroclastic origin on the Moon, Proc. Lunar Sci. Con/5th, 171–186.

Agle, D.C. (2017), New NASA radar technique finds lost Lunar spacecraft, NASA. Retrieved 10 March 2017.

Albaréde, F., 2009. Volatile accretion history of the terrestrial planets and dynamic implications. Nature 461, 1227–1233.

Alexander, J.K., Kaiser, M.L., Novaco, C., Grena, F.R., Weber, R.R., 1975. Scientific instrumentation of the Radio-Astronomy-Explorer-2 satellite. NASA. URL. https://ntrs.nasa.gov/archive/nasa/casi.ntrs.nasa.gov/19750010212.pdf.

Alexander, C.M.O., Bowden, R., Fogel, M.L., Howard, K.T., Herd, C.D.K., Nittler, L.R., 2012. The provenances of asteroids, and their contributions to the volatile inventories of the terrestrial planets. Science 337, 721–723.

Anand, M., Tartèse, R., Barnes, J.J., 2014. Understanding the origin and evolution of water in the Moon through lunar sample studies. Philos Trans A Math Phys Eng Sci 372 (2024), 20130254. doi:10.1098/rsta.2013.0254 (33 pages).

Andrews-Hanna, J.C., et al., 2013, Ancient igneous intrusions and early expansion of the Moon revealed by GRAIL gravity gradiometry. Science 339, 675–678. DOI: 10.1126/science.1231753.

Angelopoulos, M., Redman, D., Pollard, W.H., Haltigin, T.W., Dietrich, P., 2014. Lunar ground penetrating radar: minimizing potential data artifacts caused by signal interaction with a rover body. Adv. Space Res. 54, 2059–2072.

Arlin, C., 2011. Water on the Moon, I. Historical overview. Astron. Rev. 6 (8), 4–20.

Arnold, J.R., 1979. Ice in the lunar polar regions. J. Geophys. Res. 84, 5659–5668. http://doi.org/10.1029/JB084iB10p05659.

Arya, A.S., Rajasekhar, R.P., Thangjam, G., Ajai, Kiran Kumar, A.S., 2011. Detection of potential site for future human habitability on the Moon using Chandrayaan-1 data. Curr. Sci. 100 (4): 524–529.

Arya, A.S., Moorthi, S.M., Rajasekhar, R.P., Sarkar, S.S., Sur, K., Aravind, B., Gambhir, R.K., Misra, I., Patel, V.D., Srinivas, A.R., K.K.Patel, P.C, Kiran Kumar, A.S., 2015. Indian Mars-Colour-Camera captures far-side of the Deimos: a rarity among contemporary Mars orbiters. Planet. Space Sci. 117, 470–474.

Ashwal, L.D., 1993. Anorthosites. Springer, Berlin.

Azoulay, J., 2019. A proposed moon formation theory: the multiple-impact hypothesis. Astrobites. https://astrobites.org/2019/08/12/a-proposed-moon-formation-theory-the-multiple-impact-hypothesis/.

Bercovici, D., Karato, S.-I., 2003. Whole-mantle convection and the transition-zone water filter. Nature 425 (6953), 39–44.

Berczi, Sz., Jo´zsa, S., Szakma´ny, Gy., Gucsik, A., Hargitai, H., Kereszturi, A., Nagy, Sz, 2008. Studies of Solar System cumulate rocks from NASA lunar set and NIPR Martian meteorites (Abstract)European Planetary Science Congress3. Mu¨nster, Germany EPSC2008-A-002722008.

Bhandari, N., 2005. Chandrayaan-1: science goals. J. Earth Syst. Sci. 114, 699.

Bhattacharya, S., Saran, S., Dagar, A., Chauhan, P., Chauhan, M., Ajai, Kiran Kumar, A.S., 2013. Endogenic water on the Moon associated with non-mare silicic volcanism: implications for hydrated lunar interior. Curr. Sci. 105, 685–691.

Bolfan-Casanova, N., 2005. Water in the Earth's mantle. Mineral Mag. 69, 229–257.

Boyce, J.W., Liu, Y., Rossman, G.R., Guan, Y., Eiler, J.M., Stolper, E.M., Taylor, L.A., 2010. Lunar apatite with terrestrial volatile abundances. Nature 466, 466–469.

Boyce, J.W., Tomlinson, S.M., McCubbin, F.M., Greenwood, J.P., Treiman, A.H., 2014. The Lunar apatite paradox. Science 344, 400–402.

Brown, O.D.R., and Finn, J.C. (1962), The permanent lunar base—determination of biological problem areas. In: Aerospace• Medical Div., Aerospace Medical Research Labs./6570TH Wright-Patterson AFB, OH, Biologists for Space Systems Symposium, Final Report, pp. 233–241.

Brush, S.G., 1984. Early history of selenogony 1984. In: Hartmann, W.K., Phillips, R.J., Taylor, G.J. (Eds.), Origin of the Moon, Proceedings of the Conference "Origin of the Moon" Held in Kona, HI, October 13-16, 1984. Conference Supported by NASA. Houston, TX, 797 Lunar and Planetary Institute1986.

Bumsted, M.P., 1981. Research Report 20: Biocultural Adaptation Comprehensive Approaches to Skeletal Analysis. Paper 12. http://scholarworks.umass.edu/anthro_res_rpt20/12.

Bussey, D.B.J., Fristad, K.E., Schenk, P.M., Robinson, M.S., Spudis, P.D., 2005. Constant illumination at the lunar north pole. Nature 434, 842. http://doi.org/10.1038/434842a.

Bussey, D.B.J., Neish, C.D., Spudis, P., Marshall, W., Thomson, B.J., Patterson, G.W., Carter, L.M., 2011. The nature of lunar volatiles as revealed by Mini-RF observations of the LCROSS impact site. J. Geophys. Res.: Planets (E01005), 116. http://doi.org/10.1029/2010JE003647.

Butler, B.J., Muhleman, D.O., Slade, M.A., 1993. Mercury: full-disk images and the detection and stability of ice at the north pole. J. Geophys. Res. 98. 15,003–15,023. http://doi.org/10.1029/93JE01581.

Butler, B.J., 1997. The migration of volatiles on the surfaces of Mercury and the Moon. J. Geophys. Res. 102 (E8), 19283.

Cain, F., 2018. Living underground on other worlds: exploring lava tubes. Universe Today. (available at. https://www.universetoday.com/139021/living_underground_exploring_lava_tubes/.

Carlson, R.W., 2019. Analysis of lunar samples: Implications for planet formation and evolution. Science 365 (6450), 240–243. doi:10.1126/science.aaw7580.

Chambers, J.E., Wetherill, G.W., 1998. Making the terrestrial planets: N-body integrations of planetary embryos in three dimensions. Icarus 136, 304–327.

Chaussidon, M., 2008. Planetary science: the early Moon was rich in water. Nature 454 (7201), 170–172. http://doi.org/10.1038/454170a.

Ciesla, T.M., 1988. The problem of trash on the Moon (abstract), Papers Presented to the 1988 Symposium on Lunar Bases and Space Activities of the 21st Century. Lunar and Planetary Institute, Houston, p. 54.

Clark, R.N., 2009. Detection of adsorbed water and hydroxyl on the Moon. Science 326 (5952), 562–564. http://doi.org/10.1126/science.1178105.

Colaprete, A., Schultz, P., Heldmann, J., Wooden, D., et al., 2010. Detection of water in the LCROSS ejecta plume. Science 330, 463–468.

Coombs, C.R., Hawke, B.R, Wilson, L., 1987. Geologic and remote sensing studies of Rima Mozan, Proc. Lunar Planet. Sci. Conf. 18th, 339–354.

Coombs, C.R., Hawke, B.R., 1992. A search for intact lava tubes on the Moon: possible lunar base habitats. In: Mendell, W.W. (Ed.), The Second Conference on Lunar Bases and Space Activities of the 21st Century, 3166. NASA Conferences Publication, pp. 219–229 (Parts 1–4) & 2 (Parts 5–8).

Crichton, M., 1969. The Andromeda Strain. Arrow Books, London, pp. 327 pages.

Crider, D.H., Vondrak, R.R., 2000. The solar wind as a possible source of lunar polar hydrogen deposits. J. Geophys. Res. 105 (E11), 26773–26782.

Crotts, A., 2011. Water on the Moon, I. Historical overview. Astron. Rev. 6 (7), 4–20. http://doi.org/10.1080/21672857.2011.11519687.

Cruikshank, D.P., Wood, C.A., 1972. Lunar rilles and Hawaiian volcanic features: possible analogues. Moon 3, 412–447.

Ćuk, M., Stewart, S.T., 2012. Making the Moon from a fast-spinning Earth: A giant impact followed by resonant despinning. Science 338 (6110), 1047–1052. doi:10.1126/science.1225542.

Dana, J.D., 1846. On the volcanoes of the Moon. Am. J. Sci. Arts 11, 335.

Darwin, G.H., 1898. The Tides. W.H. Freeman and Company, San Francisco and London, p. 378.

Dauphas, N., 2017. The isotopic nature of the Earth's accreting material through time. Nature 541, 521–524.

Delano, J.W., 1980. Chemistry and liquidus phase relations of Apollo 15 red glass: implications for the deep lunar interior. Proc. Lunar Planet Sci. Conf. 11, 251–288.

Descartes, R. (1664), *Le Monde, ou Traite de la Lumiere*. Paris. Reprinted in Descartes (1824, vol. 4, pp. 215–332)

Díaz-Flores, A., et al., 2021. Lunar pits and lava tubes for a modern ark, 2021 IEEE Aerospace Conference. MT, USA. Big Sky.

Dixon, J.E., Leist, L., Langmuir, C., Schilling, J.-G., 2002. Recycled dehydrated lithosphere observed in plume-influenced mid-ocean-ridge basalts. Nature 420, 385.

Eke, V.R., Teodoro, L.F.A., Elphic, R.C., 2009. The spatial distribution of polar hydrogen deposits on the Moon. Icarus 200 (1), 12–18.

Eke, V.R., Bartram, S.A., Lane, D.A., Smith, D., Teodoro, L.F.A., 2014. Lunar polar craters – icy, rough or just sloping? Icarus 241, 66–78.

Elardo, S.M., Draper, D.S., Shearer, C.K., 2011. Lunar Magma Ocean crystallisation revisited: bulk composition, early cumulate mineralogy, and the source regions of the highlands Mg-suite. Geochim Cosmochim Acta 75, 3024–3045.

Elkins-Tanton, L.T., Van Orman, J.A., Hager, B.H., Grove, T.L., 2002. Reexamination of the lunar magma ocean cumulate overturn hypothesis: melting or mixing is required. Earth Planet. Sci. Lett. 196, 239–249. http://doi.org/10.1016/S0012-821X(01)00613-6.

Elkins-Tanton, L., Grove, T.L., 2011. Water (hydrogen) in the lunar mantle: results from petrology and magma ocean modeling. Earth Planet. Sci. Lett. 307, 173–180.

Elphic, R.C., Eke, V.R., Teodoro, L., Lawrence, D.J., Bussey, D.B.J., 2007. Models of the distribution and abundance of hydrogen at the lunar south pole. Geophys. Res. Lett. 34, L13204. http://doi.org/10.1029/2007GL029954.

Epstein, S., Taylor Jr., H.P., 1973. The isotopic composition and concentration of water, hydrogen, and carbon in some Apollo 15 and 16 soils and in the Apollo 17 orange soil, Proc. Fourth Lunar Sci. Conf., 2, 1559–1575 Pergamon, 1973.

Epstein, S., Taylor Jr., H.P., 1974. D/H and $^{18}O/^{16}O$ ratios of H_2O in the 'rusty' breccia 66095 and the origin of 'lunar water', Proc. Fifth Lunar Sci. Conf., 2, 1839–1854 Pergamon, 1974.

Evans, A.J., Zuber, M.T., Weiss, B.P., Tikoo, S.M., 2014. A wet, heterogeneous lunar interior: Lower mantle and core dynamo evolution. J. Geophy. Res.: Planets, 119. http://doi.org/10.1002/2013JE004494.

Faith, K., 2017. NASA finds lunar spacecraft that vanished 8 years ago. CNN Retrieved 10 March 2017.

Fang, G.-Y., Zhou, B., Ji, Y.-C., Zhang, Q.-Y., Shen, S.-X., Li, Y.-X., Guan, H.-F., Tang, C.-J., Gao, Y.-Z., Lu, W., Ye, S.-B., Han, H.-D., Zheng, J., Wang, S.-Z., 2014. Lunar penetrating radar onboard the Chang'e-3 mission. Res. Astron. Astrophys. 14, 1607–1622.

Farrell, W.M., Hurley, D.M., Zimmerman, M.I., 2015. Solar wind implantation into lunar regolith: Hydrogen retention in a surface with defects. Icarus 255, 116–126.

Feldman, W.C., et al., 1993. Redistribution of subsurface neutrons caused by ground ice on Mars. J. Geophys. Res. 98, 20855.

Feldman, W.C., et al., 1997. The neutron signature of Mercury's volatile polar deposits. J. Geophys. Res. 102, 25565.

Feldman, W.C., Maurice, S., Binder, A.B., Barraclough, B.L., Elphic, R.C., Lawrence, D.J., 1998a. Fluxes of fast and epithermal neutrons from lunar prospector: evidence for water ice at the lunar poles. Science 281, 1496–1500. http://doi.org/10.1126/science.281.5382.1496.

Feldman, W.C., Lawrence, D.J., Elphic, R.C., Barraclough, B.L., Maurice, S., Genetay, I., Binder, A.B., 2000a. Polar hydrogen deposits on the Moon. J. Geophys. Res. 105 (E2), 4175–4195. http://doi.org/10.1029/1999JE001129.

Feoktistova, E.A., Pugacheva, S.G., Shevchenko, V.V., 2018. The temperature regime of the proposed landing sites for the Luna-Glob mission in the south polar region of the Moon. Earth Moon Planets 122, 1–13. https://doi.org/10.1007/s11038-018-9518-0.

Ferl, R.J., Paul, A.L., 2010. Lunar plant biology—a review of the Apollo era. Astrobiology 10 (3), 261–274.

Garrick-Bethell, I., Weiss, B.P., Shuster, D.L., Buz, J., 2009. Early lunar magnetism. Science 323 (5912), 356–359.

Geiss, J., Rossi, A.P., 2013. On the chronology of lunar origin and evolution. Astron. Astrophys. Rev. 21, 68. https://doi.org/10.1007/s00159-013-0068-1.

Gladstone, G.R., Hurley, D.M., Retherford, K.D., Feldman, P.D., Pryor, W.R., Chaufray, J.Y., Versteeg, M., Greathouse, T.K., Steffl, A.J., Throop, H., Parker, J.W., Kaufmann, D.E., Egan, A.F., Davis, M.W., Slater, D.C., Mukherjee, J., Miles, P.F., Hendrix, A.R., Colaprete, A., Stern, S.A., 2010. LRO-LAMP observations of the LCROSS impact plume. Science 330 (6003), 472–476. http://doi.org/10.1126/science.1186474. PMID: 20966244.

Glotch, T.D., Lucey, P.G., et al., 2010. Highly silicic compositions on the Moon. Science 329 (5998), 1510–1513. doi:10.1126/science.1192148.

Gornitz, V., 1973. The origin of sinuous rilles. The Moon 6 (3-4), 337–356.

Greeley, R., 1971. Lunar Hadley rille: considerations of its origin. Science 172, 722–725.

Greeley, R., King, J.S., 1977. Volcanism of the Eastern Snake River Plain, Idaho: A Comparative Planetary Geology Guidebook. NASA CR154621, p. 308.

Greeley, R., 1987. The role of lava tubes in Hawaiian volcanoes. Volcanism in Hawaii, 1589–1602 U.S. Geol. Surv. Proc. Pap. 1350.

Greenwood, J.P., Itoh, S., Sakamoto, N., Warren, P., Taylor, L., Yurimoto, H., 2011. Hydrogen isotope ratios in lunar rocks indicate delivery

of cometary water to the moon. Nat. Geosci. 4, 79–82. http://doi.org/10.1038/ngeo1050.
Hapke, B., 1990. Coherent backscatter and the radar characteristics of outer planet satellites. Icarus 88, 407–417. http://doi.org/10.1016/0019-1035(90)90091-M.
Harmon, J.K., Slade, M.A., 1992. Radar mapping of Mercury: full-disk Doppler delay images. Science 258, 640–643. http://doi.org/10.1126/science.258.5082.640.
Hartmann, W.K., Davis, D.R., 1975. Satellite-sized planetesimals and lunar origin. Icarus 24 (4), 504–514. doi:10.1016/0019-1035(75)90070-6.
Hauri, E.H., Weinreich, T., Saal, A.E., Rutherford, M.C., Van Orman, J.A., 2011. High pre-eruptive water contents preserved in lunar melt inclusions. Science 333 (6039), 213–215.
Hayne, P.O., Greenhagen, B.T., Foote, M.C., Siegler, M.A., Vasavada, A.R., Paige, D.A, 2010. Diviner lunar radiometer observations of the LCROSS impact. Science 330 (6003), 477–479. http://doi.org/10.1126/science.1197135.
Health Canada. (2010-1005), Dietary reference intakes. http://www.hc-sc.gc.ca/fn-an/nutrition/reference/table/index-eng.php.
Heiken, G.H., McKay, D.S., Brown, P.W, 1974. Lunar deposits of possible pyroclastic origin. Geochim. Cosmochim. Acta 38, 1703–1718.
Henderson, C.W., 1962. Man and lunar housing emplacement, American Rocket Society Annual Meeting 17th, and Space Flight Exposition. Los Angeles, CA, 7 Nov. 13- 18, Paper 2688-62, ARS Paper 62-2688.
Hess, P.C., Parmentier, E.M., 1995. A model for the thermal and chemical evolution of the Moon's interior: implications for the onset of mare volcanism. Earth Planet. Sci. Lett. 134, 501–514.
Hiesinger, H., Head, J.W., 2006. New views of lunar geoscience: an introduction and overview. Rev. Mineral. Geochem 60, 1–81.
Hibbitts, C.C., Grieves, G.A., Poston, M.J., Dyar, M.D., Alexandrov, A.B., Johnson, M.A., Orlando, T.M., 2011. Thermal stability of water and hydroxyl on the surface of the Moon from temperature-programmed desorption measurements of lunar analog materials. Icarus 213 (1), 64–72.
Hirth, G., Kohlstedt, D.L., 1995. Experimental constraints on the dynamics of the partially molten upper mantle: 2. Deformation in the dislocation creep regime. J. Geophys. Res. 100 (B8) 15,441–15,449.
Hodges Jr., R.R., 1980. Lunar cold traps and their influence on argon-40. Proc. Lunar Planet. Sci. Conf. 11, 2463.
Hodges Jr., R.R., 2002. Reanalysis of lunar prospector neutron spectrometer observations over the lunar poles. J. Geophys. Res. 107 (E12), 5125.
Hood, L.L., Zuber, M.T., 2000. Recent refinements in geophysical constraints on lunar origin and evolution. In: Canup, R.M., Righter, K. (Eds.), Origin of the Earth and Moon. Univ. of Arizona Press, Tucson, pp. 397–409.
Horz, F., 1985. Lava tubes: Potential shelters for habitats. In: Mendell, W.W. (Ed.), Lunar Bases and Space Activities of the 21st Century. Lunar and Planetary Institute, Houston, pp. 405–412.
Huang, J., Xiao, Z.-Y., Flahaut, J., Martinot, M., Head, J.W., Xiao, X., Xie, M.-G., Xiao, L., 2018. Geological characteristic of Von Kármán Crater, Northwestern South Pole Aitken Basin: Chang'E 4 Landing Site Region. J. Geophys. Res. 123, 1684–1700.
Hui, H., Peslier, A.H., Zhang, Y., Neal, C.R., 2013. Water in lunar anorthosites and evidence for a wet early Moon. Nat. Geosci. 6, 177–180.
Hulme, G., 1973. Turbulent lava flow and the formation of lunar sinuous rilles. Mod. Geol. 4, 107–117.
Hulme, G., 1982. A review of lava flow processes related to the formation of lunar sinuous rilles. Geophys. Surveys 5, 245–279.
Ichimura, A.I., Zent, A.P., Quinn, R.C., Sanchez, M.R., Taylor, L.A., 2012. Hydroxyl (OH) production on all airless planetary bodies: evidence from H^+/D^+ ion-beam experiments. Earth Planet. Sci. Lett. 345, 90–94.
Ingersoll, A.P., Svitek, T., Murray, B.C., 1992. Stability of polar frosts in spherical bowl-shaped craters on the Moon, Mercury, and Mars. Icarus 100 (1), 40–47.
Jester, S., Falcke, H., 2009. Science with a lunar low-frequency array: from the dark ages of the Universe to nearby exoplanets. Cosmol. Nongalactic Astrophys. arXiv:0902.0493 [astro-ph.CO]URL. https://arxiv.org/pdf/0902.0493.pdf.
Johnston, R.S., Mason, I.A., Wooley, B.C., et al., 1975. Biomedical results of Apollo. NASA Spec. Publ. 2, 407–424.
Jolliff, B.L., Wiseman, S.A., Lawrence, S.J., et al., 2011. Non-mare silicic volcanism on the lunar farside at Compton–Belkovich. Nature Geoscience 4, 566–571. https://doi.org/10.1038/ngeo1212.
Jolliff, B.L., Floss, C., McCallum, I.S., Schwartz, J.M., 1999. Geochemistry petrology and cooling history of 141617373 a plutonic lunar sample with textural evidence of granitic-fraction separation by silicate-liquid immiscibility. Am. Min. 84, 821–837.
Kant, I. (1755), *Allgemeine Naturgeschichte und Theorie des Himmels*.
Karato, S.-I., 2010. Rheology of the deep upper mantle and its implications for the preservation of the continental roots: a review. Tectonophysics 481 (1–4), 82–98. http://doi.org/10.1016/j.tecto.2009.04.011.
Karato, S.-I., 2013. Does partial melting explain geophysical anomalies? Phys. EarthPlanet. Int. http://doi.org/10.1016/j.pepi.2013.08.006.
Kemmerer Jr., W.W., Mason, J.A., Wooley, B.C, 1969. Physical, chemical, and biological activities at the lunar receiving laboratory. BioScience 19, 712–715.
Khan, A., Pommier, A., Connolly, J.A.D., 2013. On the presence of a titanium-rich melt layer in the deep lunar interior, Paper Presented at 44th Lunar and Planetary Science Conference, LPI Contribution No. 1719. Woodlands, Tex.
Klima, R.L., Cahill, J., Hagerty, J., Lawrence, D., 2013. Remote detection of magmatic water in Bullialdus Crater on the Moon. Nat. Geosci. 6, 737–741.
Kohlstedt, D.L., 2006. The role of water in high-temperature rock deformation. Rev. Mineral. Geochem. 62, 377–396.
König, S., et al., 2011. The Earth's tungsten budget during mantle melting and crust formation. Geochimica et Cosmochimica Acta 75 (8), 2119–2136.
Langevin, Y., Arnold, J., 1977. The evolution of the lunar regolith. J. Ann. Rev. Earth Planet. Sci. 5, 449.
Laneuville, M., Wieczorek, M.A., et al., 2014. A long-lived lunar dynamo powered by core crystallization. Earth and Planetary Science Letters 401, 251–260.
Lanzerotti, L.J., Brown, W.L., Johnson, R.E., 1981. Ice in the polar regions of the Moon. J. Geophys. Res. 86 (B5), 3949.
Lawrence, D.J., Feldman, W.C., Barraclough, B.L., et al., 1998. Global elemental maps of the Moon: the Lunar Prospector gamma-ray spectrometer. Science 281 (5382), 1484–1489.
Lawrence, D.J., Feldman, W.C., Elphic, R.C., Hagerty, J.J., Maurice, S., McKinney, G.W., Prettyman, T.H., 2006. Improved modeling of Lunar Prospector neutron spectrometer data: implications for hydrogen deposits at the lunar poles. J. Geophys. Res. 111, E08001. http://doi.org/10.1029/2005JE002637.
Leyaa, I., Schönbächler, M., Wiechert, U., Krähenbühl, U., Halliday, A.N., 2008. Titanium isotopes and the radial heterogeneity of the solar system. Earth and Planetary Science Letters 266 (3–4), 233–244.

Le Bars, M., Wieczorek, M.A., Karatekin, Ö., Cebron, D., Laneuville, M., 2011. An impact-driven dynamo for the early Moon. Nature 479, 215–218.

Li, C., Xing, S., Lauro, S.E., Su, Y., Dai, S., Feng, J., Cosciotti, B., Paolo, F.D., Mattei, E., Xiao, Y., Ding, C., Pettinelli, E., 2018. Pitfalls in GPR data interpretation: false reflectors detected in Lunar radar cross sections by Chang'e-3. IEEE Trans. Geosci. Remote Sens. 56, 1325–1335.

Li, C., et al., 2020. The Moon's farside shallow subsurface structure unveiled by Chang'E-4 Lunar Penetrating Radar. Sci. Adv. 6 (9), eaay6898. http://doi.org/10.1126/sciadv.aay6898.

Lindsay, R.B., 1973. Julius Robert Mayor: Prophet of Energy. Pergamon, Oxford.

Liu, Y., Guan, Y., Zhang, Y., Rossman, G.R., Eiler, J.M., Taylor, L.A., 2012. Direct measurement of hydroxyl in the lunar regolith and the origin of lunar surface water. Nat. Geosci. 5, 779–782. http://doi.org/10.1038/ngeo1601.

Longhi, J., Ashwal, L.D., 1985. Two-stage models for lunar and terrestrial anorthosites: petrogenesis without a magma ocean. J. Geophys. Res. 90 (S02), C571–C584. http://doi.org/10.1029/JB090iS02p0C571.

Longhi, J., 1992. Experimental petrology and petrogenesis of mare volcanics. Geochim. Cosmochim. Acta 56, 2235–2251.

Lowrie, W., Kent, D.V., 2004. Geomagnetic polarity timescales and reversal frequency regimes. In: Channell, J.E.T., Kent, D.V., Lowrie, W., Meert, J. (Eds.), Timescales of the Palaeomagnetic Field. America Geophysical Union, pp. 117–129.

Lucey, P., et al., 2006. New Views of the Moon. *Mineralogical Society of America*, Chantilly, VA, pp. 83–220.

Lucey, P.G., 2009. Planetary science. A lunar waterworld. Science 326 (5952), 531–532. http://doi.org/10.1126/science.1181471.

Mangus, S., Larsen, W., 2004. Lunar Receiving Laboratory Project History. National Aeronautics and Space Administration, Washington DC NASA=CR-2004-208938.

Margot, J.L., Campbell, D.B., Jurgens, R.F., Slade, M.A., 1999. Topography of the lunar poles from radar interferometry: a survey of cold trap locations. Science 284, 1658–1660. http://doi.org/10.1126/science.284.5420.1658.

Masursky, H., Colton, G.W, El-Baz, F., 1978. Apollo Over the Moon: A View from Orbit, 255 SP-362.

Mayer, J.R. (1848), *Beitrdge zur Dynamik des Himmels in populdrer Darstellung*, Verlag von Johann Ulrich Landherr, Heilbronn. English translation in Lindsay (1973, pp. 148–196).

McLane Jr., J.C., King Jr., E.A., Flory, D.A., Richardson, K.A., Dawson, J.P., Kemmerer, W.W., Wooley, B.C., 1967. Lunar Receiving Laboratory. Science 155, 525–529.

McCord, T.B., Taylor, L.A., Combe, J.-P., Kramer, G., Pieters, C.M., Sunshine, J.M., Clark, R.N., 2011. Sources and physical processes responsible for the OH/H$_2$O in the lunar soil as revealed by the Moon Mineralogy Mapper (M^3). J. Geophys. Res. 116 (E6), E00G05. http://doi.org/10.1029/2010JE003711.

McCubbin, F.M., Steele, A., Hauri, E.H., Nekvasil, H., Yamashita, S., Hemley, R.J., 2010a. Nominally hydrous magmatism on the Moon. Proc. Natl. Acad. Sci. USA 107, 11223–11228.

McCallum, I.S., Schwartz, J.M., 2001. Lunar Mg suite: thermobarometry and petrogenesis of parental magmas. J. Geophys. Res. 106 (E11). 27,969–27,983. http://doi.org/10.1029/2000JE001397.

McCubbin, F.M., et al., 2010. Hydrous magmatism on Mars: a source of water for the surface and subsurface during the Amazonian. Earth Planet Sci. Lett. 292, 132–138.

McCubbin, F.M., et al., 2011. Fluorine and chlorine abundances in lunar apatite: Implications for heterogeneous distributions of magmatic volatiles in the lunar interior. Geochim. Cosmochim. Acta 75, 5073–5093.

Melosh, H.J., 2009. An isotopic crisis for the giant impact origin of the Moon? Meteorit. Planet. Sci. 44, A139.

Melosh, H.J., 1989. Impact Cratering: A Geologic Process. Oxford Univ. Press; Oxford, England.

Mills, R.D.J.I.S, Alexander, C.M.O'D., Wang, J., Hauri, E.H., 2016. Water in alkali feldspar: the effect of rhyolite generation on the lunar hydrogen budget. Geochem. Perspect. Lett. 3 (2). http://doi.org/10.7185/geochemlet.1712.

Mitrofanov, I.G., Sanin, A.B., Boynton, W.V., Chin, G., Garvin, J.B., Golovin, D., Evans, L.G., Harshman, K., Kozyrev, A.S., Litvak, M.L., Malakhov, A., Mazarico, E., McClanahan, T., Milikh, G., Mokrousov, M., Nandikotkur, G., Neumann, G.A., Nuzhdin, I., Sagdeev, R., Shevchenko, V., Shvetsov, V., Smith, D.E., Starr, R., Tretyakov, V.I., Trombka, J., Usikov, D., Varenikov, A., Vostrukhin, A., Zuber, M.T., 2010. Hydrogen mapping of the lunar south pole using the LRO neutron detector experiment LEND. Science 330 (6003), 483–486. http://doi.org/10.1126/science.1185696.

Moore, P., 2005. Solar System: Jupiter, Saturn and their Moons. In: Selley, R.C., Cocks, L.R.M., Pilmer, I.R. (Eds.), Encyclopedia of Geology. Elsevier Ltd.

Morbidelli, A., Chambers, J., Lunine, J.I., Petit, J.M., Robert, F., Valsecchi, G.B., Cyr, K.E., 2000. Source regions and timescales for the delivery of water to the Earth. Meteorit. Planet. Sci. 35 (6), 1309–1320.

Morgan, T.H., Shemansky, D.E., 1991. Limits to the lunar atmosphere. J. Geophys. Res. 96 (A2), 1351–1367.

Muhleman, D.O., Butler, B.J., Grossman, A.W., Slade, M.A., 1991. Radar images of Mars. Science 253, 1508–1513. http://doi.org/10.1126/science.253.5027.1508.

Mytrokhyn, O.V., Bogdanova, S.V., Shumlyanskyy, L.V., 2003. Anorthosite rocks of Fedorivskyy suite (Korosten Pluton, Ukrainian Shield). Current Problems in Geology. Kyiv National University, Kyiv, pp. 53–57.

Mytrokhyn, O.V., Bogdanova, S.V., Shumlyanskyy, L.V., 2008. Polybaricchrystalization of KorostenPlytonanortosites (Ukrainian shield). Mineral. J. 30, 36–56 In Ukrainian.

Nadia, D., 2016. Scientists may have spotted buried lava tubes on the Moon. National Geographic.

Neal, C.R., Taylor, L.A., 1992. Petrogenesis of mare basalts: A record of lunar volcanism. Geochim. Cosmochim. Acta 56, 2177–2211. http://doi.org/10.1016/0016-7037(92)90184-K.

Neumann, G., Zuber, M., Smith, D., Lemoine, F., 1996. The lunar crust: Global structure and signature of major basins. J. Geophys. Res. 101 (E7). 16,841–16,863. http://doi.org/10.1029/96JE01246.

Newell, H.E., 1972. Harold Urey and the Moon. Moon 7 (1-2), 1–5.

Noda, H., Araki, H., Goossens, S., Isihara, Y., Matsumoto, K., Tazawa, S., Sasaki, S., Kawano, N., Sasaki, S., 2008. Illumination conditions at the lunar polar regions by Kaguya (SELENE) laser altimeter. Geophys. Res. Lett. 35, L24203. http://doi.org/10.1029/2008GL035692.

Nozette, S., Lichtenberg, C., Spudis, P.D., Bonner, R., Ort, W., Malaret, E., Robinson, M., Shoemaker, E.M., 1996. The Clementine bistatic radar experiment. Science 274, 1495–1498. http://doi.org/10.1126/science.274.5292.1495.

Nozette, S., Shoemaker, E.M., Spudis, P.D., Lichtenberg, C.L., 1997. The possibility of ice on the Moon. Science 278, 144–145. http://doi.org/10.1126/science.278.5335.144.

Nozette, S., Spudis, P.D., Robinson, M., Bussey, D.B.J., Lichtenberg, C., Bonner, R., 2001. Integration of lunar polar remote-sensing data sets: evidence for ice at the lunar south pole. J. Geophys. Res. 106 (E10). 23,253–23,266. http://doi.org/10.1029/2000JE001417.

Oberbeck, V.R., Quaide, W.L., Greeley, R., 1969. On the origin of lunar sinuous rilles. Mod. Geol. 1, 75–80.

Oberbeck, V.R., Aoyagi, M., Greeley, R., Lovas, M., 1972. Planimetric shapes of lunar rilles. Apollo 16 Preliminary Science Report, Part Q 29-80 to 29-88.

Oberst, J., Nakamura, Y., 1988. A seismic risk for the lunar base (abstract), Papers Presented to the 1988 Symposium on Lunar Bases and Space Activities of the 21st Century. Lunar and Planetary Institute, Houston, p. 183.

Ohtake, M., Haruyama, J., Matsunaga, T., Yokota, Y., Morota, T., Honda, C., Team, L., 2008. Performance and scientific objectives of the SELENE (KAGUYA) Multiband Imager. Earth Planets Space 60, 257–264.

Ostro, S.J., Shoemaker, E.M., 1990. The extraordinary radar echoes from Europa, Ganymede, and Callisto: a geological perspective. Icarus 85, 335–345. http://doi.org/10.1016/0019-1035(90)90121-O.

Owen, T., Bar-Nun, A., 1995. Comets, impact, and atmospheres. Icarus 116, 155–156.

Pahlevan, K., Stevenson, D.J., Eiler, J.M., 2011. Chemical fractionation in the silicate vapor atmosphere of the Earth. Earth Planet. Sci. Lett. 301(3-4), 433–443.

Paige, D.A., Siegler, M.A., Zhang, J.A., Hayne, P.O., Foote, E.J., Bennett, K.A., Vasavada, A.R., Greenhagen, B.T., Schofield, J.T., McCleese, D.J., Foote, M.C., DeJong, E., Bills, B.G., Hartford, W., Murray, B.C., Allen, C.C., Snook, K., Soderblom, L.A., Calcutt, S., Taylor, F.W., Bowles, N.E., Bandfield, J.L., Elphic, R., Ghent, R., Glotch, T.D., Wyatt, M.B., Lucey, P.G., 2010. Diviner lunar radiometer observations of cold traps in the Moon's south polar region. Science 330 (6003), 479–482. http://doi.org/10.1126/science.1187726.

Papanastassiou, D.A., Wasserburg, G.J., 1973. Rb-Sr ages and initial strontium in basalts from Apollo 15. Earth Planet. Sci. Lett. 17, 324–337.

Pasckert, J.H., Hiesinger, H., van der Bogert, C.H., 2018. Lunar farside volcanism in and around the South Pole–Aitken basin. Icarus 299, 538–562.

Pieters, C.M., McCord, T.B., Zisk, S.H., Adams, J.B, 1973. Lunar black spots and the nature of the Apollo 17 landing area. J. Geophys. Res. 78, 5867–5875.

Pieters, C.M., McCord, T.B., Charette, M.P., Adams, J.B, 1974. Lunar surface: Identification of the dark mantling material in the Apollo 17 soil samples. Science 183, 1191–1194.

Pieters, C.M., Goswami, J.N., Clark, R.N., Annadurai, M., Boardman, J., Buratti, B., Combe, J.P., Dyar, M.D., Green, R., Head, J.W., Hibbitts, C., Hicks, M., Isaacson, P., Klima, R., Kramer, G., Kumar, S., Livo, E., Lundeen, S., Malaret, E., McCord, T., Mustard, J., Nettles, J., Petro, N., Runyon, C., Staid, M., Sunshine, J., Taylor, L.A., Tompkins, S., Varanasi, P., 2009. Character and spatial distribution of OH/H_2O on the surface of the Moon seen by M^3 on Chandrayaan-1. Science 326 (5952), 568–572. http://doi.org/10.1126/science.1178658.

Raney, R.K., 2007. Hybrid-polarity SAR architecture. IEEE Trans. Geosci. Remote Sens. 45, 3397–3404. http://doi.org/10.1109/TGRS.2007.895883.

Reynolds, J.M., 2011. An Introduction to Applied and Environmental Geophysics. John Wiley & Sons; Hoboken, New Jersey, U.S..

Richard, S., 2001. Man on the Moon: Kennedy speech ignited the dream. CNN.

Ringwood, A.E., Kesson, S.E., 1976. A dynamic model for mare basalt petrogenesis. Proc. Lunar Sci. Conf. 7, 1697–1722.

Robert, F., 2001. The origin of water on Earth. Science 293, 1056–1058. doi:10.1126/science.1064051.

Rubanenko, L., Aharonson, O., 2017. Stability of ice on the Moon with rough topography. Icarus 296, 99–109.

Rufu, R., Aharonson, O., Perets, H.B., 2017. A multiple impact origin for the Moon. Nat. Geosci. 10 (2), 89–94. http://doi.org/10.1038/ngeo2866.

Runcorn, S.K., 1996. The formation of the lunar core. Geochim. Cosmochim. Acta 60 (7), 1205–1208.

Saal, A.E., Hauri, E.H., Rutherford, M.J., Cooper, R.F., 2007. The volatile contents (CO_2, H_2O, F, S, Cl) of the Lunar picritic glasses. Lunar Planet. Sci. Conf. 38, 2148.

Saal, A.E., Hauri, E.H., Cascio, M.L., Van Orman, J.A., Rutherford, M.C., Cooper, R.F., 2008. Volatile content of lunar volcanic glasses and the presence of water in the Moon's interior. Nature 454, 192–195.

Salmon, J., Canup, R.M., 2014. Accretion of the moon from non-canonical discs. Philos. Trans. R. Soc. Lond. A: Math. Phys. Eng. Sci., 372.

Sarbadhikari, A.B., Marhas, K.K., Sameer, Goswami, J.N., 2016. Water in the lunar interior. Current Science 110 (8), 1536–1539.

Sarbadhikari, A.B., Marhas, K.K., Sameer, Goswami, J.N, 2013. Water content in melt inclusions and apatites in low-titanium lunar mare basalt, 15555, Lunar and Planetary Science Conference. Texas, 44, 2813 Abstr.

Sato, M., 1979. The driving mechanism of lunar pyroclastic eruptions inferred from the oxygen fugacity behavior of Apollo 17 orange glass. Lunar Planet. Sci. Conf. 10, 311.

Schaber, G.G., Boyce, J.M., Moore, H.J., 1976. The scarcity of mappable flow lobes on the lunar maria: unique morphology of the Imbrium flows, Proc. Lunar Sci. Conf. Seventh, 2783–2800.

Schultz, P.H., Spudis, P.D., 1983. Beginning and end of lunar mare volcanism. Nature 302, 233–236.

Schultz, P.H., Hermalyn, B., Colaprete, A., Ennico, K., Shirley, M., Marshall, W.S., 2010. The LCROSS cratering experiment. Science 330 (6003), 468–472. http://doi.org/10.1126/science.1187454.

Seth, A., Morbidelli, J.A., Raymond, S.N., O'Brien, D.P., Walsh, K.J., Rubie, D.C., 2014. Highly siderophile elements in Earth's mantle as a clock for the Moon-forming impact. Nature 508 (7494), 84. http://doi.org/10.1038/nature13172.

Shea, E.K., Weiss, B.P., Cassata, W.S., Shuster, D.L., Tikoo, S.M., Gattacceca, J., Grove, T.L., Fuller, M.D., 2012. A long-lived lunar core dynamo. Science 335 (6067), 453–456.

Snyder, G.A., Taylor, L.A., Neal, C.R., 1992. A chemical model for generating the sources of mare basalts: combined equilibrium and fractional crystallization of the lunar magmasphere. Geochim. Cosmochim. Acta 56, 3809–3823.

Solomon, S.C., Chaiken, J., 1976. Proc. Lunar Planet. Sci. Conf. 7, 541.

Spudis, P., Nozette, S., et al., 2009. Mini-SAR: an imaging radar experiment for the Chandrayaan-1 mission to the Moon. Current Science 96 (4), 533–539.

Spudis, P.D., Bussey, D.B.J., Baloga, S.M., Butler, B.J., Carl, D., Carter, L.M., Chakraborty, M., Elphic, R.C., Gillis-Davis, J.J., Goswami, J.N., Heggy, E., Hillyard, M., Jensen, R., Kirk, R.L., LaVallee, D., McKerracher, P., Neish, C.D., Nozette, S., Nylund, S., Palsetia, M., Patterson, W., Robinson, M.S., Raney, R.K., Schulze, R.C., Sequeira, H., Skura, J., Thompson, T.W., Thomson, B.J., Ustinov, E.A., Winters, H.L., 2010. Initial results for the North Pole of the Moon from Mini-SAR, Chandrayaan-1 mission. Geophys. Res. Lett. 37, L06204.

Stegman, D.R., Jellinek, A., Zatman, S., Baumgardner, J.R., Richards, M.A., 2003. An early lunar core dynamo driven by thermochemical mantle convection. Nature 421, 143–146.

Stevenson, D.J., 2003. Planetary magnetic fields. Earth Planet Sci. Lett. 208, 1–11.

Strycker, P.D., Chanover, N.J., Miller, C., Hamilton, R.T., Hermalyn, B., Suggs, R.M., Sussman, M., 2013. Characterization of the LCROSS impact plume from a ground-based imaging detection. Nat. Commun. 4, 2620. http://doi.org/10.1038/ncomms3620.

Su, Y., Fang, G.-Y., Feng, J.-Q., Xing, S.-G., Ji, Y.-C., Zhou, B., Gao, Y.-Z., Li, H., Dai, S., Xiao, Y., Li, C.-L., 2014. Data processing and initial results of Chang'E-3 lunar penetrating radar. Res. Astron. Astrophys. 14, 1623–1632.

Suavet, C., Weiss, B.P., Cassata, W.S., Shuster, D.L., Gattacceca, J., Chan, L., Garrick-Bethell, I., Head, J.W., Grove, T.L., Fuller, M.D., 2013. Persistence and origin of the lunar core dynamo. Proc. Natl. Acad. Sci. USA 110 (21), 8453–8458.

Sunshine, J.M., Farnham, T.L., Feaga, L.M., Groussin, O., Merlin, F., Milliken, R.E., A'Hearn, M., 2009. Temporal and spatial variability of lunar hydration as observed by the Deep Impact spacecraft. Science 326 (5952), 565–568. http://doi.org/10.1126/science.1179788.

Trinquiertim, A., Elliott, T., Ulfbeck, D., Coath, C., Krot, A.N., Bizzarro, M., 2009. Origin of nucleosynthetic isotope heterogeneity in the Solar protoplanetary disk. Science 324 (5925), 374–376. doi:10.1126/science.1168221.

Tartèse, R., Anand, M., 2013. Late delivery of chondritic hydrogen into the lunar mantle: insights from mare basalts. Earth Planet. Sci. Lett. 361, 480–486.

Tartèse, R., Anand, M., Barnes, J.J., Starkey, N.A., Franchi, I.A., Sano, Y., 2013. The abundance, distribution, and isotopic composition of hydrogen in the moon as revealed by basaltic lunar samples: implications for the volatile inventory of the moon. Geochim. Cosmochim. Acta 122, 58–74.

Tartèse, R., 2015. Water in the LMO. In: Cudnik, B. (Ed.), Encyclopedia of Lunar Science. Springer, Cham. https://doi.org/10.1007/978-3-319-05546-6_26-1.

Taylor, S.R., 1996. Origin of the terrestrial planets and the moon. J. R. Soc. West Aust. 79 (1), 59–65.

Taylor, S.R., Esat, T.M., 1996Earth Processes: Reading the Isotopic Code95. American Geophysical Union, Washington, DC, pp. 33–46 Geophys. Monogr.

Taylor, S.R., Pieters, C.M., MacPherson, G.J., 2006. New Views of the Moon. *Mineralogical Society of America, Chantilly, VA*, pp. 657–704.

Touboul, M., Kleine, T., Bourdon, B., Palme, H., Wieler, R., 2007. Late formation and prolonged differentiation of the Moon inferred from W isotopes in lunar metals. Nature 450, 1206–1209. https://doi.org/10.1038/nature06428.

Urey, H.C., 1952. The Planets. Yale University, New Haven, p. 245.

Urey, H.C., 1967a. Study of the Ranger Pictures of the Moon. Proc. R. Soc. Lond. Ser. A 296, 418.

Urey, H.C., 1967b. Water on the Moon. Nature 216, 1094.

Ustunisik, G., Nekvasil, H., Lindsley, D., 2011. Differential degassing of H_2O, Cl, F, and S: potential effects on lunar apatite. Am. Mineral. 96, 1650–1653.

Vaniman, D., Pettit, D., Heiken, G., 1988. Uses of lunar sulfur (abstract), Papers Presented to the 1988 Symposium on Lunar Bases and space Activities of the 21st Century. Houston. Lunar and Planetary Institute, p. 244.

Vaniman, D. et al. (1991), In: Lunar Sourcebook, a User's Guide to the Moon, G.H. Heiken, D.T. Vaniman, B.M. French, Eds. (Cambridge Univ. Press, Cambridge, 1991), pp. 5–26.

Vasavada, A.R., Paige, D.A., Wood, S.E., 1999. Near-surface temperatures on Mercury and the Moon and the stability of polar ice deposits. Icarus 141, 179–193. http://doi.org/10.1006/icar.1999.6175.

Vilas, F., Jensen, E.A., Domingue, D.L., McFadden, L.A., Runyon, C.J., Mendell, W.W., 2008. A newly-identified spectral reflectance signature near the lunar South pole & the South Pole-Aitken Basin. Earth Planets Space 60, 67.

Walkinshaw, C.H., Sweet, H.C., Venketeswaran, S., Home, W.H., 1970. Results of Apollo 11 and 12 quarantine studies on plants. Bioscience 20, 1297–1302.

Walker, D., Longhi, J., Hays, J.F., 1975, Proceedings of the Sixth Lunar Science Conference, 1975. Pergamon, pp. 1103–1120.

Walker, D., Longhi, J., Lasaga, A.C., Stolper, E.M., Grove, T.L., Hays, J.F., 1977. Slowly cooled microgabbros 15555 and 15056. Proc. Lunar Sci. Conf. 8, 1521–1547.

Wall, M., 2018. China launches relay satellite for mission to Moon's far side. Space. https://www.space.com/40646-china-queqiao-moon-relay-satellite-launch.html.

Walsh, K.J., et al., 2011. A low mass for Mars from Jupiter's early gas-driven migration. Nature 475, 206–209.

Walsh, K.J., Morbidelli, A., Raymond, S.N., O'Brien, D.P., Mandell, A.M., 2012. Populating the asteroid belt from two parent source regions due to the migration of giant planets—"The Grand Tack. Meteorit. Planet Sci. 47 (12), 1941–1947.

Ward, W.R., Cameron, A.G.W., 1978. Disc evolution within the Roche limit, Lunar and Planetary Institute Conference. Houston, Tex. Lunar and Planetary Institute, pp. 1205–1207.

Warren, P.H., Wasson, J.T., 1979. The origin of KREEP. Journal of Geophysical Research: Planets, Reviews of Geophysics, Geophysical Research Letters 17 (1), 73–88.

Watson, K., Murray, B.C., Brown, H., 1961. The behavior of volatiles on the lunar surface. J. Geophys. Res. 66, 3033.

Weber, R.C., Lin, P.-Y., Garnero, E.J., Williams, Q., Lognonne, P., 2011. Seismic detection of the lunar core. Science 331 (6015), 309–312.

Weete, J.D., Walkinshaw, C.H., 1972. Apollo 12 lunar material-effects on plant pigments. Can. J. Bot. 50, 101–104.

Wieczorek, M.A., Neumann, G.A., Nimmo, F., Kiefer, W.S., Taylor, G.J., Melosh, J.H., Phillips, R.J., Solomon, S.C., Andrews-Hanna, J.C., Asmir, S.W., Konopliv, A.S., et al., 2013. The crust of the Moon as seen by GRAIL. Science 339, 671–675.

Williams, Q., Hemley, R.J., 2001. Hydrogen in the deep earth. Annu. Rev. Earth Planet Sci. 29, 365–418.

Williams, M., 2018. Upcoming Chinese Lander Will Carry Insects and Plants to the Surface of the Moon. *Universe Today*. URL. https://www.universetoday.com/138197/upcoming-chinese-lander-will-carry-insects-plants-surface-moon/.

Wilson, L., Head, J.W., 1980. The formation of eroded depressions around the sources of lunar sinuous rilles: theory (abstract). Lunar and Planetary Science IX. Lunar and Planetary Institute, Houston, pp. 1260–1262.

Wilson, L., Head, J.W., 1981. Ascent and eruption of basaltic magma on the Earth and Moon. J. Geophys. Res. 86, 2971–3001.

Wood, J.A., 1972. Thermal history and early magmatism in the Moon. Icarus 16 (2), 229–240.

Wood, J.A., Dickey, J.S., Marvin, U.B., Powell, B.N., 1970. Lunar Anorthosites and a Geophysical Model of the Moon. Pergamon Press, New York, NY, pp. 965–988.

Wood, J.A., 1984. Moon over Mauna Loa: a review of hypotheses of formation of Earth's Moon 1984. In: Hartmann, W.K., Phillips, R.J., Taylor, G.J. (Eds.), Origin of the Moon, *Proceedings of the Conference "Origin of the Moon"* held in Kona, HI, October 13-16, 1984. Conference supported by NASA, 1986. Lunar and Planetary Institute, Houston, TX, p. 797.

Xiao, Z., 2016. Size-frequency distribution of different secondary crater populations: 1. Equilibrium caused by secondary impacts. J. Geophys. Res. 121, 2404–2425.

Xu, L., 2018. How China's lunar relay satellite arrived in its final orbit. Planet. Soc. http://www.planetary.org/blogs/guest-blogs/2018/20180615-queqiao-orbit-explainer.html.

Ye, P.J., Sun, Z.Z., Zhang, H., Li, F., 2017. An overview of the mission and technical characteristics of Change'4 Lunar Probe. Sci. Chin. Technol. Sci. 60 (5), 658–667. http://engine.scichina.com/publisher/scp/journal/SCTS/60/5/10.1007/s11431-016-9034-6?slug=full%20text.

Young, E.D., Kohl, I.E., Warren, P.H., Rubie, D.C., Jacobson, S.A., Morbidelli, A., 2016. Oxygen isotopic evidence for vigorous mixing during the Moon-forming giant impact. Science 351 (6272), 493–496. doi:10.1126/science.aad0525.

Zaets, I., Burlak, O., Rogutskyy, I., Vasilenkoa, A., Mytrokhyn, O., et al., 2011. Bioaugmentation in growing plants for lunar bases. Adv. Space Res. 47, 1071–1078.

Zhang, J., Dauphas, N., Davis, A.M., Leya, I., Fedkin, A., 2012. The proto-Earth as a significant source of lunar material. Nature Geoscience 5 (4), 251–255. www.nature.com/naturegeoscience.

Zhang, N., Parmentier, E.M., Liang, Y., 2013. A 3-D numerical study of the thermal evolution of the Moon after cumulate mantle overturn: The importance of rheology and core solidification. J. Geophys. Res. Planets 118, 1789–1804. http://doi.org/10.1002/jgre.20121.

Bibliography

Canup, R.M., 2004. Dynamics of lunar formation. Annu. Rev. Astron. Astrophys. 42 (1), 441–475.

Canup, R.M., Asphaug, E., 2001. Origin of the Moon in a giant impact near the end of the Earth's formation. Nature 412 (6848), 708–712.

Shearer, C.K., et al., 2006. Thermal and magmatic evolution of the Moon. Rev. Mineral. Geochem. 60, 365–518.

Warren, P., 1985. The magma ocean concept and lunar evolution. Annu. Rev. Earth Planet. Sci. 13, 201–240.

Chapter 10

Liquid water lake under ice in Mars's southern hemisphere—Possibility of subsurface biosphere and life

10.1 Mars' glorious past

Mars is the fourth planet from the Sun and the second-smallest planet in the Solar System, being larger than only Mercury. Unlike Earth, Mars has two moons, namely; *Phobos* and *Deimos*. The red color of Mars (Fig. 10.1) makes this planet unique in terms of external appearance. The red color we see in images of Mars is the result of iron rusting. Mars has approximately 18 wt% iron oxides in their ferrous (Fe^{2+}) and ferric (Fe^{3+}) oxidation states combined (Dreibus and Wänke, 1987), which can participate in photochemical reactions. Hydrogen peroxide reacts with catalytic ferrous iron (the standard Fenton reaction), which results in its oxidation to ferric iron and the production of hydroxyl radicals (Fenton, 1894). The Photo-Fenton reaction is a more efficient variation as it utilizes UV light to catalyze the recycling of the iron (in dissolved or in oxide form) (Oliveros et al., 1997; Tokumura et al., 2011). According to Bibring et al. (2006), liquid-water is not responsible for Mars being red.

Rocks and soil on the surface of Mars contain a kind of dust composed mostly of iron and small amounts of other elements such as chlorine and sulfur. The rocks and soil are eroded by wind and the dust is blown across the surface by ancient volcanoes. Recent evidence points to the very fine dust also being spread across the planet by water, backed up by the presence of channels and ducts across the surface of Mars.

The iron within the dust reacted with oxygen, producing a red rust color, while the sky appears red as storms carried the dust into the atmosphere. This dusty surface, which is between a few millimeters and 2 m deep, sits above hardened lava composed mostly of basalt (rock). The concentration of iron in this basalt is much higher than that on Earth, contributing to the red appearance of Mars.

Solomon et al. (2005) summarized the young Mars thus, "Mars was most active during its first billion years. The core, mantle, and crust formed within ~50 million years of Solar System formation. A magnetic dynamo in a convecting fluid core magnetized the crust, and the global magnetic field shielded a more massive early atmosphere against solar wind stripping. The Tharsis province became a focus for volcanism, deformation, and outgassing of water and carbon dioxide in quantities possibly sufficient to induce episodes of climate warming. Surficial and near-surface water contributed to regionally extensive erosion, sediment transport, and chemical alteration. Deep hydrothermal circulation accelerated crustal cooling, preserved variations in crustal thickness, and modified patterns of crustal magnetization."

According to Fairén et al. (2010)a, Mars has undergone three main climatic stages throughout its geological history, beginning with a water-rich epoch, followed by a cold and semiarid era, and transitioning into present-day arid and very cold desert conditions. These researchers consider that the just-mentioned three main global climatic eras also represent three different stages of planetary habitability: "an early, potentially habitable stage when the basic requisites for life as we know it were present (liquid water and energy); an intermediate extreme stage, when liquid solutions became scarce or very challenging for life; and the most recent stage during which conditions on the surface have been largely uninhabitable, except perhaps in some isolated niches."

The planet Mars has many Earth-like characteristics, but its evolution is different. Taking this aspect into

FIG. 10.1 Surface view of the red planet Mars. *Image courtesy of NASA. (Source: http://www.spaceanswers.com/wp-content/uploads/2012/06/Mars-red-surface.jpg).*

10.2 Mars—once a water-rich planet whose surface dried later

Just like the origin of water on Earth's Moon has been a topic of vigorous debate, the question of the origin of water on the planet Mars is also much debated, not least because it is related to the question of whether life ever existed on our red neighbor. A Mars Global Surveyor image showing Nanedi Vallis in the Xanthe Terra region of Mars, covering an area 9.8 km × 18.5 km, and revealing a canyon about 2.5 km wide (Fig. 10.2), is a clear evidence of the stable and repeated, if not persistent, flow of a low-viscosity fluid on Mars at certain times in its past history. The fluid was probably water, but the images could also suggest wind, ice, lava, even carbon dioxide or sulfur dioxide. Recently, results from the Mars Exploration Rover missions have shown that this liquid carried salts and precipitated hematite in concretions. According to McKay (2004), the case for water, we could say, is tight.

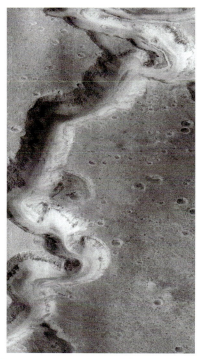

FIG. 10.2 **Water at certain times in the past history of Mars.** A Mars Global Surveyor image showing Nanedi Vallis in the Xanthe Terra region of Mars. The image covers an area 9.8 km × 18.5 km; the canyon is about 2.5 km wide. This image is the best evidence we have of liquid water at certain times in the past history of Mars. *Photo credit: NASA/ Malin Space Sciences. (Source: McKay, C.P. (2004), What is life—and how do we search for it in other worlds? PLoS Biol., 2(9):e302–e304. https://doi.org/10.1371/journal.pbio.0020302 Copyright: © 2004 Chris P. McKay. The article from which this figure is reproduced is an open-access article distributed under the terms of the Creative Commons Attribution License, which permits unrestricted use, distribution, and reproduction in any medium, provided the original work is properly cited).*

account, a European mission plan was conceived to study the planet Mars with the support of a network of Landers. Dehant et al. (2004) summarized the European mission plan thus, "An important future step in Mars' geophysics is to deploy a network of stations at the surface of Mars in order to study a wide range of properties of this planet, going from its deep interior structure to its atmosphere. Each ground station (small landers) will contain the same scientific instruments/experiments. The collected data will improve our knowledge of the Martian interior, surface and atmosphere, as well as its evolution. An important part of these objectives can only be achieved by a network of surface stations, as a network gives unique possibilities for performing studies of global scale phenomena and studies requiring simultaneous measurements from several sites."

Liquid water is widely viewed as one of the key prerequisites for life. If key nutrients such as nitrogen and phosphorus—in addition to liquid water—also were available, such locations could support life, and not necessarily just methane-producing organisms (Allen et al., 2006). Scientists have found evidence to believe that liquid water existed at least in some regions on Mars. According to Elkins-Tanton (2011), it is likely that by 4.45 Ga (giga-annum; i.e., billion years) ago, like Earth, early Mars also had developed a global 500 K ocean (or large bodies of water) enveloped in a ~100 bar, mostly carbon dioxide (CO_2) atmosphere. But nobody knows how long these oceans persisted. However, as Mars's temperature became more clement, it would open a first window of opportunity for prebiotic chemistry. On the basis of existence of liquid water on Mars sometime in the past, it is reasonably inferred that surface conditions at such locations (e.g., Meridiani) may have been inhabitable for some period of time in Martian history.

On the basis of isotopic analyses of Martian meteorites, a D/H ratio of 300×10^{-6} has been ascribed to the mantle of this planet (Leshin, 2000). This could imply a much larger contribution of cometary water on Mars than on Earth. It remains to be shown whether this scenario is compatible with the dynamical evolution of the small planets originally formed in the Solar System.

On Mars, water is photo-dissociated by the ultraviolet flux, and H, the lighter isotope of hydrogen, escapes to space at a much higher rate than D. The atmospheric D/H ratio is therefore much higher (810×10^{-6}) than that of the mantle (300×10^{-6}). The deuterium enrichment of the Martian mantle relative to that of Earth may therefore result from the recycling at depth of the deuterium-rich atmospheric water. Models of the dynamical evolution of the Martian interior can thus be constrained by the distribution of D/H ratios among different geochemical reservoirs of the planet.

It is interesting to note that Mars—the fourth planet from the Sun—has geological features like the Earth (the third planet from the Sun) and its Moon, such as craters and valleys, many of which were formed through rainfall. Simulation studies carried out by Black et al. (2017) suggests that the biggest impact craters on Mars formed very early in its history, and that later pummeling by asteroids mostly dented and dinged Mar's surface.

Mars, unlike Earth, has not undergone any active plate tectonics in its recent past. Black et al. (2017) have provided evidence that the major features of Martian topography formed very early in the history of the planet, influencing the paths of younger river systems, even as volcanic eruptions and asteroid impacts scarred the planet's surface. It is remarkable that there are three worlds (Earth, Mars, and Titan) in the Solar System where flowing rivers have carved into the landscape, either presently or in the past. On Earth, the process of plate tectonics has continuously reshaped the landscape, pushing mountain ranges up between colliding continental plates. Rivers, therefore, are constantly adapting to changes in topography, sidestepping around growing mountain ranges to reach the ocean. Whereas the upheaval of mountains by plate tectonics deflects the paths that rivers take on Earth, this telltale signature is missing from river networks on Mars (Fig. 10.3). Mars, on the other hand, is thought to have been shaped mostly during the period of primordial accretion and the so-called Late Heavy Bombardment, when asteroids carved out massive impact basins and pushed up huge volcanoes. Scientists now have well-resolved maps of river networks and topography on both Earth and Mars, along with a growing understanding of their respective histories.

Mars once harbored a huge ocean and rivers of water. Black et al. (2017) used a simulation that they previously developed, to model river erosion on Mars with different impact cratering histories. They found that the pattern of river networks on Mars today limits the extent to which

FIG. 10.3 Left to right: River networks on Mars and Earth. Mars, unlike Earth, has not undergone any active plate tectonics in its recent past and, therefore, river networks on Mars are rather linear. On the contrary, river networks on Earth (right) are deflected by tectonic activities. *Credit: Benjamin Black/NASA/Visible Earth/JPL/Cassini RADAR team. Adapted from images from NASA Viking, NASA/Visible Earth, and NASA/JPL/Cassini RADAR team (Source: https://www.sciencedaily.com/releases/2017/05/170518143817.htm).*

cratering has re-modeled the surface of Mars. Black et al. (2017) have provided evidence that the major features of Martian topography formed very early in the history of the planet, influencing the paths of younger river systems, even as volcanic eruptions and asteroid impacts scarred the planet's surface.

As just indicated, long ago, Mars harbored a huge ocean and hosted rivers of water. Those rivers scoured valleys across its now-arid surface. Although there is a growing body of evidence that there was once water on Mars, it does not rain there today. It may be noted that the only other world in the Solar System, apart from Earth where rain falls onto the surface at the present day, is Saturn's moon Titan that experiences liquid methane rainfall.

Several explanations have been proposed for the temporal differences in geologic processes associated with the modification of Martian impact craters, which occurred throughout the Noachian, and the formation of valley networks, which occurred during the Noachian/Hesperian transition. It has been proposed that the primary reason for the above-mentioned temporal differences in Mar's geological processes through ages could be the corresponding differences in Martian rainfall features. To understand how rainfall on Mars has changed over time, researchers had to consider how the Martian atmosphere has changed. It may be noted that when Mars first formed 4.5 billion years ago, it had a much more substantial atmosphere with a higher atmospheric pressure than it does now. It may be noted that the Martian atmosphere is rich in carbon dioxide (CO_2) and its atmospheric pressure ranged from 0.5 to 10 bars (in contrast, the Earth's atmospheric pressure at mean sea level [MSL] is 1 bar). Subsequently, the Mar's atmospheric pressure decreased over millions of years. The atmospheric pressure influences the size of the raindrops and how hard they fall.

According to a study carried out by Craddock and Lorenz (2017), the observed temporal differences in geologic processes, associated with the modification of Martian impact craters, could be a result of the changing nature of rainfall as the primordial atmospheric pressure on Mars waned through time. In an attempt to understand the changing nature of rainfall during the early history of Mars, Craddock and Lorenz (2017) calculated the terminal velocity and resulting kinetic energy from raindrops > 0.5 mm in diameter that would impact the surface of Mars. Note that the terminal velocity is dependent on gravity, and the lower Martian gravity results in correspondingly lower terminal velocities of the rain drops.

Craddock and Lorenz (2017)'s analyses indicate that the primordial atmospheric pressure of Mars could not have exceeded ~4.0 bars as raindrop sizes would have been limited to < 3 mm and surface erosion from rain splash and subsequent crater modification would not have occurred. At the Martian atmospheric pressures between ~3 and 4 bars, sediment transport from rain splash could occur, but surface runoff would have been limited, which could explain the modification of impact craters.

Once atmospheric pressures waned to ~1.5 bars, rainfall intensity could begin to exceed the infiltration capacity of most soils, which would be necessary to initiate Martian valley network formation. Whereas Earth's gravity is 9.807 m/s², Mars' gravity is only 3.721 m/s². Due to Mar's lower gravity, a storm on Mars that occurred in a 1 bar atmosphere could generate raindrops with a maximum diameter of ~7.3 mm compared to 6.5 mm on the Earth. However, rainfall from such a storm would only be ~70% as intense on Mars, primarily due to the lower Martian gravity and resulting lower terminal velocities of the rain drops.

The study carried out by Craddock and Lorenz (2017) implies that heavy rain on Mars reshaped the planet's impact craters and carved out river-like channels in its surface billions of years ago. They showed that changes in the atmosphere on Mars made it rain harder and harder, which had a similar effect on the planet's surface as we see on Earth. They show that there was rainfall in the past—and that it was heavy enough to change the planet's surface (see Fig. 10.4). To work this out, they used methods tried and tested here on Earth, where the erosive effect of the rain on the Earth's surface has important impacts on agriculture and the economy.

Investigations carried out by Craddock and Lorenz (2017) indicate that early on in the planet's existence, water

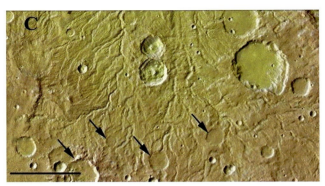

FIG. 10.4 Valley networks on Mars show evidence for surface runoff driven by rainfall. *Credit: Elsevier. (Sources: https://phys.org/news/2017-05-reveals-martian-rainfall-planet.html#nRlv; https://www.elsevier.com/about/press-releases/research-and-journals/how-hard-did-it-rain-on-mars).*

droplets would have been very small, producing something like fog rather than rain; this would not have been capable of carving out the planet we know today. As the atmospheric pressure decreased over millions of years, raindrops got bigger and rainfall became heavy enough to cut into the soil and start to alter the craters. The water could then be channeled and able to cut through the planet's surface, creating valleys.

By using basic physical principles to understand the relationship between the atmosphere, raindrop size and rainfall intensity, Craddock and Lorenz (2017) have shown that Mars would have seen some pretty big raindrops that would have been able to make more drastic changes to the surface than the earlier fog-like droplets. They showed that very early on, the atmospheric pressure on Mars would have been about 4 bars (the Earth's surface today is 1 bar) and the raindrops at this pressure could not have been bigger than 3 mm across, which would not have penetrated the soil. But as the atmospheric pressure fell to 1.5 bars, the droplets could grow and fall harder, cutting into the soil. In Martian conditions at that time, had the pressure been the same as we have on Earth, raindrops would have been about 7.3 mm—a millimeter bigger than on Earth.

Researchers admit that there could be some unknowns, such as how high a storm cloud have risen into the Martian atmosphere, but they made efforts to apply the range of published variables for rainfall on Earth. They reckon that it is unlikely that rainfall on early Mars would have been dramatically different than what is described in their paper (Craddock and Lorenz, 2017). They claim that their findings provide new, more definitive, constraints about the history of water and the climate on Mars.

10.3 Favorable position of Mars in Sun's habitable zone

Discovering life elsewhere in the Solar System and beyond is one of the great scientific challenges of our time.

Unfortunately, the Martian surface has been extremely cold, oxidizing and arid through most of its history, long before the evolution of eukaryotic life forms on Earth and for billions of years before the earliest evidence of surface life on land (Michalski et al., 2013). In terms of habitability, Earth and Mars are the only two planets that are favorably positioned in the Solar system. Have microbes ejected from Earth's atmosphere made their way to Mars? Or, on the other hand, have microbes ejected from Mars ever made their way to Earth? These are questions that have been relentlessly haunting the planetary life scientists. These issues can be better addressed by exploring Mars, a planet that shared with Earth a similar early geological history, particularly during the time when life is supposed to have appeared on Earth. Michalski et al. (2013) have argued that "the highest priority astrobiological targets on Mars should be portions of deep crust exhumed by impact and erosion, which could preserve evidence of organic chemicals from an era that is not preserved in Earth's geologic record. Alternatively, high-priority sites would also be areas where fluids have emerged from the Martian subsurface, possibly carrying clues to subsurface geochemistry and prebiotic or biotic processes. In fact, sites of groundwater upwelling might provide chemical gradients advantageous to life." Present exploration of Mars is focused on evaluating the habitability of surface environments.

The search for Martian life began in earnest when the Viking Project landed two robot space-crafts on the planet in 1976 to photograph the surface material and analyze it in detail. According to Biemann et al. (1977), one of the major goals of the Viking mission was to find out whether or not organic compounds exist on the surface of Mars and, if they do exist, to determine their structures and measure their abundances. According to them, considerable importance was attached to this mission because it was hoped that the nature of Martian organic molecules would provide a sensitive indicator of the chemical and physical environment in which they were formed. Furthermore, it was hoped that the details of their structures would indicate which of many possible biotic and abiotic syntheses are occurring on Mars.

The scientific exploration of Mars is driven by key science questions such as whether Mars was, is or can be a habitable world. This requires observations on geologic, climatic and atmospheric processes acting on Mars. Considering these requirements, five scientific payloads were selected, designed and developed and also operated from Martian orbit. Of these five instruments, two provided much information about the Martian atmosphere; the other two payloads were designed to provide information about the Martian surface and the fifth instrument was designed to measure particle environment in exosphere of Mars.

Striking a pessimistic note on the likelihood of the existence of living systems (that are similar to terrestrial biota) on Mars, Biemann et al. (1977) summarized the results of their analysis of Martian soil samples thus, "A total of four Martian samples, one surface and one subsurface sample at each of the two Viking landing sites, Chryse Planitia and Utopia Planitia, have been analyzed for organic compounds by a gas chromatograph–mass spectrometer. In none of these experiments could organic material of Martian origin be detected at detection limits generally of the order of parts per billion and for a few substances closer to parts per million. The evolution of water and carbon dioxide, but not of other inorganic gases, was observed upon heating the sample to temperatures of up to 500°C. The absence of organic compounds seems to preclude their production on the planet at rates that exceed the rate of their destruction. It also makes it unlikely that living systems that behave in a manner similar to terrestrial biota exist, at least at the two Viking landing sites."

Based on the favorable position of Mars in the Sun's habitable zone, Mars provides our second opportunity to search for traces of life. With its obvious atmosphere and polar ice caps, Mars was deemed a much more likely location in the Solar System to harbor extra-terrestrial life. Sublimation of water ice is more effective than evaporation of sorption (absorption and adsorption [the process by which a solid holds molecules of a gas or liquid or solute as a thin film] considered as a single process) water at the same temperature. Therefore, water in the form of ice must, over geologic time-scales, have left the upper Martian surface (m-scale) at mid- and low-latitudes, leaving sorption water as a possible physical form of stable subsurface water. Adsorption water is "liquid-like" at these temperatures (in the sense of a 2D-liquid). This property is the reason for the specific importance of physi-sorbed water under Martian conditions. Möhlmann (2004) showed that unfrozen adsorption water can cause numerous physical, chemical, and possibly also biological processes in the upper Martian surface and may be responsible for a number of its properties.

Vago et al. (2017) made the following observations with regard to the types of microorganisms that could have existed on early Mars. According to them, "The primordial types of microorganisms that could have existed on early Mars would have been tiny and of the order of a micron to a few microns in size. The individual cells would be too small to distinguish. However, as on Earth, their permineralized or compressed microbial colonies and biofilms would be much larger. Traces of these features may be preserved on Martian rocks as mineral-replaced structures and/or as carbonaceous remains trapped in sediments encased in mineral cement. Rover cameras and, in particular, high-resolution close-up imagers would be able to investigate many candidate microbialites similar to terrestrial thrombolites, stromatolites, layered biofilms, and abiotic/biotic organic particles and laminae (Westall, 2008; Westall et al., 2015b; Ruff and Farmer, 2016). Nevertheless, in more than 20 years of Mars surface exploration, and after having studied numerous examples of laminated sedimentary structures, there have been no claims gathering widespread support for the presence of biomediated structures."

10.4 Water-bearing minerals on Mars, its hydrologic history, and its potential for hosting life

Observations of Martian surface morphology have been used to argue that an ancient ocean once existed on Mars, and that Mars was once rich in water. Ancient fluvial networks on the surface of Mars suggest that it was warm and wet over three billion years ago. Irrespective of what happened on Mars in the past in terms of presence of water, determining whether liquid water exists at present on the Martian surface is central to understanding the hydrologic cycle and potential for extant life on Mars at present. It will be worthwhile discussing the rocky planets to better frame how their evolution may have affected the availability of liquid water; the timing of opportunities for prebiotic chemistry; and the possible emergence of life, its distribution, and its preservation record accessibility. Oyama and Berdahl (1977) found that immediate gas changes occurred when untreated Martian surface samples were humidified and/or wet by an aqueous nutrient medium in the Viking lander gas exchange experiment. The evolutions of N_2, CO_2, and Ar are mainly associated with soil surface desorption caused by water vapor. The presence of water-bearing minerals on Mars has long been discussed, but little or no data exist showing that minerals such as smectites and zeolites may be present on the surface in a hydrated state (i.e., that they could contain H_2O molecules in their interlayer or extra-framework sites, respectively).

Based on various repeated determinations of the bulk concentrations halogens in meteorites and the two-component model for the formation of terrestrial planets as proposed by Ringwood (1977, 1979), and Wanke (1981), Dreibus and Wänke (1987) argued that almost all of the water (H_2O) added to Mars during its homogeneous accretion was converted on reaction with metallic iron (Fe) to hydrogen gas (H_2), which escaped. Dreibus and Wänke (1987) found that by comparing the solubilities of H_2O and hydrogen chloride (HCl) in molten silicates, the amount of H_2O left in the mantle of Mars at the end of accretion can be related to the abundance of chlorine (Cl). In this way, H_2O content in the Martian mantle of 36 ppm was obtained, corresponding to an ocean covering the whole planet to a depth of about 130 m.

Dreibus and Wänke (1987) argued that the huge quantities of H_2 produced by the reaction of H_2O with metallic iron should also have removed other volatile species by hydrodynamic escape. Thus, they postulated that the present atmospheres of Venus, Earth, and Mars were formed by degassing the interiors of the planets, after the production of H_2 had ceased, that is, after metallic iron was no longer available. Dreibus and Wänke (1987) further postulated that the large differences in the amounts of primordial rare gases in the atmospheres of Venus, Earth, and Mars are due mainly to different loss factors.

Surface features resembling massive outflow channels provide evidence that, even more recently, the Martian crust contained the equivalent of a planet-wide reservoir of water several hundred meters deep. But arguments based on the isotopic fraction and present-day escape rate of hydrogen in the Martian atmosphere require only 0.5 m of crustal water today and about 6 m in the past. The SNC meteorites, thought to be igneous rocks from Mars, contain melt inclusions trapped at depth in early-formed crystals. Determination of the pre-eruptive water contents of SNC parental magmas from calculations of the solidification histories of these amphibole-bearing inclusions indicates that Martian magmas commonly contained 1.4% water by weight. When combined with an estimate of the volume of igneous materials on Mars, this information suggests that the total amount of water outgassed since 3.9 billion years ago corresponds to global depths on the order of 200 m. This value is significantly higher than previous geochemical estimates but lower than estimates based on erosion by floods. These results imply a wetter Mars interior than has been previously thought and support suggestions of significant outgassing before formation of a stable crust or heterogeneous accretion of a veneer (a thin covering) of cometary matter (McSween Jr and Harvey, 1993). An additional constraint on the evolution of the isotopic composition of Martian water has recently been obtained from measurements of the deuterium to hydrogen ratio (D/H ratio) of hydrous minerals in the Shergottite-Nakhlite-Chassignite (SNC) meteorites (Watson et al., 1994)—meteorites that almost certainly originated on Mars. Based on new data available on the D/H ratios in the Martian atmosphere and crust, Donahue (1995) showed that the modern crustal reservoirs of Martian water must be quite large, at least several meters in global-equivalent depth.

It has been suggested that significant quantities of water once thought to have existed on the Martian surface could have been supplied to the Martian surface through volcanic outgassing, but this suggestion is contradicted by the low magmatic water content that is generally inferred from chemical analyses of igneous Martian meteorites. However, McSween Jr. et al. (2001) reported the distributions of trace elements within pyroxenes of the Shergotty meteorite—a basalt body ejected 175 million years ago from Mars—as well as hydrous and anhydrous crystallization experiments that, together, imply that water contents of pre-eruptive magma on Mars could have been up to 1.8%. McSween Jr. et al. (2001) found that in the Shergotty meteorite, the inner cores of pyroxene minerals (which formed at depth in the Martian crust) are enriched in soluble trace elements when compared to the outer rims (which crystallized on or near to the Martian surface). This implies that water was present in pyroxenes at depth but was largely lost as pyroxenes were carried to the surface during magma ascent. McSween Jr. et al. (2001) concluded that ascending magmas possibly delivered significant quantities of water to the Martian surface in recent times, reconciling geologic and petrologic constraints on the outgassing history of Mars.

Except for gaseous species, Mars is found to be richer in volatile (halogens) and moderately volatile elements than the Earth. The resulting low release factor of ^{40}Ar for Mars is attributed to a low degree of fractionation, leading to a relatively small crustal enrichment of even the most incompatible elements like K.

Bish et al. (2003) analyzed experimental thermodynamic and X-ray powder diffraction data for smectite and the most common terrestrial zeolite, clinoptilolite, to evaluate the state of hydration of these minerals under Martian surface conditions. Thermodynamic data for clinoptilolite show that water molecules in its extra-framework sites are held very strongly, with enthalpies of dehydration for Ca-clinoptilolite up to three times greater than that for liquid water. Using these data, Bish et al. (2003) calculated the Gibbs free energy of hydration of clinoptilolite and smectite as a function of temperature and pressure. The calculations demonstrate that these minerals would indeed be hydrated under the very low partial pressure of H_2O {P (H_2O)} conditions existing on Mars, a reflection of their high affinities for H_2O. These calculations assuming the partial pressure of H_2O and the temperature range expected on Mars suggest that, if present on the surface, zeolites and Ca-smectites could also play a role in affecting the diurnal variations in Martian atmospheric H_2O because their calculated water contents vary considerably over daily Martian temperature ranges. The open crystal structure of clinoptilolite and existing hydration and kinetic data suggest that hydration/dehydration are not kinetically limited. Based on these calculations, it is possible that hydrated zeolites and clay minerals may explain some of the recent observations of significant amounts of hydrogen not attributable to water ice at Martian mid-latitudes.

Ruff (2004) found that an emissivity peak at $\sim$1630/cm observed as a spectral feature in Mars Global Surveyor Thermal Emission Spectrometer data ($\sim$1670–220/cm) of Martian surface dust is consistent with the presence of a water-bearing mineral. This spectral feature can be mapped globally and shows a distribution related to the classical bright regions on Mars that are known to be dust covered. An important spectral feature at $\sim$830/cm present in a newly derived average spectrum of surface dust likely is a transparency feature arising from the fine particulate nature of the dust. Ruff (2004) found that its shape and location are consistent with plagioclase feldspars and also zeolites, which essentially are the hydrous form of feldspar. It was found that the generally favored visible/near-infrared spectral analog for Martian dust, JSC Mars-1 altered tephra, does not display the $\sim$830/cm feature. It may be noted that zeolites commonly form from the interaction of low temperature aqueous fluids and volcanic glass in a variety of geologic settings. The combination of spectral features that are consistent with zeolites and the likelihood that Mars has (or had) geologic conditions necessary to produce them makes a strong case for recognizing zeolite minerals (the hydrous form of feldspar) as likely components of the Martian regolith.

Reports of $\sim$30 wt% of sulfate within saline sediments on Mars—probably occurring in hydrated form—suggest a role for sulfates in accounting for equatorial water (H_2O) observed in a global survey by the Odyssey spacecraft. Among salt hydrates likely to be present, those of the $MgSO_4 \cdot nH_2O$ series have many hydration states. Vaniman et al. (2004) reported the exposure of several of these phases to varied temperature, pressure and humidity to constrain their possible water contents under Martian surface conditions. They found that crystalline structure and water content are dependent on temperature–pressure history, that an amorphous hydrated phase with slow dehydration kinetics forms at <1% relative humidity, and that equilibrium calculations may not reflect the true water-bearing potential of Martian soils. Vaniman et al. (2004) found that magnesium sulfate salts can retain sufficient water to explain a portion of the Odyssey observations. Because phases in the $MgSO_4 \cdot nH_2O$ system are sensitive to temperature and humidity, they can reveal much about the history of water on Mars. However, according to Vaniman et al. (2004), the ease of transformation of $MgSO_4 \cdot nH_2O$ system implies that salt hydrates collected on Mars will not be returned to Earth unmodified, and that accurate in situ analysis is imperative.

Global mineralogical mapping of Mars by the Observatoire pour la Mineralogie, l'Eau, les Glaces et l'Activité (OMEGA) instrument on the European Space Agency's Mars Express spacecraft provides new information on Mars' geological and climatic history.

The OMEGA instrument was developed with the support of the Centre National d'Etudes Spatiales (CNES), Agenzia Spaziale Italiana (ASI), and Russian Space Agency. The scientific activity is funded by national space and research agencies and universities in France, Italy, Russia, Germany, and the United States.

The earliest observations of Mars by spacecraft showed that water had caused substantial erosion of the surface, including extensive channeling and associated transport and deposition of material (Carr, 1996). However, fundamental questions remain. Was water-driven activity on its surface transient or persistent, the latter being a prerequisite to sustain habitable surface environments? When and where did that activity take place, and when did it end? Mineralogical thermal infrared mapping from the Mars Global Surveyor

Thermal Emission Spectrometer (MGS TES) suggests that Mars must have been cold and dry over most of its history (Christensen et al. (2001a, 2001b, 201c). In a few locations, however, gray hematite was detected and was interpreted to result from aqueous processes (Christensen et al., 2001a, 2001d, 2001e). In Terra Meridiani, the Mars Exploration Rover (MER) Opportunity revealed that sulfates formed in the presence of water. Data from the Mars Express OMEGA instrument (Bibring et al., 2004a; Chicarro et al., 2004), in combination with local ground information from the MERs, provide extensive information on the aqueous history of Mars. OMEGA allows discrimination among gas, frost, ice, water absorbed, and water bound in hydrated minerals.

One of the main targets of ESA's Mars Express mission in January 2004 was to discover the presence of water in one of its chemical states. Through the initial mapping of the South polar cap on 18 January, OMEGA, the combined camera and infrared spectrometer, has already revealed the presence of water ice and carbon dioxide ice. Encouraged by recent results of NASA's year-2001 Mars Odyssey spacecraft mission and ESA's year-2004 OMEGA team (Mars Express) concerning water in equatorial latitudes between ±45° on Mars and the possible existence of hydrated minerals, Janchen et al. (2006) investigated the water sorption properties of natural zeolites and clay minerals close to Martian atmospheric surface conditions as well as the properties of Mg-sulfates and gypsum. To quantify the stability of hydrous minerals on the Martian surface and their interaction with the Martian atmosphere, the water adsorption and desorption properties of nontronite, montmorillonite, chabazite, and clinoptilolite have been investigated using adsorption isotherms at low equilibrium water vapor pressures and temperatures, modeling of the adsorption equilibrium data, thermogravimetry (TG), differential scanning calorimetry (DSC), and proton magic angle spinning nuclear magnetic resonance measurements (^{1}H MAS NMR). Mg-sulfate hydrates were also analyzed using TG/DSC methods to compare with clay mineral and zeolites. Janchen et al. (2006)'s data show that these micro-porous minerals can remain hydrated under present Martian atmospheric conditions and hold up to 2.5–25 wt% of water in their void volumes at a partial water vapor pressure of 0.001 mbar in a temperature range of 333–193 K. Results of the ^{1}H MAS NMR measurements suggest that parts of the adsorbed water are liquid-like water and that the mobility of the adsorbed water might be of importance for adsorption-water-triggered chemistry and hypothetical exobiological activity on Mars.

After 1 Martian year of operation in orbit around Mars, OMEGA has mapped 90% of the surface at a spatial sampling of 1.5–5 km and ~5% of the surface at a sampling of <0.5 km; on each pixel, the spectrum from 0.35 to 5.1 μm is acquired. OMEGA spectral data allow the identification of a variety of mafic and altered minerals, typically at a volume concentration of 5% or greater. For hydrated minerals and a few others, the sensitivity is even higher.

According to Bibring et al. (2006)'s investigations, phyllosilicates formed by aqueous alteration very early in the planet's history (the "phyllocian" era) are found in the oldest terrains; sulfates were formed in a second era (the "theiikian" era) in an acidic environment. Beginning about 3.5 billion years ago, the last era (the "siderikian") is dominated by the formation of anhydrous ferric oxides in a slow superficial weathering, without liquid water playing a major role across the planet. Two types of hydrated minerals have been identified from OMEGA data: phyllosilicates (Bibring et al., 2005; Poulet et al., 2005) and sulfates (Bibring et al., 2005; Gendrin et al.; 2005; Langevin et al., 2005; Arvidson et al., 2005), but in only a few locations (Poulet et al., 2005). Information from the Mars Express HRSC, the MGS MOC, and the Odyssey THEMIS images clearly indicates that the phyllosilicates are in rocks buried by more recent deposits; the hydrated silicate-bearing bedrock has been exposed through erosion.

Sulfates, including Mg sulfates (such as kieserite) and Ca sulfates (such as gypsum), constitute the second major class of hydrated minerals mapped by OMEGA and detected by the NASA rovers. OMEGA has shown that the sulfate-rich areas are not restricted to the gray hematite-rich regions detected by the MGS TES. Bibring et al. (2006) detected three principal types of hydrated sulfate deposits: layered deposits within Valles Marineris, extended deposits exposed from beneath younger units as in Terra Meridiani, and the dark dunes of the northern polar cap.

Martian scientists think that environments conducive to clay mineral formation may have existed at or near the surface or in the deeper subsurface. Surface or near-surface conditions would not require high-temperature conditions (hydrothermal, for example). Surface formation of these clay minerals would indicate a long-lasting wet episode, with large surface aqueous reservoirs and alkaline water resulting from this chemical alteration, occurring during the Noachian. Clay minerals could also have been formed primarily in the subsurface, by one of the three following processes (Bibring et al., 2006): hydrothermal activity; cratering, supplying subsurface water (liquid and/or ice) to the impacted minerals; or during the cooling of the mantle, if not thoroughly depleted of volatile compounds.

Sulfate mineral formation requires substantial quantities of water to account for the broad distribution of minerals seen by OMEGA. Because sulfate precipitation requires water to evaporate, it is essentially a surface process. For at least some of the sulfates identified, an acidic environment is also required. On this basis, we infer that extensive sulfate minerals formed from the late Noachian to the Hesperian, after the surface formation of phyllosilicates, which indicates a substantial change in the global aqueous chemistry of Mars.

It is believed that the period of peak volcanism on Mars was probably accompanied by a huge release of volatiles, including sulfur and water. Surface water would result from a combination of outgassing, hydrothermal activity, the rise of the water table, and massive outflows as Valles Marineris opened, over extended periods of time. Although carbonate minerals have been invoked as a possible reservoir for an early thick CO_2 atmosphere in contact with persistent liquid water, none have been detected above the OMEGA sensitivity limit of 4% in volume abundance. This result indicates that (i) the era during which surface liquid water remained stable did not last long enough to enable large amounts of CO_2 from the primordial denser atmosphere to be transformed into carbonates; or (ii) clay minerals were formed by impact or subsurface hydrothermal processes rather than by slow surface alteration within liquid water, leaving most of the primordial CO_2 in the atmosphere; or (iii) carbonates have been eliminated from the near surface by acidic weathering or decomposition; or (iv) Mars never sustained a dense CO_2-rich atmosphere.

After the formation of phyllosilicates and sulfates, the third alteration era began and has continued up to the present, in which liquid water did not play an important role. This is evidenced by the lack of hydration of the ferric oxides, in contrast with the detection of hydrated phyllosilicates and sulfates. Liquid water was probably present during transient and local events (such as the release of volatiles by impacts or the melting of ice deposits), but these episodes were too short to leave a substantial mark on the surface composition. Thus, these transient water events are not responsible for the global alteration, which has mostly been caused by surface oxidation and the production of nanophase ferric oxide, without hydration.

If indeed living organisms formed, the clay minerals could be the sites in which this biochemical development took place. The low level of the further surface alteration, in perennial cold and dry conditions, under a tenuous atmosphere, could have preserved most of the record of biological molecules, structures, or other diagnostic features in clay-rich surface or subsurface rocks. Bibring et al. (2006) consider that these areas of high habitability potential offer exciting targets for future in situ exploration.

In terms of ethical considerations, while pondering on the search for a second genesis of life on Mars, issues of contamination both with respect to the Martian environment and back contamination of the Earth are important concerns (McKay, 2016). Furthermore, it has been suggested that if there were a second genesis of life on Mars the global conditions are not favorable for it and humans may choose to intervene to improve them (McKay, 2009).

Although Earth, Venus, and Mars formed mainly from locally sourced material, the final stages of accretion blurred chemical differences among these planets by integrating contributions from elsewhere. In particular, Jupiter and Saturn's wanderings scattered objects in the region presently occupied by the asteroid belt and beyond. There are several researchers (e.g., Morbidelli et al., 2000; Albarède, 2009; Alexander et al., 2012; Marty et al., 2013; DeMeo and Carry, 2014; Hallis et al., 2015; Grazier, 2016; Meinert et al., 2016) who found reasons to believe that such objects played a role in delivering water and other volatiles (including prebiotic chemicals) not found in planetesimals formed closer to the proto-star.

Agee et al. (2013) reported data on the Martian meteorite Northwest Africa (NWA) 7034, which shares some petrologic and geochemical characteristics with known Martian meteorites of the SNC (i.e., shergottite, nakhlite, and chassignite) group, but also has some unique characteristics that would exclude it from that group. NWA 7034 is a geochemically enriched crustal rock compositionally similar to basalts and average Martian crust measured by recent Rover and Orbiter missions. It formed 2.089 ± 0.081 billion years ago, during the early Amazonian epoch in Mars' geologic history. Agee et al. (2013) found that NWA 7034 has an order of magnitude more indigenous water than most SNC meteorites, with up to 6000 parts per million extraterrestrial H_2O released during stepped heating. It also has bulk oxygen isotope values of $\Delta^{17}O = 0.58 \pm 0.05$ per mil and a heat-released water oxygen isotope average value of $\Delta^{17}O = 0.330 \pm 0.011$ per mil, suggesting the existence of multiple oxygen reservoirs on Mars.

10.5 Evidence for hydrated sulfates on Martian surface—Biological implication

Global mineralogical mapping of Mars by the OMEGA instrument on the European Space Agency's Mars Express spacecraft provides new information on Mars' geological and climatic history. Phyllosilicates formed by aqueous alteration very early in the planet's history (the "phyllocian" era) are found in the oldest terrains; sulfates were formed in a second era (the "theiikian" era) in an acidic environment.

The earliest observations of Mars by spacecraft showed that water had caused substantial erosion of the surface, including extensive channeling and associated transport and deposition of material (Carr, 1996). However, fundamental questions remain. Was water-driven activity on its surface transient or persistent, the latter being a prerequisite to sustain habitable surface environments? When and where did that activity take place, and when did it end? Mineralogical thermal infrared mapping from the Mars Global Surveyor Thermal Emission Spectrometer (MGS TES) suggests that Mars must have been cold and dry over most of its history (Christensen et al., 2001a, 2001b, 2001c). In a few locations, however, gray hematite was detected and was interpreted to result from aqueous processes (Christensen et al., 2001a, 2001d, 2001e). In Terra Meridiani, the Mars Exploration Rover (MER) Opportunity revealed that sulfates formed in the presence of water and that the hematite found by Thermal Emission Spectrometer (TES) consists of concretions weathered from the sulfate deposits (Squyres et al., 2004b). New data from the Mars Express OMEGA instrument (Bibring et al., 2004b; Chicarro et al., 2004), in combination with local ground information from the MERs, provide extensive information on the aqueous history of Mars and the mineralogical evolution of its crust.

In addition to their younger age, the sulfate-rich deposits observed by OMEGA extend over a large region. Their size requires a large source of sulfur that has been proposed to be a direct consequence of the extensive outpourings of lavas and associated degassing that primarily formed Tharsis Plateau, as well as the northern plains (in a lunar mare–like process) and Hesperian ridged plains (Phillips et al., 2001; Solomon et al., 2005). This period of peak volcanism was probably accompanied by a huge release of volatiles, including sulfur and water. This sulfur would have been rapidly oxidized in the atmosphere to form H_2SO_4, which then precipitated on the surface. Surface water would result from a combination of outgassing, hydrothermal activity, the rise of the water table, and massive outflows as Valles Mariner is opened, over extended periods of time. All the ingredients were in place for extensive sulfate deposition, either by means of acidic alteration or by weathering of both mafic minerals and, locally, their phyllosilicate alteration products.

OMEGA spectral data allowed the identification of a variety of mafic and altered minerals. OMEGA data contributed immensely to the search for the potential role that water played in that alteration: OMEGA allowed discrimination among gas, frost, ice, water absorbed, and water bound in hydrated minerals. Two types of hydrated minerals have been identified from OMEGA data: phyllosilicates (Bibring et al., 2005; Poulet et al., 2005) and sulfates (Bibring et al., 2005; Gendrin et al., 2005; Langevin et al., 2005; Arvidson et al., 2005), but in only a few locations (Poulet et al., 2005).

Sulfates, including Mg sulfates (such as kieserite) and Ca sulfates (such as gypsum), constitute a major class of hydrated minerals mapped by OMEGA and detected by the NASA rovers (Squyres et al., 2004b). OMEGA has shown that the sulfate-rich areas are not restricted to the gray hematite-rich regions detected by the MGS TES (Christensen et al., 2001a, 2001d, 2001e). The three principal types of hydrated sulfate deposits detected on Mars are: (1) layered deposits within Valles Marineris, (2) extended deposits exposed from beneath younger units as in Terra Meridiani, and (3) the dark dunes of the northern polar cap (Bibring et al., 2005; Gendrin et al., 2005; Langevin et al., 2005; Arvidson et al., 2005).

As already indicated, the Martian orbital and landed surface missions, OMEGA on Mar Express and the two Mars Explorations Rovers, respectively, have yielded evidence pointing to the presence of magnesium sulfates on the Martian surface. In situ identification of the hydration states of magnesium sulfates, as well as the hydration states of other Ca- and Fe-sulfates, are considered to be crucial in future landed missions on Mars in order to advance our knowledge of the hydrologic history of Mars as well as the potential for hosting life on Mars. Raman spectroscopy is a technique well-suited for landed missions on the Martian surface. Wang et al. (2006) reported a systematic study of the Raman spectra of the hydrates of magnesium sulfate. In their study, characteristic and distinct Raman spectral patterns were observed for each of the 11 distinct hydrates of magnesium sulfates, crystalline and noncrystalline. The unique Raman spectral features along with the general tendency of the shift of the position of the sulfate v_1 band toward higher wave-numbers with a decrease in the degree of hydration allow in situ identification of these hydrated magnesium sulfates from the raw Raman spectra of mixtures. Using these Raman spectral features, Wang et al. (2006) started the study of the stability field of hydrated magnesium sulfates and the pathways of their transformations at various temperature and relative humidity conditions. In particular they reported on the Raman spectrum of an amorphous hydrate of magnesium sulfate ($MgSO_4 \cdot 2H_2O$) that may have specific relevance for the Martian surface.

Since the Viking missions in 1976, magnesium sulfates have been predicted to exist on the surface of Mars. Recent orbital measurements suggest that Mg-sulfates are rather ubiquitous on the Martian surface. Chemical analyses by landers support the inference that Mg-sulfate hydrates may be one source of the significant quantities of equatorial near-surface hydrogen observed by the neutron and γ-ray spectrometers on the Mars Odyssey spacecraft. Chipera and Vaniman (2007) examined stability relations among the various Mg-sulfate hydrates. Using saturated salt solutions to control water-vapor pressure at temperatures of 3°C, 23°C, 50°C, 63°C, and 75°C, Mg-sulfate phases were allowed to equilibrate from 2 to 3 months to see which hydration states were formed or were stable. Starting materials consisted of hexahydrite ($6H_2O$), starkeyite ($4H_2O$), kieserite ($1H_2O$), a second monohydrate-polymorph available as a chemical reagent, and an anhydrous $MgSO_4$ reagent. Products created in this study included these minerals, along with epsomite ($7H_2O$), sanderite ($2H_2O$), amorphous $MgSO_4$ ($1-2H_2O$), several previously un-described phases, one of which was quite persistent ($2.4H_2O$), and trace amounts of pentahydrite ($5H_2O$). As expected, Mg-sulfate stability was found to be strongly dependent on water vapor pressure and temperature. Lower temperatures favor the more hydrated Mg-sulfates. However, the $MgSO_4$ system was found to be surprisingly complicated and is strongly dominated by metastability, sluggish kinetics, and reaction pathways. Unexpected results were frequently encountered, in addition to the formation of previously un-described phases. Several of the hydrates also were found to show significant metastable extensions, such that phase boundaries can only be approximated. For example, kieserite, which has been reported on Mars from OMEGA data, in addition to having a distinct stability region, is resistant to transformation and persists throughout temperature-RH space until very high relative humidities are achieved. Results of Chipera and Vaniman (2007)'s study show that $MgSO_4$ hydrates in addition to epsomite, hexahydrite, and kieserite can persist and should not be overlooked when assessing possible Mg-sulfate minerals that can occur on Mars.

Recurring slope lineae—narrow streaks of low reflectance compared to the surrounding terrain—appear and grow incrementally in the down-slope direction during warm seasons when temperatures reach about 250–300 K, a pattern consistent with the transient flow of a volatile species. Brine flows (or seeps) have been proposed to explain the formation of recurring slope lineae, yet no direct evidence for either liquid water or hydrated salts has been found. Ojha et al. (2015) analyzed spectral data from the Compact Reconnaissance Imaging Spectrometer for Mars instrument onboard the Mars Reconnaissance Orbiter from four different locations where recurring slope lineae are present. They found evidence for hydrated salts at all four locations in the seasons when recurring slope lineae are most extensive, which suggests that the source of hydration is recurring slope lineae activity. The hydrated salts most consistent with the spectral absorption features they detected are magnesium perchlorate, magnesium chlorate and sodium perchlorate. Their findings strongly support the hypothesis that recurring slope lineae form as a result of contemporary water activity on Mars. In September 2015, the Mars Reconnaissance Orbiter spectroscopically detected hydrated salts of $NaClO_4$, $Mg(ClO_4)_2$, and $Mg(ClO_3)_2$ in locations thought to be brine seeps. This may be the first direct evidence for flowing liquid water containing hydrated salts on Mars (Ojha et al., 2015).

One of the challenges to biology at Meridiani Planum on Mars deals with the persistence of water. It was found that the rocks at Eagle crater expose tens of centimeters of sulfate-rich stratigraphy. At nearby Endurance crater, the exposed thickness of similar rocks approaches 10 m. Despite this substantial thickness, the apparent prevalence of sulfate-cemented eolian sands suggest that water on Meridiani Planum may have been regionally extensive but temporally discontinuous, increasing the difficulty of biological persistence over long time intervals.

It is difficult to determine whether life was present or even possible in the waters at Meridiani, but it is clear that by the time the sedimentary rocks in Eagle crater were deposited, Mars and Earth had already gone down different environmental paths. Sample return of Meridiani rocks might well provide more certainty regarding whether life

developed on Mars. Sulfate deposits are known to preserve both chemical and morphological fossils (Bonny and Jones, 2003), and iron oxide precipitates at Rio Tinto contain beautifully preserved fossils, including minute fossils of coccoid and filamentous microorganisms (Fernández-Remolar et al., 2002). Meridiani Planum therefore can be considered an attractive candidate for further study, both by landed missions and by sample return.

10.6 Day–night variations in liquid interfacial water in Martian surface

Adsorption is the adhesion of atoms, ions or molecules from a gas, liquid or dissolved solid to a surface. This process creates a film of the adsorbate on the surface of the adsorbent. Microscopic liquid layers of water can evolve via adsorption on grain and mineral surfaces at and in the soil of the surface of Mars. The upper parts of these layers will start to freeze at temperatures clearly below the freezing point of bulk water (freezing point depression). A sandwich structure with layers of ice (top), liquid water (in between) and mineral surface (bottom) can evolve. Möhlmann (2008) described the properties of the interfacial water (of adsorption water and premelted ice) on grain surfaces by a sandwich-model of a layer of liquid-like adsorption water between the adsorbing mineral surface layer and an upper ice layer. It was shown that the thickness or number of mono-layers of the interfacial water (of adsorption water and premelted ice) depends on temperature and atmospheric relative humidity. It was found that the derived equations for the sandwich model fit well to a known phenomenological relation between thickness of the liquid layer and relative humidity, and can be a tool to estimate or to determine for appropriate materials (e.g., Hamaker's constant) for van der Waals interactions on grains and in porous media. The curvature of grain surfaces is shown to have no remarkable effects for particles in the μm-range and larger. The application of these equations to thermo-physical conditions on Mars shows that the thickness of frost-layers, which can evolve over several hours on cooling surface parts of Mars, is typically of the order or a few tenths of 1 mm or less. This is in agreement with observations. Furthermore, Möhlmann (2008) derived an equation, which relates the freezing point depression for van der Waals force govened interfacial water to the value of the Hamaker constant, to the latent heat of solidification, to the mass density of water ice, and to the thickness of the liquid-like layer. Again, this equation was found to fit well to a known phenomenological relation between freezing point depression and thickness of the liquid-like layer. The derived equation shows that the lower limiting temperature of the liquid phase can reach about 180 K under Martian conditions. An Equilibrium Moisture Content (EMC)/Equilibrium Relative Humidity (ERH) relation for the water content of Martian soil has been derived, which relates, for equilibrium conditions, soil water content and atmospheric relative humidity. This relation indicates that the content of liquid interfacial water in the upper surface of Mars can reach up to 10% by weight and more in course of saturation during night hours, and it can be of about 2% by weight during the dry daytime hours (Möhlmann, 2008).

10.7 Clues leading to existence of water at Mars' subsurface

Scientists have brought out ample evidences for the past and even present existence of water on Mars' surface. For example, the possibility of precipitation of ice at low latitudes on Mars during periods of high obliquity has been highlighted by Jakosky and Carr (1985). Head et al. (2003) found that there have even been ice ages on Mars. Forget et al. (2006) observed that formation of glaciers on Mars happened by atmospheric precipitation at high obliquity. Likewise, Christensen (2006) found that water is present at the poles and in permafrost regions of Mars. According to Clifford et al. (2010), the water periodically mobilized throughout the Martian surface through obliquity-driven climate change has probably recharged the subsurface through basal melting.

Deposits of sedimentary rocks are considered to be relics of ancient groundwater activity. According to Burt and Knauth (2003), ancient fluids that arrived in the saturated zone during periods of relatively high recharge rates on ancient Mars are likely to have become dense owing to a high concentration of dissolved solids, acquired during transport through reactive, mafic rocks. McGovern et al. (2004) have suggested that even if average surface temperatures were tens of degrees below freezing, hydrothermal conditions could have existed at depths of several kilometers during the Noachian, given reasonable estimates for ancient thermal gradients (Michalski et al., 2013). Clifford et al. (2010) have suggested that owing to its reduced gravity, Mars contains more subsurface porosity to a greater depth than Earth.

Planetary scientists consider that the Martian crust is probably hydrologically heterogeneous owing to impact fragmentation and other processes. Fortunately, based on various inputs, theoretical groundwater models provide guidance as to how and where groundwater upwelling might have occurred (see Andrews-Hanna et al., 2010). Results of such model studies have indicated that upwelling would have occurred first in deep basins (see Andrews-Hanna et al., 2010). Relying on this prediction, Michalski et al. (2013) proposed that a rising groundwater table would serve to alter and cement sediments available in those upwelling zones, resulting in deposits of sedimentary rocks, whose presence is indicative of ancient groundwater activity. Robbins and Hynek (2012) reported that the northern hemisphere of Mars contains 95 considerably large (diameter ≥40 km), deep (rim-floor depth ≥2 km) impact basins), approximately 80% of which are Noachian in age. Michalski et al. (2013) reckon that at least one of these basins contains strong evidence for groundwater activity.

Subsurface processes on Mars could be studied indirectly, either by the analysis of deep crustal rocks that have been exhumed by impact or through investigation of materials formed from subsurface fluids (Andrews-Hanna et al., 2010), where they have reached the surface. It is believed that the largest fraction of Martian water probably occurs in the subsurface as calcium-rich briny groundwater (Burt and Knauth, 2003), a deep cryosphere (Clifford et al., 2010), and water in hydrated minerals (Ehlmann et al., 2011). According to Lammer et al. (2012), water volcanically outgassed during the Pre-Noachian and Noachian periods on Mars would have ultimately become locked into the cryosphere, sequestered into the subsurface, and/or lost to space. Michalski et al. (2013) have suggested that water that infiltrated into the subsurface would have met one of several fates as pore water, ground ice, or structural and absorbed water in hydrated minerals.

Among a multitude of craters existing on Mars, McLaughlin Crater (see Fig. 10.5) on Mars is a large (diameter = 92 km), deep (2.2 km) crater located at 337.6°E, 21.9°N. It has been found that the surface layer of the McLaughlin Crater contains sulfates (Bibring et al., 2006) and layered clays (Poulet et al., 2005), as well as mobile surface sediments, snow and ice deposits, and dust (Tanaka, 2000). Edgett and Malin (2002) have suggested that this zone may occupy the uppermost hundreds to thousands of meters of the surface of many regions of Mars (see Michalski et al., 2013). According to Hurowitz et al. (2010), fluids in this zone would have been largely affected by interaction with atmospheric SO_2, Cl^- and oxidizing agents.

McLaughlin Crater presents a compelling case for groundwater upwelling. Spectra acquired of the floor of McLaughlin Crater by the Compact Reconnaissance Imaging Spectrometer for Mars (CRISM), corrected for instrument and atmospheric effects, contain absorptions at (λ =)1.4, 1.9, and 2.3 μm that are attributable to Fe–Mg-rich clay minerals (Michalski et al., 2013). Absorptions at 2.305 and 2.5–2.52 μm are also consistent with the presence of Mg-carbonates, probably mixed with clay minerals. According to Michalski et al. (2013), the deposits in McLaughlin Crater could have very high preservation potential for organic materials, in much the same manner as turbidites do on Earth (Piper, 1972). Michalski et al. (2013) argued that "the observations in McLaughlin Crater suggest the basin contained an ancient lake in one of the most likely settings where groundwater would have emerged. The evidence for alteration minerals rich in Fe and Mg, and the indication of alkaline fluids based on carbonate detections, are consistent with the expected character of fluids that would have emerged from the subsurface" from some depth zones.

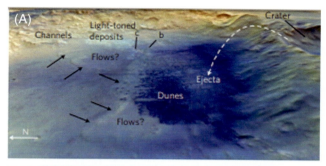
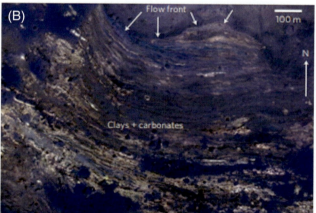

FIG. 10.5 **Geology of McLaughlin Crater.** (A), Color image data from the High-Resolution Stereo Camera (HRSC) draped onto elevation data from the same instrument show the locations of features, including possible flow fronts indicated by black arrows. The locations (B), (C) contain image data from the High-Resolution Imager for Mars (HiRISE) showing altered sediments within lobate flows (B) and layered rocks in the floor of the crater (C). *(Source: Michalski et al. (2013), Groundwater activity on Mars and implications for a deep biosphere, Nat. Geosci., doi: http://doi.org/10.1038/ngeo1706.)*

On Mars, it is not only McLaughlin Crater but Arabia Terra and elsewhere (Andrews-Hanna et al., 2010) also show evidence of groundwater activity, although more than 80 deep, ancient basins throughout the northern hemisphere do not exhibit the same evidence (see Michalski et al., 2013). In Arabia Terra and elsewhere sulfate-rich sedimentation that has been attributed to groundwater upwelling has been found. Michalski et al. (2013) have reported that some deep craters do contain interior mounds of sediments that might be attributable to groundwater-mediated sedimentation. Michalski et al. (2013) have argued that "lacustrine clay minerals and carbonates in McLaughlin Crater might be the best evidence for groundwater upwelling activity on Mars, and therefore should be considered a high-priority target for future exploration."

In a study, Lasue et al. (2013) have reviewed current estimates of the global water inventory of Mars, potential loss mechanisms, the thermo-physical characteristics of the different reservoirs that water may be currently stored in, and assess how the planet's hydrosphere and cryosphere evolved with time. According to them "the water inventory quantified from geological analyses of surface features related to both liquid water erosion, and ice-related landscapes indicate that, throughout most of Martian geologic history (and possibly continuing through to the present day), water was present to substantial depths, with a total inventory ranging from several 100 to as much as 1000 m Global Equivalent Layer (GEL)." Based on the most recent estimates of water content based on subsurface detection by orbital and landed instruments, including deep penetrating radars such as SHARAD and MARSIS, Lasue et al. (2013) showed that "the total amount of water measured so far is about 30 m GEL, although a far larger amount of water may be stored below the sounding depths of currently operational instruments." Lasue et al. (2013) have also discussed a global picture of the current state of the subsurface water reservoirs and their evolution.

10.8 Searching for presence of liquid water at the base of the Martian polar caps using ground-penetrating radar on *Mars Express* spacecraft

The polar caps of Mars have long been acknowledged to be composed of unknown proportions of water ice, solid CO_2 (dry ice), and dust (Wieczorek, 2008). The presence of liquid water at the base of the Martian polar caps was first hypothesized by Clifford (1987). According to Clifford (1987), "If ice is present throughout the cryosphere anywhere on Mars, thermodynamic considerations suggest that it is most likely at the poles. Given this condition, the deposition of dust and H_2O at the polar surface will ultimately result in a situation where the equilibrium depth to the melting isotherm has been exceeded, melting ice at the base of the cryosphere until thermal equilibrium is once again established. Should deposition persist, the polar deposits will ultimately reach a thickness where melting will occur at their actual base. At this point the cap may reach a state of equilibrium, where the deposition of any additional ice is balanced by geothermal melting." Clifford (1987)'s thermal calculations yielded basal melting thicknesses that are consistent with the inferred 4–6 km thickness of the present north polar cap; however, in the south the deposits appear sufficiently thin (1–2 km) that geothermal melting is likely to be relegated (assigned an inferior rank or position) to a depth that lies well below the regolith-polar cap interface. Clifford (1987) arrived at similar conclusions based on consideration of the polar caps' theoretical equilibrium profiles. According to Clifford (1987), "In the north a basal yield stress characteristic of ice at or near the melting point is indicated, while in the south it appears the cap has yet to achieve the necessary height for significant deformation to occur at its base." Clifford (1987)'s analysis suggests that the process of basal melting may play a key role in understanding the evolution of the Martian polar terrains and the long-term climatic behavior of water on Mars. Clifford's hypothesis on the presence of liquid water at the base of the Martian polar caps has been inconclusively debated ever since its publication in 1987.

NASA is participating in a mission of the European Space Agency and the Italian Space Agency called Mars Express, which has been exploring the atmosphere and surface of Mars from polar orbit since arriving at the red planet in 2003. Planetary scientists believed that Mars may possess a global subsurface groundwater table as an integral part of its current hydrological system. Several investigators surveyed the Martian subsurface, using radar, for more than a decade in search of evidence of liquid water, and began to wonder whether the Martian water table is hidden from radar view. The Mars Advanced Radar for Subsurface and Ionospheric Sounding (MARSIS) on Mars Express is the first sounding radar operating from orbital altitudes since the Apollo 17 Lunar Sounder flown in 1972 (Jordan et al., 2009). The radar operates from a highly elliptical orbit but acquires data only from altitudes lower than 1200 km. This radar has been successfully operating since August 2005. According to Jordan et al. (2009), "The radar is a dual channel low-frequency sounder, operates between 1.3 and 5.5 MHz (Mega Hertz) with wavelengths between 230 and 55 m in free space for subsurface sounding and between 0.1 and 5.5 MHz (wavelengths between 3000 and 55 m) for ionospheric sounding. The subsurface sounder can operate at one or two-frequency bands out of four available bands at either like or cross polarization. The subsurface sounding radar transmits radio frequency (RF) pulses of 250 µs duration through a 40 m dipole antenna. The return echoes are then converted to digital form and temporarily stored on board for some digital processing. A second antenna, a monopole, provides reception for the cross-polarized return and its data are processed by a second channel. This processing reduces the data rate produced by the instrument to rates allowed by the spacecraft communications channel. These processed returns are then sent to Earth by the telecommunications system on the spacecraft. The advances in digital data acquisition and processing, since 1972, have enabled this technique to be used in a compact spacecraft science instrument. This sounder has obtained returns from several kilometers below the surface of the Mars."

Unfortunately, the MARSIS instrument—a ground-penetrating radar (Picardi et al., 2005)—onboard the Mars Express (MEx) spacecraft could not succeed in making a definitive detection of a body of liquid water on Mars. Ground-penetrating radars use radio signals that are capable of penetrating into the ground and ice-caps, and then get reflections from the material under the surface. Water is especially reflective of radar, making the tool theoretically useful in the search for the life-sustaining liquid.

In the background of the not so encouraging experience from the MARSIS instrument in detecting subsurface water on Mars, Farrell et al. (2009) quantified the conditions that would allow a detection of a deep aquifer and argued that the lack of direct detection by MARSIS—a "null result"—does not uniquely rule out the possibility of the water table's existence.

10.9 The year-2018 discovery of an underground liquid water lake in Mars' southern hemisphere

Like Earth, Mars has ice caps at its poles. Water reaches the poles as vapor and is frozen into thin layers that build up thick deposits. Mixed with this water is dust picked up by the wind, so the caps have bright and dark layers of "clean" and "dirty" ice. Abundant water ice is also present beneath the permanent carbon dioxide ice cap at the Martian south pole and in the shallow subsurface at more temperate conditions (Bibring et al., 2004).

In an experimental investigation, Plaut et al. (2007) probed the ice-rich south polar layered deposits on Mars' southern ice cap with the Mars Advanced Radar for Subsurface and Ionospheric Sounding (MARSIS) instrument on the Mars Express spacecraft. It was found that the radar signals could penetrate deep into the deposits (more than 3.7 km). For most of the area, a reflection was detected at a time delay that was consistent with an interface such that the reflected power from this interface indicated minimal attenuation of the signal. These features were interpreted as due to the propagation of the radar signals through a very cold layer of pure water ice having negligible attenuation (Plaut et al., 2007). The results of this study on Mars' southern ice cap suggested "a composition of nearly pure water ice whose total volume was estimated to be 1.6×10^6 km^3, which is equivalent to a global water layer approximately 11 m thick" (Plaut et al. 2007).

It has been believed that large, deep craters in Mars' northern hemisphere might be candidate sites for groundwater upwelling activity, with implications for a deep biosphere.

Investigations carried out by Michalski et al. (2013a), based on spectra acquired of the floor of McLaughlin Crater by the Compact Reconnaissance Imaging Spectrometer for Mars (CRISM), provided credible indications that the crater basin contained an ancient lake in one of the most likely settings where groundwater would have emerged. According to these investigators, the evidence for alteration minerals rich in iron (Fe) and manganese (Mg), and the indication of alkaline fluids based on carbonate detections, are consistent with the expected character of fluids that would have emerged from the subsurface from some depth zones. Note that mineral alteration refers to the various natural processes that alter a mineral's chemical composition or crystallography. On Mars, it is not only McLaughlin Crater but Arabia Terra and elsewhere (Andrews-Hanna et al., 2010) also show indications of groundwater activity. However, apart from some "indications," the presence of liquid water on Mars was not confirmed.

As just indicated, presence of liquid water at the base of the Martian polar caps has long been suspected but not observed unambiguously. According to Orosei et al. (2018), radio echo sounding (RES) is a suitable technique to resolve this dispute, because low-frequency radars have been used extensively and successfully to detect liquid water at the bottom of terrestrial polar ice sheets. An interface between ice and water, or alternatively between ice and water-saturated sediments, produces bright radar reflections (Carter et al., 2007; Ashmore and Bingham, 2014). Based on radar reflection properties pertaining to subglacial lakes in East Antarctica Carter et al. (2007) found that "Definite lakes are brighter than their surroundings by at least 2 dB (relatively bright) and both are consistently reflective (specular) and have a reflection coefficient greater than −10 dB (absolutely bright). Dim lakes are relatively bright and specular but not absolutely bright, indicating nonsteady ice dynamics. Fuzzy lakes are both relatively and absolutely bright, but not specular, and may indicate saturated sediments or 'swamps.' Indistinct lakes are absolutely bright and specular but no brighter than their surroundings."

Orosei et al. (2018) surveyed a region known as the Planum Australe in the southern hemisphere of Mars (Fig. 10.6) using the MARSIS instrument on the Mars Express spacecraft. The compelling evidence for detection of a huge reservoir of liquid water buried under the base of a thick slab of polar ice in the Planum Australe region (centered at 193°E, 81°S, which is surrounded by much less reflective areas) was found based on a long and painstaking analysis of data collected between May 2012 and December 2015 using MARSIS. According to Orosei et al. (2018), "Quantitative analysis of the radar signals shows that this bright feature has high relative dielectric permittivity (>15), matching that of water-bearing materials." Orosei et al. (2018) interpret this feature as "a stable body of liquid water on Mars."

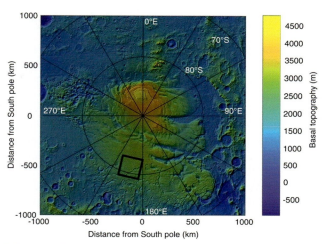

FIG. 10.6 Shaded relief map of Planum Australe, Mars, south of 75°S latitude. The map was produced using the MOLA topographic dataset. The black square outlines the study area. *(Source: Orosei et al. (2018), Radar evidence of subglacial liquid water on Mars, Science, 361 (6401): 490-493. doi: http://doi.org/10.1126/science.aar7268.) Website: https://science.sciencemag.org/content/361/6401/490/tab-figures-data.*

Orosei et al. (2018) consider that the subsurface reservoir spotted in the Planum Australe region is a sizable salt-laden liquid water lake under ice (resembling a subglacial lake on Earth) on the southern polar plain of Mars. The Planum Australe region studied by Orosei et al. (2018) is "topographically flat, composed of water ice with 10–20% admixed dust (Zuber et al., 2007; Li et al., 2012), and seasonally covered by a very thin layer of CO_2 ice that does not exceed 1 m in thickness (Aharonson et al., 2004; Litvak et al., 2007)." The location's radar profile resembled those of subglacial lakes found beneath Earth's Antarctic (southern hemisphere) and Greenland (northern hemisphere) ice sheets. The subsurface reservoir spotted in the Planum Australe region appears to be a sizable salt-laden liquid water lake under ice (resembling a subglacial lake on Earth) on the southern polar plain of Mars. Although the water in the Martian lake is below the normal freezing point, it remains liquid presumably because of the presence of high levels of dissolved salts and the pressure of the ice above.

The reservoir—roughly 12 miles (20 km) in diameter, shaped like a rounded triangle (Fig. 10.7) and located about a mile (1.5 km) beneath the ice surface—represents the first stable body of liquid water ever found on Mars. Evidence of past bodies of water and periodic flowing water has been discovered, but this is the first time a present body of water has been found on Mars (Orosei et al., 2018). Finding liquid water today on Mars, even below the surface, would greatly enhance the prospects for extant life on that planet.

The detection of liquid water by the Mars Advanced Radar for Subsurface and Ionosphere Sounding (MARSIS) at the base of the south polar layered deposits in Ultimi

Scopuli has reinvigorated the debate about the origin and stability of liquid water under present-day Martian conditions. To establish the extent of subglacial water in this region, Lauro et al. (2021) acquired new data, achieving extended radar coverage over the study area. In this study, Lauro et al. (2021) applied the traditional signal processing procedures usually applied to terrestrial polar ice sheets. Lauro et al. (2021)'s results strengthen the claim of the detection of a liquid water body at Ultimi Scopuli and indicate the presence of other wet areas nearby. These investigators suggest that the waters are hypersaline perchlorate brines, known to form at Martian polar regions and thought to survive for an extended period of time on a geological scale at below-eutectic temperatures. Note that a eutectic system is a system of a homogeneous mixture of substances that either melts or solidifies at a particular given temperature that is lower than the melting point of any of the mixture of any of the constituent elements. This particular temperature is known as the eutectic point.

10.10 Indications suggestive of Mars' subsurface harboring a vast microbial biosphere

Whether any celestial body other than Earth has ever harbored life is one of the supreme questions that remains unanswered in the history of science. Subglacial lakes provide sufficient protection from harmful radiation. Therefore, scientists like to think that this body of water could be a possible habitat for microbial life. The finding of liquid water raises the likelihood that any microbial life that arose on Mars may continue to eke out a rather bleak existence deep beneath the surface. However, planetary scientists are of the view that it could take years to verify (perhaps with a future mission, drilling through the ice to sample the water below) whether something is actually living in this body of water. However, given the fact that there are microorganisms on Earth (terrestrial organisms), which are capable of surviving and thriving even in the Antarctic glaciers (see Chapter 7), planetary scientists are looking at this discovery with a high positive note. The European Space Agency had a Mars rover in the works, set to land on our neighboring planet Mars in 2021. But that didn't happen. Subsequently, the ExoMars 2022 mission was planned for launch during a 12-day launch window starting on 20 September 2022, and scheduled to land on Mars on 10 June 2023 ("The way forward to Mars." ESA. 1 October 2020). It would have included a German-built cruise stage and Russian descent module (Clark, Stephen, 2019; "ExoMars rover leaves British factory, heads for testing in France." *Spaceflight Now*.). On 28 February 2022, the ESA announced that, as a result of sanctions related to the 2021–2022 Russo-Ukrainian crisis, a 2022 launch is "very unlikely." (Foust, Jeff, 2022, "ESA says it's 'very unlikely' ExoMars will

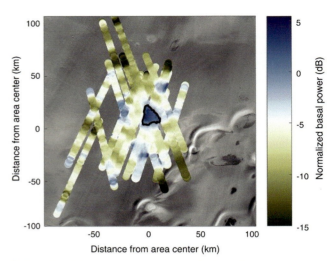

FIG. 10.7 Color-coded map of normalized basal echo power at 4 MHz, superimposed on the infrared image corresponding to the black square in Fig. 10.6. The large blue area (positive values of the normalized basal echo power) outlined in black corresponds to the main bright area; the map also shows other, smaller bright spots that have a limited number of overlapping profiles. The large blue area outlined in black—roughly 12 miles (20 km) in diameter, shaped like a rounded triangle and located about a mile (1.5 km) beneath the ice surface is a liquid-water reservoir. *(Source: Orosei et al. (2018), Radar evidence of subglacial liquid water on Mars, Science, 361 (6401), 490-493. doi: http://doi.org/10.1126/science.aar7268.) Website: https://science.sciencemag.org/content/361/6401/490/tab-figures-data.*

launch this year." *SpaceNews Now*.). On March 28th, the ExoMars rover was confirmed technically ready for launch, but confirmed the 2022 launch window for the mission is no longer possible. The next launch window will open in late 2024. The rover will be able to drill 2 m below the surface, searching for signs of life.

10.10.1 Prokaryotic life in Earth's deep subsurface offers clues to Mars' subsurface microbial biosphere

Although the investigation of life's origins on Earth will always be limited by the poor state of preservation of the earliest geologic record (>3.5 Gyr old), the prevailing general consensus among scientists suggests that the earliest record of life on Earth consists of prokaryote thermophiles (Furnes et al., 2004) in the Noachian (>4.1 Gyr ago) and into the Early Hesperian (~3.7–4.1 Gyr ago) periods. At present, prokaryotic life in the deep subsurface comprises up to 50% of the total biomass on Earth (Whitman et al., 1998). According to Heberling et al. (2010), a significant amount of diversity exists throughout the huge volume of Earth's subsurface habitable environments that may reach >5 km depth. Rothschild and Mancinelli (2001) suggested that thermophiles on Earth uniquely survived the Late Heavy Bombardment by taking refuge in the Earth's subsurface. Martin (2011) reckons that because chemoautotrophs and thermophiles are some of the oldest phyla on Earth, it is reasonable to assume that life may have originated in the Earth's subsurface by taking advantage of existing chemical gradients associated with serpentinization reactions.

Based on such evaluations pertaining to the early life on Earth, Michalski et al. (2013) argued that similar situation must have prevailed on Mars as well and, therefore, the most viable habitat for ancient, simple life forms on Mars could have been its subsurface. Thus, although the Martian surface had become extremely inhospitable by the time eukaryotic life or photosynthesis evolved on Earth, it is believed that the subsurface of Mars could potentially have contained a vast microbial biosphere. Some Martian explorers reckon that crustal fluids may have welled up from the subsurface to alter and cement surface sediments, potentially preserving clues to subsurface habitability. Taking account of these and similar arguments put forward by life-researchers, Michalski et al. (2013) suggest that the Martian subsurface is the place to search for evidence of habitability.

10.10.2 Model studies suggestive of subsurface Martian biosphere

Mars is home for large impact craters (see Ehlmann et al., 2011; Ehlmann et al., 2009; Fairén et al., 2010a). According to Michalski et al (2013), one of the most important discoveries in the exploration of Mars has been the detection of putative hydrothermal phases, including serpentine (Ehlmann et al., 2010) and phyllosilicates (Ehlmann et al., 2011), within materials exhumed from the subsurface by such large impact craters (see Fig. 10.8).

Michalski et al. (2013) reported a synthesis model of the subsurface geology of Mars, with predictions for the nature and fate of fluids in the crust and testable hypotheses for the habitability of various zones at depth. They also examined the prospects of subsurface habitability of Mars and evaluated evidence for groundwater upwelling in deep basins. They also reported evidence that crustal fluids have emerged at the surface, resulting in an alkaline lacustrine system within McLaughlin Crater. Although many ancient, deep basins lack evidence for groundwater activity, Michalski et al. (2013) found that McLaughlin Crater, one of the deepest craters on Mars, contains evidence for magnesium–iron-bearing clays and carbonates that probably formed in an alkaline, groundwater-fed lacustrine setting. According to Michalski et al. (2013), this environment strongly contrasts with the acidic, water-limited environments implied by the presence of sulfate deposits that have previously been suggested to form owing to groundwater upwelling. Michalski et al. (2013) argue that deposits formed as a result of groundwater upwelling on Mars, such as those in McLaughlin Crater, could preserve critical evidence of a deep biosphere on Mars.

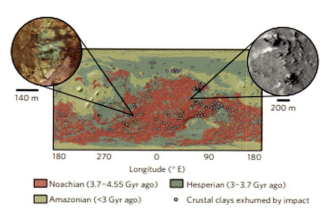

FIG. 10.8 **Distribution of exhumed deep crustal rocks on Mars.** Detections of deep crustal clays reported previously (by Ehlmann et al., 2011) are overlaid on global surface geology. Exhumed clays in Noachian terrains represent subsurface hydrothermal processes early in Mars's history. Insets show textures of two examples of exhumed crust: hydrated minerals along with mafic mineralogy exhumed from a ~2.5-km-deep unnamed crater at 306.4°E, 20.5°S (left) and Fe–Mg clays and Fe/Ca carbonates exhumed from ~6 km deep in Leighton Crater (right). *(Source: Michalski et al., (2013), Groundwater activity on Mars and implications for a deep biosphere, Nature Geoscience, doi: http://doi.org/10.1038/ngeo1706).*

10.10.3 Arguments in support of a deep biosphere on Mars

Mars possesses a lower gravity relative to that of Earth (0.375 that of Earth). This lower gravity implies less compaction of Mars' subsurface pore space. Furthermore, Mars' lower heat flow reduces the temperature constraints. These two factors together support the viability of a microbial community to be greater on Mars's subsurface than at similar depths on Earth (see Sleep and Zahnle, 1998). According to Michalski et al. (2013), the distinction of diverse geological zones (zones 1–4) within the Martian crust allows prediction of how the potential for a subsurface Martian biosphere (Fisk and Giovannoni, 1999) varies with depth.

Michalski et al. (2013) have prescribed various energy sources for the survival of microbes in the various depth zones. These investigators suggested that in the deepest zone (zone 4), a diversity of metabolic mechanisms may be expected, using hydrogen, carbon dioxide and possibly abiotic hydrocarbons, but it has been assumed that as on Earth the dominant fuel for microbes in the deep Martian subsurface would be hydrogen (Reith, 2011), possibly provided by serpentinization (Hellevang et al., 2011), radiolysis (Lin et al., 2005) or fault friction (Hirose et al., 2011). Michalski et al. (2013) suggested that the deep crust may have always been the most habitable environment on Mars. According to them Zone 4 (the deepest zone) would have been energy-rich, but potentially more nutrient-limited whereas zone 3 could have included diverse chemistry in a context with ample energy and nutrients. They proposed that, "simple life could have evolved into habitats in zones 1 and 2. Zone

2, at shallower depth and with less alteration, offers porous habitats protected from harsh surface conditions. These specifically might include vesicular basalts, which are a widely colonized habitat on Earth, and fracture systems associated with impact hydrothermal systems, also colonized in the terrestrial geological record (Parnell et al., 2010). Zone 1 offers other targets that are inhabited on Earth, including ice with brine inclusions (Mader et al., 2006), sulfates and layered clays." Note that vesicular basalt is a dark-colored volcanic rock that contains many small holes, more properly known as vesicles. A vesicle is a small cavity in a volcanic rock that was formed by the expansion of a bubble of gas that was trapped inside the lava.

According to Michalski et al. (2013), carbon for biomass could have been derived from the magmatic carbon in basalts whose occurrence is observed in Martian meteorites (Steele et al., 2012). Likewise, in the depth-zone 3, "CO_2-bearing fluids infiltrated from the surface could have reacted with the hydrogen from serpentinization to form methane, which could have fuelled methanotrophs and been mediated by methanogens. The alteration of basalt also releases cations, for example, Fe^{2+} and Mg^{2+}, which can fuel iron respiration (Holden and Feinberg, 2005) and facilitate the formation of complex organic molecules (Holm, 2012), respectively. Groundwater recharge and upwelling events provide mechanisms for replenishment of nutrients, as well as possible redox-based habitats at the interface of seepage pathways and host rocks, as is observed to be an important control on Earth (Fredrickson and Balkwill, 2006). The combination of recharging fluids and serpentinization can give rise to unusual alkaline (Ca/Mg-rich) springs with distinctive microbial communities, proposed as a potential model for Mars (Blank et al., 2009)." Several investigators have suggested that groundwater upwelling on Mars may have occurred sporadically on local scales, rather than at regional or global scales.

10.10.4 Early Mars' potential as a better place for the origin of life compared to early Earth

According to Weiss and Kirschvink (2000), it is important to compare the probable environments of the early Earth with that of early Mars in order to evaluate which of these two bodies, during the first half-billion years of the Solar System, might have produced an environment most suitable for the origin of life based on redox chemistry. Kump et al. (2001) have argued that during Archean time, both planets possessed at least local environments sufficiently reducing to allow the accumulation of prebiotic compounds. Weiss and Kirschvink (2000) argue that the question then focuses on determining which planet would have contained a better source of oxidizing atmospheric compounds capable of diffusing into this primordial soup (a solution rich in organic compounds in the primitive oceans of the Earth, from which life is thought to have originated), which Charles Darwin envisaged, to promote organic evolution.

Farquhar et al. (1998)'s isotopic studies of the ~4 Ga old Martian meteorite ALH84001 revealed that Mars' surface environment about 4 billion years ago was neutral to oxidizing, resulting in the formation of carbonates and ozone. Pavlov et al. (2001)'s studies indicated that this ozone would have shielded an early Martian biosphere from UV radiation, a benevolent protection presumably lacking on Earth during this time. Furthermore, Chang and Kirschvink (1989)'s studies of magneto-fossils and magnetization of sediments found in the Martian meteorite ALH84001 indicated the possible presence of vertical redox gradients, which magneto-tactic bacteria need for their survival. Interestingly, the oldest magneto-fossils yet identified on Earth are from the post-Snowball Gunflint Chert at about 2.1 Ga (Chang and Kirschvink, 1989), roughly coincident with the appearance of the first eukaryotic algae (Han and Runnegar, 1992).

The deuterium/hydrogen (D/H) ratio of the present-day Martian atmosphere is ~5 times that of the Earth, indicating that a large amount of H (and, by implication, H_2O) in the Martian atmosphere has been lost to space (Weiss and Kirschvink, 2000). However, the $^{18}O/^{16}O$ ratio is not much different than the terrestrial value (Owen, 1992), indicating that the protection afforded to the early Martian atmosphere by Mars' strong magnetic field at 4 Ga or earlier (Acuna et al., 1999) might have meant dramatically more loss of H than O, and so provided a cascade of oxidants to drive organic evolution on the early Mars. A recent study of Martian meteorites (Wadhwa, 2001) has shown that the mantle of Mars was more reduced than that of the Earth, while the crust of Mars is still quite oxidized. Such oxidizing conditions at the Martian surface would imply that Mars had much larger redox gradients than Earth at the same time. According to Weiss and Kirschvink (2000), both hydrogen and oxygen loss have had a much less significant effect on Earth's atmosphere both because of Earth's higher gravity and strong magnetic field (the latter reduces solar-wind-induced ionization as well as loss of ionized species due to interaction with the solar-wind magnetic field).

Considering that the last common ancestor (LCA) of living organisms on Earth is believed to have evolved on a world in which redox gradients were large, the early Mars' much larger redox gradients than early Earth's assume special significance in terms of early Mars' potential as a better place for the origin of life. Weiss and Kirschvink (2000) argue that, "at face value, all of these lines of evidence suggest that, compared to early Earth, early Mars might have had a greater supply of biologically useable energy and was perhaps, by implication, a better place for the origin of life."

10.11 Search for biosignatures and habitability on early Mars—*ExoMars* mission

Some fundamental questions that have intrigued mankind for centuries include, "Are we alone, or is there life beyond Earth? Has life ever existed on Mars?" Finding signs of past life and identifying locations where life (at least in the form of microbial organisms) could be found are the ultimate goals of extra-terrestrial missions, particularly to Mars.

10.11.1 *ExoMars* mission to Mars—An astrobiology program of the European Space Agency (ESA) and the Russian Space Agency Roscosmos

The European Space Agency (ESA) is an intergovernmental organization of 22 member states dedicated to the exploration of space. ESA was established in 1975 and it is headquartered in Paris.

The Roscosmos State Corporation for Space Activities, commonly known as Roscosmos, is a state corporation of the Russian Federation responsible for space flights, cosmonautics programs, and aerospace research. Originating from the Soviet space program founded in the 1950s, Roscosmos emerged following the dissolution of the Soviet Union in 1991. Roscosmos is headquartered in Moscow, with its main Mission Control Center in the nearby city of Korolyov, and the Yuri Gagarin Cosmonaut Training Center located in Star City in Moscow Oblast.

The first part of the program is a mission launched in 2016 that placed the Trace Gas Orbiter into Mars orbit and released the Schiaparelli EDM lander. The reported presence of methane in the atmosphere of Mars is of interest to many geologists and astrobiologists, as methane may indicate the presence of microbial life on Mars, or a geochemical process such as volcanism or hydrothermal activity.

The Trace Gas Orbiter (TGO) and a test stationary lander called Schiaparelli were launched on 14 March 2016. TGO entered Mars orbit on 19 October 2016 and proceeded to map the sources of methane (CH_4) and other trace gases present in the Martian atmosphere that could be evidence for possible biological or geological activity. The TGO features four instruments and will also act as a communications relay satellite. The Schiaparelli experimental lander separated from TGO on 16 October and was maneuvered to land in Meridiani Planum, but it crashed on the surface of Mars. The landing was designed to test new key technologies to safely deliver the subsequent Rover mission.

The second part of the program was planned to launch in July 2020, when the Kazachok lander would have delivered the Rosalind Franklin rover on the surface, supporting a science mission that was expected to last into 2022 or beyond. On 12 March 2020, it was announced that the second mission was being delayed to 2022 as a result of problems with the parachutes, which could not be resolved in time for the launch window.

The goals of ExoMars are to search for signs of past life on Mars (Vago et al., 2017), investigate how the Martian water and geochemical environment varies, investigate atmospheric trace gases and their sources and by doing so demonstrate the technologies for a future Mars sample-return mission.

ExoMars mission is an exobiology mission to the Red Planet Mars. With regard to the ExoMars Project mission, Baglioni et al. (2006) summarized the Mars exploration plans of ESA thus, "In the framework of its Aurora Exploration Program, the European Space Agency (ESA) has initiated the ExoMars Project. Aside from searching for traces of life at and near the Martian surface, ExoMars aims to characterize the Martian geochemistry and water distribution at various locations, improve the knowledge of the Mars environment and geophysics, and identify possible hazards for future missions. Beyond the ExoMars Project, there is a strong scientific and technical interest in Europe to participate in an international Mars Sample Return (MSR) mission. ESA's overall approach to MSR includes a combination of system studies, technology development work, and the identification of capability development approaches including the consideration of demonstration missions. In the longer term, the approach intends to allow the strategic down-selection of specific areas in which Europe can play a strategic role in the eventual MSR mission."

According to Vago et al. (2017), the beginnings of the ExoMars Rover mission program can be traced to 1996, when ESA tasked an exobiology science team with formulating guidelines for future search-for-life missions in the Solar System. This group was active during 1997–1998, an exciting period in Mars exploration. Success of an exploratory mission to an extra-terrestrial body such as Mars depends on various factors. One such factor is the payload. The payload of a rover, in the present context, is its carrying capacity, measured in terms of weight. Depending on the nature of the mission, the payload may include scientific instruments or experiments, or other equipment. The ExoMars Rover adorns a payload, known as Pasteur payload—which is a comprehensive suite of instruments that will characterize the Martian biological environment. The Pasteur payload is considered to be the key to the success of this scientific ques.

The mission strategy to achieve the ExoMars rover's scientific objectives is as follows (Vago et al. (2017):

- To land on an ancient location possessing high exobiological interest for past life signatures, that is, access the appropriate geological environment.
- To collect well-preserved samples (free from radiation and oxidation damage) at different sites using a rover equipped with a drill capable of reaching well into the ground and surface rocks.

- To conduct an integral set of measurements at multiple scales to achieve a coherent understanding of the geological context and, thus, inform the search for biosignatures. Beginning with a panoramic assessment of the geological environment, the rover must progress to smaller scale investigations of surface rock textures and culminate with the collection of well-selected samples to be studied in its analytical laboratory.

Vago et al. (2017) described Pasteur payload instruments thus, "The rover's Pasteur payload will produce comprehensive sets of measurements capable of providing reliable evidence for, or against, the existence of a range of biosignatures at each search location. The Pasteur payload contains panoramic instruments [cameras, an infrared (IR) spectrometer, a ground-penetrating radar, and a neutron detector]; contact instruments for studying rocks and collected samples (a close-up imager and an IR spectrometer in the drill head); a subsurface drill capable of reaching a depth of 2 m and obtaining specimens from bedrock; a sample preparation and distribution system (SPDS); and the analytical laboratory, the latter including a visual + IR imaging spectrometer, a Raman spectrometer, and a laser desorption, TV GC–MS (with the possibility to use three different derivatization agents)." Fig. 10.9 shows the sketch of ExoMars rover showing the location of the drill and the nine Pasteur payload instruments. In this sketch, ISEM (see Korablev et al., 2015, 2017 for details) is a pencil-beam IR spectrometer mounted on the rover mast that is coaligned with the PanCam high-resolution camera. ISEM is designed to record IR spectra of solar light reflected off surface targets, such as rocks and soils, to determine their bulk mineralogical composition. ISEM is considered to be a very useful tool to discriminate between various classes of minerals at a distance. This information can be employed to decide which target to approach for further studies. ISEM can also be used for atmospheric studies (Vago et al., 2017).

According to Vago et al. (2017), the WISDOM radar shown in the sketch (Ciarletti et al., 2011, 2017) will characterize stratigraphy to a depth of 3–5 m with vertical resolution of the order of a few centimeters (depending on subsurface electromagnetic properties). "WISDOM will allow the team to construct two- and three-dimensional subsurface maps to improve our understanding of the deposition environment. Most importantly, WISDOM will identify layering and help select interesting buried formations from which to collect samples for analysis. Targets of particular interest for the ExoMars mission are well-compacted, sedimentary deposits that could have been associated with past water-rich environments. This ability is fundamental to achieve the rover's scientific objectives, as deep subsurface drilling is a resource-demanding operation that can require several sols," Vago et al. (2017).

Likewise, according to Vago et al. (2017), ADRON (active detector for gamma rays and neutrons) (Mitrofanov

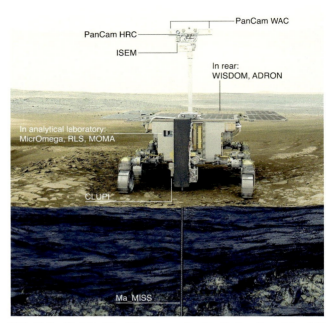

FIG. 10.9 Sketch of ExoMars rover showing the location of the drill and the nine Pasteur payload instruments; PanCam: Panoramic camera equipped with a suite of a fixed-focus, wide-angle, stereoscopic, color camera pair (WAC) complemented by a focusable, high-resolution, color camera (HRC); ISEM: a pencil-beam IR spectrometer mounted on the rover mast that is coaligned with the PanCam high-resolution camera; WISDOM: Water, Ice, and Subsurface Deposit Observations on Mars radar; ADRON: Active Detector for gamma Rays and neutrONs; CLUPI: CLose-UP Imager; Ma_MISS: Mars Multispectral Imager for Subsurface Studies; MicrOmega: very-near IR hyperspectral camera; MOMA: Mars Organic Molecule Analyzer; RLS: Raman laser spectrometer. *(Source: Vago, J.L., et al. (2017), Habitability on early Mars and the search for biosignatures with the ExoMars Rover, Astrobiology, 17(6-7): 471-510). © Jorge L. Vago et al., 2017; Published by Mary Ann Liebert, Inc. The Open Access article from which this Figure is reproduced is distributed under the terms of the Creative Commons License (http://creativecommons.org/licenses/by/4.0), which permits unrestricted use, distribution, and reproduction in any medium, provided the original work is properly credited. E-mail: jorge.vago@esa.int.*

et al., 2017) will count the number of thermal and epithermal neutrons scattered in the Martian subsurface to determine hydrogen content (present as grain adsorbed water, water ice, or in hydrated minerals) in the top 1 m. This information will complement the subsurface characterization performed by WISDOM.

CLUPI (Close-up imager) is designed to obtain high-resolution, color images to study the depositional environment (Josset et al., 2017). According to Vago et al. (2017), "By observing textures in detail, CLUPI can recognize potential morphological biosignatures, such as biolamination, preserved on surface rocks. This is a key function that complements the possibilities of PanCam when observing close targets at high magnification. CLUPI will be accommodated on the drill box and have several viewing modes, allowing the study of outcrops, rocks, soils, the fines produced during drilling, and also imaging collected samples in high resolution before delivering them to the analytical laboratory."

Vago et al. (2017) have described Ma_MISS thus, "Ma_MISS (Mars multispectral imager for subsurface studies) (De Angelis et al., 2013; De Sanctis et al., 2017) is a miniaturized IR spectrometer integrated in the drill tool for imaging the borehole wall as the drill is operated. Ma_MISS provides the capability to study stratigraphy and geochemistry in situ. This is important because deep samples may be altered after their extraction from their cold, subsurface conditions, typically of the order of −50°C at mid latitudes (Grott et al., 2007). The analysis of unexposed material by Ma_MISS, coupled with other data obtained with spectrometers located inside the rover, will be crucial for the unambiguous interpretation of rock formation conditions."

MicrOmega is a visible very-near-infrared hyperspectral microscope that is designed to characterize the texture and composition of Martian samples presented to the instrument within the ExoMars rover's analytical laboratory drawer (Bibring et al., 2017). This microscope is designed to study mineral grain assemblages in detail to try to unravel their geological origin, structure, and composition. According to Bibring et al. (2017), "The spectral range (0.5–3.65 μm) and the spectral sampling (20 cm^{-1} from 0.95 to 3.65 μm) of MicrOmega have been chosen to allow the identification of most constituent minerals, ices/frosts, and organics with astrobiological relevance within each 20 × 20 μm^2 pixel over a 5 × 5 mm^2 field of view. Such an unprecedented characterization will enable (1) identification of most major and minor phases, including the potential organics; (2) ascription of their mineralogical context, as a critical set of clues with which to decipher their formation process; and (3) location of specific grains or regions of interest in the samples, which will be further analyzed by Raman Laser Spectrometer and/or Mars Organic Molecule Analyzer." According to Vago et al. (2017), these data will be vital for interpreting past and present geological processes and environments on Mars. The rover computer can analyze a MicrOmega hyperspectral cube's absorption bands at each pixel to identify particularly interesting minerals.

MOMA (Mars Organic Molecule Analyzer) is the largest instrument in the rover, and the one that directly targets chemical biosignatures. MOMA is able to identify a broad range of organic molecules with high analytical specificity, even if present at very low concentrations (Arevalo et al., 2015; Goetz et al., 2016; Goesmann et al., 2017a).

RLS (Raman laser spectrometer) (Edwards et al., 2013; Foucher et al., 2013; Lopez-Reyes, 2015; Rull et al., 2017) provides geological and mineralogical information on igneous, metamorphic, and sedimentary processes, especially regarding water-related interactions (chemical weathering, chemical precipitation from brines, etc.). In addition, it also permits the detection of a wide variety of organic functional groups. Raman can contribute to the tactical aspects of exploration by providing a quick assessment of organic content before the analysis with MOMA.

10.11.2 Prospects of lava tube caves on Mars as an extant habitable environment

On Earth, a powerful magnetic shield, known as the magnetosphere, protects all living bodies including humans from the harsh radiation of space. Without such a powerful magnetic shield, a constant stream of electromagnetic rays would damage our cells and DNA, with dire consequences to our health (Eisler, 2012). Ionized particles, streaming through space as solar wind at the Martian atmosphere or cosmic rays add to that risk. The NASA Mars Odyssey spacecraft, equipped with a special instrument called the Martian Radiation Experiment (MARIE), detected ongoing levels of radiation that were considerably higher than what astronauts experience on the International Space Station (~200 μSv/day). Furthermore, there is always the risk that a solar flare or cosmic ray burst could expose a Martian habitat to a sudden, deadly dose.

Martian surface has witnessed occasional solar storms as well. For example, in September 2017, strong energetic particles from a coronal mass ejection were detected both in Mars orbit and on the surface (Fig. 10.10). In orbit, the Mars Atmosphere and Volatile Evolution (MAVEN) spacecraft detected energetic particles as high as 220 million E (eV), which produced radiation levels on the surface more than double any previously measured by Curiosity's

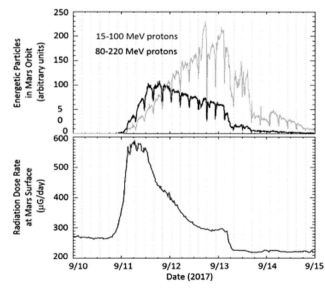

FIG. 10.10 Radiation data collected by MAVEN and RAD during the September 2017 solar storm. *(Source: NASA). Reproduced from Paris, A.J., E.T. Davies, L. Tognetti, and C. Zahniser (2019), Prospective lava tubes at Hellas Planitia: Leveraging volcanic features on Mars to provide crewed missions protection from radiation, The Journal of The Washington Academy of Sciences, arXiv:2004.13156 [astro-ph.EP].*

Radiation Assessment Detector (RAD) (Gebhardt, 2017). RAD is an energetic particle detector capable of measuring all charged particles that contribute to the radiation health risks that future crewed missions to Mars will face. During the coronal mass ejection, RAD detected radiation levels at ~600 μSv/day—over twice the normal background radiation at Gale Crater (Fig. 10.10). Studies have shown that prolonged exposure to strong energetic particles will lead to health complications—including acute radiation sickness (nausea, vomiting, weakness, headaches, purpura, hemorrhage, infections, diarrhea, leukopenia), genetic damage, and possibly death (Seedhouse, 2018).

Paris et al. (2019) summarized the health risks the explorers reaching on Mars' surface are expected to be subject to, thus: "Though astronauts in low-earth orbit are more exposed to radiation than humans on the ground, they are still protected by Earth's magnetosphere. Outside our magnetosphere, however, radiation is more problematic. Research studies of exposure to various strengths and doses of radiation provide strong evidence that degenerative diseases and/or cancer are to be expected from too much exposure to solar energetic particles and/or galactic cosmic rays (National Aeronautics and Space Administration. Why Space Radiation Matters (available at https://www.nasa.gov/analogs/nsrl/why-space-radiation-matters/)). Galactic cosmic radiation contains highly ionizing heavy ions that have large penetration power in tissue and that may produce extremely large doses, leading to early radiation sickness or death if adequate shelter is not provided (National Aeronautics and Space Administration. Space Radiation Cancer Risk Projections for Exploration Missions, JSC-29295 (available at https://pdfs.semanticscholar.org/1edf/1764e7d4cb9648e0e759ea235cac2db755ce.pdf)). During the Apollo program, for illustration, astronauts on the moon reported headaches, reported seeing flashes of light, and experienced painful cataracts (Hecht et al., 1942). These symptoms, known as Cosmic Ray Visual Phenomena, were due to radiation from cosmic rays interacting with matter and deposing its energy directly in the eyes of the astronaut (National Aeronautics and Space Administration. Apollo Flight Journal (available at http://history.nasa.gov/afj)). The Apollo missions, however, were comparatively short, and they cannot be likened to a multiyear presence on the surface of Mars."

After surveying 1500 images from NASA's Mars Reconnaissance Orbiter (MRO), investigation carried out by Paris et al. (2019) identified three candidate lava tubes in the vicinity of *Hadriacus Mons* as prospective sites for manned exploration. Their investigation concluded that terrestrial lava tubes can be leveraged for radiation shielding and, accordingly, that the candidate lava tubes on Mars (as well as known lava tubes on the lunar surface) can serve as natural radiation shelters and habitats for a prospective crewed mission to Mars. As gravity on Mars is 37% of that on Earth, lava flow, and subsequently lava tubes on Mars, are generally much larger than those found on Earth (Reference: Underground Towns on the Moon and Mars: Future Human Habitats Could Be Hidden in Lava Tubes, *Spaceflight Insider* (available at http://www.spaceflightinsider.com)).

Terrestrial analog studies carried out by Paris et al. (2019) shed much light on the feasibility of using lava tubes on Earth's Moon and the planet Mars for human settlement (despite the presence of harmful radiations on these two celestial bodies), in terms of the ability of lava tubes in substantially reducing (by ~82%) the intensity of harmful radiations inside of the lava tubes. Paris et al. (2019) reported their investigation to identify and study prospective lava tubes at *Hellas Planitia*, a plain located inside the large impact basin *Hellas* in the southern hemisphere of Mars, through the use of Terrestrial analog structures. The search for lava tubes at *Hellas Planitia* was primarily due to the low radiation environment at this particular location. Several studies by NASA spacecraft have measured radiation levels in this region at ~342 μSievert per day (μSv/day), which is considerably less than other regions on the surface of Mars (~547 μSv/day). Notwithstanding, a radiation exposure of ~342 μSv/day is still sizably higher than what human beings are annually exposed to on Earth. Paris et al. (2019) report that "By analyzing orbital imagery from two cameras onboard NASA's Mars Reconnaissance Orbiter (MRO)—the High-Resolution Imaging Science Experiment (HiRISE) and the Context Camera (CTX)—the search for lava tubes was refined by identifying pit crater chains in the vicinity of *Hadriacus Mons*, an ancient low-relief volcanic mountain along the north-eastern edge of Hellas Planitia. After surveying 1500 images from MRO, this investigation has identified three candidate lava tubes in the vicinity of Hadriacus Mons as prospective sites for manned exploration." To complement this investigation, 30 in-situ radiation monitoring experiments have been conducted at three analog lava tubes (located at Mojave, California; El Malpais, New Mexico; and Flagstaff, Arizona). Paris et al. (2019) found that on average, the total amount of solar radiation detected outside the analog lava tubes was approximately ~0.470 μSv/h, while the average inside the analog lava tubes decreased by 82% to ~0.083 μSv/hr. Based on these finding from the three terrestrial lava tubes investigated, Paris et al. (2019) infer that the candidate lava tubes identified southwest of Hadriacus Mons on Mars could be leveraged to decrease the radiation and to reduce the crew's exposure to the harmful radiation. Paris et al. (2019)'s investigation, therefore, concluded that "terrestrial lava tubes can be leveraged for radiation shielding and, accordingly, that the candidate lava tubes on Mars (as well as known lava tubes on the lunar surface) can serve as natural radiation shelters and habitats for a prospective crewed mission to the planet."

10.11.3 Mars' anticipated habitability potential

Although located in the habitable zone of our host star Sun, modern Mars is a very cold, desert-like planet. To maximize the chances of finding signs of past life, it is necessary to target the "sweet spot" in Mars' geological history—the early Noachian (periods of 4100–3700 million years ago, during the interval known as the Late Heavy Bombardment. Many of the large impact basins on Earth's Moon and Mars formed at this time). Based on available literature, Vago et al. (2017) made the case that conditions for the appearance of microbes on early Mars were similarly favorable as on our planet Earth. This provides a basis to examine how likely it is that we may find evidence, or at least some clues, of their presence. According to Vago et al. (2017), "considering the need to find landing sites suitable for pursuing our mission's science, we should establish a metric to inform us whether, how much, when, and how long a place had the capacity to host and nurture cells—the only living machine we are aware of."

There are indications suggesting that microbes could have thrived on Mars in the past (e.g., Stoker et al., 2010; McLoughlin and Grosch, 2015). However, inferences made based purely on available patchy geological information and theories pertaining to Mars' dynamic past (e.g., impacts and obliquity cycles) need to be used with caution. We, therefore, require constraints that are able to boost our confidence that microbes could have really thrived on Mars in the past. According to Vago et al. (2017), "minimum temperature and water activity thresholds have been identified below which even the hardiest known terrestrial microorganisms cannot replicate. These parameters are used to classify areas of present Mars in terms of their potential to become habitats for spacecraft delivered Earth microorganisms (Kminek and Rummel, 2015; Kminek et al., 2016; Rettberg et al., 2016)." Taking cues from fossil records found on Earth, Hull et al. (2015) suggested that species that existed over a broad area have a higher probability of being found than those that were rare or geographically restricted. The same is applicable to landed planetary missions as well.

Taking account of the probability of extensive colonization for long periods, Vago et al. (2017) proposed to categorize a candidate landing site's habitability in terms of the extent and frequency of liquid water lateral connectivity between the potential (micro) habitats. They argued that a single, short-lived meandering channel would constitute a less appealing target than a network of interconnected lakes having undergone numerous inundation episodes (wetter for longer), although both sites would have been habitable. Vago et al. (2017), therefore, concluded that, "to maximize our chances of finding signs of past life on Mars, we must target the 'sweet spot' in Mars' geological history, the one with the highest lateral water connectivity—the early Noachian—and look for large areas preserving evidence of prolonged, low-energy, water-rich environments, the type of habitat that would have been able to receive, host, and propagate microorganisms."

10.11.4 Mars landing site selection criteria

Apart from the necessary safety considerations for successful landing of the rover on Mars, the scientific characteristics of the landing site region will have the greatest effect on what the ExoMars rover will be able to discover. According to Vago et al. (2017), attributes such as (1) age; (2) nature, duration, and connectivity of aqueous environments; (3) sediment deposition, burial, diagenesis, and (4) exhumation history are decisive for the successful (or otherwise) trapping and preservation of possible chemical biosignatures. Combining scientific and engineering competence in one body was considered paramount to the success of the landing site selection process.

Vago et al. (2017) indicated that, "ESA and Roscosmos agreed a balanced sharing of responsibilities for the different elements. ESA would provide the TGO and EDM for the first mission, and the carrier and rover for the second. Roscosmos would furnish both launchers and be in charge of the second mission's descent module. NASA would also deliver important contributions to ExoMars, such as the Electra ultra-high frequency (UHF) radio package for TGO-to-Mars-surface proximity link communications, engineering support to the EDM, and a major part of Mars organic molecule analyzer (MOMA), the organic molecule characterization instrument on the rover."

There existed a general consensus among planetary scientists and astrobiologists that the ExoMars Rover mission must target a geologically diverse, ancient site interpreted to possess strong potential for past habitability and for preserving physical and chemical biosignatures (as well as abiotic/prebiotic organics).

After facing numerous difficulties (see Vago et al., 2017 for details) the first *ExoMars* mission was launched on March 14, 2016, from the Baikonur cosmodrome, in Kazakhstan (the largest country in Central Asia), and arrived at Mars on October 19, 2016. According to Vago et al. (2017)'s description, the ExoMars mission consists of two major elements: The Trace Gas Orbiter (TGO) and the Schiaparelli entry, descent, and landing demonstrator module (EDM). The objective of TGO is to conduct a detailed analysis of atmospheric gases, including methane (CH_4) and other minor constituents (Allen et al., 2006; Sherwood et al., 2006; Yung et al., 2010; Yung and Chen, 2015), and study the surface to seek signatures of possible active processes; TGO will also serve as a communications relay for surface missions until the end of 2022. The EDM's goal was to prove technologies for controlled landing and perform surface measurements. Unfortunately, the last phase of the landing sequence did not work and the lander was lost.

According to Vago et al. (2017), the second mission will deliver a rover tasked with searching for signs of past life; however, its payload also has the potential to recognize chemical indicators of extant life. The ExoMars rover will drill to depths of 2 m to collect and analyze samples that have been shielded from the harsh conditions that prevail at the surface, where radiation and oxidants can destroy organic compounds. The lander will be equipped with instruments devoted to atmospheric and geophysical investigations.

10.12 When could life have probably arisen on early Mars? Clues from early Earth environments and biosignatures

At the time when the first known life on Earth existed, Mars was habitable, but the Martian surface became hyper-arid, extremely oxidized, and inhospitable before the evolution of photosynthesis or eukaryotic life on Earth (Michalski et al., 2013). Evidence of past liquid water on the surface of Mars suggests that this world once had habitable conditions and leads to the question of life. If there was life on Mars, it would be interesting to determine if it represented a separate origin from life on Earth. To determine the biochemistry and genetics of life on Mars requires that we have access to an organism or the biological remains of one—possibly preserved in ancient permafrost. According to McKay (2010), a way to determine if organic material found on Mars represents the remains of an alien biological system could be based on the observation that biological systems select certain organic molecules over others that are chemically similar (e.g., chirality in amino acids). A "chiral" molecule is one that is not superposable with its mirror image. Like left and right hands that have a thumb, fingers in the same order, but are mirror images and not the same, chiral molecules have the same things attached in the same order, but are mirror images and not the same. Although most amino acids can exist in both left and right-handed forms, life on Earth is made of left-handed amino acids, almost exclusively. Note that many biologically active molecules are chiral, including the naturally occurring amino acids (the building blocks of proteins) and sugars. In biological systems, most of these compounds are of the same chirality: most amino acids are levorotatory (L) and sugars are dextrorotatory (D). In the case of chiral molecules (i.e., those that can exist in either of two nonidentical mirror image structures known as enantiomers), life-forms synthesize exclusively one enantiomer, for example, left-handed amino acids (L-amino acids) to build proteins and right-handed ribose (D-ribose) for sugars and the sugars within ribonucleic acid (RNA) and deoxyribonucleic acid (DNA). Opposite enantiomers (D-amino acids and L-ribose) are neither utilized in proteins nor in the genetic material RNA and DNA. The use of pure chiral building blocks is considered a general molecular property of life

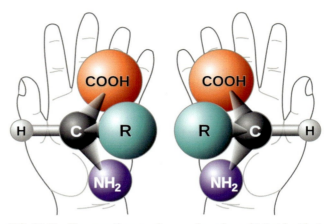

FIG. 10.11 Two enantiomers of a generic amino acid that is chiral. (Source: https://en.wikipedia.org/wiki/Chirality#/media/File:Chirality_with_hands.svg) © Public domain in the United States because it was solely created by NASA. NASA copyright policy states that "NASA material is not protected by copyright unless noted." Amino Acid Chirality chirality with hands from http://www.nai.arc.nasa.gov/.

(see Vago et al., 2017). Fig. 10.11 illustrates two enantiomers of a generic amino acid that is chiral.

There is some suggestion that early amino acids could have formed in comet dust. In this case, circularly polarized radiation (which makes up 17% of stellar radiation) could have caused the selective destruction of one chirality of amino acids, leading to a selection bias which ultimately resulted in all life on Earth being homochiral (Meierhenrich et al., 2005).

The recorded environmental conditions, however, would have posed severe challenges to development of Martian biology, particularly concerning water chemistry. Mineral assemblages similar to those in Eagle crater (a 22-m long impact crater located on the Meridiani Planum portion of the planet Mars) are found on Earth. The Opportunity Rover came to rest inside Eagle crater when it landed in 2004. Scientists were delighted that the rover landed there, as the crater contains rocky outcroppings that helped prove that Meridiani was once an ocean floor. Commonly the mineral assemblages similar to those in Eagle crater are associated with mine drainage, but in some cases, notably the Rio Tinto basin in Spain (Fernández-Remolar et al., 2003), mineral deposition predates mining activity. Precipitating waters tend to be acidic, sometimes strongly so. In addition, in the saline waters suggested by Meridiani geology, water activity could have been at least episodically prohibitive.

There are some scientists (e.g., Nisbet and Sleep, 2001; Zahnle et al., 2007) who believe that if life ever arose on the red planet, it probably did when Mars was wetter, sometime within the first half billion years after planetary formation. It has been suggested that, for a good part of its early history, Mars could have perhaps looked like a colder version of present-day Iceland—gelid (i.e., icy; extremely cold) on top, heated from below. However, Warner and Farmer

(2010) and Cousins and Crawford (2011) have suggested that the likelihood of a cold surface scenario does not constitute a serious obstacle for the possible appearance of life, as extensive subglacial, submerged, and emerged volcanic/hydrothermal activity would have resulted in numerous liquid water-rich settings. According to Westall et al. (2013) and Russell et al. (2014), the right mixture of ingredients, temperature and chemical gradients, organic molecule transport, concentration, and fixation processes could have been found just as well in a plethora of terrestrial submarine vents as in a multitude of vents under top-frozen Martian bodies of water.

Conditions on Mars within the first half billion years of its formation were similar to those when microbes gained a foothold on the young Earth. This marks Mars as a primary target to search for signs of life in our Solar System. There is a general consensus among planetary scientists and astrobiologists that the knowledge we have gathered about early Earth environments and biosignatures (Fairén et al., 2010b; Westall, 2012; Westall et al., 2013) is likely to be extremely useful in exploring the possibility of life on Mars. For example, despite certain obscurities and yet unanswered questions, life seems to have appeared on Earth as soon as the environment allowed it, sometime between 4.4 and 3.8 Ga ago. According to Thomas et al. (2006), life then continued onward more or less hampered by large impacts, a few of which could have done away with most exposed and shallow subsurface organisms. Based on these considerations and findings by several investigators (e.g., Solomon et al., 2005; McKay, 2010; Strasdeit, 2010; Yung et al., 2010), Vago et al. (2017) postulated that, although colder, conditions existed for the possible emergence of life on Mars.

Whereas microbial life quickly became a global phenomenon on Earth, the availability of transport paths between liquid water-rich environments proceeded very differently on Mars. According to Westall et al. (2013, 2015a), sometime during the late Noachian, Martian surface habitats gradually became more isolated; their lateral connectivity started to dwindle and eventually disappeared. Vago et al. (2017) have described this situation as "punctuated" habitability. Michalski et al. (2013a) have suggested that as surface conditions deteriorated, potential microbes could have found refuge in subterranean environments. Rathbun and Squyres (2002) suggested that, occasionally, impact-formed hydrothermal systems would have resulted in transient liquid water becoming available close to the surface, even if the Martian climate was cold. But Cockell et al. (2012) argued that Rathbun and Squyres (2002)'s suggestion does not necessarily mean that these later habitats could have been colonized. Fig. 10.12 illustrates the type of habitat (prolonged, low-energy, water-rich environments) that would have been able to receive, host, and propagate microorganisms.

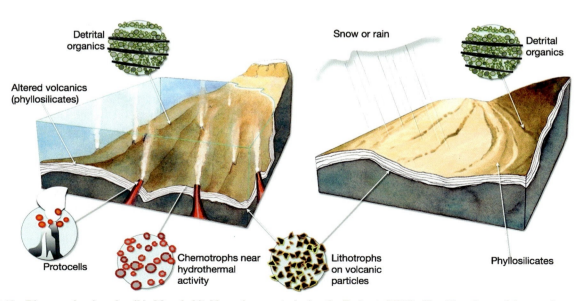

FIG. 10.12 Diagram showing plausible Mars habitable environments during the Early- to Middle-Noachian. Some of these settings may have been active long enough to witness the appearance of life (especially in the case of long-term hydrothermal activity); others could have hosted already flourishing microorganisms. (Source: Vago, J.L. et al. (2017), Habitability on early Mars and the search for biosignatures with the ExoMars Rover, Astrobiology, 17(6-7): 471-510). © Jorge L. Vago et al., 2017; Published by Mary Ann Liebert, Inc. The Open Access article from which this Figure is reproduced is distributed under the terms of the Creative Commons License (http://creativecommons.org/licenses/by/4.0), which permits unrestricted use, distribution, and reproduction in any medium, provided the original work is properly credited. Address correspondence to: Jorge L. Vago, ESA/ESTEC (SCI-S), Keplerlaan 1, 2200 AG, Noordwijk, The Netherlands, E-mail: jorge.vago@esa.int.

10.13 Search for life on Mars under the first NASA Mars Scout mission—The Phoenix mission

The unifying goal of the Mars Exploration program, as laid out in the roadmap of NASA's Mars Exploration Payload Analysis Group (MEPAG), is the search for life on Mars (http://mepag.jpl.nasa.gov/reports/index.html). To achieve this goal, it was deemed necessary to characterize the physical and chemical environment of Mars. Based on this realization, the approach recommended was first to identify habitable environments prior to conducting a mission to search for life.

The Phoenix mission is the first NASA Mars Scout mission. It is also the first mission launched with the objective of assessing the habitability of the Martian environment. The mission was designed to land in Mars' far northern hemisphere where near surface ground ice had been identified based on data from orbiting neutron detectors on the Gamma Ray Spectrometer (GRS) carried by Mars Odyssey (see Stoker et al., 2010). The area was considered important for study because of the possible periodic presence of liquid water as orbital dynamics change the regional climate. The two primary mission objectives to be achieved at the landing site were (1) study the history of water in all its phases and (2) search for habitable zones. A habitable zone, by definition, is capable of supporting living organisms with capabilities similar to terrestrial microbes. One of the major requirements for life is the presence of liquid water and the northern region of Mars, with its water ice cap and shallow subsurface ice, promised to provide the greatest likelihood for achieving both of the primary mission objectives (Stoker et al., 2010).

The Phoenix mission's key objective was to search for a habitable zone. The Phoenix mission site selection and payload were designed to gain an understanding of the habitability of the environment on Mars. Stoker et al. (2010) described the physical and chemical aspects of the Phoenix landing site and related them to habitability. They established quantitative criteria to assess the habitability of the landing site through a "habitability index" wherein a habitability probability (HI) is defined as the product of probabilities for the presence of liquid water (Plw), energy (Pe), nutrients (Pch), and a benign environment (Pb). Stoker et al. (2010) summarized the Phoenix mission thus, "Mission results are used to evaluate habitability where Phoenix landed. Observational evidence for the presence of liquid water (past or present) includes clean ice at a polygon boundary, chemical etching of soil grains, and carbonate minerals. The presences of surface and near subsurface ice, along with thermodynamic conditions that support melting, suggest that liquid water is theoretically possible. Presently, unfrozen water can form only in adsorbed films or saline brines but more clement conditions recur periodically due to variations in orbital parameters. Energy to drive metabolism is available from sunlight, when semitransparent soil grains provide shielding from UV radiation and chemical energy from the redox couple of perchlorate and reduced iron. Nutrient sources including H, O, C, N, S, and P compounds are supplied by known atmospheric sources or global dust. Environmental conditions are within growth tolerance for terrestrial microbes. Surface soil temperatures currently reach 260 K and are periodically much higher, the pH is 7.8 and is well buffered, and the water activity is high enough to allow growth when sufficient water is available. Computation of HI for the sites visited by landers yields Phoenix, 0.47; Meridiani, 0.23; Gusev, 0.22; Pathfinder, 0.05; Viking 1, 0.01; Viking 2, 0.07. HI for the Phoenix site is the largest of any site explored, but dissimilar measurements limit the comparisons' confidence."

The phyllosilicates, or sheet silicates, are an important group of minerals that includes the micas, chlorite, serpentine, talc, and the clay minerals. Phyllosilicate deposits on Mars provide an opportunity to evaluate aqueous activity and the possibility that habitable environments may have existed during the Noachian period there. According to Bishop et al. (2013), the phyllosilicate outcrops at Mawrth Vallis on Mars extend across a broad (~1000 km) region and exhibit a consistent general trend of Al-phyllosilicates and amorphous Al/Si species at the top of the clay profile and Fe/Mg-phyllosilicates on the bottom. Bishop et al. (2013) claims that "this implies either a change in water chemistry, a change in material being altered, or an alteration profile where the upper clays were leached and altered more significantly than those below." These investigators also noticed a change in iron in the phyllosilicate units such that an Fe^{2+}-bearing unit was frequently observed between the Fe^{3+}- and Mg-rich phyllosilicates below and the Al/Si-rich materials above. According to Bishop et al. (2013), abrupt changes in chemistry like this are often indicative of biogeochemical activity on Earth. The study reported by Bishop et al. (2013) analyzed the spectra obtained from the hyperspectral visible/near-infrared (VNIR) Mars Reconnaissance Orbiter (MRO) Compact Reconnaissance Imaging Spectrometer for Mars (CRISM) images from four different outcrops across the Mawrth Vallis region and evaluated the observed phyllosilicates and clay components in terms of plausible aqueous and microbial processes and the potential for retention of biosignatures, if present.

Studies carried out by these investigators showed thick, complex profiles of phyllosilicates at Mawrth Vallis, Mars that are consistent with long-term aqueous activity and active chemistry. Their inference reinforced the notion that the ancient phyllosilicates in places such as this could have served as reaction centers for organic molecules, and that phyllosilicates could have even played a role in the origin of life. Bishop et al. (2013)'s studies indicate that regardless of whether life formed on early Mars or not, evaluating the type and thickness of clay-bearing units on Mars provides insights into plausible aqueous processes and chemical conditions both during the time of formation of the phyllosilicates, but also the subsequent period following their formation.

Although many ancient, deep basins lack evidence for groundwater activity, Michalski et al. (2013) reported that McLaughlin Crater, one of the deepest craters on Mars, contains evidence for Mg–Fe-bearing clays and carbonates that probably formed in an alkaline, groundwater-fed lacustrine setting. According to them, "This environment strongly contrasts with the acidic, water-limited environments implied by the presence of sulfate deposits that have previously been suggested to form owing to groundwater upwelling. Deposits formed as a result of groundwater upwelling on Mars, such as those in McLaughlin Crater, could preserve critical evidence of a deep biosphere on Mars." Michalski et al. (2013) suggested that groundwater upwelling on Mars may have occurred sporadically on local scales, rather than at regional or global scales.

Vago et al. (2017) summarized the success stories of various missions to Mars and how those missions helped in unambiguously establishing the presence of water on Mars in the past thus, "The very successful 1996 Mars Global Surveyor and 2003 Mars Exploration Rovers (MER), which were conceived as robotic geologists, have demonstrated the past existence of wet environments (Malin and Edgett, 2000; Squyres et al., 2004a, 2004b, 2012). But perhaps it has been Mars Express 2003 and Mars Reconnaissance Orbiter 2005 that have most drawn our attention to ancient Mars, revealing many instances of finely layered deposits containing phyllosilicate minerals that could only have formed in the presence of liquid water, which reinforced the hypothesis that early Mars was wetter than today (Poulet et al., 2005; Bibring et al., 2006; Loizeau et al., 2010, 2012; Ehlmann et al., 2011; Bishop et al., 2013; Michalski et al., 2013b)."

10.14 Search for traces of life on Mars—NASA's biology experiments

As indicated earlier, one of the mission objectives of NASA's Viking missions was searching for evidence of extinct or extant life on Mars. The search for traces of life is one of the principal objectives of Mars exploration. Central to this objective is the concept of habitability, the set of conditions that allows the appearance of life and successful establishment of microorganisms in any one location.

The Viking missions to Mars in the late 1970s were the first search for life outside Earth. Each Viking conducted three incubation experiments to detect the presence of metabolism in the Martian soil. Each lander also carried a sophisticated Gas Chromatograph Mass Spectrometer for characterizing organic molecules. The results were unexpected (Klein, 1978, 1999). There was a detectable reaction in two of the incubation experiments. In the "Gas Exchange" experiment, a burst of oxygen was released when the soil was exposed to water. The "Labeled Release" experiment showed that organic material was consumed, and that carbon dioxide was released concomitantly. In the Labeled Release experiment, this reaction ceased if the soil was first heated to sterilizing temperatures, but the reaction of the Gas Exchange Experiment persisted (McKay, 2004).

McKay (2004) stated thus: "if considered alone, the Labeled Release results would be a plausible indication for life on Mars. However, the Gas Chromatograph Mass Spectrometer did not detect the presence of any organic molecules in the soil at level of one part per billion (Biemann, 1979). It is difficult to imagine life without associated organic material, and this is the main argument against a biological interpretation of the Viking results (Klein, 1999; but cf. Levin and Straat, 1981). It is also unlikely that the oxygen release in the Gas Exchange experiment had a biological explanation, because the reaction was so rapid and persisted after heating. It is generally thought that the reactivity observed by the Viking biology experiments was caused by one or more inorganic oxidants present in the soil, and was ultimately produced by ultraviolet light in the atmosphere. Consistent with the apparently negative results of the Viking biology experiments, the surface of Mars also appears to be too dry for life. Indeed, conditions are such that liquid water is rare and transient, if it occurs at all (e.g., Hecht, 2002)."

According to Westall et al. (2015a), "while environmental conditions may have been conducive to the appearance of life early in Martian history, habitable conditions were always heterogeneous on a spatial scale and in a geological time frame. This 'punctuated' scenario of habitability would have had important consequences for the evolution of Martian life, as well as for the presence and preservation of traces of life at a specific landing site." Westall et al. (2015a) hypothesized that, "given the lack of long-term, continuous habitability, if Martian life developed, it was (and may still be) chemotrophic and anaerobic. Obtaining nutrition from the same kinds of sources as early terrestrial chemotrophic life and living in the same kinds of environments, the fossilized traces of the latter serve as useful proxies for understanding the potential distribution of Martian chemotrophs and their fossilized traces. Thus, comparison with analog, anaerobic, volcanic terrestrial environments (Early Archean >3.5–3.33 Ga) shows that the fossil remains of chemotrophs in such environments were common, although sparsely distributed, except in the vicinity of hydrothermal activity where nutrients were readily available. Moreover, the traces of these kinds of microorganisms can be well preserved, provided that they are rapidly mineralized and that the sediments in which they occur are rapidly cemented." Westall et al. (2015a) evaluated the biogenicity of these signatures by comparing them to possible abiotic features. They also discussed the implications of different scenarios for life on Mars for detection by in situ exploration, ranging from its nonappearance, through preserved traces of life, to the presence of living microorganisms.

If any bioorganic compounds are detected on Mars, it will be important to show that they were not brought from Earth. Great care is being devoted during the assembly, testing, and integration of instruments and the lander's components. Strict organic cleanliness requirements apply to all parts that come into contact with the sample and to the lander's assembly process (see Vago et al., 2017).

Based on what we knew about planetary evolution in the 1970s, many scientists regarded as plausible the presence of simple micro-organisms on other planets. The twin Viking landers conducted the first in situ measurements on the Martian surface. The biggest blow was the failure of the gas chromatograph–mass spectrometer (GC–MS) to acquire evidence of organic molecules at the parts-per-billion level. Numerous attempts were made in the laboratory to simulate the reactions observed by the Viking biological package (Quinn et al., 2007). Although some reproduced certain aspects of the data, none succeeded entirely (Vago et al., 2017).

The Viking lander mission's biology package contained three experiments, all looking for signs of metabolism in soil samples (Klein et al., 1976a). One of them, the Labeled-Release Experiment, produced very provocative results (Levin and Straat, 2016). If other information had not been also obtained, these data would have been misinterpreted as proof of biological activity (Vago et al., 2017). With few exceptions, the majority of the scientific community concluded that the Viking findings did not demonstrate the presence of extant life (Klein, 1998, 1999).

Vago et al. (2017) described the importance of Viking lander mission thus, "The 1976 Viking landers can be considered the first missions with a serious chance of discovering signs of life on Mars. That the landers did not provide conclusive evidence was not because of a lack of careful preparation. In fact, these missions were remarkable in many ways, particularly taking into account the technologies available. If anything, the Viking results were a consequence of the manner in which the life question was posed, seeking to elicit signs of microbial activity from potential extant ecosystems within the Mars samples analyzed (Klein et al., 1976b)."

The Viking mission's failure to detect signs of life on Mars was a setback to the scientific community. This resulted in a kind of discouraging note from all directions. Vago et al. (2017) described this scenario thus, "The Viking program increased very significantly our knowledge of Mars; however, failure to detect organic molecules was considered a significant setback. As a consequence, our neighbor planet lost much of its allure. A multiyear gap in Mars surface exploration ensued."

10.15 Exploring Mars under United States' Viking Lander missions—Discovering volcanoes, lava plains, giant canyons, craters, wind-formed features, and seasonal dust storms

NASA's Viking mission to Mars was composed of two spacecrafts, Viking 1 and Viking 2, each consisting of an orbiter and a lander. The primary mission objectives were to obtain high-resolution images of the Martian surface, characterize the structure and composition of the atmosphere and surface, and search for evidence of life.

Viking 1 was launched on August 20, 1975 and arrived at Mars on June 19, 1976. The first month of orbit was devoted to imaging the surface to find appropriate landing sites for the Viking Landers. On July 20, 1976, the Viking 1 Lander separated from the Orbiter and touched down at Chryse Planitia (22.48° N, 49.97° W planetographic, 1.5 km below the datum [6.1 mbar] elevation). Viking 2 was launched on September 9, 1975 and entered Mars orbit on August 7, 1976. The Viking 2 Lander touched down at Utopia Planitia (47.97° N, 225.74° W, 3 km below the datum elevation) on September 3, 1976. The Orbiters imaged the entire surface of Mars at a resolution of 150–300 m, and selected areas at 8 m. The lowest periapsis (i.e., the point in the path of an orbiting body at which it is nearest to the body that it orbits) altitude for both Orbiters was 300 km. The Viking 2 Orbiter was powered down on July 25, 1978 after 706 orbits, and the Viking 1 Orbiter on August 17, 1980, after over 1400 orbits. The Orbiter images are available from NSSDCA on CD-ROM and as photographic products. These images have been converted to digital image mosaics and maps, and these are also available from NSSDCA on CD-ROM. An index giving the latitude and longitude of each Viking Orbiter image is available on the Viking image index page. The Viking Landers transmitted (sent to Earth) images of the surface, took surface samples and analyzed them for composition and signs of life, studied atmospheric composition and meteorology, and deployed seismometers. The Viking 2 Lander ended communications on 11 April 1980, and the Viking 1 Lander on 13 November 1982, after transmitting over 1400 images of the two sites. Many of these images are also available from NSSDCA on-line (https://nssdc.gsfc.nasa.gov/planetary/viking.html).

The results from the Viking experiments give our most complete view of Mars. Volcanoes, lava plains, immense canyons, cratered areas, wind-formed features, and evidence of surface water are apparent in the Orbiter images. The planet appears to be divisible into two main regions, northern low plains and southern cratered highlands. Superimposed on these regions are the Tharsis and Elysium bulges, which are high-standing volcanic areas, and Valles Marineris, a system of giant canyons near the equator. The surface material at both landing sites can best be characterized as iron-rich clay.

Measured temperatures at the landing sites ranged from 150 (−123.15°C) to 250 K (−23.15°C), with a variation over a given day of 35–50 K. Seasonal dust storms, pressure changes, and transport of atmospheric gases between the polar caps were observed. The biology experiment produced no evidence of life at either landing site (https://nssdc.gsfc.nasa.gov/planetary/viking.html). Further information on the spacecraft, experiments, and data returned from the Viking missions can be found in Flinn (1977), who discussed the following scientific aspects:

1. Orbiter imaging, taking into account some Martian volcanic features as viewed from the Viking orbiters
2. Classification and time of formation of Martian channels based on Viking data
3. Martian permafrost features, Martian impact craters and emplacement of ejecta by surface flow
4. Geology of the Valles Marineris, the geology of Chryse Planitia, geological observations in the Cydonia region of Mars from Viking
5. Martian dynamical phenomena during June-November 1976, Viking orbiter observations of atmospheric opacity during July-November 1976, Viking orbiter photometric observations of the Mars phase function July through November 1976
6. A study of variable features on Mars during the Viking primary mission and Viking observations of *Phobos* and *Deimos*, which are the two small moons of Mars.

In Flinn (1977)'s research, attention was also given to the Viking lander imaging investigation, the performance of the Viking lander cameras, spectro-photometric and color estimates of the Viking lander sites, Viking 1975 Mars lander interactive computerized video stereo-photogrammetry, the geology of the lander sites, atmospheric water detection, infrared thermal mapping, entry science, and meteorological results.

10.16 Looking for signs of primitive life on Mars

Our understanding of the potential of other planets and moons in our Solar System to host microbial life has increased considerably in the last decade. The icy moons of the outer planets no longer appear completely inhospitable, especially when viewed as warmer and more active bodies during their early history. Hypotheses concerning the possibility of life on Mars wax and wane as new data and new models related to its aqueous history appear. We tend to automatically equate the availability of water with the guarantee of conditions conducive to life. However, from a microbial point of view, the situation is very different depending on whether we are dealing with the emergence of life, with established or flourishing life, or with life in a survival or dormant mode.

10.16.1 Arguments favoring possibility of life on Mars

Similarities in the early histories of Mars and Earth suggest the possibility that life may have arisen on Mars as it did on Earth (Westall et al., 2000). Those who are inclined to favor the possibility of life on Mars argue that given the observed similarities in the early histories of Mars and Earth, early deterioration of the environment on Mars (loss of surface water, decrease in temperature) may have inhibited further evolution of life. Thus, according to their argument, life on Mars would probably be similar to the simplest form of life on Earth, the prokaryotes. Westall et al. (2000) reported a hypothetical strategy to search for life on Mars consisting of (i) identifying a suitable landing site with good exobiological potential, and (ii) searching for morphological and biogeochemical signatures of extinct and extant life on the surface, in the regolith subsurface, and within rocks. The platform to be used in this theoretical exercise is an integrated, multiuser instrument package, distributed between a lander and rover, which will observe and analyze surface and subsurface samples to obtain the following information (Westall et al., 2000): "(1) environmental data concerning the surface geology and mineralogy, UV radiation and oxidation processes; (2) macroscopic to microscopic morphological evidence of life; (3) biogeochemistry indicative of the presence of extinct or extant life; (4) niches for extant life."

There is strong evidence that Mars once had a shallow ocean and hydrothermal conditions related to volcanism, so a reasonable speculation is that microbial life could originate in hydrothermal fields on Mars and may still survive in deep subsurface brines, just as halophilic extremophiles thrive in such conditions on Earth. According to Deamer and Damer (2017), it seems likely that in the next two decades future probes and landers will provide evidence that reduces speculations to certainties.

Mars provided our second opportunity to search for traces of life. With its obvious atmosphere and polar caps, Mars was deemed a much more likely location in the Solar System to harbor extraterrestrial life. As indicated earlier, the search for Martian life began in earnest when the Viking Project landed two robot space-crafts on the planet in 1976 to photograph the surface material and analyze it in detail. These missions increased our knowledge of the physical and chemical properties of the remarkable red planet tremendously. Perhaps the most unexpected and fascinating findings, however, were returned by the two instruments designed to search for life and for life-related molecules. One instrument showed that, quite unexpectedly, the Mars soil samples contained no organic materials, not even traces of carbon from meteorites that must hit the planet's surface.

This is probably due to the destructive power of the intense ultraviolet radiation from the Sun, which destroys any exposed organic material. The second instrument, intended to detect biological reactions in the Martian soil, did not detect life on Mars either. It did, however, uncover an intriguing chemical property of the soil that, at least in part, does mimic some simple life-like reactions. Among these are the breakdown of nutrient chemicals and the synthesis of organic matter from gaseous substances.

All in all, Mars remains a planet of extreme interest to exo-biologists. Although Viking did not detect life, Mars continues to tantalize us. Why were no organic molecules detected? Were conditions in Mars' past history more favorable for prebiotic chemical syntheses than those present today? Are there other locations on Mars today where conditions are more conducive to life or to life-related chemistry? These questions and more require continued exploration of Mars. After all, the Earth is not the most favorable place to look for clues about life's origin, because life itself has altered the planet so drastically that much necessary information has been obliterated. Mars, then, provides us an essential point for comparison with Earth: an environment not extensively modified by widespread life, and perhaps still harboring secrets about the relationship of the origin of life to the origin of the Solar System.

10.16.2 Probable ancient life on Mars: Chemical arguments and clues from Martian meteorites

Evidence of biogenic activity on Mars has profound scientific implications for our understanding of the origin of life on Earth and the presence and diversity of life within the Cosmos. It is believed that primitive terrestrial life—defined as a chemical system, which is able to transfer its molecular information via self-replication and to evolve—probably originated from the evolution of reduced organic molecules in liquid water. Several sources have been proposed for the prebiotic organic molecules: terrestrial primitive atmosphere (methane or carbon dioxide), deep-sea hydrothermal systems, and extra-terrestrial meteoritic and cometary dust grains. The study of carbonaceous chondrites, which contain up to 5% by weight of organic matter, has allowed close examination of the delivery of extra-terrestrial organic material. Eight proteinaceous amino acids have been identified in the Murchison meteorite among more than 70 amino acids. L-alanine was found to have been surprisingly more abundant than D-alanine in the Murchison meteorite. Likewise, excesses of L-enantiomers have been found for nonprotein amino acids.

Thomas-Keprta et al. (2002) have briefly described the Allan Hills 84001 (ALH84001) meteorite, according to which this meteorite was discovered on 27 December 1984 by a National Science Foundation team during a snowmobile ride on the far western Allan Hills icefield, located at the eastern terminus of the Transantarctic mountain range (76°54'S, 157°01'E), (http://curator.jsc.nasa.gov/curator/antmet/mmc/mmc.htm). After 9 years' research this meteorite was identified as a Martian meteorite (Mittlefehldt, 1994; Score, 1997) based on oxygen isotope analysis (Clayton, 1993). It has a radiogenic Rb–Sr crystallization age of ~4.5 billion years (Nyquist et al., 2001), indicating that it formed shortly after the planet Mars itself was formed. Thomas-Keprta et al. (2002) recall that "ALH84001 was subsequently ejected into interplanetary space from the Martian surface ~16 million years ago (Goswami et al., 1997), presumably as a consequence of a collision of an asteroid or comet with Mars (Melosh, 1995). About 13,000 years ago it was captured by the Earth's gravity field and fell as a meteorite in Antarctica (Jull et al., 1995). With a mass of 1.94 kg, ALH84001 is primarily a volcanic, silicate-rich rock."

Possible relic biogenic activity in Martian meteorite Allan Hills 84001 (ALH84001) was proposed by McKay et al. (1996). ALH84001 has many features unlike other Martian meteorites. This ancient meteorite of 4.5 billion years old contains abundant carbonates as secondary minerals precipitated from a fluid on the Martian surface. McKay et al. (1996)'s investigations indicated that "Fresh fracture surfaces of the Martian meteorite ALH84001 contain abundant polycyclic aromatic hydrocarbons (PAHs). These fresh fracture surfaces also display carbonate globules. Contamination studies suggest that the PAHs are indigenous to the meteorite. High-resolution scanning and transmission electron microscopy study of surface textures and internal structures of selected carbonate globules show that the globules contain fine-grained, secondary phases of single-domain magnetite and Fe-sulfides. The carbonate globules are similar in texture and size to some terrestrial bacterially induced carbonate precipitates. Although inorganic formation is possible, formation of the globules by biogenic processes could explain many of the observed features, including the PAHs. The PAHs, the carbonate globules, and their associated secondary mineral phases and textures could thus be fossil remains of a past Martian biota." McKay et al. (1996) showed the following lines of evidence for the ancient life; (1) unique mineral compositions and biominerals, (2) polycyclic aromatic hydrocarbons (PAHs) in association with the carbonates, and (3) unique structures and morphologies typical of nanobacteria or microfossils. Meteorites researchers think that there is little doubt that ALH84001 as well as eleven other SNC meteorites are from Mars. However, Tsuchiyama (1996) reported that the mineralogical and biomineralogical evidence for Martian bacteria given by McKay et al. (1996) is controversial, and could be formed by nonbiogenic processes. Thus, according to Tsuchiyama (1996), further study of ALH84001 and other Martian meteorites is required. Tsuchiyama (1996) further suggested the need to consider the future Mars mission especially sample return mission.

The controversial hypothesis that the ALH84001 meteorite contains relics of ancient Martian life has spurred new findings, but the question has not yet been resolved. Organic matter probably results, at least in part, from terrestrial contamination by Antarctic ice meltwater. The origin of nanophase magnetites and sulfides, suggested, on the basis of their sizes and morphologies, to be biogenic remains contested, as does the formation temperature of the carbonates that contain all of the cited evidence for life. According to McSween Jr. (1997), the reported nonfossils may be magnetite whiskers and platelets, probably grown from a vapor. New observations, such as the possible presence of biofilms and shock metamorphic effects in the carbonates, have not yet been evaluated. However, McSween Jr. (1997) is of the opinion that regardless of the ultimate conclusion, this controversy continues to help define strategies and sharpen tools that will be required for a Mars exploration program focused on the search for life.

According to Brack and Pillinger (1998), the early histories of Mars and Earth clearly show similarities. For example, liquid water was once stable on the surface of Mars, attesting the presence of an atmosphere capable of decelerating C-rich micrometeorites (extraterrestrial microscopic particles collected at the Earth's surface (see Folco and Cordier, 2015)). Therefore, Brack and Pillinger (1998) reckon that primitive life may have developed on Mars as well; and fossilized micro-organisms may still be present in the near subsurface. Ironically, the Viking missions to Mars in 1976 did not find evidence of either contemporary or past life, but the mass spectrometer on the lander aeroshell determined the atmospheric composition, which has allowed a family of meteorites to be identified as Martian. Although these samples are essentially volcanic in origin, it has been recognized that some of them contain carbonate inclusions and even veins that have a carbon isotopic composition indicative of an origin from Martian atmospheric carbon dioxide. The oxygen isotopic composition of these carbonate-deposits allows calculation of the temperature regime existing during formation from a fluid that dissolved the carbon dioxide. As the composition of the fluid is unknown, only a temperature range can be estimated, but this falls between 0° and 90°C, which would seem entirely appropriate for life processes. It was such carbonate veins that were found to host putative microfossils. Irrespective of the existence of features that could be considered to be fossils, carbonate-rich portions of Martian meteorites tend to have material, at more than 1000 ppm, that combusts at a low temperature; that is, it is an organic form of carbon. Unfortunately, this organic matter does not have a diagnostic isotopic signature so it cannot be unambiguously said to be indigenous to the samples. However, many circumstantial arguments can be made to the effect that it is cogenetic with the carbonate and hence Martian. According to Brack and Pillinger (1998), if it could be proved that the organic matter was preterrestrial, then the isotopic fractionation between it and the carbon is in the right sense for a biological origin.

Nanocrystals of magnetite (Fe_3O_4) in a meteorite from Mars provide the strongest, albeit controversial, evidence for the former presence of extraterrestrial life. The morphological and size resemblance of the crystals from meteorite ALH84001 to crystals formed by certain terrestrial bacteria has been used in support of the biological origin of the extraterrestrial minerals. By using tomographic and holographic methods in a transmission electron microscope, Buseck et al. (2001) showed that the three-dimensional shapes of such nanocrystals can be defined, that the detailed morphologies of individual crystals from three bacterial strains differ, and that none uniquely match those reported from the Martian meteorite. In contrast to previous accounts, Buseck et al. (2001) argued that the existing crystallographic and morphological evidence is inadequate to support the inference of former life on Mars.

Martian meteorite ALH84001 preserves evidence of interaction with aqueous fluids while on Mars in the form of microscopic carbonate disks believed to have formed approx 3.9 Ga ago at beginning of the Noachian epoch. Intimately associated within and throughout these carbonate disks are nanocrystal magnetites (Fe_3O_4) with unusual chemical and physical properties, whose origins have become the source of considerable debate. Studies of sideritic carbonates conducted under a range of heating scenarios suggests that the magnetite nanocrystals in the ALH84001 carbonate disks are not the products of thermal decomposition.

One of the strongest lines of evidence that has led some investigators (e.g., McKay et al., 1996) to suggest that microbial life existed on Mars approximately 4 billion years ago is the presence of tens-of-nanometer-size magnetite (Fe_3O_4) crystals found within carbonate globules and their associated rims in the meteorite (Thomas-Keprta et al., 2000, 2001). The suggestion in 1996 that the Martian meteorite ALH84001 could contain proof of possible biologic activity in the past have generated a huge controversy. One of the most discussed evidence is the presence of magnetite crystals that resemble those produced by a particular group of bacteria, the so called magnetotactic bacteria (MTB). These microorganisms are the only known example of biologically controlled biomineralization among the prokaryotes and exert an exquisite control over the biomineralization process of intracellular magnetite that result in crystals with very unique features that, so far, cannot be replicated by inorganic means. These unique features have been used to recognize the biological origin of natural terrestrial magnetites, but the problem arises when those same biogenecity criteria are applied to extraterrestrial magnetites. Most of the problems are caused by the fact that it is not clear whether or not some of those characteristics can be reproduced inorganically.

Magnetite is a common mineral on Earth and is formed by a range of inorganic processes, including hydrothermal or volcanic precipitation from an Fe-rich fluid (Thomas-Keprta et al., 2000). Thomas-Keprta et al. (2002) recall that with the discovery by Blakemore in 1975 (Blakemore, 1975) that single-domain magnetite crystals are utilized by prokaryotic organisms called magnetotactic bacteria came the realization that magnetite might be used as a potential biosignature. Magnetotaxis is the term used to describe the passive alignment and active motility of a bacterial cell along magnetic field lines. Magnetotactic bacteria are now known to be ubiquitous in terrestrial aquatic environments, and they appear to utilize the magnetic properties of magnetite (Frankel et al., 1979; Bazylinski and Moskowitz, 1997; Bazylinski and Frankel, 2000) in conjunction with the Earth's global geomagnetic field as an orientation mechanism (Blakemore, 1975; DeLong et al., 1993). This is a process known as magnetotaxis, and when it is coupled with flagellar motility and aerotaxis, it allows an organism to locate and maintain an optimal position in vertical chemical gradients within aquatic environments by reducing a three-dimensional search problem to a one-dimensional search problem (Kirschvink, 1980; Frankel et al., 1997). Thomas-Keprta et al. (2002) recall that approximately one-quarter of these magnetite crystals have remarkable morphological and chemical similarities to magnetite particles produced by magnetotactic bacteria, which occur in aquatic habitats on Earth. Moreover, these types of magnetite particles are not known or expected to be produced by abiotic means either through geological processes or synthetically in the laboratory. Thomas-Keprta et al. (2002) therefore argued that these Martian magnetite crystals are in fact magneto-fossils (Thomas-Keprta et al., 2000, 2001). If this is true, such magneto-fossils would constitute evidence of the oldest life forms known.

Thomas-Keprta et al. (2002) examined a specific strain of a magnetotactic bacterium, designated strain MV-1. Thomas-Keprta et al. (2002) described the strain MV-1 thus: "Cells of this marine strain are gram negative, have vibrioid to helicoid cell morphology, and generally produce a single intracellular chain of ~12 well-ordered magnetite crystals, each encapsulated within a coating or membrane (Bazylinski, 1990). The chain of magnetite crystals acts much like a compass needle (Bazylinski and Moskowitz, 1997), allowing passive alignment of the bacterium along the Earth's geomagnetic field lines. All magnetotactic bacteria, including strain MV-1, produce only one morphological type of magnetite crystals, all of which fall in a narrow size distribution range from ~30 to 120 nm long (Bazylinski and Moskowitz, 1997; Bazylinski and Frankel, 2000). Particles in this size range are single magnetic domains (Bazylinski and Moskowitz, 1997). The cell appears to exert strict physical and chemical control over the biomineralization process(es) involved in magnetite synthesis (Bazylinski and Moskowitz, 1997; Bazylinski and Frankel, 2000)."

According to Thomas-Keprta et al. (2002), "It is rare for magnetite chains to remain intact after the death of the host organism. This is because a magnetite chain exists in a state of high magnetostatic potential energy and tends to spontaneously collapse into a clump, releasing the energy unless it is actively maintained by the host organism. In some instances, fortuitous fossilization can preserve a small fraction of magnetite chains, and such chains in ALH84001 carbonate have been reported by Friedmann et al. (2001)."

According to Thomas-Keprta et al. (2002), the magnetite assay for biogenicity (MAB) criteria constitute a "robust" biosignature. No inorganic magnetite population has met the MAB criteria. Thomas-Keprta et al. (2002) used the term "robust" to imply the exclusivity of the MAB, in that it even discounts many demonstrably intracellularly produced biogenic magnetites. For example, on average, ~30% of magnetites from strain MV-1 fail to meet the MAB criteria and so would not be considered biogenic (Thomas-Keprta et al., 2002). The advantage of such an exclusive biosignature is that any magnetite population meeting the MAB criteria is almost certainly a biogenic precipitate. Thomas-Keprta et al. (2002) argue that because planetary magnetic fields and magnetite crystals are not purely the provenance of Earth, the MAB biosignature is as applicable to extraterrestrial samples as it is to terrestrial samples.

Thomas-Keprta et al. (2002) summarized the conclusions of their study thus, "Perhaps the most profound implication of this study is that approximately one-quarter of the magnetite crystals embedded in the carbonate assemblages in Martian meteorite ALH84001 require the intervention of biology to explain their presence. No single inorganic process or sequence of inorganic processes, however complex, is known that can explain the full distribution of magnetites observed in ALH84001 carbonates. Under these circumstances, our best working hypothesis is that early Mars supported the evolution of Martian biota that had several traits (e.g., truncated hexa-octahedral magnetite and magnetotaxis) consistent with the traits of contemporary magnetotactic bacteria on Earth."

10.16.3 UV is not always fatal—Examples from studies of haloarchaea and spore-forming bacteria

According to the World Health Organization (WHO), the ultraviolet (UV) radiation region covers the wavelength range 100–400 nm and is divided into three bands: UVA (315–400 nm), UVB (280–315 nm), and UVC (100–280 nm) (https://www.who.int/health-topics/ultraviolet-radiation#tab=tab_1). The UV radiation range is a higher frequency and lower wavelength than visible light. The UV radiation, which comes naturally from the Sun, is ubiquitous in the Solar System and it is understood that UV radiation causes damage to cells and increases the rate of DNA mutation and eventually leads to death (Hornec, 1999).

The Martian surface is exposed to a high radiation flux in the form of ionizing radiation and solar UV-radiation due to a thin anoxic atmosphere consisting of mainly CO_2, and by the lack of a planetary magnetic field. There are many signatures of life, from biomolecules to viable microbes, and it is important to understand the ability of these biosignatures to survive and persist, as the most stable signatures may still be preserved and have the potential to be detected during future missions to the extraterrestrial worlds. Feshangsaz et al. (2020) recall that although UVB and UVC make up only 2% of the entire solar spectrum, these wavelengths are the ones primarily responsible for cell damage due to the high absorption of this wavelength range by DNA (Henning et al., 1995; Mancinelli, 2015).

10.16.3.1 Survival of some haloarchaea under simulated Martian UV radiation

Haloarchaea (also known as halophilic archaea, halophilic archaebacteria, halobacteria) are halophiles that need a high salt concentration to grow. In fact, they thrive in environments with salt concentrations approaching saturation. They are a distinct evolutionary branch of the Archaea, and are extremophiles. They are found in water bodies saturated or nearly saturated with salt.

The name "halobacteria" was wrongly assigned to this group of organisms before the existence of the domain Archaea was realized. Halobacteria are now recognized as archaea, rather than bacteria and are one of the largest groups. Halophilic archaea are generally referred to as haloarchaea to distinguish them from halophilic bacteria. These microorganisms are members of the halophile community, in that they require high salt concentrations to grow, with most species requiring more than 2.0 M NaCl for growth and survival. They are a distinct evolutionary branch of the Archaea—distinguished by possession of ether-linked lipids and the absence of murein in their cell walls. Murein is a polymer consisting of sugars and amino acids that forms a mesh-like peptidoglycan layer outside the plasma membrane of most bacteria, forming the cell wall.

Haloarchaea can grow aerobically or anaerobically. Parts of the membranes of haloarchaea are purplish in color, and large blooms of haloarchaea appear reddish, from the pigment bacteriorhodopsin, related to the retinal pigment rhodopsin, which it uses to transform light energy into chemical energy by a process unrelated to chlorophyll-based photosynthesis. Haloarchaea have a potential to solubilize phosphorus. Phosphorus-solubilizing halophilic archaea may well play a role in P (phosphorus) nutrition to vegetation growing in hypersaline soils. Haloarchaea may also have application as inoculants for crops growing in hypersaline regions (https://www.definitions.net/definition/haloarchaea).

Halophilic archaea are some of the most primitive inhabitants of hypersaline environments in Earth's history (Oren and Mana, 2002). Geological structures from the Permian and Triassic (290–206 Ma ago) have been found to contain viable archaea and bacteria. It has also been shown that halophiles can endure the extreme conditions of desiccation, starvation and radiation exposure for millions of years inside stromatolites and ancient halite (Leuko et al., 2010; Stan-Lotter and Fendrihan, 2015; Vreeland et al., 2007).

According to Feshangsaz et al. (2020), it is likely that many of the coping strategies of these cells relate to the Precambrian era when the UV-shielding ozone layer was not yet formed (Baqué et al., 2013). Particularly relevant studies of halophilic archaea have been performed which include *Halobacterium salinarium* NRC-1 under UV radiation, vacuum and desiccation (Fendrihan et al., 2009; Kottemann et al., 2005; Leuko et al., 2015), *Halococcus morrhuae, Halococcus hamelinensis, Natronorubrum* sp. strain HG-1, Hcc. Dombrowskii under UV radiation (Leuko et al., 2015; Peeters et al., 2010; Vreeland et al., 2000), *Natrialba magadii, Haloferax volcanii* under vacuum (Abrevaya et al., 2011), and *Halorubrum chaoviator* under UV radiation and vacuum (Mancinelli, 2015).

Many excellent monographs and books with extensive reviews on halophilic microorganisms exist (Rodriguez-Valera, 1988; Javor, 1989; Vreeland and Hochstein, 1993; Oren, 1999, 2002; Ventosa, 2004). The extremely halophilic archaea (also called haloarchaea or, traditionally, "halobacteria") belong to the order Halobacteriales, which contains one family, the *Halobacteriaceae* (Grant et al., 2001).

The apparent longevity of haloarchaeal strains in dry salty environments is of interest for astrobiological studies and the search for life on Mars. On Earth, microorganisms were the first life forms to emerge and were present perhaps as early as 3.8 billion years ago (Schidlowski, 1988, 2001). If Mars and Earth had a similar geological past (Schidlowski 2001; Nisbet and Sleep, 2001), then microbial life, or the remnants of it, could still be present on Mars. Since halophilic microorganisms appear to survive in dry salt over geological time scales, as several studies suggested (Norton et al., 1993; Grant et al., 1998; McGenity et al., 2000), it appears plausible to include specific searches for halophiles in the exploration of extraterrestrial samples or environments (Fendrihan et al., 2006).

According to Monk et al. (1994), apart from the present solar UV radiation stress environment, this stress factor was also very important during the Archean era as the Earth lacked an ozone layer and thus the surface was exposed to the complete spectrum of solar radiation As a result, it is hypothesized that microorganisms which evolved on the Archean surface under these conditions may retain cellular mechanisms which developed to cope with elevated levels of full-spectrum UV radiation (Wynn-Williams et al., 2001).

Feshangsaz et al. (2020) explain that extraterrestrial environments influence the biochemistry of organisms through a variety of factors, including high levels of radiation and vacuum, temperature extremes and a lack of water and nutrients. A wide variety of terrestrial microorganisms, including those counted among the most ancient inhabitants of Earth, can cope with high levels of salinity, extreme temperatures, desiccation and high levels of radiation. Key among these are the haloarchaea, considered particularly relevant for astrobiological studies due to their ability to thrive in hypersaline environments. For radiation protection, and associated damage due to hydroxyl radical production, halophilic archaea utilize protective mechanisms such as polyploidy (the heritable condition of possessing more than two complete sets of chromosomes), membrane pigments such as C_{50}, bacterioruberin and intracellular KCl (Shahmohammadi et al., 1998; Siefermann-Harms, 1987).

In a study, Feshangsaz et al. (2020) investigated the survival of two different microorganisms in simulated space and Mars-surface conditions: the halophilic archaeon *Halovarius luteus* strain DA50^T and the spore-forming bacteria *Bacillus atrophaeus*. *Bacillus atrophaeus* was selected for comparison due to its well-described resistance to extreme conditions and its ability to produce strong spore structures. Thin films were produced to investigate viability without the protective influence of cell multilayers. Late exponential phase cultures of *Hvr. luteus* and *B. atrophaeus* were placed in brine and phosphate buffered saline media, respectively. The solutions were allowed to evaporate and cells were encapsulated and exposed to radiation, desiccation and vacuum conditions, and their postexposure viability was studied by the Most Probable Number method. The protein profile using High Performance Liquid Chromatography and Matrix Assisted Laser Desorption/Ionization bench top reflector time-of-flight were explored after vacuum and UV-radiation exposure. Results showed that the change in viability of the spore-forming bacteria *B. atrophaeus* was only minor whereas *Hvr. luteus* demonstrated a range of viability under different conditions. At the peak radiation flux of 10^5 J/m^2 under nitrogen flow and after 2 weeks of desiccation, *Hvr. luteus* demonstrated the greatest decrease in viability.

Hvr. luteus strain DA50^T showed a good resistance, withstanding 30 min of Mars simulated UV radiation under air or N_2 atmosphere, high vacuum conditions, and up to 2 weeks desiccation. According to Feshangsaz et al. (2020), survival within brine inclusions may be another method by which *Hvr. luteus* strain DA50^T cells can tolerate radiation, high vacuum conditions, and desiccation. General protein profiles show a clear difference between the various extreme conditions used. Feshangsaz et al. (2020)'s study suggests that this extremely halophilic archaeon may be a suitable candidate for ongoing astrobiological research.

The result of this research illustrates the survival capacity of halophilic archaeon under simulated space and Mars conditions. It is well known that haloarchaea are excellent candidates for astrobiological studies, not only for their ability to survive salinity near saturation like on Mars (Mancinelli et al., 2004) or Jupiter's moon Europa (Marion et al., 2003), but also because of their ability to cope with a variety of extreme conditions such as desiccation, radiation, extreme pH and temperature (Abrevaya et al. 2011). An important factor that may contribute to the resistance of halophilic archaea to high levels of UV radiation, alongside with the presence of multiple genome copies, is the high concentration of intracellular halide ions including chloride and possibly bromide that play the role of chemical chaperones (Abrevaya et al. 2011; Kish et al., 2009). Note that in molecular biology, molecular chaperones are proteins that assist the conformational folding or unfolding and the assembly or disassembly of other macromolecular structures.

The capability of haloarchaea to survive in low water-activity conditions, such as evaporating environments, and their requirement for a high concentration of salt, make them suitable models to study life on Mars (Dassarma et al., 2019; Litchfield, 1998). Feshangsaz et al. (2020) claim that this study further expands our understanding of the boundary conditions of astrobiologically relevant organisms in the harsh space environment.

10.16.3.2 Survival of spores of Bacillus subtilis bacteria under simulated Martian UV radiation

Spore is a reproductive cell capable of developing into a new individual without fusion with another reproductive cell. The technique of spore formation is adopted by certain types of cells for the purpose of survival, under unfavorable conditions, often for extended periods of time. In the case of spore-forming bacteria, the technique of spore-formation allows the bacterium to produce a dormant and highly resistant cell to preserve the cell's genetic material in times of extreme stress. An endospore is a dormant, tough, nonreproductive structure produced by a small number of bacteria from the *Firmicute* family. The primary function of most endospores is to ensure the survival of a bacterium through periods of environmental stress. Endospores can survive environmental assaults that would normally kill the bacterium. The main difference between spore-forming bacteria and nonspore-forming bacteria is that the spore-forming bacteria produce highly resistant, dormant structures called spores in response to adverse environmental conditions whereas the nonspore-forming bacteria do not produce any type of dormant structures. It has been found that survival mechanisms for prokaryotic microorganisms under desiccating conditions include spore-formation (Nicholson et al., 2000) and the production of extracellular polysaccharide (Hill et al., 1997). However, these two mechanisms are not present in most halophilic archaea. Extremely resistant microorganisms, including *Bacillus subtilis* spores, have been detected in spacecraft-associated facilities (Venkateswaran et al., 2014; Checinska et al., 2015; Moissl-Eichinger et al., 2016).

It has been found that resistance of spores to extreme conditions does not rely on one single mechanism, but rather on a combination of several strategies (Setlow, 2014). Cortesão et al. (2019) explain that the first line of action is "damage prevention." The overall spore structure is composed of the core, inner membrane, cortex, coat, and crust layers, and has a wide number of properties and components that protect spores from many stress factors. Specifically, the spore core has low water content (25–55% of wet weight), due in some fashion to the spore's peptidoglycan cortex, that provides resistance to wet heat. Magge et al. (2008) explain that within the core, high levels (~25% of core dry weight) of pyridine-2,6-dicarboxylic acid—dipicolinic acid (DPA), in a 1:1 chelate with Ca^{2+} (Ca-DPA) help to protect spores from desiccation and DNA-damaging agents and maintain spore dormancy. It has been found that the core's high levels of α/β-type small, acid-soluble spore proteins (SASP) (Magge et al., 2008) that saturate spore DNA are one of the main factors protecting spores from genotoxic chemicals, desiccation, dry and wet heat, as well as UV and g-radiation (Mason and Setlow, 1986; Moeller et al., 2008). Cortesão et al. (2019) explain that "the thick proteinaceous coat and crust layers, as well as the inner membrane, function as barriers to many toxic chemicals minimizing their ability to access the spore core where DNA and most spore enzymes are located. The spore-coats also contain melanin-like pigments that absorb UV radiation, and there is evidence that such pigments can play a significant role in spore resistance to UV-B and UV-A radiation (Hullo et al., 2001; Moeller et al., 2008, 2014; Setlow, 2014)."

Several studies (e.g., Dose et al., 1995; Horneck et al., 2001, 2010; Nicholson and Schuerger, 2005; Fajardo-Cavazos et al., 2010; Moeller et al., 2012b) have indicated that despite complex stress-induced damage, spores of the gram-positive bacterium *Bacillus subtilis* have repeatedly demonstrated their resistance to many space-related extremes, becoming one of the model organisms in the field of Space Microbiology. Studies have shown *Bacillus* spores survive in extreme dryness, high levels of UV and ionizing radiation, and outer space conditions in Low Earth Orbit (LEO), where they were exposed to solar UV, high vacuum, Galactic Cosmic Radiation (GCR), and temperature fluctuations (Cortesão et al., 2019). Bacillus subtilis spores were shown to survive in Mars analog soils, confirming a potential forward contamination risk to Mars sites with liquid brines (Schuerger et al., 2017).

Cortesão et al. (2019) explain that the second line of defense is "damage repair," which takes place soon after spores germinate and begin outgrowth. *Bacillus subtilis* spores are armed with enzymes of multiple DNA repair pathways, thus marshaling multiple mechanisms that ensure spore survival. The main known mechanisms for repair of DNA damage in spores are: (1) homologous recombination (HR), (2) nonhomologous end joining (NHEJ), (3) nucleotide excision repair (NER), (4) DNA integrity scanning, (5) interstrand cross-link repair, (6) base excision repair (BER), (7) SP repair by spore photoproduct lyase (SPL), (8) mismatch repair (MMR), (9) endonuclease-dependent excision repair (UVER), and (10) error-prone translesion synthesis (TLS) (Xue and Nicholson, 1996; Rebeil et al., 1998; Duigou et al., 2005; Moeller et al., 2007b, 2012a; Lenhart et al., 2012).

In a study carried out by Cortesão et al. (2019, a systematic screening was performed to determine whether *B. subtilis* spores could survive an average day on Mars. For that purpose, spores from two comprehensive sets of isogenic *Bacillus subtilis* mutant strains, defective in DNA protection or repair genes, were exposed to 24 h of simulated Martian atmospheric environment with or without 8 h of Martian UV radiation [M(+)UV and M(−)UV, respectively]. Cortesão et al. (2019) found that when exposed to M(+)UV, the spore survival was dependent on: (1) core dehydration maintenance, (2) protection of DNA by α/β-type small acid soluble proteins (SASP), and (3) removal and repair of the major UV photoproduct (SP) in spore DNA. In turn, when exposed to M(-)UV, spore survival was mainly dependent on protection by the multilayered spore coat, and DNA double-strand breaks represent the main lesion accumulated. Cortesão et al. (2019) found that *Bacillus subtilis* spores were able to survive for at least a limited time in a simulated Martian environment, both with or without solar UV radiation. Moreover, M(−)UV-treated spores exhibited survival rates significantly higher than the M(+)UV-treated spores.

Cortesão et al. (2019)'s results indicate that, altogether, the ability of *B. subtilis* spores to maximally survive M(+/−) UV was due to the ability to prevent DNA damage through several protective components, including the coat, low core water content, Ca-DPA and α/β-type SASP. Yet, it was the spore outer coat (*cotE*) that was the most important protection mechanism, as spores lacking DPA, α/β-type SASP, and outer coat (*cotE sleB spoVF sspA sspB* spores) were 100-fold more sensitive than spores lacking DPA and α/β-type SASP, but with an intact outer coat (*sspA sspB sleB spoVF* spores). The latter observation is consistent with previous work showing that the spore coat is important for spore resistance to solar radiation, particularly UV-B and UV-A (Riesenman and Nicholson, 2000; Moeller et al., 2014). It is notable that the *B. subtilis* spore crust, the spore's outermost layer played no significant role in spore resistance to M(+/−)UV. Cortesão et al. (2019)'s finding suggests that on a real Martian surface, radiation shielding of spores (e.g., by dust, rocks, or spacecraft surface irregularities) might significantly extend survival rates.

10.17 Exploring Mars under *Rover mission* to assess past environmental conditions for suitability for life

A rover, in the present context, is a space exploration vehicle designed to move across the surface of a planet or other celestial body (wander), with the purpose of finding out information and collecting samples (e.g., dust, rocks) and even taking pictures. Rovers usually arrive at the planetary surface on a lander-style spacecraft. NASA's Mars Exploration Rover (MER) mission is a robotic space mission involving two Mars rovers, *Spirit* and *Opportunity*, exploring the planet Mars. It began in 2003 with the launch of the two rovers: MER-A *Spirit* and MER-B *Opportunity*—to explore the Martian surface and geology. Both rovers landed on Mars at separate locations in January 2004. It is a great achievement that both rovers far outlived their planned missions of 90 Martian solar days (note that a Martian day, referred to as "sol," is approximately 40 min longer than a day on Earth): MER-A *Spirit* was active until 2010, while MER-B *Opportunity* was active on Mars from 2004 until mid-2018, and holds the record for the longest distance driven by any off-Earth wheeled vehicle (https://en.wikipedia.org/wiki/Mars_Exploration_Rover).

The primary objective of the Mars Exploration Rover mission was to search for evidence in the Martian geologic record of environmental conditions that might once have been suitable for life. The Mars Exploration Rover "Opportunity" landed in Eagle crater on Meridiani Planum on 24 January 2004; 21 days after the landing of Spirit at Gusev crater. Both vehicles landed using a variant of the airbag landing system that was developed for Mars Pathfinder, deploying the rovers after the landers had come to rest on the surface. The primary scientific objective of their mission was to explore two sites on the Martian surface where water may once have been present, and to assess past environmental conditions at those sites, including their suitability for life. Squyres et al. (2004a) have provided an overview of the results from the 90-sol nominal mission of *Opportunity*.

The Mars Exploration Rover "Opportunity" carried the Athena science payload. The topography, morphology, and mineralogy of the scene around the rover have been revealed by a panoramic camera (Pancam) and a miniature thermal emission spectrometer (Mini-TES). Both instruments viewed the scene using a mast at a height of about 1.5 m above the ground. The payload also included a 5-degree-of-freedom robotic arm. The arm carried an Alpha Particle X-ray Spectrometer (APXS) that measured elemental abundances of rocks and soils and a Mössbauer Spectrometer that determined the mineralogy and oxidation state of Fe-bearing phases. It also carried a Microscopic Imager (MI) that was used to obtain high-resolution (30 μm per pixel) images of rock and soil surfaces and a Rock Abrasion Tool (RAT) having capability to remove up to ∼5 mm of material over a circular area 45 mm in diameter. Finally, the payload included seven magnets capable of attracting fine-grained magnetic materials and can be viewed by payload instruments.

The rover itself used a six-wheel rocker-bogie suspension system and onboard autonomous navigation and hazard avoidance capability, allowing traverse distances of tens of meters per sol. Navigation and hazard avoidance were aided by two monochromatic navigation cameras (Navcams) mounted on the mast and by four hazard avoidance cameras (Hazcams) mounted in fore- and aft-facing stereo pairs on the rover body.

The Meridiani Planum landing site was chosen for *Opportunity* because its smooth flat topography would favor a safe landing and because Mars Global Surveyor TES data showed it to contain ∼15–20% (by fractional area) of the mineral hematite. Hematite can form by a number of processes, many of which involve the action of liquid water. Orbital images showed that the hematite-bearing unit is the top stratum of a layered sequence about 600 m thick that overlies Noachian cratered terrain. Squyres et al. (2004a) landed on this sequence with the hope that the hematite was an indicator that aqueous processes had been involved in its formation.

The *Opportunity* has investigated the landing site in Eagle crater and the nearby plains within Meridiani Planum. The soils consist of fine-grained basaltic sand, hematite-rich spherules, spherule fragments, and other granules. In terms of Eolian processes, wind at the Opportunity landing site has eroded rock, sorted soil particles during transport, formed and changed ubiquitous ripples on the plains as well as ripples and dunes in craters and troughs, and formed wind streaks seen from orbit in MOC images. Saltation of fine-grained basaltic sand has been a particularly important eolian process, abrading clasts and outcrop rock and generating impact ripples.

Fig. 10.13 shows *Mount Sharp* on Mars, as imaged by the Curiosity rover. Underlying the thin soil layer, and

FIG. 10.13 Mount Sharp on Mars, as imaged by the Curiosity Rover. NASA/JPL-Caltech/MSSS. *(Source: Scharping (2017), Mars may be more toxic to life than we thought; http://www.astronomy.com/news/2017/07/mars-perchlorates-bacteria.)*

FIG. 10.14 Pancam image showing fine-scale laminations in the Eagle crater outcrop. Scale across the image is about 60 cm. This is an approximate true-color image, assembled using images from the 480-, 540-, and 600-nm filters. *(Source: Squyres et al., 2004a; The Opportunity Rover's Athena science investigation at Meridiani Planum, Mars, Science, 306 (5702), 1698-1703; doi: http://doi.org/10.1126/science.1106171.)*

exposed within small impact craters and troughs, are flat-lying sedimentary rocks. These rocks are finely laminated (Fig. 10.14), are rich in sulfur, and contain abundant sulfate salts. Small-scale cross-lamination in some locations provides evidence for deposition in flowing liquid water. Squyres et al. (2004a) interpreted the rocks to be a mixture of chemical and siliciclastic sediments formed by episodic inundation by shallow surface water, followed by evaporation, exposure, and desiccation. Hematite-rich spherules are embedded in the rock and eroding from them. Squyres et al. (2004a) interpreted these spherules to be concretions formed by postdepositional diagenesis (physical and chemical processes that affect sedimentary materials after deposition and before metamorphism and between deposition and weathering), again involving liquid water. The environmental conditions that the sedimentary rocks at Eagle crater in Meridiani Planum record include episodic inundation by shallow surface water, evaporation, and desiccation. The geologic record at Meridiani Planum suggests that conditions were suitable for biological activity for a period of time in Martian history.

Squyres et al. (2004b) interpreted the rocks of the Eagle crater outcrop to be a mixture of chemical and siliciclastic sediments with a complex diagenetic history. The environmental conditions that they record are episodic inundation by surface water to shallow depths, followed by evaporation, exposure, and desiccation. The festoon cross-lamination (having the appearance of a chain hung in a curve) provides evidence for inundation by water.

It is noteworthy that the environmental conditions recorded at Eagle crater may have prevailed over a large area. All of the bedrock observed during Opportunity's traverse to Endurance crater, more than half a kilometer from the landing site, showed the same characteristics as the outcrop in Eagle crater (Squyres et al., 2004b). More important, orbital data from the Mars Orbiter Camera showed that a relatively high-albedo unit that correlates with the outcrop exposures near the Opportunity landing site is at least discontinuously present across most of Meridiani Planum. It seems likely, then, that the area over which these aqueous sedimentary and diagenetic processes operated was at least tens of thousands of square kilometers in size.

In addition to geologic investigations, the remote sensing instruments on the payload have been used to observe the Martian atmosphere. Upward-looking Mini-TES spectra are diagnostic of the vertical thermal structure of the atmosphere between about 20 m and 2 km above the surface, column-integrated infrared aerosol optical depth and water vapor abundance, and aerosol particle size. In particular, temperature profiles retrieved from Mini-TES observations of the 15-μm absorption band of CO_2 provided systematic characterization of the Martian boundary layer. These observations showed the development of a near-surface super-adiabatic layer during the afternoon and a deep inversion layer at night. Upward-looking Mini-TES "stares," in which spectra have been collected every 2 s for up to an hour, showed warm and cool parcels of air moving through the Mini-TES field of view on a time scale of 30 s. Dust properties were found to have been consistent with previous measurements; specifically, cross-section mean radius is 1.5 ± 0.2 μm (Squyres et al., 2004a).

It was found that regardless of the origin of the thin surface layer of sand, the underlying sulfate-rich sedimentary rocks at Meridiani Planum clearly preserve a record of environmental conditions different from any on Mars today. Liquid water was once present intermittently at the Martian surface at Meridiani, and at times it saturated the subsurface. Because liquid water is a key prerequisite for life, Squyres et al. (2004a) infer that conditions at Meridiani may have been habitable for some period of time in Martian history.

10.18 Mangalyaan—India's Mars Orbiter Mission (MOM) spacecraft—Exploring Mars' surface features, morphology, mineralogy, and atmosphere

Apart from and subsequent to the Viking lander mission carried out by the United States (operated by NASA/JPL) and the Mars Exploration Rover mission carried out by the United States (operated by NASA/JPL), a few other missions to Mars were carried out independently by USA and India (operated by Indian Space Research Organisation (ISRO)). Mars Orbiter Mission (MOM), the interplanetary mission of ISRO to the planet Mars, launched on November 5, 2013 is a maiden and unique Indian attempt toward sending orbiters to other planets in our Solar System. It is the most cost effective interplanetary mission ever carried out globally, accomplished in record time and got inserted in the Mars orbit on September 24, 2014 in its first attempt (a unique achievement in India's MOM program). The mission has a unique and highly elliptical Martian orbit of about 261 km (Periareion) to 78,000 km (Apoareion). Among contemporary, MOM is also credited with some more laurels such as miniaturization of five heterogeneous science payloads and minimal path corrections during entire cruise phase (Arya et al., 2015a).

There was no eclipse on the MOM spacecraft during its interplanetary transfer. During the Martian phase, it started to experience eclipse shadow of Mars from the beginning. Srivastava et al. (2015) have discussed several conical shadow eclipse prediction models by accounting the effects of atmospheric dust of Mars, considering both spherical and oblate shape of the red planet. They performed a study using the results obtained by different shadow models, Systems Tool Kit (STK), and the actual telemetry data. They noticed that effects of the atmospheric dust of Mars cannot be neglected on the MOM spacecraft.

Locating the geographical position of a spacecraft is vital during launching as well as while making the interplanetary voyages. Delta Differential One-way Ranging (DDOR) is an interferometer technique used in deep space missions to calculate the angular position of a satellite with reference to a well-known source which is usually a quasar (an astronomical object of very high luminosity found in the centers of some galaxies and powered by gas spiraling at high velocity into an extremely large black hole). A transmitter has been used in Mars Orbiter Mission (MOM) for DDOR measurement to improve the accuracy of orbit determined by other techniques. The transmitter operates at S-band with carrier phase modulated by a single tone at around 4 MHz approximately. Ramamurthy et al (2015) have discussed the principle of DDOR measurement, how the specifications for the transmitter were arrived at, the design and realization using the available components and its on-orbit performance.

During the Mars solar conjunction, telecommunication and tracking between the spacecraft and the Earth degrades significantly. The radio signal degradation depends on the angular separation between the Sun, Earth and Probe (SEP), the signal frequency band, and the solar activity. All radiometric tracking data types display increased noise and signatures for smaller SEP angles. Due to scintillation (a small flash of visible or ultraviolet light emitted by fluorescence in a phosphor when struck by a charged particle or high-energy photon), telemetry frame errors increase significantly when solar elongation becomes small enough. This degradation in telemetry data return starts at solar elongation angles of around 5° at S-band, around 2° at X-band and about 1° at Ka-band. Srivastava et al. (2016) presented a mathematical model for predicting Mars superior solar conjunction for any Mars orbiting spacecraft. The described model was simulated for the Mars Orbiter Mission which experienced Mars solar conjunction during May–July 2015. It is considered that such a model may be useful to flight projects and design engineers in the planning of Mars solar conjunction operational scenarios.

On 24 September 2014 India became the first country to put an interplanetary space probe around Mars in the very first attempt. Mars Orbiter Mission, popularly known as MOM, was earlier launched from Sriharikota using a Polar Satellite Launch Vehicle (PSLV) rocket on 5 November 2013. After a cruise phase of 300 days in the heliocentric (Sun centric: measured from or considered in relation to the center of the Sun) orbit, MOM is now going around Mars in an elliptical orbit. The spacecraft is now circling Mars in an orbit whose nearest point to Mars (periapsis) is at 424 km and farthest point (apoapsis) at 77,098 km. The inclination of orbit with respect to the equatorial plane of Mars is ~150°, as intended. In this orbit, the spacecraft takes ~73 h to go around the Mars once.

The objectives of the MOM included design and realization of an interplanetary spacecraft with a capability to survive and perform Earth-bound maneuvers, cruise phase, Mars orbit insertion, and on-orbit phase around the Mars. MOM is designed to explore surface of Mars and its atmosphere.

The Alpha Particle X-ray Spectrometer on the *Opportunity* Rover determined major and minor elements of soils and rocks in Meridiani Planum. Rieder et al. (2004) found that chemical compositions differentiate between basaltic rocks, evaporite-rich rocks, basaltic soils, and hematite-rich soils. Although soils are compositionally similar to those at previous landing sites, differences in iron and some minor element concentrations signify the addition of local components. It was found that rocky outcrops are rich in sulfur and variably enriched in bromine relative to chlorine. Interestingly, the interaction with water in the past is indicated by the chemical features in rocks and soils at this site.

While exploring a planet, acquiring knowledge about its atmospheric temperature is of vital importance. In this context, Spanovich et al. (2006) modeled downward-looking spectra of the Martian surface from the Miniature Thermal Emission spectrometer (Mini-TES), onboard each of the two Mars Exploration Rovers, in order to retrieve surface and near-surface atmospheric temperatures. They determined the surface temperature and the near-surface atmospheric temperature, approximately 1.1 m above the surface, by fitting the observed radiance in the vicinity of the 15-μm CO_2 absorption feature. Spanovich et al. (2006)

used the temperatures from the first 180 sols (Martian days) of each surface mission to characterize the diurnal dependence of temperatures. It was found that the near-surface atmospheric temperatures are consistently 20 K cooler than the surface temperatures in the warmest part of each sol, which is 1300–1400 LTST (local true solar time) depending on the location. Seasonal cooling trends were seen in the data by displaying the temperatures as a function of sol. Long ground stares, 8.5 min in duration, showed as much as 8 K fluctuation in the near-surface atmospheric temperatures during the early afternoon hours when the near-surface atmosphere was unstable.

Mars Advanced Radar for Subsurface and Ionospheric Sounding (MARSIS) on the Mars Express (MEX) spacecraft has made numerous measurements of the Martian surface and subsurface. However, all of these measurements are distorted by the ionosphere and must be compensated before any analysis. Mouginot et al. (2008) developed a technique to compensate for the ionospheric distortions. This technique provides a powerful tool to derive the total electron content (TEC) and other higher-order terms of the limited expansion of the plasma dispersion function that are related to overall shape of the electron column profile. The derived parameters were fitted by using a Chapman model to derive ionospheric parameters such as n_0, electron density primary peak (maximum for solar zenith angle (SZA) equal 0), and the neutral height scale H.

The ionospheric parameters estimated by Mouginot et al. (2008) were found to be in good agreement with Mars Global Surveyor (MGS) radio-occultation data. However, because MARSIS does not have the observation geometry limitations of the radio occultation measurements, the parameters derived by Mouginot et al. (2008) extend over a large range of SZA for each Mars Express (MEX) orbit. The first results from Mouginot et al. (2008)'s technique have been discussed by Safaeinili et al. (2007).

The MARSIS radar experiment aboard the ESA Mars Express artificial satellite has recorded several unusual reflections in the Ma'adim Vallis region of Mars. These reflections display a wide variety of morphologies which are very different from those of reflections seen beneath the Polar Layered Deposits, Medusae Fossae Formation and Dorsa Argentea Formation. Their morphologies are sometimes very laterally extensive, parabolic or hyperbolic, and apparently deep, but they can also appear horizontal and shallow. Aided by a geological map of the Ma'adim Vallis region, the morphological, locational and temporal characteristics of the reflections have been studied individually by White et al. (2009) in an attempt to constrain their origin. While some may be subsurface reflections based on their shallow morphologies and correlation with the Eridania Planitia basin network, all of the reflections are ambiguous to some degree, displaying characteristics that do not allow a definite subsurface- or possibly ionospheric-sourced mechanism to be proposed for their creation. Those with more exaggerated morphologies are regarded as being much more likely to result from ionospheric distortion rather than subsurface inhomogeneity.

The SHARAD (shallow radar) sounding radar on the Mars Reconnaissance Orbiter detects subsurface reflections in the eastern and western parts of the Medusae Fossae Formation (MFF). Note that the Medusae Fossae Formation is a large geological formation of probable volcanic origin on the planet Mars. It is named for the *Medusa* of Greek mythology. "Fossae" is Latin for "trenches." The radar waves penetrate up to 580 m of the MFF and detect clear subsurface interfaces in two locations: west MFF between 150 and 155° E and east MFF between 209 and 213° E (Carter et al., 2009). Analysis of SHARAD radargrams suggests that the real part of the permittivity is ~3.0, which falls within the range of permittivity values inferred from MARSIS data for thicker parts of the MFF. The SHARAD data cannot uniquely determine the composition of the MFF material, but the low permittivity implies that the upper few hundred meters of the MFF material has a high porosity. One possibility is that the MFF is comprised of low-density welded or interlocked pyroclastic deposits that are capable of sustaining the steep-sided yardangs (streamlined protuberances carved from bedrock or any consolidated or semiconsolidated materials by the dual action of wind abrasion by dust and sand and deflation. Yardangs resemble the 'ridge and furrow' landscape of zeugen) and ridges seen in imagery. The SHARAD surface echo power across the MFF is low relative to typical Martian plains, and completely disappears in parts of the east MFF that correspond to the radar-dark Stealth region. These areas are extremely rough at centimeter to meter scales, and the lack of echo power is most likely due to a combination of surface roughness and a low near-surface permittivity that reduces the echo strength from any locally flat regions. There is also no radar evidence for internal layering in any of the SHARAD data for the MFF, despite the fact that tens-of-meters scale layering is apparent in infrared and visible wavelength images of nearby areas. These interfaces may not be detected in SHARAD data if their permittivity contrasts are low, or if the layers are discontinuous. The lack of closely spaced internal radar reflectors suggests that the MFF is not an equatorial analog to the current Martian polar deposits, which show clear evidence of multiple internal layers in SHARAD data.

10.19 Imaging of Mars' surface, its dynamical events, and its Moons *Phobos* and *Deimos* using Mars Color Camera (MCC) onboard MOM

Unlike Earth, Mars has two moons, *Phobos* and *Deimos*. Phobos (fear) and Deimos (panic) were named after the

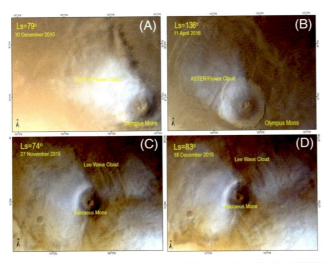

FIG. 10.15 MCC-MOM observation of Mars' surface; (A, B) ASTER/flower cloud over Olympus Mons, highest point on Mars; and (C, D) Lee-Wave clouds over Ascraeus Mons Mars. *(Source: http://www.isro.gov.in/mars-orbiter-mission-completes-1000-days-orbit.)*

horses that pulled the chariot of the Greek war god Ares, the counterpart to the Roman war god Mars. India's MOM spacecraft carried a color camera named Mars Color Camera (MCC). The MCC is an electro-optical sensor imaging the surface of Mars in three colors, varying spatial resolution between 25 m to 4 km in 16 different exposure modes, depending on its position in orbital plane and illumination conditions. The MCC onboard MOM has 16 different modes of exposures, aimed at imaging the Mars surface for morphological / structural mapping, imaging dynamic events viz. Polar Ice cap and its seasonal variations, clouds, dust storms and other opportunistic imaging. The camera has beamed several high-quality images of Mars surface. Important science objectives of MCC include studying morphology of landforms, dynamic processes such as dust storms, dust devils, clouds, and polar ice cap variability in different seasons. A higher resolution image from MCC (~300 m spatial resolution) is taken from 7300 km altitude, over the Syrtis Major region of Mars imaged on the first day itself. The image shows impact craters at the south-western edge of Syrtis Major. Syrtis Major Planum is a 'dark spot' which is considered to be a low-level shield volcano built almost entirely of lava flows and having basaltic rocks. The dark colored regions are exposures of basalts, while the red color regions show dust-covered areas. The image also reveals streaks being blown out of craters, known as wind streaks, indicating that the wind is moving in a southwest direction. Fig. 10.15 shows an MCC-MOM observation of Mars' surface, particularly ASTER/flower cloud over Olympus Mons, highest point on Mars; and Lee-Wave clouds over Ascraeus Mons. Mars Analysis of these payloads has churned the below-given interesting results (http://www.isro.gov.in/mars-orbiter-mission-completes-1000-days-orbit).

10.19.1 Atmospheric optical depth estimation in Valles Marineris—A huge canyon system

During its examination of Mars, the Viking 1 spacecraft returned images of Valles Marineris—a huge canyon system 5000 km, or about 3106 miles, long, who's connected chasma or valleys may have formed from a combination of erosional collapse and structural activity. In a study of the Valles Marineris, carried out by the Indian Space Research Organization (ISRO), the atmospheric optical depth (AOD) was estimated through an experiment involving multiview/multioptical path length images. This information was used to determine the pressure scale height of the dust (11.24 km) which is commensurate with the known GSM scale height computation (11.2–12.1 km). Note that the coefficient of attenuation of solar radiation by the atmosphere is called atmospheric optical depth, and can be calculated by measuring the intensity of direct solar radiation reaching the earth.

Mishra et al. (2016) reported analyses of bright hazes observed inside Valles Marineris formed during mid-southern spring of Mars. The analysis was performed by using data collected by MCC onboard Indian Mars Orbiter Mission on orbits 34, 49, and 52 corresponding to the observation dates of October 28, December 5, and December 13, 2014. It was found that during all these orbits the valley was hazy. On orbit 34 a thick layer of haze was observed, which became relatively thinner on orbit 49. Thick haze reappeared after 8 days on orbit 52. Mishra et al. (2016) also measured the optical depth of Martian atmosphere as a function of altitude above two opposing walls (northern and southern walls of the Valles Marineris near Coprates Chasma region) of the valley, from stereo images that were taken with MCC on December 5, 2014. The optical depth was measured from contrast comparisons of the stereo images with "stereo method." In the northern wall of Valles, Mishra et al. (2016) estimated the optical depth as a function of altitude (ranging between −6 km and 3 km) and found values between 1.7 (bottom) and 1.0 (top) in red channel and between 2.1 (bottom) and 1.2 (top) in green channel. A fit to these results yielded a scale height for the optical depth of 14.08 km and 11.24 km in red and green channel, which were more or less in good agreement to the pressure scaled height of Martian atmosphere at that time in the region as consulted from Global Circulation Model (GCM). Mishra et al. (2016) also estimated optical depth in southern wall of Valles Marineris. However, in this case optical depth remained nearly constant with decreasing altitude. Mishra et al. (2016) consulted GCM for wind direction in the region and found strong wind with direction from south-west to north-east intersecting the mountain like structure of the southern wall of Valles Marineris. Mishra et al. (2016)'s optical depth results and the wind direction suggest the presence of lee-wave cloud above the southern wall of Valles Marineris.

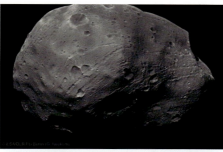

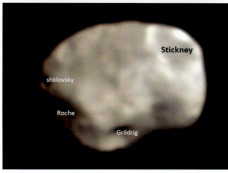

FIG. 10.16 Phobos, the closest and the biggest moon of Mars. (Top) As seen during a flyby performed by the European spacecraft Mars Express. *Credit: ESA/DLR/FU Berlin (G. Neukum) https://www.space.com/13043-mars-photos-amazing-red-planet-martian-views.html.* (Bottom) Imaged by MOM on 1st July 2020, when MOM was about 7200 km from Mars and at 4200 km from Phobos. *(Image Source: https://www.isro.gov.in/pslv-c25-mars-orbiter-mission/phobos-imaged-mom-1st-july), Image credit: ISRO.*

10.19.2 Imaging of Mars' twin Moons—*Phobos* and the far side of *Deimos*

MCC onboard Mars Orbiter Mission has imaged *Phobos* (Fig. 10.16), the closest and the biggest moon of Mars, on 1st July 2020 when MOM was about 7200 km from Mars and at 4200 km from Phobos. Spatial resolution of the image is 210 m. This is a composite image generated from 6 MCC frames and has been color corrected. Phobos is largely believed to be made up of carbonaceous chondrites. The violent phase that Phobos has encountered is seen in the large section gouged out from a past collision (Stickney crater) and bouncing ejecta. Stickney, the largest crater on Phobos along with the other craters (Shklovsky, Roche, and Grildrig) are also seen in this image.

While ISRO is not the first space agency to capture the image of Phobos, it is certainly known to be an achievement because the moon is one of the least reflective bodies in the solar system. Phobos' least reflectivity of solar light arises from the fact that this moon is made up of meteorite materials such as 'carbonaceous chondrites,' which makes it less visible compared to other celestial bodies in the Solar System.

Mars and one of its moons "Deimos" (Fig. 10.17) are tidally locked so one would always get to view the same side

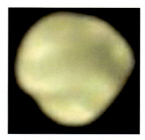

FIG. 10.17 Mars' moon "Deimos" imaged by the maiden Indian Mars Orbiter Mission (MOM). *(Sources: https://vedas.sac.gov.in/vedas/downloads/ertd/MOM/MCC_Results.pdf), Arya et al. (2015a), Mars color camera onboard mars orbiter mission: initial observations & results, 46th Lunar and Planetary Science Conference, Email: arya_as@sac.isro.gov.in.*

of Deimos from Mars. Most of the contemporary functional satellites orbiting Mars viz. Mars Reconnaissance Orbiter, Mars Express, Mars Odyssey, MAVEN, etc. are positioned between Mars and Deimos, hence they always view only one side of Deimos, that is, the Mars-side, while the far-side is always looking outward with respect to all these satellites. However, the maiden Indian Mars Orbiter Mission (MOM) is an exception as it has a very large elliptical orbit (Fig. 10.18) and goes behind the orbit of Deimos. This has enabled MOM to view the far-side of Deimos which has not been viewed for last few decades by any Mars orbiter. MCC, onboard MOM acquired four image-frames of the Deimos, the farthest of two moons of Mars, on October 14, 2014 at around 13:05 UT. These four images of MCC have been used to generate a High Dynamic Range (HDR) product so as to enhance the image details. The two well-known craters of near-side are missing in this HDR image and the shape of

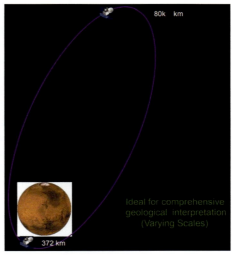

FIG. 10.18 The highly elliptical and eccentric orbit of the MOM mission is ideal for comprehensive geological interpretation and has provided the unique opportunity to view the far side of Mars' moon *Deimos*. *(Source: https://vedas.sac.gov.in/vedas/downloads/ertd/MOM/MCC_Results.pdf.)*

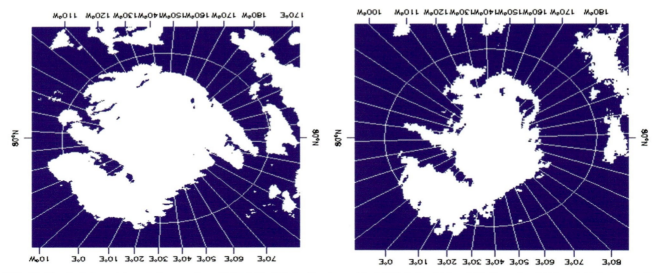

FIG. 10.19 (Left image) Area under snow/ice of Mars' North Polar Cap imaged by MCC during 24th to 26th December 2015; and (right image) 22 January 2016, showing a decrease in area from 9,52,700 km² to 6,33,825 km² due to sublimation of dry ice. *(Source: http://www.isro.gov.in/mars-orbiter-mission-completes-1000-days-orbit.)*

far-side, recently updated and reported models of Deimos proposed by other researchers, matches with the shape of Deimos in MCC images thus confirming that the far-side view of Deimos has been captured by MCC of Mars Orbiter Mission. The magnitude of Deimos has also been computed from this far-side image, which varies by a small margin from the known magnitude of Deimos. This could probably be due to possible difference in the near and far side surface characteristics, which needs further detailed investigation and more imaging of far-side in future. However, there could be reasons other than the physical characteristics of the Deimos. Thus far-side sighting has generated a scientific interest to understand it holistically. Sighting of the far-side of the Deimos is a rare event achieved by MCC after a long spell of more than three decades by any other Mars mission, including the contemporary operational Mars orbiters. This highlights the imaging capabilities of MCC-MOM and possible research to come in future. As just indicated, the highly elliptical and eccentric orbit of the MOM mission has provided the unique opportunity to view the far side of Deimos, the farthest of the two moons of Mars. This is not possible by any of the contemporary orbiters on Mars from international mission presently operational in Martian orbit. The same has been proved using orbital simulations, shape models and estimation of the apparent magnitude (Arya et al., 2015b).

10.19.3 Morphology study of *Ophir Chasma* canyon

Ophir Chasma is a canyon in the Coprates quadrangle of Mars at 4° south latitude and 72.5° west longitude. It is about 317 km long and was named after Ophir, a land mentioned in the Bible. In the Bible it was the land which King Solomon sent an expedition that returned with gold.

Ophir Chasma located in the central Valles Marineris has been imaged by MOM at a high resolution of 19.5 m, a geological map has been prepared, and various morphological units have been delineated. The various morphological features such as spur and gullies present prominently on the Chasma walls, ridges which cover the northern depression, layered domes, and dark mineral deposits were mapped. It was found that two types of layered deposits are identifiable and exposed, that is, in the canyon walls (low albedo) and Ophir Mensa (high albedo).

10.19.4 Automatic extraction, monitoring, and change detection of area under polar ice caps on Mars

It has been found that Mars' North Polar Cap suffers depletion of snow/ice due to sublimation of dry ice (see Fig. 10.19). Furthermore, long-term change detection (four decades) was accomplished by comparing snow/ice area from MCC images with Viking images. MCC showed a range 6,33,825–9,52,700 km² during imaging period, while during same imaging season of Viking mosaic (1976–1980) showed snow/ice area to be approximately 7,83,412 km² which is within the range calculated by MCC.

10.19.5 Application of reflectance data derived from a differential radiometer and the MCC on-board MOM in studying the mineralogy of Martian surface

The Methane Sensor for Mars (MSM) on-board MOM is a differential radiometer based on Fabry–Perot Etalon filters to measure columnar methane (CH_4) in the Martian

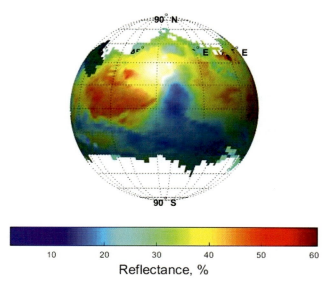

FIG. 10.20 Reflectance map of Mars at 1.65 μm derived from reference channel data of MSM, which was corrected for radiometric errors and carbon dioxide (CO_2) absorption. These data have been used to estimate the variations in albedo of Mars and it has revealed important information on seasonal changes which result in wind transport of dust (Curr. Sci., 2017, Accepted). (Source: http://www.isro.gov.in/mars-orbiter-mission-completes-1000-days-orbit.)

Physics Laboratory of Vikram Sarabhai Space Centre, measuring neutral gases in the mass range of 1–300 amu. MENCA has successfully studied the distribution of the major species in the Martian exosphere, which has helped understand the solar forcing on the Martian atmosphere.

Mars Exospheric Neutral Composition Analyser (MENCA) is a quadrupole mass spectrometer covering the mass range of 1–300 amu with mass resolution of 0.5 amu. MENCA, weighing 4 kg, would provide in-situ measurement of the neutral composition and density distribution of Martian exosphere. Technical details of MENCA can be found in Bhardwaj et al. (2015). All the five instruments have been designed to make extensive and carefully planned measurements during the expected mission time of 6 months. MCC, MSM and Thermal Infrared Imaging Spectrometer (TIS) also provide complementary information to interpret the data, for example, MCC is used for dust optical thickness estimation to correct for atmospheric scattering in MSM data for accurate estimation of methane. MCC also provides context information and TIS gives information about surface temperature to analyze MSM data.

MENCA has provided the first measurements of the low-latitude evening time exosphere of Mars (Fig. 10.21). The measured abundances of the four major Martian exospheric gases (atomic oxygen [16 amu], molecular nitrogen and carbon-monoxide [28 amu], and carbon-dioxide [44 amu]), during December 2014 showed significant orbit-to-orbit variability. These observations correspond to moderate solar activity conditions, during perihelion season (when Mars is closest to Sun) and when MOM's periapsis altitude was the lowest (~265 km). MENCA observations have shown for the first time that the abundance of oxygen exceeds that of carbon-dioxide at an altitude of ~270 ± 10 km, during the perihelion evening hours. This result

atmosphere at several parts per billion (ppb) levels. This differential signal gives a measure of columnar amount of CH_4. Though MSM could not detect any methane (above its sensitivity limit), it provided excellent reflectance data of Mars surface in the 1.65 μm region.

With 20-bit resolution and signal-to-noise ratio (SNR) > 7000, radiometric performance of MSM is extremely good. It is found that during the last 1000 days of operation, radiometric calibration of the instrument remained very stable. Fig. 10.20 gives the reflectance map of Mars generated from reference channel data of MSM which is corrected for radiometric errors and carbon dioxide (CO_2) absorption. These data together with reflectance data derived from three visible spectral bands of MCC are useful in studying the mineralogy of Mars surface. MSM data has also been used in studying the dust and cloud properties of Martian atmosphere. This is the first-time a near global albedo map of Mars has been prepared using 1.65 μm wavelength (SWIR region) of EM spectra.

10.20 Solar forcing on the Martian atmosphere and exosphere

The outermost region of a planetary atmosphere—called exosphere—holds the secrets to the atmospheric escape and evolution. This is the region being explored by Mars Exospheric Neutral Composition Analyser (MENCA) experiment aboard India's Mars Orbiter Mission (MOM), which is a quadrupole mass spectrometer-based payload, developed at India's Space

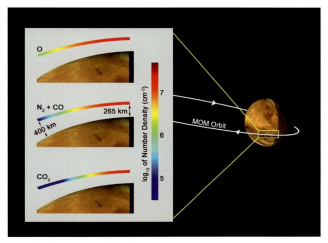

FIG. 10.21 Distribution of the major species in the Martian exosphere in the local evening sector measured by MENCA in December 2014. (Source: http://www.isro.gov.in/mars-orbiter-mission-completes-1000-days-orbit.)

indicates that the altitude where O/CO_2 ratio exceeds 1 is highly variable, is much different than at noon, and therefore it is an important input for constraining the EUV forcing in the models dealing with upper atmosphere of Mars. From the variation of the abundances of different gases with altitude, the temperature of the Martian exosphere was found to be about 270 ± 5 K, during perihelion season.

The Mars Exospheric Neutral Composition Analyser (MENCA) aboard the Indian Mars Orbiter Mission (MOM) is a quadrupole mass spectrometer which provides in situ measurement of the composition of the low-latitude Martian neutral exosphere. The altitude profiles of the three major constituents, that is, amu 44 (CO_2), amu 28 (N_2 + CO), and amu 16 (O) in the Martian exosphere during evening (close to sunset terminator) hours are reported using MENCA observations from four orbits of MOM during late December 2014, when MOM's periapsis altitude was the lowest. The altitude range of the observation encompasses the diffusively separated region much above the well-mixed atmosphere. The transition from CO_2 to O-dominated region is observed near 270 km. The mean exospheric temperature derived using these three mass numbers is 271 ± 5 K. These first observations corresponding to the Martian evening hours would help to provide constraints to the thermal escape models (Bhardwaj et al., 2016).

Another major result from MENCA is the discovery of 'hot' (supra-thermal) Argon in the exosphere of Mars (Fig. 10.22). The term 'hot' or 'supra-thermal' indicates that the atoms are more energetic compared to the thermal population, and hence their kinetic temperatures are higher.

Bhardwaj et al. (2017) reported the altitude profiles of argon-40 (Ar) in the Martian exosphere using MENCA aboard Indian Mars Orbiter Mission (MOM) from four orbits during December 2014 (Ls = 250°–257°), when MOM's periapsis altitude was the lowest. The upper limit of Ar number density corresponding to this period is $\sim 5 \times 10^5$ cm^{-3} ($\sim$250 km), and the typical scale height is $\sim$16 km, corresponding to an exospheric temperature of $\sim$275 K. However, on two orbits, the scale height over this altitude region was found to increase significantly, thereby making the effective temperature >400 K: clearly indicating the presence of supra-thermal Argon in the Martian exosphere.

Neutral Gas and Ion Mass Spectrometer observations on the Mars Atmosphere and Volatile Evolution mission also indicate that the change in slope in Ar density occurs near the upper exosphere (around 230–260 km). These observations indicate significant supra-thermal CO_2 and Ar populations in the Martian exosphere. Significant wave-like perturbations are observed but only on certain days when supra-thermal population is seen. Pickup ion-induced heating has been discussed as the other viable source.

The detection of these hot particles has important implications in the context of understanding the energy deposition in the Martian upper atmosphere, and will help understand why the Martian atmospheric escape rates are higher than what was believed previously (http://www.isro.gov.in/mars-orbiter-mission-completes-1000-days-orbit).

10.21 Emission of thermal infrared radiation from Mars

The thermal infrared imaging spectrometer (TIS) is one of the five instruments onboard Indian Mars Orbiter Mission (MOM) that measures emitted thermal Infrared radiation while orbiting around Mars in elliptical orbit. The TIS is a plane reflection grating based infrared spectrometer which uses an uncooled microbolometer detector operating in 7–13 μm wavelength range.

The TIS instrument is aimed to observe thermal emission from Mars surface to detect its temperature and hot spot regions or hydrothermal vents on Martian surface. The TIS is designed to observe emitted infrared radiation from Martian environment in 7–13 μm region of electromagnetic spectrum using micro bolometer device.

The elliptical orbit of MOM provides unique opportunity for scanning of full Mars disk from apoapsis at coarse spatial resolution and site-specific surface imaging at high spatial resolution in push broom mode from periapsis. TIS has carried out more than 90 imaging sessions over Martian surface as shown in Fig. 10.23. Observed brightness temperatures were found to be related with surface temperature, emissivity, viewing geometry and atmospheric conditions.

A scene-level analysis showed a gradual increase in binned scene-level Brightness temperature (BT) at 9.25 μm with increase in areocentric longitude (Ls). Note that areocentric latitude is the angle between the equatorial plane and a vector connecting a point on the surface and the origin of the coordinate system. Latitudes are positive in the northern hemisphere and negative in the southern hemisphere.

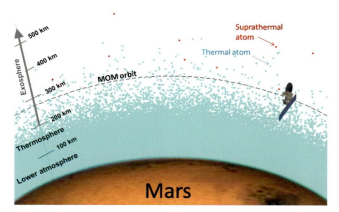

FIG. 10.22 Schematic of the MOM orbit near periapsis (drawn to scale). The blue dots represent the atmospheric gas atoms and molecules of Mars, while the red ones represent the more energetic (supra-thermal) atoms. *(Source: http://www.isro.gov.in/mars-orbiter-mission-completes-1000-days-orbit.)*

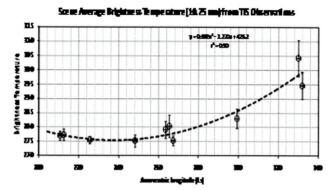

FIG. 10.23 The blue points indicate regions where TIS imaging is carried out. *(Source: http://www.isro.gov.in/mars-orbiter-mission-completes-1000-days-orbit.)*

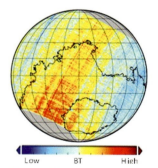 

FIG. 10.25 Brightness temperature at Ls 210.7 and corresponding MCC image. *(Source: http://www.isro.gov.in/mars-orbiter-mission-completes-1000-days-orbit.)*

10.22 D/H ratio estimation of Martian atmosphere/exosphere

The abundance and distribution of water within Mars through time plays a fundamental role in constraining its geological evolution and habitability. Studies carried out by Donahue (1995) indicate that the deuterium enrichment of the present Martian atmosphere implies that the reservoir of crustal water on ancient Mars was several hundred meters deep, consistent with the geological evidence. The isotopic composition of Martian hydrogen provides insights into the interplay between different water reservoirs on Mars. However, D/H (deuterium/hydrogen) ratios of Martian rocks and of the Martian atmosphere span a wide range of values. This has complicated the identification of distinct water reservoirs in and on Mars within the confines of existing models that assume an isotopically homogenous mantle (Barnes et al., 2020). Measurements of D/H ratio will also allow us to understand the water loss process from Mars surface through the atmosphere.

Lyman Alpha Photometer (LAP) instrument onboard MOM is an absorption cell photometer. It measures the relative abundance of deuterium (D) and hydrogen (H) from Lyman Alpha emissions in the upper atmosphere of Mars. LAP, one of the five scientific instruments of MOM spacecraft's payload suite developed at Laboratory for Electro-Optics Systems (LEOS) - ISRO, is the first Indian space-borne absorption gas cell photometer that operates on the principle of resonant scattering and resonance absorption at Lyman-A wavelengths of hydrogen (121.567 nm) and deuterium (121.534 nm), respectively. This type of instrument is best suited to measure the line-of-sight Lyman alpha intensity of hydrogen and deuterium and thereby the D/H ratio (deuterium to hydrogen ratio) estimation of a planet's atmosphere. LAP can measure the amount of deuterium compared to the amount of hydrogen in mars exosphere. Till date LAP instrument has been operated on-board successfully for more than 150 times (the 1st operation was carried out on 6th February, 2014 at 09:45:00 UT) during various phases of spacecraft's journey, namely, cruise-phase, comet-phase (Siding Spring C-2013/A, 19th October, 2014 at 18:27:13

FIG. 10.24 Areocentric longitude vs. brightness temperature. *(Source: http://www.isro.gov.in/mars-orbiter-mission-completes-1000-days-orbit.)*

Areocentric longitude increases to the east. BT were relatively higher during Ls 2600–3390 as compared to values during Ls 2040–2600. Measurements carried out during higher Sun elevation were found associated with higher BT as compared to observations from low Sun elevation angles for similar viewing geometry as shown in Fig. 10.24.

Imaging sessions were carried out (a) from apoapsis covering Martian disk, and (b) site specific imaging from periapsis. Observed BT from an altitude of 52,689 km at 12.75 μm showed warmer southern hemisphere of Mars (on Ls = 210.7°) as compared to northern region. High albedo regions of Arabia terra and Isidis, and low albedo region of Syrtis Major as seen in the synchronous image acquired from MCC onboard the MOM mission is also shown in Fig. 10.25.

Imaging in periapsis from the altitude of 386 km near Holden crater on Ls: 299.2° showed variation of Brightness temperature from 278 K to 291 K at 10.25 μm. TIS observations are draped on background of MCC data as ancillary source of information. Emissivity spectra retrieved from TIS observation near Holden crater indicated characteristic dip between 9 and 10 μm showing the basaltic surface associated with atmospheric dust. The above findings from TIS involved detailed physics-based correction procedures including instrument thermal background, atmospheric contribution, solar and viewing geometry etc.

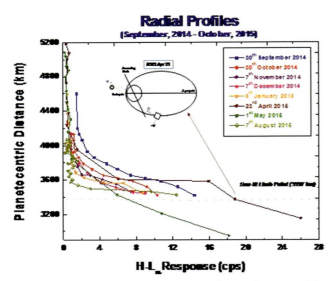

FIG. 10.26 Radial profiles of hydrogen Lyman-A measured by Lyman Alpha Photometer (LAP). *(Source: http://www.isro.gov.in/mars-orbiter-mission-completes-1000-days-orbit.)*

UT), Martian orbit phase (from 30th September, 2014) deep-space observations (6th November, 2014 at 10:19:01 UT) for assessment of payload dark count measurements and stellar source observations (3rd February, 2016 at 14:45:00 UT) to perform on-board photometric calibration.

Fig. 10.26 shows the generated radial profiles based on the first year's MOM data. Useful scientific data sets are received and are currently under analysis. Analyzed data so far has revealed successful registration of the Hydrogen Lyman-Alpha; brightness as well as clear Hydrogen Lyman-A flux absorption signatures of Martian atmosphere. Maximum Lyman-A response was recorded in the zone very close to the bright limb of Martian disk.

10.23 Success achieved by MOM

On June 19, 2017 MOM completed 1000 Earth days in its orbit, well beyond its designed mission life of 6 months. 1000 Earth days corresponds to 973.24 Mars Sols (Martian Solar days) and MOM completed 388 orbits (http://www.isro.gov.in/mars-orbiter-mission-completes-1000-days-orbit).

MOM is credited with many laurels such as cost-effectiveness, short period of realization, economical mass-budget, miniaturization of five heterogeneous science payloads etc. Satellite is in good health and continues to work as expected. Scientific analysis of the data received from the Mars Orbiter spacecraft is in progress.

Mangalyaan, India's Mars Orbiter Mission spacecraft has completed seven earth years in its orbit in 2021. According to ISRO officials, the spacecraft has successfully covered three Martian years. It was originally made to last just 6 months, but has enough punch left for more than a decade in the Martian orbit.

ISRO has also launched MOM Announcement of Opportunity (AO) programs for researchers in the country to use MOM data for research and development (R&D). The success of Mars Orbiter Mission has motivated India's student and research community in a big way. Thirty-two proposals were supported under this AO. A Planetary data analysis workshop was also conducted to strengthen the MOM-AO scientist's research interest.

First year data from MOM was released to public on September 24, 2016 through ISSDC website. Mars Atlas was prepared and made available on ISRO website.

MOM went through a communication "blackout" as a result of solar conjunction from June 2, 2015 to July 2, 2015. However, telemetry data was received during most of the conjunction period except for 9 days from June 10–18, during superior conjunction. MOM was commanded with autonomy features starting from May 18, 2015, which enabled it to survive the communication "blackout" period without any ground commands or intervention. The spacecraft emerged out of "blackout" period with auto control of the spacecraft systems successfully. This experience had enabled the mission team to program a spacecraft about 1 month in advance for all operations.

MOM spacecraft experienced the "whiteout" geometry during May 18 to May 30, 2016. A "whiteout" occurs when the Earth is between the Sun and Mars and too much solar radiation may make it impossible to communicate with the Earth. The maximum duration of "whiteout" is around 14 days. MOM spacecraft experienced the "whiteout" during May, 2016. However, MOM is built with full autonomy to take care of itself for long periods without any ground intervention. The entire planning and commanding for the "whiteout" was completed 10 days before the actual event. No commanding was carried out on the satellite in the "whiteout" period. Payload operations were suspended. Fault Detection, Isolation and Recovery were kept enabled, so as to take care of any contingency on the spacecraft. Master Recovery Sequencer was programed, to acquire the attitude of the spacecraft and ensure communication with earth even in case of loss of attitude. The spacecraft came out of "whiteout" geometry successfully on May 30, 2016 and has been normalized for regular operations.

An orbital maneuver was performed on MOM spacecraft to avoid the impending long eclipse duration for the satellite. The duration of the eclipse would have been as long as 8 h. As the satellite battery is designed to handle eclipse duration of only about 1 h 40 min, a longer eclipse would have drained the battery beyond the safe limit. The maneuvers performed on January 17, 2017 brought down the eclipse duration to zero during this long eclipse period. On the Evening of January 17, all the eight 22N-thrusters were fired for a duration of 431 s, achieving a velocity difference of 97.5 m/s. This has resulted in a new orbit for the MOM spacecraft, which completely avoided eclipse up to September 2017. About 20 kg propellant was consumed for these maneuvers leaving another 13 kg of propellant for its further mission life.

10.24 Perchlorates on Mars

Perchlorate are chemical compounds containing the perchlorate ion, ClO_4^-. Thus, a perchlorate is a negatively charged molecule made of one chlorine atom and four oxygen atoms. The perchlorate anion consists of a tetrahedral array of oxygen atoms around a central chlorine atom. Perchlorate (ClO_4^-) is widespread in Martian soils at concentrations between 0.5 and 1%. At such concentrations, perchlorate could be an important source of oxygen, but it could also become a critical chemical hazard to astronauts.

Despite the stability and lack of reactivity of perchlorate at ambient temperatures, once heated, it becomes a well-characterized and highly effective oxidizing agent, used as solid rocket fuel (Shusser et al., 2002). The ability of perchlorate to produce rocket fuel and oxygen was treated as plus points for future settlers on Mars.

10.24.1 Use of perchlorates as an energy source—Example by Antarctic microbes

In the summer of 2008, NASA's Phoenix Lander found that the soil on Mars contained 0.6 wt % perchlorate (Hecht et al., 2009; Kounaves et al. 2010b). This level of natural perchlorate (ClO_4^-) is rarely attained on Earth. With increasing evidence, the implication for Mars and Earth appears to be that given enough time, an appropriate environment, and a source of chlorine, perchlorate will accumulate.

In the past few years, it has become increasingly apparent that perchlorate (ClO_4^-) is present on all continents. Samuel et al. (2010) have reported on the discovery of perchlorate in soil and ice from several Antarctic Dry Valleys (ADVs) where concentrations reach up to 1100 μg/kg. Far from anthropogenic activity, ADV perchlorate provides unambiguous evidence that natural perchlorate is ubiquitous on Earth. The discovery has significant implications for the origin of perchlorate, its global biogeochemical interactions, and possible interactions with the polar ice sheets. The results support the hypotheses that perchlorate is produced globally and continuously in the Earth's atmosphere, that it typically accumulates in hyperarid areas, and that it does not build up in oceans or other wet environments most likely because of microbial reduction on a global scale. It is likely that perchlorate is produced globally in the stratosphere and deposited everywhere but accumulates only in arid locations because of its high aqueous solubility (Samuel et al., 2010).

According to Jackson et al. (2012), ice-covered lakes are the main sites for biological activity in the McMurdo Dry Valleys (MDVs) of Antarctica, and represent one of the most pristine aquatic ecosystems on Earth. Some of the MDV lakes contain essentially fresh water and support diverse and widespread photosynthetic microbial mats, as well as chemolithotrophic and heterotrophic processes such as nitrification, sulfur oxidation, denitrification, and sulfate reduction (Voytek et al., 1999; Green and Lyons, 2009).

Jackson et al. (2012) recall that two oxyhalides (ClO_3^- and ClO_4^-) are produced in the atmosphere by ozone (O_3) and possibly UV mediated oxidation reactions (Kang et al., 2006, 2009; Rao et al., 2010a), and are ubiquitously transferred to the surface through dry and wet deposition (Rajagopalan et al., 2009), where they are available as alternate electron acceptors to support a variety of microbial metabolisms (Coates and Achenbach, 2004, 2006). As such, the biogeochemistry of chlorine oxy-anions (ClO_x) in the MDV may provide clues regarding microbial activity and the chemical evolution of soils and lake water. Understanding the environmental limits of ClO_x metabolism is also especially important given the discovery of ClO_4^- on Mars (Hecht et al., 2009).

Based on stable isotope ratios of oxygen (O) in perchlorate (ClO_4^-), and studies carried out by several investigators (e.g., Bao and Gu, 2004; Bohlke et al., 2005; Jackson et al., 2010), it appears that ClO_4^- has at least two atmospheric production pathways: O_3 mediated oxidation of Cl species and possibly photo-oxidation of oxy-chlorine compounds (Jackson et al., 2012). Numerous investigations (e.g., Kounaves et al., 2010; Rajagopalan et al., 2009; Ericksen, 1981; Jackson et al., 2010) have revealed that ClO_4^- occurs ubiquitously throughout the world, and it tends to accumulate in arid areas including the MDV. Preferential accumulation in arid areas is due to limited liquid water and the generally oxic conditions, which prevent the infiltration into the soil and reduction of ClO_4^- by bacteria capable of using it as an electron acceptor under anoxic conditions.

During microbial reduction, ClO_4^- is sequentially reduced through ClO_3^- to ClO_2^-, which is then disproportionated by chlorite dismutase to produce O_2 and Cl^- ($ClO_4^- \rightarrow ClO_3^- \rightarrow ClO_2^- \rightarrow Cl^- + O_2$). The oxygen is consumed within the ClO_4^- reducing bacterial cell (Jackson et al., 2012). (Per)chlorate reducing organisms are facultatively ("optionally" or "discretionarily") anaerobic or micro-aerophilic and can utilize a broad range of electron donors including hydrogen, organic matter, ferrous iron, and hydrogen sulfide and have been isolated from a wide range of natural environments (Coates and Achenbach, 2004, 2006). Perchlorate (ClO_4^-) is very soluble and in oxic environments highly conserved and is not transformed by known abiotic processes under relevant environmental conditions (Jackson et al., 2012).

The permanently ice-covered lakes in the MDV are supplied by glacial melt and include streams, creeks, and the Onyx River that feed various lakes as well as ponds. Jackson et al. (2012)'s study revealed that ClO_4^- concentrations in a subset of these water bodies ranged from 0.05–8.1 μg/L but were generally less than 0.5 (45/49 samples) with an overall average of 0.19 μg/L excluding one outlier (Parera Pond).

10.24.2 Effects of perchlorate salts on the function of a cold-adapted extreme halophile from Antarctica—No significant inhibition of growth and enzyme activity

Halophiles have long been proposed as candidates for survival on Mars because they have evolved to grow in high salt concentrations and multiple extreme conditions on Earth (Landis, 2001; DasSarma, 2006). Their ability to grow in hypersaline environments requires adaptation to a number of other stressors ranging from toxic ions, periods of desiccation, and UV and ionizing radiation (Oren et al., 2014; DasSarma and DasSarma, 2017). A few halophiles, including halophilic archaea, are also adapted to cold temperatures, including subzero temperatures, with the freezing point of water depressed by high salinity (Reid et al., 2006). As a result, these extremophilic microbes may represent potential models for life on Mars where these stressors are commonly found (Laye and DasSarma, 2018).

Effects of perchlorate salts prevalent on the surface of Mars are of significant interest to astrobiology from the perspective of potential life on the Red Planet. From the perspective of possible survival on Mars, *Halorubrum lacusprofundi*—a cold-adapted polyextremophile microbe from Antarctica—is a halophilic archaeon of significant interest (Reid et al., 2006; DasSarma et al., 2017). This microorganism was isolated from Deep Lake, Antarctica, which is perennially cold and hypersaline (Franzmann et al., 1988). It has been found that *Halorubrum lacusprofundi* is able to survive under both low temperatures and hypersaline conditions, with measurable growth down to −1°C in medium containing 3.1 M NaCl and 0.4 M $MgCl_2$ (Reid et al., 2006).

To determine the effects of sodium and magnesium perchlorate Martian salts on *H. lacusprofundi*, Laye and DasSarma (2018) subjected the microbe and its enzyme to increasing concentrations of sodium and magnesium perchlorate. In their study, Laye and DasSarma (2018) found that *Halorubrum lacusprofundi* grows anaerobically in low concentrations of perchlorate. However, their study revealed that this polyextremophile's growth is inhibited at considerably higher concentrations, with more sensitivity to magnesium perchlorate compared to sodium perchlorate (50% inhibition at 0.3 M sodium perchlorate versus 0.1 M for magnesium perchlorate). Similar results were obtained for inhibition of the purified *H. lacusprofundi* b-galactosidase enzyme, with greater enzyme activity reduction for magnesium perchlorate compared to sodium perchlorate (50% inhibition with 0.88M sodium perchlorate and 0.13 M magnesium perchlorate). Interestingly, steady-state kinetic analysis showed that magnesium ions act as a competitive inhibitor for the enzyme, while perchlorate ions act as a noncompetitive inhibitor, with magnesium perchlorate acting as a mixed inhibitor.

Laye and DasSarma (2018)'s results show that this polyextremophilic halophile and its model enzyme, while exhibiting sensitivity to these ions, retain their ability to function in their presence at high concentrations far above what is likely to be encountered on the surface of our sister planet. Based on the estimated concentrations of perchlorate salts on the surface of Mars, Laye and DasSarma (2018)'s results show that neither sodium nor magnesium perchlorates would significantly inhibit growth and enzyme activity of halophiles. According to Laye and DasSarma (2018), this is the first study of perchlorate effects on a purified enzyme. Laye and DasSarma (2018)'s findings confirmed that certain haloarchaea are capable of utilizing perchlorate as a terminal electron acceptor, whereas others are not.

10.24.3 Bacteriocidal effect of UV-irradiated perchlorate

In biology, the high oxidation state of perchlorates means that they can be used as an electron acceptor by microorganisms to provide energy for growth (Coates and Achenbach, 2004). The presence of oxidants in the Martian soil was first suspected during the Viking Lander missions (Klein, 1978; Klein et al., 1972). The missions suggested low levels of reactive oxidizing substances, which were thought to explain why no evidence for organics was found (Biemann et al., 1977; Oyamaand Berdahl, 1977). The detection of chloro-hydrocarbons was initially considered to be terrestrial contamination but could be explained by the presence of oxy-chlorine species upon re-analysis (Navarro-González et al., 2010). In 2008, the NASA Phoenix Lander's onboard Wet Chemistry Lab eventually discovered the presence of perchlorate anions, at a concentration of 0.4–0.6 wt% (Hecht et al., 2009). This finding was recently supported by the Sample Analysis at Mars instrument (SAM) on the Curiosity rover (Glavin et al., 2013).

To determine if perchlorate had an effect on cell viability, Wadsworth and Cockell (2017) irradiated *Bacillus subtilis* cells in minimal media M9 in the presence of dissolved magnesium perchlorate ($Mg(ClO_4^-)_2$) at a concentration (0.6 wt%) typical of the Martian surface. Magnesium perchlorates were used because perchlorates have been detected in Martian soils directly (Hecht et al., 2009). Furthermore, they are a putative component of brine seeps, and magnesium perchlorate is thought to be a specific component (Ojha et al., 2015). However, as it is in solution, Wadsworth and Cockell (2017) were solely interested in the perchlorate ion and its effects of cell viability. Experiments were conducted under a monochromatic UV radiation source at 254 nm.

Wadsworth and Cockell (2017) argue that the mechanism of perchlorate action on cells is likely to be its degradation to deleterious reactive oxygen species. During irradiation, an increase in absorption at the expected maxima of hypochlorite (290 nm) and chlorite (260 nm) was observed. Similar photoproducts have been previously observed of perchlorate irradiated with ionizing radiation (Quinn et al., 2013; Martucci, 2012).

The chemical nature of this bacteriocidal effect was confirmed by Wadsworth and Cockell (2017) by carrying out the experiment at 4 °C, when the loss of viability is over ten times lower than at 25 °C, suggesting that lower temperatures lower the rate of the chemical reaction or the diffusion of activation products and reduce the rate of bacteriocidal effects. Nevertheless, the effect is still observable. The average surface temperature on Mars is approximately 218 K (−55°C), however the Mars Exploration Rover Opportunity measured a daily maximum of 295 K (21.85°C) (Spanovich et al., 2006). Therefore, we would expect a range of reaction rates varying with latitude and time of day.

By lowering the perchlorate concentration by one order of magnitude to below that found at the Martian surface, the loss of viability was reduced to values not statistically significant from UV irradiation alone, showing that under conditions where perchlorates are diluted, the bacteriocidal effect is mitigated. By contrast, any environment that concentrates perchlorates, such as in putative Martian brines (Navarro-González et al., 2010), will produce uninhabitable environments on account of the bacteriocidal properties of irradiated perchlorates. These properties suggest that the mere presence of liquid water seeps, thought to be good locations to search for life, does not imply environments fit for life.

When Wadsworth and Cockell (2017) combined hydrogen peroxide, hematite and perchlorates, which might represent a combination of compounds in the Martian soil, they observed the greatest loss of viability. They attributed this observation to the combined effect of UV-irradiated perchlorate-induced cell killing with Photo-Fenton-induced killing by iron oxides and hydrogen peroxides.

Although the toxic effects of oxidants on the Martian surface have been suspected for some time, Wadsworth and Cockell (2017)'s observations showed that the surface of present-day Mars is highly deleterious to cells, caused by a toxic cocktail of oxidants, iron oxides, perchlorates and UV irradiation. There has been recent research into the use of perchlorates as a potential energy source for bacteria on Mars (Oren et al., 2014; Matsubara et al., 2016) and suggestions (Levin and Straat, 2016) that such life may have been detected in the Viking Labeled Release experimental results. However, Wadsworth and Cockell (2017) showed the bacteriocidal effects of UV-irradiated perchlorates provide yet further evidence that the surface of Mars is lethal to vegetative cells and renders much of the surface and near-surface regions uninhabitable. Wadsworth and Cockell (2017)'s results show that even brine seeps, although they represent local regions of water availability, could be deleterious to cells, indigenous or contaminant if, as spectral evidence suggests, they contain perchlorates. The enhancement of the bacteriocidal properties of perchlorates by UV-irradiation suggest that these aqueous environments are even more deleterious to potential contaminants from spacecraft, and potentially less habitable, than was thought. These data have implications for planetary protection, specifically concerns about the forward contamination of Mars in both robotic and human exploration. Wadsworth and Cockell (2017)'s work focus on reporting the new finding of bacteriocidal properties of UV-irradiated perchlorate on life at ambient temperatures and under Martian conditions. Description of experiments conducted by Wadsworth and Cockell (2017) to arrive at their conclusions will enlighten the readers on the harmful effects of perchlorates in a UV-ridden Mars soil. These aspects are addressed in the below subsections.

On Earth, most UVC radiation (i.e., ultraviolet radiation with wavelengths between 200 and 290 nm) from the Sun is blocked from the Earth's surface by the ozone layer. Wadsworth and Cockell (2017) showed that when magnesium perchlorate, at concentrations relevant to the Martian surface, is irradiated under short-wave UVC radiation encountered on the Martian surface it becomes bacteriocidal. They observed this effect both in liquid culture and in a rock analog system that replicates a microenvironment within rocks. It was found that the effect is less pronounced within the rock analog system likely caused by screening within the rock, which reduces the penetration of UV radiation compared to the liquid system. The bacteriocidal effect was also replicated when using other forms of perchlorate found in the Martian regolith: calcium and sodium perchlorate. It was found that both perchlorates significantly reduce viability of cells when irradiated. Furthermore, bacterial samples in the presence of perchlorate at Martian concentrations but in the absence of UV radiation showed no loss of viability, which is consistent with the findings by Nicholson et al. (2012) that show no growth inhibition of *Bacillus subtilis 168* and *Bacillus pumilus* SAFR-032 when in the presence of perchlorate without UV exposure.

Mars is subject to UVC radiation on account of the lack of a significant oxygen concentration or ozone shield and a lower cut-off caused by CO_2. The absolute flux of radiation between 200 and 315 nm (UVC and UVB radiation) is the most damaging region of the UV radiation spectrum to DNA (Cockell et al., 2000). Wadsworth and Cockell (2017) quantified the harmful effect on viability by calculating the ratio of surviving cells ("N") with regard to the starting concentration of cells ("N_0"). They defined "viability" as any number of cells greater than zero, consequently "sterility" as zero cells. Results are shown on a log scale in all figures; therefore, cases that are strictly zero are not represented on the log plot but should be interpreted as strictly zero.

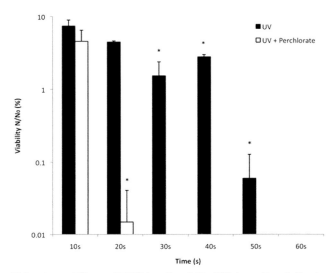

FIG. 10.27 Effects of UVC-irradiated Mg(ClO$_4$)$_2$ on *B. subtilis* viability. UV = UVC-irradiated control; Perchlorate = 0.6 wt% Mg(ClO$_4$)$_2$. $P < 0.05$ was considered statistically significant among groups; error bars are + s.d. ($n = 3$). (*Source: Wadsworth, J. and C.S. Cockell (2017), Perchlorates on Mars enhance the bacteriocidal effects of UV light, Nat. Sci. Rep., 7(4662); doi: http://doi.org/10.1038/s41598-017-04910-3.*)

Irradiated perchlorate had a significant bacteriocidal effect (Fig. 10.27). Cell viability was completely lost after 30 s exposure. By contrast, the control cells exposed to UV radiation without perchlorate took 60 s to be completely sterilized. Nonirradiated controls consisting of cells in M9, and cells in M9 in the presence of 0.6 wt% perchlorate, showed no significant difference in viability when left up to an hour.

The implications of detection of perchlorates on Mars are significant as their detection suggests the presence of other oxychlorine species, which may negatively impact the habitability of Mars and interfere with the preservation and detection of organic material (Quinn et al., 2013). As already noted, Martian soil is laced with perchlorates. Consisting of a negatively charged chloride surrounded by a tetrahedral formation of oxygen atoms, perchlorates represent the highest oxidation state of chlorine (+7) and are powerful oxidants when heated, but are stable at room temperature and lower temperatures. Perchlorates is classified as a salt, and was initially cause for celebration among extraterrestrial hopefuls because it drastically lowers the freezing point of water (Chevrier et al., 2009; Hanley et al., 2012), meaning that liquid water might conceivably exist on the surface. Lowering the freezing point of water potentially allows for a contemporary active hydrological system on Mars, which could enhance the habitability of the near-surface environment. This has prompted recent research into the use of perchlorates as a potential energy source for bacteria on Mars (Oren et al., 2014; Matsubara et al., 2016).

Wadsworth and Cockell (2017) investigated perchlorate's potential reactivity by irradiating perchlorates under UV and observed its effect on the viability of a model vegetative organism, *Bacillus subtilis*, which is a common spacecraft contaminant (La Duc et al., 2007; Puleo et al., 1977). They reported the significant bacteriocidal effects of UV-irradiated perchlorate on life at ambient temperatures and under Martian conditions.

10.24.4 Bacteriocidal effect of perchlorate salts under Martian analog conditions

Although Wadsworth and Cockell (2017) showed irradiated perchlorates are bacteriocidal to cells in liquid media, they performed the same experiment under a number of Martian analog conditions to test if this result was reproducible in a more representative environment. To simulate a rocky Martian habitat, the experiment was carried out using a simple system to more accurately simulate a rock environment ("rock analog system") in which cells were deposited within silica discs. Although the overall cell survival was higher, Fig. 10.28(A) shows a significant 9.1-fold drop in viability in the irradiated perchlorate-treated samples after 60 s, while the UV-irradiated controls show a 2-fold viability decrease after the same exposure time.

Wadsworth and Cockell (2017) then carried out experiments to investigate whether the perchlorate-sterilization effect would still be observed under the influence of other environmental parameters relevant to Mars, namely anaerobic conditions, polychromatic irradiation and low temperature, the results of which are also shown in Fig. 10.28(A). Firstly, the liquid system and rock analog systems were irradiated under anaerobic conditions. Both systems showed that perchlorate-containing samples irradiated with UV experienced a greater loss of viability than the UV-irradiated controls. In the liquid system, cells remained viable (0.12%) after 60 s under just UV irradiation, but in the presence of perchlorate, viability was completely lost after 60 s. By contrast, in the rock analog system, under UV irradiation greater cell viability was retained after 60 s (8.23%), but irradiated perchlorate still caused a significant loss of cell viability (9.7-fold decrease) compared to the UV-irradiated only control after 60 s.

Wadsworth and Cockell (2017) carried out experiments in the liquid system using polychromatic light to more accurately simulate a natural light spectrum. Under polychromatic light, cells in the presence of perchlorate showed a significant 10.8-fold decrease in viability compared to the polychromatic UV-irradiated only controls.

Wadsworth and Cockell (2017) observed an effect of low temperature on the reaction. When the experimental system was chilled to 4°C before and during monochromatic irradiation they observed after 60 s that the cell viability of the UV-irradiated perchlorate samples did not drop significantly below that of the UV-irradiated controls (Fig. 10.28(A). To test whether this was because

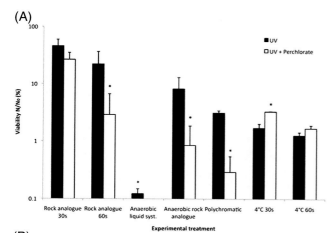
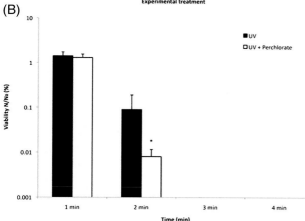

FIG. 10.28 (A) Effects of UVC-irradiated $Mg(ClO_4)_2$ in rock analogs, under anaerobic conditions, polychromatic light and low temperature. Rock analogs exposed to aerobic environment (30 and 60 s); Liquid and rock analog exposed to anaerobic environment (60 s); Liquid system exposed to polychromatic light (10 s); Liquid system chilled to 4°C while irradiated (30 and 60 s); UV = UVC-irradiated control; Perchlorate = 0.6 wt% $Mg(ClO_4)_2$. $P<0.05$ was considered statistically significant (*) among groups; error bars are + s.d. ($n=3$). (b) Effects of UVC-irradiated $Mg(ClO_4)_2$ at low temperature, 1–4-min exposure. Liquid system chilled to 4°C while irradiated. UV = UVC-irradiated control; Perchlorate = 0.6 wt% $Mg(ClO_4)_2$. $P<0.05$ was considered statistically significant (*) between or among groups; error bars are + s.d. ($n=3$). *(Source: Wadsworth, J. and C.S. Cockell (2017), Perchlorates on Mars enhance the bacteriocidal effects of UV light, Nat. Sci. Rep., 7(4662); doi: http://doi.org/10.1038/s41598-017-04910-3.)*

UV-irradiated perchlorates were no longer effective at low temperatures or whether the chemical reaction was delayed, they performed the same experiment for 10 min, measuring at each minute (Fig. 10.28B shows minutes 1–4). Once again, after 1 min of UV exposure both samples displayed no significant differences in viability. However, after 2 min of irradiation the irradiated perchlorate-treated samples showed a significant 11.4-fold decrease in viability in comparison to the UV controls; both UV and perchlorate samples were sterile after 3 min. According to Wadsworth and Cockell (2017), it is unclear if low temperature reduced perchlorate activation, reduced the diffusion of photoproducts to the cells or reduced the rate of cellular damage. Nevertheless, even at low temperatures, irradiated perchlorates proved bacteriocidal.

To confirm the production of potential biologically damaging photoproducts during perchlorate irradiation, Wadsworth and Cockell (2017) measured the absorption spectrum of an irradiated solution of perchlorate in the UV radiation range. Potential photoproducts produced during irradiation could be hypochlorite (absorbance maximum = 290 nm), chlorite (absorbance maximum = 260 nm) and chlorine dioxide (absorbance maximum = 360 nm). They observed an increase in absorbance at 260 nm and to a slightly lesser extent at 290 nm. There was a negligible increase in absorbance at 360 nm. A control sample containing nonirradiated perchlorate was also measured at the same time points and no increase in absorbance at any of the wavelengths was observed.

To determine whether altering the concentration of perchlorate affected the loss of cell viability when irradiated, Wadsworth and Cockell (2017) serially diluted the 0.6 wt% perchlorate solution to yield 0.06 wt% and 0.006 wt% solutions, which were irradiated in the presence of cells. It was found that at 0.06 wt% and 0.006 wt% there were no statistically significant differences in cell viability compared to the UV-irradiated controls.

Wadsworth and Cockell (2017) also investigated the effects of perchlorate at higher concentrations than those measured in the Martian surface regolith (Fig. 10.29). Although the regolith contains a concentration of 0.4–0.6 wt%, the spectral detection of putative perchlorate brines suggests that, in some local regions on Mars, the concentrations of this chemical could be much higher. At a perchlorate concentration of 1 wt% viability dropped over an order of magnitude after 30 s irradiation compared to results at 0.6 wt%. A complete loss of viability was observed after 60 s exposure. An increase of perchlorate concentration to 5 wt% resulted in complete loss of viability after only 30 s of irradiation.

10.24.5 Interactions of other Martian soil components

After simulating the physical effects of the Martian environment on perchlorate activity, Wadsworth and Cockell (2017) also considered the additional soil components present and their potential interactions. They undertook experiments to study whether other components of the Martian surface could affect the reactions they observed. Sulfate is an abundant component of the Martian regolith with ~30 wt% of sulfate within sediments having been reported (Vaniman et al., 2004). Wadsworth and Cockell (2017) repeated the experimental set up using perchlorate at 0.6 wt% but with the addition of sulfate at the estimated Martian concentration

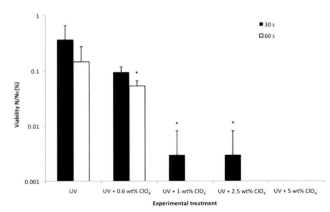

FIG. 10.29 **Influence of increased perchlorate concentration on bacteriocidal effects under UV irradiation.** $Mg(ClO_4)_2$ at representative measured Martian concentration (0.6 wt%,), 1, 2.5, and 5 wt% (30 and 60 s); UV = UVC-irradiated control (30 and 60 s). $P<0.05$ was considered statistically significant (*) between and among groups; error bars are + s.d. ($n=3$). (Source: Wadsworth, J. and C.S. Cockell (2017), Perchlorates on Mars enhance the bacteriocidal effects of UV light, Nat. Sci. Rep., 7(4662); doi: http://doi.org/10.1038/s41598-017-04910-3.)

Wadsworth and Cockell (2017) examined two additional forms of perchlorate that have been detected on Mars (Fig. 10.30). Sodium perchlorate ($NaClO_4$) was detected in the reoccurring slope lineae by Ojha et al. (2015) and calcium perchlorate ($Ca(ClO_4)_2$) is thought to be the best candidate for the oxychlorine compounds found in Rocknest (Glavin et al., 2013).

The perchlorates were both irradiated with UVC at a concentration of 0.6 wt% for comparison with the magnesium perchlorate. The calcium and sodium perchlorates showed a significantly lower cell count than the UV-irradiated controls after 30 s. The calcium perchlorate-treated samples were completely sterilized and the sodium perchlorate-treated samples showed a 15-fold drop in viability compared to the controls; all samples were sterilized after 60 s UV radiation exposure. To get a better resolution of the effect of the perchlorates, they were additionally irradiated in the same set up at a four times greater distance from the light source (16 times less irradiance). These results showed no significant difference in viability in any samples in the first 30 s of irradiation. However, after 60 s both calcium and sodium perchlorate-treated samples showed significantly lower cell counts than the UV-irradiated controls (1.9- and 1.7-fold, respectively).

Wadsworth and Cockell (2017) also examined the influence of two other components of the Martian surface environment: iron oxides and the oxidant hydrogen peroxide. They carried out experiments to determine whether these

of 30 wt% to the perchlorate at 0.6 wt% (Fig. 10.30). The results showed that there was no significant effect of sulfate on the loss of viability of the cells in the presence of UV-irradiated perchlorate, nor irradiated sulfate on its own did not differ significantly from the UV-irradiated control in terms of effects on viability.

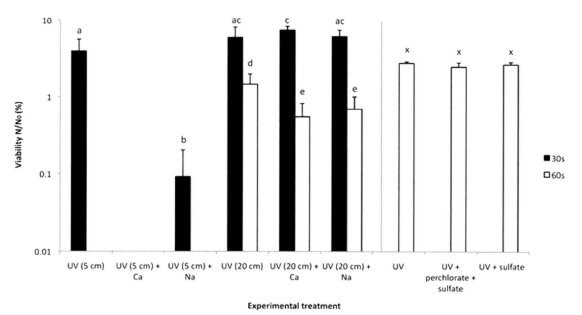

FIG. 10.30 **Effects of UVC-irradiated sodium/calcium perchlorate and sulfate on *B. subtilis* viability.** UV = UVC-irradiated control at given distance from light source; Ca = 0.6 wt% $Ca(ClO_4)_2$ at given distance from light source; Na = 0.6 wt% $NaClO_4$ at given distance from light source; perchlorate = 0.6 wt % $Mg(ClO_4)_2$; sulfate = 30 wt% $MgSO_4$. Letters shared in common between or among the groups indicate no significant difference ($P>0.05$); error bars are + s.d. ($n=3$). Vertical gray line indicates separate experiment with different control. (Source: Wadsworth, J. and C.S. Cockell (2017), Perchlorates on Mars enhance the bacteriocidal effects of UV light, Nat. Sci. Rep., 7(4662); doi: http://doi.org/10.1038/s41598-017-04910-3.)

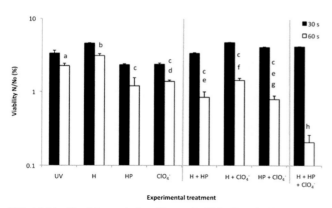

FIG. 10.31 Perchlorate-induced bacteriocidal effects in the presence of other components of the Martian surface (hematite and hydrogen peroxide) after 30- and 60-s UV exposure. UV=UV-irradiated control; H=1g/L hematite; HP=10mM hydrogen peroxide; ClO4=0.6wt% $Mg(ClO_4)_2$. Letters shared in common between or among the groups indicate no significant difference ($P>0.05$); error bars are + s.d. ($n=3$). Vertical gray lines serve as visual separation of single, double, and triple combinations. *(Source: Wadsworth, J. and C.S. Cockell (2017), Perchlorates on Mars enhance the bacteriocidal effects of UV light, Nat. Sci. Rep., 7 (4662); doi: http://doi.org/10.1038/s41598-017-04910-3).*

would act in synergy with irradiated perchlorates to make the surface of Mars inimical to life. Fig. 10.31 shows the effects of the individual components, effects of combinations of two components and the combined effect of all three components under UV irradiation.

Wadsworth and Cockell (2017) conducted experiments with the iron oxide, hematite, with a grain size of 5 μm (Sigma-Aldrich), which was added to the liquid system or rock analog system at a concentration of 1g/L. When hydrogen peroxide was used, it was added to a final concentration of 10mM. $Mg(ClO_4)_2$ was added to a concentration of 0.6wt%, as in previous experiments.

Firstly, the individual components hematite, hydrogen peroxide and perchlorate were added to M9 containing *Bacillus subtilis* cells and irradiated under the monochromatic UVC source for the indicated length of time. UV-irradiated controls containing cells in M9 served as a control. Samples in the presence of hematite showed significantly higher cell viability after 60 s exposure than cells in the UV controls after the same length of time.

Samples individually treated with hydrogen peroxide or perchlorate showed a significant drop in viability in comparison to UV-irradiated controls after 60 s. Second, the individual components were paired as follows, and irradiated for the indicated length of time: hematite and hydrogen peroxide; hematite and perchlorate; hydrogen peroxide and perchlorate. The combination of the iron oxide and hydrogen peroxide in the presence of UV radiation caused a significantly greater loss in viability than the individual components of hematite (3.7-fold) and perchlorate (1.6-fold). A decrease was also shown in comparison to samples treated only with irradiated hydrogen peroxide (1.4-fold). The iron oxide and perchlorate combination had a slight yet significant 1.04-fold increase in viability in comparison to perchlorate alone. Cells treated with perchlorate and hydrogen peroxide showed a 1.5-fold loss in viability, not significantly different to cells treated only with irradiated hydrogen peroxide; there was a significant 1.8-fold decrease in viability in comparison to samples just treated with irradiated perchlorate. When combined, Wadsworth and Cockell (2017) found that all three components resulted in the largest drop in viability. After 60s of UV radiation exposure cell viability was reduced to 0.21%, which was significantly lower than all other combinations examined.

10.24.6 Implications of perchlorate detection on Mars

Various space missions revealed the presence of perchlorates, which are known to have a high oxidizing potential in Martian regolith, at the level of 0.5%. Due to hygroscopic properties and crystallization features of perchlorate-containing solutions, assumptions leading to the possibility of the existence of liquid water in the form of brines, which can contribute to the vital activity of microorganisms, have been made. At the same time, high concentrations of perchlorates can inhibit the growth of microorganisms and cause their death. Previously performed studies have discovered the presence of highly diverse microbial communities in terrestrial perchlorate-containing soils and have also demonstrated the stability and activity of some prokaryotes cultured on highly concentrated perchlorates media (over 10%). Nevertheless, the limits of microbial tolerance to perchlorates and whether microbial communities are able to withstand the effects of high concentrations of perchlorates remain uncertain.

Identification of perchlorates on the surface of Mars prompted speculation of what their influence would be on habitability. As already indicated, Wadsworth and Cockell (2017) showed that when irradiated with a simulated Martian UV flux, perchlorates become bacteriocidal. They found that at concentrations associated with Martian surface regolith, vegetative cells of *Bacillus subtilis* (a strain of bacteria commonly found on spacecraft) in Martian analog environments lost viability within minutes. They tried this with several different kinds of perchlorates, and found similar results every time. Adding in additional environmental factors found on Mars such as low temperatures, additional minerals found on Mars and a lack of oxygen also failed to keep the bacteria alive.

This was a bit surprising for the researchers because the strain of bacteria used, *Bacillus subtilis*, belongs to a genus that actually does fine in the presence of perchlorates, as studies of the microbes in terrestrial environments have confirmed. These findings were initially good news for researchers looking for extra-terrestrial life, as they suggested that some forms of life could survive in Martian analog conditions.

Wadsworth and Cockell (2017) discovered that two other components of the Martian surface, iron oxides and hydrogen peroxide, act in synergy with irradiated perchlorates to cause a 9.8-fold increase in cell death when compared to cells exposed to UV radiation after 60 s of exposure. These data show that the combined effects of at least three components of the Martian surface, activated by surface photochemistry, render the present-day surface more uninhabitable than previously thought, and demonstrate the low probability of survival of biological contaminants released from robotic and human exploration missions.

When the researchers added in a few more Mars-like factors—UV specifically—the bacteria died in short order. They think this happens because the UV light breaks apart the perchlorate molecules into more reactive ions that wreak havoc on living cells. This hypothesis was backed up by the observation that low temperatures, which slow down chemical reactions, extended the lifespan of the bacteria in the perchlorates but still resulted in them dying. If they can't survive there, it significantly lowers our chances of finding life on Mars—life that looks similar to organisms on Earth at least.

While it is a blow to the possibility of finding life on Mars, there is at least one upside to the news: NASA regularly worries about the possibility of contaminating other planets with Earthly bacteria, even going so far as to crash probes into Saturn so that they don't hit the planet's moons. If Mars is so hostile to bacteria that they can't even make it a minute on the surface, our fears of contamination could be pretty much resolved.

While Wadsworth and Cockell (2017)'s presented data on the underlying mechanistic possibilities, a study of the exact mechanism of damage and a review of effects on multiple bacteria species would constitute interesting lines of query for follow-up studies.

Cheptsov et al. (2021) carried out a study to understand the reaction of microbial communities of hot-arid and cryo-arid soils and sedimentary rocks to the adding of a highly concentrated solution of sodium perchlorate (5%) in situ. An increase in the total number of prokaryotes, the number of metabolically active Bacteria and Archaea, and the variety of the consumed substrates were revealed. It was observed that in samples incubated with sodium perchlorate, a high taxonomic diversity of the microbial community is preserved at a level comparable to control sample. The study shows that the presence of high concentrations of sodium perchlorate (5%) in the soil does not lead to the death or significant inhibition of microbial communities.

10.25 Hydrogen peroxide—Negatively impacting the habitability of Mars?

Just like perchlorates, hydrogen peroxide (H_2O_2) also poses threat to the Martian habitability. Atreya et al. (2006) investigated a new mechanism for producing oxidants, especially hydrogen peroxide on Mars. Atreya et al. (2006) summarized the negative impact of hydrogen peroxide on habitability of Mars thus, "Large-scale electrostatic fields generated by charged sand and dust in the Martian dust devils and storms, as well as during normal saltation, can induce chemical changes near and above the surface of Mars. The most dramatic effect is found in the production of H_2O_2 whose atmospheric abundance in the 'vapor' phase can exceed 200 times that produced by photochemistry alone. With large electric fields, H_2O_2 abundance gets large enough for condensation to occur, followed by precipitation out of the atmosphere. Large quantities of H_2O_2 would then be adsorbed into the regolith, either as solid H_2O_2 'dust' or as re-evaporated vapor if the solid does not survive as it diffuses from its production region close to the surface. Atreya et al. (2006) suggested that this H_2O_2, or another superoxide processed from it in the surface, may be responsible for scavenging organic material from Mars. The presence of H_2O_2 in the surface could also accelerate the loss of methane from the atmosphere, thus requiring a larger source for maintaining a steady-state abundance of methane on Mars. The surface oxidants, together with storm electric fields and the harmful ultraviolet radiation that readily passes through the thin Martian atmosphere, are likely to render the surface of Mars inhospitable to life as we know it."

10.26 Exploring organic substances in the Martian soil

Presence of organic compounds on a celestial body could be an indication of the existence of past or present life there. While searching for extraterrestrial life, it becomes necessary to consider whether the former environment was supportive of life? If so, was it also conducive to preservation of organism remains, specifically large organic molecules? Thus, searching for organic compounds on Mars assumes great importance. In view of this, Biemann et al. (1977) analyzed a total of four Martian samples, one surface and one subsurface sample at each of the two Viking landing sites, *Chryse Planitia* and *Utopia Planitia*, for organic compounds by a gas chromatograph-mass spectrometer. Typically, mass spectrometers can be used to identify unknown compounds via molecular weight determination, to quantify known compounds, and to determine structure and chemical properties of molecules. In none of the experiments mentioned above, could organic material of Martian origin be detected at detection limits generally of the order of parts per billion and for a few substances closer to parts per million. The evolution of water and carbon dioxide, but not of other inorganic gases, was observed upon heating the sample to temperatures of up to 500°C. The absence of organic compounds seems to preclude their production on the planet at rates that exceed the rate of their destruction. Biemann et al. (1977) concluded that it is unlikely that living systems that behave in a manner similar to terrestrial biota exist, at least at the two Viking landing sites.

The 2004 arrival of the Mars Exploration Rovers (MERs) "Spirit" and "Opportunity" provided a chance to investigate ancient aqueous environments in situ and deduce their likelihood of supporting life. Initial results confirmed what data from orbiters had long suggested—that diverse aqueous environments existed on the surface of Mars billions of years ago. The success of the MERs led to the development of the Mars Science Laboratory (MSL) mission. The Curiosity rover landed at Gale crater in 2012 and was designed to specifically test whether ancient aqueous environments had also been habitable. In addition to water, did these ancient environments also record evidence for the chemical building blocks of life (H, O, C, N, S, P), as well as chemical and/or mineralogic evidence for redox gradients that would have enabled microbial metabolism, such as chemoautotrophy? Curiosity also has the capability to detect organic carbon, but it is not equipped for life detection.

The Mars Exploration Rover "Opportunity" landed at Meridiani Planum on 25 January 2004. Coincident with the 10th anniversary of this landing, Arvidson et al. (2014) reported the detection of an ancient clay-forming subsurface aqueous environment at Endeavour crater, Meridiani Planum. Though "Opportunity" does not have the ability to directly detect C or N, it has been able to establish that several of the other key factors that allow for the identification of a formerly habitable environment were in place. This potentially habitable environment stratigraphically underlies and is considerably older than the rocks detected earlier in the mission that represent acidic, hypersaline environments that would have challenged even the hardiest extremophiles. The presence of both Fe^{+3}- and Al-rich smectite clay minerals in rocks on the rim of the Noachian age Endeavour impact crater was inferred from the joint use of hyperspectral observations by the Mars Reconnaissance Orbiter and extensive surface observations by the Opportunity Rover. The rover was guided tactically by orbiter-based mapping. Extensive leaching and formation of Al-rich smectites occurred in subsurface groundwater fracture systems.

Failure to identify organic substances from a few samples from a few sites on Mars does not necessarily mean that organic substances are totally absent on Mars. Steininger et al. (2012) have described a series of TV–GC–MS experiments on powdered basalt containing known organic compounds (benzoic acid or mellitic acid) at different concentrations (500 and 0.5 ppm). With reference to the discovery of perchlorate in Martian soils by the Phoenix Wet Chemistry Laboratory (WCL) all pyrolysis (thermal) experiments were performed both in presence of 0.6 wt% magnesium perchlorate and in absence of perchlorate. Steininger et al. (2012) found that benzoic acid (BA) is detectable at both concentrations, while mellitic acid (MA) is only detectable at high concentration (500 ppm). In both cases perchlorate strongly affects type and number of organic fragments generated during pyrolysis. Note that the pyrolysis process is the thermal decomposition of materials at elevated temperatures in an inert atmosphere. It involves a change of chemical composition. The word is coined from the Greek-derived elements pyro "fire," "heat," "fever," and lysis "separating."

Ming et al. (2014) showed that H_2O, CO_2, SO_2, O_2, H_2, H_2S, HCl, chlorinated hydrocarbons, NO, and other trace gases were evolved during pyrolysis of two mudstone samples acquired by the Curiosity rover at Yellowknife Bay within Gale crater, Mars. It was found that H_2O/OH-bearing phases included 2:1 phyllosilicate(s), bassanite, akaganeite, and amorphous materials. Thermal decomposition of carbonates and combustion of organic materials are candidate sources for the CO_2. Concurrent evolution of O_2 and chlorinated hydrocarbons suggests the presence of oxychlorine phase(s). Sulfides are likely sources for sulfur-bearing species. Higher abundances of chlorinated hydrocarbons in the mudstone compared with Rocknest windblown materials previously analyzed by Curiosity suggest that indigenous Martian or meteoritic organic carbon sources may be preserved in the mudstone; however, the carbon source for the chlorinated hydrocarbons is not definitively of Martian origin. Ming et al. (2014) found that the thermal decomposition of rock powder yielded NO and CO_2, indicating the presence of nitrogen- and carbon-bearing materials. Carbon dioxide (CO_2) may have been generated by either carbonate or organic materials. Concurrent evolution of oxygen (O_2) and chlorinated hydrocarbons indicates the presence of oxychlorine species. Higher abundances of chlorinated hydrocarbons in the lake mudstones, as compared with modern windblown materials, suggest that indigenous Martian or meteoritic organic C sources are preserved in the mudstone. However, according to Ming et al. (2014), the possibility of terrestrial background sources brought by the rover itself cannot be excluded.

These results demonstrate that early Mars was habitable, but this does not mean that Mars was inhabited. Even for Earth, it was a formidable challenge to prove that microbial life existed billions of years ago—a discovery that occurred almost 100 years after Darwin predicted it, through the recognition of microfossils preserved in silica (Tyler and Barghoorn, 1954). The trick was finding a material that could preserve cellular structures. A future mission could do the same for Mars if life had existed there. "Curiosity" can help now by aiding our understanding of how organic compounds are preserved in rocks, which, in turn, could provide guidance to narrow down where and how to find materials that could preserve fossils as well. However, it is not obvious that much organic matter, of either abiologic or biologic origin, might survive degradational rock-forming and environmental processes. Our expectations are conditioned by our understanding of Earth's earliest record of life, which is very sparse (Grotzinger, 2014).

Paleontology embraces this challenge of record failure with the subdiscipline of taphonomy, through which we seek to understand the preservation process of materials of potential biologic interest. On Mars, a first step would involve detection of complex organic molecules of either abiotic or biotic origin. According to Grotzinger (2014), the point is that organic molecules are reduced and the planet is generally regarded as oxidizing, and so their preservation requires special conditions. For success, several processes must be optimized. Primary enrichment of organics must first occur, and their destruction should be minimized during the conversion of sediment to rock and by limiting exposure of sampled rocks to ionizing radiation. Of these conditions, the third is the least Earth-like (Earth's thick atmosphere and magnetic field greatly reduce incoming radiation). Curiosity can directly measure both the modern dose of ionizing cosmic radiation and the accumulated dose for the interval of time that ancient rocks have been exposed at the surface of Mars.

Hassler (2013) and Hassler et al. (2013) have quantified the present-day radiation environment on Mars that affects how any organic molecules that might be present in ancient rocks may degrade in the shallow surface (that is, the top few meters). This shallow zone is penetrated by radiation, creating a cascade of atomic and subatomic particles that ionize molecules and atoms in their path. Their measurements over the first year of Curiosity's operations provide an instantaneous sample of radiation dose rates affecting rocks, as well as future astronauts. Extrapolating these rates over geologically important periods of time and merging with modeled radiolysis data yields a predicted 1000-fold decrease in 100–atomic mass unit organic molecules in ~650 million years.

The Mars Organic Molecule Analyzer (MOMA) instrument onboard the ExoMars-2018 rover includes an additional capability of laser-desorption mass spectrometry, which may have clear advantages in diverse surface environments and for the measurement of large refractory organic molecules (Siljestrom et al., 2014; Li et al., 2015; Goesmann et al., 2017a). The search must continue relentlessly. It is heartening to note that MOMA shall apply thermal volatilization (TV), gas chromatography (GC) and mass spectrometry (MS) in order to search for organic compounds in the Martian soil.

The second ExoMars mission targets an ancient location interpreted to have strong potential for past habitability and for preserving physical and chemical biosignatures (as well as abiotic/prebiotic organics). The mission will deliver a lander with instruments for atmospheric and geophysical investigations and a rover tasked with searching for signs of extinct life. The ExoMars rover will be equipped with a drill to collect material from outcrops and at depth down to 2 m. This subsurface sampling capability will provide the best chance yet to gain access to chemical biosignatures. Using the powerful Pasteur payload instruments, the ExoMars science team planned to conduct a holistic search for traces of life and seek corroborating geological context information (Goesmann et al., 2017b). Unfortunately, on 17 March 2022, ESA suspended the second ExoMars mission due to the ongoing invasion of Ukraine by Russia. ESA expects that a restart of the mission, using a new non-Russian landing platform, is unlikely to launch before 2028.

In our efforts to characterize indigenous Martian organic matter, we must contend with a near-surface record that has been substantially altered by radiation and oxidation. Under such conditions, much of the surficial organic record on Mars may have decomposed into organic salts, which are challenging for flight instruments to conclusively identify. If organic salts are widespread on the Martian surface, their composition and distribution could offer insight into the less-altered organic record at depth and they may play an important role in near-surface carbon cycling and habitability. The organic detection techniques employed by the Mars Science Laboratory Curiosity rover include thermal extraction in combination with mass spectrometry.

Lewis et al. (2021) recall that organic matter preserved in the Martian rock record represents an invaluable resource for exploring the planet's carbon cycle, habitability, and potential biology through time. However, near-surface Martian organic matter is susceptible to transformation by ionizing radiation and oxidation (Dartnell et al., 2012; Hassler et al., 2014; Lasne et al., 2016). Flight instruments seeking to detect and characterize this near-surface organic record have utilized thermal extraction (pyrolysis) in combination with mass spectrometry (Mahaffy et al., 2012). The Sample Analysis at Mars (SAM) instrument suite on board the Curiosity rover of the Mars Science Laboratory (MSL) mission at Gale crater has detected chlorinated hydrocarbons (Szopa et al., 2020) and other chemically diverse molecular components in the pyrolysis products of multiple samples, including products released at high temperatures (>500°C) indicative of geologically refractory organic matter (Eigenbrode et al., 2018). Although the sources of these organic pyrolysis products have yet to be determined, these observations suggest preservation of organic matter over billions of years in Martian sediments (Eigenbrode et al., 2018).

Decades ago, scientists predicted that ancient organic compounds—molecules containing carbon—preserved on Mars could be breaking down into salts. These salts, they argued, would be more likely to persist on the Martian surface than big, complex molecules, such as the ones that are associated with the functioning of living things. Finding organic molecules, or their organic salt remnants, is essential in the search for life on other worlds. But this is a challenging task on the surface of Mars, where billions of years of radiation have erased or broken apart organic matter. Since arriving at Mars in 2012, NASA's Curiosity rover has drilled into rocks in search of organics. Now, Curiosity rover enabled discovering ancient organics that have been preserved in rocks for billions of years. This finding helps scientists better understand the habitability of early Mars, and it paves the way for future missions to the Red Planet.

Like an archaeologist digging up pieces of pottery, Curiosity rover collects Martian soil and rocks, which may contain tiny chunks of organic compounds, and then Sample Analysis at Mars (SAM) and other instruments identify their chemical structure. Using data that Curiosity rover beams down to Earth, scientists try to piece together these broken organic pieces. Their goal is to infer what type of larger molecules they may once have belonged to and what those molecules could reveal about the ancient environment and potential biology on Mars.

In Lewis et al. (2021)'s experiments, they used laboratory thermal extraction techniques analogous to those of the rover to examine organic salts as pure standards, as minor phases in a silica matrix, and in mixtures with O_2-evolving perchlorate salts. When Lewis et al. (2021) compared their results with flight data, they found that many of the CO_2 profiles produced by their organic salt samples were similar to the CO_2 evolutions observed by the rover. The best fits with their laboratory data included Martian materials acquired from modern eolian deposits and sedimentary rocks that had evidence for low-temperature alteration.

Scientists had already been predicting that salts found on the Martian surface could have broken from organic (carbon-containing) compounds. To reach the conclusions, Lewis et al. (2021) analyzed a range of organic salts mixed with an inert silica powder to replicate a Martian rock. They also investigated the impact of adding perchlorates to the silica mixtures. As already indicated earlier, perchlorates are salts containing chlorine and oxygen, and they are common on Mars. Scientists have long worried that perchlorates could interfere with experiments seeking signs of organic matter. Indeed, researchers found that perchlorates did interfere with their experiments, and they pinpointed how. But they also found that the results they collected from perchlorate-containing samples better matched SAM data than when perchlorates were absent, bolstering the likelihood that organic salts are present on Mars. Organic compounds and salts on Mars could have formed by geologic processes or be remnants of ancient microbial life. In a study in the search for microbial life, Lewis et al. (2021) found possible organic salts on Mars.

Lewis et al. (2021)'s laboratory experiments and analysis of data from the Sample Analysis at Mars (SAM), a portable chemistry lab inside the Curiosity rover's belly, indirectly point to the presence of organic salts. But directly identifying them on Mars is hard to do with instruments such as SAM, which heats Martian soil and rocks to release gases that reveal the composition of these samples. Investigating organic salts via thermal extraction is challenging as they produce simple pyrolysis products that can also be contributed by many other sources, such as the thermal decomposition of carbonates, the oxidation of more complex organic matter, or instrument backgrounds and contamination (Eigenbrode et al., 2018). Furthermore, additional CO_2, CO, O_2, H_2O, and catalysts in a pyrolysis oven can influence organic salt decomposition and many of the pyrolysis products can undergo side reactions. Thus, the challenge is that heating organic salts produces only simple gases that could be released by other ingredients in Martian soil.

Organic salts, such as Fe, Ca, and Mg oxalates and acetates, may be widespread radiolysis and oxidation products of organic matter in Martian surface sediments. Such organic salts are challenging to identify by evolved gas analysis but the ubiquitous CO_2 and CO in pyrolysis data from the Sample Analysis at Mars (SAM) instrument suite on the Curiosity rover indirectly points to their presence. Lewis et al. (2021) examined laboratory results from SAM-like analyses of organic salts as pure phases, as trace phases mixed with silica, and in mixtures with Ca and Mg perchlorates. They found that "Pure oxalates evolved CO_2 and CO, while pure acetates evolved CO_2 and a diverse range of organic products dominated by acetone and acetic acid. Dispersal within silica caused minor peak shifting, decreased the amounts of CO_2 evolved by the acetate standards, and altered the relative abundances of the organic products of acetate pyrolysis. The perchlorate salts scrubbed Fe oxalate CO releases and shifted the CO_2 peaks to lower temperatures, whereas with Ca and Mg oxalate, a weaker CO release was observed but the initial CO_2 evolutions were largely unchanged. The perchlorates induced a stronger CO_2 release from acetates at the expense of other products. Oxalates evolved ~47% more CO_2 and acetates yielded ~69% more CO_2 when the perchlorates were abundant." The most compelling fits between Lewis et al. (2021)'s organic salt data and SAM CO_2 and CO data included Martian samples acquired from modern eolian (relating to or arising from the action of the wind) deposits and sedimentary rocks with evidence for low-temperature alteration.

In their study, Lewis et al. (2021) used the organic detection techniques employed by the Curiosity rover including the thermal extraction in combination with mass spectrometry. A video describing the working of mass spectrometry (Fig. 10.32) is available at https://www.youtube.com/watch?v=myolF-h1kKI. Mass spectrometer is a useful tool

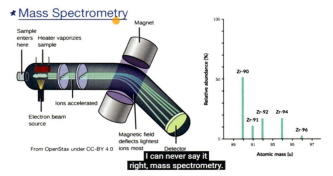

FIG. 10.32 Illustration of mass spectrometry. *(Source: https://www.youtube.com/watch?v=myolF-h1kKI.)*

that scientists rely on to find out if a sample contains certain molecules. It is an instrument that separates atoms on the basis of their mass differences. Typically, mass spectrometers can be used to identify unknown compounds via molecular weight determination, to quantify known compounds, and to determine structure and chemical properties of molecules. Lewis et al. (2021)'s study found that iron, calcium, magnesium oxalates, and acetates, affected by radiation and oxidation, may have decomposed into organic salts in Martian surface sediments. These salts could have been formed by geologic processes or be remnants of ancient microbial life. The discovery has the potential to change our understanding of the surface properties of Mars and further the search for microbial life on another planet. The researchers think that "Like shards of ancient pottery, these salts are the chemical remnants of organic compounds, such as those previously detected by the Curiosity rover." NASA said that the discovery could prove handy for future missions to Mars and support habitability in another environment, given that on Earth, some organisms can use these organic salts, such as oxalates and acetates, for energy. The researchers think that "If organic salts are widespread on the Martian surface, their composition and distribution could offer insight into the less-altered organic record and they may play an important role in near-surface carbon cycling and habitability." The researchers aim to find out what kinds of molecules may once have belonged to and what those molecules could reveal about the ancient environment and potential biology of the Red Planet, thereby trying to unravel billions of years of organic chemistry, and in that organic record there could be the ultimate prize: evidence that life once existed on the Red Planet. An important implication of Lewis et al. (2021)'s study is that "Organic matter preserved in the Martian rock record represents an invaluable resource for exploring the planet's carbon cycle, habitability, and potential biology through time." Besides adding more evidence to the idea that there once was organic matter on Mars, directly detecting organic salts would also support modern-day Martian habitability, given that on Earth, some organisms can use organic salts, such as oxalates and acetates, for energy.

While Lewis et al. (2021) used Sample Analysis at Mars (SAM), a portable chemistry lab inside Curiosity's belly, to identify the elements in a lab on Earth, identifying them directly on Mars is difficult as the rover heats Martian soil and rocks to release gases that reveal the composition of these samples. However, Lewis and his team propose that another Curiosity instrument that uses a different technique to peer at Martian soil, the Chemistry and Mineralogy instrument, or CheMin for short, could detect certain organic salts if they are present in sufficient amounts.

While NASA's *Perseverance* rover trundling in the Jazero crater is also looking for ancient microbial life on the surface (*Perseverance* rover doesn't have an instrument that can detect organic salts, but the rover is collecting samples for future return to Earth, where scientists can use sophisticated lab machines to look for organic compounds), the European Space Agency's forthcoming *ExoMars* rover, which is equipped to drill down to 2 m, will carry an instrument to analyze the chemistry of these deeper Martian layers. Meanwhile, Perseverance is storing samples to return to Earth that could help deeper analysis of the soil and biology of the surface in a cleanroom on Earth. Therefore, soon, scientists will also have an opportunity to study better-preserved soil below the Martian surface.

10.27 A ray of hope in Mars' horizon—Possibility of halophilic life on Mars

In view of the finding of perchlorate among the salts detected by the Phoenix Lander on Mars, Oren et al. (2014) investigated the relationships of halophilic heterotrophic microorganisms (archaea of the family Halobacteriaceae and the bacterium Halomonaselongata) toward perchlorate. It was found that all strains tested by Oren et al. (2014) grew well in NaCl-based media containing 0.4 M perchlorate. However, high concentrations of perchlorate appeared to be unsupportive of the growth of these microorganisms. This was attested by the fact that at the highest perchlorate concentrations, tested cells were swollen or distorted.

It was found that some species of the microorganisms (*Haloferaxmediterranei, Haloferaxdenitrificans, Haloferaxgibbonsii, Haloarculamarismortui, Haloarculavallismortis*) could use perchlorate as an electron acceptor for anaerobic growth. Although perchlorate is highly oxidizing, its presence at a concentration of 0.2 M for up to 2 weeks did not negatively affect the ability of a yeast extract-based medium to support growth of the archaeon Halobacterium salinarum. These findings show that presence of perchlorate among the salts on Mars does not preclude the possibility of halophilic life. Oren et al. (2014) are optimistic that if indeed the liquid brines that exist on Mars are inhabited by salt-requiring or salt-tolerant microorganisms similar to the halophiles on Earth, presence of perchlorate may even be stimulatory when it can serve as an electron acceptor for respiratory activity in the anaerobic Martian environment.

References

Abrevaya, X.C., Paulino-Lima, I.G., Galante, D., Rodrigues, F., Mauas, P.J., Cortón, E., Lage, C., 2011. Comparative survival analysis of Deinococcus radiodurans and the Haloarchaea *Natrialba magadii* and *Haloferax volcanii* exposed to vacuum ultraviolet irradiation. Astrobiology 11, 1034–1040.

Acuna, M., Connerney, J., Ness, N., Lin, R., Mitchell, D., Carlson, C., McFadden, J., Anderson, K., Reme, H., Mazelle, C., Vignes, D., Wasilewski, P., Cloutier, P., 1999. Global distribution of crustal magnetization discovered by the mars global surveyor MAG/ER experiment. Science 284, 790–793.

Aharonson, O., Zuber, M., Smith, D.E., Neumann, G.A., Feldman, W.C., Prettyman, T.H., 2004. Depth, distribution, and density of CO_2 deposition on Mars. J. Geophys. Res. Planets 109, E05004. http://doi.org/10.1029/2003JE002223.

Agee, C.B., et al., 2013. Unique meteorite from early Amazonian Mars: water-rich basaltic breccia Northwest Africa 7034. Science 339 (6121), 780–785. http://doi.org/10.1126/science.1228858.

Albarède, F., 2009. Volatile accretion history of the terrestrial planets and dynamic implications. Nature 461, 1227–1233.

Alexander, C.M.O., Bowden, R., Fogel, M.L., Howard, K.T., Herd, C.D.K., Nittler, L.R., 2012. The provenances of asteroids, and their contributions to the volatile inventories of the terrestrial planets. Science 337, 721–723.

Allen, M., Sherwood Lollar, B., Runnegar, B., Oehler, D.Z., Lyons, J.R., Manning, C.E., Summers, M.E., 2006. Is Mars alive? Eos Trans. Am. Geophys. Union 87, 433.

Andrews-Hanna, J.C., Zuber, M.T., Arvidson, R.E., Wiseman, S.M., 2010. Early Mars hydrology: Meridiani playa deposits and the sedimentary record of Arabia Terra. J. Geophys. Res. 115, E06002.

Arevalo, R., Brinckerhoff, W., van Amerom, F., Danell, R., Pinnick, V., Li, X., Getty, S., Hovmand, L., Grubisic, A., Mahaffy, P., Goesmann, F., Steininger, H., 2015. Designand demonstration of the Mars Organic Molecule Analyzer (MOMA) on the ExoMars 2018 rover, 2015 IEEE AerospaceConference. IEEE, pp. 1–11.

Arvidson, R.E., Poulet, F., Bibring, J.-P., Wolff, M., Gendrin, A., Morris, R.V., Freeman, J.J., Langevin, Y., Mangold, N., Bellucci, G., 2005. Spectral reflectance and morphologic correlations in eastern Terra Meridiani, Mars. Science 307 (5715), 1591–1594.

Arvidson, R.E., Squyres, S.W., Bell III, J.F., Catalano1, J.G., Clark, B.C., Crumpler, L.S., de Souza Jr., P.A., Fairén, A.G., Farrand, W.H., Fox, V.K., Gellert, R., Ghosh, A., Golombek, M.P., Grotzinger, J.P., Guinness, E.A., Herkenhoff, K.E., Jolliff, B.L., Knoll, A.H., Li, R., McLennan, S.M., Ming, D.W., Mittlefehldt, D.W., Moore, J.M., Morris, R.V., Murchie, S.L., Parker, T.J., Paulsen, G., Rice, J.W., Ruff, S.W., Smith, M.D., Wolff, M.J., 2014. Ancient aqueous environments at Endeavour crater, Mars. Science 343 (6169), 1248097. http://doi.org/10.1126/science.1248097.

Arya, A.S., et al., 2015a. Mars color camera onboard mars orbiter mission: initial observations & results, 46th Lunar and Planetary Science Conference.

Arya, A.S., Moorthi, S.M., Rajasekhar, R.P., Sarkar, S.S., Sur, K., Aravind, R., Gambhir, R.K., Misra, I., Patei, V.D., Srinivas, A.R., Patel, K.K.,
Chauhan, P., Kiran Kumar, A.S., 2015b. Indian Mars-Colour-Camera captures far-side of the Deimos: a rarity among contemporary Mars orbiters. Planet. Space Sci. 117, 470–474.

Ashmore, D.W., Bingham, R.G., 2014. Antarctic subglacial hydrology: current knowledge and future challenges. Antarct. Sci. 26, 758–773. http://doi.org/10.1017/S0954102014000546.

Atreya, S.K., Wong, A.-S., Renno, N.O., Farrell, W.M., Delory, G.T., Sentman, D.D., Cummer, S.A., Marshall, J.R., Rafkin, S.C.R., Catling, D.C., 2006. Oxidant enhancement in Martian dust devils and storms: implications for life and habitability. Astrobiology 6, 439–450.

Baglioni, P., Fisackerly, R., Gardini, B., Gianfiglio, G., Pradier, A.L., Santovincenzo, A., Vago, J.L., van Winnendael, M., 2006. The Mars exploration plans of ESA. IEEE Robot Autom. Mag. 13, 83–89.

Bao, H.M., Gu, B.H., 2004. Natural perchlorate has a unique oxygen isotope signature. Environ. Sci. Technol. 38, 5073–5077.

Baqué, M., Viaggiu, E., Scalzi, G., Billi, D., 2013. Endurance of the endolithic desert cyanobacterium *Chroococcidiopsis* under UVC radiation. Extremophiles 17, 161–169.

Barnes, J.J., et al., 2020. Multiple early-formed water reservoirs in the interior of Mars. Nat. Geosci. 13, 260–264. http://doi.org/10.1038/s41561-020-0552-y.

Bazylinski, D.A., 1990. Anaerobic production of single-domain magnetite by the marine, magnetotactic bacterium, strain MV-1. In: Frankel, R.B., Blakemore, R.P. (Eds.), Iron Biominerals. Plenum Press, New York, NY, pp. 69–77.

Bazylinski, D.A., Moskowitz, B.M., 1997. Microbial biomineralization of magnetic iron minerals: microbiology, magnetism, and environmental significance. Rev. Mineral. 35, 181–223. 6.

Bazylinski, D.A., Frankel, R.B., 2000. Biologically controlled mineralization of magnetic iron minerals by magnetotactic bacteria. In: Lovley, D.R. (Ed.), Environmental Microbe-Metal Interactions. ASM Press, Washington, DC, pp. 109–144.

Bhardwaj, A., Mohankumar, S.V., Das, T.P., Pradeepkumar, P., Sreelatha, P., Sundar, B., Nandi, A., Vajja, D.P., Dhanya, M.B., Naik, N., Supriya, G., Thampi, R.S., Padmanabhan, G.P., Yadav, V.K., Aliyas, A.V., 2015. MENCA experiment aboard India's Mars Orbiter mission. Curr. Sci. 109, 1106–1113.

Bhardwaj, A., Thampi, S.V., Das, T.P., Dhanya, M.B., Naik, N., Vajja, D.P., Pradeepkumar, P., Sreelatha, P., Supriya, G., Abhishek, J.K., Mohankumar, S.V., Thampi, R.S., Yadav, V.K., Sundar, B., Nandi, A., Padmanabhan, G.P., Aliyas, A.V., 2016. On the evening time exosphere of Mars: result from MENCA aboard Mars Orbiter mission. Geophys. Res. Lett. 43, 1862.

Bhardwaj, A., Thampi, S.V., Das, T.P., Dhanya, M.B., Naik, N., Vajja, D.P., Pradeepkumar, P., Sreelatha, P., Abhishek, J.K., Thampi, R.S., Yadav, V.K., Sundar, B., Nandi, A., Padmanabhan, G.P., Aliyas, A.V., 2017. Observation of suprathermal argon in the exosphere of Mars. Geophys. Res. Lett. 44, 2088.

Bibring, J.-P., Soufflot, A., Berthé, M., Langevin, Y., Gondet, B., Drossart, P., et al., 2004a. OMEGA: Observatoire pour la Minéralogie, l'Eau, les Glaces et l'Activité. Eur. Space Agency Spec. Publ. 1240, 37–49.

Bibring, J.-P., Langevin, Y., Poulet, F., Gendrin, A., Gondet, B., Berthé, M., Soufflot, A., Drossart, P., Combes, M., Bellucci, G., Moroz, V., Mangold, N., Schmitt, B.Omega Team, the, Erard, S., Forni, O., Manaud, N., Poulleau, G., Encrenaz, T., Fouchet, T., Melchiorri, R., Altieri, F., Formisano, V., Bonello, G., Fonti, S., Capaccioni, F., Cerroni, P., Coradini, A., Kottsov, V., et al., 2004b. Perennial water ice identified in the South Polar cap of Mars. Nature 428 (6983), 627–630.

Bibring, J.-P., Langevin, Y., Gendrin, A., Gondet, B., Poulet, F., Berthé, M., Soufflot, A., Arvidson, R., Mangold, N., Mustard, J., Drossart, P., 2005. Mars surface diversity as revealed by the OMEGA/Mars Express observations. Science 307 (5715), 1576–1581.

Bibring, J.-P., Langevin, Y., Mustard, J.F., Poulet, F., Arvidson, R., Gendrin, A., Gondet, B., Mangold, N., Pinet, P., Forget, F., Berthe´, M., Bibring, J.-P., Gendrin, A., Gomez, C., Gondet, B., Jouglet, D., Poulet, F., Soufflot, A., Vincendon, M., Combes, M., Drossart, P., Encrenaz, T., Fouchet, T., Merchiorri, R., Belluci, G., Altieri, F., Formisano, V., Capaccioni, F., Cerroni, P., Coradini, A., Fonti, S., Korablev, O., Kottsov, V., Ignatiev, N., Moroz, V., Titov, D., Zasova, L., Loiseau, D., Mangold, N., Pinet, P., Douté, S., Schmitt, B., Sotin, C., Hauber, E., Hoffmann, H., Jaumann, R., Keller, U., Arvidson, R., Mustard, J.F., Duxbury, T., Forget, F., Neukum, G., 2006. Global mineralogical and aqueous Mars history derived from OMEGA/Mars Express data. Science 312, 400–404.

Bibring, J.-P., Hamm, V., Pilorget, C., Vago, J.L., 2017. The MicrOmega investigationonboard ExoMars. Astrobiology 17, 621–626.

Biemann, K., Oro, J., Toulmin III, P., Orgel, L.E., Nier, A.O., Anderson, D.M., Simmonds, P.G., Flory, D., Diaz, A.V., Rushneck, D.R., Biller, J.E., Lafleur, A.L., 1977. The search for organic substances and inorganic volatile compounds in the surface of Mars. J. Geophys. Res. 82 (28), 4641–4658.

Biemann, K., 1979. The implications and limitations of the findings of the Viking Organic Analysis Experiment. J Mol Evol 14, 65–70.

Bish, D.L., Carey, J.W., Vaniman, D.T., Chipera, S.J., 2003. Stability of hydrous minerals on the Martian surface. Icarus 164 (1), 96–103.

Bishop, J.L., Loizeau, D., McKeown, N.K., Saper, L., Dyar, M.D., Des Marais, D.J., Parente, M., Murchie, S.L., 2013. What the ancient phyllosilicates at Mawrth Vallis cantell us about possible habitability on early Mars. Planet. Space Sci. 86, 130–149.

Black, B.A., Perron, J.T., Hemingway, D., Bailey, E., Nimmo, F., Zebker, H., 2017. Global drainage patterns and the origins of topographic relief on Earth, Mars, and Titan. Science 356 (6339), 727. http://doi.org/10.1126/science.aag0171.

Blakemore, R.P., 1975. Magnetotactic bacteria. Science 190, 377–379.

Blank, J.G., et al., 2009. An alkaline spring system within the Del Puerto Ophiolite (California, USA): a Mars analog site. Planet. Space Sci. 57, 533–540.

Bohlke, J.K., Sturchio, N.C., Gu, B.H., Horita, J., Brown, G.M., Jackson, W.A., Batista, J., Hatzinger, P.B., 2005. Perchlorate isotope forensics. Anal. Chem. 77, 7838–7842.

Bonny, S., Jones, B., 2003. Microbes and mineral precipitation, Miette Hot Springs, Jasper National Park, Alberta, Canada. Can. J. Earth Sci. 40 (11), 1483–1500. https://doi.org/10.1139/e03-060.

Brack, A., Pillinger, C.T., 1998. Life on Mars: chemical arguments and clues from Martian meteorites. Extremophiles 2 (3), 313–319.

Burt, D.M., Knauth, L.P., 2003. Electrically conducting, Ca-rich brines, rather than water, expected in the Martian subsurface. J. Geophys. Res. 108, 8026.

Buseck, P.R., Dunin-Borkowski, R.E., Devouard, B., Frankel, R.B., McCartney, M.R., Midgley, P.A., Posfai, M., Weyland, M., 2001. Magnetite morphology and life on Mars. Proc. Natl. Acad. Sci. USA 98, 13490–13495.

Carr, M.H., 1996. Water on Mars. Oxford Univ. Press, New York.

Carter, S.P., Blankenship, D.D., Peters, M.E., Young, D.A., Holt, J.W., Morse, D.L., 2007. Radar-based subglacial lake classification in Antarctica. Geochem. Geophys. Geosyst. 8, 3016. http://doi.org/10.1029/2006GC001408.

Carter, L.M., Campbell, B.A., Watters, T.R., Phillips, R.J., Putzig, N.E., Safaeinili, A., Plaut, J.J., Okubo, C.H., Egan, A.F., Seu, R., Biccari, D., Orosei, R., 2009. Shallow radar (SHARAD) sounding observations of the Medusae Fossae Formation, Mars. Icarus 199 (2), 295–302.

Chang, S.-B.R., Kirschvink, J.L., 1989. Magnetofossils, the magnetization of sediments, and the evolution of magnetite biomineralization. Ann. Rev. Earth Planet. Sci. 17, 169–195.

Checinska, A., Probst, A.J., Vaishampayan, P., White, J.R., Kumar, D., Stepanov, V.G., et al., 2015. Microbiomes of the dust particles collected from the international space station and spacecraft assembly facilities. Microbiome 3, 50. http://doi.org/10.1186/s40168-015-0116-3.

Cheptsov, V., Belov, A., Soloveva, O., Vorobyova, E., Osipov, G., Manucharova, N., Gorlenko, M., 2021. Survival and growth of soil microbial communities under influence of sodium perchlorates. Int. J. Astrobiol. 20, 36–47. https://doi.org/10.1017/S1473550420000312.

Chevrier, V., Hanley, J., Altheide, T., 2009. Stability of perchlorate hydrates and their liquid solutions at the Phoenix landing site, Mars. Geophys. Res. Lett. 36. http://doi.org/10.1029/2009GL037497.

Chicarro, A., Martin, P., Trautner, R., 2004. The Mars Express mission: an overview. Eur. Space Agency Spec. Publ. 1240, 3–13.

Chipera, S.J., Vaniman, D.T., 2007. Experimental stability of magnesium sulfate hydrates that may be present on Mars. Geochim. Cosmochim. Acta 71 (1), 241–250.

Christensen, P.R., et al., 2001a. Mars Global Surveyor Thermal Emission Spectrometer experiment: investigation description and surface science results. J. Geophys. Res. 106, 23.

Christensen, P.R., et al., 2001b. Mars Global Surveyor Thermal Emission Spectrometer experiment: investigation description and surface science results. J. Geophys. Res. 106, 823.

Christensen, P.R., 2001c. Mars Global Surveyor Thermal Emission Spectrometer experiment: investigation description and surface science results. J. Geophys. Res. 106, 871.

Christensen, P.R., Morris, R.V., Lane, M.D., Bandfield, J.L., Malin, M.C., 2001d. Global mapping of Martian hematite mineral deposits: remnants of water-driven processes on early Mars. J. Geophys. Res. 106, 873.

Christensen, P.R., Morris, R.V., Lane, M.D., Bandfield, J.L., Malin, M.C., 2001e. Global mapping of Martian hematite mineral deposits: remnants of water-driven processes on early Mars. J. Geophys. Res. 106, 885.

Christensen, P.R., 2006. Water at the poles and in permafrost regions of Mars. Elements (2), 151–155.

Ciarletti, V., Corbel, C., Plettemeier, D., Cais, P., Clifford, S.M., Hamran, S., 2011. WISDOM GPR designed for shallow and high-resolution sounding of the Martian subsurface. Proc. IEEE 99, 824–836.

Ciarletti, V., Clifford, S., Plettemeier, D., Le Gall, A., Hervé, Y., Dorizon, S., Quantin-Nataf, C., Benedix, W.-S., Schwenzer, S., Pettinelli, E., Heggy, E., Herique, A., Berthelier, J.-J., Kofman, W., Vago, J.L., Hamran, S.-E., Team, W., 2017. The WISDOM radar: unveiling the subsurface beneath the ExoMars Rover and identifying the best locations for drilling. Astrobiology 17, 565–584.

Clayton, R.N., 1993. Oxygen isotope analysis of ALH8400. Antarct. Meteorite News Lett. (JSC Curator's Office) 16, 4.

Clifford, S.M., 1987. Polar basal melting on Mars. J. Geophys. Res. 92, 9135–9152. http://doi.org/10.1029/JB092iB09p09135.

Clifford, S.M., et al., 2010. Depth of the Martian cryosphere: revised estimates and implications for the existence and detection of subpermafrost groundwater. J. Geophys. Res. 115, E07001.

Coates, J.D., Achenbach, L.A., 2004. Microbial perchlorate reduction: rocket-fuelled metabolism. Nat. Rev. Microbiol. 2, 569–580.

Coates, J.D., Achenbach, L.A., 2006. The microbiology of perchlorate reduction and its bioremediative application. In: Gu, B., Coates, J.D. (Eds.), Perchlorate, Environmental Occurrence, Interactions, and Treatment. Springer, New York, NY, pp. 279–295.

Cockell, C.S., Catling, D.C., Davis, W.L., Snook, K., Kepner, R.L., Lee, P., McKay, C.P., 2000. The ultraviolet environment of Mars: biological implications past, present, and future. Icarus 146 (2), 343–359.

Cockell, C.S., Balme, M., Bridges, J.C., Davila, A., Schwenzer, S.P., 2012. Uninhabited habitats on Mars. Icarus 217, 184–193.

Cortesão, M., Fuchs, F.M., Commichau, F.M., Eichenberger, P., Schuerger, A.C., Nicholson, W.L., Setlow, P., Moeller, R., 2019. *Bacillus subtilis* spore resistance to simulated Mars surface conditions. Front. Microbiol. 10, 333. http://doi.org/10.3389/fmicb.2019.00333.

Cousins, C.R., Crawford, I.A., 2011. Volcano-ice interaction as a microbial habitat on Earth and Mars. Astrobiology 11, 695–710.

Craddock, R.A., Lorenz, R.D., 2017. The changing nature of rainfall during the early history of Mars. Icarus 293, 172–179. http://doi.org/10.1016/j.icarus.2017.04.013 (Elsevier).

Dartnell, L.R., Page, K., Jorge-Villar, S.E., Wright, G., Munshi, T., Scowen, I.J., et al., 2012. Destruction of Raman biosignatures by ionising radiation and the implications for life detection on Mars. Anal. Bioanal. Chem. 403 (1), 131–144. https://doi.org/10.1007/s00216-012-5829-6.

DasSarma, S., 2006. Extreme halophiles are models for astrobiology. Microbe Wash. DC 1, 120–126.

DasSarma, S., DasSarma, P., 2017. Halophiles. eLS. John Wiley & Sons, Chichester, UK. http://doi.org/10.1002/9780470015902.a0000394.pub4.

DasSarma, P., Laye, V.J., Harvey, J., Reid, C., Shultz, J., Yarborough, A., Lamb, A., Koske-Phillips, A., Herbst, R., Molina, F., Grah, O., Phillips, T., DasSarma, S., 2017. Survival of halophilic archaea in Earth's cold stratosphere. Int. J. Astrobiol. 16, 321–327.

Dassarma, P., Capes, M.D., Dassarma, S., 2019. Comparative genomics of *halobacterium* strains from diverse locations. In: Das, S., Dash, H.R. (Ed.), Microbial Diversity in the Genomic Era, 1st ed. Academic Press, Cambridge, MA. USA, pp. 285–322, doi:10.1016/B978-0-12-814849-5.00017-4.

Deamer, D., Damer, B., 2017. Can life begin on Enceladus? A perspective from hydrothermal chemistry. Astrobiology 17 (9), 834–839.

De Angelis, S., De Sanctis, M.C., Ammannito, E., Di Iorio, T., Carli, C., Frigeri, A., Capria, M.T., Federico, C., Boccaccini, A., Capaccioni, F., Giardino, M., Cerroni, P., Palomba, E., Piccioni, G., 2013. VNIR spectroscopy of Mars analogues with the ExoMars-Ma_Miss instrument. Mem. Soc. Astron. Ital. Suppl. 26, 121–127.

Dehant, V., Lognonné, P., Sotin, C., 2004. Network science, NetLander: a European mission to study the planet Mars. Planet. Space Sci. 52 (11), 977–985. http://doi.org/10.1016/j.pss.2004.07.019.

DeLong, E.F., Frankel, R.B., Bazylinski, D.A., 1993. Multiple evolutionary origins of magnetotaxis in bacteria. Science 259, 803–806.

DeMeo, F.E., Carry, B., 2014. Solar System evolution from compositional mapping of the asteroid belt. Nature 505, 629–634.

De Sanctis, M.C., Altieri, F., Ammannito, E., Biondi, D., De Angelis, S., Meini, M., Mondello, G., Novi, S., Paolinetti, R., Soldani, M., Mugnuolo, R., Pirrotta, S., Vago, J.L., 2017. Ma_MISS on ExoMars: mineralogical characterization of the martian subsurface. Astrobiology 17, 612–620.

Donahue, T.M., 1995. Evolution of water reservoirs on Mars from D/H ratios in the atmosphere and crust. Nature 374 (6521), 432–434. http://doi.org/10.1038/374432a0.

Dose, K., Bieger-Dose, A., Dillmann, R., Gill, M., Kerz, O., Klein, A., et al., 1995. ERA-experiment "Space Biochemistry. Adv. Space. Res. 16, 119–129. http://doi.org/10.1016/0273-1177(95)00280-R.

Dreibus, G., Wänke, H., 1987. Volatiles on Earth and Mars: a comparison. Icarus 71 (2), 225–240.

Duigou, S., Ehrlich, S.D., Noirot, P., Noirot-Gros, M.F., 2005. DNA polymerase I acts in translesion synthesis mediated by the Y-polymerases in *Bacillus subtilis*. Mol. Microbiol. 57, 678–690. http://doi.org/10.1111/j.1365-2958.2005.04725.x.

Edgett, K.S., Malin, M.C., 2002. Martian sedimentary rock stratigraphy: outcrops and interbedded craters of northwest Sinus Meridiani and southwest Arabia Terra. Geophys. Res. Lett. 29, 2179.

Edwards, H.G.M., Hutchinson, I.B., Ingley, R., Parnell, J., Vítek, P., Jehlička, J., 2013. Raman spectroscopic analysis of geological and biogeological specimens of relevance to the ExoMars mission. Astrobiology 13, 543.

Ehlmann, B.L., et al., 2009. Identification of hydrated silicate minerals on Mars using MRO-CRISM: Geologic context near Nili Fossae and implications for aqueous Alteration. J. Geophys. Res. 114, E00D08.

Ehlmann, B.L., Mustard, J.F., Murchie, S.L., 2010. Geologic setting of serpentine deposits on Mars. Geophys. Res. Lett. 37, L06201.

Ehlmann, B.L., Mustard, J.F., Murchie, S.L., Bibring, J.-P., Meunier, A., Fraeman, A.A., Langevin, Y., 2011. Subsurface water and clay mineral formation during the early history of Mars. Nature 479, 53–60.

Eigenbrode, J.L., Summons, R.E., Steele, A., Freissinet, C., Millan, M., Navarro-González, R., et al., 2018. Organic matter preserved in 3-billion-year-old mudstones at Gale crater, Mars. Science 360 (6393), 1096–1101. https://doi.org/10.1126/science.aas9185.

Eisler, R., 2012. The Fukushima 2011 Disaster. CRC Press, Taylor & Francis Group, global network of offices at Oxford, New York, Philadelphia, Boca Raton, Boston, Melbourne, Singapore, Beijing, Tokyo, Kuala Lumpur, New Delhi and Cape Town.

Elkins-Tanton, L.T., 2011. Formation of early water oceans on rocky planets. Astrophys Space Sci 332, 359–364.

Ericksen, G.E., 1981. Geology and origin of the Chilean nitrate deposits. Geological Survey Professional Paper 1188. Government Printing Office, Reston, VA.

Fairén, A.G., Davila, A.F., Lim, D., Bramall, N., Bonaccorsi, R., Zavaleta, J., Uceda, E.R., Stoker, C., Wierzchos, J., Dohm, J.M., Amils, R., Andersen, D., McKay, C.P., 2010a. Astrobiology through the ages of Mars: the study of terrestrial analogues to understand the habitability of mars. Astrobiology 10 (8), 821–843.

Fairén, A.G., et al., 2010b. Noachian and more recent phyllosilicates in impact craters on Mars. Proc. Natl Acad. Sci. 107, 12095–12100.

Fajardo-Cavazos, P., Schuerger, A.C., Nicholson, W.L., 2010. Exposure of DNA and *Bacillus subtilis* spores to simulated martian environments: use of quantitative PCR (qPCR) to measure inactivation rates of DNA to function as a template molecule. Astrobiology 10, 403–411. http://doi.org/10.1089/ast.2009.0408.

Farquhar, J., Thiemens, M.H., Jackson, T., 1998. Atmosphere-surface interactions on Mars: delta O-17 measurements of carbonate from ALH 84001. Science 280, 1580–1582.

Farrell, W.M., Plaut, J.J., Cummer, S.A., Gurnett, D.A., Picardi, G., Watters, T.R., Safaeinili, A., 2009. Is the Martian water table hidden from radar view? Geophys. Res. Lett. 36, L15206. http://doi.org/10.1029/2009GL038945.

Fenton, H.J.H., 1894. Oxidation of tartaric acid in presence of iron. J. Chem. Soc. 65, 899–901.

Fendrihan, S., Legat, A., Pfaffenhuemer, M., Gruber, C., Weidler, G., Gerbl, F., Stan-Lotter, H., 2006. Extremely halophilic archaea and the issue of long-term microbial survival. Rev. Environ. Sci. Biotechnol. 5 (2-3), 203–218. http://doi.org/10.1007/s11157-006-0007-y.

Fendrihan, S., Bérces, A., Lammer, H., Musso, M., Rontó, G., Polacsek, T.K., Holzinger, A., Kolb, C., Stan-Lotter, H., 2009. Investigating the effects of simulated Martian ultraviolet radiation on *Halococcus dombrowskii* and other extremely halophilic archaebacteria. Astrobiology 9, 104–112.

Fernández-Remolar, D., Amils, R., Morris, R.V., Knoll, A.H., 2002, Proceedings of the 33rd Lunar and Planetary Science Conference. Houston, TX March 2002), abstr. 1226.

Fernández-Remolar, D.C., Rodriguez, N., Gómez, F., Amils, R., 2003. Geological record of an acidic environment driven by iron hydrochemistry: the Tinto River system. J. Geophys. Res. 108. http://doi.org/10.1029/2002JE001918.

Feshangsaz, N., Semsarha, F., Tackallou, S.H., Nazmi, K., Monaghan, E.P., Riedo, A., van Loon, J.J.W.A., 2020. Survival of the halophilic archaeon *Halovarius luteus* after desiccation, simulated Martian UV radiation and vacuum in comparison to *Bacillus atrophaeus*. Orig. Life Evol. Biosph. 50, 157–173. https://doi.org/10.1007/s11084-020-09597-7.

Fisk, M.R., Giovannoni, S.J., 1999. Sources of nutrients and energy for a deep biosphere on Mars. J. Geophys. Res. 104, 11805–11815.

Flinn, E.A., 1977. Scientific results of the Viking Project. J. Geophys. Res. 82, 735.

Folco, L. and Cordier, C. (2015), Micrometeorites, In: Planetary Mineralogy (pp.253-297), Edition: European Mineralogical Union, The Mineralogical Society of Great Britain and Ireland, London, Chapter: 9. Publisher: European Mineralogical Union, Editors: M.R. Lee, H. Leroux.

Forget, F., Haberle, R.M., Montmessin, F., Levrard, B., Heads, J.W., 2006. Formation of glaciers on Mars by atmospheric precipitation at high obliquity. Science 311, 368–371.

Foucher, F., Lopez-Reyes, G., Bost, N., Rull-Perez, F., Rubmann, P., Westall, F., 2013. Effect of grain size distribution on Raman analyses and the consequences for in situ planetary missions. J. Raman Spectrosc. 44, 916–925.

Frankel, R.B., Blakemore, R.P., Wolfe, R.S., 1979. Magnetite in freshwater magnetotactic bacteria. Science 203, 1355–1356.

Frankel, R.B., Bazylinski, D.A., Johnson, M.S., Taylor, B.S., 1997. Magneto-aerotaxis in marine coccoid bacteria. Biophys. J. 73, 994–1000.

Franzmann, P., Stackebrandt, E., Sanderson, K., Volkman, J., Cameron, D., Stevenson, P., McMeekin, T., Burton, H., 1988. Halobacterium lacusprofundi sp. nov., a halophilic bacterium isolated from Deep Lake, Antarctica. Syst. Appl. Microbiol. 11, 20–27.

Fredrickson, J.K., Balkwill, D., 2006. Geomicrobiological processes and biodiversity in the deep terrestrial subsurface. Geomicrobiol. J. 23, 345–356.

Friedmann, E.I., Wierzchos, J., Ascaso, C., Winklhofer, M., 2001. Chains of magnetite crystals in the meteorite ALH84001: evidence of biological origin. Proc. Natl. Acad. Sci. USA 98, 2176–2181.

Furnes, H., Banerjee, N.R., Muehlenbachs, K., Staudigel, H., de Wit, M., 2004. Early life recorded in archean pillow lavas. Science 304, 578–581.

Gebhardt, C., 2017. Year in Review, 2017 (Part II): Rovers, Orbiters Peel Away Mars' Secrets & Reveal New Mysteries. Spaceflight.com. available at: https://www.nasaspaceflight.com/2017/12/yir-2017-part-ii-rovers-mars-secrets-reveal-mysteries/.

Gendrin, A., Mangold, N., Bibring, J-P., Langevin, Y., Gondet, B., Poulet, F., Bonello, G., Quantin, C., Mustard, J., Arvidson, R., LeMouélic, S., 2005. Sulfates in Martian layered terrains: the OMEGA/Mars Express view. Science 307 (5715), 1587–1591.

Glavin, D.P., Freissinet, C., Miller, K.E., Eigenbrode, J.L., Brunner, A.E., Buch, A., Sutter, B., Archer Jr., P.D., Atreya, S.K., Brinckerhoff, W.B., Cabane, M., Coll, P., Conrad, P.G., Coscia, D., Dworkin, J.P., Franz, H.B., Grotzinger, J.P., Leshin, L.A., Martin, M.G., McKay, C., Ming, D.W., Navarro-González, R., Pavlov, A., Steele, A., Summons, R.E., Szopa, C., Teinturier, S., Mahaffy, P.R., 2013. Evidence for perchlorates and the origin of chlorinated hydrocarbons detected by SAM at the Rocknest aeolian deposit in Gale Crater. J. Geophys. Res. Planets 118, 1955–1973.

Goesmann, H.S., Goetz, W., Brinckerhoff, W., Szopa, C., Raulin, F., Westall, F., Edwards, H.G.M., Whyte, L.G., Fairén, A.G., Bibring, J-P., Bridges, J., Hauber, E., Ori, G.G., Werner, S., Loizeau, D., Kuzmin, R.O., Williams, R.M.E., Flahaut, J., Forget, F., Vago, J.L., Rodionov, D., Korablev, O., Svedhem, H., Sefton-Nash, E., Kminek, G., Lorenzoni, L., Joudrier, L., Mikhailov, V., Zashchirinskiy, A., Alexashkin, S., Calantropio, F., Merlo, A., Poulakis, P., Witasse, O., Bayle, O., Bayón, S., Meierhenrich, U., Carter, J., García-Ruiz, J.M., Baglioni, P., Haldemann, A., Ball, A.J., Debus, A., Lindner, R., Haessig, F., Monteiro, D., Trautner, R., Voland, C., Rebeyre, P., Goulty, D., Didot, F., Durrant, S., Zekri, E., Koschny, D., Toni, A., Visentin, G., Zwick, M., van Winnendael, M., Azkarate, M., Carreau, C., 2017a. Habitability on early Mars and the search for biosignatures with the ExoMars rover. Astrobiology 17 (6-7), 471–510.

Goesmann, F., Brinckerhoff, W.B., Raulin, F., Goetz, W., Danell, R.M., Getty, S.A., Siljeström, S., Mißbach, H., Steininger, H., Arevalo Jr., R.D., Buch, A., Freissinet, C., Grubisic, A., Meierhenrich, U.J., et al., 2017b. The Mars Organic Molecule Analyzer (MOMA) instrument: characterization of organic material in Martian sediments. Astrobiology 17, 655–685.

Goetz, W., Brinckerhoff, W.B., Arevalo, R., Freissinet, C., Getty, S., Glavin, D.P., Siljestrom, S., Buch, A., Stalport, F., Grubisic, A., Li, X., Pinnick, V., Danell, R., van Amerom, F.H.W., Goesmann, F., Steininger, H., Grand, N., Raulin, F., Szopa, C., Meierhenrich, U., Brucato, J.R., 2016. MOMA: the challenge to search for organics and biosignatures on Mars. Int J Astrobiol 15, 239–250.

Goswami, J.N., Sinha, N., Murty, S.V.S., Mohapatra, R.K., Clement, C.J., 1997. Nuclear tracks and light noble gases in ALH84001: pre-atmospheric size, fall characteristics, cosmic ray exposure duration and formation age. Meteoritics 32, 91–96.

Grant, W.D., Gemmell, R.T., McGenity, T.J., 1998. Halobacteria: the evidence for longevity. Extremophiles 2 (3), 279–287.

Grant, W.D., Kamekura, M., McGenity, T.J., Ventosa, A., 2001. Class III. Halobacteria class. nov. In: Boone, D.R., Castenholz, R.W., Garrity, G.M. (Eds.) second ed. Bergey's Manual of Systematic Bacteriology, vol. I Springer Verlag, New York, pp. 294–301.

Grazier, K.R., 2016. Jupiter: cosmic Jekyll and Hyde. Astrobiology 16, 23–38.

Green, W.J., Lyons, W.B., 2009. The saline lakes of the McMurdo Dry Valleys, Antarctica. Aquat. Geochem. 15, 321–348.

Grott, M., Helbert, J., Nadalini, R., 2007. Thermal structure of Martian soil and the measurability of the planetary heat flow. J. Geophys. Res. 112, E09004.

Grotzinger, J.P., 2014. Habitability, taphonomy, and the search for organic carbon on Mars. Science 343 (6169), 386–387. doi:10.1126/science.1249944.

Hallis, L.J., Huss, G.R., Nagashima, K., Taylor, G.J., Halldorsson, S.A., Hilton, D.R., Mottl, M.J., Meech, K.J., 2015. Evidence for primordial water in Earth's deep mantle. Science 350, 795–797.

Han, T.M., Runnegar, B., 1992. Megascopic eukaryotic algae from the 2.1 billion year old Negaunne iron-formation, Michigan. Science 257, 232–235.

Hanley, J., Chevrier, V., Berget, D., Adams, R., 2012. Chlorate salts and solutions on Mars. Geophys. Res. Lett. 39. http://doi.org/10.1029/2012GL051239(2012).

Hassler, D.M., et al., 2013. Mars' surface radiation environment measured with the Mars rover. Science. http://doi.org/10.1126/science.1244797.

Hassler, D.M., 2013. Mars' surface radiation environment measured with the Mars Science Laboratory's Curiosity Rover. Sciencexpress. http://www.sciencemag.org/content/early/recent/10.1126/science.1244797.

Hassler, D.M., Zeitlin, C., Wimmer-Schweingruber, R.F., Ehresmann, B., Rafkin, S., Eigenbrode, et al., 2014. Mars' surface radiation environment measured with the Mars Science Laboratory's Curiosity rover. Science 343 (6169), 1244797. https://doi.org/10.1126/science.1244797.

Head, J.W., Mustard, J.F., Kreslavsky, M.A, Milliken, R.E., Marchant, D.R., 2003. Recent ice ages on Mars. Nature 426, 797–802.

Hecht, M.H., 2002. Metastability of liquid water on Mars. Icarus 156, 373–386.

Hecht, S., Shlaer, S., Pirenne, M.H., 1942. Energy, quanta, and vision. J. Gen. Physiol. 25 (6), 819–840.

Hecht, M.H., Kounaves, S.P., Quinn, R.C., West, S.J., Young, S.M.M., Ming, D.W., Catling, D.C., Clark, B.C., Boynton, W.V., Hoffman, J., DeFlores, L.P., Gospodinova, K., Kapit, J., Smith, P.H., 2009. Detection of perchlorate and the soluble chemistry of Martian soil at the Phoenix Lander Site. Science 325, 64–67.

Hellevang, H., Huang, S.S., Thorseth, I.H., 2011. The potential for low-temperature abiotic hydrogen generation and a hydrogen-driven deep biosphere. Astrobiology 11, 711–724.

Henning, K.A., Li, L., Iyer, N., McDaniel, L.D., Reagan, M.S., Legerski, R., Schultz, R.A., Stefanini, M., Lehmann, A.R., Mayne, L.V., 1995. The Cockayne syndrome group A gene encodes a WD repeat protein that interacts with CSB protein and a subunit of RNA polymerase II TFIIH. Cell 82, 555–564.

Hill, D.R., Keenan, T.W., Helm, R.F., Potts, M., Crowe, L.M., Crowe, J.H., 1997. Extracellular polysaccharide of *Nostoc commune* (cyanobacteria) inhibits fusion of membrane vesicles during desiccation. J. Appl. Phycol. 9, 237–248.

Hirose, T., Kawagucci, S., Suzuki, K., 2011. Mechanoradical H-2 generation during simulated faulting: Implications for an earthquake-driven subsurface biosphere. Geophys. Res. Lett. 38, L1730.

Holden, J.F., Feinberg, L.F., 2005. Astrobiology and Planetary Missions. In: Hoover, R.B., Levin, G.V., Rozanov, A.Y., Gladstone, G.R. (Eds.). (International Society for Optical Engineering, 2005), 5906. Bellingham, Washington, USA.

Holm, N.G., 2012. The significance of Mg in prebiotic geochemistry. Geobiology 10, 269–279.

Horneck, G., 1999. European activities in exobiology in earth orbit: results and perspectives. Adv. Space Res. 23, 381–386.

Horneck, G., Rettberg, P., Reitz, G., Wehner, J., Eschweiler, U., Strauch, K., et al., 2001. Protection of bacterial spores in space, a contribution to the discussion on Panspermia. Orig. Life Evol. Biosph. 31, 527–547. http://doi.org/10.1023/A:1012746130771.

Horneck, G., Klaus, D.M., Mancinelli, R.L., 2010. Space microbiology. Microbiol. Mol. Biol. Rev. 74, 121–156. http://doi.org/10.1128/MMBR.00016-09.

Hull, P.M., Simon, A.F., Darroch, S.A.F., Erwin, D.H., 2015. Rarity in mass extinctions and the future of ecosystems. Nature 528, 345–351. doi:10.1038/nature16160.

Hullo, M.F., Moszer, I., Danchin, A., Martin-Verstraete, I., 2001. CotA of Bacillus subtilis is a copper-dependent laccase. J. Bacteriol. 183, 5426–5430. http://doi.org/10.1128/JB.183.18.5426-5430.2001.

Hurowitz, J.A., Fischer, W., Tosca, N.J., Milliken, R.E., 2010. Origin of acidic surface waters and the evolution of atmospheric chemistry on early Mars. Nat. Geosci. 3, 323–326.

Jackson, W.A., Bohlke, J.K., Gu, B.H., Hatzinger, P.B., Sturchio, N.C., 2010. Isotopic composition and origin of indigenous natural perchlorate and co-occurring nitrate in the Southwestern United States. Environ. Sci. Technol. 44, 4869–4876.

Jackson, W.A., et al., 2012. Perchlorate and chlorate biogeochemistry in ice-covered lakes of the McMurdo Dry Valleys, Antarctica. Geochim. Cosmochim. Acta 98, 19–30.

Jakosky, B., Carr, M.H., 1985. Possible precipitation of ice at low latitudes on Mars during periods of high obliquity. Nature 315, 559–561.

Janchen, J., Bish, D.L., Mohlmann, D.T.F., Stach, H., 2006. Investigation of the water sorption properties of Mars-relevant micro- and mesoporous minerals. Icarus 180 (2), 353–358.

Javor, B.J., 1989. Hypersaline Environments: Microbiology and Biogeochemistry. Springer Verlag, Berlin, Heidelberg, New York.

Jordan, R., Picardi, G., Plaut, J., Wheeler, K., Kirchner, D., Safaeinili, A., Johnson, W., Seu, R., Calabrese, D., Zampolini, E., Cicchetti, A., Huff, R., Gurnett, D., Ivanov, A., Kofman, W., Orosei, R., Thompson, T., Edenhofer, P., Bombaci, O., 2009. The Mars express MARSIS sounder instrument. Planet. Space Sci. 57, 1975–1986. http://doi.org/10.1016/j.pss.2009.09.016.

Josset, J.-L., Westall, F., Hofmann, B.A., Spray, J., Cockell, C., Kempe, S., Griffiths, A.D., De Sanctis, M.C., Colangeli, L., Koschny, D., Fo¨llmi, K., Verrecchia, E., Diamond, L., Josset, M., Javaux, E.J., Esposito, F., Gunn, M., Souchon-Leitner, A.L., Bontognali, T.R.R., Korablev, O., Erkman, S., Paar, G., Ulamec, S., Foucher, F., Martin, P., Verhaeghe, A., Tanevski, M., Vago, J.L., 2017. The Close-Up Imager onboard the ESA ExoMars Rover: objectives, description, operations, and science validation activities. Astrobiology 17, 595–611.

Jull, A.J.T., Eastoe, C.J., Xue, S., Herzog, G.F., 1995. Isotopic composition of carbonates in the SNC meteorites Allan Hills 84001 and Nakhla. Meteoritics 30, 311–318.

Kang, N.G., Anderson, T.A., Jackson, W.A., 2006. Photochemical formation of perchlorate from aqueous oxychlorine anions. Anal. Chim. Acta 567, 48–56.

Kang, N., Anderson, T.A., Rao, B., Jackson, W.A., 2009. Characteristics of perchlorate formation via photodissociation of aqueous chlorite. Environ. Chem. 6, 53–59.

Kiran Kumar, A.S., Chauhan, P., 2014. Scientific exploration of Mars by first Indian interplanetary space probe: Mars Orbiter Mission. Curr. Sci. 107 (7), 1096–1097.

Kirschvink, J.L., 1980. South-seeking magnetic bacteria. J. Exp. Biol. 86, 345–347.

Kish, A., Kirkali, G., Robinson, C., Rosenblatt, R., Jaruga, P., Dizdaroglu, M., Diruggiero, J., 2009. Salt shield: intracellular salts provide cellular protection against ionizing radiation in the halophilic archaeon, *Halobacterium salinarum* NRC-1. Environ. Microbiol. 11, 1066–1078.

Klein, H.P., Lederber, J., Rich, A., 1972. Biological experiments—Viking Mars Lander. Icarus 16, 139–146.

Klein, H.P., Lederber, J., Rich, A., Horowitz, N.H., Oyama vance, I., Levin, G.V., 1976a. The Viking mission search for life on Mars. Nature 262, 24–27.

Klein, H.P., Horowitz, N.H., Levin, G.V., Oyama, V.I., Lederberg, J., Rich, A., Hubbard, J.S., Hobby, G.L., Straat, P.A., Berdahl, B.J., Carle, G.C., Brown, F.S., Johnson, R.D, 1976b. The Viking biology investigation: preliminary results. Science 194 (4260), 92–105.

Klein, H.P., 1978. Viking biological experiments on Mars. Icarus 34 (3), 666–674.

Klein, H.P., 1998. The search for life on Mars: what we learned from Viking. J. Geophys. Res. 103, 28463–28466.

Klein, H.P., 1999. Did viking discover life on Mars? Orig. Life Evol. Biosph. 29, 625–631.

Kminek, G., Rummel, J.D., 2015. COSPAR planetary protection policy. Space Res. Today 193, 7–19.

Korablev, O., Ivanov, A., Fedorova, A., Kalinnikov, Y.K., Shapkin, A., Mantsevich, S., Viazovetsky, N., Evdokimova, N., Kiselev, A.V., Region, M., Measurements, R., Region, M., State, M., Gory, L., 2015. Development of a mast or robotic arm-mounted infrared AOTF spectrometer for surface Moon and Mars probes. Proc. SPIE 9608, 1–10.

Korablev, O.I., Dobrolensky, Y., Evdokimova, N., Fedorova, A.A., Kuzmin, R.O., Mantsevich, S.N., Cloutis, E.A., Carter, J., Poulet, F., Flahaut, J., Griffiths, A., Gunn, M., Schmitz, N., Martı́n-Torres, J., Zorzano, M.-P., Rodionov, D.S., Vago, J.L., Stepanov, A.V., Titov, A. Yu., Vyazovetsky, N.A., Trokhimovskiy, A.Yu., Sapgir, A.G., Kalinnikov, Y.K., Ivanov, Y.S., Shapkin, A.A., Ivanov, A.Yu, 2017. Infrared spectrometer for ExoMars: a mast-mounted instrument for the Rover. Astrobiology 17, 542–564.

Kottemann, M., Kish, A., Iloanusi, C., Bjork, S., Diruggiero, J., 2005. Physiological responses of the halophilic archaeon *Halobacterium* sp. strain NRC1 to desiccation and gamma irradiation. Extremophiles 9, 219–227.

Kounaves, S.P., Hecht, M.H., Kapit, J., Gospodinova, K., DeFlores, L., Quinn, R., Boynton, W.V., Clark, B.C., Catling, D.C., Hredzak, P., Ming, D.W., Moore, Q., Shusterman, J., Stroble, S., West, S.J., Young, S.M.M., 2010b. Wet chemistry experiments on the 2007 Phoenix Mars Scout lander mission: data analysis and results. J. Geophys. Res. 115, E00E10. http://doi.org/10.1029/2009JE003424.

Kminek, G., Hipkin, V.J., Anesio, A.M., Barengoltz, J., Boston, P.J., Clark, B.A., Conley, C.A., Coustenis, A., Detsis, E., Doran, P., Grasset, O., Hand, K., Hajime, Y., Hauber, E., Kolmasova´, I., Lindberg, R.E., Meyer, M., Raulin, F., Reitz, G., Renno´, N.O., Rettberg, P., Rummel, J.D., Saunders, M.P., Schwehm, G., Sherwood, B., Smith, D.H., Stabekis, P.E., Vago, J., 2016. COSPAR panel on planetary protection colloquium, Bern, Switzerland, September 2015. Space Res. Today 195, 42–67.

Kump, LR., Kasting, J.F., Barley, M.E., 2001. Rise of atmospheric oxygen and the "upside-down" archean mantle. Geochemistry Geophysics Geosystems 2, 2000GC000114.

La Duc, M.T., Dekas, A., Osman, S., Moissl, C., Newcombe, D., Venkateswaran, K., 2007. Isolation and characterization of bacteria capable of tolerating the extreme conditions of clean room environments. Appl. Environ. Microbiol. 73 (8), 2600–2611.

Lammer, H., et al., 2012. Outgassing history and escape of the Martian atmosphere and water inventory. Space Sci. Rev. http://dx.doi.org/10.1007/s11214-012-9943-8.

Landis, G.A., 2001. Martian water: are there extant halobacteria on Mars? Astrobiology 1, 161–164.

Langevin, Y., Poulet, F., Bibring, J.-P., Gondet, B., 2005. Sulfates in the north polar region of Mars detected by OMEGA/Mars Express. Science 307 (5715), 1584–1586.

Lasne, J., Noblet, A., Szopa, C., Navarro-González, R., Cabane, M., Poch, O., et al., 2016. Oxidants at the surface of Mars: a review in light of recent exploration results. Astrobiology 16 (12), 977–996. https://doi.org/10.1089/ast.2016.1502.

Lasue, J., Mangold, N., Hauber, E., Clifford, S., Feldman, W., Gasnault, O., Grima, C., Maurice, S., Mousis, O., 2013. Quantitative assessments of the martian hydrosphere. Space Sci. Rev. 174, 155–212. http://doi.org/10.1007/s11214-012-9946-5.

Lauro, S.E., Pettinelli, E., Caprarelli, G., et al., 2021. Multiple subglacial water bodies below the south pole of Mars unveiled by new MARSIS data. Nat. Astron. 5, 63–70. https://doi.org/10.1038/s41550-020-1200-6.

Laye, V.J., DasSarma, S., 2018. An Antarctic extreme halophile and its polyextremophilic enzyme: effects of perchlorate salts. Astrobiology 18 (4). http://doi.org/10.1089/ast.2017.1766.

Lenhart, J.S., Schroeder, J.W., Walsh, B.W., Simmons, L.A., 2012. DNA repair and genome maintenance in *Bacillus subtilis*. Microbiol. Mol. Biol. Rev. 76, 530–564. http://doi.org/10.1128/MMBR.05020-11.

Leshin, L., 2000. Geophys. Res. Lett. 27, 2017.

Leuko, S., Rothschild, L., Burns, B., 2010. Halophilic archaea and the search for extinct and extant life on Mars. J. Cosmol. 5, 940–950.

Leuko, S., Domingos, C., Parpart, A., Reitz, G., Rettberg, P., 2015. The survival and resistance of *Halobacterium salinarum* NRC-1, *Halococcus hamelinensis*, and *Halococcus morrhuae* to simulated outer space solar radiation. Astrobiology 15, 987–997.

Levin, G.V., Straat, P.A., 1981. A search for nonbiological explanation of the Viking labeled release life detection experiment. Icarus 45, 494–516.

Levin, G.V., Straat, P.A., 2016. The case for extant life on mars and its possible detection by the viking labeled release experiment. Astrobiology. 16:ast.2015.1464. http://doi.org/10.1089/ast.2015.1464.

Lewis, J.M.T., Eigenbrode, J.L., Wong, G.M., McAdam, A.C., Archer, P.D., Sutter, B., Millan, M., Williams, R.H., Guzman, M., Das, A., Rampe, E.B., Achilles, C.N., Franz, H.B., Andrejkovičová, S., Knudson, C.A., Mahaffy, P.R., 2021. Pyrolysis of oxalate, acetate, and perchlorate mixtures and the implications for organic salts on Mars. JGR Planets 126 (4), e2020JE006803.

Li, J., Andrews-Hanna, J.C., Sun, Y., Phillips, R.J., Plaut, J.J., Zuber, M.T., 2012. Density variations within the south polar layered deposits of Mars. J. Geophys. Res. Planets 117, E04006.

Li, X., Danell, R.M., Brinckerhoff, W.B., Pinnick, V.T., van Amerom, F., Arevalo, R.D., Getty, S.A., Mahaffy, P.R., Steininger, H., Goesmann, F., 2015. Detection of trace organics in Mars analog samples containing perchlorate by laser desorption/ionization mass spectrometry. Astrobiology 15, 104–110.

Lin, L.H., et al., 2005. Radiolytic H-2 in continental crust: nuclear power for deep subsurface microbial communities. Geochem. Geophys. Geosyst. 6, Q07003.

Litchfield, C.D., 1998. Survival strategies for microorganisms in hypersaline environments and their relevance to life on early Mars. Meteor. Planet. Sci. 33, 813–819.

Litvak, M.L., Mitrofanov, I.G., Kozyrev, A.S., Sanin, A.B., Tretyakov, V.I., Boynton, W.V., Kelly, N.J., Hamara, D., Saunders, R.S., 2007. Long-term observations of southern winters on Mars: estimations of column thickness, mass, and volume density of the seasonal CO_2 deposit from HEND/Odyssey data. J. Geophys. Res. Planets 112, E03S13. http://doi.org/10.1029/2006JE002832.

Loizeau, D., Mangold, N., Poulet, F., Ansan, V., Hauber, E., Bibring, J.-P., Gondet, B., Langevin, Y., Masson, P., Neukum, G., 2010. Stratigraphy in the Mawrth Vallis region through OMEGA, HRSC color imagery and DTM. Icarus 205, 396–418.

Loizeau, D., Werner, S.C., Mangold, N., Bibring, J.-P., Vago, J.L., 2012. Chronology of deposition and alteration in the Mawrth Vallis region, Mars. Planet Space Sci. 72, 31–43.

Lopez-Reyes, G., 2015. Development of Algorithms and Methodological Analyses for the Definition of the Operation Mode of the Raman Laser Spectrometer Instrument. Universidad de Valladolid, Valladolid, Spain.

Mader, H.M., Pettitt, M.E., Wadham, J.L., Wolff, E.W., Parkes, R.J., 2006. Subsurface ice as a microbial habitat. Geology 34, 169–172.

Mahaffy, P.R., Webster, C.R., Cabane, M., Conrad, P.G., Coll, P., Atreya, S.K., et al., 2012. The sample analysis at Mars investigation and instrument suite. Space Sci. Rev. 170 (1–4), 401–478. https://doi.org/10.1007/s11214-012-9879-z.

Magge, A., Granger, A.C., Wahome, P.G., Setlow, B., Vepachedu, V.R., Loshon, C.A., et al., 2008. Role of dipicolinic acid in the germination, stability, and viability of spores of *Bacillus subtilis*. J. Bacteriol. 190, 4798–4807. http://doi.org/10.1128/JB.00477-08.

Mancinelli, R., Fahlen, T., Landheim, R., Klovstad, M., 2004. Brines and evaporites: analogs for Martian life. Adv. Space Res. 33, 1244–1246.

Mancinelli, R., 2015. The effect of the space environment on the survival of Halorubrum chaoviator and Synechococcus (Nägeli): data from the space experiment OSMO on EXPOSE-R. Int. J. Astrobiol. 14, 123–128.

Malin, M.C., Edgett, K.S., 2000. Sedimentary rocks of early Mars. Science 290, 1927–1937.

Marion, G.M., Fritsen, C.H., Eicken, H., Payne, M.C., 2003. The search for life on Europa: Limiting environmental factors, potential habitats, and Earth analogues. Astrobiology 3, 785–811.

Martucci, H., 2012. Characterization of perchlorate photo-stability under simulated Martian conditions, Proc. Nat. Conf. Undergrad. Res. (NCUR), 1359–1363.

Marty, B., Alexander, C.M.O., Raymond, S.N., 2013. Primordial origins of Earth's carbon. Rev. Mineral Geochem. 75, 149–181.

Mason, J.M., Setlow, P., 1986. Essential role of small, acid-soluble spore proteins in resistance of *Bacillus subtilis* spores to UV light. J. Bacteriol. 167, 174–178. http://doi.org/10.1128/jb.167.1.174-178.1986.

Matsubara, T., Fujishima, K., Saltikov, C., Nakamura, S., Rothschild, L., 2016. Earth analogues for past and future life on Mars: isolation of perchlorate resistant halophiles from Big Soda Lake. Intl. J. Astrobiol. http://doi.org/10.1017/S1473550416000458.

McGenity, T.J., Gemmell, R.T., Grant, W.D., Stan-Lotter, H., 2000. Origins of halophilic microorganisms in ancient salt deposits. Environ. Microbiol. 2 (3), 243–250.

McGovern, P.J., et al., 2004. Correction to and 'localized gravity/topography admittance and correlation spectra on Mars: Implications for regional and global evolution'. J. Geophys. Res. 109, 1–5.

McKay, D.S., Gibson, E.K., Thomas-Keprta, K.L., Hojatollah, V., Romanek, C.S., Clemmett, S.J., Chillier, X.D.F., Maechling, C.R., Zare, R.N., 1996. Search for past life on Mars: possible relic biogenic activity in Martian meteorite ALH84001. Science 273, 924–930.

McKay, C.P., 2004. What is life—and how do we search for it in other worlds? PLoS Biol 2 (9), e302–e304. https://doi.org/10.1371/journal.pbio.0020302.

McKay, C.P., 2009. Biologically reversible exploration. Science 323, 718.

McKay, C.P., 2010. An origin of life on Mars. Cold Spring Harb. Perspect. Biol. 2, a003509.

McKay, C.P., 2016. Titan as the abode of life. Life 6, 8.

McLoughlin, N., Grosch, E.G., 2015. A hierarchical system for evaluating the biogenicity of metavolcanic- and ultramafic-hosted microalteration textures in the search for extraterrestrial life. Astrobiology 15, 901–921.

McSween Jr., H.Y., Harvey, R.P., 1993. Outgassed water on Mars: constraints from melt inclusions in SNC meteorites. Science 259 (5103), 1890–1892. http://doi.org/10.1126/science.259.5103.1890.

McSween Jr., H.Y., 1997. Evidence for life in a Martian meteorite? GSA Today 7 (7), 1–7.

McSween Jr., H.Y., et al., 2001. Geochemical evidence for magmatic water within Mars from pyroxenes in the Shergotty meteorite. Nature 409 (6819), 487–490. http://doi.org/10.1038/35054011.

Meierhenrich, U.J., Laurent, N., Christian, A., Jan, H.B., Hoffmann, S.V., Bernard, B., André, B., 2005. Asymmetric vacuum UV photolysis of the amino acid leucine in the solid state. Angew. Chem. Int. Ed. 44, 5630–5634. http://doi.org/10.1002/anie.200501311.

Meinert, C., Myrgorodska, I., deMarcellus, P., Buhse, T., Nahon, L., Hoffmann, S.V., DHendecourt, L.L.S., Meierhenrich, U.J., 2016. Ribose and related sugars from ultraviolet irradiation of interstellar ice analogs. Science 352, 208–212.

Melosh, H.J., 1995. Cratering dynamics and the delivery of meteorites to the Earth. Meteoritics Planet. Sci. 30, 545–546.

Michalski, J.R., Cuadros, J., Niles, P.B., Parnell, J., Rogers, A.D., Wright, S.P., 2013a. Groundwater activity on Mars and implications for a deep biosphere. Nat. Geosci. 6, 1–6.

Michalski, J.R., Niles, P.B., Cuadros, J., Baldridge, A.M., 2013b. Multiple working hypotheses for the formation of compositional stratigraphy on Mars: insights from the Mawrth Vallis region. Icarus 226, 816–840.

Ming, D.W., et al., 2014. Volatile and organic compositions of sedimentary rocks in Yellowknife Bay, Gale crater, Mars. Science 343 (6169), 1245267. http://doi.org/10.1126/science.1245267.

Mishra, M.K., Chauhan, P., Singh, R., Moorthi, S.M., Sarkar, S.S., 2016. Estimation of dust variability and scale height of atmospheric optical depth (AOD) in the Valles Marineris on Mars by Indian Mars Orbiter Mission (MOM) data. Icarus 264 (84).

Mitrofanov, I.G., Litvak, M.L., Nikiforov, S.Y., Jun, I., Bobrovnitsky, Y.I., Golovin, D.V., Grebennikov, A.S., Fedosov, F.S., Kozyrev, A.S., Lisov, D.I., Malakhov, A.V., Mokrousov, M.I., Sanin, A.B., Shvetsov, V.N., Timoshenko, G.N., Tomilina, T.M., Tret'yakov, V.I., Vostrukhin, A.A., 2017. The ADRON-RM instrument onboard the ExoMars Rover. Astrobiology 17, 585–594.

Mittlefehldt, D.W., 1994. ALH84001, a cumulate orthopyroxenite member of the Martian meteorite clan. Meteor. Planet. Sci. 29, 214–221.

Moeller, R., Stackebrandt, E., Reitz, G., Berger, T., Rettberg, P., Doherty, A.J., et al., 2007b. Role of DNA repair by nonhomologous-end joining in *Bacillus subtilis* spore resistance to extreme dryness, mono- and polychromatic UV, and ionizing radiation. J. Bacteriol. 189, 3306–3311.

Moeller, R., Setlow, P., Horneck, G., Berger, T., Reitz, G., Rettberg, P., et al., 2008. Roles of the major, small, acid-soluble spore proteins and spore-specific and universal DNA repair mechanisms in resistance of Bacillus subtilis spores to ionizing radiation from X rays and high-energy charged-particle bombardment. J. Bacteriol. 190, 1134–1140. http://doi.org/10.1128/JB.01644-07.

Moeller, R., Reitz, G., Li, Z., Klein, S., Nicholson, W.L., 2012a. Multifactorial resistance of *Bacillus subtilis* spores to high-energy proton radiation: role of spore structural components and the homologous recombination and nonhomologous end joining DNA repair pathways. Astrobiology 12, 1069–1077. http://doi.org/10.1089/ast.2012.0890.

Moeller, R., Schuerger, A.C., Reitz, G., Nicholson, W.L., 2012b. Protective role of spore structural components in determining *Bacillus subtilis* spore resistance to simulated mars surface conditions. Appl. Environ. Microbiol. 78, 8849–8853. http://doi.org/10.1128/AEM.02527-12.

Moeller, R., Raguse, M., Reitz, G., Okayasu, R., Li, Z., Klein, S., et al., 2014. Resistance of *Bacillus subtilis* spore DNA to lethal ionizing radiation damage relies primarily on spore core components and DNA repair, with minor effects of oxygen radical detoxification. Appl. Environ. Microbiol. 80, 104–109. http://doi.org/10.1128/AEM.03136-13.

Möhlmann, D.T.F., 2004. Water in the upper Martian surface at mid- and low-latitudes: presence, state, and consequences. Icarus 168, 318–323.

Möhlmann, D.T.F., 2008. The influence of van der Waals forces on the state of water in the shallow subsurface of Mars. Icarus 195 (1), 131–139.

Moissl-Eichinger, C., Cockell, C., Rettberg, P., 2016. Venturing into new realms? Microorganisms in space. FEMS Microbiol. Rev. 40, 722–737. http://doi.org/10.1093/femsre/fuw015.

Monk, J.D., Clavero, M.R.S., Beuchat, L.R., Doyle, M.P., Brackett, R.E., 1994. Irradiation inactivation of *Listeria monocytogenes* and *Staphylococcus aureus* in low-and high-fat, frozen and refrigerated ground beef. J. Food Prot. 57, 969–974.

Morbidelli, A., Chambers, J., Lunine, J.I., Petit, J.M., Robert, F., Valsecchi, G.B., Cyr, K.E., 2000. Source regions and timescales for the delivery of water to the Earth. Meteor. Planet Sci. 35, 1309–1320.

Mouginot, J., Kofman, W., Safaeinili, A., Herique, A., 2008. Correction of the ionospheric distortion on the MARSIS surface sounding echoes. Planet. Space Sci. 56 (7), 917–926.

Mylona, P., Pawlowski, K., Bisseling, T., 1995. Symbiotic nitrogen fixation. Plant Cell 7, 869–885.

Navarro-González, R., Vargas, E., de la Rosa, J., Raga, A.C., McKay, C.P., 2010. Reanalysis of the Viking results suggests perchlorate and organics at mid-latitudes on Mars. J. Geophys. Res. 115. http://doi.org/10.1029/2010JE003599.

Nicholson, W.L., Munakata, N., Horneck, G., Melosh, H.J., Setlow, P., 2000. Resistance of *Bacillus* endospores to extreme terrestrial and extraterrestrial environments. Microbiol. Mol. Biol. Rev. 64, 548–572.

Nicholson, W.L., Schuerger, A.C., 2005. *Bacillus subtilis* spore survival and expression of germination-induced bioluminescence after prolonged incubation under simulated Mars atmospheric pressure and composition: implications for planetary protection and lithopanspermia. Astrobiology 5, 536–544. http://doi.org/10.1089/ast.2005.5.536.

Nicholson, W.L., McCoy, L.E., Kerney, K.R., Ming, D.W., Golden, D.C., Schuerger, A.C., 2012. Aqueous extracts of a Mars analogue regolith that mimics the Phoenix landing site do not inhibit spore germination or growth of model spacecraft contaminants Bacillus subtilis 168 and Bacillus pumilus SAFR-032. Icarus 220 (2), 904–910.

Nisbet, E.G., Sleep, N.H., 2001. The habitat and nature of early life. Nature 409, 1083–1091.

Norton, C.F., McGenity, T.J., Grant, W.D., 1993. Archaeal halophiles (halobacteria) from two British salt mines. J. Gen. Microbiol. 139, 1077–1081.

Nyquist, L.E., Bogard, D.D., Shih, C.-Y., Greshake, A., Stoffler, D., Eugster, O., 2001. Ages and geologic histories of Martian meteorites. Space Sci. Rev. 96, 105–164.

Ojha, L., Wilhelm, M.B., Murchie, S.L., McEwen, A.S., Wray, J.J., Hanley, J., Massé, M., Chojnacki, M., 2015. Spectral evidence for hydrated salts in recurring slope lineae on Mars. Nat. Geosci. Lett. 8, 829–832.

Oliveros, E., Legrini, O., Hohl, M., Müller, T., Braun, A.M., 1997. Industrial waste water treatment: large scale development of a light-enhanced Fenton reaction. Chem. Eng. Process 36, 397–405.

Oren, A. (Ed.), 1999. Microbiology and Biogeochemistry of Hypersaline Environments. CRC Press, Boca Raton.

Oren, (Ed.), 2002. *Halophilic Microorganisms and Their Environments*. Kluwer Academic Publishers; Dordrecht.

Oren, A., Mana, L., 2002. Amino acid composition of bulk protein and salt relationships of selected enzymes of *Salinibacter ruber*, an extremely halophilic bacterium. Extremophiles 6, 217–223.

Oren, A., Bardavid, R.E., Mana, L., 2014. Perchlorate and halophilic prokaryotes: implications for possible halophilic life on Mars. Extremophiles 18, 75–80.

Orosei, R., Lauro, S.E., Pettinelli, E., Cicchetti, A., Coradini, M., Cosciotti, B., Di Paolo, F., Flamini, E., Mattei, E., Pajola, M., Soldovieri, F., Cartacci, M., Cassenti, F., Frigeri, A., Giuppi, S., Martufi, R., Masdea, A., Mitri, G., Nenna, C., Noschese, R., Restano, M., Seu, R., 2018. Radar evidence of subglacial liquid water on Mars. Science 361 (6401), 490–493. http://doi.org/10.1126/science.aar7268.

Owen, T., 1992. The composition and early history of the atmosphere of Mars. In: Kieffer, H.H., Jakosky, B.M., Snyder, C.W., Matthews, M.S. (Eds.), Mars. The University of Arizona Press, Tucson, pp. 818–834.

Oyama, V.I., Berdahl, B.J., 1977. The Viking gas exchange experiment results from Chryse and Utoplia surface samples. J. Geophys. Res. 82 (28), 4669–4676.

Paris, A.J., Davies, E.T., Tognetti, L., Zahniser, C., 2019. Prospective lava tubes at Hellas Planitia: leveraging volcanic features on Mars to provide crewed missions protection from radiation. J. Washington Acad. Sci. arXiv:2004.13156 [astro-ph.EP].

Parnell, J., et al., 2010. Sulfur isotope signatures for rapid colonization of an impact crater by thermophilic microbes. Geology 38, 271–274.

Pavlov, A.A., Brown, L.L., Kasting, J.F., 2001. UV shielding of NH_3 and O_2 by organic hazes in the Archean atmosphere. J. Geophys. Res. 106, 23267–23287.

Peeters, Z., Vos, D., Ten Kate, I., Selch, F., van Sluis, C., Sorokin, D.Y., Muijzer, G., Stan-Lotter, H., van Loosdrecht, M., Ehrenfreund, P., 2010. Survival and death of the haloarchaeon *Natronorubrum* strain HG-1 in a simulated martian environment. Adv. Space Res. 46, 1149–1155.

Phillips, X., et al., 2001. Science 291, 2587.

Picardi, G., Plaut, J.J., Biccari, D., Bombaci, O., Calabrese, D., Cartacci, M., Cicchetti, A., Clifford, S.M., Edenhofer, P., Farrell, W.M., Federico, C., Frigeri, A., Gurnett, D.A., Hagfors, T., Heggy, E., Herique, A., Huff, R.L., Ivanov, A.B., Johnson, W.T., Jordan, R.L., Kirchner, D.L., Kofman, W., Leuschen, C.J., Nielsen, E., Orosei, R., Pettinelli, E., Phillips, R.J., Plettemeier, D., Safaeinili, A., Seu, R., Stofan, E.R., Vannaroni, G., Watters, T.R., Zampolini, E., 2005. Radar soundings of the subsurface of Mars. Science 310, 1925–1928. http://doi.org/10.1126/science.1122165pmid:16319122.

Piper, D.J.W., 1972. Sediments of the middle Cambrian burgess shale. Lethaia 5, 169–175.

Plaut, J.J., Picardi, G., Safaeinili, A., Ivanov, A.B., Milkovich, S.M., Cicchetti, A., Kofman, W., Mouginot, J., Farrell, W.M., Phillips, R.J., Clifford, S.M., Frigeri, A., Orosei, R., Federico, C., Williams, I.P., Gurnett, D.A., Nielsen, E., Hagfors, T., Heggy, E., Stofan, E.R., Plettemeier, D., Watters, T.R., Leuschen, C.J., Edenhofer, P., 2007. Subsurface radar sounding of the south polar layered deposits of Mars. Science 316, 92–95. http://doi.org/10.1126/science.1139672pmid:17363628.

Poulet, F., Bibring, J.-P., Mustard, J.F., Gendrin, A., Mangold, N., Langevin, Y., Arvidson, R.E., Gondet, B., Gomez, C., Berthe´, M., Erard, S., Forni, O., Manaud, N., Poulleau, G., Soufflot, A., Combes, M., Drossart, P., Encrenaz, T., Fouchet, T., Melchiorri, R., Bellucci, G., Altieri, F., Formisano, V., Fonti, S., Capaccioni, F., Cerroni, P., Coradini, A., Korablev, O., Kottsov, V., Ignatiev, N., Titov, D., Zasova, L.,

Pinet, P., Schmitt, B., Sotin, C., Hauber, E., Hoffmann, H., Jaumann, R., Keller, U., Forget, F., Team, O., 2005. Phyllosilicates on Mars and implications for early martian climate. Nature 438 (7068), 623–627.

Puleo, J.R., et al., 1977. Microbiological profiles of the Viking spacecraft. Appl. Environ. Microbiol. 33, 379–384.

Quinn, R.C., Ehrenfreund, P., Grunthaner, F.J., Taylor, C.L., Zent, A.P., 2007. Decomposition of aqueous organic compounds in the Atacama Desert and in martian soils. J. Geophys. Res. Biogeosci. 112. http://doi.org/10.1029/2006JG000312.

Quinn, R.C., Martucci, H.F., Miller, S.R., Bryson, C.E., Grunthaner, F.J., Grunthaner, P.J., 2013. Perchlorate radiolysis on Mars and the origin of Martian soil reactivity. Astrobiology 13 (6), 515–520.

Rajagopalan, S., Anderson, T., Cox, S., Harvey, G., Cheng, Q.Q., Jackson, W.A., 2009. Perchlorate in wet deposition across North America. Environ. Sci. Technol. 43, 616–622.

Rao, B., Anderson, T.A., Redder, A., Jackson, W.A., 2010a. Perchlorate formation by ozone oxidation of aqueous chlorine/oxy-chlorine species: role of $Cl(x)O(y)$ radicals. Environ. Sci. Technol. 44, 2961–2967.

Rathbun, J.A., Squyres, S.W., 2002. Hydrothermal systems associated with martian impact craters. Icarus 157, 362–372.

Rebeil, R., Sun, Y., Chooback, L., Pedraza-Reyes, M., Kinsland, C., Begley, T.P., et al., 1998. Spore photoproduct lyase from *Bacillus subtilis* spores is a novel iron-sulfur DNA repair enzyme which shares features with proteins such as class III anaerobic ribonucleotide reductases and pyruvate-formate lyases. J. Bacteriol. 180, 4879–4885.

Reid, I., Sparks, W., Lubow, S., McGrath, M., Livio, M., Valenti, J., Sowers, K., Shukla, H., MacAuley, S., Miller, T., Suvanasuthi, R., Belas, R., Colman, A., Robb, F.T., DasSarma, P., Muller, J.A., Coker, J.A., Cavicchioli, R., Chen, F., DasSarma, S., 2006. Terrestrial models for extraterrestrial life: methanogens and halophiles at martian temperatures. Int. J. Astrobiol. 5, 89–97.

Reith, F., 2011. Life in the deep subsurface. Geology 39, 287–288.

Rettberg, P., Anesio, A.M., Baker, V.R., Baross, J.A., Cady, S.L., Detsis, E., Foreman, C.M., Hauber, E., Ori, G.G., Pearce, D.A., Renno, N.O., Ruvkun, G., Sattler, B., Saunders, M.P., Smith, D.H., Wagner, D., Westall, F., 2016. Planetary protection and Mars special regions—a suggestion for updating the definition. Astrobiology 16, 119–125.

Rieder, R., Gellert, R., Anderson, R.C., Brückner, J., Clark, B.C., Dreibus, G., Economou, T., Klingelhöfer, G., Lugmair, G.W., Ming, D.W., Squyres, S.W., d'Uston, C., Wänke, H., Yen, A., Zipfel, J., 2004. Chemistry of rocks and soils at Meridiani Planum from the Alpha Particle X-ray Spectrometer. Science 306 (5702), 1746–1749.

Riesenman, P.J., Nicholson, W.L., 2000. Role of the spore coat layers in *Bacillus subtilis* spore resistance to hydrogen peroxide, artificial UV-C, UV-B, and solar UV radiation. Appl. Environ. Microbiol. 66, 620–626. http://doi.org/10.1128/AEM.66.2.620-626.2000.

Ringwood, A.E., 1977. Composition of the core and implications for the origin of the earth. Geochem. J. 11, 111–135.

Ringwood, A.E., 1979. On the Origin of the Earth and Moon. Springer-Verlag, New York ISBN 978-1-4612-6167-4.

Robbins, S.J., Hynek, B.M., 2012. A new global database of Mars impact craters ≥1 km: 1. Database creation, properties, and parameters. J. Geophys. Res. 117, E05004.

Rodriguez-Valera. (Ed.), 1988. *Halophilic Bacteria*. Vols. I, II. CRC Press; Boca Raton.

Ruff, S.W., 2004. Spectral evidence for zeolite in the dust on Mars. Icarus 168 (1), 131–143.

Ruff, S.W., Farmer, J.D., 2016. Silica deposits on Mars with features resembling hot spring biosignatures at El Tatio in Chile. Nat. Commun. 7, 13554.

Rull, F., Maurice, S., Hutchinson, I., Moral, A., Perez, C., Diaz, C., Colombo, M., Belenguer, T., Lopez-Reyes, G., Sansano, A., Forni, O., Parot, Y., Striebig, N., Woodward, S., Howe, C., Tarcea, N., Rodriguez, P., Seoane, L., Santiago, A., Rodriguez-Prieto, J.-A., Medina, J., Gallego, P., Canchal, R., Santamarı́a, P., Ramos, G., Vago, J.L., 2017. The Raman Laser Spectrometer for the ExoMars Rover mission to Mars. Astrobiology 17, 627–654.

Russell, M.J., Barge, L.M., Bhartia, R., Bocanegra, D., Bracher, P.J., Branscomb, E., Kidd, R., McGlynn, S., Meier, D.H., Nitschke, W., Shibuya, T., Vance, S., White, L., Kanik, I., 2014. The drive to life on wet and icy worlds. Astrobiology 14, 308–343.

Safaeinili, et al., 2007. Estimation of the total electron content of the Martian ionosphere using radar sounder surface echoes. Geophys. Res. Lett. 34, L23204. http://doi.org/10.1029/2007GL032154.

Samuel, P., et al., 2010. Discovery of natural perchlorate in the Antarctic dry valleys and its global implications. Environ. Sci. Technol. 44, 2360–2364.

Schidlowski, M.A., 1988. 3,800 million-year old record of life from carbon in sedimentary rocks. Nature 333, 313–318.

Schidlowski, M., 2001. Search for morphologigal and biochemical vestiges of fossil life in extraterrestrial settings: utility of terrestrial evidence. In: Horneck, G., Baumstark-Khan, C. (Eds.). Astrobiology. The Quest for the Conditions of Life, 2001. Springer Verlag, Berlin, Heidelberg, New York, pp. 373–386.

Schuerger, A.C., Ming, D.W., Golden, D.C., 2017. Biotoxicity of Mars soils: 2. Survival of *Bacillus subtilis* and Enterococcus faecalis in aqueous extracts derived from six Mars analog soils. Icarus 290, 215–223. http://doi.org/10.1016/j.icarus.2017.02.023.

Score, R., 1997. Finding ALH84001. Planet. Rep. *XVII*, 5–7.

Seedhouse, E., 2018. Acute radiation sickness. Space Radiation and Astronaut Safety. Springer Briefs in Space Development, 122. Springer, Cham, Switzerland, pp. 77–86. https://doi.org/10.1007/978-3-319-74615-9.

Setlow, P., 2014. Spore resistance properties. Microbiol. Spectr. 2, 1–14. http://doi.org/10.1128/microbiolspec.TBS-0003-2012.

Shahmohammadi, H.R., Asgarani, E., Terato, H., Saito, T., Ohyama, Y., Gekko, K., Yamamoto, O., Ide, H., 1998. Protective roles of bacterioruberin and intracellular KCl in the resistance of *Halobacterium salinarium* against DNA-damaging agents. J. Radiat. Res. 39, 251–262.

Siefermann-Harms, D., 1987. The light-harvesting and protective functions of carotenoids in photosynthetic membranes. Physiol. Plant. 69, 561–568.

Sleep, N.H., Zahnle, K., 1998. Refugia from asteroid impacts on early Mars and the early Earth. J. Geophys. Res. 103, 28529–28544.

Sherwood, L.B., Lacrampe-Couloume, G., Slater, G.F., Ward, J., Moser, D.P., Gihring, T.M., Lin, L.-H., Onstott, T.C., 2006. Unravelling abiogenic and biogenic sources of methane in the Earth's deep subsurface. Chem. Geol. 226, 328–339.

Shusser, M., Culick, F., Cohen, N., 2002. Combustion response of ammonium perchlorate composite propellants. J. Propuls. Power 18, 1093–1100.

Siljestrom, S., Freissinet, C., Goesmann, F., Steininger, H., Goetz, W., Steele, A., Amundsen, H.the AMASE11 Team, 2014. Comparison of prototype and laboratory experiments on MOMA GCMS: results from the AMASE11 campaign. Astrobiology 14, 780–797.

Solomon, S.C., Aharonson, O., Aurnou, J.M., Banerdt, B.W., Carr, M.H., Dombard, A.J., Frey, H.V., Golombek, M.P., Hauck, S.A.I., Head, J.W.I., Jakosky, B.M., Johnson, C.L., McGovern, P.J., Neumann, G.A., Phillips, R.J., Smith, D.E., Zuber, M.T., 2005. New perspectives on ancient Mars. Science 307, 1214–1220.

Spanovich, N., Smith, M.D., Smith, P.H., Wolff, M.J., Christensen, P.R., Squyres, S.W., 2006. Surface and near-surface atmospheric temperatures for the Mars Exploration Rover landing sites. Icarus 180 (2), 314–320.

Squyres, S.W., Arvidson, R.E., Bell III, J.F., Brückner, J., Cabrol, N.A., Calvin, W., Carr, M.H., Christensen, P.R., Clark, B.C., Crumpler, L., Des Marais, D.J., d'Uston, C., Economou, T., Farmer, J., Farrand, W., Folkner, W., Golombek, M., Gorevan, S., Grant, J.A., Greeley, R., Grotzinger, J., Haskin, L., Herkenhoff, K.E., Hviid, S., Johnson, J., Klingelhöfer, G., Knoll, A.H., Landis, G., Lemmon, M., Li, R., Madsen, M.B., Malin, M.C., McLennan, S.M., McSween, H.Y., Ming, D.W., Moersch, J., Morris, R.V., Parker, T., Rice Jr., J.W., Richter, L., Rieder, R., Sims, M., Smith, M., Smith, P., Soderblom, L.A., Sullivan, R., Wänke, H., Wdowiak, T., Wolff, M., Yen, A., 2004a. The Opportunity Rover's Athena science investigation at Meridiani Planum, Mars. Science 306 (5702), 1698–1703. http://doi.org/10.1126/science.1106171.

Squyres, S.W., Arvidson, R.E., Bell, J.F., Brückner, J., Cabrol, N.A., Calvin, W., Carr, M.H., Christensen, P.R., Clark, B.C., Crumpler, L., Des Marais, D.J., D'Uston, C., Economou, T., Farmer, J., Farrand, W., Folkner, W., Golombek, M., Gorevan, S., Grant, J.A., Greeley, R., Grotzinger, J., Haskin, L., Herkenhoff, K.E., Hviid, S., Johnson, J., Klingelhöfer, G., Knoll, A., Landis, G., Lemmon, M., Li, R., Madsen, M.B., Malin, M.C., McLennan, S.M., McSween, H.Y., Ming, D.W., Moersch, J., Morris, R.V., Parker, T., Rice, J.W., Richter, L., Rieder, R., Sims, M., Smith, M., Smith, P., Soderblom, L.A., Sullivan, R., Wa¨nke, H., Wdowiak, T., Wolff, M., Yen, A., 2004a. The Spirit Rover's Athena science investigation at Gusev Crater, Mars. Science 305, 794–799.

Squyres, S.W., Grotzinger, J.P., Arvidson, R.E., Bell III, J.F., Calvin, W., Christensen, P.R., Clark, B.C., Crisp, J.A., Farrand, W.H., Herkenhoff, K.E., Johnson, J.R., Klingelhöfer, G., Knoll, A.H., McLennan, S.M., McSween Jr., H.Y., Morris, R.V., Rice Jr., J.W., Rieder, R., Soderblom, L.A., 2004b. In situ evidence for an ancient aqueous environment at Meridiani Planum, Mars. Science 306 (5702), 1709–1714. http://doi.org/10.1126/science.1104559.

Squyres, S.W., Arvidson, R.E., Bell, J.F., Calef, F., Clark, B.C., Cohen, B.A., Crumpler, L.A., de Souza, P.A., Farrand, W.H., Gellert, R., Grant, J., Herkenhoff, K.E., Hurowitz, J.A., Johnson, J.R., Jolliff, B.L., Knoll, A.H., Li, R., McLennan, S.M., Ming, D.W., Mittlefehldt, D.W., Parker, T.J., Paulsen, G., Rice, M.S., Ruff, S.W., Schröder, C., Yen, A.S., Zacny, K., 2012. Ancient impact and aqueous processes at Endeavour Crater, Mars. Science 336, 570–576.

Srivastava, V.K., Kumar, J., Kulshrestha, S., Kushvah, B.S., Bhaskar, M.K., Somesh, S., Roopa, M.V., Ramakrishna, B.N., 2015. Eclipse modeling for the Mars Orbiter Mission. Adv. Space Res. 56 (4), 671–679.

Srivastava, V.K., Kumar, J., Kulshrestha, S., Kushvah, B.S., 2016. Mars solar conjunction prediction modeling. Acta Astronaut. 118, 246–250.

Stan-Lotter, H., Fendrihan, S., 2015. Halophilic archaea: life with desiccation, radiation and oligotrophy over geological times. Life 5, 1487–1496.

Steele, A., et al., 2012. A reduced organic carbon component in Martian basalts. Science 337, 212–215.

Steininger, H., Goesmann, F., Goetz, W., 2012. Influence of magnesium perchlorate on the pyrolysis of organic compounds in Mars analogue soils. Planet. Space Sci. 71 (1), 9–17.

Stoker, C.R., Zent, A., Catling, D.C., Douglas, S., Marshall, J.R., Archer, D., Clark, B., Kounaves, S.P., Lemmon, M.T., Quinn, R., Renno, N., Smith, P.H., Young, S.M.M., 2010. Habitability of the phoenix landing site. J. Geophys. Res. Planets 115, 1–24.

Strasdeit, H., 2010. Chemical evolution and early Earth's and Mars' environmental conditions. Palaeodiversity 3, 107–116.

Szopa, C., Freissinet, C., Glavin, D.P., Millan, M., Buch, A., Franz, H.B., et al., 2020. First detections of dichlorobenzene isomers and trichloromethylpropane from organic matter indigenous to Mars mudstone in Gale Crater, Mars: results from the sample analysis at Mars instrument onboard the curiosity rover. Astrobiology 20 (2), 292–306. https://doi.org/10.1089/ast.2018.1908.

Tanaka, K.L., 2000. Dust and ice deposition in the Martian geologic record. Icarus 144, 254–266.

Thomas-Keprta, K.L., Bazylinski, D.A., Kirchvink, J.L., Clemett, S.J., McKay, D.S., Wentworth, S.J., Vali, H., Gibson Jr., E.K., Romanek, C.S., 2000. Elongated prismatic magnetite crystals in ALH84001 carbonate globules: potential Martian magnetofossils. Geochim. Cosmochim. Acta 64, 4049–4081.

Thomas-Keprta, K.L., Clemett, S.J., Bazylinski, D.A., Kirschvink, J.L., McKay, D.S., Wentworth, S.J., Vali, H., Gibson Jr., E.K., McKay, M.F., Romanek, C.S., 2001. Truncated hexa-octahedral magnetite crystals in ALH84001: presumptive biosignatures. Proc. Natl. Acad. Sci. USA 98, 2164–2169.

Thomas-Keprta, K.L., et al., 2002. Magnetofossils from ancient Mars: a robust biosignature in the Martian meteorite ALH84001. Appl. Environ. Microbiol. 68 (8), 3663–3672. http://doi.org/10.1128/aem.68.8.3663-3672.2002.

Thomas, P.J., Hicks, R.D., Chyba, C.F., McKay, C.P. (Eds.), 2006. Comets and the Origin and Evolution of Life, second ed., Springer Berlin Heidelberg (Advances in Astrobiology and Biogeophysics), p. 357, doi: http://doi.org/10.1007/10903490.

Tokumura, M., Morito, R., Hatayama, R., Kawase, Y., 2011. Iron redox cycling in hydroxyl radical generation during the Photo-Fenton oxidative degradation: dynamic change of hydroxyl radical concentration. Appl. Catal. B: Environ. 106, 565–576.

Tsuchiyama, A., 1996. Meteoritics and mineralogy on possible ancient Martian life. Biol. Sci. Space 10 (4), 262–270. http://doi.org/10.2187/bss.10.262.

Tyler, S.A., Barghoorn, E.S., 1954. Occurrence of structurally preserved plants in Precambrian rocks of the Canadian Shield. Science 119 (3096), 606–608.

Vago, J.L., et al., 2017. Habitability on early Mars and the search for biosignatures with the ExoMars Rover. Astrobiology 17 (6-7), 471–510.

Vaniman, D.T., Bish, D.L., Chipera, S.J., Fialips, C.I., Carey, J.W., Feldman, W.C., 2004. Magnesium sulphate salts and the history of water on Mars. Lett. Nat. 431, 663–665.

Venkateswaran, K., Vaishampayan, P., Cisneros, J., Pierson, D.L., Rogers, S.O., Perry, J., 2014. International Space Station environmental microbiome - microbial inventories of ISS filter debris. Appl. Microbiol. Biotechnol. 98, 6453–6466. http://doi.org/10.1007/s00253-014-5650-6.

Ventosa, A. (Ed.), 2004. *Halophilic Microorganisms.*Springer Verlag; Berlin, Heidelberg, New York.

Voytek, M.A., Priscu, J.C., Ward, B.B., 1999. The distribution and relative abundance of ammonia-oxidizing bacteria in lakes of the McMurdo Dry Valley, Antarctica. Hydrobiologia 401, 113–130.

Vreeland, R.H., Hochstein, L. (Eds.), 1993. *The Biology of Halophilic Bacteria.*CRC Press; Boca Raton

Vreeland, R.H., Rosenzweig, W.D., Powers, D.W., 2000. Isolation of a 250 million-year-old halotolerant bacterium from a primary salt crystal. Nature 407, 897–900.

Vreeland, R., Jones, J., Monson, A., Rosenzweig, W., Lowenstein, T., Timofeeff, M., Satterfield, C., Cho, B., Park, J., Wallace, A., 2007. Isolation of live cretaceous (121–112 million years old) halophilic Archaea from primary salt crystals. Geomicrobiol. J. 24, 275–282.

Wadhwa, M., 2001. Redox state of Mars' upper mantle from Eu anomalies in Shergottite pyroxenes. Science 291, 1527–1530. http://www.sciencemag.org/cgi/reprint/291/5508/1527.pdf.

Wadsworth, J., Cockell, C.S., 2017. Perchlorates on Mars enhance the bacteriocidal effects of UV light. Nat. Sci. Rep. 7 (4662). http://doi.org/10.1038/s41598-017-04910-3.

Wang, A., Freeman, J.J., Jolliff, B.L., Chou, I-M., 2006. Sulfates on Mars: a systematic Raman spectroscopic study of hydration states of magnesium sulfates. Geochim. Cosmochim. Acta 70 (24), 6118–6135.

Wanke, H., 1981. Constitution of terrestrial planets. Philos. Trans. Roy. Soc. Lond. Ser. A 303, 287–302.

Warner, N.H., Farmer, J.D., 2010. Subglacial hydrothermal alteration minerals in Jokulhlaup deposits of Southern Iceland, with implications for detecting past or present habitable environments on Mars. Astrobiology 10, 523–547.

Watson, L.L., Hutcheon, I.D., Epstein, S., Stolper, E.M., 1994. Water on Mars: clues from deuterium/hydrogen and water contents of hydrous phases in SNC meteorites. Science 265 (5168), 86–90. http://doi.org/10.1126/science.265.5168.86.

Weiss, B.P., Kirschvink, J.L., 2000. Life from space? Testing panspermia with Martian meteorite ALH84001. Planet. Rep. 20, 8–11.

Westall, F., Brack, A., Hofmann, B., Horneck, G., Kurat, G., Maxwell, J., Ori, G.G., Pillinger, C., Raulin, F., Thomas, N., Fitton, B., Clancy, P., Prieur, D., Vassaux, D., 2000. An ESA study for the search for life on Mars. Planet. Space Sci. 48, 181–202.

Westall, F., 2008. Morphological biosignatures in early terrestrial and extraterrestrial materials. Space Sci. Rev. 135, 95–114.

Westall, F., 2012. The early Earth. In: Impey, C., Lunine, J., Funes, J. (Eds.), Frontiers of Astrobiology. Cambridge University Press, Cambridge, pp. 331.

Westall, F., Loizeau, D., Foucher, F., Bost, N., Betrand, M., Vago, J., Kminek, G., 2013. Habitability on Mars from a microbial point of view. Astrobiology 13, 887–897.

Westall, F., Foucher, F., Bost, N., Bertrand, M., Loizeau, D., Vago, J.L., Kminek, G., Gaboyer, F., Campbell, K.A., Breheret, J.-G., Gautret, P., Cockell, C.S., 2015a. Biosignatures on Mars: what, where, and how? Implications for the search for martian life. Astrobiology 15, 998–1029.

Westall, F., Campbell, K.A., Breheret, J.G., Foucher, F., Gautret, P., Hubert, A., Sorieul, S., Grassineau, N., Guido, D.M., 2015b. Archean (3.33 Ga) microbe-sediment systems were diverse and flourished in a hydrothermal context. Geology 43, 615–618.

White, O.L., Safaeinili, A., Plaut, J.J., Stofan, E.R., Clifford, S.M., Farrell, W.M., Heggy, E., Picardi, G., Team, T.M.S., 2009. MARSIS radar sounder observations in the vicinity of Ma'adim Vallis, Mars. Icarus 201 (2), 460–473.

Wieczorek, M.A., 2008. Constraints on the composition of the martian south polar cap from gravity and topography. Icarus 196, 506–517. http://doi.org/10.1016/j.icarus.2007.10.026.

Wynn-Williams, D., Newton, E., Edwards, H., 2001. The role of habitat structure for biomolecule integrity and microbial survival under extreme environmental stress in Antarctica (and Mars?): ecology and technology. Exo-/astro-biology. Proceedings of the First European Workshop, 21 - 23 May 2001, ESRIN, Frascati, Italy. Eds.: P. Ehrenfreund, O. Angerer & B. Battrick. ESA SP-496, Noordwijk: ESA Publications Division, 2001, 225–237, ISBN 92-9092-806-9.

Xue, Y., Nicholson, W.L., 1996. The two major spore DNA repair pathways, nucleotide excision repair and spore photoproduct lyase, are sufficient for the resistance of *Bacillus subtilis* spores to artificial UVC and UV-B but not to solar radiation. Appl. Environ. Microbiol. 62, 2221–2227.

Yung, Y.L., Russell, M.J., Parkinson, C.D., 2010. The search for life on Mars. J. Cosmol. 5, 1121–1130.

Yung, Y.L., Chen, P., 2015. Methane on Mars. J. Astrobiol. Outreach 3, 3–5.

Zahnle, K., Arndt, N., Cockell, C., Halliday, A., Nisbet, E., Selsis, F., Sleep, N.H., 2007. Emergence of a habitable planet. Space Sci. Rev. 129, 35–78.

Zuber, M.T., Phillips, R.J., Andrews-Hanna, J.C., Asmar, S.W., Konopliv, A.S., Lemoine, F.G., Plaut, J.J., Smith, D.E., Smrekar, S.E., 2007. Density of Mars' south polar layered deposits. Science 317, 1718–1719. http://doi.org/10.1126/science.1146995pmid:17885129.

Bibliography

Heberling, C., Lowell, R.P., Liu, L., Fisk, M.R., 2010. Extent of the microbial biosphere in the oceanic crust. Geochem. Geophys. Geosyst. 11, Q08003.

Kounaves, S.P., Stroble, S.T., Anderson, R.M., Moore, Q., Catling, D.C., Douglas, S., McKay, C.P., Ming, D.W., Smith, P.H., Tamppari, L.K., Zent, A.P., 2010. Discovery of natural perchlorate in the Antarctic Dry Valleys and its global implications. Environ. Sci. Technol. 44, 2360–2364.

Martin, W.F., 2011. Early evolution without a tree of life. Biol. Direct. 6, 36.

Michalski, J.R., Cuadros, J., Niles, P.B., Parnell, J., Rogers, A.D., Wright, S.P., 2013. Groundwater activity on Mars and implications for a deep biosphere. Nat. Geosci., 133–138. http://doi.org/10.1038/ngeo1706.

Ramamurthy, C., et al., 2015. Delta Differential One-way Ranging (DDOR) Transmitter Onboard Mars Orbiter Mission (MOM), SPACES-2015, 458.

Rothschild, L.J., Mancinelli, R.L., 2001. Life in extreme environments. Nature 409, 1092–1101.

Whitman, W.B., Coleman, D.C., Wiebe, W.J., 1998. Prokaryotes: the unseen Majority. Proc. Natl Acad. Sci. USA 95, 6578–6583.

Chapter 11

Could near-Earth watery asteroid Ceres be a likely ocean world and habitable?

11.1 Near-Earth asteroids *Ceres* and *Vesta*—why did they become targets of NASA discovery-class mission, *Dawn*?

The near-Earth asteroids (NEAs) *Ceres* and *Vesta* are the two most massive members of the asteroid belt. At the time of its discovery on January 1, 1801, by the Italian astronomer Giuseppe Piazzi, *Ceres* was considered a planet, but upon the realization that it represented the first of a class of many similar bodies, it was reclassified as an asteroid for over 150 years. As the first such body to be discovered, Ceres was given the designation 1 Ceres or (1) Ceres under the modern system of asteroid numbering. Ceres is the smallest dwarf planet in the Solar System and the only one located in the main asteroid belt (MAB). Vesta (minor-planet designation: 4 Vesta) is one of the largest objects in the asteroid belt, with a mean diameter of 525 km (326 mi). Vesta was discovered by the German astronomer Heinrich Wilhelm Matthias Olbers on March 29, 1807.

McCord and Sotin (2005) made the following observation regarding the *Dawn* Mission initiative: "Ceres was discovered in January/February 1801 by Giuseppe Piazzi in Palermo, Italy, while looking for the planet predicted by the Titus-Bode law to be located at about 2.8 AU from the Sun (Piazzi, 1802; Encyclopædia Britannica, 2004). This was while a German team, following on the German Titus-Bode prediction, was also searching for the Ceres object. These German scientists instead found Vesta. Thus, it seems appropriate that German and Italian teams are major contributors to the DAWN mission. Ceres' history and its relevancy to the other terrestrial planets motivated NASA to select DAWN in December 2001, a mission to orbit Ceres and Vesta, as a Discovery Program mission to be flown." Summaries of the general knowledge of Ceres and its principal characteristics can be found in Parker et al. (2002) and Britt et al. (2002).

Castillo-Rogez and McCord (2010) made the following observation regarding the specialty of a class of asteroid-belt objects and why Ceres and Vesta received special attention, to the extent of warranting a spacecraft mission to these two celestial bodies, "Ceres, Vesta, and Pallas stand alone in the inner Solar System, defining a distinct class of objects (McCord et al., 2006). They are larger and more intact than asteroids, but smaller than the terrestrial planets. They are large enough to have attained a spheroid shape and experienced some planetary processes, but they are not called planets. Rather, the current name assigned to this class of objects is dwarf planets, although they often but inappropriately are referred to as asteroids or more appropriately as protoplanets. It is thought that, in the sequence of accretion from the Solar nebula, the terrestrial planets were formed from this class of objects (e.g., Canup and Agnor, 2000; Weidenschilling, 2006), but today only these three remain in the inner Solar System. Thus, Ceres and Vesta have become the targets of the NASA discovery-class mission, Dawn (Russell et al., 2006) (Pallas's inclined orbit requires too much energy for Dawn to reach)."

The interiors of asteroids *Vesta* (265-km mean radius) and Ceres (476-km mean radius) contain treasure troves of information regarding the range of processes that contributed to accretion of Solar System objects from terrestrial planets to ice-rich outer moons (see Konopliv et al., 2011). The *Dawn* mission was designed primarily to test the hypothesis about the origin and evolution of the early Solar System by visiting the largest differentiated basaltic asteroid, 4 Vesta, believed to be a survivor from the earliest times of rocky body formation (Russell et al., 2013).

Before the launch of *Dawn* mission, remote observations of these two dwarf-planet-sized asteroidal bodies have been made from ground-based telescopes and the *Hubble* Space Telescope. However, much closer observations of these asteroids for detailed studies demanded a dedicated mission. It is heartening that Dawn's decade-long journey has provided a close look at the largest objects in the main asteroid belt—Vesta and Ceres—and provided the first opportunity to investigate the interior structures of these two asteroids.

Russell et al. (2013) described Dawn and the Dawn mission thus, "Dawn is the ninth mission in the Discovery program, a series of principal-investigator-led missions exploring the Solar System. It is a unique scientific mission utilizing ion propulsion to leave the vicinity of the Earth and spiral outward to targets in the asteroid belt. When its heliocentric orbit matches that of its intended target, ion propulsion is used to slip into the body's gravitational potential well and descend until it reaches a desired mapping orbit. When it completes its mapping mission, it uses its ion thrusters to raise its orbital altitude and escape the body's gravity to achieve a heliospheric orbit around the Sun. The ion engine thrusts until reaching its next target, at which time it glides into orbit and begins mapping anew. Such a mission would be unaffordable using conventional technology, but through the power and economy of ion propulsion technology, *Dawn* is, in fact, the least expensive of the current Discovery missions." The mission is managed and operated by JPL/CALTECH (Jet Propulsion Laboratory/California Institute of Technology) with the overall mission science responsibilities from UCLA (University of California, Los Angeles).

Dawn is the first mission to orbit the most massive asteroids in MAB (Russell and Raymond, 2011; Russell et al., 2012, 2013). As just indicated, Dawn's targets were 4 Vesta and 1 Ceres. The name "Dawn" to this discovery mission originated from the belief that these asteroids are intact witnesses to the events at the "dawn" of the Solar System (Russell et al., 2013). The primary goal of this mission was to study the two minor planets (4) Vesta and (1) Ceres to understand the processes that occurred in the early Solar System (Russell and Raymond, 2011). Vesta, in particular, underwent magmatic differentiation and began to solidify as early as about 2 Myr after the first condensation of solids in the solar nebula (McSween and Huss, 2010).

The Dawn spacecraft shown in Fig. 11.1 is three-axis stabilized with large solar arrays to provide the power to run the ion engines as well as operate the flight system and payload. To map the surface of Vesta first and Ceres later, the spacecraft points its remote sensing instruments at the surface of the body. As it orbits, the spacecraft keeps its solar array pointed toward the Sun. Data are recorded on the spacecraft for later transmission to the Earth. The

FIG. 11.1 The *Dawn* **spacecraft with location of science instruments and antennas.** The spacecraft coordinate frame is shown; its origin is at the base-plate of the central ion engine. *(Source: Konopliv et al., 2011b; Space Sci. Rev., 163:461–486. DOI 10.1007/s11214-011-9794-8. Copyright info: © Springer Science+Business Media B.V. 2011)*

spacecraft has to pause in its data taking and point its high-gain antenna to Earth for this data transfer.

The spacecraft had three instruments onboard (Framing Camera [FC], Visible and Infrared Spectrometer [VIR], and Gamma Ray and Neutron Detector [GRaND]). These instruments were supplemented by the telemetry system that was used to obtain radiometric tracking data to define the gravitational field (Konopliv et al., 2011b). The Framing Camera was used to obtain observations of the surface from different viewing directions that were used to reconstruct the topography of the surface using stereophotogrammetry and stereo-photo-clinometry (Raymond et al., 2011; Jaumann et al., 2012; Preusker et al., 2012).

Of the three instruments, the Framing Camera was of particular interest and importance to the entire mission, because apart from the scientific objectives, this camera could serve the purpose of orbit navigation and control (Sierks et al., 2011).

The FC was developed and built at the Max Planck Institute for Solar System Research, Katlenburg-Lindau (now in Goettingen), with contributions from DLR/IPR (German Aerospace Centre/Institute for Planetary Science) and TUB/IDA (Technical University Braunschweig/Institute of Computer and Communication Network Engineering).

FC is equipped with seven color filters in the wavelength range from 0.4 to 1.0 µm, and a broadband clear filter (Sierks et al., 2011). The center wavelengths of the color filters are chosen to characterize major absorption bands of HED meteorites (Sierks et al., 2011). Note that HED meteorites are a clan of achondrite meteorites. HED stands for "howardite–eucrite–diogenite." While various FC color data band parameters (band curvature, band tilt, and band strength) are used for surface compositional analysis and

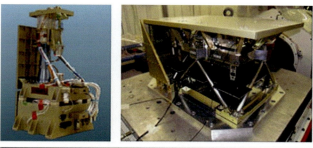
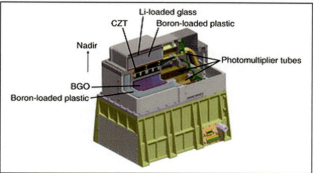

FIG. 11.2 (Upper left panel) The Framing Camera. (Upper right panel) The Visible and Infrared Mapping Spectrometer. (Bottom panel) The Gamma Ray and Neutron Detector *(Image credits, MPS/Selex Galileo/ LANL). (Source: Russell et al., 2013; Dawn completes its mission at 4 Vesta, Meteorit. Planet Sci., 48: 2076-2089). Copyright holder: © The Meteoritical Society, 2013. Publisher: Wiley-Blackwell on behalf of the Meteoritical Society.*

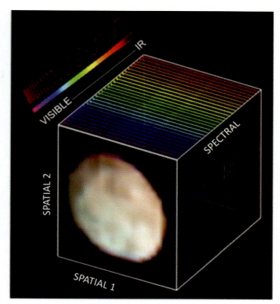

FIG. 11.3 **The VIR spectral cube**. The spectrometer obtains images at multiple wavelengths forming an image cube with two dimensions of spatial information and one dimension of spectral information. *(Source: Russell et al., 2013; Dawn completes its mission at 4 Vesta, Meteorit. Planet Sci., 48: 2076-2089). Copyright holder: © The Meteoritical Society, 2013. Publisher: Wiley-Blackwell on behalf of the Meteoritical Society.*

mapping (Reddy et al., 2012a, 2012b; Thangjam et al., 2013, 2014; Nathues et al., 2014, 2015; Le Corre et al., 2011, 2015; Schäfer et al., 2016), the clear data is used for morphological analysis and geologic mapping. The band parameters are derived by investigating the laboratory spectra. A view of the Framing Camera, the Visible and Infrared Mapping Spectrometer, and the Gamma Ray and Neutron Detector is shown in Fig. 11.2.

The VIR is a hyperspectral instrument operating in the wavelength range from near ultra-violet to infra-red (0.2–5 μm). It has two data channels, the visible channel (0.25–1.07 μm) and the infrared channel (0.95–5.1 μm) (De Sanctis et al., 2011, 2012; Ammannito et al., 2013). The main objective of VIR is to map and analyze surface mineralogy/composition, thermal behavior of the surface and their interaction with the extreme space weather in an atmosphere-free environment (De Sanctis et al., 2011). Fig. 11.3 illustrates the VIR spectral cube. The spectrometer obtains images at multiple wavelengths forming an image cube with two dimensions of spatial information and one dimension of spectral information.

It may be noted that gamma rays and neutrons are produced by the steady bombardment of the regolith of Vesta and Ceres by galactic cosmic rays. Gamma rays are also made by the decay of long-lived radio-elements such as potassium (K), thorium (Th), and uranium (U) (Prettyman et al., 2017). A portion of the gammas and neutrons escape the surface and are detected by the Gamma Ray and Neutron Detector (GraND) in orbit. GraND is also able to derive information on surface elemental abundances in oxide and silicate minerals such as O, Mg, Al, Si, K, Ca, Ti, Fe, Ni, U, Th, ices (H, C, N) and volcanic exhalation or aqueous alteration products (Prettyman et al., 2011).

The *Dawn* mission was launched in September 2007. The fourth asteroid to be discovered, *Vesta*, was explored by the *Dawn* spacecraft for more than a year (Russell et al., 2015). One of the primary scientific goals of the historic Dawn mission to (4) Vesta was indeed the realization that the study of Vesta is essential in understanding the evolution of planetary bodies in the early Solar System. Dawn spacecraft was in orbit around Vesta for more than a year (July 2011–September 2012) and imaged the entire surface.

Fig. 11.4 illustrates the comparative sizes of Ceres and Vesta in relation to those of Earth's Moon, and the planets Mercury and Mars. The investigations using Dawn spacecraft images mark a significant progress because of its higher spatial and spectral resolution data compared to all former observations.

As shown in Fig. 11.3, VIR provides an image cube of data over a broad range of wavelengths. The absorption bands are key to constraining the forms of pyroxene present and allowing mineral maps to be made (Russell et al., 2013). The spatial resolution of FC color data is about 2.3 times higher than the resolution of VIR cubes (Sierks

FIG. 11.4 Vesta and Ceres with Earth's Moon, and the planets Mercury, Mars, counter clockwise from bottom. While Vesta is found to be a much simpler body than the Moon, Mercury, and Mars, and Ceres is expected to be much simpler as well, many of the processes appearing on the larger terrestrial planets are present in primitive forms on Vesta. *(Source: Russell et al., 2013; Dawn completes its mission at 4 Vesta, Meteorit. Planet Sci., 48: 2076-2089). Copyright holder: © The Meteoritical Society, 2013. Publisher: Wiley-Blackwell on behalf of the Meteoritical Society.*

et al., 2011). Higher resolution images obtained by the onboard instruments allowed planetary researchers to analyze and study the mineralogy and geology of this body.

11.2 An overview of the dwarf planet *Ceres*

Observational campaigns for more than 3 years have made Ceres one of the best explored ice-rich bodies. These observations have revealed extensive chemical and geological activity, likely until very recent times, when the age of Occator's bright spots (called faculae) is considered (Castillo-Rogez et al., 2020). Note that Occator is an impact crater located on *Ceres*, that contains "Spot 5," the brightest of the bright spots observed by the *Dawn* spacecraft. There have been considerable controversies with regard to the status of Ceres (i.e., whether or not Ceres is a planet, an asteroid, or a dwarf planet). McCord and Sotin (2005) made the following statements on Ceres: "Ceres orbits the Sun and is large enough in size to have experienced many of the processes normally associated with planetary evolution. Therefore, it should be called a planet. Its location, in the middle of the asteroid belt, has caused it to be referred to as an asteroid. However, its size, orbit and general nature, as best we can discern it, suggest that it is much more interesting than the small pieces of larger objects (perhaps such as Ceres) that one normally thinks of when one refers to asteroids. Ceres' planet-like nature and its survival from the earliest stages of Solar System formation, when its sister/brother objects probably became the major building blocks of the Earth and the other terrestrial inner planets, makes it an extremely important object for understanding the early stages of the Solar System as well as basic planetary processes." However, Ceres is also referred to as a "dwarf planet."

Ceres is the largest object in the asteroid belt. This celestial body is classified as both a dwarf planet and an asteroid. As indicated earlier, Ceres in the inner Solar System was discovered by G. Piazzi on 1 January 1801, between the orbits of Mars and Jupiter at the heliocentric distance expected for the missing planet predicted by the Titius-Bode law (Peebles, 2000); this justifies Ceres' classification as a dwarf planet. Ceres is the largest near-Earth object in the asteroid belt; this justifies Ceres' classification as an asteroid. Ceres is not associated with a family of asteroids or known meteorites.

The Ceres' surface materials are identified as aqueous altered material such as clays and hydrated salts similar to CI and CM, that is, primitive, CC meteorites. The dominant minerals on the surface of Ceres are therefore hydrated clay minerals structurally similar to terrestrial montmorillonites (McCord and Sotin, 2005). Note that Montmorillonite is a very soft phyllosilicate group of minerals that form when they precipitate from water solution as microscopic crystals, known as clay. It is named after Montmorillon in France. It has been pointed out that the salts are products of aqueous alteration and may also be responsible for Ceres' high albedo relative to other C-type asteroids. Furthermore, the spectra of C-type asteroids in general may be reconcilable with those of carbonaceous chondrites if the asteroids' surfaces have undergone alteration by aqueous or other analogous processes (cf. Jones et al., 1988, 1990). From the spectral reflectance, the low density and the location in the Solar System, it is thought that Ceres was formed mostly of unaltered carbonaceous chondrite source material with water ice.

The typically dark surface of the dwarf planet *Ceres* is punctuated by areas of much higher albedo, most prominently in the Occator crater. These small bright areas have been tentatively interpreted as containing a large amount of hydrated magnesium sulfate, in contrast to the average surface, which is a mixture of low-albedo materials and magnesium phyllosilicates, ammoniated phyllosilicates and carbonates (King et al., 1992). In another school of thought, Ceres has been thought to be a differentiated body composed primarily of silicate materials and water ice. However, Rivkin et al. (2006) reported observational evidence that carbonates and iron-rich clays are present on the surface of Ceres. These components are also present in CI chondrites (a group of rare stony meteorites belonging to the carbonaceous chondrites) and provide a means of explaining the unusual spectrum of this object as well as providing potential insight into its evolution. It was found that some remotely observed features also indicate that

Ceres may have a composition more similar to that of the most common types of carbonaceous meteorite. The specific mineral composition of Ceres and its relationship to known meteorite mineral assemblages, however, remained uncertain (Milliken and Rivkin, 2009).

It is practically difficult for liquid water to be present on Ceres' surface. The lack of atmosphere (hence no atmospheric pressure) on Ceres means that sunlight causes liquid to evaporate away immediately (the lower the atmospheric pressure the higher the evaporation rate), leaving behind the rest of the minerals carried with it to the surface—such as carbonates. Therefore, carbonates have typically been used to infer the occurrence of liquid water. The carbonate signature in Ceres' spectrum (3.3–4.0 µm) was first detected on Earth by Rivkin et al. (2006) and subsequently confirmed by Milliken and Rivkin (2009). Based on this finding Milliken and Rivkin (2009) suggest the presence of OH- or H_2O-bearing phases on Ceres.

Studies of Ceres using ground-based and orbiting telescopes concluded that its closest meteoritic analogues are the volatile-rich CI and CM carbonaceous chondrites. Water in clay minerals, ammoniated phyllosilicates, or a mixture of brucite {$Mg(OH)_2$}, magnesium carbonate (Mg_2CO_3), and iron-rich serpentine have all been proposed to exist on the surface. In particular, brucite has been suggested from analysis of the mid-infrared spectrum of Ceres. Milliken and Rivkin (2009) showed that the spectral features of Ceres can be attributed to the presence of the hydroxide brucite, magnesium carbonates and serpentines, a mineralogy consistent with the aqueous alteration of olivine-rich materials. They therefore suggested that the thermal and aqueous alteration history of Ceres is different from that recorded by carbonaceous meteorites, and that samples from Ceres are not represented in existing meteorite collections. But the lack of spectral data across telluric (of the soil) absorption bands in the wavelength region 2.5–2.9 µm—where the OH stretching vibration and the H_2O bending overtone are found—has precluded definitive identifications.

Takir et al. (2013) found that the carbonate absorption in the Ceres spectrum is deeper than in so-far measured spectra of carbonaceous chondrites (CC), thereby suggesting a larger abundance of carbonates on Ceres than in typical CC. Note that carbonaceous chondrites are a class of chondritic meteorites comprising at least 8 known groups and many ungrouped meteorites. They include some of the most primitive known meteorites.

De Sanctis et al. (2015) reported spectra of Ceres from 0.4 to 5 µm acquired at distances from ~82,000 to 4300 km from the surface. Their measurements indicated widespread ammoniated phyllosilicates across the surface, but no detectable water ice. These investigators suggested that material from the outer Solar System was incorporated into Ceres, either during its formation at great heliocentric distance or by incorporation of material transported into the main asteroid belt. In later studies, De Sanctis et al. (2015) reconfirmed the presence of carbonate on Ceres through *Dawn* observations, showing that Ceres' average surface is an assemblage of Mg phyllosilicates, ammoniated species, dark materials, and Mg–Ca carbonates.

The NASA's *Dawn* spacecraft, which has been in orbit around Ceres since March 6, 2015, has played an important role in shedding most of the current knowledge acquired about Ceres. When the Dawn spacecraft arrived at Ceres in 2015, it found a nearly featureless world with a rocky surface. A single mountain, Ahuna Mons, rose from the surface. Scientists have found ample hints that Ahuna Mons was not just a mountain, but a cryovolcano—also known as an ice volcano. It was suggested that the many domes (tholi, montes) found on Ceres' surface (Sizemore et al., 2019) could be formed from episodes of cryovolcanic activity that lasted for several billion years up until the present (Sori et al., 2018). It is suspected that thick water ice may have risen to the surface in the past, flooding the plains with ice lava. Bright spots within craters were the only color variations on the otherwise dull, gray world. Remote sensing with telescopes at 1 astronomical unit (AU) and probing with Earth-based radar have indicated that Ceres adorns a clay-like surface (Webster et al., 1988; Rivkin et al., 2011).

The small bright areas detected on the surface of Ceres have been a focus of attention ever since their detection. De Sanctis et al. (2016) reported high spatial and spectral resolution near-infrared observations of the bright areas in the Occator crater on Ceres. According to them, "Spectra of these bright areas are consistent with a large amount of sodium carbonate, constituting the most concentrated known extraterrestrial occurrence of carbonate on kilometer-wide scales in the Solar System. The carbonates are mixed with a dark component and small amounts of phyllosilicates, as well as ammonium carbonate or ammonium chloride." De Sanctis et al. (2016) found that the compounds are endogenous and they proposed that they (the compounds) are the solid residue of crystallization of brines and entrained altered solids that reached the surface from below. De Sanctis et al. (2016) reckon that "the heat source may have been transient (triggered by impact heating). Alternatively, internal temperatures may be above the eutectic temperature of subsurface brines, in which case fluids may exist at depth on Ceres today." Note that "eutectic temperature" is the temperature at which a particular eutectic mixture (a homogeneous mixture of substances that melts or solidifies at a single temperature that is lower than the melting point of any of the constituents) freezes or melts.

A global spectrophotometric characterization of Ceres achieved through analysis of images (Schröder et al., 2017) obtained from a 5.5° × 5.5° field of view (FOV) CCD Framing Camera (FC) (Sierks et al., 2011) and observations by the onboard visible and infrared spectrometer (VIR) have revealed a dark world with little color variation and several enigmatic bright areas in the Occator crater (De Sanctis et al., 2015; Nathues et al., 2015) on Ceres. The surface shows evidence for aqueous alteration. It has been found that ammoniated phyllosilicates are widespread (Ammannito et al., 2016). More importantly, it was found that small, isolated patches of water ice lie exposed on the surface (Combe et al., 2016). The first results obtained from the analysis of data acquired by a gamma ray and neutron detector (GRaND) instrument—flown in the close proximity to the surface of Ceres in the low altitude orbit at the end of the Ceres mission—point at the widespread presence of water ice below the surface, mostly at higher latitudes (Prettyman et al., 2016). In a recent study, Raponi et al. (2018) reported the detection of a seasonal water cycle in a mid-latitude crater on Ceres.

Carrozzo et al. (2018) found that the distribution and chemical composition of the carbonates are tracers of complex and varying aqueous chemical processes on Ceres. From the global maps, Carrozzo et al. (2018) concluded that carbonates are present on the whole observed surface of Ceres, and the spectra of most of the terrains are consistent with Mg–Ca carbonates. Furthermore, according to Ammannito et al. (2016), Ceres' overall composition and the ubiquitous distribution of the OH- and NH_4-bearing species indicate a pervasive, past aqueous alteration involving ammoniated fluids and Mg phyllosilicate formation.

Using a visible and infrared mapping spectrometer (VIR)—the Dawn imaging spectrometer—Carrozzo et al. (2018) detected different carbonates on Ceres, and mapped their abundance and spatial distribution. According to Carrozzo et al. (2018), "Carbonates are abundant and ubiquitous across the surface, but variations in the strength and position of infrared spectral absorptions indicate variations in the composition and amount of these minerals. Mg–Ca carbonates are detected all over the surface, but localized areas show Na carbonates, such as nitrite (Na_2CO_3) and hydrated Na carbonates (e.g., $Na_2CO_3 \cdot H_2O$). Their geological settings and accessory NH_4-bearing phases suggest the upwelling, excavation, and exposure of salts formed from Na-CO_3-NH_4-Cl brine solutions at multiple locations across the planet." Carrozzo et al. (2018) reckon that "the presence of the hydrated carbonates indicates that their formation/exposure on Ceres' surface is geologically recent and dehydration to the anhydrous form (Na_2CO_3) is ongoing, implying a still-evolving body."

11.3 Indications favoring the presence of water on *Ceres*

The silicate and water ice mass proportions in the material available for forming bodies like Ceres early in the history of the Solar System is thought to be something like 75% and 25%, respectively (cf. Wilson et al., 1999; Grimm and McSween, 1989). This suggests that most of the water present at formation remains in some form within Ceres (McCord and Sotin, 2005).

Absorption feature observed in the spectrum of Ceres gave a credible indication of the possible presence of water on Ceres. In this respect, of great interest was the first report of a 3-μm absorption in the spectrum of Ceres (Lebofsky, 1978, 1980) interpreted to be due to OH and perhaps H_2O. Lebofsky et al. (1981) presented further observations of Ceres in the 1.7–4.2-μm range. They confirmed and elaborated on the 3-μm absorption and concluded that the strong absorption at 2.7–2.8 μm is due to structural OH groups in clay minerals. They also reported a narrow absorption feature at 3.1 μm and interpreted it as being due to a very small amount of water ice on Ceres, making this the first evidence suggested for ice on the surface of an asteroid (McCord and Sotin, 2005). Based on comparison of observational data for Ceres and laboratory measurements of some hydrated materials, Feierberg et al. (1981) inferred that the 3-μm feature is largely due to interlayer water molecules in clay minerals with a possible contribution from water molecules bound to salts.

In another study, spectroscopic observations of emissions from Ceres, obtained using the International Ultraviolet Explorer (IUE) satellite in Earth orbit to observe Ceres (A'Hearn and Feldman, 1992), were found to be consistent with the emission feature due to the O–O band of OH molecule, which is usually seen in cometary spectra. A'Hearn and Feldman (1992) calculated a possible production of H_2O from Ceres in the range of 10^{24}–10^{25}/s to explain the observations. They point out that this is consistent with the estimates of Fanale and Salvail (1989) for a flux that could be sustained over geologic time from subsurface ice layer supplied from below (see McCord and Sotin, 2005). Ceres' low density (near 2.1) indicates that considerable water is currently present, as does the association of the surface material with altered Cc meteorite types and the presence of OH and perhaps H_2O. Taking into account the density of a simple mixture of water and silicate rock, McCord and Sotin (2005) estimated that Ceres currently could be between approximately 17% and 27% water by mass.

Results obtained from theoretical models (see McCord and Sotin, 2005) suggest that Ceres likely retained much of its original water but this water became redistributed inside Ceres and a significant portion probably was taken up in the mineralization processes. Thus, Ceres may be in a similar state as is Ganymede (Jupiter's moon, which is the largest moon in our Solar System, bigger than the planet Mercury and the dwarf planet Pluto) or Callisto (Jupiter's second largest moon after Ganymede and it is the third largest moon in our Solar System) but with slightly less water. Europa (the smallest of the four Galilean moons orbiting Jupiter) and to some extent Ganymede have an additional energy source (tidal), with Europa being much more evolved. Callisto had little or no tidal heating and so may have evolved more like

Ceres, but Callisto either had more water at its origin (further out in the Solar System) or lost less of it (larger).

According to McCord and Sotin (2005), "Even if only long-lived radionuclide heating is assumed, the water ice in Ceres melts quickly and a water mantle forms, but an approximately 10-km crust does not melt. The circulating warm water would alter the silicates. As heat is lost by conduction through the frozen crust, water begins to freeze out at the base of the crust. Solid-state convection begins and transports more heat as well as perhaps material dissolved or entrained in the water to or near the surface. Ceres' water layer eventually (but perhaps not entirely) freezes, forming a layered density structure with perhaps some liquid water remaining today." According to Russell et al. (2016), "The possibility of abundant volatiles at depth is supported by geomorphologic features such as flat crater floors with pits, lobate flows of materials, and a singular mountain that appears to be an extrusive cryovolcanic dome." The size, combined with a mass estimate based on its gravitational interaction with Mars (Konopliv et al., 2011), indicate that Ceres' average density (bulk density) is about 2100 kg/m^3. It is pertinent to note that this density is between ice and rock, and planetary scientists have used this clue in inferring the presence of water on Ceres much before water has actually been detected on Ceres.

As indicated earlier, from the estimated bulk density of Ceres, thermodynamic models (McCord and Sotin, 2005; Castillo-Rogez and McCord, 2010) predicted that Ceres contained 17–27% free water by mass, and therefore it likely had fractionated into a silicate core and a water-rich mantle. Castillo-Rogez and McCord (2010) modeled Ceres' thermo-physical-chemical evolution by considering a large range of initial conditions as well as various evolutionary scenarios. Their models are constrained by available shape measurements, which point to a differentiated interior for Ceres. They addressed the role played by hydrothermal activity in the long-term evolution of Ceres and especially the evolution of its hydrosphere. Castillo-Rogez and McCord (2010) evaluated the conditions for preserving liquid water inside Ceres, a possibility enhanced by its warm surface temperature and the enrichment of its hydrosphere in a variety of chemical species.

According to Russell et al. (2016), "These inferences of a 'wet' dwarf planet were supported by a possible detection of OH with the International Ultraviolet Explorer (A'Hearn and Feldman, 1992), but could not be verified from ground-based telescopes. More recently, the Herschel Space Observatory confirmed the presence of water vapor molecules at Ceres with a source rate of about 6 kg/s (Küppers et al., 2014)." Ceres' crater distribution (Hiesinger et al., 2016) and geomorphology (Buczkowski et al., 2016) support the inference of a mechanically strong crust and upper mantle composed of rock, ice, and possibly salt hydrates that is periodically mobilized to produce extrusive features

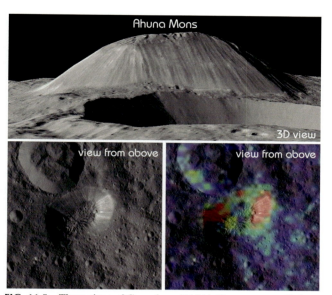

FIG. 11.5 These views of Ceres from NASA's Dawn spacecraft show Ahuna Mons (located at 316.2°E, 10.5°S)—the tallest mountain on the dwarf planet Ceres. The mountain is one of the few places on Ceres with significant amount of sodium carbonate. (*Credit: NASA/JPL-Caltech/UCLA/MPS/DLR/IDA/ASI/INAF*).

such as Ahuna Mons (Fig. 11.5), as well as flow features and bright deposits.

According to Combe et al. (2016), because Ceres has a low bulk density "its composition is expected to contain a substantial proportion of water. Internal heat (mostly from decay of radionuclides) is expected to have differentiated Ceres at least partially into a rocky core and icy mantle (McCord and Sotin, 2005; Thomas et al., 2005; Castillo-Rogez and McCord, 2010). Contact between silicates and ice or liquid water would have resulted in aqueous alteration of minerals, producing hydroxylated (OH-bearing) phases such as clays and carbonates (McCord and Sotin, 2005; Thomas et al., 2005; Rivkin et al., 2006; Milliken and Rivkin, 2009; Castillo-Rogez and McCord, 2010; De Sanctis et al., 2016; Ammannito et al., 2016), or mineral hydrates (H_2O-bearing) such as salts." In other words, it has also been hypothesized that water is likely a key component in the chemical evolution and internal activity of Ceres, possibly resulting in a layer of ice-rich material and perhaps liquid in the mantle. Another hypothesis put forward by planetary researchers was that mineral hydroxides (OH-bearing) and hydrates (H_2O-bearing), such as clays, carbonates, and various salts, would be created.

11.4 Application of reflectance spectroscopy for remote detection of water molecules on *Ceres*

Reflectance, in the present context, defines the ratio of irradiance from the surface of Ceres to the solar radiance flux.

According to Dozier et al. (2009), "The depth of photon penetration depends on the material: Photons may penetrate a few micrometers into absorbing components such as those found on Ceres or other low-albedo surfaces, whereas they may penetrate several centimeters in highly scattering materials such as pure ice and snow." Reflectance spectroscopy is known to be sensitive to surficial water and hydroxyl groups. It has been recognized that measurement of OH and H_2O absorptions near 1.5, 2.0, and 3 μm allows for remote detection of water (H_2O) molecules.

Detection and mapping of water on Ceres was one major objective of the NASA's *Dawn* spacecraft in orbit around Ceres since March 2015. Realizing this objective Dawn spacecraft indeed succeeded in remote monitoring of several features of Ceres by means of high-resolution imagery and spectrometry. The Dawn Visible and InfraRed (VIR) mapping spectrometer (De Sanctis et al., 2011) senses surface composition by reflectance spectroscopy. It covers a range of wavelengths between 0.25 and 5.1 μm with two detectors that measure radiance, overlap around 1 μm, and have spectral sampling of 1.9 nm and 9.8 nm for the visible and infrared, respectively.

According to Combe et al. (2016), "The position and shape of absorption features in VIR reflectance spectra are sensitive to the surface mineral and molecular composition. In spectroscopy, absorption bands at 2.0, 1.65, and 1.28 μm are characteristic of vibration overtones in the H_2O molecule." Based on reflectance spectroscopy measurements De Sanctis et al. (2016) observed products of aqueous activity such as bright carbonates on Ceres.

11.5 Clues leading to the presence of water on the surface of *Ceres*

The prevailing hypotheses pertaining to *Ceres* were supported by the detection of hydroxyl (OH)-rich materials, OH-bearing molecule releases, H_2O vapor molecules, and haze. Combe et al. (2016) described the reasons for optimism by planetary scientists about the possible presence of water on Ceres thus, "The state and abundance of water on the dwarf planet Ceres have been major questions ever since hydroxyl (OH)-rich materials were proposed as surface components (Chapman and Salisbury, 1973; McCord and Gaffey, 1974) and eventually observed (Lebofsky et al., 1981; Vernazza et al., 2005; Rivkin et al., 2006; Milliken and Rivkin, 2009; De Sanctis et al., 2015; Ammannito et al., 2016). The prediction of ice and perhaps liquid in the mantle (McCord and Sotin, 2005; Thomas et al., 2005; Castillo-Rogez and McCord, 2010), observations of sporadic releases of OH-bearing molecules (A'Hearn and Feldman, 1992), the detection of water vapor molecules (Küppers et al., 2014), and reports of haze (Nathues et al., 2015) support the presence of water on Ceres." However, the presence of liquid water on the surface was yet to be confirmed.

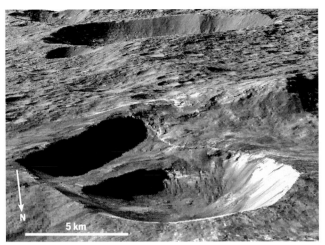

FIG. 11.6 **Perspective view of Oxo crater on Ceres (centered at 359.7°E and 42.2°N) observed by the *Dawn* Framing Camera (FC), where the two high-albedo areas right next to the scarps contain H_2O-rich materials.** Note that albedo means the proportion of the incident light or radiation that is reflected by a surface of a celestial body. *(Source: Combe et al. (2016), Detection of local H_2O exposed at the surface of Ceres, Science, 353 (6303), aaf3010; DOI: http://doi.org/10.1126/science.aaf3010)*

Combe et al. (2016)'s studies indicated the presence of water-rich surface materials in a 10-km-diameter crater named Oxo (Fig. 11.6), which exhibit all absorption bands that are diagnostic of the H_2O molecule. These spectra are most similar to those of H_2O ice, but they could also be attributable to hydrated minerals. Oxo crater has sharp rims and its floor is almost devoid of impacts, suggesting a recent exposure of surface H_2O. The high latitude and morphology of the Oxo crater protects much of the surface area from direct solar illumination for most of the cerean day, presenting favorable conditions for the stability of water ice or heavily hydrated salts (Combe et al., 2016).

Combe et al. (2016) reported VIR detections of H_2O-rich surface materials on *Ceres* from five observations of the crater Oxo, which is 10 km in diameter and is centered at 359.7°E, 42.2°N (Roatsch et al., 2016). The data were collected on 1 and 4 May 2015, with a spatial resolution of about 3.4 km per pixel, and on 6 and 9 June 2015, with a spatial resolution of about 1.1 km per pixel. To interpret the composition of the H_2O-rich area in Oxo, Combe et al. (2016) compared VIR spectra with laboratory spectra of several candidate materials by means of spectral modeling (Combe et al., 2008). According to Combe et al. (2016) an areal mixture is the simplest model, and it can be calculated by a linear combination of spectra.

According to Combe et al. (2016), "Candidate materials (McCord and Sotin, 2005; Castillo-Rogez and McCord, 2010) that exhibit absorption bands similar to VIR spectra of Oxo are free H_2O ice and H_2O-bearing minerals (mineral hydrates) such as chloride hydrates, magnesium sulfate hydrates ($MgSO_4 \cdot nH_2O$), ammonium sulfate hydrates ($NH_4SO_4 \cdot nH_2O$), sodium sulfate hydrates ($Na_2SO_4 \cdot nH_2O$), and sodium carbonate hydrates ($Na_2CO_3 \cdot nH_2O$), where n is the number of H_2O molecules per molecule of the mineral. Similar to the rest of Ceres' surface, a strong and broad absorption around 3 μm due to the fundamental vibration mode of O–H has also been observed in the study of Combe et al. (2016). However, according to Combe et al. (2016) this spectral feature alone is not unique to H_2O, as only the OH group is required. Spectral models show that H_2O-ice spectra fit VIR observations better than any spectrum of mineral hydrates." Thus, Combe et al. (2016) consider that water-ice is the most likely surface component observed in the Oxo crater.

Combe et al. (2016) provided the following clarifications and explanations with regard to the spectral features observed in their spectrometer measurements, "The spectral feature centered at 1.5 μm might also be due to O–H vibration overtone; however, it must be interpreted cautiously because of light scattering between two order-sorting filters of the instrument (De Sanctis et al., 2015), and thus it is not considered in this study. Although the diagnostic absorption feature of H_2O at 1.65 μm falls near one of these junctions, it is considered safe for interpretation. Several pieces of observational evidence indicate that instrument light scattering does not contribute significantly to the signal at this wavelength: (i) Light scattering creates spectral oscillations that stop around 1.6 μm, even for the strongest artifacts. (ii) The absorption band depths at 1.28 μm, 1.65 μm, and 2.0 μm have very similar distributions, consistent with the hypothesis that H_2O absorption is the main contributor of the 1.65-μm feature. (iii) The band depth parameter at 1.5 μm is mostly sensitive to albedo-contrasted surface features, which is radically different from the distribution pattern of absorptions attributed to H_2O."

Combe et al. (2016) explain that "Absorptions near 2 μm at the surface of Ceres require H_2O molecules, as they are attributable to an O–H stretch overtone in this molecule (Ockman, 1957; Clark, 1981) commonly seen in spectra of water ice and mineral hydrates (Clark and Lucey, 1984). Other O–H vibration overtones at 1.65 μm and 1.28 μm produce narrower yet meaningful absorption bands (Grundy and Schmitt, 1998) that are visible in the VIR spectra."

For interpreting the observations, Combe et al. (2016) took into account the following four ways to create or transport H_2O on Ceres: "(i) Exposure of near-surface H_2O-rich materials by a recent impact or an active landslide seems most consistent with the presence of both mineral hydrates and water ice. (ii) Release of subsurface H_2O may occur on Ceres, similar to release on comet nuclei, but may never recondense on the surface. (iii) Infall of ice-bearing objects is not likely to deposit water on Ceres, because the H_2O molecule likely would dissociate upon impact. (iv) Implantation of protons from the solar wind on the surface is not a probable origin of OH on Ceres because of the low flux of solar wind charged particles." Based on these considerations, Combe et al. (2016) concluded that surface water or hydrated minerals are the most plausible explanation.

According to Combe et al. (2016) pure surficial H_2O ice (i.e., exposed H_2O ice) would sublime and thus become optically undetectable within tens of years under current thermal conditions on Ceres (Fanale and Salvail, 1989; Hayne and Aharonson, 2015; Titus, 2015), where daytime surface temperatures T span the range 180–240 K (Ammannito et al., 2016), with $T > 200$ K found in Oxo crater from 11 h to 12 h local solar time; consequently, only a relatively recent exposure or formation of H_2O would explain *Dawn*'s findings. Some mineral hydrates are stable on geological time scales, but their formation would imply extended contact with ice or liquid H_2O (Combe et al., 2016).

The Gamma Ray and Neutron Detector (GRaND) (Prettyman et al., 2011) data indicate that the top meter of the regolith contains only about 10 wt% water ice (assuming 20% porosity), when averaged over broad spatial scales. Water ice is expressed in small-scale regions on Ceres' surface, in association with impact craters and mass wasting. Within Oxo crater, km-scale patches of water ice are associated with slumping regions that may have recently exposed ice from the near subsurface (Combe et al., 2017). Nine additional surface exposures of water ice (<7 km^2 total) have also been discovered at latitudes $>30°$ and in similar geologic contexts: they occur in fresh impact craters and are often associated with mass wasting features (Combe et al., 2017). Note that "mass wasting" is the movement of rock and soil down slope under the influence of gravity. Rock falls, slumps, and debris flows are all examples of mass wasting. Often lubricated by rainfall or agitated by seismic activity, these events may occur very rapidly and move as a flow. "Mass wasting" differs from other processes of erosion in that the debris transported by mass wasting is not entrained in a moving medium, such as water, wind, or ice.

Castillo-Rogez et al. (2020) recall that "Ceres' near-infrared spectrum, as observed by VIR, is best fit by a mixture of ammonium-bearing phyllosilicates, magnesium-bearing phyllosilicates, carbonates, and a dark, spectrally featureless component whose nature is unknown (De Sanctis et al., 2015). The surface mineralogy of Ceres can be reproduced qualitatively with geochemical models that include the interaction of liquid water containing carbon- and nitrogen-rich volatiles with silicates and organics of chondritic composition at temperatures below 50°C (Neveu et al., 2017), consistent with conditions expected in Ceres' early ocean (e.g., Travis et al., 2018), and a water to rock ratio >2 (Castillo-Rogez et al., 2018)."

Dawn observations have confirmed that Ceres' rocky material has been extensively processed by liquid water on a global scale (De Sanctis et al., 2015; Ammannito et al., 2016; Prettyman et al., 2017). Abundant mineral products of aqueous alteration are exposed on Ceres' surface, providing detailed insight into the chemistry of past and perhaps present liquid water environments (Castillo-Rogez et al., 2020).

11.6 Inferring presence of water in the subsurface of *Ceres*

Fanale and Salvail (1989) examined the water regime of *Ceres* and inferred that the high-latitude surface of Ceres is sufficiently cold that water ice can survive within a meter of the surface over Ceres' 4.5-billion-year lifetime. Studies carried out by McCord and Sotin (2005) and De Sanctis et al. (2015) indicated that with a measured bulk density between ice and rock, dwarf planet Ceres is expected to be rich in volatiles, with about 17–30 weight % (wt%) water ice. Based on Ceres' shape data (Castillo-Rogez and McCord, 2010), geochemical and thermodynamic considerations (Neveu and Desch, 2015), and Ceres' gravity field and shape data (Park et al., 2016), it has been inferred that internal heating generated by the decay of radioactive elements within the asteroid Ceres may have driven aqueous alteration and a partially differentiated interior, resulting in the formation of a rocky interior and an outer shell (crust) composed of a mixture of rock and ice. Park et al. (2016) further argued that physical differentiation allowed by gravity measurements (which Prettyman et al. (2017) described as "ice-rock fractionation") may have resulted in chemical fractionation, with surface regions that are not compositionally representative of the bulk. Analysis and interpretation of data obtained from orbital measurements by *Dawn*'s Visible and Infrared Mapping Spectrometer (VIR) show that Ceres' global surface contains aqueous alteration products such as ammoniated clays, serpentine, and carbonates (De Sanctis et al., 2015). Combe et al. (2016) reported detection of localized deposits of surficial water ice on Ceres but such deposits have been found to be rare.

Gamma Ray and Neutron Detector (GRaND) (Prettyman et al., 2011) is sensitive to elemental composition. Taking advantage of this capability of GRaND, Prettyman et al. (2017) used gamma-ray and neutron spectroscopy from NASA's *Dawn* spacecraft to peer below Ceres' surface and map the subsurface composition. The nuclear spectroscopy data so acquired allowed determination of the concentrations of elemental hydrogen [H], iron [Fe], and potassium [K] on Ceres. According to Prettyman et al. (2017), "measurements of hydrogen and iron provide constraints on the abundance of aqueous alteration products, assuming that the mineralogy is similar to carbonaceous chondrites. Aqueous alteration results in the separation of feed materials into a briny liquid and less-mobile solid residue. Elements such as Fe would be found primarily in the solid phase, whereas K and C are partially soluble and could be transported in the brine."

On Ceres, hydrogen is predominantly in the form of water ice and hydrated minerals (De Sanctis et al. 2015; Combe et al. 2016; Ammannito et al., 2016). Consequently, Prettyman et al. (2017) simulated changes in the hydration state of regolith materials by adding or removing hydrogen in the form of water to or from model compositions. According to Prettyman et al. (2017), "This allows hydrogen concentration to be expressed in terms of water-equivalent hydrogen (WEH). The regolith was modeled as a two-layer structure, with an ice-free upper layer covering icy soil, consistent with ice-stability models (Schorghofer, 2016). Both layers have the same nonicy composition, with the lower layer containing an additional, variable fraction of water ice." According to Prettyman et al. (2017) an estimate of [H] in the ice table can be obtained by subtracting the equatorial [H] from the values measured at the poles. Using this approach, Prettyman et al. (2017) found that Ceres' ice table contains about 10 wt% water ice, given that the ice table is within a centimeter of the surface at the poles. Prettyman et al. (2017) reckon that the water ice is most likely sourced from endogenic (i.e., formed or occurring beneath the surface of the earth) liquid not consumed by alteration processes.

According to Prettyman et al. (2017) the detection of widespread water ice at mid-to-high latitudes confirms theoretical predictions that ice can survive for billions of years within a meter of the surface. The nuclear spectroscopy data showed that surface materials were processed by the action of water within the interior. They found evidence for water ice across the dwarf planet, with water making up a larger fraction of the material near the poles than around the equator. Water vapor diffusivity and soil thermal properties determine where the ice table approaches the surface within depths sensed by GRaND. According to Schorghofer (2016), the diffusion coefficient scales with grain size and porosity. Based on forward modeling of zonally averaged thermal plus epithermal counting rates, Prettyman et al. (2017) found that water ice approaches the surface above 40° latitude in both hemispheres. Measurements of hydrogen by GRaND indicate the presence of a global, subsurface water-ice table at depths less than a few decimeters (1 dm = 10 cm) at latitudes greater than 45° (Prettyman et al., 2017).

11.7 The year-2018 discovery of a seasonal water cycle on *Ceres*

As already discussed in the previous sections in this chapter, the dwarf planet *Ceres* is known to host a significant amount of water in its interior (Russell et al., 2016), which is considered to be a key element of its evolution (McCord et al., 2011), composition (De Sanctis et al., 2015), and activity (McCord et al., 2011). Furthermore, areas of water ice on its surface were detected by the *Dawn* spacecraft. It has been noticed that water molecules from endogenic and exogenic sources can also be cold-trapped in permanent shadows at high latitudes, as happens on the Moon and Mercury. Platz et al. (2016) reported the discovery of several bright deposits in cold traps on Ceres's northern permanent shadows, and they spectroscopically identified one of the bright deposits as water ice. According to Platz et al. (2016) this detection strengthens the evidence that permanently shadowed areas have preserved water ice on airless planetary bodies.

Water ice has been estimated to be ubiquitous and abundant within a few meters from the surface (Prettyman et al., 2017). Exposed water ice was also discovered by the *Dawn* mission in a few craters on Ceres situated in the northern hemisphere above 42° latitude, indicating a strong dependence of water ice stability with the surface illumination condition and temperatures (Combe et al., 2016; Platz et al., 2016). Models predict that a substantial amount of water ice is present in Ceres' mantle and outer shell. The Herschel telescope and the Dawn spacecraft have observed the release of sporadic water vapor and hydroxyl emissions from Ceres, and exposed water ice has been detected by Dawn on its surface at mid-latitudes.

In a more up-to-date study, Raponi et al. (2018) reported the detection of water ice in a mid-latitude crater on Ceres and its variation with time. In their studies, the Dawn spectrometer data revealed "a change of water ice signatures over a period of 6 months, which is well modeled as ~2-km^2 increase of water ice." According to Raponi et al. (2018), "The observed increase, coupled with Ceres' orbital parameters, points to an ongoing process that seems correlated with solar flux. The reported variation on Ceres' surface indicates that this body is chemically and physically active at the present time." Raponi et al. (2018) noted that a notable exception is represented by Juling crater, being situated in a mid-latitude, in the southern hemisphere (35°S, 168°E), where the Dawn visible and infrared (VIR) mapping spectrometer (De Sanctis et al., 2011) detected water ice owing to the diagnostic absorption bands at 1.25, 1.5, 2.0, and 3.0 μm. In Raponi et al. (2018)'s studies, "VIR detected ice on the northern shadowed crater wall, characterized by an almost vertical rocky cliff, illuminated by a secondary light source coming from the adjacent illuminated regions. The crater floor shows evidence of the flow of ice and rock, similar to Earth's rock glacier (Schmidt et al., 2017)." According to Raponi et al. (2018), this crater has been observed by VIR several times: two times with a nominal spatial resolution of ~100 m per pixel during the Low Altitude Mapping Orbit (L1 and L2) and three times with a nominal spatial resolution of ~400 m per pixel during a phase devoted to Juling's observation: Extended Juling Orbit (E1, E2, and E3).

Raponi et al. (2018) found a growing patch of ice on Juling Crater, found in the mid-latitudes. They suspect that water from the crater floor is condensing on the wall, causing a patch of ice to grow larger. Analyzing all the acquisitions, Raponi et al. (2018) discovered changes in the spectra of the ice-rich wall. In their studies, water bands were detected in all the observations on the northern wall of Juling but with a variation of the water ice features, in particular, the 2.0-μm water ice absorption that is the most prominent band and least affected by instrumental errors. Raponi et al. (2018) found that "Intrinsic variations of the band area can be attributed to changing abundances of water ice, as well as other physical properties, such as changes in grain size." Using a model described in Raponi et al. (2018), they found that "the spectra can be modeled by an areal mixture of regolith and water ice, and the observed variation of spectral signatures can be explained by an increasing abundance of water ice as a function of time, with an effective grain diameter fixed to 100 μm."

According to Raponi et al. (2018) "The observed increasing abundance of water ice on the crater wall implies a change in temperature/pressure induced by (i) internal processes, (ii) exposure of ice present behind a regolith layer by falls, or (iii) solar input." After examining various possibilities, Raponi et al. (2018) concluded that "One possible internal process would involve subsurface displacement of brine or liquid water, which, percolating through clay layers, freezes when exposed to the shadowed cold surface of the wall." In previous studies, A'Hearn and Feldman (1992) and Küppers et al. (2014) have observed water vapor; and Villarreal et al. (2017) suggested that the observed water vapor could have been triggered by solar radiation and/or solar energetic particle events. Solar energetic protons have been indicated as possible sources of water vapor (Villarreal et al., 2017) because they could reach ice present in deeper subsurficial layers. Raponi et al. (2018) reckon that "Water vapor might originate from the rock glaciers on the crater floor below a regolith layer, and the shadowed wall of the crater may be a cold trap for the water molecules. As long as the accumulation rate of water molecules is higher than the sublimation rate from the wall itself, the ice on the wall increases." Raponi et al. (2018) used a thermophysical model to simulate the water flux coming from the crater floor, triggered by solar radiation. Assuming that the crater floor is filled by water ice at ~1 cm below the surface, the estimated vapor flux coming from the crater was found to be comparable to the total flux derived from Herschel observations. Raponi et al. (2018) studying the warmer region of *Ceres* have noticed that a patch of ice has grown larger over time. In addition, Carrozzo et al. (2018) found carbon-rich minerals on *Ceres*' surface that do not last long. Together, the new discoveries suggest that water still has a powerful presence on the tiny world.

11.8 Hint of an ocean hiding below the surface of *Ceres*

McCord and Sotin (2005) modeled several thermal evolution scenarios for *Ceres* to explore the nature of large, wet protoplanets and to predict current-day evidence that might be found by close inspection, such as by the *Dawn* mission. In their model studies, McCord and Sotin (2005) considered short- and long-lived radioactive nuclide heating during Ceres' evolution. They found that Ceres' existence and evolution depend critically on it containing water at formation. They examined differentiated (an inner core and an outer liquid H$_2$O layer) as well as homogeneous (i.e., nondifferentiated) models of Ceres. In their differentiated model of Ceres (Fig. 11.7B), McCord and Sotin (2005) considered Ceres to have an inner core (density 3.44) and an outer liquid H$_2$O layer about 100 km thick. In their homogeneous (nondifferentiated) model of Ceres (Fig. 11.7A), McCord and Sotin (2005) considered Ceres to be a homogeneous mixture of ice and silicate grains that has an initial temperature of 200 K. It was further assumed that Ceres is made of a mixture of 74% mass fraction of silicates (density 3.44) and 26% mass fraction of ice (density 1.0). Any temperature increase due to accretion was neglected, and the initial temperature profile was assumed to be isothermal.

McCord and Sotin (2005) found that in the nondifferentiated model (i.e., homogeneous distribution of silicates, ice, and liquid), Ceres would have liquid water in its interior today. They calculated that the transition between ice-silicate material and liquid water embedded in a silicate matrix today would be 130 km deep. McCord and Sotin (2005) found that this model resembles that of Callisto (the second-largest moon of the planet Jupiter) described by Nagel et al. (2004) except that Callisto being less dense (its uncompressed density is around 1.5) has a homogeneous rock-ice core covered by an H$_2$O layer. A study of the current knowledge of Ceres and the possible thermal evolution scenarios (McCord and Sotin, 2003, 2004, 2005) indicated that Ceres is likely differentiated and highly thermochemically evolved. From their measurements, Thomas et al. (2005) inferred that the body is differentiated into a core, probably with a hydrated silicate (density of 2700 kg/m^3), and an outer icy shell made of water (average density of 1000 kg/m^3). This is consistent with predictions inferred from thermal models by McCord and Sotin (2003, 2004, 2005). McCord and Sotin (2005) suggested that Ceres has probably lost little of its initial water content, and that the abundance of water in Ceres, as compared to Vesta, could be due to the difference in the time at which the two objects accreted. McCord and Sotin (2005)'s models predict that there was extensive liquid water inside Ceres during its evolution. More clearly, their model studies indicated that during its evolution Ceres maintained a layer of liquid water ocean (Fig. 11.7C) below a surface layer of conductive ice and a subsurface layer of convective ice, provided an antifreezing material (ammonia) was present. Alternatively, a fully differentiated model of Ceres with an inner iron core predict that Ceres maintained a layer of liquid water ocean (Fig. 11.7D) below a surface layer of conductive ice and a subsurface layer of convective ice even in the absence of an antifreezing material.

Castillo-Rogez and McCord (2010) continued the study of Ceres' evolution and current state, following the McCord and Sotin approach; but extending the range of possible initial conditions and updating model parameters based on recent observations and calculations. They focused on the evolution of Ceres' core and hydrosphere (the entire water layer) by comparison with other icy objects such as Europa, as well as relevant models for these bodies. Their studies showed that the long-term evolution of the hydrosphere is strongly controlled by the surface temperature. Their various models show an interior that is relatively warm and support liquid water today. For the extreme assumption that Ceres is currently in thermal equilibrium with its upper boundary condition, its temperature would be about 167 K (−106°C) on average with a maximum at the equator (180 K) and minimum at the poles (130 K).

Castillo-Rogez and McCord (2010)'s studies showed that a model assuming no silicate hydration, that is, no chemical redistribution between the core and the hydrosphere, still displays warm temperatures for the present time. According to them, if ammonia accreted in Ceres, it is also expected to play a role in the long-term evolution of Ceres

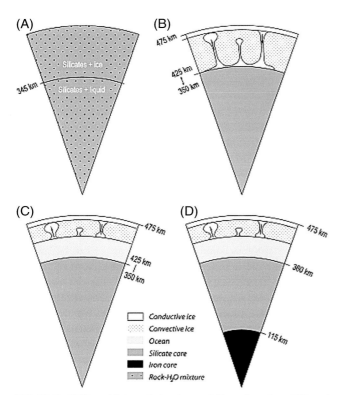

FIG. 11.7 **Different internal structures of Ceres based on different scenarios.** (A) homogeneous asteroid made of a mixture of H$_2$O and high-density silicates, (B) differentiated Ceres with high-density silicate core equivalent to Vesta (core radius of 350 km) or low-density serpentine (core radius of 425 km) and outer ice layer, (C) same as (B) but the presence of antifreezing material (ammonia) maintains a liquid layer, and (D) fully differentiated model of Ceres with an inner iron core. *(Source: McCord, T.B., and C. Sotin (2005), Ceres: evolution and current state, J. Geophys. Res., 110: E05009, doi: http://doi.org/10.1029/2004JE002244). Copyright info: © 2020 American Geophysical Union.*

by decreasing the ocean freezing temperature to about 176 K (−97°C), assuming that it has not been consumed (either decomposed into molecular nitrogen or reacted as ammonium). According to Castillo-Rogez and McCord (2010), in such case, even the coldest (and unrealistic) model of Ceres includes "a sea of liquid water enriched in ammonia and other species. This liquid layer could be as thick as 10 km at the equator, but might not exist at the poles due to the lower temperature there." These investigators further argue that, "If no ammonia accreted inside Ceres or it has been consumed, then the degree to which a deep ocean inside Ceres has been preserved over the long-term is primarily a function of: (1) the composition (in the sense that it controls the freezing point of the salt mixture), and (2) possibly the presence at the interface between the core and hydrosphere of long-lived radioisotopes due to leaching from the core."

Models of Europa's hydrosphere (the entire water layer) evolution that make use of Cc composition (Zolotov and Shock, 2001) predict that the compounds released during hydrothermal circulation in the rocky phase play a crucial antifreeze role. To strengthen the just-mentioned antifreeze role, Castillo-Rogez and McCord (2010) invoked the findings of several researchers (e.g., Kargel, 1991; Kargel et al., 2000; Wynn-Williams et al., 2001, for terrestrial analogs in hypersaline regions), which show that "A mixture of hydrates, carbonates, and chlorides can decrease water's freezing point to as low as 190 K" (−83°C). Pontefract et al. (2017) have discussed microbial diversity in a hypersaline sulfate lake (Spotted Lake [British Columbia, Canada]). Sapers et al. (2017) have addressed biological characterization of microenvironments in a hypersaline cold spring Mars analog. Castillo-Rogez and McCord (2010) argue that the evolution of the icy shell as it cools down after the primary melting phase follows the same path as described by Spaun and Head (2001) for Europa. Spaun and Head (2001) showed that Europa's hydrosphere should be stratified into an outer layer made of pure ice, and an increasing concentration of salts with depth as the ocean freezes at the salt's hydrate eutectic temperature, which ranges from 190 K (−83°C) to 253 K (−20°C) for salts expected in Ceres. Castillo-Rogez and McCord (2010) argue that a layer saturated in salt should lie over the core and separate it from a less dense water layer. According to Spaun and Head (2001)'s studies, in such conditions, and after the early phase that resulted in differentiation and silicate hydration, hydrothermal interaction at the surface of the core should be limited. Although some chemical and cosmo-chemical arguments challenge this model by demonstrating that hydrated sulfates might not be present in Europa's ocean (McKinnon and Zolensky, 2003), Castillo-Rogez and McCord (2010) invoke the findings from other chemical models involving a variety of compounds with different freezing points to support a "stratified ocean model" (e.g., Prieto-Ballesteros and Kargel, 2005).

As pointed out already, models of McCord and Sotin (2005) and Castillo-Rogez and McCord (2010) have demonstrated that there was extensive liquid water inside Ceres during its evolution. Europa or Enceladus are confirmed ocean worlds, with liquid water hiding below their outer icy crust. Considering that Ceres is smaller than Europa, but bigger than Enceladus, Castillo-Rogez and McCord (2010) reckon that it is reasonable to consider that Ceres is also likely to have liquid water ocean below its surface. These investigators argue that while Ceres does not benefit from tidal heating as a means to maintain warm temperatures at depth over the long run, its warm surface temperature helps to create warm internal temperatures at present, at least locally (i.e., equal or greater than 180 K (−93°C) at the equator instead of 80 K (−193°C) at Europa.

11.9 Possibility of hydrothermal geochemistry to take place on *Ceres*

Hydrothermal circulation can profoundly affect the evolution of Ceres. Observations and models of Ceres suggest that its evolution was shaped by interactions between liquid water and silicate rock. Hydrothermal processes in a heated core require both fractured rock and liquid. Using a new core cracking model coupled to a thermal evolution code, Neveu et al. (2015) found volumes of fractured rock always large enough for significant interaction to occur. Therefore, liquid persistence is key. It is favored by antifreezes such as ammonia, by silicate dehydration which releases liquid, and by hydrothermal circulation itself, which enhances heat transport into the hydrosphere.

It is known from terrestrial analogs that hydrothermal vents are best places for the production of organic molecules (Shock, 1992). McCord and Sotin (2005) were the first to study *Ceres* as an icy object that was subject to differentiation, hydrothermal activity, and that might host a liquid layer today. Castillo-Rogez and McCord (2010) suggested that models with times of formation shorter than about 5 My after the production of calcium–aluminum inclusions are more likely to undergo hydrothermal activity in their early history, which affects Ceres' long-term thermal evolution. Their geophysical models are fully consistent with McCord and Sotin (2005) as they predict the differentiation of a rocky core from the volatile phase under almost all circumstances of formation and evolution. Castillo-Rogez and McCord (2010)'s study expanded upon that previous work using new information and explored the core evolution and hydrosphere evolution. Castillo-Rogez and McCord (2010) showed that shape data can help constrain the amount of hydrated silicate in the core, and thus the extent of hydrothermal activity in Ceres. They discussed the importance of these results for the *Dawn* mission's arrival at Ceres in 2015.

The occurrence of the phenomenon of production of organic molecules in a small ice-rock body has been studied by Matson et al. (2007) and Glein et al. (2008) in the case of *Enceladus*. Castillo-Rogez and McCord (2010)'s modeling also indicates that conditions in *Ceres* were suitable for hydrothermal geochemistry to take place, which could yield the rich variety of organic and nonorganic molecules found in carbonaceous chondrites. For example, if ammonia accreted in *Ceres*, in hydrothermal context it dissociated into molecular nitrogen, or could react to form ammonium salts, as has been observed by King et al. (1992). The presence of clays provided optimum support for geochemical reactions to take place (e.g., Pearson et al., 2002).

The prospect for Ceres to maintain mild interior temperatures throughout its history is quantified by thermal modeling. A 2-D numerical study, of the evolution of Ceres from a "frozen mudball" to the present era, carried out by Travis et al. (2018), emphasizes the importance of hydrothermal processes in Ceres. Travis et al. (2018)'s modeling, considering both salt-free and brine fluids, suggests that Ceres's core has been warm over most of its history and is still above freezing, and convective processes are active in core and mantle to the present. They found that the addition of low eutectic solutes greatly expands the region of active convection. According to Travis et al. (2018), surface features found on Ceres, such as the bright spot in Occator crater and Ahuna Mons, could be the result of eutectic plumes. Travis et al. (2018)'s model mud mantle has a roughly 42:58 volumetric partitioning of H_2O to rock. These investigators found that particulates released as the "frozen mudball" thaws (frozen substances that become liquid or soft as a result of warming up) settle to form a roughly 290 km radius core. Hydrothermal flow is driven by radiogenic heating and serpentinization. Travis et al. (2018)'s numerical study results showed that a global muddy ocean persisted for the first 3 Gyr, and at present there may be several regional mud seas buried under a frozen crust of Ceres. Transport of interior material to the near surface occurs throughout Travis et al. (2018)'s model's history. Their study indicates that eutectic brines drive convective flow to near the surface, even breaching the surface in isolated regions, on the order of 30 km in width, similar in size to some mounds detected using the Dawn visible imaging camera (Sizemore et al., 2017).

Castillo-Rogez et al. (2020) recall that "Recent models with more elaborate physics explored a greater parameter space to address Ceres' potential for both hydrothermal activity and subsolidus (describing the region beneath the solidus in a phase diagram) convective transport in a thick ice shell (Neumann et al., 2015; Neveu et al., 2015; Travis et al., 2015; Bland and Travis, 2017). Some studies accounted for the presence of salts leached from accreted silicates following hydrothermal processing, the latter of which is inferred from Ceres' pervasively hydrated surface (Rivkin et al., 2006; Castillo-Rogez and Young, 2017). Thermal evolution models consistently yield temperatures in Ceres' shell that could remain above the eutectic temperature of relevant salts (>220 K, e.g., for chloride mixtures) until present. Therefore, modeling studies predict that pockets of concentrated brines could exist at the base of the crust today (>50km deep in Castillo-Rogez and McCord, 2010; Castillo-Rogez and Lunine, 2012; Neumann et al., 2015; Neveu et al., 2015)."

Castillo-Rogez et al. (2020) further note that "Modeling of heat transfer in a 'mud ball' even suggested that an ~300-km radius ocean loaded with silicate fines (<10s of microns) could remain until present (Neveu and Desch, 2015; Travis et al., 2015). Although not specific to Ceres, Kargel (1991) predicted that brines produced by aqueous alteration of the silicates could drive volcanic activity in large asteroids. It is remarkable that most pre-Dawn models of Ceres' thermal evolution predicted the long-term preservation of a liquid layer despite using different assumptions on the modalities of heat transfer."

11.10 Presence of organics on and inside *Ceres*

Marchi et al. (2018) found that Ceres's surface may contain up to 20 wt% of carbon, which is more than five times higher than in carbonaceous chondrites. According to their interpretation, the coexistence of phyllosilicates, magnetite, carbonates, and a high carbon content implies rock–water alteration played an important role in promoting widespread carbon chemistry. Marchi et al. (2018) argued that these findings unveil pathways for the synthesis of organic matter, with implications for their transport across the Solar System.

Based on various credible indications, Ceres is considered as a "candidate ocean world." According to Travis et al. (2018), the global ocean hiding below Ceres' surface could have been maintained for billions of years. On a young Ceres, several prebiotic pathways are pertinent to an ocean rich in fine silicate particles, as suggested by Travis et al. (2018). These may be based on hydrogen cyanide (e.g., Matthews and Minard, 2006; Patel et al., 2015) and/or formaldehyde chemistry (e.g., Saladino et al., 2012), as well as surface-catalyzed syntheses on phyllosilicates (e.g., Ertem and Ferris, 1996; Pearson et al., 2002). It is plausible that larger water-rich planetesimals, such as Ceres that hosted settings favorable to planetary differentiation, advanced serpentinization, and an ocean could have had more advanced and extensive processing of accreted organics (Castillo-Rogez et al., 2020). Note that serpentinization is the process of hydrothermal alteration that transforms Fe–Mg-silicates such as olivine, pyroxene, or amphiboles contained in ultramafic rocks into serpentine minerals. Much of the uppermost mantle in oceanic setting is so altered, as are the cumulus parts of layered intrusions. Serpentine is a soft ductile mineral and its presence in the mantle wedge lubricates subduction of the oceanic plate. Production of serpentine in the oceanic crust results in the production of hydrothermal fluids and releases gaseous methane and hydrogen, as observed along terrestrial mid-ocean ridges. The pH of the hydrothermal fluids is generally low but under some conditions, notably at low temperature, may be high enough to be favorable to life.

High concentration in organics has been found on Ceres' surface. Marchi et al. (2018) hypothesized the following three possible explanations for this: (1) Ceres formed in an accretional environment different from the carbonaceous chondrites and richer in organics; (2) Ceres hosted conditions amenable to the production of organics from simple compounds expected in an accretional environment in the outer Solar System (e.g., carbon monoxide [CO], carbon dioxide [CO_2], formaldehyde [CH_2O], and hydrogen cyanide [HCN]); and (3) the high abundance on the surface reflects concentration during Ceres' internal differentiation.

11.11 Ceres—a candidate ocean world in the asteroid belt

The prospect for long-lived oceans is a key component in the assessment of the habitability potential of large water-rich extraterrestrial icy bodies (i.e., their potential to produce and maintain an environment favorable to life). Several extraterrestrial icy moons are believed to have hosted deep oceans for at least part of their histories (e.g., Consolmagno and Lewis, 1978). The present existence of oceans has been confirmed at many such moons, for example, Europa (Carr et al., 1998; Khurana et al., 1998; Pappalardo et al., 1999; Kivelson et al., 2000; Kargel et al., 2000; Zimmer et al., 2000; Thomson and Delaney, 2001; Khurana et al., 2004; Hand et al., 2015; Lorenz, 2016; Daswani et al., 2020), Ganymede (Kivelson et al., 2002), Enceladus (Manga and Wang, 2007; Postberg et al., 2009; Tyler, 2009; Iess et al., 2014; Bouquet et al., 2015; Thomas et al., 2016), and Titan (Lunine et al., 1983; Fortes, 2000; Lorenz et al., 2008; Béghin et al., 2010; Baland et al., 2011; Iess et al., 2012; Béghin et al., 2012; Lorenz et al., 2014). Similarly, deep oceans have been proposed to occur in the dwarf planet Pluto based on geophysical grounds (e.g., McCord and Sotin, 2005; Robuchon and Nimmo, 2011; Johnson et al., 2016; Nimmo et al., 2016; Desch and Neveu, 2017). Whereas these liquid-water-rich icy moons and planets are located far beyond the asteroid belt, Ceres—the inner Solar System's only dwarf planet, and the most water-rich body in the inner Solar System after Earth—is located within the asteroid belt (i.e., among the hitherto known liquid-water-rich extraterrestrial bodies, Ceres is closest to Earth). Among planets, Mars is the closest.

The various scientific data gathered from Ceres during the near-global geological, chemical, and geophysical mapping of Ceres with the support of a large suite of remote sensing instrumentation deployed onboard the "Dawn" spacecraft (that was launched by NASA on 27 September 2007 with the mission of studying two of the three known protoplanets of the asteroid belt: Vesta and Ceres) suggest the presence of liquid inside Ceres through time, perhaps in the form of pore fluid in a silicate matrix or as a confined relict brine (or brine pockets) (see Castillo-Rogez et al., 2020). Furthermore, chemical and physical measurements obtained through the Dawn mission enabled the quantification of key parameters pertaining to Ceres. Such measurements and their quantification provided strong indication that the surface chemistry and internal structure of Ceres testify to a protracted history (i.e., extended in duration) of reactions between liquid water, rock, and likely organic compounds. Castillo-Rogez et al. (2020) reviewed the clues on chemical composition, temperature, and prospects for long-term occurrence of liquid and chemical gradients. Their review found that comparisons with giant planet moons indicate similarities both from a chemical evolution standpoint and in the physical mechanisms driving Ceres' internal evolution.

Although Ceres is located within the asteroid belt, it has been found that as per its outstanding properties, it has more resemblance to icy moons than to asteroids, in particular based on its large bulk water content and large size. McCord and Sotin (2005) pointed out that Ceres contains the right amount of both water and rock for sustained heating, resulting in the formation of a volatile-rich shell. This was supported by the early detection of hydrated materials on Ceres' surface (Lebofsky, 1978), suggesting pervasive aqueous alteration (Rivkin and Volquardsen, 2009). Castillo-Rogez et al. (2020) recall that while Ceres does not experience tidal heating, it is sufficiently close to the Sun and contains long-lived radioisotopes (~73 wt% rock) to potentially preserve brines until present (Castillo-Rogez and McCord, 2010) and meet conditions favorable for brine volcanism (Kargel, 1991). Based on the vast scientific information gathered and the results of various model studies carried out, Castillo-Rogez et al. (2020) concluded that Ceres is a "candidate ocean world" according to the definition set forth in the Roadmap for Ocean Worlds (ROW) (Hendrix et al., 2019). Ceres is thus a key piece of the overall puzzle of ocean world evolution.

11.12 Habitability potential of Ceres

An ice-dominated shell blanketing the asteroid-belt-resident dwarf planet Ceres was predicted by geophysical modeling. Ceres comprises nearly one-third of the mass of the asteroid belt. Its mean radius and bulk density, respectively, 470 km and 2162 kg/m^3 (Russell et al., 2016), are intermediate between Enceladus (252 km and 1611 kg/m^3) and Europa (1560 km and 3014 kg/m^3), which are the most important extraterrestrial bodies expected to have prospects of life, and comparable with those of many other icy moons and icy dwarf planets in the Solar System. Castillo-Rogez and McCord (2010) found that Ceres' ice-rich shell could host temperatures as warm as 240 K a few tens of kilometers deep for most of its history. Furthermore, Ceres contains about 25 wt.% water in the form of ice and/or bound to minerals, evidenced by its bulk density. Furthermore, Ceres has retained most of its original water (perhaps still in liquid form today), and underwent processes similar to those expected or observed in icy outer planet moons, especially Europa. These positive attributes led Castillo-Rogez and Lunine (2012) to propose that Ceres could be of astrobiological interest.

Dawn's arrival at Ceres—the largest world in the asteroid belt—happened in March 2015. Even ahead of the Dawn mission to Ceres, available physical properties indicated that Ceres would be an evolved, internally differentiated body with a high prospect for preserving liquid until present. For example, Ceres' bulk density was long known to be ~2100 kg/m^3 (see review by McCord and Sotin, 2005), intermediate between the values for water ice and silicates (Castillo-Rogez et al., 2020). This corresponds to a bulk fraction of water much higher than terrestrial planets and most rocky asteroids (27 wt% free water, i.e., not bound to minerals, McCord and Sotin, 2005). This suggests that Ceres is more akin to other icy moons and dwarf planets of the outer Solar System. Castillo-Rogez et al. (2020) recall that thermal evolution models of Ceres, using shape data derived from telescopic observations (e.g., Thomas et al., 2005; Carry et al., 2008; Drummond et al., 2014), indicate that, following accretion, the dwarf planet could have differentiated into a silicate core and a water-rich outer layer. These models suggest that Ceres could have harbored a global subsurface ocean for several hundred million years after its formation (e.g., McCord and Sotin, 2005; Castillo-Rogez and McCord, 2010).

The *Dawn* spacecraft gathered science data pertaining to Ceres and returned it to Earth right up to the point the spacecraft ran out of fuel. The *Dawn* spacecraft achieved many firsts until its extended mission concluded on Oct. 31, 2018, and also reinforced the idea that dwarf planets could have hosted oceans over a significant part of their history—and potentially still do. Dawn discovered that the inner Solar System's only dwarf planet was an ocean world where water and ammonia reacted with silicate rocks. As the ocean froze, salts and other minerals concentrated into deposits that are now exposed in many locations across the surface. Dawn also found organics in several locations on Ceres' surface. Furthermore, the Dawn revealed the presence of abundant carbon in Ceres' regolith (Prettyman et al., 2017; Marchi et al., 2018), as well as localized spots rich in organic matter (e.g., De Sanctis et al., 2017). Altogether, these observations not only confirm but also emphasize the astrobiological significance of Ceres. Castillo-Rogez et al. (2020) assessed the astrobiological implications of these observations and whether or not conditions within Ceres could have produced habitable environments and are amenable to advanced prebiotic chemistry. As pointed out by Castillo-Rogez et al. (2020), Ceres, the most water-rich body in the inner Solar System after Earth, has recently been recognized to have astrobiological importance. Chemical and physical measurements obtained by the Dawn mission enabled the quantification of key parameters, which helped to constrain the habitability of the inner Solar System's only dwarf planet. According to McCord and Sotin (2005), "With so much water present and the energy to distribute it in liquid form, Ceres probably experienced complex chemistry at least in its interior, probably involving organic materials. It is further likely that expressions of these processes and the materials made it to the surface, at least in places."

Castillo-Rogez et al. (2020) have provided a graphical exposition of the salient lines of evidence for the habitability of Ceres from mineralogical, elemental, geological, and geophysical observations. Such evidences include: (a) Geophysical data confirming the abundance of water ice and the need for gas and salt hydrates to explain the observed topography and crustal density, (b) Discovery of various types of carbonates and ammonium chloride in many sites across Ceres' surface (e.g., salts exposed on the floor of Dantu crater), (c) Presence of carbon species in three forms (reduced in CxHy form, oxidized in the form of carbonates, and intermediate as graphitic compounds) at Ernuter crater and its area, (d) Extensive evidence for the presence of water ice on Ceres in the form of ground ice and exposure via mass wasting and impacts (e.g., Juling crater, ~20 km), (e) Recent expressions of volcanism, pointing to the combined role of radiogenic heating and low-eutectic brines in preserving melt and driving activity (e.g., Ahuna Mons, ~4.5 km tall, ~20 km diameter), and (f) Prospects of impacts creating local chemical energy gradients in transient melt reservoirs throughout Ceres' history (e.g., Cerealia Facula, ~14 km diameter).

Castillo-Rogez et al. (2020) recall that, "The comparison with Enceladus may be particularly informative of Ceres' past (and perhaps present) habitability. Enceladus' liquids are also expected to be dominated by Na-Cl-CO$_3$ chemistry (Glein et al., 2015). At Enceladus, the bulk water to rock ratio is likely above 0.6, assuming a 10-km-thick global ocean above a 195 km rocky mantle with 25% water-filled porosity consistent with observations (Choblet et al., 2017; Neveu and Rhoden, 2019). This water to rock ratio is likely higher and closer to Ceres' water to rock ratio of ≥2 because Enceladus' ocean is thicker near the south pole (Iess et al., 2014; Thomas et al., 2016) and the mantle rock has not yet fully reacted with water (Waite et al., 2017)." The pH of Enceladus' ocean is also alkaline, having been variously constrained to 8.5–10.5 (Sekine et al., 2015), 11–12 (Glein et al., 2015), and most recently 9–11 (Waite et al., 2017; Glein et al., 2018), similar to the conditions of aqueous alteration at Ceres. Inside both worlds, the rocky interiors of density ≈2400 kg/m^3 (McKinnon, 2015; Beuthe et al., 2016; Ermakov et al., 2017) seem to be hydrated and not fully differentiated from lower density ices. Being of similar size and composition, and with related physicochemical conditions, Ceres could share similarities with Enceladus in its habitability and astrobiological potential, at least for part of its history (Castillo-Rogez et al., 2020).

NASA's discovery-class mission *Dawn* shed much light on various aspects of Ceres. However, Ceres being considered as a "candidate ocean world" having habitability potential, and located much closer to Earth compared to other known confirmed ocean worlds such as Europa, Enceladus, Titan, etc., Ceres attracts much importance in terms of future explorations. It is believed that future exploration of Ceres would reveal the degree to which liquid water and other environmental factors may have combined to make Ceres a habitable world. Confirmation of the existence of liquid inside Ceres at present, and assessment of its extent, is the natural next step when following the "Roadmaps to Ocean Worlds" (ROW), which was initiated in 2016 by NASA, and was presented in January 2019 (Hendrix et al., 2019). Another key question for any coming mission is whether there exist oxidants at the surface to help the emergence of life or at least its habitability. A favorable redox discovery would deepen the case for Ceres meeting our current definition of habitability. Another topic of major importance is about quantifying the abundance, sources and sinks, and chemical forms of CHNOPS elements, assessing the inventory and determining the origin(s) of organic compounds beyond the limited observations of *Dawn*, and determining their origin(s) (Castillo-Rogez et al., 2020).

An important aspect concerning Ceres is its representativeness as an endmember for Ocean Worlds that are not tidally heated and that accreted ammonia as in Callisto and potentially also Pluto (Nimmo et al., 2016). The prospect for a body heated only by radioisotopes to retain liquid until present is not simply a function of size; it depends on the redistribution of radioisotopes upon geochemical transfers, as well as the types of salts generated in early hydrothermal environments or brought in by accreted planetesimals. Further exploration of Ceres to assess the state of remaining liquid (extent and composition) would in turn indicate which classes of bodies (physical properties and origin) are more likely to preserve relict oceans until present (Castillo-Rogez et al., 2020). Ceres remains the smallest known dwarf planet, only 590 miles (950 km) across—roughly the size of Texas. A day on Ceres lasts a little over 9 Earth-hours, while it takes 4.6 Earth-years to travel around the Sun. Unlike Earth's Moon where the duration of day and night is 15 Earth days each, the small duration of day and night on Ceres is an added benefit in terms of habitability perspective.

References

A'Hearn, M.F., Feldman, P.D., 1992. Water vaporization on Ceres. Icarus 98, 54–60. http://doi.org/10.1016/0019-1035(92)90206-M.

Ammannito, E., de Sanctis, M.C., Palomba, E., et al., 2013. Olivine in an unexpected location on Vesta's surface. Nature 504, 122–125.

Ammannito, E., DeSanctis, M.C., Ciarniello, M., Frigeri, A., Carrozzo, F.G., Combe, J.-P., Ehlmann, B.L., Marchi, S., McSween, H.Y., Raponi, A., Toplis, M.J., Tosi, F., Castillo-Rogez, J.C., Capaccioni, F., Capria, M.T., Fonte, S., Giardino, M., Jaumann, R., Longobardo, A., Joy, S.P., Magni, G., McCord, T.B., McFadden, L.A., Palomba, E., Pieters, C.M., Polanskey, C.A., Rayman, M.D., Raymond, C.A., Schenk, P., Zambon, F., Russell, C.T., 2016. Dawn Science Team. Distribution of phyllosilicates on the surface of Ceres. Science 353, aaf4279.

Baland, R.-M., van Hoolst, T., Yseboodt, M., Karatekin, Ö., 2011. Titan's obliquity as evidence of a subsurface ocean? Astron. Astrophys. 530, A141. http://doi.org/10.1051/0004-6361/201116578.

Béghin, C., Sotin, C., Hamelin, M., 2010. Titan's native ocean revealed beneath some 45 km of ice by a Schumann-like resonance. C. R. Geosci. 342, 425–433. http://doi.org/10.1016/j.crte.2010.03.003.

Béghin, C., Randriamboarison, O., Hamelin, M., Karkoschka, E., Sotin, C., Whitten, R.C., Berthelier, J.-J., Grard, R., Simões, F., 2012. Analytic theory of Titan's Schumann resonance: constraints on ionospheric conductivity and buried water ocean. Icarus, 218, 1028–1042. http://doi.org/10.1016/j.icarus.2012.02.005.

Beuthe, M., Rivoldini, A., Trinh, A., 2016. Enceladus's and Dione's floating ice shells supported by minimum stress isostasy. Geophys. Res. Lett. 43, 10,088–10,096.

Bland, P.A., Travis, B.J., 2017. Giant convecting mud balls of the early solar system. Sci. Adv. 3, e1602514.

Bouquet, A., Mousis, O., Waite, J.H., Picaud, S., 2015. Possible evidence for a methane source in Enceladus' ocean. Geophys. Res. Lett. 42, 1334–1339. http://doi.org/10.1002/2014GL063013.

Britt, D.T., Yeomans, D., Housen, K., Consolmagno, G., 2002. Asteroid density, porosity, and structure. In: Bottke, W.F. et al (Ed.), Asteroids III. Univ. of Ariz. Press, Tucson, pp. 485–500.

Buczkowski, D.L., et al., 2016. The geomorphology of Ceres. Science 353, aaf4332.

Canup, R., Agnor, C., 2000. Accretion of the terrestrial planets and the Earth–Moon system. In: Canup, R., Righter, K. (Eds.). Origin of the Earth and Moon, 555. University of Arizona Space Science Series, pp. 115–129.

Carrozzo, F.G., De Sanctis, M.C., Raponi, A., Ammannito, E., Castillo-Rogez, J., Ehlmann, B.L., Marchi, S., Stein, N., Ciarniello, M., Tosi, F., Capaccioni, F., Capria, M.T., Fonte, S., Formisano, M., Frigeri, A., Giardino, M., Longobardo, A., Magni, G., Palomba, E., Zambon, F., Raymond, C.A., Russell, C.T., 2018. Nature, formation, and distribution of carbonates on Ceres. Sci. Adv. 4 (3), e1701645. http://doi.org/10.1126/sciadv.1701645.

Carr, M., Belton, M., Chapman, C., et al., 1998. Evidence for a subsurface ocean on Europa. Nature 391, 363–365.

Carry, B., Dumas, C., Fulchignoni, M., Merline, W.J., Berthier, J., Hestroffer, D., Fusco, T., Tamblyn, P., 2008. Near-infrared mapping and physical properties of the dwarf-planet Ceres. Astron. Astrophys. 478, 235–244.

Castillo-Rogez, J., McCord, T.B., 2010. Ceres' evolution and present state constrained by shape data. Icarus 205, 443–459. http://doi.org/10.1016/j.icarus.2009.04.008.

Castillo-Rogez, J.C., Lunine, J.I., 2012. Small worlds habitability. In: Impey, C., Lunine, J., Funes, J. (Eds.), Astrobiology: The Next Frontier. Cambridge University Press, Cambridge, United Kingdom, pp. 201–228.

Castillo-Rogez, J.C., Young, E.D., 2017. Origin and evolution of volatile-rich planetesimals. In: Elkins-Tanton, L., Weiss, B. (Eds.), Planetesimals, Early Differentiation and Consequences for Planets. *Cambridge University Press*, Cambridge, United Kingdom, pp. 92–114.

Castillo-Rogez, J.C., Neveu, M., McSween, H.Y., De Sanctis, M.C., Raymond, C.A., Russell, C.T., 2018. Insights into Ceres' evolution from surface composition. Meteor. Planet. Sci. 53, 1820–1843.

Castillo-Rogez, J.C., Neveu, M., Scully, J.E.C., House, C.H., Quick, L.C., Bouquet, A., Miller, K., Bland, M., De Sanctis, M.C., Ermakov, A., Hendrix, A.R., Prettyman, T.H., Raymond, C.A., Russell, C.T., Sherwood, B.E., Young, E., 2020. Ceres: astrobiological target and possible ocean world. Astrobiology 20 (2), 269–291. https://doi.org/10.1089/ast.2018.1999.

Chapman, C.R., Salisbury, J.W., 1973. Comparison of meteorite and asteroid spectral reflectivities. Icarus 19, 507–522. http://doi.org/10.1016/0019-1035(73)90078.

Choblet, G., Tobie, G., Sotin, C., Behounkova, M., Cadek, O., Postberg, F., Soucek, O., 2017. Powering prolonged hydrothermal activity inside Enceladus. Nat. Astron. 1, 841–847.

Clark, R.N., 1981. Water frost and ice: the near-infrared spectral reflectance 0.65-2.5 μm. J. Geophys. Res. 86, 3087–3096. http://doi.org/10.1029/JB086iB04p03087.

Clark, R.N., Lucey, P.G., 1984. Spectral properties of ice-particulate mixtures and implications for remote sensing: I. Intimate mixtures. J. Geophys. Res. 89, 6341–6348. http://doi.org/10.1029/JB089iB07p06341.

Combe, J.P., et al., 2017. Exposed H2O-rich areas on Ceres detected by Dawn, Proceedings of 48th Lunar and Planetary Science Conference. LPSC, pp. 2568 Abstract.

Combe, J.-Ph., Le Mouélic, S., Sotin, C., Gendrin, A., Mustard, J.F., Le Deit, L., Launeau, P., Bibring, J.-P., Gondet, B., Langevin, Y., Pinet, P., 2008. Analysis of OMEGA/Mars Express data hyperspectral data using a Multiple-Endmember Linear Spectral Unmixing Model (MELSUM): methodology and first results. Planet. Space Sci. 56, 951–975. http://doi.org/10.1016/j.pss.2007.12.007.

Combe, J.-P., McCord, T.B., Tosi, F., Ammannito, E., Carrozzo, F.G., De Sanctis, M.C., Raponi, A., Byrne, S., Landis, M.E., Hughson, K.H.G., Raymond, C.A., Russell, C.T., 2016. Detection of local H_2O exposed at the surface of Ceres. Science 353 (6303), aaf3010. http://doi.org/10.1126/science.aaf3010.

Consolmagno, G.J., Lewis, S.J., 1978. The evolution of icy satellite interiors and surfaces. Icarus 34, 280–293.

Daswani, M.M., Vance, S.D., Mayne, M.J., Glein, C.R., 2021. A metamorphic origin for Europa's ocean. Geophys. Res. Lett., 48 (18): e2021GL094143 (14 pp.).

De Sanctis, M.C., Coradini, A., Ammannito, E., Filacchione, G., Capria, M.T., Fonte, S., Magni, G., Barbis, A., Bini, A., Dami, M., Ficai-Veltroni, I., Preti, G., 2011. The VIR spectrometer. Space Sci. Rev. 163, 329–369. http://doi.org/10.1007/s11214-010-9668-5.

De Sanctis, M.C., et al., 2012. Detection of widespread hydrated materials on Vesta by the VIR Imaging Spectrometer on board the Dawn mission. Astrophys. J. Lett. 758, L36.

De Sanctis, M.C., Ammannito, E., Raponi, A., Marchi, S., McCord, T.B., McSween, H.Y., Capaccioni, F., Capria, M.T., Carrozzo, F.G., Ciarniello, M., Longobardo, A., Tosi, F., Fonte, S., Formisano, M., Frigeri, A., Giardino, M., Magni, G., Palomba, E., Turrini, D., Zambon, F., Combe, J.-P., Feldman, W., Jaumann, R., McFadden, L.A., Pieters, C.M., Prettyman, T., Toplis, M., Raymond, C.A., Russell, C.T., 2015. Ammoniated phyllosilicates with a likely outer Solar System origin on (1) Ceres. Nature 528, 241–244. http://doi.org/10.1038/nature16172.

De Sanctis, M.C., Raponi, A., Ammannito, E., Ciarniello, M., Toplis, M.J., McSween, H.Y., Castillo-Rogez, J.C., Ehlmann, B.L., Carrozzo, F.G., Marchi, S., Tosi, F., Zambon, F., Capaccioni, F., Capria, M.T., Fonte, S., Formisano, M., Frigeri, A., Giardino, M., Longobardo, A., Magni, G., Palomba, E., McFadden, L.A., Pieters, C.M., Jaumann, R., Schenk, P., Mugnuolo, R., Raymond, C.A., Russell, C.T., 2016. Bright carbonate deposits as evidence of aqueous alteration on (1) Ceres. Nature. 10.1038/nature18290. http://doi.org/10.1038/nature18290.

De Sanctis, M.C., Ammannito, E., McSween, H.Y., Raponi, A., Marchi, S., Capaccioni, F., Capria, M.T., Carrozzo, F.G., Ciarniello, M., Fonte, S., Formisano, M., Frigeri, A., Giardino, M., Longobardo, A., Magni, G., McFadden, L.A., Palomba, E., Pieters, C.M., Tosi, F., Zambon, F., Raymond, C.A., Russell, C.T., 2017. Localized aliphatic organic material on the surface of Ceres. Science 355, 719–722.

Desch, S.J., Neveu, M., 2017. Differentiation and cryovolcanism on Charon: a view before and after New Horizons. Icarus 287, 175–186.

Dozier, J., Green, R.O., Nolin, A.W., Painter, T.H., 2009. Interpretation of snow properties from imaging spectrometry. Remote Sens. Environ. 113, S25–S37. http://doi.org/10.1016/j.rse.2007.07.029.

Drummond, J.D., Carry, B., Merline, W.J., Dumas, C., Hammel, H., Erard, S., Conrad, A., Tamblyn, P., Chapman, C.R., 2014. Dwarf planet Ceres: ellipsoid dimensions and rotational pole from Keck and VLT adaptive optics images. Icarus 236, 28–37.

Ermakov, A.I., Fu, R.R., Castillo-Rogez, J.C., Raymond, C.A., Park, R.S., Preusker, F., Russell, C.T., Smith, D.E., Zuber, M.T., 2017a. Constraints on Ceres' internal structure and evolution from its shape and gravity measured by the Dawn spacecraft. J. Geophys. Res. Planets 122, 2267–2293.

Ertem, G., Ferris, J.P., 1996. Synthesis of RNA oligomers on heterogeneous templates. Nature 379, 238–240.

Fanale, F.P., Salvail, J.R., 1989. The water regime of asteroid (1) Ceres. Icarus 82, 97–110. http://doi.org/10.1016/0019-1035(89)90026-2.

Feierberg, M.A., Lebofsky, L.A., Larson, H.P., 1981. Spectroscopic evidence for aqueous alteration products on the surfaces of low-albedo asteroids. Geochim. Cosmochim. Acta 45, 971–981.

Fortes, A., 2000. Exobiological implications of a possible ammonia–water ocean inside Titan. Icarus 146, 444–452.

Giuseppe Piazzi; Italian astronomer; Encyclopædia Britannica (2004), Giuseppe Piazzi, Chicago, Ill., Edward F. Tedesco: The Editors of Encyclopaedia Britannica (Available at http://www.britannica.com/search?query=Giuseppe+Piazzi&ct=)

Glein, C.R., Zolotov, M., Yu., Shock, E.L, 2008. The oxidation state of hydrothermal systems on early Enceladus. Icarus 197 (1), 157–163. http://doi.org/10.1016/j.icarus.2008.03.021.

Glein, C.R., Baross, J.A., Waite, J.H., 2015. The pH of Enceladus' ocean. Geochim. Cosmochim. Acta 162, 202–219.

Glein, C.R., Postberg, F., Vance, S.D., 2018. The geochemistry of enceladus: composition and controls. In: Schenk, P.M., Clark, R.N., Howett, C.J.A., Verbiscer, A.J., Waite, J.H. (Eds.), Enceladus and the Icy Moons of Saturn. University of Arizona Press, Tucson, AZ, pp. 39–56.

Grimm, R.E., McSween, H.Y., 1989. Water and the thermal evolution of carbonaceous chondrite parent bodies. Icarus 82, 244–280.

Grundy, W.M., Schmitt, B., 1998. The temperature-dependent near-infrared absorption spectrum of hexagonal H_2O ice. J. Geophys. Res. 103, 25809–25822. http://doi.org/10.1029/98JE00738.

Hand, K.P., Carlson, R.W., 2015. Europa's surface color suggests an ocean rich with sodium chloride. Geophys. Res. Lett. 42, 3174–3178.

Hendrix, A.R., Hurford, T.A., Barge, L.M., Bland, M.T., Bowman, J.S., Brinckerhoff, W., Buratti, B.J., Cable, M.L., Castillo-Rogez, J., Collins, G.C., Diniega, S., German, C.R., Hayes, A.G., Hoehler, T., Hosseini, S., Howett, C.J.A., McEwen, A.S., Neish, C.D., Neveu, M., Nordheim, T.A., Patterson, G.W.A., Patthoff, D., Phillips, C., Rhoden, A., Schmidt, B.E., Singer, K.N., Soderblom, J.M., Vance, S.D., 2019. The NASA roadmap to ocean worlds. Astrobiology 19, 1–27.

Hayne, P.O., Aharonson, O, 2015. Thermal stability of ice on Ceres with rough topography. J. Geophys. Res. Planets 120, 1567–1584. doi:10.1002/2015JE004887.

Hiesinger, H., et al., 2016. Cratering on Ceres: implications for its crust and evolution. Science 353, aaf4759.

Iess, L., Jacobson, R.A., Ducci, M., Stevenson, D.J., Lunine, J.I., Armstrong, J.W., Asmar, S.W., Racioppa, P., Rappaport, N.J., Tortora, P., 2012. The tides of Titan. Science 337, 457–459.

Iess, L., Stevenson, D.J., Parisi, M., Hemingway, D., Jacobson, R.A., Lunine, J.I., Nimmo, F., Armstrong, J.W., Asmar, S.W., Ducci, M., Tortora, P., 2014. The gravity field and interior structure of Enceladus. Science 344, 78–80.

Johnson, B.C., Bowling, T.J., Trowbridge, A.J., Freed, A.M., 2016. Formation of the Sputnik Planum basin and the thickness of Pluto's subsurface ocean. Geophys. Res. Lett. 43, 10068–10077.

Jones, T.D., Lebofsky, L.A., Lewis, J.S., 1988. The 3 μm hydrated silicate signature on C class asteroids: Implications for origins of outer belt objects. Lunar Planet. Sci. XIX, 567–568.

Jones, T.D., Lebofsky, L.A., Lewis, J.S., Marley, M.S., 1990. The composition and origin of the C, P, and D asteroids: water as a tracer of thermal evolution in the outer belt. Icarus 88, 172–192.

Kargel, J.S., 1991. Brine volcanism and the interior structures of asteroids and icy Satellites. Icarus 94, 368–390. http://doi.org/10.1016/0019-1035(91)90235-L.

Kargel, J.S., Kaye, J.Z., Head III, J.W., Marion, G.M., Sassen, R., Crowley, J.K., Ballesteros, O.P., Grant, S.A., Hogenboom, D.L, 2000. Europa's crust and ocean: origin, composition, and the prospects for life. Icarus 148 (1), 226–265 Planetary Sciences Group, Brown University.

Khurana, K.K., Kivelson, M.G., Vasyliunas, V.M., Krupp, N., Woch, J., Lagg, A., Mauk, B.H., Kurth, W.S., 2004. The configuration of Jupiter's magnetosphere. In: Bagenal, F., Dowling, T.E., McKinnon, W.B. (Eds.). Jupiter: The Planet, Satellites and Magnetosphere Cambridge Planetary Science, Vol. 1. Cambridge University Press, Cambridge, United Kingdom, pp. 593–616, ISBN 0-521-81808-7.

King, T., Clark, R., Calvin, W., Sherman, D., Brown, R., 1992. Evidence for ammonium-bearing minerals on Ceres. Science 255, 1551–1553.

Kivelson, M.G., Khurana, K.K., Russell, C.T., Volwerk, M., Walker, R.J., Zimmer, C., 2000. Galileo magnetometer measurements: a stronger case for a subsurface ocean at Europa. Science 289 (5483), 1340–1343. https://doi.org/10.1126/science.289.5483.1340.

Kivelson, M.G., Khurana, K., Volwerk, M., 2002. The permanent and inductive magnetic moments of Ganymede. Icarus 157, 507–522.

Konopliv, A.S., Asmar, S.W., Folkner, W.M., Karatekin, Ö., Nunes, D.C., Smrekar, S.E., Yoder, C.F., Zuber, M.T., 2011. Mars high resolution gravity fields from MRO, Mars seasonal gravity, and other dynamical parameters. Icarus 211, 401–428. http://doi.org/10.1016/j.icarus.2010.10.004.

Küppers, M., O'Rourke, L., Bockelée-Morvan, D., Zakharov, V., Lee, S., von Allmen, P., Carry, B., Teyssier, D., Marston, A., Müller, T., Crovisier, J., Barucci, M.A., Moreno, R., 2014. Localized sources of water vapour on the dwarf planet (1) Ceres. Nature 505, 525–527. http://doi.org/10.1038/nature12918.

Khurana, K.K., Kivelson, M.G., Stevenson, D.J., Schubert, G., Russell, C.T., Walker, R.J., Polanskey, C., 1998. Induced magnetic fields as evidence for subsurface oceans in Europa and Callisto. Nature 395, 777–780.

Lebofsky, L.A., 1978. Asteroid 1 Ceres: evidence for water of hydration. Mon. Not. R. Astron. Soc. 182, 17–21.

Lebofsky, L.A., 1980. Infrared reflectance spectra of asteroids: a search for water of hydration. Astron. J. 85, 573–585.

Lebofsky, L.A., Feierberg, M.A., Tokunaga, A.T., Larson, H.P., Johnson, J.R., 1981. The 1.7- to 4.2-μm spectrum of asteroid 1 Ceres: evidence for structural water in clay minerals. Icarus 48, 453–459. http://doi.org/10.1016/0019-1035(81)90055-5.

Le Corre, L., Reddy, V., Nathues, A., Cloutis, E.A., 2011. How to characterize terrains on 4 Vesta using Dawn Framing Camera color bands? Icarus 216, 376–386.

Le Corre, L., et al., 2015. Exploring exogenic sources for the olivine on Asteroid (4) Vesta. Icarus 258, 483–499.

Lorenz, R.D., Stiles, B.W., Kirk, R.L., Allison, M.D., Persi del Marmo, P., Iess, L., Lunine, J.I., Ostro, S.J., Hensley, S., 2008. Titan's rotation reveals an internal ocean and changing zonal winds. Science 319, 1649–1651. http://doi.org/10.1126/science.1151639.

Lorenz, R.D., Kirk, R.L., Hayes, A.G., Anderson, Y.Z., Lunine, J.I., Tokano, T., Turtle, E.P., Malaska, M.J., Soderblom, J.M., Lucas, A., et al., 2014. A Radar map of Titan seas: tidal dissipation and ocean mixing through the throat of Kraken. Icarus 237, 9–15.

Lorenz, R.D., 2016. Europa ocean sampling by plume flythrough: astrobiological expectations. Icarus 267, 217–219.

Lunine, J.I., Stevenson, D.J., Yung, Y.L., 1983. Ethane ocean on Titan. Science 222, 1229–1230.

Manga, M., Wang, C.Y., 2007. Pressurized oceans and the eruption of liquid water on Europa and Enceladus. Geophys. Res. Lett. 34, L07202. http://doi.org/10.1029/2007GL029297.

Marchi, S., Raponi, A., Prettyman, T., De Sanctis, M.C., Castillo-Rogez, J., Raymond, C., Ammannito, E., Bowling, T., Ciarniello, M., Kaplan, H., Palomba, E., Russell, C., Vinogradoff, V., Yamashita, N., 2018. An aqueously altered carbon-rich Ceres. Nat. Astron. 3, 140–145.

Matthews, C.N., Minard, R.D., 2006. Hydrogen cyanide polymers, comets and the origin of life. Farad. Discuss. 133, 393–401.

Matson, D.L., Castillo, J.C., Lunine, J., Johnson, T.V, 2007. Enceladus' plume: compositional evidence for a hot interior. Icarus 187, 569–573. http://doi.org/10.1016/j.icarus.2006.10.016.

McCord, T.B., Gaffey, M.J., 1974. Asteroids: surface composition from reflection spectroscopy. Science 186, 352–355. http://doi.org/10.1126/science.186.4161.352.

McCord, T.B., Sotin, C., 2003. The small planet Ceres: models of evolution and predictions of current state. Bull. Am. Astron. Soc. 35, 979.

McCord, T., Sotin, C., 2004. Ceres as an exciting objective for exploration, 35th COSPAR Scientific Assembly, 18–25 July 2004, Paris, France, p. 1148.

McCord, T.B., Sotin, C., 2005. Ceres: evolution and current state. J. Geophys. Res. 110, E05009. http://doi.org/10.1029/2004JE002244.

McCord, T.B., McFadden, L.A., Russell, C.T., Sotin, C., Thomas, P.C., 2006. Ceres, vesta, and pallas: protoplanets, not asteroids. Eos.

Trans. Am. Geophys. Union 87, 105–109. http://doi.org/10.1029/2006E0100002.

McCord, T.B., Castillo-Rogez, J.C., Rivkin, A., 2011. Ceres: its origin, evolution and structure and Dawn's potential contribution. Space Sci. Rev. 163, 63–76.

McKinnon, W.B., Zolensky, M.E., 2003. Sulfate content of Europa's ocean and shell: evolutionary considerations and some geological and astrobiological implications. Astrobiology 3, 879–897.

McKinnon, W.B., 2015. Effect of Enceladus's rapid synchronous spin on interpretation of Cassini gravity. Geophys. Res. Lett. 42, 2137–2143.

McSween, H.Y., Huss, G.R., 2010. Cosmochemistry. Cambridge Univ. Press, Cambridge, pp. 565.

Milliken, R.E., Rivkin, A.S., 2009. Brucite and carbonate assemblages from altered olivine-rich materials on Ceres. Nat. Geosci. 2, 258–261. http://doi.org/10.1038/ngeo478.

Nagel, K., Breuer, D., Spohn, T., 2004. A model for the interior structure, evolution, and differentiation of Callisto. Icarus 169, 402–412.

Nathues, A., et al., 2014. Detection of Serpentine in Exogenic Carbonaceous Chondrite Material on Vesta from Dawn FC Data. Icarus 239, 222–237.

Nathues, A., et al., 2015. Exogenic olivine on Vesta from Dawn Framing Camera color data. Icarus 258, 467–482.

Nathues, A., Hoffmann, M., Schaefer, M., Le Corre, L., Reddy, V., Platz, T., Cloutis, E.A., Christensen, U., Kneissl, T., Li, J.Y., Mengel, K., Schmedemann, N., Schaefer, T., Russell, C.T., Applin, D.M., Buczkowski, D.L., Izawa, M.R., Keller, H.U., O'Brien, D.P., Pieters, C.M., Raymond, C.A., Ripken, J., Schenk, P.M., Schmidt, B.E., Sierks, H., Sykes, M.V., Thangjam, G.S., Vincent, J.B., 2015. Sublimation in bright spots on (1) Ceres. Nature 528, 237–240. http://doi.org/10.1038/nature15754.

Neumann, W., Breuer, D., Spohn, T., 2015. Modelling the internal structure of Ceres: coupling of accretion with compaction by creep and implications for the water-rock differentiation. Astron. Astrophys. 584, A117.

Neveu, M., Desch, S.J., Castillo-Rogez, J.C., 2015. Core cracking and hydrothermal circulation can profoundly affect Ceres' geophysical evolution. J. Geophys. Res. Planets 120, 123–154. http://doi.org/10.1002/2014JE004714.

Neveu, M., Desch, S.J., 2015. Geochemistry, thermal evolution, and cryovolcanism on Ceres with a muddy ice mantle. Geophys. Res. Lett. 42, 10197–10206. http://doi.org/10.1002/2015GL066375.

Neveu, M., Desch, S.J., Castillo-Rogez, J.C., 2017. Aqueous geochemistry in icy world interiors: fate of antifreezes and radionuclides. Cosmochim. Geochim. Acta 212, 324–371.

Neveu, M., Rhoden, A.R., 2019. Evolution of Saturn's midsized moons. Nat. Astron. 3, 543–552.

Nimmo, F., Hamilton, D.P., McKinnon, W.B., Schenk, P.M., Binzel, R.P., Bierson, C.J., Beyer, R.A., Moore, J.M., Stern, S.A., Weaver, H.A., Olkin, C.B., Young, L.A., Smith, K.E.and New Horizons Geology, Geophysics & Imaging Theme Team, 2016. Reorientation of Sputnik Planitia implies a subsurface ocean on Pluto. Nature 540, 94–96.

Ockman, N., 1957. Thesis. University of Michigan.

Pappalardo, R.T., et al., 1999. Does Europa have a subsurface ocean?: Evaluation of the geological evidence. J. Geophys. Res. 104, 24,015–24,055.

Park, R.S., Konopliv, A.S., Bills, B.G., Rambaux, N., Castillo-Rogez, J.C., Raymond, C.A., Vaughan, A.T., Ermakov, A.I., Zuber, M.T., Fu, R.R., Toplis, M.J., Russell, C.T., Nathues, A., Preusker, F., 2016. A partially differentiated interior for (1) Ceres deduced from its gravity field and shape. Nature 537, 515–517. http://doi.org/10.1038/nature18955.

Parker, J.W., Stern, S.A., Thomas, P.C., Festou, M.C., Merline, W.J., Young, E.F., 2002. Analysis of the first disk-resolved images of Ceres from ultraviolet observation with the Hubble Space Telescope. Astron. J. 123, 549–557.

Patel, B.H., Percivalle, C., Riston, D.J., Duffy, C.D., Sutherland, J.D., 2015. Common origins of RNA, protein and lipid precursors in a cyanosulfidic protometabolism. Nat. Chem. 7, 301–307.

Pearson, V.K., Kearsley, A.T., Sephton, M.A., Bland, P.A., Franchi, I.A., Gilmour, I., 2002. Clay mineral–organic matter relationships in the early Solar System. Meteorit. Planet. Sci. 37, 1829–1833.

Peebles, C., 2000. Asteroids: A History. Smithsonian Institution Scholarly Press, Washington, DC., vii, 280 pages.

Piazzi, G., 1802. Della scoperta del nuovo planeta Cerere Ferdinandea. Palermo Observ., Palermo, Italy.

Platz, T., Nathues, A., Schorghofer, N., Preusker, F., Mazarico, E., Schröder, S.E., Byrne, S., Kneissl, T., Schmedemann, N., Combe, J.-P., Schäfer, M., Thangjam, G.S., Hoffmann, M., Gutierrez-Marques, P., Landis, M.E., Dietrich, W., Ripken, J., Matz, K.-D., Russell, C.T., 2016. Surface water-ice deposits in the northern shadowed regions of Ceres. Nat. Astron. 1, 0007.

Pontefract, A., Zhu, T.F., Walker, V.K., Hepburn, H., Lui, C., Zuber, M.T., Ruvkun, G., Carr, C.E., 2017. Microbial diversity in a hypersaline sulfate lake: a terrestrial analog of ancient Mars. Front. Microbiol. 8, 1819. http://doi.org/10.3389/fmicb.2017.01819.

Postberg, F., Kempf, S., Schmidt, J., Brilliantov, N., Beinsen, A., Abel, B., Buck, U., Srama, R., 2009. Sodium salts in E-ring ice grains from an ocean below the surface of Enceladus. Nature 459, 1098–1101. http://doi.org/10.1038/nature08046.

Prieto-Ballesteros, O., Kargel, J.S., 2005. Thermal state and complex geology of a heterogeneous salty crust of Jupiter's satellite, Europa. Icarus 173 (1), 212–221. doi:10.1016/j.icarus.2004.07.019.

Prettyman, T.H., et al., 2016. Extensive water ice within Ceres aqueously altered regolith: Evidence from nuclear spectroscopy. Science 6765, 1–11. doi:10.1126/science.aah6765.

Prettyman, T.H., Feldman, W.C., McSween, H.Y., Dingler, R.D., Enemark, D.C., Patrick, D.E., Storms, S.A., Hendricks, J.S., Morgenthaler, J.P., Pitman, K.M., Reedy, R.C., 2011. Dawn's gamma ray and neutron detector. Space Sci. Rev. 163, 371–459. http://doi.org/10.1007/s11214-011-9862-0.

Prettyman, T.H., Yamashita, N., Toplis, M.J., McSween, H.Y., Schörghofer, N., Marchi, S., Feldman, W.C., Castillo-Rogez, J., Forni, O., Lawrence, D.J., Ammannito, E., Ehlmann, B.L., Sizemore, H.G., Joy, S.P., Polanskey, C.A., Rayman, M.D., Raymond, C.A., Russell, C.T., 2017. Extensive water ice within Ceres' aqueously altered regolith: evidence from nuclear spectroscopy. Science 355 (6320), 55–59. http://doi.org/10.1126/science.aah6765.

Prettyman, T.H., Yamashita, N., Ammannito, E., Schorghofer, N., Castillo-Rogez, J., McSween, H.Y., Toplis, M.J., Forni, O., Marchi, S., Feldman, W.C., Ehlmann, B.L., Lawrence, D.J., Sizemore, H.G., Rayman, M.D., Polanskey, C.A., Joy, S.P., Raymond, C.A., Russell, C.T., 2017. Elemental composition of Vesta and Ceres, EPSC Abstracts Vol. 11, EPSC2017-966, European Planetary Science Congress 2017.

Preusker, F., Scholten, F., Matz, K.-D., Roatsch, T., Jaumann, R., Raymond, C.A., Russell, C.T., 2012, Topography of Vesta from Dawn FC stereo images (EPSC abstract 2012-428-1). European Planetary Science Congress.

Raponi, A., De Sanctis, M.C., Frigeri, A., Ammannito, E., Ciarniello, M., Formisano, M., Combe, J-P., Magni, G., Tosi, F., Carrozzo, F.G., Fonte, S., Giardino, M., Joy, S.P., Polanskey, C.A., Rayman, M.D., Capaccioni, F., Capria, M.T., A.Longobardo, E.P, Zambon, F., Raymond, C.A., Russell, C.T., 2018. Variations in the amount of water

ice on Ceres' surface suggest a seasonal water cycle. Sci. Adv. 4 (3), eaao3757. http://doi.org/10.1126/sciadv.aao3757.
Raymond, C.A., Jaumann, R., Nathues, A., Sierks, H., Roatsch, T., Preusker, F., Scholten, F., Gaskell, R.W., Jorda, L., Keller, H.-U., Zuber, M.T., Smith, D.E., Mastrodemos, N., Mottola, S., 2011. The Dawn topography investigation. Space Sci. Rev. 163, 487–510.
Reddy, V., et al., 2012a. Color and albedo heterogeneity of Vesta from Dawn. Science 336, 700–704.
Reddy, V., et al., 2012b. Delivery of dark material to Vesta via carbonaceous chondritic impacts. Icarus 221, 544–559.
Rivkin, A.S., Volquardsen, E.L., Clark, B.E., 2006. The surface composition of Ceres: discovery of carbonates and iron-rich clays. Icarus 185, 563–567. http://doi.org/10.1016/j.icarus.2006.08.022.
Rivkin, S., McFadden, L.A., Binzel, R.P., Sykes, M., 2006. Rotationally-resolved spectroscopy of Vesta I: 2-4 μm region. Icarus 180, 464–472.
Rivkin, A.S., Volquardsen, E.L., 2009. Rotationally resolved spectra of Ceres in the 3-mm region. Icarus 206, 327–333.
Rivkin, A.S., Li, J.-Y., Milliken, R.E., Lim, L.F., Lovell, A.J., Schmidt, B.E., McFadden, L.A., Cohen, B.A., 2011. The surface composition of Ceres. Space Sci. Rev. 163, 95–116. http://doi.org/10.1007/s11214-010-9677-4.
Roatsch, Th., Kersten, E., Matz, K.-D., Preusker, F., Scholten, F., Jaumann, R., Raymond, C.A., Russell, C.T., 2016. Ceres survey atlas derived from Dawn Framing Camera images. Planet. Space Sci. 121, 115–120. http://doi.org/10.1016/j.pss.2015.12.005.
Robuchon, G., Nimmo, F., 2011. Thermal evolution of Pluto and implications for surface tectonics and a subsurface ocean. Icarus 216, 426–439.
Russell, C.T., Capaccioni, F., Coradini, A., Christensen, U., de Sanctis, M.C., Feldman, W.C., Jaumann, R., Keller, H.U., Konopliv, A., McCord, T.B., McFadden, L.A., McSween, H.Y., Mottola, S., Neukum, G., Pieters, C.M., Prettyman, T.H., Raymond, C.A., Smith, D.E., Sykes, M.V., Williams, B., Zuber, M.T., 2006. Dawn discovery mission to Vesta and Ceres: present status. Adv. Space Res. 38, 2043–2048.
Russell, C.T., Raymond, C.A., 2011. The Dawn mission to Vesta and Ceres. Space Sci. Rev. 163, 3–23.
Russell, C.T., et al., 2012. Dawn at Vesta: Testing the Protoplanetary Paradigm. Science 336 (6082), 684–686. https://doi.org/10.1126/science.1219381.
Russell, C.T., et al., 2013. Dawn completes its mission at 4 Vesta. Meteoritics & Planetary Science 48 (11), 2076–2089. doi:10.1111/maps.120912076.
Russell, C.T., McSween, H.Y., Jaumann, R., Raymond, C.A., 2015. The Dawn Mission to Ceres and Vesta. In: Asteroids IV, P., Michel, F.E., DeMeo, W.F., Bottke (Eds.), (Univ. of Arizona Press, 2015). The University of Arizona Press, Tucson, AZ, pp. 419–432.
Russell, C.T., Raymond, C.A., Ammannito, E., Buczkowski, D.L., De Sanctis, M.C., Hiesinger, H., Jaumann, R., Konopliv, A.S., McSween, H.Y., Nathues, A., Park, R.S., Pieters, C.M., Prettyman, T.H., McCord, T.B., McFadden, L.A., Mottola, S., Zuber, M.T., Joy, S.P., Polanskey, C., Rayman, M.D., Castillo-Rogez, J.C., Chi, P.J., Combe, J.-Ph., Ermakov, A., Fu, R.R., Hoffmann, M., Jia, Y.D., King, S.D., Lawrence, D.J., Li, J.-Y., Marchi, S., Preusker, F., Roatsch, T., Ruesch, O., Schenk, P., Villarreal, M.N., Yamashita, N., 2016. Dawn arrives at Ceres: exploration of a small, volatile-rich world. Science 353, 1008.
Saladino, R., Crestini, C., Pino, S., Costanzo, G., Di Mauro, E., 2012. Formamide and the origin of life. Phys. Life Rev. 9, 84–104.
Sapers, H.M., Ronholm, J., Raymond-Bouchard, I., Comrey, R., Osinski, G.R., Whyte, L.G., 2017. Biological characterization of microenvironments in a hypersaline cold spring Mars analog. Front. Microbiol. 8.

Schäfer, T., Nathues, A., Mengel, K., Izawa, M.R.M., Cloutis, E.A., Schäfer, M., Hoffmann, M., 2016. Spectral parameters for Dawn FC color data: Carbonaceous chondrites and aqueous alteration products as potential cerean analog materials. Icarus 265, 149–160. https://doi.org/10.1016/j.icarus.2015.10.005.
Schmidt, B.E., Hughson, K.H.G., Chilton, H.T., Scully, J.E.C., Platz, T., Nathues, A., Sizemore, H., Bland, M.T., Byrne, S., Marchi, S., O'Brien, D.P., Schorghofer, N., Hiesinger, H., Jaumann, R., Pasckert, J.H., Lawrence, J.D., Buzckowski, D., Castillo-Rogez, J.C., Sykes, M.V., Schenk, P.M., DeSanctis, M.-C., Mitri, G., Formisano, M., Li, J.-Y., Reddy, V., LeCorre, L., Russell, C.T., Raymond, C.A., 2017. Geomorphological evidence for ground ice on dwarf planet Ceres. Nat. Geosci. 10, 338–343.
Schorghofer, N., 2016. Predictions of depth-to-ice on asteroids based on an asynchronous model of temperature, impact stirring, and ice loss. Icarus 276, 88–95. http://doi.org/10.1016/j.icarus.2016.04.037.
Schröder, S.E., Mottola, S., Carsenty, U., Ciarniello, M., Jaumann, R., Li, J.-Y., Longobardo, A., Palmer, E., Pieters, C., Preusker, F., Raymond, C.A., Russell, C.T., 2017. Resolved spectrophotometric properties of the Ceres surface from Dawn Framing Camera images. Icarus 288, 201–225.
Sekine, Y., Shibuya, T., Postberg, F., Hsu, H.W., Suzuki, K., Masaki, Y., Kuwatani, T., Mori, M., Hong, P.K., Yoshizaki, M., Tachibana, S., 2015. High-temperature water–rock interactions and hydrothermal environments in the chondritelike core of Enceladus. Nat. Commun. 6, 8604.
Shock, E.L., 1992. Chemical environments of submarine hydrothermal systems. Origins Life Evol. Biosph. 22, 67–107. http://doi.org/10.1007/BF01808019 (Chapter 5).
Sierks, H., Keller, H.U., Jaumann, R., Michalik, H., Behnke, T., Bubenhagen, F., Büt- tner, I., Carsenty, U., Christensen, U., Enge, R., Fiethe, B., Gutiérrez Marqués, P., Hartwig, H., Krüger, H., Kühne, W., Maue, T., Mottola, S., Nathues, A., Reiche, K.-U., Richards, M.L., Roatsch, T., Schröder, S.E., Szemerey, I., Tschentscher, M., 2011. The Dawn framing camera. Space Sci. Rev. 163, 263–327. http://doi.org/10.1007/s11214-011-9745-4.
Sizemore, H.G., et al., 2017. Pitted terrains on (1) Ceres and implications for shallow subsurface volatile distribution. Geophys. Res. Lett. 44, 6570–6578. doi:10.1002/2017GL073970.
Sizemore, H.G., Schmidt, B.E., Buczkowski, D.L., Sori, M.M., Castillo-Rogez, J.C., Berman, D.C., Ahrens, C., Chilton, H.T., Hughson, K.H.G., Duarte, K., Otto, K.A., Bland, M.T., Neesemann, A., Scully, J.E.C., Crown, D.A., Mest, S.C., Williams, D.A., Platz, T., Schenk, P., Landis, M.E., Marchi, S., Schorghofer, N., Quick, L.C., Prettyman, T.H., De Sanctis, M.C., Nass, A., Thangjam, G., Nathues, A., Russell, C.T., Raymond, C.A, 2019. A global inventory of ice-related morphological features on dwarf planet Ceres: implications for the evolution and current state of the cryosphere. J. Geophys. Res. 124, 1650–1689.
Sori, M.M., Sizemore, H.G., Byrne, S., Bramson, A.M., Bland, M.T., Stein, N.T., Russell, C.T., 2018. Cryovolcanic rates on Ceres revealed by topography. Nat. Astron. 2, 946–950.
Spaun, N.A., Head, J.W., 2001. A model of Europa's crustal structure: Recent Galileo results and implications for an ocean. J. Geophys. Res. 106, 7567–7576.
Takir, D., Emery, J.P., McSween, H.Y., Hibbits, C.A., Clark, R.N., Pearson, N., Wang, A., 2013. Nature and degree of aqueous alteration in CM and CI carbonaceous chondrites. Meteorit. Planet. Sci. 48, 1618–1637.
Thangjam, G, et al., 2013. Lithologic mapping of HED terrains on Vesta using Dawn Framing Camera color data. Meteoritics & Planetary Science 48 (11), 2199–2210.

Thangjam, G., et al., 2014. Olivine-rich exposures at Bellicia and Arruntia craters on (4) Vesta from Dawn FC. Meteorit. Planet. Sci. 49, 1831–1850.

Thomas, P.C., Parker, J.W., McFadden, L.A., Russell, C.T., Stern, S.A., Sykes, M.V., Young, E.F., 2005. Differentiation of the asteroid Ceres as revealed by its shape. Nature 437, 224–226. http://doi.org/10.1038/nature03938.

Thomas, P.C., Tajeddine, R., Tiscareno, M.S., Burns, J.A., Joseph, J., Loredo, T.J., Helfenstein, P., Porco, C., 2016. Enceladus's measured physical libration requires a global subsurface ocean. Icarus 264, 37–47.

Thomson, R.E., Delaney, J.R., 2001. Evidence for a weakly stratified Europan ocean sustained by seafloor heat flux. J. Geophys. Res. 106, 12, 355–12,365.

Titus, T., 2015. Ceres: predictions for near-surface water ice stability and implications for plume generating processes. Geophys. Res. Lett. 42, 2130–2136. http://doi.org/10.1002/2015GL063240.

Travis, B.J., Bland, P.A., Feldman, W.C., Sykes, M.V., 2015. Unconsolidated Ceres model has a warm convecting rocky core and a convecting mud ocean. Lunar Planet Sci. Conf. 46, 2360.

Travis, B.J., Bland, P.A., Feldman, W.C., Sykes, M., 2018. Hydrothermal dynamics in a CM-based model of Ceres. Meteorit. Planet. Sci. 53, 2008–2032.

Tyler, R.H., 2009. Ocean tides heat Enceladus. Geophys. Res. Lett. 36 (15), L15205. http://doi.org/10.1029/2009GL038300.

Vernazza, P., Mothé-Diniz, T., Barucci, M.A., Birlan, M., Carvano, J.M., Strazzulla, G., Fulchignoni, M., Migliorini, A., 2005. Analysis of near-IR spectra of 1 Ceres and 4 Vesta, targets of the Dawn mission. Astron. Astrophys. 436, 1113–1121. http://doi.org/10.1051/0004-6361:20042506.

Villarreal, M.N., Russell, C.T., Luhmann, J.G., Thompson, W.T., Prettyman, T.H., A'Hearn, M.F., Küppers, M., O'Rourke, L., Raymond, C.A., 2017. The dependence of the cerean exosphere on solar energetic particle events. Astrophys. J. Lett. 838, L8.

Waite, J.H., Glein, C.R., Perryman, R.S., Teolis, B.D., Magee, B.A., Miller, G., Grimes, J., Perry, M.E., Miller, K.E., Bouquet, A., Lunine, J.I., 2017. Cassini finds molecular hydrogen in the Enceladus plume: evidence for hydrothermal processes. Science 356, 155–159.

Webster, W.J., et al., 1988. The microwave spectrum on asteroid Ceres. Astrophys. J. 95, 1263–1268.

Weidenschilling, S.J., 2006. Formation of planetesimals and accretion of the terrestrial planets. Space Sci. Rev. 92, 295–310.

Wilson, L., Keil, K., Browning, L.B., Krot, A.N., Bourcier, W., 1999. Early aqueous alteration, explosive disruption, and reprocessing of asteroids. Meteorit. Planet. Sci. 34, 541–557.

Wynn-Williams, D.D., Cabrol, N.A., Grin, E.A., Haberle, R.M., Stoker, C.R., 2001. Brines in seepage channels as eluants for subsurface relict biomolecules on Mars? Astrobiology 1, 165–184. http://doi.org/10.1089/153110701753198936.

Zimmer, C., Khurana, K.K., Kivelson, M.G., 2000. Subsurface oceans on Europa and Callisto: constraints from Galileo magnetometer observations. Icarus 147, 329–347.

Zolotov, M.Y., Shock, E.L., 2001. Composition and stability of salts on the surface of Europa and their oceanic origin. J. Geophys. Res. 106, 32815–32828. http://doi.org/10.1029/2000JE001413.

Bibliography

A'Hearn, M.F., Feldman, P.D., 1992. Water vaporization on Ceres. Icarus 98, 54–60.

Drummond, J.D., Fugate, R.Q., Christou, J.C., 1998. Full adaptive optics images of asteroids Ceres and Vesta: rotational poles and triaxial ellipsoid dimensions. Icarus 132, 80–99.

Fanale, F.P., Salvail, J.R., 1989. The water regime of asteroids 1 Ceres. Icarus 82, 97–110.

King, T.V.V., Clark, R.N., Calvin, W.M., Sherman, D.M., Brown, R.H., 1992. Evidence for ammonium-bearing minerals on Ceres. Science 255, 1551–1553.

Larson, H.P., Feierberg, M.A., Fink, U., Smith, H.A., 1979. Remote spectroscopic identification of carbonaceous chondrite mineralogies: applications to Ceres and Pallas. Icarus 39, 257–271.

Lebofsky, L.A., 1978. Asteroid 1 Ceres: evidence for water of hydration. Mon. Not. R. Astron. Soc. 182, 17–21.

Lebofsky, L.A., Spencer, J.R., 1989. Radiometry and thermal modeling of asteroids. In: Binzel, R.P., Gehrels, T., Matthews, J.S. (Eds.), Asteroids II. Univ. of Ariz. Press, Tucson, pp. 128–147.

Lebofsky, L.A., Feierberg, M.A., Tokunaga, A.T., Larson, H.P., Johnson, J.R., 1981. The 1.7- to 4.2-μm spectrum of asteroid 1 Ceres: evidence for structural water in clay minerals. Icarus 48, 453–459.

Millis, R.L., et al., 1987. The size, shape, density and albedo of Ceres from its occultation of BD + 8 deg 471. Icarus 72, 507–518.

Parker, J.W., Stern, S.A., Thomas, P.C., Festou, M.C., Merline, W.J., Young, E.F., 2002. Analysis of the first disk-resolved images of Ceres from ultraviolet observation with the Hubble Space Telescope. Astron. J. 123, 549–557.

Saint-Pe, O., Combes, M., Rigaut, F., 1993. Ceres surface properties by high-resolution imaging from Earth. Icarus 105, 271–281.

Viateau, B., Rapaport, M., 1998. The mass of (1) Ceres from its gravitational perturbations on the orbits of 9 asteroids. Astron. Astrophys. 334, 729–735.

Chapter 12

An ocean and volcanic seafloor hiding below the icy crust of Jupiter's Moon *Europa*—Plumes of water vapor rising over 160 km above its surface

12.1 Planet Jupiter and its water-bearing moons

Among all the planets in the solar system, Jupiter (Fig. 12.1) adorns a unique status as being the largest planet in the solar system. Fittingly, it was named after the king of the gods in Roman mythology. Jupiter helped revolutionize the way we saw the Universe and ourselves in 1610 when Galileo discovered Jupiter's four large moons—Io, Europa, Ganymede, and Callisto, now known as the *Galilean* moons. This was the first time that celestial bodies were seen circling an object other than Earth. This finding provided a major support of the Copernican view that Earth was not the center of the Universe.

12.1.1 Jupiter—the largest and the most massive planet in the solar system

Jupiter is the most massive planet in our solar system, more than twice as massive as all the other planets combined. Its atmosphere resembles that of the Sun, made up mostly of hydrogen and helium. Jupiter, the largest planet in the solar system, possesses 79 known moons orbiting around it. The immense volume of Jupiter could hold more than 1300 Earths (https://www.space.com/7-jupiter-largest-planet-solar-system.html). Planetary scientists think that Jupiter's three giant icy moons might possess liquid oceans underneath their frozen surfaces. Studies of Europa and Callisto indicate significant liquid water oceans beneath the surface (Khurana et al., 1998).

The colorful bands of Jupiter are arranged in dark belts and light zones created by strong east-west winds in the planet's upper atmosphere traveling more than 400 mph (640 km/h). The white clouds in the zones are made of crystals of frozen ammonia, while darker clouds of other chemicals are found in the belts. At the deepest visible levels are blue clouds. Far from being static, the stripes of clouds change over time. Inside the atmosphere, diamond rain may fill the skies.

The most extraordinary feature on Jupiter is undoubtedly the Great Red Spot (Fig. 12.2), a giant hurricane-like storm. The Great Red Spot is a persistent high-pressure

FIG. 12.1 **Image of Jupiter by *Pioneer 10* in 1974, showing a solid-looking spot on Jupiter's surface.** (*Source: https://commons.wikimedia.org/wiki/File:Pioneer_10_jup.jpg). This file is in the public domain in the United States because it was solely created by NASA. NASA copyright policy states that "NASA material is not protected by copyright unless noted."*

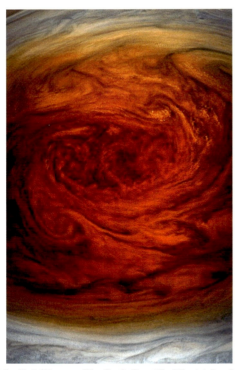

FIG. 12.2 Detail image of Jupiter's Great Red Spot, taken by the Juno spacecraft flyover (PERIJOVE 7) on July 11, 2017. *(Source: https://commons.wikimedia.org/wiki/File:Great_red_spot_juno_20170712.jpg). This file is in the <u>public domain</u> in the United States because it was solely created by NASA. NASA copyright policy states that "NASA material is not protected by copyright unless noted."*

region in the atmosphere of Jupiter, producing an anticyclonic storm, the largest in the solar system, 22° south of the planet's equator. Jupiter's Great Red Spot has been continuously observed since 1830. At its widest, the Great Red Spot is three times the diameter of the Earth, and its edge spins counter-clockwise around its center at a speed of about 225 mph (360 km/h). A wind-movement model constructed to understand the Great Red Spot succeeded in animating a mosaic image of Jupiter's Great Red Spot.

The color of the storm, which usually varies from brick red to slightly brown, may come from small amounts of sulfur and phosphorus in the ammonia crystals in Jupiter's clouds. The Great Red Spot on Jupiter is suspected to grow and shrink. Despite being long-lived, Jupiter's Great Red Spot hasn't been a model of consistency. While the massive feature has swirled across Jupiter for at least 200 years—possibly 350, if early telescope observations describe the same storm—it has been slowly shrinking. During the 19th century, and again when NASA's Voyager 1 and 2 spacecrafts sped by on the way to Saturn in 1979, the spot stretched well over two Earths wide. But Earth-based measurements today put the spot at only a third of the size measured by the Voyager probes (https://www.space.com/39066-jupiter-great-red-spot-depth-juno-spacecraft.html).

Jupiter is endowed with an ultraviolet aurora in its northern and southern polar regions. Comparisons of the northern and southern far ultraviolet (UV) auroral emissions of Jupiter from the Hubble Space Telescope (HST) or any other ultraviolet imager have mostly been made so far on a statistical basis or were not obtained with high sensitivity and resolution. Such observations are important to discriminate between different mechanisms responsible for the electron acceleration of the different components of the aurora such as the satellite footprints, the "main oval" or the polar emissions. The field of view of the Advanced Camera for Surveys (ACS) and Space Telescope Imaging Spectrograph (STIS) cameras on board HST (Fig. 12.3) is not wide enough to provide images of the full Jovian disk. In view of this lacuna, Gérard et al. (2013) compared the morphology of the north and south aurora observed 55 min apart and they pointed out similarities and differences. On one occasion HST pointed successively the two Polar Regions and auroral images were seen separated by only 3 min. This made it possible to compare the emission structure and the emitted power of corresponding regions. They found that most morphological features identified in one hemisphere have a conjugate counterpart in the other hemisphere. However, it was suspected that the power associated with conjugate regions of the main oval, diffuse or discrete equator-ward emission observed quasi-simultaneously may be different in the two hemispheres. The lack of symmetry of some polar emissions suggested that some of them could be located on open magnetic field lines.

Juno arrived at Jupiter on July 4, 2016, after a nearly 5-year flight. Since then, the spacecraft has made several science passes over the gas giant. In July 2017, it made its first close flyby of the Great Red Spot. The spacecraft's Microwave Radiometer probed the clouds surrounding the gigantic storm, measuring their depth in the atmosphere. The spot has been shrinking for quite some time, although the rate may be slowing in recent years.

NASA's Juno spacecraft is getting to the roots of Jupiter's famous Great Red Spot. New research, collected during the mission's first pass over the iconic storm, reveals that it extends far beneath the planet's surface. According to Andy Ingersoll—a Juno co-investigator—Juno found that the Great Red Spot's roots go 50–100 times deeper than Earth's oceans and are warmer at the base than they are at the top. Winds are associated with differences in temperature, and the warmth of the spot's base explains the ferocious winds experienced at the top of the atmosphere.

Jupiter is also known for its "Little Red Spot." An amazing color portrait of Jupiter's Little Red Spot (LRS) available with NASA/Johns Hopkins University Applied Physics Laboratory/Southwest Research Institute (visit https://www.space.com/16533-pluto-new-horizons-spacecraft-pictures.html) combines high-resolution images from the

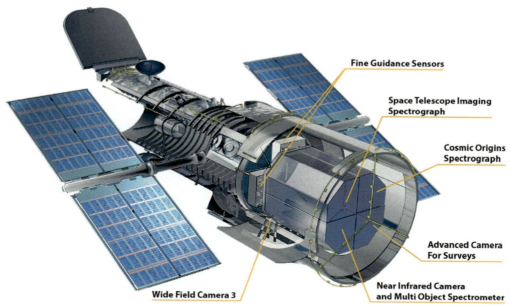

FIG. 12.3 Cutaway diagram of the Hubble Space Telescope. *Credits: NASA's Goddard Space Flight Center. (Source: https://www.nasa.gov/content/goddard/hubble-space-telescope-science-instruments).*

New Horizons probe's Long Range Reconnaissance Imager (LORRI), with color images taken nearly simultaneously by the Wide Field Planetary Camera 2 (WFPC2) on the Hubble Space Telescope.

The spacecraft also discovered two newly identified radiation bands. One lies just above the equator of the gas giant's atmosphere (Fig. 12.4) and includes hydrogen, oxygen, and sulfur ions moving at close to the speed of light (referred to as "relativistic" speeds). This radiation zone resides inside Jupiter's previously known radiation belts. The zone was identified by the mission's Jupiter Energetic Particle Detector Instrument (JEDI), enabled by Juno's unique close approach to the planet during the spacecraft's science flybys (2100 miles or 3400 km from the cloud tops) (https://www.jpl.nasa.gov/spaceimages/details.php?id=PIA22179).

FIG. 12.4 Juno detected a new radiation zone just above the atmosphere near Jupiter's equator. The diagram also indicates regions of high-energy, heavy ions at high latitudes. *(Source: https://www.jpl.nasa.gov/spaceimages/details.php?id=PIA22179). Image credit: NASA/JPL-Caltech/SwRI/JHUAPL.*

The new radiation zones could be found because Juno's unique orbit around Jupiter allows it to get really close to the cloud tops during science collection flybys, and the spacecraft literally flew through it. Juno scientists believe the particles creating this region of intense radiation are derived from energetic neutral atoms—that is, fast-moving atoms without an electric charge—coming from the tenuous gas around Jupiter's moons Io and Europa. The neutral atoms then become ions—atoms with an electric charge—as their electrons are stripped away by interaction with the planet's upper atmosphere (This discovery is discussed further by Kollmann et al., 2017).

Juno also found a second charged region around the planet's high latitudes, in realms never before explored by any spacecraft. The origin of these particles, which were detected by Juno's Stellar Reference Unit star camera, remains a mystery.

Jupiter's gargantuan (enormous) magnetic field is the strongest of all the planets in the solar system at nearly 20,000 times the strength of Earth's. It traps electrically charged particles in an intense belt of electrons and other electrically charged particles that regularly blasts the planet's moons and rings with a level of radiation more than 1000 times the lethal level for a human, damaging even heavily shielded spacecraft such as NASA's Galileo probe. The magnetosphere of Jupiter, which is composed of these fields and particles, swells out some 600,000 to 2 million miles (1 million to 3 million km) toward the Sun and tapers to a tail extending more than 600 million miles (1 billion km) behind Jupiter.

Jupiter spins faster than any other planet, taking a little under 10 h to complete a turn on its axis, compared with

24 h for Earth. This rapid spin makes Jupiter bulge at the equator and flatten at the poles, making the planet about 7 percent wider at the equator than at the poles.

Jupiter broadcasts radio waves strong enough to detect on Earth. These come in two forms—strong bursts that occur when Io, the closest of Jupiter's large moons, passes through certain regions of Jupiter's magnetic field, and continuous radiation from Jupiter's surface, and high-energy particles in its radiation belts. These radio waves could help scientists to probe the oceans on its moons (https://www.space.com/7-jupiter-largest-planet-solar-system.html).

12.1.2 NASA's Galileo mission to Jupiter and its mysterious moons

NASA's Galileo mission to Jupiter launched on October 18, 1989, from the John F. Kennedy Space Center in Florida. Galileo carried the following science instruments: Solid-state imaging camera, near-infrared mapping spectrometer, ultraviolet spectrometer, photopolarimeter radiometer, magnetometer, energetic particles detector, plasma investigation, plasma wave subsystem, dust detector, and heavy ion counter. Galileo's Jupiter arrival, and orbit insertion happened on Dec. 7, 1995. Although the primary mission was from October 1989 to December 1997, the mission was extended three times; from 1997 to 2003. Galileo orbited Jupiter 34 times during its entire mission. The total distance traveled, launch to impact, is about 2.8 billion miles (4.6 billion kilometers). More than 100 scientists from the United States, Great Britain, Germany, France, Canada, and Sweden carried out Galileo's experiments.

Galelio has to its credit, several "firsts": These include https://solarsystem.nasa.gov/missions/galileo/overview/:

- Galileo was the first spacecraft to orbit an outer planet.
- It was the first spacecraft to deploy an entry probe into an outer planet's atmosphere.
- It completed the first flyby and imaging of an asteroid (Gaspra, and later, Ida).
- It made the first, and so far, only, direct observation of a comet colliding with a planet's atmosphere (Shoemaker-Levy 9).
- It was the first spacecraft to operate in a giant planet magnetosphere long enough to identify its global structure and to investigate its dynamics.

While the aim of this mission was to study Jupiter and its mysterious moons, which it did with much success, this mission also became notable for discoveries during its journey to the gas giant. After discoveries including evidence for the existence of a saltwater ocean beneath the Jovian moon Europa's icy surface, extensive volcanic processes on the moon Io and a magnetic field generated by the moon Ganymede, Galileo plunged into Jupiter's atmosphere on September 21, 2003, to prevent an unwanted impact with Europa.

Galileo orbited Jupiter for almost 8 years and made close passes by all its major moons. Its camera and nine other instruments sent back reports that allowed scientists to determine, among other things, that Jupiter's icy moon Europa probably has a subsurface ocean with more water than the total amount found on Earth. They discovered that the volcanoes of the moon Io repeatedly and rapidly resurface the little world. They found that the giant moon Ganymede possesses its own magnetic field. Galileo even carried a small probe that it deployed and sent deep into the atmosphere of Jupiter, taking readings for almost an hour before the probe was crushed by overwhelming pressure (https://solarsystem.nasa.gov/missions/galileo/overview/).

Galileo mission executed the below 10 discoveries (https://solarsystem.nasa.gov/missions/galileo/overview/):

1. A global ocean of liquid water exists under the icy surface of Jupiter's moon Europa.
2. Galileo magnetic data provide evidence that the moons Ganymede and Callisto also likely have a liquid saltwater layer.
3. Galileo discovered the first moon around an asteroid—tiny Dactyl orbits the asteroid Ida.
4. Ganymede is the first moon known to possess a magnetic field.
5. Galileo's atmospheric probe discovered that Jupiter has thunderstorms many times larger than Earth's.
6. The probe measured atmospheric elements and found that their relative abundances were somewhat different than on the Sun, indicating Jupiter's evolution since the planet formed.
7. Io's extensive volcanic activity may be 100 times greater than that found on Earth. The heat and frequency of eruption are reminiscent of early Earth.
8. Io's complex plasma interactions in its atmosphere include support for currents and coupling to Jupiter's atmosphere.
9. Europa, Ganymede, and Callisto all provide evidence of a thin atmospheric layer known as a 'surface-bound exosphere.
10. Jupiter's ring system is formed by dust kicked up as interplanetary meteoroids smash into the planet's four small inner moons. The outermost ring is actually two rings, one embedded within the other.

12.1.3 Searching for water vapor in Jupiter's atmosphere

Microwave remote sounding from a spacecraft, flying by or in orbit around Jupiter, offers new possibilities for retrieving important and presently poorly understood properties of its atmosphere (see Karpowicz, and Steffes, 2011). Detection and measurement of atmospheric water vapor in the deep Jovian atmosphere using microwave radiometry have been discussed extensively by Janssen et al. (2005)

and de Pater et al. (2005). Janssen et al. (2005) showed that precise measurements of relative brightness temperature as a function of off-nadir emission angles, combined with absolute brightness temperature measurements, can allow planetary scientists to determine the global abundances of water and ammonia and study the dynamics and deep circulations of the atmosphere in the altitude range from the ammonia cloud region to depths greater than 30 bars in a manner which would not be achievable with ground-based telescopes.

The NASA Juno mission includes a six-channel microwave radiometer system (MWR) operating in the 1.3–50 cm wavelength range in order to retrieve water vapor abundances from the microwave signature of Jupiter (see, e.g., Matousek, 2005). In order to accurately interpret data from such observations, nearly 2000 laboratory measurements of the microwave opacity of H_2O vapor in a H_2/He atmosphere have been conducted by Karpowicz and Steffes (2011) in the 5–21 cm wavelength range (1.4–6 GHz) at pressures from 30 mbar to 101 bar and at temperatures from 330 to 525 K. It was found that the mole fraction of H_2O (at maximum pressure) ranged from 0.19% to 3.6% with some additional measurements of pure H_2O. These results have enabled development of the first model for the opacity of gaseous H_2O in an H_2/He atmosphere under Jovian conditions developed from actual laboratory data. The new model was based on a terrestrial model of Rosenkranz (1998), with substantial modifications to reflect the effects of Jovian conditions. The new model for water vapor opacity dramatically outperformed previous models and is expected to provide reliable results for temperatures from 300 to 525 K, at pressures up to 100 bars, and at frequencies up to 6 GHz. These results are expected to significantly reduce the uncertainties in the retrieval of Jovian atmospheric water vapor abundances from the microwave radiometric measurements from the NASA Juno mission, as well as provide a clearer understanding of the role deep atmospheric water vapor may play in the decimeter-wavelength spectrum of Saturn.

12.1.4 Magnetic field production and surface topography of *Ganymede*

One of the great discoveries of NASA's Galileo mission was the presence of an intrinsically produced magnetic field at Ganymede. The magnetosphere of Ganymede forms due to the interaction between its intrinsic magnetic field and Jupiter's magnetospheric environment (Kivelson et al., 1996, 1997). The internal magnetic field of Ganymede, whose equatorial surface strength is ~7 times larger than the ambient Jovian field, stands off the incident Jovian plasma at a distance of about 1 RG upstream from the moon's surface (RG, radius of Ganymede = 2634 km).

Generation of the relatively strong (750 nT) magnetic field likely requires dynamo action in Ganymede's metallic core, but how such a dynamo has been maintained into the present epoch remains uncertain. Using a one-dimensional, three-layer thermal model of Ganymede, Bland et al. (2008) found that magnetic field generation can only occur if the sulfur mass fraction in Ganymede's core is very low ($\lesssim 3\%$) or very high ($\gtrsim 21\%$), and the silicate mantle can cool rapidly (i.e. it has a viscosity like wet olivine). However, these requirements are not necessarily compatible with cosmochemical and physical models of the satellite. Bland et al. (2008) therefore investigated an alternative scenario for producing Ganymede's magnetic field in which passage through an eccentricity pumping Laplace-like resonance in Ganymede's past enables present-day dynamo action in the metallic core.

Note that in celestial mechanics, an orbital resonance occurs when orbiting bodies exert a regular, periodic gravitational influence on each other, usually because their orbital periods are related by a ratio of small integers. Most commonly this relationship is found for a pair of objects. The physical principle behind orbital resonance is similar in concept to pushing a child on a swing, where the orbit and the swing both have a natural frequency, and the other body doing the "pushing" will act in periodic repetition to have a cumulative effect on the motion. Orbital resonances greatly enhance the mutual gravitational influence of the bodies, i.e., their ability to alter or constrain each other's orbits. In most cases, this results in an unstable interaction, in which the bodies exchange momentum and shift orbits until the resonance no longer exists. Under some circumstances, a resonant system can be stable and self-correcting, so that the bodies remain in resonance. An example of orbital resonance is the 1:2:4 resonance of Jupiter's moons Ganymede, Europa, and Io. (https://en.wikipedia.org/wiki/Orbital_resonance#Laplace_resonance).

If sufficient tidal dissipation occurs in Ganymede's silicate mantle during resonance passage, silicate temperatures can undergo a runaway which prevents the core from cooling until the resonance passage ends. The rapid silicate and core cooling that follows resonance escape triggers dynamo action via thermal and/or compositional convection. To test the feasibility of this mechanism Bland et al. (2008) coupled their thermal model with an orbital evolution model to examine the effects of resonance passage on Ganymede's silicate mantle and metallic core. They found that, contrary to expectations, there are no physically plausible scenarios in which tidal heating in the silicates is sufficient to cause the thermal runaway necessary to prevent core cooling. These findings are robust to variations in the silicate rheology (the branch of physics that deals with the deformation and flow of matter, especially the non-Newtonian flow of liquids and the plastic flow of solids), tidal dissipation factor of Jupiter (Q_J), structure of the ice shell, and the inclusion of partial melting in the silicate mantle. Resonance passage, therefore, appears unlikely to explain Ganymede's magnetic field and, therefore, Bland et al. (2008) reckon that the special conditions described above should explain the presence of the field.

The internal magnetic field of Ganymede, whose equatorial surface strength is ~7 times larger than the ambient Jovian field, stands off the incident Jovian plasma at a distance of about 1 R_G upstream from the moon's surface (R_G, radius of Ganymede = 2634 km). According to Jia et al. (2009), "At Ganymede's orbit ~15 R_J (R_J, radius of Jupiter = 71,492 km), the corotating plasma overtakes the moon from the moon's trailing edge because its speed is greater than Ganymede's Keplerian speed. The typical flow speed of the ambient plasma relative to Ganymede is less than the magneto-sonic speed. Under such circumstances, the corotating plasma directly impinges on and interacts with Ganymede's magnetosphere without being modified by a bow shock such as those form upstream of planetary magnetospheres. Since Ganymede's intrinsic field is nearly antiparallel to the external field near the equator, magnetic reconnection couples the mini-magnetosphere of Ganymede with Jupiter's giant magnetosphere. In contrast to the highly fluctuating solar wind with unpredictable variations in plasma and magnetic field, the plasma at Ganymede's orbit imposes external field and plasma conditions that vary slowly (with the nearly 10.5-h synodic period of Jupiter's rotation) and the magnetic field remains in a favorable orientation (southward in this case) for reconnection. Therefore, Ganymede's magnetosphere provides us with a good opportunity to investigate the reconnection process in a relatively stable external environment."

Magnetohydrodynamics (MHD; also called magnetofluid dynamics or hydromagnetics) is the study of the magnetic properties and behavior of electrically conducting fluids. Jia et al. (2009) described a three-dimensional single-fluid MHD simulation of Ganymede's magnetosphere that accords extremely well with the Galileo particles and fields measurements. Fig. 12.5 shows the convection plot depicting the bulk motion of plasma and the projection of field lines. Fig. 12.6 shows a three-dimensional display of magnetic field lines sampled along the spacecraft trajectory (shown as white scattered dots) from the MHD simulation results.

12.1.5 Water and oxygen exospheres of *Ganymede*

The water (H_2O) and oxygen (O_2) exospheres (the outermost region of a planet's atmosphere) of Jupiter's moon Ganymede were simulated by Plainaki et al. (2015) through the application of a 3D Monte Carlo modeling technique that takes into consideration the combined effect on the exosphere generation of the main surface release processes (i.e., sputtering, sublimation and radiolysis) and the surface precipitation of the energetic ions of Jupiter's magnetosphere. In order to model the magnetospheric ion precipitation to

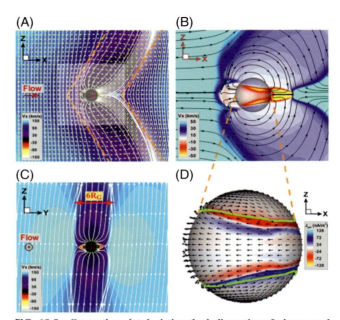

FIG. 12.5 Convection plot depicting the bulk motion of plasma and the projection of field lines. (A) Flows and the projection of field lines (white solid lines) in the XZ plane at Y = 0. Color represents the V_x contours, and unit flow vectors in yellow show the flow direction. A theoretical prediction of the Alfven characteristics (orange dashed lines) is shown for reference. The projection of the ionospheric flow is also shown as color contours on a circular disk of r = 1.08 R_G in the center. (B) A zoomed-in view of the light area in (A). Flow streamlines are superimposed on color contours of V_x. Note that the color bar differs from that in (A) and (B) in order to illustrate the relatively weak flow within the magnetosphere. (C) Same as (A) but in the YZ plane at X = 0. (D) Field-aligned current density along with unit flow vectors shown on a sphere of radius r = 1.08 R_G. *(Source: Jia et al., 2009). Journal of Geophysical Research. Copyright 2009 by the American Geophysical Union. 0148-0227/09/2009JA014375$09.00. Publisher: Wiley.*

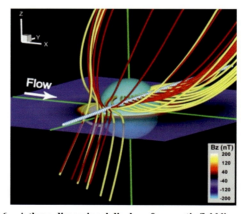

FIG. 12.6 A three-dimensional display of magnetic field lines sampled along the spacecraft trajectory (shown as white scattered dots) from the MHD simulation results. Yellow lines show the result for the case of fixed inner boundary condition, and red lines represent that of the BC2 boundary condition case. Color contours of the Bz component are shown in the equatorial plane (XY plane at Z = 0) for reference, and the centered sphere in cyan represents the inner boundary of the three-dimensional single-fluid MHD simulation of Ganymede's magnetosphere. *(Source: Jia et al., 2009). Journal of Geophysical Research. Copyright 2009 by the American Geophysical Union. 0148-0227/09/2009JA014375$09.00. Publisher: Wiley.*

Ganymede's surface, they used as an input the electric and magnetic fields from the global MHD model of Ganymede's magnetosphere (Jia et al., 2009). The exospheric model applied by Plainaki et al. (2015) is based on EGEON, a single-particle Monte Carlo model already applied for a Galilean moon (Plainaki et al., 2010; Plainaki et al., 2012; Plainaki et al., 2013); nevertheless, significant modifications have been implemented in Plainaki et al. (2015)'s investigation in order to include the effect on the exosphere generation of the ion precipitation geometry determined strongly by Ganymede's intrinsic magnetic field (Kivelson et al., 1996). Plainaki et al. (2015)'s simulation refers to a specific configuration between Jupiter, Ganymede, and the Sun in which the Galilean moon is located close to the center of Jupiter's Plasma Sheet (JPS) with its leading hemisphere illuminated.

The results of Plainaki et al. (2015)'s investigations are summarized as follows: (1) at small altitudes above the moon's subsolar point the main contribution to the neutral environment comes from sublimated H_2O; (2) plasma precipitation occurs in a region related to the open-closed magnetic field lines boundary and its extent depends on the assumption used to mimic the plasma mirroring in Jupiter's magnetosphere; (3) the spatial distribution of the directly sputtered-H_2O molecules exhibits a close correspondence with the plasma precipitation region and extends at high altitudes, being, therefore, well differentiated from the sublimated water; (4) the O_2 exosphere comprises two different regions: the first one is an homogeneous, relatively dense, close to the surface thermal-O_2 region (extending to some 100s of km above the surface) whereas the second one is less homogeneous and consists of more energetic O_2 molecules sputtered directly from the surface after water-dissociation by ions has taken place; the spatial distribution of the energetic surface-released O_2 molecules depends both on the impacting plasma properties and the moon's surface temperature distribution (that determine the actual efficiency of the radiolysis process).

12.2 An overview of *Europa*

Slightly smaller than Earth's Moon, Europa is primarily made of silicate rock and has a water-ice crust and probably an iron–nickel core. Europa is a geologically active world with evidence for recent tectonic resurfacing (Figueredo and Greeley, 2004; Greenberg et al., 1999; Pappalardo et al., 1998; Prockter et al., 2010; Squyres et al., 1983).

The four largest moons of Jupiter—Io, Europa, Ganymede, and Callisto—were first seen by Galileo Galilei in January 1610, and recognized by him as moons of Jupiter in March 1610, and therefore these four moons are named *Galilean* moons. The first reported observation of Io and Europa was made by Galileo using a 20 × -magnification refracting telescope at the University of Padua.

Europa (named after the daughter of the king of Tyre in Greek mythology) is the smallest of the four Galilean moons orbiting Jupiter, and the sixth-closest to the planet. Europa is about 3160 km (1950 miles) in diameter, or about the size of Earth's moon. Its surface is striated by cracks and streaks, but craters are relatively few. Europa is the sixth-largest moon and fifteenth-largest object in the solar system.

Ground-based spectroscopy of Europa, combined with gravity data, suggests that the moon has an icy crust roughly 150 km thick and a rocky interior (Kuiper, 1961; Morrison and Cruikshank, 1974; Anderson et al., 1997). In addition, images obtained by the Voyager spacecraft revealed that Europa's surface is crossed by numerous intersecting ridges and dark bands (called lineae) and is sparsely cratered, indicating that the terrain is probably significantly younger than that of Ganymede and Callisto (Smith et al., 1979).

The ice-rich surface of the Jovian satellite Europa is sparsely cratered, suggesting that this moon might be geologically active today. Cassen et al (1979)'s study indicated that it is possible that tidal dissipation in an ice crust on Europa preserved a liquid water layer beneath it, provided that the three-body orbital resonance for Io, Europa, and Ganymede is ancient. They suggested that the liquid water layer could be a continuing source of the observed surface frost. Images of Europa from the Galileo spacecraft show a surface with a complex history involving tectonic deformation, impact cratering, and possible emplacement of ice-rich materials and perhaps liquids on the surface (Greeley et al., 1998). Studies by Pappalardo et al. (1998) have been found to be consistent with the possibility that Europa has a liquid water ocean beneath a surface layer of ice, but suggested that further tests and observations are needed to demonstrate this conclusively. The geology and young age of Europa's surface as indicated above plus the predicted heat flow due to radiogenic decay and tidal dissipation (Cassen et al. 1979) suggests a geologically active interior and liquid water at shallow depths. Based on Galileo magnetometer measurements Kivelson et al. (2000) indicated that Europa probably contains a metallic iron core. This suggestion was supported by subsequent studies (e.g., Bhatia and Sahijpal, 2017). Its bulk density suggests that it is similar in composition to the terrestrial planets, being primarily composed of silicate rock (Kargel et al., 2000). It is suspected that Europa's surface rotates slightly faster than its interior, an effect that is possible due to the subsurface ocean mechanically decoupling Europa's surface from its rocky mantle (Hurford et al., 2007).

Europa's icy crust has an albedo (light reflectivity) of 0.64, one of the highest of all moons. The radiation level at the surface of Europa is equivalent to an amount of radiation that would cause severe illness or death in human beings exposed for a single day (The Effects of Nuclear Weapons, Revised ed., US DOD 1962, pp. 592–593). In view of this, robotic missions to Europa need to endure the high-radiation environment around itself and Jupiter.

Europa is the smoothest known object in the solar system, lacking large-scale features such as mountains and craters. There are few craters on Europa, because its surface is tectonically too active and therefore young. In addition to Earth-bound telescope observations, Europa has been examined by a succession of space probe flybys, the first occurring in the early 1970s. Studies of Voyager and Galileo images have revealed evidence of subduction on Europa's surface, suggesting that, just as the cracks are analogous to ocean ridges (Schenk and McKinnon, 1989; Kattenhorn and Prockter, 2014) so plates of icy crust analogous to tectonic plates on Earth are recycled into the molten interior. Other interesting features present on Europa are circular and elliptical lenticulae ("freckles"). It has been found that many of them are domes, some are pits and some are smooth, dark spots. Others have a jumbled or rough texture.

Since the Voyager spacecraft flew past Europa in 1979, planetary scientists have been working to understand the composition of the reddish-brown material that coats fractures and other geologically youthful features on Europa's surface. McCord et al. (1998)'s studies using Galileo's near-infrared mapping spectrometer indicated that the dark, reddish streaks and features on Europa's surface may be rich in salts such as magnesium sulfate, deposited by evaporating water that emerged from within. Carlson et al. (2005) found that sulfuric acid hydrate is another possible explanation for the contaminant observed spectroscopically. Based on the results of examination of the spectra of the ice Galilean satellites from 0.2 to 5 μm, Calvin et al. (1995) concluded that because these materials are colorless or white when pure, some other material must also be present to account for the reddish color, and sulfur compounds are suspected. A search for organics on Europa, carried out by Borucki et al., (2002) and Whalen et al. (2017) indicated that the colored regions are composed of abiotic organic compounds collectively called tholins. Studies carried out by several investigators (e.g., Coll et al., 2010; Ruiz-Bermejo et al., 2011; Trainer, 2013) have indicated that tholins bring important astrobiological implications, as they may play a role in prebiotic chemistry and abiogenesis.

Europa's most striking surface features are a series of dark streaks crisscrossing the entire globe, called lineae (Fig. 12.7). Long, dark lines seen on the surface of Europa are fractures in the crust, some of which are more than 3000 km (1850 miles) long. The larger bands are more than 20 km (12 mi) across, often with dark, diffuse outer edges, regular striations, and a central band of lighter material (Geissler, 1998). Based on pole-to-pole geological mapping investigations, Figueredo and Greeley (2004) suggested that the lineae on Europa were produced by a series of eruptions of warm ice as the Europan crust spread open to expose warmer layers beneath. According to them, the effect would have been similar to that seen in Earth's oceanic ridges. These various fractures are thought to have been caused in large part by the tidal flexing exerted by Jupiter.

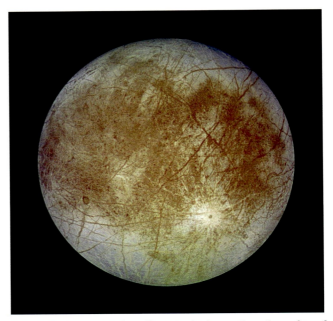

FIG. 12.7 **Long, dark lines (fractures in the crust) on the surface of Europa.** The bright feature containing a central dark spot in the lower third of the image is a young impact crater some 50 km (31 miles) in diameter. This crater has been provisionally named "Pwyll" for the Celtic god of the underworld. This image was taken on September 7, 1996, at a range of 677,000 km (417,900 miles) by the solid-state imaging television camera onboard the Galileo spacecraft during its second orbit around Jupiter. The image was processed by Deutsche Forschungsanstalt fuer Luftund Raumfahrt e.V., Berlin, Germany. (*Source: https://commons.wikimedia.org/wiki/File:Europa-moon-with-margins.jpg. This file is in the public domain in the United States because it was solely created by NASA. NASA copyright policy states that "NASA material is not protected by copyright unless noted."*)

The Galileo mission, launched in 1989, provides the bulk of current data on Europa. The Galileo mission was managed by NASA's Jet Propulsion Laboratory. Exploration of Europa began with the Jupiter flybys of Pioneer 10 and 11 in 1973 and 1974 respectively. The first close-up photos were of low resolution compared to later missions. The two Voyager probes traveled through the Jovian system in 1979, providing more-detailed images of Europa's icy surface. The images resulted in many scientists speculating about the possibility of a liquid ocean underneath. Starting in 1995, the Galileo space probe orbited Jupiter for eight years, until 2003, and provided the most detailed examination of the Galilean moons to date. It included the "Galileo Europa Mission" and "Galileo Millennium Mission," with numerous close flybys of Europa. In 2007, New Horizons probe imaged Europa, as it flew by the Jovian system while on its way to Pluto.

In 1997, the Galileo spacecraft confirmed the presence of a tenuous ionosphere (an upper-atmospheric layer of charged particles) around Europa created by solar radiation and energetic particles from Jupiter's magnetosphere

(Kliore et al., 1997), providing evidence of an atmosphere. Recent magnetic-field data from the Galileo orbiter showed that Europa has an induced magnetic field through interaction with Jupiter's, which suggests the presence of a subsurface conductive layer (Phillips and Pappalardo, 2014). This layer is likely a salty liquid-water ocean. The European Space Agency's Jupiter Icy Moon Explorer (JUICE) is a mission to Ganymede that is due to launch in 2022, and will include two flybys of Europa. NASA's planned Europa Clipper will be launched in the mid-2020s. No spacecraft has yet landed on Europa, although there have been several proposed exploration missions.

Europa orbits Jupiter in just over three and a half days, with an orbital radius of about 670,900 km. With an orbital eccentricity of only 0.009, the orbit itself is nearly circular, and the orbital inclination relative to Jupiter's equatorial plane is small, at 0.470°. Like its fellow Galilean moons, Europa is tidally locked to Jupiter, with one hemisphere of Europa constantly facing Jupiter. Because of this, there is a sub-Jovian point on Europa's surface, from which Jupiter would appear to hang directly overhead. Europa's prime meridian is a line passing through this point ("Planetographic Coordinates," Wolfram Research, 2010).

External agents have heavily weathered the visible surface of Europa. Internal and external drivers competing to produce the surface we see include, but are not limited to aqueous alteration of materials within the icy shell, initial emplacement of endogenic material by geologic activity, implantation of exogenic ions and neutrals from Jupiter's magnetosphere, alteration of surface chemistry by radiolysis and photolysis, impact gardening of upper surface layers, and redeposition of sputtered volatiles. Separating the influences of these processes is critical to understanding the surface and subsurface compositions at Europa. Recent investigations have applied cryogenic reflectance spectroscopy to Galileo Near-Infrared Mapping Spectrometer (NIMS) observations to derive abundances of surface materials including water ice, hydrated sulfuric acid, and hydrated sulfate salts. Dalton III et al. (2013) compared derived sulfuric acid hydrate ($H_2SO_4 \cdot nH_2O$) abundance with weathering patterns and intensities associated with charged particles from Jupiter's magnetosphere. They presented models of electron energy, ion energy, and sulfur ion number flux as well as the total combined electron and ion energy flux at the surface to estimate the influence of these processes on surface concentrations, as a function of location. They found that correlations exist linking both electron energy flux ($r \sim 0.75$) and sulfur ion flux ($r = 0.93$) with the observed abundance of sulfuric acid hydrate on Europa. According to Dalton III et al. (2013)'s studies, sulfuric acid hydrate production on Europa appears to be limited in some regions by a reduced availability of sulfur ions, and in others by insufficient levels of electron energy. The energy delivered by sulfur and other ions has a much less significant role. Surface deposits in regions of limited exogenic processing are likely to bear closest resemblance to oceanic composition. Dalton III et al. (2013) reckon that these results will assist future efforts to separate the relative influence of endogenic (formed or occurring beneath the surface of the earth) and exogenic (formed or occurring on the surface of the earth) sources in establishing the surface composition.

12.3 Jupiter's icy moons *Europa, Ganymede,* and *Callisto* containing vast quantities of liquid water

In January 1610, Italian astronomer Galileo Galilei discovered four of Jupiter's moons—now called *Io, Europa, Ganymede* and *Callisto*. Ever since this path-breaking discovery, these "Galilean moons" (named in honor of Galileo) have been targets of deeper investigations by astronomers. The historic landing of man on Earth's Moon on July 20, 1969—described as "One giant leap for mankind"— encouraged planetary scientists to thoroughly probe what lies on the surface and delve deep into what hides inside of the extraterrestrial planets and their moons. For example, Pilcher et al. (1972) detected water frost absorptions in the infrared reflectivity of Europa and Ganymede. These investigators determined the percentage of frost-covered surface area to be 50 to 100 percent for Europa, 20 to 65 percent for Ganymede, and possibly 5 to 25 percent for Callisto. It was found that the leading side of Ganymede has 20 percent more frost cover than the trailing side, which explains the visible geometric albedo differences between the two sides. It was further observed that the reflectivity of the material underlying the frost on Europa, Ganymede, and Callisto resembles that of silicates. Another finding made by Pilcher et al. (1972) was that the surface of Io may be covered by frost particles much smaller than those on Europa and Ganymede. Ransford et al. (1981) found that a large fraction of Europa's water content is retained as hydrated silicates.

In a study carried out by Squyres et al. (1983), Voyager images of Europa showed a bright icy surface transected by a complex pattern of long, linear fracture-like markings (known as *lineae*) with a slightly lower albedo. Very few impact craters have been observed, with sizes generally in the range 10–20 km. Indicators of surface composition and texture included deep infrared (IR) water (H_2O) absorption features (Pilcher et al., 1972), UV absorption features characteristic of sulfur in a water matrix on the trailing hemisphere (Lane et al., 1981), and a photometric function indicating much more homogeneous scattering than areas of equal albedo on Ganymede and Callisto. Several models have been suggested for the evolution and present state of Europa (Cassen et al., 1979; Ransford et al., 1981). Squyres et al. (1983) have reported arguments for resurfacing by

H$_2$O from a liquid layer, based on interpretations of spacecraft- and Earth-based observations and revised theoretical calculations.

As indicated above, it has been suggested that Europa's thin outer ice shell might be separated from the moon's silicate interior by a liquid water layer, delayed or prevented from freezing by tidal heating (Cassen et al., 1979; Squyres et al., 1983; Ross and Schubert, 1986; Ojakangas and Stevenson, 1989). In the model assuming a liquid water layer lying sandwiched between Europa's thin outer ice shell and Europa's silicate interior, the lineae clearly seen on Europa's surface could be explained by repetitive tidal deformation of the outer ice shell (McEwen, 1986). However, observational confirmation of a subsurface ocean was largely frustrated by the low resolution (>2 km per pixel) of the Voyager images (Pappalardo et al., 1996). Carr et al. (1998) reported high-resolution (54 m per pixel) Galileo spacecraft images of Europa, in which they found evidence for mobile icebergs. The detailed morphology of the terrain strongly supports the presence of liquid water at shallow depths below the surface, either today or at some time in the past. Carr et al. (1998) pointed out that lower-resolution observations of much larger regions suggest that the phenomena reported by them are widespread.

Measurements by the American robotic space probe *Galileo* (launched on 18 October 1989) of Europa's gravitational field show that Europa is strongly differentiated (with a metallic core) and that it indeed has a water-ice outer layer ~100 km thick. Anderson et al. (1998) used radio Doppler data from four encounters of the Galileo spacecraft with the Jovian moon Europa to refine models of Europa's interior. Their studies indicated that Europa is most likely differentiated into a metallic core surrounded by a rock mantle and a water-ice-liquid outer shell. They inferred that the thickness of Europa's outer shell of water ice-liquid must lie in the range of about 80–170 km.

The possibility of a liquid-water ocean beneath the icy surface of Europa has been debated for decades. In a study, Khurana et al. (1998) found that the external magnetic fields in the vicinity of both Europa and Callisto are perturbed by Jupiter's inner magnetosphere. They interpreted these perturbations as arising from induced magnetic fields, generated by the moons in response to the periodically varying plasma environment. Electromagnetic induction requires eddy currents to flow within the moons, and their calculations showed that the most probable explanation is that there are layers of significant electrical conductivity just beneath the surfaces of both moons. Khurana et al. (1998) argued that these conducting layers may best be explained by the presence of salty liquid-water oceans, for which there is already indirect geological evidence (Carr et al., 1998; Pappalardo et al., 1998) in the case of Europa.

Before the fly-bys of Ganymede by the Galileo spacecraft, viable models of the internal structure of Jupiter's largest moon (Ganymede) ranged from a uniform mixture of rock and ice to a differentiated body with a rocky core and an icy mantle (Schubert et al., 1986). Analysis of the Doppler shift of radio signals from Galileo has now shown that Ganymede is strongly differentiated with a relatively dense core surrounded by a thick shell of ice (Anderson et al., 1996a). Other instruments have revealed that Ganymede has an intrinsic magnetic field (Kivelson, et al., 1996a; Gurnett et al., 1996), which is aligned approximately antiparallel to Jupiter's magnetic field. Schubert et al. (1996) argued that these results imply that Ganymede has an outer silicate core surrounding a liquid (or partially liquid) inner core of iron or iron sulfide and that the magnetic field is generated by dynamo action within this metallic core. Io also appears to have an intrinsic magnetic field antiparallel to that of Jupiter (Kivelson et al., 1996b; Kivelson et al., 1996c), implying that it too has a metallic core (Anderson et al., 1996b) in which the field is generated either by dynamo action or by magneto-convection. Khurana et al. (1998) pointed out that Ganymede is completely differentiated, and that extensive endogenic modification of its surface and the existence of an intrinsic magnetic field (Kivelson, 1996) imply a dynamic interior in the past, and possibly also in the present (Schubert et al., 1996). Thus, it would be more plausible for Ganymede to have a subsurface liquid-water ocean. Khurana et al. (1998) concluded from an analysis of the magnetic field observations that it is very likely that both Europa and Callisto possess internal salty liquid-water oceans. Not all ocean worlds reveal their present oceans in their surface characteristics. Both Ganymede and Callisto have internal oceans, but by the present knowledge, their surfaces are currently inactive (Moore et al., 2004; Pappalardo et al., 2004). To confirm oceans on geologically active worlds, a number of geophysical measurements can be used to identify present-day oceans. In many cases, oceans can be revealed by the orbital and rotational state of a body, if it can be measured carefully enough. For systems with a strong, inclined magnetic field (*e.g.*, the Jupiter system), the electromagnetic response of the body, after correction for ionospheric background, provides a strong indication of an internal ocean, as demonstrated for Europa, Ganymede, and Callisto (Kivelson et al., 1999, 2000, 2002). With sufficient flybys, gravity data can also indicate the presence of an ocean (Iess et al., 2014; McKinnon, 2015), especially when coupled with detailed topography, as recently demonstrated for Dione (Hemingway et al., 2016).

Investigations carried out by Zimmer et al. (2000), based on magnetic field measurements during Galileo flybys of Europa and Callisto, found that these fields are close to those expected for perfectly conducting moons. This finding is interpreted as indicative of the presence of global Earth-like oceans under the surface of both moons (Europa and Callisto), provided they are at least a few kilometers thick.

The calculations, based on imaging from *Galileo* from 1995 to 1997, show Europa's ocean may be about one-fifth as salty as Earth's ocean. Thermal models predict a global, subsurface ocean on Europa (Moore and Schubert, 2000; Spohn and Schubert, 2003; Tobie et al., 2003), the existence of which is strongly supported by Galileo measurements of an induced magnetic field (Kivelson et al., 2000).

Jupiter's Galilean moons (Io, Ganymede, Europa, and Callisto) have been a subject of considerable investigations by several planetary scientists over recent years. Io most likely has a fully liquid core, while Europa, Ganymede, and Callisto are thought to have an internal global liquid water ocean beneath an external ice shell (Baland et al., 2012).

Detailed investigation of three of Jupiter's Galilean moons (Ganymede, Europa, and Callisto), which are believed to harbor subsurface water oceans, is central to elucidating the conditions for habitability of icy worlds in planetary systems in general. The study of the Jupiter system and the possible existence of habitable environments offer the best opportunity for understanding the origins and formation of the gas giants and their moon systems. The JUpiter ICy moons Explorer (JUICE) mission, selected by European Space Agency (ESA) in May 2012 to be the first large mission within the Cosmic Vision Program 2015–2025, is designed to perform detailed investigations of Jupiter and its system in all their inter-relations and complexity with particular emphasis on Ganymede as a planetary body and potential habitat. The investigations of the neighboring moons, Europa and Callisto, will complete a comparative picture of the Galilean moons and their potential habitability. In a study, Grasset et al. (2013) described the scientific motivation for this exciting new European-led exploration of the Jupiter system in the context of our current knowledge and future aspirations for exploration, and the paradigm it will bring in the study of giant (exo) planets in general.

As just indicated, Jupiter's three large icy moons (Ganymede, Callisto, and Europa) contain vast quantities of liquid water, a key ingredient for life. However, Ganymede and Callisto are weaker candidates for habitability than Europa, in part because of the model-based assumption that high-pressure ice layers cover their seafloors and prevent significant water–rock interaction. Water–rock interactions may occur, however, if heating at the rock–ice interface melts the high-pressure ice. Highly saline fluids would be gravitationally stable, and might accumulate under the ice due to upward migration, refreezing, and fractionation of salt from less concentrated liquids. To assess the influence of salinity on Ganymede's internal structure, Vance et al. (2014) used available phase-equilibrium data to calculate activity coefficients and predict the freezing of water ice in the presence of aqueous magnesium sulfate. They coupled this new equation of state with thermal profiles in Ganymede's interior—employing recently published thermodynamic data for the aqueous phase—to estimate the thicknesses of layers of ice I, III, V, and VI. It may be noted that with both cooling and pressure different types of ice exist, each being created depending on the phase diagram of ice. These are II, III, V, VI, VII, VIII, IX, … Vance et al. (2014) computed core and silicate mantle radii consistent with available constraints on Ganymede's mass and gravitational moment of inertia. Mantle radii range from 800 to 900 km for the values of salt and heat flux considered by them (4–44 mW/m^2 and 0 to 10 wt% MgSO$_4$). Ocean concentrations with salinity higher than 10 wt% have little high-pressure ice. Even in a Ganymede ocean that is mostly liquid, achieving such high ocean salinity is permissible for the range of likely S/Si ratios. However, elevated salinity requires a smaller silicate mantle radius to satisfy mass and moment-of-inertia constraints, so ice VI is always present in Ganymede's ocean. For lower values of heat flux, oceans with salinity as low as 3 wt% can co-exist with ice III. Available experimental data indicate that ice phases III and VI become buoyant for salinity higher than 5 wt% and 10 wt%, respectively. Similar behavior probably occurs for ice V at salinities higher than 10 wt%. Vance et al. (2014) concluded that flotation can occur over tens of kilometers of depth, indicating the possibility for upward 'snow' or other exotic modes of heat and material transport.

Past exploration of Jupiter's diverse moon system has forever changed our understanding of the unique environments to be found around gas giants, both in our solar system and beyond. The existence of a buried ocean sandwiched between surface ice and high-pressure (HP) polymorphs (materials which take various forms) of ice emerges as the most plausible structure for the hundreds-of-kilometers thick hydrospheres within large icy moons of the solar system (e.g., Ganymede, Callisto, Titan). The HP ice mantle is presumed to limit chemical transport from the rock component to the ocean. Choblet et al. (2017) showed that the heat power produced by radioactive decay within the rock core is mainly transported through the HP ice mantle by melt extraction to the ocean, with most of the melt produced directly above the rock/water interface. While the average temperature in the bulk of the HP ice mantle is always relatively cool when compared to the value at the interface with the rock core (∼5 K above the value at the surface of the HP ice mantle), maximum temperatures at all depths are close to the melting point, often leading to the interconnection of a melt path via hot convective plume conduits throughout the HP ice mantle. Overall, Choblet et al. (2017) predicted long periods of time during these moons' history where water generated in contact with the rock core is transported to the above ocean.

As already indicated, water-rich planetary bodies including large icy moons and ocean exo-planets may host a deep liquid water ocean underlying a high-pressure icy mantle. The latter is often considered as a limitation to the habitability of the uppermost ocean because it would limit the availability of nutrients resulting from the hydrothermal alteration of the silicate mantle located beneath the deep ice layer. In a study, Journaux et al. (2017) assessed the effects of salts on the physical properties of high-pressure ices and therefore the possible chemical exchanges and habitability inside H_2O-rich planetary bodies. It is believed that Journaux et al. (2017)'s experimental results might profoundly impact the internal dynamics of water-rich planetary bodies. For instance, an icy mantle at moderate conditions of pressure and temperature will consist of buoyant ice VI with low concentration of salt, and would likely induce an upwelling current of solutes toward the above liquid ocean. In contrast, a deep and/or thick icy mantle of ice VII will be enriched in salt and hence would form a stable chemical boundary layer on top of the silicate mantle. Such a contrasted dynamics in the aqueous–ice VI–ice VII system would greatly influence the migration of nutrients toward the uppermost liquid ocean, thus controlling the habitability of moderate to large H_2O-rich planetary bodies in our solar system (e.g., Ganymede, Titan, Calisto) and beyond.

The JUpiter ICy moons Explorer (*JUICE*)—an interplanetary spacecraft in development by the European Space Agency (ESA)—is set for launch in June 2022 and expected to reach Jupiter in October 2029 to study three of Jupiter's Galilean moons: *Ganymede*, *Callisto*, and *Europa*, all of which are thought to have significant bodies of liquid water beneath their surfaces, making them potentially habitable environments. By September 2032, the spacecraft is expected to enter the orbit around Ganymede for its close-up science mission, becoming the first spacecraft to orbit a moon other than the moon of Earth. Its period of operations will overlap with NASA's Europa Clipper mission, launching in 2024. The *JUICE* mission will characterize Ganymede's subsurface ocean, located between layers of near-surface and high-pressure ices, to better understand the formation and evolution of this ocean world. It could place bounds on communication between the subsurface ocean and the surface, energy input into the ocean layer, and the habitability of oceans separated from underlying rocky mantles. The ROW team supports the ESA JUICE mission. Because Callisto's ocean is also located between two layers of ices, Callisto studies could inform studies of Ganymede's ocean (Hendrix et al., 2019).

Planetary scientists are aware that the known ocean world *Callisto* remains to be fully characterized. Its deep subsurface ocean and its location on the edge of the Galilean moon system limits not only communication between the ocean and the surface but also vital energy input to the ocean. According to Hendrix et al. (2019), it may serve as an end member on the ocean world spectrum and help, along with *Ceres*, to characterize the limit of the ability of bodies to maintain oceans with sparse tidal input. In addition, the ROW team supports mission studies to characterize Callisto's ocean and its sustainability.

12.4 Oxygen, water, and sodium chloride on *Europa*

Europa's exosphere is a mixture of different species among which sputtered (deposited using fast ions) H_2O and H_2 dominate in the highest altitudes and O_2, formed mainly by radiolysis of ice (the decomposition of ice molecules due to ionizing radiation.) and subsequent release of the produced molecules, prevails at lower altitudes. Thus, Europa has a very thin atmosphere composed primarily of oxygen. Europa's O_2 exosphere has been demonstrated through both observation and simulation-based techniques to be spatially nonuniform. In a study, Plainaki et al. (2013) investigated Europa's exospheric O_2 characteristics under the external conditions that are likely in the Jupiter's magnetospheric environment, applying the Europa Global model of Exospheric Outgoing Neutrals (EGEON, Plainaki et al., 2012) for different configurations between the positions of Europa, Jupiter and the Sun. Plainaki et al. (2013) demonstrated for the first time that the spatial distribution of Europa's exosphere is explicitly time-variable due to the time-varying relative orientations of solar illumination and the incident plasma direction. Plainaki et al. (2013) showed also that the O_2 release efficiency depends both on solar illumination and plasma impact direction.

Europa has the smoothest surface of any known solid object in the solar system. The apparent youth and smoothness of the surface have led to the hypothesis that a water ocean exists beneath it, which could conceivably harbor extra-terrestrial life (Tritt, C.S., 2002; Possibility of Life on Europa, Milwaukee School of Engineering). Furthermore, the Hubble Space Telescope detected water vapor plumes similar to those observed on Saturn's moon *Enceladus*, which are thought to be caused by erupting cryo-geysers. In May 2018, astronomers provided supporting evidence of water plume activity on Europa, based on an updated critical analysis of data obtained from the Galileo space probe, which orbited Jupiter from 1995 to 2003. Such plume activity could help researchers in a search for life from the subsurface Europan ocean without having to land on the moon (Jia et al., 2018).

Of Jupiter's three largest icy moons, Europa, which is roughly the size of Earth's moon, is favored as having the greatest potential to sustain life. Complex and beautiful patterns adorn the icy surface of Europa.

Magnetic readings captured by NASA's Galileo spacecraft provided compelling hints that it has an ocean, and radio scans by the probe suggest a water-rich layer beneath the surface between 50 and 105 miles (80 and 170 km) thick. Recent findings even suggest its ocean could be loaded with enough oxygen to support millions of tons worth of marine life.

In a study, Plainaki et al. (2010) examined the space weathering processes on the icy surface of Europa. The heavy energetic ions of the Jovian plasma (H^+, O^+, S^+, C^+)

can erode the surface of Europa via ion sputtering (IS), ejecting up to 1000 H_2O molecules per ion. UV photons impinging the Europa's surface can also result in neutral atom release via photon-stimulated desorption (PSD) and chemical change (photolysis). Plainaki et al. (2010) studied the efficiency of the IS and PSD processes for ejecting water molecules, simulating the resulting neutral H_2O density. They also estimated the contribution to the total neutral atom release by the Ion Backscattering (IBS) process. Moreover, they estimated the possibility of detecting the sputtered high-energy atoms, in order to distinguish the action of the IS process from other surface release mechanisms. The main results of their investigation are: (1) The most significant sputtered-particle flux and the largest contribution to the neutral H_2O density come from the incident S^+ ions; (2) the H_2O density produced via PSD is lower than that due to sputtering by ~1.5 orders of magnitude; (3) in the energy range below 1 keV, the IBS can be considered negligible for the production of neutrals, whereas in the higher energy range it becomes the dominant neutral emission mechanism; (4) the total sputtering rate for Europa is 2.0×10^{27} H_2O/s; and (5) the fraction of escaping H_2O via IS is 22% of the total sputtered population, while the escape fraction for H_2O produced by PSD is 30% of the total PSD population. Because the PSD exosphere is lower than the IS one, the major agent for Europa's surface total net erosion is IS on both the nonilluminated and illuminated side. Lastly, the exospheric neutral density, estimated from the Galileo electron density measurements appears to be higher than that calculated for H_2O alone; this favors the scenario of the presence of O_2 produced by radiolysis and photolysis.

Using the Space Telescope Imaging Spectrograph (STIS), Trumbo et al. (2019) observed Europa across four Hubble Space Telescope (HST) visits, obtaining the first spatially resolved spectral dataset of the entire surface at wavelengths of 300 to 1000 nm. They observed a broad absorption near 450 nm, which corresponds well to the F-center absorption of irradiated NaCl (Hand and Carlson, 2015; Poston et al., 2017). It was found that this feature is located exclusively on the leading hemisphere and correlates with chaos terrain. Trumbo et al. (2019) argue that the low temperatures and icy environment of Europa's surface result in hydrohalite ($NaCl \cdot 2H_2O$).

According to Trumbo et al. (2019), NaCl provides an elegant explanation for the observed 450-nm feature, its geographic distribution, and previous infrared spectra interpreted to reflect endogenous (i.e., derived internally) material (Fischer et al., 2015; Fischer et al., 2017). Trumbo et al. (2019)'s dataset shows that the observed feature is concentrated in chaos, and that, of the spectra of several other irradiated salts examined by them, only NaCl is consistent with their observed feature. The presence of NaCl on Europa has important implications for our understanding of the internal chemistry and its geochemical evolution through time.

Trumbo et al. (2019) suggest that regardless of whether the observed NaCl directly relates to the ocean composition, its presence warrants a re-evaluation of our understanding of the geochemistry of Europa.

12.5 Understanding the mysteries of *Europa*

Missions to explore Europa have been imagined and conceptualized ever since the Voyager mission findings first indicated that Europa was geologically very young. Subsequently, the Galileo spacecraft supplied fascinating new insights into this moon of Jupiter. As a follow-up, an international team proposed a return to the Jupiter system and Europa, with the Europa Jupiter System Mission (EJSM). NASA and ESA engaged themselves in designing two orbiters that would explore the Jovian system and then each would settle into orbit around one of Jupiter's icy moons, Europa and Ganymede. In addition, the Japanese Aerospace eXploration Agency (JAXA) considered a Jupiter magnetospheric orbiter, and the Russian Space Agency investigated a Europa lander.

Robotic landers serve vital reconnaissance roles in the exploration of planetary surfaces, but are constrained by deliverable payload size and environment survivability. Although the Mars exploration rovers (MER) have shown incredible survivability, their solar power source limits the science output per sol. Future landers will be larger, and will incorporate more sophisticated data-collection and analysis packages, which will likely bring with them an increased demand for power. Anticipating this demand, Som et al. (2009) proposed an innovative hybrid power system combining a primary radioisotope thermoelectric generator (RTG) with a secondary alkaline fuel cell. This combination provides the opportunity to utilize more effectively the energy produced by the RTG, to produce and store O_2 and H_2 via electrolysis of melted ice, and use this obtained O_2 and H_2 in a variety of ways, including as fuel for a regenerative fuel cell. This hybrid system has applications ranging from planetary rovers and deep-space probes to human habitats.

From 2007 the Russian Academy of Sciences (RAS) and Roscosmos considered to develop a Europa surface element, in coordination with the Europa Jupiter System Mission (EJSM) international project planned to study the Jupiter system. The main scientific objectives of the Europa Lander are to search for the signatures of possible present and extinct life, in situ studies of the Europa internal structure, the surface, and the environment. The mission includes the lander, and the relay orbiter, to be launched by Proton and carried to Jupiter with electric propulsion. The mass of scientific instruments on the lander is ~50 kg, and its planned lifetime is 60 days. A dedicated international Europa Lander Workshop (ELW) was held in Moscow in February 2009. Following the ELW materials and few recent developments, Zelenyi et al. (2011) have described the planned mission, including the science goals, technical design of the mission elements, the ballistic scheme, and the synergy between the Europa Lander and the EJSM.

As already mentioned, the Europa Lander mission, an exciting part of the international Europa Jupiter System Mission (EJSM/Laplace), considered in situ planetary exploration of the moon. The distance of Europa from the Earth and the Sun asks for autonomous analytical tools that maximize the scientific return at minimal resources, demanding new experimental concepts. Pavlov et al. (2011) proposed a novel instrument, based on the atomic spectroscopy of laser generated plasmas for the elemental analysis of Europa's surface materials as far as it is in reach of the lander for example by a robotic arm or a mole, or just onboard the lander. The technique of laser-induced plasma spectrometry provides quantitative elemental analysis of all major and many trace elements. It is a fast technique, i.e. an analysis can be performed in a few seconds, which can be applied to many different types of material such as ice, dust, or rocks and it does not require any sample preparation. The sensitivity is in the range of tens of parts per million (ppm) and high lateral resolution, down to 50 μm, is feasible. In addition, it provides the potential of depth profiling, up to 2 mm in rock material and up to a few cm in more transparent icy matrices. Key components of the instrument have been developed in Germany for planetary in situ missions. This development program is accompanied by an in-depth methodical investigation of this technique under planetary environmental conditions.

The Jupiter Europa Orbiter (JEO) constituted the NASA-led portion of the EJSM; JEO addressed a very important subset of the complete EJSM science objectives, designed to function alone or in conjunction with ESA's Jupiter Ganymede Orbiter (JGO). The JEO mission concept used a single orbiter flight system that would travel to Jupiter by means of a multiple-gravity-assist trajectory and then perform a multiyear study of Europa and the Jupiter system, including 30 months of Jupiter system science and a comprehensive Europa orbit phase of 9 months.

The JEO mission was designed to investigate various options for future surface landings. The JEO mission science objectives, as defined by the international EJSM Science Definition Team, include (Clark et al., 2011):

(i) Europa's ocean: Characterize the extent of the ocean and its relation to the deeper interior.
(ii) Europa's ice shell: Characterize the ice shell and any subsurface water, including their heterogeneity, and the nature of surface–ice–ocean exchange.
(iii) Europa's chemistry: Determine global surface compositions and chemistry, especially as related to habitability.
(iv) Europa's geology: Understand the formation of surface features, including sites of recent or current activity, and identify and characterize candidate sites for future in situ exploration.
(v) Jupiter system: Understand Europa in the context of the Jupiter system.

The JEO orbital mission would provide critical measurements to support the scientific and technical selection of future landed options.

The primary challenge of a Europa mission is to perform in Jupiter's radiation environment, radiation damage being the life-limiting parameter for the flight system. Instilling a system-level radiation-hardened-by-design approach very early in the mission concept would mitigate the pervasive mission and system-level impacts (including trajectory, configuration, fault protection, operational scenarios, and circuit design) that can otherwise result in runaway growth of cost and mass.

Clark et al. (2011) have addressed the JEO mission concept developed by a joint team from JPL and the Applied Physics Laboratory to address the science objectives defined by an international science definition team formed in 2008, while designing for the Jupiter environment.

Lorenz et al. (2011) reviewed the utility of analog environments in preparations for a Europa lander mission. Such analogs are useful in the demonstration and rehearsal of engineering functions such as sample acquisition from an icy surface, as well as in the exercise of the scientific protocols needed to identify organic, inorganic, and possible biological impurities in ice. In Lorenz et al. (2011)'s studies, particular attention was drawn to Antarctic and Arctic analog sites where progress in these latter areas has been significant in recent years.

Landing site selection continues to be an important subject of investigation in planetary exploration. In this context, using data from an airborne survey of the Thwaites Glacier Catchment, West Antarctica using the High Capability Airborne Radar Sounder (HiCARS), Grima et al. (2014) demonstrated the potential for a nadir-looking radar sounder to retrieve significant surface roughness/permittivity information valuable for planetary landing site selection. In this study, the statistical method introduced by Grima et al. (2012) for surface characterization has been applied systematically along the survey flights. The coherent and incoherent components of the surface signal, along with an internally generated confidence factor, have been extracted and mapped in order to show how a radar sounder can be used as both a reflectometer and a scatterometer to identify regions of low surface roughness compatible with a planetary lander. These signal components have been used with a backscattering model to produce a landing risk assessment map by considering the following surface properties: Root mean square (RMS) heights, RMS slopes, roughness homogeneity/stationarity over the landing ellipse, and soil porosity. Comparing these radar-derived surface properties with simultaneously acquired nadir-looking imagery and laser-altimetry validated this method. The ability to assess all of these parameters with an ice-penetrating radar expands the demonstrated capability of a principle instrument in icy planet satellite science to include statistical reconnaissance of the surface roughness to identify suitable sites for a follow-on lander mission.

12.6 An approx. 100 km deep subsurface salty ocean hiding beneath *Europa*'s ice shell

Europa hosts a ~>100 km deep liquid water ocean beneath its 3–30 km ice shell (e.g., Schubert et al., 2009). Unlike the late delivery of cometary water, which has been postulated as the source of water within some extraterrestrial celestial bodies, recent studies on Europa (see Daswani et al., 2020) indicate that volatiles released from Europa's interior would have carried solutes, redox-sensitive species, and generated the entirety of Europa's ocean and potentially an early CO_2 atmosphere. According to Daswani et al. (2020), tidal dissipation from Europa's orbital interaction with Jupiter, Io, and Ganymede may have had a significant impact on early heating and differentiation of the interior, but it is difficult to disentangle the influence of tidal dissipation from other early sources of heat (Hussmann and Spohn, 2004).

Models to date propose that Europa's ocean may have evolved from a reduced NaCl-dominated composition to a more oxidized Mg-sulfate ocean today. The young surface age of the ice (40–90 Myr; Bierhaus et al., 2009) suggests periodic resurfacing and a mechanism for surface–ocean material exchange through the ice shell. Surface spectra consistent with the presence of hydrated sulfate salts were initially interpreted as indicative of a sulfate-rich ocean (McCord et al., 1998), consistent with brine evolution in CI chondrite bodies (Kargel, 1991; Kargel et al., 2000; Zolotov and Shock, 2001). But a sulfate-rich ocean in Europa has been challenged more recently because the interpretation of hydrated sulfate salts on the surface as a signature of the ocean does not appear to be consistent with more recent infrared spectroscopic observations.

Using spectra obtained with the Hubble Space Telescope, Trumbo et al. (2019) reported the detection of a 450-nm absorption indicative of irradiated sodium chloride on the surface of Europa. These recent observations favor instead chloride salts on the most geologically disrupted surfaces; surface sulfate salts and hydrated sulfuric acid are interpreted as radiolytic end-products (Brown and Hand, 2013; Ligier et al., 2016; Trumbo et al., 2019; Fischer et al., 2016). The spectral feature reported by Trumbo et al. (2019) correlates with geologically disrupted chaos terrain (quasi-circular areas of ice disruption), suggesting an interior source. The solid-state imaging (SSI) system on NASA's Galileo spacecraft provided detailed information on the complex surface features of Europa. The images so obtained indicated that Europa has the smoothest surface of any known solid object in the solar system. It was found that some of the interesting geologic features adorning Europa's icy surface include complex and beautiful patterns, ice rafts, chaos, lenticulae (dark spots), faults, domes, etc. The apparent youth and smoothness of Europa's surface and the just-mentioned interesting geologic features adorning its surface have led to the hypothesis that a layer of liquid water—with a possible depth of more than 100 km—exists several kilometers beneath Europa's icy crust. Fig. 12.8 shows artist's drawings depicting two proposed models of the subsurface structure of the Jovian moon, Europa. If a 100 km (60 mile) deep ocean existed below a 15 km (10 mile) thick Europan ice crust, it would be 10 times

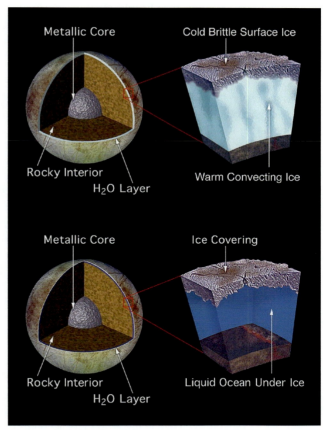

FIG. 12.8 Artist's drawings depicting two proposed models of the subsurface structure of the Jovian moon, Europa. Geologic features on the surface, imaged by the Solid State Imaging (SSI) system on NASA's Galileo spacecraft might be explained either by the existence of a warm, convecting ice layer, located several kilometers below a cold, brittle surface ice crust (top model), or by a layer of liquid water with a possible depth of more than 100 km (bottom model). *(Source: https://commons.wikimedia.org/wiki/File:EuropaInterior1.jpg). This file is in the **public domain** in the United States because it was solely created by NASA. NASA copyright policy states that "NASA material is not protected by copyright unless noted."*

deeper than any ocean on Earth and would contain twice as much water as Earth's oceans and rivers combined.

Data from various instruments on the Galileo spacecraft indicate that an Europan ocean might exist. The jet of water vapor and ice particles emanating from the surface of Europa is considered to be a sign of liquid water existing below Europa's ice crust. To date Earth is the only known place in the solar system where large masses of liquid water are located close to a solid surface. Other sources are especially interesting because water is a key ingredient for the development of life as we know it.

The predominant model suggests that heat from tidal flexing causes the ocean to remain liquid and drives ice movement similar to plate tectonics, absorbing chemicals from the surface into the ocean below. It is believed that sea salt from a subsurface ocean may be coating some geological features on Europa, suggesting that the ocean is interacting with the seafloor. This may be important in determining whether Europa could be habitable.

Under the assumption that most of the salt in Europa's shell was incorporated by freezing ocean water with a solute distribution coefficient of 2.7×10^{-3} (Gross et al., 1977), it is possible to make an estimate for the ocean salinity. Earlier studies suggested that unlike Earth's oceans, which are salty due to sodium chloride (common salt), magnesium sulfate might be a major salt component of Europa's water or ice. However, recent studies (e.g., Trumbo et al., 2019) suggest that Europa's surface is rich in NaCl, and that regardless of whether the observed NaCl directly relates to the ocean composition, its presence warrants a re-evaluation of our understanding of the geochemistry of Europa. It may be noted that Kargel et al. (2000) had predicted that, in the case of Europa, more extensive hydrothermal circulation, as on Earth, may lead to an NaCl-rich ocean. Subsequent studies strengthen this suggestion. For example, the plume chemistry of Enceladus, which is perhaps the best analog to Europa, suggests an NaCl-dominated ocean (Waite Jr. et al., 2006) and a hydrothermally active seafloor (Waite et al., 2017).

For the case of NaCl, Steinbrügge et al. (2020) estimated a value of 5.8 g/kg or, and for $MgSO_4$ they found it to be 5.1 g/kg. From Galileo magnetic induction measurements assuming a 200 km thick H_2O layer and a terrestrial ocean composition, the minimum salinity of the ocean is <1 g/kg (Zimmer et al., 2000); however, the upper limit is yet not well constrained. Geochemical models accounting for observations of Europa's atmospheric composition suggest an $MgSO_4$ dominant ocean with a salinity of 12.3 g/kg (Zolotov & Shock, 2001). Steinbrügge et al. (2020) have claimed that the value estimated by them is fully compatible with this lower limit and provides the first observational constraint on the upper bound.

Daswani et al. (2020) used geochemical and petrologic models to assess whether planetary-scale thermal processes were responsible for the build-up of Europa's ocean, and

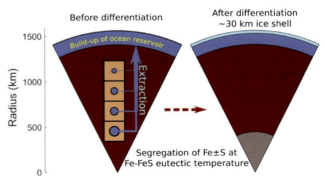

FIG. 12.9 **Schematic of the thermodynamic + extraction + structure model to simulate the build-up of Europa's ocean from exsolved volatiles.** After each heating step before differentiation, Gibbs energy minimization is carried out, resulting in an equilibrium assemblage in each cell (left figure). A portion of the fluid phase(s) is then extracted according to a specified rule, joins the ocean reservoir, and no longer affects the chemistry of the deep interior. Fe ± S is extracted from the bulk composition from the deep interior once the interior reaches the Fe–FeS eutectic temperature. Finally, Europa's structure is resolved, here assuming a 30 k mice shell, requiring a temperature of 270.8 K at the ice–ocean interface. *(Source: Daswani, M.M., S.D. Vance, M.J. Mayne, and C.R. Glein (2020), A metamorphic origin for Europa's ocean, Geophysical Research Letters (manuscript submitted)).*

whether thermal evolution of the deep interior had a significant impact on the composition of the ocean. They hypothesized that prograde metamorphism (i.e., metamorphic changes caused by increasing temperature) and associated chemical reactions in the deep interior were the driving forces behind the ocean's formation and its composition. They claim that a modern example of such behavior on Earth may be found in hydrous minerals such as chlorite and serpentine present in subducting oceanic plates, which experience dewatering with increasing pressure and temperature (e.g., Rüpke et al., 2004; Manthilake et al., 2016). They simulated the build-up of the ocean by imposing a limit on the fraction of volatiles retained in the assemblage for each heating step. That is, if fluids are thermodynamically stable, a specified portion is irreversibly fractionated from the equilibrium assemblage of the particular cell to go into the growing ocean reservoir (Fig. 12.9). Daswani et al. (2020) quantified gas saturation and mineral precipitation out of the primordial ocean, and the effects on the water column's composition, pH, and redox potential. Their thermodynamic + extraction + structure model results indicate that Europa's ocean, if derived from thermal evolution of the interior, was carbon- and sulfur-rich.

12.7 Indirect methods for detecting an ocean hidden within *Europa*

Since the very first observations of the Earth's Moon from the Earth with radar in 1946, radars are more and more frequently selected to be part of the payload of exploration missions in the solar system. They are, in fact, able to collect information on the surface structure of bodies or planets hidden by opaque atmospheres, to probe the planetsmoons' subsurface or even to reveal the internal structure of a small body comet nucleus.

Radar probing of Jovian icy moons is fundamental for understanding their origin and their thermal evolution as potential habitable environments in our solar system. One of the main objectives of proposed missions to the icy Jovian moons is to prove the existence of the postulated subsurface ocean on Europa using radar sounders. Exploration of subsurface oceans on Jupiter's (Jovian) icy moons is a key issue of the icy moons' geology as well. Electromagnetic wave propagation is the only way to probe their icy mantles from the orbit.

Radar is in fact the only sounding technique which is able to penetrate the Jovian moons' icy mantles, which can be many kilometers thick. Surface clutter, i.e. scattering of the radio waves on the rough surface, is known to be one of the most important problems of subsurface radar probing. Adequate numerical modeling of this scattering is required on all stages of subsurface radar experiment, including design of an instrument, operational strategy planning, and data interpretation. In a study, Ilyushin (2014) formulated (in general form) a computer simulation technique for numerical simulations of radar sounding of rough surfaces. As an example, Ilyushin (2014) simulated subsurface radar location of the ocean beneath Ganymedian ice with chirp radar signals. Ciarletti (2016) has provided a brief review of radars designed for the solar system planets and bodies' exploration. This review did not aim at being exhaustive but merely focused on the major results obtained. Ciarletti (2016)'s review highlighted the variety of radars that have been or are currently designed in terms of frequency or operational modes.

Two methods for detection of ocean hidden within Europa are (1) radar-based active detection, and (2) radar-based passive detection. These methods are briefly addressed below, with particular emphasis attached to practical design issues.

12.7.1 Radar-based active detection—practical constraint

Ice-penetrating radar is currently the most promising technique to directly confirm the existence of any ocean hidden within Jupiter's icy moons. Radar-based active detection works by transmitting radio signals, detecting any radio signals that reflect back, and analyzing these signals to deduce details about what they reflected off of, much like how a person might use a flashlight to illuminate objects hidden in the dark. Ice- and ground-penetrating radar systems look for signals that indicate buried objects and boundaries between layers. In Europa's case, this means looking for the boundaries between the icy crust and any hidden liquid ocean, and between such an ocean and Europa's rocky core.

To detect these oceans with ice-penetrating radar, low-frequency signals of less than 30 megahertz are needed to overcome radio wave absorption by the ice, as well as the unpredictable scattering of radio waves by the crinkled surfaces of these moons. The low-frequency radio waves that researchers would like to use are decametric, meaning they have wavelengths tens of meters long.

One problem with attempting ice-penetrating decametric radar on Jupiter's moons has to do with the powerful decametric radio bursts coming from Jupiter itself. Altogether, these signals are more than 3000 times stronger than any leaking into the solar system from the rest of the galaxy. Litvinenko et al. (2016) investigated the dynamic spectra of the Sun and Jovian decametric radiation obtained by them with the radio telescopes UTR-2 and URAN-2 (Kharkov, Poltava, Ukraine). They described briefly some mechanisms of their generation.

Jupiter's decametric waves come from clouds of electrically charged particles trapped in Jupiter's magnetic field. To overcome Jupiter's strong radio signals, a mission probing Jupiter's moons would need a relatively strong transmitter, a massive device that might be difficult to power and fit aboard the limited confines of a spacecraft.

12.7.2 Radar-based passive detection—an ingenious method

Radar instruments are part of the core payload of the two Europa Jupiter System Mission (EJSM) spacecraft: NASA-led Jupiter Europa Orbiter (JEO) and ESA-led Jupiter Ganymede Orbiter (JGO). As a part of this project, several frequency bands have been under study for radar, which ranged between 5 and 50 MHz. Part of this frequency range overlapped with that of the natural Jovian radio emissions, which are very intense in the decametric range, below 40 MHz. Radio observations above 40 MHz are free of interferences, whereas below this threshold, careful observation strategies have to be investigated.

Cecconi et al. (2012) reported a review of spectral intensity, variability, and sources of these radio emissions. It was found that as the radio emissions are strongly beamed, it is possible to model the visibility of the radio emissions, as seen from the vicinity of Europa or Ganymede. Cecconi et al. (2012) investigated Io-related radio emissions as well as radio emissions related to the auroral oval. They also reviewed the radiation belts synchrotron emission characteristics; and reported radio sources visibility products (dynamic spectra and radio source location maps, on still frames or movies), which can be used for operation planning. Their study clearly showed that a deep understanding of the natural radio emissions at Jupiter is necessary to prepare the future EJSM radar instrumentation. Cecconi et al. (2012) showed that this radio noise has to be taken into account very early in the observation planning and strategies for both JGO and JEO. They also pointed out possible synergies with RPW (Radio and Plasma Waves) instrumentations.

As already indicated, the application of active radars (to probe the suspected liquid-ocean hidden beneath Europa's icy crust) demands prohibitively strong transmitter. Fortunately, Jupiter not only hosts icy moons which could contain subsurface oceans, it is also an extremely bright radio emitter at decametric wavelengths. At these wavelengths, ice happens to be fairly transparent, providing a window to view subsurface oceans.

Instead of carrying a transmitter onboard a spacecraft to overpower Jupiter's radio signals, researchers now suggest using the giant planet's decametric radio waves to scan its moons. All the mission would then need are very low-power systems to detect radio signals reflected by the moons and any oceans lurking within them. Powerful radio signals that Jupiter generates could be used to help researchers scan its giant moons for oceans that could be home to extraterrestrial life.

Romero-Wolf et al. (2015) describe an interferometric reflectometer method for passive detection of subsurface oceans and liquid water in Jovian icy moons using Jupiter's decametric radio emission (DAM). The DAM flux density exceeds 3000 times the galactic background in the neighborhood of the Jovian icy moons, providing a signal that could be used for passive radio sounding. An instrument located between Jupiter and its icy moon could sample the DAM emission along with its echoes reflected in the ice layer of the target moon. Cross-correlating the direct emission with the echoes would provide a measurement of the ice shell thickness along with its dielectric properties. The interferometric reflectometer provides a simple solution to sub-Jovian radio sounding of ice shells that is complementary to ice-penetrating radar measurements better suited to measurements in the anti-Jovian hemisphere that shadows Jupiter's strong decametric emission. The passive nature of this technique also serves as risk reduction in case of radar transmitter failure. The interferometric reflectometer could operate with electrically short antennas, thus extending ice depth measurements to lower frequencies, and potentially providing a deeper view into the ice shells of Jovian moons.

As just indicated, the strategy that Romero-Wolf et al. (2015) developed involves placing a spacecraft between Jupiter and one of its icy moons. The probe would then monitor decametric emissions from Jupiter as well as echoes of those signals reflected off the icy moon. The technology to do this is readily available and requires no major developments. By comparing the signals from Jupiter with the echoes from its moon, the researchers can determine the thickness of the moon's icy shell and the depth of its ocean using cross-correlation method indicated earlier.

This strategy, where one analyzes both distant radio emissions and their echoes, is known as interferometric reflectometry. It was first applied by the Dover Heights radio observatory near Sydney, Australia, in the 1940s and was conceived due to the limited resources available to astronomers when the observatory first started out, like the situation faced by designers of deep space probes.

The atmospheres of Jupiter's icy moons are thin and are not expected to attenuate the decametric radio signal significantly. However, Europa does have an ionosphere, a layer of free electrons, which can distort the radio signal. Grima et al. (2015) reviewed the current state of knowledge of the Europan plasma environment, its effects on radio wave propagation, and its impact on the performance and design of future radar sounders for the exploration of Europa's ice crust. The Europan ionosphere is produced in two independently-rotating hemispheres by photo-ionization of the neutral exosphere and interaction with the Io plasma torus, respectively. This combination is responsible for temporal and longitudinal ionospheric heterogeneities not well constrained by observations. When Europa's ionosphere is active, the maximum cut-off frequency is 1 MHz at the surface. The main impacts on radar signal propagation are dispersive phase shift and Faraday rotation, both a function of the total electron content (up to $4 \times 10^{15}/m^2$) and the Jovian magnetic field strength at Europa (~420 nT). The severity of these impacts decreases with increasing center-frequency and increase with altitude, latitude, and bandwidth. Grima et al. (2015) reckon that the 9-MHz channels on the Radar for Icy Moons Exploration (RIME) and proposed Radar for Europa Assessment and Sounding: Ocean to Near-surface (REASON) will be sensitive to the Europan ionosphere. For these or similar radar sounders, the ionospheric signal distortion from dispersive phase shift can be corrected with existing techniques, which would also enable the estimation of the total electron content below the spacecraft. At 9 MHz, the Faraday fading is not expected to exceed 6 dB under the worst conditions. At lower frequencies, any active or passive radio probing of the ice shell exploration would be limited to frequencies above 1–8 MHz (depending on survey configuration) below which Faraday rotation angle would lead to signal fading and detection ambiguity. Grima et al. (2015) concluded that radar instruments could be sensitive to neutrals and electrons added in the exosphere from any plume activity if present. Fortunately, Europa's ionosphere is fairly small, and not expected to have a big impact on our ability to probe the ice layer.

Based on a study, Hartogh and Ilyushin (2016) proposed a principal concept of a passive interferometric instrument for deep sounding of the icy moons' crust. Its working principle is measuring and correlating Jupiter's radio wave emissions with reflections from the deep subsurface of the icy moons. These investigators examined a number of the functional aspects of the proposed experiment, in particular, impact of the wave scattering on the surface terrain on the instrument performance and digital sampling of the noisy signal. They have reported the results of the test of the laboratory prototype of the instrument.

The success of the proposed missions to Europa in search of the postulated subsurface ocean hidden below its icy crust will rely on the ability of the radar signals to penetrate 10 km of icy material that could potentially contain various types of impurities. In a study, Pettinelli et al. (2016) quantified the impact of magnesium sulfate hydrates on the electrical properties of water ice by performing a series of dielectric measurements on different ice/$MgSO_4 \cdot 11H_2O$ mixtures as a function of frequency and at temperatures comparable with those expected on the icy moon surfaces. Their results indicate that the salt only affects the real part of permittivity of the mixtures, whereas the imaginary part, hence the attenuation, does not significantly differ from that of pure ice. This means that in some regions signal penetration may be better than previously thought.

Romero-Wolf et al. (2016) estimated the sensitivity of a lander-based instrument for the passive radio detection of a subsurface ocean beneath the ice shell of Europa, expected to be between 3 km and 30 km thick, using Jupiter's decametric radiation. A passive technique was previously studied for an orbiter. Romero-Wolf et al. (2016) found that application of passive detection in a lander platform provides a point-measurement with significant improvements due to largely reduced losses from surface roughness effects, longer integration times, and diminished dispersion due to ionospheric effects, thereby allowing operation at lower frequencies and a wider band. Furthermore, Romero-Wolf et al. (2016) found that a passive sounder on-board a lander provides a low resource instrument sensitive to subsurface ocean at Europa up to depths of 6.9 km for high loss ice (16 dB/km two-way attenuation rate) and 69 km for pure ice (1.6 dB/km).

As already indicated, recent work has raised the potential for Jupiter's decametric radiation to be used as a source for passive radio sounding of its icy moons. Two radar sounding instruments, the Radar for Icy Moon Exploration (RIME) and the radar for Europa Assessment and Sounding: Ocean to near-surface (REASON) have been selected for ESA and NASA missions to Ganymede and Europa. In a study, Schroeder et al. (2016) revisited the projected performance of the passive sounding concept and assessed the potential for its implementation as an additional mode for RIME and REASON. They found that the Signal to Noise Ratio (SNR) of passive sounding can approach or exceed that of active sounding in a noisy sub-Jovian environment, but that active sounding achieves a greater SNR in the presence of quiescent noise and outperforms passive sounding in terms of clutter. They also compared the performance of passive sounding at the 9-MHz HF center-frequency of RIME and REASON to other frequencies within the Jovian decametric band. They concluded that the addition of a passive sounding mode on RIME or REASON stands to enhance their science return by enabling sub-Jovian HF sounding in the presence of decametric noise, but that there is not a compelling case for implementation at a different frequency.

Using the current state of knowledge of the geological and geophysical properties of Jupiter's moons Europa, Ganymede and Callisto, Heggy et al. (2017) performed a comprehensive radar detectability study to quantify the exploration depth and the lower limit for subsurface identification of water and key tectonic structural elements. To achieve these objectives, they established parametric dielectric models that reflect different hypotheses on the formation and thermal evolution of each moon of Jupiter. The models were then used for Finite Difference Time Domain (FDTD) radar propagation simulations at the 9-MHz sounding frequency proposed for both ESA JUICE and NASA Europa missions. Note that FDTD method can be used to solve Maxwell's equations when a dispersive medium can be simulated (Liu et al., 2007). Heggy et al. (2017) investigated the detectability above the galactic noise level of four predominant subsurface features: brittle-ductile interfaces, shallow faults, brine aquifers, and the hypothesized global oceans. For Ganymede, their results suggested that the brittle-ductile interface could be within radar detectability range in the bright terrains, but is more challenging for the dark terrains. Moreover, it was found that understanding the slope variation of the brittle-ductile interface is possible after clutter reduction and focusing. Furthermore, it was found that for Europa, the detection of shallow subsurface structural elements few kilometers deep (such as fractures, faults, and brine lenses) is achievable and not compromised by surface clutter. Heggy et al. (2017)'s study indicated that the objective of detecting the potential deep global ocean on Europa is also doable under both the convective and conductive hypotheses.

The surface of Europa has been hypothesized to include an ice regolith (the layer of unconsolidated solid material covering the bedrock of a planet) layer from hundreds of meters to kilometers in thickness. However, contrary to previous claims, it does not present a significant obstacle to searching for Europa's ocean with radar sounding. A study carried out by Aglyamov et al. (2017) corrected prior volume scattering loss analyses and expanded them to include observational and thermo-mechanical constraints on pore size and regolith depth. This provides a more physically realistic range of potential ice-regolith volume-scattering losses for radar-sounding observations of Europa's ice shell in the HF and VHF frequency bands. These investigators concluded that, for the range of physical processes and material properties observed or hypothesized for Europa, volume scattering losses are not likely to pose a major obstacle to radar penetration.

Planetary scientists now plan to make more detailed estimates of how well their radio strategy might detect hidden oceans in Jupiter's icy moons. For instance, they are hoping to make observations from Earth of Jupiter's decametric radio emissions as they reflect off the icy moon surfaces. The initial estimates made by scientists indicate that this may be possible—the measurements would be close to the sensitivity of current ground-based radio observatories. Although there are limitations to the technique, if the technique proves successful, it could provide valuable information about the surface properties of the moons.

12.8 Direct methods for detecting *Europa*'s Ocean—plans in the pipeline

Much of the geologic activity on Europa's icy surface has been attributed to tidal deformation, mainly due to Europa's eccentric orbit. Although the surface is geologically young (30–80 Myr), there is little information as to whether tidally driven surface processes are ongoing. However, a recent detection of water vapor near Europa's South Pole suggests that it may be geologically active. Initial observations indicated that Europa's plume eruptions are time-variable and may be linked to its tidal cycle. Saturn's moon, Enceladus, which shares many similar traits with Europa, displays tidally-modulated plume eruptions, which bolstered this interpretation. However, additional observations of Europa at the same time in its orbit failed to yield plume detection, casting doubt on the tidal control hypothesis.

Rhoden et al. (2015) carried out a study with the specific purpose of analyzing the timing of plume eruptions within the context of Europa's tidal cycle to determine whether such a link exists and examine the inferred similarities and differences between plume activity on Europa and Enceladus. To do this, they determined the locations and orientations of hypothetical tidally driven fractures that best match the temporal variability of the plumes observed at Europa. Specifically, they identified model faults that are in tension at the time in Europa's orbit when a plume was detected and in compression at times when the plume was not detected. They found that tidal stress driven solely by eccentricity is incompatible with the observations unless additional mechanisms are controlling the eruption timing or restricting the longevity of the plumes. The addition of obliquity tides, and corresponding precession of the spin pole, can generate a number of model faults that are consistent with the pattern of plume detections. The locations and orientations of these hypothetical source fractures were found to be robust across a broad range of precession rates and spin pole directions. Analysis of the stress variations across the fractures suggested that the plumes would be best observed earlier in the orbit (true anomaly ~120°). Rhoden et al. (2015)'s results indicate that Europa's plumes, if confirmed, differ in many respects from the Enceladean plumes and that either active fractures or volatile sources are rare.

Planetary scientists would like to analyze Europa's ocean directly, perhaps with missions to bore into Europa's icy shell using heat to melt through the ice, whirling blades to clear away rocks, and robot submarines (subs) to explore the ocean. However, it remains uncertain how thick this shell is, complicating any plans to penetrate it. Models of its thickness, based on the amount of heat the shell receives from the Sun and Europa itself, predict it to be roughly 18 miles (30 km) thick. In contrast, analyses of the Galileo spacecraft's data suggest the shell is no more than 9 miles (15 km) thick, and may be as little as 2.5 miles (4 km) thick (Charles Q. Choi (2014), Radio signals from Jupiter could aid search for life; Astrobiology Magazine Contributor - Jun 12, 2014; https://www.astrobio.net/news-exclusive/radio-signals-jupiter-aid-search-life/).

According to Ivanov et al. (2011), three major features make Europa a unique scientific target for a lander-oriented interplanetary mission: (1) the knowledge of the composition of the surface of Europa is limited to interpretations of the spectral data, (2) a lander could provide unique new information about outer parts of the solar system, and (3) Europa may have a subsurface ocean that potentially may harbor life, the traces of which may occur on the surface and could be sampled directly by a lander. These characteristics of Europa bring the requirement of safe landing to the highest priority level because any successful landing on the surface of this moon will yield scientific results of fundamental importance. The safety requirements include four major components: (1) A landing site should preferentially be on the anti-Jovian hemisphere of Europa in order to facilitate the orbital maneuvers of the spacecraft; (2) A landing site should be on the leading hemisphere of Europa in order to extend the lifetime of a lander and sample pristine material of the planet; (3) Images with the highest possible resolution must be available for the selection of landing sites; (4) The terrain for landing must have morphology (relief) that minimizes the risk of landing and represents a target that is important from a scientific point of view. These components severely restrict the selection of regions for landing on the surface of Europa.

After the photo-geologic analysis of all Galileo images with a resolution of better than about 70 m/pixel taken for the leading hemisphere of Europa, Ivanov et al. (2011) proposed one primary and two secondary (backup) landing sites. The primary site (51.8°S, 177.2°W) is within a pull-apart zone affected by a small chaos. The first backup site (68.1°S, 196.7°W) is also inside of a pull-apart zone and is covered by images of the lower resolution (51.4 m/pixel). The second backup site (2.4°N, 181.1°W) is imaged by relatively low-resolution images (~70 m/pixel) and corresponds to a cluster of small patches of dark and probably smooth plains that may represent landing targets of the highest scientific priority from the scientific point of view. The lack of the high-resolution images for this region prevents, however, its selection as the primary landing target.

12.9 Lakes & layer of liquid water in & beneath *Europa's* hard icy outer shell

Europa is one of the coldest moons of Jupiter. Its surface temperature averages about 110 K (−163°C) at the equator and only 50 K (−223°C) at the poles, thereby keeping Europa's icy crust as hard as granite (McFadden et al., 2006). Planetary scientists' consensus view is that a layer of liquid water exists beneath Europa's surface, and that, despite its chilly surface layer (i.e., crust), heat from tidal flexing allows the subsurface ocean to remain liquid (Greenberg, 2005).

It may be noted that the slight eccentricity of Europa's orbit, maintained by the gravitational disturbances from the other Galileans, causes Europa's sub-Jovian point to oscillate around a mean position. As Europa comes slightly nearer to Jupiter, Jupiter's gravitational attraction increases, causing Europa to elongate toward and away from it. As Europa moves slightly away from Jupiter, Jupiter's gravitational force decreases, causing Europa to relax back into a more spherical shape, and creating tides in its ocean. The orbital eccentricity of Europa is continuously pumped by its mean-motion resonance with Io (Showman and Malhotra, 1997). Thus, the tidal flexing kneads (squeezes) Europa's interior and gives it a source of heat, possibly allowing its ocean to stay liquid while driving subsurface geological processes (Showman and Malhotra, 1997). The ultimate source of this energy is Jupiter's rotation, which is tapped by Io through the tides it raises on Jupiter and is transferred to Europa and Ganymede by the orbital resonance (Showman and Malhotra, 1997; Moore, 2003).

It has been found that Europa has a tortured young surface and sustains a liquid water ocean below an ice shell of highly debated thickness. Quasi-circular areas of ice disruption called "chaos terrains" are unique to Europa, and both their formation and the ice-shell thickness depend on Europa's thermal state. Conamara Chaos is a landscape produced by the disruption of the icy crust of Europa. No model so far has been able to explain why features such as "Conamara Chaos" stand above surrounding terrain and contain matrix domes. Note that Conamara Chaos is a region of chaotic terrain on Europa. It is named after Conamara in Ireland due to its similarly rugged landscape (Pappalardo, 2009). The region consists of rafts of ice that have moved around and rotated. Surrounding these plates is a lower matrix of jumbled ice blocks which may have been formed as water, slush, or warm ice rose up from below the surface. The region is cited as evidence for a liquid ocean below Europa's icy surface. Fig. 12.10 shows "Ice rafts" in Conamara Chaos. Seen in this Figure are crustal plates ranging up to 13 km (8 miles) across, which have been broken apart and "rafted" into new positions, superficially resembling the disruption of pack-ice on polar seas during spring thaws on Earth. The size and geometry of these features suggest that motion was enabled by ice-crusted water or soft ice close to the surface at the time of disruption. The area shown is about 34 km by 42 km (21 miles by 26 miles), centered at 9.4 degrees north latitude, 274 degrees west longitude, and the resolution is 54 m (https://commons.wikimedia.org/wiki/File:Europa_Ice_Rafts.jpg). Europa's immense network of crisscrossing cracks serves as a record of the stresses caused by massive tides in its global ocean. Tidal forces are thought to generate the heat that keeps Europa's ocean liquid.

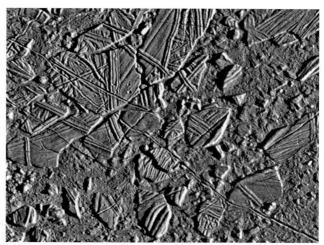

FIG. 12.10 "Ice rafts" in Conamara Chaos. *Author: By NASA/JPL - http://solarsystem.nasa.gov/multimedia/display.cfm?IM_ID=13404, Public Domain, https://commons.wikimedia.org/w/index.php?curid=24211696. (Source: https://commons.wikimedia.org/wiki/File:Europa_Ice_Rafts.jpg). This file is in the **public domain** in the United States because it was solely created by NASA. NASA copyright policy states that "NASA material is not protected by copyright unless noted."*

It has been theorized that melt-through of a thin (few-kilometre) shell is thermodynamically improbable and cannot raise the ice. It has been found that the buoyancy of material rising as either plumes of warm, pure ice called diapirs or convective cells in a thick (>10 km) shell is insufficient to produce the observed chaos heights, and no single plume can create matrix domes. Schmidt et al (2011) have reported an analysis of archival data from Europa, guided by processes observed within Earth's sub-glacial volcanoes and ice shelves. The data suggest that "chaos terrains" form above liquid water lenses perched within the ice shell as shallow as 3 km. Their results suggest that ice–water interactions and freeze-out give rise to the diverse morphologies and topography of chaos terrains. The sunken topography of Thera Macula (a region of likely active chaos production above a large liquid water lake in the icy shell of Europa) indicates that Europa is actively resurfacing over a lens comparable in volume to the *Great Lakes* in North America. Thus, evidence suggests the existence of lakes of liquid water entirely encased in Europa's icy outer shell and distinct from a liquid ocean thought to exist farther down beneath the ice shell (Schmidt et al., 2011; Airhart, 2011). If confirmed, the lakes could be yet another potential habitat for life.

It has been suspected that the ocean in Europa may have reduction-oxidation (redox) balance similar to Earth's. On Earth, low-temperature hydration of crustal olivine produces substantial hydrogen, comparable to any potential flux from volcanic activity. As part of their study in this topic, Vance et al. (2016) calculated how much hydrogen could potentially be produced in Europa's ocean as seawater reacts with rock in a process called serpentinization. In this process, water percolates into spaces between mineral grains and reacts with the rock to form new minerals, releasing hydrogen in the process. In Earth's oceanic crust, such fractures are believed to penetrate to a depth of 5–6 km (3 to 4 miles). On present-day Europa, planetary researchers expect water could reach as deep as 25 km (15 miles) into the rocky interior, driving these key chemical reactions throughout a deeper fraction of Europa's seafloor. In their study, the researchers considered how cracks in Europa's seafloor likely open up over time, as the moon's rocky interior continues to cool following its formation billions of years ago. New cracks expose fresh rock to seawater, where more hydrogen-producing reactions can take place.

Researchers compared hydrogen and oxygen production rates of the Earth system with fluxes to Europa's ocean. It was found that even without volcanic hydrothermal activity, water-rock alteration in Europa causes hydrogen fluxes 10 times smaller than Earth's. Vance et al. (2016) argue that Europa's ocean may have become reducing for a brief epoch, for example, after a thermal-orbital resonance ~2 Gyr after accretion. It was further found that the estimated oxidant flux to Europa's ocean is comparable to the estimated hydrogen fluxes. It was suggested that Europa's ice delivers oxidants to its ocean at the upper end of these estimates if its ice is geologically active, as evidence of geologic activity and subduction implies.

12.10 Europa adorning a volcanic seafloor—the year 2021 numerical model prediction

Europa is one of the rare planetary bodies that might have maintained volcanic activity over billions of years, and possibly the only one beyond Earth that has large water reservoirs and a long-lived source of energy. Volcanoes on a Jovian moon such as Europa would not be unprecedented. For example, one of Europa's fellow Galilean moons, Io, is the most volcanically active body in the solar system, and its eruptions are fuelled by the same type of gravitational tugging that Europa experiences. (Io's interior flexes more dramatically, however, because it orbits closer to Jupiter than Europa does).

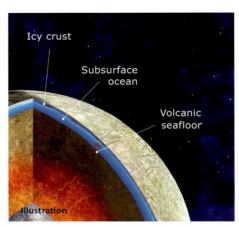

FIG. 12.11 Scientists' findings suggest that the interior of Jupiter's moon Europa may consist of an iron core, surrounded by a rocky mantle in direct contact with an ocean under the icy crust. New research models how internal heat may fuel volcanoes on the seafloor. (Image credit: NASA/JPL-Caltech/Michael Carroll). *(Source: https://www.space.com/jupiter-moon-europa-deep-sea-volcanoes?utm_source=notification).*

Běhounková et al. (2021) investigated the melting of the silicate mantle through time and its consequences for seafloor magmatism by modeling Europa's internal heat production and transfer using a three-dimensional numerical model. Based on the results gleaned from this three-dimensional numerical model, Běhounková et al. (2021) demonstrated that the magmatic activity can continue during most of Europa's history even though it progressively decays as the interior cools down. These investigators show that melt can be produced during most of Europa's history due to the limited efficiency of internal cooling by thermal convection and the presence of radiogenic heating. Their studies show that the melting rate is amplified by tidal friction, possibly leading to magmatic pulses during enhanced eccentricity periods and focusing melting to high latitudes. Běhounková et al. (2021)'s studies indicate that the volume of generated melts during magmatic episodes is comparable to those involved in Large Igneous Provinces, commonly observed on Earth, and may impact ocean chemistry. Běhounková et al. (2021) predicted that magmatic activity should be modulated in time due to change in Europa's orbit, and should focus in polar regions where the heat produced by tidal friction is the largest. Their studies further indicate that the predicted magmatic pulses should be accompanied by the release of volatiles (hydrothermal fluids such as those emitted by terrestrial submarine hydrothermal vents) that may impact the oceanic chemistry. Fig. 12.11 provides an illustration of volcanoes on the seafloor (i.e., the source of submarine hydrothermal vents) of Europa. Běhounková et al. (2021) predicted that gravity measurements, detection of anomalous H_2/CH_4, and astrometric data by upcoming spacecraft missions could confirm ongoing large-scale seafloor activity.

12.11 Plumes of water vapor erupting from Europa's surface

The high-resolution images of surface features on Europa from the Galileo Solid State Imaging (SSI) experiment (*Carr* et al., *1998*; *Greeley* et al., *1998*; *Pappalardo* et al., *1998*; *Spaun* et al., *1998, 1999*; *Greenberg* et al., *1998*; *Riley* et al., *2000*) have revealed the presence of two types of features, "chaos" (disorderly mass) and "lenticulae" (a kind of eruption visible on the surface), whose characteristics suggest that melting and disruption of the surface have occurred, even though the present temperatures on Europa average about 100 K. *Sotin* et al. (2002) pointed out that the just-mentioned confusing and difficult-to-explain observation (conundrum) with regard to the Europan surface features resulted in the development of three hypotheses to account for these features and other characteristics of the Europan crust and lithosphere: (1) the outer solid layer is less than a few km thick, and that the surface is constantly in contact with an ocean below due to tidal cracking and disruption, and small variations in the thermal gradient cause chaos and lenticulae as reported by several researchers (e.g., *Carr* et al., *1998*; *Greenberg* et al., *1998, 1999, 2001*); (2) the outer solid layer is of the order of 20–30 km thick and the lenticulae and chaos are the result of diapirism (the act of diapir formation; a major form of supply of basaltic melts into the crust of a celestial body), which delivers warmer material from depth to near the surface, causing disruption and partial melting (*Pappalardo* et al., *1998*; *Rathbun* et al., *1998*; *McKinnon, 1999*; *Pappalardo and Head, 2001*); and (3) tidal energy focused in the silicate interior at the base of an ocean causes thermal plumes to rise through the ocean and melt through an outer rigid surface layer 2–5 km thick (*Thomson and Delaney, 2001*).

A subsurface ocean hides below Europa's icy crust. In a study carried out to understand the cause of the observed chaos and lenticulae on the surface of Europa, Sotin et al. (2002) took into consideration a known geological phenomenon known as "diapirism" Diapir is a geological structure consisting of mobile material that was forced into more brittle surroundings, usually by the upward flow of material from a parent stratum. Diapir can be considered as a geologic intrusion in which more mobile and ductilly deformable materials are forced into brittle overlying harder materials. The flow may be produced by gravitational forces, tectonic forces, tidal forces, and the like. Diapirs may take the shape of domes, waves, mushrooms, teardrops, or dikes.

Sotin et al. (2002) developed a model for the rise of diapirs and the influence of tidal heating. Compared to the previous description of plumes within Europa (Rathbun et al., 1998), the present study describes plumes arising in a fully developed convective layer. In addition, these 2D numerical thermal convection models (Deschamps and Sotin, 2001) take into account the fact that ice viscosity (η) is strongly temperature-dependent. With the application of this model, Sotin et al. (2002) reported on a mechanism that can produce melting temperatures and disruption in water ice in the shallow crust of Europa. They showed that in the range of ice viscosity inferred from laboratory experiments, tidal energy can be preferentially focused, and tidal forces will heat up rising diapirs, resulting in rising plumes emanating from Europa's surface. When the plume cores reach the base of the outer cold brittle layer, they spread laterally, causing shallow melting, disruption, and formation of terrain similar to lenticulae and chaos. Sotin et al. (2002) further showed that this mechanism can readily explain the major characteristics of chaos and lenticulae even if the average outer solid ice layer thickness is larger than 20 km.

According to Sotin et al. (2002), tidal heating heats up the upwelling plume because the plume moves rapidly relative to loss of heat by lateral conduction. In Fig. 12.12, the mean temperature profile as well as the temperature in the center of the plume are shown. In that case, the upwelling velocity is so high that the temperature does not exceed the melting temperature until the plume slows down as it arrives beneath the conductive lid (around 10 km). There, tidal heating melts the ice and one can compute the amount of ice transformed into water by dividing tidal heating by latent heat.

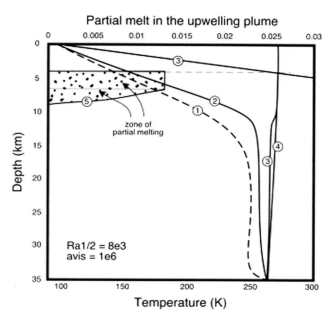

FIG. 12.12 Temperature profiles. curve 1 is the horizontally averaged temperature; curve 2 is the temperature in and above the plume core without taking into account tidal heating; and, curve 3 is the same as curve 2 but taking into account tidal heating and it matches the melting temperature (curve 4) when the plume velocity decreases and partial melt starts to be created. Curve 5 denotes the amount of partial melt in the upwelling plume and the location of partial melt is emphasized with dots. *(Source: Sotin et al., 2002, Europa: Tidal heating of upwelling thermal plumes and the origin of lenticulae and chaos melting, Geophys. Res. Lett., 29 (8): 74-1–74-4); Publisher: Wiley-Blackwell. 2020 American Geophysical Union.*

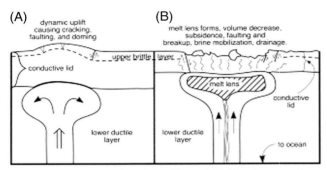

FIG. 12.13 **Interpretation of the effects of tidal heating on a plume.** Without tidal heating (A), plume rises buoyantly, impinges on overlying conductive lid, and causes doming. Plume core is tidally heated (B) and melt lens builds up in the upper part and leads to the collapse and modification seen in chaos and lenticulae. *(Source: Sotin et al., 2002, Europa: Tidal heating of upwelling thermal plumes and the origin of lenticulae and chaos melting, Geophys. Res. Lett., 29 (8): 74-1–74-4). Publisher: Wiley-Blackwell. Copyright info: 2020 American Geophysical Union.*

In the models Sotin et al. (2002) examined, 1%–10% of the ice could be melt in the upwelling warm ice. They found that this process can lead to the formation of a water-rich reservoir (melt lens) at the top of the plume (Fig. 12.12 and Fig. 12.13). It is found that the increase of heat flux induced by tidal heating increases the slope of the conductive temperature profile just above the plume by a factor 2 or 3. The conductive temperature profile intercepts the melting curve at a depth (Fig. 12.12 and Fig. 12.13) which can represent the depth where partial melt can still be present (4 km in the present example). Sotin et al. (2002) found that models with smaller values of the Rayleigh number (smaller crustal thickness) predict limited partial melting in the plume tail (1%) and more partial melt just beneath the conductive lid. Sotin et al. (2002) argue that continued pumping of warm buoyant plume material into the plume head should laterally enlarge the zone of melting and disruption. The presence of partial melt modifies the viscosity and therefore the amount of tidal heating and focuses the streamlines.

On the basis of their analysis, Sotin et al. (2002) argued that the incorporation of tidal heating into models for the ascent of diapirs from the base of a thick ice layer can produce melting temperatures in broad areas near the surface of Europa and account for many of the features observed in lenticulae and chaos. According to them, if one considers a single diapir rising in a water–ice dominated crust (Fig. 12.13). Sotin et al. (2002) explained that as the rising diapir nears the surface, it is sufficiently buoyant to cause up-bowing and fracturing of the overlying lineated plains surface to cause domes (e.g., Fig. 12.13A). With tidal heating (Fig. 12.13B), as the melting isotherm is reached in the subsurface (Fig. 12.12) and the plume head spreads laterally, a zone of partial melting and basal heating exists just below the surface.

Sotin et al. (2002) argued that the process of shallow melting indicated above has several potential consequences as explained below. Firstly, "the overlying brittle layer becomes heated, and depending on the initial thickness of the layer and the peak heat flux, can undergo partial melting and disaggregation to produce micro-chaos. Brines and other lower melting temperature impurities (Kargel et al., 2000) will enhance this effect, as will other processes such as brine mobilization (e.g. Head and Pappalardo, 1999)." Secondly, "the decrease in volume accompanying melting will result in subsidence of the overlying material, creating further surface disruption, and potentially leading to the low topography such as that associated with pits and topographic moats" (Fig. 12.13B). Thirdly, "the negative buoyancy of the resulting melt will mean that any melt will tend to migrate downward, enhancing subsidence in the surface layer and further removing support from the overlying brittle layer to produce disruption and chaos formation." Fourthly, "the thermal perturbation associated with the enhanced near-surface temperatures could lead to conditions that would enhance surface albedo change and potential discoloration through processes of surface frost sublimation, brine mobilization and exposure, and concentration of melt impurities and residues (e.g., Spencer, 1987; Pappalardo et al., 1999)." Lastly, "continued tidal deformation may over-pressurize accumulations of partial melt and propagate dikes and sills upward from the melt zone (Wilson et al., 1997; Collins et al., 2000)."

The Hubble Space Telescope acquired an image of Europa in 2012 that was interpreted to be a plume of water vapor erupting from its surface near Europa's South Pole. The image suggests the plume may be 200 km (120 mi) high, or more than 20 times the height of Mt. Everest. It has been suggested that if they exist, they are episodic and likely to appear when Europa is at its farthest point from Jupiter, in agreement with tidal force modeling predictions (Roth et al., 2013).

A Europa plume source may produce a global exosphere with complex spatial structure and temporal variability in its density and composition. To investigate this interaction Teolis et al. (2017) integrated a water plume source containing multiple organic and nitrile species into a Europan Monte Carlo exosphere model, considering the effect of Europa's gravity in returning plume ejecta to the surface, and the subsequent spreading of adsorbed and exospheric material by thermal desorption and resputtering across the entire body. Teolis et al. (2017) considered sputtered, radiolytic, and potential plume sources, together with surface adsorption, regolith diffusion, polar cold trapping, and resputtering of adsorbed materials, and examined the spatial distribution and temporal evolution of the exospheric density and composition. It was found that these models provide a predictive basis for telescopic observations and planned missions to the Jovian system by NASA and ESA.

Teolis et al. (2017) applied spacecraft trajectories to their model to explore possible exospheric compositions which may be encountered along proposed flybys of Europa to inform the spatial and temporal relationship of spacecraft measurements to surface and plume source compositions. For this preliminary study, Teolis et al. (2017) considered four cases: Case A: an equatorial flyby through a sputtered only exosphere (no plumes), Case B: a flyby over a localized sputtered 'macula' terrain enriched in nonice species, Case C: a south polar plume with an Enceladus-like composition, equatorial flyby, and Case D: a south polar plume, flyby directly through the plume.

In another study, Huybrighs et al. (2017) investigated the feasibility of detecting water molecules (H_2O) and water ions (H_2O^+) from the Europa plumes from a flyby mission. In this study, a Monte Carlo particle tracing method was used to simulate the trajectories of neutral particles under the influence of Europa's gravity field and ionized particles under the influence of Jupiter's magnetic field and the convectional electric field. As an example, mission case they investigated the detection of neutral and ionized molecules using the Particle Environment Package (PEP), which is part of the scientific payload of the future JUpiter ICy moon Explorer mission (JUICE). Huybrighs et al. (2017) considered plumes that have a mass flux that is three orders of magnitude lower than what has been inferred from recent Hubble observations. Huybrighs et al. (2017) demonstrated that the in-situ detection of H_2O and H_2O^+ from these low mass flux plumes is possible by the instruments with large margins with respect to background and instrument noise. The signal-to-noise ratio (SNR) for neutrals is up to ~5700 and ~33 for ions. They also showed that the geometry of the plume source, either a point source or 1000 km-long crack, does not influence the density distributions, and thus, their detectability. Furthermore, they discussed how to separate the plume-originating H_2O and H_2O^+ from exospheric H_2O and H_2O^+. It was inferred that the separation depends strongly on knowledge of the density distribution of Europa's exosphere.

As pointed out earlier, a huge ocean of liquid water sloshes beneath Europa's icy shell. On April 24, 1990, the space shuttle *Discovery* lifted off from Earth with the Hubble Space Telescope nestled securely in its bay. The following day, Hubble was released into orbit, ready to peer into the vast unknown of space, offering a glimpse at distant, exotic cosmic shores yet to be described. Since then, NASA's *Hubble Space Telescope* has spotted multiple times tantalizing signs of a giant water vapor plume emanating from Jupiter's moon *Europa*. NASA's *Galileo* probe, which orbited the planet Jupiter from 1995 to 2003, also detected a likely Europa plume, during a close flyby of the icy moon in 1997. An apparent plume coming from Europa's south polar region was detected in late 2012, also using Hubble. Astronomers tried repeatedly to confirm the suspected plume phenomenon but kept coming up empty—until early 2014, when Sparks et al. (2017) detected one near Europa's equator.

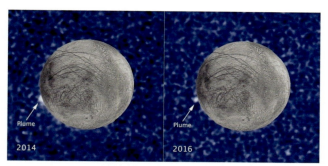

FIG. 12.14 These composite photos, which combine images by NASA's Hubble Space Telescope and Galileo spacecraft, show a suspected plume of material erupting two years apart from the same location on Jupiter's icy moon Europa. The images bolster evidence that the plumes are a real phenomenon, flaring up intermittently in the same region on the moon. *(Image credit: Science; Credit: NASA, ESA, and W. Sparks (STScI). Illustration Credit: NASA, ESA, W. Sparks (STScI), and the USGS Astrogeology Science Center). (Source: Wall, M. (2018), This May Be the Best Evidence Yet of a Water Plume on Jupiter's Moon Europa, https://www.space.com/40575-jupiter-moon-europa-plume-galileo-spacecraft.html).*

In March 2014, Hubble detected a smaller plume on Europa at a location right in the middle of an unusually warm part of Europa's surface. In February 2016, Hubble again detected a 100 km (62-mile) tall plume near Europa's equator. The March 2014 and the February 2016 detections were both accomplished using the "transit technique." As Europa passed in front of Jupiter from Hubble's perspective, the putative plumes blocked some of the ultraviolet light emitted by the giant planet. Repeat detections suggest that in a formal statistical sense, it can't happen by chance. Interestingly, the thermal imaging performed by *Galileo* two decades ago showed a "hotspot" at the location of the 2014 and 2016 plume candidates, thereby strengthening the interpretation that the "hotspot" is indicative of a warm plume. Fig. 12.14 illustrates composite photos, which combine images by NASA's Hubble Space Telescope and Galileo spacecraft, showing a suspected plume of material erupting two years apart from the same location on Europa. If the thermal anomaly and the plumes are indeed causally linked, there are two possible explanations (1) water venting through cracks in Europa's ice could be warming the surface, or (2) water from the plume may be falling back down onto the hotspot, changing the fine structure of the surface and allowing it to hold onto heat longer. Fig. 12.15 traces the location of erupting plumes observed by NASA's Hubble Space Telescope in 2014 and 2016. It is worthy of note that the location of erupting plumes observed by NASA's Hubble Space Telescope in 2014 and 2016 corresponds to a warm region on Europa's surface, as identified by the temperature map. If the eruptions are coming from the subsurface ocean, the elements could be more easily

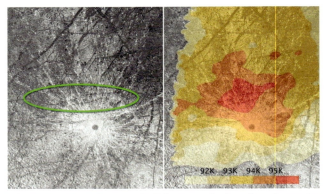

FIG. 12.15 The photo on the left traces the location of erupting plumes observed by NASA's Hubble Space Telescope in 2014 and 2016 (inside the green oval). The green oval also corresponds to a warm region on Europa's surface, as identified by the temperature map at right. The warmest area is bright red. *(Image credit: NASA, ESA, W. Sparks (STScI), and the USGS Astrogeology Science Center). (Source: Wall, 2018. This May Be the Best Evidence Yet of a Water Plume on Jupiter's Moon Europa. https://www.space.com/40575-jupiter-moon-europa-plume-galileo-spacecraft.html).*

detected by a spacecraft like the one planned for NASA's upcoming *Europa Clipper mission*.

Evidence for plumes of water on Europa has previously been found using the Hubble Space Telescope using two different observing techniques: (1) by finding line emission from the dissociation products of water, (2) by finding evidence for off-limb continuum absorption as Europa transited Jupiter.

Earth-based telescopic observations of planets and their moons have limitations in probing the details of these celestial bodies with adequate precision and resolution. However, spacecrafts are capable of approaching these celestial bodies and remotely probing them from close quarters, indeed at a substantially closer distance from the celestial body than from the far away located Earth. Spacecraft missions are, therefore, vital in the detailed studies of celestial bodies. Fortunately, two veteran NASA missions are providing new details about icy, ocean-bearing moons of Jupiter and Saturn, further heightening the scientific interest of these and other "ocean worlds" in our solar system and beyond.

In a study detailing new Hubble Space Telescope findings, Sparks et al. (2017) reported on observations of Europa from 2016, in which a consistently active plume of material was seen erupting from the moon's surface at the same location where Hubble saw evidence of a plume in 2014.

Plumes on cold celestial bodies (e.g., Saturn's moon *Enceladus*) are associated with relatively warmer regions. Therefore, after Hubble imaged this new plume-like feature on Europa, Sparks et al. (2017) examined this specific location on the Galileo thermal map, by comparison with a night-time thermal image from the Galileo Photopolarimeter-Radiometer that shows a thermal anomaly at the same location, within the measurement uncertainties. They discovered that, as expected, Europa's plume candidate is sitting right on the thermal anomaly.

These images bolster evidence that the Europa plumes could be a real phenomenon—an active cryovolcanism—that flares up intermittently in the same region on the moon's surface. The newly imaged plume rises about 62 miles (100 km) above Europa's surface, while the one observed in 2014 was estimated to be about 30 miles (50 km) high. Both correspond to the location of an unusually warm region (relative to the very cold icy crust) that contains features that appear to be cracks in the moon's icy crust, seen in the late 1990s by NASA's *Galileo* spacecraft. Researchers speculate that this could be evidence of water forcefully erupting from Europa's interior.

Sparks et al. (2017) found that the thermal anomaly has the highest observed brightness temperature on the Europa night-side. These researchers argued that if heat flow from a subsurface liquid water reservoir causes the thermal anomaly, its depth is ≈1.8–2 km, under simple modeling assumptions, consistent with scenarios in which a liquid water reservoir has formed within a thick ice shell. It was suggested that models that favor thin regions within the ice shell that connect directly to the ocean, however, cannot be excluded. Likewise, modifications to surface thermal inertia by subsurface activity also cannot be excluded. Alternatively, vapor deposition surrounding an active vent could increase the thermal inertia of the surface and cause the thermal anomaly. Researchers are hopeful that the observed active warm vent (plume) region on Europa may offer a promising location for an initial characterization of Europa's internal water and ice and for seeking evidence of Europa's habitability.

By linking the observed plumes and the warm spot, the researchers came up with the following two consequences:

(i) The high-pressure water being ejected out (vented) from beneath the moon's icy crust is warming the surrounding surface.
(ii) The high-pressure water forcefully ejected by the plume falls onto the surface as a fine mist, changing the structure of the surface grains and allowing them to retain heat longer than the surrounding landscape.

For both the 2014 and 2016 observations, the research team used Hubble's Space Telescope Imaging Spectrograph (STIS) to spot the plumes in ultraviolet light. As Europa passes in front of Jupiter, any atmospheric features around the edge of the moon block some of Jupiter's light, allowing STIS to see the features in silhouette (the dark shape and outline of something visible in restricted light against a brighter background). William Sparks, who led the Hubble plume studies in both 2014 and 2016, and his team are continuing to use Hubble to monitor Europa for additional examples of plume candidates with the hope of determining the frequency with which they appear.

As already indicated, Hubble Space Telescope (HST) observations showed evidence for active water plumes on Europa (Roth et al., 2014; Sparks et al., 2016, 2017) and a re-evaluation of the Galileo magnetometer data revealed the presence of a potential source originating from the surface during the E12 flyby of Europa (Jia et al., 2018) near Pwyll crater. However, the origin and formation mechanisms of primarily two categories of plumes (tall plumes and short plumes) remain unknown. Potential plume sources include fractures extending through the ice shell, directly feeding the eruption from the ocean (Greenberg et al., 1998), or brine reservoirs in the near surface (Fagents, 2003). The first scenario, akin to the eruption of tall plumes from the South Pole of Saturn's moon *Enceladus*, can explain the tall plumes. However, according to Manga and Wang (2007), the first scenario is unlikely because the excess pressure of Europa's ocean is insufficient for liquid water to erupt onto the surface, unless gases drive the water further upward (Crawford and Stevenson, 1988). According to another school of thought put forward by Michaut and Manga (2014) and Manga and Michaut (2017), brine reservoirs may form by injection of ocean water into the ice shell interior through basal fractures or by local melting of the ice shell. The maximum height of the plume observed from the center of the Manannán crater located near the north pole of Europa is estimated to be ~2 km only. An explanation must be found to justify the eruption of such short-height plume.

In May 2018, astronomers provided supporting evidence of water plume activity on Europa, based on an updated critical analysis of data obtained from the Galileo space probe, which orbited Jupiter during the period 1995–2003. Galileo flew by Europa in 1997 within 206 km (128 mi) of the moon's surface and the researchers suggest it may have flown through a water plume (Jia et al., 2018). As noted earlier, Hubble observed a plume on Europa. The area also corresponds to a warm region on Europa's surface. Such plume activity could help researchers in a search for life from the subsurface European ocean without having to land on the moon (Jia et al., 2018).

As already pointed out, Europa has a subsurface ocean covered by an icy shell. Despite evidence for plumes on Europa, no surface features have been definitively identified as the source of the plumes to date. Furthermore, it remains unknown whether the activity originates from near-surface water reservoirs within the ice shell or if it is sourced from the underlying global ocean. A recent study (Steinbrügge et al., 2020) indicates that rather than originating from deep within Europa's oceans, some eruptions may originate from water pockets embedded in the icy shell itself. Steinbrügge et al. (2020) investigated brine pocket migration, studied previously in the context of sea ice on Earth (described in detail by Weeks, 2010), as a process for transporting brine along thermal gradients. In this investigation, Steinbrügge et al. (2020) focused their analyses on Manannán, an

FIG. 12.16 Manannán crater on Jupiter's moon *Europa*, showing circular fractures surrounding a central depression, with linear fractures radiating from the interior. *(Source: Steinbrügge, G., J.R.C. Voigt, N.S. Wolfenbarger, C.W. Hamilton, K.M. Soderlund, D.A. Young, D.D. Blankenship, S.D. Vance, and D.M. Schroeder (2020), Brine migration and impact-induced cryovolcanism on Europa, Geophysical Research Letters, 47, e2020GL090797. https://doi.org/10.1029/2020GL090797. Copyright info: ©2020. American Geophysical Union. All Rights Reserved.*

18-mile-wide crater on Europa (Fig. 12.16) that was created by an impact with another celestial object some tens of millions of years ago. A high-resolution image showed that the immediate surroundings of the crater are dominated by inhomogeneous and knobby terrains with radiating linear streaks, whereas the interior of the crater covers ~346 km^2 and generally has a smooth floor with a central depression partially surrounded by massifs (compact groups of mountains). The central radial fracture system resembles the "spider" (araneiform) features on Mars. The clearly observed radial spider fractures are interpreted as evidence for a rigid surface subjected to pressure from below. In addition to the radial fractures, Manannán exhibits concentric normal faults marking the rim of the central depression and, thus, indicate a collapse.

Steinbrügge et al. (2020) showed that by looking at an impact crater on Europa, which was initially warm in the center and cooled inward from its colder surroundings, it is possible to study how the water migrated toward the center and formed a central water reservoir. They argued that as the final water pocket at the center of the crater started to freeze, the increasing pressure lead to a cryovolcanic eruption that emplaced brine onto the surface to form a prominent "spider" feature before the ice collapsed into the cavity below.

Using a digital terrain model of the crater and collapse feature, Steinbrügge et al. (2020) estimated how much water erupted and how salty Europa's ice shell is. Using images collected by the NASA spacecraft *Galileo*, and reasoning that a collision as indicated above would have generated a tremendous amount of heat, Steinbrügge et al. (2020) modeled how melting and subsequent freezing of a water pocket within the icy shell could have caused the water to erupt. The model they developed was used to explain how a combination of freezing and pressurization as indicated above could lead to a cryovolcanic eruption, or a burst of water rising up in the form of a plume. Fig. 12.17 illustrates a cryovolcanic eruption on Europa in which brine from within the icy shell could blast into space. After the initial impact, residual aqueous melt concentrated via brine pocket migration as the target material cooled. Freezing and over-pressurization then resulted in a cryovolcanic eruption. The volume of the emptied reservoir and the critical composition at the end of migration provide further constraints on the average salinity of Europa's ice shell. Steinbrügge et al. (2020) showed how small pockets of brine can migrate within the ice from colder areas to warmer areas. This can happen even at very low temperatures, below the point where pure water would freeze, because the water becomes saltier and saltier as it migrates.

The model indicates that as Europa's water transformed into ice during the later stages of the impact, pockets of water with increased salinity could be created in the moon's surface. The water pockets can move laterally along thermal gradients, from cold to warm, and not only in the down direction as pulled by gravity. The lateral movement of these salty water pockets through Europa's ice shell can occur by melting adjacent regions of less brackish ice, and consequently become even saltier in the process.

Steinbrügge et al. (2020) showed that the fracture system located in the center of Europa's Manannán crater is consistent with the assembly of a subsurface melt pocket (brine reservoir) via brine pocket migration. They argued that "the kinetic energy released after an impact would cause the crater to be filled rapidly with warm ice and slurry mixture composed of ice and melt water (Silber and Johnson, 2017) (Fig. 12.17A). The material would not consist entirely of either rigid ice or liquid water because dense ice would lack the mobility to fill the crater, and liquid water would quickly evaporate when exposed to vacuum or drain into a permeable substrate (Elder et al., 2012)." According to Steinbrügge et al. (2020)'s explanation, because the brine migration into the warm ice present at the center of the crater becomes slower with increasing salinity, individual smaller brine pockets can outrun the freezing front only up to a critical salt concentration that becomes the upper salinity limit of the final brine reservoir. Once this critical concentration is met, the pressure increases rapidly as the final brine reservoir freezes. Steinbrügge et al. (2020) argue that "Ultimately, this pressure exceeds the failure stress of the overlying crust and spider fractures form (Fig. 12.17B) as a conduit develops toward the surface and, subsequently, ejects a portion of the brine leaving behind the radial fractures. Finally, the ejected volume causes an empty cavity in the subsurface leading to a collapse which created the normal faults (Fig. 12.17C)."

The model predicts that when a migrating brine pocket reached the center of Manannán crater, it became stuck and began freezing, generating pressure that eventually resulted in a plume, estimated to have been over a mile (1.6 km) high. The eruption of this plume left a distinguishing mark: a spider-shaped feature on Europa's surface that was observed by Galileo imaging and incorporated in the researchers' model. According to Steinbrügge et al. (2020), the relatively small size of the plume that would form at Manannán crater indicates that impact craters probably can't explain the source of other, larger plumes on Europa that have been hypothesized based on Hubble and Galileo data.

To summarise, as just discussed, it has been pointed out (Steinbrügge et al., 2020) that features observed on Europa's surface today are consistent with the presence of current or former local melt (e.g., Craft et al., 2016; Michaut and Manga, 2014; Prockter and Schenk, 2005; Schmidt et al., 2011). Several hypotheses have been put forward to explain the possible mechanisms for melt generation. These include diapirism (Head et al., 1999; Sotin et al., 2002), shear heating (Hammond, 2019), and tidal heating (Vilella et al., 2020). In recent times, several investigators (e.g., Fagents et al., 2000; Lesage et al., 2020; Quick and Marsh, 2016) have examined the ascent of cryo-magmas to the surface from these reservoirs. However, processes of near-surface brine migration have received less attention. Steinbrügge et al. (2020) therefore discussed a scenario by which short plume is erupted from the center of Europa's Manannán crater.

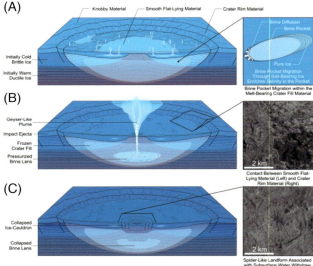

FIG. 12.17 Sketch illustrating the Manannán crater evolution. (A) Stage 1: Crater appearance shortly after the impact process. Crater is filled with refrozen material and small separated melt pockets, which begin to move toward the center of the crater through melt pocket migration. (B) Stage 2: Over-pressurization of the final water reservoir, resulting eruption of the brine to the surface and formation of the spider feature. (C) Stage 3: Collapse of the emptied reservoir and formation of the normal faults encircling the central depression and araneiform ("spider") features shown in Fig. 12.16. *(Source: Steinbrügge, G., J.R.C. Voigt, N.S. Wolfenbarger, C.W. Hamilton, K.M. Soderlund, D.A. Young, D.D. Blankenship, S.D. Vance, and D.M. Schroeder (2020), Brine migration and impact-induced cryovolcanism on Europa, Geophysical Research Letters, 47, e2020GL090797. https://doi.org/10.1029/2020GL090797. Copyright info: ©2020. American Geophysical Union. All Rights Reserved.*

12.12 Saucer-shaped sills of liquid water in *Europa*'s ice shell

Europa's surface contains numerous quasi-elliptical features called pits, domes, spots and small chaos. Manga and Michaut (2017) proposed that these features, collectively referred to as lenticulae, are the surface expression of saucer-shaped sills of liquid water in Europa's ice shell. In particular, the inclined sheets of water that surrounds a horizontal inner sill limit the lateral extent of intrusion, setting the lateral dimension of lenticulae. Furthermore, the inclined sheets disrupt the ice above the intrusion allowing the inner sill to thicken to produce the observed relief of lenticulae and to fracture the crust to form small chaos. According to Manga and Michaut (2017), scaling relationships between sill depth and lateral extent imply that the hypothesized intrusions are, or were, 1–5 km below the surface. Liquid water is predicted to exist presently under pits and for a finite time under chaos and domes.

12.13 *Europa Clipper Mission* aimed at *Europa*'s exploration

There are several inferences made on Europa and the possibility of life on this icy moon, based on telescopic and spectroscopic observations as well as model studies. The authenticity of such inferences needs to be confirmed by measurements from close quarters.

Several planetary scientists harbor questions such as whether an ice-covered ocean world like Europa could be habitable. The answer to this long-standing question boils down to whether such an ocean world can sustain a flow of electrons that might provide the energy to power life. What remains unclear is whether such icy moons could ever generate enough heat to melt rock. Planetary scientists are aware that interesting chemistry takes place within these bodies, but what reliable flow of electrons could be used by alien life to power itself in the cold, dark depths? A key aspect that makes a world "habitable" is an intrinsic ability to maintain these chemical disequilibria. Arguably, icy moons lack this ability, so this needs to be tested on any future mission to Europa. A mission to Europa to realize such goals, together with other important issues relevant to Europa, is called *Europa Clipper Mission*.

NASA's Europa Clipper mission, which is scheduled to launch in 2024, is primarily aimed at monitoring Europa's putative (i.e., generally considered or reputed to be) plume activity. Hubble's identification of a site that appears to have persistent, intermittent plume activity provides a tempting target for the Europa Clipper mission to investigate with its powerful suite of science instruments. In addition, some of Sparks' co-authors on the Hubble Europa studies are preparing a powerful ultraviolet camera to fly on Europa Clipper that will make similar measurements to Hubble's, but from thousands of times closer. And several members of the Cassini Ion and Neutral Mass Spectrometer (INMS) team are developing an exquisitely sensitive, next-generation version of their instrument for flight on Europa Clipper (https://www.nasa.gov/press-release/nasa-missions-provide-new-insights-into-ocean-worlds-in-our-solar-system).

Apart from Europa's plume activity and questions related to possible life on Europa, there are several other issues that can receive clarity based on measurements to be carried out during Europa Clipper mission. For example, contrasting with prior knowledge, Daswani et al. (2020) found that CO_2 could be the most abundant solute in the Europan ocean, followed by Ca, SO_4, and bicarbonate (HCO_3). However, gypsum ($CaSO_4 \cdot 2H_2O$) precipitation from the seafloor to the ice shell decreases the dissolved S/Cl ratio, such that Cl > S at the shallowest depths, consistent with recently inferred endogenous chlorides at Europa's surface. Gypsum would form a 3–10 km thick sedimentary layer at the seafloor. Daswani et al. (2020) have expressed optimism that the Europa Clipper mission is well-suited to test these predictions.

Steinbrügge et al. (2020) have expressed the hope that the upcoming Europa Clipper mission will be able to test their proposed geyser formation hypothesis. Due to the reprocessing of the ice on the Manannán crater on Europa (interestingly, no other similar fracture systems such as the Manannán crater have been observed on Europa), Steinbrügge et al. (2020) predicted the surroundings of the spider feature (clearly visible at the center of the Manannán crater, in association with the circular fractures surrounding a central depression, with linear fractures radiating from the interior) to have a lower salinity than the center, which would be enriched in salt associated with the erupted brine. Pettinelli et al. (2016) pointed out that the remnant melt pocket on Europa's icy surface should be observable as a radar reflector in the subsurface. Quick and Hedman (2020) noted that the surroundings of the central fracture system in the Manannán crater should also be enriched in salt and therefore observable spectroscopically. In this context, it is interesting to note that Moore et al. (2001) reported that the central pit of Manannán corresponds to a bright spot seen within the crater in a Solid-State Imager image from flyby G1 (Moore et al., 1998). According to Steinbrügge et al. (2020), this observation is compatible and provides additional evidence for their model, which is designed to understand the water particle plume emanating from the central pit of Manannán crater.

The brine migration principles advanced by Steinbrügge et al. (2020) can effectively mobilize brine over long timescales and concentrate high salinity water into a central pocket. According to them, although craters are likely a source of plumes on icy bodies, the limited plume height makes it unlikely to be the source of the plumes observed by HST or Galileo. While they offer ideal natural laboratories for investigating brine migration due to their large thermal gradients, brine migration can occur wherever thermal inhomogeneities are present. As such, it is not limited to craters but could also occur along shear heating zones or be triggered by diapirism.

According to Steinbrügge et al. (2020), because the impact rates on Europa are low, with major impactors above 1 km in size only expected every 10^6 to 10^7 years (Zahnle et al., 1998), it is less likely that impact triggered cryovolcanism is the dominant or even sole source for plumes on Europa. However, due to the directly observable spider feature with the associated collapse structure and the comparably good data quality of the region, Manannán is the ideal site for a case study (Steinbrügge et al., 2020). As such, brine pocket migration allows for the presence of plumes not sourced by the ocean and provides constraints on the shell and ocean salinity. It is therefore a critical factor to consider when studying the dynamics of the ice shell of Europa and other icy bodies within the solar system.

12.14 Prospects of extant life on *Europa*

Europa presents an internal structure similar to the inner terrestrial planets, which consist primarily of silicate rock, and appears to have an iron core (Kuskov and Kronrod, 2005). As already indicated, there is a general agreement to the presence of a subsurface water ocean, estimated to be 100 km thick (Greenberg et al., 2002), covered with an ice crust estimated to be about 20–30 km thick (Ruiz et al., 2007). A 2016 NASA study found that Earth-like levels of hydrogen and oxygen could be produced through processes related to serpentinization and ice-derived oxidants, which do not directly involve volcanism. In 2015, scientists announced that salt from a subsurface ocean may likely be coating some geological features on Europa, suggesting that the ocean is interacting with the seafloor. This may be important in determining if Europa could be habitable. The likely presence of liquid water in contact with Europa's rocky mantle has spurred calls to send a probe there. Tidal forces and deep-sea vents may provide energy to keep the ice melted and maintain a liquid water environment (Greenberg, 2002; Marion et al., 2003; Lipps, and Rieboldt, 2005).

Presence of various extremophiles under different extreme environmental conditions on Earth together with our increasing knowledge about the biology of microbes in extreme habitats has led numerous scientists to raise the possibility of finding life in various planetary bodies within the solar system, including Mars, Venus, and the moons of Jupiter (Io, Europa, Ganymede, and Callisto) and Saturn (Titan and Enceladus) (Cavicchioli, 2002; Horneck and Rettberg, 2007).

According to Worth et al. (2013), "Europa currently presents the thinnest surface ice layer, which provides less of a barrier for life to eventually find its way through, especially when considering the "chaos regions" that indicate recent partial melting. It appears that regions of the ice sheet sometimes break into large chunks separated by liquid water, which later refreezes. Any meteorites lying on top of the ice sheet in a region when this occurs would stand a chance of falling through. Additionally, the moons are thought to have been significantly warmer in the not-too-distant past."

According to the studies carried out by Marcos and Nissar (2000), the young and structurally diverse surface of Europa suggests dynamic activity in its interior. In a study. McCollom (1999) found that methanogenesis is a potential source of chemical energy for primary biomass production by autotrophic organisms in hydrothermal systems on Europa. Submarine hydrothermal vents are interesting systems, which Rampelotto (2010c) neatly reported thus, "Hydrothermal vents occur on the seafloor spreading zones and have a global distribution (Baker and German, 2004). A hydrothermal vent is a fissure in a planet's surface from which geothermally heated water emerges. Such vents are commonly found near volcanically active places, areas where tectonic plates are moving apart and where new crust is being formed. The emerging water can approach 350°C and is prevented from boiling only by the crushing pressure of the deep water. However, the water cools rapidly as it mixes with the frigid ocean around it. The water's carrying capacity drops and many of the dissolved minerals instantly precipitate out. This produces a great cloudy plume of dark particles, which gives these vents their nickname "black smokers." The effluent at black smokers is rich in dissolved transition metals, such as Fe^{2+} and Mn^{2+}, and contains high concentrations of magmatic CO_2, H_2S, and dissolved H_2, with varying amounts of CH_4 (Kelley et al., 2002). The dissolved gases and metals in black smokers fuel the microbial communities serving as the base of the food chain in these ecosystems (Takai et al., 2001)."

The high probability for the presence of hydrothermal vents on Europa as found by McCollom (1999) is encouraging. Kelley et al. (2002) found that the environments around hydrothermal vents are abundant on the floor of the Earth's oceans and are important sources of many elements and organic compounds that are transferred into the hydrosphere. Takai et al. (2001) examined the distribution of archaea in a black smoker chimney structure and found that hydrothermal vents can sustain life without input from photosynthesis and they sustain fascinating living systems with symbiotic relationships that involve lithoautotrophic microorganisms that use chemical energy to support metazoans.

Due to the presence of a subsurface ocean, Europa has been considered the most promising target in the search for extant extraterrestrial life within the solar system. Rampelotto (2010c) considers that life could exist within its sub-glacial ocean, perhaps subsisting in an environment similar to Earth's deep-ocean hydrothermal vents or the

Antarctic Lake Vostok. It may be noted that Vostok is the largest and deepest known subglacial lake and likely the oldest in Antarctica (Siegert et al., 2001). Vostok subglacial lake has been recognized as a possible analog to Europa's subsurface environment. D'Elia et al. (2008) isolated microbes from Lake Vostok accretion ice, and the assembly of microbes that were found in the Lake Vostok accretion ice samples indicates that the lake sustains a diverse population of microorganisms and potentially a complex ecosystem, although the concentrations of microbes are expected to be lower than those in most environments on Earth.

Possibility of existence of life on the surface of Europa has been debated because radiation from Jupiter can destroy molecules on Europa's surface. Material from Europa's ocean that ends up on the surface of Europa will be bombarded by radiation. The radiation breaks apart molecules and changes the chemical composition of the material, possibly destroying any biosignatures, or chemical signs that could imply the presence of life. To interpret what future space missions find on the surface of Europa we must first understand how material has been modified by radiation. Fig. 12.18 provides an artist's rendering of radiation on Europa illustrating possible effect of radiation on biosignature chemicals on Europa.

Paranicas et al. (2007) examined Europa's near-surface radiation environment and found that the Europan surface is an extremely hostile environment, as a result of constant exposure to Jupiter's intense radiation belts. However, this radiation of ions, protons, and electrons does not penetrate ice for more than about a meter and a half. While searching for potential biosignatures on Europa from surface geology observations, Figueredo et al. (2003) found that life may be found on Europa in areas that are protected from the radiation flux, such as in cracks and caves. Life on Europa could exist clustered around hydrothermal vents on the ocean floor, or below the ocean floor, where endoliths are known to inhabit on Earth. Alternatively, it could exist clinging to the lower surface of Europa's ice layer, much like algae and bacteria in Earth's polar regions, or float freely in Europa's ocean (Marion et al., 2003). If Europa's ocean is too cold, biological processes similar to those known on Earth could not take place. If it is too salty, only extreme halophiles could survive in that environment (Marion et al., 2003).

Also, studies by several researchers (e.g., Thomas and Dieckmann, 2002; Gilichinsky, 2002; Gilichinsky et al., 2005) indicated that living systems may live within the ice layer, as microbes do on Earth. Furthermore, cracks and fissures that penetrate from the bottom of the ice upwards are likely habitats for life (Greenberg, 2007). Apart from Vostok subglacial lake in Antarctica, Thomas and Dieckmann (2002) noted that Antarctic Sea ice is also a habitat for extremophiles. Based on the existence of life under extreme environmental conditions as indicated above, Rampelotto (2010c) thinks that the ice layer on Europa may provide a potential habitat for life.

Conjectures regarding extraterrestrial life have ensured a high-profile for Europa and have led to steady lobbying for future missions. The aims of these missions have ranged from examining Europa's chemical composition to searching for extraterrestrial life in its hypothesized subsurface oceans. As already indicated, evidence suggests the existence of lakes of liquid water entirely encased in Europa's icy outer shell and distinct from a liquid ocean thought to exist farther down beneath the ice shell (Schmidt et al., 2011). If confirmed, the lakes could be yet another potential habitat for life. Evidence suggests that hydrogen peroxide (H_2O_2) is abundant across much of the surface of Europa. Because hydrogen peroxide decays into oxygen (O_2) and water (H_2O) when combined with liquid water, scientists are hopeful that it could be an important energy supply for simple life forms.

So far, there is no evidence that life exists on Europa, but this Jovian moon has emerged as one of the most likely locations in the solar system for potential habitability (Hand et al., 2007). Astrobiologists think that life could exist in its under-ice ocean, perhaps in an environment similar to Earth's deep-ocean hydrothermal vents.

12.15 Future exploration of Europa in the pipeline

The European Space Agency's Jupiter Icy Moons Explorer mission, or JUICE, will study *Europa*, along with fellow moons *Callisto* and *Ganymede* and the planet Jupiter itself. JUICE, a Jupiter orbiter, is scheduled to launch in 2022. Access to the icy surface of Europa would offer scientists a much more powerful analysis of a world that they consider one of the most tantalizing prospects for finding

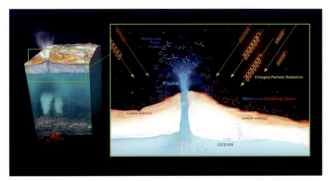

FIG. 12.18 Artist's rendering of radiation on Europa illustrating possible effect of radiation on biosignature chemicals on Europa. *(Source: https://commons.wikimedia.org/wiki/File:PIA22479-Europa-JupiterMoon-ArtistConcept-20180723.jpg). Author: NASA/JPL-Cssaltech. This file is in the **public domain** in the United States because it was solely created by NASA. NASA copyright policy states that "NASA material is not protected by copyright unless noted."*

extraterrestrial life. While pinning hopes on possibility of life on Europa, one must bear in mind the plain truth that Europa lies within Jupiter's radiation belt, so its frigid surface gets hammered by fast-moving charged particles.

Although it is possible to explore and examine habitability pretty well from Europa Clipper, it is absolutely necessary to touch the surface to actually look for biosignatures and signs of life. Touching the surface requires the support of lander and rover. The surface reconnaissance will help pave the way for a future Europa lander, which Congress has directed NASA to develop. It may be noted that there is a stark difference between knowing that the conditions for life exist and actually finding some form of life, and that difference is at the heart of the Europa Lander mission concept that NASA could one day send to Europa. The lander—just a concept at the moment, not a full-fledged mission—would dig down and hunt for signs of life about 4 inches (10 centimeters) beneath the Europan surface. Any biomolecules at such depths would be shielded from damaging radiation by overlying ice. Determining whether there is anything alive on Europa requires proper analysis of the ice dug out by the lander. Such examination requires the judicious application of a broad suite of instruments, including a microscope, a seismic package, a camera and a host of devices for chemical analysis.

The ultimate source of most biomolecules expected to be found on Europa would likely be Europa's vast ocean. Unfortunately, Europa's ice shell is a formidable barrier; the ice shell thickness being in the range 9–16 miles (15–25 km), requiring the need for a powerful drill. NASA is already trying to spur the development of such advanced tech via a program called SESAME (Scientific Exploration Subsurface Access Mechanism for Europa). SESAME ultimately envisions a nuclear-powered "penetration system" capable of getting 9 miles down within three years of operation. This drill, whose weight is capped at 440 lbs. (200 kilograms), would be delivered to the Europan surface by a lander, which would also provide a communications relay between the digger and Earth. There may well be life-hosting lakes suspended within the ice shell that such tech could access, but the ocean would be the main target (see Wall, 2020).

Wall (2020) reported that a New York-based company called *Honeybee* is developing a hybrid drill system called SLUSH (Search for Life Using Submersible Heated), which uses both thermal and mechanical means to slice through ice. SLUSH utilizes a mechanical drill to break the formation, and partially melts the fragments to enable the efficient transport of material behind the probe. Communications between the drill and lander might be tough to do via a tether once the drill gets deep enough down. Researchers are therefore investigating the possibility of deploying puck-like modems behind the drill as it descends.

References

Aglyamov, Y., Schroeder, D.M., Vance, S.D., 2017. Bright prospects for radar detection of Europa's ocean. Icarus 281, 334–337.

Airhart, M., 2011. Scientists find evidence for "Great Lake" on Europa and potential new habitat for life. Jackson School of Geosciences, Austin, Texas.

Anderson, J.D., Lau, E.L., Sjogren, W.L., Schubert, G., Moore, W.B., 1996a. Gravitational constraints on the internal structure of Ganymede. Nature 384, 541–543.

Anderson, J.D., Sjogren, W.L., Schubert, G., 1996b. Galileo gravity results and the internal structure of Io. Science 272, 709–712.

Anderson, J.D., Lau, E.L., Sjogren, W.L., Schubert, G., Moore, W.B., 1997. Europa's differentiated internal structure: inferences from two Galileo encounters. Science 276, 1236–1239.

Anderson, J.D., Schubert, G., Jacobson, R.A., Lau, E.L., Moore, W.B., Sjogren, W.L., 1998. Europa's differentiated internal structure: inferences from four Galileo encounters. Science 281 (5385), 2019–2022. doi:10.1126/science.281.5385.2019.

Baker, E.T., German, C.R., 2004. On the global distribution of hydrothermal vent fields. Geophys. Monogr. 148, 245–266.

Baland, R-M., Yseboodt, M., Hoolst, T.V., 2012. Obliquity of the Galilean satellites: the influence of a global internal liquid layer. Icarus 220 (2), 435–448.

Běhounková, M., Tobie, G., Choblet, G., Kervazo, M., Daswani, M.M., Dumoulin, C., Vance, S.D., 2021. Tidally induced magmatic pulses on the oceanic floor of Jupiter's Moon Europa. Geophys. Res. Lett. 48 (3), e2020GL090077.

Bhatia, G.K., Sahijpal, S., 2017. Thermal evolution of trans-Neptunian objects, icy satellites, and minor icy planets in the early solar system. Meteorit. Planet. Sci. 52 (12), 2470–2490.

Bierhaus, E.B., Zahnle, K., Chapman, C.R., 2009. Europa's crater distributions and surface ages. In: Pappalardo, Robert T., McKinnon, William B., Khurana, Krishan K. (Eds.), Europa. University of Arizona Press, Tucson, pp. 161–180.

Bland, M.T., Showman, A.P., Tobie, G., 2008. The production of Ganymede's magnetic field. Icarus 198 (2), 384–399.

Borucki, J.G., Khare, B., Cruikshank, D.P., 2002. A new energy source for organic synthesis in Europa's surface ice. J. Geophysical Res.: Planets 107 (E11), 24-1–24-5.

Brown, M.E., Hand, K.P., 2013. Salts and radiation products on the surface of Europa. Astronomical J. 145 (4), 110.

Calvin, W.M., Clark, R.N., Brown, R.H., Spencer, J.R., 1995. Spectra of the ice Galilean satellites from 0.2 to 5 μm: A compilation, new observations, and a recent summary. J. Geophys. Res. 100 (E9) 19, 041–19, 048.

Carlson, R.W., Anderson, M.S., Mehlman, R., Johnson, R.E, 2005. Distribution of hydrate on Europa: Further evidence for sulfuric acid hydrate. Icarus 177 (2), 461.

Carr, M.H., et al., 1998. Evidence for a subsurface ocean on Europa. Nature 391, 363–365.

Cassen, P.M., Reynolds, R.T., Peale, S.J., 1979. Is there liquid water on Europa? Geophys. Res. Lett. 6 (9), 731–734.

Cavicchioli, R., 2002. Extremophiles and the search for extraterrestrial life. Astrobiology 2, 281–292.

Cecconi, B., Hess, S., Hérique, A., Santovito, M.R., Santos-Costa, D., Zarka, P., Alberti, G., Blankenship, D., Bougeret, J.-L., Bruzzone, L., Kofman, W., 2012. Natural radio emission of Jupiter as interferences for radar investigations of the icy satellites of Jupiter. Planet. Space Sci. 61 (1), 32–45.

Choblet, G., Tobie, G., Sotin, C., Kalousová, K., Grasset, O., 2017. Heat transport in the high-pressure ice mantle of large icy moons. Icarus 285, 252–262.

Ciarletti, V., 2016. A variety of radars designed to explore the hidden structures and properties of the Solar System's planets and bodies. C.R. Phys. 17 (9), 966–975.

Clark, K., Boldt, J., Greeley, R., Hand, K., Jun, I., Lock, R., Pappalardo, R., Van Houten, T., Yan, T., 2011. Return to Europa: overview of the Jupiter Europa orbiter mission. Adv. Space Res. 48 (4), 629–650.

Coll, P.J., Poch, O., Ramirez, S.I., Buch, A., Brassé, C., and Raulin, F., 2010. Prebiotic chemistry on Titan? The nature of Titan's aerosols and their potential evolution at the satellite surface, American Geophysical Union, Fall Meeting 2010, abstract #P31C-1551.

Collins, G.C., Head, J.W., Pappalardo, R.T., Spaun, N.A., 2000. Evaluation of models for the formation of chaotic terrain on Europa. J. Geohpys. Res. 105, 1709–1716.

Craft, K.L., Patterson, G.W., Lowell, R.P., Germanovich, L., 2016. Fracturing and flow: Investigations on the formation of shallow water sills on Europa. Icarus 274, 297–313. https://doi.org/10.1016/j.icarus.2016.01.023.

Crawford, G.D., Stevenson, D.J., 1988. Gas-driven water volcanism and the resurfacing of Europa. Icarus 73 (1), 66–79. https://doi.org/10.1016/0019-1035(88)90085-1.

Dalton III, J.B., Cassidy, T., Paranicas, C., Shirley, J.H., Prockter, L.M., Kamp, L.W., 2013. Exogenic controls on sulfuric acid hydrate production at the surface of Europa. Planet. Space Sci. 77, 45–63.

Daswani, M.M., Vance, S.D., Mayne, M.J., Glein, C.R., 2021. A metamorphic origin for Europa's ocean. Geophys. Res. Lett. 48 (18), e2021GL094143. https://doi.org/10.1029/2021GL094143.

D'Elia, T., Veerapaneni, R., Rogers, S.O., 2008. Isolation of microbes from Lake Vostok accretion ice. Appl. Environ. Microbiol. 74, 4962–4965.

de Pater, I., Deboer, D., Marley, M., Freedman, R., Young, R., 2005. Retrieval of water in Jupiter's deep atmosphere using microwave spectra of its brightness temperature. Icarus 173 (2), 425–447.

Deschamps, F., Sotin, C., 2001. Thermal convection in the outer shell of large icy satellites. J. Geophys. Res. 106, 5107–5121.

Doggett, T., Greeley, R., Figuerdo, P., Tanaka, K., 2009. Europa. In: Pappalardo, Robert T., McKinnon, William B., Khurana, Krishan K. (Eds.), The University of Arizona Press, p. 137.

Elder, C.M., Bray, V.J., Melosh, H.J., 2012. The theoretical plausibility of central pit crater formation via melt drainage. Icarus 221 (2), 831–843. https://doi.org/10.1016/j.icarus.2012.09.014.

Fagents, S.A., Greeley, R., Sullivan, R.J., Pappalardo, R.T., Prockter, L.M., The Galileo, S.S.I.T, 2000. Cryomagmatic mechanisms for the formation of Rhadamanthys Linea, triple band margins, and other low-albedo features on Europa. Icarus 144 (1), 54–88. https://doi.org/10.1006/icar.1999.6254.

Fagents, S.A., 2003. Considerations for effusive cryovolcanism on Europa: the post-Galileo perspective. J. Geophys. Res. 108 (E12), 5139. https://doi.org/10.1029/2003JE002128.

Figueredo, P.H., Greeley, R., Neuer, S., Irwin, L., Schulze-Makuch, D., 2003. Locating potential biosignatures on Europa from surface geology observations. Astrobiology 3, 851–861.

Figueredo, P.H., Greeley, R., 2004. Resurfacing history of Europa from pole-to-pole geological mapping. Icarus 167 (2), 287–312. https://doi.org/10.1016/j.icarus.2003.09.016.

Fischer, P.D., Brown, M.E., Hand, K.P., 2015. Spatially resolved spectroscopy of Europa: The distinct spectrum of large-scale chaos. Astron. J. 150, 164.

Fischer, P.D., Brown, M.E., Trumbo, S.K., Hand, K.P., 2016. Spatially resolved spectroscopy of Europa's large-scale compositional units at 3–4μm with Keck NIRSPEC. Astronomical J. 153 (1), 13.

Fischer, P.D., Brown, M.E., Trumbo, S.K., Hand, K.P., 2017. Spatially resolved spectroscopy of Europa's large-scale compositional units at 3–4 μm with Keck NIRSPEC. Astron. J. 153, 13.

Geissler, P., 1998. Evolution of lineaments on Europa: Clues from Galileo multispectral imaging observations. Icarus 135 (1), 107–337.

Gérard, J.-C., Grodent, D., Radioti, A., Bonfond, B., Clarke, J.T., 2013. Hubble observations of Jupiter's north–south conjugate ultraviolet aurora. Icarus 226 (2), 1559–1567.

Gilichinsky, D.A., 2002. Permafrost model of extraterrestrial habitat. In: Horneck, G., Baumstark-Khan, C. (Eds.), Astrobiology: The Quest for the Conditions of Life. Springer, Berlin.

Gilichinsky, D.A., Rivkina, E., Bakermans, C., Shcherbakova, V., Petrovskaya, L., Ozerskaya, S., Ivanushkina, N., Kochkina, G., Laurinavichius, K., Pecheritsina, S., Fattakhova, R., Tiedje, J.M., 2005. Biodiversity of cryopegs in permafrost. FEMS Microbiol. Ecol. 53, 117–128.

Grasset, O., Dougherty, M.K., Coustenis, A., Bunce, E.J., Erd, C., Titov, D., Blanc, M., Coates, A., Drossart, P., Fletcher, L.N., Hussmann, H., Jaumann, R., Krupp, N., Lebreton, J.-P., Prieto-Ballesteros, O., Tortora, P., Tosi, F., Van Hoolst, T., 2013. JUpiter ICy moons Explorer (JUICE): An ESA mission to orbit Ganymede and to characterise the Jupiter system. Planet. Space Sci. 78, 1–21.

Greeley, R., Sullivan, R., Klemaszewski, J., Homan, K., Head, J.W., Pappalardo, R.T., …, Tufts, R., 1998. Europa: initial galileo geological observations. Icarus 135 (1), 4–24. https://doi.org/10.1006/icar.1998.5969.

Greenberg, R., Hoppa, G.V., Tufts, B.R., Geissler, P., 1998. From lenticulae to chaos: widespread melt-through to the surface of Europa? Eos 79, F540.

Greenberg, R., Hoppa, G.V., Tufts, B.R., Geissler, P., Riley, J., Kadel, S., 1999. Chaos on Europa. Icarus 141, 263–286.

Greenberg, R., Geissler, P., O'Brien, D., Hoppa, G.V., Tufts, B.R., 2001. Ocean-to-surface linkages resurface Europa: an invited review. Lunar Planet. Sci. 32, 1428.

Greenberg, R., 2002. Tides and the biosphere of Europa. Am. Sci. 90, 48–55.

Greenberg, R., Geissler, P., Hoppa, G., Tufts, B.R., 2002. Tidal-tectonic processing and their implications for the character of Europa's icy crust. Rev. Geophys. 40, 1–33.

Greenberg, R., 2005. Europa: The Ocean Moon: Search for an Alien Biosphere. (Springer Praxis Books Geophysical Sciences), Springer, pp. 395 pages, ISBN-10: 3642061265; ISBN-13: 978-3642061264.

Greenberg, R., 2007. Europa, the ocean moon: Tides, permeable ice, and life. In: Pudritz, R., Higgs, P., Stone, J. (Eds.), Planetary Systems and the Origins of Life. Cambridge University Press, Cambridge.

Grima, C., Kofmana, W., Heriquea, A., Oroseib, R., Seu, R., 2012. Quantitative analysis of Mars surface radar reflectivity at 20 MHz. Icarus 220 (1), 84–99.

Grima, C., Schroeder, D.M., Blankenship, D.D., Young, D.A., 2014. Planetary landing-zone reconnaissance using ice-penetrating radar data: Concept validation in Antarctica. Planet. Space Sci. 103, 191–204.

Grima, C., Blankenship, D.D., Schroeder, D.M., 2015. Radar signal propagation through the ionosphere of Europa. Planet. Space Sci. 117, 421–428.

Gross, G.W., Wong, P.M., Humes, K., 1977. Concentration dependent solute redistribution at the ice–water phase boundary. III. Spontaneous convection. Chloride solutions. J. Chem. Phys. 67 (11), 5264–5274.

Gurnett, D.A., Kurth, W.S., Roux, A., Bolton, S.J., Kennel, C.F., 1996. Evidence for a magnetosphere at Ganymede from plasma-wave observations by the Galileo spacecraft. Nature 384, 535–537.

Hammond, N.P. (2019). Near-surface melt on Europa: Modeling the formation and migration of brines in a dynamic ice shell, Ocean Worlds 4, held 21-22 May, 2019 in Columbia, Maryland. LPI Contribution No. 2168, id.6024. http://www.hou.usra.edu/meetings/ocenworlds2019/pdf/6024.pdf.

Hand, K.P., Carlson, R.W., Chyba, C.F., 2007. Energy, chemical disequilibrium, and geological constraints on Europa. Astrobiology 7 (6), 1006–1022.

Hand, K.P., Carlson, R.W., 2015. Europa's surface color suggests an ocean rich with sodium chloride. Geophys. Res. Lett. 42, 3174–3178.

Hartogh, P., Ilyushin, Ya.A., 2016. A passive low frequency instrument for radio wave sounding the subsurface oceans of the Jovian icy moons: An instrument concept. Planet. Space Sci. 130, 30–39.

Head, J.W., Pappalardo, R.T., 1999. Brine mobilization during lithospheric heating on Europa: Implications for formation of chaos terrain, lenticulae texture, and color variations. J. Geophys. Res. 104, 27,143–27,155.

Heggy, E., Scabbia, G., Bruzzone, L., Pappalardo, R.T., 2017. Radar probing of Jovian icy moons: Understanding subsurface water and structure detectability in the JUICE and Europa missions. Icarus 285, 237–251.

Hemingway, D.J., Zannoni, M., Tortora, P., Nimmo, F., Asmar, S.W., 2016. Diones' internal structure inferred from Cassini gravity and topography. Lunar Planet Sci Conf. 47, 1314.

Hendrix, A.R., Hurford, T.A., Barge, L.M., Bland, M.T., Bowman, J.S., Brinckerhoff, W., Buratti, B.J., Cable, M.L., Castillo-Rogez, J., Collins, G.C., Diniega, S., German, C.R., Hayes, A.G., Hoehler, T., Hosseini, S., Howett, C.J.A., McEwen, A.S., Neish, C.D., Neveu, M., Nordheim, T.A., Patterson, G.W., Patthoff, A., Phillips, D., Rhoden, C., Schmidt, A., Singer, B.E., Soderblom, K.N., J.M., Vance, S.D, 2019. The NASA Roadmap to Ocean Worlds. Astrobiology 19, 1–27. doi:10.1089/ast.2018.1955.

Horneck, G., Rettberg, P., 2007. Complete Course in Astrobiology. Wiley-VCH Verlag GmbH & Co. KGaA, Weinheim, Germany.

Hurford, T.A., Sarid, A.R., Greenberg, R., 2007. Cycloidal cracks on Europa: Improved modeling and non-synchronous rotation implications. Icarus 186 (1), 218.

Hussmann, H., Spohn, T., 2004. Thermal-orbital evolution of Io and Europa. Icarus 171 (2), 391–410.

Huybrighs, H.L.F., Futaana, Y., Barabash, S., Wieser, M., Wurz, P., Krupp, N., Glassmeier, K-H., Vermeersen, B., 2017. On the in-situ detectability of Europa's water vapour plumes from a flyby mission. Icarus 289, 270–280.

Iess, L., Stevenson, D.J., Parisi, M., Hemingway, D., Jacobson, R.A., Lunine, J.I., Nimmo, F., Armstrong, J.W., Asmar, S.W., Ducci, M., Tortora, P., 2014. The gravity field and interior structure of Enceladus. Science 344 (6179), 78–80.

Ilyushin, Y.A., 2014. Subsurface radar location of the putative ocean on Ganymede: Numerical simulation of the surface terrain impact. Planet. Space Sci. 92, 121–126.

Ivanov, M.A., Prockter, L.M., Dalton, B., 2011. Landforms of Europa and selection of landing sites. Adv. Space Res. 48 (4), 661–677.

Janssen, M.A., Hofstadter, M.D., Gulkis, S., Ingersoll, A.P., Allison, M., Bolton, S.J., Levin, S.M., Kamp, L.W., 2005. Microwave remote sensing of Jupiter's atmosphere from an orbiting spacecraft. Icarus 173 (2), 447–453.

Jia, X., Walker, R.J., Kivelson, M.G., Khurana, K.K., Linker, J.A., 2009. Properties of Ganymede's magnetosphere inferred from improved three-dimensional MHD simulations. J. Geophys. Res. 114, A09209. doi:10.1029/2009JA014375.1.

Jia, X., Kivelson, M.G., Khurana, K.K., Kurth, W.S., 2018. Evidence of a plume on Europa from Galileo magnetic and plasma wave signatures. Nature Astronomy 2 (6), 459–464.

Journaux, B., Daniel, I., Petitgirard, S., Cardon, H., Perrillat, J-P., Caracas, R., Mezouar, M., 2017. Salt partitioning between water and high-pressure ices: Implication for the dynamics and habitability of icy moons and water-rich planetary bodies. Earth Planet. Sci. Lett. 463, 36–47.

Kargel, J.S., 1991. Brine volcanism and the interior structures of asteroids and icy satellites. Icarus 94 (2), 368–390.

Kargel, J.S., Kaye, J.Z., Head III, J.W., Marion, G.M., Sassen, R., Crowley, J.K., Ballesteros, O.P., Grant, S.A., Hogenboom, D.L, 2000. Europa's crust and ocean: Origin, composition, and the prospects for life. Icarus 148 (1), 226–265 Planetary Sciences Group, Brown University.

Karpowicz, B.M., Steffes, P.G., 2011. In search of water vapor on Jupiter: Laboratory measurements of the microwave properties of water vapor under simulated jovian conditions. Icarus 212 (1), 210–223. doi:10.1016/j.icarus.2010.11.035.

Kattenhorn, S., Prockter, L., 2014. Evidence for subduction in the ice shell of Europa. Nat. Geosci. 7 (9), 762.

Kelley, D.S., Baross, J.A., Delaney, J.R., 2002. Volcanoes, fluids, and life at mid-ocean ridge spreading centers. Annu. Rev. Earth Planet Sci. 30, 385–491.

Khurana, K.K., Kivelson, M.G., Stevenson, D.J., Schubert, G., Russell, C.T., Walker, R.J., Polanskey, C., 1998. Induced magnetic fields as evidence for subsurface oceans in Europa and Callisto. Nature 395, 777–780.

Kivelson, M.G., 1996. Discovery of Ganymede's magnetic field by the Galileo spacecraft. Nature 384, 537–541.

Kivelson, M.G., Khurana, K.K., Russell, C.T., Walker, R.J., Warnecke, J., Coroniti, F.V., Polanskey, C., Southwood, D.J., Schubert, G., 1996a. Discovery of Ganymede's magnetic field by the Galileo spacecraft. Nature 384, 537–541.

Kivelson, M.G., et al., 1996b. A magnetic signature at Io: Initial report from the Galileo magnetometer. Science 273, 337–340.

Kivelson, M.G., et al., 1996c. Lo's interaction with the plasma Torus: Galileo magnetometer report. Science 274, 396–398.

Kivelson, M.G., Khurana, K.K., Coroniti, F.V., Joy, S., Russell, C.T., Walker, R.J., Warnecke, J., Bennett, L., Polanskey, C., 1997. The magnetic field and magnetosphere of Ganymede. Geophys. Res. Lett. 24 (17), 2155–2158.

Kivelson, M.G., Khurana, K., Stevenson, D.J., Bennett, L., Joy, S., Russell, C.T., Walker, R.J., Zimmer, C., Polanskey, C., 1999. Europa and Callisto: Induced or intrinsic fields in a periodically varying plasma environment. J. Geophys. Res. 104, 4609–4629.

Kivelson, M.G., Khurana, K.K., Russell, C.T., Volwerk, M., Walker, R.J., Zimmer, C., 2000. Galileo magnetometer measurements: A stronger case for a subsurface ocean at Europa. Science 289 (5483), 1340–1343.

Kivelson, M.G., Khurana, K., Volwerk, M., 2002. The permanent and inductive magnetic moments of Ganymede. Icarus 157, 507–522.

Kliore, A.J., Hinson, D.P., Flasar, F.M., Nagy, A.F., Cravens, T.E., 1997. The ionosphere of Europa from Galileo radio occultations. Science 277 (5324), 355–358.

Kollmann, P., et al., 2017. A heavy ion and proton radiation belt inside of Jupiter's rings. Geophys. Res. Lett. 44, 5259–5268. doi:10.1002/2017GL073730.

Kuiper, G.P. (1961), *Planets and Satellites (The Solar System, vol. III)*, ed. G.P. Kuiper & B. M. Middlehurst, University of Chicago Press, Chicago, pp. 575-591; ISBN 10: 0226459276; ISBN 13: 9780226459271.

Kuskov, O.L., Kronrod, V.A., 2005. Internal structure of Europa and Callisto. Icarus 177, 550–556.

Lane, A.L., Nelson, R.M., Matson, D.L., 1981. Evidence for sulfur implantation in Europa's UV absorption band. Nature 292, 38–39.

Lesage, E., Massol, H., Schmidt, F., 2020. Cryomagma ascent on Europa. Icarus 335, 113369. https://doi.org/10.1016/j.icarus.2019.07.003.

Ligier, N., Poulet, F., Carter, J., Brunetto, R., Gourgeot, F., 2016. VLT/SINFONI Observations of Europa: New insights into the surface composition. Astronomical J. 151 (6), 163.

Lipps, J.H., Rieboldt, S., 2005. Habitats and taphonomy of Europa. Icarus 177, 515–527.

Litvinenko, G.V., Shaposhnikov, V.E., Konovalenko, A.A., Zakharenko, V.V., Panchenko, M., Dorovsky, V.V., Brazhenko, A.I., Rucker, H.O., Vinogradov, V.V., Melnik, V.N., 2016. Quasi-similar decameter emission features appearing in the solar and jovian dynamic spectra. Icarus 272, 80–87.

Liu, S., Zeng, Z., Deng, Li, 2007. FDTD simulations for ground penetrating radar in urban applications. J. Geophys. Eng. 4 (3), 262–267. https://doi.org/10.1088/1742-2132/4/3/S04.

Lorenz, R.D., Gleeson, D., Prieto-Ballesteros, O., Gomez, F., Hand, K., Bulat, S., 2011. Analog environments for a Europa lander mission. Adv. Space Res. 48 (4), 689–696.

Manga, M., Wang, C.-Y., 2007. Pressurized oceans and the eruption of liquid water on Europa and Enceladus. Geophys. Res. Lett. 34, L07202. https://doi.org/10.1029/2007GL029297.

Manga, M., Michaut, C., 2017. Formation of lenticulae on Europa by saucer-shaped sills. Icarus 286, 261–269.

Manthilake, G., Bolfan-Casanova, N., Novella, D., Mookherjee, M., Andrault, D., 2016. Dehydration of chlorite explains anomalously high electrical conductivity in the mantle wedges. Sci. Adv. 2 (5), e1501631.

Marcos, R.F., Nissar, A., 2000. Possible detection of volcanic activity on Europa: Analysis of an optical transient event. Earth Moon Planets 88, 167–175.

Marion, G.M., Fritsen, C.H., Eicken, H., Payne, M.C., 2003. The search for life on Europa: Limiting environmental factors, potential habitats, and Earth analogues. Astrobiology 3, 785–811.

Matousek, S, 2005. The Juno new frontiers mission. Jet Propulsion Laboratory, California Institute of Technology, United States, pp. 8 Tech. Rep. IAC-05-A3.2. A. 04 pages.

McCollom, T.M., 1999. Methanogenesis as a potential source of chemical energy for primary biomass production by autotrophic organisms in hydrothermal systems on Europa. J. Geophys. Res. 104, 30729–30742.

McCord, T.B., Hansen, G.B., et al., 1998. Salts on Europa's surface detected by Galileo's near infrared mapping spectrometer. Science 280 (5367), 1242–1245.

McEwen, A.S., 1986. Tidal reorientation and the fracturing of Jupiter's moon Europa. Nature 321, 49–51.

McFadden, L-A., Weissman, P., Johnson, T., 2006. The Encyclopedia of the Solar System. Academic Press; Elsevier, Imprint, pp. 992 eBook ISBN: 9780080474984.

McKinnon, W.B., 1999. Convective instability in Europa's floating ice shell. Geophys. Res. Lett. 26, 951–954.

McKinnon, W.B., 2015. Effect of Enceladus' rapid synchronous spin on interpretation of Cassini gravity. Geophys. Res. Lett. 42, 2137–2143.

Michaut, C., Manga, M., 2014. Domes, pits, and small chaos on Europa produced by water sills. Journal of Geophysical Research: Planets 119, 550–573. https://doi.org/10.1002/2013JE004558.

Moore, J.M., Asphaug, E., Sullivan, R.J., Klemaszewski, J.E., Bender, K.C., Greeley, R., et al., 1998. Large impact features on Europa: Results of the Galileo nominal mission. Icarus 135 (1), 127–145. https://doi.org/10.1006/icar.1998.5973.

Moore, W.B., Schubert, G., 2000. The tidal response of Europa. Icarus 147 (1), 317–319. https://doi.org/10.1006/icar.2000.6460.

Moore, J.M., Asphaug, E., Belton, M.J.S., Bierhaus, B., Breneman, H.H., Brooks, S.M., et al., 2001. Impact features on Europa: Results of the Galileo Europa Mission (GEM). Icarus 151 (1), 93–111. https://doi.org/10.1006/icar.2000.6558.

Moore, W.B., 2003. Tidal heating and convection in Io. J. Geophys. Res. 108 (E8), 5096.

Moore, J.M., Chapman, C.R., Bierhaus, E.B., Greely, R., Chuang, F.C., Klemaszewski, J., Clarck, R.N., Dalton, J.B., Hibbits, C.A., Schenk, P.M., Spencer, J.R., Wagner, R., 2004. Callisto. In: Bagenal, F., Dowling, T.E., McKinnon, W.B. (Eds.), Jupiter. Cambridge University Press, Cambridge, pp. 397–426.

Morrison, D., Cruikshank, D.P., 1974. Physical properties of the natural satellites. Space Sci. Rev. 15, 641–739.

Ojakangas, G.W., Stevenson, D.J., 1989. Thermal state of an ice shell on Europa. Icarus 81, 220–241.

Pappalardo, R.T., Head, J.W., Greeley, R., and the Galileo Imaging Team, 1996. A Europa ocean? The (circumstantial) geological evidence, Proc. Europa Ocean Conf., 59–60.

Pappalardo, R.T., Head, J.W, Greeley, R., Sullivan, R.J., Pilcher, C., Schubert, G., Moore, W.B., Carr, M.H., Moore, J.M., Belton, M.J.S., Goldsby, D.L., 1998. Geological evidence for solid-state convection in Europa's ice shell. Nature 391, 365–368.

Pappalardo, R.T., et al., 1999. Does Europa have a subsurface ocean?: Evaluation of the geological evidence. J. Geophys. Res. 104, 24,015–24,055.

Pappalardo, R., 2009. Europa. University of Arizona Press, p. 141, ISBN 978-0-8165-2844-8.

Pappalardo, R.T., Head, J.W., 2001. The thick-shelled model of Europa's geology: Implications for crustal processes. Lunar Planet. Sci. 32, 1866.

Pappalardo, R.T., Collins, G.C., Head, J.W., Helfenstein, P., McCord, T.B., Moore, J.M., Prockter, L.M., Schenk, P.M., Spencer, J.R., 2004. The geology of Ganymede. In: Bagenal, F., Dowling, T.E., McKinnon, W.B. (Eds.), Jupiter. Cambridge University Press, Cambridge, pp. 363–396.

Paranicas, C., Mauk, B.H., Khurana, K., Jun, I., Garrett, H., Krupp, N., Roussos, E., 2007. Europa's nearsurface radiation environment. Geophys. Res. Lett. 34. doi:10.1029/2007GL030834.

Pavlov, S.G., Jessberger, E.K., Hübers, H.-W., Schröder, S., Rauschenbach, I., Florek, S., Neumann, J., Henkel, H., Klinkner, S., 2011. Miniaturized laser-induced plasma spectrometry for planetary in situ analysis – The case for Jupiter's moon Europa. Adv. Space Res. 48 (4), 764–778.

Pettinelli, E., Lauro, S.E., Cosciotti, B., Mattei, E., Paolo, F.D., Vannaroni, G., 2016. Dielectric characterization of ice/$MgSO_4 \cdot 11H_2O$ mixtures as Jovian icy moon crust analogues. Earth Planet. Sci. Lett. 439, 11–17.

Phillips, C.B., Pappalardo, R.T., 2014. Europa clipper mission concept. Eos, Transactions American Geophysical Union 95 (20), 165–167.

Pilcher, C.B., Ridgway, T.B., McCord, T.B., 1972. Galilean satellites: Identification of water frost. Science 178, 1087–1089.

Plainaki, C., Milillo, A., Mura, A., Orsini, S., Cassidy, T., 2010. Neutral particle release from Europa's surface. Icarus 210 (1), 385–395.

Plainaki, C., Milillo, A., Mura, A., Orsini, S., Massetti, S., Cassidy, T., 2012. The role of sputtering and radiolysis in the generation of Europa exosphere. Icarus 218 (2), 956–966.

Plainaki, C., Milillo, A., Mura, A., Orsini, S., Saur, J., Orsini, S., Massetti, S., 2013. Exospheric O_2 densities at Europa during different orbital phases. Planet. Space Sci. 88, 42–52.

Plainaki, C., Milillo, A., Massetti, S., Mura, A., Jia, X., Orsini, S., Mangano, V., Angelis, E.D., Rispoli, R., 2015. The H_2O and O_2 exospheres of Ganymede: The result of a complex interaction between the jovian magnetospheric ions and the icy moon. Icarus 245, 306–319.

Poston, M.J., Carlson, R.W., Hand, K.P., 2017. Spectral behavior of irradiated sodium chloride crystals under europa-like conditions. J. Geophys. Res. Planets 122, 2644–2654.

Prockter, L.M., Schenk, P., 2005. Origin and evolution of Castalia Macula, an anomalous young depression on Europa. Icarus 177 (2), 305–326. https://doi.org/10.1016/j.icarus.2005.08.003.

Prockter, L.M., Lopes, R.M.C., Giese, B., Jaumann, R., Lorenz, R.D., Pappalardo, R.T., et al., 2010. Characteristics of icy surfaces. Space Sci. Rev. 153 (1–4), 63–111. https://doi.org/10.1007/s11214-010-9649-8.

Quick, L.C., Marsh, B.D., 2016. Heat transfer of ascending cryomagma on Europa. J. Volcanol. Geotherm. Res. 319, 66–77. https://doi.org/10.1016/j.jvolgeores.2016.03.018.

Quick, L.C., Hedman, M.M., 2020. Characterizing deposits emplaced by cryovolcanic plumes on Europa. Icarus 343, 113667. https://doi.org/10.1016/j.icarus.2020.113667.

Rampelotto, P.H., 2010c. Resistance of microorganisms to extreme environmental conditions and its contribution to astrobiology. Sustainability 2, 1602–1623. doi:10.3390/su2061602.

Ransford, G.A., Finnerty, A.A., Collerson, K.D., 1981. Europa's petrological thermal history. Nature 289, 21–24.

Rathbun, J.A., Musser, G.S., Squyres, S.W., 1998. Ice diapirs on Europa: Implications for liquid water. Geophys. Res. Lett. 25, 4157–4160.

Rhoden, A.R., Hurford, T.A., Roth, L., Retherford, K., 2015. Linking Europa's plume activity to tides, tectonics, and liquid water. Icarus 253, 169–178.

Riley, J., Hoppa, G.V., Greenberg, R., Tufts, B.R., 2000. Distribution of chaotic terrain on Europa. J. Geophys. Res. 105, 22,599–22,615.

Romero-Wolf, A., Vance, S., Maiwald, F., Heggy, E., Ries, P., Liewer, K., 2015. A passive probe for subsurface oceans and liquid water in Jupiter's icy moons. Icarus 248, 463–477.

Romero-Wolf, A., Schroeder, D.M., Ries, P., Bills, B.G., Naudet, C., Scott, B.R., Treuhaft, R., Vance, S., 2016. Prospects of passive radio detection of a subsurface ocean on Europa with a lander. Planet. Space Sci. 129, 118–121.

Rosenkranz, P.W., 1998. Water vapor microwave continuum absorption: A comparison of measurement and models. Radio Sci. 33 (4), 919–928.

Ross, M., Schubert, G., 1986. Tidal heating in an internal ocean model of Europa. Nature 325, 13–134.

Roth, L., Saur, J., Retherford, K.D., Strobel, D.F., Feldman, P.D., McGrath, M.A., Nimmo, F., 2013. Transient water vapor at Europa's South Pole. Science 343 (6167), 171–174.

Roth, L., Saur, J., Retherford, K.D., Strobel, D.F., Feldman, P.D., McGrath, M.A., Nimmo, F., 2014. Transient water vapor at Europa's south pole. Science 343 (6167), 171–174. https://doi.org/10.1126/science.1247051.

Ruiz, J., Alvarez-Gómez, J.A., Tejero, R., Sánchez, N., 2007. Heat flow and thickness of a convective ice shell on Europa for grain size—dependent rheologies. Icarus 190, 145–154.

Ruiz-Bermejo, M., Rivas, L.A., Palacín, A., Menor-Salván, C., Osuna-Esteban, S., 2011. Prebiotic synthesis of protobiopolymers under alkaline ocean conditions. Orig Life Evol Biosph 41 (4), 331–345.

Rüpke, L.H., Morgan, J.P., Hort, M., Connolly, J.A., 2004. Serpentine and the subduction zone water cycle. Earth Planet. Sci. Lett. 223 (1), 17–34.

Schenk, P., McKinnon, W.B., 1989. Fault offsets and lateral plate motions on Europa: Evidence for a mobile ice shell. Icarus 79 (1), 75–100.

Schmidt, B.E., Blankenship, D.D., Patterson, G.W., Schenk, P.M., 2011. Active formation of 'chaos terrain' over shallow subsurface water on Europa. Nature 479 (7374), 502–505. doi:10.1038/nature10608.

Schroeder, D.M., Romero-Wolf, A., Carrer, L., Grima, C., Campbell, B.A., Kofman, W., Bruzzone, L., Blankenship, D.D., 2016. Assessing the potential for passive radio sounding of Europa and Ganymede with RIME and REASON. Planet. Space Sci. 134, 52–60.

Schubert, G., Spohn, T., Reynolds, R.T., 1986. Satellites In: Burns, J.A., Matthews, M.S. (Eds.), University Arizona Press, Tucson, pp. 224–292.

Schubert, G., Zhang, K., Kivelson, M.G., Anderson, J.D., 1996. The magnetic field and internal structure of Ganymede. Nature 384, 544–545.

Schubert, G., Sohl, F., Hussmann, H., 2009. Interior of Europa. In: Pappalardo, R.T., McKinnon, W.B., Khurana, K. (Eds.), Europa. University of Arizona Press, Tucson, pp. 353–367.

Showman, A.P., Malhotra, R., 1997. Tidal evolution into the Laplace resonance and the resurfacing of Ganymede. Icarus 127 (1), 93–111.

Siegert, M.J., Ellis-Evans, J.C., Tranter, M., Mayer, C., Petit, J., Salamatin, A., Priscu, J.C., 2001. Physical, chemical, and biological processes in Lake Vostok and other Antarctic subglacial lakes. Nature 414, 603–609.

Silber, E., Johnson, B., 2017. Impact crater morphology and the structure of Europa's ice shell. J. Geophys. Res. Planets 122 (2), 2685–2701. https://doi.org/10.1002/2017JE005456.

Smith, B.A., et al., 1979. The Galilean satellites and Jupiter: Voyager 2 imaging science results. Science 206, 927–950.

Som, S.M., Adam, Z.R., Vance, S., 2009. Use the water: In-situ resource technology for icy-surface landers. Acta Astronaut. 64 (9–10), 1006–1010.

Sotin, C., Head III., J.W., Tobie, G., 2002. Europa: Tidal heating of upwelling thermal plumes and the origin of lenticulae and chaos melting. Geophys. Res. Lett. 29 (8), 74-1–74-4.

Sparks, W.B., Hand, K.P., McGrath, M.A., Bergeron, E., Cracraft, M., Deustua, S.E., 2016. Probing for evidence of plumes on Europa with HST/STIS. The Astrophysical Journal 829 (2), 121. https://doi.org/10.3847/0004-637X/829/2/121.

Sparks, W.B., Schmidt, B.E., Mcgrath, M.A., Hand, K.P., Spencer, J.R., Cracraft, M., Deustua, S.E, 2017. Active Cryovolcanism on Europa? Astrophysical J. Letters 839 (2), L18. doi:10.3847/2041-8213/aa67f8.

Spaun, N.A., Head, J.W., Collins, G.C., Prockter, L.M., Pappalardo, R.T., 1998. Conamara Chaos region, Europa: reconstruction of mobile polygonal ice blocks. Geophys. Res. Lett. 25, 4277–4280.

Spaun, N.A., Prockter, L.M., Pappalardo, R.T., Head, J.W., Collins, G.C., Antman, A., Greeley, R., 1999. Spatial distribution of lenticulae and chaos on Europa. Lunar Planet. Sci. 30, 1847.

Spencer, J.R., 1987. Thermal segregation of water ice on the Galilean satellites. Icarus 69, 297–313.

Spohn, T., Schubert, G., 2003. Oceans in the icy Galilean satellites of Jupiter? Icarus 161 (2), 456–467. https://doi.org/10.1016/S0019-1035(02)00048-9.

Squyres, S.W., Reynolds, R.T., Cassen, P.M., Peale, S.J., 1983. Liquid water and active resurfacing on Europa. Nature 301 (5897), 225–226. https://doi.org/10.1038/301225a0.

Steinbrügge, G., Voigt, J.R.C., Wolfenbarger, N.S., Hamilton, C.W., Soderlund, K.M., Young, D.A., Blankenship, D.D., Vance, S.D., Schroeder, D.M., 2020. Brine migration and impact-induced cryovolcanism on Europa. Geophys. Res. Lett. 47, e2020GL090797.

Takai, K., Komatsu, T., Inagaki, F., Horikoshi, K., 2001. Distribution of archaea in a black smoker chimney structure. Appl. Environ. Microbiol. 67, 3618–3629.

Teolis, B.D., Wyrick, D.Y., Bouquet, A., Magee, B.A., Waite, J.H., 2017. Plume and surface feature structure and compositional effects on Europa's global exosphere: Preliminary Europa mission predictions. Icarus 284, 18–29.

Thomas, D.N., Dieckmann, G.S., 2002. Antarctic Sea ice—a habitat for extremophiles. Science 295, 641–644.

Thomson, R.E., Delaney, J.R., 2001. Evidence for a weakly stratified Europan ocean sustained by seafloor heat flux. J. Geophys. Res. 106, 12,355–12,365.

Tobie, G., Choblet, G., Sotin, C., 2003. Tidally heated convection: Constraints on Europa's ice shell thickness. J. Geophys. Res. 108 (E11), 5124. https://doi.org/10.1029/2003JE002099.

Trainer, M.G., 2013. Atmospheric prebiotic chemistry and organic hazes. Curr. Org. Chem. 17 (16), 1710–1723. doi:10.2174/13852728113179990078.

Tritt, C.S., 2002; Possibility of Life on Europa, Milwaukee School of Engineering, 1025 N Broadway, Milwaukee, WI 53202, USA.

Trumbo, S.K., Brown, M.E., Hand, K.P., 2019. Sodium chloride on the surface of Europa. Sci. Adv. 5 (6), eaaw7123. doi:10.1126/sciadv.aaw7123.

Vance, S., Bouffard, M., Choukroun, M., Sotin, C., 2014. Ganymede's internal structure including thermodynamics of magnesium sulfate oceans in contact with ice. Planet. Space Sci. 96, 62–70.

Vance, S.D., Hand, K.P., Pappalardo, R.T., 2016. Geophysical controls of chemical disequilibria in Europa. Geophys. Res. Lett. 43 (10), 4871–4879. doi:10.1002/2016GL068547.

Vilella, K., Choblet, G., Tsao, W.-E., Deschamps, F., 2020. Tidally heated convection and the occurrence of melting in icy satellites: Application to Europa. J. Geophysical Res.: Planets 125, e2019JE006248. https://doi.org/10.1029/2019JE006248.

Waite, J.H. Jr., Combi, M.R., Ip, W.-H., Cravens, T.E., McNutt, R.L. Jr., Kasprzak, W., Yelle, R., Luhmann, J., Miemann, H., Gell, D., Magee, B., Fletcher, G., Lunine, J., Tseng, W.-L, 2006. Cassini ion and neutral mass spectrometer: enceladus plume composition and structure. Science 311, 1419–1422.

Waite, J.H., Glein, C.R., Perryman, R.S., Teolis, B.D., Magee, B.A., Miller, G., Grimes, J., Perry, M.E., Miller, K.E., Bouquet, A., Lunine, J.I., Brockwell, T., Bolton, S.J., 2017. Cassini finds molecular hydrogen in the Enceladus plume: Evidence for hydrothermal processes. Science 356, 155–159.

Wall, M., 2018. This May Be the Best Evidence Yet of a Water Plume on Jupiter's Moon Europa, https://www.space.com/40575-jupiter-moon-europa-plume-galileo-spacecraft.html

Wall, M., 2020. Alien-life hunters are eyeing icy ocean moons Europa and Enceladus, https://www.space.com/alien-life-ocean-moons-europa-enceladus.html

Whalen, K., Lunine, J.I., Blaney, D.L., 2017. MISE: A search for organics on Europa. American Astronomical Society AAS Meeting #229, id.138.04.

Wilson, L., Head, J.W., Pappalardo, R.T., 1997. Eruption of lava flows on Europa: Theory and application to Thrace Macula. J. Geophys. Res. 102, 9263–9272.

Worth, R.J., Sigurdsson, S., House, C.H., 2013. Seeding life on the moons of the outer planets via lithopanspermia. Astrobiology 13 (12), 1155–1165. doi:10.1089/ast.2013.1028.

Zahnle, K., Dones, L., Levison, H.F., 1998. Cratering rates on the Galilean satellites. Icarus 136 (2), 202–222. https://doi.org/10.1006/icar.1998.6015.

Zelenyi, L., Korablev, O., Martynov, M., Popov, G.A., Blanc, M., Lebreton, J.P., Pappalardo, R., Clark, K., Fedorova, A., Akim, E.L., Simonov, A.A., Lomakin, I.V., Sukhanov, A., Eismont, N., the Europa Lander Team, 2011. Europa Lander mission and the context of international cooperation. Adv. Space Res. 48 (4), 615–628.

Zimmer, C., Khurana, K.K., Kivelson, M.G., 2000. Subsurface oceans on Europa and Callisto: Constraints from Galileo magnetometer observations. Icarus 147 (2), 329–347. https://doi.org/10.1006/icar.2000.6456.

Zolotov, M.Y., Shock, E.L., 2001. Composition and stability of salts on the surface of Europa and their oceanic origin. J. Geophysical Res.: Planets 106 (E12), 32815–32827.

Chapter 13

Salty ocean and submarine hydrothermal vents on Saturn's Moon *Enceladus* — Tall plume of gas, jets of water vapor & organic-enriched ice particles spewing from its south pole

13.1 Saturn and its glaring similarities and dissimilarities with Earth

Saturn is the sixth planet from the Sun and the second-largest in the solar system, after Jupiter. Saturn is about 75,000 miles (120,000 km) across. One day on Saturn takes only 10.7 h (the time it takes for Saturn to rotate or spin around once), and Saturn makes a complete orbit around the Sun (a year in Saturnian time) in about 29.4 Earth years (10,756 Earth days). Saturn is flattened at the poles because of its very rapid rotation. Saturn travels at an average speed of 21,637 miles per hour (34,821 km/h) in its orbit around the Sun. Saturn is a gas giant [a giant planet composed mainly of hydrogen and helium (D'Angelo and Lissauer, 2018)] with an average radius about nine times that of Earth. Apart from hydrogen and helium, heavier elements make up between 3 and 13 percent of the mass (Guillot et al., 2004). Saturn is thought to consist of an outer layer of molecular hydrogen surrounding a layer of liquid metallic hydrogen, with probably a molten rocky core. Note that metallic hydrogen is a phase of hydrogen in which it behaves like an electrical conductor. This phase was predicted in 1935 on theoretical grounds by Eugene Wigner and Hillard Bell Huntington (Wigner and Huntington, 1935). According to Guillot et al. (2004), at high pressure and temperatures, metallic hydrogen can exist as a liquid rather than a solid, and researchers think it might be present in large quantities in the hot and gravitationally compressed interiors of Jupiter, Saturn, and in some exoplanets. Liquid metallic hydrogen (LMH) is a fundamental system in condensed matter sciences and the main constituent of gas giant planets. Because of exceptional challenges in experimentation and theory, its transport properties remained poorly understood.

In the Saturnian system, the large, 250-km radius moon *Enceladus* is considered to be the creator and sustainer of Saturn's gorgeous E ring. A system of rings around a planet is akin to a miniature solar nebula, in which the planet is the giant mass at the center of a rotating system, much as the early Sun was, and the rings are the material rotating around it. Material is taken from moons to make rings. Geysers on the recently discovered, geologically active south-polar region of *Enceladus* are now recognized as the dominant source of material in Saturn's E ring. The rings are made of dusty water ice, in the form of boulder-sized and smaller chunks that gently collide with each other as they orbit around Saturn. The rings, as shown in Fig. 13.1, have a slight pale reddish color due to the presence of organic material mixed with the water ice. Once formed, the ring does not remain forever: forces from radiation, meteoroid impacts, and drag from the outer parts of the planet's atmosphere (the exosphere) begin to erode the ring. Saturn's gravitational field constantly disrupts these ice chunks, keeping them spread out and preventing them from combining to form a moon. Looking at the relatively small lifetime of even dense rings (a few hundreds of thousands or millions of years), they are merely a fleeting and transient phenomenon in the history of the solar system. The Saturn system is indeed a dynamic place, with the Enceladus plumes creating the E ring and loading the magnetosphere with water which interacts with Titan and the other Saturnian moons.

584 Water worlds in the solar system

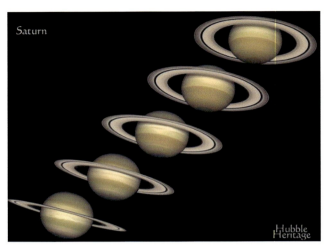

FIG. 13.1 Hubble space telescope images of Saturn's rings, captured from 1996 to 2000, opening up from just past edge-on to nearly fully open as it moves from autumn toward winter in its Northern Hemisphere. *(Source: https://solarsystem.nasa.gov/resources/14621/a-change-of-seasons-on-saturn/). Credit: NASA and The Hubble Heritage Team (STScI/AURA).*

It was once thought that Saturn's rings are incredibly thin, with a thickness of only about 30 feet (10 m). The ring was traditionally thought to span the region between 3 and 8 R_S, where R_S is the radius of Saturn. However, new in situ dust measurements indicate that the density of small grains might continuously extend far beyond these boundaries, and the E ring could reach even beyond the orbit of Titan (20.3 R_S). Horányi et al. (2008) reported on the modeling results of the long-term evolution of dust particles comprising the E ring to show that grains from *Enceladus* could indeed reach the outskirts of Saturn's magnetosphere.

The planet Saturn is similar to Earth in some ways. For example, Saturn's equatorial plane is tilted relative to its orbital plane by 27 degrees, very similar to the 23-degree tilt of the Earth. As Saturn moves along its orbit, first one hemisphere, then the other is tilted toward the Sun. This cyclical change causes seasons on Saturn, just as the changing orientation of Earth's tilt causes seasons on our planet. But on Saturn, one season lasts seven Earth years. Fig. 13.2 shows a diagram of Saturn, to scale. Unlike Earth, having just one moon orbiting around it, Saturn (apart from its beautiful rings) has 62 confirmed moons, of which 9 are waiting to be officially named. Saturn's moons have mythological names such as Mimas, Enceladus, Tethys, etc. Titan has its own atmosphere (a very thick atmosphere which is mostly nitrogen), which is very unusual for a moon. Among its moons, *Enceladus* has been found to harbor a subsurface ocean and *Titan* is believed to harbor an internal ocean.

13.2 Exploring Saturn and its two major moons (*Enceladus* and *Titan*)—importance of Cassini spacecraft mission

Cassini is a sophisticated robotic spacecraft, which orbited the ringed planet Saturn and helped studying the Saturnian system in detail. The Cassini mission to Saturn is one of the most ambitious efforts in planetary space exploration ever mounted. It is a joint endeavor and a cooperative mission of the United States of America's National Aeronautics and Space Administration (NASA); the European Space

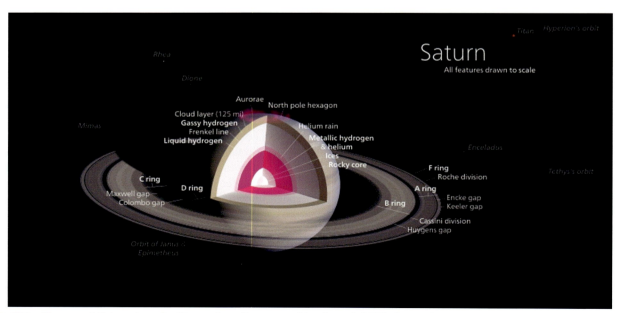

FIG. 13.2 Diagram of Saturn, to scale. *(Source: https://commons.wikimedia.org/wiki/File:Saturn_diagram.svg; https://commons.wikimedia.org/w/index.php?curid=32219154). Copyright info: This file is licensed under the Creative Commons Attribution-Share Alike 3.0 Unported. By Kelvinsong - Own work, CC BY-SA 3.0.*

Agency (ESA); and the Italian space agency, Agenzia Spaziale Italiana (ASI), with participation by hundreds of scientists and engineers from Europe and the US. The Jet Propulsion Laboratory, a division of the California Institute of Technology, manages the Cassini mission for NASA.

With the launch of Cassini spacecraft on Oct. 15, 1997, a seven-year journey (1997–2004) to Saturn began (Fig. 13.3). Cassini also carried a probe called *Huygens* to explore the surface of Saturn's largest moon, *Titan*. After a nearly 2.2-billion mile journey, the spacecraft arrived in the Saturn system on June 30, 2004. In its 7-year historical cruise to Saturn on a gravity-assist trajectory, Cassini spacecraft carried out two swing-bys of Venus, one of Earth and one of Jupiter to give the spacecraft the boost needed to reach Saturn. The Cassini spacecraft has been orbiting Saturn since mid-2004. Cassini spacecraft's final orbit happened on 15 September 2017, with its fall into Saturn's atmosphere, ending its extraordinary mission of discovery.

Note that in orbital mechanics and aerospace engineering, gravity assist maneuver, swing-by, or gravitational slingshot is the use of the relative movement (e.g., orbit around the Sun) and gravity of a planet or other astronomical object to alter the path and speed of a spacecraft, typically to save propellant and reduce expense. The gravity assist maneuver was first used in 1959 when the Soviet probe Luna 3 photographed the far side of Earth's Moon and it was used by interplanetary probes from Mariner 10 onwards, including the two Voyager probes' notable flybys of Jupiter and Saturn.

On June 30, 2004, Cassini passed between the F and G rings of Saturn and allowed itself to be captured as a moon of Saturn. Having been released on December 24, 2004, the Huygens probe descended (parachuted) through Titan's nitrogen atmosphere, landed on Titan, and relayed data to Earth on January 14, 2005. During an eventful four years of orbital tour (2004–2008), Cassini made 45 Titan encounters, 10 icy moon encounters, and 76 orbits of Saturn. In the next two years (2008–2010) Cassini succeeded in making 64 orbits around Saturn, 28 Titan encounters, 8 Enceladus encounters, three encounters with other smaller icy moons; and carried out equinox crossing in August 2009. The next seven years (2010–2017) also proved to be as eventful as the earlier years, with several success stories such as 155 orbits, 54 Titan encounters, 11 Enceladus encounters, and 5 other icy moon encounters.

The 10th anniversary (June 30, 2014) of Cassini spacecraft in Saturn's orbit has been hailed as a decade of discoveries. As of this historic date, the stalwart spacecraft has beamed back to Earth more than 500 gigabytes of scientific data through NASA's Deep Space Network (DSN), enabling the publication of more than 3000 scientific papers in subsequent years. Note that the NASA DSN is a worldwide network of U.S. spacecraft communication facilities, located in the United States, Spain, and Australia, that supports NASA's interplanetary spacecraft missions. The DSN also provides radar and radio astronomy observations that improve our understanding of the solar system and the larger universe.

The Grand Finale of the Cassini–Huygens space-research mission to Saturn (15 October 1997–15 September 2017) culminated in the Cassini spacecraft's final orbit, in which Cassini spacecraft fell into Saturn's atmosphere, ending its extraordinary 20-year mission of discovery. It would be interesting to skim through the voyage of discovery made possible by the Cassini spacecraft mission. Fig. 13.4 shows Cassini at the planet *Saturn*. Fig. 13.5 shows an overview of Cassini spacecraft orbiting around Saturn.

The Cassini Spacecraft Mission had several interesting science objectives that pertained to not only the planet Saturn but also its magnificent rings and some of the icy moons. Understanding Saturn's magnetosphere turned out to be an important mission objective. Studies pertaining to Saturn's magnetosphere included particle composition, sources and sinks, dynamics of the magnetosphere, and interaction with solar wind, moons, and rings. Understanding Saturn's

FIG. 13.3 Launch of Cassini Spacecraft to Saturn on Oct. 15, 1997. *(Image source: NASA: https://www.nasa.gov/image-feature/oct-15-1997-launch-of-cassini-to-saturn).*

FIG. 13.4 Spacecraft "Cassini" at planet *Saturn*. *(Image Source: https://www.nasa.gov/mission_pages/cassini/whycassini/index.html).*

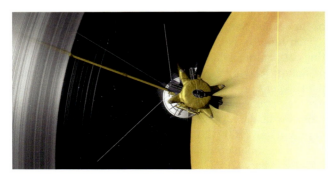

FIG. 13.5 An overview of "Cassini" spacecraft orbiting around Saturn. *(Image source: https://www.nasa.gov/mission_pages/cassini/whycassini/index.html).*

FIG. 13.6 Cassini spacecraft orbiting in the close vicinity of the planet Saturn's rings. *(Image source: NASA: https://saturn.jpl.nasa.gov/legacy/files/Cassini_Grand_Finale_Fact_Sheet_508.pdf).*

hazy moon Titan's interaction with solar wind and magnetosphere was of special interest. Other studies aimed at understanding Saturn included its cloud properties and atmospheric composition; winds and temperatures; internal structure and rotation; ionosphere; and origin and evolution. Studies aimed at understanding Saturn's rings included the rings' structure and composition; dynamical processes; inter-relation of rings and moons; dust and micro-meteoroid environment. Fig. 13.6 shows the Cassini spacecraft orbiting in the close vicinity of Saturn's rings. Studies of Saturn's hazy moon *Titan* were one of the Cassini mission objectives. These included Titan's atmospheric constituent abundances, distribution of trace gases and aerosols, winds and temperatures, surface state and composition, and upper atmosphere. Because Saturn's icy moon *Enceladus* was believed to harbor a subsurface ocean and *Titan* is believed to harbor an internal ocean, much importance has been attached to achieve as much understanding as possible about these two icy moons. Saturn's other icy moons were also explored during the Cassini mission. Studies of Saturn's icy moons included characteristics and geological histories, mechanisms of surface modification, surface composition and distribution, bulk composition and internal structure, and interaction with magnetosphere.

Cassini mission can boast of several achievements to its credit. During this mission, Cassini spacecraft completed more than 200 orbits of Saturn, carried out 132 close flybys of Saturn's moons, and discovered seven new moons. In this chapter, weightage is given to Saturn's icy moon *Enceladus* because of its value as an organic-rich ocean world having prospects of harboring an ecosystem based on microbial populations and adorning a favorable scenario for life's emergence.

13.3 Presence of a global subsurface ocean on *Enceladus*—inferences from gravitational field measurements and forced physical wobbles in its rotation

The early flybys of the Cassini–Huygens mission to the Saturnian system revealed the incredible nature of Enceladus. It was found that Enceladus is an active, dynamic world, including cryovolcanic plumes ejecting ice grains and vapor from faults near its south pole (Waite et al., 2006; Thomas et al., 2016), as reviewed by Dougherty and Spilker (2018). Mission modifications led to Cassini flying close to Enceladus 23 times over a decade, in order to obtain more details about the moon.

Iess et al. (2014) determined the quadrupole gravity field (the quadrupole represents how stretched-out along some axis the mass is) of Enceladus and its hemispherical asymmetry using Doppler data from three spacecraft flybys. Their results indicate the presence of a negative mass anomaly in the south-polar region, largely compensated by a positive subsurface anomaly compatible with the presence of a south polar subsurface ocean of about 10 km thickness located beneath an ice crust 30 to 40 km thick and above a rocky core and extending up to south latitudes of about 50°. Thomas et al. (2016) found that the subsurface ocean is not merely a regional one but rather a 26–31 km deep global salty water ocean.

Detection of sodium-salt-rich ice grains emitted from the plume of Enceladus suggests that the grains formed as frozen droplets from a liquid water reservoir that is or has been, in contact with rock. This finding implies rock–water interactions in regions surrounding the core of Enceladus. The resulting chemical 'footprints' are expected to be preserved in the liquid and subsequently transported upwards to the near-surface plume sources, where they eventually would be ejected and could be measured by a spacecraft (Hsu et al., 2015).

Cassini scientists analyzed more than seven years' worth of images of Enceladus taken by the spacecraft, which has been orbiting Saturn since mid-2004. They carefully mapped the positions of features on Enceladus—mostly craters—across hundreds of images, in order to measure changes in the moon's rotation with extreme precision. As a result, they found Enceladus has a tiny, but measurable wobble as it orbits Saturn. Because the icy moon is not perfectly spherical—and because it goes slightly faster and slower during different portions of its orbit around Saturn—the giant planet subtly rocks Enceladus back and forth as it rotates.

The research team plugged their measurement of the wobble, called a *libration*, into different models for how Enceladus might be arranged on the inside, including ones in which the moon was frozen from surface to core. They argued that if the surface and core were rigidly connected, the core would provide so much dead weight that the wobble would be far smaller than they observe it to be.

Thomas et al. (2016) used measurements of control points across the surface of Enceladus accumulated over seven years of spacecraft observations to determine the moon's precise rotation state, and found a forced physical wobble (called *libration*) of $0.120 \pm 0.014°$ (2σ). This value is too large to be consistent with Enceladus's core being rigidly connected to its surface, and thus implies the presence of a global ocean rather than a localized polar sea. In simple language, after detailed analysis of more than seven years measurements of wobbles in Enceladus's rotation, researchers found that the magnitude of the moon's very slight wobble, as it orbits the planet Saturn, can only be accounted for if the crust is moving separately from the rocky core, meaning that there must be a widespread layer of liquid between them (Thomas et al., 2016). If the solid icy crust and a solid rocky core of the moon is loosely connected as inferred above, then it implies that a global ocean must be present underlying the moon's south polar region (see Fig. 13.7), rather than a lens-shaped body of water sandwiched between the icy layers of the solid crust as in the case of Jupiter's moon Europa, or a sea, as previously thought.

13.4 Liquid water ocean hiding below the icy crust of *Enceladus*—inference drawn from plumes of water vapor & salty ice grain jutting out from the surface of *Enceladus*

The Cassini spacecraft passed within 168.2 km of the surface above the southern hemisphere at 19:55:22 universal time coordinate (UTC) on 14 July 2005 during its closest approach to *Enceladus*. Fig. 13.8 provides a view of Enceladus, showing surface features and the Cassini ground track during the flyby on 14 July 2005. Cassini's encounters with Enceladus led to the discovery of an active south pole and a south polar hot spot (Spencer et al., 2006). In the year 2005, a water-vapor and ice-grain plume jetting out from near the south polar region of Enceladus was detected (Hansen et al., 2006). Fig. 13.9 shows NASA's Cassini spacecraft closing in on Saturn's icy moon *Enceladus*. It was found that more than 100 individual geysers blast water ice, organic molecules and other material into space from the tiger stripes in south polar region of Enceladus (Fig. 13.10). This close-up view of Enceladus shows a distinctive pattern of continuous, ridged, slightly curved and roughly parallel faults within the moon's southern polar latitudes. These surface features have been informally referred to by imaging

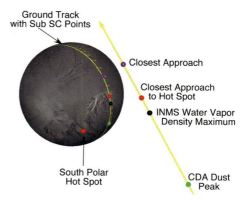

FIG. 13.8 **View of Enceladus showing surface features and the Cassini ground track during the flyby on July 14, 2005.** The south polar hot spot is shown in red, amidst the surface feature known as the tiger stripes. The spacecraft trajectory is shown in yellow. The colors of the points along the trajectory represent Cassini's closest approach to Enceladus (purple), the closest approach to the southern polar hot spot (red), the point along the track where INMS saw the maximum water vapor density (black), and the point along the track where the CDA saw the peak in dust particle density (green). The direction of motion of the spacecraft (ram direction) is represented by the arrowhead on the trajectory. SC, spacecraft. *(Source: Waite Jr. et al. (2006), Cassini ion and neutral mass spectrometer: Enceladus plume composition and structure, Science, 311 (5766), 1419-1422. DOI: 10.1126/science.1121290). Copyright info: Copyright 2006, American Association for the Advancement of Science.)*

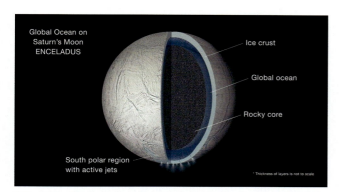

FIG. 13.7 **Illustration of the interior of Saturn's moon *Enceladus* showing a global liquid water ocean located between its rocky core and icy crust.** Thickness of layers shown here is not to scale. *Credits: NASA/JPL-Caltech. (Source:https://www.nasa.gov/press-release/cassini-finds-global-ocean-in-saturns-moon-enceladus).*

FIG. 13.9 NASA's Cassini spacecraft closing in on Saturn's icy moon *Enceladus*. *(Source: https://www.nasa.gov/mission_pages/cassini/main/index.html).*

scientists as "tiger stripes" due to their distinctly stripe-like appearance when viewed in false color.

Clearly, one of the spectacular discoveries of the Cassini spacecraft was the plume of water vapor and icy particles (dust) originating near the south pole of Enceladus, particularly in the region of long, parallel crevasses dubbed "tiger stripes" (see Porco et al., 2006; Brown et al., 2006). Tian et al. (2006) used Monte Carlo simulations to model the July 14, 2005, Ultraviolet Imaging Spectrograph (UVIS) stellar occultation observations of the water vapor plumes on Enceladus. These simulations indicate that the observations can be best fit if the water molecules ejected along the Tiger Stripes in the South Polar Region of Enceladus have a vertical surface velocity of 300–500 m/s at the surface. According to Tian et al. (2006), the high surface velocity suggests that the plumes on Enceladus originate from some depth beneath the surface. The Monte Carlo simulations indicated that the total escape rate of water molecules is 120–180 kg/s—more than 100 times the estimated mass escape rate for ice particles.

Enceladus's tall water vapor plume aroused considerable interest among planetary and life scientists, resulting in the commencement of various studies. For example, Kieffer et al. (2006) put forward a clathrate (a compound in which molecules of one component are physically trapped within the crystal structure of another. In the present case, methane molecule is trapped/caged within water molecules. Nimmo et al. (2007) also suggested that the plume characteristics and local high heat flux are ascribable either to the presence of liquid water within a few tens of meters of the surface, or the decomposition of clathrates (Fig. 13.11). Observations of Enceladus's South Pole revealed large localized fractures or rifts in the crust, informally called 'tiger stripes', which exhibit higher temperatures than the surrounding terrain and are probably sources of the observed eruptions (Hurford et al., 2007). McKay et al. (2008) suggested that the jets of icy particles and water vapor issuing from the south pole of Enceladus are evidence for activity driven by some geophysical energy source. The vapor has also been shown to

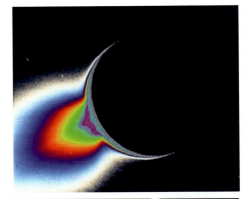

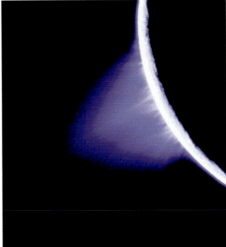

FIG. 13.10 (top) Saturn's moon Enceladus backlit by the Sun shows the fountain-like sources of the fine spray of material that towers over the South Polar Region. *Image credit: NASA/JPL/Space Science Institute. (Source: https://saturn.jpl.nasa.gov/science/overview/).* (middle) The Enceladus plume, sourced by a potentially habitable subsurface ocean. Credit: NASA/JPL/Space Science Institute. *(Source: Hendrix et al. (2019), The NASA Roadmap to Ocean Worlds, Astrobiology, 19(1): 1–27). Amanda R. Hendrix and Terry A. Hurford et al., 2018; Published by Mary Ann Liebert, Inc.* (bottom) Illustration showing NASA's Cassini spacecraft diving through a geyser plume on Saturn's moon Enceladus in 2015. *Credit. NASA/JPL-Caltech. (Source: https://www.space.com/13531-photos-enceladus-saturn-moon-cassini.html).*

contain simple organic compounds, and the south polar terrain is bathed in excess heat coming from below. Based on the July 14, 2005, ultraviolet imaging spectrograph (UVIS) stellar occultation observations of the water vapor plumes on Enceladus, Tian et al. (2006) carried out Monte Carlo

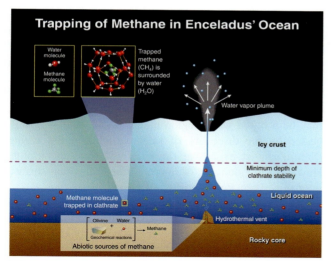

FIG. 13.11 **Trapping of Methane in Enceladus' Ocean.** This illustration depicts potential origins of methane found in the plume of gas and ice particles that sprays from Saturn's moon, Enceladus, based on research by scientists working with the Ion and Neutral Mass Spectrometer on NASA's Cassini mission. http://photojournal.jpl.nasa.gov/catalog/PIA19059. *Copyright info: This file is in the **public domain** in the United States because it was solely created by NASA. NASA copyright policy states that "NASA material is not protected by copyright **unless noted**". (Source: https://commons.wikimedia.org/wiki/File:PIA19059-SaturnMoon-Enceladus-OceanMethane-20150311.png).*

simulations to better understand the Enceladen water vapor plume. These simulations indicate that the observations can be best fit if the water molecules ejected along the Tiger Stripes in the South Polar region of Enceladus have a vertical surface velocity of 300–500 m/s at the surface. According to these researchers, the high surface velocity suggests that the plumes on Enceladus originate from some depth beneath the surface. It was also found that the total escape rate of water molecules is $4-6 \times 10^{27}$ molecules/s, or 120–180 kg/s, and more than 100 times the estimated mass escape rate for ice particles. It was also found that the average deposition rate in the South Polar region is on the order of $10^{11}/cm^2/s$, yielding a resurfacing rate as high as 3×10^{-4} cm/yr. It would be interesting to note that the globally averaged deposition rate of water molecules is about one order of magnitude lower.

The imaging science subsystem (ISS) observed a young surface and evidence of recent tectonic activity in the southern hemisphere (Porco et al. 2006). Their estimates from cratering models place the age of the surface at less than 10–100 million years. Interestingly, the Visible and Infrared Mapping Spectrometer (VIMS) instrument saw that the ice in and around the "tiger stripes" is crystalline in nature, whereas it is amorphous elsewhere on the moon (Brown et al., 2006). According to Parkinson et al. (2007), the high degree of crystallinity seen in the coldest areas of the "tiger stripes" is consistent with fresh material, perhaps as young as a few decades.

Manga and Wang (2007) argued that pressurized subsurface oceans are responsible for the observed eruption of liquid water on Europa and Enceladus. Chemical analysis of the plume emanating from near the south pole of *Enceladus* indicates that the interior of this Saturnian moon is hot (Spencer and Grinspoon, 2007). They wondered whether it could have been hot enough for complex organic molecules to be made? The energy budget of Enceladus holds a number of perplexing and puzzling questions. It is indeed surprising that although this icy moon orbits Saturn in the frozen outer reaches of the solar system, it is volcanically and tectonically active, with giant geysers of water vapor and ice erupting into space through fissures in its surface. No wonder, with a young, highly reflective surface and vigorous geological activity, Enceladus has begun to be considered as one of the most mysterious planetary bodies in the solar system.

Hurford et al. (2007) suggested that the eruptions emanating from Enceladus's south pole arise from tidally controlled periodic openings of rifts on the tiger stripes on Enceladus. The prolific activity and presence of a plume on Enceladus offer us a unique opportunity to sample the interior composition of an icy moon, and to look for interesting chemistry and possible signs of life. The discovery of a plume of water vapor and ice particles emerging from warm fractures (tiger stripes) in Enceladus raised the question of whether the plume emerges from a subsurface liquid source or from the decomposition of ice. To shed light on this interesting question, Schmidt et al. (2008) examined the nature of motion of dust in Enceladus' plume in relation to that of the vapor in this plume. They found that the observed data imply considerably smaller velocities for the grains than for the vapor, which has been difficult to understand. They argued that the gas and dust are too dilute in the plume to interact, so the difference must arise below the surface. Schmidt et al. (2008) reported a model for grain condensation and growth in channels of variable width. They showed that repeated wall collisions of grains, with re-acceleration by the gas, induce an effective friction, offering a natural explanation for the reduced grain velocity. In their study, Schmidt et al. (2008) derived particle speed and size distributions that reproduce the observed and inferred properties of the dust plume. They found that the gas seems to form near the triple point of water. Note that triple point is the temperature and pressure at which a substance can exist in equilibrium in the liquid, solid, and gaseous states. The triple point of pure water is at 0.01°C (273.16K, 32.01°F) and 4.58 mm (611.2Pa) of mercury. According to Schmidt et al. (2008), gas densities corresponding to sublimation from ice at temperatures less than 260 K are generally too low to support the measured particle fluxes. They argue that this in turn suggests liquid water below Enceladus' South Pole. According to Schmidt et al. (2008), the flow through these tiger stripes is supersonic and choked. McKay et al. (2008)

predicted that it is possible that a liquid water environment exists beneath the south polar cap, which may be conducive to life. Fig. 13.12 indicates fresh ice that has been deposited on the surface at the south pole of Enceladus.

McKay et al. (2008) also expressed confidence that Cassini's instruments may detect plausible evidence for life by analysis of hydrocarbons in the plume during close encounters. The water ice particles are 1–10 µm in diameter, and the plume composition shows H_2O, CO_2, CH_4, NH_3, Ar, and evidence that more complex organic species might be present (Tsou et al., 2012).

Investigations carried out by Postberg et al. (2009a,b) showed that the sodium salts detected in Saturn's E-ring ice grains originated from an ocean below the surface of Enceladus. Identifying the relatively large source of heat in the south polar region of this tiny icy moon became a curious topic for research. To explain the observed heat, planetary scientists began to think that the extraordinary activity at Enceladus' warm South Pole could be an indication of the presence of an internal global or local reservoir of liquid water beneath the surface. In this respect, Tyler (2009) showed that if the spin axis of Enceladus is tilted with respect to its orbital plane by at least 0.05 degree then strong tidal flow will be generated with enough dissipative heating to explain the observed heat flux. Tyler (2009) further showed that in an alternative case of a shallow (10 km or less) ocean, comparable flow velocities and heating may be obtained by eccentricity tidal forces.

The plume characteristics and local high heat flux have been ascribed either to the presence of liquid water within a few tens of meters of the surface, or the decomposition of clathrates (i.e., gas hydrates existing in ice-like solid form under suitable combination of pressure and temperature). There is reason to believe that the source of the jets of water ice from surface fractures near the south pole of Enceladus may be a liquid water region under the ice shell—as suggested by the discovery of salts particles in the E-ring (Saturn's most expansive ring) derived from the plume. However, how delivery of internal heat to the near-surface is sustained remained poorly understood.

As a further evidence to support the notion of the presence of liquid water below the surface of Enceladus, it was found that some liquid escapes into space through cracks in the ice, which is the source of one of Saturn's rings. Cassini confirmed that the plume feeds particles into the E ring. The spacecraft has come as close as 25 km (15 miles) from the moon's icy surface during its investigation, revealing the presence of a global subsurface ocean that might have conditions suitable for life.

Waite Jr. et al. (2009) reported that ammonia is present in the plume, along with various organic compounds, deuterium and, very probably, ^{40}Ar. The presence of ammonia provides strong evidence for the existence of at least some liquid water, given that temperatures in excess of 180 K have been measured near the fractures from which the jets emanate. Waite Jr. et al. (2009) concluded, from the

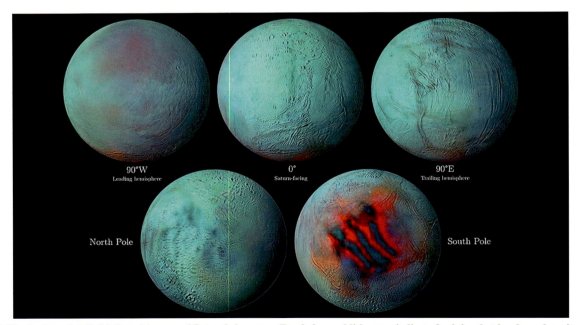

FIG. 13.12 In these detailed infrared images of Saturn's icy moon Enceladus, reddish areas indicate fresh ice that has been deposited on the surface. *Image Credit: NASA/JPL-Caltech/University of Arizona/LPG/CNRS/University of Nantes/Space Science Institute. Copyright info: This file is in the **public domain** in the United States because it was solely created by NASA. NASA copyright policy states that "NASA material is not protected by copyright **unless noted**". (Source: https://commons.wikimedia.org/wiki/File:PIA24023-16-SaturnMoon-Enceladus-FreshIce-20200918.jpg). Author: NASA/JPL-Caltech/University of Arizona/LPG/CNRS/University of Nantes/Space Science Institute. Dated: September 18, 2020.*

overall composition of the material from the jet, that the plume derives from both a liquid reservoir (or from ice that in recent geological time has been in contact with such a reservoir) as well as from degassing, volatile-charged ice, reinforcing the earlier suggestion that Enceladus holds a subsurface ocean. Based on the observation of ammonia and ^{40}Ar in the plume, Waite, Jr, et al. (2009) argued that the Enceladus's subsurface ocean is formed by liquid water.

As already noted, the observation that water plumes erupt from cracks on Enceladus has fueled and fired speculation about a possible subsurface ocean. The finding that a global ocean is located beneath the icy crust of geologically active Enceladus implies that the fine spray of water vapor, icy particles, and simple organic molecules Cassini has observed coming from fractures near the moon's South Pole is being fed by this vast liquid water reservoir.

Deciphering, this new knowledge was a hard problem that required years of observations, and calculations involving a diverse collection of disciplines, but scientists finally managed to get it right. Planetary scientists think that this discovery is a major step beyond what we understood about this moon before. This discovery further demonstrates the kind of deep-dive discoveries that can be made with long-lived orbiter missions (such as Cassini) to other planets.

Postberg et al. (2011) found that the plume on Enceladus is compositionally stratified, and that the existence of a saltwater reservoir below the tiger stripes is responsible for this stratification. In a contemporary study, Hansen et al. (2011) determined the composition and structure of the Enceladus plume. Based on Cassini observations and models, Nimmo et al. (2014) put forward the idea of tidally modulated eruptions on Enceladus.

The searches for sodium salts point to the existence of a subsurface ocean on Enceladus (see Spencer, 2009). Do the spectacular plumes of water vapor and ice particles seen on Enceladus come from liquid water just below its frigid surface? That is the fascinating question addressed by Postberg et al. (2009) using data from the Saturn-orbiting Cassini spacecraft, and by Schneider et al. (2009) using ground-based telescopes. It is an accepted view that the plume of water vapor and ice particles emitted from the south pole of Enceladus are the dominant source of Saturn's E ring.

An in-situ analysis of these particles had already concluded that the minor organic or siliceous components, identified in many ice grains, could be evidence of interaction between Enceladus' rocky core and liquid water. It was not clear, however, whether the liquid is still present today or whether it has frozen. Postberg et al. (2009) reported the identification of a population of E-ring grains that are rich in sodium salts (~0.5–2% by mass), which can arise only if the plumes originate from liquid water. The abundance of various salt components in these particles, as well as the inferred basic pH, exhibits a compelling similarity to the predicted composition of a subsurface Enceladus ocean in contact with its rock core. The plume vapor is expected to be free of atomic sodium. Thus, the absence of sodium from optical spectra is in good agreement with the results obtained by Postberg et al. (2009).

The presence of water in the plume hinged on either of the two sources: (1) an internal ocean, and (2) ice (warmed, melted, or crushed by tectonic motions). Sodium chloride (that is, common salt) is expected to be present in a long-lived ocean in contact with a rocky core. Schneider et al. (2009) reported a ground-based spectroscopic search for atomic sodium near Enceladus that places an upper limit on the mixing ratio in the vapor plumes orders of magnitude below the expected ocean salinity. The low sodium content of escaping vapor, together with the small fraction of salt-bearing particles, argues against a situation in which a near-surface geyser is fueled by a salty ocean through cracks in the crust. The lack of observable sodium in the vapor is consistent with a wide variety of alternative eruption sources, including a deep ocean, a freshwater reservoir, or ice. However, Schneider et al. (2009) cautioned that the existing data might be insufficient to distinguish between these hypotheses.

As just mentioned, Cassini had spotted jets of liquid spewing from the moon's surface and scientists followed it up with a series of discoveries about the materials gushing from the warm fractures near its south pole. Based on initial analysis of Cassini data, in 2014, scientists announced strong evidence for a lens-shaped underground regional sea only near the moon's South Polar Region. However, gravity data collected during the spacecraft's several close passes over the South Polar Region lent support to the possibility that the sea might be global. Subsequent new results, derived using an independent line of evidence based on Cassini's images, confirmed this inference to be true.

Cooper et al. (2009) have provided an explanation for forces that could generate the Enceladus geysers that spew water vapor into space. According to them the water molecules and other associated molecules are possibly reaching the Enceladus surface through some form of "cryovolcanism." Note that cryovolcanism is some kind of eruptive or intrusive processes analogous to volcanism occurring at low temperatures in the icy crusts of certain planets and moons, typically involving liquid or gaseous water, methane, nitrogen, etc. To be more specific, "cryovolcanism" is the icy volatile mechanics which can produce subsurface volatile gases that can explode outward and create plumes as a result of chemical reaction between chemically altered icy grains and icy contaminants such as ammonia, methane, and other hydrocarbons (see Cooper et al., 2009). Energetic particles raining down from Saturn's magnetosphere—at Enceladus, mostly electrons from Saturn's radiation belts—can break up molecules within the surface. This process is called radiolysis. Like a process called photolysis, in which sunlight can break apart molecules in the atmosphere, energetic radiation from charged particles that hit an icy surface, like that of Enceladus, can cause damage to molecules within the ice. These damaged molecules can get buried deeper and deeper under the surface by the perpetual churning forces that can repave the icy surface. Meteorites constantly crashing into the surface and splashing out material might also be burying the molecules (https://www.nasa.gov/mission_pages/cassini/whycassini/saturn201 00701.html). When chemically altered icy grains come into contact beneath the surface with icy contaminants such as ammonia, methane and other hydrocarbons, they can produce volatile gases that can explode outward. Such gases can create plumes of the size seen by Cassini. Cooper and colleagues call such icy volatile mechanics "cryovolcanism."

Cooper et al. (2009) have reported a model (the "Old Faithful" model, after the Old Faithful geyser in Yellowstone National Park) on how chemically driven cryo-volcanism might contribute to episodic outgassing at the icy moon Enceladus and potentially elsewhere including Europa and Kuiper Belt Objects. What is unique about the "Old Faithful" model is that it is a model for cryo-volcanism that is based on not only liquid water, but also requires the production of gases by the radiolytic chemistry observed at Enceladus. These investigators explain that "exposed water ices can become oxidized from radiolytic chemical alteration of near-surface water ice by space environment irradiation. In contact with primordially abundant reductants such as NH_3, CH_4, and other hydrocarbons, the product oxidants can react exothermically to produce volatile gases driving cryo-volcanism via gas-piston forces on any subsurface liquid reservoirs. Radiolytic oxidants such as H_2O_2 and O_2 can continuously accumulate deep in icy regoliths and be conveyed by rheological flows to subsurface chemical reaction zones over million-year time scales indicated by cratering ages for active regions of Enceladus and Europa. Surface blanketing with cryovolcanic plume ejecta would further accelerate regolith burial of radiolytic oxidants. Episodic heating from transient gravitational tides, radioisotope decay, impacts, or other geologic events might occasionally accelerate chemical reaction rates and ignite the exothermic release of cumulative radiolytic oxidant energy." Note that in this context, "rheological flows" means viscoelastic flows.

In the model experiments carried out by Cooper et al. (2009), gas pressure slowly builds up inside Enceladus, then gets released occasionally in geyser-like eruptions. Unlike terrestrial geysers, or even geyser-like forces on Jupiter's moon Io, the model proposed by Cooper et al. (2009) shows that charged particle radiation raining down from Saturn's magnetosphere can create the forces from below the surface that is required to eject gaseous jets. According to Cooper et al. (2009), the time history for the suggested model of radiolytic gas-driven cryo-volcanism at Enceladus and elsewhere consists of long periods of chemical energy accumulation punctuated by much briefer episodes of cryovolcanic activity.

The Ion and Neutral Mass Spectrometer (INMS) aboard the Cassini spacecraft made measurements of the gaseous composition of the plume at altitudes of greater than 175 km and the Infrared Mapping Spectrometer (VIMS) determined that Enceladus' surface is composed mostly of nearly pure water ice, except near its south pole, where there are light organics, CO_2, and water ice, particularly in the "tiger stripes" region (Brown et al., 2006). Besides H_2O, the INMS on-board Cassini probe detected methane (CH_4), carbon dioxide (CO_2), ammonia (NH_3), molecular nitrogen (N_2), and molecular hydrogen (H_2) in the plume (Waite et al., 2017). In addition, carbon monoxide (CO) and ethane (C_2H_4) were found among other substances with moderate ambiguity (Waite et al., 2006; Waite et al., 2009; Bouquet et al., 2015; Magee and Waite Jr., 2017).

The inferred heat fluxes emanating from the tiger stripes were estimated to have been about 100 mW/m^2 for the same region (Tyler, 2009). Previous compositional analyses of particles injected by the plume into Saturn's diffuse E ring have already indicated the presence of liquid water. Postberg et al. (2011) reported an analysis of the composition of freshly ejected particles close to the sources. They found that salt-rich ice particles dominate the total mass flux of ejected solids (more than 99%) but they are depleted in the population escaping into Saturn's E ring. It was found that ice grains containing organic compounds are more abundant in dense parts of the plume.

Although the mechanisms driving the plume emission continued to be debated, Postberg et al. (2011)'s investigations compelled them to conclude that a salt-water reservoir with a large evaporating surface hiding beneath the tiger stripes is the source of a compositionally stratified plume on Enceladus. Their studies strongly suggest that such a salt-water reservoir provides nearly all of the matter in the plume.

Studies on Enceladus, carried out by several investigators (e.g., Postberg et al., 2009; Tyler, 2009; Tyler, 2011; Iess et al., 2014; Chen et al., 2014; Tyler, 2014; Bouquet et al., 2015), established the presence of a south polar subsurface ocean of about 10 km thickness located beneath an ice crust 30 to 40 km thick and above a rocky core and extending up to south latitudes of about 50°. Postberg et al. (2011) found that Enceladus's subsurface ocean is a salt-water reservoir, and that such a reservoir is the source of a compositionally stratified plume on Enceladus. Observational data from the Cassini spacecraft are used to make a chemical model of ocean water on Enceladus. Glein et al. (2015) reported the pH of Enceladus' ocean; their model suggests that Enceladus' ocean is a Na–Cl–CO_3 solution with an alkaline pH of ~11–12. Thomas et al. (2016) theorized that Enceladus's measured physical libration requires that Enceladus's subsurface ocean is not merely a regional one below its tiger stripes, but rather a 26–31 km deep global salty water ocean.

It was found that slow dust is generated in Enceladus' plume from condensation and wall collisions in tiger stripe fractures (Schmidt et al., 2008). Spectra of Cassini's dust detector CDA (Postberg et al., 2009) allowed determination of the dust constituent in the plume. Kempf et al. (2010) also found that the Enceladus plume is dusty, and provided an explanation as to how the dust plume feeds Saturn's E-ring. Detection and measurement of ice grains and gas distribution in the Enceladus plume were carried out by Cassini's ion neutral mass spectrometer (Teolis et al., 2010). Ingersoll and Pankine (2010) reported subsurface heat transfer on Enceladus, and the conditions under which melting occurs. Matson et al. (2012) reported a hypothesis for bringing both heat and chemicals to the surface of Enceladus. Perry et al. (2015) determined Enceladus plume density using Cassini ion and neutral mass spectrometer (INMS) measurements (Waite, Jr et al., 2004). Based on model parameter insights from Cassini-INMS, Magee and Waite Jr. (2017) determined the neutral gas composition of Enceladus' plume.

Since these discoveries, planetary- and life-scientists expended much focus on this tiny icy moon, as a result of which Enceladus became well-known for sustained eruptions consisting of tall plumes of gas, jets of water vapor, and organic-enriched salty ice particles spewing from the tiger stripes on its south pole (see Hansen et al., 2011; Dong et al., 2011; Spitale et al., 2015). Spitale et al. (2015) described the gas plumes and jets of water vapor and ice particles forcibly and copiously gushing out from Enceladus' south-polar terrain as "curtain eruptions". Yeoh et al. (2015) examined and attempted to understand the physics of the Enceladus south polar plume via numerical simulation. Looking from an astrobiological perspective and considering the implications for future missions to Enceladus, Porco et al. (2017) suspected that the sustained plume detected on the south pole of Enceladus could be snowing microbes on Enceladus. Based on an assessment of the Enceladen plume, McKay et al. (2014) pointed to the possibility of the habitability of Enceladus.

Each of the four eruptive fissures is flanked by < 1-km-wide belts of endogenic (i.e., occurring beneath the surface of the earth) thermal emission (10^4 W/m for the ~500-km total tiger stripe length), a one-to-one correspondence indicating a long-lived internal source of water and energy (Nimmo and Spencer, 2013). The tiger stripe region is tectonically resurfaced, suggesting an underlying mechanism accounting for both volcanism and resurfacing, as on Earth (Kite and Rubin, 2016). Enceladus' tiger stripes have been erupting continuously since their discovery in 2005 (Hansen et al., 2011; Dong et al., 2011; Porco et al., 2014; Spitale et al., 2015). The spray of icy particles from the surface jets forms a towering plume. Eruptions on Enceladus provide access to materials from Enceladus' ocean. Enceladus' 30 ± 10-km-thick ice shell is probably underlain by an ocean or sea of liquid water; and Enceladus' plume samples a salty liquid water reservoir containing ^{40}Ar, ammonia, nanosilica, and organics (Postberg et al., 2011; WaiteJr. et al., 2009; Iesset al., 2014; Hsuet al., 2015). According to McKay et al. (2014), the sustainability of water eruptions on Enceladus affects the moon's habitability, as well as astrobiology, and therefore assumes special interest.

Spitale et al. (2015) showed that much of the eruptive activity can be explained by broad, curtain-like eruptions. According to them, optical illusions in the curtain eruptions resulting from a combination of viewing direction and local fracture geometry produce image features that were probably misinterpreted previously as discrete jets. Spitale et al. (2015) reported maps of the total emission along the fractures, rather than just the jet-like component, for five times during an approximately 1-year period in 2009 and 2010. It was suggested that an accurate picture of the style, timing, and spatial distribution of the south-polar eruptions is crucial to evaluating theories for the mechanism controlling the eruptions.

Despite a decade of intense research the mechanical origin of the tiger-stripe fractures (TSF) and their geologic relationship to the hosting South Polar Terrain (SPT) of Enceladus remain poorly understood. Yin and Pappalardo (2015) showed via a systematic photo-geological mapping that the semi-squared SPT is bounded by right-slip, left-slip, extensional, and contractional zones on its four edges. Discrete deformation along the edges in turn accommodates translation of the SPT as a single sheet with its transport direction parallel to the regional topographic gradient. According to Yin and Pappalardo (2015)'s interpretation, this parallel relationship implies that the gradient of gravitational potential energy drove the SPT motion. In map view, internal deformation of the SPT is expressed by distributed right-slip shear parallel to the SPT transport direction. The broad right-slip shear across the whole SPT was facilitated by left-slip bookshelf faulting along the parallel tiger-stripe fractures (TSF). Yin and Pappalardo (2015) suggested that the flow-like tectonics, to the first approximation across the SPT on Enceladus, is best explained by the occurrence of a transient thermal event, which allowed the release of gravitational potential energy via lateral viscous flow within the thermally weakened ice shell.

Tidal heating has been described as one of the reasons for the presence of a liquid water subsurface ocean on Enceladus (Meyer and Wisdom, 2007). Tyler (2011) found that tidal dynamical considerations constrain the state of an ocean on Enceladus. Investigations carried out by Běhounková et al. (2012) indicated that tidally-induced melting events are responsible for the origin of south-pole activity on Enceladus. Based on Cassini observations and models, Nimmo et al. (2014), showed that the eruptions on Enceladus is tidally modulated. Based on theoretical investigations in which Enceladus's subsurface ocean is modeled as a constant-pressure bath, Kite and Rubin (2016) showed that the erupted flux from Enceladus varies on diurnal timescales, which they attribute to daily flexing of the source fissures from tidal stresses induced by Saturn in Enceladus's subsurface ocean. Just like the primary driving force for the generation of tides in Earth's oceans arise from tidal stresses induced by Earth's Moon (and partly from the distant Sun), the primary driving force for the generation of tides in Enceladus's subsurface ocean is derived from Saturn, around which Enceladus orbits. It may be noted that Enceladus is tidally locked with Saturn, keeping the same face toward the planet. It completes one orbit every 32.9 h within the densest part of Saturn's E Ring. Because the orbit of Enceladus around Saturn is considerably (in relative terms) far from Sun, the tide-generating force acting on Enceladus is primarily contributed by Saturn. It may be noted that in Earth's oceans, "flood" is the tidal phase during which the tidal current is flowing inland (flood current), and "ebb" is the tidal phase during which the tidal current is flowing seaward (ebb current). In short, "flood" is the tidal phase during which the water level is rising (flooding the seacoast), and "ebb" is the tidal phase during which the water level is falling. Because the upper boundary (i.e., sea surface) of Earth's oceans is a highly compressible atmosphere, the tides in Earth's oceans are free to rise and fall as dictated by the tide-generating force. However, the upper boundary of Enceladus's subsurface ocean is a rather rigid and noncompressible icy crust, and therefore the tides in Enceladus's subsurface ocean are constrained to rise and fall through the vertical slots in the tiger stripes. Based on the just-mentioned reasons, the tidal flexing encountered by Enceladus's subsurface ocean is compelled to drive vertical flow in slots underneath the source fissures (see Figs. 13.9 and 13.10). According to Kite and Rubin (2016), such a vertical flow generates heat through viscous dissipation. This heat helps to maintain the slots against freeze-out despite strong evaporitic cooling by vapor escaping from the water table. The vapor ultimately provides heat (via condensation) for the envelope of warm surface material bracketing the tiger stripes.

Kite and Rubin (2016)'s theoretical investigation reveal that although the amplitude of the cycle in water table height is reduced when the slot is hydrologically connected to the ocean relative to a hypothetical situation where the slot is isolated from the ocean, the flow velocity, driven by the deviation of the water table from its equilibrium elevation, is very much larger than in the hydrologically isolated case.

Kite and Rubin (2016) explain that the observed eruptions arise from turbulent dissipation in the tiger stripes, which extend to the water table lying below Enceladus's south pole. They argue that in tune with the rhythmic rise and fall of water level induced by the tidal cycle in the ocean below, the water table lying below the tiger stripes rises and falls periodically. Thus, depending on the phase of the tidal cycle in Enceladus's subsurface ocean, the water in the tiger stripes is flushed outwards and inwards periodically (the slots of the tiger stripes widen during flood tide, and become narrow during ebb tide). Thus, the eruptions through the tiger stripes get enhanced and diminished periodically in tune with the tidal cycle (but the eruptions never cease).

13.5 Maintaining liquid oceans inside cold planets and moons—role of tidal heating

Tiny, icy Enceladus, at a mere 500 km in diameter, is planet Saturn's sixth-largest moon. *Enceladus* is Saturn's second nearest major satellite. Of the hitherto known icy moons in the solar system (Europa, Ganymede, Callisto, Enceladus, and Titan), Enceladus allows for the deepest ocean where the transition from obliquity to eccentricity tidal flow dominance occurs (Tyler, 2009). Furthermore, Enceladus is one of only three solid planetary bodies in the solar system with an internal heat source large enough to have been detected by remote sensing (Spencer et al., 2006; Porco, 2006). The other two—Earth, and Jupiter's moon Io—are much larger than Enceladus. With a diameter of only 500 km, the volume of Enceladus is so small that a radiogenic source for the observed heat flux is ruled out (Tyler, 2009). *Enceladus* shows higher heat loss than expected and a wide range of surface ages. Numerical simulations indicate that occasional catastrophic overturn events could be responsible for both observations by recycling portions of the icy lid to the interior, which would cause transiently enhanced heat loss.

The usual calculations of tidal heat depend on the orbital eccentricity of the moon, but the eccentricity of Enceladus (0.0047) is relatively small (Tyler, 2009). According to Meyer and Wisdom (2007), the neighboring moon Mimas has a larger eccentricity (0.020) and for similar composition should receive about eleven times as much tidal heating as Enceladus. Mimas, however, is cool and geologically dead. An ocean layer under the ice of Enceladus has long been suspected as the source of Saturn's E-ring.

Several proposals have been put forth to describe how tidal stresses on an uncoupled ice shell can, under restrictive assumptions, lead to elevated heat fluxes similar to those observed (Nimmo et al., 2007), but a consensus has not been reached and the heat source of Enceladus remains a mystery. An explanation must provide enhanced heat fluxes over the broad region south of about 55° latitude, an average of about 100 W/m^2 south of 65°, and peak local heat fluxes should associate with the four prominent troughs near the south pole dubbed the "tiger stripes" (Tyler, 2009). The South Pole of the icy moon Enceladus is anomalously warm, geologically youthful, and cryo-volcanically active. Episodic convective overturn explains how the moon's modest sources of internal heat can be channeled into intense geological activity (see Helfenstein, 2010).

Data from recent space missions have added strong support for the idea that there are subsurface liquid oceans on several icy moons of the outer planets in the solar system. But given the extremely cold surface temperatures and meagre radiogenic heat sources of these moons, how these oceans remain liquid was a matter of debate for quite some time in the past. On Earth, about 25%–30% of the total tidal dissipation occurs in the deep ocean. In the modern view, significant tidal dissipation in Earth's oceans is generated by deep-ocean tidal flow as it crosses rough bathymetry (seafloor topography) and transfers energy to internal waves (see Joseph, 2016) that eventually lose this energy to smaller-scale dissipative processes (heat).

Nimmo et al. (2007) showed that a likely explanation for the heat and vapor production is shear heating by tidally driven lateral (strike-slip) fault motion with the displacement of approx. 0.5 m over a tidal period. The vapor produced by this heating may escape as plumes through cracks reopened by the tidal stresses (tidal flexing). It has been estimated that the ice shell thickness needed to produce the observed heat flux is at least 5 km. It was further found that the tidal displacements required imply a Love number of $h_2 > 0.01$, suggesting that the ice shell is decoupled from the silicate interior by a subsurface ocean. Note that the Love numbers h, k, and l are dimensionless parameters that measure the rigidity of a planetary body and the susceptibility of its shape to change in response to a tidal potential.

Parkinson et al. (2008) discussed evidence for surface/ocean material exchange on Enceladus based on the amounts of silicate dust material present in the Enceladus' plume particles. Microphysical cloud modeling of Enceladus' plume carried out by these investigators shows that the particles originate from a region of Enceladus' near surface where the temperature exceeds 190 K. According to them this could be consistent with a shear-heating origin of Enceladus' tiger stripes, which would indicate extremely high temperatures (approximately 250–273 K) in the subsurface shear fault zone, leading to the generation of subsurface liquid water, chemical equilibration between surface and subsurface ices, and crustal recycling on a time scale of 1 to 5 Myr.

Another school of thought pertaining to interior heating is that when an object is in an elliptical orbit, tidal forces that flex the solid moon (rock plus ice) during its eccentric orbit generate internal friction which heats its interior. The prevailing conjecture is that this heat entering the ocean does not rapidly escape because of the insulating layer of ice over the ocean surface. However, Tyler (2008) came forward with another mechanism in which the moon's obliquity (spin axial tilt of the moon with respect to its orbital plane) plays an important additional role in generating heat. Tyler (2008) showed that a subdominant and previously unconsidered tidal force due to obliquity has the right form and frequency to resonantly excite large-amplitude Rossby waves in these oceans and that the strong tidal dissipation in the liquid oceans arising from such large-amplitude waves can generate sufficient heat to prevent the liquid oceans from solidifying. It is believed that tidal heating is responsible for the geologic activity of the most volcanically active body in the solar system.

As indicated above, tidal forces on the outer moons in the solar system come in at least two different types. The one considered in all the early work involving tidal-heating effects is due to the eccentricity of the moon's orbit around the planet. A second tidal force is due to the moon's obliquity (the axial tilt of the moon's spin axis relative to its orbital plane). According to Tyler (2009), the reason for the previous focus on the eccentricity tidal forces, and the consequent neglect of obliquity, has surely been because the eccentricity tidal forces are typically much larger. Tyler (2009) further explains that the response of the dynamical system (the ocean) is quite different in each case, and that "Technically, the important difference is that the obliquity tidal forces include a component with a degree-two, order-one spherical-harmonic spatial form propagating westward around the moon with a diurnal frequency; while the eccentricity tidal forces do not have such a component."

Tyler (2008; 2009) suggested that obliquity tides could drive large-scale flow in the oceans of Europa and Enceladus, leading to significant heating. A critical unknown in previous work is what the tidal quality factor, Q, of such an ocean should be. The corresponding tidal dissipation spans orders of magnitude depending on the value of Q assumed.

To address the issue of tidal heating, Chen et al. (2014) adopted an approach employed in terrestrial ocean modeling, where a significant portion of tidal dissipation arises due to bottom drag, with the drag coefficient O (0.001) being relatively well-established. From numerical solutions to the shallow-water equations including nonlinear bottom drag, they obtained scalings for the equivalent value of Q as a function of this drag coefficient. In addition, they provided new scaling relations appropriate for the inclusion of ocean tidal heating in thermal–orbital evolution models. Their approach is appropriate for situations in which the ocean bottom topography is much smaller than the ocean thickness.

Using these novel scalings, Chen et al. (2014) calculated the ocean contribution to the overall thermal energy budgets for many of the outer solar system moons. Although uncertainties such as ocean thickness and moon obliquity remain, they found that for most moons it is unlikely that ocean tidal dissipation is important when compared to either radiogenic heating or solid-body tidal heating. Chen et al. (2014) reckon that of known moon, Neptune's moon Triton is the most likely icy moon to have ocean tidal heating play a role in its present-day thermal budget and long-term thermal evolution.

In another study, Matsuyama (2014) extended previous theoretical treatments for ocean tidal dissipation by taking into account the effects of ocean loading, self-attraction, and deformation of the solid regions. It was found that these effects modify both the forcing potential and the ocean thicknesses for which energy dissipation is resonantly enhanced, potentially resulting in orders of magnitude changes in the dissipated energy flux. According to Matsuyama (2014), "Assuming a Cassini state obliquity, Enceladus' dissipated energy flux due to the obliquity tide is smaller than the observed value by many orders of magnitude. On the other hand, the dissipated energy flux due to the resonant response to the eccentricity tide can be large enough to explain Enceladus' observed heat flow."

It has been shown in previous studies that ocean tides, if resonantly forced, can supply heat at or exceeding the rates necessary for maintaining liquid oceans on icy moons in the outer solar system. Tyler (2014) extended from the previous work and sought to examine the full set of dynamically-consistent ocean tidal solutions to describe why do some of the icy moons in the outer solar system have oceans and others do not? Tyler (2014)'s study found that even with no other sources of heat, a liquid ocean on many of these moons would be maintained by ocean tidal heat because the process of freezing (which changes the thickness of the remaining liquid ocean and thereby the Eigen modes) would push the ocean into a resonant configuration, with the associated increase in heat production preventing further freezing and stabilizing the configuration. Tyler (2014)'s study indicated that an ocean on Io (Jupiter's moon; the most volcanically active world in the solar system) or Mimas (the smallest and innermost of Saturn's major moons) would suffer extreme tides (with heat generated exceeding 1 W/m^2) unless an implausibly large volume of water were present to lift the eigen modes of the configuration out of resonance with the tidal forces.

Baland et al. (2016) investigated the influence of an internal subsurface ocean and of tidal deformations of the solid layers on the obliquity of Enceladus. Their Cassini state model takes into account the external torque exerted by Saturn on each layer of the moon and the internal gravitational and pressure torques induced by the presence of the liquid layer. As a new feature, their model also includes additional torques that arise because of the periodic tides experienced by the moon. They found that the upper limit for the obliquity of a solid Enceladus is 4.5×10^{-4} degrees and is negligibly affected by elastic deformations. They found that the presence of an internal ocean decreases this upper limit by 13.1%, elasticity attenuating this decrease by only 0.5%. Because the obliquity of Enceladus cannot reach Tyler's requirement, obliquity tides are unlikely to be the source of the large heat flow of Enceladus. According to Baland et al. (2016), more likely, the geological activity at Enceladus' South Pole results from eccentricity tides. They determined that even in the most favorable case, the upper limit for the obliquity of Enceladus corresponds to about 2 m at most at the surface of Enceladus. This is well below the resolution of Cassini images. Baland et al. (2016) suggested that control point calculations cannot be used to detect the obliquity of Enceladus, let alone to constrain its interior from an obliquity measurement.

The thermal and mechanical evolution of icy moons—the physics behind creating and sustaining a subsurface water ocean—depends almost entirely on the mechanical dissipation of tidal energy in ice to produce heat, the mechanism(s) of which remain poorly understood. The general consensus among planetary scientists is that dissipation of tidal energy is an important mechanism for the evolution of outer solar system moons, several of which are known to contain subsurface or internal oceans. The power generated in the core from tidal friction creates high temperatures and thermal gradients in the ocean (Choblet et al., 2017). Although tidal heating was suspected to play a role in maintaining a subsurface liquid ocean on several icy moons, it remained unclear whether tidal dissipation in a global liquid ocean can represent a significant additional heat source.

13.6 Eruptions in the vicinity of Enceladus' south pole—role of tidally driven lateral fault motion at its south polar rifts

While Cassini spacecraft had a close encounter with Enceladus (see Fig. 13.8) on July 14, 2005, planetary scientists first detected remarkable water-rich active, vast plumes of vapor erupting from geysers near the moon's south pole. This was one of the most unexpected discoveries from this mission. Observations of the South Pole revealed localized large rifts in the south-polar terrain, named Alexandria, Baghdad, Cairo, and Damascus Sulci, informally called "tiger stripes."

Parkinson et al. (2007) reviewed the probable reasons for the higher temperatures exhibited by the "tiger stripes" region (which was found to be the source of the observed eruptions) relative to the surrounding terrain. It had been noted that the global temperature maps made by the Composite Infrared Spectrometer (CIRS) show that the "tiger stripes" are as much as 25 K warmer than the surrounding regions (Spencer et al., 2006). This warmth is anomalously higher than predictions from solar heating models, consistent with heat escaping from an internal heating source in this region (Burger et al., 2005). It was found that a thermal evolution model, developed by Matson et al. (2005) for Enceladus as a function of its time of formation, using radiogenic species and tidal dissipation can raise the core temperature of Enceladus in the vicinity of 1000 K. This allows for the presence of a liquid layer at the interface between the rocky core and an icy mantle.

The "tiger stripes" fractures have been found to be the sources of the observed jets of water vapor and icy particles. Subsequent observations have focused on obtaining close-up imaging of this region to better characterize these emissions. Those newer datasets have been examined; and triangulation of discrete jets has been used to produce maps of jetting activity at various times. Evidence from multiple instruments onboard the spacecraft show that there is a large plume of water vapor and particles emanating from Enceladus' South Polar Region. The moon's weak gravitational field prevents the retention of an atmosphere, which indicates that this gas is likely the result of some currently active venting geothermal process (Parkinson et al., 2007).

Hurford et al. (2007) reported a mechanism in which temporal variations in tidal stress open and close the tiger-stripe rifts, governing the timing of eruptions. It was found that during each orbit, every portion of each tiger stripe rift spends about half the time in tension, which allows the rift to open, exposing volatiles, and allowing eruptions. In a complementary process, periodic shear stress along the rifts also generates heat along their lengths, which has the capacity to enhance eruptions. According to Hurford et al. (2007), plume activity is expected to vary periodically, affecting the injection of material into Saturn's E ring and its formation, evolution, and structure. Hurford et al. (2007) argue that "the stresses controlling eruptions imply that Enceladus' icy shell behaves as a thin elastic layer, perhaps only a few tens of kilometers thick."

In another study of the same topic, Nimmo et al. (2007) showed that the most likely explanation for the heat and vapor production is shear heating by tidally driven lateral (strike-slip) fault motion with displacement of approximately 0.5 m over a tidal period. They argued that the vapor produced by this heating may escape as plumes through cracks reopened by the tidal stresses. They estimated that the ice shell thickness needed to produce the observed heat flux is at least 5 km. According to their calculation, the tidal displacements required imply a Love number of $h_2 > 0.01$, suggesting that the ice shell is decoupled from the silicate interior by a subsurface ocean. Nimmo et al. (2007) predicted that the tiger-stripe regions with the highest relative temperatures will be the lower-latitude branch of Damascus, Cairo around 60°W longitude, and Alexandria around 150°W longitude.

13.7 Presence of macromolecular organic compounds in the subglacial water-ocean of Enceladus

"Do conditions favorable to the formation of life occur elsewhere in the solar system?" is a query every space enthusiast throws. A hydrological cycle (i.e., processes that describe the constant movement of water above, on, and below Enceladus' surface) governing the weathering of rocks by liquid water at the rock/liquid-water interface and any concomitant radioactive emissions or other geothermal energy sources are possible incipient conditions for life (Hartman et al., 1993). Parkinson et al. (2007) have discussed the evidence for this on Enceladus, based on the materials that gush out from Enceladus's interior through the plume jetting out from the tiger stripes existing on the south pole of Enceladus. During the hydrologic cycle, the water undergoes a continual change of state between liquid, solid, and gas in four distinct processes. These processes are: (1) evaporation and venting; (2) precipitation; (3) infiltration and resurfacing; and (4) runoff (Parkinson et al., 2007).

Parkinson et al. (2007) have discussed the search for signatures of species and organics in the Cassini Ultraviolet Imaging Spectrograph (UVIS) spectra of the plume and implications for the possible detection of life. Parkinson et al. (2007) estimated the flux of H_2O to be about 2000 times that of meteoritic material that impinge on Enceladus's icy surface. According to their calculations, the flux of water must exceed 2×10^{-13} g/cm^2/s. Subsequent resurfacing of the southern polar region due to plume material venting from the "tiger stripes" will eventually cause sinking of micrometeorite material down to the rock/ice interface.

Hydrogen peroxide (H_2O_2) is a chemical compound, which in its pure form, is a very pale blue liquid, slightly more viscous than water (H_2O). Hydrogen peroxide is used as an oxidizer, bleaching agent, and antiseptic. Some hydrogen peroxide (H_2O_2) is being produced in the surface ice of Enceladus. Additionally, one can introduce the notion of Saturn's E-ring as a "chemical processor" to provide additional H_2O_2 (Parkinson et al., 2007). On Enceladus, the H_2O sputtering flux (Jurac et al., 2002; Richardson and Jurac, 2004) is $\sim 10^9$ molecules/cm^2/s, implying that the production rate of H_2O_2 at Enceladus would be of the same magnitude as that on Europa ($\approx 10^{11}$ molecules/cm^2/s). Hydrothermal alteration and weathering could drive the equilibration of a liquid water and silicate crust chemistry (Gaidos et al., 1999).

Water is being ejected from Enceladus in sufficient quantities to keep this ring in existence and fairly stable over time (Hansen et al., 2006). This frozen water in the ring will be exposed to energetic particles and UV irradiance over time and some of the water will be converted to H_2O_2 which can be picked up again by the moon as it sweeps through the ring while it is moving in its orbit. In this manner, the E-ring can be seen as acting as an extended "surface" providing a new oxidant source for Enceladus (Parkinson et al., 2007).

Organic compounds can be identified using mass spectrometry (Silverstein et al., 2005; Dass, 2007). Organic compounds and other minerals present in Saturn's E-ring and Enceladus's plume were analyzed using time-of-flight mass spectra of ions in plasmas produced by hypervelocity impacts of organic and mineralogical microparticles on a cosmic dust analyzer (CDA) (Srama et al., 2004; Hillier et al., 2006; Srama et al., 2011) onboard Cassini spacecraft which conducted deep dives through Saturn's E-ring and Enceladus's plumes (see Goldsworthy et al., 2003; Srama et al., 2009). An interesting finding was the presence of High Mass Organic Cations (HMOC), which are organic species with an increasing number of carbon atoms (C_7–C_{15}), in Saturn's E-ring. The first of HMOC spectra were already detected in 2004 and early 2005, before Cassini's first close flyby of Titan.

Parkinson et al (2007) suspected that the extraterrestrial micrometeoroids flux, similar to those found at Saturn, will form a source of organic material for Enceladus. Cassini's mission ended in September 2017, when scientists sent it plummeting into Saturn, but researchers are still poring over the data it beamed back to Earth. Data from the NASA-ESA Cassini-Huygens mission indicate that Enceladus has complex organic molecules floating in its global ocean. A new analysis of those data finds that Enceladus not only has small organic molecules but also larger, more complex carbon compounds bursting from its ocean.

The CDA records the Time-of-Flight (TOF) mass spectra of cations generated by high-velocity impacts of individual grains onto a rhodium target with a mass resolution of $m/\Delta m \approx$ 20–5014. Postberg et al. (2009) noted that the detection probability increases with impact speed, increasing the available energy for fragmentation and ionization. The maximum mass nominally covered by CDA is at about 200u. Previous CDA measurements by several investigators (e.g., Postberg et al., 2008; Postberg et al., 2009; Postberg et al., 2011) showed that about 25% of E-ring ice grain spectra, so-called Type 2 spectra, exhibit organic material.

With the support of two mass spectrometers aboard the Cassini spacecraft, the Ion and Neutral Mass Spectrometer (INMS) and the Cosmic Dust Analyzer (CDA), Kempf et al. (2010) made compositional in situ measurements inside both the Enceladus's plume and Saturn's E-ring; the latter's formation is by virtue of ice grains escaping Enceladus' gravity. The INMS identifies compounds based on mass-to-charge ratio. The CDA can detect compounds that are more than 10 times more massive than the INMS can.

The mechanism by which the ascending gas bubbles in Enceladus's subsurface ocean, which lies beneath the tiger stripes at Enceladus's south pole, efficiently transport water vapor, organic aerosols (Porco et al., 2017), and salty water droplets into water-filled cracks in the south polar ice crust is an interesting subject matter. Organics ultimately concentrate in a thin organic layer on top of the water table located inside the icy vents. When gas bubbles burst, they form aerosols made of insoluble organic material that later serve as efficient condensation cores for an icy crust from water vapor thereby forming HMOC-type particles. In parallel, larger, pure salt-water droplets form (blue), which freeze and are later detected by CDA as salt-rich Type 3 ice particles in the plume (Postberg et al., 2009; Postberg et al., 2011).

By performing INMS and CDA measurements during Cassini spacecraft's Enceladus plume dives and detailed analyses of materials emerging from the subsurface of Enceladus, Postberg et al. (2018) reported discovery of complex organic molecules in the plumes emanating from the subglacial water-ocean of Enceladus. Grains of organic species with an increasing number of carbon atoms have been detected in the plume (Fig. 13.13) and in the E ring at a wide range of Saturnian distances, except in its outermost fringe (Srama et al., 2011). However, the fraction of such grains in the E ring increases closer to Enceladus orbit.

Analyses by Postberg et al. (2018) indicated that an Enceladus origin of these macromolecules is evident. During Cassini's E17 flyby one freshly ejected HMOC type ice grain was observed inside the plume. Furthermore, the Cosmic Dust Analyzer (CDA), which was a mass spectrometer aboard the Cassini spacecraft, observed many HMOC at the locations close to Enceladus' orbit, where most grains have been ejected from the plume only a few days to few months ago (Postberg et al., 2018). The chemical contents found in Saturn's E-ring, as well as during Cassini spacecraft's deep Enceladus plume dives, shed light on the chemical composition of the plume material gushing out from the tiger stripes on Enceladus's south pole, and therefore the chemical composition of the subsurface ocean lying below Enceladus's south pole. Detailed analyses of the spectral data gathered during Cassini's deep Enceladus plume dives enabled Postberg et al. (2018) to the deduction of an organic enriched layer at the Enceladean water table.

According to Postberg et al. (2018), as a simple possibility, one can consider a primordial origin (i.e., inherited from the interstellar medium, and partly modified due to early

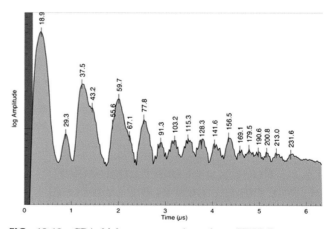

FIG. 13.13 **CDA high mass organic cations (HMOC) spectrum recorded in Enceladean plume. The impact speed was 8.6 km/s and the particle was about 2 μm in radius.** Note that organic species with an increasing number of carbon atoms (C_7 – C_{15}) is referred to as high mass organic cations (HMOC). *(Source: Postberg et al. (2018), Macromolecular organic compounds from the depths of Enceladus, Nature, 558: 564–568).*

hydrothermal processing on carbonaceous chondrite parent bodies) of Enceladan organic materials. These investigators hypothesise that "It may be the case that Enceladus's organic matter is similar to chondritic insoluble organic matter (IOM)" and argue that "If the rocky materials that were accreted by Enceladus were analogous to CI/CM/CR chondrites or refractory cometary solids, then a substantial organic inventory is inescapable". IOM in CI, CM, and CR classes is considered to be the most primitive IOM type. Postberg et al. (2018) further suggest that "a large accreted inventory of IOM-like material could establish the potential for rocks in Enceladus's core to serve as the basis of a more evolved organic factory, similar to oil and gas generating sedimentary basins on Earth. Hydrothermal activity could facilitate this process."

It has been suggested that the analogy to petroleum geochemistry on Earth implies that thermal processing of organic materials will inevitably produce some CH_4 accompanying more complex organics (Tissot and Welte, 1984). At Guaymas Basin, hydrothermal fluids have very high concentrations of CH_4 (~60 mmol/kg) (Von Damm et al., 2005). If the rocky core of Enceladus is also organic-rich and heated sufficiently, then this becomes a plausible scenario (Postberg et al., 2018).

The chemical analyzer (CA) is the subsystem of the CDA (Srama et al., 2004) that provides chemical information about an impacting dust particle. If a dust particle impacts the central rhodium target (chemical analyzer target—CAT) with sufficient energy the particle is totally vaporized and partly ionized, forming an impact plasma of target and particle ions, together with electrons and neutral molecules and atoms. The instrument separates this plasma. Depending on the operation mode, cations or anions, which pass through a skimmer, are analyzed in a reflectron Time of Flight (TOF) mass spectrometer. The CDA detected large organic molecules in particles from the plume, and these organics most likely contain dozens of carbon atoms. The researchers realized that the fast fly-bys had pulverized these molecules in the INMS, which appeared more clearly during a slower fly-by. Postberg et al. (2018) found that "low-speed flybys result in minimal fragmentation of the organic compounds within the ice grains. In contrast, high-speed impacts lead to fragmentation of large organic parent molecules beyond the INMS mass range of 99 u". Note that u represents atomic mass units. Postberg et al. (2018) postulated that "Species with molecular masses above 99 u in the plume are expected to be extremely depleted in the gas phase and have to reside in or on ice grains".

The spectrometer is sensitive to positive ions only. The mass resolution (m/Δm), derived from laboratory experiments with the instrument, depends on the atomic masses of the ions. Recording can be triggered up to several μs after the actual impact by the arrival of an abundant ion species. For water-dominated particles these generally are hydrogen cations (H^+) or hydronium ions (H_3O^+). The spectra are logarithmically amplified, digitized at 8 bit resolution, and sampled at 100 MHz for a period of 6.4 μs after the trigger. Because the TOF is proportional to the square root of the mass to charge ratio of ions, its spectrum in an ideal case also represents a mass spectrum for identical ion-charges (Postberg et al., 2018).

Postberg et al. (2018) found that a subgroup (≈ 3%) of Type 2 spectra is characterized by a sequence of repetitive peaks beyond 80u usually separated by mass intervals of 12u to 13u in most cases (Fig. 13.14).

It is quite natural that one might be curious about the organic molecules that are dissolved, dispersed, or suspended in the Enceladan aqueous body. Organic layers, whether liquid oils floating on the ocean's surface or organic solids sinking to the rocky core, would undergo chemical

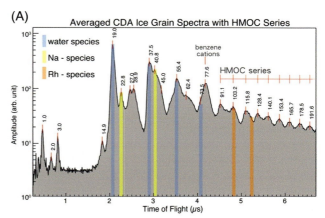

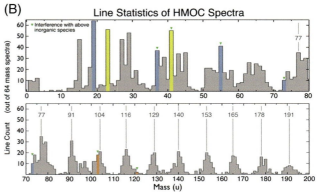

FIG. 13.14 Co-added CDA HMOC spectrum and mass line histogram (note that u represents atomic mass units). (A) Time-of-flight spectrum representing the average of 64 high quality spectra. The amplitudes of individual spectra were normalized and co-added. The spectrum provides a representation of the average abundance of cation species. (B) Occurrences ('counts') of resolved mass lines and "flank peaks" in 64 high quality HMOC spectra. In contrast to the spectrum shown in a), the histogram makes no distinction of the peak amplitude and just shows the frequency of occurrences of resolved mass lines. The overall pattern suggests hydrocarbons or hydrocarbon fragments. *(Source: Postberg et al. (2018), Macromolecular organic compounds from the depths of Enceladus, Nature, 558: 564–568).*

alteration and differentiation, interacting with core minerals and metal ions to generate new chemical compositions. With regard to Fig. 13.14A, "the broad and irregular shape of the peaks indicates that they are composed of multiple unresolved overlapping mass lines" (Postberg et al., 2018). According to Postberg et al. (2018), the mass intervals between the peaks suggest organic species with an increasing number of carbon atoms (C_7–C_{15}), which they refer to as High Mass Organic Cations (HMOC). In an experimental scenario, HMOC is a subgroup of spectra with a pattern of repetitive nonwater peaks above 80u (Fig. 13.14).

According to Postberg et al. (2018)'s interpretation, fragmentation of macromolecular parent molecules or networks creates the HMOC. In other words, the HMOC pattern is indicative of highly unsaturated organic cations. They further suggest that "while an interval of 14u would indicate the addition of a saturated CH_2 group to an organic 'backbone', the actual average mass difference of 12.5u (note that u represents atomic mass units) indicates the presence of predominately unsaturated carbon atoms." From Fig. 13.14, it is seen that the HMOC sequence always appears together with intense nonwater signatures below 80u, offering further compositional and structural constraints. Organic mass lines below 45u show up preferentially at uneven masses (Fig. 13.14B) indicating a typical cationic hydrocarbon fragmentation fingerprint.

Postberg et al. (2018) found that each spectrum also exhibits water cluster cations of the form $H_3O(H_2O)n^+$, typical for water ice impacts (Postberg et al. 2008; 2009). Postberg et al. (2018) further found that an ice/organic mixture constitutes the bulk composition of these particles. According to these investigators, ion abundances are indicative of an organic fraction in ice grains. It was found that particle radii are mostly between 0.2 and 2 μm. In Postberg et al. (2018)'s studies, Na^+ ions and sodium/water cluster $(Na(H_2O)n^+)$ appear in most HMOC spectra.

Cassini's INMS has measured the integrated composition of the Enceladan plume at several flyby speeds (Waite, Jr et al., 2017). In contrast to CDA, which records cations forming upon impact, INMS simultaneously measured the composition of neutral gas entering the instrument aperture and volatile neutral molecules that are generated upon ice grain impact onto the instrument's antechamber (Teolis et al., 2010). Postberg et al. (2018) found that "there is a striking over-abundance of organic species in spectra obtained at high flyby speeds (14–18 km/s) compared to those obtained at slower velocities (7–8 km/s)" that they attribute to "fragmentation of large organic parent molecules beyond the upper INMS mass limit at 99u." In Postberg et al. (2018)'s studies, a high abundance of aromatic structures is further supported by mass lines at 63u - 65u and 39u that can be interpreted as coincident unsaturated benzene fragments (Dass, 2007). Postberg et al. (2018)'s studies found that also aliphatic fragmentation is indicated by strong peaks at 27u–29u, 41u–43u, and less pronounced at 15u. It was further found that mass lines at 55u–57u are in agreement with aliphatic C_4 species, whereas aliphatic structures with more than 4 C atoms are generally absent (Postberg et al., 2018). It has been found that although signatures in agreement with tropylium ions at ≈ 91u are present in every HMOC spectrum, they are on average about 3 times less abundant than the energetically favorable phenyl cation peak (Fig. 13.14A). It has also been found that many of the more abundant low-mass species are in agreement with aromatic fragmentation and oxygen-bearing parent molecules. According to Postberg et al. (2018), "INMS compositional analysis of signals extracted explicitly from ice grain impacts provides further evidence that CO is the dominant fragment species at 28u and that an N-bearing fragment (C_2H_3N) might be present."

In spite of the relatively low mass range and resolution of the Cassini mass spectrometers, the measurements lead to the following key constraints (Postberg et al., 2018): "The HMOC-pattern in CDA spectra and the extended spectra argue for the presence of organic molecules with masses clearly above 200u. Prominent benzene-like species in CDA and INMS spectra indicate the presence of abundant sub-structures of isolated benzene rings. From INMS spectra the features cannot originate from benzene itself and are fragments of larger molecules. From the suppressed formation of tropylium cations in CDA spectra it can be concluded that the rings are either connected to functional groups without C-atoms or to dehydrogenated C-atoms. Unsaturated species at low masses are in close agreement with aromatic fragmentation. Aliphatic cations indicate saturated aliphatic structures with 1 to 4 C atoms in parallel to the unsaturated (aromatic) structures. Oxygen-bearing species indicated in both CDA and INMS spectra likely originate from hydroxyl, ethoxy and/or carbonyl functional groups."

In the data analyzed by Postberg et al. (2018), they found the presence of unsaturated carbon and the benzene ring in particular. There is a multitude of speculative options to explain the presence of complex organic materials in an icy moon such as Enceladus in the outer solar system. The two general categories of origin are accretion of primordial material and endogenic synthesis. In the former hypothesis, the organic carbon on Enceladus would predate the formation of the moon, and Enceladus would have acquired an organic inventory via its building blocks (icy planetesimals) (Postberg et al., 2018). The latter hypothesis would employ hydrothermal systems inside Enceladus's porous rocky core (Choblet et al., 2017) to produce complex organic molecules from small molecule precursors. Vance et al. (2007) examined means for driving hydrothermal activity in extraterrestrial oceans on planets and moons of less than one Earth mass, with implications for sustaining a low level of biological activity over geological timescales. Based on a

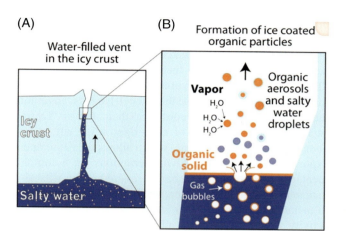

FIG. 13.15 **Schematic on the formation of organic condensation cores from a refractory organic film.** Ascending gas bubbles in the ocean efficiently transport organic material into water filled cracks in the south polar ice crust (A). Organics ultimately concentrate in a thin organic layer (orange) on top of the water table located inside the icy vents. When gas bubbles burst, they form aerosols made of insoluble organic material that later serve as efficient condensation cores for an icy crust from water vapor thereby forming HMOC-type particles (B). In parallel, larger, pure salt-water droplets form (blue), which freeze and are later detected by CDA as salt-rich Type 3 ice particles in the plume. *(Source: Postberg et al. (2018), Macromolecular organic compounds from the depths of Enceladus, Nature, 558: 564–568).*

model for cooling-induced thermal cracking, they found that in Europa and Enceladus, tidal flexing may drive hydrothermal circulation.

In order to deduce the mechanism that enables the presence of an organic enriched layer at the Enceladean water table located inside the water-filled cracks (tiger stripes) in the Enceladean ice crust, Postberg et al. (2018) invoked a scenario, well known from ice cloud formation over polar waters on Earth (Wilson et al., 2015). There, organic aerosols of mostly biogenic origin (deLeeuw et al., 2011) thrown up by bubble bursting serve as highly efficient nucleation seeds. Fig. 13.15 shows the schematic on the formation of organic condensation cores from a refractory organic film. Ascending gas bubbles in the ocean efficiently transport organic material (Porco et al., 2017) into water-filled cracks in the south polar ice crust. Organics ultimately concentrate in a thin organic layer (orange in Fig. 13.15) on top of the water table located inside the icy vents. When gas bubbles burst, they form aerosols made of insoluble organic material that later serve as efficient condensation cores for an icy crust from water vapor thereby forming HMOC-type particles. In parallel, larger, pure salt-water droplets form (blue in Fig. 13.15), which freeze and are later detected by CDA as salt-rich Type 3 ice particles in the plume (Postberg et al., 2009; 2011).

According to the argument put forward by Postberg et al. (2018) "When bubbles burst on Earth's ocean an organic-free sea spray forms in parallel with pure organic aerosols and mixed-phase organic-bearing sea spray (Gaston et al., 2011; Gantt and Meskhidze, 2013). The organic mass fraction of sea-spray aerosol has been consistently shown to be inversely related to aerosol size (Keene et al., 2007; Facchini et al., 2008; Gantt and Meskhidze, 2013) and is mostly water-insoluble (Facchini et al., 2008; deLeeuw et al., 2011). The purely organic endmembers are found preferentially in the smallest aerosols (Leck and Bigg, 2008; deLeeuw et al., 2011)".

Postberg et al. (2018) further argue that "nearly pure water ice grains (Type 1) can form from condensation of supersaturated vapor inside and above the ice vents (Schmidt et al., 2008; Postberg et al., 2009; Yeoh et al., 2015). The HMOC-producing organic material is detected in salt-poor ice grains (Type 2) and thus cannot have formed directly from the salty ocean spray that preserves the liquid composition upon flash freezing (Postberg et al., 2011). Salt-rich ice grains (Type 3), however, are thought to be frozen ocean spray (Postberg et al., 2009; Postberg et al., 2011), generated when bubbles of volatile gas (CO_2, CH_4, or H_2) reach the oceanic water table and burst (deLeeuw et al., 2011)."

Aerosol formation on Earth also provides a plausible analog mechanism for the simultaneous production of salty ocean spray and organic aerosols. Postberg et al. (2018) offered the following argument in support of the role of bubbles in lifting organic aerosols and bringing them in the Enceladean jet: "If the droplets are smaller than a few micrometers they can be thermally supported in the water vapor with gas densities slightly below the triple point against Enceladus' gravity. They don't fall back to the liquid, but are carried upwards through the ice vents by gas from evaporating water following the pressure gradient into space. In parallel to the formation of smaller organic aerosols, larger salty ocean droplets form, later detected by CDA as salt rich Type 3 particles in the plume and in the E-ring." Postberg et al. (2018) suggest that the analog from Earth would also predict some larger mixed-phase particles that carry both ocean salts and complex organics but these have not been identified by CDA. According to them, "one explanation could be that the macromolecular HMOC parent substance might be hydrophobic and in that way, would naturally avoid forming mixed phase organic/sea-water aerosols."

Based on detailed chemical scrutiny, Postberg et al. (2018) invoked primordial or endogenously synthesized carbon-rich monomers (<200 u) and polymers (up to 8000 u). However, according to Postberg et al., 2018), "it is still an open question whether we are seeing features that are representative of bulk organic materials in the subsurface. Processes that could fractionate organic compounds from their hypothesized source region to our instruments include expulsion from the core, (bio)degradation, hydrophobic phase separation in the ocean, plume outgassing, and impacts during high-speed flybys."

The Cassini spacecraft's Ion and Neutral Mass Spectrometer (INMS) detected volatile, gas phase, organic species in the plume and the Cosmic Dust Analyser (CDA) discovered high-mass, complex organic material in a small fraction of ice grains. Khawaja et al. (2019) reported a broader compositional analysis of CDA mass spectra from organic-bearing ice grains. Through analog experiments, they found spectral characteristics attributable to low-mass organic compounds in the Enceladean ice grains: nitrogen-bearing, oxygen-bearing, and aromatic. By comparison with INMS results, Khawaja et al. (2019) identified low-mass amines [particularly (di)methylamine and/or ethylamine] and carbonyls (with acetic acid and/or acetaldehyde most suitable) as the best candidates for the N- and O-bearing compounds, respectively. It was found that inferred organic concentrations in individual ice particles vary but may reach tens of mmol levels. The low-mass nitrogen- and oxygen-bearing compounds are dissolved in the ocean, evaporating efficiently at its surface and entering the ice grains via vapor adsorption. The potentially partially water-soluble, low-mass aromatic compounds may alternatively enter ice grains via aerosolization (the process or act of converting some physical substance into the form of particles small and light enough to be carried on the air, i.e., into an aerosol). These amines, carbonyls, and aromatic compounds could be ideal precursors for mineral-catalyzed Friedel–Crafts hydrothermal synthesis of biologically relevant organic compounds in the warm depths of Enceladus' ocean. As indicated earlier, the Enceladean plumes have been sampled by the Cassini spacecraft, and both low- and high-mass organics have been observed (Postberg et al., 2018; Khawaja et al., 2019).

13.8 Mechanisms driving the cryovolcanic plume emission from the warm fractures in *Enceladus*

Spacecraft observations suggest that the cryovolcanic plumes of Saturn's icy-moon Enceladus draw water from a subsurface ocean. But the mechanism that drives and sustains the eruptions on Enceladus, and the sustainability of conduits linking ocean and surface remained poorly understood. Furthermore, it is also not known what sets the rate of volcanism.

Initiation of ocean-to-surface conduits on ice moons including Enceladus and Europa is suspected to be related to ice shell disruption during a past epoch of high orbital eccentricity. It is believed that such disruption could have created partially water-filled conduits with a wide variety of apertures, and evaporative losses caused by tiger stripe activity would ensure that only the most dissipative conduits endure to the present day.

Models of the ultimate interior source for the eruptions have been considered by some researchers. It was found that models of an expanding plume require eruptions from discrete sources, as well as less voluminous eruptions from a more extended source, to match the observations. However, physical mechanism that matches the observations has not been identified to control these eruptions. Hurford et al. (2007) reported a mechanism in which temporal variations in tidal stress open and close the tiger-stripe rifts, governing the timing of eruptions. According to their studies, "During each orbit, every portion of each tiger stripe rift spends about half the time in tension, which allows the rift to open, exposing volatiles, and allowing eruptions. In a complementary process, periodic shear stress along the rifts also generates heat along their lengths, which has the capacity to enhance eruptions." Their studies suggested that plume activity would vary periodically, affecting the injection of material into Saturn's E ring and its formation, evolution, and structure. According to them, the stresses controlling eruptions imply that Enceladus' icy shell behaves as a thin elastic layer, perhaps only a few tens of kilometres thick.

Studies carried out by several investigators (e.g., Hurford et al., 2009; Porco et al., 2014; Nimmo et al., 2014; Spitale et al., 2015) have found that the plume's peak eruptive output anomalously lags peak tidal extension (by 5.1 ± 0.8 h relative to a fiducial model of the tidal response), and fissure eruptions continue from all four tiger stripes at Enceladus' periapsis (the point in the path of an orbiting body at which it is nearest to the body that it orbits) when all tidal crack models predict that eruptions should cease. In an attempt to understand the sustainability of the eruptions, Kite and Rubin (2016) asked the following questions, "How can eruptions continue throughout the tidal cycle? How can the liquid water conduits obtain the energy to stay open—as needed to sustain eruptions—despite evaporitic cooling and viscous ice inflow? Why is the total power of the system ~5 GW (not ~0.5 GW or ~50 GW)? Do tiger stripe mass and energy fluxes drive ice shell tectonics, or are the tiger stripes a passive tracer of tectonics?"

Kite and Rubin (2016) found that "a simple model of the fissures as open conduits can simultaneously explain both the maintenance of Enceladus' eruptions throughout the tidal cycle and the sustainability of eruptions on 10^{-1}-y to 10^{1}-y timescales, while predicting that eruptions are sustained over 10^{6}-y timescales." In Kite and Rubin (2016)'s investigation, fissures are modeled as parallel rectangular slots with length L ≈ 130 km, depth Z = 35 km, stress-free half-width W_0, and spacing S = 35 km. Slots are connected to vacuum at the top, and open to an ocean at the bottom (Fig. 13.16). In the model, the slots are mostly filled with water, and Saturn tides drive turbulent water flow in the slots whose dissipation produces enough heat to keep slots open. In turn, long-lived water-filled slots drive a volcano-tectonic feedback that buffers the rate of volcanism to approximately the observed value. Kite and Rubin (2016)'s results suggest that "a model of the tiger stripes as tidally flexed slots that puncture the ice shell can simultaneously explain

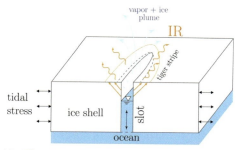

FIG. 13.16 The erupted flux from Enceladus (blue arrows) varies on diurnal timescales, which is attributed to daily flexing (dashed lines) of the source fissures by Saturn tidal stresses (horizontal arrows). Such flexing would also drive vertical flow in slots underneath the source fissures (vertical black arrow), which through viscous dissipation generates heat. This heat helps to maintain the slots against freeze-out despite strong evaporitic cooling by vapor escaping from the water table (downward-pointing triangle). The vapor ultimately provides heat (via condensation) for the envelope of warm surface material bracketing the tiger stripes (orange arrows; "IR" corresponds to infrared cooling from this warm material). *(Source: Kite, E.S., and A.M. Rubin (2016), Sustained eruptions on Enceladus explained by turbulent dissipation in tiger stripes, Proc Natl Acad Sci U S A., 113(15): 3972–3975; doi: 10.1073/pnas.1520507113).*

the persistence of the eruptions through the tidal cycle, the phase lag, and the total power output of the tiger stripe terrain, while suggesting that eruptions are maintained over geological timescales." Kite and Rubin (2016)'s studies revealed that the ocean–surface connection on Enceladus may be sustained on million-year timescales.

Solving the coupled equations for elastic deformation of the icy shell with turbulent flow of water within the tiger stripes, Kite and Rubin (2016) found that sustained eruptions on Enceladus can be explained by turbulent dissipation in tiger stripes. According to them, "Turbulent liquid water flow into and out of the slots generates heat. Water temperature is homogenized by turbulent mixing, allowing turbulent dissipation to balance water table losses and prevent icing over. Ice forming at the water table is disrupted by aperture variations and vertical pumping; water cooled by evaporation, if sufficiently saline, will sink and be replaced by warmer water from below. A long-lived slot must satisfy the heat demands of evaporitic cooling at the water table (about 1.1 × the observed IR emission) plus heating and melt-back of ice driven into the slot (Cuffey and Patterson, 2010) by the pressure gradient between the ice and the water in the slot (Gogue net al., 2013)."

Since the 2005-discovery of the plumes emanating from the tiger stripes of Enceladus, observations show eruptions from "tiger stripe" fissures are strongly tidally variable but sustained over the tidal cycle throughout each orbit. It was found that peak plume flux lags peak tidal extension by ~1 rad, suggestive of resonance. According to Kite and Rubin (2016)'s studies, turbulent dissipation of diurnal tidal flows explains the observed phase lag and the diurnal-to-decadal sustainability of liquid-water–containing tiger stripes, and the coupling between tiger stripes and the ice shell forces a $> 10^6$-y geologic cycle that buffers Enceladus' power to approximately the Cassini-era value.

Kite and Rubin (2016) are of the view that "the tiger stripes are the loci of sustained emission because other fractures are too short (L < 100 km) for sustained flow. Because sloshing homogenizes water temperatures along stripe strike, the magnitude of emission should be relatively insensitive to local tiger stripe orientation, a prediction that distinguishes the slot model from all crack models". Kite and Rubin (2016) claim that their model is more easily reconciled with curtain eruptions (Spitale et al., 2015) than jet eruptions (Porco et al., 2014) and it can provide a physical underpinning for curtain eruptions. However, Kite and Rubin (2016) suggest that localized emission might still occur, for example near Y junctions. They further suggest that "The pattern of spatial variability in orbit-averaged activity should be steady, in contrast to bursty hypotheses, and vapor flux should covary with ice-grain flux". Kite and Rubin (2016)'s studies have indicated that "The delay associated with flushing and refilling of O(1)-m-wide slots with ocean water causes erupted flux to lag tidal forcing and helps to buttress slots against closure, while tidally pumped in-slot flow leads to heating and mechanical disruption that staves off slot freeze-out. Much narrower and much wider slots cannot be sustained." Kite and Rubin (2016)'s model shows how to open connections to an ocean can be reconciled with, and sustain, long-lived eruptions. Another inference gleaned from this model is that "Turbulent dissipation in long-lived slots helps maintain the ocean against freezing, maintains access by future Enceladus missions to ocean materials, and is plausibly the major energy source for tiger stripe activity." Kite and Rubin (2016) caution that "eccentricity variations on $> 10^7$-y timescales may also be required if the ocean is to be sustained for the $> 10^7$-y timescales that are key to ocean habitability (Meyer and Wisdom, 2007; Tyler, 2011; Thomas et al., 2016). Furthermore, ocean longevity could be affected by heat exchange with self-sustained slots in the ice shell". It is believed that the mass and heat fluxes associated with long-lived slots (Howett et al., 2014) would drive regional tectonics. Kite and Rubin (2016) claim that their basic model might be used as a starting point for more sophisticated models of Enceladus coupling fluid and gas dynamics (Ingersoll and Pankine, 2010), as well as the tectonic evolution and initiation of the tiger stripe terrain (e.g., Běhounková et al., 2012).

The mechanisms that might have prevented Enceladus' ocean from freezing remain a mystery. Thomas et al. (2016) suggest a few ideas for future study that might help resolve the question, including the surprising possibility that tidal flexing suffered by Enceladus due to Saturn's gravity could be generating much more heat within Enceladus than previously thought.

13.9 Hydrothermal vents on the seafloor of *Enceladus*—possibility for harboring an ecosystem based on microbial populations

Origin of life researchers consider that primordial organic molecules may have evolved in seafloor hydrothermal systems by mineral catalysis. Williams et al. (2005) simulated experimentally the role of clay minerals in the abiotic synthesis of organic molecules near seafloor spreading centers. Although most organic compounds decompose in >300°C vent fluid, Williams et al. (2005)'s experiments simulating seafloor hydrothermal conditions showed that smectite-type clays can protect and promote development of diverse organic compounds that may be precursors to biomolecules. It may be noted that smectite is defined in clay mineralogy as a 2:1 clay—consisting of an octahedral sheet sandwiched between two tetrahedral sheets. Clays are common hydrothermal alteration products of volcanic glass and due to their nanoscale crystal size, provide extensive and variably charged surfaces that interact with aqueous organic species. It was found that organic products extracted with dichloromethane from the two expandable smectite clays (montmorillonite, saponite) contained a variety of complex organic molecules including: alkanes, alkyl-benzenes, alkyl-naphthalenes, alkyl-phenols, alkyl-naphthols, alkyl-anthrols, methoxy, and alkyl-methoxy-phenols, methoxy and alkyl-methoxy-naphthols, and long-chain methyl esters (Williams et al., 2011). Volcanic gases H_2 and CO_2 have been shown to react on magnetite surfaces to form methanol, a primary organic molecule, under hydrothermal conditions (Williams et al., 2011). Smectites may swell by the uptake of water in the clay interlayer spacing. Williams et al. (2005)'s studies indicated that smectite provides a safe haven for the synthesis of organic molecules, essentially like a "primordial womb."

The Enceladus plume was interrogated by two mass spectrometry-based instruments on the Cassini spacecraft. Measurements by the Ion and Neutral Mass Spectrometer and the Cosmic Dust Analyzer confirmed the presence of H_2, CH_4, and silica nanograins, which indicate hydrothermal processes. Large organic molecules and ammonia were also observed.

As already indicated, it was found that Enceladus has a rocky core, surrounded by a global ocean encrusted by a thick sheet of ice. NASA's Cassini spacecraft discovered a plume of material erupting from cracks in the ice. What Cassini has discovered in the plumes must reflect the chemical and physical conditions of the subsurface ocean and its dynamics over time. Based on initial analysis of Cassini data, in 2015, scientists shared results that suggest hydrothermal activity is taking place on the ocean floor. For example, silicon-rich nanoparticles observed in the plumes have suggested the existence of active hydrothermal vents at the water-rock interface, where temperatures are about 90°C, and alkaline pH of 8.5–10.5 and pressures of 10–80 MPa prevail (Hsu et al., 2015). Molecular hydrogen, detected in the plumes also likely reflects hydrothermal activity at the core (Waite et al., 2017). Fig. 13.17 illustrates how water interacts with rock at the bottom of the ocean of Enceladus, producing hydrogen gas.

Several investigators (e.g., Spahn et al., 2006; Porco et al., 2006; Waite Jr. et al., 2006; Hansen et al., 2006) found that through warm cracks in the crust (Spencer et al., 2006) a cryo-volcanic plume ejects ice grains and vapor into space. It was found that the plume contains chemical signatures of water-rock interaction between the ocean and a rocky core and that the plume contains materials that originate from the ocean (Postberg et al., 2009; Postberg et al., 2011). Testing habitability on Enceladus ultimately requires access to ocean materials. In realization of this requirement, in October 2015, the Cassini spacecraft flew directly through the plume of escaping material and sampled its chemical composition.

As indicated earlier, detection of sodium-salt-rich ice grains emitted from the plume of Enceladus suggests that the grains formed as frozen droplets from a liquid water

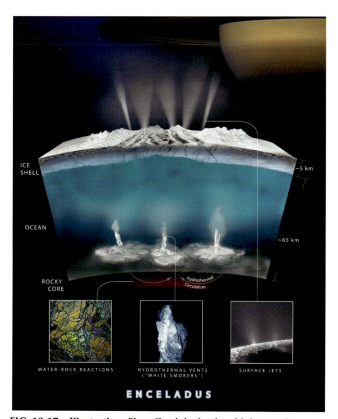

FIG. 13.17 Illustration of how Cassini scientists think water interacts with rock at the bottom of the ocean of Saturn's icy moon *Enceladus*, producing hydrogen gas. *Credits: NASA/JPL-Caltech. (Source: https://www.nasa.gov/press-release/nasa-missions-provide-new-insights-into-ocean-worlds-in-our-solar-system).*

reservoir that is or has been, in contact with rock. This finding implies rock–water interactions in regions surrounding the core of Enceladus. The recent discovery of silica nanoparticles (so-called stream particles) derived from Enceladus shows the presence of ongoing hydrothermal reactions in the interior. Hsu et al. (2015) reported an analysis of silicon-rich, nanometre-sized dust particles that stand out from the water-ice-dominated objects characteristic of Saturn. They interpret these grains as nanometre-sized SiO_2 (silica) particles, initially embedded in icy grains emitted from Enceladus' subsurface waters and released by sputter erosion in Saturn's E ring. According to Hsu et al. (2015), "The composition and the limited size range (2 to 8 nm in radius) of stream particles indicate ongoing high-temperature (>90°C) hydrothermal reactions associated with global-scale geothermal activity that quickly transports hydrothermal products from the ocean floor at a depth of at least 40 km up to the plume of Enceladus."

Sekine et al. (2015) reported results from detailed laboratory experiments to constrain the reaction conditions. They found that to sustain the formation of silica nanoparticles, the composition of Enceladus' core needs to be similar to that of carbonaceous chondrites. They further showed that the presence of hydrothermal reactions would be consistent with NH_3- and CO_2-rich plume compositions. Sekine et al. (2015) suggested that high reaction temperatures (>50°C) are required to form silica nanoparticles whether Enceladus' ocean is chemically open or closed to the icy crust. According to Sekine et al. (2015), "such high temperatures imply either that Enceladus formed shortly after the formation of the solar system or that the current activity was triggered by a recent heating event. Under the required conditions, hydrogen production would proceed efficiently, which could provide chemical energy for chemoautotrophic life."

In another study, using ion neutral mass spectrometer on-board the Cassini spacecraft, Waite et al. (2017) detected molecular hydrogen in the plume of gas and icy particles spraying from Enceladus. By using the instrument's open-source mode, background processes of hydrogen production in the instrument were minimized and quantified, enabling the identification of a statistically significant signal of hydrogen native to Enceladus. Waite et al. (2017) found that the most plausible source of molecular hydrogen (H_2), in the plume is on-going hydrothermal reactions of rock containing reduced minerals and organic materials. This drives the ocean out of chemical equilibrium, in a similar way to water around Earth's hydrothermal vents, potentially providing a source of chemical energy. Waite et al. (2017) also found that the relatively high hydrogen abundance in the plume signals thermodynamic disequilibrium that favors the formation of methane from CO_2 in Enceladus' ocean.

Hydrothermal activity was suspected to occur deep inside the porous core (Hsu et al., 2015; Sekine et al., 2015; Waite, Jr et al., 2017) powered by tidal dissipation (Choblet et al., 2017). The discovery of hydrothermal vents on the seafloor of Enceladus (see Waite et al., 2017) also provides further evidence that warm, mineral-laden water is pouring into the ocean from vents on the seafloor.

Simple organic compounds with molecular masses mostly below 50 atomic mass units (amu) have been observed in plume material (Waite. Jr et al., 2006; Postberg et al., 2008; Waite. Jr. et al., 2009). Postberg et al. (2018) reported observations of emitted ice grains containing concentrated and complex macromolecular organic material with molecular masses above 200 atomic mass units. It has been argued that, as an example of endogenic synthesis, relatively oxidizing hydrothermal conditions may promote the conversion of simple primordial organics into reactive unsaturated compounds, such as quinones, (poly)phenols, or aldehydes, which may polymerize in turn (possibly in the presence of catalytic minerals) to form relatively hydrogen-poor macromolecules. According to Postberg et al. (2018), macromolecules at Enceladus containing aromatic units with connecting short aliphatic chains that include more or less oxidized functional groups may resemble some humic substances on Earth.

Based on their findings, the Enceladan researchers hypothesize that hydrothermal fluids rising from the seafloor deliver insoluble organic compounds to Enceladus' ocean. Currents carry them to the surface, where they float as a film. Eruptions then force the organic molecules, along with seawater, out into space. The organic compounds become encased in water ice, emerge from Enceladus' plume, and finally escape to form Saturn's ghostly E ring. Enceladan researchers think that discovery of organic molecules in the Enceladan plume could be the tip of the "organic iceberg," because the instruments deployed on the Cassini spacecraft did not have the capability to detect other organics such as amino acids, which are organic molecules related to life-forms.

Deamer et al. (2017) recall that soon after submarine hydrothermal vents ("black smokers") were discovered at several locations on the seafloor of Earth, it was proposed that not only do they harbor an ecosystem based on microbial populations using hydrogen sulfide and hydrogen as energy sources, but they may also have been sites conducive for the origin of life (Corliss et al., 1981; Baross and Hoffman, 1985; Russell et al., 1993; Russell and Hall,1997; Zierenberg et al., 2000; Kelley et al., 2001). It may be noted that black smokers are associated with volcanism in which heat from an underlying magma plume brings circulating seawater to temperatures exceeding 400°C. When the heated water emerges from the vent, the dissolved minerals precipitate as a black cloud of iron sulfide and form chimneys. It was found that on Earth, such hydrothermal vents support thriving communities of life in complete isolation from sunlight (see Joseph, 2016). Enceladus now appears likely to have all three of the ingredients scientists think life needs: liquid water, a source of energy (such as sunlight or chemical energy), and the right chemical ingredients (such as carbon, hydrogen, nitrogen, oxygen). According to Kahana et al. (2019), it is important to assess the capacity of Enceladus' aqueous organic content to jump-start life. This complements other legitimate questions regarding the similarity of the moon's environments to those proposed by some as essential for terrestrial life's origin—hydrothermal vents and volcanic fields (Deamer and Damer, 2017).

13.10 Resemblance of *Enceladus's* organics-rich ocean to earth's primitive prebiotic ocean—a favorable scenario for life's emergence on *Enceladus*

The origin of life is a long-standing and controversial subject concerned with how the first known single-cell organisms called prokaryotes probably originated in the Archean period (4-2.5 BYA) and about 3.8 BYA in the oceans. The "origin of life" researchers consider that an organic solution has formed in the Earth's ocean during Earth's first few billion years, presumably resulting from the carbon in-fall from dust, meteorites, and comets (Chyba and Sagan, 1992). Thus, during Earth's first few billion years, the chemical composition of the Earth's ocean and the atmosphere was very different from what it is today (see Dostal et al., 2009). The just mentioned concentrated organic solution in the primitive Earth's ocean—known as the "Primordial Soup" or "Prebiotic Broth" (Oparin, 1962)—is considered to supply the required amount of chemical components essential for life's emergence.

An important candidate mechanism involves lightning in the early atmosphere and the consequent production of amino acids that, when combined in long chains (polymers), provided the basic constituent of life. Thus, Earth's primitive prebiotic oceans have played a vital role in the origin of life and in the biological evolution on Earth (see review by Joseph, 2016). Based on this notion, the prebiotic soup is considered to provide a conceivable milieu for early steps in life's origin.

According to Kahana et al. (2019, "A prebiotic soup is by definition a water body that carries dissolved or stably dispersed organic molecules, the latter being in the form of micelles, vesicles, or emulsions. The three major sources for soup compounds are infalling water-dispersible organics, infalling water-nondispersible organics that become solubilized by abiotic differentiation, and very simple compounds that undergo abiotic synthesis to both dispersible and nondispersible compounds." In general, a prebiotic soup is any water body containing organic compounds from a variety of supply sources. The just-mentioned "soup" and "broth" is akin to the "warm little pond" first envisaged by Charles Darwin (1871). Arguments against the existence of a prebiotic soup are not strong enough to negate its likely existence and role in life's origin.

One possible path enabling the supply of organic molecules is the terrestrial endogenous synthesis of organic compounds from very simple carbon-containing molecules, which necessitates a free energy source, such as light, electric discharge, or concentration gradients. A groundbreaking example of artificially producing prebiotic soup was the Miller-Urey experiment originally performed with hydrogen-rich reducing gases such as methane (CH_4), ammonia (NH_3), and hydrogen (H_2) (Miller, 1953), "which demonstrated the generation of soup-worthy organic compounds from simple reducing gases energized by spark discharge. This experiment motivated several decades of scrutiny in which life-related molecules were shown to be abiotically synthesized from different precursors and based on electrical, thermal, chemical, and photochemical energy as reviewed (Ferris and Hagan, 1984; Maden, 1995; Cleaves, 2012; Bada, 2013). This mode of soup generation is represented as the abiotic path from simple primordial compounds to dispersible and nondispersible organics" (Kahana et al. (2019). While terrestrial endogenous syntheses quite strongly dominate the origin-of-life literature as noted above, exogenous in-fall mechanisms are also important (Ehrenfreund et al., 2011), with sources including interplanetary dust particles, meteorites, and comets (Chyba and Sagan, 1992; Kaiser et al., 2013).

The Enceladus observations teach us by direct measurements that an extraterrestrial organic-rich soup exists even in the absence of an atmosphere. Such a soup is likely fed by primordial accretion, in-fall, and subsequent abiotic chemical modifications and syntheses, the combination of which may increase its diversity (Kahana et al., 2019).

The jets of icy particles and water vapor emanating from the tiger stripe region in the vicinity of the south pole of Enceladus are evidence for activity driven by some geophysical energy source. The vapor has also been shown to contain simple and complex organic compounds, and the south polar terrain is bathed in excess heat coming from below. There are strong indications that a liquid water environment exists beneath the south polar cap, which may be conducive to life. Several theories for the origin of life on Earth would apply to Enceladus, which include (McKay et al. 2008): (1) origin in an organic-rich mixture, (2) origin in the redox gradient of a submarine vent, and (3) panspermia.

Ribonucleic acid (RNA) is considered as a primordial molecule that can store genetic information, and thus RNA has been referred to as an "informational polymer." It is believed that at some stage in the origin of life, such a molecule must have arisen by purely chemical means (see Powner et al., 2009). It has been discovered that some RNA molecules also carry out chemical reactions. In other words, this single polymer molecule might have been able to handle all the basic tasks required for life. The hypothesis that RNA preceded DNA (Deoxyribonucleic acid) as genetic material (*RNA World hypothesis*) is supported by several lines of evidence (see Oro et al., 1990).

Kahana et al. (2019) recall that "The RNA World scenario calls for yet another abiotic synthesis—the generation of replicating biopolymers from specific monomers in the prebiotic soup, so as to jump-start an evolutionary process (Cafferty and Hud, 2014). In contrast, the Lipid World scenario posits that replication, selection, and evolution can transpire at the level of monomers, by compositional inheritance mediated by a mutually catalytic network (Lancet et al., 2018)." The Lipid World is hypothesized to initially rest on pure compositional information.

With regard to the validity of the Lipid World scenario, some seminal questions are often posed; such as the source of prebiotically formed lipids. In this context, it may be recalled that prebiotic lipids can be formed via high-temperature aqueous Fischer-Tropsch reaction from oxalic acid (Rushdi and Simoneit, 2001). Lipids have several properties. For example, based on the Lipid First GARD model, it has been shown that lipids can exert catalysis (Segré et al., 2000; Segré and Lancet, 2000; Segre et al., 2001a,b; Shenhav et al., 2003, 2005; Markovitch and Lancet, 2012, 2014; Gross et al., 2014). The GARD model is based on the storage and transmission of compositional information, which is different from but equivalent to sequence-based information. Because lipids in present-day life obey specific structural and functional constraints (Georgiou and Deamer, 2014), lipid structure evolution is also expected to occur en route to more elaborate life-forms.

Phospholipids are the primary components of most cell membranes and contain two carboxylic acids attached by ester bonds to glycerol phosphate, which in turn is esterified to head groups like choline, ethanolamine, serine, and glycerol. When dispersed in dilute solutions of ionic solutes, carboxylic acids and phospholipids spontaneously self-assemble into microscopic membranous vesicles (small fluid-filled bladders, sacs, cysts, or vacuoles within the body) bounded by bilayer structures (Fig. 13.18).

Kahana et al. (2019)'s comparison between the RNA First and Lipid First scenarios points to the possibility that Lipid First nonequilibrium, catalysis-based dynamics could serve as a pre-RNA first step on a long road to full-fledged proto-cells. Such portrayal addresses the widely expressed doubts regarding the probability of molecules as complex as self-copying covalently strung RNA to emerge directly from random chemical mixtures (Shapiro, 2006).

Early chemical analyses of Enceladus's cryovolcanic plumes, and of Saturn's E-ring generated from their ejecta, established compositions of organic molecules in an H_2O-dominated context (Waite et al., 2006; Postberg et al., 2008, 2011). These observations suggested organic compounds with 2–4 carbon atoms, potentially derived from

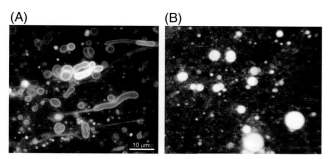

FIG. 13.18 (A) Membranous vesicles spontaneously form when a mixture of fatty acid and fatty alcohol is dispersed in water. If the same mixture is exposed to a single evaporation-rehydration cycle in the presence of short strands of duplex DNA stained with acridine orange, a fluorescent dye, the concentrated polymer is encapsulated within membrane-bounded vesicles (B). *(Source: Deamer, D., and Damer, B. (2017), Can Life Begin on Enceladus? A Perspective from Hydrothermal Chemistry, Astrobiology, 17(9):834–839).*

larger aromatic entities. A study by Postberg et al. (2018) has extended these observations, providing a much more detailed picture of the Enceladan ocean's organic repertoire, suggesting that some of the ice grains in Enceladus's cryovolcanic plumes contain concentrated and much more complex organic material. Kahana et al. (2019) found that the mass spectrometry observations extend the organics' size range from <50 u (atomic mass units) to ≤200 u. Kahana et al. (2019) suggest that this higher range, defined as high-mass organic cations (HMOCs), likely represents molecules with 7–15 carbon atoms.

Enceladus appears to represent the first reported extraterrestrial organics-rich water body. A detailed chemical scrutiny of the complex organic molecules in the plumes emanating from the subglacial water ocean of Enceladus, as reported in the landmark discovery by Postberg et al. (2018), prompted Kahana et al. (2019) to invoke primordial or endogenously synthesized carbon-rich monomers (<200 u) and polymers (up to 8000 u). In the search for finding an answer to the question "which origin-of-life scenario appears more consistent with the reported molecular configurations on Enceladus?" Kahana et al. (2019)'s investigations revealed that the monomeric organics observed in the Enceladan plumes are carbon-rich unsaturated molecules that are quite chemically akin to simple lipids. These investigators proposed that the organic polymers found in the Enceladen plumes resemble terrestrial insoluble kerogens (solid, insoluble organic matter in sedimentary rocks) and humic substances (i.e., substances relating to or consisting of organic component formed by the decomposition of leaves and other plant material by soil microorganisms), as well as refractory organic macromolecules found in carbonaceous chondritic meteorites. Interestingly, the composition of Enceladus' core bears similarity to that of carbonaceous chondrites (Sekine et al., 2015; Postberg et al., 2018).

It may be noted that carbonaceous meteorites are a group of asteroid fragments containing between 2% and 8% carbon (Kerridge, 1985), including kerogen-like macromolecules (Kerridge et al., 1987), as well as numerous simpler compounds (Cronin, 1998). They therefore may carry a record of early carbon chemistry, and provide clues to the organic compounds on planetary and sub-planetary bodies, including on ocean worlds, pointing to the compositions that might have led to the onset of life (Pizzarello and Shock, 2017).

Results gleaned from numerous studies lend strong support to the earlier observation that components in a Murchison extract form vesicular boundary structures (Deamer, 1985). In focused analysis aimed at lifelike small molecules, Cronin and Chang (1993) and Cronin (1998) found that the Murchison carbonaceous meteorite extracts contain a complex mixture of hundreds of monomeric compounds, including carboxylic acids, amino acids, hydroxy acids, phosphonic acids, amines, amides, nitrogen heterocycles (including purines and pyrimidines), alcohols, as well as aliphatic, aromatic, and polar hydrocarbons. Based on several studies the meteorite appears to contain millions of specific molecular structures yet to be fully resolved (Schmitt-Kopplin et al., 2010).

There has been suggestion that, if some forms of life does not exist now or did not exist earlier in the Enceladan ocean, the Enceladan subglacial ocean could be the first observationally discovered concrete example of what may well be a primeval soup (see Kahana et al., 2019).

Kahana et al. (2019) further posit that the organic polymers found in the Enceladen plumes "upon long-term hydrous interactions, might break down to micelle-forming amphiphiles." In support of this notion, they invoked the already available detailed analyses of the Murchison chondrite, which are dominated by an immense diversity of likely amphiphilic monomers. In this context, Deamer et al (2017) elaborated on the importance of "amphiphilic compounds" thus, "An essential requirement for the origin of cellular life is the availability of amphiphilic compounds and conditions that permit such compounds to assemble into the boundary membranes of primitive cells. Amphiphile is a general term referring to molecules composed of a nonpolar hydrophobic moiety, usually a hydrocarbon chain, with a polar hydrophilic group at one end of the molecule. Because the hydrophobic portion of an amphiphile is insoluble in water and the hydrophilic group interacts strongly with water, certain amphiphiles are able to assemble into bilayer structures that form the boundaries of all cellular life today. The simplest amphiphiles are carboxylic acids, basically a hydrocarbon chain with a carboxyl group at one end. Carboxylic acids up to 13 carbons in length are present in carbonaceous meteorites and have been demonstrated to assemble into membranous vesicles (see Deamer et al., 2002, for a review). For this reason, they are often used in simulations of prebiotic protocellular compartments (Budin et al., 2014)."

Kahana et al. (2019)'s the specific quantitative amphiphile-based model for compositionally reproducing lipid micelles, which benefits from a pronounced organic diversity, support the possibility of a pre-RNA Lipid World scenario for life's origin on Enceladus. It is important to note that lipid assemblies are considerably resistant to extreme conditions.

Kahana et al. (2019) tersely summarize their argument on the possible resemblance of Enceladus's organics-rich ocean to Earth's primitive prebiotic ocean thus, "The cumulative evidence from the enceladan ocean thus suggests a chemically rich and physically dynamic water body, as befits a prebiotic soup (Wollrab et al., 2016). Further, that the enceladan ocean resides under a thick layer of ice and still becomes rich in increasingly complex organic compounds opposes the notion that atmospheric sparking is crucial for the emergence of a prebiotic soup (Maden, 1995)." Kahana et al. (2019) further note that "alternative free radical reactions could supplement this missing synthetic pathway."

It is interesting that the Enceladus data are consistent with a soup replete with highly diverse organic compounds that are conducive for the chemical evolution leading to life's origin, and which could serve as chemical supply for the formation of early lifelike entities. According to Kahana et al. (2019), in an RNA-centric view, evolution cannot begin before several life component types join forces. However, an advantage of the GARD Lipid World view is that rudimentary evolution becomes possible at the lipid-only stage, and RNA, proteins, and metabolism may then emerge in the realm of a long, gradual Darwinian evolutionary process, in a set of molecularly describable routes (Lancet et al., 2018). However, Kahana et al. (2019) caution that more extensive evidence from astrobiological research, chemical experimentation, and computer analyses would be necessary for the ultimate arbitration between origin-of-life scenarios.

13.11 Science goals and mission concept for future exploration of *Enceladus*

Enceladus is one of the solar system's most enigmatic (mysterious) bodies. According to McKay et al. (2008), much importance is attached to the study of Enceladus because (1) there is ample evidence that liquid water exists under ice coverage in the form of active geysers in the "tiger stripes" area of the southern Enceladus hemisphere; (2) it is an active cryovolcanic moon, wherein sustained eruptions in its ocean floor break through its South Pole ice and produce the plumes of frozen vapor mixtures (H_2O, CO_2, CH_4, NH_3, Ar, and other compounds) that were detected by Cassini; (3) the active plume on Enceladus enables the sampling of fresh materials from the interior that may originate from a liquid water source; and (4) Enceladus is such a scientifically compelling target that provides a great opportunity to understand organic chemistry in an ocean beyond Earth's, with great potential to provide at least some new clues about the origin of life and the potential for life elsewhere in the Universe. Because life on Earth exists whenever liquid water, organics, and energy coexist, understanding the chemical components of the emanating ice particles could indicate whether life is potentially present on Enceladus. The icy worlds of the outer planets are testing grounds for some of the theories for the origin of life on Earth.

Future ground-based and space-based observations as well as future space missions are needed to advance our understanding of Enceladus. These include (1) long-term observations to establish whether the plumes are transient or in steady-state; (2) searching for molecules of astrobiological significance, such as NH_3; (3) searching for the presence of photo-pigments (e.g. chlorophyll); (4) identifying the presence of oxidants such as H_2O_2 on the surface, and (5) mapping the North-South gradient in surface properties to quantify the rate of impact erosion and resurfacing (see Parkinson et al., 2007).

Laboratory studies are needed, (1) to identify the chemical species responsible for 3.4 μm absorption feature in the ice in the "tiger stripes"; (2) to study the chemical evolution of organics in ice in the presence of energetic photons and particles; and (3) to quantify the rate of production of oxidants such as H_2O_2 for conditions appropriate for Enceladus. Finally, modeling studies are needed to link the observations and laboratory experiments to the evolution of the hydrological, chemical redox, and geochemical cycles on Enceladus.

Because the Cassini mission detected plumes of ice particles erupting hundreds of kilometers into space through the ice cover of Enceladus, it has been proposed that Enceladus may have environments resembling oceanic hydrothermal vents (Hsu et al., 2015). It has been found that the plumes contain compounds consistent with the presence of hydrothermal sites on the ocean floor that are powered by chemical reactions. If so, the effluents are likely to be similar in properties (pH, temperature, composition, and origin) to the *Lost City* hydrothermal vents on Earth's ocean floor, with energy made available by serpentinization (Kelley et al., 2001). It is essential to have more detailed information about the properties of hydrothermal conditions and their potential to promote physical and chemical processes related to the origin of life.

On Enceladus, the presence of Si-containing nanoparticles in the plume, and the vent conditions inferred from it, suggest serpentinization as a possible source of chemical energy. In the absence of chemical energy, life could not be sustained even if molecular systems capable of growth, replication, and metabolism assembled from organic compounds are present in the ocean. It has also been found that the Enceladan Ocean is salty and contains organic compounds that resemble the mixture of organic compounds in carbonaceous meteorites.

According to Deamer and Damer (2017), the submarine hydrothermal vents and the land-based volcanic hydrothermal fields that have been proposed as alternative sites for the origin of life on Earth can guide the development of future flight missions designed to sample icy plumes of Enceladus, and future landers on Enceladus or Europa. Furthermore, given the assumption that life can begin in hydrothermal vent conditions, a reasonable speculation is that life could also originate on these ocean worlds and may still exist. If so, depending on the abundance of microbial life in the Enceladan Ocean, it might be possible to detect biosignatures in plume samples.

Although Enceladus may not look very hospitable, data from NASA's Cassini spacecraft suggest that Enceladus has all the raw ingredients necessary for life as we know it. No wonder, Enceladus is one of the prime targets for future space exploration. Enceladen researchers have discussed the possibility of sending a spacecraft to Enceladus in the future, armed with more sophisticated mass spectrometers that could discern the full range of individual compounds in the Enceladus plumes. Enceladen researchers are of the view that such an investigation could even determine isotopic ratios, which would provide a powerful means to probe sources of the organic molecules. Furthermore, sending spacecraft through the Enceladus plumes would allow the collection of microgram amounts of plume components and then analyzing the sample using ultrasensitive biosensors to identify biosignatures, if any present. Or, a lander could visit the moon and scoop up a sample from its snowy surface to scan for potential biomolecules such as proteins. According to Taubner et al. (2018) future, lander missions could easily collect physical evidence for the presence of autotrophic, hydrogenotrophic methanogenic life on Enceladus.

Kahana et al. (2019) is of the view that there is a likelihood of saltwater grains containing monomeric organic material (i.e., material of low molecular weight capable of reacting with identical or different materials of low molecular weight to form a polymer), which stays below the sensitivity threshold of the Cassini instruments. Future missions might be able to characterize such organic molecules, by extending the geophysical characterization toward the search for life-worthy organic molecules (McKay et al., 2014). It may be noted that a monomer is a molecule of any class of organic compounds that can bind chemically to the same molecules or other compounds to form a large molecule or polymer containing large number of repeating units of monomers. Polysaccharides, polypeptides, and polynucleotides are some of the biopolymers in living cells. According to Kahana et al. (2019) there are indications that "hydrothermal pyrolysis leads to the formation of diverse polar compounds, including alkanoic acids and alcohols, isoprenoid ketones, and alkanoate esters, in the C9–C33 range (Rushdi and Simoneit, 2011). All these results are consistent with the possible generation of monomeric organics from insoluble polymers under conditions similar to those prevailing on Enceladus. Whether this actually happens on Enceladus could only be verified by future missions and analyses."

The astrobiological exploration of other worlds in our solar system is moving from initial exploration to more focused astrobiology missions. In this context, McKay et al. (2014) present the case that the plume of Enceladus currently represents the best astrobiology target in the solar system. Analysis of the plume by the Cassini mission indicates that the steady plume derives from a subsurface liquid water reservoir that contains organic carbon, biologically available nitrogen, redox energy sources, and inorganic salts. Furthermore, samples from the plume jetting out into space are accessible to a low-cost flyby mission. No other world has such well-studied indications of habitable conditions. Thus, according to McKay et al. (2014), the science goals that would motivate an Enceladus mission are more advanced than for any other solar system body. They argue furthermore that the goals of such a mission must go beyond further geophysical characterization, extending to the search for biomolecular evidence of life in the organic-rich plume. This will require improved in situ investigations and a sample return.

Tsou et al. (2012) proposed an Enceladen mission concept known as "LIFE: Life Investigation For Enceladus," a low-cost sample return mission to Enceladus, a body with high astrobiological potential. The LIFE mission concept is envisioned in two parts: first, to orbit Saturn (in order to achieve lower sampling speeds, approaching 2 km/s, and thus enable a softer sample collection impact than Stardust, and to make possible multiple flybys of Enceladus); second, to sample Enceladus' plume, the E ring of Saturn, and the Titan upper atmosphere. With new findings from these samples, NASA could provide detailed chemical and isotopic and, potentially, biological compositional context of the plume. Tsou et al. (2012) consider that because the duration of the Enceladus plume is unpredictable, it is imperative that these samples are captured at the earliest flight opportunity. The proponents of LIFE mission are optimistic that this concept offers "science returns comparable to those of a Flagship mission but at the measurably lower sample return costs of a Discovery-class mission."

13.12 Implementable mission concepts to further explore Enceladus in the near future—Enceladus life finder (ELF) mission

Enceladus is unique amongst ocean worlds in our solar system primarily because the contents of its internal ocean are continuously emitted to space by its present-day activity, and some of these materials are redeposited on the surface. As indicated earlier, Enceladus is a high-priority target for astrobiology as it harbors a subsurface ocean that erupts into space and is likely habitable. Searching for life at Enceladus is a top priority, near-term goal of the "NASA Roadmap to Ocean Worlds" exploration (e.g., Hendrix et al., 2019, The NASA Roadmap to Ocean Worlds, and references therein). Several mission concepts to search for life in Enceladus' plume have been studied including the proposed *Discovery* and *New Frontiers* missions.

NASA's *Discovery* program gives scientists a chance to dig deep into their imaginations and find new ways to unlock the mysteries of our solar system. When it began in 1992, the program represented a breakthrough in the way NASA explores space. Discovery invites scientists and engineers to assemble a team to design exciting, focused planetary science missions that deepen what we know about the solar system and our place in it. The *Enceladus Life Finder (ELF) mission* was first proposed in 2015 for *Discovery* Mission 13 funding, and then it was proposed in May 2017 to NASA's *New Frontiers* program Mission, but it was not selected. A single flagship-level mission that would search for life on Enceladus' surface, within its ice shell and in the subsurface ocean may be the most comprehensive opportunity for astrobiology in the coming decade (Hofgartner et al., 2019). The ELF scientists think that a life-hunting mission to the geyser-spewing moon would deliver impressive "bang for the buck" (the worth of one's money or exertion) astrobiologically, allowing humanity to take a solid crack at perhaps the biggest mystery facing humanity.

If selected at another future opportunity, the ELF mission would search for biosignature and biomolecules in the geysers of Enceladus. Cassini spacecraft has flown through the Enceladen plume multiple times, but that spacecraft was not equipped to search for life. ELF, on the other hand, would probe the habitability of Enceladus' ocean and hunt for evidence of biological activity. ELF would carry two mass spectrometers; one would be optimized to study gaseous plume molecules, whereas the other would focus on solid grains. These instruments would study amino acids (the building blocks of proteins), fatty acids, methane and other molecules, allowing mission scientists to perform three separate tests for life. Positive results for all three would strongly argue for life within Enceladus. ELF brings the most compelling question in all of space science within reach of NASA's *Discovery Program*, providing an extraordinary opportunity to discover life elsewhere in the solar system in a low-cost program.

The baseline concept calls for ELF to launch aboard a United Launch Alliance Atlas V rocket and endure a 9.5-year-long journey to Saturn (though the trip would be much shorter if NASA's Space Launch System megarocket, which is currently in development, were used). ELF would enter orbit around Saturn, then fly through Enceladus' plume eight to 10 times over the course of three years. These sampling sojourns would bring the robotic probe within about 31 miles (50 km) of Enceladus' surface. ELF is a logical follow-on from Cassini and leverages much of the older mission's heritage.

Dozens of geysers detected on the tiger stripes region in the south pole region of Enceladus blast water ice, organic molecules, and other material into space from these cracks. These geysers are powerful and prolific; the plume they create makes up Saturn's E ring. The geyser material is likely coming from Enceladus' sea, so the tiger stripes could be a portal, potentially allowing a probe to get into the water without drilling through miles of ice. And some scientists and engineers are developing technology that could exploit this convenient loophole.

For example, a group at JPL is working on the Exobiology Extant Life Surveyor (EELS), an autonomous, 13-foot-long (4 m) snakelike robot that would spiral its way down the tiger stripe cracks until it reached liquid water. A long tether would connect EELS to a surface lander, which would provide power to the ocean explorer. A recent internal JPL study envisions EELS and the lander as part of a larger overall mission, which would also include a melt probe and an Enceladus orbiter, which would relay data from the surface craft back to Earth.

As a consequence of the plumes emitted by Enceladus, this tiny moon of Saturn presents a golden opportunity to directly measure the composition of the ocean and seek evidence for habitability (including past or extant life), either by collecting and analyzing plume particles as previously proposed by *Discovery* and *New Frontiers* mission concepts, or via more ambitious mission concepts that involve landing, surface sampling and analysis, and potential deployment of subsurface probes to reach the ocean itself (Hofgartner et al., 2019). Is a mission that accesses the ocean and directly searches for life in unaltered ocean samples feasible in the coming decade? Hofgartner et al. (2019) reported the results of an internal study at the Jet Propulsion Laboratory for an ambitious yet implementable mission concept to Enceladus' ocean. This innovative mission would include autonomous landing at Enceladus' south pole and penetration through the ice shell via both an intelligent melt probe and an adaptable multiterrain robot (robotic eel) that navigates down an erupting vent. Both the eel and melt probe would be instrumented, able to search for evidence of life in samples acquired during their descent, and capable of manoeuvring within the ocean.

The low surface gravity (1% of Earth's) and extreme cryogenic conditions in the South Pole regions ($\sim$ 50 K, away from the Tiger Stripes) raise questions: how to best sample the upper $\sim$ 1 cm of the surface around a lander, made of most freshly deposited plume materials? What are the expected properties of these materials, i.e. how fast does sintering proceed and how strong would these materials be as function of their exposure age? Hodyss et al. (2019) have provided answers to these questions via a two-pronged approach.

Firstly, Hodyss et al. (2019) surveyed experimentally the time evolution of mechanical strength of large samples of ice spherules at several temperatures. A custom sample preparation system has been developed by them to synthesize ice spheres with a grain size distribution of mean $\sim$ 12 microns. The samples were subsequently held at temperatures of -30, -50, and -80°C, over extended periods of time (several months), and their strength was tested at frequent intervals using cone penetration tests. The data obtained suggest that the observed temperature dependence of the strength evolution is commensurate with expectations from vapor diffusion.

Secondly, Hodyss et al. (2019) developed a new sampling system that enables rapid sampling and transfer of surface materials into receptacles. Those receptacles can then deposit the sampled materials into the inlet of an instrument dedicated to analyzing the chemical composition of these materials and seek tracers of past or extant life. The geometry of the system and principles of operation have been established and validated by experimental tests, as well as dynamical simulations.

Enceladus' plume enables ocean sampling with a flyby spacecraft, but flyby measurements at hypervelocity (>1 km/s) present a challenge. To obtain the most useful and detailed information, it is imperative to understand hypervelocity sampling. Specifically, it is unknown at what velocities plume constituents will volatilize or fragment. Furthermore, understanding how hypervelocity impacts induce fragmentation is important as mass spectral patterns might be sensitive to impact velocity.

To better understand hypervelocity sampling, a lab-based instrument that creates hypervelocity species is being characterized. Waller et al. (2019) described the relevant instrumentation and the technicalities involved therein thus: "The Hypervelocity Ice Grain System (HIGS) generates ions and neutrals by laser-induced desorption (LID) from a water jet. Charged LID products are extracted into a time-of-flight mass spectrometer, accelerated to >1 km/s, and analyzed by mass-to-charge ratio. Currently, research efforts focus on characterizing LID products and probing how salts and pH affect amino acid and polypeptide ion distributions. Ultimately, a mass spectrometer under development at JPL (the quadrupole ion trap mass spectrometer; QITMS) will be coupled to HIGS, and vaporization and fragmentation thresholds of ions with known composition and velocity will be determined. Methods to soften ionization after a hypervelocity impact in the QITMS are also being developed. Note that other mass spectrometers/instruments sampling ice grains at hypervelocity could be validated using HIGS. These efforts represent a strong start to understanding hypervelocity sampling, which will inform design, planning, and implementation of future mission concepts to Enceladus."

The icy jets emanating from the tiger stripes in the south polar region of Enceladus are considered to be in contact with Enceladus' underground ocean, which offers a rare and tantalizing opportunity—gathering samples from a potentially habitable alien environment without even touching down. (Furthermore, the oceans of Europa and Enceladus lie beneath miles of ice, which could make sampling by a landed mission tough). Current ELF plans call for including a technology-demonstration instrument designed to determine the chirality, or "handedness," of amino acids. All Earth life uses left-handed amino acids rather than right-handed ones; according to astrobiologists a similar preference found in an extraterrestrial sample would be a strong indication of alien life.

The baseline mission architecture would include a solar-powered orbiter to identify safe, compelling landing zones and relay data to Earth as well as a Radioisotope Thermoelectric Generator (RTG)-powered lander that would execute a precision landing, deploy the melt and eel probes, relay data to the orbiter, and also search for evidence of life in samples of surface material. Cassini spacecraft was also powered by three RTGs, which convert the heat of plutonium-238′s radioactive decay into electricity. But ELF would be solar-powered, because NASA, concerned about its dwindling stockpile of plutonium-238, prohibited the use of nuclear fuel for this *Discovery* mission. Demonstrating the utility of solar power at Saturn is an important goal in itself, because nuclear fuel will always be in relatively short supply and therefore reserved for future missions that cannot do without it. Examples of plutonium-dependent missions include efforts to explore the surface or atmosphere of Saturn's huge, haze-shrouded moon Titan or probes that journey to extremely faraway destinations such as Neptune. The ELF scientists want to push the boundaries for solar power so that, for missions in orbit around Saturn, we don't need to use that valuable inventory of radio-isotopic fuel that is going to be needed for these other missions.

A new surface sampling system has been developed to enable the latter search, based on experimentally-derived surface strength expectations (Hodyss et al., 2019). The mission design would not be inhibited by winter darkness at Enceladus' south pole; all operations could be accomplished at any Enceladus season. The concept would utilize the state-of-the-art in astrodynamics techniques, systems architecture, optimization, and advanced manufacturing. A single Flagship-level mission that would search for life on Enceladus' surface, within its ice shell, and in the subsurface ocean may be the most comprehensive opportunity for astrobiology in the coming decade.

References

Bada, J.L., 2013. New insights into prebiotic chemistry from Stanley Miller's spark discharge experiments. Chem Soc Rev. 42, 2186–2196.

Baland, R-M., Yseboodt, M., Hoolst, T.V., 2016. The obliquity of Enceladus. Icarus 268, 12–31.

Baross, J.A., Hoffman, S.E., 1985. Submarine hydrothermal vents and associated gradient environments as sites for the origin and evolution of life. Orig. Life Evol. Biosph. 15, 327–345.

Běhounková, M., Tobie, G., Choblet, G., Čadek, O., 2012. Tidally-induced melting events as the origin of south-pole activity on Enceladus. Icarus 219, 655–664.

Bouquet, A., Mousis, O., Waite, J.H., Picaud, S., 2015. Possible evidence for a methane source in Enceladus' ocean. Geophys. Res. Lett. 42, 1334–1339. doi:10.1002/2014GL063013.

Brown, R.H., Clark, R.N., Buratti, B.J., et al., 2006. Composition and physical properties of Enceladus' surface. Science 311, 1425.

Budin, I., Prwyes, N., Zhang, N., Szostak, J.W., 2014. Chain-length heterogeneity allows for the assembly of fatty acid vesicles in dilute solutions. Biophys. J. 107, 1582–1590.

Burger, M.H., Sittler, E.C., Johnson, R.E., Shematovich, V.I., Tokar, R.L., 2005, The surprising, humid atmosphere at Enceladus, AGU Fall Meeting, abstract #P32A-08. San Francisco, CA Dec. 2005.

Cafferty, B.J., Hud, N.V., 2014. Abiotic synthesis of RNA in water: a common goal of prebiotic chemistry and bottom-up synthetic biology. Curr. Opin. Chem. Biol. 22, 146–157.

Chen, E.M.A., Nimmo, F., Glatzmaier, G.A., 2014. Tidal heating in icy satellite oceans. Icarus 229, 11–30.

Choblet, G., et al., 2017. Powering prolonged hydrothermal activity inside Enceladus. Nat.Astronomy 1, 841–847.

Chyba, C., Sagan, C., 1992. Endogenous production, exogenous delivery and impact-shock synthesis of organic molecules: an inventory for the origins of life. Nature 355, 125–132.

Cleaves, H.J., 2012. Prebiotic chemistry: what we know, what we don't. Evolution: Education Outreach 5, 342–360.

Cooper, J.F., Cooper, P.D., Sittler, E.C., Sturner, S.J., Rymer, A.M., 2009. Old Faithful model for radiolytic gas-driven cryovolcanism at Enceladus. Planet. Space Sci. 57 (13), 1607–1620.

Corliss, J.B., Baross, J.A., Hoffman, S.E., 1981. An hypothesis concerning the relationship between submarine hot springs and the origin of life on Earth. Oceanolog. Acta 4, 59–69.

Cronin, J.R., Chang, S., 1993. Organic matter in meteorites: molecular and isotopic analyses of the Murchison meteorite. In: Greenberg J.M., Mendoza-Gomez C.X., Pirronellos V. (Eds.), The Chemistry of Life's Origins. Springer: Dordrecht, pp. 209–258.

Cronin, J.R., 1998. Clues from the origin of the Solar System: meteorites. In: Brack, A. (Ed.), The Molecular Origin of Life: Assembling Pieces of the Puzzle. Cambridge University Press: Cambridge, pp. 119–146.

Cuffey, K., Patterson, W., 2010. The Physics of Glaciers, 4th Ed. Elsevier, Burlington, MA.

Darwin, C., 1871. Letter no. 7471, Darwin Correspondence Project. Cambridge University Library: Cambridge. http://www.darwinproject.ac.uk/DCP-LETT-7471.

Dass, C., 2007. Fundamentals of Contemporary Mass Spectrometry, 1st edn. John Wiley and Sons, Hoboken, pp. 210–238.

D'Angelo, G., Lissauer, J.J., 2018. Formation of giant planets. In: Deeg, H., Belmonte, J. (Eds.), Handbook of Exoplanets. Springer International Publishing AG, part of Springer Nature, pp. 2319–2343.

Deamer, D.W., 1985. Boundary structures are formed by organic compounds of the Murchinson carbonaceous chondrites. Nature 317, 792–794.

Deamer, D., Dworkin, J.P., Sandford, A., Bernstein, M.P., Allamandola, L.J., 2002. The first cell membranes. Astrobiology 2, 371–382.

Deamer, D., Damer, B., 2017. Can life begin on enceladus? a perspective from hydrothermal chemistry. Astrobiology 17 (9), 834–839.

deLeeuw, G., et al., 2011. Production flux of sea spray aerosol. Rev. Geophys. 49, RG2001.

Dong, Y., Hill, T.W., Teolis, B.D., Magee, B.A., Waite, J.H., 2011. The water vapor plumes of Enceladus. J. Geophys. Res. 116, A10204.

Dostal, J., Murphy, J.B., Nance, R.D., 2009. History of the Earth, Volume 2 of Earth Systems: History and Natural Variability, Encyclopedia of Life Support Systems. Developed Under the Auspices of the UNESCO. Eolss Publishers, Oxford.

Dougherty, M., and Spilker, L., 2018. Review of Saturn's icy moons following the Cassini mission, Reports on Progress in Physics, 81 (6), ISSN: 0034-4885. DOI:10.1088/1361-6633/aabdfb.

Ehrenfreund, P., Spaans, M., Holm, N.G., 2011. The evolution of organic matter in space. Philos. Trans. A. Math. Phys. Eng. Sci. 369, 538–554.

Facchini, M.C., et al., 2008. Primary submicron marine aerosol dominated by insoluble organic colloids and aggregates. Geophys. Res. Lett. 35, L17814.

Ferris, J.P., Hagan Jr., W.J., 1984. HCN and chemical evolution: the possible role of cyano compounds in prebiotic synthesis. Tetrahedron 40, 1093–1120.

Gantt, B., Meskhidze, N., 2013. The physical and chemical characteristics of marine primary organic aerosol: a review. Atmos. Chem. Phys. 13, 3979–3996.

Gaston, C.J., et al., 2011. Unique ocean-derived particles serve as a proxy for changes in ocean chemistry. J. Geophys. Res. 116, D18310.

Georgiou, C.D., Deamer, D.W., 2014. Lipids as universal biomarkers of extraterrestrial life. Astrobiology 14, 541–549.

Glein, C.R., et al., 2015. The pH of Enceladus' ocean. Geochim. Cosmochim. Acta 162, 202–219.

Goguen, J.D., et al., 2013. The temperature and width of an active fissure on Enceladus measured with Cassini VIMS during the 14 April 2012 South Pole flyover. Icarus 226, 1128–1137.

Gross, R., Fouxon, I., Lancet, D., Markovitch, O., 2014. Quasispecies in population of compositional assemblies. BMC Evol Biol. 14. doi:10.1186/s12862-014-0265-1.

Guillot, T., Stevenson, D.J., Hubbard, W.B., Saumon, D., 2004. The interior of Jupiter. In: Bagenal, F., Dowling, T.E., McKinnon, W.B. (Eds.), Jupiter: The Planet, Satellites and Magnetosphere. Cambridge University Press, Cambridge, UK; New York: ISBN 978-0-521-81808-7.

Hansen, C.J., et al., 2006. Enceladus water vapor plume. Science 311, 1422–1425.

Hansen, C.J., et al., 2011. The composition and structure of the Enceladus plume. Geophys. Res. Lett. 38, L11202.

Hartman, H., Sweeney, M.A., Kropp, M.A., Lewis, J.S., 1993. Carbonaceous chondrites and the origin of life. Origin Life Evolution Biosphere 23, 221–227.

Helfenstein, P., 2010. Planetary science: tectonic overturn on Enceladus. Nat. Geosci. 3, 75–76. doi:10.1038/ngeo763.

Hendrix, A.R., Hurford, T.A., Barge, L.M., Bland, M.T., Bowman, J.S., Brinckerhoff, W., Buratti, B.J., Cable, M.L., Castillo-Rogez, J., Collins, G.C., Diniega, S., German, C.R., Hayes, A.G., Hoehler, T., Hosseini, S., Howett, C.J.A., McEwen, A.S., Neish, C.D., Neveu, M., Nordheim, T.A., Patterson, G.W., A. Patthoff, D., Phillips, C., Rhoden, A., Schmidt, B.E., Singer, K.N., Soderblom, J.M., Vance, S.D, 2019. The NASA roadmap to ocean worlds. Astrobiology 19, 1–27.

Hofgartner, J.D., Choukroun, M., Brophy, J.R., Casillas, R.P., Cooley, P., Fleurial, J-P., Parness, A., Wilcox, B.H., Cable, M.L., Carpenter, K., Chmielewski, A.B., Cutts, J.A., Landau, D., Reh, K.R., 2019. P34C-06 - Feasibility of a mission to Enceladus' subsurface ocean for the next planetary science decadal survey, AGU 100 Fall Meeting. San Francisco, CA, 9–13.

Horányi, M., Juhász, A., Morfill, G.E., 2008. Large-scale structure of Saturn's E-ring. Geophys. Res. Lett. 35, L04203.

Howett, C., et al., 2014. Enceladus' enigmatic heat flow, Proceedings of the 46th Meeting of the Division for Planetary Sciences. American Astronomical Society, Washington, DC Abstract 405.02.

Hodyss, et al., 2019. A new sampling system tailored to experimentally-derived mechanical properties of icy analogs for evolved Enceladus surface plume deposits, AGU Fall Meeting 2019, Advancing Earth and Space Science, Dec 9-13, 2019. Moscone Center, San Francisco, northern California, U.S.

Hsu, H.W., Postberg, F., Sekine, Y., Shibuya, T., Kempf, S., Horányi, M., Juhász, A., Altobelli, N., Suzuki, K., Masaki, Y., et al., 2015. Ongoing hydrothermal activities within Enceladus. Nature 519 (7542), 207–210.

Hurford, T.A., Helfenstein, P., Hoppa, G.V., Greenberg, R., Bills, B.G., 2007. Eruptions arising from tidally controlled periodic openings of rifts on Enceladus. Nature 447 (7142), 292–294.

Hurford, T.A., et al., 2009. Geological implications of a physical libration on Enceladus. Icarus 203, 541–552.

Iess, L., Stevenson, D.J., Parisi, M., Hemingway, D., Jacobson, R.A., Lunine, J.I., Nimmo, F., Armstrong, J.W., Asmar, S.W., Ducci, M., Tortora, P., 2014. The gravity field and interior structure of Enceladus. Science 344 (6179), 78–80.

Ingersoll, A.P., Pankine, A.A., 2010. Subsurface heat transfer on Enceladus: conditions under which melting occurs. Icarus 206, 594–607.

Joseph, A., 2016. Investigating Seafloors and Oceans: From Mud Volcanoes to Giant Squid. Elsevier Science & Technology Books Publishers, New York, p. 581 p., ISBN: 978-0-12- 809357-3.

Jurac, S., McGrath, M.A., Johnson, R.E., et al., 2002. Geophys. Res. Lett. 29, 2172.

Kahana, A., Schmitt-Kopplin, P., Lancet, D., 2019. Enceladus: first observed primordial soup could arbitrate origin-of-life debate. Astrobiology 19 (10), 1263–1278.

Kaiser, R., Stockton, A., Kim, Y., Jensen, E., Mathies, R., 2013. On the formation of dipeptides in interstellar model ices. Astrophys. J., 765. doi:10.1088/0004-637X/765/2/111.

Keene, W.C., et al., 2007. Chemical and physical characteristics of nascent aerosols produced by bursting bubbles at a model air–sea interface. J. Geophys. Res. 112, D21202.

Kelley, D.S., Karson, J.A., Blackman, D.K., Früh-Green, G.L., Butterfield, D.A., Lilley, M.D., Olson, E.J., Schrenk, M.O., Roe, K.K., Lebon, G.T., 2001. An off-axis hydrothermal vent field near the Mid-Atlantic Ridge at 30° N. Nature 412, 145–149.

Kempf, S., Beckmann, U., Schmidt, J., 2010. How the Enceladus dust plume feeds Saturn's E ring. Icarus 206, 446–457.

Kerridge, J.F., 1985. Carbon, hydrogen and nitrogen in carbonaceous chondrites: abundances and isotopic compositions in bulk samples. Geochim. Cosmochim. Acta 49, 1707–1714.

Kerridge, J.F., Chang, S., Shipp, R., 1987. Isotopic characterisation of kerogen-like material in the Murchison carbonaceous chondrite. Geochim. Cosmochim. Acta 51, 2527–2540.

Khawaja, N., Postberg, F., Hillier, J., Klenner, F., Kempf, S., Nölle, L., Reviol, R., Zou, Z., and Srama, R., 2019. Low-mass nitrogen-, oxygen-bearing, and aromatic compounds in Enceladean ice grains. Mon. Notices Royal Astron. Soc. 489 (4), 5231–5243.

Kite, E.S., Rubin, A.M., 2016. Sustained eruptions on Enceladus explained by turbulent dissipation in tiger stripes. Proc. Natl. Acad. Sci. U S A. 113 (15), 3972–3975. doi:10.1073/pnas.1520507113.

Lancet, D., Zidovetzki, R., Markovitch, O., 2018. Systems protobiology: origin of life in lipid catalytic networks. J. R. Soc. Interface 15. doi:10.1098/rsif.2018.0159.

Leck, C., Bigg, E.K., 2008. Comparison of sources and nature of the tropical aerosol with the summer high arctic aerosol. Tellus B 60, 118–126.

Maden, B.E.H., 1995. No soup for starters? Autotrophy and the origins of metabolism. Trends Biochem. Sci 20, 337–341.

Magee, B.A., Waite Jr, J.H., 2017. Neutral gas composition of Enceladus' plume – model parameter insights from Cassini-INMS, 48th Lunar and Planetary Science Conference abstr.2974.

Markovitch, O., Lancet, D., 2012. Excess mutual catalysis is required for effective evolvability. Artif. Life 18, 243–266.

Markovitch, O., Lancet, D., 2014. Multispecies population dynamics of prebiotic compositional assemblies. J. Theor. Biol. 357, 26–34.

Matson, D.L., Castillo, J.C., Johnson, T.V., Lunine, J.I., McCord, T.B., Sotin, C., Thomas, P.C., and Turtle, E.P., 2005. Thermal evolution models for Enceladus defining the context for the formation of the south pole thermal anomaly. Eos 86 (Fall Suppl.). Abstract #P32A-05.

Matsuyama, I., 2014. Tidal dissipation in the oceans of icy satellites. Icarus 242, 11–18.

McKay, C.P., Porco, C.C., Altheide, T., Davis, W.L., Kral, T.A., 2008. The possible origin and persistence of life on Enceladus and detection of biomarkers in the plume. Astrobiology 8, 909–919. doi:10.1089/ast.2008.0265.

McKay, C.P., Anbar, A.D., Porco, C., Tsou, P., 2014. Follow the plume: The habitability of Enceladus. Astrobiology 14 (4), 352–355.

Meyer, J., Wisdom, J., 2007. Tidal heating in Enceladus. Icarus 188, 535–539.

Miller, S.L., 1953. A production of amino acids under possible primitive Earth conditions. Science 117, 528–529.

Nimmo, F., Spencer, J.R., Pappalardo, R.T., Mullen, M.E., 2007. Shear heating as the origin of the plumes and heat flux on Enceladus. Nature 447, 289–291. doi:10.1038/nature05783.

Nimmo, F., Spencer, J.R., 2013. Enceladus: An active ice world in the Saturn system. Annu. Rev. Earth Planet. Sci. 41, 693–717.

Nimmo, F., et al., 2014. Tidally modulated eruptions on Enceladus: Cassini observations and models. Astron. J. 148, 48.

Oparin, A., 1962. Origin and evolution of metabolism. Comp. Biochem. Physiol. 4, 371–377.

Oro, J., Miller, S.L., Lazcano, A., 1990. The origin and early evolution of life on Earth. Annu. Rev. Earth Planet Sci. 18, 317–356.

Parkinson, C.D., Liang, M.-C., Hartman, H., Hansen, C.J., Tinetti, G., Meadows, V., Kirschvink, J.L., Yung, Y.L., 2007. Enceladus: Cassini observations and implications for the search for life (Research Note). Astron. Astrophys. 463, 353–357. doi:10.1051/0004-6361:20065773.

Parkinson, C.D., Liang, M.C., Yung, Y.L., Kirschivnk, J.L., 2008. Habitability of enceladus: planetary conditions for life. Origins of Life and Evolution of the Biosphere: the Journal of the International Society for the Study of the Origin of Life 38 (4), 355–369. doi:10.1007/s11084-008-9135-4 PMID: 18566911.

Pizzarello, S., Shock, E., 2017. Carbonaceous chondrite meteorites: the chronicle of a potential evolutionary path between stars and life. Orig Life Evol. Biosph. 47, 249–260.

Porco, C.C., et al., 2006. Cassini observes the active south pole of Enceladus. Science 311, 1393–1401.

Porco, C., DiNino, D., Nimmo, F., 2014. How the jets, heat and tidal stresses across the south polar terrain of Enceladus are related. Astron. J. 148, 45.

Porco, C.C., Dones, L., Mitchell, C., 2017. Could it be snowing microbes on Enceladus? Assessing conditions in its plume and implications for future missions. Astrobiology 17, 876–901.

Postberg, F., et al., 2008. The E ring in the vicinity of Enceladus. II. Probing the moon's interior—the composition of E-ring particles. Icarus 193, 438–454.

Postberg, F., Kempf, S., Schmidt, J., Brilliantov, N., Beinsen, A., Abel, B., Buck, U., Srama, R., 2009a. Sodium salts in E-ring ice grains from an ocean below the surface of Enceladus. Nature 459, 1098–1101. doi:10.1038/nature08046.

Postberg, F., et al., 2009b. Discriminating contamination from particle components in spectra of Cassini's dust detector CDA. Planet. Space Sci. 57, 1359–1374.

Postberg, F., Schmidt, J., Hillier, J., Kempf, S., Srama, R., 2011. Salt-water reservoir as the source of a compositionally stratified plume on Enceladus. Nature 474 (7353), 620–622.

Postberg, F., Khawaja, N., Abel, B., Choblet, G., Glein, C.R., Gudipati, M.S., Henderson, B.L., Hsu, H-W., Kempf, S., Klenner, F., Moragas-Klostermeyer, G., Magee, B., Nölle, L., Perry, M., Reviol, R., Schmidt, J., Srama, R., Stolz, F., Tobie, G., Trieloff, M., Waite, J.H., 2018. Macromolecular organic compounds from the depths of Enceladus. Nature 558, 564–568.

Powner, M.W., Gerland, B., Sutherland, J.D., 2009. Synthesis of activated pyrimidine ribonucleotides in prebiotically plausible conditions. Nature 459, 239–242.

Richardson, J.D., Jurac, S., 2004. Geophys. Res. Lett. 31, L24803.

Rushdi, A.I., Simoneit, B.R., 2001. Lipid formation by aqueous Fischer-Tropsch-type synthesis over a temperature range of 100 to 400 degrees C. Orig. Life Evol. Biosph. 31, 103–118.

Rushdi, A.I., Simoneit, B.R., 2011. Hydrothermal alteration of sedimentary organic matter in the presence and absence of hydrogen to tar then oil. Fuel 90, 1703–1716.

Russell, M.J., Daniel, R.M., Hall, A., 1993. On the emergence of life via catalytic iron-sulphide membranes. Terra Nova 5, 343–347.

Russell, M.J., Hall, A.J., 1997. The emergence of life from iron monosulphide bubbles at a submarine hydrothermal redox and pH front. J. Geol. Soc. London 154, 377–402.

Schmidt, J., Brilliantov, N., Spahn, F., Kempf, S., 2008. Slow dust in Enceladus' plume from condensation and wall collisions in tiger stripe fractures. Nature 451 (7179), 685–688.

Schmitt-Kopplin, P., Gabelica, Z., Gougeon, R.D., Fekete, A., Kanawati, B., Harir, M., Gebefuegi, I., Eckel, G., Hertkorn, N., 2010. High molecular diversity of extraterrestrial organic matter in Murchison meteorite revealed 40 years after its fall. Proc. Natl. Acad. Sci. USA. 107, 2763–2768.

Schneider, N.M., Burger, M.H., Schaller, E.L., Brown, M.E., Johnson, R.E., Karge, J.S., Dougherty, M.K., Achilleos, N.A., 2009. No sodium in the vapour plumes of Enceladus. Nature 459, 1102–1104. doi:10.1038/nature08070.

Segré, D., Lancet, D., 2000. Composing life. EMBO Rep. 1, 217–222.

Segré, D., Ben-Eli, D., Lancet, D., 2000. Compositional genomes: prebiotic information transfer in mutually catalytic noncovalent assemblies. Proc. Natl. Acad. Sci, USA. 97, 4112–4117.

Segré, D., Ben-Eli, D., Deamer, D.W., Lancet, D., 2001a. The lipid world. Orig. Life Evol. Biosph. 31, 119–145.

Segré, D., Shenhav, B., Kafri, R., Lancet, D., 2001b. The molecular roots of compositional inheritance. J. Theor. Biol. 213, 481–491.

Sekine, Y., Shibuya, T., Postberg, F., Hsu, H.W., Suzuki, K., Masaki, Y., Kuwatani, T., Mori, M., Hong, P.K., Yoshizaki, M., et al., 2015. High-temperature water-rock interactions and hydrothermal environments in the chondrite-like core of Enceladus. Nat. Commun. 6, 8604.

Shapiro, R., 2006. Small molecule interactions were central to the origin of life. Q. Rev. Biol. 81, 105–126.

Shenhav, B., Segrè, D., Lancet, D., 2003. Mesobiotic emergence: molecular and ensemble complexity in early evolution. Adv. Complex Syst. 6, 15–35.

Shenhav, B., Bar-Even, A., Kafri, R., Lancet, D., 2005. Polymer GARD: computer simulation of covalent bond formation in reproducing molecular assemblies. Orig. Life Evol. Biosph. 35, 111–133.

Spahn, F., et al., 2006. Cassini dust measurements at Enceladus and implications for the origin of the E ring. Science 311, 1416–1418.

Spencer, J., Pearl, J., Segura, M., Flasar, F., 2006. Cassini encounters Enceladus: Background and the discovery of a south polar hot spot. Science 311, 1401–1405.

Spencer, J., and Grinspoon, D., 2007. Planetary science: inside Enceladus. Nature 445 (7126), 376–7, doi: 10.1038/445376b.

Spencer, J., 2009. Planetary science: Enceladus with a grain of salt. Nature 459, 1067–1068. doi:10.1038/4591067a.

Spitale, J.N., Hurford, T.A., Rhoden, A.R., Berkson, E.E., Platts, S.S., 2015. Curtain eruptions from Enceladus' south-polar terrain. Nature 521 (7550), 57–60.

Srama, R., et al., 2004. The Cassini cosmic dust analyzer. Space Sci. Rev. 114, 465–518.

Srama, R., et al., 2011. The cosmic dust analyzer onboard Cassini: ten years of discoveries. CEAS Space Jour 2, 3–16.

Taubner, R.S., Pappenreiter, P., Zwicker, J., Smrzka, D., Pruckner, C., Kolar, P., Bernacchi, S., Seifert, A.H., Krajete, A., Bach, W., Peckmann, J., Paulik, C., Firneis, M.G., Schleper, C., Rittmann, S.K.R., 2018. Biological methane production under putative Enceladus-like conditions. Nat. Commun. 9 (1), 748.

Teolis, B.D., Perry, M.E., Magee, B.A., Westlake, J., Waite, J.H., 2010. Detection and measurement of ice grains and gas distribution in the Enceladus plume by Cassini's ion neutral mass spectrometer. J. Geophys. Res. 115, A09222.

Thomas, P.C., Tajeddine, R., Tiscareno, M.S., Burns, J.A., Joseph, J., Loredo, T.J., Helfenstein, P., Porco, C., 2016. Enceladus's measured physical libration requires a global subsurface ocean. Icarus 264, 37–47.

Tian, F., et al., 2006. Monte Carlo simulations of the water vapor plume on Enceladus. Icarus 188, 154–161.

Tsou, P., Brownlee, D.E., McKay, C.P., Anbar, A.D., Yano, H., Altwegg, K., Beegle, L.W., Dissly, R., Strange, N.J., Kanik, I., 2012. LIFE: Life Investigation For Enceladus: A Sample Return Mission Concept in Search for Evidence of Life. Astrobiology 12 (8), 730–742.

Tyler, R.H., 2008. Strong ocean tidal flow and heating on moons of the outer planets. Nature 456, 770–772.

Tyler, R.H., 2009. Ocean tides heat Enceladus. Geophys. Res. Lett. 36 (15), L15205. doi:10.1029/2009GL038300.

Tyler, R., 2011. Tidal dynamical considerations constrain the state of an ocean on Enceladus. Icarus 211 (1), 770–779. doi:10.1016/j.icarus.2010.10.007.

Tyler, R., 2014. Comparative estimates of the heat generated by ocean tides on icy satellites in the outer Solar System. Icarus 243, 358–385.

Vance, S., Harnmeijer, J., Kimura, J., Hussmann, H., Demartin, B., Brown, J.M., 2007. Hydrothermal systems in small ocean planets. Astrobiology 7 (6), 987–1005.

Waite Jr, J.H., et al., 2006. Cassini ion and neutral mass spectrometer: Enceladus plume composition and structure. Science 311, 1419–1422.

Waite Jr, J.H., Lewis, W.S., Magee, B.A., Lunine, J.I., McKinnon, W.B., Glein, C.R., Mousis, O., Young, D.T., Brockwell1, T., Westlake, J., Nguyen, M.-J., Teolis, B.D., Niemann, H.B., McNutt Jr., R.L., Perry, M., Ip, W.-H., 2009. Liquid water on Enceladus from observations of ammonia and ^{40}Ar in the plume. Nature 460, 487–490. doi:10.1038/nature08153.

Waite J.H.Jr., C.R.G, Perryman, R.S., Teolis, B.D., Magee, B.A., Miller, G., Grimes, J., Perry, M.E., Miller, K.E., Bouquet, A., Lunine, J.I., Brockwell, T., Bolton, S.J., 2017. Cassini finds molecular hydrogen in the Enceladus plume: Evidence for hydrothermal processes. Science 356 (6334), 155–159. doi:10.1126/science.aai8703.

Waller, S.E., Tallarida, N., Lambert, J.L., Belousov, A.E., Madzunkov, S.M., Darrach, M., Hodyss, R.P., Malaska, M., Hofmann, A., Charvat, A., Abe, B., Postberg, F., Lunine, J.I., Cable, M.L., 2019. P24A-07 - Analyzing Enceladus' plume constituents: First steps to experimentally simulating hypervelocity impacts, AGU 100 Fall Meeting. San Francisco, CA., 9–13.

Wigner, E., Huntington, H.B., 1935. On the possibility of a metallic modification of hydrogen. J. Chem. Phys. 3 (12), 764.

Williams, L.B., Canfield, B., Voglesonger, K.M., et al., 2005. Organic molecules formed in a "primordial womb. Geology 33, 913–916.

Williams, L.B., et al., 2011. Birth of biomolecules from the warm wet sheets of clays near spreading centers. In: Golding, S., Glikson, M. (Eds.), Earliest Life on Earth: Habitats, Environments and Methods of Detection. Springer, Dordrecht.

Wilson, T.W., et al., 2015. A marine biogenic source of atmospheric ice-nucleation particles. Nature 525, 234–238.

Wollrab, E., Scherer, S., Aubriet, F., Carré, V., Carlomagno, T., Codutti, L., Ott, A., 2016. Chemical analysis of a "Miller-type" complex prebiotic broth part I: chemical diversity, oxygen and nitrogen based polymers. Orig. Life Evol. Biosph. 46, 149–169.

Yeoh, S.K., Chapman, T.A., Goldstein, D.B., Varghese, P., Trafton, L.M., 2015. On understanding the physics of the Enceladus south polar plume via numerical simulation. Icarus 253, 205–222.

Yin, A., Pappalardo, R.T., 2015. Gravitational spreading, bookshelf faulting, and tectonic evolution of the South Polar Terrain of Saturn's moon Enceladus. Icarus 260, 409–439.

Zierenberg, R.A., Adams, M.W.W., Arp, A., 2000. Life in extreme environments: hydrothermal vents. Proc. Natl. Acad. Sci. USA. 97, 12961–12962.

Bibliography

Gaidos, E.J., Nealson, K.H., Kirschvink, J.L., 1999. Life in ice-covered oceans. Science 284, 1631–1633.

Goldsworthy, B.J., et al., 2003. Time of flight mass spectra of ions in plasmas produced by hypervelocity impacts of organic and mineralogical microparticles on a cosmic dust analyser. Astron. Astrophys. 409, 1151–1167.

Hillier, J.K., McBride, N., Green, S.F., Kempf, S., Srama, R., 2006. Modelling CDA mass spectra. Planet, Space Sci. 54, 1007–1013.

Kieffer, S.W., et al., 2006. A clathrate reservoir hypothesis for Enceladus' south polar plume. Science 314, 1764–1766.

Manga, M., Wang, C.Y., 2007. Pressurized oceans and the eruption of liquid water on Europa and Enceladus. Geophys. Res. Lett. 34, L07202. doi:10.1029/2007GL029297.

Matson, D.L., et al., 2012. Enceladus: A hypothesis for bringing both heat and chemicals to the surface. Icarus 221, 53–62.

Perry, M.E., et al., 2015. Cassini INMS measurements of Enceladus plume density. Icarus 257, 139–162.

Silverstein, R.M., Webster, F.X., Kiemle, D.J., 2005. Spectrometric Identification of Organic Compounds, 7th edn. John Wiley and Sons, Hoboken, pp. 1–70.

Srama, R., et al., 2009. Mass spectrometry of hyper-velocity impacts of organic micro grains. Rapid Commun. Mass Spectrom. 23, 3895–3906.

Tissot, B.P., Welte, D.H., 1984. Petroleum Formation and Occurrence, 2nd edn. Springer, Berlin 1984.

Von Damm, K.L., et al., 2005. The Escanaba Trough, Gorda Ridge hydrothermal system: temporal stability and subseafloor complexity. Geochim. Cosmochim. Acta 69, 4971–4984.

Waite Jr, J.H., et al., 2004. The Cassini ion and neutral mass spectrometer (INMS) investigation. Space Sci. Rev. 114, 113–231.

Chapter 14

Hydrocarbon lakes and seas & internal ocean on *Titan* — Resemblance with primitive earth's prebiotic chemistry

14.1 Titan—an earth-like system in some ways

Titan—the giant planet Saturn's largest moon—is the second largest moon in the solar system, outranked only by Jupiter's moon Ganymede. Titan's diameter of 5150 km makes it bigger than Mercury and only 25% smaller than Mars. Titan is small compared to Earth, with a surface gravity of 1/7 that of Earth. Titan is locked in a synchronous orbit around Saturn (i.e., an orbit in which Titan has a period equal to the average rotational period of Saturn, and having the same direction of rotation of Saturn) with a period of 16 Earth-days (the cycle of light and dark). The tilt of Saturn's spin axis to its orbit plane is ~27°, resulting in seasonal changes in the position of the Sun in Titan's sky over the ~30-year period of Saturn's orbit around Sun. Compared to most moons in our solar system, Titan's surface is relatively smooth, with few craters pock-marking its surface.

Much importance is attached to the study of Titan because this moon is considered to be an Earth-like system. Planetary scientists recognize that Titan provides an analog for many processes relevant to the Earth, more generally to outer solar system bodies, and a growing host of newly discovered icy exoplanets. Processes represented include atmospheric dynamics, complex organic chemistry, meteorological cycles (with methane as a working fluid), astrobiology, surface liquids and lakes, geology, fluvial (river-related) and aeolian (wind-induced) erosion, and interactions with an external plasma environment. Aeolian processes play a dominant role in shaping Titan's landscape. An analysis of material transport directions indicates that aeolian transport is the dominant transport mechanism at the equator and midlatitudes (Malaska et al., 2016).

Titan's special importance lies in the primitive chemically-reducing nature of its atmosphere. To date, Titan has been a focus of a number of spacecraft missions, as well as numerous Earth-based telescopic observations. The collected data have provided global observations of Titan's atmosphere and surface at a range of spatial and spectral resolutions. Pioneer 11 was the first spacecraft to encounter Saturn, and acquired the first images of Titan in 1979 (Tomasko, 1980). The Pioneer 11 mission was followed by the Voyager mission (an American scientific program that employs two robotic probes, Voyager 1 and Voyager 2, launched in 1977 to study the planetary systems of Jupiter, Saturn, Uranus, and Neptune; and even the interstellar space). The Voyager 1 flew by Saturn and Titan in 1980, and Voyager 2 in 1981. The Voyager missions returned important information about Titan's atmospheric chemistry (e.g., Hanel et al., 1981; Kunde et al., 1981; Maguire et al., 1981; Yung et al., 1984), but the cameras on Voyager were unable to resolve any of the fine details of the surface (Richardson et al., 2004).

Based on measurements from Cassini's magnetometer, Titan does not currently have an internal magnetic field (Backes et al., 2005). Titan is around four billion years old, about the same age as the rest of the solar system.

Titan is one of the solar system's most mysterious bodies and is a prime target for future space exploration. The information gathered by Huygens probe and various instruments deployed onboard Cassini spacecraft have revealed many details of a surprisingly Earth-like world and raised fascinating new questions for future study. Titan offers abundant complex organics on the surface of a water-ice-dominated ocean world, making it an ideal destination to study prebiotic chemistry and to document the habitability of an extraterrestrial environment. Having a methane cycle instead of a water cycle, Titan is a unique natural laboratory to investigate prebiotic chemistry and to search for signatures of hydrocarbon-based life instead of water-based terrestrial life. Titan has the potential for organics to interact with liquid water near or at the surface, furthering the potential for the progression of prebiotic chemistry as well as the search for signatures of water-based life (Turtle et al., 2017). *Titan*'s carbon-rich surface is shaped not only by impact craters and by winds that sculpt drifts of aromatic organics into long linear dunes but also by methane rivers and possible eruptions of liquid water ("cryovolcanism"). Titan is a compelling astrobiology target because its surface contains abundant complex carbon-rich chemistry and because both liquid water and liquid hydrocarbons can occur on its surface, possibly forming a prebiotic primordial soup.

14.2 Pre-Cassini mission knowledge of Titan

Cassini spacecraft mission—which carried out observation and measurements between mid-June 2004 and mid-September 2017 (a period exceeding 13 years)—has played an important role in achieving a substantially clearer understanding of Titan. However, much prior to this extraordinary mission of discovery, initial studies of Titan were carried out using Earth-based observations. Such studies provided a basic understanding of Titan. The basic knowledge of Titan thus achieved prior to the Cassini mission is briefly addressed in a few Sections below.

14.2.1 Titan's dense atmosphere

Titan is the only moon with a substantial atmosphere, the only other thick N_2 atmosphere besides Earth's, the site of extraordinarily complex atmospheric chemistry that far surpasses any other solar system atmosphere, and the only other solar system body with stable liquid currently on its surface.

Titan has an atmosphere chiefly made up of nitrogen (N_2) and methane (CH_4) and including many organics. This atmosphere also partly consists of hazes (photochemically generated involatile particles in Titan's atmosphere) and aerosol particles (particles suspended in air) which shroud the surface of this moon, giving it a reddish appearance (Coll et al., 1998). The aerosols observed in Titan's atmosphere are thought to be synthesized at high altitudes (>300 km) and fall to the surface. Varying with temperature profiles, condensation phenomena take place in the lower atmosphere, about 100 km below. These solid particles, often called "tholins," have been investigated for many years by laboratory scientists and physics modelers.

Titan's surface temperature is estimated to be about 94 K (−179°C) and its rivers run with liquid methane and ethane but it is a weirdly Earth-like place, even with this exotic combination of materials and temperatures. It may be noted that long ago, Mars also hosted rivers, which scoured valleys across its now-arid surface.

One of the most noticeable features of Titan is the orange blanket of haze that hides its surface. Because of Titan's large distance from the Sun (10 AU) and the haze in the atmosphere, the maximum level of sunlight on the surface of Titan is about 0.1% that of the overhead Sun on Earth's surface (McKay et al., 1991). Titan is a world with lakes and reservoirs composed of liquid methane and ethane near its poles, with vast, arid regions of hydrocarbon-rich dunes girdling its equator. Outside of Earth, Titan is so far the only other planetary body in the solar system that is known to have liquid bodies (lakes, reservoirs, seas) on its surface, with actively flowing rivers, though they are fed by liquid methane instead of water (Raulin, 2008).

Titan with its massive atmosphere is a prime planetary object in the outer solar system to perform both remote and in-situ investigations. Hörst (2017) suggested that because a large number of exoplanets appear to have an aerosol absorber in their atmospheres (see, e.g., Kreidberg et al., 2014; Knutson et al., 2014a, 2014b), which may be photochemically produced (Marley et al., 2013), there is increasing interest in using Titan as an exoplanet analog (Lunine, 2010; Bazzon et al., 2014; Robinson et al., 2014). The current understanding of Titan is forged from the powerful combination of Earth-based observations, remote sensing and in situ spacecraft measurements, laboratory experiments, and models.

Long before ESA's Huygens probe arrived at Titan, planetary scientists knew that the moon's dense atmosphere was mainly composed of nitrogen, with some methane, but the atmosphere's structure—its temperature and pressure at different altitudes—was poorly understood. The optical properties of the haze in the 1 to 3μm spectral region and the implications for the visibility of the surface were considered to be probably the most pressing research questions before the Cassini mission. Other key questions in the pre-Cassini period included the nature of the high-altitude detached haze layer, altitude and seasonal changes in the composition of the haze, the role of haze particles as condensation nuclei for clouds, and the nature of any condensate clouds.

Titan's atmosphere is composed primarily of N_2, with a little methane, argon, hydrogen, simple hydrocarbons and nitriles (i.e., organic compounds containing a cyanide group), carbon monoxide, and carbon dioxide. According to Owen (1987), sources of nitrogen may be a product of the photo-dissociation of ammonia (NH_3) or trapped in the ices that formed Titan. But theoretical models designed to study the formation of N_2 and organics on primordial Titan suggest that the initial form of nitrogen in Titan's atmosphere may have been NH_3. McKay et al. (1988) investigated the possible importance of strong shocks produced during high-velocity impacts accompanying the late states of accretion as a method for converting NH_3 to N_2. To simulate the effects of an impact in Titan's atmosphere they used the focused beam of a high-power laser, a method that has been shown to simulate shock phenomena. For mixtures of 10%, 50%, and 90% NH_3 (balance CH_4) they obtained yields of 0.25, 1, and 6×10^{17} molecules of N_2 per joule, respectively. They also found that the yield of hydrogen cyanide (HCN) is comparable to that for N_2. McKay et al. (1988) found that, in addition, several other hydrocarbons are produced, many with yields in excess of theoretical high-temperature-equilibrium models. It was found that, the above yields, when combined with models of Titan's accretion, result in a total N_2 production comparable to that present in Titan's atmosphere and putative ocean (i.e., an ocean assumed to exist).

Based on an assessment of D-to-H ratio in methane (i.e., the ratio of CH_3D to CH_4) in the stratosphere of Titan, Pinto et al. (1986) found that Titan's atmosphere is enriched in deuterium by a factor of > or = 3 relative to Jupiter and Saturn. On examination of probable potential causative factors for this enrichment, Pinto et al. (1986) found that fractionation occurring over a hypothetical methane-ethane (CH_4-C_2H_6) ocean and between the ocean and the clathrate crust beneath could be one of the reasons for the observed D-to-H ratio enrichment on Titan.

Although Titan was a target of studies for several decades, a detailed understanding of Titan was missing because of the presence of a thick haze enveloping it. Thus, Titan remained to be a mysterious solar system body. In fact, the importance of carrying out a detailed study of this mysterious moon encouraged the launch of Cassini Spacecraft Mission. The inspiration behind the Cassini Mission, supported by a dedicated probe (known as Huygens probe), was the limited knowledge already gained on this moon. Studies carried out on Titan before the launch of Cassini Mission are examined in the following subsections.

14.2.2 Size and shape of aerosol particles in Titan's atmosphere—results from voyager 1 and 2 missions

The size and shape of the aerosols in Titan's atmosphere initially perplexed researchers. Measurements of polarized scattered light from below 300 km, carried out during the *Pioneer* mission (NASA's first mission to the outer planets launched in 1972) and the *Voyager* program mentioned earlier suggested that the particles have an average size of $0.15\,\mu m$ (Tomasko and Smith, 1982). On the other hand, the observed high-phase angle scattering of the main haze layer required particles of larger size, close to $0.5\,\mu m$ (Rages et al., 1983). Gaining insight into the size and shape of aerosol particles in Titan's atmosphere is of paramount importance in understanding the peculiarities of Titan's unique hazy atmosphere. In this context, Rages et al. (1983) derived the limits on the physical properties of the scattering haze near the top of Titan's atmosphere from the data obtained from seven high-phase-angle images from Voyager 1 and 2. From the ratio of the intensities observed at two different high phase angles, an estimate can be made of the forward scattering lobe of the single-scattering phase function. Comparing the forward scattering estimate with diffraction lobes from particles of different radii, Rages et al. (1983) concluded that the average radius of the particles found in the upper few tenths of an optical depth exceeds $0.19\,\mu m$. They further argued that judging from the data observed at four different phase angles, the haze particles probably have a refractive index near 1.6 and a mean size of $\sim 0.5\,\mu m$, if the widths of their diffraction peaks are close to those for equal volume spheres. However, the highly polarizing nature of the particles over a broad wavelength bandpass (Tomasko and Smith, 1982) combined with their forward scattering behavior makes it very unlikely that the particles are spherical. It was suggested that the nonsphericity contributes to the uncertainty about the radii of the particles, but it is thought that the average radius is several tenths of a micron.

Owen and Gautier (1989)'s analyses of Voyager spectra of Titan have led to improvements in the determination of abundances of minor constituents as a function of latitude and altitude. Ground-based microwave observations have extended the Voyager results for hydrogen cyanide (HCN), and have demonstrated that carbon monoxide (CO) is mysteriously deficient in the stratosphere. While the origin of CH_4, CO, and N_2 in Titan's atmosphere remained unresolved, it has been found that both primordial and evolutionary sources are compatible with the available evidence.

Calculations of the optical properties of aggregate particles are able to resolve a persistent problem in understanding the shape and size of haze aerosols in the atmospheres of planets and moons (e.g., Jupiter and Titan). West and Smith (1991) found that most of the photometric and polarimetric observations for Titan can be explained by the presence of aggregate particles whose mean projected area is equal to that of a sphere with radius $0.14\,\mu m$, containing monomers (molecules that can be bonded to other identical molecules to form polymers) with mean radii near $0.06\,\mu m$. An additional mode of smaller particles is needed to fit ultraviolet data. Knowledge of the size and shape of the particles will allow for more precise estimates of the sedimentation rates and provide a key constraint on the coupled surface/atmosphere evolution of Titan.

In order to reconcile the two different sizes, West and Smith (1991) suggested that the aerosols in the main haze layer are aggregates composed of small spherical particles (the primary particles), with the latter responsible for the observed high degree of polarization and the former providing the scattering at visible wavelengths. Rannou et al. (1995) reported that theoretical calculations for the optical properties of aggregate particles are consistent with the observed geometric albedo (the proportion of the incident light or radiation that is reflected by the surface of a celestial body) of Titan's disk, providing further support to this scenario. Rannou et al. (2003) reported that the aggregates in these calculations have fractal properties characterized by a fractal dimension of two. The formation of aggregate particles is considered to take place outside the aerosol production region. Because the main aerosol layer is in the stratosphere (the second layer of the atmosphere above the troposphere, which is the lowest layer), most of the previous aerosol models assumed a production region at a higher altitude, usually at 400 km (right above the detached haze layer observed at 350 km in *Voyager* images).

In another related study, Israël et al. (2005) reported an in situ chemical analysis of Titan's aerosols by pyrolysis at 600 °C. In this study, NH_3 and HCN have been identified as the main pyrolysis products. This finding clearly shows that the aerosol particles include a solid organic refractory core. According to Israël et al. (2005), NH_3 and HCN are gaseous chemical fingerprints of the complex organics that constitute this core, and their presence demonstrates that carbon and nitrogen are in the aerosols.

Aerosols in Titan's atmosphere play an important role in determining its thermal structure. They also serve as sinks for organic vapors and can act as condensation nuclei for the formation of clouds, where the condensation efficiency will depend on the chemical composition of the aerosols (Israël et al., 2005).

14.2.3 Understanding the mechanisms of aerosol particle building in Titan's atmosphere

Planetary scientists had intuitive thinking that some kind of collisional mechanisms of Titan's aerosols may lead to a fractal structure (infinitely complex patterns that are self-similar across different scales, created by repeating a simple process over and over in an ongoing feedback loop) in which the aerosols are built by the aggregation of spherical submicrometer particles (monomers). In an initial study of the problem, Rannou et al. (1995) modeled the optical behavior of these aggregates, assuming that each monomer radiates a dipole field (a field between two charges of the same but opposite strength) in response to the incident radiation including the radiated fields of all the other elements in the aggregate. This dipole approximation, valid if the monomer radius is smaller than the wavelength, was used to calculate the scattering and extinction efficiencies of such aerosol particles, which were assumed to be composed of tholins (organic-rich aerosols). By applying the two-stream approximation for radiative transfer to the vertical distribution of aerosols obtained by microphysical modeling, Rannou et al. (1995) computed the geometric albedo of Titan. Computed values and observational values of the albedo were compared for wavelengths from 0.22 to 1.0 μm, and the effects of parameters, such as the fractal dimension of aerosols, their formation altitude or mass production rate, and, in addition, the methane abundance, were investigated. It was found that the hypothesized fractal structure of particles can explain both the visible and the UV albedos. In previous models, these measurements could only be matched simultaneously under the assumption of a bimodal population. Furthermore, for a fractal dimension $D_f \approx 2$ in the settling region, corresponding to a growth governed by cluster-cluster aggregation, the computed albedo in the near-UV range matched the observations. A good fit between measurement and calculated albedo was obtained, for a formation altitude $Z_0 = 535$ km, over the whole wavelength range by adjusting the absorption coefficient of the particles within a factor of two from that of tholins. Lower formation altitudes, such as the preferred case, $Z_0 = 385$ km chosen by Rannou et al. (1995), could not be investigated in the UV range due to limitations of the dipolar approximation, but this case is expected to give the same behavior.

As already indicated, a prominent feature of Titan's atmosphere is a thick haze region that acts as the end product of hydrocarbon and nitrile chemistry. Using a one-dimensional photochemical model, Wilson and Atreya (2003) carried out an investigation into the chemical mechanisms responsible for the formation of this haze region. The model-derived profiles for Titan's atmospheric constituents were found to be consistent with observations. The model demonstrated that the growth of polycyclic aromatic hydrocarbons (PAHs) throughout the lower stratosphere plays an important role in furnishing the main haze layer, with nitriles (organic compounds containing a cyanide group) playing a secondary role. It was found that the peak chemical production of haze layer ranges from 140 to 300 km, peaking at an altitude of 220 km, with a production rate of 3.2×10^{-14} g/cm^2/s. Wilson and Atreya (2003) have discussed the possible mechanisms for polymerization and copolymerization and suggestions for further kinetic study, along with the implications for the distribution of haze in Titan's atmosphere.

In another study, Rannou et al. (2004) developed a coupled general circulation model (GCM) of Titan's atmosphere in which the aerosol haze is treated with a microphysical model and is advected by the winds. Note that the radiative transfer accounts for the nonuniform haze distribution and, in turn, drives the dynamics. They analyzed the GCM results, especially focusing on the difference between a uniform haze layer and a haze layer coupled to the dynamics. It was found that in the coupled simulation the aerosols tend to accumulate at the poles, at latitudes higher than ±60°. During winter, aerosols strongly radiate at thermal infrared wavelengths enhancing the cooling rate near the pole. Because this effect tends to increase the latitudinal gradients of temperature the direct effect of this cooling excess, in contrast to the uncoupled haze case, is to increase the strength of the meridional cells as well as the strength of the zonal winds and profile. This is a positive feedback of the haze on dynamics. It was found that the coupled model reproduces observations about the state of the atmosphere better than the uniform haze model, and in addition, the northern polar hood and the detached haze are qualitatively reproduced.

14.2.4 Chemical transition of simple organic molecules into aerosol particles in Titan's atmosphere

Chemical transition of simple organic molecules into aerosol particles contributes significantly to the observed thick haze on Titan's atmosphere. In a study of the photochemistry of Titan's atmosphere, Yung et al. (1984) used updated chemical schemes and estimates of key rate coefficients C-, H-, and O-atom containing simple molecules, according to a model incorporating exospheric boundary conditions, vertical transport, and condensation processes at the tropopause (the interface between the troposphere and the stratosphere). Based on their studies, Yung et al. (1984) suggested that the composition, climatology, and evolution of the Titan atmosphere are controlled by five major processes, namely: (1) CH_4 photolysis (the decomposition or separation of molecules by the action of light radiation) and photosensitized dissociation; (2) H-to-H_2 conversion and hydrogen escape; (3) higher hydrocarbon synthesis; (4) nitrogen and hydrocarbon coupling; and (5) oxygen and hydrocarbon coupling. The model accounted for the minor species concentrations observed by Voyager instruments. As part of this study, Yung et al. (1984) have briefly discussed the implications of abiotic organic synthesis on Titan for the origin of life on earth.

In another study, Lara et al. (1999) suggested that the incorporation of HCN in Titan's atmospheric haze may be an important process affecting the HCN profile and the carbon/nitrogen (C/N) ratio in Titan's haze. Titan's haze is optically thick in the visible band of the electromagnetic spectrum. The haze varies with latitude in a seasonal cycle and has a detached upper layer. Microphysical models, photochemical models, and laboratory simulations all imply that the production rate of the haze is in the range of 0.5–2×10^{-14} g/cm^2/s. Given the rate of sedimentation, the total mass loading is about 250 mg/m^2. The transparency of the haze is high for wavelengths above 1µm because the haze material becomes almost purely scattering and the optical depth decreases with increasing wavelength. Note that "optical depth" is the natural logarithm of the ratio of incident to transmitted radiant power through a material. Thus, the larger the optical depth, the smaller the amount of transmitted radiant power through the material. The particles in the main haze deck are probably fractal in structure with an equivalent volume radius of 0.2µm. The haze material is organic and, if similar to laboratory tholin (organic-rich aerosol), has a C/N ratio in the range of 2–4 and a C/H ratio of about unity.

According to McKay et al. (2001), condensate clouds (collection of particles that form by condensation of species in Titan's atmosphere; and may be solid or liquid phase) of ethane or methane, if present, are thin, patchy, or transient. Stratospheric clouds of condensed nitriles (organic compounds containing cyanide groups) and possibly hydrocarbons appear to be associated with, though not contained entirely in, the polar shadow, suggesting abundances may vary with the season. Precipitating condensate particles from the stratosphere probably act as nucleating centers for the formation and rapid growth of methane ice particles in the troposphere, where the gas phase appears to be highly supersaturated. Once formed, fallout times for these hailstones are ~2h or less. According to McKay et al. (2001), melting and possible subsequent fragmentation of methane raindrops should occur at ~12km and below. Almost complete evaporation should occur just above the surface. A thin residue of ethane-enriched fog particles would then slowly settle to the surface, steadily modifying an existing surface or subsurface residue of liquid hydrocarbons.

Lebonnois et al. (2002) investigated the chemical transition of simple molecules such as acetylene (C_2H_2) and hydrogen cyanide (HCN) into aerosol particles in the context of Titan's atmosphere. They found that experiments that synthesize analogs (tholins) for these aerosols can help illuminate and constrain these polymerization mechanisms. Using information available from these experiments, Lebonnois et al. (2002) suggested chemical pathways that can link simple molecules to macromolecules, which will be the precursors to aerosol particles: polymers of acetylene and cyanoacetylene, polycyclic aromatics, polymers of HCN and other nitriles, and polyynes. Note that polyynes are any organic compound with alternating single and triple bonds; that is, a series of consecutive alkynes, $(-CC-)_n$. with n greater than 1. The simplest example is diacetylene or buta-1,3-diyne, H−CC−CC−H. Although Lebonnois et al. (2002)'s goal was not to build a detailed kinetic model for this transition, they proposed parameterizations to estimate the production rates of these macromolecules, their C/N and C/H ratios, and the loss of parent molecules (C_2H_2, HCN, HC_3N and other nitriles, and C_6H_6) from the gas phase to the haze. Lebonnois et al. (2002) used a one-dimensional photochemical model of Titan's atmosphere to estimate the formation rate of precursor macromolecules. They found a production zone slightly lower than 200 km altitude with a total production rate of 4×10^{-14} g/cm^2/s and a C/N$\simeq$4. They compared the results with experimental data, and to microphysical model requirements. The planetary researchers in the pre-Cassini/Huygens mission eagerly waited to see a detailed picture of the haze distribution and properties, which was expected to be a great challenge for our understanding of these chemical processes.

14.2.5 Understanding particle size distribution in Titan's hazy atmosphere

Particle size distribution in the aerosol above Titan's surface determines the thickness and haziness of its atmosphere. With a view to understanding this distribution, Mitchell and Frenklach (2003) modeled particle aggregation with simultaneous surface growth using a dynamic Monte Carlo method. In their experiment, the Monte Carlo algorithm began in the particle inception zone and constructed aggregates via ensemble-averaged collisions between spheres and deposition of gaseous species on the sphere surfaces. Simulations were conducted using four scenarios. The first, referred to as scenario 0, was used as a benchmark and simulated aggregation in the absence of surface growth. Scenario 1 forced all balls to grow at a uniform rate, while scenario 2 only permitted them to grow once they have collided and stuck to each other. The last one is a test scenario constructed to confirm conclusions drawn from scenarios 0–2. Mitchell and Frenklach (2003) further investigated the transition between the coalescent and the fully developed fractal aggregation regimes using shape descriptors to quantify particle geometry. They were used to define the transition between the coalescent and fractal growth regimes. The simulations demonstrated that the morphology (a particular form, shape, or structure) of aggregating particles is intimately related to both the surface deposition and particle nucleation rates.

14.2.6 Gaining Insight on the vertical distribution of Titan's atmospheric haze

To achieve an adequate insight on the observed haze on Titan's surface, it is necessary to understand its vertical distribution. In this context, Rannou et al. (2003) used Titan's geometric albedo to constrain the vertical distribution of the haze. It was found that microphysical models incorporating fractal aggregates do not readily fit the methane features at 0.62 μm band and the dark 0.88 μm of the albedo spectrum simultaneously. Rannou et al. (2003) took advantage of this apparent discrepancy to constrain the haze vertical profile. They used the geometric albedo and several results and constraints from other works to better constrain the vertical haze extinction profile, especially in the low stratosphere. The objective of this model was to give a solution that simultaneously fits the main constraints known to apply to the haze.

Rannou et al. (2003) found that the haze extinction increases with decreasing altitude with a scale height about equal to the atmospheric scale height down to 100 km. Below this altitude, extinction must decrease down to 30km. According to Rannou et al. (2003) this is necessary in order to have enough haze to sustain a relatively high albedo (0.076) in the dark 0.88μm methane band and to show the 0.62μm band in the haze continuum. These investigators set the haze production rate around 7×10^{-14}kg/m²/s, and the aerosols production altitude around 400km (or at pressure 1.5Pa). Rannou et al. (2003) admit that the physical processes which generate such a profile are not clear. However, purely one-dimensional effects such as condensation, sedimentation, and rainout can be ruled out, and it is believed that this relative clearing in Titan's troposphere and lower stratosphere is due to particle horizontal transport by the mean circulation.

14.2.7 Greenhouse and antigreenhouse effects on Titan

There are many parallels between the atmospheric thermal structure of Titan and the terrestrial greenhouse effect. These parallels provide a comparison for theories of the heat balance of Earth. According to McKay et al. (1991), Titan's atmosphere has a greenhouse effect caused primarily by pressure-induced opacity of nitrogen (N_2), methane (CH_4), and hydrogen (H_2). Hydrogen is a key absorber because it is primarily responsible for the absorption in the wave number 400/cm to 600/cm "window" region of Titan's infrared spectrum. McKay et al. (1991) explain that the concentration of CH_4, also an important absorber, is set by the saturation vapor pressure and hence is dependent on temperature. According to McKay et al. (1991), in this respect there is a similarity between the role of H_2 and CH_4 on Titan and that of CO_2 and H_2O vapor on Earth.

McKay et al. (1991) found that Titan also has an antigreenhouse effect that results from the presence of a high-altitude haze layer that is absorbing at solar wavelengths but transparent in the thermal infrared. It has been found that winds and haze significantly affect the thermal balance of Titan, causing an antigreenhouse effect that cools the surface by 9 K. Interestingly, whereas the antigreenhouse effect on Titan reduces the surface temperature by 9 K, the greenhouse effect increases it by 21 K. Therefore, the net effect is that the surface temperature (94 K) is 12 K warmer than the effective temperature (82 K). McKay et al. (1991) estimated that "if the haze layer were removed, the antigreenhouse effect would be greatly reduced, the greenhouse effect would become even stronger, and the surface temperature would rise by over 20 K."

14.2.8 Inferring clues on Titan's surface

In addition to methane, a number of products of Titan's atmospheric photochemistry are also liquid at Titan surface conditions; of these, ethane is the most abundant. Based on photochemical considerations, Lunine et al. (1983) argued that Titan's surface might be covered in vast seas or oceans of ethane. Based on model studies, Yung et al. (1984) had predicted that photolysis of N_2 and CH_4 abundantly available in Titan's immensely thick atmosphere can produce a plethora of organic molecules that would end up as liquid and solid sediments on Titan's surface. Based on this study, it was inferred that Titan's surface might have two distinct components, an icy bedrock and the atmospheric-derived organic sediments. However, concrete observational evidence was lacking. In several studies prior to Cassini–Huygens mission (e.g., Muhleman et al., 1990; Lemmon et al., 1993; Griffith, 1993; Lemmon et al., 1995; Smith et al., 1996), Earth-based measurements revealed that Titan's surface is heterogeneous.

14.3 Role of Cassini spacecraft and Cassini–Huygens probe in understanding Titan better

Although several studies of Titan have been carried out prior to the Cassini spacecraft mission, they were based on earth-based observations and models. A much more detailed study of Titan could be achieved only through much closer observations and measurements carried out by the Cassini Spacecraft, which orbited Titan several times, and the Cassini–Huygens probe that pierced through Titan's hazy atmosphere and landed on Titan's surface.

14.3.1 Cassini spacecraft mission to explore planet Saturn and some of its icy moons

With the launch of Cassini spacecraft on Oct. 15, 1997, a seven-year journey (1997–2004) to Saturn began. Because of the fact that Titan is swathed in a dense, hazy atmosphere that obscures its surface to telescopes and cameras, Cassini carried a probe called *Huygens* to explore the surface of Saturn's largest and hazy moon, *Titan*. After a nearly 2.2-billion-mile journey, the spacecraft arrived in the Saturn system on June 30, 2004. Fig. 14.1 shows an artist's concept of Cassini during the Saturn Orbit Insertion (SOI) maneuver. The Cassini spacecraft has been orbiting Saturn since mid-2004. Its final orbit happened on September 15, 2017, with its fall into Saturn's atmosphere, ending its extraordinary mission of discovery.

On June 30, 2004 Cassini passed between the F and G rings of Saturn and allowed itself to be captured as a moon of Saturn. Having been released on December 24, 2004, the Huygens probe descended (parachuted) through Titan's hazy atmosphere, landed on Titan and relayed data to Earth on January 14, 2005. During an eventful four years of its initial orbital tour (2004–2008), Cassini made 45 Titan encounters, 10 other icy moon encounters and 76 orbits of Saturn. In the next two years (2008–2010) Cassini succeeded in making 64 orbits around Saturn, 28 Titan encounters, 8 Enceladus encounters, 3 encounters with other smaller icy moons; and carried out equinox crossing in August 2009. The next seven years (2010–2017) also proved to be as eventful as

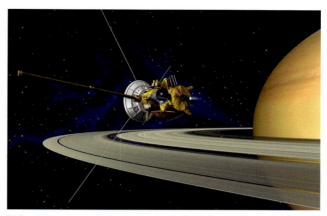

FIG. 14.1 An artist's concept of Cassini spacecraft during the Saturn Orbit Insertion (SOI) maneuver, just after the main engine has begun firing. The spacecraft is moving out of the plane of the page and to the right (firing to reduce its spacecraft velocity with respect to Saturn) and has just crossed the ring plane. The SOI maneuver, which is approximately 90 min long, allowed Cassini to be captured by Saturn's gravity into a 5-month orbit. Cassini's close proximity to the planet after the maneuver offers a unique opportunity to observe Saturn and its rings at extremely high resolution. *Credit: NASA/JPL. (Source: https://commons.wikimedia.org/wiki/File:Cassini_Saturn_Orbit_Insertion.jpg. Copyright: This file is in the public domain in the United States because it was solely created by NASA. NASA copyright policy states that "NASA material is not protected by copyright unless noted.")*

the earlier years, with several success stories such as 155 orbits, 54 Titan encounters, 11 Enceladus encounters, and 5 other icy moon encounters.

14.3.2 The European space agency's Huygens probe to explore Titan's hazy atmosphere and its surface

The Cassini–Huygens mission is a cooperative project of NASA, the European Space Agency and the Italian Space Agency. The Jet Propulsion Laboratory, a division of the California Institute of Technology in Pasadena, manages the Cassini–Huygens mission for NASA's Science Mission Directorate, Washington, D.C. The Cassini orbiter was designed, developed and assembled at JPL. The radio science team is based at Wellesley College, Wellesley, Massachusetts.

Until the Cassini mission, little was known about Titan, except that it is a hazy orange ball about the size of Mercury, and that Titan's surface is veiled beneath a thick, nitrogen-rich atmosphere—the only known world with a dense nitrogen atmosphere besides Earth. But what might lie beneath the smoggy clouds was still largely a mystery. But Cassini mapped Titan's surface, studied its atmospheric reactions, discovered liquid seas there (although it was unknown what the liquid is), and even sent a probe to the moon's surface, completely rewriting our understanding of this remarkably

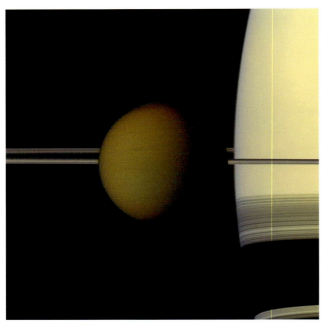

FIG. 14.2 **The colorful globe of Titan, the largest of Saturn's 53 moons, passing in front of the planet Saturn and its rings in this true color snapshot from NASA's Cassini spacecraft.** The rings appear as a thin band, and their striped shadows encircle the planet Saturn's face. Credit: NASA/JPL-Caltech/Space Science Institute. (Source: https://saturn.jpl.nasa.gov/science/Titan/https://saturn.jpl.nasa.gov/science/Titan/ (https://solarsystem.nasa.gov/resources/15441/Titan-up-front/).

FIG. 14.3 **The European Space Agency's Huygens Probe — a unique, advanced spacecraft and a crucial part of the overall Cassini mission to explore Saturn.** The probe was about 9 feet wide (2.7 m) and weighed roughly 700 pounds (318 kilograms). It was built like a shellfish: a hard shell protected its delicate interior from high temperatures during the 2.25 h descent through the atmosphere of Saturn's giant moon Titan (Source: https://saturn.jpl.nasa.gov/science/Titan/).

Earth-like world. Data from Cassini–Huygens revealed Titan has lakes and seas of liquid methane and ethane, replenished by rainfall from hydrocarbon clouds. The mission also provided evidence that Titan is hiding an internal liquid ocean beneath its surface, likely composed of saltwater and ammonia. Fig. 14.2 shows Titan passing in front of the planet Saturn and its rings in a true color snapshot from NASA's Cassini spacecraft.

The European Space Agency's Huygens probe (Fig. 14.3) was designed to land on the surface of Titan during Cassini spacecraft's flyby around Saturn. The probe was a unique, advanced spacecraft and a crucial part of the overall Cassini mission to explore Titan in detail. The probe was about 9 feet wide (2.7 m) and weighed roughly 700 pounds (318 kilograms). It was built like a shellfish. A hard shell protected its delicate interior from high temperatures during a 2-h and 27-min descent through the atmosphere of Titan.

The probe had two parts: An Entry Assembly Module and the Descent Module. The Entry Assembly Module carried the equipment to control Huygens after separation from Cassini, and a heat shield that acted as a brake and as thermal protection. The Descent Module contained the scientific instruments and three different parachutes that were deployed in sequence to control Huygens' descent to the surface of Titan (https://saturn.jpl.nasa.gov/science/Titan/).

The scientific payload onboard Huygens probe consisted of several instruments for measuring the physical, electrical (permittivity and the distribution of ions), and thermal properties of the atmosphere, its electromagnetic wave activity, wind gusts, etc. The Huygens Atmospheric Structure Instrument (HASI) made the first in situ measurements of Titan's atmosphere, such as the atmospheric temperature, pressure, and density from an altitude of 1400 km down to the surface.

Another instrument known as Descent Imager/Spectral Radiometer (DISR) made a range of imaging and spectral observations using several sensors and fields of view. By measuring the upward and downward flow of radiation, the radiation balance (or imbalance) of the thick Titan atmosphere was measured. Solar sensors measured the light intensity around the Sun as received at Titan's surface due to scattering by aerosols in Titan's atmosphere. These measurements permitted the calculation of the size and number density of the suspended particles. Two imagers (one visible, and the other infrared) observed Titan's surface during the latter stages of the descent and, as the probe slowly rotated, built up a mosaic of pictures around the landing site. There was also a side-view visible imager that obtained a horizontal view of the horizon and the underside of the cloud deck. A lamp switched on shortly before landing augmented the weak sunlight. This artificial illumination permitted spectral measurements of Titan's surface from close quarters.

Apart from these measurements, an instrument known as a gas chromatograph mass spectrometer (GCMS) identified and measured chemicals in Titan's atmosphere. During descent, the GCMS analyzed pyrolysis products (i.e., samples altered by artificial heating in an oven) passed to it from another instrument known as Aerosol Collector Pyrolyser (ACP). The process of pyrolysis vaporized volatiles and decomposed the complex organic materials. The GCMS also measured the composition of Titan's surface after Huygens' safe landing.

Another suite of sensors enabled the determination of the physical properties of Titan's surface at the point of impact. These sensors could also determine whether the surface was solid or liquid. An acoustic sounder, which could be activated during the last 100 m (328 feet) of the descent, continuously determined the distance to the surface, measuring the rate of descent and the surface roughness. During descent, measurements of the speed of sound provided information on atmospheric composition and temperature, and an accelerometer accurately recorded the deceleration profile at impact, providing information on the hardness and structure of the surface. A tilt sensor measured any pendulum motion during the descent and indicated the probe attitude after landing.

14.3.3 Landing of Huygens probe on Titan's surface

The Huygens probe landed on Titan on January 14, 2005. The probe support equipment (PSE) remained attached to the orbiting spacecraft. The support equipment included the electronics necessary to track the probe, recover the data gathered during its descent, and process and deliver the data to the orbiter. The data was then transmitted or down-linked from the orbiter to Earth. The Huygens probe payload also consisted of six scientific instruments; each designed to perform a different function as the probe descended through Titan's murky atmosphere. It may be borne in mind that the Huygens probe was designed with essentially no information about Titan's surface and was not guaranteed to survive impact. As a result, it was not capable of precision landing near a site of astrobiological interest, such as an impact crater or cryovolcano.

As the Huygens Probe descended for two and a half hours, Huygens took measurements of Titan's atmospheric composition and pictures of its surface. The hardy probe not only survived the descent and landing, but continued to transmit data for more than an hour on Titan's frigid surface, until its batteries were drained.

It has been found that Titan's surface is darker than originally expected, consisting of a mixture of water and hydrocarbon ice. There is also evidence of erosion at the base of these objects, indicating possible fluvial activity. The landing site itself resembled a dried-up riverbed. Rounded cobbles, 10 cm to 15 cm in diameter and probably made of hydrocarbons and water ice, rested on a darker granular surface.

No evidence of surface liquid was found at the time of the landing. However, according to ESA, it seems likely that, from time to time, the entire dark region is inundated by floods of liquid methane and ethane. If the darker region is a dry lakebed, it is too large to have been caused by the creeks and channels visible in the images. It may have been created by other larger river systems or some large-scale catastrophic event, which predates deposition by the rivers seen in the images. It has been believed that the dark material that covers the plains may have been carried along by the flows and could be made up of photochemical deposits rained down from above (http://sci.esa.int/cassini-huygens/55229-science-highlights-from-huygens-8-dry-river-beds-and-lakes/).

It is heartening to every space enthusiast that the Huygens probe provided detailed information about Titan's atmospheric profile and chemistry (Fulchignoni et al., 2005; Niemann et al., 2005). Although the Huygens probe was able to image Titan's surface at the meter scale from an altitude of 10 km, surface spectra could not be obtained outside of a few specific spectroscopic windows (Tomasko et al., 2005). This is because, at these altitudes, there is little solar illumination for the surface to reflect, since much of the sunlight has been absorbed or scattered by the overlying atmosphere (Tomasko et al., 2005). Thus, remotely identifying biomolecules on Titan's surface from above or within Titan's atmosphere would be difficult, even with an infrared camera that has finer spatial and spectral resolution and wider spectral range than the Visual and Infrared Mapping Spectrometer (VIMS) (Neish et al., 2018).

McDonald et al. (2015) modeled the effect of methane absorption with altitude and found a slight widening of the spectral windows at altitudes closer to the surface. However, they neglected to include the effects of atmospheric scattering, and thus judge that the broadening they observed is at best an upper limit. As a result, an airplane or balloon would provide little, if any, improvement in the wavelengths available for spectroscopy over an orbiter. Given these constraints, it would be difficult for a remote spectrometer to identify spectral features associated with common biological molecules on Titan.

14.4 Gaining better understanding of Titan based on data gleaned from Cassini spacecraft mission

For many years, Titan's thick, methane- and nitrogen-rich opaque atmosphere precluded astronomers from deciphering what lies beneath. Titan appeared through telescopes as a hazy orange spherical object, in contrast to other heavily cratered moons in the solar system. Since 2011, Cassini has caught glimpses of the transition from fall to winter at Titan's South Pole (https://saturn.jpl.nasa.gov/science/Titan/). Taking account of the fact that each Titan season lasts about 7.5 Earth years, having been able to see for the first time the onset of a Titan winter, and watching as summer came to the north was indeed a matter of pride to the entire humanity.

14.4.1 Titan's clouds, storms, and rain

Clouds have been observed on Titan, through the thick haze, using near-infrared spectroscopy and images near the south pole and in temperate regions near 40°S. Telescope and Cassini orbiter observations have provided an insight into cloud climatology. Interestingly, unlike Earth, only two kinds of clouds have been detected on Titan: large storms near the south pole and long clouds predominantly at 40°S latitude. It has been found that the south polar clouds reside at the altitude of neutral buoyancy, which is indicative of convection.

By means of a three-dimensional general circulation model employed to investigate the influence of Saturn's gravitational tide on the atmosphere of Titan, Tokano (2002) showed that unlike atmospheric tides on terrestrial planets, Saturn's tide on Titan has a large impact on the dynamic meteorology down to the surface.

In a study carried out by Griffith et al. (2005), spectra from Cassini's Visual and Infrared Mapping Spectrometer revealed that the horizontal structure, height, and optical depth of Titan's clouds are highly dynamic. In addition, vigorous cloud centers have been seen to rise from the middle to the upper troposphere within 30 min and dissipate within the next hour. According to Griffith et al. (2005), the development of such vigorous cloud centers indicates that "Titan's clouds evolve convectively; dissipate through rain; and, over the next several hours, waft downwind to achieve their great longitude extents. These and other characteristics suggest that temperate clouds originate from circulation-induced convergence, in addition to a forcing at the surface associated with Saturn's tides, geology, and/or surface composition."

Titan and Earth are the only worlds in the solar system where rain reaches the surface. Hueso and Sánchez-Lavega (2006) reported that on Titan raindrops of 1–5 mm in radius produce precipitation rainfalls on the surface as high as 110 kg/m^2 and are comparable to flash flood events on Earth (Maddox, 1980). However, the atmospheric cycles of water and methane are expected to be very different (Lorentz et al., 1997). Based on several studies (e.g., Porco et al., 2005; Lebreton et al., 2005), the nitrogen atmosphere of Titan appears to support a methane meteorological cycle that sculptures the surface and controls its properties. Hueso and Sánchez-Lavega (2006) reported three-dimensional dynamical calculations showing that severe methane convective storms accompanied by intense precipitation may occur in Titan under the right environmental conditions. Their calculations show that the strongest storms grow when the methane relative humidity in the middle troposphere is above 80%, producing updrafts with maximum velocities of 20 m/s, able to reach altitudes of 30 km before dissipating in 5–8 h.

To study clouds, Rannou et al. (2006) developed a general circulation model of Titan that includes cloud microphysics. They identified and explained the formation of several types of ethane and methane clouds, including south polar clouds and sporadic clouds in temperate regions and especially at 40° in the summer hemisphere. The locations, frequencies, and composition of these cloud types are essentially explained by the large-scale circulation. The presence of dry fluvial river channels and the intense cloud activity in the south pole of Titan (Roe et al., 2002; Porco et al., 2005; Tomasko et al., 2005) suggest the presence of methane rain (Hueso and Sánchez-Lavega, 2006). Titan shows landscapes with fluvial features, suggestive of hydrology based on liquid methane. Post-Cassini efforts in understanding Titan's methane hydrological cycle have focused on occasional cloud outbursts near the south pole or cloud streaks at southern mid-latitudes and the mechanisms of their formation. It is not known, however, if the clouds produce hydrocarbon rain or if there are also nonconvective clouds, as predicted by several models. A study by Tokano et al. (2006) forecasts light drizzle over much of Titan's surface for the next few years. This is consistent with the damp surface observed by the Huygens probe at its landing site (at 10° S latitude, 192° W longitude). But such drizzle is too delicate to create the fluvial features that cut through Titan's hard surface (Griffith, 2006).

With regard to the detailed nature of Titan's storms, Hueso and Sánchez-Lavega (2006) reported models of the formation and evolution of Titan's storms with a resolution of 0.5 km. These models describe the small-scale structure of discrete clouds that are formed by updrafts of air. These researchers found that small temperature perturbations of 0.5°C and updrafts of 1 m/s initiate the formation of Titan's high clouds, and explain their altitudes and morphologies. These models also predict unexpectedly heavy rainfall, equivalent to severe terrestrial storms (Griffith, 2006). Hueso and Sánchez-Lavega (2006)'s study suggests how these river valleys may have formed. Models of Titan's storms predict torrential rain strong enough to carve the rugged washes. Yet no storms were seen at the Huygens landing site, nor was there any indication of how storms might develop in the calm equatorial atmosphere of that area (Tokano et al., 2006.

Griffith (2006) explained how a picture emerges of the pole-to-pole migration of rainstorms (Hueso and Sánchez-Lavega, 2006; Rannou et al., 2006) that shapes river valleys and creates hydrocarbon lakes. According to this researcher, the leftover rivers and metre-deep lakes evaporate during the dry seasons, into an atmosphere that can hold enough methane to cover Titan's surface to a depth of 4 m. With regard to the physical mechanism of how storms form in Titan's tropics, Griffith (2006) suggested that perhaps atmospheric tides from Saturn as illustrated by Tokano (2002) cause stronger local updrafts to occur during the equinox than are currently predicted. In another study, Tokano (2005) reported a systematic investigation of the seasonal and spatial variation of the surface temperature and air temperature in the lower troposphere by a 3-dimensional general circulation model for different putative surface types. Based on these study results, Griffith (2006) suggested that atmospheric tides from Saturn, possibly in combination with others, such as solar warming of a porous surface (Tokano, 2005), may cause tropical storms at the equinox (Hueso and Sánchez-Lavega, 2006).

Griffith (2006) has noted that just like stratiform clouds develop over Earth's Arctic in the summer, as warm, low-latitude air flows over the cooler ice shelves, stratiform clouds exist on Titan as well (Tokano et al., 2006). Griffith (2006) further noted that "measurements taken by the Huygens probe (Niemann et al., 2005; Fulchignoni et al., 2005) show that Titan has a complicated methane distribution over the tropics. Here, the abundance of methane is mysteriously pegged at around 80% humidity, assuming pure methane condensation, at altitudes of 8–16 km, and rises to 100% humidity at around 21 km." Griffith (2006) recalls that investigators studying Titan's weather suggest that "below 16 km, where the clouds consist of liquid droplets, it is not pure methane that condenses; a liquid solution of nitrogen dissolved in methane condenses instead. The atmosphere is found to be saturated at altitudes of 8–16 km with respect to methane–nitrogen liquid condensation, and saturated at around 21–30 km for pure methane-ice condensation. These wet conditions indicate the existence of stratiform methane clouds at 21–30 km, as seen by Huygens, and tenuous methane–nitrogen clouds at 8–16 km, not yet detected. The highly stable conditions in Titan's tropics suggest that the resulting drizzle is typical of this region over most of Titan's year, except possibly near the equinox." Griffith (2006) is of the view that "although clouds on both Titan and Earth may, on average, be married to their atmospheric circulations, Titan's weather is enticingly alien."

Schaller et al. (2009) found a storm from ground-based observations that was largely missed by Cassini. These were low-latitude storms, including one that resulted in extensive alteration of the surface, presumably caused by large amounts of methane rainfall (Turtle et al., 2011b). Studies by several investigators (e.g., Hueso and Sánchez-Lavega, 2006; Rafkin and Barth, 2015; Charnay et al., 2015) indicated that these convective storms may produce fast surface winds that increase the lifetime of the storm and enhance precipitation.

Turtle et al. (2011) have reported the detection by Cassini's Imaging Science Subsystem of a large low-latitude cloud system early in Titan's northern spring and extensive surface changes (spanning more than 500,000 km^2) in the wake of this storm. They found that the changes are most consistent with widespread methane rainfall reaching the surface, which suggests that the dry channels observed at Titan's low latitudes are carved by seasonal precipitation.

Turtle et al. (2011) observed rain in equatorial latitudes. Williams et al. (2012) suggested that moist surface conditions could persist for 5 to 50 days after a rain event, depending on moisture level. Based on analysis of spectral data, Rannou et al. (2016) concluded that the ground at low latitudes is persistently wet, resulting either from a subsurface rich in liquid or from frequent enough rains to maintain moist surface conditions against evaporation.

Much like Earth, Titan exhibits a number of types of clouds resulting from different formation processes and conditions. Unlike Earth, Titan's clouds form from a number of different volatiles. Tsai et al. (2012) divided clouds into three categories: convective methane clouds, stratiform ethane clouds, and high-altitude cirrus clouds (HCN, HC$_3$N, etc.). In addition to those clouds, there is currently a large, high-altitude HCN ice cloud, first observed in May 2012, that formed in the polar vortex at the south pole (de Kok et al., 2014). (Hörst, 2017) found that, in general, convective methane clouds are observed at the summer pole and midlatitudes, while the other types of clouds are observed at the winter pole.

14.4.2 Detection of tall sand dunes in the equatorial regions of Titan

The area around the Huygens landing site (Fig. 14.4) turned out to be a huge plain of dirty water ice over which lay blankets of organic (carbon-bearing) deposits. Although dark, longitudinal dunes form vast "sand seas" throughout Titan's optically dark equatorial regions, Huygens descended over a region of bright and dark units that was free of the pervasive dune fields found elsewhere.

Titan's surface possesses a variety of features such as lakes, mountains, and dunes, as shown in Fig. 14.5 **(top) (middle, left)**. Rippling sand dunes, such as those in Earth's Arabian Desert, can be seen in the dark equatorial regions of Titan. Data from Cassini RADAR revealed widespread regions of equatorial dunes (±30°) covering 10–20% of Titan's surface (Elachi et al., 2005; Lorenz et al., 2006b; Radebaugh et al., 2008; Lorenz and Radebaugh, 2009; Le Gall et al., 2011; Rodriguez et al., 2014). Scientists had thought that the dunes are not made of silicates as on Earth, but of solid water ice coated with hydrocarbons that fall from the atmosphere. It was believed that the dunes on Titan are probably composed of sand-sized hydrocarbon and/or nitrile (an organic compound containing a cyanide

FIG. 14.4 Titan's surface at the Huygens landing site, 10.2°S, 192.4°W. There are at least eight rocks visible in the image—numbered in red with size indicated for two of them. Distance from the lander is shown in blue and the horizon at 88.5° is labeled in green. The small rocks are thought to be H_2O ice mostly coated by organic solid material. The rounded nature of these rocks suggests past fluvial activity. The right panel shows, approximately, the true color of the scene. Direct measurements by the Huygens probe indicated that the ground at the landing was moist with methane and ethane at the time of the landing, 14 January 2005. Image from ESA. *(Source: McKay, C.P. (2016), Titan as the abode of life, Life, 6:8.). Copyright info: 2016 by the author; licensee MDPI, Basel, Switzerland. This article from which this Figure is reproduced here is an open access article distributed under the terms and conditions of the Creative Commons by Attribution (CC-BY) license (http://creativecommons.org/licenses/by/4.0/).*

group—CN bound to an alkyl group) grains mixed with lesser amounts of water ice. The particles rained down from above onto the surface and were subsequently eroded and moved by surface and aeolian processes, such as liquid methane runoff and wind erosion. In order for the sand to migrate across the surface under the influence of Titan's weak surface winds, a process called saltation, scientists have concluded that the dune material are likely composed of 100–300 μm particles in diameter (Lorenz et al., 2006b).

Analyses of RADAR and visible and infrared mapping spectrometer (VIMS) data indicate that the dunes on Titan's surface are composed of pure organic or organic coated materials (McCord et al., 2006; Soderblom et al., 2007b; Barnes et al., 2008; Clark et al., 2010; Le Gall et al., 2011; Hirtzig et al., 2013; Rodriguez et al., 2014). It was found that the dune particles are much larger than the aerosols measured by Descent Imager/Spectral Radiometer (DISR) near the surface (Hörst, 2017).

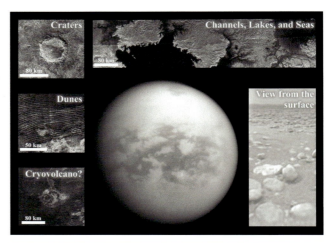

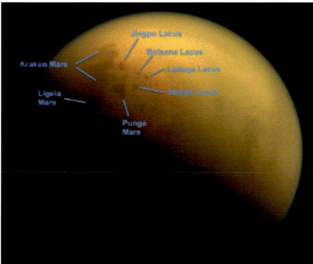

FIG. 14.5 (top) Shown here are examples of some of the features of Titan's surface. The crater shown at the top left is Sinlap, which has a diameter of 79 km (PIA16638, NASA/JPL-Caltech/ASI/GSFC). An example of Titan's equatorial dunes from the Shangri-la dunes is shown at the middle left (PIA12037, NASA/JPL-Caltech/ASI). Doom Mons and Sotra Patera, a putative cryovolcanic region, are shown at the bottom left (PIA09182, NASA/JPL-Caltech/ASI). At the top is Kraken Mare (left) and Ligeia Mare (right) and associated channel systems (PIA09217, NASA/JPL-Caltech/ASI). Shown at the bottom right is the view from the surface taken by the Descent Imager Spectral Radiometer carried by Huygens. The cobbles in the foreground are 10 to 15 cm (PIA06440, ESA/NASA/JPL/University of Arizona). The middle image was taken at 938 nm (CB3 filter), which sees down to the surface. The dark equatorial regions are Titan's expansive dune fields, while the dark regions at the north pole are lakes and seas (PIA14584, NASA/JPL-Caltech/Space Science Institute). *(Source: Hörst, S.M (2017), Titan's atmosphere and climate, Journal of Geophysical Research Planets, 122(3): 432-482); Wiley Online Library.* (bottom) Titan as seen from NASA's Cassini spacecraft shows the largest hydrocarbon lakes, including the largest sea Kraken Mare. Credit: NASA/JPL-Caltech/Space Science Institute. *(Source: https://www.space.com/12638-amazing-photos-Titan-saturn-moon.html).*

Images show Titan's icy dunes are gigantic, eastward propagating longitudinal dunes around 300 feet (100 m) tall, 0.6 to 1.2 miles (1–2 km) wide, and on average 30–50 km long (Lorenz et al., 2006b; Radebaugh et al., 2008; Barnes et al., 2008; Neish et al., 2010b; Le Gall et al., 2011). The observed 1–3 km spacing (Lorenz et al., 2006b; Radebaugh et al., 2008; Savage et al., 2014) is approximately the same height as the boundary layer indicating that the dunes are mature and have therefore stopped growing (Griffith et al., 2008; Lorenz et al., 2010; Lorenz, 2014b). Studies carried out by Lopes et al. (2019) show that Titan's surface is dominated by sedimentary or depositional processes with a clear latitudinal variation, with dunes at the equator, plains at mid-latitudes and labyrinth (i.e., a complicated irregular network) of terrains and lakes at the poles.

14.4.3 Understanding Titan's ionosphere using Cassini plasma spectrometer (CAPS)

Titan's ionosphere contains a rich positive ion population including organic molecules. However, using CAPS electron spectrometer data from sixteen Titan encounters, Coates et al. (2007) revealed the existence of negative ions. It has been found that these ions, with densities up to ~100/cm^3, are in mass groups of 10–30, 30–50, 50–80, 80–110, 110–200 and 200 + amu/charge. During one low encounter, negative ions with mass per charge as high as 10,000 amu/q have been observed. Researchers believe that due to their unexpectedly high densities at ~950 km altitude, these negative ions must play a key role in the ion chemistry and they may be important in the formation of organic-rich aerosols (tholins) eventually falling to the surface.

Multiple Titan encounters by the Cassini spacecraft have shown that ion chemistry in Titan's upper atmosphere is much more complex than previously thought. As well as showing a great variety of species present below 100 amu, they also include the detection of negative ions and of large abundances of ions above 100 amu. Crary et al. (2010) used data from two Cassini instruments, the Cassini plasma spectrometer's ion beam sensor (CAPS/IBS) and the ion and neutral mass spectrometer (INMS) during fourteen Cassini encounters with Titan's upper atmosphere. By simultaneous analysis of the combined data, they determined the ion temperature, one component of the wind speed, and spacecraft potential. Using these derived quantities, they extended the analysis of CAPS/IBS data to quantify the abundance of ions above 100 amu and to statistically estimate their composition.

14.4.4 Confirming the existence of strong winds in Titan's atmosphere

As indicated earlier, winds and haze significantly affect the thermal balance of Titan, causing an antigreenhouse effect that cools the surface by 9 K. Even before the Cassini spacecraft mission to Titan, McKay et al. (2001) suggested that Titan's faintly banded appearance could be indicative of strong zonal winds in Titan's lower stratosphere. Amazingly, McKay et al. (2001)'s prediction was confirmed later during the Cassini mission to the planet Saturn.

The winds in Titan's atmosphere result primarily from solar forcing, which varies seasonally (Hörst, 2017). There are a number of different ways to estimate wind speeds on Titan. Direct measurements of the wind speeds come from a few methods, such as measurements from the Doppler Wind Experiment (DWE) carried by the Huygens probe; measuring Doppler shifts in the emission lines of atmospheric constituents like ethane; and cloud tracking.

Although spacecraft observations had indicated that strong zonal (east-west) winds may exist in Titan's atmosphere, the first direct measurements were made by the Doppler Wind Experiment on ESA's Huygens probe. The DWE provided measurements at one place and time from 145 km to the surface (Hörst, 2017). By measuring the Doppler shift of the radio signal from Huygens and studying panoramic mosaics from the onboard imager to work out the descent trajectory, it was possible to create a high-resolution vertical profile of Titan's winds, with an estimated accuracy of better than 1 m/s.

Huygens found that the zonal winds were prograde (i.e., in the same direction as Titan's rotation) during most of the atmospheric descent. The probe generally drifted east, driven by remarkably strong westerly winds which peaked at roughly 120 m/s (430 km/h) at an altitude of about 120 km.

Down to a height of 60 km, large variations in the Doppler measurements were observed—evidence that Huygens endured a rough ride as a result of significant vertical wind shear. Wind speeds then decreased toward the surface, dropping from 30 m/s (108 km/h) at an altitude of 55 km to 10 m/s (36 km/h) at a height of 30 km, eventually slowing to 4 m/s (14 km/h) at 20 km. The winds dropped to zero and then reversed direction at around 7 km.

The large prograde wind speeds measured between 45 km and 70 km altitude and above 85 km were much faster than Titan's equatorial rotation speed. It was the first in situ confirmation of the predicted super-rotation of the moon's atmosphere, even though the speed observed was slightly lower than expected. Fig. 14.6 shows measurements from various techniques, revealing that although wind speeds are relatively low in the troposphere, there are super-rotating winds in the stratosphere. In this Figure, black circles are the measurements from the Doppler Wind Experiment (DWE) (Bird et al., 2005; Folkner et al., 2006). Cloud-tracking measurements from Porco et al. (2005) are shown as filled diamonds; the altitude for these clouds is not well constrained, and shown here is a presumed upper limit. The measurements made from Doppler shifts in

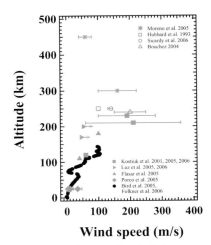

FIG. 14.6 Measurements from various techniques reveal that wind speeds are relatively low in the troposphere but in the stratosphere there are super-rotating winds. *(Source: Hörst, S.M (2017), Titan's atmosphere and climate, Journal of Geophysical Research Planets, 122(3): 432-482); Wiley Online Library.*

the emission lines of atmospheric constituents are shown as filled squares (Kostiuk et al., 2001, 2005, 2006), filled right-pointing triangles (arrows indicate these measurements are lower limits) (Luz et al., 2005, 2006), and asterisks (Moreno et al., 2005). Hubbard et al. (1993) (open square), Sicardy et al. (2006) (open circle), and Bouchez (2004) (open triangle) used stellar occultations to estimate the stratospheric wind speeds. Wind speeds calculated from the thermal wind equation using Cassini CIRS temperature measurements near the Huygens' landing site are shown as triangles (Flasar et al., 2005). Note that although there is some spatial and temporal overlap, this plot does not necessarily represent a snapshot of Titan's atmosphere at one place and time (Hörst, 2017).

A layer with surprisingly slow wind, where the sideways velocity decreased to near zero, was detected at altitudes between 60 km and 100 km. During the last 15 min of the descent, Huygens headed west-northwest at a speed of approximately 1 m/s. The wind speed on the surface was between 0.3 m/s and 1 m/s. Over the duration of the descent, the probe drifted eastward a distance of 165.8 km with respect to the surface of Titan.

14.4.5 Knowing more on an unusual atmosphere surrounding Titan

The arrival of the Cassini–Huygens mission to the Saturn system ushered in a new era in the study of Titan. Carrying a variety of instruments capable of remote sensing and in situ investigations of Titan's atmosphere and surface, the Cassini Orbiter and the Huygens Probe have provided a wealth of new information about Titan and have finally allowed humankind to visualize the surface realistically.

High-energy photons, electrons, and ions initiate ion–neutral chemistry in Titan's upper atmosphere by ionizing the major neutral species (nitrogen and methane). The ion and neutral mass spectrometer (INMS) onboard the Cassini spacecraft performed the first composition measurements of Titan's ionosphere. The INMS revealed that Titan has the most compositionally complex ionosphere in the solar system, with roughly 50 ions at or above the detection threshold. Modeling of the ionospheric composition constrains the density of minor neutral constituents, most of which cannot be measured with any other technique. The species identified with this approach include the most complex molecules identified so far on Titan. This confirms the long-thought idea that a very rich chemistry is actually taking place in this atmosphere. In a study, Vuitton et al. (2007) have discussed the production and loss reactions for the ions and how this affects the neutral densities. They compared their results to neutral densities measured in the stratosphere by other instruments, to production yields obtained in laboratory experiments simulating Titan's chemistry, and to predictions of photochemical models. They have suggested neutral formation mechanisms and highlighted the need for new experimental and theoretical data.

Rodriguez et al. (2009) have reported that the global spatial cloud coverage on Titan is in general agreement with the models, confirming that cloud activity is mainly controlled by the global circulation. According to Rodriguez et al. (2009), the nondetection of clouds at latitude ~40°N and the persistence of the southern clouds while the southern summer is ending are, however, both contrary to predictions. These researchers argue that this observation suggests that Titan's equator-to-pole thermal contrast is over-estimated in the models and that its atmosphere responds to the seasonal forcing with a greater inertia than expected.

The prominent aerosol layers on Titan result from the complex photochemistry in the moon's atmosphere. Dissociation of the main atmospheric gas composition, dominantly by energetic photons (note that a photon is a particle representing a quantum of light or other electromagnetic radiation, in which the said particle carries energy proportional to the radiation frequency but has zero rest mass) and photoelectrons, initiates the growth of organic molecules, the increasing complexity of which eventually yields the observed aerosols (particles dispersed in air). These particles interact strongly with the solar radiation field. But the most prominent aerosol signature is the almost featureless appearance of Titan's disk in visible wavelengths, due to the screening of Titan's surface by the aerosols.

Interestingly, it has been found that Saturn's icy moons Titan and Enceladus are chemically connected by the flow of material through the Saturn system (Cooper et al., 2009). Based on detailed analyses of the data gathered by the

FIG. 14.7 Shown here are examples of some of Titan's distinctive atmospheric features. At the left, a natural color view of Titan from 2012 (PIA14925; 25 July 2012) showing the detached haze layer and the newly formed vortex at the south pole (shown also at the bottom middle, PIA14919; 27 June 2012). The north-south asymmetry in the haze is shown at the top middle (PIA14610; 31 January 2012). Titan's north polar hood stands out in the image shown top right (PIA08137; 27 January 2006). At the bottom right, extensive methane clouds at a number of different latitudes, including the equator, show the post-equinox shift in cloud location (PIA12810; 18 October 2010). *All image credits are NASA/JPL/Space Science Institute. (Source: Hörst, S.M (2017), Titan's atmosphere and climate, Journal of Geophysical Research Planets, 122(3): 432-482). Wiley Online Library.*

instruments onboard the Cassini Orbiter and the Huygens Probe, it was found that perhaps more so than anywhere else in the solar system, Titan's atmosphere and surface are intimately linked (Hörst, 2017).

The surface of Titan is hidden beneath a thick atmosphere (Fig. 14.7). The Huygens Atmospheric Structure Instrument (HASI) data showed that the upper atmosphere (the thermosphere) is generally warmer and denser than expected. Titan's atmosphere was also found to be highly stratified. Above 500 km, the average temperature was approximately minus 100°C but strong variations of 10–20°C were detected due to inversion layers and other phenomena, such as, gravity waves and tides. The mesosphere (the region of the atmosphere above the stratosphere and below the thermosphere) was virtually absent, in contrast with theoretical predictions.

Below 500 km, the temperature increased quite rapidly, reaching a maximum of minus 87°C at the top of the stratosphere, at an altitude of 250 km. The temperature then decreased steadily throughout the stratosphere, reaching a minimum of minus 203°C at an altitude of 44 km. This marked the boundary between the stratosphere and the troposphere. The temperature increased again as the probe neared the surface, rising to a chilly minus 180°C at the landing site. The surface pressure was 1.47 times that on Earth (http://sci.esa.int/cassini-huygens/55222-science-highlights-from-huygens-1-profiling-the-atmosphere-of-Titan).

14.4.6 Making the first direct identification of bulk atmospheric nitrogen and its abundance on Titan

Titan and Earth are the only worlds in our solar system that have thick nitrogen atmospheres. Although data from the Voyager mission had implied that nitrogen (N_2) was the main atmospheric gas, the gas chromatograph mass spectrometer (GCMS) on ESA's Huygens probe made the first direct identification of bulk atmospheric nitrogen and its abundance. Other GCMS atmospheric measurements provided clues about where this atmosphere came from.

During its descent to the surface, the GCMS measured isotopic ratios and trace species in the atmosphere. One of the objectives for the GCMS was to search for heavy, noble gases such as argon-36 (^{36}Ar), argon-38 (^{38}Ar), krypton (Kr), and xenon (Xe) (http://sci.esa.int/cassini-huygens/55225-science-highlights-from-huygens-4-the-origin-of-Titans-nitrogen-atmosphere/)

These primordial gases have been detected and measured in several celestial bodies such as meteorites; in the atmospheres of Earth, Mars, Venus (to some extent); and also, in Jupiter's atmosphere. Differing patterns of relative abundances and isotopic ratios of the gases provide insights into the origin and evolution of these objects. As a result, their measurements in the atmosphere of Titan were eagerly anticipated.

Scientists had theorized that these noble gases were present throughout the solar nebula, and should therefore have been incorporated into both Saturn and Titan during the early stages of planet formation. In the context of the origin of nitrogen, ^{36}Ar is of particular importance, and the GCMS found that the ratio of ^{36}Ar to nitrogen was about one million times less than is found in the Sun.

It is believed that direct condensation of gases in the young Titan would have resulted in the capture of ^{36}Ar, as well as nitrogen, in solar proportions. However, the depleted ratio detected by the GCMS on Huygens implies that the nitrogen was captured as ammonia (NH_3) or in other nitrogen-bearing compounds. The rarity of noble gases on Earth has long been viewed as strong support for the atmosphere having been formed by the impacts of gas-rich planetesimals, and the near absence of noble gases from Titan provides more support for this hypothesis.

14.4.7 Obtaining evidence for formation of Tholins in Titan's upper atmosphere and understanding the process

Photochemically produced aerosols are common among the atmospheres of our solar system and beyond. Observations and models have shown that photochemical aerosols have direct consequences on atmospheric properties as well as important astrobiological ramifications (Lavvas et al.,

2013). Titan's lower atmosphere has long been known to harbor tholins (organic-rich aerosols) presumed to have been formed from simple molecules, such as methane (CH_4) and nitrogen (N_2). For long, it has been assumed that tholins were formed at altitudes of several hundred kilometers by processes as yet unobserved. Using measurements from a combination of mass/charge and energy/charge spectrometers on the Cassini spacecraft, Waite et al. (2007) obtained evidence for tholin formation at high altitudes (~1000 km) in Titan's atmosphere. The observed chemical mix strongly implies a series of chemical reactions and physical processes that lead from simple molecules (CH_4 and N_2) to larger, more complex molecules (80 to 350 daltons) to negatively charged massive molecules (~8000 daltons), which Waite et al. (2007) identified as tholins. Note that dalton is a unit used in expressing the molecular weight of proteins, equivalent to atomic mass unit. That the process involves massive negatively charged molecules and aerosols was completely unexpected.

14.4.8 Detection of benzene (C_6H_6) in Titan's atmosphere

Benzene is an organic chemical compound with the molecular formula C_6H_6. The benzene molecule is composed of six carbon atoms joined in a ring with one hydrogen atom attached to each carbon atom. As it contains only carbon and hydrogen atoms, benzene is classed as a hydrocarbon. The Cassini spacecraft detected benzene high in Titan's atmosphere. Vuitton et al. (2008) reported a study of the formation and distribution of benzene (C_6H_6) on Titan. Analysis of the Cassini ion and neutral mass spectrometer (INMS) measurements of benzene densities on 12 Titan passes showed that the benzene signal exhibits an unusual time-dependence, peaking ~20 s after closest approach, rather than at the closest approach. Vuitton et al. (2008) showed that this behavior can be explained by recombination of phenyl radicals (C_6H_5) with H atoms on the walls of the instrument and that the measured signal is a combination of (1) C_6H_6 from the atmosphere and (2) C_6H_6 formed within the instrument. In parallel, these investigators examined Titan benzene chemistry with a set of photochemical models. A model for the ionosphere predicts that the globally averaged production rate of benzene by ion-molecule reactions is ~10^7/cm^2/s, of the same order of magnitude as the production rate by neutral reactions of ~4×10^6/cm^2/s. Vuitton et al. (2008) showed that benzene is quickly photolyzed (caused to undergo chemical decomposition under the influence of light) in the thermosphere, and that C_6H_5 radicals, the main photo-dissociation products, are ~3 times as abundant as benzene. This result is consistent with the phenyl/benzene ratio required to match the INMS observations. Loss of benzene occurs primarily through reaction of phenyl with other radicals, leading to the formation of complex aromatic species. These species, along with benzene, diffuse downward, eventually condensing near the tropopause. Vuitton et al. (2008) found a total production rate of solid aromatics of ~10^{-15} g/cm^2/s, corresponding to an accumulated surface layer of ~3 m.

14.4.9 Direct measurements of carbon-based aerosols in Titan's atmosphere and deciphering their chemical composition

Cassini followed up Huygens' measurements from space, detecting chemicals in Titan's atmosphere. Analysis of the Cassini Ultraviolet Imaging Spectrograph (UVIS) stellar and solar occultations at Titan include chemical species such as N_2 (nitrogen), CH_4 (methane), C_2H_2 (acetylene), C_2H_4 (ethylene), C_2H_6 (ethane), C_4H_2 (diacetylene), C_6H_6 (benzene), C_6N_2 (dicyanodiacetylene), C_2N_2 (cyanogen), poisonous HCN (hydrogen cyanide), HC_3N (cyanoacetylene), and aerosols (**tiny particles**) distinguished by a structureless continuum extinction (absorption plus scattering) of photons in the Extreme Ultraviolet (EUV) (Liang et al., 2007). Fig. 14.8 shows a summary of the major production and loss pathways for 10 of the most abundant photochemically produced molecules in Titan's atmosphere (see extensive discussion in Vuitton et al. [2014]).

Aerosols in Titan's atmosphere have long been suspected to play an important role in determining its thermal structure and atmospheric processes. However, until the

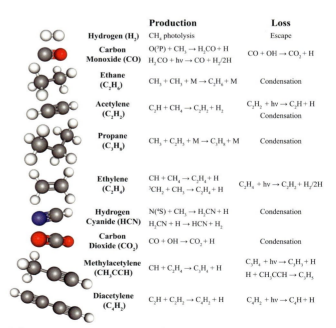

FIG. 14.8 A summary of the major production and loss pathways for 10 of the most abundant photochemically produced molecules in Titan's atmosphere in approximately decreasing order of abundance. *(Source: Hörst, S.M (2017), Titan's atmosphere and climate, Journal of Geophysical Research Planets, 122(3): 432-482).*

Huygens mission, no direct measurements had been made of the chemical composition of these particles. One set of measurements was made by the Gas Chromatograph and Mass Spectrometer (GCMS) and the Aerosol Collector and Pyrolyser (ACP) experiment. The collected aerosol particles were heated in the ACP oven in order to vaporize all volatile components, and the composition of the gases released by each sample was then analyzed by the GCMS (Source: http://sci.esa.int/cassini-huygens/55228-science-highlights-from-huygens-7-Titans-tiny-aerosols/).

Two atmospheric samples were obtained during the descent of Huygens. One was taken at 130-35 km (the middle stratosphere) and the other at 25-20 km (the middle troposphere). Ammonia (NH_3) and hydrogen cyanide (HCN) were identified as the main gases released in the oven, confirming that carbon and nitrogen are major constituents of the aerosols. Interestingly, no substantial difference was found between the two samples, suggesting that the aerosols' composition was the same at both altitudes. This supports the idea that they have a common source in the upper atmosphere, where ultraviolet sunlight photo-chemically alters gases such as methane.

Meanwhile, the Descent Imager/Spectral Radiometer (DISR) characterized the optical properties of the photochemical aerosols from 150 km altitude to the surface. They were found to match the properties of "tholins," materials created in laboratories by sending electrical discharges into mixtures of nitrogen and methane.

It was found that the aerosols' optical properties can be reproduced by the condensation of hydrogen cyanide close to 80 km, ethane condensation close to the tropopause (44 km), and methane condensation from the tropopause down to 8 km. Fig. 14.9 illustrates the mechanism of aerosols formation in Titan's haze. This mechanism, constructed based on the simulations described by Lavvas et al. (2011), shows the various steps that lead to the formation of the aerosols that make up the haze on Titan.

When sunlight or highly energetic particles from Saturn's magnetosphere hit the layers of Titan's atmosphere above 1000 km, the nitrogen and methane molecules there are broken up. This results in the formation of massive positive ions and electrons, which trigger a chain of chemical reactions that produce a variety of hydrocarbons. Many of these hydrocarbons have been detected in Titan's atmosphere, including Polycyclic Aromatic Hydrocarbons (PAHs), which are large carbon-based molecules that form from the aggregation of smaller hydrocarbons. Some of the PAHs detected in the atmosphere of Titan also contain nitrogen atoms.

PAHs are the first step in a sequence of increasingly larger compounds. Models show how PAHs can coagulate and form large aggregates, which tend to sink, due to their greater weight, into the lower atmospheric layers. The higher densities in Titan's lower atmosphere favor the

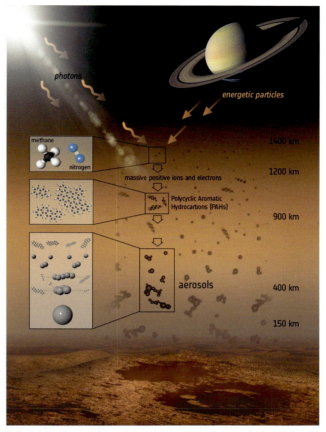

FIG. 14.9 Illustration of the mechanism of aerosols formation in Titan's haze. *Credit: ESA / ATG medialab. (Source: http://sci.esa.int/cassini-huygens/55228-science-highlights-from-huygens-7-Titans-tiny-aerosols/).*

further growth of these large conglomerates of atoms and molecules. These reactions eventually lead to the production of carbon-based aerosols, large aggregates of atoms and molecules that are found in the lower layers of the haze that enshrouds Titan, well below 500 km (http://sci.esa.int/cassini-huygens/55224-science-highlights-from-huygens-3-methane-mystery/).

Surprisingly, Titan is the only moon in the solar system with a dense atmosphere, which produces a surface pressure (see Sotin, 2007). The substantial atmosphere, which obscures the surface, was poorly understood before the Cassini mission, leading to intense speculation about Titan's nature. Porco et al. (2005) have reported observations of Titan from the imaging science experiment onboard the Cassini spacecraft that address some of these issues.

Convective clouds are found to be common near Titan's south pole, and the motion of mid-latitude clouds consistently indicates eastward winds, from which scientists infer that the troposphere is rotating faster than the surface. A detached haze at an altitude of 500 km is 150–200 km higher than that observed by Voyager.

In the investigation of Titan's hazy atmosphere, caused by aerosol particles in its atmosphere, Liang et al. (2007) found that di-cyano-di-acetylene is condensable at ~650 km, where the atmospheric temperature minimum is located. Liang et al. (2007) found that this species is the simplest molecule identified to be condensable. However, they cautioned that observations are needed to confirm the existence and production rates of di-cyano-di-acetylene.

The descent imager/spectral radiometer (DISR) instrument aboard the Huygens probe measured the brightness of sunlight using a complement of spectrometers, photometers, and cameras that covered the spectral range from 350 to 1600 nm, looked both upward and downward, and made measurements at altitudes from 150 km to the surface. The DISR measurements from the upward-looking visible and infrared spectrometers are described in Tomasko et al. (2008a). Fig. 14.10 shows Huygens's view of Titan from five altitudes, taken by the DISR on board the Huygens probe, during its descent to Titan's surface on 14 January 2005.

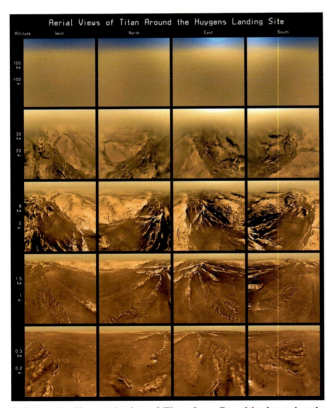

FIG. 14.10 Huygens's view of Titan from five altitudes, taken by the descent imager/spectral radiometer (DISR) on board the Huygens probe, during its descent to Titan's surface on 14 January 2005. The views from the probe, in the four cardinal directions (West, North, East, South), were taken at 5 different altitudes (top to bottom): 150 km, 30 km, 8 km, 1.5 km and 300 m. (*Source: http://sci.esa.int/cassini-huygens/39211-huygens-view-of-Titan-from-five-altitudes/). Copyright: ESA/NASA/JPL/University of Arizona.*

Tomasko et al. (2008a) have very briefly reviewed the measurements by the violet photometers, the downward-looking visible and infrared spectrometers, and the upward-looking solar aureole (SA) camera. Taken together, the DISR measurements constrain the vertical distribution and wavelength dependence of opacity, single-scattering albedo, and phase function of the aerosols in Titan's atmosphere.

Comparison of the inferred aerosol properties with computations of scattering from fractal aggregate particles indicates the size and shape of the aerosols. Tomasko et al. (2008a) found that the aggregates require monomers (i.e., molecules that can be bonded to other identical molecules to form polymers) of radius 0.05 μm or smaller and that the number of monomers in the loose aggregates is roughly 3000 above 60 km. Furthermore, the single-scattering albedo of the aerosols above 140 km altitude was found to be similar to that predicted for some tholins measured in laboratory experiments, although Tomasko et al. (2008a) found that the single-scattering albedo of the aerosols increases with depth into the atmosphere between 140 and 80 km altitude, possibly due to condensation of other gases on the haze particles. The number density of aerosols was found to be about $5/cm^3$ at 80 km altitude, and decreases with a scale height of 65 km to higher altitudes. It was also found that the aerosol opacity above 80 km varies as the wavelength to the −2.34 power between 350 and 1600 nm.

It was found that between 80 and 30 km the cumulative aerosol opacity increases linearly with increasing depth in the atmosphere. The total aerosol opacity in this altitude range varies as the wavelength to the −1.41 power. The single-scattering phase function of the aerosols in this region is also consistent with the fractal particles found above 60 km.

Tomasko et al. (2008a) found that in the lower 30 km of the atmosphere, the wavelength dependence of the aerosol opacity varies as the wavelength to the −0.97 power, much less than at higher altitudes. This suggests that the aerosols here grow to still larger sizes, possibly by incorporation of methane into the aerosols. Here the cumulative opacity also increases linearly with depth, but at some wavelengths the rate is slightly different than above 30 km altitude.

Another interesting finding gleaned from Tomasko et al. (2008a)'s studies is the possibility that either the brightest aerosols near 30 km altitude contain significant amounts of methane, or there may be spherical particles in the bottom few kilometers of the atmosphere.

In another study, Tomasko et al. (2008b) very briefly reviewed the measurements by the violet photometers, the downward-looking visible and infrared spectrometers, and the upward-looking solar aureole (SA) camera. Taken together, the DISR measurements constrain the vertical distribution and wavelength dependence of opacity, single-scattering albedo, and phase function of the aerosols in Titan's atmosphere. Comparison of the inferred aerosol properties with computations of scattering from fractal aggregate particles indicates the size and shape of the aerosols. The studies provided an indication that the brightest aerosols near 30 km altitude contain significant amounts of methane, and that the decreasing albedo at lower altitudes may reflect the evaporation of some of the methane as the aerosols fall into dryer layers of the atmosphere.

Much before the Cassini mission, it was known that there exists a high-altitude detached haze layer in Titan's atmosphere. For example, *Voyager* images have indicated a detached haze layer at 350 km in Titan's hazy atmosphere. According to Hörst (2017), the origin of the detached haze layer, shown in Fig. 14.7, is not yet well understood. The Imaging Science Subsystem (ISS) and the Ultraviolet Imaging Spectrograph (UVIS) have detected a detached haze layer at 500 km, along with less pronounced layers above the main detached haze layer (Porco et al., 2005; Liang et al., 2007; Koskinen et al., 2011). This location is about 150 km higher than the detached haze layer observed by Voyager.

By comparing observations from the Cassini imaging system, UV spectrometer, and Huygens atmospheric structure instrument, Lavvas et al. (2009) pointed out that the detached haze layer in Titan's atmosphere is coincident with a local maximum in the measured temperature profile and showed that the temperature maximum is caused by absorption of sunlight in the detached haze layer. This finding rules out condensation as the source of the layer. Because the aerosol size and mass flux derived for the detached layer agreed with those determined for the main layer, Lavvas et al. (2009) have suggested that the main haze layer in Titan's stratosphere is formed primarily by sedimentation and coagulation of particles in the detached layer. Based on this finding, Lavvas et al. (2009) argued that high-energy radical and ion chemistry in the thermosphere is the main source of haze on Titan.

Given several difficulties in explaining various aspects related to the detached haze layer on Titan, it seems likely that both dynamics and microphysics play a role in determining the structure and temporal evolution of the detached haze layer (Hörst, 2017). From 2007 to 2010, the location of the detached haze layer dropped from 500 km to 380 km (West et al., 2011), returning to the same altitude at which it was observed during the Voyager era at the same point in Titan's year. According to Hörst (2017), the apparent seasonal evolution in the presence and location of the detached haze layer strengthens the argument that atmospheric dynamics play a role in its origin. The drop in altitude was most rapid at equinox. Starting in late 2012 the detached haze layer was not detectable until early 2016 when it reappeared, with very low contrast, near 500 km (West et al., 2016).

14.4.10 Understanding the role of nitrogen and methane in generating the orange blanket of haze in Titan's atmosphere

One of the most noticeable features of Titan is the orange blanket of haze (mist) that hides its surface. However, no one knew exactly whether the haze extended to the surface until ESA's Huygens probe landed on the icy moon. The measurements of the Descent Imager/Spectral Radiometer (DISR) on the Huygens probe provided in situ information on the optical properties, size, and density of the haze particles. The observations showed that there was a significant amount of haze at all altitudes throughout the descent, extending all the way down to the surface. With decreasing altitude, the haze particles became brighter, and the particle sizes increased, due to collisions which resulted in a "snowball" effect, as well as condensation of methane, ethane, and hydrogen cyanide gases onto small aerosol nuclei at lower levels (http://sci.esa.int/cassini-huygens/55227-science-highlights-from-huygens-6-hazy-Titan/).

Huygens detected three distinct haze regions (region I above 80 km; region II between 80 and 30 km; and region III between 30 km and the surface), based on the density and optical properties of the atmosphere. Before the Huygens mission, it was generally believed that the tiny haze particles slowly sink through the stratosphere, eventually acting as condensation nuclei for lower-level clouds. Some scientists theorized that the haze might clear below an altitude of 50 km to 70 km due to condensation of gases such as methane. However, the probe's Descent Imager/Spectral Radiometer (DISR) showed that Huygens began to emerge from the haze only in the troposphere, 30 km above the surface.

Another thin layer of methane haze was detected at an altitude of 21 km, where the local temperature was minus 197°C and the pressure was 450 mbar. Researchers think that this feature may be an indication of methane condensation. When combined with ground-based measurements, the data suggest an upper methane ice cloud (or haze) between approximately 20 km and 30 km and a liquid methane-nitrogen cloud layer between 8 km and 16 km, perhaps with a gap in between. Fig. 14.11 shows stereographic (fish-eye) images of Titan's surface, taken during Huygens's descent by the Descent Imager/Spectral Radiometer (DISR) on 14 January 2005. This view shows the surface from 6 different altitudes from 150 km down to 200 m. It also shows the haze layer at 20–21 km altitude.

It may be noted that two of the key questions about Titan are the origin of the nitrogen and methane in its atmosphere, and the mechanisms by which methane levels are maintained. The Huygens probe made the first direct measurements of Titan's lower atmosphere. Data returned by the probe included altitude profiles of the gaseous constituents, isotopic ratios, and trace gases (including organic

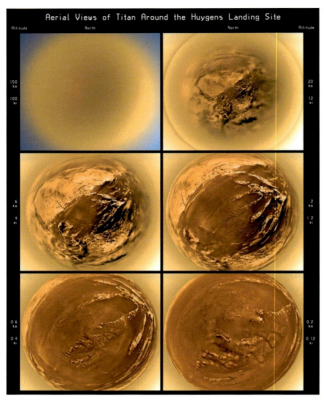

FIG. 14.11 Stereographic (fish-eye) images of Titan's surface, taken during Huygens's descent by the Descent Imager/Spectral Radiometer (DISR) on 14 January 2005. This view shows the surface from 6 different altitudes from 150 km down to 200 m. It also shows the haze layer at 20–21 km altitude *(Source: http://sci.esa.int/cassini-huygens/39213-aerial-stereographic-views-of-Titan/); Copyright: ESA/NASA/JPL/University of Arizona.*

compounds). Huygens also directly sampled aerosols in the atmosphere and confirmed that carbon and nitrogen are their major constituents.

From Cassini mission results, the primary constituents of Titan's atmosphere were confirmed to be nitrogen and methane. For the last part of the descent, methane amounts remained relatively constant until the probe touched down on the surface. A sudden, 40% increase in the methane signal after landing, while the nitrogen count rate remained constant, suggested the presence of liquid methane on the surface. Measurements of the carbon isotopes in the methane provide no support for suggestions that it is generated by active micro-organisms on Titan. The methane was probably accreted by Titan during the moon's formation, and large quantities of liquid methane are now trapped in ices beneath the surface, possibly reaching the surface through some form of cryovolcanism. It may be noted that a number of potential cryovolcanic features have been identified on Titan's surface through the use of Cassini's RADAR and VIMS measurements. This activity would replace the methane that is lost as a result of photochemistry in the atmosphere. The spectra taken on the surface also showed signatures characteristic of more complex hydrocarbons, such as ethane, cyanogen and benzene.

It is interesting to note that Titan has a massive nitrogen atmosphere containing up to 5% methane near its surface. The variety of chemicals observed in Titan's atmosphere indicates a rich and complex chemistry originating from nitrogen and methane; and evolving into complex molecules, eventually forming the smog that surrounds the icy moon. It is believed that methane and ethane rain down from clouds in Titan's atmosphere, but the ultimate source of the methane is still unclear.

Tobie et al. (2006) have shown that episodic out-gassing of methane stored as clathrate hydrates within an icy shell above an ammonia-enriched water ocean is the most likely explanation for Titan's atmospheric methane. On the basis of their models, Tobie et al. (2006) have predicted that future fly-bys should reveal the existence of both a subsurface water ocean and a rocky core, and should detect more cryo-volcanic edifices.

Lavvas et al. (2008a) introduced a one-dimensional (1D) coupled Radiative/Convective-Photochemical-Microphysical model for a planetary atmosphere and applied it to Titan. The model incorporated detailed radiation transfer calculations for the description of the short-wave and long-wave fluxes which provide the vertical structure of the radiation field and temperature profile. These were then used for the generation of the photochemistry inside the atmosphere from the photolysis of Titan's main constituents, nitrogen (N_2) and methane (CH_4). The resulting hydrocarbons and nitriles were used for the production of the haze precursors, whose evolution is described by the microphysical part of the model. The calculated aerosol and gas opacities were iteratively included in the radiation transfer calculations in order to investigate their effect on the resulting temperature profile and geometric albedo (the proportion of the incident light or radiation that is reflected by the surface of a celestial body). The main purpose of their model was to help in the understanding of the missing link between the gas production and particle transformation in Titan's atmosphere. Lavvas et al. (2008a) described the basic physical mechanisms included in the model. The final results regarding the eddy mixing profile, the chemical composition and the role of the different haze precursors suggested in the literature are presented in Lavvas et al. (2008b) along with the sensitivity of the results to the molecular nitrogen photoionization scheme and the impact of galactic cosmic rays in the atmospheric chemistry.

Lavvas et al. (2008b) applied the one-dimensional radiative–convective/photochemical/microphysical model described in Lavvas et al. (2008a) to the study of Titan's atmospheric processes that lead to haze formation. Their model generated the haze structure from the gaseous species photochemistry. Lavvas et al. (2008b) have presented the model results for the species vertical concentration profiles, haze formation and its radiative properties, vertical temperature/density profiles, and geometric albedo. These were validated against Cassini/Huygens observations and other ground-based and space-borne measurements. The model reproduced well most of the latest measurements from the Cassini/Huygens instruments for the chemical composition of Titan's atmosphere and the vertical profiles of the observed species. For the haze production Lavvas et al. (2008b) included pathways that are based on pure hydrocarbons, pure nitriles and hydrocarbon/nitrile copolymers. From these, the nitrile and copolymer pathways provided the stronger contribution, in agreement with the results from the Aerosol Collector Pyrolyser (ACP) instrument, which supported the incorporation of nitrogen in the pyrolized (thermally decomposed) haze structures. Note that pyrolysis is the thermal decomposition of materials at elevated temperatures in an inert atmosphere. It involves a change of chemical composition.

Lavvas et al. (2008b)'s model results have been compared with the DISR-retrieved haze extinction profiles and have been found to be in very good agreement. They have also incorporated in their model heterogeneous chemistry on the haze particles that converts atomic hydrogen to molecular hydrogen. The resultant H_2 profile was found to be closer to the ion and neutral mass spectrometer (INMS) measurements, while the vertical profile of the diacetylene formed was found to be closer to that of the Composite Infrared Spectrometer (CIRS) profile when this heterogenous chemistry was included.

Tomasko et al. (2008b) reported new low-temperature methane absorption coefficients pertinent to the Titan environment as derived from the Huygens DISR spectral measurements combined with the in-situ measurements of the methane gas abundance profile measured by the Huygens Gas Chromatograph/Mass Spectrometer (GCMS). The GCMS instrument on the probe measured the methane mixing ratio throughout the descent. The DISR measurements are the first direct measurements of the absorbing properties of methane gas made in the atmosphere of Titan at the path-lengths, pressures, and temperatures that occur there. Tomasko et al. (2008b) used the DISR spectral measurements to determine the relative methane absorptions at different wavelengths along the path from the probe to the Sun throughout the descent. These transmissions as functions of methane path length were fit by exponential sums and used in a haze radiative transfer model to compare the results to the spectra measured by DISR. Tomasko et al. (2008b) also compared the laboratory measurements of methane absorption at low temperatures (Irwin et al., 2006) with the DISR measurements. Tomasko et al. (2008b) found that the strong bands formed at low pressures on Titan act as if they have roughly half the absorption predicted by the laboratory measurements, while the weak absorption regions absorb considerably more than suggested by some extrapolations of warm measurements to the cold Titan temperatures. Tomasko et al. (2008b) gave factors as a function of wavelength that can be used with the published methane coefficients between 830 and 1620 nm to give agreement with the DISR measurements. They also gave exponential sum coefficients for methane absorptions that fit the DISR observations. They found that the DISR observations of the weaker methane bands short-ward of 830 nm agree with the methane coefficients given by Karkoschka (1994). Tomasko et al. (2008b) also discussed the implications of their results for computations of methane absorption in the atmospheres of the outer planets.

14.4.11 Examining the contribution of polycyclic aromatic hydrocarbons (PAHs) in producing organic haze layers in Titan's atmosphere

Polycyclic aromatic hydrocarbons (PAHs) are believed to be responsible for the formation of organic haze layers in Titan's atmosphere, but the nature of PAHs existing on Titan and their formation and growth mechanisms are not well understood. The Cassini spacecraft detected the presence of large mass positive and negative ions. Previous work has suggested that these large mass ions could be composed of fused-ring polycyclic aromatic hydrocarbon (PAH) compounds. These fused-ring PAHs, such as naphthalene (a white crystalline solid having molecules composed of two benzene rings) and anthracene (having a crystalline structure, pale yellow in color, and exhibiting a weak aromatic odor (see Navrátil and Minařík, 2002), are usually the result of high-temperature processes that may not occur in Titan's thin, cold, upper thermosphere. Delitsky and McKay (2010) have suggested that a different class of aromatic compounds, polyphenyls, may be a better explanation. It may be noted that polyphenyls can grow to be large polymeric structures and could condense to form the aerosols seen in Titan's cloud and hazes. They have similar properties to fused-ring PAHs (for example, electron affinity, ionization potential) and could be the negative ion species seen in the Cassini Plasma Spectrometer (CAPS) instrument data from the Cassini spacecraft.

Mebel et al. (2008) established a novel ethynyl addition mechanism (EAM) (see Fig. 14.12) computationally as a practicable alternative to high-temperature hydrogen-abstraction-C_2H_2-addition (HACA) sequences to form polycyclic aromatic hydrocarbon (PAH)-like species under low-temperature conditions in the interstellar medium and also in hydrocarbon-rich low-temperature atmospheres of planets and their moons such as Titan. Referring to Fig. 14.12, Mebel et al. (2008) have claimed that if the final ethynyl addition to 1,2-diethynylbenzene is substituted by a barrier-less addition of a cyano (CN) radical, this newly proposed mechanism can even lead to the formation

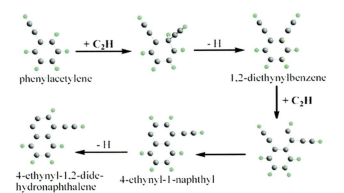

FIG. 14.12 A computationally novel ethynyl addition mechanism (EAM) as a practicable alternative to high-temperature hydrogen-abstraction-C_2H_2-addition (HACA) sequences to form polycyclic aromatic hydrocarbon (PAH)-like species under low-temperature conditions in the interstellar medium and in hydrocarbon-rich atmospheres of planets and their moons. *(Source: Mebel et al. (2008); J. Am. Chem. Soc.).*

of cyano-substituted naphthalene cores in the interstellar medium and in planetary atmospheres.

Raj et al. (2009a) reported a theoretical study on the physical interaction between polycyclic aromatic hydrocarbons (PAHs) and their clusters of different sizes in laminar premixed flames. They employed two models for this study: a detailed PAH growth model, referred to as the kinetic Monte Carlo—aromatic site (KMC-ARS) model (Raj et al., 2009b); and a multivariate PAH population balance model, referred to as the PAH—primary particle (PAH-PP) model. Both the models were solved by kinetic Monte Carlo methods. In their investigation, PAH mass spectra were generated using the PAH-PP model, and they compared it to the experimentally observed spectra for a laminar premixed ethylene flame. In this study, the position of the maxima of PAH dimers in the spectra and their concentrations were found to depend strongly on the collision efficiency of PAH coagulation. In their study, Raj et al. (2009a) examined the variation in the collision efficiency with various flame and PAH parameters to determine the factors on which it may depend. They proposed a correlation for the collision efficiency by comparing the computed and the observed spectra for an ethylene flame. With this correlation, a good agreement between the computed and the observed spectra for a number of laminar premixed ethylene flames was found.

Considering the high abundance of nitrogen in Titan's atmosphere, it is likely that the haze layers hold not only pure hydrocarbon PAHs but also their nitrogenated analogs, N-containing polycyclic aromatic compounds (N-PACs) with "hetero" N atoms in aromatic rings. Laboratory studies of Titan's tholins also support the hypothesis that, together with pure PAHs and their cations (positively charged ions), N-PACs may be the fundamental building blocks of microphysical tholin particles.

In another study, Landera and Mebel (2010) carried out ab initio quantum chemical calculations of potential energy surfaces for various reaction mechanisms of incorporation of nitrogen atoms into aromatic rings of polycyclic aromatic compounds, which may lead to the formation of N-PACs under the low-temperature and low-pressure conditions of Titan's atmosphere.

In yet another study, Totton et al. (2010) reported investigations of clusters assembled from polycyclic aromatic hydrocarbon (PAH) molecules similar in size to small soot particles. The clusters studied were comprised of coronene ($C_{24}H_{12}$) or pyrene ($C_{16}H_{10}$) molecules and represent the types of soot precursor molecule typically found in flame environments. A stochastic (which represents the element of chance, or probability) "basin-hopping" global optimization scheme was used to locate low-lying local minima on the potential energy surface of the molecular clusters.

Transmission electron microscopy (TEM) is a major analytical method employed in the physical, chemical and biological sciences. Transmission electron microscopes are capable of imaging at a significantly higher resolution than light microscopes, owing to the smaller de Broglie wavelength of electrons. This enables the instrument to capture fine detail—even as small as a single column of atoms, which is thousands of times smaller than a resolvable object seen in a light microscope. In Totton et al. (2010)'s studies, TEM-style projections of the resulting geometries showed similarities with those observed experimentally in TEM images of soot particles. The mass densities of these clusters have also been calculated and were found to be lower than bulk values of the pure crystalline PAH structures. It was found that they are also significantly lower than the standard value of 1.8 g/cm^3 used in their soot models. Consequently, they varied the mass density between 1.0 g/cm^3 and 1.8 g/cm^3 to examine the effects of varying soot density on their soot model and observed how the shape of the particle size distribution changes. According to these investigators, based on similarities between nascent soot particles and PAH clusters a more accurate soot density is likely to be significantly lower than 1.8 g/cm^3. Totton et al. (2010) recommend that, for modeling purposes, the density of nascent soot should be taken to be the value obtained for their coronene cluster of 1.12 g/cm^3. It is recognized that many aspects of the nitrogen fixation process by photochemistry in the Titan atmosphere are not fully understood. Although the Cassini mission revealed organic aerosol formation in the upper atmosphere of Titan, it is not clear how much and by what mechanism nitrogen is incorporated in Titan's organic aerosols. Using tunable synchrotron radiation at the Advanced Light Source, Imanaka and Smith (2010) demonstrated the first evidence of nitrogenated organic aerosol production by extreme ultraviolet–vacuum ultraviolet irradiation of an N_2/CH_4 gas mixture. In their investigations, the ultrahigh-mass-resolution study with laser desorption ionization-Fourier transform-ion cyclotron resonance mass spectrometry of N_2/CH_4 photolytic solid products at 60 and 82.5 nm indicated the predominance of highly nitrogenated compounds. They argued that the distinct nitrogen incorporations at the elemental abundances of H_2C_2N and HCN, respectively, are suggestive of the important roles of H_2C_2N/HCCN and HCN/CN in their formation. In Imanaka and Smith (2010)'s studies, the efficient formation of unsaturated hydrocarbons was observed in the gas phase without abundant nitrogenated neutrals at 60 nm, and this was confirmed by separately using ^{13}C and ^{15}N isotopically labeled initial gas mixtures. Imanaka and Smith (2010) argue that these observations strongly suggest a heterogeneous incorporation mechanism via short-lived nitrogenated reactive species, such as HCCN radical, for nitrogenated organic aerosol formation, and imply that substantial amounts of nitrogen is fixed as organic macromolecular aerosols in Titan's atmosphere.

14.4.12 Deciphering the particle size distribution in Titan's hazy atmosphere

In a study conducted with a view to understanding particle size distribution in Titan's hazy atmosphere, Morgan et al. (2007) made use of a detailed particle model and stochastic numerical methods to simulate the particle size distributions of soot particles formed in laminar premixed flames. The model was claimed to have been able to capture the evolution of mass and surface area along with the full structural detail of the particles. The model was validated against previous models for consistency and then used to simulate flames with bimodal and unimodal soot particle distributions. The change in morphology between the particles from these two types of flames provided further evidence of the interplay among nucleation, coagulation, and surface rates. The results confirmed the previously proposed role of the strength of the particle nucleation source in defining the instant of transition from coalescent to fractal growth of soot particles.

In a different contemporary study, Patterson and Kraft (2007) introduced new bivariate models for soot particle structure to qualitatively replicate observed particle shapes; and these models have been found to offer quantitative improvements over older single-variable models. These investigators have described models for the development of particle shape during surface growth and for particle collision diameters. They implemented their model along with two models taken from other published work. Using a stochastic approach, bivariate soot particle distributions have been calculated for the first time. Distributions calculated for the new models have been found to be insensitive to the collision diameter model used for coagulation. The total mass of soot produced in a laminar premixed flame was found to vary by no more than 20% as the model for the geometric effects of chemical reactions on the surface of particles was changed. Patterson and Kraft (2007) analyzed the histories of individual particles and showed the limitations of collector particle techniques for predicting the evolution of aggregate shape descriptors.

Various studies have demonstrated that the decomposition of nitric oxide (NO) on a soot molecule forms surface nitrogen and oxygen. The surface nitrogen can be recombined to gaseous nitrogen (N_2) while the surface oxygen desorbs from the soot molecule as carbon monoxide (CO). Sander et al. (2009a) investigated this noncatalytic conversion of gaseous NO into N_2 using density functional theory, transition state theory, and a kinetic Monte-Carlo (kMC) simulation. They validated the results against experiments. They also explored a mechanism for the conversion of NO to N_2 on a soot surface. The geometries of the intermediate stable species as well as the transition states were optimized to identify the different reaction steps. The forward and backward reaction rate of each intermediate reaction was calculated applying transition state theory. A kMC simulation using the current rates and intermediate species demonstrated feasible mechanisms for the conversion of NO to N_2 on a soot surface. It is also suggested that a portion of NO is trapped on the soot surface and this increases during the reaction and blocks the active carbon sites inhibiting further reactions. By combining different theoretical techniques in a multi-scale model, Sander et al. (2009a) were able to describe the conversion of soot in the presence of NO accurately.

14.4.13 Investigating the role of organic haze in Titan's atmospheric chemistry

In Titan's atmosphere, consisting primarily of nitrogen (N_2) and methane (CH_4), large amounts of atomic hydrogen are produced by photochemical reactions during the formation of complex organics. This atomic hydrogen may undergo heterogeneous reactions with organic aerosol in the stratosphere and mesosphere of Titan. Note that heterogeneous reaction is any of a class of chemical reactions in which the reactants are components of two or more phases (solid and gas, solid and liquid, two immiscible liquids) or in which one or more reactants undergo chemical change at an interface, for example, on the surface of a solid catalyst. In order to investigate both the mechanisms and kinetics of the heterogeneous reactions, Sekine et al. (2008a) irradiated atomic deuterium (one form of naturally occurring pure hydrogen) onto Titan tholin (organic-rich aerosol) formed from nitrogen and methane gas mixtures at various surface-temperatures of the tholin ranging from 160 to 310 K. In their experiments, the combined analyses of the gas species and the exposed tholin indicated that the interaction mechanisms of atomic deuterium with the tholin are composed of three reactions; (1) abstraction of hydrogen from tholin resulting in gaseous HD formation (HD recombination), (2) addition of D atom into tholin (hydrogenation), and (3) removal of carbon and/or nitrogen (chemical erosion). Sekine et al. (2008a) have reported the reaction probabilities of HD recombination and hydrogenation. It was found that the chemical erosion process is very inefficient under the conditions of temperature range of Titan's stratosphere and mesosphere. Under Titan conditions, the rates of hydrogenation > HD recombination ≫ chemical erosion. Sekine et al. (2008a)'s measured HD recombination rate about 10 times (with an uncertainty of a factor of 3–5) the prediction of the previous theoretical model. These results imply that organic aerosol can remove atomic hydrogen efficiently from Titan's atmosphere through the heterogeneous reactions and that the presence of aerosol may affect the subsequent organic chemistry.

One of the key components controlling the chemical composition and climatology of Titan's atmosphere is the removal of reactive atomic hydrogen from the atmosphere. A proposed process of the removal of atomic hydrogen is the heterogeneous reaction with organic aerosol. In another study, Sekine et al. (2008b) investigated the effect of heterogeneous reactions in Titan's atmospheric chemistry using new measurements of the heterogeneous reaction rate (Sekine et al., 2008a) in a one-dimensional photochemical model. Their results indicate that 60%–75% of the atomic hydrogen in the stratosphere and mesosphere are consumed by the heterogeneous reactions. This result implies that the heterogeneous reactions on the aerosol surface may predominantly remove atomic hydrogen in Titan's stratosphere and mesosphere. The results of Sekine et al. (2008b)'s calculation also indicate that a low concentration of atomic hydrogen enhances the concentrations of unsaturated complex organics, such as C_4H_2 and phenyl radical, by more than two orders in magnitude around 400 km in altitude. Such an increase in unsaturated species may induce efficient haze production in Titan's mesosphere and upper stratosphere. These results imply a positive feedback mechanism in haze production in Titan's atmosphere. The increase in haze production would affect the chemical composition of the atmosphere, which might induce further haze production. According to Sekine et al. (2008b), such a positive feedback could tend to dampen the loss and supply cycles of CH_4 due to an episodic CH_4 release into Titan's atmosphere.

14.4.14 Understanding the processes responsible for the evolution of aerosols in Titan's atmosphere

The application of stochastic approach is customary in many studies involving a multitude of small particles. In order to simulate the coagulation of the particles in tandem with the impact of surface chemistry, Lavvas et al. (2011) used a stochastic approach for the purpose of understanding the production and evolution of aerosols in Titan's atmosphere. The stochastic approach allows tracking the evolution of a particle assembling from the initial chemical species to the final aggregate structures (Sander et al. 2009a, 2009b). The starting point of the simulation is the benzene molecules, which are detected at high abundance in Titan's thermosphere (Vuitton et al., 2008). These are allowed to grow to polycyclic aromatic compounds (PACs) through a chemical reaction with radicals. There are two types of particles considered in the calculations: primary particles that are spherical and aggregates of primary particles (the specific surface area of an aggregate is smaller than the sum of its original primary particles!). Coagulation (the process of a liquid changing to a solid or semi-solid state) among primary particles provides aggregates, while surface chemistry can transform an aggregate to a larger primary particle. In Lavvas et al. (2011)'s investigation, the stochastic approach was designed to follow the evolution of particles through coagulation and surface chemistry, by simulating all growth stages from benzene molecules to PACs, to primary particles, and eventually aggregates. It is known that the stochastic particle model is able to simulate a multivariate particle population (Balthasar and Kraft, 2003). It is also known that a high dimensional state space can be used to describe the particles and consequently detailed chemical and structural information of each particle can be stored (Celnik et al., 2009; Totton et al., 2010). Consequently, the rounding of the particles can be simulated on a very detailed level.

In Lavvas et al. (2011)'s studies, the simulation initiated from the benzene molecules observed in the thermosphere and subsequently followed their evolution to larger aromatic structures through reaction with gas-phase radical species. The different processes included in the model are summarized in Fig. 14.13. Aromatics were allowed to collide and provided the first primary particles, which further grew to aggregates through coagulation. In Lavvas et al. (2011)'s experiment, they considered the collision of two PACs to create a spherical primary particle with the mass of the two colliding PACs. The collision of two PACs is not always successful and therefore they used an empirical formula that depends on the mass of the colliding PACs to describe the sticking efficiency (Appel et al., 2000; Raj et al., 2009b). Lavvas et al. (2011) assumed a typical mass density of 1 g/cm^3 for the primary particles. The coagulation of two primary particles provided an aggregate. The two colliding primary particles were in point contact directly after the coagulation event, but the neck at the touching point disappeared and the aggregate got rounder due to surface growth from the interaction with gas-phase molecules and PAC deposition. The aggregates could further grow by coagulation with other primary particles or aggregates, and by surface growth. The simulation speed is increased by taking advantage of the majorant rate approach (Eibeck and Wagner, 2000; Goodson and Kraft, 2002) and the

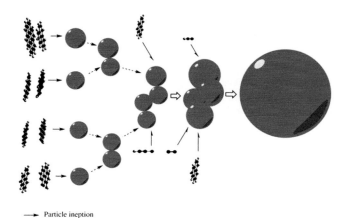
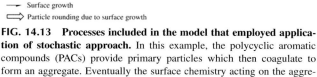

FIG. 14.13 **Processes included in the model that employed application of stochastic approach.** In this example, the polycyclic aromatic compounds (PACs) provide primary particles which then coagulate to form an aggregate. Eventually the surface chemistry acting on the aggregate provides a new, larger primary particle. *(Source: Lavvas, P., M. Sander, M. Kraft, and H. Imanaka (2011), Surface chemistry and particle shape: processes for the evolution of aerosols in TTitan's atmosphere, The Astrophysical Journal, 728(2). Copyright info: © 2011. The American Astronomical Society. All rights reserved).*

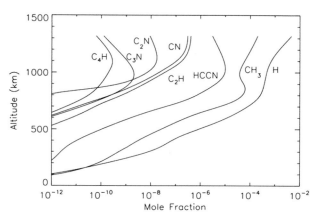

FIG. 14.14 **Radical abundances in Titan's atmosphere from the Lavvas et al. (2008b) photochemical model.** *(Source: Lavvas, P., M. Sander, M. Kraft, and H. Imanaka (2011), Surface chemistry and particle shape: processes for the evolution of aerosols in TTitan's atmosphere, The Astrophysical Journal, 728(2). Copyright info: 2011. The American Astronomical Society. All rights reserved).*

linear process deferment algorithm developed by Patterson et al. (2006). A similar model has already been successfully applied to model the sintering of silica nanoparticles (Sander et al., 2009b) and the formation of soot in flames.

Lavvas et al. (2011) also considered for the first time the contribution of heterogenous processes at the surface of the particles, which they described by the deposition of the formed aromatic structures on the surface of the particles, and also through the chemical reaction with radical species. Their results demonstrated that the evolution of aerosols in terms of size, shape, and density is a result of competing processes between surface growth, coagulation, and sedimentation. Furthermore, their simulations clearly demonstrated the presence of a spherical growth region in the upper atmosphere followed by a transition to an aggregate growth region below. Lavvas et al. (2011)'s studies showed that the transition altitude ranges between 500 and 600 km based on the parameters of the simulation.

At the low-temperature conditions of Titan's atmosphere, reactions among closed cell molecules are inhibited by high energy barriers, and chemical growth proceeds through reactions with radical species. In Lavvas et al. (2011)'s calculations, they considered the C_2H, CN, and HCCN radical species. Although these are just a few of the anticipated radicals in Titan's atmosphere (Fig. 14.14), they started their calculations with these three for two main reasons: based on theoretical calculations the first two are expected to have a major role in the growth of PACs, while laboratory experiments suggest that the last one has a dominant role in the aerosol production and evolution.

At high temperatures PACs form through the acetylene addition mechanisms, while at the low-temperature conditions of the interstellar medium (ISM) aromatic structures grow through ion–molecule reactions (see Bauschlicher and Ricca, 2000 and references therein). Although the low-temperature conditions in Titan's atmosphere do not favor the acetylene addition mechanism, the role of ion–molecule reactions could be important as demonstrated for the formation of benzene in Titan's thermosphere (Vuitton et al., 2008). On the other hand, though, the contribution of ions will be constrained over a narrow altitude region.

Based on laboratory measurements it has been known that the addition of C_2H radicals on benzene is barrierless and proceeds readily at the low-temperature conditions of Titan's atmosphere (Goulay and Leone, 2006). Furthermore, theoretical calculations suggest that multiple ethynyl additions to a benzene ring lead to the formation of larger aromatic structures (Mebel et al., 2008). Landera and Mebel (2010) reckon that the same is expected for CN addition leading to heterogeneous aromatic structures. Ion chemistry might contribute to the production of aromatic structures from benzene, but the large abundance of the C_2H and CN radicals in Titan's upper atmosphere, combined with their high reaction rate with benzene molecules, suggests that any further growth will be dominated by neutral reactions. Nevertheless, according to Lavvas et al. (2011), the contribution of ion chemistry in the overall aerosol growth should be investigated in the future. Lavvas et al. (2011)'s results demonstrate that the evolution of aerosols in terms of size, shape, and density is a result of competing processes between surface growth, coagulation, and sedimentation. As

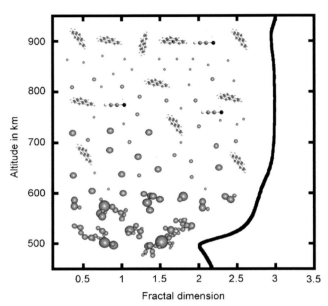

FIG. 14.15 **Shape of the particles at different altitudes.** *(Source: Lavvas, P., M. Sander, M. Kraft, and H. Imanaka (2011), Surface chemistry and particle shape: processes for the evolution of aerosols in TTitan's atmosphere, The Astrophysical Journal, 728(2). Copyright info: © 2011. The American Astronomical Society. All rights reserved).*

illustrated in Fig. 14.15, the particle evolution can be separated into the following three stages (Lavvas et al., 2011):

1) "The benzene molecules produced in the thermosphere react with radical species and grow to bigger aromatic structures. The PACs coagulate and generate the first primary particles, while deposition of the former on the surface of the latter preserves the spherical shape of the particles.
2) Once the abundance of PACs is consumed, the produced particles aggregate and the average fractal dimension of the system reduces to values close to 2. At the pressure conditions of their formation region, aggregates sediment with a velocity close to the sedimentation velocity of their primary particles. This velocity is significantly smaller than the sedimentation velocity of the same mass spherical particle, thus, aggregates fall slowly. Consequently, aggregates grow rapidly over a narrow altitude region.
3) As the produced aggregates slowly sediment to lower altitudes, gas-phase radicals are interacting with their surface. Depending on the abundance and sticking efficiency of the radicals, the surface of the aggregates is once again smoothed toward a more spherical shape."

14.5 Organic compounds on Titan's surface

Despite providing an enormous amount of new information about Titan's atmosphere, Titan's thick photochemical haze and abundant methane prevented the instruments carried by Pioneer 11 and the Voyager spacecraft from seeing Titan's surface. Images beamed back to Earth as the spacecraft sped through the Saturnian system showed only a featureless orange ball, providing no hint of the incredible landscape below. With hardly any clues available on Titan's surface conditions and surface materials, this icy and hazy moon remained shrouded in mystery until the arrival of the Cassini–Huygens mission in 2004. It has been found that, as predicted by Yung et al. (1984), Titan's surface composition is uniquely tied to its atmosphere. Chemical and physical processes occurring in Titan's atmosphere result in the formation of organic liquids and solids that are eventually deposited on its surface.

Only one spacecraft has so far acquired in situ information about Titan's surface. In January 2005, the Cassini–Huygens probe became the first (and only) probe to descend through Titan's atmosphere and safely land on its surface (Lebreton et al., 2005). As already indicated, images with at least close to desirably high-resolution were not obtained until the Cassini–Huygens mission entered orbit around Saturn in 2004. During Cassini's 13-year orbiting around Saturn, covering its moons including Titan, the Cassini RADAR instrument managed to image approximately two-thirds of Titan's surface at resolutions of 350–2000 m. Combined measurements carried out by the radar instrument, Visual and Infrared Mapping Spectrometer (VIMS), and Imaging Science Subsystem (ISS) instruments deployed onboard Cassini spacecraft have provided our first detailed view of the surface of Titan (Barnes et al., 2005; Elachi et al., 2005; Porco et al., 2005). The uncertainty that hitherto shrouded Titan changed in 2004 when Huygens probe penetrated Titan's haze, providing planetary scientists with their first detailed images of Titan's surface as well as its surface properties (Niemann et al., 2005; Tomasko et al., 2005; Zarnecki et al., 2005).

Only 11% of the sunlight incident at the top of Titan's atmosphere reaches the surface (Tomasko et al., 2005), which means there is very little solar energy available for further chemical processing on the surface. However, the molecules produced in the atmosphere, in particular, those that possess triple bonds (like acetylene), carry chemical energy to the surface that could potentially be released (Lunine and Hörst, 2011).

The spectral windows through which Titan's surface can be observed include those near 1.08, 1.28, 1.58, 2.01, 2.67, and 2.78 μm, but the 5-μm window is the widest and least affected by aerosol scattering (Sotin et al., 2005; Barnes et al., 2005; McCord et al., 2006; Soderblom et al., 2009). The Huygens probe firmly identified methane and ethane, and tentatively identified cyanogen (C_2N_2), benzene (C_6H_6), and carbon dioxide (CO_2) on the surface of Titan (Niemann et al., 2010).

Measurements from Cassini RADAR provide constraints on the dielectric constant of the surface and Janssen et al. (2016) found that the radar dark regions are consistent with organics (see Hörst, 2017). Grard et al. (2006) and Hamelin et al. (2016) observed that the measured values of dielectric constant (2.5 ± 0.3) and conductivity (1.2 ± 0.6 nS/m) are consistent with photochemically produced organic material in the first meter of the surface.

Some suggested surface constituents from interpretation of VIMS data include CO_2 (Barnes et al., 2005; McCord et al., 2008), C_6H_6, C_2H_6, and HC_3N (Clark et al., 2010). As already indicated, our ability to constrain the composition of Titan's surface is fundamentally limited by the resolution of visual and infrared mapping spectrometer (VIMS) and the small wavelength windows where it is possible to observe the surface from orbit. VIMS data can be used to identify distinct units on the surface (Barnes et al., 2007; Soderblom et al., 2007a), but conclusive identification of the composition of the units is probably not possible without another mission given the constraints of the instrument (Clark et al., 2010).

Clark et al. (2010) reported the identification of compounds on Titan's surface by spatially resolved imaging spectroscopy methods through Titan's atmosphere, and set upper limits to other organic compounds. They have reported evidence for surface deposits of solid benzene (C_6H_6), solid and/or liquid ethane (C_2H_6), or methane (CH_4), and clouds of hydrogen cyanide (HCN) aerosols using diagnostic spectral features in data obtained from the Cassini VIMS. Clark et al. (2010) identified benzene (an aromatic hydrocarbon) in larger abundances than expected by some models. However, they failed to detect acetylene (C_2H_2), expected to be more abundant on Titan according to some models than benzene. Clark et al. (2010) reckon that solid acetonitrile (CH_3CN) or other nitriles might be candidates for matching other spectral features in some Titan spectra. According to them, an as yet unidentified absorption at 5.01 μm indicates that yet another compound exists on Titan's surface. Clark et al. (2010) placed upper limits for liquid methane and ethane in some locations on Titan and found local areas consistent with millimeter path lengths. Clark et al. (2010)'s investigations indicate that except for potential lakes in the southern and northern Polar Regions, most of Titan appears "dry." Their studies also indicate that there is little evidence for exposed water ice on the surface. According to Clark et al. (2010), water ice, if present, must be covered with organic compounds to the depth probed by 1–5-μm photons: a few millimeters to centimeters.

It has been suggested that both acetylene and benzene should be produced from methane by photochemical processes in Titan's upper atmosphere and incorporated into aerosols by condensation. Once aerosols form, they carry the condensates, and eventually fall to the surface, which is well-shielded from ultraviolet radiation. Vuitton et al. (2008) predicted surface acetylene to be 125 times that of benzene. They reported models that indicate benzene is efficiently produced by ion chemistry in the upper atmosphere but would be quickly photolyzed in the ionosphere leading to downward diffusions of complex aromatic species. Their model indicates an accumulated surface layer of ~3 m of solid aromatic hydrocarbons, including a ~20 cm benzene thickness. According to Clark et al. (2010), possible mechanisms for altering the ratio of these two hydrocarbons on the surface are considered to be: (1) chemical conversion of acetylene to benzene aided by radiolysis from cosmic rays; (2) spontaneous conversion at the surface of acetylene to benzene; (3) preferential transport of one or the other hydrocarbon; and (4) sequestering acetylene in the subsurface.

Clark et al. (2010) argued that a methane + ethane rain may mix with a surface coating of hydrocarbons and nitriles to produce an organic-rich slurry that is preferentially removed from the bright highlands to accumulate in channels and lakes, at the base of slopes, and on low-lying plains. These low-lying plains may remain muddy at or just below the surface, as indicated at the Huygens landing site and more extensively in the VIMS data. Erosional processes initially concentrate certain compounds like benzene, followed by later erosional processes that expose those accumulated deposits of benzene and other organic compounds. This implies a dynamic surface with a continued history of erosion and deposition.

According to Clark et al. (2010), "while many organic compounds have now been detected in the atmosphere and on Titan's surface, we still have only a small spectral database of organic compounds to compare to spectra of Titan. Spectra of additional compounds are needed, along with higher spatial and spectral resolution data on Titan's surface in order to better understand the full compositional range of compounds on Titan."

14.6 Hydrocarbon reservoirs, seas, lakes, and rivers on Titan

Hidden beneath an all-embracing blanket of haze, Titan's surface remained a mystery until the Descent Imager/Spectral Radiometer (DISR) on ESA's Huygens probe sent back a series of unique, spectacular images. The DISR took several hundred visible-light images with its three cameras during its 2-h 27-min descent, including several sets of stereo image pairs which enabled scientists to construct digital terrain models. It was found that hundreds of lakes and seas are spread across the surface and subsurface of Titan.

14.6.1 Hydrocarbon seas on Titan's surface

The surface of the heavily haze-shrouded Titan has long been proposed to have oceans or lakes, on the basis of the stability of liquid methane at the surface. There are features similar to terrestrial lakes and seas, and widespread evidence for fluvial erosion, presumably driven by precipitation of liquid methane from Titan's dense, nitrogen-dominated atmosphere. On the surface of Titan there is a coexistence of organic lakes and water-ice rocks (Tokano, 2009). Landforms common to Earth are found across Titan, and include lakes and seas (Stofan et al., 2007; Hayes et al., 2008; Hayes, 2016), river valleys (Lorenz et al., 2008b), fans and deltas (Witek and Czechowski, 2015; Radebaugh et al., 2016; Birch et al., 2016).

Several studies (e.g., Stofan et al., 2007; Hayes et al., 2008) have identified hundreds of lakes and seas on Titan's surface. Indirect inferences gleaned from many studies (e.g., Elachi et al., 2004; Brown et al., 2008; Lunine and Lorenz, 2009) indicated that the liquid in Titan's seas are ethane and methane. The surprising ability to use Cassini RADAR to measure the depths of the lakes and seas has resulted in the production of the first extraterrestrial bathymetric maps (Mastrogiuseppe et al., 2014; Le Gall et al., 2016; Hayes, 2016). During Cassini flybys, the details of which are given in Mastrogiuseppe et al. (2014), the Cassini RADAR mapped Ligeia Mare, located at 79°N latitude and 250°W longitude and spanning an area of 126,000 km^2. Ligeia Mare's surface area is second only to that of Kraken Mare (Mastrogiuseppe et al., 2014). The International Astronomical Union (IAU) has given these and one other large feature the designation of "mare," or sea, rather than a "lacus," or lake.

Although several studies have identified hundreds of lakes and seas on Titan's surface, unlike freshwater lakes on Earth, the lakes on Titan are hydrocarbon lakes. Neish et al. (2018) recalls that there is an almost complete absence of craters near Titan's poles, which may be indicative of marine impacts into a former ocean in these regions (Neish and Lorenz, 2014) or an increased rate of fluvial erosion (Neish et al., 2016).

On Earth, water rains down from clouds and fills rivers, lakes and oceans. On Titan, clouds spew hydrocarbons such as methane and ethane (which are gases on Earth) in liquid form due to Titan's frigid climate. Rainfall occurs everywhere on Titan, but the equatorial regions are drier than the poles. Titan is the only solar system object other than Earth boasting stable liquids on the surface, with lakes and seas dominating the Polar regions. Plains (covering 65% of the surface) and dunes (covering 17% of the surface) made up of frozen bits of methane and other hydrocarbons dominate Titan's mid-latitudes and equatorial regions, respectively.

Indirect inferences gleaned from many studies (e.g., Elachi et al., 2004; Brown et al., 2008; Lunine and Lorenz, 2009) indicated that the liquid in Titan's seas are ethane and methane. For example, Lunine and Lorenz (2009) have discussed liquid methane rainfall, and the presence of liquid methane rivers and lakes on Titan. On the other hand, the liquid identified in Titan's Ontario Lacus (lake) is ethane (Brown et al., 2008). Tokano et al (2014) carried out numerical simulation of tides and oceanic angular momentum of Titan's hydrocarbon seas. The discovery of seas and northern lakes on Titan (Hayes, 2016) provides extraterrestrial explorers reason for cheer. Titan's lakes and seas cover ~1% of Titan's surface (Hayes et al., 2008; Hayes, 2016) and contain liquids that represent a global ocean depth of about 1 m (Hayes, 2016).

Birch et al. (2016) have reported a geomorphologic map of Titan's polar terrains. The map was generated from a combination of Cassini Synthetic Aperture Radar (SAR) and Imaging Science Subsystem imaging products, as well as altimetry, SAR Topo and radargrammetry topographic datasets. It was found that at the lowest elevations of each polar region, there are large seas (Fig. 14.16), which are currently liquid methane/ethane filled at the north and empty at the south (Birch et al., 2016). With an atmospheric pressure at the surface of 1.5 bars (compare this with Earth's atmospheric pressure at the mean sea level of 1 bar) and a surface temperature of 91–95 K, methane and ethane are both able to condense out of the atmosphere and rain to the surface (Atreya, 2006), where the fluid runoff concentrates, incises channels and transports sediment (Collins, 2005; Burr et al., 2006).

Lakes and seas are of particular atmospheric interest because waves can be used to constrain the wind speeds, which are very difficult to measure directly (Hörst, 2017). The size of the wave depends on wind speed and liquid viscosity, both of which, in Titan's case, are ultimately determined by the atmosphere. Lorenz and Hayes (2012) showed that the 2 m/s maximum wind speeds predicted by GCMs would produce 1 m waves.

Laboratory measurements of the dielectric properties of liquid nitrogen (Smith et al., 1991) and liquid alkanes (Dagg and Reesor, 1972; Gelsthorpe and Bennett, 1978; Smith et al., 1991) imply that the Cassini RADAR could probe depths of up to few hundred meters through Titan's lakes and seas in a mode in which the radar is pointed straight downward, rather than to the side when used as imager (Mastrogiuseppe et al., 2014). Mastrogiuseppe et al. (2014) constructed the depth profile—the bathymetry—of Titan's large sea Ligeia Mare from Cassini RADAR data collected during the 23 May 2013 (T91) nadir-looking altimetry flyby. Fig. 14.17 provides the bathymetric profile across Ligeia Mare. The greatest depth was found to be about 160 m (occurring close to 79° of latitude). Another observation with regard to the bathymetry of Ligeia Mare is that its seabed slope is gentler toward the northern shore. However, seabed slope in the southern region was found to be very different, where steep hills and flooded valleys are found. Whereas Ligeia Mare is up to 160 m deep as already indicated, Ontario Lacus is 90 m deep. Titan's lakes and seas cover ~1% of Titan's surface (Hayes et al., 2008;

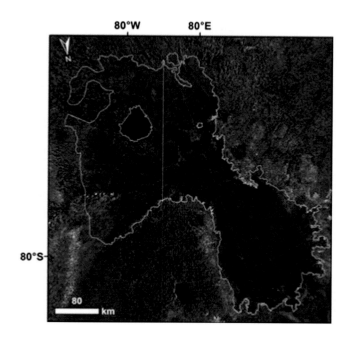

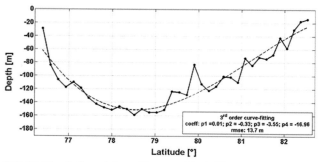

FIG. 14.17 **Bathymetric profile across Ligeia Mare.** The deepest point of the sea approach to 160 m close to 79° of latitude. The dashed lines indicate the third order polynomial fitting the depths values with an RMS error of about 14 m. *(Figure Source: Mastrogiuseppe et al. (2014); The bathymetry of a Titan sea, Geophys. Res. Lett., 41: 1432–1437).*

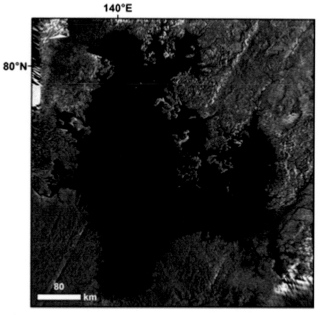

FIG. 14.16 (Top) Empty sea (*Em*) around Titan's south pole with the boundary marked in yellow. These basins occupy the lowest elevations at the south. Evidence of valley incision is seen along their perimeters; (Bottom) Filled sea (Ligeia Mare) at the north shows a similar shoreline morphology to the southern empty sea. The largest valley networks drain into the seas at the north, while the southern empty seas also show evidence for large drainage systems. *(Source: Birch et al. (2016), Geomorphologic mapping of Titan's polar terrains: constraining surface processes and landscape evolution, Icarus, 282: 1–23. (Elsevier publication).*

Hayes, 2016) and contain liquids that represent a global ocean depth of about 1 m (Hayes, 2016).

Low radio signal attenuation through the sea demonstrated that the liquid is remarkably transparent, requiring a nearly pure methane-ethane composition, and further that microwave absorbing hydrocarbons, nitriles, and suspended particles be limited to less than the order of 0.1% of the liquid volume.

14.6.2 Likelihood of finding transient liquid water environments on Titan's surface

Liquid water is both a crucial source of oxygen and a useful solvent for the generation of biomolecules on Titan's surface. Thus, according to Neish et al. (2018), if we wish to identify molecular indicators of prebiotic chemistry on Titan, we need to determine where liquid water is most likely to have persisted. Although Titan's average surface temperature of ~94 K precludes the existence of bodies of liquid water over geological timescales (unless there is an active hotspot), it does not rule out the presence of water on the surface for short periods of time. According to Neish et al. (2018), "We are likely to find transient liquid water environments on the surface of Titan in two distinct geological settings: (1) cryovolcanic lavas and (2) melt in impact craters. In addition, Titan's deep interior has a liquid water layer, perhaps hundreds of kilometers thick, which may also contain biomolecules (Fortes, 2000; Iess et al., 2012). Samples of this ocean may be transported to the surface through cryovolcanic processes before eventually freezing. Thus, if we wish to find biomolecules on the surface of Titan, we should focus our search in and around cryovolcanoes and impact craters."

14.6.3 Subsurface hydrocarbon reservoirs on Titan

Titan's subsurface reservoirs (see, for example, Fig. 14.18) contain hydrocarbons, including methane. While most of the liquid in the lakes is thought to be replenished by hydrocarbon rainfall from clouds in Titan's atmosphere, the

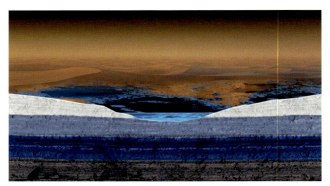

FIG. 14.18 Titan's subsurface reservoir containing hydrocarbons, including methane. Credit: ESA/ATG medialab. (Sources: http://sci.esa.int/cassini-huygens/55224-science-highlights-from-huygens-3-methane-mystery/; http://sci.esa.int/cassini-huygens/54589-Titans-subsurface-reservoirs-unannotated/). Copyright: ESA/ATG medialab.

cycling of liquid between the subsurface, surface, and atmosphere is still not well understood.

Scientists have modeled how a subsurface reservoir ("alkanofer") of liquid hydrocarbons, filled with rainfall runoff, would diffuse throughout Titan's porous icy crust. They found that this diffusion could cause a new reservoir—formed from clathrates—to form where the bottom of the original reservoir meets layers of nonporous ice. Note that clathrates are compounds that form a crystal structure with small cages that trap other substances like methane and ethane. Titan's subsurface clathrate reservoirs would interact with and fractionate (separate) the liquid phase within the original underground hydrocarbon lake, slowly changing its composition. Eventually, subsurface lakes that had come into contact with the clathrate layer would mainly be composed of either propane or ethane, depending on the type of clathrate that had formed. Importantly, this would continue up to Titan's surface. Lakes fed by these propane or ethane subsurface reservoirs would show the same kind of composition, whereas those fed by the primarily hydrocarbon rainfall would be different and contain methane, nitrogen, and trace amounts of argon and carbon monoxide. The composition of the lake would indicate what was happening deep underground (http://sci.esa.int/cassini-huygens/55224-science-highlights-from-huygens-3-methane-mystery/).

14.6.4 Hydrocarbon lakes on Titan and their astrobiological significance

Given the presence of a dense nitrogen-methane atmosphere and the likelihood of liquid hydrocarbons, Titan has long been considered a natural laboratory of prebiotic chemistry. The Cassini RADAR instrument has imaged approximately two-thirds of the surface of Titan, producing views of the landscape with resolutions as good as 350 m. The radar imaging pole-wards of 70° north shows more than 75 circular-to-irregular radar-dark patches, in a region where liquid methane and ethane are expected to be abundant and stable on the surface. As the Huygens Probe descended through Titan's atmosphere, images taken by DISR revealed a network of channels flowing into a plain reminiscent of flood plains (Tomasko et al., 2005). The complex dendritic (having a branched form resembling a tree) drainage systems observed are likely indicative of a distributed source and possibly formed by rapid erosion that creates deeply incised valleys (Perron et al., 2006; Soderblom et al., 2007a).

Although the Huygens Probe only found evidence of past fluvial activity, RADAR data obtained during the Cassini Radar flyby of Titan on 22 July 2006 (T_{16}) revealed numerous lakes and seas near the north pole (Stofan et al., 2007). Stofan et al. (2007) have provided definitive evidence for the presence of lakes on the surface of Titan, obtained during T_{16}. While orbiting Saturn, the Cassini spacecraft spotted lakes on Titan containing ethane. The radar-dark patches are interpreted as lakes on the basis of their very low radar reflectivity and morphological similarities to lakes, including associated channels and location in topographic depressions. Some of the lakes do not completely fill the depressions in which they lie, and apparently dry depressions are present. Stofan et al. (2007) interpreted this finding to indicate that lakes are present in a number of states, including partly dry and liquid-filled. These northern-hemisphere lakes constitute the strongest evidence yet that a condensable-liquid hydrological cycle is active in Titan's surface and atmosphere, in which the lakes are filled through rainfall and/or intersection with the subsurface "liquid methane" table. When the Cassini spacecraft found no methane ocean swathing Titan, it was a blow to the proponents of an Earth-like world existing on Titan (see Sotin, 2007). However, the discovery of northern lakes on Titan gives them reason for cheer. For a review of the lakes and seas, see Hayes (2016).

The nature of the lakes, the processes that fill them, and the factors that set their distribution are still not fully understood. Likewise, the geological processes that form the lakes, both large and small, on Titan are still unclear. Cornet et al. (2015) proposed that formation of hydrocarbon lakes on Titan results from the dissolution of solid organics by the liquid methane and ethane. With the exception of Ontario Lacus, the hydrocarbon lakes on Titan occur in the northern hemisphere (these are shown in Fig. 14.19) and 97% of the lakes on Titan are clustered in a region of 900 by 1800 km—about 2% of Titan's surface area (McKay, 2016). According to McKay (2016), the large lakes and small lacustrine depressions seen in Fig. 14.19 can be roughly grouped into two principle modes based on size. According to several investigators (Lorenz et al., 2008c; Lorenz et al., 2014; Marco et al., 2014), the large lakes (several hundred kilometers in width) are thought

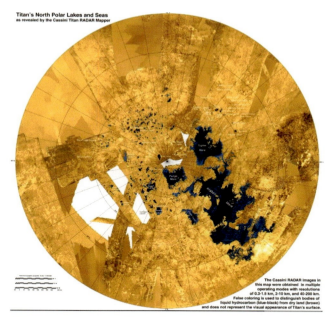

FIG. 14.19 Lakes on Titan. False color map of the Northern Hemisphere down to 50°N shows Titan's lakes in blue and areas that are not lakes in brown. *The map is based on data from the radar on the Cassini spacecraft, taken during flybys between 2004 and 2013. Kraken Mare, the largest known lake on Titan is shown to the right and below the pole in this image. Above and right of the pole is Ligeia Mare, the second largest known lake. Image from NASA. (Source: McKay, C.P. (2016), Titan as the abode of life, Life, 6:8.). Copyright info: © 2016 by the author; licensee MDPI, Basel, Switzerland. This article from which this Figure is reproduced here is an open access article distributed under the terms and conditions of the Creative Commons by Attribution (CC-BY) license (http://creativecommons.org/licenses/by/4.0/).*

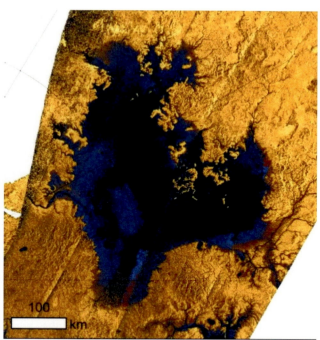

FIG. 14.20 Close-up radar image of Ligeia Mare, showing coastline features and rivers. The lake is ~400 km across. *Image from ESA based on Cassini Data. (Source: McKay, C.P. (2016), Titan as the abode of life, Life, 6:8.). Copyright info: 2016 by the author; licensee MDPI, Basel, Switzerland. This article from which this Figure is reproduced here is an open access article distributed under the terms and conditions of the Creative Commons by Attribution (CC-BY) license (http://creativecommons.org/licenses/by/4.0/).*

to be deep (up to hundreds of meters in depth). Several researchers (Stofan et al., 2007; Sotin et al., 2012; Wasiak et al., 2013) have observed that the hydrocarbon lakes discovered in Titan's northern hemisphere possess fractal shorelines (i.e., having irregular geometric structure) and are connected to fluvial channels (e.g., Ligeia Mare, Fig. 14.20). However, exceptions have also been found. For example, Hayes et al. (2008) and Lorenz et al. (2013) reported that the small lacustrine depressions appear to be more shallow depressions with rounded shorelines and show a gradient with elevation. McKay (2016) reported that empty depressions, which appear similar to the small lakes are found at elevations about 250 m above the similarly shaped small lakes (as discussed by Cornet et al., 2015). According to McKay (2016), this could be indicative of the presence of aquifers and subsurface connectivity establishing defined liquid levels (Cornet et al., 2015; Hayes et al., 2008; Cornet et al., 2012).

As noted earlier, although Titan's entire surface has not been mapped, there appears to be a north-south asymmetry in the distribution of the lakes; with indications of the southern hemisphere almost entirely devoid of currently filled lakes (see Hörst, 2017). It may be noted that, from Titan GCMs results, methane is efficiently transported away from the equator, drying out the low latitudes and midlatitudes and depositing liquid at the poles (Mitchell et al., 2006; Mitchell, 2008; Mitchell, 2009; Schneider et al., 2012; Mitchell, 2012; Lora et al., 2014, 2015).

Titan researchers have found that the vast majority of Titan's lakes sit in sharp-edged depressions that "look like you took a cookie cutter and cut out holes in Titan's surface." The so-called "cookie cutter" lakes found near the north pole have steep walls and are hundreds of meters deep (Hayes et al., 2008; Stiles et al., 2009), morphology that is consistent with dissolution karst (Cornet et al., 2015)—a topography formed from the dissolution of soluble rocks such as limestone, dolomite, or gypsum. According to (Hörst, 2017), "Since methane does not dissolve water ice but it can dissolve organics like solid acetylene (Glein and Shock, 2013; Malaska and Hodyss, 2014), the presence of dissolution karst would indicate that there are hundreds of meters of organics at the north pole, which has implications for atmospheric chemical and dynamical processes that would result in such a significant deposit of organics."

McKay (2016) pointed out that the spectral data in the regions inside and around some polar (and equatorial) lacustrine (lake-like) depressions indicate the presence of various hydrocarbons and nitriles (Clark et al., 2010; Moriconi et al., 2010) and are not compatible with the presence of water ice exposed on the surface (Barnes et al., 2009). Rannou et al. (2016) found that the surface spectrum at the Huygens landing site indicates a layer of water-ice grains overlaid by a moist layer of weakly compacted haze particles. The spectrum is inconsistent with dry haze particles.

Cassini radar detected several putative liquid hydrocarbon lakes in the polar region of Titan. These lakes are likely to contain organic sediments deposited from the atmosphere. According to Tokano (2008, 2009), such liquid hydrocarbon lakes may contain organic sediments deposited from the atmosphere that would promote prebiotic-type chemistry driven by cosmic rays, the result of which could be the production of more complex molecules such as nitrogen-bearing organic polymer or azides (i.e., compounds containing the group N_3 combined with an element or radical). Stevenson et al. (2015) suggested that nitrogen compounds could act as a lipid bilayer in the lakes. In addition to solubility, density also plays a role in the fate of materials in the lakes and seas. The likelihood or otherwise of materials sinking/floating in Titan's lakes has been a matter of debate among some Titan researchers. For example, whereas Tokano (2009) suggested that hydrocarbon solids are generally denser than their corresponding liquids and, therefore, will sink forming deposits on the bottom of Titan's lakes. However, Roe and Grundy (2012) and Hofgartner and Lunine (2013) pointed out the likelihood of ice formed during the winter to float in Titan's hydrocarbon lakes. Based on experimental simulations of haze formation, Hörst and Tolbert (2013) found the possibility that the haze particles would also float. However, the findings of Le Gall et al. (2016), who reported evidence of the bottom of the Ligeia Mare being coated in a layer of nitriles, strengthened Tokano (2009)'s suggestion.

As indicated earlier, Titan researchers are generally of the view that Titan's lakes and seas represent a region of particular astrobiological interest. According to Hörst (2017), "Some fraction of the complex organics produced in the atmosphere will end up in the lakes and seas either directly through sedimentation from the atmosphere (dry removal), washout from rain (wet removal), or from subsequent transport processes once they are deposited on the surface. Some of the compounds are soluble in liquid hydrocarbons, but others are not; thus, the lakes will partition the organics produced in the atmosphere potentially resulting in subsequent chemistry." Based on the finding that the methane (CH_4) and ethane (C_2H_6) liquids are widespread on Titan in lakes and in the moist surface, McKay (2016) suggests that if life can exist in such liquids, life should be widespread on Titan. According to Hörst (2017), an improved understanding of the chemistry of the lakes requires experimental investigations and additional measurements from a future mission.

14.6.5 Ammonia-enriched salty liquid–water inner-ocean hiding far beneath Titan's frozen surface

Based on credible indications, a putative ethane-ocean below the icy crust of Titan was suspected by several planetary researchers. For example, Lunine et al. (1983) predicted an ethane ocean on Titan. Based on Titan's rotation, Lorenz et al. (2008) predicted an internal ocean hiding below Titan's crust as well as changing zonal winds on Titan. In later studies, the presence of a salty liquid–water inner ocean hiding far beneath Titan's frozen surface has been confirmed based on evidence gleaned from a deformable interior, Schumann resonance, and cryo-volcanism. Based on a Schumann-like resonance on Titan driven by Saturn's magnetosphere as revealed by the Huygens Probe carried by NASA's Cassini spacecraft (Béghin et al., 2007), Béghin et al. (2010) found that Titan possesses a native ocean beneath some 45 km of ice. The presence of a global subsurface ocean of liquid water beneath an icy outer shell was inferred by radio science data collected by the Cassini spacecraft.

The presence of an electrically conductive ocean of water and ammonia which is buried at a depth of 55–80 km below a nonconducting, icy crust has been indicated from theoretical studies. Further theoretical studies also indicated that Titan's observed Schumann resonance is indicative of Titan's buried water ocean (Béghin et al., 2012). Thus, data from Cassini–Huygens mission provided evidence that Titan is hiding an internal liquid ocean beneath its surface, likely composed of saltwater and ammonia. Fortes (2000) examined the exobiological implications of a possible ammonia–water ocean inside Titan.

Mitri et al. (2008) showed that ammonia–water mixtures may erupt from a subsurface ocean on Titan through the ice shell, leading to cryo-volcanism. Based on theoretical considerations, Baland et al. (2011) showed that Titan's obliquity could be considered as evidence of a subsurface ocean. Observations by the Cassini mission have confirmed the presence of a liquid water subsurface ocean (Iess et al., 2012). Based on the discovery of Schumann resonance on Titan, Beghin et al. (2012) estimated that the lower layer of a salty, subsurface ocean lies 55–80 km below Titan's surface. Tokano et al. (2014) carried out numerical simulation of tides and oceanic angular momentum of Titan's hydrocarbon seas. Lorenz et al. (2014) reported a radar map of Titan seas, illustrating tidal dissipation and ocean mixing through the throat of Kraken. Apart from investigations such as these, other analyses of Titan's geophysical features have been found to be consistent with an ice shell of this thickness overlying a relatively dense subsurface ocean (Nimmo and Bills, 2010; Mitri et al., 2014a,b).

14.6.6 Rivers and drainage networks on Titan—comparison and contrast with those on Earth and Mars

Earth, Mars, and Titan have all hosted rivers at some point in their histories. Rivers have eroded the topography of Earth, Mars, and Titan, creating diverse landscapes. However, the dominant processes that generated topography on Titan (and to some extent on early Mars) are not well known. Rivers erode the landscape, leaving behind signatures that depend on whether the surface topography was in place before, during, or after the period of liquid flow. Titan is viewed as a sibling of Earth, as both bodies have rainy weather systems and landscapes formed by rivers. But as we study these similarities, Titan emerges as an intriguingly foreign world.

Cassini's Titan Radar Mapper imaged the surface of Titan on its February 2005 flyby (denoted T3), collecting high-resolution synthetic-aperture radar and larger-scale radiometry and scatterometry data. These data provided the first definitive identification of impact craters on the surface of Titan, networks of fluvial channels and surficial dark streaks that may be longitudinal dunes. Elachi et al. (2006) have described this great diversity of landforms. They concluded that much of the surface thus far imaged by radar of the haze-shrouded Titan is very young, with persistent geologic activity. Images of Titan, taken during the joint NASA/European Space Agency Cassini–Huygens mission, invoke a sense of familiarity: river channels meander downhill to damp lake-beds, where icy, rounded stones, resembling river cobbles, litter the ground; massive cumulus clouds form and quickly dissipate, suggestive of rain; and dark oval regions resemble lakes (Griffith, 2006).

The Cassini spacecraft images revealed intricate surface albedo features that suggest aeolian, tectonic and fluvial processes; they also showed a few circular features that could be impact structures. These observations imply that substantial surface modification has occurred over Titan's history. Initial experimentations failed to directly detect liquids on the surface. However, observation of clouds tempted scientists to the possibility of rainfall and the resulting flow of liquid methane over Titan's surface, and the existence of liquid methane rivers and lakes.

The surface of the haze-shrouded Titan has long been proposed to have oceans or lakes, on the basis of the stability of liquid methane at the surface. Initial visible and radar imaging failed to find any evidence of an ocean, although abundant evidence was found that flowing liquids existed on the surface.

The presence of dry fluvial river channels and the intense cloud activity in the south pole of Titan suggest the presence of methane rain. Planetary scientists think that the nitrogen atmosphere of Titan, therefore, appears to support a methane meteorological cycle that sculptures the surface and controls its properties. Interestingly, Titan and Earth are the only worlds in the solar system where rain reaches the surface, although the atmospheric cycles of water and methane are expected to be very different. Hueso and Sánchez-Lavega (2006) reported three-dimensional dynamical calculations showing that severe methane convective storms accompanied by intense precipitation may occur in Titan under the right environmental conditions. The strongest storms grow when the methane relative humidity in the middle troposphere is above 80%, producing updrafts with maximum velocities of 20 m/s, able to reach altitudes of 30 km before dissipating in 5–8 h. Raindrops of 1–5 mm in radius produce precipitation rainfalls on the surface as high as 110 kg/m^2 and are comparable to flash flood events on Earth.

There are features similar to terrestrial lakes and seas, and widespread evidence for fluvial erosion, presumably driven by precipitation of liquid methane from Titan's dense, nitrogen-dominated atmosphere. Brown et al. (2008) have reported infrared spectroscopic data, obtained by the Visual and Infrared Mapping Spectrometer (VIMS) on board the Cassini spacecraft, that strongly indicate that ethane, probably in liquid solution with methane, nitrogen, and other low-molecular-mass hydrocarbons, is contained within Titan's Ontario Lacus (lake).

Titan is the only other place in the solar system hitherto known to have an Earth-like cycle of liquids flowing across its surface as the planet cycles through its seasons. Radar images revealed an icy terrain carved out over millions of years by rivers of liquid methane, similar to how rivers of water have etched into Earth's rocky continents. Several impact craters have been found on Titan's surface, the most important of which include Sinlap, Selk, and Menrva craters. The craters that have been investigated in great detail, including Sinlap and Selk craters, show evidence of fluvial and aeolian erosion (Le Mouélic et al., 2008; Soderblom et al., 2010; Wood et al., 2010). Neish et al. (2013) found that craters on Titan are shallower than comparable size craters on Ganymede. These researchers argue that the craters have been filled in from aeolian transport. Soderblom et al. (2010) and Neish et al. (2015) found evidence for large river channels in the ejecta blankets of both Selk and Sinlap respectively. Menrva is also characterized by many large fluvial networks (Lorenz et al., 2008b; Wood et al., 2010; Williams et al., 2011), which likely expose impact melt deposits in the channel walls and as riverbed sediments (Hörst, 2017).

Radar images of Titan's surface provided clear indications of fluvial features that cut through Titan's hard surface. The Huygens probe observed the damp surface at its landing site (at 10° S latitude, 192° W longitude). Images of Titan's surface taken by the Huygens probe, which descended into Titan's atmosphere in January 2005 showed indications of river channels flowing from the hills into a damp lake-bed, which is riddled with "cobbles" (Fig. 14.21). Rounded

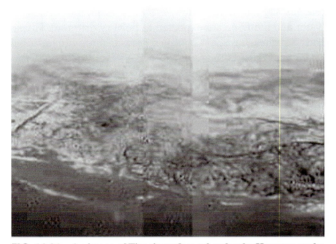

FIG. 14.21 An image of Titan's surface taken by the Huygens probe, which descended into Titan's atmosphere in January 2005. River channels flow from the lightercoloured hills into a damp lake-bed, which is riddled with "cobbles." *(Source: Griffith, C.A. (2006), Titan's exotic weather, Nature (Planetary Science), 442: 362-363).*

cobbles at the landing site (see Fig. 14.4) provided additional evidence of fluvial erosion (Tomasko et al., 2005). Channel morphology can be used to constrain formation mechanisms. For example, the presence of dry valleys, found only at midlatitudes, indicates strong episodic events (Langhans et al., 2012).

The variety of features on Titan's surface, including equatorial dunes; hydrocarbon rivers, lakes, and seas, with liquid methane playing the role on Titan that water plays on Earth, has surprised and delighted scientists and the enthusiastic general public alike. Surprisingly, Titan's surface doesn't look as old and weather-beaten as it should. The rivers have caused surprisingly little erosion and there are fewer impact craters than would be expected. Planetary scientists are curious to know the secret to Titan's youthful complexion. While images of Titan have revealed its present landscape, very little is known about its geologic past.

Apart from lakes and reservoirs as already indicated, the cameras revealed a plateau with a large number of dark channels cut into it, forming drainage networks that bore many similarities to those on Earth. Brighter regions north of the Huygens Probe's landing site on Titan displayed two different drainage patterns: (1) bright highlands with rough topography and deeply incised branching (dendritic) drainage networks with dark-floored valleys that indicated erosion by methane rainfall; and (2) short, stubby channels that followed linear fault patterns, forming canyon-like features suggestive of spring sapping by liquid methane. The narrow channels converged into broad rivers, which drained into a broad, dark, lowland region. The ravines (deep, narrow gorge with steep sides) cut by the rivers are approximately 100 m deep and their valley slopes are very steep, which suggest rapid erosion due to sudden, violent flows. The spatial distribution of different channel types provides information about spatial variation in the strength and frequency of rainfall (Hörst, 2017).

Initially thought to be rocks or ice blocks, the rock-like objects are more pebble-sized (see Fig. 14.22). The topographic data showed that the bright highland terrains are extremely rugged, often with slopes of up to 30 degrees. These drain into relatively flat, dark lowland terrains.

Rivers develop on planetary bodies from the interplay of climate and topography (Burr et al., 2017). In planetary contexts, rivers and their effects are used to constrain short- and long-term climate models. However, rivers are also controlled by the short- and long-wavelength topography of the planetary surface. Rivers thus provide a means to constrain the generation of topography.

Black et al. (2017) developed two metrics to measure how well river channels align with the surrounding large-scale topography. Earth's plate tectonics introduce features such as mountain ranges that cause rivers to divert, processes that clearly differ from those found on Mars and Titan. Black et al. (2017) compared topography at a range of scales with mapped river drainages to provide new

FIG. 14.22 Titan's surface viewed by Descent Imager/Spectral Radiometer (DISR). This image was returned 14 January 2005, by ESA's Huygens probe after its successful descent to land on Titan. This colored view, following processing to add reflection spectra data, gives an indication of the actual colour of the surface. *Credit: ESA/NASA/JPL/ University of Arizona. (Sources: http://sci.esa.int/cassini-huygens/36378-disr-image-of-Titan-s-surface/; http://sci.esa.int/cassini-huygens/55229-science-highlights-from-huygens-8-dry-river-beds-and-lakes/). Copyright: ESA/NASA/JPL/University of Arizona.*

insights into the topography-generating mechanisms on Earth, Mars, and Titan.

Black et al. (2017) analyzed drainage patterns on all three bodies (Earth, Mars, and Titan) and found that large drainages, which record interactions between deformation and erosional modification, conform much better to long-wavelength topography on Titan and Mars than on Earth. Black et al. (2017) used a numerical landscape evolution model to demonstrate that short-wavelength deformation causes drainage directions to diverge from long-wavelength topography, as observed on Earth. They attributed the observed differences to ancient long-wavelength topography on Mars, recent or ongoing generation of long-wavelength relief on Titan, and the creation of short-wavelength relief by plate tectonics on Earth.

It has been found that the history of Titan's landscape resembles that of Mars, not Earth. Rivers on three worlds (Earth, Mars, and Titan) tell different tales. Scientists have found that Titan, like Mars but unlike Earth, has not undergone any active plate tectonics in its recent past. The upheaval of mountains by plate tectonics deflects the paths that rivers take. The research team found that this telltale signature was missing from river networks on Mars and Titan.

The environment on Titan may seem surprisingly familiar to that on Earth: clouds condense and rain down on the surface, feeding rivers (though with liquid methane, instead of water) that flow into lakes. Black et al. (2017) have found that despite these similarities, the origins of topography, or surface elevations, on Mars and Titan are very different from that on Earth. According to these investigators, while the processes that created Titan's topography are still enigmatic, this rules out some of the mechanisms we are most familiar with on Earth. Fig. 14.23 illustrates a comparison of the river networks on Mars, Earth, and Titan. Maps of Earth are sharp in detail, as are those for Mars, showing mountain peaks and impact basins in high relief. By contrast, due to Titan's thick, hazy atmosphere, the global map of Titan's topography is extremely fuzzy, showing only the broadest features. Fig. 14.24 illustrates methane river networks draining into lakes in Titan's North Polar Region, gleaned through the images obtained from the Cassini mission.

Titan is around four billion years old, roughly the same age as the rest of the solar system. But the low number of impact craters found on Titan put estimates of its surface at only between 100 million and one billion years old. Black et al (2017) analyzed images of Titan's river networks and suggested two possible explanations: either erosion on Titan is extremely slow, or some recent phenomena have wiped out older surface features. According to these researchers, Titan's surface should have eroded much more than what we are seeing today, if the river networks have been active for a long time. It raises some very interesting questions

FIG. 14.23 **Comparison of the river networks on Mars, Earth, and Titan.** From top to bottom, images span ~100 km on Mars, ~2000 km on Earth, and ~400 km on Titan. *Credit: Benjamin Black, adapted from images from NASA Viking, NASA/Visible Earth, and NASA/JPL/Cassini RADAR team. (Source: https://phys.org/news/2017-05-history-Titan-landscape-resembles-mars.html).*

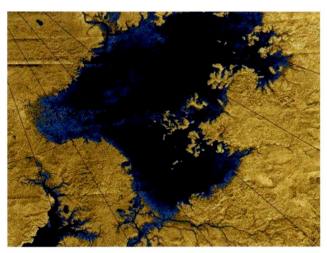

FIG. 14.24 **Methane river networks draining into lakes in Titan's North Polar Region, gleaned through the images obtained from the Cassini mission.** *Credit: NASA/JPL/USG. (Source: https://phys.org/news/2012-07-beneath-mask-Titan-surprisingly-smooth.html#jCp).*

about what has been happening on Titan in the last billion years.

Black et al. (2017) have suggested that geological processes on Titan may be like those we see here on Earth, where plate tectonics, erupting volcanoes, advancing glaciers, and river networks reshaped our planet's surface over billions of years. Based on this logic, planetary scientists think that on Titan, similar processes—tectonic upheaval, cryovolcanic eruptions (i.e., icy lava eruptions), erosion, and sedimentation by rivers—may be at work and altering the surface.

Unfortunately, identifying which of these geological phenomena may have modified Titan's surface is a significant challenge. Researchers lament that the images generated by the Cassini spacecraft are like aerial photos but with much coarser resolution. Because the images are flat, depicting terrain from a bird's-eye perspective, no information about a landform's elevation or depth can be gleaned. Consequently, discovering which processes are at work is not that easy.

Black et al. (2017) determined the extent to which river networks may have renewed Titan's surface. The team analyzed images taken from Cassini–Huygens, and mapped 52 prominent river networks from four regions on Titan. The researchers compared the images with a model of river network evolution. This model depicts the evolution of a river over time, given variables such as the strength of the underlying material and the rate of flow through the river channels. As a river erodes slowly through the ice, it transforms from a long, spindly thread into a dense, treelike network of tributaries. They compared their measurements of Titan's river networks with the model, and found that Titan's rivers most resembled the early stages of a typical terrestrial river's evolution. The observations indicate that rivers in some regions have caused very little erosion, and hence very little modification of Titan's surface.

Going a step further, Black et al. (2017) compared Titan's images with recently renewed landscapes on Earth, including volcanic terrain on the island of Kauai and recently glaciated landscapes in North America. It was found that the river networks in those locations are similar in form to those on Titan, thereby suggesting that geologic processes may have reshaped Titan's icy surface in the recent past.

14.7 Presence of a salty liquid water inner ocean hiding far beneath Titan's frozen surface—evidence gleaned from a deformable interior, Schumann resonance, and cryovolcanism

Titan was once thought to have a global ocean of light hydrocarbons on its surface, but after 40 close flybys of Titan by the Cassini spacecraft, it has become clear that no global ocean exists on Titan's surface. However, there are hints to believe that deep below the surface, Titan harbors a large internal ocean.

Tobie et al. (2005) showed that theoretical models of Titan's formation and evolution predict the present-day existence of a substantial liquid water layer in its interior, provided a sufficient amount of ammonia is present in the ocean. Observations by the Cassini mission have confirmed the presence of a liquid water subsurface ocean. For example, Iess et al. (2012) showed that measurements of the tidal Love number by the Radio Science experiment require that Titan's interior is deformable over its orbital period, consistent with a global ocean at depth. Furthermore, the permittivity, wave, and altimetry instrument on ESA's Huygens probe detected an electric current in Titan's ionosphere, consistent with a *Schumann resonance* between two conductive layers (Beghin et al., 2012).

Schumann resonance: The phenomenon of *Schumann resonance* can be explained as follows, with the help of a terrestrial example. Between Earth's surface and its ionosphere, thousands of lightning flashes take place every second, and each bolt generates a radio "crackle." This means that Earth's atmosphere is continuously generating extremely low frequency (ELF) radio signals. Schumann resonances (SR) are a set of spectrum peaks in the extremely low frequency (ELF) portion of the Earth's electromagnetic field spectrum. Schumann resonances are global electromagnetic resonances, generated and excited by lightning discharges in the "cavity resonator" formed between Earth's surface (an electric conductor) and the ionosphere (a region of electrically charged particles in Earth's upper atmosphere). A cavity resonator is an excitable device or system that exhibits resonance or resonant behavior. That is, it naturally oscillates with greater amplitude at some frequencies, called resonant frequencies, than at other frequencies. The oscillations in a resonator can be either electromagnetic or mechanical (e.g., acoustic resonators in musical instruments that produce sound waves of specific tones; quartz crystals used in electronic devices to produce oscillations of very precise frequency). It had long been considered that the existence of Schumann resonances on other planets would make it possible to reveal the presence of both storm activity and a conductive ground.

Scientists had wondered whether lightning might be generated in Titan's atmosphere, so Huygens was equipped for Permittivity, Wave and Altimetry (PWA) experiment to detect tell-tale radio signals. One of the most surprising discoveries of ESA's Huygens mission was the detection of an unusual source of electrical excitation in Titan's atmosphere. Although no lightning or thunderstorms were detected, the PWA did detect an unusual ELF signal at a frequency of around 36 Hertz. Huygens also discovered a lower ionospheric layer between 140 km and 40 km from Titan's surface, with electrical conductivity peaking near 60 km (http://sci.esa.int/cassini-huygens/55230-science-highlights-from-huygens-9-schumann-like-resonances-hints-of-a-subsurface-ocean/).

In order to explain the unique pattern of ELF signals observed on Titan using the PWA equipment on the Huygens probe, scientists have proposed that Titan's atmosphere behaves like a giant electrical circuit. The electrical currents are generated in the ionosphere when it interacts with Saturn's magnetosphere. This results in a dynamo effect as plasma trapped in the magnetosphere corotates with the planet every 10 h or so.

The lower boundary of Titan's "cavity resonator," which reflects the radio signals, is thought to be a conductive ocean of water and ammonia which is buried at a depth of 55-80 km below a nonconducting, icy crust. Huygens' discovery of this unique Schumann resonance is seen as key supporting evidence for the existence of such a subsurface ocean, hidden far beneath the moon's frozen surface. Fig. 14.25 shows artist's concept illustrating a possible model of Titan's internal structure that incorporates data from NASA's Cassini spacecraft. In this model, Titan is fully differentiated, which means the denser core of the moon has separated from its outer parts. This model proposes a core consisting entirely of water-bearing rocks and a subsurface ocean of liquid water. The mantle, in this image, is made of icy layers, one that is a layer of high-pressure ice closer to the core and an outer ice shell on top of the subsurface ocean.

To support the presence of a conductive ocean of water and ammonia which is buried at a depth of 55-80 km below a nonconducting, icy crust, planetary scientists have invoked the principles of radioactive decay and cryo-volcanism. It may be noted that one of the trace gases detected by the gas chromatograph mass spectrometer (GCMS) on ESA's Huygens probe was radiogenic argon-40 (^{40}Ar). This isotope offers a window to the interior of the giant moon (http://sci.esa.int/cassini-huygens/55226-science-highlights-from-huygens-5-radioactive-decay-and-cryovolcanism/).

Radiogenic argon was detected by the GCMS below 18 km. This detection was important because ^{40}Ar originates solely from the decay of potassium-40 (^{40}K), a radioactive isotope of potassium found in rocks. The only possible source of this ^{40}Ar is rocks that exist deep in Titan's interior, below its mantle of hydrocarbon and water ice.

Planetary scientists think that the presence of ^{40}Ar at the levels seen by Huygens is a strong indication of geological activity on Titan, and consistent with periodic replenishment of atmospheric methane. The apparent evidence for cryo-volcanism observed by the Cassini orbiter – involving water or a mixture of water and ammonia – provides one possible process for the release of both gases from the interior. Fig. 14.26 illustrates a possible scenario for the internal structure of Titan (i.e., concept sketch of the interior of Titan), which includes a global subsurface ocean beneath an

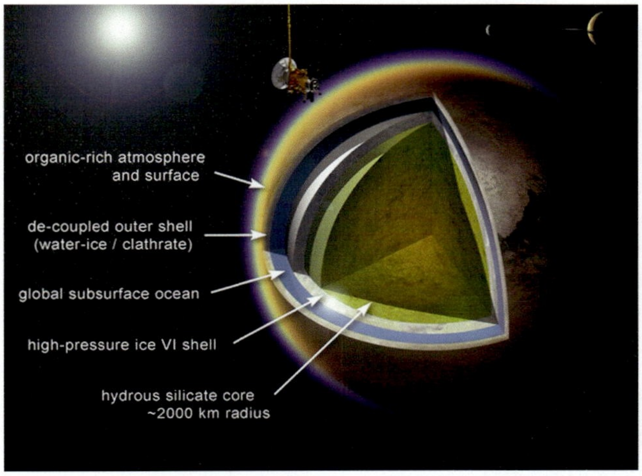

FIG. 14.25 **Artist's concept illustrating a possible model of Titan's internal structure that incorporates data from NASA's Cassini spacecraft.** In this model, Titan is fully differentiated, which means the denser core of the moon has separated from its outer parts. This model proposes a core consisting entirely of water-bearing rocks and a subsurface ocean of liquid water. The mantle, in this image, is made of icy layers, one that is a layer of high-pressure ice closer to the core and an outer ice shell on top of the subsurface ocean. A model of Cassini is shown making a targeted flyby over Titan's cloud-tops, with Saturn and Enceladus appearing at upper right. *(Source: http://sci.esa.int/cassini-huygens/50128-layers-of-Titan/). Copyright: A.D. Fortes/UCL/STFC.*

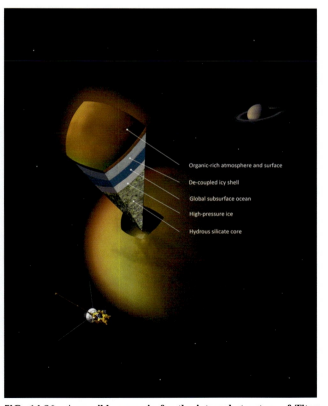

FIG. 14.26 **A possible scenario for the internal structure of Titan (i.e., concept sketch of the interior of Titan), which includes a global subsurface ocean beneath an icy outer shell, as inferred by radio science data collected by the Cassini spacecraft.** *Credit: Angelo Tavani. (Sources: http://sci.esa.int/cassini-huygens/55226-science-highlights-from-huygens-5-radioactive-decay-and-cryovolcanism/); http://sci.esa.int/cassini-huygens/50518-inside-Titan/).*

icy outer shell, as inferred by radio science data collected by the Cassini spacecraft.

Based on the discovery of Schumann resonance on Titan, Beghin et al. (2012) estimated that the lower layer of a salty, subsurface ocean lies 55–80 km below Titan's surface. Apart from these findings, other analyses of Titan's overall shape, topography, and gravity field have been found to be consistent with an ice shell of this thickness overlying a relatively dense subsurface ocean (Nimmo and Bills, 2010; Mitri et al., 2014).

Cryo-lavas have been found on some pocket areas on Titan. On Titan, lavas are generally referred to as cryo-lavas, because they involve the eruption of substances that are considered volatiles on the surface of Earth (water, water–ammonia mixtures, etc.) (Neish et al., 2018). Features suggested to be caused by cryo-volcanism were first discovered on the icy moons during the Voyager missions (e.g., Jankowski and Squyres, 1988; Showman et al., 2004). Two conditions must be satisfied for cryovolcanic flows to be present on a surface: first, liquids must be present in the interior and those liquids must then migrate to the surface (Neish et al., 2018). The second requirement for cryo-volcanism is for liquid to be transported from the interior to the surface. One plausible way to transport lava from the interior to the surface is through fluid-filled cracks. Mitri et al. (2008) proposed a model in which ammonia–water pockets are formed through cracking at the base of the ice-I shell on Titan. As these ammonia–water pockets undergo partial freezing, the ammonia concentration in the pockets would increase, thereby decreasing the negative buoyancy of the ammonia–water mixture. Although these pockets could not easily become buoyant on their own (given the difference in density of ~20–30 kg/m^3), they are sufficiently close to the neutral buoyancy point that large-scale tectonic stress patterns (tides, nonsynchronous rotation, moon volume changes, solid state convection, or subsurface pressure gradients associated with topography) could enable the ammonia–water to erupt effusively onto the surface. Studies carried out by Cook-Hallett et al. (2015) and Liu et al. (2016) found substantial evidence for the existence of such stress patterns on Titan.

14.8 Gaining insight into Titan's overall similarities and dissimilarities to Earth

In terms of active or recent surface-shaping processes, Titan is one of the most earthlike worlds in the solar system, often being referred to as the Earth of the outer solar system (Lopes et al., 2013). There are many analogies between Titan and the Earth, in spite of the much lower temperature in the Saturn system, the complex organic chemistry in the atmosphere, from the gas to the aerosol phases, but also the potential organic chemistry on Titan's surface, and in its possible internal water ocean (Raulin et al., 2012). The atmosphere of Titan contains methane, which, like water on Earth, can exist as a gas, ice, and liquid. Despite the differences in materials, temperatures and gravity fields between Earth and Titan, many of their surface features are similar and can be interpreted as products of the same geologic processes (Malaska et al., 2016).

Earth's seeming familiarity with Titan arose primarily in the wake of images of Titan taken during the joint NASA/European Space Agency Cassini–Huygens mission. Spectra and high-resolution images obtained by the Huygens Probe Descent Imager/Spectral Radiometer instrument in Titan's atmosphere reveal the traces of once-flowing liquid. Two years before the Cassini spacecraft reached the Saturn system (Cassini observation time: Mid-June 2004 to mid-September 2017), Brown et al. (2002) found that the atmospheric conditions on Titan allow the possibility that it could possess a methane condensation and precipitation cycle with many similarities to Earth's hydrological cycle. They reported images and spectra of Titan that show clearly transient clouds concentrated near the south pole. The discovery of these clouds demonstrates the existence of condensation and localized moist convection in Titan's atmosphere. Their location suggests that methane cloud formation is controlled seasonally by small variations in surface temperature, and that the clouds will move from the south to the north pole on a 15-year timescale. As just indicated, Titan's methane cycle is similar to the terrestrial hydrological cycle, involving methane clouds, rain, and surface liquids (Griffith et al., 2005). The methane-based hydrologic cycle on Titan is an extreme analog to Earth's water cycle. Titan is the only planetary body in the solar system, other than Earth, that is known to have an active hydrologic cycle. With a surface pressure of 1.5 bar and temperatures of 90 to 95 K, methane and ethane condense out of a nitrogen-based atmosphere and flow as liquids on the moon's surface. Exchange processes between atmospheric, surface and subsurface reservoirs produce methane and ethane cloud systems, as well as erosional and depositional landscapes that have strikingly similar forms to their terrestrial counterparts (Hayes et al., 2018).

The Cassini mission provided a wealth of information on Titan. It was found that surprisingly like Earth, the brighter highland regions show complex systems draining into flat, dark lowlands (Tomasko et al., 2005). These images showed that river channels meander downhill to damp lake-beds, where icy, rounded stones, resembling river cobbles, litter the ground (Elachi et al., 2005). It was found that massive cumulus clouds form and quickly dissipate, suggestive of rain (Griffith et al., 2000; Porco et al., 2005; Griffith et al., 2005); and dark oval regions resemble lakes. Taking these similarities into account, Titan is viewed as a sibling of Earth, as both bodies have rainy weather systems and landscapes formed by rivers.

Although Titan is tantalizingly similar to Earth in many ways, there are glaring dissimilarities as well. For example, Titan's atmosphere is ten times thicker, and much cooler, so that it takes longer than a Titan year (29.5 Earth years) for the atmosphere to respond radiatively to seasonal heating changes (Griffith, 2006). In terms of another dissimilarity, although both Earth and Titan cycle liquid between their surfaces and atmospheres, in Titan's cool atmosphere (minus 179°C at the surface) it is methane, rather than water, that exists as a gas, liquid, and ice.

As an example of another dissimilarity between these two celestial bodies, Titan's weather differs markedly from Earth's. Evidence for this is given by the location of Titan's large clouds, which presently reside either in a narrow band at 40° S latitude or within 30° of the south pole (Brown et al., 2002; Roe et al., 2002; Roe et al., 2005). In contrast, terrestrial clouds visit all latitudes. Yet they too exhibit striking latitudinal patterns—they occur preferentially near the Equator, but are scarce at latitudes of 15–35° N and S (regions that contain many deserts). Clouds also dominate the 40–60° northern and southern skies, covering northern Europe, for example (Griffith, 2006).

14.9 Titan's resemblance to prebiotic Earth

Titan has gained widespread acceptance in the origin of life field as a model for the types of evolutionary processes that could have occurred on prebiotic Earth. The surface and atmosphere of Titan constitute a system which is potentially as complex as that of the Earth, with occurrence of precipitation, surface erosion due to liquids, chemistry in large surface or subsurface hydrocarbon reservoirs, surface expressions of internal activity, and occasional major impacts leading to crustal melting (see Lunine and McKay, 1995).

With a dense N_2-CH_4 atmosphere rich in organics, both in gas and aerosol phases, and with the presence of hydrocarbons lakes and rivers, Titan appears as a natural laboratory to study chemical evolution toward complex organic systems, in a planetary environment and over a long time-scale. Thanks to many analogies with planet Earth, it provides a unique way to look at the various physical and chemical processes, and their couplings which may have been involved in terrestrial prebiotic chemistry (Raulin et al., 1995).

Both Titan and Earth possess significant atmospheres (1.5 and 1 bar, respectively at their surfaces) composed mainly of molecular nitrogen with smaller amounts of more reactive species. Both of these atmospheres are processed primarily by solar ultraviolet light with high energy particle interactions contributing to a lesser extent. The products of these reactions condense or are dissolved in other atmospheric species (aerosols/clouds) and fall to the surface.

There these products may have been further processed on Titan and the primitive Earth by impacting comets and meteorites (Clarke and Ferris, 1997).

Even prior to the Cassini mission, Clarke and Ferris (1997) have reported direct comparisons between the conditions on present-day Titan and those proposed for prebiotic Earth. It has been found that Titan has all the ingredients needed to produce "life as we know it." When exposed to liquid water, organic molecules analogs to those found on Titan produce a range of biomolecules such as amino acids. Titan thus provides a natural laboratory for studying the products of prebiotic chemistry.

Indeed, analogies with the Earth have a limit because Titan's temperatures (approximately 72-180 K) are much lower than those on the Earth; and this limitation preclude the presence of permanent liquid water on Titan's surface. However, it has been suggested that tectonic activity or impacts by meteors and comets could produce liquid water pools on Titan's surface for thousands of years. Clarke and Ferris (1997) suggested that hydrolysis (chemical breakdown of a compound due to reaction with water) and oligomerization reactions (chemical processes that convert monomers to macromolecular complexes through a finite degree of polymerization) in these pools might form chemicals of prebiological significance.

According to Clarke and Ferris (1997), if one considers the exobiological significance of selected extraterrestrial bodies in our solar system, Titan is the most similar to primitive Earth in terms of its basic physical properties (liquid at or near the surface, atmospheric density), energy sources (both energetic electrons and solar UV) and evolutionary processes (complex chemical reactions, precipitation, erosion, volcanism, impact processing) expected to have occurred throughout its history (Raulin et al., 1995; Lunine and McKay, 1995).

Mitri et al. (2008) showed that ammonia–water mixtures may erupt from a subsurface ocean on Titan through the ice shell, leading to cryo-volcanism. The possibility that cryo-volcanism, or at least outgassing, is still active on Titan has been proposed by Nelson et al. (2009a, 2009b). Mitri et al. (2008) proposed that cryo-volcanism may be related to fracturing in the ice crust overlying the ocean, which together with convection may lead to upward transport of ammonia–water fluid to the quiescent near-surface crust, where it refreezes and primes the crust for later episodes of volcanism. If eruption of ammonia–water occurs, ammonia–water cryo-volcanism would likely behave much as basaltic volcanism does on Earth, with comparable construction of low-profile volcanic shields and other constructional volcanoes and flow fields (Kargel, 1992).

Titan is the only celestial body in the solar system with a dense atmosphere. Mainly composed of dinitrogen (N_2) with several % of methane, this atmosphere experiences complex organic processes, both in the gas and aerosol phases, which are of prebiotic interest and within an environment of astrobiological interest (Raulin et al., 2012). Cassini–Huygens mission provided an observational tool to planetary researchers, and gave a new astrobiologically oriented vision of Titan which is now available by coupling three scientific approaches, namely, observation, theoretical modeling, and experimental simulation (Raulin et al., 2012).

14.10 Possibility of finding biomolecules on Titan

Observations made initially with the support of Voyager missions (Hanel et al., 1981; Kunde et al., 1981; Maguire et al., 1981) and subsequently with the support of Cassini Huygens mission (Niemann et al., 2005; Lavvas et al., 2008b; Janssen et al., 2016) have indicated that ultraviolet photons and charged particles dissociate the methane and nitrogen in Titan's atmosphere to produce a suite of carbon, hydrogen, and nitrogen containing products (CxHyNz), which eventually settle onto Titan's surface. Fortes (2000) proposed that Titan's subsurface ocean may contain biomolecules, or even simple life forms. If so, evidence of such biology could be found frozen in the cryovolcanic lavas on the surface of Titan.

Although Titan's surface temperature on average (~94 K) is too low for liquid water to exist, studies carried out by several researchers (e.g., Thompson and Sagan, 1992; O'Brien et al., 2005; Neish et al., 2006) have found that transient liquid water environments may be available in special locations such as impact melts and cryo-lavas. Several studies (e.g., Neish et al. 2008, 2009, 2010; Poch et al., 2012; Cleaves et al., 2014) have shown that when exposed to liquid water, organic molecules analogous to those found on Titan produce a range of biomolecules such as amino acids and possibly nucleobases. Titan thus provides a natural laboratory for studying the products of prebiotic chemistry (Neish et al., 2018). According to Neish et al., 2018), "These products provide crucial insight into what may be the first steps toward life in an environment that is rich in carbon and nitrogen, as well as water. It is even possible that life arose on Titan and survived for a short interval before its habitat froze. Alternatively, life may have developed in Titan's subsurface ocean, and evidence of this life could be brought to the surface through geophysical processes such as volcanism (Fortes, 2000)."

Neish et al. (2018) determined the ideal locales to search for biomolecules on Titan, and suggested future Titan mission scenarios to test the hypothesis that the first steps toward life have already occurred there. In this scenario, Neish et al. (2018) have considered "a substantial presence of biomolecules (i.e., compounds that are essential to life as we know it) as either a compelling indicator of an advanced prebiotic environment or as a possible sign of extinct (or more speculatively, extant) life."

It may be recalled that Neish et al. (2018) had suggested that impact melts and cryo-lavas of different volumes—and hence, different freezing timescales (O'Brien et al., 2005; Davies et al., 2010)—give us a unique window into the extent to which prebiotic chemistry can proceed over different timescales. Although Neish et al. (2018) proposed cryo-lavas and impact melt deposits to be two possible geological settings where prebiotic molecules could be found, they think that the aqueous chemistry in cryo-lavas may not have sufficient time or energy to produce more complicated prebiotic molecules. They further think that given the uncertain presence of biomolecules in the subsurface ocean, and the challenges inherent in transporting material to the surface, Fortes (2000)'s proposal may not be feasible. Based on these considerations, Neish et al. (2018) suggested that the priority for exploration should focus on another geological setting—impact melt deposits—where biomolecules are more likely to be present. Neish et al. (2018) determined that the best sites to identify biological molecules are deposits of impact melt on the floors of large, fresh impact craters, specifically Sinlap, Selk, and Menrva craters. These researchers consider that determining the extent of prebiotic chemistry within these melt deposits would help us to understand how life could originate on a world very different from Earth.

Clastic rocks are composed of fragments, or clasts, of pre-existing minerals and rock. A clast is a fragment of geological detritus, chunks, and smaller grains of rock broken off other rocks by physical weathering. In impact craters on Earth, impact melt often incorporates large amounts of clastic material from nonmelted, but shocked target rocks (Osinski et al., 2018), suggesting that there would be efficient mixing between liquid water and organic clasts on Titan. On this basis, Neish et al. (2018) argue that impact melts could provide "oases" for prebiotic chemistry to occur on Titan's surface. According to them, impact melts would provide an excellent medium for aqueous chemistry on Titan. Neish et al. (2018) are of the view that melted crustal rock (as opposed to water extruded from depth, as in the case of cryo-lavas resulting from cryo-volcanism) is more likely to yield a water-rich composition, with temperatures near the water liquidus (273 K). Given the large amounts of energy available from an impact, super-heating of several hundred Kelvins in impact melts [such as that observed on Earth (El Goresy, 1965) and the Moon (Simonds et al., 1976)] could increase the temperature of the melt above the liquidus, accelerating the chemistry occurring in the melt ponds on Titan.

Taking account of all the factors indicated above, and considering the fact that Titan's dense nitrogen–methane atmosphere supports a rich organic photochemistry (Horst, 2017), Neish et al. (2018) concluded that Titan has all the ingredients needed to produce "life as we know it" (i.e., carbon-based life that uses water as a solvent).

14.11 Practicability of photosynthesis on *Titan*'s surface

Life on Earth employs primarily photosynthesis, the only exception being life in the vicinity of submarine hydrothermal vents. On Earth, photosynthesis is used to produce organic material primarily from CO_2 and H_2O. Models of light penetration in Titan's atmosphere (McKay et al., 1989), and direct measurements by the Huygens Probe (Tomasko et al., 2008) provide a good understanding of the availability of sunlight in the atmosphere and at the surface of Titan. Despite a thick haze covering the entire Titan, the distribution of solar flux with wavelength roughly follows the solar spectrum, with the peak flux occurring at about 0.6 μm. This low-light level is more than adequate for photosynthesis on Earth. Studies carried out by Raven et al. (2000) and McKay (2014) have indicated that the process of photosynthesis on Earth can utilize light levels as low as 10^{-6} of the Earth noonday solar flux. Thus, photosynthesis should be possible on Titan with pigments that would not be dissimilar to those used on Earth. Even with attenuation by the atmospheric haze and the large distance from the Sun, the energy in sunlight reaching the surface of Titan is orders of magnitude larger than the chemical energy in hydrocarbons descending from the upper atmosphere (McKay, 2016).

On Titan, presumably, photosynthesis should produce organic material from CH_4; and H_2 would be a by-product. If at all life exists on Titan, it does not need to fix nitrogen from N_2 because N is available in organic compounds produced photochemically. Because of the low surface temperature (approximately 95 K) on Titan, hydrogen bonds provide enough binding strength to form useful structures.

14.12 Active cycling of liquid methane and ethane on *Titan*

On Titan, photochemical reactions beginning with the dissociation of the atmospheric nitrogen (N_2) and methane (CH_4) create an array of organic molecules in Titan's atmosphere and produce a solid organic haze in the upper atmosphere that obscures the lower atmosphere and surface. The organic haze particles eventually settle on the surface. A major product of the photochemistry is ethane, which accumulates on the surface and mixes with liquid methane. There are large hydrocarbon lakes of methane and ethane on Titan (Griffith et al., 2012), ranging in size up to ~1000 km across. Observations of the Huygens lander indicated moisture (methane and ethane) in the ground at the landing site.

On Earth, the extensive surface area of water and the resulting evaporation ensures that there is an active hydrological cycle, continually producing rain and snow. On Titan liquids are composed of three major components: methane, ethane, and dissolved atmospheric nitrogen. The solubility of N_2 in the methane and ethane liquids is significant, up to 20% (Graves et al., 2008). Ethane is nonvolatile compared to methane and nitrogen and is left behind as the liquid evaporates. Nitrogen and methane (and perhaps ethane to a smaller extent) are present in the atmosphere and can equilibrate with surface liquids. Rain on Titan is expected to be a mixture of these gases (Graves et al., 2008) and the fraction of methane and ethane depending on the mixing ratio of ethane in the lower atmosphere, which is not known (Williams et al., 2012). It has been shown (Tan et al., 2013; 2015; and Luspay-Kuti et al., 2015) that the mixture of three components of differing volatility creates unusual exotic behavior during evaporation and condensation, compared to a single component fluid. They also show that the liquid density decreases with pressure, very much unlike water on Earth. Furthermore, ethane is ~20 times better than methane regarding solubility of small organic molecules (Tiffin et al., 1979). These properties affect the global circulation of fluids on Titan. Thus, the cycling of liquid on Titan, and its various compositions, is complex when compared to the cycling of water on Earth (McKay, 2016). Apart from these fascinating characteristics of liquid cycling on Titan, the composition of the northern hydrocarbon lakes on Titan appears to be dominated by methane (Luspay-Kuti et al., 2015), whereas ethane has been identified as a major component in the southern-hemisphere Ontario Lacus (Brown et al., 2008; Luspay-Kuti et al., 2015), which may indicate that it is a remnant evaporitic basin (Cornet et al., 2015), analogous to the Dead Sea on Earth (McKay, 2016). Based on these results, there are credible reasons to believe that liquid methane and ethane are widespread and actively cycled on Titan. These findings pertaining to Titan led to the query "Is a 'follow the methane' strategy plausible? (McKay, 2016)."

14.13 Cassini spacecraft's retirement in September 2017 after successful 13 years' orbiting around the Saturn system

An eventful conclusion of the Cassini spacecraft mission on September 15, 2017, after Cassini spacecraft's successful thirteen years' orbiting around the Saturn system (30 June 2004 to 15 September 2017), is a memorable occasion for everyone having an interest in matters concerning extraterrestrial exploration. Cassini spacecraft mission has to its credit Cassini's cameras having taken amazing, ultra-close images of Saturn's rings and clouds, thereby shedding light on the vertical structures in Saturn's rings. The mission revealed that Saturn's rings are active and dynamic, and that they could serve the purpose of a laboratory to investigate how planets form. The Cassini measurements vastly improved our knowledge of how much material is in the rings, bringing us closer to understanding their origins. The Huygens probe (carried by Cassini Spacecraft for the specific purpose of close investigation of Titan's hazy atmosphere and its unseen surface) making the first landing on a moon (Titan) in the outer solar system was another major achievement, which paved the way in clearing the mystery that has been shrouding Titan's orange coloured atmosphere. Discovery of giant hurricanes at both poles of Saturn was another achievement of the Cassini mission. Fig. 14.27 provides pictorial representations of some selected findings related to the planet Saturn, captured by Cassini spacecraft.

Images from Cassini's flyby in the region of Saturn's large, icy moons with a series of three close encounters with Enceladus, which started on October 14, 2015 provided the first opportunity for a close-up look at its north polar region. Since Cassini's 2005-discovery of continually-erupting fountains of icy material on Enceladus, Titan has become

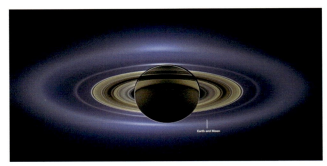

FIG. 14.28 **With Saturn sheltering Cassini from the Sun's glare, Cassini spacecraft captured 141 images to create a panoramic mosaic of Saturn.** *(Source: NASA: https://saturn.jpl.nasa.gov/legacy/files/Cassini_Grand_Finale_Fact_Sheet_508.pdf).*

one of the most promising places in the solar system to search for present-day habitable environments.

Cassini probed the water-rich plume of the active geysers on the planet's intriguing moon *Enceladus*. With Saturn sheltering Cassini from the Sun's glare, the spacecraft captured 141 images to create a panoramic mosaic of Saturn (Fig. 14.28).

After a last targeted Titan flyby, the spacecraft then hopped the Saturn's rings and dived between Saturn's uppermost atmosphere and its innermost ring 22 times (Fig. 14.29). As Cassini plunged past Saturn, the spacecraft collected some incredibly rich and valuable information far beyond the mission's original plan, including making detailed maps of Saturn's gravity and magnetic fields, revealing how the planet is composed on the inside, determining ring mass, sampling the atmosphere and ionosphere, and making the last views of the Saturnian moon *Enceladus* (https://www.nasa.gov/press-release/nasa-s-cassini-spacecraft-ends-its-historic-exploration-of-saturn).

FIG. 14.27 **Pictorial representation of some selected findings related to the planet Saturn, captured by Cassini spacecraft.** 1. This huge northern hemisphere storm wrapped around the entire planet. 2. This view of Saturn's rings in the ultraviolet indicates there is more ice toward the outer part of the rings. 3. Cassini captured this false-color view of the tiny moon Hyperion in 2005. The color differences may indicate differences in surface material composition. 4. Color was used in this simulated image to represent information about ring particle sizes based on measurements of three radio signals sent to Earth by Cassini. 5. Vertical structures rise from the edge of the B ring, perhaps formed as moonlets disturb the ring particles streaming nearby. 6. Radar imaging data from Cassini indicate the presence of large bodies of liquid on Titan's surface. 7. Three of Saturn's moons are seen against the darkened nightside of the planet. Dione (at left) is partly obscured by Saturn. 8. The final part of Cassini's mission—the Grand Finale—took the spacecraft to previously unexplored regions of the Saturn system. *(Image source: NASA: https://saturn.jpl.nasa.gov/legacy/files/Cassini_Grand_Finale_Fact_Sheet_508.pdf).*

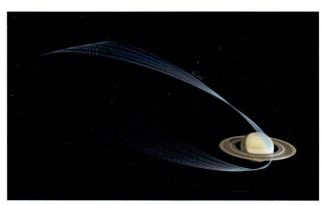

FIG. 14.29 **Cassini ended its historic mission with 22 daring loops passing through the gap between Saturn and the innermost ring.** *(Source: NASA: https://saturn.jpl.nasa.gov/legacy/files/Cassini_Grand_Finale_Fact_Sheet_508.pdf).*

A thrilling epoch in the exploration of our solar system came to a close on Sept. 15, 2017, as NASA's Cassini spacecraft made a fateful plunge into the atmosphere of Saturn, ending its 13-year tour of the ringed planet.

14.14 Is existence of life on Titan possible? Significance of carbon-rich and oxygen-loaded fullerenes in introducing oxygen to Titan's surface chemistry

Titan is the only world we know, other than Earth, that has lakes and seas presently on its surface, although the liquids in them are methane and ethane, instead of water. Titan also has a thick atmosphere composed of nitrogen and methane. An external flux of O^+ ions (Hartle et al., 2006) flowing into the top of Titan's atmosphere is believed to be most likely from Enceladus. Complex interactions between Saturn and its moons have led scientists to a comprehensive model that could explain how oxygen may end up on the surface of Titan. Sittler Jr. et al. (2009) have described a unique new process in which oxygen that circulates in the upper atmosphere of Titan can be carried all the way to its surface without further chemical contamination by being encased in carbon cages called "fullerenes," which are hollow, soccer-ball shaped shells made of carbon atoms. The work draws upon previous and subsequent investigations carried out by Sittler and others (Sittler Jr. et al., 2005, 2006, 2008, 2010) that model the dynamics of how particles, including water molecules, travel from Enceladus to Titan. In Sittler Jr. et al. (2009)'s model, the fullerenes condense into larger clusters that can attach to polycyclic aromatic hydrocarbons. The fullerene clusters form even larger aerosols that travel down to Titan's surface. This process protects the trapped oxygen from Titan's atmosphere, which is saturated with hydrogen atoms and compounds that are capable of breaking down other molecules. Otherwise, the oxygen would combine with methane in Titan's atmosphere and form carbon monoxide or carbon dioxide. As the oxygen-rich aerosols fall to Titan's surface, they are further bombarded by products of galactic cosmic ray interactions with Titan's atmosphere. Cosmic rays bombarding the oxygen-stuffed fullerenes could produce more complex organic materials, such as amino acids, in the carbon-rich and oxygen-loaded fullerenes. Amino acids are considered important for prebiological chemistry.

The presence of these oxygen atoms could potentially provide the basis for prebiological chemistry. The O^+ ions observed by Cassini Plasma Spectrometer (CAPS) onboard Cassini spacecraft are deposited near 1100 km altitude, in the same region as the very heavy ions observed by CAPS. The conditions in Titan's atmosphere (very complicated chemistry, started by sunlight), together with Titan's mildly reducing atmosphere, are favorable for formation of thick smog-like involatile particles (called haze) made of carbon, hydrogen, nitrogen, and oxygen (sometimes called "organic" particles). These organic molecules in Titan's atmosphere ultimately land on Titan's surface where they are moved by wind and rain to form dunes, lakes, and seas. Until recently, scientists have not been able to explain how oxygen fits into the picture of the dynamics and chemistry of Saturn and its moons. Scientists have been able to couple the new models that describe the generation of plumes at Enceladus and oxygen ion capture in fullerenes near the top of Titan's atmosphere to existing theories of the transport of oxygen across the magnetosphere. Taken together, the findings embodied in Sittler Jr. et al. (2009) on magnetospheric introduction of free oxygen into Titan's atmosphere and in Cooper et al. (2009) on radiolytic gas-driven cryo-volcanism at Enceladus suggest a chemical pathway that allows the oxygen to be introduced to Titan's surface chemistry.

Titan is a world with vast, arid regions of hydrocarbon-rich dunes girdling its equator. There is a growing school of thought in the planetary scientific community, which believe that the organic haze present in Titan's atmosphere when combined with some oxygen-bearing molecules (CO, CO_2, and H_2O) present in its atmosphere may result in the production of molecules of prebiotic interest. CO is a remarkably stable molecule, and its discovery in Titan's atmosphere (Lutz et al., 1983) led to investigations into whether the observed abundance is a primordial remnant. According to Hörst (2017), the addition of an oxygen source to complex organic compounds leads to the exciting possibility that molecules of prebiotic interest may form. Experiments indicate that amino acids and nucleobases may form under Titan atmospheric conditions (Hörst et al., 2012), but the instruments carried by Cassini–Huygens do not have a large enough mass range or sufficiently high resolution to detect such molecules.

The Titan life-searchers consider two fundamentally diverse possibilities for life to evolve or exist on Titan; (1) water-based life, and (2) liquid methane/ethane-based life. They justify their arguments based on the facts that (1) the limited level of solar energy received on the surface of Titan (about 0.1% that of the overhead Sun on Earth's surface; McKay et al., 1991) is more than adequate for photosynthesis on Earth—which process can utilize light levels as low as 10^{-6} of the Earth noonday solar flux (Raven et al., 2000; McKay, 2014), (2) availability of nutrients and water molecules, (3) availability of small hydrocarbons and N_2, both of which are readily soluble in liquid methane (Cornet et al., 2015). Some N- and O-containing molecules may be present in the liquid as well. According to McKay (2016), "Organic compounds are common on Titan, including N containing organics. Thus, C, H, and N are available in multiple compounds. The surface contains H_2O as ice but no other compounds with O atoms are common at the surface. Thus, life on Titan may have a restricted set of elements as nutrients, compared to the wide variety available and used by life on Earth."

The presence of both a subsurface liquid water ocean (Béghin et al., 2010; Baland et al., 2011; Bills and Nimmo, 2011; Béghin et al., 2012; Iess et al., 2012) and extensive hydrocarbon seas on the surface make Titan a prime target for NASA's renewed interested in exploring "Ocean Worlds." Planetary researchers are optimistic that biomolecules similar to those found on Earth are likely present on Titan. According to Lunine (2009), Titan's diverse oceans serve as a test for the ubiquity and diversity of life. The combination of organics and liquid, in the form of water in a subsurface ocean and methane/ethane in the surface lakes and seas, means that Titan may be the ideal place in the solar system to test ideas about habitability, prebiotic chemistry, and the ubiquity and diversity of life in the Universe (Lunine, 2009).

Schulze-Makuch et al. (2013) suggested that a critical problem for life on Titan would be access to inorganic elements, such as Fe, Cu, Mn, Zn, Ni, S, Ca, Na, K, etc., which life on Earth can access through their solubility in water. These inorganic elements are used, often in trace levels, in various life functions. Of particular interest is the use of metals in the active sites of enzymes (McKay, 2016).

A possible source of inorganic elements is the influx of meteorites and comets in the upper atmosphere. This source is known to be responsible for CO, CO_2, and H_2O in Titan's atmosphere (De Kok et al., 2007). Thus, a small but possibly biologically useful flux of inorganic elements is descending from the upper atmosphere with the haze and may be adequate for life forms that carefully recycle and reuse these elements (McKay, 2016). According to McKay (2016), based on the present knowledge, it can be concluded that the question of Titan as an abode of life remains unresolved but indications so far warrant further investigation.

14.15 Viability of a nonwater liquid capable of sustaining life on *Titan*

The only so far known solar system celestial body having the presence of a liquid on its surface is Titan. However, the liquid on Titan's surface is not water, but methane and ethane, both of which are hydrocarbons. This interesting fact led to a rethinking among life-hunting planetary researchers in terms of the viability of a nonwater liquid capable of sustaining life. In recent years, an idea which is gaining foothold among these researchers is examining the suitability of the liquid present on the surface of Titan in serving as a solvent for life, and thereby supporting and sustaining life. The area related to the habitability of Titan that has received the most detailed attention is the possibility of chemical energy sources.

McKay (2016) attempted to develop an approach to characterizing Titan as a possible abode of life by examining the possibilities of physical environments created by the presence and cycling of Titan's liquid, the biochemistry of carbon in that liquid, and the ecological systems that life might develop in that liquid. The role of water as a solvent for life on Earth depends on its chemical reactivity and on its physical properties, and the two are closely related. Thus, like water, considering other liquids requires addressing the needs life has for chemical functionality, both in terms of physicochemical and reaction-chemistry-related requirements (McKay, 2016). According to McKay (2016), if we turn to the possibility of life based on another liquid, each of these background assumptions must be re-examined.

An inquisitive scientific query as to whether life is possible in liquid methane/ethane medium instead of water medium has begun to arise among a growing school of thought comprising the search-for-life planetary researchers, primarily from Titan's primordial soup-like surface characteristics and its atmospheric conditions. In this context, it is fascinating to note that the only other world we know that has a liquid on its surface is Titan, where we find liquids of methane and ethane (Brown et al., 2008). The atmosphere is dominated by N_2 (95%) with CH_4 (5%) and H_2 (0.1%) and various trace organic compounds (McKay and Smith, 2005). Although its surface temperature is very low (95 K), its surface atmospheric pressure is sufficiently high (about 1.5 times that of Earth). The lower atmosphere has an active hydrological cycle of liquid methane, including convective clouds and rain, both of which vary with the seasons (McKay and Smith, 2005).

Life-hunting researchers think that perhaps there could be life in Titan's liquids. According to McKay (2016), "The case of water and life on Earth is the only guide we have for Titan as an abode of life, and so we must try to separate those features of habitability on Earth that are specific to Earth from those features that might be generalized to other liquids on other worlds."

Life on Earth is based on carbon biochemistry in the medium of liquid water. Carbon biochemistry in liquid water on Earth is able to create the following (McKay, 2016): "compartmentalization for autonomy and reproduction, information storage molecules and a way to duplicate them, structural molecules and a way to build them." The postulated life on the surface of Titan is also carbon biochemistry, but in a medium of liquid methane and ethane, rather than water. Many of the key biochemical structures that life on Earth uses are specific to solvents with properties like liquid water. These include, for example, lipid bilayers as membrane components, amino acids and their collection into folded proteins, and the shape of DNA. For life on Titan, structures similar to those on Earth must be created in a solution of methane and ethane.

It has been shown that Titan's very low surface temperature (approximately 95 K) need not be an impediment for the genesis and sustenance of life on this icy moon.

For example, Schulze-Makuch and Grinspoon (2005) and McKay and Smith (2005) noted that the photochemically produced organics in Titan's atmosphere would produce energy if reacted with atmospheric H_2, and that this could be a source of biological energy. McKay and Smith (2005) quantified the energy released from such reactions. Hydrogenation of C_2H_2 provided a particularly energetic reaction, with 334 kJ per mole of C_2H_2 consumed. McKay (2016) found that this can be compared to the minimum energy required to power methanogen growth on Earth of ~40 kJ mole^{-1}, determined by Kral et al. (1998), or the energy from the reaction of O_2 with CH_4, which produces ~900 kJ mole^{-1}. The reactions indicated above are exothermic (i.e., reactions accompanied by the release of heat) but kinetically inhibited at Titan temperatures. This is ideal for biology. For example, the reaction of O_2 with CH_4 is kinetically inhibited at Earth temperatures, but methane-oxidizing microorganisms present in the environment catalyze the reaction and thus derive energy from it (McKay, 2016).

According to McKay (2016), given that redox couples (i.e., reducing species and their corresponding oxidizing forms) are produced in Titan's atmosphere, and that these small molecules will be widespread on Titan and are readily soluble in liquid ethane and methane (Cornet et al., 2015), there may not be any need for photosynthesis on Titan. McKay and Smith (2005) pointed out that H_2 is the most promising atmospheric constituent on Titan to show a biological effect. In addition to H_2, McKay and Smith (2005) suggested that methane-based life on Titan could consume acetylene and ethane. If this is true, depletion of acetylene and ethane could be used as an indicator of biological effect manifesting on Titan. McKay (2016) gathered available evidences to support depletion of acetylene and ethane on Titan, and reported thus: "The data that suggest that there is less ethane on Titan than expected are well established (Lorenz et al., 2008a). Photochemical models have predicted that Titan should have a layer of ethane sufficient to cover the entire surface to a thickness of many meters but Cassini has found no such layer. Clark et al. (2010) find a lack of acetylene on the surface despite its expected production in the atmosphere and subsequent deposition on the ground. There was also no evidence of acetylene in the gases released from the surface after the Huygens Probe landing (Niemann et al., 2005; Lorenz et al., 2006a). Thus, the evidence for less ethane and less acetylene than expected seems clear." According to McKay (2016), these depletions may not be just due to a lack of production but are due to some kind of chemical reaction at the surface, which is more significant in the astrobiological sense. McKay (2016) further argues that "one could also speculate that the ethane-enhanced southern hemisphere of Titan is biologically consuming more H_2 than the methane-dominated northern hemisphere due to the superior properties of ethane as a solvent." The depletion of hydrogen, acetylene, and ethane, could be due to a new type of liquid methane/ethane-based life form, as predicted in several studies (e.g., Benner et al., 2004; Schulze-Makuch and Grinspoon, 2005; McKay and Smith, 2005).

14.16 Science goals and mission concept for future exploration of *Titan*

Titan is one of the solar system's most mysterious bodies and is a prime target for future space exploration. Neish et al. (2018) recalls that "Even before Cassini reached the outer solar system, it was recognized that a post-Cassini scientific priority, especially for astrobiology, would be to access surface material for detailed investigation (Chyba et al., 1999; Lorenz, 2000). More recently, identifying "Planetary Habitats" was included as one of the three cross-cutting themes of the National Research Council's 'Visions and Voyages for Planetary Science in the Decade 2013–2022'. In addition, Titan is currently listed as one of six potential mission themes for NASA's next New Frontiers mission (such a mission could be specifically designed to identify the products of prebiotic chemistry on Titan's surface)."

The information gathered by Huygens probe and various instruments deployed onboard Cassini spacecraft have revealed many details of a surprisingly Earth-like world and raised fascinating new questions for future study. Planetary scientists and astrobiologists recognize that in the exploration and deep study of Titan, it is important to address its origin and evolution. These requirements call for measuring the mineralogy and chemistry of bulk crust, excavated impact ejecta, and cryo-lava or ash. It is also important to understand the atmospheric photochemistry and climate history by measuring the mineralogy and chemistry of organic sediments found in various environments. All the more important is to understand Titan's astrobiological evolution.

Various instruments have been developed with a view to fulfilling the just-mentioned requirements. For example, Fortes et al. (2009) have described the scientific case for and preliminary design of an instrument whose primary goal is to determine the chemistry (element abundance) and mineralogy (compound identity and abundance) of Titan's surface using a combination of energy dispersive X-ray fluorescence spectroscopy (EDXRF) and X-ray diffraction (XRD). The latter is capable of identifying any crystalline substance present on Titan's surface at relative abundances greater than ~1 wt%, allowing unambiguous identification of, for example, structure I and II clathrates (even in the presence of ice), and various organic solids, which may include C_2H_2, C_2H_4, C_4H_2, HCN, CH_3CN, HC_3N, and C_4N_2.

The instrument will be able to address the origin and evolution of Titan by measuring the mineralogy and chemistry of bulk crust (water ice/clathrates), excavated impact ejecta (possibly including XRD-identifiable high-pressure polymorphs of ice), and cryo-lava or ash (distinguishing between ammonia–methanol and sulfate–carbonate magmas). The instrument has the ability to address atmospheric photochemistry and climate history by measuring the mineralogy and chemistry of organic sediments found in various environments, such as dunes, fluvial and peri-lacustrine areas (including possible evaporites), and possibly exhumed from deep reservoirs by mud volcanism. The instrument has the capacity to address Titan's astrobiological evolution by enabling detection of more complex organic materials (crystalline carboxylic/amino acids or sugars) in areas where sediments have interacted with aqueous solutions.

Tobie et al. (2014) have described some of the science goals and key measurements to be performed by a future exploration mission involving a Saturn–Titan orbiter and a Titan balloon, which was proposed to ESA in response to the call for definition of the science themes of the next large-class mission. The mission scenario for the study of Titan is built primarily around the expectation that the chemistry of Titan may provide clues for the origin of life. The measurements envisaged would provide a step change in our understanding of planetary processes and evolution, with many orders of magnitude improvement in temporal, spatial, and chemical resolution over that which is possible with Cassini–Huygens. This mission concept builds upon the successes of Cassini–Huygens and takes advantage of previous mission heritage in both remote sensing and in situ measurement technologies.

Mitri et al. (2014b) suggested that fundamental questions involving the origin, evolution, and history of both Titan and the broader Saturnian system can be answered by exploring this moon from an orbiter and also in situ. They reported the science case for an exploration of Titan and one of its lakes from a dedicated orbiter and a lake probe. According to them, observations from an orbit-platform can improve our understanding of Titan's geological processes, surface composition and atmospheric properties. Further, they suggest, combined measurements of the gravity field, rotational dynamics and electromagnetic field can expand our understanding of the interior and evolution of Titan. According to Mitri et al. (2014b), an in-situ exploration of Titan's lakes provides an unprecedented opportunity to understand the hydrocarbon cycle, investigate a natural laboratory for prebiotic chemistry and habitability potential, and study meteorological and marine processes in an exotic environment. They have briefly discussed possible mission scenarios for a future exploration of Titan with an orbiter and a lake probe.

Hörst (2017) listed the following select outstanding questions that remain to be answered:

1. What are the very heavy ions in the ionosphere, how do they form, and what are the implications for complexity of prebiotic chemistry?
2. What is the connection between the plumes of Enceladus and Titan's atmosphere?
3. What is the composition of the haze and how does it vary spatially and temporally?
4. How do the organic compounds produced in the atmosphere evolve once reaching the surface?
5. What are the dynamics of Titan's troposphere and how does that affect the evolution of the surface?
6. How variable is Titan's weather from year to year and how variable is the climate over longer timescales?
7. How old is Titan's current atmosphere?
8. What happened on Titan 300–500 Myrs ago?
9. Is Titan's atmospheric methane cyclic and/or episodic, and if so, what are the implications of those timescales on habitability?
10. Where is the origin of Titan's methane and what is the fate of the photochemically produced ethane?
11. What is controlling Titan's H_2 profile and potential spatial variations?
12. What is the composition of the surface and on what scales is it spatially variable?
13. What is the composition of the lakes and seas and what chemistry occurs there?
14. What is the circulation in the lakes and seas and how is it affected by the atmospheric dynamics?
15. What is the composition of the dune particles and how are they produced?
16. Does cryovolcanism occur on Titan?

According to Hörst (2017), finding answers to many of these questions will require future missions to Titan.

To identify biological molecules on Titan, it will be necessary to obtain more detailed data than are currently available from the past ground- and space-based observations. Neish et al. (2018) have explained that remote sensing data sets lack the spatial and spectral resolution to make definitive conclusions about the composition of Titan's surface. According to them, "compositional information regarding the potential presence of biological molecules could be obtained from in situ observations, but only if (a) the associated instrumentation is designed for such a task, and (b) the surface material can be obtained from the targeted regions."

Neish et al. (2018) examined the ideal locales to search for evidence of, or progression toward, life on Titan. According to them, "Impact craters are preferred over cryovolcanoes for a number of reasons. Chief among them is the temperature of the aqueous medium: higher temperatures at impact craters will increase reaction rates exponentially, increasing the likelihood of forming complex biomolecules. Determining the extent of prebiotic chemistry within these melt deposits would help us to understand how life could originate on a world very different from Earth, and shed light on prebiotic synthesis more generally." They found that it is not possible to identify biomolecules on Titan through remote sensing, but rather through in situ measurements capable of identifying a wide range of biological molecules. They consider that to identify and characterize the biomolecules would require in situ measurements of Titan's surface material, obtained through precision targeting of a lander, equipped with instrumentation capable of measuring a wide range of biological molecules. According to Neish et al. (2018), the ideal landing sites would be the floors of Titan's largest freshest impact craters, where mass wasting and fluvial erosion expose fresh deposits of impact melt for sampling. Note that in geography, mass wasting, also known as slope movement or mass movement, is the geomorphic process by which soil, sand, regolith, and rock move downslope typically as a solid, continuous or discontinuous mass, largely under the force of gravity, frequently with characteristics of a flow as in debris flows and mudflows.

Neish et al. (2018) consider that, given the nonuniformity of impact melt exposures on the floor of a weathered impact crater, the ideal lander would be capable of precision targeting. This would allow it to identify the locations of fresh impact melt deposits, and/or sites where the melt deposits have been exposed.

A lander-based future exploration of Titan with a view to discovering biological molecules, if present on its surface, has been advocated by several life-hunting investigators. For example, according to Neish et al. (2018), "The benefit to accessing melt deposits at the bottom of river valleys is that no drilling would be needed to reach an unaltered melt sample. Since liquid hydrocarbons do not react chemically with water ice (Lorenz and Lunine, 1996), even samples exposed to erosion and weathering in the Titan environment would remain relatively pristine. We would also not expect any major alteration due to high-energy electromagnetic radiation and/or charged particles, since ultraviolet radiation and galactic cosmic rays do not penetrate all the way to the surface of Titan (Horst, 2017). Thus, any biological molecules present would be trapped inside the chemically inert water ice, and so should be accessible when the sample is ingested into a lander. Therefore, if we can identify river valleys on the floors of Sinlap, Selk, and Menrva impact craters, these would be ideal landing sites."

Neish et al. (2018) have further suggested that future mission planners may wish to differentiate between those biomolecules formed by abiotic processes and those formed by biotic processes. They have suggested that there are several indicators that may be able to differentiate between biomolecules of biotic origins from those of abiotic origins. For example, one may use isotopic signatures to differentiate between the two; life on Earth preferentially utilizes the lighter isotope of carbon, ^{12}C, over the heavier isotope, ^{13}C (Cockell, 2015). More clearly, when organisms ingest carbon, they preferentially use ^{12}C over ^{13}C. Note that ^{14}C is radioactive, and thus won't remain over a long time period. Carbon with a high ratio of ^{12}C compared to ^{13}C is, therefore, an indicator of living processes. Another suggestion is to look for an abundance of molecules with a single chirality. Note that in chemistry, a molecule or ion is called chiral if it cannot be superposed on its mirror image by any combination of rotations and translations. This geometric property is called chirality. A chiral molecule or ion must have at least one chiral center or stereocenter. McKay (2016) pointed out that life on Earth uses only the l-stereoisomer of amino acids, and not their mirror image, the d-stereoisomer. Assuming that life on both Earth and Titan has similar characteristics, the difference observed on terrestrial stereoisomer of amino acids could be used as an indicator to differentiate between those biomolecules formed by abiotic processes and those formed by biotic processes. Neish et al. (2018) further suggest that one could also consider the broader suite of molecules present in the melt pond on Titan's surface. McKay (2004) indicated that abiotic processes typically produce smooth distributions of organic material, whereas biological processes select a highly specific set of molecules. According to him, homochirality is a special and powerful example of such biology selection.

References

Appel, J., Bockhorn, H., Frenklach, M., 2000. Kinetic modeling of soot formation with detailed chemistry and physics: laminar premixed flames of C_2 hydrocarbons. Combust. Flame 121 (1–2), 122–136.

Atreya, S.K., Adams, E.Y., Niemann, H.B., Demick-Montelara, J.E., Owen, T.C., Fulchignoni, M., Ferri, F., Wilson, E.H., 2006. Titan's methane cycle. Planet. Space Sci. 54, 1177–1187.

Backes, H., Neubauer, F.M., Dougherty, M.K., Achilleos, N., André, N., Arridge, C.S., Bertucci, C., Jones, G.H., Khurana, K.K., Russell, C.T., Wennmacher, A., 2005. Titan's magnetic field signature during the first Cassini encounter. Science 308, 992–995. doi:10.1126/science.1109763.

Baland, R.-M., van Hoolst, T., Yseboodt, M., Karatekin, Ö., 2011. Titan's obliquity as evidence of a subsurface ocean? Astron. Astrophys. 530, A141. doi:10.1051/0004-6361/201116578.

Balthasar, A., Kraft, M., 2003. A stochastic approach to calculate the particle size distribution function of soot particles in laminar premixed flames. Combust. Flame 133 (3), 289–298.

Barnes, J.W., Brown, R.H., Turtle, E.P., McEwen, A.S., Lorenz, R.D., Janssen, M., Schaller, E.L., Brown, M.E., Buratti, B.J., Sotin, C., Griffith, C., Clark, R., et al., 2005. A 5-micron-bright spot on Titan: evidence for surface diversity. Science 310, 92–95.

Barnes, J.W., Brown, R.H., Soderblom, L., Buratti, B.J., Sotin, C., Rodriguez, S., Le Mouèlic, S., Baines, K.H., Clark, R., Nicholson, P., 2007. Global-scale surface spectral variations on Titan seen from Cassini/VIMS. Icarus 186, 242–258. doi:10.1016/j.icarus.2006.08.021.

Barnes, J.W., et al., 2008. Spectroscopy, morphometry, and photoclinometry of Titan's dunefields from Cassini/VIMS. Icarus 195, 400–414. doi:10.1016/j.icarus.2007.12.006.

Barnes, J.W., Soderblom, J.M., Brown, R.H., Buratti, B.J., Sotin, C., Baines, K.H., Clark, R.N., Jaumann, R., McCord, T.B., Nelson, R., et al., 2009. VIMS spectral mapping observations of Titan during the Cassini prime mission. Planet. Space Sci. 57, 1950–1962.

Bauschlicher Jr., C.W., Ricca, A., 2000. Mechanisms for polycyclic aromatic hydrocarbon (PAH) growth. Chem. Phys. Lett. 326 (3–4), 283–287.

Bazzon, A., Schmid, H.M., Buenzli, E., 2014. HST observations of the limb polarization of Titan. Astron. Astrophys. 572, A6. doi:10.1051/0004-6361/201323139.

Béghin, C., et al., 2007. A Schumann-like resonance on Titan driven by Saturn's magnetosphere possibly revealed by the Huygens Probe. Icarus 191 (1), 251–266. doi:10.1016/j.icarus.2007.04.005.

Béghin, C., Sotin, C., Hamelin, M., 2010. Titan's native ocean revealed beneath some 45 km of ice by a Schumann-like resonance. C. R. Geosci. 342, 425–433. doi:10.1016/j.crte.2010.03.003.

Beghin, C.B., Randriamboarison, O., Hamelin, M., Karkoschka, E., Sotin, C., Whitten, R.C., Berthelier, J.-J., Grard, R., Simoes, F., et al., 2012. Analytic theory of Titan's Schumann resonance: constraints on ionospheric conductivity and buried water ocean. Icarus 218, 1028–1042.

Benner, S.A., Ricardo, A., Carrigan, M.A., 2004. Is there a common chemical model for life in the universe? Curr. Opin. Chem. Biol. 8, 672–689.

Bills, B.G., Nimmo, F., 2011. Rotational dynamics and internal structure of Titan. Icarus 214, 351–355. doi:10.1016/j.icarus.2011.04.028.

Birch, et al., 2016. Geomorphologic mapping of Titan's polar terrains: constraining surface processes and landscape evolution. Icarus 282, 1–23.

Bird, M.K., et al., 2005. The vertical profile of winds on Titan. Nature 438, 800–802. doi:10.1038/nature04060.

Black, B.A., Perron, J.T., Hemingway, D., Bailey, E., Nimmo, F., Zebker, H., 2017. Global drainage patterns and the origins of topographic relief on Earth, Mars, and Titan. Science 356 (6339), 727–731. doi:10.1126/science.aag0171.

Bouchez, A.H., 2004. PhD thesis. California Inst. of Technol., Pasadena, CA.

Brown, M.E., Bouchez, A.H., Griffith, C.A., 2002. Direct detection of variable tropospheric clouds near Titan's south pole. Nature 420, 795–797.

Brown, R.H., Soderblom, L.A., Soderblom, J.M., Clark, R.N., Jaumann, R., Barnes, J.W., Sotin, C., Buratti, B., Baines, K.H., Nicholson, P.D., 2008. The identification of liquid ethane in Titan's Ontario Lacus. Nature 454, 607–610.

Burr, D.M., Emery, J.P., Lorenz, R.D., Collins, G.C., Carling, P.A., 2006. Sediment transport by liquid surficial flow: application to Titan. Icarus 181, 235–242.

Burr, D., et al., 2017. Defining the topography of a planetary body. Science 356 (6339), 708. doi:10.1126/science.aan2719.

Celnik, M.S., Sander, M., Raj, A., West, R.H., Kraft, M., 2009. Modelling soot formation in a premixed flame using an aromatic-site soot model and an improved oxidation rate. Proc. Combust. Inst. 32 (1), 639–646.

Charnay, B., Barth, E., Rafkin, S., Narteau, C., Lebonnois, S., Rodriguez, S., Courrech Du Pont, S., Lucas, A., 2015. Methane storms as a driver of Titan's dune orientation. Nat. Geosci. 8, 362–366. doi:10.1038/ngeo2406.

Chyba, C.F., McKinnon, W.B., Coustenis, A., Johnson, R.E., Kovach, R.L., Khurana, K., Lorenz, R., McCord, T.B., McDonald, G.D., Pappalardo, R.T., Race, M., Thomson, R., 1999. Europa and Titan: preliminary recommendations of the campaign science working group on prebiotic chemistry in the outer solar system, 30th Lunar and Planetary Science Conference Abstracts. Houston, TX.

Clark, R.N., Curchin, J.M., Barnes, J.W., Jaumann, R., Soderblom, L., Cruikshank, D.P., Brown, R.H., Rodriguez, S., Lunine, J., Stephan, K., et al., 2010. Detection and mapping of hydrocarbon deposits on Titan. J. Geophys. Res., 115.

Clarke, D.W., Ferris, J.P., 1997. Chemical evolution on Titan: comparisons to the prebiotic earth. Orig. Life Evol. Biosph. 27 (1-3), 225–248.

Cleaves, H.J., Neish, C., Callahan, M.P., Parker, E., Fernandez, F.M., Dworkin, J.P., 2014. Amino acids generated from hydrated Titan tholins: comparison with Miller-Urey electric discharge products. Icarus 237, 182–189.

Coates, A.J., Crary, F.J., Lewis, G.R., Young, D.T., Waite Jr., J.H., Sittler Jr., E.C., 2007. Discovery of heavy negative ions in Titan's ionosphere. Geophys. Res. Lett. 34 (22), L22103.

Cockell, C.S., 2015. Astrobiology: Understanding Life in the Universe. John Wiley & Sons, Ltd., Sussex, pp. 449.

Coll, P., Coscia, D., Gazeau, M.C., Guez, L., Raulin, F., 1998. Review and latest results of laboratory investigations of Titan's aerosols. Orig Life Evol Biosph 28 (2), 195–213.

Collins, G.C., 2005. Relative rates of fluvial and bedrock incision on Titan and Earth. Geophys. Res. Lett. 32, L22202.

Cook-Hallett, C., Barnes, J.W., Kattenhorn, S.A., Hurford, T., Radebaugh, J., Stiles, B., Beuthe, M., 2015. Global contraction/expansion and polar lithospheric thinning on Titan from patterns of tectonism. J. Geophys. Res. Planets 120, 1220–1236.

Cooper, J.F., Cooper, P.D., Sittler, E.C., Sturner, S.J., Rymer, A.M., 2009. Old Faithful model for radiolytic gas-driven cryovolcanism at Enceladus. Planet. Space Sci. 57 (13), 1607–1620.

Cornet, T., Bourgeois, O., le Mouélic, S., Rodriguez, S., Gonzalez, T.L., Sotin, C., Tobie, G., Fleurant, C., Barnesf, J.W., Brown, R.H., et al., 2012. Geomorphological significance of Ontario Lacus on Titan: Integrated interpretation of Cassini VIMS, ISS and RADAR data and comparison with the Etosha Pan (Namibia). Icarus 218, 788–806.

Cornet, T., Cordier, D., Le Bahers, T., Bourgeois, O., Fleurant, C., Le Mouélic, S., Altobelli, N., 2015. Dissolution on Titan and on Earth: Towards the age of Titan's karstic landscapes. J. Geophys. Res.: Planets 120, 1044–1074.

Crary, F.J., Magee, B.A., Mandt, K., Waite Jr., J.H., Westlake, J., Young, D.T., 2010. Heavy ions, temperatures and winds in Titan's ionosphere: Combined Cassini CAPS and INMS observations. Space Sci 57 (14-15), 1847–1856.

Dagg, I.R., Reesor, G.E., 1972. Dielectric loss measurements on nonpolar liquids in the microwave region from 18 to 37 GHz. Can. J. Phys. 50, 2397–2401.

Delitsky, M.L., McKay, C.P., 2010. The photochemical products of benzene in Titan's upper atmosphere. Icarus 207 (1), 477–484.

Davies, A.G., Sotin, C., Matson, D.L., Castillo-Rogez, J., Johnson, T.V., Choukroun, M., Baines, K.H., 2010. Atmospheric control of the cooling rate of impact melts and cryolavas on Titan's surface. Icarus 208, 887–895.

de Kok, R., Irwin, P.G.J., Teanby, N.A., Lellouch, E., Bézard, B., Vinatier, S., Nixon, C.A., Fletcher, L., Howett, C., Calcutt, S.B., et al., 2007. Oxygen compounds in Titan's stratosphere as observed by Cassini CIRS. Icarus 186, 354–363.

de Kok, R.J., Teanby, N.A., Maltagliati, L., Irwin, P.G.J., Vinatier, S., 2014. HCN ice in Titan's high-altitude southern polar cloud. Nature 514, 65–67. doi:10.1038/nature13789.

Eibeck, A., Wagner, W., 2000. Approximative solution of the coagulation–fragmentation equation by stochastic particle systems. Stochastic Anal. Appl. 18 (6), 921–948.

Elachi, C., et al., 2004. RADAR: The Cassini Titan Radar Mapper. Space Sci. Rev. 115, 71–110.

Elachi, C., Wall, S., Allison, M., Anderson, Y., Boehmer, R., Callahan, P., Encrenaz, P., Flamini, E., Franceschetti, G., Gim, Y., Hamilton, G., Hensley, S., Janssen, M., Johnson, W., Kelleher, K., Kirk, R., Lopes, R., Lorenz, R., Lunine, J., Muhleman, D., Ostro, S., Paganelli, F.,

Picardi, G., Posa, F., Roth, L., Seu, R., Shaffer, S., Soderblom, L., Stiles, B., Stofan, E., Vetrella, S., West, R., Wood, C., Wye, L., Zebker, H., 2005. Cassini radar views the surface of Titan. Science 308, 970–974. doi:10.1126/science.1109919.

Elachi, C., Wall, S., Janssen, M., Stofan, E., Lopes, R., Kirk, R., Lorenz, R., Lunine, J., Paganelli, F., Soderblom, L., Wood, C., Wye, L., Zebker, H., Anderson, Y., Ostro, S., Allison, M., Boehmer, R., Callahan, P., Encrenaz, P., Flamini, E., Francescetti, G., Gim, Y., Hamilton, G., Hensley, S., Johnson, W., Kelleher, K., Muhleman, D., Picardi, G., Posa, F., Roth, L., Seu, R., Shaffer, S., Stiles, B., Vetrella, S., West, R., 2006. Titan radar mapper observations from Cassini's T3 fly-by. Nature 441, 709–713. doi:10.1038/nature04786.

El Goresy, A., 1965. Baddeleyite and its significance in impact glasses. J. Geophys. Res. 70, 3453–3456.

Flasar, F.M., et al., 2005. Titan's atmospheric temperatures, winds, and composition. Science 308, 975–978. doi:10.1126/science.1111150.

Folkner, W.M., et al., 2006. Winds on Titan from ground-based tracking of the Huygens probe. J. Geophys. Res. 111, E07S02. doi:10.1029/2005JE002649.

Fortes, A., 2000. Exobiological implications of a possible ammonia–water ocean inside Titan. Icarus 146, 444–452.

Fortes, A.D., Wood, I.G., Dobson, D.P., Fewster, P.F., 2009. An icy mineralogy package (IMP) for in-situ studies of Titan's surface. Adv. Space Res. 44 (1), 124–137.

Fulchignoni, M., Ferri, F., Angrilli, F., Ball, A.J., Bar-Nun, A., Barucci, M.A., Bettanini, C., Bianchini, G., Borucki, W., Colombatti, G., Coradini, M., Coustenis, A., Debei, S., Falkner, P., Fanti, G., et al., 2005. In situ measurements of the physical characteristics of Titan's environment. Nature 438, 785–791.

Gelsthorpe, R.V., Bennett, R.G., 1978. Some measurements on the temperature variation of loss angle of a group of low-loss liquid dielectrics at microwave frequencies. J. Phys. D Appl. Phys. 11, 717. doi:10.1088/0022-3727/11/5/015.

Glein, C.R., Shock, E.L., 2013. A geochemical model of non-ideal solutions in the methane-ethane-propane-nitrogen-acetylene system on Titan. Geochim. Cosmochim. Acta 115, 217–240. doi:10.1016/j.gca.2013.03.030.

Goodson, M., Kraft, M., 2002. An efficient stochastic algorithm for simulating nano-particle dynamics. Comput. Phys. 183 (1), 210–232.

Goulay, F., Leone, S.R., 2006. Low-temperature rate coefficients for the reaction of ethynyl radical (C_2H) with benzene. J. Phys. Chem. A 110 (5), 1875–1880.

Grard, R., et al., 2006. Electric properties and related physical characteristics of the atmosphere and surface of Titan. Planet. Space Sci. 54, 1124–1136. doi:10.1016/j.pss.2006.05.036.

Graves, S.D.B., McKay, C.P., Griffith, C.A., Ferri, F., Fulchignoni, M., 2008. Rain and hail can reach the surface of Titan. Planet. Space Sci. 56, 346–357.

Griffith, C.A., 1993. Evidence for surface heterogeneity on Titan. Nature 364, 511–514. doi:10.1038/364511a0.

Griffith, C.A., Hall, J.L., Geballe, T.R., 2000. Detection of daily clouds on Titan. Science 290, 509–513.

Griffith, C.A., Penteado, P., Baines, K., Drossart, P., Barnes, J., Bellucci, G., Bibring, J., Brown, R., Buratti, B., Capaccioni, F., Cerroni, P., Clark, R., Combes, M., Coradini, A., Cruikshank, D., Formisano, V., Jaumann, R., Langevin, Y., Matson, D., McCord, T., Mennella, V., Nelson, R., Nicholson, P., Sicardy, B., Sotin, C., Soderblom, L.A., Kursinski, R., 2005. The evolution of Titan's mid-latitude clouds. Science 310, 474–477.

Griffith, C.A., 2006. Planetary science: Titan's exotic weather. Nature 442, 362–363. doi:10.1038/442362a.

Griffith, C.A., McKay, C.P., Ferri, F., 2008. Titan's tropical storms in an evolving atmosphere. Astrophys. J. 687, L41–L44. doi:10.1086/593117.

Griffith, C.A., Lora, J.M., Turner, J., Penteado, P.F., Brown, R.H., Tomasko, M.G., Doose, L., See, C., 2012. Possible tropical lakes on Titan from observations of dark terrain. Nature 486, 237–239.

Hamelin, M., et al., 2016. The electrical properties of Titan's surface at the Huygens landing site measured with the PWA-HASI Mutual Impedance Probe. New approach and new findings. Icarus 270, 272–290. doi:10.1016/j.icarus.2015.11.035.

Hanel, R., Conrath, B., Flasar, F.M., Kunde, V., Maguire, W., Pearl, J., Pirraglia, J., Samuelson, R., Herath, L., Allison, M., Cruikshank, D., Gautier, D., Gierasch, P., Horn, L., Koppany, R., Ponnamperuma, C., 1981. Infrared observations of the saturnian system from voyager 1. Science 212, 192–200.

Hartle, R.E., et al., 2006. Preliminary interpretation of Titan plasma interaction as observed by the Cassini Plasma Spectrometer: Comparisons with Voyager 1. Geophys. Res. Lett. 33, L08201. doi:10.1029/2005GL024817.

Hayes, A., Aharonson, O., Callahan, P., Elachi, C., Gim, Y., Kirk, R., Lewis, K., Lopes, R., Lorenz, R., Lunine, J., et al., 2008. Hydrocarbon lakes on Titan: Distribution and interaction with a porous regolith. Geophys. Res. Lett. 35 (9): L09204 (6 p), https://doi.org/10.1029/2008GL033409.

Hayes, A.G., 2016. The lakes and seas of Titan. Annu. Rev. Earth Planet. Sci. 44 (1), 57–83. doi:10.1146/annurev-earth-060115-012247.

Hayes, A.G., Lorenz, R.D., Lunine, J.I., 2018. A post-Cassini view of Titan's methane-based hydrologic cycle. Nat. Geosci. 11 (5), 306–313.

Hirtzig, M., et al., 2013. Titan's surface and atmosphere from Cassini/VIMS data with updated methane opacity. Icarus 226, 470–486. doi:10.1016/j.icarus.2013.05.033.

Hofgartner, J.D., Lunine, J.I., 2013. Does ice float in Titan's lakes and seas? Icarus 223, 628–631. doi:10.1016/j.icarus.2012.11.028.

Hörst, S.M., et al., 2012. Formation of amino acids and nucleotide bases in a Titan atmosphere simulation experiment. Astrobiology 12, 809–817. doi:10.1089/ast.2011.0623.

Hörst, S.M., Tolbert, M.A., 2013. In situ measurements of the size and density of Titan aerosol analogs. Astrophys. J. 770, L10. doi:10.1088/2041-8205/770/1/L10.

Horst, S.M., 2017. Titan's atmosphere and climate. J. Geophys. Res. Planets 122, 432–482.

Hubbard, W.B., et al., 1993. The occultation of 28 SGR by Titan. Astron. Astrophys. 269, 541–563.

Hueso, R., Sánchez-Lavega, A., 2006. Methane storms on Saturn's moon Titan. Nature 442, 428–431. doi:10.1038/nature04933.

Iess, L., Jacobson, R.A., Ducci, M., Stevenson, D.J., Lunine, J.I., Armstrong, J.W., Asmar, S.W., Racioppa, P., Rappaport, N.J., Tortora, P., 2012. The tides of Titan. Science 337, 457–459.

Imanaka, H., Smith, M.A., 2010. Formation of nitrogenated organic aerosols in the Titan upper atmosphere. Proc. Natl. Acad. Sci. 107 (28), 12423–12428. doi:10.1073/pnas.0913353107.

Irwin, P.G.J., Sromovsky, L.A., Strong, E.K., Sihra, K., Teanby, N.A., Bowles, N., Calcutt, S.B., Remedios, J.J., 2006. Improved near-infrared methane band models and k-distribution parameters from 2000 to 9500 cm^{-1} and implications for interpretation of outer planet spectra. Icarus 181, 309–319.

Israël, G., Szopa, C., Raulin, F., Cabane, M., Niemann, H.B., Atreya, S.K., Bauer, S.J., Brun, J.F., Chassefière, E., Coll, P., et al., 2005. Complex organic matter in Titan's atmospheric aerosols from in situ pyrolysis and analysis. Nature 438 (7069), 796–799.

Jankowski, D., Squyres, S., 1988. Solid-state ice volcanism on the satellites of Uranus. Science 241, 1322–1325.

Janssen, M.A., Le Gall, A., Lopes, R.M., Lorenz, R.D., Malaska, M.J., Hayes, A.G., Neish, C.D., Solomonidou, A., Mitchell, K.L., Radebaugh, J., Keihm, S.J., Choukroun, M., Leyrat, C., Encrenaz, P.J., Mastrogiuseppe, M., 2016. Titan's surface at 2.18-cm wavelength imaged by the Cassini RADAR radiometer: results and interpretations through the first ten years of observation. Icarus 270, 443–459.

Kargel, J.S., 1992. Ammonia-water volcanism on icy satellites: Phase relations at 1 atmosphere. Icarus 100, 556–574.

Karkoschka, 1994. Spectrophotometry of the Jovian planets and Titan at 300- to 1000-nm wavelength: the methane spectrum. Icarus 111, 174–192.

Knutson, H.A., Dragomir, D., Kreidberg, L., Kempton, E.M.-R., McCullough, P.R., Fortney, J.J., Bean, J.L., Gillon, M., Homeier, D., Howard, A.W., 2014a. Hubble space telescope near-ir transmission spectroscopy of the super-Earth HD 97658b. Astrophys. J. 794, 155. doi:10.1088/0004-637X/794/2/155.

Knutson, H.A., Benneke, B., Deming, D., Homeier, D., 2014b. A featureless transmission spectrum for the Neptune-mass exoplanet GJ436b. Nature 505, 66–68. doi:10.1038/nature12887.

Koskinen, T.T., Yelle, R.V., Snowden, D.S., Lavvas, P., Sandel, B.R., Capalbo, F.J., Benilan, Y., West, R.A., 2011. The mesosphere and lower thermosphere of Titan revealed by Cassini/UVIS stellar occultations. Icarus 216, 507–534. doi:10.1016/j.icarus.2011.09.022.

Kostiuk, T., Fast, K.E., Livengood, T.A., Hewagama, T., Goldstein, J.J., Espenak, F., Buhl, D., 2001. Direct measurement of winds on Titan. Geophys. Res. Lett. 28, 2361–2364. doi:10.1029/2000GL012617.

Kostiuk, T., Livengood, T.A., Hewagama, T., Sonnabend, G., Fast, K.E., Murakawa, K., Tokunaga, A.T., Annen, J., Buhl, D., Schmülling, F., 2005. Titan's stratospheric zonal wind, temperature, and ethane abundance a year prior to Huygens insertion. Geophys. Res. Lett. 32, L22205. doi:10.1029/2005GL023897.

Kostiuk, T., et al., 2006. Stratospheric global winds on Titan at the time of Huygens descent. J. Geophys. Res. 111, E07S03. doi:10.1029/2005JE002630.

Kral, T.A., Brink, K.M., Miller, S.L., McKay, C.P., 1998. Hydrogen consumption by methanogens on the early Earth. Orig. Life Evol. Biosph. 28, 311–319.

Kreidberg, L., Bean, J.L., Désert, J.-M., Benneke, B., Deming, D., Stevenson, K.B., Seager, S., Berta-Thompson, Z., Seifahrt, A., Homeier, D., 2014. Clouds in the atmosphere of the super-Earth exoplanet GJ1214b. Nature 505, 69–72. doi:10.1038/nature12888.

Kunde, V.G., Aikin, A.C., Hanel, R.A., Jennings, D.E, 1981. C_4H_2, HC_3N and C_2N_2 in Titan's atmosphere. Nature 292, 686–688.

Landera, A., Mebel, A.M., 2010. Mechanisms of formation of nitrogen-containing polycyclic aromatic compounds in low-temperature environments of planetary atmospheres: a theoretical study. Farad. Discuss. 147, 479.

Langhans, M.H., et al., 2012. Titan's fluvial valleys: Morphology, distribution, and spectral properties. Planet. Space Sci. 60, 34–51. doi:10.1016/j.pss.2011.01.020.

Lara, L.-M., Lellouch, E., Shematovich, V., 1999. Titan's atmospheric haze: the case for HCN incorporation. A&A 341, 312.

Lavvas, P.P., Coustenis, A., Vardavas, I.M., 2008a. Coupling photochemistry with haze formation in Titan's atmosphere, Part I: Model description. Planet. Space Sci. 56 (1), 27–66.

Lavvas, P.P., Coustenis, A., Vardavas, I.M., 2008b. Coupling photochemistry with haze formation in Titan's atmosphere, Part II: Results and validation with Cassini/Huygens data. Planet. Space Sci. 56, 67.

Lavvas, P., Yelle, R.V., Vuitton, V., 2009. The detached haze layer in Titan's mesosphere. Icarus 201 (2), 626–633.

Lavvas, P., Sander, M., Kraft, M., Imanaka, H., 2011. Surface chemistry and particle shape: processes for the evolution of aerosols in TTitan's atmosphere. The Astrophysical Journal 728 (2), 11 pp. DOI:10.1088/0004-637X/728/2/80.

Lavvas, P., et al., 2013. Aerosol growth in Titan's ionosphere. Proc. Natl Acad. Sci. USA 110, 2729–2734.

Lebonnois, S., Bakes, E.L.O., McKay, C.P., 2002. Transition from gaseous compounds to aerosols in Titan's atmosphere. Icarus 159 (2), 505–517.

Lebreton, J.-P., Witasse, O., Sollazzo, C., Blancquaert, T., Couzin, P., Schipper, A.-M., Jones, J.B., Matson, D.L., Gurvits, L.I., Atkinson, D.H., Kazeminejad, B., Perez-Ayucar, M., 2005. An overview of the descent and landing of the Huygens probe on Titan. Nature 438, 758–764.

Le Gall, A., et al., 2011. Cassini SAR, radiometry, scatterometry and altimetry observations of Titan's dune fields. Icarus 213, 608–624. doi:10.1016/j.icarus.2011.03.026.

Le Gall, A., et al., 2016. Composition, seasonal change, and bathymetry of Ligeia Mare, Titan, derived from its microwave thermal emission. J. Geophys. Res. Planets 121, 233–251. doi:10.1002/2015JE004920.

Le Mouélic, S., et al., 2008. Mapping and interpretation of Sinlap crater on Titan using Cassini VIMS and RADAR data. J. Geophys. Res. 113, E04003. doi:10.1029/2007JE002965.

Lemmon, M.T., Karkoschka, E., Tomasko, M., 1993. Titan's rotation—Surface feature observed. Icarus 103, 329–332. doi:10.1006/icar.1993.1074.

Lemmon, M.T., Karkoschka, E., Tomasko, M., 1995. Titan's rotational light-curve. Icarus 113, 27–38. doi:10.1006/icar.1995.1003.

Liang, M.-C., Yung, Y.L., Shemansky, D.E., 2007. Photolytically generated aerosols in the mesosphere and thermosphere of Titan. Astrophysical J. 661 (2), L199–L202.

Liu, Z.Y.C., Radebaugh, J., Christiansen, E.H., Harris, R.A., Neish, C.D., Kirk, R.L., Lorenz, R.D., 2016. The tectonics of Titan: global structural mapping from Cassini RADAR. Icarus 270, 14–29.

Lopes, R.M.C., Kirk, R.L., Mitchell, K.L., LeGall, A., Barnes, J.W., Hayes, A., Kargel, J., Wye, L., Radebaugh, J., Stofan, E.R., Janssen, M.A., Neish, C.D., Wall, S.D., Wood, C.A., Lunine, J.I., Malaska, M.J., 2013. Cryovolcanism on Titan: new results from Cassini RADAR and VIMS. J. Geophys. Res. Planets 118, 1–20.

Lopes, R.M.C., Malaska, M.J., Schoenfeld, A.M., Solomonidou, A., Birch, S.P.D., Florence, M., Hayes, A.G., Williams, D.A., Radebaugh, J., Verlander, T., Turtle, E.P., Le Gall, A., Wall, S.D., 2019. A global geomorphologic map of Saturn's moon Titan. Nature Astronomy. doi:10.1038/s41550-019-0917-6.

Lora, J.M., Lunine, J.I., Russell, J.L., Hayes, A.G., 2014. Simulations of Titan's paleoclimate. Icarus 243, 264–273. doi:10.1016/j.icarus.2014.08.042.

Lora, J.M., Mitchell, J.L., 2015. Titan's asymmetric lake distribution mediated by methane transport due to atmospheric eddies. Geophys. Res. Lett. 42, 6213–6220. doi:10.1002/2015GL064912.

Lorentz, R.D., McKay, C.P., Lunine, J.I., 1997. Photochemically driven collapse of Titan's atmosphere. Science 275, 642–645.

Lorenz, R.D., Lunine, J.I., 1996. Erosion on Titan: Past and present. Icarus 122, 79–91.

Lorenz, L.D., Niemann, H.B., Harpold, D.N., Way, S.H., Zarnecki, J.C., 2006a. Titan's damp ground: Constraints on Titan surface thermal

properties from the temperature evolution of the Huygens GCMS inlet. Meteorit. Planet. Sci. 41, 1705–1714.
Lorenz, R.D., et al., 2006b. The sand seas of Titan: Cassini RADAR observations of longitudinal dunes. Science 312, 724–727. doi:10.1126/science.1123257.
Lorenz, R.D., Stiles, B.W., Kirk, R.L., Allison, M.D., del Marmo, P.P., Iess, L., Lunine, J.I., Ostro, S.J., Hensley, S., 2008a. Titan's rotation reveals an internal ocean and changing zonal winds. Science 319, 1649–1651.
Lorenz, R.D., Lopes, R.M., Paganelli, F., Lunine, J.I., Kirk, R.L., Mitchell, K.L., Soderblom, A., Stofan, E.R., Ori, G., Myers, M., Miyamoto, H., Radebaugh, J., Stiles, B., Wall, S.D., Wood, C.A., Team, C.R., 2008b. Fluvial channels on Titan: initial Cassini RADAR observations. Planet. Space Sci. 56, 1132–1144.
Lorenz, R.D., Mitchell, K.L., Kirk, R.L., Hayes, A.G., Aharonson, O., Zebker, H.A., Paillou, P., Radebaugh, J., Lunine, J.I., Janssen, M.A., et al., 2008c. Titan's inventory of organic surface materials. Geophys. Res. Lett. 35 (2): L02206 (6 p). doi:10.1029/2007GL032118.
Lorenz, R.D., Radebaugh, J., 2009. Global pattern of Titan's dunes: Radar survey from the Cassini prime mission. Geophys. Res. Lett. 36, L03202. doi:10.1029/2008GL036850.
Lorenz, R.D., Claudin, P., Andreotti, B., Radebaugh, J., Tokano, T., 2010. A 3 km atmospheric boundary layer on Titan indicated by dune spacing and Huygens data. Icarus 205, 719–721. doi:10.1016/j.icarus.2009.08.002.
Lorenz, R.D., Hayes, A.G., 2012. The growth of wind-waves in Titan's hydrocarbon seas. Icarus 219, 468–475. doi:10.1016/j.icarus.2012.03.002.
Lorenz, R.D., Stiles, B.W., Aharonson, O., Lucas, A., Kirk, R.L., Zebker, H.A., Turtle, E.P., Neish, C.D., Stofan, E.R., et al., 2013. A global topographic map of Titan. Icarus 225, 367–377.
Lorenz, R.D., Kirk, R.L., Hayes, A.G., Anderson, Y.Z., Lunine, J.I., Tokano, T., Turtle, E.P., Malaska, M.J., Soderblom, J.M., Lucas, A., et al., 2014. A Radar map of Titan seas: Tidal dissipation and ocean mixing through the throat of Kraken. Icarus 237, 9–15.
Lorenz, R.D., 2014b. Physics of saltation and sand transport on Titan: a brief review. Icarus 230, 162–167. doi:10.1016/j.icarus.2013.06.023.
Lorenz, R.D., 2000. Post-Cassini exploration of Titan: Science rationale and mission concepts. J. Br. Interplanet. Soc. 53, 218–234.
Lunine, J.I., Stevenson, D.J., Yung, Y.L., 1983. Ethane ocean on Titan. Science 222, 1229–1230.
Lunine, J.I., McKay, C.P., 1995. Surface-atmosphere interactions on Titan compared with those on the pre-biotic Earth. Adv. Space Res. 15, 303–311.
Lunine, J.I., 2009. Saturn's Titan: A strict test for life's cosmic ubiquity. Proc. Am. Philos. Soc. 153 (4), 403–418.
Lunine, J.I., Lorenz, R.D., 2009. Rivers, lakes, dunes and rain: Crustal processes in Titan's methane cycle. Ann. Rev. Earth Planet. Sci. 37, 299–301.
Lunine, J.I., 2010. Titan and habitable planets around M-dwarfs. Farad. Discuss. 147, 405–418. doi:10.1039/c004788k.
Lunine, J.I., Hörst, S.M., 2011. Organic chemistry on the surface of Titan. Rendiconti Lincei 22 (3), 183–189. doi:10.1007/s12210-011-0130-8.
Luspay-Kuti, A., Chevrier, V.F., Cordier, D., Rivera-Valentin, E.G., Singh, S., Wagner, A., Wasiak, F.C., 2015. Experimental constraints on the composition and dynamics of Titan's polar lakes. Earth Planet. Sci. Lett. 410, 75–83.
Lutz, B.L., de Bergh, C., Owen, T., 1983. Titan—Discovery of carbon monoxide in its atmosphere. Science 220, 1374–1375.

Luz, D., Civeit, T., Courtin, R., Lebreton, J.-P., Gautier, D., Rannou, P., Kaufer, A., Witasse, O., Lara, L., Ferri, F., 2005. Characterization of zonal winds in the stratosphere of Titan with UVES. Icarus 179, 497–510. doi:10.1016/j.icarus.2005.07.021.
Luz, D., et al., 2006. Characterization of zonal winds in the stratosphere of Titan with UVES: 2. Observations coordinated with the Huygens Probe entry. J. Geophys. Res. 111, E08S90. doi:10.1029/2005JE002617.
Maddox, R.A., 1980. Mesoscale convective complexes. Bull. Am. Meteorol. Soc. 61, 1374–1387.
Maguire, W.C., Hanel, R.A., Jennings, D.E., Kunde, V.G., 1981. C_3H_8 and C_3H_4 in Titan's atmosphere. Nature 292, 683–686.
Malaska, M.J., Hodyss, R., 2014. Dissolution of benzene, naphthalene, and biphenyl in a simulated Titan lake. Icarus 242, 74–81. doi:10.1016/j.icarus.2014.07.022.
Malaska, M.J., et al., 2016. Geomorphological map of the Afekan Crater region, Titan: terrain relationships in the equatorial and mid-latitude regions. Icarus 270, 130–161.
Marco, M., Valerio, P., Alexander, H., Ralph, L., Jonathan, L., Giovanni, P., Roberto, S., Enrico, F., Giuseppe, M., Claudia, N., et al., 2014. The bathymetry of a Titan sea. Geophys. Res. Lett. 41, 1432–1437.
Marley, M.S., Ackerman, A.S., Cuzzi, J.N., Kitzmann, D., 2013. Clouds and Hazes in Exoplanet Atmospheres. Univ. of Ariz., Tucson, pp. 367–391.
Mastrogiuseppe, M., Poggiali, V., Hayes, A., Lorenz, R., Lunine, J., Picardi, G., Seu, R., Flamini, E., Mitri, G., Notarnicola, C., Paillou, P., Zebker, H., 2014. The bathymetry of a Titan sea. Geophys. Res. Lett. 41, 1432–1437.
McCord, T.B., et al., 2006. Composition of Titan's surface from Cassini VIMS, Planet. Space Sci. 54, 1524–1539. doi:10.1016/j.pss.2006.06.007.
McCord, T.B., et al., 2008. Titan's surface: Search for spectral diversity and composition using the Cassini VIMS investigation. Icarus 194, 212–242. doi:10.1016/j.icarus.2007.08.039.
McDonald, G.D., Corlies, P., Wray, J.J., Horst, S.M., Hofgartner, J.D., Liuzzo, L.R., Buffo, J., Hayes, A.G., 2015. Altitude-dependence of Titan's methane transmission windows: Informing future missions, 46th Lunar and Planetary Science Conference Abstracts. The Woodlands, TX, 2307 LPI Contribution No. 1832.
McKay, C.P., Scattergood, T.W., Pollack, J.B., Borucki, W.J., Van Ghyseghem, H.T., 1988. High-temperature shock formation of N_2 and organics on primordial Titan. Nature 332 (6164), 520–522.
McKay, C.P., Pollack, J.B., Courtin, R., 1989. The thermal structure of Titan's atmosphere. Icarus 80, 23–53.
McKay, C.P., Pollack, J.B., Courtin, R., 1991. The greenhouse and anti-greenhouse effects on Titan. Science 253, 1118–1121.
McKay, C.P., Coustenis, A., Samuelson, R.E., Lemmon, M.T., Lorenz, R.D., Cabane, M., Rannou, P., Drossart, P., 2001. Physical properties of the organic aerosols and clouds on Titan. Planet. Space Sci. 49 (1), 79–99.
McKay, C.P., 2004. What is life—and how do we search for it in other worlds? PLoS Biol. 2, e302–e304.
McKay, C.P., Smith, H.D., 2005. Possibilities for methanogenic life in liquid methane on the surface of Titan. Icarus 178, 274–276.
McKay, C.P., 2014. Requirements and limits for life in the context of exoplanets. Proc. Natl. Acad. Sci. USA 111, 12628–12633.
McKay, C.P., 2016. Titan as the abode of life. Life 6, 8.
Mebel, A.M., Kislov, V.V., Kaiser, R.I., 2008. Photo-induced mechanism of formation and growth of polycyclic aromatic hydrocarbons in low-

temperature environments via successive ethynyl radical additions. J. Am. Chem. Soc. 130 (41), 13618–13629.

Mitchell, P., Frenklach, M., 2003. Particle aggregation with simultaneous surface growth. Phys. Rev. E 67, 061407.

Mitchell, J.L., Pierrehumbert, R.T., Frierson, D.M.W., Caballero, R., 2006. The dynamics behind Titan's methane clouds. Proc. Natl. Acad. Sci. USA 103, 18,421–18,426. doi:10.1073/pnas.0605074103.

Mitchell, J.L., 2008. The drying of Titan's dunes: Titan's methane hydrology and its impact on atmospheric circulation. J. Geophys. Res. 113, E08015. doi:10.1029/2007JE003017.

Mitchell, J.L., 2009. Coupling convectively driven atmospheric circulation to surface rotation: Evidence for active methane weather in the observed spin rate drift of Titan. Astrophys. J. 692, 168–173. doi:10.1088/0004-637X/692/1/168.

Mitchell, J.L., 2012. Titan's transport-driven methane cycle. Astrophys. J. 756, L26. doi:10.1088/2041-8205/756/2/L26.

Mitri, G., Showman, A., Lunine, J., Lopes, R., 2008. Resurfacing of Titan by ammonia-water cryomagma. Icarus 196, 216–224.

Mitri, G., Meriggiola, R., Hayes, A., Lefevre, A., Tobie, G., Genova, A., Lunine, J.I., Zebker, H., 2014a. Shape, topography, gravity anomalies and tidal deformation of Titan. Icarus 236, 169–177.

Mitri, G., et al., 2014b. The exploration of Titan with an orbiter and a lake probe. Planet. Space Sci. 104, 78–92. doi:10.1016/j.pss.2014.07.009.

Moreno, R., Marten, A., Hidayat, T., 2005. Interferometric measurements of zonal winds on Titan. Astron. Astrophys. 437, 319–328. doi:10.1051/0004-6361:20042117.

Morgan, N., Kraft, M., Balthasar, M., Wong, D., Frenklach, M., Mitchell, P., 2007. Numerical simulations of soot aggregation in premixed laminar flames. Proc. Combust. Inst. 31 (1), 693–700.

Moriconi, M.L., Lunine, J.I., Adriani, A., D'Aversa, E., Negrao, A., Filacchione, G., Coradini, A., 2010. Characterization of Titan's Ontario Lacus region from Cassini/VIMS observations. Icarus 210, 823–831.

Muhleman, D.O., Grossman, A.W., Butler, B.J., Slade, M.A., 1990. Radar reflectivity of Titan. Science 248, 975–980. doi:10.1126/science.248.4958.975.

Navrátil, T., Minařík, L., 2002. Trace elements and contaminants. In: Cílek, V., Smith, R.H. (Eds.),. Earth's System: History and Natural Variability EOLSS-UNESCO. Encyclopedia of Life Support Systems (EOLSS). Oxford.

Neish, C.D., Lorenz, R.D., O'Brien, D.P., Team, C.R., 2006. The potential for prebiotic chemistry in the possible cryovolcanic dome Ganesa Macula on Titan. Int. J. Astrobiol. 5, 57–65.

Neish, C.D., Somogyi, A., Imanaka, H., Lunine, J.I., Smith, M.A., 2008. Rate measurements of the hydrolysis of complex organic macromolecules in cold aqueous solutions: implications for prebiotic chemistry on the early earth and Titan. Astrobiology 8, 273–287.

Neish, C.D., Somogyi, A., Lunine, J.I., Smith, M.A., 2009. Low temperature hydrolysis of laboratory tholins in ammonia-water solutions: implications for prebiotic chemistry on Titan. Icarus 201, 412–421.

Neish, C.D., Somogyi, A., Smith, M.A., 2010a. Titan's primordial soup: formation of amino acids via low-temperature hydrolysis of tholins. Astrobiology 10, 337–347.

Neish, CD., Lorenz, R.D., Kirk, R.L., Wye, L.C., 2010b. Radarclinometry of the sand seas of Africa's Namibia and Saturn's moon Titan. Icarus 208, 385–394. doi:10.1016/j.icarus.2010.01.023.

Neish, C.D., Kirk, R.L., Lorenz, R.D., Bray, V.J., Schenk, P., Stiles, B.W., Turtle, E., Mitchell, K., Hayes, A., Team, C.R., 2013. Crater topography on Titan: Implications for landscape evolution. Icarus 223, 82–90. doi:10.1016/j.icarus.2012.11.030.

Neish, C.D., Lorenz, R.D., 2014. Elevation distribution of Titan's craters suggests extensive wetlands. Icarus 228, 27–34.

Neish, C.D., Barnes, J.W., Sotin, C., MacKenzie, S., Soderblom, J.M., Le Mouélic, S., Kirk, R.L., Stiles, B.W., Malaska, M.J., Le Gall, A., Brown, R.H., Baines, K.H., Buratti, B., Clark, R.N., Nicholson, P.D, 2015. Spectral properties of Titan's impact craters imply chemical weathering of its surface. Geophys. Res. Lett. 42, 3746–3754.

Neish, C.D., Molaro, J.L., Lora, J.M., Howard, A.D., Kirk, R.L., Schenk, P., Bray, V.J., Lorenz, R.D., 2016. Fluvial erosion as a mechanism for crater modification on Titan. Icarus 270, 114–129.

Neish, C.D., Lorenz, R.D., Turtle, E.P., Barnes, J.W., Trainer, M.G., Stiles, B., Kirk, R., Hibbitts, C.A., Malaska, M.J., 2018. Strategies for detecting biological molecules on Titan. Astrobiology 18 (5), 571–585. https://doi.org/10.1089/ast.2017.1758.

Nelson, R.M., et al., 2009a. Saturn's Titan: Surface change, ammonia, and implications for atmospheric and tectonic activity. Icarus 199, 429–441. doi:10.1016/j.icarus.2008.08.013.

Nelson, R.M., et al., 2009b. Photometric changes on Saturn's moon Titan: Evidence for cryovolcanism. Geophys. Res. Lett. 36, L04202. doi:10.1029/2008GL036206.

Niemann, H.B., Atreya, S.K., Bauer, S.J., Carignan, G.R., Demick, J.E., Frost, R.L., Gautier, D., Haberman, J.A., Harpold, D.N., Hunten, D.M., Israel, G., Lunine, J.I., Kasprzak, W.T., Owen, T.C., Paulkovich, M., Raulin, F., Raaen, E., Way, S.H., 2005. The abundances of constituents of Titan's atmosphere from the GCMS instrument on the Huygens probe. Nature 438, 779–784.

Niemann, H.B., Atreya, S.K., Demick, J.E., Gautier, D., Haberman, J.A., Harpold, D.N., Kasprzak, W.T., Lunine, J.I., Owen, T.C., Raulin, F., 2010. Composition of Titan's lower atmosphere and simple surface volatiles as measured by the Cassini-Huygens probe gas chromatograph mass spectrometer experiment. J. Geophys. Res. 115, E12006.

Nimmo, F., Bills, B.G., 2010. Shell thickness variations and the long-wavelength topography of Titan. Icarus 208, 896–904.

O'Brien, D.P., Lorenz, R.D., Lunine, J.I., 2005. Numerical calculations of the longevity of impact oases on Titan. Icarus 173, 243–253.

Osinski, G.R., Grieve, R.A.F., Bleacher, J.E., Neish, C.D., Pilles, E.A., Tornabene, L.L., 2018. Igneous rocks formed by hypervelocity impact. J. Volcanol. Geotherm. Res. 353, 25–54.

Owen, T., 1987. How primitive are the gases in Titan's atmosphere? Adv. Space Res. 7 (5), 51–54.

Owen, T., Gautier, D., 1989. Titan: some new results. Adv. Space Res. 9 (2), 73–78.

Patterson, R.I.A., Singh, S., Balthasar, M., Kraft, M., 2006. The Linear Process Deferment Algorithm: A new technique for solving population balance equations. SIAM J. Sci. Comput. 28 (1), 303–320.

Patterson, R.I.A., Kraft, M., 2007. Models for the aggregate structure of soot particles. Combust. Flame 151 (1–2), 160–172.

Perron, J.T., Lamb, M.P., Koven, C.D., Fung, I.Y., Yager, E., Ádámkovics, M., 2006. Valley formation and methane precipitation rates on Titan. J. Geophys. Res. 111, E11001. doi:10.1029/2005JE002602.

Pinto, J.P., Lunine, J.I., Kim, S.J., Yung, Y.L., 1986. D to H ratio and the origin and evolution of Titan's atmosphere. Nature 319 (6052), 388–390.

Poch, O., Coll, P., Buch, A., Ramírez, S.I., Raulin, F., 2012. Production yields of organics of astrobiological interest from H_2O–NH_3 hydrolysis of Titan's tholins. Planet. Space Sci. 61, 114–123.

Porco, C.C., Baker, E., Barbara, J., Beurle, K., Brahic, A., Burns, J.A., Charnoz, S., Cooper, N., Dawson, D.D., DelGenio, A.D., Denk, T., Dones, L., Dyudina, U., Evans, M.W., Fussner, S., Giese, B., Grazier, K., Helfenstein, P., Ingersoll, A.P., Jacobson, R.A., Johnson, T.V., McEwen, A.,

Murray, C.D., Neukum, G., Owen, W.M., Perry, J., Roatsch, T., Spitale, J., Squyres, S., Thomas, P., Tiscareno, M., Turtle, E.P., Vasavada, A.R., Veverka, J., Wagner, R., West, R., 2005. Imaging of Titan from the Cassini spacecraft. Nature 434, 159–168. doi:10.1038/nature03436.

Radebaugh, J., et al., 2008. Dunes on Titan observed by Cassini radar. Icarus 194, 690–703. doi:10.1016/j.icarus.2007.10.015.

Radebaugh, J., Ventra, D., Lorenz, R., Farr, T., Kirk, R., Hayes, A., Malaska, M., Birch, S., Liu, Z., Lunine, J., Barnes, J., Le Gall, A., Lopes, R.M.C., Stofan, E., Wall, S., Paillou, P., 2016. Alluvial and fluvial fans on Saturn's moon Titan reveal materials and regional geology From: Ventra, D. & Clarke, L. E. Geology and Geomorphology of Alluvial and Fluvial Fans: Terrestrial and Planetary Perspectives. Geological Society. Special Publications, London, pp. 440.

Rafkin, S.C.R., Barth, E.L., 2015. Environmental control of deep convective clouds on Titan: The combined effect of CAPE and wind shear on storm dynamics, morphology, and lifetime. J. Geophys. Res. Planets 120, 739–759. doi:10.1002/2014JE004749.

Rages, K., Pollack, J.B., Smith, P.H., 1983. Size estimates of Titan's aerosols based on Voyager high-phase-angle images. J. Geophys. Res. 88 (A11), 8721–8728.

Raj, A., Sander, M., Janardhanan, V., Kraft, M., 2009a. A study on the coagulation of polycyclic aromatic hydrocarbon clusters to determine their collision efficiency. Combust. Flame 157 (3), 523–534.

Raj, A.M., et al., 2009b. A statistical approach to develop a detailed soot growth model using PAH characteristics. Combust. Flame 156, 896–913.

Rannou, P., Cabane, M., Chassefiere, E., Botet, R., McKay, C.P., Courtin, R., 1995. Titan's Geometric Albedo: role of the fractal structure of the aerosols. Icarus 118 (2), 355–372.

Rannou, P., McKay, C.P., Lorenz, R.D., 2003. A model of Titan's haze of fractal aerosols constrained by multiple observations. Planet. Space Sci. 51 (14–15), 963–976.

Rannou, P., Hourdin, F., McKay, C.P., Luz, D., 2004. A coupled dynamics-microphysics model of Titan's atmosphere. Icarus 170 (2), 443–462.

Rannou, P., Montmessin, F., Hourdin, F., Lebonnois, S., 2006. The latitudinal distribution of clouds on Titan. Science 311, 201–205.

Rannou, P., Toledo, D., Lavvas, P., D'Aversa, E., Moriconi, M.L., Adriani, A., Le Mouélic, S., Sotin, C., Brown, R., 2016. Titan's surface spectra at the Huygens landing site and Shangri-La. Icarus, 270: 291–306. https://doi.org/10.1016/j.icarus.2015.09.016.

Raulin, F., Bruston, P., Coll, P., Coscia, D., Gazeau, M-C., Guez, L., de Vanssay, E., 1995. Exobiology on Titan: a reference laboratory for studying prebiotic chemistry on a planetary scale. J. Biol. Phys. 20, 39–53.

Raulin, F., 2008. Planetary science: Organic lakes on Titan. Nature 454, 587–589. doi:10.1038/454587a.

Raulin, F., Brassé, C., Poch, O., Coll, P., 2012. Prebiotic-like chemistry on Titan. Chem. Soc. Rev. 41 (16), 5380–5393.

Raven, J.A., Kübler, J.E., Beardall, J., 2000. Put out the light, and then put out the light. J. Mar. Biolog. Assoc. UK 80, 1–25.

Richardson, J., Lorenz, R.D., McEwen, A., 2004. Titan's surface and rotation: new results from Voyager 1 images. Icarus 170, 113–124.

Robinson, T.D., Maltagliati, L., Marley, M.S., Fortney, J.J., 2014. Titan solar occultation observations reveal transit spectra of a hazy world. Proc. Natl. Acad. Sci. USA 111, 9042–9047. doi:10.1073/pnas.1403473111.

Rodriguez, S., Mouélic, S.L., Rannou, P., Tobie, G., Baines, K.H., Barnes, J.W., Griffith, C.A., Hirtzig, M., Pitman, K.M., Sotin, C., Brown, R.H., Buratti, B.J., Clark, R.N., Nicholson, P.D., 2009. Global circulation as the main source of cloud activity on Titan. Nature 459, 678–682. doi:10.1038/nature08014.

Rodriguez, S., et al., 2014. Global mapping and characterization of Titan's dune fields with Cassini: correlation between RADAR and VIMS observations. Icarus 230, 168–179. doi:10.1016/j.icarus.2013.11.017.

Roe, H.G., de Pater, I., Macintosh, B.A., McKay, C.P., 2002. Titan's clouds from Gemini and Keck adaptive optics imaging. Astrophys. J. 581, 1399–1406.

Roe, H.G., Bouchez, A.H., Trujillo, C.A., Schaller, E.L., et al., 2005. Discovery of temperate latitude clouds on Titan. Astrophys. J. Lett. 618, L49–L52.

Roe, H.G., Grundy, W.M., 2012. Buoyancy of ice in the CH_4-N_2 system. Icarus 219, 733–736. doi:10.1016/j.icarus.2012.04.007.

Sander, M., Raj, A., Inderwildi, O.R., Kraft, M., Kureti, S., Bockhorn, H., 2009a. The simultaneous reduction of nitric oxide and soot in emissions from diesel engines. Carbon 47 (3), 866–875.

Sander, M., West, R.H., Celnik, M.S., Kraft, M., 2009b. A detailed model for the sintering of polydispersed nanoparticle agglomerates. Aerosol Sci. Technol. 43 (10), 978–989.

Savage, C.J., Radebaugh, J., Christiansen, E.H., Lorenz, R.D., 2014. Implications of dune pattern analysis for Titan's surface history. Icarus 230, 180–190. doi:10.1016/j.icarus.2013.08.009.

Schaller, E.L., Roe, H.G., Schneider, T., Brown, M.E., 2009. Storms in the tropics of Titan. Nature 460, 873–875. doi:10.1038/nature08193.

Schneider, T., Graves, S.D.B., Schaller, E.L., Brown, M.E., 2012. Polar methane accumulation and rainstorms on Titan from simulations of the methane cycle. Nature 481, 58–61. doi:10.1038/nature10666.

Schulze-Makuch, D., Grinspoon, D.H., 2005. Biologically enhanced energy and carbon cycling on Titan? Astrobiology 5, 560–567.

Schulze-Makuch, D., Fairén, A.G., Davila, A., 2013. Locally targeted ecosynthesis: a proactive in situ search for extant life on other worlds. Astrobiology 13, 674–678.

Sekine, Y., Imanaka, H., Matsui, T., Khare, B.N., Bakes, E.L.O., McKay, C.P., Khare, B.N., Sugita, S., 2008a. The role of organic haze in Titan's atmospheric chemistry: I. Laboratory investigation on heterogeneous reaction of atomic hydrogen with Titan tholin. Icarus 194 (1), 186–200.

Sekine, Y., Lebonnois, S., Imanaka, H., Matsui, T., Bakes, E.L.O., McKay, C.P., Khare, B.N., Sugita, S., 2008b. The role of organic haze in Titan's atmospheric chemistry: II. Effect of heterogeneous reaction to the hydrogen budget and chemical composition of the atmosphere. Icarus 194 (1), 201–211.

Showman, A.P., Mosqueira, I., Head III., J.W., 2004. On the resurfacing of Ganymede by liquid–water volcanism. Icarus 172, 625–640.

Sicardy, B., et al., 2006. The two Titan stellar occultations of 14 November 2003. J. Geophys. Res. 111, E11S91. doi:10.1029/2005JE002624.

Simonds, C.H., Warner, J.L., Phinney, W.C., 1976. Thermal regimes in cratered terrain with emphasis on the role of impact melt. Am. Mineral. 61, 569–577.

Sittler Jr., E.C., Hartle, R.E., Vinas, A.F., Johnson, RF., Smith, H.T., Miueller-Wodarg, I., 2005. Titan interaction with Saturn's magnetosphere: Voyager I results revisited. J. Geophys. Res. 110, A09302 IGA(129,2004JA010759.

Sittler Jr., E.C., Johnson, R.E., Smith, H.T., Richardson, J.D., Jurac, S., Moole, M., Cooper, J.F., Mauk, B.H., Michael, M., Paranicus, C., Armstrong, T.P., Tsufultani, B., 2006. Energetic nitrogen ions within the inner magnetosphere of Saturn. J. Geophys. Res. 111, A09223. doi: 10.1029,2004JA010509.

Sittler Jr., E.C., et al., 2008. Ion and neutral sources and sinks within Saturn's inner magnetosphere: Cassini results. Planet. Space Sci. 56, 3–18.

Sittler Jr., E.C., Ali, A., Cooper, J.F., Hartle, R.E., Johnson, R.E., Coates, A.J., Young, D.T., 2009. Heavy ion formation in Titan's ionosphere: Magnetospheric introduction of free oxygen and a source of Titan's aerosols? Planet. Space Sci. 57, 1547–1557.

Sittler Jr., E.C., Hartle, R.E., Johnson, R.E., Cooper, J.F., Lipatov, A.S., Bertucci, C., Coates, A.J., Szego, K., Shappirio, M., Simpson, D.G., Wahlund, J.-E., 2010. Saturn's magnetospheric interaction with Titan as defined by Cassini encounters T9 and T18: New results. Planet. Space Sci. 58, 327–350.

Smith, P.A., Davis, L.E., Button, T.W., Alford, R.M., 1991. The dielectric loss tangent of liquid nitrogen. Supercond. Sci. Technol. 91, 128–129.

Smith, P.H., Lemmon, M.T., Lorenz, R.D., Sromovsky, L.A., Caldwell, J.J., Allison, M.D., 1996. Titan's surface, revealed by HST imaging. Icarus 119, 336–349. doi:10.1006/icar.1996.0023.

Soderblom, L.A., et al., 2007a. Topography and geomorphology of the Huygens landing site on Titan. Planet, Space Sci. 55, 2015–2024. doi:10.1016/j.pss.2007.04.015.

Soderblom, L.A., et al., 2007b. Correlations between Cassini VIMS spectra and RADAR SAR images: Implications for Titan's surface composition and the character of the Huygens Probe landing site. Planet. Space Sci. 55, 2025–2036. doi:10.1016/j.pss.2007.04.014.

Soderblom, L.A., Barnes, J.W., Brown, R.H., Clark, R.N., Janssen, M.A., McCord, T.B., Niemann, H.B., Tomasko, M.G., 2009. Composition of Titan's surface, In: *Titan from Cassini-Huygens*, R. H. Brown et al. (Eds.), chap. 6, pp. 141–175, Springer, New York, NY.

Soderblom, J.M., Brown, R.H., Soderblom, L.A., Barnes, J.W., Jaumann, R., Le Mouélic, S., Sotin, C., Stephan, K., Baines, K.H., Buratti, B.J., Clark, R.N., Nicholson, P.D., 2010. Geology of the Selk crater region on Titan from Cassini VIMS observations. Icarus 208, 905–912.

Sotin, C., et al., 2005. Release of volatiles from a possible cryovolcano from near-infrared imaging of Titan. Nature 435, 786–789. doi:10.1038nature03596.

Sotin, C., 2007. Planetary science: Titan's lost seas found. Nature 445, 29–30. doi:10.1038/445029a.

Sotin, C., Lawrence, K.J., Reinhardt, B., Barnes, J.W., Brown, R.H., Hayes, A.G., le Mouélic, S., Rodriguez, S., Soderblom, J.M., Soderblom, L.A., et al., 2012. Observations of Titan's northern lakes at 5 microns: Implications for the organic cycle and geology. Icarus 221, 768–786.

Stevens, M.H., Evans, J.S., Lumpe, J., Westlake, J.H., Ajello, J.M., Bradley, E.T., Esposito, L.W., 2015. Molecular nitrogen and methane density retrievals from Cassini UVIS dayglow observations of Titan's upper atmosphere. Icarus 247, 301–312. doi:10.1016/j.icarus.2014.10.008.

Stiles, B.W., et al., 2009. Determining Titan surface topography from Cassini SAR data. Icarus 202 (2), 584–598.

Stofan, E.R., Elachi, C., Lunine, J.I., Lorenz, R.D., Stiles, B., Mitchell, K.L., Ostro, S., Soderblom, L., Wood, C., Zebker, H., Wall, S., Janssen, M., Kirk, R., Lopes, R., Paganelli, F., Radebaugh, J., Wye, L., Anderson, Y., Allison, M., Boehmer, R., Callahan, P., Encrenaz, P., Flamini, E., Francescetti, G., Gim, Y., Hamilton, G., Hensley, S., Johnson, W.T.K., Kelleher, K., Muhleman, D., Paillou, P., Picardi, G., Posa, F., Roth, L., Seu, R., Shaer, S., Vetrella, S., West, R., 2007. The lakes of Titan. Nature 445, 61–64.

Tan, S.P., Kargel, J.S., Marion, G.M., 2013. Titan's atmosphere and surface liquid: New calculation using Statistical Associating Fluid Theory. Icarus 222, 53–72.

Tan, S.P., Kargel, J.S., Jennings, D.E., Mastrogiuseppe, M., Adidharma, H., Marion, G.M., 2015. Titan's liquids: Exotic behavior and its implications on global fluid circulation. Icarus 250, 64–75.

Thompson, W.R., Sagan, C., 1992. Organic chemistry on Titan: surface interactions, Proceedings of the Symposium on Titan, September 9–12, 1991, SP-338, ESA. Toulouse, 167–176.

Tiffin, D.L., Kohn, J.P., Luks, K.D., 1979. Solid hydrocarbon solubility in liquid methane-ethane mixtures along three-phase solid-liquid-vapor loci. J. Chem. Eng. Data 24, 306–310.

Tobie, G., Grasset, O., Lunine, J.I., Mocquet, A., Sotin, C., 2005. Titan's internal structure inferred from a coupled thermal-orbital model. Icarus 175, 496–502.

Tobie, G., Lunine, J.I., Sotin, C., 2006. Episodic outgassing as the origin of atmospheric methane on Titan. Nature 440, 61–64. doi:10.1038/nature04497.

Tobie, G., Teanby, N.A., Coustenis, A., Jaumann, R., Raulin, F., Schmidt, J., Carrasco, N., Coates, A.J., Cordier, D., De Kok, R., Geppert, W.D., Lebreton, J.-P., Lefevre, A., Livengood, T.A., Mandt, K.E., Mitri, G., Nimmo, F., Nixon, C.A., Norman, L., Pappalardo, R.T., Postberg, F., et al., 2014. Science goals and mission concept for the future exploration of Titan and Enceladus. Planet. Space Sci. 104 (Part A), 59–77.

Tokano, T., 2002. Tidal winds on Titan caused by Saturn. Icarus 158, 499–515.

Tokano, T., 2005. Meteorological assessment of the surface temperatures on Titan: constraints on the surface type. Icarus 173, 222–242.

Tokano, T., McKay, C.P., Neubauer, F.M., Atreya, S.K., Ferri, F., et al., 2006. Methane drizzle on Titan. Nature 442, 432–435.

Tokano, T., 2008. Limnological structure of Titan's hydrocarbon lakes and its astrobiological implication, European Planetary Science Congress 2008.

Tokano, T., 2009. Limnological structure of Titan's hydrocarbon lakes and its astrobiological implication. Astrobiology 9, 147–164.

Tokano, T., Lorenz, R.D., Van Hoolst, T., 2014. Numerical simulation of tides and oceanic angular momentum of Titan's hydrocarbon seas. Icarus 242, 188–201. doi:10.1016/j.icarus.2014.08.021.

Tomasko, M.G., 1980. Preliminary results of polarimetry and photometry of Titan at large phase angles from Pioneer 11. J. Geophys. Res. Space Phys. 85, 5937–5942.

Tomasko, M.G., Smith, P.H., 1982. Photometry and polarimetry of Titan: Pioneer 11 observations and their implications for aerosol properties. Icarus 51 (1), 65–95.

Tomasko, M.G., Archinal, B., Becker, T., Bézard, B., Bushroe, M., Combes, M., Cook, D., Coustenis, A., de Bergh, C., Dafoe, L.E., Doose, L., Douté, S., Eibl, A., Engel, S., Gliem, F., Grieger, B., Holso, K., Howington-Kraus, E., Karkoschka, E., Keller, H.U., Kirk, R., Kramm, R., Küppers, M., Lanagan, P., Lellouch, E., Lemmon, M., Lunine, J., McFarlane, E., Moores, J., Prout, G.M., Rizk, B., Rosiek, M., Rueffer, P., Schröder, S.E, Schmitt, B., See, C., Smith, P., Soderblom, L., Thomas, N., West, R., 2005. Rain, winds and haze during the Huygens probe's descent to Titan's surface. Nature 438 (7069), 765–778.

Tomasko, M.G., Doose, L., Engel, S., Dafoe, L.E., West, R., Lemmon, M., Karkoschka, E., See, C., 2008a. A model of Titan's aerosols based on measurements made inside the atmosphere. Planet. Space Sci. 56, 669–707.

Tomasko, M.G., Bézard, B., Doose, L., Engel, S., Karkoschka, E., 2008b. Measurements of methane absorption by the descent imager/spectral radiometer (DISR) during its descent through Titan's atmosphere. Planet. Space Sci. 56 (5): 624–647. DOI: 10.1016/j.pss.2007.10.009.

Tomasko, M.G., Doose, L., Engel, S., Dafoe, L.E., West, R., Lemmon, M., Karkoschka, E., See, C., 2008c. Limits on the size of aerosols from measurements of linear polarization in Titan's atmosphere. Planet. Space Sci. 56, 669–707.

Totton, T.S., Chakrabarti, D., Misquitta, A.J., Sander, M., Wales, D.J., Kraft, M., 2010. Modelling the internal structure of nascent soot particles. Combust. Flame 157 (5), 909–914.

Tsai, I.-C., Liang, M.-C., Chen, J.-P., 2012. Methane-nitrogen binary nucleation: A new microphysical mechanism for cloud formation in Titan's atmosphere. Astrophys. J. 747, 36. doi:10.1088/0004-637X/747/1/36.

Turtle, E.P., Perry, J.E., Hayes, A.G., Lorenz, R.D., Barnes, J.W., McEwen, A.S., West, R.A., Del Genio, A.D., Barbara, J.M., Lunine, J.I., Schaller, E.L., Ray, T.L., Lopes, R.M.C., Stofan, E.R., 2011a. Rapid and extensive surface changes near Titan's equator: evidence of April showers. Science 331 (6023), 1414–1417.

Turtle, E.P., Perry, J.E., Hayes, A.G., Lorenz, R.D., Barnes, J.W., McEwen, A.S., West, R.A., del Genio, A.D., Barbara, J.M., Lunine, J.I., et al., 2011b. Rapid and extensive surface changes near Titan's equator: Evidence of April showers. Science 331, 1414–1417.

Turtle, E.P., Barnes, J.W., Trainer, M.G., Lorenz, R.D., MacKenzie, S.M., Hibbard, K.E., Adams, D., Bedini, P., Langelaan, J.W., Zacny, K., 2017. Dragonfly: Exploring Titan's prebiotic organic chemistry and habitability, and the Dragonfly Team Lunar and Planetary Science Conference. https://www.hou.usra.edu/meetings/lpsc2017/eposter/1958.pdf.

Vuitton, V., Yelle, R.V., McEwan, M.J., 2007. Ion chemistry and N-containing molecules in Titan's upper atmosphere. Icarus 191 (2), 722–742.

Vuitton, V., Yelle, R.V., Cui, J., 2008. Formation and distribution of benzene on Titan. J. Geophys. Res. 113 (E5), E05007.

Vuitton, V., Dutuit, O., Smith, M.A., Balucani, N., 2014. Chemistry of Titan's Atmosphere, In: Titan, I. Müller-Wodarg et al. (Eds.), pp. 224–271, Cambridge Univ. Press, Cambridge.

Waite Jr., J.H., Young, D.T., Cravens, T.E., Coates, A.J., Crary, F.J., Magee, B., Westlake, J., 2007. The process of Tholin formation in Titan's upper atmosphere. Science 316 (5826), 870–875. doi:10.1126/science.1139727.

Wasiak, F.C., Androes, D., Blackburn, D.G., Tullis, J.A., Dixon, J., Chevrier, V.F., 2013. A geological characterization of Ligeia Mare in the northern polar region of Titan. Planet. Space Sci. 84, 141–147.

West, R.A., Smith, P.H., 1991. Evidence for aggregate particles in the atmospheres of Titan and Jupiter. Icarus 90 (2), 330–333.

West, R.A., Balloch, J., Dumont, P., Lavvas, P., Lorenz, R., Rannou, P., Ray, T., Turtle, E.P., 2011. The evolution of Titan's detached haze layer near equinox in 2009. Geophys. Res. Lett. 38, L06204. doi:10.1029/2011GL046843.

West, R.A., Del Genio, A.D., Barbara, J.M., Toledo, D., Lavvas, P., Rannou, P., Turtle, E.P., Perry, J., 2016. Cassini imaging science subsystem observations of Titan's south polar cloud. Icarus 270, 399–408. doi:10.1016/j.icarus.2014.11.038.

Williams, D.A., Radebaugh, J., Lopes, R.M.C., Stofan, E., 2011. Geomorphologic mapping of the Menrva region of Titan using Cassini RADAR data. Icarus 212, 744–750.

Williams, K.E., McKay, C.P., Persson, F., 2012. The surface energy balance at the Huygens landing site and the moist surface conditions on Titan. Planet. Space Sci. 60, 376–385.

Wilson, E.H., Atreya, S.K., 2003. Chemical sources of haze formation in Titan's atmosphere. Planet. Space Sci. 51 (14–15), 1017–1033.

Williams, K.E., McKay, C.P., Persson, F., 2012. The surface energy balance at the Huygens landing site and the moist surface conditions on Titan. Planet. Space Sci. 60, 376–385.

Witek, P.P., Czechowski, L., 2015. Dynamical modelling of river deltas on Titan and Earth. Planet. Space Sci. 105, 65–79.

Wood, C.A., Lorenz, R., Kirk, R., Lopes, R., Mitchell, K., Stofan, E., 2010. Impact craters on Titan. Icarus 206, 334–344.

Yung, Y.L., Allen, M., Pinto, J.P., 1984. Photochemistry of the atmosphere of Titan — comparison between model and observations. Astrophysical J. Supplement Series (ISSN 0067-0049) 55, 465–506.

Zarnecki, J.C., Leese, M.R., Hathi, B., Ball, A.J., Hagermann, A., Towner, M.C., Lorenz, R.D., McDonnell, J.A.M., Green, S.F., Patel, M.R., Ringrose, T.J., Rosenberg, P.D., et al., 2005. A soft solid surface on Titan as revealed by the Huygens surface science package. Nature 438, 792–795.

Chapter 15

A likely ocean world fostering a rare mixing of CO and N₂ ice molecules on Neptune's Moon *Triton*

15.1 General features of *Triton*

Triton is planet Neptune's principal moon. Its radius is approximately 1353 km, with an average density of approximately 2065 kg/m^3 (McKinnon and Kirk, 2007). Models of Triton's internal structure indicate that Triton likely has a large silicate core, estimated to be approximately 950 km in radius. Based on the density of Triton, and the likely abundance of radiogenic elements in its core, Brown et al. (1991) estimated approximately 0.75–1.5×10^{11} W of heat flow is generated by radioactive decay in the silicate part of Triton's core. Brown and Kirk (1994) demonstrated that this radiogenic heat can drive lateral heat transport and explain the surface distribution of volatiles.

Triton was the last solid object visited by the Voyager 2 spacecraft on its epic 10-year tour of the outer solar system. This moon was one of the most unusual and surprising bodies observed as part of the Voyager mission. During its distant flyby in 1989, Voyager 2 captured a series of images, mostly of the southern, sub-Neptune hemisphere, establishing Triton as one of a rare class of solar system bodies with an active geology. The Voyager mission imaged about 40% of Triton's surface. From this sampling of the surface, it was found that three types of terrains can be distinguished – volcanic plains, dynamic "cantaloupe" terrain, and polar caps (McKinnon and Kirk, 2007). Triton's infamous cantaloupe terrain (an organized cellular pattern of noncircular dimples) most likely formed when the icy crust of Triton underwent wholesale overturn, forming large numbers of rising blobs of ice (diapirs). The numerous irregular mounds found on the cantaloupe terrain are a few hundred meters (several hundred feet) high and a few kilometers (several miles) across and formed when the top of the crust buckled during overturn. Candidate endogenic features include a network of tectonic structures, including most notably long linear features which appear to be similar to Europa's double ridges (Prockter et al., 2005); several candidate cryovolcanic landforms (Croft et al., 1995); widespread cantaloupe terrain unique to Triton; and several particulate plumes and associated deposits (see Prockter et al., 2019).

McKinnon and Kirk (2007) suggested that Triton is differentiated given surface features indicative of melting and the distribution of various icy phases over the surface. Triton's surface appears to possess two different types of ices, one composed by the volatile ices and the second one formed by water and carbon dioxide ices.

The Raz Fossae (two prominent, ≈15 km wide, troughs en echelon; note that en echelon is characterized by an approximately parallel formation at an oblique angle to a particular direction) is among the most outstanding tectonic features of Triton (Fig. 15.1). These structures have been interpreted as grabens (Croft et al., 1995). Note that a graben is an elongated block of a celestial body's crust lying between two faults and displaced downwards relative to the blocks on either side, as in a rift valley. Through simple geometric concepts, the maximum depth of faulting can be estimated from the width of a graben.

Triton has several interesting features. For example, if the upper atmosphere and ionosphere of Triton are controlled by precipitation of electrons from Neptune's magnetosphere as previously proposed, Triton could have the only ionosphere in the solar system not controlled by solar radiation (Lyons et al., 1992). Furthermore, a myriad of exotic ices is observed spectroscopically on Triton (e.g., Quirico et al., 1999), but water ice or ammonia hydrate ice is thought to be the "bedrock ice" in order to preserve topographic relief (Croft et al., 1995; Cruikshank et al., 2000). Triton is considered as a candidate ocean world based on

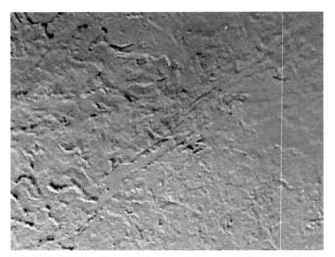

FIG. 15.1 **The Raz Fossae, Triton, seen in a mosaic of images taken by the Voyager 2 spacecraft.** These features are ≈ 15 km wide and are centered on 8° N, 21.5° E. *(Source: Ruiz, J. (2003), Heat flow and depth to a possible internal ocean in Triton, Icarus, 166: 436–439). Elsevier publication.*

hints from limited spacecraft observations. The NASA's ROW team supports the creation of an exploration program that studies the full spectrum of ocean worlds, that is, not just the exploration of known ocean worlds such as Europa but candidate ocean worlds such as Triton as well. Triton is the highest priority candidate ocean world to target in the near term.

15.2 Origin of *Triton*—its uniqueness among all large moons in the solar system

A simple hypothesis on the origin of Triton is that Triton, like Pluto, is an independent representative of large outer Solar System planetesimals (bodies which could come together with many others under gravitation to form planets). Triton's starting point and initial composition were that of a Dwarf Planet originating in the Kuiper Belt Objects (KBO). Triton is by far the largest moon (its mass is ~40% greater than that of Pluto) in our Solar System that did not form in situ around its host planet. Triton's relatively high-density of 2065 kg/m^3 (McKinnon and Kirk, 2007) implies that a substantial part of its interior is likely composed of rock and possibly metal (Gaeman et al., 2012).

Triton orbits Neptune, the eighth planet from the Sun, ~2.7 billion miles from Earth—at the cold outer fringe of our Solar System's major planet zone. Triton is unique among all large moons in the Solar System for its inclined and retrograde orbit around its planet (i.e., it orbits in a direction opposite to its host planet Neptune's rotation). The peculiar motion of Triton (i.e., orbiting "backwards" or in the opposite direction to its host planet's rotation) suggests that Triton is a captured trans-Neptunian object from the Kuiper Belt – a region of leftovers from the Solar System's early history, which is why it shares several features with the dwarf planet Pluto and Eris.

Triton's inclined orbit lies between a group of small inner prograde moons (i.e., those moons orbiting in the direction of rotation of Neptune) and a number of exterior irregular moons with both prograde and retrograde orbits. This unusual configuration has also led to the belief that Triton originally orbited the Sun before being captured in orbit around Neptune (McCord, 1966; McKinnon, 1984).

Prockter et al. (2005) suggested that following the primordial Triton's capture into a highly eccentric (i.e., highly elliptical) orbit, tidal dissipation would have reduced Triton's semi-major axis a, and eccentricity e (i.e., Triton's orbit approaching a circular orbit), with a timescale dependent on the dissipation factor Q and Love number k of Triton. At each passage of periapse (i.e., when Triton comes nearest to Neptune), Neptune's gravity raised tides at Triton's surface, dissipating energy within Triton, and circularizing its orbit. Ross and Schubert (1990) showed that this circularization process took place over 800 Ma since its capture. In an alternative approach, Cuk and Gladman (2005) suggested that Triton's capture strongly perturbed the preexisting moons, leading to collision and the formation of a debris disk. Subsequently, Triton's orbit evolved rapidly (6105 yr) by interaction with this debris disk.

In modeling the circularization of Triton's retrograde orbit, Ross and Schubert (1990) showed that there is a peak in tidal dissipation during a brief period when the semi-major axis is reduced but eccentricity remains relatively high. According to them, following capture, Triton's orbit likely circularized on a timescale of ~1 billion years (By). As already noted, Triton's orbital history likely involves capture from a binary system by Neptune, followed by a period of circularization. Gaeman et al. (2012) investigated Triton's coupled thermal and structural evolution past its circularization, driven by tidal dissipation and radiogenic heating.

Some planetary researchers (e.g., Goldreich et al., 1989; Benner and McKinnon, 1995) are of the opinion that if Triton's capture occurred relatively early in solar system history, it might have been enabled by gas drag; alternatively, more recent capture may have been facilitated by collision with an original Neptunian moon. Thus, Triton probably formed in the protoplanetary nebula as an icy dwarf planet and may have a composition similar to other large primitive bodies such as Pluto, Eris, Sedna, Ceres, and Vesta.

Craig et al. (2006) reported that a three-body gravitational encounter between a binary system (of ~10^3-kilometre-sized bodies) and Neptune is a likely explanation for Triton's capture. Their model predicts that Triton was once a member of a binary with a range of plausible characteristics, including ones similar to the Pluto–Charon pair. According to them, one possible outcome of gravitational encounters between a binary system and a planet is an exchange reaction, where one member of the binary is expelled and its place taken by the planet. Triton's duality as both captured dwarf planet and large icy moon that has experienced extreme collisional and tidal processing, make it a unique lens for understanding two of the Solar System's principal constituencies and the fundamental processes that govern their evolution.

According to Agnor et al. (2009), "Due to its unique dynamical history, Triton possesses both properties known to drive interesting geology and chemistry on icy worlds: hydrocarbons and a history of tidal dissipation. Thus, Triton holds the key to understanding the evolution of the entire spectrum of icy objects in the solar system, from large icy moons to small KBOs."

Some potentially spectacular consequences that resulted from Triton's capture include runaway melting of interior ices and release to the surface of clathrated methane (CH_4), carbon monoxide (CO) and nitrogen (N_2). Condensed remnants of this proto-atmosphere could account for features in Triton's unique spectrum (McKinnon, 1984).

15.3 Triton's surface temperature and pressure

Instruments on board Voyager 2 showed that the surface of Triton is very cold, about 38 Kelvins (−235°C). Interestingly, Triton is having the coldest surface temperature of any solar system object, including Pluto, which is much farther from the Sun. Because the surface temperatures of Triton hover near absolute zero, the common compounds we know as gases on Earth freeze into ices on Triton. The temperature on Triton varies wildly with the seasons. This means that during its "year" (165 Earth years long), icy material should be repeatedly shifted from one pole to the other.

Significant seasonal changes undergone by Triton would reveal themselves as changes in its mean frost temperature (Spencer, 1990; Hansen and Paige, 1992; Spencer and Moore, 1992). But whether this temperature should at the present time be increasing, decreasing or constant depends on a number of parameters (such as the thermal properties of the surface, and frost migration patterns) that are unknown. Elliot et al. (1998) reported observations of a stellar occultation by Triton which, when combined with earlier results, show that Triton has undergone a period of global warming since 1989. Their most conservative estimates of the rate of temperature and surface-pressure increase during this period imply that the atmosphere is doubling in bulk every 10 years—significantly faster than predicted by any published frost model for Triton (Hansen and Paige, 1992; Spencer and Moore, 1992). Elliot et al.'s result suggest that permanent polar caps on Triton play a dominant role in regulating seasonal atmospheric changes.

The Voyager 2 mission indicated that Triton has a thin atmosphere of gaseous nitrogen. Triton's atmospheric pressure is only 15 microbars (note that Earth's standard atmospheric pressure at mean sea level, by definition, is 1.01325 bars). It is of interest to note that Triton is one of the very few moons with an atmosphere and one of only four solar system objects to have an atmosphere consisting largely of nitrogen. Triton is one of the brightest objects in the solar system, with an albedo of about 0.7. According to Brown et al. (1991), indications of recent global albedo change on Triton suggest that Triton's surface temperature and pressure may not now be in a steady state, further suggesting that atmospheric pressure on Triton was as much as ten times higher in the recent past. Arguing in favor of the unsteady state of Triton's surface temperature and pressure, Brown et al. (1991) made the following observation, "Internal heat flow from radioactive decay in Triton's interior along with absorbed thermal energy from Neptune total 5 to 20 percent of the insolation absorbed by Triton; thus comprising a significant fraction of Triton's surface energy balance. These additional energy inputs can raise Triton's surface temperature between approximately 0.5 and 1.5 K above that possible with absorbed sunlight alone, resulting in an increase of about a factor of approximately 1.5 to 2.5 in Triton's basal atmospheric pressure. If Triton's internal heat flow is concentrated in some areas, as is likely, local effects such as enhanced sublimation with subsequent modification of albedo could be quite large."

According to Prockter et al. (2019) although Triton's atmosphere is thin, "it is sufficiently substantial to be a major sink for volatiles, and sufficiently dynamic to play a role in the movement of surface materials. Its youthful age implies a highly dynamic environment, with surface atmosphere volatile interchange, and potentially dramatic climate change happening over obliquity and/or season timescales. An extensive south polar cap, probably mostly consisting of nitrogen which can exchange with the atmosphere, was observed."

15.4 Chemical composition of Triton's atmosphere & surface

Triton has a remarkable but poorly understood atmosphere and surface that hint strongly at ongoing geological activity, suggesting an active interior and a possible subsurface ocean. Even without the presence of a surface ocean and endogenic activity, Triton remains one of the most compelling targets in the solar system. Triton's atmosphere is ~70,000 times less dense than Earth's and is primarily composed of nitrogen, methane, and carbon monoxide. How this information about a very distant moon in the Solar System has been obtained is of much interest. In situations where in-situ sample collection is practically impossible (e.g., very remotely located celestial bodies), information on the chemical composition of the surface of most of the celestial bodies is obtained through a thorough analysis of the remotely obtained absorption spectra (visible, near-infrared, etc. spectral region) from such bodies. Through analyses of the near-infrared spectrum of Triton, Cruikshank et al. (1993) identified ices of nitrogen, methane, carbon monoxide, and carbon dioxide on Triton, of which nitrogen is the dominant component. Their studies suggest that carbon dioxide ice may be spatially segregated from the other more volatile ices, covering about 10 percent of Triton's surface. Triton's hydrogen and nitrogen atmosphere is a main source of Neptune's magnetospheric plasma, dominating the middle-magnetosphere (e.g., Summers and Strobel, 1991; Eviatar et al., 1995).

Krasnopolsky and Cruikshank (1995) found that the photochemistry of Triton's atmosphere below 50 km is driven mostly by photolysis of methane by the solar and interstellar medium Lyman-alpha photons, producing hydrocarbons C_2H_4, C_2H_6, and C_2H_2 which form haze particles. According to the studies of these investigators, the chemistry above 200 km is driven by the solar EUV radiation (wavelength <1000 angstroms) and by precipitation of magnetospheric electrons. Note that angstrom is a unit of length equal to equal to 10^{-10} m, used mainly to express wavelengths and interatomic distances. The most abundant photochemical species are N, H_2, H, O, and C. They found that atomic species are transported to a region of 50–200 km and drive the chemistry there. It was found that ionospheric chemistry explains the formation of an E region at 150–240 km with HCO^+ as a major ion, and of an F region above 240 km with a peak at 320 km and C^+ as a major ion. The ionosphere above 500 km was found to consist of almost equal densities of C^+ and N^+ ions. Krasnopolsky and Cruikshank (1995)'s model profiles were found to agree with the measured atomic nitrogen and electron density profiles. Although a number of other models with varying rate coefficients of some reactions, differing properties of the haze particles (chemically passive or active), etc., were developed, all of them showed that there are four basic unknown values that have strong impacts on the composition and structure of the atmosphere and ionosphere.

It is generally accepted that Triton's icy layer is mainly composed of water ice (Smith and team, 1989; McKinnon et al., 1995), though it has been argued using cosmochemical concepts, that ammonia–water system ices (e.g., ammonium dehydrate ($NH_3 \cdot 2H_2O$)) might be a closer approximation for an icy moon (e.g., Hogenboom et al., 1997).

Molecular nitrogen is thought to have been the most common type of nitrogen available when the Solar System was forming. Its abundance in the outer Solar System is an important key to life's origins, as it is an important part of the building blocks of life. Analysis of Voyager mission data has provided a glimpse of Triton's chemical constituents, but not a detailed analysis of its composition. Observations of surface features and spectral analysis indicate that the surface composition of Triton is predominantly H_2O with trace amounts of volatiles: N_2, CO, CO_2, and CH_4 (Brada and Clarke, 1997). About half of Triton's surface is composed of N_2 in solid solution with a minute amount of CO and CH_4. The other half is comprised of sections of either CO_2 or H_2O.

The Voyager spacecraft passed quickly by Triton, so could not adequately characterize the interaction of Triton's atmosphere and ionosphere with Neptune's magnetic field under a variety of magnetospheric configurations. Based on more recent reports, Triton is mainly covered by N_2, CO, CO_2, CH_4, and H_2O in the solid state and, except for H_2O and CO_2, these species are also present in the gas phase (Merlin et al., 2018). Sublimation and recondensation of the volatile species may lead to geographical and temporal variation on the surface composition, and could participate to the formation of complex chemical compounds from photochemistry occurring on Triton. The presence of methane in the atmosphere, and possibly on the surface, makes possible a wide range of "hot atom" chemistry allowing higher-order organic materials to be produced in a similar albeit slower manner to Titan. The presence of such materials is of potential importance to habitability, especially if conditions exist where they come into contact with liquid water (Prockter et al., 2019)."

15.5 Triton's nitrogen deposits—interesting consequences

Triton's polar caps are considered to be permanent nitrogen deposits hundreds of meters thick. This icy moon's clouds appear to be made of frozen nitrogen. The principal ion in the ionosphere of Triton is N^+. Energetic electrons of magnetospheric origin are the primary source of ionization, with a smaller contribution due to photoionization. To explain the topside plasma scale height, Yung and Lyons (1990) postulated that N^+ ions escape from Triton. With reference to the postulated N^+ ions escape, these investigators have discussed the implications for the magnetosphere of Neptune and Triton's evolution.

Complex temperature variations on Triton's surface induce reversible transitions between the cubic and hexagonal phases of solid nitrogen, often with two coexisting propagating transition fronts. Duxbury and Brown (1993) calculated subsurface temperature distributions using a two-dimensional thermal model with phase changes. Their calculations revealed that the phase changes fracture the upper nitrogen layer, increasing its reflectivity and thus offering an explanation for the surprisingly high southern polar cap albedo (approximately 0.8) seen during the Voyager 2 flyby. It was found that the model has other implications for the phase transition phenomena on Triton, such as a plausible mechanism for the origin of geyser-like plume vent areas and a mechanism of energy transport toward them.

15.6 Triton's surface—among the youngest surfaces in the solar system

Triton is one of the most fascinating and enigmatic bodies in the solar system. Among its numerous interesting traits, Triton appears to have far fewer craters than would be expected if its surface was primordial. From the low density (i.e., small number) of visible impact craters, it may be inferred that the surface of Triton is relatively young (Smith and team, 1989). Triton shows evidence of recent geologic activity based on observations of its surface (Prockter et al., 2005). These results imply that Triton almost certainly has the youngest surface age of any planetary body in the solar system, with the exception of the violently volcanic world, Io (Prockter et al., 2019).

Deformation via tidal activity commonly occurs in planetary bodies orbiting at large eccentricities. Taking account of the fact that Triton's orbit in the past exhibited large eccentricities, Prockter et al. (2005) suggested that one possible mechanism for Triton's past activity is tidal diurnal stress, modulated by the eccentricity of Triton's orbit. Triton's current orbit is nearly circular (Ross and Schubert, 1990; McKinnon and Kirk, 2007), implying that tidal stresses are less likely to significantly deform the surface at present. Conversely, tidal heating likely played a significant role early in Triton's history (Gaeman et al., 2012), when it was orbiting at large eccentricities. Taking account of Triton's present likely lesser tidal heating resulting from the current nearly circular orbit of Triton, several investigators (e.g., Ross and Schubert, 1990; Brown et al., 1991; Brown and Kirk, 1994; Ruiz, 2003) suggested that sustained internal activity, fueled by the combined action of tidal dissipation and radiogenic heating may provide an alternative mechanism for the current surface activity on Triton.

While the ages of many of the documented features of Triton remain unknown, the surface age of Triton is approximated between 10 and 100 Myr old (Schenk and Zahnle, 2007). More specifically, the estimated cratering rate on Triton by ecliptic comets is used to put an upper limit of ~50 Myr on the age of the more heavily cratered terrains, and of ~6 Myr for the Neptune-facing cantaloupe terrain (Fig. 15.2). According to Schenk and Jackson (1993), the cantaloupe terrain of the Tritonian Bubembe Regio may have formed via diapiric overturn of a layered crust ~20 km thick. Note that diapirism is an anticlinal fold (a type of fold that is an arch-like shape and has its oldest beds at its core) in which a mobile core has pierced through the more brittle overlying material.

Recent age estimates for Triton give 300–350 Myr (Stern and McKinnon, 2000; Zahnle et al., 2003; Schenk et al., 2004). However, Triton's young surface seems to

FIG. 15.2 High resolution "Cantaloupe" Terrain on Triton. Image Credit: NASA/JPL/Universities Space Research Association/Lunar & Planetary Institute. (Source: https://www.nasa.gov/mission_pages/voyager/pia12186.html).

indicate that this distant moon is currently active (Stern and McKinnon 1999, Schenk and Sobieszczyk, 1999). In a study, Stern and McKinnon (2000) combined the best available crater count data for Triton with improved estimates of impact rates by including the Kuiper Belt as a source of impactors. They found that the population of impactors creating the smallest observed craters on Triton must be sub-km in scale and that this small-impactor population can be best fit by a differential power-law size index near -3. Based on the modern, Kuiper Belt and Oort Cloud impactor flux estimates, Stern and McKinnon (2000) recalculated the estimated ages for several regions of Triton's surface imaged by Voyager 2, and found that Triton was probably active on a time scale no greater than 0.1-0.3 Gyr ago (indicating Triton was still active after some 90% to 98% of the age of the solar system), and perhaps even more recently.

Stern and McKinnon (2000) estimated that, even for their conservative age estimates, Triton has the highest resurfacing rates of the outer Solar System after Io and Europa, being of the same order as those calculated for Venus and the Earth's intraplate zones. These high rates of geological activity are consistent with the high surface heat flows necessary for putting the brittle–ductile transition at the base of Raz Fossae (interpreted as grabens). It is believed that such heat flows might be the remains of the intense heat generated by the capture of a heliocentric Triton by Neptune.

According to Stern and McKinnon (2000), the time-averaged volumetric resurfacing rate on Triton implied by these results, 0.01 km^3/yr or more, is likely second only to Io and Europa in the outer solar system, and is within an order of magnitude of estimates for Venus and for the Earth's intra-plate zones. Based on these findings Stern and McKinnon (2000) suggested that Triton likely remains a highly geologically active world at present, some 4.5 Gyr after its formation.

Though the event of diapiric overturn of a layered crust ~20 km thick precedes the opening of the Raz Fossae at the Neptune-facing cantaloupe terrain (Croft et al., 1995), this thickness is comparable to the depth to the internal ocean proposed by Ruiz (2003). The results of Ruiz (2003)'s investigations also suggest that the cantaloupe terrain may have formed by diapiric overturn — but an overturn of the whole ice layer rather than of a layered crust ~20 km thick as proposed by Schenk and Jackson (1993). According to Ruiz (2003), it is feasible that Triton has retained part of the heat generated during its capture by Neptune and is able to sustain an internal ocean at a depth as low as 20 or 30 km below its surface.

Schenk and Zahnle (2007) proposed that the vast majority of cratering on Triton is by planeto-centric debris. If this is correct, then the surface everywhere is probably less than 10 Myr old. According to Schenk and Zahnle (2007), although the uncertainty in these cratering ages is at least a factor 10, it seems likely that Triton's surface is among the youngest surfaces in the Solar System, a candidate ocean moon, and an important target for future exploration.

15.7 Spectral features of *Triton*'s water–ice

Solid H_2O at low pressure can exist in at least four phases: amorphous ice (not crystalline, or not apparently crystalline; and like glass in appearance or physical properties) in a high-density configuration (Iah) and a low-density configuration (Ial), and the crystalline cubic (Ic) and crystalline hexagonal (Ih) phases (see Jenniskens et al., 1995). The phase depends upon the temperature and other conditions of formation, the thermal annealing history, and the elapsed time. Water–ice that forms at T < 100 K is amorphous, and when formed at T < 30 K it is Iah (see Cruikshank et al., 2000). The transformation Iah → Ial occurs at about 38 K (Jenniskens and Blake, 1996), and in the absence of irradiation (e.g., electrons or UV) is irreversible. The transition Ial → Ic occurs quickly at T~150K, but in time it will occur at somewhat lower temperature (Cruikshank et al., 2000).

Electron diffraction studies of vapor-deposited water–ice have characterized the dynamical structural changes during crystallization that affect volatile retention in cometary materials. Crystallization is found to occur by nucleation of small domains, while leaving a significant part of the amorphous material in a slightly more relaxed amorphous state that coexists metastably with cubic crystalline ice. The amorphous component can effectively retain volatiles during crystallization if the volatile concentration is ~10% or less. For higher initial impurity concentrations, a significant amount of impurities is released during crystallization, probably because the impurities are trapped on the surfaces of micro-pores. Jenniskens and Blake (1996) have described a model for crystallization over long timescales that can be applied to a wide range of impure water ices under typical astrophysical conditions if the fragility factor D, which describes the viscosity behavior, can be estimated.

It has been found that the crystallization time of amorphous ice is less than 10 years at T = 110 K, and about 1000 years at T = 100 K (Schmitt et al., 1989a, 1989b, 1992). In the absence of irradiation this transition is also irreversible. Ice formed at T > ~190 K is hexagonal, and transitions to Ic or the amorphous phases do not occur, even if the ice is cooled to lower temperatures (Cruikshank et al., 2000).

The absorption spectrum of ice changes with temperature in several ways. With higher temperature, the shapes of absorption bands become more smoothed, the strengths of some absorption bands decrease, the absorption in continuum wavelengths increases, and the band centers of some absorption bands shift to shorter wavelengths. In a study, Grundy and Schmitt (1998) reported the new absorption coefficient spectra along with an examination of the different temperature effects. According to them, these data should prove extremely valuable for analysis of near-infrared reflectance spectra of low-temperature icy surfaces, such as those of outer solar system moons, Kuiper Belt objects, Pluto and Charon, comet nuclei, the polar caps of Mars, and terrestrial snow-and-ice-covered regions.

One of the spectral features of H_2O ice of special interest in planetary spectroscopic studies occurs at 1.65 μm. It is present in the spectra of all of the icy moons of Jupiter, Saturn, and Uranus (e.g., Cruikshank et al., 1998a). This feature is unique to crystalline H_2O ice, and its strength and band center are strongly temperature-dependent (Fink and Larson, 1975; Grundy and Schmitt, 1998).

According to Cruikshank et al. (2000), the spectroscopic signatures of the various phases of ice in the wavelength region of interest in the case of water–ice on Triton have not been fully explored, although there is considerable information on the hexagonal phase and its spectral variation with temperature (Grundy and Schmitt, 1998). Because of technical difficulties in preparing thick samples of amorphous and cubic ice, the near-infrared spectrum of the amorphous phase has been studied only in thin films (Fink and Sill, 1982; Schmitt et al., 1998).

The spectral reflectance of Triton in the photo-visual region is known to be variable (e.g., Smith and team, 1989; Brown et al., 1995). Buratti et al. (1998, 1999) have reported an abrupt change from a relatively neutral and colorless reflectance (in the region 0.35-0.95 μm) to a distinct red color, and then back to neutral over an interval of about seven months.

Using data from the cooled grating array spectrometer (CGS4) of the United Kingdom Infrared Telescope (UKIRT) in May 1995, and October 1998, Cruikshank et al. (2000) discussed the spectroscopic detection of H_2O ice on Triton, evidenced by the broad absorption bands in the near infrared at 1.55 μm and 2.04 μm. Although crystalline H_2O ice has a distinctive spectral band at 1.65 μm, and Cruikshank et al. (2000)'s new models slightly favor the presence of this phase, their investigation of spectra obtained for the same longitude on Triton, over an interval of nearly 3.5 years, did not provide any unambiguous results indicating whether Triton's water–ice is crystalline or amorphous. Their results suggested that both phases might be present, and special conditions in the surface microstructure may affect the spectroscopic signature of water ice in such a way that crystalline ice is present and its 1.65 μm spectral band is masked. For lack of precise information, Cruikshank et al. (2000) suggested that continued monitoring of Triton's spectrum appears to be warranted.

15.8 Ridges on *Triton*

Triton displays a variety of distinctive curvilinear ridges and associated troughs. Images from Triton show linear ridges with similar morphologies that are not observed on any other planetary bodies. Prockter et al. (2005) proposed that ridges on Triton may have formed by diurnal tidal stresses, in a manner similar to that proposed for ridge formation on Europa. Triton ridges have remarkably similar morphological forms to those on Europa (Smith and team, 1989), and include single isolated troughs and double ridges. Triple and multi-crested ridges are also present but are rare. In general, Triton's ridges tend to be morphologically subdued compared to Europa's ridges, but are much larger in overall scale (Prockter et al., 2005). Despite differences in horizontal scale, the topographic profiles across ridges on Europa and Triton both clearly show well-pronounced central troughs and flanking ridge crests. Note that flank of a ridge is the lateral part or side of a ridge. Prockter et al. (2005) examined the shear heating model of Gaidos and Nimmo (2000) and Nimmo and Gaidos (2002) in which heating along cracks from diurnal tidally driven strike-slip motion causes upwelling of warm ice to form ridges, with possible associated partial melting. Employing the stress model of Stempel et al. (2004), Prockter et al. (2005) determined diurnal stresses on Triton through time, based on the coupled evolution of Triton's semi-major axis *a*, and eccentricity *e*, after its capture as modeled by Ross and Schubert (1990). The model assumes an ice shell that deforms above a global liquid ocean, which, planetary scientists think, is likely considering the great amount of tidal dissipation following capture by Neptune (McKinnon et al., 1995). In a related further study involving Triton's eccentricity, Gaeman et al. (2012) examined the rate of ice shell growth as a function of different orbital eccentricities, in the presence of radiogenic heating. They found that "tidal dissipation in the ice shell, proportional to orbital eccentricity squared, concentrates heating near the base, reducing the basal heat flux. As the growth of the ice shell is proportional to the basal heat flux, increased tidal heating creates a blanketing effect, reducing the rate of ice shell growth.

Radiogenic heating from Triton's core is the other, more dominant, source of heat to the shell."

Prockter et al. (2005)'s investigations revealed that diurnal stressing is a plausible deformation mechanism on Triton. According to them, significant effects from diurnal stresses are expected only for a moon with an elliptical orbit, periapse close to its primary planet, and an ice shell that is decoupled from the interior by a global ocean. Based on model-study results, Prockter et al. (2005) came up with the following observation, "Because the stress maximum occurs after the dissipation maximum, the surface we see today, including its ridges and troughs, is likely a relic of Triton's waning geological activity. Triton's ridges themselves are not necessarily young, but the young surface age and lack of impact craters imply that they formed relatively recently." In the view of Nimmo and Spencer (2015), Triton's geological activity is driven by obliquity tides, which arise because of its inclination.

According to Prockter et al. (2005), it is reasonable that double ridges are preferentially located within Triton's cantaloupe terrain, because that region is believed to be associated with vigorous vertical (diapiric) overturn of mobile crustal materials (Schenk and Jackson, 1993). It is believed that depressions alongside major ridges may have formed when buoyant material was depleted from these regions by flow laterally toward the ridge axis, analogous to rim synclines (trough or fold of stratified rocks in which the strata slope upwards from the axis) of terrestrial diapirs (e.g., Trusheim, 1960). Prockter et al. (2005)'s model-based studies gave an indication that outside of the cantaloupe region, frosts and icy volcanic deposits tend to overprint older terrains, presumably hiding ridges and troughs (Croft et al., 1995).

15.9 Dust devils-like tall plumes of gas and dark material rising through *Triton*'s atmosphere

In the summer of 1989, NASA's Voyager 2 spacecraft became the first spacecraft to observe the planet Neptune, its final planetary target. Passing about 4,950 kilometers (3,000 miles) above Neptune's North Pole, Voyager 2 made its closest approach to any planet 12 years after leaving Earth in 1977. Voyager 2 remains the only spacecraft to visit Triton. When Voyager 2 whizzed by Neptune, it was the planet's moon Triton that stole the show. According to Kerr (1990), "Ice-lava lakes, cantaloupe-skin terrains, and subtle peach hues made Triton a celebrity, but Voyager's most remarkable discovery may have been 8-km-tall plumes of gas and debris rising through Triton's atmosphere." At least four active geyser-like eruptions were discovered in Voyager 2 images of Triton. The radii of the rising columns appear to be in the range of several tens of meters to a kilometer (Soderblom et al., 1990). The two best-documented eruptions occurred as columns of dark material rising to an

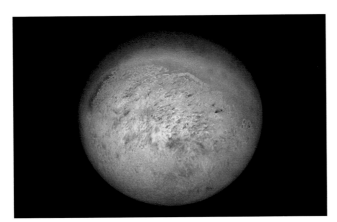

FIG. 15.3 Voyager 2 image of planet Neptune's icy moon Triton showing the south polar region with dust devil-like dark streaks visible on the icy surface. *Credit: NASA/JPL. (Source: https://www.gemini.edu/pr/neptune-s-moon-triton-fosters-rare-icy-union).*

altitude of about 8 kilometers where dark clouds of material are left suspended to drift downwind over 100 kilometers. The mass flux in the trailing clouds is estimated to consist of up to 10 kilograms of fine dark particles per second or twice as much nitrogen ice and perhaps several hundred or more kilograms of nitrogen gas per second. Each eruption may last a year or more, during which on the order of a tenth of a cubic kilometer of ice is sublimed (Soderblom et al., 1990). The geyser-like eruptions are dark in color because of the carbon-rich materials rising from beneath the surface of Triton. Fig. 15.3 shows Voyager 2 image of Triton showing the southern polar region with dark strokes produced by geysers visible on the icy surface.

Although Triton is a frigid moon, subsurface hot spots are formed on Triton when sunlight penetrates the translucent icy surface. Such subsurface hot spots behave like trapped greenhouses below the Triton's surface. One model for the mechanism to drive the plumes involves heating of nitrogen ice in a subsurface greenhouse environment; nitrogen gas pressurized by the solar heating explosively vents to the surface carrying clouds of ice and dark particles into the atmosphere. A temperature increase of less than 4 Kelvins above the ambient surface value of 38 ± 3 Kelvins is more than adequate to drive the plumes to an 8-kilometer altitude. Kirk et al. (1990) examined a variety of models for the storage of solar energy in a sub-greenhouse layer and for the supply of gas and energy to a geyser. Solid-state greenhouse calculations carried out by Kirk et al. (1990) have shown that solar energy (sunlight) can generate substantially elevated subsurface temperatures, leading to the observed active geyser-like eruptions and related features close to the sub-solar latitude on Triton. According to Kirk et al. (1990)'s calculations, the inferred active geyser lifetime is ∼ five Earth years.

Beyond Earth and Mars, large dust devils have been observed on *Triton*. Because Triton has such a low surface pressure, extremely unstable layers could develop during the day. One may ask, on Triton, how could an ephemeral (lasting for a very short time), hurriedly moving dust devil stay in one place long enough to create a straight 150-kilometer long streamer or the dark streaks? Ingersoll and Tryka (1990) explained that dust devils on Triton would not stray from their warm patches and they could reform every Triton day (5.877 Earth days).

Ingersoll and Tryka (1990) disagreed with the theory that nitrogen gas pressurized by the solar heating explosively venting to the surface, carrying clouds of ice and dark particles into the atmosphere, is the reason for the generation of the observed geyser-like plumes on Triton. They argued that the observed structure of the plume suggests that the plumes are an atmospheric rather than a surface phenomenon. They proposed that the plumes might be swirling funnels of dust, like dust devils on Earth. It may be noted that on Earth, dust devils are atmospheric vortices originating in the unstable layer close to the ground. Dust devils form if there is a very unstable layer of air. Such layers occur in deserts because often there is a large temperature difference between the ground and the air only a short distance above it. Large dust devils have also been observed on Mars, where they can rise to a height of about 6 kilometres in its thin atmosphere. Because Triton has such a low surface pressure, extremely unstable layers could develop during the day.

According to Kerr (1990), "The dust devil idea has a certain appeal because, if correct, it would mean that nothing—neither heat nor gas nor "dust"—would have to come from beneath Triton's surface. That would sidestep a problem confronting proponents of plumes as geysers. Those researchers have to explain how subsurface heat sources, which have been waning for billions of years, can produce such dynamic plumes from an icy surface hovering just a few tens of degrees above absolute zero."

Ingersoll and Tryka pointed out that the frigid 38 Kelvin temperature at the bottom of the atmosphere could be maintained by the chilling effects of the nitrogen frost thought to cover much of Triton's surface. Assuming that there could be small frost-free areas (e.g., unfrosted ground near the sub-solar point on Triton), Ingersoll and Tryka proposed that the feeble sunlight could make these locations perhaps 10 Kelvin or higher warmer than surrounding areas. Then, just as happens on some parts of Earth and Mars, the warmed layer of atmosphere overlying a frost-free area could begin swirling, forming funnels that carry dust and heat upward, thereby establishing the unusual temperature gradient observed by the Voyager radio science team. Assuming that velocity scales as the square root of temperature difference times the height of the mixed layer, Ingersoll and Tryka (1990) derived a velocity of 20 meters per second for the strongest dust devils on Triton. According to them, winds of this speed are sufficient enough to raise particles on Triton.

Kerr (1990) further noted that "Ingersoll and Tryka came up with the dust devil idea after Leonard Tyler of Stanford University and his colleagues reported that the temperature of Triton's atmosphere behaves most peculiarly. The Tyler group's analysis of the changes in Voyager's radio signals as they passed through Triton's vanishingly thin atmosphere indicated that the atmosphere gets warmer with increasing altitude instead of cooling, as Earth's atmosphere does."

15.10 Likely existence of a 135–190 km thick inner ocean at a depth of ~20–30 km beneath Triton's icy surface

Assuming that Triton possesses a homogeneous internal structure, Ross and Schubert (1990) developed a parameterized thermal–orbital evolution model of Triton, incorporating tidal dissipation and radiogenic heating. Tidal dissipation in their model generated sufficient energy to cause large-scale melting and the formation of a global ocean early in Triton's history. It is now generally agreed among planetary scientists that deformation incurred by the primordial Triton's orbital evolution would have dissipated a large amount of heat within Triton's interior, melting Triton's icy layer into a global ocean. McKinnon and Kirk (2007) approximated the dissipation of energy to be $\sim 10^4$ kJ/kg, which is enough to melt the icy moon entirely.

Considering water or ammonia dehydrate ($NH_3 \cdot 2H_2O$) as possible components of Triton's lithosphere and a feasible range of strain rates, it was estimated that surface heat flow is greater than that inferred from radiogenic heating, especially for a lithosphere dominated by water (Ruiz, 2003). According to Ruiz (2003)'s calculations, "an internal ocean could lie at a depth of only ~20 km beneath the surface". According to this investigator, "The presence over the surface of an insulating layer of ice of low thermal conductivity (e.g., nitrogen) or of regolith would only substantially alter these estimates if the effective surface temperature were considerably higher than the observed value of 38 K."

Ruiz (2003) suggested that if Triton's ice mantle contains significant amounts of ammonia (a potent anti-freezing agent), there could be a liquid-water-dominated ocean within Triton. Although the thermal conductivity, k, of water–ice is dependent on its temperature according to the expression: $k = k_0/T$ (where T is the absolute temperature; and k_0 is the thermal conductivity at absolute zero temperature); Ruiz (2003) used a constant value of 1 W/m/K for the thermal conductivity of ammonia dehydrate, and in this case the heat flow, F, is given by (Ruiz, 2003):

$$F = \frac{k_a (T_{BDT} - T_s)}{z} \quad (15.1)$$

where k_a is the thermal conductivity of ammonia dehydrate ($NH_3 \cdot 2H_2O$); T_{BDT} is the temperature at the level of brittle–ductile transition, T_s is the surface temperature, and z is the maximum possible depth of the grabens (an elongated block of Triton's crust lying between two faults and displaced downwards relative to the blocks on either side, as in a rift valley) at Raz Fossae (a selected location on Triton).

In the absence of convection (the movement caused within a fluid by the tendency of hotter and therefore less dense material to rise, and colder, denser material to sink under the influence of gravity, which consequently results in the transfer of heat), the temperature profile of the outer ice layer is purely conductive (in the present case, having the property of conducting heat). If one assumes an $NH_3 \cdot 2H_2O$ ice layer, then the depth to the ocean top would be given by (Ruiz, 2003):

$$z_{ocean} = \frac{k_a (T_m - T_s)}{F} \quad (15.2)$$

where T_m is the melting temperature of $NH_3 \cdot 2H_2O$ ice, ~176 K (e.g., Kargel, 1992), a value that varies slightly with pressure (e.g., Hogenboom et al., 1997). In the case of a water–ice layer (of variable thermal conductivity), the depth of the ocean top is given by (Ruiz, 2003):

$$z_{ocean} = \frac{k_0}{F} \ln\left(\frac{T_m}{T_s}\right) \quad (15.3)$$

Although T_m is pressure-dependent, given the low pressures of Triton's lithosphere, a constant value of 273 K was considered for T_m.

Ross and Kargel (1998) reported that, on Triton, ices other than water–ice show low thermal conductivity. Fig. 15.4A shows surface heat flow deduced from the depth of Raz Fossae as a function of effective surface temperature. Ruiz (2003) noted that heat flow diminishes with T_s because the difference between T_s and T_{BDT} becomes smaller. Ruiz (2003) further noted that for surface heat flow to be consistent with the predicted value of radiogenic heating, the effective surface temperature needs to be increased to at least ~90 K (for $NH_3 \cdot 2H_2O$) or to ~120–130 K (for H_2O).

Similarly, Fig. 15.4B shows the depth to an internal ocean as a function of effective surface temperature. The increase in T_s raises the interior temperature, but this effect is offset by the reduction in the value of F necessary to put the brittle–ductile transition to the base of the Raz Fossae. Hence, for a conductive ice shell, the depth to a possible ocean increases with T_s. Given the overall thickness of Triton's ice layer is estimated at ~350–400 km (Smith and team, 1989; McKinnon et al., 1995), Ruiz (2003) argued that the existence of an internal ocean seems inevitable in the absence of convection in the outer ice layer. Moreover, according to Ruiz (2003), if the effective surface temperature is not increased by the insulating layer to the extent of several dozen degrees, the ocean would lie at a depth of ~20–30 km beneath the surface.

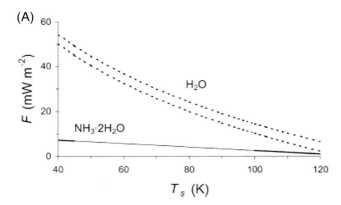
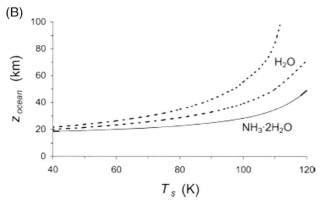

FIG. 15.4 (A) Surface heat flow necessary for putting the brittle–ductile transition at the base of Raz Fossae, as a function of the effective surface temperature for a strain rate of 10^{-15}/s. The dotted lines indicate upper and lower limits for an H_2O lithosphere and the solid line a $NH_3 \cdot 2H_2O$ lithosphere. (B) Depth to a possible internal ocean underlying a conductive outer ice layer as a function of the effective surface temperature. The increase in T_s raises the interior temperature but this effect is offset by the diminished value of F. The dotted lines indicate upper and lower limits for an H_2O lithosphere and the solid line a $NH_3 \cdot 2H_2O$ lithosphere. *(Source: Ruiz, J. (2003), Heat flow and depth to a possible internal ocean in Triton, Icarus, 166: 436–439). Elsevier publication.*

The steady-state model (i.e., assuming steady-state internal structure for Triton) of Hussmann et al. (2006) indicates that a 135–190 km thick ocean likely exists in Triton's interior. According to Gaeman et al. (2012), if an ocean on Triton formed approximately 4 Ga ago, the presence of volatiles may contribute significantly to whether or not the ocean can be sustained until present.

According to Agnor et al. (2009), "Determining whether Triton has an ocean, and whether any of its oceanic chemistry is expressed on its surface or in its atmosphere should be a high scientific priority. Triton's geysers (identified in Voyager 2 flyby data) and significant nitrogen and hydrogen atmosphere put Triton into the rare class of moons with a substantial atmosphere and active geology. Studying Triton's geysers and atmosphere would provide clues on interior composition, and provide a key comparison to Pluto, and to other KBOs that may have atmospheres. Finally, Triton's internal energy source and possible liquid ocean make it an attractive astrobiological target for exploration." Based on appropriate calculations; these investigators consider that when the Raz Fossae formed, it is probable there was an ocean on Triton lying at a depth of ∼15–30 km.

A scientific question that plagued the minds of several planetary researchers was whether an ocean formed early during Triton's evolution can be sustained until present within Triton's interior? Gaeman et al. (2012) suggested that "to determine if an ocean has survived, it is necessary to explore the coupled thermal–structural evolution." Gaeman et al (2012) argued that the evolution of tidal dissipation due to an increase in the thickness of the ice shell and variation of orbital eccentricity may exert important influence in the evolution of the icy moons in general, and Triton in specific. Gaeman et al (2012) further suggested that tidal dissipation within a coupled ice-shell–ocean system, with a moving boundary or crystallization front varies spatially and temporally. Therefore, it was found necessary that to study the retention of an ocean in Triton's interior it is required to quantify the heat transfer in the coupled system with a robust description of tidal dissipation within Triton's interior. Gaeman et al (2012)'s theoretical investigations addressed these issues in the coupled ice-shell–ocean model with a time-dependent ice shell thickness. With a view to modeling the evolution of an ocean and to identify the conditions under which the ocean can be sustained over Triton's history, Gaeman et al (2012)'s theoretical work coupled the thermal, structural, and orbital evolution of Triton, assuming tidal dissipation occurs entirely within the ice shell, while radiogenic heating from the silicate fraction of the core is transmitted to the base of the ice shell by the convecting ocean. Gaeman et al (2012) summarized their study results pertaining to Triton's inner ocean thus, "Despite being several orders of magnitude higher than the tidal dissipation, radiogenic heating alone fails to sustain an ocean within Triton over 4.5 Ga. For orbital eccentricities of 5×10^{-7} and 3×10^{-5} it takes approximately 2 Ga and 3 Ga, respectively, to completely freeze the ocean. For higher values of orbital eccentricities, an ocean can be sustained in Triton's interior over 4.5 Ga. If Triton's history past circularization involves a slow decrease in orbital eccentricity to the current value, a thin, possibly NH_3-rich ocean exists beneath Triton's icy shell". Note that orbital circularization of a celestial body simply means that its orbital eccentricity has become zero.

Nimmo and Spencer (2015) proposed that radiogenic heating is sufficient to maintain a long-lived ocean beneath the conductive ice shell of a celestial body such as Triton. Their investigations revealed that Triton's high inclination likely causes a significant (≈0.7°) obliquity, resulting in large heat fluxes due to tidal dissipation in its subsurface ocean. Their calculations took into account a rather conflicting requirement of convection placing an upper bound on the ice shell viscosity, while the requirement for yielding imposes a lower bound. Nimmo and Spencer (2015) found that both bounds can be satisfied with an ocean temperature ≈240 K for their nominal temperature-viscosity relationship, suggesting the presence of antifreeze such as ammonia (NH_3) in Triton's subsurface ocean to prevent solidification. Note that the presence of antifreeze in Triton's subsurface ocean has been suggested by Ruiz (2003) as well. According to Prockter et al. (2019), "confirmation of the presence of an ocean would establish Triton as arguably the most exotic and probably the most distant ocean world in the solar system, potentially expanding the habitable zone."

If an inner ocean is present within the ice shell of Triton, Prockter et al. (2019) sought to determine its properties and whether the ocean interacts with the surface environment. To address these questions, they proposed another spacecraft mission to Triton, with an onboard focused instrument suite consisting of:

(1) A magnetometer, primarily for detection of the presence of an induced magnetic field which would indicate compellingly the presence of an ocean.
(2) A camera, for imaging of the mostly unseen anti-Neptune hemisphere, and repeat imaging of the sub-Neptune hemisphere to look primarily for signs of change.
(3) An infrared imaging spectrometer with spectral range up to 5 μm, suitable for detection and characterization of surface materials at the scales of Triton's features. In addition to the Triton science, opportunistic Neptune science data will also be acquired (Prockter et al., 2019).

To summarize, Neptune's moon *Triton* is considered to be the highest priority candidate ocean world to target in the near term. This 135–190 km thick inner ocean is existing at a depth of ~20-30 km beneath Triton's icy surface (Ross and Schubert, 1990; Ruiz, 2003; Hussmann et al., 2006; Agnor et al., 2009; Gaeman et al., 2012; Nimmo and Spencer, 2015; Prockter et al., 2019).

15.11 The Year-2019 discovery of *Triton* fostering a rare mixing of CO and N_2 ice molecules—shedding more light on Triton's geysers

Carbon monoxide (CO) and nitrogen (N_2) are the most volatile species on Triton and Pluto {methane (CH_4) is a distant third}, and so dominate seasonal volatile transport across their surfaces. It is entirely reasonable to expect CO and N_2 molecules to mix throughout the ice. CO and N_2 molecules have similar sizes, shapes, and masses. They have similar volatility (N_2 is somewhat more volatile than CO), and they are unusual in that they are fully miscible in one another, in both liquid and solid phases.

While spectroscopically monitoring Triton (Grundy et al., 2010) and Pluto-Charon pair (Grundy et al., 2016) with a view to searching for changes on the volatile ices CH_4, CO, N_2, etc. on their surface as witnessed by changes in the infrared absorption bands of their surface ices, it has been suspected that the CO molecules are intimately mixed in the N_2 ice (i.e., CO and N_2 ice coexist) on these distant icy celestial bodies. However, clear spectroscopic evidence validating this suggestion was missing. Finding spectroscopic evidence first requires an understanding of the physical structure of pure N_2 ice and pure CO ice. In this context, Tegler et al. (2020) made the following observation. "Both species undergo a solid-solid phase transition between a higher temperature β-phase with an orientationally disordered hexagonal crystal structure and a lower temperature α-phase with an orientationally ordered cubic structure. In pure N_2, this α–β transition occurs at 35.61 K (Scott, 1976), while in pure CO it occurs at a much warmer 61.6 K (Barrett and Meyer, 1965). Since the two are fully miscible, the transition temperature can be expected to vary as a function of composition, and a phase diagram published by Angwin & Wasserman (1966) from x-ray diffraction shows exactly that."

While carrying out an extensive laboratory-based spectroscopic study of a sample of CO (4%) diluted in solid α-N_2 ice Quirico and Schmitt (1997) first detected a new and unidentified band at wavenumber 4467.3/cm (wavelength 2.239 μm). Considering the importance of ice phase on the spectroscopic signature of a CO/N_2 ice mixture, Tegler et al. (2020) decided to revisit the phase diagram, but instead of using transmission spectroscopy, they opted to use Raman spectroscopy. Raman spectroscopy was chosen in Tegler et al. (2020)'s studies because CO and N_2 both have strong Raman bands (Cahil and Leroi, 1969), so use of Raman spectroscopy to track phase changes in N_2/CO mixtures obviates difficulties associated with the weakness of N_2 absorption.

Tegler et al.'s laboratory results show that the band is strongest in samples (of CO/N_2 mixture) with near equal amounts of CO and N_2 and is not present in either pure CO or pure N_2 samples. In addition, they found that summing the frequencies of the CO (0-1) fundamental and the N_2 (0-1) fundamental agreed with the frequency of the band under study (i.e., wavenumber 4467.3/cm). Tegler et al. carried out additional laboratory-based isotope experiments involving different samples of CO/N_2 ice mixtures — one of $^{13}C^{16}O/^{14}N$ and the other of $^{12}C^{18}O/^{14}N$. In all samples, they found summing the frequencies of the CO and N_2 fundamentals agreed with the frequencies of the band under study. According to Tegler et al. (2020) these experiments indicate that photons are exciting adjacent CO and N_2 molecules to produce the observed band.

Tegler et al. (2020)'s experiments showed that the "unidentified band at 4467.3/cm" mentioned in Quirico and Schmitt (1997)'s studies "is due to the simultaneous vibrational excitation of a CO molecule and an adjacent N_2

molecule in the ice by a single photon". Tegler and coresearchers termed such a band a "two-molecule combination band," which is also known in the literature as a "dimol absorption band," dimol meaning "two-molecule."

Thus, Tegler et al. (2020) succeeded in demonstrating a very specific wavelength of infrared light absorbed when carbon monoxide (CO) and nitrogen (N_2) molecules join together and vibrate in unison. Individually, carbon monoxide and nitrogen ices each absorb their own distinct wavelengths of infrared light, but the tandem vibration of an ice mixture absorbs at an additional, distinct wavelength identified in this study. What Tegler et al. (2020) demonstrated was evidence of a combination band—one where a single photon simultaneously excites two adjacent molecules in a CO/N_2 ice. In particular, Tegler et al. (2020) presented near-infrared spectra of laboratory CO/N_2 ice samples where they identified a band at wavenumber 4467.5/cm (wavelength 2.239 μm) that results from single photons exciting adjacent pairs of CO and N_2 molecules (Fig. 15.5A). In more technical terms, "the combination bands result from photons that simultaneously excite the fundamentals of two adjacent molecules, N_2 and CO, rather than photons that excite two fundamentals in a single molecule (Fig. 15.5B)."

In addition to their laboratory experiments, Tegler and coresearchers obtained an 80-minute spectrum of Triton using the 8-meter Gemini Observatory Telescope in Chile (South America). Key to the discovery was the high-resolution spectrometer called IGRINS (Immersion Grating Infrared Spectrometer) which was built as collaboration between the University of Texas at Austin and the Korea Astronomy and Space Science Institute (KASI). Tegler and coresearchers recorded this same unique infrared signature showing clear evidence for the $CO-N_2$ two-molecule combination band at wavelength λ=2.239 μm (wavenumber = 4466.5/cm) on Triton.

Merlin et al. (2018) also had found the same band in their VLT spectrum of Triton. However, they did not identify the band. Tegler et al. (2019) argue that the existence of the band in a spectrum of Triton indicates that CO and N_2 molecules are intimately mixed in the surface ice rather than existing as separate spatially distinct regions of pure CO and pure N_2 deposits. Tegler et al. (2019)'s finding attains special significance because CO and N_2 are the most volatile species on Triton and so dominate seasonal volatile transport across its surface. The discovery team is optimistic that their result will place constraints on the interaction between the surface and atmosphere of Triton. According to Tegler, who led this study, "While the icy spectral fingerprint we uncovered was entirely reasonable, especially as this combination of ices can be created in the lab, pinpointing this specific wavelength of infrared light on another world is unprecedented" (Gemini Observatory Press Release, May 22, 2019).

Whereas carbon monoxide and nitrogen molecules exist as gases, not ices, in the Earth's atmosphere, on distant

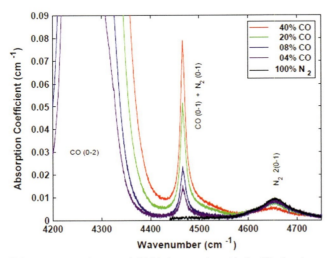

FIG. 15.5a (A) Spectra of CO/N_2 ice samples with the CO abundance ranging from 0% to 40% and all taken at T = 60 K. The N_2 ice is in the β-phase. The spectra show the new band near 4467/cm. The new band is not present in the pure N_2 sample (black line) and increases in strength with increasing CO abundance. The saturated band at ~4252/cm is CO first vibrational overtone (0-2), that is, CO (0-2) and the weak, broad band at ~4654/cm is N_2. *(Source: Tegler et al. (2020), A new two-molecule combination band as diagnostic of carbon monoxide diluted in nitrogen ice on Triton, Astronomical Journal).*

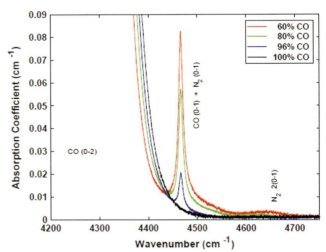

FIG. 15.5b Spectra of CO/N_2 ice samples with the CO abundance ranging from 60% to 100% and all taken at T = 60 K. The N_2-ice is in the β-phase. The spectra show the new band near 4467/cm. Its band strength at a CO abundance of 60% is nearly the same as its strength at 40% (see panel 15.5a) and then decreases in strength with increasing CO abundance. The band is not present in the pure CO ice sample. A maximum strength for samples with nearly equal amounts of CO and N_2 and absence in pure N_2 and pure CO reinforces the idea that the band is due to photons exciting adjacent pairs of CO and N_2 molecules. *(Source: Tegler et al. (2020), A new two-molecule combination band as diagnostic of carbon monoxide diluted in nitrogen ice on Triton, Astronomical Journal).*

Triton carbon monoxide and nitrogen freeze and exist as solid ices. They can form their own independent ices, or can condense together in the icy mix detected in the Gemini data.

It is already known that despite Triton's large distance from the Sun and the extremely cold temperatures existing on Triton, the weak sunlight is enough to drive strong seasonal changes on Triton's surface and atmosphere (Kirk et al., 1990). Seasons are known to progress slowly on Triton (a season on Triton lasts a little over 40 years), as Neptune takes 165-Earth years to orbit the Sun.

Subsequent to the discovery of geyser-like particle plumes on Triton, several theories have been proposed to explain the possible sources of the erupted material. There exists a thin layer of volatile ice on Triton's Sun-facing surface, potentially involving the mixed carbon monoxide and nitrogen ice revealed by the Gemini observation. The researchers associated with Gemini Observatory study argue that the just-mentioned CO/N_2 icy mix on Triton's surface could be involved in Triton's iconic geysers first seen in Voyager 2 spacecraft images as dark, windblown streaks on Triton's surface. They further argued that the geysers may erupt when the summertime Sun heats this thin layer of volatile ice on Triton's surface, potentially involving the mixed carbon monoxide and nitrogen ice revealed by the Gemini observation. Furthermore, that ice mixture could also migrate around the surface of Triton in response to seasonally varying patterns of sunlight. The discovery made by the Gemini Observatory study team is considered to offer insights into how the volatile mixture of carbon monoxide and nitrogen ices can transport material across the moon's surface via geysers, trigger seasonal atmospheric changes, and provide a context for conditions on other distant, icy worlds.

The Gemini Observatory study team recalls astronomers having suspected that the mixing of carbon monoxide and nitrogen ice exists not only on Triton, but also on Pluto, where the *New Horizons* spacecraft found the two ices coexisting. This Gemini finding is claimed to be the first direct spectroscopic evidence of these ices mixing and absorbing this type of light on either world. According to the research team, "This work demonstrates the power of combining laboratory studies with telescope observations to understand complex planetary processes in alien environments so different from what we encounter every day here on Earth."

15.12 *Triton* in the limelight as a high-priority target under NASA's "Ocean Worlds" program

Triton is deemed the highest priority target to address as part of an Ocean Worlds Program mooted by NASA. This priority is given based on the extraordinary hints of

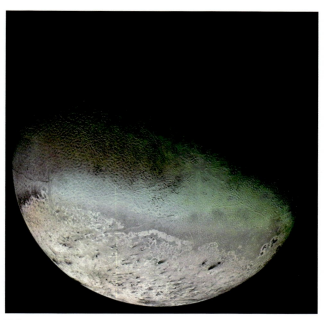

FIG. 15.6 **Voyager 2 image of Neptune's moon Triton.** *Image credit: NASA. (Source: Hendrix, A.R., T.A. Hurford, L.M. Barge, M.T. Bland, J.S. Bowman, W. Brinckerhoff, B.J. Buratti, M.L. Cable, J. Castillo-Rogez, G.C. Collins, S. Diniega, C.R. German, A.G. Hayes, T. Hoehler, S. Hosseini, C.J.A. Howett, A.S. McEwen, C.D. Neish, M. Neveu, T.A. Nordheim, G. W. Patterson, D.A. Patthoff, C. Phillips, A. Rhoden, B.E. Schmidt, K.N. Singer, J.M. Soderblom, and S.D. Vance (2019), The NASA Roadmap to Ocean Worlds, Astrobiology, 19(1): 1–27). Copyright info: © Amanda R. Hendrix and Terry A. Hurford et al., 2018; Published by Mary Ann Liebert, Inc.*

activity shown by the Voyager spacecraft (e.g., plume activity; smooth, walled plains units; the cantaloupe terrain suggestive of convection) (Fig. 15.6) and the potential for ocean-driven activity given by Cassini results at Saturn's icy moon Enceladus. Triton ocean confirmation and characterization are now in the horizon. Although the source of energy for Triton's activity remains unclear, all active bodies in the Solar System are driven by endogenic heat sources, and Triton's activity coupled with the young surface age makes investigation of an endogenic source important (Hendrix et al., 2019). Planetary scientists think that great science is possible with a Neptune-Triton flyby. Description of such missions are given in the Argo white paper by Hansen et al. (2009) and the Ice Giants SDT report by Hofstadter et al. (2017).

15.13 Forthcoming mission to map *Triton*, characterize its active processes, and determine the existence of the predicted subsurface ocean—TRIDENT Mission

NASA's Voyager 2 mission showed that Triton has active resurfacing—generating the second youngest surface in the solar system—with the potential for erupting plumes and an atmosphere. Coupled with an ionosphere that can create organic snow and the potential for an interior ocean, Triton is an exciting exploration target to understand how habitable worlds may develop in our solar system and others (Hendrix et al., 2019).

Hendrix et al. (2019) have suggested that similar to KBO and Pluto studies, Triton investigations will benefit from improved understanding of the rheology of N_2 ice and water–ice mixtures (water-ammonia, methane clathrates) at the temperatures expected on these targets. In addition, lab work to understand the spectral signatures of mixtures of ices in solid solution would be beneficial as well. Hendrix et al. (2019) have further suggested that, specifically, phase diagrams for mixtures of species such as N_2, CH_4, CO, and CO_2 are needed as well as optical constants in the UV and IR.

Because Triton is in the limelight as a high-priority target to address as part of NASA's "Ocean Worlds Program", it is necessary to unambiguously determine whether Triton has an internal ocean. According to Hendrix et al. (2019), this could be accomplished by looking for the magnetic induction signature and/or gravity field measurements and/or searching for libration with high-resolution images and/or LIDAR. It has been suggested that a Triton orbiter or Neptune orbiter with multiple Triton flybys (with magnetometer, gravity, thermal imagery, high-resolution imagery) would address the goals of a Triton ocean mission. It is also necessary to determine whether Triton's plumes sample a subsurface liquid layer. To achieve this goal, Hendrix et al. (2019) suggested that high-resolution images to look for ongoing eruptions are the minimal data set, but additional information on composition is important. They further suggested that the composition of the plumes could be assessed by using UV spectroscopic imaging of solar occultations from a spacecraft at the Neptune–Triton system. Furthermore, the UV spectra can be used to derive density profiles of different species in the atmosphere. If the solar occultation probed a plume, then we would have direct information about the composition of the plume and therefore the subsurface source. Mass spectrometry could also be useful (Hendrix et al. (2019).

Trident is a mission under consideration by NASA that would send a spacecraft to explore Triton. The Trident concept was proposed in March 2019 to NASA's Discovery Program. The mission concept is supported by NASA's Ocean Worlds Exploration Program and it is intended to help answer some of the questions generated by Voyager 2's flyby in 1989 (Hendrix et al., 2019). Trident would explore Triton, to understand pathways to habitable worlds at tremendous distances from the Sun. Using a single flyby, Trident would map Triton, characterize active processes, and determine whether the predicted subsurface ocean exists. Fig. 15.7 provides a pictorial illustration of the proposed Trident Mission concept. Trident aims to

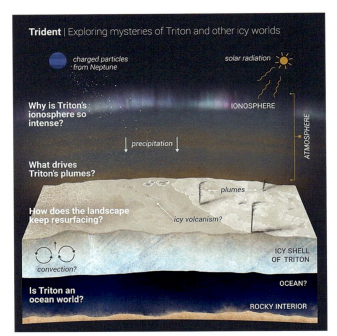

FIG. 15.7 Pictorial illustration of the proposed Trident Mission to Neptune and Triton PIA23874: Trident: Exploring the Mysteries of Triton https://photojournal.jpl.nasa.gov/catalog/PIA23874. *(Source: https://en.wikipedia.org/wiki/Trident_(spacecraft)). Author: NASA/JPL-Caltech Copyright information: This file is in the **public domain** in the United States because it was solely created by NASA. NASA copyright policy states that "NASA material is not protected by copyright **unless noted.**" (See Template: PD-USGov, NASA copyright policy page or JPL Image Use Policy.) (Copyright information source: https://commons.wikimedia.org/wiki/File:PIA23874-NeptuneMoonTriton-TridentMission-20200616.jpg).*

answer the questions outlined in the graphic illustration given in Fig. 15.7. The questions fall under Trident's three main goals. The first goal is to explore the factors that lead to a solar system body having the necessary ingredients—including water—to be habitable. A second goal is to explore vast, unseen lands. Most of what we know of the moon came from Voyager 2 data. But we have only seen 40% of the moon's surface; Trident would map most of the remainder. Trident's third major goal is to understand how Triton's surface keeps renewing itself. The surface is remarkably young, geologically speaking (possibly only 10 million years old in a 4.6-billion-year-old solar system), and has almost no visible craters.

References

Agnor, C., Barr, A. Bierhaus, B. et al., 2009. The exploration of Neptune and Triton, Submitted to the NRC 2009 planetary science decadal survey; NASA Science, Solar System Exploration; https://solarsystem.nasa.gov/studies/141/the-exploration-of-neptune-and-triton/.

Angwin, M.J., Wasserman, J., 1966. Nitrogen—carbon monoxide phase diagram. J. Chem. Phys. 44, 417–418.

Barrett, C.S., Meyer, L., 1965. Phase diagram of argon—carbon monoxide. J. Chem. Phys. 43, 3502.

Benner, L., McKinnon, W.B., 1995. Orbital evolution of captured satellites: the effect of solar gravity on Triton's post capture orbit. Icarus 114, 1–20.

Brada, M.P., Clarke, D.R, 1997. Thermodynamic approach to the wetting and dewetting of grain boundaries. Acta Mater 45, 2501–2508.

Brown, R.H., Johnson, T.V., Goguen, J.D., Schubert, G., Ross, M.N., 1991. Triton's global heat budget. Science 251 (5000), 1465–1467.

Brown, R.H., Kirk, R.L., 1994. Coupling of volatile transport and internal heat flow on Triton. J. Geophys. Res. 99, 1965–1981.

Brown, R.H., Cruikshank, D.P., Veverka, J., Helfenstein, P., Eluszciewicz, J., 1995. Surface composition and photometric properties of Triton. In: Cruikshank, D.P. (Ed.), Neptune and Triton. Univ. of Arizona Press, Tucson, pp. 991–1030.

Buratti, B., Hicks, M.D., Tryka, K.A., Newburn Jr., R.L., 1998. Observation of a reddening of Triton's spectrum. Bull. Am. Astron. Soc. 30, 1107.

Buratti, B., Hicks, M.D., Newbum Jr., R.L., 1999. Does global warming make Triton blush? Nature 397, 219.

Craig, B., Agnor, C., Hamilton, D.P., 2006. Neptune's capture of its moon Triton in a binary–planet gravitational encounter. Nature 441, 192–194.

Croft, S.K., et al., 1995. Geology of Triton. In: Cruikshank, D.P., Matthews, M.S. (Eds.), Neptune and Triton. Univ. of Ariz. Press, Tucson, pp. 879–948.

Croft, S.K., Kargel, J.S., Kirk, R.L., Moore, J.M., Schenk, P.M., and Strom, R.G. (1995), The geology of Triton. In: Cruikshank, D.P. (Ed.), Neptune and Triton. Univ. of Arizona Press, Tucson, pp. 879–947.

Cruikshank, D.P., Roush, T.L., Owen, T.C., Geballe, T.R., de Bergh, C., Schmitt, B., Brown, R.H., Bartholomew, M.J., 1993. Ices on the surface of Triton. Science 261, 742–745.

Cruikshank, D.P., Brown, R.H., Calvin, W.M., Roush, T.L., Bartholomew, M.J., 1998. Ices on the satellites of Jupiter, Saturn, and Uranus. In: Schmitt, B., Festou, M., deBergh, C. (Eds.), Solar System Ices. Kiuwer Academic Pub., Dordrecht, pp. 579–606.

Cruikshank, D.P., Schmitt, B., Roush, T.L., Owen, T.C., et al., 2000. Water ice on Triton. Icarus 147, 309–316.

Cuk, M., Gladman, B.J., 2005. Constraints on the orbital evolution of Triton. Astrophys. J. 626, L113–L116.

Duxbury, N.S., Brown, R.H., 1993. The phase composition of Triton's Polar caps. Science 261 (5122), 748–751.

Elliot, J.L., Hammel, H.B., Wasserman, L.H., Franz, O.G., McDonald, S.W., Person, M.J., Olkin, C.B., Dunham, E.W., Spencer, J.R., Stansberry, J.A., Buie, M.W., Pasachoff, J.M., Babcock, B.A., McConnochie, T.H., 1998. Global warming on Triton. Nature 393, 765–767.

Eviatar, A., Vasyliūnas, V.M., Richardson, J.D., 1995. Plasma temperature profiles in the magnetosphere of Neptune. J. Geophys. Res. 100 (A10), 19551–19557.

Fink, U., Larson, H.P., 1975. Temperature dependence of the water-ice spectrum between 1 and 4 microns: Application to Europa, Ganymede and Saturn's rings. Icarus 24, 411–420.

Fink, U., Sill, G.T., 1982. The infrared spectral properties of frozen volatiles. In: Wilkening, L.L. (Ed.), Comets. Univ. of Arizona Press, Tucson, pp. 164–202.

Gaeman, J., Hier-Majumder, S., Roberts, J.H., 2012. Sustainability of a subsurface ocean within Triton's interior. Icarus 220 (2), 339–347.

Gaidos, E., Nimmo, F., 2000. Tectonics and water on Europa. Nature 405, 637.

Goldreich, P., et al., 1989. Neptune's story. Science 245, 500.

Grundy, W.M., Schmitt, B., 1998. The temperature-dependent near-infrared absorption spectrum of hexagonal H_2O ice. J. Geophys. Res. 103 (25), 809-25,822.

Grundy, W.M., Young, L.A., Stansberry, J.A., et al., 2010. Near-infrared spectral monitoring of Triton with IRTF/SpeX II: Spatial distribution and evolution of ices. Icarus 205, 594.

Grundy, W.M., Binzel, R.P., Buratti, B.J., et al., 2016. Surface compositions across Pluto and Charon. Science 351, 9189.

Hansen, C.J., Paige, D.A., 1992. A thermal model for the seasonal nitrogen cycle on Triton. Icarus 99, 273–288.

Hansen, C.J., Hammel, H.B., and Spilker, L.J., 2009. Argo: Exploring the Neptune System and Beyond, European Planetary Science Congress 2009, 14–18 September in Potsdam, Germany, p.796. http://meetings.copernicus.org/epsc2009

Hendrix, A.R., Hurford, T.A., Barge, L.M., Bland, M.T., Bowman, J.S., Brinckerhoff, W., Buratti, B.J., Cable, M.L., Castillo-Rogez, J., Collins, G.C., et al., 2019. The NASA Roadmap to Ocean Worlds. Astrobiology 19 (1), 1–27. doi:10.1089/ast.2018.1955.

Hofstadter, M., Simon, A., Atreya, S., Banfield, D., Fortney, J., Hayes, A., Hedman, M., Hospodarsky, G., Mandt, K., Masters, A., Showalter, M., Soderlund, K., Turrini, D., Turtle, E.P., Elliott, J., and Reh, K. (2017), A vision for ice giant exploration. Planetary Science Vision 2050 Workshop, 27–28, February and 1 March, 2017 in Washington, DC. LPI Contribution No. 1989, id.8115.

Hogenboom, D.L., Kargel, J.S., Consolmagno, G.J., Holden, T.C., Lee, L., Buyyounouski, M., 1997. The ammonia–water system and the chemical differentiation of icy satellites. Icarus 126, 171–180.

Hussmann, H., Sohl, F., Spohn, T., 2006. Subsurface oceans and deep interiors of medium-sized outer planet satellites and large trans-neptunian objects. Icarus 185, 258–273.

Ingersoll, A.P., Tryka, K.A., 1990. Triton's plumes: the dust devil hypothesis. Science 250 (4979), 435–437.

Jenniskens, P., Blake, D.F., Wilson, M.A., Pohorille, A., 1995. High-density amorphous ice, the frost on interstellar grains. Ap. J. 455, 389–401.

Jenniskens, P, Blake, D.F., 1996. Crystallization of amorphous water ice in the solar system. Ap. J. 473, 1104–1113.

Kargel, J.S., 1992. Ammonia–water volcanism on icy satellites: phase relations at 1 atmosphere. Icarus 100, 556–574.

Kerr, R.A., 1990. Geysers or dust devils on Triton?. Science 250 (4979), 377. doi:10.1126/science.250.4979.377.

Kirk, R.L., Brown, R.H., Soderblom, L.A., 1990. Subsurface energy storage and transport for solar-powered geysers on Triton. Science 250 (4979), 424–429.

Krasnopolsky, V.A., Cruikshank, D.P., 1995. Photochemistry of Triton's atmosphere and ionosphere. J Geophys Res 100 (E10), 21271–21286.

Lyons, J.R., Yung, Y.L., Allen, M., 1992. Solar control of the upper atmosphere of Triton. Science 256, 204–206.

McCord, T.B., 1966. Dynamical evolution of the Neptunian system. Astron. J. 71, 585–590.

McKinnon, W.B., and Kirk, R.L., 2007. Triton. In: Lucy-Ann McFadden, L.A., Weissman, P., Johnson, T. (Eds.), Encyclopedia of the Solar System. Academic Press, 483–502.

McKinnon, W.B., Lunine, J.I., Banfield, D., 1995. Origin and evolution of Triton. In: Cruikshank, D.P., Matthews, M.S. (Eds.), Neptune and Triton. Univ. of Ariz. Press, Tucson, pp. 807–877.

Merlin, F., Lellouch, E., Quirico, E., Schmitt, B., 2018. Triton's surface ices: distribution, temperature and mixing state from VLT/SINFONI observations. Icarus 314, 274.

Nimmo, F., Gaidos, E., 2002. Strike-slip motion and double ridge formation on Europa. J. Geophys. Res. 107 (E4), 5021. doi:10.1029/2000JE001476.

Nimmo, F., Spencer, J.R., 2015. Powering Triton's recent geological activity by obliquity tides: implications for Pluto geology. Icarus 246, 2–10.

Prockter, L.M., Nimmo, F., Pappalardo, R.T., 2005. A shear heating origin for ridges on Triton. Geophys. Res. Letters 32 (14) L14202 (4 pages), doi:10.1029/2005GL022832.

Prockter, L.M., Mitchell, K.L., Howett, C.J.A., Smythe, W.D., Sutin, B.M., Bearden, D.A., Frazier, W.E., 2019. Exploring Triton with Trident: A Discovery Class Mission. In: Proceedings of the 50th Lunar and Planetary Science Conference, LPI, Woodlands, TX, USA, 18–22 March 2019..

Quirico, E., Schmitt, B., 1997. A spectroscopic study of CO diluted in N_2 ice: applications for Triton and Pluto. Icarus 128, 181.

Quirico, E., et al., 1999. Composition, physical state, and distribution of ices at the surface of Triton. Icarus 139, 159–178.

Ross, M.N., Schubert, G., 1990. The coupled orbital and thermal evolution of Triton. Geophys. Res. Lett. 17, 1749–1752.

Ross, R.G., Kargel, J.S., 1998. Thermal conductivity of ices with special reference to martian polar caps. In: Schmitt, B., de Bergh, C., Festou, M. (Eds.), Solar System Ices. Kluwer Academic, Dordrecht, pp. 33–62.

Ruiz, J., 2003. Heat flow and depth to a possible internal ocean in Triton. Icarus 166, 436–439.

Schenk, P.M., Jackson, M.P.A., 1993. Diapirism on Triton: a record of crustal layering and instability. Geology 21, 299–302.

Schenk, P., Sobieszczyk, S., 1999. Cratering asymmetries on Ganymede and Triton: from the sublime to the ridiculous. Bull. Am. Astron. Soc. (4), 31. Abstract 70.02.

Schenk, P.M., et al., 2004. Ages and interiors: the cratering record of the Galilean satellites. In: Bagenal, F. (Ed.), Jupiter: The Planet, Satellites and Magnetosphere. Cambridge Univ. Press, New York, NY, pp. 427–456.

Schenk, P.M., Zahnle, K., 2007. On the negligible surface age of Triton. Icarus 192 (1), 135–149.

Schmitt, B., Grim, R.J.A., Greenberg, J.M., 1989a. Spectroscopy and physicochemistry of $CO:H_2O$ and $CO_2:H_2O$ ices. In: Infrared Spectroscopy in Astronomy. Proc. 22 Eslab Symposium, Salamanca. ESA Spec. Pub. SP-290, 213–219.

Schmitt, B., Espinasse, S., Grim, R.J.A., Greenberg, J.M., Klinger, J., 1989b. Laboratory studies of cometary ice analogues. Proc. of International Workshop on Physics and Mechanics of Cometary Materials, Munster. ESA Spec. Pub. SP-302, 65–69.

Schmitt, B., Grim, R.J.A., Greenberg, J.M., Klinger, J., 1992. Crystallization of water rich amorphous mixtures. In: N., Maeno, Hondoh, T. (Eds.), Physics and Chemistry of Ice. Hokkaido Univ. Press, Sapporo, pp. 344–348.

Schmitt, B., Quirico, E., Trotta, F., Grundy, W.M., 1998. Optical properties of ices from UV to infrared. In: Schmitt, B., Festou, M., deBergh, C. (Eds.), Solar System Ices. Kluwer Academic Pub., Dordrecht, pp. 199–240.

Scott, T.A., 1976. Solid and liquid nitrogen. Phys. Rep. 27, 89.

Smith, B.A., the Voyager Imaging Team, 1989. Voyager 2 at Neptune: imaging science results. Science 246, 1422–1449.

Soderblom, L.A., Kieffer, S.W., Becker, T.L., Brown, R.H., Cook 2nd, A.F., Hansen, C.J., Johnson, T.V., Kirk, R.L., Shoemaker, E.M., 1990. Triton's geyser-like plumes: discovery and basic characterization. Science 250 (4979), 410–415.

Spencer, J.R., 1990. Nitrogen frost migration on Triton: a historical model. Geophys. Res. Lett. 17, 1769–1772.

Spencer, J.R., Moore, J.M., 1992. The influence of thermal inertia on temperatures and frost stability on Triton. Icarus 99, 261–272.

Stempel, M., et al., 2004. Combined effects of diurnal and nonsynchronous surface stresses on Europa. Proc. Lunar. Planet. Sci. Conf., [CD-ROM] 35, 2061.

Stern, S.A., McKinnon, W.B., 1999. Triton's surface age and impactor population revisited (evidence for an internal ocean), Proc. Lunar Planet. Sci. Conf. 30th. Abstract, 1766.

Stern, S.A., McKinnon, W.B., 2000. Triton's surface age and impactor population revisited in light of Kuiper Belt fluxes: Evidence for small Kuiper Belt objects and recent geological activity. Astron. J. 119, 945–952.

Summers, M.E., Strobel, D.F., 1991. Triton's atmosphere—a source of N and H for Neptune's magnetosphere. Geophys. Res. Lett. 18 (12), 2309–2312.

Tegler, S.C., Stufflebeam, T.D., Grundy, W.M., Hanley, J., Dustrud, S., Lindberg, G.E., Engle, A., Dillingham, T.R., Matthew, D., Trilling, D., Roe, H., Llama, J., Mace, G., Quirico, E., 2019. A new two-molecule combination band as a diagnostic of carbon monoxide diluted in nitrogen ice on Triton. The Astronomical Journal 158 (8pp), 17. https://doi.org/10.3847/1538-3881/ab199f.

Trusheim, F., 1960. Mechanism of salt migration in northern Germany. Assoc. Pet. Geol. Bull. 44, 1519–1540.

Zahnle, K., Schenk, P., Levison, H., Dones, L., 2003. Cratering rates in the outer solar system. Icarus 163, 263–289.

Bibliography

Agnor, C.B., Hamilton, D.P., 2006. Neptune's capture of its moon Triton in a binaryplanet gravitational encounter. Nature 441, 192–194.

Brown, R.H., Cruikshank, D.P., 1997. Determination of the composition and state of icy surfaces in the outer solar system. Annu. Rev. Earth Planet Sci. 25, 243–277.

Choukroun, M., Grasset, O., 2007. Thermodynamic model for water and high-pressure ices up to 2.2 GPa and down to the metastable domain. J. Chem. Phys. 127, 124506-1–124506-11.

Choukroun, M., Grasset, O., 2010. Thermodynamic data and modeling of the water and ammonia–water phase diagrams up to 2.2 GPa for planetary geophysics. J. Chem. Phys. 133, 144502-1–144502-13.

Clark, R.N., F.P. Fanale, and M.J. Gaffey (1986), Surface Composition of Satellites. In Satellites (J. Burns and M. S. Matthews, Eds.), pp 437-491, Univ. of Arizona Press, Tucson.

Craig, B., Agnor, 1., Hamilton, D.P., 2006. Neptune's capture of its moon Triton in a binary–planet gravitational encounter. Nature 441, 192–194.

Cruikshank, D.P., Brown, R.H., Clark, R.N., 1984. Nitrogen on Triton. Icarus 58, 293–305.

Cruikshank, D.P., Roush, T.L., Owen, T.C., Quirico, E., deBergh, C., 1998. The surface compositions of Triton, Pluto, and Charon. Ices on the satellites of Jupiter, Saturn, and Uranus. In: Schmitt, B., Festou, M., deBergh, C. (Eds.), Solar System Ices. Kluwer Academic Pub., Dordrecht, pp. 655–684.

Davies, J.K., Roush, T.L., Cruikshank, D.P., Geballe, T.R., Bartholomew, M.J., Owen, T.C., 1997. Detection of ice on comet C/1995 O1 (Hale-Bopp). Icarus 127, 238–245.

Elliot, J.L., Stansberry, J.A., Olkin, C.B., Agner, M.A., Davies, M.E., 1997. Triton's distorted atmosphere. Science 278, 436–439.

Elliot, J.L., Hammel, H.B., Wasserman, L.H., Franz, O.G., McDonald, S.W., Person, M.J., Olkin, C.B., Dunham, E.W., Spencer, J.R., Stansberry, J.A., Buie, M.W., Pasachoff, J.M., Babcock, B.A., Mc-

Connochie, T.H., 1998. Global warming on Triton. Nature 393, 765–767.

Eluszkiewicz, J., 1991. On the microphysical state of the surface of Triton. J. Geophys.Res. 96 (19), 217-19,229.

Ingersoll, A.P., 1990. Dynamics of Triton's atmosphere. Nature 344, 315–317.

Lellouch, E., Crovisier, J., Lim, T., Bockel6e-Morvan, D., Leech, K., Hanner, M.S., Altieri, B., Schmitt, B., Trotta, F., Keller, H.U., 1998. Evidence for water ice and estimate of dust production rate in comet Hale-Bopp at 2.9 AU from the Sun. Astron. Astrophys 339, L9-L 12.

Marouf, E.A., Tyler, G.L., Eshleman, V.R., Rosen, P.A., 1991. Voyager radio occultation of Triton: surface topography and radius. Bull. Am. Astron. Soc. 23, 1207.

Olkin, C.B., et al., 1997. The structure of Triton's atmosphere: Results from the entire ground-based occultation data set. Icarus 129, 178–201.

Quirico, E., Douté, S., Schmitt, B., de Bergh, C., Cruikshank, D.P., Owen, T.C., Geballe, T.M., Roush, T.L., 1999. Composition, physical state, and distribution of ices at the surface of Triton. Icarus 139, 159–178.

Rubincam, D.P., 2003. Polar wander on Triton and Pluto due to volatile migration. Icarus 163, 469–478.

Schenk, P., Jackson, P.A, 1993. Diapirism on Triton: a record of crustal layering and instability. Geology 21, 299–302.

Schenk, P.M., Zahnle, K., 2007. On the negligible surface age of Triton. Icarus 192, 135–149.

Stern, S.A., McKinnon, W.B., 2000. Triton's surface age and impactor population revisited in light of Kuiper belt fluxes: evidence for small Kuiper belt objects and recent geological activity. Astron. J. 119, 945–952.

Tegler, S.C., Stufflebeam, T.D., Grundy, W.M., Hanley, J., Dustrud, S., Lindberg, G.E., Engle, A., Dillingham, T.R., Matthew, D., Trilling, D., Roe, H., Llama, J., Mace, G., Quirico, E., 2019. A new two-molecule combination band as a diagnostic of carbon monoxide diluted in nitrogen ice on Triton. Astronomical J. 158 (8pp), 17. https://doi.org/10.3847/1538-3881/ab199f.

Trafton, L., 1984. Large seasonal variations on Triton. Icarus 58, 312–324.

Tsui, K.H., 2002. Satellite capture in a four-body system. Planet. Space Sci. 50, 269–276.

Tyler, G.L., et al., 1989. Voyager radio science observations of Neptune and Triton. Science 246, 1466–1473.

Chapter 16

Subsurface ocean of liquid water on Pluto

16.1 Pluto and its five moons

Dwarf planet Pluto may be the best known of the larger objects in the Kuiper Belt. The Kuiper belt is a disc-shaped region beyond Neptune that extends from about 30–55 astronomical units (compared to Earth which is one astronomical unit, or AU, from the Sun). This distant region is probably populated with hundreds of thousands of icy bodies larger than 100 km (62 miles) across and an estimated trillion or more comets.

Comets from the Kuiper Belt take less than 200 years to orbit the Sun and travel approximately in the plane in which most of the planets orbit the Sun. Objects in the Kuiper Belt are presumed to be remnants from the formation of the Solar System about 4.6 billion years ago. Kuiper belt is named after the astronomer Gerard Kuiper, who published a scientific paper in 1951 that speculated about objects beyond Pluto (see Kuiper, 1951).

Telling the tale of Pluto's discovery, Cruikshank and Sheehan (2018) recount the grand story of our unfolding knowledge of the outer Solar System, from William Herschel's serendipitous discovery of Uranus in 1781, to the mathematical prediction of Neptune's existence, and to Percival Lowell's studies of the wayward motions of those giant planets leading to his prediction of another world farther out. Lowell's efforts led to Clyde Tombaugh's heroic search and discovery of Pluto—then a mere speck in the telescope—at Lowell Observatory in 1930. Thus, Pluto—the first known Kuiper belt object (KBO) in the Solar System—was discovered by Clyde W. Tombaugh in 1930 and was originally considered to be the ninth planet from the Sun. Pluto has been a speck of light barely visible to the largest ground-based telescopes available in 1930, Tombaugh (Fig. 16.1) discovered this tiny celestial body by examining time-series photographs of stars for tiny changes; and he identified Pluto when its position on Jan 23, 1930 changed in a photograph taken a week later. Eleven-year-old Venetia Burney, on hearing of the discovery, suggested that the newly discovered planet should be named "Pluto" (Fig. 16.2) after the god of the underworld. It wasn't until 1988 that a chance alignment gave

FIG. 16.1 The astronomer Clyde Tombaugh, discoverer of Pluto here shown with his homemade 9-inch telescope. *(Source: https://commons.wikimedia.org/wiki/File:Clyde_W._Tombaugh.jpeg). This work is in the public domain because it was published in the United States between 1924 and 1963 and although there may or may not have been a copyright notice, the copyright was not renewed.*

astronomers a lucky break, when Pluto happened to cross in front of a distant star. As the starlight filtered through Pluto's atmosphere, scientists were able to disentangle the molecules there (including nitrogen, carbon monoxide and methane). Pluto was finally recognized as the premier body in the Kuiper Belt, the so-called third zone of our solar system. The first zone contains the terrestrial planets (Mercury through Mars) and the asteroid belt; the second, the gas-giant planets Jupiter through Neptune. The third

FIG. 16.2 **High-resolution MVIC image of Pluto in enhanced color to bring out differences in surface composition.** *(Source: https://commons.wikimedia.org/wiki/File:Pluto-01_Stern_03_Pluto_Color_TXT.jpg), Copyright info: Public Domain NASA / Johns Hopkins University Applied Physics Laboratory / Southwest Research Institute - http://www.nasa.gov/sites/default/files/thumbnails/image/crop_p_color2_enhanced_release.png (Converted to JPEG).*

zone, holding Pluto and the rest of the Kuiper Belt, is the largest and most populous region of the Solar System. As on 13-Dec-2021, *New Horizons* spacecraft is in the Kuiper Belt, where it will continue to collect data on Kuiper Belt objects and faraway worlds like Neptune and Uranus for the foreseeable future. Now well beyond Pluto, *New Horizons* spacecraft continues to wend its lonely way through the galaxy, transmitting data, thereby inspiring future generations to uncover more secrets of Pluto, the Solar System, and the Universe (Cruikshank and Sheehan, 2018).

During the early days of planetary research, there has been much speculation that Pluto is an escaped moon of Neptune. However, studies and modeling of the physical characteristics and composition of Pluto (McKinnon and Mueller, 1988; Tancredi and Fernandez, 1991) have led to the conclusion that it accumulated in a heliocentric orbit of ice-rich planetesimals in the outer reaches of the Solar Nebula rather than in a circum-planetary disk (Malhotra, 1995). In consonance with this, Malhotra (1993) proposed that Pluto formed in a near-circular coplanar heliocentric orbit beyond the orbits of the giant planets, but they differ in the physical and dynamical mechanisms that placed Pluto in its unusual orbit.

A phenomenon known as "orbital resonance" plays an important role in determining the orbit and inclination of some pairs or groups of celestial bodies, which are under mutual gravitational influences exerted by these bodies. In celestial mechanics, orbital resonance occurs when orbiting bodies exert regular, periodic gravitational influence on each other, usually because their orbital periods are related by a ratio of small integers. Most commonly this relationship is found for a pair of objects. Orbital resonances greatly enhance the mutual gravitational influence of the bodies (i.e., their ability to alter or constrain each other's orbits). Under some circumstances, a resonant system can be self-correcting and thus stable. The 2:3 resonance between the planets Pluto and Neptune (i.e., a resonance phenomenon in which for every two orbits that Pluto makes around the Sun, Neptune makes exactly three orbits around the Sun) is a simple example of orbital resonance interaction that controls their motions.

It is now recognized that Pluto is a dwarf planet in the Kuiper belt, which is a ring of bodies beyond Neptune. Malhotra (1995) examined the implications of "resonance capture" scenario for the origin of Pluto's orbit to understand the architecture of the Solar System beyond Neptune, that is, the "Kuiper Belt" of comets orbiting around the Sun approximately between Neptune's orbit and 50 AU. The numerical experiments reported by Malhotra (1995) indicate that the dynamical structure of the Kuiper Belt is dominated by concentrations of objects trapped in orbital resonances with Neptune, particularly at the 3:2 resonance and the 2:1 resonance.

Because Pluto rotates about its axis much more slowly than Earth does about its axis, a day on Pluto is approximately 6.4 Earth days or 153.3 h long (much longer than a day on Earth). Pluto's journey around the Sun is also slow (one orbit of Pluto around the Sun takes 248 Earth-years).

Pluto, orbiting in the outer periphery of the Solar System, is so distant from Earth that radio signals traveling at the speed of light (3×10^5 km/s) need more than 4 h and 20 min to reach Earth from Pluto. In contrast to Earth's surface gravity of 9.8 m/s^2, Pluto has a considerably smaller surface gravity (0.617 m/s^2; Stern et al., 2015). Pluto's mass is only two-thousandths of the Earth's mass (Malhotra, 1993).

Pluto's density is in the range 1.84 - 2.14 g/cm^3. Pluto is thus very rock-rich, with a rock/(rock + H_2O-ice) mass ratio of approximately 0.68–0.80, much greater than those of the icy moon Ganymede, Callisto or Titan. Pluto is so rock-rich that spontaneous un-mixing of rock and ice phases in its convecting interior is just possible. Pluto's closest structural cousin is Europa (McKinnon and Mueller, 1988)—the smallest of the four Galilean moons orbiting the planet Jupiter, and the sixth-closest to the planet of all the 79 known moons of Jupiter.

In 1996, the Hubble Space Telescope (which orbits around Earth with the ability to record images in wavelengths of light spanning from ultraviolet to near-infrared) helped scientists to finally see surface details of Pluto at a resolution of 500 km. Although fuzzy, the images revealed a world—at that time still defined as a planet—that had more large-scale contrast than any other in the Solar System, except Earth. It was found that Pluto's icy crust is wrinkled with mountains that are dusted with methane snow. Much of Pluto's terrain looks like snakeskin, rippling with gray and reddish-brown creases and pits.

The dwarf planet Pluto has five moons (Fig. 16.3) down to a detection limit of about 1 km in diameter (Kenyon et al.,2019). In order of distance from Pluto, they are Charon, Styx, Nix, Kerberos, and Hydra. The innermost and the largest of the five moons, Charon, is mutually tidally locked with Pluto, and is massive enough that Pluto–Charon is sometimes considered a double dwarf planet.

Charon was discovered by James Christy on 22 June 1978, nearly half a century after Pluto was discovered. Two additional moons (Nix and Hydra) were imaged on 15 May 2005 by astronomers of the Pluto Companion Search Team preparing for the *New Horizons* mission and working with the Hubble Space Telescope; but this discovery was confirmed only in 2006 (Weaver et al., 2006). Kerberos was discovered in 2011 (Showalter et al., 2011) using images from the Hubble Space Telescope (HST). Kerberos's discovery, announced on 20 July 2011, was accidently made while searching for Plutonian rings. Follow-up observations in 2012, while looking for potential hazards for New Horizons spacecraft during its flyby of Pluto, led to the discovery of the still smaller moon Styx (Showalter et al., 2012).

Planetary scientists consider that Charon was probably formed by a large impact into Pluto (see Canup, 2005), and the outer moons accreted from the leftover debris. Pluto's five moons show a tantalizing orbital configuration: the ratios of their orbital periods are close to 1:3:4:5:6 (Buie et al., 2006; Showalter et al., 2011; Showalter et al., 2012; Buie et al., 2013). This configuration is reminiscent of the Laplace resonance at Jupiter, where the moons Io, Europa, and Ganymede have periods in the ratio 1:2:4 (Showalter and Hamilton, 2015). As just indicated in the cases of Jupiter's moons Io, Europa, and Ganymede, a Laplace resonance occurs when three or more orbiting bodies have a simple integer ratio between their orbital periods. This is an example of stable orbital resonance in the Solar System.

Pluto's four small moons—Styx, Nix, Kerberos, and Hydra—follow near-circular, near-equatorial orbits around the central 'binary planet' comprising Pluto and its large moon, Charon (Showalter and Hamilton, 2015). Note that a binary system is a system of two astronomical bodies which are close enough that their gravitational attraction causes them to orbit each other around a common center-of-mass (barycenter). More restrictive definitions require that this common center-of-mass is not located within the interior of either object, in order to exclude the typical planet–moon systems and planetary systems. The most common binary systems are binary stars and binary asteroid, but brown dwarfs, planets, neutron stars, black holes and galaxies can also form binaries. Charon is about half the diameter of Pluto and is massive enough (nearly one-eighth of the mass of Pluto) that the system's barycenter lies between them, approximately 960 km above Pluto's surface.

Pluto's four small moons orbit Pluto at two to four times the distance of Charon, ranging from Styx at 42,700 km to Hydra at 64,800 km from the barycenter of the system. They have nearly circular prograde orbits in the same orbital plane as Charon. Note that prograde or direct motion in astronomy is, in general, orbital or rotational motion of an object in the same direction as the primary (central object) rotates. Pluto's four small moons' near orbital resonances with Charon (Showalter and Hamilton, 2015) suggest that they formed closer to Pluto than they are at present and migrated outward as Charon reached its current orbit.

Among all the five moons of Pluto, the largest moon Charon is particularly important for various reasons. As

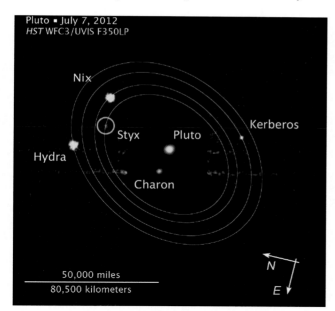

FIG. 16.3 Discovery image of Styx, overlaid with orbits of the satellite system, Author: NASA, ESA, and L. Frattare (STScI). *(Source: http://hubblesite.org/newscenter/archive/releases/2012/32/image/c/), This image, taken by NASA's Hubble Space Telescope, shows five moons orbiting the distant, icy dwarf planet Pluto. The diagram shows that Styx is in a 58,000-mile-diameter circular orbit around Pluto that is assumed to be co-planar with the other moons in the system. Though Charon (discovered in 1978) is an even closer moon to Pluto, some astronomers consider the Pluto-Charon pair a "double planet" because Charon's mass is 12 percent of Pluto's mass (by comparison, Earth's Moon is 1.2 percent Earth's mass). This image was taken with Hubble's Wide Field Camera 3 on July 7. Other observations that collectively show the moon's orbital motion were taken on June 26, 27, and 29, 2012 and July 9, 2012. The original NASA image has been modified by replacing "P4" and "P5" with newer designations "Kerberos" and "Styx," respectively. Copyright info: This file is in the **public domain in the United States** because it was solely created by NASA. NASA copyright policy states that "NASA material is not protected by copyright unless noted." https://commons.wikimedia.org/wiki/File:Pluto_moon_P5_discovery_with_moons%27_orbits.jpg.*

Charon approached its current distance from Pluto, the obliquity of Pluto stabilized to something like its current value, and Charon locked into synchronous orbit above a particular spot on Pluto's surface. The dynamical state of the Pluto–Charon binary is distinctive in several respects including its well-known position in the Neptune 3:2 mean motion resonance, its librating argument of perihelion, and its high heliocentric eccentricity and inclination (Levison and Stern, 1995). Note that the eccentricity (e) of an ellipse is, most simply, the ratio of the distance c between the center of the ellipse and each focus to the length of the semi-major axis a. That is, $e = \dfrac{c}{a}$. When the eccentricity is 0 the foci coincide with the center point and the figure is a circle. As the eccentricity tends toward 1, the ellipse gets a more elongated shape.

According to Ward and Canup (2006), if Charon had a large initial eccentricity, then its co-rotation resonances could lock material into the 1:3:4:5:6 relationship. Kenyon and Bromley (2014) suggest that as Charon's eccentricity damped, the resonant strengths waned, but the moons were left with periods close to these integer ratios. However, several investigators (e.g., Lithwick and Wu, 2008; Lithwick and Wu, 2013; Levison and Walsh, 2013) have suggested that this seemingly appealing model has numerous shortcomings. For example, the presence of a strong Laplace-like resonance places a new constraint on formation models. Additionally, future models must account for the non-zero eccentricities and inclinations of the small moons (Showalter and Hamilton, 2015); for example, these might imply that the system was excited in the past by resonances that are no longer active (Zhang and Hamilton, 2007; Zhang and Hamilton, 2008).

According to Hamilton et al. (2016), "As seen from Pluto, Charon is an extremely large and extremely close satellite, appearing fully seven times larger than the full Moon in Earth's sky, which itself is larger than any other planetary satellite seen from its primary. Moreover, owing to its tidally evolved state, Charon hovers over one point on Pluto's equator, continuously illuminating the same hemisphere." Interestingly, gravitational torques from Charon on the tidal bulge raised on Pluto (by Charon) cause a monotonic decrease in the spin rate of Pluto. Furthermore, Charon's phases vary over a Pluto year, always oscillating about, but rarely departing appreciably from, half full. Hamilton et al. (2016) reported that "More than half of the sunlight incident on Charon is reprocessed into thermal infrared, which delivers energy to ices far more efficiently than does visible light. Finally, more indirect energy from Charon is delivered to the night side of Pluto than to the dayside." Buie et al. (1992) examined the effect of sunlight reductions to Pluto due to eclipses by Charon that occur every 124 years. Hamilton et al. (2016) found that when averaged over a Pluto year, this effect is smaller, but not substantially smaller, than reflected sunlight and thermal emission from Charon.

Showalter and Hamilton (2015) have listed the mean disk-integrated photometry for each moon of Pluto. According to them, to infer the sizes of these bodies, we also require their geometric albedos p_v. It has been found that "Charon is a relatively bright, with $p_v \approx 38\%$. Kuiper Belt objects (KBOs) exhibit a large range of albedos, but the smallest KBOs tend to be dark; $p_v \approx 4$–8% is common (Grundy et al., 2005; Lykawka and Mukai, 2005; Stansberry et al., 2008; Lacerda et al., 2014)."

Showalter and Hamilton (2015) have pointed out that Kerberos seems to be very different from the other moons of Pluto. According to them, "The dynamical inference that its mass is ~ 1/3 that of Nix and Hydra, yet it reflects only ~5% as much sunlight, implies that it is very dark. This violates our expectation that the moons should be self-similar, due to the ballistic exchange of regolith (Stern, 2009). Such heterogeneity has one precedent in the Solar System: at Saturn, Aegaeon is very dark ($p_v < 15\%$), unlike any other satellite interior to Titan, and even though it is embedded within the ice-rich G ring (Hedman et al., 2011). The formation of such a heterogeneous satellite system is difficult to understand."

According to the studies of Showalter and Hamilton (2015), "Nix has an unusually large axial ratio of ~2:1, comparable to that of Saturn's extremely elongated moon, Prometheus. Hydra is also elongated, but probably less so. Also, Nix's year-by-year variations are the result of a rotation pole apparently turning toward the line of sight; this explains both its brightening trend and also the decrease in its variations during 2010–2012. Pluto's sub-Earth latitude is 46°, so Hydra's measured pole is nearly compatible with the system pole. Nix's pole was ~20° misaligned in 2010 but may have reached alignment by 2012."

16.2 Pluto's highly eccentric orbit—an oddity in the general scheme

In the widely accepted paradigm for the formation of the Solar System, all the planets accumulated in a highly dissipative disk of dust and gas orbiting the proto-Sun, and most planets formed in near-circular and nearly coplanar orbits (Malhotra, 1995). Pluto's orbital plane is off by 17° relative to the coplanar plane of the Solar System. In the pre-New Horizons spacecraft mission, our knowledge of Pluto has mainly been from telescopic observations. From such observations, it has been found that the outermost planet/dwarf planet, Pluto, is an oddity in the general scheme, in the sense that its orbit is highly eccentric (eccentricity, e = 0.25). Note that the orbital eccentricity of an astronomical object is a dimensionless parameter that determines the amount by which its orbit around another body deviates from a perfect circle.

The origin of Pluto's unusual orbit—the most eccentric of all the planets—has long been a puzzle. A possible explanation has been proposed by Malhotra (1993) which suggests that the extraordinary orbital properties of Pluto (highly eccentric and resonance-locked orbit) may be a natural consequence of the formation and early dynamical evolution of the outer Solar system.

According to Malhotra (1993), "One consequence of Neptune's orbital expansion is that its orbital resonances would have swept across a range of heliocentric distances comparable to its radial migration. During this resonance sweeping, a small body such as Pluto, initially in a nearly circular orbit beyond Neptune, would have a significant probability of being captured into a resonance. Resonant perturbations from Neptune would transfer sufficient angular momentum to the body to keep it trapped in the resonance and to expand its orbit in concert with Neptune. A by-product of such a resonance capture is that Pluto's orbital eccentricity would have been excited to a high value."

Pluto's large orbital eccentricity means that Pluto crosses the orbit of Neptune, and it traverses a very large region of space from just inside the orbit of Neptune at 30 AU to almost 50 AU. Although it was initially suspected that the dynamical lifetime of Pluto could be short because of the possibility of its close encounter with Neptune (Lyttleton, 1936), it was later found that a dynamical protection mechanism exists that prevents close encounters between Pluto and Neptune. In a study based on a 120,000 yr orbit integration of the outer planets, Cohen and Hubbard (1965) showed that Pluto is locked in a 3:2 orbital resonance with Neptune which maintains a large longitude separation between the planets at orbit crossing and causes Pluto's perihelion to librate about a center ±90° away from Neptune. Thus, Cohen and Hubbard (1965) showed that although the orbits of Pluto and Neptune overlap, close approaches of these two planets are prevented by the existence of a resonance condition.

Subsequent studies carried out by several researchers (e.g., Williams and Benson, 1971; Applegate et al., 1986; Sussman and Wisdom, 1988; Milani et al., 1989), based on orbit integrations of increasingly longer times, showed that at perihelion Pluto is close to its maximum excursion above the mean plane of the solar system. This mechanism has the effect of increasing the minimum approach distance between Pluto and Neptune and between Pluto and Uranus than would otherwise be the case (see Malhotra, 1995).

It was long suspected that Pluto's orbit would become Neptune-crossing because of Pluto's high eccentricity of 0.25. In order to test the hypothesis more regorously, Malhotra (1993) carried out a numerical integration of the orbit evolution of the four jovian planets and Pluto. Her theory does not invoke any catastrophic collisions, and is possibly compatible with the standard paradigm for planet formation {see Levy and Lunine (1993) for reviews of planet formation theory}. In this model, "an initially low-inclination, nearly circular orbit of Pluto beyond the orbits of the giant planets evolves into its Neptune-crossing but resonance-protected orbit as a result of early dynamical evolution of the outer solar system." As Pluto's mass is several orders of magnitude smaller than those of the jovian planets, Pluto was treated as a massless "test particle." The numerical method used by Malhotra (1993) was a modified version of the mixed-variable-symplectic method (Wisdom and Holman, 1991; Saha and Tremaine, 1992), which is a very fast integrator particularly suited to Solar System integrations. According to Malhotra (1993), as Neptune moved outwards, a small body like Pluto in an initially circular orbit could have been captured into the 3:2 resonance, following which its orbital eccentricity would rise rapidly to its current Neptune-crossing value.

Malhotra (1993) showed that Pluto could have acquired its current orbit during the late stages of planetary accretion, when the jovian planets (i.e., Jupiter, Saturn, Uranus, and Neptune) underwent significant orbital migration as a result of encounters with residual planetesimals (Fernandez and Ip, 1984). "In particular, Neptune's orbit may have expanded considerably, and its exterior orbital resonances would have swept through a large region of trans-Neptunian space. During this resonance sweeping, Pluto could have been captured into the 3:2 orbital period resonance with Neptune and its eccentricity (as well as inclination) would have been pumped up during the subsequent evolution."

The scheme outlined by Malhotra (1993) for the origin of Pluto's highly eccentric, Neptune-crossing orbit is considered to be quite robust in that the probability of capturing Pluto in the 3:2 Neptune resonance from an initially circular orbit is very high. According to Malhotra (1993), the role of possible planetesimal collisions with Pluto during its evolution in the 3:2 Neptune resonance needs to be evaluated. This may have an important bearing on the origin and properties of the Pluto–Charon binary.

Levison and Stern (1995) reported a suite of numerical simulations of bodies in the outer Solar System which demonstrate that Pluto's high-heliocentric eccentricity and high-inclination states can both result directly from objects initially on low-inclination, nearly-circular orbits. In the model proposed by Levison and Stern (1995), the orbital configuration of the planets is taken as observed today, except that a "test Pluto" is placed in an initially low-eccentricity, low-inclination orbit near the 3:2 Neptune resonance. With some fine-tuning of initial conditions, orbit integrations showed that such an orbit has its eccentricity and inclination pumped up to values comparable to those of the real Pluto in a timescale of about 10^7 yr. However, the orbit remains chaotic during this evolution. According to Levison and Stern (1995), the evolution of Pluto from an initially low-heliocentric eccentricity, low-inclination orbit to a high-heliocentric eccentricity and high-inclination orbital states occurs entirely due to gravitational interactions with the giant planets in their present orbits and causes objects to be trapped in both the Neptune 3:2 mean motion resonance and a perihelion libration (oscillation) similar to Pluto's, but with a 3:2 libration amplitude that is much larger than that of the Pluto–Charon binary. Levison and Stern (1995)'s studies revealed that in order to achieve a complete scenario for the evolution of Pluto into its present dynamical state, it is also necessary that Pluto's 3:2 libration-amplitude be damped by some dissipative event or events. Levison and Stern (1995) showed that "there are several mechanisms that can achieve this dissipation including (i) a single giant impact (which may have formed the binary itself), or (ii) a large number of physical collisions and gravitational interactions with the primordial Kuiper belt population." Levison and Stern (1995) then proposed that Pluto was "knocked" into the stable 3:2 resonance libration region by one or more dissipative collisions with a neighboring small body or bodies.

Malhotra (1995) found that a resonance capture mechanism is possible during the clearing of the residual icy planetesimal debris and the formation of the Oort Cloud—which is a roughly isotropic distribution of comets surrounding the planetary system at distances in excess of approximately 10^4 AU. According to her, "if this mechanism were in operation during the early history of the planetary system, the entire region between the orbit of Neptune and approximately 50 AU would have been swept by first-order mean motion resonances. Thus, resonance capture could occur not only for Pluto, but quite generally for other trans-Neptunian small bodies. Some consequences of this evolution for the present-day dynamical structure of the trans-Neptunian region are: (1) most of the objects in the region beyond Neptune and up to ~50 AU exist in very narrow zones located at orbital resonances with Neptune (particularly the 3:2 and the 2:1 resonances), and (2) these resonant objects would have significantly large eccentricities." According to the studies reported by Malhotra (1993,1994,1995), the unusual properties of Pluto's orbit may be a natural consequence—and a signature—of the early dynamical evolution in the outer Solar System. Malhotra (1995)'s studies indicate that the resonant objects in the Kuiper Belt move "on highly eccentric orbits, with a significant fraction on Neptune-crossing orbits; the inclinations of most of the objects remain low (less than 10°), but a small fraction (up to approximately 10%) are in the 15°–20° range. Libration of the argument of perihelion (about ±90°) is not an uncommon occurrence among the resonant objects." Thus, it can be safely inferred that Pluto's orbital peculiarities are akin to those of many other Kuiper Belt objects.

16.3 Long-period perturbations in the chaotic motion of Pluto

An examination of the orbital periods of Uranus, Neptune, and Pluto by Williams and Benson (1971) showed that they are very nearly in the ratio of 1:2:3. According to them,

"Such commensurabilities or near-commensurabilities of the periods give rise to perturbations of the motion which have long periods and moderately large amplitudes. Long-period perturbations in the motion of Pluto are expected from both Uranus and Neptune." Sussman and Wisdom (1988) performed an integration of the motion of the outer planets for 845 million years. This integration indicates that the long-term motion of the planet Pluto is chaotic. Wisdom and Holman (1991) generalized the mapping method of Wisdom (1982) to encompass all gravitational n-body problems with a dominant central mass. The method is used to compute the evolution of the outer planets for a billion years. This calculation provides independent numerical confirmation of the result of Sussman and Wisdom (1988) that the motion of the planet Pluto is chaotic. Long-term orbit integrations by several researchers (e.g., Williams and Benson, 1971; Applegate et al., 1986; Sussman and Wisdom, 1988; Milani et al., 1989) have uncovered some subtle resonances and near-resonances, and indicate that Pluto's orbit is chaotic yet remains macroscopically stable over billion-year time-scales.

According to Levison and Duncan (1993), the suggestion by Sussman and Wisdom (1988) and some other investigators that the orbit may have evolved purely by chaotic dynamics appears unlikely in light of subsequent orbital stability studies, unless one appeals to a well-timed collision to place Pluto in its stable orbit (Levison and Stern, 1993). However, Levison and Duncan (1993)'s view has been refuted by Malhotra (1995), who reported that Pluto's orbit is confined in a very narrow region of relative orbital stability near the 3:2 Neptune resonance, a region in phase space that is bounded by highly chaotic orbits [see Malhotra and Williams (1994) for a review of Pluto's orbital dynamics].

It has been found that apart from the chaotic long-term motion experienced by Pluto as outlined above, Pluto's moon Nix also has been found to exhibit chaotic motion. Dynamical simulations carried out by Showalter and Hamilton (2015) have indicated that a binary planet tends to drive its moons into chaotic rotation. This is illustrated in Fig. 16.4 showing the simulated rotation period and orientation of Nix vs. time. It is found that "the moon has a tendency to lock into near-synchronous rotation for brief periods, but these configurations do not persist. At other times, the moon rotates at a period entirely unrelated to its orbit. For example, it shows occasional pole flips, a phenomenon consistent with the observed changes in Nix's orientation." Although the torques acting on a less elongated body such as Hydra are weaker, Showalter and Hamilton (2015)'s integrations support chaos.

In this context, Showalter and Hamilton (2015) have pointed out that "Nearly every moon in the solar system rotates synchronously; the only confirmed exception is Hyperion, which is driven into chaotic rotation by

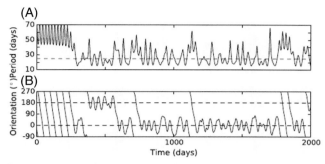

FIG. 16.4 **Numerical simulations of Nix's rotation.** (A) The instantaneous rotation period is compared to the synchronous rate (dashed line). (B) The orientation is described by the angle between Nix's long axis and the direction toward the barycenter. Nix librates about 0° or 180° for periods of time, but it jumps out of these states frequently. *(Source: Showalter, M.R., and Hamilton, D.P. (2015), Resonant interactions and chaotic rotation of Pluto's small moons, Nature, 522, 45–49).*

a resonance with Titan (Wisdom et al., 1984; Klavetter, 1989). Neptune's highly eccentric moon Nereid may also rotate chaotically (Dobrovolskis, 1995), but observational support is lacking (Buratti et al., 1997; Grav et al., 2003)." Showalter and Hamilton (2015) have suggested that "Chaotic dynamics makes it is less likely to find rings or additional moons of Pluto. Within the Styx-Hydra region, the only stable orbits are co-orbitals of the known moons."

16.4 Pluto's complex crater morphology

Moons in the outer solar system frequently exhibit impact basins having basin diameter comparable to the moon radius (Kamata and Nimmo, 2014). Pluto, like most other icy planetary bodies, is likely to possess large impact basins. As noted by Nimmo and Matsuyama (2007), Pluto's slow rotation rate makes it rotationally unstable. Thus, formation of a large impact basin could cause significant prompt reorientation (true polar wander), potentially generating global tectonic stress patterns. Subsequent slow basin relaxation (the dominant factor controlling which is radiogenic heat produced in the rocky core) would result in a slow reversal of the initial true polar wander path. Further tectonic features, both from this reorientation and the relaxation itself, would likely result. Thus, the tectonic consequences of large impact basins on Pluto are likely to prove interesting (Kamata and Nimmo, 2014).

As an icy body, Pluto's craters are expected to be similar in morphology to those on icy moons. These include lesser depth-diameter ratios (d/D), shallower wall slopes, and the development of central uplifts in craters of smaller rim-to-rim diameter than craters on rocky bodies of similar gravity. Bray and Schenk (2015) investigated the pristine impact crater morphology on Pluto, and made some suggestions on the expectations for *New Horizons* mission. Their studies combined previous cratering studies and numerical modeling of the impact process at different impact velocities to predict crater morphology on Pluto.

Impactors on Pluto are expected to have relatively modest impact velocities, because of Pluto's modest (heliocentric) orbital velocity of 4.7 km/s. Zahnle et al. (2003) give a mean impact velocity of 1.9 km/s (barely supersonic). According to Bray and Schenk (2015), the low impact velocity of the Pluto system (~2 km/s) might cause deviation from the generalization that Pluto's craters are expected to be similar in morphology to those on the icy moons. Note that "impact velocity" is essentially the "effective velocity" at which an object is traveling at a moment in which it impacts another object. When other forces on the two objects are considered, such as the force of gravity, friction, wind resistance and similar outside forces, then the value of the velocity at the moment of impact is significantly more complicated. Impact velocity can be determined for any two objects that come into contact with each other.

Bray and Schenk (2015)'s simulations suggested that decreasing impact velocity from 10 km/s to 2 km/s results in deeper craters (larger d/D) and a simple-to-complex transition diameter at larger crater sizes than predicted based on gravity scaling alone (D > 6 km). Another interesting result obtained from Bray and Schenk (2015)'s simulations suggests that decreasing impact velocity from 2 km/s to 300 m/s produced smaller d/D, akin to the lower d/D noted for secondary craters. According to Bray and Schenk (2015), this complex relationship between impact velocity and d/D suggests that there might be a larger range of "pristine" simple crater depths on Pluto than on bodies with higher mean impact velocity. These investigators suggested that "The low impact velocities and correspondingly low volumes of impact melt generated at Pluto might prevent the occurrence, or limit the size, of floor-pits if their formation involves impact melt water. The presence, or not, of central floor-pit craters on Pluto will thus provide a valuable test of floor-pit formation theories. The presence of summit-pits or concentric craters on Pluto is plausible and would indicate the presence of layering in the near sub-surface. Palimpsests, multi-ring basins, and other crater morphologies associated with high heat flow are not expected and would have important implications for Pluto's thermal history if observed by New Horizons."

16.5 Controversy over Pluto's planetary status—recent arguments

Although Pluto's average distance from the Sun is 6×10^9 km (i.e., 40 AU), the Sun's gravitational sphere of influence extends much further, out to $\sim 2 \times 10^{13}$ km. This space is occupied by the Oort cloud (Oort, 1950), comprising 10^{12} to 10^{13} cometary nuclei, formed in the primordial solar nebula. Observations and computer modeling have contributed to a detailed understanding of the structure and dynamics of the

Oort cloud, which is thought to be the source of the long-period comets and possibly comet showers (Weissman, 1990). The Oort cloud is completely unobservable from Earth with the current technology and thus escapes the scrutiny of observers. However, the largest members of the Kuiper belt (a region of the outer Solar System extending from a distance of ~ 40 AU [roughly the inner edge of the Kuiper belt] to approximately 50 AU from the Sun) are within the reach of available optical telescopes.

Pluto's status as a planet until the year 1992 was questioned following the discovery of several objects of similar size in the Kuiper belt (Kuiper, 1951). As already indicated, the Kuiper belt is accessible to ground-based studies and, as a fossil from the formation of the Solar System, is an important reservoir of primordial material (Luu, 1993). This reservoir awaits exploration and may yield important information on early Solar System conditions. Unfortunately, the dynamics of the Kuiper belt remain largely unexplored till date. Recent numerical simulations suggest that chaos and gravitational scattering by large belt members play a role in the delivery of comets from the belt to the inner planetary region. Kuiper belt objects are expected to retain primordial irradiation mantles, produced by prolonged irradiation of ices by cosmic rays.

In 2005, Eris, a dwarf planet in the scattered disc which is 27% more massive than Pluto, was discovered. This led the International Astronomical Union (IAU)—a global group of astronomy experts—to define the term "planet" formally in 2006, during their 26th General Assembly. That definition excluded Pluto and reclassified it as a dwarf planet.

The exclusion of Pluto from the planet league was primarily because Pluto's gravity is affected by its adjacent planet Neptune, and Pluto shares its orbit with objects in the Kuiper belt and frozen gases. It is often claimed that asteroids' sharing of orbits is the reason they were reclassified from planets to non-planets. However, a recent research reported by Metzger et al (2018) claims that the reason due to which Pluto lost its status as a planet is invalid. According to Metzger et al (2018), the evidence obtained from a critical review of the literature from the 19th Century to the present demonstrates that the consensus among planetary scientists for re-classification of asteroids from planets to non-planets formed on the basis of geophysical differences between asteroids and planets, and not the sharing of orbits. Metzger et al (2018) suggested that "attempts to build consensus around planetary taxonomy not rely on the nonscientific process of voting, but rather through precedent set in scientific literature and discourse, by which perspectives evolve with additional observations and information, just as they did in the case of asteroids." The study team recommends classifying a celestial body as a planet only if it is large enough that its gravity allows it to become spherical in shape. A celestial body becoming spherical in shape is an important milestone in the evolution of a planetary body because apparently this process is accompanied by initiation of active geology in the body. Furthermore, the special case of 1:1 orbital resonance (between bodies with similar orbital radii) causes large Solar System bodies to "clear out" the region around their orbits by ejecting nearly everything else around them. Because no planetesimals (i.e., bodies that could come together with many others under gravitation to form a planet) are found orbiting around Pluto, it can be argued that Pluto might have performed the above-mentioned "clearing out" process in its history. This could be used as an argument to categorize Pluto as a large Solar System body, with the status of a planet. In any case, the blue-skied Pluto is still a planet in the heart of many planetary scientists as well as the lay people, though not in official registers.

16.6 Studies on Pluto system and Pluto's physical and geological features prior to the launch of "New Horizons" spacecraft

Pluto and its moon, Charon, are the most prominent members of the Kuiper belt, and their existence holds clues to outer solar system formation processes. Several planetary researchers have investigated Pluto much before NASA's "New Horizons" Spacecraft approached Pluto on July 14, 2015 for observing its surface. As a result, researchers had obtained some new information about this distant celestial body in our Solar System.

16.6.1 Dynamics of Pluto and its largest moon Charon

Pluto and its largest moon "Charon" form a twin-body system. They both orbit around a common center-of-mass, which is one-eighth of the distance from the center-of-mass of Pluto to Charon, on the side toward Charon and along a straight line connecting the center-of-masses of Pluto and Charon. This positioning of the common center-of-mass of the Pluto–Charon system is because Pluto has eight times the mass of Charon. In analogy, Earth and its moon also form a twin-body system. However, because the Earth is 81 times more massive than its moon, the common center-of-mass of the Earth-Moon system lies within the Earth at a distance of approximately 1718 km beneath the Earth's surface, on the side toward the Moon and along a straight line connecting the center-of-masses of the Earth and the Moon. Because Pluto and Charon are two constituents of a coupled rotating system with a common center-of-mass, Charon always shows the same face to Pluto, the way the Earth's moon is locked in the same direction toward Earth. The mutual motion of the "binary planet" comprising Pluto and Charon creates a time-variable and distinctly asymmetric gravity field. According to Lee and Peale (2006), this induces wobbles in the outer moons' orbits and also drives much slower apsidal precession {i.e., precession of the line of apsides, that is, of the orientation of the major axes of the orbits (see Greenberg, 1981)} and nodal regression (i.e., shift of the orbit's line of nodes; that is, shift in the intersection of the orbital plane and some reference plane, usually the ecliptic).

Charon is comparable in size, density and composition to the 11 classical midsize moons of Saturn and Uranus (Schenk et al., 2018). Volatile-rich Pluto, on the other hand, is comparable in size and density to Triton; both bodies are transitional between the smaller midsize icy moons and the larger ice-rich worlds of Ganymede, Callisto, and Titan.

The dynamics of Pluto and Charon has been a subject of interest among planetary scientists. In this context, Dobrovolskis (1989) reviewed the dynamics of the Pluto-Charon system from a historical perspective, and observed that although Pluto's orbit crosses Neptune's orbit in an apparently chaotic manner, an intricate system of nested resonances keeps these planets apart. Interestingly, Pluto always keeps the same face turned toward Charon, and vice versa. It was found that tides play an important role by way of damping Charon's orbital eccentricity and inclination. Another observation made in this review is that the precession of Pluto's orbital plane causes Pluto's obliquity to vary periodically from formally prograde (i.e., proceeding from west to east) to retrograde (moving backwards). Planetary scientists think that Pluto is probably an original member of the Solar system, but not an escaped moon of Neptune. Dobrovolskis (1989) expressed optimism that the Voyager II encounter with Neptune, the final Pluto-Charon mutual events, and the next generation of telescopes are bound to reveal some surprises.

Canup (2005) used hydrodynamic simulations to demonstrate that the formation of Pluto–Charon coupled system by means of a large collision is quite plausible. Fig. 16.5 shows time-series of a potential Pluto–Charon–forming impact yielding a planet-disk system. The Pluto–Charon pair shares key traits with the Earth–Moon system. Canup (2005)'s studies shed some light on a giant impact origin of Pluto and Charon. He showed that such an impact probably produced an intact Charon, although it is possible that a disk of material orbited Pluto from which Charon later accumulated. According to Canup (2005), these findings suggest that collisions between 1000-km-class objects occurred in the early inner Kuiper belt. That the outermost parts of the Solar system may be populated by primordial icy planetesimals has been conjectured on both theoretical and observational grounds. For example, Kuiper (1951) suggested this on the basis of theoretical considerations of the genesis of the planetary system from the primordial Solar Nebula. With regard to the possible mechanism of the shaping/carving of the Kuiper Belt, Malhotra (1995) has

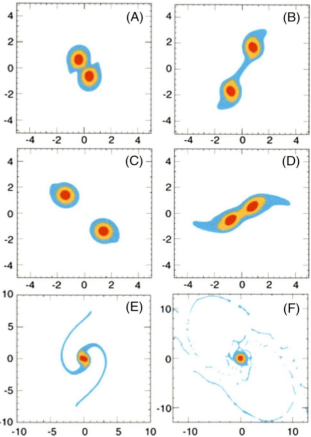

FIG. 16.5 Time series of a potential Pluto-Charon–forming impact yielding a planet-disk system. Units shown are distance in 10^3 km. Color indicates material type (blue, water ice; orange, dunite; red, iron), with all of the particles in the 3D simulation over-plotted in order of increasing density. After an initially oblique impact in the counter-clockwise sense (A), the two objects separate (B and C) before recolliding. After the second collision, the denser cores migrate toward the center, as a bar-type mode forms in the rapidly rotating merged objects (D). From each end of the bar emanate spiral structures (D and E), whose self-gravity acts to transport angular momentum from inner to outer portions. The arms wrap up on themselves and finally disperse to yield a ring of material (whose differential motion would on a longer time scale produce a disk), together with the central planet (F). *(Source: Canup, R.M. (2005), A giant impact origin of Pluto-Charon, Science, 307: 546–550), Copyright info: Copyright 2005 by the American Association for the Advancement of Science; all rights reserved.*

discussed a "dynamical sculpting" mechanism that would have occurred due to the early orbital evolution of the outer planets (during the late stages of their formation) as predicted by the "resonance capture theory" for the origin of Pluto's orbit. The results of this study indicate that "the Kuiper Belt would have been "sculpted" into a highly non-uniform distribution early in solar system history, and this structure would be largely preserved to the present epoch: the region beyond Neptune's orbit and up to approximately 50 AU heliocentric distance should have most of the primordial small bodies locked in orbital resonances with Neptune, particularly the 3:2 and the 2:1 orbital resonances which are located at semi-major axes of approximately 39.4 and 47.8 AU, respectively." Cheng et al. (2014) developed Pluto–Charon tidal evolution models and with their application, they investigated the tidal evolution of Pluto and its largest moon Charon.

16.6.2 Three-body orbital resonant interactions among Pluto's small moons Styx, Nix, and Hydra

Orbital resonances are ubiquitous in the Solar system. A general resonance involves an angle Φ (the libration angle) and its time-derivative $\dot{\Phi}$. A resonance is recognized by $\dot{\Phi}$'s coefficients that sum to zero and produce a very small value of $\dot{\Phi}$; In addition, the resonant argument Φ usually librates (oscillates) around either 0° or 180° (Showalter and Hamilton, 2015).

Orbital resonances play a decisive role in the long-term dynamics, and in some cases the physical evolution, of the planets and of their natural moons, as well as the evolution of small bodies (including dust) in the planetary system (Malhotra, 1994). In celestial mechanics, orbital resonance is a gravitational phenomenon that occurs when orbiting bodies exert regular, periodic gravitational influence on each other, usually because their orbital periods are in a specific pattern, related by a ratio of small integers. The Solar System is known for resonances of various kinds. The phenomenon of capture into resonance as a result of some slow dissipative forces is common in nature, and there exists a large body of literature devoted to its study. A well-developed solar system example is the formation of orbit-orbit resonances among the moons of the giant planets by the action of slow tidal dissipation {see Peale (1986) and Malhotra (1994) for reviews}.

Orbital resonances greatly enhance the mutual gravitational influence of the orbiting bodies. In most cases, this results in an unstable interaction, in which the orbiting bodies exchange momentum and shift orbits until the resonance no longer exists. For small bodies, destabilization is actually far more likely. For example, unstable resonances with Saturn's inner moons give rise to gaps in the rings of Saturn. It may be noted that in 1675, Giovanni Domenico Cassini determined that Saturn's ring was composed of multiple smaller rings with gaps between them; the largest of these gaps was later named the "Cassini Division." In the rings of Saturn, the "Cassini Division" is a 4,800-km-wide gap between the inner B-Ring and the outer A-Ring that has been cleared by a 2:1 resonance with the moon Mimas. For every two orbits made by particles at this distance from Saturn, Mimas makes one orbit. The moon's repeated gravitational tugs force ring particles away from this region. More specifically, the site of the resonance is the "Huygens Gap," which bounds the outer edge of the B-Ring. Likewise, in the asteroid belt within 3.5 AU from the Sun, the major mean-motion resonances with Jupiter are locations of gaps in the asteroid distribution, the Kirkwood gaps (Wisdom, 1982) {most notably at the 3:1, 5:2, 7:3, and 2:1 resonances}. Kirkwood gaps are gaps in the distribution of asteroid orbital periods; and these gaps are caused by the gravitational interaction between Jupiter and asteroids with these orbital periods. Asteroids have been ejected from these almost empty lanes by repeated perturbations.

Under some circumstances, a resonant system can be stable and self-correcting, so that the orbiting bodies remain in resonance. Stabilization occurs when the orbiting bodies move in such a synchronized fashion that they never closely approach. In general, capture into a stable (long-lived) orbit-orbit resonance is possible when the orbits of two bodies approach each other as a result of the action of some dissipative process. Cohen and Hubbard (1965) computed the orbits of the five outer planets by special perturbations over 120,000 years. There was revealed a remarkable libration of the close approaches of Pluto to Neptune such that the distance between the bodies is never less than 18 AU. Cohen and Hubbard (1965) concluded that the orbit of Pluto is safe from very close approaches to Neptune and no particular instability results from the fact that the radius of perihelion of Pluto is less than the radius of the orbit of Neptune. The study revealed that the period of the oscillation is about 19,670 years and the amplitude is about 76 degrees.

Malhotra (1995) proposed that Pluto may have been captured into the 3:2 resonance with Neptune during the late stages of planet formation, when Neptune's orbit expanded outward as a result of angular momentum exchange with residual planetesimal debris. The simplest case of stable orbital resonance in the Solar System is the 2:3-resonance between the planets Pluto and Neptune, in which for every two orbits that Pluto makes, Neptune makes exactly three orbits. This is why it is impossible for the two bodies ever to collide, despite the fact that their orbits cross. This is the only known stable resonance in the Solar System involving two planetary bodies (although Pluto is a dwarf planet). However, there exists a dynamical group of small planet-like bodies orbiting the Sun in the region of the Kuiper belt and in 2:3 mean-motion resonance with Neptune (called Trans-Neptunian objects; also called Plutinos). Pluto and the Plutinos are in stable orbits, despite crossing the orbit of the much larger Neptune. This is because a 2:3 resonance keeps them always at a large distance from it. Other (much more numerous) Neptune-crossing bodies that were not in resonance were ejected from that region by strong perturbations due to Neptune.

Results derived from Cheng et al. (2014)'s Pluto–Charon tidal evolution models revealed that the orbits of Pluto's four small moons (Styx, Nix, Kerberos, and Hydra) are nearly circular and coplanar with the orbit of the largest moon Charon, with orbital periods nearly in the ratios 3:1, 4:1, 5:1, and 6:1 with Charon's orbital period. According to Cheng et al. (2014), these properties suggest that the small moons mentioned above were created during the same impact event that placed Charon in orbit and had been pushed to their current positions by being locked in mean-motion resonances with Charon as Charon's orbit was expanded by tidal interactions with Pluto. It was found that the test particle used in the simulation exercise has significant orbital eccentricity at the end of the tidal evolution of Pluto–Charon in almost all cases. Based on results of simulations with finite but minimal masses of Nix and Hydra, Cheng et al. (2014) concluded that the placing of the small moons at their current orbital positions by resonant transport is extremely unlikely.

New observational details of Pluto's moons have emerged following the discoveries of Kerberos and Styx. Using the orbital elements and their uncertainties, Showalter and Hamilton (2015) performed an exhaustive search for strong resonances in the Pluto system. In continuation of studies involving Pluto and its moons, they investigated the resonant interactions and chaotic rotation of Pluto's small moons. Showalter and Hamilton (2015)'s investigations indicated that Styx, Nix, and Hydra are tied together by a three-body resonance, which is reminiscent of the Laplace resonance linking Jupiter's moons Io, Europa and Ganymede. It was found that if λ denotes the mean longitude and Φ the libration angle, then the resonance can be formulated as:

$$\Phi = 3\lambda_{Styx} - 5\lambda_{Nix} + 2\lambda_{Hydra} \approx 180°$$

As with the Laplace resonance of the Galilean moons of Jupiter, triple conjunctions never occur. Φ librates (oscillates) about 180° with an amplitude of at least 10° (Showalter and Hamilton, 2015). They found that the time-derivative of the libration angle, $\dot{\Phi}$ = -0.007 ± 0.001°/day and that the libration angle Φ decreased from 191° to 184° during 2010–2012; this is all consistent with a small libration about 180°.

Showalter and Hamilton (2015) identified the harmonics of a second three-body resonance:

$$\Phi' = 42\lambda_{Styx} - 85\lambda_{Nix} + 43\lambda_{Kerberos} \approx 180°$$

This was the second strongest resonance found in their search; at the orbit of Styx, the two resonances are separated by just 4 km. According to Showalter and Hamilton (2015) this is reminiscent of the Uranus system, where chains of near-resonances drive the chaos in that system.

Showalter and Hamilton (2015)'s studies showed that perturbations by the other bodies inject chaos into this otherwise stable configuration. In the case of Pluto's moons, it has been proposed that the present near-resonances are relics of a previous precise resonance that was disrupted by tidal damping of the eccentricity of Charon's orbit. A particularly interesting observation obtained from Showalter and Hamilton (2015)'s studies is that whereas Nix and Hydra have bright surfaces similar to that of Charon, Kerberos may be much darker, raising questions about how a heterogeneous moon system might have formed. Their investigations further revealed that Nix and Hydra rotate chaotically, driven by the large torques of the Pluto–Charon binary.

16.6.3 Insolation and reflectance changes on Pluto

The two primary topics of interest in studying Pluto's history of insolation (i.e., the amount of solar radiation reaching a given area) are: (1) the variations in insolation patterns when integrated over different intervals, and (2) the evolution of diurnal insolation patterns over the last several decades. Naturally, fundamental questions such as the insolation at Pluto and insolation changes on it caused by orbital element variations have been subjects of scientific investigations.

16.6.3.1 Insolation changes on Pluto caused by orbital element variations

In the context of understanding insolation changes on Pluto, Van Hemelrijck (1982) carried out calculations of the daily solar radiation incident at the top of Pluto's atmosphere and its variability with latitude and season for three literature-cited fixed values (60, 75, and 90°) of Pluto's obliquity ε and for the presently adopted values of the eccentricity (e) and the longitude of the perihelion (λ_p). Note that "perihelion" is the point in the orbit of a planet, asteroid, or comet at which it is closest to the Sun. Although Pluto's obliquity was very poorly known till then, the results obtained through Van Hemelrijck (1982)'s study illustrated fairly well the sensitivity of Pluto's insolation to changes in ε. The obliquity of Pluto changes the frost pattern facing the Sun and finally the heat flow from or to the substrate. It was found that Pluto receives nearly three times less sunlight at aphelion (the point in the orbit of Pluto at which it is farthest from the Sun) than perihelion. Pluto's varying heliocentric distance results in considerable insolation changes during its 248-year orbit. According to Olkin et al. (2015), over the course of a Pluto-year (248 Earth-years), changes in global insolation drives the migration of 1 m of frost, therefore, seasonal changes in frost distribution are likely.

Near the perihelion equinox, the southern hemisphere surface is warm, ~42 K. Approaching and after equinox (at perihelion), the southern hemisphere receives less sunlight, and radiatively cools slowly due to high thermal inertia. In the near future, the year 2080 is the period when the southern pole starts to be illuminated by the Sun given the obliquity of Pluto (see Olkin et al., 2015).

16.6.3.2 Latitudinal variations of Pluto's insolation and reflectance

The amount of solar radiation reflected from the surface of Pluto and its moons was a topic that planetary scientists showed much interest in investigating. Albedo is a parameter commonly used in such investigations. The albedo is defined as the ratio between the reflected energy and the incident energy over a unit area. In this context, Buie et al. (1992) prepared single-scattering albedo maps (i.e., texture maps defining the color and pattern of diffused light) of the surface of Pluto and its largest moon Charon, based primarily on mutual event observations. Buie et al. (1992) applied the technique of maximum entropy image reconstruction to invert the light-curves, thus revealing surface maps of single-scattering albedo. The dataset contained 3374 photometric observations that cover 15 different moon transit events, 14 moon eclipse events, and other out-of-eclipse photometry spanning 1954 to 1986. The maps consisted of a 59 × 29 grid of tiles for each body. The study revealed that a south polar cap is present on Pluto. It was also found that Pluto's North Polar Region is brighter than its equatorial regions but is not as bright as the South Pole. Buie et al. (1992) reported that "Single-scattering albedos range from 0.98 in the south polar cap to a low near 0.2 at longitudes corresponding to the light-curve minimum. The map of Charon is somewhat darker with single-scattering albedos as low as 0.03."

16.6.4 Pluto's atmosphere

Continuing observations of Pluto's atmospheric pressure on decadal timescales constrain thermal inertia, providing insight into deeper layers of the surface that are not visible in imaging (Olkin et al., 2015). An interesting finding that Pluto receives nearly three times less sunlight at aphelion than perihelion prompted early modelers to predict that Pluto's atmosphere would expand and collapse over its 248-year orbit (Stern and Trafton, 1984). According to Olkin et al. (2015), definitive detection of Pluto's atmosphere in 1988 (Millis et al., 1993) and the discovery of nitrogen (N_2) as the dominant volatile in the atmosphere and on the surface (Owen et al., 1993) prompted the development of more sophisticated models in the 1990s (Hansen and Paige, 1996). Stellar occultations, where a body such as Pluto passes between an observer and a distant star, provide the most sensitive method for measuring Pluto's changing atmospheric pressure.

Pluto was predicted to occult a 14th magnitude (R filter) star on May 4, 2013 (Assafin et al., 2010), and the predicted stellar occultation indeed happened on May 4, 2013. This was one of the most favorable Pluto occultations of 2013 because of the bright star, slow shadow velocity (10.6 km/s at Cerro Tololo), and shadow path near large telescopes (Olkin et al., 2015). Olkin et al. (2015) fit the three Las Cumbres Observatory Global Telescope Network (LCOGT) light curves simultaneously using a standard Pluto atmospheric model reported by Elliot and Young (1992). According to Olkin et al. (2015), "this model was developed after the 1988 Pluto occultation, which showed a

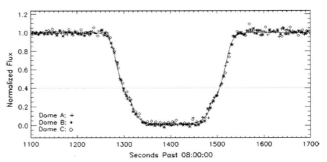

FIG. 16.6 The observed occultation light curves overlaid with the best fitting model. Time plotted is seconds after 2013 May 04 08:00:00 UTC. The line at normalized flux of 0.4 corresponds to 1275 km in Pluto's atmosphere. The transition from the upper atmosphere to the lower atmosphere in the fitted model occurs at a flux level of ∼0.25 in these data. All three telescopes are 1.0-m telescopes located at the Cerro Tololo LCOGT node. The WGS 84 Coordinates of the three telescopes are (1) Dome A: Latitude: 30°.167383S, Longitude: 70°.804789W, (2) Dome B: Latitude: 30°.167331S, Longitude: 70°.804661W and (3) Dome C: Latitude: 30°.167447S, Longitude: 70°.804681W. All telescopes are at an altitude of 2201 m. The Dome A telescope used a 2-s integration time; Dome B used a 3-s integration time and Dome C used a 5-s integration time. (*Source: Olkin et al. (2015), Evidence that Pluto's atmosphere does not collapse from occultations including the 2013 May 04 event, Icarus, 246: 220–225). Copyright info: © 2014 The Authors. Published by **Elsevier Inc**. This is an **open access** article under the CC BY license (http://creativecommons.org/licenses/by/3.0/).*

distinct kink, or change in slope, in the light curve indicating a difference in the atmosphere above and below about 1215 km from Pluto's center." Based on this interesting observation, this model separates the atmosphere into two domains: a clear upper atmosphere with at most a small thermal gradient, and a lower atmosphere that potentially includes a haze layer. In particular, the lower atmosphere can be described with either a haze layer, or by a thermal gradient as reported by Eshleman (1989), Hubbard et al. (1990), and Stansberry et al. (1994); or a combination of the two to match the low flux levels in the middle of the occultation light curves (Olkin et al., 2015). Fig. 16.6 shows the LCOGT light curves and the best fitting model with a pressure of 2.7 ± 0.2 microbar and a temperature of 113 ± 2 K for an isothermal atmosphere at 1275 km from Pluto's center. The lower atmosphere was fit with a haze onset radius of 1224 ± 2 km, a haze extinction coefficient at onset of 3.2 ± 0.3 × 10^3/km and a haze scale height of 21 ± 5 km (see Elliot and Young, 1992, for details). This atmospheric pressure extends the trend of increasing surface pressure with temperature since 1988.

Based on the volatile transport model runs in Young (2013), Olkin et al. (2015) found that the models can produce solutions with a bright southern pole. With regard to the source of the bright south pole at equinox, Olkin et al. (2015) suggested that "the south pole could be bright due to methane (CH_4) ice at the south pole and this would not be reflected in the volatile-transport models because the models only consider the dominant volatile, N_2."

According to Olkin et al. (2015), the permanent northern volatile (PNV) model has a high thermal inertia, such that the atmosphere does not collapse over the course of a Pluto year with typical minimum values for the surface pressure of roughly 10 microbar. At this surface pressure the atmosphere is collisional (meaning that the gas molecules in the atmosphere collide with one another before traveling any appreciable distance) and present globally, and Olkin et al. (2015) concluded that Pluto's atmosphere does not collapse at any point during its 248-year orbit. Olkin et al. (2015) further consider that "an atmosphere has not collapsed if it is global, collisional, and opaque to UV radiation. An atmosphere that is global and collisional can efficiently transport latent heat over its whole surface. The cutoff for a global atmosphere is ~0.06 microbars (Spencer et al., 1997) or more than 2 orders of magnitude smaller than the typical minimum pressure for PNV models." Combining stellar occultation observations probing Pluto's atmosphere from 1988 to 2013, and models of energy balance between Pluto's surface and atmosphere, Olkin et al (2015) found that the preferred models are consistent with Pluto retaining a collisional atmosphere throughout its 248-year orbit.

Unfortunately, it is not possible to measure the atmospheric pressure at the surface of Pluto from the ground. To circumvent this lacuna, it becomes necessary to use the pressure at a higher altitude as a proxy for the surface pressure. Olkin et al. (2015) investigated the validity of this proxy measurement. They started with synthetic occultation light curves derived from GCM models (Zalucha and Michaels, 2013) at a variety of different methane column abundances and surface pressures ranging from 8 to 24 mbars. They fit the synthetic light curves with the Elliot and Young (1992) model to derive a pressure at 1275 km. They found that the ratio of the pressure at 1275 km to the surface pressure was a constant within the uncertainty of the model fit (0.01 mbar). Because of this, they concentrated on those occultations for which the pressures at 1275 km have been modeled by fitting Elliot and Young (1992) models, which is a subset of the occultation results presented in Young (2013).

Olkin et al. (2015) explained the overall state of Pluto's atmosphere thus: "Pluto's atmosphere is protected from collapse because of the high thermal inertia of the substrate. The mechanism that prevents the collapse is specific to Pluto because it relies on Pluto's high obliquity and the coincidence of equinox with perihelion and aphelion. In the PNV model, volatiles are present on both the southern and northern hemispheres of Pluto just past aphelion. Sunlight absorbed in the southern hemisphere (the summer hemisphere) from aphelion to perihelion powers an exchange of volatiles from the southern hemisphere to the northern (winter) hemisphere. Latent heat of sublimation cools the southern hemisphere and warms the northern hemisphere, keeping the N_2 ice on both hemispheres the same temperature. This exchange of volatiles continues until all the N_2 ice on the southern hemisphere sublimates and is condensed onto Pluto's northern hemisphere. However, the thermal inertia of the substrate is high, so the surface temperature on the northern hemisphere does not cool quickly. The ice temperature drops by only a few degrees K before the N_2-covered areas at mid-northern latitudes receive insolation again, in the decades before perihelion." Olkin et al. (2015) argued that "as Pluto travels from perihelion to aphelion, N_2 ice is always absorbing sunlight on the northern hemisphere keeping the ice temperatures relatively high throughout this phase and preventing collapse of Pluto's atmosphere."

According to Olkin et al. (2015), "the current epoch is a time of significant change on Pluto. Most of the PNV models show a maximum surface pressure between the years 2020 and 2040. Regular observations over this time period will constrain the properties of Pluto's substrate and the evolution of its atmosphere."

16.6.5 Presence of a subsurface ocean on Pluto—inference based on numerical studies

Determining whether or not Pluto possesses, or once possessed, a subsurface ocean is crucial to understanding its astrobiological potential. Astronomers have long theorized that Pluto's icy exterior could act to insulate vestiges of warmer moisture below the surface. Astronomers think there is a water–ocean between the inner rock (rock-ball core) and the icy crust. As already noted, our knowledge of Pluto prior to the *New Horizons* mission was mainly from telescopic observations. Its size and mass are constrained with a relatively small error (e.g., Person et al., 2006), while spectral information pertaining to composition (Owen et al., 1993) and surface color (Buie et al., 2010) have also been obtained. These observations, however, are insufficient to determine the interior structure and thermal history of Pluto. Numerical studies are therefore necessary. Most such studies have either assumed conduction or treated convection using a parameterized approach as discussed in McKinnon et al. (2017), Hussmann et al. (2006), Desch et al. (2009), and Barr and Collins (2014).

Scientific evidences are fast emerging to believe that Pluto is a prime candidate for the presence of a subsurface ocean. Robuchon and Nimmo (2011) discussed thermal evolution of Pluto and implications for surface tectonics and a subsurface ocean. They used a 3D convection model to investigate Pluto's thermal and structural evolution, and the conventional observational consequences of different evolutionary pathways. These investigators tested the sensitivity of their model results to different initial temperature profiles, initial spin periods, silicate potassium concentrations and ice reference viscosity (i.e., the viscosity of ice at its melting point). Robuchon and Nimmo (2011) found that the evolution depends mainly on the reference viscosity of ice and the amount of radiogenic material present. The ice reference viscosity is considered to be the primary factor controlling whether or not an ocean develops and whether that ocean survives to the present day. In most of the models present-day Pluto consists of a convective ice shell without an ocean. However, if the reference viscosity is higher than 5×10^{15} Pascal-second (Pa. s), the icy shell is conductive, and a subsurface ocean develops. On the other hand, if the reference viscosity is lower than that value, the icy shell is convective, and a subsurface ocean does not develop. A typical ice reference viscosity is $\sim 10^{14}$ Pa. s, though it depends on many factors, such as grain size of ice (e.g., Goldsby and Kohlstedt, 2001) and the temperature of the subsurface ocean, if present. A global pattern of tectonic features, if observed on Pluto's surface, would provide constraints on its evolution. For example, thickening of an ice shell above an ocean would result in recent extension, while cooling of a shell in the absence of a subsurface ocean would generate compression (Robuchon and Nimmo, 2011).

Robuchon and Nimmo (2011) formulated the case for determining whether or not an ocean develops and whether that ocean survives to the present day based on the following logic: "For the nominal potassium concentration the present-day ocean and conductive shell thickness are both about 165 km; in conductive cases, an ocean will be present unless the potassium content of the silicate mantle is less than 10% of its nominal value. If Pluto never developed an ocean, predominantly extensional surface tectonics should result, and a fossil rotational bulge will be present. For the cases which possess, or once possessed, an ocean, no fossil bulge should exist. A present-day ocean implies that compressional surface stresses should dominate, perhaps with minor recent extension. An ocean that formed and then re-froze should result in a roughly equal balance between (older) compressional and (younger) extensional features." Robuchon and Nimmo (2011)'s investigations shed light on the thermal evolution of Pluto and implications for surface tectonics and a subsurface ocean.

In continuation of Robuchon and Nimmo (2011)'s studies, Kamata and Nimmo (2014) re-examined the interior structure model of Pluto, which is adopted from Robuchon and Nimmo (2011) (see Fig. 16.7). In Kamata and Nimmo (2014)'s studies, they assumed a differentiated Pluto, consisting of an H_2O layer overlying a silicate core. According to Kamata and Nimmo (2014), "The radius of the core is assumed to be 850 km (McKinnon et al., 2017). Here the silicate mass fraction ~ 0.67 is assumed, though it might range from 0.5 to 0.7 (McKinnon et al., 2017). Since this uncertainty leads to a change in the core radius by only 50 km, different core radii would not change our results significantly. The key point is that only cold, near-surface ice can support topographic loads for billions of years; thus, the total thickness of the ice shell is of only secondary importance. The thickness of the H_2O layer depends on the thickness of a subsurface ocean. This is because the total mass of Pluto

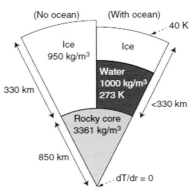

FIG. 16.7 **Interior structure model of Pluto, after Robuchon and Nimmo (2011), adopted by Kamata and Nimmo (2014).** The radius of the rocky core is fixed to 850 km (McKinnon et al., 2017). The thickness of the H_2O layer depends on the thickness of a subsurface ocean. When there is no liquid water layer, the thickness of the H_2O layer is 330 km. A development of a subsurface ocean reduces the thickness of the H_2O layer to conserve the total mass. *(Source: Kamata, S., and F. Nimmo (2014), Impact basin relaxation as a probe for the thermal history of Pluto, J. Geophys. Res. Planets, 119: 2272–2289, doi:10.1002/2014JE004679). Copyright info: ©2014. American Geophysical Union. All Rights Reserved.*

needs to be conserved. Specifically, the thickness of the H_2O layer is assumed to be 330 km when it is completely frozen, and the presence of a subsurface ocean leads to a decrease in the thickness."

The modelers seemed to be unanimous in the suggestion that inferences and predictions made by planetary researchers based purely on numerical studies with regard to the possible presence of a subsurface ocean on Pluto would need to be tested by observations obtained from a spacecraft that orbited or flew past Pluto. The *New Horizons* mission helped realizing this much sought-after requirement.

16.7 "New Horizons"—the spacecraft that brilliantly probed the distant Pluto from close quarters

New Horizons is an interplanetary space probe (Fig. 16.8) that was launched as a part of NASA's New Frontiers program. The spacecraft was launched on Jan. 19, 2006 on a 9-year journey to the Solar System's distant dwarf planet, Pluto. The primary mission was to perform a flyby study of the Pluto system in 2015, and a secondary mission to fly by and study one or more other Kuiper Belt objects (KBOs) in the decade to follow.

16.7.1 Powering the New Horizons spacecraft

The New Horizons probe is powered by a Radio-isotope Thermo-electric Generator (RTG) functioning as a "nuclear battery" (Fountain et al., 2008). This generator has no moving parts. RTGs have been used as power sources in artificial satellites, space probes, and un-crewed remote facilities. RTGs are usually the most desirable power source for

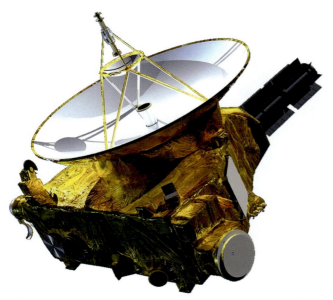

FIG. 16.8 Image of the *New Horizons* spacecraft. Source: "PEPSSI Instrument Tastes Pluto's Atmosphere" from the Applied Physics Laboratory New Horizons website Public Domain: This file is in the public domain in the United States because it was solely created by NASA. NASA copyright policy states that "NASA material is not protected by copyright unless noted." (Source: https://commons.wikimedia.org/wiki/File:New_Horizons_Transparent.png).

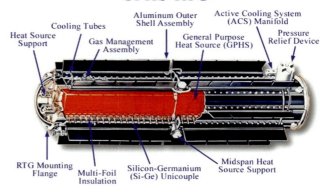

FIG. 16.9 A cut-drawing of an GPHS-RTG that are used for Galileo, Ulysses, Cassini-Huygens and New Horizons space probes. (Source: http://saturn.jpl.nasa.gov/spacecraft/safety.cfm) Copyright info: This file is in the public domain in the United States because it was solely created by NASA. NASA copyright policy states that "NASA material is not protected by copyright unless noted." https://commons.wikimedia.org/wiki/File:Cutdrawing_of_an_GPHS-RTG.jpg.

FIG. 16.10 A pellet of ^{238}PuO$_2$ as used in the RTG. This photo was taken after insulating the pellet under a graphite blanket for several minutes and then removing the blanket. The pellet is glowing red hot because of the heat generated by radioactive decay (primarily α). Author: The original uploader was Deglr6328 at English Wikipedia. (*Source: Transferred from en.wikipedia to Commons., Attribution, https://commons.wikimedia.org/w/index.php?curid=11198885*) This image comes from Los Alamos National Laboratory, a national laboratory privately operated under contract from the United States Department of Energy by Los Alamos National Security, LLC betweeen October 1, 2007 and October 31, 2018. LANL allowed anyone to use it for any purpose, provided that the copyright holder is properly attributed. Redistribution, derivative work, commercial use, and all other use is permitted. LANL requires the following text be used when crediting images to it: (link) Unless otherwise indicated, this information has been authored by an employee or employees of the Los Alamos National Security, LLC (LANS), operator of the Los Alamos National Laboratory under Contract No. DE-AC52-06NA25396 with the U.S. Department of Energy. The U.S. Government has rights to use, reproduce, and distribute this information. The public may copy and use this information without charge, provided that this Notice and any statement of authorship are reproduced on all copies. Neither the Government nor LANS makes any warranty, express or implied, or assumes any liability or responsibility for the use of this information. https://commons.wikimedia.org/wiki/File:Radioisotope_thermoelectric_generator_plutonium_pellet.jpg.

unmaintained situations that need a few hundred watts (or less) of power for durations too long for fuel cells, batteries, or generators to provide economically, and in places where solar cells are not practical. Safe use of RTGs requires containment of the radioisotopes long after the productive life of the unit. This nuclear battery (Fig. 16.9) releases electric power by means of the heat generated by the radio-active decay of 24 pounds (11 kg) of plutonium-238 oxide pellets (Fig. 16.10). Each pellet is clad in iridium, then encased in a graphite shell. A pellet of plutonium-238 isotope glows red with the heat it gives off as atomic nuclei decay (Karl Tate, NASA, Johns Hopkins Applied Physics Laboratory). Similar generators were used by the Voyager space probes and on the Earth's Moon by Apollo astronauts. The RTG provided 245.7 W of power at launch, and was predicted to drop approximately 3.5 W every year, decaying to 202 W by the time of its encounter with the Plutonian system in 2015 and will decay too far to power the transmitters in the 2030s. There are no onboard batteries because RTG output is predictable, and load transients are handled by a capacitor bank and fast circuit breakers.

The science goal of the mission was to understand the formation of the Plutonian system, the Kuiper belt, and the transformation of the early Solar System. Some of the questions the mission attempts to answer are: What is Pluto's atmosphere made of and how does it behave? What does its surface look like? Are there large geological structures? How do solar wind particles interact with Pluto's atmosphere?

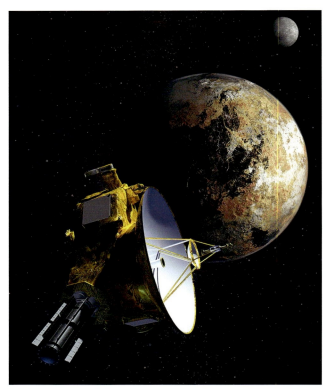

FIG. 16.11 Artist's concept of NASA's *New Horizons* spacecraft as it passes Pluto and Pluto's largest moon, Charon, in July 2015. *Copyright info: Public Domain: This file is in the public domain in the United States because it was solely created by NASA. NASA copyright policy states that "NASA material is not protected by copyright unless noted." Author: NASA/JHU APL/SwRI/Steve Gribben (Source: https://commons.wikimedia.org/wiki/File:15-011a-NewHorizons-PlutoFlyby-ArtistConcept-14July2015-20150115.jpg).*

The spacecraft collected data on the atmospheres, surfaces, interiors, and environments of Pluto and its moons. It will also study other objects in the Kuiper belt. On July 14, 2015, the probe approached Pluto and spent the next three months observing its surface before completing the flyby and continuing on to the Kuiper belt.

16.7.2 New Horizons spacecraft and its science payloads

Launched in 2006, the *New Horizons* probe is the first spacecraft to fly past Pluto (see Fig. 16.11). New Horizons' primary structure includes an aluminum central cylinder that supports the spacecraft body panels, supports the interface between the spacecraft and its RTG power source, and houses the propellant tank. It also served as the payload adapter fitting that connected the spacecraft to the launch vehicle. In its size/structure (see Fig. 16.12), the New Horizons probe is 27 inches (0.7meter) tall, 83 inches (2.1 m) long, and 108 inches (2.7 m) at its widest. The spacecraft is comparable in size and general shape to

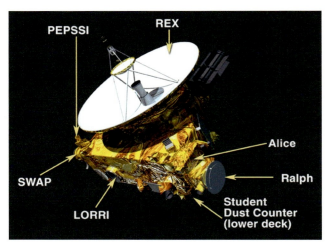

FIG. 16.12 New Horizons spacecraft's' primary structure. (*Source: http://pluto.jhuapl.edu/Mission/Spacecraft.php*).

a grand piano and has been compared to a piano glued to a cocktail bar-sized satellite dish (Moore, 2010). Its main antenna is 83 inches (2.1 m) in diameter. A telescopic camera known as LORRI (Long Range Reconnaissance Imager) provides high-resolution photos and geologic data. LORRI is a long-focal-length imager designed for high resolution and responsivity at visible wavelengths. The instrument is equipped with a 1024×1024 pixel by 12-bits-per-pixel monochromatic CCD imager giving a resolution of 5 μrad (~1 arcsec). A dust-detector known as VENITIA—built by students—measures the dust encountered in space throughout the mission of the New Horizons probe. The dust counter is named for Venetia Burney, who first suggested the name "Pluto" at the age of 11.

New Horizons carries seven instruments: three optical instruments, two plasma instruments, a dust sensor and a radio science receiver/radiometer. The instruments are to be used to investigate the global geology, surface composition, surface temperature, atmospheric pressure, atmospheric temperature, and escape rate of Pluto and its moons. The rated power is 21 watts, though not all instruments operate simultaneously (Guo and Farquhar, 2006). In addition, New Horizons has an Ultrastable Oscillator subsystem, which may be used to study and test the Pioneer anomaly toward the end of the spacecraft's life (Nieto, 2008).

The payload is incredibly power efficient—with the instruments collectively drawing less than 28 watts—and represents a degree of miniaturization that is unprecedented in planetary exploration. The instruments were designed specifically to handle the cold conditions and low light levels in the Kuiper Belt.

An instrument known as PEPSSI (Pluto Energetic Particle Spectrometer Science Investigation) measures the composition and density of ions escaping from Pluto's atmosphere. PEPSSI is a time of flight ion and electron sensor (which measures particles of up to 1 MeV) that makes

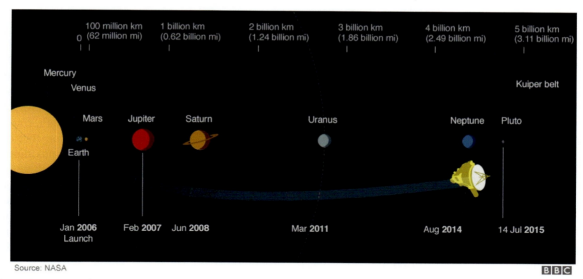

FIG. 16.13 Illustration of the *New Horizons* spacecraft encountering various celestial bodies during its long space travel ever since it lifted off from Cape Canaveral Air Force Station, Florida, at 19:00 UTC on January 19, 2006 until it is on the verge of speeding past Pluto on 14 July, 2015 in its valiant onward journey, piercing through the Kuiper belt and beyond. (*Source*: NASA).

up one of the two instruments comprising New Horizons' plasma and high-energy particle spectrometer suite (PAM), the other being SWAP (Solar Wind Around Pluto) that measures the rate of escape of Pluto's atmosphere. SWAP measures particles of up to 6.5 keV and, because of the tenuous solar wind at Pluto's distance, the instrument is designed with the largest aperture of any such instrument ever flown.

An ultraviolet imaging spectrometer, named Alice, was employed to study Pluto's atmosphere. It resolves 1,024 wavelength bands in the far and extreme ultraviolet (from 50 to 180 nm), over 32 view fields. Its goal was to determine the composition of Pluto's atmosphere. The New Horizons probe was also equipped with a visible and infra-red imager and spectrometer, named Ralph. The Ralph telescope is 75 mm in aperture.

The Venetia Burney Student Dust Counter (VBSDC), built by students at the University of Colorado Boulder, is operating periodically to make dust measurements. The detector contains fourteen polyvinylidene difluoride (PVDF) panels, twelve "science" panels and two "reference" panels, which generate voltage when impacted. The VBSDC is always turned on, measuring the masses of the interplanetary and interstellar dust particles (in the range of nano- and pico-grams) as they collide with the PVDF panels mounted on the New Horizons spacecraft. The measured data is expected to greatly contribute to the understanding of the dust spectra of the Solar System. The dust spectra can then be compared with those from observations of other stars, giving new clues as to where Earth-like planets can be found in the universe.

An instrument known as REX (Radio science EXperiment) uses a high-gain antenna as a passive radiometer for measuring the atmospheric composition and temperature.

16.7.3 Discoveries by New Horizons during its long travel to Pluto and beyond

Ever since the New Horizons Spacecraft lifted off from Cape Canaveral Air Force Station, Florida, at 19:00 UTC on January 19, 2006, the probe has been engaged in encountering various celestial bodies during its long travel to Pluto (Fig. 16.13) and even beyond the Solar System. After leaving the Earth, New Horizons took only 9 h to pass the Earth's Moon's orbit. On April 7, 2006, the spacecraft passed the orbit of Mars, moving at roughly 76,000 km/h (47,000 mph) away from the Sun at a solar distance of 243 million kilometers.

Because of the need to conserve fuel for possible encounters with Kuiper belt objects subsequent to the Pluto flyby, intentional encounters with objects in the asteroid belt were not planned. However, in May 2006 the New Horizons team discovered that the probe would pass close to the tiny asteroid 132524 APL on June 13, 2006. Closest approach occurred at 4:05 UTC at a distance of 101,867 km (63,297 mi). The asteroid was imaged by Ralph, which gave the team a chance to make observations of the asteroid's composition. The asteroid was estimated to be 2.5 km (1.6 mi) in diameter. The spacecraft successfully tracked the rapidly moving asteroid over June 10–12, 2006.

After crossing the asteroid belt located between Mars and Jupiter, New Horizons took its first photographs of Jupiter on September 4, 2006, from a distance of 291 million kilometers (181 million miles). More detailed exploration of the system began in January 2007 with an infrared image of Callisto (the second-largest moon of Jupiter), as well as several black-and-white images of Jupiter itself. New Horizons received a gravity assist from Jupiter, with its closest approach at 05:43:40 UTC on February 28, 2007, when it was 2.3 million kilometers (1.4 million miles) from Jupiter. Note that in orbital mechanics and aerospace engineering, gravity assist manoeuvre is the use of the relative movement and gravity of an astronomical object to alter the path and speed of a spacecraft, typically to save propellant and reduce expense. This mechanism increased New Horizons' speed by 14,000 km/h (9,000 mph), accelerating the probe to a velocity of 83,000 km/h (51,000 mph) relative to the Sun and shortening its voyage to Pluto by three years.

During its encounter with Jupiter, the New Horizons probe observed and measured heat-induced lightning strikes in Jupiter's Polar Regions and "waves" that indicate violent storm activity. The Little Red Spot, spanning up to 70% of Earth's diameter, was imaged from up close for the first time. New Horizons took detailed images of Jupiter's faint ring system based on recording from different angles and illumination conditions. Such images resulted in the discovery of debris left over from recent collisions within the rings or from other unexplained phenomena. The search for undiscovered moons within the rings showed no results.

Traveling through Jupiter's magnetosphere, *New Horizons* collected valuable particle readings. "Bubbles" of plasma that are thought to be formed from material ejected by Jupiter's innermost Galilean moon, Io (whose active volcanoes shoot out tons of material into Jupiter's magnetosphere, and further), were noticed in the magnetotail. An eruption event observed by New Horizons reached an altitude of up to 330 km (210 mi), giving scientists an unprecedented look into the structure and motion of the rising plume and its subsequent fall back to the surface.

New Horizons' measurement of the surface of Callisto (the second-largest moon of Jupiter, after Ganymede) enabled gaining an insight of how lighting and viewing conditions affect infrared spectrum readings of its surface water ice. Furthermore, New Horizons' measurements allowed refinement of the orbit solutions of Jupiter's minor moons such as Amalthea. The cameras determined their positions, acting as "reverse optical navigation."

Jupiter's icy moon Europa was imaged by New Horizons space probe on February 27, 2007, from a distance of 3.1 million km (1.9 million mi). The first Europa image returned by New Horizons was taken with the spacecraft's Long-Range Reconnaissance Imager (LORRI) camera.

After passing Jupiter, New Horizons spent most of its journey toward Pluto in hibernation mode: redundant components as well as guidance and control systems were shut down to extend their life cycle, decrease operation costs and free the Deep Space Network for other missions.

New Horizons crossed the orbit of Saturn on June 8, 2008, and Uranus on March 18, 2011. While in hibernation mode in July 2012, New Horizons started gathering scientific data. The first set of data was transmitted in January 2013 during a three-week activation from hibernation.

The first images of Pluto from New Horizons were acquired on September 21–24, 2006, during a test of LORRI. They were released on November 28, 2006. The images, taken from a distance of approximately 4.2 billion km (2.6 billion mi; 28 AU), confirmed the spacecraft's ability to track distant targets, critical for maneuvering toward Pluto and other Kuiper belt objects. Note that the astronomical unit (or AU) is a unit of length, roughly the distance from Earth to the Sun. However, that distance varies as Earth orbits the Sun, from a maximum (aphelion) to a minimum (perihelion) and back again once a year. Originally conceived as the average of Earth's aphelion and perihelion, since 2012 it has been defined as exactly 149597870700 m, or about 150 million km (93 million miles). The astronomical unit is used primarily for measuring distances within the Solar System or around other stars.

Images from July 1 to 3, 2013, by LORRI were the first by the New Horizons probe to resolve Pluto and its moon Charon as separate objects. During July 19–24, 2014, New Horizons managed to gather 12 images of Charon revolving around Pluto, covering almost one full rotation at distances ranging from about 429 to 422 million kilometers (267,000,000 to 262,000,000 mi). High-precision measurements of Pluto's location and orbit around the Sun using the Atacama Large Millimeter/submillimeter Array (ALMA) allowed New Horizons spacecraft to accurately home in on Pluto. A command signal sent by the mission controllers for the craft to "wake up" from its final Pluto-approach hibernation and begin regular operations was positively responded by the craft. The craft began its intended regular operations from December 7, 2014, 02:30 Coordinated Universal Time (UTC), which is equivalent to the Greenwich Mean Time (GMT). Fig. 16.14 shows photograph of Pluto and its largest and the nearest moon, Charon, taken by the Ralph color imager aboard NASA's New Horizons spacecraft on April 9, 2015 and downlinked to Earth the following day.

16.7.4 New Horizons spacecraft performing the first ever flyby of Pluto

The *New Horizons* spacecraft probe managed to uncover most of the secrets of Pluto, which is 40 times more distant from the Sun than Earth and warms itself from within using radioactive decay of elements in its rock-ball core. On July 14, 2015, this spacecraft performed the first-ever flyby of the far-away dwarf planet, zooming within 12,500 kilometers (7,800 miles) of its frigid surface and sent reams of amazing data uncovering its hidden secrets. The close encounter is giving researchers their first up-close looks at Pluto, which has remained mysterious since its discovery on January 23, 1930. Fig. 16.15 shows color photos of Charon (left) and Pluto (right) taken by *New*

16.8 Understanding Pluto and its moons through the eyes of *New Horizons* spacecraft

Although several preliminary studies about Pluto have been carried out based on information gleaned from Earth-based telescopes and the Earth-orbiting Hubble Space Telescope much before the launch of NASA's New Horizons spacecraft, several relatively close-view photographs of Pluto sent to Earth from New Horizons during its flyby of Pluto shed considerably more information on this distant planet. The New Horizons spacecraft provided near-global observations of Pluto that far exceed the resolution of Earth-based data sets. While results of previous research, based primarily on Earth-based observations and theoretical studies, have provided context for observations of the Pluto system by the New Horizons spacecraft in July 2015, the voluminous amount of data transmitted by New Horizons spacecraft probe has enabled planetary researchers to shed much more light on Pluto and its five moons. The 2015 New Horizons flyby has produced the first high-resolution maps of the morphology and topography of Pluto and Charon, the most distant objects so mapped. Global integrated mosaics of Pluto were produced using both LORRI framing camera and MVIC line scan camera data, showing the best resolution data obtained for all areas of the illuminated surface, ~78% of the body (Schenk et al., 2018).

16.8.1 Shedding light on the consequences of Pluto's high orbital eccentricity and high obliquity

Nimmo et al. (2016) estimated the mean radius and shape of Pluto and its largest moon Charon based on New Horizons images of Pluto and Charon. These measurements and subsequent calculations revealed that Pluto has a diameter of 2370 kilometers, which is 18.5 percent that of Earth, while Charon has a diameter of 1208 km, which is 9.5 percent that of Earth. Binzel et al. (2016) determined the climate zones on Pluto and Charon. Once thought to be a planet as massive as Earth, Pluto is now known to be an icy body only two-thirds the diameter of Earth's Moon.

It has been suspected that Pluto's high orbital eccentricity and high obliquity lead to complex changes in surface insolation over a Pluto-year (248 Earth-years), and, therefore, in surface temperatures (Dobrovolskis and Harris, 1983; Olkin et al., 2015). The change in Pluto's tilt is the largest single factor that affects the strength of incident sunlight. Hamilton et al. (2016) reported that the 1.4-million-year swing from $\varepsilon = 104°$ to $\varepsilon = 127°$ causes a change of roughly 20% in the solar flux absorbed at Pluto's equator and poles. In the current epoch, Pluto's poles receive about 10% more solar energy than does its equator, but this ratio varies from about 45% more at $\varepsilon = 104°$ to near parity at $\varepsilon = 127°$ (Hamilton et al., 2016).

FIG. 16.14 Photograph of Pluto and its largest and the nearest moon, Charon, taken by the Ralph color imager aboard NASA's New Horizons spacecraft on April 9, 2015 and downlinked to Earth the following day. It is the first color image ever made of the Pluto system by a spacecraft on approach. The image is a preliminary reconstruction. *(Source: http://pluto.jhuapl.edu/Multimedia/Science-Photos/pics/20150414_First_Color_Image_Ralph.png) (Source: https://commons.wikimedia.org/wiki/File:PlutoCharon-1stColorImage-NewHorizons-Ralph-20150409.png) Copyright info: Public Domain: This file is in the public domain in the United States because it was solely created by NASA. NASA copyright policy states that "NASA material is not protected by copyright unless noted."*

FIG. 16.15 Photo of Pluto and Charon taken by *New Horizons* spacecraft on approach to Pluto. It was taken at 06:49 UT on July 14, 2015, 5 h before Pluto's closest approach, from a range of 150,000 miles (250,000 kilometers), with the spacecraft's Ralph instrument. In order to fit Pluto and Charon in the same frame in their correct relative positions, the image has been rotated so the North Pole on both Pluto and Charon is pointing toward the upper left. *(Source: http://photojournal.jpl.nasa.gov/jpeg/PIA19856.jpg) Author: NASA/Johns Hopkins University Applied Physics Laboratory/Southwest Research Institute Copyright info: This file is in the public domain in the United States because it was solely created by NASA. NASA copyright policy states that "NASA material is not protected by copyright unless noted." https://commons.wikimedia.org/wiki/File:PIA19856-PlutoCharon-NewHorizons-Color-20150714.jpg.*

Horizons spacecraft shortly before its closest approach to the dwarf planet. New Horizons completed the historic flyby of Pluto successfully on July 14, 2015. The image highlights the contrasting appearance of the two worlds: Charon is mostly grey, with a dark reddish polar cap, while Pluto shows a wide variety of subtle color variations, including yellowish patches on the north polar cap and subtly contrasting colors for the two halves of Pluto's "heart," informally named Tombaugh Regio, seen in the upper right quadrant of the image. The image was made with the blue, red, and near-infrared color filters of Ralph's Multi-color Visible Imaging Camera, and shows colors that are similar, but not identical, to what would be seen with the human eye, which is sensitive to a narrower range of wavelengths.

The distant location of Pluto's orbit in the Solar System is where conditions are such that exotic compounds such as N_2 and CH_4 can exist in abundance as solids on the surface (White et al., 2017). Because Pluto is lying on its side, its poles get more sunlight than its equator. As it slowly journeys around the Sun (one orbit taking 248 Earth-years), nitrogen and other gases freeze in the permanently shadowed areas, then evaporate into gases again, and then re-condense. According to planetary researchers, this nitrogen snow might pile up over billions of years, and eventually, a heavy nitrogen glacier could overwhelm the planet's shape. Analysis carried out by Hamilton et al. (2016) shows that variations in Pluto's obliquity and albedo are the strongest drivers of sublimation and deposition, with weaker but noticeable effects from orbital eccentricity and even radiation from the large moon Charon.

According to White et al. (2017), the exotic compounds mentioned above are nevertheless "volatile enough such that they can be readily mobilized by the radiogenic heat that still emanates from Pluto's interior (in the case of the N_2 ice in Sputnik Planitia) or by the profound insolation and climatic changes that take place across Pluto's 2.8 Myr obliquity cycle (in the case of sublimation and re-deposition, and glacial advance and retreat, of N_2 ice)."

16.8.2 Achieving confirmation on Pluto's small Moons' rotation, obliquity, shapes, and color & the absence of a predicted ring system surrounding Pluto

Pluto researchers compiled a series of images of the moons Nix and Hydra taken from January 27 through February 8, 2015, beginning at a range of 201 million kilometers (125,000,000 mi). Starting May 11, a hazard search was performed, by looking for unknown objects that could be a danger to the spacecraft, such as rings or more moons, which were possible to avoid by a course change.

The Pluto system was visited by the *New Horizons* spacecraft in July 2015. Images with resolutions of up to 330 m per pixel were returned of Nix and up to 1.1 kilometers per pixel of Hydra. Still lower-resolution images were returned of Styx and Kerberos. Pluto's four small moons (Styx, Nix, Kerberos, and Hydra) are much smaller than Charon (Fig. 16.16).

Prior to the New Horizons mission, Pluto's small moons (Nix, Hydra, Styx, and Kerberos) were predicted to rotate chaotically or tumble (i.e., experience a sudden or headlong fall) (see Showalter and Hamilton, 2015; Correia et al., 2015), just like Saturn's moon Hyperion, which is known to tumble (Wisdom et al., 1984). However, New Horizons imaging found that they had not tidally spun down to near a spin synchronous state where chaotic rotation or tumbling would be expected (Weaver, 2016; Quillen et al., 2017).

FIG. 16.16 **Pluto's small moons to approximate scale, compared to Pluto's largest moon Charon.** This composite image shows a small part of Pluto's largest moon, Charon, and all four of Pluto's small moons, as resolved by the Long Range Reconnaissance Imager (LORRI) onboard the New Horizons spacecraft. All the moons are displayed with a common intensity stretch and spatial scale (see scale bar). Charon is by far the largest of Pluto's moons, with a diameter of 751 miles (1212 km). Nix and Hydra have comparable sizes, approximately 25 miles (40 kilometers) across in their longest dimension above. Kerberos and Styx are much smaller and have comparable sizes, roughly 6–7 miles (10–12 kilometers) across in their longest dimension. *(Source: http://www.nasa.gov/feature/last-of-pluto-s-moons-mysterious-kerberos-revealed-by-new-horizons). Author: NASA/JHUAPL/SwRI Copyright info: This file is in the public domain in the United States because it was solely created by NASA. NASA copyright policy states that "NASA material is not protected by copyright unless noted." https://commons.wikimedia.org/wiki/File:Nh-pluto_moons_family_portrait-truecolor.png.*

In terms of obliquity, New Horizons imaging found that all 4 small moons of Pluto are at high obliquity (Weaver, 2016). Quillen et al. (2017) suggested that either they were born that way, or they were tipped by a spin precession resonance. It was further suggested that Styx may be experiencing intermittent and chaotic obliquity variations.

Nix and Hydra, the two larger, are roughly 42 and 55 kilometers on their longest axis respectively, and Styx and Kerberos are 7 and 12 kilometers respectively. Images taken by the New Horizons spacecraft clearly showed that all four small moons are irregularly shaped, having highly elongated features—a characteristic thought to have been typical of small bodies in the Kuiper Belt. However, the nearly circular orbits of the smaller moons suggest that, just like Charon, the smaller moons were also created in a massive collision, similar to the "Theia impact event" thought to have created the Earth's moon (Canup, 2005; Stern et al., 2006), rather than being captured Kuiper Belt objects.

An intense search conducted by *New Horizons* confirmed that no moons larger than 4.5 km in diameter exist at the distances up to 180,000 km from Pluto (Stern et al., 2015). It was found that their grey color is different from that of Pluto, one of the reddest bodies in the Solar System. This is thought to be due to a loss of volatiles during the impact or subsequent coalescence, leaving the surfaces of the moons dominated by water ice.

The discovery of Pluto's small moons Nix and Hydra had given rise to the suggestion that Pluto could have a ring system. This is because small-body impacts can create debris that can form into a ring system. However, data from a deep-optical survey by the Advanced Camera for Surveys on the Hubble Space Telescope, by occultation studies (Pasachoff et al., 2006) suggested that no ring system is present. Pasachoff et al. (2006)'s findings were substantiated by the data obtained from New Horizons (Kenyon and Bromley, 2019).

16.8.3 Runaway Albedo effect on Pluto

Ice is very reflective, therefore some of the solar or stellar radiation/energy is reflected back to space. Ice–albedo feedback is a positive feedback process where a change in the area of ice caps and glaciers alters the albedo and surface temperature of a planet. The runaway albedo effect describes the possible runaway cooling of the planetary surface. Decreasing surface temperature leads to an increase of snow and ice cover, which increases the surface albedo. Hence, more incoming solar/stellar radiation is reflected back to space, which in turn decreases surface temperature.

To illustrate the runaway albedo effect on Pluto, Hamilton et al. (2016) developed a code to track the individual sizes of several spatially separated ice deposits through Pluto's seasonal cycles. The instantaneous flux (F) of sunlight that is absorbed by a unit surface element on Pluto is given by the expression:

$$F = \frac{L}{4\pi r^2}(1-A)\cos(\gamma) \quad (16.1)$$

where L is the solar luminosity, r is the Pluto–Sun distance, A is Pluto's albedo and γ is the angle between the Sun–Pluto line and the normal to the surface element. Referring to equation (16.1), Hamilton et al. (2016) assumed that a fraction $(1 - A)$ of incident sunlight is absorbed by the ices. They further assumed that a fraction f of absorbed sunlight converts ice to vapor during sunlight hours and that all of this vapor redeposits onto the ice caps at night. Note that Pluto takes 6 1/2 Earth days/nights to rotate, so one day on Pluto is about 6 1/2 days/nights on Earth. The rest of the incident energy goes to heating of the surface and subsurface layers, and re-deposition is assumed to be in proportion to the exposed surface areas of the ice caps. Over a single summer day, ice preferentially sublimates into the atmosphere, whereas in winter the converse is true. Hamilton et al. (2016) considered only the longer annual cycle, assuming that the vapor content of the atmosphere remains constant and using it only as a conduit for communication from one ice deposit to another. In their simulation, Hamilton et al. (2016) set the number of ice caps, and their sizes, depths and albedos as initial conditions, and tracked changes to the deposits over many annual cycles. Hamilton et al. (2016) considered ice caps near a configuration in which each spot receives the same annual solar insolation. The total mass in ice was set to approximate that believed to be in the ice cap today, and was preserved over the course of the simulation. Hamilton et al. (2016) found that mass is rapidly removed from the darker two ice caps, owing to their more efficient absorption of solar radiation and consequent copious production of vapor. Simultaneously, the brighter two ice caps both grow in mass and spatial extent, with their larger surface areas attracting more of the available vapor each annual cycle. Ultimately, and on relatively short timescales, only one ice cap remains.

Fig. 16.17 shows a typical run with four large ice caps of initially equal size, but different albedos. Hamilton et al. (2016) recognizes that the four-ice-cap situation in Fig. 16.17 is unlikely to have ever occurred on Pluto, but serves as an effective demonstration of the power of the runaway albedo effect. As suggested by McKinnon et al. (2016), this simple model does not capture all of the relevant physics, for example, it does not include the expected brightening of the ice cap once its size exceeds that necessary for solid-state convection. However, according to Hamilton et al. (2016), this simple model effectively demonstrates that very small albedo differences are magnified to the point where only a single ice cap survives.

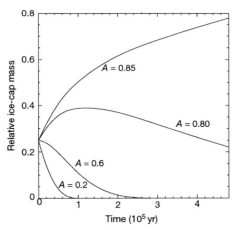

FIG. 16.17 **The runaway albedo effect.** This simulation tracks the time evolution of four ice caps with initially equal mass, each with an initial radius of 800 km and an average depth of 500 m, but with different albedos A. In each seasonal cycle of 248 Earth years, it has been assumed that a fraction $f = 0.5$ of absorbed solar energy converts ice to vapor, which is then redeposited on the ice caps in proportion to their exposed surface areas. As ice caps gain or lose mass, their areal coverage and average depth evolve as well, such that the ratio of diameter to depth remains constant. The darker two ice caps ($A = 0.2$ and $A = 0.6$) efficiently absorb solar energy and sublimate away rapidly, while the brighter two ($A = 0.8$ and $A = 0.85$) initially grow. Eventually, the brighter two ice caps compete more directly for the available vapor, and the one with the higher albedo captures all of the available ice, growing to a depth of 700 m and a radius of 1270 km. Finally, the depth-to-diameter ratio of the surviving ice cap will grow as the lower-albedo margins preferentially sublimate in a process analogous to that considered by Hamilton et al. (2016). *(Source: Hamilton, D.P., S.A. Stern, J.M. Moore, L.A. Young, and the New Horizons Geology, Geophysics & Imaging Theme Team (2016), The rapid formation of Sputnik Planitia early in Pluto's history,* Nature, *540, 97-99.) Copyright info: 2016 Macmillan Publishers Limited, part of Springer Nature. All rights reserved.*

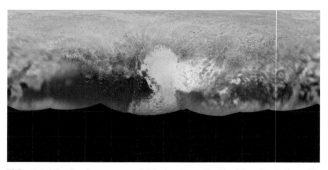

FIG. 16.18 **Surface map of Pluto.** Sputnik Planitia, the informally named western lobe of the white heart-shaped feature, is approximately 1,000 km across and is centered on a latitude of 25° N and longitude of 175°, with the zero of longitude defined to run directly underneath Pluto's nearest and the largest moon Charon. Owing to Pluto's slow rotation and the approach vector of New Horizons, the highest-resolution images were of Sputnik Planitia, and most other regions were imaged at much lower resolution from farther away. Consequently, this map is a mosaic of multiple images of differing resolution. Faint grid lines of latitude and longitude are spaced by 30°. Regions south of about 30°S were not sunlit and, hence, were not imaged, because they are currently experiencing polar night. *(Source: Hamilton, D.P., S.A. Stern, J.M. Moore, L.A. Young, and the New Horizons Geology, Geophysics & Imaging Theme Team (2016), The rapid formation of Sputnik Planitia early in Pluto's history, Nature, 540, 97-99.) Copyright info: © 2016 Macmillan Publishers Limited, part of Springer Nature. All rights reserved.*

Pluto's surface albedo determines the fraction of incident energy that is actually absorbed by Pluto. Hamilton et al. (2016)'s analysis indicated that dark areas in the surface map of Pluto derived from the images obtained from the New Horizons cameras (Fig. 16.18) absorb about 85% of the incident sunlight ($A = 0.15$ in eq. (16.1)), whereas Grundy et al. (2016) found that the bright ices of Sputnik Planitia reflect a comparable fraction of incident sunlight back into space ($A \approx 0.85$). Hamilton et al. (2016) offered the following explanation with regard to the bright and dark regions seen in the surface map of Pluto depicted in Fig. 16.18: "Owing to diminished absorption of sunlight, temperatures over bright icy regions should be much lower than over darker regions, thereby enhancing the deposition rate of volatiles and inhibiting sublimation (Earle et al., 2016). The strong effect of albedo can drive longitudinal variations in ice cover. The dark equatorial regions of Pluto remain dark by discouraging deposition of bright frost layers, whereas the bright areas attract additional frost. The equally marked albedo variations on Saturn's moon Iapetus are driven by similar processes (Spencer and Denk, 2010). Acting in concert with variations in incident sunlight, albedo differences will cause volatiles to be preferentially deposited on the brightest terrain in the ±30° latitude bands." Hamilton et al. (2016) suggested that "perhaps the most likely trigger for the formation of Sputnik Planitia was bright terrain formed from refreezing of melt produced by a mid-sized impactor or an early period of cryo-volcanism. Shadowing by mountains or in craters are also possibilities, although these are probably weaker effects. In any case, once ices begin to be deposited in one area, they raise the albedo and reflect more sunlight back to space. Thus, the absorption of energy decreases, sublimation is inhibited and deposition of more ices is enhanced, leading to a runaway process." Fig. 16.17 provides support to these conclusions. It is believed that the runaway albedo effect, acting over time, probably concentrated the ice into the deep convecting structure that we now observe. Convection works to keep the surface bright by continually exposing fresh ice and reprocessing dirty ice, protecting the ice cap against additional absorption of sunlight. Hamilton et al. (2016) suggest that the runaway ice-cap formation on Pluto is the reverse of the accelerating melting that is now ongoing in Earth's north polar ice pack.

16.8.4 Knowing more on Pluto's insolation and reflectance in the light of new data from New Horizons mission

Since previous long-term insolation modeling in the early 1990s, new atmospheric pressure data, and increased computational power have generated new motivation and increased capabilities for the study of Pluto's complex long-term (million-years) insolation history. Earle and Binzel (2015) carried out studies on Pluto's insolation history, with particular reference to latitudinal variations and effects on atmospheric pressure. In their studies, Earle and Binzel (2015) found latitudinal dichotomies (contrasts) when comparing average insolation over timescales of days, decades, centuries, and millennia, where all timescales they considered are short relative to the predicted timescales for Pluto's chaotic orbit. These researchers suggested that depending on the timescales of volatile migration, some consequences of these insolation patterns may be manifested in the surface features revealed by New Horizons. Earle and Binzel (2015) found that the Maximum Diurnal Insolation (MDI) at any latitude is driven most strongly when Pluto's obliquity creates a long arctic summer (or "midnight sun") beginning just after perihelion (the point in the orbit of a planet, asteroid, or comet at which it is closest to the Sun). Earle and Binzel (2015) found that Pluto's atmospheric pressure, as measured through stellar occultation (defined as the passage of one celestial body in front of another, thus hiding the other from view) observations during the past three decades, shows a circumstantial correlation with this midnight sun scenario as quantified by the MDI parameter.

Nadeau and McGhee (2018) invented a simple method for calculating a planet's mean annual insolation by latitude. The inventers used a sixth-order Legendre series expansion to approximate the mean annual insolation by latitude of a planet with obliquity angle β, leading to faster computations with little loss in the accuracy of results. They have discussed differences between their method and selected computational results for insolation found in the literature.

16.8.5 Ice-laden "Heart-Shaped" region on Pluto's surface—formation and stability of Sputnik Planitia crater

When NASA's New Horizons spacecraft sped past Pluto on July 14, 2015, the craft was moving at 52,000 km per hour, so fast that it was able to capture close-ups of only one side of Pluto—the hemisphere that the Sun illuminated at the encounter time (described as "near-side"). The other side was temporarily shrouded in shadow.

In one part of Pluto's surface, the spacecraft's cameras captured a giant heart-shaped feature (see Fig. 16.18). Following the passage on July 14, 2015 of Pluto by the New Horizons spacecraft, the "Cold Heart of Pluto" was named Tombaugh Regio, after Clyde W. Tombaugh who discovered Pluto on Jan 23, 1930. The western lobe of the vast, deep, volatile-ice-filled basin inside Pluto's heart-shaped feature is provisionally called Sputnik Planitia (SP), formerly known as Sputnik Planum. According to Hamilton et al. (2016), SP is a bright, roughly circular feature that resembles a polar ice cap. It is approximately 1,000 km across and is centered on latitude of 25° North and a longitude of 175°, almost directly opposite the side of Pluto that always faces its nearest and the closest moon Charon as a result of tidal locking (Stern et al., 2015). In this context it may be noted that "tidal locking," in the most well-known case, occurs when an orbiting astronomical body always has the same face toward the object it is orbiting. This is known as synchronous rotation: *the tidally locked body takes just as long to rotate around its own axis as it does to revolve around its partner*. It may be noted that the zero of Pluto's longitude is defined to run directly underneath Pluto's nearest and the largest moon Charon. According to Hamilton et al. (2016), "Owing to the strength of even a thin lithosphere, the initial ice deposits would have had positive topography contributing to a global longitudinal asymmetry. Such asymmetry would be subject to tidal torques from Charon that would act to rotate the ice cap toward the Pluto–Charon line. This probably occurred very early, within a few million years after Pluto and Charon were formed by the impact of two large Kuiper belt objects (Canup, 2005)." Once the basin was established, ice would naturally have accumulated there (Bertrand and Forget, 2016). Then, provided that the basin was a positive gravity anomaly, true polar wander could have moved the feature toward the Pluto–Charon tidal axis, on the far side of Pluto from Charon (Nimmo et al., 2016; Keane et al., 2016). Once locked in, the ice cap will remain near its current equilibrium longitude of 180° against all but the most extreme changes.

McKinnon et al. (2016) found that SP is a $\sim$ 900,000 km^2 oval-shaped unit of high-albedo plains set within a topographic basin at least 2–3 km deep. By analogy with other large basins in the Solar System (Zuber et al., 2012), it is believed that the "heart-shaped" basin was formed more than 4 billion years ago by a cosmic impact (Schenk et al., 2015), presumably as a result of an object from the Kuiper Belt striking Pluto. The basin's scale, depth and ellipticity, and rugged surrounding mountains, also suggest an origin as a huge impact. Scientists now know that Sputnik Planitia exerts an extraordinary influence over Pluto's activity. As sunlight warms the frozen plain, a pulse of ice sublimates into vapor that wafts (gently moves) upwards, before dropping back down at the end of the day. The heart might have even knocked Pluto on its side.

Hamilton et al. (2016) reported modeling that shows that "ice quickly accumulates on Pluto near latitudes of 30 degrees north and south, even in the absence of a basin, because, averaged over its orbital period, those are Pluto's coldest regions." Hamilton et al. (2016) argued that "within a million years of Charon's formation, ice deposits on Pluto concentrate into a single cap centered near a latitude of 30 degrees, owing to the runaway albedo effect." According to Hamilton et al. (2016), "This accumulation of ice causes a positive gravity signature that locks, as Pluto's rotation slows, to a longitude directly opposite Charon. Once locked, Charon raises a permanent tidal bulge on Pluto, which greatly enhances the gravity signature of the ice cap. Meanwhile, the weight of the ice in Sputnik Planitia causes the crust under it to slump, creating its own basin, as has happened on Earth in Greenland (Spada et al., 2012). Even if the feature is now a modest negative gravity anomaly, it remains locked in place because of the permanent tidal bulge raised by Charon. Any movement of the feature away from 30 degrees latitude is countered by the preferential re-condensation of ices near the coldest extremities of the cap." Therefore, Hamilton et al. (2016)'s modeling suggests that Sputnik Planitia formed shortly after Charon did and has been stable, albeit (though) gradually losing volume, over the age of the solar system.

In the solar system, calamitous events generally do not spell the end of worlds. A planet or moon can take a hit from an asteroid or comet. Wrenched off its previous trajectory, it might falter for a time, or tilt on its axis, or experience a dramatic reorganization of its exterior. But things will eventually stabilize. Titanic changes like these are happening at Pluto today, largely because of the iconic "heart" on its surface. Pluto's orientation in space is controlled by heavy ice in this heart, and probably by a massive global subsurface ocean. The crater retention age of SP is very young, no more than $\sim$10 Myr based on models of the impact flux of small Kuiper belt objects onto Pluto (Greenstreet et al., 2015). According to McKinnon et al. (2016), this indicates renewal, burial or erosion of the surface on this timescale or shorter.

The SP basin has been filling with nitrogen ice over the years and now contains huge amounts of nitrogen. Because of the fact that nitrogen deposition is latitude-dependent (Binzel et al., 2016), it is argued that nitrogen loading and reorientation may have exhibited complex feedbacks (Keane et al., 2016). There are indications to believe that SP's present nitrogen ice may be up to 10 km (6 miles) thick. It has been suggested that the two bluish-white 'lobes' that extend to the southwest and northeast of the 'heart' may represent exotic ices being transported away from SP. McKinnon et al. (2016) have suggested that the apparent flow lines around obstacles in northern SP and the pronounced distortion of some fields of pits in southern SP are evidence for the lateral, advective flow of SP ices (Stern et al., 2015; Moore et al., 2016).

It is interesting that the deep nitrogen-covered SP basin is located in a particularly interesting spot, named "tidal axis," on Pluto's surface such that SP always faces away from Pluto's tidally locked largest moon *Charon*. In terms of planetary jargon, Pluto and Charon are "tidally locked," meaning these two celestial bodies always show the same face to each other, just as Earth's moon always shows just one side to Earth. "Tidal axis" is a term used in describing the gravitational relationship (locking) between a planet and its moon. In the case of Pluto and Charon, "tidal axis" is the line along which the gravitational pull experienced by Pluto from Charon is the strongest. In the case of Pluto, the "heart-shaped" geological feature (SP) is aligned nicely with (i.e., located very close to the longitude of) Pluto's "tidal axis" (Moore et al., 2016). If a straight line is drawn from the center of Charon through the center of Pluto and out the other side, this straight line would pass very close to the location where SP is located today. The alignment is so precise that it is as if Charon floats over the area directly opposite SP.

Pluto researchers unanimously agree that the location of SP is no accident. They consider that Pluto reoriented itself to make sure the formation ended up where it did in a process called "true polar wander." The phenomenon known as true polar wander happens when something very catastrophic takes place (e.g., forming a new feature causing changes in the distribution of a planet's mass that changes its balance of mass). On a spinning body, extra mass migrates toward the equator, and areas with lower mass end up closer to the poles.

Rubincam (2003) had discussed polar wander on Triton and Pluto due to volatile migration. If one area suddenly gains more mass due to some reason, the celestial body will reorient itself so that the more massive feature is closer to the equator. If a part of the celestial body suddenly loses a lot of mass, that feature will drift toward the poles. What sort of mechanism brought SP to its current location, directly opposite Charon, became a scientific curiosity that needed to be resolved. Interestingly, SP is a crater-like depression (a negative topography) in the ground. The fact that SP is located near Pluto's equator indicates that it has comparatively more mass than other parts of Pluto. It is a little counter-intuitive, however, because SP is essentially a negative topography in the ground.

According to planetary researchers investigating Pluto, its polar wander started with the suspected SP impact (see Keane et al., 2016). In a study Hamilton (2015) examined the icy cold heart of Pluto; and found that the dwarf planet's skin is riven with cracks and faults (see Keane et al., 2016). Surprisingly, the pattern of faults existing on Pluto's surface skin was found to match what one might expect to see during simulations of true polar wander (see Keane et al., 2016). The Pluto researchers found that if most of the material ejected from the crater during the impact ultimately landed along the sides of the crater, and if a disproportionate amount of volatile ices such as methane, nitrogen, and carbon got deposited in the impact basin due to seasonal snowfall, it is possible for SP to gain what is called a positive mass anomaly, inducing true polar wander (see Keane et al., 2016).

Keane et al. (2016) modeled what happened as nitrogen ice accumulated in SP. They found that once enough nitrogen ice has piled up, an extra mass is formed in SP. If there is an excess of mass in one spot on Pluto, there is a tendency for it to drift to the equator. Eventually, over millions of years, it will drag the whole planet over. This tumble brought SP to the southeast, until the plain-faced directly away from Charon as it does today. Keane et al. (2016)'s models predicted this very orientation: In this spot, Pluto's reorientation would have stressed Pluto's crust. These findings indicate that this "heart-shaped feature" caused Pluto to roll over the eons, and this reorientation probably wouldn't have been possible without a subsurface ocean.

According to several researchers (e.g., Rubincam, 2003; Nimmo and Matsuyama, 2007) reorientation of SP, arising from tidal and rotational torques, can explain the basin's present-day location, but requires the feature to be a positive gravity anomaly (Keane et al., 2016), despite its negative topography. According to Hamilton et al. (2016), "any equator-ward displacement of the ice cap, or indeed even an initially equatorial ice cap, would be affected by many thousands of annual sublimation and deposition cycles that would move the ice cap slowly pole-ward toward the latitudes with the least orbit-averaged flux of sunlight. The north–south orientation of Sputnik Planitia may provide some evidence for such motions. Eventually conditions stabilized, leaving Pluto with a single dominant ice cap centered on a latitude of 25° determined by a minimum of solar illumination and a longitude of 175° set by long-ago tidal forces from Charon. Therefore, Pluto should be in one of four possible end states: with an ice cap centered near 30° or −30° latitude and near 0° or 180° longitude. Positions within 10° of these end states cover just 3% of Pluto's surface, making Sputnik Planitia's location near one of them particularly noteworthy."

16.8.6 Gaining insights on Pluto's SP crater basin and its surroundings

Pluto's positive pole is defined by Archival (2011a, b) and points in the direction of the angular momentum vector. Pluto's prime or 0° meridian is the sub-Charon longitude. Informally, we refer to the positive pole direction as "north," and to the direction of increasing longitude as "east" (Schenk et al., 2018). The longitude and latitude on Pluto are defined according to the right-hand rule and following the recommendations of Zangari (2015). Owing to Pluto's slow rotation around the Sun (it's one orbit around the Sun takes 248 Earth-years) and the approach vector of New Horizons probe, the highest-resolution images were of Sputnik Planitia (SP), and most other regions were imaged at much lower resolution from farther away. Consequently, the available topographic map is a mosaic of multiple images of differing resolutions. The best mapping and stereo imaging of Pluto covered the illuminated anti-Charon hemisphere, the hemisphere observed during closest approach on July 14, 2015. An important characteristic of the Pluto encounter was that the phase angle remained essentially constant at ~15° until the final hours, facilitating production of a global map with generally uniform illumination quality. During approach, LORRI imaging of Pluto (and Charon) was acquired every ~15° of longitude during the last Pluto rotation before encounter, thereby providing continuous longitudinal mapping at increasing resolution as the two bodies rotated under the approaching spacecraft (Schenk et al., 2018).

The stunning images captured by the New Horizons spacecraft in July 2015 could be used by planetary researchers to detect subtle differences in Pluto's surface, and confirmed the notion that Pluto's surface is endowed with a "heart-shaped" geological feature. SP is the most prominent geological feature on Pluto revealed by NASA's New Horizons mission; and this basin is considered to be central to Pluto's geological activity (Stern et al., 2015; Moore et al., 2016). The bright volatile-ice-rich deposit forming SP is broadly flat-lying and forms a roughly 1500 × 900-km-wide pear-shaped oval with its smaller end to the south (e.g., Moore et al., 2016; McKinnon et al., 2016).

Some of the available images of Pluto are actually a combination of observations from the main camera on New Horizons Long-Range Reconnaissance Imager (LORRI) and the probe's visible/infrared imager (known as Ralph) which provided data for the colors in the view. The enhanced colors allow planetary researchers to identify differences in the composition and texture of Pluto's surface. During its flyby of Pluto, New Horizons collected measurements of surface features, including the dimensions of Pluto's bright, heart-shaped region. From the probe's measurements, Binzel et al. (2016) determined the size and depth of SP, which is similar in proportional size to the largest basins on Mercury and Mars. SP covers 5% of Pluto's surface. The massive basin also appears extremely bright relative to the rest of the planet; and the reason, the New Horizons data suggest, is that it is composed of a mixture of nitrogen, methane and carbon monoxide ices (Grundy et al., 2016; Protopapa et al., 2016)—substances that are volatile at the temperatures expected on Pluto (about 40 K). Its location at 30°N—a latitude that is temperate on Earth—currently receives substantially less solar energy per Pluto-year than all other latitudes. SP is essentially a vast, frozen sea, one in which convective turnover (now, and even more vigorously in the past) continually refreshes the surface volatile ice inventory (McKinnon et al., 2016).

An unexpected bonus of the encounter was that scattering of light by atmospheric haze illuminated areas on Pluto otherwise in darkness. This illumination permits mapping of haze-lit terrains in the three highest resolution MVIC scans at 0.650, 0.475 and 0.315 km/pixel resolutions and in stereo. Haze-illuminated imaging near longitude 150° extends down to −56° latitude, the southernmost imaging obtained of Pluto, and extend useful imaging 200–250 km into what would have been dark regions at the time of closest approach on the encounter hemisphere. Thus, a unique feature of the Pluto imaging data set is the observation of terrains illuminated only by light scattered from atmospheric haze, allowing scientists to map terrains in the southern hemisphere that would otherwise have been in darkness (Schenk et al., 2018). However, the brightness of haze-illuminated areas was < 1% that of the solar illuminated terrains, and the dynamic range and signal-to-noise quality were significantly poorer than in the sun-lit regions (Schenk et al., 2018). Fig. 16.19 illustrates the global image mosaic of Pluto produced at 300 m/pixel. Cylindrical map projection centered on 180° longitude. Black areas in all global map products were unilluminated during the 2015 encounter. The encounter hemisphere is in map center, with approach imaging to the left starting at ~35 km/pixel resolution and increasing in quality to the west. Area extending downward from the general east-west boundary between the illuminated and unilluminated areas are those illuminated by light scattered by atmospheric haze, processed to have similar brightness properties as areas illuminated by sunlight. The resulting global map product shows the illuminated surface of Pluto down to ~35°S.

The global digital elevation model (DEM) of Pluto (Fig. 16.20) clearly indicating Sputnik Planum (SP) on Pluto's surface. It is clear that this deep basin overwhelmingly dominates the topography of Pluto. Based on the morphology and preliminary topographic data indicating a depression several kilometers deep, Moore et al. (2016) concluded that this basin was a large highly degraded impact structure subsequently partially filled with volatile ices. Schenk et al. (2018) reported concurrence with this conclusion and explored the topographic properties of this large structure in

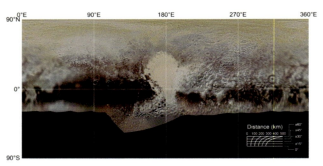

FIG. 16.19 Global image mosaic of Pluto produced at 300 m/pixel. Cylindrical map projection centered on 180° longitude. Black areas in all global map products were unilluminated during the 2015 New Horizons encounter. The encounter hemisphere is in map center, with approach imaging to the left starting at ~35 km/pixel resolution and increasing in quality to the west. Area extending downward from the general east-west boundary between the illuminated and unilluminated areas are those illuminated by light scattered by atmospheric haze, processed to have similar brightness properties as areas illuminated by sunlight. *(Source: Schenk, P.M., R.A. Beyer, W.B. McKinnon, J.M. Moore, J.R. Spencer, O.L. White, K. Singer, F. Nimmo, C. Thomason, T.R. Lauer, S. Robbins, O.M. Umurhan, W.M. Grundy, S.A. Stern, H.A. Weaver, L.A. Young, K.E. Smith, C. Olkin, the New Horizons Geology and Geophysics Investigation Team (2018), Basins, fractures and volcanoes: Global cartography and topography of Pluto from New Horizons, Icarus, 314:400–433)* **Elsevier Publication**.

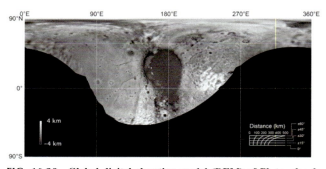

FIG. 16.20 Global digital elevation model (DEM) of Pluto, clearly indicating Sputnik Planum (SP) on Pluto's surface. Cylindrical map projection centered on 180° longitude. Dark areas were unilluminated or do not have resolvable stereogrammetric data from the 2015 encounter. Note the low resolution and in some areas noisy DEM data sets at upper left and right, derived from imaging data acquired 5–10 h prior to encounter. *(Source: Schenk, P.M., R.A. Beyer, W.B. McKinnon, J.M. Moore, J.R. Spencer, O.L. White, K. Singer, F. Nimmo, C. Thomason, T.R. Lauer, S. Robbins, O.M. Umurhan, W.M. Grundy, S.A. Stern, H.A. Weaver, L.A. Young, K.E. Smith, C. Olkin, the New Horizons Geology and Geophysics Investigation Team (2018), Basins, fractures and volcanoes: Global cartography and topography of Pluto from New Horizons, Icarus, 314:400–433)* **Elsevier Publication**.

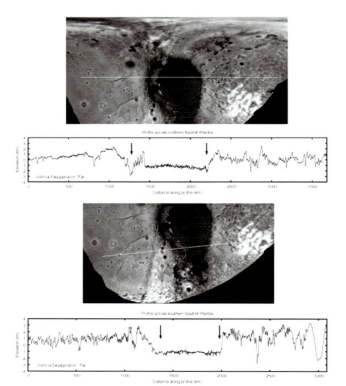

FIG. 16.21 Profiles across the northern (top) and southern (bottom) halves of Sputnik Planitia and surrounding terrains. Arrows indicate depressed moats and raised margins along the outer edges of the low-lying ice sheet in SP. *(Source: Schenk, P.M., R.A. Beyer, W.B. McKinnon, J.M. Moore, J.R. Spencer, O.L. White, K. Singer, F. Nimmo, C. Thomason, T.R. Lauer, S. Robbins, O.M. Umurhan, W.M. Grundy, S.A. Stern, H.A. Weaver, L.A. Young, K.E. Smith, C. Olkin, the New Horizons Geology and Geophysics Investigation Team (2018), Basins, fractures and volcanoes: Global cartography and topography of Pluto from New Horizons, Icarus, 314:400–433).* **Elsevier Publication**.

greater detail. The SP bright ice deposit is enclosed within a larger pear-shaped topographic basin, defined by an eroded and modified broadly arched raised ridge 125–175 km wide and variable in elevation, rising locally up to ~1 km above surrounding plains and 2.5–3.5 km above the surface of the Sputnik Planitia ice sheet (Fig. 16.21). The DEM also reveals that the outer 10–20 km of the ice sheet is depressed a few hundred meters relative to the interior of the ice sheet along its contact with the eroded rim escarpment (Schenk et al., 2018). With a basin-to-planet diameter ratio of ~0.67, the SP basin is among the largest in size known with respect to the host planet, larger than the Odysseus basin on Tethys, which has a ratio of ~0.4, though not as large as Rheasilvia on Vesta, which has a ratio of ~0.95 (Schenk et al., 2018).

16.8.7 Subtle topography of ice domes and troughs of cellular plains within Sputnik Planitia crater

Following its flyby of Pluto in July 2015 (Stern et al., 2015), NASA's *New Horizons* spacecraft has returned high quality images that cover the entire encounter hemisphere of the planet, and which have revealed a highly (and unexpectedly) diverse range of terrains, implying a complex geological history. The centerpiece of the encounter hemisphere is the prominent Sputnik Planitia (SP) crater, and it is on this feature and its immediate environs that initial geological mapping of Pluto's surface has been focused. McKinnon

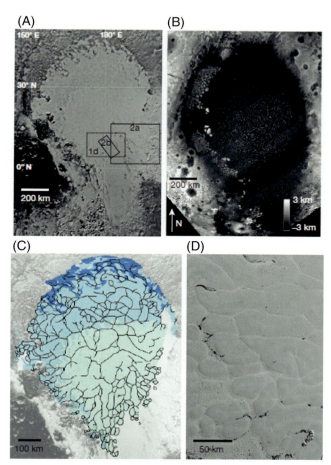

FIG. 16.22 **Topographic and map views of Sputnik Planum (SP) on Pluto's surface.** (A) Base map showing locations of some figure panels. (B) Stereo-derived topography, showing that SP lies within a kilometre-deep basin (depth coded on greyscale, see key at bottom right). Southwest–northeast banding and central basin 'speckle' are artefacts or noise; elevations are relative. (C) Map of troughs (black lines), which define cell boundaries (note enlarged scale compared with (A) and (B)). Cell size increases and/or becomes less well connected toward SP center, consistent with a thickened N_2 ice layer there. Aquamarine shading indicates 'bright cellular plains', within which troughs are topographically defined. (D) 350 m per pixel Multispectral Visible Imaging Camera (MVIC) image (position shown in (A)) that shows cellular/polygonal detail (north is to right). *(Source: McKinnon et al. (2016), Convection in a volatile nitrogen-ice-rich layer drives Pluto's geological and atmospheric vigour, Nature, https://c, 82–85). Copyright info: 2016 Macmillan Publishers Limited. All rights reserved.*

et al. (2016) carried out a detailed study of SP and its volatile nitrogen-ice-rich layer. They also carried out mapping of the SP region and surroundings (Fig. 16.22) using the New Horizons LORRI data. The LORRI basemap in Fig. 16.22A was created from the 5 × 4 mosaic sequence P_LORRI (890 m per pixel), taken by LORRI. Mapping of cell/polygon boundaries (Fig. 16.22C) was carried out in ArcGIS using this mosaic and additional images from P_LORRI_Stereo_Mosaic (390 m per pixel). Fig. 16.22A–C shows simple cylindrical projections, so the scale bars are approximate. Locations of Fig. 16.22A,B,D are shown as insets in Fig. 16.22A.

Based on the hitherto unknown information obtained from New Horizons, McKinnon et al. (2016) made the following observation with regard to the variety of ices found in SP, "From New Horizons spectroscopic mapping, N_2, CH_4 and CO ice all concentrate within Sputnik Planum (Grundy et al. 2016). All three ices are mechanically weak, van der Waals bonded molecular solids and are not expected to be able to support appreciable surface topography over any great length of geological time (Eluszkiewicz and Stevenson, 1990; Yamashita et al., 2010; Moore et al., 2015; Stern et al., 2015), even at the present surface ice temperature of Pluto (37 K) (Stern et al., 2015)." According to McKinnon et al. (2016), this is consistent with the overall smoothness of SP over hundreds of km (Fig. 16.22B). Convective overturn that reaches the surface would also eliminate impact and other features.

According to McKinnon et al. (2016)'s studies, quantitative radiative transfer modeling of the relative surface abundances of N_2, CH_4 and CO ices within SP (Protopapa et al., 2016) shows that N_2 ice dominates CH_4 ice, especially in the central portion of the planum. Water ice has been identified in the rugged mountains that surround SP (Grundy et al., 2016). N_2 and CO ice have nearly the same density (close to 1.0 g/cm^3), whereas CH_4 ice is half as dense as this (Moore et al., 2016). Hence water-ice blocks can float in solid N_2 or CO, but not in solid CH_4 (McKinnon et al., 2016).

McKinnon et al. (2016) found that the surface of SP is at least 2–3 km below the surrounding terrain (Fig. 16.22B). With regard to Fig. 16.22B, McKinnon et al. (2016) provide the following description: "Pluto was assumed to be a sphere of 1187-km radius (Stern, et al., 2015), and elevations were determined using an automated stereo photogrammetry method based on scene-recognition algorithms (Schenk et al., 2004). Spatial resolutions are controlled by the lower resolution MVIC scan and, using this method, are further reduced by a factor of three to five. Vertical precisions can be calculated through standard stereo technique from $m_{rp}(\tan e_1 + \tan e_2)$, where m is the accuracy of pixel matching (0.2–0.3), rp is pixel resolution, and e_1 and e_2 are the emission angles of the stereo image pair."

McKinnon et al. (2016) found that the central and northern regions of SP display a distinct cellular/polygonal pattern (Fig. 16.22C). It was found that in the bright central portion, the cells are bounded by shallow troughs locally up to 100 m deep (Fig. 16.22D), and the centers of at least some cells are elevated by ∼ 50 m relative to their edges (Moore et al., 2016). The southern region and eastern margin of SP do not display cellular morphology, but instead show featureless plains and dense concentrations of km-scale pits (Moore et al., 2016).

In further explaining Fig. 16.22B, McKinnon et al. (2016) states that "the precision is about 230 m, well suited for determining elevations of Pluto's mountains and deeper craters as well as the rim-to-floor depth of the SP basin. It is not sufficient to determine planum cell/polygon elevations."

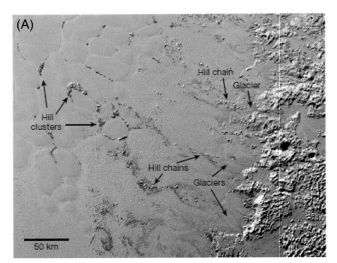

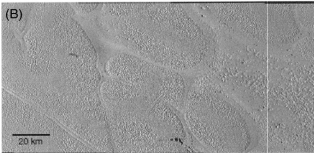

FIG. 16.23 High-resolution images of cellular terrain within SP. (A) Kilometre-scale hills appear to emanate from uplands to the east (at right), and are probably darker water-ice blocks and methane-rich debris (arrows) that have broken away and are being carried by denser, N_2-ice-dominated glaciers into SP, where they become subject to the convective motions of SP ice, and are pushed to the downwelling edges of the cells at left. (B) Part of the highest-resolution image sequence taken by New Horizons (80 m per pixel); surface texture (for example, pitting) concentrates toward cell boundaries and in regions apparently unaffected by convection (such as at right). *(Source: McKinnon et al. (2016), Convection in a volatile nitrogen-ice-rich layer drives Pluto's geological and atmospheric vigour,* Nature, *534, 82–85). Copyright info: 2016 Macmillan Publishers Limited. All rights reserved.*

It has been suggested that in the planum center, the dearth of sufficient frequency topography inhibits closure of the stereo algorithm, hence the noise in the center of SP in Fig. 16.22B. McKinnon et al. (2016) determined the subtle topography of the raised cells within SP from a preliminary photoclinometric (shape from shading) analysis (for example, Schenk, 2002), and is subject to further refinement of the photometric function for the bright cellular plains. McKinnon et al. (2016) observed that "Photoclinometry offers high-frequency topographic data at spatial scales of image resolution, but can be poorly controlled over longer wavelengths. Photoclinometry is sensitive to inherent albedo variations, but can be especially useful for investigating features with assumed symmetry, such as impact craters, which allows a measure of topographic control. Fig. 16.23 illustrates the high-resolution images of cellular terrain within SP. The ovular domes and bounding troughs of the bright cellular plains within SP are such symmetric features, and intrinsic albedo variations are muted in the absence of dark knobs or blocks, so photoclinometry is well-suited to determining elevations across individual cells within the bright cellular plains" see Fig. 16.22D and Fig. 16.23B. McKinnon et al. (2016) found slight topographic dimples over down-wellings in some of their calculations, which they interpret to be related to trough formation at cell edges (Fig. 16.23B). According to them, the troughs themselves, however, are likely to be finite amplitude topographic instabilities of the sort seen on icy moons elsewhere (Bland and McKinnon, 2015).

Gladstone et al. (2016) found that Pluto's atmospheric nitrogen escape rate is much lower than previously estimated. The volatile nitrogen-ice-rich layer of SP is composed of molecular nitrogen, methane, and carbon monoxide ices (Grundy et al. 2016), but dominated by N_2-ice. For Pluto, SP acts as an enormous glacial catchment or drainage basin, the major topographic trap for Pluto's surficial, flowing N_2 ice.

Judging from Figs. 16.22 and 16.23, McKinnon et al. (2016) found that the volatile nitrogen-ice-rich layer of SP is organized into cells or polygons, typically ~10–40 km across, that resemble the surface manifestation of solid-state convection (Stern et al., 2015; Moore *et al.,* 2016). Taking account of the realization that the cells/polygons on SP are the surface expression of convective cells, McKinnon et al. (2016) described the convection mechanism thus, "Convection in a layer occurs if the critical Rayleigh number (Ra_{cr}) is exceeded. The Rayleigh number, the dimensionless measure of the vigor of convection, for a power-law fluid heated from below is given by (Solomatov, 1995).

$$Ra = \frac{\rho g \alpha A^{1/n} \Delta T D^{(2+n)/n}}{K^{1/n} \exp\left(E^* / nRT\right)} \quad (16.2)$$

where D is the thickness of the convecting layer, κ is the thermal diffusivity, g is the acceleration due to gravity, ρ the ice layer density, α the volume thermal expansivity, ΔT the super-adiabatic temperature drop across the layer, and A is the pre-exponential constant in the relationship between stress and strain-rate, E^* is the activation energy of the dominant creep mechanism, and R is the gas constant." The critical Rayleigh number depends on the temperature drop and the associated change in viscosity (Solomatov, 1995), as deformation mechanisms are thermally activated processes. For a given ΔT, the Ra_{cr} implies a critical or minimum layer thickness, D_{cr}, below which convection cannot occur. The surface temperature of the ices on Pluto at the time of the New Horizons encounter was 37 K (Stern et al., 2015; Gladstone et al., 2016) and the melting temperature of N_2 ice was 63.15 K (Eluszkiewicz, 1991). McKinnon et al. (2016) assumed

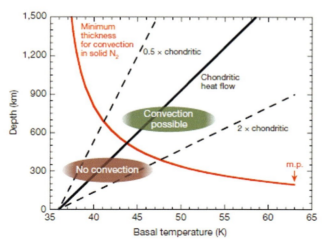

FIG. 16.24 Minimum thickness for convection in a layer of solid N_2 ice on Pluto, as a function of basal temperature. Convection can occur above the solid red curve provided a sufficient perturbation exists (area labelled "convection possible"). Limit is based on numerical and laboratory experiments & theory, and creep measurements for nitrogen ice. Basal temperatures due to conductive heat flow (3 mW/m^2) from Pluto are shown for comparison (solid black line), along with variations of a factor of 2 in heat flow (dashed black lines). For approximately present-day chondritic heat flows, basal temperatures exceed the convective threshold for layer thicknesses in excess of about 500 m. In contrast, the minimum thickness for convection by volume diffusion creep would plot off the graph to the upper right. *(Source: McKinnon et al. (2016), Convection in a volatile nitrogen-ice-rich layer drives Pluto's geological and atmospheric vigour,* Nature, *534, 82–85). Copyright info: 2016 Macmillan Publishers Limited. All rights reserved.*

an average ice surface temperature of 36 K set by vapor-pressure equilibrium over an orbital cycle (Stansberry and Yelle, 1999), and an upper limit on the basal temperature set by the N_2 ice melting temperature of 63 K (Scott, 1976). From their studies depicted in Fig. 16.24, McKinnon et al. (2016) concluded that convection in solid nitrogen on Pluto is a facile process: critical thicknesses are generally low, less than 1 km, as long as the necessary temperatures at depth are achieved.

According to McKinnon et al. (2016), if the cells/polygons on SP are the surface expression of convective cells, then cell diameters of 20–40 km imply depths to the base of the N_2 ice layer in SP of about 10–20 km. Rheological studies (i.e., studies of the flow of matter in any state in response to an applied force) are sometimes employed in gaining a better understanding of the processes taking place on the surfaces of celestial bodies. Based on available rheological measurements (Yamashita et al., 2010), McKinnon et al. (2016) reported that solid layers of N_2 ice approximately greater than 1 km thick should convect for estimated present-day heat flow conditions on Pluto. According to McKinnon et al. (2016), in sluggish lid convection, the surface is in motion and transports heat, but moves at a much slower pace than the deeper, warmer subsurface. They consider that a defining characteristic of this regime—depending on Ra$_b$ (the Rayleigh number defined with the basal viscosity) and Arrhenius viscosity ratio ($\Delta\eta$)—is convection cells with large aspect ratios (widths of convective cells divided by layer depth).

McKinnon et al. (2016) carried out numerical convection calculations with the well-benchmarked fluid dynamics finite element code CitCom (Moresi and Solomatov, 1995). {CitCom is freely available, in the version CitComS, released under a General Public License and downloadable from the Computational Infrastructure for Geodynamics (http://geodynamics.org)}. CitCom solves the equations of thermal convection of an incompressible fluid in the Boussinesq approximation and at infinite Prandtl number using an Arrhenius viscosity or an exponential law (the Frank–Kamenetskii approximation). McKinnon et al. (2016) used the latter approximation, for both Newtonian (stress-independent) and non-Newtonian viscosities, to best compare their results with those in the literature (Moresi and Solomatov, 1995; Solomatov, 1995; Solomatov and Moresi, 1997; Hammond and Barr, 2014).

In McKinnon et al. (2016)'s studies, numerical simulations were carried out in terms of dimensionless parameters, without pre-supposing any particular values for the depth of the SP volatile ice layer or Pluto's heat flow, and so on. It was found that convection in a km-thick N_2 layer within Pluto's SP basin emerges as a compelling explanation for the remarkable appearance of the Planum surface (Fig. 16.22). Larger Kuiper belt objects are known to be systematically brighter (more reflective) than their smaller cousins in the Kuiper belt (Brown, 2008). According to McKinnon et al. (2016)'s interpretation, "convective renewal of volatile ice surfaces, as in a basin or basins similar to SP, may be one way in which the dwarf planets of the Kuiper belt maintain their youthful appearance."

McKinnon et al. (2016)'s studies indicate that "The transition from cellular to non-cellular plains could reflect several things, including shallowing of the volatile ice layer, lower heat flow, and in the case of non-Newtonian flow, an insufficient initial temperature perturbation (Solomatov, 1995; Barr and McKinnon, 2007; Solomatov and Barr, 2007)." According to McKinnon et al. (2016), "The simplest explanation, however, for smaller cell sizes with distance from the center of SP (Fig. 16.23C), and then a transition to level plains (no cells) toward the south (for example, Fig. 16.23B), is that the SP basin is shallower toward its margins, and particularly shallow toward its southern margin." McKinnon et al. (2016) argued that their explanation as indicated above is consistent with the expected basin topography created by an oblique impact to the SSW (Elbeshausen et al., 2013). They further suggested that "The less well-defined cellular structure in the very center of SP may, in contrast, reflect the deeper center of the basin, implying a larger Ra for the N_2 ice layer there and more chaotic, time dependent convection."

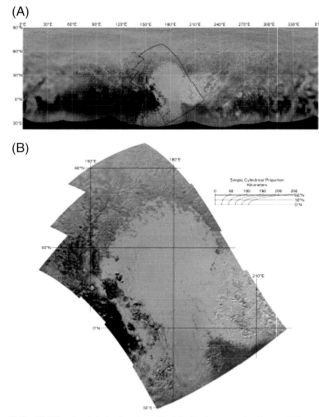

FIG. 16.25 A global, simple cylindrical, photometrically equilibrated mosaic of LORRI images of Pluto is shown in (A). Latitudes south of ~30°S were in darkness during the New Horizons flyby. The mapping area (comprising a mosaic of 12 LORRI images obtained at 386 m/pixel) is highlighted by the black boundary and is expanded in (B). *(Source: White et al. (2017), Geological mapping of Sputnik Planitia on Pluto, Icarus, 287: 261–286)* Copyright info: © 2017 **Elsevier Inc.** All rights reserved.

In a subsequent study, White et al. (2017) carried out geological mapping of SP, and the uplands terrain surrounding SP, and documented its geology and stratigraphy through geologic mapping at 1:2,000,000 scale. They reported mapping within an area covered by a mosaic of 12 images obtained by New Horizons' Long-Range Reconnaissance Imager (LORRI) (Cheng et al., 2008) that includes all of Sputnik Planitia and portions of the surrounding terrain (shown in Fig. 16.25, along with a global context map). Because in the early stages of analysis of New Horizons images, SP was recognized to be a highly complex entity that displays a wide range of surface textures and patterns and which is undergoing rapid surface renewal (Stern et al., 2015; Moore et al., 2016; McKinnon et al., 2016), the purpose of White et al. (2017)'s study was focused to "specifically map, describe, and interpret units within the Planitia itself, as well as any units in the surrounding terrain that are considered to be presently directly affected by, or which are directly affecting, Sputnik Planitia, and to determine the sequence of events that formed and modified them."

White et al. (2017) found that "All units that have been mapped are presently being affected to some degree by the action of flowing N_2 ice. The N_2 ice plains of Sputnik Planitia display no impact craters, and are undergoing constant resurfacing via convection, glacial flow and sublimation. Condensation of atmospheric N_2 onto the surface to form a bright mantle has occurred across broad swathes of Sputnik Planitia, and appears to be partly controlled by Pluto's obliquity cycles. The action of N_2 ice has been instrumental in affecting uplands terrain surrounding Sputnik Planitia, and has played a key role in the disruption of Sputnik Planitia's western margin to form chains of blocky mountain ranges, as well in the extensive erosion by glacial flow of the uplands to the east of Sputnik Planitia."

16.8.8 Latitudinal variations of solar energy flux on Pluto—theoretical investigations

The incident solar energy flux (i.e., the amount of solar radiation reaching a given area) on a celestial body is known as insolation. The axial tilt and orbital eccentricity of Pluto evolve on million-year timescales (Dobrovolskis and Harris, 1983; Dobrovolskis et al., 1997), leading to important changes in the annual energy flux to Pluto and to climate variations that are analogous to ice-age cycles on Earth. Such variations have been addressed by several investigators (e.g., van Hemelrijck, 1982; van Hemelrijck, 1985; Binzel, 1992; Spencer, et al., 1997; Lissauer et al., 2012; Earle and Binzel, 2015; Hamilton, 2015; Earle et al., 2016; Binzel et al., 2016; Nadeau and McGhee, 2015). In a different approach, Hamilton et al. (2016) investigated these variations by considering the instantaneous flux of sunlight that is absorbed by a unit surface element on Pluto (see Eqn: 16.1). Hamilton et al. (2016) averaged the above equation (Eq. 16.1) over one orbit of Pluto about the Sun (see Fig. 16.26), assuming a perfect absorber (albedo [A] = 0) to obtain the solar energy flux to different latitudes on Pluto. The daily average of the energy flux from sunlight to a given latitude ly on Pluto over the course of one Pluto rotation (the integration is over time with P_{day} equal to one Pluto day or approximately 6.39 Earth days) is given by the integral of the equation (16.3) with $A = 0$ (Hamilton et al., 2016):

$$\overline{F}_{day} = \frac{L}{4\pi r^2} \frac{1}{P_{day}} \int_0^{P_{day}} \cos(\gamma) dt$$
$$= \frac{L}{4\pi r^2} \frac{1}{\pi} \int_0^{\beta_{max}} \cos(\gamma) d\beta \quad (16.3)$$

where L is the solar luminosity and r is the instantaneous Pluto–Sun distance, which is assumed to remain approximately constant over one Pluto day. The angle γ is the angular distance between the Sun and the point of interest on Pluto's surface

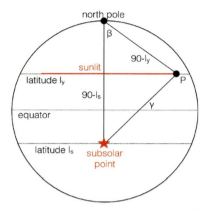

FIG. 16.26 Geometry for solar energy deposition on Pluto. The large circle represents Pluto, with our view centered on the intersection of the equator and the noon meridian. At the location of the red star, the Sun is directly overhead, and it tracks along the latitude *ls* over the course of a full rotation of Pluto. The thick horizontal red line segment shows the regions along latitude *ly* that are currently illuminated by sunlight. The angle γ is the angular distance between the Sun and the point of interest on Pluto's surface (black dot labelled "P" at latitude *ly*), as measured from the center of Pluto. The Sun is on the horizon when $\gamma = 90°$, which Hamilton et al. (2016) defined to occur at the meridional angle of $\beta = \beta_{max}$. The spherical triangle formed by γ and the two meridians connecting the North Pole to the above-mentioned points defines γ in terms of the other variables and simplifies derivations of average energy fluxes. *(Source: Hamilton, D.P., S.A. Stern, J.M. Moore, L.A. Young, and the New Horizons Geology, Geophysics & Imaging Theme Team (2016), The rapid formation of Sputnik Planitia early in Pluto's history, Nature, 540, 97-99.) Copyright info: 2016 Macmillan Publishers Limited, part of Springer Nature. All rights reserved.*

(Fig. 16.26); γ changes continuously in time owing to Pluto's rotation. Hamilton et al. (2016) used the fact that β—the hour angle measured from the north pole (Fig. 16.26)—increases uniformly as Pluto rotates so that $\beta/(2\pi) = t/P_{day}$. Finally, β_{max} is the largest β such that the Sun is still visible ($\gamma = 90°$). Eq. (16.3) shows that the daily average flux is highest when both β_{max} and $\cos(\gamma)$ are maximized (long sunlit periods with the Sun nearly overhead). According to Hamilton et al. (2016), this situation is approximately realized at two locations along the orbit of Uranus where that planet's spin axis points nearly at the Sun.

The integral is evaluated by expanding the angle γ with the law of cosines for the spherical triangle shown in Fig. 16.26 (Hamilton et al., 2016), and the following result is obtained (Cross, 1971; Ward, 1974):

$$\overline{F}_{day} = \frac{L}{4\pi r^2}\frac{1}{\pi}\left[\beta_{max}\sin(l_s)\sin(l_y) + \sin(\beta_{max})\cos(l_s)\cos(l_y)\right] \quad (16.4)$$

where l_s is the subsolar latitude, l_y is the latitude of interest and $\cos(\beta_{max}) = -\tan(l_s)\tan(l_y)$, with $\beta_{max} \in [0, \pi]$. Hamilton et al. (2016) have found that Eq. (16.4) can be used to show that, on the summer solstice, Pluto's summer pole receives nearly 5.5 times the solar energy than its equator does. According to Hamilton et al. (2016)'s estimates, Pluto receives more daily sunlight at its north pole than at its equator whenever $l_s > 17.66°$. For higher latitudes, the more direct sunlight to the equator is more than offset by the greater amount of time that the Polar Regions spend in sunlight. For a given position of the Sun north of Pluto's equator, the greatest daily flux of energy is always to a more northerly latitude.

To determine the annually averaged energy flux, Hamilton et al. (2016) employed the technique of integrating Eq. (16.4) over a full Pluto year:

$$\overline{F}_{year} = \frac{L}{4\pi}\frac{1}{\pi}\frac{1}{P_{year}}$$
$$\int_0^{P_{year}} \frac{1}{r^2}\left[\beta_{max}\sin(l_s)\sin(l_y) + \sin(\beta_{max})\cos(l_s)\cos(l_y)\right]dt$$

Hamilton et al. (2016) used Kepler's second law

$$\frac{dv}{dt} = \frac{h}{r^2} \quad (16.5)$$

to replace integration over time with an integration over the angle v. Here

$$h = \sqrt{GM_\odot a(1-e^2)}$$

is Pluto's constant orbital angular momentum per unit mass, a, e and v are Pluto's orbital semi-major axis, eccentricity and true anomaly, respectively, G is the gravitational constant and $M_\odot$ is the solar mass. The true anomaly is the angle between Pluto and its orbital pericenter as measured from the Sun. Hamilton et al. (2016) have recognized that the r^2 introduced by this change of variables conveniently cancels the r^{-2} from the definition of flux. Taking all these into account, Hamilton et al. (2016) found that:

$$\overline{F}_{year} = \frac{L}{8\pi^3}\frac{1}{a^2}\frac{1}{\sqrt{1-e^2}}$$
$$\int_0^{2\pi}\left[\beta_{max}\sin(l_s)\sin(l_y) + \sin(\beta_{max})\cos(l_s)\cos(l_y)\right]dv$$

(16.6)

Because the integrand is a complicated function of the true anomaly, the integral in equation (16.6) must be evaluated numerically (for details, see Hamilton et al., 2016). According to Hamilton et al. (2016)'s investigations, the solution of Eq. (16.6) shows that cold polar regions as existing on Earth or Mars require obliquities of less than 45° or more than 135°, whereas deep equatorial minima occur for all planets tilted between 66° and 114°. Pluto's obliquity varies around the 114° boundary, spending about 1.3 million years with an equatorial minimum followed by 1.5 million years with minima at low to mid-latitudes (Fig. 16.27).

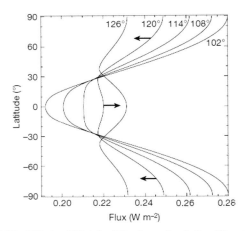

FIG. 16.27 **Effects of Pluto's obliquity on insolation.** The orbit-averaged incident solar energy flux, or insolation, is plotted as a function of latitude on Pluto for five different values of Pluto's axial tilt ε. All curves were computed assuming the present-day value of Pluto's eccentricity, $e = 0.25$. The change in Pluto's tilt is the largest single factor that affects the strength of incident sunlight; the 1.4-million-year swing from $ε = 104°$ to $ε = 127°$ causes a change of roughly 20% in the flux absorbed at Pluto's equator and poles. Latitudes near ± 30° have a far more muted response to changes in Pluto's obliquity. The black arrows highlight the present-day obliquity of Pluto and show how the obliquity will change in the future; the minimum insolation at the equator occurred about 0.85 million years ago and the peak is due in about 0.55 million years. *(Source: Hamilton, D.P., S.A. Stern, J.M. Moore, L.A. Young, and the New Horizons Geology, Geophysics & Imaging Theme Team (2016), The rapid formation of Sputnik Planitia early in Pluto's history,* Nature, *540, 97-99.)* Copyright info: 2016 Macmillan Publishers Limited, part of Springer Nature. All rights reserved.

16.9 The current state of Pluto's atmosphere

Pluto's atmosphere expanded and increased in mass (and thus surface pressure) by a factor of 2 between the 1988 and 2002 occultations (Elliot and Kern, 2003; Elliot et al., 2007). Observations made during the New Horizons flyby provide a detailed snapshot of the current state of Pluto's atmosphere. Gladstone et al. (2016) found that "Molecular nitrogen (N_2) dominates the atmosphere (at altitudes of less than 1800 kilometers or so), whereas methane (CH_4), acetylene (C_2H_2), ethylene (C_2H_4), and ethane (C_2H_6) are abundant minor species and likely feed the production of an extensive haze that encompasses Pluto. The cold upper atmosphere shuts off the anticipated enhanced-Jeans, hydrodynamic-like escape of Pluto's atmosphere to space." However, Gladstone et al. (2016) caution that it is unclear whether the current state of Pluto's atmosphere is representative of its average state—over seasonal or geologic time scales.

Images from New Horizons revealed that Pluto's mainly nitrogen atmosphere is hazy and foggy. Apart from nitrogen, the other gases that adorn Pluto's atmosphere include hydrocarbon particles as well. Gladstone et al. (2016) investigated the atmosphere of Pluto as observed by New Horizons. The more we learn about Pluto, the more interesting the little world becomes. According to a study, hydrocarbon particles in Pluto's atmosphere are responsible for Pluto's surprisingly low temperatures. Pluto is therefore the only planetary body known whose temperatures are driven more by haze particles than by gas molecules. Scientists suspect that Pluto's atmosphere is gradually vanishing through an escaping process, driven by solar wind (the supersonic outflow of electrically charged particles from the Sun) and Pluto's weak gravity. Bagenal et al. (2016) examined Pluto's interaction with its space environment: in particular solar wind, energetic particles, and dust. Some of the atmosphere's molecules possess enough energy to overcome Pluto's weak gravity and escape into space.

Photographs taken by the New Horizons probe showed that Pluto's surface is ornamented with a multitude of snakeskin ridges, which may have been shaped by surface winds. It is suspected that Pluto's atmosphere could have contributed to the dwarf planet's unusual features. A zoomed-in view of a near-sunset scene photo sent by the New Horizons probe, taken on July 14, 2015, when it was just 11,000 miles (18,000 km) from Pluto, showed rugged ice mountains up to 11,000 feet (3500 m) high and wide, flat plains.

In an image from NASA's New Horizons spacecraft, which made the first-ever flyby of the dwarf planet in July 2015, a black-and-white view revealed dozens of ringed craters (NASA describes these formations as "haloed") strewn across the dark landscape of Vega Terra, a region in the far western reaches of the hemisphere photographed by New Horizons during its flyby. The craters have been found to have bright walls and rims, making them stand out from their darker surroundings. While the haloed craters are eye-catching, what has really stumped scientists is what these features are made of. The discovery of strange halo-like craters on Pluto has raised a new mystery about how the odd scars formed on the icy world. NASA scientists believe that the halo-like craters on Pluto's surface display a puzzling distribution of methane ice and water ice (https://www.space.com/32710-pluto-halo-craters-mystery-new-horizons.html).

An interesting achievement obtained from New Horizons is the discovery of flowing ice (probably frozen nitrogen) from Pluto's mountains through valleys on to the plains. McKinnon et al. (2016) found that convection in a volatile nitrogen-ice-rich layer drives Pluto's geological and atmospheric vigor. The range of surface features on Pluto is worthy of mention. Some of the images from NASA's New Horizons spacecraft indicated wide areas on Pluto with very varied surface reflectivity and geological landforms. Bertrand and Forget (2016) investigated the observed glacier and volatile distribution on Pluto from atmosphere–topography processes. Trowbridge et al. (2016) found that vigorous convection is responsible for Pluto's polygonal terrain.

New Horizons' Ralph instrument revealed evidence of frozen carbon monoxide (carbon monoxide ice) on Pluto, in the western part of the region known presently as Tombaugh Regio (Tombaugh Region), the highly visible "heart of Pluto." The contours that overlain on the image show that the concentration of frozen carbon monoxide increases toward the center of the "bull's eye."

The orbit of Pluto is extremely tilted in regards to the other planets. According to a newly fleshed out theory, Pluto could have been scarred when a smaller body smashed into it producing a crater on the planet's surface, prompting Pluto to roll over, perhaps as much as 60 degrees. Note that the axis a planet spins on is not random. The planets like to spin in such a way that they minimize the amount of energy in the spin. That means they like to spin around the shortest axis.

As already indicated, Pluto's surface is covered in numerous CH_4 ice deposits, that vary in texture and brightness, as revealed by the New Horizons spacecraft as it flew by Pluto in July 2015. These observations suggest that CH_4 on Pluto has a complex history, involving reservoirs of different composition, thickness, and stability controlled by volatile processes occurring on different timescales. In order to interpret these observations, Bertrand et al. (2019) used a Pluto volatile transport model, which is able to simulate the cycles of N_2 and CH_4 ices over millions of years. By assuming fixed solid mixing ratios, they explored how changes in surface albedos, emissivities, and thermal inertias impact volatile transport. Results show that "bright CH_4 deposits can create cold traps for N_2 ice outside Sputnik Planitia, leading to a strong coupling between the N_2 and CH_4 cycles." In Bertrand et al. (2019)'s simulations, the massive and perennial CH_4 ice reservoirs at the same equatorial latitudes where the Bladed Terrain Deposits are observed, or at mid-latitudes (25°–70°) are not in an equilibrium state and either one can dominate the other over long timescales, depending on the assumptions made for the CH_4 albedo. According to Bertrand et al. (2019), this suggests that long-term volatile transport exists between the observed reservoirs. They further showed that Pluto's atmosphere always contained, over the last astronomical cycles, enough gaseous CH_4 to absorb most of the incoming Lyman-α flux.

16.10 Indirect detection of subsurface ocean of liquid water inside Pluto—in support of pre-New Horizons numerical studies

Even before the *New Horizons* spacecraft flew past Pluto and made detailed measurements from close quarters, presence of a subsurface ocean on Pluto has been inferred based on numerical studies (Robuchon and Nimmo, 2011). It is fascinating to think that Pluto—orbiting in the outer periphery of the Solar System and hugging the Kuiper belt—is harboring a hidden ocean buried beneath its frozen heart located on its frozen surface crust. It is considered that Pluto's subsurface ocean potentially contains as much water as all of Earth's seas. Based on an analysis of images and data collected by NASA's *New Horizons* spacecraft, which flew past Pluto and its entourage of moons in July 2015, it was found that Pluto's subsurface ocean is laden with ammonia. It was also found that Pluto's subsurface ocean, which is likely slushy with ice, lies 93 to 124 miles (150 to 200 km) beneath Pluto's icy surface and is about 62 miles (100 km) deep.

When the *New Horizons* probe conducted its fly-by of Pluto in July 2015, one of the most striking features the spacecraft saw was a 1600-km-wide heart-shaped plain on the surface of the celestial body. A 1000-km region within the heart-shaped plain known as Sputnik Planitia has become the focus of the research, with scientists believing it is home to a massive hidden ocean. Shortly after the first images of the near-side arrived at Earth, it was realized that Sputnik Planitia was in a strange place: it is aligned almost exactly opposite Pluto's largest moon, Charon. Mathematical models suggest that when the basin formed, an underground ocean began to well up into the chasm (a deep fissure in the surface). Afterwards, nitrogen gas in Pluto's atmosphere condensed and froze in the frigid basin. The weight of the new water and ice created a heavy load that tipped Pluto into its current alignment (Nimmo et al., 2016).

There was a school of thought that Pluto's ocean had a 'cold start', in that the ocean was frozen when the dwarf planet formed. It was further believed that the initially frozen subsurface ocean that came into existence on Pluto during its formation would have melted subsequently under the heat provided by decaying radioactive elements in its rocky core. In this scenario, the ice would have contracted as it melted—leading to wrinkles on the surface. Then, the ice would have expanded as it refroze when the radiogenic heat produced in the rocky core dwindled over a time—resulting in the formation of cracks on the surface. If this scenario is correct, images of Pluto's surface should reveal older wrinkles and newer cracks. But *New Horizons* photographed only cracks, suggesting that Pluto's subsurface ocean began as a liquid and has partially frozen over time (Nimmo et al., 2016).

Johnson et al. (2016) examined the formation of the SP basin and the thickness of Pluto's subsurface ocean. The collapse of the huge crater is believed to have lifted Pluto's subsurface ocean and the dense water—combined with dense surface nitrogen ice that fills in the hole—formed a huge mass excess that caused Pluto to tip over, reorienting itself with respect to its largest moon Charon. But the ocean uplift won't last if warm water ice at the base of the covering ice shell can flow and adjust in the manner of glaciers on Earth. If enough ammonia is added to the water, it can chill to incredibly cold temperatures (down to minus 145 Fahrenheit) and still be liquid, even if quite viscous. It is worth noting that New Horizons detected ammonia as a compound on Pluto's largest moon, Charon, as well as its smaller moons Nix and Hydra, as a broad absorption band in the 2.2-μm spectral region (Buie and Grundy, 2000; Brown and Calvin, 2000; Cook et al., 2007; Cook et al., 2018). It is, therefore, considered that ammonia is almost certainly present inside Pluto. At these chill temperatures made possible by the availability of sufficient quantity of ammonia, water ice is rigid, and the uplifted surface ocean becomes permanent.

Based on a study involving Pluto's heart-shaped ice-accumulated impact-crater driven feature named Sputnik Planitia (SP), Nimmo et al. (2016) argued that Pluto might be harboring a subsurface ocean. As indicated earlier, the SP region always faces away from Pluto's largest moon Charon; and the alignment is so precise that it is as if Charon floats over the area directly opposite SP. This suggests that there is extra mass in SP, and it forced Pluto to roll over to balance itself between its own mass and that of its sister moon. However, because of the very fact that SP is located in a crater in the ground, there ought to be less mass, not more. If this is right, there ought to be a way of hiding that extra mass. Nimmo et al. (2016) argued that the just-mentioned extra mass (i.e., positive mass anomaly) gained by SP is probably caused by an underground ocean that moved closer to the surface in the area of the impact-crater—an observation in concurrence with that of Johnson et al. (2016). To explain this more clearly, when an enormous impactor pummelled Pluto, it would have excavated some of the planet's ice shell. The ocean beneath the now-thinned crust would well up, filling the void. Because liquid water is denser than ice, Pluto's mass would now be unevenly spread out. The entire body of Pluto would be unbalanced, as though it were heavier on one side. Over time, this would reorient Pluto's spin until it eventually balanced itself again. According to Nimmo et al. (2016), that would be what brought SP to its current location, directly opposite Charon. According to them, the pattern of faults existing on Pluto's surface skin also bolster the idea of a subsurface ocean.

Nimmo et al. (2016) argued that the giant impact that is suspected to have created Pluto's enormous impact crater "Sputnik Planitia" was followed by a subsequent upwelling of a dense interior ocean. Once the basin was established, ice would naturally have accumulated there (Bertrand and Forget, 2016). Then, provided that the basin was a positive gravity anomaly (with or without the ocean), true polar wander could have moved the feature toward the Pluto–Charon tidal axis, on the far side of Pluto from Charon (Nimmo et al., 2016; Keane et al., 2016). Hamilton et al. (2016)'s modeling studies suggest that Sputnik Planitia formed shortly after Charon did and has been stable, albeit gradually losing volume, over the age of the Solar System.

Reorientation of Sputnik Planitia (Rubincam, 2003; Nimmo and Matsuyama, 2007; Keane et al., 2016), arising from tidal and rotational torques, can explain the basin's present-day location, but requires the feature to be a positive gravity anomaly (Keane et al., 2016), despite its negative topography. Nimmo et al. (2016) have argued that if Sputnik Planitia did indeed form as a result of an impact and if Pluto possesses a subsurface ocean, the required positive gravity anomaly would naturally result because of shell thinning and ocean uplift, followed by later modest nitrogen deposition. Without a subsurface ocean, a positive gravity anomaly requires an implausibly thick nitrogen layer (exceeding 40 km). The Sputnik Planitia basin has been filling with nitrogen ice over the years and now contains huge amounts of nitrogen. To prolong the lifetime of such a subsurface ocean to the present day (Robuchon and Nimmo, 2011) and to maintain ocean uplift, a rigid, conductive water-ice shell is required. Because of the fact that nitrogen deposition is latitude-dependent (Binzel et al., 2016), it is argued that nitrogen loading and reorientation may have exhibited complex feedbacks (Keane et al., 2016). There are indications to believe that Sputnik Planitia's present nitrogen ice may be up to 10 km (6 miles) thick.

Keane et al. (2016) modeled what happened as nitrogen ice accumulated in Sputnik Planitia. They found that once enough nitrogen ice has piled up, may be a hundred meters thick, it starts to overwhelm the planet's shape, which dictates the planet's orientation. If there is an excess of mass in one spot on the planet, it wants to go to the equator. Eventually, over millions of years, it will drag the whole planet over. This tumble brought Sputnik Planitia to the southeast, until the plain faced directly away from Charon as it does today. Keane et al. (2016)'s models predicted this very orientation: In this spot, Pluto's reorientation would have stressed its crust. Based on these findings, scientists consider that this "heart-shaped feature" caused the dwarf planet to roll over the eons, and this reorientation probably wouldn't have been possible without a subsurface ocean.

As already indicated, Pluto and its largest moon, *Charon*, are tidally locked, and the deep nitrogen-covered Sputnik Planitia basin on Pluto is aligned nicely with (located very close to the longitude of) Pluto's "tidal axis" (Moore et al., 2016). And that is probably no coincidence, according to the studies of Hamilton et al. (2016) and Nimmo et al. (2016). According to these studies, the additional mass located near the tidal axis of Pluto causes the least wobble in Pluto's spin.

A network of faults and fractures on Pluto's surface, spotted by the New Horizons' probe, and the characteristics of this network matching those predicted by Nimmo et al., (2016)'s model, which assumes that Pluto harbors a subsurface ocean of liquid water, provide convincing indication that Pluto's "wandering heart" indeed harbors a subsurface ocean. It has been argued that Pluto's surface fractures may be the result of this ocean gradually freezing over time.

Water expands as it freezes, which would lead to stresses in the overlying rock or ice. Nimmo et al. (2016)'s studies indicate that all these lines of argument are pointing in the same direction. Researchers had indeed anticipated that slow refreezing of this ocean would conceivably crack the planet's shell—a scenario consistent with photos taken by New Horizons.

The research findings suggest that the impact that created the basin weakened the crust overlying a buried ocean, causing some of the water to rise close to the surface. Nimmo et al. (2016) argued that this action, along with the deposition of nitrogen ice in Sputnik Planitia, would have created enough of a "positive mass anomaly" to roll the dwarf planet. It is indeed a tremendous achievement of NASA's *New Horizon* space mission program that, using computer models along with topographical and compositional data culled from the *New Horizon* spacecraft's July 2015 flyby of Pluto, planetary scientists managed to discover that Sputnik Planitia's churning nitrogen ice layer hides a subsurface liquid water ocean.

It is fascinating to think that Pluto is harboring a hidden ocean buried beneath its frozen heart located on its frozen surface crust—potentially containing as much water as all of Earth's seas. However, queries have been raised as to how Pluto managed to avoid freezing up entirely over the past 4.5 billion years of its history if Pluto does have an ocean? It has been argued that Pluto is big enough that it may have retained a substantial amount of radiogenic internal heat. Furthermore, as indicated earlier, Pluto's subsurface ocean water may contain significant amounts of ammonia or other substances that act as an antifreeze.

In a nutshell, beneath the heart-shaped region on Pluto, known as Sputnik Planitia, there lies an ocean laden with ammonia. Pluto's subsurface ocean, which is likely slushy with ice, lies 93 to 124 miles (150 to 200 km) beneath Pluto's icy surface and is about 62 miles (100 km) deep. The discovery was made through an analysis of images and data collected by NASA's *New Horizons* spacecraft, which flew past Pluto and its entourage of moons in July 2015.

Pluto researchers have found that despite being about 40 times farther from the Sun than Earth, Pluto has enough radioactive heat left over from its formation 4.6 billion years ago to keep the subsurface water liquid. They are of the view that Pluto has enough rock that there is quite a lot of heat being generated through radioactive decay, and an ice shell a few hundred kilometers thick is quite a good insulator. So, a deep subsurface ocean is not too surprising, especially if the ocean contains ammonia, which acts like an anti-freeze. It is heartening that sheer scientific curiosity has given rise to an important discovery of a liquid ocean hiding beneath the petite planet Pluto's icy surface. Interestingly, scientists made this discovery while they were trying to figure out why a 621-mile (1,000-km) wide impact basin known as Sputnik Planitia, which contains the curious

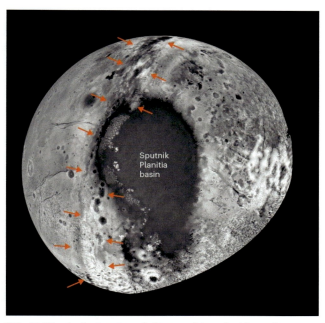

FIG. 16.28 **A giant system of ridges and troughs (shown by orange arrows) encircles Pluto.** This global crack could have formed as a subsurface ocean froze. *Credit: NASA/JHUAPL/SwRI.*

heart-shaped region, was located in its present position (on tidal axis) near Pluto's equator. The existence of a subsurface ocean is believed to have likely solved a longstanding puzzle: A thick, heavy ocean, may have served as a "gravitational anomaly," which would factor heavily in Pluto and Charon's gravitational tug-of-war. It is believed that over millions of years, Pluto would have spun around, aligning its subsurface ocean and the heart-shaped region above it, almost exactly opposite along the line connecting Pluto and Charon.

Although the idea of a subsurface ocean has existed for some time, the far-side (the hemisphere facing Charan) images have helped to support/strengthen this idea. Some of the strongest evidence comes from a feature known as chaotic terrain—a muddled mess of ridges, cracks and plains on the exact opposite side of Pluto from Sputnik Planitia. Some geological oddities found on Pluto—in particular, a large number of cracks discovered on the far-side—add support to the notion of existence of a hidden ocean beneath the icy crust of Pluto, and even shed light on how it formed. Such cracks, though in relatively limited number, distributed along a path extending from one pole to the other, are found even on the near-side (the hemisphere antipodal to the hemisphere that faces Charan) of Pluto (Fig. 16.28).

16.11 Distinct topographic signatures on Pluto's "Near-Side" and "Far-Side"

The flyby of Pluto by NASA's New Horizons spacecraft revolutionized knowledge of Pluto in many ways, including via the moderate and high-resolution imaging of its "near-side" hemisphere (the hemisphere that the Sun was illuminating at the time the New Horizons zipped past Pluto; also called the "encounter hemisphere" of Pluto). However, owing to the fact that Pluto is a slow rotator with a 6.3872-Earth-day period, a single spacecraft's fast flyby like that of New Horizons could only observe one hemisphere of the planet closely.

In the first few years after the New Horizons sped past Pluto in July 2015, after having sent streams of high-resolution data from the sun-illuminated "near-side" and low-resolution data from the temporarily dark "far-side" of Pluto, planetary scientists have been focussing on scientifically scrutinizing the "near-side" close-up images of Pluto. As a result, most Pluto New Horizons analysis has initially been focused on the encounter hemisphere of Pluto (i.e., the anti-Charon hemisphere containing Sputnik Planitia).

The captured close-up images showed a world that was much more dynamic than anyone had imagined. It was found that the dwarf planet hosts icy nitrogen cliffs that resemble the rugged coast of Norway, and giant skyscraper-like shards (sharp-edged broken pieces) of methane ice protrude out from Pluto's surface. Equally interestingly, cracks deeper than the Grand Canyon were found scaring the surface, while icy volcanoes rise taller than Mount Everest.

16.11.1 Deep depression enclosing Sputnik Planitia ice sheet & north-south running complex ridge-trough system along 155° meridian

The *New Horizons* spacecraft passed through the Pluto system on July 14, 2015, and executed a series of observations designed to map the surface morphology, color, and topography of both Pluto and its largest moon, Charon (Stern et al., 2015; Moore et al., 2016; Grundy et al., 2016). In a study, Schenk et al. (2018) produced Digital elevation models (DEMs) over ~42% of Pluto using combinations of MVIC hemispheric scans and LORRI mosaics, from which slopes at scales of ~1 km could be determined.

Studies by various researchers as indicated above revealed that Pluto can be divided into regions each with distinct topographic signatures, corresponding with major physiographic terrain types. They found that large areas of Pluto are comprised of low-relief moderately cratered plains units. An interesting feature revealed in their study is deeply pitted and glaciated plains east of Sputnik Planitia, in which the said plains are elevated ~0.7 km. According to Schenk et al. (2018), "The most dominant topographic feature on Pluto is the 1200-by-2000-km wide depression enclosing the bright Sputnik Planitia ice sheet, the surface of which is 2.5-to-3.5 km deep (relative to the rim) and ~2 km deep relative to the mean radius." They found that the partial ring of steep-sided massifs, several of which are more than 5 km high, along the western margins of Sputnik Planitia produce some of the locally highest and steepest relief on Pluto, with slopes of 40–50°. The second major topographic feature on Pluto, revealed in Schenk et al. (2018)'s studies, is a complex, eroded, ridge-trough system ~300–400 km wide and at least 3200 km long extending north-to-south along the 155° meridian. They found that this enormous structure has several kilometers of relief. Note that "relief" is the term used for the differences in height from place to place on the land's surface and it is greatly affected by the underlying geology. Relief relies on the hardness, permeability, and structure of a rock. Schenk et al. (2018) reckon that this enormous structure may predate the large impact event forming the basin, though some post-Sputnik Planitia deformation is evident.

16.11.2 Bladed terrain on Pluto's "Near-Side"

Bladed Terrain observed in the encounter hemisphere (sometimes termed "near-side") of Pluto consists of deposits of massive CH_4, occurring within latitudes 30° of the equator and are found almost exclusively at the highest elevations (> 2 km above the mean radius). When the images from New Horizons first reached Earth, scientists noticed a bizarre terrain that consisted of skyscraper-sized shards of ice (sharp-edged pieces of ice) along the near-side's easternmost region. These evenly spaced ridges are only a few kilometres apart, yet rise sharp and knife-like into the sky, occasionally soaring as high as 1 km. They can be as long as 30 km. Moore et al. (2018) found that well-developed blades are typically spaced ~3–7 km crest-to-crest, have a typical local relief of ~300 m, and flank slopes of ~20° (as determined from photoclinometry & photogrammetry). They found that blades dominantly display a N-S orientation (Fig. 16.29), but those near the equator additionally exhibit a more rectilinear pattern. They also found that the blades are located on broad ridges averaging ~100 km wide, separated by troughs that appear to be of structural origin. Analysis carried out by Moore et al. (2018) indicated that the deposits of massive CH_4 preferentially precipitate at low latitudes where net annual solar energy input is lowest. These investigators found that CH_4 and N_2 will both precipitate at low elevations. However, because there is much more N_2 in Pluto's atmosphere than CH_4, the N_2 ice will dominate at these low elevations.

It was found that available observations do not readily point to a single simple analogous terrestrial or planetary process or landform. Therefore, Moore et al. (2018) separately considered the origin of the Bladed Terrain Deposits (BTD), and the bladed textures on their surface. Interestingly, the blades have been found to occur on ridges rather than in depressions, as is commonly seen for these processes on other planetary surfaces. Moore et al. (2018)

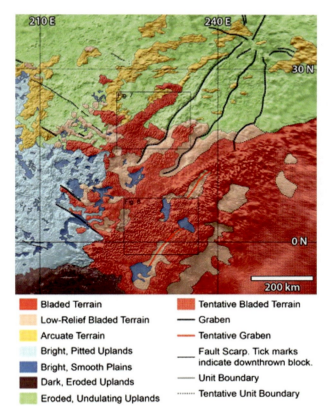

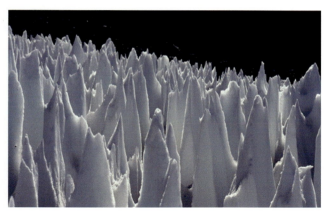

FIG. 16.30 **Field of Penitentes on the Upper Rio Blanco, Central Andes of Argentina.** The blades are between 1.5 and 2m in height, slightly tilted northwards, or more exactly about 11°, the approximate position of the Sun at noon at this latitude and time of the year. *(Source: https://en.wikipedia.org/wiki/Penitente_(snow_formation)#/media/File:Penitentes_Upper_Rio_Blanco_Argentine.jpg) (Permission details for reusing this file: Permission is granted to copy, distribute and/or modify this document under the terms of the GNU Free Documentation License, Version 1.2 or any later version published by the Free Software Foundation; with no Invariant Sections, no Front-Cover Texts, and no Back-Cover Texts. A copy of the license is included in the section entitled GNU Free Documentation License. This file is licensed under the Creative Commons Attribution-Share Alike 3.0 Unported license.*

FIG. 16.29 **Geological map of the encounter hemisphere region that includes the Bladed Terrain.** The ridges tend to be oriented N-S. They are clearly identifiable across most of the mapping area, but the white stippled pattern in the lower-right portion of the map indicates where imaging degrades to the extent that their presence is inferred here rather than directly observed. *(Source: Moore et al. (2018), Bladed Terrain on Pluto: Possible origins and evolution, Icarus, 300: 129–144)* **Elsevier Publication.**

argued that the strong correlation of BTD occurrence with high elevation suggests an atmospheric temperature control and source for their presence and modification. It may be recalled that during the time of New Horizons' encounter with Pluto, with the exception of the 1 km-thick boundary layer exclusively above Sputnik Planitia, Pluto's lower atmosphere temperature profile displayed an increase with altitude. The consequence of warmer air temperatures at higher altitudes is that the condensation of N_2 ice is suppressed, while the formation of CH_4 ice is currently promoted at higher elevations. Simply stated, at high elevations the atmosphere is too warm for N_2 to precipitate; therefore, only CH_4 can do so (Moore et al., 2018).

Moore et al. (2018) further argued that since the time the BTD were emplaced, there have been sufficient excursions in Pluto's climate to partially erode these deposits into the blades we see today. The blades themselves are partially analogous to terrestrial penitents (snow humps or ridges in an ice field) (Fig. 16.30). More specifically, penitents are snow formations—found at low-latitude, high-elevation ice fields—that take the form of elongated, thin blades of hardened snow or ice, closely spaced and pointing toward the general direction of the Sun (Moore et al., 2016; Moores et al., 2017). Moore et al. (2018) recognised that the processes that contribute to, and control the amplitude and spacing of, the blades are not yet fully understood. For instance, according to Moore et al. (2018) these Plutonian blades are at least two orders of magnitude larger than terrestrial penitentes. They suggest that Plutonian blades may be entirely erosional, like terrestrial penitentes, or may form by erosion at the base and condensation at the crests, which is marginally permitted in Pluto's current climate at those altitudes. According to Moore et al. (2018), following the time of massive CH_4 emplacement, there have been sufficient excursions in Pluto's climate to partially erode these deposits via sublimation into the blades we see today. These investigators argue that blades composed of massive CH_4 ice implies that the mechanical behavior of CH_4 can support at least several hundred meters of relief at Pluto surface conditions. Moore et al. (2018) conclude that Bladed Terrain, along with other deposits of volatiles in Tombaugh Regio proper (including Sputnik Planitia), represents an active response of the landscape to current and past climates, and very likely a major terrain type on Pluto.

16.11.3 Bladed terrain on Pluto's "Far-Side"

Based on color data and the CH_4 band depth map of the non-encounter hemisphere, Moore et al. (2018), who carried out a comprehensive study of the Bladed Terrain on

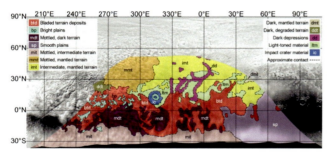

FIG. 16.31 Geological map of Pluto's far side showing geological units identified by analysis of New Horizons imaging, spectral, and limb topography data. This map is overlain on photometrically equilibrated LORRI imaging ranging in pixel scale from 2.2 km/pixel (at the western boundary) to 40.6 km/pixel (at the eastern boundary), and which is surrounded by higher resolution, near-side hemisphere imaging. Given the low resolution of imaging covering Pluto's far side, the study team treat all boundary contacts as approximate. *(Source: Stern, S.A., O.L. White, P.J. McGovern, J.T. Keane, J.W. Conrad, C.J. Bierson, C.B. Olkin, P.M. Schenk, J.M. Moore, K.D. Runyon, H.A. Weaver, L.A. Young, K. Ennico, The New Horizons Team (2020), Pluto's far side, Earth and Planetary Astrophysics (astro-ph.EP), arXiv:1910.08833, (published by Cornell University).*

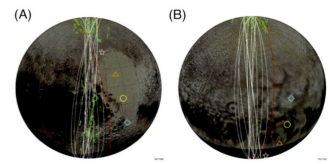

FIG. 16.32 Comparison of Pluto's near-side (left image) and far-side (right image) in relation to the Sputnik Planitia basin-related locations and the antipodes of the corresponding locations, respectively. The left image indicates spherical orthographic projection of the global mosaic of Pluto (Schenk et al., 2018) focusing on the near- side hemisphere. Center of projection is 150°E, 15°N. Tectonic lineations are mapped with various colors. The green lineations indicate the tectonic system dubbed the "great north-south ridge-trough system," or RTS (Schenk et al., 2018). The white lines represent great-circle paths fit to individual lineation segments of RTS. The red line represents a great-circle path fit to the divergent north-western segments of RTS. The star, triangle, circle, and diamond symbols represent a potential first contact point for an impact incidence from the NNW, the center of the deep portion of the Sputnik basin, a southern extension of SP basin-filling materials, and the potential first-contact point for an impact incidence from the SSE, respectively. The right image indicates global mosaic, as in the left image, but focusing on Pluto's far-side. Center of projection 330°E, 30°N. Green lineations at the top of the figure represent the northernmost elements of the RTS. The star, triangle, circle and diamond symbols indicate the antipodes of the corresponding Sputnik Planitia basin-related locations in the left image. *(Source: Stern, S.A., O.L. White, P.J. McGovern, J.T. Keane, J.W. Conrad, C.J. Bierson, C.B. Olkin, P.M. Schenk, J.M. Moore, K.D. Runyon, H.A. Weaver, L.A. Young, K. Ennico, The New Horizons Team (2020), Pluto's far side, Earth and Planetary Astrophysics (astro-ph.EP), arXiv:1910.08833, (published by Cornell University)).*

Pluto's near-side, provided a hint that Bladed Terrain may extensively occur within the entire ± 30° latitude band of Pluto (i.e., including the far-side of Pluto). These investigators had also suggested that these putative Bladed Terrain regions presumably also occur at high elevations. Furthermore, based on spectral observations, Moore et al. (2018) have argued that Bladed Terrain deposits may be widespread in the low latitudes of the poorly seen sub-Charon hemisphere (i.e., Pluto's far-side). They further suggested that if these locations are indeed Bladed Terrain deposits, they may mark heretofore unrecognized regions of high elevation.

Once the analyzes of the near-side of Pluto has been accomplished and shed considerable new knowledge on Pluto, as briefly described above, planetary scientists went on to analyze the other half, which the spacecraft photographed from far, days before it shot past Pluto, speeding at 14km per second (50,400 km/h), in its valiant onward journey of discovery to the Kuiper belt and beyond. In a comprehensive study, Stern et al. (2020) summarized and interpreted data on the far side (i.e., the non-encounter hemisphere), providing the first integrated New Horizons overview of the far-side terrains. Fig. 16.31 illustrates the geological map of Pluto's far-side showing geological units identified by analysis of New Horizons imaging, spectral, and limb topography data. Fig. 16.32 illustrates a comparison of Pluto's near-side and far-side in relation to the Sputnik Planitia basin-related locations and the antipodes of the corresponding locations.

The far-side measurements have generated a number of mysteries. Although the terrain map created based on the images from the far-side of Pluto is blurry (with a resolution too poor to actually see the individual ridges themselves), it was found that the bladed terrain wraps all the way around the far-side and pops out again on the western edge of the near-side in a region that was previously overlooked (Stern et al., 2020). On the far-side, they cover an area that is 3.5 times larger than their extent on the near-side—making them one of the biggest mysteries on Pluto. It now seems to run across the North Pole and back down toward the south pole on the far-side, thus wrapping around the entire dwarf planet (Stern et al., 2020).

Interestingly, images of Pluto's "far-side" reveal a giant crack that stretches up the "near-side" of Pluto. The giant crack observed on Pluto's surface—probably a scar from the freezing and ever-expanding ocean—has been likened to the East African Rift System, which is cleaving that continent in two.

Stern et al. (2020) found strong evidence for an impact crater about as large as any on the near-side hemisphere, evidence for complex linear markings (lineations) approximately antipodal to Sputnik Planitia that may be causally related, and evidence that the far-side maculae (large/irregular marks) are smaller and more structured than the encounter hemisphere maculae.

Pluto's surface is covered in numerous CH_4 ice deposits, that vary in texture and brightness, as revealed by the

New Horizons spacecraft as it flew past Pluto in July 2015. These observations suggest that CH_4 on Pluto has a complex history, involving reservoirs of different composition, thickness, and stability controlled by volatile processes occurring on different timescales. In order to interpret these observations, Bertrand et al. (2019) used a Pluto volatile transport model able to simulate the cycles of N_2 and CH_4 ices over millions of years. Climate model-based studies carried out by Bertrand et al. (2019) indicated that the icy shards on Pluto's landscape cannot be understood without closely studying its weather. Their model results revealed that methane accumulates at higher altitudes, whereas nitrogen builds up in the low atmosphere—explaining why the Sputnik Planitia basin is rich in nitrogen ice, but the bladed terrain is dominated by methane ice.

16.12 Is life possible on Pluto?

Determining whether or not Pluto possesses, or once possessed, a subsurface ocean is crucial to understanding its astrobiological potential. Based on theoretical studies carried out by several investigators (e.g., Robuchon and Nimmo, 2011; Kamata and Nimmo, 2014; Nimmo et al., 2016; Johnson et al., 2016; Bertrand and Forget, 2016; Keane et al., 2016; Hamilton et al., 2016), as discussed earlier, it has been inferred that Pluto possesses a subsurface ocean.

Astrobiologists reckon that the subsurface ocean existing beneath the frozen crust of Pluto might be ripe for the existence of life. Observations of water that had probably gushed out of the ocean on Pluto's "near-side" show that it is red (Fig. 16.33)—hinting that it is stained with organic molecules. Although the possibility of existence of organic molecules might seem impossible on a sunlight-limited world such as Pluto, laboratory experiments have shown that radiation similar to solar wind or cosmic rays can create complex organic matter that is reddish brown (Miller, 1953). Furthermore, if ammonia is present, it is possible to form molecules that are crucial for life, including amino acids and the bases that are present in RNA and DNA.

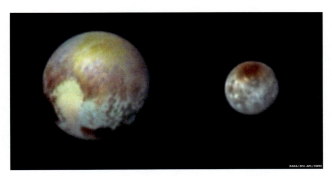

FIG. 16.33 Stretched color images of Pluto (L) and its largest and nearest moon Charon (R). The stretching amplifies surface differences. *Credit: NASA/JHU-APL/SWRI.*

Cruikshank et al. (2019) carried out studies pertaining to the astrobiological potential of Pluto and found strong indications for the presence of complex organic molecules, such as amino acids and nucleobases, formed by abiotic processes on the surface and in near-subsurface regions of Pluto. It is believed that in the early stages of the evolution of life, RNA specifically may have served as both a functional catalyst and information-bearing molecule that permitted early cells to reproduce themselves. Because of their potential involvement in the emergence of life, understanding the abiotic formation of nucleobases is of great interest. Cruikshank et al. (2019) found that Pluto's surface is tinted with a range of non-ice substances with colors ranging from light yellow to red to dark brown. It was found that these colors match those of laboratory organic residues called tholins (broadly characterized as complex, macromolecular organic solids consisting of a network of aromatic structures connected by aliphatic bridging units). Photolysis and radiolysis of a mixture of ices relevant to Pluto's surface composition (N_2, CH_4, CO) have produced strongly colored, complex organics with a significant aromatic content having a high degree of nitrogen substitution similar to the aromatic heterocycles pyrimidine and purine.

The synthesis of tholins in planetary atmospheres and in surface ices has been explored in numerous laboratory experiments, and both gas- and solid-phase varieties are found on Pluto (Cruikshank et al., 2019). According to them, a third variety of tholins, exposed at a site of tectonic surface fracturing called Virgil Fossae, appears to have come from a reservoir in the subsurface. Cruikshank et al. (2019) suggest that eruptions of tholin-laden liquid H_2O from a subsurface aqueous repository appear to have covered portions of Virgil Fossae and its surroundings with a uniquely colored deposit that is geographically correlated with an exposure of H_2O-ice that includes spectroscopically detected NH_3.

Nucleobases are the informational subunits of RNA and DNA, and are essential to all known forms of life. Nucleobases play an important role in life on Earth, and their function as genetic information storage constitutes a major event in the evolution of biological organisms. Biological nucleobases fall into two main families: pyrimidines (uracil, cytosine, and thymine) and purines (adenine and guanine). It has been demonstrated that uracil, cytosine, and thymine (pyrimidine-based compounds) and other non-biological, less common nucleobases can form abiotically from the ultraviolet (UV) photo-irradiation of pyrimidine in simple astrophysical ice analogs containing combinations of H_2O, NH_3, and CH_4. In a study, Materese et al. (2017) focused on the UV photo-irradiation of purine mixed with combinations of H_2O and NH_3 ices to determine whether or not the full complement of biological nucleobases can be formed abiotically under astrophysical conditions. Experiments with pyrimidines and purines frozen in

730 Water worlds in the solar system

H_2O-NH_3 ice resulted in the formation of numerous nucleobases, including the biologically relevant guanine, cytosine, adenine, uracil, and thymine (Materese et al.2017).

Cruikshank et al. (2019) found that Pluto's near-side ices were both red and laced with ammonia—a crucial sign that Pluto might be well-supplied with organic molecules. They reckon that the red material associated with the H_2O-ice on Pluto's surface may contain nucleobases resulting from energetic processing on Pluto's surface or in the interior. According to them, "some other Kuiper Belt objects also exhibit red colors similar to those found on Pluto and may therefore carry similar inventories of complex organic materials. The widespread and ubiquitous nature of similarly complex organic materials observed in a variety of astronomical settings drives the need for additional laboratory and modeling efforts to explain the origin and evolution of organic molecules. Pluto observations reveal complex organics on a small body that remains close to its place of origin in the outermost regions of the Solar System."

These investigators agree that this finding, which has gained plenty of support and interest among planetary scientists, doesn't necessarily mean that life got a start on Pluto. However, if life were introduced to Pluto, it could survive. Results gleaned from Pluto's "far-side" strengthen the theory of an astrobiological potential of Pluto. The results gleaned from the New Horizons spacecraft's flyover of Pluto reveal a red band, suspected to be of organic material, that stretches across the equator—corresponding to an area with the availability of most sunlight and the most temperate climate (Stern et al., 2020). The most vital ingredients required to support life are liquid water, organics and a source of energy. According to the astrobiologists, given that there is an increasingly strong case for liquid water and a reasonably trustworthy case for organics, the third astrobiologically important requirement to be met on Pluto is a source of energy.

16.13 Expectations from future explorations of Pluto

While some researchers are still examining the data stream received from Pluto, some others are already dreaming of what we might do next. In science, inferences made purely based on theoretical considerations are accepted as truth only when such inferences are validated through actual measurements and observations. For example, in the case of Pluto, it will be hard to know for certain if there is an ocean unless we send another spacecraft out there, and that is unlikely to happen anytime soon. According to McKinnon, who is one of the researchers who theorized about the presence of a subsurface ocean on Pluto, all of these ideas about an ocean inside Pluto are credible, but they are just inferences, not direct detections. If we want to confirm that such an ocean exists, we will need gravity measurements or subsurface radar sounding, all of which could be accomplished by a future orbiter mission to Pluto.

The exploration of the binary Pluto–Charon and its small moons during the New Horizons flyby in July 2015 revealed not only widespread geologic and compositional diversity across Pluto, but surprising complexity, a wide range of surface unit ages, evidence for widespread activity stretching across billions of years to the near-present, as well as numerous atmospheric puzzles, and strong atmospheric coupling with its surface. Robins et al. (2021) briefly reviewed the findings made by New Horizons and the case for a follow up mission to investigate the Pluto system in more detail. According to them, "As the next step in the exploration of this spectacular planet-satellite system, we recommend an orbiter to study it in considerably more detail, with new types of instrumentation, and to observe its changes with time. We further call for the in-depth study of Pluto orbiter missions as a precursor to the 2023 Planetary Science Decadal Survey."

According to Stern et al. (2020), "Future progress on far side geology, geophysics, and compositional studies would benefit tremendously from a Pluto orbiter. In particular, such an orbiter could address the key science questions including:

(i) What is the global distribution of Pluto's volatile units, and how does this relate to Pluto's climatic history?
(ii) What is the nature of Pluto's great ridge-trough system (RTS), and is it truly global?
(iii) How did Sputnik Planitia form, and did its formation play a significant role in shaping the far side geology of Pluto?
(iv) Are the FS maculae being exhumed by volatile transport?"

Planetary scientists studying Pluto hope that if some other probe would be sent someday, the probe would be equipped with a radar instrument, which could be used to peer beneath Pluto's crust, and into its ocean. It is planetary scientists' dream that far into the future, they could send an orbiter, or may even be a pair, to carefully map Pluto's gravity. A spacecraft sailing overhead would be able to study the layers of nitrogen ice in Sputnik Planitia's glacier, and the ice that makes up its crust. It would be able to observe the slow turning of Pluto's seasons. Furthermore, such a mission would tell what really lies beneath the ice, and how over the eons, in the face of whatever the solar system has thrown at it, this distant world has been able to remake itself.

Planetary scientists are optimistic that another probe will eventually be launched to Pluto, in a multi-year cruise. Indeed, many argue that it is almost inevitable given the numbers of questions that New Horizons left in its wake. To that end, NASA recently awarded scientists the opportunity to study the feasibility of an orbiter, which would allow them to map all of Pluto in detail and even to watch it change over time. This study is a long way from being an actual mission, which may take decades to come to fruition.

Stern et al. (2020) suggested that because the arrival of any such orbiter is unfortunately at least two decades away, we must in the nearer term rely on advances that can be obtained from observatories with new capabilities on or near Earth, and laboratory and modeling efforts. According to these researchers, "most notably among those will be the new generation of 25-to-30 m telescopes coming on line in the 2020s. These devices will have diffraction limited resolutions ~10x better than Hubble Space Telescope, offering to obtain panchromatic, color, and even compositional spectroscopic maps of Pluto with resolutions of 30-50 km on a regular basis." Note that panchromatic maps would provide surface details of Pluto in all visible colors of the spectrum. Such maps would be useful in identifying astrobiologically important information (e.g., presence of diverse organic compounds relevant to biochemistry)." Stern et al. (2020) recognize that the resolution of such maps will not surpass the New Horizons FS mapping, but will allow many more colors to be explored, and will also permit studies of surface albedo time variability as Pluto's complex orbital-obliquity seasons advance in the coming decades. Stern et al. (2020) further suggested that "In addition to yielding new knowledge about Pluto itself, the comparison of these datasets to similar datasets obtained on Triton and other Kuiper Belt dwarf planets will inform our understanding of the range of surface variation and variegation (appearance of differently colored zones) on such bodies as a class. Time-dependent Atacama Large Millimeter Array (ALMA) thermal mapping at crudely similar resolution will also be possible in the coming decades. The study of these small planets as a group with individually changing insolation and seasonal effects will better anchor volatile transport models used to understand them as a class."

References

Applegate, J.H., Douglas, M.R., Gursel, J., Sussman, G.J., Wisdom, J, 1986. The outer solar system for 200 million years. Astr. J. 92, 176–194.

Archival, B., 2011a. Report of the IAU working group on cartographic coordinates and rotational elements: 2009. Celest. Mech. Dynam. Astros. 109, 101–135.

Archival, B., 2011b. Erratum to: report of the IAU working group on cartographic coordinates and rotational elements: 2006 & 2009. Cel. Mech. Dyn. Ast. 109, 401–403.

Assafin, M., et al., 2010. Precise predictions of stellar occultations by pluto, charon, nix, and hydra for 2008–2015. Astron. Astrophys. 515 (A32), 1–14.

Bagenal, F., et al., 2016. Pluto's interaction with its space environment: solar wind, energetic particles, and dust. Science 351, aad9045.

Barr, A.C., Collins, G.C., 2014. Tectonic activity on pluto after the charon-forming impact. Icarus. doi:10.1016/j.icarus.2014.03.042.

Barr, A.C., McKinnon, W.B., 2007. Can enceladus' ice shell convect? Geophys. Res. Lett. 34, L09202.

Bertrand, T., Forget, F., 2016. Observed glacier and volatile distribution on pluto from atmosphere–topography processes. Nature. doi:10.1038/nature19337.

Bertrand, T., Forget, F., Umurhan, O.M., Moore, J.M., et al., 2019. The CH_4 cycles on pluto over seasonal and astronomical timescales. Icarus 329, 148–165.

Binzel, R.P., 1992. 1991 Urey prize lecture: physical evolution in the solar system—present observations as a key to the past. Icarus 100, 274–287.

Binzel, R.P., et al., 2016. Climate zones on pluto and charon. Icarus. doi:10.1016/j.icarus.2016.07.023.

Bland, M.T., McKinnon, W.B., 2015. Forming ganymede's grooves at smaller strain: toward a self-consistent local and global strain history for ganymede. Icarus 245, 247–262.

Bray, V.J., Schenk, P.M., 2015. Pristine impact crater morphology on pluto—expectations for new horizons. Icarus 246, 156–164.

Brown, M. E., 2008. The compositions of Kuiper belt objects, In: The Solar System Beyond Neptune (eds. Barucci, M. A., Boehnhardt, H., Cruikshank, D. & Morbidelli, A.), 335–344 (Univ. Arizona Press).

Brown, M.E., Calvin, W.M., 2000. Evidence for crystalline water and ammonia ices on pluto's satellite charon. Science 287, 107–109.

Buie, M.W., et al., 2006. Orbits and photometry of pluto's satellites: charon, S/2005 P1, and S/2005 P2. Astron. J. 132, 290–298.

Buie, M.W., Grundy, W.M., 2000. The distribution and physical state of H_2O on charon. Icarus 148, 324–339.

Buie, M.W., Grundy, W.M., Tholen, D.J., 2013. Astrometry and orbits of nix, kerberos, and hydra. Astron. J. 146 (6): 152–164. DOI:10.1088/0004-6256/146/6/152.

Buie, M.W., Grundy, W.M., Young, E.F., Young, L.A., Stern, S.A., 2010. Pluto and charon with the hubble space telescope. II. resolving changes on pluto's surface and a map for charon. Astron. J. 139 (3), 1128–1143. doi:10.1088/0004-6256/139/3/1128.

Buie, M.W., Tholen, D.J., Horne, K., 1992. Albedo maps of pluto and charon: initial mutual event results. Icarus 97, 211–227.

Buratti, B.J., Gougen, J.D., Mosher, J.A., 1997. No large brightness variations on nereid. Icarus 126, 225–228.

Canup, R.M., 2005. A giant impact origin of pluto-charon. Science 307, 546–550.

Cheng, A.F., et al., 2008. Long-range reconnaissance imager on new horizons. Space Sci. Rev. 140, 189–215. doi:10.1007/s11214-007-9271-6.

Cheng, W.H., Lee, M.H., Peale, S.J., 2014. Complete tidal evolution of pluto–charon. Icarus 233, 242–258.

Cohen, C.J., Hubbard, E.C., 1965. Libration of the close approaches of pluto to neptune. Astr. J. 70, 10–13.

Cook, J.C., Desch, S.J., Roush, T.L., Trujillo, C.A., Geballe, T.R., 2007. Near-infrared spectroscopy of charon: possible evidence for cryovolcanism on kuiper belt objects. Astrophys. J. 663, 1406–1419.

Cook, J.C., Dalle Ore, C.M., Protopapa, S., Binzel, R.P., Cartwright, R., Cruikshank, D.P., Earle, A., Grundy, W.M., Ennico, K., Howett, C., Jennings, D.E., Lunsford, A.W., Olkin, C.B., Parker, A.H., Philippe, S., Reuter, D., Schmitt, B., Stansberry, J.A., Alan Stern, S., Verbiscer, A., Weaver, H.A., Young, L.A., 2018. Composition of pluto's small satellites: analysis of new horizons spectral images. Icarus 315, 30–45.

Correia, A.C.M., Leleu, A., Rambaux, N., Robutel, P., 2015. Spin-orbit coupling and chaotic rotation for circumbinary bodies: application to the small satellites of the pluto-charon system. Astron. Astrophys. 580, L7.

Cross, C.A., 1971. The heat balance of the Martian polar caps. Icarus 15, 110–114.
Cruikshank, D.P., Materese, C.K., Pendleton, Y.J., et al., 2019. Prebiotic chemistry of pluto. Astrobiology 19, 831–848.
Cruikshank, D.P., Sheehan, W., 2018. Discovering Pluto: Exploration at the Edge of the Solar System. University of Arizona Press, Tucson, p. 502.
Desch, S.J., Cook, J.C., Doggett, T., Porter, S.B., 2009. Thermal evolution of kuiper belt objects, with implications for cryovolcanism. Icarus 202 (2), 694–714. doi:10.1016/j.icarus.2009.03.009.
Dobrovolskis, A.R., 1989. Dynamics of pluto and charon. Geophys. Res. Lett. 16 (11), 1217–1220.
Dobrovolskis, A.R., 1995. Chaotic rotation of nereid? Icarus 118, 181–198.
Dobrovolskis, A.R., Harris, A.W., 1983. The obliquity of pluto. Icarus 55, 231–235.
Dobrovolskis, A.R., Peale, S.J., and Harris, A.W., 1997. Dynamics of the Pluto-Charon binary, In: Pluto and Charon (eds., Stern, S.A., and Tholen, D.J.) 159–190 (Univ. Arizona Press, 1997).
Dombard, A.J., McKinnon, W.B., 2006. Elastoviscoplastic relaxation of impact crater topography with application to ganymede and callisto. J. Geophys. Res. 111, E01001. doi:10.1029/2005JE002445.
Earle, A.M., et al., 2016. Long-term surface temperature modeling of pluto. Icarus. doi:10.1016/j.icarus.2016.09.036.
Earle, A.M., Binzel, R.P., 2015. Pluto's insolation history: latitudinal variations and effects on atmospheric pressure. Icarus 250, 405–412.
Elbeshausen, D., Wünnemann, K., Collins, G.S., 2013. The transition from circular to elliptical impact craters. J. Geophys. Res. 118, 2295–2309.
Elliot, J.L., Kern, S.D., 2003. Pluto's atmosphere and a targeted-occultation search for other bound kbo atmospheres. Earth, Moon Planets 92, 375–393.
Elliot, J.L., Person, M.J., Gulbis, A.A.S., Souza, S.P., Adams, E.R., Babcock, B.A., Gangestad, J.W., Jaskot, A.E., Kramer, E.A., Pasachoff, J.M., Pike, R.E., Zuluaga, C.A., Bosh, A.S., Dieters, S.W., Francis, P.J., Giles, A.B., Greenhill, J.G., Lade, B., Lucas, R., Ramm, D.J., 2007. Changes in pluto's atmosphere: 1988-2006. Astronomical J. 134 (1), 1–13.
Elliot, J.L., Young, L.A., 1992. Analysis of stellar occultation data for planetary atmospheres. I. model fitting, with application to pluto. Astron. J. 103, 991–1015.
Eluszkiewicz, J., 1991. On the microphysical state of the surface of triton. J. Geophys.Res. 96, 19217–19229.
Eluszkiewicz, J., Stevenson, D.J., 1990. Rheology of solid methane and nitrogen: application to triton. Geophys. Res. Lett. 17, 1753–1756.
Eshleman, V.R., 1989. Pluto's atmosphere: models based on refraction, inversion, and vapor–pressure equilibrium. Icarus 80, 439–443.
Fernandez, J.A., Ip, W.H., 1984. Some dynamical aspects of the accretion of uranus and neptune: the exchange of orbital angular momentum with planetesimals. Icarus 58, 109–120.
Fountain, G.H., et al., 2008. The new horizons spacecraft. Space Sci. Rev. 140 (1–4), 23–47.
Gladstone, G.R., Stern, S.A., Ennico, K., Olkin, C.B., Weaver, H.A., Young, L.A., Summers, M.E., Strobel, D.F., Hinson, D.P., Kammer, J.A., Parker, A.H., Steffl, A.J., Linscott, I.R., Parker, J.W., Cheng, A.F., Slater, D.C., Versteeg, M.H., Greathouse, T.K., Retherford, K.D., Throop, H., Cunningham, N.J., Woods, W.W., Singer, K.N., Tsang, C.C., Schindhelm, E., Lisse, C.M., Wong, M.L., Yung, Y.L., Zhu, X., Curdt, W., Lavvas, P., Young, E.F., Tyler, G.L.New Horizons Science Team, 2016. The atmosphere of pluto as observed by new horizons. Science 351 (6279), aad8866. doi:10.1126/science.aad8866.
Goldsby, D.L., Kohlstedt, D.L., 2001. Superplastic deformation of ice: experimental observations. J. Geophys. Res. 106 (B6), 11017–11030. doi:10.1029/2000JB900336.
Grav, T., Holman, M.J., Kavelaars, J.J., 2003. The short rotation period of nereid. Astrophys. J. Lett. 591, L71.
Greenberg, R., 1981. Apsidal precession of orbits about an oblate planet. Astron. J. 86, 912–914.
Greenstreet, S., Gladman, B., McKinnon, W.B., 2015. Impact and cratering rates onto pluto. Icarus 258, 267–288.
Grundy, W.M., et al., 2016. Surface compositions across pluto and charon. Science 351, aad9189.
Grundy, W.M., Noll, K.S., Stephens, D.C., 2005. Diverse albedos of small transneptunian objects. Icarus 176, 184–191.
Guo, Y., Farquhar, R.W., 2006. Baseline design of new horizons mission to pluto and the kuiper belt. Acta Astronaut. 58 (10), 550–559.
Hamilton, D.P., 2015. The icy cold heart of pluto, In: 47th Division For Planetary Sciences (DPS) Meeting (American Astronomical Society, 2015), American Geophysical Union, Fall Meeting, 2015, abstract id. P51A-2036, DPS meeting #47, id.200.07; Bibcode: 2015AGUFM.P51A2036H: The 47th DPS meeting was held 8-13 November 2015 in Washington, DC.
Hamilton, D.P., Stern, S.A., Moore, J.M., Young, L.A., Geology, the New HorizonsGeophysics & Imaging Theme Team, 2016. The rapid formation of sputnik planitia early in pluto's history. Nature 540, 97–99. doi:10.1038/nature20586.
Hammond, N.P., Barr, A.C., 2014. Formation of ganymede's grooved terrain by convection-driven resurfacing. Icarus 227, 206–209.
Hammond, N.P., Barr, A.C., Parmentier, E.M., 2016. Recent tectonic activity on pluto driven by phase changes in the ice shell. Geophys. Res. Lett. 43, 6775–6782.
Hansen, C.J., Paige, D.A, 1996. Seasonal nitrogen cycles on pluto. Icarus 120, 247–265.
Hedman, M.M., Burns, J.A., Thomas, P.C., Tiscareno, M.S., Evans, M.W., 2011. Physical properties of the small moon aegaeon (Saturn LIII). EPSC Abstracts 6, 531.
Hubbard, W.B., Yelle, R.V., Lunine, J.I., 1990. Nonisothermal pluto atmosphere models. Icarus 84, 1–11.
Hussmann, H., Sohl, F., Spohn, T., 2006. Subsurface oceans and deep interiors of medium-sized outer planet satellites and large transneptunian objects. Icarus 185 (1), 258–273. doi:10.1016/j.icarus.2006.06.005.
Johnson, B.C., Bowling, T.J., Trowbridge, A.J., Freed, A.M., 2016. Formation of the sputnik planum basin and the thickness of pluto's subsurface ocean. Geophys. Res. Lett. 43, 10068–10077.
Kamata, S., et al., 2013. Viscoelastic deformation of lunar impact basins: implications for heterogeneity in the deep crustal paleo-thermal state and radioactive element concentration. J. Geophys. Res. Planets 118, 398–415. doi:10.1002/jgre.20056.
Kamata, S., Nimmo, F., 2014. Impact basin relaxation as a probe for the thermal history of pluto. J. Geophys. Res. 119, 2272–2289.
Keane, J.T., Matsuyama, I., Kamata, S., Steckloff, J.K., 2016. Reorientation and faulting of pluto due to volatile loading within sputnik planitia. Nature 540, 90–93. doi:10.1038/nature20120.
Kenyon, S.J., Bromley, B.C., 2014. The formation of pluto's low-mass satellites. Astron. J 147, 8–24.
Kenyon, S.J., Bromley, B.C., 2019. A pluto-charon sonata: the dynamical architecture of the circumbinary satellite system. Astrophysical J. 157 (2), 79.
Klavetter, J.J., 1989. Rotation of hyperion. I-Observations. Astron. J. 97, 570–579.

Kuiper, G.P., 1951. On the origin of the solar system. In: Hynek, J.A. (Ed.), Astrophysics. McGraw-Hill, New York, NY, p. 357–427.

Lacerda, P., et al., 2014. The albedo–color diversity of transneptunian objects. Astrophys. J. Lett. 793, L2.

Lee, M.H., Peale, S.J., 2006. On the orbits and masses of the satellites of the pluto-charon system. Icarus 184, 573–583.

Levison, H.F., Duncan, M.J., 1993. The gravitational sculpting of the kuiper belt. Astrophys. J. 406, L35–L38.

Levison, H.F., and Stern, S.A (1993), The early dynamical history of the pluto-charon system, In: Pluto-Charon Conf., 6-9 July 1993, Flagstaff, Arizona.

Levison, H.F., Stern, S.A., 1995. Possible origin and early dynamical evolution of the pluto-charon binary. Icarus 116 (2), 315–339.

Levison, H.F., and Walsh, K. (2013), Forming the small satellites of pluto, DPS meeting #45, #503.05

Levy, E.H., Lunine, J.I., 1993. Protostars & Planets III. The University of Arizona Press, Tucson.

Lissauer, J.J., Barnes, J.W., Chambers, J.E., 2012. Obliquity variations of a moonless earth. Icarus 217, 77–87.

Lithwick, Y., and Wu, Y. (2008), The effect of charon's tidal damping on the orbits of pluto's three moons. arXiv0802.2939L

Lithwick, Y., Wu, Y., 2013. On the Origin of Pluto's minor Moons. Nix and Hydra arXiv:0802.2951.

Edited by , J.Luu, 1993. The kuiper belt, asteroids, comets, meteors, proceedings of the 160th symposium of the international astronomical union, held in belgirate, italy, june 14-18, 1993. In: Milani, Andrea, Di Martino, Michel, Cellino, A. (Eds.), International Astronomical Union. Kluwer Academic Publishers, Dordrecht, p. 31 Symposium no. 160.

Lykawka, P.S., Mukai, T., 2005. Higher albedos and size distribution of large transneptunian objects. Plan. Space Sci. 53, 1319–1330.

Lyttleton, R.A., 1936. On the possible results of an encounter of pluto with the neptunian system. Mon. Not. R. Astron. Soc. 97 (2), 108–115.

Malhotra, R., 1993. The origin of pluto's peculiar orbit. Nature 365, 819–821.

Malhotra, R., 1994. Nonlinear resonances in the solar system. *Physica D*: Nonlinear Phenomena. Elsevier.

Malhotra, R., 1995. The origin of pluto's orbit: implications for the solar system beyond neptune. Astronomical J. 110, 420–429.

Malhotra, R., and Williams, J.G. (1994), In: *Pluto & Charon*, edited by D. Tholen and S. A. Stern (University of Arizona Press, Tucson)

Materese, C.K., Nuevo, M., Sandford, S.A., 2017. The formation of nucleobases from the ultraviolet photo-irradiation of purine in simple astrophysical ice analogs. Astrobiology 17 (8), 761–770.

McKinnon, W.B., Mueller, S., 1988. Pluto's structure and composition suggest origin in the solar, not a planetary, nebula. Nature 335, 240–243.

McKinnon, W.B., Nimmo, F., Wong, T., Schenk, P.M., White, O.L., Roberts, J.H., Moore, J.M., Spencer, J.R., Howard, A.D., Umurhan, O.M., Stern, S.A., Weaver, H.A., Olkin, C.B., Young, L.A., Smith, K.E. The New Horizons Geology, Geophysics and Imaging Theme Team, 2016. Convection in a volatile nitrogen-ice-rich layer drives pluto's geological and atmospheric vigour. Nature 534, 82–85.

McKinnon, W.B., et al., 2017. Origin of the Pluto–Charon system: Constraints from the New Horizons fly, Icarus, 287, 2–11. https://doi.org/10.1016/j.icarus.2016.11.019.

Metzger, P.T., Sykes, M.V., Stern, A., Runyon, K., 2018. The reclassification of asteroids from planets to non-planets. Icarus. https://doi.org/10.1016/j.icarus.2018.08.026.

Milani, A., Nobili, A.M., Carpino, M., 1989. Dynamics of pluto. Icarus 82, 200–217.

Miller, S., 1953. A production of amino acids under possible primitive earth conditions. Science 117, 528–529.

Millis, R.L., et al., 1993. Pluto's radius and atmosphere: results from the entire 9 june 1988 occultation data set. Icarus 105, 282–297.

Mohit, P.S., Phillips, R.J., 2007. Viscous relaxation on early mars: a study of ancient impact basins. Geophys. Res. Lett. 34, L21204. doi:10.1029/2007GL031252.

Moore, J.M., et al., 2014. Geology before pluto: pre-encounter considerations. Icarus. doi:10.1016/j.icarus.2014.04.028.

Moore, J.M., et al., 2016. The geology of pluto and charon through the eyes of new horizons. Science 351, 1284–1293.

Moore, J.M., Howard, A.D., Umurhan, O.M., White, O.L., et al., 2018. Bladed terrain on pluto: possible origins and evolution. Icarus 300, 129–144.

Moore, P., 2010. The Sky at Night. Springer, p. 35 ISBN 978-1-4419-6408-3.

Moores, J.E., Smith, C.L., Toigo, A.D., Guzewich, S.D., 2017. Penitentes as the origin of the bladed terrain of tartarus dorsa on pluto. Nature 541, 188–190. doi:10.1038/nature20779.

Moresi, L.-N., Solomatov, V.S., 1995. Numerical investigation of 2D convection with extremely large viscosity variations. Phys. Fluids 7, 2154–2162.

Nadeau, A., McGhee, R.A., 2015. Preprint At.

Nadeau, A., McGhee, R.A., 2018. Preprint At.

Nieto, M.M., 2008. New horizons and the onset of the pioneer anomaly. Phys. Lett. B 659 (3), 483–485.

Nimmo, F., et al., 2016. Mean radius and shape of pluto and charon from new horizons images. Icarus. http://dx.doi.org/10.1016/j.icarus.2016.06.027.

Nimmo, F., Hamilton, D.P., McKinnon, W.B., Schenk, P.M., Binzel, R.P., Bierson, C.J., Beyer, R.A., Moore, J.M., Stern, S.A., Weaver, H.A., Olkin, C.B., Young, L.A., Smith, K.E., 2016. Reorientation of sputnik planitia implies a subsurface ocean on pluto. Nature 540, 94–96. doi:10.1038/nature20148.

Nimmo, F., Matsuyama, I., 2007. Reorientation of icy satellites by impact basins. Geophys. Res. Lett. 34, L19203.

Olkin, C.B., Young, L.A., Borncamp, D., Pickles, A., Sicardy, B., Assafin, M., Bianco, F.B., Buie, M.W., Dias de Oliveira, A., Gillon, M., French, R.G., Ramos Gomes Jr., A., Jehin, E., Morales, N., Opitom, C., Ortiz, J.L., Maury, A., Norbury, M., Braga-Ribas, F., Smith, R., Wasserman, L.H., Young, E.F., Zacharias, M., Zacharias, N., 2015. Evidence that pluto's atmosphere does not collapse from occultations including the 2013 may 04 event. Icarus 246, 220–225.

Oort, J.H., 1950. The structure of the cloud of comets surrounding the solar system and a hypothesis concerning its origin. Bull. Astr. Inst. Neth. 11, 91.

Owen, T.C., Roush, T.L., Cruikshank, D.P., Elliot, J.L., Young, L.A., de Bergh, C., Schmitt, B., Geballe, T.R., Brown, R.H., Bartholomew, M.J., 1993. Surface ices and the atmospheric composition of pluto. Science 261 (5122), 745–748. doi:10.1126/science.261.5122.745.

Parmentier, E., Head, J., 1981. Viscous relaxation of impact craters on icy planetary surfaces: determination of viscosity variation with depth. Icarus 47 (1), 100–111. doi:10.1016/0019-1035(81)90095-6.

Pasachoff, J.M., Babcock, B.A., Souza, S.P., et al., 2006. A search for rings, moons, or debris in the pluto system during the 2006 july 12 occultation. Bulletin Am. Astronomical Society 38 (3), 523.

Peale, S.J. (1986), In: *Satellites*, edited by J. Burns and M. Matthew, (University of Arizona Press, Tucson), p. 159

Person, M.J., Elliot, J.L., Gulbis, A.A.S., Pasachoff, J.M., Babcock, B.A., Souza, S.P., Gangestad, J., 2006. Charon's radius and density from the combined data sets of the 2005 july 11 occultation. Astron. J. 132 (4), 1575–1580. doi:10.1086/507330.

Protopapa, S., et al., 2016. First look at global pluto's surface composition through pixel-by-pixel radiative scattering model of new horizons ralph/leisa data. Icarus. (in the press); preprint at. https://arxiv.org/abs/1604.08468.

Quillen, A.C., Nichols-Fleming, F., Chen, Y.-Y., Noyelles, B., 2017. Obliquity evolution of the minor satellites of pluto and charon. Icarus 293, 94–113.

Robbins, S., Stern, A., Binzel, R., Grundy, W., Hamilton, D., Lopes, R., McKinnon, B., Olkin, C., 2021. Pluto system follow on missions: Background, rationale, and new mission recommendations, https://baas.aas.org/pub/2021n4i193/release/1?readingCollection=7272e5bb.

Robuchon, G., Nimmo, F., 2011. Thermal evolution of pluto and implications for surface tectonics and a subsurface ocean. Icarus 216, 426–439.

Robuchon, G., Nimmo, F., Roberts, J., Kirchoff, M., 2011. Impact basin relaxation at iapetus. Icarus 214 (1), 82–90. doi:10.1016/j.icarus.2011.05.011.

Rubincam, D.P., 2003. Polar wander on triton and pluto due to volatile migration. Icarus 163, 469–478.

Saha, P., Tremaine, S, 1992. Symplectic integrators for solar system dynamics. Astr. J. 104 (4), 1633–1640.

Schenk, P.M., 2002. Thickness constraints on the icy shells of the galilean satellites from a comparison of crater shapes. Nature 417, 419–421.

Schenk, P.M., et al., 2015. A large impact origin for sputnik planum and surrounding terrains, pluto? Div. Planet. Sci. Meet. 47 abstr. 200.06.

Schenk, P.M., Beyer, R.A., McKinnon, W.B., et al., 2018. Basins, fractures and volcanoes: global cartography and topography of pluto from new horizons. Icarus 314, 400–433. doi:10.1016/j.icarus.2018.06.008.

Schenk, P.M., Wilson, R.R., Davies, A.G., 2004. Shield volcano topography and the rheology of lava flows on IO. Icarus 169, 98–110.

Scott, T.A., 1976. Solid and liquid nitrogen. Phys. Rep. (Phys. Lett. C) 27, 89–157.

Showalter, M.R. et al. (2011), New satellite of (134340) Pluto: S/2011 (134340). IAU Circular 9221, #1 (2011). Central Bureau Electronic Telegrams, No. 2769, #1 (2011). Edited by Green, D. W. E: Bibcode: 2011IAUC.9221....1S.

Showalter, M.R. et al. (2012), New satellite of (134340) Pluto: S/2012 (134340). IAU Circular 9253: IAU Circ., No. 9253, #1 (2012). Edited by Green, D. W. E.: Bibcode: 2012IAUC.9253....1S: © The SAO/NASA Astrophysics Data System.

Showalter, M.R., Hamilton, D.P., 2015. Resonant interactions and chaotic rotation of pluto's small moons. Nature 522, 45–49.

Solomatov, V.S., 1995. Scaling of temperature- and stress-dependent viscosity convection. Phys. Fluids 7, 266–274.

Solomatov, V.S., Barr, A.C., 2007. Onset of convection in fluids with strongly temperature-dependent, power-law viscosity: 2. dependence on the initial perturbation. Phys. Earth Planet. Inter. 165, 1–13.

Solomatov, V.S., Moresi, L.-N., 1997. Three regimes of mantle convection with non-Newtonian viscosity and stagnant lid convection on the terrestrial planets. Geophys. Res. Lett. 24, 1907–1910.

Solomon, S.C., Stephens, S.K., Head, J.W., 1982. On venus impact basins: viscous relaxation of topographic relief. J. Geophys. Res. 87 (B9), 7763–7771. doi:10.1029/JB087iB09p07763.

Spada, G., et al., 2012. Greenland uplift and regional sea level changes from ICES at observations and GIA modelling. Geophys. J. Int. 189, 1457–1474.

Spencer, J.R., Denk, T., 2010. Formation of iapetus' extreme albedo dichotomy by exogenically triggered thermal ice migration. Science 327, 432–435.

Spencer, J.R., Stansberry, J.A., Trafton, L.M., Young, E.F., Binzel, R.P., Croft, S.K., 1997. Volatile transport, seasonal cycles, and atmospheric dynamics on Pluto, In: Pluto and Charon, ed. by S.A. Stern, D.J. Tholen (Univ. of Arizona Press, Tucson), pp. 435–473.

Stansberry, J., Grundy, W., Brown, M., Cruikshank, D., Spencer, J., Trilling, D., et al., 2008. Physical properties of kuiper belt and centaur objects: constraints from spitzer space telescope. In: Barucci, M.A., Boehnhardt, H., Cruikshank, D.P., and Morbidelli, A., (Ed.). The Solar System Beyond Neptune, 2008. University of Arizona Press, Tucson, 592, p. 161–179.

Stansberry, J.A., Lunine, J.I., Hubbard, W.B., Yelle, R.V., Hunten, D.M., 1994. Mirages and the nature of pluto's atmosphere. Icarus 111, 503–513.

Stansberry, J.A., Yelle, R.V, 1999. Emissivity and the fate of pluto's atmosphere. Icarus 141, 299–306.

Stern, S.A., 2009. Ejecta exchange and satellite color evolution in the pluto system, with implications for KBOs and asteroids with satellites. Icarus 199, 571–573.

Stern, S.A., Bagenal, F., Ennico, K., et al., 2015. The pluto system: initial results from its exploration by new horizons. Science 350 (6258) aad1815-1- aad1815-8.

Stern, S.A., Trafton, L., 1984. Constraints on bulk composition, seasonal variation, and global dynamics of pluto's atmosphere. Icarus 57, 231–240.

Stern, S.A., Weaver, H.A., Steff, A.J., Mutchler, M.J., Merline, W.J., Buie, M.W., Young, E.F., Young, L.A., Spencer, J.R., 2006. A giant impact origin for pluto's small moons and satellite multiplicity in the kuiper belt. Nature 439 (7079), 946–948.

147 Stern, S.A., O.L. White, P.J. McGovern, J.T. Keane, J.W. Conrad, C.J. Bierson, C.B. Olkin, P.M. Schenk, J.M. Moore, K.D. Runyon, H.A. Weaver, L.A. Young, K. Ennico, The new horizons team (2020), pluto's far side, Earth and Planetary Astrophysics (astro-ph.EP), arXiv:1910.08833

Sussman, G.J., Wisdom, J, 1988. Numerical evidence that the motion of pluto is chaotic. Science 241, 433–437.

Tancredi, G., Fernandez, J.A., 1991. The angular momentum of the pluto-charon system: considerations about its origin. Icarus 93, 298.

Thomas, P.J., Squyres, S.W., 1988. Relaxation of impact basins on icy satellites. J. Geophys. Res. 93 (B12), 14919–14932. doi:10.1029/JB093iB12p14919.

Trowbridge, A.J., et al., 2016. Vigorous convection as the explanation for pluto's polygonal terrain. Nature 534, 79–81.

van Hemelrijck, E., 1982. The insolation at pluto. Icarus 52, 560–564.

van Hemelrijck, E., 1985. Insolation changes on pluto caused by orbital element variations. Earth Moon Planets 33, 163–177.

Ward, W.R., 1974. Climatic variations on mars: 1. astronomical theory of insolation. J. Geophys. Res. 79, 3375–3386.

Ward, W.R., Canup, R.M., 2006. Forced resonant migration of pluto's outer satellites by charon. Science 313, 1107–1109.

Weaver, H., et al., 2006. Discovery of two new satellites of pluto. Nature 439, 943–945.

Weaver, H.A., 2016. The small satellites of pluto as observed by new horizons. Science 351 (6279), 1281.

Weissman, P.R., 1990. The oort cloud. Nature 344, 825–830.

White, O.L., Moore, J.M., McKinnon, W.B., Spencer, J.R., Howard, A.D., Schenk, P.M., Beyer, R.A., Nimmo, F., Singer, K.N., Umurhan, O.M.,

Stern, S.A., Ennico, K., Olkin, C.B., Weaver, H.A., Young, L.A., Cheng, A.F., Bertrand, T., Binzel, R.P., Earle, A.M., Grundy, W.M., Lauer, T.R., Protopapa, S., Robbins, S.J., Schmitt, B.the New Horizons Science Team, 2017. Geological mapping of sputnik planitia on pluto. Icarus 287, 261–286.

White, O.L., Schenk, P.M., Dombard, A.J., 2013. Impact basin relaxation on rhea and iapetus and relation to past heat flow. Icarus 223 (2), 699–709. doi:10.1016/j.icarus.2013.01.013.

Williams, J.G., Benson, G.S., 1971. Resonance in the neptune-pluto system. Astron. J. 76, 167.

Wisdom, J., 1982. The origin of the kirkwood gaps—A mapping for asteroidal motion near the 3/1 commensurability. Astronomical J. 87 (3), 577–593.

Wisdom, J., Holman, M., 1991. Symplectic maps for the n-body problem. Astr. J. 102 (4), 1528–1538.

Wisdom, J., Peale, S.J., Mignard, F., 1984. The chaotic rotation of hyperion. Icarus 58 (2), 137–152.

Yamashita, Y., Kato, M., Arakawa, M., 2010. Experimental study on the rheological properties of polycrystalline solid nitrogen and methane: implications for tectonic processes on triton. Icarus 207, 972–977.

Young, L.A., 2013. Pluto's seasons: new predictions for new horizons. Astrophys. J. 766, L22–L28.

Zahnle, K., Schenk, P., Levison, H., Dones, L., 2003. Cratering rates in the outer solar system. Icarus *163* (2), 263–289. doi:10.1016/S0019-1035(03)00048-4.

Zalucha, A., Michaels, T., 2013. A 3D general circulation model for pluto and triton with fixed volatile abundance and simplified surface forcing. Icarus 223, 819–831.

Zangari, A., 2015. A meta-analysis of coordinate systems and bibliography of their use on pluto from charon's discovery to the present day. Icarus 246, 93–145.

Zhang, K., Hamilton, D.P., 2007. Orbital resonances in the inner neptunian system I. the 2:1 proteus-larissa mean-motion resonance. Icarus 188, 386–399.

Zhang, K., Hamilton, D.P., 2007. Orbital resonances in the inner neptunian system I. the 2:1 proteus-larissa mean-motion resonance. Icarus 188, 386–399.

Zuber, M.T., et al., 2012. Topography of the northern hemisphere of mercury from messenger laser altimetry. Science 336, 217–220.

Bibliography

Arakawa, M., Maeno, N., 1994. Effective viscosity of partially melted ice in the ammonia-water system. Geophys. Res. Lett. 21 (14), 1515–1518. doi:10.1029/94GL01041.

Balcerski, J.A., Hauck, S.A., Dombard, A.J., Turtle, E.P., 2010. The influence of local thermal anomalies on large impact basin relaxation. Proc. Lunar Planet. Sci. Conf. 41st, Abstract #2535.

Barr, A.C., Collins, G.C., 2014. Tectonic activity on pluto after the charon-forming impact. Icarus. doi:10.1016/j.icarus.2014.03.042.

Bray, V.J., Schenk, P.M., 2014. Pristine impact crater morphology on pluto—expectations for new horizons. Icarus. doi:10.1016/j.icarus.2014.05.005.

Binzel, R.P., 1991. Urey prize lecture: physical evolution in the solar system—present observations as a key to the past. Icarus 100, 274–287 (1992).

Brown, M.E., 2012. The compositions of kuiper belt objects. Annu. Rev. EarthPlanet. Sci. 40, 467–494.

Buie, M.W., Grundy, W.M., Young, E.F., Young, L.A., Stern, S.A., 2010. Pluto and charon with the hubble space telescope. II. resolving changes on pluto's surface and a map for charon. Astron. J. 139 (3), 1128–1143. doi:10.1088/0004-6256/139/3/1128.

Canup, R.M., 2005. A giant impact origin of pluto-charon. Science 307 (5709), 546–550. doi:10.1126/science.1106818.

Choblet, G., Cadek, O., Couturier, F., Dumoulin, C., 2007. ŒDIPUS: a new tool to study the dynamics of planetary interiors, ˇ geophys. J. Int. 170 (1), 9–30. doi:10.1111/j.1365-246X.2007.03419.x.

Dobrovolskis, A.R., Peale, S.J., and Harris, A.W. (1997), In: Pluto and Charon (eds Stern, S. A. & Tholen, D. J.) 159–190 (Univ. Arizona Press, 1997)

Hamilton, D.P., 2015. The icy cold heart of pluto, 47th Division For Planetary Sciences Meeting abstr. 200.07. American Astronomical Society.

Kargel, J.S., 1992. Ammonia water volcanism on icy satellites—phase relations at 1-atmosphere. Icarus 100, 556–574.

Lissauer, J.J., Barnes, J.W., Chambers, J.E., 2012. Obliquity variations of a moonless earth. Icarus 217, 77–87.

Matsuyama, I., Mitrovica, J.X., Manga, M., Perron, J.T., Richards, M.A., 2006. Rotational stability of dynamic planets with elastic lithospheres. J. Geophys. Res., Planets 111, E02003.

Matsuyama, I., Nimmo, F., 2007. Rotational stability of tidally deformed planetary bodies. J. Geophys. Res. 112, E11003.

McKinnon, W.B., Simonelli, D.P., and Schubert, G. (1997), In: Pluto and Charon (eds Stern, S. A. & Tholen, D. J.) 295–346 (Univ. Arizona Press, 1997).

Murray, C.D., Dermott, S.F., 2000. Solar System Dynamics, 577pp. Cambridge University Press, United States of America.

Mohit, P.S., Phillips, R.J., 2007. Viscous relaxation on early mars: a study of ancient impact basins. Geophys. Res. Lett. 34, L21204. doi:10.1029/2007GL031252.

Nimmo, F., Matsuyama, I., 2007. Reorientation of icy satellites by impact basins. Geophys. Res. Lett. 34, L19203.

Nimmo, F., Spencer, J.R., 2014. Powering triton's recent geological activity by obliquity tides: implications for pluto geology. Icarus. doi:10.1016/j.icarus.2014.01.044.

Owen, T.C., Roush, T.L., Cruikshank, D.P., Elliot, J.L., Young, L.A., de Bergh, C., Schmitt, B., Geballe, T.R., Brown, R.H., Bartholomew, M.J., 1993. Surface ices and the atmospheric composition of pluto. Science 261 (5122), 745–748. doi:10.1126/science.261.5122.745.

Parmentier, E., Head, J., 1981. Viscous relaxation of impact craters on icy planetary surfaces: determination of viscosity variation with depth. Icarus 47 (1), 100–111. doi:10.1016/0019-1035(81)90095-6.

Person, M.J., Elliot, J.L., Gulbis, A.A.S., Pasachoff, J.M., Babcock, B.A., Souza, S.P., Gangestad, J., 2006. Charon's radius and density from the combined data sets of the 2005 July 11 occultation. Astron. J. 132 (4), 1575–1580. doi:10.1086/507330.

Robuchon, G., Nimmo, F., 2011. Thermal evolution of pluto and implications for surface tectonics and a subsurface ocean. Icarus 216 (2), 426–439. doi:10.1016/j.icarus.2011.08.015.

Robuchon, G., Nimmo, F., Roberts, J., Kirchoff, M., 2011. Impact basin relaxation at iapetus. Icarus 214 (1), 82–90. doi:10.1016/j.icarus.2011.05.011.

Rubin, M.E., Desch, S.J., Neveu, M., 2014. The effect of rayleigh-taylor instabilities on the thickness of undifferentiated crust on kuiper belt objects. Icarus 236, 122–135.

Scott, T.A., 1976. Solid and liquid nitrogen. Phys. Rep. 27, 89–157.

Spencer, J. R. et al. (1997), In: Pluto and Charon (eds Stern, S.A. & Tholen, D.J.) 435–473 (Univ. Arizona Press, 1997).

Spencer, J.R., Denk, T., 2010. Formation of iapetus' extreme albedo dichotomy by exogenically triggered thermal ice migration. Science 327, 432–435.

Turcotte, D.L., et al., 1981. Role of membrane stresses in the support of planetary topography. J. Geophys. Res. 86, 3951–3959.

White, O.L., Schenk, P.M., Dombard, A.J., 2013. Impact basin relaxation on Rhea and Iapetus and relation to past heat flow. Icarus 223, 699–709.

Willemann, R.J., 1984. Reorientation of planets with elastic lithospheres. Icarus 60, 701–709.

Zahnle, K., Schenk, P., Levison, H., Dones, L., 2003. Cratering rates in the outer solar system. Icarus 163 (2), 263–289. doi:10.1016/S0019-1035(03)00048-4.

Zhong, S., Zuber, M.T., 2000. Long-wavelength topographic relaxation for self-gravitating planets and implications for the time-dependent compensation of surface topography. J. Geophys. Res. 105 (E2), 4153–4164. doi:10.1029/1999JE001075.

Chapter 17

Hunting for environments favorable to life on planets, moons, dwarf planets, and meteorites

17.1 A new frontier in planetary simulation

Astronomers and cosmologists have found that subsurface oceans exist on the moons of some planets (e.g., Saturn moons *Enceladus* and *Titan*). Apart from these celestial bodies, evidence have been found suggesting that subsurface water exists also on some other moons (e.g., the Jupiter moons *Europa, Ganymede,* and *Callisto*); and a number of other bodies in the solar system. There are probably also many more planetary moons that have not yet been discovered. Indeed, buried water bodies, akin to that of Pluto, may be abundant in the *Kuiper Belt*. According to researchers conducting studies of the solar system, one of the lessons of the last 20 years is that oceans pop up in all kinds of unexpected places. It is heartening that the recent Cassini discovery of water vapor plumes ejected from the vicinity of the south pole of the Saturnian moon *Enceladus* presents a unique window of opportunity for the detection of extant life in our solar system.

The Voyager 1 and 2 missions have shown that most of the ice moons of Jupiter and Saturn have had some form of tectonic or internal activity. Understanding the tectonics and building reasonable models of the interior of these planets depend on our knowledge of the properties and viscosity of high-pressure and low-temperature forms of ice (see Poirier, 1982).

Based on studies of the potential habitability of Jupiter's moon Europa, icy moon oceans can be habitable if they are chemically mixed with the overlying ice shell on Myr time scales. According to Parkinson et al (2007), with Enceladus's significant geothermal energy source propelling its plumes from the surface of the moon and the ensuing large temperature gradient with the surrounding environment, it is possible to have the weathering of rocks by liquid water at the rock/liquid interface. Parkinson et al (2007) have argued that on Enceladus, the weathering of rocks by liquid water and any concomitant radioactive emissions are possible incipient conditions for life. If there is CO, CO_2, and NH_3 present in the spectra obtained from the plume, then this is possible evidence that amino acids could be formed at the rock/liquid interface of Enceladus. The combination of a hydrological cycle, chemical redox gradient, and geochemical cycle give favorable conditions for life. Parkinson et al. (2008) hypothesized that Enceladus' plume, tectonic processes, and liquid water ocean may create a complete and sustainable geochemical cycle that may allow it to support life.

Investigations of other planetary bodies, including Mars and icy moons such as Enceladus and Europa, show that they may have hosted aqueous environments in the past and may do so even today. Therefore, a major challenge in astrobiology is to build facilities that will allow us to study the geochemistry and habitability of these extra-terrestrial environments. Martin and Cockell (2015) have described a simulation facility (PELS: Planetary Environmental Liquid Simulator) with the capability for liquid input and output that allows for the study of such environments. The facility, containing six separate sample vessels, allows for statistical replication of samples. Control of pressure, gas composition, UV irradiation conditions, and temperature allows for the precise replication of aqueous conditions, including subzero brines under Martian atmospheric conditions. A sample acquisition system allows for the collection of both liquid and solid samples from within the chamber without breaking the atmospheric conditions, enabling detailed studies of the geochemical evolution and habitability of past and present extra-terrestrial environments. The facility Martin and Cockell (2015) described represents a new frontier in planetary simulation—continuous flow-through simulation of extra-terrestrial aqueous environments.

17.2 Role of tidally heated oceans of giant planets' moons in supporting an environment favorable to life

Tidal dissipation in the moons of giant planets may provide sufficient heating to maintain an environment favorable to life on the moon's surface or just below a thin ice layer. Reynolds et al. (1987) recalled that in our own solar system, Europa, one of the Galilean moons of Jupiter, could have a liquid ocean that may occasionally receive sunlight through cracks in the overlying ice shell. In such cases, sufficient solar energy could reach liquid water so that organisms similar to those found under Antarctic ice could grow. In other solar systems, larger moons with more significant heat flow could represent environments that are stable over an order of Aeons and in which life could perhaps evolve. In this context, Reynolds et al. (1987) have defined a zone around a giant planet in which such moons could exist as a tidally-heated habitable zone. According to Reynolds et al. (1987), this zone can be compared to the habitable zone which results from heating due to the radiation of a central star.

Ice-covered oceans are primarily heated from below, creating convection that could transport putative microbial cells and cellular cooperatives upward to congregate beneath an ice shell, potentially giving rise to a highly focused shallow biosphere. According to Russell et al. (2017), "it is here where electron acceptors, ultimately derived from the irradiated surface, could be delivered to such life-forms through exchange with the icy surface. Such zones would act as 'electron disposal units' for the biosphere, and occupants might be transferred toward the surface by buoyant diapirs and even entrained into plumes."

In considering ocean worlds, there are several with confirmed oceans: Does life originate and take hold in some ocean worlds and not others, and why? Thus, the Roadmaps to Ocean Worlds (ROW) team operating under NASA supports the creation of a program that studies the full spectrum of ocean worlds; if only one or two ocean worlds are explored and life is discovered (or not), we will not fully understand the distribution of life, its origin and variability, and the repeatability of its occurrences in the Solar System.

Planetary scientists have considered that Enceladus, Europa, Titan, Ganymede, and Callisto have known subsurface oceans, as determined from geophysical measurements by the Galileo and Cassini spacecraft. These are confirmed/known ocean worlds. Hendrix et al. (2019) have described the scientific content and priorities for investigations that are needed for the exploration of ocean worlds. Such investigations would be carried out by a robotic flight program that would measure needed quantities at ocean worlds, and by research efforts to characterize important physical processes potentially at work on ocean worlds.

Before sending spacecraft to a target body to search for life within the ocean, we must first demonstrate that an ocean exists. There are several questions that can be addressed to determine the presence of an ocean. For the confirmed ocean worlds (Europa, Enceladus, Titan, Ganymede, and Callisto), these questions have already been answered—or enough of the questions have been answered that the presence of an ocean is (reasonably) certain (Hendrix et al., 2019).

17.3 Prospects of life on *Mars* and Jupiter's Moon *Europa*

According to McKay (2004), based on a categorization of structures that comprise terrestrial life (ecological, biochemical, and chemical), we can speculate about how life might be different on Mars or Europa. A reasonable assumption is that life is composed of matter. Carbon and liquid water are the next levels in terms of life. This makes Mars and Europa likely candidates, because they have carbon and have, or have had, liquid water.

17.3.1 Role of biomineralization in providing bacteria with an effective UV screen on Mars

Investigation of the habitability potential of Mars requires serious consideration of the harmful radiation reaching the surface of Mars. It may be noted that Earth's magnetosphere, arising from its magnetic field—generated because of the circulating lava in its mantle (magnetic dynamo effect)—protects the inhabitants of Earth from harmful solar and galactic cosmic radiation. However, approximately 4.2 billion years ago, due to either rapid cooling in its central core or a massive impact from an asteroid or comet, the magnetic dynamo effect on Mars stopped, and its magnetosphere weakened dramatically (Matt, 2016). As a result, over the course of the next 500 million years, the Martian atmosphere was gradually stripped away by the solar wind.

The absence of protective components in the early Mars atmosphere forced any possible primordial life forms to deal with high doses of UV radiation. A similar situation occurred on the primitive Earth during the development of early life in the Archean (Berkner and Marshall, 1965; Kasting, 1993). It is known that some cellular and/or external components can shield organisms from damaging UV radiation or quench its toxic effects (Olson and Pierson, 1986; García-Pichel, 1998; Cockell et al., 2003). In a study carried out by Phoenix et al. (2001), cyanobacteria, isolated from the Krisuvik hot spring, Iceland, were mineralized in an iron-silica solution and irradiated with high levels of ultraviolet light. Analysis of the rates of photosynthesis, chlorophyll-a content, and phycocyanin autofluorescence revealed that these mineralized bacteria have a marked resistance to UV compared to nonmineralized bacteria.

Naturally occurring sinters (a mixture of iron ore and other materials) composed of iron-silica biominerals collected from the Lysúholl hot spring, and made into wafers of 150–250 μm thickness, also provided cyanobacteria with an effective UV screen. Analysis of the UV-absorbing capacity of these wafers showed that they absorbed an order of magnitude more UV than photosynthetically active light (required for photosynthesis). From these results, it is evident that both natural and experimental biomineralization provide bacteria with an effective UV screen through the passive precipitation of iron-enriched silica crusts. The UV-shielding capacity of iron-bearing silicate biominerals may have been important for early life forms. Phoenix et al. (2001) proposed that the biomineralization of Archean bacteria similarly provided protection from the high-intensity UV present at that time, and hence allowed colonization and bacterial diversification of shallow-water environments.

In another study, Gómez et al. (2007) demonstrated the effect of iron in solution, particularly the protective effect of soluble ferric iron against UV radiation on acidophilic photosynthetic microorganisms. These results offer an interesting alternative means of protection for life on the surface of early Mars and Earth, especially in light of the geochemical conditions in which the sedimentary minerals, jarosite, and goethite, reported by the MER missions, were formed (Squyres et al., 2004; Klingelhöfer et al., 2004). Thus, Gómez et al. (2007)'s study offers a clear indication that soluble ferric iron is an effective protective agent against the UV radiation present on Mars; and that the presence of vast quantities of iron minerals present on the Martian surface has vast implications for early life on Mars.

In yet another study, with the aim of evaluating this possibility Gómez et al. (2010) studied the viability of two microorganisms under different conditions in a Mars simulation chamber. An acidophilic chemolithotroph isolated from Río Tinto belonging to the Acidithiobacillus genus and Deinococcus radiodurans, a radiation resistant microorganism, were exposed to simulated Mars conditions under the protection of a layer of ferric oxides and hydroxides, a Mars regolith analog. Samples of these microorganisms were exposed to UV radiation in Mars atmospheric conditions at different time intervals under the protection of 2- and 5-mm layers of oxidized iron minerals. Viability was evaluated by inoculation on fresh media and characterization of their growth cultures. Gómez et al. (2010) have reported the survival capability of both bacteria to simulated Mars environmental conditions.

17.3.2 Redox gradients may support life and habitability on Mars

Measurements of methane (CH_4) by the Mars Science Laboratory (MSL) have revealed a baseline level of CH_4 (~0.4 parts per billion by volume [ppbv]), with seasonal variations, as well as greatly enhanced spikes of CH_4 with peak abundances of ~7 ppbv. Do these CH_4 revelations with drastically different abundances and temporal signatures represent biosignature? Discerning how CH_4 generation occurs on Mars may shed light on the potential habitability of Mars. There is no evidence of life on the surface of Mars today, but microbes might reside beneath the surface. In this case, the carbon flux represented by CH_4 would serve as a link between a putative subterranean biosphere on Mars and what we can measure above the surface.

Missions such as ExoMars Trace Gas Orbiter may provide mapping of the global distribution of CH_4. To discriminate between abiotic and biotic sources of CH_4 on Mars, Yung et al. (2018) suggest that future studies should use a series of diagnostic geochemical analyses, preferably performed below the ground or at the ground/atmosphere interface, including measurements of CH_4 isotopes, methane/ethane ratios, H_2 gas concentration, and species such as acetic acid. Advances in the fields of Mars exploration and instrumentation will be driven, augmented, and supported by an improved understanding of atmospheric chemistry and dynamics, deep subsurface biogeochemistry, astrobiology, planetary geology, and geophysics. Future Mars exploration programs will have to expand the integration of complementary areas of expertise to generate synergistic and innovative ideas to realize breakthroughs in advancing our understanding of the potential of life and habitable conditions having existed on Mars. Yung et al. (2018) further suggested future exploration of the Martian subsurface and enhancement of spatial tracking of key volatiles, such as CH_4.

Arguing in favor of the potential habitability of Mars, Yung et al. (2018) made the following observation: "The Martian atmosphere and surface are an overwhelmingly oxidizing environment, and life requires pairing of electron donors and electron acceptors, that is, redox gradients, as an essential source of energy. Therefore, a fundamental and critical question regarding the possibility of life on Mars is, 'Where can we find redox gradients as energy sources for life on Mars?' Hence, regardless of the pathway that generates CH_4 on Mars, the presence of CH_4, a reduced species in an oxidant-rich environment, suggests the possibility of redox gradients supporting life and habitability on Mars."

17.3.3 Probable life on Mars—speculations driven by biochemistry

In the absence of any firm/credible indications of life on Mars, scientists have come up with speculations. For example, Pace (2001) has argued that alien biochemistry will turn out to be the same as biochemistry on Earth, because there is one best way to do things and that natural selection will ensure that life everywhere discovers that way. However, only observation will tell if there is one possible biochemistry, or many.

McKay (2004) argued that other worlds may have a different chemical baseline for life. The usual speculation in this area is that the presence of ammonia and silicon, rather than water and carbon, might be preconditions for life on other planets. Such speculation has yet to lead to any specific suggestions for experiments, or to new ways to search for such life, but this may just reflect a failure of human imagination rather than a fundamental limitation on the nature of life.

McKay (2004) argued that life on Mars is also likely to be the same at the ecological level. According to him, primary production in a Martian ecosystem is likely to be phototrophic (i.e., tendency of an organism to grow in response to a light stimulus), using carbon dioxide and water. Heterotrophs (i.e., organisms that cannot produce their own food, instead taking nutrition from other sources of organic carbon, mainly plant or animal matter) are likely to be present to consume the phototrophs and in turn to be consumed by predators. McKay (2004) is of the view that on Mars, Darwinian evolution would result in many of the same patterns we see in ecosystems on Earth. While it may be similar at the ecological and chemical levels, life on Mars could be quite alien in the realm of biochemistry.

McKay (2004) points out that on Mars, there is ice-rich ground in the cratered southern polar regions (Feldmann et al., 2002), which presumably overlies deeper, older ice. The surprise discovery of strong magnetic fields in the southern hemisphere of Mars (Acuña et al., 1999; Connerney et al., 1999) indicates that the area may be the oldest undisturbed permafrost on that planet. Organic material of biological origin will eventually lose its distinctive pattern when exposed to heat and other types of radiation, (examples of this include the thermal racemization of amino acids), but at the low temperatures in the Martian permafrost, calculations suggest that there has been no thermal alteration (Kanavarioti and Mancinelli, 1990). Note that in chemistry, racemization is a conversion, by heat or by chemical reaction, of an optically active compound into a racemic form. Half of the optically active substance becomes its mirror image referred as racemic mixtures.

While McKay (2004) agreed that like the mammoths [a large extinct elephant of the Pleistocene epoch (the geological epoch that lasted from about 2,580,000 to 11,700 years ago, spanning the earth's most recent period of repeated glaciations), typically hairy with a sloping back and long curved tusks] extracted from the ice in Siberia, any Martian microbes found in this ice would be dead, but their biochemistry would be preserved. He is optimistic that from these biological remains, it would then be possible to determine the biochemical composition of, and the phylogenetic relationship between, Earth life and Martian life.

17.3.4 Best possible hideout on Mars—a string of lava tubes in the low-lying Hellas Planitia

Current surface conditions (strong oxidative atmosphere, UV radiation, low temperatures, and very dry conditions) on Mars are considered extremely challenging for life. The question is whether there are any features on Mars that could exert a protective effect against the sterilizing conditions detected on its surface. Potential habitability in the subsurface would increase if the overlaying material played a protective role (Gómez et al., 2010).

The data from Mars Atmosphere and Volatile Evolution (MAVEN) spacecraft, Martian Radiation Experiment (MARIE), and Radiation Assessment Detector (RAD) show that the solar wind and other violent solar activity, such as solar flares (brief eruptions of intense high-energy radiation from the Sun's surface, associated with sunspots) and coronal mass ejections (events in which a large cloud of energetic and highly magnetized plasma erupts from the solar corona into space), continue to strip away the Martian atmosphere—but more importantly, they make the northern latitudes less desirable for human exploration (see Paris et al., 2019).

Between the loss of its magnetic field and its atmosphere, the surface of Mars is now exposed to much higher levels of solar and cosmic radiation than Earth. In addition to regular exposure to solar energetic particles and galactic cosmic rays, Mars receives intermittent harmful blasts that occur from strong solar flares, as well as the bombardment of meteors [National Aeronautics and Space Administration. Apollo Flight Journal (available at http://history.nasa.gov/afj/]. A crewed mission to the surface of Mars, consequently, will introduce the crew and its critical life-support equipment to a harmful radiation environment. The Martian crew will be at risk of absorbing sudden fatal radiation doses, as well as assuredly suffering cellular and DNA (the hereditary material in humans and almost all other organisms) damage from chronic high background radiation, which will lead to cancer (see Paris et al., 2019). Additionally, there is a risk of an unpredictable cosmic ray burst or a meteor shower capable of critically damaging the crew's life-support equipment [National Aeronautics and Space Administration. Why Space Radiation Matters (available at https://www.nasa.gov/analogs/nsrl/why-space-radiation-matters/)].

The net result of these findings is that there appears to be no safe place to camp out on Mars. But a team of researchers has identified what could be future Martian explorers' best possible hideout: a string of lava tubes in the low-lying Hellas Planitia—an impact basin blasted into Mars's surface by ancient meteor impacts.

Mars' surface is arid, starved of oxygen, and blasted daily with unrelenting, unfiltered solar radiation. Any future Martian explorers will put their lives in peril when they embark. NASA has decades of experience hauling oxygen, food, and water beyond Earth. But that last killer, the radiation, is a harder problem to tackle.

Mars is currently at the center of intense scientific study aimed at potential human colonization. Consequently, there has been increased curiosity in the identification and study of lava tubes (details on lava tubes are given in Chapters 7–9) for deriving information on the paleo-hydrological (the study of ancient use and handling of water; e.g., as in irrigation or urban water supplies), geomorphological (of or relating to the form or surface features of any celestial body), geological (relating to the study of a celestial body's physical structure and substance), and potential biological (relating to living organisms) history of Mars, including the prospect of present microbial life on the planet.

Hunting through images taken from probes in Mars' orbit, planetary researchers identified several pit crater-chains and other evidence of old lava flows that burrowed into the Martian crust around Hadriacus Mons. For example, an analysis of HiRISE and CTX imagery of the Hellas Planitia basin, specifically southwest of Hadriacus Mons, carried out by Paris et al. (2019) identified pit crater chains consistent with known lava tube morphology. Although the internal structural conditions of the candidate lava tubes remain largely unknown, a close examination of the satellite surface imagery suggests that sections of the pit crater-chains have not collapsed, and therefore that lava tubes below the surface could be internally intact (Paris et al., 2019). Furthermore, the candidate lava tubes identified in Paris et al. (2019)'s investigation are positioned in a region of Mars that regularly has lower radiation exposure than other regions on Mars (Fig. 17.1). Note that Sievert (Sv) is the unit of radiation absorption in the International System of Units (SI). Though a background radiation environment of ~342.46 microsieverts per day (μSv/day) is still significantly high, the terrestrial analog experiments conducted by Paris et al. (2019) at three locations (Mojave, CA, El Malpais, NM, and Flagstaff, AZ) concluded the lava tubes southwest of Hadriacus Mons could reduce the crew's exposure to radiation to ~61.64 μSv/day. The results of Paris et al. (2019)'s investigation, therefore, indicate that the proposed lava tubes southwest of Hadriacus Mons can and should be utilized to serve as natural shelters for a crewed mission to Mars. Paris et al. (2019) are optimistic that these natural caverns would provide the crew protection from excessive radiation exposure, shelter them from the bombardment of micrometeorites, reduce their exposure to hazardous perchlorates in the Martian regolith, and provide them a degree of protection from extreme temperature fluctuations, solar energetic particles, and unpredictable high-energy cosmic radiation (i.e., gamma-ray bursts). The candidate lava tubes identified on Mars, moreover, can serve as important locations for direct observation and study of Martian geology and geomorphology, as well as potentially uncovering any

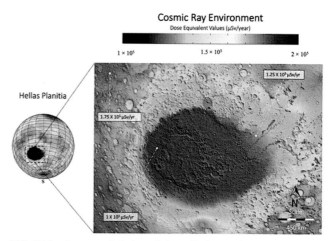

FIG. 17.1 An expanded view of the radiation environment at Hellas Planitia, which experiences lower radiation exposure than other regions on Mars. *(Source: Center for Planetary Science). (Source: Paris, A.J., E.T. Davies, L. Tognetti, and C. Zahniser (2019), Prospective lava tubes at Hellas Planitia: Leveraging volcanic features on Mars to provide crewed missions protection from radiation, The Journal of The Washington Academy of Sciences, arXiv:2004.13156 [astro-ph.EP].*

evidence for the development of microbial life early in the natural history of Mars (Paris et al., 2019).

17.3.5 Europa—one of most promising places in the solar system where possible extra-terrestrial life forms could exist

Planetary scientists have figured out how the subsurface ocean on Jupiter's moon *Europa* may have formed and determined that this vast expanse of water may have been able to support microbial life in the past. Europa, with an ocean hidden beneath a thick shell of ice, long has been viewed as a potential habitat for extra-terrestrial life in our solar system, alongside other candidates such as Mars and Saturn's moon Enceladus. On Europa, the cracks and icebergs on the surface of the ice, the presence of NaCl, and plumes of water particles gushing out from the surface indicate water beneath the ice. Present water on Europa is also indicated by the magnetic disturbance Europa makes as it moves through Jupiter's magnetic field. It has been found that Europa's inside is holding a large conductor, and this is most likely a global, salty layer of water.

On Europa, the organic material for testing in the search for life might be collected right from the dark regions on the surface. According to McKay (2004), an interesting question, as yet unanswered, is how long organic material frozen into the surface ice of Europa would retain a biological signature in the strong radiation environment.

It is believed that a water ocean existing beneath Europa's icy crust could conceivably harbor extraterrestrial life. Astrobiologists want to investigate Europa, Ganymede, and Callisto for extra-terrestrial life, as there is life virtually wherever there is liquid water on Earth. Europa is one of the most promising places in our Solar System where possible extra-terrestrial life forms could exist either in the past or even presently.

17.4 In the hope of finding evidence of past life on Mars—arrival of NASA's "Mars 2020 Perseverance Rover" at Mars

The United States has succeeded in sending its fifth rover—the Mars 2020 Perseverance Rover (Fig. 17.2)—NASA's most capable ever, in the hope of finding evidence of past life on Mars and collecting a set of rocks and soil samples that will one day be the first samples flown back to Earth. Mars 2020 is the next step in NASA's robotic exploration of Mars, a primary target of astrobiology research in the Solar System. Many researchers supported by elements of the Astrobiology Program are involved in the design and development of the Mars 2020 mission and its scientific goals.

Previous missions, from Pathfinder to Curiosity, have helped astrobiologists determine that habitable environments were present on Mars in the planet's ancient past. For example, NASA's twin Viking landers did look for extant Mars life after they touched down in 1976. Likewise, Curiosity has been assessing the habitability of ancient Mars and investigating the planet's long-ago transition from relatively warm and wet to extremely cold and dry. However, while the Martian environments may have been habitable, we do not know if they were inhabited (i.e., if life was ever present). The Mars 2020 mission will take the next step, actively hunting for signs of life on the ancient Red Planet. Perseverance will not be looking for organisms living on Mars today. However, the rover is collecting data that could be used to identify biosignatures of ancient microbial life. Even if Perseverance does not discover any signs of past life, it paves the way for human life on Mars someday.

The car-sized Perseverance rover landed inside the Red Planet's Jezero Crater on Feb. 18, 2021, tasked with searching for signs of ancient Mars life and collecting dozens of samples for future return to Earth. What really sets the "Mars 2020 Perseverance rover mission" apart from the previous missions to Mars is that Perseverance will be collecting a suite of samples to be returned to Earth via a Mars Sample Return mission through a joint NASA-European Space Agency campaign. Once the pristine Mars material is on the ground, scientists in labs around the world can scrutinize it using far more powerful and precise equipment than a single rover can carry to the Red Planet. The return samples from another planet will allow our researchers to interrogate them with all of the sophistication and thoroughness of Earth-based instrumentation, and will provide an immense opportunity to dramatically advance our scientific understanding of Mars.

Missions before Perseverance have found that liquid water existed on Mars in the ancient past and they explored the planet's "habitability." For example, at its landing site in Gale crater, the Curiosity rover found the chemical building blocks of life and energy sources that microbes could have used, and established that Mars indeed had regions that could have been friendly to life in the ancient past.

Perseverance is a familiar-looking rover—basically a copy of the Curiosity rover that has been exploring Gale Crater on Mars since Aug. 5, 2012. The US rover is a car-sized vehicle, packed with seven instruments. Perseverance is based upon the Mars Science Laboratory heritage architecture, and as such is a step-wise improvement on tested technology. The instruments on Perseverance rover are similar to the Curiosity rover, some are more powerful upgrades of previous instruments and others have completely different, and new capabilities. NASA's four previous Mars rovers—1997's Sojourner; 2004's Spirit and Opportunity; and 2012's Curiosity—were all about exploration.

Perseverance was planned to take the next natural step in Mars exploration—to look for places in Jezero Crater (Fig. 17.3) that were habitable in the distant past, ask whether life ever existed there, and look for signs of ancient life. Jezero Crater was once home to a lake and a river delta (Fig. 17.4). That ancient delta offers a rich variety of geological landscapes—where Perseverance could collect many samples that might contain signs of past life. The rover Perseverance is providing important data relevant to astrobiology research, along with a vast amount of geological information about the landing site and the planet at large that will help put the astrobiological data into context.

One of the mission's primary goal is to explore the geology of Jezero Crater in order to assess past habitability. Jezero Crater is a geologically rich terrain, with many features and minerals formed by water that may date back as far as 3.6 billion years ago. Studying the geology and mineralogy of this site will provide a window into the planet's climate history, and allows astrobiologists to determine if

FIG. 17.2 Artist's concept depicting NASA's Mars 2020 rover exploring Mars. *(Source: https://mars.nasa.gov/resources/nasas-mars-2020-rover-artists-concept-1/) Image Credit: NASA/JPL-Caltech.*

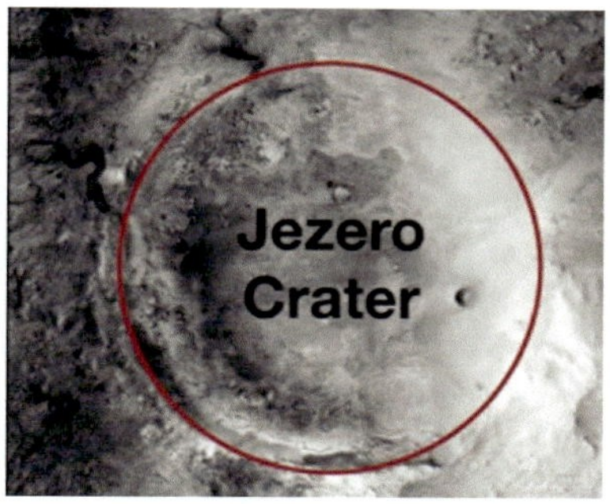

FIG. 17.3 Jezero Crater that is considered to have been habitable in the distant past. *(Source: https://mars.nasa.gov/mars2020/mission/overview/).*

FIG. 17.5 A broad picture of the Perseverance rover's landing site on Mars, covering the Jezero Crater. *(Source: https://mars.nasa.gov/mars2020/mission/overview/).*

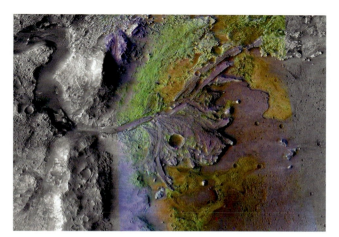

FIG. 17.4 The Mars 2020 Perseverance Rover will explore an ancient river delta where water once flowed on Mars's surface. Credit: NASA/JPL/JHUAPL/MSSS/Brown University, *(Source: https://www.nature.com/articles/d41586-020-01861-0?utm_source=Nature±Briefing&utm_campaign=13e014bf84-briefing-dy-20200707&utm_medium=email&utm_term=0_c9dfd39373-13e014bf84-44566193).*

The Perseverance Rover has a suite of several scientific instruments for poking and probing the Martian surface and atmosphere for hunting for evidence of life. The rover carries advanced versions of some of Curiosity's sensors, including a chemical analyzer that blasts rocks with a laser to identify the atoms and molecules they are made of, and a sharp-eyed camera system that can zoom in on areas of interest to produce stereo and 3D pictures. Perseverance also sports an experiment that will try to produce oxygen from Mars's carbon dioxide-rich atmosphere, as a test of ways to support future human explorers. The rover has X-ray and ultraviolet spectrometers for analyzing mineralogy in detail—and, for a bit more novelty, microphones for listening to Martian sounds, plus a squat, solar-powered helicopter.

The six-wheeled, 3-m-long rover has a drill to extract samples of Martian rock and soil, then store them in sealed tubes that a future mission will one day pick up and bring back to Earth for detailed analysis, possibly by 2031. It would be the first-ever sample return from Mars. Perseverance is designed to study Mar's weather and geology, hunt for water, produce oxygen from carbon dioxide, record sounds for the first time, and test a solar-powered helicopter. Perseverance Rover is supposed to fulfill its mission in one Mars year—nearly two Earth years (about 687 Earth days). Whatever the rover picks up will help to shape the course of Mars science for decades to come.

Perseverance carries 43 tubes in its belly. When it encounters a rock that mission scientists want to sample, the rover will reach out its 2.1-m-long robotic arm (Fig. 17.6) and drill a sample about the size of a penlight: 60 mm long and 13 mm across. The sample goes into a tube. The arm delivers the sample to a carousel, which moves the sample to the rover's underside (Fig. 17.7). A second robotic arm carries the sample to different instruments for initial measurements, then seals the tube (Fig. 17.8). Eventually, once Perseverance has filled at least 20 of its tubes, it will cache them on the surface of Mars until some future, yet-to-be-funded robot arrives to retrieve them. NASA currently plans

sites like these were persistent habitable environments that could have supported life in the past (https://astrobiology.nasa.gov/missions/2020-mars-rover/).

Perseverance will also study the evolution of Mars' climate, surface, and the interior of the planet. The rover will test out technologies for future human exploration of Mars. In addition to these science goals, Perseverance has a unique goal of collecting and caching samples of Mars material for possible future return. Fig. 17.5 shows a broad picture of the Perseverance rover's landing site on Mars, covering the Jezero Crater.

Because other missions before it did not collect and cache samples, they were focused solely on studying the surface to answer science questions. For the Perseverance rover, on the other hand, the goal of collecting and caching samples changes the team's approach to science exploration. The scientific exploration and analysis that are expected to be carried out support the unique goal of caching samples that best represent Mars as a planet.

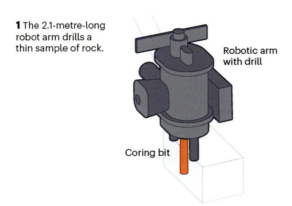

FIG. 17.6 Robotic arm on "Mars 2020 Perseverance Rover." *(Source: https://www.nature.com/articles/d41586-020-01861-0?utm_source=Nature±Briefing&utm_campaign=13e014bf84-briefing-dy-20200707&utm_medium=email&utm_term=0_c9dfd39373-13e014bf84-44566193), Copyright: Nature.*

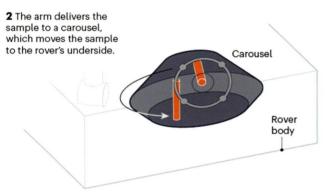

FIG. 17.7 Delivery of sampled filled tube to a carousel, which moves the sample to the Mars 2020. Perseverance Rover's underside. *(Source: https://www.nature.com/articles/d41586-020-01861-0?utm_source=Nature±Briefing&utm_campaign=13e014bf84-briefing-dy-20200707&utm_medium=email&utm_term=0_c9dfd39373-13e014bf84-44566193), Copyright: Nature.*

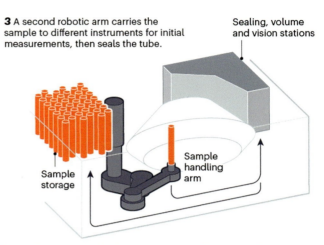

FIG. 17.8 A second robotic arm of the Mars 2020 Perseverance Rover carries the collected sample to different instruments for initial measurements, and then seals the tube. *(Source: https://www.nature.com/articles/d41586-020-01861-0?utm_source=Nature±Briefing&utm_campaign=13e014bf84-briefing-dy-20200707&utm_medium=email&utm_term=0_c9dfd39373-13e014bf84-44566193), Copyright: Nature.*

to work with the European Space Agency (ESA) to launch a mission in 2026 that would return the rocks to Earth in 2031.

One of the several scientific instruments on the Perseverance Rover is Scanning Habitable Environments with Raman & Luminescence for Organics & Chemicals (SHERLOC), which illuminates rocks with an ultraviolet laser and records spectra of the luminescence and reflectance. It can identify the signal of organic molecules and minerals that formed in watery environments (Fig. 17.9). SHERLOC is attached at the end of the rover's arm, and is meant to work in tandem with a second instrument named WATSON (Wide Angle Topographic Sensor for Operations and eNgineering), a camera that can take close-up pictures of rock textures. Together the two instruments will be used to map the presence of certain minerals and organic molecules on the surfaces of rocks. Mineral maps can also be combined with data from additional instruments, including PIXL (Planetary Instrument for X-ray Litho-chemistry).

17.5 Looking for signs of primitive life on Jupiter & Saturn and their Moons

Observations of Jupiter and Saturn had shown that they contained ammonia and methane, and large amounts of hydrogen were inferred to be present there (this inference was confirmed later, and it is now known that hydrogen is the major atmospheric component of these planets). These "chemically reducing" atmospheres of the giant planets were regarded as captured remnants of the solar nebula; and by analogy the atmosphere of the early Earth was assumed to have been similar. Thus, the experimental conditions in the Miller-Urey experiment were assumed to simulate those on primitive Earth. This assumption was found to have been realistic as attested in a subsequent study, which supports an early "reducing" atmosphere.

The significance of these experiments to the origin of life has been debated, with no firm conclusions emerging, except that the generation of biogenic molecules under astronomical conditions is not particularly difficult, and can be achieved in a variety of ways. Therefore, it is interesting that essentially all the starting gases of the various Miller-Urey experiments are now found in the interstellar gas, under conditions that are radically different from those in planetary atmospheres and oceans, with densities lower by at least 15 orders of magnitude (see Thaddeus, 2006).

Some of the end products of the Miller-Urey experiment are found in space as well; for example, formaldehyde and cyano-acetylene. Note that cyano-acetylene is a major

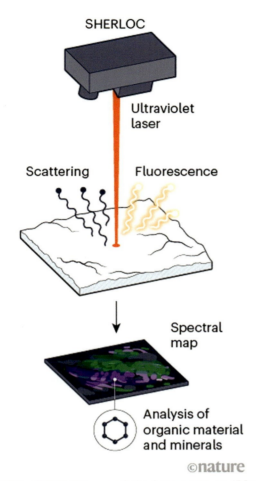

FIG. 17.9 SHERLOC, one of the instruments on "Mars 2020 Perseverance Rover," which illuminates rocks with an ultraviolet laser and records spectra of the luminescence and reflectance to identify the signal of organic molecules and minerals that formed in watery environments. *(Source: https://www.nature.com/articles/d41586-020-01861-0?utm_source=Nature±Briefing&utm_campaign=13e014bf84-briefing-dy-20200707&utm_medium=email&utm_term=0_c9dfd39373-13e014bf84-44566193), Copyright: Nature.*

nitrogen-containing product of the action of an electric discharge on a mixture of methane and nitrogen. It reacts with simple inorganic substances in an aqueous solution to give products including aspartic acid, asparagine, and cytosine (Sanchez et al., 1966). It can be reasonably suspected that many more end-products of the Miller-Urey experiment are currently lurking just below our current level of sensitivity, amino-acids especially.

17.5.1 Prospects of life on Europa

The prospects of habitability of Europa's subsurface ocean is conditioned by heat released from the deep interior and by the intensity of magmatic activity. As indicated earlier, a recent study carried out by Běhounková et al. (2021) suggests that active volcanoes may lurk on the seabed of the 1940-mile-wide (3120 km) Europa, which harbors a huge ocean of salty water beneath its icy shell. Such volcanoes could power deep-sea hydrothermal systems, environments rich in chemical energy that potential Europa lifeforms could exploit. Jupiter's icy moon Europa harbors underneath a tectonically modified ice shell and a salty ocean in direct contact with a rocky interior. Such an oceanic environment makes Europa a primary target for the search of a habitable world outside the Earth. The occurrence of magmatic activity on the seafloor is essential to determine if it constitutes an environment hospitable to life.

Planetary scientists and astrobiologists are of the general view that Europa is one of our best chances of finding life in our Solar System. The putative ocean of Europa has focused considerable attention on the potential habitats for life on Europa, which are considered to be the ice layer, the brine ocean, and the seafloor environment. Beneath its icy crust, Europa hosts a salty, liquid-water ocean in contact with a rocky seafloor (Anderson et al., 1998; Kivelson et al., 2000; Zimmer et al., 2000), making it an exciting place to explore habitability in the Solar System. However, the ocean's potential to support life relies heavily on its composition and chemical energy budget (Chyba and Hand, 2001; Hand et al., 2009), which remain largely unconstrained. Chemical exchange between the ocean and the rocky interior may have been significant in the past so potential life may have been able to use chemical energy to survive.

Scientists are well aware that an unambiguous observation of a subsurface ocean or liquids in the ice of Europa is only the first step toward identifying the possibility for life. What the scientists are proposing will not be able to tell the world community whether there are living organisms in Europa, but it could provide strong evidence for that possibility. By generally clement Earth standards, these Europan habitats are likely to be extreme environments. Model simulations carried out by Marion et al. (2003) demonstrated that "hypothetical oceans could exist on Europa that is too cold for biological activity (T < 253 K). These simulations also demonstrate that salinities are high, which would restrict life to extreme halophiles. An acidic ocean (if present) could also potentially limit life. Pressure, per se, is unlikely to directly limit life on Europa. But indirectly, pressure plays an important role in controlling the chemical environments for life."

According to Marion et al. (2003), "The dual stresses of lethal radiation and low temperatures on or near the icy surface of Europa preclude the possibility of biological activity anywhere near the surface. Only at the base of the ice layer could one expect to find the suitable temperatures and liquid water that are necessary for life. An ice layer turnover time of 10 million years is probably rapid enough for preserving in the surface ice layers dormant life forms originating from the ocean." The six factors likely to be most important in defining the environments for life on Europa and the focus for future work are liquid water, energy, nutrients, low temperatures, salinity, and high pressures (Marion et al., 2003).

Evidence suggests the existence of lakes of liquid water entirely encased in Europa's icy outer shell and distinct from a liquid ocean thought to exist farther down beneath the ice shell (Schmidt et al., 2011; Airhart, 2011). If confirmed, the lakes could be yet another potential habitat for life.

The geologically young, and extensively disrupted chaos terrain found on Europa's surface is potentially indicative of locations of subsurface upwelling or melt-through (O'Brien et al., 2002; Sotin et al., 2002; Collins and Nimmo, 2009). The color captured in Galileo images of Europa's surface (e.g., Geissler et al., 1998) resembles the color-centers that form in sodium chloride (NaCl) and NaCl brine evaporates under Europa-like surface conditions (Hand and Carlson, 2015; Poston et al., 2017). To investigate the hypothesis that Europa's endogenous units contain chloride salts, Trumbo et al. (2019) used the Hubble Space Telescope (HST) to search for signatures of these color-centers on Europa's surface.

The key to understanding the past and present habitability of Europa is its composition and evolution (Daswani et al., 2021). Water, solutes, and possible oxidants needed to carry out metabolic processes (Gaidos et al., 1999; Hand et al., 2007) in Europa's ocean were delivered through some combination of Europa's accreted materials, release by chemical reactions, and subsequently by meteoritic or Io-genic influx. The potential habitability of Europa's subsurface ocean depends on its chemical composition, which may be reflected in that of Europa's geologically young surface. The observed active warm vent (plume) region on Europa may offer a promising location for an initial characterization of Europa's internal water and ice; and for seeking evidence of Europa's habitability. It is believed that large fluxes of surface-derived oxidants must have been delivered into the Europan ocean through geodynamic overturning of the icy lithosphere (Hand et al., 2007; Pasek and Greenberg, 2012). Alternatively, the Europan ocean may have remained reduced and sulfidic if volatile escape to space was limited (McKinnon and Zolensky, 2003). On the basis of thermodynamic equilibrium and extensive water-rock interaction between the Europan ocean and its seafloor, Zolotov and Kargel (2009) predicted a "low pH" fluid that rapidly ($\sim 10^6$yr) evolved to a reduced and basic primordial ocean (pH = 13–13.6) rich in H_2, Na^+, K^+, Ca^+, OH^-, and Cl^-. According to Daswani et al. (2021), the CO_2-rich ocean delivered by metamorphism may facilitate life's emergence by contributing to the generation of a proton gradient between acidic ocean water and alkaline hydrothermal fluids (Camprubí et al., 2019), if the latter is present in Europa.

17.5.2 Strategies for detection of life on Europa

Europa is strongly believed to hide a deep ocean of salty liquid water beneath its icy shell. Whether the Jovian moon has the raw materials and chemical energy in the right proportions to support biology is a topic of intense scientific interest. The answer may hinge on whether Europa has environments where chemicals are matched in the right proportions to power biological processes. Life on Earth exploits such niches.

Planned future exploration missions to the Jovian moon Europa have a strong astrobiological motivation. Characterization of the potential habitability of the liquid water environments, and searching for life signals are the main astrobiological objectives of these missions. To meet these objectives, specific strategies and instrumentation are required. In this context, Irwin and Schulze-Makuch (2003) described a general strategy for modeling ecosystems on other worlds. They modeled two alternative biospheres beneath the ice surface of Europa, based on analogous ecosystems on Earth in potentially comparable habitats, with reallocation of biomass quantities consistent with different sources of energy and chemical constituents. The first ecosystem models a benthic (ecological region at the lowest level of a body of water such as an ocean, lake, or stream, including the sediment surface and some subsurface layers) biosphere supported by chemoautotrophic producers. Note that chemoautotrophs are organisms that obtain their energy from a chemical reaction (chemotrophs) but their source of carbon is the most oxidized form of carbon, carbon dioxide (CO_2). In the second model, two concentrations of biota at the top and bottom of the subsurface water column supported by energy harvested from transmembrane ionic gradients. Irwin and Schulze-Makuch (2003)'s calculations indicated the plausibility of both ecosystems, including small macroorganisms at the highest trophic levels, with ionotrophy supporting a larger biomass than chemoautotrophy.

Prieto-Ballesteros et al. (2011) discussed some scenarios for the development of Europa potential biospheres. These scenarios are based on assumptions of the life similarity concept and knowledge about terrestrial life in extreme environments. Because the potential habitable environments on Europa are in the interior of the moon, it is not possibly to directly detect life. However, there are processes that link aqueous subsurface environments with the near-surface environment, such as tectonism or magmatism. Therefore, by analyzing endogenous materials that arise from the interior it is possible to make predictions about what is in the subsurface. Prieto-Ballesteros et al. (2011) proposed some measurements and instrumentation for future missions to detect biosignatures on the upper layers of Europa, including the simple physico-chemical traces of metabolism to complex biomolecules or biostructures. Prieto-Ballesteros et al. (2011) proposed that Raman spectroscopy or biosensor technologies are the future for in situ exploration of the Solar System.

NASA is currently formulating a mission to explore Europa and investigate in depth whether the icy moon might be habitable. Sometime in the 2020s, NASA would send a highly capable, radiation-tolerant spacecraft into a long, looping orbit around Jupiter to perform repeated close flybys of Europa. During these flybys, the mission would take high-resolution images; determine the composition of the icy moon's surface and faint atmosphere; and investigate its ice shell, ocean and interior.

17.5.3 Anticipation of an Earth-like chemical balance of Europa's ocean

A recent NASA study, modeling conditions in the ocean of Jupiter's moon Europa, suggests that the necessary balance of chemical energy for life could exist there, even if the moon lacks volcanic hydrothermal activity. This new model is part of a large body of evidence that is guiding the development of the future mission to Europa. In this study, scientists at NASA's Jet Propulsion Laboratory, Pasadena, California, compared Europa's potential for producing hydrogen and oxygen with that of Earth, through processes that do not directly involve volcanism. The balance of these two elements is a key indicator of the energy available for life. The study found that the amounts would be comparable in scale; on both worlds, oxygen production is about 10 times higher than hydrogen production.

The study draws attention to the ways that Europa's rocky interior may be much more complex and possibly earth-like than people typically think. NASA's Jet Propulsion Laboratory is studying an alien ocean using methods developed to understand the movement of energy and nutrients in Earth's own systems (Vance et al., 2016). The cycling of oxygen and hydrogen in Europa's ocean will be a major driver for Europa's ocean chemistry and any life there, just as it is on Earth. Ultimately, Europa research scientists want to also understand the cycling of life's other major elements in the ocean: carbon, nitrogen, phosphorus, and sulfur.

As part of their study, the researchers calculated how much hydrogen could potentially be produced in Europa's ocean as seawater reacts with rock, in a process called serpentinization. In this process, water percolates into spaces between mineral grains and reacts with the rock under high temperature and pressure to form new minerals, releasing hydrogen in the process. The researchers considered how cracks in Europa's seafloor likely open up over time, as Europa's rocky interior continues to cool following its formation billions of years ago. New cracks expose fresh rock to seawater, where more hydrogen-producing reactions can take place.

In Earth's oceanic crust, such fractures are believed to penetrate to a depth of 3 to 4 miles (5 to 6 kilometers). On present-day Europa, the researchers expect water could reach as deep as 15 miles (25 km) into the rocky interior, driving these key chemical reactions throughout a deeper fraction of Europa's seafloor.

The rest of Europa's chemical-energy-for-life equation would be provided by oxidants—oxygen and other compounds that could react with the hydrogen—being cycled into the Europan ocean from the icy surface above. Europa is bathed in radiation from Jupiter, which splits apart water ice molecules to create these materials. Scientists have inferred that Europa's surface is being cycled back into its interior, which could carry oxidants into the ocean. Europa researchers consider that the oxidants from the ice are like the positive terminal of a battery, and the chemicals from the seafloor, called reductants, are like the negative terminal. Whether or not life and biological processes complete the circuit is part of what motivates the researchers' exploration of Europa.

Europa's rocky, neighboring Jovian moon, Io, is the most volcanically active body in the Solar System, due to heat produced by the stretching and squeezing effects of Jupiter's gravity as Io orbits Jupiter. Scientists have long considered it possible that Europa might also have volcanic activity, as well as hydrothermal vents, where mineral-laden hot water would emerge from the seafloor.

Researchers previously speculated that volcanism is paramount for creating a habitable environment in Europa's ocean. If such activity is not occurring in its rocky interior, the thinking goes, the large flux of oxidants from the surface would make the ocean too acidic, and toxic, for life. Researchers reckon that if the rock is cold, it is easier to fracture. This allows for a huge amount of hydrogen to be produced by serpentinization that would balance the oxidants in a ratio comparable to that in Earth's oceans.

17.5.4 Proposal for probing subglacial ocean of enceladus in search of life—orbiter and lander missions

Detection of silica-rich dust particles as an indication for on-going hydrothermal activity, and the presence of water and organic molecules observed in the plume of Enceladus, have made this icy moon of Saturn a hot spot in the search for potential extraterrestrial life. Enceladus is one of the Solar System's most enigmatic bodies and is a prime target for future space explorations. Exploring Enceladus over multiple targeted flybys will give us a unique opportunity to further study the most active icy moon in our Solar System as revealed by Cassini, and to analyze in situ its active plume with highly capable instrumentation addressing its complex chemistry and dynamics. Enceladus' plume likely represents the most accessible samples from an extra-terrestrial liquid water environment in the Solar system, which has far-reaching implications for many areas of planetary and biological science (Tobie et al., 2014). Enceladus, with its active plume, is a prime planetary object in the Outer Solar System to perform in situ investigations. Understanding the chemistry of Enceladus is hoped to provide clues for the origin of life.

A plume of water vapor and ice spews from Enceladus's South Polar Region. Cassini data suggest that this plume, sourced by a liquid reservoir beneath the moon's icy crust, contain organics, salts, and water–rock interaction derivatives. The plumes emanating from the South Pole region of Enceladus and the unique chemistry found in them have fueled speculations that Enceladus may harbor life. The presumed aquiferous fractures from which the plumes emanate would make a prime target in the search for extra-terrestrial life and would be more easily accessible than the moon's subglacial ocean.

Thus, Enceladus offers a unique opportunity in the search for life and habitable environments beyond Earth, a key theme of the National Research Council's 2013–2022 Decadal Survey. The ingredients for life as we know it – liquid water, chemistry, and energy sources – are available in Enceladus's subsurface ocean. We have only to sample the plumes to investigate this hidden ocean environment.

A lander mission that is equipped with a subsurface maneuverable ice melting probe will be most suitable to assess the existence of life on Enceladus. A lander would have to land at a safe distance away from a plume source and melt its way to the inner wall of the fracture to analyze the plume subsurface liquids before potential biosignatures are degraded or destroyed by exposure to the vacuum of space. A possible approach for the in-situ detection of biosignatures in such samples can be based on the hypothesis of universal evolutionary convergence, meaning that the independent and repeated emergence of life and certain adaptive traits is wide-spread throughout the cosmos. In view of this, Konstantinidis et al. (2015) conceived a hypothetical evolutionary trajectory leading toward the emergence of methanogenic chemoautotrophic micro-organisms as the baseline for putative biological complexity on Enceladus. To detect their presence, several instruments are proposed that may be taken aboard a future subglacial melting probe.

The "Enceladus Explorer" (EnEx) project funded by the German Space Administration (DLR), aims to develop a terrestrial navigation system for a subglacial research probe and eventually test it under realistic conditions in Antarctica using the EnEx-IceMole, a novel maneuverable subsurface ice melting probe for clean sampling and in situ analysis of ice and subglacial liquids. As part of the EnEx project, an initial concept study is foreseen for a lander mission to Enceladus to deploy the IceMole near one of the active water plumes on the moon's South-Polar Terrain, where it will search for signatures of life.

The general mission concept is to place the Lander at a safe distance from an active plume. The IceMole would then be deployed to melt its way through the ice crust to an aquiferous fracture at a depth of 100 m or more for an in-situ examination for the presence of micro-organisms.

The driving requirement for the mission is the high energy demand by the IceMole to melt through the cold Enceladan ices. This requirement is met by a nuclear reactor providing 5 kW of electrical power. The nuclear reactor and the IceMole are placed on a pallet lander platform. An Orbiter element is also foreseen, with the main function of acting as a communications relay between Lander and Earth.

After launch, the Lander and Orbiter will perform the interplanetary transfer to Saturn together, using the onboard nuclear reactor to power electric thrusters. After Saturn orbit insertion, the Combined Spacecraft will continue using Nuclear Electric Propulsion to reach the orbit of Enceladus. After orbit insertion at Enceladus, the orbiter will perform a detailed reconnaissance of the South-Polar Terrain. At the end of the reconnaissance phase, the Lander will separate from the Orbiter and an autonomously guided landing sequence will place it near one of the active vapor plumes. Once landed, the IceMole will be deployed and start melting through the ice, while navigating around hazards and toward a target subglacial aquiferous fracture.

Konstantinidis et al. (2015) have given an initial estimation of the mission's cost, as well as recommendations on the further development of enabling technologies. They have also addressed the planetary protection challenges posed by such a mission.

In another effort in search of life on Enceladus, MacKenzie et al.(2016) conceived a New Frontiers class, solar-powered Enceladus orbiter—Testing the Habitability of Enceladus's Ocean (THEO). Developed by the 2015 Jet Propulsion Laboratory Planetary Science Summer School student participants under the guidance of TeamX, this mission concept includes remote sensing and in situ analyses with a mass spectrometer, a sub-mm radiometer–spectrometer, a camera, and two magnetometers. These instruments were selected to address four key questions for ascertaining the habitability of Enceladus's ocean within the context of the moon's geological activity: (1) How are the plumes and ocean connected? (2) Are the abiotic conditions of the ocean suitable for habitability? (3) How stable is the ocean environment? (4) Is there evidence of biological processes? By taking advantage of the opportunity Enceladus's plumes offer, THEO represents a viable, solar-powered option for exploring a potentially habitable ocean world of the outer solar system.

17.5.5 Possibility of life on Enceladus

Over the past few decades, numerous techniques have been developed for assessing whether subsurface oceans are present on icy worlds. In some cases, investigation of a moon's surface is sufficient to infer the possible presence of an ocean below (Pappalardo et al., 1999). Recent or ongoing geological activity, such as a young surface shaped by tectonics, hotspots, and plumes, is indicative of a warm interior that can potentially sustain an ocean. For example, the plume of Enceladus, along with the young, crater-free terrain and warm fractures from which it emanates are strong indicators of an ocean, even in the absence of other geophysical data (Porco et al., 2006). Likewise, surface change could indicate ongoing geological activity, again requiring a warm interior. Enceladus is a target of future missions designed to search for existing life or its precursors.

Observations of stellar occultations are a sensitive method for looking for the presence of tenuous atmospheres, particularly at ultraviolet wavelengths, where many gases have strong absorptions. The Cassini spacecraft flew close to Enceladus three times in 2005. The Ultraviolet Imaging Spectrograph (UVIS) on-board Cassini observed stellar occultations on two flybys and confirmed the existence, composition, and regionally confined nature of a water vapor plume in the south polar region of Enceladus. This plume provides an adequate amount of water to resupply losses from Saturn's E ring and to be the dominant source of the neutral OH and atomic oxygen that fill the Saturnian system (Hansen et al., 2006). The UVIS high-speed photometer recorded a decrease in signal intensity ~24 s prior to the occultation of the star by the hard limb, consistent with the presence of an atmosphere (see Parkinson et al., 2007).

Within the context of Cassini observations of a tall plume of water vapor and other organic-rich gases & sodium-salt-rich ice grains spewing from the tiger stripes in *Enceladus'* South Pole terrain, the stage is set for understanding some of the underlying mysterious life-related processes possibly happening in the subsurface of Enceladus. Life as we know it requires three primary ingredients: liquid water; a source of energy for metabolism; and the right chemical ingredients, primarily carbon, hydrogen, nitrogen, oxygen, phosphorus, and sulfur. It has been seen from analysis of the spectra obtained from the plume during the Cassini flyby of Enceladus that many required species such as CO, CO_2, and NH_3 may be present, and that there is possible evidence that amino acids could be formed at the rock/liquid interface of Enceladus, where clay formations should exist due to rock weathering from liquid water or from micrometeorite accumulation due to resurfacing (Parkinson et al., 2007). Power is not a problem because resonance with Dione (a moon of Saturn) forces tidal heating and there is radiogenic heating as well.

For Enceladus, it appears that we have the necessary hydrological cycle (i.e., processes that describe the constant movement of water above, on, and below Enceladus' surface) as well as geologically active, energy-generating reactions sufficient to create a redox gradient favorable for life (see Fig. 17.10), as well as the possibility of peroxide-laden ice from the E-ring of Saturn re-accreting to the surface (Parkinson et al., 2007). The most abundant oxidant on icy moons is most likely hydrogen peroxide (H_2O_2), which has been detected on Europa (Carlson et al., 1999). According

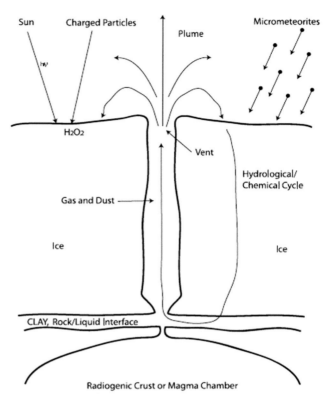

FIG. 17.10 Schematic illustration the hydrological, geochemical and chemical redox cycles on Enceladus. *(Source: Parkinson, C.D., M.-C. Liang, H. Hartman, C.J. Hansen, G. Tinetti, V. Meadows, J.L. Kirschvink, and Y.L. Yung (2007), Enceladus: Cassini observations and implications for the search for life (Research Note); Astronomy and Astrophysics, 463:353-357. DOI: 10.1051/0004-6361:20065773), Copy right: ESO 2007, Article published by EDP Sciences and available at http://www.aanda.org or http://dx.doi.org/10.1051/0004-6361:20065773.*

to Parkinson et al. (2007), on Enceladus, however, there is a more active hydrological cycle transporting the oxidants down to the rock/water interface (see Fig. 17.10). The combination of a hydrological cycle, chemical redox gradient, and geochemical cycle give favorable conditions for life on Enceladus. In addition, the weathering of rocks by liquid water at the rock/liquid interface and any concomitant radioactive emissions (that some researchers have proposed) are possible incipient conditions for life (Hartman et al., 1993).

Flybys of Enceladus by the Cassini probe have confirmed the existence of a long-lived global ocean laced with organic compounds and biologically available nitrogen. This immediately suggests the possibility that life could have begun and may still exist on Enceladus (Deamer et al., 2017). It has been hypothesized (see Parkinson et al., 2008) that Enceladus' plume, tectonic processes, and possible liquid water ocean may create a complete and sustainable geochemical cycle that may allow it to support life. Fly-bys of Enceladus has revealed liquid water, hydrogen gas, and small carbon compounds, which together could potentially support microbial cells. From a perspective from hydrothermal chemistry, Deamer and Damer (2017) consider that life beginning on Enceladus could be feasible. Thus, Cassini's discovery of ocean worlds at Titan and Enceladus provides food for optimism about surprising places to search for potential life beyond Earth in the Solar System.

Chemical analysis of the plume emanating from near the south pole of *Enceladus* indicates that the interior of this Saturnian moon is hot (Spencer and Grinspoon, 2007). These investigators suspect that the moon's interior could be hot enough for complex organic molecules to be made. Studies suggest that hot water is interacting with rock beneath the sea, and that the rock could be reacting chemically to produce hydrogen. NASA's Cassini spacecraft indeed detected hydrogen in the plume of gas and icy material spraying from Enceladus during its last, and deepest, dive through the plume on Oct. 28, 2015. Cassini also sampled the plume's composition during flybys earlier in the mission. From these observations scientists have determined that nearly 98 percent of the gas in the plume is water, about 1 percent is hydrogen and the rest is a mixture of other molecules including carbon dioxide, methane, and ammonia.

Planetary scientists have attached great importance to the discovery of hydrogen in the plume of gas and icy material jetting out from the tiger stripes on the south pole terrain of Enceladus. Considering the fact that microorganisms on Earth can survive in the deep, cold ocean in the absence of light energy; and that for these life-forms, the reducing power of hydrogen sulfide or dissolved hydrogen gas is used as a source of electrons, the presence of hydrogen gas in the Enceladan plume assumes special significance. An electron transport chain allows the electrons to fall through the redox potential to an acceptor, thereby releasing chemical energy that drives metabolism. The acceptor is usually molecular oxygen, a product of photosynthesis, but can also be any of a variety of other acceptors such as ferric iron, nitrate, sulfate, and in the case of methanogens, carbon dioxide. Given an energy source, together with sources of carbon and nitrogen, it seems possible that microorganisms could survive in the oceans of Enceladus and Europa (Deamer et al., 2017).

It may be noted that among the proposed sites for the origin of life on Earth are (1) hydrothermal vents on the ocean floor, and (2) hydrothermal volcanic fields at the land surface. Deamer et al. (2017) investigated whether similar conditions could have fostered the origin of life on Enceladus. According to Deamer et al. (2017), "A hydrothermal vent origin would allow life to begin in the Enceladus Ocean, but if the origin of life requires freshwater hydrothermal pools undergoing wet-dry cycles, the Enceladus ocean could be habitable but lifeless. These arguments also apply directly to Europa and indirectly to early Mars."

The discovery of hydrothermal vents on the seafloor of Enceladus (see Spencer and Grinspoon, 2007; Waite et al., 2017) also provides further evidence that warm, mineral-laden water is pouring into the ocean from hydrothermal vents in the seafloor. On Earth, such hydrothermal vents support thriving communities of life in complete isolation from sunlight. Investigations carried out by Waite et al. (2017) indicated that the presence of ample hydrogen in the moon's ocean means that microbes – if any exist there – could use it to obtain energy by combining the hydrogen with carbon dioxide dissolved in the water. This chemical reaction, known as "methanogenesis" because it produces methane as a by-product, is at the root of the tree of life on Earth, and could even have been critical to the origin of life on our planet.

Enceladus now appears likely to have all three of the ingredients scientists think life needs: liquid water, a source of energy, and the right chemical ingredients (such as carbon, hydrogen, nitrogen, oxygen). According to Taubner et al. (2018), methanogenic archaea are among the organisms that could potentially thrive under the predicted conditions on Enceladus, considering that both molecular hydrogen (H_2) and methane (CH_4) have been detected in the plume. Taubner et al. (2018) showed that a methanogenic archaeon, *Methanothermo coccuso kinawensis*, can produce CH_4 under physicochemical conditions extrapolated for Enceladus. They found that up to 72% carbon dioxide to CH_4 conversion is reached at 50 bar pressure in the presence of potential inhibitors. Furthermore, kinetic and thermodynamic computations of low-temperature serpentinization carried out by Taubner et al. (2018) indicated that there may be sufficient H_2 gas production to serve as a substrate for CH_4 production on Enceladus. Taubner et al. (2018) concluded that some of the CH_4 detected in the plume of Enceladus might, in principle, be produced by methanogens. They recalled that methanogens produce distinct and lasting biosignatures, in particular lipid biomarkers such as ether lipids and isoprenoid hydrocarbons.

According to Parkinson et al. (2007), the presence of organics and the probable presence of clay minerals alongside an ample (and possibly huge) source of oxidants on Enceladus make it a plausible venue for the origin life, as has been suggested for similar system on Earth (Russell et al.,2002). They further suggested that unless another as-yet undiscovered abiotic pathway exists for ribose synthesis, life on Enceladus would most likely have a fundamentally different chemistry than that on Earth. Panspermia, the transfer of life through space on meteorites (e.g., Kirschvink and Weiss, 2002) is essentially impossible because of the high impact velocity of particles falling into Saturn's high gravitational well, and the lack of an atmosphere to decelerate them without sterilizing the interior.

Although life has not yet been detected on Enceladus, with the discovery of hydrothermal vents it is clear that there is a food source there for life to exist. Astrobiologists think that the seafloors of Enceladus would be like a candy store for microbes and chemosynthetic life.

With the finding of liquid water and the right chemical ingredients provided by hydrothermal vents, it is now clear that Enceladus—a small, icy moon a billion miles farther from the Sun than Earth—has nearly all of the ingredients necessary for habitability. This finding further highlights how Enceladus could be one of the most likely places for extra-terrestrial life. Cassini has not shown that phosphorus and sulfur are present in Enceladus's ocean, but scientists suspect them to be, because the rocky core of Enceladus is thought to be chemically similar to meteorites that contain the two elements.

Astrobiologists think that confirmation of the existence of the chemical energy for life within the ocean of a small moon of Saturn is an important milestone in the human search for habitable worlds beyond Earth. The measurement was made using Cassini's Ion and Neutral Mass Spectrometer (INMS) instrument, which sniffs gases to determine their composition.

INMS was designed primarily to sample the upper atmosphere of Saturn's moon *Titan*. However, after Cassini's surprising discovery of a towering plume of icy spray in 2005, emanating from hot cracks near the South Pole of Enceladus, scientists turned its detectors toward this small moon. Cassini was not designed to detect signs of life in the Enceladus plume – indeed; scientists didn't know the plume existed until after the spacecraft arrived at Saturn. The unfolding story of Enceladus has been one of the great triumphs of the NASA spacecraft Cassini's long mission at Saturn.

Cassini was not able to detect life, and has found no evidence that Enceladus is inhabited. But if life is there, that means life is probably common throughout the cosmos; if life has not evolved there, it would suggest life is probably more complicated or unlikely than we have thought. Either way the implications are profound. There have been suggestions that planetary protection issues aside, if life does not yet exist on Enceladus, artificial introduction of terrestrial life to this environment would be an interesting, and most likely successful experiment. Future missions to this icy moon may shed light on its habitability.

17.5.6 The anticipated nature of life on Saturn's Moon Titan

Cassini's numerous gravity measurements of Titan revealed that this moon is hiding an internal, liquid water and ammonia ocean beneath its surface. The Huygens probe also measured radio signals during its descent that strongly suggested the presence of an ocean 35 to 50 miles (55 to 80 kilometers) below the moon's surface. The discovery of a global ocean of liquid water adds Titan to the handful of worlds in our Solar System that could potentially contain habitable environments.

Titan, with its reducing atmosphere, rich organic chemistry, and heterogeneous surface, moves into the astrobiological spotlight. Environmental conditions on Titan and Earth were similar in many respects 4 billion years ago, the approximate time when life originated on Earth. Schulze-Makuch and Grinspoon (2005) proposed that life may have originated on Titan during its warmer early history and then developed adaptation strategies to cope with the increasingly cold conditions. According to them, if organisms originated and persisted, metabolic strategies could exist that would provide sufficient energy for life to persist, even today. Metabolic reactions might include the catalytic hydrogenation of photochemically produced acetylene, or involve the recombination of radicals created in the atmosphere by ultraviolet radiation. They consider that metabolic activity may even contribute to the apparent youth, smoothness, and high activity of Titan's surface via biothermal energy.

Studies of Titan indicate significant liquid water oceans beneath the surface (Lorenz et al., 2008). Titan currently has a roughly 50 km thick crust, but the moon only cooled enough to form this shell after 4 Gyr (Trobie et al., 2006), before which it had only a few kilometers of methane clathrate over the surface, allowing a significant time in which life could have more easily penetrated into the liquid water ocean.

In a study, McKay (2016) considered Titan as an abode of life based, of necessity, on aspects and the phenomenon of Earth and of life on Earth. The presence of a widespread liquid on its surface, the abundance of light and chemical energy, and the continual production of organic gases and solids all encourage the concept of indigenous life on Titan. According to McKay (2016), "Light levels on the surface of Titan are more than adequate for photosynthesis, but the biochemical limitations due to the few elements available in the environment may lead only to simple ecosystems that only consume atmospheric nutrients. Life on Titan may make use of the trace metals and other inorganic elements produced by meteorites as they ablate in its atmosphere. It is conceivable that H_2O molecules on Titan could be used in a biochemistry that is rooted in hydrogen bonds in a way that metals are used in enzymes by life on Earth."

It is generally thought that the ability to create a boundary layer between the interior of a living system and the external environment is a requirement for life (Davies, 2000; Koshland, 2002). On Earth, this is accomplished using a lipid bilayer, where the nature of the membrane is the result of the interaction of the bi-polar lipids with liquid water (McKay, 2016). Based on computational chemical analysis, Stevenson et al. (2015) proposed a type of membrane which they termed "azotosome" capable of forming and functioning in liquid methane at cryogenic temperatures. The membrane is composed of small organic nitrogen compounds, such as acrylonitrile. The structural integrity of the membrane results from the attraction between polar heads of short-chain molecules that are rich in nitrogen and the interlocking nitrogen and hydrogen atoms that reinforce the structure. Although the azotosome structures may be difficult to construct in the laboratory, the structures provide a persuasive example of a possible membrane system for life on Titan (McKay, 2016).

Although theoretical simulations have indicated a possible membrane capable of forming and functioning in Titan liquids, the search for a plausible information-containing molecule for life in Titan liquids remains an open research topic—poly-ethers have been considered and shown to be insoluble at Titan temperatures (McKay, 2016). However, laboratory investigations have failed to suggest a plausible information containing molecule. There exist several biopolymers (i.e., polymeric substances occurring in living organisms) such as DNA, protein, or cellulose. Benner et al. (2004) have observed that polymeric molecules (i.e., very large molecules, or macromolecules, composed of many repeating subunits as occurring in a polymer) suitable for information storage must have the property that the shape of the molecule does not depend on the information stored in it. The shape of the biopolymer DNA is insensitive to the sequence encoded by that DNA. In contrast, the shape of a protein, another biopolymer, can be radically altered by the change of even a single amino acid in the sequence. Thus, DNA is a suitable information-storage molecule and proteins are not (McKay, 2016).

According to McKay (2016), the search for a plausible information molecule for life in Titan liquids remains an open research topic. For the reasons discussed above, if such a Titan information molecule is discovered, it is likely that the binding of the information bits is achieved via hydrogen bonds. In chemistry, polarity is a separation of electric charge leading to a molecule or its chemical groups having an electric dipole moment, with a negatively charged end and a positively charged end. Polar molecules must contain polar bonds due to a difference in electronegativity (the tendency of an atom participating in a covalent bond to attract the bonding electrons) between the bonded atoms. A possible approach for the molecular structure of the information system applicable in Titan surface liquids (methane and ethane) is a two-letter code involving hydrogen bonding with polar molecules containing O and N—both of which form hydrogen bonds and are available on Titan (O in H_2O, and N in HCN, as well as in other nitriles). Another possible class of polymers that could store information is conducting polymers, such as polypyrrole and polyaniline, which are currently of much interest in the nanomaterials field (Abdulla, and Abbo, 2012). These polymers are composed of C, N, and H only and can transition between stable redox states—possibly a basis for information encoding.

According to McKay (2016), "if life on Titan utilizes a set of biomolecules compatible with low-temperature methane and ethane liquids, then it is plausible that the same sort of ecological communication and exchange that occurs on Earth can occur in communities of Titan life forms. Signaling molecules could be low-molecular weight hydrocarbons that would be mobile in the Titan liquid."

The premier example of biochemical selectivity is chirality. A chiral molecule is one that does not contain a plane of symmetry, and thus cannot be superimposed on its mirror image. It is now known that chiral molecules contain one or more chiral centers, which are almost always tetrahedral (sp^3-hybridized) carbons with four different substituents. The chirality property is most well known in the amino acids used in proteins. Life on Earth uses only the L version of amino acids in proteins, not the mirror image, the D version. If life on Titan also uses molecules with chiral centers then detection of homochirality is a powerful indication of life. The simplest and most common chirality center is an atom that has four different groups bonded to it in such a manner that it has a nonsuperimposable mirror image. With N atoms added to hydrocarbons, chiral centers can be expected to form. Chirality can be determined by chromographic and optical methods, and thus may provide a robust and easily determinable method to test whether organic material in sediments on Titan is of biological origin (McKay, 2016).

According to McKay (2016), "Possible search strategies for life on Titan include looking for unusual concentrations of certain molecules reflecting biological selection. Homochirality is a special and powerful example of such biology selection."

Based on what we know so far, we can list the following fundamental challenges to life on Titan, in approximate order of severity (McKay, 2016):

1. The small diversity of elements available in the environment.
2. The low temperature of solution and the resulting negligible solubility.
3. The nonpolar nature of the methane and ethane solvent, further lowering solubility.
4. The limited diversity of hydrocarbon structural molecules (compared to proteins).

Taking account of these limitations, McKay (2016) proposed that "it may be that if there is life in the liquids on Titan's surface it may be simple, heterotrophic, slow to metabolize, and slow to adapt with limited genetic and metabolic complexity. The simple molecules needed for metabolism may be widespread in the environment and in the methane/ethane liquids, but the complex organics needed for structural or genetic systems may be hard to obtain or synthesize. The communities formed may be ecologically simple—perhaps analogous to the microbial ecosystems found in extreme cold and dry environments on Earth." It may be noted that nonpolar solvents contain bonds between atoms with similar electronegativities, such as carbon and hydrogen. Bonds between atoms with similar electronegativities will lack partial charges; it is this absence of charge which makes these molecules "nonpolar." Life in the liquids on Titan's surface is expected to be largely consisting of heterotrophic organisms (i.e., those organisms obtaining carbon for growth and energy from complex organic compounds) because the liquids available on Titan's surface are hydrocarbons (methane and ethane) and the nutrients needed to sustain life on Titan are derived from the hydrocarbon rain on Titan. McKay (2016) suggests that "if there is methane- or ethane-based life on Titan, the global environment is probably quite suitable for it, and it will not need human help for many billions of years."

McKay (2016) proposed that the advantages of the Titan environment for life may include several aspects such as availability of free food from the sky in the form of organics (principally C_2H_2), the chemically gentle nature of the nonpolar solvent in terms of attacking biomolecules (in contrast to water), the lack of ultraviolet or ionizing radiation at the surface, and low rates of thermal decomposition due to the low temperature. McKay (2016) further argued that the simple low-temperature life-forms and communities envisaged would have very low energy demands and would grow slowly. If genetic material on Titan is based on soluble polymers as suggested by Stevenson et al. (2015), then these also could be mobile in Titan liquid. McKay (2016) surmised thus: "If life on Titan had genetics, and were thus Darwinian, then what a wonderful life it would be: a second genesis different enough from Earth life to suggest that our Universe is full of diverse and wondrous life forms."

According to McKay (2016), given how different carbon-based life in liquid methane and ethane must be from carbon-based life in liquid water, it will be a challenge to implement a search strategy for life on Titan. However, some general principles that have emerged in the search for life may be applicable. Based on the several considerations indicated above, it may be that any planned missions to search for life on Titan will have to consider whether any biological material found represents an independent origin, rather than another branch in the family tree populated by Earth life. Life-hunting planetary researchers think that further measurements of hydrogen, as well as of acetylene and ethane abundances in the lower atmosphere of Titan, may be the most promising immediate strategy for a life search on that world.

In a nutshell, although Titan possesses a large subsurface ocean, it also has an abundant supply of a wide range of organic species and surface liquids, which are readily accessible and could harbor more exotic forms of life. Further, Titan may have transient surface liquid water such as impact melt pools and fresh cryovolcanic flows in contact with both solid and liquid surface organics. These environments present unique and important locations for investigating prebiotic chemistry and, potentially, the first steps toward life (Hendrix et al., 2019).

17.5.7 Possibility of amino acids production in Titan's haze particles

In a study of considerable astrobiological interest, Cleaves et al. (2014) produced in the laboratory various analogs of Titan haze particles (termed "tholins"). They found that in certain geologic environments on Titan, these haze particles may come into contact with aqueous ammonia (NH_3) solutions, hydrolyzing them into molecules of astrobiological interest. A Titan tholin analog hydrolyzed in aqueous NH_3 at room temperature for 2.5 years was analyzed by Cleaves et al. (2014) for amino acids. In this analysis they made use of highly sensitive ultra-high-performance liquid chromatography coupled with fluorescence detection and time-of-flight mass spectrometry (UHPLC-FD/ToF-MS) analysis after derivatization with a fluorescent tag. They compared the amino acids produced from this reaction sequence with those generated from room-temperature Miller-Urey (MU) type electric discharge reactions. Cleaves et al. (2014) found that most of the amino acids detected in low-temperature MU $CH_4/N_2/H_2O$ electric discharge reactions are generated in Titan simulation reactions, as well as in previous simulations of Triton chemistry. According to Cleaves et al. (2014), this result argues that many processes provide very similar mixtures of amino acids, and possibly other types of organic compounds, in disparate environments, regardless of the order of hydration. Cleaves et al. (2014) argued that "although it is unknown how life began, it is likely that given reducing conditions, similar materials were available throughout the early Solar System and throughout the Universe to facilitate chemical evolution".

17.6 Astrobiological potential of the dwarf planet Pluto

The most vital ingredients required to support life are liquid water, organics, and a source of energy. Numerical studies carried out during the pre-New Horizons era as well as a variety of measurements obtained from the New Horizons spacecraft when it sped past Pluto in July 2015 provided trustworthy indications that Pluto possesses a liquid–water subsurface ocean.

Pluto's surface was observed at high spatial resolution by the near-infrared spectral imager, known as Linear Etalon Imaging Spectral Array (LEISA) with a resolution of ~2700 m per pixel, and charge-coupled device camera, known as multispectral Visible Imaging Camera (MVIC) with a resolution of ~650 m per pixel system (Reuter et al., 2008) on the New Horizons spacecraft during the flyby in 2015 (Stern et al., 2015; Grundy et al., 2016; Moore et al.,

2016). In the narrow geographic region of Pluto scanned at highest resolution as just indicated, Cthulhu is a central feature, with a dark red-brown color, indicating a substantial nonice composition rich in materials processed both in the atmosphere and on the surface and regarded as tholins (Grundy et al., 2016). Note that "tholins" are a wide variety of organic compounds formed by solar ultraviolet or cosmic rays irradiation of simple carbon-containing compounds such as carbon dioxide, methane, or ethane, often in combination with nitrogen or water. In the northern part of Cthulhu, two geographic features stand out: Elliot crater and the adjacent Virgil Fossae troughs, in which the latter has a unique red color as seen in the color-enhanced MVIC image (Fig. 17.11B).

Most recently, researchers used the New Horizons data to find a strong sign of ammonia on Pluto's surface. The red stains covering large regions of Pluto's surface are considered to be a clear indication that Pluto's icy surface is not all water ice, but contains other elements. Virgil Fossae was one of the particularly red features in Pluto's so-called Cthulhu region, the dark area to the left of Pluto's bright and famous heart. The Virgil Fossae complex is part of a tectonic pattern radiating away from Sputnik Planitia and the basin in which it lies (Moore et al., 2016). The ejection of fluid onto the surface through faults or cracks could be propelled by fluid pressure due to ocean or crustal reservoir freezing (Manga and Wang, 2007) and or by gas pressure due to exsolution in a cryovolcanic event. A few small craters nearby Virgil Fossae troughs share this distinctive coloration. The red-colored region in and around the fossa (a shallow depression or hollow) is spatially coincident with a prominent exposure of H_2O ice that is seen in only a few other regions of Pluto's surface (Cook et al., 2016; Olkin et al., 2017; Protopapa et al., 2017; Schmitt et al., 2017). Cryoclastic material was deposited from one or more erupting fountains. In this background the researchers looked at spectral information for that region, to learn what kinds of materials are present. According to Dalle Ore et al. (2019), "if cryovolcanism is involved in the creation of the ammonia-rich spectral signature described here, then it is further suggestive of both ongoing tectonism and escape of ammonia-bearing aqueous fluids presumably derived from an internal ocean or from a crustal reservoir ultimately sourced from the ocean."

A number of investigations of Pluto's largest moon Charon, as well as its smaller moons Nix and Hydra, have reported the presence of ammonia as a compound or an ammoniated species, revealed as a broad absorption band in the 2.2-μm spectral region (Buie and Grundy, 2000; Brown and Calvin, 2000; Cook et al., 2007; Cook et al., 2018). It is, therefore, considered that ammonia is almost certainly present inside Pluto.

Dalle Ore et al. (2019) reported the detection of ammonia (NH_3) on Pluto's surface in spectral images obtained

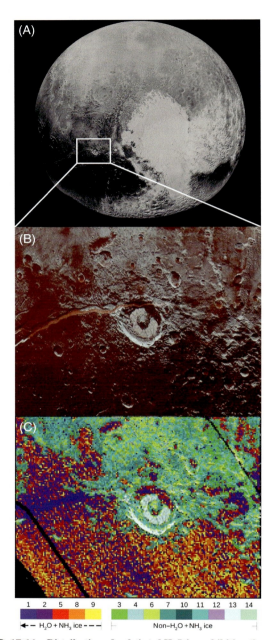

FIG. 17.11 Distribution of red-tinted H_2O ice exhibiting the spectral signature of NH_3 in Virgil Fossae and surrounding terrain. (A) Pluto's encounter hemisphere as seen by New Horizons during the 14 July 2015 flyby. (B) Selected region of interest (ROI) in the MVIC image illustrating the uniquely bright red coloring of Virgil Fossa. (C) Geographical distribution of the 14 clusters where clusters 1 (dark blue), 2 (purple), 5 (red), 8 (orange), and 9 (yellow) show the $H_2O + NH_3$–rich clusters in a gradation from maximum to minimum (indicated by the arrow direction). *Image credit: NASA, Johns Hopkins University, Southwest Research Institute. This figure is reproduced from an open-access article distributed under the terms of the **Creative Commons Attribution-NonCommercial license**, which permits use, distribution, and reproduction in any medium, so long as the resultant use is **not for commercial advantage** and provided the original work is properly cited. Copyright © 2019 The Authors, some rights reserved; exclusive licensee American Association for the Advancement of Science. No claim to original U.S. Government Works. Distributed under a Creative Commons Attribution NonCommercial License 4.0 (CC BY-NC).*

with the New Horizons spacecraft that show absorption bands at 1.65 and 2.2 μm. It is believed that Virgil Fossae and surroundings represent the site of cryovolcanic activity in the fractured crust of Pluto in a region of planet-wide stress induced by the formation of Sputnik Planitia (Keane et al., 2016; Nimmo et al., 2016) and subsequent events. Dalle Ore et al. (2019) found that the ammonia signature is spatially coincident with a region of past extensional tectonic activity (Virgil Fossae) where the presence of H_2O ice is prominent. They found that the ammonia lines up with a cracked region called "Virgil Fossae," which has mounds of water ice and shows signs of past tectonic activity. Because ammoniated ices are believed to be easily broken up by the Sun's ultraviolet light and charged particles, and by cosmic rays, and therefore it is eventually destroyed in environments like Pluto's, the researchers are taking this as a sign that the activity might be recent. According to Dalle Ore et al. (2019), in the present case, the areal distribution is more suggestive of cryovolcanic emplacement, adding to the evidence for ongoing geological activity on Pluto and the possible presence of liquid water at depth today. According to Nimmo et al. (2016), the liquid H_2O–NH_3 mixture could be part of a subsurface ocean or a more localized crustal reservoir. Dalle Ore et al. (2019) reckon that cracks or conduits in the icy crust could be routes of egress (the process of emerging from water) for the liquid H_2O–NH_3 that, upon reaching the vacuum and cold at the surface, both freezes and boils, forming fountains that shower the Fossa surroundings with icy particles. The ammonia leaked to the surface might provide an explanation for how icy Pluto would be keeping liquid reservoirs underground. Cruikshank et al. (2019) found that Pluto's near-side ices were both red and laced with ammonia—a crucial sign that Pluto might be well-supplied with organic molecules. They reckon that the red material associated with the H_2O-ice on Pluto's surface may contain nucleobases resulting from energetic processing on Pluto's surface or in the interior.

Astrobiologists reckon that the subsurface ocean existing beneath the frozen crust of Pluto might be ripe for the existence of life. Observations of water that had probably gushed out of the ocean on Pluto's "near-side," or a reservoir in Pluto's subsurface, show that the gushed out water is tinted with a range of nonice substances with colors ranging from light yellow to red/dark brown—hinting that the observed reddish brown colored regions of Pluto's surface are the store house of complex organic molecules, which are probably produced under radiation from solar wind or cosmic rays.

The possible presence of ammonia in the liquid water body of Pluto's subsurface ocean suggests the potential of formation of molecules that are crucial for life, including amino acids and the bases that are present in RNA and DNA. Indeed, strong indications for the presence of complex organic molecules, such as amino acids and nucleobases, on the surface and in near-subsurface regions of Pluto have been found in the studies of Cruikshank et al. (2019). Apart from the probable presence of biologically important organic compounds in Pluto's ammonia-rich subsurface ocean waters, Pluto's surface composition consisting of N_2, CH_4, and CO has produced strongly colored, complex organics having a high degree of nitrogen substitution similar to the aromatic heterocycles pyrimidine and purine. As indicated earlier, given that there is an increasingly strong case for liquid water and a reasonably trustworthy case for organics, the third astrobiologically important requirement to be met is a source of energy.

17.7 Is there a prospect for life to arise on the asteroid belt resident dwarf planet ceres?

The *Dawn* mission to Vesta and Ceres orbiting in the Asteroid Belt (located between the orbits of Mars and Jupiter) has revolutionized our understanding of the dwarf planet *Ceres* in the same decade that has also seen the rise of ocean worlds as a research and exploration focus. In a comprehensive review, Castillo-Rogez et al. (2020) assessed the implications of observational results from the Dawn mission for Ceres' astrobiological significance: the prospect for the preservation of liquid until present; possible active and passive mechanisms driving internal activity; constraints on Ceres' early ocean composition and the prospect for the maintenance of chemical gradients until present; and possible origins for the organics found on Ceres' surface as well as the prospect for life to arise on Ceres. These researchers used this information to assess the place of Ceres in the ROW and justify why it was identified as a "candidate" ocean world (Hendrix et al., 2019). For the benefit of the members of the research community, who are interested in the astrobiological aspects of Ceres, which is located relatively closer to our Earth than the other extraterrestrial bodies possessing astrobiological importance, Castillo-Rogez et al. (2020) provided a summary of the current state of knowledge of Ceres in the ROW framework in a tabular form.

Castillo-Rogez et al. (2020) recalled that arguably, the most significant finding from the Dawn mission is unambiguous evidence for extensively, aqueously altered material on Ceres' surface, as well as features indicative of recent cryo-volcanism. In the setting described by Travis et al. (2018), thermal convection in a long-lived ocean triggers convective upwelling in the crust, which would be responsible for the observed domes (Castillo-Rogez et al., 2020). On Europa, the existence, spacing, and morphology of pits, spots, and domes have long been used to infer locations of enhanced local heating and the possibility of local liquid pockets in the ice shell (e.g., Pappalardo et al., 1998; Michaut and Manga, 2014). Recent work has also suggested that some of Europa's domes may have formed from eruptions of briny cryo-lava in areas of enhanced local heating (Quick et al., 2017). Diapirism offers another possible mechanism for material extrusion (Castillo-Rogez et al., 2020).

The mineralogy observed on the surface of Ceres implies that its material went through a phase of advanced aqueous alteration (De Sanctis et al., 2015). Conditions in Ceres' ocean that led to the currently observed mineralogy are addressed at length by Neveu et al. (2017) and Castillo-Rogez et al. (2018). Several studies on Ceres' mineralogy suggest that the aqueously altered material on Ceres' surface reached chemical equilibrium as a result of advanced alteration. Modeling by Vance et al. (2016) also suggests that the majority of Ceres' rock could be serpentinized. Furthermore, the globally homogeneous distribution of the aqueously altered material on Ceres' surface suggests production in an ocean, presumably early on in Ceres' history before internal differentiation (Ammannito et al., 2016; Park et al., 2016).

Dawn's observations confirmed earlier predictions for a volatile-rich crust encompassing the bulk of a former ocean, now frozen, and provide hints for a weak interior that may reflect the presence of a relict liquid layer or brine pockets. A long-term supply of (both endo- and exo-genic) bioessential elements in liquid water in subsurface brines inferred from the Dawn observations suggests that Ceres may have been habitable at some point in its history (Castillo-Rogez et al., 2020). Terrestrial organisms can grow at temperatures ranging between 261 and 395 K and pressures in excess of 100 MPa (Jones and Lineweaver, 2010; Harrison et al., 2013). Pikuta et al. (2007) have reported an overview of barophilic bacteria. Inside Ceres, these pressures are encountered in the outer 170–220 km. Thermal models indicate that at these depths, temperatures in the above range could have prevailed for much of Ceres' history (Castillo-Rogez et al., 2020). In these regions deprived of sunlight, the dominant energy sources are chemical. While serpentinization is geologically rapid (e.g., McCollom et al., 2016), it may have been episodically restarted by impact events. In addition, photo- and/or radiolysis (dissociation of molecules by ionizing radiation) may have prolonged subdued redox gradients to the present day (see Castillo-Rogez et al., 2020 for a detailed discussion).

According to Ruiz et al. (2007), the diapirs found on Ceres could conceivably create transient habitable zones and/or reactivate dormant ones by warming the surrounding ice for hundreds of thousands of years. The same could be said for sills and fractures containing briny cryomagmas (Castillo-Rogez et al., 2020). It is perhaps possible that the brine pockets on Ceres could offer a propitious environment for halophiles. These and similar other observations led to Ceres' classification as a "candidate" ocean world in the ROW (Hendrix et al., 2019). Current knowledge indicates that Ceres once had water, organic building blocks for life, energy sources, and redox gradients, and perhaps still does today. Perhaps more importantly, Ceres' astrobiological value comes from its potential for continuous habitability, commencing directly after accretion with a global ocean in which advanced chemical differentiation developed (Castillo-Rogez et al., 2020).

Ruesch et al. (2019a) proposed that brine mobility could serve as an important exchange process on Ceres, promoting the transfer of volatiles and organics between the subsurface and surface, potentially throughout Ceres' history if helped by large impacts and/or via diapirism. Impact mixing is another mechanism that could promote recycling of material in the first kilometers of Ceres' crust (Castillo-Rogez et al., 2020). Judging from its size and water abundance, Ceres belongs to a class of objects that could host a high fugacity (quickly fading or disappearing) of hydrogen, organic molecules, and alkaline conditions, as was suggested for Europa (e.g., McKinnon and Zolensky, 2003) and inferred from Cassini observations of Enceladus (Postberg et al., 2011; Marion et al., 2012).

With regard to the question of whether there is a prospect for life to arise on Ceres, relying on the strength of the hitherto available information, Castillo-Rogez et al. (2020) put forward the following arguments: "In light of a likely scarce long-term energy supply whether life could initiate and persist on Ceres remains uncertain. Recent studies have provocatively argued that life could have been present on Ceres (Houtkooper, 2011; Sleep, 2018), having emerged in situ and/or been transported throughout the inner solar system (e.g., Gladman et al., 2005; Warmflash and Weiss, 2005; Worth et al., 2013). In the latter case, however, by the time Earth had a thriving biosphere, Ceres's near-surface environments should have become less suitable for the implantation of transported organisms due to internal cooling. Likewise, the above considerations make it premature to conclude that life could have emerged on Ceres. Consequently, any relationship between Ceres and the emergence of life on Earth remains speculative. Castillo-Rogez et al. (2020) further argued that "If life was ever present on Ceres, signs of it might still be detectable today. Among potential biosignatures likeliest to survive to the present day, many lipid biomarkers are stable on billion-year timescales (Georgiou and Deamer, 2014). Others, such as particular biological amino acids (e.g., Dorn et al., 2003), are stable on shorter geologic timescales when frozen in ice (Kanavarioti and Mancinelli, 1990). Nucleic acid chains are recoverable only up to about 1 Ma from ice and permafrost (Willerslev et al., 2004). Because Ceres is airless, the lifetime of all potential chemical biosignatures in near-surface material could be significantly further shortened due to radiation (Pavlov et al., 2012)."

It is known that radiolysis is an important energy source in subsurface environments on Earth, as it maintains redox gradient availability to support microbial life. Taking a cue from the knowledge that, in general, throughout Earth's history, the production of redox gradients driven by radiolysis appears to be a significant source of energy for subsurface microbial life, Castillo-Rogez et al. (2020) argued thus with a tint of optimism: "On early Ceres, radiolysis could augment other chemical redox available and could maintain habitability of such environment long after the initial chemical gradients would have been exhausted (Onstott et al., 1997; Pedersen, 2000; Lin et al., 2005; Onstott et al., 2019)."

17.8 Recent and upcoming missions to the extraterrestrial worlds in the solar system in search of ingredients for life

By studying how humans, animals, plants, and microbes survive and thrive on Earth, scientists have identified key ingredients that appear to be essential for life to evolve. For generations, scientists have been scouring our galaxy for evidence of life on other planets. They are searching for a specific set of circumstances and chemicals to come together in the right place, at the right time. The ingredients essential for life include water, carbon, nitrogen, sulfur, and phosphorus.

Dr. Anne Jungblut, a specialist in life living in extreme conditions, and Dr. Paul Kenrick, a specialist on the early evolution of life, explain what they are looking for (https://www.nhm.ac.uk/discover/eight-ingredients-life-in-space.html). Apart from water (which functions as a solvent, allowing key chemical reactions to take place), other characteristics that make water a good habitat for life are its heat conduction, surface tension, high boiling and melting points, and its ability to let light penetrate it. Astrobiologists consider that because water plays such an essential role in life on Earth, the presence of water has been vital in the search of other habitable planets, dwarf planets, and moons.

Carbon molecules are strong and stable, so they are perfect to build a body with. Carbon is one of the most abundant chemical elements on Earth and a major part of all living organisms. Therefore, one working hypothesis is that life on other celestial bodies might also be carbon-based.

Carbon is a fundamental component of organic compounds. The complex proteins required for life are built up from smaller compounds called amino acids—simple organic compounds that contain nitrogen. Finding biochemically usable nitrogen could be a big clue for biological activity on another celestial body. The phosphate group acts like glue in DNA, so the bodies of living organisms would not work without it.

Some micro-organisms are able to grow under extreme conditions such as permanently frozen lakes, deep-sea hydrothermal vents, high radioactive radiation and hypersalinity. They expand our understanding of the capability of some life forms to resist extreme stress. This helps us to understand how habitable other celestial bodies might be.

In the case of Earth, having all the right chemicals on the same planet seems fortunate. Earth—a tiny planet in the middle of an enormous Universe—has been a lucky planet, with all the ingredients of life coming together simultaneously; lucky to have enough of the right chemicals to support a vast abundance of life. Over time, major catastrophes such as impact by asteroids and massive volcanic eruptions have wiped out many species. However, the gaps created afforded opportunities for the survivors to flourish. These accidents along the road mean chance has a huge role in shaping our destinies.

The development of complex life takes billions of years, and there is no shortcut in the journey from single-celled organisms to complex life. Earth is 4.5 billion years old, but in its earliest stages it was far too hot to support life. The oldest fossil evidence of life comes from rocks that are 3.4 billion years old. It took a long time to evolve plants and animals from single-celled organisms. It is possible that life exists on other celestial bodies—but it is likely such life would have a lot of evolutionary catching up to do.

Earth falls into the Goldilocks zone, meaning it is just the right distance from the Sun: not too hot or too cold to have liquid water on the surface. Life needs an energy source to power growth—either the right amount of light from a star or chemically generated energy. Life also needs protection from certain wavelengths of solar radiation. Exposure to ultraviolet B damages DNA, but this wavelength is mostly absorbed by the ozone layer.

In the search for life in the Solar System, one strategy is to follow the water. Liquid water has been found to exist below the dry surface of Mars, the frozen surface of Jupiter's moon Europa, and Saturn's moons Enceladus and Titan. The search for life has broadened to consider worlds far from the Sun.

17.8.1 Upcoming missions to Earth's Moon

Some new nations have recently begun to enter the space mission programs. Their initiatives and mission goals are briefly addressed below.

17.8.1.1 America's plans to explore the far-side of Earth's Moon in search of water and other sustaining minerals

With China successfully roving on the Moon, the United States has plans to send a robot to the lunar surface in late 2023 to hunt for ice and other resources. The robot that goes up with the Artemis program will look for signs of water both over and under the surface of the Earth's Moon. Dubbed as Volatiles Investigating Polar Exploration Rover (VIPER), the robot will map resources at the lunar South Pole that could one day be harvested for long-term human exploration at the Moon. A video demonstrating the working of VIPER is available at the following YouTube link: https://youtu.be/S9Y6n1G5hhc. NASA aims to explore the far-side of Earth's Moon in search of critical water molecules and other sustaining minerals and resources.

According to NASA's Planetary Science Division, the data received from VIPER has the potential to aid scientists in determining precise locations and concentrations of ice on the Moon and will help evaluating the environment and potential resources at the lunar south pole in preparation for Artemis astronauts. During its 100-Earth-day mission, the VIPER rover will roam several miles and use its four science instruments to sample various soil environments. The rover will explore lunar craters using a specialized set of wheels and a suspension system to cover a variety of inclines and soil types. It will operate in the permanently shadowed regions of the Moon that haven't seen sunlight in billions of years and are some of the coldest spots in the Solar System.

The four scientific instruments onboard rover will include the Regolith and Ice Drill for Exploring New Terrains (TRIDENT) hammer drill, the Mass Spectrometer Observing Lunar Operations (MSolo) instrument, the Near-Infrared Volatiles Spectrometer System (NIRVSS), and the Neutron Spectrometer System (NSS).

17.8.1.2 ESA's "Moonlight Initiative"—raising a network of satellites around Earth's Moon to enhance telecommunications and navigation services

When the Soviet Union first sent a satellite in Earth's orbit in 1957, no one had imagined the vast network of thousands of big and small satellites that will encircle the planet within decades. Now, plans are afoot to create a constellation of satellites around Earth's Moon as humans return to the lunar surface, not for fun but to stay. The European Space Agency (ESA) has proposed to create a commercially viable constellation of lunar satellites to aid future missions with better communication and navigational capabilities. The Moonlight initiative was released as part of Europe's bid to reach the Moon in the coming years. ESA is going to Earth's Moon together with its international partners including NASA. Dozens of international, institutional, and commercial teams are sending missions to the Moon that envisage a permanent lunar presence. These will become regular trips to Earth's natural satellite rather than one-off expeditions.

As part of the initiative, ESA in partnership with global space agencies will raise a network of satellites in Earth's Moon's orbit to enhance telecommunications and navigation services that would allow missions to land wherever they wanted. Radio astronomers could set up observatories on the far side of the Moon. Rovers could trundle over the lunar surface more speedily. It could even enable the teleoperation of rovers and other equipment from Earth.

The shared services will benefit nations across the world by reducing the design complexity of individual missions and make them lighter, freeing up space for more scientific instruments and cargo, making each individual mission more cost-efficient. The agency aims to lower ticket prices to lunar exploration and empower member states to launch their own national lunar missions even on low budgets.

According to the ESA, a lasting link with Earth's Moon enables sustainable space exploration for all of the ESA's international partners, including commercial space companies. By using ESA-backed telecommunications and navigation service for the Moon, explorers will be able to navigate smoothly and to relay to Earth all the knowledge gained from these lunar missions.

17.8.1.3 Turkey's plans to send a rover to Earth's Moon by the Year 2030

In a major milestone, Turkey, which launched its space agency [Turkey Space Agency (TUA)] in 2018 amid an economic crisis, joins the list of countries hoping to land a rover on Earth's Moon by the late 2020s using a domestically built rocket engine that will first fly to the Moon in a test mission in 2023. The 10-year space program envisages making contact with the celestial body in 2023, sending citizens into space, working with other countries on building a spaceport, and creating a global brand in satellite technology. According to the current plan, the rover, which will be launched in 2028 or 2029, will land softly on the Moon and collect scientific data on its surface (Pultarova, 2021).

17.8.2 New missions to Mars initiated in 2021

United Arab Emirates, China, and India have initiated space programs aimed at exploring Mars in the coming years.

17.8.2.1 United Arab Emirates (UAE)'s "Hope Mission" to Mars

In a bold undertaking by a young nation, the United Arab Emirates (UAE)'s optimistically named *Hope* probe—launched from the Tanegashima Space Center near Minamitane, Japan, on 20 July 2020—successfully entered the Martian orbit on 9 February 2021 after a seven-month journey, and joined an elite club of nations that have successfully sent missions to Mars: the United States, the Soviet Union, Europe, and India.

It may be recalled that Mars once had a thick atmosphere and a significant amount of liquid water on its surface, but much of the atmosphere has leaked away over billions of years. In 2015, NASA's MAVEN mission showed that the solar wind helps erode the Martian atmosphere. Hope will probe the link between processes in the lower atmosphere, which contains most of the Martian atmosphere's water vapor, and the escape of hydrogen and oxygen from the upper atmosphere.

During its two-year mission, Hope will track daily weather variations and changing seasons. As well as helping to prepare for future human missions, it should reveal how atmospheric conditions cause hydrogen and oxygen to escape into space. This could help scientists to understand Mars's climate and how it lost its once-thick atmosphere. The Hope is hoping to study the entire Martian atmosphere, creating the first global weather map by observing every region of Mars at every time of day.

17.8.2.2 China's Tianwen-1 Mission to Mars

In China's first successful pioneering mission to Mars, *Tianwen-1* spacecraft—designed to explore Mars—departed Earth from an island in southern China in July 2020 and arrived at the planet successfully on 10 February 2021 and started orbiting Mars. The Tianwen-1 mission included an orbiter, a lander and a rover, packed with 13 scientific instruments. Just one day after the United Arab Emirates (UAE)'s *Hope* probe successfully entered Martian orbit; China's Tianwen-1 successfully entered Martian orbit. With this success, China became the sixth player to successfully reach Mars, following the US, Russian, European, Indian, and UAE space agencies.

China has the credit of being the first nation to achieve all three (i.e., carrying an orbiter, lander, and rover) in one mission. The Tianwen-1 program envisaged that in three months' time, it will drop a lander and rover on the planet's northern hemisphere. The mission, named "Tianwen-1," which means "quest for heavenly truth," is China's deepest probe into space.

China's Zhurong rover now joins several other active Mars missions. NASA's Perseverance rover, which arrived on 18 February 2021, is several hundred kilometres away from the landing site, and NASA's Curiosity rover has been poking around the planet since 2012. Several spacecrafts are also circling Mars.

Two cameras are fitted on a mast to take images of nearby rocks while the Zhurong rover is stationary; these will be used to plan the journeys that it takes. A multispectral camera placed between these two navigation imagers will reveal the minerals present in these rocks. The rover's subsurface radar instrument will be used to study geological structures below the surface.

During the rover's mission, the orbiter will act as a communication link, and then will move into closer orbit to survey the planet for an entire Martian year. The orbiter and rover will explore the geology and soil characteristics of Mars, including searching for water and ice. Europe's Mars Express, an orbiter that has been circling Mars since 2003, previously found water lurking below the surface of the southern polar cap. Mars Express, which has been in operation for almost two decades, NASA's Mars Reconnaissance Orbiter, and the Mars Atmosphere and Volatile Evolution orbiter, known as MAVEN could be nearing the end of their lifetimes. Planetary scientists are of the view that continuous monitoring of Mars will benefit the scientific community at a time when many other space agencies will be busy building sample-return missions. In fact, China has its own plans to collect and bring back samples from Mars by 2030.

The Tianwen-1 mission aims to conduct a comprehensive survey of the Martian atmosphere, surface environment, and internal structures—including searching for the presence of water and signs of life. The Tianwen-1 mission will explore whether an ancient ocean ever existed in the northern region of Mars, and study the geological evolution of volcanoes there. The mission expects to reveal new geological information.

Like NASA's Perseverance rover, China's Zhurong rover has ground-penetrating radar. As it winds its way across the basin, this will reveal the geological processes that led to the formation of the regions through which the rover travels. Knowing how deep this lies, and its general characteristics, could offer insights into more recent climate changes on Mars, and reveal the fate of ancient water that could have once soaked the surface.

17.8.2.3 India's Mars Orbiter Mission 2

Following the successful insertion of the Mars Orbiter Mission (MOM; also called *Mangalyaan*) into Martian orbit, the Indian Space Research Organisation (ISRO) announced its intent to launch a second mission to Mars by 2025. Mars Orbiter Mission 2 (MOM 2), also called Mangalyaan-2, is India's second interplanetary mission planned by ISRO. The proposed launch vehicle for this campaign is the GSLV Mk III, which flew for the first time on 5 June 2017, which might be powerful enough to place MOM 2 on a direct-to-Mars trajectory alongside carrying a much more heavier satellite, unlike the lighter Mars Orbiter Mission 1 (MOM 1), which used a less powerful PSLV rocket. The architecture for mission is yet to be finalized and may also have a lander and rover, but no timeline was announced.

In February 2021, ISRO called for 'Announcement of Opportunities' on MOM 2. The total science payload mass is estimated at 100 kg (220 lb). One of the science payloads under development is an ionosphere plasma instrument named ARIS. It is being developed by the Space Satellite Systems and Payloads Centre (SSPACE), which is part of the Indian Institute of Space Science and Technology (IIST). The engineering model and high vacuum test have been completed.

17.8.3 Titan's exploration planned to launch in 2026—NASA's dragonfly mission

In 2005, the European Space Agency's Huygens lander acquired some atmospheric and surface measurements on

Titan, detecting tholins, which are a mix of various types of hydrocarbons (organic compounds) in the atmosphere and on the surface. Because Titan's atmosphere obscures the surface at many wavelengths, the specific compositions of solid hydrocarbon materials on Titan's surface remain essentially unknown. Measuring the compositions of materials in different geologic settings will reveal how far prebiotic chemistry has progressed in environments that provide known key ingredients for life, such as pyrimidines (bases used to encode information in DNA) and amino acids, the building blocks of proteins. Areas of particular interest are sites where extra-terrestrial liquid water in impact melt or potential cryovolcanic flows may have interacted with the abundant organic compounds.

NASA Ocean Worlds Science Objectives for the Titan Habitability Mission Theme are (1) Understand the organic and methanogenic cycle on Titan, especially as it relates to prebiotic chemistry, and (2) Investigate the subsurface ocean and/or liquid reservoirs, particularly their evolution and possible interaction with the surface (Turtle et al., 2017).

The major post-Cassini knowledge gap concerning Titan is in the composition of its diverse surface, and in particular how far its rich organics may have ascended up the "ladder of life." The NASA *New Frontiers 4* solicitation sought mission concepts addressing Titan's habitability and methane cycle. A team led by the Johns Hopkins University Applied Physics Laboratory (APL) proposed a revolutionary lander that uses rotors to land in Titan's thick atmosphere and low gravity and can repeatedly transit to new sites, multiplying the mission's science value from its capable instrument payload. For Titan, the science objectives (listed without priority) of the Ocean Worlds mission theme are (Lorenz et al., 2018):

- Understand the organic and methanogenic cycle on Titan, especially as it relates to prebiotic chemistry.
- Investigate the subsurface ocean and/or liquid reservoirs, particularly their evolution and possible interaction with the surface.

According to Turtle et al. (2017), "Mobility is key to accessing material in different settings, and Titan's dense atmosphere provides the means to explore different geologic settings 10s to 100s km apart using an aerial vehicle. Multiple landers could address Titan's surface chemical diversity but would require multiple copies of instrumentation and sample acquisition equipment. Most efficient approach is to convey a single highly capable instrument suite to multiple locations on a lander with aerial mobility. Several airborne strategies have been considered for in situ Titan exploration, including helicopter (Lorenz, 2000), helium, or hydrogen airship (Levine and Wright, 2005; Hall et al., 2006), Montgolfière hot-air balloon (Reh et al., 2007; Leary et al., 2008; Coustenis et al., 2011), and airplane (Levine and Wright, 2005; Barnes et al., 2012)."

Turtle et al. (2017) recall that environments that offer the most likely prospects for chemical evolution similar to that on Earth occur on Titan's land. For example, dune sands may represent a 'grab bag' site of materials sourced from all over Titan (Leary et al., 2008), much as the rocks at the Mars Pathfinder landing site were intended to collect samples from a wide area (Golombek et al., 1997). On Titan, sites of particular interest include impact melt sheets and potential cryovolcanic flows where transient liquid water may have interacted with the abundant (but oxygen-poor) photochemical products that litter the surface (Thompson and Sagan, 1992).

Mobility is key for in situ measurements. Compositions of solid materials on Titan's surface are still largely unknown. Measuring the composition of materials in different geologic settings can reveal how far prebiotic chemistry has progressed in environments that provide known key ingredients for life.

Heavier-than-air mobility is highly efficient at Titan (Lorenz, 2000; Lorenz, 2001). Titan's atmosphere is four times denser than Earth's, reducing the wing/rotor area required to generate a given amount of lift – this makes all forms of aviation easier (lighter-than air as well as heavier-than-air). Titan's gravity is 1/7th Earth's, reducing the required magnitude of lift—this is a strong factor in favor of a heavier-than-air vehicle. Modern control electronics make a multirotor vehicle (Langelaan et al., 2017) mechanically simpler than a helicopter (cf. proliferation of terrestrial quadcopter drones). Multirotor vehicles offer improved flight control authority and surface sampling capability, redundancy, and failure tolerance; moreover, the system is straightforward to test on Earth and to package in an entry vehicle (Turtle et al., 2017).

Dragonfly is a revolutionary mission concept, providing the capability for in situ exploration of diverse locations to characterize the habitability of Titan's environment, investigate how far prebiotic chemistry has progressed, and search for chemical signatures indicative of water-based and/or hydrocarbon-based life (Turtle et al., 2017). The Dragonfly mission builds on several earlier studies of Titan mobile aerial exploration, including the 2007 Titan Explorer Flagship study, which advocated a Montgolfier balloon for regional exploration, and AVIATR, an airplane concept considered for the Discovery program (Lorenz et al., 2018). The concept of a rotorcraft lander that flew on battery power, recharged during the 8-Earth-day Titan night from a radio-isotope power source, was proposed by Lorenz (2000). More recent discussion has included a 2014 Titan rotorcraft study by Larry Matthies, at the Jet Propulsion Laboratory, that would have a small rotorcraft deployed from a lander or a balloon. The hot-air balloon concepts would have used the heat from a radioisotope thermoelectric generator (RTG). NASA announced the selection of Dragonfly on 27 June 2019, for development under the *New Frontiers* program *Mission 4*, and it will be built and launched by June 2027.

Hunting for environments favorable to life on planets, moons, dwarf planets, and meteorites **Chapter | 17** 761

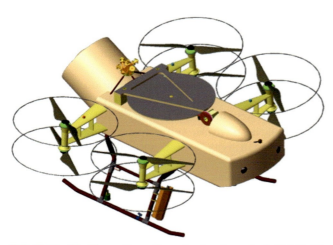

FIG. 17.12 The Dragonfly configuration for atmospheric flight [with the gray circular high-gain antenna (HGA) stowed flat]. Note the aerodynamic fairing in front of the HGA gimbal. The cylinder at rear is the multimission radioisotope thermoelectric generator (MMRTG). A sampling drill mechanism is visible in the nearside skid leg, and forward-looking cameras are recessed into the tan insulating foam forming the rounded nose of the vehicle. The rotor wing section and planform are designed for the Titan atmosphere. *(Source: Lorenz et al. (2018), Dragonfly: A rotorcraft lander concept for scientific exploration at Titan,* Johns Hopkins APL Technical Digest, *34(3): 374-387; www.jhuapl.edu/techdigest).*

Dragonfly is designed taking into account all the concerns and requirements echoed and indicated above. The dragonfly configuration is shown in Fig. 17.12. Dragonfly has introduced a revolutionary new paradigm in planetary exploration by demonstrating a detailed implementation proposal for unparalleled regional mobility. Having laid out this concept, Lorenz et al. (2018) predict that henceforth it may be difficult to imagine a Titan lander mission that does not exploit this capability. The initial descent of the Dragonfly is illustrated in Fig. 17.13. The Dragonfly mission concept is illustrated in Fig. 17.14.

Accordingly, Dragonfly is capable of undertaking two kinds of measurements, namely, (1) Surface Measurements, and (2) In-flight Measurements. Dragonfly surface measurements consist of the following operations (Turtle et al., 2017):

- Sample surface material into a mass spectrometer to identify chemical components available and processes at work to produce biologically relevant compounds.
- Measure bulk elemental surface composition with a neutron-activated gamma-ray spectrometer
- Monitor atmospheric and surface conditions with meteorology sensors and remote-sensing instruments, including spatial and diurnal variations.
- Characterize geologic features via remote-sensing observations, which also provide context for samples and scouting for scientific targets.
- Perform seismic studies to detect subsurface activity and structure.

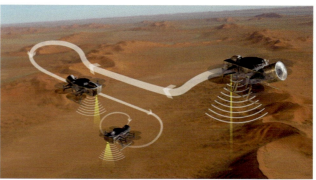

FIG. 17.13 **(top)** This illustration shows NASA's Dragonfly rotorcraft-lander approaching a site on Saturn's exotic moon, Titan. Taking advantage of Titan's dense atmosphere and low gravity, Dragonfly will explore dozens of locations across the icy world, sampling and measuring the compositions of Titan's organic surface materials to characterize the habitability of Titan's environment and investigate the progression of prebiotic chemistry. Credits: NASA/JHU-APL, **(below)** Initial descent of Dragonfly. After release from the entry system and parachute, the vehicle can traverse many kilometers at low altitude using sensors to identify the safest landing site. The schematic is shown against an aerial image of the Namib sand sea, a geomorphological analogue of the Titan landing site, with ~100-m-high dunes spaced by several kilometers, *(Source: Lorenz et al. (2018), Dragonfly: A rotorcraft lander concept for scientific exploration at Titan,* Johns Hopkins APL Technical Digest, *34(3): 374-387; www.jhuapl.edu/techdigest).*

Dragonfly In-flight Measurements consist of the following operations (Turtle et al., 2017):

- Atmospheric profiles, including diurnal and spatial variations.
- Aerial observations of surface geology, also to provide sampling context and identify sites of highest scientific potential for characterizing prebiotic chemistry, Titan's environment, and its habitability to inform prioritization of activities.

The technicalities of the Dragonfly operations can be summarized as follows:

- Flight durations of up to a few hours—ranges of 10s of kilometers—are possible using power from a battery, recharged by an MMRTG between flights and science activities (Lorenz, 2000).

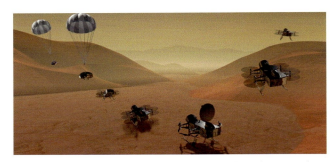

FIG. 17.14 **Dragonfly mission concept.** After delivery from space in an aeroshell and parachute descent, the vehicle lands under rotor power and deploys a high-gain antenna for direct-to-Earth (DTE) communication. Powered by a radioisotope power supply that provides heat and trickle-charges a large battery, the vehicle can operate nearly indefinitely as a conventional lander but can also make periodic brief battery-powered rotor flights to new locations. *(Source: Lorenz et al. (2018), Dragonfly: A rotorcraft lander concept for scientific exploration at Titan,* Johns Hopkins APL Technical Digest, *34(3): 374-387; www.jhuapl.edu/techdigest).*

- In situ operations strategies similar to those proven by Mars rovers (Squyres et al., 2004; Grotzinger et al., 2010) at a more relaxed pace with 16-day Titan-sols.
- Direct-to-Earth (DTE) communication.

Dragonfly will be a rotorcraft lander. Flight on Titan is aerodynamically benign as Titan has low gravity and little wind, and its dense atmosphere allows for efficient rotor propulsion. The radioisotope thermoelectric generator (RTG) power source has been proven in multiple spacecraft, and the extensive use of quad drones on Earth provides a well-understood flight system that is being complemented with algorithms to enable independent actions in real-time. The craft will be designed to operate in a space radiation environment and in temperatures averaging 94 K (−179.2°C).

Titan's dense atmosphere and low gravity mean that the flight power for a given mass is a factor of about 40 times lower than on Earth (Lorenz et al., 2018). The atmosphere has 1.45 times the pressure and about four times the density of Earth's, and local gravity (13.8% of Earth's) will make it easier to fly, although cold temperatures, lower light levels and higher atmospheric drag on the airframe will be challenges.

Dragonfly will be able to fly several kilometers, powered by a lithium-ion battery, which will be recharged by a multimission radioisotope thermoelectric generator (MMRTG) during the night (Lorenz, 2000). MMRTGs convert the heat from the natural decay of a radioisotope into electricity (Lorenz et al., 2018). The rotorcraft will be able to travel ten kilometers on every battery charge and stay aloft for a half hour each time. The vehicle will use sensors to scout new science targets, and then return to the original site until new landing destinations are approved by mission controllers.

The Dragonfly rotorcraft will be approximately 450 kg (990 lb), and packaged inside a 3.7 m (12 ft) diameter heatshield (Lorenz et al., 2018). Regolith samples will be obtained by two sample acquisition drills and hoses, one on each landing skid, for delivery to the mass spectrometer instrument (Lorenz et al., 2018).

The craft will remain on the ground during the Titan nights, which last about 8 Earth days or 192 h (Lorenz et al., 2018). Activities during the night may include sample collection and analysis, seismological studies like diagnosing wave activity on the northern hydrocarbon seas (Stähler et al., 2019), meteorological monitoring, and local microscopic imaging using LED illuminators as flown on Phoenix lander and Curiosity rover (Lorenz et al., 2018). The craft will communicate directly to Earth with a high-gain antenna (Lorenz et al., 2018).

Leveraging proven rotorcraft systems and technologies, Dragonfly will use a multirotor vehicle to transport its instrument suite to multiple locations to make measurements of surface composition, atmospheric conditions, and geologic processes (Langelaan J. W. et al., 2017: *Proc. Aerospace Conf. IEEE*). Dragonfly will provide the capability to explore diverse locations to characterize the habitability of Titan's environment, investigate how far prebiotic chemistry has progressed, and search for biosignatures indicative of life based on water as solvent and even hypothetical types of biochemistry (Turtle et al., 2017). During its 2.7-year baseline mission, Dragonfly will explore diverse environments from organic dunes to the floor of an impact crater where liquid water and complex organic materials key to life once existed together for possibly tens of thousands of years. Its instruments will study how far prebiotic chemistry may have progressed. They also will investigate the moon's atmospheric and surface properties and its subsurface ocean and liquid reservoirs. Additionally, instruments will search for chemical evidence of past or extant life.

17.9 Astrobiologists' thoughts on collection of samples from plumes emitted by Enceladus and Europa

As already noted, evidence suggests that Saturn's icy moon Enceladus has a subsurface ocean that sources plumes of water vapor and ice vented to space from its south pole terrain. The material particles vent into space via surface cracks, producing plumes (Spencer et al., 2006). It was found that tidal stresses likely play a role in opening and closing fractures at Europa's surface. Enceladus is erupting a plume of gas and ice grains from its south pole terrain. Linked directly to the moon's subsurface global ocean, plume material travels through cracks in the icy crust and is ejected into space. In situ analyses of this material by the Cassini spacecraft have shown that the ocean contains key ingredients for life (elements H, C, N, O, and possibly S; simple and complex organic compounds; chemical disequilibria at water-rock interfaces; clement temperature, pressure, and pH). The Cassini discoveries make Enceladus' interior a prime locale for life detection beyond Earth. Scant material exchange with the inner Solar System makes it likely that such life would have emerged independently of life on Earth. Thus, its discovery would illuminate life's universal characteristics (Neveu et al., 2020). According to Neveu et al. (2020), the alternative result of an upper bound on a detectable biosphere in an otherwise habitable environment would considerably advance our understanding of the prevalence of life beyond Earth.

Roth et al. (2014) analyzed spectral images taken by the Hubble Space Telescope that show ultraviolet emissions from Europa's atmosphere, and reported a statistically significant emission signal extending above Europa's southern hemisphere. This emission is consistent with two 200-km-high plumes of water vapor. The subsurface ocean is believed to be in contact with the rocky core, with ongoing hydrothermal activity present (see Shematovich, 2018, or Hendrix et al., 2019, for reviews of the ocean worlds). It is presently unknown whether the plumes carry any microorganisms because the instruments borne by the Cassini spacecraft, which flew through the Enceladus plume, were designed for chemical analysis of the plume particles, and not biological detection.

Unlike terrestrial oceans, collecting samples from the subsurface oceans of the very remotely orbiting icy moons in the solar system is difficult because penetrating through the thick icy crust of such moons with an unmanned device and reaching the subsurface ocean with conventional sampling devices is a Herculean task. Fortunately, Jupiter's icy moon *Europa* and Saturn's icy moon *Enceladus* emit high-velocity tall plumes from their surfaces, carrying organic-rich particles and ice grains thought to be sourced by these moons' hydrothermally active subsurface oceans.

Neveu et al. (2020) have outlined the rationale for returning vented ocean samples, accessible from Enceladus' surface or low altitudes, to Earth for life detection. Returning samples allows analyses using laboratory instruments that cannot be flown, with decades or more to adapt and repeat analyses. Neveu et al. (2020) have described an example set of measurements to estimate the amount of sample to be returned and discuss possible mission architectures and collection approaches. Neveu et al. (2020) have also addressed the challenges of preserving sample integrity and implementing planetary protection policy.

Traspas and Burchell (2021) recalled that the encounter speed with a space vehicle (e.g., Cassini spacecraft) during a flyby is high; Cassini data was obtained at encounter speeds above 5 km/s. Lower speeds in the 3.5–4.5 km/s range are possible if the Saturnian orbit is optimized (Tsou et al., 2012), but this will still result in shock pressures on solid collectors of significantly greater than 1 GPa.

However, use of a porous material such as aerogel can be considered for collection of materials from the plume (Burchell et al., 2006). Note that aerogel is an ultra-low-density material that can be used to capture small particles incident upon it at speeds in excess of 1 km/s. This permits capture of cosmic dust in space where the high speeds usually result in destructive impact events. Based on the observed performance of aerogel in laboratory impact tests, it has been found that completely intact capture is rare; most studies show that between 10% and 100% of the incident particle's mass is captured. However, in all cases, unaltered domains were found in the particles captured in the laboratory at speeds up to 6 or 7 km/s. Several analytic techniques can be applied in-situ to particles captured in aerogel, yielding data on the pre-impact composition of the particle. Extraction techniques for removing small particles from aerogel have been described by Burchell et al. (2006), and after extraction, handling, and analysis in the laboratory can proceed as for any small-sized particle. Coupled with the survival of intact regions in the captured particles, this allows detailed identification of the composition of the dust.

If, as suggested by Tsou et al. (2012), a porous material such as aerogel were used as the collector for the plume materials, peak shock pressures have been calculated to be less than 1 GPa even at speeds of 6 km/s (Trigo-Rodríguez et al., 2008). At 5 or 6 km/s, for impactors of 100 μm diameter and density 2500 kg/m^3, approximately 1.5 cm thickness of aerogel would contain an impact for aerogel of density 50 kg/m^3 (Burchell et al., 2009). This, however, rises to 15 cm depth of aerogel for a 1 mm impactor. The required aerogel depth would be decreased if the impact speed were lowered but would increase again for lower aerogel densities. These depths of aerogel, however, are not implausible and would permit capture of material at shock pressures below those that kill tardigrades (Traspas and Burchell, 2021).

Traspas and Burchell (2021) have suggested that as an alternative to a flyby of an icy moon, an orbiter could be employed. If the encounter is determined by the spacecraft orbital speed (and any contribution from the motion of the material in the plume is ignored), then the impact speed depends on the altitude of the orbit. These speeds can readily be predicted at both Europa and Enceladus (Fig. 17.15A). If we assume a range of metal collectors (as suggested by Neveu et al. (2020), for collecting organic particles during a transit of Enceladus's plume), the peak impact shock pressures can be found by using the Planar Impact Approximation (PIA) (Melosh, 2013). Traspas and Burchell (2021) have modeled aluminum, indium, copper, gold, and silver, as suggested in the work of Neveu et al. (2020), with the necessary linear wave speed coefficients. The resulting shock pressures versus orbital altitude are shown in Fig. 17.15B. These values are well within survival limits for tardigrades at Enceladus, but they are too great for survival at Europa. Traspas and Burchell (2021) have suggested that, at the higher altitudes at Enceladus (and

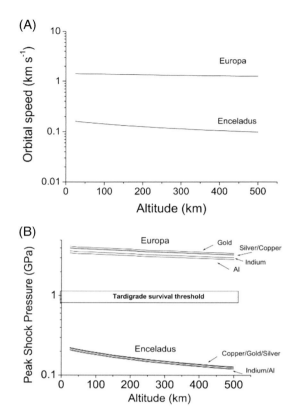

FIG. 17.15 (A) Orbital speed vs. altitude at Europa and Enceladus. (B) Peak shock pressure on various metals as a function of altitude (note that within the resolution of the graphs the curves for several metals overlap). For Europa, all impacts produce pressures above the tardigrade survival limit, whereas for Enceladus impacts on all the metal surfaces are below the limit. *(Source: Traspas, A., and M.J. Burchell (2021), Tardigrade survival limits in high-speed impacts —Implications for panspermia and collection of samples from plumes emitted by ice worlds, Astrobiology, https://doi.org/10.1089/ast.2020.2405) Open Access; Open Access license Website: https://www.liebertpub.com/doi/full/10.1089/ast.2020.2405.*

lower impact speeds), the main problem may be rebound of the impactor rather than the impactor sticking to the target, in which case a funnel-like arrangement may be needed to direct rebounding grains into a detector. According to Traspas and Burchell (2021), it is reasonable to envisage that, in a plume at Enceladus or Europa, a flyby mission could feasibly collect viable small animals such as tardigrades if an aerogel collector was used, and at Enceladus an orbiter could successfully use a solid metal target as well.

According to Traspas and Burchell (2021), collection of material from plumes also depends on the content of the plume and surface area of the collector. For example, it is estimated that the amount of material ejected in the Enceladus plume is 150–300 kg/s, which, at a height of 80 km (the plume expands sideways as it rises up), produces 1 ice particle per m^3 (Tsou et al., 2012). This suggests that an orbiter, as opposed to a flypast, would yield more samples. Further, the method by which material is entrained into the plume may not be straightforward, and the height to which larger (100 μm to millimeter scale) objects are ejected could be limited. Nevertheless, it does appear that, if larger objects could get into a plume, then animals such as tardigrades could survive capture by a passing spacecraft. Whether such a spacecraft should perform an analysis in situ or conduct a sample return to Earth is an open question. As noted in the work of Tsou et al. (2012), the lack of sample material and low sensitivity of in-situ analysis tools would militate against in-situ analysis. However, the cost of implementing planetary protection protocols for a sample return mission to a place potentially harboring life would dwarf the purely spacecraft-related mission costs (Traspas and Burchell, 2021).

It may be noted that no orbiting mission is currently planned for Enceladus. However, upcoming NASA and European flyby missions will swoosh (move with a rushing sound) by the latter at high speeds of several kilometers per second. But perhaps one day far in the future an orbiter might be in the cards, with an ability to detect life at gentler speeds. Planetary scientists, who have studied plume impacts on spacecraft, are aware that if the collected life-forms happened to have died on impact, there is now way to know whether they have been dead for millions of years. However, if it becomes really possible to collect microscopic life and if they were found moving around, one can say they are alive.

References

Abdulla, H.S., Abbo, A.I., 2012. Optical and electrical properties of thin films of polyaniline and polypyrrole. Int J. Electrochem. Sci. 7, 10666–10678.

Acuña, M.H., Connerney, J.E.P., Ness, N.F., Lin, R.P., Mitchell, D., et al., 1999. Global distribution of crustal magnetism discovered by the Mars Global Surveyor MAG/ER experiment. Science 284, 790–793.

Airhart, M., 2011. Scientists Find Evidence for "Great Lake" on Europa and Potential New Habitat for Life, Texas Geosciences, The University of Texas at Austin. Jackson School of Geosciences, https://scienceprojects.lv/Projects/Solar/objects/Europa.php.

Ammannito, E., De Sanctis, M.C., Ciarniello, M., Frigeri, A., Carrozzo, F.G., Combe, J.-Ph., Ehlmann, B.E., Marchi, S., McSween, H.Y., Raponi, A., Toplis, M.J., Tosi, F., Castillo-Rogez, J.C., Capaccioni, F., Capria, M.T., Fonte, S., Giardino, M., Jaumann, R., Longobardo, A., Joy, S.P., Magni, G., McCord, T.B., McFadden, L.A., Palomba, E., Pieters, C.M., Polanskey, C.A., Rayman, M.D., Raymond, C.A., Schenk, P.M., Zambon, F., Russell, C.T., 2016. Distribution of ammoniated magnesium phyllosilicates on. Ceres. Sci. 353, 1–5 aaf4279.

Anderson, J.D., Schubert, G., Jacobson, R.A., Lau, E.L., Moore, W.B., Sjogren, W.L., 1998. Europa's differentiated internal structure: Inferences from four Galileo encounters. Science 281, 2019–2022.

Anglada-Escude, G., Amado, P.J., Barnes, J., Berdinas, Z.M., Butler, R.P., Coleman, G.A.L., de La Cueva, I., Dreizler, S., Endl., M., Giesers, B., Jeffers, S.V., Jenkins, J.S., Jones, H.R.A., Kiraga, M., Kurster, M., Lopez-Gonzalez, M.J., Marvin, C.J., Morales, N., Morin, J., Nelson, R.P., Ortiz, J.L., Ofir, A., Paardekooper, S.-J., Reiners, A., Rodriguez, E., Rodriguez-Lopez, C., Sarmiento, L.F., Strachan, J.P., Tsapras, Y., Tuomi, M., Zechmeister, M, 2016. A terrestrial planet candidate in a temperate orbit around Proxima Centauri. Nature 536 (7617), 437–440. doi:10.1038/nature19106.

Ball, P., 2008. Water as an active constituent in cell biology. Chem. Rev. 108 (1), 74–108.

Běhounková, M., Tobie, G., Choblet, G., Kervazo, M., Daswani, M.M., Dumoulin, C., Vance, S.D., 2021. Tidally induced magmatic pulses on the oceanic floor of Jupiter's Moon Europa. Geophys. Res. Lett. 48 (3), e2020GL090077.

Benner, S.A., Ricardo, A., Carrigan, M.A., 2004. Is there a common chemical model for life in the universe?. Curr. Opin. Chem. Biol. 8 (6), 672–689.

Berkner, L.V., Marshall, L.C., 1965. On the origin and rise of oxygen concentration in the Earth's atmosphere. J. Atmos. Sci. 22 (3), 225–261.

Billings, L., 2016. $100-million plan will send probes to the nearest star. ScientificAmerican. https://www.scientificamerican.com/article/100-million-plan-will-send-probes-to-the-nearest-star1/.

Boyajian, T.S., et al., 2012. Stellar diameters and temperatures. II: Main-sequence K- and M-stars. Astrophys. J. 757, 112.

Brown, M.E., Calvin, W.M., 2000. Evidence for crystalline water and ammonia ices on Pluto's satellite Charon. Science 287, 107–109.

Brown, J.H., Gillooly, J.F., Allen, A.P., Savage, V.M., West, G.B., 2004. Toward a metabolic theory of ecology. Ecology 85 (7), 1771–1780.

Brownlee, D.E., 1981. Extraterrestrial components in deep sea sediments. In: Emiliani, E. (Ed.). The Sea, 7. Wiley, New York, NY, pp. 733–762.

Brownlee, D.E., Bates, B., Schramm, L., 1997. The elemental composition of stony cosmic spherules. Meteoritics Planetary Sci. 32, 157–175.

Buie, M.W., Grundy, W.M., 2000. The distribution and physical state of H_2O on Charon. Icarus 148, 324–339.

Burchell, M.J., Graham, G, Kearsley, A., 2006. Cosmic dust collection in aerogel. Annu. Rev. Earth Planet Sci. 34, 385–418.

Burchell, M.J., Fairey, S.A.J., Foster, N.J., et al., 2009. Hypervelocity capture of particles in aerogel: dependence on aerogel properties. Planet Space Sci. 57, 58–70.

Camprubí, E., de Leeuw, J.W., House, C.H., Raulin, F., Russell, M.J., Spang, A., …, Westall, F., 2019. The emergence of life. Space Sci. Rev. 215 (8), 56.

Carlson, R.W., 1999. Hydrogen peroxide on the surface of Europa. Science 283, 2062–2064.

Castillo-Rogez, J.C., Neveu, M., McSween, H.Y., De Sanctis, M.C., Raymond, C.A., Russell, C.T., 2018. Insights into Ceres' evolution from surface composition. Meteorit Planet Sci 53, 1820–1843.

Castillo-Rogez, J.C., Neveu, M., Scully, J.E.C., House, C.H., Quick, L.C., Bouquet, A., Miller, K., Bland, M., De Sanctis, M.C., Ermakov, A., Hendrix, A.R., Prettyman, T.H., Raymond, C.A., Russell, C.T., Sherwood, B.E., Young, E., 2020. Ceres: Astrobiological target and possible ocean world. Astrobiology 20 (2), 269–291. https://doi.org/10.1089/ast.2018.1999.

Chyba, C.F., Hand, K.P., 2001. Life without photosynthesis. Science 292, 2026–2027.

Chyba, C.F., Hand, K.P., 2005. Astrobiology: The study of the living universe. Annu. Rev. Astron. Astrophys. 43, 31–74.

Cleaves II, H.J., Neish, C., Callahan, M.P., Parker, E., Fernández, F.M., Dworkin, J.P, 2014. Amino acids generated from hydrated Titan tholins: comparison with Miller–Urey electric discharge products. Icarus 237, 182–189.

Cockell, C., Rettberg, P., Horneck, G., Scherer, K., Stokes, M.D., 2003. Measurements of microbial protection from ultraviolet radiation in polar terrestrial microhabitats. Polar Biol. 26, 62–69.

Cockell, C.S., Bush, T., Bryce, C., Direito, S., Fox-Powel, M., Harrison, J.P., Lammer, H., Landenmark, H., Martin-Torres, J., Nicholson, N., Noack, L., O'Malley-James, J., Payler, S.J., Rushby, A., Samuels, T., Schwendner, P., Wadsworth, J., Zorzano, M.P., 2016. Habitability: a review. Astrobiology 16 (1), 89–117.

Cockell, C.S., Schwendner, P., Perras, A., Rettberg, P., Beblo-Vranesevic, K., Bohmeier, M., Rabbow, E., Moissl-Eichinger, C., Wink, L., Marteinsson, V., Vannier, P., Gomez, F., Garcia-Descalzo, L., Ehrenfreund, P., Monaghan, E.P., Westall, F., Gaboyer, F., Amils, R., Malki, M., Pukall, R., Cabezas, P., Walter, N., 2017. Anaerobic microorganisms in astrobiological analogue environments: from field site to culture collection. Int. J. Astrobiol. 43, 1–15.

Cook, J.C., Dalle Ore, C.M., Protopapa, S., Binzel, R.P., Cartwright, R., Cruikshank, D.P., Earle, A., Grundy, W.M., Ennico, K., Howett, C., Jennings, D.E., Lunsford, A.W., Olkin, C.B., Parker, A.H., Philippe, S., Reuter, D., Schmitt, B., Stansberry, J.A., Alan Stern, S., Verbiscer, A., Weaver, H.A., Young, L.A., 2018. Composition of Pluto's small satellites: analysis of *New Horizons* spectral images. Icarus 315, 30–45.

Cook, J.C., Desch, S.J., Roush, T.L., Trujillo, C.A., Geballe, T.R., 2007. Near-infrared spectroscopy of Charon: possible evidence for cryovolcanism on Kuiper Belt objects. Astrophys. J. 663, 1406–1419.

Collins, G., and Nimmo, F. (2009), Chaotic terrain on Europa, In: R.T. Pappalardo, W.B. McKinnon, & K. Khurana (Eds.), *Europa* (pp. 259–282). Tucson: University of Arizona Press.

Connerney, J.E.P., Acuna, M.H., Wasilewski, P., Ness, N.F., Reme, H., et al., 1999. Magnetic lineations in the ancient crust of Mars. Science 284, 794–798.

Cook, J.C., Cruikshank, D.P., Dalle Ore, C.M., Ennico, K., Grundy, W.M., Olkin, C.B., Protopapa, S., Stern, S.A., Weaver, H.A., Young;, L.A.New Horizons surface composition theme team, 2016. The identification and distribution of Pluto's non-volatile inventory. 47th Lunar Planetary Science Conference. The Woodlands, Texas 21 to 26 March 2016.

Cruikshank, D.P., Materese, C.K., Pendleton, Y.J., et al., 2019. Prebiotic chemistry of Pluto. Astrobiology 19, 831–848.

Dalle Ore, C.M., Cruikshank, D.P., Protopapa, S., Scipioni, F., McKinnon, W.B., Cook, J.C., Grun, W.M., et al., 2019. Detection of ammonia on Pluto's surface in a region of geologically recent tectonism. Sci. Adv. 5 (5), eaav5731. doi:10.1126/sciadv.aav5731.

Daswani, M.M., Vance, S.D., Mayne, M.J., Glein, C.R., 2021. A metamorphic origin for Europa's ocean. Geophys. Res. Lett., 48 (18): e2021GL094143. https://doi.org/10.1029/2021GL094143.

Davies, P.C.W., 2000. The Fifth Miracle: The Search for the Origin and Meaning of Life. Simon and Schuster, New York, NY.

Deamer, D., Damer, B., 2017. Can Life Begin on Enceladus? A Perspective from Hydrothermal Chemistry. Astrobiology 17 (9), 834–839.

De Sanctis, M.C., Ammanito, E., Raponi, E., Marchi, S., Mcord, T.B., McSween, H.Y., Capaccioni, F., Capria, M.T., Carrozzo, F.G., Ciarniello, M., Longobardo, A., Tosi, F., Fonte, S., Formisano, M., Frigeri, A., Giardino, M., Magni, G., Palomba, E., Turrini, D., Zambon, F., Combe, J.P., Feldman, W., Jaumann, R., McFadden, L.A., Pieters, C.M., Prettyman, T., Toplis, M., Raymond, C.A., Russell, C.T., 2015. Ammoniated phyllosilicates with a likely outer Solar system origin on (1) Ceres. Nature 528, 241–244.

Dillon, M.E., Wang, G., Huey, R.B., 2010. Global metabolic impacts of recent climate warming. Nature 467 (7316), 704–706.

Dittmann, J.A., Irwin, J.M., Charbonneau, D., Bonfils, X., Astudillo-Defru, N., Haywood, R.D., Berta-Thompson, Z.K., Newton, E.R., Rodriguez, J.E., Winters, J.G., Tan, T.-G., Almenara, J.-M., Bouchy, F., Delfosse, X., Forveille, T., Lovis, C., Murgas, F., Pepe, F., Santos, N.C., Udry, S., Wunsche, A., Esquerdo, G.A., Latham, D.W., Dressing, C.D., 2017. A

temperate rocky super-Earth transiting a nearby cool star. Nature 544 (7650), 333–336.
Dorn, E.D., McDonald, G.D., Storrie-Lombardi, M.C., Nealson, K.H, 2003. Principal component analysis and neural networks for detection of amino acid biosignatures. Icarus 166, 403–409.
Dressing, C.D., Charbonneau, D, 2015. The occurrence of potentially habitable planets orbiting M dwarfs estimated from the full Kepler dataset and an empirical measurement of the detection sensitivity. Astrophys. J. 807 (1), 45.
Feldman, W.C., Boynton, W.V., Tokar, R.L., Prettyman, T.H., Gasnault, O., et al., 2002. Global distribution of neutrons from Mars: results from Mars Odyssey. Science 297, 75–78.
Gaidos, E.J., Nealson, K.H., Kirschvink, J.L., 1999. Life in ice-covered oceans. Science 284 (5420), 1631.
García-Pichel, F., 1998. Solar ultraviolet and the evolutionary history of cyanobacteria. Origins Life Evol. 28, 321–347 B.
Georgiou, C.D., Deamer, D.W., 2014. Lipids as universal biomarkers of extraterrestrial life. Astrobiology 14, 541–549.
Gillon, M., Jehin, E., Lederer, S.M., Delrez, L., de Wit, J., Burdanov, A., Van Grootel, V., Burgasser, A.J., Triaud, A.H.M.J., Opitom, C., Demory, B.-O., Sahu, D.K., BardalezGagliuffi, D., Magain, P., Queloz, D, 2016. Temperate Earth-sized planets transiting a nearby ultracool dwarf star. Nature 533 (7602), 221–224.
Gillon, M., Triaud, A.H.M.J., Demory, B.-O., Jehin, E., Agol, E., Deck, K.M., Lederer, S.M., de Wit, J., Burdanov, A., Ingalls, J.G., Balmont, E., Leconte, J., Raymond, S.N., Selsis, F., Turbet, M., Barkaoui, K., Burgasser, A., Burleigh, M.R., Carey, S.J., Chaushev, A., Copperwheat, C.M., Delrez, L., Fernandes, C.S., Holdsworth, D.L., Kotze, E.J., Van Grootel, V., Almleaky, Y., Benkhaldoun, Z., Magain, P., Queloz, D, 2017. Seven temperate terrestrial planets around the nearby ultracool dwarf star TRAPPIST-1. Nature 542 (7642), 456–460.
Gladman, B., Dones, L., Levison, H.F., Burns, J.A., 2005. Impact seeding and reseeding in the inner solar system. Astrobiology 5, 483–496.
Gohd, C. (2021), NASA's perseverance rover takes its first test drive on Mars. https://www.space.com/perseverance-rover-first-mars-test-drive?utm_source=notification.
Gómez, F., Aguilera, A., Amils, R, 2007. Soluble ferric iron as an effective protective agent against UV radiation: implications for early life. Icarus 191, 352–359. doi:10.1016/j.icarus.2007.04.008.
Gómez, F, Mateo-Martí, E, Prieto-Ballesteros, O, Martín-Gago, J, Amils, R, 2010. Protection of chemolithoautotrophic bacteria exposed to simulated Mars environmental conditions. Icarus 209, 482–487. doi:10.1016/j.icarus.2010.05.027.
Grundy, W.M., et al., 2016. Surface compositions across Pluto and Charon. Science 351, aad9189.
Güdel, M., Audard, M., Reale, F., Skinner, S.L., Linsky, J.L., 2004. Flares from small to large: X-ray spectroscopy of Proxima Centauri with XMM-Newton. Astron. Astrophys. 416, 713–732.
Hand, K., Carlson, R., Chyba, C., 2007. Energy, chemical disequilibrium, and geological constraints on Europa. Astrobiology 7 (6), 1006–1022.
Hand, K.P., Chyba, C.F., Priscu, J.C., Carlson, R.W., Nealson, K.H (2009), Astrobiology and the potential for life on Europa. In: Pappalardo, R.T., McKinnon, W.B., Khurana, K.K., (Eds.), Europa, Univ. Arizona Press, Tucson, pp. 589-630.
Hand, K.P., Carlson, R.W., 2015. Europa's surface color suggests an ocean rich with sodium chloride. Geophys. Res. Lett. 42, 3174–3178.
Hansen, C.J., Esposito, L., Stewart, A.I.F., Colwell, J., Hendrix, A., Pryor, W., Shemansky, D., West, R., 2006. Enceladus' water vapor plume. Science 311, 1422–1425.
Harrison, J.P., Gheeraert, N., Tsigelnitskiy, D., Cockell, C.S., 2013. The limits for life under multiple extremes. Trends Microbiol 21, 204–212.
Hartman, H., Sweeney, M.A., Kropp, M.A., Lewis, J.S., 1993. Carbonaceous chondrites and the origin of life. Origin of Life Evolution Biosphere 23, 221–227. https://doi.org/10.1007/BF01581900.
Hatzes, A.P., 2016. Astronomy: Earth-like planet around Sun's neighbour. Nature 536, 408–409. doi:10.1038/536408a.
Hendrix, A.R., Hurford, T.A., Barge, L.M., Bland, M.T., Bowman, J.S., Brinckerhoff, W., Buratti, B.J., Cable, M.L., Castillo-Rogez, J., Collins, G.C., Diniega, S., German, C.R., Hayes, A.G., Hoehler, T., Hosseini, S., Howett, C.J.A., McEwen, A.S., Neish, C.D., Neveu, M., Nordheim, T.A., Patterson, G.W., A. Patthoff, D., Phillips, C., Rhoden, A., Schmidt, B.E., Singer, K.N., Soderblom, J.M., Vance, S.D, 2019. The NASA roadmap to ocean worlds. Astrobiology 19, 1–27.
Houtkooper, J.M., 2011. Glaciopanspermia: seeding the terrestrial planets with life?. Planet Space Sci 59, 1107–1111.
Hsu, H.-W., Postberg, F., Sekine, Y., Shibuya, T., Kempf, S., Horányi, M., Juhász, A., Altobelli, N., Suzuki, K., Masaki, Y., Kuwatani, T., Tachibana, S., Sirono, S.I., Moragas-Klostermeyer, G., Srama, R., 2015. Ongoing hydrothermal activities within Enceladus. Nature 519, 207–210.
Irwin, L.N., Schulze-Makuch, D, 2003. Strategy for modeling putative multilevel ecosystems on Europa. Astrobiology 3 (4), 813–821.
Jones, E.G., Lineweaver, C.H., 2010. To what extent does terrestrial life "follow the water"?. Astrobiology 10, 349–361.
Kanavarioti, A., Mancinelli, R.L., 1990. Could organic matter have been preserved on Mars for 3.5 billion years?. Icarus 84, 196–202.
Kasting, J.F., 1993. Earth's early atmosphere. Science 259 (5097), 920–926.
Kasting, J.F., Catling, D, 2003. Evolution of a habitable planet. Annu. Rev. Astron. Astrophys. 41 (1), 429–463.
Kasting, J.F., Kopparapu, R., Ramirez, R.M., Harman, C.E., 2014. Remote life-detection criteria, habitable zone boundaries, and the frequency of Earth-like planets around M and late K stars. Proc. Natl. Acad. Sci. USA 111 (35), 12641–12646.
Kingsolver, J.G., Huey, R.B., 2008. Size, temperature, and fitness: three rules. Evol. Ecol. Res. 10 (2), 251–268.
Kiraga, M., Stepien, K., 2007. Age–rotation–activity relations for M dwarf stars. Acta Astron. 57, 149–172.
Kirschvink, J.L., Weiss, B.P., 2002. Mars, Panspermia, and the Origin of Life: Where did it all begin?. Palaeontologia Electronica v.4 (2). http://palaeo-electronica.org/paleo/2001_2/editor/mars.htm.
Kivelson, M.G., Khurana, K.K., Russel, C.T., Volwerk, M., Walker, R.J., Zimmer, C., 2000. Galileo magnetometer measurements: a stronger case for a subsurface ocean at Europa. Science 289, 1340–1343.
Klingelhöfer, et al., 2004. Jarosite and hematite at Meridiani Planum from Opportunity's Mössbauer spectrometer. Science 306, 1740–1745.
Koeberl, C., Hagen, E.H., 1989. Extraterrestrial spherules in glacial sediment from the Transantarctic Mountains, Antarctica: Structure, mineralogy, and chemical composition. Geochim. Cosmochim. Acta 53, 937–944.
Kopparapu, R.K., et al., 2013. Habitable zones around main-sequence stars: new estimates. Astrophys. J. 765, 131.
Konstantinidis, K., Martinez, C.L.F., Dachwald, B., Ohndorf, A., Dykta, P., Bowitz, P., Rudolph, M., Digel, I., Kowalski, J., Voigt, K., Förstner, R., 2015. A lander mission to probe subglacial water on Saturn's moon Enceladus for life. ActaAstronautica 106, 63–89.
Koshland, D.E., 2002. The seven pillars of life. Science 295, 2215–2216.
Lammer, H., Bredehoft, J.H., Coustenis, A., Khodachenko, M.L., Kaltenegger, L., Grasset, O., Prieur, D., Raulin, F., Ehrenfreund, P., Yamauchi,

M., Wahlund, J.-E., Griebmeier, J.-M., Stangl, G., Cockell, C.S., Kulikov, Y.N., Grenfell, J.L., Rauer, H., 2009. What makes a planet habitable?. Astron. Astrophys. Rev. 17 (2), 181–249.

Lin, L.H., Slater, G.F., Lollar, B.S., Lacrampe-Couloume, G., Onstott, T.C., 2005. The yield and isotopic composition of radiolytic H2, a potential energy source for the deep subsurface biosphere. Geochim. Cosmochim. Acta. 69, 893–903.

Lingam, M., Loeb, A., 2017a. Enhanced interplanetary panspermia in the TRAPPIST-1 system. Proc. Natl. Acad. Sci. USA 114 (26), 6689–6693.

Lingam, M., Loeb, A., 2017b. Physical constraints on the likelihood of life on exoplanets. Int. J. Astrobiol. 816, 1–11.

Lorenz, R.D., Stiles, B.W., Kirk, R.L., Allison, M.D., del Marmo, P.P., Iess, L., Lunine, J.I., Ostro, S.J., Hensley, S., 2008. Titan's rotation reveals an internal ocean and changing zonal winds. Science 319, 1649–1651.

Lorenz, R.D., Turtle, E.P., Barnes, J.W., Trainer, M.G., Adams, D.S., Hibbard, K.E., Sheldon, C.Z., Zacny, K., Peplowski, P.N., Lawrence, D.J., Ravine, M.A., McGee, T.G., Sotzen, K.S., MacKenzie, S.M., Langelaan, J.W., Schmitz, S., Wolfarth, L.S., Bedini, P.D., 2018. Dragonfly: A rotorcraft lander concept for scientific exploration at Titan. Johns Hopkins APL Technical Digest 34 (3), 374–387 www.jhuapl.edu/techdigest.

Love, S.G., Brownlee, D.E., 1993. A direct measurement of the terrestrial mass accretion rate of cosmic dust. Science 256, 550–553.

MacKenzie, S.M., Caswell, T.E., Phillips-Lander, C.M., Stavros, E.N., Hofgartner, J.D., Sun, V.Z., Powell, K.E., Steuer, C.J., O'Rourke, J.G., Dhaliwal, J.K., Leung, C.W.S., Petro, E.M., Wynne, J.J., Phan, S., Crismani, M., Krishnamurthy, A., John, K.K., DeBruin, K., Budney, C.J., Mitchell, K.L., 2016. THEO concept mission: testing the Habitability of Enceladus's Ocean. Adv. Space Res. 58 (6), 1117–1137.

Manga, M., Wang, C.-Y., 2007. Pressurized oceans and the eruption of liquid water on Europa and Enceladus. Geophys. Res. Lett. 34, L07202.

Martin, D., Cockell, C.S., 2015. PELS (Planetary Environmental Liquid Simulator): A new type of simulation facility to study extra-terrestrial aqueous environments. Astrobiology 15, 111–118.

Marion, G.M., Fritsen, C.H., Eicken, H., Payne, M.C., 2003. The search for life on Europa: limiting environmental factors, potential habitats, and Earth analogues. Astrobiology 3 (4), 785–811.

Marion, G.M., Kargel, J.S., Catling, D.C., Lunine, J.I., 2012. Modeling ammonia-ammonium aqueous chemistries in the solar system's icy bodies. Icarus 220, 932–946.

Matt. W. (2016), How bad is the radiation on Mars? Universe Today. https://www.universetoday.com/14979/mars-radiation1/).

Maurette, M., Hammer, C., Brownlee, D.E., Reeh, N., 1986. Traces of cosmic dust in blue ice lake Greenland. Science 233, 869–872.

Maurette, M., Jehanno, C., Robin, E., Hammer, C., 1987. Characteristics and mass distribution of extraterrestrial dust from the Greenland ice cap. Nature 328, 699–702.

Maurette, M., Olinger, C., Michel-Levy, Ch.M., Kurat, G., Purchet, M., Brandstatter, F., Bourot-Denise, M., 1991. A collection of diverse micrometeorites recovered from 100 tonnes of Antarctic blue ice. Nature 351, 44–47.

Maurette, M., Immel, G., Perreau, M., Porchet, M., Vincent, C., Kurat, G., 1992. The 1991 EUROMET collection of micrometeorites at Cap Prudhomme, Antarcica: Discussion of possible collection biases, Proceedings of the Lunar Planetary Science Conference 33rd., 859–860.

McKay, C.P., 2004. What is life—and how do we search for it in other worlds?. PLoS Biol 2 (9), e302–e304. https://doi.org/10.1371/journal.pbio.0020302.

McKay, C.P., 2016. Titan as the abode of life. Life 6, 8.

McKinnon, W.B., Zolensky, M.E., 2003. Sulfate content of Europa's ocean and shell: evolutionary considerations and some geological and astrobiological implications. Astrobiology 3 (4), 879–897.

Melosh, H.J., 2013. The contact and compression stage of impact cratering. In: Osinski, G.R., Pierazzo, E (Eds.), Impact Cratering: Processes and Products. Wiley-Blackwell, Hoboken, NJ, pp. 32–42.

Michaut, C., Manga, M., 2014. Domes, pits, and small chaos on Europa produced by water sills. J Geophys Res 119, 550–573.

Millard, H.T., Finkelman, R.B., 1970. Chemical and mineralogical compositions of cosmic and terrestrial spherules from a marine sediment. J. Geophys. Res. 75, 2125–2133.

Moore, J.M., et al., 2016. The geology of Pluto and Charon through the eyes of New Horizons. Science 351, 1284–1293.

Neveu, M., Desch, S.J., Castillo-Rogez, J.C., 2017. Aqueous geochemistry in icy world interiors: fate of antifreezes and radionuclides. Cosmochim Geochim Acta 212, 324–371.

Neveu, M., Anbar, A.D., Davila, A.F., Glavin, D.P., MacKenzie, S.M., Phillips-Lander, C.M., Sherwood, B., Takano, Y., Williams, P., Yano, H., 2020. Returning Samples from Enceladus for Life Detection. *Front Astron Space Sci.*, 7, 26. DOI: 10.3389/fspas.2020.00026.

Nimmo, F., et al., 2016. Reorientation of Sputnik Planitia implies a subsurface ocean on Pluto. Nature 540, 94–96.

O'Brien, D.P., P. Geissler, and R. Greenberg (2002), A melt-through model for chaos formation on Europa, Icarus, 156:152–161

Olson, J.M., Pierson, B.K., 1986. Photosynthesis 3.5 thousand million years ago. Photosynth. Res. 9, 251–259.

Onstott, T.C., Tobin, K., H.Dong, DeFlaun, M.F., Fredrickson, J.K., Bailey, T., Brockman, F.J., Kieft, T.L., Peacock, A., White, D.C., Balkwill, D., Phelps, T.J., Boone, D.R., 1997. The deep gold mines of South Africa: windows into the subsurface biosphere. In: Hoover, R.B. (Ed.), SPIE (International Society for Optics and Photonics) Proceedings, Instruments, Methods, and Missions for the Investigation of Extraterrestrial Microorganisms, Vol. 3111. Optical Science, Engineering and Instrumentation '97, 1997, San Diego, CA, pp. 344–357.

Onstott, T.C., Ehlmann, B.L., Sapers, H., Coleman, M., Ivarsson, M., Marlow, J.J., Neubeck, A., Niles, P, 2019. Paleo-rock-hosted life on Earth and the search for life on Mars: a review and strategy for exploration. Astrobiology 19, 1230–1262.

Pace, N.R., 2001. The universal nature of biochemistry. Proc. Natl. Acad. Sci. USA 98 (3), 805–808.

Pappalardo, R.T., Head, J.W., Greeley, R., Sullivan, R.J., Pilcher, C., Schubert, G., Moore, W.B., Carr, M.H., Moore, J.M., Belton, M.J.S., Goldsby, D.L., 1998. Geological evidence for solid-state convection in Europa's ice shell. Nature 391, 365–368.

Paris, A.J., Davies, E.T., Tognetti, L., Zahniser, C., 2019. Prospective lava tubes at Hellas Planitia: Leveraging volcanic features on Mars to provide crewed missions protection from radiation. J. Washington Acad. Sci., 105(3), 13–36. DOI:10.13140/RG.2.2.23401.03689.

Phoenix, V.R., Konhauser, K.O., Adams, D.G., Bottrell, S.H., 2001. Role of biomineralization as an ultraviolet shield: Implications for Archean life. Geology 29 (9), 823–826.

122 Geological evidence for solid-state convection in Europa's ice shell, Nature, 391:365–368.

Park, R., Konopliv, A.S., Bills, B.G., Rambaux, N., CastilloRogez, J.C., Raymond, C.A., Vaughan, A.T., Ermakov, A.I., Zuber, M.T., Fu, R.R., Toplis, M.J., Russell, C.T., Nathues, A., 2016. Interior structure of dwarf planet Ceres from measured gravity and shape. Nature 537, 515–517.

Parkinson, C.D., Liang, M.-C., Hartman, H., Hansen, C.J., Tinetti, G., Meadows, V., Kirschvink, J.L., Yung, Y.L., 2007. Enceladus: Cassini observations and implications for the search for life (Research Note). Astron. Astrophys. 463, 353–357. doi:10.1051/0004-6361:20065773.

Parkinson, C.D., Liang, M.C., Yung, Y.L., Kirschivnk, J.L., 2008. Habitability of enceladus: planetary conditions for life. Origins Life Evolution Biosphere 38 (4), 355–369. doi:10.1007/s11084-008-9135-4PMID 18566911.

Pasek, M.A., Greenberg, R., 2012. Acidification of Europa's subsurface ocean as a consequence of oxidant delivery. Astrobiology 12 (2), 151–159.

Pavlov, A.A., Vasilyev, G., Ostryakov, V.M., Pavlov, A.K., Mahaffy, P., 2012. Degradation of the organic molecules in the shallow subsurface. Geophys. Res. Lett. 39, L13202.

Pedersen, K., 2000. The hydrogen driven intra-terrestrial biosphere and its influence on the hydrochemical conditions in crystalline bedrock aquifers. In: Stober, I., Bucher, K. (Eds.), Hydrogeology of Crystalline Rocks. Springer, Dordrecht, pp. 249–259 edited by.

Pikuta, E.V., Hoover, R.B., Tang, J., 2007. Microbial extremophiles at the limits of life. Crit. Rev. Microbiol. 33, 183–209.

Postberg, F., Schmidt, J., Hillier, J., Kempf, S., Srama, R., 2011. A saltwater reservoir as a source of compositionally stratified plume on Enceladus. Nature 474, 620–622.

Poston, M.J., Carlson, R.W., Hand, K.P., 2017. Spectral behavior of irradiated sodium chloride crystals under europa-like conditions. J. Geophys. Res. Planets 122, 2644–2654.

Prieto-Ballesteros, O., Vorobyova, E., Parro, V., Manfredi, J.A.R., Gómez, F., 2011. Strategies for detection of putative life on Europa. Adv. Space Res. 48 (4), 678–688.

Protopapa, S., et al., 2017. Pluto's global surface composition through pixel-by-pixel Hapke modeling of New Horizons Ralph/LEISA data. Icarus 287, 218–228.

Quick, L.C., Glaze, L.S., Baloga, S.M., 2017. Cryovolcanic emplacement of domes on Europa. Icarus 284, 477–488.

Reuter, D.C., Stern, S.A., Scherrer, J., Jennings, D.E., Baer, J.W., Hanley, J., Hardaway, L., Lunsford, A., McMuldroch, S., Moore, J., Olkin, C., Parizek, R., Reitsma, H., Sabatke, D., Spencer, J., Stone, J., Throop, H., van Cleve, J., Weigle, G.E., Young, L.A., 2008. Ralph: A visible/infrared imager for the New Horizons Pluto/Kuiper Belt mission. Space Sci. Rev. 140, 129–154.

Ritchie, R.J., Anthony, W.D., Ribas, L.I., 2017. Could photosynthesis function on Proxima Centauri b?. Int. J. Astrobiol. doi:10.1017/S1473550417000167 Published online: 18 July 2017.

Roth, L., Saur, J., Retherford, K.D., et al., 2014. Transient water vapor at Europa's south pole. Science 343, 171–174.

Ruesch, O., Genova, A., Neumann, W., Quick, L.C., CastilloRogez, J.C., Zuber, M., Raymond, C.A., Russell, C.T., 2019a. Slurry extrusion on ceres from a convective mudbearing mantle. Nat Geosci 12, 505–509.

Ruiz, J., Montoya, L., López, V., Amils, R, 2007. Thermal diapirism and habitability of the Icy Shell of Europa. Orig Life Evol Biosphere 37, 287–295.

Russell, M.J., Hall, A.J., 2002. From Geochemistry to Biochemistry: Chemiosmotic coupling and transition element clusters in the onset of life and photosynthesis. Geochemical News 113, 6.

Sanchez, R.A., Ferris, J.P., Orgel, L.E., 1966. Cyanoacetylene in prebiotic synthesis. Science 154, 784–785.

Scalo, J., Kaltenegger, L., Segura, A,G., Fridlund, M., Ribas, I., Kulikov, Y.N., Grenfell, J.L., Rauer, H., Odert, P., Leitzinger, M., Selsis, F., Khodachenko, M.L., Eiroa, C., Kasting, J., Lammer, H., 2007. M Stars as targets for terrestrial exoplanet searches and biosignature detection. Astrobiology 7 (1), 85–166.

Schmidt, B.E., Blankenship, D.D., Patterson, G.W., Schenk, P.M., 2011. Active formation of 'chaos terrain' over shallow subsurface water on Europa. Nature 479 (7374), 502–505. doi:10.1038/nature10608.

Schmitt, B., et al., 2017. Physical state and distribution of materials at the surface of Pluto from New Horizons LEISA imaging spectrometer. Icarus 287, 229–260.

Schulze-Makuch, D., Grinspoon, D.H., 2005. Biologically enhanced energy and carbon cycling on Titan?. Astrobiology 5, 560–567.

Schulze-Makuch, D., Guinan, E., 2016. Another Earth 2.0? Not so fast. Astrobiology 16 (11), 817–821.

Shematovich, V.I., 2018. Ocean worlds in the outer reaches of the Solar System: a review. Sol. Syst. Res. 52, 371–381.

Shields, A.L., Ballard, S., Johnson, J.A, 2016. The habitability of planets orbiting M-dwarf stars. Phys. Rep. 663 (1), 1–38.

Sleep, N.H., 2018. Geological and geochemical constraints on the origin and evolution of life. Astrobiology 18, 1199–1219.

Sotin, C., Head III., J.W., Tobie, G., 2002. Europa: Tidal heating of upwelling thermal plumes and the origin of lenticulae and chaos melting. Geophys. Res. Lett. 29 74-1–74-4.

Spencer, J.R., et al., 2006. Cassini encounters Enceladus: background and the discovery of a south polar hot spot. Science 311, 1401–1405.

Spencer, J., Grinspoon, D., 2007. Planetary science: Inside Enceladus. Nature 445, 376–377. doi:10.1038/445376b.

Squyres, S.W., et al., 2004. In situ evidence for an ancient aqueous environment at Meridiani Planum, Mars. Science 306, 1698–1703.

Squyres, S.W., Knoll, A.H., 2005. Sedimentary rocks at Meridiani Planum: Origin, diagenesis, and implications for life on Mars. Earth Planet. Sci. Lett. 240, 1–10.

Stein, G.M., 2015. Discovering the North-West Passage: The Four-Year Arctic Odyssey of H.M.S. Investigator and the McClure Expedition. McFarland & Company, Inc., Publishers, Jefferson, NC, pp. 376.

Stern, S.A., et al., 2015. The Pluto system: Initial results from its exploration by New Horizons. Science 350, aad1815.

Stevenson, J., Lunine, J., Clancy, P., 2015. Membrane alternatives in worlds without oxygen: creation of an azotosome. Sci. Adv. 1, e1400067.

Tarter, J.C., Backus, P.R., Mancinelli, R.L., Aurnou, J.M., Backman, D.E., Basri, G.S., Boss, A.P., Clarke, A., Deming, D., Doyle, L.R., Feigelson, E.D., Freund, F., et al., 2007. A reappraisal of the habitability of planets around M dwarf stars. Astrobiology 7 (1), 30–65.

Tasker, E., Tan, J., Heng, K., Kane, S., Spiegel, D., Brasser, R., Casey, A., Desch, S., Dorn, C., Hernlund, J., Houser, C., Laneuville, M., Lasbleis, M., Libert, A.-S., Noack, L., Unterborn, C., Wicks, J., 2017. The language of exoplanet ranking metrics needs to change. Nat. Astron. 1, 0042.

Taubner, R.S., Pappenreiter, P., Zwicker, J., Smrzka, D., Pruckner, C., Kolar, P., Bernacchi, S., Seifert, A.H., Krajete, A., Bach, W., Peckmann, J., Paulik, C., Firneis, M.G., Schleper, C., Rittmann, S.K.R., 2018. Biological methane production under putative Enceladus-like conditions. Nat. Commun. 9 (1), 748.

Thaddeus, P., 2006. The prebiotic molecules observed in the interstellar gas. Phil. Trans. R. Soc. B 361, 1681–1687.

Tobie, G., Teanby, N.A., Coustenis, A., Jaumann, R., Raulin, F., Schmidt, J., Carrasco, N., Coates, A.J., Cordier, D., De Kok, R., Geppert, W.D., Lebreton, J.-P., Lefevre, A., Livengood, T.A., Mandt, K.E., Mitri, G., Nimmo, F., Nixon, C.A., Norman, L., Pappalardo, R.T., Postberg, F., et al., 2014. Science goals and mission concept for the future exploration of Titan and Enceladus. Planet. Space Sci. 104 (Part A), 59–77.

Tosi, N., Godolt, M., Stracke, B., Ruedas, T., Grenfell, J.L., Höning, D., Nikolaou, A., Plesa, A.-C., Breuer, D., Spohn, T., 2017. The habitability of a stagnant-lid Earth. Astronomy Astrophys. 605, A71.

Traspas, A., Burchell, M.J., 2021. Tardigrade survival limits in high-speed impacts—Implications for panspermia and collection of samples from plumes emitted by ice worlds. Astrobiology. doi:10.1089/ast.2020.2405.

Travis, B.J., Bland, P.A., Feldman, W.C., Sykes, M., 2018. Hydrothermal dynamics in a CM-based model of Ceres. Meteorit. Planet Sci. 53, 2008–2032.

Trigo-Rodríguez, J., Domínguez, G., Burchell, M.J., et al., 2008. Bulbous tracks arising from hypervelocity capture in aerogel. Meteorit. Planet Sci. 43, 75–86.

Trobie, G., Lunine, J.I., Sotin, C., 2006. Episodic outgassing as the origin of atmospheric methane on Titan. Nature 440, 61–64.

Trumbo, S.K., Brown, M.E., Hand, K.P., 2019. Sodium chloride on the surface of Europa. Sci. Adv. 5 (6), eaaw7123. doi:10.1126/sciadv.aaw7123.

Tsou, P., Brownlee, D.E., McKay, C.P., Anbar, A.D., Yano, H., Altwegg, K., Beegle, L.W., Dissly, R., Strange, N.J., Kanik, I., 2012. LIFE: life investigation for enceladus: a sample return mission concept in search for evidence of life. Astrobiology 12 (8), 730–742.

vanLeeuwen, F., 2007. Validation of the new Hipparcos reduction. Astron. Astrophys. 474, 653–664.

Vance, S.D., Hand, K.P., Pappalardo, R.T., 2016. Geophysical controls of chemical disequilibria in Europa. Geophys. Res. Lett. 43 (10), 4871–4879. doi:10.1002/2016GL068547.

Waite, J.H., Glein, C.R., Perryman, R.S., Teolis, B.D., Magee, B.A., Miller, G., Grimes, J., Perry, M.E., Miller, K.E., Bouquet, A., Lunine, J.I., Brockwell, T., Bolton, S.J., 2017. Cassini finds molecular hydrogen in the Enceladus plume: evidence for hydrothermal processes. Science 356 (6334), 155–159. doi:10.1126/science.aai8703.

Wall, M (2021), NASA's Mars rover Perseverance landing: Everything you need to know. https://www.space.com/mars-rover-perseverance-landing-explained.

Warmflash, D., Weiss, B., 2005. Did life come from another world? Sci Am 293, 64–71.

Whitmire, D.P., 2017. Implication of our technological species being first and early. Int. J. Astrobiol. https://doi.org/10.1017/S1473550417000271 (Published online: 03 August 2017).

Willerslev, E., Hansen, A.J., Rønn, R., Brand, T.B., Barnes, I., Wiuf, C., Gilichinsky, D., Mitchell, D., Cooper, A, 2004. Long-term persistence of bacterial DNA. Curr Biol 6, R9–R10.

Winn, J.N., Fabrycky, D.C., 2015. The occurrence and architecture of exoplanetary systems. Annu. Rev. Astron. Astrophys. 53, 409–447.

Worth, R.J., Sigurdsson, S., House, C.H., 2013. Seeding life on the moons of the outer planets via lithopanspermia. Astrobiology 13, 1155–1165.

Yung, Y.L., et al., 2018. Methane on Mars and Habitability: Challenges and Responses. Astrobiology 18 (10), 1221–1242. doi:10.1089/ast.2018.1917.

Zimmer, C., Khurana, K.K., Kivelson, M.G., 2000. Subsurface oceans on Europa and Callisto: constraints from Galileo magnetometer observations. Icarus 147, 329–347.

Zolotov, M.Y., Kargel, J.S., 2009. On the chemical composition of Europa's icy shell, ocean, and underlying rocks. In: Pappalardo, R.T., McKinnon, W.B., Khurana, K. (Eds.), Europa. University of Arizona Press, Tucson, AZ, pp. 431.

Bibliography

Aigrain, S., Pont, F., Zucker, S., 2012. A simple method to estimate radial velocity variations due to stellar activity using photometry. Mon. Not. R. Astron. Soc. 419, 3147–3158.

Anglada-Escudé, G., Butler, R.P., 2012. The HARPS-TERRA Project. I: description of the algorithms, performance, and new measurements on a few remarkable stars observed by HARPS. Astrophys. J. Suppl. Ser. 200, 15.

Arriagada, P., et al., 2013. Two planetary companions around the K7 dwarf GJ 221: a hot super-Earth and a candidate in the sub-Saturn desert range. Astrophys. J. 771, 42.

Anglada-Escudé, G., et al., 2013. A dynamically-packed planetary system around GJ 667C with three super-Earths in its habitable zone. Astron. Astrophys. 556, A126.

Baliunas, S.L., et al., 1995. Chromospheric variations in main-sequence stars. Astrophys. J. 438, 269–287.

Baluev, R.V., 2009. Accounting for velocity jitter in planet search surveys. Mon. Not. R. Astron. Soc. 393, 969–978.

Baluev, R.V., 2013. The impact of red noise in radial velocity planet searches: only three planets orbiting GJ 581?. Mon. Not. R. Astron. Soc. 429, 2052–2068.

Barnes, J.R., et al., 2012. Red Optical Planet Survey: a new search for habitable Earths in the southern sky. Mon. Not. R. Astron. Soc. 424, 591–604.

Barnes, J.R., et al., 2014. Precision radial velocities of 15 M5–M9 dwarfs. Mon. Not. R. Astron. Soc. 439, 3094–3113.

Berdiñas, Z.M., Amado, P.J., Anglada-Escudé, G., Rodríguez-López, C., Barnes, J., 2016. High-cadence spectroscopy of M-dwarfs. I: analysis of systematic effects in HARPS-N line profile measurements on the bright binary GJ 725A+B. Mon. Not. R. Astron. Soc. 459, 3551B.

Berger, J.O. (1980), Statistical Decision Theory and Bayesian Analysis, Section 3.3 (Springer, 1980), Springer New York, NY. XVI, 618 pp. Hardcover ISBN: 978-0-387-96098-2. Softcover ISBN: 978-1-4419-3074-3: eBook ISBN: 978-1-4757-4286-2. https://doi.org/10.1007/978-1-4757-4286-2.

Bolmont, E., et al., 2016, Water loss from Earth-sized planets in the habitable zones of ultracool dwarfs: implications for the planets of TRAPPIST-1. Preprint at http://arxiv.org/abs/1605.00616.

Bonfils, X., et al., 2007. The HARPS search for southern extra-solar planets. X: A $m \sin i = 11 M_\oplus$ planet around the nearby spotted M dwarf GJ 674. Astron. Astrophys. 474, 293–299.

Bonfils, X., et al., 2013. The HARPS search for southern extra-solar planets. XXXI: the M-dwarf sample. Astron. Astrophys. 549, A109.

Brown, T.M., et al., 2013. Las Cumbres observatory global telescope network. Publ. Astron. Soc. Pacific 125, 1031–1055.

Butler, R.P., et al., 1996. Attaining Doppler precision of 3 m s^{-1}. Publ. Astron. Soc. Pacif. 108, 500–509.

Cabrol, N.A., 2016. Alien mindscapes—A perspective on the search for extraterrestrial intelligence. Astrobiology 16 (9), 661–676.

Collins, K.A., Kielkopf, J.F., Stassun, K.G., 2016. Image processing and photometric extraction for ultra-precise astronomical light curves. Astro Image J. Preprint at. http://arxiv.org/abs/1601.02622.

Cumming, A., 2004. Detectability of extrasolar planets in radial velocity surveys. Mon. Not. R. Astron. Soc. 354, 1165–1176.

Davila, A.F., Schulze-Makuch, D., 2016. The last possible outposts for life on Mars. Astrobiology, 159–168.

Dawson, R.I., Fabrycky, D.C., 2010. Radial velocity planets de-aliased: a new, short period for super-Earth 55 Cnc e. Astrophys. J. 722, 937–953.

Delfosse, X., et al., 2000. Accurate masses of very low mass stars. IV. Improved mass-luminosity relations. Astron. Astrophys. 364, 217–224.

Deutsch, C.A., Tewksbury, J.J., Huey, R.B., Sheldon, K.S., Ghalambor, C.K., Haak, D.C., Martin, P.R., 2008. Impacts of climate warming on terrestrial ectotherms across latitude. Proc. Natl. Acad. Sci. U.S.A 105 (18), 6668–6672.

Donati, J.-F., Brown, S.F., 1997. Zeeman–Doppler imaging of active stars. V: sensitivity of maximum entropy magnetic maps to field orientation. Astron. Astrophys 326, 1135–1142.

Endl, M., Kürster, M., 2008. Toward detection of terrestrial planets in the habitable zone of our closest neighbor: Proxima Centauri. Astron. Astrophys. 488, 1149–1153.

Ferraz-Mello, S., 1981. Estimation of periods from unequally spaced observations. Astron. J. 86, 619–624.

Filippidou, S., Wunderlin, T., Junier, T., Jeanneret, N., Dorador, C., Molina, V., Johnson, D.R., Junier, P., 2016. A combination of extreme environmental conditions favor the prevalence of endospore-forming firmicutes. Front. Microbiol. 7, 1707; doi: 10.3389/fmicb.2016.01707. PMID: 27857706; PMCID: PMC5094177.

Gaboyer, F., Le Milbeau, C., Bohmeier, M., Schwendner, P., Vannier, P., Beblo-Vranesevic, K., Rabbow, E., Foucher, F., Gautret, P., Guégan, R., Richard, A., Sauldubois, A., Richmann, P., Perras, A.K., Moissl-Eichinger, C., Cockell, C.S., Rettberg, P., Marteinsson, E.Monaghan, Ehrenfreund, P., Garcia-Descalzo, L., Gomez, F., Malki, M., Amils, R., Cabezas, P., Walter, N., Westall, F., 2017. Mineralization and preservation of an extremo-tolerant bacterium isolated from an early Mars analog environment. Sci. Rep. 7 (1).

Gomes da Silva, J., et al., 2012. Long-term magnetic activity of a sample of M-dwarf stars from the HARPS program. II: activity and radial velocity. Astron. Astrophys. 541, A9.

Grazier, K.R., 2016. Jupiter: Cosmic Jekyll and Hyde. Astrobiology, 16 (1), 23–38. doi: 10.1089/ast.2015.1321.

Haario, H., Saksman, E., Tamminen, J., 2001. An adaptive Metropolis algorithm. Bernouilli 7, 223.

Haario, H., Laine, M., Mira, A., Saksman, E., 2006. Dram: efficient adaptive MCMC. Stat. Comput. 16, 339–354.

Harrison, J.P., Angel, R., Cockell, C.S., 2017. Astrobiology as a framework for investigating antibiotic susceptibility: a study of *Halomonashydrothermalis*. J. R. Soc., Interface 14 (126), 20160942.

Hastings, W.K., 1970. Monte Carlo sampling methods using Markov chains and their applications. Biometrika 57, 97–109.

Horneck, G., Walter, N., Westall, F., Grenfell, J.L., Martin, W.F., Gomez, F., Leuko, S., Lee, N., Onofri, S., Tsiganis, K., Saladino, R., Pilat-Lohinger, E., Palomba, E., Harrison, J., Rull, F., Muller, C., Strazzulla, G., Brucato, J.R., Rettberg, P., Capria, M.T., 2016. AstRoMap European astrobiology roadmap. Astrobiology, 16 (3), 201–243. doi: 10.1089/ast.2015.1441.

Jehlička, J., Culka, A., Nedbalová, L., 2016. Colonization of snow by micro-organisms as revealed using miniature Raman spectrometers—Possibilities for detecting carotenoids of psychrophiles on Mars?. Astrobiology 16 (12), 913–924.

Jenkins, J.S., et al., 2006. An activity catalogue of southern stars. Mon. Not. R. Astron. Soc. 372, 163–173.

Jenkins, J.S., et al., 2008. Metallicities and activities of southern stars. Astron. Astrophys. 485, 571–584.

Kopparapu, R.K., et al., 2016. The inner edge of the habitable zone for synchronously rotating planets around low-mass stars using general circulation models. Astrophys. J. 819, 84.

Kürster, M., et al., 2003. The low-level radial velocity variability in Barnard's star (= GJ 699): secular acceleration, indications for convective redshift, and planet mass limits. Astron. Astrophys. 403, 1077–1087.

Lingam, M., Loeb, A., 2017. Natural and artificial spectral edges in exoplanets. Monthly Notices Royal Astronomical Soc. Letters 470 (1), L82–L86.

Lomb, N.R., 1976. Least-squares frequency analysis of unequally spaced data. Astrophys. Space Sci. 39, 447–462.

Lubin, P., 2016. Preprint at.

Metropolis, N., Rosenbluth, A., Rosenbluth, M., Teller, A., Teller, E., 1953. Equations of state calculations by fast computing machines. J. Chem. Phys. 21, 1087–1092.

Moissl-Eichinger, C., Cockell, C., Rettberg, P., Albers, S.-V., 2016. Venturing into new realms? Microorganisms in space. FEMS Microbiol. Rev. 40 (5), 722–737.

Newton, M.A., Raftery, A.E., 1994. Approximate Bayesian inference with the weighted likelihood bootstrap. J. R. Stat. Soc. B. 56, 3–48.

Noack, L., Snellen, I., Rauer, H., 2017. Water in extra-solar planets and implications for habitability. Space Sci. Rev.

Ofir, A., 2014. Optimizing the search for transiting planets in long time series. Astron. Astrophys. 561, A138.

Pascual-Granado, J., Garrido, R., Suárez, J.C., 2015. Limits in the application of harmonic analysis to pulsating stars. Astron. Astrophys. 581, A89.

Pepe, F., et al., 2011. The HARPS search for Earth-like planets in the habitable zone. I. Very low-mass planets around HD 20794, HD 85512, and HD 192310. Astron. Astrophys 534, A58, DOI: https://doi.org/10.1051/0004-6361/201117055. (Number of page(s): 16).

Press, W.H., Teukolsky, S.A., Vetterling, W.T., Flannery, B.P., 1992, 2nd edn Numerical Recipes in FORTRAN. The Art of Scientific Computing 1992. Cambridge Univ. Press Section 4.1 1003 pages: ISBN 0-521-43064-X.

Queloz, D., et al., 2001. No planet for HD 166435. Astron. Astrophys 379, 279–287.

Rajpaul, V., Aigrain, S., Roberts, S., 2016. Ghost in the time series: no planet for Alpha Cen B. Mon. Not. R. Astron. Soc. 456, L6–L10.

Ranjan, S., Sasselov, D.D., 2016. Influence of the UV environment on the synthesis of prebiotic molecules. Astrobiology 16 (1), 68–88, https://doi.org/10.1089/ast.2015.1359.

Reiners, A., Basri, G., 2008. The moderate magnetic field of the flare star Proxima Centauri. Astron. Astrophys. 489, L45–L48.

Reynolds, R.T., McKay, C.P., Kasting, J.F., 1987. Europa, tidally heated oceans, and habitable zones around giant planets. Adv Space Res 7 (5), 125–132.

Robertson, P., Mahadevan, S., Endl, M., Roy, A, 2014. Stellar activity masquerading as planets in the habitable zone of the M dwarf Gliese 581. Science 345, 440–444.

Russell, M.J., Murray, A.E., Hand, K.P., 2017. The possible emergence of life and differentiation of a shallow biosphere on irradiated icy worlds: The example of Europa. Astrobiology 17 (12), 1265–1273.

Scargle, J.D., 1981. Studies in astronomical time series analysis. I: modeling random processes in the time domain. Astrophys. J. Suppl. Ser. 45, 1–71.

Scargle, J.D., 1982. Studies in astronomical time series analysis. II: statistical aspects of spectral analysis of unevenly spaced data. Astrophys. J. 263, 835–853.

Scharf, C., Virgo, N., Cleaves II, H.J., Aono, M., Aubert-Kato, N., Aydinoglu, A., Barahona, A., Barge, L.M., Benner, S.A., Biehl, M., Brasser, R., Butch, C.J., Chandru, K., Cronin, L., Danielache, S., Fischer, J., Hernlund, J., Hut, P., Ikegami, T., Kimura, J., Kobayashi, K., Mariscal, C., McGlynn, S., Menard, B., Packard, N., Pascal, R., Pereto, J., Rajamani, S., Sinapayen, L., Smith, E., Switzer, C., Takai, K., Tian, F., Ueno, Y., Voytek, M., Witkowski, O., Yabuta, H., 2015. A strategy for origins of life research. Astrobiology 15, 1031–1042.

Schuerger, A.C., Ming, D.W., Golden, D.C., 2017. Biotoxicity of Mars soils: 2. Survival of *Bacillus subtilis* and *Enterococcus faecalis* in aqueous extracts derived from six Mars analog soils. Icarus 290, 215–223.

Shields, A.L., Ballard, S., Johnson, J.A., 2016. The habitability of planets orbiting M-dwarf stars. Phys. Rep. 663, 1–38.

Sicardy, B., et al., 2011. A Pluto-like radius and a high albedo for the dwarf planet Eris from an occultation. Nature 478, 493–496.

Southworth, J., et al., 2014. High-precision photometry by telescope defocussing. VI: WASP-24, WASP-25 and WASP-26. Mon. Not. R. Astron. Soc. 444, 776–789.

Snellen, I., et al., 2015. Combining high-dispersion spectroscopy with high contrast imaging: probing rocky planets around our nearest neighbours. Astron. Astrophys. 576, A59.

Tanaka, H., Takeuchi, T., Ward, W.R., 2002. Three-dimensional interaction between a planet and an isothermal gaseous disk. I: corotation and Lindblad torques and planet migration. Astrophys. J. 565, 1257–1274.

Tuomi, M., 2012. Evidence for nine planets in the HD 10180 system. Astron. Astrophys. 543, A52.

Tuomi, M., Anglada-Escudé, G., 2013. Up to four planets around the M dwarf GJ 163: sensitivity of Bayesian planet detection criteria to prior choice. Astron. Astrophys 556, A111.

Tuomi, M., et al., 2013. Signals embedded in the radial velocity noise: periodic variations in the τCeti velocities. Astron. Astrophys. 551, A79.

Tuomi, M., Jones, H.R.A., Barnes, J.R., Anglada-Escudé, G., Jenkins, J.S., 2014. Bayesian search for low-mass planets around nearby M dwarfs—estimates for occurrence rate based on global detectability statistics. Mon. Not. R. Astron. Soc. 441, 1545–1569.

Tuomi, M., 2014. A new cold sub-Saturnian candidate planet orbiting GJ 221. Mon. Not. R. Astron. Soc. 440, L1–L5.

Vago, J.L., Westall, F.Pasteur Instrument Teams, Landing Site Selection Working Group, and Other Contributors, Coates, A.J., Jaumann, R., Korablev, O., Ciarletti, V., Mitrofanov, I., Josset, J.-L., De Sanctis, M.C., Bibring, J.-P., Rull, F., Goesmann, F., Steininger, H., Goetz, W., Brinckerhoff, W., Szopa, C., Raulin, F., Westall, F., Edwards, H.G.M., Whyte, L.G., Fairén, A.G., Bibring, J.-P., Bridges, J., Hauber, E., Ori, G.G., Werner, S., Loizeau, D., Kuzmin, R.O., Williams, R.M.E., Flahaut, J., Forget, F., Vago, J.L., Rodionov, D., Korablev, O., Svedhem, H., Sefton-Nash, E., Kminek, G., Lorenzoni, L., Joudrier, L., Mikhailov, V., Zashchirinskiy, A., Alexashkin, S., Calantropio, F., Merlo, A., Poulakis, P., Witasse, O., Bayle, O., Bayón, S., Meierhenrich, U., Carter, J., García-Ruiz, J.M., Baglioni, P., Haldemann, A., Ball, A.J., Debus, A., Lindner, R., Haessig, F., Monteiro, D., Trautner, R., Voland, C., Rebeyre, P., Goulty, D., Didot, F., Durrant, S., Zekri, E., Koschny, D., Toni, A., Visentin, G., Zwick, M., van Winnendael, M., Azkarate, M., Carreau, C.the ExoMars Project Team, 2017. Habitability on early Mars and the search for biosignatures with the ExoMars Rover. Astrobiology 17 (6-7), 471–510.

Vidotto, A.A., et al., 2013. Effects of M dwarf magnetic fields on potentially habitable planets. Astron. Astrophys. 557, A67.

Vilella, K., Kaminski, E., 2017. Fully determined scaling laws for volumetrically heated convective systems, a tool for assessing habitability of exoplanets. Physics of the Earth andPlanetary Interiors 266, 18–28.

Weidenschilling, S.J., 1977. Aerodynamics of solid bodies in the solar nebula. Mon. Not. R. Astron. Soc. 180, 57–70.

Westall, F., Foucher, F., Bost, N., Bertrand, M., Loizeau, D., Vago, J.L., Kminek, G., Gaboyer, F., Campbell, K.A., Bréhéret, J.-G., Gautret, P., Cockell, C.S., 2015. Biosignatures on Mars: What, where, and how? Implications for the search for Martian life. Astrobiology 15, 998–1029.

Wright, J.T., Howard, A.W., 2009. Efficient fitting of multiplanetKeplerian models to radial velocity and astrometry data. Astrophys. J. Suppl. Ser. 182, 205–215.

Zechmeister, M., Kürster, M., Endl, M., 2009. The M dwarf planet search programme at the ESO VLT + UVES: a search for terrestrial planets in the habitable zone of M dwarfs. Astron. Astrophys. 505, 859–871.

Zuluaga, J.I., Bustamante, S., Cuartas, P.A., Hoyos, J.H., 2013. The influence of thermal evolution in the magnetic protection of terrestrial planets. Astrophys. J. 770, 23.

Appendix

Chemical names and their chemical formulae

1. Acetonitrile: CH_3CN
2. Acetylene: C_2H_2
3. Adenine: $C_5H_5N_5$
4. Ammonia: NH_3
5. Ammonia dehydrate: $NH_3 \cdot 2H_2O$
6. Ammonium sulfate hydrates: $NH_4SO_4\text{-}nH_2O$
7. Apatite: $Ca_5(PO_4)_3(F,Cl,OH)$: This is a hydrogen-rich calcium phosphate mineral
8. Benzene: C_6H_6
9. Bicarbonate: HCO_3
10. Biotite: $K(Mg, Fe)_3[AlSi_3O_{10}](OH,F)_2$
11. Brucite: $Mg(OH)_2$
12. Calcite (Calcium carbonate): $CaCO_3$
13. Calcium chloride: $CaCl_2$
14. Calcium sulfate: $CaSO_4$
15. Carbon monoxide: CO
16. Carbon dioxide: CO_2
17. Carbonic acid: H_2CO_3
18. Coronene: $C_{24}H_{12}$
19. Cyanoacetylene: HC_3N
20. Cyanogen: C_2N_2
21. Cytosine: $C_4H_5N_3O$
22. Deuterium: D_2
23. Diacetylene: C_4H_2
24. Dicyanodiacetylene: C_6N_2
25. Dimethyl sulfide: C_2H_6S
26. Dolomite (Calcium magnesium carbonate): $CaMg(CO_3)_2$
27. Ethane: C_2H_6 (This is a colorless, odorless gas at standard temperature and pressure)
28. Ethylene: C_2H_4
29. Formaldehyde: CH_2O
30. Guanine: $C_5H_5N_5O$
31. Gypsum: $CaSO_4 \cdot 2H_2O$
32. Helium: He
33. Hematite: Fe_2O_3 (This is a common iron oxide compound widely found in rocks and soils)
34. Hydrogen: H_2
35. Hydrogen chloride/Hydrocloric Acid: HCl
36. Hydrogen cyanide: HCN
37. Hydrogen fluoride: HF
38. Hydrogen peroxide: H_2O_2
39. Hydrogen sulphide: H_2S
40. Hydronium: H_3O
41. Hydronium jarosite: $KFe_3(SO_4)_2(OH)_6$
42. Hydrous ferric sulfate: $Fe_2(SO_4)_3 \cdot 9H_2O$
43. Ilmenite: $FeTiO_3$
44. Jarosite: $KFe_3(SO_4)_2(OH)_6$
45. Lithium chloride: $LiCl$
46. Magnesium bromide: $MgBr_2$
47. Magnesium carbonate: Mg_2CO_3
48. Magnesium chloride: $MgCl_2$
49. Magnesium sulfate salts: $MgSO_4 \cdot nH_2O$
50. Magnetite: Fe_3O_4
51. Methane: CH_4
52. Methyl chloride: CH_3Cl
53. Nitrogen: N_2
54. Nitric oxide: NO
55. Nitrous oxide: N_2O
56. Olivine: $(Mg,Fe)_2SiO_4$
57. Opal: SiO_2
58. Orthoclase: $K[AlSi_3O_8]$
59. Orthopyroxenes: $(Mg,Fe,Ca)(Mg,Fe,Al)(Si,Al)_2O_6$
60. Oxygen: O_2
61. Ozone: O_3
62. Pyrene: $C_{16}H_{10}$
63. Pyroxene: $(Mg,Fe)SiO_3$
64. Schwertmannite: $Fe_8O_8(OH)_6(SO_4) \cdot nH_2O$
65. Silica: SiO_2
66. Sodium carbonate: Na_2CO_3

67. Sodium carbonate hydrates: $Na_2CO_3 \cdot nH_2O$, where n is the number of H_2O molecules per molecule of the mineral
68. Sodium chloride: NaCl
69. Sodium sulfate hydrates: $Na_2SO_4 \cdot nH_2O$
70. Sulfur dioxide: SO_2
71. Sulfuric acid hydrate: $H_2SO_4 \cdot nH_2O$
72. Thymine: $C_5H_6N_2O_2$
73. TNT (Trinitrotoluene): $C_6H_2(NO_2)_3CH_3$
74. Uracil: $C_4H_4N_2O_2$
75. Water: H_2O

Water Worlds in the Solar System: Exploring Prospects of Extraterrestrial Habitability & Life

Short forms and their expansion

1. 1D: One-dimensional
2. ABRF MGRG: Association of Biomolecular Resource Facilities Metagenomics Research Group
3. ACP: Aerosol Collector Pyrolyser
4. ACS: Advanced Camera for Surveys
5. ADRON: Active Detector for gamma Rays and neutrONs
6. ADV: Antarctic Dry Valley
7. ALH84001 meteorite: Allan Hills 84001 Martian meteorite
8. ALMA: Atacama Large Millimeter/submillimeter Array
9. ALS system: Advanced life support system
10. AMD: Acid and metalliferous drainage
11. AMD: Acid mine drainage
12. AMO: Anaerobic methane oxidation
13. AMS: Accelerator mass spectrometry
14. ANME: ANaerobic MEthanotrophic Archaea
15. AO: Announcement of opportunity
16. AOD: Atmospheric optical depth
17. AOGCM: Atmosphere–Ocean General Circulation Model
18. AOT: Aerosol optical thickness
19. APL: Applied Physics Laboratory
20. APXS: Alpha Particle X-ray Spectrometer
21. ARC: Ames Research Center
22. ARD: Acid rock drainage
23. ARISA: Automated Approach for Ribosomal Intergenic Spacer Analysis
24. ASI: Agenzia Spaziale Italiana (the Italian space agency)
25. ATP: Adenosine triphosphate. ATP is a nucleotide that consists of three main structures: the nitrogenous base (adenine); the sugar (ribose); and a chain of three phosphate groups serially bonded to ribose. ATP is the energy-carrying molecule found in the cells of all living things. ATP captures chemical energy obtained from the breakdown of food molecules and releases it to fuel other cellular processes
26. AU: Astronomical Unit (an astronomical unit (AU) is the average distance between Earth and the Sun, approximately 150×10^6 km. The speed of light is about 3.0×10^8 m/s)
27. BA: Benzoic acid
28. BAS: British Antarctic Survey
29. BER: Base excision repair
30. BP: Before Present
31. BPC: Biomass production chamber
32. BT: Brightness temperature
33. BTD: Bladed terrain deposits
34. CA: Chemical Analyser
35. CAPS: Cassini Plasma Spectrometer
36. CAPS/IBS: Cassini Plasma Spectrometer's Ion Beam Sensor
37. CAST: China Aerospace Science and Technology Corp
38. CAT: Chemical Analyser Target
39. CBD: Convention on biological diversity
40. CBL: Convective boundary layer
41. CC: Carbonaceous chondrite
42. CCD: Charge-coupled device
43. CCD: Carbonate compensation depth
44. CDA: Cosmic dust analyzer (which was a mass spectrometer aboard the Cassini spacecraft, and discovered high-mass, complex organic material in a small fraction of ice grains in the plume emitted by Saturn's moon *Enceladus*)
45. CDA: Chilika Development Authority
46. CELSS: Controlled Ecological Life Support System
47. CGS: Cooled Grating array Spectrometer
48. CHACE: Chandra's Altitudinal Composition Explorer
49. CheMin: Chemistry and Mineralogy instrument
50. CHZ: Circum-stellar Habitable Zone
51. CIRS: Composite Infrared Spectrometer
52. C1XS: Chandrayaan-1 imaging X-ray Spectrometer

53. CLUPI: CLose-UP Imager
54. CORK: Circulation Obviation Retrofit Kit (a sub-seafloor observatory, installed at North Pond (an isolated sediment pond on the western flank of the Mid-Atlantic Ridge) by the Integrated Ocean Drilling Program in 2011)
55. CMB: Core–Mantle boundary
56. CNSA: Chinese National Space Administration
57. CPR: Circular polarization ratio
58. CRISM: Compact Reconnaissance Imaging Spectrometer for Mars
59. CSI: Conserved Signature Insert
60. CTX: Context Camera
61. DAM: Decametric radio emission
62. D-CB: Devonian–Carboniferous boundary
63. DD: Dust Devil
64. DDOR: Delta Differential One-way Ranging
65. DEM: Digital Elevation Map
66. DEM: Digital Elevation Model
67. DHU: Data Handling Unit
68. DISR: Descent Imager/Spectral Radiometer
69. DMS: Dimethyl sulfide
70. DNA: Deoxyribonucleic acid
71. DOM: Deep Ocean Mission
72. DSC: Differential scanning calorimetry
73. DSN: Deep Space Network
74. DSW: Dead Sea Water
75. DTE: Direct-To-Earth
76. DTM: Digital Terrain Model
77. DWE: Doppler Wind Experiment
78. EAM: Ethynyl Addition Mechanism
79. EBSA: Ecologically or Biologically Significant marine Area
80. EDX: Energy dispersive X-ray spectroscopy
81. EDXRF: Energy dispersive X-ray fluorescence spectroscopy
82. EELS: Exobiology Extant Life Surveyor
83. EGEON: Europa Global model of Exospheric Outgoing Neutrals
84. EJSM: Europa Jupiter System Mission
85. ELF: Enceladus Life Finder
86. ELF: Extremely low frequency
87. ELW: Europa Lander Workshop
88. EMC: Equilibrium moisture content
89. EML points: Earth–Moon Liberation points (also called Lagrange points)
90. EMP: Earth Microbiome Project
91. EnEx project: Enceladus Explorer project
92. EPS: Extracellular polymeric substances (high molecular weight carbohydrate polymers that make up a substantial component of the extracellular polymers surrounding most microbial cells in the harsh environment)
93. EPS: Exopolysaccharides (high molecular weight carbohydrate polymers that make up a substantial component of the extracellular polymers surrounding most microbial cells in the harsh environment)
94. ERH: Equilibrium relative humidity
95. ESA: European Space Agency
96. ESCRT: Endosomal Sorting Complex Required for Transport
97. ESTs: Expressed sequence tags
98. EUV: Extreme Ultra Violet
99. FAPs: Filamentous anoxygenic phototrophs
100. FC: Framing camera
101. F-F boundary: Frasnian–Famennian boundary
102. FOV: Field of view
103. FDTD: Finite difference time domain
104. Ga: Giga-annum (10^9 years); i.e., billion years
105. GC: Gas chromatography
106. GC/MS: Gas chromatography/mass spectrometry
107. GCM: General circulation model
108. GCM: Global circulation model
109. GCMS: Gas chromatograph/mass spectrometer
110. GCR: Galactic cosmic radiation
111. GDGT: Glycerol di-alkyl glycerol tetraether
112. GEL: Global equivalent layer
113. GFDL: Geophysical Fluid Dynamics Laboratory
114. GMT: Greenwich Mean Time
115. GO: Gene ontology
116. GOE: Great oxygenation event (also called great oxidation event)
117. GPMS: Galileo Probe Mass Spectrometer
118. GPR: Ground Penetrating Radar
119. GRAIL: Gravity Recovery and Interior Laboratory
120. GRaND: Gamma Ray and Neutron Detector
121. GRS: Gamma-Ray Spectrometer
122. Gy (gray): A derived unit of ionizing radiation dose in the International System of Units (SI). It is defined as the absorption of one joule of radiation energy per kilogram of matter
123. HACA: Hydrogen-abstraction-C_2H_2-addition
124. HAMO: High altitude mapping orbit
125. HASI: Huygens Atmospheric Structure Instrument
126. Hazcam: Hazard avoidance camera
127. HDR: High dynamic range
128. HERCULES: Highly eccentric rotating concentric U (potential) layers equilibrium structure. This is a computer software that mathematically solves for the equilibrium structure of planets as a series of overlapping constant-density spheroids
129. HEX: High Energy X-ray Spectrometer
130. HGA: High-gain antenna
131. HiCARS: High capability airborne radar sounder
132. HIGS: Hypervelocity ice grain system
133. HiRISE: High-Resolution Imaging Science Experiment
134. HMOC: High mass organic cations
135. HOCNSP: Abbreviation of biologically important chemical elements hydrogen (H), oxygen (O), carbon (C), nitrogen (N), sulfur (S), and phosphorous (P)

136. HP: High pressure
137. HR: Homologous recombination
138. HRC: High-resolution camera
139. HRSC: High-Resolution Stereo Camera (this camera was deployed on board ESA's Mars Express spacecraft)
140. HST: Hubble Space Telescope
141. HT: Hyper-thermophilic
142. HZ: Habitable zone
143. IAU: International Astronomical Union
144. IBS: Ion BackScattering
145. IBS: Ion beam sensor
146. ICR: Institute of Cotton Research
147. IDPs: Interplanetary dust particles
148. IETM: Initial Eocene Thermal Maximum (also called PETM)
149. IGRINS: Immersion Grating Infrared Spectrometer
150. IIST: Indian Institute of Space Science and Technology
151. IM: Induced Magnetosphere
152. Indel: Insertion/deletion polymorphism
153. INMS: Ion and Neutral Mass Spectrometer. The INMS was deployed in Cassini spacecraft for detection of volatile, gas phase, and organic species in the plume gushing out from the "tiger stripes" of Saturn's moon *Enceladus*
154. IODP: Integrated Ocean Drilling Program
155. IOM: Insoluble organic matter
156. IPBSL: Iberian Pyrite Belt Subsurface Life Detection
157. IPCC: Intergovernmental Panel on Climate Change
158. IR: Infrared
159. IS: Ion sputtering
160. ISM: Inter stellar medium
161. ISRO: Indian Space Research Organization
162. ISS: International Space Station
163. ISS: Imaging Science Subsystem
164. ITS: Internal Transcribed Spacer
165. IUCN: International Union for Conservation of Nature
166. IUE: International Ultraviolet Explorer
167. JAXA: Japanese Aerospace eXploration Agency
168. JEB: Jovian Early Bombardment
169. JEDI: Jupiter Energetic Particle Detector Instrument
170. JEO: Jupiter Europa Orbiter
171. JFCs: Jupiter-Family Comets
172. JGA: Judea Group Aquifers
173. JGO: Jupiter Ganymede Orbiter
174. JPL/CALTECH: Jet Propulsion Laboratory/ California Institute of Technology
175. JPS: Jupiter's Plasma Sheet
176. JSC: Johnson Space Center
177. JUICE: JUpiter ICy moons Explorer (an interplanetary spacecraft in development by the European Space Agency (ESA) and set for launch in June 2022 and expected to reach Jupiter in October 2029 to study three of Jupiter's Galilean moons: Ganymede, Callisto, and Europa, all of which are thought to have significant bodies of liquid water beneath their surfaces, making them potentially habitable environments)
178. KASI: Korea Astronomy and Space Science Institute
179. KBO: Kuiper Belt Object
180. KMC: Kinetic Monte-Carlo
181. KMC-ARS: Kinetic Monte Carlo—Aromatic Site (a polycyclic aromatic hydrocarbons (PAHs) growth model)
182. KPBD: Cretaceous/Paleogene (K–Pg) boundary deposit
183. KREEP: Potassium (K), rare earth elements (REE), and phosphorus (P)
184. KSC: Kennedy Space Center
185. LADEE: Lunar Atmosphere and Dust Environment Explorer
186. LAMP: Lyman Alpha Mapping Project
187. LAP: Lyman Alpha Photometer
188. LC/MS: Liquid chromatography/mass spectrometry
189. LCA: Last Common Ancestor
190. LCHF: Lost City Hydrothermal Field
191. LCOGT: Las Cumbres Observatory Global Telescope Network
192. LCM: Lost City Methanosarcinale
193. LCROSS: Lunar CRater Observation and Sensing Satellite
194. LEISA: Linear Etalon Imaging Spectral Array
195. LEND: Lunar Exploration Neutron Detector
196. LEO: Low Earth Orbit
197. LEOS: Laboratory for Electro-Optics Systems (at ISRO, India)
198. LET: Linear Energy Transfer
199. LFS: Low Frequency Spectrometer
200. LHB: Late Heavy Bombardment
201. LID: Laser-Induced Desorption
202. LIFE: Life Investigation For Enceladus
203. LIP: Large Igneous Province
204. LLRI: Lunar Laser Ranging Instrument
205. LMH: Liquid metallic hydrogen
206. LNA: Locked nucleic acid
207. LOME: Late Ordovician Mass Extinction
208. LORRI: Long Range Reconnaissance Imager
209. LP: Lunar Prospector
210. LP-NS: Lunar Prospector Neutron Spectrometer
211. LPR: Lunar Penetrating Radar
212. LPTM: Latest Paleocene Thermal Maximum, which occurred roughly 55 million years ago
213. LRL: Lunar Receiving Laboratory
214. LRO: Lunar Reconnaissance Orbiter
215. LRS: Little Red Spot
216. Ls: Areocentric longitude
217. LTST: Local True Solar Time
218. LUCA: Last Universal Common Ancestor

219. LUCA: Last Universal Cellular Ancestor
220. MA: Mellitic acid
221. MAB: Main Asteroid Belt
222. MAB: Magnetite Assay for Biogenicity
223. Ma_MISS: Mars Multispectral Imager for Subsurface Studies
224. MAR: Mid-Atlantic Ridge
225. MARIE: Martian Radiation Experiment
226. MARSIS: Mars Advanced Radar for Subsurface and Ionospheric Sounding
227. MARTE: Mars Astrobiology Research and Technology Experiment
228. MAVEN: Mars Atmosphere and Volatile Evolution (spacecraft)
229. MBSF: Meters Below Seafloor
230. MCC: Mars Color Camera
231. MCD: Mars Climate Database
232. MDI: Maximum diurnal insolation
233. MDVs: McMurdo Dry Valleys
234. MENCA: Mars Exospheric Neutral Composition Analyser
235. MEPAG: Mars Exploration Payload Analysis Group
236. MER: Mars Exploration Rover
237. MEX: Mars Express
238. MFF: Medusae Fossae Formation
239. MGS: Mars Global Surveyor
240. MGS TES: Mars Global Surveyor Thermal Emission Spectrometer
241. MHD: Magneto Hydro Dynamics
242. MI: Multiband Imager
243. MI: Microscopic Imager
244. Mini-TES: Miniature Thermal Emission Spectrometer
245. MIP: Moon Impact Probe
246. MISS: Microbially Induced Sedimentary Structures
247. MMR: MisMatch Repair
248. MMRTG: Multi-Mission Radioisotope Thermoelectric Generator
249. MOC: Mars Orbiter Camera
250. MOLA: Mars Orbiter Laser Altimeter
251. MOM: Modular Ocean Model
252. MOM: Mars Orbiter Mission (of India)
253. MOMA: Mars Organic Molecule Analyzer
254. MMSN: Minimum Mass Solar Nebula
255. MOR: Mid Ocean Ridge
256. MRO: Mars Reconnaissance Orbiter
257. mRNA: messenger RNA
258. MS: Mass spectrometry
259. MSL: Mars Science Laboratory
260. MSL: Mean sea level
261. MSM: Methane Sensor for Mars
262. MSolo: Mass Spectrometer Observing Lunar Operations
263. MSR mission: Mars Sample Return mission
264. MTB: Magneto Tactic Bacteria
265. MU: Miller-Urey
266. MVIC: Multispectral Visible Imaging Camera
267. MWR: Micro wave radiometer system
268. Mya: Million years ago
269. Myr: Million years
270. NASA: National Aeronautics and Space Administration (an independent agency of the US federal government responsible for the civilian space program, as well as aeronautics and space research)
271. Navcam: Navigation camera
272. NEAs: Near-Earth Asteroids
273. NER: Nucleotide Excision Repair
274. NGS: Next-Generation Sequencing
275. NHEJ: Non-Homologous End Joining
276. NIMS: Near-Infrared Mapping Spectrometer
277. NIR: Near-Infra Red
278. NIRVSS: Near-Infrared Volatiles Spectrometer System
279. NM: New Mexico
280. nm: Nanometer
281. N-PACs: N-containing polycyclic aromatic compounds
282. NSF: National Science Foundation (of USA)
283. NSS: Neutron Spectrometer System
284. OMEGA: Observatoire pour la Mineralogie, l'Eau, les Glaces et l'Activité (an instrument on the European Space Agency's Mars Express spacecraft, which allows discrimination among gas, frost, ice, water absorbed, and water bound in hydrated minerals)
285. Orbital triggering: Events (e.g., extreme global warming events) occurring during orbits with a combination of high eccentricity and high obliquity in Earth's orbital chronology
286. ODP: Ocean Drilling Program
287. ORP: Oxido-Reduction Potential
288. OSS: Outer Solar System
289. OWEP: Ocean Worlds Exploration Program (conceived by NASA). OWEP envisages to explore ocean worlds in the outer Solar System that could possess subsurface oceans to assess their habitability and to seek biosignatures of simple extraterrestrial life
290. PAC: Polycyclic aromatic compound
291. PAH: Polycyclic aromatic hydrocarbon
292. PAH-PP: Polycyclic aromatic hydrocarbon - primary particle (a multivariate PAH population balance model)
293. P.A.L: Present atmospheric level
294. PBL: Planetary boundary layer
295. PCR: Polymerase chain reaction
296. PELS: Planetary environmental liquid simulator
297. PEP: Particle Environment Package
298. PEPSSI: Pluto Energetic Particle Spectrometer Science Investigation
299. PETM: Paleocene–Eocene Thermal Maximum
300. PIA: Planar impact approximation
301. PIDD: Primary ion dose density

302. PIXL: Planetary Instrument for X-ray Litho-chemistry
303. PLFs: Pingo-like forms
304. PMA: Propidium monoazide
305. PNA: Peptide nucleic acid
306. PNG: Portable Network Graphics
307. PNV model: Permanent Northern Volatile model
308. Ppb: parts per billion
309. Ppbv: Parts per billion by volume
310. PPMV: Parts per million by Volume
311. PRC: People's Republic of China
312. PSD: Photon-stimulated desorption
313. PSE: Probe support equipment
314. PSLV: Polar Satellite Launch Vehicle
315. PVDF: Poly vinylidene di fluoride
316. PWA: Permittivity, wave, and altimetry
317. QITMS: Quadrupole Ion Trap Mass Spectrometer
318. qPCR: Quantitative polymerase chain reaction
319. RAD: Radiation Assessment Detector
320. RADOM: Radiation Dose Monitor
321. RAS: Russian Academy of Sciences
322. RAT: Rock Abrasion Tool
323. REASON: Radar for Europa Assessment and Sounding: Ocean to Near-surface
324. Redox: Reduction–oxidation/oxidation–reduction
325. RES: Radio Echo Sounding
326. REX: Radio science EXperiment
327. RIME: Radar for Icy Moons Exploration
328. RLS: Raman Laser Spectrometer
329. RMS: Root mean square
330. RNA: Ribonucleic acid
331. ROI: Region of interest
332. ROV: Remotely Operated Vehicle
333. ROW: Roadmaps to Ocean Worlds
334. RPW: Radio and plasma waves
335. rRNA: ribosomal RNA
336. RNAi: RNA interference
337. RSL: Recurring Slope Lineae
338. RTG: Radioisotope Thermoelectric Generator
339. RTS: Ridge–Trough system
340. tRNA: transfer RNA
341. SA: Solar Aureole (a camera aboard Huygens probe dropped to Titan's surface)
342. SAC: Space Applications Centre
343. SAM: Sample Analysis at Mars instrument; a portable chemistry lab inside the Curiosity rover's belly
344. SAR: Synthetic Aperture Radar
345. SASP: Small acid soluble proteins
346. SCOR: Scientific Committee on Oceanographic Research
347. s.d.: standard deviation
348. SEM: Scanning Electron Microscopy
349. SEP: Sun, Earth, and Probe
350. SESAME: Scientific Exploration Subsurface Access Mechanism for Europa
351. SETI: Search for Extraterrestrial Intelligence
352. SHARAD: SHAllow RADar (sounding radar)
353. SHERLOC: Scanning Habitable Environments with Raman & Luminescence for Organics & Chemicals (an instruments on the Perseverance Rover)
354. SHS: Submarine hydrothermal system
355. SIM: Southwest Iberian Margin
356. SIMS: Secondary ion mass spectrometry
357. SLUSH: Search for Life Using Submersible Heated (a hybrid drill system, which uses both thermal and mechanical means to slice through ice)
358. SNC meteorites: Shergottite–Nakhlite–Chassignite meteorites—a basalt body ejected 175 million years ago from Mars
359. SNDM: Solar Nebular Disk Model
360. SNR: Signal-to-noise ratio
361. SOI: Saturn Orbit Insertion
362. SP: Sputnik Planitia, formerly known as Sputnik Planum (planet Pluto's heart-shaped feature)
363. SPA: South Pole-Aitken
364. SPC: Strelley Pool Chert
365. SPDS: Sample preparation and distribution system (for astrobiology investigation of Mars)
366. SPL: Spore Photoproduct Lyase
367. SPT: South Polar Terrain
368. SR: Schumann Resonance
369. SSI: Solid-state imaging
370. SSPACE: Space Satellite Systems and Payloads Centre
371. SST: Sea surface temperature
372. ss rRNA: Small subunit ribosomal RNA
373. STIS: Space Telescope Imaging Spectrograph
374. STK: Systems Tool Kit
375. SWAP: Solar Wind Around Pluto (an instrument that measures the rate of escape of Pluto's atmosphere)
376. Sv: Sievert (the unit of radiation absorption in the International System of Units (SI))
377. SZA: Solar Zenith Angle
378. TACK: Thaumarchaeota, Aigarchaeota, Crenarchaeota, and Korarchaeota (the first groups of archaea discovered)
379. TDPs: Tardigrade-specific intrinsically disordered proteins. (TDPs form noncrystalline amorphous solids (vitrify) upon desiccation, and this vitrified state mirrors their protective capabilities against desiccation)
380. TDS: Total dissolved salts
381. TEC: Total electron content
382. TEM: Transmission electron microscopy
383. TES: Thermal Emission Spectrometer
384. TEX: Tumor-derived EXosomal (biomarkers)
385. TIS: Thermal Infrared Imaging Spectrometer
386. TG: Thermogravimetry
387. TGO: Trace Gas Orbiter (entered Mars orbit on 19 October 2016 and proceeded to map the sources of methane (CH_4) and other trace gases present in the Martian atmosphere that could be evidence for

possible biological or geological activity. It crashed on the surface of Mars)
388. THEMIS: Thermal Emission Imaging Spectrometer
389. THEO: Testing the Habitability of Enceladus's Ocean
390. TNA: Threose nucleic acid
391. TNT: Trinitrotoluene (a chemical compound with the formula $C_6H_2(NO_2)_3CH_3$). This pale yellow, solid organic nitrogen compound is used chiefly as an explosive, and is prepared by stepwise nitration of toluene
392. TO: Total organisms (dead and live)
393. ToF-SIMS: Time-of-flight secondary ion mass spectrometry
394. TPD: Temperature-programmed desorption
395. T-RFLP analysis: Terminal restriction fragment length polymorphism analysis
396. TRIDENT: The Regolith and Ice Drill for Exploring New Terrains
397. TSBs: Taxon-specific biomarkers
398. TSF: Tiger–Stripe fractures
399. TT: Thumbprint Terrains (on planet Mars)
400. TUB/IDA: Technical University Braunschweig/Institute of Computer and Communication Network Engineering
401. TV: Thermal volatilization
402. TW: Terawatt (10^{12} W)
403. UCLA: University of California, Los Angeles
404. UHF: Ultra-high frequency
405. UHPLC-FD/ToF-MS: Ultra-high-performance liquid chromatography coupled with fluorescence detection and time-of-flight mass spectrometry
406. UKIRT: United Kingdom Infrared Telescope
407. UNA: Unlocked nucleic acid
408. UTC: Universal Time Coordinate or Coordinated Universal Time
409. UV: Ultra violet
410. UVIS: Ultraviolet imaging spectrograph
411. VBSDC: Venetia Burney Student Dust Counter
412. VIMS: Visible and Infrared Mapping Spectrometer
413. VIPER: Volatiles Investigating Polar Exploration Rover
414. VIR: Visible and Infrared Spectrometer
415. VIRTIS: Visible and Infrared Thermal Imaging Spectrometer
416. VNIR: Visible/near-infrared
417. VO: Viable organisms (live)
418. WATSON: Wide Angle Topographic Sensor for Operations and eNgineering (a camera that can take close-up pictures of rock textures)
419. WCL: Wet chemistry laboratory
420. WEH: Water-equivalent hydrogen
421. WFPC: Wide Field Planetary Camera
422. WHO: World Health Organization
423. WHOI: Woods Hole Oceanographic Institution
424. WISDOM: Water, Ice, and Subsurface Deposit Observations on Mars radar
425. wt %: weight %
426. XMP: Extreme Microbiome Project
427. XRD: X-ray diffraction
428. XRF: X-ray fluorescence
429. 16S rRNA: 16S ribosomal RNA

Definition/Meaning

1. Abiogenic: Not produced or brought about by living organisms; of inorganic origin.
2. Abiotic methane: Methane (CH_4) formed by chemical reactions which do not directly involve organic matter; e.g., methane produced by high-temperature magmatic processes in volcanic and geothermal areas, or via low-temperature (<100°C) gas–water–rock reactions in continental settings.
3. Accretion: Growth by the gradual accumulation of additional layers or matter.
4. Acetyl-CoA: A molecule that participates in many biochemical reactions in protein, carbohydrate, and lipid metabolism. Its main function is to deliver the acetyl group to the citric acid cycle (Krebs cycle) to be oxidized for energy production.
5. Acetyl-CoA (acetyl coenzyme A) pathway: The acetyl-CoA pathway utilizes carbon dioxide as a carbon source and often, hydrogen as an electron donor to produce acetyl-CoA.
6. Achondrite: A stony meteorite containing no small mineral granules (chondrules).
7. Acidophilic: Growing best in acidic conditions.
8. Adenosine triphosphate (ATP): An organic compound and hydrotrope (a compound that solubilizes hydrophobic compounds in aqueous solutions by means other than micellar solubilization) that provides energy to drive many processes in living cells, such as muscle contraction, nerve impulse propagation, condensate dissolution, and chemical synthesis.
9. Adsorption: The process by which a solid holds molecules of a gas or liquid or solute as a thin film. In other words, adsorption is the adhesion of atoms, ions or molecules from a gas, liquid or dissolved solid to a surface. Holding as a thin film on the outside surface or on internal surfaces within the material.
10. Aeolian: Wind-induced.
11. Aerobes: Microorganisms which grow in the presence of air or require oxygen for growth.
12. Aerosol particles: Particles suspended in air.
13. Aerosolization: The process or act of converting some physical substance into the form of particles small and light enough to be carried on the air, i.e., into an aerosol.
14. Areocentric latitude: The angle between the equatorial plane and a vector connecting a point on the surface and the origin of the coordinate system. Latitudes are positive in the northern hemisphere and negative in the southern hemisphere. Areocentric longitude increases to the east.
15. Aerosol: A suspension of fine solid or liquid particles dispersed in air or gas.
16. Agglutinates: Materials clumped together.
17. Albedo: The proportion of the incident light or radiation that is reflected by a surface of a celestial body.
18. Algae: Simple, non-flowering, and typically aquatic plants of a large group that includes the seaweeds and many single-celled forms that contain chlorophyll but lack true stems, roots, leaves, and vascular tissue.
19. Alluvium: Loose, soil or sediment that has been eroded, reshaped by water in some form, and redeposited in a non-marine setting.
20. ALMA: Atacama Large Millimeter/submillimeter Array.
21. ALS system: Advanced Life Support system.
22. Amazonian epoch: A geologic system and time period on the planet Mars, thought to have begun around 3 billion years ago and continues to the present day, characterized by low rates of meteorite and asteroid impacts and by cold, hyperarid conditions broadly similar to those on Mars today.
23. Amino acids: Organic compounds that contain amino and carboxyl functional groups, along with a side chain specific to each amino acid. Amino acids are biologically important organic compounds composed of amine (—NH_2) and carboxylic acid (—COOH) functional groups, along with a side chain specific to each amino acid. Thus, the key elements of an amino acid are carbon, hydrogen, oxygen, and nitrogen, although other elements are found in the side chains of certain amino acids. Amino acids are natural molecules that form a very large network of molecules, known as proteins, by a process of polymerization; that is, by chemical binding to other molecules. Thus, proteins are simply amino acid polymers.
24. Amor Asteroids: Earth approaching asteroids with orbits that lie between the Earth and Mars.
25. Amorphous: Without a clearly defined shape or form.
26. Amphidromic point, also called tidal node: A geographical location which has zero tidal amplitude for one harmonic constituent of the tide. The tidal range

(the peak-to-peak amplitude, or height difference between high tide and low tide) for that harmonic constituent increases with distance from this point.
27. Amplicon: A piece of DNA or RNA that is the source and/or product of amplification or replication events. It can be formed artificially, using various methods including polymerase chain reactions or ligase chain reactions, or naturally through gene duplication.
28. Amplicon sequencing: A highly targeted approach that enables researchers to analyze genetic variation in specific genomic regions.
29. Amylase: An enzyme, found chiefly in saliva and pancreatic fluid, that converts starch and glycogen into simple sugars.
30. Anabiosis or Cryobiosis: see Cryobiosis.
31. Anadromous fish: Fish born in freshwater who spend most of their lives in saltwater and return to freshwater to spawn (e.g., salmon and some species of sturgeon).
32. Anaerobes: Organisms requiring an absence of free oxygen.
33. Anaerobic: Requiring an absence of free oxygen.
34. Angular momentum (L): The quantity of rotation of a body, which is the product of its mass (m), velocity (v), and radius (r). L = mvr; In physics, angular momentum is the rotational analog of linear momentum. It is an important quantity in physics because it is a conserved quantity—the total angular momentum of a closed system (a physical system that does not allow transfer of matter in or out of the system) remains constant. Angular momentum has both a direction and a magnitude, and both are conserved.
35. Anhydrobiosis: The phenomenon of desiccation tolerance, which involves the ability of an organism to survive the loss of almost all cellular water without sustaining irreversible damage.
36. Anhydrobiotic state: "Life without water", referring to the remarkable ability of some organisms to survive the loss of all, or almost all, water and enter into a state of suspended animation in which their metabolism comes reversibly to a standstill.
37. Anoxia: An absence of oxygen.
38. Anoxic: Oxygen-depleted; oxygen-free.
39. Anoxygenic phototrophic bacteria: A diverse collection of organisms that are defined by their ability to grow using energy from light without generating oxygen.
40. Anthropocentric: Regarding humankind as the central or most important element of existence, especially as opposed to God or animals.
41. Apex predators: Predators that are at the top of their food chain without natural predators.
42. Aphelion: The point in the orbit of a planet, asteroid, or comet at which it is farthest from the Sun.
43. Apoapsis: The point of greatest separation (e.g., in a two-body system, when an object is in an elliptical orbit).
44. Apollo Asteroids: Earth crossing asteroids with semi-major axes larger than 1 AU.
45. A priori: From logical point of view rather than from observation or experience.
46. Aptamers: Oligonucleotide or peptide molecules that bind to a specific target molecule.
47. Araneiform: "Spider" features seen on some spots on the icy crusts of Mars and Europa.
48. Archean Eon: Geologic eon, 4000–2500 million years ago (period in the range 4–2.5 billion years ago).
49. Archetypical: Representing or constituting an original type after which other similar things are patterned.
50. Aragonite: Calcium carbonate.
51. Archaea: Microorganisms which are similar to bacteria in size and simplicity of structure but radically different in molecular organization.
52. Artemia: Brine shrimp; a genus of crustaceans found in salt lakes and the brines of saltworks.
53. Artesian pressure: Natural pressure developed in water-bearing strata lying at an angle producing a constant supply of water with little or no pumping; natural pressure producing a constant supply of water with little or no artificial intervention.
54. Arthropods: Invertebrate animals having an exoskeleton, a segmented body, and paired jointed appendages.
55. Asteroid: A hydrated rock in orbit generally between Mars and Jupiter. Asteroids are leftovers from the formation of our Solar System about 4.6 billion years ago. Sometimes these rocky materials get bounced towards Earth.
56. Astrobiology: A branch of biology which studies the origin, evolution, and distribution of life in the Universe.
57. Astronomical unit (AU): An astronomical unit (AU) is the average distance between Earth and the Sun, defined as exactly 149,597,870,700 meters, or about 150 million kilometers (150×10^6 km) (93 million miles). Astronomical units are usually used to measure distances within our Solar System.
58. Atmospheric Optical Depth (AOD): The coefficient of attenuation of solar radiation by the atmosphere. AOD can be calculated by measuring the intensity of direct solar radiation reaching the Earth.
59. Atolls: Ring-shaped reefs, or chain of islands formed of coral.
60. Autotrophs: Organisms that manufacture their own food from inorganic substances, such as carbon dioxide and ammonia.
61. Autotrophy: The process of making food from inorganic substances, using photosynthesis. By oxidizing sulfides, especially hydrogen sulfide (H_2S), as well as

other reduced substrates such as hydrogen (H_2), iron (Fe^{2+}), or manganese (Mn^{2+}) released from vents, the microbes obtain energy to synthesize organic compounds from carbon dioxide (CO_2) in seawater; this process is known as autotrophy.
62. Autotrophic: Requiring only carbon dioxide or carbonates as a source of carbon and a simple inorganic nitrogen compound for metabolic synthesis of organic molecules (such as glucose).
63. Autotrophic cells: Living thing that can make its own food from simple chemical substances such as carbon dioxide.
64. Avifauna: Birds of a given region, considered as a whole.
65. Axial tilt: (also known as obliquity) is the angle between a celestial body's rotational axis and its orbital axis, or, equivalently, the angle between its equatorial plane and orbital plane.
66. Azides: Compounds containing the group N_3 combined with an element or radical.
67. Bacteria: A type of biological cells constituting a large domain of prokaryotic microorganisms (single-celled organism that lacks a nucleus, and other membrane-bound organelles).
68. Bacteroidetes: A phylum of rod-shaped, Gram-negative bacteria that are commonly found in the environment, including in soil, seawater, and in the GI tract and on the skin of animals. Members of this genus are among the so-called good bacteria, because they produce favorable metabolites, including SCFAs, which have been correlated with reducing inflammation.
69. Band curvature: In spectroscopy, band curvature is the spectral curvature due to the 1-μm absorption feature at around 0.9 μm.
70. Band strength: In spectroscopy, band strength—a proxy for 1-μm band depth—corresponds to the abundance of ferrous iron in a mineral. Band strength is a ratio of 0.75- and 1-μm filters.
71. Band tilt: In spectroscopy, band tilt is the reflectance ratio of 0.9- and 1-μm filters. Band curvature and band tilt are sensitive to mineral chemistry (Fe-rich/Ca-rich). Higher band curvature and lower band tilt are indicative of orthopyroxene, and vice versa for clinopyroxene.
72. Banded iron formations: A type of rock consisting of thin alternating layers of chert (a fine-grained form of silica) and iron oxides, magnetite, and hematite.
73. Barophilic: Relatively high pressure (up to 110 MPa).
74. Barotrauma: Injury to one's body because of changes in barometric or water pressure.
75. Basalt: A dark gray to black dense volcanic rock.
76. Basaltic: Referring to basalt.
77. Bathyal zone: The zone of the sea between the continental shelf and the abyssal zone.
78. Beater beads: An effective method of cell lysis used to disrupt virtually any biological sample by rapidly agitating samples with a lysing matrix, sometimes referred to as grinding media or beads, in a bead beater.
79. Benthic zone: Ecological region at the lowest level of a body of water such as an ocean, lake, or stream, including the sediment surface and some sub-surface layers.
80. Binary system: A system of two astronomical bodies which are close enough that their gravitational attraction causes them to orbit each other around a common center-of-mass (barycenter).
81. Biochemicals: Compounds/substances that contain carbon and are found in living things; e.g., carbohydrates, proteins, lipids (fats), and nucleic acids.
82. Biofilms: Thin but robust layers of viscous secretions adhering to solid surfaces and containing highly organized microbial communities of bacteria and other microorganisms.
83. Biogenic: Produced or brought about by living organisms.
84. Biogenic molecules: Molecules that are present in living organisms, including large macro-molecules such as proteins, polysaccharides, lipids, and nucleic acids (DNA and RNA), as well as small molecules such as primary metabolites, secondary metabolites, and so forth.
85. Bioleaching: The process of extracting metals from ores or waste by using microorganisms to oxidize the metals, producing soluble compounds.
86. Biological: Relating to living organisms.
87. Biomarker: Another name of biosignature.
88. Biome: A large naturally occurring community of flora and fauna occupying a major habitat.
89. Biominerals: Natural composite materials based upon biomolecules (such as proteins) and minerals produced by living organisms via processes known as biomineralization, yielding materials with impressive mechanical properties such as bones, shells, and teeth.
90. Biomolecules: see Biogenic molecules.
91. Biopolymers: Natural polymers produced by the cells of living organisms (e.g., DNA, protein, or cellulose).
92. Biosignature: Any substance, element, molecule, or feature that can be used as evidence (or some clue) for past or present life.
93. Biosphere: The regions of the surface, atmosphere, and hydrosphere of celestial bodies occupied by living organisms.
94. Biotechnology: The branch of biology involving the exploitation of biological processes for industrial and other purposes, especially the genetic manipulation of microorganisms to produce antibiotics, hormones, etc.
95. Biotic methane: Methane (CH_4) formed by biotic processes (e.g., microbial).

96. Biotopes: The region of a habitat associated with a particular ecological community.
97. Bolide: A large meteor which explodes in the atmosphere.
98. Breccia: Lithified sedimentary rock consisting of angular fragments.
99. Bridgmanite: A magnesium-silicate mineral, $MgSiO_3$, the most abundant mineral on Earth, making up around 70% of the lower mantle.
100. Brines: Water strongly impregnated with salt.
101. Brucite: Hydrated magnesium hydroxide.
102. BS: Band strength.
103. Butte: An isolated flat-topped hill.
104. Calcite (chemical formula: $CaCO_3$): A rock-forming mineral. Calcite is the principal constituent of limestone and marble. Calcite is the most stable form of calcium carbonate crystal.
105. CAI: A Calcium–Aluminum-rich Inclusion or Ca–Al-rich inclusion is a submillimeter- to centimeter-sized light-colored calcium- and aluminum-rich inclusion found in carbonaceous chondrite meteorites. CAIs consist of minerals that are among the first solids condensed from the cooling protoplanetary disk. They are thought to have formed as fine-grained condensates from a high temperature (>1300 K) gas that existed in the protoplanetary disk at early stages of Solar System formations. Some of them were probably remelted later resulting in distinct coarser textures. The most common and characteristic minerals in CAIs include anorthite, melilite, perovskite, aluminous spinel, hibonite, calcic pyroxene, and forsterite-rich olivine. As CAIs are the oldest dated solids, this age is commonly used to define the age of the Solar System (4567.30 ± 0.16 Myr).
106. Caldera: A large volcanic crater, especially one formed by a major eruption leading to the collapse of the mouth of the volcano.
107. Cambrian period: Approx. 500 million years ago, when the first complex animals were evolving.
108. Cantaloupe terrain: An organized cellular pattern of noncircular dimples, most likely formed when the top of the icy crust buckled during wholesale overturn. Widespread cantaloupe terrains are unique to the planet Neptune's largest moon Triton.
109. Carbonaceous: Containing carbon or its compounds.
110. Carbonaceous chondrites (C chondrites): A class of chondritic meteorites comprising at least 8 known groups and many ungrouped meteorites. They include some of the most primitive known meteorites. A rich variety of organic and non-organic molecules are found in carbonaceous chondrites.
111. Carbonaceous chondrite meteorites: A class of chondritic meteorites comprising at least 8 known groups and many ungrouped meteorites, representing only a small proportion (4.6%) of meteorite falls.
112. Carbonate compensation depth (CCD): The depth in the oceans below which the rate of supply of calcite (calcium carbonate) lags behind the rate of solvation (i.e., interaction of solvent with dissolved molecules), such that no calcite is preserved. Shells of animals therefore dissolve and carbonate particles may not accumulate in the sediments on the seafloor below this depth.
113. Carrying capacity: The maximum sustainable population and socioeconomic scale that the water environment can support in a specific region for some period of time without obvious adverse effect on the local water environment.
114. Cation: Positively charged ion.
115. Caviar: Pickled eggs of fish or shellfish, eaten as a delicacy.
116. Cellulase: An enzyme that converts cellulose into glucose.
117. Cenozoic era: Era covering the period from 66 million years ago to the present day. Cenozoic era is notorious in Earth's history because this era has seen the extinction of the dinosaurs. However, the Cenozoic era is equally famous in the sense that the genus *Homo* (i.e., the genus that comprises the species *Homo sapiens*, which includes modern humans, as well as several extinct species classified as ancestral to or closely related to modern humans, most notably *Homo erectus*) began to appear on earth subsequent to the extinction of dinosaurs.
118. Cephalopod: An active predatory mollusk of the large class Cephalopoda, such as an octopus or squid.
119. Chaos terrains: Quasi-circular areas of ice disruption on the surface of a celestial body (for example, "chaos terrains" are found on Jupiter's moon *Europa*). There are indications suggesting that ice-water interactions and freeze-out below the surface give rise to the diverse morphologies and topography of chaos terrains.
120. Chaotropes: Disorder-makers; One of two or more solutes in the same solution that can disrupt the hydrogen bonding network between water molecules and reduce the stability of the native state of proteins.
121. Chaotropic activity: An activity that disrupts the structure of, and denatures, macromolecules such as proteins and nucleic acids (e.g., DNA and RNA).
122. Chaotropicity: The entropic disordering of lipid bilayers and other biomacromolecules, caused by substances dissolved in water.
123. Chasmoendolithic: Colonizing fissures and cracks in the rock.

124. Chasmoendolithic hyphae: Long, branching filamentous structured organisms, growing in the interior of rocks, inhabiting fissures and cracks of the rock.
125. Chasmoendolithic habitats: Habitats found in fissures and cracks within rocks.
126. Chemoautotrophs: Organisms that obtain their energy from a chemical reaction (chemotrophs) in which their source of carbon is carbon dioxide (CO_2).
127. Chondritic meteorites: Asteroidal fragments that retain records of the first few million years of Solar System history.
128. Chromatin: A substance within a chromosome consisting of DNA and protein.
129. Chromatography: A technique for the separation of a mixture by passing it in solution or suspension through a medium in which the components move at different rates.
130. Chemical fossil: Another name of Biosignature.
131. Chemiosmosis: also known as Chemiosmotic coupling: The process of moving ions (e.g., protons) to the other side of the membrane, resulting in the generation of an electrochemical gradient; movement of ions across a semipermeable membrane bound structure, down their electrochemical gradient.
132. Chemisorption: An adsorption process that involves a chemical reaction and that implies the formation of a covalent bond between the molecule and one or more atoms on the surface.
133. Chemolithoautotrophic: Obtaining the necessary carbon for metabolic processes from carbon dioxide in the environment.
134. Chemolithotroph: An organism that is able to use inorganic reduced compounds as a source of energy. These microbial communities commonly colonize on and in the vicinity of terrestrial submarine hydrothermal vents.
135. Chemolithotrophic microbial communities: Organisms that are able to use inorganic reduced compounds as a source of energy.
136. Chemolithotrophy: A mode of metabolism in which some organisms, known as chemolithotrophs, are able to use inorganic reduced compounds as a source of energy.
137. Chemosynthesis: The use of energy released by inorganic chemical reactions to produce food.
138. Chemotrophic: Deriving energy from the oxidation of organic (chemo-organotrophic) or inorganic (chemo-lithotrophic) compounds; said of bacteria.
139. Chert: A fine-grained form of silica.
140. Chondrite: Primitive material from the solar nebulae; A stony meteorite containing small mineral granules (chondrules).
141. Chondritic: (of stone) of meteoric origin characterized by chondrules and consisting of some of the oldest solid material in the solar system.
142. Chondritic meteorites: Asteroidal fragments that retain records of the first few million years of Solar System history.
143. Chondrites: The most primitive meteorites; fragments of main-belt asteroids.
144. Chemolithotrophy: A type of metabolism where energy is obtained from the oxidation of inorganic compounds.
145. Chromosome: A threadlike structure of nucleic acids and protein found in the nucleus of most living cells, carrying genetic information in the form of genes.
146. Ciliate: A single-celled animal of a large and diverse phylum distinguished by the possession of short microscopic hair-like vibrating structure found in large numbers on the surface of certain cells, either causing currents in the surrounding fluid, or, in some protozoans and other small organisms, providing propulsion.
147. CI chondrites: or C1 chondrites: A group of rare stony meteorites belonging to the carbonaceous chondrites. Samples have been discovered in France, Canada, India, and Tanzania.
148. Clast: A fragment of particles of rock derived from pre-existing rock through processes of weathering and erosion (detritus).
149. Clathrates: Compounds that form a crystal structure with small cages that trap other substances like methane and ethane. Gas hydrates (methane hydrates) existing in ice-like solid form under suitable combination of pressure and temperature are examples of clathrates.
150. Coacervates: Spherical aggregation of lipid molecules making up a colloidal inclusion which is held together by hydrophobic forces.
151. Coagulation: The process of a liquid changing to a solid or semi-solid state.
152. Coccolithopores: Unicellular, eukaryotic phytoplankton belonging either to the kingdom Protista, according to Robert Whittaker's Five kingdom classification, or clade Hacrobia, according to the newer biological classification system.
153. Cofactor: A non-protein chemical compound or metallic ion that is required for an enzyme's activity as a catalyst. Cofactors can be considered "helper molecules" that assist in biochemical transformations.
154. Condensation phenomena: The process where water vapor becomes liquid. It is the reverse of evaporation, where liquid water becomes a vapor. Condensation happens one of two ways: Either the air is cooled to its dew point or it becomes so saturated with water vapor that it cannot hold any more water.
155. Conglomerate: Lithified sedimentary rock consisting of rounded fragments greater than 2 millimeters (0.08 inch) in diameter.

156. Coagulants: Substances that cause fluids to coagulate, i.e., cause to change to a solid or semi-solid state.
157. Conamara Chaos: A region of chaotic terrain on Europa. It is named after Conamara in Ireland due to its similarly rugged landscape.
158. Coronal mass ejection: An event in which a large cloud of energetic and highly magnetized plasma erupts from the solar corona into space. It causes radio and magnetic disturbances on the Earth.
159. Comet: A chunk of solid body, usually around 1–10 km across and made of ices, dust, and rock originating from the outer Solar System (Uranus–Neptune region and the Kuiper Belt region).
160. Commensalism: A long-term biological interaction in which members of one species gain benefits while those of the other species neither benefit nor are harmed.
161. Condensate clouds: Collection of particles that form by condensation of species in the atmosphere of a celestial body; and may be solid or liquid phase.
162. Condensation: The conversion of a vapor or gas to a liquid.
163. Corioli's force: Deflecting force of the Earth's rotation.
164. Coronal mass ejections: Events in which a large cloud of energetic and highly magnetized plasma erupts from the solar corona into space.
165. Cosmic pluralism: Philosophical thinking that, apart from Earth, the planets of even other stars (exoplanets) might foster life of their own.
166. Cretaceous period: Period in the range from 140 million to 65 million years ago. The Cretaceous period is characterized by the advent of flowering plants and modern insects.
167. Cretaceous–Paleogene (K–Pg) extinction event: also known as the Cretaceous–Tertiary (K–T) extinction: Mass extinction of some three-quarters of the plant and animal species on Earth that occurred over a geologically short period of time, approximately 66 million years ago. This is the famous event which killed the dinosaurs at the end of the Cretaceous period. A widely accepted theory is that a large asteroid or comet about 10–15 kilometers in diameter hit the Earth about 65.5 million years ago, creating a large crater (about 180–200 km in diameter), known as the Chicxulub crater.
168. Cretaceous–Tertiary (K–T) extinction: see Cretaceous–Paleogene (K–Pg) extinction event.
169. Cretaceous/Paleogene (K-Pg) Boundary Deposit (KPBD): A "cocktail" of mass transport deposits and debris widely recognized as resulting from the impact of a large bolide 66 million years ago (Ma) creating the Chicxulub impact crater on the northwestern corner of the Yucatan Peninsula of Mexico.
170. Cryobiology: The branch of biology which deals with the properties of organisms and tissues at low temperatures.
171. Cryobiosis or anabiosis: A metabolic state of life entered by an organism in response to adverse environmental conditions such as desiccation, freezing, and oxygen deficiency. In the cryptobiotic state, all measurable metabolic processes stop, preventing reproduction, development, and repair. When environmental conditions return to being hospitable, the organism will return to its metabolic state of life as it was prior to the cryptobiosis.
172. Cryo-environments/cryogenic environments: Low-temperature environments in which the temperature range is below the point at which permanent gases begin to liquefy.
173. Cryogenic: Relating to very low temperature range; below the point at which permanent gases begin to liquefy.
174. Cryophiles: Organisms that grow best at temperatures close to freezing.
175. Cryophilic: Preferring or thriving at low temperatures.
176. Cryoprotectant: A substance used to protect biological tissue from freezing damage. Arctic and Antarctic insects, fish and amphibians create cryoprotectants in their bodies to minimize freezing damage during cold winter periods.
177. Cryosphere: Those portions of a celestial body's surface where water is in solid form, including sea ice, lake ice, river ice, snow cover, glaciers, ice caps, ice sheets, and frozen ground.
178. Cryostatic pressure: The pressure exerted on soil when freezing occurs.
179. Cryovolcanic eruptions: Icy lava eruptions.
180. Cryovolcano: A type of volcano that erupts volatiles such as water, ammonia, or methane, instead of molten rock. Collectively referred to as cryomagma, cryolava or ice-volcanic melt, these substances are usually liquids and can form plumes, but can also be in vapor form.
181. Cryovolcanism: The act of erupting volatiles as in a cryovolcano.
182. Cryptobiosis: A physiological state in which metabolic activity is reduced to an undetectable level without disappearing altogether. It is known in certain plant and animal groups adapted to survive periods of extremely dry conditions.
183. Cryptoendolithic: One of the three subclasses in which "Endolithic" microorganisms are classified. Cryptoendolithic microorganisms are those able to colonize the empty spaces or pores inside a rock with the connotation of being hidden.
184. Crystal: A solid material whose constituents are arranged in a highly ordered microscopic structure, forming a crystal lattice that extends in all directions.

185. Crystalline: Having the structure and form of a crystal.
186. Cyanobacteria: An ancient but advanced form of life. Cyanobacteria are microbes that live primarily in seawater. Cyanobacteria is the only microorganisms capable of oxygenic photosynthesis, and oxygenic photosynthesis is the largest source of O_2 in the atmosphere today.
187. Cyclomorphosis: Occurrence of cyclic or seasonal changes in the phenotype of an organism through successive generations.
188. Cyclostratigraphy: A subdiscipline of stratigraphy that studies astronomically forced climate cycles within sedimentary successions.
189. Cyst: A thin-walled hollow organ or cavity in an animal or plant, containing a liquid secretion; a sac, vesicle, or bladder.
190. Cytoplasm: A thick solution that fills each cell and is enclosed by the cell membrane. It is mainly composed of water, salts, and proteins. All of the organelles in eukaryotic cells, such as the nucleus, endoplasmic reticulum, and mitochondria, are located in the cytoplasm.
191. Dalton: A unit used in expressing the molecular weight of proteins, equivalent to atomic mass unit.
192. Decametric radio waves: Radio waves having wavelengths tens of meters long.
193. Deep Space Network (DSN): NASA's DSN is a worldwide network of U.S. spacecraft communication facilities, located in the United States, Spain, and Australia, that supports NASA's interplanetary spacecraft missions. The DSN also provides radar and radio astronomy observations that improve our understanding of the Solar System and the larger Universe.
194. Degassing: Removal of dissolved gases.
195. Deglaciation: Melting and retreat of glaciers and ice sheets; the disappearance of ice from a previously glaciated region.
196. Deliquesce effect: The habit of becoming liquid by absorbing moisture from the air, as certain salts.
197. Demersal fish: Fish that lives on or near the bottom.
198. Desiccation: Removal of moisture.
199. Detrital sediments: Sedimentary rocks that form from transported solid material.
200. Detritus (in geology): Particles of rock derived from pre-existing rock through processes of weathering and erosion. Detrital particles can consist of lithic fragments, or of monomineralic fragments.
201. Detritus (in biology): Dead organisms and fecal material, typically hosting communities of microorganisms that colonize and decompose them.
202. Deuterium (or hydrogen-2, also known as heavy hydrogen): One of two stable isotopes of hydrogen (the other being protium, or hydrogen-1). The nucleus of a deuterium atom, called a deuteron, contains one proton and one neutron, whereas the far more common protium has no neutrons in the nucleus.
203. Devolatilization: Removal of volatile substances from a solid.
204. Devonian-Carboniferous (D-C) transition: A time (359 Ma) of extreme climate and faunal change in Earth's history, associated with the end-Devonian biodiversity crisis, characterized by transgressive/regressive cycles, which culminated in the onset of widespread ocean anoxia (the Hangenberg Black Shale event) and a remarkable sea-level fall close to the Devonian-Carboniferous boundary.
205. Diagenesis: Physical and chemical processes that affect sedimentary materials after deposition and before metamorphism and between deposition and weathering. The effects of diagenetic processes on rock properties such as porosity and the degree of lithification are progressive.
206. Diagenetic minerals: Minerals formed by water-rock interactions.
207. Diagenetically altered: Changes and alterations that have taken place on biological material.
208. Diapause: A period of suspended development in an insect, other invertebrate, or mammal embryo, especially during unfavorable environmental conditions.
209. Diapir: Material buoyantly rising as either plumes of warm, pure ice etc. on to the surface of a celestial body. Diapirs may take the shape of domes, waves, mushrooms, teardrops, or dikes. The surface of Jupiter's moon *Europa* is famous for the presence of diapirs.
210. Diapirism: A major form of supply of basaltic melts into the crust of a celestial body.
211. Diatoms: A major group of algae, specifically microalgae, having a cell wall of silica, and found in the oceans, waterways, and soils of the world. Diatoms are a type of plankton called phytoplankton, the most common of the plankton types. Diatoms are unicellular: they occur either as solitary cells or in colonies, which can take the shape of ribbons, fans, zigzags, or stars.
212. Dielectric constant (in physics): A quantity measuring the ability of a substance to store electrical energy in an electric field.
213. Dike: (In geology) A large slab of rock that cuts through another type of rock; (In hydrology) A barrier used to regulate or hold back water from a river, lake, or even the ocean.
214. Dilational bands: A tabular zone of new crustal material that intruded between the progressively dilating walls of a tension fracture.
215. Diogenites: Igneous rocks, having solidified slowly enough deep within the asteroid Vesta's crust to form

crystals which are larger than in the eucrites. These crystals are primarily magnesium-rich orthopyroxene, with small amounts of plagioclase and olivine.
216. Dipole (in chemistry): The separation of electric charges within a molecule between two covalently bonded atoms.
217. Dipole field: An electric field between two charges of the same but opposite strength.
218. Dipole moment: A measurement of the separation of two oppositely charged charges.
219. DNA: The hereditary material in humans and almost all other organisms. Nearly every cell in a person's body has the same DNA. The information in DNA is stored as a code made up of four chemical bases: adenine (A), guanine (G), cytosine (C), and thymine (T).
220. Dolostone: Sedimentary rock whose primary component is Dolomite (Calcium magnesium carbonate: $CaMg(CO_3)_2$).
221. Dolomite (calcium magnesium carbonate), whose chemical formula is $CaMg(CO_3)_2$: A common rock-forming mineral. Dolomite is the primary component of the sedimentary rock known as dolostone and the metamorphic rock known as dolomitic marble.
222. Dolomitic marble: Metamorphic form of Dolostone.
223. Dosimetry: Determination and measurement of the amount or dosage of radiation absorbed by a substance or living organism by means of a dosimeter.
224. Dropstones: Isolated fragments of rock—ranging in size from small pebbles to boulders—found within finer-grained water-deposited sedimentary rocks.
225. Dry ice: Solid Carbon Dioxide.
226. Dust Devils (DDs): Whirlwinds that result from solar warming of the ground, prompting convective air to rise into the atmosphere.
227. Earth-like planet: A planet with about the same size and mass as Earth, with oceans and continents, a thin N_2-CO_2-O_2 atmosphere, and a radiation environment similar to that of Earth's.
228. East: (of a celestial body): The direction of increasing longitude.
229. Eccentric orbit: Elliptical orbit.
230. Eccentricity: (of a celestial body's orbit): A measure of the amount by which its orbit deviates from a circle. Eccentricity (e) of an ellipse is, most simply, the ratio of the distance c between the center of the ellipse and each focus to the length of the semimajor axis a. That is, $e = \dfrac{c}{a}$. When the eccentricity is 0 the foci coincide with the center point and the figure is a circle. In a different description, eccentricity is found by dividing the distance between the focal points of the ellipse by the length of the major axis.
231. Ecdysozoa: A group of protostome animals, including Arthropoda Nematoda, and several smaller phyla.
232. Ecdysozoan animals: A morphologically heterogeneous group of animals which have a cuticle and grow by molting.
233. Eclogite: A metamorphic rock containing granular minerals, typically garnet (a precious stone consisting of a deep red vitreous silicate mineral) and pyroxene (a group of important rock-forming inosilicate minerals found in many igneous and metamorphic rocks).
234. Edgeworth–Kuiper belt: see Kuiper belt.
235. Efflorescence: Migration of a salt to the surface of a porous material, where it forms a coating. The essential process involves the dissolving of an internally held salt in water, or occasionally in another solvent.
236. Electronegativity: The tendency of an atom participating in a covalent bond to attract the bonding electrons. The tendency of an atom in a molecule to attract the shared pair of electrons towards itself. It basically indicates the net result of the tendencies of atoms in different elements to attract the bond-forming electron pairs.
237. Emergence bottleneck: Low probability for the emergence of life.
238. Emulsifiers: Substances that stabilize an emulsion, in particular additives used to stabilize processed foods.
239. Endemic: (of a plant or animal) native and restricted to a certain place.
240. Endogenic: Formed or occurring beneath the surface of a celestial body (e.g., Earth, Europa).
241. Endolith: An organism (archaeon, bacterium, fungus, lichen, algae, or amoeba) that lives inside rock, coral, animal shells, or in the pores between mineral grains of a rock.
242. Endolithic environment: The pore space in rocks.
243. Endolithic microbes: Organisms such as archaeon, bacterium, fungus, lichen, algae, or amoeba that are able to acquire the necessary resources for growth in the inner part of a rock, mineral, coral, animal shells, or in the pores between mineral grains of a rock.
244. Endorheic: (of a lake) having no outlet besides evaporation.
245. EnEx-IceMole: A novel maneuverable subsurface ice melting probe for clean sampling and in situ analysis of ice and sub-glacial liquids.
246. Enstatite meteorites: A diverse group of strange rocks that contain little or no oxidized iron—a rare occurrence in the Solar System.
247. Encystment: The process of becoming enclosed by a cyst.
248. Engrailed gene: A "selector" gene that controls the expression of other genes to confer a "posterior identity" on groups of cells that are related to each other by lineage.
249. Entropy: A thermodynamic quantity representing the unavailability of a system's thermal energy for

conversion into mechanical work, often interpreted as the degree of disorder or randomness in the system. A terminology indicating lack of order or predictability; gradual decline into disorder. The second law of thermodynamics says that entropy always increases with time.
250. Enzyme: A substance produced by a living organism which acts as a catalyst to bring about a specific biochemical reaction. Enzymes are proteins that act as biological catalysts that accelerate chemical reactions.
251. Eocene Epoch: The interval lasting from 56 to 33.9 million years ago.
252. Eolian: Relating to or arising from the action of the wind.
253. Epithermal: Having energy above that of thermal agitation.
254. Epilithic organisms: Organisms growing on the surface of rock.
255. E-ring: Planet Saturn's most expansive ring.
256. Escape velocity: In astronomy and space exploration, escape velocity is the velocity that is sufficient for a body to escape from a gravitational center of attraction without undergoing any further acceleration. To escape from a celestial body, an object must exceed the escape velocity v, given by the expression, $v = \sqrt{\dfrac{2Gm}{R}}$, where G is the universal constant of gravitation; m is the mass of the celestial body; and R is nominally the distance from the center of the celestial body to the top of the atmosphere, if there is any. From the above expression, it is seen that escape velocity decreases with altitude and is equal to the square root of 2 (or about 1.414) times the velocity necessary to maintain a circular orbit at the same altitude. At the surface of the Earth, if atmospheric resistance could be disregarded, escape velocity would be about 11.2 km (6.96 miles) per second. The velocity of escape from the less massive Moon is about 2.4 km per second at its surface. A planet (or satellite) cannot long retain an atmosphere if the planet's escape velocity is low enough to be near the average velocity of the gas molecules making up the atmosphere.
257. Estuary: Tidal mouth of a large river, where the tide meets the stream.
258. Euendolithic habitats: Habitats formed by active boring/penetration by micro-organisms.
259. Eustatic sea level: The distance from the center of the earth to the sea surface. Eustatic (global) sea level refers to the volume of Earth's oceans. This is not a physical level but instead represents the sea level if all the water in the oceans were contained in a single basin.
260. Eutardigrades: A class of tardigrades without lateral appendices. Primarily freshwater bound, some species have secondarily gained the ability to live in marine environments. By cryptobiosis many species are able to live temporarily in very dry environments. More than 700 species have been described.
261. Euxinia: Conditions occurring when water is both anoxic (no oxygen) and sulfidic (raised level of free hydrogen sulfide).
262. Eucrites: Eucrites are achondritic stony meteorites, many of which originate from the surface of the asteroid 4 Vesta and as such are part of the HED meteorite clan. They are the most common achondrite group with well over 100 distinct finds at present.
263. Euryhaline: (of an aquatic organism) able to tolerate a wide range of salinity.
264. Eutectic system: A system of a homogeneous mixture of substances that either melts or solidifies at a particular given temperature that is lower than the melting point of any of the mixture of any of the constituent elements. This particular temperature is known as the eutectic point.
265. Eutectic temperature: The temperature at which a particular eutectic mixture (a homogeneous mixture of substances that melts or solidifies at a single temperature that is lower than the melting point of any of the constituents) freezes or melts.
266. Eutectic-point: The temperature at which a particular eutectic (homogeneous) mixture of substances freezes or melts; this single temperature is lower than the freezing/melting point of any of the constituents.
267. Evaporation: The process where liquid water becomes a vapor.
268. Evaporite: A water-soluble sedimentary mineral deposit that results from concentration and crystallization by evaporation from an aqueous solution.
269. Exfoliation: The process of removing dead skin cells from the outer layer of the skin.
270. Exogenic: Formed or occurring on the surface of a celestial body including the Earth.
271. Exo-planets: Planets which orbit stars outside the Solar System.
272. Exosphere: The outermost region of a planet's or moon's atmosphere.
273. Exothermic reactions: Reactions accompanied by the release of heat.
274. Extremophilic bacteria: Bacteria requiring severe conditions for growth.
275. Extremotolerant bacteria: Bacteria that are tolerant to extreme conditions.
276. Extremozymes: Enzymes, often created by archaea, which are known prokaryotic extremophiles, that can function under extreme environments.
277. Fahrenheit: Conversion of temperature (T) between degrees Fahrenheit (°F) and degrees Celsius (°C) is given by the expression: T(°F) = {T(°C) × 1.8} + 32.

278. False positive: An error in data reporting in which a test result improperly indicates presence of a condition, when in reality it is not present; i.e., the positive test result is false.
279. False negative: An error in which a test result improperly indicates no presence of a condition (the result is negative), when in reality it is present; i.e., the negative test result is false.
280. Fecundity: Fertility, the natural capability to produce offspring, measured by the number of eggs, seed set, etc.
281. Felsic rocks: Rocks which are rich in silicon, sodium, potassium, calcium, aluminum, and lesser amounts of iron and magnesium.
282. Feldspar: An abundant rock-forming mineral typically occurring as colorless or pale-colored crystals and consisting of aluminosilicates of potassium, sodium, and calcium.
283. Fjord: a long, narrow, deep inlet of the sea between high cliffs, as in Norway.
284. Flippers: Broad flat limb without fingers, used for swimming by various sea animals such as seals, whales, and turtles.
285. Fluvial: River-related.
286. Foliated: Composed of layers of sheet-like planar structures.
287. Foraminifera: Forams for short, single-celled organisms, characterized by an external shell for protection, that live in the open ocean, along the coasts and in estuaries, either floating in the water column (planktonic) or living on the seafloor (benthic).
288. Fossa: A shallow depression or hollow (e.g., Virgil Fossae troughs on Pluto's surface).
289. Fractal structure: Infinitely complex patterns that are self-similar across different scales, created by repeating a simple process over and over in an ongoing feedback loop.
290. Frasnian-Famennian (F-F) global event: One of the five largest biotic crises of the Phanerozoic period.
291. Frustule: The silicified cell wall of a diatom, consisting of two valves or overlapping halves.
292. Fullerene: Soccer-ball shaped hollow cage/shell made of carbon atoms.
293. Fumarole: A vent in the surface of the Earth or other rocky planet from which hot gases and vapors (fumes) are emitted, without any accompanying liquids or solids.
294. Gaian regulation: The emergence of life's abilities to modify its environment and regulate initially abiotic positive feedback mechanisms rather than abiotic negative feedback.
295. Galilean moons: The four largest moons of planet Jupiter (named the *Galilean moons* in honor of the Italian astronomer Galileo, who observed them in 1610). The Galilean Moons are *Io, Europa, Ganymede*, and *Callisto*.
296. Gamete: A mature haploid male or female germ cell which is able to unite with another of the opposite sex in sexual reproduction to form a zygote.
297. Gardening: Continuous bombardment of celestial bodies by minor and larger impacts has led to a mechanical reworking commonly known as "gardening". Impact gardening is the process by which impact events stir the outermost crusts of moons and other celestial objects with no atmospheres. In the particular case of the Earth's Moon, this is more often known as lunar gardening.
298. Gas giants: Giant, low-density, planets composed mainly of hydrogen, helium, methane, and ammonia in either gaseous or liquid state—Jupiter, Saturn, Uranus, and Neptune.
299. Geological: Relating to the study of a celestial body's physical structure and substance.
300. Geology: The science which deals with the physical structure and substance of a celestial body such as the earth, their history, and the processes which act on them.
301. Geomorphological: Of or relating to the form or surface features of any celestial body.
302. Geomorphology: Scientific study of the origin and evolution of topographic and bathymetric features created by physical, chemical or biological processes operating at or near the Earth's surface.
303. Genome: An organism's complete set of DNAs, including all of its genes—the units of heredity. Each genome of an organism contains all of the information needed to build and maintain that organism. The human genome contains about 3 billion base pairs that spell out the instructions for making and maintaining a human being.
304. Geometric albedo: The proportion of the incident light or radiation that is reflected by the surface of a celestial body.
305. Geysers: Springs characterized by intermittent discharge of water ejected turbulently and accompanied by steam, as in hydrothermal vents.
306. Glacial isostasy: A concept concerned with a planet's response to the changing surface loads of ice and water during the waxing and waning of large ice sheets.
307. Gneiss (plural noun: gneisses): A metamorphic rock with a banded or foliated structure, typically coarse-grained and consisting mainly of feldspar, quartz, and mica.
308. Goethite: A dark or yellowish-brown mineral consisting of hydrated iron oxide, occurring typically as masses of fibrous crystals.
309. Graben: An elongated block of a celestial body's crust lying between two faults and displaced downwards

relative to the blocks on either side, as in a rift valley. The maximum depth of faulting can be estimated from the width of a graben.
310. Gram-positive bacteria: Bacteria that give a positive result in the Gram stain test, which is traditionally used to quickly classify bacteria into two broad categories according to their type of cell wall.
311. Gravity Assist: In orbital mechanics and aerospace engineering, gravity assist maneuver is the use of the relative movement and gravity of an astronomical object to alter the path and speed of a spacecraft, typically to save propellant and reduce expense. This mechanism increased *New Horizons* spacecraft's speed by 14,000 km/h (9000 mph), accelerating the probe to a velocity of 83,000 km/h (51,000 mph) relative to the Sun and shortening its voyage to the planet Pluto by three years.
312. Great Oxidation Event, sometimes also called the Great Oxygenation Event: A time period when the Earth's atmosphere and the shallow ocean first experienced a rise in oxygen, approximately 2.4–2.0 Ga during the Paleoproterozoic era.
313. Greenhouse effect: The trapping of the Sun's warmth in a planet's lower atmosphere, due to the greater transparency of the atmosphere to visible radiation from the Sun than to infrared radiation emitted from the planet's surface.
314. Gully: A ravine (deep, narrow gorge with steep sides) formed by the action of water.
315. Habitable Zone (HZ): The zone around a star where a rocky planet with a thin atmosphere, heated by its star, may have liquid water on its surface.
316. Hadean Eon: A geologic eon of the Earth, which began with the formation of the Earth about 4.6 billion years ago and ended 4 billion years ago (period in the range 4.6 to 4 billion years ago). Hadean Eon is named after the mythological underworld ruler—Hades—because during most of the Hadean period (4.6 to 4 billion years ago) the surface of the Earth must have been like our image of Hell.
317. Halobacterium (plural halobacteria): A genus in the family *Halobacteriaceae*. The genus Halobacterium consists of several species of Archaea with an aerobic metabolism which requires an environment with a high concentration of salt; many of their proteins will not function in low-salt environments.
318. Halophiles: Organisms, especially microorganisms, that grow in or can tolerate saline conditions.
319. Halophilic: Capable of flourishing in a salty environment.
320. Halotolerant: Having ability to grow at salt concentrations higher than those required for growth.
321. Haze: Photochemically generated involatile particles in Titan's atmosphere.
322. Heavy metals: Metals with relatively high densities, atomic weights, or atomic numbers (e.g., As, Cd, Pb, Cr, Cu, Hg, and Ni).
323. HED (Howardite–Eucrite–Diogenite) meteorites: A clan of achondrite meteorites. A type of regolith breccia (rock consisting of angular fragments of stones cemented by finer calcareous material) formed by impact-induced mixing on the surface of the asteroid Vesta.
324. Heliocentric: Sun centric: measured from or considered in relation to the center of the Sun.
325. Hematite: A common iron oxide compound with the formula, Fe_2O_3 and is widely found in rocks and soils.
326. HERCULES: A computer software, named Highly Eccentric Rotating Concentric U (potential) Layers Equilibrium Structure, that mathematically solves for the equilibrium structure of planets as a series of overlapping constant-density spheroids, and a smoothed particle hydrodynamics software.
327. Hesperian Eon: An eon beginning 3.7 Ga—a time of ephemeral (lasting a very short time) lakes, resulting in the widespread deposition of sulfate and chloride salts observed today on the Martian surface.
328. Heterogeneous: Diverse in character or content.
329. Heterogeneous reaction: Any of a class of chemical reactions in which the reactants are components of two or more phases (solid and gas, solid and liquid, two immiscible liquids) or in which one or more reactants undergo chemical change at an interface, e.g., on the surface of a solid catalyst.
330. Heterotardigrades: A class of tardigrades that have cephalic appendages and legs with four separate but similar digits or claws on each.
331. Heterotroph: An organism that cannot produce its own food, instead taking nutrition from other sources of organic carbon, mainly plant or animal matter. In the food chain, heterotrophs are primary, secondary, and tertiary consumers, but not producers.
332. Heterotrophic: Requiring complex organic compounds of nitrogen and carbon (such as that obtained from plant or animal matter) for metabolic synthesis.
333. Heterotrophy: The quality or condition of being capable of utilizing only organic materials as a source of food.
334. High Mass Organic Cations (HMOC): Organic species with an increasing number of carbon atoms ($C_7 - C_{15}$); Organic species representing molecules with 7–15 carbon atoms.
335. Holocene: The current geological epoch, which began approximately 11,650 years before present, after the last glacial period.
336. Holomictic: Having a uniform temperature and density from top to bottom at a specific time during the year, which allows the water body to completely mix.

337. Homo: The genus that comprises the species *Homo sapiens*, which includes modern humans, as well as several extinct species classified as ancestral to or closely related to modern humans, most notably *Homo erectus*.
338. Homogenization: The process of making things uniform or similar.
339. Homolog: A gene related to a second gene by descent from a common ancestral DNA sequence. The term, homolog, may apply to the relationship between genes separated by the event of speciation or to the relationship between genes separated by the event of genetic duplication.
340. Hoofed animals: Those animals having horny covering encasing their foot, as the ox and horse.
341. Hot Jupiters: A class of gas giant exoplanets that are inferred to be physically similar to Jupiter but that have very short orbital periods (P < 10 days). The close proximity to their stars and high surface-atmosphere temperatures resulted in their informal name "hot Jupiters".
342. Howardites: Howardites are achondritic stony meteorites that originate from the surface of the asteroid 4 Vesta, and as such are part of the HED meteorite clan. There are about 200 distinct members known.
343. Human Genome Project: An international scientific research project with the goal of determining the base pairs that make up human DNA, and of identifying, mapping, and sequencing all of the genes of the human genome from both a physical and a functional standpoint.
344. Human microbiome: The genes of the 10–100 trillion symbiotic microbial cells harbored by each human being, primarily bacteria in the gut.
345. Humic substances: Substances relating to or consisting of organic component formed by the decomposition of leaves and other plant material by soil microorganisms.
346. Hyaloclastite: A volcanoclastic accumulation or breccia consisting of glass fragments formed by quench fragmentation of lava flow surfaces during submarine or subglacial extrusion.
347. Hydrogenotrophic: That converting hydrogen to other compounds as part of their metabolism.
348. Hydrophobic: Tending to repel or fail to mix with water.
349. Hydrophobic effect: The observed tendency of nonpolar substances to aggregate in an aqueous solution and exclude water molecules.
350. Hydrolase: A class of enzyme that commonly perform as biochemical catalysts that use water to break a chemical bond, which typically results in dividing a larger molecule into smaller molecules.
351. Hydrolysis: Chemical breakdown of a compound due to reaction with water.
352. Hydrophilic: Having a tendency to mix with, dissolve in, or be wetted by water (e.g., hydrophilic amino acids).
353. Hydrosphere: The entire water layer.
354. Hyperthermals: Extreme global warming events.
355. Hyperthermophiles: Particularly extreme thermophiles for which the optimal temperatures are above 80°C.
356. Hyper-thermophilic: "Superheat-loving" (from 60°C upwards) bacteria and archaea, which are found within high-temperature environments, representing the upper temperature border of life.
357. Hypolith: In Arctic and Antarctic ecology, a hypolith is a photosynthetic organism, and an extremophile, that lives underneath rocks in climatically extreme deserts such as Cornwallis Island and Devon Island in the Canadian high Arctic. The community itself is the hypolithon.
358. Hypolithic habitats: Habitats found on underside of rocks.
359. Hypoxic: Not having enough oxygen in the tissues to sustain bodily functions.
360. Ice Age: A period of colder global temperatures and recurring glacial expansion capable of lasting hundreds of years. There have been at least five significant ice ages in Earth's history, with approximately a dozen epochs of glacial expansion occurring in the past 1 million years. The most recent Ice Age occurred about 11,700 years ago. The global temperature of the most recent Ice Age was only about 5°C below the current average.
361. Igneous rock: Rock having solidified from lava or magma.
362. Igneous differentiation, or magmatic differentiation: An umbrella term used for the various processes by which magmas undergo *bulk chemical change* during the partial melting process, cooling, emplacement, or eruption.
363. Ionized gas: Atoms or molecules which have one or more orbital electrons stripped, thus attaining positive electrical charge or, rarely, an extra electron attached, thus attaining negative electrical charge.
364. Ionosphere: An upper-atmospheric layer of charged particles; a region of electrically charged particles in Earth's upper atmosphere.
365. Impact velocity: Essentially the "effective velocity" at which an object is traveling at a moment in which it impacts another object.
366. Insertion/deletion polymorphism (Indel): A type of genetic variation in which a specific nucleotide sequence is present (insertion) or absent (deletion). Indels are widely spread across the genome.

367. Insolation: The incident solar energy flux (i.e., the amount of solar radiation reaching a given area) on a celestial body.
368. Interferometric reflectometry: The strategy of analyzing both distant radio emissions and their echoes to glean useful information (e.g., thickness of Europa's ice shell).
369. Isolates: Culture of microorganisms isolated for study.
370. Isostasy: A concept that describes the response of a planet to a change in surface load.
371. Isotope: Any of two or more forms of a chemical element, having the same number of protons in the nucleus (i.e., the same atomic number), but having different numbers of neutrons in the nucleus.
372. Jarosite: A basic hydrous sulfate of potassium and ferric iron (Fe-III) with a chemical formula of $KFe_3(SO_4)_2(OH)_6$.
373. Jovian Early Bombardment (JEB): The formation of Jupiter resulted in the triggering of a sudden spike in the flux of impactors in the early history of the Solar System. This event is named the Jovian Early Bombardment (JEB in short).
374. Jovian planets: Jupiter, Saturn, Uranus, and Neptune.
375. Kerogens: Solid, insoluble organic matter in sedimentary rocks.
376. Keystone species: A species which has a disproportionately large effect on its natural environment relative to its abundance.
377. Kimberlite: An ultrabasic rock, which means it does not contain any quartz or feldspar, the two most common rock-forming minerals. Kimberlite is composed of at least 35% olivine, together with other minerals such as mica, serpentine, and calcite.
378. Kirkwood gaps: Gaps in the distribution of asteroid orbital periods; these gaps are caused by the gravitational interaction between Jupiter and asteroids with these orbital periods. Asteroids have been ejected from these almost empty lanes by repeated perturbations.
379. Knot: A unit of speed equal to one nautical mile per hour, exactly 1.852 km/h.
380. Komatiite: A type of ultramafic mantle-derived volcanic rock defined as having crystallized from a lava of at least 18 wt% MgO.
381. Kosmotrope: Order-maker; stabilizing ion.
382. Kuiper Belt: A region of leftovers from the Solar System's early history. Occasionally called the Edgeworth–Kuiper belt, is a circumstellar disc in the outer Solar System, extending from the orbit of Neptune to approximately 50 AU from the Sun. It is similar to the asteroid belt, but is far larger – 20 times as wide and 20–200 times as massive.
383. Labyrinth: A complicated irregular network.
384. Lacustrine: Relating to or associated with lakes.
385. Late Heavy Bombardment; or Lunar cataclysm: A hypothesized bombardment event thought to have occurred approximately 4.1 to 3.8 billion years (Ga) ago.
386. Lava: Molten rock that is ejected by a volcano during an eruption.
387. Lava flow: An outburst of lava that moves during a non-explosive effusive eruption.
388. Lava tube/pyroduct: A tunnel under a celestial body's surface, formed by an intense flow of molten rock during a volcanic explosion. Lava tubes manifest themselves as natural subsurface caves or caverns. Lava tubes, believed to have potential for human settlement, have been spotted on Earth's Moon and the planet Mars.
389. Lenticulae: Quasi-elliptical features called pits, domes, spots, and small chaos; a kind of eruption (dark spots) visible on Europa's icy surface. Their similar sizes and spacing suggest that Europa's icy shell may be churning away like a lava lamp, with warmer ice moving upward from the bottom of the ice shell while colder ice near the surface sinks downward.
390. Leucocytes: A type of blood cell that is made in the bone marrow and found in the blood and lymph tissue. Leukocytes are part of the body's immune system. They help the body fight infection and other diseases.
391. Lichen: A very small grey or yellow plant that spreads over the surface of rocks, walls and trees and does not have any flowers.
392. Ligand: A molecule that binds to another (usually larger) molecule.
393. Limnic: Relating to bodies of water with low salt concentration, such as lakes and ponds.
394. Limno-terrestrial: Being or inhabiting a moist terrestrial environment that is subject to periods of both immersion and desiccation.
395. Lineaments: Linear features on a planet's surface, such as a fault.
396. Lipase: A type of protein, made by the pancreas, which helps the body to digest fats.
397. Lipid: A substance of biological origin that is soluble in nonpolar solvents (e.g., liquid hydrocarbons).
398. Liquid metallic hydrogen (LMH): A fundamental system in condensed matter sciences and the main constituent of gas giant planets.
399. Lithic: relating to stone/rock.
400. Lithify: Transformed into stone.
401. Lithification: Conversion to rock.
402. Lithoautotrophs: A type of lithotrophs (a diverse group of organisms using an inorganic substrate to obtain reducing equivalents for use in biosynthesis or energy conservation via aerobic or anaerobic respiration).

403. Lithosphere: The rocky outer part of the Earth. It is made up of the brittle crust and the top part of the upper mantle. The lithosphere is the coolest and most rigid part of the Earth.
404. Lithopanspermia: Movement of rocks that contain life from one planetary surface to another; life's launch into space from the surface of a planet on impact ejecta and subsequent arrival on a new body. Lithopanspermia postulates that meteors could be the transfer vehicles for life through space.
405. Lithophytic: Capable of, or habituated to, growing on rock.
406. Lineae: Complex pattern of long, linear fracture-like markings.
407. Locked nucleic acid (LNA): A structurally rigid modification that increases the binding affinity of a modified-oligonucleotide.
408. Love numbers: Love numbers h, k, and l are dimensionless parameters that measure the rigidity of a planetary body and the susceptibility of its shape to change in response to a tidal potential.
409. LUCA: The Last Universal Common Ancestor or Last Universal Cellular Ancestor, also called the Last Universal Ancestor (LUA), is the most recent common ancestor of all current life on Earth. LUCA was the common ancestor of bacteria and archaea.
410. Lunar synodic period: The time required for the Moon to return to the same or approximately the same position relative to the Sun as seen by an observer on the Earth.
411. Lysis: The disintegration of a cell by rupture of the cell wall or membrane.
412. Lysozyme: An enzyme which catalyzes the destruction of the cell walls of certain bacteria, and occurs notably in tears and egg white.
413. Macromolecules: Very large molecules, such as proteins, composed of thousands of covalently bonded atoms.
414. Mafic: Relating to or denoting a group of dark-colored, mainly ferromagnesian minerals such as pyroxene and olivine.
415. Mafic Rock/Mineral: A mafic mineral or rock is silicate mineral or igneous rock rich in magnesium and iron. Most mafic minerals are dark in color, and common rock-forming mafic minerals include mainly ferromagnesian minerals such as pyroxene and olivine, and amphibole, and biotite. Common mafic rocks include basalt, diabase, and gabbro.
416. Magma: An extremely hot liquid and semi-liquid rock located under Earth's surface. This magma can push through holes or cracks in the crust, causing a volcanic eruption. When magma flows or erupts onto Earth's surface, it is called lava. Like solid rock, magma is a mixture of minerals.
417. Magmatism: The motion or activity of magma.
418. Magmatic differentiation or igneous differentiation: An umbrella term used for the various processes by which magmas undergo *bulk chemical change* during the partial melting process, cooling, emplacement, or eruption.
419. Magnesiowüstite: A mineral composing 20% of the lower mantle. It is a cubic phase of composition (Mg,Fe)O.
420. Magnetohydrodynamics (MHD): Also called magneto-fluid dynamics or hydromagnetics): Study of the magnetic properties and behavior of electrically conducting fluids.
421. Main Asteroid Belt (MAB): The asteroid belt located between the orbits of Mars and Jupiter (a relatively narrow annulus ~2.1–3.3 AU from the Sun).
422. Main sequence stars: Stars that fuse hydrogen atoms to form helium atoms in their cores. About 90 percent of the stars in the Universe, including the Sun, are main sequence stars.
423. Mangalyaan: India's Mars Orbiter Mission spacecraft.
424. Mare: Large, dark, basaltic plains on Earth's Moon, formed by ancient asteroid impacts. They were dubbed maria (Latin for "seas"), by early astronomers who mistook them for actual seas.
425. Mass spectrometer: An instrument that separates atoms on the basis of their mass differences. Typically, mass spectrometers can be used to identify unknown compounds via molecular weight determination, to quantify known compounds, and to determine structure and chemical properties of molecules.
426. Mass wasting: "Mass wasting", also known as slope movement or mass movement, is the movement of rock and soil down slope under the influence of gravity. Rock falls, slumps, and debris flows are all examples of mass wasting. Often lubricated by rainfall or agitated by seismic activity, these events may occur very rapidly and move as a flow. "Mass wasting" differs from other processes of erosion in that the debris transported by mass wasting is not entrained in a moving medium, such as water, wind, or ice.
427. Massif: A compact group of mountains.
428. Medusae Fossae Formation: A large geological formation of probable volcanic origin on the planet Mars. It is named for the *Medusa* of Greek mythology. "Fossae" is Latin for "trenches".
429. Megafauna: The large or giant animals of an area, habitat, or geological period.
430. Melt inclusion: A small parcel or "blobs" of melt(s) that is entrapped by crystals growing in magma and eventually forming igneous rocks. Melt inclusions tend to be microscopic in size and can be analyzed for volatile contents that are used to interpret trapping pressures of the melt at depth.

431. Meromictic lake: A lake which has layers of water that do not intermix.
432. Mesophile: An organism that grows best in moderate temperature, neither too hot nor too cold, typically between 20°C and 45°C (68°F and 113°F).
433. Mesophilic: Growing best at moderate temperatures, between 20°C and 45°C.
434. Mesosphere: The region of the atmosphere above the stratosphere and below the thermosphere.
435. Mesozoic Era: An era lasting from about 252 to 66 million years ago (an era characterized by the development of flying reptiles, birds, and flowering plants and by the appearance and extinction of dinosaurs).
436. Metabolic pathways: Linked series of chemical reactions occurring within a cell.
437. Metamorphism: (In geology) Alteration of the composition or structure of a rock by heat, pressure, or another natural agency.
438. Metamorphosis: A conspicuous and relatively abrupt change in a living organism's body structure through cell growth and differentiation; e.g., the metamorphosis in the life of a butterfly.
439. MetaPolyZyme: A novel multi-enzyme extraction protocol.
440. Metatranscriptome: Microbes within natural environments.
441. Metazoans: All animals having the body composed of cells differentiated into tissues and organs and usually a digestive cavity lined with specialized cells.
442. Meteoroid: A small rocky or metallic body in outer space.
443. Meteorite: A solid piece of debris from an object, such as a comet, asteroid, or meteoroid, that originates in outer space and survives its passage through the atmosphere to reach the surface of a planet or moon.
444. Metabolism: Chemical reactions in the body's cells that change food into energy. Our bodies need this energy to do everything from moving to thinking to growing. Specific proteins in the body control the chemical reactions of metabolism.
445. Metabolite: An intermediate or end product of metabolism.
446. Metagenome: The genetic content of any group of microorganisms.
447. Metagenomics: Study of a collection of genetic material (genomes) from a mixed community of organisms recovered directly from environmental samples.
448. Metatranscriptomic: Study of gene expression of microbes within natural environments, i.e., the metatranscriptome.
449. Metazoans: All animals having the body composed of cells differentiated into tissues and organs and usually a digestive cavity lined with specialized cells.
450. Methanogens: Methane-producing organisms possessing the capability to chemically reduce carbon dioxide to methane.
451. Methanogenesis: Production of methane as a part of the metabolism in methanogens. It is usually the final step in the decomposition of biomass. Production of methane as a by-product of the chemical reaction between hydrogen and carbon dioxide dissolved in water.
452. Methanogenic: Producing methane as a by-product of energy metabolism.
453. Methanogenic bacteria: Bacteria that produce methane.
454. Methanotrophic: Methane-utilizing.
455. Microaerophilic: Requiring little free oxygen, or oxygen at a lower partial pressure than that of atmospheric oxygen.
456. Microbialites: Rock-like underwater structures that look like reefs but are made entirely of millions of microbes.
457. Microbial streamers: Bacterial biofilms degenerated in the form of filamentous structures.
458. Microbiome: The community of microorganisms that can usually be found living together in any given reasonably well-defined habitat which has distinct physio-chemical properties.
459. Microbiota: The microorganisms of a particular site, habitat, or geological period.
460. Macrofauna: Animals that are one centimeter or more long but smaller than an earthworm.
461. Mid-oceanic ridge: Long, narrow submarine hilltop, mountain range.
462. Mineral alteration: Various natural processes that alter a mineral's chemical composition or crystallography.
463. Mineralogy: A subject of geology specializing in the scientific study of the chemistry, crystal structure, and physical properties of minerals and mineralized artifacts.
464. Mitotic activity: Having to do with the presence of dividing (proliferating) cells. Cancer tissue generally has more mitotic activity than normal tissues.
465. Mixolimnion: The freely circulating upper layer of a meromictic lake.
466. Moiety: A part of a molecule that is given a name because it is identified as a part of other molecules as well.
467. Molecular fossil: Another name of Biosignature.
468. Molting: The process of shedding the old outer covering of the body to make way for a new growth.
469. Monomer molecule: Molecules that can be bonded to other identical molecules to form polymers. One of many repeating subunits of a polymer. Small molecules, mostly organic, that can join with other similar molecules to form very large molecules, or polymers.

All monomers have the capacity to form chemical bonds to at least two other monomer molecules.

470. Monomeric organic material: Material of low molecular weight capable of reacting with identical or different materials of low molecular weight to form a polymer.

471. Monte Carlo Simulations: Simulations that are often used to model the probability of different outcomes in a process that cannot easily be predicted due to the intervention of random variables (the use of random variables is most common in probability and statistics, where they are used to quantify outcomes of random occurrences). Monte Carlo methods, or Monte Carlo experiments, are a broad class of computational algorithms that rely on repeated random sampling to obtain numerical results. The underlying concept is to use randomness to solve problems that might be deterministic in principle.

472. Montmorillonite: A very soft phyllosilicate group of minerals that form when they precipitate from water solution as microscopic crystals, known as clay. It is named after Montmorillon in France.

473. Morphology: Form, shape, or structure. Morphology is a branch of biology dealing with the study of the form and structure of organisms and their specific structural features. This includes aspects of the outward appearance (shape, structure, color, pattern, size), i.e., external morphology, as well as the form and structure of the internal parts such as bones and organs, i.e., internal morphology.

474. Morphotypes: Any of a group of different types of individuals of the same species in a population.

475. M stars (also called Type M stars, M dwarfs or Red dwarfs): Relatively cool (characterized primarily by a surface temperature of 3000 kelvins or less), red stars, which make up about 75% of all stars in the Galaxy. They are extremely long-lived, and because they are much smaller in mass than the Sun (between 0.5 M and 0.08 M (Sun)), their temperature and stellar luminosity are low and peaked in the red. Because the luminosity of these dwarf stars peaks in the red, these stars are also called "Red dwarfs". These diminutive stars, much smaller and dimmer than our own Sun are not bright enough to see with the naked eye.

476. Multispectral band parameters: The three multispectral band parameters are band curvature, band strength, and band tilt.

477. Murchison meteorite: A meteorite that fell in near Murchison, Victoria, Australia in 1969. It belongs to a group of meteorites rich in organic compounds. Murchison meteorite, which have been determined to be 7 billion years old, about 2.5 billion years older than the 4.54-billion-year age of the Earth and the Solar System.

478. Mush: A soft, wet, pulpy mass.

479. Mutation: The change in the structure of a gene (DNA), resulting in a variant form which may be transmitted to subsequent generations. The changing of the structure of a gene is caused by the alteration of single base units in DNA, or the deletion, insertion, or rearrangement of larger sections of genes or chromosomes. Mutation is, ultimately, the only way in which new variation enters the species.

480. Nauplii: The first larval stage of many crustaceans, having an unsegmented body and a single eye.

481. Near-Earth Asteroids (NEAs): Asteroids whose orbits bring them relatively close to the Earth {perihelon distances of less than 1.3 astronomical unit (AU)} are known as Near Earth Asteroids (NEAs). Some NEAs' orbits intersect Earth's orbit so they pose a collision danger. NEAs only survive in their orbits for 10 million to 100 million years.

482. Nebula: A distinct body of interstellar clouds (which can consist of cosmic dust, hydrogen, helium, molecular clouds; possibly as ionized gases).

483. Nematodes: Roundworms.

484. Neoproterozoic era: About 750 to 635 million years ago. This is an era during which the Earth went through a difficult time when the entire Earth became fully covered with thick ice. According to this hypothesis (known as "Snowball Earth" hypothesis), Earth's surface became either entirely or nearly entirely frozen at least once.

485. Neovolcanic zones: Zones of recent volcanic activity; from the Tertiary period (up to about 65 million years ago) to the present time (contrasted with palaeovolcanic).

486. New Frontiers program: A series of space exploration missions being conducted by NASA with the purpose of furthering the understanding of the Solar System.

487. Newtonian fluid: A fluid in which the viscous stresses arising from its flow are at every point linearly correlated to the local strain rate—the rate of change of its deformation over time. Stresses are proportional to the rate of change of the fluid's velocity vector.

488. Nitrifiers: Organisms that change ammonia or ammonium into nitrite or change nitrite into nitrate as part of the nitrogen cycle.

489. Nitriles: Organic compounds containing a cyanide group —CN bound to an alkyl group.

490. Noachian: A geologic system and early time period (4.1 to 3.7 billion years ago) on the planet Mars characterized by high rates of meteorite and asteroid impacts and the possible presence of abundant surface water.

491. Noble gases (historically also the inert gases): A class of chemical elements with similar properties; under standard conditions, they are all odorless, colorless,

monatomic gases with very low chemical reactivity. The six naturally occurring noble gases are helium (He), neon (Ne), argon (Ar), krypton (Kr), xenon (Xe), and the radioactive radon (Rn).
492. Non-avian dinosaurs: Members of the dinosaur descendants other than birds and their immediate ancestors.
493. Non-Newtonian fluid: A fluid that does not follow Newton's law of viscosity, i.e., constant viscosity independent of stress. In non-Newtonian fluids, viscosity can change when under force to either more liquid or more solid.
494. North (Positive pole): (of a celestial body) Direction of the angular momentum vector.
495. Nucleotide: A subunit of DNA or RNA that consists of a nitrogenous base, a phosphate molecule, and a sugar molecule.
496. Nucleic acid: An acid that is enclosed in the nucleus of every biological cell.
497. Obliquity: (also known as axial tilt): The angle between a celestial body's rotational axis and its orbital axis, or, equivalently, the angle between its equatorial plane and orbital plane.
498. Occator: An impact crater located on Ceres, that contains "Spot 5", the brightest of the bright spots observed by the Dawn spacecraft.
499. Oceana: The largest international advocacy group dedicated entirely to ocean conservation.
500. Oceanic cyanobacteria (also called *blue-green algae*): The first organisms on Earth to perform oxygenic photosynthesis.
501. Oligomerization reactions: Chemical processes that convert monomers to macromolecular complexes through a finite degree of polymerization.
502. Oligotrophic: Relatively poor in plant nutrients and containing abundant oxygen in the deeper parts.
503. Olivine: A magnesium iron silicate with the chemical formula $(Mg, Fe)_2SiO_4$. Olivine being the primary component of the Earth's upper mantle, it is a common mineral in Earth's subsurface, but weathers quickly on the surface. Olivine is a rock-forming mineral found in dark-colored igneous rocks.
504. Ontogeny: The study of the entirety of an organism's lifespan.
505. Ophicalcite: Crystalline limestone or marble spotted with greenish serpentine, like a serpent or snake.
506. Ophiolite: A section of Earth's oceanic crust and the underlying upper mantle that has been uplifted and exposed above sea level and often emplaced onto continental crustal rocks.
507. Opportunistic feeders: Predators exhibiting the practice of taking advantage of circumstances, guided primarily by self-interested motives.
508. Optical depth: Natural logarithm of the ratio of incident to transmitted radiant power through a material. Thus, the larger the optical depth, the smaller the amount of transmitted radiant power through the material.
509. Order-of-magnitude: An order-of-magnitude estimate of a variable, whose precise value is unknown, is an estimate rounded to the nearest power of ten.
510. Ordovician period: Period from 485 to 444 million years ago, which is a time of dramatic changes for life on Earth.
511. Organelles: Specialized structures that perform various jobs inside cells. The term literally means "little organs." In the same way organs, such as the heart, liver, stomach, and kidneys, serve specific functions to keep an organism alive, organelles serve specific functions to keep a cell alive.
512. Organism: an individual animal, plant, or single-celled life form.
513. Orphan biomarkers: Biomarkers with no known biological source.
514. Oscillatoria: A genus of filamentous cyanobacterium which is named after the oscillation in its movement.
515. Outgroup: A group of organisms not belonging to the group whose evolutionary relationships are being investigated. Such a group is used for comparison, to assess which characteristics of the group being studied are more widely distributed and may therefore be older in origin.
516. Oviparous animals: Female animals that lay their eggs, with little or no other embryonic development within the mother. This is the reproductive method of most fish, amphibians, most reptiles, and dinosaurs (including birds).
517. Oxic: Having enough oxygen in the tissues to sustain bodily functions; Of a process or environment in which oxygen is involved or present.
518. Oxidase: An enzyme which promotes the transfer of a hydrogen atom from a particular substrate to an oxygen molecule, forming water or hydrogen peroxide. Any of a group of enzymes that bring about biological oxidation.
519. Pahoehoe lava: Extremely fluid smooth, unbroken lava.
520. Paleohydrological study: The study of ancient use and handling of water; e.g., as in irrigation or urban water supplies.
521. Paleolakes: Ancient lakes, especially those that no longer exist.
522. Palaeontologists: Scientists who study fossils as a way of getting information about the history of life on Earth and the structure of rocks.
523. Paleoproterozoic era: An era spanning the time period from 2500 to 1600 million years ago. This era is the first of the three sub-divisions of the Proterozoic Eon. The Paleoproterozoic is also the longest era of the

Earth's geological history. It was during this era that the continents first stabilized.
524. Paleosols: Fossil soils.
525. Panarthropoda: A proposed animal clade containing the extant phyla Arthropoda, Tardigrada and Onychophora.
526. Pangaea (also spelled Pangea): A supercontinent that incorporated almost all the landmasses on Earth. Its name is derived from the Greek pangaia, meaning "all the Earth".
527. Panspermia: The theory that life on Earth originated from microorganisms or chemical precursors of life present in outer space and able to initiate life on reaching a suitable environment. According to this hypothesis, life exists throughout the Universe, distributed by space dust, meteoroids, asteroids, comets, planetoids, and also by spacecraft carrying unintended contamination by microorganisms.
528. Parthenogenesis: A natural form of asexual reproduction in which growth and development of embryos occur without fertilization by sperm. In animals, parthenogenesis means development of an embryo from an unfertilized egg cell.
529. Patera (plural: Paterae): A broad, shallow bowl-shaped feature on the surface of a planet.
530. Pedogenic: Relating to or denoting processes occurring in soil or leading to the formation of soil.
531. Pelagic fish: Fish living neither close to the bottom nor near the shore.
532. Peptidase: An enzyme which breaks down peptides into amino acids.
533. Peptide: A compound consisting of two or more amino acids linked in a chain, the carboxyl group of each amino acid being joined to the amino group of the next by a bond of the type -OC-NH-.
534. Peptide Nucleic Acid (PNA): An artificially synthesized polymer similar to DNA or RNA.
535. Periapsis: The point of closest approach (e.g., in a two-body system, when an object is in an elliptical orbit).
536. Peridotite: A dense, coarse-grained plutonic rock (an igneous rock of holocrystalline granular texture regarded as having solidified at considerable depth below the surface) containing a large amount of olivine, believed to be the main constituent of the Earth's mantle.
537. Periglacial terrain: Terrain subject to repeated freezing and melting.
538. Perihelion: The point in the orbit of a planet, asteroid, or comet at which it is closest to the Sun.
539. Petagram: 10^{12} kg.
540. PETM/IETM: Palaeocene-Eocene Thermal Maximum (PETM), also called Initial Eocene Thermal Maximum (IETM), is a relatively short-lived (in geological terms) transient climatic event of maximum temperature lasting approximately 100,000–200,000 years during the late Paleocene and early Eocene epochs. The PETM period is believed to be the most rapid and extreme natural global warming event of the last 56 million years. The global temperature increase was estimated to have been in the range 5–9°C.
541. Petrology: The branch of science concerned with the origin, structure, and composition of rocks.
542. Petrogenesis: Branch of science dealing with the origin and formation of rocks.
543. pH: A quantitative measure of the acidity or basicity of aqueous or other liquid solutions. The values of the concentration of the hydrogen ion—which ordinarily ranges between about 1 and 10^{-14} gram-equivalents per litre—into numbers between 0 and 14.
544. Phylogenetically: In a way that relates to the evolutionary development and diversification of a species or group of organisms.
545. Phanerozoic Eon: The current geologic eon in the geologic time scale, and the one during which abundant animal and plant life has existed. It covers 541 million years to the present, and began with the Cambrian Period when diverse hard-shelled animals first appeared.
546. Photolysis: The decomposition or separation of molecules by the action of solar light radiation.
547. Photon: A particle representing a quantum of light or other electromagnetic radiation, in which the said particle carries energy proportional to the radiation frequency but has zero rest mass.
548. Photolyze: Cause to undergo chemical decomposition under the influence of light.
549. Photospheric surface: Outer shell from which light is radiated.
550. Photosynthesis: Harvesting light to produce energy and oxygen. Photosynthesis is the process by which green plants and some other organisms use sunlight to synthesize nutrients from carbon dioxide and water. Photosynthesis in plants generally involves the green pigment chlorophyll and generates oxygen as a by-product.
551. Photosynthetic bacteria: Bacteria that produce oxygen gas.
552. Photosynthetic microorganisms: Microorganisms that are capable of converting light energy (from the Sun) into chemical energy through a process known as photosynthesis.
553. Phototrophs: Organisms that carry out photon capture—i.e., use the energy from light to carry out various cellular metabolic processes—to produce complex organic compounds and acquire energy.
554. Phototrophic: Tendency of an organism to grow in response to a light stimulus.

555. Phylogeny: The branch of biology that deals with phylogenesis; i.e., the evolutionary development and diversification of a species or group of organisms, or of a particular feature of an organism.
556. Phylogenetic: Relating to the evolutionary development and diversification of a species or group of organisms, or of a particular feature of an organism.
557. Phyllosilicate: Compound with a structure in which silicate tetrahedrons—a triangular pyramid composed of four triangular faces, six straight edges, and four vertex corners—(each consisting of a central silicon atom surrounded by four oxygen atoms at the corners of a tetrahedron) are arranged in sheets. Examples are talc and mica.
558. Phylum: A principal taxonomic category that ranks above class and below kingdom, equivalent to the division in botany.
559. Physiology: Normal functions of living organisms and their parts.
560. Piezophiles: Organisms with optimal growth under high hydrostatic pressure.
561. Pillow basalt: A type of rock formed during an underwater eruption.
562. Pillow lava: Lava that contains characteristic pillow-shaped structures that are attributed to the extrusion of the lava underwater, or subaqueous extrusion. Pillow lavas in volcanic rock are characterized by thick sequences of discontinuous pillow-shaped masses, commonly up to one meter in diameter.
563. Pit crater: Circular or elliptical depression shaped by the collapsing or sinking of the surface lying above a void or hollow cavity; e.g., pit craters noticed above the lava tubes lying below.
564. Planet: An astronomical body orbiting a star.
565. Planetesimal: A body which could come together with many others under gravitation to form a planet.
566. Plant-bacteria microcosms: Plant-bacteria regarded as encapsulating in miniature the characteristics of something much larger.
567. Plasma: An ionized gas consisting of positive ions and free electrons in proportions resulting in more or less no overall electric charge, typically at low pressures (as in the upper atmosphere and in fluorescent lamps) or at very high temperatures (as in stars and nuclear fusion reactors).
568. Plate tectonics: A scientific theory describing the large-scale motion of the plates making up the Earth's lithosphere since tectonic (building) processes began on Earth between 3.5 and 3.3 billion years ago.
569. Playa: Salt flat.
570. Pleistocene epoch: The geological epoch that lasted from about 2,580,000 to 11,700 years ago, spanning the Earth's most recent period of repeated glaciations.
571. Plio-Pleistocene: Geological pseudo-period, which begins about 5 million years ago and, drawing forward, combines the time ranges of the formally defined Pliocene and Pleistocene epochs—marking from about 5 Mya to about 12 kya.
572. Plutonic rock: An igneous rock (as granite) of holocrystalline granular texture regarded as having solidified at considerable depth below the surface.
573. Podzol soil: An infertile acidic soil characterized by a white or grey subsurface layer resembling ash, from which minerals have been leached into a lower dark-colored stratum.
574. Polymeric molecules: Very large molecules, or macromolecules, composed of many repeating subunits as occurring in a polymer.
575. Polyphyletic origin: (of a group of organisms) derived from more than one common evolutionary ancestor or ancestral group and therefore not suitable for placing in the same taxon.
576. PolyZyme: A novel multi-enzyme extraction protocol involving substituting a novel multi-enzyme blend.
577. Portable Network Graphics (PNG) analysis: A tool that has been employed for detecting structural microbial biosignatures.
578. Polar (in biology): A separation of electric charge leading to a molecule or its chemical groups having an electric dipole.
579. Polar wander: A phenomenon which happens on a spinning body, in which extra mass migrates toward the equator, and areas with lower mass end up closer to the poles. If one area suddenly gains more mass due to some reason, the spinning celestial body will reorient itself so that the more massive feature is closer to the equator. If a part of the celestial body suddenly loses a lot of mass, that feature will drift toward the poles.
580. Polymer: A substance or material consisting of very large molecules, or macromolecules, composed of many repeating subunits. Due to their broad spectrum of properties, both synthetic and natural polymers play essential and ubiquitous roles in everyday life.
581. Polymerization: A process of reacting monomer molecules (small molecules, mostly organic, that can join with other similar molecules to form very large molecules) together in a chemical reaction to form polymer chains or three-dimensional networks.
582. Polynucleotide: A biopolymer composed of 13 or more nucleotide monomers covalently bonded in a chain. DNA and RNA are examples of polynucleotides with distinct biological function.
583. Polysaccharides, or polycarbohydrates: The most abundant carbohydrate found in food. They are long chain polymeric carbohydrates composed of monosaccharide units bound together by glycosidic

linkages. This carbohydrate can react with water using amylase enzymes as catalyst, which produces constituent sugars.
584. Polyyne: Any organic compound with alternating single and triple bonds; that is, a series of consecutive alkynes, $(-C\equiv C-)_n$. with n greater than 1. The simplest example is diacetylene or buta-1,3-diyne, $H-C\equiv C-C\equiv C-H$.
585. Pore-water: Water contained in pores in soil or rock.
586. Precession: A gravity-induced slow and continuous change in the orientation of Earth's rotational axis (wobble), in which the Earth's rotational axis completes a rotation about another line intersecting it so as to describe a cone in approximately 26,000 years.
587. Primordial Soup: A solution rich in organic compounds in the primitive oceans of the Earth, from which life is thought to have originated, which Charles Darwin envisaged, to promote organic evolution.
588. Prograde metamorphism: Metamorphic changes caused by increasing temperature.
589. Prograde moon: The moon that orbits in the direction of rotation of its host planet.
590. Prokaryote: A microscopic single-celled organism which has neither a distinct nucleus with a membrane nor other specialized organelles. Examples are bacteria and cyanobacteria.
591. Proteins: Macromolecular polypeptides—i.e., very large molecules (macromolecules) composed of many peptide-bonded amino acids. Proteins are simply amino acid polymers. In other words, amino acids are the building blocks of proteins. A complex network of more than 200 essential proteins is found in today's most elementary biological cells.
592. Proterozoic Eon: The Proterozoic Eon, meaning "earlier life," is the eon of time after the Archean eon and ranges from 2.5 billion years old to 541 million years old. During this time, most of the central parts of the continents had formed and the plate tectonic process had started.
593. Protist: A single-celled organism of the kingdom Protista, such as a protozoan or simple alga.
594. Protium: The basic hydrogen atom—a single proton circled by a single electron; Isotope of hydrogen with atomic weight of approximately 1; its nucleus consists of only one proton. Ordinary hydrogen is made up almost entirely of protium.
595. Proto-Earth: A still-forming Earth. The Earth at an early stage of its development; i.e., an early-stage fluid-Earth.
596. Proto-planet: A large body of matter in orbit around the Sun or a Star and believed to be developing into a planet.
597. Protoplanetary disk: A protoplanetary disk is a rotating circumstellar disc (a ring-shaped accumulation of matter composed of gas, dust, planetesimals, asteroids, or collision fragments in orbit around a star) of dense gas and dust surrounding a young newly formed star.
598. Proto-solar nebula: The rotating, flattened disk of gas and dust from which the Solar System originated ~4.6 Ga (billion years ago).
599. Protozoa: A phylum or grouping of phyla which comprises the single-celled microscopic animals.
600. Psychrophiles: Organisms that grow best at temperatures close to freezing.
601. Psychrophilic: Extremophilic cold-loving.
602. Putative ocean: An ocean assumed to exist.
603. Pyroclastic: Fragments of rock erupted by a volcano.
604. Pyro-ducts: Collapsed lava tubes.
605. Pyrolysis process: Thermal decomposition of materials at elevated temperatures in an inert atmosphere. It involves a change of chemical composition.
606. Pyrolyzed: Thermally decomposed.
607. Quasar: An astronomical object of very high luminosity found in the centers of some galaxies and powered by gas spiraling at high velocity into an extremely large black hole.
608. Racemic mixture: A mixture that has equal amounts of left- and right-handed enantiomers of a chiral molecule. The first known racemic mixture was racemic acid, which Louis Pasteur found to be a mixture of the two enantiomeric isomers of tartaric acid.
609. Radioactive decay: The process by which an unstable atomic nucleus loses energy by radiation.
610. Radiogenic: Produced by radioactivity (e.g., "a radiogenic isotope", "radiogenic heat" in a celestial body).
611. Radiolarians: Predatory single-celled organisms of the kingdom Protista, such as a protozoan or simple alga.
612. Radiolysis: Dissociation of molecules by ionizing radiation.
613. Radiometric dating: Dating by use of radioactive decay.
614. Red beds: Red-colored sandstones that are coated with hematite.
615. Redox: Oxidation and reduction considered together as complementary processes.
616. Redox couples: Reducing species and their corresponding oxidizing forms.
617. Redox gradient: Pairing of electron donors and electron acceptors, an essential source of energy that life requires.
618. Redox reaction: A type of chemical reaction in which the oxidation states of atoms are changed. Redox reactions are characterized by the actual or formal transfer of electrons between chemical species, most often with one species undergoing oxidation while another species undergoes reduction. Many important

biological processes involve redox reactions. Before some of these processes can begin iron must be assimilated from the environment.
619. Reef fish: Fish that are associated with coral reefs.
620. Refractory elements: A group of metallic elements that are highly resistant to heat and wear. It is generally accepted that tungsten, molybdenum, niobium, tantalum and rhenium best fit most definitions of refractory metals.
621. Regolith: The layer of loosely arranged solid material covering the bedrock of a planet or moon. It includes dust, broken rocks, and other related materials and is present on Earth, Earth's Moon, Mars, and some asteroids. Regolith is considered to be a biologically active medium and a key component in plant growth.
622. Retrograde orbit: Orbiting "backwards" or in the opposite direction to its host planet's rotation.
623. Reverse gyrase: An enzyme that induces positive supercoiling in closed circular DNA in vitro. It is unique to thermophilic organisms and found without exception in all microorganisms defined as hyperthermophiles, that is, those having optimal growth temperatures of 80°C and above.
624. Rheic Ocean: An ocean which separated two major palaeo-continents, Gondwana and Laurussia (Laurentia-Baltica-Avalonia).
625. Rheology: The study of the flow of matter, primarily in a liquid or gas state, but also as "soft solids" or solids under conditions in which they respond with plastic flow rather than deforming elastically in response to an applied force.
626. Rheological studies: Studies of the flow of matter in any state in response to an applied force.
627. Rheological flows: Viscoelastic flows.
628. Rhythmite: Layers of sediment or sedimentary rock which are laid down with an obvious periodicity and regularity.
629. Ribozyme: A ribonucleic acid (RNA) molecule capable of acting as an enzyme.
630. Rille: A fissure or narrow channel on the surface of a celestial body.
631. RNA: Ribonucleic acid, a nucleic acid present in all living cells. Its principal role is to act as a messenger carrying instructions from DNA for controlling the synthesis of proteins, although in some viruses RNA rather than DNA carries the genetic information. RNA and DNA are nucleic acids.
632. Roche limit: The minimum distance that the Moon (or any satellite held together only by gravity) can be from the Earth (or any large body) without breaking up. If the Moon comes any closer than the Roche limit, it will break apart because of tidal forces. Thus, no planet can have a moon that lies within the Roche limit – it can have only rings.
633. Rocky planet: see Terrestrial planet.
634. Roe: Fully ripe internal egg masses in the ovaries, or the released external egg masses of fish (usually used for preparing pickles).
635. Rover: A space exploration vehicle designed to move across the surface of a planet or other celestial body, with the purpose of finding out information and collecting samples (e.g., dust, rocks) and even taking pictures.
636. Runaway greenhouse effect: The complete evaporation of the oceans.
637. Runaway glaciation effect: Lowering the temperature and/or water activity to levels not conducive to life, like Mars.
638. Salar: Salt flat.
639. Salares: Salar ecosystems.
640. Salinity: The total concentration of dissolved salts.
641. Saltern: A set of pools in which seawater is left to evaporate to make salt.
642. Scanning Electron Microscopy (SEM): A form of microscopy in which a focused beam of accelerated electrons is scanned across the surface of a specimen, generating a number of signals that yield information about its morphology, elemental composition, and, when outfitted with appropriate detectors, crystalline microstructure or other features.
643. Scintillation: A small flash of visible or ultraviolet light emitted by fluorescence in a phosphor when struck by a charged particle or high-energy photon.
644. Secondary-Ion Mass Spectrometry (SIMS): A technique used to analyze the composition of solid surfaces and thin films by sputtering the surface of the specimen with a focused primary ion beam and collecting and analyzing ejected secondary ions.
645. Secular change: A linear long-term trend.
646. Selenogony: The study of the origin of Earth's Moon.
647. Serpentinization: A process that takes place at depth in the seafloor, leading to significant changes in topography, focused micro-seismic activity as a result of continuous cracking, and significant heat generation. Geologically, serpentinization is the process of hydrothermal alteration that transforms Fe-Mg-silicates such as olivine, pyroxene, or amphiboles contained in ultramafic rocks into serpentine minerals. Chemically, serpentinization is the hydration of the olivine $((Mg,Fe)_2SiO_4)$ and orthopyroxene minerals that mainly constitute the upper mantle. In the process of serpentinization, the low-temperature (150–400°C) hydrolysis and transformation of ultramafic rocks produces H_2. The hydrogen thus produced reacts with simple oxidized carbon compounds, such as CO_2 and CO, under reducing conditions to produce and release methane (CH_4) and other organic molecules through Fischer–Tropsch-type synthesis. Thus,

serpentinization is considered to be responsible for methane (CH_4) generation.
648. Serpentine: Serpentine is a soft ductile mineral and its presence in the mantle wedge lubricates subduction of the oceanic plate. Production of serpentine in the oceanic crust results in the production of hydrothermal fluids and releases gaseous methane and hydrogen, as observed along terrestrial mid-ocean ridges. The pH of the hydrothermal fluids is generally low but under some conditions, notably at low temperature, may be high enough to be favorable to life.
649. Sessile: Permanently attached or established: not free to move about.
650. Sexual dimorphism: The condition where the sexes of the same species exhibit different characteristics, particularly characteristics not directly involved in reproduction.
651. Shales: Soft finely stratified sedimentary rock that formed from consolidated mud or clay and can be split easily into fragile plates.
652. Shield volcanoes: A type of volcano named for its low profile, resembling a warrior's shield lying on the ground.
653. Siliceous: Containing or consisting of silica.
654. Silicic: Rich in silica.
655. Silicified: Converted into or impregnated with silica.
656. Sievert (Sv): The unit of radiation absorption in the International System of Units (SI). Accordingly, one sievert is generally defined as the amount of radiation roughly equivalent in biological effectiveness to one gray (or 100 rads) of gamma radiation.
657. Silhouette: The dark shape and outline of something visible in restricted light against a brighter background.
658. Silicate hydration: Chemical redistribution between the core and the hydrosphere of a celestial body. Silicate dehydration releases liquid.
659. Sinuous: Having many curves and turns.
660. Sinuous rilles: Curved paths having the shape and features of a mature river, and are commonly thought to be the remains of collapsed lava tubes or extinct lava flows.
661. Skylights or skylight holes: Light-transmitting windows that form part of the roof space of a dwelling/building for daylighting and ventilation purposes. "Skylights" associated with lava tubes are dark, nearly rounded features that are hypothesized as entrances to lava tubes. At this point, light from the Sun enters into the permanent darkness of the lava tube from above, forming a skylight. Many lava tubes on Earth, for instance, have been identified through the discovery of skylights. Natural skylights have been spotted on the lava tubes on Earth's Moon and the planet Mars.
662. Smectite: A clay mineral which undergoes reversible expansion on absorbing water. With no net force pushing the change in one direction or the other, the change is said to be reversible or to occur reversibly.
663. Snowline: The boundary between a snow-covered and snow-free surface.
664. Sol: A Martian day (approximately 40 minutes longer than a day on Earth).
665. Solar flare: A brief eruption of intense high-energy radiation from the Sun's surface, associated with sunspots. Powerful flares are often, but not always, accompanied by a coronal mass ejection. It causes radio and magnetic disturbances on the Earth.
666. Solar wind: A thin stream of electrically conducting gas (chiefly protons and electrons together with nuclei of heavier elements in smaller numbers) ejected from the surface of the Sun; the supersonic outflow of electrically charged particles from the Sun.
667. Solidus: The maximum temperature at which all components of a mixture, such as an alloy, can be in a solid state.
668. Solstice: The time or date (twice each year) at which the Sun reaches its maximum or minimum declination, marked by the longest and shortest days (about 21 June and 22 December, respectively).
669. Sorption: Absorption and adsorption (the process by which a solid holds molecules of a gas or liquid or solute as a thin film) considered as a single process.
670. Specific weight, γ: Weight of a substance per unit volume in absolute units; $\gamma = \rho g$, where ρ is the density, and g is the standard gravity).
671. Speleogenetic processes: Those processes which operate on or close to the surface of the Earth and which involve weathering, mass movement, fluvial, aeolian, glacial, periglacial, and coastal processes. The term is normally used in contrast to the endogenetic processes, whose origin is within the Earth.
672. Speleological exploration: Exploration of caves.
673. Spherule: A small sphere.
674. Spore: A reproductive cell capable of developing into a new individual without fusion with another reproductive cell. Spores are produced by bacteria, fungi, algae, and plants.
675. Staging area: Sites that attract large concentrations (many thousands) of birds.
676. Star sensor: A device used for determining a spacecraft's orientation in space.
677. Stellar occultation: The passage of one celestial body in front of another, thus hiding the other from view.
678. Stenohaline: (of an aquatic organism) able to tolerate only a narrow range of salinity.
679. Stereochemistry: The branch of chemistry concerned with the three-dimensional arrangement of atoms and molecules and the effect of this on chemical reactions.

680. **Stochastic:** Representing the element of chance, or probability. Having a random probability distribution or pattern that may be analyzed statistically but may not be predicted precisely. Stochastic refers to the property of being well described by a random probability distribution. Although stochasticity and randomness are distinct in that the former refers to a modeling approach and the latter refers to phenomena themselves, these two terms are often used synonymously.
681. **Stratigraphy:** The study of the composition, relative positions, etc., of rock strata in order to determine their geological history: Branch of geology concerned with the study of rock layers and layering.
682. **Stratosphere:** The layer of gases surrounding the Earth at a height of between 15 km and 50 km. More specifically, the second layer of atmosphere above the troposphere, which is the lowest layer.
683. **Strelley Pool Chert (SPC):** A sedimentary rock formation containing laminated structures of probable biological origin (stromatolites). SPC has been discovered from Pilbara Craton, Australia.
684. **Stromatolites:** Dome-shaped structures consisting of alternating layers of carbonate or silicate sediments and fossilized algal mats, produced over geologic time by the trapping, binding, or precipitating of sediment by groups of micro-organisms, primarily cyanobacteria. Stromatolites are commonly defined as "organo-sedimentary structures built by microbes".
685. **Subsolidus:** Describing the region beneath the solidus in a phase diagram.
686. **Sulfate reducers:** Organisms which can perform anaerobic respiration utilizing sulfate as terminal electron acceptor, reducing it to hydrogen sulfide.
687. **Sulfide/Sulfur oxidizers:** Organisms having the capability to oxidize various reduced inorganic sulfur compounds with high efficiency to obtain electrons for their autotrophic growth.
688. **Super-Earths:** Planets of less than 10 Earth masses—the so-called 10 M_{Earth}.
689. **Super-Earth exoplanets:** Exoplanets that are larger and more massive than Earth, but smaller and less massive than Neptune (about 17 times Earth's mass). Surface blisters: Small to large, broken, or unbroken bubbles, which are under or within a coating.
690. **Surface clutter:** Scattering of the radio waves on the rough surface.
691. **Surfactant:** A substance which tends to reduce the surface tension of a liquid in which it is dissolved.
692. **SYBR Safe:** A cyanine dye used as a nucleic acid stain in molecular biology. This dye binds to DNA. The resulting DNA-dye-complex absorbs blue light (λ_{max} = 509 nm) and emits green light (λ_{max} = 524 nm).
693. **Symbiosis:** Any type of a close and long-term biological interaction between two different biological organisms, be it mutualistic, commensalistic (one benefits and the other derives neither benefit nor harm), or parasitic. The organisms, each termed a symbiont, must be of different species.
694. **Symbiotic relationship:** Mutually beneficial relationship or interaction between two or more different organisms living in close physical association.
695. **Synestia:** A giant doughnut-shaped object consisting of a mass of hot, vaporized rock spinning around a molten mass left over from a planetary collision. Scientists think that about 4.6 billion years ago, synestia formation was a painfully common occurrence as the Solar System formed from the giant disc of dust, gas, and debris that orbited the young Sun.
696. **Synodic period:** The time required for a body within the Solar System, such as a planet, the Moon, or an artificial Earth satellite, to return to the same or approximately the same position relative to the Sun as seen by an observer on the Earth.
697. **Syntrophic communities:** Cross-feeding communities in which the growth of one partner depends on the nutrients, growth factors, or substrates provided by the other partner.
698. **Talik:** An unfrozen zone within a permafrost; An area of unfrozen ground surrounded by permafrost.
699. **Talus:** A slope formed especially by an accumulation of rock debris.
700. **Taphonomic changes:** Changes taking place during the process of fossilization.
701. **Tardigrades:** Microscopic, normally semi-aquatic, creatures affectionately known as "water bears". The name "water-bear" comes from the way they walk, reminiscent of a bear's gait.
702. **Taxon, plural Taxa:** Any unit used in the science of biological classification, or taxonomy. Taxa are arranged in a hierarchy from kingdom to subspecies, a given taxon ordinarily including several taxa of lower rank.
703. **Taxonomic:** Concerned with the classification of things, especially organisms.
704. **Taxonomy:** Scientific study of naming, defining, and classifying groups of biological organisms based on shared characteristics.
705. **Taxon-specific biomarkers (TSBs):** Complex biosynthetic molecules (biomarkers) that are utilized or synthesized by one specific group of organisms. Thus, they are signature compounds with demonstrated efficacy for tracing evolutionary history and the early development of the Earth's biosphere.
706. **Telluric:** Of the earth as a planet; of the soil.
707. **Telluric planet:** see Terrestrial planet.

708. Terrestrial planets: Planets that are earth-like in nature, meaning those that are similar in composition {silicate rocks or metals} and structure to that of the Earth; Terrestrial planet (also called telluric planet, or rocky planet) is a planet that is composed primarily of silicate rocks or metals. Within the Solar System, the terrestrial planets are the inner planets closest to the Sun, i.e., Mercury, Venus, Earth, and Mars.
709. Thaws: Frozen substances that become liquid or soft as a result of warming up.
710. Thaw lake: A body of freshwater, usually shallow, that is formed in a depression formed by ice-rich permafrost becoming liquid or soft as a result of warming up.
711. Theia: A Mars-sized planetesimal (Impactor), which is believed to have impacted the late-stage accreting Earth, giving rise to the formation of Earth's Moon.
712. Thera Macula: A region of likely active chaos production above a large liquid water lake in the icy shell of Europa.
713. Thermophile: A bacterium or other microorganism that grows best at higher than normal temperatures.
714. Thermophilic: Thriving at relatively high temperatures, between 41°C and 122°C.
715. Tholin: Organic-rich aerosol.
716. Throese Nucleic Acid (TNA): An artificial genetic polymer in which the natural five-carbon ribose sugar found in RNA has been replaced by an unnatural four-carbon throese sugar.
717. Tidal axis: The line along which the gravitational pull experienced by a planet (e.g., Pluto) from a tidally locked moon (e.g., Charon) is the strongest. In terms of planetary jargon, Pluto and Charon are "tidally locked", meaning these two celestial bodies always show the same face to each other, just as Earth's moon always shows just one side to Earth.
718. Tidal friction: Strain produced in a celestial body that undergoes cyclic variations in gravitational attraction as it orbits, or is orbited by, a second body.
719. Tidal node: (see Amphidromic point).
720. Tidal quality/dissipation factor (Q): A measure of a body's response to tidal distortion. In planetary science, Q is a parameter quantifying the tidal dissipation in celestial bodies. The Q factor of a planet is a complex function of the interior structure of the planet. Whereas Earth has $Q \approx 12$, the quality factor of other terrestrial planets in our solar system lie in the range $10 \leq Q \leq 190$. Rocky bodies tend to have Q-values near 100, giant planets and stars have Q-values near 10^6. The Q factor of a planet is critical to driving the dynamics and heating in these systems.
721. Tidal range: Difference in sea level elevation between successive high water and low water.
722. Time-of-Flight Secondary Ion Mass Spectrometry (ToF-SIMS): A technique designed to analyze the composition and spatial distribution of molecules and chemical structures on surfaces.
723. Transcription: The process of copying a segment of DNA into RNA.
724. Transcriptome: The sum of all the messenger RNA molecules expressed from the genes of an organism. More specifically, the transcriptome is the set of all RNA transcripts, including coding and non-coding, in an individual or a population of cells. The term can also sometimes be used to refer to all RNAs, or just mRNA, depending on the particular experiment.
725. Translation: The process of making proteins. Translation is the process in which ribosomes in the cytoplasm or endoplasmic reticulum synthesize proteins after the process of transcription of DNA to RNA in the cell's nucleus.
726. Translucence: The quality or state of being semi-transparent.
727. Trehalose: A sugar of the di-saccharide class produced by some fungi, yeasts, and similar organisms.
728. Trilobite: A fossil marine arthropod that occurred abundantly during the Palaeozoic era, with a carapace over the forepart, and a segmented hind part divided longitudinally into three lobes, hence the name trilobite.
729. Tropopause: The interface between the troposphere and the stratosphere.
730. Ultramafic: Relating to or denoting igneous rocks composed chiefly of mafic minerals.
731. Ultramafic rocks: Magmatic rocks with very low silica content and rich in minerals such as hypersthene, augite, and olivine.
732. Universal evolutionary convergence: The hypothesis that the independent and repeated emergence of life and certain adaptive traits is wide-spread throughout the cosmos.
733. Universal Time Coordinate (UTC): The primary time standard by which the world regulates clocks and time. It is within about 1 second of mean solar time at 0° longitude such as UT1 and is not adjusted for daylight saving time. It is effectively a successor to Greenwich Mean Time (GMT).
734. Uranium–Lead dating (abbreviated U–Pb dating): One of the oldest and most refined of the radiometric dating schemes. It can be used to date rocks that formed and crystallized from about 1 million years to over 4.5 billion years ago with routine precisions in the 0.1–1 percent range.
735. UVC radiation: Ultraviolet radiation with wavelengths between 200 nm and 290 nm.
736. Vent ecosystem communities: Organisms that utilize only inorganic and/or abiotic simple molecules for their carbon and energy sources so that they do not

rely on other living organisms to feed, develop, and multiply.
737. Vesicle: A structure within or outside a biological cell, consisting of liquid or cytoplasm enclosed by a lipid bilayer. Small fluid-filled bladders, sacs, cysts, or vacuoles within the body.
738. Vesicular basalt: A dark-colored volcanic rock that contains many small holes, more properly known as vesicles. A vesicle is a small cavity in a volcanic rock that was formed by the expansion of a bubble of gas that was trapped inside the lava.
739. Vitrification: Conversion into a glassy substance by heat and fusion.
740. Volatiles: Substances (elements and compounds, including water) that easily evaporate/vaporize at low temperatures.
741. Volcanism: The phenomenon of eruption of molten rock onto the surface of the Earth or a solid-surface planet or moon, where lava, pyroclastics and volcanic gases erupt through a break in the surface called a vent.
742. Volcano: A mountain or hill, typically conical, having a crater or vent through which lava, rock fragments, hot vapor, and gas are or have been erupted from the earth's crust.
743. Volcano-speleogenetic processes: Those processes which operate on or close to the surface of the Earth and which involve weathering, mass movement, fluvial, aeolian, glacial, periglacial, and coastal processes.
744. Water activity (*aw*) of a food: The ratio between the vapor pressure of the food itself, when in a completely undisturbed balance with the surrounding air media, and the vapor pressure of distilled water under identical conditions.
745. Water Cycle: The water cycle describes how water evaporates from the surface of the earth, rises into the atmosphere, cools, and condenses into rain or snow in clouds, and falls again to the surface as precipitation. The cycling of water in and out of the atmosphere is a significant aspect of the weather patterns on Earth.
746. Weakly reducing atmosphere: An atmosphere of N_2 and CO_2 with trace amounts of H_2O, CH_4, carbon monoxide (CO), and hydrogen (H_2). Such an atmosphere contains practically no oxygen.
747. Weasel family: A family of carnivorous mammals.
748. Yardang: A streamlined protuberance ('ridge and furrow' landscape) carved from bedrock or any consolidated or semi-consolidated material by the dual action of wind abrasion by dust and sand and deflation.
749. Zircon: A mineral occurring as prismatic crystals, typically brown but sometimes in translucent forms of gem quality. It consists of zirconium silicate and is the chief ore of zirconium.
750. Zygote: Fertilized egg cell that results from the union of a female gamete (egg, or ovum) with a male gamete (sperm).
751. 16S rRNA (16S ribosomal RNA): A component of the prokaryotic ribosome 30S subunit. The 16S rRNA gene is the DNA sequence corresponding to rRNA encoding bacteria, which exists in the genome of all bacteria. 16S rRNA is highly conserved and specific, and the gene sequence is long enough.

Index

Page numbers followed by "*f*" and "*t*" indicate, figures and tables respectively.

A

Abitibi Greenstone Belt, 360–361
Acid and metalliferous drainage, 231
Acidic waters, 320
Acid mine drainage (AMD) habitats, 321
Acidophilic green alga, 321
Acid rock drainage (ARD), 231
Adenine, 168
Adenosine triphosphate (ATP), 127, 157, 206
    synthesis, 123–124
Advanced Camera for Surveys (ACS), 546
Advanced Life Support (ALS) system, 313
Alcyonium glomeratum, 720*f*
Algae, 694–695, 705*f*
Alkaline hydrothermal systems, 128
American robotic space probe, 554
Amino acids, 151, 156–157, 167
    hydrophobicity of, 152
Aminoacyl-tRNA synthetases, 145, 152
AMISTAD expedition, 229
Ammonochemistry, 62
Amoebae, 148
Amor Asteroids, 14
Antarctic, 217
    ice sheet, 216, 217*f*
    lakes, 692
    marine environment, 210–211
    micrometeorites, 192, 193–194
    Ross Desert, 177, 321, 322
    sessile benthic community, 217, 218*f*
Antarctic Dry Valleys (ADVs), 499
Antigreenhouse effects, 622
Apollo Asteroids, 14
Apollo mission, 407*f*, 407, 408, 408*f*, 409*f*, 410*f*, 437
    existence of water, 405, 412
        analysis of hydrogen (H) and hydroxyl, 412–413, 415
    missions 11, 12, 14, 15, 16, and 17, 409–410
    plant biology experiments, 436–437
Apsidal precession, 96
Archaea, 149, 227, 328
    in acid mine drainage, 232*f*
    cell membranes of, 234
    discovery of, 227, 228, 229
    extreme halophilic, 231
    extremophile, 231
    general features, 230
    halophilic, 485, 486
    hyperthermophilic, 231
    search for life on extraterrestrial worlds, 235
    sizes and shapes exhibited, 230–231
    thermoplasmata, 321
    unique features, 230
Archaeal halophiles, 231
Archaebacteria, 228–229, 703
Archean Eon, 55, 77
Archean hydrothermal vents, 126
Arctic and Antarctic thawing, 94–95
Aristotle, 45
Assam meteorite, 12
Association of Biomolecular Resource Facilities Metagenomics Research Group (ABRF MGRG), 318, 319
Asteroid and comet-impacts on Earth, 213
Asteroids, 13, 14
    discovery and characterization of, 13
Astrobiology, 201
    definition, 201
    interdisciplinary nature of, 201
Atacama Desert, 218, 726
    brines of, 295
    habitability of, 318
    microbial diversity, 211
    similarity with Mars surface, 318
Atacama Large Millimeter/submillimeter Array (ALMA), 4
Aten Asteroids, 14
Athalassic environment, 691, 692
Athalassic water bodies on Earth, 692
    chemical composition of, 693
    life in brackish water, 708, 709, 710, 711
Atlantic Blue Crab, 706*f*
Atlantic Lobster, 707*f*
Atlantis Massif, 339*f*, 339, 341–342
Atmospheric chemical disequilibrium, 192
Aurorae, 391
Automated Approach for Ribosomal Intergenic Spacer Analysis (ARISA), 717
Autotrophic cells, 127
Autotrophy, 124, 348
Axel Heiberg Island, 330
    cryo-environments on, 331
    hypersaline cold springs on, 330, 332, 333
    saline spring systems on, 331
Axial precession, 96

B

Bacillus species, 723
    colonial morphology, 723
    endospore formation, 723–724
Bacteria, 149, 344
    Archaebacteria, 228–229, 703
    extremophilic and extremotolerant, 206, 207, 207, 208, 209
    gram-positive bacteria, 234, 723
    in Great Salt Lake, 719
    green sulfur, 730–731
    halophilic, 730
    iron-reducing bacteria (FeRB), 692
    in Lost City Hydrothermal Field, 346–347
    magnetotactic bacteria, 483, 484
    oxygen-producing, 56
    phototrophic, 730
Baltic Sea, 704, 729*f*
    fauna of, 704
    salinity of, 704
Barberton Greenstone Belt, 80
Basal mound smoker, 338
Bay of Bengal, 693–694
Bay of Pigs, 693–694
Bdelloid rotifers, 226
Beehive Smoker, 121*f*, 121–122
Beluga sturgeon fish, 710
Beresheet, 240
Bernoulli, David, 41–42, 69
Bioactive peptides, 120
Biogenic isotopic anomalies, 204
Bioleaching, 319–320
Biological bet hedging, 223–224
Biomarkers, 175–176
    of cryptoendolithic activity, 177
    detection, 171–172
    identified on Venus, 177
    lipid-derived, 176, 178
    as metabolites, 175–176
    methods for identification, 181
Biosignatures, 190, 311–312
    carbon chemistry, 170, 170, 171
    characterisation using time-of-flight secondary ion mass spectrometry, 181
    chemical, 170
    chemical identification of, 178
    definition, 169
    under diverse environmental conditions, 178
    false positives and false negatives, 189, 189, 190
    gases, detection of, 186, 187
        extrasolar, 188, 189, 194
    means of studying, 180
    microbes, 169–170
    of microorganisms, 170
    mineralogical, 176

molecular, 181
    measurable attributes of, 191
morphological, 172, 173, 177
    Noachian period, 172
remote sensing of, 187
type-I, 190
vs bioindicators, 175
Black smoker hydrothermal vent, 115, 121–122, 336–337, 337*f*, 337, 337, 338, 339, 344
Blue catfish, 713
Blue-footed booby *(Sula nebouxii)*, 702, 728*f*
Boivin, André, 142
Bottlenose dolphins, 708
Brain Coral, 700, 721*f*
Breadboard Project was the Biomass Production Chamber (BPC), 313
Brenner, Sydney, 139
Brown-headed Gull, 706

C

Calcite, 91
Calderas, 379
Callisto, 553, 556
Cancer Genome Atlas project, 141
Cantaloupe terrain, 335
Carbon, 157, 170
    $^{13}C/^{12}C$ ratio, 171
    inorganic, 170–171
    organic, 170–171
Carbonaceous chondrites, 12, 128
Carbonate–silicate cycle, 90, 91
    global climate and, 91
Carbon chemistry, 176
Carbonyl sulfide, 120
Carboxylic acid, 167
Caspian Sea, 692*f*, 692, 708, 709
    animals living in and around, 708, 709
    biodiversity, 708
    fish species, 708–709
Caspian seal, 709
Caspian sturgeon, 710
Cassini-Huygens mission, 623, 656
Cassini plasma spectrometer, 629, 637
Cassini spacecraft mission, 586, 618, 623, 624, 636, 649, 654
Cassini Synthetic Aperture Radar, 644
Cellular respiration, 715
Cellulosimicrobium cellulans, 217
Cephalopods, 695
Ceres, 526*f*, 530, 532, 535, 536–537
    internal structures, 534*f*
    views, 529*f*
Ceustigma, 321
Chaos terrains, 312
Chaotropes, 328
Chasmoendolithic habitats, 212
Chelyabinsk event, 2013, 14
Chemiosmosis, 123–124, 127
Chemiosmotic coupling, 127
Chemolithotrophic microbial communities, 313–314, 346
Chemosynthesis, 344
Chesapeake Bay, 693–694
Chicxulub crater, 88, 89, 90
Chicxulub impact tsunami, 88

Chilika Lake, 704–705
    birds, 706
    fauna, 706
    flora, 706
Chirality, 476
Chloroplast, 715
Chondrites, 118
Chondritic meteorites, 12
    carbonaceous, 12
    classes, 13
Circumstellar habitable zone, 203, 204
Coastal tides, ecological implications, 76
    on animals, 76
    on intertidal species, 76
    lunar periodicity and, 76
    phytoplankton aggregations, 76
Coccolithus pelagicus ssp. *braarudii*, 704*f*
Colonization on rocks, 208
Compact Reconnaissance Imaging Spectrometer for Mars (CRISM), 326
Conamara Chaos, 312
Conserved signature inserts and deletions (CSIs), 230
Continuous planetary habitability, 202
Controlled Ecological Life Support System (CELSS), 313
Cool early Earth hypothesis, 67
Copernicus, Nicolaus, 45–46
Corals, 699, 700
    algae of, 700
Corioli's force, 72–73, 75
Cornwallis Island, 208
Cosmic dust, 193
Cosmic rays, 659
Crenarchaeol, 175
Cretaceous–Paleogene extinction (K–T extinction). *See* K–Pg extinction
Crustaceans, 695
Cryoconite holes, 216–217
Cryo-environments/cryogenic environments, 330
Cryosectioning, 185
Cryptobiosis, 223–224
Cryptoendoliths, 177, 178
    habitats, 212
    microorganisms, 322
Curiosity rover mission, 507, 508
    ionizing cosmic radiation, 508
    search of organics, 508, 509
    working of mass spectrometry, 509–510, 510*f*
Cuttlefish or cuttles, 695
Cyanoacetylene, 118
Cyanobacteria, 78*f*, 78, 82, 153–154, 321, 328, 694, 714, 715
    anoxygenic photosynthesis and, 79
    multicellular forms, 79
    oxygenic photosynthesis and, 79
    role of solar ultraviolet radiation (UVR) on, 78–79
Cytosine, 168

D

Danube sturgeon. *See* Ossetra sturgeon
Darwinism, 126

Dawn mission, 523, 524*f*, 524, 525, 531–532
Dawn spacecraft, 525, 532–533, 538
Dawn spectrometer data, 533
Dawn Visible and InfraRed (VIR) mapping spectrometer, 530
Dead Sea, 714
    archae, 714
    bacteria and cyanobacteria, 715, 717
    chloroplasts, 715
    diatoms, 718–719
    Halobacteria, 714
    hydrogen sulfide (H_2S) content, 714, 717
    microbial life, 718, 719
    mitochondria, 715
    oxygen level, 714
    underwater springs, 715
Deep-ocean cold gas seeps, 210*f*
Deep-sea hydrothermal vents, 120, 121, 122, 173*f*, 209*f*, 211
    benthic communities, 345
    biological and geological processes in, 345
    geothermal energy, 348
Deoxyribonucleic acid (DNA), 129, 130*f*, 156, 159, 206, 319
    base pairs, 141, 168
    basis of grouping organisms, 148
    compaction, 234–235
    dark, 147, 148
    discovery of, 131
    double helix, 132*f*, 134, 136*f*, 136, 137*f*, 229
    genetic role of, 134
    pairing rule, 136–137
    re-modeling, 235
    as a repository of genetic information, 139–140
    sequencing, 134, 144*f*
        nanopore-based, 141
Descent Imager/Spectral Radiometer (DISR), 635
Devon Island, 207, 208, 213
Diapirs, 312
Diatoms, 694
Dimethyl sulfide (DMS), 179
Dinoflagellates, 694
Discovery program, 524
Dolomite, 91
Dolomitic marble, 91
Dolphins, 697, 710*f*
Don Juan Pond, 692
    cyanobacteria, 77
    diatoms, 721
Donohue, Jerry, 136
Doppler shift of radio signals, 554
"Dry Moon Paradigm", 436
Dugong *(Dugong dugon)*, 706
Dust devils and vortices, 335–336, 370
    on Aeolis Mons (volcanoes), 372
    aircraft accidents, 371
    electric fields, 370, 371
    impact on climate and environment, 372, 372, 373
    phenomenon of, 370*f*, 370
    role in dust cycle, 373
    terrestrial, 371
    on Venus, 373, 374

E

Early Eocene hyperthermals, 94–95
Earth, 169, 202, 359
   analog ecosystems, 313–314
   asteroid and comet-impacts on, 207–208
   atmospheric composition and climate, 365, 366, 369f
      "Late Heavy Bombardment" period, 366
   based measurements, 546
   based radar, 527
   biosphere, 336
   bubbles bursting in, 334, 335
   centric universe, 45
   clement surface conditions on, 77
   core of, 32–33
   diameter of, 33
   distance from Sun, 374, 375
   early climate, 55–56
   equatorial plane, 71
   first ice age, 56
   geodynamics, 360–361
   gravity, 376
   inclination and orbit, 34
   influence of frequent hits/collisions, 32
   interior, 362, 363
   liquid water formation, 65
   magnetic field, 391
   mass, 376
   mass extinction events, 83
   Microbiome Project, 318
   microbores, 77
   ocean, 41f, 41, 42
      chemical composition, 41
      geological evidence, 66
      Hadean oceanic crust, 57
      heating of surface, 40–41
      hydrostatic pressure, 41
      IWs, 42
      oceanic surface currents, 40–41
      Panthalassic, 67
      sources, 66
      thermohaline conveyor belt circulation, 40–41, 41f, 41
      tidal forces, 41, 42
      water density, 41
   oxygen, 23
   paleo-rotation and revolution, 33
      balancing of opposing torques, 33
      gradual rotational deceleration, 34
      sedimentary cyclic rhythmites of tidal origin, 33–34
   plants, 56
   size and composition of, 362
   spin, 33
   stromatolites on, 77, 78, 78, 79
   surface, 33
      core, 11, 363
      crust, 11, 33, 55, 363
      mantle, 11, 363
      molybdenum (Mo) isotope ratios, 58
      noble gas isotope ratios, 58
      "Snowball Earth" hypothesis, 83
      Xe/Kr ratios, 58
   tectonic evolution, 359, 359, 360
      radiogenic heating, effect of, 361, 362f
      scale of, 361
      surface velocities, 361
      temperatures, 360
      terrain, 362
   volcanism and surface features, 378f, 378, 378, 379 See also (Lava tubes)
   water content in, 34, 35
   water vapor on, 65
Earth's Moon, 25f, 314, 315, 405
   anomalies, 23–24
   Bullialdus crater, 429
   from cataclysmic impact, 24
   cold traps, 420
   from collision impact, 28
   early-forming apatite, 426
   evolution of, 412
   existence of water, 404, 405, 406, 412, 429, 430, 431 See also (LRO-LCROSS mission)
      lunar convective core dynamo considerations, 427–428
      magmatic water, 425–426
      water molecules trapped in rocks, 411, 431
   "giant impact" model, 426
   Giant-impact theory on origin (See Giant-impact theory on origin)
   growth of plants on, 317
   hydration, 419, 420
   hydrodynamic simulations, 26
   impact on Earth's environmental conditions, 31, 32
   impactors of, 27
   isotopic homogeneity, 23, 24
   lava tubes, 315, 316, 432, 433
   mineral content, 418
   Moonlet formation, 24f, 27, 28
   multi-impact hypothesis, 24, 25, 27
   numerical simulations analysis, 25
   from off-axis energetic impact, 25
   orbit, 375
   phase space analysis, 25–26
   Polar Regions, 64
   polar regions, 418, 419–420
   siderophile elements, 28–29
   study of SPA, 438
   underground caves and tunnel complexes, 433
   volatile zoning, 426
   Wolfram (Tungsten), 26–27
Earth system, 566
Eccentricity, 95f, 95
Echiniscoides sigismundi, 222
Ecliptic, 1
Eextremophiles, 211
Einstein's General Theory of Relativity, 159
Electromagnetic induction, 554
Elliptical orbit, 1
Enceladus, 11, 13, 42, 61
   energy of, 45
Enceladus, 583, 584f, 584, 587–588, 589, 594
   CH_4 under, 334
   ecologies of, 334
   gravity conditions, 335
   Saturn's moon, 333, 334, 335
   subglacial water ocean, 335
   transport of organics, 334, 334, 335
Enceladus' south polar terrain, 312
Endangered Species Act, 696
Endolith, 322
Endolithic organisms, 212
Endosomal Sorting Complex Required for Transport (ESCRT)-III proteins, 235
Energy sources, 44, 45
Eocene Epoch, 90
Epilithic habitats, 212
Epsilonproteobacteria, 714
Escape velocity, 26
Escherichia coli, 146, 147
Euendolithic habitats, 212
Eukarya, 233–234
Eukaryotes, 148, 149, 321
Europa, 312, 551, 552, 553, 557, 566, 571, 741, 745, 746
   chaos terrains, 312
   chaotic terrains of, 350
   ground-based spectroscopy, 551
   hydrothermal sulfate reduction within, 350
   ice shell, 312, 313
   subsurface ocean, 350
Europa Clipper mission, 42, 553, 570, 573
Europa Global model of Exospheric Outgoing Neutrals, 556
Europa Jupiter System Mission (EJSM) spacecraft, 561
Europan ionosphere, 562
European Space Agency (ESA), 744
Euryhaline, 693
Eutectic-point, 312
ExoMars mission, 171, 471, 472f, 472, 473, 508, 510
   beginnings, 471
   exobiology mission, 471
   goal, 471
   strategy to achieve objectives, 471–472
Exons, 145
Exoplanet Biosignatures Workshop Without Walls (EBWWW), 188, 189–190
Exoplanets, 187, 205, 336
Exopolysaccharides (EPSs), 209
   as biosurfactants, 211
   biosynthesis, 210
   commercial application of, 211
   cryoprotective role, 210
   marine EPS-producing microorganisms, 210, 211
   microbial, 210
Extraterrestrial life, 156, 157, 159, 167–168
   biospheres, absorption features of, 179, 180
   habitability, 202
   Kumar's hypothesis, 158, 159
Extraterrestrial ocean worlds, 43
Extreme Microbiome Project (XMP), 318, 319
Extremophiles, 205, 207, 209, 218, 322, 691
   archaea, 231
   habitats, 236
   mapping of, 318
   taxonomic classification of, 318
Extremozymes, 211

F

Fabaceae, 706
Fatty acids, 167

Felix Hoppe-Seyler's laboratory, 132
Filamentous anoxygenic phototrophs (FAPs), 178
Fischer-Tropsch-type synthesis, 174
Formaldehyde, 118
Frost line, 7, 36
Fucus serratus, 705f

G

Gaian bottleneck, 204, 205
Gaian regulation, 204
Galilei, Galileo, 45–46
Galileo, 555, 571–572
Galileo magnetometer data, 570
Galileo mission, 548
Galileo Probe Mass Spectrometer (GPMS), 8, 9
Galileo Solid State Imaging, 567
Galileo spacecraft, 560
Gamma ray, 525
Gamma Ray and Neutron Detector (GRaND), 532
Ganymede, 553
Ganymede's silicate mantle and metallic core, 549–550
Gas Chromatograph and Mass Spectrometer (GCMS), 633
Gas chromatograph mass spectrometer (GCMS), 625
Genome sequencing, 147
Giant-impact theory on origin, 14–15
    distribution of moonlet-Earth impact angles, 28f
    multiple, 22, 23, 28–29
        measurements of isotopes, 23
        oxygen content, 23
    single, 2, 15f, 15, 16, 22
        corotation limit (CoRoL), 17
        formation of Moon, 16, 21–22
        Lagrange point, 18
        proto-Earth, 19–20, 21
        Roche lobe, Roche sphere, and *Roche limit*, 17–18, 18f, 18, 19, 20, 21
        smoothed particle hydrodynamics method (SPH), 20, 21
Giant planet formation
    core accretion mechanism, 8
    core accretion model, 8, 9
    diffusion equation, 7
    disk instability model, 8
    enhancement of abundance of solid material, 7
    formation of Uranus and Neptune, 9
    frost line, 7
    gas giants, 7
    ice giants, 7
    Jupiter's atmosphere, 8–9
    snow line, 8
    solid accretion rate, 8
    water vapor condensation, role of, 7
Giant white clams *(Calyptogena magnifica)*, 345f
Glacial isostasy, 323–324
Global climate change, 56, 92
Glycerol, 167
Glycerol di-alkyl glycerol tetraether (GDGT) lipids, 175

Gneissic banding, 213f, 213
Goldilocks zone, 34, 202
Gram-positive bacteria, 234, 723
Great Oxidation Event, 81, 82
Great Oxygenation Event (GOE), 79
Great Red Spot, 545–546
Great Salt Lake, 692, 693f, 719
    algae, 719, 720
    bacteria and protozoa, 719
    brine flies (genus *Ephydra*), 720
    brine shrimp, 720, 721
    cysts, 288
    migratory birds, 719, 721
        avian (relating to birds) sex roles, 721
        Wilson's Phalaropes, 721
    salinity of, 692, 719
Greenhouse effects, 55, 56f, 366, 622
    carbon dioxide-induced, 57
    effect on Earth, 55
Greenhouse gas, 55
    in Earth's atmosphere, 55
Greenland's icecap, 216–217
Guanine, 168
Gypsum Hill spring, 330f, 331, 332f, 332–333
    microbial community of, 331f, 331, 332
Gypsum Hill summer flood plain, 332

H

Habitability, 201–202, 311–312 *See also* Mars
    abiotic habitable zones, 205–206
    in Antarctic ice sheet, 216, 217f
    in Arctic and Antarctic polar deserts, 208
    in astrobiology, 202–203
    availability of liquid water, 205
    beneath seafloor, 214, 214, 215, 215, 216
    continuous planetary, 202
    definition, 201–202
    determining, 202
    at driest desert, 218
    ExoGaia model, 204–205
    extraterrestrial, 202
    of extremophilic and extremotolerant bacteria, 206, 207, 207, 208, 209
    factors contributing to enhancement, 207–208
    Gaian bottleneck model, 204, 205
    habitable conditions, 202
    in high-latitude polar deserts, 207
    on and inside rocks, 211, 212, 212f, 212, 213, 214
    interior liquid water worlds, 202
    planet, 204, 205
    prerequisites and ingredients for life, 204
    of saline and hypersaline environments, 328
    surface liquid water worlds, 202
Habitable conditions, 202
Habitable planets, 202
Habitable zone (HZ), 202
    terrestrial planets, 187
Hadean Eon, 55, 56f
Halites, 302
Haloarchaea, 230
Halobacterium sp., 231
Halophiles, 211, 231, 500

Haloquadratum walsbyi, 232f
Halorubrum lacusprofundi, 500
Haughton crater, 335
Helicopter catfish, 706
Helio-centrism, 45–46
Helium, 34
Hellas Planitia, 315–316
Hellas Planitia lava tubes, 381, 382
Hematite, 320–321
High-energy photons, 630
High Mass Organic Cations (HMOC), 335
High- Resolution Imaging Science Experiment (HiRISE), 326
High-Resolution Stereo Camera (HRSC), 326
HOCNSP compounds, 157, 188–189
Holley, Robert, 147
Homo erectus, 153–154, 154f
Homo sapiens, 153–154, 155
Hubble space telescope, 524, 546, 547f, 559–560, 568, 569, 584f, 692–693
Hubble's Space Telescope Imaging Spectrograph, 570
Hudson Bay, 693–694
Huygens Probe, 625
Hydrolases, 211
Hydrothermal activity, 125
Hydrothermal circulation, 535
Hydrothermal Vent Fields Ecologically or Biologically Significant Marine Area (EBSA), 347
Hypersaline brines, 328
Hyperthermophiles, 211
Hyperthermophilic (HT) organisms, 232, 233, 234
    archaebacteria, 703
    chemolithoautotrophic, 234f, 234
Hypolith, 208
Hypolithic habitats, 212
Hypsibius dujardini, 236, 240–241

I

Ice Ages, 97
Ice-covered oceans, 738
Ice on the Earth, 311
Ice-penetrating radar, 561
Icy Moons Exploration (RIME), 562
Icy worlds, 43
Imaging science subsystem (ISS), 589
India's Chandrayaan-1 Lunar Probe, 417f, 417, 417, 418
    achievements, 419
    ESA Payload, 418–419
    last phase, 425, 431
    mapping of landing sites of Apollo 15 and Apollo 17, 418
India's National Centre for Polar and Ocean Research (NCPOR), 217
Infrared Mapping Spectrometer, 592
Instantaneous habitability, 202
Integrated Ocean Drilling Program (IODP), 215–216
Intergovernmental Panel on Climate Change (IPCC), 57
Interior liquid water worlds, 202

Internal waves (IW), 42
International Ultraviolet Explorer (IUE), 528
Interplanetary dust particles (IDPs), 60
Introns, 145
IPBSL (Iberian Pyrite Belt Subsurface Life Detection), 321
Iron-reducing bacteria (FeRB), 692
Irrawaddy dolphins *(Orcaella brevirostris)*, 708
Isua Greenstone Belt, 66, 203, 204, 204f, 212

J

James Webb Space Telescope (JWST), 187
Jellyfish, 701
Jezero Crater, 743f
Jezero lake, 389f, 389
Jordan Dead Sea Rift, 692
    holomictic nature of, 692
    microbial ecosystem in, 692
    salt concentration in, 692
    spring system, 692
Jovian satellite Europa, 551
Jovian system, 557
Juan de Fuca Ridge, 338
Judea Group Aquifers (JGA), 717–718
Jupiter, 545f, 545, 562
    colorful bands, 545
    core, 3
    gravitational forces, 60
    Great Red Spot, 64, 546f
    Moon, 39–40, 42, 311–312
Jupiter broadcasts radio waves, 548
Jupiter Energetic Particle Detector Instrument (JEDI), 547
Jupiter Europa Orbiter (JEO), 561
Jupiter's atmosphere, 548–549
Jupiter's gargantuan (enormous) magnetic field, 547

K

Kant, Immanuel, 1, 2f
Kelvin, Lord, 41–42, 69
Kendrew, John, 139
Kepler, 1
    three laws on planetary motion, 46
Kodiak butte, 388
Komatiite, 80
Kuiper Belt, 59–60, 60f, 60, 691, 692, 698, 706, 710, 737
    Objects, 175

L

Laboratory of Molecular Biology (LMB), 139
Lagoons in Australasia, 79
Lake Assal, 692–693, 697f, 724
    bacteria, 724
    halophiles, 725
    minerals, 724
    salinity level, 724
    watershed area of, 724
Lake Texoma, 711
    species, 711
Lake Vostok, 333f, 333
    microbial community, 333

Laplace, Marquis de, 41–42, 69
Laplace, Pierre-Simon, 1–2
Laplace-like resonance, 549
Laplace resonance, 701
Laplacian nebular model, 2
Laser-induced desorption, 594
Last common ancestor (LCA), 148–149
Last Universal Common Ancestor (LUCA), 128, 149, 153
Late Heavy Bombardment (LHB), 79–80
Lava tubes, 314, 315
    Earth's, 315, 316
    Hellas Planitia, 381, 382
    as human habitats, 317
    of Kilauea Volcano, 316
    near Hadriacus Mons, 382f, 382, 383f, 383, 384f
Libration, 587
Life-forms, classification of, 148
    cells and viruses, 148, 149, 150
    cellular domains, 148–149
    three-domain classification, 149
Life Marker Chip, 171
Life on Earth, 62, 63, 64, 65, 314, 476
    biological evolution, 129
        adenine, 129, 130f
        A-DNA, 129–130
        B-DNA, 129–130
        cytosine, 129, 130f
        deoxyribonucleic acid (DNA) and ribonucleic acid (RNA), 129, 130f, 131
        guanine, 129, 131f
        thymine, 129, 131f
        uracil, 129, 131f
        Z-DNA, 129–130
    building blocks of life, 158, 169
    chemical elements for, 63–64
    communities of microorganisms, 63
    diversity of minerals, 170
    in Earth's lithosphere, 175
    evolution, 153, 154f, 154, 155
    influence of thermodynamic disequilibrium, 155, 156
    role of topographical gradients, 155
    water vapor transport, 155
    interpretations, 152, 153
    isotopic biomarkers, 170
    level of metabolism, 63
    lightning in early atmosphere, 116
    liquid-water environment for, 63, 64
    Miller-Urey experiment, 116f, 116, 116, 117, 117, 118, 119, 120
    at molecular level, 156, 156, 157
    origin of life (OoL), 115
        breakdown of carbonyl sulfide (COS), 120
        from carbon and hydrocarbons of comets and meteorites, 128, 129
        chemical evolution, 150
        chemical processes at submarine volcanic vents, 120
        chemolithotrophic microbial communities, 125
        deep-ocean alkaline hydrothermal systems, 126–127

        hydrothermal vents, importance of, 121, 122
        importance of organic molecules, 150
        microenvironments, 125–126
        Oparin's tenets, 115
        of organic compounds, 122, 123
        role of hydrothermal vent, 123, 123, 124, 125f, 125, 126
        serpentinization process, 121
        spontaneous, 115–116
    prebiotic chemical evolution, 153
    production of amino acids, 116
    use of proteins, 63
Life on Mars, 176, 470
    biominerals, 176
    measurable attributes of molecular biosignatures, 176
    microbial signatures in extraterrestrial rocks, 178
    signs of primitive life, 481
        arguments for, 481
        environmental conditions, 488
        organic compounds on celestial body, 506, 507, 507, 508, 508, 509, 509, 510
        probable ancient life, 482, 483
        relic biogenic activity, 482
        spores of *Bacillus subtilis bacteria*, 486, 487
        studies of Haloarchaea and spore-forming bacteria, 484, 485, 486
Lithium chloride-dominated hypersaline salt flats and ponds, 725
    effect on microbial cells, 725–726
    microbial diversity, 726
Lithium Triangle Zone, 725
"Little Red Spot.", 546–547
Locked nucleic acid (LNA), 140–141
Long Range Reconnaissance Imager (LORRI), 547
Lost City Hydrothermal Field (LCHF), 339–340, 343f
    active carbonate structures, 341
    bacteria and Archaea, 346–347
    fissure-filling deposits, 342f
    geochemistry, 341
    hydrothermal geochemistry, 341
    inactive structures, 342f
    IODP Expedition, 347
    Methanosarcinales phylotype, 346–347
    mineralogy, 342–344
    organisms supported by, 346, 346, 347, 347, 348
    petrographic and geochemical features, 342–344, 344f, 344
    serpentinization reactions, 341
Lost City hydrothermal field (LCHF), 121–122, 122f, 123, 126, 127
    alkalinity, 123–124
    chimneys at, 123–124
Lost City hydrothermal vent structures, 339, 340
LRO-LCROSS mission, 420–421, 422
    design, 421
    detection of thermal emission, 421
Lunar magma ocean (LMO), 29f, 29, 30
Lunar Receiving Laboratory (LRL), 437
Lyman-alpha photons, 676

M

Magma chambers, 337
    fed hydrothermal vents, 344
        animal communities, 345
        bacteria and archaea, 344
        benthic communities, 345
        ecosystems, 344–345
Magnetic field
    Earth, 391
    Mars, 392, 393
    Venus, 391, 392
Magnetite, 484
Magnetohydrodynamics, 550
magnetotactic bacteria (MTB), 483, 484
Magnetotaxis, 484
Main asteroid belt (MAB), 523
Mangalyaan mission, 489, 490, 491
    Delta Differential One-way Ranging (DDOR) measurements, 490
    D/H ratio estimation, 497, 498
    elliptical and eccentric orbit of, 493f, 496
    imaging of Mars' surface, 491–492, 492f
    Mars Exospheric Neutral Composition Analyser (MENCA) altitude profiles, 495f, 495, 496
    Mars Orbiter Mission (MOM), 489, 490
        image of Phobos and Deimos, 493f, 493, 494
    Methane Sensor for Mars (MSM), 494, 495
    minerology analysis, 494f, 494–495, 495f, 495
    morphology study of Ophir Chasma, 494
    Neutral Gas and Ion Mass Spectrometer observations, 496
    objectives, 490
    radial profiles, 498f, 498
    SHARAD data, 491
    success of, 498, 499
    thermal emission, study of, 496, 497f
MarineLab, 313
Mars, 315, 359, 363, 364f
    Anatolia trough system, 333f
    Arabia Terra, 389, 390f
    Aster clouds, 380
    atmospheric composition and climate, 368, 369f
        atmospheric pressure, 330–331, 455, 456
        deuterium/hydrogen (D/H) ratio, 470
        raindrops, 456
        temporal differences, 456
        terminal velocity, 456
    clay mineral formation, 460, 461
    climatic stages, 453
    craters, 455, 456, 464, 466–467
        NASA's *Perseverance* mission, 510
    crust, 364, 365, 464
    dark pits, 381
    diameter, 365
    difference in color between Earth and, 365
    distance from Sun and Earth, 376f, 376
    dust on, 364–365, 368, 369, 370, 459
        primary effect of, 369
    energetic particles, 473–474
    escape velocities, 238
    existence of primordial ocean, 385, 386, 386f, 386, 387
    evidence of tsunami waves, 389, 390
    existence of water, 464
    galactic cosmic rays, 474
    gravity on, 364, 469, 470
    habitability, 172f, 456–457, 470, 475
        environmental conditions, 476, 477f
        presence of hydrogen peroxide (H_2O_2) and, 506
        Viking mission, 457
    landing site selection criteria, 475
    landslides, 365
    lava tube caves on, 473, 474 See also (Lava tubes)
        at *Hellas Planitia*, 474
        near Hadriacus Mons, 474
    liquid water presence on, 330–331, 454f, 454, 455, 457, 475
        Clifford's hypothesis, 465–466
        day-night variations, 463–464
        Mars Express (MEx) spacecraft missions, 466, 489
        MARSIS expedition, 467–468
        at polar caps, 467
        underground lake, 466
    magnesium sulfate salts, 329–330
    magnetosphere of, 392, 393, 474
    Mare Acidalium quadrangle, 389
    mass of, 364
    meteorites of, 458, 461, 470
    methane content, 239
    microbial life, 468, 475, 477
        microbial reduction, 499
        model studies, 469
        prokaryotic life, 468, 469
    mineral compositions of, 239, 385, 458, 458, 459
        clinoptilolite, 459
        ESA's Mars Express mission, 460
        Mg-Fe-bearing clays, 479
        mineralogical mapping, 459
        phyllosilicate deposits, 478
        phyllosilicates, 461
        sulfates, 460, 461, 462, 463
        zeolites, 459
    mountains and valleys, 365
    Oasis on Alexander Island, 208
    open-lake systems, 388, 388, 389
    Opportunity mission, 330f
    origin of life (*see* Life on Mars)
    paleolakes, 330
    perchlorates on, 499
        bacteriocidal effect of UV-irradiated, 500, 504f
        as energy source, 499, 500
        impact of, 505, 506
        induced bacteriocidal effects, 504–505, 505f
        interactions with other soil components, 503, 504, 505
        salts, 500
    pingos, 325, 326, 327f, 327–328
    polar caps of, 465–466
        Planum Australe region, 467f, 467
    position in Solar System, 456–457
    possibility of halophilic life on, 510, 511
    presence of organic substances, 506–507
        Mars Exploration Rover missions, 454, 507
    presence of saline bodies of water, 330
    rainfall features, 455–456
    river networks on, 455f, 455
    river plains, 387, 387, 388
    rocks and soil, 453, 457–458
    saline sediments, 459
    sedimentary rocks, 464
    solar storms, 473–474
    subsurface processes on, 464
    surface, 454f
    tectonic evolution, 455
    temperature, 454
    types of microorganisms, 457–458
    Utopia Planitia region, 326f
    valley networks, 456f
    volcanic regions, 314
    volcanism and surface features, 379, 379, 380 See also (Lava tubes)
Mars Advanced Radar for Subsurface and Ionospheric Sounding (MARSIS) experiment, 491
Mars Orbiter Camera (MOC), 326
Mars Orbiter Laser Altimeter (MOLA), 326
Mars Science Laboratory (MSL) mission, 507
MARTE (Mars Astrobiology Research and Technology Experiment), 321
Matagorda Bay, 694
Maximum Diurnal Insolation (MDI), 712
Max Planck Institute for Solar System Research, 524
Maxwell, James Clerk, 2
McCarty, Maclyn, 133f, 133–134
McLaughlin Crater, 464, 465f, 465
McMurdo Dry Valleys of Antarctica, 212
Medusae Fossae Formation (MFF), 491
mega-ripples, 88–89, 89f, 89
    size and orientation, 89, 89, 90
Mercury, 364
Meridiani Planum rocks, 320
Metagenome, 226
Metagenomics, 347
Meteorites, 61, 66
    Chelyabinsk event, 2013, 14
    iron, 14
Methane (CH_4), 55, 56, 174–175, 190, 314, 334
    abiotic, 57
Methane river networks, 651f
Methanogenic archaea, 230
Methanopyrus kandleri Strain 116, 231–232
Microalgae, 694
Microbes, 228–229
Microbes in subseafloor sediment layers, 703
Microbial ecosystems on Earth, 314
Microbial life on and inside rocks, 211
    diversity of habitats, 211–212
    escape from environmental extremes, 212–213
    on impact-shocked rocks, 213
    rock porosity, 212
Microbially Induced Sedimentary Structures (MISS), 172–173, 174

Microbores, 77
Microhabitats, 207–208
Micrometeorites, 193–194
Microphysical models, 621
Microwave remote sounding, 548–549
Mid-oceanic ridges, 337
Mid Ocean Ridge (MOR) system, 125
Miescher, Johannes Friedrich, 131–132, 132*f*
 notion of DNA, 133
Milankovitch cycles, 95, 96, 97
Miller-Urey experiment, 744
Miller-Urey "prebiotic soup" experiment, 116*f*, 116, 116, 117, 117, 118, 119, 120
Milnesium tardigradum, 237
Minimum Mass Solar Nebula (MMSN), 35, 36
Mitochondria, 715
Mobile Bay, 694
Models of Europa's hydrosphere, 535
Monte Carlo algorithm, 622
Monte Carlo model, 551
Murchison meteorite, 192, 193, 194*f*

N

NASA discovery-class mission, 523
NASA Juno mission, 549
NASA's Galileo mission, 548, 549
NASA's Juno spacecraft, 546
National Human Genome Research Institute (NHGRI) Genome Technology Program, 141
Neanderthals, 153–154, 154*f*, 154
Near Earth asteroids (NEAs), 9–10, 60–61, 61*f*
Near-Earth object (NEO), 13, 14
 types of, 14
"*Nebular*" or "*Disk instability*" Hypothesis, 1, 2, 4, 5
Neptune, 612
Nitrogen, 157
Nitrous oxide (N₂O), 191
Nix's rotation, 697*f*
Nonpolar solvents, 169
Nonterrestrial biochemistry, 62
Nucleic acids, 118
Nucleobases, 729–730
Nuvvuagittuq Greenstone Belt, 66

O

Ocean
 fishes, 695–696
 marine mammals, 696, 697, 698, 699, 700, 701, 702, 703
 sea turtles and reptiles, 696
Oceana, 696
Ocean-atmosphere chemistry on Earth, 191
Ocean Drilling Program (ODP), 703
Oceanic crust, 215
Oceanic crustal aquifer, rock-hosted, 216
Oceans, 694, 699*f*
 photosynthesis in, 694
Ocean Worlds Exploration Program (OWEP), NASA, 44
Octopuses, 695
Octopus Spring, 208–209, 209*f*
Olivine, 11, 12

Olympus Mons, 365
Oman ophiolite, 174–175
"Orbital resonance", 692
Orbital resonances, 549, 700
Oscillatoria filaments, 289
Osmo-regulators and osmo-conformers, 693
Ossetra sturgeon, 710
Oxo crater, 530*f*
Oxygenic photosynthesis, 23
Oxygen on Earth, evolution of, 81, 82
Oxygen-producing bacteria, 56

P

Pacific sea nettle, 725*f*
Palade, George E, 139
Paleocene–Eocene thermal maximum event, 92–93, 99, 100
 changes in eccentricity, 95
 consequences, 92
  carbonate compensation depth, 92
  sea surface temperature (SST), 92
  weather change, 92
 dissociation hypothesis, 99
 fate of excess carbon released during, 100
 intermediate-water warming, 100
 Methane hydrates emission, 93, 94
 methane hydrates emission, 93, 93, 94
 Milankovitch cycles, 94, 95, 96
 obliquity, 95, 96
 precession, 96
 volcanic eruptions and seaquakes, 92, 93, 97
Panspermia hypothesis, 128–129, 237, 238, 239
Panthalassa, 67
Panthalassic Ocean floor, 67
Paramecia, 148
Patearoa saline pond, 693
Pauling, Linus, 134–136
Pelagic fish, 696
Pelican *(Pelecanus occidentalis)*, 702, 728*f*
Penguin, 701, 702, 727*f*
Peptide Nucleic acids (PNA), 118
Peptides, 151
Periglacial rock sorting, 208
Permanent northern volatile (PNV) model, 703
Petersen, Johann Friedrich, 1
Phoenix mission, 478, 479
 findings, 478, 479
 objective, 478
Phosphorus, 157
Photoautotrophs, 207
Photochemical models, 621
Photolysis, 591–592
Photosynthesis, 179
Photosynthetic life, 155
Photosynthetic organisms, 63, 207
Phylogenetic tree, 228*f*
Phytoplankton community, 694, 702*f*, 721
Picrophilus torridus, 231–232
Piezophiles, 211
Pillow basalt, 338
Pillow lavas, 66*f*, 66
Pingos, 322*f*, 322, 325*f*
 closed-system, 323, 324, 325
 features of, 323

 forms, 323
 freezing cold regions, 323
 ice, 323–324
 on Mars, 325, 326
 open-system, 323*f*, 323, 324, 324*f*, 324–325
Pit craters, 314
Planetary embryos, 6
Planetary formation and evolution, 1, 438
 age dating of meteorite, 3
 asteroids, meteorites, and chondrites, 9*f*, 9, 10, 12
 astrophysical constraints, 3
 bridgmanite and magnesiowüstite, 12
 comets, 11
 of denser regions, 5–6
 dynamics of dust, 3
 electrostatic and gravitational interactions, 6
 giant, 7, 8, 8, 9
 magnitude of growth, 3
 nebular hypothesis, 1, 2, 4, 5
 from rotating protoplanetary disks, 4, 5*f*
 Safronov (nebular) hypothesis, 2
 SNDM model, 2, 3
 terrestrial, 6, 6, 7
Planetesimals, 203
Plate tectonics, 79–80
 breakup of supercontinent Pangaea, 80*f*
 first proto-subduction, 80
 lithospheric damage, 80–81
 process of, 80
Pluto, 1, 60, 691
 interior structure model, 704*f*
Pluto-Charon system, 699
Plutonic rocks, 317
Pluto system, 710
Polycyclic aromatic hydrocarbons, 620
Polymerase chain reaction (PCR), 206
Poly-peptides, 151
Polyps, 700
Porcelain white brachyuran crabs, 345*f*
Poseidon, 342
Potassium (K), rare earth elements (REEs) and phosphorus (P) (KREEP), 30, 31
Prebiotic soup theory, 122
Precambrian euendolithic, 212
Precambrian rocks, 173–174
Prokaryotes, 148, 150
Prokaryotic extremophiles, 211
Proterozoic Eon, 23, 55, 77
Protists, 148
Protoplanetary disks, 5*f*, 35
Psychrophiles, 211
Pythagoras, 45

R

Radar, 561
Radar-based active detection, 561
Radar-based passive detection, 561
Radar for Icy Moon Exploration (RIME), 563
Radiative-convective climate model, 57
Radiogenic heating, 44
Radiolytic oxidants, 592
Ramazzottius varieornatus, 225
Rayleigh number, 718–719

Red River, 711
Reflectance spectroscopy, 529–530
Ribonucleic acid (RNA), 129, 130f, 142
    Crick's idea on, 142, 143f
    form, 142
    functional, 142
    functions, 145
    messenger, 142, 142, 144
        pre-mRNA, 145
        prokaryotic, 145
        relationship between DNA and, 142–144, 143f
    microsomal particles, 142
    molecule, 151
    production of proteins, 144, 152
    ribosomal, 142, 145, 227, 233–234
        phylogenetic tree, 233f
    16rRNA, 175, 228
    sequencing, 145, 146, 319
    soluble, 142
    16S rRNA, 148
    transfer, 145, 151
    virus, 142
RNA-mediated RNA interference (RNAi), 237
"Roadmaps to Ocean Worlds" (ROW), 539
Roche, Edouard, 18f
Rocky planet formation, 203
Roscosmos State Corporation for Space Activities, 471
Runaway albedo effect, 694
Russian sturgeon. See Ossetra sturgeon

S

Safronov, Victor, 2
Safronov (nebular) hypothesis, 2
Salar de Atacama (Chilean salt flat), 729–730
    brines, 255
    green sulfur bacteria, 730–731
    halophilic bacteria, 730
    lithium concentrations, 729–730
    lithium-tolerant bacterial isolates, 301
    microbial communities, 281
    microbial life in, 730
    phototrophic bacteria, 730
        phototrophic Gammaproteobacteria, 730
    saline systems in, 730
    water variability, salt concentrations, and light conditions, 731
Salar de Hombre Muerto, 302
Salar del Hombre Muerto, 726
    James's Flamingos, 294
    Llamas, 294
    microbial life, 726
    salt flat, 295
Salar de Uyuni, 726
    brine pool communities of, 729
    climatic conditions, 726
    geological and hydrochemical superlatives, 727
    hypersalinity, 727
    lithium concentrations, 728
    microbial communities of polyextremophilic organisms, 728
    microbial diversity in, 729
    salt flat, 726–727, 728
    sulfate-rich habitats, 728–729
Saline and hypersaline habitats, 691
    ocean, 691f
    survival and growth of living organisms in, 691
    thalassic environment, 692
Salt Lake, Sutton, 693
San Francisco Bay, 693–694
Sanger, Frederick, 135f, 146, 147
Saturn, 583
Saturnian system, 586
Saturn's moon, 61
    Enceladus, 333, 334
        deep subsurface ocean on, 349, 350
        habitability of, 349–350
        silica nanoparticles, 349
        submarine hydrothermal vents on, 349, 350
Saturn's moons, 39, 40, 42, 43
Scalloped topography, 326–327
Seabirds, 701
Sea otter, 698, 716f
Sea slug, 701, 727f
Seastar *(Fromia monilis)*, 722f
Sea turtles and reptiles, 696
Sea urchins, 716f
Seaweed species, 694–695
Secondary-ion mass spectrometry (SIMS), 181
    dynamic, 183
Serpentinite-hosted carbonate chimneys, 339, 340
    aragonite and brucite, 340
    fluid circulation in, 340
    formation, 340
    hydrothermal circulation in, 340
Serpentinization, 174, 339
    along MORs, 174–175
    by-product of, 174–175
    Fischer-Tropsch-type synthesis, 174
    hydrogen production by, 174, 175
    hydrolysis and transformation of ferromagnesian minerals, 174
    as weathering process, 175
Serpentinized mantle rocks, 347–348
Sessile benthic community, 217, 218f
Sevruga sturgeon, 710
Sharks and rays, 698, 717f
    African frilled shark *(Chlamydoselachus africana)*, 699
    frilled shark *(Chlamydoselachus anguineus)*, 699, 718f
Shergottite-Nakhlite-Chassignite (SNC) meteorites, 458, 459
Signal to Noise Ratio (SNR), 563
Signal-to-noise ratio (SNR), 569
Sinuous lunar rilles, 432
Six-channel microwave radiometer system, 549
Slack water, 75
Smith, John, 146
Smoothed particle hydrodynamics method (SPH), 20, 21, 25–26, 28
Snow line, 8
Solar nebula, 15
Solar nebular disk model (SNDM), 2
Solar System, 1, 46f, 63, 238, 239–240, 242, 359, 362, 363, 523, 526, 698
    age, 3–4
    comets and asteroids, 60, 61–62
    four regions, 59–60
    Main Asteroid Belt (MAB), 9–10, 10f
    Nice model, 11
    ocean worlds, 43
    origin, 1
    primordial evolution of, 61
    reservoirs, 68
    sources of, 61
    Sun-centric model of, 45–46
Solid-state imaging (SSI) system, 560
South Polar Terrain, 593
South-Pole Aitken (SPA) basin, 437–438
Space missions
    soft-landings on the lunar surface see also specific entries, 399
Space Telescope Imaging Spectrograph (STIS), 546
Spatial resolutions, 717
Spotted Lake, 328, 329, 329f
    analogous sulfate-rich closed-basin paleolakes, 330
    concentration of sulfates, 329
    habitability, 329
    microbial structure in, 328–329
    salt concentrations in, 329
Squid *Euprymna scolopes*, 705f
Sri Aurobindo, 153
Starfish, 701
Stenohaline, 693
Stingrays, 699
Stratovolcano, 379
Strecker condensation, 117
Strelley Pool Chert (SPC), 173–174
Strelley Pool Formation, 174
Striped bass *(Morone saxatilis)*, 712, 713
Stromatolites, 78f, 78, 173, 174, 203
    definition, 77–78
    identification of, 180–181
    sedimentary rock formations, 78
Sturgeons' caviar, 710
Submarine hydrothermal systems, 126, 127–128
Submarine hydrothermal vents, 336, 338f
    diffuse, 338
    on *Enceladus*, 349
    on *Europa*, 350
    features of, 336
    magma-chambers-fed black smoker and white smoker vents, 336, 336, 337, 338
    off-axis vents, 340–341
    water's carrying capacity, 339
Subsurface hydrocarbon reservoirs, 645–646
Suizhou meteorite, 11–12
Sulfate- and hematite-rich sedimentary rocks, 320
Sulfate-reducing firmicutes, 124
Sulfolobus, 228–229, 232, 234
Sulfur, 157
Sun's habitable zone, 202
Super-Earths, 186
Surface liquid water worlds, 202

Sutton Salt Lake, 722
  fauna and chemical composition, 722, 723
  rainfall in, 722
Synestia, 16

T

Tardigrades, 218, 219f, 219, 236, 237, 240f, 240, 240, 241
  antioxidant production, 219
  body plan, 225f
  cryptobiotic, 223–224
  desiccation tolerance in, 220, 221
  dormancy strategies in, 223
  effect of extreme environmental stresses, 224
  eggs, 242
  hydrostatic pressure tolerance of, 224
  limno-terrestrial, 222
  as potential model organisms in space research, 225
  radiation tolerance in, 221, 222, 223, 223f, 223
  specific intrinsically disordered proteins (TDPs), 221
  structure, 219
  survival high-speed impact shocks, 239, 240
  temperature tolerance in, 219
Terrestrial analogs, 311, 336
Terrestrial life
  carbon chemistry of living systems, 169
  characteristic of living thing, 167, 168
  co-evolution of life, 168
  evidence of microbial life, 168
  molecules associated with living organisms, 168
Terrestrial microbes, 311–312
Terrestrial-planet formation
  basic architecture, 6–7
  cosmochemistry evidence, 6
  Earth's Moon, 6
  gravitational collapse of molecular cloud spanning, 7
  Kyoto Model, 6
  modification of planetary orbits, 7
  oligarchic growth, 6
  planetary embryos, 6
  runaway growth, 6
  stages of growth, 6
  stochastic nature of, 6–7
  time scale for lunar formation, 6
Thalassic environment, 691–692
Thalassic water bodies
  brackish water bodies, 703–704
  life in, 693
  saline and hypersaline waters, 692
Thaw lakes, 324–325
Theory of lithopanspermia, 238
Thermal Emission Imaging Spectrometer (THEMIS), 326
Thermal Emission Spectrometer (TES), 326
Thermal metamorphism, 13
Thermal model of Ganymede, 549
Thermocrinis ruber, 208–209
Thermophiles, 231–232
Thermophoresis, 128

Thiomicrospira, 332
Thiotrichaceae, 714
Thrace Macula, 312–313
Threose nucleic acid (TNA), 118
Tidal bore, 73, 75
  cultural heritage of, 76–77
  impact on ecosystems, 77
Tidal currents, 75
  acceleration, 375–376
  across-valley, 76
  in coastal water bodies, 75
  current rose, 75
  cycle, 75
  ebb current, 75
  flood current, 75
  in Juan de Fuca plates, 76
  neap and apogean, 75–76
  reversing, 75
  in ridge valleys, 76
  speed, 75
    magnitude of, 75–76
  spring and perigean, 75–76
  strength, 75
Tidal information, 76
Tidal rhythms, 69, 70
  diurnal inequalities, 71f, 71
  effect of Moon on, 71, 375–376
  form factor, 70f, 70
  high tidal level, 71
  long-period, 72
  mixed and diurnal tides, 70
  neap tides, 71
  at perigee, 71–72
  regimes, 70
  semidiurnal, 70–71
  spring tides, 71, 73–75
  tidal amplitude, 70–71
  tidal node, 70–71
  topographical influences, 72, 73f
    Chilika Lake, India, 72–73, 73f
    flood tide, 73
    geometrical amplification, 72, 73
    Gulf of Kachchh and Gulf of Khambhat, India, 72, 73f
    Kochi backwaters, 72–73, 74f
Tidal-stress-induced heating, 44
  of Jupiter's moon, 44
Tidal variations, 41–42
Time-of-flight secondary ion mass spectrometry (ToF-SIMS), 181, 182, 183f, 184f, 186
  analysis of membrane-derived lipids, 185
  analysis of pristine biomarkers, 186
  analyzer, 183, 184
  area of analysis, 183
  benefits/advantages, 182, 183, 186
  cell interior for, 185
  data interpretation, 185
  generic scheme of, 182f, 182
  lipid analysis, 185
  potential of, 181, 185
  primary ion dose density (PIDD) of, 182–183
  secondary ion yield, 184
  sensitivity of, 181–182, 185
  static, 181
  study of lipids, 185, 186

ToF analysis, 183
  topographic effect, 183–184
Timpie Springs, 719
Titan, 612, 617, 618, 624f, 650f, 651
Titan, 586
Titan Radar Mapper, 649
Titus-Bode law, 523
Todd, Alexander, 134–136
Trehalose, 221
Trinitrotoluene (TNT), 14
Triton, 335, 673, 677, 679, 680
  cantaloupe terrain, 335
  dust devil formation, 335–336
Tunguska events, 14
Tyndall, John, 55

U

Ukrainian rocks, 317
Ultraviolet imaging spectrometer, 707
Uni-genes, 237
United Nations Office for Outer Space Affairs (UNOOSA), 13–14
Unlocked nucleic acid (UNA), 140–141
Uranus's moons, 43
Urey cycle of CO_2, 91
Utopia Planitia, 326–327, 327f, 327–328
Utopia Planitia region of Mars, 326f, 326–327
  identification of PLFs, 327

V

Valles Marineris system of valleys, 365
Vapor-deposited water-ice, 678
Venetia Burney Student Dust Counter (VBSDC), 707
Vent ecosystem communities, 313–314
Venus, 359
  atmospheric composition and climate, 366, 366, 367, 368
  distance from Sun, 376
  dry surface, 391
  magnetic field, 391, 392
  similarties with Earth, 363, 364
  size and composition of, 362
    clouds, 363
    crater population, 364
    crust, 363
    exoplanet, 364
    interior, 363
    seismology, 363
    surface, 363
    volcanic edifices, 364
  southern hemisphere, 377f
  temperature profiles, 377
  volcanism and surface features, 377, 378
Vesta, 526f
Vibrio fischeri, 695
Viking missions, 479, 479, 480
  biological interpretation of the Viking results, 479
  detection of bioorganic compounds, 480
  failure of, 480
  incubation experiments, 479, 480
  observation of Valles Marineris, 492
  scientific aspects, 480–481

Visible and infrared mapping spectrometer
    (VIMS), 642
        data, 628
        instrument, 589
Visible and infrared mapping spectrometer
    (VIR), 528
Volcano eruptions, 97, 98f, 98
    drifting of Greenland and North America, 92–93
Volga River Delta, 708–709
Voyager 2 mission, 675
Voyager spacecraft, 552

W

Water, 157
    abundant celestial bodies, 39, 40
    based life, 62
    condensation of, 35
    in Earth, 34, 35, 38–39, 57–58
        by comets and asteroids, 59, 60
        lost to space, 58–59
        by photolysis, 58–59
        through collisions, 58
        through mantle evolution, 59
    as good solvent, 62
    from meteors, 68f, 68
    molecule, 62f, 62, 65
        deuterium to hydrogen ratio, 67, 68, 69
    in protoplanetary disks, 35, 36, 61
    of terrestrial planets, 36
Water-equivalent hydrogen (WEH), 532
Water-rich planetary bodies, 555–556
Water-rock interactions, 555
Watson, James D., 135f, 136, 137
Wegener, Alfred, 67
Whales, 697, 711f, 712f
White and Thompson Glaciers, 331
White smokers, 336, 337, 338
Wiedemann–Franz law, 427–428
Wilkins, Maurice, 134–136, 137
Wilson, Alexander, 721
Woese, Carl, 227–228, 228f, 233–234,
    235–236
World Line, 158

Y

Yellowstone hot springs, 186

Z

Zambales ophiolite, 174–175
Zircons, 67
Zygote, 721